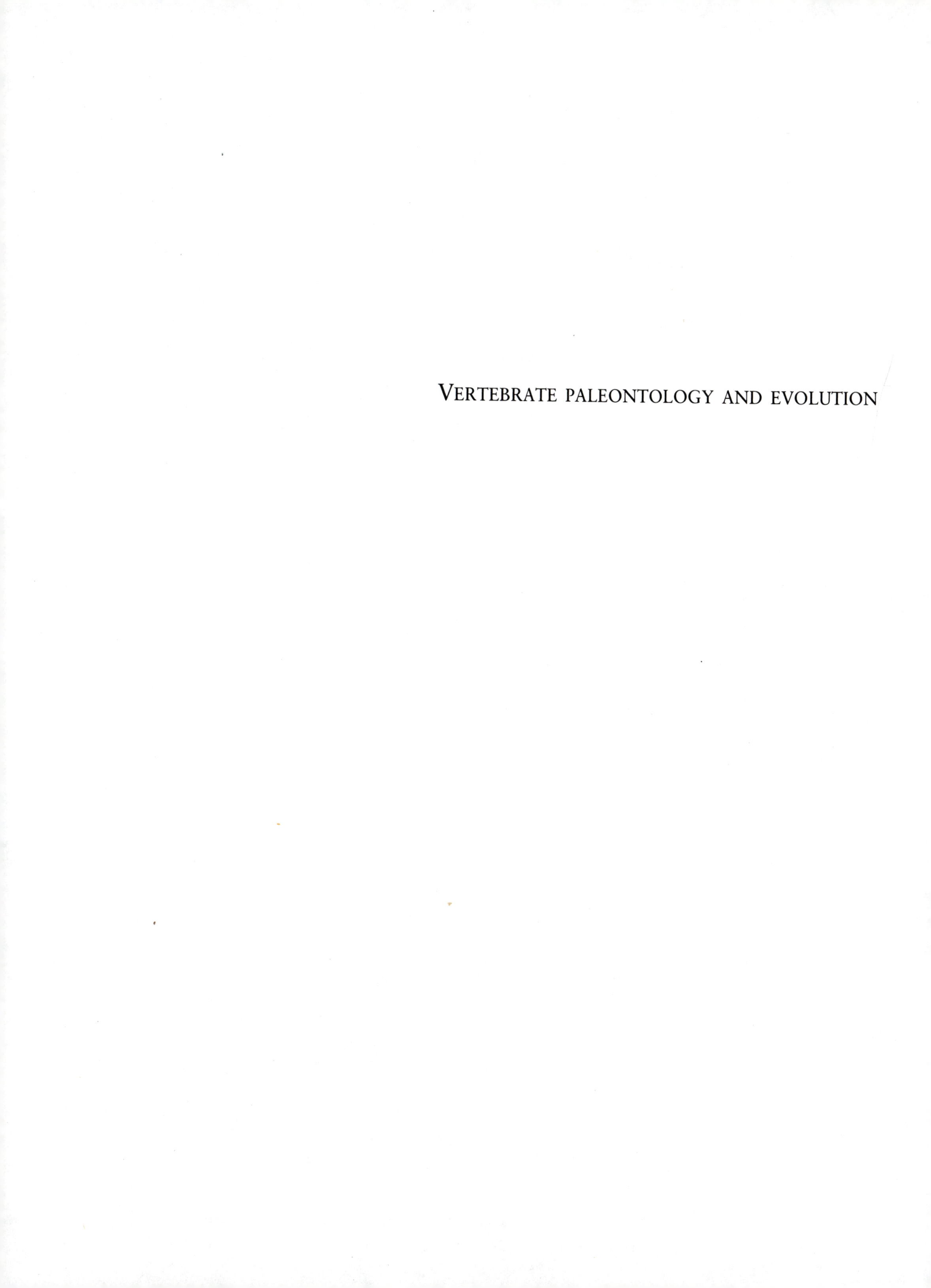

Vertebrate paleontology and evolution

VERTEBRATE PALEONTOLOGY AND EVOLUTION

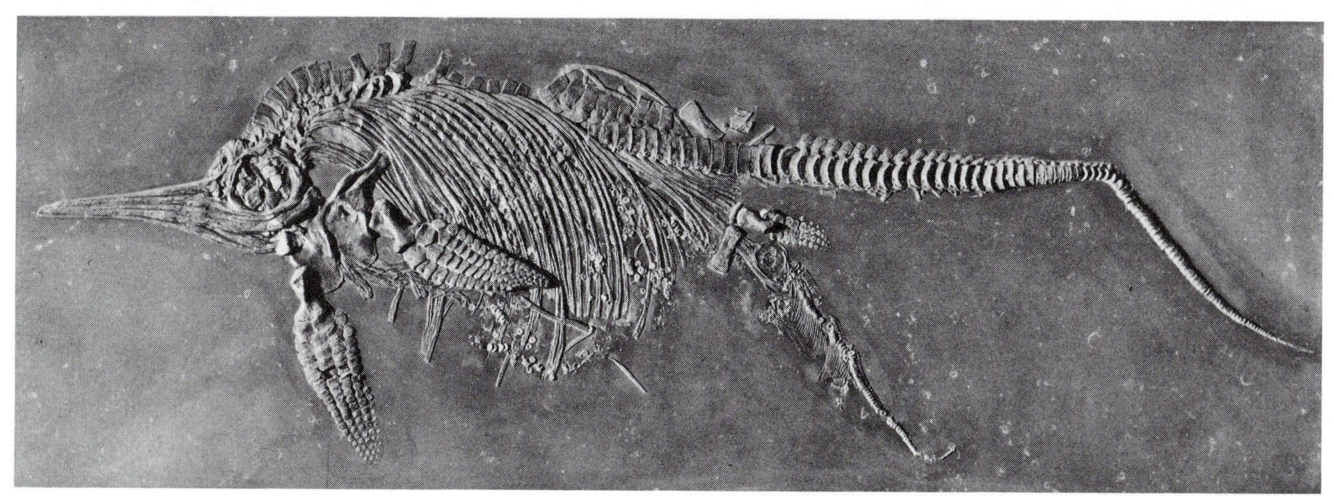

ROBERT L. CARROLL

Redpath Museum
McGill University

W. H. FREEMAN AND COMPANY | NEW YORK

Library of Congress Cataloging-in-Publication Data

Carroll, Robert L.
Vertebrate paleontology and evolution.
Includes index.
1. Vertebrates, Fossil. 2. Vertebrates—Evolution. I. Title.
QE841.C254 1987 566 86-31808
ISBN 0-716-71822-7

Printed in the United States of America.
1 2 3 4 5 6 7 8 9 0 HL 6 5 4 3 2 1 0 8 9 8

Dedicated to the
memory of my father,
John H. Carroll
who introduced me to
the joys of paleontology

CONTENTS

Last night in the museum's hall
The fossils gathered for a ball.
There were no drums or saxophones,
But just the clatter of their bones,
A rolling, rattling carefree circus
Of mammoth polkas and mazurkas.
Pterodactyls and brontosauruses
Sang ghostly prehistoric choruses.
Amid the mastodonic wassail
I caught the eye of one small fossil.
Cheer up, sad world, he said, and winked—
It's kind of fun to be extinct.

Ogden Nash
"Carnival of the Animals"*

P R E F A C E

Vertebrate Paleontology and Evolution is intended as a comprehensive text for university courses in vertebrate biology and as a reference for biologists and geologists who want to become familiar with the pattern of vertebrate evolution. I hope that it also provides the necessary background for research into more specific details of the anatomy, relationships, and distribution of both fossil and living vertebrates.

During the past 20 years, our knowledge of fossil vertebrates has increased immensely. Entirely new groups of jawless fish, sharks, amphibians, and dinosaurs have been discovered, and the major transitions between amphibians and reptiles, reptiles and mammals, and dinosaurs and birds have been thoroughly studied. Evidence from both paleontology and molecular biology provides much new information on the initial radiation of both birds and placental mammals. This book integrates knowledge of the anatomy of extinct vertebrates to serve as the basis for understanding the origin and relationships of the modern orders.

The past 20 years have also seen a revolution in the methodology of establishing relationships, and scientists have proposed many new theories regarding the factors controlling the rate and direction of evolution. The use of the fossil record in determining the polarity and homology of characters is emphasized throughout the book as a necessary step in establishing phylogenies. I describe numerous patterns of radiation that can be used to test current hypotheses regarding the processes and constraints that govern the course of evolution.

This book reflects the current direction of paleontological research by emphasizing functional anatomy and the interactions between extinct organisms and their environment. In addition to considering the anatomy and way of life of each of the major vertebrate groups, I have paid special attention to transitions between groups. I have also focused on periods of vertebrate history during which the fossil record is sparse and the available evidence has led to conflicting hypotheses of relationship.

I hope that this book will not only provide knowledge of our current understanding of the history of vertebrates but will also serve to stimulate further research into the many remaining problems of relationships and evolutionary processes.

Acknowledgments

In a discipline as broad as vertebrate paleontology, one worker can be personally familiar with only a very small portion of the current research. My own studies have focused on a few families of Paleozoic and early Mesozoic amphibians and reptiles. For information on the remainder of vertebrate phylogeny, I am indebted to a host of friends and colleagues from this and previous generations.

I would like to offer special thanks to my graduate students, past and present: Donald Brinkman, Philip Currie, Stephen Godfrey, the late Malcolm Heaton, Robert Holmes, Robert Reisz, Olivier Rieppel, Denis Walsh, and Carl Wellstead. Their work has contributed substantially to the chapters on early amphibians and reptiles.

Many other scientists have provided suggestions and critical comments on sections of the manuscript that deal with groups with which they are most familiar. In addition, several have allowed me to make use of unpublished manuscripts and illustrations so that this text could reflect the most recent research in this field. To all these colleagues I express my gratitude. David Bardack, Gerry Case, the late Robert Denison, David Dineley, Beverly Halstead, Karel F. Liem, Richard Lund, Roger Miles, Alex Ritchie, Bobb Schaeffer, Hans-Peter Schultze, Barbara Stahl, and Rainer Zangerl all provided assistance in the fish chapters. Donald Baird, Angela and Andrew Milner, E. C. Olson, Alec Panchen, Tim Smithson, Zdenek Spinar, Keith Thomson, and Peter Vaughn have shared generously the results of their research into the ancestry and relationships of Paleozoic tetrapods. Alan Charig, Sankar Chatterjee, Ned Colbert, Peter Dodson, Richard Estes, Gene Gaffney, Peter Galton, Carl Gans, Cris Gow, Wann Langston, John MacIntosh, John Ostrom, Dale Russell, and Rupert Wild have provided literature, information from unpublished studies, and many helpful suggestions as to the anatomy and relationships of the various reptile groups. The chapter on flight benefited greatly from the assistance of Alan Feduccia, Peter Houde, Storrs Olson, John Ostrom, and Kevin Padian.

Mike Archer, Bill Clemens, A. W. Crompton, Richard Fox, Jim Hopson, Nick Hotton, Zofia Kielan-Jaworowska, Farish Jenkins, Jason Lilligraven, and Hans-Dieter Sues provided much help in documenting the origin of mammals and their Mesozoic radiation. Assistance in the preparation of the chapters chronicling the radiation of placental mammals was provided by Mary Dawson, Philip Gingerich, Robert Hoffmann, Leonard Krishtalka, Malcolm McKenna, Nancy Neff, Michael Novacek, Ken Rose, the late George Gaylord Simpson, Jean Sudre, David Webb, and John Wible. Larry Barnes, Daryl Domning, Ed Mitchell, and Clayton Ray helped in establishing relationships among the marine mammals.

The final chapter on evolution benefited greatly from suggestions offered by Niles Eldredge, Andrew Knoll, and David Webb. I am particularly grateful to Hans-Dieter Sues and Denis Walsh for reading the entire manuscript and offering suggestions as to its improvement. Additional reviews of part or all of the manuscript were provided by Leo Laporte, Everett Lindsay, Robert Emry, Milton Hildebrand, Michael Williams, and Lance Grande.

The spirit of this text owes much to the work of Alfred S. Romer, whose outstanding contribution to our understanding of the evolution of Paleozoic amphibians and reptiles and sheer enthusiasm for paleontology serve as examples for all workers in the field. The evolutionary concepts that are elaborated here were developed primarily from the writings and teaching of George Gaylord Simpson, who continues to stand as the major figure in the integration of the fossil record and evolutionary theory.

The greatest assistance in preparing this text has been provided by Pamela Gaskill, who drafted nearly all the illustrations. Heide Hanson carried the major task of manuscript preparation and the technical aspects of the figures and bibliography. Aruna Ashtakala provided assistance in preparing the generic lists. Marie La Ricca and Delise Alison assisted with typing the text and the generic lists. Access to the paleontological literature was greatly facilitated by the help of the staff of Blacker-Wood Library, Eleanor MacLean, Ann Habbick, and Jennifer Adams.

Production of this book benefited greatly from the efforts of the staff of W. H. Freeman and Company, notably the project editor, Scott Amerman; designer, Lynn Pieroni; art director, Mike Suh; and Julia DeRosa, coordinator, composition and manufacturing. The senior editor, Jerry Lyons, offered advice at every stage.

Anna Di Turi has provided encouragement and support throughout the preparation of this book.

C H A P T E R **I**

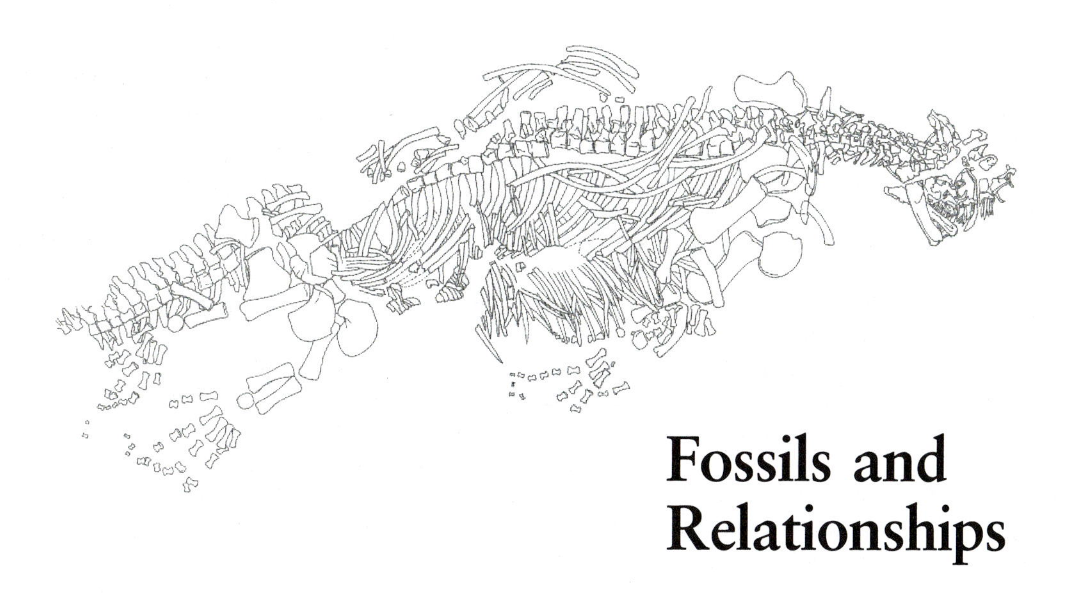

Fossils and Relationships

The history of vertebrates spans more than 500 million years, from their appearance in the Cambrian seas to the present rich and varied fauna. Their history is recorded in a sequence of fossils that documents their skeletal anatomy, distribution, and evolutionary change.

From the fossil record we can determine the interrelationships of the modern species and trace the origin of the skeletal features that characterize each major group. The pattern of vertebrate evolution provides the best available evidence for determining the major processes of evolution: the origin of new structures, physiological processes, and adaptive strategies. Fossils document the rate of evolution and may enable us to establish the degree to which environmental, developmental, or other factors influence its direction.

This text will review the anatomy, relationships, and evolutionary patterns of the major groups of vertebrates from their differentiation in the early Paleozoic through their last major radiation in the early Cenozoic. I will emphasize the way in which vertebrates have adapted to diverse environments and the challenges faced during major adaptive shifts. Particular attention will be paid to the earlier and previously less well-known forms that are fundamental to understanding the origin of the anatomical and adaptive patterns of the modern groups.

Vertebrate evolution is an interdisciplinary subject, involving aspects of geology, comparative anatomy, physiology, and evolutionary theory.

Knowledge of geological processes helps one to understand how vertebrates are preserved as fossils and why animals from different environments have different likelihoods of being preserved. Evidence of changes in the patterns of the major land masses, mountain chains, oceans, and shallow seas helps us to explain the distribution of extinct vertebrates and to reconstruct the climates in which they lived.

Knowledge of the skeletal anatomy of modern vertebrates is necessary for the study of the bones and teeth, which provide most of our information regarding extinct animals. Direct evidence of the soft anatomy of extinct genera is rarely preserved; nevertheless, an understanding of the basic patterns and functions of the muscular, circulatory, respiratory, reproductive, and nervous systems is necessary to appreciate their way of life.

In addition to knowledge of adult anatomy, an understanding of embryological structures and developmental processes will help in interpreting the possible direction of evolutionary changes and the origin of new structures. These geological and biological subjects will be covered briefly in Chapters One and Two.

The term **evolution** as it appears in the title covers two distinct concepts. Throughout the text, vertebrate evolution is treated in the historical sense of changes in structure, function, and adaptation. Evolution may also be considered from the viewpoint of the underlying genetic changes and processes of selection and population dynamics that explain how these changes occur. The final chapter will be devoted to a consideration of structural, developmental, and environmental factors that help to explain the known patterns and rates of vertebrate evolution.

The next twenty chapters will describe vertebrate history as established from the fossil record and through our knowledge of living species. This story is by no means complete or entirely coherent. The early stages of vertebrate history are poorly known, and significant gaps still separate many major groups. This text will document our current knowledge of each group and discuss the problems of establishing their relationships and ways of life. Many aspects of vertebrate evolution remain contentious, and alternative viewpoints will be considered.

Technical papers describing fossil vertebrates appear in many journals. Short notes on exceptionally important discoveries appear in *Science* and *Nature*. Longer descriptive papers are published in *Journal of Vertebrate Paleontology, Journal of Paleontology, Palaeontology, Palaeontographica, Zoological Journal of the Linnean Society, Philosophical Transactions of the Royal Society,* and the publications of many universities and public museum, for example, *Bulletin of the American Museum, Bulletin of the Museum of Comparative Zoology, Harvard.* All literature through 1983 is cited in the *Bibliography of Fossil Vertebrates,* published by the Society of Vertebrate Paleontology, whose membership includes professional vertebrate paleontologists from all countries. The office of this society is currently associated with the Natural History Museum, 900 Exposition Boulevard, Los Angeles, California, 90007.

THE NATURE OF THE FOSSIL RECORD

Our understanding of the relationships and ways of life of extinct vertebrates depends on knowledge of their geological age and the nature of the deposits in which they were buried. The geological time scale is shown in Figure 1-1, together with an outline of vertebrate phylogeny. Detailed subdivisions of the geological periods and diagrams of relationships within each of the vertebrate groups accompany their descriptions.

The processes of fossilization limit the type of information we can gain from extinct organisms. The bodies of most animals are consumed or scattered by predators and scavengers soon after death, and their bones are broken up and decompose. Perhaps no more than one in a million are so quickly buried that they may become fossilized. The flesh almost invariably decays, but the bones may be infiltrated by water carrying sediments and soluble minerals that fill up the large cavities and precipitate in smaller channels once occupied by cells and blood vessels. In most vertebrate fossils, mineral components of the bone retain their integrity, so that histological details and chemical composition are little altered even after hundreds of millions of years. A fossil bone is still bone, but it contains a hard and heavy infilling of other minerals as well.

Occasionally, entire animals have been enclosed in sediments that hardened to give an impression of the complete body surface. Fossils of dinosaurs that became mummified after death have been preserved in this manner. Impressions of other animals have been preserved in lava flows and amber. Flattened, coalified remains showing body outlines are found in fine-grained sediments that were presumably deposited in an anaerobic environment, free from decomposing organisms. Occasionally, the soft tissues may be preserved by waters bearing pyrite or silica. Histological details of muscle fibers and kidney tubules are preserved in fossil sharks from the Upper Devonian Cleveland shale.

Footprints are relatively common trace fossils that provide information about the posture, gait, and other aspects of behavior of extinct animals, and they may pro-

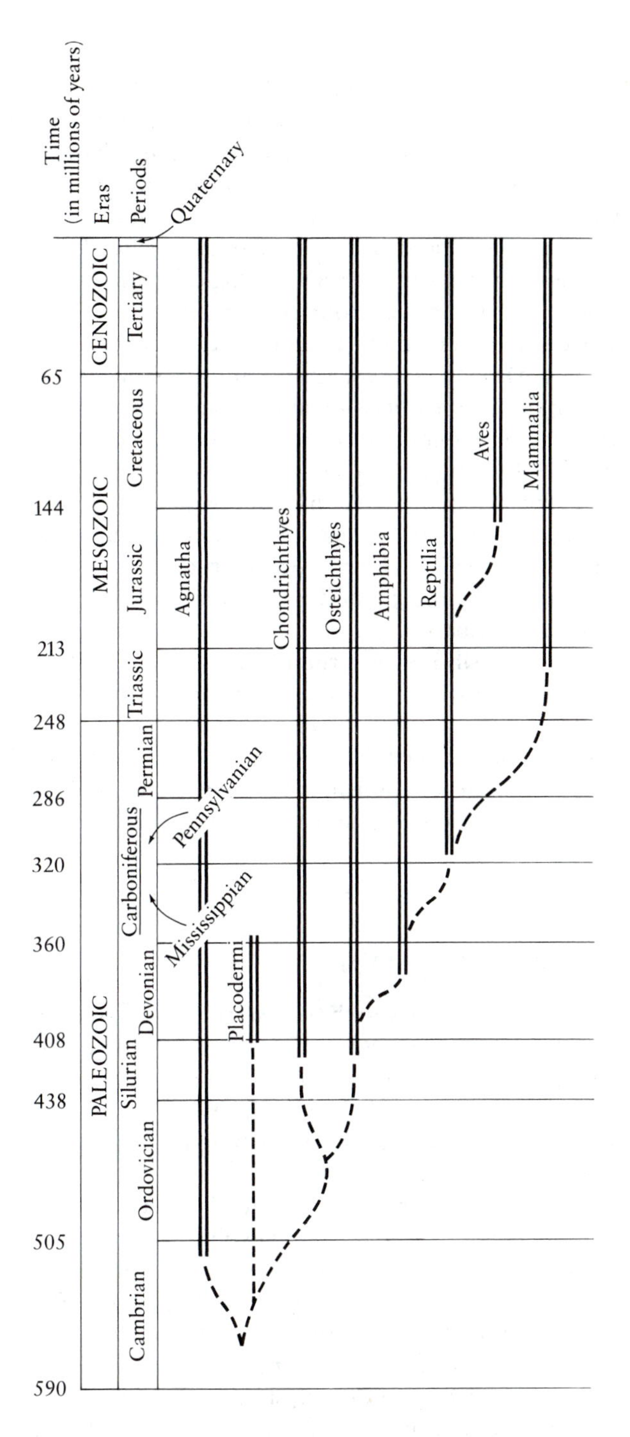

Figure 1-1. TEMPORAL DISTRIBUTION OF VERTEBRATE CLASSES AND THEIR PROBABLE RELATIONSHIPS. The geological time scale used in this and other phylogenies is from Harland et al., 1982.

vide information about distribution of groups not locally represented by skeletons. Coprolites, or fossilized feces, may reflect the shape of the internal surface of the intestines (notably the spiral valve of sharks) and the diet of the animal that produced them. In some cases, coprolites contain small bones of animals that are not otherwise found in a given fauna.

Although the actual tissue of the nervous system is never fossilized, we can learn about the surface features of the brain from impressions of the inner surface of the cranial bones. In some cases, natural infillings of the cranial cavities, or **endocasts,** are preserved and exposed when the overlying bone is eroded or intentionally removed. In relatively large animals, we can remove the matrix filling the skull and make an artificial mold of the inside surface of the bone. We can also study the inside of small and complex skulls through the use of serial sections.

Endocasts of mammals are especially informative and can provide a close approximation of the original shape and volume of the brain. In contrast, the brains of fish, amphibians, and reptiles do not fully occupy the cranial cavity, and neither the surface features nor the volume can be accurately reconstructed on the basis of the surrounding bones (Jerison, 1973).

Behrensmeyer and Hill (1980) consider other aspects of the relationships between fossils and the sediments in which they are buried in "Fossils in the Making." Rixon (1976) and the journal *Curator* discuss collection and preparation of fossil vertebrates.

Not all animals have an equal chance of fossilization, which results in significant biases in the fossil record. Generally, animals that live in or near the water have a much greater chance of being quickly buried and preserved than animals that live in strictly terrestrial environments. On the other hand, fossils of animals that lived in the depths of the ocean are not usually found, because this part of the earth's crust is rarely exposed on the land surface. The shallower waters of continental margins and inland seas, in contrast, are ideal areas for the preservation and subsequent exposure of fossils. The fauna may be rich and the flow of sediments from the land copious. These areas are frequently subject to tectonic activity that can thrust them up into mountains, where sediments erode to expose the enclosed fossils.

The fossil record is also rich in deltaic and lagoonal environments closer to shore, but it may be difficult to determine whether the animals actually lived in this area, inhabited a more truly marine environment, or washed in from freshwater streams or lakes. We also find large numbers of animals living at the margin of the sea—in the zone of the surf and tide—but their remains tend to be fragmented by the action of the sea.

Fossils of animals living in fresh water are rarer simply because throughout geological history, areas of fresh water have been less extensive and served as basins of deposition for shorter periods of time. Nevertheless, large lake systems have yielded rich faunas, which may record very short time intervals over wide areas.

Many localities record the fauna of coal swamps during the Upper Carboniferous. However, the area occupied by the productive localities is infinitesimal compared with the millions of square miles of coal deposits. Hook and Ferm (1985) provide a useful model for the special conditions that were responsible for the preservation of vertebrates in coal swamps. They attribute their concentration at Linton, Ohio, to reducing anaerobic conditions associated with deep abandoned river channels that were dominated by spores and degraded plant remains. For most of the Mesozoic, the expected fauna of coal swamps is missing from the fossil record.

Large river systems, especially where they form deltas on the margins of large lakes or oceans, provide an excellent source of fossils from a spectrum of aquatic and semiaquatic environments. The famous Lower Permian redbeds of north central Texas are an example of such a deposit.

Low-lying flood plains beyond the banks of large river channels preserve some of the richest fossil accumulations. Deposits from this environment in southern Africa range in age from the Upper Permian throughout the Triassic; included is a magnificent fauna of mammal-like reptiles that documents the transition between reptiles and mammals. The vast assemblage of Cenozoic mammals from western North America was preserved under similar conditions.

Further from major areas of sedimentation the fossil record is drastically reduced. Small bodies of water, ponds, and streams are only temporary features in the geological time scale. Above the basins of deposition, they are likely to be eroded, rather than being preserved in the sedimentary record. Under exceptional conditions such higher land may be preserved as the result of major faults, where large areas are suddenly dropped hundreds or thousands of feet and become basins of deposition. The peculiar fauna of the Upper Permian of Madagascar appears to be preserved in this manner (Currie, 1981).

Even more unusual but locally rich sources of fossils have been found in cave and fissure fillings (Robinson, 1962) and within the upright tree stumps of the lycopod genus *Sigillaria* (Carroll et al., 1972). Such modes of preservation tend to accumulate animals not ordinarily found in aquatic deposits and partially correct the bias against fully terrestrial vertebrates.

The discovery of fossils depends not only on their initial burial and preservation but also on present-day conditions of exposure and erosion. Deserts and semideserts provide excellent areas for the discovery of fossils, since they are exposed but not immediately weathered away or covered by vegetation. In contrast, the wet tropics are among the least profitable areas for collecting, since rapid chemical weathering destroys the bones before they are ever exposed and what remains is quickly covered with vegetation.

We know little about the fossil history of large areas of the tropics. Fossils have been collected at the margins of the Arctic and in Antarctica, but large areas remain obscured by ice.

The changing positions of the continents and areas of tectonic activity have also led to varying distributions of depositional environments. In addition, some deposits have been completely removed by subsequent erosion, and others are so deeply buried that their fauna may never be exposed. Thus, while some geological periods are represented by a wide spectrum of different environments, others have a more limited record. There are very few nonmarine deposits of the Lower Carboniferous or the upper part of the Lower Jurassic anywhere in the Northern Hemisphere. Consequently, the knowledge of semiaquatic, amphibious, and terrestrial vertebrates of these ages is extremely limited. In the Paleozoic, the fossil record remains a patchwork—certain geological horizons, geographical areas, and environments of deposition are well represented, while others are completely unknown.

The hazards of preservation and subsequent exposure impose another bias—against groups of animals that were rare or geographically restricted. This bias is particularly unfortunate, since most major evolutionary changes probably occurred in small, isolated populations that were subject to stringent selection pressure (Dobzhansky et al., 1977; Mayr, 1963; Simpson, 1953). Where information regarding transitional forms is most eagerly sought, it is least likely to be available. We have no intermediate fossils between rhipidistian fish and early amphibians or between primitive insectivores and bats; only a single species, *Archaeopteryx lithographica* represents the transition between dinosaurs and birds. On the other hand, certain genera of fish, amphibians, and reptiles are known from thousands upon thousands of fossils from every continent.

EVOLUTION

Although fossil vertebrates were very incompletely known in the nineteenth century, remains of dinosaurs, giant marine reptiles, and mammals without modern descendants provided Darwin and other biologists with irrefutable evidence of extinction and, less directly, of the process of evolution.

Evolution might have been accepted without fossil evidence, but fossils now seem inexplicable without evolution. Yet the foremost vertebrate paleontologists of the nineteenth century—Cuvier, Owen, and Agassiz—did not accept evolution as put forward by Darwin but argued for a succession of creations and extinctions. Until the 1940s, vertebrate paleontologists remained outside the mainstream of evolutionary thought. Simpson (1944) and Jepsen (Jepsen, Mayr, and Simpson, 1949) made the first important contributions to the modern evolutionary synthesis (Mayr and Provine, 1980).

Perhaps we should not be surprised that vertebrate paleontologists did not support the prevailing view of slow, progressive evolution but tended to elaborate theories involving saltation, orthogenesis, or other vitalistic hypotheses. Most of the evidence provided by the fossil record does *not* support a strictly gradualistic interpretation, as pointed out by Eldredge and Gould (1972), Gould and Eldredge (1977), Gould (1985), and Stanley (1979, 1982).

Few contemporary paleontologists would deny that natural selection controls the direction of evolution, but many would seek additional factors to account for the rapid evolution that characterizes the early diversification and radiation of groups and the early stages in the elaboration of major new structures. The great longevity of many groups and the minor evolutionary changes they exhibited pose another problem. I will discuss these and other general aspects of vertebrate evolution in the final chapter.

ESTABLISHING RELATIONSHIPS

One of the most challenging aspects of vertebrate paleontology is establishing relationships—between extinct genera and possible living counterparts and among totally extinct groups. Taxonomists since the time of the ancient Greeks have based relationships and systems of classification primarily on overall similarities. Linnaeus and other early taxonomists thought these similarities resulted from the orderly creation of life by God. Yet, these similarities so soundly reflected basic, inherited attributes of the organisms that the hierarchical system of classification developed by Linnaeus could be accepted, with little change, by evolutionary biologists.

Until recently, few paleontologists were seriously concerned with the methods of classification; nevertheless, they developed a workable and relatively consistent system that integrates well with the classification of living vertebrates.

HENNIGIAN SYSTEMATICS

During the past 20 years, the writings of Willi Hennig (1965, 1966, 1981) have aroused increasing interest in the methodology of determining relationships and their expression in systems of classification. Hennig sought a more objective method of establishing relationships and a pattern of classification that was not prey to the subjective approach of individual taxonomists. He also sought to make classification reflect phylogenetic patterns as closely as possible. This system is referred to as **phylogenetic systematics, cladism,** or **Hennigian systematics.** Wiley (1981) provides the most effective recent review of this method.

Paleontologists, including Eldredge and Cracraft (1980), Gaffney (1979), and Patterson (1981), have been among the most active supporters of phylogenetic systematics. Despite the difficulty of applying some of the conventions of Hennigian classification to extinct groups, the methods of determining relationships provide an objective basis that greatly assists in establishing reliable phylogenies.

The most important aspect of Hennigian methodology is the emphasis on specialized or derived characters in establishing relationships, in contrast with the use of general similarities, as has been the common practice for many taxonomists. Two groups may share a large number of attributes in common, but only those that are specialized relative to other more distantly related groups demonstrate close relationships. For example, all fish share a great number of features, but sharks and chimaeras (holocephalians or ratfish) are unique in having claspers and prismatic calcification of their cartilage, which suggests that sharks and chimaeras are more closely related to one another than either are to other fishes.

Hennig coined several terms to emphasize the unique aspects of this approach. The term **apomorphy** refers to a derived or specialized trait; **plesiomorphy** refers to a primitive trait. An **autapomorphy** is a specialization unique to one group, and a **synapomorphy** is a specialized trait shared by two or more groups. A **symplesiomorphy** is a shared primitive trait. All these terms must be defined in relation to a particular taxonomic level. The presence of a hemipenes (a paired, eversible penis) is an autapomorphy of the Squamata, the single group that includes lizards and snakes. It is a synapomorphy of the Lacertilia (lizards) and the Ophidia (snakes). Within either of these groups, the presence of the hemipenes is a primitive character or a symplesiomorphy.

An autapomorphy has no value in establishing relationships with other groups that lack this character; the presence of a hemipenes does not contribute to our understanding of the specific relationship between the Squamata and other groups of reptiles that lack this trait. Nor is the presence of this structure valuable in establishing relationships *among* the various lizard or snake groups, all of which possess it.

The term **character** may be applied to any recognizable attribute of an organism. The total number of characters of any organism may be nearly infinite, but when classifying, we need to consider only the characters that vary among the taxa being studied. The term **character state** refers to the presence or absence of a particular attribute or to a series of alternative ways in which a trait may be expressed. For example, teeth may be present or absent, or their attachment may be to the side of the jaw (pleurodont), to the edge of the jaw (acrodont), or in sockets (thecodont). The term **morphocline** refers to characters that vary quantitatively within a group, for instance, total body size or proportionate length of limbs.

Hennig applied the term **sister groups** to two groups united by the presence of one or more synapomorphic characters. Hennig argued that a system of classification could be based entirely on such pairs of taxonomic groups (Figure 1-2). This depends on the assumption that both of the sister groups will inherit one or more uniquely derived characters from their immediate ancestors. (However, evolution may not always occur according to this pattern. A conservative lineage might give rise to a series of daughter species as a result of geographical separation of small portions of the population. In this case, divergent

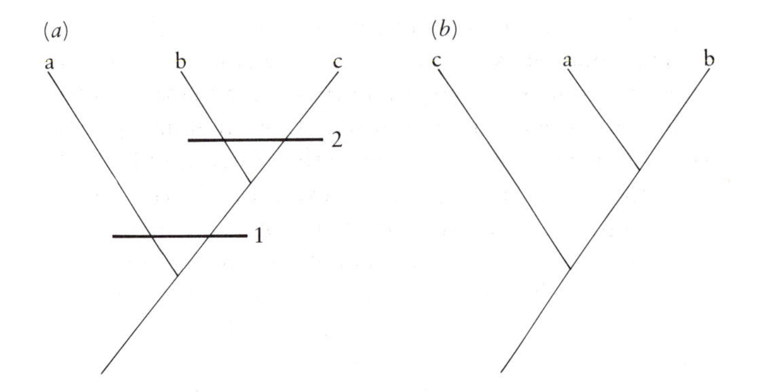

Figure 1-2. FOUR ALTERNATIVE AND MUTUALLY EXCLUSIVE HYPOTHESES OF PHYLOGENETIC RELATIONSHIP AMONG THREE TAXA. In (*a*) a, b, and c belong to a monophyletic group if they all express some derived characters (1), not shared with other taxa. Taxa b and c are considered to be more closely related to each other than either are to a if b and c share other derived characters (2) not found in a. Taxon a is considered to be the sister group of b and c. Taxa b and c are sister groups of each other. (*d*) An unresolved tricotomy in which no unique characters are shared by any two taxa to the exclusion of the third.

species and their descendants will not necessarily share any unique derived features with the conservative lineage. Their affinities can then be judged only on the basis of overall similarities.)

The term sister group can apply to any taxonomic level. Most taxonomists agree that the origin of major groups, like birds or mammals, can ultimately be traced to a particular speciation event. Thus, we may initially apply the concept of sister groups at the level of the species. Higher taxonomic categories, for example, placental and marsupial mammals, may also be considered as sister groups. In other cases, a single species might be identified as the sister group of a larger taxonomic assemblage such as a family or an order.

MONOPHYLY

A major objective of classification is to unite species sharing a common ancestry. Such groups are termed **monophyletic.** It would be desirable if we could demonstrate that all taxonomic groups evolved from individual species, but this demonstration is rarely if ever possible. Monophyletic groups can be identified by the presence of unique, derived characters. All mammals, for instance, can be recognized by the presence of hair, mammary glands, and three ear ossicles, which are not present in any other vertebrates.

Groups are sometimes recognized that are later found to have evolved from two or more distinct ancestors. Such groups are termed **polyphyletic.** The suborder Pinnipedia was long used to refer to both seals (Phocoidea) and the sea lions and walruses (Otarioidea). We now recognize that each group evolved independently from separate families of terrestrial carnivores, rather than having a single common ancestor. Taxonomists try to avoid naming polyphyletic groups, but they cannot always differentiate characters that are uniquely evolved from those that have arisen convergently in two or more groups.

HOMOLOGY AND CONVERGENCE

In order for characters to be useful in establishing monophyletic groups and relationships between groups, we must first demonstrate that the characters are truly comparable. Long before the acceptance of evolutionary theory, scientists recognized that particular elements of the skeleton of all vertebrates are fundamentally similar to one another. This similarity is clearly evident among terrestrial vertebrates. The bones of the forelimb are readily recognizable whether the limb is used for walking (as in mammalian carnivores), grasping (as in primates), flying (as in bats), or swimming (as in secondarily aquatic forms such as whales). **Homology** refers to the fundamental similarity of individual structures that belong to different species within a monophyletic group. Their homologous nature originates from a common ancestral pattern.

The homology of a structure, such as the humerus, can be demonstrated on the basis of several criteria:

1. Position—The configuration of the humerus may change significantly, but its position relative to other structures (the pectoral girdle and the lower limb) remains constant.
2. Development—The embryological tissue from which the humerus develops is the same in all tetrapods, and the pattern of its ontogeny is similar in all groups in which the bone occurs.
3. Evolution—The change in shape of the humerus as revealed by the fossil record can be traced from one group to another with no significant breaks, and the most primitive condition observed in sister groups is the most similar.

Not all homologous characters have such a similar structure and function. Some homologies would be almost impossible to recognize without information from either developmental patterns or the fossil record. One of the most striking cases is provided by the mammalian ear ossicles (see Chapter 17). Only a single ear ossicle, the stapes, is present in reptiles, but three bones are present in mammals. The two additional bones—the malleus and incus—are homologous with bones that form the jaw articulation in reptiles—the articular and quadrate. This homology was first established as a result of observed changes in position and function of these bones in the embryos of modern mammals and, more recently, was demonstrated in a sequence of fossils from the reptilian-mammalian transition.

Not only are some characters that are very different in structure and function homologous, but other characters that appear very similar may be nonhomologous. For example, a structurally similar patella, or knee cap, is present in birds, mammals, and some lizards, yet it was not present in the ancestors of any of these groups and certainly evolved separately in each.

Characters that are similar in structure and function but have arisen separately rather than from a common ancestor are termed **convergent.** Convergence is common in groups that become adapted to a similar habitat or way of life; the similar body shape of whales, ichthyosaurs, and mosasaurs—all of which had different terrestrial ancestors—resulted from secondary adaptation to an aquatic way of life.

A second evolutionary pattern that results in the development of nonhomologous characters, **parallelism,** is frequently recognized. Simpson (1961) distinguished these patterns as follows: Convergence is the development of similar characters separately in two or more lineages without a common ancestry pertinent to the similarity. Parallelism is the development of similar characters separately in two or more lineages of common ancestry on the basis of, or channeled by, characteristics of that ancestry.

These two terms may be useful in characterizing evolutionary patterns, but in practice they are difficult to distinguish. When we attempt to determine specific relationships between groups, parallelism may be considered a special case within the broader category of convergence. **Homoplasy** is frequently used as a synonym for convergence. A major problem in establishing relationships is distinguishing characters that are homologous from those that have arisen as a result of convergence. We can establish directly whether or not a character observed in two groups is homologous only through knowledge of the groups' immediate common ancestor. If this form can be recognized and has this character, it is homologous. If the ancestor can be recognized on the basis of other criteria but lacks this character, we may assume it evolved convergently. If a common ancestor is not known, convergence might be detected from knowledge of the immediate ancestors of either group, but homology cannot be demonstrated in this manner.

The fossil record is the final arbitrator in determining homology for traits that are capable of being preserved, but this excludes almost all aspects of soft anatomy and physiology. Even among groups with a good fossil record, ancestors are frequently not known, and might not be recognized without knowledge from other sources regarding the possible homology of structures.

Several other lines of evidence may be used to distinguish homologous characters from features that have arisen convergently. All are based on probabilities. If structures or processes are complex and similar, it is statistically unlikely that they would have evolved as a result of convergence. The two major groups of vertebrates, the Agnatha (or jawless vertebrates) and the jawed vertebrates, have basically similar patterns of the brain, cranial nerves, and associated sense organs. Even without direct knowledge of the common ancestor of all vertebrates, we may safely assume that it had a comparable pattern. Similarly, complex developmental changes are unlikely to be matched exactly in groups with different ancestral patterns.

A common, rather than multiple, origin of a trait may also be considered more likely if many related forms exhibit the same condition. For example, all placental mammals have a fundamentally similar pattern of reproduction, involving a complex system of endocrine control and many similar features of the placenta. Since these occur consistently in all modern mammalian orders, it seems much more likely that this pattern developed in a common ancestral stock, rather than evolving separately in each group.

Similar lines of reasoning can be used to support the homology of elements known in fossil groups where specific ancestry has not been established. Patterson (1982) advanced an extension of these arguments, suggesting that any relationships established on the basis of other criteria may be used to determine the homologous nature of particular characters. If monophyletic groups reflect common ancestry, then any characters they exhibit are likely to be homologous. This argument is based only on probability and is much less significant in establishing homology than evidence from the fossil record, because it provides no specific information regarding individual characters.

PARSIMONY

All means of establishing homology other than those involving developmental patterns and the fossil record are based on the principle of **parsimony**—that it is more logical to accept a hypothesis that depends on a small number of processes rather than one based on a large number of independent processes. Cladists have applied this principle broadly to judge alternative relationships on the basis of the relative number of derived characters that they share. In general, this principle is logical and to some degree underlies all scientific thinking. However, we may question the degree to which it is applicable to establishing phylogenies. In the case of phylogenetic analysis, the use of parsimony is based on the assumption that most characters evolved only once and that convergence is rare. Surprisingly, supporters of this doctrine have never tested this assumption.

In contrast, biologists working with both modern and extinct groups argue that convergence is very common (Cain, 1982; Carroll 1982). Arguments for the close relationship of groups based only on the common presence of derived features are of little value, if convergence is equally or more common than the unique origin of derived characters.

There is no simple correlation between elapsed time and the amount of evolutionary change, but in general, more changes are likely to have taken place in groups that share only a very distant common ancestry than in those that diverged from a common ancestry more recently. The number of features resulting from convergence will almost inevitably increase during evolutionary history, while the number of characters that they share as a result of their common ancestry cannot increase, and may decrease, if synapomorphies are lost or altered. Hence, arguments based on parsimony are increasingly less likely to be correct the more distant the proposed common ancestor is in time.

Attempts to determine the relationships of groups on the basis of characters exhibited only in their modern representatives may result in phylogenies that are far different from those established on the basis of the fossil record. Two striking examples are provided by the recent works of Gardiner (1982) and Rosen et al. (1981).

When attempting to establish relationships of any group with a fossil record, we must emphasize the earliest known members, because they have had the shortest amount of time to evolve new characters since their initial divergence. Hence, they should provide us with the best opportunity to identify the derived features that they share with their closest sister group. Characters that evolve within a group, rather than being present in its most primitive members, have no value in establishing its relationship with a possible sister group.

POLARITY

If we emphasize the use of derived, rather than primitive, characters in establishing relationships, we must devise a means for determining the direction of evolutionary change of those characters. Within a series of related species, the number of teeth may differ widely. Is the number of teeth increasing or decreasing? Is a small or a large number of teeth likely to be the primitive condition? The direction of evolutionary change is referred to as **polarity.**

Determining the direction of the polarity of character states or morphoclines is very important to all evolutionary biologists, and many authors have written on this subject in recent years (Kluge, 1977; Patterson, 1981; Wiley, 1981).

STRATIGRAPHIC SEQUENCE

The most obvious way a paleontologist can determine polarity is by the order in which characters appear in the fossil record. If the fossil record is relatively complete or at least representative of the actual pattern of evolution within a group, the direction of evolutionary change should be evident from the sequence of fossils. Paleozoic sharks are in general more primitive than the living shark groups and among mammals, the character states exhibited by Paleocene genera are consistently more primitive than those of animals from the later Tertiary. Butler (1982) recently commented on the question of polarity as applied to the configuration of mammalian molars:

> Cope (1883) found that in the oldest Tertiary mammalian fauna, from the Lower Palaeocene, ... 38 out of 41 species had upper molars with three principal cusps, and 35 of these were triangular, whereas many Eocene and later mammals had quadrate molars with four cusps. In this way he was able to recognize the tritubercular type as the ancestral form of upper molar.

On the other hand, paleontologists have noted the difficulty of relying exclusively on the fossil record to determine the sequence of anatomical changes (Schaeffer, Hecht, and Eldredge, 1972), and cladists in general have been suspicious of the fossil record and minimize its importance in establishing polarity (Hennig, 1981; Patterson, 1981; Wiley, 1981).

The degree of completeness of the fossil record differs widely from group to group and from one horizon to another in the history of any particular group. Because many species and genera have relatively long time spans, we cannot assume that the earliest appearance of a particular character state indicates that it is the most primitive. However, in groups with an extensive fossil record, it is probable that the most primitive character states will be expressed in the older fossils and for derived characters to be more common in later forms. As the fossil record improves, it will better reflect the actual sequence of evolution (Fortey and Jefferies, 1982). Despite the difficulties of interpreting the fossil record, it is the final arbitrator in establishing the antiquity of groups and the distribution of character states (Paul, 1982).

However, determining the direction of polarity from the fossil record requires some degree of prior knowledge of the relationships of the groups being investigated. A particular character or character state may have evolved at different times in different lineages. The early appearance of a character in one group does not necessarily indicate that it represents the ancestral condition in another group that appears later in the fossil record. The earliest known vertebrates have an extensive external carapace, but this does not necessarily indicate that the presence of a carapace was also primitive for sharklike fish, which appear at a later time in the fossil record. This problem can be minimized if the monophyletic nature of groups can be established first on the basis of other characters. However, in practice most significant taxonomic problems involve assemblages in which both their monophyletic nature and the polarity of important characters are in question.

ONTOGENY

Since the recognition of the evolutionary relationship of organisms, it has been evident that the process of embryological development generally parallels phylogeny. Terrestrial vertebrates go through a stage in which they have fishlike gill slits, and mammals go through stages in which elements of their circulatory and excretory systems resemble those seen in the early development of fish, amphibians, and reptiles before they achieve the definitive mammalian pattern. Haeckel (1866) summarized this phenomenon with the expression "Ontogeny recapitulates phylogeny."

In some cases, this phenomenon enables us to determine the relationships of an animal on the basis of early growth stages, when the adult is so specialized that its affinities are not readily recognized. The relationship of tunicates (urochordates) with vertebrates would be almost impossible to guess on the basis of the adult structure, but the larvae exhibit a notochord and dorsal hollow nerve cord that demonstrate their affinities with other chordates.

Gould (1977) reviewed the relationships between ontogeny and phylogeny, pointing out problems in a literal interpretation of Haeckel's phrase while discussing other aspects of development that may be significant in understanding phylogenetic change.

Fink (1982) stressed the importance of ontogenetic change as a means of establishing the direction of evolution. The study of ontogeny is not limited to living species, because changes of the skeleton during growth are frequently seen in fossils. Among early amphibians and reptiles, the pattern of fusion of tarsal elements, that can be followed ontogenetically, closely parallels the same process when viewed phylogenetically and provides convincing evidence of the homology of elements in the two classes.

The process of neoteny or paedomorphosis, in which juvenile characters are retained in sexually mature individuals, results in an altered ontogenetic sequence relative to that of related groups, but such changes can usually be recognized by the polarity of other traits.

OUTGROUP COMPARISON

Cladists generally stress outgroup comparison as one of the most important single means of establishing the polarity of traits. This method is usually discussed in terms of contemporary, living groups, but it is equally or even more convincingly applied to groups with a fossil record. Within a particular group, characters or character states are judged to be derived if they do not appear in other, closely related groups (the outgroups) and primitive if they do. The presence of hair in mammals is clearly a derived character, since it is not possessed by any other vertebrate class. The presence of marsupial bones in both marsupials and monotremes suggests that this feature may be primitive for mammals; the absence of these bones is a specialization of placentals.

To establish polarity, the ideal outgroup would be the ancestors of the group being studied. The polarity of character change among early amphibians can be established on the basis of the anatomy of their putative ancestors, the rhipidistian fish. On the other hand, the specific nature of the interrelationships among the various fossil hominids is difficult to assess because we lack knowledge of their immediate ancestors from the Pliocene and thus have difficulty determining which features of the various australopithecine species are likely to be primitive or specialized relative to those of our own genus, *Homo*.

CHARACTER ANALYSIS WITHIN GROUPS

It has been suggested that a trait that is rarely found within a group is probably derived, rather than primitive. A high degree of aquatic specialization among the mosasaurs is certainly a specialization among lizards. On the other hand, only two living mammals lay eggs, yet the echidna and platypus are unquestionably regarded as primitive mammals and the egg-laying trait is almost certainly a primitive feature inherited from their reptilian ancestors. On its own, rarity is not a good criterion for judging whether a character is primitive or advanced.

Among the modern lizard infraorders, the presence of amphicoelous vertebrae is rare—encountered only in some genera of the Gekkota. To judge by the otherwise specialized nature of the genera with this character, it is probably derived relative to the procoelous condition of other modern lizard groups. On the other hand, amphicoelous vertebrae are also present in the most closely related living outgroups of lizards, the genus *Sphenodon*, and the extinct eosuchians. In addition, the most primitive fossil lizards also have amphicoelous centra that are nearly identical with those of the modern Gekkota. Thus, the amphicoelous condition is both a primitive and an advanced character among lizards. Only knowledge of the remaining aspects of the anatomy and phylogenetic relationships of the groups involved enable us to determine the polarity of this character in each instance.

CORRESPONDENCE WITH OTHER CHARACTERS

It is tempting to assume that a particular character found in a species that otherwise displays a large number of primitive character traits is primitive as well. There are many examples where this assumption is not correct, however. For instance, the great anterior extent of the notochord in amphioxus is unlikely to be a primitive condition relative to that of vertebrates, although many other features of amphioxus almost certainly are.

ANATOMICAL PROGRESSION

Even without evidence from the fossil record, ontogeny, or outgroup comparison, certain structural morphoclines usually proceed in only one direction. Once lost, limbs are rarely, if ever, regained. Loss of digits and loss or fusion of carpals and tarsals are much more common than gain, although polydactyly and polyphalangy are common derived features in aquatic groups. The presence of many sacral ribs is generally derived relative to the presence of only one pair, but the reduction of sacral ribs may occur in groups that are secondarily adapted to an aquatic way of life.

No single method of establishing polarity is dependable in all cases, and confirmation should be sought through the use of alternative evidence.

CLADOGRAMS

The evolutionary history of vertebrate groups has long been represented by phylogenetic trees. The horizontal axis represents the relative amount of anatomical divergence between taxa, and the vertical axis shows their temporal distribution. Phylogenetic trees may reflect a broad range of anatomical, behavioral, and physiological evidence, but they rarely include reference to the specific traits used to establish relationships.

Emphasis on the use of derived characters has led to the development of a different type of diagram, the **cladogram,** to represent relationships. A cladogram specifically shows the derived characters used to recognize the monophyletic nature of each taxonomic group (Figure 1-3). A cladogram is frequently presented as a hypothesis of possible relationships that can be tested and either corroborated or refuted on the basis of additional data. The sequence of branching in a cladogram reflects the relative times of the divergence of each group, but unlike phylogenetic trees, cladograms are not associated with an absolute time scale.

Vertebrate paleontologists are using cladograms with increasing frequency. Their advantage lies in the specific presentation of all data that form the basis for postulated relationships. Clearly established relationships are indicated by simple dichotomies. Where three or more lineages appear to radiate from a single point, there is not yet sufficient information available to establish specific sister-group relationships. Eventually, it may be possible to represent all relationships among vertebrates by cladograms. However, at present, the fossil record is still too incomplete to do this in a systematic manner.

The most difficult taxonomic problems involve determining specific interrelationships among the many separate lineages that evolved in the early stages of major adaptive radiations. Most of the major groups of vertebrates appear to have radiated very rapidly, giving rise within 10 to 20 million years to many different lineages, each of which is characterized by conspicuous specialized or apomorphic characters in addition to the synapomorphies by which the larger group is recognized. This pattern is particularly evident in the radiation of the placoderms in the late Siliurian and early Devonian, the Chondrichthyes during the Devonian, primitive tetrapods in the late Devonian and early Carboniferous, the amniotes in the late Carboniferous, the archosaurs in the late Permian, the dinosaurs in the late Triassic, and the teleosts, therian mammals, and birds in the late Cretaceous and early Tertiary.

It has been difficult to recognize shared derived characters that would allow us to determine specific sister-group relationships within these radiations. As the fossil record has become better known, some interrelationships have been established, but it is still not possible to provide well-documented cladograms to demonstrate the specific sequence of dichotomous branching that occurred during the origin of these groups.

We have a great deal to learn about the polarity and homology of traits in the primitive members of most major vertebrate groups. Until these facts are established, it is very difficult to draft cladograms that do not give the appearance of being better documented than they actually are.

One of the most conspicuous aspects of cladograms is the hierarchical arrangement of monophyletic groups. Many authors who make use of cladistic methodology urge that systems of classification should follow a similar, hierarchical pattern, with the generally accepted convention that sister groups be given equivalent taxonomic rank. Hence, among vertebrates, the Agnatha would have the same rank as all other vertebrates, the placoderms should have the same rank as all other gnathostomes, and the crocodiles would have the same rank as birds. All tetrapods would be included within a taxonomic rank much below that of the few living members of the Agnatha.

This system's advantage is that taxonomic rank would directly reflect the relative time of origin of all taxonomic groups and their specific position in a hierarchical system. However, it would require a potentially enormous number of taxonomic ranks and the introduction of a plethora of unfamiliar group names (Panchen, 1982).

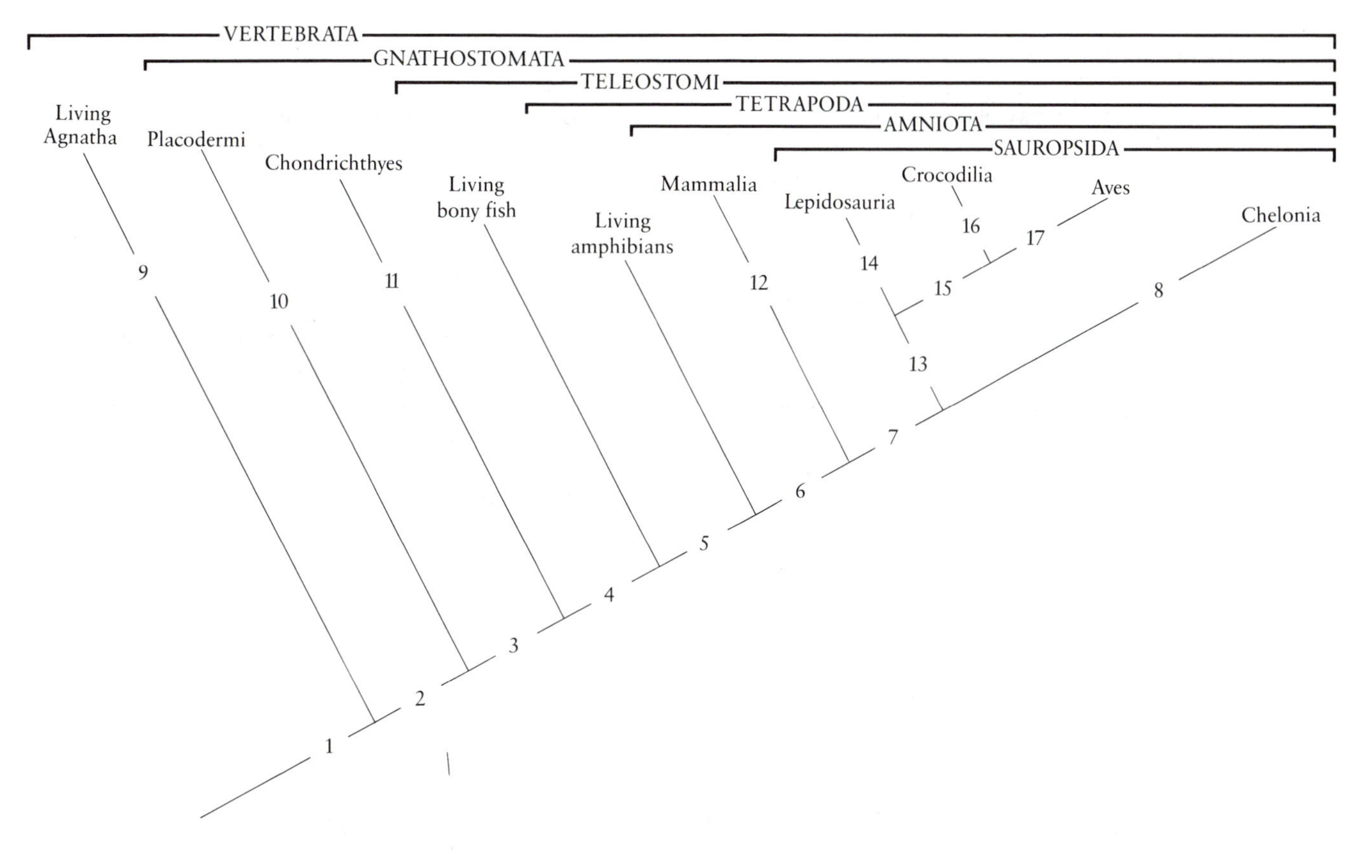

Figure 1-3. A CLADOGRAM SHOWING A NESTED SERIES OF MONOPHYLETIC GROUPS AMONG THE VERTEBRATES. The derived characters that define the nodes of the cladogram and characterize each group are as follows: 1. Brain, specialized paired sense organs, ability to form bone. 2. Jaws, gill filaments lateral to gill supports. 3. Regular tooth replacement, palatoquadrate medial to adductor jaw musculature. 4. Bone a regular constituent of the endochondral skeleton. Swimbladder or lung. 5. Paired limbs used for terrestrial locomotion. 6. Extraembryonic membranes, allantois, chorion, and amnion, direct development without an aquatic larval stage. 7. Loss of medial centrale of pes. 8. Loss of ectopterygoid bone, closely integrated dorsal carapace and ventral plastron. 9. Gill filaments medial to gill supports. 10. Palatoquadrate lateral to adductor jaw musculature. 11. Prismatic cartilage. 12. Fur, mammary glands. 13. Dorsal and lateral temporal openings, suborbital fenestra. 14. Large sternum on which the scapulocoracoid rotates. 15. Hooked fifth metatarsal, foot directed forward for much of stride. 16. An akinetic skull with extensive pneumatization, elongate coracoid. 17. Feathers. Neither osteichthyes nor modern amphibians can be defined on the basis of unique derived characters.

As Charig (1982) emphasized, phylogenetic trees and cladograms provide different kinds of information than do Linnean systematics and there is no particular value in attempting to combine the two into a single, extremely complex system of formal classification. Linnean classification provides a relatively simple way of indicating the taxa included in each major group. Phylogenetic trees or cladograms indicate the nature of the interrelationships among the included taxa.

Another problem with attempting to derive a formal system of classification from a cladogram is that any change in the classification of the most primitive groups alters the systematic position of all other taxa. The earliest points of divergence are typically the most poorly represented in the fossil record, and the living members of these groups have had the longest period of evolution during which they may have lost or modified primitive characters. Hence, the portions of the cladograms most vital in establishing a system of classification are the most difficult to document. Therefore, the classification used in this text will follow the familiar Linnean pattern.

Paraphyly and Holophyly

The great emphasis on the use of derived rather than primitive features in establishing relationships led Hennig to modify the definition of the term monophyly in a way that has been accepted by all subsequent workers using cladistic methodology. All taxonomists agree that monophyletic groups are defined on the basis of having a common ancestry. Hennig further specified that a monophyletic group should include an ancestral species and all its descendants. This definition appears logical, but it differs from that used by most earlier taxonomists who did not specify the inclusion of all descendants. Hennig coined the term **paraphyletic** for groups that have a common ancestry, but from which one or more descendant groups have been excluded. Cladists in general discourage the recognition of paraphyletic groups. Patterson (1981, 1982) refers to them as unnatural and no more than an artifact of taxonomists.

Earlier definitions of monophyly include both paraphyletic and monophyletic groups (of cladistic usage). To alleviate the problem of having two definitions for the same term, Ashlock (1971) coined the term **holophyly** for Hennig's use of the term monophyly. Unfortunately, the term holophyly has not achieved general usage, and most cladists continue to use the term monophyly in reference to groups that are not paraphyletic. In this text the term monophyletic will be used to refer to the origin of groups without consideration of their descendants.

Paraphyletic groups cannot be defined in a strictly cladistic manner because they do not possess uniquely derived characters. All the derived characters that they exhibit are shared with their descendants. The class Reptilia is frequently cited as a prime example of a paraphyletic group. Reptiles, birds, and mammals together make up a holophyletic group, the Amniota, but reptiles can only be defined on the basis of the absence of features that characterize birds and mammals.

The classes Amphibia, Osteichthyes, and Agnatha (including fossil members) are also paraphyletic groups. The fossil record of the earliest members of these classes are too poorly known to establish how they might be assigned to holophyletic groups. This is a particularly difficult problem in the case of the early tetrapods.

The problem of paraphyletic groups may be analyzed more closely at other taxonomic levels, beginning with species. Early papers by Hennig (1966) and other cladists (e.g., Bonde, 1975) introduced the convention that all speciation events should be considered as resulting in the extinction of the parental species. This convention follows logically from the prohibition of paraphyletic groups, since the ancestral species could only be recognized as a monophyletic group if it included all of its descendants. If one daughter lineage is recognized as a separate species, the other would have to be considered distinct as well. Wiley (1981) recognized that this procedure cannot be justified biologically. As Mayr (1963) and others have emphasized,

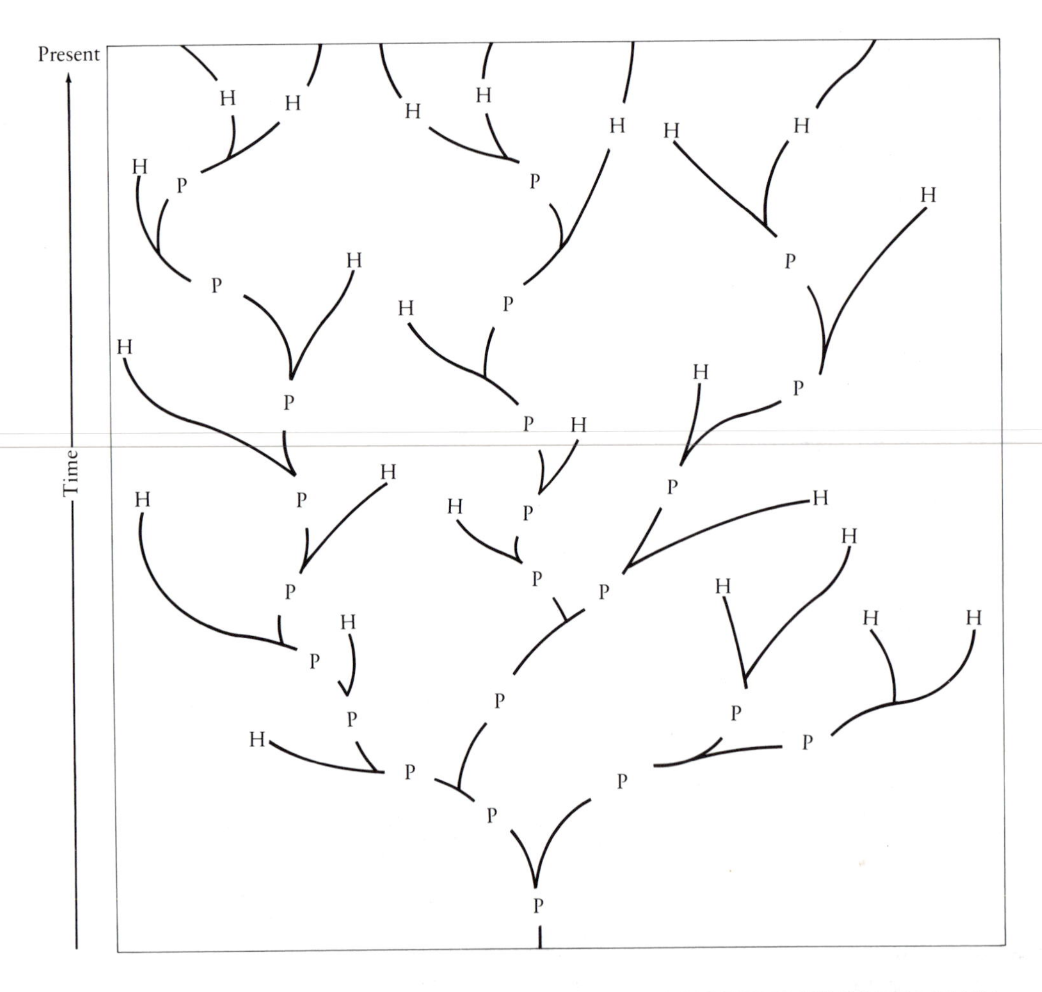

Figure 1-4. PATTERN OF SPECIATION WITHIN A GROUP, SHOWING THE PREVALENCE OF PARAPHYLETIC TAXA. Each dichotomy represents a speciation event. Only those species that are living today or have become extinct without issue are holophyletic. The remainder are paraphyletic. P, paraphyletic species. H, holophyletic species.

most and probably all speciation events among vertebrates occur subsequent to geographical separation of the parental population. This separation may result in a relatively equal subdivision of daughter populations, but the division is usually unequal, resulting in the parent population occupying most of its original territory and one daughter population having a much smaller range and population size. According to most current thinking on speciation (Dobzhansky et al., 1977), the smaller population is more likely to undergo rapid evolutionary change and the larger population may continue essentially unaltered. We cannot justify the assumption that the parental species becomes extinct. An extreme case can be envisioned in which a few individuals (perhaps a single pregnant female) become isolated from the rest of the species. The parental population is no more affected if she founds a new species than if she dies without issue.

If we generate a random pattern of branching—comparable to a series of speciation events—combined with extinction, we may conclude that approximately half of the species that have ever existed are paraphyletic (Figure 1-4). The only species that are not paraphyletic are those living today and those that became extinct without issue. There is no inherent difference between species that eventually become extinct, those that undergo subsequent speciation, and those living at the present time. We cannot ignore paraphyletic species or treat them in a different manner taxonomically from those that are holophyletic.

The problem of paraphyly is most contentious in the case of taxa that have been recognized as including the ancestors of major groups. Rhipidistian fish are thought to include the ancestors of tetrapods, the therapsids include the ancestors of mammals, and the theropod dinosaurs almost certainly include the ancestors of birds. By definition, rhipidistians, therapsids, and theropod dinosaurs are all paraphyletic groups. But does this require that they be classified differently than groups that are living today or that have become extinct without issue?

Presumably, the ancestry of tetrapods, mammals, and birds may each be traced to a single lineage that diverged from a single species within the ancestral group. Were it not for this speciation event, the parental group would be holophyletic and defined by derived characters alone. There is no reason to believe that the ancestral group was otherwise altered by this speciation event. Rhipidistians, therapsids, and theropod dinosaurs all survived the divergence of their descendant groups; it is not practical to subdivide them into a holophyletic assemblage that existed after the dichotomy and a paraphyletic taxon that existed before.

The problem of classifying these particular paraphyletic groups might be diminished by including them within the taxon of their holophyletic descendants, but the problem is only extended to the next earlier taxonomic group. If theropod dinosaurs are classified with birds, than the more primitive saurischian dinosaurs that gave rise to the theropods become a paraphyletic group. If they in turn are included with the birds, other, more primitive archosaurs become paraphyletic, ad infinitum (Figure 1-5).

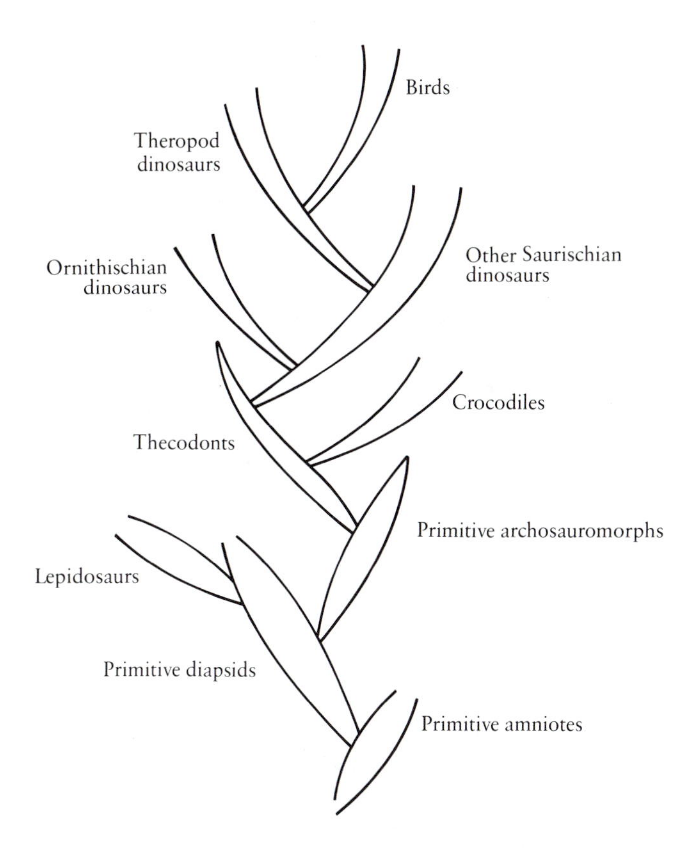

Figure 1-5. SUCCESSION OF PARAPHYLETIC GROUPS LEADING FROM PRIMITIVE AMNIOTES TO BIRDS.

The existence of paraphyletic groups is an inevitable result of the process of evolution. Their existence requires that we define them in terms of the absence of the characters of their descendants as well as the presence of characters absent in their ancestors.

SUMMARY

The history of vertebrate evolution is revealed by fossils buried in sediments that have accumulated during the past 500 million years. Most fossils consist of bones infiltrated by suspended sediments and dissolved minerals. Footprints, coprolites, endocasts, and, occasionally, soft parts preserved as impressions or as the result of replacement by pyrite or silica provide further evidence of the structure and behavior of extinct vertebrates.

Geological processes strongly influence the relative likelihood of preservation of animals from different environments. Shallow margins of the ocean and other large bodies of water, deltas, and coastal planes receive copious sediments and are subject to subsequent tectonic activity. These processes contribute to a rich fossil record. In contrast, the oceanic depths and terrestrial environments beyond the margins of large water bodies provide few vertebrate fossils. The relative position of the continents, the extent of shallow inland seas, and the amount of tectonic activity have varied throughout geological history and influence the probability of preservation and recovery of fossils from different environments during different periods of time. The fossil record has the potential for documenting the rate and pattern of evolution, but its irregularities have made these data difficult to interpret.

Hennig emphasized that relationships are most effectively demonstrated on the basis of shared derived characters. Only strictly homologous characters are significant in establishing relationships. Homology is established with certainty only by the discovery of comparable structures in the common ancestor of the groups in question. The longer the period of time since the divergence of two groups, the more likely that similar characters have evolved separately, rather than from a common ancestor. The most certain relationships are based on knowledge of the earliest fossils of the groups in question.

The use of derived characters in establishing relationships requires prior determination of the direction of evolutionary change, or polarity. Polarity can be established on the basis of the fossil record, ontogenetic sequence, and by outgroups comparison.

Cladograms represent relationships established on the basis of shared derived characters. Eventually, we may be able to represent all affinities within vertebrates in this manner, but as yet the fossil record of most major radiations is too incomplete to establish the specific sequence of successive dichotomous branchings.

Hennig and other cladists apply the term monophyletic to groups that have a common ancestry and include all their descendants. They use the term paraphyletic for groups, like reptiles, that have a common ancestry but do not include all their descendants. Approximately half the recognized taxonomic groups, ranging from species to classes, are paraphyletic. Paraphyletic groups can be defined on the basis of derived features not present in their ancestors and the absence of derived features present in their descendants.

REFERENCES

Ashlock, P. H. (1971). Monophyly and associated terms. *Syst. Zool.*, **20**: 63–69.

Behrensmeyer, A. K., and Hill, A. P. (eds.). (1980). Fossils in the making: Vertebrate taphonomy and paleontology. In *Prehistoric Archeology and Ecology Series, July 1979, Burg Wartenstein, Austria.* University of Chicago Press, Chicago.

Bonde, N. (1975). Origin of "Higher Groups": Viewpoints of phylogenetic systematics. *Coll. Intl. C.N.R.S.*, **218**: 293–324.

Butler, P. M. (1982). Directions of evolution in the mammalian dentition. In K. A. Joysey and A. E. Friday (eds.), *Problems of Phylogenetic Reconstruction. Systematics Assoc. Spec. Vol.*, **21**: 235–244.

Cain, A. J. (1982). On homologies and convergence. In K. A. Joysey and A. E. Friday (eds.), *Problems of Phylogenetic Reconstruction. Systematics Assoc. Spec. Vol.*, **21**: 1–19.

Carroll, R. L. (1982). Early evolution of reptiles. *Ann. Rev. Ecol. Syst.*, **13**: 87–109.

Carroll, R. L., Belt, E. S., Dineley, D. L., Baird, D., and McGregor, D. C. (1972). Vertebrate paleontology of Eastern Canada. *Intl. Geol. Congr., Excursion 59, Guidebook*, 24th International Geological Congress, Ottawa, p. 113.

Charig, A. J. (1982). Systematics in biology: A fundamental comparison of some major schools of thought. In K. A. Joysey and A. E. Friday (eds.), *Problems of Phylogenetic Reconstruction. Systematics Assoc. Spec. Vol.*, **21**: 363–440.

Cope, E. D. (1883). Note on the trituberculate type of superior molar and the origin of the quadrituberculate. *Am. Nat.*, **17**: 407–408.

Currie, P. (1981). *Hovasaurus boulei*, an aquatic eosuchian from the Upper Permian of Madagascar. *Palaeont. afr.*, **24**: 99–168.

Dobzhansky, Th., Ayala, F. J., Stebbins, G. L., and Valentine, J. W. (1977). *Evolution.* W. H. Freeman, New York.

Eldredge, N., and Cracraft, J. (1980). *Phylogenetic Patterns and the Evolutionary Process.* Columbia University Press, New York.

Eldredge, N., and Gould, S. J. (1972). Punctuated equilibria: An alternative to phyletic gradualism. In T. J. M. Schopf (ed.), *Models in Paleobiology*, pp. 82–115. Freeman, Cooper and Co., San Francisco.

Fink, W. L. (1982). The conceptual relationship between ontogeny and phylogeny. *Paleobiology*, **8**(3): 254–264.

Fortey, R. A., and Jefferies, R. P. S. (1982). Fossils and phylogeny—a compromise approach. In K. A. Joysey and A. E. Friday (eds.), *Problems of Phylogenetic Reconstruction. Systematics Assoc. Spec. Vol.*, **21**: 197–234.

Gaffney, E. S. (1979). An introduction to the logic of phylogeny reconstruction. In J. Cracraft and N. Eldredge (eds.), *Phylogenetic Analysis and Paleontology*, pp. 79–111. Columbia University Press, New York.

Gardiner, B. G. (1982). Tetrapod classification. *Zool. J. Linn. Soc.*, **74**(3): 207–232.

Gould, S. J. (1977). *Ontogeny and Phylogeny.* Harvard University Press, Cambridge.

Gould, S. J. (1985). The paradox of the first tier: An agenda for paleobiology. *Paleobiology*, **11**(1): 2–12.

Gould, S. J., and Eldredge, N. (1977). Punctuated equilibria: The tempo and mode of evolution reconsidered. *Paleobiology*, **3**: 115–151.

Haeckel, E. (1866). *Generelle Morphologie der Organismen. II.* Georg Reiner, Berlin.

Harland, W. B., Cox, A. V., Llewellyn, P. G., Pickton, C. A. G., Smith, A. G., and Walters, R. (1982). *A Geologic Time Scale.* Cambridge University Press, Cambridge.

Hennig, W. (1965). Phylogenetic systematics. *Ann. Rev. Entomol.*, **10**: 97–116.

Hennig, W. (1966). *Phylogenetic Systematics.* University of Illinois Press, Urbana.

Hennig, W. (1981). *The Phylogeny of Insects.* Pitman Press, Bath.
Hook, R. W., and Ferm, J. C. (1985). A depositional model for the Linton tetrapod assemblage (Westphalian D, Upper Carboniferous) and its palaeoenvironmental significance. *Phil. Trans. Roy. Soc. Lond. B,* **311**: 101–109.
Jepsen, G. L., Mayr, E., and Simpson, G. G. (eds.). (1949). *Genetics, Paleontology, and Evolution.* Princeton University Press, Princeton, N.J.
Jerison, H. J. (1973). *Evolution of the Brain and Intelligence.* Academic Press, New York.
Kluge, A. G. (1977). *Chordate Structure and Function.* (2d ed.). Chapter 1. Macmillan, New York.
Mayr, E. (1963). *Animal Species and Evolution.* Belknap Press of Harvard University Press, Cambridge.
Mayr, E., and Provine, W. B. (eds.). (1980). *The Evolutionary Synthesis. Perspectives on the Unification of Biology.* Harvard University Press, Cambridge.
Panchen, A. L. (1982). The use of parsimony in testing phylogenetic hypotheses. *Zool. J. Linn. Soc.,* **74**(3): 305–328.
Patterson, C. (1981). Significance of fossils in determining evolutionary relationships. *Ann. Rev. Ecol. Syst.,* **12**: 195–223.
Patterson, C. (1982). Morphological characters and homology. In K. A. Joysey and A. E. Friday (eds.), *Problems of Phylogenetic Reconstruction. Systematics Assoc. Spec. Vol.,* **21**: 21–74.
Paul, C. R. C. (1982). The adequacy of the fossil record. In K. A. Joysey and A. E. Friday (eds.), *Problems of Phylogenetic Reconstruction. Systematics Assoc. Spec. Vol.,* **21**: 75–117.
Rixon, A. E. (1976). *Fossil Animal Remains. Their Preparation and Conservation.* Athlone Press, University of London.
Robinson, P. L. (1962). Gliding lizards from the Upper Keuper of Great Britain. *Proc. Geol. Soc. Lond.,* **1601**: 137–146.
Rosen, D. E., Forey, P. L., Gardiner, B. G., and Patterson, C. (1981). Lungfishes, tetrapods, paleontology and plesiomorphy. *Bull. Am. Mus. Nat. Hist.,* **167**: 159–276.
Schaeffer, B., Hecht, M. K., and Eldredge, N. (1972). Paleontology and phylogeny. *Evol. Biol.,* **6**: 31–46.
Simpson, G. G. (1944). *Tempo and Mode in Evolution.* Columbia University Press, New York.
Simpson, G. G. (1953). *The Major Features of Evolution.* Columbia University Press, New York.
Simpson, G. G. (1961). *Principles of Animal Taxonomy.* Columbia University Press, New York.
Stanley, S. M. (1979). *Macroevolution: Pattern and Process.* W. H. Freeman, New York.
Stanley, S. M. (1982). Macroevolution and the fossil record. *Evolution,* **36**: 460–473.
Wiley, E. O. (1981). *Phylogenetics. The Theory and Practice of Phylogenetic Systems.* Wiley, New York.

CHAPTER II

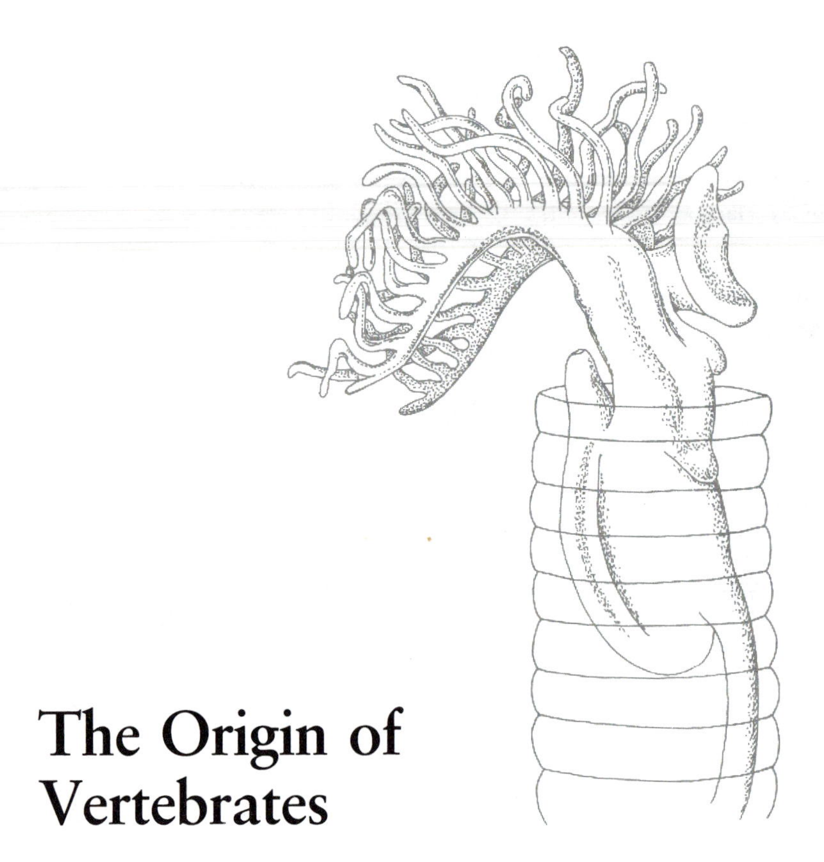

The Origin of Vertebrates

Figure 1-1 shows the broad pattern of vertebrate evolution. We recognize eight classes of vertebrates on the basis of distinct anatomical and physiological characteristics. Each appears to have a monophyletic origin. In cladistic terminology, placoderms, Chondrichthyes, birds, and mammals are holophyletic, while the Agnatha, Osteichthyes, Amphibia, and Reptilia are paraphyletic.

Various groupings of these classes are recognized. The Agnatha, Placodermi, Chondrichthyes, and Osteichthyes may be referred to informally as fish. The term *Gnathostome* distinguishes all the jawed vertebrates from the Agnatha. The term *Tetrapoda* recognizes the common origin of the terrestrial classes; the *Amniota* includes reptiles, birds, and mammals.

We know very little about the early stages in the evolution of vertebrates. Both the earliest vertebrates and their immediate ancestors were almost certainly soft-bodied animals that were nearly incapable of fossilization.

The early vertebrates were capable of being preserved only after the development of bone, scales, or teeth. The capacity to deposit bone apparently evolved considerably after the initial radiation of vertebrates. The earliest known members of the major groups are so different from one another that their interrelationships are difficult to establish. Each group exhibits a mosaic of primitive and derived features, and none are close to the anatomical pattern that would be expected of primitive ancestors of the other groups.

THE BODY PLAN OF PRIMITIVE VERTEBRATES

Because of the absence of relevant fossils, we must establish the relationship of vertebrates to the other major groups of multicellular animals largely on the basis of living forms. We will begin with a brief description of an idealized primitive vertebrate. This model can serve simultaneously as a basis for considering the origin of this group and as a starting point for the derivation of more advanced vertebrates. We can deduce the probable pattern of the earliest vertebrates on the basis of features common to primitive members of the living fish groups and evidence from the skeletal anatomy of Paleozoic fossils.

We may picture the early vertebrates as small animals, having a generally fishlike appearance, with a fusiform body and the head closely integrated with the trunk (Figure 2-1). The most conspicuous structural features are associated with active swimming. A series of segmental muscles, the myotomes, extend along both sides of the body. Their fibers run longitudinally between transverse septae. These septae are anchored medially to the notochord, a supporting rod that runs the length of the body. It is capable of lateral flection but resists longitudinal compression and so bends from side to side with contraction of the myotomes.

Swimming results from waves of muscle contraction passing posteriorly down the trunk on alternate sides of the body. The most primitive vertebrates lacked paired fins, making it difficult to control movements in the vertical plane, but they may have had fleshy dorsal and caudal fins.

The dorsal hollow nerve cord, which runs the length of the body dorsal to the notochord, controls and integrates movements. Segmental ventral motor nerves extend to the muscle blocks; dorsal sensory nerves extend laterally between the muscles. Anteriorly, the nerve cord expands as a brain, associated with specialized paired sensory structures for smell, sight, and balance. Lateral line organs, extending over the surface of the body, detect changes in pressure produced by movement through the water.

Primitive members of most modern fish groups are also sensitive to electrical currents, which are detected by organs associated with the lateral line system (Bodznick and Northcutt, 1981). Sharks and other fish use these organs to feed in dark waters and beneath the mud by detecting the prey's electrical activity resulting from nerve transmissions and muscle contractions (Boord and Campbell, 1977). Structures with a similar shape and, one presumes, a comparable function are common in many groups of early fossil vertebrates (Thomson, 1977), which indicates that electrical sensitivity was a primitive attribute of the group.

In early developmental stages of all vertebrates, the pharynx is pierced laterally by a series of gill slits. In all living fish groups, they are associated with thin, lamellar gills that provide the main surface for gas exchange. Contraction of the muscles of the mouth and pharynx enable fish to feed and breathe. Jaws were absent in the most primitive vertebrates, but they may have fed actively on small prey that could be engulfed whole (Jollie, 1982; Northcutt and Gans, 1983).

The circulatory system of all aquatic vertebrates includes a ventral heart that pumps the blood anteriorly and dorsally through a series of aortic arches running

(a)
Myosepta
Myotomes
Eye
Nostril
Pharyngeal slits
Mouth

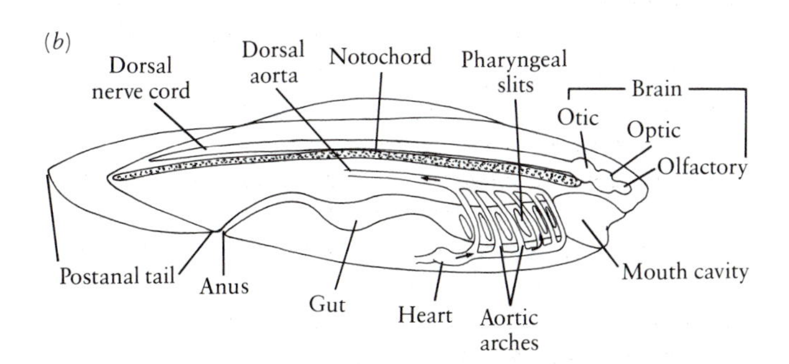

Figure 2-1. SCHEMATIC VIEWS OF AN ANCESTRAL VERTEBRATE. (*a*) External view showing conspicuous >-shaped myotomes separated by myosepta, pharyngeal slits, and paired sensory structures of the head. The mouth is not supported by jaws and there are no paired fins. Filimentous gills were probably absent. The estimated size is 10 centimeters or less. (*b*) Sagittal section, showing presence of brain with expanded olfactory, optic and otic areas, notochord, elements of the circulatory system associated with the pharynx, and digestive tract.

between the gill slits. Blood flows posteriorly through the major dorsal vessel. Capillary beds are present in the gills, and in association with the digestive system, muscles and kidneys, for maximum efficiency in exchanging respiratory gases, products of digestion, and metabolic wastes between the animal and its environment.

The early developmental stages of the excretory system in living vertebrates may reflect the primitive condition for the group (Romer and Parsons, 1977). The kidney tubules, or nephrons, are anteriorly located and segmentally arranged. They drain the coelomic fluids and may or may not be associated with a tuft of capillary tissue. They function both to control the concentration of body fluids and to remove metabolic wastes. Ducts that drain the tubules extend posteriorly to the cloaca. In the primitive state, they are not associated with the gonads, although in more advanced vertebrates they are associated with the transport of sperm and eggs.

The sexes in primitive vertebrates were almost certainly separate. The paired gonads, as in the lamprey, probably expelled both the sperm and eggs into the coelom, from which they emptied into the cloacal area via genital pores. Northcutt and Gans (1983) argue that primitive vertebrates may have had a distinct larval stage like some members of modern fish groups, with a planktonic filter feeding habit in contrast with the more active, predatory adult.

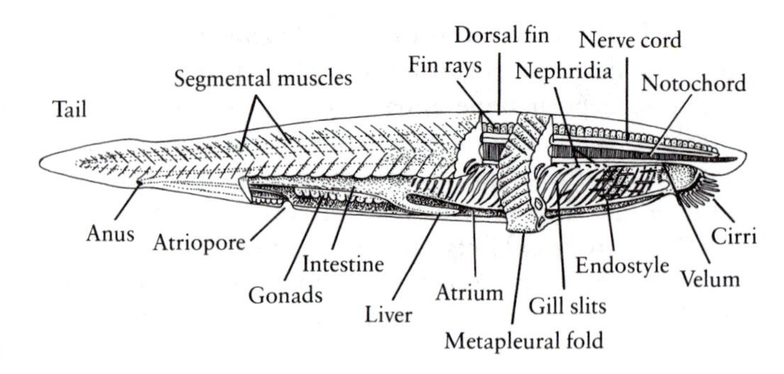

Figure 2-2. THE CEPHALOCHORDATE *BRANCHIOSTOMA* (AMPHIOXUS). This living animal provides an excellent model for the ancestors of vertebrates that lived approximately 600 million years ago. It is clearly more primitive than vertebrates because of the absence of a brain and paired sensory organs of the head. The nephridea are paired anterior excretory structures unlike those of any known vertebrates. Natural size was about 5 centimeters. *From Storer and Usinger, 1965. Reprinted with permission of McGraw-Hill Book Company.*

INVERTEBRATE CHORDATES

We can recognize vertebrate remains as early as the Ordovician on the basis of evidence of a brain, paired sensory organs, and bone. As far as we can determine from Paleozoic fossils, most of the organ systems of early vertebrates already approach the pattern seen in living species. The origin of these systems logically lies with even more primitive animals.

Several significant structures evident in primitive vertebrates—the notochord, dorsal hollow nerve cord, perforation of the pharynx, and the arrangement of the segmental muscle masses—also occur in more primitive metazoans that are collectively termed invertebrate chordates. Living representatives of this group include the cephalochordates, typified by amphioxus, urochordates (the tunicates or ascidians), and the hemichordates (pterobranchs and acorn worms). Amphioxus (*Branchiostoma*) most closely resembles primitive vertebrates in its general appearance and way of life (Figure 2-2).

Except for the graptolites (tiny, colonial animals common in the Paleozoic that may be related to the pterobranchs), few invertebrate chordates are known as fossils and none provide specific information relevant to the origin of vertebrates. *Pikaia* (Conway-Morris, 1979) from the Middle Cambrian is a vaguely fishlike animal that may eventually provide more information on early chordates, but it has not yet been described in detail. Oelofsen and Loock (1981) have described a cephalochordate from the early Permian of South Africa.

Although vertebrates and invertebrate chordates must have diverged from one another considerably more than 500 million years ago, amphioxus appears to have retained the primitive pattern in most of its organ systems and may be used as a model of the immediately prevertebrate condition.

Amphioxus is widely distributed in shallow, nearshore marine waters and spends most of its life partially buried in the sediments, feeding by filtering particles out of the water. Despite its broadly fishlike appearance, amphioxus is distinctly more primitive than fish, lacking a brain and all of the specialized sense organs that characterize the vertebrates. Posteriorly, the dorsal hollow nerve cord resembles that of the most primitive vertebrates but lacks the ganglia that in more advanced forms join the ventral and dorsal nerves. What appear as ventral motor nerves, in fact, are actually muscle cells extending from the myotomes.

The notochord of amphioxus differs from that of vertebrates by extending to the extreme anterior end of the body, but it resembles that of primitive and embryonic vertebrates in consisting of a tough fibrous sheath of connective tissue surrounding cells that are vacuolated to produce a rigid structure.

In contrast with vertebrates, the segmental muscles of the body wall in amphioxus extend around the pharynx and are joined ventrally by connective tissue, producing a cavity external to the pharynx, the atrium. The atrium opens posteriorly via the atriopore just anterior to the anus. The ventral edge of the muscles lateral to the pharynx forms ridges, the metapleural folds, which have been compared with the paired fins in vertebrates. However, their position and composition are entirely different.

Figure 2-3. LARVAL STAGE OF THE LAMPREY, TERMED AN AMMOCOETE. It is about the size and general appearance of amphioxus and feeds by filtering particles out of the water. The pharynx is muscular, in contrast with that of amphioxus, and the brain has the same basic plan as other vertebrates. At metamorphosis, its endostyle is converted into a thyroid gland, and the configuration of the pharynx changes dramatically in relationship to the predatory feeding habits of the adult. The eye becomes fully differentiated and the nasal sacs migrate from a ventral to a dorsal, median position.

There are no gills in the pharynx. Because of the small body size of amphioxus, exchange of respiratory gases through the general body surface is sufficient to maintain its relatively low metabolic rate (Ruben and Bennett, 1980).

Very little musculature is present in the gut, pharynx, or mouth of cephalochordates; movement of water and food particles is produced primarily by the action of cilia. Large amounts of water which carry food particles are drawn into the mouth. Most of the water is discharged through the slits in the wall of the pharynx. The endostyle, a trough-shaped structure extending the length of the ventral surface of the pharynx, secretes mucus, which traps particles of food that are swept back into the gut. This mode of feeding provides important evidence for the specific relationship of invertebrate chordates to vertebrates; the larval stage of the lamprey exhibits a nearly identical structure of the pharynx and a similar feeding behavior (Figure 2-3).

Cephalochordates do not have a heart; blood flow is maintained by contractile tissue at the base of the aortic arches. The general geometry of the circulatory system (Figure 2-4) is similar to that of primitive vertebrates, but there are no capillary beds, and both red blood cells and hemoglobin are missing from the blood. Oxygen is carried in solution in the body fluids.

Amphioxus has peculiar excretory structures situated in segmental fashion in the dorsal portion of the pharynx. Their association with both the coelom and the circulatory system resemble functionally the nephrons of primitive vertebrates. Associated structures have been compared with the flame cells or solenocytes of flatworms, but a more recent study indicates that these cells are similar to those that line Bowman's capsule of the vertebrate kidney (Welsch, 1975). Developmentally, however, vertebrate kidney tissue appears to be strictly mesodermal in origin, while that of amphioxus is thought to be ectodermal (Romer and Parsons, 1977).

The gonads of amphioxus, in contrast to those in vertebrates, are numerous, segmental structures. When the gametes are fully developed, the gonads rupture and discharge sperm or eggs into the atrium, from which they are released into the water. The eggs have little yolk and the young quickly hatch out as simple larvae that swim actively in the plankton. When the young reach maturity, they sink to the bottom and take up an almost completely sedentary mode of life.

The most important chordate characters—the notochord, dorsal hollow nerve cord, and segmentally arranged muscles—appear to be specifically evolved for an active way of life, including open-water swimming. The inactivity of the adult stage of amphioxus, together with the oddly asymmetrical early development of the mouth and pharynx and the nature of the kidney, may be specializations peculiar to the modern cephalochordates.

THE ORIGIN OF CHORDATE CHARACTERS

The hemichordates, urochordates, and cephalochordates can be interpreted as representing a morphological series that shows the sequential evolution of chordate characteristics. The most primitive hemichordates, the tiny colonial pterobranchs (Figure 2-5), have no definitive chordate features. They feed by filtering food particles out of the water, but they do so with tentacles, rather than through a series of pharyngeal slits. Acorn worms, including *Balanoglossus*, have many pharyngeal slits but no structures that can be confidently identified as either a notochord or a dorsal hollow nerve cord.

The adult of most tunicates has a huge pharynx with an endostyle and feeds like amphioxus but lacks other chordate features (Figure 2-6). In contrast, the larva is motile, with both a notochord and a dorsal nerve cord.

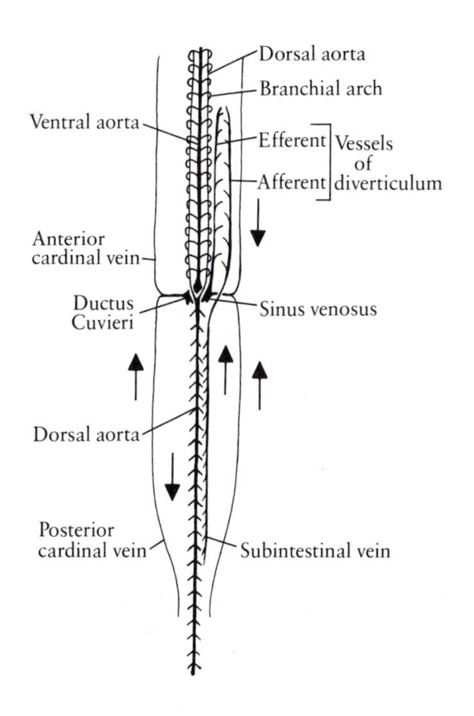

Figure 2-4. THE CIRCULATION OF AMPHIOXUS. The basic pattern is similar to that of primitive vertebrates but lacks a large muscular heart, capillaries, blood cells, and blood pigment. *From Young, 1981 (after Grobben and Zarnik). Reproduced with permission of Oxford University Press.*

(a)

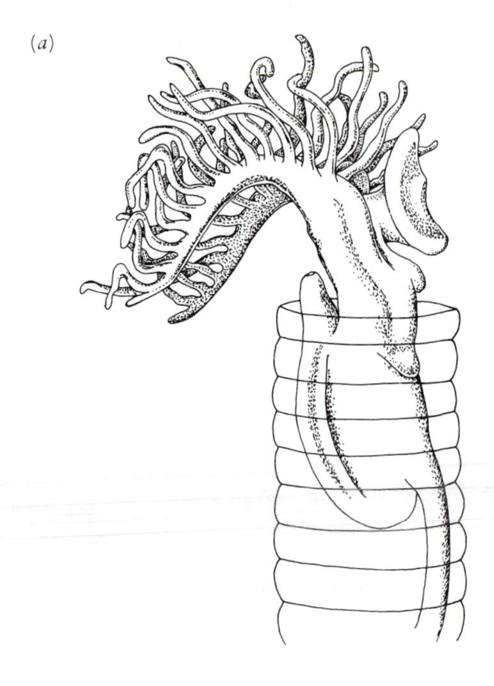

(b)

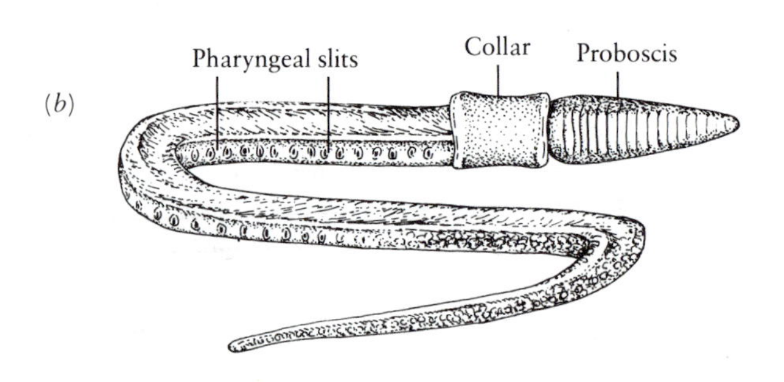

Figure 2-5. HEMICHORDATES. (*a*) Single zooid of *Rhabdopleura*, a pterobranch that feeds via tenticles. It has none of the characteristic chordate features. A related genus, *Cephalodiscus*, has a single gill slit. *Rhabdopleura* is a tiny animal that lives in colonies connected by tubular structures. *From Borradaile et al., 1959.* (*b*) An acorn worm, which is related to pterobranchs on the basis of similar larval stages. It has many pharyngeal slits but lacks a notochord and dorsal hollow nerve cord.

In most tunicates, these structures are lost after a few days when the larva undergoes a complex metamorphosis. Berrill (1955) elaborated the hypothesis that the notochord and dorsal nerve cord evolved within the urochordates to facilitate habitat selection via an actively swimming larval stage. Habitat selection is clearly a critical problem in a group in which most members are attached to the substrate during their adult life. Berrill further suggested that certain urochordates may have prolonged the larval stage so that they reached sexual maturity while retaining juvenile locomotor characters. Further elaboration of the notochord, nerve cord, and associated musculature would result in the structures seen in amphioxus and primitive vertebrates. Such prolongation of the larval features, a process termed **neoteny,** occurs in some groups of urochordates, although in other ways these particular forms do not closely resemble cephalochordates or vertebrates.

Although Berrill's argument is plausible, Carter (1957) and Jollie (1982) have argued that this complex of characters might have evolved in a free-swimming ancestor of both urochordates and cephalochordates and that the attached adult stage of most urochordates is a specialized rather than primitive feature of that group. The origin of the major chordate characters remains unclear.

THE RELATIONSHIPS OF CHORDATES

Early anatomists and some paleontologists proposed that vertebrates may have evolved from specialized members of various invertebrate groups including the annelids, arthropods, and cephalopods. Fundamental differences in developmental patterns are now known to distinguish vertebrates and invertebrate chordates from molluscs and the advanced segmented invertebrates, while demonstrating affinities with echinoderms. These differences involve the geometrical pattern of the early cell divisions, the developmental potential of the embryonic cells, and the way in which the coelom is formed.

(a)

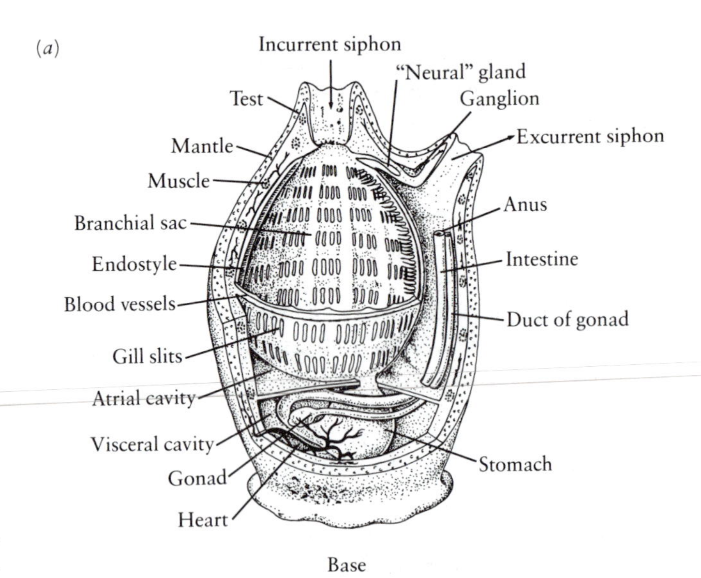

(b)

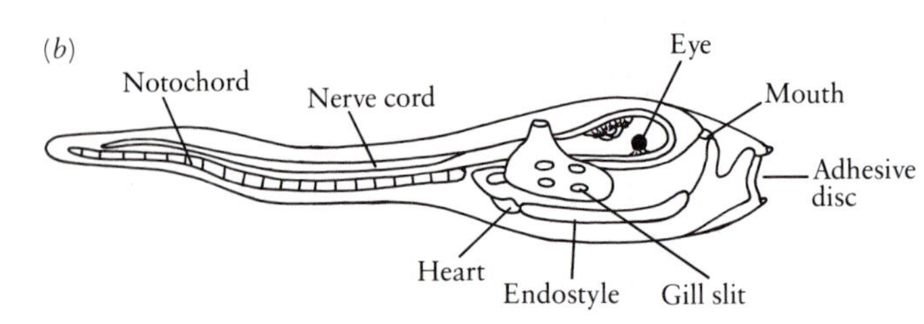

Figure 2-6. TUNICATES. (*a*) Structure of an adult tunicate. Tunic, mantle, and upper half of branchial sac removed on left side. Arrows indicate path of water currents. It has an enormous pharynx, pierced with many rows of slits, but no structure equivalent to the notochord or dorsal hollow nerve cord. (*b*) Diagram of an ascidian larva, showing typical chordate characters. Tail muscles are continuous rather than being divided into segments. *From Storer and Usinger, 1965. Reprinted with permission of McGraw-Hill Book Company.*

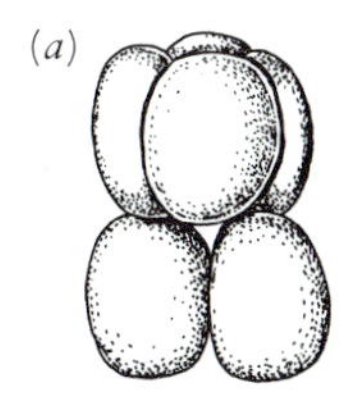

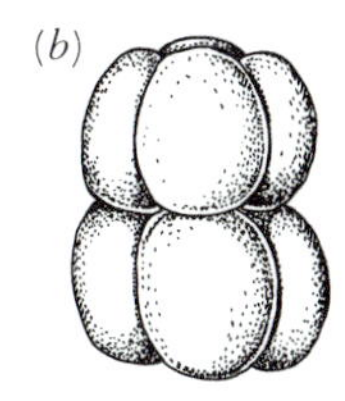

Figure 2-7. (*a*) Geometrical arrangement of cells characteristic of segmented invertebrates that results from spiral cleavage. Lateral view. (*b*) Arrangement of cells characteristic of chordates and echinoderms that results from radial cleavage.

A discussion of these differences not only enables us to appreciate the relationships of the chordates to other groups but is also useful in outlining the basic processes of vertebrate development that are necessary in understanding their structure and evolution.

In the annelid-mollusc assemblage, early cell division occurs so that cells in successive rows are nested between one another, a process known as **spiral cleavage** (Figure 2-7). In this group, the fate of each cell is predetermined, so that if one is removed, a particular part of the organism fails to develop. This is referred to as **determinate cleavage.** In the group including chordates and echinoderms, early cell division results in **radial cleavage,** in which the cells are directly above one another. If a particular cell is removed at an early stage, adjacent cells can compensate for its loss, and the embryo develops normally (**indeterminate cleavage**).

In primitive members of both groups in which the eggs have relatively little yolk, successive cell divisions produce a simple ball of cells, the blastula. Cells at one end grow inward to form a second layer of tissue. The external layer, the ectoderm, forms the outer surface of the body. The inner layer, the endoderm, forms the lining of the gut. In the annelid-mollusc assemblage, the mesoderm, a third embryonic tissue that will form many of the internal structures, develops as a clump of cells between the ectoderm and endoderm. The mesoderm splits to form the coelom, the major internal body cavity. This pattern of coelom development is termed **schizocoelic.** In primitive chordates and echinoderms, the mesoderm develops as an outpocketing of the gut, a pattern termed **enterocoelic.**

In most groups of multicellular animals, the opening that occurs in the blastula as the result of the infolding of the endoderm is retained as the mouth. Animals with this pattern are called **protostomes.** This group includes all the animals with spiral cleavage and schizocoelic development of the coelom. In echinoderms and chordates, the opening in the blastula becomes the anus, and the mouth develops as a separate structure. This group is named the **deuterostomes,** indicating the secondary nature of the mouth. Echinoderms and hemichordates also have a similar type of larva, the tornaria, in contrast with the trochophore larva of the annelids and molluscs. The patterns of early development clearly separate the chordates and echinoderms from the other major groups of metazoans.

Jefferies (1968, 1973, 1975) and Jefferies and Lewis (1978) have proposed particularly close affinities between one group of animals typically classified as echinoderms (the Calcichordata) and the vertebrates. Most of the similarities they cite are based on reconstructions of the soft anatomy of the calcichordates, a group known only from the early Paleozoic. However, there are no convincing similarities if comparisons are limited to the known hard skeleton of calcichordates and early vertebrates (Philip, 1979; Jollie, 1982).

THE ORIGIN OF VERTEBRATE CHARACTERS

How can we account for the origin of definitive vertebrate features such as the brain, specialized paired sensory structures, and bone from a pattern like that of amphioxus? No existing fossils illustrate an intermediate condition. The earliest adequately preserved vertebrates show a pattern that is fundamentally similar to that of modern genera.

Northcutt and Gans (1983) recently tackled this problem. They point out that most of the differences between cephalochordates and vertebrates can be associated with three types of tissue. The most important are the neural crest cells and the ectodermal placodes. The third is the muscular hypomere, the ventral continuation of mesodermal tissue beneath the segmental myotomes. This tissue accounts for the muscularization of the mouth, pharynx, and gut, and the adjacent elements of the vascular system. This contributes to the development of a muscular heart and more active transport of food and water used for gas exchange. All are necessary for the higher metabolic rate associated with the active vertebrate way of life. The elaboration of the hypomere is not difficult to account for, since a few muscle fibers are present in this area in the invertebrate chordates. Evolution of the vertebrate condition requires only amplification of tissues already present in more primitive chordates.

In contrast, the brain and special sensory structures of vertebrates do not have obvious homologues in invertebrate chordates. Northcutt and Gans argue that scattered epidermal sensory structures, common in primitive chordates, which have the capacity for chemical, tactile, light, and electrical sensitivity, have become specialized and localized in the form of the ectodermal placodes characteristic of all vertebrates. These placodes develop in the head region and form the major portions of the olfactory tissue, the eyes (except for the retina, which develops as an outgrowth of the diencephalon of the forebrain), the sensory structures of the inner ear, and the lateral line organs. A median placode forms the adenohypophysis. Taste buds are formed from a series of ventral placodes.

The peripheral epidermal and perivisceral sensory tissues may also have been the predecessors of the neural crest cells. These cells are unique to vertebrates; they proliferate at the margin of the neural tube, between it and the surrounding ectoderm (Figure 2-8). They then spread ventrally in the head region and migrate throughout the body. Either the neural crest cells or other tissues whose development they influence contribute to many very different vertebrate structures, including:

Dermal skeleton
Cartilage, bone, and muscles of the jaws and visceral arch skeleton
Some portions of the sensory capsules
Trabeculae (paired rods that underlie the anterior portion of the brain in jawed vertebrates)
Sensory ganglia of cranial nerves V, VII, IX, and X
Visceral motor neurons innervating the heart and muscles of the aortic arch
Pigment cells

Neural crest tissue and/or placodes also contribute to the formation of the portion of the brain anterior to the diencephalon.

Clearly, evolution of the neural crest and placode tissue was a major factor in the origin of vertebrates. Northcutt and Gans attribute the development of the brain and special sensory structures to selection for a more active way of life including predatory feeding, in contrast to the nearly passive filtering of food practiced by the primitive chordates. However, the specific manner in which these tissues became elaborated remains uncertain.

THE ORIGIN OF SKELETAL TISSUES

Bone and teeth are unique to vertebrates and vital to their sustained success. It is important to consider how they may have originated.

Bone was almost certainly not present in the most primitive vertebrates, but evolved separately in several lineages. We can deduce this from an examination of the fundamentally different patterns of the exoskeleton in each major group when they first appear in the fossil record. On the other hand the developmental processes and selective pressures associated with the origin of bone may have been similar in all these groups.

In the earliest fossil vertebrates, bony tissue forms only a superficial covering of the body. The earliest adequately known calcified tissue is that of the middle Ordovician genus *Astraspis* (see Figure 3-4). It is a complex material with features that resemble both bone and teeth in modern vertebrates. It consists of a basal layer of laminar tissue that resembles bone except for the absence of lacunae, which in later vertebrates mark the position of bone cells, or **osteocytes.** Because of its acellular nature, this primitive bony tissue is given a distinct name, **aspidin.** More superficially, the skeletal material consists of tissue that is histologically similar to the dentine and enamel of teeth and develops around openings that resemble pulp cavities. Ørvig (1968, 1977) uses the term **odontode** to refer to this toothlike complex of tissues.

The formation of skeletal tissues in modern vertebrates is a complicated process involving specialized cells, enzymes, mineral salts, and fibrous proteins (Moss, 1964 and 1968). Developmentally, the cells associated with skeletal tissue can be traced to the neural crest cells. Bone is deposited as a result of the activity of osteoblast cells which appear in the embryonic connective tissue of the ectoderm and mesoderm. They deposit a noncellular matrix containing mucopolysaccharides as well as collagen fibers that are capable of being impregnated by calcium phosphate. This material crystallizes as a hydrated form of the mineral apatite. Controversy still exists over the manner in which mineralization of the bone occurs. The process may be mediated by an enzyme, or it may occur spontaneously when the appropriate concentration of calcium and phosphate salts is present (Halstead, 1974).

The deposition of dentine and enamel-like material in the scales and teeth of sharks provides an example of how tissues resembling the superficial layers of the skeleton of early vertebrates may be formed (Moss, 1977). In sharks, bony fish, and terrestrial vertebrates, teeth are formed through the integrated activity of both ectodermal and underlying mesodermal tissue. The enamel is produced primarily by the ameloblast cells of the ectoderm. The ectodermal cells also induce formation of the dentine and tooth pulp by the underlying odontoblasts.

It is difficult to understand how such a complex association of specialized cells in different tissue layers could have evolved among primitive vertebrates. However, most of the chemical constituents were available even in their primitive chordate ancestors. Collagen (similar to that in many invertebrates) may have been present in vertebrates

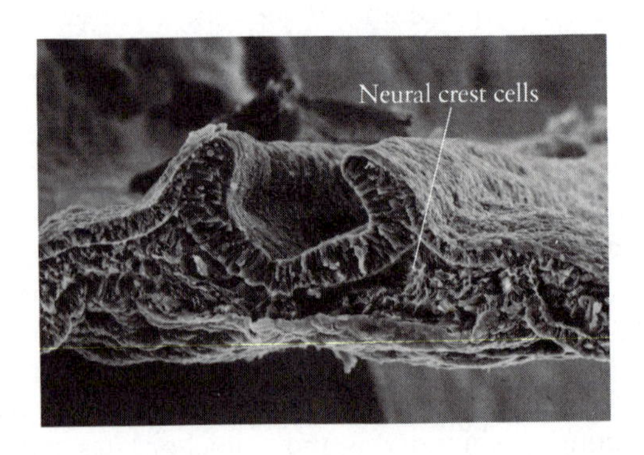

Figure 2-8. CROSS SECTION OF A VERTEBRATE EMBRYO. Photograph shows the proliferation of neural crest cells between the ectoderm and the dorsal hollow nerve cord. *From Tosney, 1982.*

prior to the appearance of bone. Calcium and phosphate salts are elements of the body chemistry of all metazoans; these salts are involved with energy transfer, transmission of nervous impulses, osmotic regulation, and other metabolic processes. However, localization and concentration of the salts would have had to be significantly altered before they could accumulate as hard tissues. The ancestors of vertebrates may not have had to evolve totally new factors to form bone, but they brought together preexisting systems so as to cause its deposition (Hall, 1975).

Ørvig (1968) described aggregates of globular structures within the bony tissue of primitive vertebrates and suggested that they may have preceded laminar bone in evolution. These structures may have been deposited without the presence of collagen fibers or specific bone cells. In contrast, Northcutt and Gans (1983) attribute the formation of even the earliest ossified tissues to particular functional complexes that would require specifically controlled deposition from the very beginning.

Our understanding of how ossified tissues evolved depends on recognition of their original functions. Structurally, bone is more effective as a skeletal material than calcium carbonate, which is present in many other groups of metazoans. It is much stronger and harder, so that a smaller amount can be used with the same effect. However, it is difficult to visualize how selection would have acted to elaborate this tissue preferentially in the earliest vertebrates.

Tarlo (1964) argued that bone originally developed not for protection, but as a phosphate reserve. Phosphate is both a vital component for energy storage and transfer in all vertebrates and a substance that may be available in fluctuating amounts in the natural environment. The amount of phosphate is frequently a limiting factor in population growth. Superficial bone in some Paleozoic fish shows changes in its extent that are thought to be seasonal and may correspond to the periodic deposition and resorption of phosphate. Bone also acts as a calcium reserve in all vertebrates.

Northcutt and Gans (1983) point out that dentine and enamel would serve to insulate the electrosensory organs present in many primitive vertebrates. They suggest that these may have been the first tissues to ossify. The underlying lamellar bone may have been added later. This tissue is not as hard as dentine and enamel, but it is much less brittle and would better serve to support the sensory tissue.

We can integrate these ideas into a single scheme. Calcium phosphate may have been initially concentrated by early vertebrates as a metabolic reserve. With changes in the nature of previously existing collagen fibers, this material could then crystallize near electrosensory organs, thus better insulating them from currents generated within the body and giving them better directional resolution. Once the capacity to form hard tissues had developed, a more complete protective covering could then have been formed.

None of the early vertebrate groups shows evidence of a bony internal skeleton. The capacity to ossify tissues within the body certainly evolved later than did the exoskeleton. Unossified cartilage may have preceded bone as an internal supporting tissue, but this is difficult to demonstrate since cartilage is incapable of fossilization. Calcified cartilage has been described in *Eriptychius*, a poorly known jawless fish that accompanies *Astraspis* in the middle Ordovician (Denison, 1956).

By the early Devonian, many major groups of primitive vertebrates had evolved the capacity to ossify the internal skeleton. In modern vertebrates, ossification of the internal skeleton occurs in two stages. Initially, the bone is preformed in cartilage. Cartilage is deposited by cells from the mesenchyme (the chondroblasts) similar to those that deposit bone. Instead of forming collagen fibers, they secrete a matrix of sulfated polysaccharide that forms a firm gel through which is spread a network of connective tissue fibers. The cartilage may remain as such or be replaced by bone. If bone is to be present in the adult, the area of cartilage is invaded by blood vessels and the cells that had formerly been elaborating cartilage now act to break it down. Osteoblasts now begin laying down bony tissue.

Such preformation in cartilage enables jointed bones of the internal skeleton (vertebrae, ribs, girdles, and limbs) to function during growth. The ends of the bones remain cartilaginous until maturity, and bone is formed in the more central part of the structure. More superficial bone can grow at the edges and does not need to be preformed in cartilage.

The terms **dermal** and **endochondral** (or **cartilage-replacement bone**) are used to distinguish the ossified tissue that forms in the superficial layers of the body from those that develop more deeply. Once formed, however, they are histologically indistinguishable.

Although bone, cartilage, dentine and enamel are structurally distinct, all are formed by a single class of cells, the scleroblast; its manifestation as an osteoblast, chondroblast, odontoblast, or ameloblast may be altered by changes in the physical or chemical environment of the surrounding tissue (Hall, 1975), which suggests a considerable degree of plasticity in the expression of skeletal tissues throughout their evolutionary history.

THE ENVIRONMENT OF EARLY VERTEBRATES

It was long thought that the origin of vertebrates was associated with movement by their ancestors from the sea into fresh water. Romer was the most influential advocate of this point of view (Romer, 1968; Romer and Grove, 1935). He argued that evolution of the brain, specialized sensory organs, and improved swimming abilities resulted from selection in animals that had to swim against the currents of rivers and streams flowing into the ocean. Analysis of the structure and function of the kidney (Smith, 1932) seemed to support the assumption that the earliest vertebrates had evolved in fresh water rather than salt water.

Paleozoic vertebrates are found in many depositional environments, ranging from marine through estuarine to freshwater. Romer argued that the majority were freshwater, and that the marine occurrences resulted from the remains of fish being swept out to sea from the mouths of rivers. However, if only the earliest fossil vertebrates, from the late Cambrian and Ordovician, are considered, all the available evidence now points to a marine origin. A recent review by Darby (1982) shows that all early vertebrate remains are preserved in shallow, near-shore marine deposits. There is no evidence from more typical, deeper water deposits that early vertebrates lived in the open ocean, nor is there any evidence of their presence in freshwater deposits prior to the early Devonian. Smith's arguments regarding kidney structure are no longer thought to preclude the origin of vertebrates in salt water (Northcutt and Gans, 1983).

The consensus among all persons now working with early vertebrates is that this group evolved in a marine environment. However, this is partially based on negative evidence. Although it is true that no fossil vertebrates have been described from freshwater deposits earlier than the Lower Devonian, no demonstrable freshwater deposits of this age have been described. Perhaps we could recognize early Paleozoic freshwater deposits by the presence of typical freshwater invertebrates (such as some ostracods and gastropods, which are known later in the Paleozoic), but none have so far been reported (Copeland, 1983).

One may still argue that the oldest of all vertebrates evolved in fresh water and that the early fossils from marine deposits represent a secondary invasion of salt water. This hypothesis becomes less likely as primitive members of more and more fish groups are found in progressively earlier marine deposits. We can also argue that vertebrates evolved in a marine environment rather than having reinvaded it, for their ultimate origin certainly lies among the strictly marine invertebrate chordates.

SUMMARY

Because of the absence of bone, there is no fossil record of the earliest vertebrates or their immediate ancestors. Based on the anatomy of primitive living fish and fossils from the early Paleozoic, we can infer that ancestral vertebrates were probably small, soft-bodied animals, lacking jaws and paired fins, but possessing a brain, specialized paired sense organs for smell, sight, and balance, and electrosensory structures. The pattern of the circulatory, reproductive, and digestive systems probably resembled those of primitive living fish. The adult was probably an actively swimming predator, but there may have been a filter-feeding larval stage.

The presence of a notochord, dorsal hollow nerve cord, and pharyngeal perforations, as well as segmental body musculature and a postanal tail, are features that all vertebrates share with invertebrate chordates. The modern cephalochordate amphioxus resembles vertebrates in its general body form, but it is a sedentary filter feeder throughout its adult life. Hemichordates (pterobranchs and acorn worms) and urochordates (tunicates) show progressive elaboration of chordate features, but the nature of the origin of the notochord, dorsal hollow nerve cord, and pharyngeal perforations has not been established.

Chordates, like echinoderms, are characterized by radial cleavage, indeterminant (or equipotential) development, and enterocoelic formation of the coelom, in contrast with advanced segmented invertebrates, in which cleavage is spiral, development determinant, and coelom formation schizocoelic.

Vertebrates are more advanced than invertebrate chordates by virtue of features associated with three distinctive tissues, the ventral hypomere, the placodes, and the neural crest cells. The ventral hypomere contributes to the muscularization of the mouth and pharynx, the heart, and the gut. The placodes contribute to the paired sense organs of the head and the lateral-line canal organs. Neural crest cells form melanocytes and contribute to the development of the visceral arches, skull, and other elements of the skeleton, as well as the ganglia of cranial and spinal nerves. Placodes and neural crest cells may have evolved from scattered epidermal sensory structures common in invertebrate chordates, but no evidence exists to show how this may have occurred.

Bone was not present in the earliest vertebrates, but it evolved separately in many distinct lineages. Bony tissue may have evolved initially as a metabolic reserve for phosphate and calcium. It may have served to insulate the electrosensory organs in early vertebrates and later became elaborated to form a protective covering over the entire body.

The transition between invertebrate chordates and vertebrates almost certainly occurred in a shallow, near-shore marine environment, after which many groups became adapted to life in fresh water.

REFERENCES

Berrill, N. J. (1955). *The Origin of Vertebrates.* Oxford University Press, Oxford.

Bodznick, D., and Northcutt, R. G. (1981). Electroreception in lampreys: Evidence that the earliest vertebrates were electroreceptive. *Science*, **212**: 465–467.

Boord, R. L., and Campbell, C. B. G. (1977). Structural and functional organization of the lateral line system of sharks. *Amer. Zool.*, **17**: 431–441.

Borradaile, L. A., Eastman, L. E. S., Potts, F. A., and Saunders, J. T. (1959). *The Invertebrata* (3d ed.). Cambridge University Press, Cambridge.

Carter, G. S. (1957). Chordate phylogeny. *Syst. Zool.*, **6**: 187–192.

Conway-Morris, S. (1979). The Burgess Shale (Middle Cambrian) fauna. *Ann. Rev. Ecol. Syst.*, **10**: 327–349.

Copeland, M. (1983). Personal communication. Geological Survey of Canada.

Darby, D. G. (1982). The early vertebrate *Astraspis*, habitat based on a lithologic association. *J. Paleont.*, **56**: 1187–1196.

Denison, R. H. (1956). A review of the habitat of the earliest vertebrates. *Fieldiana, Geol.*, **11**: 359–457.

Hall, B. K. (1975). Evolutionary consequences of skeletal differentiation. *Amer. Zool.*, **15**: 329–350.

Halstead, L. B. (1974). *Vertebrate Hard Tissues.* Wykeham Publications Ltd., London.

Jefferies, R. P. S. (1968). The subphylum Calcichordata (Jefferies 1968)—primitive fossil chordates with echinoderm affinities. *Bull. Brit. Mus. Nat. Hist. (Geol.)*, **16**: 243–339.

Jefferies, R. P. S. (1973). The Ordovician fossil *Lagynocystis pyramidalis* (Barrande) and the ancestry of amphioxus. *Phil. Trans. Roy. Soc. Lond., B*, **265**: 409–469.

Jefferies, R. P. S. (1975). Fossil evidence concerning the origin of the chordates. *Symp. Zool. Soc. Lond.*, **36**: 253–318.

Jefferies, R. P. S., and Lewis, D. N. (1978). The English Silurian fossil *Placocystites forbesianus* and the ancestry of the vertebrates. *Phil. Trans. Roy. Soc. Lond. B*, **282**: 205–323.

Jollie, M. (1982). What are the 'Calcichordata'? and the larger question of the origin of Chordates. *Zool. J. Linn. Soc.*, **75**: 167–188.

Moss, M. L. (1964). The phylogeny of mineralised tissues. *Int. Rev. Gen. Exp. Zool.*, **1**: 297–331.

Moss, M. L. (1968). The origin of vertebrate calcified tissues. In T. Ørvig (ed.), *Current Problems of Lower Vertebrate Phylogeny*, pp. 359–372. Interscience, New York.

Moss, M. L. (1977). Skeletal tissue in sharks. *Am. Zool.*, **17**: 335–342.

Northcutt, R. G., and Gans, C. (1983). The genesis of neural crest and epidermal placodes: A reinterpretation of vertebrate origins. *Quart. Rev. Biol.*, **58**(1): 1–28.

Oelofsen, B. W., and Loock, K. (1981). A fossil cephalochordate from the early Permian Whitehill Formation of South Africa. *South Afr. Jour. Sci.*, **77**: 178–180.

Ørvig, T. (1968). The dermal skeleton: General considerations. In T. Ørvig (ed.), *Current Problems of Lower Vertebrate Phylogeny*, pp. 373–397. Almquist and Wiksell, Stockholm.

Ørvig, T. (1977). A survey of odontodes ("dermal teeth") from developmental, structural, functional, and phyletic points of view. In S. M. Andrews, R. S. Miles, and A. D. Walker (eds.), *Problems in Vertebrate Evolution. Linn. Soc. Symp. Ser.*, **4**: 53–75.

Philip, G. M. (1979). Carpoids—Echinoderms or chordates? *Biol. Rev.*, **54**: 439–471.

Romer, A. S. (1968). *Notes and Comments on Vertebrate Paleontology.* University of Chicago Press, Chicago.

Romer, A. S., and Grove, B. H. (1935). Environment of the early vertebrates. *Amer. Mid. Nat.*, **16**: 805–856.

Romer, A. S., and Parsons, T. S. (1977). *The Vertebrate Body* (5th ed.). W. B. Saunders, Philadelphia.

Ruben, J. A., and Bennett, A. F. (1980). Antiquity of the vertebrate pattern of activity metabolism and its possible relation to vertebrate origin. *Nature*, **286**: 886–888.

Smith, H. M. (1932). Water regulation and its evolution in the fishes. *Quart. Rev. Biol.*, **7**: 1–26.

Storer, T. I., and Usinger, R. L. (1965). *General Zoology* (4th ed.). McGraw-Hill, New York.

Tarlo, L. B. Halstead. (1964). The origin of bone. Discussion. In H. J. J. Blackwood (ed.), *Bone and Tooth: Proceedings of the First European Symposium, Oxford, April 1963*, pp. 3–17. Macmillan, New York.

Thomson, K. S. (1977). On the individual history of cosmine and possible electroreceptive function of the pore-canal system in fossil fishes. In S. M. Andrews, R. S. Miles, and A. D. Walker (eds.), *Problems in Vertebrate Evolution. Linn. Soc. Symp. Ser.*, **4**: 247–270.

Tosney, K. W. (1982). The segregation and early migration of cranial neural crest cells in the avian embryo. *Developmental Biol.*, **89**: 13–24.

Welsch, U. (1975). The fine structure of the pharynx, cyrtopodocytes and digestive caecum of amphioxus (*Branchiostoma lanceolatum*). *Symp. Zool. Soc. Lond.*, **36**: 17–41.

Young, J. Z. (1981). *The Life of Vertebrates* (3d ed.). Clarendon Press, Oxford.

C H A P T E R **III**

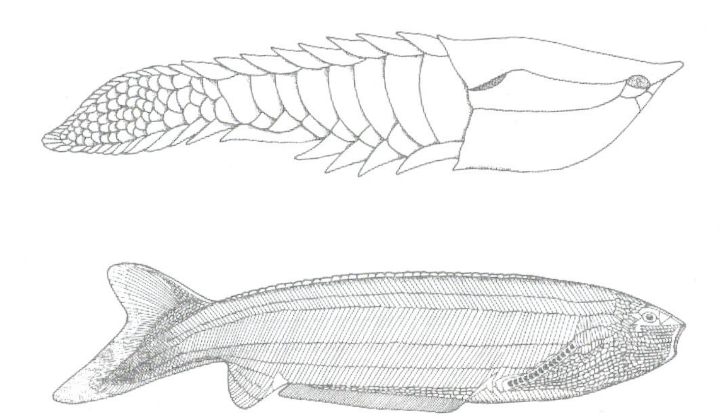

The Diversity of Jawless Fish

Ostracoderms

Ancestral vertebrates may have broadly resembled amphioxus but had a complex brain and associated sense organs of smell, sight, and balance. They were probably only a few centimeters long but may have been fairly active swimmers, despite the absence of paired fins.

The radiation of vertebrates began in the late Precambrian or early Cambrian periods. One or more lines not represented in the fossil record until the early Silurian developed jaws and became effective predators. Several other lineages retained the primitive jawless condition (Figure 3-1). These fish, collectively termed the Agnatha, were the first vertebrates to achieve dominance.

The fossil record of jawless vertebrates began when they evolved the capacity to form a bony skeleton. The early agnathans are commonly termed ostracoderms (shell skinned) because of their extensive exoskeleton, in the form of a solid carapace, large bony plates, or scales. The presence of a bony covering complicated growth in early

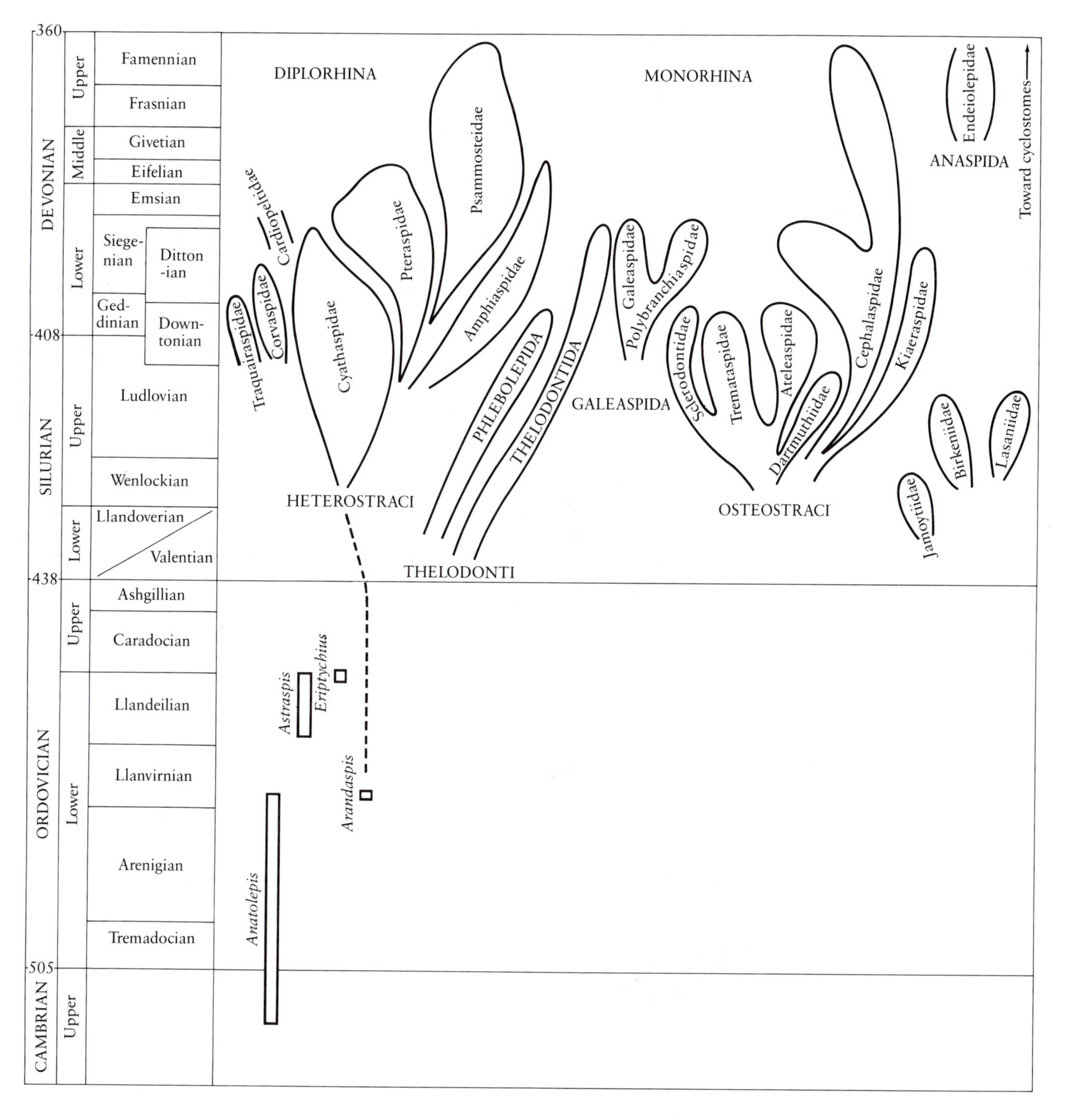

Figure 3-1. PHYLOGENY OF THE PALEOZOIC AGNATHAN FISHES.

vertebrates. One of the most important changes we see among the ostracoderms is the evolution of different ways in which an exoskeleton forms early in development to allow for continued growth. We may conclude from the distinctive nature of the armor that bone probably evolved separately in each of the major groups of ostracoderms (Halstead, 1982).

Some ostracoderms with a light flexible covering of small scales and a fusiform body may have lived in the open water. Others, with heavier armor and a dorsoventrally flattened body, were presumably benthonic.

The first adequately known ostracoderms are from the Middle Ordovician. They continued as a numerous and diverse assemblage into the Silurian and early Devonian, but after the Devonian they are no longer represented in the fossil record. The living jawless fish, hagfishes and lampreys, evolved from one or more groups of Paleozoic Agnatha.

Two major groups of Paleozoic Agnatha are recognized. The Cephalaspidomorphi, or Monorhina, are characterized by a nasal sac and narial opening that are medial in position and confluent with the hypophyseal duct. The

Pteraspidomorphi, or Diplorhina, retain what is usually considered a more primitive condition in which the nasal sacs and, apparently, the external narial openings are paired. Members of the Pteraspidomorphi are the first group of vertebrates to appear in the fossil record.

PTERASPIDOMORPHI

Cambrian and Ordovician

The oldest known fossils that may be remains of vertebrates are phosphatic fragments of late Cambrian and early Ordovician age from widely scattered areas in North America, Greenland, and Spitsbergen (Figure 3-2) (Repetski, 1978). These fragments are associated with benthonic trilobites, conodonts, and brachiopods. They are only a few square millimeters in area and less than one-tenth of a millimeter thick. The surface shows scalelike ornamentation but gives no evidence of the overall shape of the animal. The chemical composition of the material is apatite, the mineral constituent of bone, but the histology is not readily comparable with that of later, but still very primitive vertebrates. These fragmentary remains are known as *Anatolepis*.

Far more complete remains are known of the genus *Arandaspis* from the lower Middle Ordovician of Australia (Figure 3-3). The bone itself is not preserved in these specimens but is represented by molds in the rock showing detailed impressions of both internal and external surfaces. Large dorsal and ventral plates cover the head and pharyngeal region and smaller lateral (branchial) plates cover the area of the gills. Small scales covered the anterior trunk region and probably extended posteriorly to cover the remainder of the body and tail. The nature of the caudal fin is unknown, and paired fins were not present. The anterior plates are 6 to 7 centimeters long; we estimate that the length of the entire fish was 12 to 14 centimeters. The body is oval in cross section, with the ventral plate broadly rounded. Ritchie and Gilbert-Tomlinson (1977) suggested that *Arandaspis* lived just above the bottom, feeding on organic debris and microorganisms carried by currents immediately above the surface of the sediments.

Lateral notches for the eyes are near the anterior extremities of the dorsal plate. The specimens do not show nasal capsules, a pineal opening, or the inner ear. Open lateral line canal grooves extend longitudinally along both the dorsal and ventral plates. The branchial plates may have had as many as 15 rectangular subdivisions, representing the position of the gill pouches. They appear not to have opened to the exterior; there were probably common external branchial ducts leading to a pair of posteriorly placed openings, as in better-known Silurian relatives.

The external plates are less than $\frac{1}{10}$ millimeter thick. They show no evidence of growth or of fusion from initially smaller units, indicating that they formed only after the animal had reached maximum size. The external surface is ornamented with tiny scales like those of *Anatolepis*. The internal surface of the plates shows little evidence of superficial body structures, except for the gills pouches. The bone appears to have been almost completely superficial to the soft anatomy of the body and would have served primarily as a protective covering.

A second genus, *Porophoraspis*, accompanies *Arandaspis* but is represented only by fragments of dermal armor showing a distinct pattern of ornamentation (Figure 3-3*d*). There is no evidence of an internal skeleton in these early vertebrates.

Bony fragments of two other ostracoderm genera, *Astraspis* and *Eriptychius*, from western North America appear just slightly later in the Ordovician than *Arandaspis* (Denison, 1967). Thousands of fragments were found in a series of localities within a narrow band of sediments termed the Harding Sandstone, which extends along the foothills of the Rocky Mountains in the western United States and Canada. These sediments were almost certainly deposited at the margin of the sea and contain marine invertebrates as well as fish fragments. Only two specimens of *Astraspis* have been found that show a significant portion of the carapace (Figure 3-4). The dorsal shield is incomplete anteriorly and shows no evidence of sensory structures except for a few short segments of open lateral line canals. The most striking difference from *Arandaspis* is the presence of many distinct plates called **tesserae** that make up the shield. Large tubercles at the center of each plate represent the point at which ossification was initiated in small individuals of *Astrapis*. The

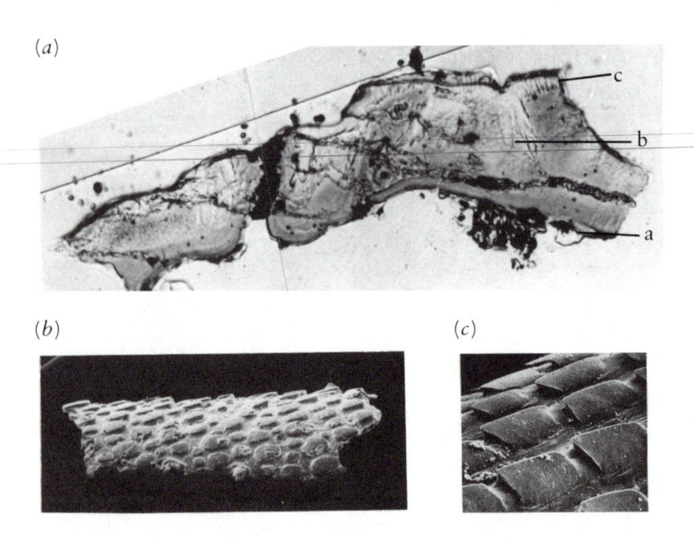

Figure 3-2. *ANATOLEPIS.* Specimens identified as bony fragments of fish from the Upper Cambrian and Lower Ordovician. (*a* and *b*) From the Upper Cambrian. (*a*) Cross section of plate showing: a, basal lamellar layer; b, intermediate cavernous layer; c, superficial layer; approximately 175 times natural size (×175). *From Repetski, 1978.* (*b*) Fragment showing outer surface, ×40. *Courtesy of Dr. Bockelie.* (*c*) Bony plate with scales from the Lower Ordovician of Spitsbergen, ×250. *From Bockelie and Fortey, 1976. Reprinted by permission from* Nature, **260**:*36–38. Copyright © 1976, Macmillan Journals Ltd.*

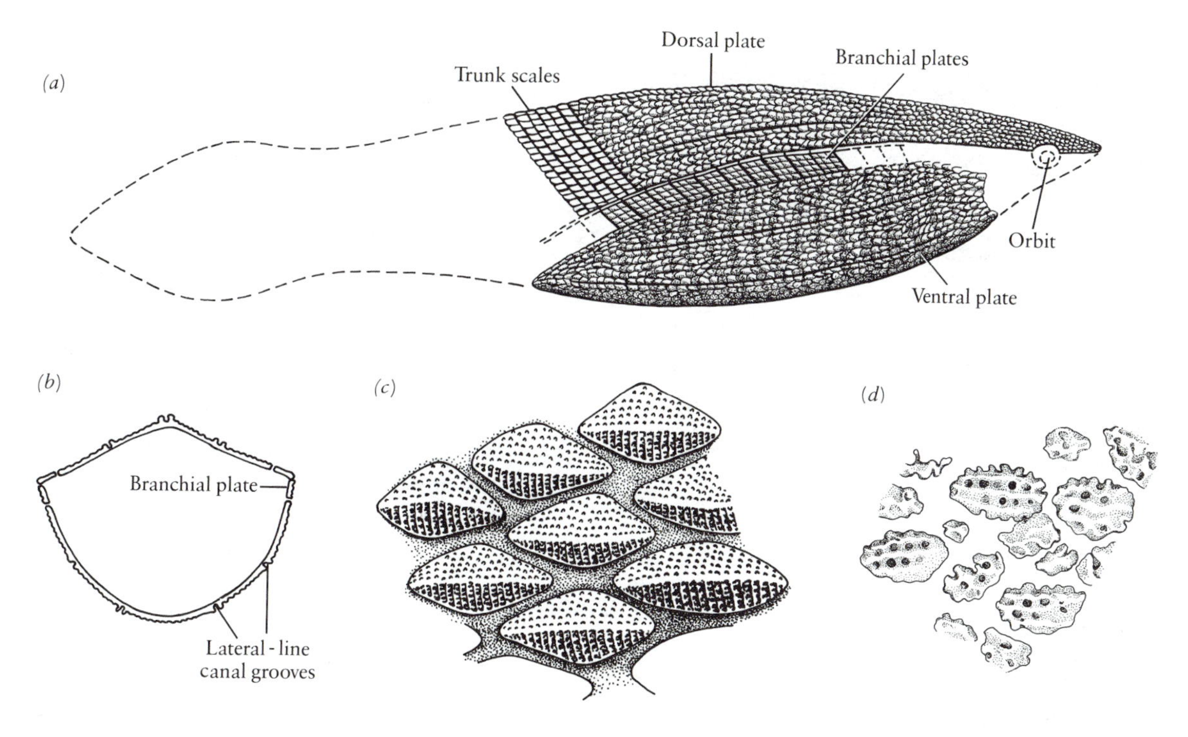

Figure 3-3. *ARANDASPIS,* A MIDDLE ORDOVICIAN HETEROSTRACAN OSTRACODERM FROM AUSTRALIA. (*a*) Restoration of body in lateral view. Original length estimated to be 12 to 14 centimeters. (*b*) Cross section of cephalothoracic shield. (*c*) Detail of dermal ornament. (*a–c*) *From Ritchie and Gilbert-Tomlinson, 1977.* (*d*) *Porophoraspis,* dermal ornament of a second ostracoderm found with *Arandaspis,* ×17. *Redrawn from photograph in Ritchie and Gilbert-Tomlinson, 1977.*

Figure 3-4. INCOMPLETE DORSAL SHIELD OF *ASTRASPIS,* ×2. *Astraspis* is a heterostracan ostracoderm from the Middle Ordovician of western North America. The shield is made up of numerous tesserae that ossify as discrete units and become fused when growth is complete. *Courtesy of Dr. Ørvig.*

tesserae grew through the addition of smaller tubercles at their margins. The shield could thicken by developing additional layers of tubercles.

The histology of the numerous isolated plates has been studied in great detail, and its pattern is taken as representative of primitive vertebrates (Figure 3-5). The plates are relatively thicker than those of *Anatolepis* and *Arandaspis* and consist of three distinct layers—a basal laminar layer, a cancellous vascular layer, and a superficial layer bearing tubercles, each formed around a pulp cavity. A layer of dentine-like material immediately surrounds the pulp cavity, and contains many minute tubules extending from it. A clear, dense material resembling enamel caps the tubercles.

The bony material does not show the lacunae associated with bone cells in modern vertebrates. As in the formation of dentine, the cells apparently escaped rather than becoming entrapped in the hard tissue. Ruben and Bennett (1980) suggest that the acellular nature of bone in primitive vertebrates might have been a response to high levels of lactic acid in the blood, resulting from a high rate of fermentative metabolism. The skeletal material of *Astraspis,* like that of more advanced vertebrates, shows evidence of local dissolution and redeposition, indicating remodeling for growth and repair.

The second genus from the Harding Sandstone, *Eriptychius,* has calcified cartilage in the endoskeleton, which is not found in other pteraspidomorphs (Figure 3-5*b*). Unfortunately, the fragments are too incomplete to provide information of the overall structure of this fish.

(a)

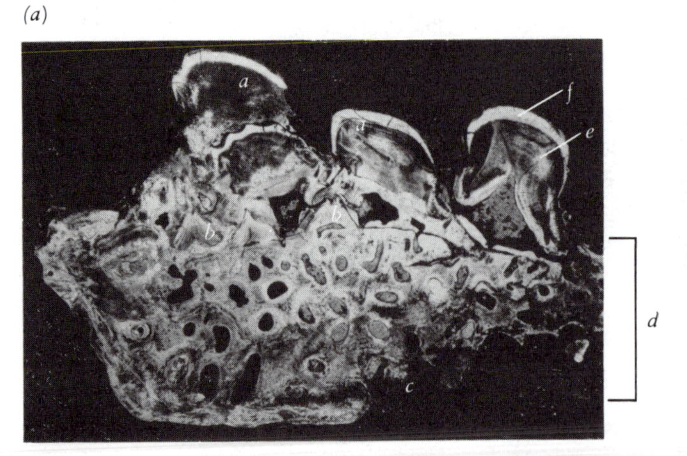

(b)

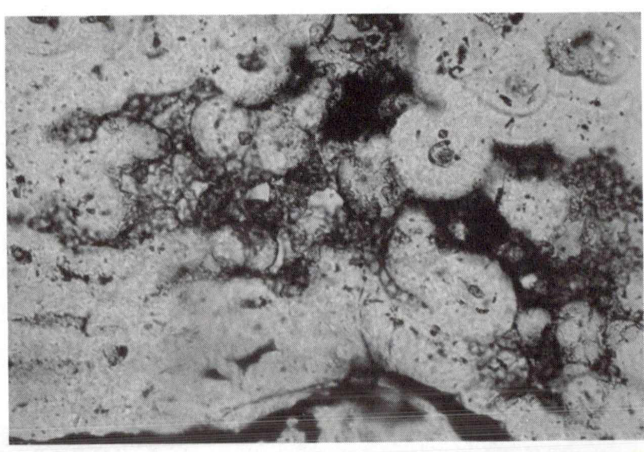

Figure 3-5. (*a*) *Astraspis,* vertical section through dermal plate. Several generations of mushroom-shaped tubercles, a, have grown over small stellate tubercles, b; c, laminar layer; d, vascular layer; e, dentine; f, superficial, enamel-like material, ×20.

(*b*) *Eriptychius,* ostracoderm found with *Astraspis.* Cross section showing globular calcified cartilage, ×180. *Courtesy of Robert Denison. Fieldiana,* Geology, **16.** *By permission of Field Museum of Natural History, Chicago.*

The five genera just discussed are all the named vertebrate remains from the Cambrian and Ordovician. The fossil record remains very incomplete until toward the end of the Silurian, at which time a variety of vertebrate groups are well represented.

The few fish genera known from the Cambrian and Ordovician probably represent only a small portion of the fauna that was present at that time, judging by the absence of ancestors for many groups that appear in the Silurian and early Devonian. Some of the early vertebrates may not have been ossified and so were not readily fossilized. Others may have lived in environments (including freshwater) where preservation was less likely to occur.

Cyathaspididae

Of the diverse agnathan families present in the middle and late Silurian, the cyathaspids are the most closely related to the known Ordovician genera. They are also the most primitive of adequately known vertebrates. The pattern of their dermal armor may have been derived from that of the much earlier Australian genus *Arandaspis,* and the histology of the bone resembles that of *Astraspis.*

We recognize nearly 20 genera of cyathaspids, ranging from the base of the Middle Silurian to the top of the Lower Devonian, a period of approximately 50 million years. The early specimens are entirely from marine deposits, but there is a gradual transition through marginal and brackish water to largely freshwater sediments in which the Lower Devonian genera are found (Denison, 1964).

The entire body is enclosed by bony plates and scales (Figures 3-6 and 3-7). The head and thorax are covered by extensive dorsal and ventral plates. The upper margin of the long, narrow branchial plates is notched by a single pair of excurrent gill openings. Smaller plates border the orbits ventrally. A series of plates making up the ventral surface of the mouth were probably connected by soft tissue to form a flexible scoop.

Paleontologists have long presumed that these animals fed by drawing in water and mud and filtering out the food particles in a manner comparable to amphioxus. The pharynx as a whole could not be expanded, and muscles associated with the gills or a velum must have pumped the water and mud. Northcutt and Gans (1983) doubt these assumptions, pointing out that these and other ostracoderms might have been capable of feeding on soft-bodied invertebrates. They question whether enough food could have been gained by filter feeding to support the metabolic processes of a relatively large and potentially active vertebrate.

The body is oval in cross section, with a tendency toward dorsoventral flattening. The caudal fin is laterally compressed and nearly symmetrical, although the main axis may bend somewhat ventrally. There are no dorsal, anal, or paired fins. These fish probably swam with only limited control over directional movements and may have spent much time resting on the bottom. The openings for the eyes are at the side of the dorsal shield, as in *Arandaspis.* Tubes of the lateral line canals within the cancellous layer of the bony plates open to the outside by a series of pores.

The endoskeleton is neither ossified nor calcified in cyathaspids. The dermal skeleton shows a more intimate association with the underlying soft anatomy than was evident in *Arandaspis,* and superficial structures are reflected as impressions on its internal surface (Figure 3-6). Some structures are interpreted differently by different authors, but the presence of a series of seven gill pouches with external openings leading to a pair of excurrent ducts is clearly shown. Paired depressions on the inner surface of the dorsal shield just posterior to the maxillary brim may have housed nasal sacs, and paired notches on the

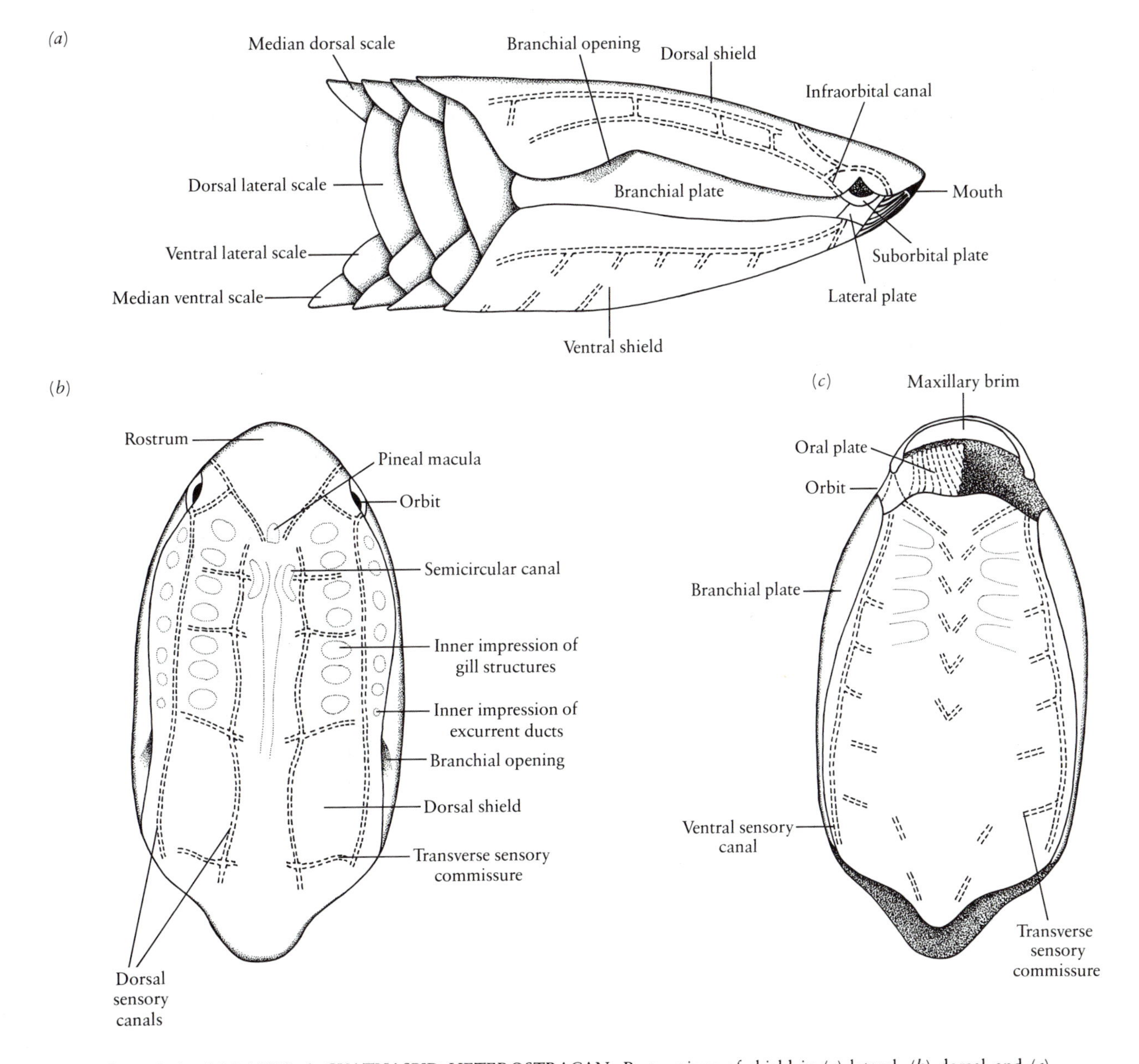

Figure 3-6. *PORASPIS,* A CYATHASPID HETEROSTRACAN. Restorations of shield in (*a*) lateral, (*b*) dorsal and (*c*) ventral views. Impressions on the inner surface of the shield indicated with dotted lines, ×2. *From Moy-Thomas and Miles, 1971. By permission from Chapman and Hall Ltd.*

maxillary brim may have been nostrils. More posteriorly, there are impressions of the brain stem and the pineal organ, as well as the anterior and posterior vertical semicircular canals. There is no direct evidence of the horizontal canal, which is absent in living agnathans, and it is usually presumed not to have been present. However, because of its deeper position, it would not have left any mark on the dermal shield, and it is possible that the cyathaspids had the typical vertebrate complement of three pairs of semicircular canals. Impressions on the dorsal surface of the ventral plate may show the position of the endostyle (Halstead, 1973).

The superficial ornamentation of the dermal armor of cyathaspids typically consists of a series of rounded ridges of dentine running parallel with the margins of the plates (Figure 3-8*a*). In *Tolypelepis* (Figure 3-8*b*), the ridges are broken up in a pattern somewhat reminiscent of the tesserae of *Astraspis*. These units are termed **epitega.** Unlike the tesserae of *Astraspis,* they are superficial only, and the internal surface of the carapace shows no evidence of growth. Cyathaspids apparently ossified the superficial layers of their armor early in ontogeny by the progressive deposition of dentinal ridges. Once adult size was reached, the deeper layers of the carapace ossified rapidly and growth was terminated.

Amphiaspididae, Pteraspididae, and Psammosteidae

The cyathaspids apparently include the ancestors of a wide variety of more specialized ostracoderms that radiated throughout the Devonian. A family of highly mod-

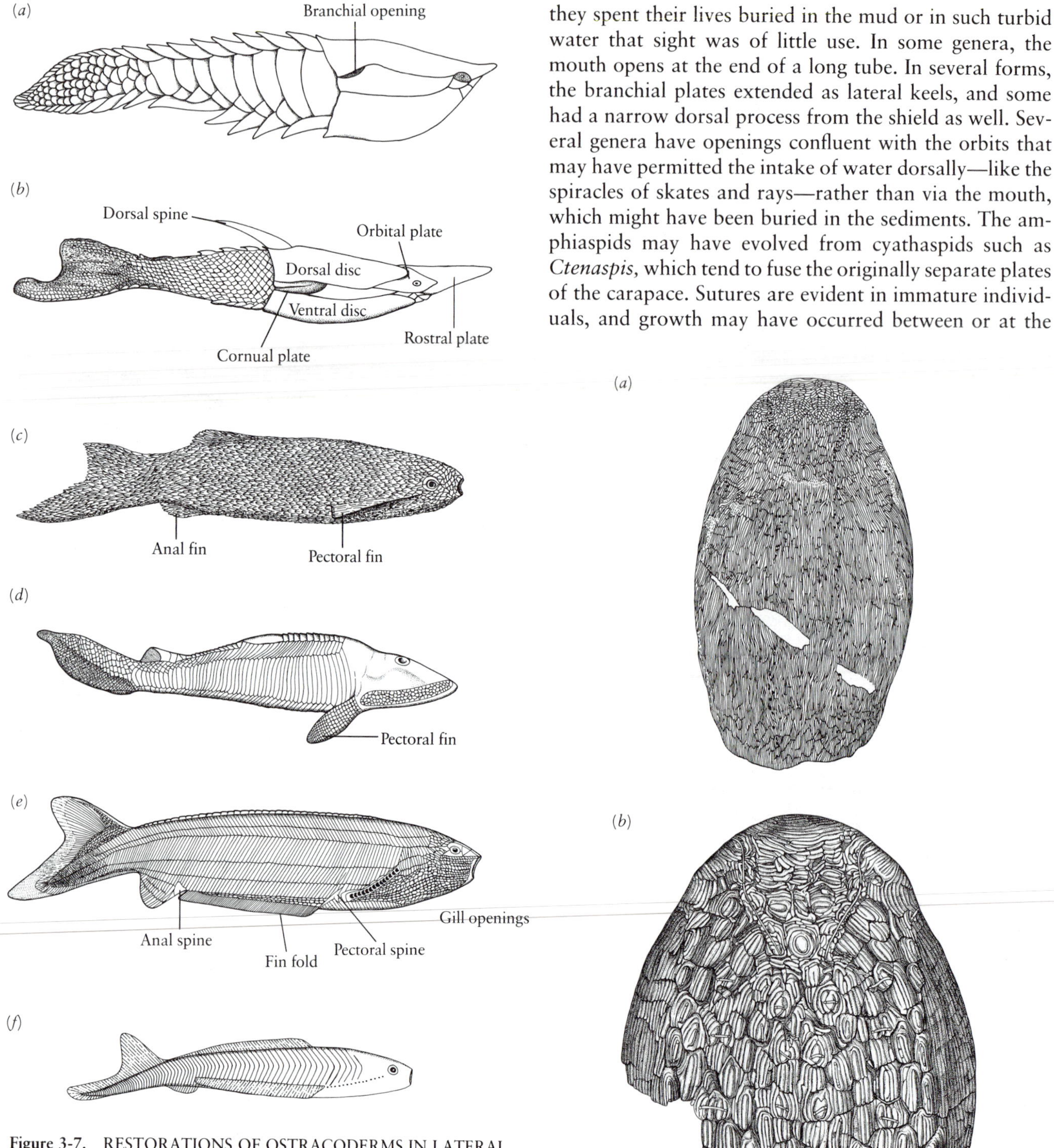

Figure 3-7. RESTORATIONS OF OSTRACODERMS IN LATERAL VIEW. (*a*) *Anglaspis*, $\times 1\frac{1}{2}$. (*b*) *Pteraspis*, $\times\frac{1}{3}$. (*c*) *Phlebolepis* (some specimens show a row of gill openings beneath the pectoral fin), $\times 1$. (*d*) *Hemicyclaspis*, $\times\frac{1}{3}$. (*e*) *Pharyngolepis*, $\times\frac{1}{3}$. (*f*) *Jaymoytius*, $\times\frac{1}{3}$. *From Moy-Thomas and Miles, 1971. By permission from Chapman and Hall Ltd.*

Figure 3-8. PATTERNS OF ORNAMENTATION OF DERMAL ARMOR IN HETEROSTRACANS. (*a*) *Ptomaspis*, $\times 1$. *From Denison,* Fieldiana, Geology, **16**, 1964. *By permission of Field Museum of Natural History, Chicago.* (*b*) *Tolypelepis*, $\times 2$. *From Stensiö, 1964.*

ified descendants, the Amphiaspididae (Figure 3-9*a*, *b*), are known from the Lower and early Middle Devonian of central Siberia and arctic Canada. In a recent review of the group, Novitskaya (1971) recognized nearly 20 genera. All show nearly complete fusion of the carapace and reduction or complete loss of the eyes. Apparently they spent their lives buried in the mud or in such turbid water that sight was of little use. In some genera, the mouth opens at the end of a long tube. In several forms, the branchial plates extended as lateral keels, and some had a narrow dorsal process from the shield as well. Several genera have openings confluent with the orbits that may have permitted the intake of water dorsally—like the spiracles of skates and rays—rather than via the mouth, which might have been buried in the sediments. The amphiaspids may have evolved from cyathaspids such as *Ctenaspis*, which tend to fuse the originally separate plates of the carapace. Sutures are evident in immature individuals, and growth may have occurred between or at the

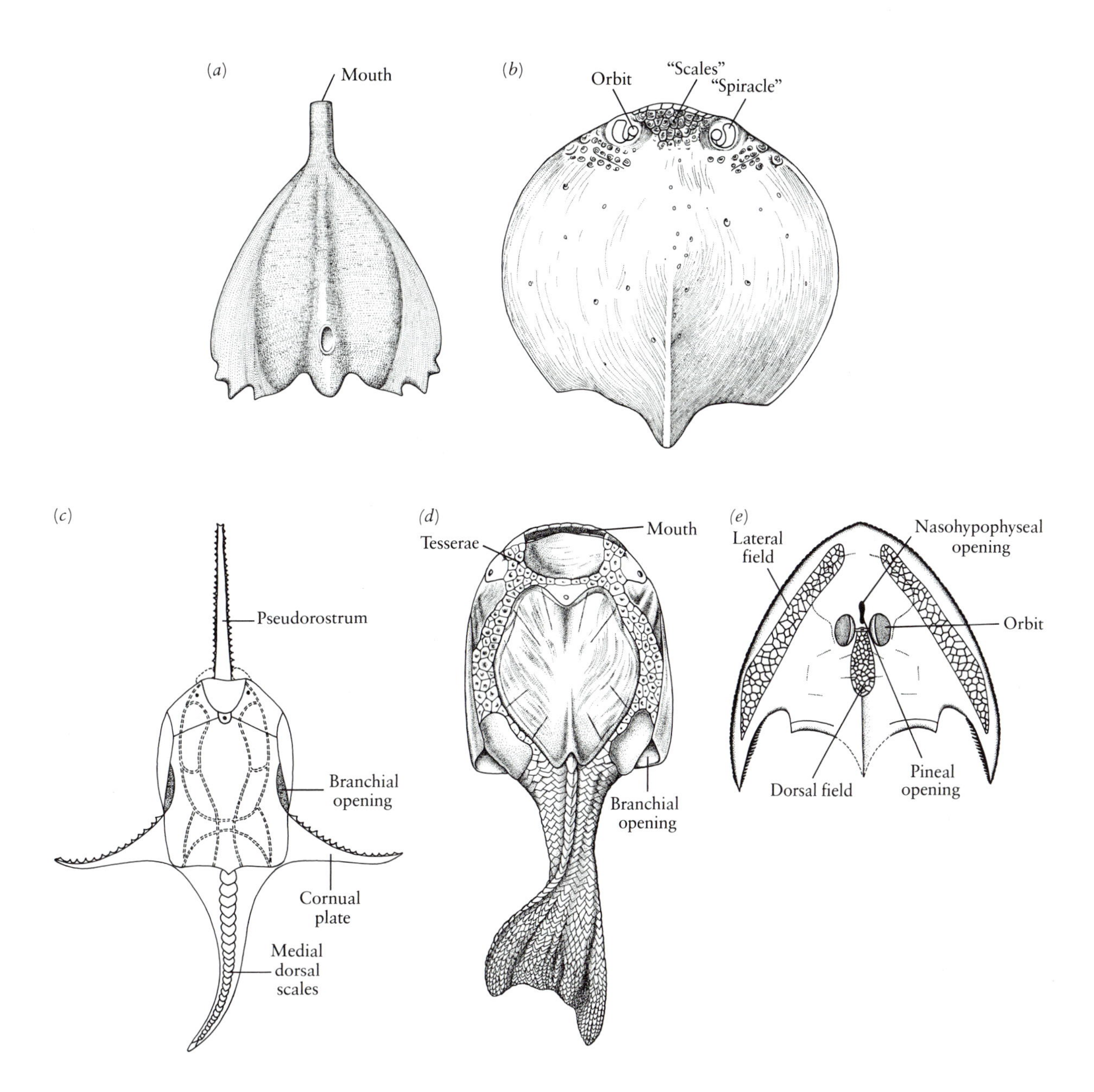

Figure 3-9. DORSAL VIEW OF THE HEAD SHIELD OR CARAPACE OF OSTRACODERMS. (*a*) and (*b*), representatives of the family Amphiaspididae. (*a*) *Eglonaspis*. The eyes are lost and the mouth opens at the end of a long tube, $\times\frac{1}{7}$. (*b*) *Gabreyaspis*. Openings lateral to the orbits may be similar to spiracles of skates and rays, $\times\frac{1}{3}$. (*c*) *Doryaspis*, a peculiar pteraspidomorph, $\times 5$. (*d*) The psammosteid *Drepanaspis*, $\times\frac{1}{4}$. (*e*) *Cephalaspis*, $\times\frac{3}{4}$. *From Moy-Thomas and Miles, 1971. By permission from Chapman and Hall Ltd.*

margins of the plates. However, a continuous layer of superficial material covers the plates in adults.

More typical cyathaspids were succeeded in time and anatomical complexity by the pteraspids (Figure 3-7*b*), animals of a basically similar body pattern and way of life. Pteraspids are among the best known, most numerous, and taxonomically diverse groups of ostracoderms (Denison, 1970; Blieck, 1984). The dermal shield shows more individual plates than are recognized in the cyathaspids; the plates show evidence of growth in most members of the group. In primitive pteraspids, the plates are relatively small. Like the cyathaspids, these animals elaborated their armor only when of nearly adult size. White (1958) demonstrated a range of plate sizes in later pteraspids (Figure 3-10). The larger plates show growth rings, indicating the initiation of ossification at a progressively smaller body size.

Although true paired fins never developed, the pteraspids did evolve laterally and dorsally projecting spines that may have served as protection, increased their stability in the water, and fixed their position in the mud. The trunk and caudal region is covered with smaller, more numerous scales than the cyathaspids, giving them more flexibility for swimming. The ventral lobe of the tail is considerably longer than the dorsal lobe. Many pteraspids developed an anterior rostral plate, extending far in front of the mouth. The genus *Doryaspis* (Figure 3-9*c*) evolved an apparently movable pseudorostrum, resembling that of a sawfish, together with greatly extended lateral cornual plates.

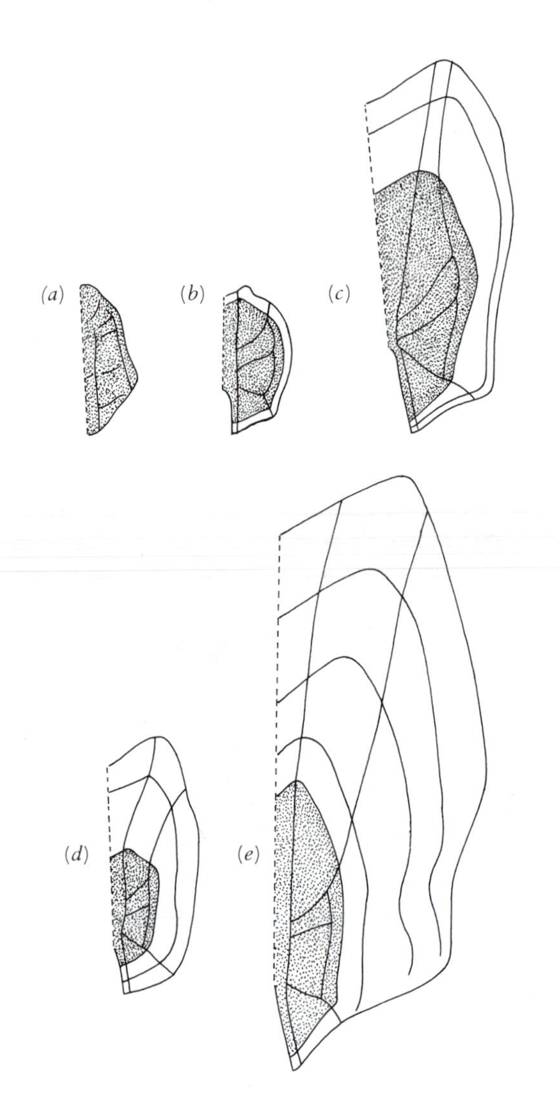

Figure 3-10. "GROWTH STAGES" IN THE ARMOR OF PTERASPIDS REPRESENTED BY A SERIES OF SPECIES FROM PROGRESSIVELY LATER DEPOSITS. Stippled area indicates the extent of the armor when it first ossifies; subsequent growth lines are parallel to its margin. Other lines show the course of the lateral line canals. *From White, 1958.*

In time and degree of specialization, the pteraspids are succeeded by the family Psammosteidae. This group has the same basic pattern of dermal armor, with the addition of a postorbital plate. They become more conspicuously dorsoventrally flattened, with both the branchial fenestrae and the mouth opening dorsally. The earliest Lower Devonian psammosteids are from river deposits, but by the late Lower Devonian, the best-known genus *Drepanaspis* (Figure 3-9*d*) is found in strictly marine sediments. The major radiation in the Middle and Upper Devonian occurred principally in the area which is now the Baltic, but with later distribution throughout northern Europe and North America. Obruchev and Mark-Kurik (1968) envisage early psammosteids as living at the bottom of rapidly flowing streams, supported by greatly elongated branchial plates and specialized ventral plates, some in the form of sled runners. However, it is difficult to see how their dorsally placed mouth would have functioned in this position. Most genera are known only from fragments, but isolated plates suggest a length of up to 2 meters. Fields of tesserae between the major plates developed in areas of particularly rapid growth in later psammosteids.

The arandaspids, astraspids, cyathaspids, amphiaspids, pteraspids, and psammosteids may all be grouped in a single order, the Heterostraci. This group is characterized by the presence of a single pair of external gill openings, paired nasal sacs, and an essentially solid carapace over the head and pharynx. They lack endoskeletal ossification and bone cells; paired fins never developed.

Thelodonti

The coelolepids or thelodonts are frequently allied with the heterostracans (Figure 3-7*c*). Their anatomy is not well known. The entire body is covered with tiny rhomboidal, nonimbricating scales. Without a solid exoskeleton or any internal ossification, these animals are nearly always preserved in a flattened condition, making restoration and establishment of anatomical details very difficult. Ritchie (1968b) made the most complete recent study of the body form, and Gross (1967) and Karatajute-Talimaa (1978) published extensive studies of the histology of their scales and proposed classifications of the known genera.

The body is fusiform in profile but somewhat dorsoventrally flattened. The mouth is nearly terminal and bordered by several rows of transversely oriented scales. A medial indentation immediately posterior to the dorsal scales bordering the mouth may represent the position of the pineal opening. The orbits are laterally placed, each surrounded by two crescentic scales. There is no evidence of either lateral narial openings or of a medial opening. Exceptionally well-preserved specimens show a straight row of approximately eight external gill openings, in contrast to the single pair of gill openings in all adequately known heterostracans. The gill openings are located beneath fragile, laterally oriented flaps of tissue identified as pectoral fins. In addition, there are small rounded dorsal and anal fins; the caudal fin is **hypocercal**, that is, the main axis is angled ventrally.

The scales (Figure 3-11) have a crown of dentinal tissue over an open pulp cavity. The basal region has no bone cells. Two subgroups are recognized on the basis of scale structures; the Thelodontida have one to three pulp cavities, and the Phlebolepidida have many pulp cavities. Some genera show a slightly different pattern of scales covering the head and trunk, while others have a more uniform appearance. The early thelodonts are known from marine waters, but later genera are found in both freshwater and marine deposits.

Adequately known specimens are limited to the Upper Silurian and Lower Devonian. Isolated scales with a similar structure are known as early as the Lower Silurian (Karatajute-Talimaa, 1978).

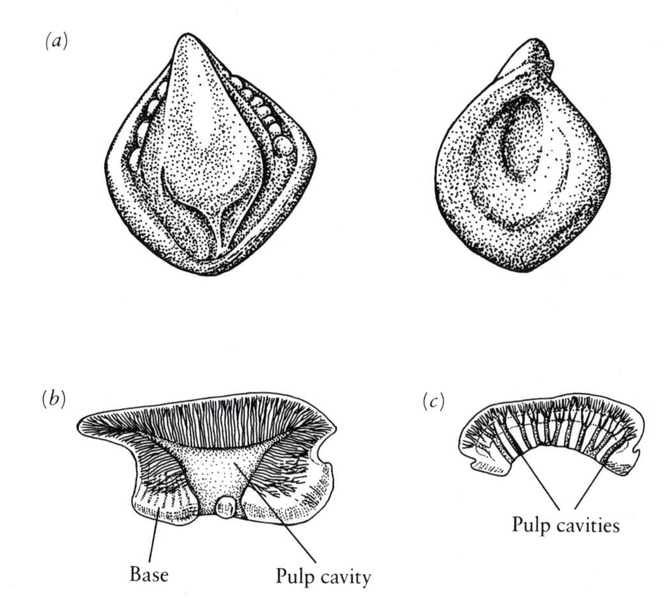

Figure 3-11. THELODONT SCALES. (*a*) *Logania* scale in dorsal and ventral view, ×40. *From Turner, 1973. By permission of Blackwell Scientific Publications Ltd.* (*b*) Cross section of *Thelodus* scale, showing simple pulp cavity characteristic of the Thelodontida. (*c*) Cross section of *Katoporus* scale, showing multiple pulp cavities characteristic of the Phlebolepidida, ×25. (*b–c*) *From Moy-Thomas and Miles, 1971. By permission from Chapman and Hall Ltd.*

Thelodonts have been classified with heterostracans based on the absence of a median nasal opening, a feature that clearly differentiates both groups from the cephalaspidomorphs. Since paired narial openings, multiple external gill openings, and the absence of bone cells are apparently primitive vertebrate characteristics, they do not indicate specific affinities between thelodonts and other groups of primitive fish. The appearance of the armor of *Lepidaspis*, from the Lower Devonian, suggests that thelodont denticles may have been derived from superficial dermal ornamentation such as is present in primitive heterostracans (Dineley and Loeffler, 1976).

CEPHALASPIDOMORPHI

Although there is some question as to whether the heterostracans and thelodonts should be classified together, a further assemblage of agnathans, the cephalaspidomorphs, are more obviously related to one another by the common presence of specialized traits. In most cephalaspidomorphs, both the narial opening and the nasal sac are single and medial, rather than paired and lateral, as in other vertebrate groups.

A detailed study of the brain and cranial nerves in fossil cephalaspidomorphs with heavily ossified endocrania demonstrates a striking similarity to the living lamprey. The lamprey also has a single medial narial opening that is known to develop ontogenetically from paired lateral structures (see Figure 3-17). We presume that the medial nostril of cephalaspidomorphs developed similarly and that they evolved from more primitive vertebrates with paired nasal capsules.

Not only are the nasal capsule and the narial opening uniquely medial in position, but they are also closely associated with a duct leading from the pituitary, the hypophyseal duct. In most cephalaspidomorphs and the lamprey, both emerge from the top of the head in a common nasohypophyseal opening. These changes are related to a reorganization of the entire anterior portion of the head and development of a rostral region that is not directly comparable with that of other vertebrates.

Two groups of ostracoderms that are otherwise very different in their basic anatomy exhibit the specialized monorhine condition. The anaspids (Figure 3-7*e* and *f*) are fusiform and similar to thelodonts in body outlines, but they are covered with larger, overlapping scales. The osteostracans (Figures 3-7*d* and 3-9*e*), often called cephalaspids, are heavily armored like the cyathaspids, and most are strongly adapted for a benthonic way of life. The dorsal position of the narial opening may have an adaptive advantage for the specialized benthonic osteostracans, but this is not obvious for the anaspids, which appear to be adapted to open-water swimming. At present, we cannot explain the original shift in the nasal capsules and the hypophyseal duct in the ancestors of the cephalaspidomorphs.

Osteostraci

The Osteostraci have the best-known anatomy of all Paleozoic Agnatha and are frequently used to initiate discussions of this assemblage. However, they are neither typical nor primitive ostracoderms and are not known in the fossil record until the Upper Silurian, as much as 100 million years after the appearance of the early heterostracans. They exhibit three advanced features that were independently evolved in jawed vertebrates: paired pectoral fins, an ossified endoskeleton, and lacunae for bone cells.

The earliest Osteostraci are known from marginal marine and lagoon deposits, but the majority of the later genera are found in stream, lake, and delta sediments. Their sudden appearance in the fossil record is presumably associated with the initial development of dermal ossification. In most, if not all osteostracans, growth must have been nearly complete prior to the ossification of the exoskeleton.

The tremataspids (Figure 3-12) are the most primitive osteostracans. They are known primarily from the Upper Silurian, are small, and have a solid carapace covering the body as far back as the base of the tail (Denison, 1951; Janvier, 1985b). The body is dorsoventrally compressed, but the ventral surface is rounded rather than flattened, as it is in the more advanced cephalaspids. The configuration of the armor is reminiscent of the cyathaspids, but the gills (approximately 10 in number) open separately to the exterior along the anteroventral margin of the carapace. The branchial chamber is compressed anteriorly,

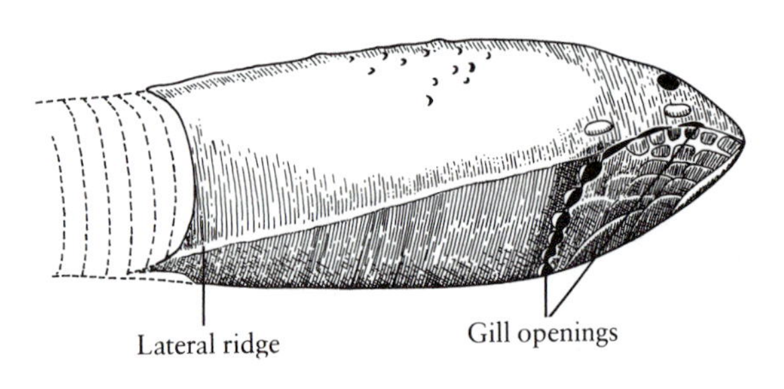

Figure 3-12. CEPHALOTHORACIC SHIELD OF THE PRIMITIVE OSTEOSTRACAN, *TREMATASPIS*, $\times 1\frac{1}{2}$. *From Stensiö, 1964.*

like that of other ostracoderms. The dermal skeleton shows no evidence of subdivision into separate areas of ossification, except for the presence of a mosaic of small plates ventrally, between the rows of gill openings. The head and pharynx are ossified as a unit, incorporating the braincase and the gill supports. Small scales cover the caudal region. There is no evidence of paired, anal, or dorsal fins.

In the late Silurian and Devonian, the tremataspids are succeeded by the better-known cephalaspids, which Janvier (1985a) suggests are an unnatural, polyphyletic, assemblage. They are more completely committed to a benthonic way of life to judge by their broad, flat, ventral surface. Nevertheless, they alone among the Agnatha evolved muscular paired fins (Figure 3-13). Early tremataspids show no trace of fins. In later genera, the bony carapace is reduced and the scale-covered portion of the body extends further anteriorly. Lateral ridges develop on the head shield, which may extend as horns, or cornua. Openings on their posterior margin mark the position of the emergence of the nerves, muscles, and blood vessels associated with the limb. The pectoral fins are not known to have had an internal skeleton. Pelvic fins never developed in the Agnatha.

In osteostracans, the notochordal axis of the tail is tilted upward to give the **heterocercal** configuration that is common to sharks and other primitive jawed fish. A ventrally directed caudal fin, such as is present in most other ostracoderms, would ill suit the benthonic osteostracans.

The eyes are crowded toward the midline, nearly adjacent to the pineal and nasohypophyseal openings. A striking feature of the osteostracans is the presence of paired and medial areas of polygonal plates (Figure 3-9*e*), one behind the nasohypophyseal opening and others on the lateral margin of the head shield. The polygons are formed of very thin layers of superficial bone, and a narrow space separates them from the underlying exoskeleton. The terminal branches of the large nerves of the acousticofacialis complex, or fluid-filled tubes extending from this region of the brain, ramify within the endoskeleton beneath the plates (Figure 3-14*a*). These areas have been termed "electric fields," because of their superficial resemblance to the muscles that generate current in electric fish. Their relationship to the underlying tubes from the vestibular area of the brain suggests that they may have been pressure-sensitive areas.

The high degree of ossification of the endoskeleton among the Osteostraci provides detailed evidence of the structure of the brain, cranial nerves, and mouth and gill region. This evidence is valuable in evaluating the origin of jaws and the ancestry of higher vertebrates. In the fossils, the cavities once occupied by the brain, cranial nerves, and blood vessels are filled with matrix and can be revealed by natural weathering, mechanical removal of the surrounding bone, or grinding serial sections at close intervals. The configuration of the endocranial cavities indicates a similar brain structure to that of other primitive vertebrates, with particular resemblances to the lamprey, *Petromyzon*. Canals leading from the posterior portion of the brain stem indicate that the left and right spino-occipital nerves exit alternately and that the dorsal and ventral rami do not join to form ganglia—as is the case in higher vertebrates—but remain distinct. The semicircular canals appear very large relative to the narrow brain stem. Two pairs of vertical canals are present, but the horizontal canal is absent. Large canals extend toward the lateral sensory fields from the vestibular area.

On the roof of the mesencephalon, the area occupied by the right habenular ganglion (associated with the sense of smell) is more expanded than the left, as is the case with modern lampreys. The anterior portion of the brain stem is expanded laterally to form the telencephalon, above which is the unpaired nasal cavity leading to the external nasohypophyseal opening. Ventrally, the large hypophyseal duct ends bluntly in the area of the hypophysis.

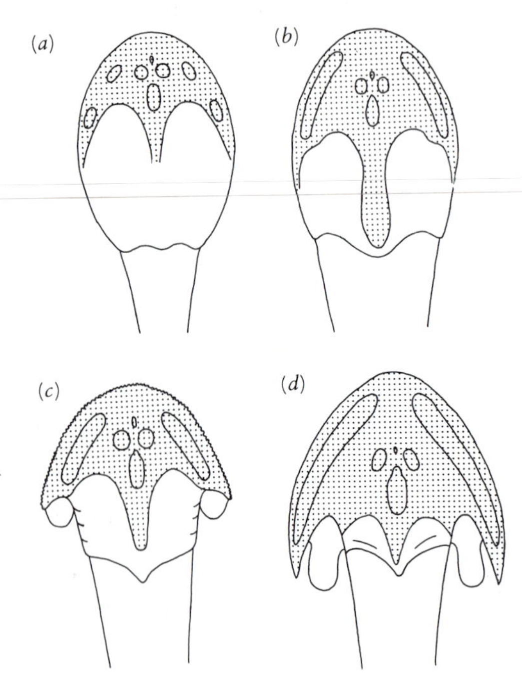

Figure 3-13. OSTEOSTRACANS. Diagram shows progressive changes in the extent of the carapace and the development of pectoral fins. Stippled areas indicate extent of endochondral braincase. *After Westoll, 1958.*

(a)

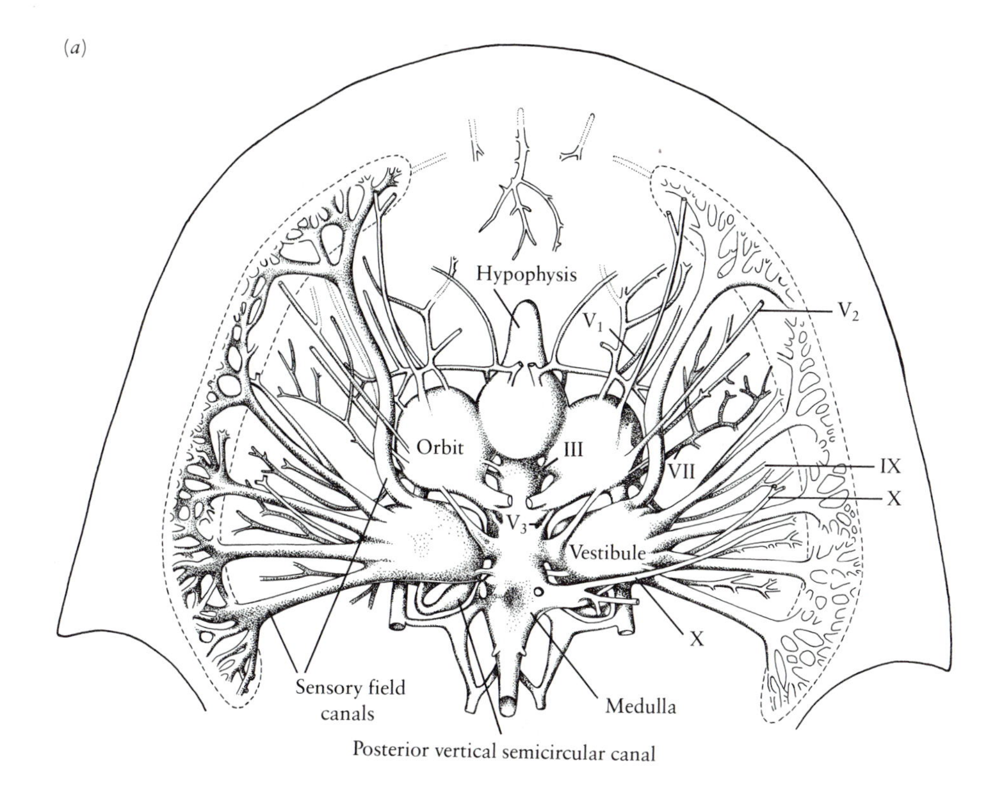

(b)

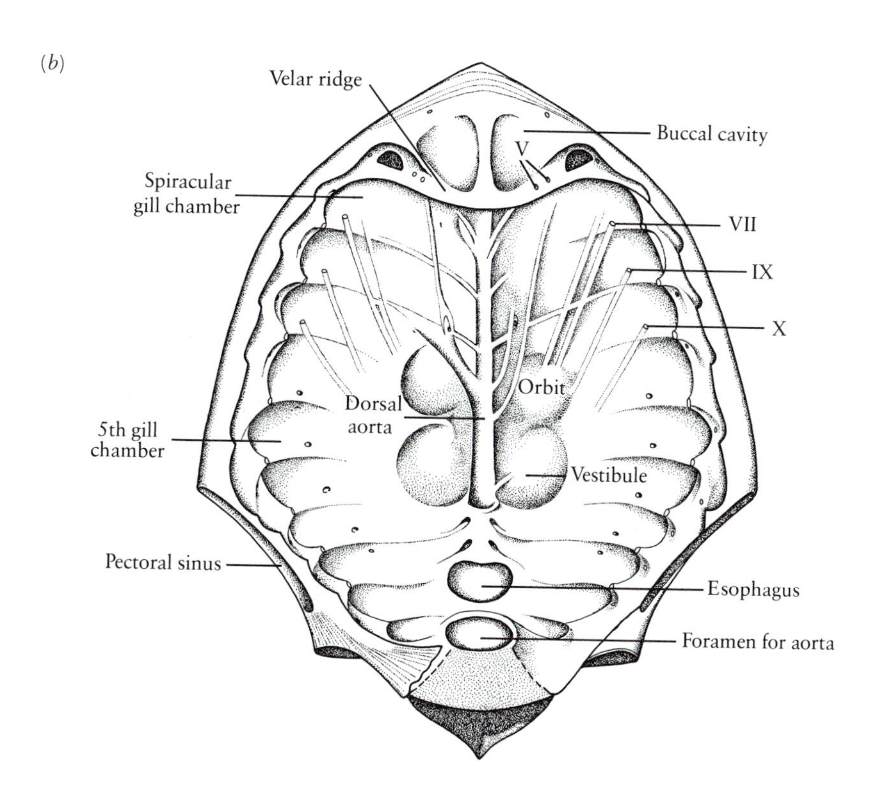

Figure 3-14. INTERNAL STRUCTURE OF THE HEAD OF OSTEOSTRANS. (a) *Kiaeraspis*. Reconstruction of the endocranial cavities containing the brain and associated sense organs, ventral view. Roman numerals identify cranial nerves. *From Stensiö, 1963*. (b) *Nectaspis*. Ventral view of orobranchial chamber. *From Moy-Thomas and Miles, 1971. By permission from Chapman and Hall Ltd.*

We can expose the oropharyngeal cavity by removing the polygonal plates between the gill openings (Figure 3-14*b*). The dorsal surface of the cavity typically bears a ridge between the mouth cavity and the pharynx, which presumably marks the position of a velum. This ridge is associated with the Vth or trigeminal nerve, which innervates the jaws in higher vertebrates. The first gill chamber is behind the ridge. It is considered comparable with the spiracular cleft and is innervated by the VIIth or facial nerve. More posterior chambers are associated with nerves IX and X.

The osteostracans emerged in the late Silurian, diversified in the early Devonian, and were extinct by the end of that period.

Galeaspida

During the past 15 years, the discovery in southern China of a host of unusual genera from the Lower Devonian demonstrates the presence of an entirely new order of jawless vertebrates, the Galeaspida (Figure 3-15). Members of this order have recently been reviewed by Halstead, Liu, and P'an (1979), Janvier (1984), and Pan Jiang (1984). Early descriptions suggested that some of the genera might be Osteostraci and that others were related to the Heterostraci, but additional material indicates that all belong to a single distinct group of cephalaspidomorphs. This group evolved in isolation, but it paralleled other ostracoderm orders in many ways.

The galeaspids resemble the osteostracans in having a heavy carapace and a well-ossified endochondral braincase. In some specimens, infillings of the cranial cavities provide considerable structural detail. Like the osteostracans, the galeaspids have two vertical semicircular canals but lack the horizontal canal. Surprisingly, they have paired nasal sacs that are associated with a median nasohypophyseal cavity. As in osteostracans, the ventral surface of the carapace consists of a mosaic of small plates that lie between the multiple gill openings.

Dorsally, toward the anterior end of the carapace is a large opening whose shape differs from genus to genus. In *Polybranchiaspis* (Figure 3-15*a*), it is a transversely elongate oval, but in *Galeaspis*, it is a longitudinal slit. In *Polybranchiaspis*, it was originally interpreted as the mouth, but that opening is apparently in the typical position on the ventral surface. The dorsal opening may be related to the nasohypophyseal organ or another sensory structure. The pineal opening does not pierce the dorsal shield but is represented by a depression in its ventral surface, as is the case in heterostracans.

In *Polybranchiaspis*, the dorsal surface of the skull is covered by stellate tubercles and has a pattern of sensory canal grooves that more closely resembles that of heterostracans than osteostracans. Dorsal sensory fields, so conspicuous in the osteostracans, are not characteristic of galeaspids but have tentatively been reported in one genus (Halstead, 1982).

The axis of the tail in galeaspids is inclined ventrally, in contrast with that of osteostracans, and no specimens show any evidence of paired fins.

Anaspida

The anaspids (Figure 3-7*e* and *f*) are entirely different from the Osteostraci in general body form but have long been considered closely related because they have a median nasohypophyseal opening. From their first occurrence in the Upper Silurian, anaspids are fusiform animals with perhaps the most normal fishlike appearance of all ostracoderms. However, their fin structure sets them strongly apart from early jawed fish. The axis of the caudal fin bends ventrally rather than dorsally, giving them a reversed heterocercal tail. There is usually a stout spine in the shoulder region. No genera have normal pectoral or pelvic fins, but some forms have long lateral fin folds running most of the length of the body that superficially resemble the metapleural folds of amphioxus. While some authors suggest that paired fins in fish evolved by the subdivision of such primitively continuous folds, the earliest known osteostracans lack any trace of lateral fins or fin folds.

Typically, the body is covered with small overlapping scales composed of laminar bone, resembling the basal layer of the dermal plates in other ostracoderms. The scale rows appear to correspond with the body segments. Dorsal and lateral scale rows are in the pattern of an anteriorly facing *V*, suggesting the outline of myotomes in amphioxus, in contrast with the *W*-shaped pattern in living Agnatha and jawed fish. The scales lack bone cells; growth occurs at the margins of the scales, and some species show a considerable range of body sizes.

The bony plates covering the head are small but distinctly shaped. The orbits are more laterally placed than in the Osteostraci, and the pineal opening pierces the

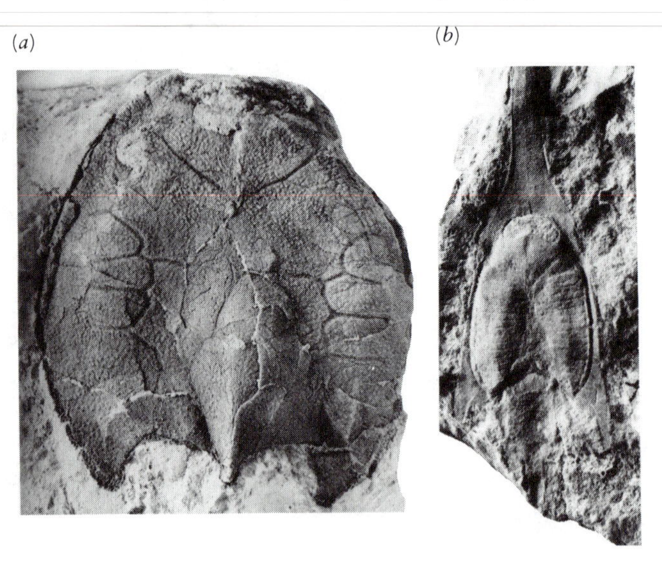

Figure 3-15. GALEASPID OSTEOSTRACANS FROM THE DEVONIAN OF CHINA. (*a*) *Polybranchiaspis*, $\times\frac{3}{4}$. (*b*) *Sangiaspis*, $\times 1\frac{1}{4}$. *Courtesy of Dr. Y.-H. Liu.*

scales just behind the nasohypophyseal duct. The mouth is terminal or nearly so. There are 6 to 15 gill openings running diagonally down the flank. No anaspids show more than traces of the endochondral skeleton.

The genera *Birkenia, Paryngolepis, Pterygolepis,* and *Rhyncholepis* from the Upper Silurian all have substantial bony scales. Other anaspids, ranging from the Upper Silurian *Lasanius* and *Jaymoytius* to *Endeiolepis* and *Euphanerops* in the Upper Devonian, have little if any skeletal ossification.

Anaspids are known primarily from freshwater deposits, except for the oldest known genus, *Jaymoytius,* which is from near-shore marine or brackish water deposits (Ritchie, 1968a).

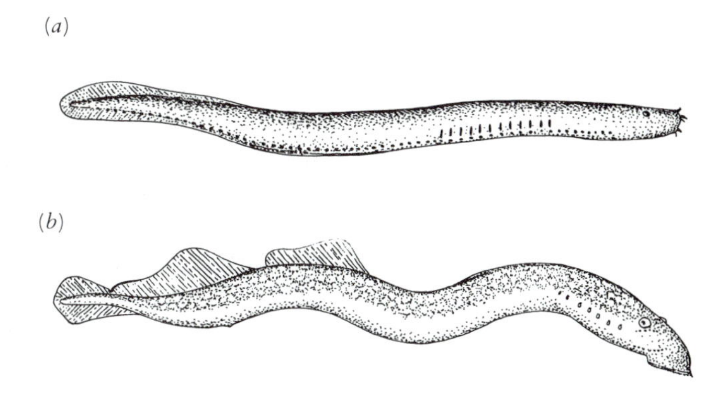

Figure 3-16. LIVING CYCLOSTOMES. (*a*) The hagfish, *Polistotrema.* (*b*) The lamprey, *Petromyzon. From Storer and Usinger, 1965. Reprinted with permission of McGraw-Hill Book Company.*

LIVING JAWLESS VERTEBRATES

Most groups of jawless vertebrates became extinct by the end of the Devonian, presumably as a result of competition and predation from the jawed fish that diversified greatly within the Devonian.

However, one or more lineages survived and gave rise to the modern agnathous forms. Two distinct groups of jawless fish are recognized today, the Petromyzontoidea, which includes the lampreys, and the Myxinoidea—hagfish and slime hags (Figure 3-16). Together, the modern jawless fish are termed cyclostomes because of the circular shape of the mouth. Both groups differ from the Paleozoic agnathans in the total absence of bone and scales. They have an elongate, eel-shaped body, without a trace of paired fins. Both lack the horizontal semicircular canal.

The gills are primarily endodermal in origin and are located in pouches, medial to the muscular wall of the pharynx. We assume that the gills of the ostracoderms were also medial to the wall of the pharynx and the gill supports. The visceral skeleton, which is well developed in the lamprey, is a basketlike structure, quite unlike the jointed visceral arches of gnathostomes.

Although both lampreys and hagfish have a medial nasal opening confluent with the hypophyseal duct, the relative position of the opening is quite different in the two groups. It is dorsal in the lamprey, more or less between the eyes, while in the Myxinoidea it opens anteriorly and is continuous with a duct that leads into the pharynx. This passage allows water to be drawn into the gill pouches when the mouth is occupied with feeding. A muscular pump developed from the velum moves water for respiration. In lampreys, water can be pumped directly in and out of the gill pouches. Muscles in the pharyngeal wall compress the visceral arch skeleton as the gill pouches are emptied. They expand as a result of elastic recoil of the cartilage (Northcutt and Gans, 1983).

The gill pouches of the lamprey open directly to the exterior. The myxinoids' gill pouches exit via posteriorly directed branchial ducts that lead to a variable number of excurrent openings. The presence of only a single pair of excurrent openings in some hagfish led Stensiö (1964) to ally them with the heterostracans. The variability of this feature among the myxinoids suggests that it probably evolved within this group, rather than being inherited from ancestors in which this trait was always expressed. Its development among the myxinoids is attributed to their burrowing habits.

Lampreys and hagfish are also divergent in other structures and habits, suggesting a long and separate history whatever their specific ancestry.

Hagfish are entirely marine and live in burrows in soft sediments on the continental shelf. Their sensory organs are poorly differentiated, which may reflect a combination of their primitive state and degeneration associated with life in the mud. The eyes show most of the basic components common to other vertebrates, but their development is equivalent to that of the larval lamprey. They are covered with thick but translucent tissue. Their limited sense of vision is supplemented by the presence of sensory tentacles surrounding the mouth. The two pairs of dorsal semicircular canals are poorly differentiated from one another, which may be a primitive feature.

Hagfish have horny plates on either side of a protrusible tongue, which can be spread apart or folded together like pincers to grasp and bite into prey. They feed mostly on polychaete worms but also attack dead and dying fish whom they may enter by the mouth and consume the soft tissues from the inside. The young hatch out as miniatures of the adult, without a larval stage (Brodal and Fänge, 1963).

Most lampreys spend their adult life in the sea and come into fresh water to reproduce. The larval stage, termed the **ammocoete,** is so different from the adult that it was originally considered to belong to a distinct taxon (see Figure 2-3). The similarities of the pharynx and feeding behavior to those of amphioxus were emphasized in Chapter 2. The ammocoete differs significantly in using the muscular walls of the pharynx to pump water in and out, rather than relying on cilia as do the cephalochordates. The filter-feeding larval stage may last for several years. The entire head and pharynx are reorganized at metamorphosis. The nostril and the hypophyseal duct migrate from the primitively ventral position to the top of the head, behind an enlarged rostral area (Figure 3-17). The mouth is transformed into a circular sucker that is used by the adult lamprey to attach to fish on which it feeds. Rows of epidermal structures resembling teeth line the sucker and cover the protrusible tongue. The endostyle is transformed into a thyroid gland, and the pharynx becomes displaced so that the esophagus passes directly into the gut. The possibility of such an extensive metamorphosis among some ostracoderms provides a new way of looking at the differences between the major groups (Halstead, 1982).

Because of the medial dorsal position of the nasal opening and its association with the hypophyseal duct, many authors believe that lampreys have affinities with the cephalaspidormorphs. The closest resemblance lies with the anaspids, which have fusiform bodies and relatively poorly developed fins. In particular, genera such as *Jaymoytius* (Figure 3-18*a*) and *Endeiolepis,* which have reduced body armor, approach the condition of the boneless lampreys and hagfish. As in lampreys, the gill pouches of *Jaymoytius* are circular and an annular cartilage supports the mouth.

Recent discoveries of fossil lampreys from the Upper Mississippian of Montana (Janvier and Lund, 1983) and the Middle Pennsylvanian of Mazon Creek, Illinois (Bardack and Zangerl, 1968, 1971; Bardack and Richardson, 1977) provide evidence of the specific origin of cyclostomes. The similarities in position and configuration of the cartilaginous elements of the head in the fossil *Mayomyzon* (Figure 3-18*b*) and the modern lamprey are striking. Epidermal teeth are not evident in *Mayomyzon,* but the general structure of the mouth indicates a mode of feeding similar to the lamprey. The olfactory, optic, and otic capsules are similarly supported and are com-

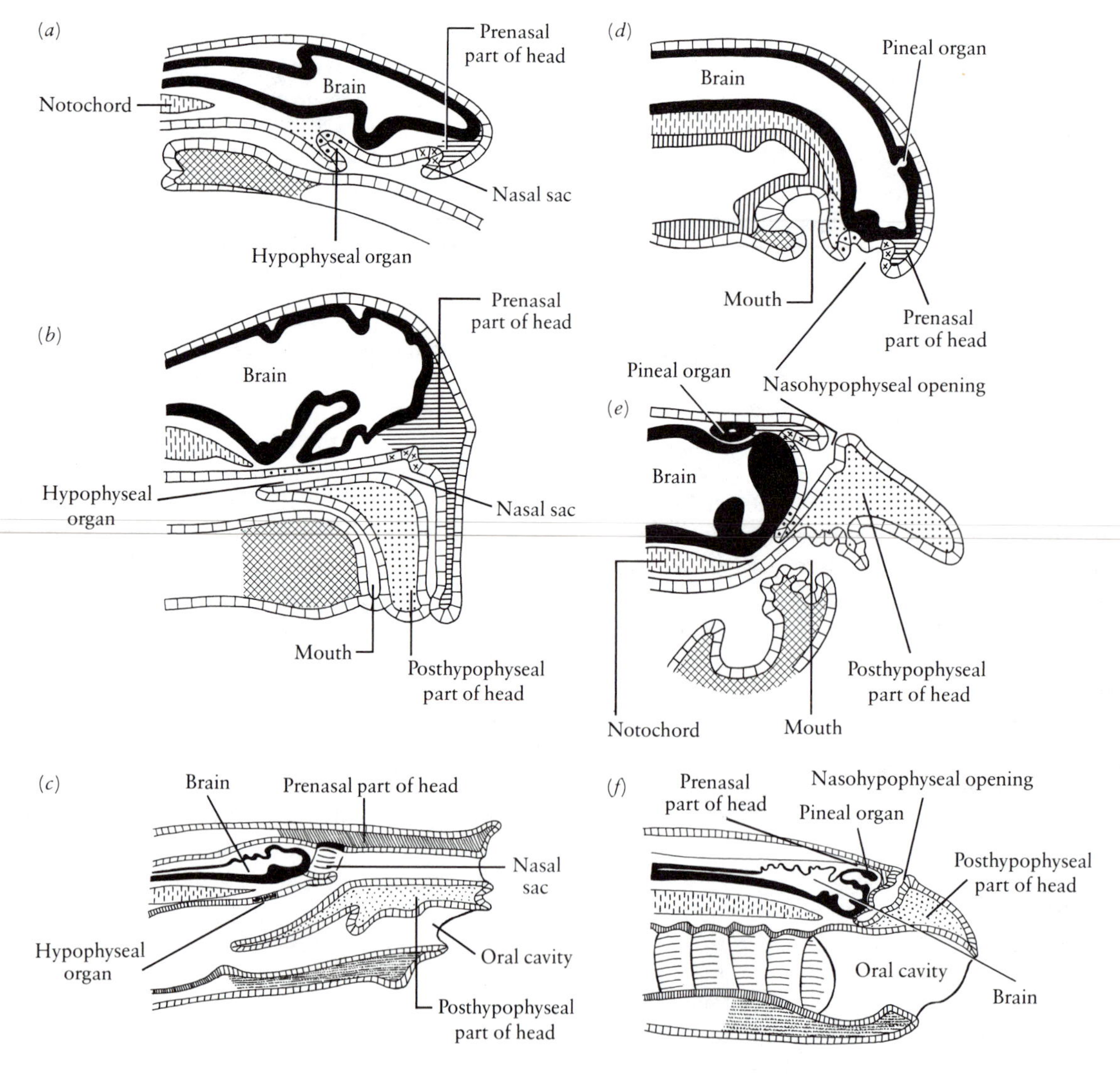

Figure 3-17. Comparison of the developmental stages of the recent Myxinida (*a–c*) and Petromyzontida (*d–f*). *From Heintz, 1963. Reprinted by permission of Norwegian University Press.*

parably positioned relative to the remainder of the head. In both, there are seven external gill openings, posterior to which a perichordal cartilage anchors the piston cartilage to the tongue. *Mayomyzon* is similar to the living lamprey but is placed in a distinct family because of the absence of rasping "teeth" and the simplicity of the branchial skeleton. *Hardistiella* from the Upper Mississippian retains a hypocercal tail and an anal fin with fin rays, in common with both anapsids and thelodonts.

In contrast with their similarities to the lamprey, *Mayomyzon* and *Hardistiella* show no specialized features of the second living cyclostome groups, the hagfishes. Stensiö (1964) argued that the hagfishes evolved from the heterostracans and suggested that the nasal sacs in hagfishes were derived from paired structures, comparable to those of the pteraspids. Since the nasal sacs in all vertebrates appear to be derived embryologically from paired placodes, this argument has little weight. Nevertheless, biochemical as well as structural evidence points to an early divergence of hagfishes from the anaspid-lamprey stock. Certainly the two lineages must have separated long before the appearance of *Mayomyzon*—a definite lamprey—in the Carboniferous. The structural peculiarities of the hagfishes may result from their divergence prior to the appearance of any of the known fossil monorhines.

Bardack (1986) has recently reported a fossil hagfish of essentially modern morphology from the same locality that yielded *Mayomyzon*. It provides no specific evidence as to the origin of this group.

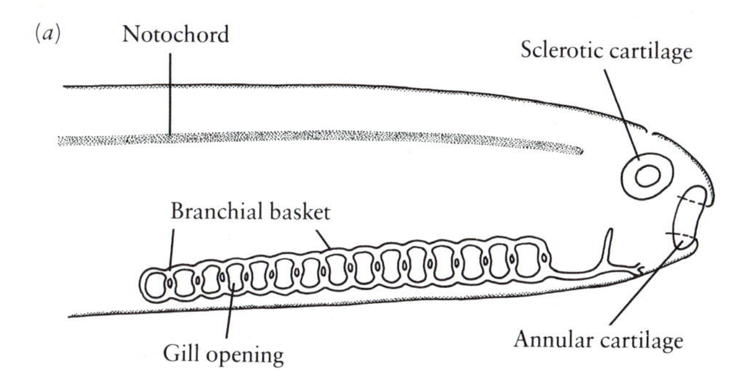

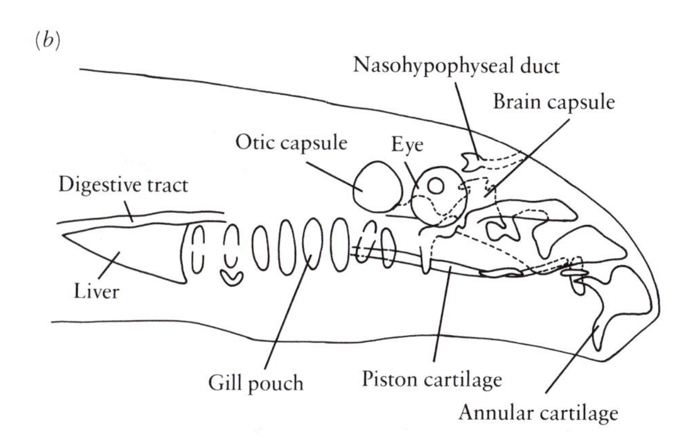

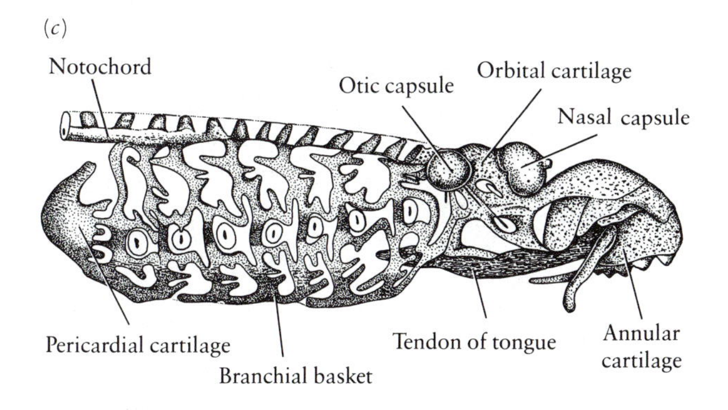

Figure 3-18. ANASPIDS AND CYCLOSTOMES. (*a*) The oldest anaspid, *Jaymoytius*, showing internal structure of the head region, × 1¼. (*b*) *Mayomyzon*, a lamprey from the *Middle Pennsylvanian*, × 4. *(a–b) From Moy-Thomas and Miles, 1971. By permission from Chapman and Hall Ltd.* (*c*) *Petromyzon*, a modern lamprey. *From Young, 1950. Reproduced with permission of Oxford University Press.*

INTERRELATIONSHIPS OF PRIMITIVE VERTEBRATES

The nature of the interrelationships of the major groups of jawless vertebrates remains uncertain. It is possible that additional fossils will be found in the late Cambrian and early Ordovician, but the sudden appearance of several ostracoderm groups, each with a distinct pattern of dermal armor, indicates that they may have had a long history without a bony skeleton. In its absence, the chances of fossilization are limited. Additional information from later fossils and more complete understanding of the structure and developmental processes of living jawless fish may contribute to establishing their interrelationships, but this seems a distant goal.

Several recent papers deal with the interrelationships of primitive vertebrates. All stress the use of derived characters to establish affinities and make use of essentially the same characters that have been discussed here. Differences in the interpretations of polarity and emphasis on different characters have led to divergent conclusions.

Janvier (1981) argues that the Myxinoidea and the heterostracans are relatively closely related to one another but not to other vertebrates; he maintains that pteromyzontoids, osteostracans, and jawed vertebrates share a common ancestry.

Halstead (1982) suggests that the lampreys are closely related to the cephalaspidomorphs but have no close affinities to hagfish. He allies the heterostracans with the jawed vertebrates, primarily on the assumption that paired nasal openings are derived, rather than a primitive feature of vertebrates. He argues that the cyclostomes are primitive because of their total absence of armor and paired appendages.

Schaeffer and Thomson (1980), who emphasize the modern genera in establishing relationships, stress the close affinities of the two groups of cyclostomes. They argue that the pattern of development and the adult structure of the gills clearly separate the modern Agnatha from gnathostomes. In both cyclostome groups, the gills are primarily endodermal in origin and are medial to the visceral skeleton and the muscular wall of the pharynx. In jawed vertebrates, the gills are ectodermal in origin and lateral to the gill arches. Neither condition is plausibly antecedent to the other, indicating that the gills were little developed in the primitive vertebrates that were ancestral to both the Agnatha and to gnathostomes.

Yalden (1985) recognized 11 synapomorphies that unite the feeding structures of lampreys and hagfish and concluded that the cyclostomes are a natural group. He included the Myxinoidea within the Cephalaspidomorphi and emphasized the fundamental differences between both groups of cyclostomes and gnathostomes.

In marked contrast, Forey (1984) argues that lampreys are the sister group of gnathostomes, while Janvier (1984) finds evidence that the osteostracans are most closely related to the gnathostomes. These papers stress particular derived characters that are shared by one or another group of agnathans and jawed vertebrates, but neither author establishes that these characters are homologous. They provide no direct evidence that the characters they use to unite the gnathostomes with either the lampreys or the osteostracans were present in the immediate common ancestors of these groups.

Summary

Primitive members of the class Agnatha are known from the late Cambrian to the end of the Devonian. All have a covering of scales or an external bony carapace. They are referred to collectively as ostracoderms but are formally divided into two subclasses, the Pteraspidomorphi and the Cephalaspidomorphi.

The earliest remains of possible vertebrates are small fragments of phosphatic plates from the late Cambrian. Five pteraspidomorph genera are known in the Ordovician. *Arandaspis,* from the middle Ordovician of Australia, is represented by remains of most of the carapace. *Astraspis* provides our earliest knowledge of the histology of bony tissue. Unlike most modern vertebrates, it lacks lacunae for bone cells.

The Cyathaspididae range in age from the middle Silurian into the early Devonian and are the earliest vertebrates for which the entire body form is known. The head and trunk were completely encased in bony plates that ossified when they reached adult size, precluding further growth. The tail was symmetrical and covered with overlapping scales. Paired fins were absent. There was no internal ossification, but impressions of the inner surface of the carapace show the outline of the dorsal surface of the brain, two pairs of vertical semicircular canals, and a series of gill pouches that had a common external exit. The eyes were laterally placed and the nasal sacs paired.

Cyathaspids may include the ancestors of the Amphiaspididae, Pteraspididae, and Psammosteidae, which were common in the Devonian. All are included in the order Heterostraci.

The thelodonts, known from the early Silurian through the early Devonian, were small, fusiform, and completely covered by rhomboidal, nonimbricating scales. Like the Heterostraci, the orbits were lateral and there is no trace of an internal skeleton. However, the gill pouches had individual external openings, and there are fleshy lateral structures that have been identified as pectoral fins. Thelodonts retain many features found in the most primitive vertebrates, but almost nothing is known of their internal anatomy and their phylogenetic position remains subject to speculation.

Osteostracans, galeaspids, and anaspids are members of the Cephalaspidomorphi; they are distinguished from the Pteraspidomorphi because they possess a medial nasohypophyseal opening in contrast with the paired lateral narial openings of other vertebrates. Osteostracans, including the cephalaspids, are the best known of all ostracoderms. Their internal skeleton is heavily ossified, showing the configuration of the brain and other cranial structures in great detail. They definitely lack a horizontal semicircular canal. The head is characterized by paired and medial dorsal sensory fields. Pectoral fins evolved within the group. In most genera, the body is strongly dorsoventrally flattened and they were probably benthonic.

The galeaspids are common in the Lower Devonian of China. Like the osteostracans, they have an ossified endocranium and medially placed orbits. They differ in having paired nasal capsules, although they are associated with a medial nasohypophyseal passage. The anaspids were fusiform and presumably actively swimming fish, with small overlapping scales and a reverse heterocercal tail.

Both groups of modern agnathans, the lamprey and hagfish, appear in the Carboniferous. They completely lack bones, scales, and paired fins. The lampreys may have evolved from anaspid cephalaspidomorphs such as *Jaymoytius* and *Endeiolepis* that appear to have reduced or lost their body scales. There is no specific evidence to establish the origin of the hagfish. Much remains to be learned concerning the relationships among the early jawless vertebrates.

References

Bardack, D. (1986). Manuscript in preparation.

Bardack, D., and Richardson, E. S. (1977). New agnathous fishes from the Pennsylvanian of Illinois. *Fieldiana, Geol.,* 33: 489–510.

Bardack, D., and Zangerl, R. (1968). First fossil lamprey: A record from the Pennsylvanian of Illinois. *Science*, **162**: 1265–1267.

Bardack, D., and Zangerl, R. (1971). Lampreys in the fossil record. In M. W. Hardisty and I. C. Potter (eds.), *The Biology of Lampreys*, pp. 67–84. Academic Press, New York.

Blieck, A. (1984). *Les Hetérotracés Pteraspidiformes, Agnathes du Silurien-Dévonien du Continent Nord—Atlantique et des Blocs Avoisiments: Révision systematique, phylogénie, biostratigraphie, biogéographie*. Editions du Centre National de la Recherche Scientifique, Paris.

Bockelie, T., and Fortey, R. A. (1976). An early Ordovician vertebrate. *Nature*, **260**: 36–38.

Brodal, A., and Fänge, R. (eds.) (1963). *The Biology of Myxine*. Universitetsforlaget, Oslo.

Denison, R. H. (1951). Evolution and classification of the Osteostraci. *Fieldiana, Geol*, **11**: 157–196.

Denison, R. H. (1964). The Cyathaspididae: A family of Silurian and Devonian jawless vertebrates. *Fieldiana, Geol.*, **13**: 309–473.

Denison, R. H. (1967). Ordovician vertebrates from western United States. *Fieldiana, Geol*, **16**: 131–192.

Denison, R. H. (1970). Revised classification of Pteraspididae with description of new forms from Wyoming. *Fieldiana, Geol.*, **20**: 1–41.

Dineley, D. L., and Loeffler, E. J. (1976). Ostracoderm faunas of the Delorme and associated Siluro-Devonian Formations, North West Territories, Canada. *Spec. Papers Palaeont.*, **18**: 1–214.

Forey, P. L. (1984). Yet more reflections on agnathan-gnathostome relationships. *J. Vert. Pal.*, **4**(3): 330–343.

Gross, W. (1967). Über Thelodontier-Schuppen. *Palaeontographica, Abt. A*, **127**: 1–67.

Halstead, L. B. (1973). The heterostracan fishes. *Biol. Rev.*, **48**: 279–332.

Halstead, L. B. (1982). Evolutionary trends and the phylogeny of the Agnatha. In K. A. Joysey and A. E. Friday (eds.), *Problems of Phylogenetic Reconstruction. Systematics Assoc. Spec. Vol.*, **21**: 159–196.

Halstead, L. B., Liu, Y.-H., and P'an, K. (1979). Agnathans from the Devonian of China. *Nature*, **282**: 831–833.

Heintz, A. (1963). Phylogenetic aspects of myxinoids. In A. Brodal and R. Fänge (eds.), *The Biology of Myxine*, pp. 9–21. Universitetsforlaget, Oslo.

Janvier, P. (1981). The phylogeny of the Craniata, with particular reference to the significance of fossil "agnathans". *J. Vert. Paleont.*, **1**(2): 121–159.

Janvier, P. (1984). The relationships of the Osteostraci and Galeaspida. *J. Vert. Paleont.*, **4**(3): 344–358.

Janvier, P. (1985a). *Les Céphalaspides du Spitsberg*. Cahiers de Paléontologie, Section Vertébrés, Ed. CNRS, Paris.

Janvier, P. (1985b). Les Thyestidiens (Osteostraci) du Silurien de Saaremaa (Estonia) 2e partie: Analyse phylogénétique, répartition stratigraphique, remarques sur les genres *Auchenaspis, Timanaspis, Tyriaspis, Didymaspis, Sclerodus* et *Tannuaspis*. *Annales de Paleontologie*, **71**: 187–216.

Janvier, P., and Lund, R. (1983). *Hardistiella montanensis* n. gen. et sp. (Petromyzontida) from the Lower Carboniferous of Montana, with remarks on the affinities of lampreys. *J. Vert. Paleont.*, **2**: 407–413.

Karatajute-Talimaa, V. (1978). *Silurian and Devonian Thelodonts of the USSR and Spitsbergen*. Mokslas Publishers, Vilnius.

Moy-Thomas, J. A., and Miles, R. S. (1971). *Palaeozoic Fishes* (2d ed.). Chapman and Hall, Ltd., London.

Northcutt, R. G., and Gans, C. (1983). The genesis of neural crest and epidermal placodes: A reinterpretation of vertebrate origins. *Quart. Rev. Biol.*, **58**: 1–28.

Novitskaya, L. I. (1971). Les amphiaspides (Heterostraci) du Dévonien de la Sibérie. *Cahiers de Paléontologie, Paris* **1971**: 1–130.

Obruchev, D. V., and Mark-Kurik, E. (1968). On the evolution of the psammosteids (Heterostraci). *Toimetised Eesti NSV Teaduste Akad. ser. chem., geol.*, **17**: 279–284.

Pan Jiang. (1984). The phylogenetic position of the Eugaleaspida in China. *Proc. Linn. Soc. N.S.W.*, **107**(3): 309–319.

Repetski, J. E. (1978). A fish from the Upper Cambrian of North America. *Science*, **200**: 529–531.

Ritchie, A. (1968a). New evidence of *Jaymoitius kerwoodi* White, an important ostracoderm from the Silurian of Lanarkshire, Scotland. *Palaeontology*, **11**: 21–39.

Ritchie, A. (1968b). *Phlebolepis elegans* Pander, an Upper Silurian thelodont from Oesel, with remarks on the morphology of thelodonts. In T. Ørvig (ed.), *Current Problems of Lower Vertebrate Phylogeny. Proc. 4th Nobel Symp.*, pp. 81–88. Almquist and Wiksell, Stockholm.

Ritchie, A., and Gilbert-Tomlinson, J. (1977). First Ordovician vertebrates from the Southern Hemisphere. *Alcheringa*, **1**: 351–368.

Ruben, J. A., and Bennett, A. F. (1980). Antiquity of the vertebrate pattern of activity metabolism and its possible relation to vertebrate origin. *Nature*, **286**: 886–888.

Schaeffer, B., and Thomson, K. S. (1980). Reflections on agnathan-gnathostome relationships. In L. L. Jacobs (ed.), *Aspects of Vertebrate History*, pp. 19–33. Museum of Northern Arizona Press, Flagstaff.

Stensiö, E. (1963). The brain and the cranial nerves in fossil, lower craniate vertebrates. *Skrifter utgitt av Det Norske Videnskaps-Akademi i Oslo. I. Mat.-Naturv. Kl. Ny Ser.*, **13**: 1–120.

Stensiö, E. (1964). Les Cyclostomes fossiles ou Ostracodermes. In J. Piveteau (ed.), *Traité de Paléontologie*, Tome IV, vol. 1: 96–382. Masson S.A., Paris.

Storer, T. I., and Usinger, R. L. (1965). *General Zoology*. (4th ed.). McGraw-Hill, New York.

Turner, S. (1973). Siluro-Devonian thelodonts from the Welsh Borderland. *J. Geol. Soc. Lond.*, **129**: 557–584.

Westoll, T. S. (1958). The lateral fin-fold theory and the pectoral fins of ostracoderms and early fishes. In T. S. Westoll (ed.), *Studies on Fossil Vertebrates*, pp. 180–211. Athlone Press, London.

White, E. I. (1958). Original environment of the craniates. In T. S. Westoll (ed.), *Studies on Fossil Vertebrates*, pp. 212–234. Athlone Press, London.

Yalden, D. W. (1985). Feeding mechanisms as evidence for cyclostome monophyly. *Zool. J. Linn. Soc.*, **84**: 291–300.

Young, J. Z. (1950). *The Life of Vertebrates*. Clarendon Press, Oxford.

CHAPTER IV

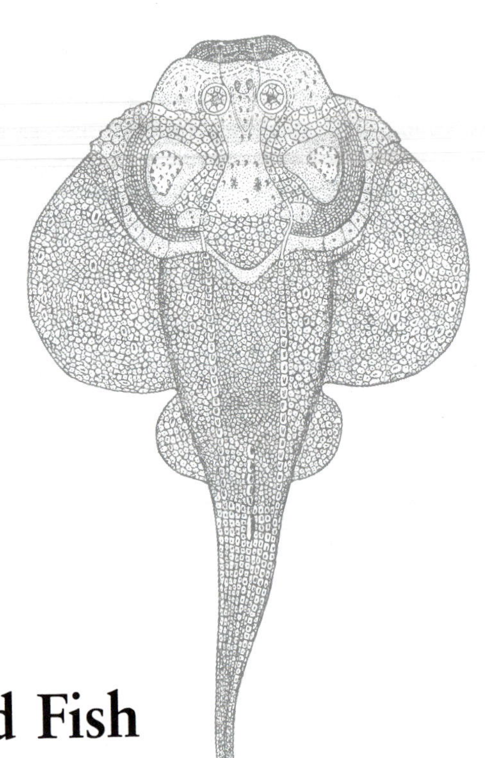

Primitive Jawed Fish

THE ORIGIN OF JAWS

One of the most important advances among primitive vertebrates was the evolution of jaws, which allowed the development of a truly predatory way of life. The early jawed vertebrates were also advanced over the Agnatha in having well-developed pelvic and pectoral fins and a braincase resembling that of modern jawed fish.

Jawed fish are first found in the Lower Silurian, some 100 million years after the appearance of the ostracoderms. None of the known agnathans are likely ancestors. Differences in the structure of the braincase and gill supports indicate that the two groups diverged prior to the appearance of the earliest fossil agnathans in the late Cambrian. Differences in the position and development of the gill filaments indicate that they evolved separately in the two groups (Schaeffer and Thomson, 1980) (Figure 4-1). The medial position of the gill supports in jawed fish may have been a response to the need to protect the gill filaments from large food particles passing down the throat.

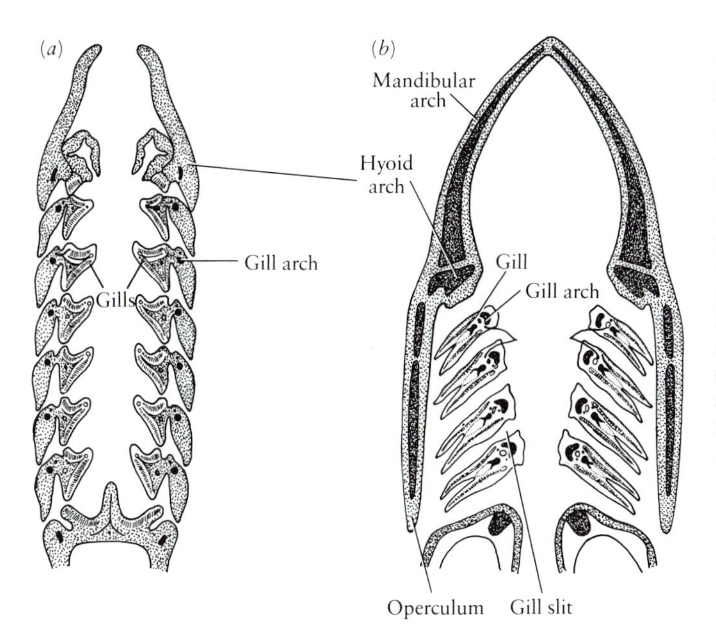

Figure 4-1. COMPARISON OF THE POSITION OF THE GILL FILAMENTS AND GILL SUPPORTS IN AGNATHA AND JAWED VERTEBRATES. In the modern lamprey (left), the gill filaments are medial to the gill supports; in the jawed fish, they are lateral. Because the filamentous gills differ in both their position and their pattern of development in the two groups, they probably evolved independently in Agnatha and gnathostomes. The most primitive vertebrates were probably so small that specialized surfaces for gas exchange were not necessary. The sequence of the cranial nerves can be used to identity the relative position of the gill arches in the two groups. This evidence indicates that the division between the mouth and pharynx is homologous in jawed and jawless fish. There is no convincing evidence in either group that there were once gill pouches in the area of the mouth, as would be the case if the palatoquadrate and Meckel's cartilage evolved from gill supports in ancestral vertebrates. *From Moy-Thomas and Miles, 1971. By permission from Chapman and Hall Ltd.*

In many Paleozoic agnathans, the gill supports are integrated with the cranium; in the gnathostomes, they are separate, jointed elements. In both agnathans and jawed vertebrates, the gill supports are derived from neural crest cells rather than from the mesoderm (the embryonic tissue that gives rise to most of the skeleton), but the evolution of the visceral arch skeleton clearly followed different pathways in the two groups (Figure 4-2).

The resemblance between the jaws and gill supports in jawed fish, especially primitive sharks (Figure 4-3), suggests that jaws may have evolved through the specialization of a more anterior pair of visceral arches. The jaws,

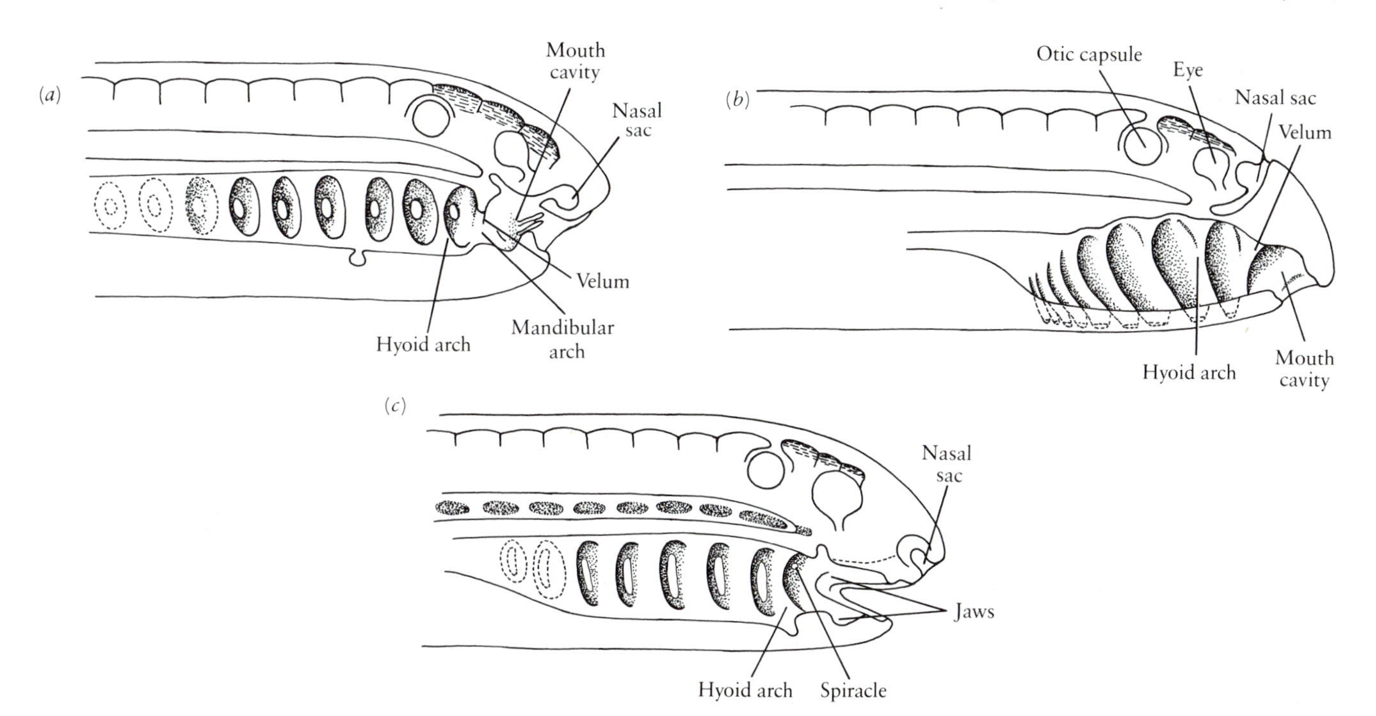

Figure 4-2. Reconstruction of the possible sequence of changes in the structure of the mouth and pharynx from ancestral vertebrates to primitive members of the agnatha and primitive jawed fish. (*a*) Hypothetical ancestral vertebrate. The mouth cavity is a distinct structure anterior to the pharynx, which is pierced by a series of paired gill slits. No skeletal tissue is associated with the mouth or pharynx, and gill filaments are absent. The olfactory sac, eyes, and otic capsule are paired sensory structures. The olfactory capsule is ventral. (*b*) Agnathans, as exemplified by cephalaspidomorphs and cyclostomes, in which the olfactory sac has migrated to a dorsal position and has a single, medial opening. In these agnathans, the gill filaments are medial to the gill supports and the gill supports are closely integrated with the head. (*c*) Ancestral jawed vertebrates. The nasal capsule has maintained its primitively paired and ventral condition. The gill supports and the palatoquadrate and Meckel's cartilage have evolved essentially simultaneously by the utilization of tissue generated from the neural crest cells. The trabeculae, which are paired structures that underlie the front of the braincase, develop similarly from neural crest cells. *From Moy-Thomas and Miles, 1971. By permission from Chapman and Hall Ltd.*

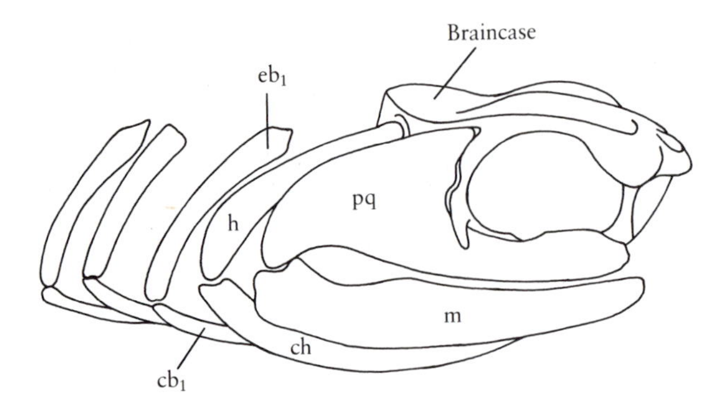

Figure 4-3. THE BRAINCASE, JAWS, AND GILL SUPPORTS OF THE CARBONIFEROUS SHARK *COBELODUS*. This general pattern is common to primitive sharks and early bony fish. The similarity in postion and structure of the jaws to the gill supports has led to the hypothesis that jaws evolved from structures that originally functioned as gill supports. The hyomandibular serves simultaneously as a gill support and as a link between the jaws and braincase. Like the jaws, the gill supports have a hingelike articulation between the dorsal and ventral units. *From Zangerl and Williams, 1975*. Abbreviations: cb_1, first ceratobranchial; ch, ceratohyal; eb_1, first epibranchial; h, hyomandibular; m, Meckel's cartilage; pq, palatoquadrate.

like the gill supports, have hinged dorsal and ventral components that are bent forward. The muscles that close the jaws are comparable to those that constrict the gill slits. The dorsal portion of the first gill support, the hyomandibular, is enlarged and links the jaws with the braincase.

In addition to the jaws and gill arches, the trabeculae (paired cartilages that support the anterior portion of the braincase) also develop from neural crest cells in jawed vertebrates. Because the trabeculae are positioned anterior to the jaws, some authors suggest that they evolved from another, premandibular gill arch. This assumption implies that there may have been yet another pair of gill pouches in the position of the mouth in the ancestors of jawed vertebrates. Evidence from agnathans seems to contradict this hypothesis. In both cyclostomes and osteostracan ostracoderms, the Vth nerve (which innervates the jaws in gnathostomes) is associated with the ridge or velum that separates the mouth from the pharynx (Figure 3-14*b*). No ostracoderm shows evidence of a gill pouch anterior to the mandibular arch, nor is there convincing evidence for their previous existence in jawed vertebrates, which suggests that the mouth was distinct from the pharynx in all vertebrates and is in a comparable position in Agnatha and gnathostomes.

Based on this reasoning, we now think that the palatoquadrate and Meckel's cartilage evolved originally in association with the mouth and were never gill supports. In the ancestors of the modern jawed fish, the hyomandibular apparently had a dual role from the time of its origin, serving as a gill support and linking the braincase with the jaws. According to Schaeffer (1975), the trabeculae evolved as discrete structures to support the anterior portion of the braincase when the integrative centers of the brain in early jawed fish were enlarged as both a predatory mode of feeding and more sophisticated swimming patterns developed.

Several groups of jawed vertebrates are first represented in the fossil record in the late Silurian and early Devonian. The major classes of modern fishes, the bony fish and sharks, first appear at this time, as do two archaic groups, the acanthodians and the placoderms. The acanthodians show a combination of features that suggest a relationship to modern jawed fish.

In contrast, placoderms differ in many features of their anatomy, particularly in the structure of their jaws and teeth. In modern jawed fish, the endochondral element of the upper jaw, the palatoquadrate, is medial to the major jaw-closing muscles. In primitive placoderms, the palatoquadrate is closely integrated with the cheek and thus is lateral to the space occupied by the jaw muscles. Most placoderms have large bony plates attached to the margins of the jaws (instead of teeth), and none show the pattern of regular tooth replacement common to other jawed fish. These differences suggest that placoderms evolved from a lineage of primitive jawed vertebrates other than those that gave rise to the Osteichthyes and Chondrichthyes.

PLACODERMS

Placoderms are a peculiar assemblage of heavily armored jawed fish. They appeared first in the Lower Devonian and radiated throughout that period, but only one or two genera survived into the Lower Carboniferous (Figure 4-4). Not only do placoderms lack any living descendants, but because of their massive external armor, they are also without modern analogues, making their ways of life particularly difficult to interpret.

Some of the earliest known Lower Devonian placoderms resemble the early heterostracan ostracoderms in that the entire head and trunk region is encased in bone, with only the short scaly tail exposed posteriorly (Figure 4-5). In most genera, the exoskeleton is distinctive in having a clear separation between the armor covering the head and trunk. Each unit is further subdivided into smaller plates that are capable of growth at their margins. Primitively, the armor was similar histologically to that of osteostracans in having three distinct layers and lacunae for bone cells.

Most placoderms are dorsoventrally compressed, and some are very flattened. Together with their heavy armor, this suggests a basically benthonic way of life. The braincase is heavily ossified in primitive forms, and the elements of the upper jaw are attached directly to the undersurface of the cheek and braincase. Some genera have small tooth-like structures, but most have one or two large bony plates in the upper jaw and a single pair in the lower. Unlike

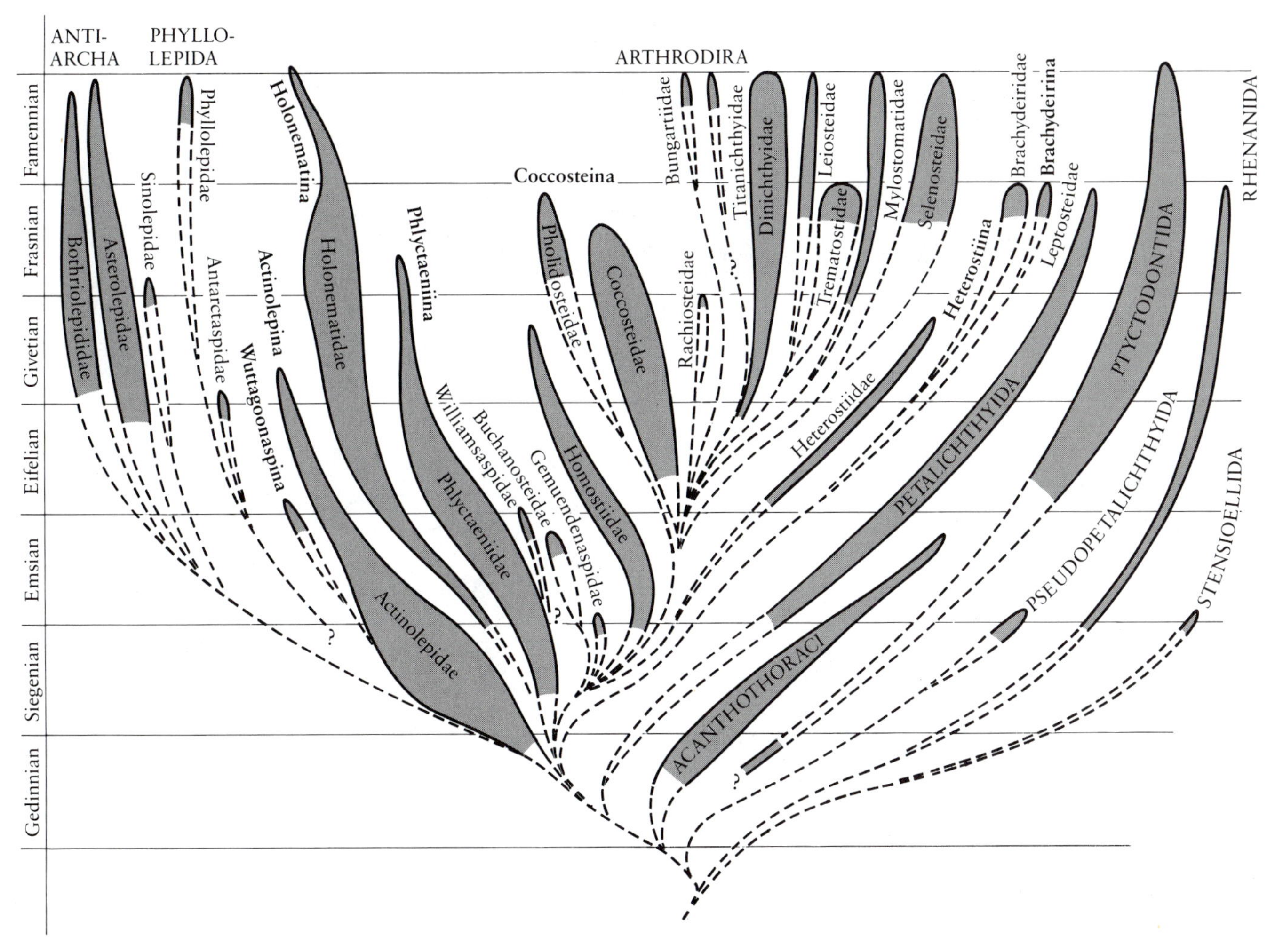

Figure 4-4. GRAPHIC REPRESENTATION OF THE RADIATION OF PLACODERMS. Solid lines represent the known fossil record of each group and dotted lines indicate probable range extensions and possible relationships The fossil record of this group is almost completely limited to the Devonian. The nature of their relationships to bony fish and sharks, both of which appear in the Silurian, remains uncertain. Only a few fossils of placoderms are known in the Carboniferous. The placoderms are the only totally extinct vertebrate class. *From Denison, 1975.*

the ancestors of the modern jawed fish, placoderms lack a spiracle. The water from the gills was apparently expelled through the gap between the cranial and trunk shields.

In contrast with the ostracoderms, ossified neural and haemal arches reinforce the notochord in several groups. To facilitate raising the head, nearly all placoderms have the anterior elements fused into a composite structure, the synarcual, which articulates with the occipital condyles. The paired articulating surfaces ensure that movement is restricted to the vertical plane.

The presence of jaws probably required that the head was movable relative to the trunk, at least in benthonic genera, suggesting that the separation between the cranial and trunk armor developed when the exoskeleton first evolved.

All genera had pectoral and pelvic fins, but none are known to have had an anal fin. The tail is typically heterocercal. During the Devonian, several groups of placoderms adaptated toward a more open-water (nectonic) habit, with a reduction in the thickness and extent of the dermal armor and a loss of endoskeletal ossification resulting in a loss of weight and a freeing of the trunk and tail for more active swimming. Other placoderms intensified their benthonic habits and some spread from the marine environment into fresh water.

There are two schools of thought regarding the history of ossification within the Placodermi. Stensiö (1963), Moy-Thomas and Miles (1971), Miles and Young (1977), Young (1980), and Goujet (1984b) hold that the most primitive members of the group were heavily armored and that reduction of the skeleton was a feature common to several orders. In contrast, Gross (1963) and Denison (1975, 1983) suggest that the most primitive pattern is exhibited by genera such as *Stensioella* (see Figure 4-20), in which there are no large bony plates but a mosiac of tesserae—small superficial dermal elements without a basal layer of laminar bone.

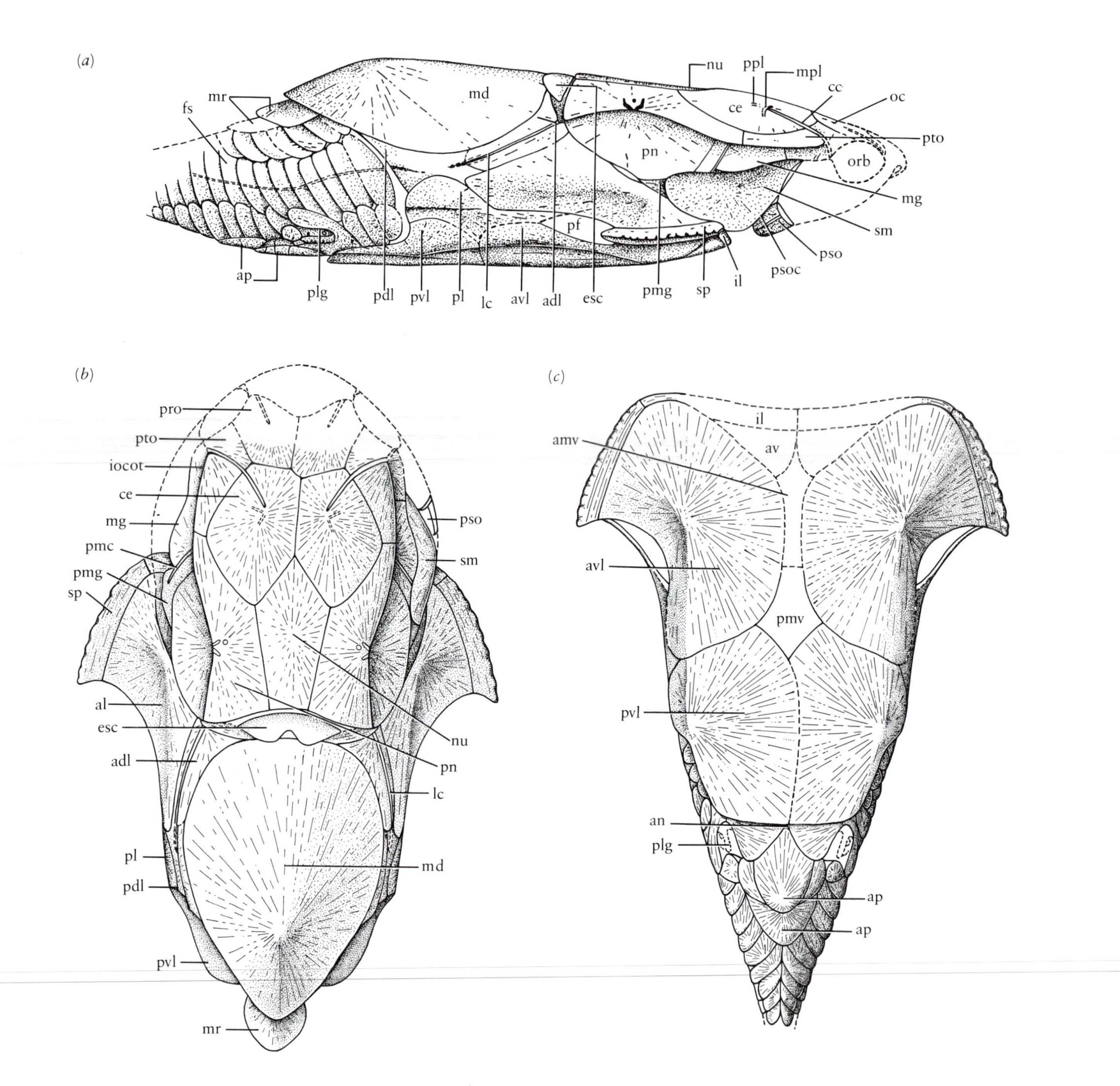

Figure 4-5. THE EARLY ARTHRODIRE *SIGASPIS*, × 1½. (*a*) Lateral view of body; neither the paired fins nor the end of the tail is known in this genus. (*b*) Dorsal view of head and thoracic shield. (*c*) Ventral view of thoracic shield and base of tail. *From Denison, 1978. Sigaspis* shows the complex pattern of dermal armor that is common to many placoderms. Such an extensive external skeleton is probably a primitive feature of arthrodires. This condition probably evolved rapidly within the early placoderms, which may initially have been covered with small denticles or tesserae as in the genus *Stensioella* (see Figure 4-20). The pattern of the dermal plates in placoderms is not comparable with that of the dermal bones in bony fish. The names used here are unique to this group. The general arrangement of the dermal plates is similar in most placoderm orders. Differences in the configuration of individual plates and their apparent fusion, breakup, and loss are important features in evaluating possible relationships within this group. Abbreviations as follows:

adl, anterior dorso-lateral plate; al, anterior lateral plate; amv, anterior medio-ventral plate; an, anus; ap, anal plate; asg, anterior superognathal; av, antero-ventral plate; avl, anterior ventro-lateral plate; cc, central sensory line; ce, central plate; dend, endolymphatic duct; exc, extrascapular plate; fs, flank scales; hm, hyomandibula; lc, main lateral line canal; Il, intero-lateral plate; iocot, infraorbital sensory line, otic branch; md, median dorsal plate; mg, marginal plate; mpl, middle pit line; mr, dorsal ridge plate; ncav, nasal cavity; nu, nuchal plate; oc, occipital cross commissure; orb, orbit; pdl, posterior dorso-lateral plate; pf, pectoral fenestra; pi, pineal plate; pl, posterior lateral plate; plg, pelvic girdle; pmc, postmarginal sensory line; pmg, postmarginal plate; pmv, posterior medio-ventral plate; pn, paranuchal plate; pna, postnasal plate; ppl, posterior pit line; pq, palatoquadrate; pro, preorbital plate; psg, posterior superognathal; pso, postsuborbital plate; psoc, postsuborbital sensory line; psp, parasphenoid; ptn, postnasal plate; pto, postorbital plate; pvl, posterior ventro-lateral plate; qu, quadrate; ro, rostral plate; scl, sclerotic plate; sm, submarginal plate; so, suborbital plate; sp, spinal plate; VII, seventh cranial nerve; IX, ninth cranial nerve; X, tenth cranial nerve.

Although remains of both cartilaginous and bony fish are known from the Upper Silurian, no fossils of placoderms are recognized from beds older than the Devonian. This fact supports Denison's hypothesis that the ancestors of placoderms did not have extensive armor but were either naked or covered with isolated denticles or tesserae that would render fossilization unlikely and identification difficult. Denison (1983) suggests that the Rhenanida, Stensioellidae, and paraplesiobatids (see Figures 4-18 and 4-20) are primitive representatives of the ancestral placoderm condition.

On the other hand, we may infer from the similarity of the dermal armor in the numerous groups that are known from the early Devonian that fairly extensive armor must have evolved prior to the differentiation of the orders Acanthothoraci, Ptyctodontida, Phyllolepida, Arthrodira, and Antiarchi.

Unfortunately, the placoderms that lack extensive dermal armor are not well known, but features of their jaws and dentition appear specialized in a different manner than those of other placoderms, rather than representing a primitive condition. Because of the continuing uncertainty regarding the nature of primitive placoderms, the interrelationships of the various orders remain subject to controversy. This description will begin with the arthrodires because they are both the best known and the most numerous of all placoderms.

ARTHRODIRA

Although the primitive structural pattern for placoderms as a whole cannot be confidently reconstructed the sequence of fossils and the probable direction of anatomical change within the Arthrodira indicate that small, extensively ossified forms were the most primitive members of this extremely diverse order (Goujet, 1984a).

PRIMITIVE ARTHRODIRES: ACTINOLEPIDAE AND PHLYCTAENIIDAE

The general body form of primitive arthrodires is exemplified by *Sigaspis* (see Figure 4-5), a member of the family Actinolepidae. The head and almost all of the trunk are protected by a nearly continuous covering of large plates. The cranial and trunk shields slide past one another without a specific joint. The pectoral fin is surrounded posteriorly by bony plates and protected anteriorly by a short lateral extension of the body armor. The trunk shield extends to the level of the anus. The pelvic girdle is surrounded by large overlapping scales that cover the base of the tail and surround the anus. The nature of the caudal fin is not known; all we know about the paired fins is that they were narrow based.

The dermal plates of the skull and trunk have no counterparts in higher, bony fish and consequently have a unique terminology (see Figure 4-5). The pattern is sufficiently consistent with other placoderms that it forms a convenient basis for evaluating relationships. In primitive forms, the dermal ossification of the cranial shield has two or three movable units. Typically, a single unit covers most of the dorsal and lateral surface. The cheek area, associated internally with the hyomandibular, forms a movable operculum.

The angle between the cranial and trunk shields is approximately 45 degrees. The dorsal surface of the skull is much longer than the ventral, and the jaws are no more than half the length of the skull roof. The gill chamber lies entirely beneath the back of the skull roof.

Most of the early arthrodires were probably poor swimmers, with their very short tail region enclosed in heavy scales, and their head and trunk region heavily ossified and ventrally flattened. Despite the presence of pectoral and pelvic fins, these animals were probably as restricted to the sea bottom as the heavily armored osteostracans and heterostracans.

The braincase is extremely well ossified in actinolepids, as exemplified by *Kujdanowiaspis* (Figure 4-6). It is formed by a single low and wide ossification (a configuration termed **platybasic**). The otic region is long and includes three semicircular canals. The hypophysis is far anterior and opens into the mouth through the buccohypophyseal foramen. This opening is surrounded by a small dermal plate termed the parasphenoid. Anterior to the ossified braincase is an area that was occupied by unossified nasal capsules. The back of the braincase has paired occipital condyles that provide a well-defined articulation between the head and the trunk. The gill supports are unossified in most placoderms and little is known of the jaws in actinolepids.

Arthrodires underwent an extensive radiation above the actinolepid level, giving rise to some 20 families (see Figure 4-4). Arthrodire evolution may also be considered in terms of a series of grades of development that show increasing sophistication in the structure of the feeding and locomotor apparatus. Miles (1969) termed these grades the phlyctaenaspid, the coccosteomorph, and the pachyosteomorph levels. The structure of the upper jaws is first known in the Phlyctaeniidae (Figure 4-7). The palatoquadrate is ossified as two units, the anterior autopalatine and the posterior quadrate. Both units have extensive contact with the lateral wall of the cheek, beneath the suborbital and postsuborbital plates. The only space available for the muscles that close the jaws is medial to the palatoquadrate, a condition quite unlike that in higher fish, in which the main mass of jaw musculature is lateral to the palatoquadrate. The hyomandibular appears to be incorporated with the cheek and is not part of a jaw support mechanism, as it is in higher jawed fish. One of two pairs of dermal elements, the posterior superognathal, is attached to the autopalatine. The anterior superogna-

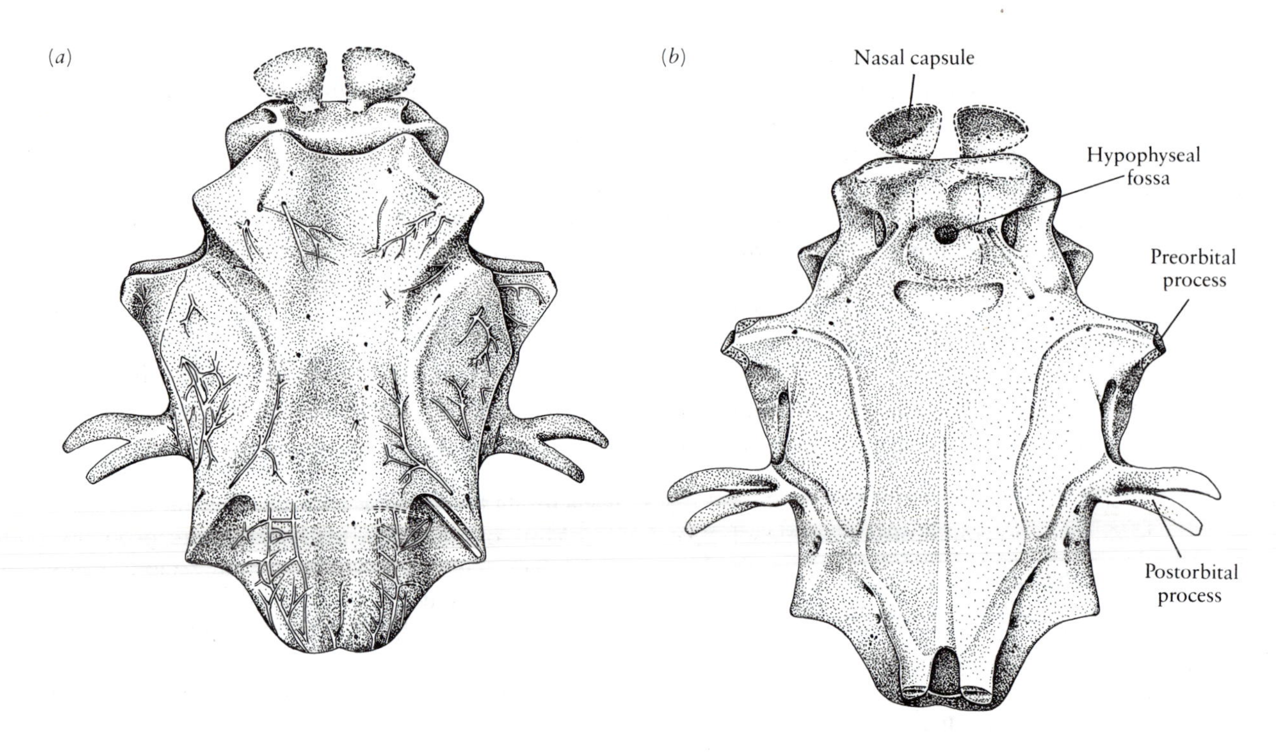

Figure 4-6. DORSAL AND VENTRAL VIEWS OF THE BRAINCASE OF THE LOWER DEVONIAN ARTHRODIRE *KUJDANOWIASPIS*, ×1. It is formed from a single ossification. The nasal capsules are restored. Grooves indicate the position of blood vessels. *From Stensiö, 1969.*

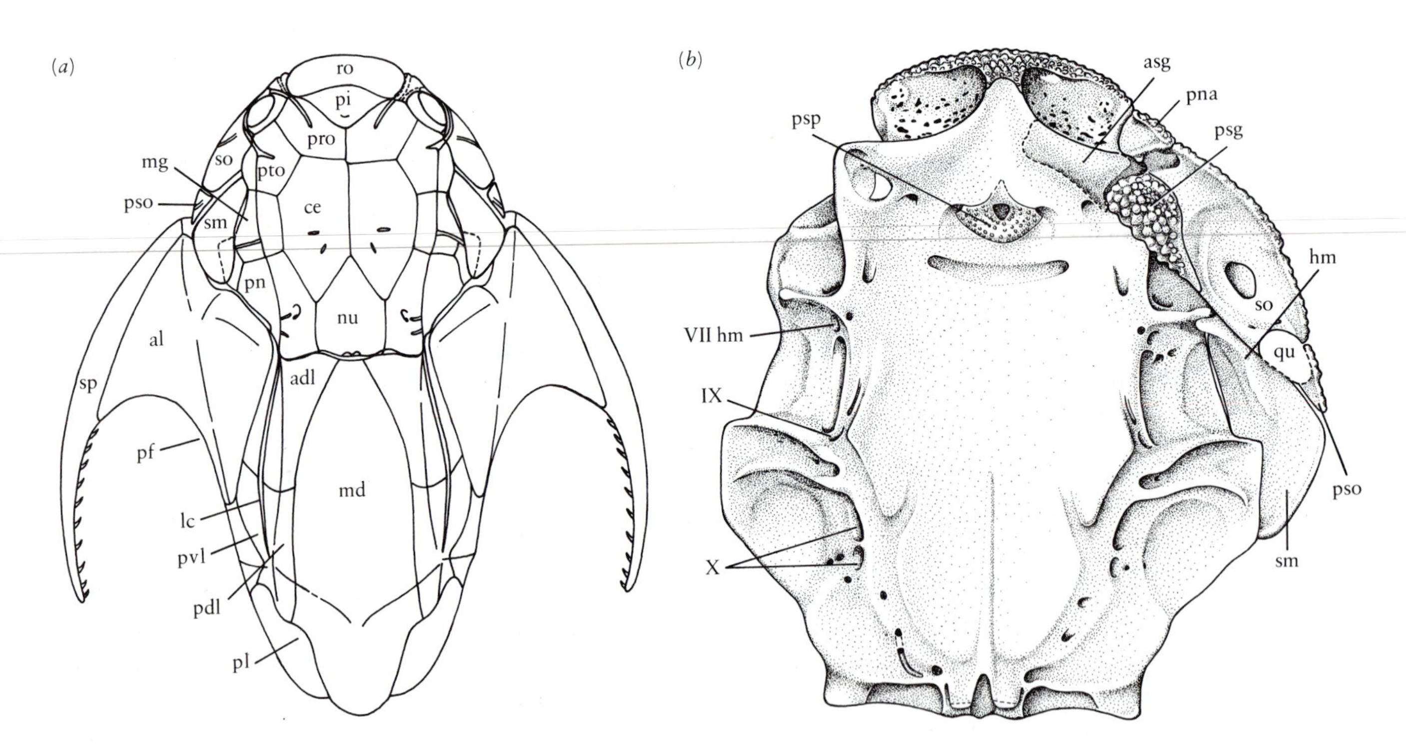

Figure 4-7. (*a*) Dorsal view of the dermal skeleton of the primitive arthrodire *Dicksonosteus*, natural size. The long cornua anterior to the pectoral fins resemble those of advanced cephalaspids. See Figure 4-5 for a key to abbreviations. (*b*) Reconstruction of the skull in ventral view, twice natural size. The articulation for the lower jaw is far forward, and the elements of the upper jaw are attached to the base of the braincase and the inside surface of the cheek plate. *From Goujet, 1975.*

thal is attached directly to the endocranium. These dermal elements are studded with denticles and serve as the teeth of the upper jaw. The jaw articulation is far forward.

The lower jaw is well known in more advanced forms but has not been described in the most primitive genera. Where known, it consists of a perichondrally ossified Meckelian bone that may be divided into as many as three elements. A large dermal element, the inferognathal, is attached to the endochondral lower jaw and bears denticles or a cutting or crushing surface that serves the function of teeth in more typical jawed vertebrates.

In the Phlyctaeniidae, a true joint between the head and trunk shields has evolved, with a recessed area, or glenoid fossa, on the paranuchals and a knoblike condyle on the anterodorsal laterals. The spines anterior to the pectoral fins are much more extensive than in *Sigaspis* and look like those of contemporary cephalaspids (Figure 4-7).

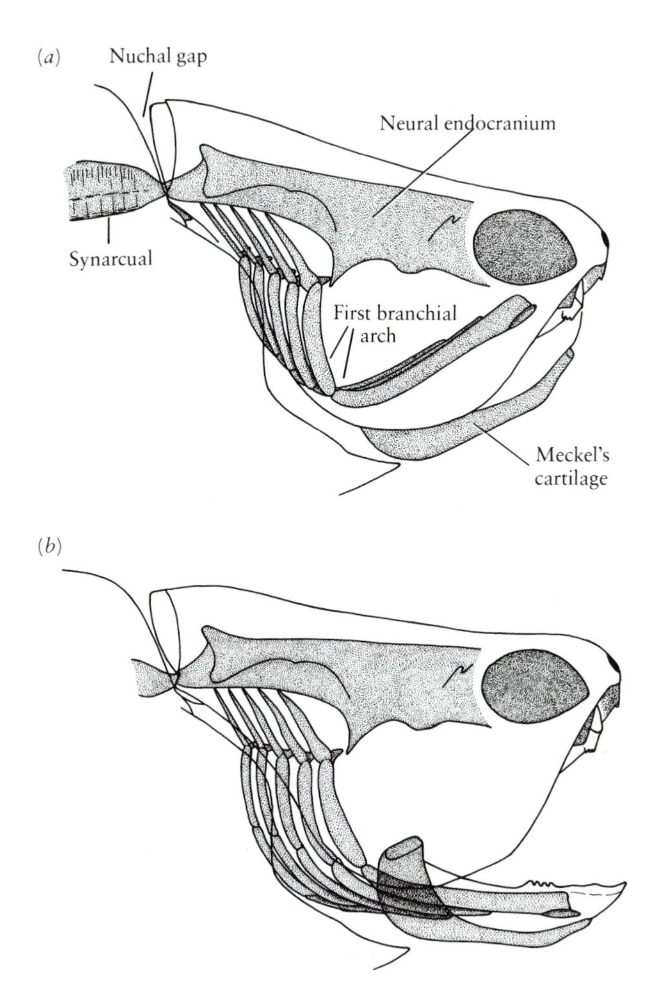

Figure 4-8. RESPIRATORY AND FEEDING MOVEMENTS IN *COCCOSTEUS*. Drawn as if dermal skeleton were transparent. (*a*) Head and visceral skeleton in position of rest and mouth closed. (*b*) Head raised, branchial skeleton expanded, and mouth wide open. *From Miles, 1969.*

ADVANCED ARTHRODIRES

As a result of work by Miles and Westoll (1968), the level of evolution exemplified by *Coccosteus* is particularly well-known. The jaws and associated respiratory structures in this genus serve as a model for arthrodires in general and as a basis for explaining the structural changes in more advanced forms (Figure 4-8).

With the exception of one highly specialized genus (see Figure 4-9*b*), all placoderms have a well-defined joint between the cervical vertebrae and the braincase. The arthrodires also elaborated the exoskeletal joint between the cranial and trunk shields. We can determine the range of movements by the geometry of the shields and the position of the joint. The joint is relatively high on the head, so that the jaws move forward as the head is tilted upward. This movement also opens the operculum and spreads the gill supports apart.

In primitive arthrodires, the angle between the open jaws is large but the jaws themselves are short, so that the total gape is small and precludes feeding on large prey. In more advanced arthrodires (Figure 4-9*a*), the potential for raising the head is increased by widening the nuchal gap—the area between the head and trunk shields above the craniothoracic joint. Raising the height of the joint also increases anterior movement of the jaw. The length of the jaws is increased as well, with the angle between the head and trunk shield becoming nearly vertical. The inferognathals are enlarged and differentiated into an anterior toothed portion and a posterior plate for attachment of the jaw-closing musculature. In some genera, the inferognathal extends dorsally as a coronoid process, and the level of jaw articulation may be lowered to increase the leverage on the lower jaws. Arthrodires have many different tooth plate patterns, suggesting a range of diets or ways of handling prey; some crush, some slash, and other stab with long picks. Among the diverse advanced arthrodires of the Upper Devonian Cleveland shale, the genera *Dunkleosteus, Heintzichthys,* and *Titanichthys* reached a length of well over 2 meters and were the dominant marine predators of their age. However they may have been quite clumsy compared to contemporary sharks.

The structure of the feeding mechanism in arthrodires is an adaptation to a basically benthonic way of life. Nearly all of these fish may have lived on, or very near, the bottom. It was impractical to lower the mandible in this position, and so the head was raised while the mouth simultaneously opened and thrust forward toward the prey. In most arthrodires, this feeding pattern is accompanied by a dorsoventrally flattened body. Two advanced arthrodire families, the Leptosteidae and the Brachydeiridae, have laterally compressed bodies and heads. They were probably actively swimming, open-water feeders. The mouth can open ventrally without the head shield being raised. In these families, the nuchal gap is restricted and the cheek is fused to the remainder of the skull. In one genus, *Synauchenia* (Figure 4-9*b*), the classic placoderm structure of a separate head and trunk shield is absent, and the two units are completely fused. The specialized joint between the braincase and the anterior vertebrae, which is present in all other placoderms, is also lost.

At the same time as feeding mechanics were evolving, changes in the dermal armor affected swimming (Figure 4-10). Primitive arthrodires were barely advanced above the level of early ostracoderms. Most of the body was encased in heavy armor, with only the tail protruding behind the trunk shield. Small paired fins suggest limited maneuverability.

In the actinolepid and phlyctaenaspid levels, the opening for the pectoral fin is small and entirely surrounded by dermal bone. An endoskeletal scapulocoracoid with a short site for the attachment of the fin lies under the anterior lateral plate. The endoskeletal pelvic girdle is similarly surrounded with dermal ossifications. The pelvic girdle is a small plate of bone, which is adequately known only in later genera. To judge by the limited area of attachment, the pelvic fin must have been quite small in early arthrodires and frequently appears to be little more than a thin flap of tissue.

In more advanced arthrodires, Miles's coccosteomorph and pachyosteomorph levels, the entire exoskeleton is lighter because of the reduced thickness of dermal bone that resulted from the loss of superficial layers. The endochondral ossification of the head region is reduced, with the braincase and the jaw supports largely cartilaginous. Reduction in the area of the trunk shield saves weight and, more importantly, frees much of the trunk for more effective use of the swimming muscles and opens up the posterior margin of the pectoral fenestra. The pectoral spine is reduced and finally lost. The articular crest of the scapulocoracoid becomes much longer, indicating a larger pectoral fin with as many as eleven basal elements. In more advanced pachyosteomorph arthrodires, some basals may be fused to form a sharklike metapterygium (see Chapter 5). In *Coccosteus* (Figure 4-10), the endochondral pelvic girdle is formed by a pair of ventral plates that meet at the midline. They articulate with fin elements termed radials.

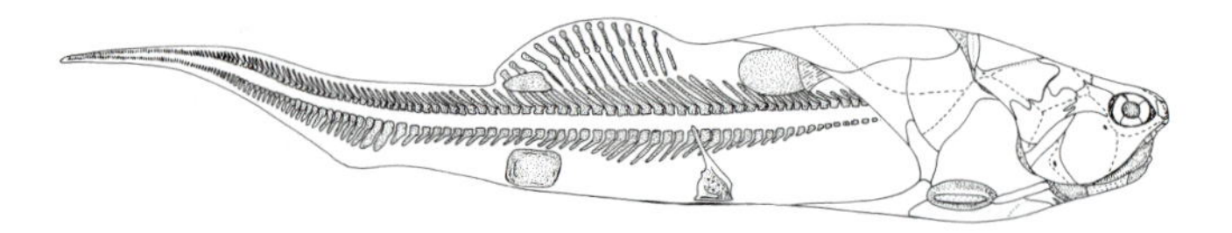

Figure 4-10. *COCCOSTEUS*, A SMALL MIDDLE DEVONIAN ARTHRODIRE. Original is about 35 centimeters long. The dermal skeleton is much less extensive than that of primitive arthrodires. The greater flexibility of the trunk presumably resulted in much more effective swimming. *From Miles and Westoll, 1968.*

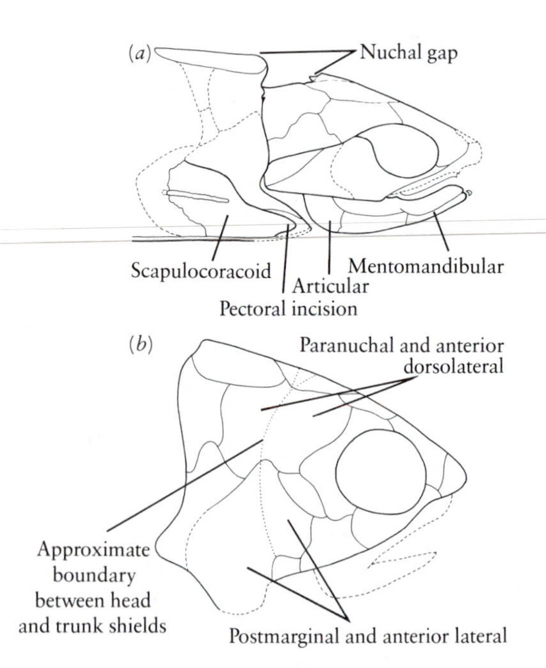

Figure 4-9. HEAD AND TRUNK ARMOR OF ADVANCED ARTHRODIRES. (*a*) *Erromenosteus*. Note the long nuchal gap. The angle between the head and thoracic shield is almost vertical. The craniothoracic joint is very high; the lower jaw is much longer than in primitive arthrodires, × ¼. *From Denison, 1975* (*b*) *Synauchenia*. The head and trunk shields are fused to one another. × ¼. *From Moy-Thomas and Miles, 1971. By permission from Chapman and Hall Ltd.*

The notochord supplies the main longitudinal support in arthrodires, and there is no development of ossified vertebral centra. However, there are frequently more superficial ossifications in the form of partial or complete rings derived from the neural and haemal arches. In *Coccosteus*, there are neural and haemal arches throughout the length of the column, but the notochord remains unconstricted.

The arthrodires radiated throughout the Devonian, becoming ever more diverse and specialized until the end of the period, when they quickly became extinct. None of the other placoderm orders approached the success of the arthrodires, and most remained restricted to a single adaptive zone. The interrelationships of other placoderm orders are the subject of continuing controversy. Most authors ally the Antiarchi, which are known throughout the Devonian, with the arthrodires.

ANTIARCHI

Young (1984) recently published a comprehensive review of the phylogeny and biogeography of the antiarchs. While arthrodires show a progressive reduction of armor and become strongly predatory, antiarchs (Figure 4-11) emphasize their dermal ossification and retain a benthonic way of life that is confined primarily to fresh water. Like the primitive arthrodires, they have a very long trunk shield with ossification behind the pectoral fin and distinct posterior lateral and posterior medioventral plates. The pattern of the bones of the dorsal head shield differs considerably from the arthrodires as a result of the dorsal migration of both the orbits and narial openings. Most genera are dorsoventrally flattened. The mouth is

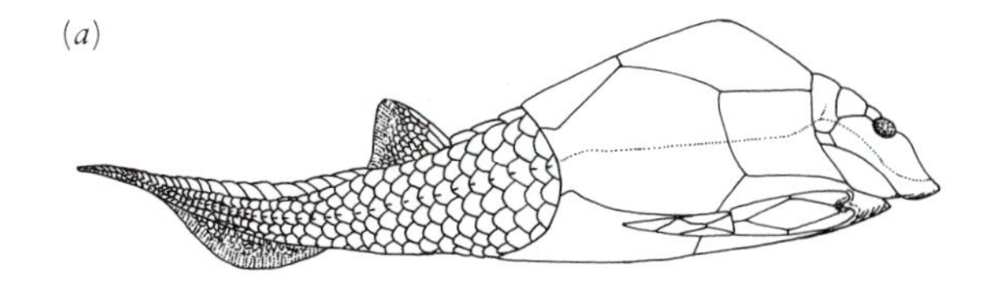

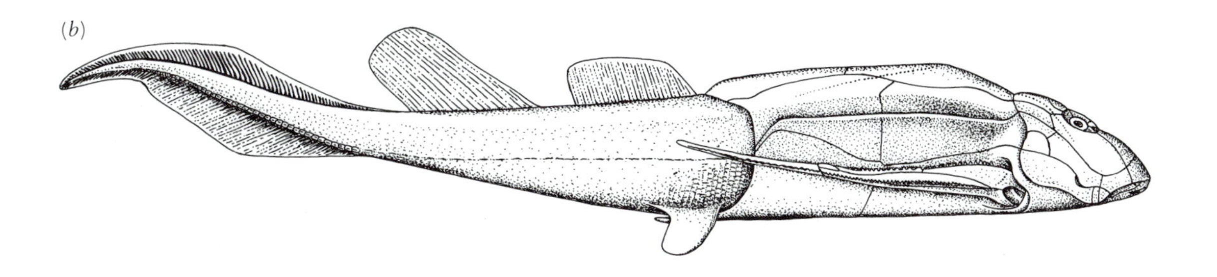

Figure 4-11. ANTIARCHS, THE MOST EXTENSIVELY ARMORED PLACODERMS. (*a*) *Pterichthyodes*, a scale-covered Middle Devonian genus; original about 15 centimeters long. (*b*) *Bothriolepis* from the Upper Devonian, about 40 centimeters long. Both genera are freshwater, bottom-dwelling fish. *From Stensiö, 1969.*

terminal but opens ventrally. There is only one pair of superognathals, rather than two. None of the internal skeleton is ossified. The head shield in most antiarchs is much smaller than the trunk shield, although they are nearly equal in the genus *Sinolepis,* from the Upper Devonian of China.

The pectoral fins are completely enclosed in bone. This ossification may have evolved from the spinal plate but appears as a distinct structure articulating via a complex joint with the anterior margin of the trunk shield. In most antiarchs, the fin is jointed midway along its length, but this joint is absent in the Upper Devonian genus *Remigolepis,* which has rather short fins compared with those of other antiarchs. The pectoral fins may have served as props or holdfasts but could not have functioned in swimming. The pelvic fins are represented by small flaps of tissue just behind the trunk shield. There may be either one or two dorsal fins. In the Middle Devonian genus *Pterichthyodes,* the tail is covered with overlapping scales, but in the well-known Upper Devonian genus *Bothriolepis,* it is almost naked. Specimens of *Bothriolepis* have been found by the thousands in the shallow water deposits of the Escuminac formation, outcropping on the Bay of Chaleur on the south side of the Gaspé Peninsula in eastern Canada. Denison (1978) has described the soft anatomy of this genus based on differential infillings of the body cavities. He interpreted diverticula from the esophagus as lungs, but they probably did not function in either gas exchange or buoyancy control since neither role is suggested by the way of life of this placoderm.

Some authors suggest that antiarchs may have evolved from primitive arthrodires, but no known genera appear intermediate between the two groups. Antiarchs are rare in the Lower Devonian, where they are represented by only a few poorly known specimens from China. They become widespread and common in the Middle and particularly the Upper Devonian, and some material has even been found in Lower Mississippian sediments.

PHYLLOLEPIDIDA

Like the arthrodires and antiarchs, the phyllolepids have extensive dermal armor. They are known principally from a single widespread genus from freshwater deposits of the Upper Devonian. *Phyllolepis* (Figure 4-12) is extremely flattened dorsoventrally. Only the central portions of the head and thoracic shields are adequately known. The plates are marked by a conspicuously concentric pattern of ridges that permits ready identification. The plates are like those of arthrodires, particularly in the common presence of a posterior ventrolateral plate, but the pectoral fenestrae are not closed posteriorly. The proportions differ greatly, with the medial plates much larger than those at the margins. Simple flanges form the craniothoracic joint.

Recently described material from the Upper Devonian of Australia has led Long (1984) to ally the phyllolepids specifically with the actinolepoid arthrodires.

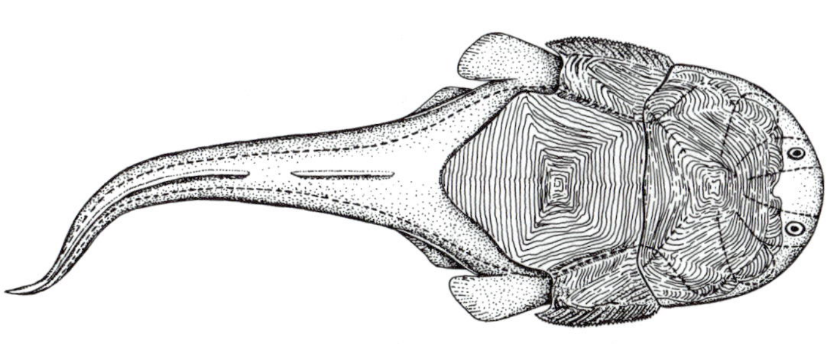

Figure 4-12. RESTORATION OF *PHYLLOLEPIS.* The dorsoventrally flattened late Devonian placoderm is shown in dorsal view. (Anterior part of head and much of body conjectural) $\times\frac{1}{4}$. *From Stensiö, 1969.*

Petalichthyida

The petalichthyids include eight genera known from marine deposits in the Lower to Upper Devonian. They broadly resemble the primitive arthrodires but appear to have led a strictly benthonic way of life, with the orbits facing dorsally and the body dorsoventrally compressed. The cranial and trunk shields are generally comparable to the primitive arthrodires, but they differ in a number of consistent features. The trunk shield is short and there are never plates behind the pectoral fin. Denison (1983) suggests that this pattern is primitive among the better-known arthrodires. Where known, the spinal plates are very long and resemble the cornua of cephalaspids. In the earliest genera, there is no specific area of articulation between the head and thoracic shields, and when it develops, it is oriented longitudinally rather than transversely as in arthrodires. In some genera, the anterior margin of the head shield is not formed by large plates but by a covering of small scales. In *Lunaspis* (Figure 4-13*a*), the trunk is long and covered with small overlapping scales. The braincase of *Macropetalichthys* (Figure 4-13*b*) is heavily ossified and has been described in detail by Stensiö. Like that of early arthrodires, it is very long and wide, but has a particularly elongate occipital area. Surprisingly, nothing at all is known of the jaw structure of this group.

Ptyctodontida

In contrast with the placoderms we have discussed, the ptyctodonts (Figures 4-14 and 4-15) bear a superficial resemblance to modern fishes. The armor is limited to the back of the head and the thoracic shield is very short, freeing the trunk for active anguilliform swimming. The trunk is long, slim, and scaleless, and the tail is reduced to a narrow filament. The ptyctodonts have the general appearance and many of the structural specializations of modern chimaeroids or holocephalians (which are discussed in the next chapter), and it has been repeatedly claimed (most recently by Ørvig (1985)) that they are related to that group. However, Miles and Young (1977) argue that the similarities are the result of convergence and do not reflect close relationship.

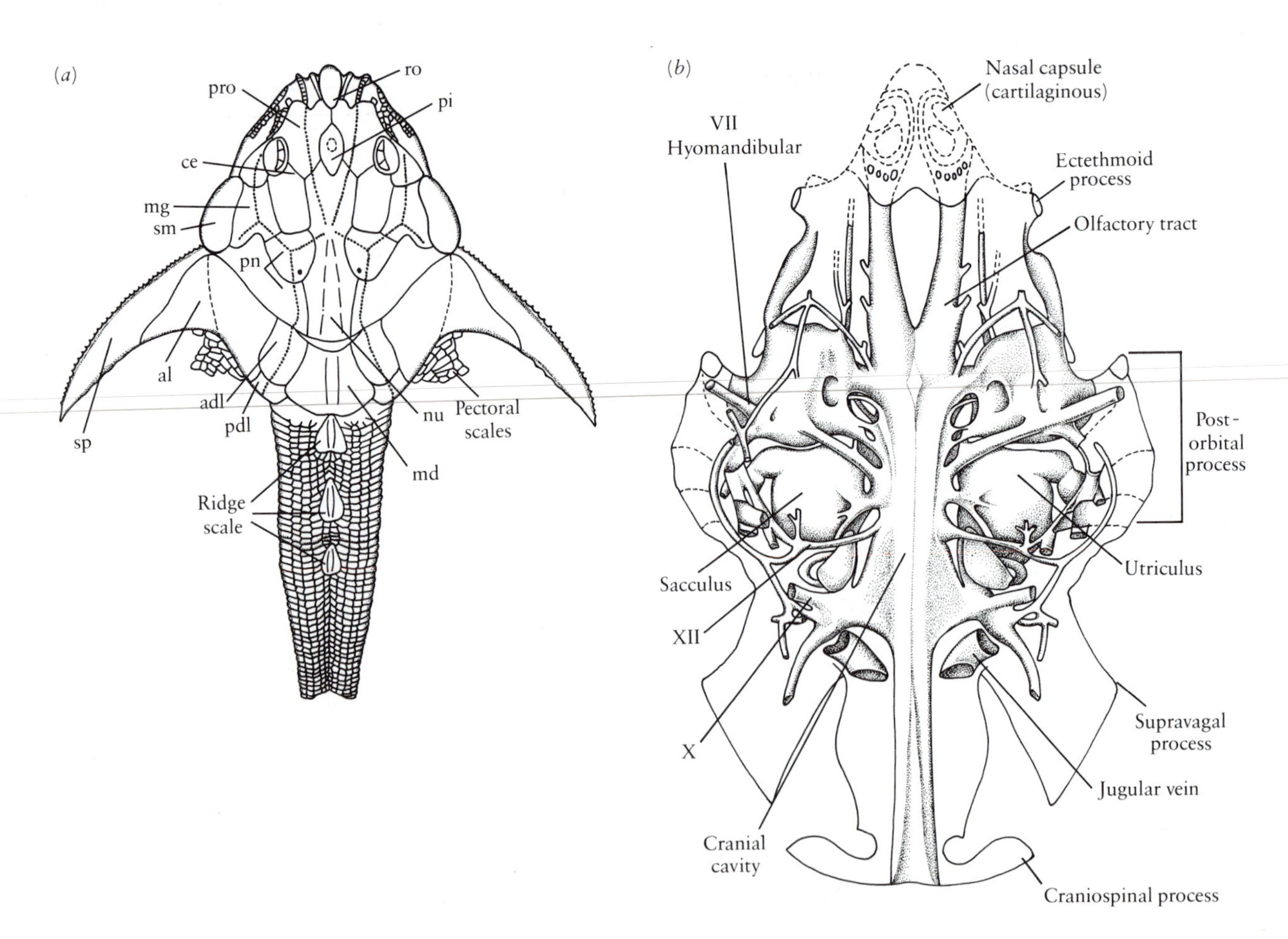

Figure 4-13. PETALICHTHYIDS. (*a*) *Lunaspis*, Dorsal view, $\times\frac{1}{2}$. *From Moy-Thomas and Miles, 1971.* (*b*) *Marcropetalichthys*. Restoration of brain cavity and canals of neurocranium in ventral view, $\times\frac{1}{2}$. *From Moy-Thomas and Miles, 1971. By permission from Chapman and Hall Ltd.*

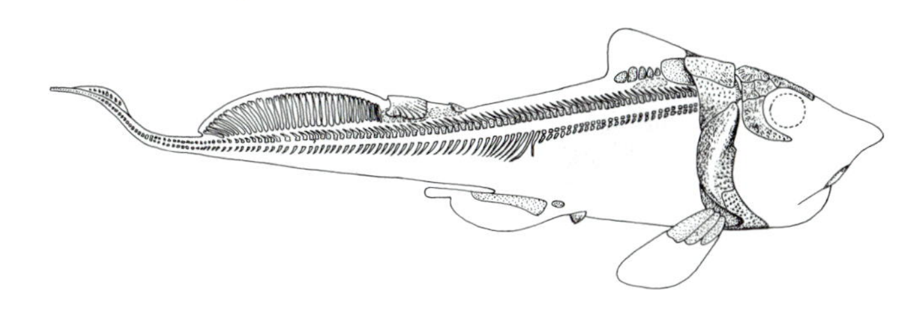

Figure 4-14. RESTORATION OF THE PTYCTODONT *CTENURELLA*, ×⅔. The general body form and details of the tooth plate and jaws resemble those of holocephalians (see Chapter 5), but the pattern of the dermal armor is similar to other placoderms. *From Stensiö, 1969.*

A single family of ptyctodonts is recognized and includes approximately ten genera known from the Middle and Upper Devonian. Most are less than 20 centimeters long. They are primarily, but not exclusively, marine and several genera have a very widespread distribution.

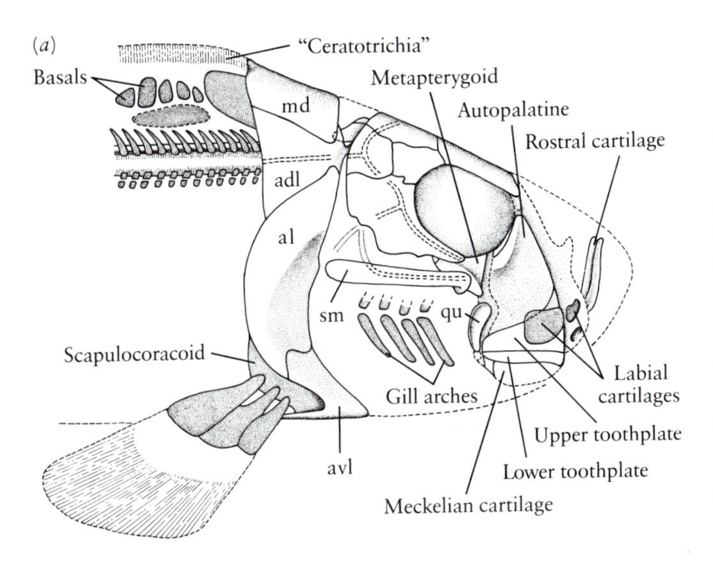

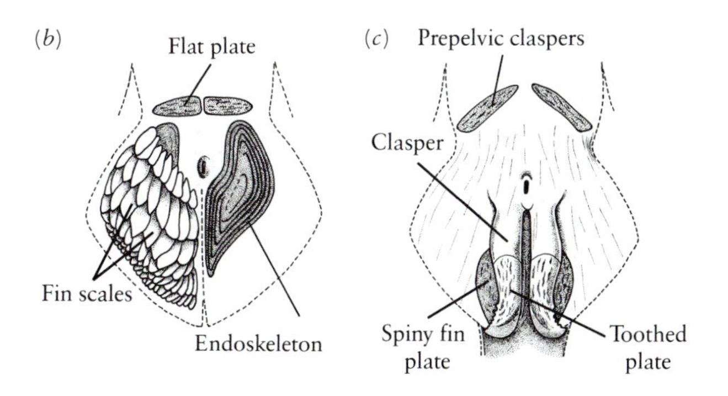

Figure 4-15. PTYCTODONTS. (*a*) *Ctenurella*. Restoration of head in lateral view, ×2. (*b*) Restoration of pelvic fin of *Rhamphodopsis* female. (*c*) Restoration of pelvic fin of *Rhamphodopsis* male, showing claspers. Both (*b*) and (*c*) are ventral views, ×2. *From Moy-Thomas and Miles, 1971. By permission from Chapman and Hall Ltd.*

The general body form, and particularly the reduction of armor, suggests a way of life that is fundamentally different from that of other placoderms. They are described as benthonic feeders, but they must have been capable of much more active swimming than any but the most specialized arthrodires.

Ptyctodonts are noted for the presence of a single pair of highly distinctive upper and lower tooth plates. Crushing or shearing areas within the teeth are formed by a particularly hard, hypermineralized tissue called secondary dentine. The upper jaw is formed by three ossifications—the quadrate, autopalatine, and metapterygoid—apparently united by cartilage in the living animal. These elements are firmly attached to the braincase. Traces of four gill supports are present more posteriorly, but there is no evidence of a distinct hyomandibular. Rostral and labial cartilages associated with the mouth resemble those of sharks and chimaeras. The craniothoracic joint is not as fully developed as in arthrodires. In some genera, the trunk shield extends dorsally as a spine that supports an anterior dorsal fin. Posterior to this is a second dorsal fin supported by a large number of fin rays.

A striking specialization noted in many ptyctodonts is the presence of claspers (Figure 4-15*c*) associated with the pelvic fin. Structurally, they distinguish ptyctodonts from all other placoderms. As in sharks and holocephalians, they may be associated with the practice of internal fertilization. The ptyctodont claspers differ structurally from those of the Chondrichthyes, and they are not considered homologous.

The taxonomic position of the ptyctodonts remains uncertain. Miles and Young (1977) contend that the presence of claspers separates them from all other placoderms, but such a specialized feature is an inadequate basis on which to establish their relationship with other groups. Denison (1975) placed them close to the Acanthothoraci and Rhenanida because of the short trunk shield. They may have evolved from primitive arthrodires by a reduction in their armor (for the pattern of the remaining bones appears similar), or they may have evolved from a more primitive stage in which the armor was well defined but less extensive.

ACANTHOTHORACI

Other placoderm orders differ still more from the arthrodires and are even more difficult to classify. A group called alternatively the Palaeacanthaspidoidei by Moy-Thomas and Miles and Acanthothoraci by Denison is known from eight exclusively marine genera from the Lower Devonian. The head shield is similar to that of the actinolepids, but the nasal openings are dorsal, as in antiarchs. The rostral region shows extensive endochondral ossification. The head shield in *Romundina* (Figure 4-16), the earliest known genus, is tiny and appears as a single

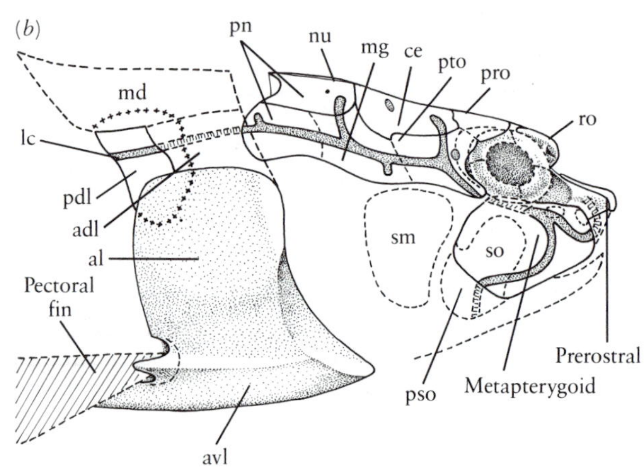

Figure 4-16. THE ACANTHOTHORACID *ROMUNDINA*. (*a*) Head in dorsal view, ×3. (*b*) Restoration of head and trunk armor in lateral view, ×2. *From Ørvig, 1975.*

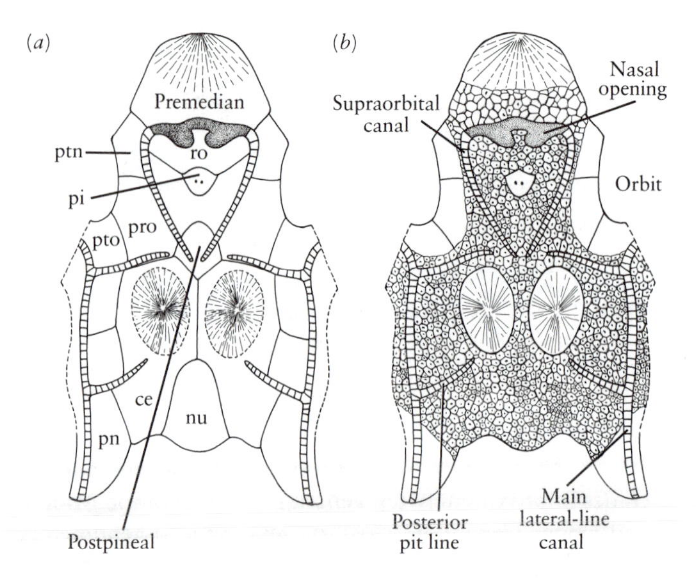

Figure 4-17. *RADOTINA [HOLOPETALICHTHYS]* SHOWING DERMAL ARMOR. (*a*) Armor expressed as large plate. (*b*) Armor expressed as a mosaic of tesserae, ×½. *From Moy-Thomas and Miles, 1971, and Ørvig, 1975. By permission from Chapman and Hall Ltd.*

unit in the adult. It is clear from the size range of isolated elements that the plates grew at their margins until at some stage they became indistinguishably fused. They are ornamented with tiny stellate tubercles. The later genus *Radotina* (*Holopetalichthys*) (Figure 4-17) shows a peculiar pattern of dermal armor that alternatively exhibits a solid exoskeleton or one covered with tesserae. The tesserae may be a secondary feature in this group. Acanthothoracids have a platybasic braincase like arthrodires and petalichthyids. As in arthrodies, the palatoquadrate is fused to the cheek. There is no joint between the dermal bones of the head and the short trunk shield.

Little is known of the gill structures or paired fins. With the exception of the dorsal migration of the nares and the well-ossified endochondral rostrum, they are relatively close to the pattern of primitive arthrodires.

Goujet (1984b) argues that the anatomy of the Acanthothoraci may be close to the primitive pattern for all placoderms. He contends that this order may be a stem group and its constituent genera are sister groups to a number of more derived orders.

RHENANIDA

The orders so far discussed are related to each another by one common characteristic: a relatively complete covering of dermal bones over the head and thorax. Where known, the jaws bear large gnathal plates. In the remaining placoderms, there are some large plates covering the head and thorax, but the dermal ossification is primarily a mosaic of small tesserae or denticles. The jaws are studded with discrete denticles rather than large gnathal plates. The order Rhenanida, as defined by Denison (1975), includes only three well-known genera, one each from the Lower, Middle, and Upper Devonian. The group is rare but worldwide in distribution and entirely marine. Their anatomy is well known and the feeding apparatus differs markedly from that of the previously discussed placoderms.

Gemuendina (Figure 4-18) is considered typical of the order. Specimens may be over 100 centimeters long. The body is extremely dorsoventrally flattened. Orbits and external nares face dorsally and are close to the midline. The pectoral fin resembles that of skates and rays and extends from the level of the orbit to the pelvic fin. Internally, it is supported by a single wide basal element and a series of long jointed radials. The head shield has several large plates, and the surface between them is covered by many polygonal ossicles. The trunk shield is very narrow anteroposteriorly and has typical placoderm plates, although their surface is partially covered by tesserae. Behind a short synarcual, the vertebrae are formed from fused neural and haemal arches, giving the appearance of centra.

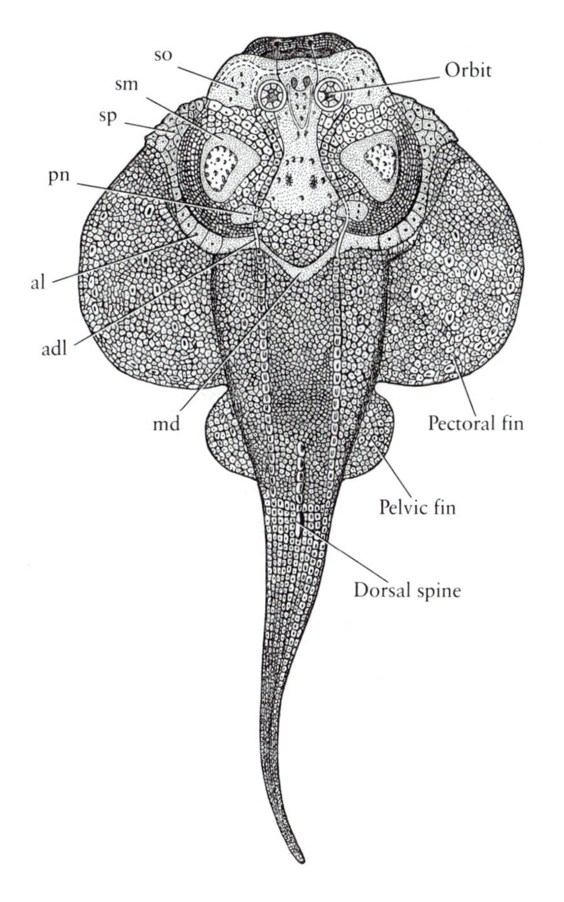

Figure 4-18. THE RHENANID *GEMUENDINA*. Restoration in dorsal view, $\times \frac{1}{2}$. *From Moy-Thomas and Miles, 1971. By permission from Chapman and Hall Ltd.*

The endoskeleton of the head is well known in *Jagorina* (Figure 4-19) and is similar in *Gemuendina*. Only the medial portion of the braincase is ossified. Its appearance differs markedly from the wide flat structure of primitive arthrodires and petalichthyids, although some advanced arthrodires also have a narrow braincase. The structure of the jaws is a more important feature. Unlike arthrodires, the palatoquadrates do not appear to be as strongly integrated with the cheek region, which is not ossified as large plates but as a mosaic of smaller ossicles. In contrast with other placoderms, the palatoquadrates approach one another anteriorly. There are no gnathal plates, but the surface of the palatoquadrate is armed with stellate tubercles. Posteriorly, the palatoquadrate is linked to the braincase via a very long, narrow hyomandibular. The configuration and relationship of this bone is entirely unlike that encountered in other placoderms, but it generally resembles that of advanced sharks and even more closely that of modern rays. Six more posterior gill arches are present according to Stensiö's restorations. The mouth is terminal and protrusible.

Both adequately known genera, *Gemuendina* in the Lower Devonian and especially *Jagorina* in the Upper Devonian, are conspicuously tesserate. This scanty evidence suggests that tesserae replaced armor in this group. We may postulate that this occurred rapidly in the early evolution of rhenanids and at a slower rate in later forms. In contrast, Denison (1983) proposed that the tesserate condition in rhenanids may be primitive for placoderms, with the elaboration of large bony plates a derived condition. The configuration of the paired fins and the dorsal position of the narial openings in rhenanids are almost certainly specializations of this group, which suggests that the tesserate condition of the exoskeleton is also derived.

Miles and Young (1977) suggest that the specialized dorsal position of the narial opening in the Acanthothoraci and Rhenanida indicates a relatively close relationship. Like the pseudopetalichthyids, they also lack an anterior median ventral plate in the trunk shield.

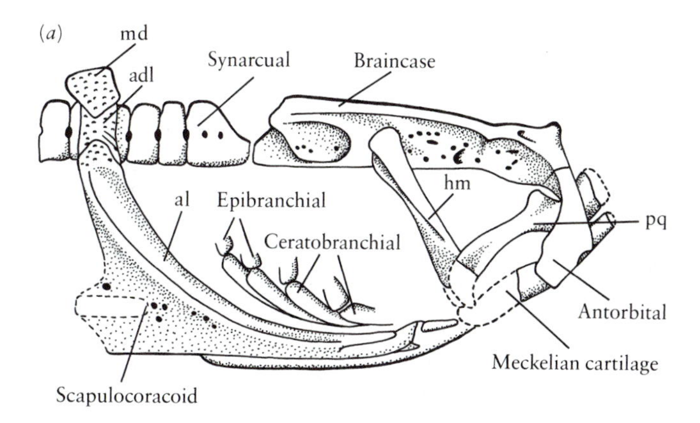

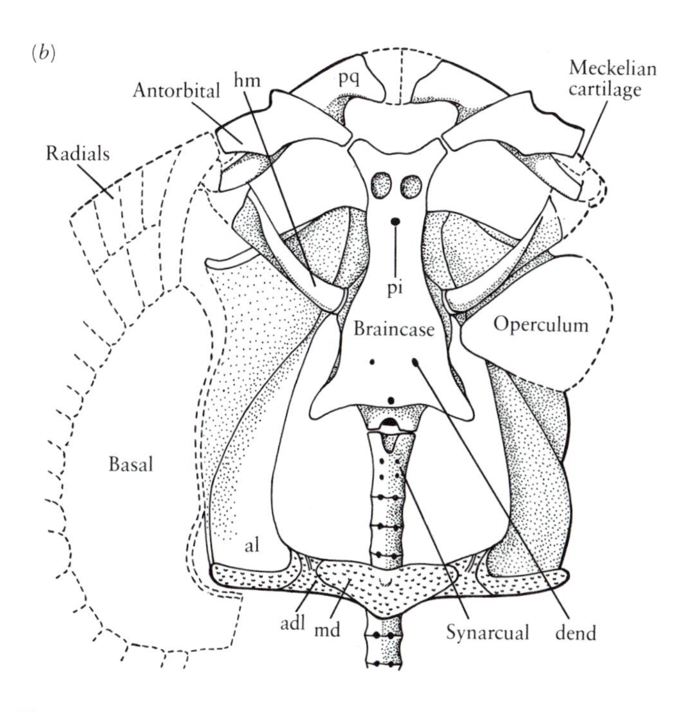

Figure 4-19. THE ENDOSKELETON OF THE HEAD REGION OF THE RHENANID *JAGORINA*. (*a*) Lateral view, $\times \frac{1}{2}$. (*b*) Dorsal view, $\times \frac{1}{2}$. The large hyomandibular linking the braincase and the back of the jaws resembles the pattern in sharks and differs greatly from that in arthrodires. *From Denison, 1978, after Stensiö.*

STENSIOELLIDAE AND PARAPLESTOBATIDAE

The final assemblage of placoderms includes three poorly known genera. Denison (1975, 1983) places them in two orders: the Stensioellida, including only the genus *Stensioella*, and the Pseudopetalichthyida, including *Paraplesiobatis* and *Pseudopetalichthys*. Miles and Young (1977) recognize only one group, the pseudopetalichthyids. These authors differ greatly in the significance they assign to these genera. Moy-Thomas and Miles (1971) relegate all three genera to *incertae sedis*, indicating that they are too poorly known to evaluate their phylogenetic position. Denison suggests that they are relicts of an early stage in placoderm evolution and are very important to our understanding of the group's history.

All are from marine Lower Devonian deposits in Germany. They are small and dorsoventrally flattened. Most of the body is covered with denticles or small scales, but the head and shoulder region may have some larger bony plates. In *Stensioella* (Figure 4-20), the braincase is poorly ossified except posteriorly where the surface for articulation with the vertebrae is well developed. The jaws are short and covered with small pointed denticles, rather than large gnathal plates. This pattern is similar to that of rhenanids and differs from other placoderms. We may draw a parallel between the predominance of bony plates on the body and the presence of gnathal plates and the occurence of tesserae or denticles on the body and denticulate teeth in the mouth. We do not know whether the common presence of these characteristics indicates a close relationship. In *Pseudopetalichthys* (Figure 4-20*b*), the thoracic armor resembles that of more orthodox placoderms, while the fin supports resemble those of advanced sharks.

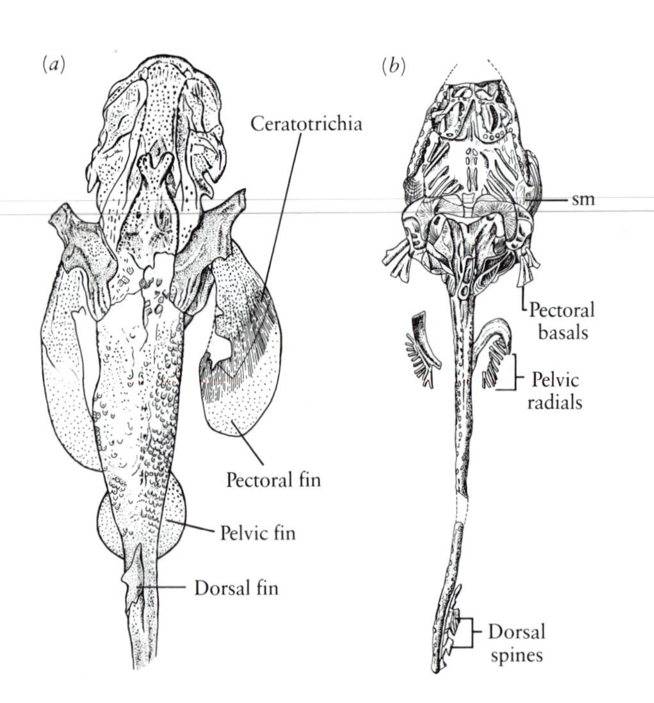

Figure 4-20. (*a*) *Stensioella* in dorsal view, $\times \frac{1}{3}$. (*b*) *Pseudopetalichthys* in ventral view, one-half natural size. The low degree of consolidation of the dermal skeleton may be a primitive feature of placoderms. *From Moy-Thomas and Miles, 1971. By permission from Chapman and Hall Ltd.*

Placoderms became extinct near the end of the Devonian, when they appear to have been near the height of their diversity. The suddenness of their extinction may be partially a geological artifact, since their remains are particularly common in the widespread black shales at the end of the Devonian in North America and Europe. The overlying Mississippian sediments are mostly clastic and represent an environment in which placoderms may have been uncommon even if they had survived to this time. Nevertheless, many placoderm lineages apparently became extinct over a relatively short period of time, presumably as a result of competition with the more advanced jawed fish, the Osteichthyes and Chondrichthyes.

INTERRELATIONSHIPS OF PLACODERMS

Since there is no fossil record of either plausible ancestors or a closely related sister group, it has been very difficult to establish the polarity of character states among primitive placoderms. Thus, the interrelationships of the various orders are subject to continuing debate. Figure 4-21 shows four alternative cladograms based on recent analyses of this group.

SUMMARY

Jawed vertebrates first appear in the Lower Silurian. Jaws, gill supports, and trabeculae all develop from neural crest cells. The division between the mouth and pharynx is comparable in jawed fish and agnathans, but the skeletal supports and gill filaments evolved separately in the two groups.

The placoderms are the most primitive jawed vertebrates but are not closely related to either cartilaginous or bony fish. They were very diverse within the Devonian but became extinct during the early Carboniferous. Most placoderms have a massive exoskeleton that is jointed between the head and thorax. Most appear to have been benthonic and were probably poor swimmers, despite the presence of pectoral and pelvic fins. Most placoderms had bony plates on the margin of their jaws instead of true teeth.

The ancestors of placoderms may have been naked or covered with small, isolated denticles or tesserae. However, most placoderm orders have a similar pattern of

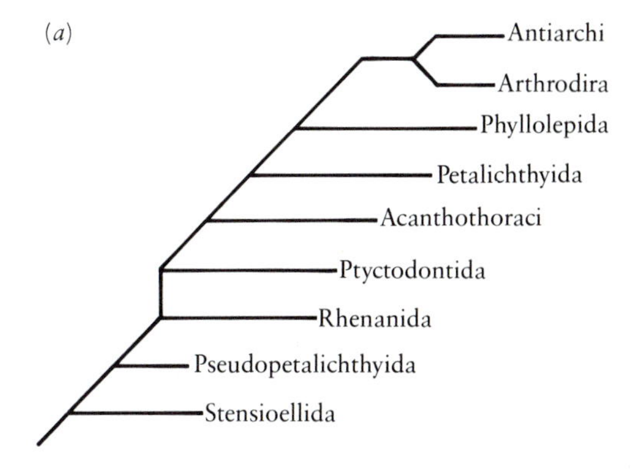

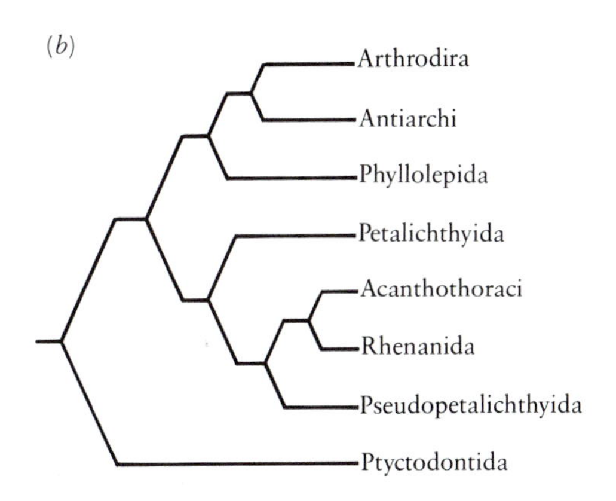

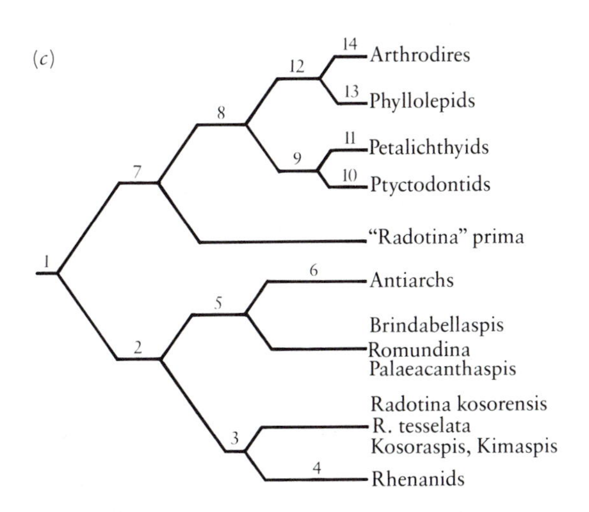

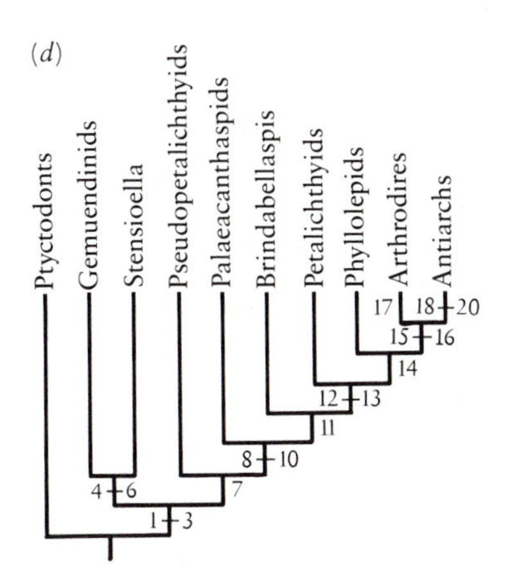

Figure 4-21. FOUR ALTERNATIVE CLADOGRAMS SHOWING POSSIBLE INTERRELATIONSHIPS OF THE PLACODERM ORDERS. (*a*) *From Denison, 1978.* (*b*) *From Miles and Young, 1977. With permission from the Linnean Society Symposium Series 4. Copyright 1977 by the Linnean Society of London.* (*c*) *From Goujet, 1984.* (*d*) *From Gardiner, 1984.* Characters used by Goujet (somewhat shortened from original) to establish interrelationships are as follows:

Characters assumed to be basic placoderm apomorphies are:

1. Body armour forming a complete ring surrounding the pectoral girdle; double cervical joint with an endoskeletal component and a dermal component; omega-shaped palatoquadrate closely associated with the anterior cheek plates, with the adductor mandibulae muscle attached on the ventral face of the metapterygoid and the medial face of the suborbital plate.

Apomorphies acquired at the following numbered stages are:

2. Premedian plate, and backward shifting of the nasal capsules.
3. Tesserae on the skull roof, particularly in the pineal and orbital regions.
4. Rhenanid apomorphies; pseudobatoid body shape with very flattened skull, loss of the dermal cervical joint, fragmented (or tessellated) anterolateral and spinal plates, strong reduction in body armor length, and loss of the posterior dorsolateral and posterolateral plates.
5. Large premedian plate with a preorbital depression, dermal cervical joint with a fossa on the articular flange of the anterior dorsolateral plate, external endolymphatic pore placed medially relative to the endocranial opening, and strong cohesion of the plates covering the endoskeletal shoulder girdle to form a pectoral unit (anterolateral, spinal, anterior ventrolateral plates).
6. Antiarchan apomorphies; a second (posterior) median dorsal plate incorporated in the body armor, pectoral unit represented by a single plate (anterior ventrolateral) enclosing the small pectoral fenestra, pectoral fin modified into a slender appendage covered by plates derived from modified scales, antiarch skull roof pattern with the orbits laterally enclosed by large lateral plates, and semilunar plates.
7. Dermal bone ornamented with evenly distributed tubercles of pallial semidentine.
8. Anterior and posterior median ventral plates added to the body armor, and endolymphatic duct opening on the same paranuchal plate as the confluence of the posterior pit line and the main lateral line of the skull.
9. Sensory lines enclosed in tubes projecting below the visceral surface of the dermal skull plates and opening by pores, central sensory line lost, *X* pattern of the sensory lines on the middle plate of the skull (nuchal or postpineo-nuchal), fragmentation of the suborbital and postsuborbital plates, and similar structure of the cervical joint.
10. Ptyctodontid apomorphies; loss of rostral plate, endolymphatic duct closed, loss of suborbital and postsuborbital plates, loss of posterior dorsolateral, posterolateral, posterior ventrolateral, posterior median ventral plates, presence of bony pelvic claspers.
11. Petalichthyid apomorphies; laterodorsal orbits, bounded laterally by pre- and postorbital plates and medially by the central plate.
12. Shortened preorbital region of the skull and separate interolateral plate with a deep transverse ventral groove.
13. Phyllolepid apomorphies; flattened body, much expanded nuchal plate, loss of rostral plate, loss of posterior dorsolateral and posterolateral plates, and loss of semidentine on the superficial ornament.
14. Arthrodire apomorphies; two pairs of upper tooth plates, large endocranial posterior postorbital processes, wide and short rostral plate, and pectoral fenestra closed posteriorly by the posterior ventrolateral plates, unless secondarily transformed into a deep pectoral emargination in some brachythoracids.

(*continued on page 60*)

Figure 4-21. *(Continued)*
Characters used by Gardiner to establish interrelationships are as follows:

1. Long, anteromedially directed coracoid process and a prepectoral process that enters spinal plate versus coracoid process not medially directed and prepectoral process absent. (Spinal plate absent in *Ctenurella* and possibly missing in *Stensioella* and antiarchs).
2. Synarcual encloses nerve cord and notochord versus synarcual formed of fused neural arches surrounding nerve cord only.
3. Ventral fossa of pectoral girdle closed anteriorly versus ventral fossa open. (Fossa absent altogether in more advanced placoderms).
4. Large, elongate hyomandibular versus small, squat hyomandibular.
5. Infraorbital sensory canals joined anteriorly and also joined to supraorbital versus supraorbital and infraorbital not united.
6. Small polygonal scales clothe the dermal shoulder girdle versus no polygonal scales.
7. Interolateral bone present versus interolateral absent. (Interolaterals are said to be missing in antiarchs, but the paired semilunars of asterolepids could be their homologues).
8. Posterior dorsolateral bone present versus posterior dorsolateral absent. (Posterior dorsolateral absent in phyllolepids).
9. Omega-shaped palatoquadrate fused to covering dermal bones versus palatoquadrate not "omega shaped" and free of dermal bones.
10. Occipital region elongate, with many spino-occipital nerve foramina versus occipital region behind vagus short.
11. Ginglymoid articulation in which condyles of anterior dorsolaterals rotate in glenoid fossae on the paranuchals; joint supported by the "craniospinal" process versus dermal neck joint absent or without glenoid fossa.
12. Posterior ventrolateral bone present versus posterior ventrolateral absent.
13. Posterior median ventral bone present versus posterior median ventral absent. (There are some specimens of *Lunaspis* and *Phyllolepis* where the posterior median ventral is present but the anterior median ventral is absent).
14. Great width of skull roof across posterolateral angles and greatly enlarged median dorsal bone with branch of lateral-line canal versus a skull roof longer than broad and a shoulder girdle with a short median dorsal.
15. Postmarginal bone present versus postmarginal absent from lateral margin of skull roof.
16. Posterior lateral bone present versus posterior lateral bone absent.
17. Two superognathals present versus one superognathal.

Additional hypotheses of placoderm relationships are discussed by Forey and Gardiner (1986) and Young (1986).

large bony plates, which indicates that they share a common ancestor that had a fairly extensive dermal skeleton.

The Arthodira is the most diverse order of placoderms. Advanced members of this group reduced the exoskeleton, freeing the trunk for more active swimming. Antiarchs, phyllolepids, petalichthyids, and acanthoracids are less diverse orders, each of which is specialized in body form, the nature of the limbs, and the dermal armor pattern. All were primarily benthonic. The antiarchs were particularly common in freshwater deposits, but most placoderms lived in the ocean. The ptyctodonts reduced the external armor and may have been fairly active swimmers. Their body form and dentition resemble the pattern of holocephalians among modern cartilaginous fish; the males had claspers.

Instead of large armor plates, the rhenanids, stensioellids, and paraplesiobatids had a covering of small tesserae, perhaps reflecting the primitive pattern for placoderms. The rhenanids had very large pectoral fins, giving them the appearance of skates and rays. In this group, the hyomandibular supported the jaws like that of modern sharks. The interrelationships of the placoderm orders have not been established.

References

Denison, R. H. (1975). Evolution and classification of placoderm fishes. *Breviora*, **432**: 1–24.

Denison, R. H. (1978). *Placoderms. Handbook of Paleoichthyology*. 2. Gustav Fischer Verlag, Stuttgart.

Denison, R. H. (1983). Further consideration of placoderm evolution. *J. Vert. Paleont.*, **3**: 69–83.

Forey, P. L. and Gardiner, B. G. (1986). Observations on *Ctenurella* (Ptyctodontida) and the classification of placoderms fishes. *Zool. J. Linn. Soc.*, **86**: 43–74.

Gardiner, B. G. (1984). The relationship of placoderms. *J. Vert. Paleont.*, **4**: 379–395.

Goujet, D. (1975). *Dicksonosteus*, un nouvel arthrodire du Dévonien du Spitsberg. *Coll. Intl. C.N.R.S., Problèmes actuels de paléontologie*, **218**: 81–99.

Goujet, D. (1984a). *Les poissons placodermes du Spitsberg. Arthrodires Dolichothoraci de la Formation de Wood Bay (Dévonien Inférieur)*. Cahiers de Paléontologie. Section vertébrés. Editions du Centre National de la Recherche Scientifique, Paris.

Goujet, D. (1984b). Placoderm interrelationships: A new interpretation with a short review of placoderm classification. *Proc. Linn. Soc. N.S.W.*, **107**(3): 211–243.

Gross, W. (1963). *Gemuendina stuertzi* Traquair, Neuuntersuchung. *Hessisches Landesamt für Bodenforschung, Wiesbaden, Notizblatt*, 91: 36–73.

Long, J. A. (1984). New phyllolepids from Victoria and the relationships of the group. *Proc. Linn. Soc. N.S.W.*, **107**(3): 263–308.

Miles, R. S. (1969). Features of placoderm diversification and the evolution of the arthrodire feeding mechanism. *Trans. Roy. Soc. Edinburgh*, 68: 123–170.

Miles, R. S., and Westoll, T. S. (1968). The placoderm fish *Coccosteus cuspidatus* Miller ex Agassiz from the Middle Old Red Sandstone of Scotland. Part 1. Descriptive morphology. *Trans. Roy. Soc. Edinburgh*, **67**: 373–476.

Miles, R. S., and Young, G. C. (1977). Placoderm interrelationships reconsidered in the light of new ptyctodontids from Gogo, Western Australia. In S. M. Andrews, R. S. Miles, and A. D. Walker (eds.), *Problems in Vertebrate Evolution, Linn. Soc. Symp. Ser.*, **4**: 123–198. Academic Press, London.

Moy-Thomas, J. A., and Miles, R. S. (1971). *Palaeozoic Fishes* (2d ed.). Chapman and Hall, Ltd. London.

Ørvig, T. (1975). Description, with special reference to the dermal skeleton of a new radotinid arthrodire from the Gedinnian of the Arctic Canada. *Coll. Intl. C.N.R.S. Problèmes actuels de paléontologie (évolution des vertébrés)*, **218**: 41–71.

Ørvig, T. (1985). Histologic studies of ostracoderms, placoderms and fossil elasmobranchs. 5. Ptyctodontid tooth plates and their bearing on holocephalan ancestry: The condition of chimaerids. *Zoologica Scripta*, **14**(1): 55–79.

Schaeffer, B. (1975). Comments on the origin and basic radiation of the gnathostome fishes with particular reference to the feeding mechanism. *Coll. Intl. C.N.R.S. Problèmes actuels de paléontologie (évolution des vertébrés)*, **218**: 101–109.

Schaeffer, B., and Thomson, K. S. (1980). Reflections on agnathan-gnathostome relationships. In L. L. Jacobs (ed.), *Aspects of Vertebrate History; Essays in Honor of Edwin Harris Colbert*, pp. 19–33. Museum of Northern Arizona Press, Flagstaff.

Stensiö, E. A. (1963a). Anatomical studies on the arthrodiran head. Part I. *Kungl. Svenska Vetenskapakad. Handlingar*, ser. 4, **9**: 1–419.

Stensiö, E. A. (1963b). The brain and the cranial nerves in fossil, lower craniate vertebrates. *Skrifter utgitt av Det Norske Vidlenskap-Akad: Oslo. I. Mar.-Naturv. Klasse. Ny Serie*, **13**: 1–120.

Stensiö, E. A. (1969). Elasmobranchiomorphi. Placodermata. Arthrodires. In J. Piveteau (ed.), *Traité de Paléontologie*, **4**(2): 71–692. Masson S.A., Paris.

Young, G. C. (1980). A new early Devonian placoderm from New South Wales, Australia, with a discussion of placoderm phylogeny. *Palaeontographica, A*, **167**: 10–76.

Young, G. C. (1984). Comments on the phylogeny and biogeography of antiarchs (Devonian placoderm fishes), and the use of fossils in biogeography. *Proc. Linn. Soc. N.S.W.*, **107**(3): 443–473.

Young, G. C. (1986). The relationships of placoderm fishes. *Zool. J. Linn. Soc.*, **88**: 1–57.

Zangerl, R., and Williams, M. E. (1975). New evidence on the nature of the jaw suspension in Palaeozoic anacanthous sharks. *Palaeontology*, **18**: 333–341.

C H A P T E R V

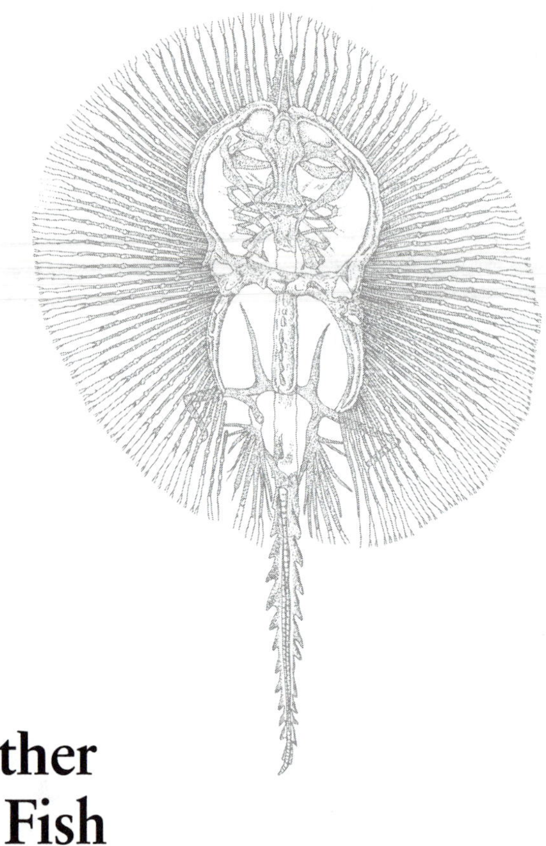

Sharks and Other Cartilaginous Fish

ADVANCED JAWED FISH

Except for the jawless cyclostomes, all living fish belong to two major classes, the Chondrichthyes, or cartilaginous fish, and the Osteichthyes, or bony fish. They differ from one another in many respects, but the structure of the jaws and the pattern of tooth replacement indicate that they share a common ancestry that is separate from that of the placoderms.

As Schaeffer pointed out (1975), in the earliest known members of both groups the palatoquadrate extends most of the length of the braincase and has a distinctive, cleaver shape. The narrow anterior "handle" supports the teeth, and the posterior "blade" forms the medial surface of the chamber for the jaw musculature (Figure 5-1). Teeth are formed along an infolding of the epithelium of the mouth's mucous membrane, called the **dental lamina.** Unlike the gnathal plates of placoderms, the teeth are continuously replaced; they develop on the medial surface of the jaw and rotate into a more lateral position as they mature.

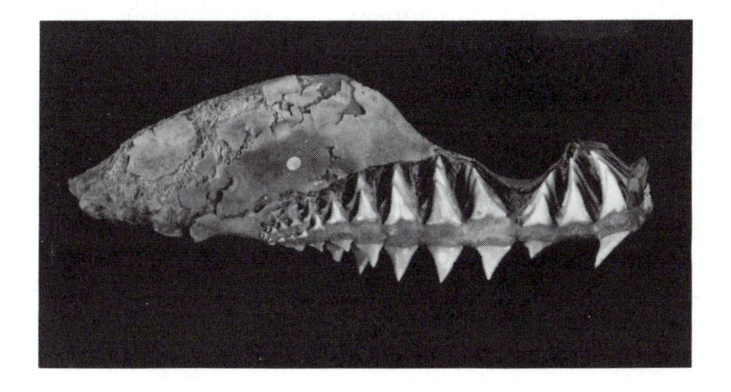

Figure 5-1. THE JAW OF A MODERN SHARK. Note the "cleaver" shape of the palatoquadrate characteristic of all primitive, jawed vertebrates, except the placoderms. Among both bony and cartilaginous fish, the teeth develop on the medial surface of the jaw and rotate to the lateral margin as they mature. The teeth that occupy each position along the jaw belong to one tooth "family." The size of the teeth progressively increases from older to younger to accommodate the increase in jaw size. The superficial nature of the prismatic calcification may be seen on the broken surface. *Courtesy of the Redpath Museum, McGill University, Montreal, Canada.*

In contrast with ostracoderms and most placoderms, there is no evidence that an extensive, inflexible exoskeleton was present in the ancestors of sharks and bony fish. Remains of both groups are known from the Upper Silurian. The sharks are represented by tiny denticles, and the bony fish by small, overlapping scales. Both groups, as far back as they can be traced, were active swimmers, with a fusiform body and well-developed paired fins. A swim bladder evolved among primitive Osteichthyes but not among the Chondrichthyes. Modern sharks have a large, oil-filled liver that greatly reduces their specific gravity; Paleozoic sharks may also have had this feature. The internal skeleton in Chondrichthyes, in contrast with the Osteichthyes, never ossifies; although bone may be present at the base of the scales in some genera, it is never an important constituent of the exoskeleton. The cartilage of Chondrichthyes is frequently strengthened by calcification which occurs as a superficial layer of prismatic granules.

Two groups of Chondrichthyes are present in the modern seas: the Elasmobranchii, including the sharks, skates, and rays; and the Holocephali, composed of the less-common chimaeras or ratfish. The two groups differ significantly in body form and dentition, but the similarities in the calcification of the cartilaginous skeleton and the specializations of their reproductive pattern suggest that they share a common ancestry.

All modern Chondrichthyes practice internal fertilization; most genera are ovoviviparous or viviparous, but others lay large, yolky eggs enclosed in thick horny egg cases. In all living Chondrichthyes and most fossil genera, the male has paired structures, termed claspers (see Figure 5-5), that are supported by the skeleton of the pelvic fin and facilitate internal fertilization. Internal fertilization is associated with the production of a small number of eggs, each of which has a high chance of survival, in contrast with the reproductive strategy of typical bony fish who produce many eggs, each of which has only an infinitesimal chance of maturing.

The earliest remains of chondrichthyan fishes are found in sediments deposited in marine waters; the group has remained predominantly marine. However, one Paleozoic order was common in freshwater, and other genera, both living and fossil, seem to be able to tolerate considerable changes in salinity.

The elasmobranchs and holocephalians each have a long fossil history, extending back into the Devonian. In a recent review of Paleozoic chondrichthyans, Zangerl (1981) suggests that the most fundamental distinction between these groups lies in the proportions of the gill region. In most jawed fish, the gills are close behind and somewhat beneath the back of the braincase, and the shoulder girdle lies a short distance behind the occiput. Zangerl thinks that this pattern is primitive for jawed fish and has been retained in the holocephalians. In sharks, there is considerable space between the gill supports and they are primarily behind the braincase; the shoulder girdle is well separated from the occiput. One consequence of the length of the gill region is that sharks did not evolve an operculum, which is a typical feature of holocephalians and bony fish.

Maisey (1984b) presents an alternative hypothesis that chimaeroids diverged from among the Paleozoic sharklike chondrichthyans and occupy a phylogenetic position between the primitive cladoselachian and symmoriid sharks and the more advanced assemblage including *Ctenacanthus* and modern sharks.

PALEOZOIC ELASMOBRANCHS

On the basis of the known fossil record, paleontologists think that elasmobranchs may have undergone two major episodes of adaptive radiation (Figure 5-2). The first began during the late Silurian and Devonian, resulting in sharks that broadly resembled modern genera but had a more primitive skeleton. Most of the archaic shark groups became extinct by the end of the Triassic. The Ctenacanthida, which we recognize in the late Devonian, apparently gave rise to all modern sharks as well as skates and rays. These groups began their radiation in the Jurassic and achieved an essentially modern appearance by the early Cretaceous.

Throughout their history, elasmobranchs have maintained a dominant role as marine predators. Many genera have retained a streamlined shape and a piercing and slashing dentition, but there has been repeated evolution of groups adapted toward a benthonic way of life, with the development of crushing dentition that enables them

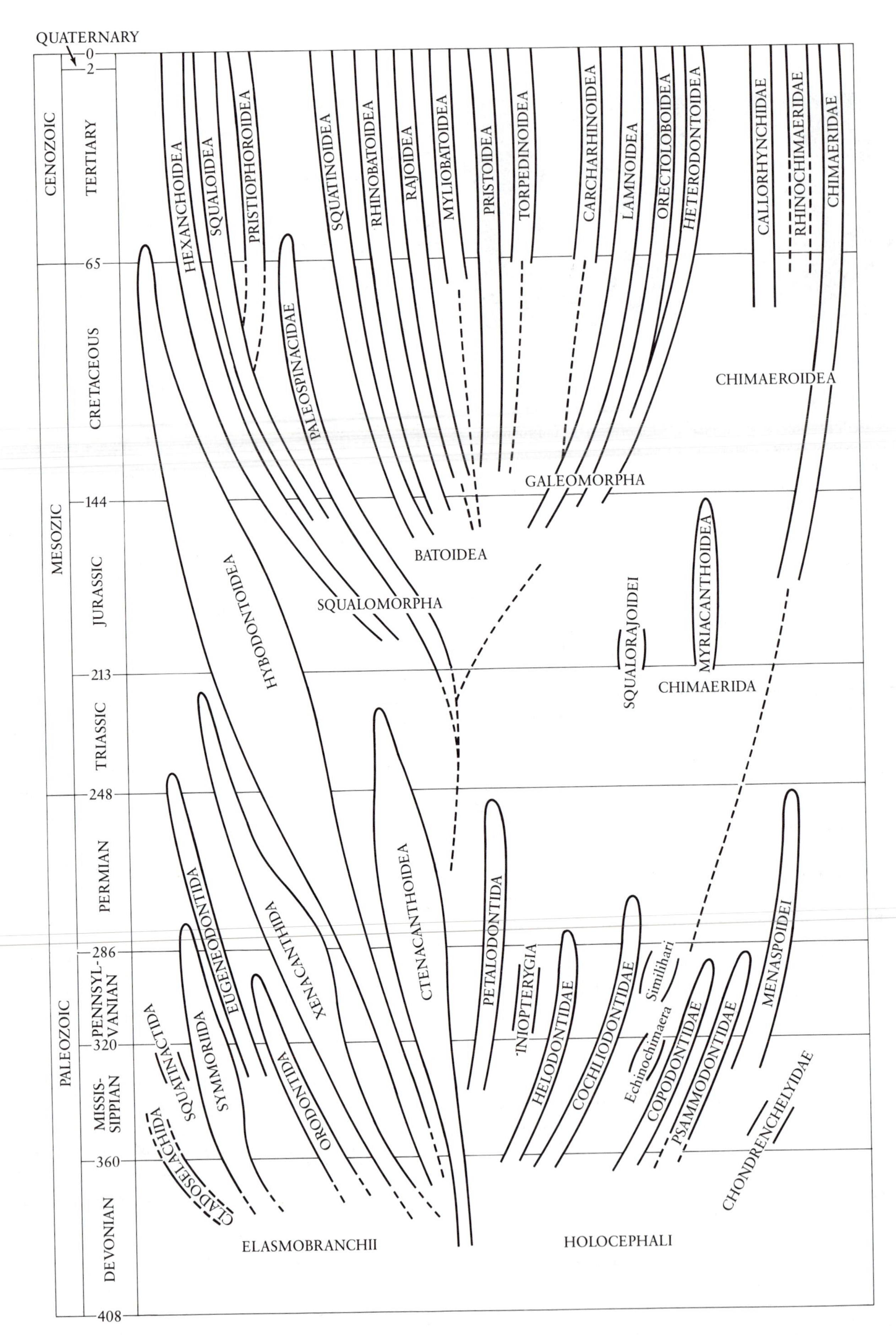

Figure 5-2. PHYLOGENY OF CHONDRICHTHYAN FISHES. Classification primarily from Zangerl (1981) and Compagno (1977). Note the presence of two successive radiations among the elasmobranchs. All modern elasmobranchs can be traced to a single ancestral group. The relationships among the Paleozoic orders have not been established, and their grouping is largely arbitrary.

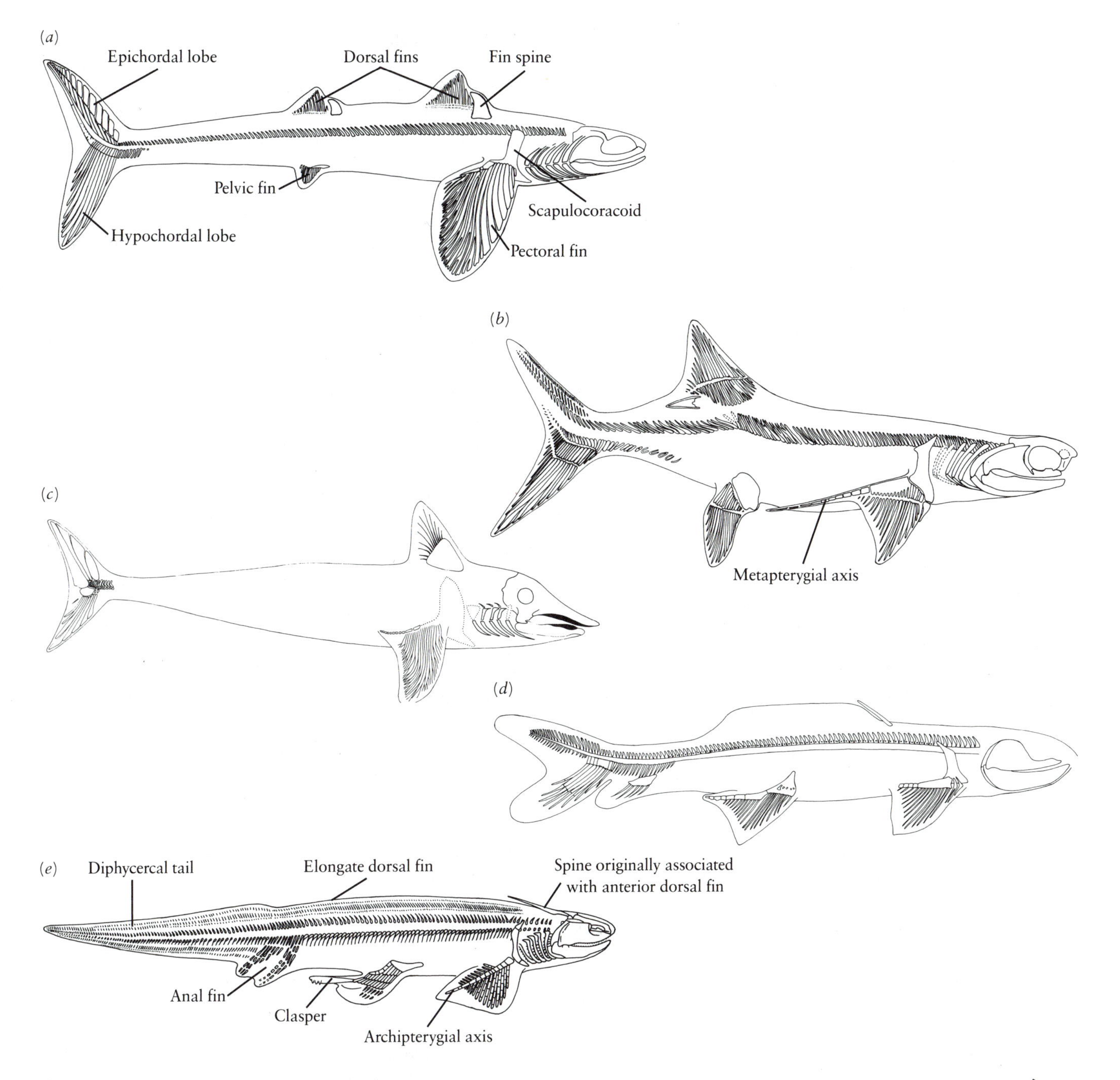

Figure 5-3. THE BODY SHAPE OF PALEOZOIC SHARKS. (*a*) The Devonian Cladoselachida, *Cladoselache*, $\times\frac{1}{5}$. *From Zangerl, 1981.* (*b*) The Carboniferous Symmoriida, *Denaea*, $\times\frac{1}{3}$. *From Schaeffer and Williams, 1977.* (*c*) The Upper Carboniferous Eugeneodontida, *Fadenia*, $\times\frac{1}{5}$. *From Zangerl, 1981.* (*d*) The lower Carboniferous xenacanth *Diplodoselache*, $\times\frac{1}{7}$. *From Dick, 1981.* (*e*) The lower Permian Xenacanthida, *Xenacanthus*, $\times\frac{1}{7}$. *From Schaeffer and Williams, 1977.*

to feed on a wide variety of hard-shelled invertebrates as well as vertebrates.

Although the structure of their fin supports and teeth vary greatly, most Paleozoic sharks (Figure 5-3) resemble common living genera such as the dogfish. The body is streamlined and fusiform, with the mouth nearly terminal. There may be one or two dorsal fins. The anal fin is absent primitively. The notochordal axis is directed dorsally to give a tail that is technically heterocercal, although the ventral supports are commonly arranged to give a superficially symmetrical appearance. The paired fins are fully developed and, usually, broadly triangular.

The braincase of all sharks (Figure 5-4) is an undivided structure that is broad and low like that of primitive placoderms. Three areas may be recognized: ethmoid, orbital, and otico-occipital. The anterior ethmoid region varies greatly, with an extremely long rostrum in some genera (see Figures 5-8 and 5-12). The orbital region is relatively uniform in its proportions, with a fairly wide interorbital septum and protective shelves above and below the eyeball. Behind the orbits, the postorbital processes extend far laterally to support the otic process of

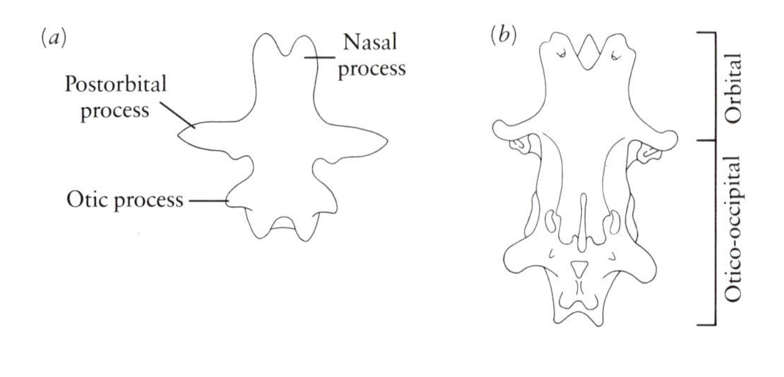

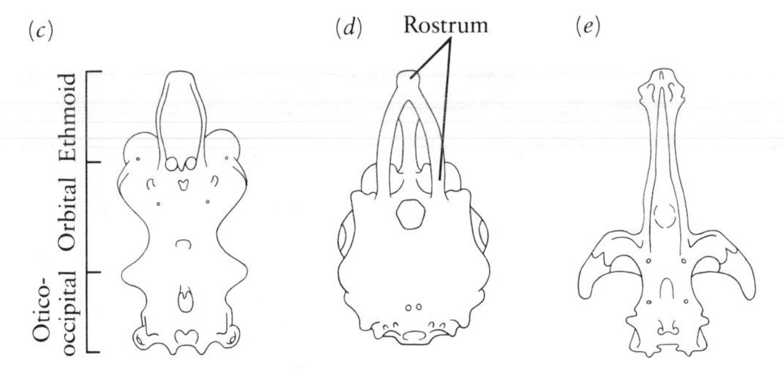

Figure 5-4. THE BRAINCASES OF SHARKS. (*a*) The Upper Devonian *Cladoselache*, showing prominent postorbital processes and otico-occipital region of moderate length. (*b*) *Tamiobatis*, a Lower Carboniferous xenacanth with strongly elongated otico-occipital region. (*c*) The squaloid *Squalus* showing the trough-shaped ethmoid region. (*d*) The galeoid *Isurus* with a tripod-shaped ethmoid region. (*e*) *Rhinobatos*, with a greatly elongated ethmoid region. *All from Schaeffer, 1967.* Schaeffer (1981) and Maisey (1983) recently published detailed descriptions of the braincases of Paleozoic and Mesozoic sharks.

the palatoquadrate on their posterior surface. The relative length of the otico-occipital portion of the braincase varies among primitive sharks. The fossil record is not complete enough to indicate whether a long or short otico-occipital region is more primitive. Most groups of primitive sharks have a cleaver-shaped palatoquadrate. The narrow "handle" bearing the teeth is typically connected to the neurocranium beneath the subocular shelf. The extensive lateral surface of the otic process provides the origin for the adductor jaw musculature.

The gill arches are frequently not well preserved in primitive sharks but appear to adhere to a fairly consistent pattern. The hyomandibular articulates with the braincase posterior to the postorbital process. In most groups, it is supported by the dorsomedial surface of the otic process of the palatoquadrate and helps connect the upper jaw with the neurocranium. We refer to this double support of the palatoquadrate (directly to the postorbital process of the braincase and via the hyomandibular) as **amphistylic.**

Two evolutionary trends in the nature of jaw support occur among Chondrichthyes. One is the closer integration of the palatoquadrate with the braincase, resulting among most holocephalians in a fusion of the two elements, a condition termed **holostylic** or **autostylic.** The other is a clear separation of the palatoquadrate from the braincase, which evolves among advanced elasmobranchs, with the hyomandibular forming a mobile connection between the two units (see Figure 5-14). This condition is termed **hyostylic** and is characteristic of advanced sharks, skates, and rays.

The ventral portion of the hyoid arch, the ceratohyal, lies between the lower jaws. Five additional gill arches lie behind the hyoid. As in modern sharks, each presumably consisted of four elements: a dorsal pharyngobranchial, an epibranchial, a ceratobranchial, and a basibranchial.

Vertebral centra are not evident in early sharks; noncalcified segmental structures may have been present, but the notochord probably remained unrestricted. Neural and haemal arches are generally calcified and outline the axial skeleton, as in some placoderms. Calcified ribs are present in some Paleozoic sharks but not in the most primitive forms.

The skeleton of the pectoral and pelvic girdles is relatively constant in primitive sharks. The scapulocoracoid forms a broad arc extending dorsally to the level of the notochord and curving ventrally toward the midline. The two sides may meet ventrally, but they never fuse, as is the case in modern sharks. The pelvic girdle consists of small paired oval or triangular plates. In the primitive state, the pectoral and pelvic fin both may have had a number of parallel supports that extended from the girdle into the fin. The supports are typically divided into proximal elements, termed **basals,** and distal elements, termed **radials.** The radials in primitive sharks extend to the margins of the fins, a condition termed **plesodic.** In the more advanced sharks of the Mesozoic and Cenozoic, the radials are supplanted by more flexible, horny supports called **ceratotrichia.** In both the pectoral and pelvic fins, the basals tend to fuse to one another and to form a more effective articulation with the girdle.

Various trends are evident in the evolution of the pectoral fin. *Cladoselache* (Figure 5-3*a*) displays a primitive condition, in which the anterior radials articulate directly with the girdle. There are a series of small basals that articulate with the more posterior surface of the scapulocoracoid and support the more posterior radials. The basals are arranged as an axis extending along the proximal margin of the fin. The large anterior basal is termed the **metapterygium.** In several other Paleozoic groups (see Figures 5-3*b* and 5-11), the axis may extend posteriorly beyond the limit of the radials. In some xenacanths (see Figure 5-3*e*), radials are borne on both the postaxial and preaxial surface of the axis. The pattern of the pectoral fin seen in ctenacanthid sharks (see Figure 5-11*b*) appears to have developed by the fusion of the anterior basals to form a tripartite fin support that resembles the pectoral fin of Mesozoic and Cenozoic sharks.

The structure of the pelvic fin is less well known but appears to remain relatively simple, except for some fusion of the basals and the elaboration of the claspers in the males (Figure 5-5). Their structure differs from one group to another, and the individual components are not strictly homologous.

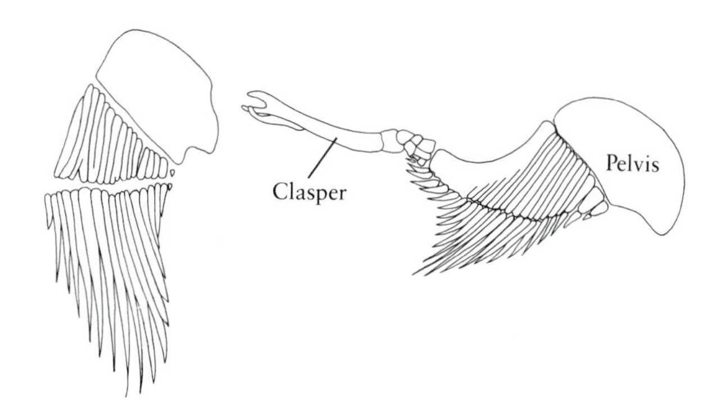

Figure 5-5. PELVIC FINS OF THE SYMMORIIDA, *COBELODUS*. Left, female; right, male showing clasper. *From Zangerl and Case, 1976.*

Although scales (or denticles) are reduced or missing in some genera, they are usually an important feature of chondrichthyan fishes. The simplest type of scales seen in Paleozoic sharks consists of a cone of dentine developed over a pulp cavity and usually capped with a hard, shiny enameloid substance (Figure 5-6*a*). We call such elementary scales **lepidomoria,** and they probably represent the ancestral pattern for sharks. More commonly, the scales are composite structures that developed by the fusion of many lepidomoria. Increase in size occurs by the progressive addition of lepidomorial units, a growth pattern termed **cyclomorial** (Figure 5-6*c*).

In modern sharks and some Paleozoic genera (including remains from the Upper Silurian), the scales show no evidence of growth or formation by fusion of lepidomoria. Their definitive size is achieved when they first form. We apply the term **placoid** to these scales. Small scales are present early in ontogeny and are replaced by larger scales as the animal grows. Placoid scales have apparently evolved many times in different lineages and may be directly derived from lepidomorial units.

The teeth of sharks are clearly derived from scales that have become elaborated along the margins of the jaws. Primitively, they have an open pulp cavity surrounded by a cone of dentine with a hard, shiny, enamel-like substance at the surface. They are attached to the jaws by connective tissue. Shark teeth are first recognizable in Lower Devonian sediments and already show considerable diversity in cusp pattern. Most Devonian and Carboniferous shark teeth have a pattern termed **cladodont** (Figure 5-7*a*), with a single major cusp and smaller lateral cusps above a broad base. *Diplodus* teeth have two lateral cusps that are much longer than the central cusp. Many shark groups have blunt, broad cusps, which may form a crushing pavement dentition. Although shark teeth vary greatly in their form, they may be poor indicators of taxonomic identity. A single shark may have teeth of many shapes, and an individual tooth pattern may be present in genera that are very distantly related on the basis of other anatomical features.

Fin spines are a prominent feature of both ancient and modern sharks. Spines, like teeth, were apparently derived from single denticles that became greatly enlarged. Spines typically consist of an inner trunk of dentine, the base of which becomes deeply inserted in the body in advanced genera, and an outer mantle of ornamented enameloid material. Most sharks have fin spines associated with each of two dorsal fins. According to Young (1982), primitive elasmobranchs and holocephalians probably had only a single dorsal fin and accompanying spine. He regards the absence of spines among Paleozoic sharks as a specialization.

We know at least seven major groups of sharks in the Paleozoic. Although it was once thought that they could be arranged in a simple phyletic sequence or a series

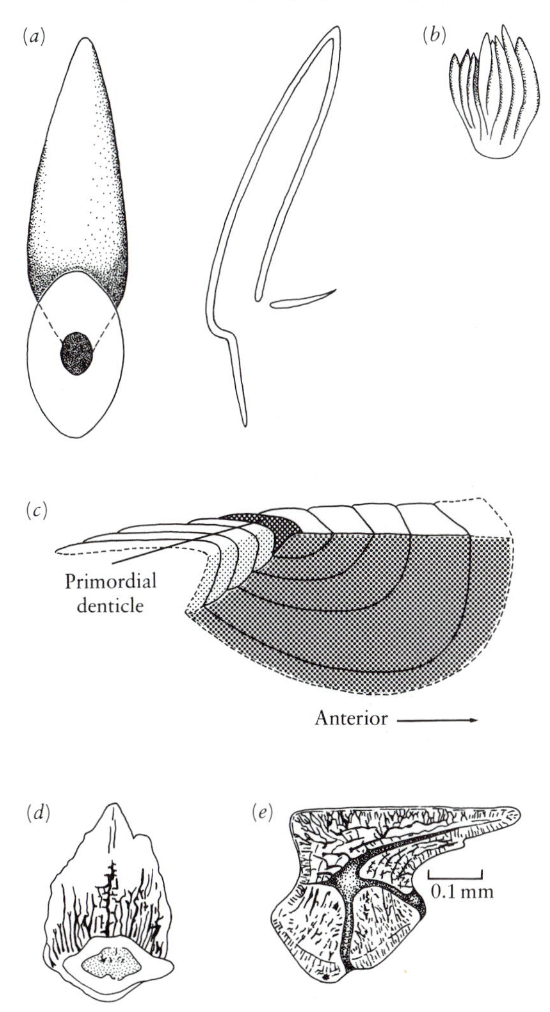

Figure 5-6. SCALES OF ELASMOBRANCHS. (*a*) Single lepidomorial unit of *Eugeneodus* in ventral view and in longitudinal section. This exemplifies the most primitive type of shark scale. It also illustrates the pattern from which shark teeth and spines are thought to have evolved. (*b*) Aggregate of lepidomoria in *Eugeneodus*. (*c*) Diagram to illustrate the mode of growth of a cyclomorial scale of the type found in *Orodus*. The first fingerlike denticle is a simple lepidomorium. Additional fingerlike elements develop posteriorly, and massive denticles develop anteriorly. The latest additions to the scale are dotted. The fused base of the fingerlike elements is lightly stippled; the fibrous cushion base of the massive denticles is more darkly stippled. (*d* and *e*) Placoid denticles of *Elegestolepis*, a possible elasmobranch from the Upper Silurian, in ventral view and in longitudinal section. *From Zangerl, 1981.*

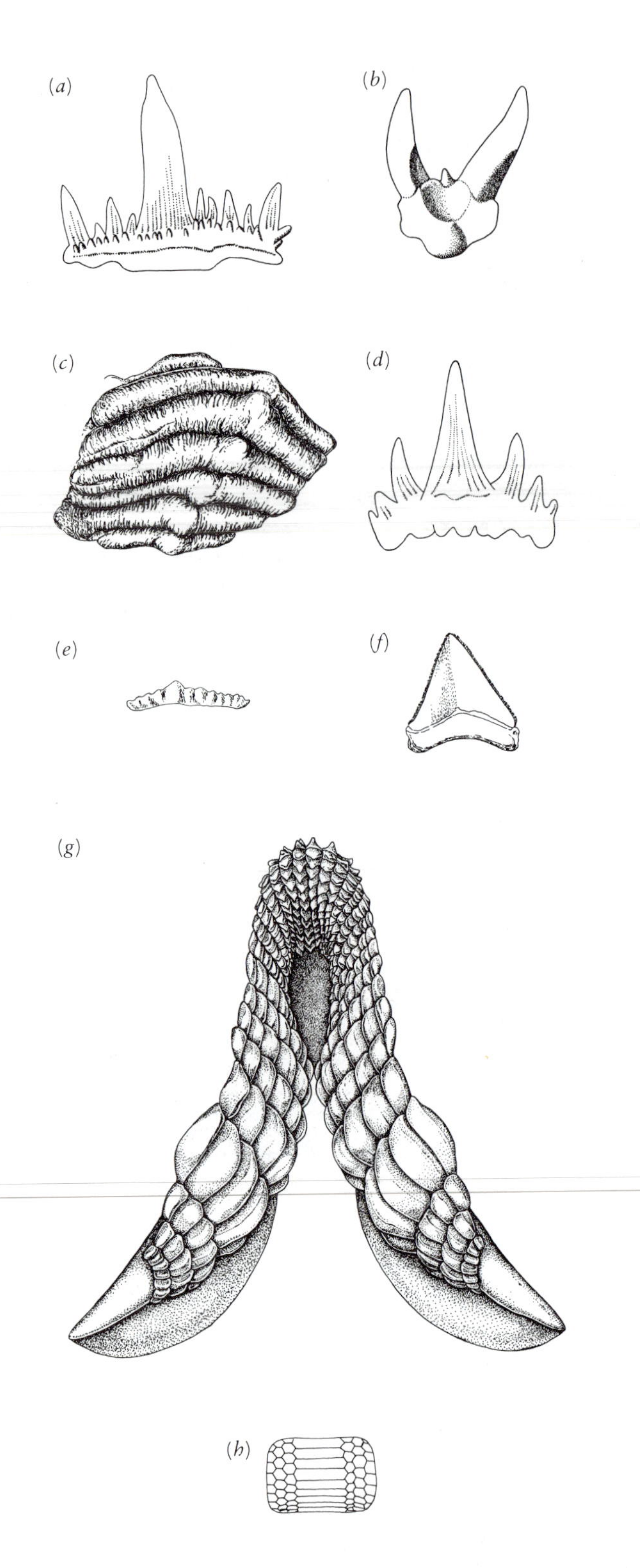

Figure 5-7. TEETH OF ELASMOBRANCHS. (*a*) *Cladodus* or cladodont pattern, common to the Cladoselachida, Symmoriida, and Ctenacanthoids, ×2. *From Schaeffer, 1967.* (*b*) Diplodus pattern, illustrated by *Xenacanthus*, ×1. *From Schaeffer, 1967.* (*c*) Crown view of tooth series of *Orodus*, a pavement toothed shark, ×¾. *From Zangerl, 1981.* (*d* and *e*) Two types of teeth borne by hybodont sharks, ×2. *From Schaeffer, 1967 and Gregory, 1951, respectively.* (*f*) The modern galeoid shark *Carcharodon*, ×¼. *From Gregory, 1951.* (*g*) Jaw of *Heterodontus* showing range of tooth patterns, ×1. *From Redpath Museum, McGill University.* (*h*) Part of the dental battery of the ray *Myliobatis*, ×½. *From Gregory, 1951.*

of levels of development, we now recognize that each group has a long history that extends back into the Lower or Middle Devonian. Each group exhibits a mosaic of primitive, advanced, and specialized characteristics. None clearly represent an ancestral anatomical pattern. Figure 5-9 shows a scheme of possible relationships.

CLADOSELACHIDA

The earliest recognized remains of sharks are isolated scales from the Upper Silurian of Central Asia (see Figure 5-6*d*) (Karatajute-Talimaa, 1973). A variety of teeth known from the Middle Devonian provide a veiled glimpse of the early stages of shark diversification. By the end of the Devonian, seven orders are recognized, but most are represented by little more than teeth and spines. As a result of fortuitous conditions of preservation, one group of Upper Devonian sharks, the Cladoselachida, is represented by a large number of specimens that show not only the skeletal remains but the body outline and such elements of the soft anatomy as muscle fibers and kidney tubules. *Cladoselache*, the only adequately known representative of the order Cladoselachida, is preserved in nodules collected from the Cleveland shale in the area of Cleveland, Ohio.

Like other primitive sharks, *Cladoselache* has a fusiform body outline, amphistylic jaw suspension, and cladodont dentition. The notochord is unrestricted and the ventral supports for the notochord and ribs are either absent or not calcified. The anal fin is absent, and the notochord bends sharply dorsally to support a heterocercal caudal fin. The paired and dorsal fins are largely supported by radials that extend to their margins. The edges of the fins are further strengthened by ceratotrichia. The base of the pectoral fin is narrow and presumably allowed considerable maneuverability.

The shape of the trunk and the configuration of the fins compare closely with those of modern fast-swimming pelagic sharks and bony fish such as the tuna. As in these forms, there are small, laterally directed keels at the base of the tail. As in the tuna and sailfish, the caudal fin is lunate in outline, which produces an effective forward thrust during swimming.

Other features that distinguish *Cladoselache* include the presence of two dorsal fins and unusual fin spines. Unlike those of contemporary ctenacanth sharks and their modern derivatives (see Figure 5-13), the fin spines of *Cladoselache* consist entirely of trabecular dentine and lack the superficial, enamel-like material comparable to the surface of the denticles from which they were presumably derived. This absence suggests that the spines were developed entirely within the dermis and were not exposed superficially. Most of the body surface was free of scales. Scales are restricted to the margins of the fins, and a circle of specialized denticles surrounds the eye. In contrast with all other Chondrichthyes, this genus has no claspers.

SYMMORIIDA

No adequately known members of the Cladoselachida have been found above the Devonian. Within the Carboniferous, ten elasmobranch orders are known from fairly complete remains. Several groups have recently been recognized as a result of work by Dr. Zangerl and his colleagues on the black shales in the North American midcontinent. Use of X-rays has enabled detailed study of many specimens.

Among the newly characterized groups, members of the order Symmoriida (see Figure 5-3*b*) share many of the primitive features noted in the Cladoselachida but differ in the nature of the fins and the absence of orthodox fin spines. The teeth have the primitive cladodont pattern. Like *Cladoselache*, the anal fin is absent and the caudal fin is symmetrical and deeply forked. Symmoriids differ in having only a single dorsal fin above the pelvic area. The pectoral fin has a long, jointed metapterygial axis that ends in a free whip. The postorbital processes of the braincase extend ventrally to support the otic process of the palatoquadrate laterally as well as anteriorly. The body surface has few scales, except along the lateral line canals. Claspers (see Figure 5-5) are well developed but not comparable in detail with those of other sharks.

Two families are recognized. The Symmoriidae (in which three genera have been described) lacks fin spines of any kind. The Stethacanthidae, known from a single genus, has a very curious complex above the pectoral girdle consisting of a wide spine formed of trabecular dentine and, posteriorly, a bushy-looking structure topped with denticles. Large denticles also cover the dorsal surface of the head. Zangerl (1984) suggests that these two areas of denticles may have formed a threat display, giving the appearance of an enormous mouth.

EUGENEODONTIDA

The dentition of the sharks so far mentioned follows a single broad pattern that is similar to that of living sharks, with an emphasis on sharp-cusped marginal teeth that pierce and shear prey. An assemblage of forms known from the Upper Devonian into the Triassic—grouped by Zangerl (1981) as the order Eugeneodontida—possessed a different pattern (Figure 5-8). The teeth at the symphysis of the lower jaw are very large, and replacement teeth accumulate to form a complex spiral or tooth whorl. The lateral teeth form a crushing pavement, functionally comparable with that of modern skates and rays. Zangerl recognizes two suborders, the Caseodontoidea, in which the symphyseal teeth are transversely crested or swollen, and the Edestioidea, in which the symphyseal teeth are even more prominent but are laterally compressed to form sharp cutting blades. In some genera, the lateral teeth resemble those of the Caseodontoidea or they may be much reduced or completely lost. The edestoids (Figures

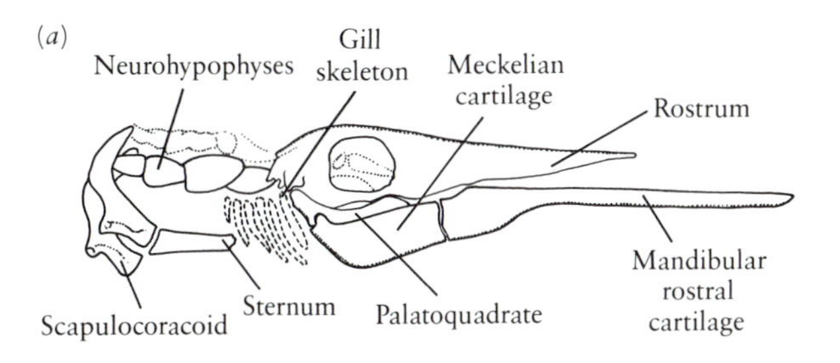

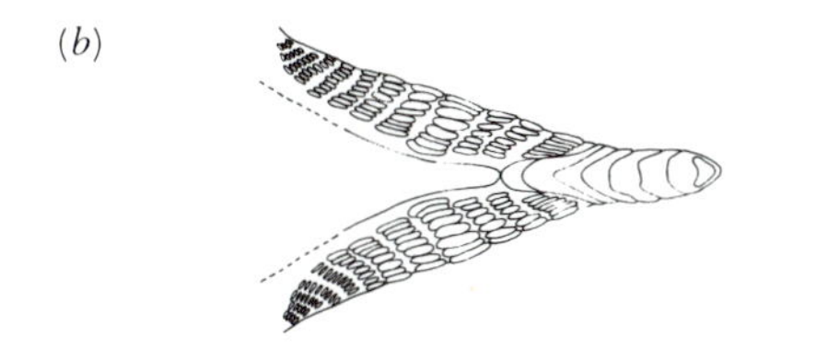

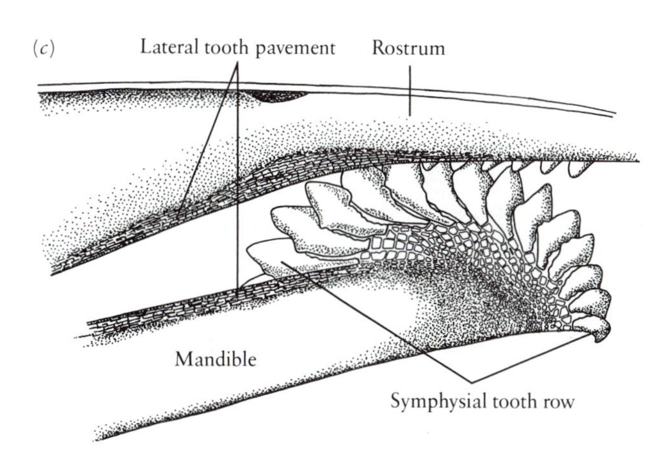

Figure 5-8. REMAINS OF THE EUGENEODONTIDA, A PALEOZOIC SHARK GROUP CHARACTERIZED BY THE POSSESSION OF PROMINENT SYMPHYSEAL TEETH. (*a*) *Ornithoprion*, restoration of skull and shoulder region in lateral view, $\times \frac{1}{3}$. *From Moy-Thomas and Miles, 1971. By permission from Chapman and Hall, Ltd.* (*b*) Lower jaw dentition of *Fadenia* showing symphyseal tooth whorl, $\times \frac{1}{5}$. *From Zangerl, 1981.* (*c*) *Sarcoprion*, restoration of anterior region of skull with symphyseal dentition in lateral view (the most complete material yet known of an edestoid shark), $\times \frac{1}{5}$. *From Moy-Thomas and Miles, 1971. By permission from Chapman and Hall, Ltd.* (*d*) The symphyseal tooth whorl of *Helicoprion*, $\times \frac{1}{3}$. *From Bendix-Almgreen, 1966.*

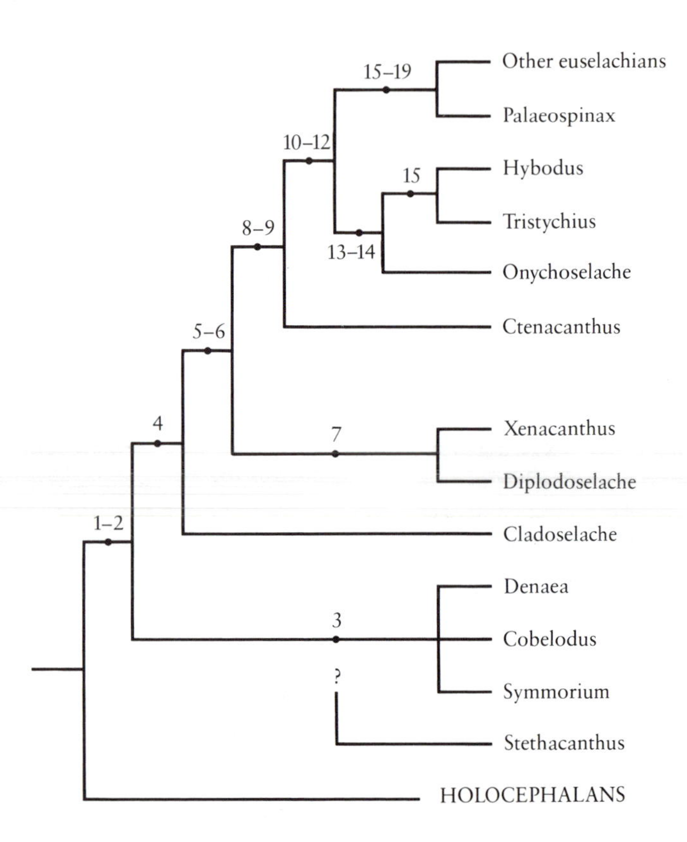

Figure 5-9. INTERRELATIONSHIPS OF CERTAIN ELASMOBRANCH GROUPS. Cladogram, slightly modified from Young (1982), showing the distribution of advanced (or derived) characters thought to link a series of nested sister groups. The characters are as follows:

1. Prominent metapterygial axis in the pectoral fin; palatoquadrates with palatine processes and symphyseal connection.
2. Loss of pharyngohyal and development of an articulation between the hyomandibula and the braincase (assuming the unmodified hyoid arch of holocephalans is primitive).
3. Loss of dorsal fin-spine and development of a short, deep neurocranium and large orbits.
4. Two dorsal fins, each with an unornamented spine lacking an outer orthodentine coat.
5. Ctenacanthiform type of fin-spine consisting mainly of osteodentine with little or no lamellar tissue and an ornamented outer coat of orthodentine. Spine cross-section is triangular with a posteriorly placed central cavity, a thinner, flat-to-concave posterior wall, and no posterior ornament or denticles.
6. Evolution of an anal fin.
7. Braincase with elongated otic region and large semi-circular canals.
8. Fin-spines deeply embedded between the myotomes.
9. Tribasal pectoral fin.
10. Caudal fin not lunate.
11. Calcified ribs.
12. Pectoral metapterygial axis reduced or lost.
13. Hybodontiform type of fin-spine, oval in cross-section, with a centrally placed cavity and posterior denticle rows.
14. Teeth lacking a lingual torus.
15. Aplesodic pectoral fin (radials supplanted by ceratotrichia).
16. Calcified centra.
17. Jaws sub-terminal and hyostylic.
18. Right and left halves of pectoral and pelvic girdles fused ventrally.
19. Smooth fin-spines with clearly defined mantle and trunk components, the latter made up largely or completely of lamellar tissue.

5-8*c, d*) are as yet known from little more than the dentition and tooth-bearing elements of the skull. The remainder of the skeleton was apparently poorly calcified, making preservation very difficult.

The skeleton is better known among the caseodontoids, but even here our knowledge is limited. *Fadenia* (see Figure 5-3*c*) resembles the Cladoselachida and Symmoriida in having a fusiform body with an externally symmetrical caudal fin that is stiffened by large cartilaginous plates. The anal fin is absent, and there is only a single, spineless dorsal fin located above the shoulder girdle. A large triangular cartilaginous plate supports the radials. The pectoral fin has an elongate, posteriorly directed axis, but it is not as extensive as that found in the Symmoriida. The pelvic girdle and limb are entirely missing in this and other genera.

The braincase is long and laterally compressed. The otico-occipital portion is very short, but the ethmoid region is extended as a long rostrum. The gill arches are primarily behind the braincase, as in typical elasmobranchs. In *Fadenia,* the upper teeth are borne directly on the neurocranium, and it was assumed that the palatoquadrate was fused to the braincase in the manner of holocephalians. In *Ornithoprion* (Figure 5-8*a*), the palatoquadrate is recognizable as a short element that still articulates freely beneath the orbit and with the postorbital process as in other Paleozoic elasmobranchs. Zangerl (1981) suggests that *Fadenia* lost the palatoquadrate entirely rather than fusing it to the braincase. Although it is apparently primitive in retaining a distinct palatoquadrate, *Ornithoprion* is specialized in having a long mandibular rostral cartilage that articulates with the symphysis of the lower jaw and extends far beyond the rostrum.

SQUATINACTIDA, PETALODONTIDA, AND ORODONTIDA

Although most Paleozoic elasmobranchs have a similar fusiform shape, a few approach the modern skates and rays in their habitus and others are elongate with reduced fins. *Squatinactis* (Figure 5-10*a*) has a skatelike body plan but a cladodont dentition, with sharp-cusped teeth. This genus, the only known representative of the order Squatinactida, is from the Upper Carboniferous of North America. The pectoral fins are greatly expanded like the modern genus *Squatina* (see Figure 5-16*a*), and the long straight tail has a spine near its end like that of stingrays. Nothing of the anatomy of this genus indicates special affinities with any other Paleozoic elasmobranchs.

The petalodonts, which are usually classified with the holocephalians, also have obscure affinities. This assemblage is known primarily on the basis of isolated teeth of a variety of patterns, which form a pavement dentition (Lund, 1977 a,b). Some 17 genera are recognized, ranging from the Lower Carboniferous through the Permian. Only *Janassa* (Figure 5-10*b*) and *Heteropetalus* show a well-

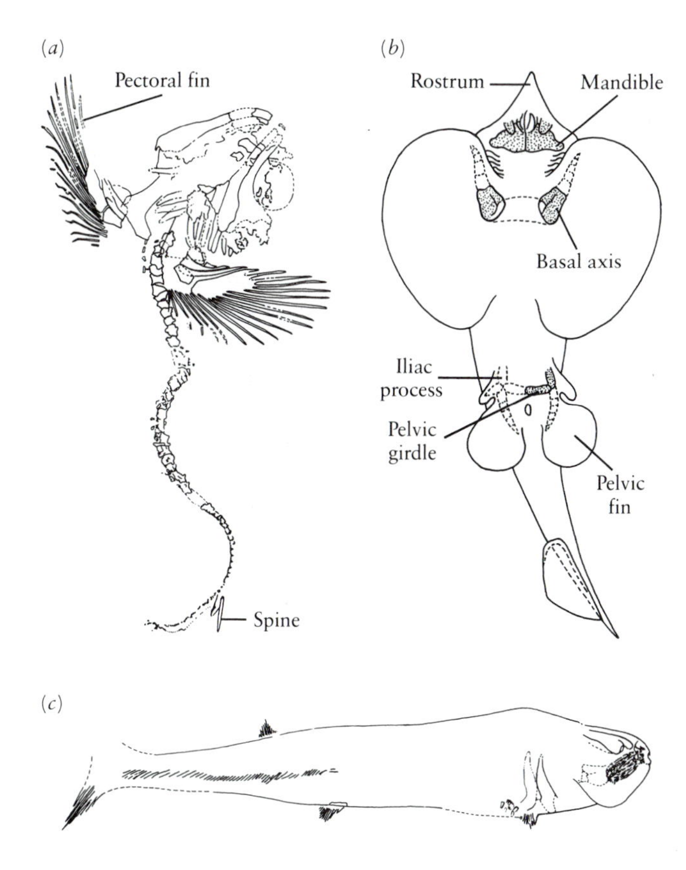

Figure 5-10. PALEOZOIC ELASMOBRANCHS WITH SPECIALIZED BODY FORM. (*a*) *Squatinactis*, the only known member of the Squatinactida, × 1. *From Lund and Zangerl, 1974.* (*b*) *Janassa*, the only elasmobranch possessing a petalodent dentition for which the entire skeleton is known, × $\frac{1}{3}$. *From Moy-Thomas and Miles, 1971, by permission from Chapman and Hall, Ltd. After Malzahn.* (*c*) *Orodus*, × $\frac{1}{12}$. A greatly elongate form with tiny fins. *From Zangerl, 1981.*

preserved skeleton in association with the teeth. The body of *Janassa* is strikingly similar to that of skates, with greatly expanded pectoral fins, and the pelvic girdle has an anteriorly directed iliac process. The gills are well behind the neurocranium, indicating affinities with elasmobranchs rather than holocephalians. In this genus, the teeth are not replaced but accumulate to form a platform upon which rests the most recently formed tooth (see Figure 5-19*c*). *Heteropetalus* (1977b) has a more sharklike body form.

Teeth that are referred to as *Orodus* (see Figure 5-7*c*) have long been recognized in Upper Paleozoic marine sediments. They are wide based with a broad, ridged surface rather than sharp cusps, and they form a pavement dentition. Symphyseal teeth develop, but they are not greatly enlarged as is the case among the Eugeneodontida. In the past, *Orodus* teeth have been associated with various better-known groups of sharks, including edestoids and hybodonts. Recently discovered material from the Upper Carboniferous of North America demonstrates that this tooth pattern is characteristic of a distinct group of sharks having very long bodies and relatively small fins (Figure 5-10*c*). They are among the largest of early sharks, with a length approaching 4 meters. Like the Symmoriida, they have only a single dorsal fin, no fin spines, and no anal fin. The body is covered with composite denticles formed from modified lepidomoria. Orodont teeth are known as early as the late Devonian, which indicates that this group was established nearly as early as more orthodox sharks.

XENACANTHIDA

In contrast with most Paleozoic elasmobranchs whose remains are found primarily in marine deposits, the Xenacanthida (see Figures 5-3*d* and *e*) was a predominantly freshwater group. Xenacanths are immediately recognized by their distinctive dentition. The teeth (termed "*Diplodus*") have two laterally directed cusps and a little button at the base that articulates with the next tooth in the jaw (see Figure 5-7*b*). Such teeth are known from the Lower Devonian to the end of the Triassic. Skeletal remains are found primarily in the Upper Carboniferous and Lower Permian. The pattern of the jaws resembles that of the Cladoselachida and Symmoriida, but the fin structure is highly specialized. The dorsal fin extends from just behind the head to the level of the caudal fin, from which it is differentiated by a slight notch. The notochordal support for the caudal fin is only slightly bent, and the supporting radials are arranged symmetrically above and below it to form a **diphycercal** caudal fin. There are two stalked anal fins, one behind the other. The pectoral fin has a long segmented central axis with radials diverging from both the pre- and postaxial borders. This pattern, termed **archipterygial,** had been considered primitive for fish, but probably evolved from a pattern like that seen in the Symmoriida, with the addition of radials posterior to the metapterygial axis. There is a single dorsal fin spine that may articulate with either the shoulder girdle or the occipital surface of the braincase. The pattern of the fins of xenacanths appears to be so different from that of other elasmobranchs that until recently the xenacanths were thought to be only distantly related to other Paleozoic sharks (Moy-Thomas and Miles, 1971; Schaeffer and Williams, 1977).

In 1981, Dick described a newly recognized genus, *Diplodoselache* from the Lower Carboniferous (see Figure 5-3*d*). This form is immediately identifiable as a xenacanth by its dentition and long dorsal fin, but the paired fins, the anal fin, and the caudal fin all resemble those of the ctenacanth sharks, which are described in the next section. This early genus demonstrates that xenacanths are closely related to one of the major groups of Paleozoic sharks and that all the peculiarities of the fins have evolved within the group. Dick attributes the elongation of the dorsal fin and the specialization of the anal and caudal fins to an accentuation of sinusoidal movements of the body, which may have provided a more effective mode of locomotion in the shallow, vegetation-choked, Carboniferous coal swamps.

CTENACANTHIDA

In our discussion of Paleozoic elasmobranchs, we have left to the last the phylogenetically most important group, the Ctenacanthida, which almost certainly includes the ancestors of all modern sharks, skates, and rays. Ctenacanthids broadly resemble the Cladoselachida and Symmoriida but are readily distinguished by the structure of the fins and the nature of the fin spines (Maisey, 1982a). The spines differ in having an external covering or mantle of enameloid substance (see Figure 5-13). Like the outer layer of the scales and teeth, this material (when compared with modern sharks) was probably formed below the epidermis, which indicates that the distal end of the spine was exposed at the surface. The spine is deeply set between the myotomes and is supported by a large triangular cartilage that fits into its base. An anal fin is present and the base of the pectoral fin has at least an incipient development of the tribasal structure common to modern sharks. In front of the metapterygium, two new fin supports known as the **mesopterygium** and **propterygium** have developed. The body is covered with placoid denticles or aggregates of modified lepidomorial scales. The ribs are typically calcified.

We recognize two groups of Ctenacanthida in the late Devonian. The ctenacanthoids (Figure 5-11*a*) are the more primitive and apparently represent a basal stock from which both the second group of Ctenacanthida, the hybodontoids, and the modern, neoselachian elasmobranchs were derived. The ctenacanthoids survived into the Triassic but are primarily a Carboniferous group. None of the genera are adequately known. The braincase and jaws were weakly calcified; the notochord remained unrestricted, and the ribs were uncalcified. The dentition is cladodont. The pectoral fin resembles that of the symmoriids; the anterior basals show some tendency toward formation of distinct pro- and mesopterygia, and the caudal fin is externally symmetrical and bilobate. In contrast

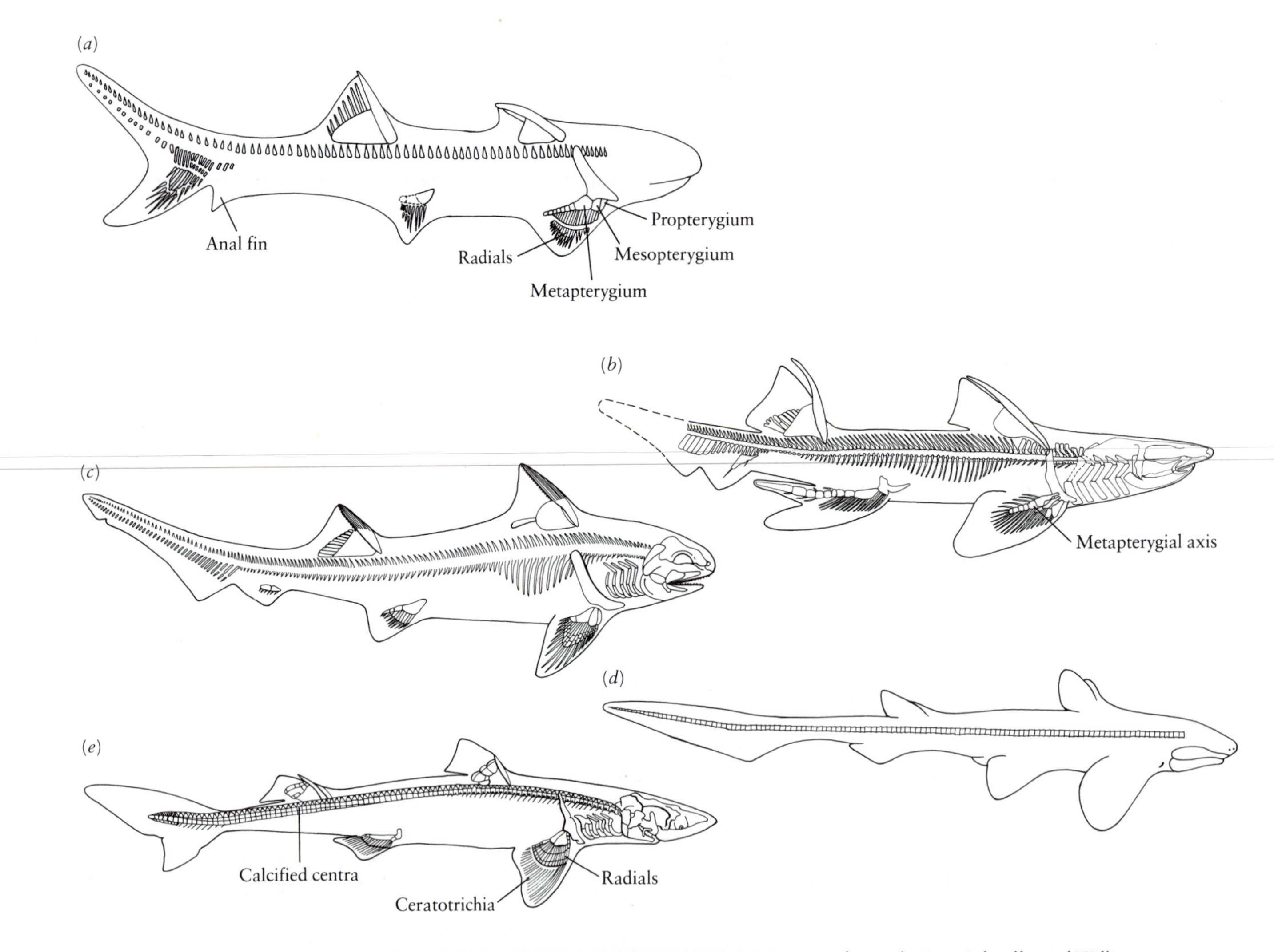

Figure 5-11. CTENACANTHIDA AND NEOSELACHIAN SHARKS. (*a*) *Ctenacanthus*, ×⅕. *From Schaeffer and Williams, 1977.* (*b*) The Lower Mississippian hybodont *Tristychius*, ×⅕. *From Dick, 1978.* (*c*) The Mesozoic hybodont *Hybodus*, ×⅐. *From Schaeffer and Williams, 1977.* (*d*) The Lower Jurassic neoselachian *Palaeospinax*, ×¼. *From Schaeffer and Williams, 1977.* (*e*) The modern squaloid shark *Squalus*, ×⅙. *From Schaeffer and Williams, 1977.*

to genera of normal proportions, *Bandringa* (Figure 5-12), a small form known from both freshwater and marine deposits, has a greatly elongated rostrum.

The hybodontoids (Figures 5-11*b* and *c*) are common from the Carboniferous into the Cretaceous and are the early Mesozoic successors of the ctenacanthoids as the dominant predaceous sharks (Maisey (1982b, 1983). Their fins have some advanced features that resemble the pattern of modern sharks. In the Mesozoic, the tribasal nature of the pectoral fin is clearly established, the area of articulation with the scapulocoracoid is narrow, and the posterior axis is lost. The radials do not extend to the margins of the fins but are replaced distally by more flexible ceratotrichia. A continuous puboischiadic bar joins the two sides of the pelvic fin. The caudal fin is clearly heterocercal, with a limited expression of the ventral lobe. Hybodonts are also advanced over the ctenacanthoids in the absence of articulation between the quadrate portion of the palatoquadrate and the postorbital process of the braincase and the elaboration of contact between these bones in the ethmoid region.

Tristychius and *Onchoselache* from the Lower Carboniferous show primitive features that demonstrate the affinities of early hybodonts with ctenacanths. In *Onchoselache,* the radials extend to the fin margin, and in *Tristychius* the metapterygial axis is still long and, like that of most xenacanths, bears radials on both the pre- and postaxial surfaces. The otic region of the braincase is relatively long in both these genera.

According to Maisey (1975), the configuration of the fin spines provides a consistent basis for distinguishing hybodontoids and ctenacanthoids. In ctenacanthoids, the posterior surface of the base of the spine is concave, thinly walled, and typically not ornamented. Among hybodonts, the posterior surface is convex, thick walled, and ornamented (Figure 5-13). Both patterns of spines can be recognized as early as the late Devonian. The individual teeth of hybodonts (see Figures 5-7*d* and *e*) tend to have blunter cusps than those of ctenacanthoids. There is also a greater variety of tooth shapes within the mouth, suggesting that hybodonts fed on a broad range of prey.

The hybodonts were common in the Triassic and Jurassic, but their numbers were greatly reduced by the late Cretaceous. The advanced hybodonts were contemporaries of the early neoselachians and paralleled them in many features that were more advanced than the early

Figure 5-12. THE UPPER CARBONIFEROUS CTENACANTH *BANDRINGA*, $\times\frac{1}{2}$. It is known both from freshwater and marine deposits. *From Baird, 1978. Courtesy of the Library Services Department, American Museum of Natural History.*

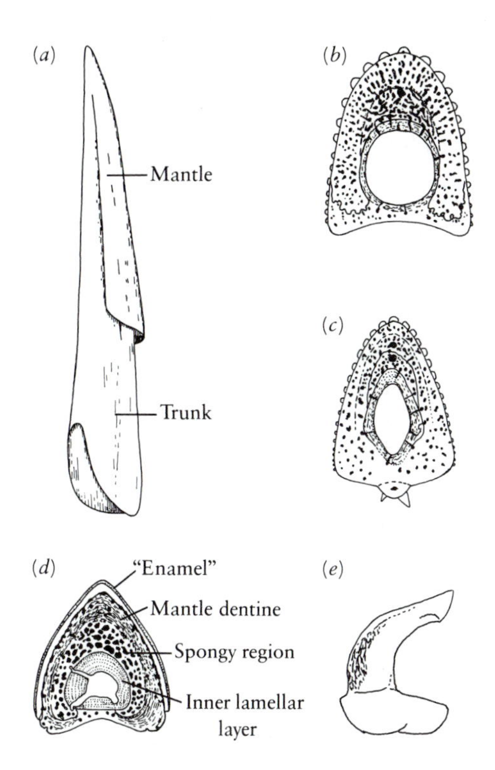

Figure 5-13. (*a*) Fin spine of *Squalus* showing mantle and trunk portions. *After Zangerl, 1981.* (*b–d*) Diagrammatic transverse sections of the fin spines of (*b*) ctenacanth, (*c*) hybodont, (*d*) *Palaeospinax. From Maisey, 1977. With permission from the Zoological Journal of the Linnean Society. Copyright 1977 by the Linnean Society of London.* (*e*) Head spine of a hybodont shark. *From Maisey, 1982.*

ctenacanthid level. Because of the similar body form and the specific configuration of the fins, paleontologists once thought that the Mesozoic hybodonts were ancestral to one or more groups of modern elasmobranchs. Maisey (1982b) pointed out a number of specialized characteristics of hybodonts that are not present in the neoselachians. The most important feature that defines the Mesozoic hybodonts is the presence of one or two pairs of hook-shaped cephalic spines (Figure 5-13*e*). Details of the jaw suspension and the histology of the teeth and scales may also be unique to this assemblage.

Our knowledge of Paleozoic and early Mesozoic elasmobranchs is still very limited. Many genera are known only from teeth or spines, and several groups that are recognized on the basis of very distinct anatomical patterns are known from very few specimens, a single geological horizon, or a limited geographical area.

Despite the obvious deficiencies of the fossil record, Paleozoic sharks represent a much wider adaptive spectrum than do the genera living today. In addition to the habitus occupied by modern sharks, there were analogues of the modern skates and rays, as well as groups with a variety of patterns of pavement dentition that are difficult to compare with those of any fish living today. Several orders probably occupied habitats that were taken over after the Triassic by various groups of bony fish.

NEOSELACHII

The Ctenacanthida are the only Paleozoic sharks that survived beyond the early Mesozoic. Sometime in the Triassic or possibly as early as the Permian, members of this order gave rise to a distinct group, the Neoselachii (modern sharks, skates, and rays), which dominated the seas of the late Mesozoic and Cenozoic. The earliest adequately known neoselachians are from the early Jurassic and already show several significant advances over the early Ctenacanthida in structures associated with feeding and locomotion.

The vertebral centra are strongly calcified to resist compressional forces associated with swimming. The pelvic girdle becomes fused at the midline, and the two halves of the pectoral girdle either articulate with each other or are fused. The basal elements that support the fins are further concentrated to articulate more effectively with the girdles. The radials are reduced, and the more flexible ceratotrichia become the major elements that support the distal portion of the fins. The fin spines are reduced and lose their ornamentation.

Among early neoselachians, the relationship between the braincase and the jaws is modified to form a more maneuverable feeding apparatus. The postorbital process of the braincase and the otic portion of the palatoquadrate are reduced so that the upper jaw moves freely. The hyomandibular serves as a movable link between the back of the braincase and the area of articulation between the upper and lower jaws, allowing the jaws to be raised and lowered relative to the braincase and move fore and aft (Figure 5-14). In sharks that feed on large prey, the jaws can dig into the body and gouge out large pieces. In genera feeding on smaller prey, the mobile jaws form an effective suction device. Many of the neoselachians have an elongate rostrum that extends beyond the neurocranium. As a result, the mouth opens ventrally rather than terminally. The increased size of the nasal capsules leads us to believe that the sense of smell was augmented in early neoselachians.

In conjunction with the increased sophistication of feeding, locomotion, and enhanced sensory input, neoselachians have larger brains than their Paleozoic ancestors and other jawed fish. Recent work by Northcutt (1977) shows that large predatory sharks and advanced rays have the highest known brain weight to body weight ratio of any group of anamniotic vertebrates and even exceed that of many birds and mammals. The telencephalon is enlarged, and both the forebrain and the cerebellum become enfolded like those of advanced mammals.

The earliest neoselachians include *Palaeospinax* (Figure 5-11*d*) from the Lower Jurassic and some vertebrae and spines from the Triassic that may represent more ancient members of the lineage (Maisey, 1977). Early neoselachians retain many features that link them with more primitive sharks; the teeth have a basically cladodont pattern with small lateral cusps and the jaws are long and have a wide gape. The two dorsal fin spines and an anal fin of the ctenacanthid pattern are retained.

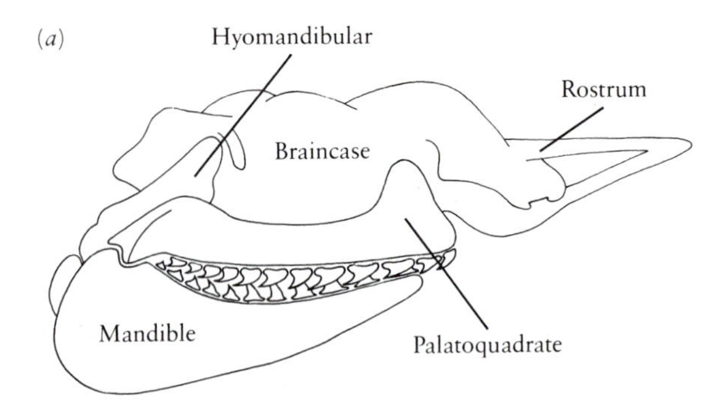

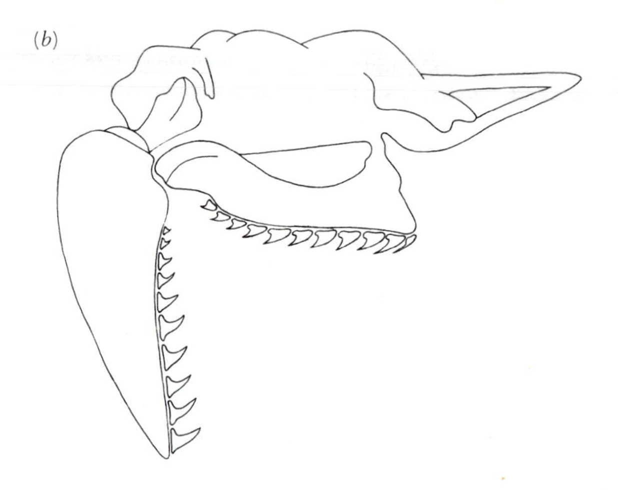

Figure 5-14. BRAINCASE AND JAWS OF THE MODERN SHARK *CARCHARHINUS*. Diagram showing movement of the palatoquadrate during feeding. The distal end of the hyoid and the posterior end of both jaws move laterally as the jaw is opened. *From Moss, 1972. By permission of the Zoological Society of London.*

We have not established the specific origin of the neoselachians. Similarities of fin structure to the Mesozoic hybodonts were apparently achieved by convergence. Dick (1978) and Young (1982) suggest that the neoselachians may be derived from Paleozoic hybodonts similar to *Tristychius* prior to the origin of the specialized pattern of the hybodont fin spine. Maisey (1975) argued that the pattern of the neoselachian fin spine indicates affinities with more primitive ctenacanthoid sharks. These authors think that the lineage leading to the neoselachians must have been distinct from both major ctenacanthid groups since the early Carboniferous, which would indicate that the specific ancestors of the modern shark groups have no known fossil record from the early Mississippian to the end of the Triassic.

We recognize three major groups of neoselachians from the Jurassic. The sharks themselves are divided into two orders, the galeoids (or galeomorphs) and the squaloids (or squalomorphs). The skates and rays constitute a third order, the Batoidea. Compagno (1973, 1977) re-

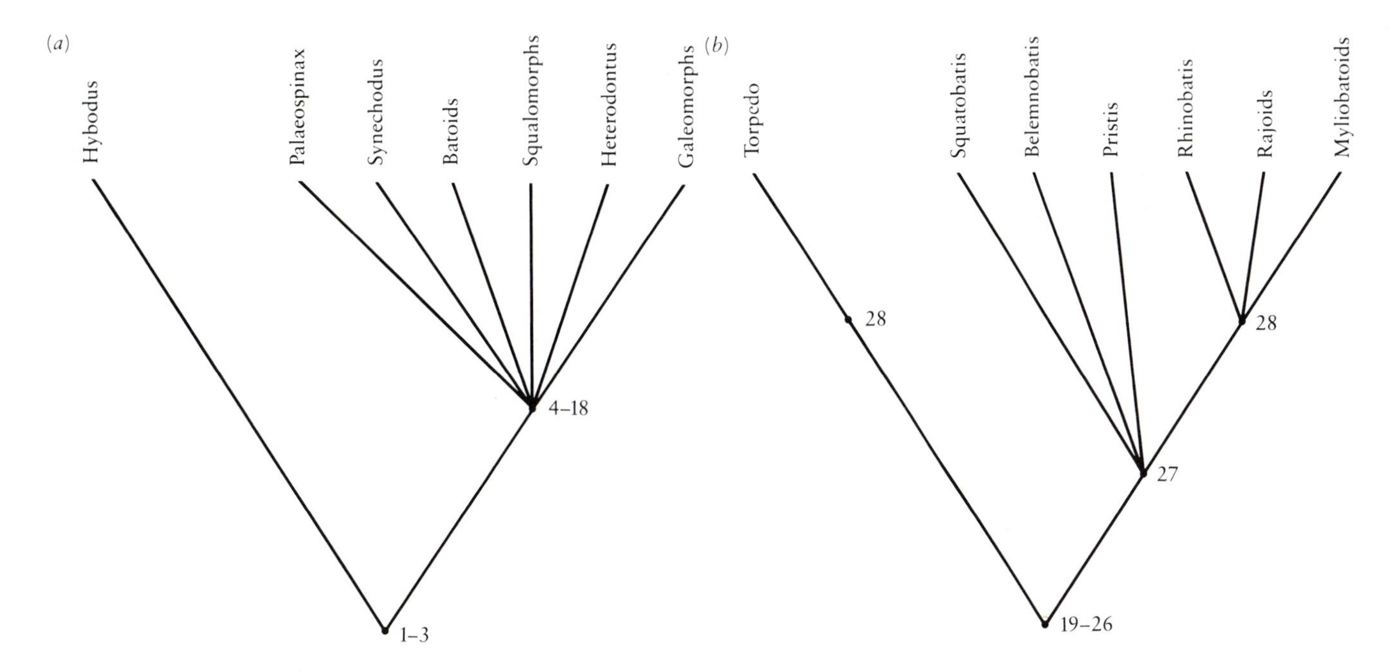

Figure 5-15. HYPOTHESIS OF HIGHER ELASMOBRANCH MONOPHYLY. Recent elasmobranchs are not distinguished as a monophyletic group unless fossil taxa *Palaeospinax* and *Synechodus* are included. Numbers refer to the following derived characters.

1. Fusion of pelvic half-girdles into transverse puboischiadic bar.
2. Presence of synchronomorial (nongrowing) scales.
3. Hypotic lamina fused to floor of otic capsule, leaving the glossopharyngeal nerve enclosed by a canal.
4. Notochordal sheath segmented and calcified between the *membrana elastica interna* and *externa.*
5. Notochord becomes constricted except intervertebrally.
6. Fin spines have shiny enameloid mantle, lack posterior denticles.
7. Scales have simple pulp cavity and single basal canal.
8. Occipital condyles present (except for some sharks).
9. Occipital hemicentrum incorporated into occiput (not batoids).
10. Basicranial arteries arranged into a broad bell-shaped circuit.
11. Adult hypophyseal duct closed externally.
12. Three-layered enameloid ultrastructure in teeth.
13. Only one or two intermediate pelvic segments between the metapterygium and clasper.
14. Long metapterygium spans pelvic fin base (unknown in *Synechodus*).
15. No median ventral keel in the ethmoid region (unknown in *Palaeospinax*).
16. Postorbital process reduced ventrally (unknown in *Palaeospinax*).
17. Complete series of perichondrally calcified radials in the anterior dorsal fin (unknown in *Palaeospinax* or *Synechodus;* does not apply to hexanchoids or *Pentanchus*).
18. Haphazardly fibered enameloid in teeth (except *Palaeospinax, Synechodus, Squalicorax;* many taxa not yet sampled).

b. Hypothesis of batoid relationships. Forms having "rhinobatoid" habitus (e.g., *Rhinobatus, Spathobatis, Belemnobatis*) are considered paraphyletic. Numbers refer to the following derived characters.

19. Cornea attached directly to eye; upper eyelid absent.
20. Anterior vertebrae fused to form synarcual.
21. Suprascapulae joined over vertebral column, articulating with column or synarcual or fused to synarcual.
22. Pseudohyal takes over function of ceratohyal.
23. Pectoral fins expanded, anterior lobes continuous.
24. Last hypobranchial articulates or is fused with the shoulder girdle.
25. Antorbital cartilages present.
26. Palatoquadrate lacks articulations with the neurocranium.
27. Loss of ceratohyal-hyomandibular articulation.
28. Propterygium contacts chondrocranium antorbitally. *From Maisey, 1984a.*

cently reviewed the relationships between these groups. Figure 5-15 illustrates relationships of the major groups of modern elasmobranchs.

Although both shark groups are of equal antiquity and include a variety of primitive and highly specialized genera, the squaloids are more primitive in their general anatomy, in particular the small brain size. They are also geographically restricted. Galeoids have three times the number of genera and are the dominant marine carnivores in the tropical oceans. Squaloids tend to dwell in cold, deep water.

SQUALOMORPHA

The central squaloid stock includes the familiar laboratory animal, the spiny dogfish *Squalus*. Being typical of the squalomorphs, *Squalus* has fin spines and lacks the anal fin. The braincase has a distinctive trough or scoop-shaped rostrum (see Figure 5-4*c*). This group shows some specializations toward bottom-dwelling habits. The spiracle is usually large, but the body is well rounded and the pectoral fins are shaped normally. Some authors suggest that *Protospinax* (Figure 5-16*a*) from the Upper Jurassic is close to the ancestry of the modern squalomorphs. However, it also shows features expected in batoid ancestors, and Maisey (1976) suggests a close relationship between these groups. Several modern squaloid genera are already known in the Upper Cretaceous.

The pristiophoroids, or saw sharks, are specialized derivatives of the early squaloids, with a very long rostrum armed with laterally directed toothlike structures. The fin spines are lost in this group. Saw sharks are not recognized in the fossil record until the Upper Cretaceous, at which time they are represented by the living genus *Pristiophorus*.

GALEOMORPHA

The galeomorphs are active sharks with fusiform bodies. The spiracle is small or absent, and the anal fin is always present. They are represented by several genera in the Upper Jurassic from Bavaria and France (Saint-Seine, 1949); *Crossorhinus, Phorcynis,* and *Corysondon* appear to be primitive members of the group which includes the blind, nurse, zebra, wobbegong, and whale sharks of the modern fauna—the Orectoloboidea. The dominant predaceous sharks of modern seas are grouped in two superfamilies. We know the lamnoids, which include the sand tiger, crocodile, goblin, thresher, basking, mackerel, porbeagle, mako, and great white sharks, from the Lower Cretaceous. We first recognize the carcharhinoids, including the cat, false cat, hound, leopard, soupfin, tiger, gray, sharpnose, blue, lemon, and hammerhead sharks, in the Eocene. Together, these groups account for more than 60 percent of the modern shark species. The lamnoids and carcharhinoids are characterized by a tripod-shaped rostrum (Figure 5-4*d*) and the loss of fin spines. Possibly in correlated with the loss of fin spines, the dorsal fins are supported by a series of basals and radials rather than the

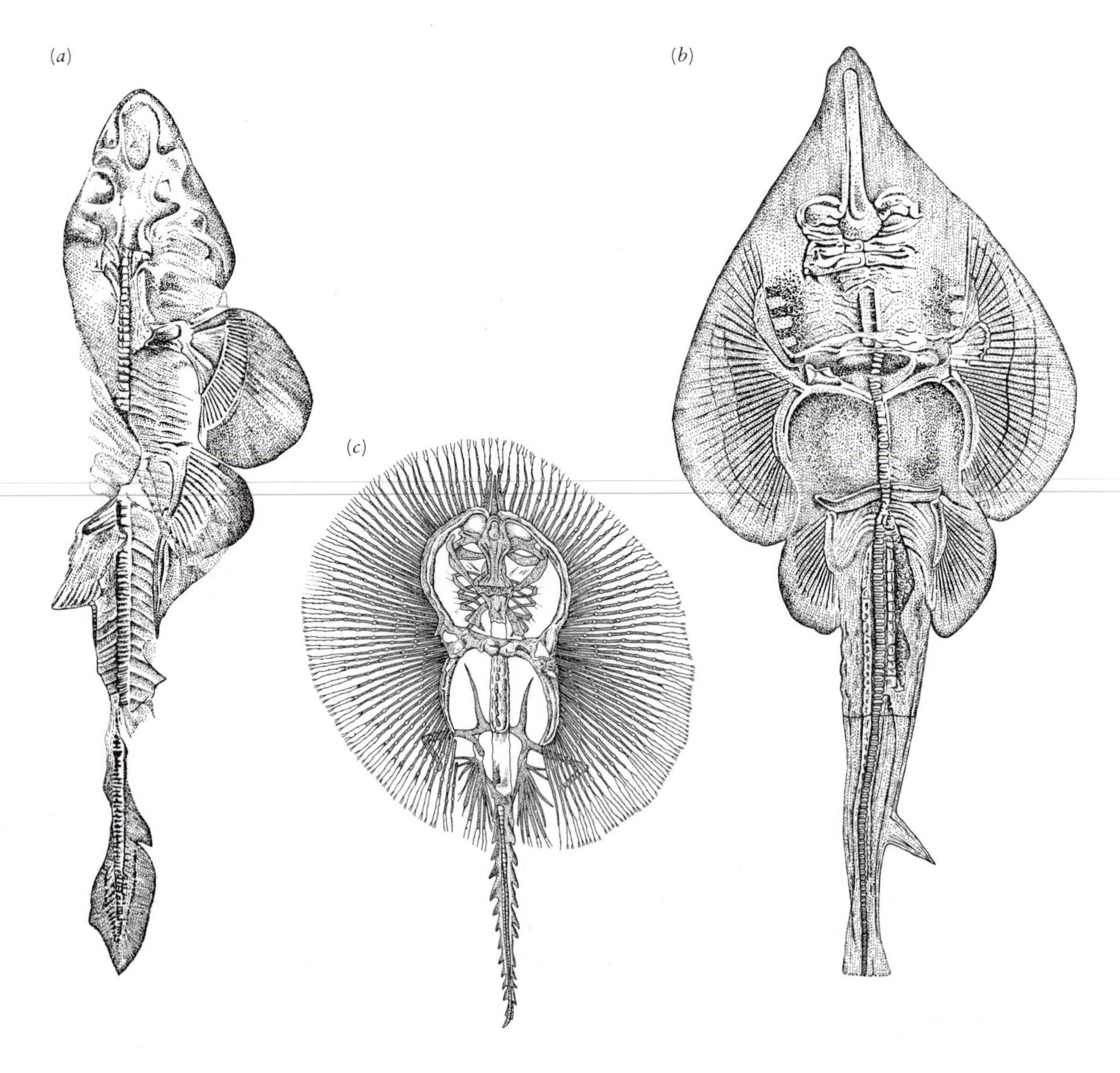

Figure 5-16. MESOZOIC BATOIDS. (*a*) *Protospinax*, from the Upper Jurassic, ×$\frac{1}{7}$. *Spathobatis*, Upper Jurassic, ×$\frac{1}{2}$. (*c*) *Cyclobatis*, Upper Cretaceous, ×$\frac{3}{4}$. *From de Saint-Seine, Devillers, and Blot, 1969.*

broad basal plate that is characteristic of both the ctenacanthids and the squaloids.

HEXANCHOIDS AND HETERODONTOIDS

Two other groups of living sharks, the hexanchoids and heterodontoids, are typically distinguished from the more common galeomorphs and squalomorphs. They are quite different from one another anatomically, but both are significantly more primitive than the remaining sharks and have been thought to represent surviving lineages of the hybodont radiation.

The hexanchoids include four living genera, *Hexanchias, Heptranchias, Notorhynchus,* and *Chlamydoselachus*. They are specialized in the loss of the anterior dorsal fin and both fin spines, but the nature of the jaw suspension has been considered to be particularly primitive. Like the Paleozoic sharks, they retain a strong postorbital process on the neurocranium, which suggests that an amphistylic jaw suspension is retained. They also lack fully calcified vertebral centra.

Recent studies by Maisey (1980, 1982b) indicate that the palatoquadrate is capable of considerable dorsoventral movement and that the sharp distinction made between amphistylic and hyostylic jaw suspension in living sharks is not justified. The absence of calcified vertebral centra in these genera may be a specialized (and not a primitive) feature, because the notochord shows septate subdivisions.

Based on other anatomical features, these sharks may be allied with the most primitive group of neoselachians, the Squalomorpha. There is no fossil evidence to justify the assumption that they were directly derived from the major hybodont radiation of the Triassic and Jurassic. Hexanchoids were present in the Jurassic, but *Chlamydoselachus* is represented by fossils only during the Tertiary.

The presence of six or seven gill slits in hexanchoids is almost certainly a specialization rather than a retention of a primitive condition, since no Paleozoic sharks are known to have more than five. The teeth of hexanchoids are sharp cusped but quite unlike those of the older selachians; they are peculiar sawtooth structures in *Hexanchias;* they are three pronged in *Chlamydoselachus*.

Heterodontus, known from the Upper Jurassic, has a crushing dentition (see Figure 5-7*g*) that generally resembles that of some hybodontoids but was apparently separately evolved. The two dorsal fins have fin spines, but this is an attribute of primitive neoselachians and does not imply a more ancient origin (Maisey, 1982c). The centra are well developed. The palatoquadrate is not braced by the postorbital process but retains a more medial, anteroventral support. Heterodontoids share many features of their soft anatomy and jaw structure with the orectoloboids. These features suggest a relationship to the galeomorphs.

BATOIDS

The living batoids are among the most readily definable group of chondrichthyans and evolved from a common ancestor in the early Jurassic. A series of interrelated changes occurred as a result of their bottom-dwelling habit and the great elaboration of the pectoral fins. The body is very dorsoventrally flattened and the anal fin is lost. The eyes are dorsally placed. The spiracle, used for intake of water, has shifted to a dorsal position and is much enlarged; the remaining gills have shifted to a ventral position. The anterior vertebrae are fused into a rigid synarcual (as in placoderms) to facilitate dorsoventral movement of the head on the trunk. The anterior basal element of the pectoral fin, the propterygium, extends anteriorly to the front of the head and articulates with a new element, the antorbital cartilage. The fin fuses to the head above the gill openings, and the pectoral girdle either articulates with or has a fixed attachment to the vertebral column. The jaws become even more mobile than in sharks; the palatoquadrate loses the suborbital articulation and in most genera the ventral element of the hyoid arch, the ceratohyal, becomes separate from the hyomandibular.

Swimming is accomplished by an up-and-down motion of the enormously expanded pectoral fins; the tail tends to be reduced and is often a mere whiplash; the dorsal fins are reduced or absent. Skates and rays are omnivorous; the teeth are usually flattened and often form a solid crushing pavement (see Figure 5-7*h*), enabling them to feed on hard-shelled invertebrates.

The most primitive of the modern raylike fishes, *Squatina* (Figure 5-17*a*) is separated taxonomically from the remainder of the batoids because few of the specialized skeletal features that characterize the group are fully developed. The body is very much dorsoventrally flattened, and the pectoral fins are greatly expanded but do not fuse with the head. The spiracles and eyes have shifted dorsally, but the gill openings are still ventrolateral. There is only a trace of a synarcual. *Squatina* is known as early as the Upper Jurassic. It has retained a morphological pattern very close to that of the oldest batoids. The pectoral fin resembles that of *Protospinax* (see Figure 5-16*a*).

The earliest known rays are the guitarfishes, or Rhinobatoidea, represented in the Upper Jurassic by *Spathobatis* (Figure 5-16*b*). They are primitive in retaining dorsal fin spines and relatively short propterygia but specialized in having a long rostrum.

All other batoids may have evolved from the early rhinobatoids, but no intermediates between the various well-established lines have been found. The pristoids, or sawfish, retain the basically primitive features of the rhinobatoids. The synarcual is extremely short and the scapulocoracoid is not fused to the neural arches. However, the rostrum is greatly extended and resembles that of the sawsharks in having laterally directed toothlike structures. The pristoids do not appear in the fossil record until the Lower Cretaceous.

Torpedinoids, the torpedo rays, also retain a number of primitive features, such as well-defined dorsal fins and a connection between the hyomandibular and the ceratohyal. They do not appear in the fossil record until the Paleocene. Their most notable specialization is the huge electric organs developed from the muscles of the pectoral fin.

The rajoids, or skates, are readily derived from the primitive rhinobatoids. They have reduced dorsal and caudal fins, and the pelvic fins are nearly supplanted by the enlarged pectorals. The pectoral-synarcual complex is enlarged and strengthened. The rajoids appear in the fossil record in the Lower Cretaceous, represented by the modern genus *Raja*. *Cyclobates* (Figure 5-16*c*) is a well-known Upper Cretaceous genus.

The myliobatoids are known from the Paleocene and are the most numerous of the batoids. There are 18 to 20 living genera, which include the stingray and the butterfly, eagle, cownosed, and devil rays. The suprascapula is fused to the side of the synarcual to form a socket for the articulation of the dorsal tip of the pectoral girdle. A second synarcual is present behind the first. The head is short and has no rostrum. The myliobatoids were probably derived from rhinobatoids of essentially modern morphology. The fossil record indicates that radiation of all the major types of batoids occurred by the early Cre-

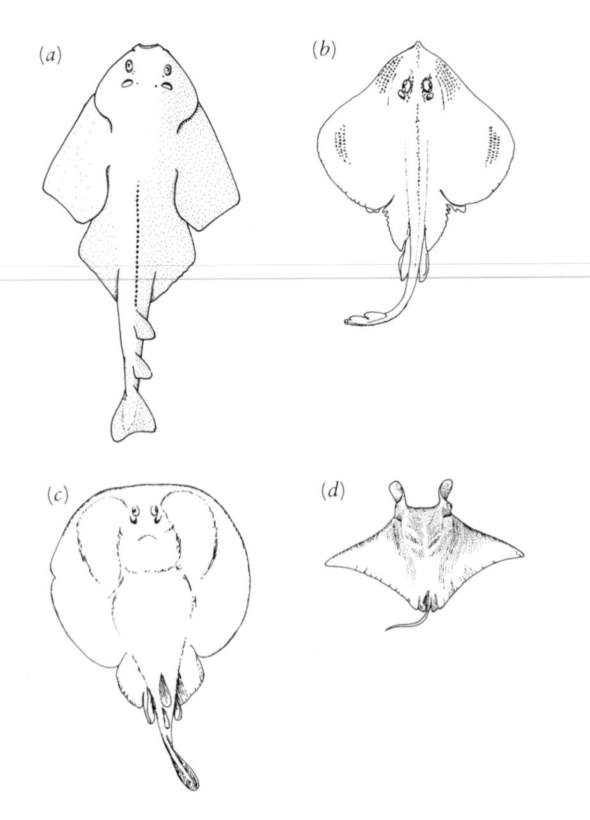

Figure 5-17. MODERN SKATES AND RAYS. (*a*) Angel "shark" *Squatina*. *From Compagno, 1977.* (*b*) *Raja*, 50 centimeters long. (*c*) *Torpedo*, 80 centimeters long. (*d*) Devil ray, *Manta*, 4 meters wide. *(b–d) From Bigelow and Schroeder, 1953.*

Holocephali

CHIMAERIFORMES

Although we have not determined specific relationships among the Paleozoic sharks, the general pattern of evolution of the elasmobranchs appears relatively well established. This is not the case among the animals grouped as holocephalians. Rather than tracing this assemblage from its origin in the Paleozoic, we must begin with the living genera and try to trace their ancestry among a confusing assemblage of Paleozoic groups whose general relationships have not been determined.

The modern holocephalians—the chimaeras or rat fish—differ from modern and Paleozoic sharks in many respects. The palatoquadrate is short and fused with the neurocranium. The gills lie beneath the back of the braincase and are covered by an operculum. A spiracle appears as a transitory structure during development but is absent in the adult. The first gill arch is not recognizable as a supporting hyomandibular. The identity of a separate dorsal element as a pharyngohyal led to the suggestion that the hyoid arch never had a supporting role in the ancestors of chimaeras. Maisey (1984a) argues convincingly that this element is not a pharyngohyal and that the hyoid arch is comparable with that of elasmobranchs. The dentition is reduced to two pairs of tooth plates in the upper jaw and one in the lower jaw, which are not replaced but grow at the margin during ontogeny. The notochord is little restricted, but in most living genera, there are several ring-shaped perichordal calcifications to each body segment. The anterior vertebrae unite to form a synarcual to which an anterior dorsal spine articulates. The caudal fin is not obviously heterocercal but is reduced to a narrow whip. As in primitive sharks, the pelvic girdle is paired, but the pectoral girdle has fused at the midline. In addition to having claspers on the pelvic fins, the males have clasping organs on the "forehead." The sensory canals are in open grooves formed by crescentic scales.

The affinities of modern holocephalians with sharks are evidenced by similar fin structures, the prismatic calcification of the cartilage, and similarities of their reproductive system.

The six living genera of chimaeroids are all marine and are placed in three families. Within the Chimaeridae, the genus *Chimaera* is known from the Lower Eocene and several extinct genera are present in the Middle Jurassic. The modern genera of the Rhinochimaeridae are unknown as fossils, but genera from the Lower Tertiary and Cretaceous can be included within the family. The only genus in the family Callorhynchidae is first known in the Upper Cretaceous. Although never diverse, the modern

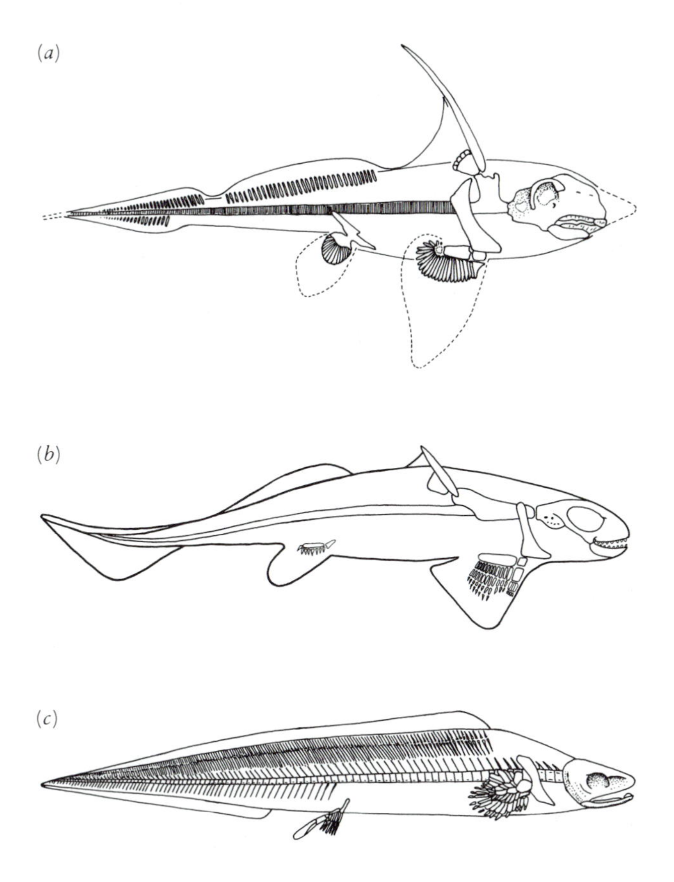

Figure 5-18. HOLOCEPHALIANS. (*a*) The chimaeroid *Ischyrodus*, from the Middle Jurassic, $\times \frac{1}{4}$. *From Patterson, 1965.* (*b*) *Helodus*, Upper Carboniferous, $\times \frac{1}{5}$. *From Moy-Thomas and Miles, 1971.* (*c*) *Chondrenchelys*, Lower Carboniferous, $\times \frac{1}{3}$. *From Moy-Thomas and Miles, 1971.* (*b* and *c*). *By permission from Chapman and Hall, Ltd.*

families show a very persistent pattern, and the genus *Ischyrodus* (Figure 5-18) from the Upper Jurassic is essentially like members of the living fauna. These three families are included in the suborder Chimaeroidei within the order Chimaeriformes. Several other genera from the Jurassic show broad similarities with the living chimaeroids but are placed in distinct suborders—the Squalorajoidei and the Myriacanthoidei (Patterson, 1965).

Recently, fossils have been discovered that suggest that a nearly modern chimaeroid pattern was achieved by the Carboniferous. *Similihariotta* (Figure 5-20*a*), from the Middle Pennsylvanian, has a body form similar to that of the modern chimaeroid genus *Hariotta* (Zangerl, 1979). The only definitely established difference is the absence of a synarcual spine. Unfortunately, the single specimen on which this genus is based is very immature. None of the skeleton is calcified and detailed comparison is not possible.

The late Mississippian genus *Echinochimaera* (Figure 5-20*b*) is better known. While it is specialized in some respects that preclude it from actual ancestry of the living families, *Echinochimaera* can be included in the same order. It is primitive in retaining a complete covering of placoid denticles and the fin spine is ornamented. The lateral line canals are enclosed by small denticles rather than by large crescentic scales, and the head has no claspers (Lund, 1977c).

These specimens indicate that chimaeroids are much more ancient than any groups of modern elasmobranchs. What is known of the anatomy of *Echinochimaera* appears so similar to that of the living families that it supplies little information about the ultimate ancestry of the group beyond their general chondrichthyan affinities.

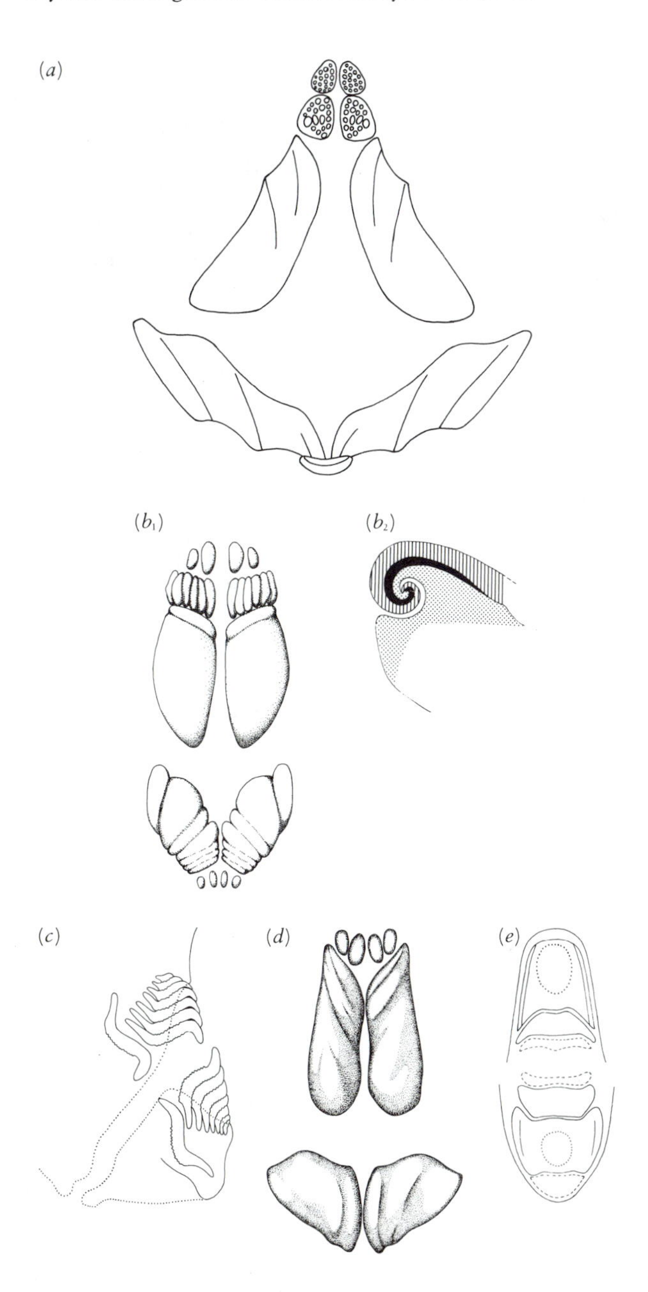

Figure 5-19. DENTITION OF CHIMAEROIDS AND "BRADYODONTS." (*a*) The Jurassic chimaeroid *Myriacanthus*, upper and lower dentition, symphyseal plate of lower jaw links mandibular plates, $\times \frac{1}{3}$. Seen as if looking directly into an open mouth. *From Patterson, 1965.* (b_1) Upper and lower dentition of *Cochliodus*. (b_2) Section shows curling of margin of plate associated with continuous growth, $\times \frac{1}{4}$. *From Patterson, 1968.* (*c*) *Janassa*, a petalodont, anterior dentition in vertical section. *From Zangerl, 1981.* (*d*) *Deltoptychius*, a Lower Carboniferous menaspoid, $\times \frac{1}{4}$. *From Patterson, 1965.* (*e*) *Copodus*, possible reconstruction of the symphyseal dentition, $\times \frac{1}{2}$. *From Zangerl, 1981.*

(a)

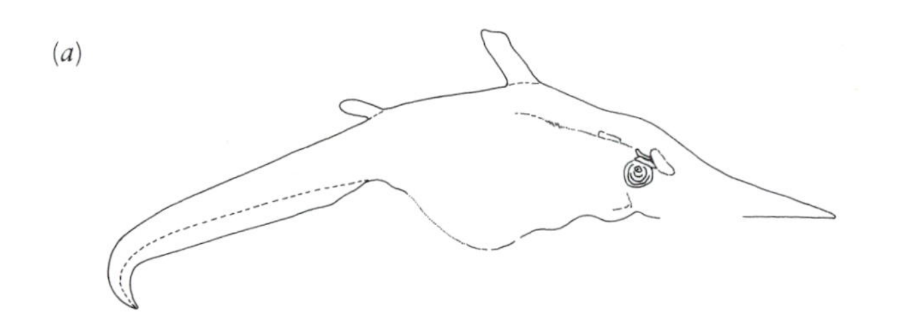

(b)

Figure 5-20. (*a*) *Similihariotta*, a possible chimaeroid from the Upper Carboniferous, approximately natural size. *From Zangerl, 1979.* (*b*) *Echinochimaera*, a possible chimaeroid from the Lower Carboniferous, approximately natural size. *Photograph courtesy of Richard Lund.*

No skeletons of chimaeroids are known prior to the late Mississippian. Tooth plates known as cochliodonts (Figure 5-19*b*), which broadly resemble those of chimaeroids, are known as early as the late Devonian. Articulated specimens show a pair of large posterior tooth plates in the upper jaw and two or three large plates in the lower jaws, as well as several smaller anterior plates that resemble shark teeth. The large plates grew at the margins in the manner of modern chimaeroids.

Other patterns of tooth plates known from the Paleozoic might be from animals related to the cochliodonts, but there is no strong evidence. The copodonts (Figure 5-19*e*) apparently have only a single functional tooth in the upper and lower jaws. The psammodonts have several rectangular plates arranged in two rows that meet at the midline. These and other types of tooth plates have been collectively referred to as **bradyodonts** on the assumption that they grew and were replaced very slowly.

In addition to these isolated tooth plates, the Helodontidae provides indirect evidence of an early stage in the evolution of chimaeroids. This family is known from the Upper Devonian into the Permian. The general configurations of the skull and postcranial skeleton in the Upper Carboniferous genus *Helodus* (Figures 5-18*b* and 5-21) resemble those of the chimaeroids, in particular the structure of the synarcual and holostylic jaw suspension. The dorsal fin spine is unornamented and articulates with the synarcual, as in modern chimaeroids. The dibasal pectoral fin has a structure that is almost identical to that of modern chimaeroids, although the pectoral girdle is not fused at the midline. However, *Helodus* is strikingly different in lacking large tooth plates. There are a series of smaller elements (about 10 in each jaw ramus) that appear to result from the partial fusion of flattened, sharklike teeth. As in primitive sharks, there are no calcified centra, and the tail is definitely heterocercal. The scales are placoid. *Helodus* is known only from freshwater deposits and is much too late in time to be an actual ancestor of chimaeroids, but its anatomy is what would be expected in a Devonian ancestor if chimaeroids did evolve from sharks.

MENASPOIDS, CHONDRENCHELYS, AND INIOPTERYGIANS

In addition to the chimaeroids and the Helodontidae, a number of Paleozoic chondrichthyans are known from more or less complete skeletons that resemble the modern holocephalians in either the nature of their dentition and/or the fusion of the palatoquadrate to the neurocranium. These genera all have gills that are closely arranged beneath the back of the braincase and a shoulder girdle that is close to the occiput, but the remainder of their anatomy is very different from that of the chimaeroids.

Menaspoids are a Carboniferous and Permian group in which the dentition (see Figure 5-19*d*) resembles that of the chimaeroids, but what we know of the remainder

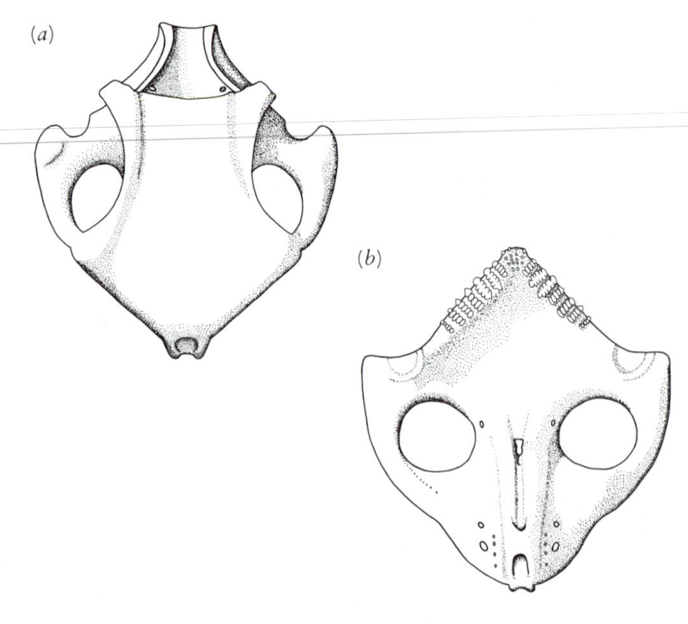

Figure 5-21. NEUROCRANIUM OF *HELODUS*, A POSSIBLE RELATIVE OF THE CHIMAEROIDS FROM THE UPPER CARBONIFEROUS. (*a*) Dorsal view. (*b*) Ventral view, ×¾. Marginal teeth resemble those of sharks, but tooth families are in the process of fusion. *From Moy-Thomas and Miles, 1971. By permission from Chapman and Hall, Ltd.*

of the anatomy is specialized in quite a different way. The body is dorsoventrally flattened and the denticles have coalesced to form a nearly continuous dermal armor. There are no dorsal fins, but one genus has three pairs of spines articulating with the head. Menaspoids, which were last reviewed by Patterson (1968), may be very divergent chimaeroids or a quite unrelated group. *Chondrenchelys* (Figure 5-18*c*) and *Harpagofututor* (Lund, 1982) represent another group from the Carboniferous that has a somewhat chimaeroid dentition. *Chondrenchelys* has four pairs of tooth plates in the upper jaw. The elongate posterior plates lie side by side. The two small anterior pairs are arranged longitudinally. There are three pairs of tooth plates in the lower jaw, a small anterior plate and, as in the upper jaw, two elongate posterior plates side by side.

The skull generally resembles that of chimaeroids, and the palatoquadrate is indistinguishably fused to it. In contrast, the postcranial skeleton of this fish is grossly similar to that of the xenacanth sharks. The dorsal fin is continuous with a diphycercal caudal fin. The two halves of the pectoral girdle are not fused at the midline, and the pectoral fin is an archipterygium. The pelvic fin bears a well-developed clasper. No synarcual or dorsal fin spine is present, but the notochordal sheath is calcified as segmented rings. We do not know whether the Chondrenchelyiformes are holocephalians that have converged on the xenacanth pattern of the fins or representatives of a distinct line of chondrichthyan fish.

The recently described iniopterygians from the Pennsylvanian of central North America (Figure 5-22)—first recognized by Zangerl and Case (1973)—are the most bizarre of the well-documented chondrichthyans. They resemble the chimaeroids in having gills under the back of the skull that are covered by an operculum. The head is disproportionately large, the body is stocky, and the tail is rounded except at the tip, which is expanded as an oval plate.

Of the two otherwise similar families, the Iniopterygidae have a separate palatoquadrate, but this bone is fused to the neurocranium in the Sibyrhynchidae (Stahl, 1980). The dentition of the Iniopterygidae consists of many delicate teeth of the elasmobranch pattern; those at the symphysis are arranged in a whorl. In the Sibyrhynchidae, the marginal dentition consists of a series of tooth whorls formed by the fusion of the bases of the teeth in six tooth families in each jaw ramus. There is an additional symphyseal whorl in the lower jaw. The mucous membrane of the roof and floor of the mouth is covered by stellate complexes formed from fused denticles. A large protruding tubercle is present at the symphysis of both the upper and lower jaws, and a pair of lateral tubercles flanks the medial tubercle in the upper jaw. Polygonal dermal denticles form an incomplete armor atop the skull. There is no synarcual or dorsal fin spine in either family. The notochord was surmounted by segmental neural arches and supported ventrally by calcifications in the notochordal sheath.

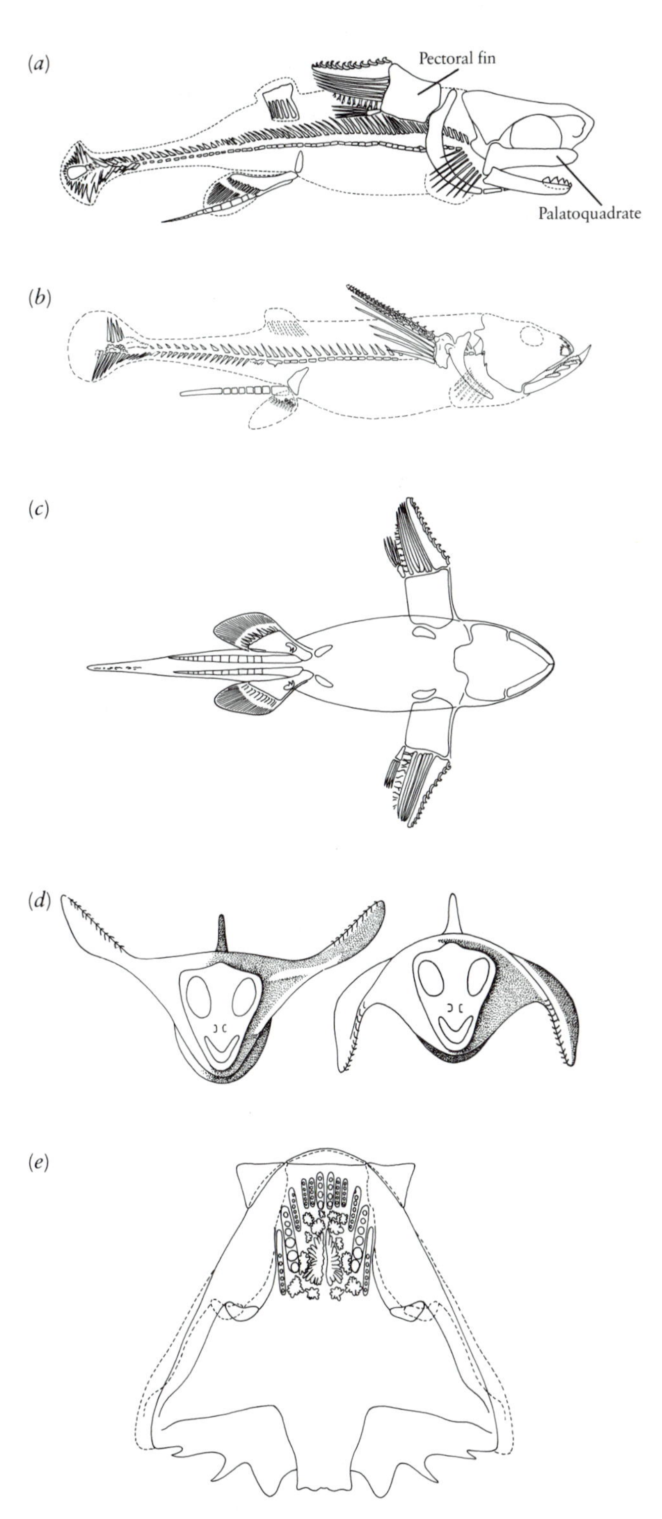

Figure 5-22. INIOPTERYGIANS FROM THE CARBONIFEROUS OF NORTH AMERICA. (*a*) Restoration of *Iniopteryx* in which the palatoquadrate is not fused to the braincase, approximately 20 centimeters long. *From Zangerl, 1981.* (*b*) Restoration of *Sibyrhynchus* in which the palatoquadrate is fused to the neurocranium, original about 30 centimeters long. *From Zangerl and Case, 1973.* (*c*) Ventral view of *Iniopteryx, From Zangerl and Case, 1973.* (*d*) Assumed swimming position of the pectoral fins of *Iniopteryx. From Zangerl and Case, 1973.* (*e*) *Sibyrhynchus*, neurocranium with snout tubercles, tooth whorls and mouth plates. *From Zangerl and Case, 1973.* (*b–e*) *Fieldiana Geological Memoirs. By permission from the Field Museum of Natural History, Chicago.*

The position and configuration of the pectoral fin are the strangest features of these animals. The fin articulates high on the lateral edge of the scapulocoracoid and apparently moved primarily in a dorsoventral direction. The base of the fin is formed from a large quadrangular block of cartilage to which a series of radials articulated. The first radial is by far the largest and is especially massive in males, where it bears a series of large hooks. The single dorsal fin is supported by a series of radials that are fused both distally and proximally. The pelvic girdle is a simple rod of cartilage that articulated with a massive basal bearing a large tenacular hook. There are many short radials that are succeeded by ceratotrichia. The claspers are in the form of long segmented rods.

If Zangerl (1981) is correct in assuming that the location of the gills beneath the back of the braincase and the anterior position of the shoulder girdle are primitive features of Chondrichthyes, there are no specialized characteristics of the iniopterygians that indicate close relationship with the holocephalians, as exemplified by the Chimaeriformes. The dentition of the iniopterygians can be derived from the pattern known in primitive elasmobranchs. The specializations of the appendicular skeleton show no significant similarities to the pattern of any early elasmobranchs or holocephalians.

Rather than attempt to classify iniopterygians with either of the major groups of chondrichthyans, we may place them in a separate assemblage until more is known of their possible relationships.

SUMMARY

The configuration of the jaws and regular pattern of tooth replacement in Chondrichthyes and Osteichthyes indicate that they share a common ancestry above the level of the placoderms. The Chondrichthyes include two major groups, the elasmobranchs and holocephalians, which are characterized by prismatic calcification of the cartilage and internal fertilization.

The elasmobranchs underwent two major radiations, one in the late Silurian and early Devonian, and a second in the early Mesozoic. The interrelationships of the major groups of Paleozoic elasmobranchs remain uncertain. The ancestry of the modern elasmobranchs lies among the Ctenacanthida, which first appears in the late Devonian. The neoselachians may have diverged from the Ctenacanthida as early as the Carboniferous, but well-preserved remains are not known until the Lower Jurassic. The three modern orders, Galeoidea, Squaloidea, and Batoidea, share a common ancestry in the early Mesozoic, characterized by calcification of the vertebral centra and a hyostylic jaw suspension. All the modern elasmobranch families had differentiated by the early Cretaceous.

Holocephalians of the modern order Chimaeriformes are known from the early Carboniferous. They differ from elasmobranchs in the close association of the gill supports with the back of the skull, the presence of an operculum, and the fusion of the palatoquadrate with the braincase. The dentition consists of a small number of large, continuously growing tooth plates. Numerous chondrichthyan orders are known in the Paleozoic which share one or more of these features with the modern holocephalians, but their interrelationships have not been established.

REFERENCES

American Society of Zoologists. (1977). Proceedings of symposium: Recent advances in the biology of sharks. *Amer. Zoologist,* **17**: 282–515.

Baird, D. (1978). Studies on Carboniferous freshwater fishes. *Amer. Mus. Novitates,* **2641**: 1–22.

Bendix-Almgreen, S. E. (1966). New investigations on *Helicoprion* from the Phosphoria Formation of south-east Idaho, U.S.A. *Biol. Skr. Dan. Vid. Selskab,* **14**(5): 1–54.

Bigelow, H. B., and Schroeder, W. C. (1953). Fishes of the Gulf of Maine. *Fishery Bull. Fish & Wildlife Serv.,* **53**: 1–577.

Compagno, L. J. V. (1973). Interrelationships of living elasmobranchs. In P. H. Greenwood, R. S. Miles, and C. Patterson (eds.), *Interrelationships of Fishes. Suppl. No. 1, Zool. J. Linn. Soc.,* **53**: 15–61.

Compagno, L. J. V. (1977). Phyletic relationships of living sharks and rays. *Amer. Zoologist,* **17**: 303–322.

Dick, J. R. F. (1978). On the Carboniferous shark *Tristychius arcuatus* Agassiz from Scotland. *Trans. Roy. Soc. Edinburgh,* **70**: 63–109.

Dick, J. R. F. (1981). *Diplodoselachi woodi* gen. et sp. nov., an early Carboniferous shark from the Midland Valley of Scotland. *Trans. Roy. Soc. Edinburgh, Earth Sci.,* **72**: 99–113.

Gregory, W. K. (1951). *Evolution Emerging.* Macmillan, New York.

Karatajute-Talimaa, V. (1973). *Elegestoleps grossi* gen. et sp. nov., ein neuer Typ der Placoidschuppe aus dem Oberen Silur der Tuwa. *Gross-Festschrift, Palaeontographica A,* **143**: 35–50.

Lund, R. (1977a). New information on the evolution of the bradyodont Chondrichthyes. *Fieldiana, Geol.,* **33**: 521–539.

Lund, R. (1977b). A new petalodont (Chondrichthyes, Bradyodonti) from the Upper Mississippian of Montana. *Ann. Carnegie Mus.,* **46**: 129–155.

Lund, R. (1977c). *Echinochimaera meltoni,* a new genus and species (Chimaeriformes) from the Mississippian of Montana. *Ann. Carnegie Mus.,* **46**: 195–221.

Lund, R. (1982). *Harpagofututor volsellorhinus* new genus and species (Chondrichthyes, Chondrenchelyiformes) from the Namurian Bear Gulch Limestone, *Chondrenchelys problematica Traquair* (Visean), and their sexual dimorphism. *Jour. Paleon.,* **56**: 938–958.

Lund, R., and Zangerl, R. (1974). *Squatinactis caudispinatus,*

a new elasmobranch from the Upper Mississippian of Montana. *Ann. Carnegie Mus.*, **45**: 43–55.
Maisey, J. G. (1975). The interrelationships of phalacanthous selachians. *Neues Jahrb. Geol. Paläont. Mh.*, **9**: 553–567.
Maisey, J. G. (1976). The Jurassic selachian fish *Protospinax* Woodward. *Palaeontology*, **19**: 733–747.
Maisey, J. G. (1977). The fossil selachian fishes *Palaeospinax* Egerton, 1872 and *Nemascanthus* Agassiz, 1837. *Zool. J. Linn. Soc.*, **60**: 259–273.
Maisey, J. G. (1980). An evaluation of jaw suspension in sharks. *Amer. Mus. Novitates*, **2706**: 1–17.
Maisey, J. G. (1982a). Studies on the Paleozoic selachian genus *Ctenacanthus* Agassiz: No. 2 *Bythiacanthus* St. John and Worthen, *Amelacanthus*, new genus, *Eunemacanthus* St. John and Worthen, *Sphenacanthus* Agassiz and *Wodnika* Münster. *Amer. Mus. Novitates*, **2722**: 1–24.
Maisey, J. G. (1982b). The anatomy and interrelationships of Mesozoic hybodont sharks. *Amer. Mus. Novitates*, **2724**: 1–48.
Maisey, J. G. (1982c). Fossil hornshark finspines (Elasmobranchi; Heterodontidae) with notes on a new species (*Heterodontus tuberculatus*). *Neues Jahrb. Geol. Paläont. Abh.*, **164**: 393–413.
Maisey, J. G. (1983). Cranial anatomy of *Hybodus basanus* Egerton from the Lower Cretaceous of England. *Amer. Mus. Novitates*, **2758**: 1–64.
Maisey, J. G. (1984a). Higher elasmobranch phylogeny and biostratigraphy. *Zool. J. Linn. Soc.*, **82**: 33–54.
Maisey, J. G. (1984b). Chondrichthyan phylogeny: A look at the evidence. *J. Vert. Paleont.*, **4**: 359–371.
Moss, S. A. (1972). The feeding mechanism of sharks of the family Carcharhinidae. *J. Zool. London*, **167**: 423–436.
Moy-Thomas, J. A., and Miles, R. S. (1971). *Palaeozoic Fishes*. Chapman and Hall, London.
Northcutt, R. G. (1977). Elasmobranch central nervous system organization and its possible evolutionary significance. *Am. Zoologist*, **17**: 411–429.
Patterson, C. (1965). The phylogeny of the chimaeroids. *Phil. Trans. Roy. Soc. London, B*, **249**: 101–219.
Patterson, C. (1968). *Menaspis* and the bradyodonts. In T. Ørvig (ed.), *Current Problems of Lower Vertebrate Phylogeny. Nobel Symp.* **4**: 171–205. Almquist and Wiksell, Stockholm.
Saint-Seine, P. de. (1949). Les poissons des calcaires lithographiques de Cerin (Ain). *Nouvelles Archives du Musée d'Histoire Naturelle de Lyon*, **2**: 1–357.
Saint-Seine, P. de, Devillers, C., and Blot, J. (1969). Holocéphales et élasmobranches. In J. Piveteau (ed.), *Traité de Paléontologie*, Tome IV: 693–776. Masson S. A., Paris.
Schaeffer, B. (1967). Comments on elasmobranch evolution. In P. W. Gilbert, R. F. Mathewson, and D. P. Rall (eds.), *Sharks, Skates and Rays*, pp. 1–35. Johns Hopkins Press, Baltimore.
Schaeffer, B. (1975). Comments on the origin and basic radiation of the gnathostome fishes with particular reference to the feeding mechanism. *Coll. Intl. C.N.R.S. Probl. Actuel de Paléont.* (*Evol. des Vert.*), **218**: 101–109.
Schaeffer, B. (1981). The xenacanth shark neurocranium, with comments on elasmobranch monophyly. *Bull. Am. Mus. Nat. Hist.*, **169**: 1–66.
Schaeffer, B., and Williams, M. (1977). Relationships of fossil and living elasmobranchs. *Amer. Zoologist*, **17**: 293–302.
Stahl, B. J. (1980). Non-autostylic Pennsylvanian iniopterygian fishes. *Palaeontology*, **23**: 315–324.
Young, G. C. (1982). Devonian sharks from south-eastern Australia and Antarctica. *Palaeontology*, **25**: 817–843.
Zangerl, R. (1979). New Chondrichthyes from the Mazon Creek fauna (Pennsylvanian) of Illinois. In M. H. Nitecki (ed.), *Mazon Creek Fossils*, pp. 449–500. Academic Press, New York.
Zangerl, R. (1981). Chondrichthyes I. Paleozoic elasmobranchs. In H. P. Schultze (ed.), *Handbook of Paleoichthyology, Vol. 3. Elasmobranchi*. Gustav Fischer Verlag, Stuttgart.
Zangerl, R. (1984). On the microscopic anatomy and possible function of the spine "brush" complex of *Stethacanthus* (Elasmobranchii: Symmoriida). *J. Vert. Paleont.*, **4**: 372–378.
Zangerl, R., and Case, G. R. (1973). Iniopterygia, a new order of chondrichthyan fishes from the Pennsylvanian of North America. *Fieldiana, Geol. Mem.*, **6**: 1–67.
Zangerl, R., and Case, G. R. (1976). *Cobelodus aculeatus* (Cope), an anacanthous shark from Pennsylvanian black shales of North America. *Palaeontographica A*, **154**: 107–157.

C H A P T E R VI

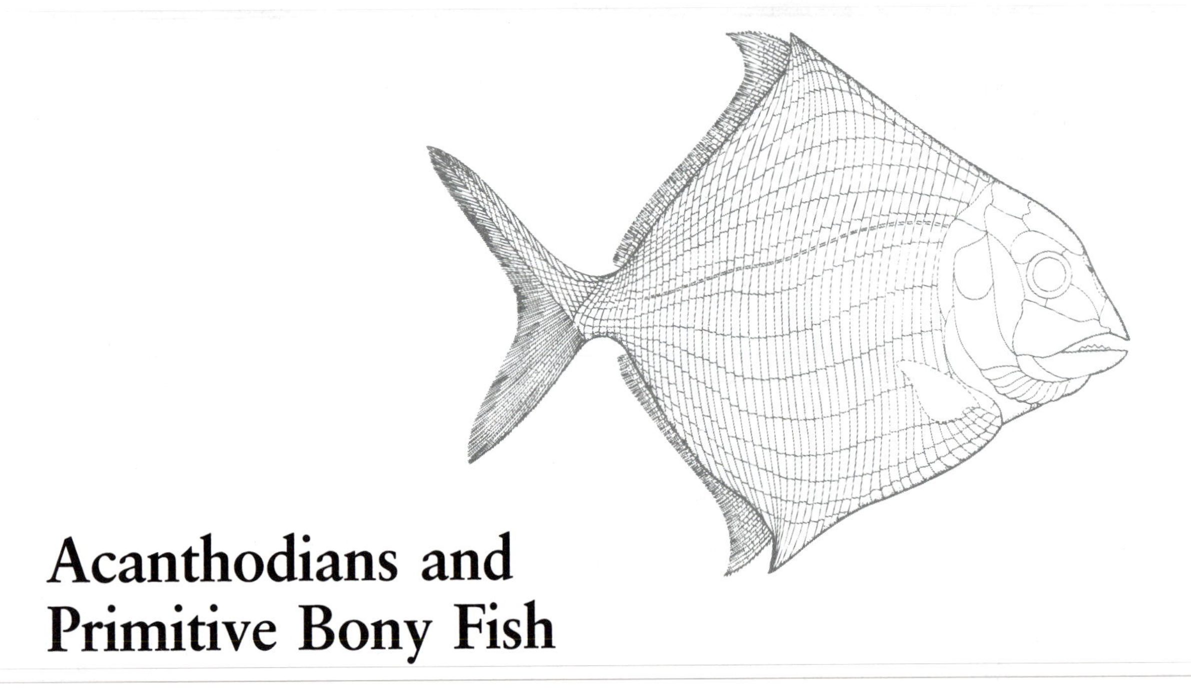

Acanthodians and Primitive Bony Fish

Most modern fishes belong to a single, extremely diverse group, the Osteichthyes, or bony fish, which have dominated marine and freshwater environments since the Middle Paleozoic. In contrast with the Chondrichthyes, the Osteichthyes have an endoskeleton that is well ossified and dermal bones and scales that cover the body. Bony fish are also characterized by the presence of a swim bladder.

Living bony fish are divided into two major groups, the actinopterygians, or ray-finned fishes (such as the trout and the perch and more than 25,000 other living species), and the much less common lobe-finned fishes, or sarcopterygians (including the lungfish and coelacanth). Modern actinopterygians are the most diverse of all vertebrate groups, while Paleozoic sarcopterygians are phylogenetically significant because they include the ancestors of all terrestrial vertebrates.

Acanthodians

Before we discuss these advanced subclasses, we must consider an enigmatic group of early bony fish whose relationships are difficult to establish: the acanthodians (Figures 6-1 to 6-6), known from the Silurian to the end of the Lower Permian. At first glance they resemble modern bony fish except for the markedly heterocercal caudal fin and stout spines that support all the fins except the caudal. Although the palatoquadrate resembles that of Chondrichthyes and orthodox bony fish, the teeth have a different histology (without an enamel-like surface) and they do not show evidence of regular replacement. This feature takes on major importance in establishing the relationship of the acanthodians, because the pattern of tooth replacement is so important in differentiating Chondrichthyes and the major groups of Osteichthyes from placoderms. We may need to recognize acanthodians as a distinct class separate from both Osteichthyes and Chondrichthyes. However, the similarities between acanthodians and Osteichthyes make it more convenient to discuss this group with the true bony fish. Denison (1979) recently reviewed the acanthodians; Miles (1965, 1968, 1973a, b), Jarvik (1977), Ørvig (1973), and Watson (1937) published important earlier accounts.

Acanthodians are the first jawed fish to appear in the fossil record. They are represented by scales, teeth, and fin spines in near-shore marine deposits of the Upper Silurian and by isolated spines as early as the Lower Silurian. More complete material is known from the Lower Devonian, and the group continues in diminishing numbers into the Lower Permian. Most were small fusiform fish, but the largest reached a length of over 2 meters. They may have possessed a swim bladder, although there is no direct evidence.

The conspicuously heterocercal tail broadly resembles that of sharks. However, a similarly shaped tail is also evident in many primitive bony fish. There may be one or two dorsal fins. Acanthodians are distinguished from all other primitive fish by the presence of conspicuous fin spines on the anal and paired fins as well as the dorsal fins, which must have provided a very effective defense against predators. Many genera also have a number of intermediate spines between the pectoral and pelvic

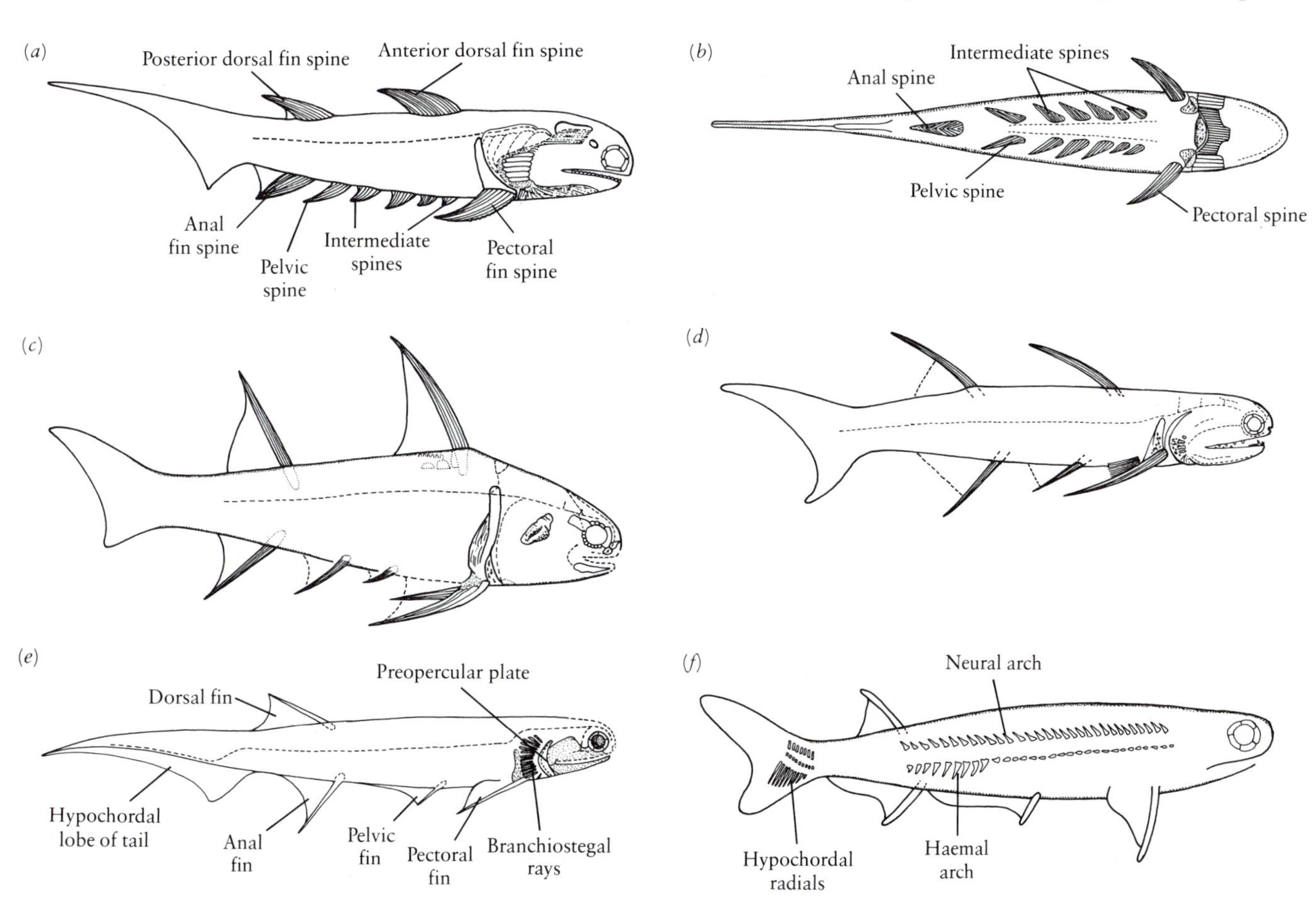

Figure 6-1. ACANTHODIAN FISHES. This group is characterized by conspicuous fin spines anterior to the pelvic, pectoral, anal, and dorsal fins. Primitive genera have several pairs of spines without accompanying fin structures between the pectoral and pelvic fins. (*a*) *Climatius*, family Climatiidae, in lateral view, $\times\frac{3}{5}$. (*b*) *Euthacanthus*, family Climatiidae, in ventral view, $\times\frac{1}{2}$. (*c*) *Diplacanthus*, family Diplacanthidae, $\times 1$. (*d*) *Ischnacanthus*, family Ischnacanthidae, $\times\frac{4}{5}$. (*e*) *Homalacanthus*, family Acanthodidae, $\times 1$. (*f*) *Acanthodes* showing internal skeleton, $\times\frac{1}{4}$. *From Moy-Thomas and Miles, 1971. By permission from Chapman and Hall, Ltd.*

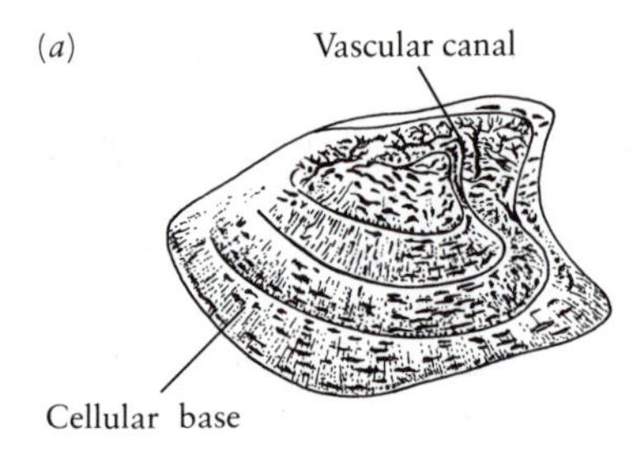

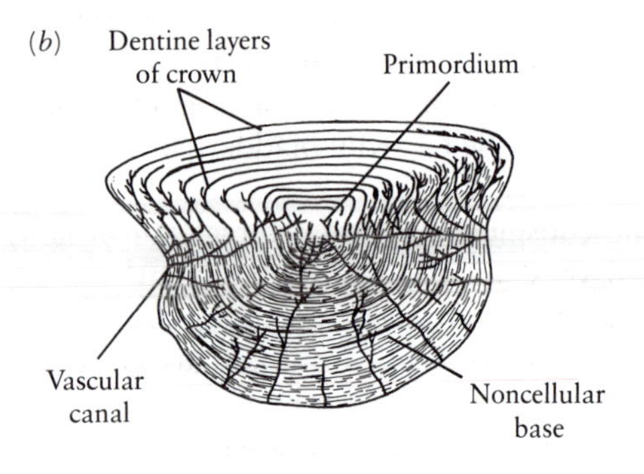

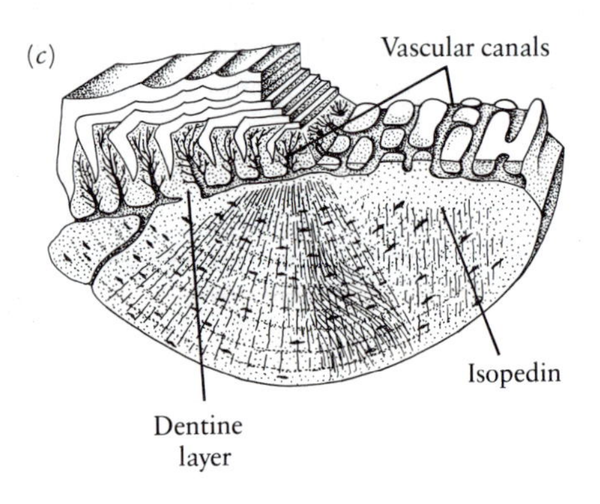

Figure 6-2. SCALES OF PRIMITIVE BONY FISHES. (*a*) The pattern of primitive acanthodians exemplified by *Nostolepis,* in which the bony base is cellular, ×40. (*b*) *Acanthodes,* which lacks lacunae for bone cells, ×20. (*c*) The primitive osteichthyan fish *Cheirolepis,* ×30. *From Moy-Thomas and Miles, 1971. By permission from Chapman and Hall, Ltd.*

girdles. These are most numerous (up to six pairs) in the more primitive genera and may be lost entirely in the Permian forms. The intermediate spines have no accompanying fin structure.

Primitive acanthodians are completely covered with small, thin, nonoverlapping scales that generally resemble those of elasmobranchs. However, the absence of a pulp cavity distinguishes them from the denticles of primitive sharks, as does the pattern of growth that occurs through the addition of concentric layers of tissue. The scales are formed of a dentine or dentine-like superficial portion and a bony base. In primitive acanthodians such as *Nostolepis* (Figure 6-2), the scales have lacunae for bone cells, but these are lost in later genera such as *Acanthodes.* In some acanthodians, the scales ossify slowly and the body is partially naked during development. The fins are covered with scales that are similar, albeit smaller, than those on the general body surface. They are arranged in parallel rows and are comparable in structure to the fin scales or **lepidotrichia** in early osteichthyan fish, although they also resemble those of osteostracan ostracoderms.

The dorsal fin of the acanthodian *Diplacanthus,* the pelvic fin of *Cheiracanthus,* and the pectoral fins of other genera have a second, deeper layer of fin supports. They resemble lepidotrichia in size and orientation but are not jointed, although they may have periodic nodes. They have been compared with ceratotrichia in sharks.

The head is covered with large dermal scales that vaguely resemble the dermal bones of the Osteichthyes, although the individual elements cannot be homologized (Figure 6-3). The gill region is protected by a variable covering of enlarged scales. We cannot recognize a clearly defined spiracular gill slit. There usually is a large anterior gill cover supported by a number of elongated scales that extend posteriorly from the hyoid arch. In primitive genera, smaller ancillary gill covers protect the more posterior gill openings. In advanced genera, the hyoid gill cover extends posteriorly to form a complete operculum.

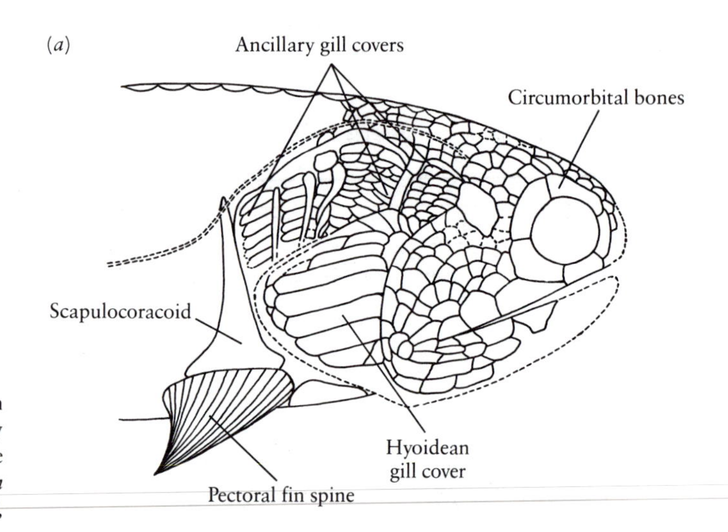

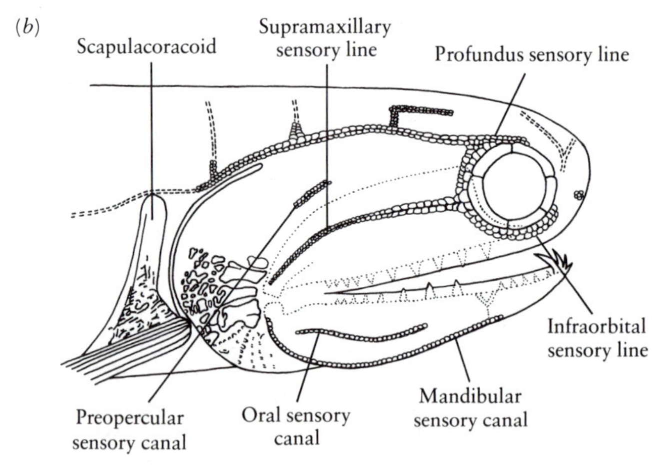

Figure 6-3. RESTORATIONS OF THE HEADS OF ACANTHODIANS. (*a*) *Brachyacanthus,* family Climatiidae, ×3.3. (*b*) *Ischnacanthus,* with scales outlining the course of the lateral line canals, ×2.8. The small scales that cover the head cannot be directly compared with the dermal skull bones of ostheichthyan fishes. *From Denison, 1979.*

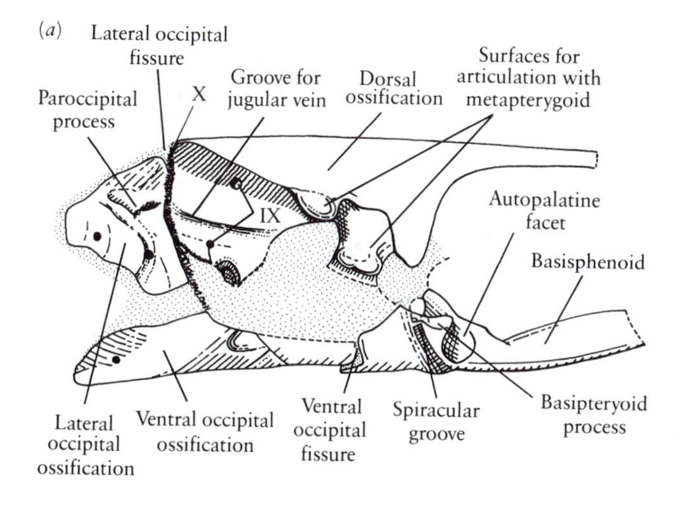

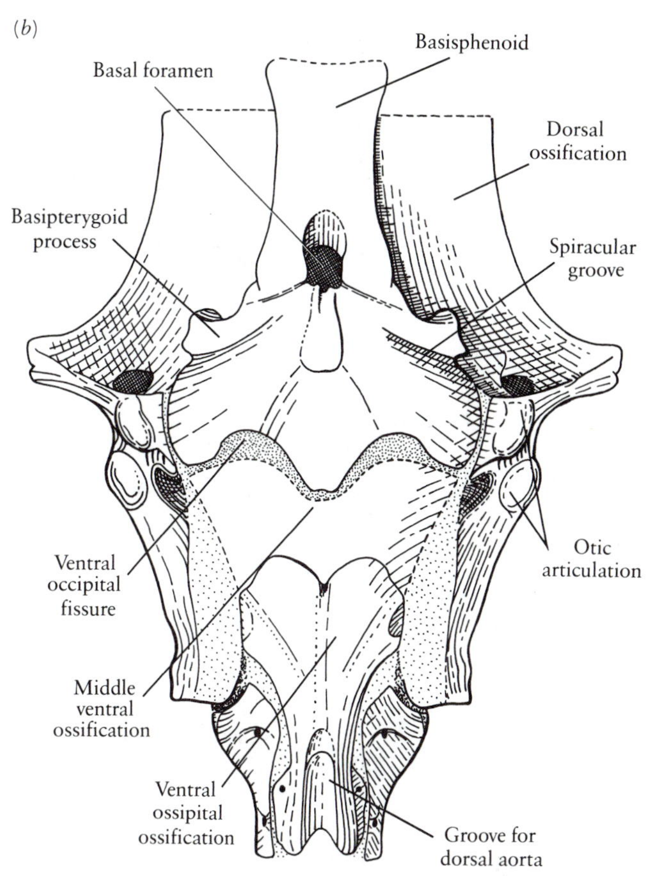

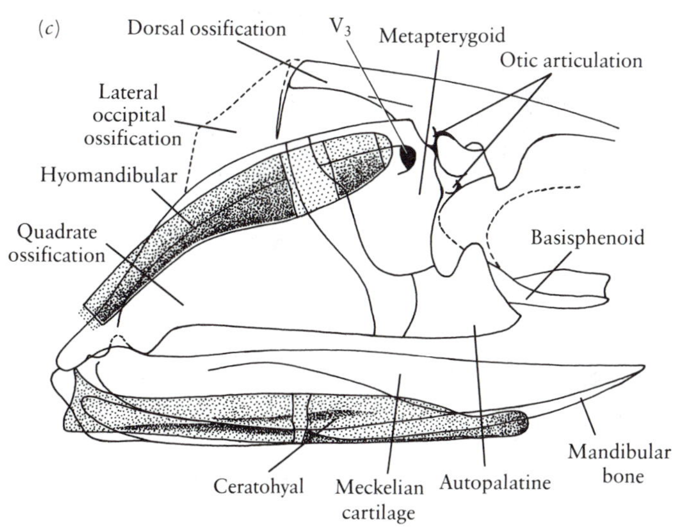

Figure 6-4. ENDOSKELETAL ELEMENTS OF THE HEAD OF THE LOWER PERMIAN ACANTHODIAN *ACANTHODES.* (*a*) Braincase in lateral view. (*b*) Braincase in ventral view. (*a* and *b*) *From Miles, 1973b. With permission from R. S. Miles. Copyright 1973 by the Linnean Society of London.* (*c*) Lateral view of visceral arch elements in articulation with braincase. Heavy stippling indicates hyomandibula and ceratobranchial; lighter stippling indicates cartilage. *From Moy-Thomas and Miles, 1971. By permission from Chapman and Hall, Ltd.*

The endoskeleton of the head region is variably ossified in acanthodians. It is best known in the late and obviously specialized genus *Acanthodes*, and the braincase is adequately known only in this genus (Figure 6-4). It is at least superficially and possibly fundamentally similar to that of actinopterygians. Its multipartite structure consists of a number of paired and medial bones held together by extensive cartilage. A large dorsal ossification, incorporating the otic capsules posteriorly, forms much of the roof of the braincase. Dorsally, it is pierced by an anterior fontanelle that may have housed the pineal organ. Behind the broad recesses for the eyes, lateral extensions provide surface for articulation with the palatoquadrate. Paired lateral occipital ossifications behind the dorsal ossification are separated from the otic capsules by a gap in ossification, the **lateral cranial fissure.** This gap, an important feature in actinopterygian fish, provided passage for the Xth nerve.

Two ventral ossifications complete the braincase. Posteriorly, the ventral occipital ossification is separated from the lateral occipital ossifications by wide, cartilage-filled gaps. Dorsally, this ossification is grooved for the notochord, which extends far forward into the braincase. The anterior basal ossification completes the floor of the braincase. It is comparable in position and structure to a complex of two bones in osteichthyans—the dermal parasphenoid and the endoskeletal basisphenoid. Lateral processes from this ossification articulate with the medial surface of the palatoquadrate. The ethmoid region, including the olfactory capsules, is not ossified.

The basic pattern of the endoskeletal jaw supports in acanthodians corresponds closely with that of primitive sharks and osteichthyans (Figure 6-4). Both the large cleaver-shaped palatoquadrate and Meckel's cartilage extend posteriorly behind the back of the braincase; the palatoquadrate does not extend anteriorly as far as the Meckel's cartilage but stops at the level of the orbit. In primitive acanthodians, each bone ossifies as a single unit, but in the late and specialized genus *Acanthodes,* both form as a series of ossifications united by cartilage.

In *Acanthodes,* each of the three areas of palatoquadrate ossification articulates with other parts of the

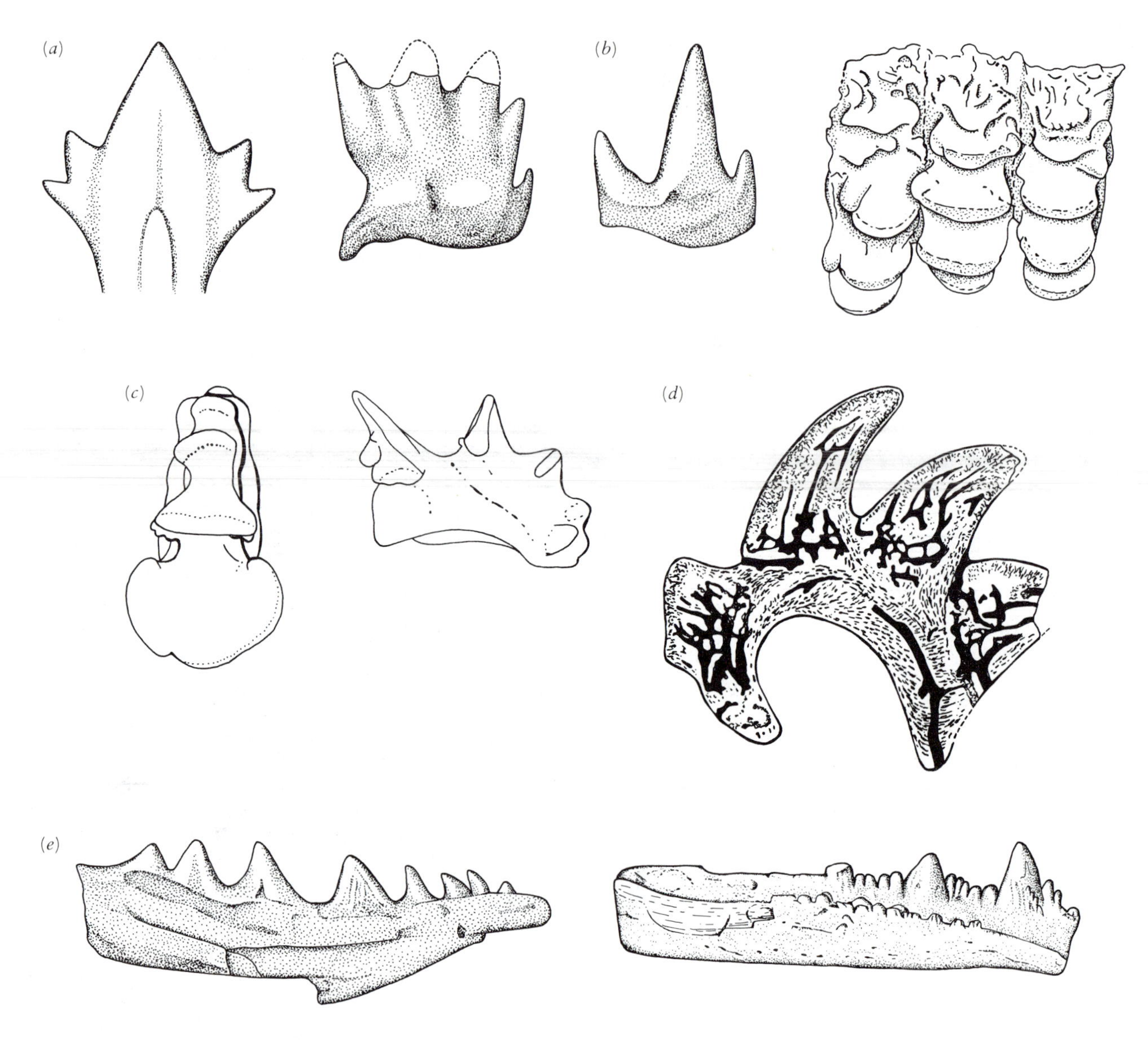

Figure 6-5. ACANTHODIAN TEETH. (*a*) Individual teeth of *Climatius*, much enlarged. (*b*) *Ptomacanthus*, family Climatiidae, tooth whorls of the upper jaw, ×8.5. (*c*) *Nostolepis*, family Climatiidae, tooth whorl in occlusal and lateral view, ×32 and ×37. (*d*) *Nostolepis*, section through tooth whorl, ×40. (*e*) Dentigerous jaw bones of members of the Ischnacanthidae, ×7.9 and ×3.2. Individual teeth are not regularly replaced among the acanthodians. Teeth that are fused to the jaws are most heavily worn posteriorly and are added progressively at the front. *From Denison, 1979.*

head. The posterior, or quadrate, unit forms the jaw articulation and is closely integrated with the hyomandibular. The dorsal, metapterygoid ossification articulates with the postorbital process of the braincase, and the antero-ventral element, the autopalatine, articulates with the basipterygoid process of the anterior basal ossification. A trough-shaped dermal ossification, the mandibular bone, sheaths the ventral surface of Meckel's cartilage throughout its length.

The dentition of acanthodians differs significantly from that of other bony fish. Although many acanthodians have no teeth at all, we can recognize three types of teeth in those genera that do: isolated elements, tooth whorls, and teeth fused to the surface of the jaw bones (Figure 6-5). Some species show only a single type of tooth, while others possess both tooth whorls or isolated teeth and teeth fused to the jaws. Unlike primitive sharks and most bony fish, none of the three types of teeth is known to have been regularly replaced.

Tooth wear increases posteriorly in teeth that are fused to the jaw, which suggests that teeth were added progressively at the front of the mouth. The tooth whorls resemble the fused tooth families of some sharks. They are aligned transverse to the jaws and may have been formed by the progressive addition of new teeth to the medial surface.

The absence of a regular pattern of tooth replacement among the acanthodians suggests that it was not present in their immediate ancestors. If the acanthodians are close to the ancestry of other bony fish (as is indicated by the similarity of the braincase and scales), we may assume that advanced bony fish and the Chondrichthyes achieved

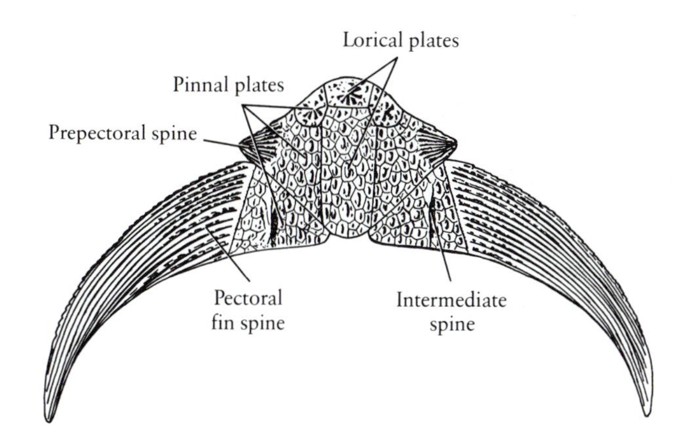

Figure 6-6. DERMAL SHOULDER GIRDLE OF *CLIMATIUS* IN VENTRAL VIEW, ×1.5. According to Miles, the presence of dermal elements associated with the shoulder girdle is a primitive feature of acanthodians. Denison feels that this condition has evolved within the group. *From Denison, 1979.*

similar patterns of tooth replacement separately. Alternately, if Chondrichthyes and Osteichthyes inherited their pattern of tooth replacement from a common ancestor, acanthodians would not be closely allied to either of these advanced classes, and their skeletal similarities to Osteichthyes must have evolved independently.

The hyomandibular is well known in *Acanthodes,* where it ossifies in two or more units. It articulates with the back of the braincase at two points and rests in a groove on the posteromedial surface of the palatoquadrate. Five gill arches are posterior to the hyoid.

As in other groups of primitive fish, vertebral centra in acanthodians are not ossified, but neural and haemal arches support the notochord dorsally and ventrally for most of the column length. There are no ribs.

In many early acanthodians, the shoulder girdle is a complex combination of dermal and endochondral units closely associated with a series of dermal spines, as can be clearly seen in *Climatius* (Figure 6-6). The most conspicuous element is a large pectoral spine that extends laterally far beyond the body wall. Its base is firmly attached to both a large dermal plate and the base of the scapulocoracoid. The most anterior of the intermediate spines extends from the same dermal plate. In other acanthodians, the plates may be fused to one another in various ways, and one or more of the spines may be lost. In advanced genera, the pectoral spine has a movable articulation with the dermal shoulder girdle. Miles (1973a) suggests that the presence of dermal elements is primitive for acanthodians. Denison (1983) argues that they evolved within the group.

The Lower Permian genus *Acanthodes* belongs to an order that never showed evidence of dermal elements, even among the Lower Devonian members of the group. Among primitive relatives of *Acanthodes,* the pectoral spine is firmly fixed to the scapulocoracoid. In *Acanthodes,* it articulated with the girdle and was presumably erected and depressed by muscles.

The acanthodians' scapulocoracoid appears to be solid superficially but is composed of a thin layer of perichondral bone surrounding a large space originally filled with cartilage. It forms as a single unit in primitive acanthodians, but three distinct elements are evident in *Acanthodes.* The pelvic spine does not articulate with the girdle.

Most acanthodians may be grouped in three orders: the Climatiida, the Ischnacanthida, and the Acanthodida. However, many genera are based on isolated spines, scales, or teeth that cannot be assigned to particular orders. Both the Climatiida and the Ischnacanthida are present in the late Silurian and survive into the late Carboniferous. The Acanthodida are not known until the early Devonian but persist into the early Permian.

The Climatiida have two dorsal fins and a complex of dermal plates associated with the pectoral girdle. Most genera have tooth whorls, but the Diplacanthidae, which appear as early as the Lower Devonian, are toothless. The genus *Gyracanthus* is a common form in the Carboniferous and is represented primarily by fin spines that are up to 40 centimeters long.

The Ischnacanthida have a relatively longer and lighter body than the Climatiida, with thin scales and no dermal bones associated with the shoulder girdle. The fins are deeply inserted, and there are no ancillary gill covers or intermediate spines. They have both tooth whorls and tooth-bearing dermal bones on the jaw margins. The dentition of the Climatiida and early Ischnacanthida suggests that they were relatively unspecialized microphageous feeders. Later Ischnacanthida may have been effective predators on larger prey.

From their first appearance in the fossil record, the Acanthodida lack teeth and dermal plates associated with the pectoral girdle. There is a single dorsal fin. This group is adapted toward filter feeding. The palatoquadrate is articulated so that it can swing far laterally to increase the gape greatly. Gill rakers and the entire gill region are considerably elongated.

The Climatiida, Ischnacanthida, and Acanthodida represent clearly defined taxonomic assemblages, but several progressive changes have occurred as a result of convergence in each of these orders. These changes include loss of ancillary gill covers and the spread of the hyoid gill cover to form a large operculum, deeper insertion of spines, development of articulation between the pectoral spine and the girdle, loss of dermal elements of the shoulder girdle, loss of intermediate and prepectoral spines, loss of cell spaces in the bones, and loss of teeth.

OSTEICHTHYES

Both actinopterygians and sarcopterygians are known from the early Devonian. Unlike acanthodians, they have no fin spines but do have a regular pattern of tooth replacement. The braincase is more completely ossified, and the palate is covered with dermal bones. The dermal bones of the skull and shoulder girdle effectively integrate the functions of feeding and respiration. Primitive members of both groups are fusiform, actively swimming, predaceous fish of small to moderate size. The earliest fossils are from sediments deposited in marine waters, but both groups quickly spread into brackish and freshwater environments.

Isolated scales from the Upper Silurian have been attributed to actinopterygians but tell us little about the origin of the group. Early actinopterygians resemble the acanthodians more closely than do the sarcopterygians in the structure of the scales, fins, and braincase and may represent the pattern from which all higher bony fish evolved.

The sarcopterygians and actinopterygians are clearly distinct from one another when they first appear in the fossil record, and four very distinct groups of sarcopterygians had already evolved by the Middle Devonian.

ACTINOPTERYGIANS

Early actinopterygians possess a number of features, many of them primitive, that differentiate them from the sarcopterygians. The term "ray-finned" refers to the fact that the paired and median fins are supported by a large number of parallel endoskeletal fin rays and contain relatively little intrinsic musculature. Movements of the fins are controlled almost entirely by muscles within the body wall and the fins are moved as a unit except in advanced genera. The endoskeletal fin supports of sarcopterygians are condensed into a thick bony axis that is moved by muscles within the fin (see Chapter 8). Primitive actinopterygians have a single dorsal fin, in contrast with two in early sarcopterygians. The caudal fin in primitive actinopterygians is fully heterocercal with no epaxial lobe above the notochord; the flexible portion of the tail is formed entirely by the hypaxial lobe. All sarcopterygians have at least a small epaxial lobe, although the tail is primitively heterocercal. Actinopterygians lack the specialized layer of tissue termed **cosmine** that characterizes the surface of the scales and dermal bones of sarcopterygians.

Patterson (1982) listed a number of additional characters thought to be specializations that further distinguish actinopterygians from other bony fish. These include a pectoral propterygium, the absence of a squamosal bone (an important element that makes up the cheek region in sarcopterygians), the presence of a dermohyal (a bone located superficial to the dorsal extremity of the hyomandibular), and the fact that the hyomandibular has only a single point of articulation above the jugal canal (rather than two points of articulation, above and below the canal, as in sarcopterygians and acanthodians).

We may divide actinopterygians into two major groups, the chondrosteans and the neopterygians, which represent successive radiations (Figure 6-7). The chondrosteans are a primarily Paleozoic and Triassic assemblage, with several genera that survive to the present day. The neopterygians are first known in the late Permian, and their primary radiation occurred during the Triassic and Jurassic. During the later Mesozoic, one lineage gave rise to a further radiation of much greater magnitude, which resulted in the diversification of the majority of modern bony fish groups. Advanced neopterygians, including almost all the living bony fish, are termed the Teleostei. In the past, more primitive neopterygians, including the majority of Jurassic genera and a few surviving forms, have been included in a distinct group, the "Holostei". These fish do not appear to form a natural group, and their taxonomic position remains uncertain.

PRIMITIVE ACTINOPTERYGIANS: THE PALAEONISCIFORMES

Skeletal anatomy

Although first known in the late Silurian, the chondrosteans show little diversity until the end of the Devonian. Most of the Paleozoic actinopterygians belong to a single order, the Palaeonisciformes. Primitive members of this order include the genera *Moythomasia, Mimia,* and *Cheirolepis* (Figures 6-8 and 6-13) from the early part of the late Devonian. *Cheirolepis* approaches $\frac{1}{2}$ meter in length, but the other genera were considerably smaller. The body is fusiform with a strongly heterocercal caudal fin. Like the anal and paired fins, the single dorsal fin is broadly triangular in shape.

The body is covered with small, overlapping rhomboidal scales (Figure 6-9). In most genera, the scales articulate with one another by a dorsal peg that fits into a socket on the bottom of the scale above it. Scales from the Silurian, which Schultze (1977) attributed to chondrosteans, lack this peg-and-socket articulation. The scales at the leading edge of the caudal fin, termed fulcral scales, are enlarged to act as a cutwater. The scales on the fins are smaller and somewhat more elongated than those covering the trunk, but they are basically similar in configuration. As in acanthodians, they run in closely set rows parallel to the orientation of the fin. These jointed scale rows are referred to as dermal fin rays and are more numerous in the primitive state than the underlying endoskeletal fin supports. The scales consist of a basal layer of laminar bone, a middle layer of dentine containing vascular canals, and a superficial layer of enamel-like material termed ganoine (see Figure 6-2*c*). Growth occurs by the addition of concentric layers of tissue, both dorsally and ventrally. This type of scale, termed **ganoid,** could have been derived from primitive acanthodians.

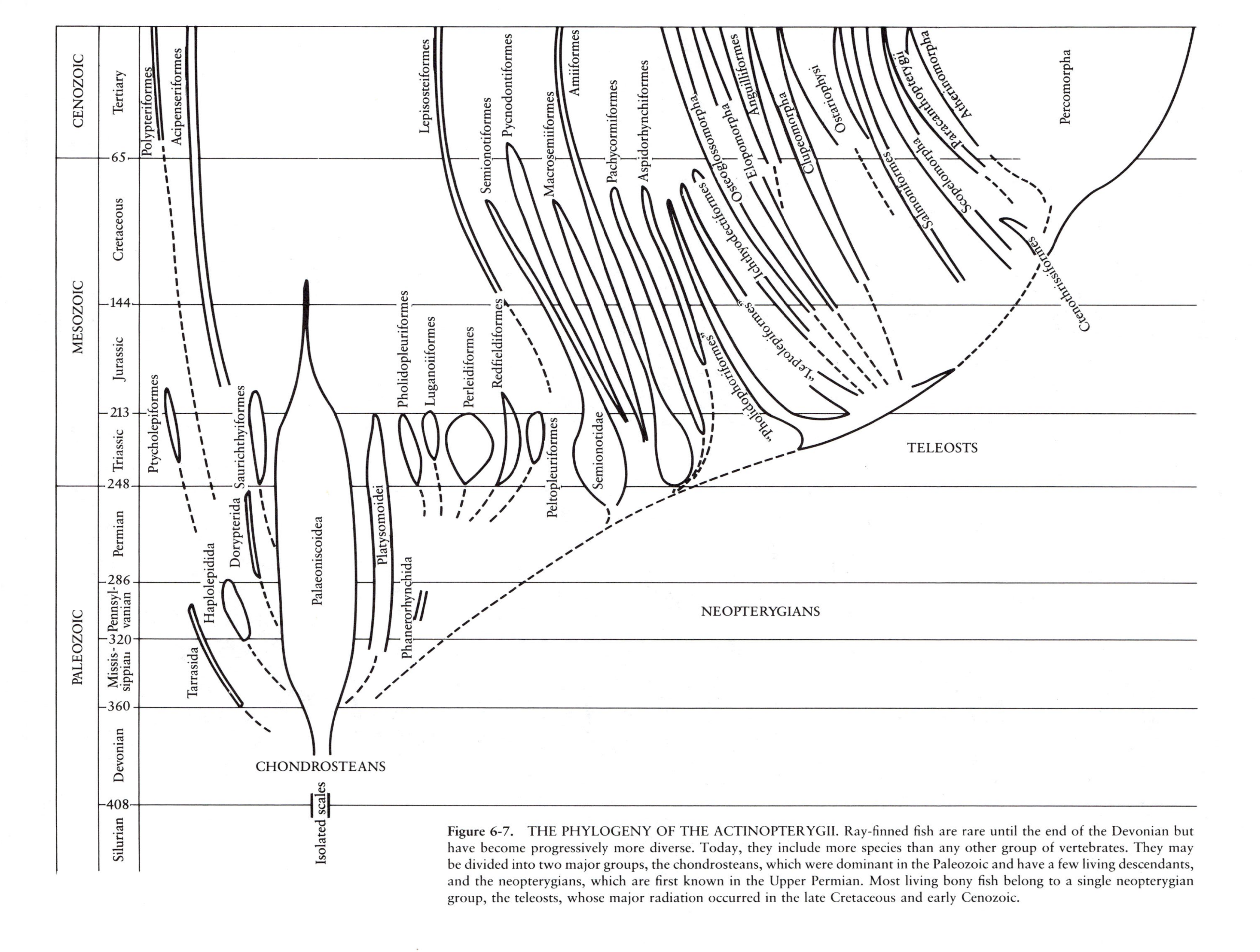

Figure 6-7. THE PHYLOGENY OF THE ACTINOPTERYGII. Ray-finned fish are rare until the end of the Devonian but have become progressively more diverse. Today, they include more species than any other group of vertebrates. They may be divided into two major groups, the chondrosteans, which were dominant in the Paleozoic and have a few living descendants, and the neopterygians, which are first known in the Upper Permian. Most living bony fish belong to a single neopterygian group, the teleosts, whose major radiation occurred in the late Cretaceous and early Cenozoic.

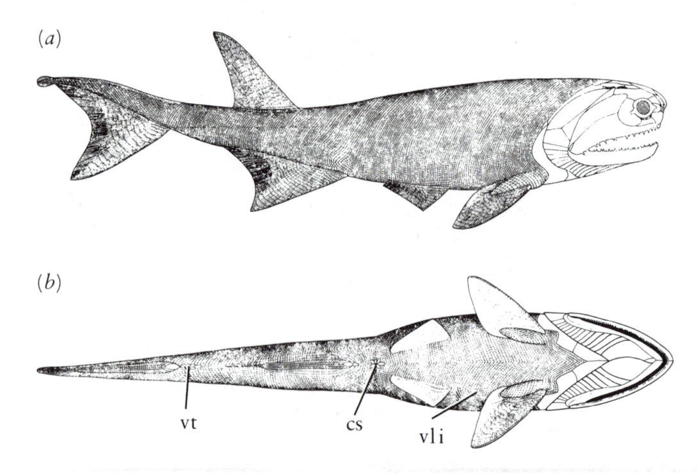

Figure 6-8. THE UPPER DEVONIAN PALAEONISCIFORM *CHEIROLEPIS* IN LATERAL AND VENTRAL VIEWS. The small size of the scales and the oblique jaw suspension may illustrate the primitive condition for actinopterygians, $\times \frac{1}{4}$. Abbreviations are as follows: cs, circumcloacal scales; vt, ventral caudal fulcra; vli, ventrolateral scale inversion. *From Pearson and Westoll, 1979.*

The notochord in primitive actinopterygians is unrestricted and extends to the very tip of the heterocercal tail. Paired neural spines lie dorsal to the notochord throughout the trunk and into the base of the tail, and ventral haemal arches support it in the posterior portion of the trunk and tail. One or two additional rows of endochondral supports extend dorsally from the neural arches to support the dorsal fin, and one row extends ventrally from the haemals to support the anal fin and the hypaxial lobe of the caudal fin (Figure 6-10).

A series of parallel endochondral rods, termed radials, support the pectoral fin and articulate with the scapulocoracoid. They extend only a short distance into the fin. While ceratotorichia are not present primitively, they do evolve in some specialized groups. The scapulocoracoid is a small ossification attached to the medial surface of the cleithrum (Figure 6-10). It has a short blade, in contrast with that of sharks and acanthodians which lack the lateral support of the dermal shoulder girdle. The pelvic fin is also supported by a row of radials. The halves of the girdle are small triangular ossifications.

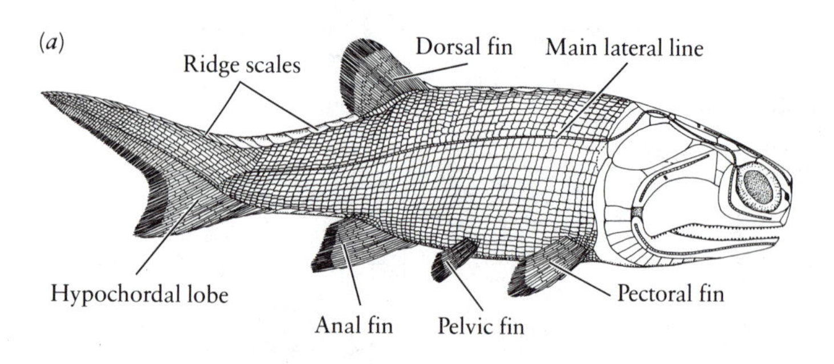

Figure 6-9. THE UPPER DEVONIAN PALAEONISCOID *MOYTHOMASIA*, $\times 1$. This genus shows the well-developed fulcral scales on the caudal fin that are characteristic of chondrostean fishes. *From Moy-Thomas and Miles, 1971, after Jessen. By permission from Chapman and Hall, Ltd.*

We will discuss the structure and function of the head region of the palaeonisciforms in considerable detail, since these fish exhibit a pattern that is retained among later ray-finned fishes and also serves as a basis for understanding the structure of sarcopterygian fish and their descendants, the terrestrial vertebrates.

The internal skeleton of the head in primitive actinopterygians resembles that of sharks and acanthodians in its most general features (Figures 6-11 and 6-12). The braincase is a nearly unitary structure to which the long palatoquadrate and the hyomandibular articulate. However, there are few detailed similarities between the almost completely ossified braincase of early actinopterygians and the cartilaginous structure of sharks. The braincase of actinopterygians differs in having a narrow interorbital septum that accommodates relatively large eyes. It could have evolved from a pattern like that of acanthodians with few fundamental changes, except for an increase in ossification. Two unossified areas are retained between elements that form separately during embryological development. As in acanthodians and some sharks, the occipital plate is separated from the otic capsule by the lateral cranial fissure. This fissure is confluent dorsally with the posterior dorsal fontanelle. A second, **ventral fissure** lies between areas that are recognizable in later actinopterygians as the basioccipital and basisphenoid. In embryos of modern fish, these bones develop from the areas formed by the parachordal cartilages and the trabeculae. In actinopterygians, we see no evidence that the

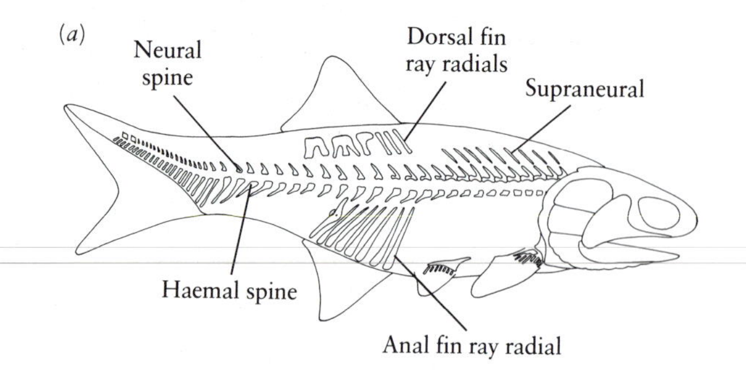

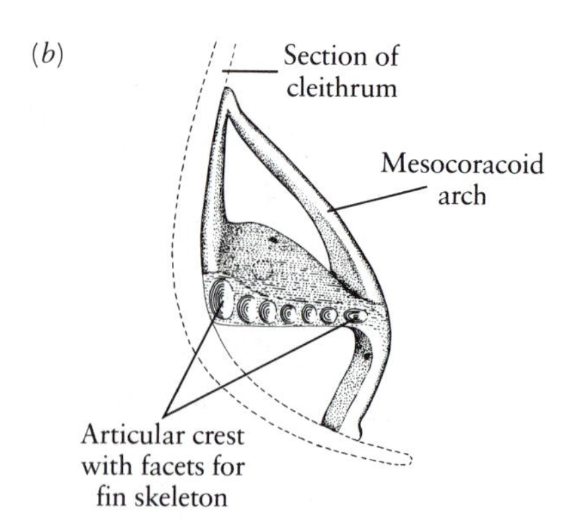

Figure 6-10. (*a*) The palaeoniscoid *Mimia*, showing the endoskeleton. *From Gardiner, 1984.* (*b*) Scapulocoracoid of *Moythomasia* in posterior view, $\times 6$. *From Moy-Thomas and Miles, 1971. By permission from Chapman and Hall, Ltd.*

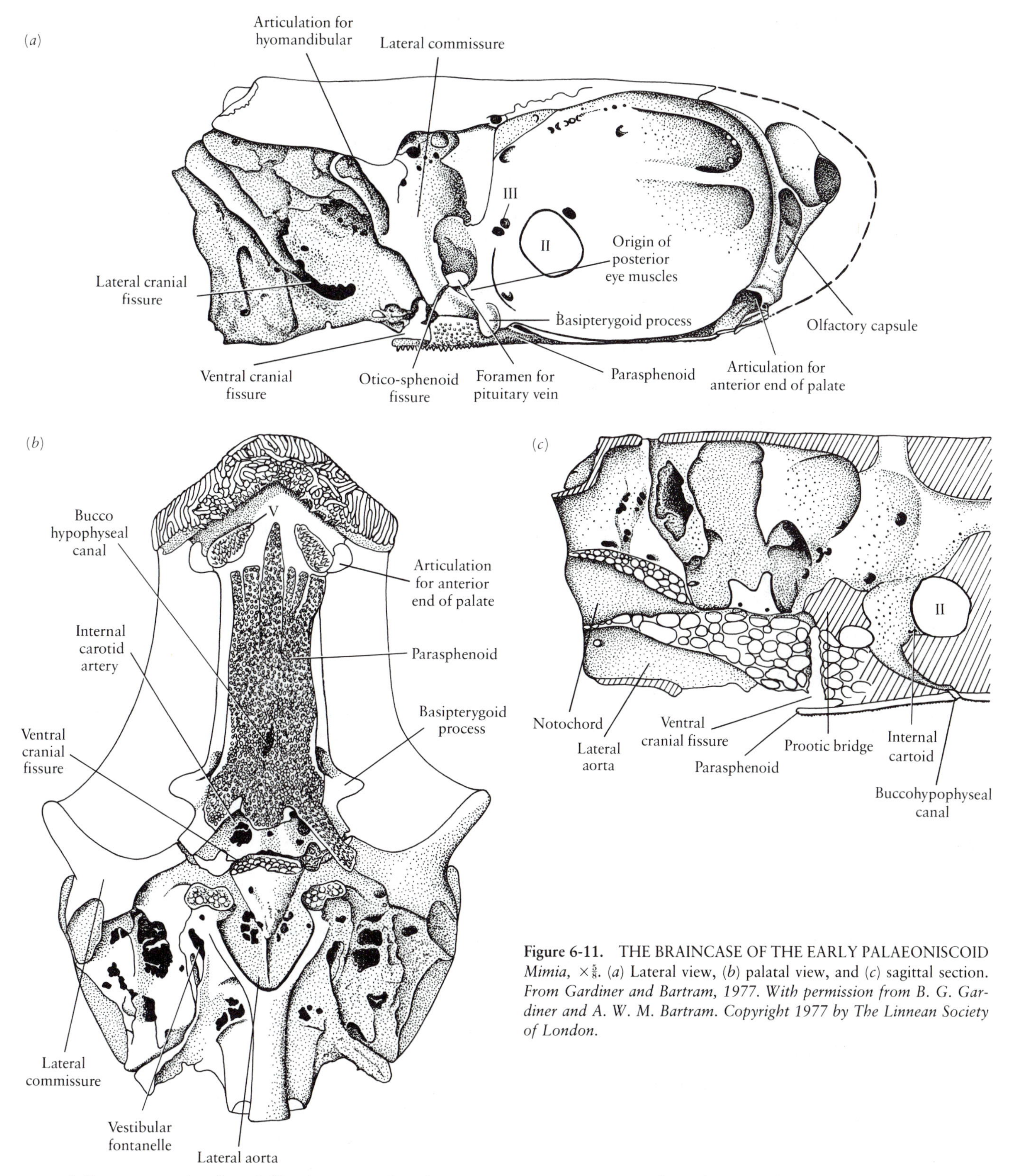

Figure 6-11. THE BRAINCASE OF THE EARLY PALAEONISCOID *Mimia*, ×⅝. (*a*) Lateral view, (*b*) palatal view, and (*c*) sagittal section. *From Gardiner and Bartram, 1977. With permission from B. G. Gardiner and A. W. M. Bartram. Copyright 1977 by The Linnean Society of London.*

ventral fissure permitted mobility between the elements of the braincase, as is the case in crossopterygian sarcopterygians. The back of the braincase is pierced by the notochord, which extends anteriorly nearly to the pituitary fossa.

In acanthodians, there is a large unossified area ventral and anterior to the otic capsule. In actinopterygians and sarcopterygians, this area is ossified as the **lateral commissure.** In the Palaeonisciformes, it supports a single large facet for articulation of the hyomandibular and bears a vertical groove for the spiracular gill slit. Anteriorly, a narrow interorbital septum extends to the small, ossified nasal capsules.

The shape of the palatoquadrate (Figure 6-12) resembles that of primitive sharks, but the otic process does not articulate with the braincase. Instead, there is a well-

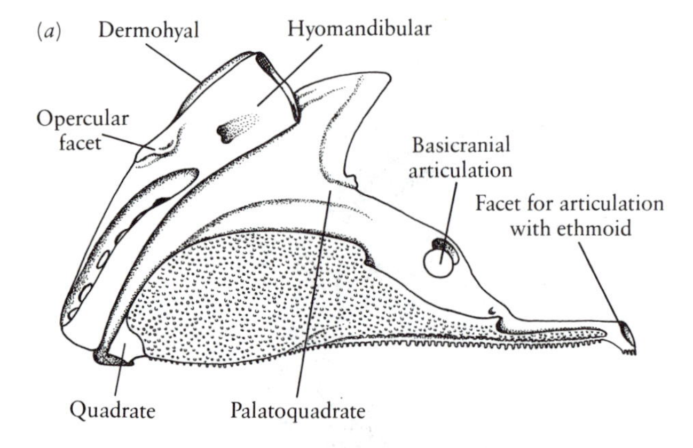

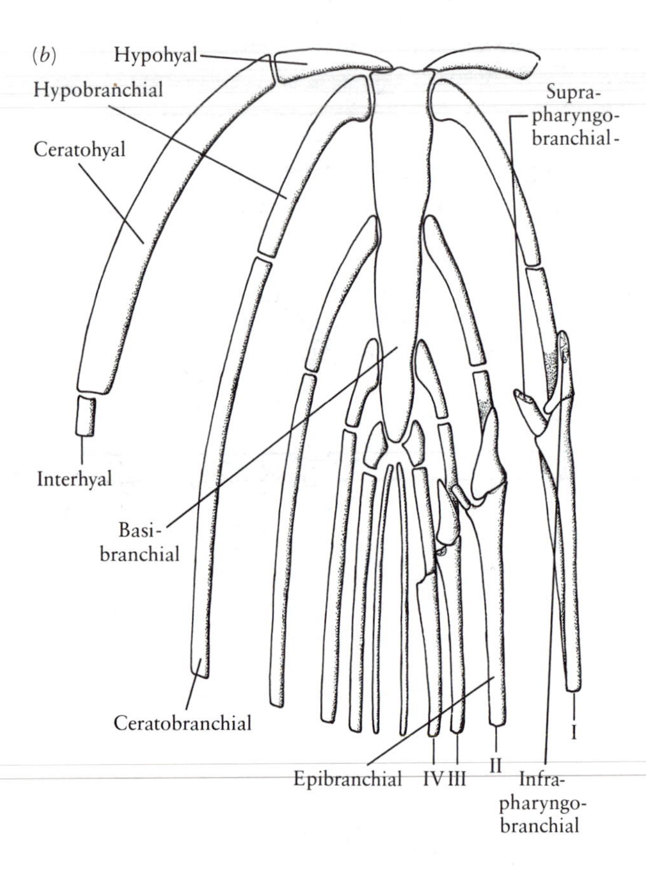

Figure 6-12. VISCERAL ARCH ELEMENTS OF *MIMIA*. (*a*) Hyomandibular and palatoquadrate in medial view, ×½. (*b*) Branchial arches in dorsal view, ×⅗. *From Gardiner, 1973.*

developed articulation with the ventral surface of the braincase near the midline (the basicranial articulation), as in acanthodians. There is a second, more anterior articulation with the ethmoid region behind the nasal capsule. The palatoquadrate in acanthodians does not extend this far forward. Marginal teeth are not borne on the palatoquadrate and Meckel's cartilage, as is the case in Chondrichthyes, but on the dermal bones at the margin of the skull roof and lower jaw. Meckel's cartilage is almost completely supplanted by dermal bones. It ossifies as a small element, the articular, at the posterior end of the jaw, where it forms the surface for articulation with the palatoquadrate. The hyomandibular articulates with the lateral commissure of the braincase and rests distally in a groove on the quadrate portion of the palatoquadrate. The ceratohyal lies between the lower jaws. Four additional gill arches lie behind the hyoid (Figure 6-12*b*).

The dermal bones of the actinopterygian skull (Figure 6-13) form an external covering surrounding the endoskeletal units, the sense organs, jaw muscles, mouth, and pharynx. Openings are present dorsally for the pineal and laterally for the large orbits. The incurrent and excurrent nasal openings notch the nasal and rostral bones. The spiracular cleft is anterior to the operculum. The dermal skull is not rigid but consists of several units that move relative to one another to facilitate feeding and respiration. These functions are closely associated with a particular pattern and configuration of the bones that can be readily traced from the Devonian to modern genera. The bones bear no significant similarities to the large scales covering the head of acanthodians, but they do generally correspond with those of sarcopterygian fish and with the dermal bones of tetrapods.

The bones forming the central portion of the skull roof from the snout to the occiput include the tooth-bearing premaxilla, one or more rostrals, the nasal, frontal, parietal, intertemporal, supratemporal, and a series of bones above the orbit; all are closely attached to one another and more or less fixed to the neurocranium. Laterally, the bones surrounding the posterior and ventral margin of the orbit and the cheek make up a second unit whose posterior margin can swing laterally and somewhat ventrally as the mouth is opened. This unit includes the lacrimal, jugal, maxilla, preopercular, dermohyal, and quadratojugal. The maxilla has an ill-defined anterior articulation with the premaxilla. Internally, the maxilla and preopercular are attached to the palate, and the dermohyal is attached to the hyomandibular.

The gill arches extend from behind the palatoquadrate, ventrally and forward between the lower jaws. They are covered by a flexible arrangement of dermal bones—the large gulars and branchiostegal rays between the jaws and the opercular series behind the cheek. The bones are spread apart laterally and ventrally as the jaws are opened and the branchial chamber is expanded (Figures 6-14 and 6-15).

The dermal shoulder girdle serves a series of interrelated functions. The cleithrum is a particularly important element that forms the back of the gill chamber, supports the scapulocoracoid, and serves for the insertion of major hypaxial muscles that open the mouth and expand the gill chamber. The supracleithrum connects the cleithrum to the back of the cheek and, via the posttemporal, articulates with the extrascapular bone at the back of the skull table. The ventral clavicle extends forward between the branchiostegal rays.

Dermal ossifications also cover the ventral surface of the braincase and the roof of the mouth. Two small vomers, which are studded with denticles, underlie the ethmoid region of the braincase. The medial parasphenoid

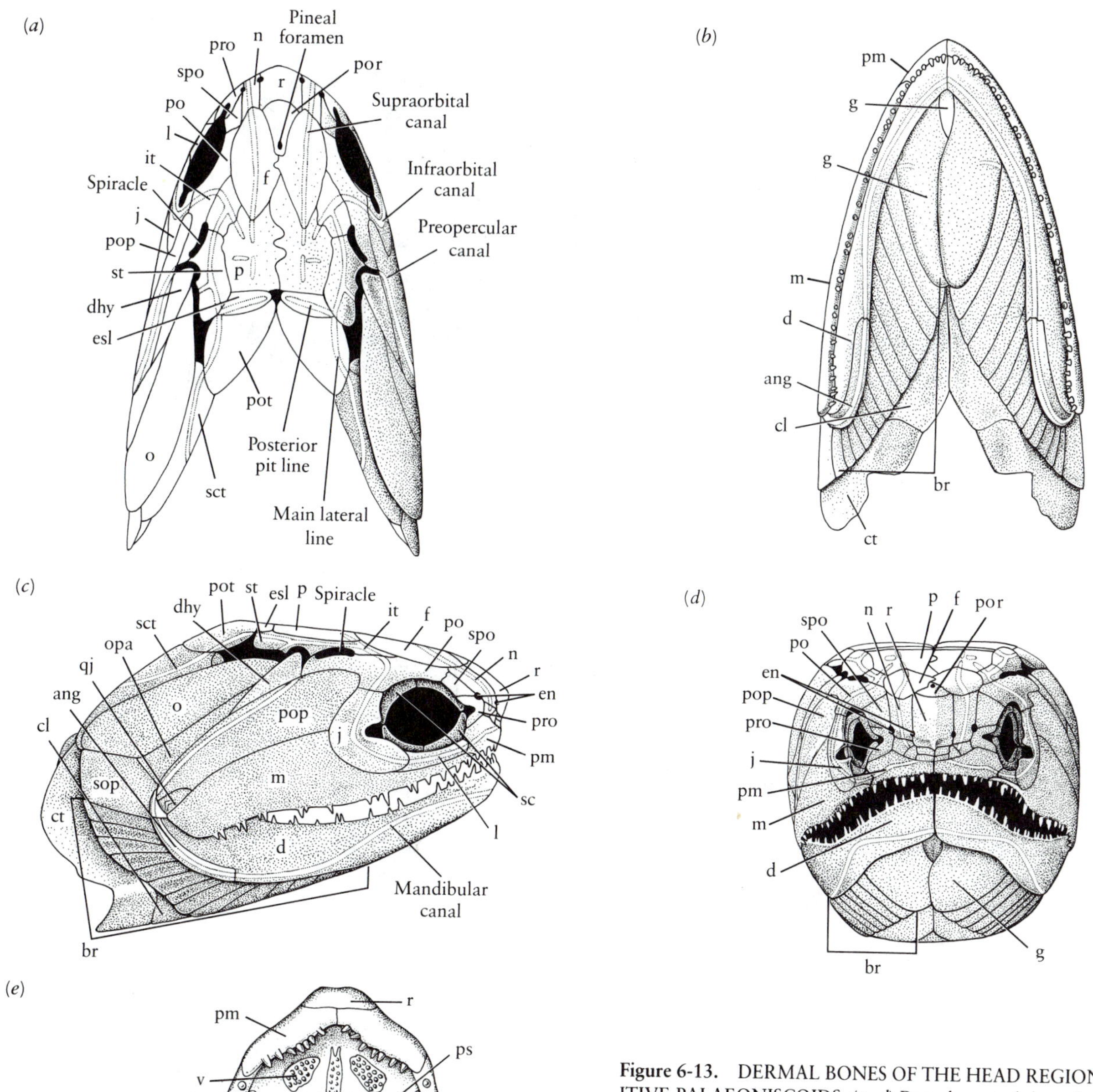

Figure 6-13. DERMAL BONES OF THE HEAD REGION OF PRIMITIVE PALAEONISCOIDS. (*a–d*) Dorsal, ventral, lateral, and anterior views of the skull of *Cheirolepis. From Pearson and Westoll, 1979.* (*e*) Palate of *Mimia. From a sketch provided by Brian Gardiner.* The general pattern of the bones resembles that of rhipidistians, tetrapods, and more advanced bony fish. Abbreviations as follows: a, articular; ang, angular; bo, basioccipital region of braincase; br, branchiostegal rays; bs, basisphenoid region of braincase; cl, clavicle; ct, cleithrum; d, dentary; dhy, dermohyal; dpt, dermopterotic; ec, ectopterygoid; en, external naris; ep, epipterygoid; es, ethmosphenoid portion of braincase; esl, lateral extrascapular; esm, medial extrascapular; f, frontal; g, gulars; hy, hyomandibular; ifo, infraorbital; in, internal naris; iop, interopercular; it, intertemporal; j, jugal; l, lacrimal (infraorbital); m, maxilla; n, nasal; nc, nasal capsule; ntc, notochord; o, opercular; oo, otico-occipital portion of braincase; opa, accessory opercular; ot, otic region of braincase; p, parasphenoid; pa, parietal (frontal); pf, postfrontal (supraorbital); pl, palatine; pm, premaxilla; pn, postnarial; po, postorbital; pop, preopercular; por, postrostral or internasal; pos, postsplenial; pot, posttemporal; pp, postparietal (parietal); pro, preorbital; ps, parasphenoid; pt, pterygoid; pv, vomer (prevomer); q, quadrate; qj, quadratojugal; r, rostral; rar, retroarticular process; s, suprapterygoid(s); sa, surangular; sc, sclerotic ring; sct, supracleithrum; sm, supramaxilla; smp, symplectic; so, suborbital(s); soc, supraoccipital; sop, subopercular; sp, spiracular cleft; spo, supraorbital; sq, squamosal; st, supratemporal (dermal sphenotic); t, tabular (supratemporal-intertemporal); v, vomer.

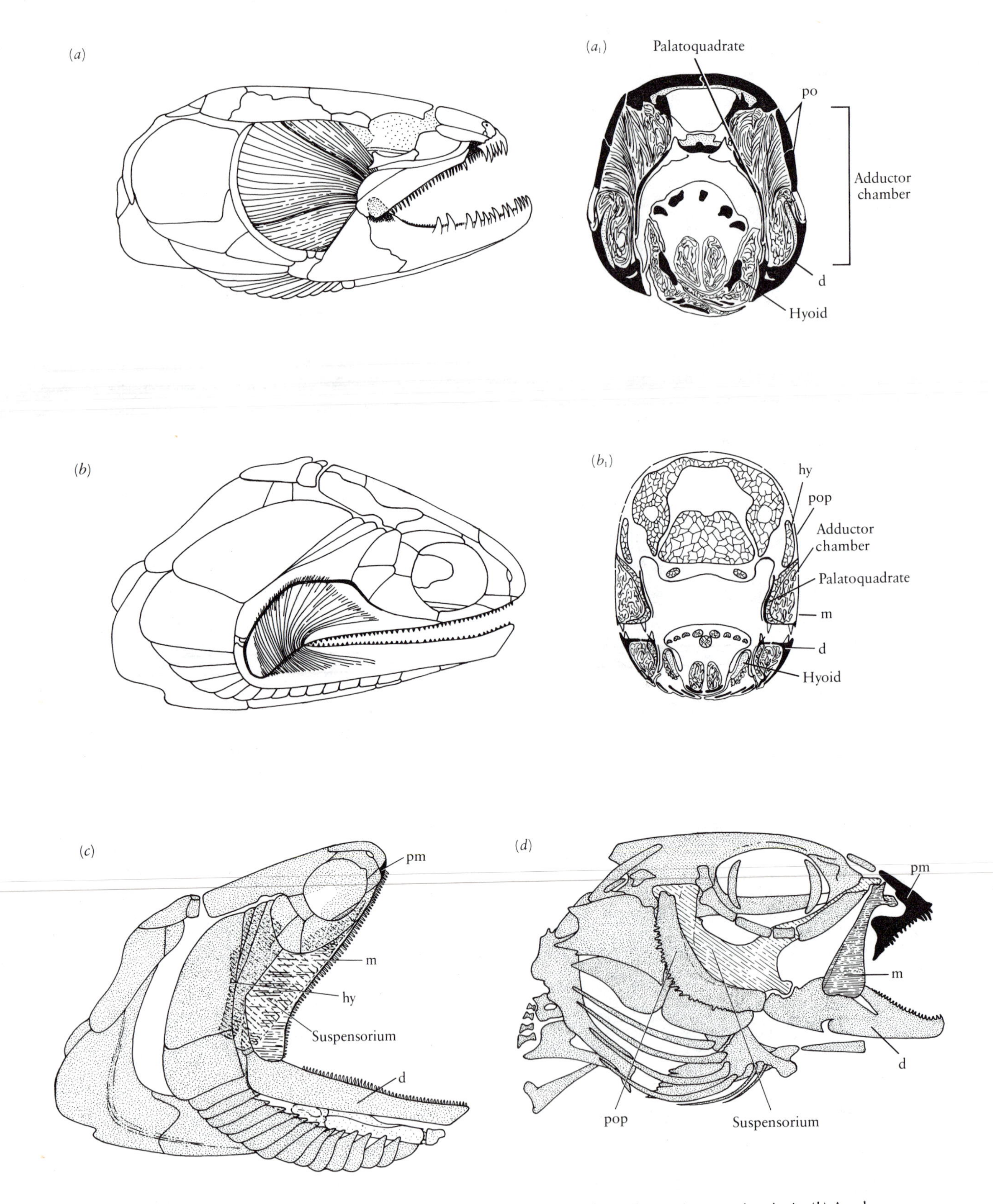

Figure 6-14. JAW BONES AND MUSCULATURE OF BONY FISH. (*a*) The modern actinopterygian *Amia*. (*b*) A palaeoniscoid. *Amia* was drawn as if the bones of the cheek were removed, and the palaeoniscoid was drawn as if the dermal bones were transparent. In *Amia*, the musculature has spread dorsally and posteriorly out of the originally closed adductor chamber. (a_1 and b_1) Cross-sections through the back of the skull showing the extent of the jaw musculature of *Amia* and palaeoniscoid. (*c* and *d*) Diagrams of a palaeoniscoid and a neopterygian in which the jaws are widely open. For abbreviations see Figure 6-13. Premaxilla black, maxilla horizontally shaded, suspensorium diagonally shaded. The cheek is covered with bones in palaeoniscoids but is widely open in the neopterygian. *From Schaeffer and Rosen, 1961.*

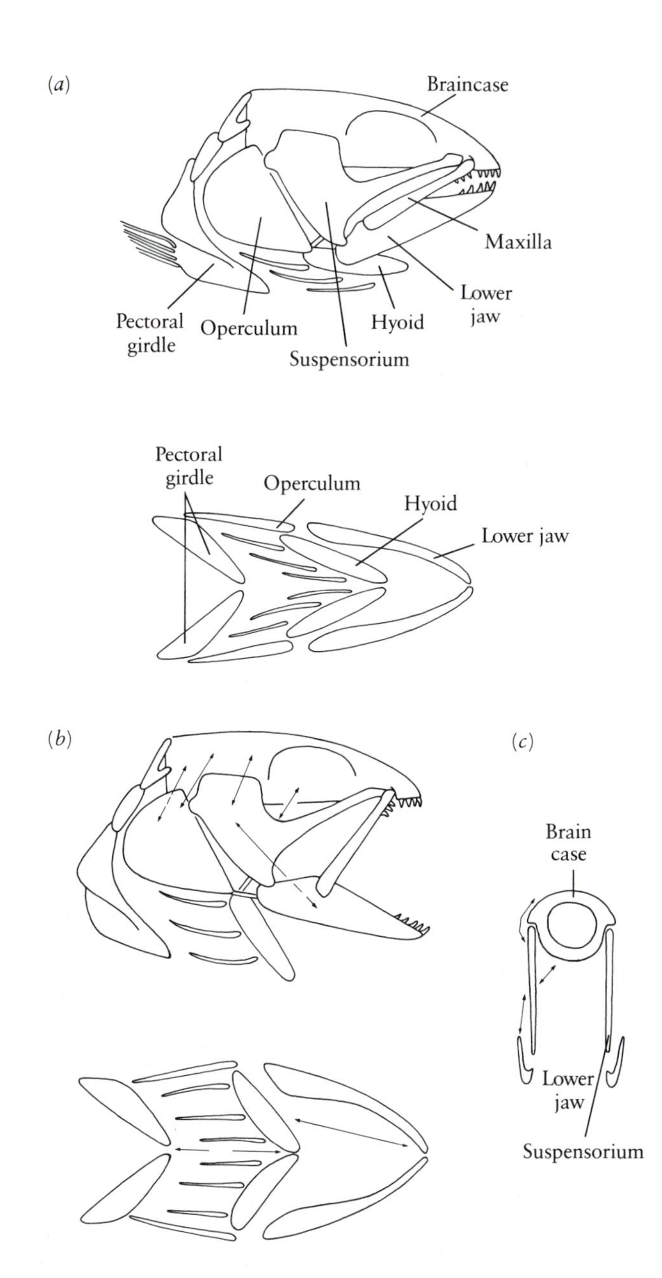

Figure 6-15. MECHANICS OF THE SKULL AND SHOULDER GIRDLE IN RAY-FINNED FISH. Based on teleosts, but except for the mobility of the maxilla, applicable to chondrosteans as well. (*a*) Lateral and ventral views when mouth is closed and (*b*) when the mouth is open. (*c*) Transverse section demonstrates mobility of suspensorium in relation to braincase. *Modified from Alexander, 1970.*

extends forward from the ventral fissure. Paired parotic plates lay in the tissue of the roof of the mouth, just behind the fissure. The palatoquadrate is covered by a number of distinct but contiguous dermal bones. A series of dermopalatines, which bear large teeth, lie parallel with the maxilla. The remainder of the surface is covered by the entopterygoids, ectopterygoid, and dermometapterygoids, which are comparable with the pterygoids in tetrapods.

The dermal bones of the lower jaw include the large, tooth-bearing dentary, the angular and surangular below the jaw articulation, and the splenial, prearticular, and coronoids on the medial surface.

Functional anatomy of feeding and respiration

We attribute the great success achieved by actinopterygians in the Mesozoic and Cenozoic to the evolution of a particularly effective feeding apparatus (Lauder, 1982). The palaeonisciforms show an early stage in the development of this system (Figure 6-14).

We can reconstruct the major jaw muscles of primitive actinopterygians on the basis of those in living ray-finned fish. The high degree of ossification of the skull in palaeonisciforms clearly delimits the areas that the muscles could occupy. The adductor mandibulae—the principal muscles for closing the jaws—originate in a small, closed chamber, formed by the palatoquadrate dorsally and medially, and by the dermal bones of the cheek, the maxilla and preopercular, laterally. They pass through the palate immediately anterior to the jaw articulation and insert in the Meckelian fossa at the back of the lower jaw. Attachment of the muscles at the very back of the jaw provides a wide gape and gives a fast, relatively weak bite.

The jaw-closing mechanism in early actinopterygians is easy to understand because it is mechanically similar to our own way of biting. In mammals, the opening of the jaws is relatively passive and can be accomplished by little more than relaxing the muscles that close the jaw. However, there is contribution from both the hyoid muscles running between the jaws and the diagastric muscle originating at the back of the skull. The jaw-opening muscles in fish are much more important since they must counteract the viscosity of the water. As in mammals, the hyoid musculature in fish is an important element of the jaw-opening apparatus. In addition, the major axial muscles of the body are involved (Figure 6-15). Fish lack a clear-cut division between head and trunk. Both the dorsal (epaxial) and ventral (hypaxial) musculature of the trunk contribute to opening the mouth. The epaxial muscles attach directly to the back of the braincase and pull it back to raise the snout. The hypaxial muscles attach to the dermal shoulder girdle, particularly the cleithrum. When they contract, the jaws are pulled backward and downward via the intervening hyoid apparatus. The mouth and gill chamber are also capable of *lateral* expansion, which is important in both feeding and repiration.

Bony fish respire by means of a double pump that produces a continuous flow of water through the gills. The oral cavity forms the anterior pump that expands both vertically and laterally as the jaws are lowered and the skull is raised. As the jaws are closed, the water is forced through the gills, which form a partial division between the two pumps. The space lateral to the gills and beneath the operculum forms the second pump. It is expanded and contracted by mediolateral movements of the operculum and dorsoventral movement of the area covered by the gular plates and branchiostegal rays.

The lateral expansion of the orobranchial chamber is produced by linkage between the hyomandibular, operculum, palatoquadrate, and lower jaw. In opening the mouth, the hyoid and hypaxial musculature pull the area of the jaw symphysis posteriorly. Posterior movement at the jaw articulation is restricted by the continuous covering of dermal bones of the cheek, operculum, and shoulder girdle, so that the back of the jaws and cheeks are thrust laterally. This lateral movement is facilitated by the mobility of the palatoquadrate and hyomandibular on the braincase.

In primitive palaeonisciforms, the oblique orientation of the hyomandibular limited the movement of the cheek to a short anterolateral arc in the horizontal plane. This movement was important in opening and closing the operculum and provided a small increase in the size of the oral cavity, but it was very limited compared with that of later actinopterygians.

Major changes in the jaw structure separate the higher actinopterygians from the Palaeonisciformes. These include reorientation of the hyomandibular to increase the amount of lateral movement, freeing of the maxilla from the cheek and palate, and a great increase in the size and complexity of the adductor jaw musculature.

DIVERSITY OF CHONDROSTEAN FISH

We include approximately 80 genera of primitive actinopterygian fish in the order Palaeonisciformes. They range from Lower Devonian to Lower Cretaceous. The suborder Palaeoniscoidea encompasses more than a score of families, including the earliest known actinopterygians as well as members of a conservative lineage extending into the Mesozoic (Figure 6-16). Most of the Palaeoniscoidea are small, but some reach 1 meter in length. The scales are thick and rhomboidal, the dorsal and anal fins are triangular, and the heterocercal caudal fin is deeply cleft. All fins are supported by numerous jointed dermal fin rays. The jaws are long and usually articulate behind the back of the braincase. The suborder evolved rapidly in the Carboniferous from the three or four genera that are adequately known in the Devonian. It retained a dominant position until the end of the Permian, when more specialized actinopterygians evolved.

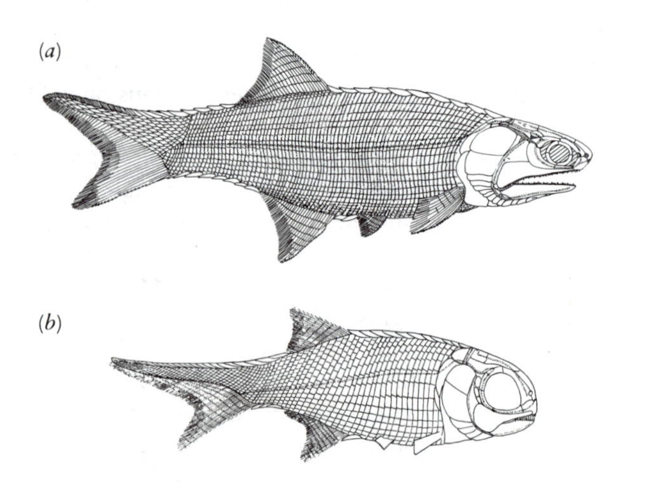

Figure 6-16. PALAEONISCOIDS. (*a*) *Mimia*, Upper Devonian. *From Gardiner, 1984.* (*b*) *Canobius*, Carboniferous, showing an upright jaw suspensorium. About natural size. *From Moy-Thomas and Dyne, 1938.*

Mimia and *Moythomasia,* included in the family Stegotrachelidae, characterize primitive actinopterygians. Several other families within the suborder Palaeoniscoidea possess a different scale pattern or fin structure but otherwise retain a similar body plan. In the Cheirolepidae (see Figure 6-8), the scales are extremely tiny, like those of acanthodians, and lack the peg-and-socket articulation common to other genera. Those of the Lower Carboniferous Cryphiolepididae are cycloid rather than rhomboid. In their contemporaries, the Carbovelidae, most of the scales are lost. In other families, the scales are conservative, but the fin structure is specialized. In the Holuriidae, the dorsal and anal fins become long based, rounded, and lose their fulcral scales. In contrast, the fulcral scales in the Styracopteridae are much emphasized.

We see greater changes from the primitive body form in the suborder Platysomoidei (Figure 6-17), in which the body is laterally compressed and oval or rhomboidal in lateral view. In modern fish, such a deep body is associated with life in quiet waters. Dorsal and anal fins may be extremely long based. One family, the Chirodontidae (Amphicentridae), has developed a crushing dentition.

The jaw suspension in *Canobius* (Figure 6-16*b*) from the Lower Carboniferous and *Aeduella* from the Lower Permian has become almost vertical, in contrast with its oblique orientation in primitive stegotrachelids. The jaw articulation lies in nearly the same transverse plane as the articulation between the hyomandibular and the braincase. This feature would be expected in the ancestors of the higher neopterygian fish. In other respects, these genera resemble their Paleozoic contemporaries.

Other chondrosteans are similar to palaeonisciforms in some primitive features but differ so greatly in other characters that they are placed in distinct orders (Figures 6-17 and 6-18). The Upper Permian *Dorypterus* is the only known representative of the order Dorypterida. This genus is deep bodied like the Platysomoidei, with comparably primitive fins, but the dermal ossification of the skull is reduced and most of the scales are lost, as are the teeth.

The Phanerorhynchida is known from a single Upper Carboniferous genus in which the body scales are greatly enlarged, the number of dermal fin rays is reduced, and their primitive jointed nature is lost. The rostrum is elongate.

The Upper Carboniferous order Haplolepidida is more numerous and diverse than the preceding orders, but the body form is more conservative. The jaw suspension is upright, and the number of branchiostegal rays is reduced. The dermal fin rays, although jointed, are also reduced in number.

The Tarrasiida (Figure 6-18*b*) are among the most highly modified of primitive actinopterygians but are rep-

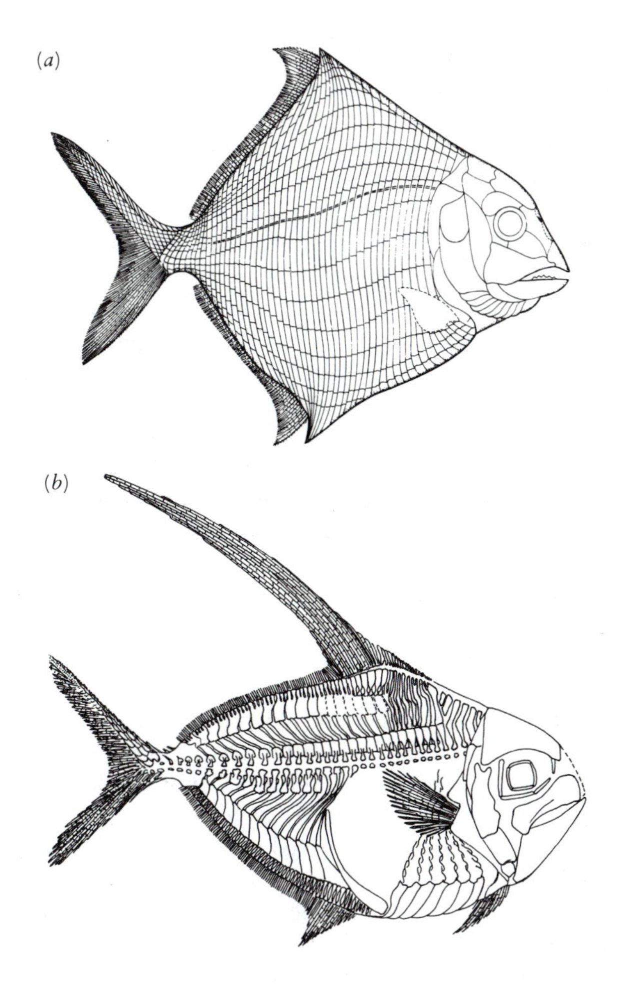

Figure 6-17. DEEP-BODIED CHONDROSTEAN FISH. (*a*) *Cheirodus* [*Amphicentrum*], a palaeonisciform of the suborder Platysomoidei from the Carboniferous, $\times \frac{1}{2}$. *From Smith-Woodward, 1891–1901.* (*b*) *Dorypterus* of the order Dorypterida, from the Upper Permian, $\times \frac{1}{2}$. *From Westoll, 1941.*

resented by only one or two genera from the Carboniferous. The caudal fin is symmetrical (diphycercal) and confluent with both dorsal and anal fins. The pelvic fin is lost and the pectoral fin resembles that of sarcopterygians. Scales are greatly reduced. Despite these radical changes in the postcranial skeleton, the skull pattern resembles that of typical palaeoniscoids.

The Saurichthyiformes, including *Saurichthys* and *Acidorhynchus* (Figure 6-18*c*) are common and widespread in the Triassic and Lower Jurassic. They are advanced in having a superficially symmetrical caudal fin and specialized in reducing the scales. Their elongated body and posteriorly placed dorsal fin (symmetrical with the anal) resemble the pattern of the modern pike, which suggests a similarly predaceous way of life. The upper and lower jaws are extended as a long, toothed beak.

We have not determined the specific affinities of the orders Dorypterida, Haplolepidida, Phanerorhynchida, Tarrasiida, and Saurichthyiformes. They are currently being studied by Schaeffer and Gardiner. These diverse and specialized chondrosteans are known from the late Paleozoic into the Jurassic. Two other orders that retain many primitive characteristics, the Polypteriformes and the Acipenseriformes, survive today.

The genera *Polypterus* and *Erpetoichthys* [*Calamoichthys*] constitute a phylogenetically isolated group of modern fish, alternately designated the Polypteriformes, Cladistia, or Brachiopterygii. They inhabit the streams and lakes of tropical Africa. Both are elongated; *Erpetoichthys* approaches the proportions of an eel and has lost the pelvic fins. Both are characterized by a long dorsal fin that is divided into a series of shorter elements, each bearing a sharp anterior spine (Figure 6-19). The caudal fin is nearly symmetrical. The histology of the thick, rhomboidal scales and the pattern of the bones of the skull roof resemble those of early palaeoniscoids. The "swim bladder" is in the form of paired, ventrally situated lungs that presumably represent the primitive pattern for all bony fish. *Polypterus* will drown if prevented from using atmospheric oxygen. The fleshy appearance of the pectoral fin is apparently a specialization of the group, although it was once thought to demonstrate affinities with the sarcopterygians. Recent authors, including Patterson (1982), argue that these genera are the most primitive of living actinopterygians. Unfortunately, their meager fossil record, which goes back to the Eocene, does not demonstrate their specific relationships among earlier chondrosteans. Bjerring (1985) argues that the Cladistia diverged from the base of the osteichthyan radiation, and lack specific affinities with the actinopterygians.

The Acipenseriformes are a more familiar order of living chondrosteans and include the Acipenseridae and

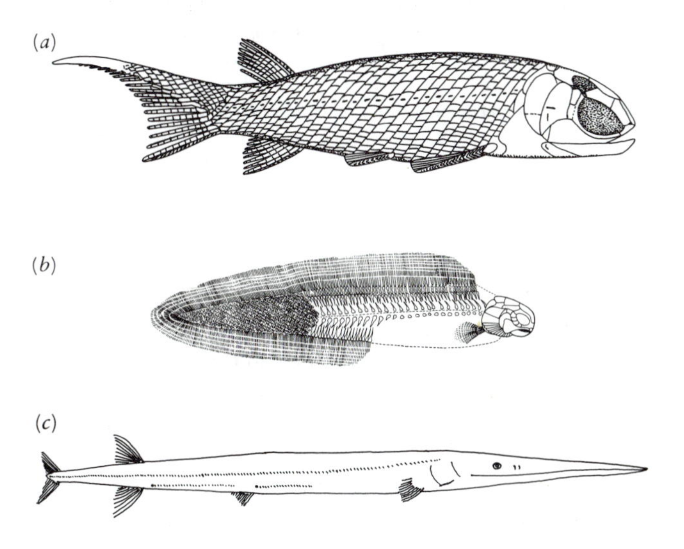

Figure 6-18. SPECIALIZED CHONDROSTEANS. (*a*) The Upper Carboniferous genus *Pyritocephalus*, order Haplolepidida. The jaw suspension is vertical and the number of dermal fin rays is reduced, about $\times 2$. (*b*) *Tarrasius*, order Tarrasida, from the Lower Carboniferous. Most of the scales are lost and the median fins are fused into a continuous structure, $\times \frac{2}{3}$. *From Moy-Thomas and Miles, 1971.* (*a* and *b*) *By permission from Chapman and Hall, Ltd.* (*c*) *Saurichthys*, a common and widespread genus from the Triassic that is characterized by a long, slender body with dorsal and anal fins that are located far posteriorly. Large individuals reach approximately 1 meter in length. *From Rieppel, 1985.*

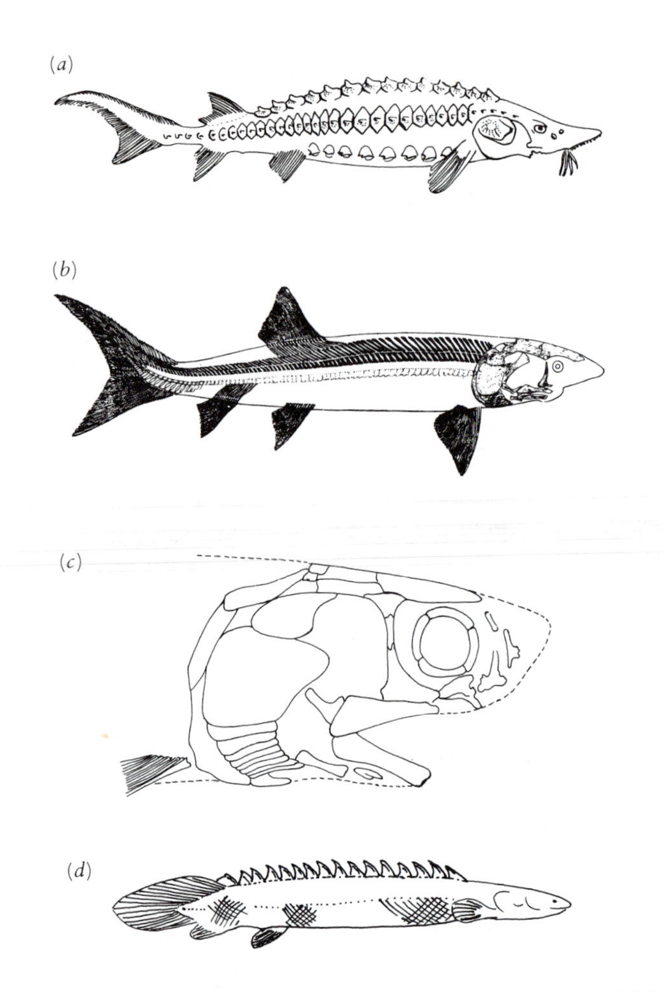

Figure 6-19. MODERN GROUPS OF BONY FISH BELONGING TO THE CHONDROSTEI. (*a*) The sturgeon *Acipenser,* the source of much commercial caviar. The scales are limited to five rows of large scutes. There is a long rostrum that overhangs very short jaws. The tail is fully heterocercal, with the notochord extending to the very tip. *From Bond, 1979.* (*b*) Skeleton of the Lower Jurassic genus *Chondrosteus,* the earliest recognized member of the Acipenseriformes. *From Smith-Woodward, 1891–1901.* (*c*) Skull of *Chondrosteus. From Watson, 1925.* (*d*) *Polypterus.*

the Polyodontidae. The sturgeons *Acipenser* and *Huso,* which produce most commercial caviar, and *Scaphirhynchus* are all predators. The paddlefish *Polyodon* from eastern North America is an open-water filter feeder using long gill rakers to trap plankton. *Psephurus* is a similar form from China.

Like the Paleozoic chondrosteans, the Acipenseriformes retain a markedly heterocercal tail, fulcral scales, and numerous jointed fin rays. The notochord is unrestricted, the clavicle is retained, and there is no myodome (a recess for the eye muscles that is highly developed in neopterygians). The maxilla is in contact with the palate in primitive fashion.

This order differs significantly in other respects from primitive bony fish. Scales are reduced in both living families. In the sturgeon, they are limited to five rows of large scutes (Figure 6-19); in the paddlefish, there are only a few scales at the base of the tail. The endocranium, which is heavily ossified in palaeoniscoids, is almost totally cartilaginous. The dermal bones of the snout and cheek are reduced, with the preopercular lateral line canal supported by a series of small ossicles. The skull roof shows a covering of small elements. Jollie (1980) suggests that their irregular pattern is associated with regression of the dermal skeleton. The rostrum is greatly elongated and the jaws are very short. One or two pairs of sensory barbels are just anterior to the mouth. The premaxillae are lost, and the palatoquadrates meet at the midline. The hyomandibular forms a movable jaw support resembling that of advanced sharks (see Figure 5-14).

It was once thought that the configuration of the mouth and the low degree of ossification of the endoskeleton allied the Acipenseriformes with sharks. When they were first recognized as being related to the bony fish, a low degree of ossification was judged to be a primitive feature of that group. Only with the discovery of the Paleozoic bony fish did we realize that the early actinopterygians were heavily ossified and that the loss of ossification in the Acipenseriformes was a specialization.

The Acipenseridae and Polyodontidae can both be traced back into the late Cretaceous, at which time they are already clearly distinct from one another. Because the Lower Jurassic genus *Chondrosteus* (Figure 6-19) shares a number of important derived features with the modern families (more specifically with the sturgeons)—particularly the configuration of the jaws—it is included in the same order. The Lower Triassic genus *Errolichthys* shows a similar reduction of the dermal bones, but the mouth is terminal and the palatoquadrates do not meet at the midline. We have not established the specific origin of acipenseriforms among the Paleozoic actinopterygians. However, Patterson (1982) places them closer to the neopterygians than he does the Cladistia.

The Lepisosteidae—the gars—have also been classified among the chondrosteans. This family includes two genera, *Lepisosteus* and the very similar *Atractosteus.* The gars are elongated predatory fish with posteriorly and symmetrically placed dorsal and anal fins that resemble the Mesozoic Saurichthyiformes. We attribute the similar body forms of these groups to a similar mode of swimming rather than being indicative of close relationship, since their detailed anatomy is very different.

Both groups have very long snouts, but the gars's jaw suspension is anterior to the orbit (Figure 6-20). In the adult, the upper jaw is formed by a series of tooth-bearing elements termed lachrymo-maxillaries. During development, another bone that initially bears teeth is present in the position normally occupied by the maxilla. This element later loses its teeth. The gars's postcranial skeleton retains such primitive characters as a heterocercal tail and thick ganoid scales with peg-and-socket articulation, but this group is somewhat advanced over the palaeoniscoids in the reduction of the number of dermal fin rays and the loss of their segmental structure; fulcral scales remain prominent.

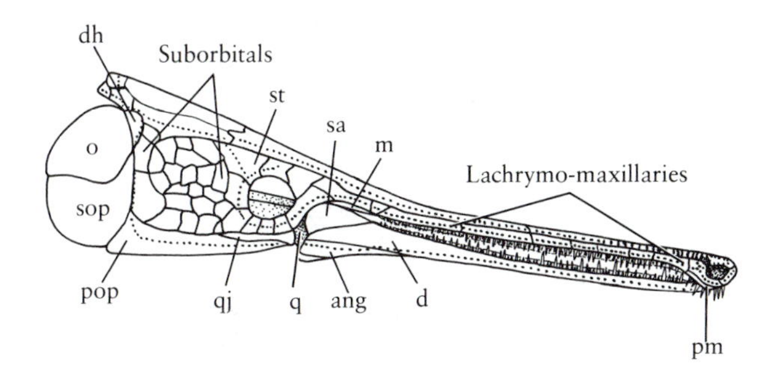

Figure 6-20. SKULL OF THE MODERN GENUS *LEPISOSTEUS*. This genus was classified as a holostean but lacks most of the characteristics of neopterygians. For abbreviations, see Figure 6-13. *From Patterson, 1973.*

From the mid-nineteenth to the mid-twentieth century, *Lepisosteus* was united with *Amia* in the Holostei, which was thought to represent an intermediate level of actinopterygian evolution between chondrosteans and teleosts. Patterson (1973) and Wiley (1976) showed that there are few significant derived characters that are common to *Amia* and gars. *Amia* shares many derived features with teleosts that are not evident in *Lepisosteus*. Wiley discovered a number of derived characters that *Lepisosteus* shares with neopterygians, including loss of the clavicle, the fact that the basipterygoid process is formed entirely by the parasphenoid, and several features of the visceral arch apparatus. On this basis, Wiley placed the gars among the neopterygians. He did not find any characters that united gars with the living chondrostean orders Polypteriformes or Acipenseriformes. However, he made no comparisons with the vast array of fossil chondrosteans.

Gars remain primitive in the absence of the interopercular and supramaxilla that form important elements in the specialized feeding mechanism of neopterygians. The posttemporal fossa and the myodome (discussed in the next section) are not developed. McAllister (1968) placed the gars among the chondrosteans. In contrast, Olsen (1984) believes that the structure of the jaw suspension indicates that gars are more closely related to teleosts than is *Amia*.

Both gar genera are represented by fossils from the Upper Cretaceous that are essentially modern in character. Fragmentary remains from the Lower Cretaceous provide no additional evidence regarding the ultimate ancestry of the group. Gars are now limited to North and Central America and Cuba, but they were present in Europe, Africa, and India during the Cretaceous and early Tertiary. Wiley (1976) suggests that they evolved within the primitive land mass of Pangea, which has progressively split apart since the late Paleozoic.

In addition to the conservative chondrosteans of the Paleozoic and Mesozoic and their living survivors, several late Paleozoic and early Mesozoic orders evolved toward the pattern of more advanced actinopterygians. These include the ancestors of neopterygians as well as a number of groups that became extinct without advancing far beyond the level of palaeoniscoids. Six orders—the Perleidiformes (Figure 6-21), Redfieldiiformes, Pholidopleuriformes, Peltopleuriformes, Luganoiiformes, and Cephaloxeniformes—may have a common ancestry distinct from that of other palaeonisciform derivatives. They are primarily Triassic in age and include both freshwater and marine genera. Most are small fusiform fish without strong specializations in fin or scale structure. The Cleithrolepididae (Figure 6-21*c*) has a deep body, reminiscent of the platysomids. Other genera with a crushing dentition were apparently bottom feeders.

The skulls of all members of this assemblage are advanced in having a vertical hyomandibular and a tall, narrow preopercular. However, the maxilla remains fixed to the cheek, a supramaxilla is absent, and there is no other evidence of modernization of the jaw structure. The fins are somewhat advanced, with reduction and fusion of dermal fin rays. Most genera have a **hemiheterocercal**

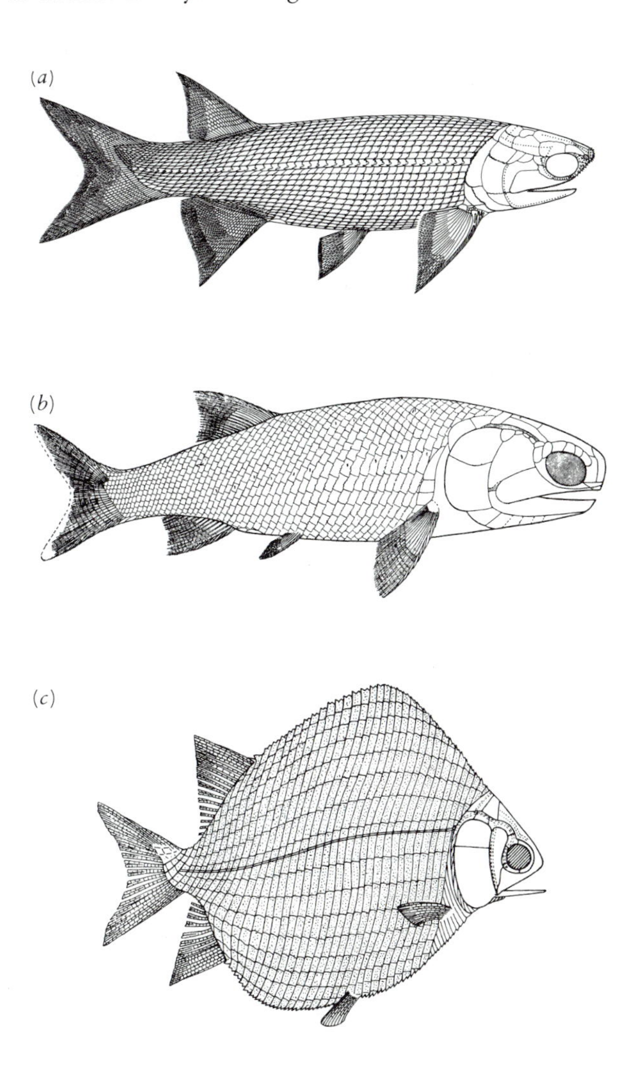

Figure 6-21. PROGRESSIVE TRIASSIC CHONDROSTEANS. (*a*) *Perleidus*, order Perleidiformes, $\times \frac{1}{2}$. *From Lehman, 1966.* (*b*) *Redfieldia* [*Catopterus*], order Redfieldiiformes, $\times \frac{2}{3}$. *From Schaeffer, 1978.* (*c*) The deep-bodied genus *Cleithrolepis*, $\times \frac{1}{2}$. *From Lehman, 1966.*

tail, that is, the notochordal axis does not extend to the tip, but the fin rays are still asymmetrically arranged. In some genera, the tail has become superficially symmetrical and the notochord is restricted to its base. The pholidopleuriforms are the most advanced in this respect. In contrast with the teleosts, the neural and haemal arches to which the dermal fin rays are attached assume almost symmetrical proportions.

NEOPTERYGIAN CHARACTERISTICS

Feeding apparatus

We may include all actinopterygians other than the chondrosteans in a single vast assemblage, the neopterygians, that are first recognized in the late Paleozoic.

A host of anatomical changes related to feeding, respiration, and locomotion, as well as modifications in the structure of the braincase, occurred between primitive palaeoniscoids and advanced neopterygians. Changes in the feeding apparatus that facilitated great expansion of the oral cavity laid the basis for the wide radiation of neopterygians within the Mesozoic. Readily observed differences that allow us to distinguish the neopterygians from their chondrostean ancestors include the separation of the maxilla from the cheek and the development of a distinct element, the interopercular, between the subopercular and the branchiostegal rays (Figure 6-22). To understand the significance of these changes, we must consider other modifications of the feeding apparatus.

In early palaeoniscoids, the jaw-closing musculature was confined to a small space surrounded by the dermal bones of the cheek and the palatoquadrate. Among advanced palaeoniscoids, this chamber opened dorsally and

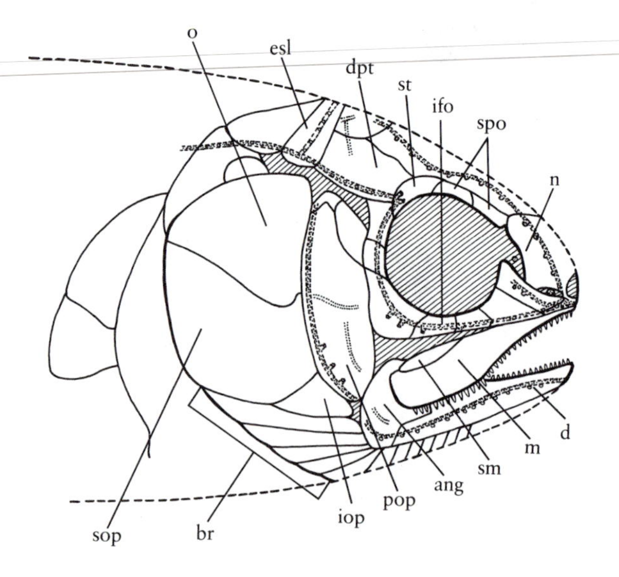

Figure 6-22. AN EARLY NEOPTERYGIAN, *PARASEMIONOTUS* FROM THE LOWER TRIASSIC. The maxilla is free from the cheek and both a supramaxilla and an interopercular bone are evident, ×2. For abbreviations see Figure 6-13. *From Lehman, 1952.*

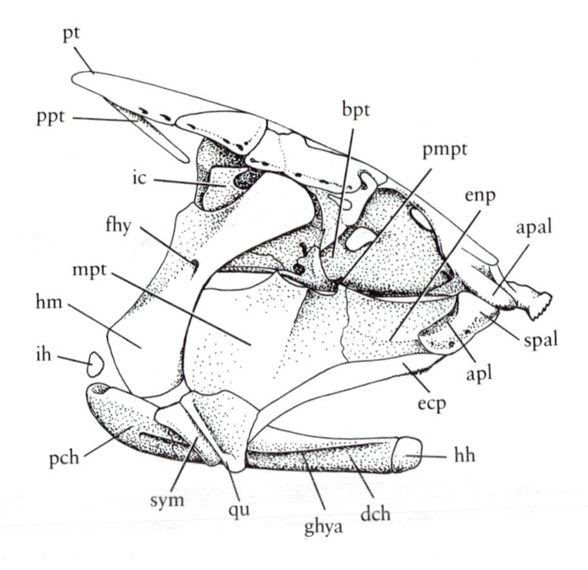

Figure 6-23. THE SKULL OF *WATSONULUS,* A PARASEMIONOTID. The dermal bones of the head are removed to show the braincase and jaw suspension. Abbreviations as follows: apal, articular facet for palatoquadrate complex; apl, autopalatine; bpt, basipterygoid process; dch, distal ceratohyal; ecp, ectopterygoid; enp, entopterygoid; fhy, foramen for hyomandibular nerve; ghya, groove for afferent hyoidean artery; hh, hypohyal bone; hm, hyomandibular; ic, intercalar; ih, interhyal; mpt, metapterygoid; pch, proximal ceratohyal; pmpt, process on metapterygoid for ligamentous attachment to basipterygoid process; ppt, posttemporal process; pt, posttemporal; qu, quadrate; spal, surface on autopalatine for articulation with ethmoidal region; sym, symplectic. *From Olsen, 1984.*

the adductor musculature extended upward to originate on the lateral wall of the braincase. The chamber also expanded laterally. The greater length of the muscle fibers allowed the mouth to open wider, and their greater volume resulted in a stronger bite (see Figure 6-14).

Primitive neopterygians have shorter jaws that articulate beneath the point at which the hyomandibular articulates with the back of the braincase. Lateral movement of the distal end of the hyomandibular becomes more extensive, allowing much greater mediolateral movement of the cheek and operculum. With the increased mechanical stress placed on the hyomandibular and its greater lateral extent, it was supplemented by a second endoskeletal bone, the symplectic, that was not originally present in chondrosteans (Figure 6-23). The symplectic forms a strong link between the hyomandibular and the quadrate portion of the palatoquadrate and later connects with the overlying preopercular. Together, these bones form a specialized jaw support, or **suspensorium.** As the scope of movement of the palatoquadrate increases in later neopterygians, its posterior articulation with the base of the braincase (the basicranial articulation) is lost, but the more anterior ethmoid articulation increases in importance and size. The quadratojugal, which is a separate element in chondrosteans, becomes fused to the quadrate in teleosts.

With the greater lateral movement of the jaw suspension in early neopterygians, the maxilla becomes separated from the bones of the cheek and the opercular series. A new joint develops anteriorly. A medial peg from the maxilla that fits into the palate behind the premaxilla allows the back of the maxilla to swing ventrally and anteriorly. This development is very important because it allows the shape of the mouth to be altered significantly.

We presume that primitive palaeoniscoids fed by engulfing their prey. Since mouth opening was accompanied by a considerable increase in the volume of the orobranchial chamber, the potential existed for sucking in prey with the water. However, the simple, scissorslike jaw action made this potential of little practical importance. With the greater mobility of the maxilla, the mouth cavity could assume more tubular proportions, and effective suction could be produced as it expanded. This mode of feeding is basic to many neopterygians.

An additional small bone, the supramaxilla, developed above the maxilla as it separated from the cheek. One supramaxilla is present in most primitive neopterygians, but a second develops in early teleosts.

Changes in the lower jaw also occur in the transition between chondrosteans and neopterygians. The articulation drops below the level of the tooth row, and a coronoid process develops so that a portion of the jaw musculature inserts at a considerable distance from the jaw articulation, which increases the force of the jaw musculature without reducing the gape of the jaw.

The presence of the interopercular bone between the subopercular and the branchiostegal rays in early neopterygians is associated with a new mechanism for depressing the lower jaw. This bone directs the force of a ligament from the opercular series to a point on the lower jaw below and behind the area of articulation with the quadrate. The opercular series is in turn elevated by a muscle originating on the braincase.

Braincase

Changes in the structure of the braincase accompany the evolution of the jaw structure (Patterson, 1975). In primitive actinopterygians, the adult braincase is ossified with few if any traces of subdivisions, except for the ventral and lateral fissures. In more advanced fish, we find two trends that appear contradictory. The conspicuous fissures close or are overgrown by surrounding bones, but the braincase as a whole is composed of many smaller units that are separated by sutures that persist in the adult (compare Figures 6-11 and 6-24). Closure of the fissures is related to the greater role played by the braincase as a unit of the feeding apparatus and because of the elaboration of the muscles controlling eye movements.

We find it more difficult to explain the separation of the adult braincase into many distinct areas of ossification. It results developmentally from a delay in the fusion of elements that were distinct in early growth stages. Specimens of juvenile palaeoniscoids show many separate elements that become indistinguishably fused in the adult. The delay in ossification among neopterygians may have been necessary because of the greater complexity of the units and the greater length of time necessary for their separate development.

Changes in the areas for the origin of the eye muscles have a significant effect on the evolution of the actinopterygian braincase. Jawed vertebrates have six muscles that control eye movements. Two of these, the superior and inferior oblique, arise from the anterior part of the eye socket. The four posterior muscles (superior, inferior, medial, and external rectus) arise from the back of the eye socket.

In primitive actinopterygians, muscle scars indicate that the probable areas of origin for the superior, inferior, and medial (internal) rectus muscles on the basisphenoid pillar—behind and above the basicranial articulation—were comparable to their position in the primitive living neopterygian *Amia*. There is no direct evidence for the origin of the external rectus. In later actinopterygians, we find distinct sockets lateral to the pituitary cavity. Comparable structures in *Amia* serve as areas for origin of the external rectus. In most advanced actinopterygians, these sockets form deep cavities, the myodomes. In modern teleosts, they house both external and internal rectus muscles.

The ventral cranial fissure migrates posteriorly as the myodome deepens, until it becomes confluent with the lateral fissure. The confluence of these fissures is a structural weakness that is compensated for among neopterygians by filling the lateral fissure with cartilage and/or by extensions of bone from the adjacent centers of ossification.

The penetration of epaxial musculature into the occipital surface is another important change in the structure of the braincase. Even in palaeoniscoids, these muscles are important in pulling back on the braincase and lifting the snout as the mouth is opened. The occipital surface is relatively flat in these fish. Most early neopterygians develop deep posterior pits, the posttemporal fossae, to accommodate these muscles. In early teleosts, they are very conspicuous and become confluent with a second pair of more primitive openings on the occipital surface, the fossae bridgei, through which pass branches of the facial, trigeminal, and glossopharyngeal nerves.

The evolution of the posttemporal fossae correlates with the elaboration of a medial bone, the supraoccipital, which is a particularly prominent component in the skull of most higher teleosts (see Figure 7-19).

Scales, vertebrae, and fins

The bony elements of the locomotor system, the scales, vertebral column, and fins, can be easily seen in fossils and enable us to appreciate the major patterns of actinopterygian evolution.

The most conspicuous superficial feature of Paleozoic and early Mesozoic actinopterygian fish is their heavy shiny scales. Due to the enamel-like surface of these scales, the term *ganoid fish* was long applied to these and other archaic Osteichthyes. Although these heavy, overlapping, interlocking scales certainly would have provided a large

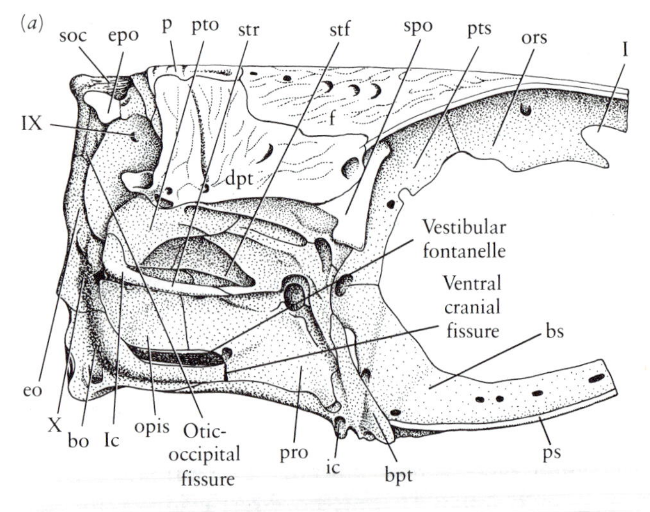

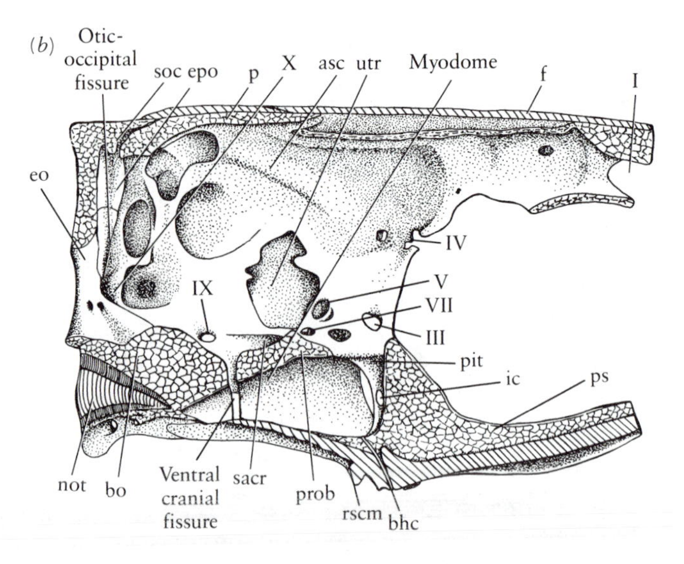

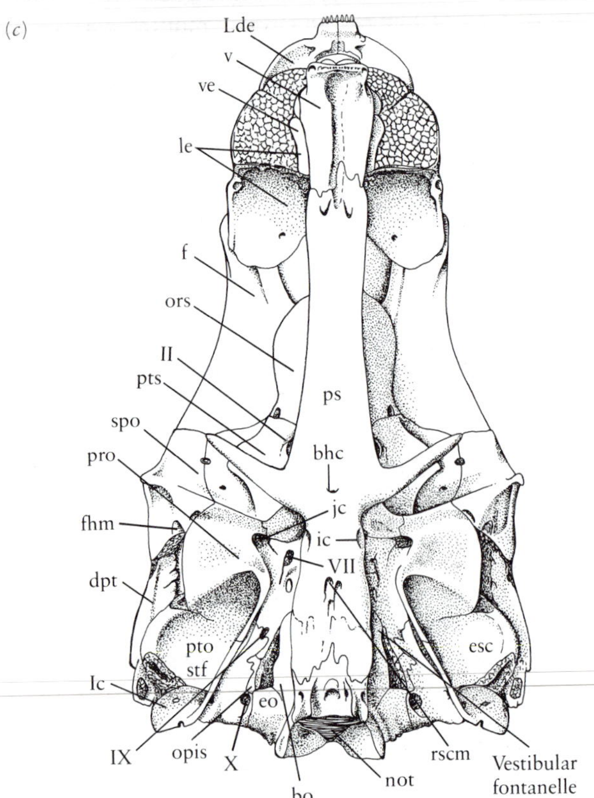

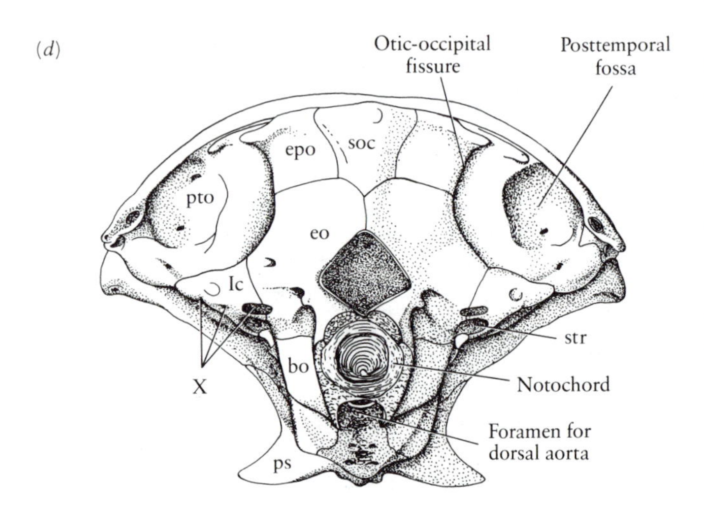

Figure 6-24. BRAINCASE OF THE PRIMITIVE TELEOST *PHOLIDOPHORUS.* (*a*) Lateral, (*b*) sagittal, (*c*) palatal, and (*d*) occipital views. We can see more individual centers of ossification here than in primitive actinopterygian fish (see Figure 6-11). Abbreviations as follows: asc, ridge over anterior semicircular canal; bhc, bucco-hypophysial canal; bo, basioccipital; bpt, basipterygoid process; bs, basisphenoid; dpt, dermopterotic; eo, exoccipital; epo, epioccipital; esc, ridge over external semicircular canal; f, frontal; fhm, hyomandibular facet; ic, foramina for internal carotid; Ic, intercalar; jc, jugular canal; L de, lateral dermethmoid; le, lateral ethmoid; not, notochordal calcification in notochordal pit; opis, opisthotic; ors, orbitosphenoid; p, parietal; pit, pituitary fossa; pro, prootic; prob, prootic bridge; ps, parasphenoid; pto, pterotic; pts, pterosphenoid; rscm, recess on parasphenoid housing origin of subcephalic muscles; sacr, saccular recess; soc, supraoccipital; spo, sphenotic; stf, subtemporal fossa; str, prootic or intercalar portion of strut across subtemporal fossa; utr, utricular recess; v, vomer; ve, ventral ethmoid; I–X, foramina of cranial nerves. *From Patterson, 1975.*

measure of protection, they must have significantly restricted mobility and would have added considerable weight to the fish.

Since the Devonian, the scales have been gradually reduced in most lines, although the living polypteriforms and gars may have as thick a covering as primitive palaeoniscoids. The central (dentine and vascular) layer of the scales tends to be lost first. In the ancestors of teleosts, the superficial layer of ganoine is also reduced and finally lost. However, it still persists in the tail region in early teleosts. There is a concomitant tendency for the shape of the scales to be modified from a diamond-shaped or rhombic configuration to a circular or cycloid scale. These changes occur at different rates in various lineages and some groups, even among the Palaeoniscoidea, totally lost their scales.

In the evolution between chondrosteans and later neopterygians, the reduction of scales is accompanied by gradual ossification of the vertebral centra. These changes may be related to the mode of swimming. In general, the palaeonisciforms were relatively long bodied; this presumably facilitated an anguilliform mode of swimming, with broad and relatively slow undulations passing from the head to the tail. The heavy scalation probably precluded rapid or sharp-angled bending of the body. Consequently, the notochord did not have to be reinforced to resist strong muscular contraction.

With reduction of the scales to thin, flexible elements, the body could be bent more rapidly and powerfully. The primitive notochord was presumably not strong enough to resist the greater tensile and compressive forces acting on the column. Some actinopterygians develop cylindrical vertebral centra as early as the Lower Permian, but in general they are present only in advanced neopterygians. Centra are only just beginning to form among early teleosts of the Triassic. Ossification of the centra is particularly important in establishing the form of the skeletal supports of the tail. These supports provide a key to understanding the evolution of teleosts and will be discussed with the caudal fin.

The fins in palaeoniscoids are covered by scales comparable to those on the trunk but arranged in long, jointed rows termed *dermal fin rays*. They strengthen and protect the fins but limit their mobility. In later actinopterygians, the number of fin rays is reduced and those that remain are solidified by fusion of the scales. In teleosts, the number of dermal fin rays of the dorsal and anal fins corresponds with the number of endochondral supports with which they articulate basally.

In contrast with the rigid structure of the primitive fins, the reduced number of rays in teleosts allows the whole fin to be raised or lowered, and portions of it can be swung from side to side. Many of these changes occurred within the early neopterygians. Further advances in fin structure occur within the teleosts themselves and will be discussed with the diversification of that group.

The caudal fin

The shape of the caudal fin is one of the most obvious differences in the general appearance of chondrosteans and teleosts. It is completely heterocercal in palaeonisciforms with the notochord extending to the tip of its dorsal margin. In teleosts, it is superficially symmetrical, or **homocercal,** and the vertebrae extend only as far as the base (Figure 6-25).

The asymmetrical shape of the tail in primitive palaeoniscoids may have been necessary to produce a strongly asymmetrical thrust to counteract the weight of a body that was covered by heavy bony scales. We may relate the achievement of a symmetrical tail by early teleosts to a reduction of body weight resulting from thinning of the scales and the greater effect of the swim bladder in achieving neutral buoyancy.

Evolution first affected the more superficial aspects of the tail, so that the surface that was in contact with the water became symmetrical relatively rapidly. It took much longer for the endoskeletal supports to approach a symmetrical appearance. Structurally, the elements supporting the ventral and dorsal lobes of the tail remain different, even in the most advanced teleosts, as a result of a disparity between the epichordal and the hypochordal elements of the tail that are inherited from palaeoniscoids.

In palaeoniscoids, the notochord is supported dorsally and ventrally by neural and haemal arches, but caudal lepidotrichia are entirely confined to the hypochordal lobe. In several lines of advanced chondrosteans, the notochordal axis is shortened and the dermal fin rays extend to the dorsal margin of the caudal fin.

In *Amia,* a living representative of the primitive neopterygian level of evolution, the caudal fin is formed primarily by redistributing the fin rays of the hypaxial lobe while retaining a relatively long, upturned axial skeleton (Figure 6-25*b*). The axial skeleton is reduced in the ancestors of the teleosts, and both epaxial and hypaxial elements contribute to the caudal fin.

In considering the origin of the teleost pattern, we may start with the parasemionotiforms of the Lower Triassic (Figure 6-25*a*). They may not be phylogenetic ancestors of the teleosts, but many of their skeletal features are almost ideal structural antecedents. There are no central ossifications. The notochordal axis is upturned and extends nearly to the end of the tail. Dorsally, the base of the tail is supported by nine skeletal elements, the **epurals,** articulating with the last neural arches. Ventrally, the tail is supported by haemal arches, the last five of which lie medial to the haemal arteries and veins. These arches are termed **hypurals** and are the primary supports for the dermal fin rays.

The notochordal region in pholidophorids (Figures 6-25*c, d*), the most primitive group included in the teleosts, still shows little ossification, except for small elements between the neural and haemal arches, the intercalaries. The arches themselves have grown around the notochord and develop as calcifications in its sheath. There is a clear division ventrally between the anterior haemal arches and the hypural elements of the tail. The first two hypurals form the primary support for all the fin rays of the ventral lobe of the tail. The dorsal lobe is supported by nine or ten additional hypurals, termed **dorsal hypurals.** They support seven to eight principal caudal fin rays. Dorsally, the caudal or ural neural arches, the **uroneurals,** show the initiation of specializations characteristic of early teleosts. They are paired, rather than medial, as are the anterior neural arches. All but the most anterior uroneurals are elongated, so that they overlap one another and the more posterior neural arches. There are seven to eight pairs of uroneurals above which are four to five epurals. There are 22 to 24 principal caudal fin rays.

The general appearance of the caudal supports at the next level of teleost evolution is very distinct, but it can be interpreted as having arisen directly from pholidophoroids. Although generally somewhat smaller fish than the pholidophorids, the leptolepids (Figure 6-25*e*) have solidly ossified vertebral centra in the form of close-fitting cylinders. The articulation of the first two hypurals to a single centrum is a simple, diagnostic feature of teleosts that is evident at this evolutionary level. This centrum

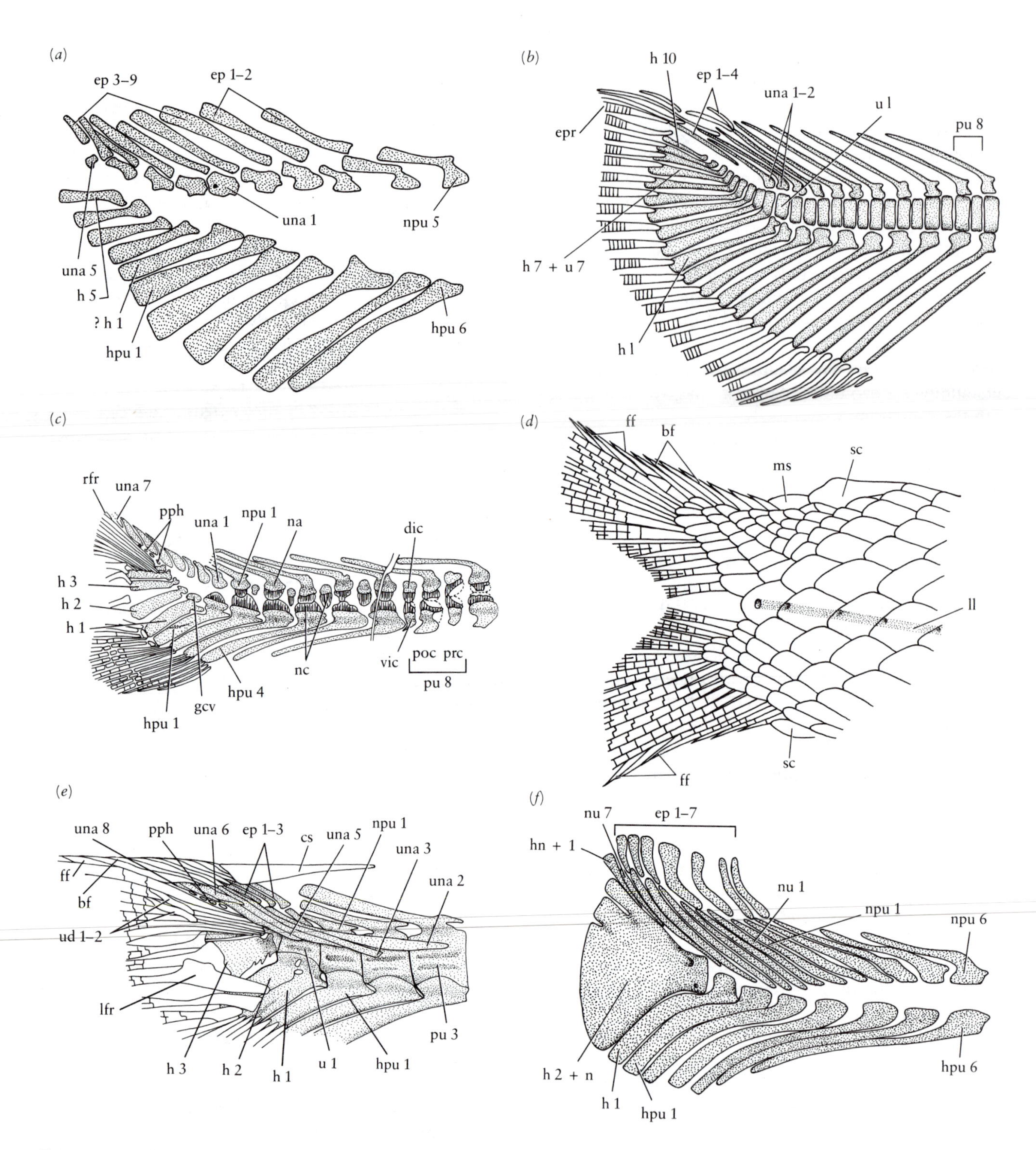

Figure 6-25. CAUDAL FIN SUPPORTS OF NEOPTERYGIAN FISH. (*a*) Undetermined parasemionotid from the early Triassic of Greenland showing primitive neopterygian condition, about ×5. *From Patterson, 1973.* (*b*) *Amia.* In contrast with teleosts, the centra are perichordally ossified and diplospondylous (two centra to a segment), approximately ×1. *From Patterson, 1973.* (*c*) *Pholidophorus bechei,* vertebral centra are just beginning to form, ×2⅓. *From Patterson, 1968.* (*d*) *Pholidophorus bechei,* external surface of caudal fin showing the symmetrical arrangement of scales and dermal fin rays. Fulcral scales are still retained, ×⅔. *From Patterson, 1968.* (*e*) *Tharsis* [*Leptolepis*] *dubius,* a primitive teleost, representative of the leptolepid level of development. Note two hypurals articulate with one ural centrum, ×3½. *From Patterson, 1968.* (*f*) *Pachycormus,* an Upper Jurassic neopterygian of uncertain taxonomic affinities, ×3. *From Patterson, 1968.* Abbreviations as follows: bf, epaxial basal fulcra; cs, caudal scute; dic, dorsal intercalary; ep, epurals; epr, lowermost epaxial fin ray; ff, fringing fulcra; gcv, groove for caudal vein and artery on hypural; h, hypurals; h 2 + n, hypural plate, comprising second and an unknown number of succeeding hypurals; h n + 1, free hypural; hpu, preural haemal arches and spines; ll, course of lateral line; 1fr, expanded base of lowermost fin ray of upper caudal lobe; ms, median dorsal scale; na, neural arch; nc, notochordal calcifications (hemicentra); npu, preural neural arch; nu, ural neural arch; poc, postcentrum of preural vertebra; pph, postero-dorsal processes on heads of upper hypurals; prc, precentrum of preural vertebra; pu, preural centrum; rfr, reduced uppermost hypaxial fin ray; sc, upper and lower caudal scutes; u, ural centrum; ud, urodermals; una, ural neural arches; vic, ventral intercalary; vra, ventral caudal radials.

probably represents the fusion of elements from two segments. There is only one additional ural centrum. Five dorsal hypurals are present. The uroneurals are more elongated than in early pholidophoroids, and their origin as neural arches is only barely recognizable. The number of epurals is reduced to three. As in the basal members of all major teleost groups, there are 19 principal caudal fin rays. The caudal structure of leptolepids forms an ideal ancestral pattern for the derivation of advanced teleosts.

We will discuss additional changes in the structure of the endochondral supports of the caudal fin in relation to the differentiation of the modern teleost groups of the late Mesozoic in Chapter 7.

PRIMITIVE NEOPTERYGIANS

Many families of Mesozoic fish show one or more of the advanced features that characterize neopterygians. Among these families are the ancestors of the teleosts. However, most Mesozoic groups are not closely related to the lineage that gave rise to the dominant late Mesozoic and Cenozoic actinopterygians. Several of these are very distinct when they first appear in the fossil record, and we have not yet determined their relationships. In the past, this assemblage of Mesozoic actinopterygians was placed in a distinct infraclass, the Holostei. Until we have a better understanding of their affinities, they will be considered simply as primitive neopterygians.

Semionotidae

The earliest known neopterygian, the Upper Permian genus *Acentrophorus* (Figure 6-26), is distinguished from chondrosteans by the presence of an interopercular and the separation of the maxilla from the opercular series. However, no supramaxilla is yet developed above it. As in advanced chondrosteans, the lower jaw is short and the jaw suspension is nearly vertical. The notochordal axis of the tail is shortened, so that the caudal fin is not so strikingly heterocercal. The number of dermal fin rays is reduced and their joints are lost.

We have not yet established the specific ancestry of *Acentrophorus*. It is older than the advanced chondrosteans of the Triassic and presumably arose directly from among the palaeoniscoids. Carboniferous genera such as *Canobius* (see Figure 6-16) and *Aeduella* have an upright jaw suspension like that of the early neopterygians but show few other advanced features in either the skull or the postcranial skeleton.

Acentrophorus is classified among the Semionotidae, which includes some 20 other genera from the Mesozoic (Figure 6-26). *Semionotus* from the Triassic and *Lepidoteus* ranging from the Upper Triassic into the Lower Cretaceous are typical representatives. *Dapedium*'s deep-bodied form is analogous to genera seen among the earlier actinopterygians. The short mouth is armed with peglike teeth, indicating that they feed on some type of hard in-

(*a*)

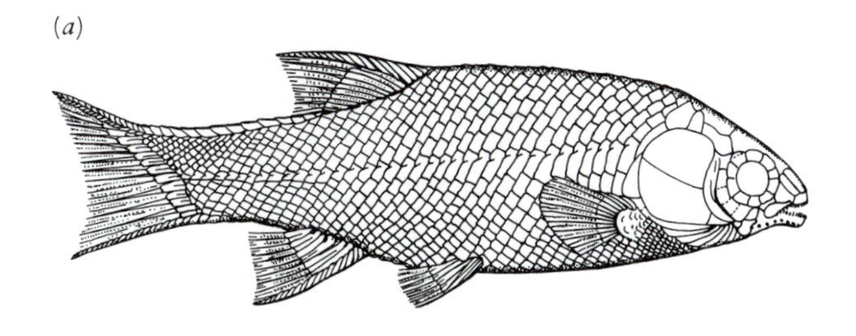

(*b*)

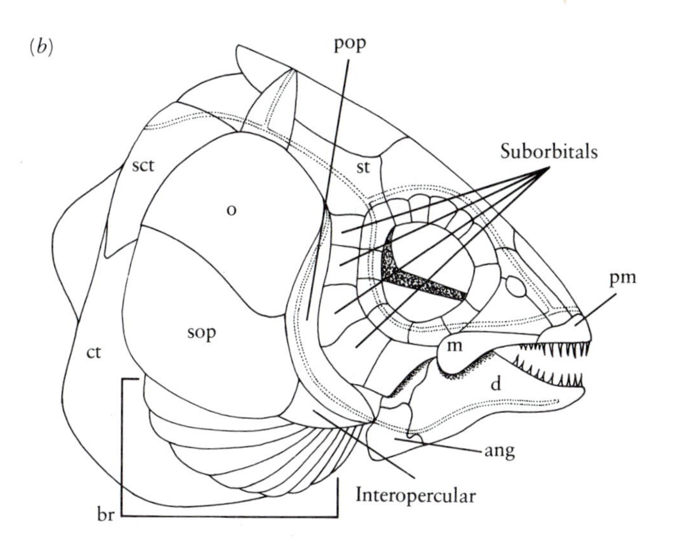

(*c*)

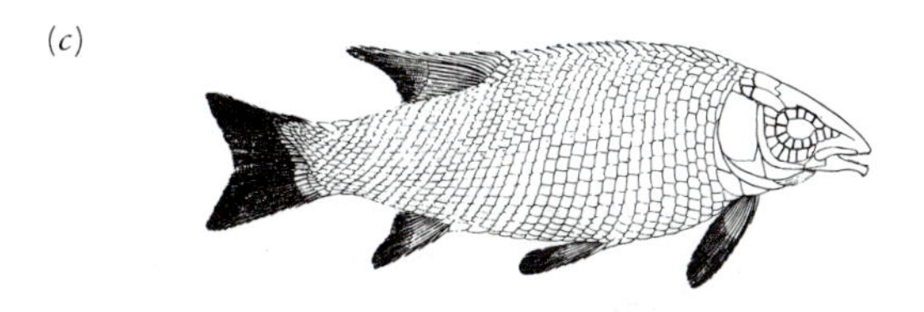

(*d*)

Figure 6-26. SEMIONOTID NEOPTERYGIANS. (*a*) Restoration of *Acentrophorus*, the oldest known neopterygian from the Upper Permian, approximately natural size. *From Moy-Thomas and Miles, 1971.* (*b*) Skull of *Acentrophorus*, × 4. *From Moy-Thomas and Miles, 1971.* (*a* and *b*) *By permission from Chapman and Hall, Ltd.* (*c*) *Lepidotes minor* from the Jurassic, approximately 30 centimeters long. *From Smith-Woodward, 1891–1901.* (*d*) *Dapedium*, a deep-bodied genus, about 35 centimeters long. *Photograph courtesy of Dr. Wild.*

vertebrate prey. The scales in this group retain a heavy layer of ganoine.

The quadratojugal remains distinct and the vomers are paired among early semionotids. In later genera in which the braincase is well known, the lateral fissure is closed, whereas it is still open in early teleosts. The later genera have also developed a large myodome and a post-temporal fossa. Despite their early appearance, we do not think that semionotids were ancestral to teleosts. They may not in fact represent a natural group (Patterson, 1973).

Halecomorphi: Amiidae, Caturidae, and Parasemionotidae

The term halecomorph is applied to the largest natural group of advanced actinopterygians not included among the teleosts. This group includes the living *Amia*, a score of closely related Jurassic and Cretaceous genera, and the Triassic parasemionotids (Figure 6-27). They are characterized by the ventral elaboration of the symplectic that articulates with the lower jaw along with the quadrate. This feature is not seen in teleosts.

In typical neopterygian fashion, the amiids are advanced in having a mobile maxilla, a supramaxilla, and a well-developed interopercular. The posttemporal fossa is confluent with the fossa bridgei. They are more primitive than the early teleosts in that the myodome does not extend into the basioccipital. In *Amia*, it accommodates only the external rectus; the other eye muscles originate on the basisphenoid. There is only a single supramaxilla, no supraoccipital ossification, and the caudal neural arches are not specialized like the uroneurals in pholidophorids.

The genus *Amia* is known as early as the Upper Cretaceous. Nine other members of the family Amiidae are known from the Upper Jurassic into the Cretaceous. The caturids from the Upper Triassic to the Cretaceous precede the amiids in both time and degree of anatomical specialization. They include *Heterolepidotus* and seven other genera. In turn, the Caturidae may be derived from the Parasemionotidae of the Lower Triassic. This latter group shows the initiation of all the important neopterygian characters, and at least one genus shows the peculiarities of the symplectic-lower jaw articulation characteristic of caturids and amiids. Among most halecomorphs, the lateral cranial fissure is closed either by cartilage, endochondral bone, or a membranous overgrowth. However, in the parasemionotid *Watsonulus*, it is still open in the manner of chondrosteans (Olsen, 1984).

We know about a dozen genera of parasemionotids, primarily from eastern Greenland and Madagascar, where the marine Lower Triassic is particularly well exposed. These genera show considerable variations in the anatomy of the cheek region, suggesting that they represent a grade of development, rather than a well-defined taxonomic group. The fossa bridgei and posttemporal fossa are still distinct. Parasemionotids are primitive relative to other members of the Neopterygii in this and most other fea-

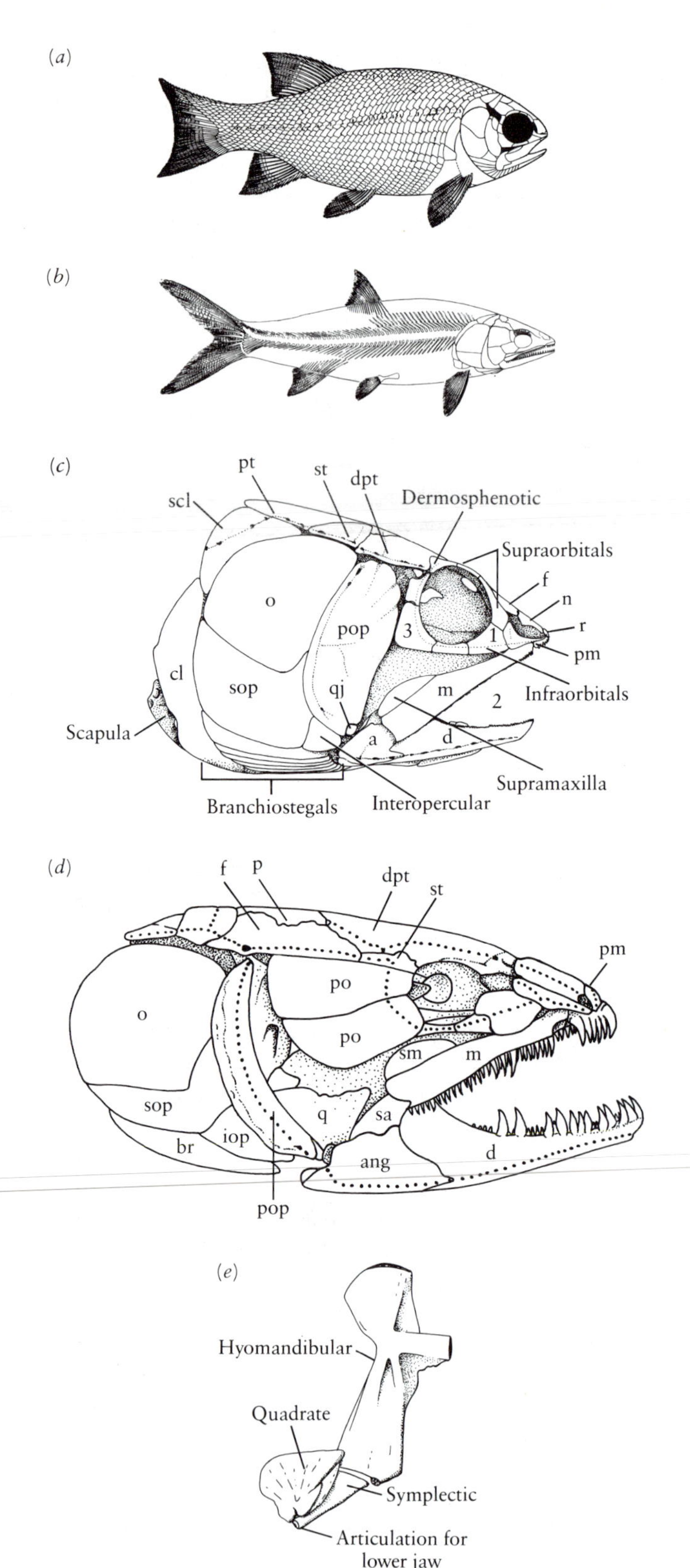

Figure 6-27. PROGRESSIVE CHANGES AMONG HALECOMORPH FISHES. (*a*) The Lower Triassic parasemionotid *Parasemionotus*. *After Lehman, 1966.* (*b*) *Caturus* (Caturiidae) from the Jurassic. *From Smith-Woodward, 1891–1901.* (*c*) Skull of the Lower Triassic parasemionotid *Watsonulus*. Abbreviations as in Figure 6-13. *From Olsen, 1984.* (*d*) The modern genus *Amia*. *From Patterson, 1973.* (*e*) Relationships of the symplectic and quadrate, both of which articulate with the lower jaw in halecomorphs. *From Patterson, 1973.*

tures of the skeleton, but unquestionably advanced above the chondrostean level. In most skeletal features, particularly the modification of the cheek region related to neopterygian feeding patterns, parasemionotids are nearly ideal ancestors for the pholidophorids. Patterson (1973) argues that extension of the symplectic to the articular is a derived feature that excludes parasemionotids from the ancestry of the teleosts. In contrast, Olsen (1984) suggests that this character may have been a primitive feature of the ancestors of teleosts.

Pycnodonts

The pycnodonts (Figure 6-28) are a diverse order that includes at least six families. All known members have many specialized features. They are characterized by a deep, narrow body; the dorsal and anal fins are long and the caudal fin homocercal. Reduced scales are a peculiar feature of many pycnodonts. Only a few anterior oblique rows of scale bars (the keels of chondrostean peg-and-socket scales) persisted in advanced forms such as *Pycnodus*.

The snout is elongated and terminates in a small mouth bearing peglike or incisiform premaxillary and dentary teeth, with a pavement of crushing teeth on the vomer and inner surface of the lower jaw. If we compare the pycnodonts with living teleosts that have a similar body shape and dentition, we can see that this group probably lived in quiet reef waters and fed on molluscs, echinoids, and other hard-bodied prey.

The pycnodonts appeared first in the late Triassic, flourished in the Jurassic and Cretaceous, and apparently died out in the Eocene. Their ancestry is difficult to establish. The general body shape resembles that of the palaeonisciform *Cheirodus* (see Figure 6-17*a*) and the advanced chondrostean *Cleithrolepis* (see Figure 6-21*c*). We can interpret these similarities as the result of convergence toward a similar way of life rather than an indication of close relationship.

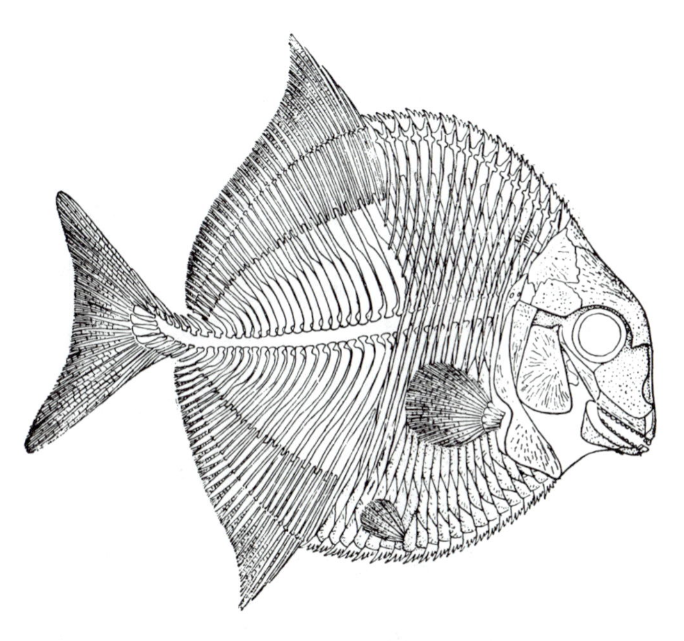

Figure 6-28. *PROSCINETES [MICRODON]*, A JURASSIC PYCNODONT, $\times \frac{1}{2}$. Most of the scales are lost, leaving only a lattice work formed from the keels of the archaic peg-and-socket articulation. *From Smith-Woodward, 1916.*

(*a*)

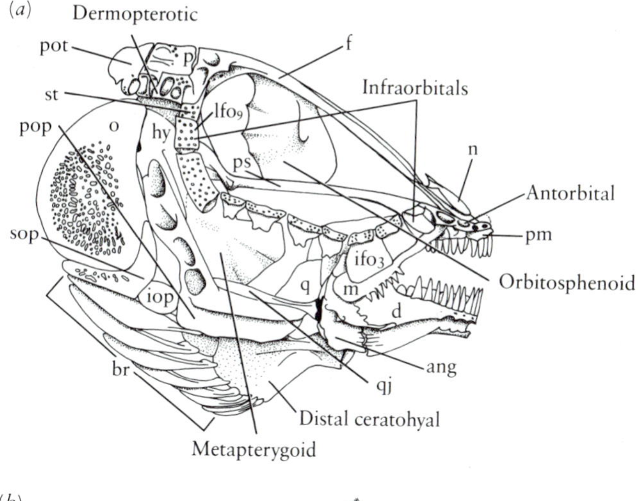

(*b*)

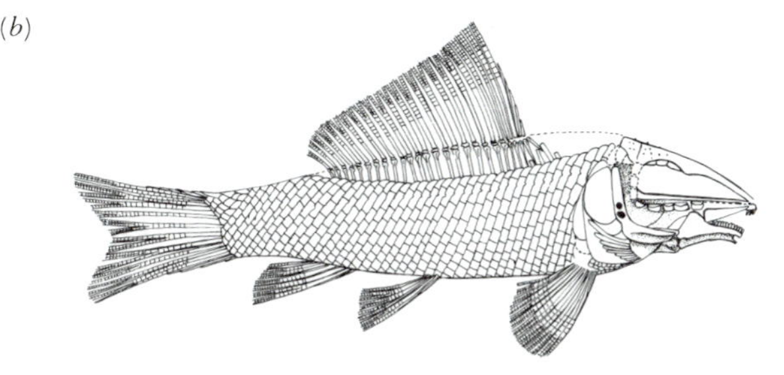

Figure 6-29. REPRESENTATIVES OF THE JURASSIC NEOPTERYGIAN FAMILY MACROSEMIIDAE. (*a*) Skull of *Macrosemius*, $\times 1\frac{1}{4}$. (*b*) Skeleton of *Legnonotus*, about natural size. *From Bartram, 1977.*

The mobility of the maxilla and the presence of a myodome suggest that pycnodonts evolved from the base of the neopterygian assemblage. Unlike other early neopterygian orders, they lack both the supramaxilla and the interopercular. We assume that the absence of these bones is a specialization rather than a primitive feature of the group. The symplectic articulates with the lower jaw as is the case with amiids and caturids, which indicates at least a distant relationship to the halecomorphs (Patterson, 1973).

Macrosemiidae

The macrosemiids, which Bartram (1977) recently reviewed, are a clearly monophyletic family that include six genera ranging from the Upper Triassic into the Lower Cretaceous. They are immediately differentiated from all other neopterygians by the presence of seven peculiar scroll-shaped infraorbital bones and two others with a tubular configuration behind the eye (Figure 6-29). The interopercular is small and remote from the mandible. Some genera have a very long dorsal fin and may lose the scales on either side of the dorsal midline. There is no supramaxilla, and the quadratojugal is retained as a free element. The symplectic lacks the specialization of the halecomorphs. We do not know the condition of the myodome and the posttemporal fossa. No genera are known that link macrosemiids with other Mesozoic actinopterygians and they share no important specialized features with any group. They may have arisen from a very primitive level of neopterygian evolution.

Pachycormidae and Aspidorhynchidae

The Pachycormidae and Aspidorhynchidae are only known from specialized Jurassic and Cretaceous genera whose relationships to more primitive actinopterygians is not obvious. The Pachycormidae (Figure 6-30) include 11 genera that are characterized by elongated scythe like pectoral fins and reduction or loss of the pelvic fins. All have an elongate bony rostrum; in the Cretaceous genus *Protosphyraena*, it resembles that of the swordfish. The caudal fins have some teleost characteristics. They possess structures that can be interpreted as uroneurals that are otherwise present only in pholidophorids and their derivatives. The first hypural is separate, while the second hypural is fused with an undetermined number of more posterior hypurals into a broad triangular plate (see Figure 6-25*f*). This pattern is quite different from the pholidophorid's but can be derived from it. There is a small, movable lateral premaxilla, as in teleosts, but only a single supramaxilla. The vomers are paired. The symplectic does not reach the articular, and we have no other reason to associate this family with the halecomorphs, as has been a common practice.

Aspidorhynchus (Figure 6-31) and *Belonostomus* of the Middle Jurassic to Upper Cretaceous are primitive neopterygians that resemble gars in both the retention of thick ganoid scales and in body form that relates to fast swimming and predaceous feeding habits. However, their distinctive cranial structures make a close relationship improbable. The lower jaws are moderately elongated, and there is a long, typically toothless rostrum. The skeleton of the caudal fin has rudimentary uroneurals comparable with primitive teleosts. More precise assignment awaits study of the braincase.

Most of the primitive neopterygian groups were extinct before the end of the Mesozoic. A few pycnodonts survived into the Eocene, but only a single genus, *Amia*, is alive today.

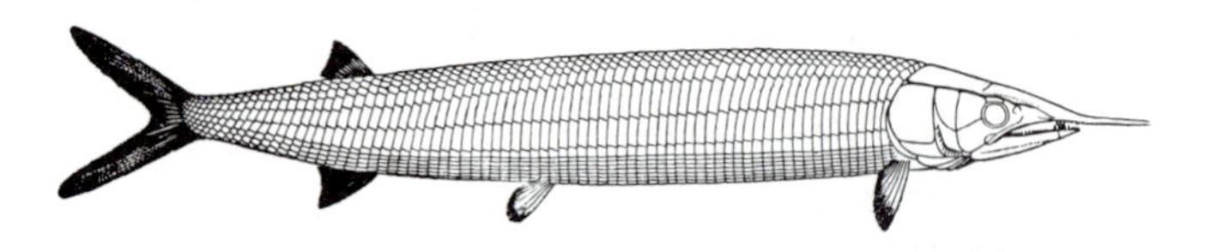

Figure 6-31. *ASPIDORHYNCHUS*. The predaceous Jurassic neopterygian, possibly closely related to the teleosts, is about 60 centimeters long. *From Smith-Woodward, 1891–1901.*

SUMMARY

Acanthodians are primitive jawed fish that are first known in the Lower Silurian and persist into the Lower Permian. The structure of the jaws resembles the pattern of both Chondrichthyes and Osteichthyes, but they lack a regular pattern of tooth replacement and may

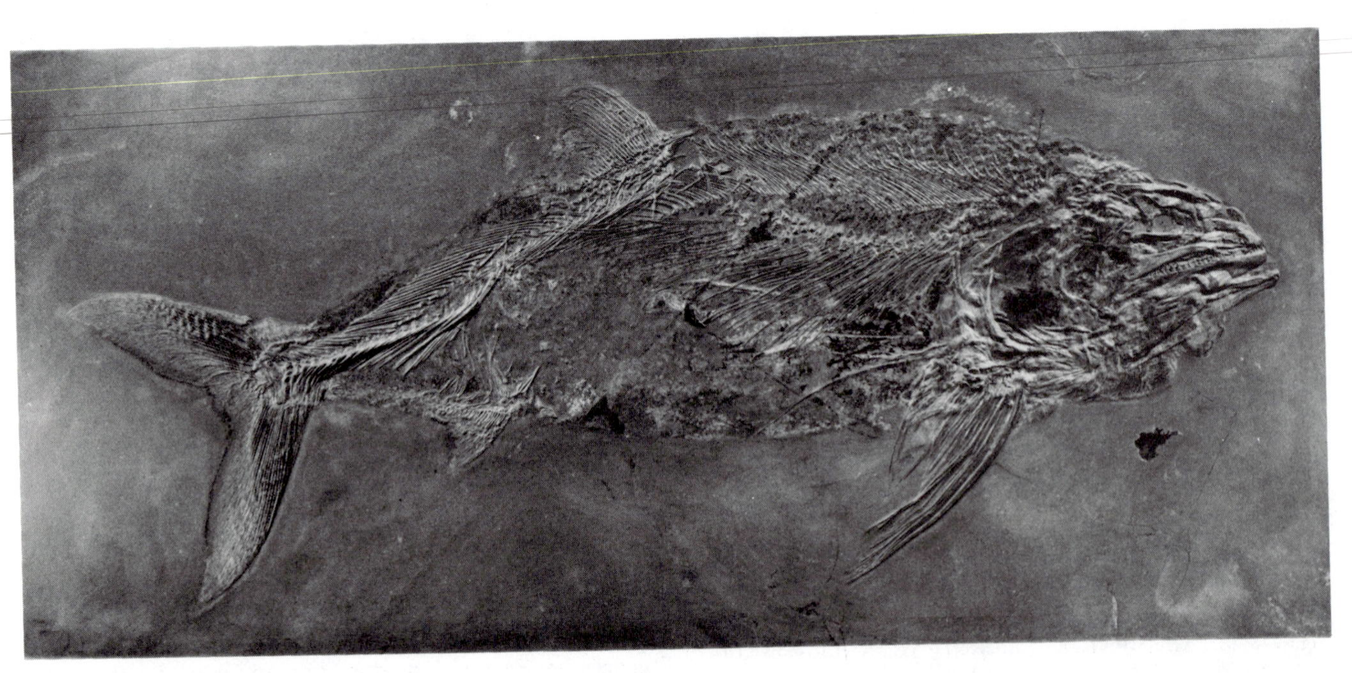

Figure 6-30. THE PACHYCORMID *PACHYCORMUS*, THOUGHT TO BE CLOSELY RELATED TO THE TELEOSTS, $\times \frac{1}{20}$. *Photograph courtesy of Dr. Wild.*

not be closely related to either of the major groups of jawed fish. Acanthodians are characterized by large spines supporting the dorsal, anal, and paired fins. Primitively they had additional paired spines between the girdles. We recognize three orders: Climatiida, Ischnacanthida, and Acanthodida.

Ostseichthyes are distinguished from Chondrichthyes by the ossification of the skeleton and the presence of a swim bladder. The bony exoskeleton of the head and shoulder girdle contribute to an effective integration of feeding and respiration. Isolated scales of osteichthyan fish are first known from the end of the Silurian. Two major groups, the actinopterygians and the sarcopterygians, have dominated marine and fresh waters from the beginning of the Devonian. Actinopterygians are characterized by multiple parallel endochondral fin supports controlled by muscles within the body wall and the absence of an epichordal lobe of the caudal fin.

Actinopterygians are divided into two large groups, the chondrosteans, which radiated primarily in the Devonian and Carboniferous, and the neopterygians, which emerged at the end of the Paleozoic, radiated extensively in the Mesozoic and early Cenozoic, and constitute the vast majority of living fish.

The Palaeonisciformes represent the central stock of Paleozoic chondrosteans and continued as a dominant group into the early Mesozoic. They exhibit heavy ganoid scales, a strongly heterocercal tail, and jointed dermal fin rays. The maxilla is long and closely integrated with the cheek; the vertebral centra are not ossified. Two groups of chondrosteans, the Polypteriformes and the Acipenseriformes, have survived to the present. We have not established the specific ancestry of these orders. The neopterygians also evolved from among the Paleonisciformes.

The evolution of actinopterygian fish during the Mesozoic is marked by progressive changes in the jaws and hyoid apparatus that increased the size and speed of opening of the oropharyngeal chamber. The maxilla becomes freed from the cheek and three new bones, the symplectic, supramaxilla, and interopercular, evolve. The braincase is modified for more effective attachment of the eye muscles and the epaxial muscles, which lift the skull to open the mouth. The scales are lightened and the vertebral centra have ossified to facilitate rapid swimming. The tail becomes more symmetrical as the swim bladder becomes more effective in achieving neutral buoyancy.

These changes occurred in a number of distinct lineages during the Mesozoic, one of which gave rise to the modern teleost fish. Other neopterygian lineages common in the Mesozoic were the semionotids, halecomorphs, pycnodonts, macrosemiids, pachycormids, and aspidorhynchids.

References for Chapters 6 and 7 are combined on pages 133–135.

CHAPTER VII

Advanced Bony Fish: The Teleosts

Almost all living actinopterygian fish—ranging from the familiar perch and goldfish to such peculiar forms as sea-horses, elephant-snout fish, flatfish, and deep-sea scaly dragonfish—belong to a single division, the Teleostei. Many approaches can be taken in discussing such a diverse group, including emphasizing functional anatomy and environmental adaptation, zoogeography, or reproductive and developmental strategies. In this chapter, we will concentrate on the phylogenetic relationships of the major groups to provide a framework that is necessary to study a host of evolutionary processes such as the direction and rates of morphological change, the frequency of convergence, parallelism and reversal, and the patterns of geographical dispersal.

TELEOST CLASSIFICATION

Because of their great numbers and wide range of anatomical diversity, it is a monumental problem to establish phylogenetic relationships among the teleosts. Nelson (1984) classified the over 20,000 living species in 35 orders and 409 families. The challenge of understanding the relationships within this assemblage has attracted many workers in the past 20 years, including the most knowledgeable taxonomic innovators. This work has resulted in a continuing flood of papers and a rapid reorganization of the taxonomy of teleosts. The work of Greenwood, Rosen, Weitzman, and Myers (1966) represents a watershed as the earliest and most comprehensive of the recent revisions. Many details of classification have since been modified, but their paper provided an important impetus for further work, especially because of its emphasis on establishing strictly monophyletic groups based on unique derived characters.

Earlier classifications of the teleosts had been based on the recognition of a series of evolutionary grades, starting from a primitive level broadly resembling the modern salmon, through intermediate steps, toward the most advanced forms of which the perch is the best-known example. In addition, a number of highly specialized groups, such as the Anguilliformes (eels) and the Ostariophysii (catfish and carps), have long been recognized. We now realize that most of the major groups of teleosts have a very long history going back to the late Mesozoic. The early members of each modern group are very similar to one another. Their relationships with living teleosts are based on specific characters of the caudal fin and details of the jaw and pharyngeal apparatus. Several taxa that have been accepted for many years—most notably the primitive level represented by the salmon, the Protacanthopterygii—were found to be unnatural assemblages including the ancestors of many distinct modern lineages.

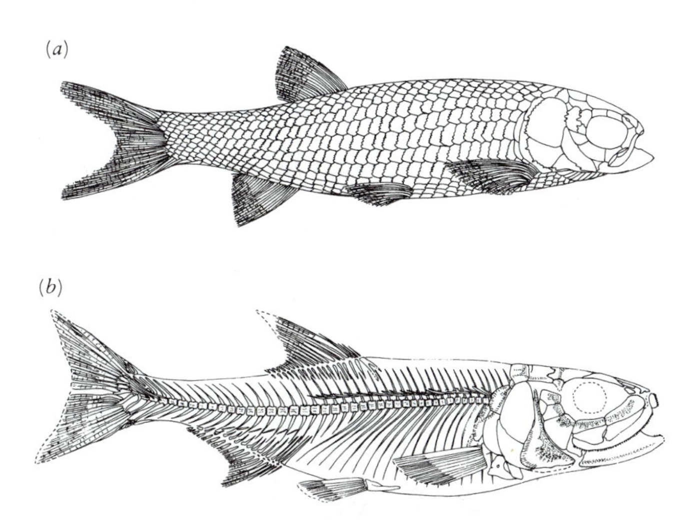

Figure 7-1. REPRESENTATIVES OF THE PHOLIDOPHORID AND LEPTOLEPID LEVELS OF EARLY TELEOST EVOLUTION. (*a*) *Oreochima*, a pholidophorid from the Lower Jurassic of Antarctica; $\times 1\frac{1}{2}$. *From Schaeffer, 1972. Courtesy of the Library Services Department, American Museum of Natural History.* (*b*) *Varasichthys griasi*, an Upper Jurassic representative of the "leptolepid" level of organization, $\times\frac{1}{3}$. This fish represents the pattern from which all advanced teleosts evolved. *From Arratia, 1981.*

Lauder and Liem (1983) provide the most recent review and appraisal of teleost relationships. Their analysis is based on works by Patterson (1973), Rosen (1973), Patterson and Rosen (1977), Rosen and Patterson (1969), Fink and Weitzman (1982), Forey (1973a and b), Greenwood (1977), Nelson (1969a and b, 1973), Weitzman (1974), and Fink and Fink (1981).

More major changes in teleost classification are expected within the next decade. Areas that are likely to be subject to the most significant changes include the specific interrelationships among the major groups and the classification of perciform fishes, which embrace over 7000 species in 150 families and 21 suborders.

ANCESTRAL TELEOSTS

PHOLIDOPHORIDAE AND LEPTOLEPIDAE

The earliest known teleosts are the late Triassic and early Jurassic pholidophorids (Figures 7-1*a* and 7-2*a*). Most are relatively small fusiform fish but some were as much as 40 centimeters long. Many have large teeth, which suggests that they had active predaceous habits. The tail is superficially symmetrical and the number of dermal fin rays is reduced. Most species retain rhomboidal scales with a thin layer of ganoine.

Patterson and Rosen (1977) recognized the pholidophorids as teleosts by the presence of the following features that are characteristic of early members of all the late Mesozoic teleost groups:

1. Two supramaxillae
2. Extension of the myodome into the basioccipital
3. Ural neural arches modified as uroneurals
4. Premaxilla divided into a lateral, mobile, tooth-bearing portion and a more medial bone termed the lateral dermethmoid
5. Quadratojugal fused with quadrate
6. Fused vomers

Early pholidophorids are primitive in that the lateral cranial fissure is not completely closed. The vertebral centra in this group are weakly ossified.

We have not established the specific ancestry of pholidophorids. The parasemionotids resemble them in general appearance but some are more specialized in the closure of the lateral cranial fissure. Other parasemionotids might be ancestral to pholidophorids, or they may have evolved directly from palaeoniscoids.

The pholidophorids are succeeded in the Jurassic by the closely related leptolepids (Figures 7-1*b* and 7-2*b*, *c*). Leptolepids are more similar to modern teleosts in having their vertebrae fully ossified. The scales are cycloid, which is a condition that is achieved in some pholidophorids. Their small size (as little as 5 centimeters long), general skull proportions, and tiny teeth resemble modern teleosts that are effective planktonic feeders (Nybelin, 1974). Close similarities of particular features of the skull and caudal skeleton indicate that some species typically included among the leptolepids are closely related to the major teleost groups that become fully differentiated in the Cretaceous.

Patterson and Rosen (1977) do not consider the groups referred to as pholidophorids and leptolepids to be valid taxonomic units. Several species once assigned to the family Leptolepidae are now classified with more advanced teleost groups; these include the type species of the genus *Leptolepis*. Pholidophorids are also difficult to define since various species may be included in more advanced groups. The terms pholidophorid and leptolepid do remain convenient however, in referring to early stages in the development of teleost characteristics.

The radiation that led to the modern teleost groups began among the Jurassic leptolepids. Five major groups (designated as cohorts) are recognizable within the Mesozoic: the Osteoglossomorpha, Elopomorpha, Clupeomorpha, and Euteleostei, which constitute the modern fauna, and the Ichthyodectiformes, which were prominent in the Cretaceous but are now extinct.

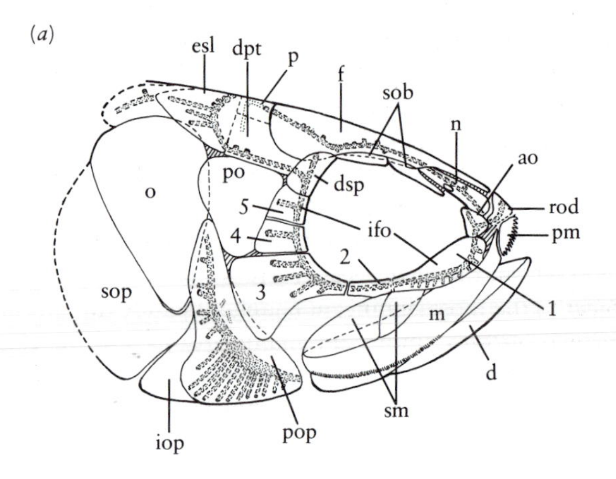

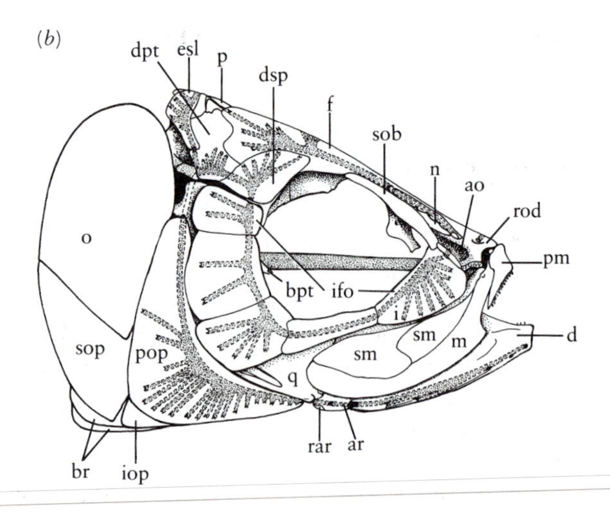

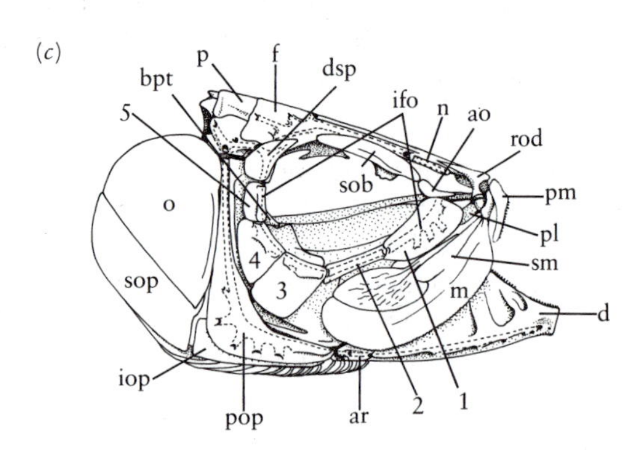

Figure 7-2. THE SKULLS OF PRIMITIVE TELEOSTS. (*a*) *Pholidophorus bechei*, Lower Jurassic, $\times 2$. *From Nybelin, 1966.* (*b*) *Tharsis* [*Leptolepis*] *dubius*, Upper Jurassic, $\times 1\frac{1}{2}$. *From Patterson and Rosen, 1977.* (*b* and *c*) *Courtesy of the Library Services Department, American Museum of Natural History.* (*c*) A primitive Upper Jurassic clupeocephalan, *Leptolepides sprattiformis*, $\times 5$. *From Patterson and Rosen, 1977.* For abbreviations see Figure 6-13.

ICHTHYODECTIFORMES

The Ichthyodectiformes are known from approximately a dozen genera, typified by the Cretaceous genus *Xiphactinus* (Figure 7-3). Most are large predaceous fish with a marginal dentition composed of long conical teeth. In contrast, *Allothrissops* from the late Jurassic was a small form with a microphagous dentition. Two peculiarities unite these genera and distinguish them from other teleosts. The anterior uroneurals are expanded and cover the lateral faces of the preural centra (Figure 7-4*a*), and a uniquely shaped endoskeletal ethmo-palatine bone comprises part of the floor of the nasal capsule. Most members of the group are marine, but some fossils are found in brackish or even freshwater deposits. The Ichthyodectiformes probably arose from among the genera that are

Figure 7-3. *Xiphactinus* [*Portheus*], a representative of the primitive teleost order Ichthyodectiformes that was prominent in the Cretaceous, about 4 meters long. The skeleton of another ichthyodectiform, *Gillicus*, is within the body cavity. *From Bardack, 1965.*

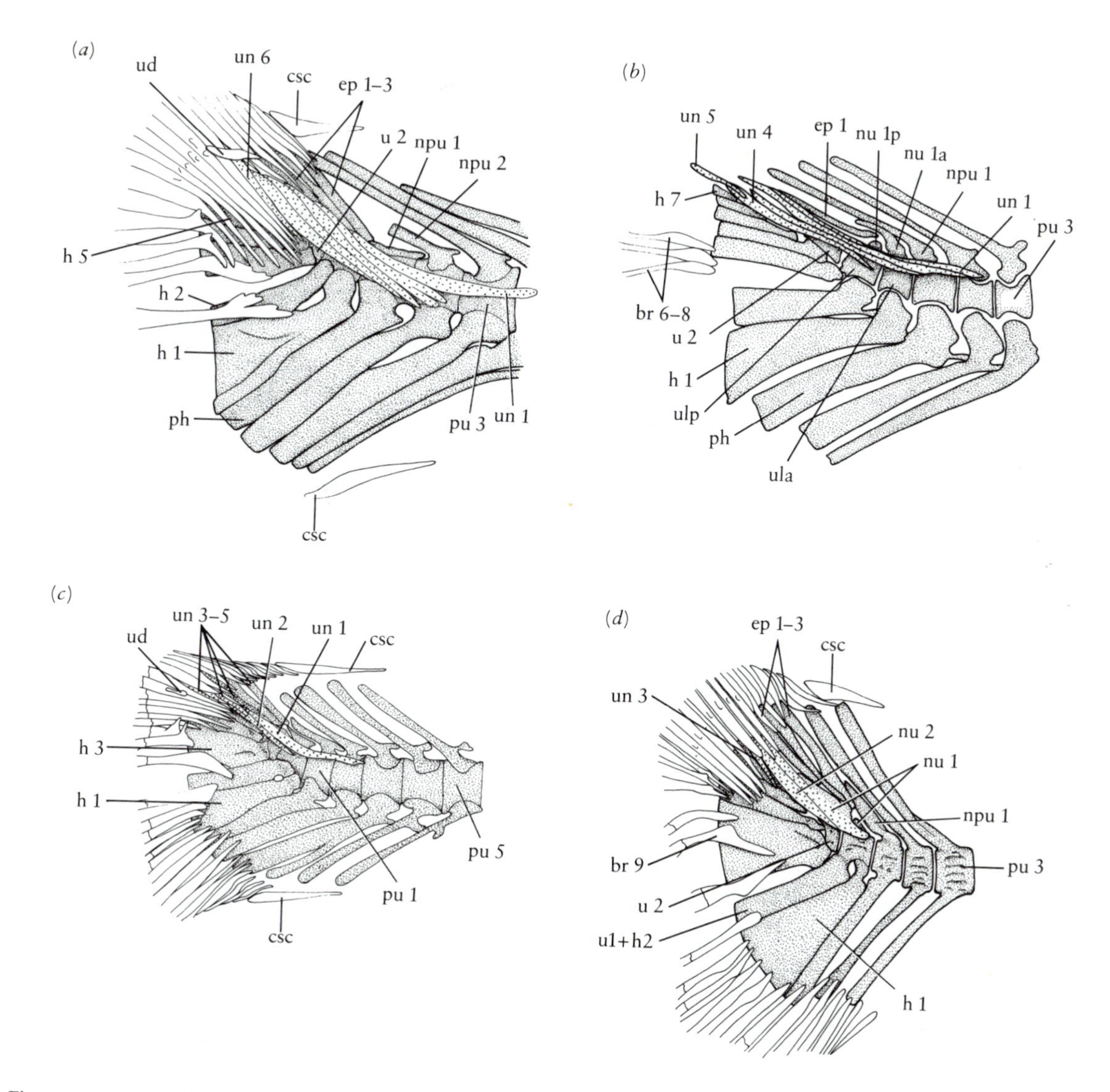

Figure 7-4. CONFIGURATION OF THE CAUDAL SUPPORTS IN PRIMITIVE TELEOSTS. (*a*) The ichthyodectiform *Thrissops* from the Upper Jurassic, approximately ×2. (*b*) The osteoglossomorph *Lycoptera*, Upper Jurassic, approximately ×8. (*c*) The primitive clupeocephalan *Leptolepides sprattiformis* from the Upper Jurassic, approximately ×7. (*d*) The early clupeomorph *Diplomystus* from the Lower Cretaceous, ×3½. Abbreviations as follows: br, branched caudal fin rays of upper lobe, numbered from top downward; csc, caudal scute; ep, epurals; h, hypurals; npu, neural arch or spine of numbered preural centrum; nula, nulp, anterior and posterior neural arches of first ural centrum, ph, parhypural bone; pu, preural centra; u, ural centra; ud, urodermals; ula, ulp, anterior and posterior parts of first ural centrum; ul+h2, fused first ural centrum and second hypural, characteristic of clupeomorphs; un, uroneural bones. *From Patterson and Rosen, 1977. Courtesy of the Library Services Department, American Museum of Natural History.*

typically considered as leptolepids, at a slightly more primitive level than that of the ancestors of the living teleost groups.

Living Teleost Groups

The interrelationships of the four living teleost cohorts have been much debated in recent years, and almost all possible taxonomic combinations have been suggested, depending on which anatomical characters are emphasized. The anatomy of the Tertiary and Recent members of these groups is very distinct, but their early Cretaceous and late Jurassic predecessors were extremely similar in overall morphology. Many species now classified as basal members of the advanced groups were included in the family Leptolepidae.

According to Patterson and Rosen (1977), the structure of the lower jaws, gill arches, and caudal skeleton indicate that the closest relatives of the euteleosts (which include most living teleosts) are the clupeomorphs. The clupeomorphs and euteleosts together constitute the Clupeocephala. They argue that the Clupeocephala shared a common ancestry with the Elopomorpha. The Clupeocephala and Elopomorpha are together designated as the Elopocephala. The Elopocephala probably shared a common ancestry with the Osteoglossomorpha, which are

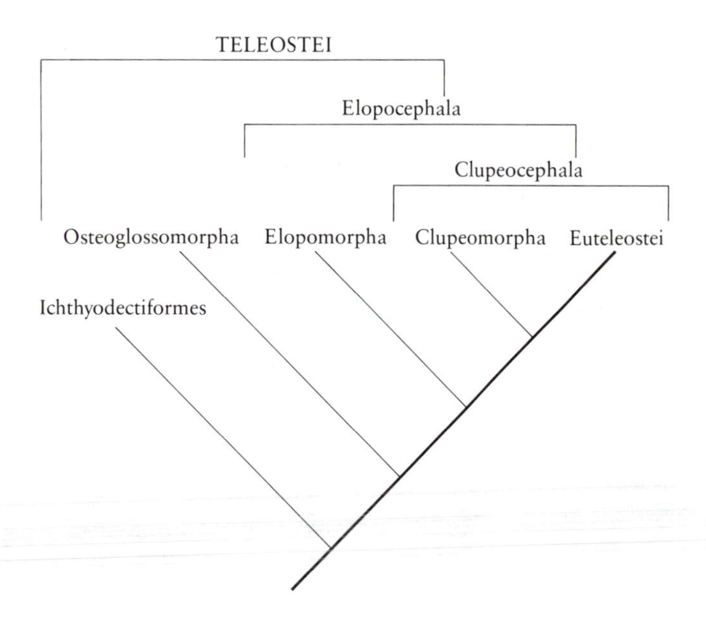

Figure 7-5. A cladogram showing hypotheses of relationships among the major groups of teleosts that can be recognized by the end of the Mesozoic. *From Patterson and Rosen, 1977. Courtesy of the Library Services Department, American Museum of Natural History.*

considered to be the sister group of all living teleosts. These relationships are graphically illustrated in Figure 7-5.

Early members of the primitive living teleost groups may have resembled the living salmon and trout in general appearance (although only the primitive euteleosts possess a fatty posterior dorsal fin). They were larger than the leptolepids but with a similarly elongate body. The fins had no spines (which are a feature of more advanced teleosts), and the pelvic fin was posterior in position. The maxilla was separate from the cheek, but it remained toothed. Some elopomorphs, most clupeomorphs, and the primitive euteleosts retain this general pattern, but the living osteoglossomorphs show extensive modification of their body proportions and fin structure.

OSTEOGLOSSOMORPHA

Modern osteoglossomorphs are freshwater fish that are common in South America, Africa, Southeast Asia, and Australia. Only the genus *Hiodon* occurs in North America. Most are predaceous. Although the maxilla is usually toothed, the primary bite is between the parasphenoid and the tongue which is supported by toothed basihyals and the glossohyal. The basihyal teeth shear against the pterygoquadrate teeth.

Osteoglossum from Brazil, Egypt, and the East Indies and the South American genus *Arapaima*—the largest of all primary freshwater fish—are members of the order Osteoglossiformes, which is conservative in body form except for the posterior position of the dorsal and anal fins. A few members of this order are known from the early Tertiary. The elephant-snout fish *Mormyrus* and several other genera from Africa represent a related order, the Mormyriformes, that is characterized by the presence of electric organs in the caudal region. This group has no fossil record.

The Osteoglossomorpha can be differentiated from more advanced teleosts by the retention of 18 principal caudal fin rays, a full neural spine on the first preural centrum, and (in Recent genera) the coiling of the gut so that the intestine passes to the left of the stomach. We can recognize early fossils by the presence of three or four straplike uroneurals that extend forward beyond the second ural centrum. The genus *Lycoptera* (Figure 7-4*b*) from the Upper Jurassic of China is the oldest recognized osteoglossomorph. Some individuals of this genus still retain two centers of ossification where even the leptolepids already show a single centrum in association with hypurals 1 and 2. There are seven or fewer hypurals compared with nine or more in leptolepids. The caudal axis turns sharply upward at the level of the first preural centrum, rather than gradually from preural centra 3 to 5.

ELOPOMORPHA

Patterson and Rosen (1977) used the term Elopocephala to designate all more advanced teleosts. We recognize ancestors of this assemblage from the late Jurassic. Species from Bavaria attributed to the genus *Anaethalion* show an advancement in caudal structure over the condition in osteoglossomorphs. The number of uroneurals that extend forward beyond the second ural centrum is reduced to two. As in higher teleosts, epipleural intermuscular bones (see Figure 7-13) are developed. Species with this morphology would be appropriate ancestors for both the elopomorphs and the clupeocephalans.

The Lower Cretaceous species *Anaethalion vidali* from Montsech, Spain, shows specific characters of the elopomorphs in the formation of a compound neural arch developed above the first ural and first preural centra (Figure 7-6). The fusion of the angular and retroarticular bones at the back of the lower jaw is another derived character of this group. This species cannot be assigned to any of the modern elopomorph orders, but other fossils indicate that the major elopomorph groups were all distinguishable by the Upper Cretaceous.

Greenwood (1977) and Forey (1973b) recently reviewed the Elopomorpha. This group is represented in the modern fauna by a heterogenous assemblage of fish including the tarpons, eels, and bizarre deep-sea forms. We would not have recognized the interrelationships of this assemblage were it not for the fact that all members have a highly specialized larval stage that is unlike that of any other fish groups (Figure 7-7). This larval type, termed the leptocephalus, is leaf or ribbon shaped and is so thin and deficient in pigments that it is translucent. The head is small but may be conspicuously toothed. The leptocephalus larva can migrate thousands of miles and can

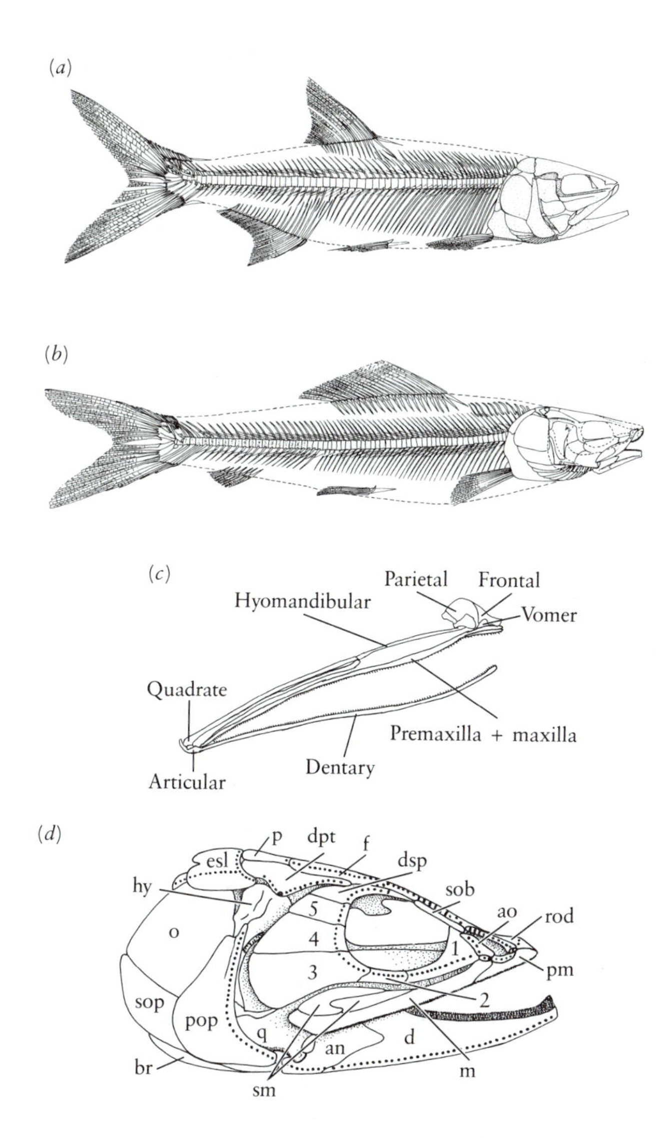

Figure 7-6. REPRESENTATIVES OF THE ELOPOMORPHA. (*a*) *"Anaethalion" vidali*, an elopomorphon of uncertain affinities from the Upper Jurassic, ×¼. *From Forey, 1973b.* (*b*) The Upper Cretaceous albuloid *Lebonichthyes gracilis*, ×¼. Both the modern albuloids and the eels may have evolved from a genus with a similar morphology. *From Forey, 1973b.* (*c*) Skull of the modern saccopharyngoid eel *Gastrostomus. From Gregory, 1933.* (*d*) Skull of the modern genus *Elops*, the ten pounder. *With permission from Patterson, 1973. Interrelationships of holosteans, In P.H. Greenwood, R.S. Miles, and C. Patterson (eds.),* Interrelationship of Fishes. *Copyright 1973 by the Limnean Society of London.* For abbreviations, see Figure 6-13.

tolerate great changes in the salinity, oxygen content, and temperature of the water. We know of a somewhat similar larval stage in the clupeomorphs, but it is not nearly as highly specialized.

Three major types of elopomorphs may be recognized. *Elops* (the ten-pounder) and the tarpon *Megalops* somewhat resemble the modern salmon in their body configuration, except for the absence of a second, adipose dorsal fin. According to Forey (1973b), the elopoids have remained virtually unchanged since the Upper Jurassic.

The Albuloidei includes genera such as *Albula*, the bonefish, in which the general body shape resembles that of the elopids and megalopids, but the specialized nature of the lateral line canals and associated ossifications in the snout region demonstrate affinities with the specialized deep-sea families Notacanthidae (spiny eels) and Halosauridae. These forms have a long, tapering trunk that ends in a much reduced caudal fin. There is a posteriorly directed spine on the dorsal edge of the maxilla. The anal fin is elongate and merges with the tail, but the dorsal fin is short and anterior in position. Some possess photophores. The genus *Osmeroides* represents the Albuloidei in the Lower Cretaceous and links this group to the base of the Elopidae. Both the Notacanthidae and Halosauridae are known from the Upper Cretaceous.

We think that primitive albuloids are closely related to the ancestry of eels. The Anguilloidea, with more than 600 species in 19 families, are the most diverse of the elopomorphs. All living eels are specialized in the loss of the pelvic girdle and the elongation of the dorsal and anal fins along with the general elongation of the body. The jaws are modified by the fusion of the premaxillae, vomers, and ethmoid into a single unit. Fossils of eels are known as early as the Upper Cretaceous. *Anguillavus* and *Enchelurus* still retain the pelvic fin but are otherwise nearly as specialized as the living species and give little evidence of affinities with other teleosts. The living genus *Anguilla* is known from the Eocene. Modern species breed in the Sargasso Sea and the larvae drift to North America and Europe where they mature in freshwater streams.

The Saccopharyngoidei, a group of eel derivatives, are even more modified. They live in the dark waters at the depths of the major ocean basins. Their bodies are eel shaped, but the enormous head and distensible pharynx allow them to swallow extremely large prey. In addition to the pelvic fins that are already missing in eels, they lose their scales, ribs, swim bladder, caudal fin, and many of the bones of the skull (see Figure 7-6). Nelson (1984) suggests that they may be the most anatomically modified of any vertebrate group. Unfortunately, there is no fossil record of the saccopharyngoids, and we do not know how long this high degree of specialization took to evolve.

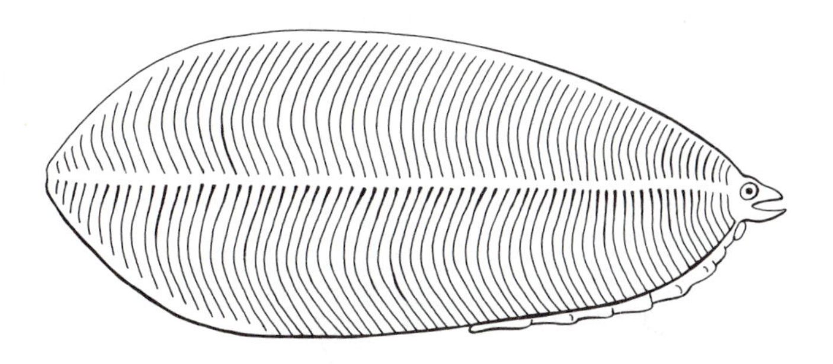

Figure 7-7. The leptocephalus larva that is common to elopoids and eels. *From* Biology of Fishes *by Carl E. Bond. Copyright © 1979 by Saunders College Publishing/Holt, Rinehart and Winston. Reprinted by permission of Holt, Rinehart and Winston, CBS College Publishing.*

CLUPEOMORPHA

Patterson and Rosen (1977) placed *Leptolepides sprattiformis* (see Figure 7-4*c*) from the Upper Jurassic and *Sombroclupeoides bahiaensis* from the Lower Cretaceous (both of which were formerly assigned to the Leptolepidae) at the base of the Clupeocephala, the assemblage that includes the ancestors of the clupeomorphs and the euteleosts. These genera are too primitive to be assigned specifically to either of these cohorts. In clupeocephalans, the retroarticular is excluded from the jaw joint and the angular is fused to the articular rather than to the retroarticular, as is the case in elopomorphs. As in other higher teleosts, the neural arch over the first ural centrum is reduced or absent and the number of hypurals is reduced to six.

Clupeomorphs are distinguished from other teleosts in having the swim bladder penetrate the exoccipital and extend into the proötic. The Lower Cretaceous genus *Spratticeps* already shows this condition. The tail is distinctive in having the second hypural fused with ural vertebra 1, while the first hypural is free proximally (see Figure 7-4*d*).

In their skeleton, the clupeomorphs are nearly as primitive as the elopoids, but with 4 families, 80 genera, and 300 species they are far more successful in the Cenozoic. They include many small fish, typical of which are the herring, *Clupea,* as well as the shad, sardines, and anchovies. Most clupeomorphs are plankton feeders with numerous long gill rakers; *Ornategulum* (Figure 7-8) was present in the late Cretaceous, Forey (1973c) and Grande (1982a and b) described many additional genera from the Tertiary.

EUTELEOSTS

The term euteleost refers to all remaining teleosts. They can be divided into three unequal groups, the primitive salmonlike fish, the specialized Ostariophysi, and a vast assemblage termed the neoteleosts, or spiny teleosts. According to Patterson and Rosen, all arose ultimately from a common ancestry with clupeomorphs, but their specific interrelationships are not clearly established.

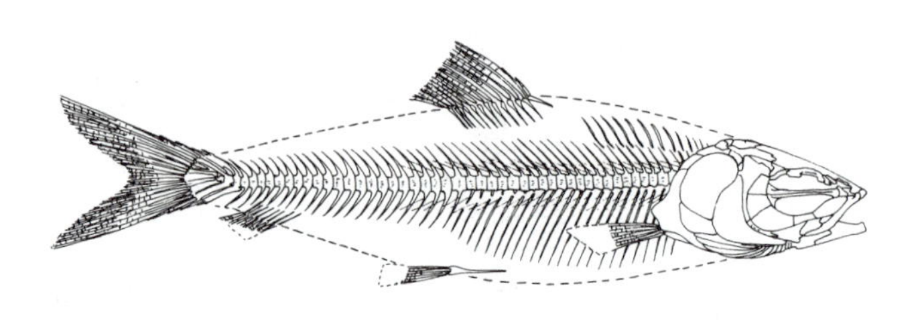

Figure 7-8. THE UPPER CRETACEOUS CLUPEOMORPH *ORNATEGULUM*, $\times\frac{1}{3}$. *From Forey, 1973c.*

(*a*)

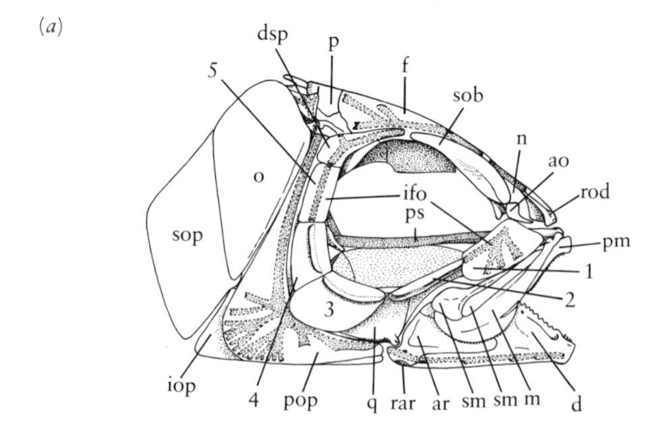

(*b*)

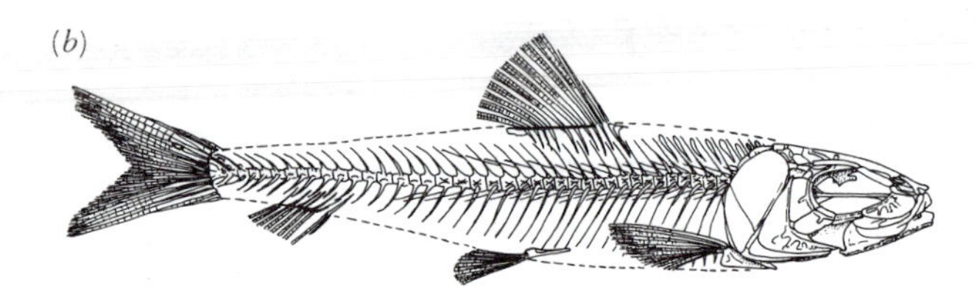

(*c*)

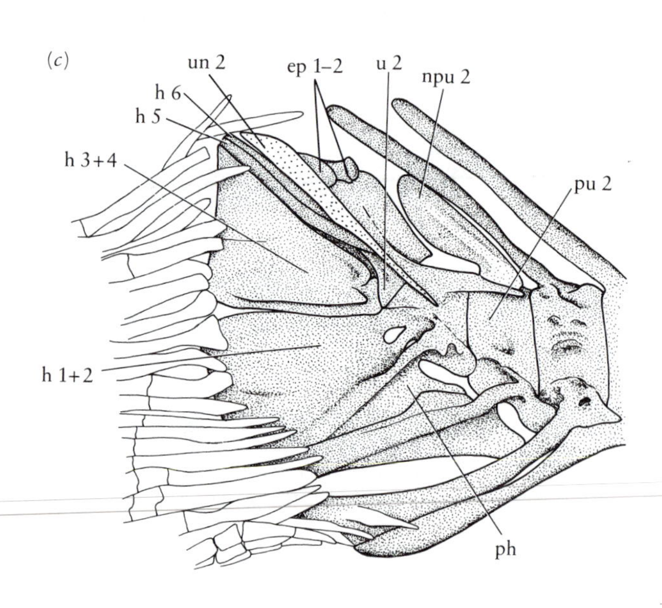

Figure 7-9. PRIMITIVE EUTELEOSTS FROM THE LATE CRETACEOUS. (*a*) Skull of *Humbertia*, $\times 2\frac{1}{2}$. For abbreviations, see Figure 6-13. (*b*) The skeleton of *Gaudryella*, approximately natural size. (*c*) Caudal supports of *Gaudryella*, approximately $\times 13$. Abbreviations as follows: ep, epural; h, hypural; npu 2, neural spine of second preural centrum; ph, parhypural; pu, preural centrum; u, ural centrum; un, uroneural. Note fusion of first preural and first ural centrum. *From Patterson, 1970.*

The euteleosts possess few derived features that are unique to that group. Patterson and Rosen (1977) recognize euteleosts on the basis of an adipose fin and nuptial tubercles. Primitive genera are unique in having a structure termed a stegural that is formed by a membranous anterior extension of the first ural neural arch, which may grow forward beneath the epurals to the level of preural centrum 1 or 2 (Figure 7-9*c*). A similar structure is also present in clupeomorphs, although it may not have been present in the common ancestors of the two groups. In elopoids, the neural arches of the first preural and the first ural centra fuse to form a platelike structure. This is functionally, but not developmentally, equivalent to the structure in euteleosts.

Most euteleosts are characterized by the possession of acellular bone, whereas the bone is cellular in osteoglossomorphs, most clupeomorphs, and some elopomorphs. However, bone cells are present in most ostariophysans and some salmoniforms, so the evolution of acellular bone presumably occurred many times in parallel after the initial diversification of the euteleosts. Few euteleosts are known prior to the Upper Cretaceous. But by the Cenomanian, at the base of the Upper Cretaceous, we find evidence of a considerable diversity of advanced teleosts. (Unlike most geological periods, there is no Middle Cretaceous.)

PRIMITIVE EUTELEOSTS

The body form common to modern salmon and trout, which are included in the order Salmoniformes, is primitive for euteleosts. They are generally long bodied with a second, adipose dorsal fin. This fin has no skeletal support and is only rarely evident in fossils. It is difficult to characterize this assemblage further except by primitive features that are lost or modified in higher teleosts. The fins lack spines. Most species have a large, toothed maxilla. The pelvic fins are posterior in position and the pectoral fin is low on the flank.

Greenwood and his coauthors (1966) included a number of other primitive euteleost groups—the alepocephaloid, myctophid, neoscopelid and questionably the ostariophysan fishes (which will be discussed shortly)—with the salmoniforms in a single superorder, the Protacanthopterygii. Rosen (1973) removed all but the salmoniforms from this assemblage, since there is no evidence that they shared a common ancestry separate from that of other euteleosts. Fink and Weitzman (1982) demonstrated that the salmoniforms themselves were a polyphyletic group. Lauder and Liem (1983) recognized that the term Protacanthopterygii no longer has any taxonomic significance and suggested that it be abandoned.

The following groups have been embraced within the Salmoniformes:

Esocidae	pikes and pickerels
Umbridae	mudminnows
Argentinoidea	smeltlike fish living in deep marine waters
Osmeridae	smelts, sweetfish, and icefishes
Galaxiidae	smeltlike fish from the southern hemisphere
Salmonidae	trout, salmon, white fish, char

Fink and Weitzman contend that the Esocidae and Umbridae represent a single assemblage that is more primitive than any other euteleosts. Unlike the higher teleosts, they retain a toothplate on the fourth basibranchial but lack an adipose fin that is common to most other "salmoniforms." The pike's dorsal fin is far posterior, so that it appears symmetrical with the anal, as in saurichthyids and gars.

The argentinoids, osmeroids, and galaxiids (all of which have a smeltlike body form) can be recognized as a monophyletic assemblage based on the fusion of the posterior caudal neural arch with either the uroneural or the first ural vertebra.

Present evidence indicates that the Ostariophysi, the smeltlike fish, and the higher teleosts (including the Salmonidae) share a common ancestry above the level of the pikes and their close relatives. The Salmonidae appears to have evolved separately from the other "salmoniform" fish and is closer to the ancestry of the more advanced neoteleosts. Both the exoccipital and basioccipital articulate with the first cervical vertebra, while only the basioccipital articulates in more primitive teleosts, and there is a single medial cartilage between the ethmoid and the premaxillae. We have not found any unique characters that distinguish all salmonids from other euteleosts.

Recent taxonomic revisions of primitive teleosts have concentrated on living forms, with little consideration for their fossil relatives. Among primitive euteleosts, complete, well-preserved esocoids are known from the Paleocene of Canada (Wilson, 1980, 1984). We know argentinoids from the Eocene and galaxiids from the Oligocene. *Gaudryella* and *Humbertia* (Figure 7-9) from the early Upper Cretaceous appear close to the argentinoids but are more advanced in the presence of a stegural and the loss of cellularity of the bone. While we cannot demonstrate the existance of an adipose fin, its presence is suggested by the anterior position of the dorsal. According to Patterson (1970), there are broad similarities between *Gaudryella* and *Humbertia* and some species of the genus *Clupavus* from the Upper Jurassic and Lower Cretaceous that have been assigned to the Leptolepidae, indicating the early divergence of the line leading to modern argentinoids. The salmonids are not known prior to the Tertiary.

OSTARIOPHYSI

The Ostariophysi have more than 6000 species, including the carp, goldfish, characoids, cyprinoids, and catfish. They constitute the majority of modern freshwater fish. Their origin apparently lies near the base of euteleost evolution, perhaps just prior to the emergence of the salmonids. Fink and Fink (1981) recently reviewed the relationships of the Ostariophysi.

Four major groups are recognized in the modern fauna: Gonorynchiformes (milk fishes), Cypriniformes (carp, minnows, goldfish, suckers, and loaches), Characiformes (tetras and piranhas) and Siluriformes (catfish and gymnotid "eels"). This assemblage is unquestionably monophyletic, united by a striking specialization of the anterior cervical vertebrae, ribs, and neural arches, which serve as a series of movable units (termed the Weberian ossicles; Figure 7-10) that transmit vibrations from an anterior portion of the swim bladder to the inner ear. Physiological studies indicate that the Weberian ossicles serve to improve sensitivity to high-frequency sounds.

The Ostariophysi are also unique in possessing special epidermal alarm cells that produce a substance which results in a fright reaction when a fish is injured. Other ostariophysan fish in the vicinity, whether of the same species or not, immediately scatter and head for the bottom.

The Gonorynchiformes, fusiform fish once placed with the Clupeiformes, show the lowest degree of development of the Weberian ossicles, but the rib of the third vertebra is expanded and associated with the anterior chamber of the swim bladder. This order is distinguished from other ostariophysans by the extensive joint between the exoccipital and neural arch one. Gonorynchiform fossils are known from the Lower Cretaceous, and modern genera occur as early as the Eocene. More advanced ostariophysans have a fully developed Weberian apparatus and show several derived features of the caudal fin supports. The earliest genus in which the Weberian apparatus is known in detail is *Chanoides* from the Eocene of Monte Bolca, Italy (Patterson, 1984). This genus is considered to be a sister group of all the Recent ostariophysans.

The Cypriniformes differ from most teleosts in having the majority of their species specialized as herbivores. The marginal teeth are lost, but the pharyngeal "jaws" form an effective grinding and crushing apparatus. They have a highly specialized mechanism that protrudes the jaws, as do the spiny teleosts, but it evolved separately in the two groups (Alexander 1967b). Eighty percent of the species in the Cypriniformes are included in the Cyprinidae. This family is common in North America, Africa, Europe, and Asia but is not present in South America. Members of this group are known as early as the Eocene, but most fossils are from the middle to late Tertiary and represent living genera.

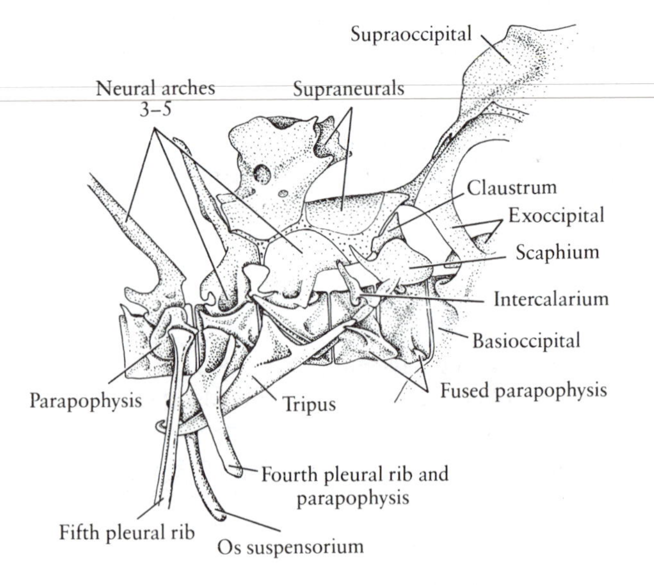

Figure 7-10. WEBERIAN APPARATUS. Specialized elements of cervical vertebrae and ribs that transmit vibrations from the swim bladder to the inner ear in osterophysian fish. *From Fink and Fink, 1981. With permission from the Zoological Journal of the Limnean Society, Vol. 72. Copyright 1981 by the Limnean Society of London.*

Characiforms, with at least 1000 species, have radiated most extensively in South America, making up nearly half the fish species in the Amazon Basin. There are 16 families that show a wide diversity in body form and diet. We do not know of any Characiformes before the Eocene.

There are approximately 3000 species of catfish (Siluroidei) in 31 families, three of which are primarily marine. The group is characterized by having a spiny anterior ray on the pectoral fin, which is controlled by muscles and may be locked in an erect position. The group lacks true scales, but some species are covered with bony plates which gives them the appearance of heavily ossified ostracoderms. Catfish have one or more pairs of barbels that are tactile and chemically sensitive. The earliest fossil catfish are late Cretaceous (Wenz, 1968), and the group was fairly diverse by the end of the Eocene (Lundberg, 1975).

The eel-like Gymnotoidei of Central and South America are closely related to the catfish. These fish generate weak electrical currents from either nervous or muscular tissue that are used in navigation and intraspecific communication. This group has no fossil record.

NEOTELEOSTS

We may include all euteleosts other than the salmoniforms and ostariophysans in a single large assemblage termed the neoteleosts, reviewed by Rosen (1973). Most are characterized by stiff fin spines and modifications in the position of the pectoral and pelvic fins and body proportions that can be associated with advances in locomotor patterns. Other, somewhat less highly advanced teleosts, including several groups adapted to the extreme conditions of deep-sea life, are more difficult to classify. According to Rosen, these groups may be classified with the more advanced neoteleosts based on the common presence of a specialized muscle, the retractor arcuum branchialium (RAB), that extends from the anterior vertebrae and inserts variously on the pharyngobranchials, epibranchials, and the upper pharyngeal tooth patches (Figure 7-11). Lauder and Liem (1983) call this muscle the retractor dorsalis (RD). It assists in prey manipulation and swallowing. The RAB presumably evolved from the longitudinal esophageal musculature that is present, but poorly developed, in primitive teleosts. There is no trace of the RAB in ostariophysans or salmoniforms and the pharyngeal teeth are poorly developed in these fish.

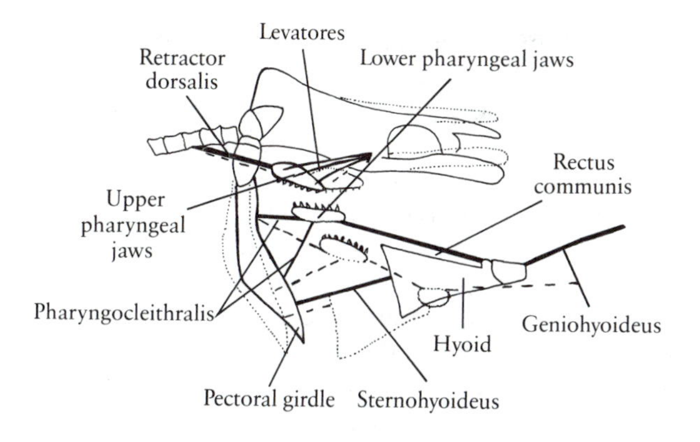

Figure 7-11. THE RETRACTOR DORSALIS (OR RETRACTOR ARCUUM BRANCHIALIUM—RAB) MUSCLE IN EUTELEOSTS. The diagram is simplified. Solid lines show position of bones during retraction of the pharyngeal jaws. Bony elements in dotted lines represent positions during protraction. *From Lauder and Liem, 1983.*

STOMIIFORMES

The Stomiiformes are the most primitive of modern fish that can be assigned to the neoteleost assemblage. They possess the retractor arcuum branchialium, but most features of their anatomy have led in the past to their assignment to the salmoniform assemblage. The modern members of the group are all moderately to highly specialized deep-sea fish—large-mouthed, voracious carnivores. We recognize eight families that have such colorful names as hatchetfishes, viperfishes, and deep-sea scaly dragonfishes. The fossil record of undoubted stomiatoids extends only as far back as the Miocene. We find otoliths (ossifications within the inner ear) similar to those of modern genera as early as the Eocene, which demonstrates the divergence of the highly specialized modern groups. They may have originated within the Cretaceous, but no specific affinities can be established on the basis of the current fossil evidence. Fink and Weitzman (1982) recently reviewed the stomiiformes. They emphasize the structure of the photophores as being particularly important in demonstrating the monophyletic nature of this group.

AULOPIFORMES

No stomiiforms are known from the Upper Cretaceous, but two other groups of primitive neoteleosts, the Myctophiformes and Aulopiformes, are common. They are represented today by a large assemblage of deep-sea fish. Rosen (1973) defines the Aulopiformes on the basis of unique pharyngobranchial characteristics. The second pharyngobranchial is greatly elongated posterolaterally, so that it extends away from the third pharyngobranchial. The uncinate process of the second epibranchial is drawn out to bridge the gap between its origin and its contact with the third pharyngobranchial. The Aulopiformes have a primitive RAB pattern, and they retain a toothed maxilla and supratemporal bones; the posttemporal fossa is not open dorsally, and the ural and preural centra are not fused.

The living aulopiform fish are extremely varied in habits and body form. The modern Synodontidae (lizard fish) have an essentially salmonid body form with an elongate trunk and an adipose fin. The giganturids are deep-sea fish with an enormous gape, no pelvic fin, and an extremely elongate ventral lobe of the caudal fin.

Numerous Cretaceous fossils described by Goody (1969) may be included among the Aulopiformes; most of these were once classified as myctophoids. These include fish with essentially normal body proportions such a *Phylactocephalus* (Figure 7-12), genera with greatly

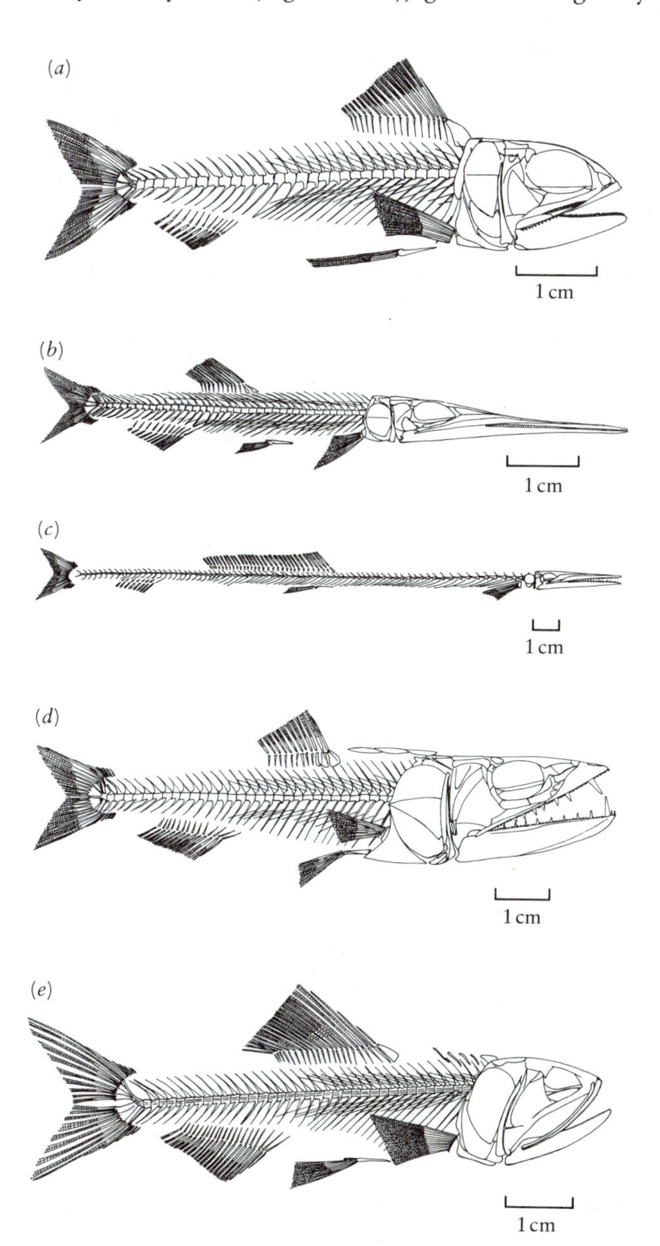

Figure 7-12. CRETACEOUS TELEOSTS AT A LEVEL BETWEEN PRIMITIVE EUTELEOSTS AND SPINY TELEOSTS. Rosen (1973) includes all five genera shown among the Aulopiformes. (*a*) *Phylactocephalus*, about ×1. (*b*) *Ichthyotringa*, ×1. (*c*) *Dercetis*, $\times\frac{3}{4}$. (*d*) *Eurypholis*, $\times\frac{2}{3}$. (*e*) *Sardinius*, $\times\frac{3}{4}$. *From Goody, 1969.*

elongated heads included in the Family Ichthyotringidae, the Dercetidae, which have long narrow bodies, and the Enchodontidae, which have large heads and greatly elongated teeth. Rosen suggests that the Upper Cretaceous genus *Sardinius*, a primitive member of the Synodontidae differing only in details from the modern lizard fish, may represent the central stock of this assemblage.

MYCTOPHIFORMES

The next level of neoteleost evolution is exemplified by the myctophiforms, represented in the Recent fauna by the families Myctophidae (lanternfishes) and Neoscopelidae. These include about 150 modern species of small, deep-sea fish of relatively normal body proportions that undergo extensive vertical migration daily. The body bears numerous photophores that are capable of being preserved in the fossils. Rosen (1973) defined the group on the basis of specialization of the pharyngeal dentition and musculature. The early myctophoids are primitive in the retention of an adipose fin but specialized in the exclusion of the maxilla from the gape.

The Upper Cretaceous genus *Sardinioides* is one of the earliest known forms that can definitely be assigned to the Myctophiformes. *Nematonotus* (Figure 7-13*a*), which is also from the Upper Cretaceous, might be classified either with this group or with the Aulopiformes. It is advanced in the structure of the fin supports and in the configuration of the upper jaw. The first pectoral and pelvic fin rays are stout and closely segmented, approaching the condition of spiny teleosts. A few maxillary teeth are retained, but the well-developed articulation between the maxilla and premaxilla is comparable to that of primitive acanthopterygians and paracanthopterygians, although it is not as elaborate. The myctophoids were once placed in the Protacanthopterygii with the Salmoniformes, but they are distinctly more advanced in the initiation of spine development—a hallmark of advanced teleosts—as well as the specialization of the mouth parts and the presence of a well-developed retractor arcuum branchialium. Nelson (1984) unites the orders Aulopiformes and Myctophiformes in the Scopelomorpha.

(*a*)

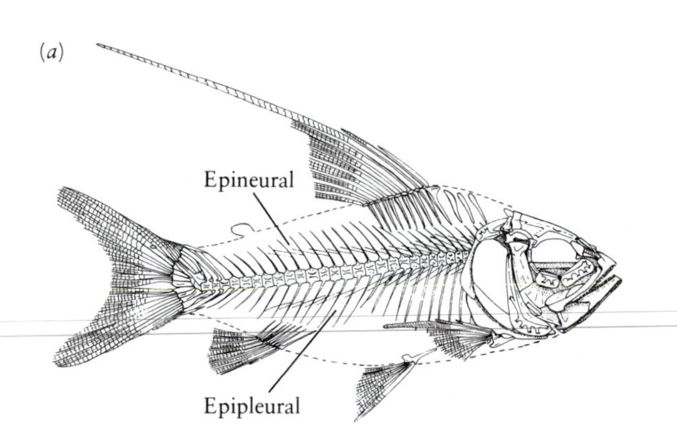

(*b*)

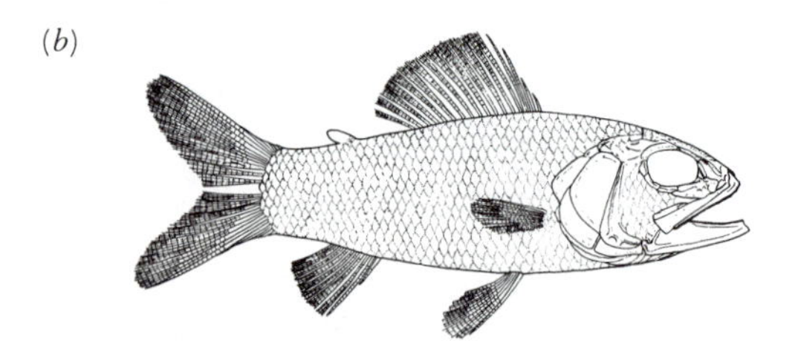

Figure 7-13. (*a*) *Nematonotus.* The aulopiform, close to the ancestry of the myctophids, from the lower part of the Upper Cretaceous. Note the riblike epipleural and epineural bones, which are characteristic of primitive euteleosts. They are functionally replaced by zygopophyses in spiny teleosts. *From Rosen and Patterson, 1969. Courtesy of the Library Services Department, American Museum of Natural History.* (*b*) *Sardinioides crassicaudus,* a myctiphoid from the Upper Cretaceous, not unlike Recent neoscopelids. *From Patterson, 1964.*

SPINY TELEOSTS— THE ACANTHOMORPHA

Although the teleost level of evolution was reached by the end of the Triassic, the group only began to diversify significantly in the Upper Cretaceous, and the number of modern groups represented in the Mesozoic fossil record is still very small (Patterson, 1964). The explosive radiation that led to the establishment of the modern families is first evident in the early Cenozoic. This is dramatically demonstrated in the rich Eocene fauna from Monte Bolca in northern Italy that Blot (1969, 1984) recently reviewed. This radiation, which is comparable to that which occurred among placental mammals at about the same time, may have been related to global changes in climate and ecosystems that led to the extinction of most archaic actinopterygian groups. On the other hand, it may be attributed to anatomical and physiological changes in the teleosts themselves, highlighted by advances in the feeding and locomotor apparatus. Modern teleosts that resulted from this dramatic radiation are classified among the Acanthomorpha—the spiny teleosts.

CHARACTERISTICS OF SPINY TELEOSTS

Patterson (1964) outlined a series of changes in locomotion, protection, and feeding that differentiate spiny teleosts from primitive euteleosts. The most pervasive advances are in the locomotor system.

Locomotion

In primitive teleosts, swimming occurs by sinuous movements of the entire trunk. In neoteleosts, the trunk is short and rigid and nearly all the swimming force is concentrated in rapid, large amplitude movements of the caudal fin. Stiffening of the trunk is assisted by the presence of zygapophyses, which functionally replace the intermuscular epineural and epipleural bones of more primitive teleosts. Muscles throughout the trunk are contracted, but nearly all the force is transmitted to the base of the tail by tendons. As a result of the more effective concentration of force, a large neoteleost such as the tuna can swim at speeds up to 70 kilometers per hour, compared with 5 kilometers per hour in primitive teleosts such as the trout.

In primitive spiny teleosts, there are only 24 or 25 vertebrae, 10 precaudal and 14 or 15 caudals, whereas in the primitive teleost *Elops,* there are almost twice as many precaudal as caudal vertebrae. The body of typical spiny teleosts is deep and laterally compressed, which helps in turning and resisting rolling and facilitates lateral contraction of the trunk musculature.

The posterior position of the anal fin and posterior portion of the dorsal fin counteract the thrust of the tail and reduce lateral undulation in the rear part of the body. The anterior portion of the dorsal fin functions most effectively as a rudder when located anterior to the center of gravity.

We can also see important changes in the position and function of the pectoral and pelvic fins. In primitive bony fish, the pectoral fins are low on the flank and serve as hydrofoils to lift the front of the body against the force of gravity. Thinning of the scales and the development of a homocercal tail in early teleosts reduced the importance of this function. In spiny teleosts, buoyancy is more effectively controlled by a closed (physoclistous) swim bladder and the pectoral fins are used primarily for braking and turning. In the primitive ventral position, the body would tend to somersault if the pectoral fins were suddenly spread out as brakes. They assume a higher position on the flank in advanced teleosts that equalizes the force along the horizontal axis.

The pelvic fins move forward in advanced teleosts (in some cases anterior to the pectorals) and become attached to the cleithrum. This anterior movement can be explained by the fins' role in countering the tendency of the fish to rise in the water when the pectoral fins are extended as brakes. In the primitive posterior position, they would only lower the tail region. The anterior fin spine on the pelvic fins is controlled by strong muscles that affect the orientation of the entire fin. The number of soft rays is reduced to five. The shortening of the entire trunk places both the pelvic and pectoral fins close to the center of gravity. The structure of the caudal fin is also modified in higher teleosts. This modification occurs in different ways in two major groups and will be discussed separately.

Protection

The term spiny teleost refers to the fin spines that most members of this group possess. These are particularly conspicuous on the dorsal fin but also occur in the anal and paired fins. The stiff spines result from fusion of the two halves of primitively paired and jointed dermal fin rays into a median, unjointed structure. The spine loses its flexibility and is effectively moved as a unit by muscles attached to its base.

When fully elaborated, the spines serve a defensive function. The importance of the anterior fin spines in controlling the orientation of both the dorsal and pelvic fins suggests that selection may have acted first to modify the fin rays in relation to their role in locomotion. Advanced teleosts also have spines extending from the operculum, preoperculum, lacrimal, and even from the individual scales. These spines are in the form of comblike teeth on the posterior margin, which is why the scales are termed **ctenoid.**

Feeding

Feeding structures of all spiny teleosts are significantly modified from the pattern of salmoniform fish. Within the structural complex already considered, the attachment of the retractor arcuum branchialium to the pharyngobranchials tends to shift anteriorly and centers on the third, rather than the fourth, pharyngobranchial. In addition, the pharyngeal dentition is emphasized and the tooth patches that primitively occur on the palate and opposing hyoid elements are reduced.

More conspicuously, the maxilla in all higher teleosts tends to lose its role as a tooth-bearing bone at the margin of the jaw and serves primarily as a lever manipulating the premaxilla. The tooth-bearing ramus of the premaxilla (which extends beneath the maxilla) lengthens to approach the angle of the jaw. In spiny teleosts as a group, the premaxilla and maxilla function together to form a nearly circular mouth opening that facilitates suction feeding. The manner of manipulation of the maxilla and the structure and function of the premaxilla differ in two major living groups, the paracanthopterygians, including the cod and haddock, and the acanthopterygians, including most living species (Rosen and Patterson, 1969).

In paracanthopterygians (Figure 7-14*a*), the premaxilla moves relatively little and serves primarily as a pivot around which the maxilla rotates. There is little protrusion. The posterior end of the maxilla is highly mobile and can swing anteriorly, so that the bone approaches a vertical orientation. In advanced paracanthopterygians, the movement of the maxilla is principally controlled by an internal portion of the adductor mandibulae, designated $A_1\beta$, which inserts far forward on the maxilla. The more posterior, external head of the adductor mandibulae, A_1, which is common in more primitive teleosts, is small or absent (however, primitive paracanthopterygians retain a fairly large A_1). The decreased mass of this muscle allows free movement of the posterior end of the maxilla.

In acanthopterygians (Figure 7-14*b*), the premaxilla moves anteriorly as the mouth is opened. This mobility results from a longitudinal twisting of the maxilla whereby a long internal process is rotated laterally and anteriorly as a cam. In acanthopterygians, the $A_1\beta$ is not an important muscle, but the A_1, or external head of the adductor mandibulae, is large and the posterior end of the

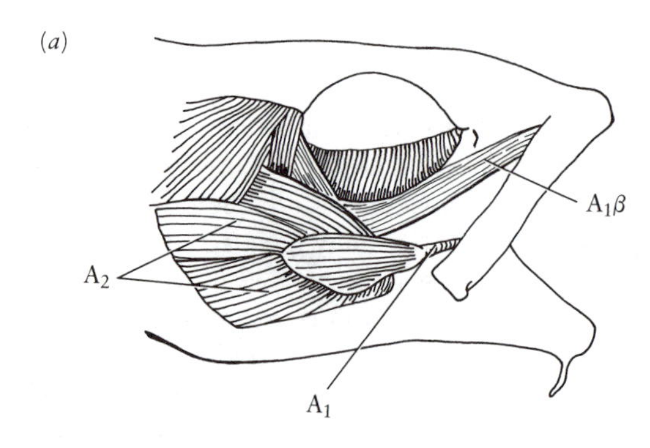

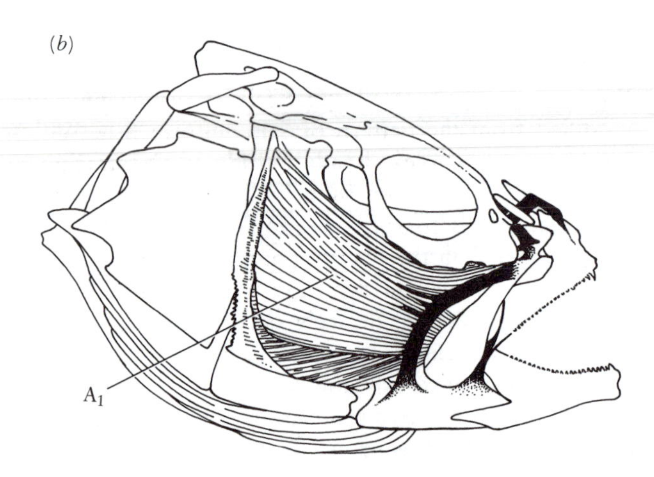

Figure 7-14. JAW MUSCULATURE OF SPINY TELEOSTS. (*a*) The paracanthopterygian *Microgadus,* or tomcod. *From Rosen and Patterson, 1969. Courtesy of the Library Services Department, American Museum of Natural History.* (*b*) The acanthopterygian *Epinephelus. From Schaeffer and Rosen, 1961.* In the paracanthopterygian, the premaxilla protrudes only slightly, but the maxilla swings forward. In the acanthopterygian, twisting of the maxilla causes the premaxilla to protrude. An internal process at the anterior end of the maxilla acts as a cam to move the ascending process forward. Abbreviations as follows: A_1, external head of adductor mandibulae; $A_1\beta$, internal portion of adductor mandibulae; A_2, posterior portion of adductor mandibulae.

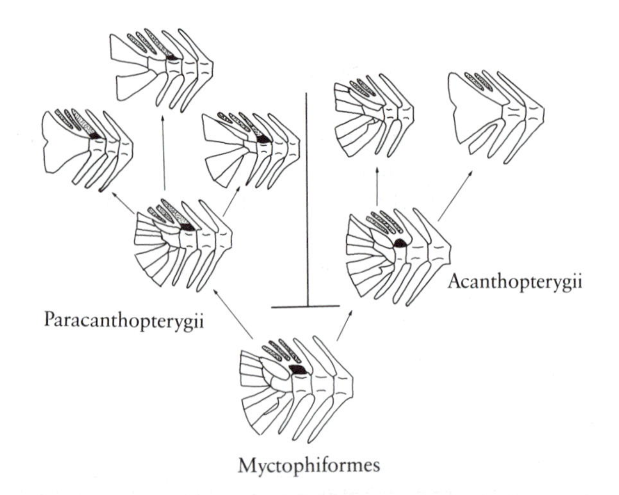

Figure 7-15. EVOLUTION OF THE CAUDAL SKELETON IN ADVANCED NEOTELEOSTS. The arrows connecting the different patterns indicate possible structural changes, not phyletic lineages. The epurals are stippled, the second preural neural spine crest is black. The primitive configuration of the caudal skeleton is exhibited in the Myctophiformes. In the Paracanthopterygii, the most anterior epural fuses with the second preural neural spine crest. Further specializations involve the fusions of the hypurals into platelike elements. In the Acanthopterygii, fusions occur between two preural vertebrae, of which one has a complete neural spine and the other has a reduced neural spine crest. The result of this fusion is a caudal skeleton configuration that converges with that of the Paracanthopterygii. Among more specialized Acanthopterygii, fusions of the hypurals result in the formation of hypural plates. *From Lauder and Liem, 1983.*

maxilla cannot extend forward as much as in the paracanthopterygians. Protrusibility is important to capture and manipulate prey and also to maintain a low pressure within the oral cavity as the mouth begins to close.

Caudal skeleton

Researchers have used the structure of the caudal fin to distinguish teleosts from other neopterygians and euteleosts from other teleosts. It may also be used to differentiate the major groups of spiny teleosts. We can see the primitive pattern in the Upper Cretaceous genus *Nematonotus* (Figure 7-15). It has a free second ural centrum, six hypurals, three epurals, and two uroneurals. The first ural and first preural centra are fused into a compound centrum carrying the first preural haemal arch as well as the two lower hypurals.

This pattern is modified in different ways in acanthopterygians and paracanthopterygians. In acanthopterygians, the second ural centrum is fused with the preceding compound centrum, and the number of hypurals is reduced to five. Three epurals are retained in the primitive state. In paracanthopterygians, the second ural centrum does not fuse with the more anterior caudal centrum but with the dorsal hypurals. The number of epurals is reduced to two and there are 16 branched caudal fin rays.

These caudal features are highly modified in specialized genera and are not absolutely consistent within the two groups, but they do provide a relatively effective means of differentiating primitive lineages.

PRIMITIVE ACANTHOMORPHS—THE CTENOTHRISSOIDS

The ctenothrissoids, *Ctenothrissa* (Figure 7-16) and *Aulolepis,* accompany the aulopiforms and myctophiforms in the early Upper Cretaceous (Cenomanian). The short, deep body of these fishes greatly resembles that of generalized acanthopterygians and primitive paracanthopterygians. There are 29 to 35 vertebrae. The pectoral fins have moved far up the side of the shoulder region, and the pelvic fins have migrated forward to a position in the same transverse plane. As in primitive teleosts, there are still two supramaxillae, but the maxilla is excluded from the margin of the mouth. There are no true fin spines,

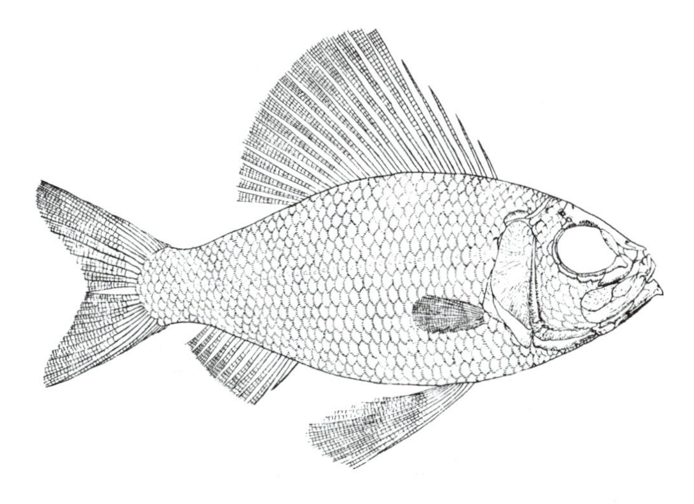

Figure 7-16. *Ctenothrissa,* a Cretaceous fish that closely approaches the structure of the spiny finned teleosts, ×⅓. *From Patterson, 1964.*

but the scales are ctenoid. There is no adipose dorsal fin. The caudal fin retains six hypurals and three epurals. There is a large third pharyngobranchial.

The ctenothrissoids have an anatomical pattern that could be ancestral to both major groups of spiny teleosts.

PARACANTHOPTERYGIANS

It has been thought that acanthopterygians and paracanthopterygians evolved in parallel from a salmoniform level and separately developed fin spines and complex protrusible mouthparts. But Rosen and Patterson (1969) demonstrated that the earliest paracanthopterygians closely resemble the early acanthopterygians and that they subsequently lost primitive acanthopterygian features. Apparently the groups diverged from each other at a level slightly advanced above that represented by the ctenothrissoids. The early paracanthopterygians appear later in the fossil record than the first acanthopterygians.

We recognize approximately 30 families and 200 to 250 genera of living paracanthopterygians. They are grouped in five orders (Percopsiformes, Gadiformes, Batrachoidiformes, Lophiiformes, and Gobiesociformes), most of which have fossil records dating from at least as early as the Eocene. Lauder and Liem (1983) point out that it is uncertain that these orders all share a single common ancestry since no unique derived characters have been recognized.

Modern percopsiforms are entirely freshwater and include the pirate and trout "perch" and the blind cave fish *Amblyopsis.* They resemble salmonids in their general body outline and the position of the pelvic fins and have a relatively small number of fin spines. Our knowledge of earlier percopsiforms indicates that these features result from a reversal of evolutionary trends. The ancestral genera had a greater number of spines and more anteriorly placed pelvic fins. Even the number of vertebrae appears to have increased between primitive and advanced Percopsiformes.

The earliest known and only Cretaceous paracanthopterygian is the percopsiform *Sphenocephalus* (Figure 7-17) from the Campanian, near the end of the Upper Cretaceous. It greatly resembles the polymixioid beryciforms (classified among the percomorph fishes) and supports the close relationships of acanthopterygians and paracanthopterygians. It shows less specialization in cranial proportions than do later percopsiforms but has more medial fin spines and more fin rays in the pelvic fin.

Such important commercial fish as the cod, haddock, and whiting belong to the Gadiformes. Members of this group tend to have elongate anal and dorsal fins, and the latter may be divided in two or three parts. The pelvic fins are usually anterior to the pectoral, but fin spines are very reduced. Eocene gadiforms closely resemble the early percopsids.

The batrachoidiform lineage includes three orders of bottom dwelling and/or deep-sea fishes. In this assemblage, the skull is greatly flattened so that the parasphenoid either approaches or is sutured to the frontals. In some forms, the pelvic fins are modified for "walking" along the bottom. The Batrachoidiformes, or toad fishes, live at the bottom of shallow coastal waters. Some can move about on land and utter loud grunts and growls by use of the swim bladder. Only a single fossil is known, a Miocene member of the modern genus *Batrachoides.*

The Lophiiformes have 18 families, including some of the most striking deep-sea fish, the anglers (Figure 7-18). The European angler *Lophius* is known from the Eocene. Like other members of this group, the anterior dorsal fin ray is elongate and hangs over the mouth to serve as a lure to attract prey. This lure has the technical name **illicium.** They have wide skulls and mouths capable of engulfing large prey. In at least four families of deep sea anglers, the males are much smaller than the females. Following metamorphosis from a shallower water larval stage, the male attaches to the female and feeds from her

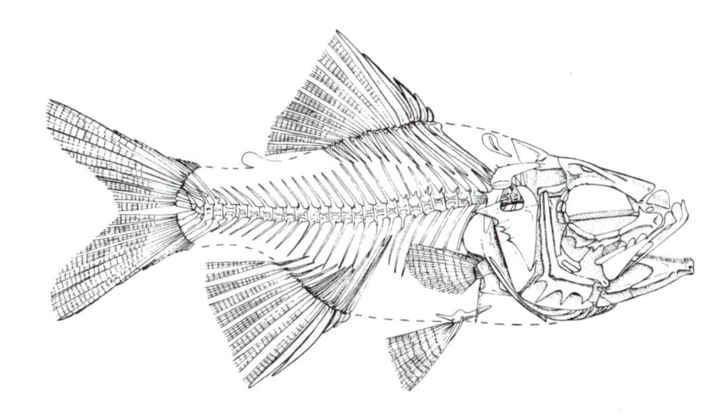

Figure 7-17. *Sphenocephalus fissicaudus* from the Upper Cretaceous, the oldest fossil paracanthopterygian. *From Rosen and Patterson, 1969. Courtesy of the Library Services Department, American Museum of Natural History.*

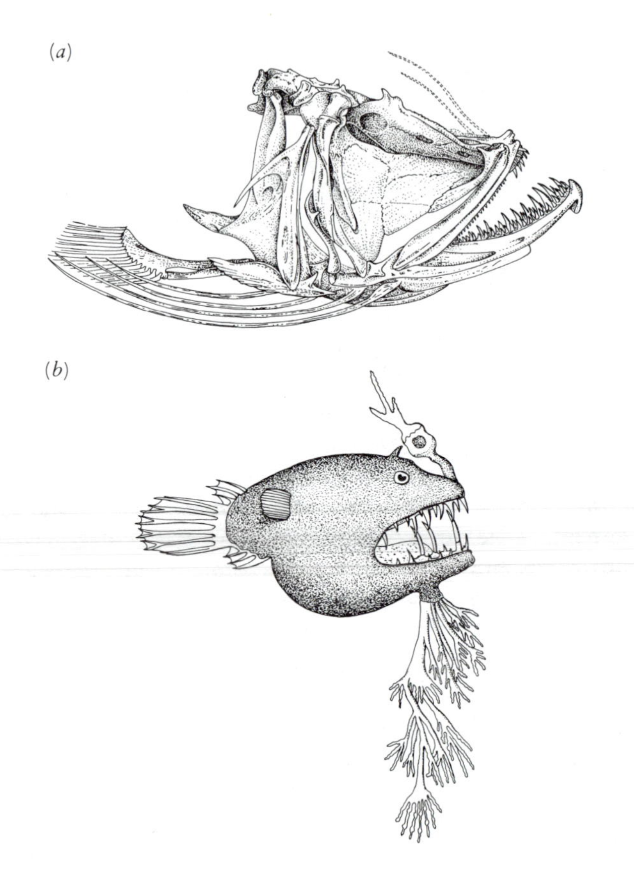

Figure 7-18. ANGLER FISH. (*a*) The living genus *Lophius*, which has fossil representatives from the Eocene. (*b*) The deep-sea angler *Linophryne arborifer*. *From Gregory, 1951. Courtesy of the Library Services Department, American Museum of Natural History.*

bloodstream, which ensures he will never have to search for a mate in the dark and cold of the bottom of the sea. This reproductive strategy is unique among vertebrates.

The Gobiesociformes, or cling fish, are dorsoventrally flattened, with the pelvic fin and surrounding tissue forming a ventral sucker to hold the fish in place in the fast-moving water of tidal zones and coastal streams. One genus, which is questionably associated with this group, has been described from the Miocene.

ACANTHOPTERYGIANS

The largest single group of teleosts are the acanthopterygians, which include some 8000 living species. This group can be recognized by the elaboration of the mechanism for protruding the upper jaw. The development of a long, ascending process on the premaxilla facilitates its anterior extension. The pharyngeal jaws are also modified, with the retractor dorsalis (RAB) attaching principally or entirely on the third pharyngobranchial. The second and third epibranchials are enlarged for support of the upper pharyngeal jaws. Acanthopterygians can be divided into two unequal groups, the atherinomorphs and the percomorphs. Both have elaborated the capacity to protrude the upper jaws, but the mechanisms in the two groups are different (Alexander, 1967a, 1967b).

ATHERINOMORPHA

The Atherinomorpha is a heterogenous assemblage of approximately 1000 genera that Rosen and Parenti (1981) and Parenti (1981) group in 20 families. The mechanism for protruding the upper jaw differs from that of the other acanthopterygian fish in that the medial process of the maxilla does not articulate directly with the premaxilla but rather articulates with the median rostral cartilage that is linked to the premaxilla (Figure 7-19). The ascending process of the atherinomorphs' premaxilla has the unique ability to slide forward on the median rostral cartilage as the skin over the snout is pulled taut by the rotation of the maxilla. Alexander (1967b) thinks that it is likely that jaw protrusion evolved within the atheriniform fish rather than being a primitive characteristic of this assemblage, since the most primitive members of each of the major subgroups lack this capacity. The Atherinomorpha is also distinguished by the loss of infraorbital bones 3, 4 and 5 and the fourth pharyngobranchial, which are retained in primitive percomorphs. The killifishes (Cyprinodontidae), silversides and grunions (Atherinidae), halfbeaks (Hemiramphidae), marine flying fishes (Exocoetidae), and the guppies (Poeciliidae) are among the better-known atherinomorphs. Rosen and Parenti have defined each family on the basis of differences in the structure of the dorsal gill arches.

There are fossil representatives of living genera of the families Atherinidae, Exocoetidae, and Hemiramphidae from the Eocene.

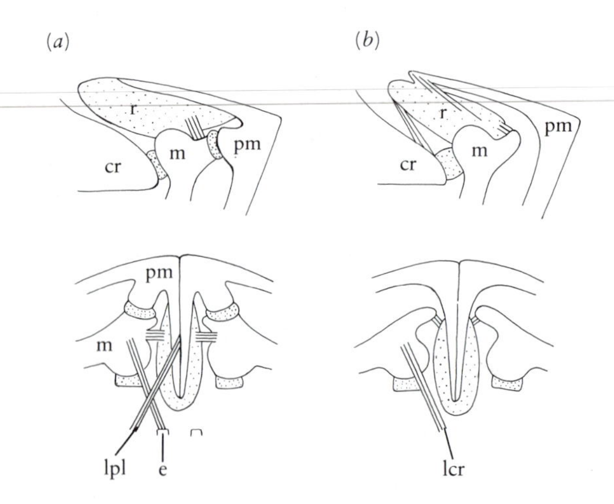

Figure 7-19. JAW PROTRUSION IN SPINY TELEOSTS. (*a*) Percomorph fish. (*b*) Atheriniformes. Snout regions in lateral and dorsal views. Cartilage is stippled. In Atheriniformes, the inner process of the maxilla articulates with the median ethmoid cartilage; in percomorphs, it articulates directly with the premaxilla. Abbreviations as follows: cr, cranium; e, ethmoid bone; lcr, ligament to cranium; lpl, ligament to palatine; m, maxilla; pm, premaxilla; r, rostral cartilage. *From Alexander, 1967b. By permission of the Zoological Society of London.*

PERCOMORPHA

Although we can define the Atherinomorpha and the Acanthopterygi as a whole on the basis of derived characters common to all the major subgroups, we cannot distinguish the percomorphs as readily. Some orders, including the Pleuronectiformes and Tetraodontiformes, can be clearly defined, but the great assemblage of perciform fishes (the most diverse and numerous of all vertebrate orders) and their primitive relatives are in need of general study to determine what monophyletic groups exist. The enormity of this assemblage makes any comprehensive revision extremely difficult.

Primitive percomorphs

According to Lauder and Liem (1983), three orders of percomorphs may have branched off the main stem prior to the appearance of the perciforms and their close relatives. These are the Lampridiformes, Gasterosteiformes, and Dactyopteriformes. We know the Gasterosteiformes from fossils as early as the Paleocene, but the other orders have a very limited fossil record.

The Lampridiformes are mainly deep-water fish, of which the best known genus, the opah *Lampris,* is known from the Miocene. This order also includes a bizarre assemblage of deep-sea forms, most of which have no fossil record. Lampridiforms lack fin spines and have a unique mechanism of jaw protrusion in which the maxilla slides in and out with the premaxilla rather than being attached to the ethmoid. Unlike other percomorphs, the pelvic girdle is not attached to the cleithrum.

The fossil record is better known among the Gasterosteiformes (encompassing the Syngnathiformes and Pegasiformes), many of which have large bony plates covering much of the body. The Syngnathiformes, which include the pipefish, sticklebacks, seahorses, and trumpetfish, are characterized by long, narrow bodies. The sea moths have greatly expanded pectoral fins that enable them to glide over the bottom. As is typical of percomorphs, the spinous and soft parts of the dorsal fin are separated (unless secondarily lost), and in some genera the pelvic girdle is attached to the cleithrum.

In the Dactylopteriformes (flying Gurnard), the pectoral fin is greatly expanded so that the fish can glide above the water like members of the Exocoetidae. They can also crawl along the bottom with their pelvic fins. No fossils are known, and we have no hint of the origin of the single genus within this order.

The remaining percomorphs all appear to form a single vast assemblage that shares a common ancestry in the Upper Cretaceous. The earliest known and most primitive percomorphs are grouped in the order Beryciformes, which includes some 15 families, 11 of which are present in the modern fauna. The polymixiids, berycids, holocentrids, and stephanoberycoids are particularly important beryciform groups (Figures 7-20 and 7-21). They resemble modern perciform fish in the anterior position of the pelvic fin, deep body, and the beginning of fin spines. Approximately 15 beryciform genera are known in the Upper Cretaceous.

The beryciforms are defined primarily by primitive characters such as the retention of an orbitosphenoid. However, they have lost basibranchial and endopterygoid teeth. There are 18 or 19 fin rays in the tail, whereas more advanced forms never have more than 17. Three epurals are present.

The beryciforms and perciforms probably share an immediate common ancestry that is characterized by the attachment of the pelvic to the pectoral girdle and the presence of one spine and five soft rays on the pelvic fin. Patterson (1964) argues that the similarities of the otoliths (ossification within the sacculus of the inner ear) in beryciforms and zeiforms support a common ancestry of these groups following the divergence of the perciforms. Zeiformes (dories and boarfish, which resemble the perch) appear in the fossil record in the Paleocene. They are specialized in the reduced number of vertebrae, deepening of the body, and loss of the orbitosphenoid, all of which are common tendencies among the percomorphs.

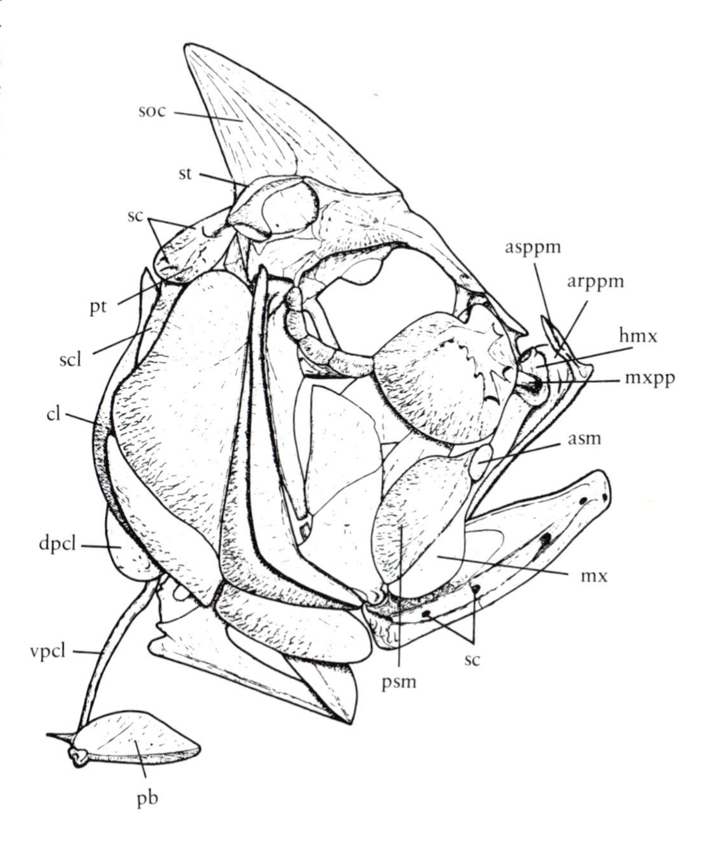

Figure 7-20. SKULL OF THE CRETACEOUS BERYCIFORM *BERYCOPSIS.* Abbreviations as follows: arppm, articular process of premaxilla; asm, anterior supramaxilla; asppm, ascendary process of premaxilla; cl, cleithrum; dpcl, dorsal postcleithrum; hmx, articular head of maxilla; mx, maxilla; pb, pelvic bone; psm, posterior supramaxilla; pt, posttemporal; sc, foramen of sensory canal; scl, supracleithrum; soc, supraoccipital; st, supratemporal; vpcl, ventral postcleithrum. *From Patterson, 1964.*

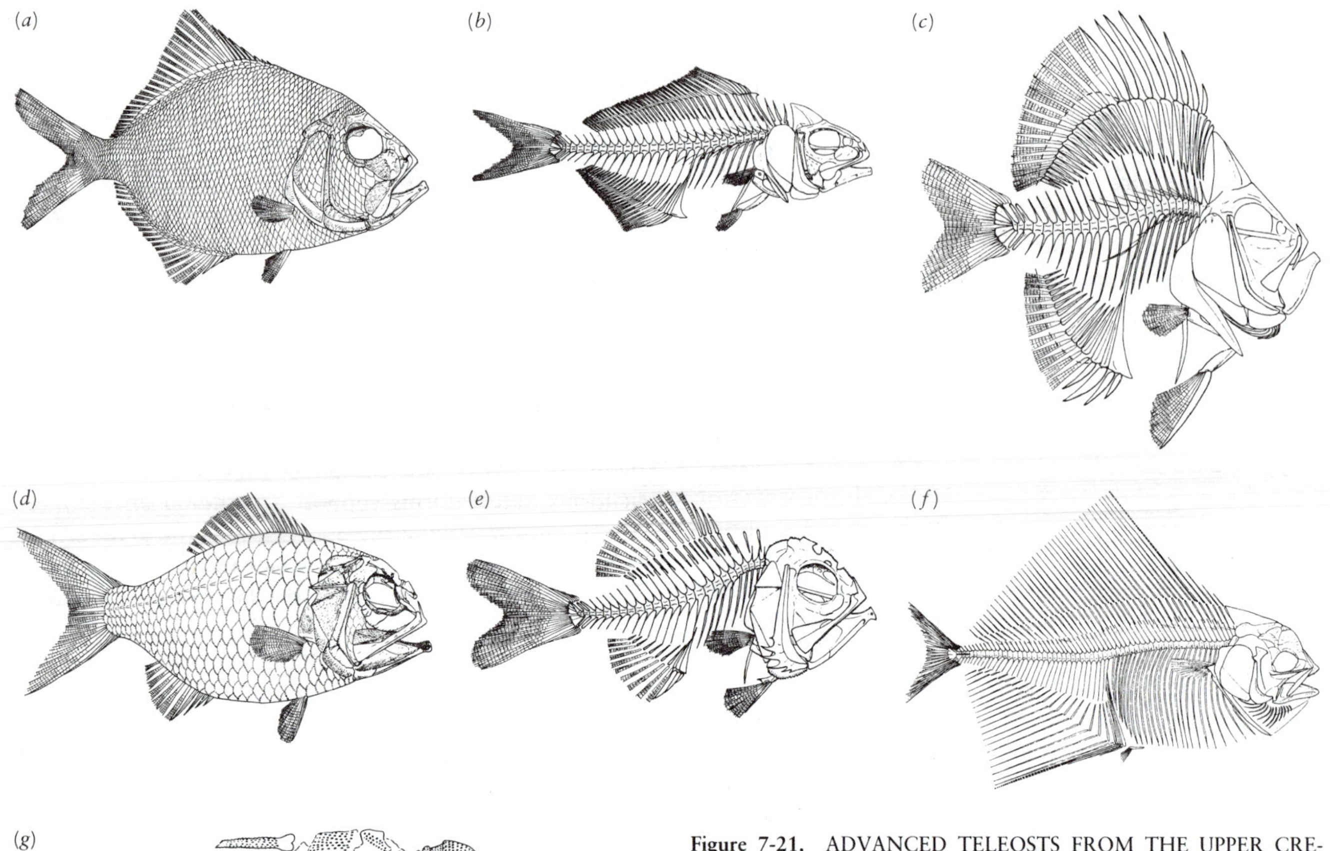

(g)

Figure 7-21. ADVANCED TELEOSTS FROM THE UPPER CRETACEOUS. (*a–e*) Members of the order Beryciformes. (*a* and *b*) *Berycopis* (English Chalk) and *Omosoma* (Upper Senonian) family Polymixiidae. (*c*) *Pycnosteroides* (Upper Cenomanian), family Pycnosteroididae. (*d* and *e*) *Hoplopteryx* (Upper Senonian) and *Acrogaster* (Lower Cenomanian), Trachithyidae. (*f*) *Tselfatia* (Lower Cenomanian), a possible member of the Exocoetoidei within the Antherinomorpha. (*g*) *Protriacantha* (Lower Cenomanian), uncertain position, but possibly allied with the Gasterosteiformes. *From Patterson, 1964.*

Lauder and Liem (1983) recognized several specialized features that unite an assemblage including the most advanced of all the teleost fishes. The Perciformes, Scorpaeniformes, Tetraodontiformes, Pleuronectiformes, Synbranchiformes, and Channiformes are all advanced over other acanthopterygians in the loss or fusion of the second ural centrum, reduction of the number of principal fin rays to 17, and the presence of 5 hypurals.

Perciformes

As currently classified, the Perciformes have the largest number of species of any vertebrate order (about 7800 grouped in 150 families) and show the greatest amount of morphological diversity. Unfortunately, this assemblage cannot be rigorously defined taxonomically. We cannot recognize a single derived character or combination of characters that uniquely distinguishes the perciforms. The Perciformes are probably not monophyletic but may have evolved from more than one lineage of acanthopterygians at the beryciform level. Perciforms are probably also paraphyletic, for they are generally thought to have given rise to several more specialized percomorph orders.

The perciforms are currently understood as including all advanced acanthopterygians that do not belong to the orders Scorpaeniformes, Tetraodontiformes, Pleuronectiformes, Synbranchiformes, and Channiformes. As such, this assemblage includes some 20 suborders. The definition of these suborders is also subject to dispute. Lauder and Liem (1983) discuss some of these problems. Nelson (1984) briefly describes and illustrates each of the suborders and families.

Figure 7-24 shows a general view of perciform diversification. The pattern of evolution of this assemblage is far different from that of their contemporaries among the mammals (see Chapters 20 and 21). Following a period of rapid adaptive radiation in the late Cretaceous and early Cenozoic, most mammalian groups show a gradual change in their morphology throughout the Tertiary. Many archaic groups have arisen and become extinct, and the modern fauna appears only within the last 10 million years. The percoid fish must have undergone a period of extremely rapid anatomical evolution in the late Cretaceous and early Cenozoic, because most of the groups have already achieved an essentially modern appearance by the Eocene (Figures 7-22 to 24). Of the 20

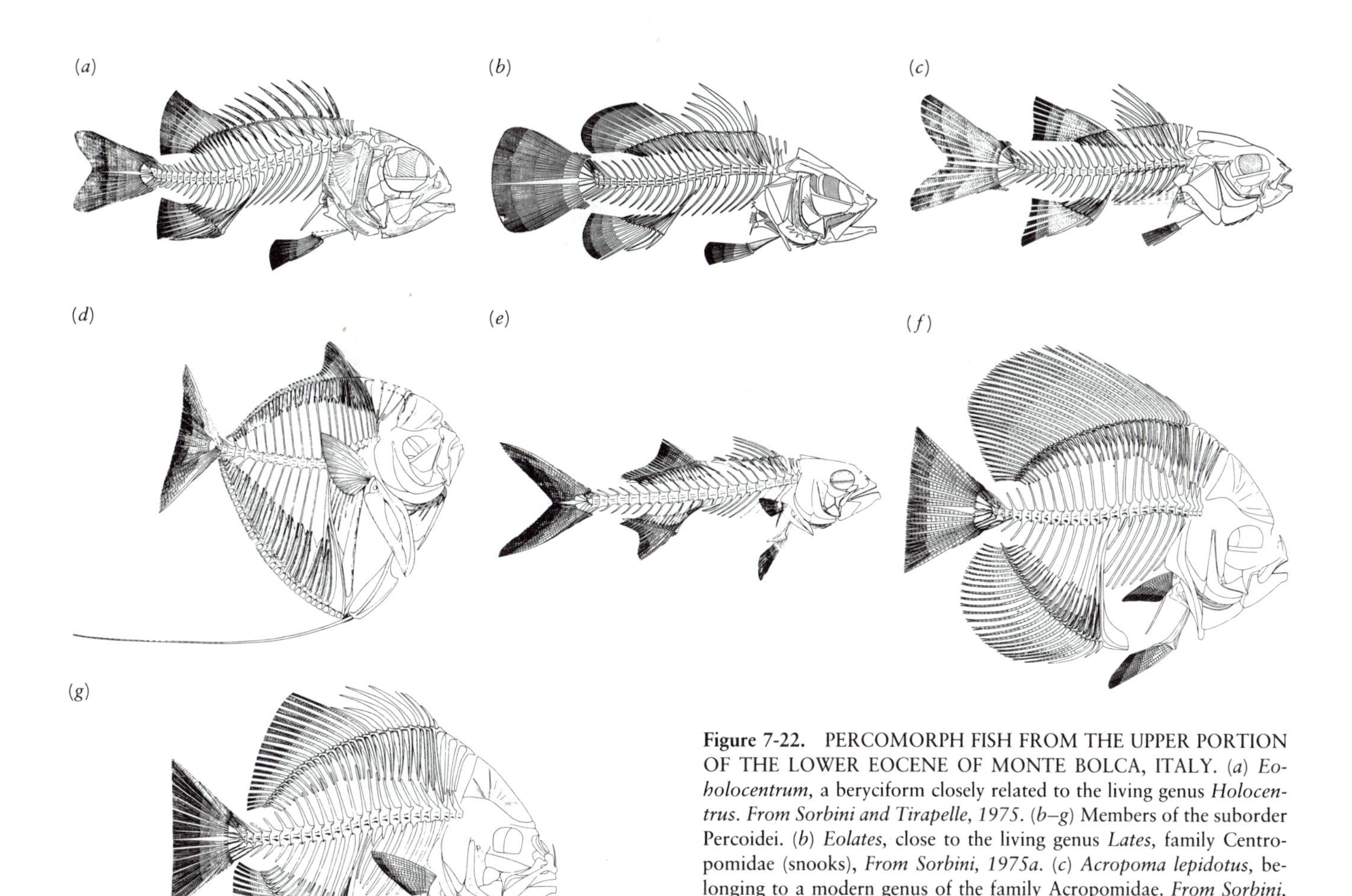

Figure 7-22. PERCOMORPH FISH FROM THE UPPER PORTION OF THE LOWER EOCENE OF MONTE BOLCA, ITALY. (*a*) *Eoholocentrum*, a beryciform closely related to the living genus *Holocentrus*. *From Sorbini and Tirapelle, 1975.* (*b–g*) Members of the suborder Percoidei. (*b*) *Eolates*, close to the living genus *Lates*, family Centropomidae (snooks), *From Sorbini, 1975a.* (*c*) *Acropoma lepidotus*, belonging to a modern genus of the family Acropomidae. *From Sorbini, 1975b.* (*d*) *Mene rhombea*, belonging to the only genus of the living family Menidae. *From Blot, 1969.* (*e*) *Seriola prisca*, belonging to a modern genus in the family Carangidae. *From Blot, 1969.* (*f*) *Psettopsis*, a member of the family Monodactylidae, close to the living moonfish *Psettias*. *From Blot, 1969.* (*g*) *Archaephippus*, a member of the family Ephippidae, close to the modern spadefishes. *From Blot, 1969.* This and the following figure demonstrate the great range of variability within the suborders Percoidei as well as the early establishment of modern genera.

suborders, 6 have no fossil record, but these include only 14 of the 1367 living genera. Of the remaining 14 suborders, 11 are represented by living genera in the Paleocene or Eocene, and the others are first represented by living genera by the Oligocene or the Miocene. Examples of modern genera present in the early Tertiary include such highly specialized fish as the tuna, swordfish, barracudas, anglefish, remoras, as well as the common perch. Only one or two families have been described that lack living representatives.

One may see this pattern as indicative of the slow rate of morphological change since the early Tertiary, but one can also look at it as evidence of very rapid evolution from primitive acanthopterygians in the late Mesozoic and early Cenozoic.

Only about 20 million years is available for the essentially modern and enormously diverse percomorph fauna to evolve from the beryciforms of the late Cretaceous. On the other hand, if beryciforms and perciforms are representatives of a single radiation, the differentiation of the modern "percomorph" lineages may have begun as early as the beginning of the Upper Cretaceous. This would allow perhaps 50 million years for the "modern" genera to evolve from primitive, spiny finned fishes.

"PERCIFORM" DERIVATIVES

The remaining percomorph orders are thought to have diverged from the perciform assemblage. Each is highly specialized and readily defined. One lineage is made up of two very small orders, the Channiformes (snakeheads), with one modern genus also known from Pliocene fossils, and the Synbrachiformes, the swamp eels and rice eels, with four living genera but no fossil record. Both show a peculiar specialization of the brain—fusion of the forebrain hemispheres—and modification of the circulatory system associated with air breathing. Specific ancestry of these orders within the perciform assemblage has not been established.

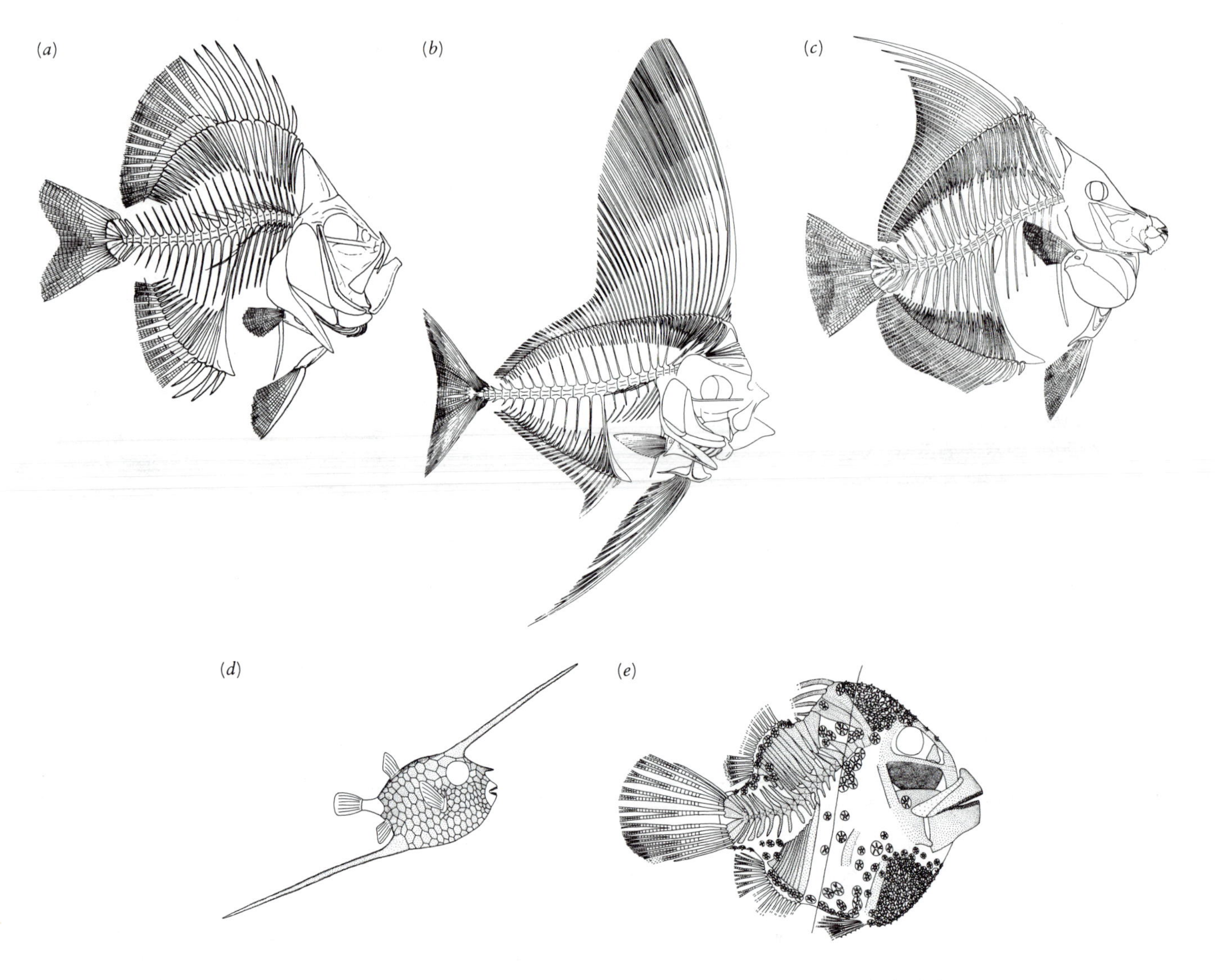

Figure 7-23. PERCOMORPH FISH FROM MONTE BOLCA. (*a*) *Scatophagus frontalis,* belonging to a modern genus of the family Scatophagidae. *From Blot, 1969.* (*b*) *Exellia,* the only member of the Eocene family Exellidae, suborder Percoidei. *From Blot, 1969.* (*c*) *Eozanclus,* closely related to the modern moorish idol *Zanclus,* suborder Acanthuroidei. *From Blot and Voruz, 1975.* (*d*) *Eolactoria,* a member of the family Ostraciontidae, within the order Plectognathi. *From Tyler, 1975a.* (*e*) *Eoplectus,* a member of the family Triacanthodidae, within the order Plectognathi. *From Tyler, 1975b.*

Tetraodontiformes

The Tetraodontiformes (Plectognathi) are much more diverse. They include such well-known forms as puffers, porcupine fish, boxfish, filefishes, and sunfish. We recognize more than 300 species that are grouped in 8 families, all but one of which are represented by fossils from the Eocene (Figure 7-23). All have rounded bodies. The scales are usually modified as spines or plates. The hyomandibular and palatine are fused to the skull to support the heavy teeth, and the maxilla and premaxilla are solidly attached to one another. They are further characterized by the loss of the suborbital bones, parietals, and nasals. They may have evolved from the Acanthuroidei within the perciforms (Patterson 1964), but even the early Eocene tetraodontiforms look modern and have no obvious affinities with other groups.

Pleuronectiformes

Flat fish—flounders, sole, and halibut—are grouped in the order Pleuronectiformes. They are unique among vertebrates in having both eyes on one side of the head, a condition achieved during development from larvae that have symmetrical, laterally compressed bodies. The dorsal fin is extremely long; in all but one family it extends onto the head. The dorsal, anal, and pelvic spines are progressively lost within the group. Most of the eight families are known from Eocene fossils, including genera such as *Solea* and *Bothus* that are living today. Only 15 of the approximately 117 living genera are known from fossils, but most extend back to the early Tertiary. Only nine fossil genera have been described that are not represented by living species. This evidence strongly supports the notion that flat fish, like other percomorph groups, under-

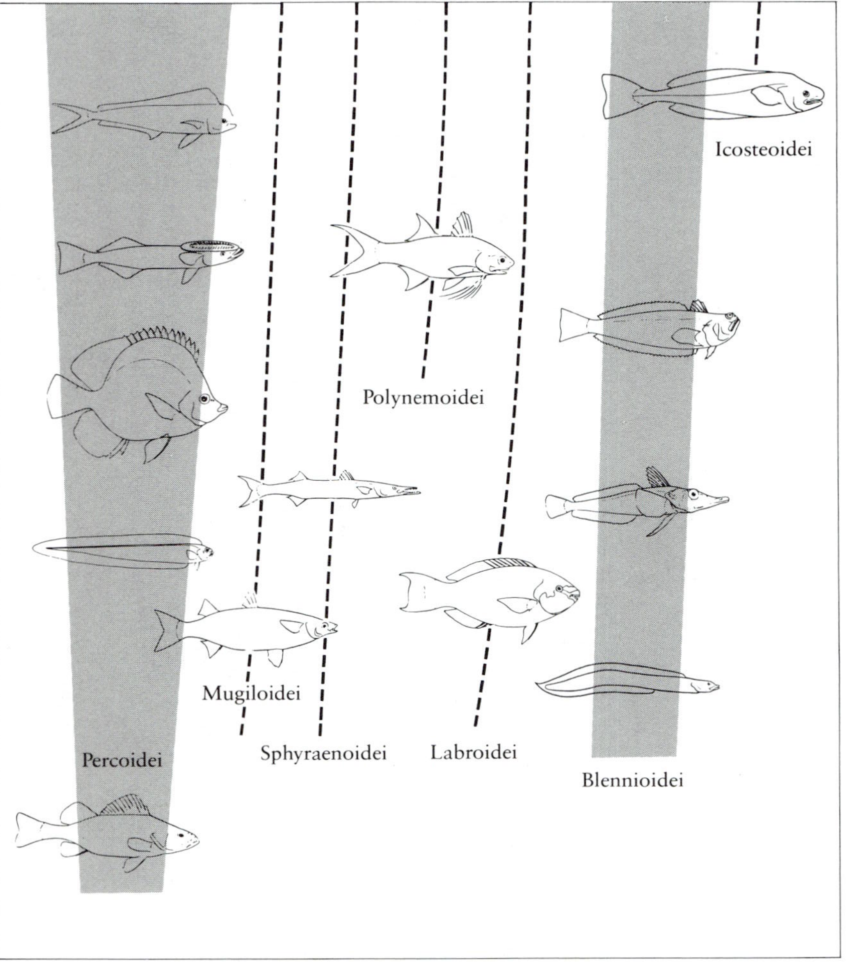

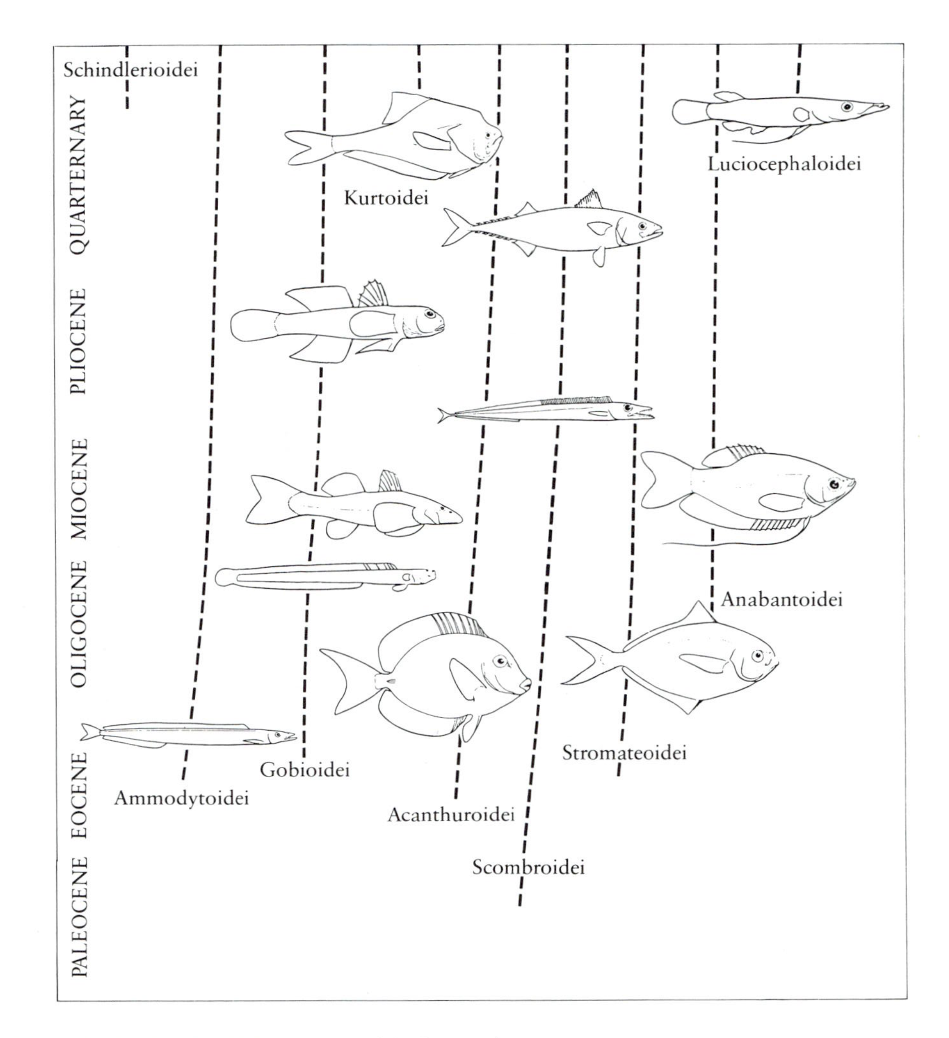

Figure 7-24. DIVERSITY AND LONGEVITY OF THE PERCIFORM SUBORDERS. All the fish illustrated are living genera; their vertical position has no significance. The Percoidei is the largest group, with 72 families. Only the infraorders Icostedidei, Schindlerioidei, Kurtoidei, and Luciocephaloidei have no fossil record. *Individual fish redrawn from* Fishes of the World *by T.S. Nelson (1984). Copyright © 1976. By permission of John Wiley & Sons, Inc.*

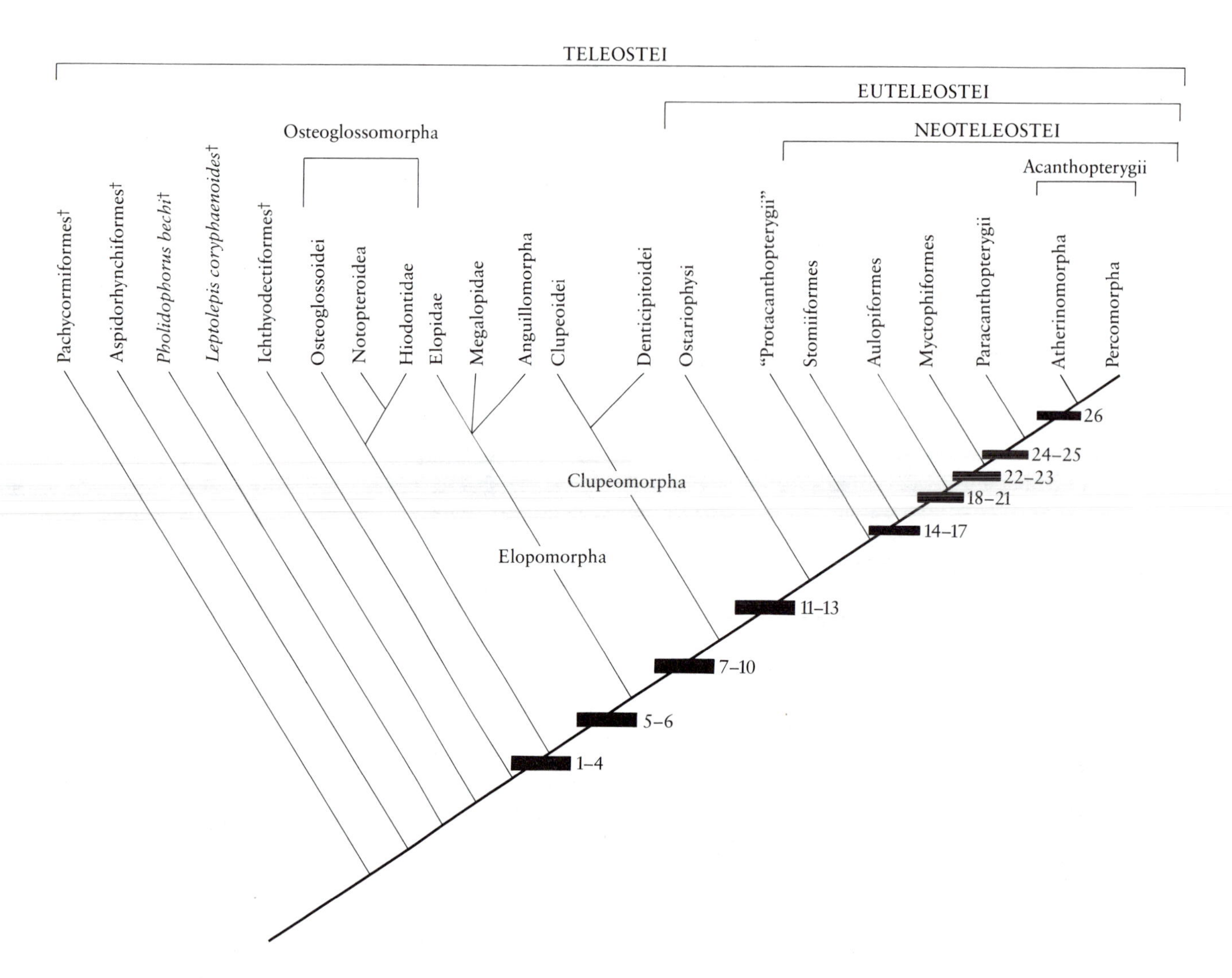

Figure 7-25. PHYLOGENY OF THE TELEOSTEI. Taxa known only from fossils are indicated with a dagger. The characters are: (1) presence of an endoskeletal basihyal; (2) four pharyngobranchials; (3) three hypobranchials present; (4) medial toothplates overlay basibranchial and basihyal cartilages; (5) two uroneurals extend anteriorly over the second ural centrum; (6) epipleural intermuscular bones developed throughout the abdominal and anterior caudal regions; (7) retroarticular bone excluded from the quadratomandibular joint surface; (8) toothplates fused with endoskeletal gill arch elements; (9) neural arch on ural centrum one reduced or absent; (10) articular bone coossified with the angular; (11) presence of an adipose fin; (12) presence of nuptial tubercles; (13) presence of an anterior membraneous outgrowth of the first uroneural that does not meet its antimere in the midline; (14) presence of a retractor dorsalis muscle; (15) a rostral cartilage; (16) tooth attachment to bone is Type 4; (17) Aw division of adductor mandibulae has a posterior tendinous insertion on the quadrate, preoperculum, or operculum; (18) reduction of second preural neural spine to a half spine; (19) retractor dorsalis with a tendinous insertion on the third pharyngobranchial; (20) loss of the primitive mandibulohyoid ligament and the presence instead of an interoperculo-hyoid ligament; (21) fusion of a toothplate to the third epibranchial. (In 1973, Rosen listed several additional characters at this level, but these have been studied in more detail and are not found to be uniquely derived features); (22) pharyngohyoideus (the primitive teleostean rectus communis) inserts on the urohyal; (23) reduction or loss of pharyngobranchial four and the main insertion of the retractor dorsalis onto the third pharyngobranchial; (24) the presence of well-developed ctenoid scales (however, ctenoid scales are also present in more primitive clades); (25) expansion of ascending and articular premaxillary processes. This level on the cladogram is not well defined. (26) insertion of the retractor dorsalis onto the third pharyngobranchial only, and enlargement of epibranchials two and three. Various other features of the upper jaw mechanism are discussed by Rosen (1973, 1982) but none are unique to the Acanthopterygii and this group remains poorly defined. *From Lauder and Liem, 1983.*

went one basic radiation in the earliest Tertiary with very rapid establishment of essentially modern body plans.

The specific origin of this group remains unknown. Amaoka (1969) suggests that they may have evolved from more than a single percoid lineage.

Scorpaeniformes

Scorpaeniformes include the stonefish, sculpins, searobins, and approximately 1000 additional species. Their ancestry remains unknown and the interrelationships of the 21 constituent families are uncertain. Members of this order have a narrow process from the third circumorbital bone that extends across the cheek to the premaxilla and two platelike hypurals. There are four suborders, two of which are represented by modern genera as early as the Eocene. In many genera, the dorsal, anal, and pelvic spines carry dangerous toxins. The genus *Sebastes* has 100 species which bear live young.

Figure 7-25 is a cladogram that summarizes the relationships among the Teleostei.

SUMMARY

Most of the 20,000 species of modern bony fish belong to a single taxonomic group, the Teleostei, which is first recognized in the late Triassic. The earliest teleosts, the pholidophorids and leptolepids, are distinguished from more primitive neopterygians by derived features of the jaws and braincase that presumably improved their capacity to capture and manipulate prey and by changes in the caudal fin that led to more effective locomotion.

The ancestry of all advanced teleost groups may be traced to the late Triassic and Jurassic leptolepids, which were small fish that resemble modern plankton feeders. Teleosts are divided into five major groups whose origins can be separately traced to the leptolepid assemblage. The Ichthyodectiformes are typified by large predatory genera. They are limited to the Mesozoic. The osteoglossomorphs, elopomorphs, and clupeomorphs evolved from progressively more advanced leptolepids.

More advanced teleosts, the euteleosts, which first appear at the base of the Upper Cretaceous, share a common ancestry with the clupeomorphs. Like the modern salmonids, the most primitive euteleosts have relatively long bodies, a second adipose dorsal fin, and a caudal fin possessing a stegural. The Ostariophysi and Salmoniformes shared a close common ancestry with the more advanced neoteleosts.

Neoteleosts are characterized by changes in fin position and body form and by elaboration of the pharyngeal region for crushing and manipulating prey. They possess a large muscle, the retractor arcuum branchialium, that runs from the anterior vertebrae to the dorsal elements of the pharyngeal apparatus. Stomiiformes, Aulopiformes and Myctophiformes are the most primitive teleosts in which this muscle is developed. They were common nectonic fish in the early Cretaceous, but their living descendants are primarily deep-sea forms.

The Myctophiformes may include the ancestors of the most advanced teleosts, the Acanthomorpha or spiny teleosts, which are characterized by the elaboration of spiny fin rays and ctenoid scales. The Acanthomorpha represent the culmination of a series of trends among the teleosts toward improved locomotion. In the central stock of higher teleosts, the body is laterally compressed and has fewer trunk vertebrae. The tail produces most of the propulsive force, and the trunk is nearly rigid. The pectoral fins are high on the shoulder, and the pelvic fins have moved anteriorly and are attached to the cleithrum.

The vast assemblage of higher teleosts radiated from the primitive Acanthomorpha, which are represented in the early Upper Cretaceous by the ctenothrissoids. The great adaptive radiation of the group may have resulted largely from their improved capacity to capture and manipulate prey. Most members of this assemblage can protrude their jaws to maintain reduced pressure as the oropharyngeal chamber is closed. The paracanthopterygians, which are represented by the cod, can move the maxilla extensively, but the premaxilla remains relatively immobile. In the acanthopterygians, the premaxilla protrudes as a result of the rotation of a camlike process of the maxilla. Atherinomorphs have separately evolved the capacity to protrude the premaxilla. They have a cam that moves the median rostral cartilage, rather than being attached directly to the premaxilla.

Most acanthopterygians belong to a single assemblage, the Percomorpha. This group is represented early in the Upper Cretaceous by the beryciforms, which may constitute a primitive grade of evolution within a larger assemblage that includes the perciform fishes and their derivatives. Most of the modern percomorph families are known from the early Cenozoic, and numerous living genera are represented in the Eocene. This vast assemblage apparently evolved relatively rapidly during the Upper Cretaceous and Paleocene and had achieved a modern appearance by the Eocene.

REFERENCES*

Alexander, R. McN. (1967a). The functions and mechanisms of the protrusible upper jaws of some acanthopterygian fish. *J. Zool., Lond.*, **151**: 43–64.

Alexander, R. McN. (1967b). Mechanisms of the jaws of some atheriniform fish. *J. Zool., Lond.*, **151**: 233–255.

Alexander, R. McN. (1970). *Functional Design in Fishes*. Hutchison, London.

Amaoka, K. (1969). Studies on the sinistral flounders found in the waters around Japan—taxonomy, anatomy, and phylogeny. *J. Shimonoseki Univ. Fish.*, **18**(2): 1–340.

Arratia, G. (1981). *Varasichthys ariasi* n. gen. et sp. from the Upper Jurassic of Chile (Pisces, Teleostei, Varasichthyidae n. fam.). *Palaeontographica, Abt. A*, **175**: 107–195.

Bardack, D. (1965). Anatomy and evolution of chirocentrid fishes. *Paleont. Contr. Univ. Kansas, Vertebrata*, **10**: 1–88.

Bartram, A. W. H. (1977). The Macrosemiidae, a Mesozoic family of holostean fishes. *Bull. Brit. Mus. (Nat. Hist.) Geol.*, **29**: 137–234.

Bjerring, H. C. (1985). Facts and thoughts on piscine phylogeny. In R. E. Forman, A. Gorbman, J. M. Dodd, and R. Olsson (eds.), *Evolutionary Biology of Primitive Fishes*. Plenum Press, New York.

Blot, J. (1969). Les poissons fossiles du Monte Bolca classés jusqu'ici dans les familles des Carangidae-Menidae-Ephippidae-Scatophagidae. *Studi e Ricerche sui Giacimenti Terziari di Bolca. I. Mem. Mus. Civ. Storia Nat. Verona;* Mem. out of normal ser. no. 2, Tome I (text), 526 pp., Tome II (plates), Museo Civico di Storia Naturale di Verona, Verona.

Blot, J. (1984). Actinopterygii. Ordre des Scorpaeniformes? Famille des Pterygocephalidae Blot 1980. In *Studi e Ricerche sui Giacimenti Terziari di Bolca. IV. Miscellanea Paleontologica*. II., pp. 265–299. Museo Civico di Storia Naturale di Verona, Verona.

*These references are for chapters 6 and 7.

Blot, J., and Voruz, C. (1975). La famille de Zanclidae. In *Studi e Ricerche sui Giacimenti Terziari di Bolca. II. Miscellanea Paleontologica*, pp. 233–278. Museo Civico di Storia Naturale di Verona, Verona.

Bond, C. E. (1979). *Biology of Fishes*. Saunders, Philadelphia.

Denison, R. H. (1979). *Acanthodii. Handbook of Paleoichthyology*, Vol. 5. Gustav Fischer Verlag, Stuttgart.

Denison, R. H. (1983). Further consideration of placoderm evolution. *J. Vert. Paleont.*, 3(2): 69–83.

Fink, S. V., and Fink, W. L. (1981). Interrelationships of the ostariophysan teleost fishes. *Zool. J. Linn. Soc.*, **72**: 297–353.

Fink, W. L., and Weitzman, S. H. (1982). Relationships of the stomiiform fishes (Teleostei), with a description of *Diplophos*. *Bull. Mus. Comp. Zool.*, **150**: 31–93.

Forey, P. L. (1973a). Relationships of elopomorphs. In P. H. Greenwood, R. S. Miles, and C. Patterson (eds.), *Interrelationships of Fishes*, pp. 351–368. Suppl. No. 1, *Zool. J. Linn. Soc.* Vol. 53. Academic Press, London.

Forey, P. L. (1973b). A revision of the elopiform fishes, fossil and Recent. *Bull. Brit. Mus. (Nat. Hist.) Geol. Suppl.*, **10**: 1–222.

Forey, P. L. (1973c). A primitive clupeomorph fish from the Middle Cenomanian of Hekel, Lebanon. *Can. J. Earth Sci.*, **10**: 1302–1318.

Gardiner, B. G. (1973). Interrelationships of teleostomes. In P. H. Greenwood, R. S. Miles, and C. Patterson (eds.), *Interrelationships of Fishes*. pp. 105–135. Suppl. No. 1, *Zool. J. Linn. Soc.* Vol. 53. Academic Press, London.

Gardiner, B. G. (1984). The relationships of the palaeoniscid fishes, a review based on new specimens of *Mimia* and *Moythomasia* from the upper Devonian of Western Australia. *Bull. Brit. Mus. (Nat. Hist.) Geol.*, **37**: 173–428.

Gardiner, B. G., and Bartram, A. W. H. (1977). The homologies of ventral cranial fissures in osteichthyans. In S. M. Andrews, R. S. Miles, and A. D. Walker (eds.), *Problems in Vertebrate Evolution. Linn. Soc. Symp. Ser.*, **4**: 227–245.

Goody, P. C. (1969). The relationships of certain Upper Cretaceous teleosts with special reference to the myctophoids. *Bull. Brit. Mus. (Nat. Hist.) Geol. Suppl.*, **7**: 1–255.

Grande, L. (1982a). A revision of the fossil genus *Diplomystus* with comments on the interrelationships of the clupeomorph fishes. *Amer. Mus. Novitates*, **2728**: 1–34.

Grande, L. (1982b). A revision of the fossil genus *Knightia*, with a description of a new genus from the Green River formation (Teleostei, Clupeidae). *Amer. Mus. Novitates* **2731**: 1–22.

Greenwood, P. H. (1977). Notes on the anatomy and classification of elopomorph fishes. *Bull. Brit. Mus. (Nat. Hist.) Zool.*, **32**: 65–102.

Greenwood, P. H., Rosen, D. E., Weitzman, S. H., and Myers, G. S. (1966). Phyletic studies of teleostean fishes, with a provisional classification of living forms. *Bull. Am. Mus. Nat. Hist.*, **131**: 339–456.

Gregory, W. K. (1933). Fish skulls. A study of the evolution of natural mechanisms. *Trans. Am. Phil. Soc.*, **23**: 75–481.

Gregory, W. K. (1951). *Evolution Emerging*. Vols. 1 and 2. Macmillan, New York.

Jarvik, E. (1977). The systematic position of acanthodian fishes. In S. M. Andrews, R. S. Miles, and A. D. Walker (eds.), *Problems in Vertebrate Evolution. Linn. Soc. Symp. Ser.*, no. 4: 199–225.

Jollie, M. (1980). Development of head and pectoral girdle skeleton and scales in *Acipenser*. *Copeia*, **1980**(2): 226–249.

Lauder, G. V., and Liem, K. F. (1983). The evolution and interrelationships of the actinopterygian fishes. *Bull. Mus. Comp. Zool.*, **150**(3): 95–197.

Lehman, J.-P. (1952). Étude complémentaire des poissons de L'Eotrias Madagascar. Kungl. Svenska Vetenskapsakad. Handlingar. 2: 1–201.

Lehman, J.-P. (1966). Actinopterygii. In J. Piveteau (ed.), *Traité de Paléontologie*, **4**: 1–242. Masson S. A., Paris.

Lundberg, J. G. (1975). The fossil catfishes of North America. *Mus. Paleont., Univ. Michigan, Papers on Paleont.*, **11** (Claude W. Hibbard Memorial Volume 2): 1–51.

McAllister, D. E. (1968). The evolution of branchiostegals and associated opercular, gular, and hyoid bones and the classification of teleostome fishes, living and fossil. *Bull. Natl. Mus. Canada*, **221**: 1–239.

Miles, R. S. (1965). Some features in the cranial morphology of acanthodians and the relationships of Acanthodi. *Acta Zoologica*, **46**: 233–255.

Miles, R. S. (1968). Jaw articulation and suspension in *Acanthodes* and their significance. *Proc. 4th Nobel Symp., Stockholm*, pp. 109–127.

Miles, R. S. (1973a). Articulated acanthodian fishes from the Old Red Sandstone of England, with a review of the structure and evolution of the acanthodian shoulder-girdle. *Bull. Brit. Mus. (Nat. Hist.) Geol.*, **24**: 113–213.

Miles, R. S. (1973b). Relationships of acanthodians. In P. H. Greenwood, R. S. Miles, and C. Patterson (eds.), *Interrelationships of Fishes*, pp. 63–103. Suppl. No. 1, *Zool. J. Linn. Soc.*, Vol. 53. Academic Press, London.

Moy-Thomas, J. A., and Dyne, B. M. (1938). Actinopterygian fishes from the Lower Carboniferous of Glencartholm, Eskdale, Dumfriesshire. *Trans. Roy. Soc. Edinburgh*, **59**: 437–480.

Moy-Thomas, J. A., and Miles, R. S. (1971). *Palaeozoic Fishes*. Chapman and Hall, London.

Nelson, G. J. (1969a). Gill arches and the phylogeny of fishes, with notes on the classification of vertebrates. *Bull. Am. Mus. Nat. Hist.*, **141**: 475–552.

Nelson, G. J. (1969b). Infraorbital bones and their bearing on the phylogeny and geography of osteoglossomorph fishes. *Amer. Mus. Novitates*, **2394**: 1–37.

Nelson, G. J. (1973). Relationships of clupeomorphs, with remarks on the structure of the lower jaw in fishes. In P. H. Greenwood, R. S. Miles, and C. Patterson (eds.), *Interrelationships of Fishes*. pp. 333–349. Suppl. No. 1, *Zool. J. Linn. Soc.* Vol. 53. Academic Press, London.

Nelson, J. S. (1984). *Fishes of the World*. (2d ed.) Wiley, New York.

Nybelin, O. (1966). On certain Triassic and Liassic representatives of the family Pholidophoridae s. str. *Bull. Brit. Mus. (Nat. Hist.) Geol.*, **11**: 351–432.

Nybelin, O. (1974). A revision of the leptolepid fishes. *Acta Regiae Societatis scientiarum et litterarum gothoburgensis, (Zoologica)*, **9**: 1–202.

Olsen, P. E. (1984). The skull and pectoral girdle of the parasemionotid fish *Watsonulus eugnathoides* from the Early Triassic Sakamena Group of Madagascar, with comments on the relationships of the holostean fishes. *J. Vert. Paleont.*, **4**(3): 481–499.

Ørvig, T. (1973). Acanthodian dentition and its bearing on the relationships of the group. *Palaeontographica, Abt. A*, **143**: 119–150.

Parenti, L. R. (1981). A phylogenetic and biogeographic analysis of cyprinodontiform fishes (Teleostei, Atherinomorpha). *Bull. Am. Mus. Nat. Hist.*, **168**: 335–557.

Patterson, C. (1964). A review of Mesozoic acanthopterygian fishes, with special reference to those of the English chalk. *Phil. Trans. Roy. Soc. London*, **247** (B): 213–482.

Patterson, C. (1968). The caudal skeleton in Lower Liassic pholidophorid fishes. *Bull. Brit. Mus. (Nat. Hist.) Geol.*, **16**: 201–239.

Patterson, C. (1970). Two Upper Cretaceous salmoniform fishes from the Lebanon. *Bull. Brit. Mus. (Nat. Hist.)Geol.*, **19**(5): 205–296.

Patterson, C. (1973). Interrelationships of holosteans. In P. H. Greenwood, R. S. Miles, and C. Patterson (eds.), *Interrelationships of Fishes*, pp. 233–305. Supplement no. 1, *Zool. J. Linn. Soc.*, vol. 53. Academic Press, London.

Patterson, C. (1975). The braincase of pholidophorid and leptolepid fishes, with a review of the actinopterygian braincase. *Phil. Trans. Roy. Soc. London (B)*, **269**: 275–579.

Patterson, C. (1982). Morphology and interrelationships of primitive actinopterygian fishes. *Amer. Zool.* **22**: 241–259.

Patterson, C. (1984). *Chanoides*, a marine Eocene otophysan fish (Teleostei: Ostariophysi). *J. Vert. Paleont.*, **4**(3): 430–456.

Patterson, C., and Rosen, D. E.(1977). Review of ichthyodectiform and other Mesozoic teleost fishes and the theory and practice of classifying fossils. *Bull. Am. Mus. Nat. Hist.*, **158**: 81–172.

Pearson, D. M. (1982). Primitive bony fishes, with especial reference to *Cheirolepis* and palaeonisciform actinopterygians. *Zool. Jour. Linn. Soc.*, **74**: 35–67.

Pearson, D. M., and Westoll, T. S. (1979). The Devonian actinopterygian *Cheirolepis* Agassiz. *Trans. Roy. Soc. Edinburgh*, **70**: 337–399.

Rieppel, O. (1985). Die Triasfauna der Tessiner Kalkalpen XXV. Die Gattung *Saurichthys* (Pisces, Actinopterygii) aus der mittleren Trias des Monte San Giorgio, Kanton Tessin. *Schweizerische Paläontologische Abhandlungen.*, **108**: 1–103.

Rosen, D. E. (1973). Interrelationships of higher euteleostean fishes. In P. H. Greenwood, R. S. Miles, and C. Patterson (eds.), *Interrelationships of Fishes*, pp. 397–513. Suppl. No. 1, *Zool. J. Linn. Soc.*, Vol. 53. Academic Press, London.

Rosen, D. E. (1982). Teleostean interrelationships, morphological function and evolutionary inference. *Amer. Zool.*, **22**: 261–273.

Rosen, D. E., and Parenti, L. R. (1981). Relationships of *Oryzias*, and the groups of atherinomorph fishes. *Amer. Mus. Novitates*, **2719**: 1–25.

Rosen, D. E., and Patterson, C. (1969). The structure and relationships of the paracanthopterygian fishes. *Bull. Am. Mus. Nat. Hist.*, **141**: 357–474.

Schaeffer, B. (1972). A Jurassic fish from Antarctica. *Amer. Mus. Novitates.*, **2495**: 1–17.

Schaeffer, B. (1973). Interrelationships of chondrosteans. In P. H. Greenwood, R. S. Miles, and C. Patterson (eds.), *Interrelationships of Fishes*, pp. 207–226. Suppl. No. 1, *Zool. J. Linn. Soc.* Vol. 53. Academic Press, London.

Schaeffer, B. (1978). Redfieldiid fishes from the Triassic-Liassic Newark Supergroup of Eastern North America. *Bull. Am. Mus. Nat. Hist.*, **159**: 133–173.

Schaeffer, B., and Rosen, D. E. (1961). Major adaptive levels in the evolution of the actinopterygian feeding mechanism. *Amer. Zool.* **1**: 187–204.

Schultze, H.-P. (1977). Ausgangsform und Entwicklung der rhombischen Schuppen der Osteichthyes (Pisces). *Paläont. Z.*, **51**: 152–168.

Sorbini, L. (1975a). Evoluzione e distribuzione del genere fossile *Eolates* e suoi rapporti con il genere attuale *Lates* (Pisces—Centropomidae). In *Studi e Ricerche sui Giacimenti Terziari di Bolca. II. Miscellanea Paleontologica*, pp. 1–54. Museo Civico di Storia Naturale di Verona, Verona.

Sorbini, L. (1975b). Studio paleontologico di *Acropoma lepidotus* (Agassiz). Pisces, Acropomidae. In *Studi e Ricerche sui Giacimenti Terziari di Bolca. II. Miscellanea Paleontologica*, pp. 177–204. Museo Civico di Storia Naturale di Verona, Verona.

Sorbini, L., and Tirapelle, R. (1975). Gli Holocentridae di Monte Bolca. I: *Eoholocentrum*, nov. gen. *Eoholocentrum macrocephalum* (de Blainville) (Pisces—Actinopterygii). In *Studi e Ricerche sui Giacimenti Terziari di Bolca. II. Miscellanea Paleontologica*, pp. 205–232. Museo Civico di Storia Naturale di Verona, Verona.

Tyler, J. C. (1975a). A new species of Boxfish from the Eocene of Monte Bolca, Italy, the first unquestionable fossil record of the Ostraciontidae. In *Studi e Ricerche sui Giacimenti Terziari di Bolca. II. Miscellanea Paleontologica*, pp. 103–126. Museo Civico di Storia Naturale di Verona, Verona.

Tyler, J. C. (1975b). A new species of triaconthid fish (Plectognathi) from the Eocene of Monte Bolca, Italy, representing a new subfamily ancestral to the Triodontidae and the other gymnodonts. In *Studi e Ricerche sui Giacimenti Terziari di Bolca. II. Miscellanea Paleontologica*, pp. 127–156. Museo Civico di Storia Naturale di Verona, Verona.

Watson, D. M. S. (1925). The structure of certain palaeoniscids and the relationships of that group with other bony fish. *Proc. Zool. Soc. London*, **1925**: 815–870.

Watson, D. M. S. (1937). The acanthodian fishes. *Phil. Trans. Roy. Soc. London, B*, **228**: 49–146.

Weitzman, S. H. (1974). Osteology and evolutionary relationships of the Sternoptychidae, with a new classification of stomiatoid families. *Bull. Am. Mus. Nat. Hist.*, **153**: 329–478.

Wenz, S. (1968). Note préliminaire sur la faune ichthyologique du Jurassique supérieur du Montsech (Espagne). *Bull. Soc. Géol., France*, **10**: 116–119.

Westoll, T. S. (1941). The Permian fishes *Dorypterus* and *Lekanichthys*. *Proc. Zool. Soc., Series B*, **111**: 39–58.

Wiley, E. O. (1976). Phylogeny and biogeography of fossil and Recent gars (Actinopterygii: Lepisosteidae). *Univ. Kansas, Mus. Nat. Hist. Misc. Publ.*, **64**: 1–111.

Wilson, M. V. H. (1980). Oldest known *Esox* (Pisces: Esocidae), part of a new Paleocene teleost fauna from western Canada. *Can. J. Earth Sci.*, **17**: 307–312.

Wilson, M. V. H. (1984). Osteology of the Palaeocene teleost *Esox tiemani*. *Palaeontology*, **27**: 597–608.

Woodward, A. Smith (1891–1901). *Catalogue of the Fossil Fishes in the British Museum*. Parts 1–4. British Museum, London.

Woodward, A. Smith (1916). The fossil fishes of the English Wealden and Purbeck Formations. Part II. *Publ. Palaeont. Soc. London*, **70**: 49–104.

CHAPTER VIII

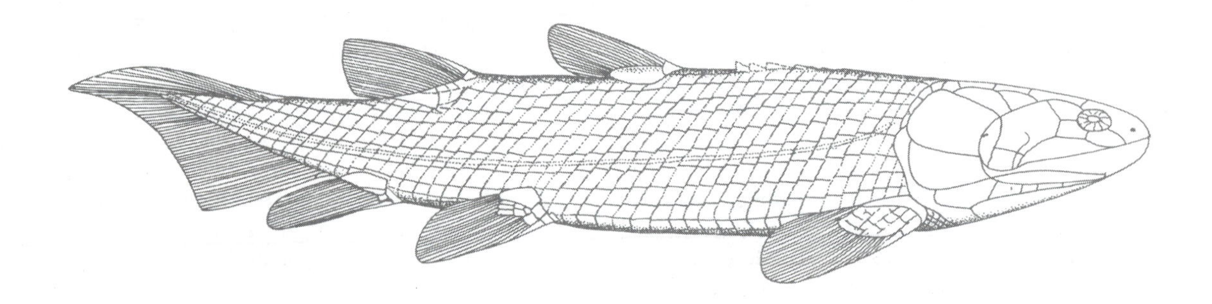

Sarcopterygian Fish

THE ANATOMY OF LOBE-FINNED FISH

In their great numbers and degree of anatomical diversity, the modern ray-finned fishes may be considered the most successful of all vertebrates. In contrast, the lobe-finned fishes are today reduced to only four genera. However, sarcopterygians are of great phylogenetic importance, since they include the ancestors of all land vertebrates.

We recognize three major groups of lobe-finned fishes: the lungfish and coelacanths, which are present in the modern fauna, and the Paleozoic rhipidistians. Lungfish and rhipidistians are known from the earliest Devonian, and the coelacanths (also known as Actinistia) appear in the middle Devonian. Rhipidistians and coelacanths are classified as separate suborders of the order Crossopterygii. Lungfish are placed in a distinct order, the Dipnoi. Fossils of early sarcopterygians are found in both freshwater and marine sediments, but both rhipidistians and Dipnoi were most common in shallow freshwater where the presence of lungs may have been especially important.

The modern lungfish and coelacanths are highly specialized derivatives of groups that were common in the Paleozoic and early Mesozoic. The early members of these groups were much more similar to one another and to the Paleozoic rhipidistians than are the living genera; the crossopterygians and Dipnoi probably shared a common ancestry in the late Silurian.

Like the chondrosteans, early lungfish and rhipidistians were elongate, fusiform fish with heavy rhomboidal scales. The sarcopterygians were generally larger, and the histology of their scales and dermal bones differed significantly. Other features that are common to the early sarcopterygians and distinguish them clearly from the early ray-finned fish include the presence of two dorsal fins, fleshy lobed fins, and an epichordal lobe in the caudal fin.

The muscular nature of the fins in sarcopterygians is a specialization relative to other groups of jawed fish that may have evolved as an adaptation to life on or near the bottom where they could be used to push against the substrate or obstacles. The presence of an epichordal lobe of the caudal fin may be a retention of a more primitive stage than we find in early chondrosteans if, as suggested by the condition in early heterostracans, a diphycercal tail is primitive for vertebrates.

A particularly important feature that is common to early dipnoans and crossopterygians is the presence of a special sensory system associated with the superficial layers of the dermal skeleton. A material termed **cosmine** covers the spongy and laminar bone of the skull and scales in early rhipidistians and Dipnoi. Cosmine, which Thomson (1975, 1977) described in detail, is a complex of bony and soft tissues including a layer of dentine that encloses a mosaic pore-canal system consisting of flask-shaped pore-cavities and a complicated network of interconnecting canals (Figure 8-1). The dentine and pore-canal system are covered with an enamel layer pierced by small sensory pores. The flask-shaped pore-cavities resemble the ampullary canals of Lorenzini in sharks and the ampullary and tuberous organs in teleosts. Thomson (1977) suggests that they contain comparable sensory organs. In the modern groups, these organs are electroreceptive and are used to locate prey and avoid predators and obstacles in water where vision is limited (Boord and Campbell, 1977). In sharks and teleosts, the sensory system is supported by soft tissue rather than bone. The importance of these sensory structures in early sarcopterygians may indicate an early adaptation toward life in deep or turbid waters. The relatively small size of the orbits and the large olfactory sacs in early lungfish and rhipidistians, (compared with those of early chondrosteans) reinforces this notion.

The cosmine layer that includes these sensory structures covers not only the scales and surface of the individual dermal bones of the head and shoulder girdle but may extend across the sutures between the skull bones to

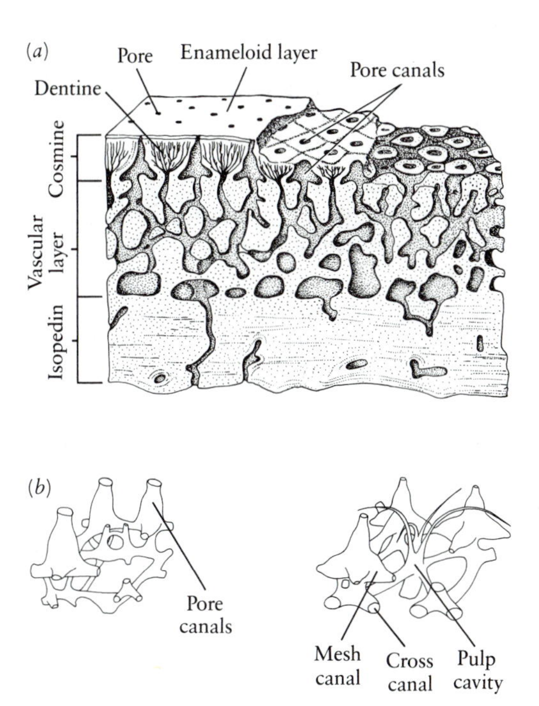

Figure 8-1. (*a*) The dermal exoskeleton of the rhipidistian *Megalichthys* showing the superficial cosmine layer, including the pore-canal system. *From Moy-Thomas and Miles, 1971. By permission from Chapman and Hall, Ltd.* (*b*) Greatly enlarged restoration of segments of the pore-canal system of the lungfish *Dipterus* and the rhipidistian *Ectosteorhachis*. Similar structures supported by soft tissues in modern sharks and teleosts are electrosensitive. *From Thomson, 1975.*

form an unbroken covering. It would be difficult for normal growth to have occurred at the margins of the dermal bones that lay beneath this continuous layer. According to Thomson (1975), growth was accomplished during periods of resorption of the cosmine. These periods are well documented in the fossil record of both dipnoans and rhipidistians. Cosmine may constitute up to 10 percent of the calcified tissue of the body; its resorption and redeposition must have required a major metabolic effort in these sarcopterygians. Cosmine is reduced in several groups of later rhipidistians, completely lost in modern lungfish, and is not known in any coelacanths. The presence of organs sensitive to electrical currents in divergent groups of modern fish and the occurrence of a similar pore-canal system in some ostracoderms suggest that this sensory faculty may have been a widespread heritage among fish. Although it is not unique to sarcopterygians, the pore-canal system is greatly emphasized within this group.

Similarities of the postcranial anatomy and the common presence of cosmine suggest that crossopterygians and dipnoans share a common ancestry separate from that of actinopterygians. In contrast, the pattern of the dermal bones of the skull, the structure of the braincase, and the configuration of the jaw musculature are strikingly different in these two sarcopterygian groups.

No fossils are known of an immediate common ancestor that might have given rise to crossopterygians and Dipnoi, nor are any actinopterygian fish known by more than isolated scales prior to the appearance of sarcopterygians. The broad similarities in the relationship between the braincase and the endoskeletal jaw supports and the probable position of the main jaw-closing muscles in primitive sharks, acanthodians, and actinopterygians suggest that this pattern is primitive for jawed fish (other than placoderms). Both lungfish and rhipidistians show marked specializations that exemplify different ways of accommodating a greater mass of muscles to close the mouth than we have seen in early actinopterygians.

In primitive actinopterygians, the adductor chamber is small and enclosed dorsally by the palatoquadrate and the dermal bones of the cheek (Figure 8-2*a*). Among early crossopterygian fish, the adductor chamber remains closed dorsally and mobility between the braincase, palate, hyomandibular, and cheek is retained (Figure 8-2*b*). However, the configuration of the braincase is strikingly different than in actinopterygians. Instead of being a single, essentially unitary structure, it is divided into two parts that articulate with one another. This pattern has long been recognized in Paleozoic rhipidistians, but its function only became clear with the discovery of the modern coelacanth, in which the soft anatomy could be studied. In the living genus, a pair of subcephalic muscles extends from the posterior (otico-occipital) unit of the braincase to the anterior (ethmoid) unit. Since the front part of the braincase is attached to the palate and marginal bones bearing the dentition, contraction of this muscle forces the teeth down into the prey. Thomson (1967) described the function of this system in both coelacanths and rhipidistians.

Elaboration of this feeding apparatus may have been particularly important in feeding on the heavily armored bony fish of the Paleozoic. From the Devonian into the early Permian, the rhipidistians are among the most important freshwater predators.

Similarities in the relationship between the palate, cheek, and operculum in primitive actinopterygians and rhipidistians may explain the similar patterns of the dermal bones covering these areas (see Figures 6-13 and 8-3).

We may attribute the much different pattern of the dermal bones in lungfish (see Figures 8-15 and 8-16) to a radical reorganization of the adductor jaw musculature. Even among the earliest lungfish, the adductor chamber opened dorsally so that the jaw musculature could spread medially over the braincase toward the midline (see Figures 8-2*c* and 8-15*c*). We may associate the great increase in the jaw musculature with the fusion of the palatoquadrate to the braincase to form a holostylic jaw suspension. The capacity for lateral expansion of the cheek, which is common to rhipidistians and actinopterygians, is absent in Dipnoi. The dermal bones of the cheek do not have a specific role in forming the adductor chamber and so lack the consistent pattern seen in actinopterygians and rhipidistians. The palatal dentition is emphasized, and most lungfish lack marginal teeth, which probably accounts for their lack of marginal jaw bones comparable with the maxilla and premaxilla. The highly specialized pattern of both the dermal and endoskeletal elements of the skull that is characteristic of most Dipnoi evolved soon after the group first appeared in the early Devonian; essentially the same configuration is retained in the living lungfish.

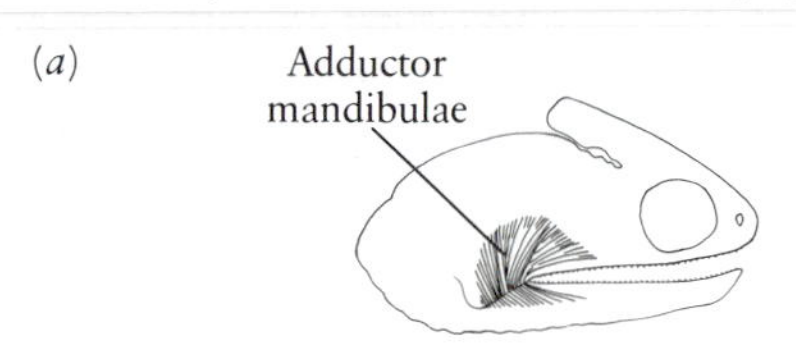

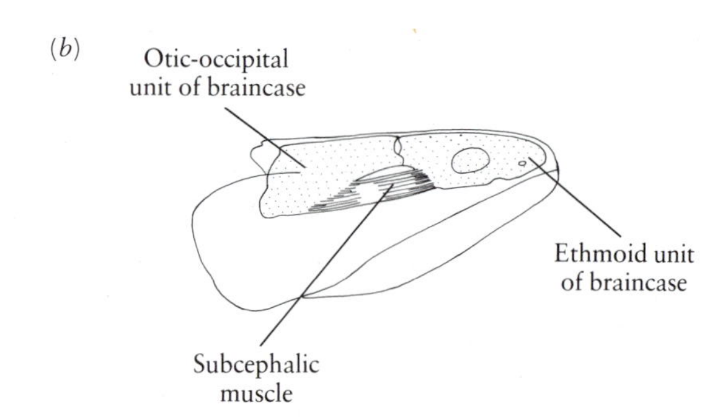

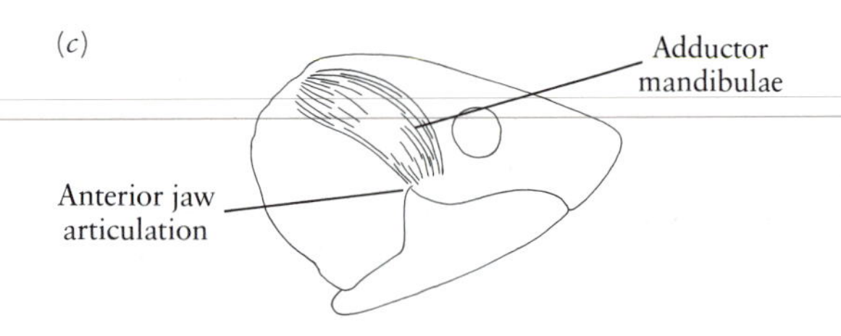

Figure 8-2. SPECIALIZATIONS OF THE JAW MUSCULATURE IN SARCOPTERYGIAN FISH. (*a*) The adductor jaw musculature in a primitive actinopterygian fish, which may represent the pattern ancestral to lobe-finned fish. The adductor chamber is small and enclosed, with little space for the jaw-closing musculature. *Modified from Schaeffer and Rosen, 1961.* (*b*) The rhipidistian fish *Ectosteorhachis*, in which the subcephalic muscle links the anterior and posterior units of the braincase. Contraction of this muscle lowers the front of the skull and pulls the large teeth of the palate and the marginal dentition of the maxilla and premaxilla down on the prey. The amount of movement is not great, but the added strength of this muscle could drive the teeth through the thick scaly covering of primitive fish. *Modified from Thomson, 1967.* (*c*) The Devonian lungfish *Chirodipterus*, in which the adductor chamber has greatly expanded above the braincase. The jaw articulation is well anterior to the occiput, and the palatoquadrate is fused to the braincase so that the muscles can exert a very great force on the jaws. *Based on Miles, 1977.*

RHIPIDISTIANS

Rhipidistians were the dominant freshwater predators among the bony fish during the late Paleozoic. They became extinct by the end of the Lower Permian, but sometime in the Devonian they gave rise to the amphibians. They are divided into two major groups, the Porolepiformes, which appear at the very base of the Devonian and continue to the end of that period, and the Osteolepiformes, which appear in the middle Devonian and survive into the early Permian. These groups probably diverged from a common ancestor in the late Silurian.

Many specializations of the rhipidistians may be attributed to adaptation to a predatory way of life in shallow water. We will discuss their anatomy in considerable detail since this group shows a pattern that is basic to that of primitive tetrapods.

SKULL

Externally, the skull of rhipidistians resembles that of early actinopterygians in its general features. The cheek and operculum are movable relative to the skull roof, and the dermal shoulder girdle appears as an extension of the skull. A major difference is that the bones of the skull roof hinge on one another above the division between the anterior and posterior units of the braincase, except in the early genus *Powichthys* (Figure 8-4*a*).

The pattern of dermal bones in osteolepiform rhipidistians such as *Eusthenopteron* (Figure 8-3) is very similar to that of early amphibians (see Figure 9-3), and the names correspond very closely. The names of the bones in actinopterygian fish are mostly based on those of tetrapods as well, but the homology of the elements is more questionable. A major difference in terminology involves the bones between the orbits that border the pineal opening. They are typically termed "frontals" in actinopterygians but "parietals" in rhipidistians. Anterior to the parietals in rhipidistians, the snout is made up of a mosaic of small bones that occupy the position of the rostrals in actinopterygians. Lateral to the parietals in the rhipidistians are the pre- and postfrontals. The posterior border of the parietals and the intertemporals articulate with the more posterior portion of the skull roof formed by the postparietals, supratemporals, and tabulars. Behind them,

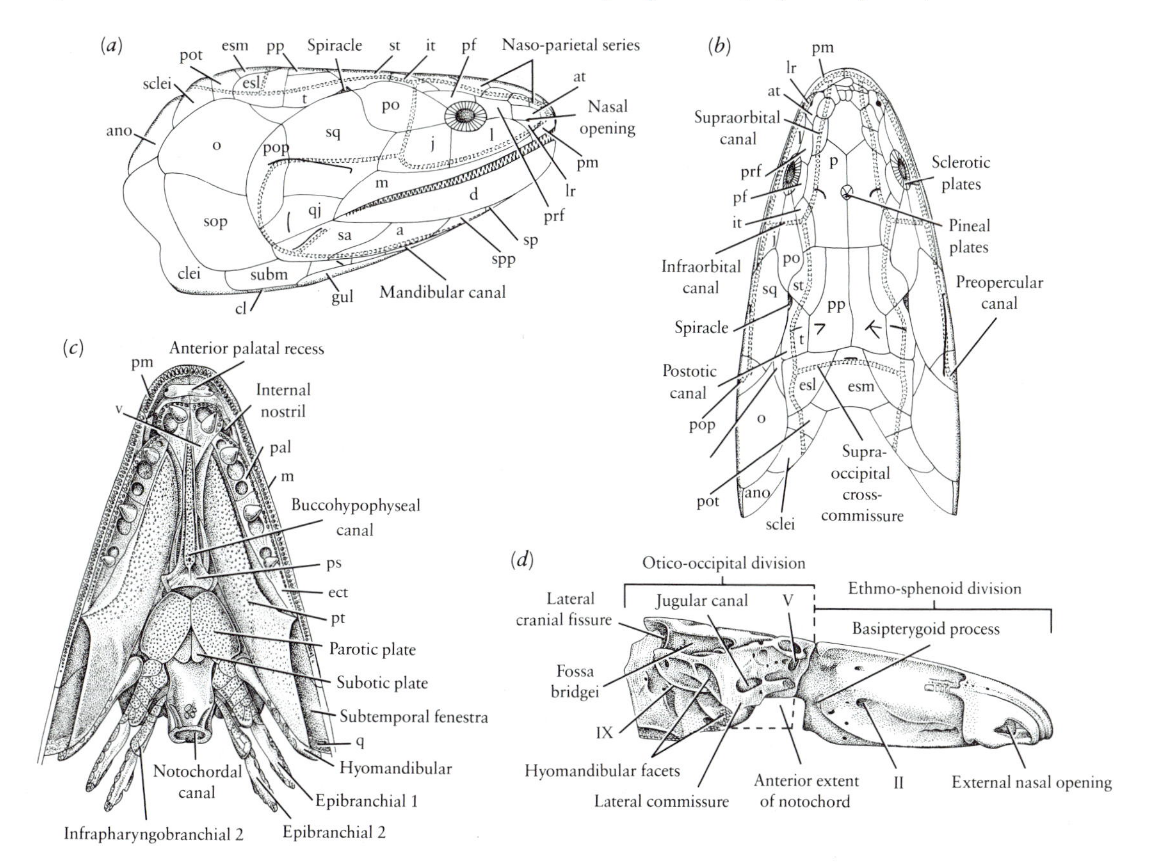

Figure 8-3. THE SKULL OF THE OSTEOLEPIFORM RHIPIDISTIAN *EUSTHENOPTERON.* (*a*) Lateral, (*b*) dorsal, and (*c*) palatal views. (*d*) Braincase in lateral view, $\times\frac{1}{2}$. Abbreviations (which will be used extensively for tetrapods) as follows: a, angular; ano, anocleithrum; art, articular; at, anterior tectal; bo, basioccipital; bs, basisphenoid; cl, clavicle; clei, cleithrum; cor, coronoid; d, dentary; ect, ectopterygoid; eo, exoccipital; ept, epipterygoid; esl, lateral extrascapular; esm, medial extrascapular; f, frontal; gul, gular plate; ina, internasal; inf, interfrontal; it, intertemporal; j, jugal; l, lacrimal; lr, lateral rostral; m, maxilla; n, nasal; o, opercular; opis, opisthotic; p, parietal; pal, palatine; part, prearticular; pf, postfrontal; pm, premaxilla; po, postorbital; pop, preopercular; pos, postspiracular; pot, posttemporal; pp, postparietal; prf, prefrontal; pro, proötic; ps, parasphenoid; pt, pterygoid; ptf, posttemporal fossa; q, quadrate; qj, quadratojugal; sa, surangular; sclei, supracleithrum; sm, septomaxilla; so, supraoccipital; sop, subopercular; sp, splenial; sph, sphenethmoid; spp, postsplenial; sq, squamosal; st, supratemporal; subm, submandibular (includes also the more anterior plates in this series); t, tabular; v, vomer; I–XII, cranial nerves. *From Moy-Thomas and Miles, 1971, after Jarvik. By permission from Chapman and Hall, Ltd.*

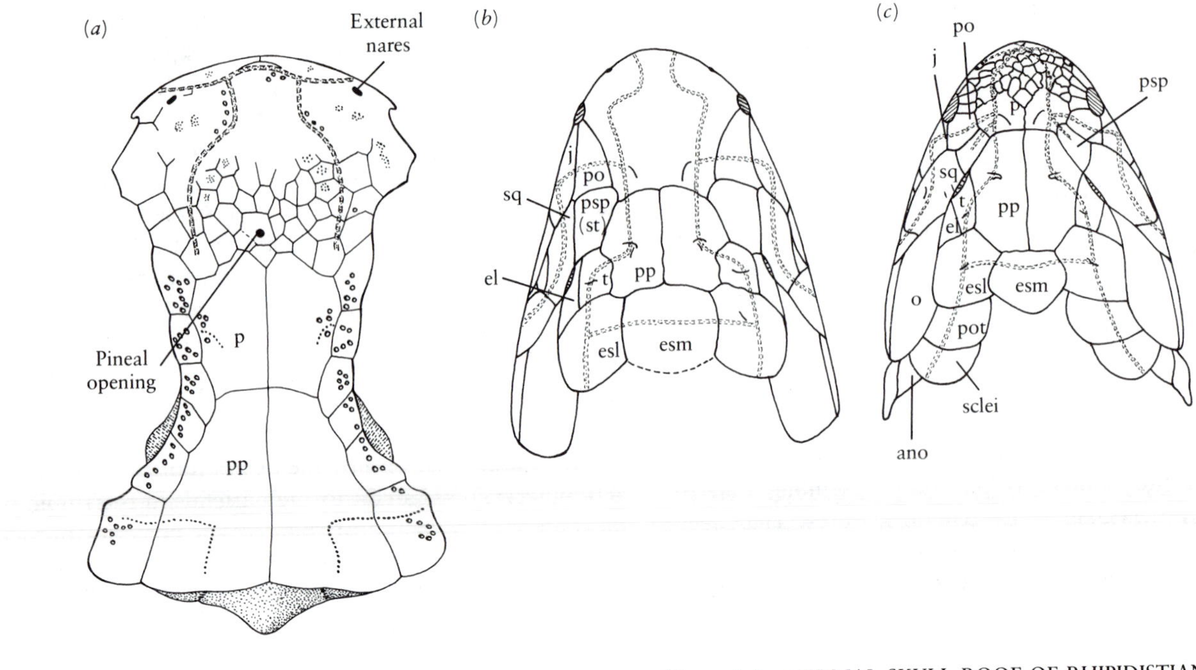

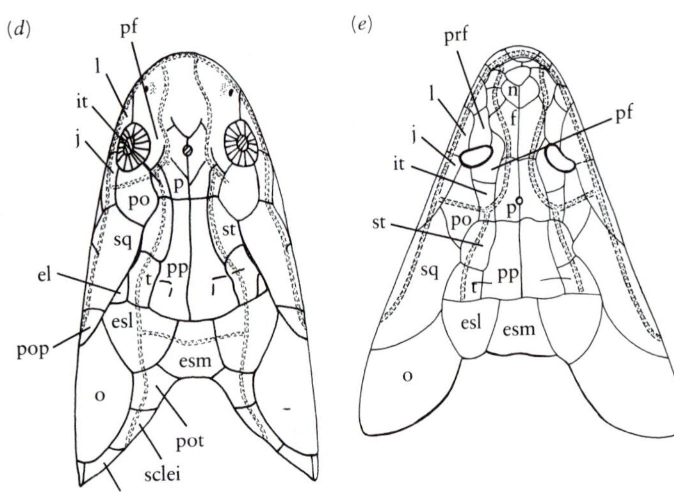

Figure 8-4. DERMAL SKULL ROOF OF RHIPIDISTIAN FISH. (*a*) *Powichthys*, early Devonian, × 1⅓. *From Jessen, 1980.* (*b* and *c*) The Porolepiformes *Porolepis* and *Holoptychius* from the Lower and Upper Devonian. *From Jarvik, 1972.* (*d* and *e*) The Osteolepiformes *Osteolepis* and *Panderichthys*, from the Middle Devonian. (*d*) *From Jarvik, 1972* and (*e*) *From Vorobyeva, 1977,* respectively. Abbreviations as in Figure 8-3, plus: el, extratemporal; psp, prespiracular. *Powichthys* differs from all other rhipidistians in lacking a break in the skull roof between the parietal and the postparietal above the fissure between the anterior and posterior elements of the braincase. The series of small bones above the orbits vaguely resembles those in lungfish. The Upper Devonian genus *Eusthenopteron*, not the lower Devonian porolepiforms, shows the greatest similarity to the pattern of dermal bones of early actinopterygians.

the extrascapulars are linked to the posttemporal and supracleithrum of the dermal shoulder girdle, as in actinopterygians.

The cheek in rhipidistians resembles that of actinopterygians, except for the large size of the quadrotojugal and the presence of two or more centers of ossification, including a large squamosal, in the area occupied by the preopercular in actinopterygians (Figure 8-5).

The orbit is surrounded by up to 30 sclerotic plates, which is similar to the number in amphibians but in contrast to the situation in actinopterygians, in which there are usually only four.

As in actinopterygians, the principal bones making up the operculum are termed the opercular and the subopercular. Extending forward and between the jaws are bones that are roughly equivalent to the branchiostegal rays and gulars, although their particular configuration differs among the rhipidistian genera. The structure and function of the dermal shoulder girdle are comparable in the two groups, although rhipidistians have an additional medial element, the interclavicle.

The pattern of the rhipidistian palate is broadly comparable to that of primitive actinopterygians, except for the very important presence of a choana, or internal naris. It lies medial to the suture separating the premaxilla and maxilla and is bordered on the palatal surface by the vomer and palatine. Rosen, Forey, Gardiner, and Patterson (1981) suggest that this opening could not function as an internal nostril, in part because the margins of the premaxilla and maxilla serve as a joint associated with the lateral expansion of the cheek. The arrangement of the bones surrounding the opening is otherwise nearly identical in osteolepiform rhipidistians and amphibians (see Chapter 9). We are justified in thinking that it served as an air passage in both groups. Posterior to the palatine is the ectopterygoid. Most of the palatal surface is composed of the pterygoids, which sheath the surface of the palatoquadrates. They surround the subtemporal fenes-

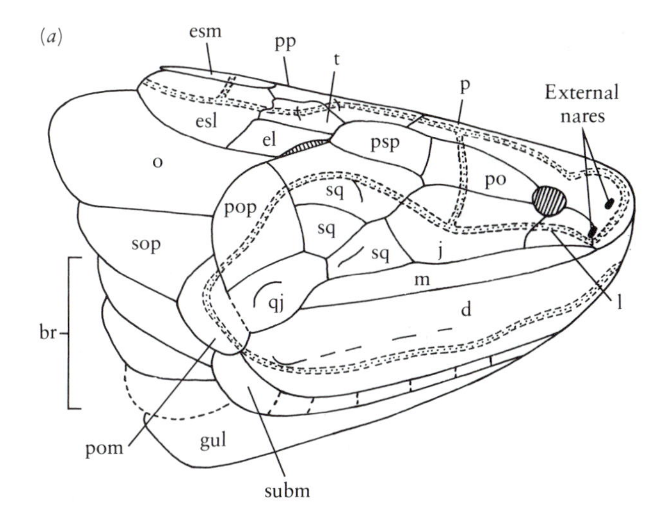

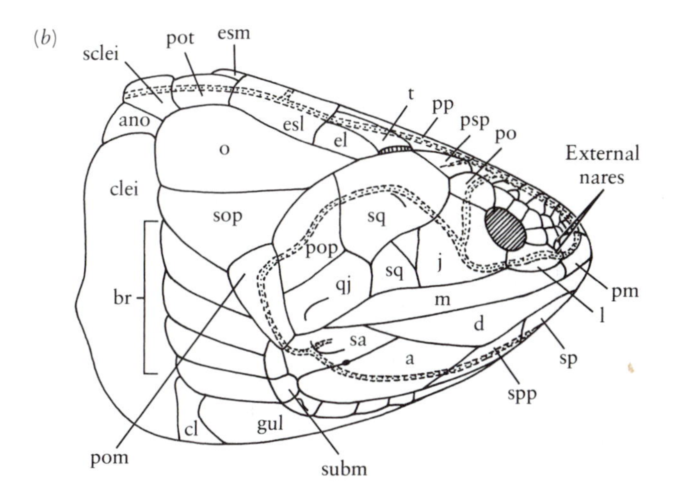

Figure 8-5. LATERAL VIEW OF THE SKULLS OF *POROLEPIS* AND *HOLOPTYCHIUS.* The cheek region and operculum generally resemble those of the early actinopterygians but are very distinct from those of lungfish (see Figure 8-15). Abbreviations as in Figures 8-3 and 8-4, plus: br, branchiostegal rays; pom, preoperculosubmandibular. *From Jarvik, 1972.*

trae, through which the adductor jaw muscles emerge, and extend anteriorly nearly to the choanae. As in primitive actinopterygians, the parasphenoid covers the ventral surface of the sphenethmoid from between the vomers anteriorly to the level of the ventral cranial fissure posteriorly. It is pierced by the buccohypophyseal canal. Paired parotic plates underlie the posterior portion of the braincase as they do in early actinopterygians.

All three of the marginal bones of the palate—the vomer, palatine, and ectopterygoid—bear large teeth. Like the marginal teeth, they are replaced rapidly, and a conspicuous pit usually lies adjacent to each fang where a tooth has been lost. Primitive labyrinthodont amphibians retain a similar pattern. In addition to the large fangs, a row of small teeth is present on the margin of the lateral palatal bones just medial to the marginal dentition.

As in the early actinopterygians, Meckel's cartilage is largely supplanted by the dermal bones of the lower jaw, principally the tooth-bearing dentary. The ventral surface of the lower jaw is formed by a series of smaller bones, termed from back to front: surangular, angular, postsplenial, and splenial. Medially, the surface is completed by the large prearticular and a series of coronoids just medial to the marginal dentition that, like the ectopterygoid, palatine, and vomer, bear large fangs.

The teeth of rhipidistians are characterized by complex folding of the dentine. When seen in cross-section (Figure 8-6) the infolding makes the teeth appear like a maze, which is why we call these teeth labyrinthodont. This pattern is retained in many groups of primitive tetrapods.

We have noted that the braincase in rhipidistians (and the related coelacanths) is ossified in two large units in relation to the specialized feeding mechanism. We can infer that this division resulted from the dorsal extension of the ventral fissure between the basioccipital and basisphenoid that characterizes early actinopterygians and reflects the embryological division between the parachordal and trabecular regions of the braincase.

In the rhipidistians, the articulation between the anterior, ethmoid unit (including the nasal capsules) and the posterior otico-occipital occurs just anterior to the opening for the Vth (trigeminal) nerve (see Figure 8-3). The otico-occipital portion is closely comparable to that of primitive actinopterygians. The lateral occipital fissure is still evident. A large lateral commissure, which is external to the jugular canal, supports two facets for articulation of the hyomandibular. A large opening above the commissure is functionally comparable with the posttemporal fossa in teleosts, which serves for the attachment of the epaxial musculature that raises the head.

In primitive actinopterygians, the notochordal canal pierces the otico-occipital plate and extends anteriorly to the proötic bridge, which separates it from the pituitary

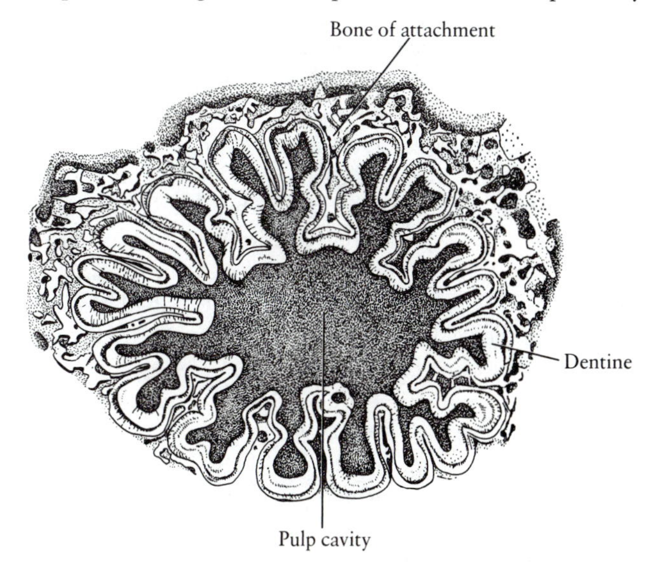

Figure 8-6. Great enlarged cross section of a tooth of *Eusthenopteron* to show labyrinthodont infolding of dentine. *From Moy-Thomas and Miles, 1971, after Schultze.*

fossa. The proötic bridge is missing in rhipidistians, and the notochord abuts against the back of the ethmoid ossification. Paleontologists postulated, on the basis of the bony configuration of the otico-occipital element, that rhipidistians had a great anterior extension of the notochord; this was confirmed by the discovery of the notochord in exactly this position in the living coelacanth *Latimeria*.

The ethmoid portion of the braincase is more extensively ossified than it is in actinopterygians in relation to its role in feeding. The orbits are relatively smaller and the interorbital septum is wider. The nasal capsules are large and heavily ossified. In *Eusthenopteron* and other osteolepiform rhipidistians, the capsules have three openings: a ventral opening that leads toward the choana, an external opening above the suture between the maxilla and premaxilla, and a posterior opening that leads toward the orbit via a short tube. This tube may be comparable with the lacrimal duct in tetrapods. The external narial opening in rhipidistians is probably homologous with the anterior or incurrent narial opening in actinopterygians, and the nasolacrimal duct is comparable with the posterior, or excurrent opening. Panchen (1967) considers this problem in detail. The presence of the nasolacrimal duct is logically associated with the adaptation of later rhipidistians to life near the surface of the water, where the eyes and nasal openings are out of the water so often that they need to be bathed in secretions from a lacrimal gland. We may attribute the initial development of the internal narial opening to the advantage of having a sense of smell within the oral cavity. Its function in respiration was probably secondary.

The structure of the hyomandibular and palatoquadrate are fundamentally similar to those of primitive actinopterygians, although their attachments to the braincase are modified in relation to the articulation within the braincase and the peculiarities of the feeding mechanism. The basicranial articulation and adjacent structures, including the pituitary fossa and areas of origin of the rectus eye muscles, are relatively further forward than in primitive actinopterygians, and the entire otico-occipital portion of the braincase appears longer. The hyomandibular has a double articulation with the braincase, precluding dorsoventral movement but facilitating anterior movement of the distal extremity as the cheek is expanded. As in actinopterygians, the palatoquadrate articulates anteriorly with the back of the nasal capsule and laterally with the maxillary bone of the cheek.

We can reconstruct the mechanics of feeding movements on the basis of the pattern in primitive actinopterygians and the living coelacanth (Figure 8-7). As in actinopterygians, the contraction of the axial and hyoid musculature that open the jaws forces the cheeks and operculum to move laterally. Because of their linkage with the cheek, the distal end of the hyomandibular and the attached palatoquadrate move anteriorly. The anterior movement of the palatoquadrate lifts the anterior end of

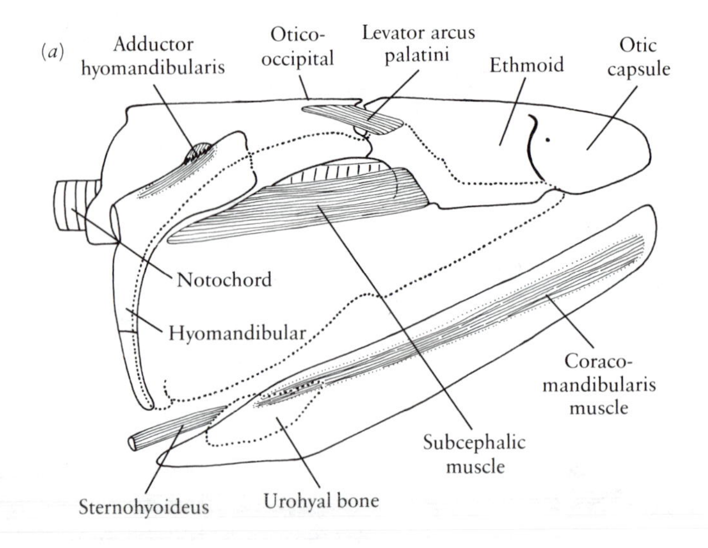

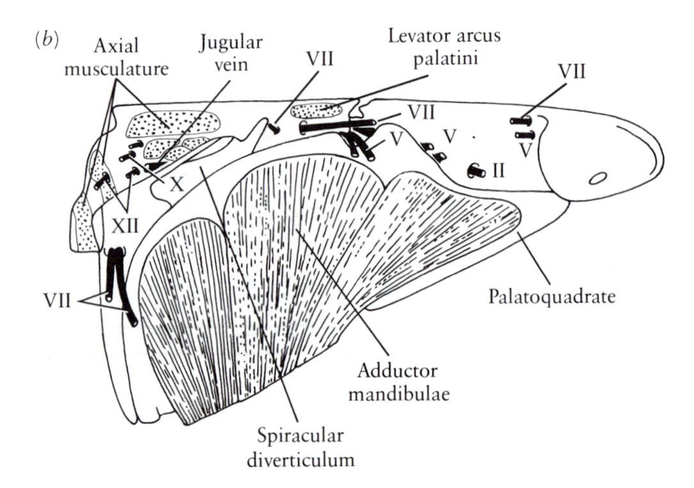

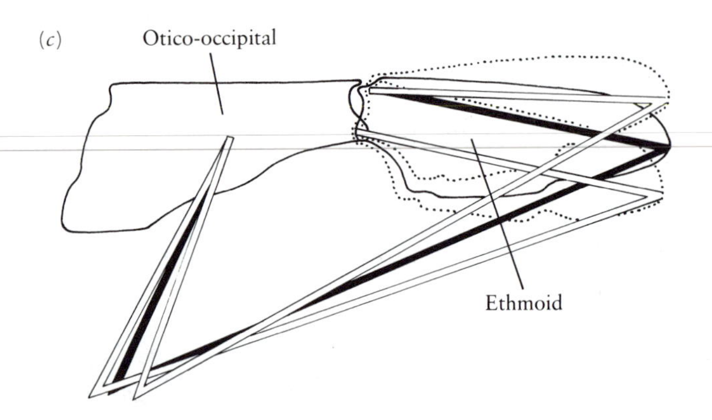

Figure 8-7. CRANIAL KINESIS OF RHIPIDISTIANS. (*a*) Outline of braincase, lower jaw, and hyomandibular showing subcephalic muscles. Palatoquadrate indicated in dotted lines. (*b*) Braincase with palatoquadrate and adductor jaw musculature. (*c*) Relative movement of skull elements during feeding. Straight lines indicate changes in orientation of the hyomandibular and palatoquadrate and dotted lines show outline of ethmoid. As the jaws are opened, the distal end of the hyomandibular swings anteriorly. The palatoquadrate acts as a link between this bone and the ethmoid region of the braincase, which is lifted as the palatoquadrate moves forward. The subcephalic muscle lowers the ethmoid and the palate and dermal bones of the snout to which it is attached. *From Thomson, 1967. With permission from the Zoological Journal of the Linnean Society. Copyright 1967 by the Linnean Society of London.*

the ethmoid portion of the braincase and the dermal bones bearing the marginal dentition. The adductor mandibulae close the jaws as in actinopterygians, but their force is greatly augmented by the large subcephalic muscles that depress the front of the braincase, the palate, and the tooth-bearing maxilla and premaxilla. The extent of movement of the anterior unit of the braincase was not great but would have been sufficient to drive the large fangs of the palatal dentition through the heavy dermal armor of Paleozoic bony fish, which may have constituted a large portion of the diet of rhipidistians.

The hyoid apparatus (Figure 8-8) follows the general pattern of acanthodians and primitive actinopterygians, with a series of medial ventral supports, the basibranchials, and a jointed lateral series supporting the gills—the hypobranchials, ceratobranchials, epibranchials, infrapharyngobranchials, and suprapharyngobranchials. There are four or five gill arches. The first two are attached dorsally to the neurocranium. The internal surfaces of these elements are covered with tooth plates.

VERTEBRAE AND RIBS

As in most primitive fish, rhipidistians have a notochord that is large and little restricted, but in contrast with early actinopterygians, all species have ossified centra. These centra are usually relatively thin walled, which indicates that the diameter of the notochord is large, but they may surround it more or less completely. The pattern of the centra differs considerably from genus to genus (Figures 8-9 and 8-10). This variability does not follow the major taxonomic divisions of the group based on the pattern of the dermal skull roof, dentition, and other features. The structure of the centra probably evolved in relation to changes in locomotor patterns that occurred in each of the major taxonomic groups. Despite detailed descriptions of the various elements, we are not certain of their specific functions. Cartilage was certainly present in the vertebral column as well, but its extent, relative to that of the notochord, cannot be accurately judged from the fossils (Andrews and Westoll, 1970a, 1970b).

The pattern of the centra seen in *Eusthenopteron* and *Osteolepis* appears structurally similar to those of primitive labyrinthodont amphibians, and similar names have been applied to the elements. However, the proportions and functional relationships differ somewhat between the members of these two groups, which has led to controversy over their specific homology (Thomson and Vaughn, 1968). In *Eusthenopteron* and *Osteolepis* (Figure 8-9), there are two successive elements in each segment: the pleurocentrum is dorsal, posterior, and always paired; the intercentrum is ventral, anterior, and paired throughout much of the trunk region but fused medially in both the cervical and the posterior trunk and caudal regions. In the tail, the intercentrum is continuous with the haemal arch, a configuration that is retained in most primitive tetrapods.

In *Eusthenopteron*, the central elements and the neural arches were apparently not in contact with one another but separated by cartilage. The neural arches themselves may be paired in the trunk region, and their bases are separated well above the level of the neural tube to allow passage of the dorsal longitudinal ligament. This opening persists in primitive labyrinthodonts. In *Osteolepis*, the

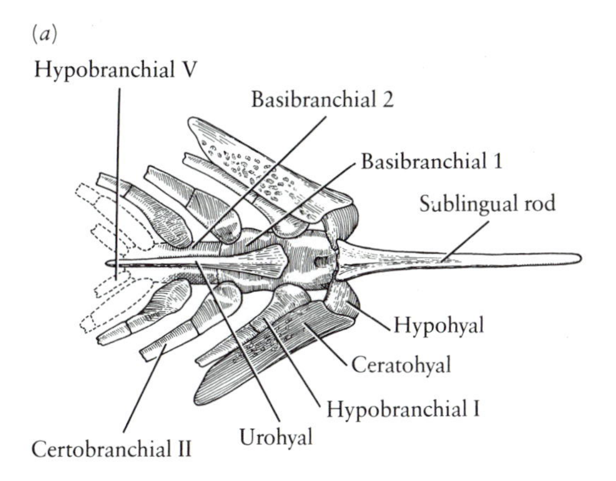

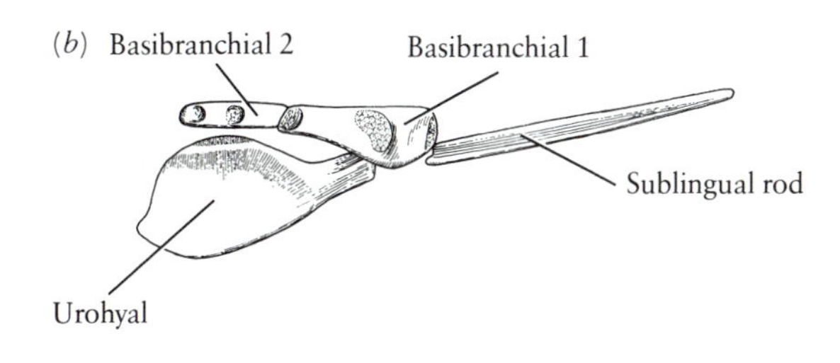

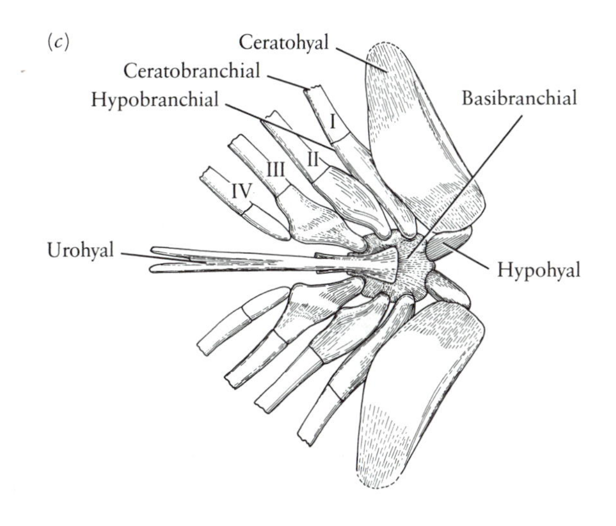

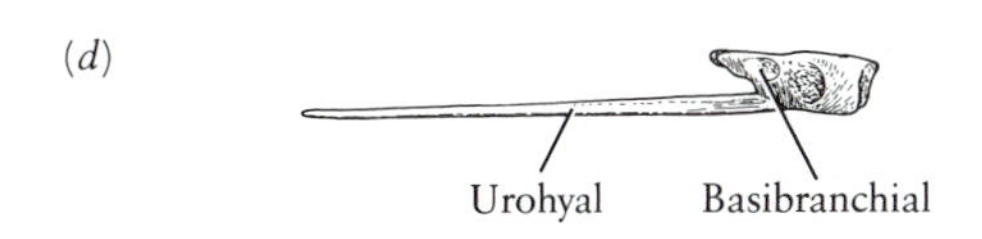

Figure 8-8. HYOID APPARATUS OF RHIPIDISTIANS. The osteolepiform *Eusthenopteron* (*a*) in ventral view and (*b*) lateral view of ventral elements. (*c* and *d*) Comparable views of the porolepiform *Glyptolepis*. The structure is basically similar except for the very long sublingual rod in *Eusthenopteron*. *From Jarvik, 1963.*

(a)

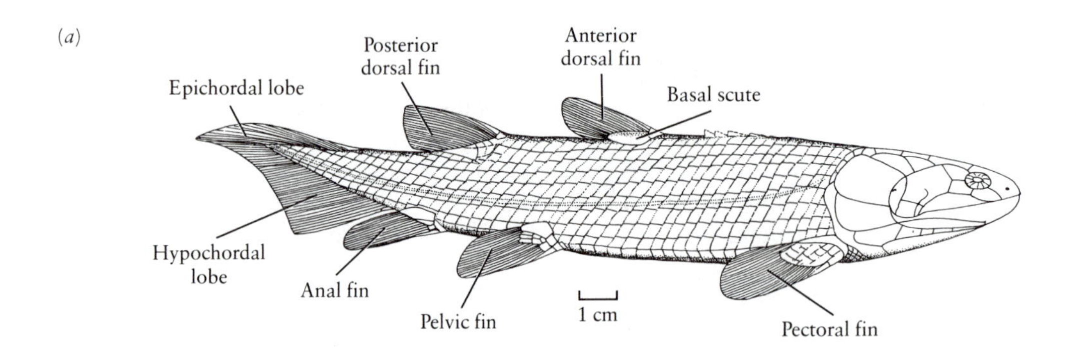

(b)

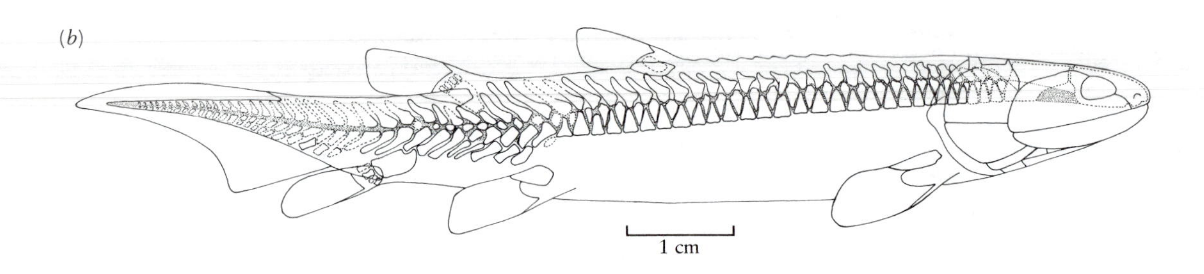

(c)

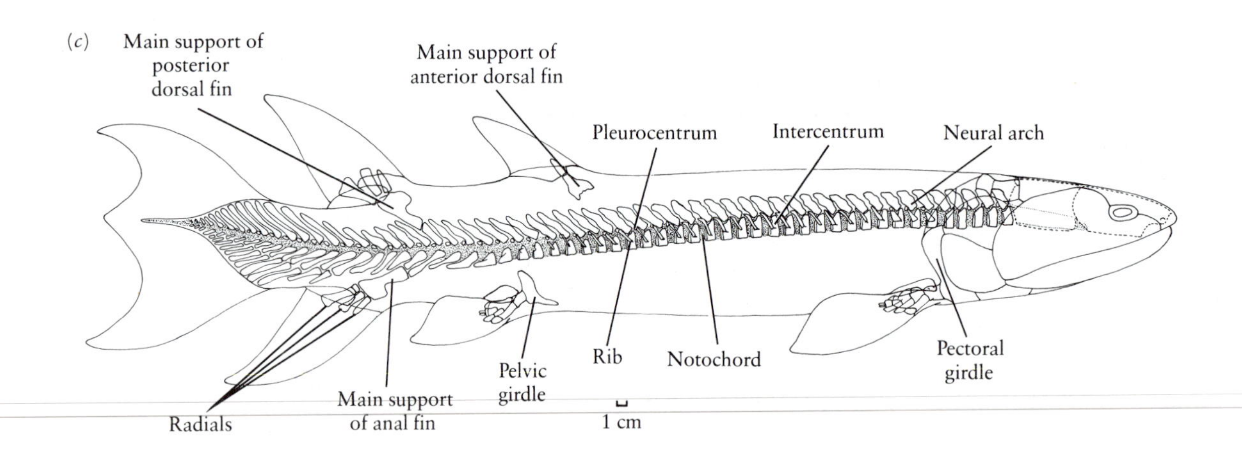

(d)

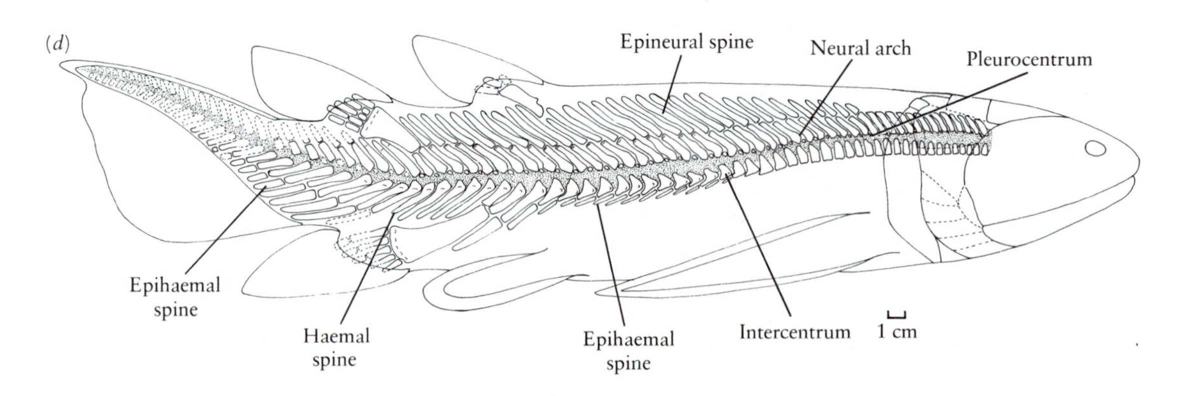

Figure 8-9. RHIPIDISTIAN FISHES. (*a*) *Osteolepis* restoration with scales. *From Moy-Thomas and Miles, 1971. By permission from Chapman and Hall, Ltd.* (*b*) *Osteolepis* reconstruction of endoskeleton. *From Andrews and Westoll, 1970b.* (*c*) *Eusthenopteron. From Andrews and Westoll, 1970a.* (*d*) *Glyptolepis. From Andrews and Westoll, 1970b.* (*a–c*) are osteolepiforms and (*d*) is a porolepiform.

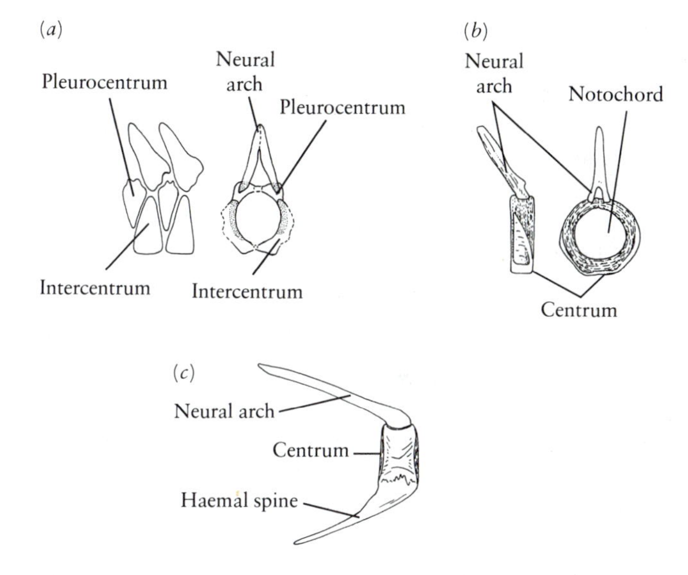

Figure 8-10. RHIPIDISTIAN VERTEBRAE. (*a*) *Osteolepis* trunk vertebrae in lateral and anterior views. This pattern is similar to that of primitive Paleozoic amphibians. (*b*) *Megalichthys* trunk vertebrae in lateral and anterior views. The ring centra may have developed from smaller elements, like those of *Osteolepis*. (*c*) *Megalichthys* caudal vertebrae in lateral view. *From Andrews and Westoll, 1970b.*

centra are extensive and nearly cover the notochord laterally. *Glyptolepis* (Figure 8-9*d*), a porolepiform rhipidistian, appears to show basically the same pattern as *Osteolepis,* although the elements are relatively smaller. Other rhipidistians of both major groups show an apparent consolidation of pleurocentra and intercentra into complete cylinders that occupy most of the segment. In *Thursius,* the pattern of fusion from distinct elements is still evident. It is less so in *Megalichthys* (Figure 8-10*b*), in which the centra, although never as long, appear like the unitary structure seen in some early lungfish. In the genera in which the centra are complete cylinders, the neural arches articulate with them. In the genera with small paired pleurocentra, they appear to remain separate. There are never well-developed zygapophyses linking successive neural arches, although there are traces of surfaces that may have been in contact with one another.

Eusthenopteron and *Osteolepis,* which have multipartite centra also have ribs, but genera with cylindrical centra do not. In *Eusthenopteron,* the ribs are short rods linking the area of the arch and pleurocentrum with the intercentrum. The intercentrum bears a distinct notch for their attachment, but the development of an articulating surface on the arch is questionable. A very short segment that extends dorsally and posteriorly corresponds with the shaft of the rib in amphibians and other tetrapods.

Despite the general similarities of the vertebrae to those of early amphibians, rhipidistians show little of the specialization of the anterior cervicals to articulate with the occiput in the manner of the atlas-axis complex of tetrapods. This is presumably a result of the strong, but not rigid, attachment of the skull to the trunk via the notochord that extends deeply into the occiput.

FINS

The fleshy basal portion of the fins of rhipidistians is surrounded by rhomboidal or rounded scales similar to those that cover the trunk (Figure 8-9). On the distal portion or fin web, these scales are succeeded by elongate lepidotrichia like those of actinopterygians. Osteolepiformes have large scutes at the base of all but the caudal fins.

The anterior dorsal fin has an expanded basal support that is not in contact with the neural arches and one or more radials. The endoskeletal supports of the second dorsal and anal fins articulate directly with the neural and haemal arches. The basals have a large distal portion that supports three or more radials and a narrower base that articulates with one or two neural or haemal spines. Whether the tail is heterocercal or diphycercal, the primary supports are provided by the neural and haemal arches and additional ventral elements, the epihaemal spines. The development of a symmetrical tail in rhipidistians appears to involve little more than straightening the notochordal axis and compensation in the length of the endoskeletal supports. As in actinopterygians, these changes can be associated with reduction in the weight of the scales. They may also be associated with an increase in the effectiveness of the lung as a hydrostatic organ.

We are especially interested in the paired fins of rhipidistians (Figures 8-9 and 8-11) because they form the basis of the structure of the limbs in amphibians and all higher tetrapods. As in the actinopterygians, the shoulder girdle consists of both dermal and endochondral elements. The dermal bones are a direct continuation of the back of the opercular series and the skull roof. The largest ventrolateral element is the cleithrum, which, as in palaeoniscoids, supports the scapulocoracoid. Large paired clavicles and a small median interclavicle lie ventrally. Dorsally, the cleithrum is succeeded by the anocleithrum, the supracleithrum, and the posttemporal, which is in contact with the extrascapulars. The scapulocoracoid is a small bone in *Eusthenopteron* and has a triradiate attachment to the cleithrum.

We can directly compare elements of both the forelimbs and hind limbs of osteolepiform rhipidistians and primitive amphibians (see Figures 9-8 through 9-10). In the forelimb, the proximal element is clearly comparable to the tetrapod humerus and articulates with a long bladelike radius and a stout posterior equivalent of the ulna. The ulna in turn articulates with an anterior intermedium and posterior ulnare. The ulnare articulates with two or more distal elements that we may compare with the more distal carpals of amphibians. More distally, there are no large endoskeletal supports for the fin and one must suppose that the metacarpals and phalanges of tetrapods developed as almost, if not entirely, new structures.

The pelvic girdle consists of a single pair of ossifications that may have been joined at the midline by cartilage. A large acetabulum faces almost directly poste-

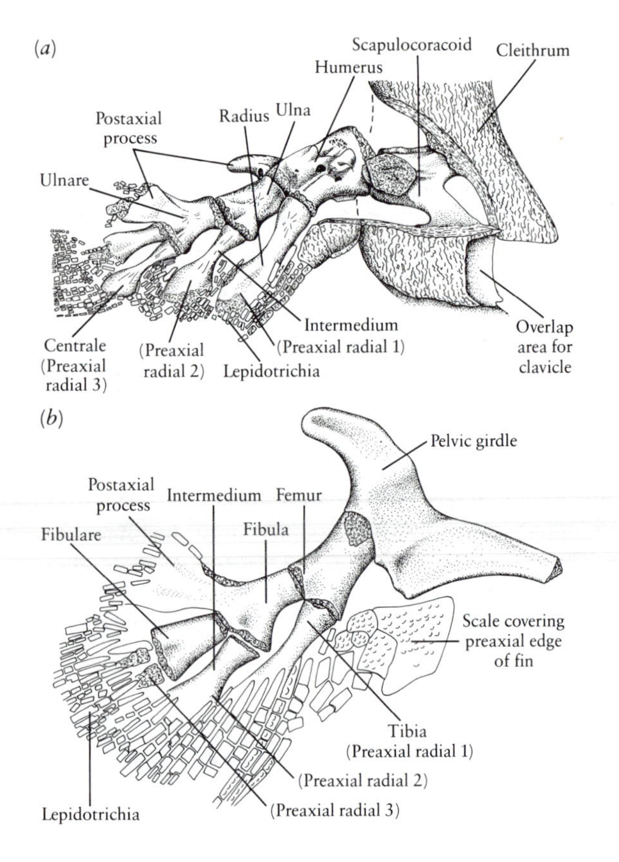

Figure 8-11. ENDOSKELETON OF PAIRED FINS OF *EUSTHENOPTERON*. (*a*) Pectoral girdle and fin. (*b*) Pelvic girdle and fin. *From Andrews and Westoll, 1970a.*

riorly. The proximal element of the hindlimb is clearly equivalent to the femur of tetrapods. It is succeeded by homologues of the tibia anteriorly and the fibula posteriorly. The fibula is succeeded by equivalents of the fibulare and intermedium, and the fibulare in turn is succeeded by two or more endoskeletal ossifications. As in the forelimb, there is no evidence of more distal elements, nor of anything functionally equivalent to an ankle or foot. We will consider the function of the paired fins in rhipidistians in the next chapter.

RHIPIDISTIAN DIVERSITY

Despite the fact that most rhipidistians conform to a single, relatively specialized anatomical pattern based on a predatory habit and a distinctive type of cranial kinesis, we have long recognized that at least two major phylogenetic groups could be distinguished: the rare and generally more primitive Porolepiformes, and the more numerous and advanced Osteolepiformes (see Figures 8-4, 8-5, 8-8, and 8-9).

The Porolepiformes are represented by three well-known genera: *Porolepis* from the Lower Devonian, *Glyptolepis* from the Middle Devonian, and *Holoptychius* from the Upper Devonian (Jarvik, 1972). In contrast with the Osteolepiformes, the pineal usually does not pierce the skull roof, and the anterior portion of the skull is especially short, with small orbits and two external narial openings. The progressive shortening of the anterior portion of the braincase within this group indicates a greater reliance on cranial kinesis and the use of the subcephalic muscle in feeding. Porolepiformes have several tooth whorls lateral to the symphysis, in contrast with the patches of small teeth or large fangs common to osteolepiforms. We know little of the endoskeletal fin supports in this group, but the pectoral fin may be very long and slender. Infolding of the enamel is even more complex than in osteolepiforms (Schultze, 1970).

Porolepis, *Powichthys*, and *Youngolepis* from the Lower Devonian are the earliest known rhipidistians, and many of the distinctive features of these genera may be primitive (Chang, 1982). These features include a mosaic of small bony plates covering extensive areas of the skull roof, the failure of the vomers to meet at the midline, the absence of basal scutes on the fins, and the presence of several bones in the position of the squamosal and quadratojugal. *Powichthys* had a pineal foramen but lacks division between the anterior and posterior elements of the dermal skull roof (Jessen, 1980).

The osteolepiforms are typified by the Upper Devonian genus *Eusthenopteron*, whose anatomy has served as the basis for description of most features common to rhipidistians as well as for comparison with amphibians (Jarvik, 1980). The earlier osteolepiform *Osteolepis* from the Middle Devonian is a member of a conservative family that extends into the Lower Permian, where it is represented by the genus *Ectosteorhachis*. This family retains a heterocercal tail and thick rhomboidal scales covered with cosmine. *Eusthenopteron* represents a more progressive family that is characterized by a symmetrical, three-lobed tail and thin, extensively overlapping cycloid scales. Vorobyeva (1975, 1980) has recently discussed other aspects of evolution among the osteolepiforms.

STRUNIIFORMES

Four poorly known genera are usually allied with the rhipidistians: *Strunius* (Figure 8-12), *Onychodus*, *Grossius*, and *Quebecius* from the Middle and Upper Devonian (Jessen, 1966; Schultze, 1973). *Strunius* is a small fish known from complete but poorly preserved material; it is widely distributed geographically and occurs in both marine and fresh waters. *Onychodus* is a larger form that is known primarily from disarticulated elements.

The possession of a bipartite braincase and kinesis of the dermal skull roof demonstrate the affinities of these genera with rhipidistians. The bones within the two units of the braincase are not solidly fused to one another but may be distinguished by sutures or separated by large amounts of cartilage. Like other sarcopterygians, Struniiformes possess two dorsal fins and numerous sclerotic plates. In the genus *Quebecius*, the tail is heterocercal, but in *Strunius*, it is trifid like that of both *Eusthenopteron*

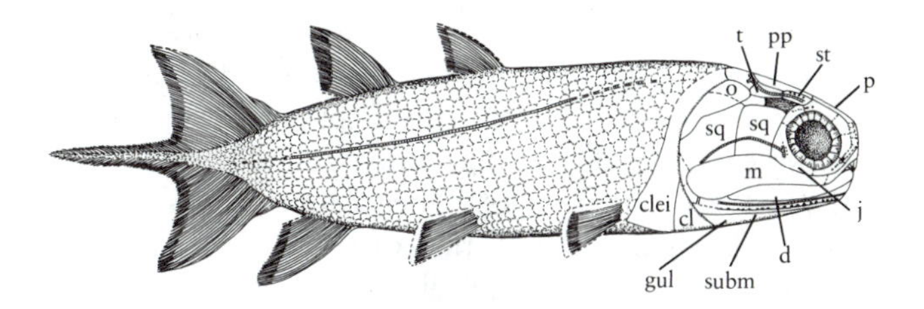

Figure 8-12. *STRUNIUS*, × 1½. Like actinopterygians, the struniiforms have large orbits and lack conspicuously lobate fins. However, the braincase ossifies in two sections like the crossopterygians. Abbreviations as in Figure 8-3. *From Moy-Thomas and Miles, 1971. By permission from Chapman and Hall, Ltd.*

and coelacanths. Unlike osteolepiform rhipidistians, there are no internal nares, but like the more primitive rhipidistians and actinopterygians, there are two external nares. There is no pineal opening or interclavicle. Tooth whorls are present on either side of the symphysis, as in some sharks, acanthodians, and porolepiforms.

Because of its relatively large orbits, small nasal capsules, and cheek bone configuration, *Strunius* resembles actinopterygians, but these features can be attributed to convergence. The paired and dorsal fins appear to be superficially like those of ray-finned fish. We may tentatively place the Struniiformes (or Onychodontiformes) within the crossopterygians as a taxon equivalent to the rhipidistians. Andrews (1973) suggests closer affinities with the Osteolepiformes than with the Porolepiformes.

COELACANTHIFORMES (ACTINISTIA)

Rhipidistians are not represented in the fossil record after the Lower Permian. By the Middle Devonian, they had given rise to two derived groups that survive to the present: tetrapods, which will be discussed in the next chapter, and coelacanths (Figure 8-13). We have not established the specific origin of coelacanths, although Andrews (1973) contends that their closest affinities may lie with the porolepiforms. The primary reason for considering that coelacanths are closely related to the rhipidistians and including them in a single order, Crossopterygii, is the great similarity in the structure of the braincase and its peculiar kinesis.

Although several families are recognized, coelacanths are remarkably consistent in their general morphology and even in details of bony structure (Jarvik, 1980). Coelacanths are already specialized in skull proportions, greatly reduced marginal dentition, and absence of cosmine when we first encounter them in the Middle Devonian. Unlike osteolepiforms, there is no choana, but two external nares are present. The body is generally short; the tail is typically a trifid structure resembling that of *Eusthenopteron*, except in the Upper Devonian genus *Miguashaia*, in which it is heterocercal (Schultze, 1973).

The structure and proportions of the braincase are similar to those of rhipidistians; however, the opening for the Vth nerve passes through the joint rather than via a separate opening behind it. The proportions of the remainder of the skull and the distinctive orientation and configuration of the hyomandibular contribute to significant differences in cranial mechanics, as Thomson (1967) described. The skull is shorter and the quadrate is almost at the level of the intracranial joint. The hyomandibular is not directly linked with the palatoquadrate but articulates via a symplectic with an elongate retroarticular process behind the articulation between the quadrate and the lower jaw. Elevation of the front of the skull is accompanied by a forward movement of the mandibles, but unlike the rhipidistians, the movement is not necessarily accompanied by lateral expansion of the cheeks. Feeding and respiration are mechanically distinct, although both involve movements of the hyomandibular.

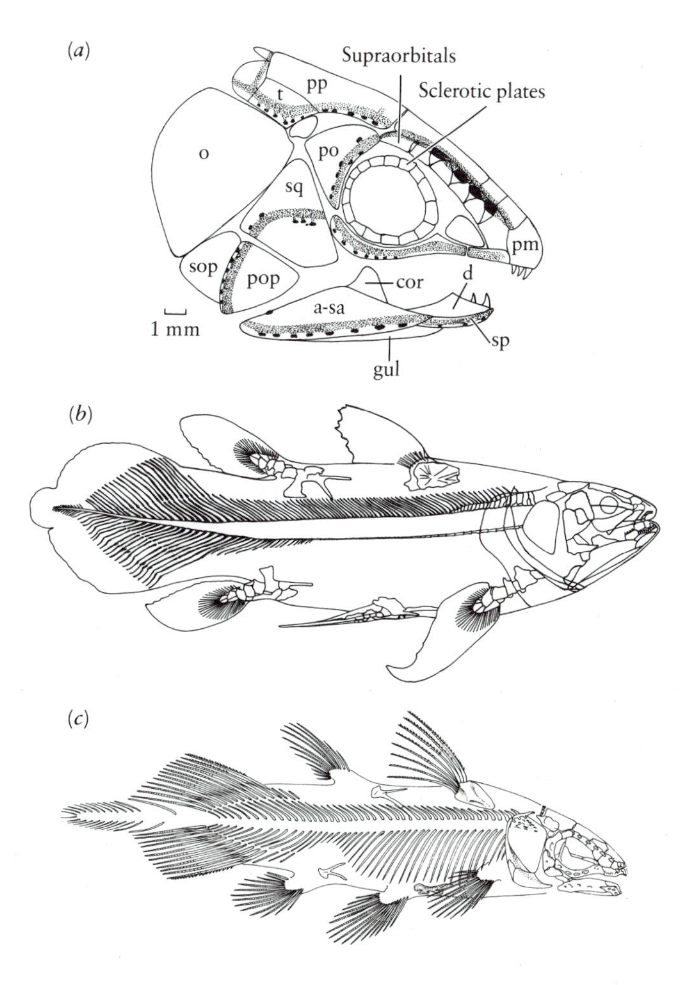

Figure 8-13. COELACANTHS. (*a*) Skull of the Carboniferous coelacanth *Rhabdomerma*. Abbreviations as in Figure 8-3. *From Moy-Thomas and Miles, 1971. By permission from Chapman and Hall, Ltd.* (*b*) Skeleton of the modern genus *Latimeria*. *From Thomson, 1969.* (*c*) The Triassic coelacanth *Diplurus*. *From Schaeffer, 1952. Courtesy of the Library Services Department, American Museum of Natural History.*

The notochord remains persistent throughout the group, with only tiny central elements recognizable. The fleshy lobe tends to be more restricted to the base of the fins, and the first dorsal in *Latimeria* is supported primarily by lepidotrichia. By the Triassic, the skeleton had achieved an essentially modern appearance. According to Schaeffer (1952), most features of the modern genus had been established by the late Devonian. The Upper Devonian coelacanths are known from both marine and fresh water. Mesozoic coelacanths were largely, but not entirely, marine. In the late Paleozoic and Mesozoic, the group had a worldwide distribution.

There is no fossil record of coelacanths beyond the Cretaceous, at which time they are known from shallow marine deposits. In 1938, a living coelacanth was discovered in the Indian Ocean near the coast of Madagascar. Dozens of specimens have been collected since, providing the most direct evidence of the soft anatomy of the other, long extinct crossopterygians.

Millot and Anthony (1958, 1965, 1978) described the anatomy of the living coelacanth in detail, Andrews (1977) described the axial skeleton, and McCosker and Lagios (1979) discussed various aspects of the anatomy and physiology. Northcutt, Neary, and Senn (1978) showed that the brain to body weight ratio in *Latimeria* is comparable to that of the most primitive living sharks and bony fish and is similar to that of rhipidistians that were measured on the basis of endocasts.

The long period of separate evolution between rhipidistians and coelacanths makes it unwise for us to extrapolate uncritically the specific features of the living genus *Latimeria* to the Paleozoic groups. The presence of subcephalic muscles and the extension of the notochord to the level of the pituitary, which occur in coelacanths, were previously suspected features of the rhipidistians. Modern coelacanths give birth to live young, but this was almost certainly not a feature of the ancestors of amphibians. In *Latimeria,* the lungs are filled with fat, but they appear to be calcified in Paleozoic and Mesozoic coelacanths. In contrast, they served for respiration and buoyancy control in Paleozoic rhipidistians.

LUNGFISH

THE ANATOMY OF PRIMITIVE GENERA

Nearly all Dipnoi are highly modified from the pattern of other bony fish in the fusion of the palatoquadrate with the braincase, loss of marginal tooth-bearing bones, and emphasis on the palatal dentition. The pattern of the bones of the skull roof is not directly comparable with that of either rhipidistians or actinopterygians. So different are typical lungfish from sarcopterygians that Jarvik (1980) allied them with holocephalians among the chondrichthyes rather than with bony fish.

In contrast with all later genera, the earliest-known lungfish, *Diabolichthyes* from the Lower Devonian of China, shows a remarkable mosaic of characters considered typical of lungfish and primitive rhipidistians (Chang and Yu, 1984) (Figure 8-14). The pattern of the skull roof is clearly comparable with that of later lungfish, but the ventral surface of the skull shows that the basicranial articulation was still mobile and that the palatoquadrate was not fused to the braincase. The posterior portion of the braincase has not been described, but comparison with the primitive rhipidistian *Youngolepis* suggests that the ethmoid and otic-occipital elements were separately ossified. The premaxilla is recognizable as a distinct tooth-bearing bone of the skull margin that separates the anterior narial opening from the mouth cavity. The vomer occupies a position that is comparable to that of primitive rhipidistians, and the parasphenoid is a long, toothed element extending anteriorly between the pterygoids. In contrast with all later lungfish, the dentary retains marginal teeth. On the other hand, the teeth on the premaxilla do not form a marginal row but were exposed primarily within the mouth cavity. The teeth covering the pterygoid and prearticular are densely packed and arranged in a radiating pattern, as in later lungfish, but are not fused to form definite tooth plates.

Diabolichthys is clearly allied with later lungfish in the emphasis on the palatal dentition and in the pattern of the dermal skull roof, but it retains many features that reflect an ancestry among the crossopterygians. The postcranial skeleton of *Diabolichthys* has not been described, but that of the slightly younger genus *Uranolophus* resembles that of rhipidistians (Denison, 1968a, b).

Diabolichthys comes from a facies that is transitional between marine and continental. Other early Devonian lungfish are known from freshwater and marine deposits.

All lungfish other than *Diabolichthys* are distinguished by ossification of the braincase as a single unit to which the palatoquadrate is fused, the loss of the premaxillae, and the great reduction in the anterior extent of the parasphenoid. The advanced pattern of the palate and braincase are already evident in *Uranolophus* from the Lower Devonian of North America (Figure 8-15). Unlike most later lungfish, *Uranolophus* lacks tooth plates. The pteryogoids are elongate triangular bones that cover most of the palate. Both they and the prearticulars bear tooth ridges on their margins.

Well-defined tooth plates are a hallmark of more advanced lungfish. In the middle Devonian genus *Dipterus* (Figure 8-16*d*), there are large paired plates with radiating rows of denticles that occupy much of the surface of the pterygoids and smaller plates that developed from the vomers. Another pair are borne on the prearticulars.

Typically, the bones of the snout of primitive lungfish are turned under as a "lip" that may bear sharp ridges and tubercles. The ventral margin of the "lip" is notched for paired narial openings, and there is space posteriorly for a large nasal capsule. Modern lungfish have a second

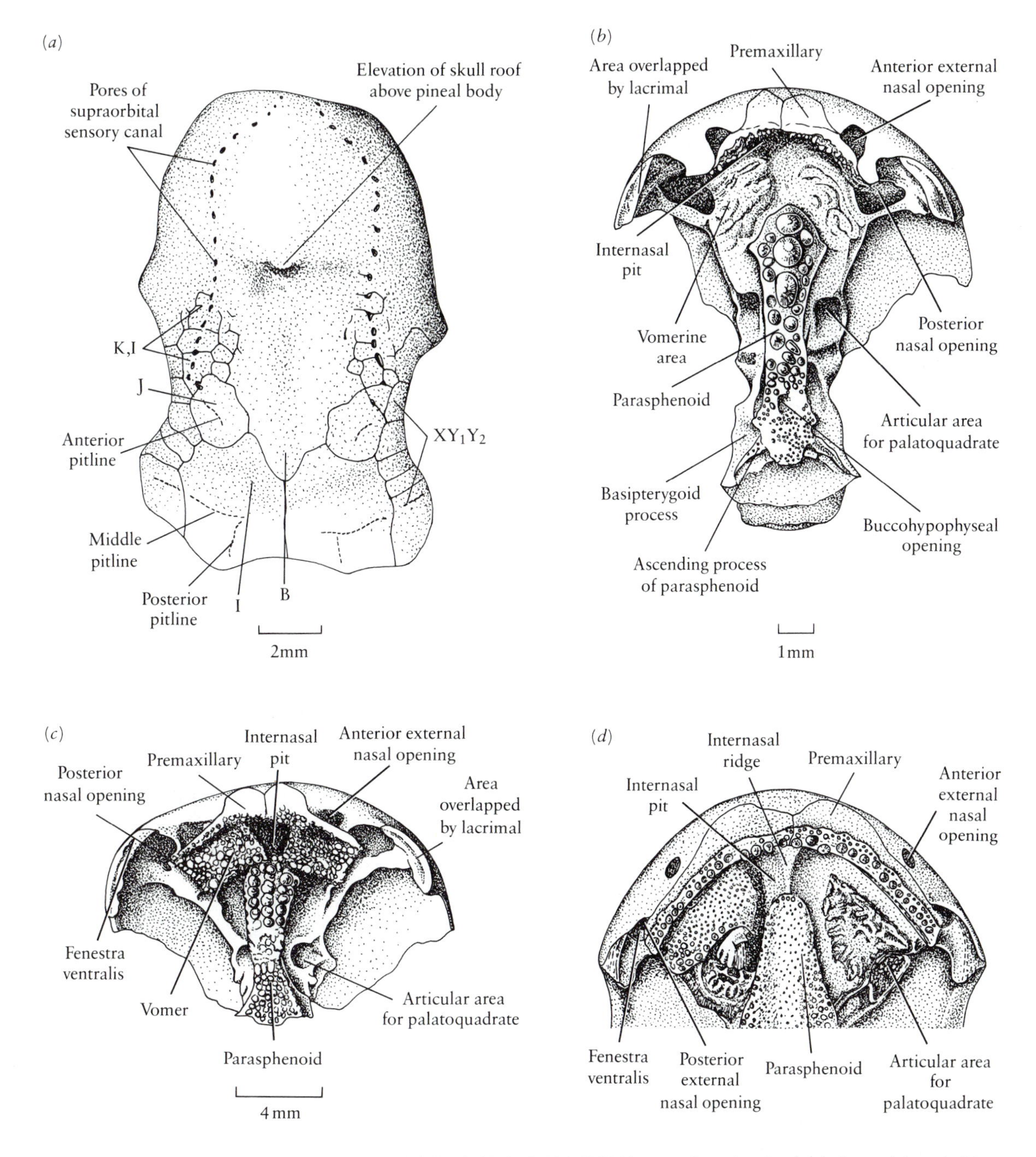

Figure 8-14. LOWER DEVONIAN SARCOPTERYGIANS FROM CHINA. (*a–c*) Dorsal and palatal views of the primitive lungfish *Diabolichthys*. The skull roof closely resembles the pattern of later lungfish, but the palate is primitive in that it shows an articulating surface for the palatoquadrate, which indicates that the palate and braincase were not fused as is the case in all other lungfish. The premaxilla remains a separate area of ossification, and the external narial opening is at the margin of the mouth. Numbers and letters are used to identify the dermal bones of the lungfish skull because most do not appear to be directly homologous with those of other bony fish or amphibians. (*d*) Anterior portion of the palate of the very primitive rhipidistian *Youngolepis* showing a pattern very similar to that of *Diabolichthys*. *From Chang and Yu, 1984.*

opening from the nasal sac into the mouth cavity. Some authors believe that this opening is homologous either with the internal nostril of rhipidistians or the posterior external naris of actinopterygians. This controversy remains unresolved.

Rosen, Forey, Gardiner, and Patterson (1981) identify an opening in the palate of *Griphognathus*, an upper Devonian lungfish that lacks tooth plates, as an internal naris. Although this opening may have served as an air passage into the nasal capsule, it is difficult to homologize with the internal naris of rhipidistians and tetrapods because of the absence of marginal tooth-bearing bones (Campbell and Barwick, 1984). No other lungfish possesses such an opening.

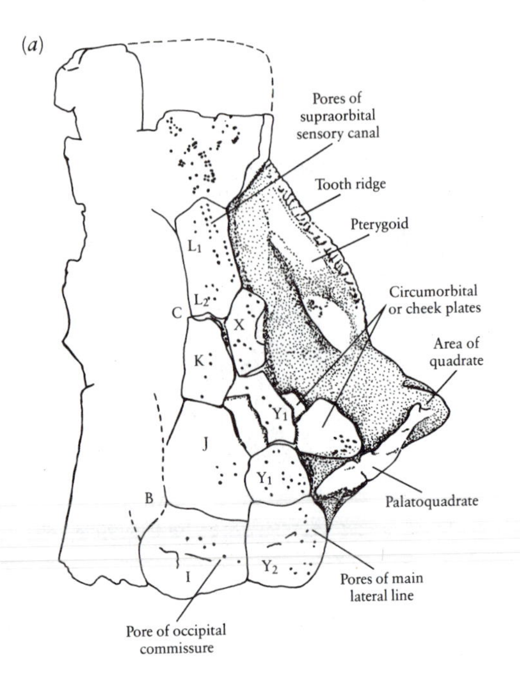

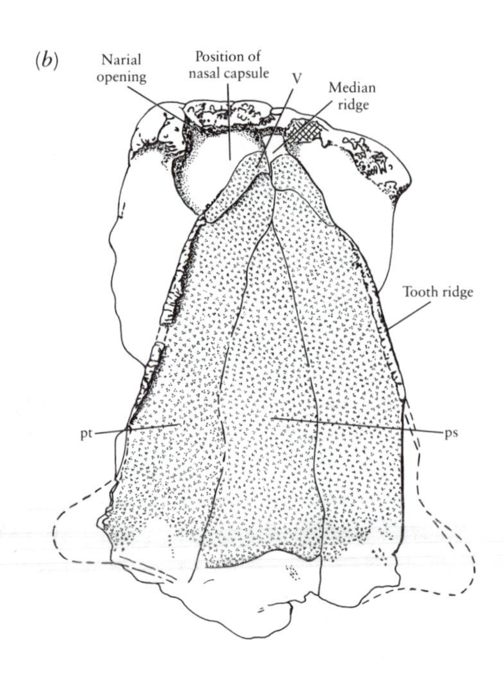

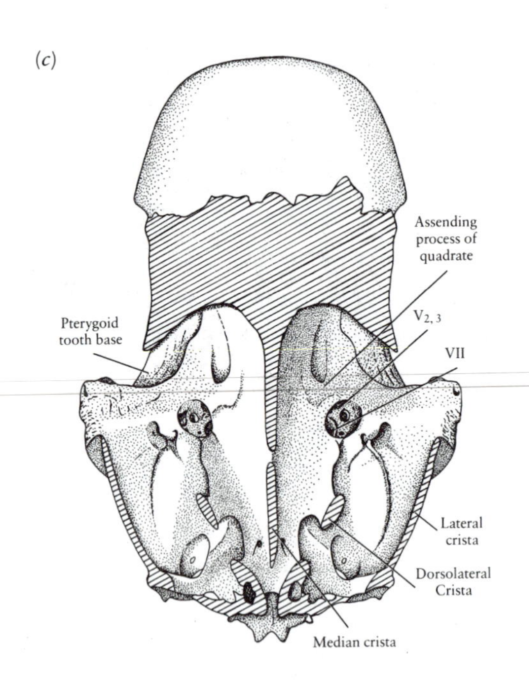

Figure 8-15. SKULLS OF FOSSIL DIPNOI. Skull roof of an early Devonian lungfish *Uranolophis*. (*a*) Dorsal view, $\times\frac{2}{3}$. (*b*) Ventral view, $\times 1$. The area labelled ps (parasphenoid) is probably a broken piece of the pt (pterygoid). No lungfish other than *Diabolichthys* has a long parasphenoid, as in rhipidistians. *From Denison, 1968a. By permission from the Field Museum of Natural History, Chicago.* (*c*) Dorsal view of the braincase of the lungfish *Chirodipterus* with the skull roof removed to show the extent of chambers for the adductor jaw musculature. The jaw muscles occupy almost all of the space above the braincase and extend anteriorly toward the short lower jaws. *From Miles, 1977. With permission from the Zoological Journal of the Linnean Society. Copyright 1977 by the Linnean Society of London.*

Figures 8-14 to 8-17 show the dermal skull roofs of several Dipnoi. The pattern is not directly comparable with those of either rhipidistians or actinopterygians, and a unique terminology was established using numbers and letters. The number of separate centers of ossification tends to be reduced in later genera. The general homology of the major units can be established on the basis of their association with the lateral line canals. Ventral to the operculum and between the jaws there are subopercular, gular, and submandibular plates that are broadly but not specifically comparable with bones in these positions in actinopterygians and rhipidistians.

In Upper Devonian lungfish, the hyomandibular is recognizable as a distinct element that articulates with the posterior surface of the palatoquadrate and the braincase (Campbell and Barwick, 1982). Its movement may affect the operculum, but it plays no role in the connection between the palatoquadrate and the braincase. The hyomandibular retains the same position and configuration in *Neoceratodus*, although it is smaller in the living genus.

The surface of the lower jaw is usually covered with a continuous layer of cosmine, but in some specimens sutures divide a series of bones. Paired dentary bones such as those of rhipidistians and actinopterygians are not evident, and there are no marginal teeth except in *Diabolichthys*. The jaw symphysis is formed by a large medial element that may incorporate the anterior portion of the area of ossification that forms the dentary in other bony fish. A sequence of more posterior and ventral bones are topographically comparable to the splenial, postsplenial, angular, and surangular of rhipidistians. Most of the medial surface is formed by the prearticular, which bears the tooth plate in advanced genera. This bone is usually fused at the symphysis. Coronoid bones, which are present in both actinopterygians and rhipidistians, are missing in lung fish, presumably in conjunction with the absence of a normal marginal dentition.

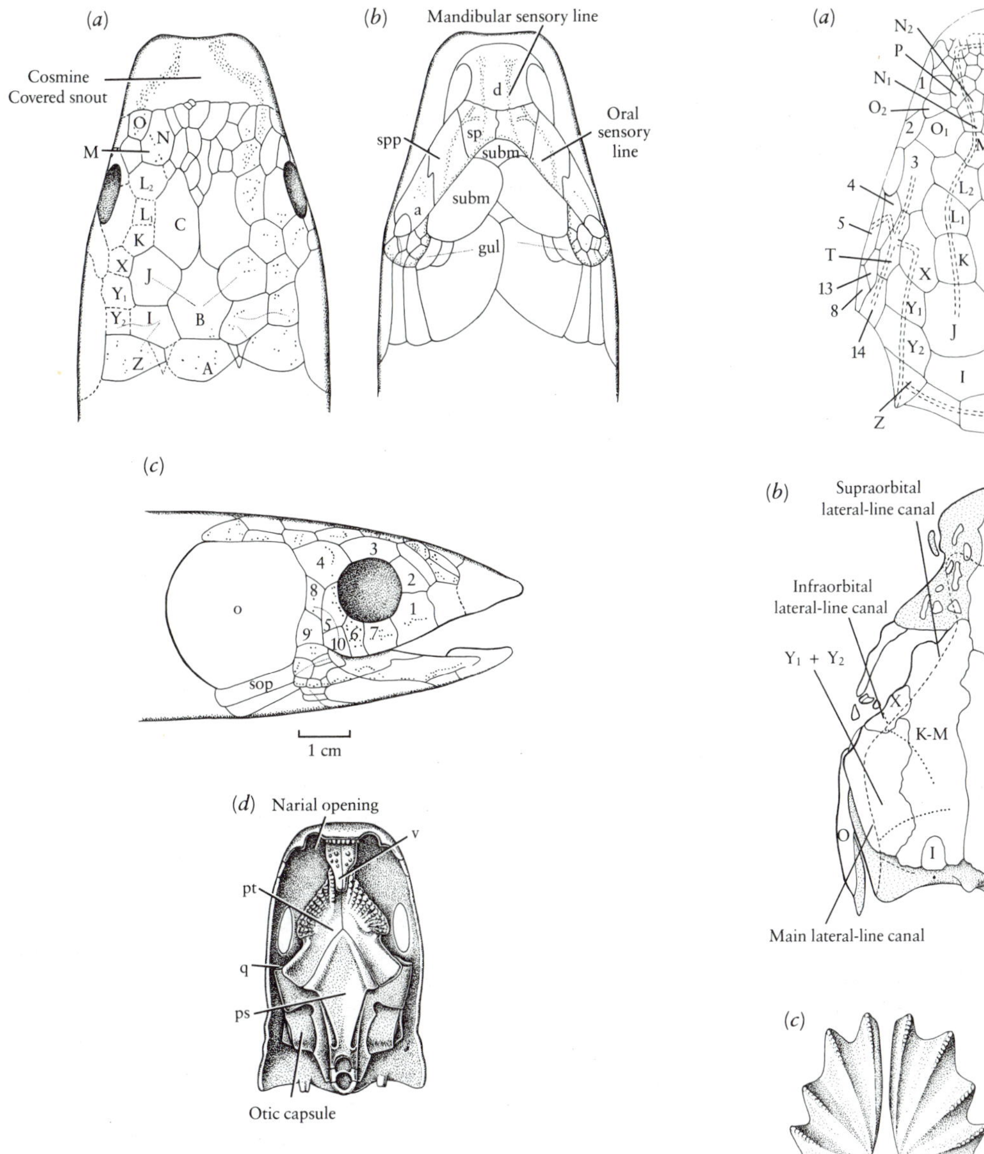

Figure 8-16. SKULL OF THE MIDDLE DEVONIAN LUNGFISH *DIPTERUS*. (*a*) Dorsal, (*b*) ventral, (*c*) lateral, and (*d*) palatal views. *(a–c) From Moy-Thomas and Miles, 1971. By permission from Chapman and Hall, Ltd. (d) From Westoll, 1949. Copyright 1949 by Princeton University Press, copyright renewed by Princeton University Press. Reprinted by permission of Princeton University Press.*

Like primitive rhipidistians and actinopterygians, the rhomboidal scales of *Uranolophus* have an anterior dorsal process that articulates with the next scale above. There are two posteriorly placed dorsal fins. The tail is heterocercal in *Uranolophus* and other early lungfish.

The shoulder girdle of primitive lungfish resembles that of rhipidistians in being formed by a clavicle, a small interclavicle, and a cleithrum bearing an ossified scapulocoracoid. It is unlike that of either rhipidistians or actinopterygians in lacking a succession of large bones dorsal to the cleithrum that bind it to the skull table. The supracleithrum and the anocleithrum (which are present in later lungfish) may have evolved separately in actinopterygians and sarcopterygians.

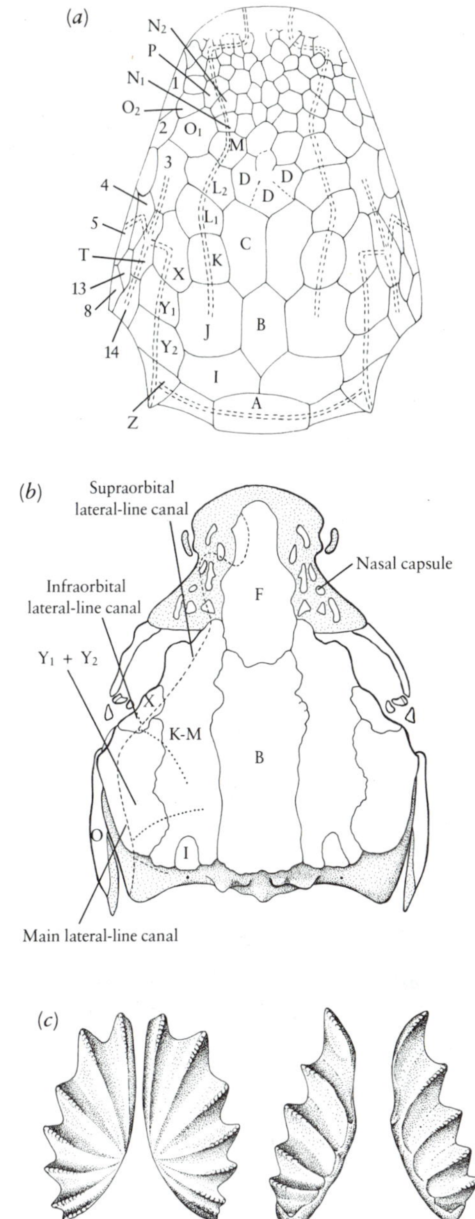

Figure 8-17. (*a*) Skull roof of the Lower Devonian lungfish *Dipnorhynchus. From Thomson and Campbell, 1971.* (*b*) Skull roof of the modern lungfish *Neoceratodus. From Miles, 1977, after Goodrich. With permission from the Zoological Journal of the Linnean Society. Copyright 1977 by the Linnean Society of London.* (*c*) Pterygoid and lower jaw tooth plates of the late Paleozoic lungfish *Sagenodus. From Gregory, 1951. Courtesy of the Library Services Department, American Museum of Natural History.*

Uranolophus vertebrae have well-developed neural arches and one crescentric centrum in each segment, the intercentrum, commonly the largest unit in primitive rhipidistians. In contrast to *Uranolophus*, the centra of a few later Devonian lungfish are ossified as complete rings and the notochord is much reduced. After the Devonian, however, there is almost no evidence of ossification of the centra in this group. In contrast with other early bony fish, Dipnoi's pleural ribs are very well developed.

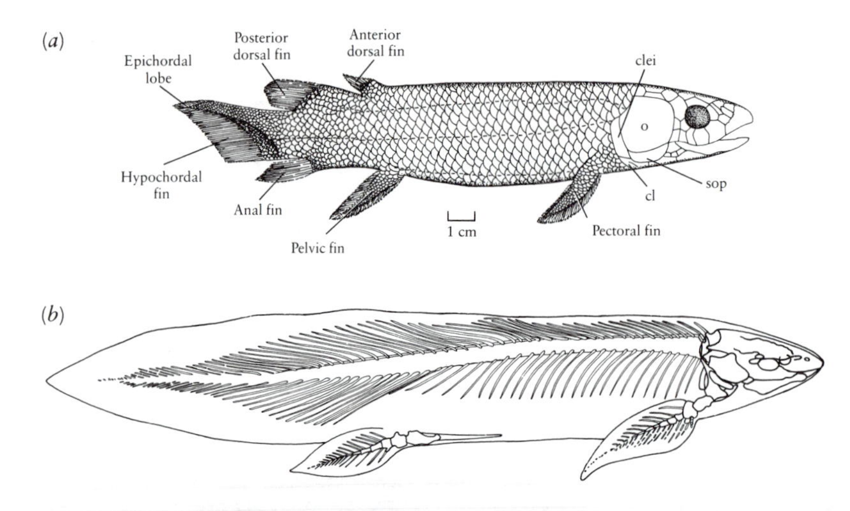

Figure 8-18. (*a*) The Upper Devonian lungfish *Dipterus*. *From Moy-Thomas and Miles, 1971. By permission from Chapman and Hall, Ltd.* (*b*) The modern lungfish *Neoceratodus*. *From Thomson, 1969.*

Endoskeletal supports for the paired fins are described in very few primitive lungfish. In the Carboniferous genus *Conchopoma,* they are arranged in an archipterygial pattern comparable to that of the living genus *Neoceratodus* (Figure 8-18). There is no evidence that any lungfish developed the large preaxial radials characteristic of osteolepiform rhipidistians.

EVOLUTIONARY TRENDS

The feeding apparatus of lungfish evolved very rapidly during the early Devonian. Further consolidation of the palate and braincase and experimentation in the nature of the dentition continued to the end of that period. By the late Devonian, the basic pattern of the head region was established and its general structure has remained almost unaltered to the present day.

The fusion of the palatoquadrate to the braincase and the increase in size of the adductor chamber would have given lungfish a very strong bite, which is necessary for feeding on hard-shelled prey, but would also have been effective for feeding on a wide range of foods, which is the habit among living lungfish.

The capacity to bite down very hard on the food can be associated with the development of tooth plates among early Dipnoi. Their elaboration almost certainly occurred after the modification of the braincase and upper jaw, since several of the Devonian lungfish lack tooth plates. However, an isolated plate has been found in the same deposit as *Uranolophus,* indicating that some lungfish achieved this feature very early in their evolution. Structural intermediates have been described between the primitive pattern seen in *Uranolophus* and fully developed tooth plates such as occur in the middle Devonian genus *Dipterus*. *Fleurantia* (from the late Devonian) has enlarged denticles that are arranged in a radiating pattern within a field of smaller denticles.

Denison (1974) described progressive changes that occurred during the evolution of lungfish that made the tooth plates more effective. In *Uranolophus,* as in other primitive bony fish, the individual palatal denticles are regularly replaced. In the early stages of tooth plate development, the ability to replace the denticles is lost and the plates may be severely worn. In more advanced species, a pulp chamber develops between the dentine of the teeth and the underlying spongy bone and surface wear is compensated by the continuous growth of the dentine.

In early lungfish, the histology of the individual denticles does not differ from that of other early bony fish. In later genera, the surface is more resistant to wear because of the elaboration and hypermineralization of tubular dentine. Another specialized, highly mineralized tissue, termed petrodentine, is present in the more advanced living lungfish family, the Lepidosirenidae. Areas of petrodentine are separated by areas of trabecular dentine that are less dense and more rapidly worn. This arrangement gives rise to a sharp cutting edge between the two tissues.

Like some surviving members of the most primitive group of ray-finned fish, modern lungfish have greatly reduced the ossification of their internal skeleton. After the Upper Devonian, the braincase and vertebral centra became largely cartilaginous. The dermal bones of the skull became reduced in number and extent, lost their cosmine, and sank beneath the surface (see Figure 8-17*b*). The scales became cycloid, lost their cosmine, and, in modern species, overlap extensively. In the primitive state, the fins were covered by bony lepidotrichia. These are lost and the fin webs of modern genera are supported by horny camptotrichia (the equivalent of ceratotrichia in sharks). Progressive changes also occur in the medial fins. The two dorsal fins move posteriorly and become confluent with the caudal. The caudal fin itself becomes symmetrical and tapering and the anal becomes confluent with it ventrally.

These trends are far advanced in the Upper Devonian genus *Phaneropleuron,* with only the anal fin remaining separate.

The Dipnoi appear to have undergone several successive episodes of adaptive radiation. The specific relationships of the many families are still subject to controversy. Thomson and Campbell (1971) and Miles (1977) recently discussed the problems of classifying lungfish.

Several distinct lineages that are characterized by differences in the palate and its dentition appear at the base of the Devonian. Later in the period the Rhynchodipteridae appear. They include the genera *Griphognathus, Soederberghia,* and *Rhynchodipterus,* which are specialized in the elongation of the snout and the development of ring centra. Two additional Upper Devonian families are represented by *Fleurantia* and *Scaumenacia,* which are specialized in having a deep, laterally compressed body that is associated with fast swimming and predatory habits. *Dipterus* may represent a main line of dipnoan evolution from which have diverged various specialized lineages, including both living families and a variety of Carboniferous genera.

After the Carboniferous, the diversity of lungfish declined dramatically, with only two recognizable lineages persisting in the Mesozoic and Cenozoic: the Ceratodidae and the Lepidosirenidae. The Ceratodidae represent a continuation of the main stream of Paleozoic dipnoan adaptation represented by *Dipterus* and *Sagenodus.* The tooth plates are large and well developed but have relatively low tooth ridges (see Figure 8-17). The body is thick and round, and the paired fins are short and stout. The genus *Ceratodus* is known from the Lower Triassic into the Upper Cretaceous of nearly all continents. The modern genus *Neoceratodus* [*Epiceratodus*] is known as early as the Upper Cretaceous in Australia; it is now restricted to a few rivers on the east side of that continent.

We know the Lepidosirenidae from early Cenozoic remains in both Africa and South America, where the living genera *Protopterus* and *Lepidosiren* persist. The tooth plates are distinguished from those of the Ceratodidae in having a small number of very high ridges that are strengthened by petrodentine. The body is long and slim, with the paired fins reduced to tentaclelike appendages.

The modern African and South American lungfish live in areas of seasonal drought and have the ability to aestivate. Members of the genus *Protopterus* burrow into the mud where they can remain dormant for more than a year, surviving at a reduced metabolic rate. The early Permian genus *Gnathorhiza* is also known to aestivate. Numerous specimens are preserved in burrows, still waiting for the next rainy season. Empty burrows, attributed to lungfish, are known from the Carboniferous and, questionably from the Devonian. All three living genera of lungfish can make use of atmospheric oxygen, and members of the Lepidosirenidae can drown without access to the air.

ALTERNATIVE RELATIONSHIPS

Despite extensive study of exceptionally well-preserved fossils of sarcopterygian fish near the beginning of their radiation in the Devonian, considerable controversy remains regarding their relationships. Lagios (1979) suggests that coelacanths have close affinities with sharks, while Wiley (1979) maintains that they are only distantly related to the rhipidistians among the bony fish. The position of the Dipnoi appears especially contentious. Jarvik (1968, 1980) contends that the absence of marginal dentition and fusion of the palatoquadrate to the braincase support affinities with holocephalians among the chondrichthyes, rather than to bony fish. Compagno (1979) argues that the complete absence of dermal bone in any of the chondrichtyan groups demonstrates an almost insurmountable difficulty in establishing close affinities between either sharks or holocephalians and the dipnoi.

The affinities of the sarcopterygians to amphibians are also subject to different interpretations. Most authors think that the ancestry of all tetrapods can be traced to the osteolepiform rhipidistians. Jarvik (1942, 1963, 1980) has argued repeatedly that urodeles are unique in having arisen from the porolepiforms, while all other tetrapods have evolved from osteolepiforms. His arguments have been countered by Thomson (1968), Vorobyeva (1975), and Schultze (1970).

Rosen, Forey, Gardiner, and Patterson (1981) have attempted to demonstrate that lungfish, not rhipidistians, are close to the ancestry of tetrapods. The numerous, detailed similarities between amphibians and the well-known rhipidistian *Eusthenopteron* are dismissed as being plesiomorphies, rather than indicative of close relationship, and the many, highly specialized features of the dipnoan skull that are evident in even the earliest genera are not considered as precluding a close relationship to the early amphibians. A number of similarities between dipnoans (fossil and living) and amphibians (Paleozoic and modern) are discussed, but there is little evidence that any of these are unique to the earliest amphibians and Paleozoic lungfish. Holmes (1985) points out many procedural and factual errors in the cladistic analysis by Rosen and his colleagues.

Although there is not a continuous sequence of genera linking rhipidistians and amphibians, nearly all aspects of the skeleton of primitive amphibians can be traced directly to comparable features in well-known rhipidistians. In no dipnoans, from the Devonian to the Recent, is the structure of the skull or appendicular skeleton similar to that of any Paleozoic amphibian.

In cladistic terms, rhipidistians are a paraphyletic group. They almost certainly include the ancestors of amphibians, and they may have given rise to coelacanths and struniiforms as well. The host of shared derived characters that link amphibians with osteolepiform rhipidistians could be used to ally them in a single taxonomic group that is distinct from porolepiforms, coelacanths, and struniiforms. Anatomically and biologically, osteolepiforms nevertheless were more similar to these fish groups than they are to amphibians (Thomson, 1969). This similarity is reflected in their continued classification among the crossopterygian fish.

Summary

Sarcopterygian fish include four living genera, the coelacanth *Latimeria,* and the lungfish *Neoceratodus, Protopterus,* and *Lepidosiren.* These fish are relics of an extensive Paleozoic assemblage that also includes the rhipidistians, which gave rise to all tetrapods. Sarcopterygians are distinguished from actinopterygians by the possession of fleshy lobed fins, two dorsal fins, and an epichordal lobe of the caudal fin. Early rhipidistians and lungfish have a specialized superficial tissue termed *cosmine* that overlies the laminar and vascular bone of the head and scales. Cosmine supports a mosaic pore-canal system that resembles electrosensory structures in modern sharks and teleosts. The fleshy lobed fins and emphasis on nonvisual sensory structures in early sarcopterygians suggest that this group was initially specialized toward life near the bottom of dark or turbid water.

Rhipidistians and dipnoans specialized their skull and jaw musculature in divergent ways that resulted in a more forceful bite than that of primitive actinopterygians. In rhipidistians and their descendants the coelacanths, the braincase is ossified in two units that are joined ventrally by a large subcephalic muscle. Contraction of this muscle lowers the anterior unit and the attached bones of the palate and marginal dentition to force the teeth into the prey. The braincase in lungfish is a unitary structure to which the palate is fused. They have a very large adductor chamber extending above the braincase that accommodates a much larger muscle mass than that of primitive actinopterygians and rhipidistians. Specializations of the feeding apparatus in rhipidistians and lungfish are evident from the very base of the Devonian; these specializations are associated with very different patterns of the dermal bones of the skull.

Most of the skeletal features of rhipidistians are comparable to those of the early amphibians. The pattern of the skull roof and palate are nearly identical, and the proximal bones of the paired limbs of tetrapods are already evident in rhipidistians although they retain the basic function of fins.

We recognize three groups of rhipidistians: the primitive porolepiforms, the osteolepiforms, which bear the closest resemblance to amphibians, and the poorly known struniiforms. Coelacanths first appear in the middle Devonian and achieved an essentially modern appearance by the end of that period. Lower Devonian lungfish had already achieved the highly specialized skull structure typical of all members of that group. Development of tooth plates and changes in the configuration of the fins occurred rapidly during the Devonian and at a much slower rate in the lineages leading to the modern families. The habit of aestivation practiced by modern African and South American lungfish is evidenced by fossils as early as the Lower Permian.

References

Andrews, S. M. (1973). Interrelationships of crossopterygians. In P. H. Greenwood, R. S. Miles, and C. Patterson (eds.), *Interrelationships of Fishes,* pp. 137–177. Supplement No. 1, *Zool. J. Linn. Soc.,* Vol. 53. Academic Press, London.

Andrews, S. M. (1977). The axial skeleton of the coelacanth *Latimeria.* In S. M. Andrews, R. S. Miles, and A. D. Walker (eds.), *Problems in Vertebrate Evolution,* pp. 271–288. *Linn. Soc. Symp. Series,* No. 4. Academic Press, London.

Andrews, S. M., and Westoll, T. S. (1970a). The postcranial skeleton of *Eusthenopteron foordi* Whiteaves. *Trans. Roy. Soc. Edinburgh,* **68**: 207–329.

Andrews, S. M., and Westoll, T. S. (1970b). The postcranial skeleton of rhipidistian fishes excluding *Eusthenopteron. Trans. Roy. Soc. Edinburgh,* **68**: 391–489.

Boord, R. L., and Campbell, C. B. G. (1977). Structural and functional organization of the lateral line systems of sharks. *Am. Zoologist,* **17**: 431–442.

Campbell, K. S. W., and Barwick, R. E. (1982). The neurocranium of the primitive dipnoan *Dipnorhynchus sussmilchi. J. Vert. Paleont.,* **2**(2): 286–327.

Campbell, K. S. W., and Barwick, R. E. (1984). The choana, maxillae, premaxillae and anterior palatal bones of early dipnoans. *Proc. Linn. Soc. N.S.W.* **107**: 147–170.

Chang, M.-M. (1982). The braincase of *Youngolepis,* a Lower Devonian crossopterygian from Yunnan, South-Western China. Department of Geology University of Stockholm, and section of Palaeozoology, Swedish Museum of Natural History, 3–113.

Chang, M.-M. and Yu, X. (1984). Structure and phylogenetic significance of *Diabolichthys speratus* gen. et sp. nov., a new dipnoan-like form from the Lower Devonian of Eastern Yunnan, China. *Proc. Linn. Soc. N.S.W.,* **107**: 171–184.

Compagno, L. J. V. (1979). Coelacanths: Shark relatives or bony fishes? In J. E. McCosker and M. D. Lagios (eds.), The biology and physiology of the living coelacanth. *Calif. Acad. Sci. Occ. Papers,* **134**: 45–55.

Denison, R. H. (1968a). Early Devonian lungfishes from Wyoming, Utah and Idaho. *Fieldiana: Geol.,* **17**: 353–413.

Denison, R. H. (1968b). The evolutionary significance of the earliest known lungfish. *Uranolophus. Nobel. Symp.* No. 4: 247–257.

Denison, R. H. (1974). The structure and evolution of teeth in lungfishes. *Fieldiana: Geol.*, **33**: 31–58.

Gregory, W. K. (1951). *Evolution Emerging*. Macmillan, New York.

Holmes, E. B. (1985). Are lungfishes the sister group of tetrapods? *Biol. J. Linn. Soc.*, **25**(4): 379–397.

Jarvik, E. (1942). On the structure of the snout of crossopterygians and lower gnathostomes in general. *Zool. Bidr. Uppsala*, **21**: 235–675.

Jarvik, E. (1963). The composition of the intermandibular division of the head in fish and tetrapods and the diphyletic origin of the tetrapod tongue. *Kungliga Svenska Vetenskapsakademiens Handlingar*, **9**: 1–74.

Jarvik, E. (1968). The systematic position of the Dipnoi. *Nobel Symp.* No. 4: 223–245.

Jarvik, E. (1972). Middle and Upper Devonian Porolepiformes from East Greenland with special reference to *Glyptolepis groenlandica* n. sp. *Meddelelser om Grønland*, **187**: 1–178.

Jarvik, E. (1980). *Basic Structure and Evolution of Vertebrates*. Academic Press, London.

Jessen, H. (1966). Struniiformes. In J. P. Lehman (ed.), *Traité de Paléontologie*, Tome 4, Vol. 3: 387–398, Masson S. A. Paris.

Jessen, H. (1980). Lower Devonian Porolepiformes from the Canadian Arctic with special reference to *Powichthys thorsteinssoni* Jessen. *Palaeontographica, Abt. A.*, **167**: 180–214.

Lagios, M. D. (1979). The Coelacanth and the Chondrichthyes as sister groups: A review of shared apomorph characters and a cladistic analysis and reinterpretation. In J. E. McCosker and M. D. Lagios (eds.), The biology and physiology of the living coelacanth. *Calif. Acad. Sci. Occ. Papers*, **134**: 25–44.

McCosker, J. E. and Lagios, M. D. (eds.). (1979). The biology and physiology of the living coelacanth. *Calif. Acad. Sci. Occ. Papers*, **134**: 1–175.

Miles, R. S. (1977). Dipnoan (lungfish) skulls and the relationships of the group: A study based on new species from the Devonian of Australia. *Zool. J. Linn. Soc.*, **61**: 1–328.

Millot, J. and Anthony, J. (1958, 1965, 1978). *Anatomie de Latimeria chalumnae*. C. N. R. S., Paris.

Moy-Thomas, J. A. and Miles, R. S. (1971). *Palaeozoic Fishes* (2d ed.). Chapman and Hall, London.

Northcutt, R. G., Neary, T. J., and Senn, D. G. (1978). Observations on the brain of the coelacanth *Latimeria chalumnae:* External anatomy and quantitative analysis. *J. Morph.*, **155**: 181–192.

Panchen, A. L. (1967). The nostrils of choanate fishes and early tetrapods. *Biol. Rev.*, **42**: 374–420.

Rosen, D. E., Forey, P. L., Gardiner, B. G., and Patterson, C. (1981). Lungfishes, tetrapods, paleontology and plesiomorphy. *Bull. Am. Mus. Nat. Hist.*, **167**: 163–275.

Schaeffer, B. (1952). The Triassic coelacanth fish *Diplurus*, with observations on the evolution of the Coelacanthini. *Bull. Am. Mus. Nat. Hist.*, **99**: 25–78.

Schaeffer, B., and Rosen, D. E. (1961). Major adaptive levels in the evolution of the actinopterygian feeding mechanism. *Am. Zoologist*, **1**: 187–204.

Schultze, H-P. (1970). Folded teeth and the monophyletic origin of tetrapods. *Am. Mus. Novit.*, **2408**: 1–10.

Schultze, H-P. (1973). Crossopterygier mit heterozerker Schwanzflosse aus dem Oberdevon Kanadas, nebst einer Beschreibung von Onychodontida—Resten aus dem Mitteldevon Spaniens und aus dem Karbon der USA. *Palaeontographica, Abt. A*, **143**: 188–208.

Thomson, K. S. (1967). Mechanisms of intracranial kinetics in fossil rhipidistian fishes (Crossopterygii) and their relatives. *J. Linn. Soc.* (Zool.), **46**: 223–253.

Thomson, K. S. (1968). A critical review of certain aspects of the diphyletic theory of tetrapod relationships. *Nobel Symp.*, **4**: 285–305.

Thomson, K. S. (1969). The biology of the lobe-finned fishes. *Biol. Rev.*, **44**: 91–154.

Thomson, K. S. (1975). On the biology of cosmine. *Peab. Mus. Nat. Hist. Bull.*, **40**: 1–58.

Thomson, K. S. (1977). On the individual history of cosmine and a possible electroreceptive function of the pore-canal system in fossil fishes. In S. M. Andrews, R. S. Miles, and A. D. Walker (eds.), *Problems in Vertebrate Evolution. Linn. Soc. Symp. Ser.* No. 4: 247–270. Academic Press, London, New York.

Thomson, K. S., and Campbell, K. S. W. (1971). The structure and relationships of the primitive Devonian lungfish—*Dipnorhynchus sussmilchi* (Etheridge). *Bull. Peabody Mus. Nat. Hist.*, **38**: 1–109.

Thomson, K. S., and Vaughn, P. P. (1968). A new pattern of vertebral structure in a fossil rhipidistian fish (Crossopterygii). *Postilla Peab. Mus. Nat. Hist. Yale Univ.*, **127**: 1–19.

Vorobyeva, E. I. (1975). Formenvielfalt und Verwandtschaftsbeziehungen der Osteolepidida (Crossopterygii, Pisces). *Paläont. Z.*, **49**: 44–55.

Vorobyeva, E. I. (1977). Morphology and the features of the evolution of crossopterygian fish. *Trudy Paleont. Inst.*, **163**: 1–240. (In Russian).

Vorobyeva, E. I. (1980). Observations on two rhipidistian fishes from the Upper Devonian of Lode, Latvia. *Zool. J. Linn. Soc.*, **70**: 191–201.

Westoll, T. S. (1949). On the evolution of the Dipnoi. In G. L. Jepsen, G. G. Simpson, and E. Mayr (eds.), *Genetics, Paleontology and Evolution*, pp. 121–184. Princeton University Press, Princeton, NJ.

Wiley, E. O. (1979). Ventral gill arch muscles and the phylogenetic relationships of Latimeria. In J. E. McCosker and M. D. Lagios (eds.), The biology and physiology of the living coelacanth. *Calif. Acad. Sci. Occ. Papers*, **134**: 56–67.

CHAPTER IX

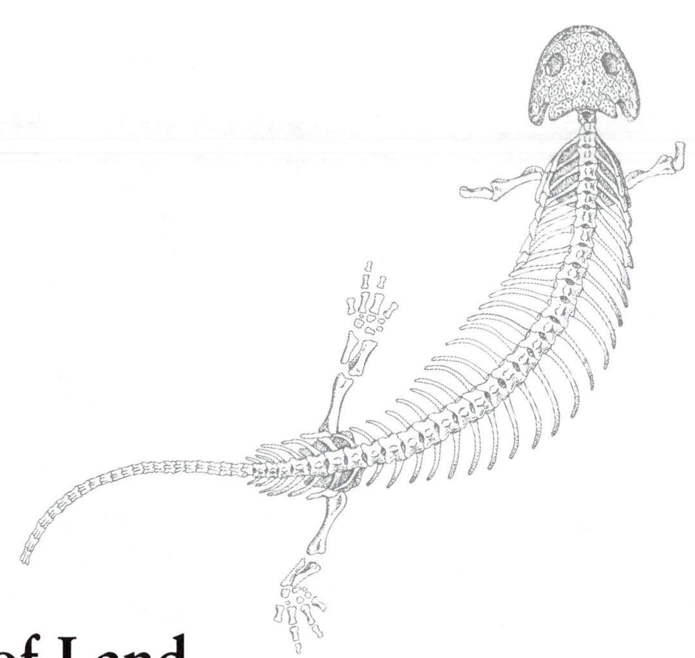

The Conquest of Land and the Radiation of Amphibians

Amphibians are biological intermediates between bony fish and reptiles. They are basically terrestrial animals, but many genera lay their eggs in the water and have an aquatic larval stage; none have extra-embryonic membranes. We include two anatomically distinct groups of tetrapods among the Amphibia: the living frogs, salamanders, and caecilians and their late Mesozoic and Cenozoic ancestors, and a much more diverse assemblage known from the late Devonian into the early Mesozoic (Figure 9-1) (Frost, 1985; Carroll, 1977).

Paleozoic amphibians gave rise to both the living amphibians and, via the reptiles, to all other groups of terrestrial vertebrates. Unfortunately, we have not established the specific nature of these relationships and thus will consider the Paleozoic amphibians as a unit.

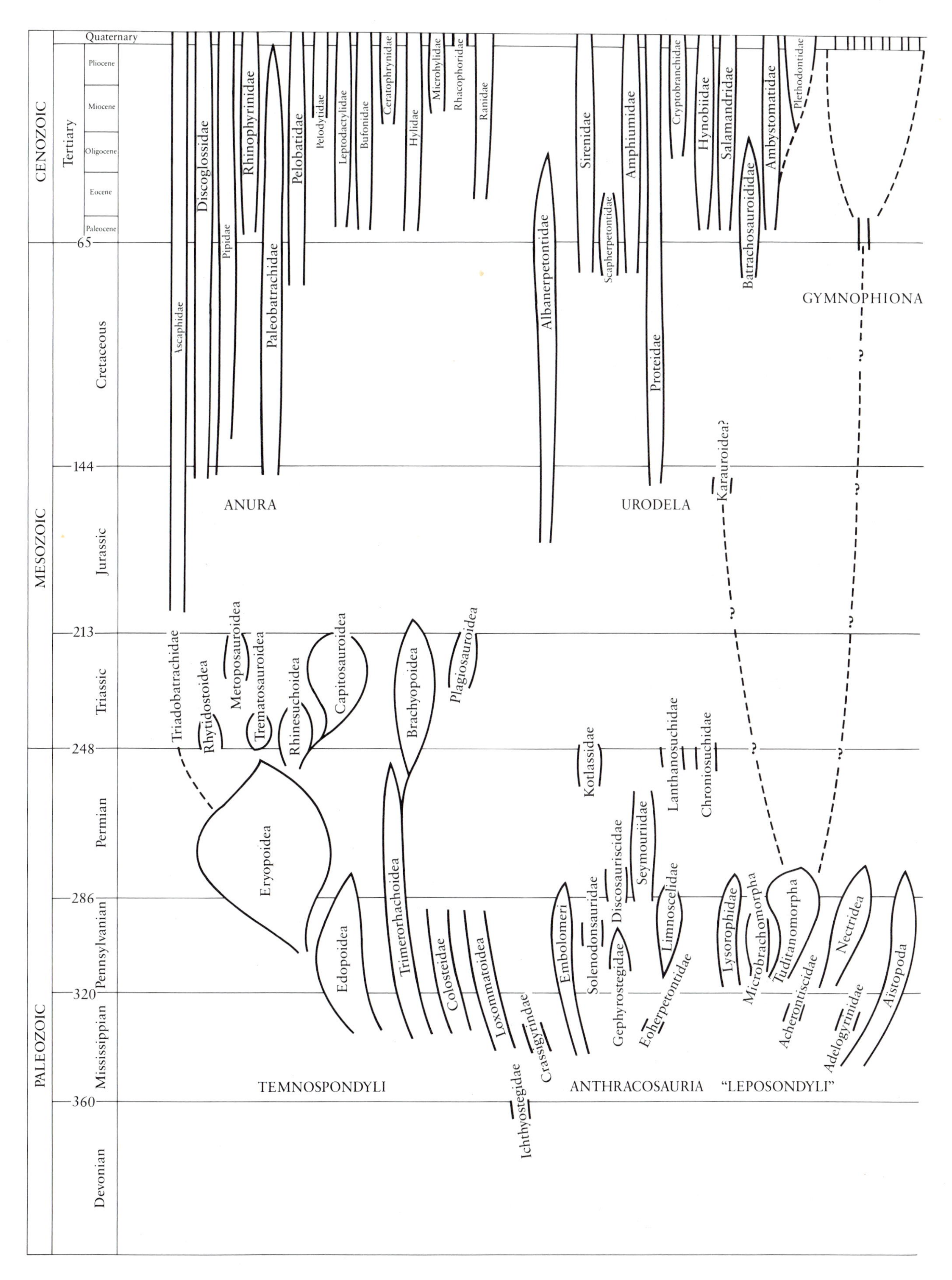

Figure 9-1. PHYLOGENY OF THE AMPHIBIA. All of the modern families, with the exception of minor anuran groups without a fossil record, are indicated. Because of the much greater diversity of Paleozoic and Triassic amphibians, it is not practical to represent each family separately. For example, we recognize 12 families of eryopoids and 11 families of tuditanomorph and microbrachomorph microsaurs.

Paleozoic amphibians

The earliest amphibians were ponderous and clumsy-looking animals, over a meter long with large skulls, short trunks, and stocky limbs (Figure 9-2*a*). These animals, the ichthyostegids, are found in late Devonian sediments from east Greenland (Jarvik, 1980). Their skeletal pattern is basic to all terrestrial vertebrates, while the proximal limb bones, girdles, vertebrae, and dermal bones of the skull are all directly comparable with those of osteolepiform rhipidistians. We have not found any fossils that are intermediate between such clearly terrestrial animals and the strictly aquatic rhipidistians described in the previous chapter.

We recognize many distinct amphibian lineages within the Carboniferous. Their basic anatomical patterns can be derived from the ichthyostegids, but we have not established their specific interrelationships.

THE STRUCTURE OF PRIMITIVE AMPHIBIANS

Because they are close to the ancestry of all later land vertebrates, the ichthyostegids will be considered in some detail. Most of the skeletal features that differentiate early amphibians from rhipidistians can be associated with adaptation to a terrestrial way of life. The following challenges to life on land are particularly important:

1. Support
2. Feeding
3. Respiration
4. Locomotion
5. Water balance
6. Sensory input
7. Reproduction

Support—structure of the axial skeleton

The body of rhipidistians and other fish is supported against the force of gravity by the surrounding water. Out of the

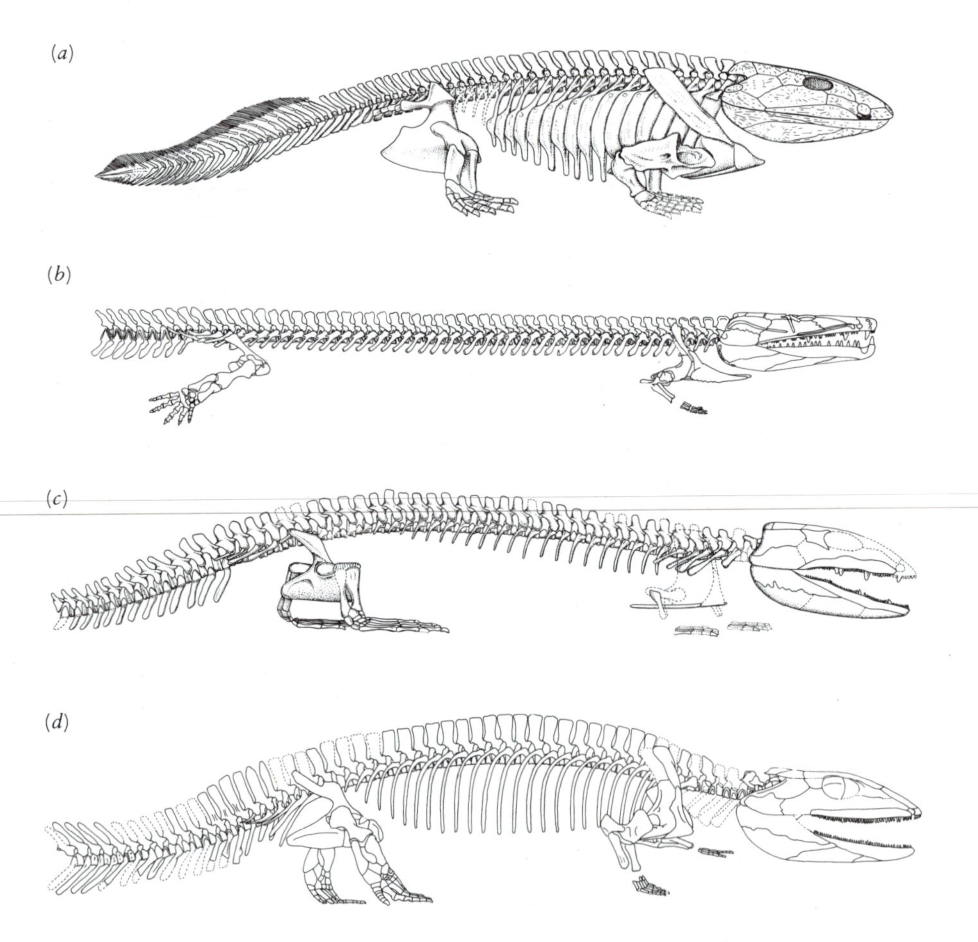

Figure 9-2. RECONSTRUCTION OF THE SKELETONS OF DEVONIAN AND MISSISSIPPIAN AMPHIBIANS. (*a*) *Ichthyostega*, a primitive labyrinthodont from the Upper Devonian of East Greenland. Original about 1 meter long. *From Jarvik, 1955. By permission from Scientific Monthly.* (*b*) *Greererpeton*, from the Upper Mississippian of West Virginia. Length about 1.5 meters. *From Godfrey, 1986.* (*c*) *Caerorhachis*, a terrestrial temnospondyl from the Upper Mississippian of Scotland. Approximately 30 cm long. *From Holmes and Carroll, 1977.* (*d*) *Proterogyrinus*, a primitive anthracosaur from the Upper Mississippian of West Virginia. Approximately 1 meter long. *From Holmes, 1984.*

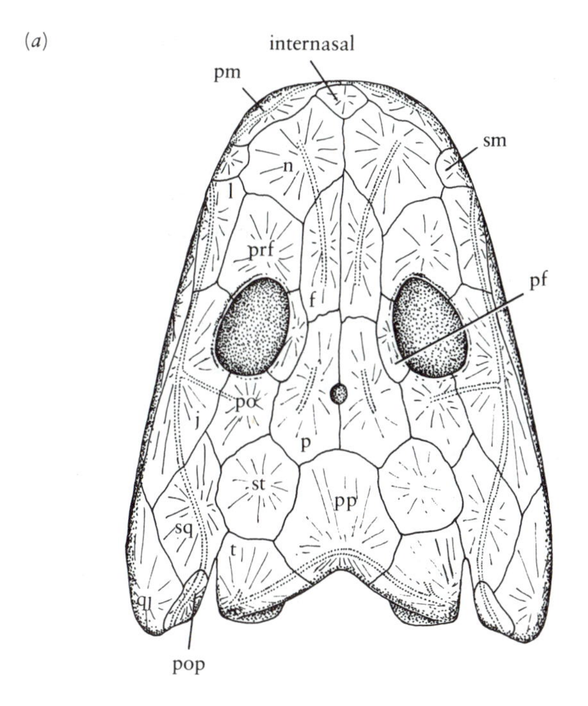

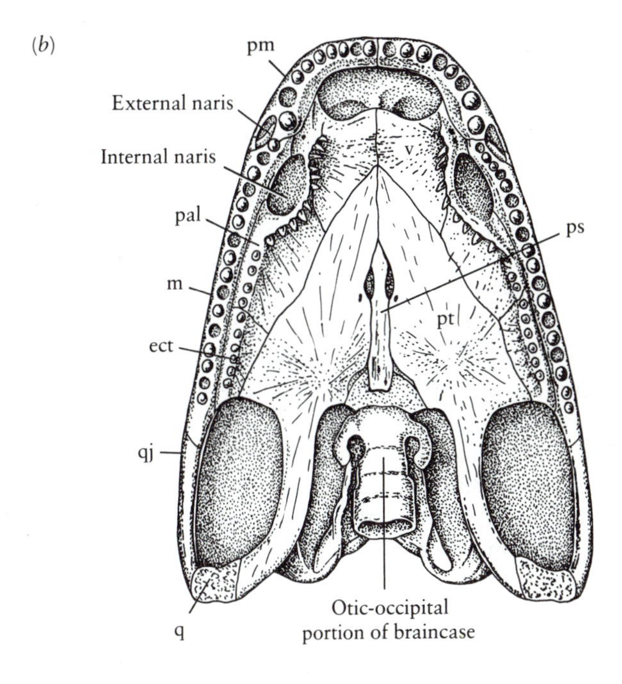

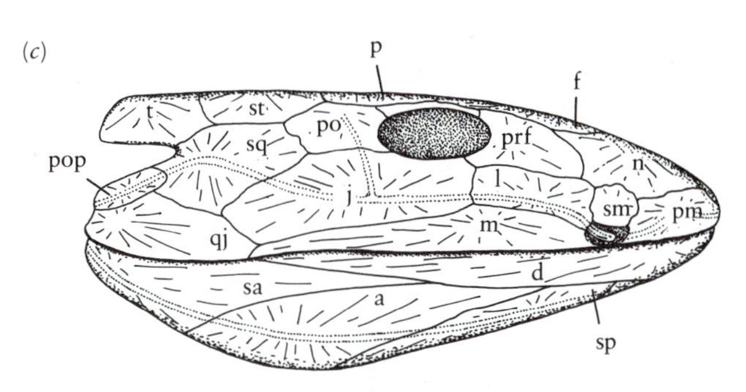

Figure 9-3. SKULL OF THE LATE DEVONIAN AMPHIBIAN *ICHTHYOSTEGA*. (*a*) Dorsal, (*b*) ventral, and (*c*) lateral views. Original about 20 cm long. For abbreviations, see Figure 8-3. *From Jarvik, 1980.* The actual pattern of sculpturing resembles that of *Dendrerpeton* (Figure 9-17).

water, elaboration of the vertebral column, ribs, and musculature of the body wall is required to prevent collapse of the body on the gut and lungs and to support the head above the ground.

The notochord in ichthyostegids was still a major structural element that was little constricted by the vertebral centra. As in rhipidistians, it extended through the otico-occipital portion of the braincase toward the pituitary fossa. The pattern of the centra resembles that of *Eusthenopteron* and *Osteolepis*, with large crescentic intercentra and small paired pleurocentra; neither were suturally attached to the neural arches (see Figure 9-13*a*,*b*). The only major advance we see in the ichthyostegids is the development of well-defined zygapophyses that link the neural arches with one another. The anterior cervical vertebrae are not described in ichthyostegids. Evolution of the atlas-axis complex occurred separately in several groups of primitive tetrapods following their evolution from rhipidistian fish (see Figure 9-12).

In *Eusthenopteron*, the ribs are short rods that connect the neural arches and intercentra. In most tetrapods, they extend from the vertebrae to reinforce the body wall. In ichthyostegids, extremely massive ribs form a solid body wall that supports and protects the lungs and viscera and supplements the vertebral column in supporting the trunk. Later tetrapods strengthen the vertebral column and reduce the ribs. Ribs are attached to all the trunk vertebrae and the first several caudal vertebrae in primitive tetrapods. Beneath the ilia, sacral ribs are specialized to connect the pelvic girdle with the vertebral column. In *Ichthyostega*, the neural and haemal arches of the tail support a fishlike caudal fin, which indicates that aquatic locomotion must have remained important in this genus no matter how well-developed the paired limbs were.

Support of the head above the substrate is necessary in tetrapods for feeding and respiration as well as locomotion. In rhipidistians, the skull is a complex of mobile units associated with specialized cranial kinesis. The skull roof is in four pieces that are attached to the bipartite braincase. When the head was supported by the water and connected posteriorly with the operculum and shoulder girdle, these loose attachments created no problems. In amphibians, the operculum is lost and the head is freed from the shoulder girdle for effective terrestrial feeding and locomotion. One of the major changes in the skull between rhipidistians and amphibians is a solidification of the dermal bones of the skull and their closer integration with the braincase.

In all known amphibians, mobility of the skull roof between the parietals and postparietals is lost and the supratemporal extends anteriorly across the former line of mobility. In ichthyostegids, the skull is further strengthened by the solid attachment of the skull roof to the cheek (Figure 9-3). However, these units remain movable relative to one another in some groups of early amphibians.

In all tetrapods, the medial and lateral extrascapulars at the back of the skull table and the major portion of the opercular series are lost, and the number of separate centers of ossification in the snout region is much reduced from the pattern in rhipidistians. Two remnants of the opercular series remain in ichthyostegids: a small preopercular that forms part of the back of the cheek and a small subopercular that was apparently suspended in soft tissue behind the quadratojugal.

The skull proportions in most early tetrapods are distinct from those of typical rhipidistians. The area anterior to the eyes is much longer and the skull table, particularly the postparietals, is much shorter. Several groups of early amphibians lose the intertemporal bone. The space that this bone occupied in rhipidistians may be incorporated into an adjacent bone—in some groups the supratemporal and in others the postorbital or parietal.

Aside from these changes, there are relatively few differences between the pattern of the dermal bones of the skull of early amphibians and rhipidistians such as *Osteolepis* and *Eusthenopteron* (Figure 8-3). Nearly all the bones can be directly and unquestionably homologized.

Initially, the palate differs little from that in rhipidistians. The dentition is simplified in *Ichthyostega;* there are two more or less parallel rows of teeth, one on the marginal bones and a second, including larger fangs like those in rhipidistians, on the ectopterygoid, palatine, and vomer. In rhipidistians, the palatoquadrate forms a nearly continuous endoskeletal ossification on the dorsal surface of the palate. This ossification is progressively reduced in amphibians where it becomes concentrated in the areas of the epipterygoid and quadrate that articulate with the braincase and lower jaw, respectively. In primitive amphibians, the palate is open between the vomers and premaxillae beneath the olfactory capsules.

In early amphibians, solid attachment of the braincase to the dermal skull roof and integration of the otico-occipital and ethmoid elements of the braincases lag behind the solidification of the dermal skull roof. In ventral view, the otico-occipital element of the braincase in ichthyostegids resembles that of rhipidistians. The ventral cranial fissure is still open, and the top of the braincase appears to be loosely attached to the skull roof. Other early tetrapods evolved a variety of different ways of integrating the occipital and otic elements with the skull roof. The hyomandibular, which links the cheek and the braincase in rhipidistians, remains a large unit (termed the stapes in tetrapods) that helps to support the braincase against the palatoquadrate (Figure 9-4).

The ethmoid unit of ichthyostegids resembles that of rhipidistians in bearing the basicranial articulation. In all other early tetrapods, this articulation with the palate becomes associated with the otico-occipital unit (Smithson, 1982). The parasphenoid extends posteriorly, bridges the ventral cranial fissure, and underlies the posterior portion of the braincase.

Despite the mechanical necessity of integrating the skull in early amphibians, the degree of ossification of the braincase is markedly reduced. There is no ossification of the nasal capsules, and the otic capsules are largely cartilaginous in several groups. Perhaps the need to reduce weight in these areas was more important than the requirements for strength. The space between the ethmoid and otico-occipital regions in the area of the Vth nerve is also unossified in most groups.

As in teleosts, the early amphibians show a division of the braincase into a number of distinct areas of ossification that are not apparent in early bony fish. The occipital arch can be differentiated into the ventral basioccipital (filling the opening occupied by the notochord in rhipidistians and ichthyostegids), lateral exoccipitals,

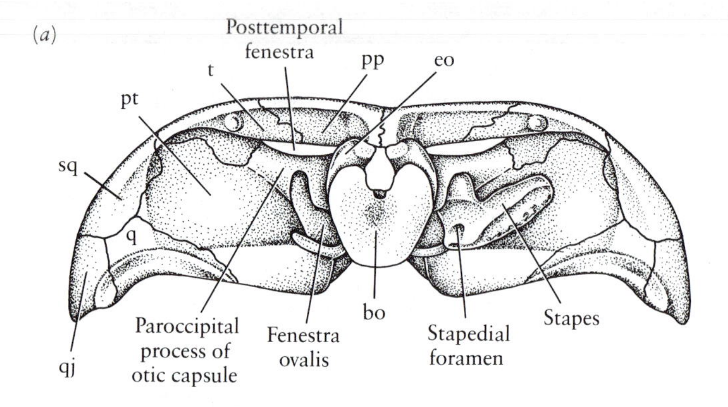

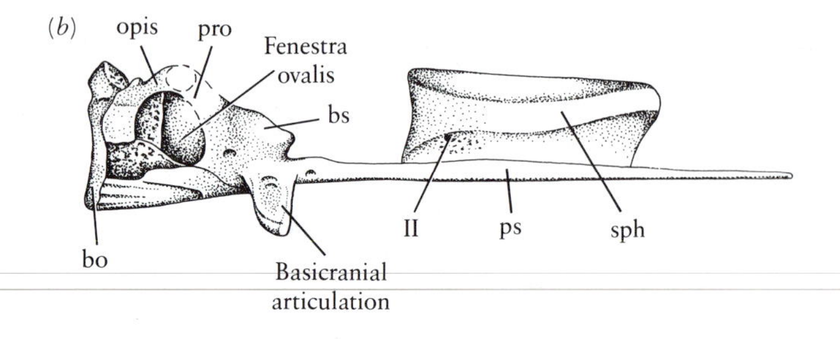

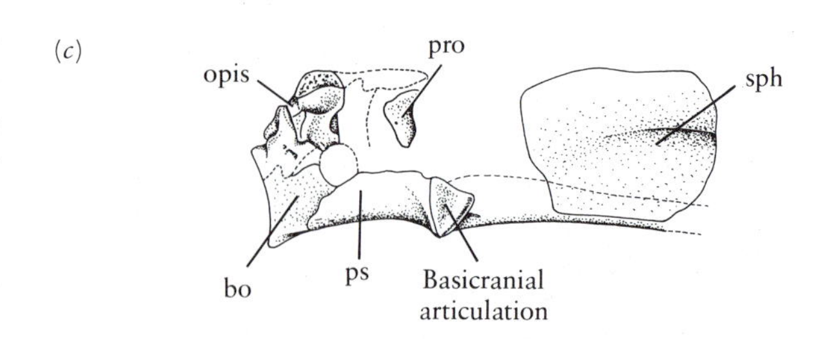

Figure 9-4. (*a*) Occipital view of the skull of *Greererpeton*, an aquatic temnospondyl from the Lower Carboniferous, showing the large stapes that supports the back of the braincase on the pterygoid. *From Smithson, 1982.* (*b*) Lateral view of the braincase. Natural size. *From Smithson, 1982.* (*c*) Lateral view of the braincase of the Lower Carboniferous anthracosaur *Proterogyrinus. From Holmes, 1984.* Abbreviations as follows: bo, basioccipital; bs, basisphenoid; eo, exoccipital; opis, opisthotic; pp, postparietal; pro, proötic; ps, parasphenoid; pt, pterygoid; q, quadrate; qj, quadratojugal; sph, sphenethmoid; sq, squamosal; t, tabular; II, second cranial nerve.

and in some forms, a dorsal median supraoccipital. The exoccipitals and basioccipital form a variably developed occipital condyle for articulation with the first cervical vertebra. The exoccipital shows passages for the XIIth cranial nerve. Between the exoccipital and otic capsule, there is a remnant of the lateral cranial fissure through which the Xth cranial nerve passes.

The otic capsule ossifies as a posterior opisthotic and anterior proötic. Ventromedial to the otic capsule, the braincase floor is formed by the basisphenoid, which extends laterally as the basicranial processes that articulate with the palate. This bone is closely associated with the otico-occipital elements of the braincase in most amphibians, rather than with the ethmoid element as it is in rhipidistians and ichthyostegids. The ethmoid area of the braincase ossifies as a single unit, the sphenethmoid.

Feeding and respiration

Because both rhipidistians and most early amphibians have large, fanglike palatal teeth and a wide gape, we may assume that they were active predators on relatively large prey. Rhipidistians presumably relied on the water to provide the buoyancy to bring prey into their mouths. Marked behavioral, if not structural, changes would have been necessary to accommodate terrestrial feeding. Early amphibians would have had to use the jaws to lift up the prey and practice inertial feeding to force it back into the throat. Since the tongue is not highly modified in the primitive members of any of the modern amphibian groups, it was almost certainly not specialized to capture or manipulate prey in the Paleozoic amphibians (Regal and Gans, 1976).

The functional kinesis of the rhipidistian skull was certainly lost early in the origin of amphibians. Elongation of the snout would lead to a gradual loss in the effectiveness of kinetic movements, so that they would have been negligible in a skull with the proportions of early amphibians (Thomson, 1967). There is no evidence for the retention of the subcephalic muscles in amphibians. The increased size of the adductor chamber and its relatively more anterior position would have augmented the force of the adductor muscles in early tetrapods.

The early amphibians may have fed in the water or on fish stranded at the margins of lakes and swamps. There were also several groups of large terrestrial invertebrates in the late Devonian and Carboniferous that may have provided an adequate food supply (Rolfe, 1980).

We assume that rhipidistians, like other groups of primitive bony fish, had functional lungs. However, the large size of the hyoid apparatus and the operculum indicate that gill respiration remained well developed. The use of lungs would have been of great advantage for fish living in bodies of quiet, shallow water that might be poor in oxygen, particularly if they were warm and filled with decaying vegetation. Internal nares would have allowed the fish to breathe without opening the mouth, although this is not necessary for air breathing.

Both rhipidistians and early tetrapods probably drew air into the mouth by depressing the hyoid apparatus to expand the oral cavity. Elevating the floor of the mouth while the mouth and nostrils were closed would force the air back into the lungs. Air could be forced out of the lungs by contraction of the elastic tissue in their walls. Expansion and contraction of the thoracic cavity in most tetrapods (other than modern frogs and salamanders that have lost their ribs) results from movement of the ribs, but in ichthyostegids, the ribs may have been too closely integrated to function in respiration.

Locomotion—structure and function of the appendicular skeleton

The most conspicuous differences between rhipidistians and amphibians are in the appendicular skeleton. These differences are associated with the evolution of terrestrial locomotion. In rhipidistians, as in most fish, swimming occurred primarily by lateral undulation of the trunk, and the main muscles for locomotion were concentrated in the body, not in the fins. The paired fins in rhipidistians were used primarily for steering and braking. Because they extended ventrally beyond the body, they could have been used to push the fish along the bottom when the body was submerged and supported by the buoyancy of the water.

The rhipidistians were unique among Paleozoic fish in having thick, lobate paired fins with a pattern of proximal bones that is directly comparable with that of tetrapods. However, their function, as Andrews and Westoll (1970a,b) determined was very different. Both the glenoid and acetabulum face posteriorly and only slightly laterally. The humerus and femur were held close to the body, which restricted lateral or rotational movement. The distal ends of these bones may have been capable of moving in a dorsoventral arc of 15 to 20 degrees. The preaxial margin of both fins was directed ventrolaterally (Figure 9-5).

The ulna and radius articulate with the distal end of the humerus. This joint could bend freely and accounted for most of the fin's mobility. The more distal elements of the forelimb were capable of some movement on the ulna and radius, but there is no major joint comparable to the carpus in tetrapods. Beyond the "elbow," the fin functioned as a relatively stiff paddle. The orientation and movement of the rearlimb appear very similar to the forelimb. Bending, and to a lesser extent, twisting, occurred at the "knee" joint, and the remainder of the fin was stiff.

In amphibians, the shoulder girdle is functionally separate from the skull. The dorsal supracleithrum and anocleithrum are lost, but the lateral cleithrum and clavicle are retained and the interclavicle is elaborated as a major midventral unit. The endoskeletal shoulder girdle in rhipidistians is attached to the inner surface of the cleithrum. The same relationship is retained in ichthyostegids, although the scapula is relatively larger and a ventral, coracoid area is added (Figure 9-6). The orientation of both

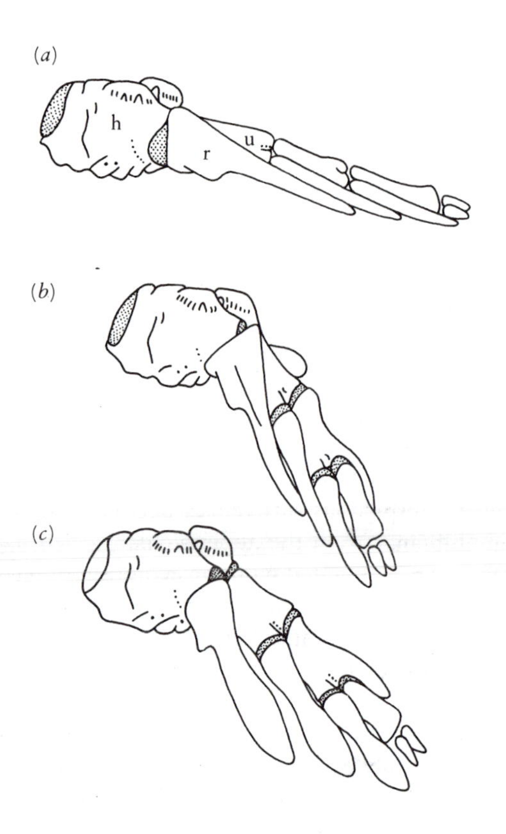

Figure 9-5. RANGE OF MOVEMENTS OF THE PECTORAL FIN OF THE RHIPIDISTIAN *EUSTHENOPTERON*. Anterior view of the left forelimb (*a*) with the fin skeleton in extended position, (*b*) with the elbow joint flexed, and (*c*) with the elbow flexed and the ulna and radius rotated preaxially. Abbreviations as follows: h, humerus; r, radius; u, ulna. *From Andrews and Westoll, 1970a.*

the glenoid and acetabulum is significantly altered. Both are much larger and face primarily laterally rather than posteriorly, so that the humerus and femur extend at nearly right angles to the trunk.

In the early amphibians, the structure of the limb joints and the distal portion of the limb are considerably remodeled. The major joints of the forelimb and hind limb differ from one another in their structure and function in tetrapods (Figure 9-7). The forelimb is bent and twisted at the elbow in a manner presaged in the rhipidistians, so that the lower extremity of the ulna and radius face forward. The carpus forms a simple hinge joint between the ulna and radius and the manus. In the rear limb, the knee forms a hinge and the lower end of the tibia and fibula face more or less laterally. The ankle must both twist and hinge to bring the foot into an anteriorly facing orientation.

In primitive tetrapods, the limbs are extended forward and away from the trunk during locomotion and the axial musculature pulls the body toward the extended limb, following the same pattern of undulatory movements as in fish. The side of the body toward the extended limb is raised and then lowered as the limb is retracted.

The function of the limbs in early tetrapods has been interpreted by comparing them with those in living amphibians and reptiles. The primitive features of locomotion common to both groups can logically be attributed to their ancestors in the Paleozoic. We can also study the pattern of limb movements in early tetrapods on the basis of numerous fossil footprints that are known as early as the Devonian.

Holmes (1980) shows that the specific configuration of the articulating surface of the glenoid, humerus, and ulna results in closely controlled, stereotyped movements (Figure 9-8). As the humerus is pulled forward at the initiation of the stride, it is twisted so that the ulna and radius assume their most anterior position. At this point, the entire limb is nearly horizontal and the body is closest to the ground. As the humerus is retracted (and the body pulled forward), it is twisted so that the ulna and radius become vertical. The body is then raised by the muscles that depress the distal end of the humerus. As the humerus achieves its most posterior position, the movement of its proximal end across the glenoid results in raising the distal end so that the body again approaches the ground. The reduction in the angle made by the humerus on the ulna and radius move the body laterally to bring it over the hand.

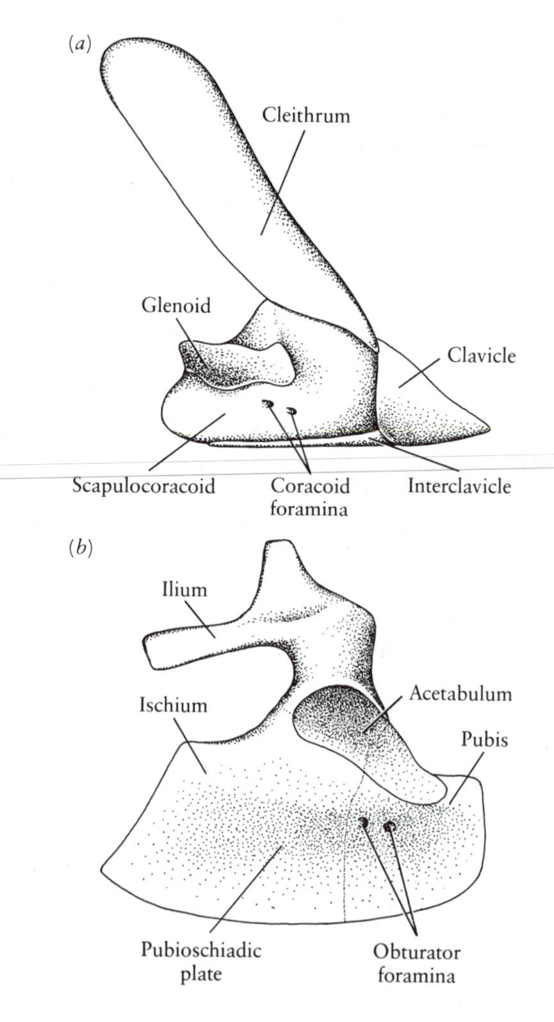

Figure 9-6. (*a*) Pectoral and (*b*) pelvic girdles of *Ichthyostega*, $\times \frac{1}{2}$. *From Jarvik, 1955, (by permission from Scientific Monthly) and 1965,* respectively.

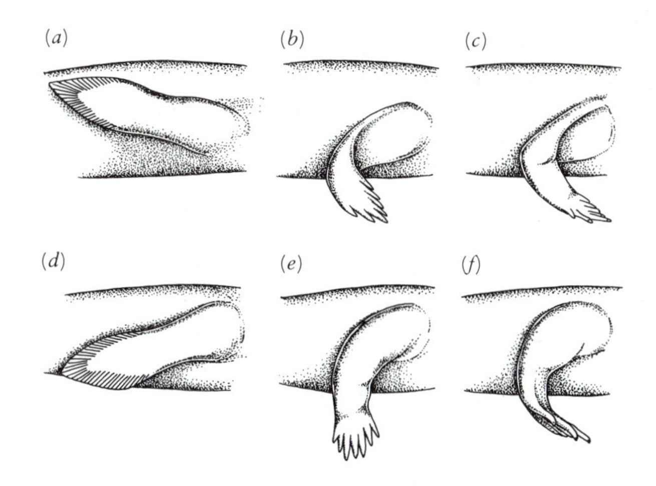

Figure 9-7. DIAGRAM SHOWING THE POSTURAL SHIFT IN THE PAIRED LIMBS IN THE TRANSITION BETWEEN FISH AND AMPHIBIANS. (*a–c*) pectoral limb; (*d–f*) pelvic limb. (*a* and *d*) Fish position; (*b* and *e*) transitional stage; (*c* and *f*) amphibian position. *From Romer, 1949. Copyright © 1977 by W. B. Saunders Company. Reprinted by permission of CBS College Publishing.*

The wrist joint in amphibians evolved from the distal end of the rhipidistian fin. The long, overlapping elements in the rhipidistians appear designed to resist bending in this area. The long bones may have been "broken up" developmentally to form the mosaic of carpals and metacarpals and elaborated distally as phalanges. The carpals and manus are not known in *Ichthyostega,* but they are presumably comparable in their stage of evolution to the tarsus and pes which already show the basic features of later amphibians.

The general configuration of the carpus in amphibians and early reptiles is sufficiently similar to suggest that they diverged from a common ancestral pattern (Figure 9-9). From the fish, they directly inherit the ulnare and intermedium, which articulate with the end of the ulna. A small lateral element, the pisiform, articulates with the distal end of the ulna and the margin of the ulnare. Fish have no bone in this position. The pisiform apparently evolved when the carpal joint formed in association with muscles that flex and extend the carpus and extend the lower limb laterally. One might think of the radius, a much longer element in the fish, as ossifying in two units in amphibians—the radius proximally and a separate distal radiale that remains functionally integrated with the radius in most primitive tetrapods. One or more bones, termed centralia, occur lateral to the radiale. A row of distal carpals, probably five in the primitive state, complete the carpus.

Among primitive amphibians, two patterns of the digits of the hand are common. There may be five digits with a phalangeal count of 2, 3, 4, 5, 3, or four digits with a phalangeal count of 2, 3, 3, 3. The larger number appears in several groups and may represent the primitive condition.

The pelvis in *Ichthyostega* and other tetrapods is formed from three centers of ossification, the pubis, ilium, and ischium, that are suturally attached to one another (see Figure 9-6). The acetabulum develops at the center where the three bones meet. The left and right halves of the girdle are joined in a symphysis that is formed primarily by the pubes in primitive genera. Contact between the ischia is strengthened in later tetrapods. The ilium extends dorsally to be supported by the sacral rib. The pubis and ischium form a large platelike surface that faces ventrolaterally to which muscles that raise the body on the femur are attached.

The femur in early tetrapods differs little from that in rhipidistians. The proximal articulating surface is terminal and more or less hemispherical. Much more freedom of rotation is possible at this joint than in the pectoral limb. Again as in rhipidistians, the tibia and fibula have a hingelike articulation with the end of the femur that allows little rotation at this joint. Rotation, as well as bending, must occur at the ankle joint, and the foot as a whole is rotated through a considerable arc during locomotion, in contrast with the hand. In general, the structure of the joint surfaces exert much less specific control in the hindlimb than in the forelimb. Two groups of muscles—a large mass that is attached ventrally and extends

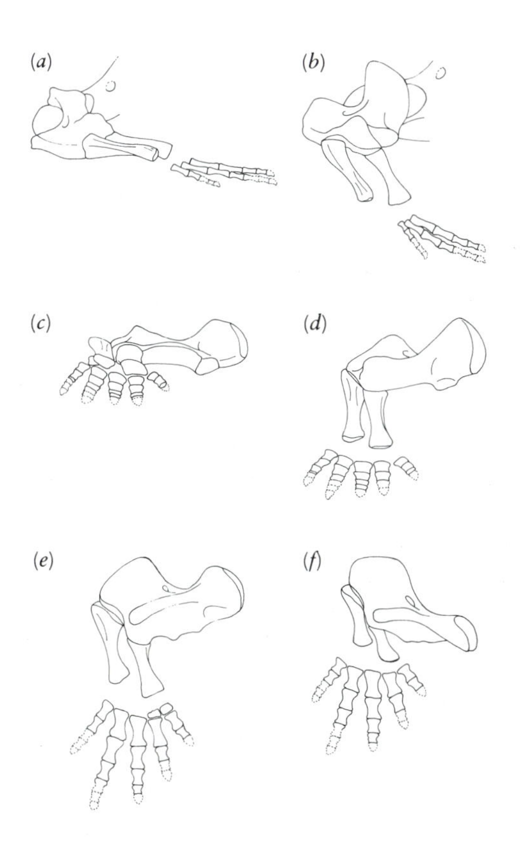

Figure 9-8. DIAGRAM OF THE RIGHT FORELIMB OF THE PRIMITIVE AMPHIBIAN *PROTEROGYRINUS* TO SHOW MOVEMENT DURING LOCOMOTION. (*a* and *b*) Lateral views; (*c* and *d*) anterior views; (*e* and *f*) dorsal views. (*a, c,* and *e*) Most anterior (extended) position; (*b, e,* and *f*) most posterior (retracted) position. Movement of the limbs in primitive tetrapods is largely controlled by the shape of the articulating surfaces. As in many early tetrapods, the carpus is poorly ossified. *From Holmes, 1980.*

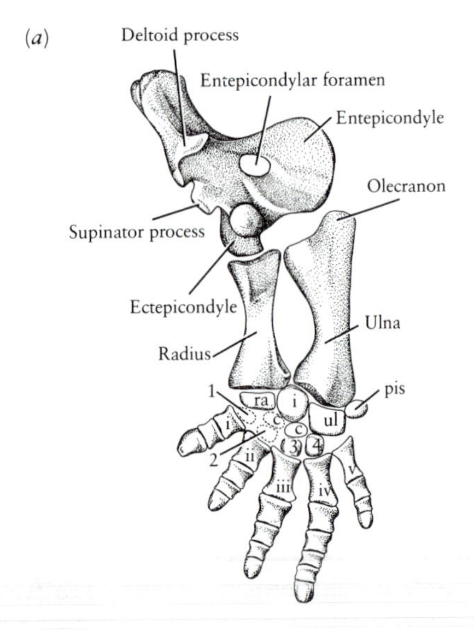

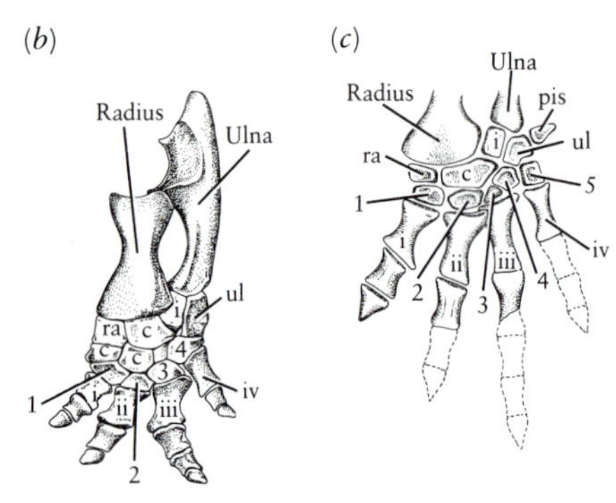

Figure 9-9. THE CARPUS OF EARLY TETRAPODS. (*a*) The anthracosaur *Limnoscelis*. *From Williston, 1925.* (*b*) The temnospondyl *Eryops*. *From Williston, 1925. Reprinted by permission of Harvard University Press.* (*c*) The microsaur *Pantylus*. *From Carroll and Gaskill, 1978.* All from the Lower Permian. Abbreviations as follows: c, centrale; i, intermedium; pis, pisiform; ra, radiale; ul, ulnare; 1–5, distal carpals; i–v, metacarpals.

from the puboischiadic plate to the femur and a group that extends anteriorly from the base of the tail—control the movement of the hindlimb. The body is not lifted far from the ground by the stride of the rear limb.

The tarsus and pes in *Ichthyostega* presage the conditions in all other early tetrapods (Figure 9-10). The fibulare and intermedium, which articulate with the distal end of the fibula, are directly inherited from rhipidistians. The tibiale is developed distal to the tibia, as are several centralia. As in most primitive tetrapods, *Ichthyostega* has five distal tarsals and five digits. Like several groups of primitive tetrapods, this genus has a phalangeal count of 2, 3, 3, 3, 3. Other Paleozoic amphibians and primitive reptiles have a phalangeal count of 2, 3, 4, 5, 4. Reduction in the number of phalanges is common in tetrapods, and the larger number may be primitive for amphibians, although both patterns may have evolved separately when the foot evolved from the rhipidistian fin. It is also conceivable that additional phalanges were added from a configuration such as that in *Ichthyostega*.

Water balance

As long as they were in the water, rhipidistians had no problem with dessication, and those genera living in fresh water probably excreted dilute urine to counter the intake of excess water into the tissues. Once on dry land, early amphibians would have been subject to to serious water loss through the mouth, lungs, and general body surface. Water would also have been lost through the gills, if they were retained in semiterrestrial forms. The heavy scales in most primitive genera may have slowed water loss over the body surface, but initially the skin was probably no less permeable than that of their fish ancestors. If cutaneous respiration were of importance in Paleozoic amphibians, water loss would have been a particularly serious problem as it is in modern amphibians.

Sense organs

The physical and chemical differences between water and air required reorganization of many sensory structures between rhipidistians and fully terrestrial vertebrates. The lateral line canals of fish would not function on land because of the thousandfold difference in the density of

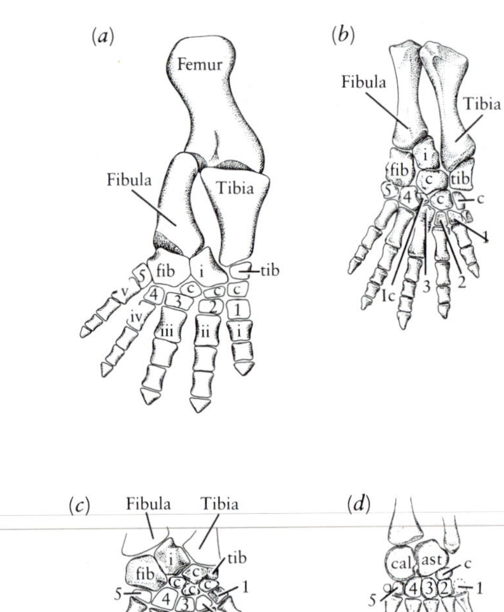

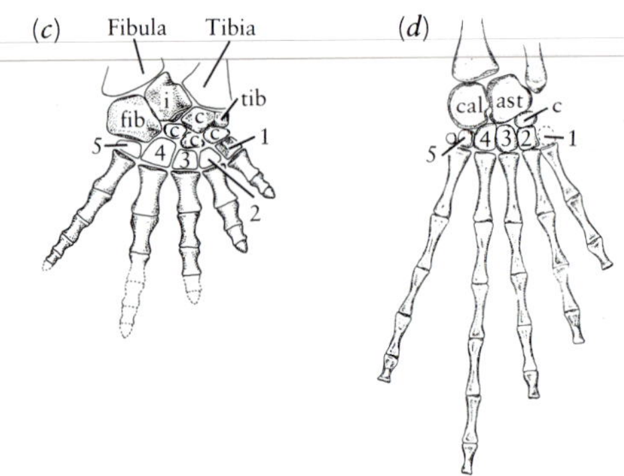

Figure 9-10. THE TARSUS OF EARLY TETRAPODS. (*a*) *Ichthyostega*. *From Jarvik, 1952.* (*b*) The temnospondyl *Trematops*. *From Gregory, 1951.* (*c*) The anthracosaur *Proterogyrinus*. *From Holmes, 1984.* (*d*) The microsaur *Tuditanus*. *From Carroll and Gaskill, 1978.* In the microsaur, the tibiale, intermedium, and proximal centrale have become coossified to form an astragalus. In *Proterogyrinus*, the tibiale is in the process of fusing with a proximal centrale. Abbreviations as follows: ast, astragalus; c, centrale; cal, calcaneum; fib, fibulare; i, intermedium; tib, tibiale; 1–5, distal tarsals; i–v, metatarsals.

air and water and would have been an additional source of water loss. The lateral line canals of both rhipidistians and ichthyostegids are surrounded by the bones of the skull, but they are conspicuously exposed in many Paleozoic amphibians that are secondarily aquatic. In more terrestrial amphibians, they are completely lost.

The difference between the refractive index of air and water required a change in the effective focal length of the lens of the eye but did not require a fundamental change in other structures associated with vision. The presence of a nasolacrimal duct in rhipidistians presumably indicates that both the eye and the olfactory epithelium of the nostril were bathed in water and so could continue to function in the air. However, airborne odors probably had much different properties than those encountered by most fish and may have required long-term modification of the olfactory epithelium.

The semicircular canals are much larger in rhipidistians and other fish than in tetrapods. Bernacsek and Carroll (1981) suggest that this difference may be related to the mechanical separation of the head and trunk in tetrapods. The dimensions of the semicircular canals are proportional to the mass of the animal within each group, but their size in the tetrapods is logically associated with the head alone, since the canals are of little significance in measuring the orientation of the remainder of the body.

In reptiles, mammals, and birds, stretch receptors in the muscles of the trunk and limbs monitor the orientation of the body and limbs. Fish in general lack muscle stretch receptors. The lateral line canals on the body may have served a similar function in monitoring the body movements associated with locomotion. Stretch receptors are poorly developed in modern salamanders and caecilians, which suggests that they evolved slowly within primitive amphibians (Bone, Ridge, and Ryan, 1976). We may correlate the probable absence of muscle stretch receptors in the earliest tetrapods with their stereotyped pattern of locomotion, which is controlled largely by the configuration of the joint surfaces.

Many modern fish have specialized structures (such as the Weberian ossicles in the Ostariophysi) to detect airborne vibrations. Thomson (1966) suggested that rhipidistians used their operculum as a tympanum to detect airborne sounds in the manner of modern reptiles, birds, and mammals. If we consider the fact that both reptiles and mammals later evolved very different structures to detect airborne vibrations and evaluate the structure of the middle ear region in rhipidistians, it seems probable that their ability to hear airborne sounds must have been limited to those of very low frequency and high intensity (Wever, 1965). A tympanum and a lightweight stapes evolved independently in several lines of primitive tetrapods (see Chapters 11 and 17), but sensitivity to high-frequency airborne sounds was probably little developed among early amphibians.

In most groups of early tetrapods, the stapes was a stout, ventrally directed rod. Because of its large size, it would not have served effectively to transmit vibrations but probably functioned primarily to support the braincase against the cheek (see Figure 9-4) (Carroll, 1980). The embayment above the squamosal in *Ichthyostega* and some other early labyrinthodonts may represent the area that was occupied by the operculum in rhipidistians. In the aquatic stage, the embayment may have surrounded a spiracle rather than being an area for support of a tympanum as has frequently been assumed (Panchen, 1985).

Reproduction

The early amphibians probably retained a mode of reproduction like that of most actinopterygian fish, which have external fertilization and lay large numbers of small eggs in the water. Fossils of larval stages with external gills are known among various genera of Paleozoic amphibians (Figure 9-11). Like some lungfish and primitive

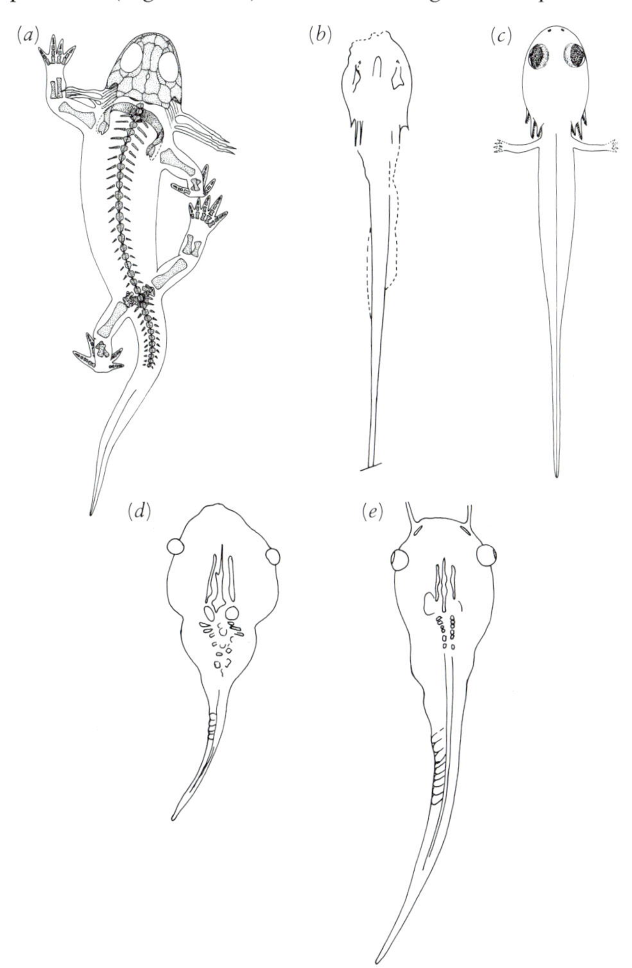

Figure 9-11. LARVAL AMPHIBIANS. (*a*) Reconstruction of "*Branchiosaurus*," a larval labyrinthodont from the Lower Permian of Europe. Some species with this general pattern mature into terrestrial forms, while others are apparently neotenic, reaching sexual maturity without undergoing metamorphosis. Approximately natural size. *From Boy, 1971.* (*b*) Much earlier larval stage of a Paleozoic amphibian, ×2. *From Heyler, 1975.* (*c*) Diagrammatic reconstruction of an intermediate stage in development. *From Boy, 1974.* (*d*) Larva of the Lower Cretaceous pipid frog, *Shomronella*, ×1½. *From Estes, Spinar, and Nevo, 1978.* (*e*) Larva of a modern species of *Pipa*, ×1½. *From Estes, Spinar, and Nevo, 1978.*

living actinopterygians, rhipidistians may also have had external gills in early developmental stages, but there is no fossil evidence.

Early tetrapods probably emerged onto the land only when a basically adult morphology was achieved. There is no reason to believe that Paleozoic amphibians had a marked metamorphosis such as that which characterizes modern frogs.

THE ORIGIN OF AMPHIBIANS

We may interpret the absence of any fossils intermediate between rhipidistians and amphibians as evidence of either very rapid evolution or of evolution over a relatively long period of time, during which intermediate forms were rare or lived in environments in which fossilization was unlikely. A skeletal pattern appropriate to the ancestors of amphibians is evident in middle Devonian rhipidistians such as *Osteolepis* and *Gyroptychius.* At least 20 million years may have been available for the evolution of the amphibian appendicular skeleton before the appearance of the ichthyostegids in the late Devonian. However, the requirements of locomotion and support probably necessitated dramatic and rapid change. It is difficult to envisage long-term adaptation to an environment intermediate between that of rhipidistians and tetrapods.

Romer (1957b) suggested that the rhipidistians ancestral to amphibians elaborated the limbs to move from pond to pond in areas of seasonal drought. The availability of food on land and freedom from competition with more conservative rhipidistians may also have contributed to the selective pressures favoring the evolution of tetrapod limbs.

Of the adequately known rhipidistians, the osteolepiforms seem more appropriate ancestors than do the porolepiforms, although the structure of the limbs is too poorly known in the latter group for comparison. The genus *Panderichthys* (Vorobyeva, 1980, Schultze and Arsenault, 1985) appears particularly close to amphibians in the loss of mobility between the postparietals and parietals and the elongation of the snout. Unfortunately the appendicular skeleton remains poorly known.

We can recognize many amphibian lineages within the Carboniferous. It has long been assumed that these groups had a common ancestor that was already a tetrapod. Several groups from among the rhipidistians may have approached the amphibian level of organization. By analogy, we have seen that several lineages within the Chondrostei approached the level of the neopterygians. However, there is no definite evidence indicating that tetrapods had multiple origins. Other aspects of the origin and early radiation of terrestrial vertebrates are discussed in *The Terrestrial Environment and the Origin of Land Vertebrates,* edited by Panchen (1980a).

In the absence of adequate fossil evidence, two models of early amphibian radiation may be considered. The early differentiation of amphibians might have had a pattern that was similar to mammalian or teleost evolution in the early Tertiary, with a sudden wide adaptive radiation and a great increase in numbers. On the other hand, if one takes into account the challenges encountered by the descendants of the rhipidistians in the new terrestrial environment, it is possible that only a few highly specialized groups were present, as is the case in the modern fauna of arctic mammals and deep-sea fish.

We know from the presence of numerous groups in the late Mississippian and early Pennsylvanian that several lineages must have been present in the early Carboniferous, but individuals may have been rare and groups little differentiated. The similarity of amphibians on the two sides of what is now the Atlantic Ocean and from several localities in Great Britain suggests that there were only a few genera that were fairly widespread geographically and may have had broad ecological tolerance (Smithson, 1980).

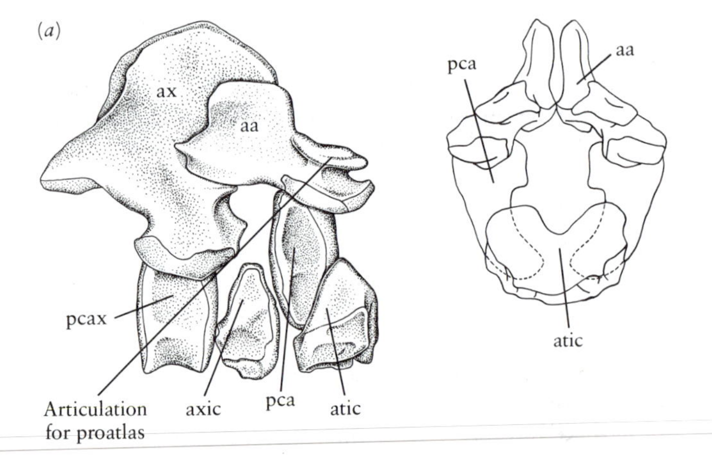

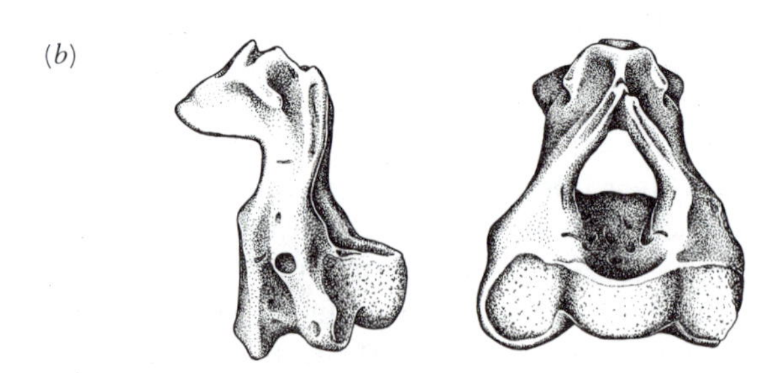

Figure 9-12. (*a*) Atlas-axis complex of the early anthracosaur *Proterogyrinus* showing the multipartite structure common to most labyrinthodonts; lateral and anterior views, ×2. A further small paired element, the proatlas, links the atlas arch with the exoccipital. *From Holmes, 1984.* (*b*) First cervical vertebra of the microsaur *Euryodus* showing pattern common to lepospondyls, in which the first vertebra is a unitary structure, ×4.5. The articulating surface with the skull is very broad, limiting lateral movement of the head. *From Carroll and Gaskill, 1978.* Abbreviations as follows: aa, atlas arch; ax, axis arch; atic, atlas intercentrum; axic, axis intercentrum; pca, pleurocentrum of atlas; pcax, pleurocentrum of axis.

THE RADIATION OF PALEOZOIC AMPHIBIANS

Two major groups of Paleozoic amphibians have long been recognized, the labyrinthodonts and the lepospondyls. The labyrinthodonts, characterized by genera such as *Ichthyostega,* apparently evolved directly from the rhipidistians and may have a single common ancestor. The lepospondyls are a heterogeneous assemblage of groups, each of which may have evolved independently from early labyrinthodonts. Many of the features that characterize labyrinthodonts are primitive attributes retained from the rhipidistians: labyrinthine infolding of the dentine (from which the name of the groups is derived); conspicuous fangs on the marginal bones of the palate, each of which is associated with a large replacement pit; and vertebral centra composed of more than one element per segment. The presence of well-developed limbs clearly differentiates labyrinthodonts from rhipidistians, but the basic pattern is shared with all other tetrapods.

An important advanced feature of most labyrinthodonts, with the exception of the ichtyostegids, is a specialized atlas-axis complex composed of several units that forms a ringlike attachment with the occipital condyle (Figure 9-12).

Many labyrinthodonts have an embayment of the posterior margin of the cheek, which is termed an otic notch and has been interpreted as supporting a tympanum. The embayment may have been retained from the rhipidistians, but the presence of a tympanum is almost certainly not a primitive feature of labyrinthodonts. Primitive labyrinthodonts are relatively large (up to a meter or more in length) compared with most rhipidistians and lepospondyls.

Lepospondyls usually lack labyrinthine infolding of the dentine. None had conspicuous fangs on the palate associated with replacement pits. They typically have only a single central element per segment, but intercentra and/or haemal arches are reported in some groups. Most are small, and none had an otic notch supporting a tympanum. In this assemblage, the first cervical vertebra is functionally a single ossification that either fits into a recess at the back of the skull or has a broad strap-shaped surface for articulation with widely spaced exoccipitals (Figure 9-12*b*).

LABYRINTHODONTS

We may divide the labyrinthodonts into two large groups—the temnospondyls and the anthracosaurs (or batrachosaurs). The pattern of the vertebral centra is one of the most important features that differentiates them from one another. Most temnospondyls have large crescentic intercentra and small, paired pleurocentra (Figure 9-13). Anthracosaurs are typified by the elaboration of the pleurocentrum as the major element of support and the reduction of the intercentrum.

The skull roof is solidly attached to the cheek in temnospondyls, and the tabular is small and separated from the parietal by the supratemporal (see Figure 9-14). The manus has four toes and a phalangeal count of 2, 3, 3, 3, and the pes has five toes with a count of 2, 3, 3, 3, 3. The skull is marked by a regular pattern of rounded pits or grooves that are separated by anastomosing ridges (Romer, 1947).

The skull roof of anthracosaurs is loosely attached to the cheek, except in specialized genera. The tabular is large and attaches to the parietal. It frequently extends posteriorly beyond the occiput. The manus has five toes and a phalangeal count of 2, 3, 4, 5, 3, and the pes has one or more extra phalanges on the fifth digit. The skull is usually marked by fine radiating grooves.

Unfortunately, we cannot use broader biological criteria to differentiate these groups. Both temnospondyls and anthracosaurs radiated extensively in aquatic, semiaquatic, and terrestrial habitats in the late Paleozoic. The most significant differences between temnospondyls and anthracosaurs may be in locomotor patterns that are related to the elaboration of different vertebral elements. However, we have not established the functional differences between these vertebral patterns.

Temnospondyls and anthracosaurs both exhibit a mosaic of skeletal characteristics. Some resemble the condition in rhipidistians and others are more specialized. We cannot determine which group is generally more primitive. The vertebrae in anthracosaurs are more specialized, as is the configuration of the tabular. However, the mobility of the skull table and cheek seems to be retained from the rhipidistian condition, as is a closed palate with the pterygoids approaching the midline. Temnospondyls develop an open palate with large interpterygoid vacuities between the parasphenoid and the pterygoid.

The development of a more solid attachment of the braincase to the dermal bones of the skull appears to have occurred separately in the two groups. Anthracosaurs evolved a bony attachment between the otic capsule and the skull table at an early stage. Temnospondyls rely more on the palate and cheek for braincase support, and the otic capsules were slow to ossify.

Some early labyrinthodonts do not fit into these two categories. The ichthyostegids have been placed in a separate order because they retain such primitive cranial characteristics as the persistence of the notochord within the occipital element of the braincase and a parasphenoid that was restricted to the area anterior to the ventral cranial fissure. They are specialized in the loss of the intertemporal and the medial fusion of the postparietals, which preclude them from ancestry to any of the more advanced groups of labyrinthodonts. Except for the structure of the hand (which is unknown) they fit most of the criteria used to identify temnospondyls and they may rep-

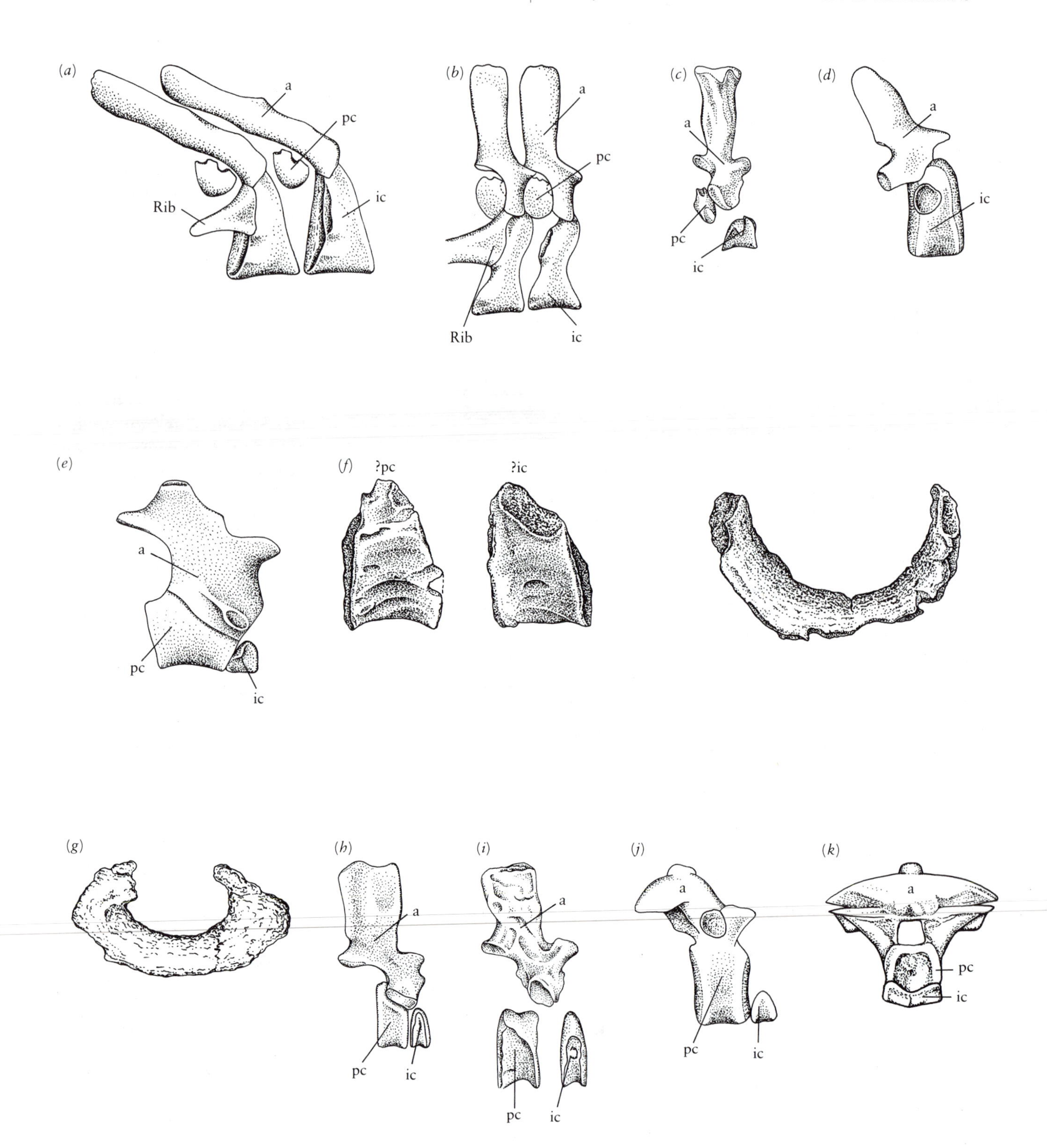

Figure 9-13. LABYRINTHODONT VERTEBRAE. (*a*) *Eusthenopteron*, from the Upper Devonian, a rhipidistian illustrating the pattern that apparently gave rise to that of early amphibians. *From Jarvik, 1955. By permission from Scientific Monthly.* (*b*) *Ichthyostega*, from the Upper Devonian. *From Jarvik, 1955. By permission from Scientific Monthly.* (*c*) *Eryops*, which illustrates the typical temnospondyl pattern. Note the small paired pleurocentra and large crescentic intercentra. *From Moulton, 1974.* (*d*) *Mastodonsaurus*, which shows the pattern typical of "stereospondyls," in which the intercentrum has become the dominant element. *From Panchen, 1967.* (*e*) *Doleserpeton*, a small temnospondyl in which a large crescentic pleurocentrum is the major central element. *From Bolt, 1969. With permission from Science. Copyright 1969 by The American Association for the Advancement of Science.* (*f*) Central elements of *Crassigyrinus* in lateral and anterior views. *From Panchen, 1980.* (*g*) Vertebra possibly associated with loxommatid. *From Panchen, 1980.* (*h*) *Proterogyrinus*, a primitive anthracosaur; pleurocentra and intercentra are both crescentic. *From Holmes and Carroll, 1977.* (*i*) The embolomere *Eogyrinus*; both intercentra and pleurocentra are complete cylinders. *From Panchen, 1966. By permission of the Zoological Society of London.* (*j* and *k*) *Seymouria* in lateral and anterior views. Note the widely expanded neural arches. *From Panchen, 1967, and Gregory, 1951, respectively. Courtesy of the Library Services Department, American Museum of Natural History.* Abbreviations as follows: a, arch; ic, intercentrum; pc, pleurocentrum.

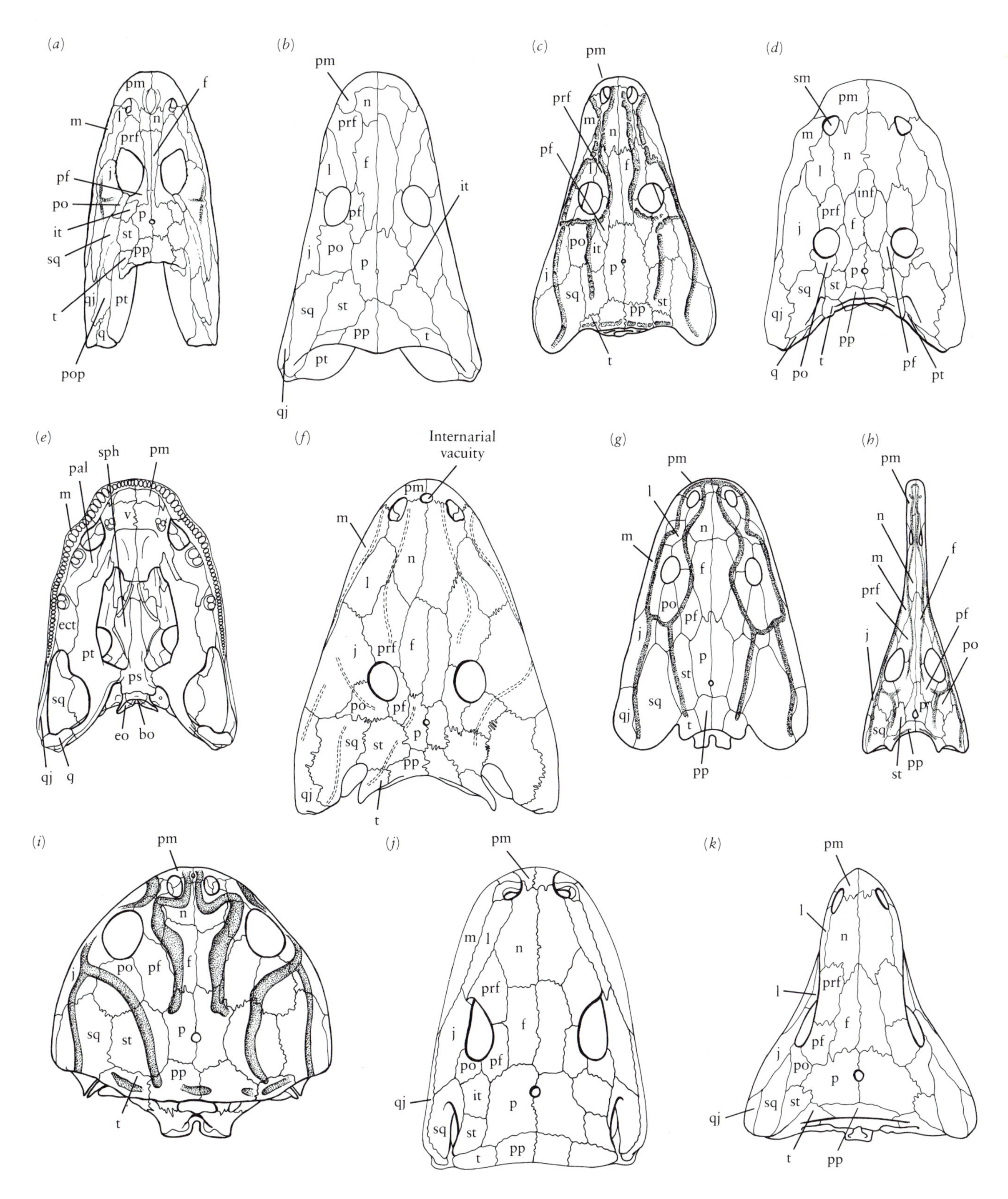

Figure 9-14. SKULLS OF LABYRINTHODONTS. (*a*) *Crassigyrinus* of uncertain relationships. *After Panchen, 1985.* (*b*) The primitive, secondarily aquatic temnospondyl *Greererpeton. From Smithson, 1982.* (*c*) *Neldasaurus*, a Lower Permian aquatic temnospondyl. *From Chase, 1965.* (*d* and *e*) The eryopid *Eryops.* The skull roof and palate show absence of the intertemporal bone and the development of sutural attachment between the palate and base of the braincase. *From Sawin, 1941.* (*f*) *Rhinesuchus*, an Upper Permian rhinesuchoid. *After Watson, 1962.* (*g*) *Metoposaurus*, an Upper Triassic metoposaur. *From Colbert and Imbrie, 1956.* (*h*) *Wantzosaurus*, a trematosauroid. *After Lehman, 1961.* (*i*) *Batrachosuchus*, a brachiopoid. *From Watson, 1956.* (*j*) *Seymouria*, an anthracosaur in which the cheek is solidly attached to the skull roof. *From White, 1939.* (*k*) *Limnoscelis*, an anthracosaur in which the postorbital has taken over the area once occupied by the intertemporal. *From Romer, 1946.* Abbreviations as in Figure 8-3.

resent an early offshoot from that assemblage. If so, the loss of the anterior extension of the notochord must have occurred separately in temnospondyls and anthracosaurs.

Crassigyrinus (Figure 9-14) from the early Carboniferous is more difficult to classify (Panchen, 1985). The cheek is movable on the skull roof, as in anthracosaurs, but the tabular does not extend as far forward as the parietal. The sculpturing resembles that of most temnospondyls, but the vertebral column appears to consist of intercentra and pleurocentra that ossify as crescents of roughly equal size (see Figure 9-13*f*). The skull is large, with the jaw articulation far to the rear. The orbits are very close to the midline, in contrast with all other early tetrapods. This animal appears to share almost equally the characteristics of temnospondyls and anthracosaurs and may represent another distinct lineage that did not have an extensive radiation.

Lower Carboniferous temnospondyls

In the late Mississippian, temnospondyls are represented by several distinct groups that cover a broad range of adaptive types. The best known are the colosteids, represented by *Pholidogaster* and *Greererpeton* (see Figures 9-2*b*, 9-4). These animals have very elongated bodies, up to 40 presacral vertebrate, and relatively small limbs. The skull is flattened, lacks an otic notch, and has open grooves for the lateral line canals. Were it not for the earlier appearance of the more terrestrial ichthyostegids, we might think that these genera represent an early stage in the development of a terrestrial way of life. Instead, this group was probably secondarily aquatic, with the limbs reduced and the trunk elongated relative to more primitive tetrapods. The resulting body proportions and the probable pattern of aquatic locomotion are quite different from the rhipidistians.

Greererpeton (see Figure 9-4*a*) retains a large stapes that supports the braincase on the palatoquadrate. The otic capsule is incompletely ossified, and the occipital arch is weakly attached to the skull roof. This pattern may represent a more primitive stage in the consolidation of the braincase and dermal skull than is shown by any amphibians other than the ichthyostegids (Smithson, 1982). *Colosteus,* from the Upper Pennsylvanian, is a late surviving member of this group (Hook, 1983).

The trimerorhachids (Figure 9-14*c*) and saurerpetontids of the late Pennsylvanian and Permian resemble the colosteids in general body proportions, but they may represent a separate group of secondarily aquatic temnospondyls in which the intertemporal bone is retained and shallow otic notches are present.

A further group of aquatic labyrinthodonts is represented by the late Mississippian and Pennsylvanian loxommatoids, which Beaumount (1977) recently described (Figure 9-15). We know almost nothing of the postcranial skeleton of this group. The skulls show a combination of primitive and specialized characteristics. The palate is closed as in rhipidistians and anthracosaurs. As in temnospondyls, the tabular is small and does not reach the parietal, and the cheek and skull roof are suturally attached. As in anthracosaurs, the otic capsule extends to the tabular, which provides a strong attachment between the back of the braincase and the skull. Loxommatoids are unique among Paleozoic tetrapods in the anterior extension of the orbital margin which results in a keyhole-shaped opening. They have well-developed "otic notches" and conspicuous lateral line canal grooves. Vertebrae that may be associated with loxommatoids resemble those of *Crassigyrinus,* with crescent-shaped intercentra and pleurocentra. Although loxommatoids are usually classified among the temnospondyls, they may represent a distinct group of labyrinthodonts that, like the ichthyostegids and the crassigyrinids, show only a modest radiation.

Caerorhachis from the Upper Mississippian is the sole representative of a third group (see Figure 9-2*c*). Holmes and Carroll (1977) show that it was a more terrestrial form than were the colosteids, with body proportions like those of *Ichthyostega.* It has 32 presacral vertebrae and well-developed limbs. Lateral line canals are missing. The sculpturing and pattern of the skull are typical of later temnospondyls, except for the apparent absence of an otic notch. Surprisingly, the dominant vertebral elements are the pleurocentra, rather than the intercentra. Large pleurocentra in an animal that would otherwise be classified as a temnospondyl suggests that vertebral patterns vary in response to locomotor adaptations and may not be a reliable guide to taxonomic affinities.

A further group of terrestrial temnospondyl amphibian with small paired pleurocentra have recently been reported (but not yet described) from the Lower Carboniferous of Scotland (Milner, Smithson, Milner, Coates, and Rolfe, 1986). They appear to be close to the ancestry of a host of genera with this vertebral pattern in the late Carboniferous and Lower Permian (see Figure 9-13*c*). These advanced temnospondyls are further characterized by having well-developed otic notches.

The impedance-matching ear

Interpretations of the function of the otic notch and middle ear have played an important part in discussions of the relationships of early amphibians and reptiles. We long assumed that the stapes of the earliest amphibians was an important element in transmitting airborne vibrations to the inner ear. The embayment of the posterior margin of the cheek, which is present in most labyrinthodonts, has been interpreted as having evolved to support a tympanum.

Capranica (1976) and Wever (1978) performed physiological studies of the function of the ear in modern amphibians and reptiles that provided much more information on the specific function of the stapes and tympanum. As a result, we must reevaluate their role in Paleozoic tetrapods.

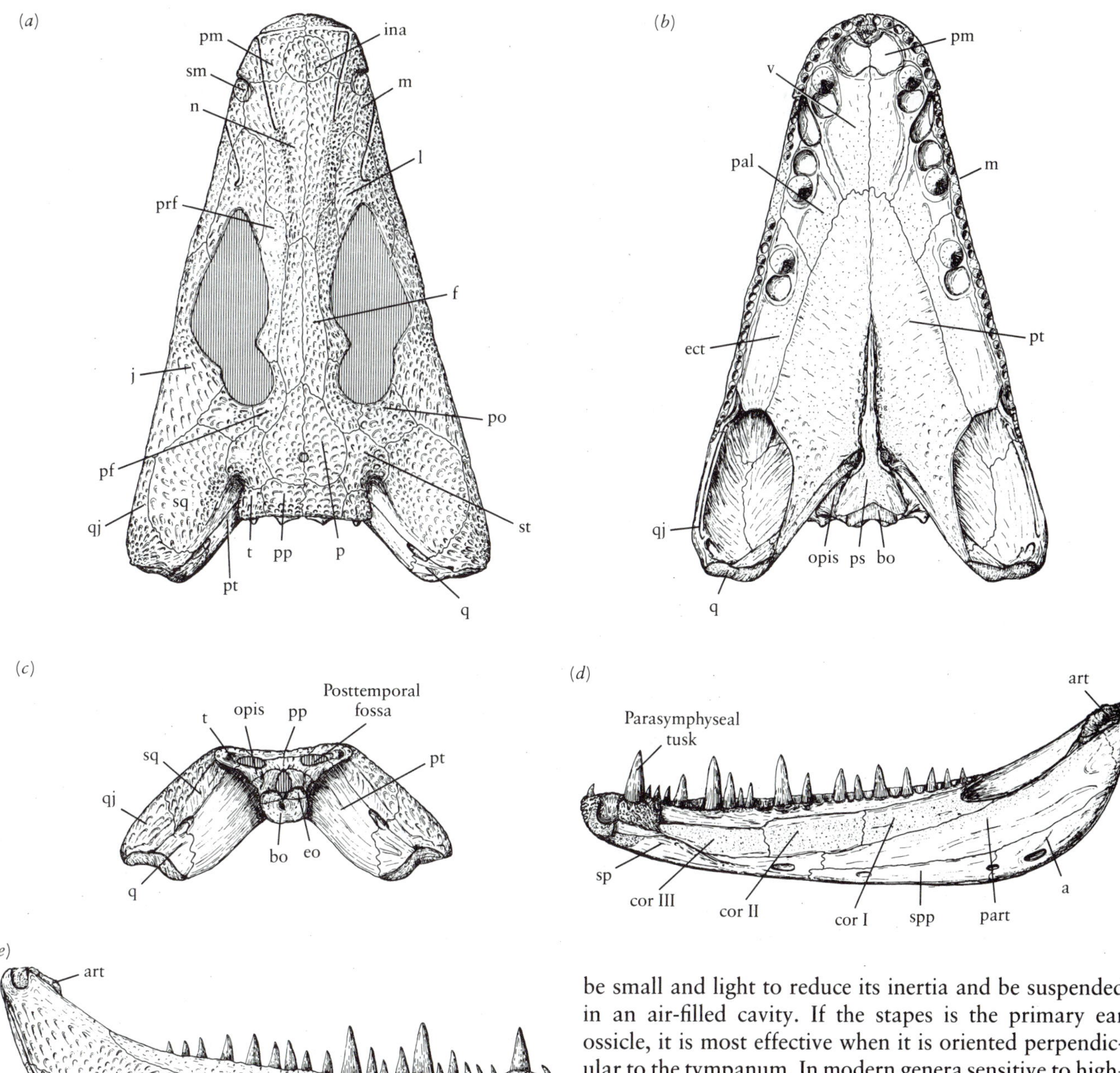

Figure 9-15. *Megalocephalus,* a loxommatoid skull in (*a*) dorsal, (*b*) palatal, and (*c*) occipital views, (*d*) medial and (*e*) lateral view of lower jaw, ×¼. Abbreviations as in Figure 8-3. *From Beaumont, 1977.*

In modern frogs, reptiles, and mammals, the middle ear functions as an impedance-matching mechanism to amplify the force that impinges on the tympanum and compensate for the thousandfold difference in density between the air outside the ear and the fluid within the inner ear (Figure 9-16). This force is magnified by an amount proportional to the difference in area between the tympanum and the foot plate of the stapes. The stapes must be small and light to reduce its inertia and be suspended in an air-filled cavity. If the stapes is the primary ear ossicle, it is most effective when it is oriented perpendicular to the tympanum. In modern genera sensitive to high-frequency sounds (above 1000 hertz), the ratio of the area of the tympanum to the stapedial foot plate is approximately 20 to 1.

The massive platelike stapes of *Greererpeton* and early anthracosaurs would not have functioned effectively as part of such an impedance-matching system, nor would the stapes in early reptiles or any of the lepospondyl groups. Upper Carboniferous temnospondyls such as *Amphibamus* (Bolt, 1977) are the earliest known vertebrates with a stapes that was small enough to have functioned in this manner. These genera also have an otic notch that could have supported a tympanum large enough to contribute effectively to an impedance-matching system.

Advanced temnopondyls

Romer (1947) recognized two superfamilies of Upper Carboniferous and Lower Permian temnospondyls—the edopoids and the eryopoids. The edopoids, such as *Den-*

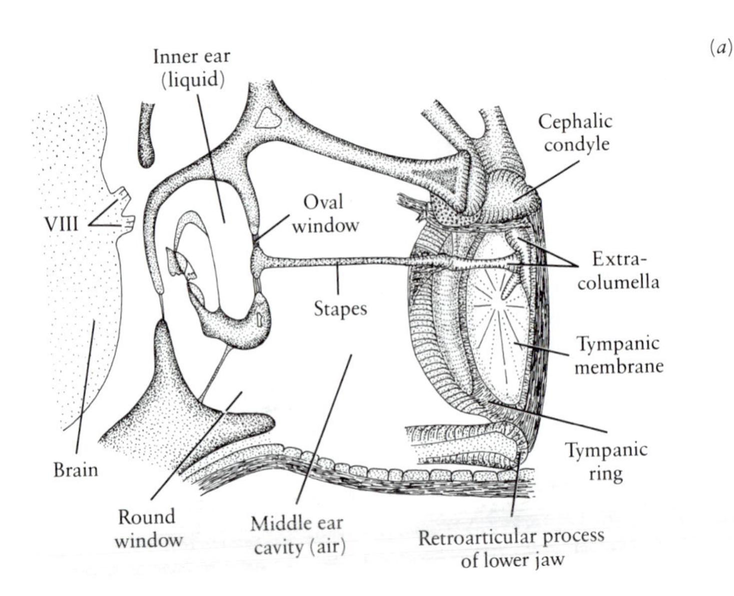

Figure 9-16. Diagrammatic representation of the ear of a lizard that illustrates the basic components of an impedance matching system. The large tympanum transmits vibrations via the small, light columella (suspended in an air-filled chamber) to the oval window of the middle ear. The greater area of the tympanum relative to the foot plate of the stapes mechanically amplifies the force impinging on the ear drum to compensate for the differential in density between the fluid of the inner ear and the outside air. *From Robinson, 1973 (after Wever).*

drerpeton (Figure 9-17), are more primitive. They retain an intertemporal bone and a movable articulation between the base of the braincase and the pterygoid. Most eryopoids evolved a sutural attachment between the parasphenoid and the pterygoid, and all have lost the intertemporal bone. They are the dominant temnospondyl amphibians in the later Pennsylvanian and early Permian and occupied a spectrum of aquatic, semiaquatic, and semiterrestrial habitats. The large semiaquatic genus *Eryops* (Figure 9-18*a*) is one of the most common and best-known early Permian amphibians. Other genera such as *Cacops* (Figure 9-18*b*) were probably among the most terrestrial of all amphibians. This genus is a member of the extremely diverse family Dissorophidae, many members of which have plates of dermal bone protecting the trunk region (Carroll, 1964; DeMar, 1968).

We include among the eryopoids a number of animals known mostly from Europe that are small, poorly ossified, and have external gills (see Figure 9-11). They were once thought to constitute a distinct order of amphibians, the branchiosaurs. Boy (1972) identified some branchiosaurs as early growth stages of larger terrestrial genera. Others appear to have been neotenic, that is, they reached their maximum size and presumably sexual maturity without metamorphosing into more terrestrial forms.

The Lower Permian may be considered the high-water mark for terrestrial labyrinthodonts. Later temnopondyls were primarily aquatic and typically had small, poorly ossified limbs and flat skulls. The reduction of the number of terrestrial amphibians after the Lower Permian may be attributed to the increasing dominance of large terrestrial reptiles.

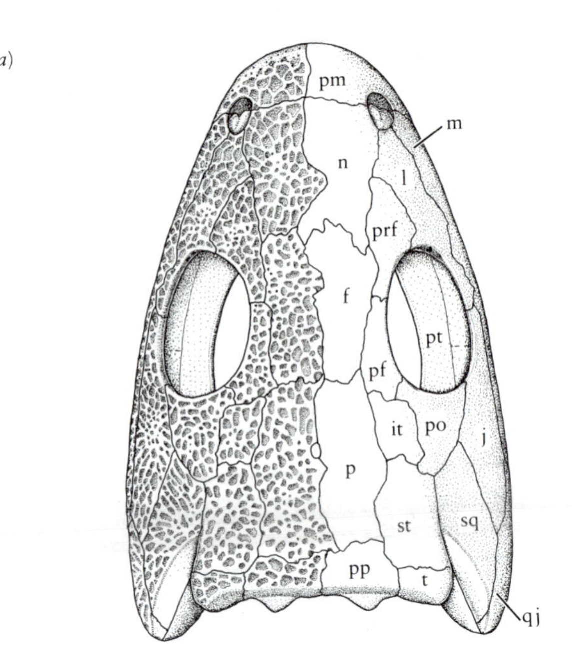

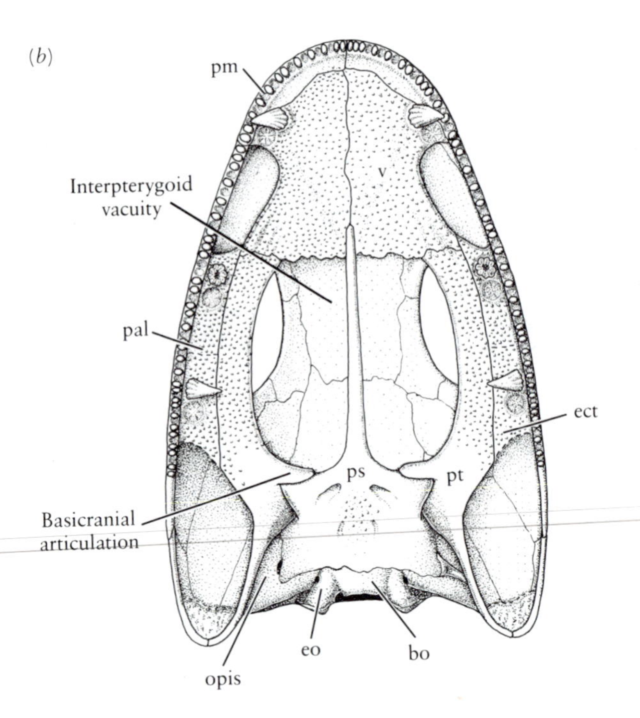

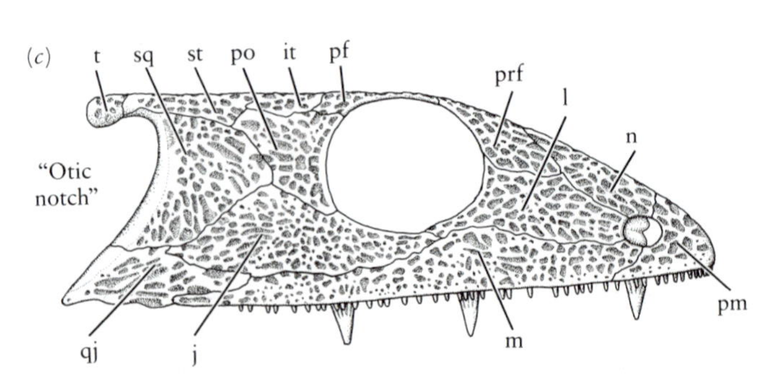

Figure 9-17. The edopoid temnospondyl *Dendrepeton* showing a movable articulation between the palate and the braincase and the retention of the intertemporal bone (*a*) dorsal, (*b*) palatal, (*c*) lateral views, natural size. *From Carroll, 1967.* Abbreviations as in Figure 8-3.

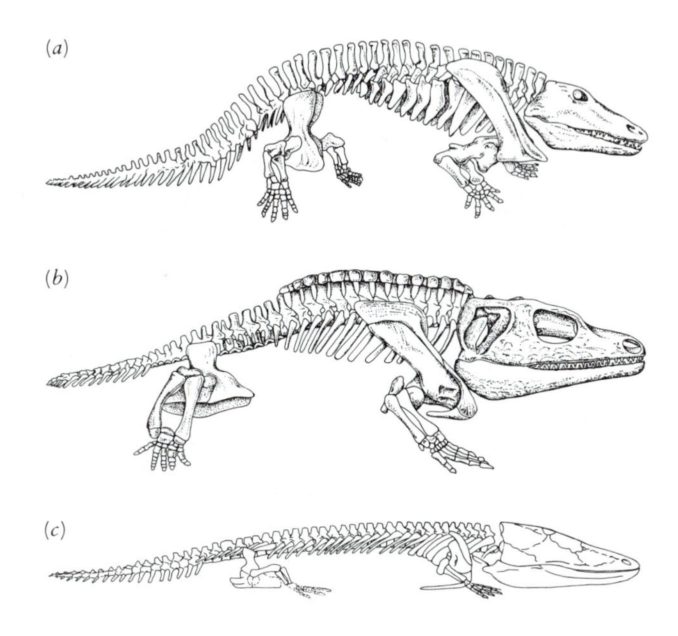

Figure 9-18. Permian and Triassic temnospondyl amphibians. (*a*) *Eryops*, original nearly 2 meters long. *From Gregory, 1951.* (*b*) *Cacops*, a dissorophid, 40 centimeters long. *From Williston, 1910.* (*c*) the Triassic "stereospondyl" *Paracyclotosaurus*, over 2 meters long. *From Watson, 1958. Redrawn with permission by the Trustees of the British Museum (Natural History).*

In several groups of Triassic amphibians, the intercentra are ossified as complete cylinders and the pleurocentra are small or totally missing (see Figure 9-13*d*). The term stereospondyl has been applied to these animals on the assumption that they constituted a unified taxonomic assemblage with a common ancestry. We now recognize that the late surviving temnospondyls are not a single taxonomic group. Some almost certainly evolved from advanced eryopoids, but others probably arose from other groups of early temnospondyls.

The problem of establishing the relationships of the late Permian and Triassic temnospondyls results partly from the hiatus in the fossil record at the end of the Lower Permian, when the relatively complete record of Upper Carboniferous and Lower Permian vertebrates in Europe and North America is interrupted by geological events and changes in climate. The record of later stages in amphibian evolution resumes in the Middle and Upper Permian of Russia and South Africa, when we encounter several distinct groups for the first time.

The Rhinesuchoides, known from the Upper Permian and Lower Triassic of South Africa, are an assemblage of medium- and large-sized semiaquatic forms that evolved from among the Lower Permian eryopoids. They are succeeded in time by the capitosauroids (see Figure 9-14*e*, *f*), with which they form an evolutionary continuum. The capitosauroids are limited to the Triassic but span the length of that period and have a cosmopolitan distribution. In this group, the skulls are very much flattened, the limbs small, and ossification reduced. Most were obligatorily aquatic. They reached the greatest dimensions of any labyrinthodonts, with skulls approaching a meter in length (Paton, 1974).

Metoposauroids resemble the capitosauroids in size and body proportions but are differentiated by the anterior position of the orbits (see Figures 9-14*g*). They are known from both eastern and western North America, Europe, and India. We sometimes find large numbers of metoposauroid remains in what were presumably the last remnants of shallow lakes. This group is common in the Upper Triassic, but their antecedents have not been recognized.

The Trematosauroidea is another phylogenetically isolated group that is restricted to the Lower Triassic but is common in such distant regions as Greenland, Spitsbergen, Madagascar, western North America, South Africa, Australia, and Russia. Trematosauroids are typically long-snouted forms with well-developed lateral line canal grooves and rather high narrow skulls (see Figure 9-14*h*). They retain the vertebral pattern of most primitive temnospondyls with the persistence of paired pleurocentra. The nature of the deposits in which they are found indicates that some were marine in habits—the only group of amphibians to invade that environment. Trematosaurs may have evolved from long-snouted aquatic forms such as *Archegosaurus* from the Lower Permian, but there are no known intermediates.

Branchyopoids, which Welles and Estes (1969) reviewed, are a rare group that extends throughout the Triassic. They may have evolved from among the Lower Permian trimerorhachoids. The skull is short and flat (see Figure 9-14*i*). Their remains have been described from North and South America, Australia, Asia, South Africa, Europe, and Antarctica. This group includes the last of the labyrinthodonts, with one genus surviving into the Middle Jurassic in China (Dong, 1985).

Plagiosaurs are the most peculiar of the temnospondyls, with extremely short, wide skulls that had pustular ornamentation. Some, such as *Plagiosuchus*, retain external gills in what are apparently adults (Figure 9-19).

Figure 9-19. Dorsal view of the specialized late Triassic temnospondyl *Plagiosuchus*, original about 1 meter long. *Photograph courtesy of Dr. Wild.*

They are represented by rare fossils throughout the Triassic of Europe. Although they resemble the brachyopoids in skull proportions, the vertebrae are complete cylinders, which Panchen (1959) has homologized with the pleurocentra, in contrast with the condition in most other temnospondyls.

The rarity of appropriate deposits in the Lower Jurassic may conceal the last phase in the evolution of aquatic temnospondyls, but it is unlikely that any lineages survived long into the later Mesozoic without leaving some fossil record. The late surviving labyrinthodonts certainly left no descendants.

Anthracosaurs

The anthracosaurs were never as numerous or as diverse as the temnospondyls. The best known of the early anthracosaurs is *Proterogyrinus* (see Figure 9-2*d*) from the late Mississippian of North America (Holmes, 1984). As in the related Pennsylvanian genus *Palaeoherpeton,* the skull is primitive in retaining a line of weakness between the cheek and the skull table, and the palate is closed (Figure 9-20). The skull table is advanced in the forward extension of the tabular to the parietal. The braincase is solidly fixed to the skull roof, with the otic capsule attached to the underside of the tabular. From their first appearance in the fossil record, the anthracosaurs have an embayment in the posterior margin of the cheek that was generally considered to have supported a tympanum. Clack (1983) described a large, bladelike stapes in early genera that is quite different from the narrow rodlike structure usually found in animals known to have an impedance matching ear.

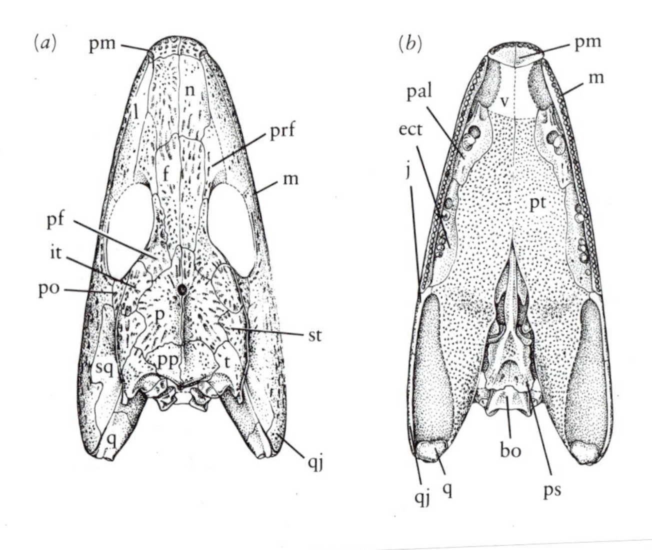

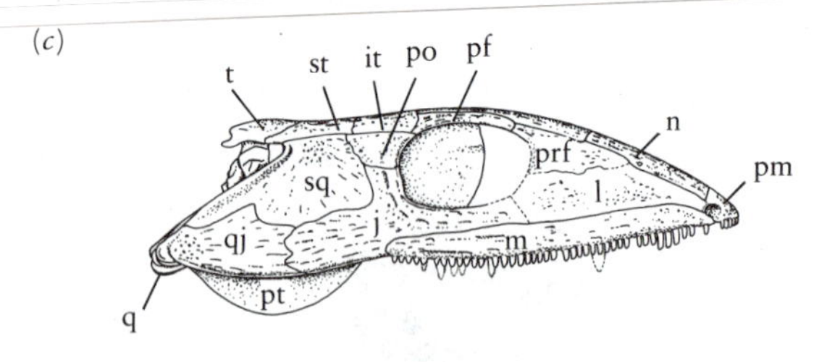

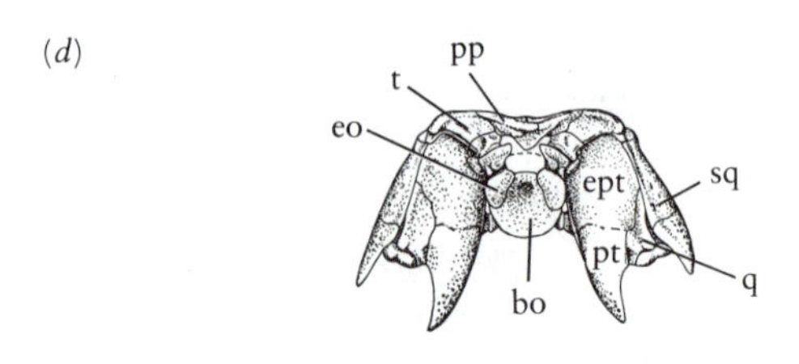

Figure 9-20. SKULL OF *PROTEROGYRINUS.* (*a*) Dorsal, (*b*) palatal, (*c*) lateral, (*d*) occipital views. Original about 20 centimeters long. Abbreviations as in Figure 8-3. *From Holmes, 1984.*

The postcranial skeleton of *Proterogyrinus* is typical of terrestrial animals with well-developed limbs and a relatively short trunk. There are 32 presacral vertebrae, each consisting of a crescentic intercentrum and a large horseshoe-shaped pleurocentrum supporting the neural arch (see Figure 9-13*h*). This condition is clearly advanced over typical temnospondyls. We may assume that the pleurocentra, which were paired in rhipidistians and primitive temnospondyls, have grown ventrally to meet at the midline. The girdles and limbs are similar to those of primitive temnospondyls, which suggests derivation from a similar common ancestor. Several other anthracosaur lines evolved from animals similar to *Proterogyrinus.* Some of them became secondarily aquatic, and others evolved toward a more terrestrial way of life.

The best known anthracosaurs are the embolomeres, reviewed by Panchen (1970). They became specialized for aquatic locomotion by elongation of the trunk region, with approximately 40 presacral vertebrae (Figure 9-21). The intercentrum and the pleurocentrum of each vertebra (see Figure 9-13*i*) were elaborated as subequal cylinders, which probably permitted a great deal of flexibility in the column and facilitated undulatory swimming. However, the limbs are similar to their more terrestrial antecedents (Romer, 1957a). Some genera reevolved a dorsal caudal fin. The embolomeres are best represented in the Carboniferous of Great Britain, where they apparently inhabited fairly deep bodies of water. However, in the Lower Permian, their remains are restricted to deltaic deposits in southwestern North America. The group did not survive beyond the Lower Permian.

The remaining Carboniferous anthracosaurs were apparently more terrestrial. The gephyrostegids are relatively small forms, with no more than 24 trunk vertebrae and a relatively great limb-to-trunk-length ratio. The earliest known genus, *Bruktererpeton* (Figure 9-22*a*), has body proportions approaching those of primitive reptiles.

Forms allied to the known gephyrostegids or to the Lower Carboniferous *Eoherpeton* (Figure 9-23) may have given rise to the seymouriamorphs of the Permian. Unlike other anthracosaurs (but like temnospondyls), the skull and cheek region are solidly attached and the dermal bones are marked by distinct pits and grooves. The otic notch is very deep and the stapes is reduced to a narrow rod.

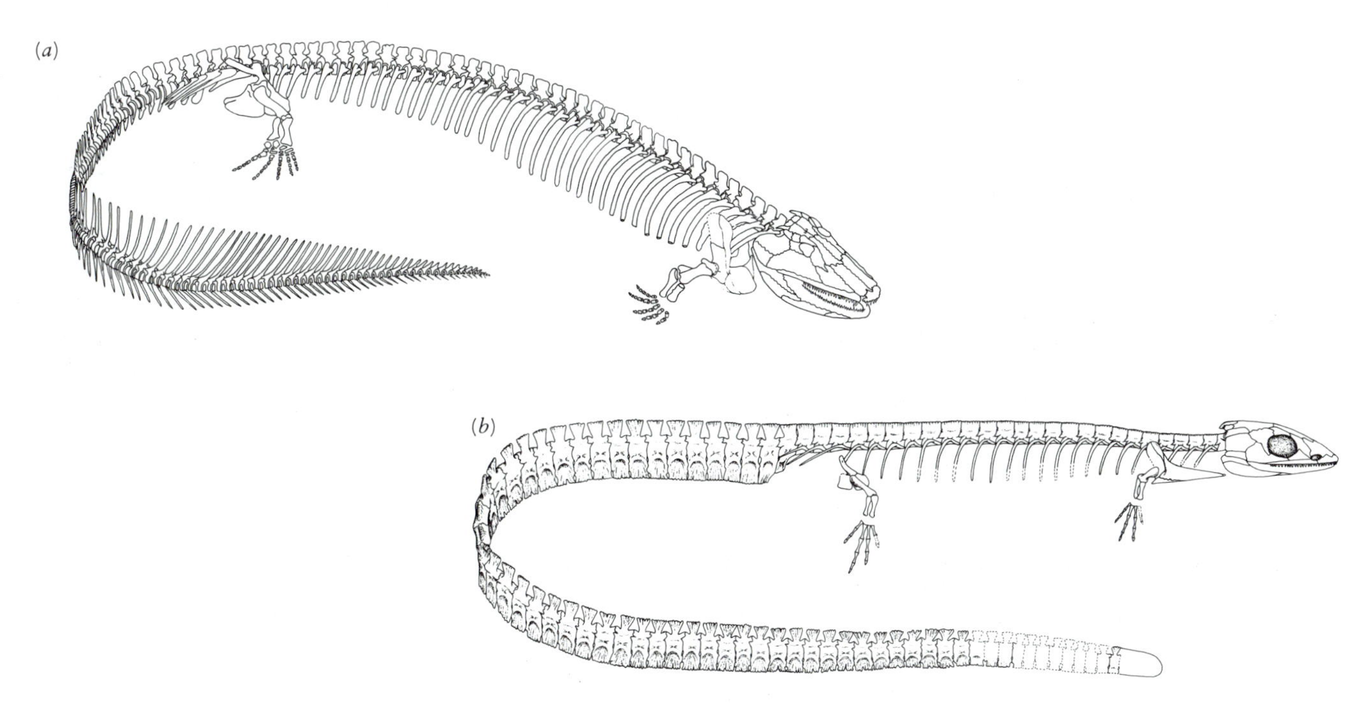

Figure 9-21. (*a*) Reconstruction of the skeleton of the embolomere *Eogyrinus*. Original about 2 meters long. Length of trunk and finlike tail are restored, based on the Lower Permian embolomere *Archeria*. *From Panchen, 1972.* (*b*) Restoration of the skeleton of the nectridean *Ptyonius marshii*, which shows a contrasting mode of aquatic specialization, about 20 centimeters long. *From Bossy, 1976.*

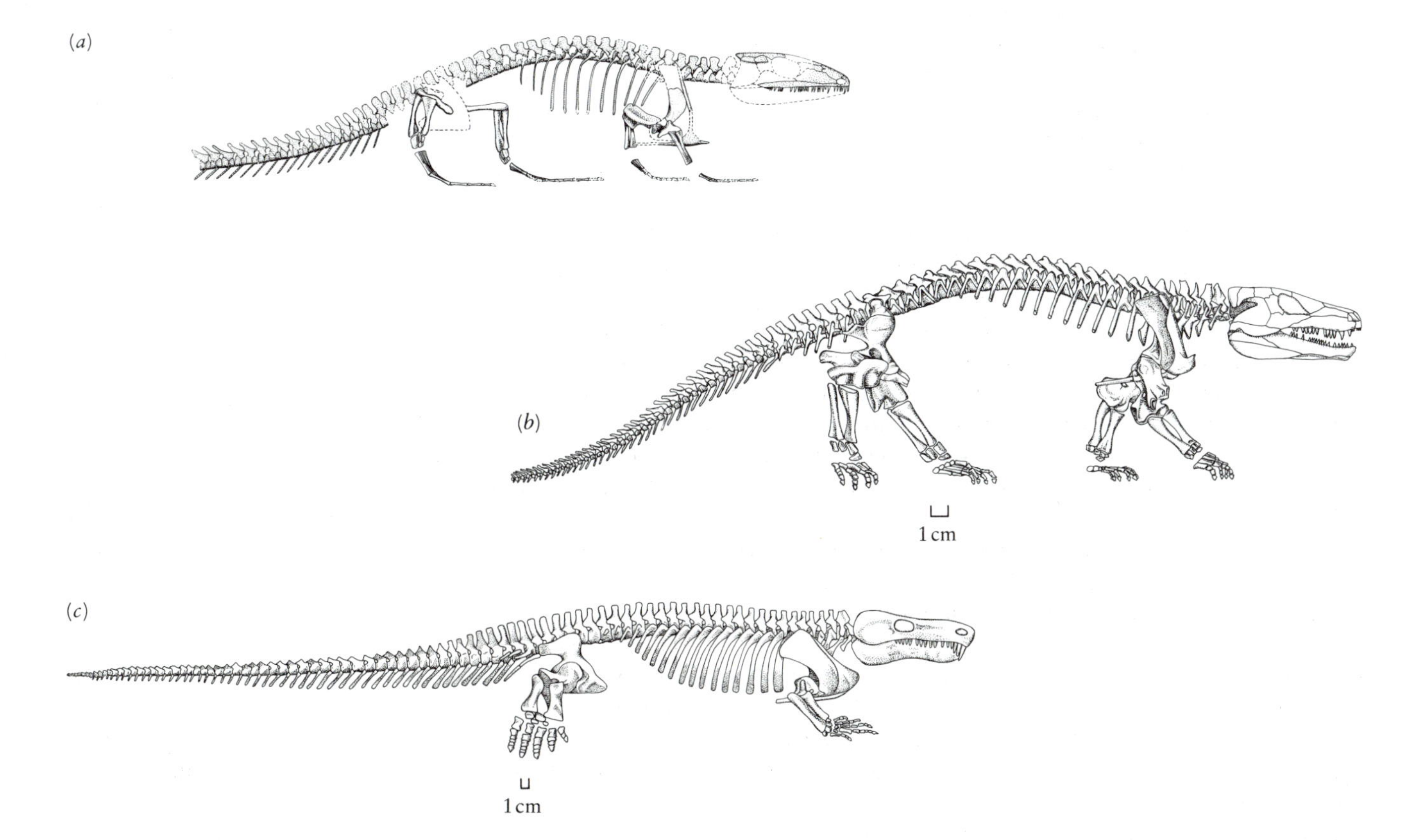

Figure 9-22. TERRESTRIAL ANTHRACOSAURS. (*a*) The gephyrostegid *Bruktererpeton*. *From Boy and Bandel, 1973.* (*b*) *Seymouria*. *From White, 1939.* (*c*) *Limnoscelis*. *From Williston, 1912.*

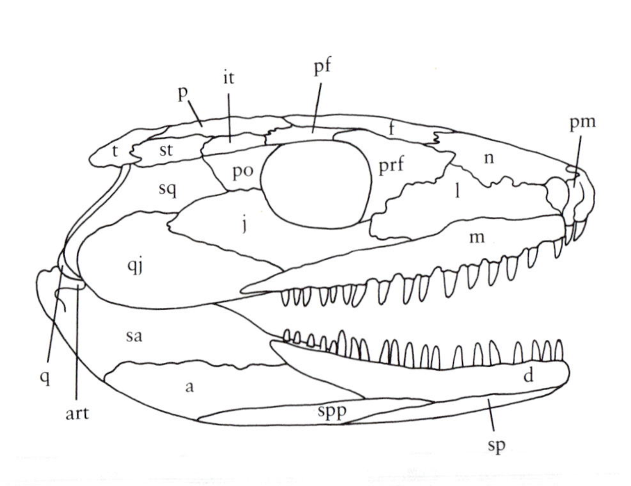

Figure 9-23. *Eoherpeton*, a primitive anthracosaur; lateral view of the skull. Abbreviations as in Figure 8-3. A slightly modified illustration of *Eoherpeton* is presented by Smithson (1985) in a new description of this genus. *From Panchen, 1980b.*

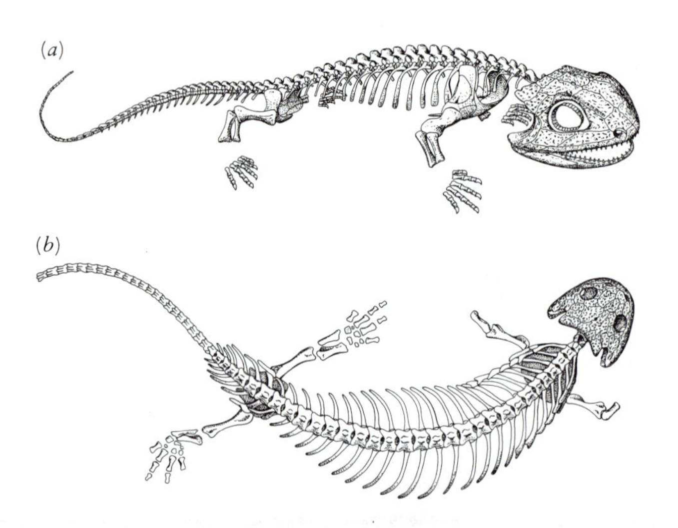

Figure 9-24. AQUATIC ANTHRACOSAURS. (*a*) *Discosauriscus*, a small early Permian seymouriamorph, slightly less than natural size. *From Spinar, 1952.* (*b*) The skeleton of the Upper Permian seymouriamorph *Kotlassia*, $\times \frac{1}{12}$. The manus is unknown. *From Bystrow, 1944.*

The Seymouriidae, Discosauriscidae, and Kotlassiidae (Figures 9-22 and 9-24) apparently belong to a single assemblage with relatively consistent skeletal patterns but different ways of life. The seymouriids are primarily North American, with questionably assigned material from Russia. Their limb proportions indicate they were the most terrestrial of anthracosaurs. Discosauriscids, from central and eastern Europe and recently reported in China (Zhang, Li, and Wang, 1984) are known only from larval or neotenic forms. The kotlassiids, known only from the Upper Permian of Russia, reverted to an aquatic way of life, like the terminal temnospondyls. In common with some dissorophids and plagiosaurs, they developed dermal armor covering the trunk region.

Other anthracosaurs from the Upper Permian of Russia are difficult to associate with any of the previously mentioned groups. We have classified them among the seymouriamorphs because they may have diverged from the same general stock, but they show few similarities with the other families. Members of the Chroniosuchidae have very long, narrow skulls that are rather similar to those of some embolomeres, and large plates of dermal armor. The lanthanosuchids have wide, low skulls with a large opening above the area of the jaw musculature (Tatarinov, 1972).

None of these Upper Permian groups show any evidence of continuing into the Triassic, although beds of this age in Russia have very abundant remains of temnospondyls.

A number of families from the Upper Carboniferous and Lower Permian may be affiliated with the Anthracosauria, but we have difficulty assessing their specific position. They approach the reptilian grade of development in some skeletal features but retain others that are typically amphibian (Carroll, 1970; Heaton, 1980). Members of the Limnoscelidae (see Figure 9-22*c*), Solenodonsauridae, and Tseajaiidae may not be particularly closely related to one another, but all have an otic notch and a primitive configuration of the occiput and temporal region. All are relatively large, and their skeleton suggests that they were primarily terrestrial, at least as adults.

LEPOSPONDYLS

In addition to the generally large, heavy bodied labyrinthodonts, a number of smaller amphibians in the Paleozoic vaguely resemble a variety of modern tetrapods, including newts and other salamanders, lizards, and snakes. These animals have been grouped together in a single assemblage, the Lepospondyli, that is most notably characterized by the possession of spool-shaped centra (Figure 9-25). Other features of the vertebrae and skull are common to these animals, but the general body proportions and the patterns of the dermal skull roof differ markedly among the various groups.

We recognize six major anatomical patterns and group them in three diverse orders—the Aïstopoda, Nectridea, and Microsauria, and three isolated families—the Lysorophidae, Adelogryinidae, and Acherontiscidae. We have not established their specific ancestry. Except for the nectrideans, which are known in North Africa, their remains have been found only in North American and Europe.

Aïstopods

The most specialized of the lepospondyls are the snakelike aïstopods. They have a very long body with up to 230 vertebrae and lack limbs and girdles. Most of our information on the group comes from the Upper Carboniferous families, Ophiderpetontidae and Phlegethontiidae (McGinnis, 1967). The skulls of these forms are highly specialized, which implies that each family has a long history separate from the other as well as from any labyrinthodonts or rhipidistians. In the most specialized genus, *Phlegethontia* (Figure 9-26*a*, *b*, and *c*), the skull is fenestrated in a manner somewhat resembling snakes. In the more primitive genus *Ophiderpeton* (Figure 9-26*d*), dermal bones of the skull are more extensive and their pattern somewhat resembles that of labyrinthodonts.

The vertebrae of aïstopods have only a single ossification per segment, a condition termed **holospondylous.** The neural arches are fused to the centra; separate haemal arches are absent in the caudal region. The atlas resembles

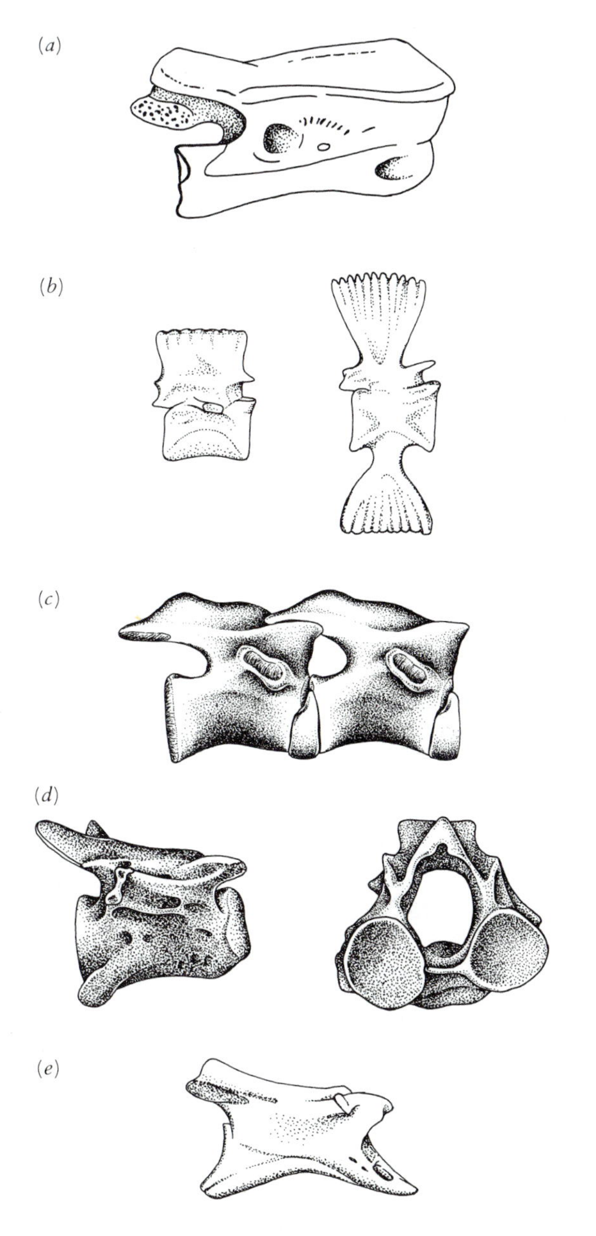

Figure 9-25. VERTEBRAE OF LEPOSPONDYL AND MODERN AMPHIBIANS. (*a*) Trunk vertebrae of an aïstopod. *From McGinnis, 1967.* (*b*) trunk and caudal vertebrae of the nectridean *Sauropleura. From Bossy, 1976.* (*c*) The trunk vertebrae of the microsaur *Eryodus* showing intercentra. *From Carroll and Gaskill, 1978.* (*d*) Trunk and cervical vertebrae of the Cretaceous salamander *Opisthotriton. From Estes, 1976.* (*e*) trunk vertebra of the caecilian *Hypogeophis. From Carroll and Currie, 1975. With permission from the Zoological Journal of the Linnean Society. Copyright 1975 by the Linnean Society of London.*

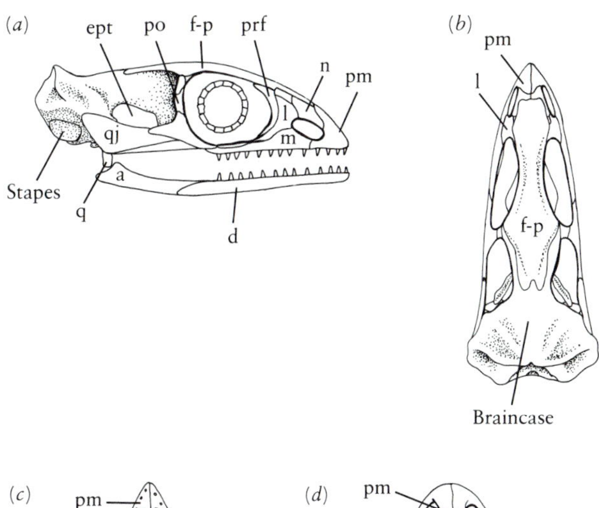

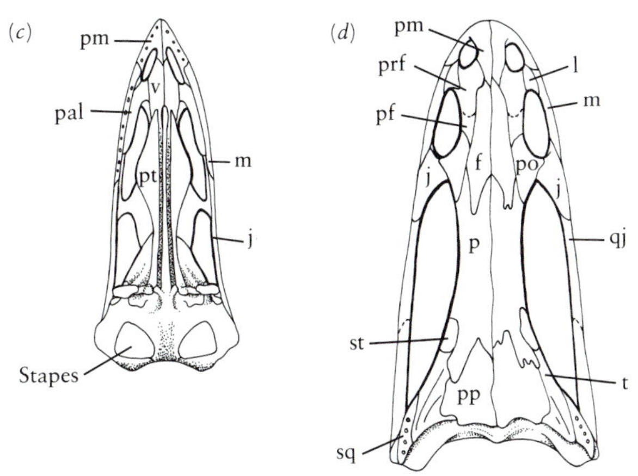

Figure 9-26. THE SKULL OF AISTOPODS. (*a–c*) Lateral, dorsal, and ventral views of the skull of *Phlegethontia*, ×3. *From Gregory, 1948.* (*d*) Dorsal view of the skull of *Ophiderpeton*, ×3. *From Bossy, 1976.* Abbreviations as in Figure 8-3.

the remaining trunk vertebrae and fits against the rim of a rounded concavity in the occiput. The heads of the ribs are described as **K**-shaped.

A single specimen from the early Lower Carboniferous has been referred to this order by Wellstead (1982). The skull is specialized in the manner of *Ophiderpeton* with the orbits far forward; the cheek region has no ossification (Figure 9-27*a*). This animal is interpreted as an ancestor of the Upper Carboniferous aïstopods, but there are no features of this genus (the oldest of all known lepospondyls) that suggest a close relationship with other groups of Carboniferous tetrapods.

Nectrideans

Nectrideans are an entirely aquatic group. Many are newtlike in appearance and have a very long, laterally compressed tail that serves as the principle swimming organ. Like the aïstopods, none has more than a single central ossification, but all have well-differentiated, although relatively small, paired limbs.

Members of this order can be distinguished from all other Paleozoic amphibians by the symmetrical appearance of the neural and haemal arches of the tail (see Figure

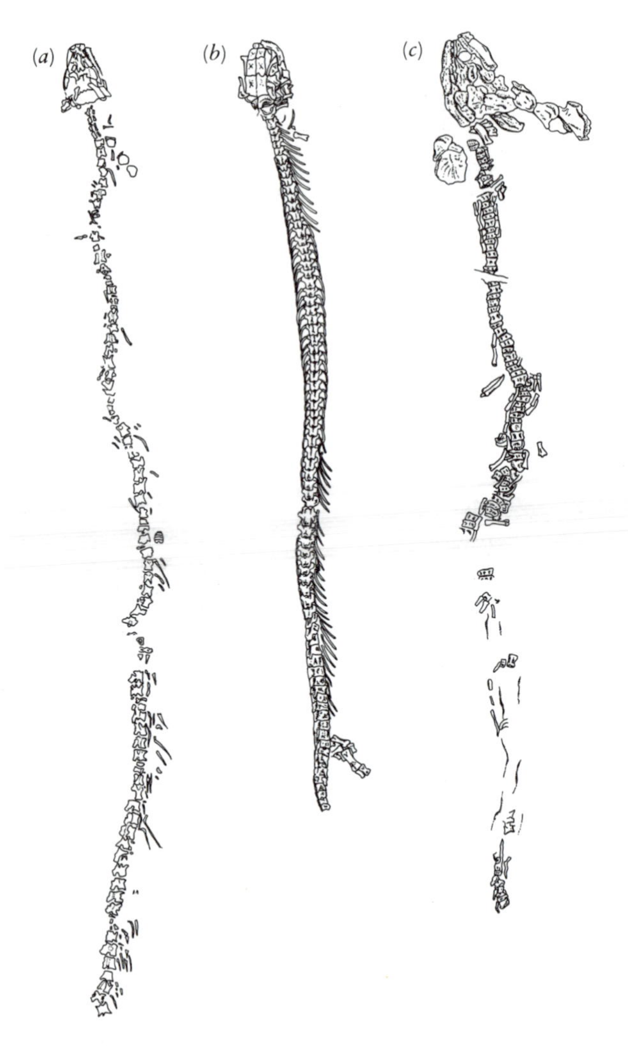

Figure 9-27. LONG-BODIED "LEPOSPONDYLS." (*a*) The aïstopod *Lethiscus*, from the Lower Carboniferous, $\times\frac{1}{3}$. *From Wellstead, 1982.* (*b*) The lysorophid *Cocytinus*, $\times 1$. *From Carroll and Gaskill, 1978.* (*c*) *Acherontiscus*, $\times 1$. *From Carroll and Gaskill, 1978.* Skeletons of *Lethiscus* and *Acherontiscus* have been straightened for convenience of illustration.

9-25*b*). The neural spines in the trunk are also laterally compressed plates that interdigitate by accessory articulating surfaces. Twisting of the trunk would be minimal and lateral undulation would be limited. Aquatic locomotion is almost certainly a secondary adaptation of the group since the swimming mechanism is unlike that of rhipidistians or primitive labyrinthodonts.

In one of the earliest and most primitive nectrideans, *Urocordylus*, the tail is more than twice as long as the trunk and the limbs are well developed, with five toes on both the manus and pes. In more advanced urocordylid nectrideans, the degree of ossification of the limbs and girdles is reduced, the rear limb is comparatively enlarged, and the number of digits in the manus is reduced to four. In contrast with aquatic temnospondyls, anthracosaurs, and many microsaurs, the number of trunk vertebrae in nectrideans is not increased but ranges from 15 to 26 (Bossy, 1976).

The pattern of the skull roof of primitive nectrideans appears to be closest of all the lepospondyl groups to that of labyrinthodonts (Figure 9-28*a*). As in embolomeres, the skull table is narrow and movable on the cheek. Of the primitive elements, only the intertemporal has been lost. The area that it occupied may have been appropriated by the elongated postorbital. The supratemporal is a long, narrow bone, lost in the more specialized families. The elongated parietal extends back to the tabular. A line of weakness, like that of rhipidistians, extends anterior to the orbit and runs between the lacrimal and the prefrontal. In these primitive nectrideans, the cheeks can be flared and the snout raised and lowered at joints that permit the frontal to slide over the parietal and the nasal. Primitively, the skull is short with the eyes relatively far forward, but in more specialized species the snout becomes greatly elongated.

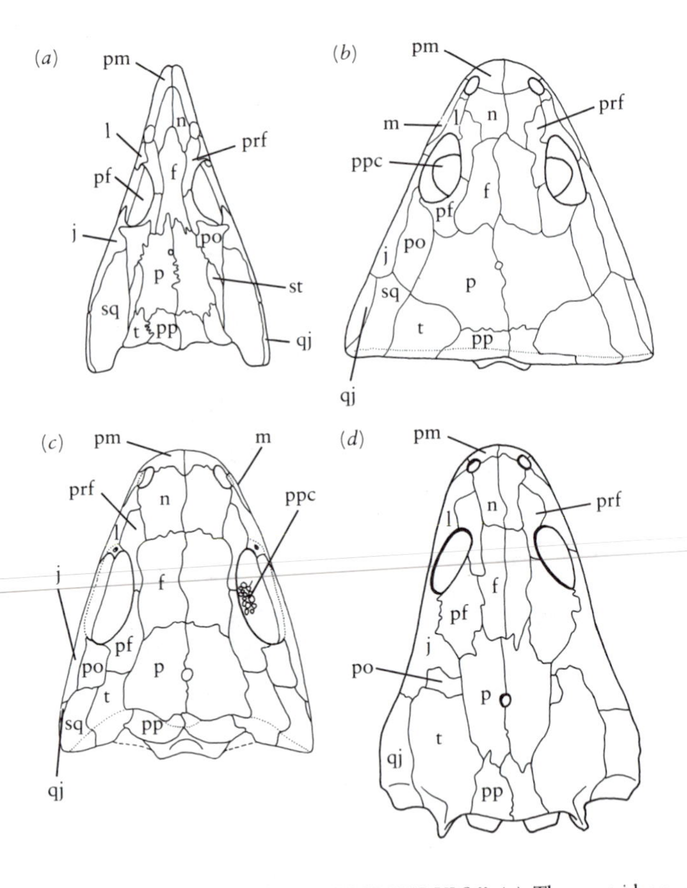

Figure 9-28. SKULLS OF "LEPOSPONDYLS." (*a*) The nectridean *Sauropleura*, which retains the most primitive pattern of the skull table seen among "lepospondyls." *From Bossy, 1976.* (*b*) The microsaur *Microbrachis*, in which the space once occupied by the intertemporal has apparently been incorporated into the parietal. *From Carroll and Gaskill, 1978.* (*c*) The microsaur *Asaphestra*, in which the space once occupied by the intertemporal may be incorporated into the postfrontal. In all microsaurs, there is a single bone that occupies the position of the tabular and supratemporal in labyrinthodonts. *From Carroll and Gaskill, 1978.* (*d*) *Adelospondylus*, in which the squamosal, supratemporal, and tabular are represented by a single ossification. *From Carroll and Gaskill, 1978.* Abbreviations as in Figure 8-3 plus: ppc, palpebral cup, an ossification of the eyelid not uncommon in primitive tetrapods.

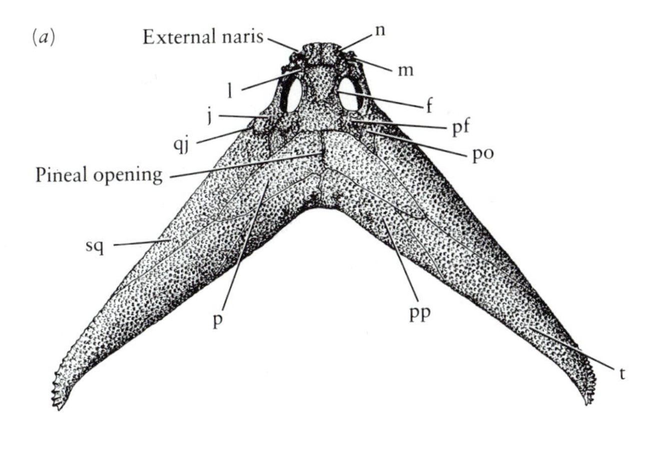

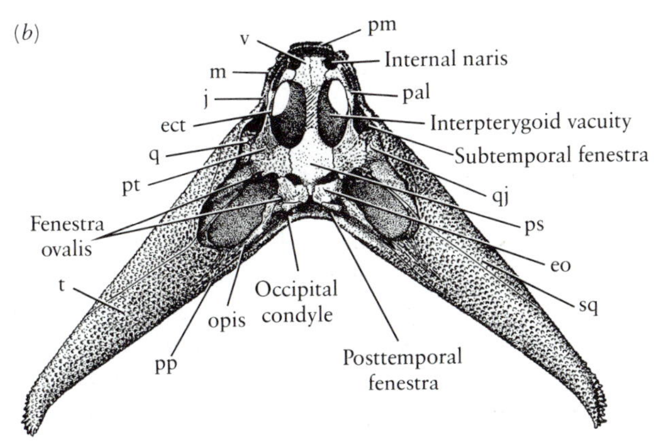

Figure 9-29. (*a*) Dorsal and (*b*) ventral views of the skull of *Diploceraspis*, a "horned" nectridean of the early Permian, ×⅖. Abbreviations as in Figure 8-3. *From Beerbower, 1963.*

The Diplocaulidae, which are represented in the Lower Permian by the genera *Diploceraspis* (Figure 9-29) and *Diplocaulus*, are more specialized nectrideans in which kinesis has been lost. The skull is dorsoventrally flattened, and the tabulars and squamosals are drawn out into long processes (Milner, 1977). Like the brachyopids, plagiosaurs, and other late surviving temnospondyls, the diplocaulids have fused the palate to the base of the braincase and developed vacuities between the pterygoids. The trunk and tail are short, and some authors suggest that they swam by dorsoventral undulation. This family occurs in North Africa as well as North America and Europe.

Microsaurs

Microsaurs (Figure 9-30) are the most varied of lepospondyls. Carroll and Gaskill (1978) recognize 11 families, several of which appear in the Lower Pennsylvanian. The major diversification of the order may have been completed by that time, although other families are known only from the later Pennsylvanian or Lower Permian. Their habits range from perenni-branchiate to lizardlike, with a variety of large and small burrowing genera.

We recognize two patterns of the bones of the dermal skull roof, which probably resulted from the incorporation of the intertemporal into different adjacent bones. In the Suborder Tuditanomorpha, the area of the intertemporal is apparently incorporated into the postfrontal; in the Microbrachimorpha, it was apparently incorporated into the parietal (Figure 9-28*b*, *c*). In both groups, there is only one remaining temporal bone, the tabular, which

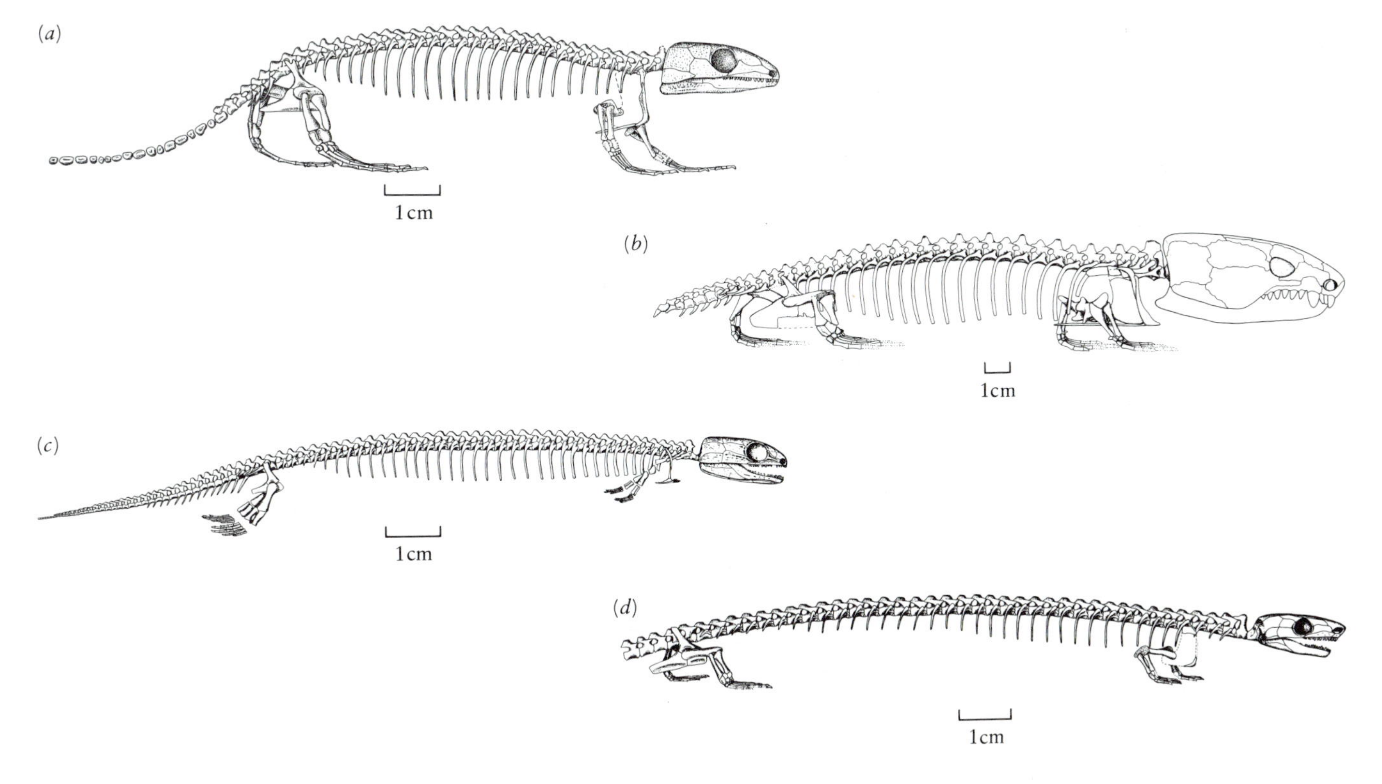

Figure 9-30. SKELETONS OF MICROSAURS WITH DIVERSE BODY PROPORTIONS. (*a*) *Tuditanus.* (*b*) *Pantylus.* (*c*) *Microbrachis.* (*d*) *Rhynchonkos. From Carroll and Gaskill, 1978.*

may incorporate the supratemporal. In primitive genera, the cheek region is completely roofed, but it is deeply emarginated in representatives of three families in a manner somewhat resembling that of primitive salamanders.

Microsaurs exhibit great variety in body proportions, with the vertebral count ranging from 19 to 44. The limbs may be small, but they are always present, and the tail is never specialized in the manner of nectrideans. Several microsaurs have intercentra in the trunk region in addition to spool-shaped pleurocentra. Trunk intercentra have never been reported in nectrideans and aïstopods.

Other lepospondyls

Two additional families are usually allied with microsaurs: the lysorophids and the adelogyrinids. In both, the pattern of the skull appears to preclude close affinities with typical microsaurs. The lysorophids (Figures 9-27 and 9-31) have a very open skull with large orbits; their maxilla and premaxilla are quite freely movable. There are up to 99 presacral vertebrae and very small limbs. The short tail exhibits haemal arches, but there are never trunk intercentra. The family is most common in the Upper Pennsylvanian and Lower Permian of North America. Forms that are possibly related occur in the Lower Pennsylvanian in Ireland (Wellstead, 1985).

The adelogyrinids (Figure 9-28*d*) are much more restricted in diversity and geography, with four genera described from the Upper Mississippian of Scotland. The skull is solidly roofed but lacks one of the elements present in typical microsaurs. The eyes are very far forward. The trunk is apparently quite long, but the limbs are well developed (Brough and Brough, 1967).

The Acherontiscidae is represented by a single specimen described by Carroll (1969) from the Mississippian of Scotland (see Figure 9-27). The centra are formed by two subequal cylinders that grossly resemble the pleurocentra and intercentra of embolomeres. The skull is poorly known but has a solid roof and lateral line canal grooves. The trunk region is long and the limbs little developed.

Lysorophids, adelogyrinids, and acherontiscids appear like the isolated orders of Paleozoic sharks; neither their time of origin nor phylogenetic position is apparent. They hint at an even greater diversity of small amphibians in the Paleozoic.

Unlike the labyrinthodonts, the lepospondyls show no obvious affinities with the rhipidistians. The characteristics by which they differ from both the rhipidistians and labyrinthodonts—the lack of labyrinthodont infolding of the dentine, the otic notch, and the pairs of palatine fangs—may all be related to their small size and do not necessarily indicate close relationship to each other. The presence of cylindrical vertebral centra in all groups of lepospondyls may also be related to their relatively small size. The various lepospondyl groups may each have arisen from different stocks of labyrinthodonts during the Upper Devonian and Lower Carboniferous.

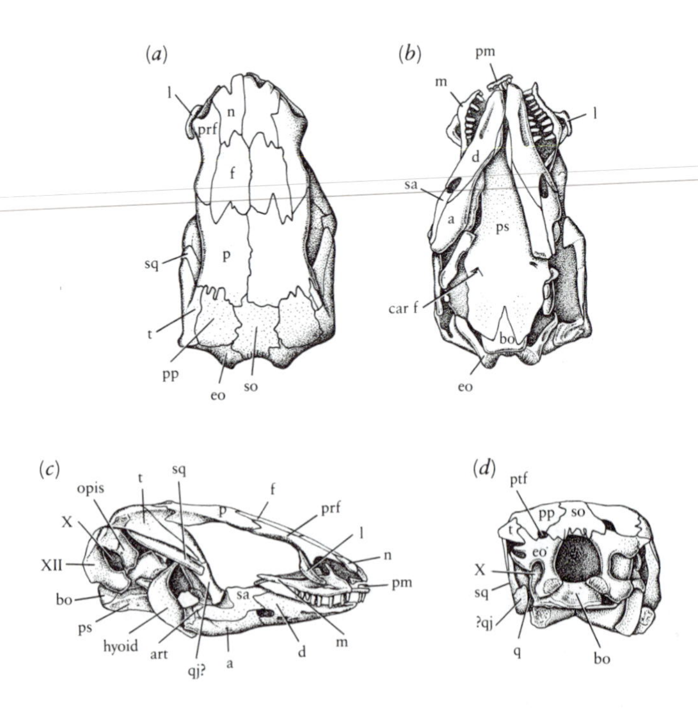

Figure 9-31. *Lysorophus*, from the Lower Permian. Skull in (*a*) dorsal, (*b*) palatal, (*c*) lateral, and (*d*) occipital views, ×2. Abbreviations same as Figure 8-3, plus: carf, carotid foramen; ptf, posttemporal foramen. *From Carroll and Gaskill, 1978.*

MODERN AMPHIBIANS

STRUCTURE AND BIOLOGY OF FROGS, SALAMANDERS AND GYMNOPHIONA

We have not found any lepospondyl fossils later than the Lower Permian. There is a surprising gap in the record of small amphibians until the appearance of frogs and salamanders within the Jurassic. The survivors of the small Paleozoic amphibians may have been extremely rare from the late Permian into the middle Mesozoic. Or perhaps this stage of amphibian evolution remains obscure because appropriate conditions for the preservation of small amphibians were uncommon during this time or restricted to regions where little collecting has been done.

When they first appear in the fossil record during the Jurassic, both frogs and salamanders appear essentially modern in their skeletal anatomy (Figures 9-32 and 9-33). The described fossil record of gymnophionans (caecilians) is limited to isolated vertebrae from the Upper

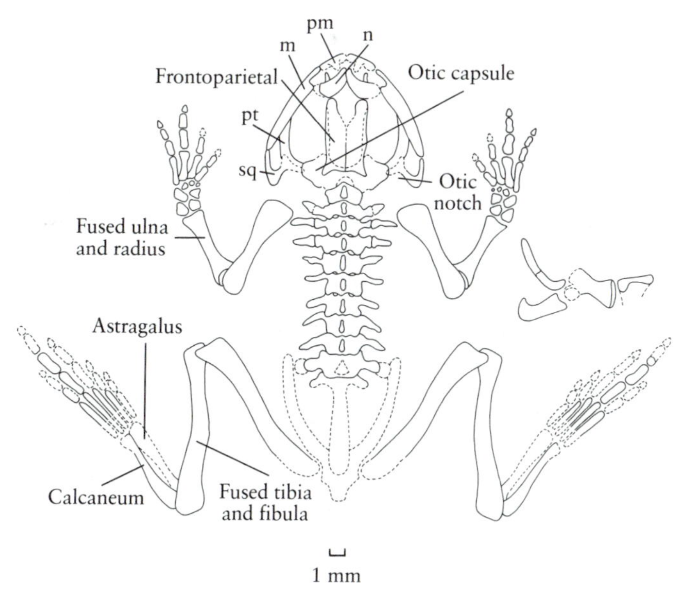

Figure 9-32. The oldest known frog, *Vieraella,* from the Lower Jurassic of South America. Shoulder girdle in ventral view drawn to right of skeleton. Abbreviations same as Figure 8-3. *From Estes and Reig, 1973. By permission of the University of Missouri Press. Copyright 1983 by the curators of the University of Missouri.*

Cretaceous and Paleocene that are very similar to those of modern genera (Rage, 1986).

Frogs, salamanders, and caecilians share a number of derived features that have led herpetologists to ally them in a single group, the Lissamphibia. Parsons and Williams (1963) are the most forceful advocates of this arrangement. All are small animals that live in damp environments and depend on cutaneous respiration. All have spool-shaped vertebrae. Most members of all three groups have a peculiar tooth structure in which the base and crown are separated by a zone of fibrous tissue. Such teeth are termed **pedicellate.**

Most frogs and salamanders have a very open skull, in contrast with that of Paleozoic amphibians. Many terrestrial frogs and salamanders possess a protrusible tongue and an extra ear ossicle called the operculum adjacent to the base of the stapes. The operculum has a thin muscular attachment to the shoulder girdle.

Despite these similarities, frogs, salamanders, and caecilians are very different from one another in skeletal structure and ways of life, both now and throughout their known fossil history.

Frogs are the most readily characterized, because they all adhere to a single skeletal pattern that is among the most specialized found in any vertebrate order. The vertebral column is greatly reduced, with only five to nine trunk vertebrae and no caudal vertebrae in the adult stage. The region immediately posterior to the sacrum is in the form of a longitudinal rod, the urostyle. The rear limbs are emphasized and both terrestrial and aquatic locomotion is typically produced by their symmetrical movement. The skull is typically open, in contrast with the labyrinthodonts. Its main support is provided by the braincase which forms a stout longitudinal bar. Posteriorly, the otic capsules extend laterally to reach the dorsal portion of the cheek. There is a bony connection between the maxilla and the squamosal and quadrate, although the jugal is lost. The palate exhibits very large interpterygoid vacuities. In most genera, there is a large tympanum supported by a tympanic annulus set into a notch in the squamosal. Vibrations of the tympanum are transmitted to the inner ear via a thin stapes, or columella. This condition is almost certainly primitive for frogs and could readily have been derived from the pattern seen in advanced terrestrial temnospondyls (Bolt and Lombard, 1985).

Both frogs and salamanders evolved specialized means to protrude the tongue, but the mechanism differs in the two groups. Regal and Gans (1976) showed that in advanced frogs the muscles at the front of the lower jaw

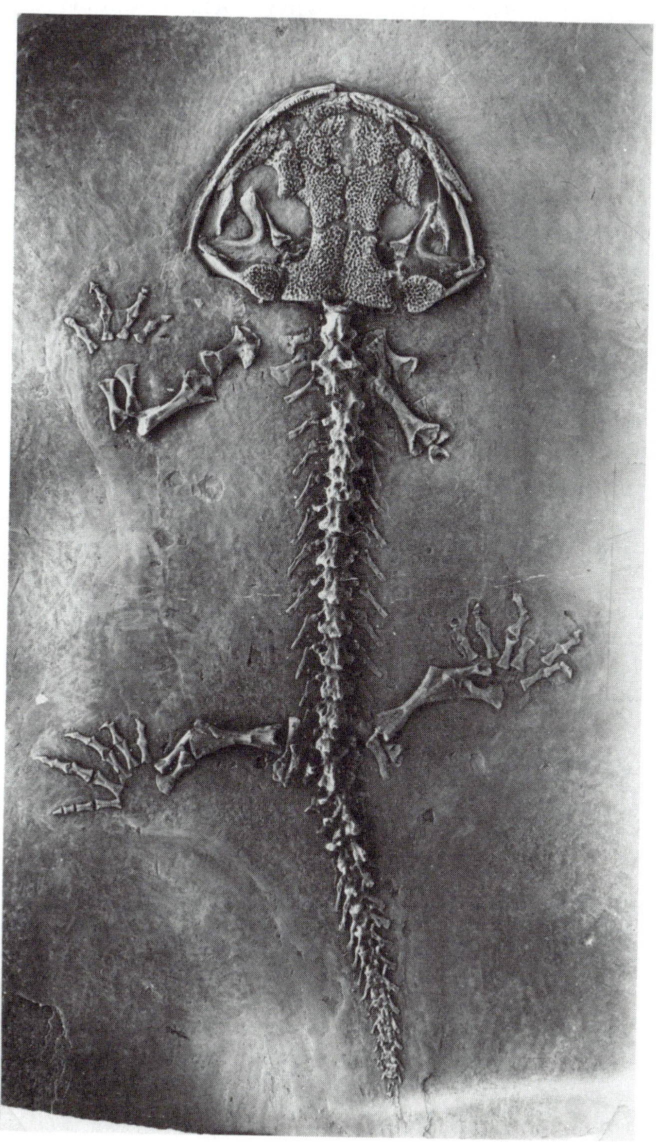

Figure 9-33. The oldest known salamander, *Karaurus,* from the Upper Jurassic of Russia, slightly less than natural size. *Photograph courtesy of M. F. Ivachnenko.*

and within the tongue flip the back of the tongue forward and out of the mouth. In contrast with salamanders, the hyoid apparatus is not involved in tongue protrusion.

One of the most striking specialization of frogs is the evolution of a larval stage that is radically different in structure and biology from the adult. Tadpoles characterize all frog groups. Even in genera in which direct development occurs, early stages retain the appearance of a tadpole. They differ from the adults not only in structure but also in diet; most are herbivorous suspension feeders and algae grazers, in contrast with the carnivorous adults. Fossil tadpoles are known as early as the Lower Cretaceous (see Figure 9-11). Despite the high degree of specialization of the larval stage, nearly all frogs are primitive in having external fertilization.

We may consider salamanders to be the least specialized of living amphibians because of their conservative body proportions. Most forms have small limbs, and most of the power for locomotion is generated by the axial musculature. Locomotor specializations within the group include loss of the pelvic limbs in the Sirenidae and elongation of the trunk and limb reduction in the Amphiumidae and some plethodontids.

The skull of salamanders is modified in a manner analogous with that of frogs, with the main support provided by the braincase; the otic capsules attach posterolaterally to the dorsal margin of the cheek. In contrast with most frogs, the maxilla is separated from the squamosal and quadrate by a long gap. In primitive genera, the squamosal, quadrate, and pterygoid form a suspensorium that is moveable on the braincase in a manner somewhat analogous with that of primitive teleosts. There is never an otic notch, tympanum, or middle ear cavity. The stapes is a relatively massive element, primitively linking the movable cheek to the braincase.

Lombard and Wake (1976) described how terrestrial salamanders protrude their tongue using the hyoid apparatus. Muscles attached to the ceratohyals pull them forward and thrust the more anterior portion of the hyoid apparatus, to which the tongue is attached, out of the mouth.

The reproductive pattern of primitive salamanders probably resembles that of Paleozoic amphibians. External fertilization is the rule, and the larvae are similar to the adults in general anatomy, except for the presence of external gills. In advanced aquatic and terrestrial groups, the male deposits a spermatophore that is picked up by the cloacal lips of the females, so that fertilization is internal. Complicated courtship coordinates the sexual behavior of the males and females. In one advanced salamander family, the Plethodontidae, many genera lay their eggs on land and development is direct; the young are miniature replicas of the adults with no specifically larval features. The European salamandrid *Salamandra atra* gives birth to live young.

Caecilians are the least-well-known of the living amphibians; all are restricted to the damp tropics and are aquatic or burrowing in habit. They have no trace of limbs or girdles, and although the trunk region may have more than 200 vertebrae, the tail is very short or absent. They are unlike frogs and salamanders in having well-developed ribs throughout the column, and in most genera the skull is solidly roofed as in Paleozoic amphibians (Figure 9-34). The orbital openings are small and may be completely covered with bone. A specialized tactile or chemosensory organ, the tentacle, extends from the skull anterior to the area of the orbit. Caecilians all practice internal fertilization. In contrast with frogs and salamanders, the males possess a copulatory organ.

All the structural and physiological differences between frogs, salamanders, and apodans may be attributed to divergence. These differences do not in themselves preclude evolution from a single ancestral group. However, we have found no fossil evidence of any possible antecedents that possessed the specialized features common to all three modern orders. Some of the features shared by the modern orders may have evolved separately in each, and others may have been inherited from a pattern common to all early amphibians.

Although cutaneous respiration may have been a common attribute of rhipidistians and primitive tetrapods, its emphasis in living amphibians is certainly a specialization. The efficiency of cutaneous respiration depends on a large surface-to-volume ratio and might be elaborated in any small vertebrate that did not evolve an impervious skin.

One of the most convincing arguments to support the derivation of frogs, salamanders, and caecilians from a common ancestor is the presence of pedicellate teeth. We have not found pedicellate marginal teeth in any Paleozoic amphibians, which suggests that they may have been uniquely evolved in the ancestors of the modern orders.

Larsen and Guthrie (1975) argued that terrestrial salamanders evolved pedicellate teeth in association with their use of the tongue to capture and manipulate prey. If the teeth pierce the prey, they can be broken off readily at the jaw margin so that they do not interfere with further tongue manipulation. This argument may be extended to terrestrial frogs, which also rely on the tongue in feeding. The great differences in the mechanism of tongue protrusion in frogs and salamanders indicate that it evolved separately in the two groups. This suggests that pedicellate teeth may have evolved independently as well. Estes (1981) described nonpedicellate teeth in several genera of primitive fossil salamanders which indicates that this specialization may have arisen within the urodeles rather than being retained from an ancestral condition. Larsen and Guthrie's argument does not account for the presence of pedicellate teeth in caecilians, in which all but the tip of the teeth are embedded in a thick layer of spongy tissue.

Another specialized feature of frogs and salamanders (but not caecilians) is the presence of an opercularis muscle that links the shoulder girdle with the otic capsule. It

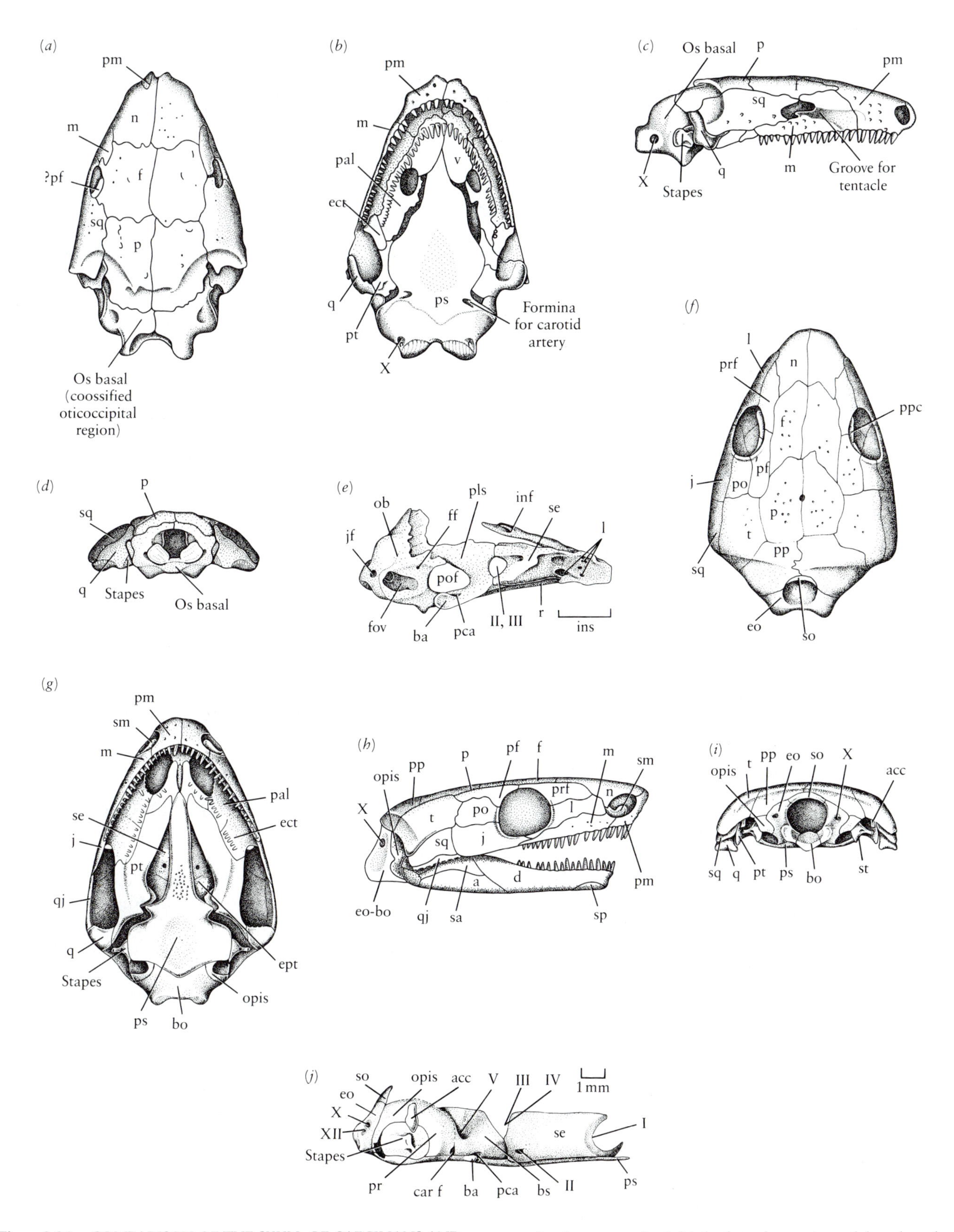

Figure 9-34. COMPARISON OF THE SKULL OF CAECILIANS AND THE MICROSAUR *RHYNCHONKOS.* (*a–c*) The modern caecilian *Grandisonia* in dorsal, palatal, and lateral views, ×5. (*d*) Occiput of the caecilian *Hypogeophis.* (*e*) Braincase of *Oscaecilia* in lateral view. (*f–j*) *Rhynchonkos* in comparable views. Abbreviations as in Figure 8-3 plus: acc, accessory ear ossicle; ba, basicranial articulation; carf, carotid foramen; eo-bo, exoccipital-basioccipital complex; ff, facial foramen; fov, fenestra ovalis; inf, infra frontal extensions of the sphenethmoid; ins, internasal septum; jf, jugular foramen; ob, os basale; opis, opisthotic; pca, palatine canal; pls, pleurosphenoid; pof, proötic foramen; ppc, palpebral cup; pr, proötic; r, rostral process of parasphenoid; se, sphenethmoid; soc, supraoccipital; st, stapes. *From Carroll and Currie, 1975. With permission from the Zoological Journal of the Linnean Society. Copyright 1975 by the Linnean Society of London.*

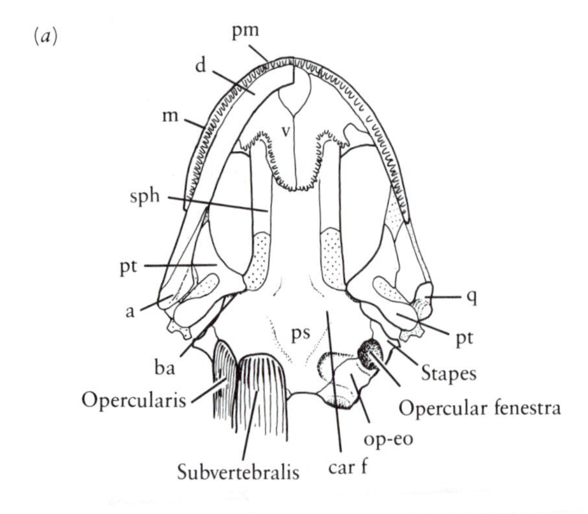

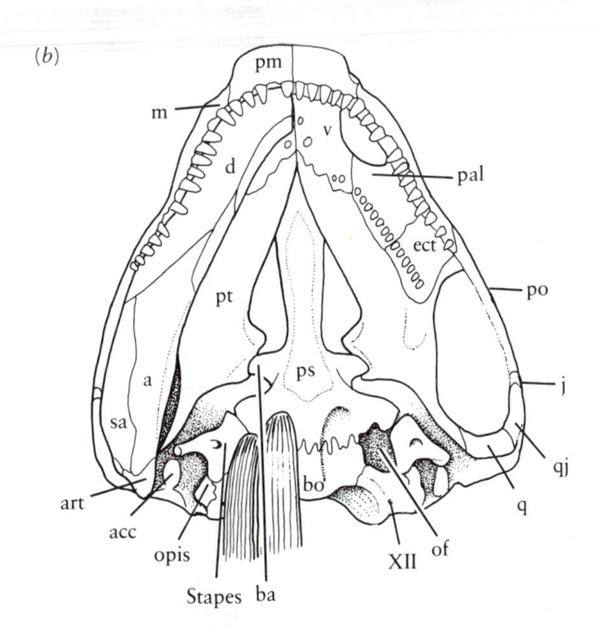

Figure 9-35. (*a*) Position of the opercularis muscle in the primitive salamander *Hynobius*. The operculum is not ossified in this genus. (*b*) The microsaur *Micraroter* with the opercularis muscle restored in a comparable position relative to an unossified portion of the otic capsule and adjacent axial musculature. Abbreviations as in Figure 8-3, plus op-eo, fused otic capsule and exoccipital; of, opercular fenestra. *From Carroll and Holmes, 1980. With permission from the Zoological Journal of the Linnean Society. Copyright 1980 by the Linnean Society of London.*

is usually attached to a small bone, the operculum, at the base of the stapes (Figure 9-35*a*). These structures have long been assumed to be associated with hearing, but Capranica's (1976) description of the function of the inner ear in frogs indicates that displacement of a segment of the wall of the otic capsule by contraction of the opercularis muscle would interfere with reception of sound conducted by the stapes.

Experimental evidence reported by Baker (1969) indicates that severing this muscle interferes with balance in frogs. Equivalent experiments have not been done with salamanders. The orientation of the muscle in frogs and salamanders would make it an effective means of determining the degree of twisting of the body relative to the head. Such a system does not appear to be necessary in advanced tetrapods, which have well-developed stretch receptors in all the somatic muscles. As indicated earlier in the chapter, stretch receptors probably were not present in ancestral amphibians; they are absent in the axial musculature of all three groups of modern amphibians (Bone, Ridge, and Ryan, 1976). The operculum-opercularis system may have been a primitive sensory system that was evolved by early amphibians and has been retained in both frogs and salamanders but lost in more advanced tetrapod groups. We attribute its absence in caecilians to their lack of a shoulder girdle.

An unossified area in the ventral surface of the otic capsule in several microsaurs may be an opening for the operculum (Figure 9-35*b*), but there is no evidence of this bone in other primitive tetrapods. The opercularis muscle may initially have attached to the stapes but shifted to a separate ossicle when the stapes became important in sound conduction.

In the absence of fossil evidence that frogs, salamanders, and caecilians evolved from a close common ancestor, we must consider the possibility that each of the modern orders evolved from a distinct group of Paleozoic amphibians.

THE ORIGIN AND DIVERSIFICATION OF FROGS

The only fossil that provides a link between any of the modern orders and Paleozoic amphibians is *Triadobatrachus* from the Lower Triassic of Madagascar (Figure 9-36) (Estes and Reig, 1973). The skull is frog-like with a median fronto-parietal, and the squamosal is recessed to support a tympanum. There are 14 trunk and 6 caudal vertebrae. The ilium is specialized like that of frogs, but to a lesser degree. Their closest similarity to Paleozoic amphibians is with the dissorophid temnospondyls, particularly the genus *Doleserpeton* (Figure 9-37) (Bolt, 1977). In the family Dissorophidae, the otic notch is particularly large, and the number of trunk vertebrae and the length of the tail is reduced (see Figure 9-18). There are no known features that would preclude the dissorophids from having given rise to frogs. One important feature that immature dissorophids share with salamanders and caecilians, as well as frogs, is the bicuspid configuration of the teeth, as Bolt (1977) demonstrated. This configuration may have evolved to assist in holding, but not piercing, the prey. However, the remainder of the anatomy of dissorophids provides no evidence of the initiation of trends leading to either salamanders or caecilians.

Modern frogs are classified in 23 families according to Frost (1985), most of which have a long, but somewhat discontinuous, fossil record (see Figure 9-1) (Vial, 1973).

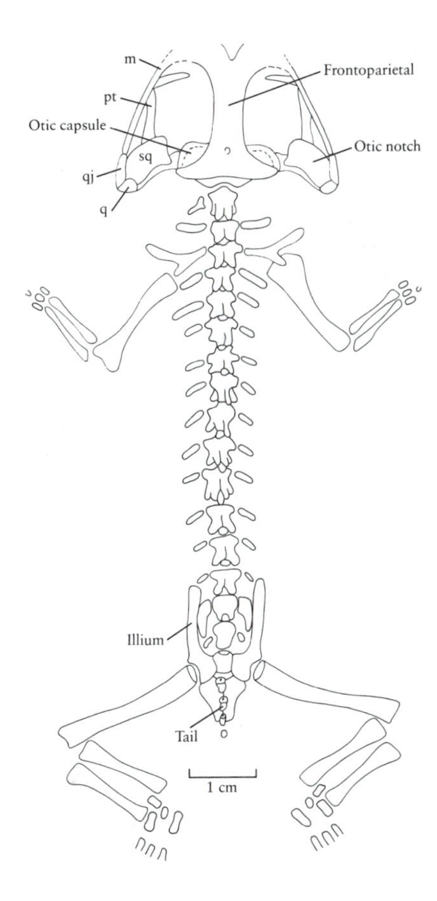

Figure 9-36. *Triadobatrachus* from the Lower Triassic of Madagascar. This genus provides a plausible link between Lower Permian dissorophids and primitive frogs. Abbreviations as in Figure 8-3. *From Estes and Reig, 1973. By permission of the University of Missouri Press. Copyright 1983 by the curators of the University of Missouri.*

Families that have been considered primitive on the basis of their skeletal morphology—Ascaphidae, Discoglossidae, Pipidae, and the extinct Palaeobatrachidae—have a long fossil record that goes back to the Jurassic or the base of the Cretaceous. Rhinophrynids, which are also considered primitive on the basis of their tadpole morphology, appear in the Paleocene. Pelobatids, which are generally accepted as intermediate in terms of the morphology of the living genera, have a record that extends to the top of the Cretaceous. Microhylidae, although specialized in some features, may have diverged from the other frog families at a comparatively early stage, but their fossils are not known prior to the Lower Miocene.

Leptodactylids, bufonids, and hylids—all of which are structurally more modern frogs—appear in the Paleocene, where they are represented by members of modern species groups. These families were probably all present in the Cretaceous and may have diverged as early as the Jurassic.

We have no record of the family Ranidae before the Eocene, and almost all fossils can be placed in the genus *Rana*. Modern species of *Rana* appear at the base of the Miocene in North America (Carroll, 1977).

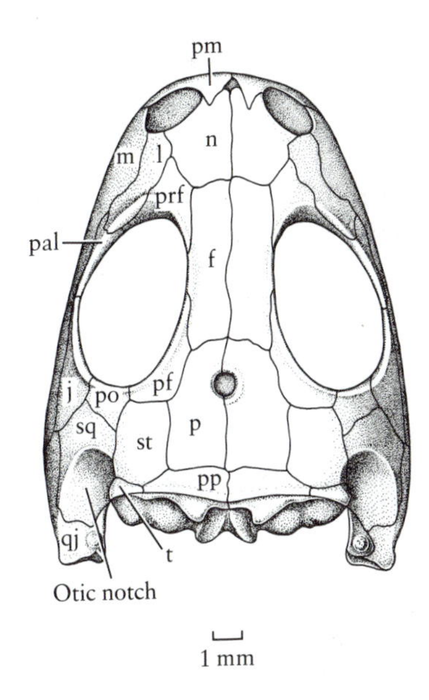

Figure 9-37. Skull roof of the dissorophid *Doleserpeton*, which may be close to the ancestry of frogs. Abbreviations as in Figure 8-3. *From Bolt, 1977.*

SALAMANDER ANCESTRY AND INTERRELATIONSHIPS

It has been assumed that the similarities of the skulls of frogs and salamanders were the result of derivation from a common ancestor that had already reduced the extensive ossification characteristic of Paleozoic amphibians. Carroll and Holmes (1980) have described consistent differences in the configuration of the skull and adductor jaw musculature between all living frogs and salamanders that suggest that reduction in dermal ossification proceeded separately in the ancestors of the two groups.

In amphibians and most reptiles, the adductor jaw musculature may be subdivided into three major units on the basis of their relationships to branches of the Vth nerve (Figure 9-38). The adductor mandibulae internus lies be-

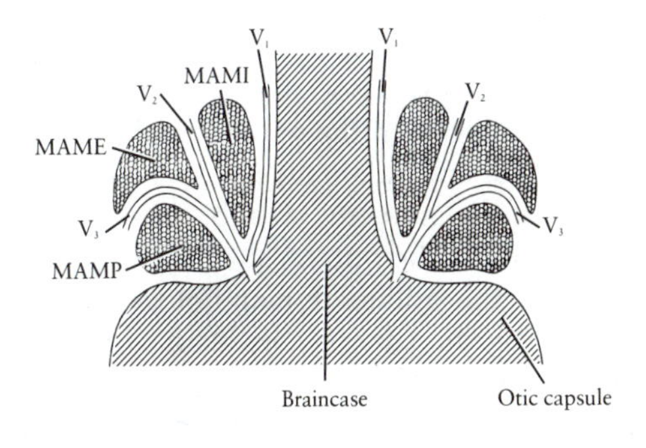

Figure 9-38. Diagram showing the major units of the adductor jaw muscles of amphibians and reptiles. Drawn as a horizontal section through the back of the skull seen in dorsal view. Area of diagonal shading represents the braincase. Cross sections of the muscles are indicated by polygonal shading. Abbreviations: V_1, V_2, V_3, branches of the Vth nerve. Abbreviations for muscles as in Figure 9-39. *From Carroll and Holmes, 1980. With permission from the Zoological Journal of the Linnean Society. Copyright 1980 by the Linnean Society of London.*

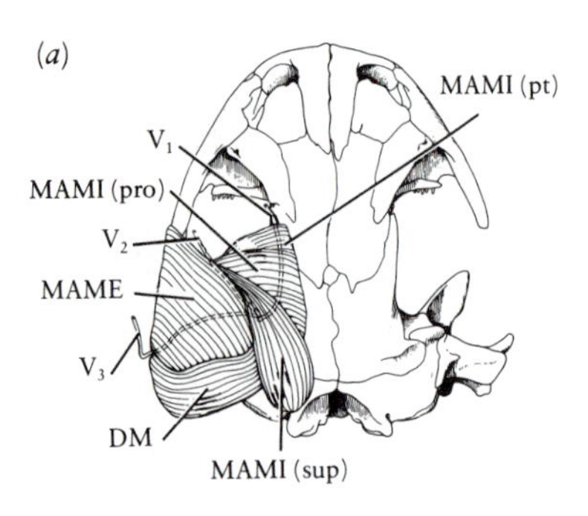

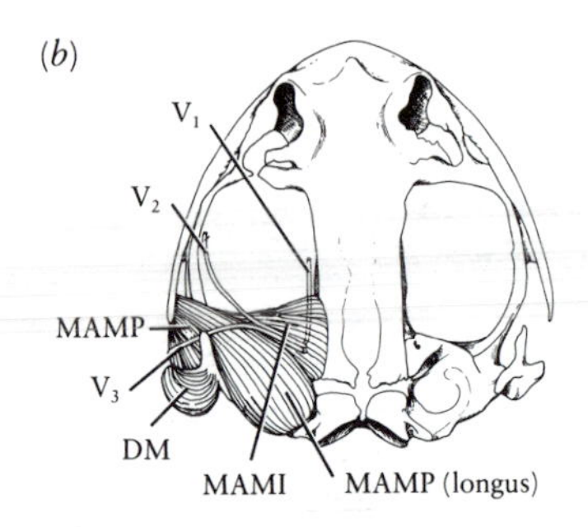

Figure 9-39. PATTERN OF JAW MUSCULATURE. (*a*) *Ambystoma*, a representative salamander and (*b*) *Ascaphus*, a primitive frog. Abbreviations: V_1, V_2, V_3, branches of the Vth nerve; DM, depressor mandibulae; MAME, adductor mandibulae externus (large in salamanders and missing in most frogs); MAMI, adductor mandibulae internus; MAMI (pro), profundus head of MAMI; MAMI (pt), pterygoideus head of MAMI; MAMI (sup), superficialis head of MAMI (large in salamanders, but missing in frogs); MAMP, adductor mandibulae posterior (elaborated in frogs, but not in salamanders); MAMP (longus), longus head of MAMP. *With permission from the Linnean Society. Copyright 1980 by the Linnean Society of London.*

tween the ophthalmic and maxillary branches of the Vth nerve, the adductor mandibulae externus lies between the maxillary and mandibular branches, and the adductor mandibulae posterior lies behind the mandibular branch.

From this basic pattern, frogs and salamanders show divergent specializations (Figure 9-39). Because of the prominence of the otic notch in frogs, the area for origin of the adductor mandibulae externus is much reduced, and this muscle is missing in most genera. In contrast, this is one of the principal muscles in salamanders. Its elaboration is probably associated with the absence of a bony connection between the maxilla and jugal.

In both frogs and salamanders, a major muscle has expanded out of the adductor chamber and originates on the dorsal surface of the otic capsule. In frogs, this is the longus head of the adductor mandibulae posterior; in salamanders, it is the superficialis head of the adductor mandibulae internus. The spread of this muscle onto the upper surface of the braincase could not have occurred prior to the loss of the dermal bone behind the orbit. It seems unlikely that different muscles would have spread onto this area if frogs and salamanders had an immediate common ancestor that had already developed a very open skull. Both the frog and salamander patterns of the jaw musculature could have originated from that which is assumed to be primitive for all tetrapods, but there is little likelihood that the pattern in either modern group could have given rise to that of the other.

Because of the conspicuous otic notch, large orbit, and solid cheek in small temnospondyl amphibians such as the dissorophids, their adductor chamber resembles that of frogs (Figure 9-40). We may logically assume a similar distribution of the jaw musculature. Some aspects of the skull pattern typical of salamanders are seen in a number of genera of microsaurs that have an embayment of the cheek, with the maxilla not connected with the quadrate (Figure 9-41). Some species show a groove on the lateral surface of the lower jaw for the insertion of a very large adductor mandibulae externus. The microsaurs lack an otic notch, and the distribution of the adductor mandibulae internus and posterior would be expected to follow the pattern of salamanders rather than that of frogs. However, the adductor mandibulae internus superficialis would not be able to extend over the otic capsule, since the skull roof is still complete. An early stage in the posterior migration of this muscle is seen in the Upper Jurassic salamander *Karaurus* (Figure 9-42), in which the occipital margin of the otic capsule is still covered by ornamented dermal bone.

Even the earliest fossil salamanders have an essentially modern morphology, and we do not know any fossils that connect them with microsaurs. Nevertheless, the general similarity of the skull between the most primitive salamanders and microsaurs is much closer than it is between frogs and salamanders or between salamanders and other groups of Paleozoic amphibians.

Salamanders probably evolved during the later Permian and Triassic. Naylor (1980) showed that the radiation of the modern salamander families must have occurred by the late Mesozoic. We know seven of the eight living families by the Paleocene. The plethodontids are not known until the Lower Miocene but sirenids and amphiumids are known in the late Cretaceous. Proteiids are questionably reported from the Upper Jurassic (Estes, 1981).

In terms of their cranial morphology, hynobiids are the most primitive of living salamanders and closest to the pattern of Paleozoic microsaurs. We consider cryptobranchids to be closely related since both groups practice external fertilization. The basic cranial anatomy of primitive ambystomatids and salamandrids closely resembles that of the hynobiids. Plethodontids probably evolved from ambystomatids early in the Cenozoic. The proteids, amphiumids, and sirenids are all very highly modified in their cranial anatomy from the pattern of hynobiids, but in divergent ways that give no evidence of specific relationships. The extremely modified anatomy of these families may be attributed to major changes in their pattern of early development.

Gymnophionans (caecilians)

Carroll and Currie (1975) argued that the cranial anatomy of recent caecilians is quite similar to that of one family

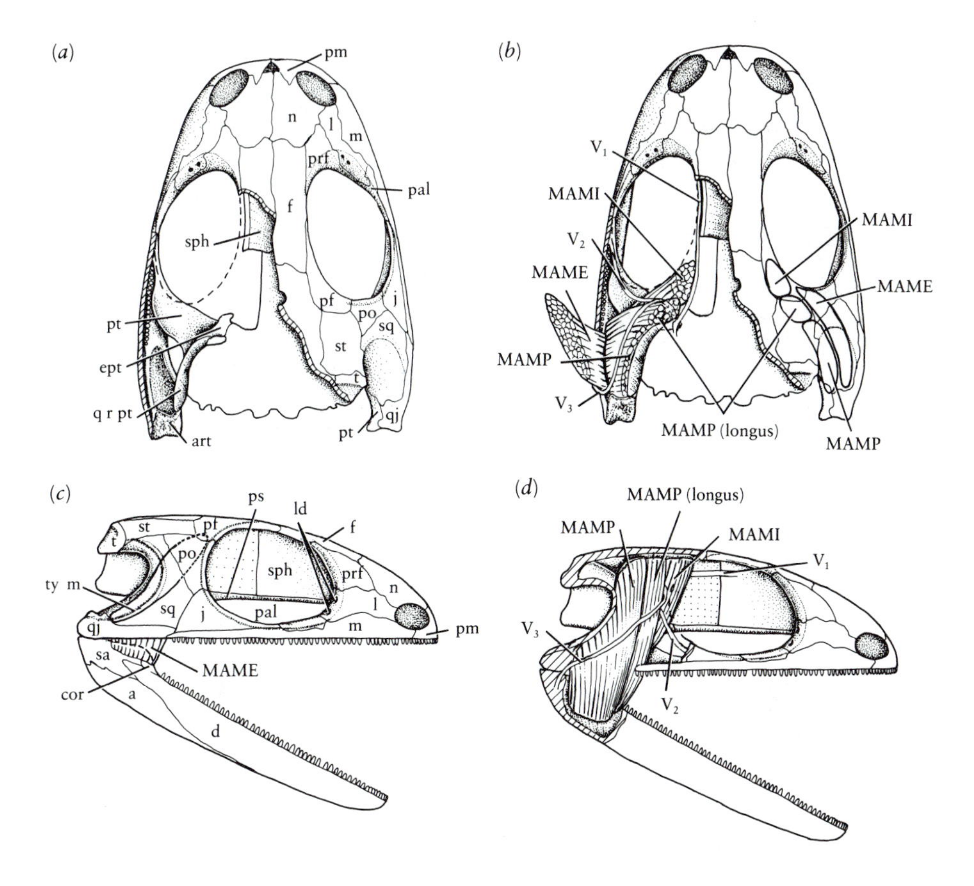

Figure 9-40. RESTORATION OF THE JAW MUSCULATURE OF THE TEMNOSPONDYL *DOLESERPETON*, BASED ON SKELETAL SIMILARITIES TO MODERN FROGS. (*a*) Cutaway view of skull to show dorsal surface of palate. (*b*) Restoration of jaw musculature in dorsal view. (*c*) Lateral view of skull. (*d*) Cutaway view of skull in lateral view to show jaw muscles. Abbreviations as in Figures 8-3 and 9-39, plus ld, lacrimal duct; q r pt, quadrate ramus of pterygoid; ty m, line of attachment for tympanic membrane. *From Carroll and Holmes, 1980. With permission from the Zoological Journal of the Linnean Society. Copyright 1980 by the Linnean Society of London.*

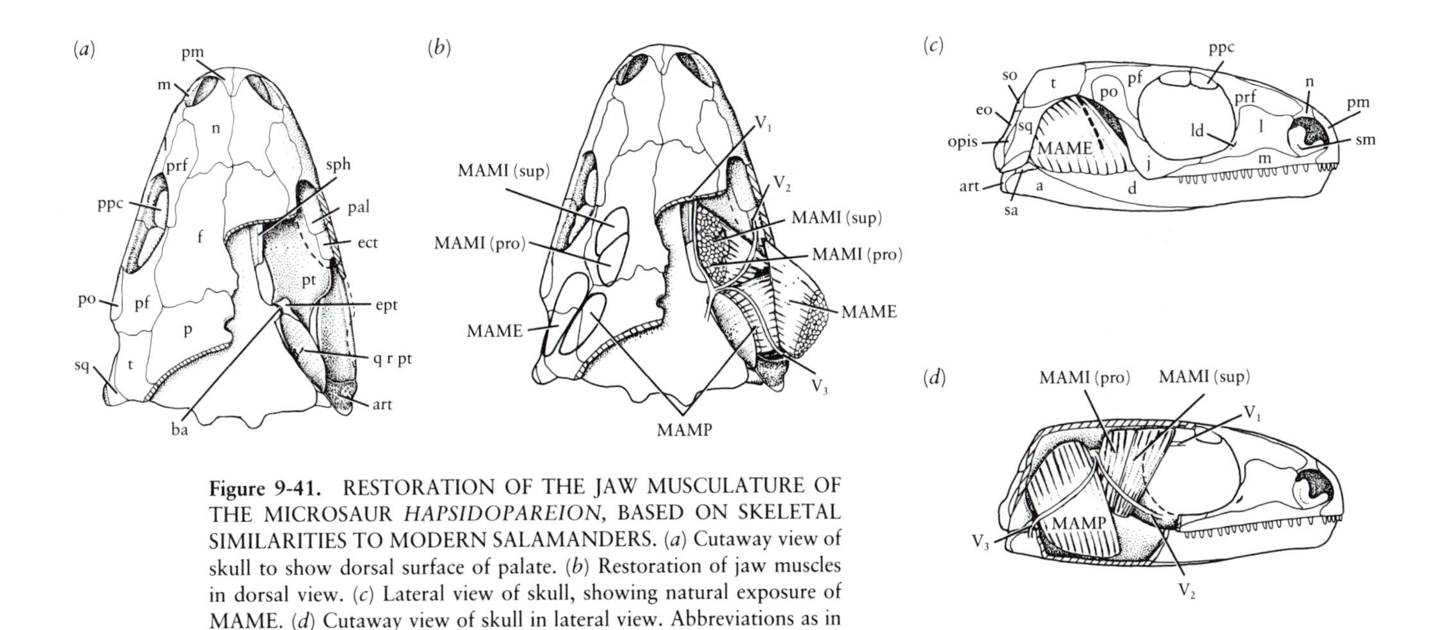

Figure 9-41. RESTORATION OF THE JAW MUSCULATURE OF THE MICROSAUR *HAPSIDOPAREION*, BASED ON SKELETAL SIMILARITIES TO MODERN SALAMANDERS. (*a*) Cutaway view of skull to show dorsal surface of palate. (*b*) Restoration of jaw muscles in dorsal view. (*c*) Lateral view of skull, showing natural exposure of MAME. (*d*) Cutaway view of skull in lateral view. Abbreviations as in Figure 9-40. *From Carroll and Holmes, 1980. With permission from the Zoological Journal of the Linnean Society. Copyright 1980 by the Linnean Society of London.*

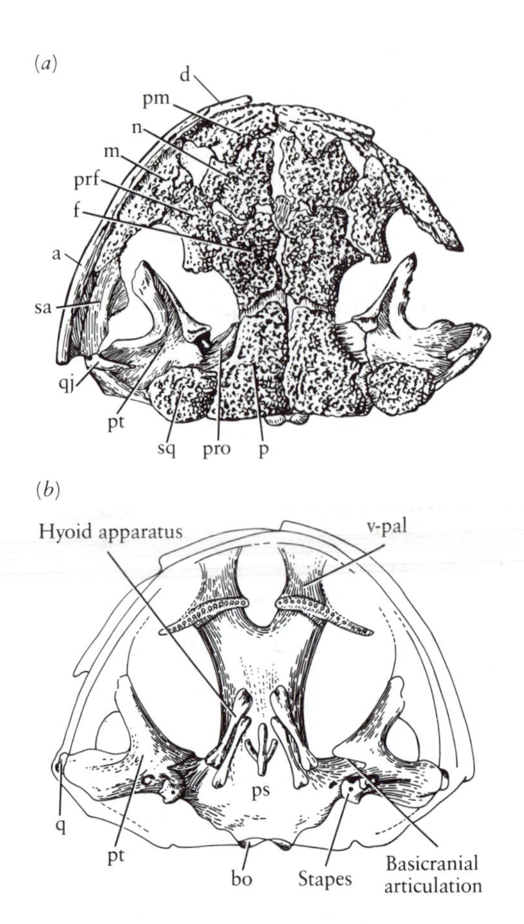

Figure 9-42. (*a*) Dorsal and (*b*) ventral views of the Upper Jurassic salamander *Karaurus*, $\times 1\frac{1}{2}$. The right lower jaw has been omitted from the drawing to emphasize the separation of the suspensorium from the maxilla. Abbreviations as in Figure 8-30, plus v-pal, coossified vomer and palatine. *From Ivachnenko, 1978.*

of microsaurs, the Goniorhynchidae (see Figure 9-34). In both groups, the skull is fully roofed, the jaw articulation is well anterior to the occiput, the snout overhangs the tooth row, and both the palate and the lower jaw have a medial row of teeth parallel with the marginal dentition. In both groups, the stapes extends from the otic region to the quadrate in a similar manner. The caecilians have a solid lateral wall to the braincase in the area of the opening for the Vth nerve, which is unique among modern amphibians. In *Rhynchonkos* [*Goniorhynchus*], this area is walled by a specialized extension from the basisphenoid.

The particular pattern of the trunk vertebrae in caecilians could have arisen by the fusion of the intercentrum of goniorhynchids with the anterior margin of the pleurocentrum. *Rhynchonkos* shows a greater degree of limb reduction than other microsaurs and has approximately 36 trunk vertebrae (see Figure 9-30*d*).

We may attribute many of the anatomical similarities between *Rhynchonkos* and caecilians to adaptation to a burrowing way of life in both groups. However, in the absence of any more positive evidence of their ancestry, it is as reasonable to assume that caecilians evolved from microsaurs with a generally similar morphology as it is to postulate that they are derived from the ancestors of frogs or salamanders, which are known to have adapted to entirely different ways of life.

Wake and Hanken (1982) reached an entirely different conclusion in a recent study of the embryological development of the skull in the caecilians *Dermophis*. They argued that the extensive dermal ossification in this group developed secondarily from a more open configuration and that they could not be derived directly from any of the known groups of Paleozoic amphibians. We will have difficulty establishing the phylogenetic position of the Gymnophiona with any assurance until relevant fossils are found in the early or middle Mesozoic.

Modern caecilians are classified in four families, with a total of approximately 34 genera, none of which has a fossil record (Taylor, 1968; Nussbaum, 1977). The geographical distribution of the families does not conform with the modern arrangement of the continents. This may reflect an original distribution prior to the establishment of the modern continental pattern, which suggests an early Mesozoic or late Paleozoic origin for the group.

SUMMARY

Paleozoic amphibians form a phylogenetic link between the rhipidistian fish and more advanced tetrapods. The earliest amphibians, the Upper Devonian ichthyostegids, retain many primitive features of the skull, vertebrae, and girdles from their fish ancestors but have fully developed limbs for terrestrial locomotion.

Changes in the skeleton between rhipidistian fish and early amphibians are related to the requirements of support, feeding, and locomotion on land. Changes in the physiology of water balance and the structure of the sense organs were also necessary to adapt to life out of the water. Most amphibians lay their eggs in the water or in damp places on land, and many have a long aquatic larval stage.

The fossil record of amphibians in the Lower Carboniferous is very incomplete and little is known of the specific interrelationships of the numerous lineages. We recognize two major groups of Paleozoic amphibians, the labyrinthodonts, which retain many of the features of their rhipidistian ancestors, and the smaller, more specialized lepospondyls.

Labyrinthodonts were typically large animals, up to a meter in length, with an anatomy that was quite unlike modern amphibians. They occupied a spectrum of terrestrial, semiaquatic, and secondarily aquatic habitats. The most common labyrinthodonts were the temnospondyls, which are typified by multipartite vertebral centra that retain large intercentra. They were represented in the late Carboniferous and early Permian by a number of terrestrial families, but in the later Permian and Triassic, they are known primarily by secondarily aquatic groups.

The anthracosaurs, which are characterized by large pleurocentra, include the aquatic embolomeres and the more terrestrial gephyrostegids and seymouriids. The solenodonsaurids, limnoscelids, and tseajiids share several derived features with primitive reptiles. We have not established their specific relationships.

Several very distinct groups of Paleozoic amphibians are grouped as lepospondyls. Their origin and interrelationships remain unknown. They include the snake-like aïstopods, the newt-like nectrideans, and the microsaurs. All have spool-shaped pleurocentra and lack the palatine fangs and otic notches that characterize labyrinthodonts. These features may all be associated with small body size and do not necessarily support common ancestry.

Most groups of Paleozoic amphibians became extinct by the end of the Triassic; one genus survives into the Middle Jurassic.

The radiation of the modern amphibian groups began in the early Mesozoic. A possible ancestor of the frogs, *Triadobatrachus,* is known in the early Triassic. Members of several modern frog families are present in the Jurassic. Salamanders are known since the late Jurassic and caecilian vertebrae are reported from the Upper Cretaceous and Paleocene.

Triadobatrachus provides a plausible link between Paleozoic labyrinthodonts and frogs. The origin and relationships of salamanders and caecilian remain uncertain. The presence of pedicellate teeth in all three groups and the operculum-opercularis complex in frogs and salamanders suggest that the modern amphibian groups share a common ancestry from among the Paleozoic amphibians. However, no fossils are known that support this hypothesis, and all three groups may have evolved separately from distinct ancestral groups.

Frogs almost certainly evolved from temnospondyl labyrinthodonts, with which they share an impedance-matching ear. Salamanders and caecilians may have evolved from different groups of microsaurs. The separate origin of the three groups is supported by the different pattern of the jaw musculature in modern frogs, salamanders, and caecilians which suggests that each evolved separately from ancestors that retained a solidly roofed temporal region.

REFERENCES

Andrews, S. M., and Westoll, T. S. (1970a). The postcranial skeleton of *Eusthenopteron foordi* Whiteaves. *Trans. Roy. Soc. Edinb.,* **68**: 207–329.

Andrews, S. M., and Westoll, T. S. (1970b). The postcranial skeleton of rhipidistian fishes excluding *Eusthenopteron. Trans. Roy. Soc. Edinb.,* **68**: 391–489.

Baker, M. C. (1969). The effect of severing the opercularis muscle on body orientation of the leopard frog *Rana pipiens. Copeia,* **3**: 613–617.

Beaumont, E. H. (1977). Cranial morphology of the Loxommatidae (Amphibia: Labyrinthodontia). *Phil. Trans. Roy. Soc. Lond., B,* **280**: 29–101.

Beerbower, J. R. (1963). Morphology, paleoecology and phylogeny of the Permo-Pennsylvanian amphibian *Diploceraspis. Bull. Mus. Comp. Zool.,* **130**: 31–108.

Bernacsek, G. N., and Carroll, R. L. (1981). Semicircular canal size in fossil fishes and amphibians. *Can. J. Earth Sci.,* **18**: 150–156.

Bolt, J. R. (1969). Lissamphibian origins: Possible protolissamphibians from the Lower Permian of Oklahoma. *Science,* **166**: 888–891.

Bolt, J. R. (1977). Dissorophid relationships and ontogeny, and the origin of the Lissamphibia. *J. Paleont.,* **51**: 235–249.

Bolt, J. R., and Lombard, R. E. (1985). Evolution of the amphibian tympanic ear and the origin of frogs. *Biol. J. Linn. Soc.,* **24**: 83–99.

Bone, Q., Ridge, R. M. A. P., and Ryan, K. P. (1976). Stretch receptors in urodele limb muscles. *Cell Tiss. Res.,* **165**: 249–266.

Bossy, K. V. (1976). Morphology, paleoecology, and evolutionary relationships of the Pennsylvanian urocordylid nectrideans (Subclass Lepospondyli, Class Amphibia). Unpublished thesis, Yale University.

Boy, J. A. (1971). Zur Problematik der Branchiosaurier (Amphibia, Karbon-Perm). *Palaeont. Z.,* **45**: 107–119.

Boy, J. A. (1972). Die Branchiosaurier (Amphibia) des saarpfaelzischen Rotliegenden (Perm, SW-Deutschland). *Abh. Hess. Landesamt, Bodenforschung,* **65**: 1–137.

Boy, J. A. (1974). Die Larven der rhachitomen Amphibien (Amphibia: Temnospondyli; Karbon-Trias). *Palaeont. Z.,* **48**: 236–268.

Boy, A. J., and Bandel, K. (1973). *Bruktererpeton fiebigi* n. gen. n. sp. (Amphibia: Gephrostegida) der erste Tetrapode aus dem rheinisch-westfaelischen Karbon (Namur B; W. Deutschland). *Palaeontographica, A,* **145**: 39–77.

Brough, M. C., and Brough, J. (1967). Studies on early tetrapods. *Phil. Trans. Roy. Soc. Lond., B,* **252**: 107–165.

Bystrow, A. P. (1944). *Kotlassia prima* Amalitzky. *Bull. Geol. Soc. America,* **55**: 379–416.

Capranica, R. R. (1976). Morphology and physiology of the auditory system. In R. Llinas and W. Precht (eds.), *Frog Neurobiology,* pp. 551–575. Springer-Verlag, Berlin-Heidelberg.

Carroll, R. L. (1964). Early evolution of the dissorophid amphibians. *Bull. Mus. Comp. Zool.,* **131**: 161–250.

Carroll, R. L. (1967). Labyrinthodonts from the Joggins Formation. *J. Paleont.,* **41**: 11–142.

Carroll, R. L. (1969). A new family of Carboniferous amphibians. *Palaeontology,* **12**(4): 537–548.

Carroll, R. L. (1970). The ancestry of reptiles. *Phil. Trans. Roy. Soc. Lond., B,* **257**: 267–308.

Carroll, R. L. (1977). Patterns of amphibian evolution: An extended example of the incompleteness of the fossil record. In A. Hallam (ed.), *Patterns of Evolution,* pp. 405–437. Elsevier Scientific Publications, Amsterdam.

Carroll, R. L. (1980). The hyomandibular as a supporting element in the skull of primitive tetrapods. In A. L. Panchen (ed.), *The Terrestrial Environment and the Origin of Land*

Vertebrates, pp. 293–317. Academic Press, London.

Carroll, R. L., and Currie, P. J. (1975). Microsaurs as possible apodan ancestors. *Zool. J. Linn. Soc.,* **57**: 229–247.

Carroll, R. L., and Gaskill, P. (1978). The Order Microsauria. *Mem. Am. Phil. Soc.,* **126**: 1–211.

Carroll, R. L., and Holmes, R. (1980). The skull and jaw musculature as guides to the ancestry of salamanders. *Zool. J. Linn. Soc.,* **68**: 1–40.

Chase, J. N. (1965). *Neldasaurus wrightae,* a new rhachitomous labyrinthodont from the Texas Lower Permian. *Bull. Mus. Comp. Zool.,* **133**: 153–225.

Clack, J. A. (1983). The stapes of the Coal Measures embolomere *Pholiderpeton scutigerum* Huxley (Amphibia: Anthracosauria) and otic evolution in early tetrapods. *Zool. J. Linn. Soc.,* **79**: 149–179.

Colbert, E. H., and Imbrie, J. (1956). Triassic metoposaurid amphibians. *Bull. Am. Mus. Nat. Hist.,* **110**: 399–452.

DeMar, R. E. (1968). The Permian labyrinthodont amphibian *Dissorophus multicinctus,* and adaptations and phylogeny of the family Dissorophidae. *J. Paleont.,* **42**: 1210–1242.

Dong, Z. (1985). A Middle Jurassic labyrinthodont (*Sinobrachyops placenticephalus* gen. et sp. nov) from Dashaupu, Zigong, Sichuan Province. *Vertebrata Palasiatica,* **23**: 301–307.

Estes, R. (1976). Middle Paleocene lower vertebrates from the Tongue River Formation, Southeastern Montana. *J. Paleont.,* **50**: 500–520.

Estes, R. (1981). Gymnophiona, Caudata. In O. Kuhn (ed.), *Handbuch des Palaeoherpetologie.* 2. Gustav Fischer Verlag, Stuttgart.

Estes, R., and Reig, O. (1973). The early fossil record of frogs: A review of the evidence. In J. Vial (ed.), *Evolutionary Biology of the Anurans,* pp. 11–63. University of Missouri Press, Columbia.

Estes, R., Spinar, Z. V., and Nevo, E. (1978). Early Cretaceous pipid tadpoles from Israel (Amphibia: Anura). *Herpetologica,* **34**: 374–393.

Frost, D. R. (ed.) (1985). *Amphibian Species of the World.* Allen Press and Association of Systematic Collections, Lawrence, Kansas.

Godfrey, S. (1986). The skeletal anatomy of *Greererpeton burkemorani* Romer 1969, an Upper Mississippian temnospondyl amphibian. Ph.D. thesis, Department of Biology, McGill University, Montreal.

Gregory, J. T. (1948). A new limbless vertebrate from the Pennsylvanian of Mazon Creek, Illinois. *Am. J. Sci.,* **246**: 636–663.

Gregory, W. K. (1951). *Evolution Emerging.* Macmillan, New York.

Heaton, M. J. (1980). The Cotylosauria: A reconsideration of a group of archaic tetrapods. In A. L. Panchen (ed.), *The Terrestrial Environment and the Origin of Land Vertebrates,* pp. 497–551. Academic Press, London.

Heyler, D. (1975). Sur les "Branchiosaurus" et autres petits amphibiens apparentes de la Sarre et du Bassin d'Autun. *Bull. de la Societe d'histoire naturelle d'Autun,* **75**: 15–27.

Holmes, R. (1980). *Proterogyrinus scheelei* and the early evolution of the labyrinthodont pectoral limb. In A. L. Panchen (ed.), *The Terrestrial Environment and the Origin of Land Vertebrates,* pp. 351–376. Academic Press, London.

Holmes, R. (1984). The Carboniferous amphibian *Proterogyrinus scheelei* Romer, and the early evolution of tetrapods. *Phil. Trans. Roy. Soc. Lond., B,* **306**: 431–527.

Holmes, R., and Carroll, R. L. (1977). A temnospondyl amphibian from the Mississippian of Scotland. *Bull. Mus. Comp. Zool.,* **147**: 489–511.

Hook, R. W. (1983). *Colosteus scutellatus* (Newberry), a primitive temnospondyl amphibian from the Middle Pennsylvanian of Linton, Ohio. *Amer. Mus. Novitates,* **2770**: 1–41.

Ivachnenko, K. F. (1978). Urodelans from the Triassic and Jurassic of Soviet Central Asia. *Paleontological Journal,* **12**(3): 362–368.

Jarvik, E. (1952). On the fish-like tail in the ichthyostegid stegocephalians. *Meddelelser Om Grønland,* **114**(12): 1–90.

Jarvik, E. (1955). The oldest tetrapods and their forerunners. *Sci. Monthly,* **80**: 141–154.

Jarvik, E. (1965). Specializations in early vertebrates. *Ann. Soc. Royale Zool. Belgique,* **94**: 11–95.

Jarvik, E. (1980). *Basic Structure and Evolution of Vertebrates.* Vols. 1 and 2. Academic Press, London.

Larsen, J. H., Jr., and Guthrie, D. J. (1975). The feeding system of terrestrial tiger salamanders (*Ambystoma tigrum melanostictum* Baird). *J. Morph.,* **147**: 137–154.

Lehman, J.-P. (1961). Les stégocéphales du Trias de Madagascar. *Ann. de Paléont.,* **47**: 111–154.

Lombard, R. E., and Wake, D. B. (1976). Tongue evolution in the lungless salamanders, Family Plethodontidae. *Jour. Morph.* **148**: 265–286.

McGinnis, H. J. (1967). The osteology of *Phlegethontia,* a Carboniferous and Permian aistopod amphibian. *Univ. Calif. Publ. Geol. Sci.,* **71**: 1–47.

Milner, A. C. (1977). Morphology and taxonomy of Carboniferous Nectridea (Amphibia). Thesis, University of Newcastle Upon Tyne.

Milner, A. R., Smithson, T. R., Milner, A. C., Coates, M. I., and Rolfe, W. D. I. (1986). The search for early tetrapods. *Modern Geology,* **10**: 1–28.

Moulton, J. M. (1974). A description of the vertebral column of *Eryops* based on the notes and drawings of A. S. Romer. *Breviora,* **428**: 1–44.

Naylor, B. (1980). Radiation of the Amphibia Caudata: Are we looking too far into the past? *Evol. Theory,* **5**: 119–126.

Nussbaum, R. (1977). Rhinatrematidae, a new family of Caecilians (Amphibia, Gymnophione). *Mus. Zool., Univ. Mich., Occ. Papers,* **682**: 1–30.

Panchen, A. L. (1959). A new armoured amphibian from the Upper Permian of East Africa. *Phil. Trans. Roy. Soc. Lond., B,* **242**: 207–281.

Panchen, A. L. (1966). The axial skeleton of the labyrinthodont *Eogyrinus attheyi. J. Zool. Lond.,* **150**: 199–222.

Panchen, A. L. (1967). The homologies of the labyrinthodont centrum. *Evolution,* **21**: 24–33.

Panchen, A. L. (1970). Anthracosauria. In O. Kuhn (ed.), *Handbuch der Palaeoherpetologie,* Part 5a, pp. 1–84. Gustav Fischer Verlag, Stuttgart.

Panchen, A. L. (1972). The skull and skeleton of *Eogyrinus attheyi* Watson (Amphibia: Labyrinthodontia). *Phil. Trans. Roy. Soc. Lond., B,* **263**: 279–326.

Panchen, A. L. (ed.). (1980a). *The Terrestrial Environment and the Origin of Land Vertebrates.* Academic Press, London.

Panchen, A. L. (1980b). The origin and relationships of the anthracosaur Amphibia from the late Paleozoic. In A. L. Panchen (ed.), *The Terrestrial Environment and the Origin of Land Vertebrates,* pp. 319–350. Academic Press, London.

Panchen, A. L. (1985). On the amphibian *Crassigyrinus scoticus*

Watson from the Carboniferous of Scotland. *Phil. Trans. Roy. Soc. Lond., B,* **309**: 505–568.

Parsons, T., and Williams, E. (1963). The relationship of modern Amphibia: A re-examination. *Quart. Rev. Biol.*, **38**: 26–53.

Paton, R. L. (1974). Capitosauroid labyrinthodonts from the Trias of England. *Palaeontology,* **17**: 253–289.

Rage, J.-C. (1986). Le plus ancien Amphibien apode (Gymnophiona) fossile. *C. R. Acad. Sc. Paris, t. 302, Serie II,* **16**: 1033–1036.

Regal, P. L., and Gans, C. (1976). Functional aspects of the evolution of frog tongues. *Evolution,* **30**(4): 718–734.

Robinson, P. L. (1973). A problematic reptile from the British Upper Trias. *J. Geol. Soc. Lond.*, **129**: 457–479.

Rolfe, I. (1980). Early invertebrate terrestrial faunas. In A. L. Panchen (ed.), *The Terrestrial Environment and the Origin of Land Vertebrates,* pp. 117–157. Academic Press, London.

Romer, A. S. (1946). The primitive reptile *Limnoscelis* restudied. *Am. J. Sci.*, **244**: 149–188.

Romer, A. S. (1947). Review of the Labyrinthodontia. *Bull. Mus. Comp. Zool.*, **99**: 1–368.

Romer, A. S. (1949). *The Vertebrate Body.* Saunders, Philadelphia and London.

Romer, A. S. (1957a). The appendicular skeleton of the Permian embolomerous amphibian *Archeria. Contr. Mus. Pal. Univ. Mich.*, **13**: 103–159.

Romer, A. S. (1957b). Origin of the amniote egg. *Sci. Monthly,* **85**: 57–63.

Romer, A. S. (1964). The skeleton of the Lower Carboniferous labyrinthodont *Pholidogaster pisciformis. Bull. Mus. Comp. Zool.*, **131**: 129–159.

Sawin, H. J. (1941). The cranial anatomy of *Eryops megacephalus. Bull. Mus. Comp. Zool.*, **88**: 407–463.

Schultze, H.-P., and Arsenault, M. (1985). The panderichthyid fish *Elpistostege:* A close relative of tetrapods? *Palaeontology* **28**: 293–309.

Smithson, T. R. (1980). An early tetrapod fauna from the Namurian of Scotland. In A. L. Panchen (ed.), *The Terrestrial Environment and the Origin of Land Vertebrates,* pp. 407–438. Academic Press, London.

Smithson, T. R. (1982). The cranial morphology of *Greererpeton burkemorani* Romer (Amphibia: Temnospondyli). *Zool. J. Linn. Soc.*, **76**: 29–90.

Smithson, T. R. (1985). The morphology and relationships of the Carboniferous amphibian *Eoherpeton watsoni* Panchen. *Zool. J. Linn. Soc.*, **85**: 317–410.

Spinar, Z. V. (1952). Revision of some Moravian Discosauriscidae. *Roz. Ustred. Ustad. Geol.*, **15**: 1–159. (In Czech)

Tatarinov, L. P. (1972). *Seymouria morphen* aus der Fauna der U.S.S.R. In O. Kuhn (ed.), *Handbuch der Palaeoherpetologie,* **5B**: 70–80. Gustav Fischer Verlag, Stuttgart.

Taylor, E. H. (1968). *The Caecilians of the World: A Taxonomic Review.* University of Kansas Press, Lawrence.

Thomson, K. S. (1966). The evolution of the middle ear in the rhipidistian amphibian. *Amer. Zoologist,* **6**: 379–397.

Thomson, K. S. (1967). Mechanisms of intracranial kinetics in fossil rhipidistian fishes (Crossopterygii) and their relatives. *J. Linn. Soc. (Zool).*, **178**: 223–253.

Vial, J. L. (ed.). (1973). *Evolutionary Biology of the Anurans.* University Missouri Press, Columbia.

Vorobyeva, E. I. (1980). Observations on two rhipidistian fishes from the Upper Devonian of Lode, Latvia. *Zool. J. Linn. Soc.*, **70**(2): 191–201.

Wake, M. H., and Hanken, J. (1982). Development of the skull of *Dermophis mexicanus* (Amphibia: Gymnophiona), with comments on skull kinesis and amphibian relationships. *J. Morph.*, **173**: 203–223.

Watson, D. M. S. (1956). The brachiopid labyrinthodonts. *Bull. Brit. Mus. (Nat. Hist.), Geol.*, **2**(8): 318–391.

Watson, D. M. S. (1958). A new labyrinthodont (*Paracyclotosaurus*) from the Upper Trias of New South Wales. *Bull. Brit. Mus. (Nat. Hist.), Geol.*, **3**: 233–263.

Watson, D. M. S. (1962). The evolution of the labyrinthodonts. *Phil. Trans Roy. Soc. Lond., B,* **245**: 219–265.

Welles, S. P., and Estes, R. (1969). *Hadrokkosaurus bradyi* from the Upper Moenkopi Formation of Arizona: With a review of the brachyopid labyrinthodonts. *Univ. Cal. Publ., Geol. Sci.*, **84**: 1–56.

Wellstead, C. F. (1982). A Lower Carboniferous aistopod amphibian from Scotland. *Palaeontology,* **25**: 193–208.

Wellstead, C. F. (1985). Taxonomic revision of the Permo-Carboniferous lepospondyl amphibian families Lysorophidae and Molgophidae. Unpublished Ph.D. thesis, McGill University, Montreal.

Wever, E. G. (1965). Structure and function of the lizard ear. *J. Audit. Res.*, **5**: 331–371.

Wever, E. G. (1978). *The Reptile Ear.* Princeton University Press, Princeton.

White, T. E. (1939). Osteology of *Seymouria baylorensis* Broili. *Bull. Mus. Comp. Zool.*, **85**(5): 325–409.

Williston, S. W. (1910). *Cacops, Desmospondylus:* New genera of Permian vertebrates. *Bull. Geol. Soc. America,* **21**: 249–284.

Williston, S. W. (1912). Restoration of *Limnoscelis,* a cotylosaur reptile from New Mexico. *Am. J. Sci.*, **34**: 457–468.

Williston, S. W. (1925). W. K. Gregory (ed.), *The Osteology of the Reptiles.* Harvard University Press, Cambridge, MA.

Zhang, F., Li, Y., and Wang, X. (1984). A new occurrence of Permian seymouriamorphs in Xinjiang, China. *Vertebrata Palasiatica,* **22**(4): 294–304. (In Chinese)

CHAPTER X

Primitive Amniotes and Turtles

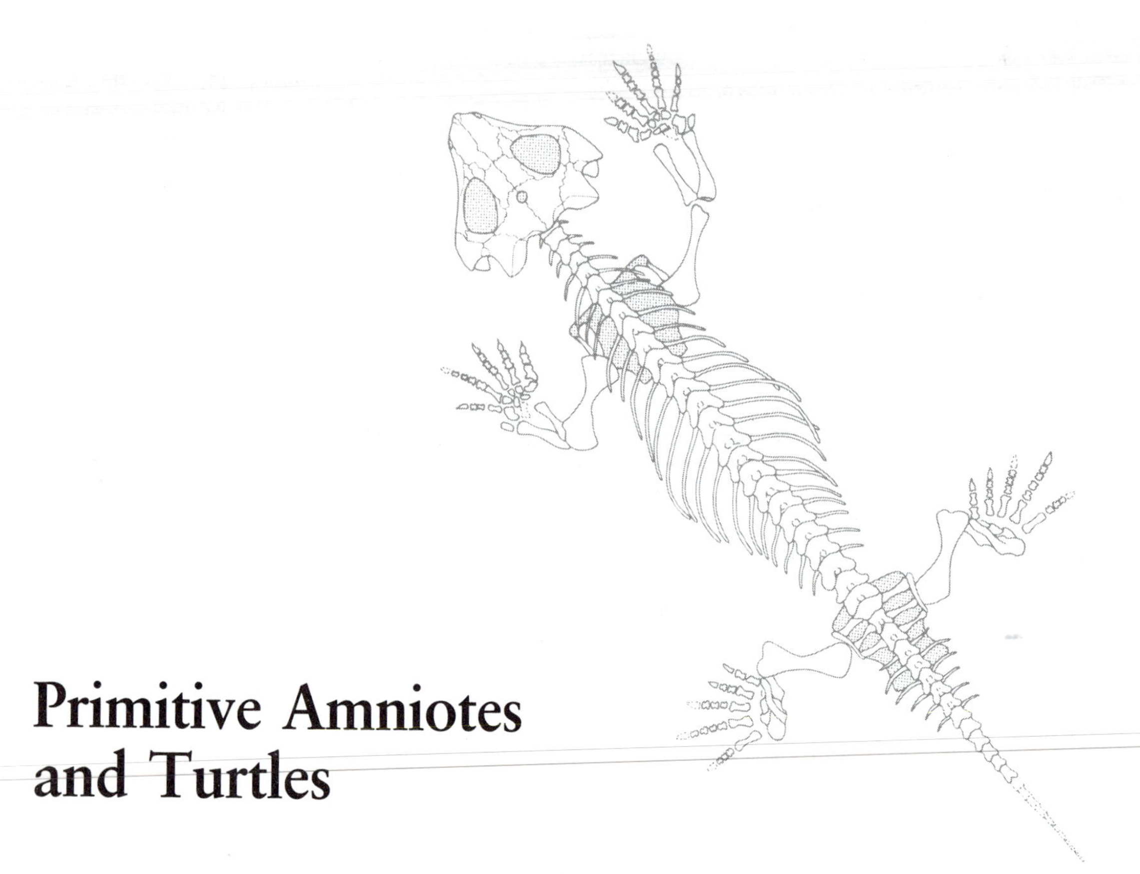

Reptiles, birds, and mammals constitute a single assemblage, the Amniota, that is distinguished from amphibians by the evolution of a reproductive pattern free of standing water. Many amphibians lay their eggs in the water, and the young typically hatch out at an immature or larval stage that depends on the water for support, gas exchange, and food. The eggs of modern genera are no more than 10 millimeters in diameter and have only a limited supply of yolk and no protective membranes or shell. The eggs of all other tetrapods develop extraembryonic membranes—the allantois, chorion, and amnion—that serve to retain water and provide for gas exchange, support, and protection so that the embryo may achieve an advanced stage of development before it is hatched or born.

Relationships of Amniotes

Amniotes appear to be a monophyletic group that evolved from a single stock of primitive tetrapods during the early Carboniferous. By the Upper Carboniferous, the amniotes had diverged into three major lineages: one that gave rise to mammals, a second to turtles, and a third to the majority of other reptilian groups and to the birds.

A simplified phylogeny (Figure 10-1) shows that the groups customarily referred to as reptiles are a phylogenetically heterogeneous assemblage. Within the modern fauna, crocodiles share a more recent common ancestry with birds than they do with lizards, snakes, or turtles. The ancestors of mammals are commonly classified as reptiles, yet they are phylogenetically closer to modern mammals than to any of the modern reptilian orders.

Based on phylogenetic considerations, two major groups of amniotes can be recognized: the mammals and their ancestors, and all other amniotes. Nonmammalian amniotes may in turn be divided into two groups, the turtles and their immediate ancestors and an assemblage including crocodiles, birds, and lepidosaurs (lizards, snakes, and *Sphenodon*). The relative times of divergence of these groups is fairly well established, and all important extinct orders can be fitted into one or the other of these categories with reasonable assurance.

We may use the term reptile to refer informally to turtles, crocodiles, lepidosaurs, and primitive amniotes, but we cannot define the Reptilia rigorously except as amniotes that are not birds or mammals.

In sharp contrast with the fossil record of amphibians, modern amniotes are linked to their Paleozoic ancestors by a relatively complete sequence of intermediate forms. The earliest known amniotes are immediately recognizable as members of this assemblage because of the similarities of their skeleton to those of primitive living lizards.

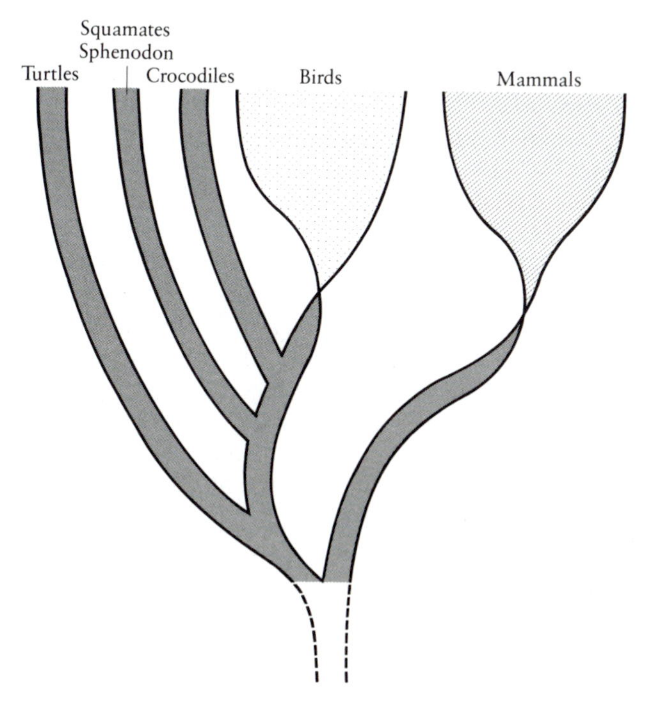

Figure 10-1. SIMPLIFIED PHYLOGENY OF AMNIOTES. The class Reptilia, as customarily defined, includes all the darkly stippled lineages. Reptiles have a common ancestry but are considered paraphyletic since they do not include all their descendants. Birds and mammals can each be defined on the basis of unique, shared derived characters. Reptiles can be defined as amniotes that lack the specialized characters of birds and mammals.

Primitive Amniotes

Remains of the earliest known amniotes are found in two localities of early and middle Pennsylvanian age in Nova Scotia, eastern Canada (Carroll, Belt, Dineley, Baird, and McGregor, 1972). These fossils are not found in normal coal-swamp deposits, such as those from which the majority of Carboniferous tetrapods have been found, but rather within the upright stumps of the giant lycopod *Sigillaria*. These trees grew in areas that were subject to periodic flooding, which resulted in the burial of the trees in several meters of sediments. The trees died and the central portion rotted out, but the bark was stronger and retained the cylindrical shape of the stump. After the withdrawal of the water, animals living on the newly developed land surface would occasionally fall into the hollow stumps. Eventually they died and were covered with sediments and fossilized. This unusual series of events led to the preferential preservation of the first truly terrestrial vertebrates.

The best-known primitive amniotes from the early Pennsylvanian are *Hylonomus* (Figure 10-2) and *Paleothyris* (Figure 10-3). We include these genera and other genera from the later Pennsylvanian (Carroll and Baird, 1972) and Lower Permian (Clark and Carroll, 1973) within the family Protorothyridae (Romeriidae) in the order Captorhinida.

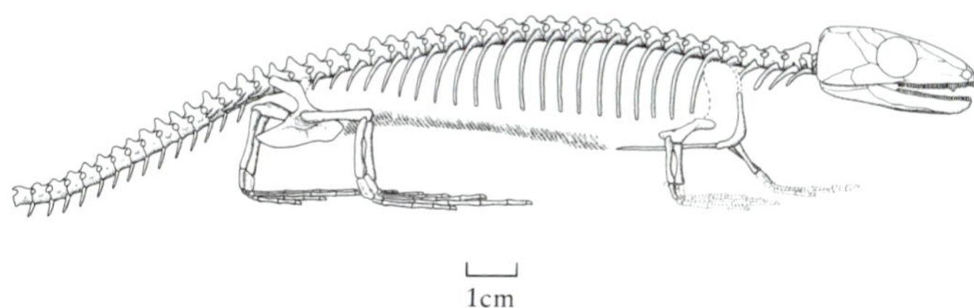

Figure 10-2. SKELETON OF ONE OF THE EARLIEST KNOWN AMNIOTES, *HYLONOMUS LYELLI* FROM THE EARLY PENNSYLVANIAN OF JOGGINS, NOVA SCOTIA. The remains were found within the upright stump of the giant lycopod *Sigillaria. From Carroll and Baird, 1972.*

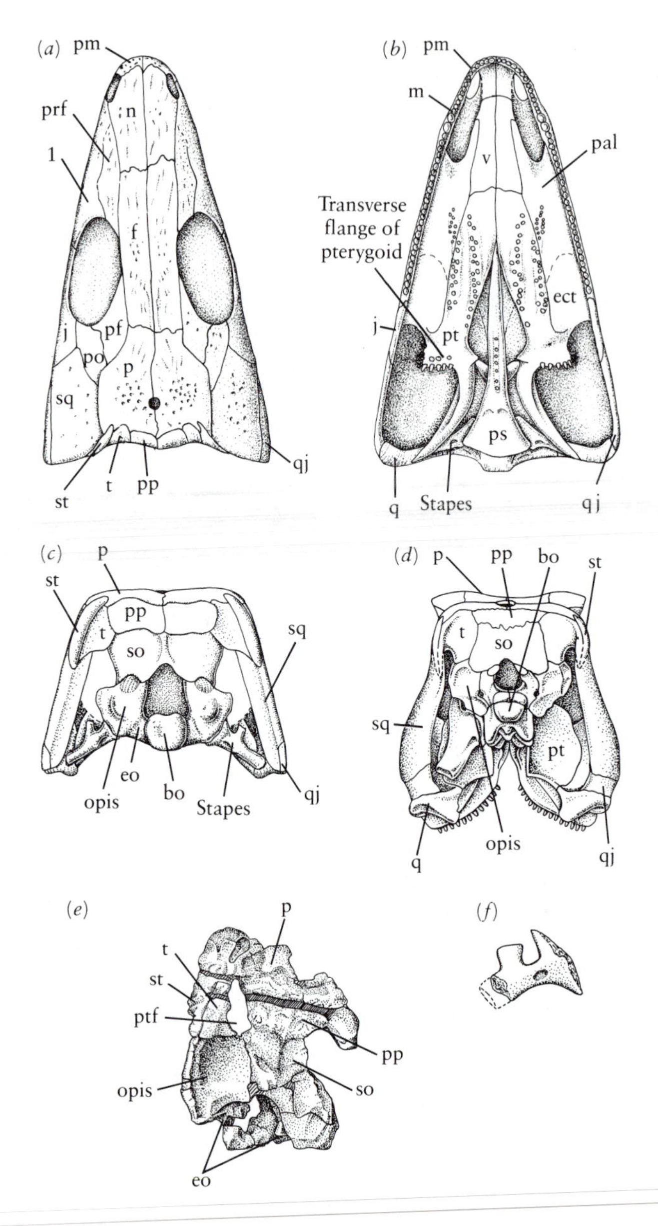

Figure 10-3. THE SKULL OF PRIMITIVE AMNIOTES. The skull of *Paleothyris* in (*a*) dorsal, (*b*) palatal, and (*c*) occipital views. *From Carroll and Baird, 1972.* (*d*) Occiput of the primitive mammal-like reptile *Ophiacodon. From Romer and Price, 1940. With permission of the Geological Society of America.* (*e*) Occiput of an immature specimen of *Desmatodon,* a carboniferous relative of *Diadectes. From Vaughn, 1972.* In all these skulls the supraoccipital is a broad plate of bone that links the exoccipitals and otic capsules with the dermal bones at the back of the skull table. (*f*) The stapes of *Hylonomus,* which is characteristic of primitive amniotes. *From Carroll, 1969b.* Abbreviations as in Figure 8-3.

SKELETAL ANATOMY OF EARLY AMNIOTES

The skeleton of early protorothyrids provides us with a good insight into the anatomy of the ancestral amniotes as well as a basis for evaluating their origin. The body was small—approximately 100 millimeters from the snout to the base of the tail—and the skeleton was very well ossified. Their general appearance would have closely resembled that of modern lizards.

Skull

The skull roof, like that of Paleozoic amphibians, forms a nearly complete dermal covering, with openings for the eyes, pineal, and nostrils. The pattern of the bones resembles that of small anthracosaurs such as the gephyrostegids, except for the absence of an otic notch and the intertemporal bone. The area that was occupied by that bone in primitive amphibians appears to have been taken over by the extensive parietal, which separates the postorbital from the supratemporal. The postparietal, tabular, and supratemporal are reduced in size and their exposure is limited to the occipital surface.

The pattern of the occiput, which is unique among early tetrapods, indicates the divergence of amniote ancestors from the other major groups prior to the evolution of a solid attachment between the back of the braincase and the skull roof. Protorothyrids have a large, platelike supraoccipital bone linking the dermal bones of the skull table with the exoccipital and otic capsule. In the primitive state, there appears to have been a sliding contact between the supraoccipital and the postparietal and tabular. The paroccipital processes, which extend from the otic capsules to the quadrates in modern lizards, were poorly developed and did not link the braincase to the cheek.

Because the back of the braincase was initially not firmly attached to either the skull roof or the cheek, the stapes retained an important role in its support. The stapes is a large element, with an expanded foot plate, a thick shaft that extends obliquely and ventrolaterally toward the quadrate, and a short dorsal process that articulates with the otic capsule. The large size of the stapes and particularly its broad foot plate precludes its effective participation in an impedance-matching system such as was described previously in Chapter 9. However, a fenestra ovalis is present.

It was long assumed that early amniotes had a tympanum, but they lack an otic notch and a tympanum would have had to be very large to activate a stapes of this size. In later amniotes, either the squamosal, quadrate, or angular became specialized to support a large tympanum. Hearing in early amniotes may have occurred by bone conduction, as in salamanders, caecilians, and those modern reptiles without a tympanum. In these groups, hearing is limited to relatively low-frequency, high-intensity sounds (Wever, 1978).

The most significant feature of the palate in early amniotes is the presence of a transverse flange on the pterygoid. The portion of this bone that is lateral to the basicranial articulation is angled ventrally into the mouth cavity. In modern lizards, the transverse flange of the pterygoid serves as the origin of one of the largest of the jaw-closing muscles, the pterygoideus. The orientation of this muscle, which is at nearly right angles to the other adductor jaw muscles, enables it to exert its maximum force when the jaws are wide open. There is little evidence at the existence of a large pterygoideus muscle in any primitive amphibian.

In contrast with labyrinthodonts, the palate of primitive amniotes does not have large fangs. In most genera there are three rows of denticles, one on the transverse flange of the pterygoid, a second along the margin of the small interpterygoid vacuities, and a third extending across the palatine bone. Some early amniotes with large marginal teeth show a trace of labyrinthine infolding, but this infolding is absent in most genera. Many early amniotes have two pairs of large marginal teeth near the anterior end of the maxilla that are broadly comparable with the mammalian canine teeth.

The general structure of the braincase of early amniotes resembles that of modern lizards (Figure 10-4). The occipital plate and otic capsule are well ossified, but the more anterior portion is largely cartilaginous, in contrast with the condition in most labyrinthodonts. The area ossified in labyrinthodonts as the massive sphenethmoid may be represented by short, narrow plates of bone. Between the sphenethmoid and the otic capsule, the lateral wall of the braincase is not ossified. A movable basicranial articulation links the base of the braincase with the palate.

The basioccipital and base of the exoccipitals form a prominent medial occipital condyle that fits into a ring formed by the intercentrum and the arches of the atlas vertebra like the ball of a ball-and-socket joint. The otic capsule consists of a posterior opisthotic and anterior proötic. Together with the supraoccipital, they enclose the semicircular canals.

In early amniotes, the elements of the palatoquadrate, which are still extensive in primitive amphibians, are reduced to the relatively small epipterygoid and the quadrate. The epipterygoid has a wide base that is in contact with the upper surface of the pterygoid and forms the basicranial articulation with the basisphenoid. Dorsally, the epipterygoid extends nearly to the skull roof as a narrow rod lateral to the area of the Vth nerve. The quadrate has a broad base for articulation with the articular bone of the lower jaw and a narrow dorsal flange that is supported by the squamosal and the quadrate ramus of pterygoid.

The lower jaw (Figure 10-5) lacks parasymphyseal tusks, which are present in early labyrinthodonts, and has one or two coronoids and a splenial. Otherwise, the elements resemble those of most primitive amphibians.

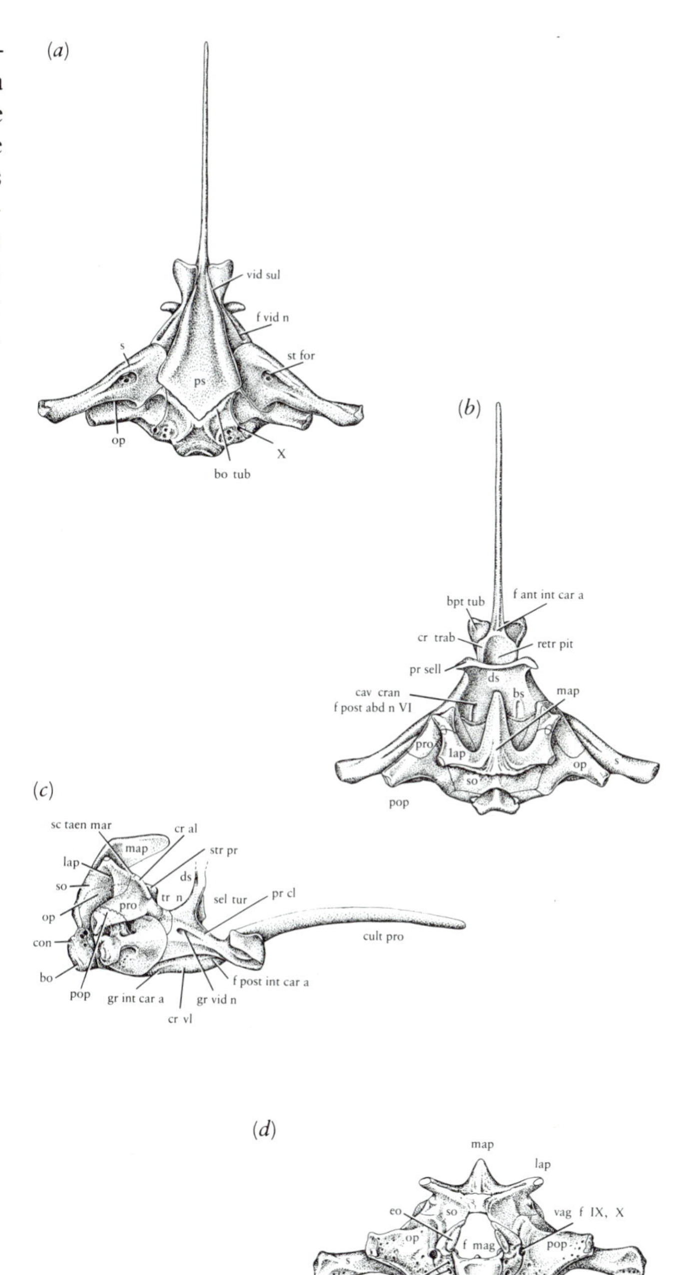

Figure 10-4. BRAINCASE OF PRIMITIVE AMNIOTES. (*a*) Ventral, (*b*) dorsal, (*c*) lateral, and (*d*) occipital views of the captorhinid *Eocaptorhinus*. Abbreviations as follows: bo tub, basioccipital tubercle; bpt tub, basipterygoid tubercle; bs, basisphenoid; cav cran, cavum cranii; con, occipital condyle; cr al, crista alaris; cr trab, crista trabecularis; cr vl, crista ventrolateralis; cult pro, cultriform process; ds, dorsum sella, f ant int car a, foramen anterior for internal carotid artery canal; f mag, foramen magnum; f post abd n VI, foramen posterior of abducens (VI) nerve canal; f post int car a, foramen posterior of internal carotid artery canal; f vid n, foramen for vidian (VII) nerve; gr int car a, groove for internal carotid artery; gr vid n, groove for vidian (VII) nerve; hyp f XII, hypoglossal (XII) nerve foramina; lap, lateral ascending process of supraoccipital; map, median ascending process; op, opisthotic; pop, paroccipital process; pr cl processus clinoideus; pro, proötic; pr sell, processus sellaris; ps, parasphenoid; retr pit, retractor pit; s, stapes sc taen mar, scar for attachment of taenia marginalis; sel tur, sella turcica; so, supraoccipital; st for, stapedial foramen; str pr, supratrigeminal process; tr n, trigeminal notch; vag f IX, X, vagus foramen (IX, X); vid sul, vidian sulcus. *From Heaton, 1979.*

Postcranial skeleton

The postcranial skeleton of early amniotes generally resembles that of the primitive living reptile *Sphenodon*. The vertebral centra consist of large spool-shaped pleurocentra, with deep recesses at both ends for the notochord, and small crescentic intercentra. In the primitive state, the neural arches are narrow and the zygapophyses are close to the midline (Figure 10-6). A suture between the neural arch and the pleurocentrum may remain in the trunk vertebrae. The elements of the atlas-axis complex are basically similar to those of primitive anthracosaurs, except for the greater degree of consolidation, with the

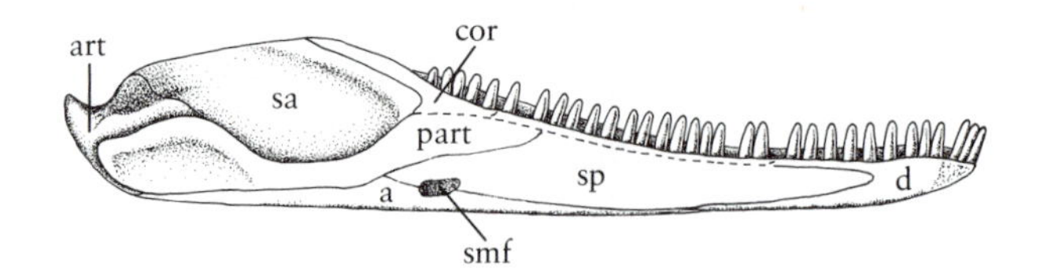

Figure 10-5. MEDIAL VIEW OF THE LOWER JAW OF *PROTOROTHYRIS*, A PRIMITIVE AMNIOTE FROM THE LOWER PERMIAN. Abbreviations as in Figure 8-3, plus: smf, submeckelian fossa. *From Clark and Carroll, 1973.*

axis arch and centrum fused. The earliest amniote in which an accurate count can be made has 32 presacral vertebrae, but other early genera have 24 or 25. Most early amniotes have two sacral vertebrae and a very long tail. Ribs are present throughout the trunk and at the base of the tail.

The basic structure of the girdles and limbs resembles that of early anthracosaurs, although their proportions differ, suggesting more agile locomotion (Figures 10-7 and 10-8). The size of the clavicles and cleithra is reduced, and the interclavicle has a distinctive *T* shape. In the primitive state, the adult scapulocoracoid is ossified as a single unit with a screw-shaped glenoid. In immature specimens, we can see that it is formed from three centers of ossification: the vertical scapula and separate anterior and posterior coracoids. The humerus has a relatively long narrow shaft from which the extremities are expanded at right angles to one another. As in many primitive Paleozoic amphibians, there is an entepicondylar foramen. A supinator process, which is proximal to the ectepicondyle, may be a primitive feature for amniotes. The carpus has 11 well-ossified, tightly fitting elements, including a pisiform (Figure 10-8). The tarsus is specialized over that of most Paleozoic amphibians in having the tibiale, intermedium, and proximal centrale fused into a single bone, the astragalus. Peabody (1951) documented this fusion ontogenetically. The fibulare is enlarged and is now termed the calcaneum. As in anthracosaurs, the phalangeal count of the manus is 2, 3, 4, 5, 3 and that of the pes is 2, 3, 4, 5, 4. Based on the similarity of the joints of the girdles and limbs and the evidence of foot prints, we assume that the general pattern of locomotion in early amniotes probably was broadly similar to that of primitive amphibians and has been little altered in primitive living lizards (Holmes, 1977).

Dorsal dermal scales are lost, but there are small scales extending between the pectoral and pelvic girdles that form a chevron pattern.

REPRODUCTION AND DEVELOPMENT

The most important characteristic that unites amniotes—the nature of the egg—cannot be directly determined from the skeleton. However, two lines of indirect evidence suggest the evolution of the amniote egg before the appearance of protorothyrids in the fossil record.

Extraembryonic membranes develop in a similar way in all three living groups of amniotes. It is more parsimonious to assume that their common ancestor had already developed this feature than to propose that it had evolved separately two or three times. The small size of the early fossils also suggests that they may already have achieved a pattern of reproduction approaching that of modern amniotes.

Among salamanders, there is a close correlation between the size of the egg and the size of the adult within three broad categories—those laying eggs in still water, those laying eggs in streams, and those laying eggs on land (Salthe, 1969). The closest comparison to amniotes occurs among plethodontid salamanders, which lay their eggs on land and whose young hatch out as miniatures of the adults. The eggs in this family are the largest in relationship to body size of any amphibian group. The large yolk sac contains enough nutrients for the young to develop

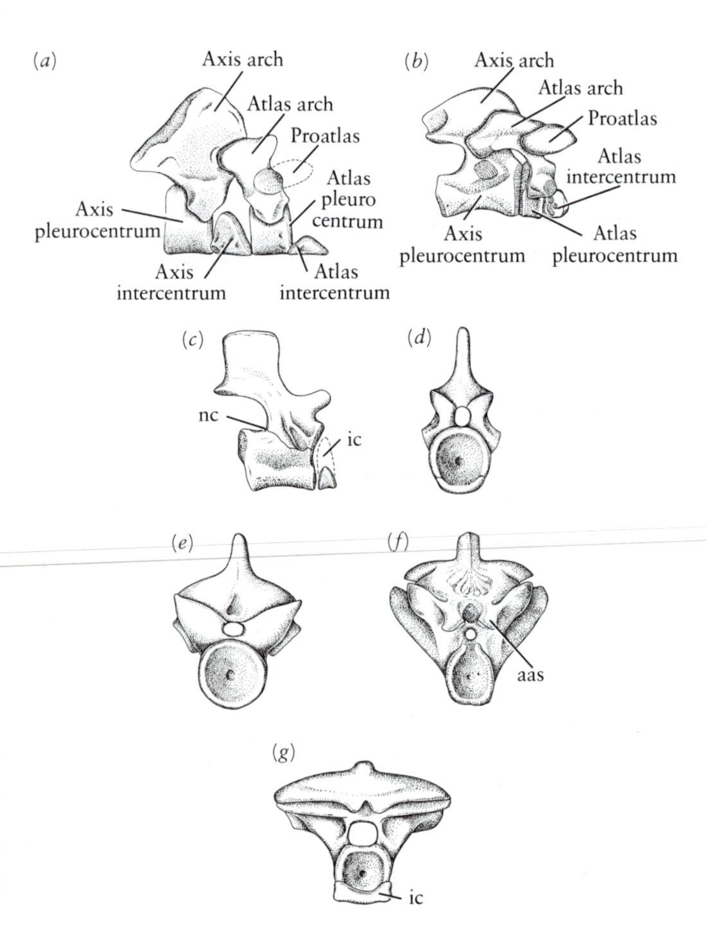

Figure 10-6. VERTEBRAE OF PRIMITIVE AMNIOTES AND ANTHRACOSAURS. (*a*) Atlas-axis complex of the anthracosaur *Gephyrostegus*. *From Carroll, 1970.* (*b*) Atlas-axis of the protorothyrid *Paleothyris*. *From Carroll, 1969b.* (*c* and *d*) Trunk vertebra of a primitive amniote based on the Carboniferous specimen number 1901.1378 from the Humboldt Museum, Berlin. (*c*) *From Carroll, 1970.* Anterior views of three primitive tetrapods with laterally expanded neural arches: (*e*) the limnoscelid *Limnostygis*, (*f*) *Diadectes*, and (*g*) *Seymouria*. Abbreviations as follows: aas, accessory articulating surfaces; ic, intercentra; nc, neurocentral suture.

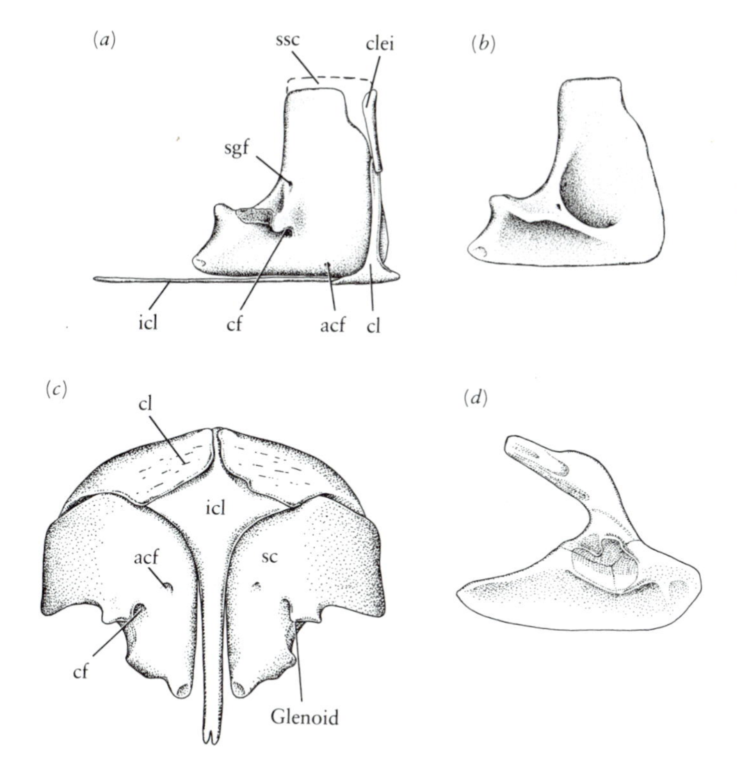

Figure 10-7. GIRDLE OF PRIMITIVE AMNIOTES. Pectoral girdle of *Protorothyris* in (*a*) lateral, (*b*) medial, and (*c*) ventral views. *From Clark and Carroll, 1973.* (*d*) Lateral view of the pelvis of *Hylonomus. From Carroll, 1969a. With permission of Cambridge University Press.* Abbreviations as follows: acf, anterior coracoid foramen; cf, coracoid foramen; cl, clavicle; clei, cleithrum; icl, interclavicle; sc, scapula; sgf, supraglenoid foramen; ssc, suprascapular cartilage.

to maturity before hatching. The absence of extraembryonic membranes is probably the factor that limits the size of the egg in all amphibians. Without the capacity for effective exchange of gases between the air and the embryo and retention of water for development, a larger size could not be achieved.

The largest plethodontid salamanders that undergo direct development on land are no more than 100 millimeters long from their snout to the base of the tail. Lizards of this body length lay eggs that are similar in size to those of the larger plethodontids.

By analogy, one can argue that the origin of amniotes proceeded via a stage in which nonamniotic eggs were laid on land and development was direct. Prior to the evolution of extraembryonic membranes, adult size would have been restricted as it was in plethodontids, so that the largest eggs that could develop without extra support and protection would contain enough nutrients for the young to develop fully before hatching. Very large amphibians, such as most labyrinthodonts, probably hatched out of similarly small eggs, but they had a protracted larval period in which they depended on the water for support, food, and exchange of respiratory gases.

According to Szarski (1968), the first of the extraembryonic membranes to evolve would have been the allantois. In modern amniotes, it develops as a bladder to retain water that is otherwise lost from the embryo. Solutes within the urine render its contents hypertonic relative to fresh water. If the egg is laid on damp ground, water is drawn into the allantois by osmosis. In modern reptiles such as many lizards and turtles that have permeable egg shells, the water content within the egg may more than double during development because of absorption into the allantois.

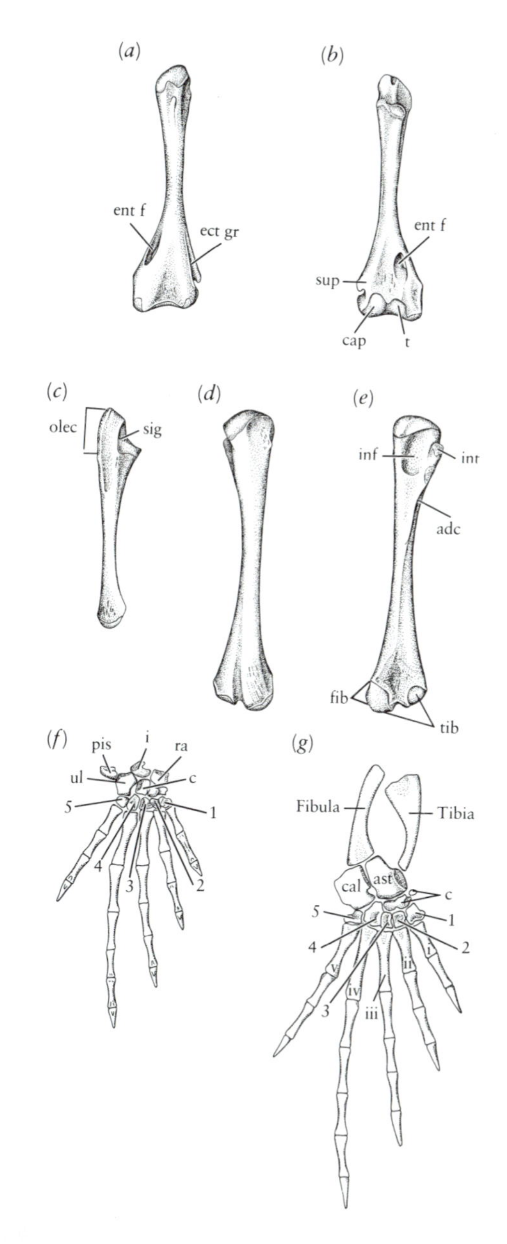

Figure 10-8. LIMBS BONES OF PRIMITIVE AMNIOTES. Humerus in (*a*) dorsal and (*b*) ventral views. (*c*) Ulna in anterior view. Femur in (*d*) dorsal and (*e*) ventral views, based on a protorothyrid from the Lower Permian, $\times 1\frac{1}{2}$. *From Reisz, 1980.* (*f*) Carpus and manus of *Paleothyris. From Carroll, 1969a.* (*g*) Lower limbs, tarsus and pes of *Paleothyris*, $\times 1\frac{1}{2}$. *From Carroll, 1969a.* (*d* and *e*) *With permission of Cambridge University Press.* Abbreviations as in Figure 9-10 plus: adc, adductor crest; cap, capitellum; ect gr, ectepicondylar groove; ent f, entepicondylar foramen; fib, condyle for fibula; in f, intertrochanteric, fossa; in t, internal trochanter; olec, olecranon; sig, sigmoid notch; sup, supinator process; t, trochlea; tib, condyle for tibula.

With the development and enlargement of the allantois, the more superficial layers of tissue that covered the embryo in the primitive state were forced out from their original position and reflected over the body. These layers form the chorion, which together with the surface of the allantois serves for gas exchange. A further membrane, the amnion, forms a fluid-filled chamber that surrounds the embryo. In order to fertilize eggs laid on land, internal fertilization probably evolved early in amniote evolution. Since copulatory organs are absent in the primitive living genus *Sphenodon* and have apparently evolved separately in the ancestors of lizards and crocodiles, they were almost certainly absent in the early amniotes.

A permeable membranous shell presumably evolved in conjunction with the extraembryonic membranes. An object that was presumed to be the oldest amniote egg was described from the Lower Permian of Texas, but it lacks a calcareous shell and the evidence that it is an egg is equivocal (Hirsch, 1979).

The absence of palatal fangs, labyrinthine infolding of the enamel, and the posterior emargination of the cheek (otic notch) are also features of lepospondyl amphibians. Since these structures all become elaborated during growth in labyrinthodonts, their absence in all these groups may be associated with maturation at a small body size. Early amniotes also resemble lepospondyls in having spool-shaped vertebral centra, which also may be associated with achieving a higher degree of ossification at a small body size than is the case with most labyrinthdonts. The association of these characters suggests that the origin of amniotes may be related to a major change in developmental processes.

FEEDING AND LOCOMOTION IN EARLY AMNIOTES

We may associate the small body size in early amniotes with dietary specialization as well as reproductive patterns. The large size of the body and disproportionately large head and long sharp teeth in early labyrinthodonts suggest that they were most effective in preying on other large vertebrates. In contrast, the early amniotes closely resemble primitive living lizards in their small body size and proportionately small skull. The structure of the teeth and probable arrangement of the jaw musculature in early amniotes resemble those of living lizards that feed almost exclusively on insects and other small arthropods (Pough, 1973). This association suggests that a major factor in the emergence of early amniotes was their adaptation to feeding on small terrestrial arthropods that were becoming increasingly diverse during the Carboniferous.

Early amniotes have a more fully ossified skeleton than most Paleozoic amphibians, and the proportions of the limbs suggest that they were more agile. All living amniotes have a much more complete and effective system of muscle stretch receptors than do amphibians. An advance in the coordination of their locomotion probably was another important factor in the success of the early amniotes.

THE ANCESTRY OF AMNIOTES

Prior to the Lower Pennsylvanian, no deposits are known that yield a truly terrestrial fauna, and there is no earlier record of either reptiles or their immediate ancestors. The early amniotes are sufficiently distinct from all Paleozoic amphibians that their specific ancestry has not been established. In their body proportions, they resemble microsaurs such as *Tuditanus* (see Figure 9-30), but they differ from all lepospondyls in having a multipartite atlas-axis complex. Early amniotes differ from temnospondyls in the primitive absence of a solid union between the skull table and the cheek, in the large size of the pleurocentrum and higher phalangeal count.

Among Paleozoic amphibians, only the anthracosaurs share significant derived characters with early amniotes. These features include contact between the tabular and parietal and the dominance of the pleurocentrum. Although a multipartite atlas-axis complex is a feature of primitive labyrinthodonts, only in anthracosaurs is the pleurocentrum the major element that supports the atlas and axis arches as it does in early amniotes.

Amniotes also share with anthracosaurs a five-toed manus and a phalangeal count of 2, 3, 4, 5, 3. However, we are not certain that these features are unique to anthracosaurs among primitive Paleozoic amphibians.

Another reason for relating amniotes with anthracosaurs is the presence of several groups of late Carboniferous and early Permian tetrapods—including the limnoscelids, tseajaiids, and solenodonsaurids—that resemble primitive anthracosaurs in their anatomy, particularly in the structure of the occiput, but that have some characteristics of primitive amniotes. The largest forms are more than 1 meter long; their postcranial skeleton is specialized for terrestrial locomotion with stout limbs and well-developed pleurocentra. The most important feature that these groups share with primitive amniotes is the transverse flange of the pterygoid. *Limnoscelis* has lost the intertemporal bone, but the postorbital rather than the parietal has grown into the area it once occupied.

Tseajaia and limnoscelids both have greatly expanded neural arches, a feature that is also observed among some Permian amniotes (see Figure 10-6). This specialization was once thought to be a primitive character of amniotes. However, it is not present in the earliest amniote groups and may have evolved convergently within several later lineages in association with large body size.

Some of the advanced features seen in these anthracosaurs and early amniotes were apparently achieved separately. Nevertheless, the tendency toward more effective terrestrial locomotion and similar changes in the palate

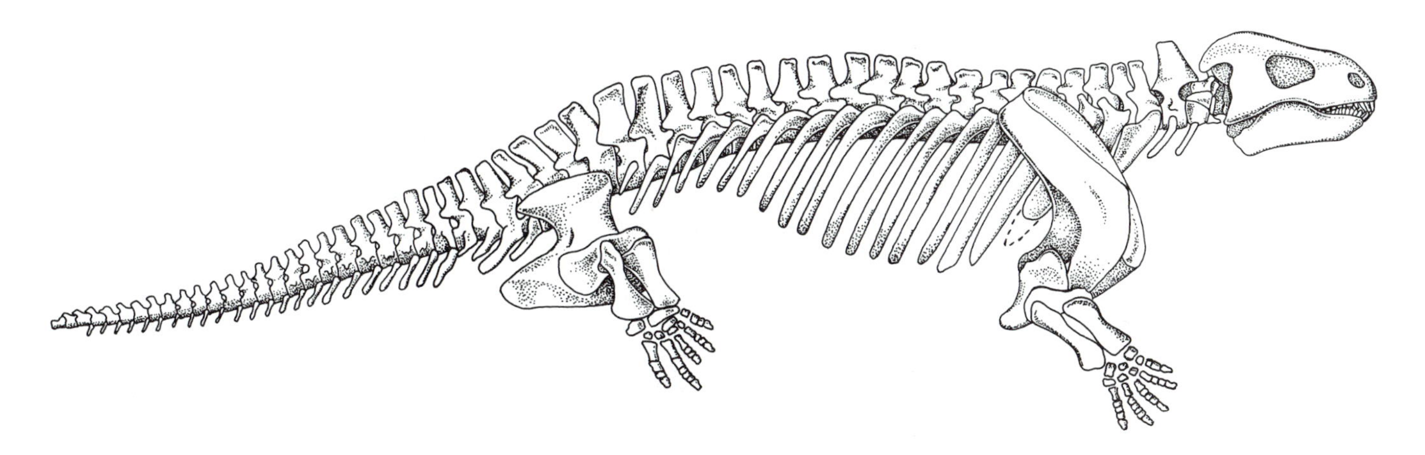

Figure 10-9. RESTORATION OF THE SKELETON OF *DIADECTES*, A PRIMITIVE TETRAPOD OF UNCERTAIN AFFINITIES (about 3 meters long). Most of the postcranial skeleton resembles that of primitive amniotes if adjustment is made for its great size. *From Carroll, 1969c.*

may indicate a closer relationship between these two groups than between amniotes and any other orders of Paleozoic tetrapods.

The taxonomic position of another family, the Diadectidae, remains in doubt. *Diadectes* (Figures 10-9 and 10-10), one of the largest of Lower Permian vertebrates, reaches 3 meters in length. Most of the postcranial skeleton resembles that of primitive amniotes, except for features associated with its large size. The skull is highly specialized in having a partial secondary palate and molariform cheek teeth. These features suggest a diet of plants or hard-shelled invertebrates. *Diadectes* has an otic notch, which had led some to compare this genus with anthracosaurs, but the notch is formed by the quadrate, as in

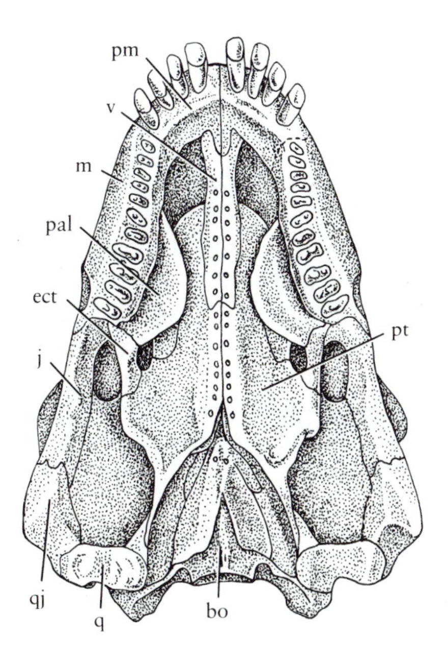

Figure 10-10. THE PALATE OF *DIADECTES*. The cheek teeth are expanded transversely to form a large crushing area. The palatine bone forms a partial secondary palate. The skull roof is formed of thick spongy bone, unlike that of other early tetrapods. Abbreviations as in Figure 8-3. *From Carroll, 1969c.*

advanced amniotes, not by the squamosal, as in amphibians. The presence of a large supraoccipital in the early diadectids suggests that they evolved from the base of the same lineage that gave rise to the protorothyrids, but they may have evolved separately from the primitive anthracosaurs.

The term Cotylosauria, which is commonly used in reference to the primitive reptilian stock, was originally coined for the genus *Diadectes*. It became associated with a wide range of primitive tetrapods, including the seymouriamorphs and other terrestrial anthracosaurs as well as a host of primitive amniotes. The presence of swollen neural arches was one of the most conspicuous features by which this group was recognized (Heaton, 1980). Since the term cotylosaur has been used to include both amphibians and reptiles and because the phylogenetic position of the Diadectidae remains contentious, an alternate name, Captorhinida, is used here for the order that includes the most primitive amniotes.

TEMPORAL OPENINGS AND THE CLASSIFICATION OF AMNIOTES

From their basic adaptation to a terrestrial way of life as small, agile insectivores, the early amniotes diversified during the Carboniferous into several distinct adaptive zones. Some increased their body size and developed large slashing teeth in relation to a truly carnivorous diet. Other lineages developed a crushing or grinding dentition as an adaptation toward feeding on hard-shelled prey or plant material. By the middle Permian, at least one group had become secondarily adapted to an aquatic way of life.

Amniotes can be broadly classified on the basis of the pattern of openings in the dermal skull roof behind the orbits (Figure 10-11). In primitive amniotes, this area is completely covered with bone. This pattern is termed

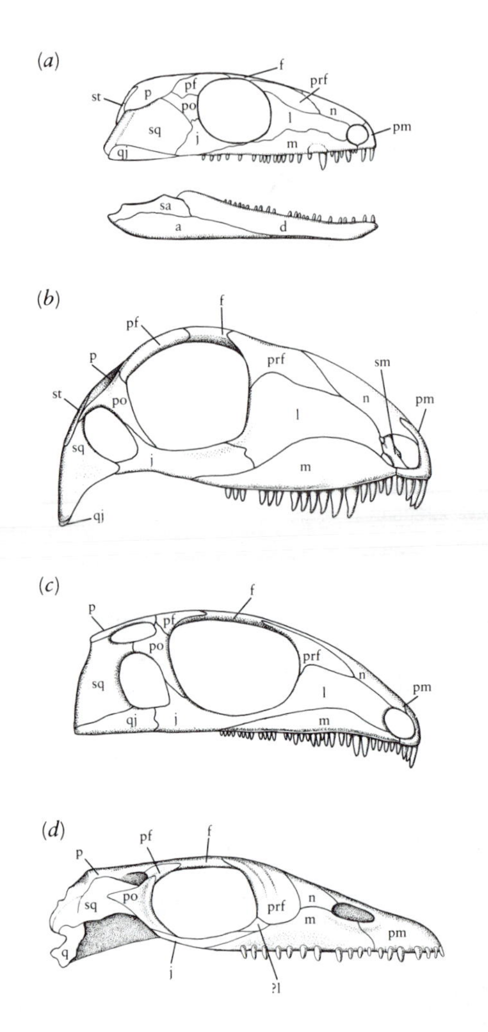

Figure 10-11. SKULLS OF EARLY AMNIOTES SHOWING THE PATTERN OF TEMPORAL OPENINGS THAT DISTINGUISH THE MAJOR GROUPS. (*a*) The anapsid condition, illustrated by the protorothyrid *Paleothyris*. (*b*) The synapsid condition, exemplified by the early mammal-like reptile *Haptodus*. (*c*) The diapsid condition, shown by *Petrolacosaurus*. (*d*) The nothosaur *Neusticosaurus*, illustrating the parapsid or euryapsid condition. The diapsid and synapsid configurations are thought to have evolved separately from the anapsid condition. The euryapsid pattern has evolved from the diapsid pattern by loss of the lower temporal bar. Abbreviations as in Figure 8-3.

anapsid and forms the basis for the subclass Anapsida, including the Captorhinida as a primitive stock. The turtles are typically included among the anapsids since they retain a continuous bony covering behind the orbits.

Other amniote groups developed one or more pairs of temporal openings. The first group to diverge from the ancestral stock were the **synapsids** (Figure 10-12). This group, also termed the mammal-like reptiles, includes the ancestors of mammals. They have a single pair of openings located low in the cheek and bordered by the jugal, squamosal, and postorbital. The synapsids were the first group of amniotes to become abundant and diversified. Members had achieved large body size and differentiated into a range of habitats by the late Carboniferous. Their subsequent evolution is discussed in Chapter 17.

Late in the Pennsylvanian, a second major group diverged from the basal anapsid stock: the **Diapsida.** Diapsids are characterized by the presence of two pairs of temporal openings—one pair located ventral to the postorbital, like that of the synapsids, and a second pair dorsal to the postorbital and squamosal and lateral to the parietal. The major groups of Mesozoic amniotes, including the dinosaurs and the pterosaurs, arose from the base of the diapsid stock. This stock also gave rise to all living reptiles other than turtles.

The diapsids are so diverse that it is convenient to recognize two major subgroups: the Lepidosauromorphs, which are represented in the modern fauna by sphenodontids, lizards, and snakes, and the Archosauromorphs, which include the dinosaurs and their kin as well as the living crocodiles.

Two major groups of Mesozoic marine reptiles, the plesiosaurs and ichthyosaurs, have an upper temporal opening like that of the diapsids but lack a clearly defined lateral opening. The postorbital and squamosal form a wide cheek. This pattern has been termed **parapsid** or **euryapsid** and was thought to have arisen directly from an anapsid configuration. The pattern seen in plesiosaurs can be derived from that of early diapsids by elimination of the lower temporal bar and thickening of the postorbital and squamosal (Carroll, 1981). An intermediate stage is illustrated by the Triassic nothosaurs. Nothosaurs and plesiosaurs are classified in the subclass Sauropterygia, which is discussed in Chapter 12.

Ichthyosaurs may also have evolved from the diapsid stock, but no intermediate forms are recognized. The ichthyosaurs have diverged so far from the basic reptilian body plan that they may be placed in a distinct subclass, the Ichthyopterygia.

Development of temporal openings was not limited to these major subclasses but also occurred among several minor groups that are included among the Anapsida on the basis of their otherwise conservative anatomy. The nature of the temporal openings nevertheless provides a practical basis for distinguishing the major groups of amniotes. Surprisingly, the reasons for their initial development are still not adequately established. When fully elaborated, their configuration can be related to particular patterns of the adductor jaw musculature, but this function does not account for their initiation as small openings in the otherwise relatively smooth skull roof.

The development of temporal openings may be attributed to two major factors: differential concentration of mechanical stress in the skull and concentration of areas of muscle attachment. From studies of living vertebrates, we know that during growth bone is deposited most thickly along lines of stress; between areas of stress the bone may be thin or absent altogether. The particular shape and proportions of the skull in primitive amniotes determine where stress was concentrated. The fact that muscles can be more strongly attached to ridges and edges of bone than to flat surfaces would account for the con-

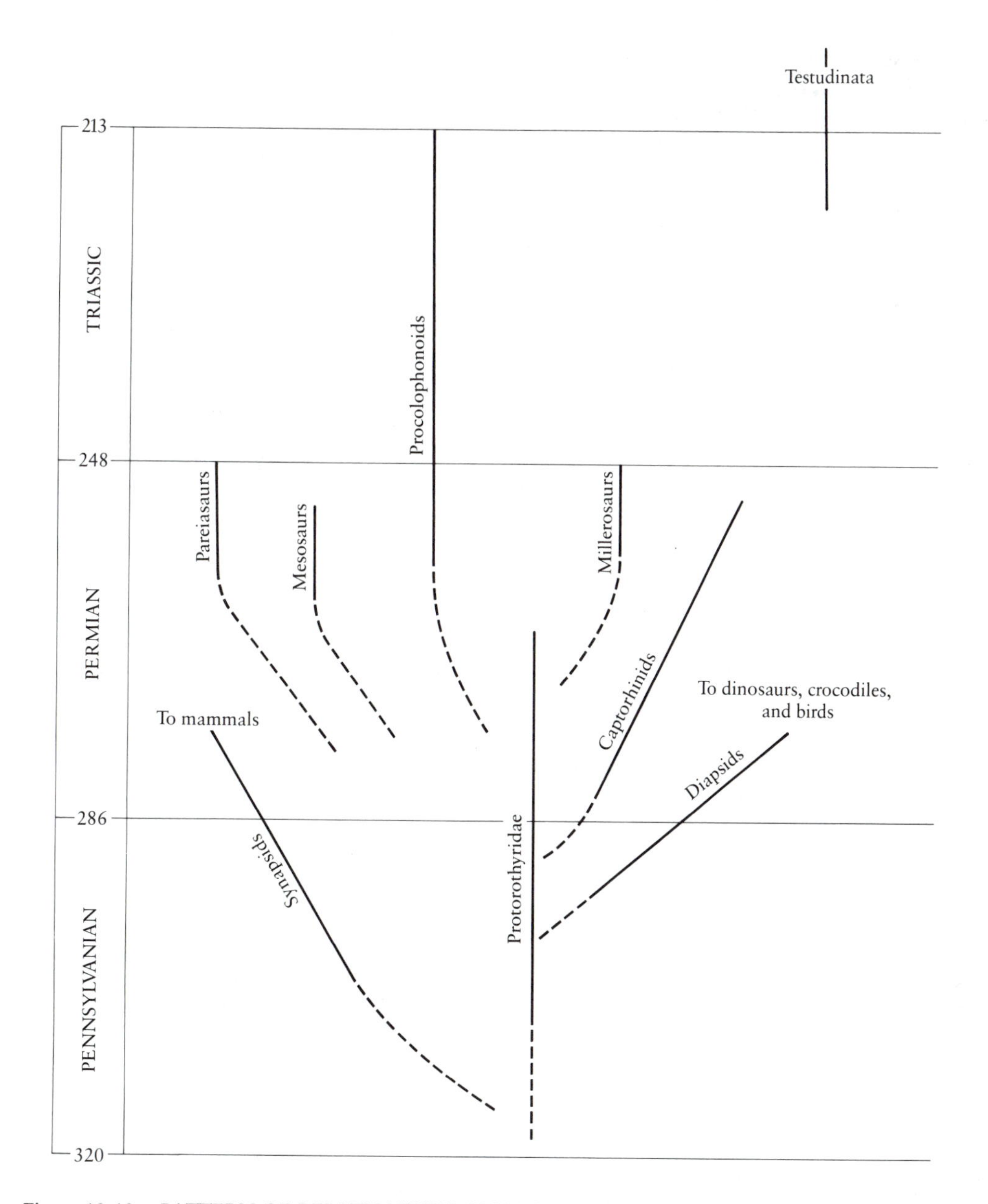

Figure 10-12. PATTERN OF RELATIONSHIPS AMONG EARLY AMNIOTES. The skeletal anatomy of protorothyrids remains relatively constant from the Lower Pennsylvanian through the Lower Permian. During this time, a series of other amniote groups appear in the fossil record. Each may have independently evolved from the protorothyrid pattern. None show close relationships with one another. *From Carroll, 1982.*

centration of muscle origins along the thickened areas. The forces applied by the muscles would result in further thickening.

Very thin areas of bone between the thickened ridges are subject to cracking, especially where they are crossed by sutures, such as the area of the cheek in primitive amniotes where the squamosal, postorbital, and jugal join. During feeding, force is concentrated on the sutures. This stress could be dissipated by the development of a larger opening with rounded margins in which the force is distributed around the periphery. (In metal work and bone surgery, rounded openings may be made to dissipate force concentrated by sharp cracks, see Frost, 1967). This factor may account for the initial fenestration seen in synapsids and early diapsids. The different proportions of the skull in the specific ancestors of these groups may account for the particular position of the temporal openings.

DIVERSITY AMONG PRIMITIVE ANAPSIDS

CAPTORHINOMORPHS

We may include most of the primitive anapsids from the Pennsylvanian and Lower Permian in the suborder Captorhinomorpha, within which several families may be recognized. The Protorothyridae, which includes the earliest and most primitive amniotes, appears to be a very conservative group; except for a slight increase in overall size and a disproportionate increase in the size of the skull, there are few anatomical changes evident from the Lower Pennsylvanian into the Lower Permian (Clark and Carroll, 1973). This family appears to represent a basal stock from which a succession of more specialized groups evolved. These groups include several suborders within the Captorhinida as well as the ancestors of all the advanced amniote groups (Carroll, 1982).

Each of these derivative groups is already well differentiated when it first appears in the fossil record, and the specific times of their derivation has not been established. Some groups do not appear in the fossil record until the Middle Permian, but their ancestors may have diverged any time after the early Pennsylvanian.

The Acleistorhinidae and Bolosauridae are among the minor groups that are recognized as distinct families (Figure 10-13). Both have lower temporal openings but show no other similarities with the synapsids. *Bolosaurus* has molariform cheek teeth with a specific occlusal pattern. The coronoid process of the lower jaw is elevated to provide a more powerful bite.

The captorhinids are one of the most successful immediate descendants of the protorothyrids (Figure 10-14). They diversified in the Lower and Middle Permian and

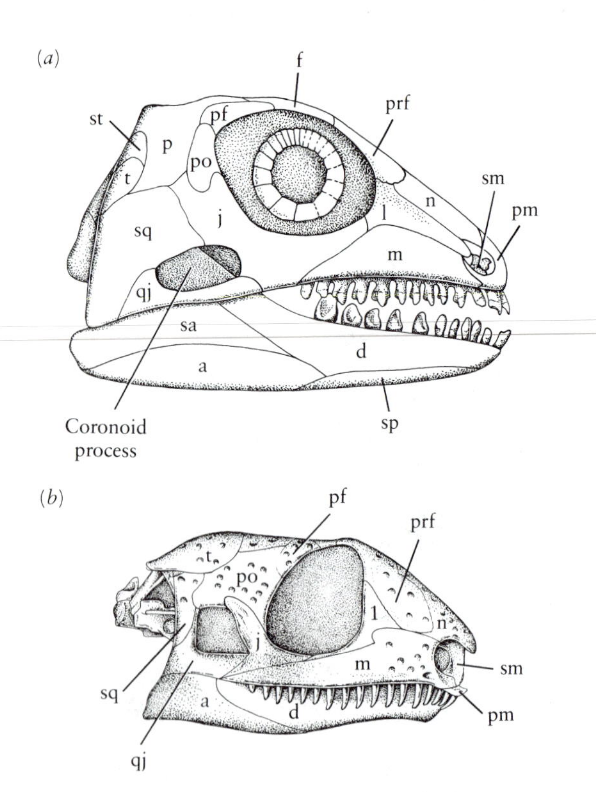

Figure 10-13. SKULLS OF TWO PRIMITIVE AMNIOTES FROM THE LOWER PERMIAN SHOWING LATERAL TEMPORAL OPENINGS. They show no close affinities with other groups. (*a*) *Bolosaurus*, in which the cheek teeth are expanded and show precise occlusion; note also the high coronoid process. (*b*) *Acleistorhinus*. Abbreviations as in Figure 8-3. *From Daly, 1969.*

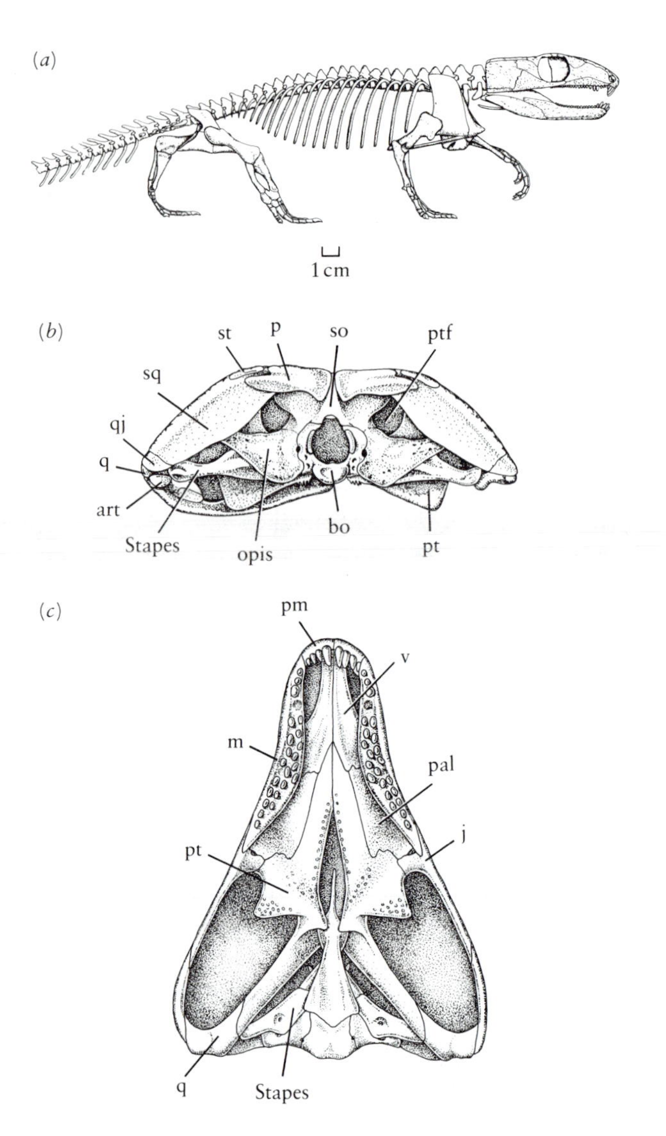

Figure 10-14. REPRESENTATIVES OF THE FAMILY CAPTORHINIDAE. (*a*) Reconstruction of the skeleton of *Eocaptorhinus*. *From Heaton and Reisz, 1980.* (*b*) Occipital view of *Eocaptorhinus*. *From Heaton, 1979.* Note the massive stapes and large posttemporal fossa. (*c*) Palate of *Captorhinus*, showing multiple tooth row. The ectopterygoid bone is missing; its position is replaced by a medial process of the jugal. *From Clark and Carroll, 1973.* Abbreviations as in Figure 8-3.

are known in Africa, Russia, India, China, and North America. Heaton (1979), Gaffney and McKenna (1979), de Ricqlès and Taquet (1982), and Kutty (1972) have published recent studies of this group. Most captorhinids have multiple rows of marginal teeth. This may have come about by a delay in tooth loss at the time new teeth were added. As many as 12 rows were functional at a time, with the earlier tooth rows "drifting" laterally across the jaw surface (de Ricqlès and Bolt, 1983). Such a dentition might have been an adaptation to feeding on plant material or hard-shelled invertebrates for which a wide tooth plate might be more effective than just a single row of marginal teeth.

Captorhinids are common reptiles in the early Permian, and we can recognize many genera on the basis of the number and arrangement of the tooth rows. The supratemporals are reduced and later lost, and the tabulars are missing. In advanced genera, the braincase is supported by the paroccipital processes of the otic capsules, which extend to the cheek. The early members of the group were as small as the protorothyrids, but some of the later survivors had skulls up to 40 centimeters long. In relation to their large size, the neural arches are swollen and the zygapophyses laterally placed, like those of diadectids, seymouriamorphs, and limnoscelids. Captorhinids are not known to survive the Permian, but early members of the group show some characteristics that are expected in the ancestors of turtles, including the loss of the ectopterygoid bone (Gaffney and McKenna, 1979).

MILLEROSAURS

The millerosaurs (Figure 10-15) which are known from the middle and late Permian of South Africa, may be considered as a logical extension of the way of life seen in primitive captorhinomorphs. They have small bodies, proportionately small skulls, and simple conical teeth, which suggest an insectivorous diet. As Gow (1972) described, the skull roof is primitive in the retention of large postparietals, tabulars, and supratemporals, but other cranial features are significantly specialized. There is typically (but not always) a lateral temporal opening; the lower temporal bar may not be complete, and the quadrate, squamosal, and quadratojugal may have been movable relative to both the maxilla and the skull roof. The jaw articulation is anterior to the occipital condyle, so that the posterior margin of the cheek is in line with the fenestra ovalis. The stapes is a small, light rod that extends directly laterally. The squamosal and quadratojugal are embayed posteriorly like an otic notch to support a tympanum. This pattern is expected for the structure of an impedance matching ear and is apparently the first manifestation of such a structure in amniotes. Most advanced diapsids and turtles support the tympanum by the quadrate. In millerosaurs, the quadrate retains the primitive configuration seen in captorhinomorphs.

The millerosaurs were succeeded in time by true lizards, which probably were more successful in exploiting the same general way of life. Some authors have suggested that millerosaurs were ancestral to lizards, but it is now clear that lizards (as discussed in the next chapter) evolved from primitive diapsids with a dorsal as well as a lateral temporal opening.

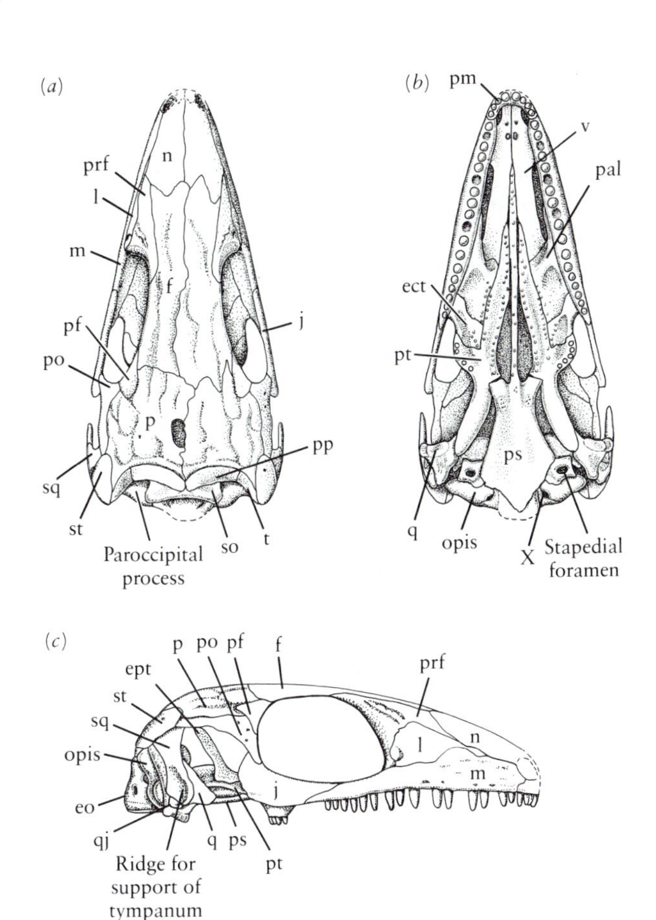

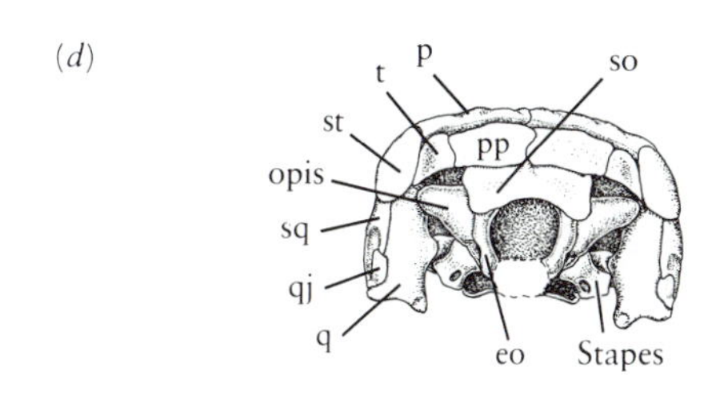

Figure 10-15. SKULL OF THE MILLEROSAUR *MILLEROSAURUS*, ×2. (*a*) Dorsal view. (*b*) Palatal view. (*c*) Lateral view. (*d*) Occipital view. Abbreviations as in Figure 8-3. *Based on specimens in the Bernard Price Institute, University of Witwatersrand, South Africa.*

PROCOLOPHONOIDS

Another group that evidently evolved from the primitive captorinomorph stock are the procolophonoids (Figures 10-16 and 10-17). They are first recognized in the late Permian and are common and diverse throughout the world until the end of the Triassic. We have had difficulty establishing their ancestry because of the specialized anatomy of the later members of this group, but recently described specimens from the Upper Permian of Russia and Madagascar link them with the basic protorothyrid stock (Ivachnenko, 1979).

Procolophonoids retain what appears to be a very primitive pattern of the postcranial skeleton. Like protorothyrids, but in contrast to all other reptiles from beds above the Lower Permian, the caudal ribs are not fused to the centra and extend posteriorly to parallel the axis of the tail. Within the group (and like the larger captorhinids), the neural arches widen and the zygapophyses flatten to facilitate lateral undulation. The limbs show a sprawling posture. The endochondral shoulder girdle is ossified in three units that are distinct in the adults: the dorsal scapula and the anterior and posterior coracoids. These elements are evident in immature specimens of cap-

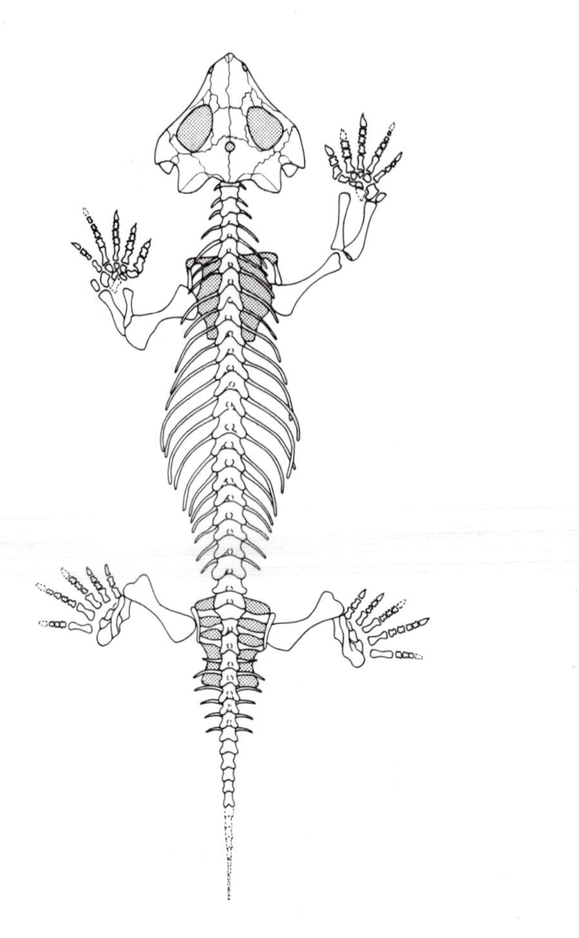

Figure 10-16. Skeleton of the Lower Triassic procolophonoid *Procolophon*, ×⅓. *From Colbert and Kitching, 1975. Courtesy of the Library Services Department, American Museum of Natural History.*

torhinids and pelycosaurs and may be primitive for amniotes, but in most early genera they coossify at an early stage so that they are indistinguishable in the adult. The attachment of the pelvic girdle is enhanced in all procolophonoids by the incorporation of a third sacral rib.

The skull of procolophonoids is specialized in having the orbital margin embayed posteriorly to expose the area of the jaw musculature as a pseudotemporal opening. As in millerosaurs, the jaw articulation is anterior to the occipital condyle and there is a narrow, laterally directed stapes that probably participated in an impedance-matching system. As in millerosaurs, the tympanum would have been supported by the squamosal rather than by the quadrate, as was the case in most diapsids and turtles.

The dentition of early procolophonoids from the Upper Permian of Russia, Madagascar, and South Africa is primitive. It consists of a large number of small, peglike teeth. The skull roof is thin and fragile. In the widespread and diverse Triassic procolophonoids, the number of teeth is much reduced and each is a bulbous, transversely expanded structure that was possibly associated with a herbivorous diet. The skull is enlarged and the bones variably thickened (Colbert, 1946; Carroll and Lindsay, 1985).

In the Middle and Upper Triassic, the late members of the group reach a relatively large size and the skulls

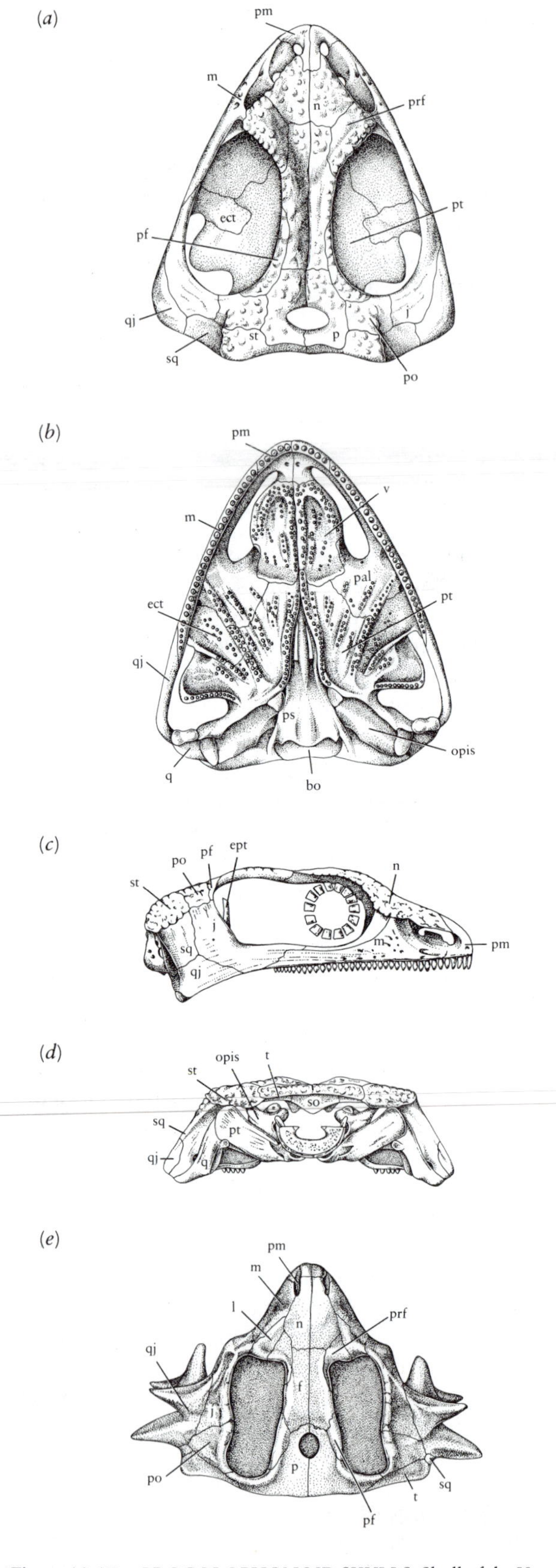

Figure 10-17. PROCOLOPHONOID SKULLS. Skull of the Upper Permian procolophonoid *Nyctiphruetus* in (*a*) dorsal, (*b*) palatal, (*c*) lateral, and (*d*) occipital view, ×1. *From Ivachnenko, 1979.* (*e*) The Upper Triassic genus *Hypsognathus*, ×⅓ *From Colbert, 1946. Courtesy of the Library Services Department, American Museum of Natural History.* Abbreviations as in Figure 8-3.

become specialized in a bizarre manner by the elaboration of long processes from the lower cheek, which probably protected animals that, to judge by their postcranial skeleton, must have been clumsy and slow moving.

PAREIASAURS

The largest members of the Captorhinida are the pareiasaurs from the Middle and Upper Permian of Africa, western Europe, Russia, and China (Figures 10-18 and 10-19). They were elephantine animals that approached a length of 3 meters. The limbs were stocky but held in a more upright pose than in other primitive amniotes to support the massive trunk. The scapula is much longer than in other primitive tetrapods, and the pelvis has an almost mammalian configuration, with the pubis and ischium small and rotated posteriorly behind the elongate ilium. This configuration may have helped accommodate muscles that moved the rear limbs in a manner approaching the fore and aft gait of mammals. The feet were short, had a reduced number of phalanges, and faced anteriorly.

The skulls are short, massive, and laterally expanded. The jaw articulation is well anterior to the occipital condyle, which increased the mechanical advantage of the jaw musculature while decreasing the gape. The palate is strongly integrated with the base of the braincase and the margins of the skull. The teeth have laterally compressed leaf-shaped crowns that are similar in shape to the teeth of modern herbivorous lizards. Together with the massive trunk region, they suggest that the pareiasaurs had a herbivorous diet. Most genera had small bony plates embedded in the skin of the trunk region.

Although pareiasaurs resemble both the diadectids and procolophonids in adapting to a herbivorous diet, the dentition and most aspects of the cranial anatomy are very different, which discourages any suggestion that they might be closely related. The origin of the pareiasaurs remains speculative. Wild (1985) provides the latest information bearing on their possible relationships.

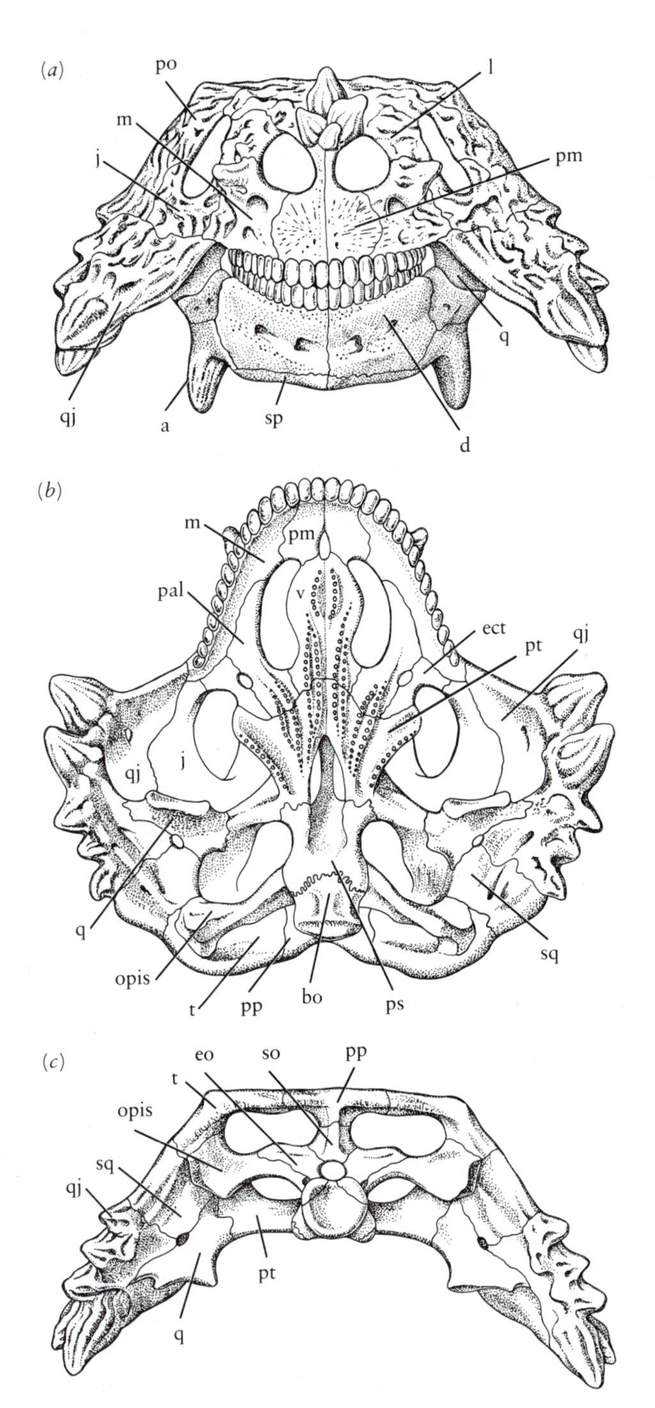

Figure 10-19. SKULL OF THE PAREIASAUR *SCUTOSAURUS*. (*a*) Anterior, (*b*) palatal, and (*c*) occipital views. Specimen is from the Upper Permian of Russia. Width about 50 centimeters. Abbreviations as in Figure 8-3. *From Kuhn, 1969.*

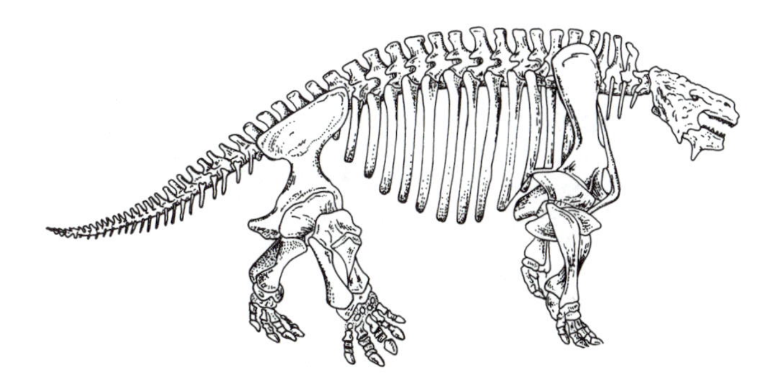

Figure 10-18. THE PAREIASAUR *SCUTOSAURUS* FROM THE LATE PERMIAN OF RUSSIA; ORIGINAL 2 METERS LONG. *From Gregory, 1951. Courtesy of the Library Services Department, American Museum of Natural History.*

MESOSAURS

Although the basic adaptation of amniotes was toward a terrestrial way of life, again and again divergent groups have become specialized for life in the water. The earliest-known amniotes that were fully committed to an aquatic way of life were the mesosaurs (Figures 10-20 and 10-21) from the Permian of southern Africa and eastern South America. Conservative features of their skeleton resemble

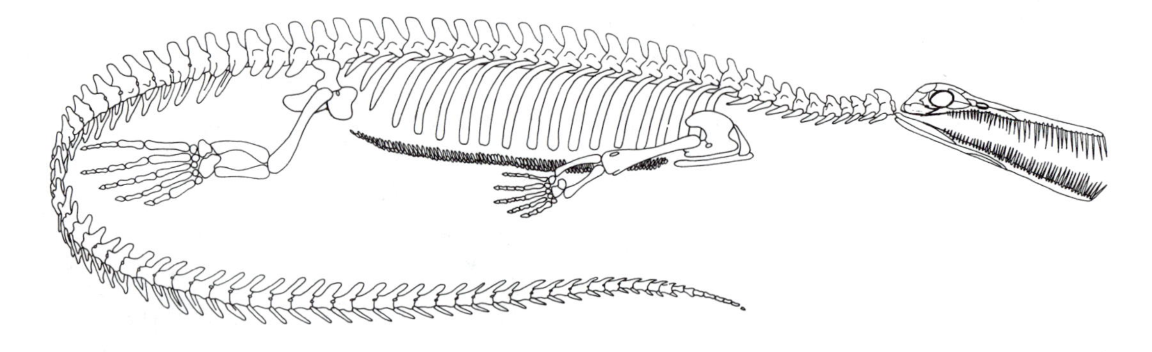

Figure 10-20. *MESOSAURUS* SKELETON, ABOUT 1 METER LONG. This aquatic genus is known from the Middle Permian of South America and South Africa. *Modified from von Huene, 1941, and McGregor, 1908.*

the primitive captorhinomorphs, but the body proportions were altered in adapting to swimming and feeding in the water.

The snout was greatly elongated, a common feature of many secondarily aquatic tetrapods. The very long, slender teeth may have formed a straining device that enabled the mesosaurs to feed on the small crustaceans that abound in the same deposits. The skull has long been thought to have a lateral temporal opening, as in synapsids, but specimens being described by Olafson and Reisz show that the cheek is not fenestrated.

The neck is much longer than that of the captorhinomorphs—another feature common to secondarily aquatic reptiles—with 10 vertebrae anterior to the shoulder girdle. This condition may have been achieved by a posterior shift of the shoulder girdle or by the addition of extra cervical vertebrae.

The long and laterally compressed tail probably served for aquatic propulsion. The ribs at the base of the tail are fused to the caudal vertebrae, unlike those of more primitive amniotes, to permit more solid attachment of the muscles. Surprisingly, the more posterior vertebrae show evidence of caudal autotomy like those of the terrestrial captorhinids.

The neural arches of the trunk are widely expanded, which limited twisting of the column but facilitated lateral bending. The trunk ribs are thickened, approaching the shape of bananas, in contrast with the slim ribs of most primitive amniotes. The same pattern is seen in modern sirenians, which show a similar degree of aquatic adaptation. Thickening of the ribs and an increase in their internal ossification, a condition termed **pachyostosis,** would have increased the weight of the animals so that they could stay submerged without muscular effort.

The girdles are somewhat smaller than their counterparts in protorothyrids and captorhinids, and the blade of the scapula is notably shorter. The ulna and radius and the tibia and fibula are shorter relative to the humerus and femur. The rear foot is a large, paddle-shaped structure, with the fifth toe more elongated than in other early amniotes.

The presence of mesosaurs on both sides of the Atlantic basin was used as an argument for the juxtaposition of Africa and South America long before continental drift became an accepted theory in geology. The mesosaurs may have been restricted to a single, large body of water. If we judge from its size and the presence of numerous marine invertebrates, this body of water was apparently a salt-water basin. No amniotes other than mesosaurs are known from the Carboniferous or Lower Permian in South America or southern Africa. There is no evidence for the relationship of mesosaurs to any subsequent group of aquatic reptiles. McGregor (1908) and von Huene (1941) were the last to describe mesosaurs. Additional study might clarify their taxonomic position.

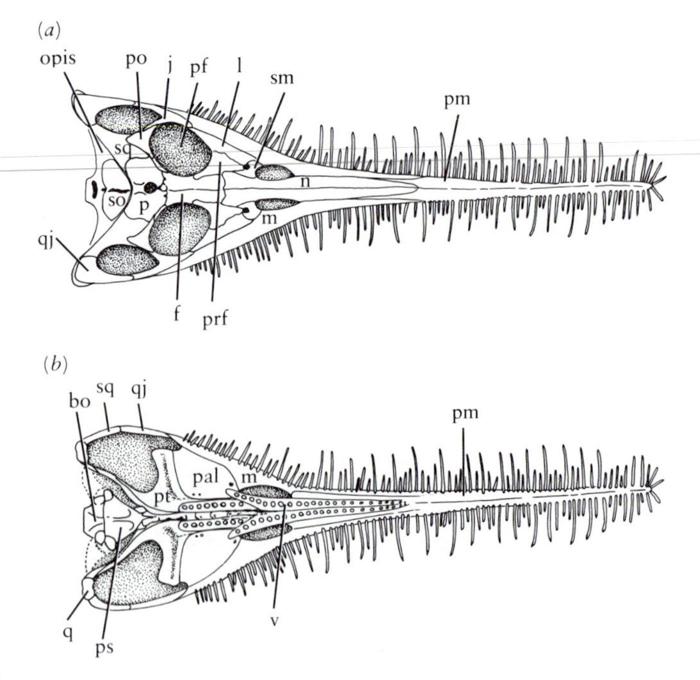

Figure 10-21. SKULL OF *MESOSAURUS.* (*a*) Dorsal and (*b*) palatal view. Recent work indicates that the cheek was solidly ossified, without a lateral temporal opening. Abbreviations as in Figure 8-3. *From von Huene, 1941.*

EUNOTOSAURUS

Some authors have suggested that *Eunotosaurus*, from the middle Permian of southern Africa, was a link between primitive amniotes and the ancestors of turtles. The post-cranial skeleton shows widely expanded ribs, like those of turtles, and there are only 10 trunk vertebrae with greatly elongate centra (Figure 10-22). The ilium is very high and attached by only a single sacral rib. In other features of the postcranial skeleton, *Eunotosaurus* remains primitive. There is no trace of dermal armor, and the blade of the scapula is external to the ribs. What is known of the limbs resembles the pattern of captorhinomorphs.

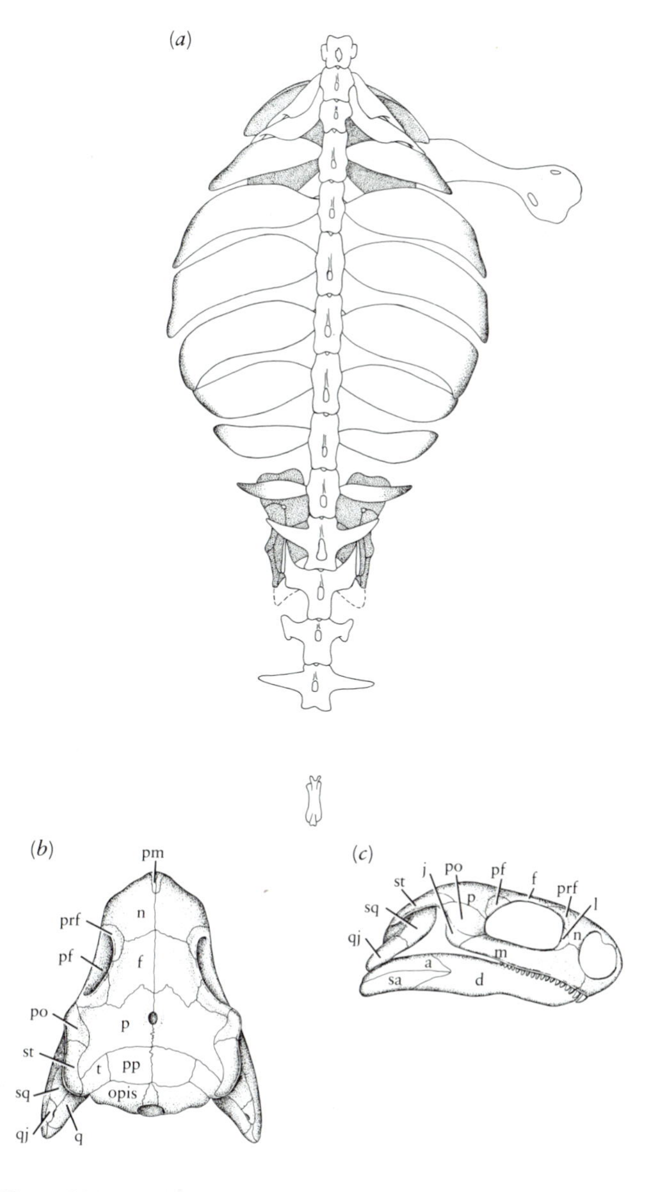

Figure 10-22. *EUNOTOSAURUS* FROM THE UPPER PERMIAN OF SOUTHERN AFRICA. (*a*) Postcranial skeleton. *From Cox, 1969.* (*b* and *c*) Skull. The pattern of the trunk and ribs is similar to that of turtles, but the skull shows no evidence of affinities with that group. *From Keyser and Gow, 1981.* Abbreviations as in Figure 8-3.

Until recently, very little was known of the skull. Keyser and Gow (1981) demonstrate that *Eunotosaurus* shares no important derived features of the cranium with turtles. The presence of a platelike occiput, which is formed by very large postparietals that reach the opisthotic would appear to preclude development of the pattern of the adductor jaw musculature that is characteristic of all Chelonia. In turtles, the musculature extends posteriorly through the occiput via the posttemporal fossae. The cheek of *Eunotosaurus* is deeply emarginated ventrally, with the quadrate and quadratojugal being separated from the jugal. There is a normal marginal dentition.

Eunotosaurus is classified in a distinct suborder of the Captorhinida but has no obvious affinities with other members of this group.

TESTUDINES

All of the groups so far discussed became extinct by the end of the Triassic. All the advanced amniotes evolved from among the early captorhinomorphs. The ancestors of the mammals and diapsids were already distinct within the Pennsylvanian. A third major group of modern amniotes, the turtles, may also have evolved from the early Captorhinida. Because they did not develop orthodox temporal openings, turtles are usually included within the subclass Anapsida and distinguished only as a separate order, the Testudines or Chelonia.

However, the assignment of the turtles to the Anapsida is somewhat arbitrary, since their ancestors probably diverged from the primitive captorhinomorph stock later than either the synapsids or the diapsids. In addition, their overall anatomy and way of life is so different from that of the early anapsids that it is more logical to place them in a distinct subclass.

The earliest turtles are found in Upper Triassic sediments in Germany. They are immediately recognizable from the pattern of the shell, which is closely comparable to that of modern genera. No trace of earlier or more primitive turtles has been described, although turtle shells are readily fossilized and even small pieces are easily recognized. Apparently the earlier stages in the evolution of the shell either occurred very rapidly or took place in an environment or part of the world where preservation and subsequent discovery were unlikely.

In relationship to the evolution of the shell, the postcranial skeleton of even the earliest turtles is so altered from that of primitive amniotes that it gives few clues of their specific relationships. The skulls of late Triassic turtles are also highly specialized. There is no evidence for the prior existence of temporal openings, which precludes their close relationship to early synapsids or diapsids. The presence of large posttemporal fossae and the absence of the ectopterygoid bone in the palate are features that are shared with members of the Family Captorhinidae, but the skull is otherwise so much altered that we cannot

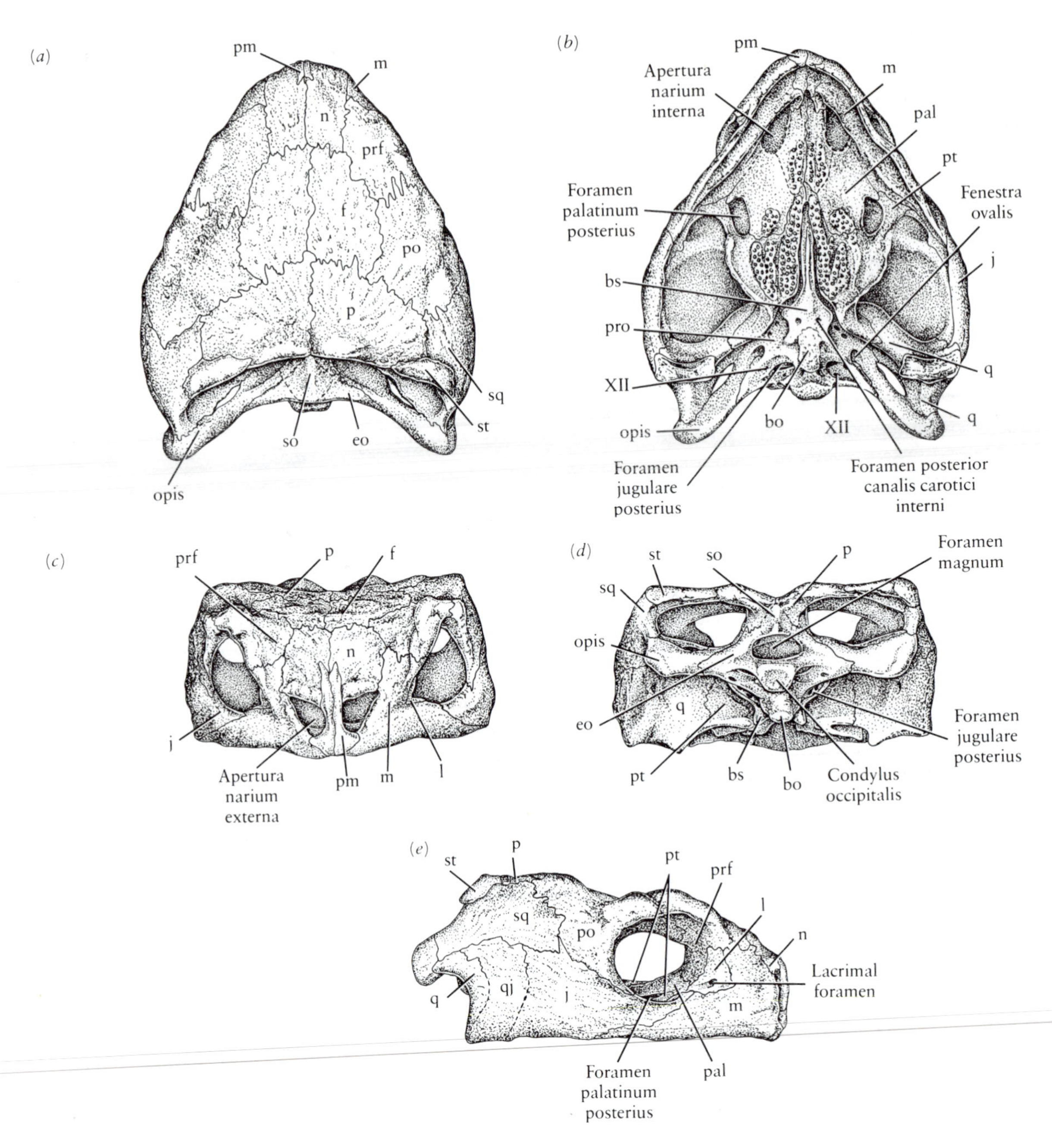

Figure 10-23. SKULL OF ONE OF THE EARLIEST KNOWN TURTLES, *PROGANOCHELYS* FROM THE UPPER TRIASSIC OF GERMANY. $\times\frac{1}{2}$ (*a*) Dorsal view. (*b*) Palatal view. (*c*) Anterior view. (*d*) Occipital view. (*e*) Lateral view. Abbreviations as in Figure 8-3. *From Gaffney and Meeker, 1983.*

establish with assurance that turtles evolved from that family rather than from some other, as yet unrecognized, lineage of early anapsids.

TRIASSIC TURTLES

Many superbly preserved skeletons of turtles are found in the late Triassic of Germany. Other material has recently been reported from the Upper Triassic of Thailand (de Broin, Ingavat, Jauvier, and Sattayarak, 1982) and in beds of early Jurassic age in North America and southern Africa (Gaffney, 1987).

Most of the specimens from Germany belong to the genus *Proganochelys,* which represents an early stage in the evolution of chelonians (Figures 10-23 to 10-25). The specimens are large for turtles, approaching 1 meter in length. In their high arched shell and large body size, they resemble modern tortoises that are terrestrial in habits and herbivorous in diet.

The skull, which Gaffney and Meeker (1983) recently described, already has many features that resemble those in modern turtles. Marginal teeth are lost, and the surface of the premaxilla, maxilla, and dentary resemble the areas in living turtles that are covered with a horny beak. The temporal region that surrounds the jaw muscles, is solidly roofed, as in primitive amniotes. The tabular and postparietal bones of the skull table are lost, but the supratemporal, lacrimal bone, and lacrimal duct, which are missing in later turtles, are retained. In contrast with most advanced reptiles, the quadratojugal remains a large element in *Proganochelys* and other turtles. A pineal opening no longer pierces the skull roof.

The quadrate is embayed posteriorly to form a broad otic notch for support of a tympanum, a condition not achieved in any members of the Captorhinomorpha. Dorsally, the quadrates are attached to the paroccipital processes, which form a strong bar beneath the posttemporal fossae. These openings are large and separated by a narrow supraoccipital, as in the captorhinids and later turtles.

The palate is primitive in retaining rows of denticles on the pterygoid and vomer and in having a loose attachment with the base of the braincase, but the basicranial articulation has shifted posteriorly toward the level of the jaw articulation and the chamber for the adductor jaw musculature is closer to the occipital surface than in more primitive amniotes.

The shell of *Proganochelys* broadly resembles that of modern terrestrial turtles (Figure 10-24). There is a broad arched carapace dorsally and a flat plastron protecting the ventral surface of the body. The plastron is attached to the carapace laterally by an area termed the bridge and is broadly open anteriorly for the head and forelimbs and posteriorly for the rear limbs and tail.

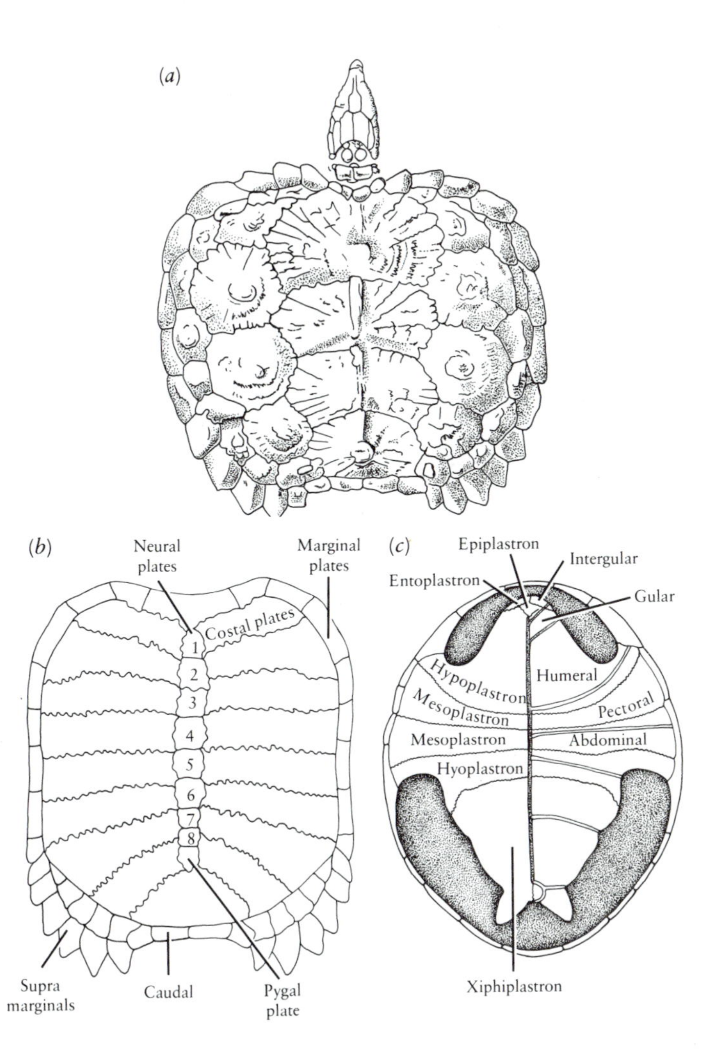

Figure 10-24. (*a*) Dorsal view of the carapace of the Triassic turtle *Proganochelys*, showing the pattern of the epidermal scutes. *From Młynarski, 1976.* (*b*) Carapace of *Proganochelys* drawn so as to emphasize the pattern of the dermal bones. This pattern is common to all turtles. *From Jaekel, 1915.* (*c*) Plastron of the Upper Triassic pleurodire *Proterocheris*. On the left side, the single lines show the division between the dermal elements; the double lines on the right side outline the epidermal scutes. *From Młynarski, 1976.*

The carapace is composed entirely of bone that forms as new centers of ossification within the dermis. Bony scutes are present in the skin of pareiasaurs and crocodiles as well as in several Mesozoic groups, but they never develop such a solid interconnected armor as that of turtles.

The plates that make up the carapace compare closely with those of most later turtles. Along the midline there are eight neural plates, each fused to the neural spines of the underlying vertebrae. Further medial plates, including an anterior nuchal and one or more pygal plates posteriorly, are not fused to the vertebrae. Eight pairs of pleural or costal plates that are fused to the surfaces of the ribs make up the bulk of the carapace. More laterally there are 11 pairs of marginal plates. These complete the shell in modern turtles, but *Proganochelys* also has supramarginal plates forming the periphery of the carapace.

The plastron of *Proganochelys* appears modern in its general configuration but differs from that of most living turtles in having a greater number of separate elements. Like the carapace, most of the bones of the plastron form from newly developed centers of ossification. In contrast, the anterior plates, the paired epiplastron and median entoplastron, are derived from the expanded ventral surfaces of the clavicles and interclavicle. In most modern turtles, the central portion of the plastron, including the bridge, is formed from two large paired plates, the hypoplastra and hyoplastra. In *Proganochelys* and other primitive turtles, there are one or more pairs of additional bones, the mesoplastra, between the hypoplastra and the hyoplastra. As in modern turtles, the posterior margin is composed of paired xiphiplastra.

In addition to the dermal elements, living turtles have a regular pattern of large epidermal scutes overlapping the bones of the carapace and plastron. These scutes do not fossilize, but they are closely integrated with the underlying bones and impressions of their margins indicate that they were already present in *Proganochelys*.

The vertebral column beneath the shell is greatly modified in all turtles. Only 10 trunk vertebrae are present, and the individual centra are considerably elongated. Both the neural arches and the ribs are attached between

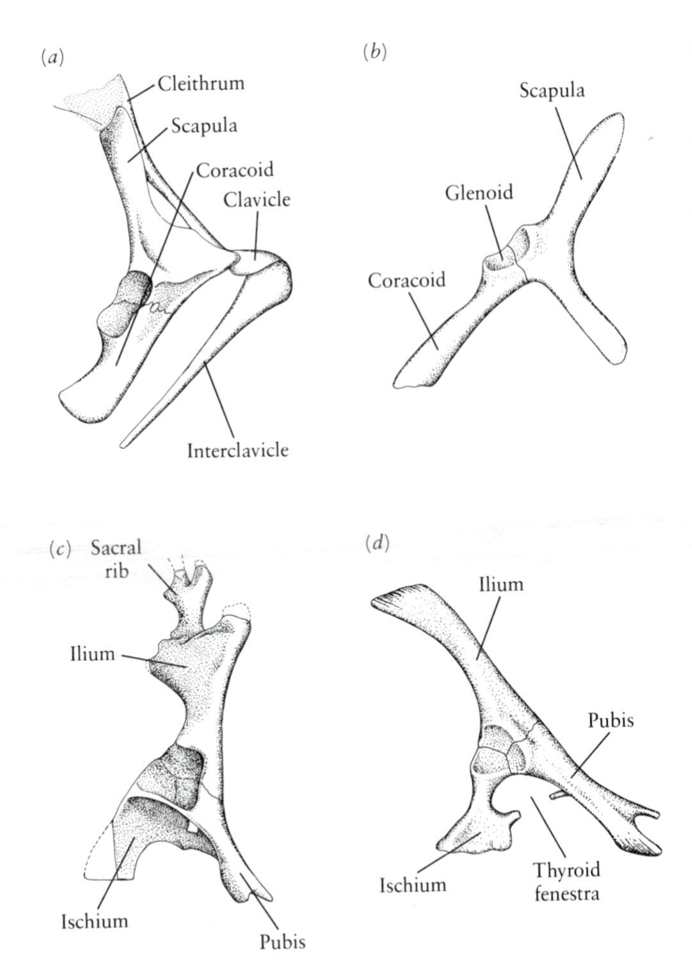

Figure 10-25. PECTORAL AND PELVIC GIRDLES OF TURTLES. *(a* and *c) Proganochelys. From Jaekel, 1915–1916. (b* and *d)* A modern turtle.

the centra. The eight cervical vertebrae in *Proganochelys* show no specializations that would enable the head to be retracted, as is the case in the modern turtles. The neck is protected by overlying dermal ossifications. This genus has only a single pair of sacral ribs, but most turtles have two.

Turtles are unique among terrestrial vertebrates in having the ribs external to the girdles. In modern turtles, the ribs assume an unusually superficial position during embryological development. A broad disk of tissue in which they and the carapace ossify spreads out laterally beyond the girdles as a result of rapid growth. However, this developmental specialization does not provide any hint as to the way in which this pattern evolved phylogenetically.

The limbs and girdles of all turtles are greatly modified to accommodate the shell. Despite their partial integration in the plastron, the clavicles and interclavicle are still clearly recognizable in *Proganochelys* (Figure 10-25). In the primitive state, the cleithrum (which is missing in later turtles) is a stout element that extends dorsally and expands beneath the surface of the carapace. The scapula and coracoid are separated by a suture running through the broad, widely open glenoid. In later turtles, an anterior process from the base of the scapula gives the primary girdle a triradiate shape. The pelvic girdle is more conservative. In most turtles, it retains the triradiate pattern of primitive reptiles, although a wide opening, the thyroid fenestra, develops between the ventral portion of the pubis and ischium.

The configuration of the shell and its relationship to the girdles restricts the movement of the limbs and confines their posture to the sprawling gait of primitive captorhinomorphs. The humerus and femur are advanced over the primitive configuration in having a longer, more slender shaft and a hemispherical head, which presumably enabled them to move more freely in the otherwise confined space within the shell. The feet are much modified, with the phalangeal count reduced to 2, 2, 2, 2, 2 in *Proganochelys*. Later turtles more commonly show a count of 2, 3, 3, 3, 3.

MODERN TURTLES

Proganochelys represents a primitive stock, the suborder Proganochelida, from which all other turtles may have evolved. More advanced turtles may all be classified in one of two suborders, the Cryptodira and the Pleurodira. These names refer to the manner in which the living members of these groups retract their necks. The pleurodires do so by lateral flexure of the cervical vertebrae and the cryptodires by vertical flexure. The fossil record shows that changes in the configuration of the cervical vertebrae that enabled the head to be retracted did not occur before the late Cretaceous. Other differences in the anatomy of the skull and the nature of the shell characterize each of these groups as far back as they can be traced in the fossil record. The pattern that is typical of the cryptodire skull is evident by the Upper Jurassic. We do not find typical pleurodire skulls until the Lower Cretaceous, but shells with peculiarities common to the later genera are known as early as the Upper Triassic.

Although the basic pattern of the shell common to living turtles was achieved by the late Triassic, subsequent changes have occurred in both of the advanced suborders. Nearly all advanced turtles have lost the extramarginal plates. Most pleurodires retain the mesoplastra, which are lost in all modern cryptodires. Pleurodires are specialized in having the pelvic girdle fused to the plastron and suturally attached to the carapace.

The skull changes significantly between the proganochelids and more modern turtles. These changes involve the ear region, the adductor chamber, and the area of attachment between the base of the braincase and the palate. The resulting pattern of the jaw musculature is somewhat analogous with that of lungfish (Gaffney, 1975).

In primitive amniotes, such as protorothyrids and early captorhinids, the adductor jaw musculature is vertically oriented and originates from the inside surface of the cheek, the underside of the skull roof, and the lateral surface of the braincase. The otic capsule is a small structure separated from the adductor chamber by the quadrate ramus of the pterygoid. The braincase is loosely attached to the palate by the basicranial articulation and to the back of the skull roof via the supraoccipital. The paroccipital process does not reach the cheek, and the stapes braces the otic capsule against the quadrate (Figure 10-26).

In *Proganochelys,* as in several groups of Triassic diapsids, the paroccipital processes have extended to the cheek and are suturally attached to the top of the quadrates. This change frees the stapes from its supporting role, enabling it to become a light rod that is associated with an impedance-matching system. The quadrate ramus of the pterygoid is low and close to the otic capsule so that the jaw muscles could have extended posteriorly to develop areas of origin around the margins of the large posttemporal fenestrae.

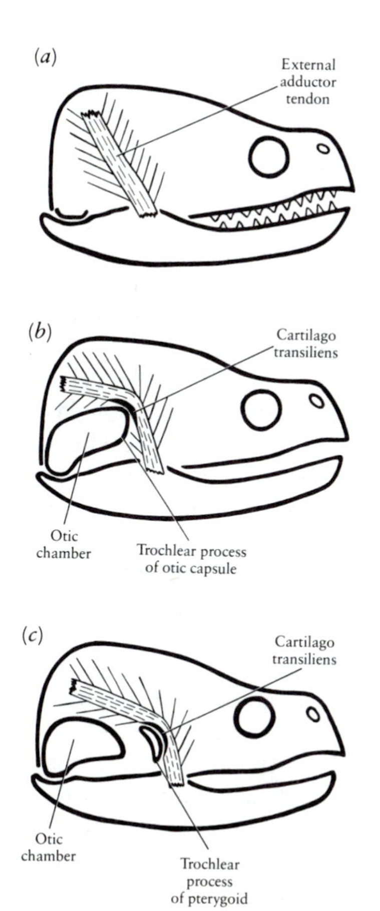

Figure 10-26. EVOLUTION OF THE ADDUCTOR CHAMBER IN ANAPSIDS. (*a*) The condition in primitive anapsids, such as the protorothyrids and captorhinids. This basic pattern is retained in the earliest turtle, *Proganochelys.* (*b* and *c*) The chelonian condition. (*b*) Cryptodires, in which the adductor mandibulae presses over a pulleylike surface on the otic capsule. (*c*) The pleurodire condition, in which the jaw musculature passes over the surface of the pterygoid. *From Gaffney, 1975. Courtesy of the Library Services Department, American Museum of Natural History.*

In bony fish and modern frogs and salamanders, we have seen that the jaw musculature spread out of the primitive confines of the adductor chamber formed by the cheek and palatoquadrate. An analogous process occurred in turtles. Here it is modified by the presence of a very large otic capsule. In all turtles above the level of the proganochelids, the paracapsular network, a fluid-filled sinus in the middle ear, is greatly enlarged compared with most other amniotes and is enclosed in a chamber formed from the palatoquadrate and elements of the braincase.

The great expansion of the otic capsule results in a significant modification of the space occupied by the jaw musculature. Instead of being oriented primarily in a vertical direction, the muscles bend over the otic capsule and extend posteriorly toward the posttemporal fossae. In all advanced turtles, the muscles pass over a pulleylike structure, the **trochlear process** (Figure 10-26 and 10-27). However, this structure develops differently in the two groups of modern turtles. In cryptodires, it forms on the anterior surface of the otic capsule. In pleurodires, it is formed by a lateral process of the pterygoid. This difference provides a clear way of distinguishing between all members of these two groups.

Other differences are evident in the structure of the palate. In *Proganochelys,* the base of the braincase is only loosely attached to the palate. In more advanced turtles a solid sutural attachment develops. In cryptodires, the pterygoid is suturally attached to the basisphenoid and extends out laterally toward the quadrate. In pleurodires, a process of the quadrate extends medially to reach the base of the braincase. Pleurodires also differ from cryptodires in having lost the epipterygoid bone.

Młynarski (1976) classifies all species of fossil turtles, and Gaffney (1979) provides detailed descriptions of the skulls of both fossil and living genera that provide a basis for establishing the phylogenetic relationships of the major groups.

Pleurodires

The recently discovered early Jurassic turtles have not yet been described. By the end of that period, both modern groups were clearly established. Turtle shells that resemble those of modern pleurodires in having the pelvis firmly attached to the carapace and plastron are known from the Upper Triassic and Upper Jurassic. The specimen from the Upper Jurassic is the only recognized representative of the family Platychelyidae.

All other adequately known pleurodires are placed in two families—both with living representatives—the Pelomedusidae and the Chelidae. These families are now limited to the southern continents. The oldest pelomedusid is early Cretaceous in age. In the Upper Cretaceous and Lower Tertiary, this group had a nearly worldwide distribution. The genus *Taphrosphys,* for example, is known from North America, Africa, and South America. Other genera were common in Europe and Asia. Today the group is limited to three genera, *Pelomedusa* and *Pelusios* in

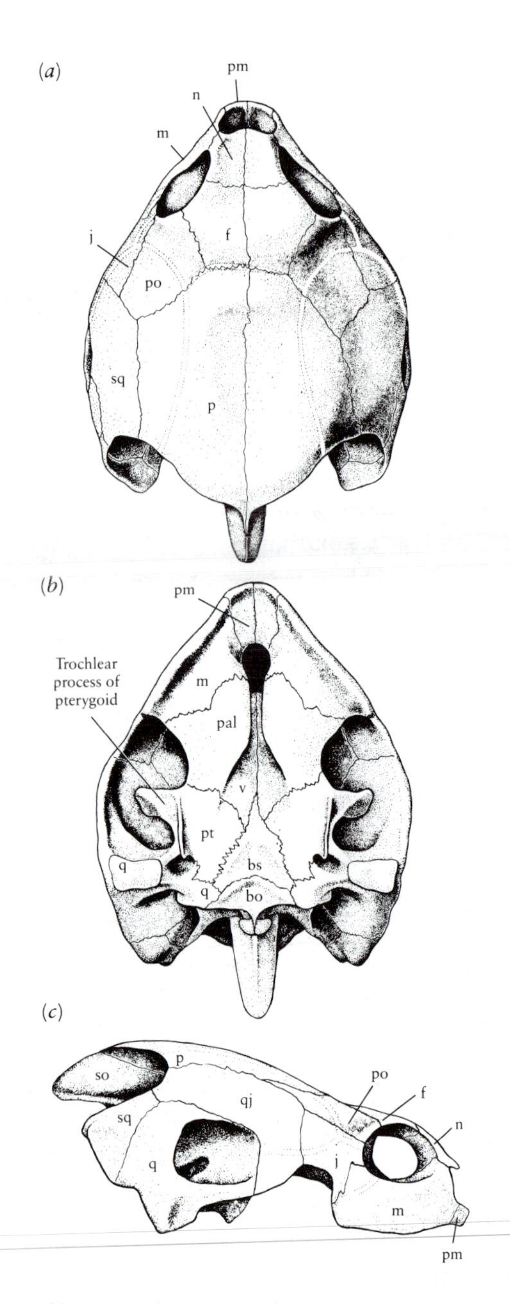

Figure 10-27. THE PELOMEDUSID PLEURODIRE *SHWEBOEMY* FROM THE LATE EOCENE OF EGYPT. (*a*) Dorsal, (*b*) palatal, and (*c*) lateral views. Note the different pattern of attachment between the base of the braincase and the palate in this pleurodire and the cryptodire illustrated in Figure 10-28. Abbreviations as in Figure 8-3. *From Gaffney, 1979. Courtesy of the Library Services Department, American Museum of Natural History.*

Africa and Madagascar and *Podocnemis* in Madagascar and South America. The Pliocene pelomedusid *Stupendemys* from Venezuela is the largest known turtle, fossil or living, with the carapace over 2 meters long (Wood, 1976).

The chelids are advanced over the pelomedusids in the loss of the mesoplastra. The oldest known fossil, from the Miocene, has an essentially modern anatomy (Gaffney, 1979). Chelids are now represented by seven genera in South America, Australia (where they are the dominant freshwater turtles), and New Guinea.

While all living pleurodires inhabit fresh water, extinct specimens include forms with high arched shells (which in living cryptodires is indicative of a terrestrial habit) and others that may have lived in a marine environment.

Cryptodires

Cryptodires are much more diverse in the modern fauna than are the pleurodires. The antiquity of the two groups appears to be comparable, but cryptodire remains have not been described from beds below the Upper Jurassic. Cryptodires apparently underwent a rapid diversification within the Jurassic that culminated in the development of several major groups by the end of that period. The ancestors of the modern groups were probably all distinct by that time. In addition, there were several families that were common in the late Mesozoic and Tertiary but are now extinct.

The extinct cryptodire groups resemble one another in the retention of primitive features. None show evidence of specialization of the cervical vertebrae that would enable them to retract the head, and all retain mesoplastra. The presence of these features was once used to group them (and other primitive genera) in a distinct suborder, the Amphichelida, but we now recognize that their retention of primitive features does not form an adequate basis for establishing close relationships between these groups. The structure of the skull and the probable nature of the jaw musculature allies them with living cryptodires, but we have not established specific affinities with any of the modern superfamilies.

The Baenidae represent a relatively conservative assemblage of genera that is common in North America from the early Cretaceous to the late Eocene. Gaffney (1972) showed that they are more primitive than living cryptodires in having the opening for the internal carotid arteries located midway along the length of the pterygoid suture (Figure 10-28). The internal carotid arteries enter above and behind the pterygoids in modern cryptodires. The primitive condition is shared by a second family, the Glyptopsidae, which we know from the Upper Jurassic of North America and the early Cretaceous of England.

The Meiolaniidae is a further isolated group of turtles known from the early Tertiary and possibly late Cretaceous of South America and the Pleistocene of the Australian region. They are generally similar to living cryptodires but cannot be specifically allied with any other groups within the suborder. The skulls are up to 50 centimeters wide and characterized by bony protruberances that resemble those of the procolophonoids.

Modern cryptodires may be grouped in three superfamilies: the Testudinoidea, including the tortoises and most freshwater turtles; the Chelonioidea, the sea turtles, with the limbs specialized as flippers and the shell (particularly the plastron) reduced; and the Trionychoidea, the soft-shelled turtles. Each of these groups is characterized by details of the basicranial anatomy and the pat-

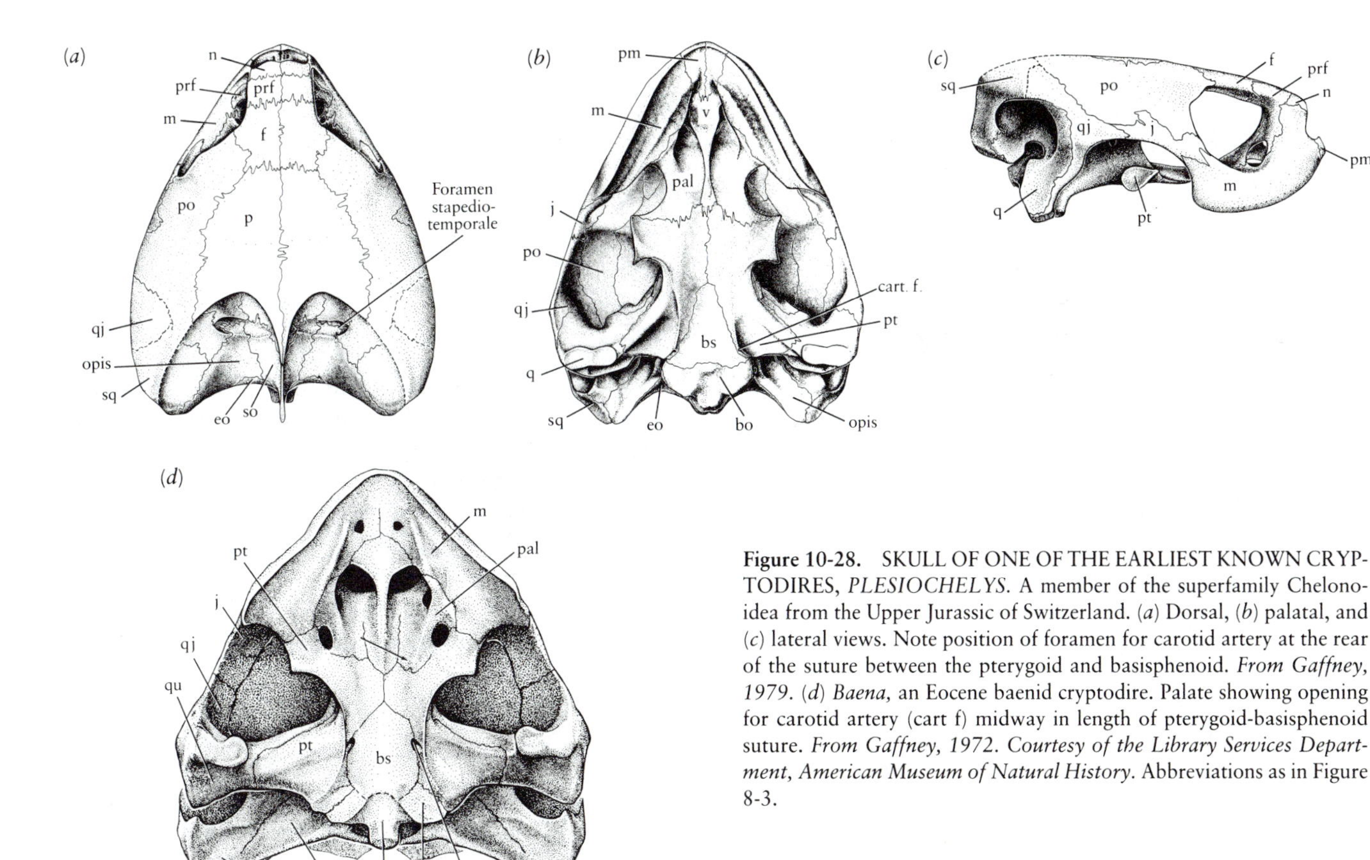

Figure 10-28. SKULL OF ONE OF THE EARLIEST KNOWN CRYPTODIRES, *PLESIOCHELYS*. A member of the superfamily Chelonoidea from the Upper Jurassic of Switzerland. (*a*) Dorsal, (*b*) palatal, and (*c*) lateral views. Note position of foramen for carotid artery at the rear of the suture between the pterygoid and basisphenoid. *From Gaffney, 1979.* (*d*) *Baena*, an Eocene baenid cryptodire. Palate showing opening for carotid artery (cart f) midway in length of pterygoid-basisphenoid suture. *From Gaffney, 1972. Courtesy of the Library Services Department, American Museum of Natural History.* Abbreviations as in Figure 8-3.

tern of the cranial arteries. According to Gaffney (1975), the Chelonioidea and the Trionychoidea can be recognized as monophyletic, but the Testudinoidea has no unique specializations and may represent a primitive grade of cryptodire evolution rather than a well-defined taxonomic group.

The Chelonioidea are the first of the modern superfamilies to appear in the fossil record and are represented by members of the family Plesiochelyidae in the Upper Jurassic. The Cretaceous genera *Protostega* and *Archelon* (Figure 10-29*a*) show the manner of reduction of the shell that is common to this group. The limbs have become highly modified as paddles that propel the body through the water like the wings of penguins. The family Chelonidae, which is known from the Lower Cretaceous, is represented today by four genera, the best known of which is *Chelonia*. *Dermochelys,* the leatherback turtle, belongs to a related family. The plastron and carapace are reduced to a layer of small, irregularly shaped bones that are embedded in leathery skin. This genus is striking in its ability to maintain a body temperature at least 18 degrees above that of the water (Greer, Lazell, and Wright, 1973). Like mammals and birds, the high body temperature is achieved by muscular activity and is retained by an external layer of insulating fat and by countercurrent heat exchangers in the limbs.

The Trionychoidea, which are typified by the family Trionychidae, are primarily freshwater carnivores. They are characterized by the loss of the external horny covering of the shell that is replaced by a soft, leathery skin. The carapace is low and rounded. The peripheral bones are lost in most genera (Figure 10-29*b*). *Trionyx,* which lives today in North America, Asia, and Africa, is known from the Lower Cretaceous and possibly from the Upper Jurassic. We may also include the three families Kinosternidae, Carettochelyidae, and Dermatemydidae within the Trionychoidea.

The Testudinoidea is the last of the major turtles groups to appear in the fossil record, with each of the three modern families in this group occurring first in the early Tertiary. The group lacks the specializations of the shell and limbs seen in the other modern superfamilies and has been considered to represent the general anatomical pattern from which all other cryptodires have evolved. The Chelydridae, which include the snapping turtles, are known as early as the Paleocene. The Emydidae include a large assemblage of freshwater and amphibious genera such as the terrapins, pond, and box turtles. They are common in all continents except Australia. The Testudinidae encompass the most terrestrial of all turtles, the tortoises, which are characterized by high arched shells. They are almost exclusively vegetarian in diet, and the limbs are elephantine. They include the largest of all land turtles. The giant tortoise, *Geochelone* from the Galapagos Islands, is also known on the mainland of Asia, Africa, and South America. Tortoises are known as early as the Eocene and are common elements in Tertiary deposits.

(a)

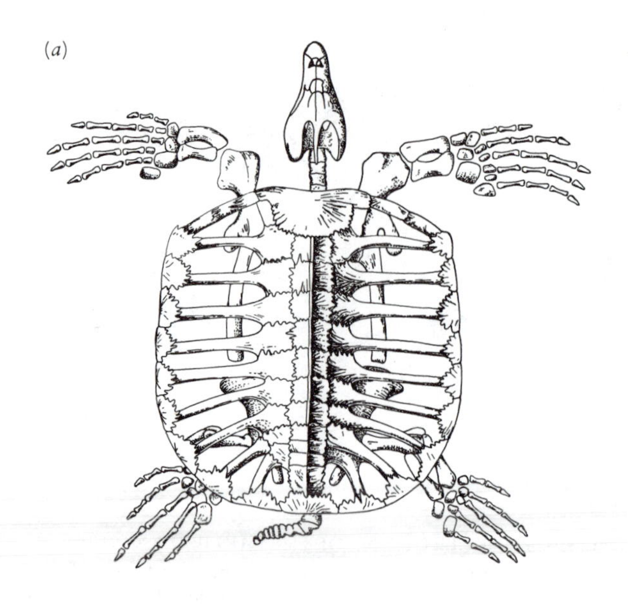

(b)

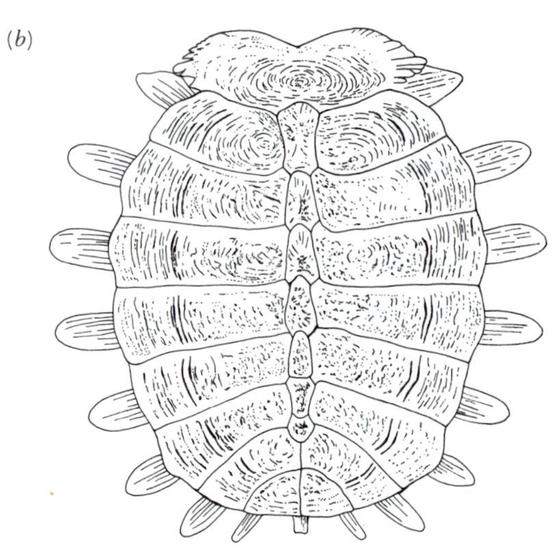

Figure 10-29. (*a*) Dorsal view of the Cretaceous marine reptile *Archelon*. (*b*) The carapace of a trionychid turtle, showing the loss of marginal elements. *From Młynarski, 1976.*

SUMMARY

The amniotes achieve a fully terrestrial way of life. The females lay their eggs on land or retain them within their bodies, and the young develop directly without an aquatic larval stage. This assemblage, which includes reptiles, birds, and mammals, first appears in the fossil record during the early Pennsylvanian. In their general appearance, the early amniotes resembled primitive living lizards. They were distinguished from Paleozoic amphibians by improvements of jaw mechanics, which suggest specialization for an insectivorous diet. The stapes served to connect the braincase to the skull roof and was too massive to have contributed to an impedance-matching system. The body may have been less than 100 centimeters in snout-to-vent length, but the skeleton was extremely well ossified. The limbs were slender, suggesting agile locomotion. Two very important aspects of the soft anatomy, the presence of extraembryonic membranes and stretch receptors associated with the axial and appendicular muscles, are presumed to have been present in early amniotes, although these structures are not reflected in the skeletal anatomy. Early amniotes probably practiced internal fertilization but lacked copulatory organs.

We have not established the specific origin of amniotes, although they do share some derived features with anthracosaurs. The family Protorothyridae, a primitive lineage of small-sized, lizardlike amniotes, persisted throughout the Upper Carboniferous and into the Permian. All other amniotes may have evolved from this common stock. Three distinct patterns of cranial anatomy are recognized which serve as a basis for the classification of amniotes. Primitive amniotes retain a pattern common to Paleozoic amphibians, in which the skull is solidly roofed above the chamber for the adductor jaw musculature. This configuration, termed anapsid, is retained in the turtles. The mammal-like reptiles and their descendants, the mammals, are characterized by the presence of a single pair of openings low on the cheek. This is termed the synapsid condition. All other amniotes, including birds and all living reptiles other than turtles, have two pairs of temporal openings (the diapsid condition) or show modification from this pattern.

The lineage leading to mammals was the first group to diverge from the primitive amniote stock. The diapsids appeared in the late Pennsylvanian. Other early groups include the captorhinids, which are typified by multiple tooth rows, and the lizardlike millerosaurs, which were the first amniotes to develop an impedance-matching middle ear. The pareiasaurs and procolophonoids are divergent groups of herbivores that appear in the middle and late Permian. The mesosaurs are unique among early amniotes in adapting to a secondarily aquatic way of life.

We have not established the specific ancestry of the Testudines, although it may lie among the family Captorhinidae. Primitive turtles are represented by complete skeletons from the Upper Triassic. The structure of the shell is similar to that of modern turtles, except for the possession of additional centers of ossification in both the carapace and plastron. The skull shows the presence of a horny beak but is primitive in the loose attachment of the braincase and the lack of specialization of the adductor chamber. The Upper Triassic genus *Proganochelys* is the only adequately known member of the suborder Proganochelida. All more advanced turtles can be assigned to either the Cryptodira or Pleurodira on the basis of specializations of the skull and shell. In all advanced turtles, the jaw musculature passes over a pulleylike structure, the trochlear process. The position of the trochlear process and the manner of attachment of the braincase to the palate distinguish the two advanced groups. The specialization of the cervical vertebrae that enabled the head to retract also evolved separately in the two groups and was not elaborated until the late Cretaceous. The two modern families of pleurodires appear in the early Cretaceous and Miocene. Modern cryptodires are grouped in three superfamilies. The Chelonoidea, the sea turtles, and the Trionychoidea, the soft-shelled turtles both appear in the Upper Jurassic. The Testudinoidea, including the tortoises and most freshwater turtles, appear to represent the central stock of the modern cryptodires but do not appear in the fossil record until the early Tertiary.

REFERENCES

Carroll, R. L. (1969a). Problems of the origin of reptiles. *Biol. Reviews*, **44**: 393–432.

Carroll, R. L. (1969b). A Middle Pennsylvanian captorhinomorph and the interrelationships of primitive reptiles. *J. Paleont.*, **43**: 151–170.

Carroll, R. L. (1969c). Origin of reptiles. In C. Gans (ed.), *Biology of the Reptilia*, pp. 1–44. Academic Press, London.

Carroll, R. L. (1970). The ancestry of reptiles. *Phil. Trans. Roy. Soc., London*, **257**: 267–308.

Carroll, R. L. (1981). Plesiosaur ancestors from the Upper Permian of Madagascar. *Phil. Trans. Roy. Soc., London*, **293**: 315–383.

Carroll, R. L. (1982). Early evolution of reptiles. *Ann. Rev. Ecol. Syst.*, **13**: 87–109.

Carroll, R. L., and Baird, D. (1972). Carboniferous stem-reptiles of the family Romeriidae. *Bull. Mus. Comp. Zool.*, **143**: 321–363.

Carroll, R. L., Belt, E. S., Dineley, D. L., Baird, D., and McGregor, D. C. (1972). Vertebrate paleontology of Eastern Canada, Excursion A 59. *24th Intl. Geol. Conference*, Ottawa: 1–113.

Carroll, R. L., and Lindsay, W. (1985). The cranial anatomy of the primitive reptile *Procolophon*. *Can. J. Earth Sci.*, **22**: 1571–1587.

Clark, J., and Carroll, R. L. (1973). Romeriid reptiles from the Lower Permian. *Bull. Mus. Comp. Zool.*, **147**: 353–407.

Colbert, E. H. (1946). *Hypsognathus*, a Triassic reptile from New Jersey. *Bull. Am. Mus. Nat. Hist.*, **86**: 225–274.

Colbert, E. H., and Kitching, J. W. (1975). The Triassic reptile *Procolophon* in Antarctica. *Amer. Mus. Novitates*, **2566**: 1–23.

Cox, C. B. (1969). The problematic Permian reptile *Eunotosaurus*. *Bull. Brit. Mus. (Nat. Hist.)*, **18**: 165–196.

Daly, E. (1969). A new procolophonoid reptile from the Lower Permian of Oklahoma. *J. Paleont.*, **43**(3): 676–687.

De Broin, F., Ingavat, R., Janvier, P., and Sattayarak, N. (1982). Triassic turtle remains from northeastern Thailand. *J. Vert. Paleont.*, **2**: 41–46.

de Ricqlès, A. and Bolt, J. R. (1983). Jaw growth and tooth replacement in *Captorhinus aguti* (Reptilia: Captorhinomorpha): A morphological and histological analysis. *J. Vert. Paleont.*, **3** (1): 7–24.

de Ricqlès, A. and Taquet, P. (1982). La faune de vertébrés du Permien Supérieur du Niger I. Le captorhinomorphe *Moradisaurus grandis* (Reptilia, Cotylosauria). *Ann. de Paléont. (Vert.-Invert.)*, **68**, fasc. 1: 33–106.

Frost, H. M. (1967). *An Introduction to Biomechanics*. Charles C Thomas, Springfield, Illinois.

Gaffney, E. S. (1972). The systematics of the North American family Baenidae (Reptilia, Cryptodira). *Bull. Am. Mus. Nat. Hist.*, **147**: 241–320.

Gaffney, E. S. (1975). A phylogeny and classification of the higher categories of turtles. *Bull. Am. Mus. Nat. Hist.*, **155**: 387–436.

Gaffney, E. S. (1979). Comparative cranial morphology of Recent and fossil turtles. *Bull. Am. Mus. Nat. Hist.*, **164**: 65–370.

Gaffney, E. S. (1987). Triassic and early Jurassic turtles. In K. Padian (ed.), *The Beginning of the Age of Dinosaurs: Faunal Changes Across the Triassic-Jurassic Boundry*, pp. 183–187. Cambridge University Press, Cambridge.

Gaffney, E. S., and McKenna, M. C. (1979). A late Permian captorhinid from Rhodesia. *Amer. Mus. Novitates*, **2688**: 1–15.

Gaffney, E. S., and Meeker, L. J. (1983). Skull morphology of the oldest turtles: A preliminary description of *Proganochelys quenstedti*. *J. Vert. Paleont.*, **3** (1): 25–28.

Gow, C. E. (1972). The osteology and relationships of the Millerettidae (Reptilia: Cotylosauria). *J. Zool., London*, **167**: 219–264.

Greer, A. E., Lazell, J. D., and Wright, R. M. (1973). Anatomical evidence for a countercurrent heat exchanger in the leatherback turtle (*Dermochelys coriacea*). *Nature*, **244**(5412): 181.

Gregory, W. K. (1951). *Evolution Emerging*. Macmillan, New York.

Heaton, M. J. (1979). Cranial anatomy of primitive captorhinid reptiles from the late Pennsylvanian and early Permian Oklahoma and Texas. *Univ. Oklahoma, Geol. Survey, Bull.*, **127**: 1–81.

Heaton, M. J. (1980). The Cotylosauria: A reconsideration of a group of archaic tetrapods. In A. L. Panchen (ed.), *Terrestrial Environment and the Origin of Land Vertebrates*, pp. 497–551. Academic Press, London, New York.

Heaton, M. J., and Reisz, R. R. (1980). A skeletal reconstruction of the early Permian captorhinid reptile *Eocaptorhinus laticeps* (Williston). *J. Paleont.*, **54**: 136–143.

Hirsch, K. F. (1979). The oldest vertebrate egg? *J. Paleont.*, **53** (5): 1068–1084.

Holmes, R. (1977). The osteology and musculature of the pectoral limb of small captorhinids. *J. Morphol.*, **152**: 101–140.

Huene, F. von (1941). Osteologie und systematische Stellung von *Mesosaurus*. *Palaeontographica, A*, **92**: 45–58.

Ivachnenko, M. F. (1979). The Permian and Triassic procolophonians from the Russian platform. *Akad. Nauk. SSSR, Paleont. Inst., Tr.*, **164**: 1–80. In Russian.

Jaekel, O. (1915–1916). Die Wirbeltierfunde aus dem Keuper von Halberstadt. *Paläont. Zeitschrift*, **2**: 88–214.

Keyser, A. W., and Gow, C. E. (1981). First complete skull of the Permian reptile *Eunotosaurus africanus* Seeley. *South Afr. J. Sci.*, **77**: 417–420.

Kuhn, O. (1969). Cotylosauria. *Handbuch der Paläoherpetologie*. Part 6. Gustav Fischer Verlag, Stuttgart.

Kutty, T. S. (1972). Permian reptilian fauna from India. *Nature*, **237**: 462–463.

McGregor, J. H. (1908). On *Mesosaurus brasiliensis* no. sp. from the Permian of Brazil. *Comm. Estud, Minas Carvao Pedra Brasil*: 303–336.

Młynarsky, M. (1976). Testudines. In P. Wellnhofer (ed.), *Handbuch der Paläoherpetologie*. Part 7. Gustav Fischer Verlag, Stuttgart.

Peabody, F. E. (1951). The origin of the astragalus of reptiles. *Evolution*, **5**: 339–344.

Pough, F. H. (1973). Lizard energetics and diet. *Ecology*, **54**: 837–844.

Reisz, R. R. (1980). A protorothyrid captorhinomorph reptile from the Lower Permian of Oklahoma. *Roy. Ont. Mus., Life Sci. Contr.*, **121**: 1–16.

Romer, A. S., and Price, L. I. (1940). Review of the Pelycosauria. *Geol. Soc. Am., Spec. Papers*, **28**: 1–538.

Salthe, S. N. (1969). Reproductive modes and the numbers and sizes of ova in the urodeles. *Am. Midl. Nat.*, **81**: 467–490.

Szarski, H. (1968). The origin of vertebrate foetal membranes. *Evolution*, **22** (1): 211–214.

Vaughn, P. P. (1972). More vertebrates, including a new microsaur, from the Upper Pennsylvanian of Central Colorado. *Contr. Sci., Nat. Hist. Mus. Los Angeles County*, **223**: 1–30.

Wever, E. G. (1978). *The Reptile Ear*. Princeton University Press, Princeton.

Wild, R. (1985). Ein Schädelrest von *Parasaurus geinitzi* H. v. Meyer (Reptilia, Cotylosauria) aus dem Kupferschiefer (Perm) von Richelsdorf (Hessen). *Geol. Bl. NO-Bayern (Gedenkschrift B. v. Freyberg)*, **34/35**: 897–920.

Wood, R. C. (1976). *Stupendemy geographicus*, the world's largest turtle. *Breviora*, **436**: 1–31.

CHAPTER XI

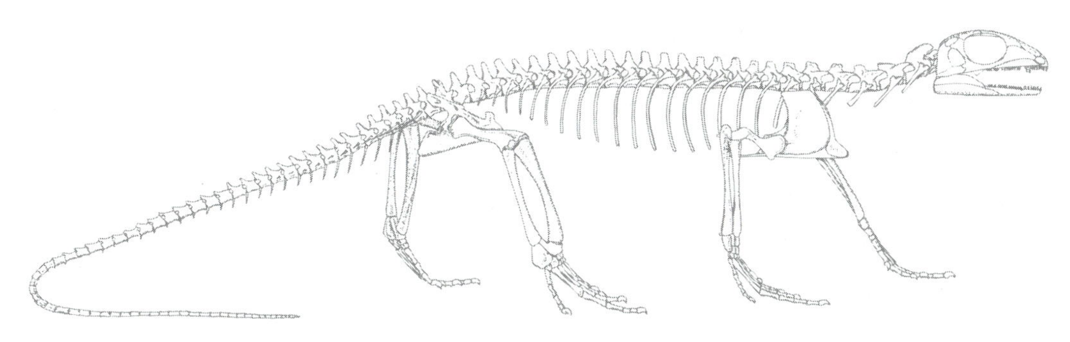

Primitive Diapsids and Lepidosaurs

All living reptiles, except for turtles, are members of the diapsid assemblage. The ancestors of crocodiles, lizards, snakes, and *Sphenodon* can be traced back to the early Mesozoic. Other extinct diapsid groups include the dinosaurs and pterosaurs and a variety of less-well-known genera that go back to the late Carboniferous (Figure 11-1).

EARLY DIAPSIDS

The earliest known diapsid is *Petrolacosaurus* from the Upper Pennsylvanian of Kansas (Reisz, 1981) (Figure 11-2). In most features of the skeleton it resembles the protorothyrids *Hylonomus* and *Paleothyris*. The body is somewhat larger, with an adult length of approximately 20 centimeters, not including a very long tail. The limbs

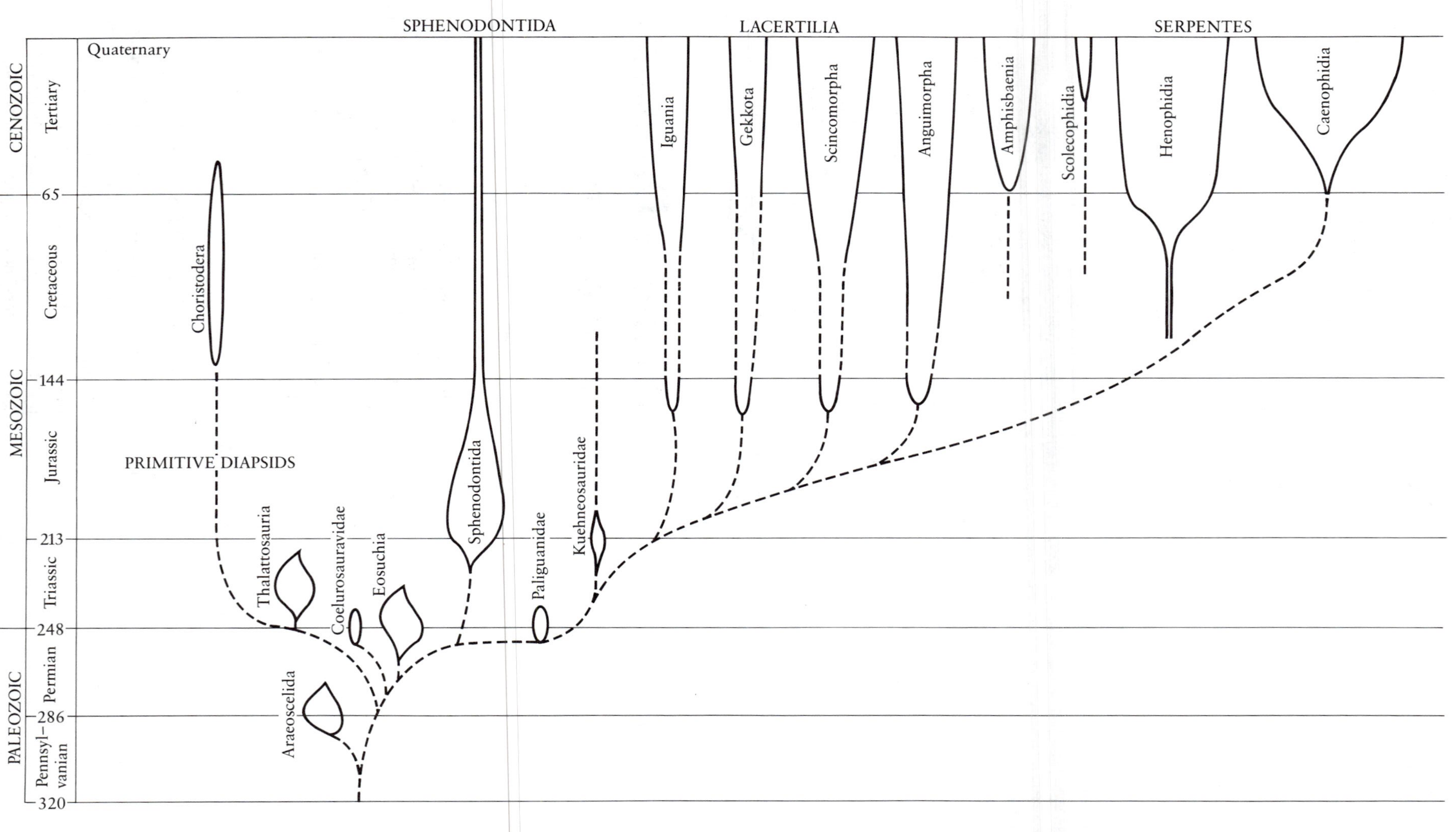

Figure 11-1. PHYLOGENY OF PRIMITIVE DIAPSIDS AND LEPIDOSAURS. *Data on lizards primarily from Estes, 1983a; data on snakes from Rage, 1984.*

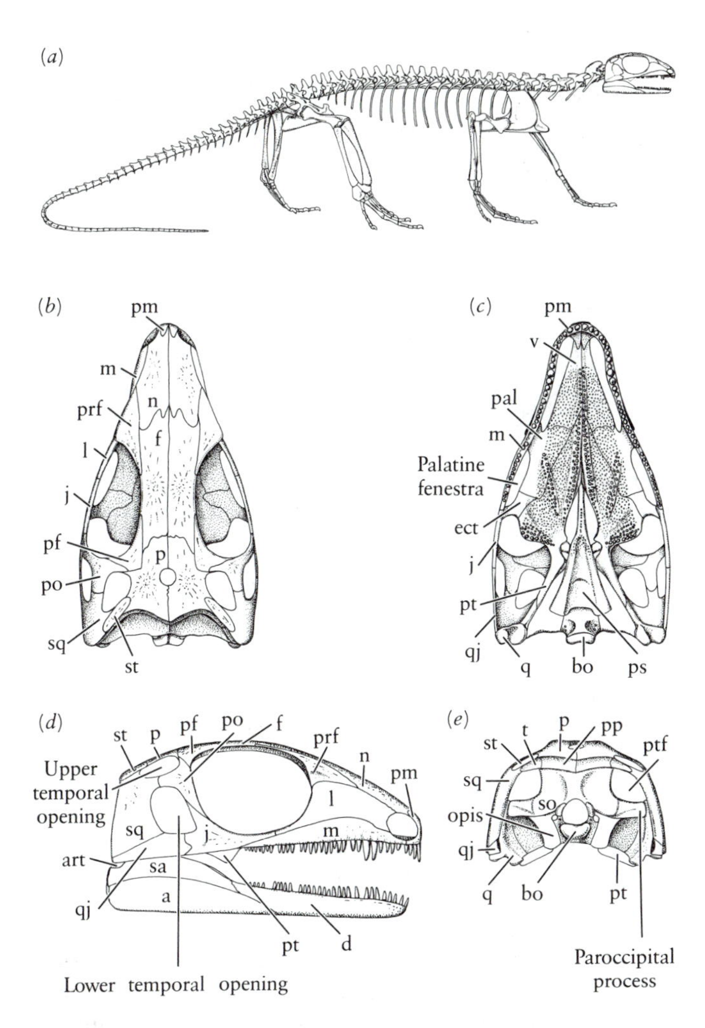

Figure 11-2. THE EARLIEST AND MOST PRIMITIVE KNOWN DIAPSID, *PETROLACOSAURUS* FROM THE UPPER CARBONIFEROUS OF KANSAS. (*a*) Skeleton, about 20 centimeters long, not including the tail. Skull in (*b*) dorsal, (*c*) palatal, (*d*) lateral, and (*e*) occipital views. Approximately $\times \frac{2}{3}$. Abbreviations as in Figure 8-3. *From Reisz, 1981.*

are proportionately longer which gives a more gracile appearance. The skull of *Petrolacosaurus* resembles that of early protorothyrids in its proportions and small size relative to the length of the body. A small supratemoral, postparietal, and tabular are still present, and a pineal opening has been retained. Like *Hylonomus,* the dentition of *Petrolacosaurus* is composed of many small marginal teeth, two of which are recognizable as canines. The teeth are set in a shallow groove at the edge of the jaw. Like the most primitive amniotes, the early diapsids were probably basically insectivorous but were able to feed on larger prey than did the protorothyrids.

Like other primitive amniotes, the early diapsids lack evidence of an impedance-matching middle ear. The cheek is not specialized for attachment of a tympanum and the paroccipital processes are not securely attached to the cheek.

Petrolacosaurus is distinguished by the presence of dorsal and lateral temporal openings in the dermal bones that cover the jaw muscles. The postorbital and squamosal form a narrow bar between these openings and a ventral bar is formed by the jugal and quadratojugal. This genus is also distinguished from its anapsid ancestors in having a relatively large opening in the palate, called the suborbital or palatine fenestra, between the ectopterygoid, palatine and maxilla, as in most later diapsids. Judging by the configuration of the adductor chamber, and comparison with modern diapsids, there seems to have been no significant change in the pattern of the jaw musculature associated with the initial development of temporal openings. They may have been elaborated as a more effective way of distributing stress while lightening the skull.

Petrolacosaurus has only 25 presacral vertebrae, but the shoulder girdle has shifted posteriorly so that 6 vertebrae may be recognized as cervicals. Each of these vertebrae is elongate, so that the neck is very long compared with that of other primitive reptiles. The configuration of the girdles and limbs is basically the same as in protorothyrids, and the bones of the carpus and tarsus have a similar pattern but different proportions. The stride would have been increased by the elongation of the distal limb elements, but the mechanics of limb movement appear no different from those of protorothyrids.

The closest known relative of *Petrolacosaurus* is *Araeoscelis* (Figure 11-3) from the Lower Permian. The body proportions resemble those of the Upper Pennsylvanian genus and the postcranial skeleton is similar, but the skull differs in lacking a lateral temporal opening and in the presence of a smaller number of somewhat bulbous cheek teeth. The absence of a lateral temporal opening might be a primitive feature, indicating that a dorsal opening had developed first in this group, but it is more probable that the opening was primitively present and secondarily lost in *Araeoscelis* to strengthen the skull against the more powerful bite implied by the bulbous teeth. The posterior margin of the jugal is bifurcated like that of *Petrolacosaurus,* in which the bone forms the anterior and ventral margins of the temporal opening.

Although most features of the skeleton of *Petrolacosaurus* and *Araeoscelis* are primitive relative to those of later diapsids, the great elongation of the neck and distal limb elements and details of vertebral structure are specializations that allow them to be recognized as a monophyletic assemblage, the Araeoscelida, that may be considered a sister group of all later diapsids.

There is no fossil evidence to elucidate the pattern of diapsid evolution during the middle part of the Permian. By the Upper Permian and Lower Triassic, the group had radiated into a number of distinct lineages whose interrelationships have been difficult to establish.

Most diapsids since the Permian may be grouped in one of two large assemblages, the Archosauromorpha or the Lepidosauromorpha. The archosauromorphs include the living crocodiles, the extinct dinosaurs and pterosaurs, and a variety of late Permian and Triassic families. The lepidosauromorphs include *Sphenodon,* lizards, snakes and their extinct relatives, as well as two major groups of

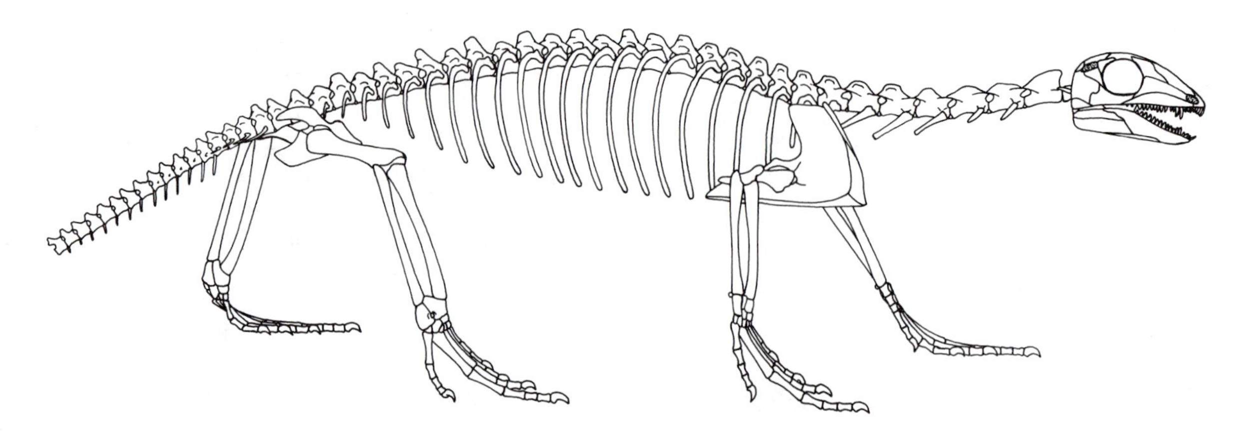

Figure 11-3. SKELETON OF *ARAEOSCELIS* FROM THE LOWER PERMIAN. *Araeoscelis* was long thought to be only distinctly related to the diapsids because of the absence of a lateral temporal opening. The great similarity of the postcranial skeleton to that of *Petrolacosaurus* (see Figure 11-2) suggests that the two genera share a close common ancestry. The lateral temporal opening may have been secondarily closed to strengthen the skull in relationship to the more massive dentition. *From Reisz, Berman, and Scott, 1984.*

aquatic reptiles. The archosauromorphs and lepidosauromorphs can each be defined on the basis of a complex of shared derived characters.

Two highly specialized groups from the Upper Permian and Triassic fit in neither of these categories but apparently represent an earlier radiation of more primitive diapsids. Their specific taxonomic position cannot be established at present.

The coelurosauravids are the most highly specialized of the primitive diapsids (Figures 11-4 and 11-5). This family includes two or three genera from the Upper Permian whose ribs are greatly elongated and were presumably connected by tissue to form a gliding surface that was closely comparable to that of the living lizard *Draco* (Carroll, 1978; Evans, 1982). The skull is also highly modified, with the elaboration of a squamosal frill and the loss of the lower temporal bar. We know coelurosauravids from Madagascar, Germany, and Great Britain.

The marine thalattosaurs from the Triassic of Switzerland and western North America. *Askeptosaurus* (Figure 11-6) is the best-known member of this group. The neck and trunk are greatly elongated, as in many aquatic groups, and the nostrils are set well back on the long snout. The limbs are short but not specialized as paddles.

Characteristics of Advanced Diapsids

We can differentiate archosauromorphs and lepidosauromorphs on the basis of distinct patterns of posture and locomotion. Lepidosauromorphs retain the sprawling posture and mediolateral excursion of the limbs common to primitive tetrapods. Lateral undulation of the vertebral column is emphasized as an important component of locomotion and reaches its extreme expression

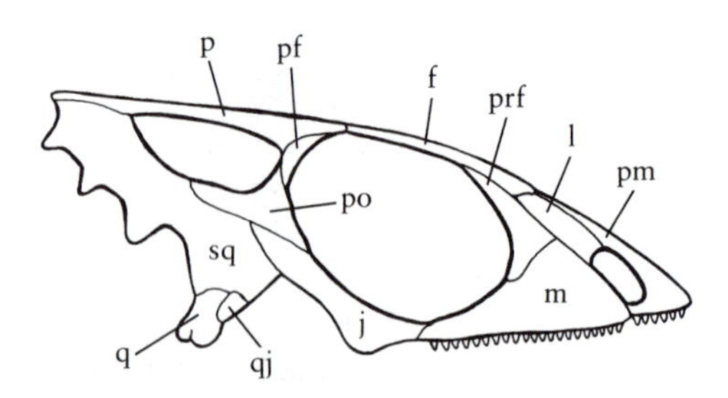

Figure 11-4. SKULL OF *COELUROSAURAVUS*, A SPECIALIZED, DIAPSID FROM THE UPPER PERMIAN OF MADAGASCAR, ×2. The lower temporal bar is lost and the squamosal forms a wide frill. Abbreviations as in Figure 8-3. *From Evans, 1982. With permission from The Zoological Journal of The Linnean Society. Copyright © 1982 by The Linnean Society of London.*

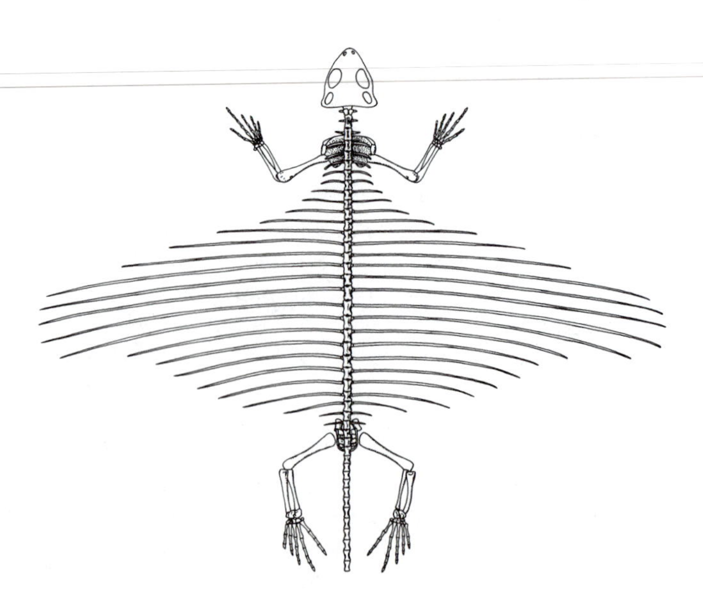

Figure 11-5. SKELETON OF *COELUROSAURUS* FROM THE UPPER PERMIAN OF MADAGASCAR. The greatly elongated ribs may have supported a gliding membrane like that of the living lizard *Draco*. "Wing" span approximately 30 centimeters. *From Carroll, 1978.*

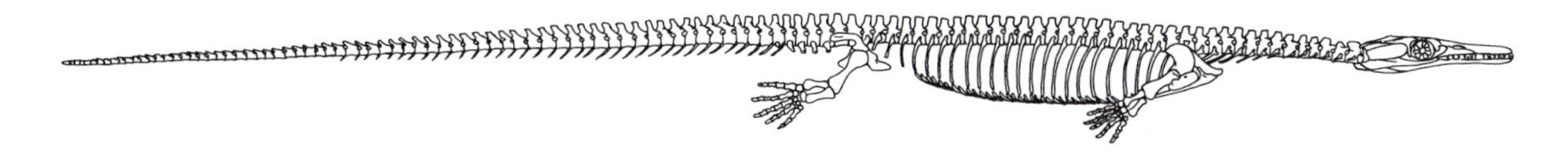

Figure 11-6. *ASKEPTOSAURUS,* A PRIMITIVE AQUATIC DIAPSID FROM THE MIDDLE TRIASSIC OF SWITZERLAND. Approximately 2 meters long. *From Kuhn-Schnyder, 1974.*

in snakes. In contrast, archosauromorphs reduce and, in some forms, eliminate lateral flexure of the vertebral column and bring the limbs more directly under the body. This pattern reaches its highest expression in dinosaurs and pterosaurs, in which the rear limbs move in a parasagittal plane. The initiation of these trends is evident in late Permian fossils, which show divergent specializations in the skeleton of the girdles and limbs.

Early lepidosauromorphs are most clearly distinguished from primitive diapsids and archosauromorphs by the development of a large sternum. A bone with this name occurs in some archosaurs, birds, and mammals, but it is certainly not a homologous structure since it was not present in the common ancestor of these groups and has a very different structure and function in each. The configuration of the sternum in Upper Permian lepidosauromorphs is very similar to that of primitive living lizards and *Sphenodon* (Figure 11-7). Jenkins and Goslow (1983) recently described the function of the sternum in lizards on the basis of x-ray cinematography and electromyography (Figure 11-8). The sternum prevents the scapulocoracoid from moving posteriorly when the forelimb is strongly retracted. In addition, the coracoids can rotate horizontally along the anterior margin of the sternum, which increases the arc of movement of the forelimb by approximately 35 degrees. This is very important in increasing the length of the stride. The movement of the shoulder girdle and forelimb in early lepidosaurs is associated with the lateral undulation of the trunk, which is an important aspect of locomotion in primitive tetrapods. The carpus and manus of the lepidosauromorphs retain the general features of earlier diapsids. In the primitive state, the tarsus also retains the pattern of primitive diapsids, in contrast with early changes in the archosaur assemblage that are associated with a more upright posture (see Chapter 13).

EOSUCHIANS

The position of the eosuchians has been crucial to understanding the pattern of diapsid evolution. Romer (1966) classified eosuchians among the lepidosaurs but included a number of groups that are now recognized as relatives of the archosaurs. Other authors have enlarged the concept of the Eosuchia to include the most primitive diapsids. As in the case of the leptolepids among the early teleosts, we can now identify many of the genera that were once included within the eosuchians as primitive members of several more advanced groups.

Broom (1914) first used the name eosuchian to refer to the genus *Youngina,* from the Upper Permian of South Africa. The tarsus and pes of this genus are primitive, but it possesses a sternum and short transverse processes of the vertebrae that demonstrate its affinities with the lepidosauromorphs. In most skeletal characters, *Youngina* and closely related genera (the Younginoidea) are almost ideal ancestors of the more specialized lepidosaur groups, including the squamates (lizards and snakes) and *Sphenodon* of the modern fauna.

The skull of *Youngina* (Figure 11-9) adheres to the same basic pattern as that of *Petrolacosaurus.* The margins of the temporal openings are thicker, and the lateral edge of the parietal bone extends ventrally, medial to the adductor chamber, and may have provided origin for some of the jaw musculature. As in *Petrolacosaurus* and other Carboniferous amniotes, the quadrate is not embayed posteriorly and the stapes is still a massive element. A gap remains between the paroccipital process and the cheek.

Details of the postcranial skeleton are not well known in *Youngina,* but they are fully described in the related

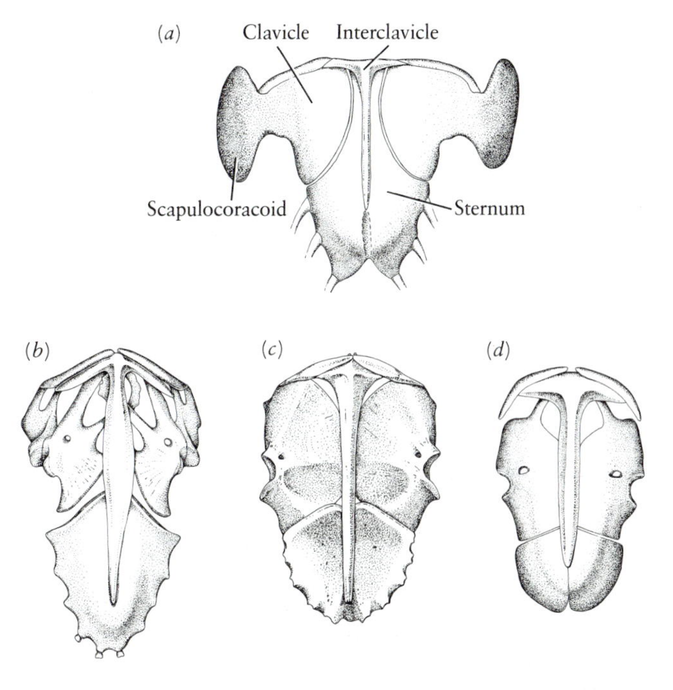

Figure 11-7. STERNA OF LEPIDOSAURS. (*a*) *Sphenodon.* (*b*) A modern iguanid lizard. (*c*) An Upper Permian eosuchian. (*d*) The Upper Permian lizard *Saurosternon.*

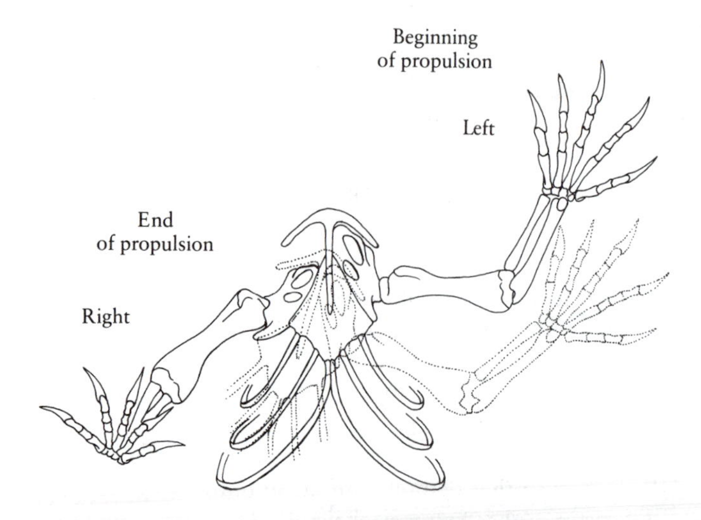

Figure 11-8. THE EFFECT OF MOVEMENT BETWEEN THE CORACOIDS AND STERNUM IN A VARANID LIZARD, REVEALED BY X-RAY CINEMATOGRAPHY. The coraco-sternal joint and limbs are depicted in ventral view, with the right forelimb in a posture typical of the end of propulsion, the left forelimb in a posture typical of the beginning of propulsion. Translation at the coraco-sternal joint advances the trunk relative to the glenoid and simultaneously rotates the contralateral shoulder ahead of the propulsive phase shoulder (dark lines). With a fixed coraco-sternal joint (dashed lines), the same limb excursion results in a shorter step length. *From Jenkins and Goslow, 1983.*

genus *Thadeosaurus* (Figure 11-10). In this genus and all other early lepidosauromorphs, the neck and the distal limb elements are much shorter than those of *Petrolacosaurus*. However, aside from these particular features, most skeletal characters of the early lepidosauromorphs can be directly derived from those of *Petrolacosaurus*. *Thadeosaurus* has five cervical vertebrae; the centra are recessed at both ends (a condition termed **amphicoelous**); intercentra are present throughout the trunk and continue for two or three segments beyond the sacrum. The transverse processes that support the ribs in the trunk region are very short, in sharp contrast with those of many early archosauromorphs. Unlike primitive diapsids, the caudal ribs are fused to the centra and extend directly laterally. This feature has evolved convergently in many groups of late Permian amniotes.

In the adult, the scapulocoracoid appears as a single unit, without a trace of suture. However, early growth stages show that it forms from two centers of ossification, with only a single coracoid element. This element seems to occupy the same position as the two coracoids in primitive amniotes. The anterior margin of the scapulocoracoid is thin but forms a continuous arc. The cleithrum, which is lost in advanced lepidosaurs, is retained in younginoids. The supinator process of the humerus has extended distally to enclose the radial nerve and blood vessels in an ectepicondylar foramen.

The only skeletal feature of these younginoids that appears to preclude them from the ancestry of later lepidosauromorphs is the pattern of the carpus. In both primitive diapsids and advanced lepidosaurs, the lateral centrale is in contact with the fourth distal carpal. In the younginoids, the medial centrale reaches the fourth distal carpal, separating it from the lateral centrale (Figure 11-11). The presence of this specialized feature allows us to recognize the younginoids as a monophyletic group, to which the name Eosuchia can be restricted.

The younginoids include not only the terrestrial genera *Youngina* and *Thadeosaurus* but also aquatic forms. *Tangasaurus* and *Hovasaurus* from the late Permian of Madagascar and East Africa resemble other eosuchians in most skeletal features but possess a very long, paddle-shaped tail (Figure 11-12). Currie (1981) showed that most specimens of *Hovasaurus* have many small stones in the area of the abdominal cavity that were presumably used as ballast to facilitate diving or underwater feeding.

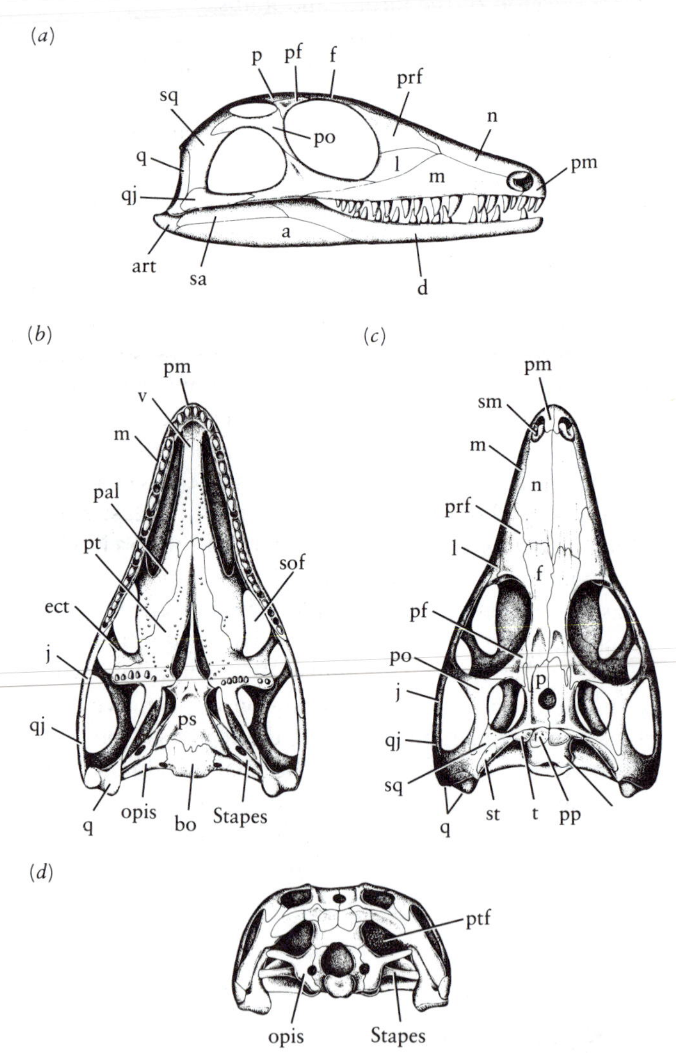

Figure 11-9. SKULL OF THE CHARACTERISTIC EOSUCHIAN *YOUNGINA* FROM THE UPPER PERMIAN OF SOUTHERN AFRICA. (*a*) Lateral, (*b*) palatal, (*c*) dorsal, and (*d*) occipital views. The quadrate is exposed laterally, but there is no embayment for support of a tympanum. The stapes is massive and oriented obliquely to the surface of the skull. There is no evidence that it participated in an impedance-matching mechanism at this stage in diapsid evolution. Approximately natural size. Abbreviations as in Figure 8-3, plus: sof suborbital fenestra. *From Carroll, 1981.*

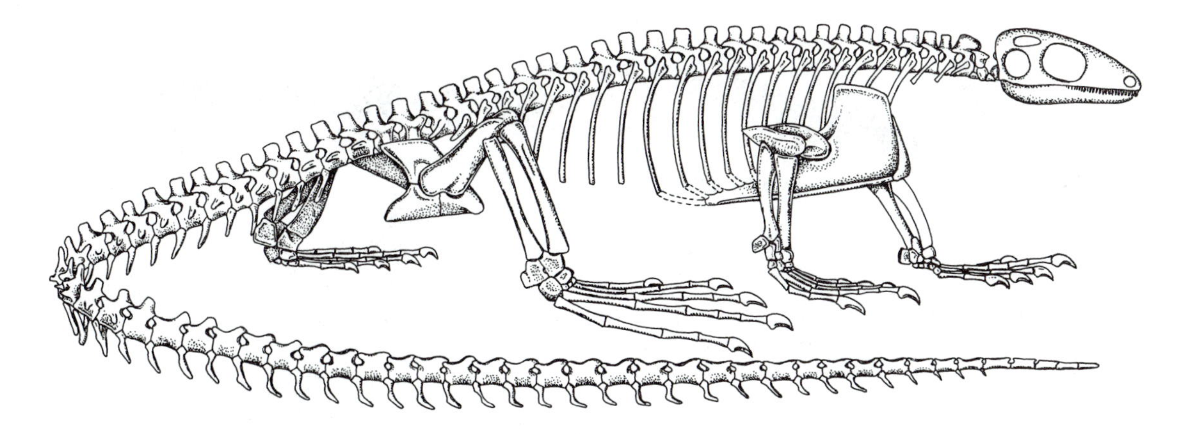

Figure 11-10. *THADEOSAURUS,* AN EOSUCHIAN REPTILE FROM THE UPPER PERMIAN OF MADAGASCAR. The pattern of the skeleton is almost ideal for an ancestor of the modern lepidosaurs. Length of skeleton without tail about 20 centimeters. *From Currie and Carroll, 1984.*

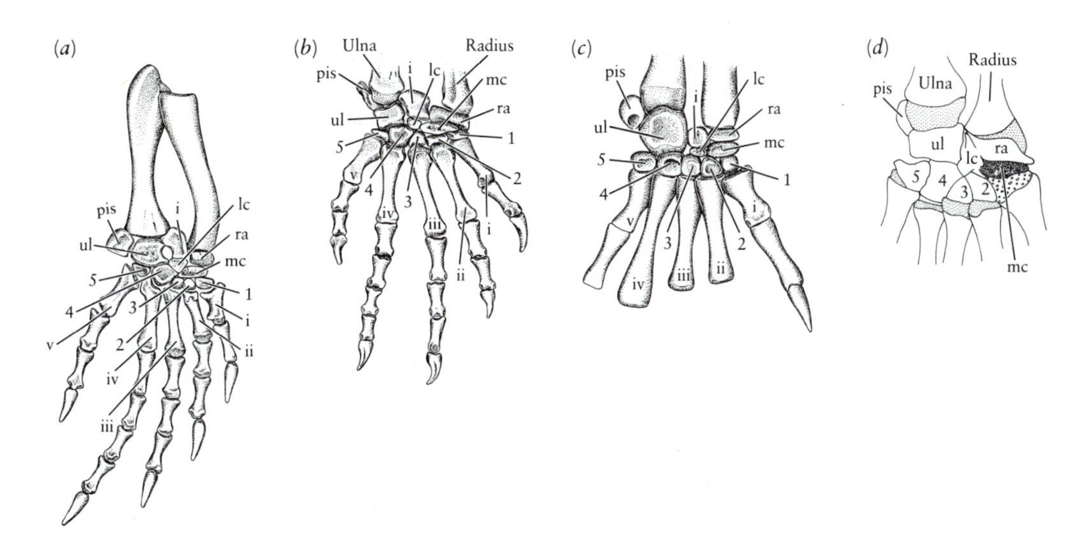

Figure 11-11. PATTERNS OF THE CARPUS IN LEPIDOSAURS. (*a*) The eosuchian *Thadeosaurus. From Carroll, 1981.* (*b*) *Sphenodon.* (*c*) The Permo-Triassic lizard *Saurosternon. From Carroll, 1977.* (*d*) A modern iguanid lizard. *From Carroll, 1977.* In contrast with archosauromorphs, the lepidosauromorphs retain most of the elements of the primitive diapsid carpus, including the pisiform. Eosuchians are specialized in the contact of the medial central and the fourth distal tarsal, which may separate the lateral centrale from the third distal tarsal. Both sphenodontids and lizards retain a more primitive pattern. Lizards are advanced over sphenodontids in the reduction or loss of the intermedian, which is large in eosuchians. In modern lizards, the epiphyses (dotted elements) are important in the formation of the carpus. A bone that has been thought to be the first distal carpal (open circles) may be a large epiphyseal element. Abbreviations as follows: i, intermedium; lc, lateral centrale; mc, medial centrale; pis, pisiform; ul, ulnare; ra, radiale; 1–5, distal carpals; i–v, metacarpals.

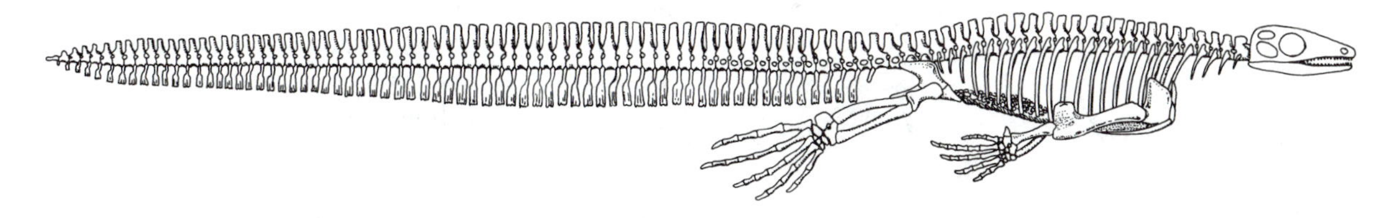

Figure 11-12. *HOVASAURUS,* AN AQUATIC EOSUCHIAN FROM THE UPPER PERMIAN OF MADAGASCAR. Approximately 50 centimeters long. Most of the specimens of this genus have small stones in the abdominal cavity that would have served as ballast. *From Currie, 1981.*

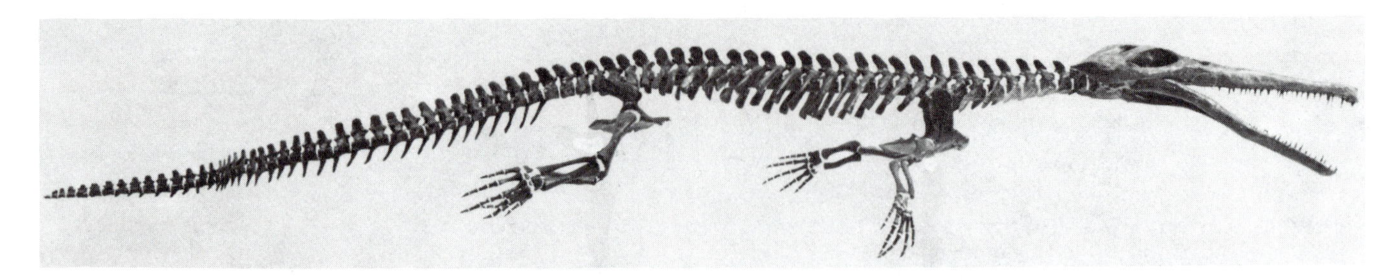

Figure 11-13. SKELETON OF *CHAMPSOSAURUS* FROM THE UPPER CRETACEOUS OF WESTERN NORTH AMERICA. It is an aquatic animal with the body proportions and size of a modern crocodile. It may have evolved from early lepidosauromorphs. *Photo courtesy of Bruce Erickson.*

GEPHYROSAURUS AND CHAMPSOSAURS

In addition to the eosuchians, several other groups diverged from the basic lepidosauromorph stock. Lizards and sphenodontids, which have a well-developed sternum, certainly evolved from this lineage, as did an assemblage of highly specialized Mesozoic aquatic reptiles that we will discuss in the next chapter. *Gephyrosaurus* from the Lower Jurassic appears to be a relatively conservative derivative that retains a younginoid body pattern but has some features of the skull that resemble those of lizards and sphenodontids (Evans, 1981).

Champsosaurus and *Simoedosaurus* from the Cretaceous and Lower Tertiary resemble eosuchians in many primitive features, but they are specialized toward a crocodilelike aquatic way of life (Figures 11-13 and 11-14). The temporal bars are greatly extended laterally to accommodate a massive jaw musculature. The snout is elongate, the limbs poorly ossified, and the ribs, like those of mesosaurs, are pachyostotic. They lack a sternum, but the functional significance of this structure would be lost if they swam like modern crocodiles and marine iguanids, which propel themselves by lateral undulation of the trunk and tail while holding the forelimbs against the side of the body. Erickson (1972) and Sigogneau-Russell (1979) recently described representatives of this group. We have found no fossils that link advanced champsosaurs with Permian diapsids, and their taxonomic position remains uncertain.

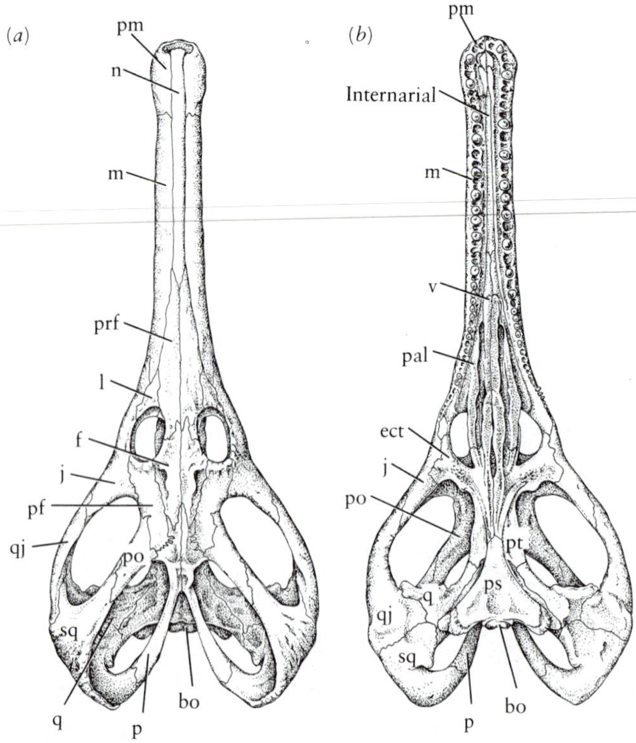

Figure 11-14. SKULL OF *CHAMPSOSAURUS.* (*a*) Dorsal and (*b*) palatal view. The narial opening is terminal and there is an extensive secondary palate. Abbreviations as in Figure 8-3. *From Erickson, 1972.*

LEPIDOSAURS

Within the larger assemblage of lepidosauromorphs, lizards, snakes, and *Sphenodon* constitute the Lepidosauria. Lepidosaurs are by far the most diverse of modern reptiles, including nearly 6000 species of lizards and snakes. Snakes appear only in the Cretaceous, but lizards are one of the earliest of the advanced reptile groups to appear in the fossil record, with specimens known in the Upper Permian. Sphenodontids are now restricted to a single species, but they have a fossil record going back to the Triassic and exhibited a modest radiation within the middle Mesozoic.

Lizards and sphenodontids may share a close common ancestry. Throughout their history both groups have included primarily small, terrestrial animals that have a basically similar skeletal structure. One of the most important features they share (in contrast with eosuchians and all other reptiles) is the presence of epiphyses—separate centers of ossification that form the articulating surfaces of the limb bones. As in mammals, growth is terminated when the cartilaginous plate that separates the ends of the long bones and the epiphyses becomes fully ossified. In lizards and sphenodontids, the epiphyses also facilitated the development of more effective limb joints.

Lizards and sphenodontids are also advanced over the younginoids in developing a solid attachment between the paroccipital process and the quadrate and the presence of a thin, rodlike stapes. Most lizards use the stapes in an impedance-matching system that is capable of detecting high-frequency airborne vibrations (Wever, 1978). The modern genus *Sphenodon* has a lizardlike stapes but lacks a tympanum and middle ear cavity. It is difficult to explain the nature of this stapes except as a remnant of a once functional impedance-matching system. As in many burrowing lizards, the loss of the tympanum and middle ear cavity in *Sphenodon* may be specializations. Fossil relatives of *Sphenodon* have a similar stapes and a quadrate that may have supported a tympanum, but the configuration of the bone is not sufficient in itself to demonstrate whether or not a tympanum was present. On the evidence of the stapes, we can assume that the common ancestor of sphenodontids and lizards had probably evolved an impedance-matching middle ear, in contrast with primitive lepidosauromorphs.

Other advanced features are shared by most lizards and sphenodontids but are not fully developed in the early lizards. These characteristics may have evolved in parallel in the two groups.

Both advanced lizards and sphenodontids have a large opening (the thyroid fenestra) between the pubis and ischium (Figure 11-15). Similar fenestration of the pelvis also evolved separately in archosaurs and turtles. In most lizards and *Sphenodon,* the anterior tail vertebrae are specialized to facilitate caudal autonomy.

Advanced lizards and sphenodontids also have a similar tarsal and foot structure. The proximal tarsals, including the calcaneum, astragalus, and centrale, are fused into a single unit, the astragalocalcaneum, which is closely integrated with the tibia and fibula (Figure 11-16). There is a hingelike articulation between these proximal elements and the more distal bones of the foot that are closely integrated with one another. The foot can be strongly flexed on the ankle as a result of contraction of the gastrocnemius muscle, which is attached to tubercles on the ventral surface of the fifth metatarsal. This bone is sharply bent (or hooked) to articulate with the fourth distal tarsal and serves as a lever for more effective action of this muscle. A hingelike joint between the proximal and distal tarsals is developing in early lizards, but the proximal elements are not fused to one another. The fifth metatarsal is short and diverges from the other metatarsals, as in modern lizards, but is not hooked.

According to Robinson (1973), the initial divergence between the ancestors of lizards and sphenodontids may have been associated with different feeding strategies. In

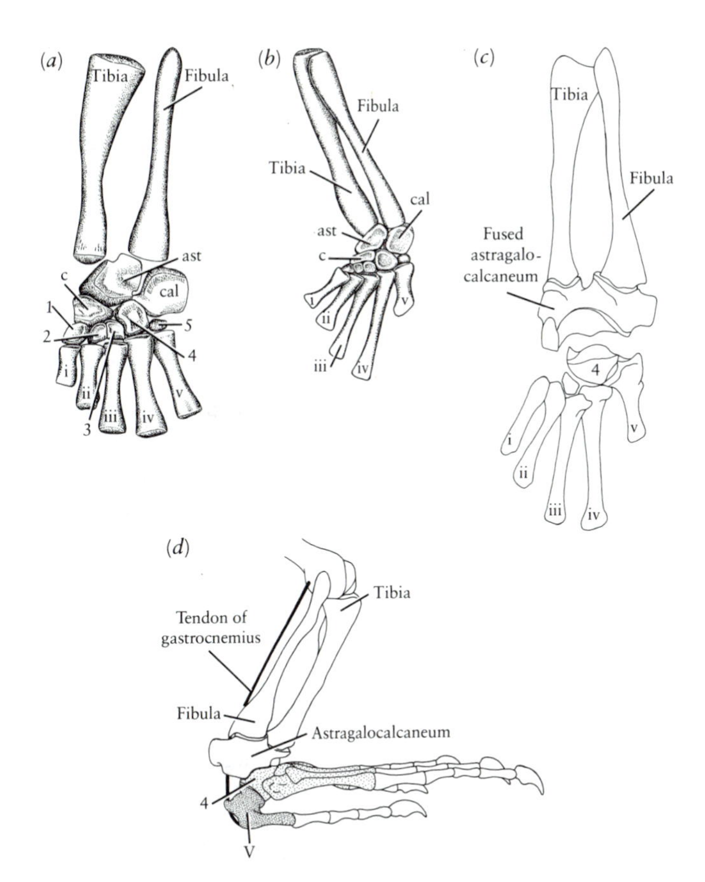

Figure 11-16. LEPIDOSAUR TARSUS. (*a*) Lower rear limb of the primitive diapsid *Galesphyrus. From Carroll, 1977.* (*b*) The Permo-Triassic lizard *Saurosternon. From Carroll, 1977.* (*c*) The modern lizard *Varanus* with the proximal and distal tarsals separated to show their articulating surfaces. *From Robinson, 1975.* (*d*) Lower limb of *Varanus* in lateral view, showing the course of the gastrocnemius muscle. *From Robinson, 1975.* In *Saurosternon,* the centrale has become closely integrated with the astragalus and a line of flexure has evolved between the medial and distal tarsals. The fifth metatarsal is much shorter than in primitive diapsids and may have served as a level to ventroflex the foot. In *Varanus,* the calcaneum, astragalus, and centrale have fused into a single element. Flexure of the foot is limited to a complex joint surface between the fourth distal tarsal and the astragalocalcaneum. The gastrocnemeus attaches to tubercles on the ventral surface of the fifth metatarsal (closely spaced dots), which is sharply angled or "hooked" proximally. The other metatarsals (open dots) move as a unit. Abbreviations: ast, astragalus; c, centrale; cal, calcaneum; 1–5 distal tarsals; i–v, metatarsals.

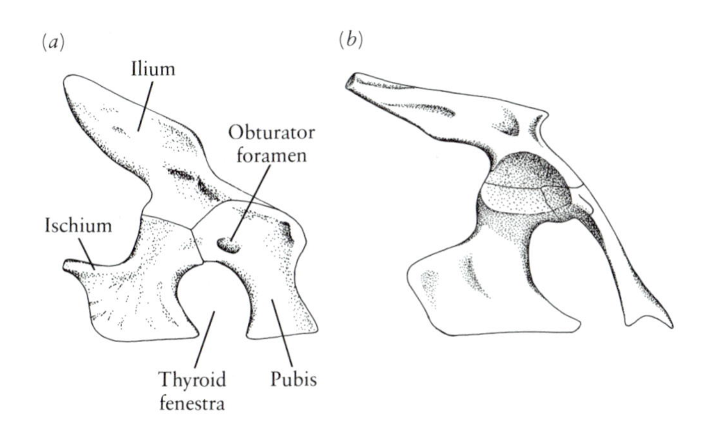

Figure 11-15. PELVES OF LEPIDOSAURS. (*a*) Pelvis of the Upper Triassic sphenodontid *Planocephalosaurus. From Fraser and Walkden, 1984.* (*b*) Pelvis of *Iguana.* In both sphenodontids and modern lizards, there is a large opening ventrally between the pubes and ischia, the thyroid fenestra. This opening may have evolved in the common ancestor of these two groups or independently after their divergence. A thyroid fenestra evolves separately in several other groups of diapsid reptiles.

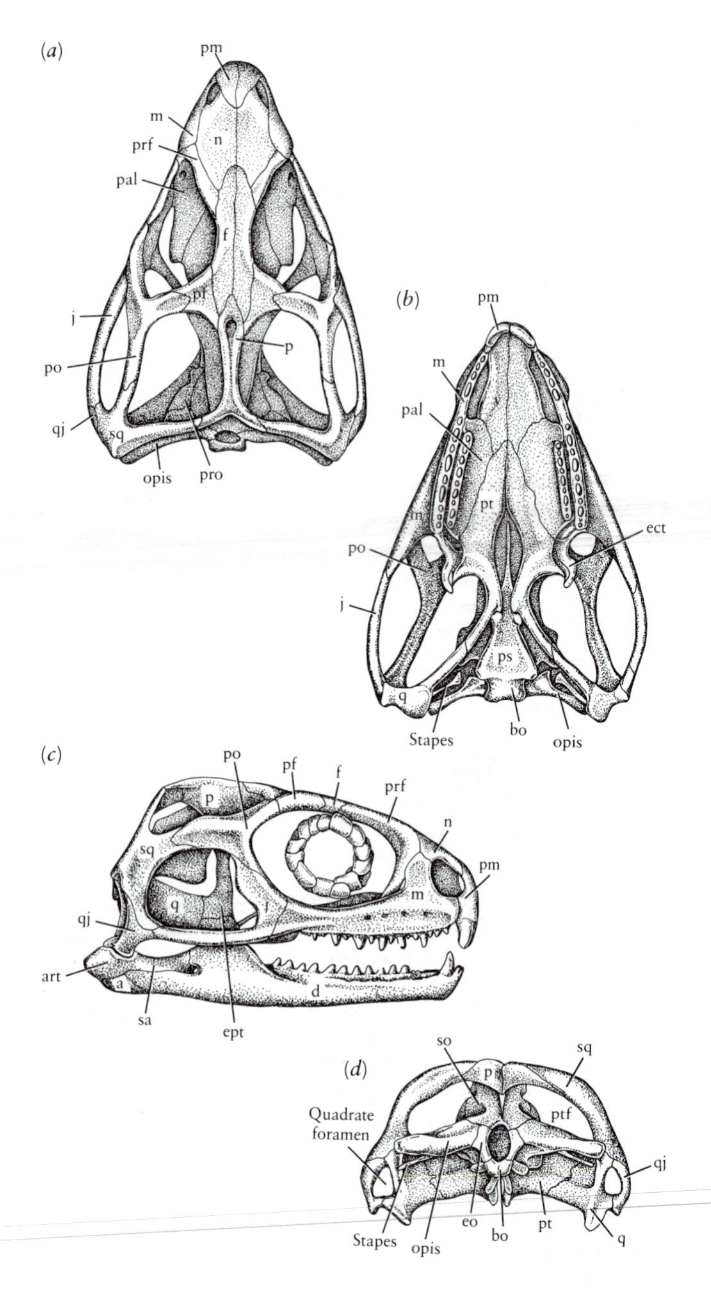

Figure 11-17. SKULL OF THE MODERN GENUS *SPHENODON*. (*a*) Dorsal, (*b*) palatal, (*c*) lateral, and (*d*) occipital views. Like other members of the Sphenodontidae, the teeth are acrodont, the palatine bears a row of denticles that are comparable in size to those of the maxilla, and the lower temporal bar is complete but bowed laterally, which allows passage of the external portion of the adductor mandibulae to the lateral surface of the lower jaw. Abbreviations as in Figure 8-3. *Drawn from a specimen in the collection of Redpath Museum.*

sphenodontids (Figure 11-17), the adductor chamber is relatively larger than in primitive eosuchians, indicating that the adductor jaw musculature had a proportionately larger mass and a stronger but slower bite. The dentition is specialized, with a relatively small number of marginal teeth fused to the edge of the jaw (the **acrodont** condition). The teeth are not replaced but are added at the back of the jaw as growth proceeds. In contrast, the adductor chamber in early lizards is relatively shorter than in eosuchians, implying a weaker but faster bite (see Figure 11-22). Although some advanced lizards have evolved acrodont tooth implantation, most have teeth that are loosely attached to the jaws, which facilitates rapid replacement. The differences in jaw mechanics may be associated with differences in diet. Living sphenodontids have a broad range of food and can bite the heads off young birds. Most lizards are predominantly insectivorous.

Sphenodontids have a rigid quadrate, as do the younginid eosuchians, which is related to the elaboration of a strong bite. The most significant feature of lizards, which is evident in the earliest genera, was the evolution of a movable quadrate.

SPHENODONTIDA

Jaws resembling those of modern sphenodontids in having acrodont teeth have been reported from the Lower Triassic (Malan, 1963). Unfortunately, they are not sufficiently diagnostic to be certain that they belong to this group. Complete skeletons, unquestionably related to the living *Sphenodon* are known from the Upper Triassic of Europe (Figure 11-18). They show considerable diversity in the pattern of the dentition, the configuration of the skull table, and the proportions of the limbs. The upper and lower teeth of *Glevosaurus* are triangular and shear against one another as the jaw is closed. This wear pattern requires precise occlusion that is only possible if the quadrate is rigid (Robinson, 1973). Most early sphenodontids show accommodation for the lateral expansion of the adductor jaw musculature so that the superficial portion can insert on the lateral surface of the lower jaw. In *Glevosaurus* and *Planocephalosaurus* (Fraser, 1982), this accommodation is achieved by the development of a gap in the lower temporal bar. In *Brachyrhinodon* and *Polysphenodon* (Carroll, 1985), it occurs by a lateral bowing of the bar. This pattern is retained in *Sphenodon*. In the Upper Jurassic genus *Sapheosaurus*, which some authors place in a separate family, the dentition is entirely lost.

Sphenodontids remain diverse throughout the early Mesozoic and their remains are common in the Upper Jurassic of Europe (Farbre, 1981). Some sphenodontids are known from the Cretaceous, but the group has no fossil record in the Cenozoic. This absence may reflect the general decline of the group relative to the lizards that became progressively more common and diverse since the Upper Jurassic.

Sphenodontids are now limited to a single species of the genus *Sphenodon* that lives on small islands off the coast of New Zealand. The modern species shows no significant differences in the postcranial skeleton from the well-known Upper Jurassic genera. The skull differs in having the row of large palatine teeth parallel with the maxillary dentition, in contrast with the more primitive oblique orientation seen in the Triassic and Jurassic genera. In contrast with the shearing occlusion in *Glevosaurus*, the lower jaw of *Sphenodon* moves in an anterior-

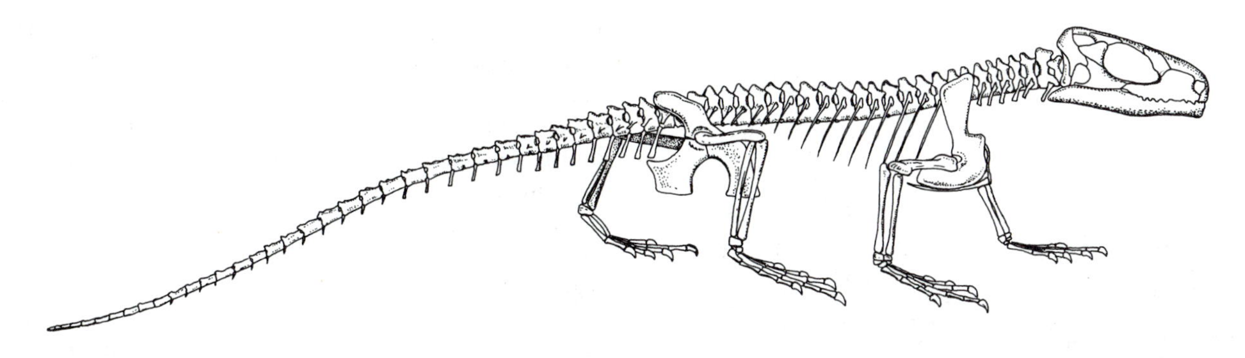

Figure 11-18. *PLANOCEPHALOSAURUS,* A SPHENODONTID FROM THE UPPER TRIASSIC OF EUROPE. The lower temporal bar is not complete, and the palatine tooth row is oblique to that of the maxilla. Most of the remainder of the skeleton is similar to that of *Sphenodon. From Fraser and Walkden, 1984.*

posterior direction as the jaw is closed (**propalinal movement**), producing a sawlike action between the upper and lower teeth (Gorniak, Rosenberg, and Gans, 1982).

Throughout their known history, which goes back into the Triassic, terrestrial sphenodontids have resembled primitive lizards in most aspects of their skeleton and small body size. The basic feeding adaptations of the sphenodontids remain relatively stereotyped, while those of lizards have diversified extensively. Their conservative skull structure and perhaps differences in the physiology and behavior may explain why sphenodontids are now restricted to a single geographically limited species, while there are some 3000 living species of lizards that are distributed throughout the world.

PLEUROSAURS

Although the sphenodontids have remained extremely conservative for some 200 million years, they gave rise to one divergent group early in their history. The pleurosaurs are an assemblage of aquatic diapsids that we know from the early Jurassic into the early Cretaceous (Figure 11-19). The late members of this group are greatly elongated, with up to 57 trunk vertebrae and very long tails. The limbs are greatly reduced, although they retain the pattern of terrestrial diapsids. The skull is elongate, with the nostrils posterior in position. Most authors have allied them with sphenodontids on the basis of their acrodont dentition, but the lower temporal bar is incomplete. Recently discovered specimens from the Lower Jurassic (Carroll, 1985) show an intermediate condition between the advanced forms and the early sphenodontids. The trunk is somewhat elongate but has only 37 vertebrae. The skull is also somewhat elongate but resembles the sphenodontids in many features, although the lower temporal bar is in the process of reduction (Figure 11-20). The pleurosaurs show progressive elongation of the trunk and reduction of the limbs, especially the forelimbs, throughout the Jurassic and into the Cretaceous.

We may classify the sphenodontids and pleurosaurids together within the Order Sphenodontida. This name is used in preference to Rhynchocephalia (which has been the common practice) because of possible confusion with the rhynchosaurs (see Chapter 13), which have long been classified among the lepidosaurs but are more appropriately allied with the archosauromorphs.

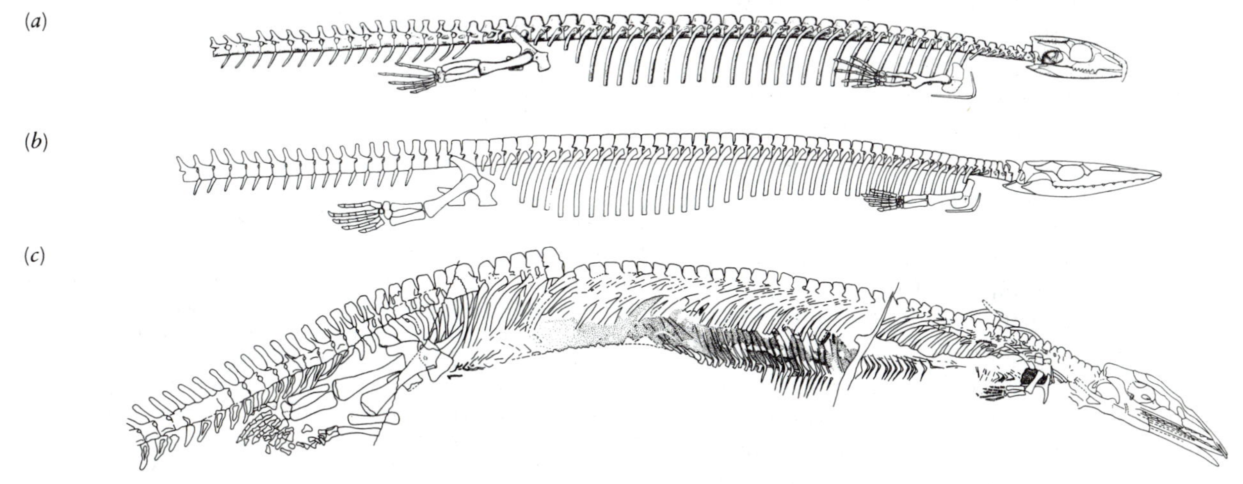

Figure 11-19. PROGRESSIVE CHANGES IN BODY PROPORTIONS WITHIN THE FAMILY PLEUROSAURIDAE. (*a*) *Paleopleurosaurus* from the Lower Jurassic. (*b*) *Pleurosaurus goldfussi* from the Upper Jurassic. (*c*) *Pleurosaurus ginsburgi* of the Lower Cretaceous. *From Carroll, 1985.*

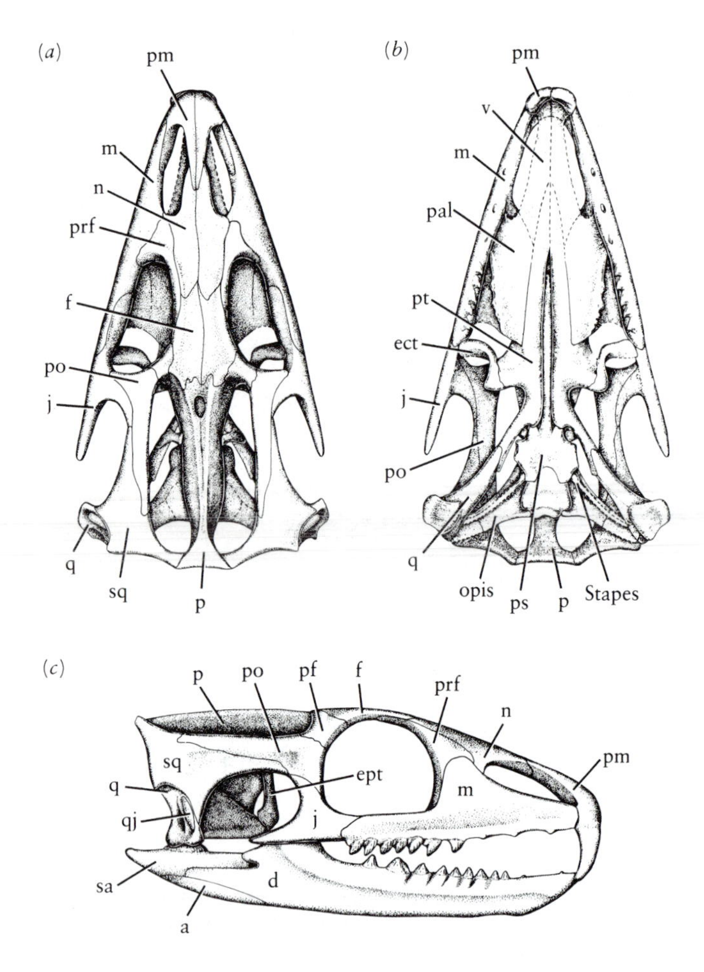

Figure 11-20. SKULL OF THE PRIMITIVE PLEUROSAUR *PALEOPLEUROSAURUS*. (*a*) Dorsal, (*b*) palatal, and (*c*) lateral views. Note similarity to *Sphenodon* (see Figure 11-17). Abbreviations as in Figure 8-3. *From Carroll, 1985.*

LIZARDS

Lizards and snakes are the most successful of modern reptiles in terms of the number of species and their wide geographical distribution. All can be traced to fossils from the late Permian and early Triassic. These early genera differ most significantly from sphenodontids and eosuchians in the loss of the lower temporal bar and the development of a joint between the dorsal extremity of the quadrate and the squamosal. Mobility of the quadrate at this joint is referred to as **streptostyly;** it appears to be a major factor in the success of lizards.

Based on her study of the jaw muscles in modern lizards, Smith (1980) determined that the main function of streptostyly is to increase the force of the pterygoideus muscle (typically the largest of the jaw adductors in lizards) when the jaw is nearly closed (Figure 11-21). In general, the effective force of a muscle is proportional to the sine of the angle between the muscle and the element to which it is inserted. The greatest force is delivered when the angle is 90 degrees, and the least when it approaches zero. If the normal jaw joint, between the quadrate and the mandible, is the only one that is considered, the effective angle between the pterygoideus muscle and the mandible is very small when the jaw is nearly closed and so its resultant force is weak. If the quadrate-mandibular joint is held stable and the contact between the quadrate and squamosal acts as the functional jaw joint, the angle of the pterygoideus is close to 90 degrees, which gives the muscle a much greater effective force. In even the earliest lizard *Paliguana* (Figure 11-22), the quadrate extends to the skull roof, forms a joint with the squamosal, and is embayed posteriorly to support a large tympanum in the manner of modern lizards.

The combination of more effective hearing, jaw mechanics, and locomotion (which is facilitated by the development of epiphyseal joint structures) may have enabled the lizards to usurp the habitats that were previously occupied by other lizardlike tetrapods, including the millerosaurs, early procolophonoids, and small eosuchians.

Small body size is apparently an important attribute of lizards as a group and is characteristic of the early members, which barely exceed 10 centimeters from the tip of the skull to the base of the tail. We can attribute the limitation of body size to the development of epiphyses. Pough (1973) associated the small size in modern lizards with their almost exclusively insectivorous diet.

Several lizard genera have been described from the Upper Permian and Lower Triassic of South Africa (Carroll, 1977) (Figure 11-23), and others are known from the Upper Triassic of Great Britain and North America (Colbert, 1970; Robinson, 1962). Many aspects of their skeletal anatomy are more primitive than those of modern lizard groups that appear in the late Jurassic, and they are placed in a distinct group, the Eolacertilia. The proximal tarsals are not fused to one another; the fenestration

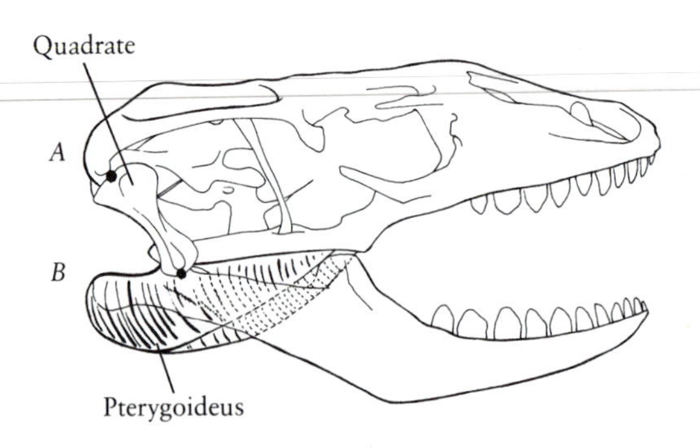

Figure 11-21. SKULL OF A MODERN LIZARD SHOWING THE ADVANTAGE OF STREPTOSTYLY. The pterygoideus muscle lies nearly parallel with the long axis of the skull. In most reptiles, the maximum force of this muscle is achieved when the jaw is widely open, at which time the muscle acts at nearly right angles to the axis of the jaw. When the jaw is nearly closed, the force of the pterygoideus is nearly parallel to the long axis of the jaw and has little effect in rotating the jaw about the quadrate-mandibular articulation (*B*). Since the force of the pterygoideus acts at nearly right angles to the quadrate, a joint between this bone and the skull (at *A*) enables this muscle to deliver much more effective force when the jaw is nearly closed. *From Smith, 1980. Redrawn by permission from Nature. Copyright © 1980, Macmillan Journal, Ltd.*

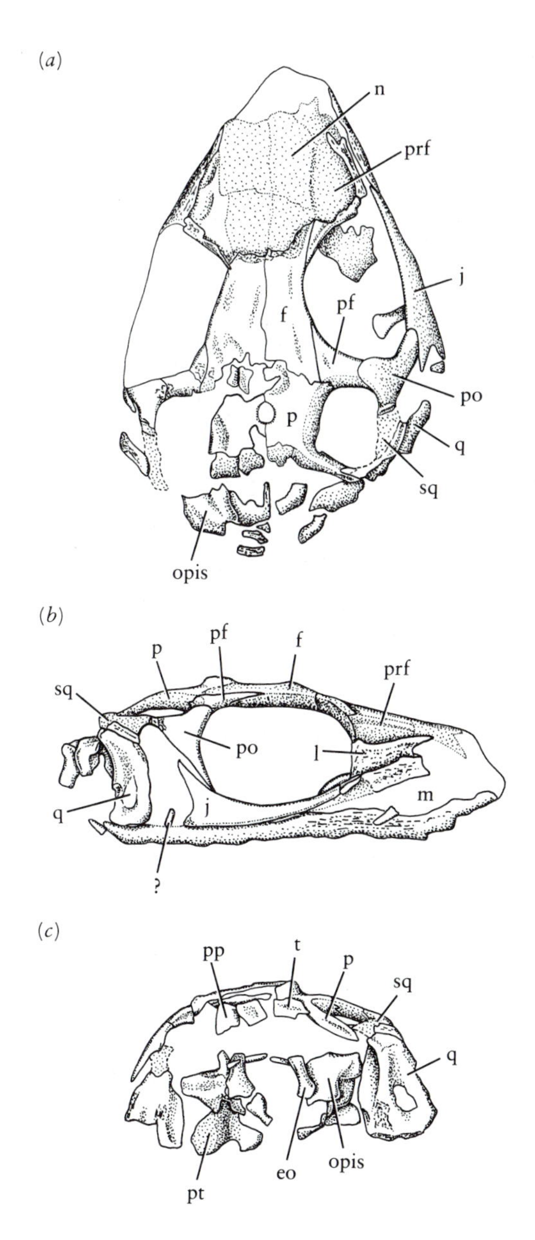

Figure 11-22. SKULL OF THE EARLIEST KNOWN LIZARD *PALIGUANA*. From the Upper Permian of southern Africa. (*a*) Dorsal, (*b*) lateral, and (*c*) occipital views, ×2. Abbreviations as in Figure 8-3. *From Carroll, 1977.*

of the scapulocoracoid that is common to advanced lizards is poorly developed and thyroid fenestration of the pelvis is not evident. The vertebral centra are still amphicoelous. As in most primitive tetrapods, the teeth are attached to a shallow groove at the edge of the jaw.

Although the eolacertilians are primitive in these anatomical features, they underwent a considerable adaptive radiation that paralleled that of several groups of modern lizards. Most genera had body proportions that were similar to those of the primitive eosuchians, but one genus had relatively short limbs, as is the case in many of the modern skinks. Another had short forelimbs and long hind limbs, suggesting that it was capable of facultatively bipedal locomotion like the modern genera *Crotaphytus* and *Basiliscus* (Carroll and Thompson, 1982). The greatest degree of specialization occurs in the family Kuehneosauridae (Figure 11-24), in which the ribs were greatly elongated like those of the coelurosauravids and the modern agamid *Draco;* in that genus, they support a broad membrane that enables it to glide up to 30 meters from tree to tree.

The fossil record of lizards is very incomplete for much of the Jurassic. At the end of that period, deposits in China and in several localities in Europe include primitive members of most of the major groups of modern lizards. Because all modern lizard groups share many features that are advanced over those of the eolacertilians, we assume that they can be traced to a single ancestral stock, but we have not established their specific interrelationships.

Nearly all modern lizards have a transverse joint between the frontals and parietals that enables the snout to be raised and lowered through a small arc. Especially in the larger carnivorous lizards, this joint may function to improve the orientation of the long curving teeth that penetrate prey. In association with this mobility, the frontals and/or parietals are fused at the midline in most advanced lizards. The paroccipital process and the supra-

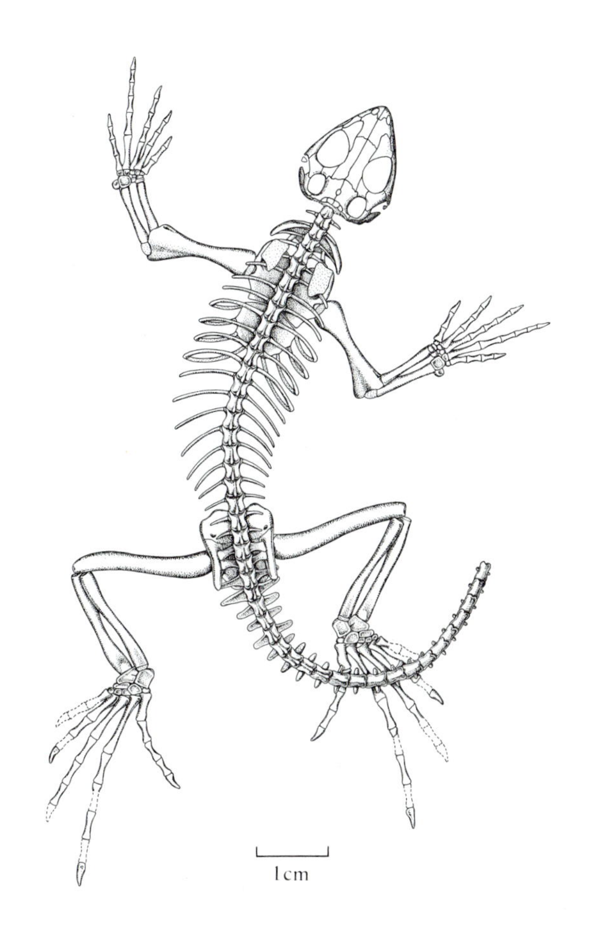

Figure 11-23. RESTORATION OF THE SKELETON OF A PRIMITIVE LIZARD. Based on the skeletons of *Saurosternon* and *Palaeagama* and the skull of *Paliguana* from the Upper Permian and Lower Triassic of southern Africa. *From Carroll, 1977.*

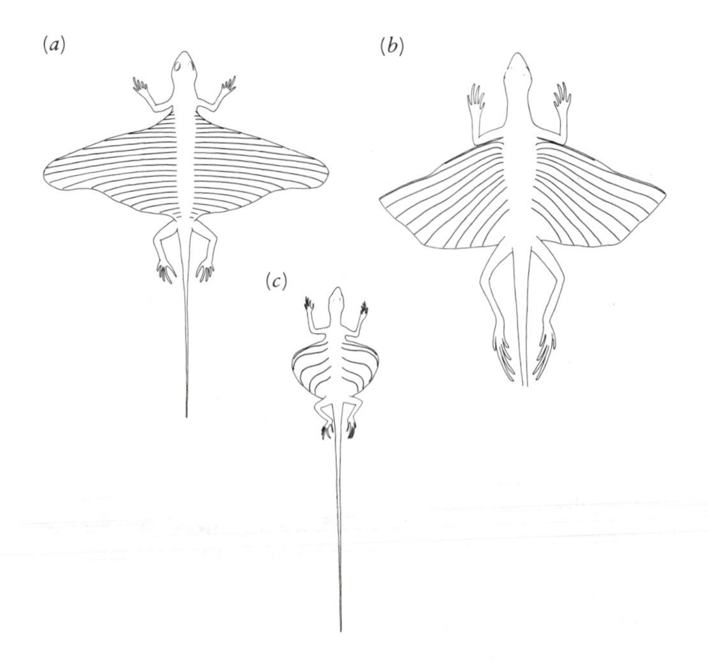

Figure 11-24. GLIDING REPTILES. Flesh restorations of reptiles from the Upper Permian, Upper Triassic, and Recent that use their ribs to support a gliding membrane. (*a*) *Coelurosauravus,* a primitive diapsid from the Upper Permian of Madagascar. (*b*) *Kuehnosaurus,* a primitive lizard from Great Britain. (*c*) *Draco,* a living agamid lizard. This specialization evolved separately in each of these groups. Note progressive reduction in the number of ribs. *From Carroll, 1978.*

temporal contribute to support the quadrate and the streptostylic joint is further perfected.

Advanced lizards also specialize the nature of tooth implantation. In most groups, the jaw margin has extended to cover a large portion of the base of the teeth. The teeth are attached largely by their lateral surface, a pattern termed **pleurodont.** In two families, the Agamidae and Chamaeleontidae, the teeth are fused to the edge of the jaw as in sphenodontids. In one extinct group of large lizards, the mosasaurs, the teeth are set in deep sockets.

The presence of specialized joints between the vertebral centra is an important characteristic of most advanced lizard groups. A socket develops in the anterior surface into which fits a condylar process that extends from the posterior surface of the next anterior centrum. This condition is termed **procoely** and is common to all modern lizard groups with the exception of many genera within the Gekkota (Figure 11-25).

We recognize four major lizard stocks in the late Mesozoic and Cenozoic, the Iguania, the Gekkota, an assemblage including both the Scincomorpha and the Anguimorpha, and the Amphisbaenia. All had apparently diverged by the end of the Jurassic, although the Iguania is not known with certainty until the end of the Cretaceous, and the Amphisbaenia appear only in the Paleocene. No fossil lizards have been described from the early Cretaceous, which greatly hampers the establishment of specific interrelationships of the modern families. Estes (1983a,b) comprehensively reviewed the fossil record of all these squamate groups.

Despite their late appearance in the fossil record, members of the Iguania retain many primitive skeletal features. In primitive genera, the skull remains little modified from that of the eolacertilians. We recognize three families. The Iguanidae is primarily restricted to North and Central America, but with living genera also known in Madagscar and Fiji. The oldest fossils are from the Upper Cretaceous of South America. The Agamidae is widespread in the Old World, with fossils as early as the Upper Cretaceous. This group is characterized by acrodont implantation of the marginal teeth. *Arretosaurus,* from the Upper Eocene of central Asia, has a superficially agamidlike skull and pleurodont or subpleurodont tooth implantation like that of iguanids. *Euposaurus,* from the Upper Jurassic of France, has been tentatively assigned to the Iguania on the basis of the configuration of the upper temporal opening and the presence of acrodont teeth. These features suggest a relationship to the Agamidae, but we do not know any specimens in sufficient detail to substantiate this assignment.

The chameleons are among the most specialized of modern lizards. The tongue is highly protrusible, the tail prehensile, and the digits of the hands and feet are greatly modified for grasping. The group is primarily arboreal. There are only six living genera, but they are divided into 109 species. Like the agamids, the teeth are fused to the jaws. Fossils are known as early as the Paleocene in eastern Asia. Intermediate forms are not known, but the common presence of acrodont tooth implantation supports a close relationship between chameleons and agamids. The modern genus *Chamaeleo* is known from the Miocene of Europe and Africa.

Members of the Gekkota are among the earliest known and most divergent of the modern lizard groups (Figure 11-26). Unlike all other advanced lizards, most Gekkota have amphicoelous vertebrae. This condition is known as early as the Upper Jurassic but may be a reversion to a primitive condition resulting from the suppression of structures that are expressed late in embryological devel-

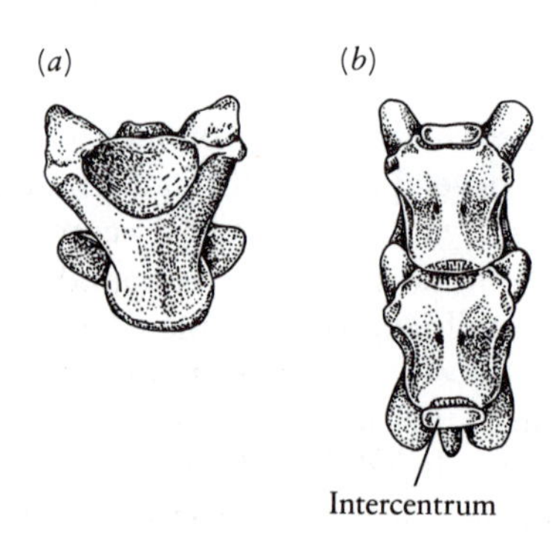

Figure 11-25. LIZARD VERTEBRAE. (*a*) The procoelus pattern of most groups of advanced lizards. (*b*) The trunk vertebrae of a gecko. They are amphicoelous, as in eosuchians, but this condition may be secondarily evolved within the geckos rather than being retained from their more primitive ancestors. *Drawn from specimens in Redpath Museum.*

Figure 11-26. SKELETON OF *ARDEOSAURUS.* A gecko from the Upper Jurassic of southern Germany, approximately ½ natural size. *From Mateer, 1982.*

opment in other lizard groups (Kluge, 1967). We also know procoelous gekkonids in the late Jurassic, and the condition is present in some living genera that are otherwise relatively primitive in their anatomy.

Gekkotans are specialized in the loss of the dorsal process of the squamosal, the down growth of the frontal bones below the brain, and the presence of the facial artery anterior to the stapes. The karyotype is very different from that of other lizards. The eyes of modern geckos are adapted to feeding under conditions of limited light. The diverse family Gekkonidae is known from the Upper Eocene. The Pygopodidae are rare burrowing lizards, without a fossil record, which are apparently related to the geckos.

Most remaining lizards apparently had an ancestry that was separate from the Iguania and Gekkota. Members of the Scincomorpha and Anguimorpha are both known in the Upper Jurassic. Like the gekkotans, the dorsal process of the squamosal is lost in most genera, and the remainder of the bone in the shape of a long, curving rod. However, a dorsal process is present in the Teiidae, which suggests that it was lost within the scincomorphs. There is a tendency for the body to be elongate and the limbs to be reduced in many members of these two major groups. Many have developed osteoscutes that cover the head and trunk.

The Scincomorpha includes the modern families Lacertidae, Xantusiidae, Scincidae, Cordylidae, Teiidae, and Gymnophthalmidae. The Paramacellodidae is known only from the Upper Jurassic.

The teiids are the most primitive, with a skull that closely resembles that of the iguanids (Figure 11-27). This family is common today in South and Central America and was quite diverse in the Upper Cretaceous of Central Asia and North America. The Gymnophthalmidae are an associated group without a fossil record.

More advanced scincomorphs tend to fill in the area of the upper temporal opening with extensions from one or more of the surrounding bones and to develop osteoscutes. The Lacertidae, which is widespread in the Old World, is known from the late Paleocene. The cordylids, which are known from the Upper Eocene, may be derived from the late Jurassic Paramacellodidae, which have similar scutes. The xantusids have a fossil record that goes back to the Paleocene. Some authors suggested that they are allied with the gekkotans, but they are more commonly associated with the skinks. The skinks, which are among the most diverse of modern lizards, appear in the Upper Cretaceous, with possible forebears in the Upper Jurassic. Many genera show marked limb reduction.

We can place most remaining lizards in a single group, the Anguimorpha, which includes a considerable variety of body forms. The group is represented in the Upper Jurassic by incomplete remains that are placed in the family Dorsetisauridae. They show some features of the living families Xenosauridae and Anguidae. The Anguidae, which appears in the late Cretaceous, encompasses a few genera with normally developed limbs, and others have snakelike proportions, including the "glass snake" *Ophisaurus,* and the European "slow worm" *Anguis.* Osteoderms are highly developed in the anguids, and they are especially prominent in a number of early Tertiary genera including *Glyptosaurus,* in which they form a massive armor (Figure 11-28). The "Gila monster," *Heloderma,* with antecedents from the Upper Cretaceous, belongs to an allied family, *Helodermatidae,* that is associated with more advanced anguimorphs. It is the only venomous lizard.

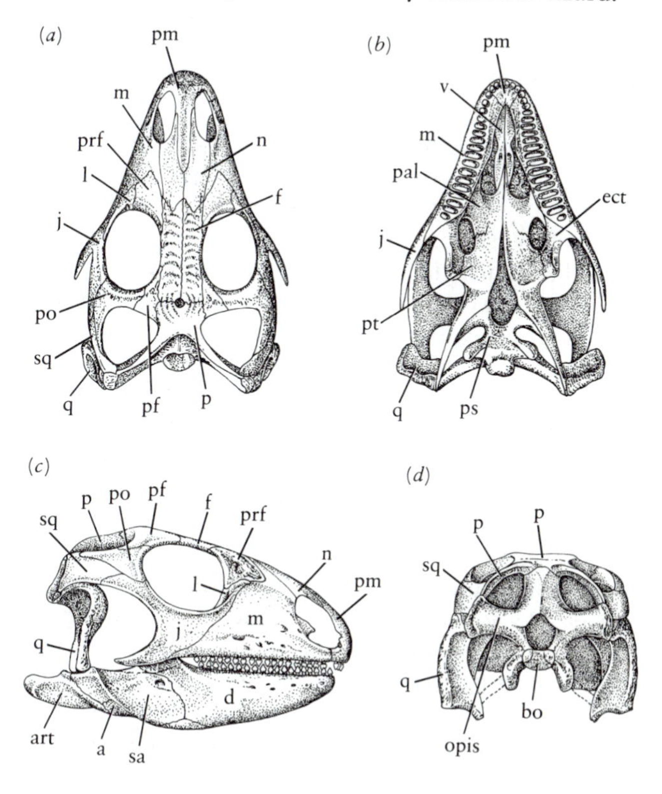

Figure 11-27. THE UPPER CRETACEOUS TEIID LIZARD *POLYGLYPHANODON.* (*a*) Dorsal, (*b*) palatal, (*c*) lateral, and (*d*) occipital views, ×½. The pattern of the skull root is typical of modern lizards. A transverse suture separates the frontals and parietals, which permits raising the snout in some genera. The pineal opening lies along this suture rather than further posteriorly as in more primitive lepidosaurs. The parietal and premaxilla are both fused at the midline. Abbreviations as in Figure 8-3. *From Gilmore, 1942. By permission of Smithsonian Institution Press.*

Figure 11-28. *HELODERMOIDES,* AN ARMORED ANGUID LIZARD FROM THE OLIGOCENE. *Photograph courtesy of Robert Sullivan.*

The varanoids (which are also termed platynotans) are the most advanced of all lizards in achieving large size and an active, predaceous way of life. The modern Komodo dragon of Indonesia exceeds 3 meters and preys on deer and pigs. *Megalania,* from the Pleistocene of Australia, was perhaps twice the size. The Komodo dragon is one of 24 living species of the genus *Varanus,* which are widely distributed in the Old World tropics. Characteristics that are partially developed in other anguimorphs reach their full expression in this genus. The tongue is a highly specialized structure with a bifid sensory tip that can be retracted into an elastic posterior portion. Replacement teeth are located between, rather than medial to, mature teeth. The mandible is hinged in the middle. At least 29 presacral vertebrae are present, nine of which contribute to a long neck. We know members of the family Varanidae no earlier than the late Cretaceous, but rather similar forms may have been present as early as the late Jurassic to account for the origin of several derived groups within the early Cretaceous.

The family Necrosauridae, which is known from the Upper Cretaceous into the Oligocene, is structurally intermediate between primitive members of the Anguioidea and the Varanoidea. Necrosaurs retain osteoscutes and the jaw hinge is only partially developed.

Aquatic varanoid lizards

Three other families of varanoid lizards, most recently reviewed by Russell (1967), are known from the late Jurassic and Cretaceous. The dolichosaurs were long-bodied, short-limbed lizards from the Middle Cretaceous of Europe. The long neck, with 11 vertebrae, and short skull clearly distinguish them from the other families. Haas (1980) suggested that dolichosaurs may be related to the ancestry of snakes. Unfortunately, their cranial anatomy is too poorly known for detailed comparison.

We know the aigialosaurs from one incomplete skull from the Upper Jurassic of Southern Germany and several skeletons from the middle Cretaceous (Cenomanian-Toronian) of Yugoslavia. The neck, with eight cervical vertebrae, resembles that of terrestrial varanids; the limbs are reduced but not structurally altered. In *Opetiosaurus* (Figure 11-29), the tail is laterally compressed and the tip is bent ventrally, as in marine crocodiles and advanced ichthyosaurs. The body, including the tail, is a little more than 1 meter long. By themselves, the aigialosaurs show only a modest radiation and limited specialization toward an aquatic way of life. Their importance lies in their close affinities with the most specialized of marine lizards, the mosasaurs, judged by the near identity of the skull structure (Figure 11-30).

The mosasaurs were the most spectacular of all lizards. We know these large, predaceous, fishlike forms only from the Upper Cretaceous. Nearly 20 genera are recognized, including *Clidastes, Tylosaurus,* and *Plotosaurus,* the largest specimens of which exceed 10 meters (Figure 11-31).

The body and tail are long and slender, suggesting an anguilliform mode of swimming, rather than the more specialized carangiform mode of advanced ichthyosaurs (see Chapter 12) and dolphins. The limbs are modified as fins for steering rather than for propulsion. Neither the ears nor the braincase show specializations comparable to those found in modern marine mammals in association with the capacity for deep diving. Some species have cancellous bone in the ribs that may have contributed to buoyancy.

The deposits in which most mosasaurs are found indicate that they were common in shallow, near-shore, marine waters. We know hundreds of extremely well-preserved specimens, but none shows a trace of young within the body cavity. Small, juvenile specimens are rarely found with the large adults. These observations suggest that mosasaurs may have laid eggs on land, like other varanoid lizards, although it must have been extremely difficult for the females to move without support of the water.

The skull is basically similar to that of modern varanids, but the closest resemblance is to the aigialosaurs. Mosasaurs and aigialosaurs differ from varanids in the fusion of the frontals, the extension of the premaxillae to the frontals, and the reduction of the nasals. There is a well-developed transverse joint between the angular and splenial, midway in the length of the lower jaw. There is some mobility between these bones in varanids, but the joint is already fully developed in aigialosaurs. As in aigialosaurs, the quadrate of mosasaurs is oval in lateral view and extremely massive. The tympanum is calcified in several genera.

In most genera, the teeth are large, sharp cones, set in sockets rather than fused to the side of the jaw. In the genus *Globidens,* the crowns of the teeth are hemispherical, as might be expected in forms feeding on molluscs or other hard-shelled prey. Several ammonites have been described which show a pattern of tooth marks comparable to that of the marginal dentition of *Prognathodon overtoni* (Kauffman and Kesling, 1960).

The neck is relatively short, with seven cervical vertebrae, but would have allowed considerable flexibility of

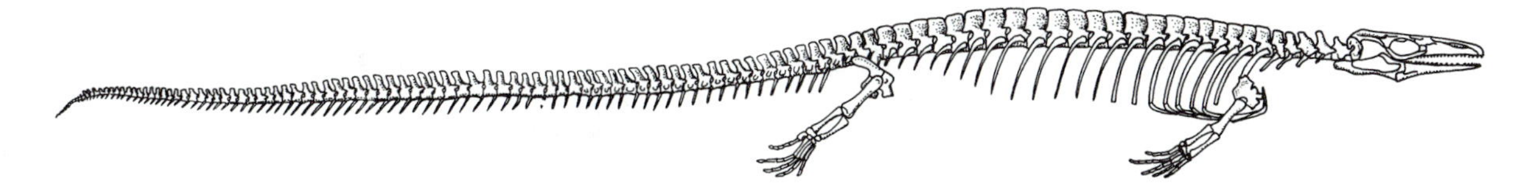

Figure 11-29. SKELETON OF THE AIGIALOSAUR *OPETIOSAURUS*. From the mid-Cretaceous of Yugoslavia. Approximately 1 meter long. *Restoration based on a specimen described by Kornhuber (1901).*

the head. Differences in body proportions among the many species suggest different degrees of maneuverability and maximum speed, which may have been associated with specializations for different types of prey.

The trunk is long and the number of presacral vertebrae ranges from 29 to 51. The tail is approximately as long as the presacral region; in some genera it is expanded distally, and in others it tapers gradually to a point. It never assumes the high aspect ratio of ichthyosaurs and fast-swimming fish. There is no connection between the vertebral column and the ilium. The rib cage is expanded anteriorly, with the first five thoracic ribs being attached to the sternum as in primitive terrestrial lizards. The posterior trunk ribs are much reduced. The manus and pes are extended by **hyperphalangy** (multiplication of the units in each digit) but do not exhibit **hyperdactyly** (multiplication of digits). The phalangeal count of the manus ranges from approximately 4, 5, 5, 5, 3 to 10, 10, 10, 7, 2.

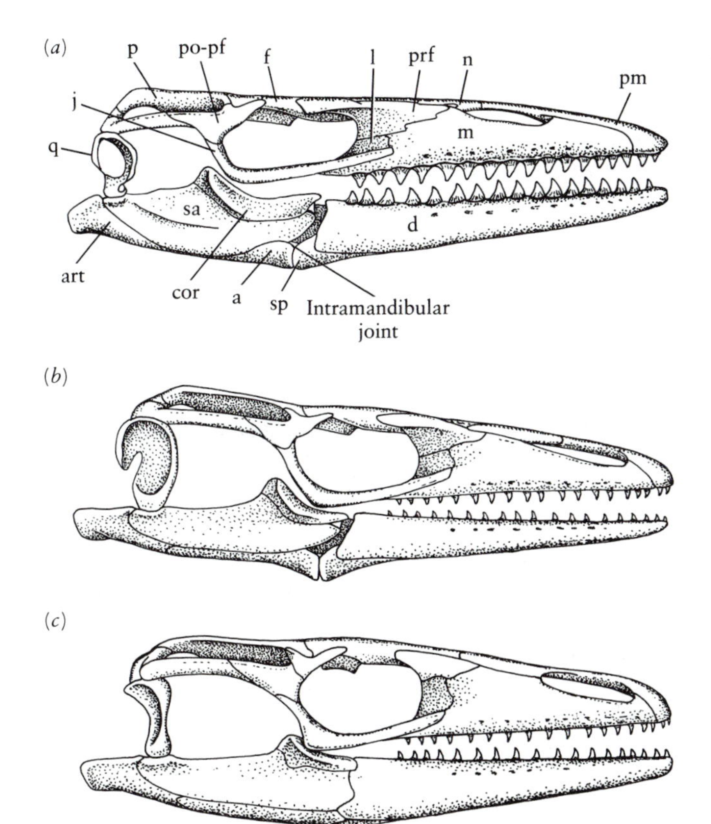

Figure 11-30. VARANOID LIZARD SKULLS. (*a*) The mosasaur *Clidastes*. (*b*) The aigialosaur *Opetiosaurus*. (*c*) Hypothetical primitive platynotan. *From Russell, 1967*. Abbreviations as in Figure 8-3.

In contrast with the similarity of the skull of aigialosaurs and mosasaurs, the postcranial skeleton is very different, which indicates that mosasaurs underwent a marked change early in their evolution that was associated with marine adaptation. The size is greatly increased and the limbs and tail are modified for aquatic propulsion and maneuverability. Unfortunately, we cannot establish the length of time during which these changes occurred. We know of no intermediate forms from the time of appearance of the first recognized aigialosaurs in the late Jurassic until the first appearance of highly specialized mosasaurs in the early Upper Cretaceous, approximately 60 million years later.

Once the basic pattern of mosasaur anatomy was achieved, the group remained relatively constant in body form until its extinction at the end of the Mesozoic, although details of vertebral number, fin configuration, and skull morphology allow us to recognize some 40 species that are distributed throughout the world.

Amphisbaenia (Annulata)

Although we have not established the specific interrelationships of the modern lizard infraorders, their anatomy shows some degree of convergence as the fossil record is traced back into the Mesozoic, and they probably share a common ancestry above the level of the eolacertilians. A further group of squamates, the Amphisbaenia, shows a very distinct anatomical pattern throughout its fossil history and has no significant characters in common with any one of the lacertilian infraorders. The anatomy and taxonomy of this group has been extensively studied by Gans (1978), who feels that they should be accorded taxonomic rank equivalent to that of lizards and snakes.

Except for the genus *Bipes*, which retains well-developed forelimbs, all amphisbaenids lack limbs and girdles and have an elongate, snakelike body. In strong contrast with most snakes and lizards, the skull is solidly constructed, with the bones closely interlocking and surrounding the braincase (Figure 11-32). All amphisbaenids use the skull as a digging tool and are exclusively subterranean in habits. The postorbital portion of the skull is elongate, with a large mass of temporal musculature. The temporal arcade is lost in most genera.

The jaw articulation is far forward, and there is a large extra-columella but no tympanum or fenestra rotundum. Amphisbaenids are unique in having a median tooth in the fused premaxillae. The vertebral number ranges from 80 to 175, with a very short tail that is not capable

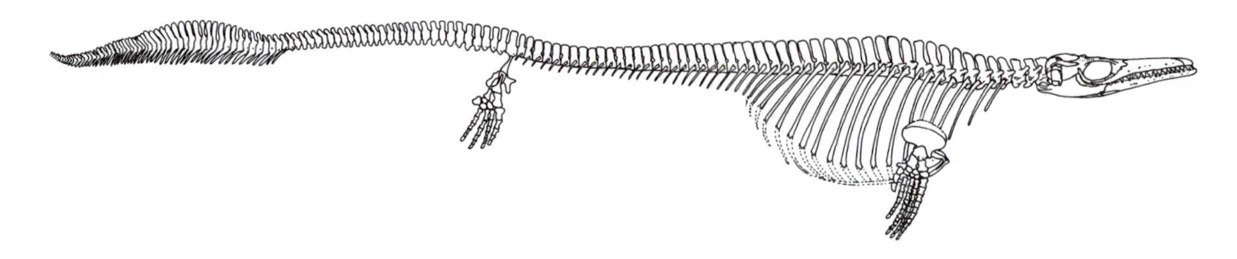

Figure 11-31. SKELETON OF THE MOSASAUR *PLOTOSAURUS*. Genus is a giant marine lizard from the Upper Cretaceous, approximately 10 meters long. *From Russell, 1967.*

of regeneration. The living species are widespread in Africa, the Middle East, and South and Central America.

We recognize four modern families: the Trogonophidae and Bipedidae, which have no fossil record; the Rhineuridae, with only a single living genus but a rich fossil record in the early and middle Tertiary of North America; and the Amphisbaenidae, the most common of living groups, but with a limited fossil record. Estes (1983a) recognizes two additional families that are known only by fossils, the Oligodontosauridae, which we know primarily from isolated jaws of Paleocene age, and the Hyporhinidae from the Oligocene.

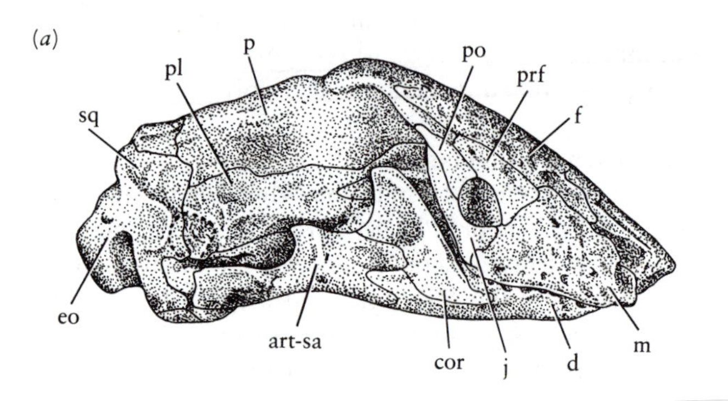

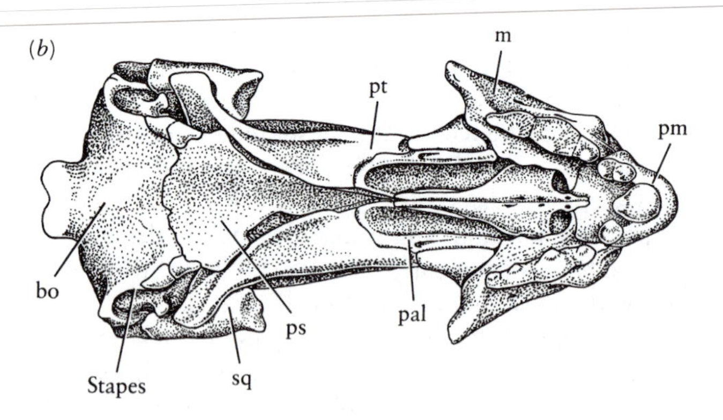

Figure 11-32. AMPHISBAENID LIZARD SKULLS. (*a*) Lateral view of the skull of *Spathorhynchus* from the early Oligocene, ×3. The temporal arcade common to other lizards is lost. The bones of the snout and braincase are very solidly integrated; the lateral wall of the braincase is formed by a large pleurosphenoid (pl). Dashed line indicates position of fenestra ovalis. *From Berman, 1977.* (*b*) Palate of the modern genus *Trogonophis*, ×6. *From Gans, 1960.* The presence of a median premaxillary tooth is a unique feature of this group. Abbreviations as in Figure 8-3.

According to Gans, most features of the amphisbaenids are closer to the pattern of lizards than they are to the snakes. Their well-developed procoely unites amphisbaenids with advanced lizards above the level of the eolacertilians. Unless we can demonstrate that amphisbaenids evolved earlier than any of the other groups now classified as lizards, there is no phylogenetic justification for recognizing them as a separate taxonomic group. The known members have a very distinct skeletal anatomy, but this is the case for mosasaurs as well and yet they are universally accepted as no more than a family among the Varanoidea.

SNAKES

Structure and origin

Snakes are the most rapidly evolving group of reptiles. Most modern genera belong to families whose major radiation has occurred since the beginning of the Miocene. The skull structure and feeding behavior of many snakes seems specifically adapted toward warm-blooded prey, which suggests that the recent evolution of snakes has been closely tied with the diversification of mammals.

Snakes have exaggerated the tendency that is seen in many lizard groups toward elongation of the trunk and reduction of the limbs. No snakes show even a trace of the pectoral girdle or forelimb, although many primitive genera retain vestiges of the pelvic girdle and rear limb. The number of precaudal vertebrae ranges from 120 to 454. The vertebrae (which are often the only elements known for many fossil snakes) are characterized by the presence of accessory articulating facets above the neural canal, the anterior zygosphene, and posterior zygantrum (Figure 11-33). Some lizards have similar structures, but their specific configuration allows us to distinguish the vertebrae of the two groups.

In most snakes, the elements of the upper and lower jaw and the palate are highly mobile in association with feeding on large prey that is swallowed whole (Figure 11-34). The upper temporal arch is lost and the quadrate is even more mobile than in lizards. It serves a role that is comparable to that of the hyomandibular in advanced sharks (see Figure 5-14) by contributing to a great mobility of the jaws that are not strongly attached at the

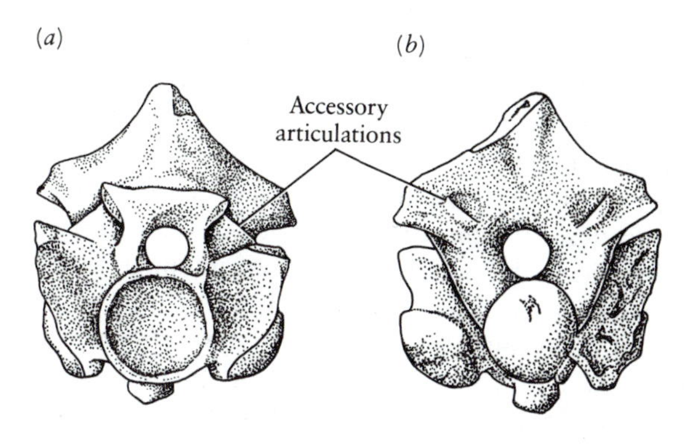

Figure 11-33. VERTEBRA OF THE PRIMITIVE CAENOPHIDIAN SNAKE *NIGEROPHIS* FROM THE PALEOCENE OF NIGER. (*a*) Anterior and (*b*) posterior views showing accessory articulating surfaces that are diagnostic of snakes, ×6. *From Rage, 1975. By permission of Comptes Rendus de l'Academie des Sciences de Paris, Sen. D, 281.*

symphysis. The teeth are long, laterally compressed, and recurved. They are newly elaborated on the pterygoid and palatine but may be lost from the premaxilla. The temporal region is elongated, which provides for a great increase in the area of jaw muscles, and the eyes are far forward. The feeding mechanisms of snakes have been described by Pough in a recent symposium (1983).

To protect the braincase from struggling prey, the frontals and parietals have grown ventrally so that they completely surround the forebrain. In contrast with lizards and other reptiles, the braincase is low and flat. Embryologically, the trabeculae remain paired in the interorbital region, whereas they fuse medially in all other amniotes.

The oldest snake in which the skull is adequately known, *Dinilysia* from the Upper Cretaceous of South America, shows a fairly advanced stage of evolution of these features (Figure 11-35).

In less specialized aspects of their anatomy, snakes closely resemble lizards, and it has long been accepted that they evolved from within that group. The most important derived feature that all species of snakes and lizards possess is the hemipenes (paired, eversible copulatory organs). However, there is continued debate as to whether

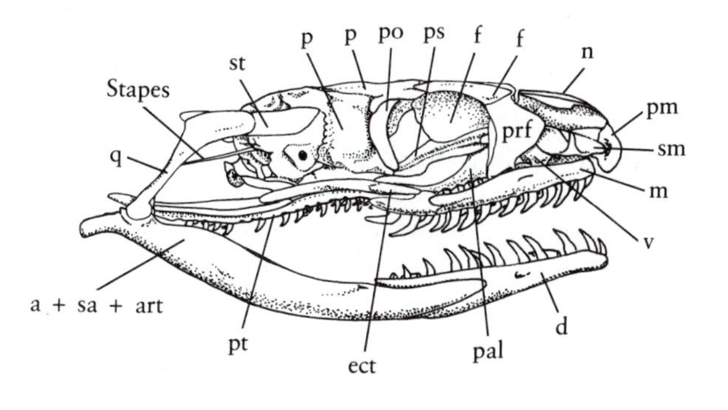

Figure 11-34. SKULL OF A MODERN COLUBRID SNAKE, *PTYAS*. The loose connection of the upper and lower jaw is clearly evident. Abbreviations as in Figure 8-3. *From Parker and Grandison, 1977. © Trustees of The British Museum (Natural History) 1977. Used by permission of the publisher, Cornell University Press.*

any of the modern lizard families are closely related to snakes, and there are also conflicting views as to the environment in which snakes evolved.

Snakes share several features (including their most important attributes, trunk elongation and limb reduction) with burrowing lizards. Like these groups, snakes have lost the tympanum and the middle ear cavity. Even more striking similarities are found in the eye, which some authors have interpreted as indicating that snakes passed through a burrowing stage in their evolution during which the eye underwent marked degeneration. It is hypothesized that the eye became reconstituted when modern snakes again became adapted to living on the surface of the ground.

As in some burrowing lizards, snakes have modified the movable eyelids into a fixed spectacle. The retinal cells appear to have been simplified, and both intrinsic and extrinsic eye muscles have been reduced. All snakes lack oil droplets and most lack a fovea, which is present in advanced lizards. The absence of chondrified and ossified supports of the sclera, which are retained even in burrowing lizards provides further evidence of degeneration of the eye (Walls, 1942).

Despite these arguments, there are several problems in accepting a burrowing stage in the origin of snakes. The specializations that occurred in the skulls of burrowing lizards and amphisbaenids (see Figure 11-32) are very different than those that characterize early snakes such as *Dinilysia*. The bones of the skull are tightly knit into a solid structure, in contrast with the extremely mobile skull of most snakes. Modern burrowing snakes, like burrowing lizards, have upper jaws that are strongly attached to the skull and tend to reduce the length of the lower jaws. These genera appear to have specialized their jaw structure from a more mobile antecedent condition, but the skull is otherwise quite distinct from that of burrowing lizards. The earliest known genera that have been identified as snakes or intermediates between lizards and snakes are not burrowers but were aquatic and have a highly mobile skull.

Several of the characteristics of snakes that have been attributed to burrowing may have a different basis. The loss of the tympanum and middle ear cavity has occurred in several groups of nonburrowing lizards and can be associated in snakes with the great mobility of the quadrate, which would have made support of a tympanum extremely difficult (Berman and Regal, 1967). Several of the changes in eye structure that have been cited as indicating a fossorial stage in the origin of snakes also occurred in the origin of mammals, although in association with nocturnal vision rather than with burrowing. Other changes in the eye can be attributed to life on the ground and movement in dense vegetation and loose soil and were not necessarily associated with a strictly subterranean or burrowing way of life.

Among modern lizards, the most significant similarities to snakes are found among the varanoids. These include the nature of the tongue, the manner of tooth

(*a*)

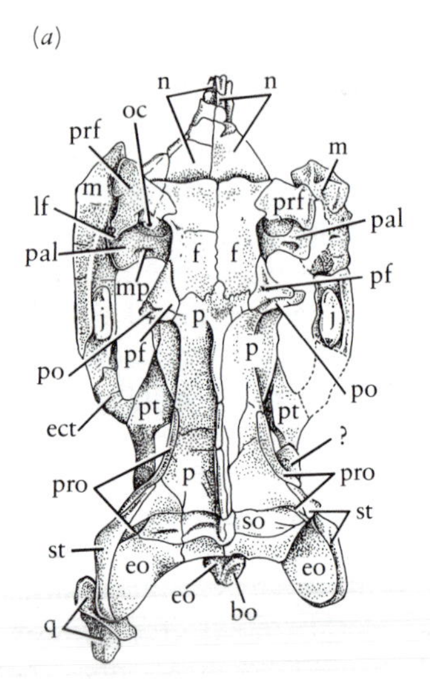

(*b*)

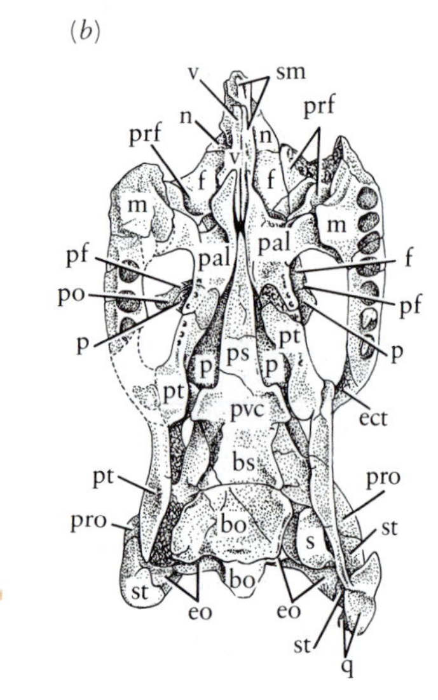

Figure 11-35. SKULL OF THE EARLIEST ADEQUATELY KNOWN SNAKE *DINILYSIA*. From the Upper Cretaceous of South America, in dorsal and palatal views, ×1. Abbreviations as in Figure 8-3 plus: lf, lacrimal foramen; mp, maxillopalatine foramen; oc, orbitonasal canal; pvc, posterior opening of vidian canal; s, stapes. *From Estes, Frazetta, and Williams, 1970.*

replacement, and the presence of an intramandibular joint. In both groups, Jacobson's organ is an important sensory structure. No one feature is unique to varanoids, but no other group of lizards possesses all of them. Rieppel (1980, 1983) questions varanoid affinities and suggests that the pattern of kinesis of both the skull roof and the lower jaws differ from that of all modern lizard groups and probably evolved separately from the eolacertilian level.

In contrast with members of several other lizard groups, modern varanoids show little tendency toward limb reduction or trunk elongation. However, there are several early Cretaceous genera that combine the cranial features of varanoids (including great mobility of the jaw bones and quadrate) with an elongate body and reduced limbs. These genera include the dolichosaurs, which have already been mentioned, as well as two genera that Haas (1979, 1980) recently described from the early Upper Cretaceous of Israel (Figure 11-36). The latter are clearly snakelike in general body form and the total absence of forelimbs and girdles. *Ophiomorphus* has well-developed remnants of the rear limb and girdle, but they are much reduced in *Pachyrhachis*. Both are well over 1 meter long, with more than 100 precaudal vertebrae. *Pachyrhachis* has pachyostotic ribs, which suggests aquatic adaptation. Both show vertebral similarities with previously named members of the family Simoliophidae, which have been classified among both the varanoids and the snakes. These genera seem to be almost ideal intermediates between the two groups.

The ancestors of snakes probably diverged from the lizards before the end of the Jurassic. Although the evidence is not conclusive, their affinities are probably close to the base of the varanoid stock, from which their primary specialization was toward great elongation of the body and limb reduction. The early members may have been relatively large terrestrial carnivores that became further specialized by an increase in jaw mobility that enabled them to swallow prey whole. Great elongation of the body and a relative reduction of its girth would have preadapted snakes for both aquatic locomotion and burrowing, which have been elaborated in divergent lines many times during their evolutionary history (Gans, 1975).

The diversity of snakes

The fossil record of terrestrial snakes in the Cretaceous and Lower Tertiary remains very incomplete. The interrelationships of the modern families is based almost entirely on the anatomy of living species.

We can divide modern snakes into three groups: the infraorders Scolecophidia, Henophidia, and Caenophidia (Parker and Grandison, 1977). The Scolecophidia includes two families of burrowing snakes, the Typhlopidae and the Leptotyphlopidae, which are today widely distributed in Europe, southeast Asia, Africa, South America, and Australia. They are primitive in retaining rudiments of the pelvic girdle and in having only a single opening for the trigeminal nerve. However, they are highly specialized in their cranial anatomy in a pattern unlike that of other snakes, modern or fossil. Their skull structure may be associated with a burrowing way of life and adaptation to feeding on small invertebrate prey. Most of the skull bones are solidly joined together, the jaws are shortened, and the dentary and pterygoid may lack teeth. Vertebrae that are assignable to this group are known from the Eocene.

The Henophidia, or Booidea, includes a large number of genera that are relatively primitive but lack the specializations of the Scolecophidians. Primitive features that are retained in many genera include vestiges of the pelvic girdle and hind limb, coronoid bones in the lower jaw, and teeth on the premaxilla. In contrast with more ad-

Figure 11-36. *PACHYRHACHIS.* This genus from the Lower Cretaceous of Israel may be intermediate between varanoid lizards and snakes. Approximately 1 meter long. *Photograph courtesy Dr. Tchernov.*

vanced snakes, the left and right common carotids are both retained. Three families of burrowing snakes and the Boidae are included in this group. The Aniliidae, Uropeltidae, and Xenopeltidae (sometimes united in a single family) are all medium-sized snakes that feed on small vertebrates and invertebrates. They are specialized in the solid attachment of the upper jaw. The Uropeltidae and Xenopeltidae, which are restricted to India, Ceylon, and southeast Asia, have no fossil record. The Aniliidae, which have modern genera in South America and southeast Asia, are known in North America from the Upper Cretaceous through the Middle Tertiary.

Although the Boidae includes some burrowing genera, members of this family show much greater jaw mobility, with lengthening of the squamosal and quadrate. The Boidae includes the well-known boas and pythons, which may reach a length of 10 meters, as well as many smaller forms. Some boids are fossorial, while others are terrestrial, arboreal, or semiaquatic. We find fossils that belong to the family Boidae as early as the Upper Cretaceous. *Dinilysia* (see Figure 11-35), one of the few fossil snakes for which nearly the entire skeleton is known, appears close to the base of the stock from which all later henophidians evolved.

The Caenophidia includes all the more advanced snakes—the vipers, cobras, and other poisonous forms, as well as the extremely diverse colubrid assemblage. They are differentiated from more primitive snakes by a further increase in jaw mobility. The basipterygoid processes are lost so that the pterygoid is more freely movable relative to the base of the braincase. The teeth on the premaxillae are lost, as are the coronoid bones in the lower jaw. The pelvic vestiges are lost, which increases the capacity of the abdominal region to accept large prey. Further technical distinctions from henophidians include the loss of the right common carotid artery and the wider spacing of the intercostal arteries.

The Colubridae is the most diverse of all snake families, with over 300 genera and 1400 species and a worldwide distribution. Most are terrestrial, but some have adapted to fresh water and others to marine habitats. A few are venomous, with fixed fangs at the back of the mouth. They are the earliest of the advanced Caenophidian groups to appear in the fossil record, with isolated vertebrae reported from the late Eocene of Europe.

The major groups of venomous snakes, the Viperidae and the Elapidae, both appear first in the Lower Miocene of Europe and enter the North American fauna by the end of the epoch. Both are now nearly worldwide in distribution. The Elapidae, which include the cobras, coral snakes, mambas, and kraits, are characterized by the fixed nature of the fangs. Although not known as fossils, the sea snakes or Hydrophidae, appear to be closely related to the Elapidae and have been placed in the same family. They are widespread in warm oceans.

In the Viperidae, the maxillae can be rotated so that the fangs are folded back during feeding. The subfamily Viperinae includes the adders and true vipers of the Old World. The Crotalinae or pit vipers are characterized by the presence of a heat-sensitive pit between the eye and nostril. This adaptation seems especially well suited for hunting warm-blooded mammals in the night and in dark tunnels and crevices. This subfamily includes rattlesnakes, water moccasins, copperheads, the bushmasters, and fer-de-lance.

Modern colubroid snakes are clearly distinct from most henophidians, but several families appear to occupy an intermediate position. The living Acrochordidae, which are typically placed with the boids, show many caenophidan features—especially in the mobility of the jaws—

but these may have been achieved in parallel. The Archaeophidae, Anomalophidae, and Palaeophidae of the Upper Cretaceous and Lower Cenozoic may be related to the ancestry of the caenophidians, but they are poorly known and, like the Acrochordidae, primarily aquatic.

SUMMARY

Lepidosaurs include the modern lizards, snakes, *Sphenodon,* and their ancestors going back to the late Paleozoic. Lepidosaurs and archosaurs (including the modern crocodiles and the dinosaurs) belong to a larger assemblage, the Diapsida, that is differentiated from other reptile groups by the presence of two pairs of temporal openings and a suborbital fenestra. The oldest known genus to have these features is *Petrolacosaurus,* from the Upper Carboniferous. *Petrolacosaurus* closely resembles the protorothyrids in other, more primitive features of the skeleton.

Petrolacosaurus has an elongate neck and distal limb elements that unite it with *Araeoscelis* as a distinct group of early diapsids, the Araeoscelida. Other primitive diapsids include the coelurosauravids, in which the ribs are greatly expanded to form a gliding membrane, and the thalattosaurs, which have become secondarily aquatic.

We may group more advanced diapsids into two large assemblages, the lepidosauromorphs and the archosauromorphs, which are distinguished by basic differences in locomotion. Lepidosauromorphs retain the primitive pattern of sinusoidal movements of the trunk that are common to early tetrapods, but the force and stride of their forelimbs is increased through the development of a massive sternum that forms a base for the rotation of the coracoids.

The most primitive lepidosauromorphs are the Upper Permian and Lower Triassic eosuchians. *Youngina* and *Thadesoaurus* are almost ideal ancestors of later lepidosaurs, except for specialization of the carpus. The crocodilelike champsosaurs of the late Cretaceous and early Tertiary may have evolved from primitive eosuchians, but we cannot establish this connection with certainty.

Lizards and sphenodontids may share a common ancestry with the eosuchians. The modern lepidosaur groups are unique among reptiles in having separate epiphyseal ossifications that allow the formation of specialized joint surfaces and permit controlled termination of growth. Early squamates and sphenodontids have a light, elongate stapes, and they probably had an impedance-matching middle ear.

We know the sphenodontids from the Upper Triassic. Most features of their skeleton already resemble the living species. This group is characterized by the presence of an acrodont marginal dentition. The palatine bears a row of large denticles that in *Sphenodon* are parallel with those of the maxilla. The lower temporal bar is complete in most genera but is bowed laterally to allow passage of the external adductor to the lateral surface of the lower jaw. In some early genera, the lower temporal bar is incomplete. The pleurosaurs were an early offshoot of the sphenodontids that became adapted to an aquatic way of life. The trunk region is greatly elongate, and the limbs are greatly reduced.

Lizards, which appear in the fossil record at the end of the Permian, are distinguished from sphenodontids in the mobility of the quadrate (streptostyly), which enables the pterygoideus musculature to exert a much greater force when the jaw is nearly closed. In even the earliest genus, *Paliguana,* the quadrate is deeply embayed posteriorly to support a tympanum as in modern lizards.

Modern lizard infraorders had all evolved by the end of the Jurassic. Most living families have a fossil record going back to the later Cretaceous or early Tertiary, and we know many modern genera from the early Tertiary.

The most spectacular of fossil lizards are the mosasaurs which were large marine carnivores of the late Cretaceous. They share a common ancestry with the modern varanoids.

Elongate, snakelike forms from the early Cretaceous support affinities of this group with the varanoid lizards. The evolution of the advanced snake families is closely associated with that of the Tertiary mammals on which they preyed.

REFERENCES

Berman, D. S. (1977). *Spathorhynchus natronicus,* a new species of rhineurid amphisbaenian (Reptilia) from the early Oligocene of Wyoming. *J. Paleont.,* **51**: 986–991.

Berman, D. S., and Regal, P. J. (1967). The loss of the ophidian middle ear. *Evolution,* **21**(3): 641–643.

Broom, R. (1914). A new thecodont reptile. *Proc. Zool. Soc. Lond.*: 1072–1077.

Carroll, R. L. (1977). The origin of lizards. In S. M. Andrews, R. S. Miles, and A. D. Walker (eds.), *Problems in Vertebrate Evolution. Linn. Soc. Symp. Ser.,* **4**: 359–396. Academic Press, London and New York.

Carroll, R. L. (1978). Permo-Triassic "Lizards" from the Karoo System. Part II. A gliding reptile from the Upper Permian of Madagascar. *Palaeont. afr.,* **21**: 143–159.

Carroll, R. L. (1981). Plesiosaur ancestors from the Upper Permian of Madagascar. *Phil. Trans. Roy. Soc. Lond., B,* **293**: 315–383.

Carroll, R. L. (1985). A pleurosaur from the Lower Jurassic and the taxonomic position of the Sphenodontida. *Palaeontographica, A,* **189**: 1–28.

Carroll, R. L., and Thompson, P. (1982). A bipedal lizardlike reptile from the Karoo. *J. Paleont.,* **56**: 1–10.

Colbert, E. H. (1970). The Triassic gliding reptile *Icarosaurus. Bull. Amer. Mus. Nat. Hist.,* **143**: 85–142.

Currie, P. J. (1981). *Hovasaurus boulei,* an aquatic eosuchian from the Upper Permian of Madagascar. *Palaeont. afr.,* **24**: 99–168.

Currie, P. J., and Carroll, R. L. (1984). Ontogenetic changes in the eosuchian reptile *Thadeosaurus. J. Vert. Paleont.,* **4**(1): 68–84.

Erickson, B. R. (1972). The lepidosaurian reptile *Champsosaurus* in North America. *Monograph Sci. Mus. Minn.,* **1**: 1–91.

Estes, R. (1983a). *Sauria terrestria, Amphisbaenia.* In P. Wellnhofer (ed.), *Handbuch der Palaeoherpetologie.* Part 10A. Gustav Fischer Verlag, Stuttgart.

Estes, R. (1983b). The fossil record and early distribution of lizards. In A. G. J. Rhodin and K. Miyata (eds.), *Advances in Herpetology and Evolutionary Biology: Essays in Honor of Ernest E. Williams.* Museum of Comparative Zoology, Cambridge, Massachusetts.

Estes, R., Frazzetta, T. H., and Williams, E. E. (1970). Studies on the fossil snake *Dinilysia patagonica* Woodward: Part 1. Cranial morphology. *Bull. Mus. Comp. Zool.,* **140**(2): 25–73.

Evans, S. E. (1981). The postcranial skeleton of the Lower Jurassic eosuchian *Gephyrosaurus bridensis. Zool. J. Linn. Soc.,* 73: 81–116.

Evans, S. E. (1982). The gliding reptiles of the Upper Permian. *Zool. J. Linn. Soc.,* **76**: 97–123.

Farbre, J. (1981). *Les rhynchocéphales et les ptérosaurien à crête pariétale du Kiméridgien supérieur—Berriasien d'Europe occidentale. Les gisement de Canjuers (Var-France) et ses abords.* Editions de la Fondation Singer-Polignac, Paris.

Fraser, N. C. (1982). A new rhynchocephalian from the British Upper Trias. *Palaeontology,* **25**(4): 709–725.

Fraser, N. C., and Walkden, G. M. (1984). The postcranial skeleton of the Upper Triassic sphenodontid *Planocephalosaurus robinsonae. Palaeontology,* **27**(3): 575–595.

Gans, C. (1960). Studies on amphisbaenids (Amphisbaenia, Reptilia). 1. A taxonomic revision of the Trogophinae, and a functional interpretation of the amphisbaenid adaptive pattern. *Bull. Amer. Mus. Nat. Hist.,* **119**: 129–204.

Gans, C. (1975). Tetrapod limblessness: Evolution and functional corollaries. *Amer. Zool.,* **15**: 455–467.

Gans, C. (1978). The characteristics and affinities of the Amphisbaenia. *Trans. Zool. Soc. Lond.,* **34**: 347–416.

Gilmore, C. W. (1942). Osteology of *Polyglyphanodon,* an Upper Createaceous lizard from Utah. *Proc. U.S. Nat. Mus.,* **92**: 229–265.

Gorniak, G. C., Rosenberg, H. I., and Gans, C. (1982). Mastication in the Tuatara, *Sphenodon punctatus* (Reptilia: Rhynchocephalia): Structure and activity of the motor system. *J. Morph.,* **171**: 321–353.

Haas, G. (1979). On a new snakelike reptile from the Lower Cenomanian of Ein Jabrud, near Jerusalem. *Bull. Mus. Natn. Hist. Nat.,* Paris 4e sér., **1**: 51–64.

Haas, G. (1980). Remarks on a new ophiomorph reptile from the Lower Cenomanian of Ein Jabrud, Israel. In L. L. Jacobs (ed.), *Aspects of Vertebrate History,* pp. 177–192. Museum of Northern Arizona Press, Flagstaff, Arizona.

Jenkins, F. A., Jr., and Goslow, G. E., Jr. (1983). The functional anatomy of the shoulder of the Savannah monitor lizard (*Varanus exanthematicus*). *J. Morph.,* **175**: 195–216.

Kauffman, E. G., and Kesling, R. V. (1960). An Upper Cretaceous ammonite bitten by a mosasaur. *Contr. Mus. Paleont., Univ. Michigan,* **15**(9): 193–248.

Kluge, A. (1967). Higher taxonomic categories of gekkonid lizards and their evolution. *Bull. Amer. Mus. Nat. Hist.,* **135**: 1–59.

Kornhuber, A. G. (1901). *Opetiosaurus bucchichi.* Eine neue fossile Eidechse aus der unteren Kreide von Lesina in Dalmatien. *Abh. geol. Reichsanst. Vienna,* **17**: 1–24.

Kuhn-Schnyder, E. (1974). Die Triasfauna der Tessiner Kalkalpen. *Neujahrsblatt, Naturf. Ges. Zuerich,* **176**: 1–119.

Malan, M. E. (1963). The dentition of the South African Rhynchocephalia and their bearing on the origin of the rhynchosaurs. *S. Afr. J. Sci.,* **59**: 214–219.

Mateer, N. J. (1982). Osteology of the Jurassic lizard *Ardeosaurus brevipes* (Meyer). *Palaeontology,* **25**(3): 461–469.

Parker, H. W., and Grandison, A. G. C. (1977). *Snakes—A Natural History* (2d ed.). British Museum and Cornell University Press, London and Ithaca, New York.

Pough, H. F. (1973). Lizard energetics and diet. *Ecology,* **54**: 837–844.

Pough, H. F. (ed.). (1983). Adaptive radiation within a highly specialized system: The diversity of feeding mechanisms of snakes. *Amer. Zool.,* **23**: 338–460.

Rieppel, O. (1983). A comparison of the skull of *Lanthanotus borneensis* (Reptilia: Varanoidea) with the skull of primitive snakes. *Z. Zool. Systematik und Evolutionsforschung,* **21**(2): 142–153.

Robinson, P. L. (1962). Gliding lizards from the Upper Keuper of Great Britain. *Proc. Geol. Soc. Lond.,* **1601**: 137–146.

Robinson, P. L. (1973). A problematic reptile from the British Upper Trias. *J. Geol. Soc.,* **129**: 457–479.

Robinson, P. L. (1975). The functions of the hooked fifth metatarsal in lepidosaurian reptiles. In *Problemes actuels de paléontologie; evolution des vertebres. Colloq. Int. C.N.R.S.,* **218**: 461–483.

Romer, A. S. (1966). *Vertebrate Paleontology.* University of Chicago Press, Chicago.

Russell, D. A. (1967). Systematics and morphology of American mosasaurs (Reptilia, Sauria). *Peabody Mus. Nat. Hist. Bull.,* **23**: 1–237.

Rage, J.-C. (1975). Un serpent du Paléocene du Niger. Etude préliminaire sur l'origine des Caenophidiens (Reptilia, Serpentes). *C.R., Acad. Sci. (Paris), Ser. D,* **281**: 515–518.

Rage, J.-C. (1984). Serpentes. In P. Wellnhofer (ed.), *Handbuch der Palaeoherpetologie,* **11**: 1–80. Gustav Fischer Verlag, Stuttgart.

Reisz, R. (1981). A diapsid reptile from the Pennsylvanian of Kansas. *Univ. Kansas, Mus. Nat. Hist., Spec. Publ.,* **7**: 1–74.

Reisz, R., Berman, D. S., and Scott, D. (1984). The anatomy and relationships of the Lower Permian reptile *Araeoscelis. J. Vert. Paleont.,* **4**: 57–67.

Rieppel, O. (1980). The phylogeny of anguinomorph lizards. *Denkschriften Schweiz, Naturf. Ges.,* **94**: 1–86.

Sigogneau-Russell, D. (1979). Les champsosaures Européens: Mise au point sur le Champsosaure d'Erquelinnes (Landénien Inferieur, Belgique). *Ann. Pal. Vert.,* **65**: 93–154.

Smith, K. K. (1980). Mechanical significance of streptostyly in lizards. *Nature,* **283**: 778–779.

Walls, G. L. (1942). The vertebrate eye and its adaptive radiation. *Bull. Cranbrook Instit. Sci.,* **19**: 1–785.

Wever, E. G. (1978). *The Reptile Ear.* Princeton University Press, Princeton, New Jersey.

CHAPTER XII

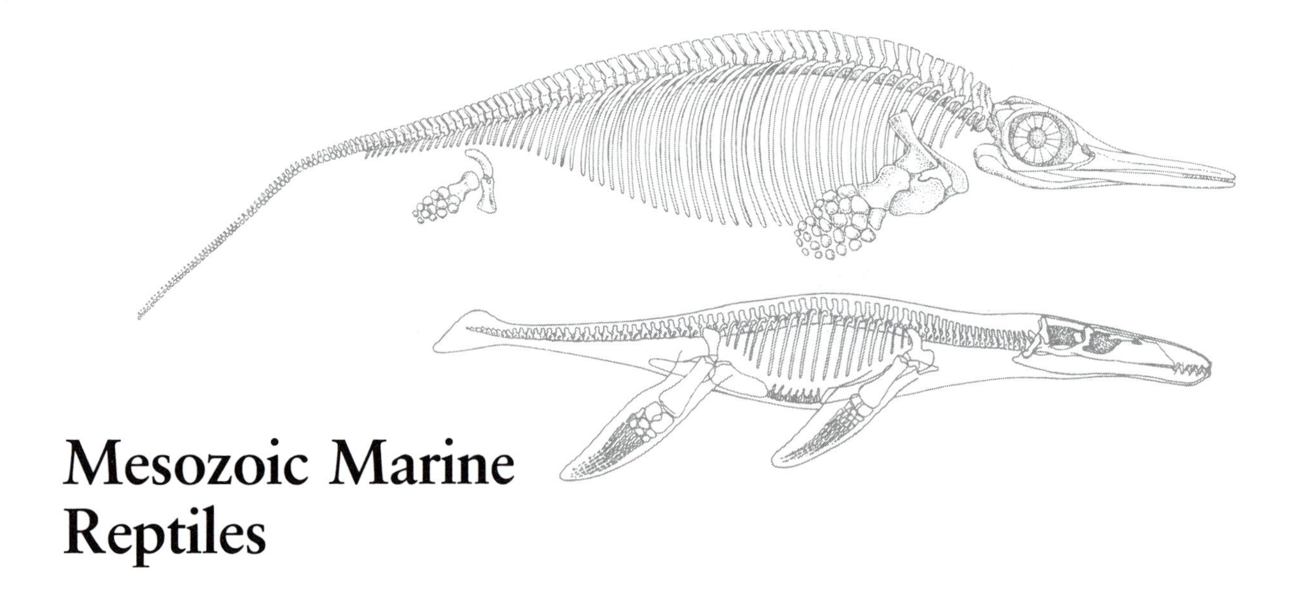

Mesozoic Marine Reptiles

Reptiles are a primarily terrestrial assemblage, but during the Mesozoic several groups also dominated the marine environment. Among the most conspicuous marine reptiles were the ichthyosaurs, which were comparable in size and shape to large pelagic sharks and dolphins, and the plesiosaurs, which are characterized by their long necks, short trunks, and large paddlelike limbs. Ichthyosaurs appeared in the early Triassic, were most numerous in the early Jurassic, but died out well before the end of the Cretaceous. The plesiosaurs were diverse throughout the Jurassic and Cretaceous. Less well known are the nothosaurs and placodonts of the Triassic, which show lesser degrees of aquatic adaptation but are already clearly distinct from all groups of earlier terrestrial reptiles. The pattern of radiation of these groups is shown in Figure 12-1.

The specific ancestry and relationships of these groups have long been subject to dispute. There is still little evi-

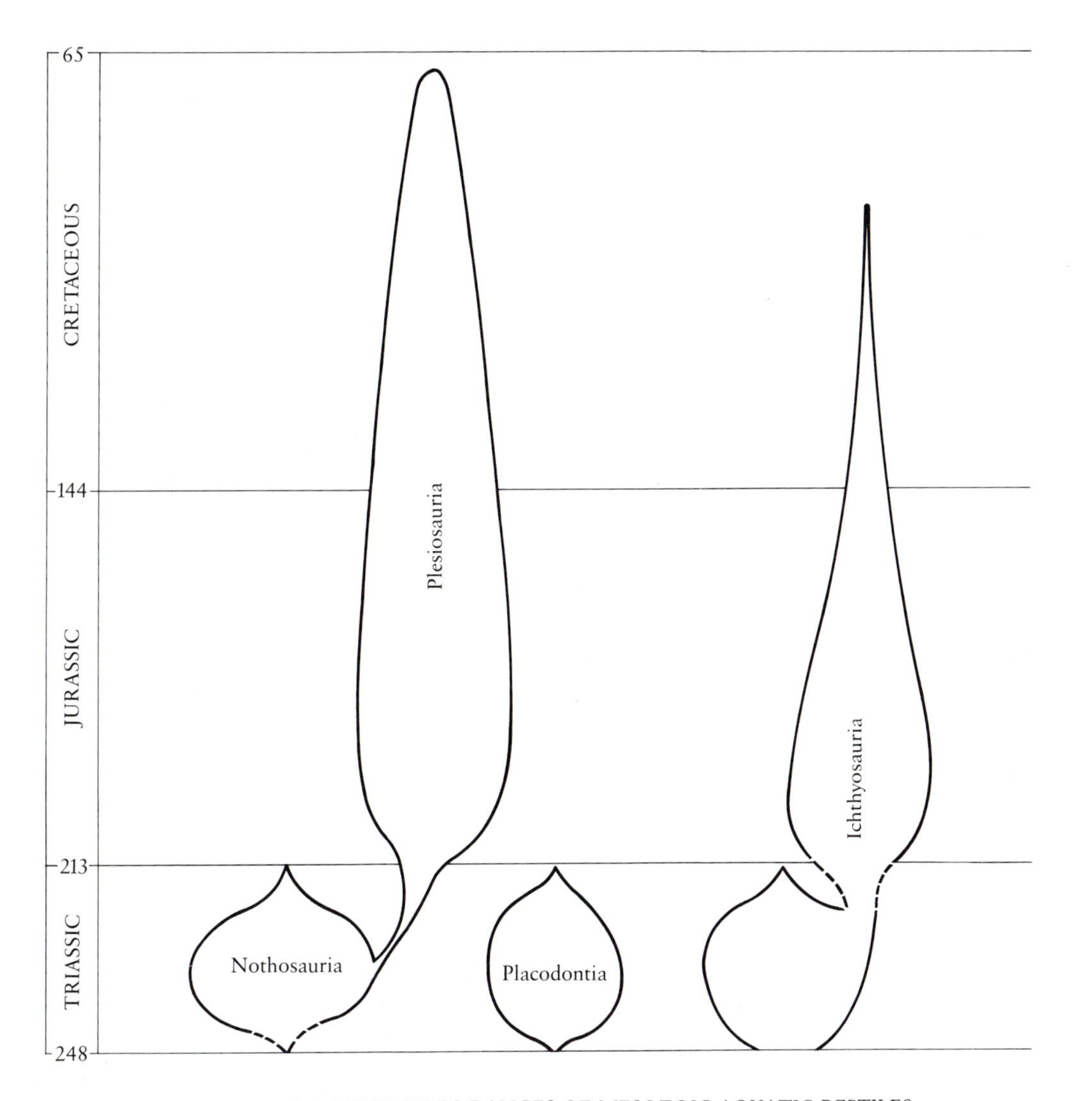

Figure 12-1. STRATIGRAPHIC RANGES OF MESOZOIC AQUATIC REPTILES.

dence to establish the affinities of the ichthyosaurs, which are distinguished as a separate reptilian subclass, the Ichthyopterygia. The placodonts appear to be derived from among early diapsids, but we cannot be more specific about their ancestry at present. Nothosaurs and plesiosaurs are united by unique derived features of the skull and shoulder girdle that support their inclusion in a distinct order, Sauropterygia, which may be a sister group of the lepidosaurs.

Aquatic adaptation is a common phenomenon among reptiles, as exemplified by the mesosaurs among the Captorhinida; the turtle families Plesiochelyidae, Chelonidae, and Dermochelyidae; the pleurosaurs; and several families among the Squamata. Seymour (1982) pointed out that aquatic adaptation is particularly easy among primitive amniotes because of their low metabolic rate, tolerance of anoxia and low body temperature, and great capacity to make use of fermentative metabolism for muscle activity. If we judge from modern marine iguanids, we see that adaptation to locomotion and feeding in the water does not necessarily require any structural or physiological specializations (Dawson, Bartholomew, and Bennett, 1977). Aquatic locomotion requires only one-fourth the metabolic expenditure of terrestrial locomotion in the iguana. Reptiles may have become secondarily aquatic whenever the balance between terrestrial and aquatic food sources and predators favored life in the water.

SAUROPTERYGIANS

The sauropterygians provide the most complete evidence of the sequence of events that leads to a specialized aquatic way of life. The radiation of primitive diapsids includes a number of aquatic lineages. Among the younginoids, the families Younginidae and Tangasauridae are similar in most skeletal features, but the tail in tangasaurids is laterally compressed with long neural and haemal spines to form an effective swimming structure (see Figure 11-12). The abdominal cavity is filled with small stones that would have served for ballast. Tangasaurids are precluded from the ancestry of later aquatic diapsids by the specialization of the carpus, in which the

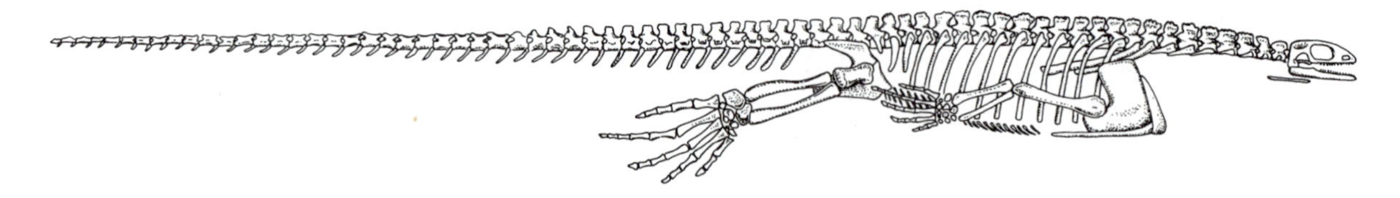

Figure 12-2. SKELETON OF *CLAUDIOSAURUS*. This diapsid from the Upper Permian of Madagascar shows an early stage of aquatic adaptation. Joint surfaces of the girdles and limbs are slow to ossify, and the small size of the head and long neck would have facilitated feeding in the water. Approximately 60 centimeters long. *From Carroll, 1981.*

lateral centrale is separated from contact with the fourth distal carpal by the large medial centrale. However, the remainder of their anatomy is close to that of primitive sauropterygians. Neither the front nor hind limbs are specialized for aquatic propulsion, nor are they reduced in size. As in modern aquatic lizards and crocodiles, the forelimbs were probably held against the side of the body to reduce drag.

CLAUDIOSAURUS

Contemporary with these eosuchians in the Upper Permian of Madagascar are the remains of a further aquatic genus, *Claudiosaurus*, which may be closer to the ancestry of the later sauropterygians (Figures 12-2 and 12-3). Most skeletal features of *Claudiosaurus* resemble those of early terrestrial eosuchians, although the carpus retains the primitive diapsid pattern in which the lateral centrale reaches the fourth distal tarsal. The most conspicuous difference is the presence of a long neck, which results from the posterior displacement of the shoulder girdle. The presacral vertebral count remains 25. The third, rather than the fourth, digit of the manus is the longest, which gives the hand a more paddlelike appearance.

The most important derived feature that *Claudiosaurus* shares with nothosaurs and plesiosaurs is the loss of the lower temporal bar. The remainder of the skull closely resembles that of the early eosuchian *Youngina* (see Figure 11-9). The general similarity of the skulls of eosuchians, *Claudiosaurus*, and nothosaurs (see Figure 12-5) demonstrates that the configuration of the cheek that is typical of sauropterygians evolved from that of primitive diapsids through the loss of the lower temporal bar rather than from genera that primitively lacked a lateral temporal opening (Carroll, 1981). In contrast with squamates, which have also lost the lower temporal bar, the quadrate is solidly supported by the pterygoid in *Claudiosaurus* and later sauropterygians.

The palate of *Claudiosaurus* also resembles that of nothosaurs and plesiosaurs in the loss of the transverse flange of the pterygoid and the reduction in the size of the suborbital fenestrae and interpterygoid vacuities.

In further contrast with terrestrial lepidosaurs, the sternum is not calcified or ossified in this genus, although an impression of an apparently cartilaginous structure is evident in most specimens between the gastralia and the posterior margin of the coracoids. The loss of ossification of the sternum may be attributed to a change in function of the forelimbs. In lizards, the sternum serves as a surface for the rotation of the coracoid to extend the stride. However, the alternating movements of the forelimbs that are so important in terrestrial locomotion would be disadvantageous in the water, since they would force the head and front of the trunk from side to side. If such alternating movements of the limbs were not advantageous in aquatic reptiles, selection would quickly act to reduce the sternum.

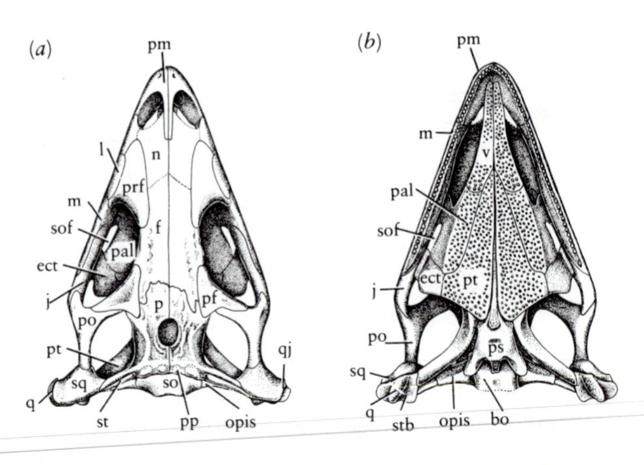

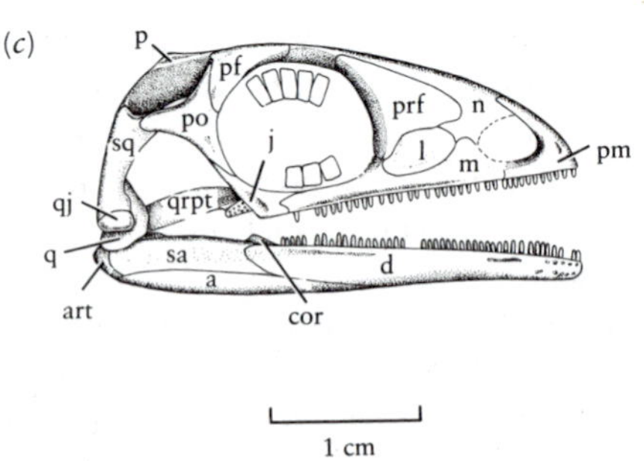

Figure 12-3. SKULL OF *CLAUDIOSAURUS*. (*a*) Dorsal, (*b*) palatal, and (*c*) lateral views showing structures intermediate between those of eosuchians (compare with Figure 11-9) and nothosaurs (see Figure 12-5). The lower temporal bar is lost, but the quadrate is strongly supported by the squamosal and pterygoid. The transverse flange of the pterygoid is lost, and the size of the suborbital and interpterygoid vacuities are greatly reduced as would be expected in an ancestor of sauropterygians, but the skull roof remains primitive. Abbreviations as in Figure 8-3 plus: stb, stapedal boss; qrpt, quadrate ramus of pterygoid; sof, suborbital fenestra. *From Carroll, 1981.*

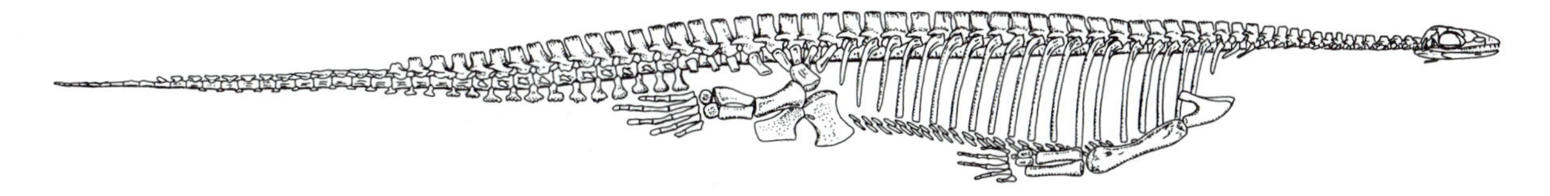

Figure 12-4. SKELETON OF *PACHYPLEUROSAURUS*, A NOTHOSAUR FROM THE MIDDLE TRIASSIC OF SWITZERLAND. The joint surfaces of the limbs and girdles are poorly ossified and would probably not support the body for continuous terrestrial locomotion. The limbs are otherwise little specialized for aquatic locomotion. The primary propulsive force was probably produced by the tail, in which the anterior neural spines and haemal arches are expanded dorsally and ventrally. The adult specimen is approximately 120 centimeters long. *From Carroll and Gaskill, 1985.*

In contrast with tangasaurids, the tail of *Claudiosaurus* is slender and shows no specializations for aquatic locomotion.

To judge by proportions of the trunk, tail, and limbs, this genus probably swam like modern aquatic lizards and crocodiles, with the front limbs folded alongside the body and the propulsive force produced by lateral undulation of the rear portion of the trunk and tail. The configuration of the joints of the girdles and limbs indicates that *Claudiosaurus* was probably still capable of terrestrial locomotion, but the large amount of cartilage suggests that it may have depended on the buoyancy of the water for support most of the time.

NOTHOSAURS

The nothosaurs, which are known primarily from the Middle Triassic of Europe and China, represent a more advanced stage in aquatic adaptation. Their limbs are reduced relative to primitive terrestrial reptiles but not highly modified for aquatic propulsion. Ossification of the girdles, carpals, and tarsals is greatly reduced, and the structure of the joints shows no features that would facilitate support and movement on land.

Nothosaurs range from fewer than 20 centimeters to more than 4 meters long, and show considerable variation in the proportions of the head and neck. However, the structure and proportions of the remainder of the body are relatively uniform, which indicates a similar manner of aquatic propulsion throughout the group.

Pachypleurosaurus (Figures 12-4 and 12-5) from the Middle Triassic of the Alpine region is a well-known genus that belongs to the family Pachypleurosauridae (Carroll and Gaskill, 1985). It reached a little more than 1 meter in length, including a moderately long neck and long tail. The skull is particularly small and its relative size decreases significantly during growth.

Apart from its small size, the skull is typical of primitive nothosaurs. As in eosuchians and *Claudiosaurus*, the upper temporal opening is smaller than the orbit. There is a strong upper temporal bar that is important in supporting the cheek in the absence of a lower temporal bar. The quadrate of pachypleurosaurids is embayed posteriorly like that in lizards and may have supported a tympanum. The presence of an impedance-matching ear is confirmed by the structure of the stapes, which is a slender rod that extends directly lateral toward the margin of the quadrate. It is suprising that a primarily aquatic group would have specialized the middle ear to be sensitive to airborne vibrations, especially since there is no evidence for an impedance-matching middle ear in primitive eosuchians and *Claudiosaurus*. However, most primarily aquatic modern frogs have a well-developed middle ear.

The palate of *Pachypleurosaurus* is typical of other nothosaurs in the broad extension of the pterygoids posteriorly and medially, so that the interpterygoid vacuities are completely closed and the base of the braincase is

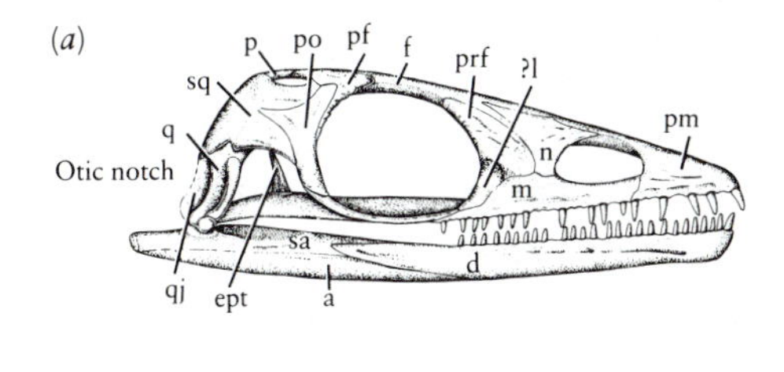

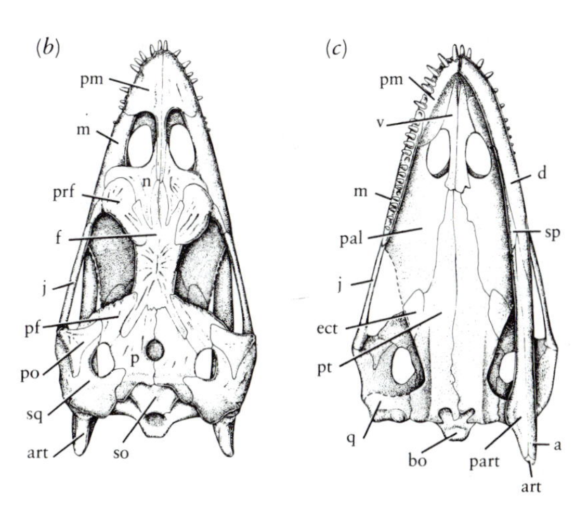

Figure 12-5. SKULL OF THE PRIMITIVE NOTHOSAUR *PACHYPLEUROSAURUS*. (*a*) Lateral, (*b*) dorsal, and (*c*) palatal views. The skull roof retains the features of eosuchians, except for the posterior movement of the external nares, but the palate is greatly modified with the complete closure of the suborbital vacuities. The pterygoids meet at the midline beneath the braincase. The quadrate is emarginated posteriorly like that of lizards and may have supported a tympanum. Abbreviations as in Figure 8-3. *From Carroll and Gaskill, 1985.*

covered ventrally. This can be interpreted as an extension of the changes already observed in *Claudiosaurus*.

The neck is long in all nothosaurs. *Pachypleurosaurus* has approximately 18 cervical vertebrae. The number in the trunk ranges from 20 to 21. Intercentra are lost, except anterior to the atlas and axis and at the base of the tail. There are three sacral vertebrae—one more than in eosuchians. Other nothosaurs may have as many as six. The sacral ribs have very small surfaces for attachment to the ilium, which is itself very reduced. The sacral attachment appears very weak in all nothosaurs, and it is difficult to explain why the number of sacral ribs has increased.

In *Pachypleurosaurus*, the neural and haemal spines near the base of the tail are expanded dorsally and ventrally to form an effective surface for sculling.

In many nothosaurs, the ribs and vertebrae are greatly thickened or pachyostotic, as were those of mesosaurs. This thickening would have increased their specific gravity, enabling them to remain submerged with a minimum of effort. This characteristic is particularly evident in smaller genera, including *Neusticosaurus*.

The configuration of the pectoral girdles in nothosaurs is unique among reptiles (Figure 12-6). The bones are relatively poorly ossified, with little definition of the glenoid. The base of the scapulae, clavicular blades, and interclavicle form a strong anteroventral bar. Behind this bar is a very large medial opening that is bordered posteriorly by very elongated coracoids. The blade of the scapula is greatly reduced relative to terrestrial reptiles and located well anterior to the glenoid. It would have provided little area for the attachment of muscles from the trunk that support the body on the girdle in terrestrial reptiles. The stem of the clavicle is fused to the scapula. Ventrally, the scapula expands beneath the clavicle, in contrast with the relationship in most reptiles, in which the blade of the clavicle is external to the scapula.

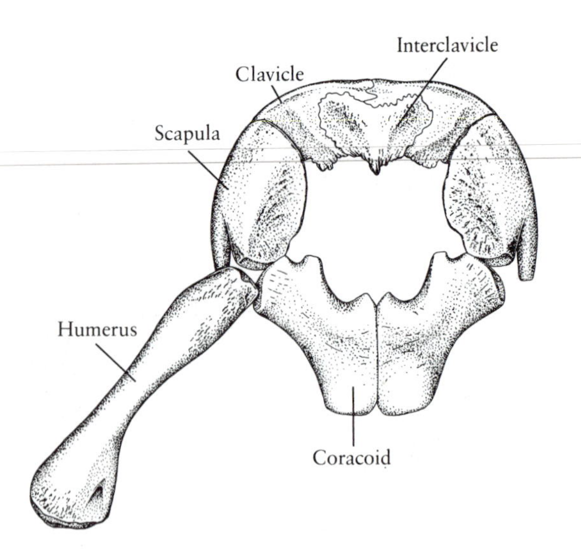

Figure 12-6. PECTORAL GIRDLE OF THE NOTHOSAUR *PACHYPLEUROSAURUS*. The clavicles and interclavicles form a solid transverse bar anteriorly, which is characteristic of this group. The central portion of the girdle is open and the coracoids extend posteriorly and meet broadly at the midline. Like the plesiosaurs (but unlike other amniotes), the anterior margin of the scapula extends superficially beneath the clavicles and the interclavicle underlies the clavicles. *From Carroll and Gaskill, 1985.*

In all nothosaurs, the distal limb elements are reduced relative to the humerus, which is a particularly stout element in the larger species. The elbow joint is very poorly defined. The olecranon of the ulna is not ossified, and articulating surfaces for the ulna and radius on the ventral surface of the humerus are absent. The lower limb was probably capable of being extended directly laterally, in contrast with its habitually flexed position in quadrupedal reptiles. All nothosaurs retain five distinct digits. In most genera the primitive phalangeal formula of 2, 3, 4, 5, 3 is retained, but *Pachypleurosaurus* shows a loss of phalanges while *Ceresiosaurus* exhibits hyperphalangy (Figure 12-7).

The pubis and ischium resemble those of primitive reptiles except for the development of a thyroid fenestra. The blade of the ilium is very narrow and provides little area of support for the sacral ribs. In *Pachypleurosaurus*, the rear limb is considerably shorter than the anterior, but the foot is broad and may have been important in steering.

Aquatic locomotion in nothosaurs

Nothosaurs have long been thought to be intermediate in their degree of aquatic locomotion between the pattern of primitive terrestrial reptiles and plesiosaurs. The limbs appear to be suitable for paddling but not for more sophisticated aquatic locomotion.

The restoration of *Pachypleurosaurus* in lateral view shows that the vertical expansion at the base of the tail would have provided the largest surface for aquatic propulsion (see Figure 12-4). The primitive pattern of nothosaur locomotion was probably an extension of that seen in aquatic lizards and crocodiles, in which lateral undulation of the trunk and tail played the primary role in propulsion. The limbs in all nothosaurs are reduced relative to the length of the trunk and may initially have been held to the sides of the body to reduce drag. The reduction of the size and degree of ossification of the limbs in nothosaurs probably dates from a primitive stage in their adaptation to life in the water. In some nothosaurs, the ventral portion of the shoulder girdle is extremely reduced, with neither the clavicles nor the scapulae expanded ventrally. This reduction may reflect the most primitive condition in the group.

The configuration in other genera, in which the ventral portion of the scapula and the clavicular blade are expanded but have a relationship that is reversed from that of most reptiles, is certainly a specialization of nothosaurs. It may be associated with selection for more active use of the forelimbs in propulsion. All nothosaurs are characterized by large coracoids that are greatly extended posteriorly. We can associate this extension with

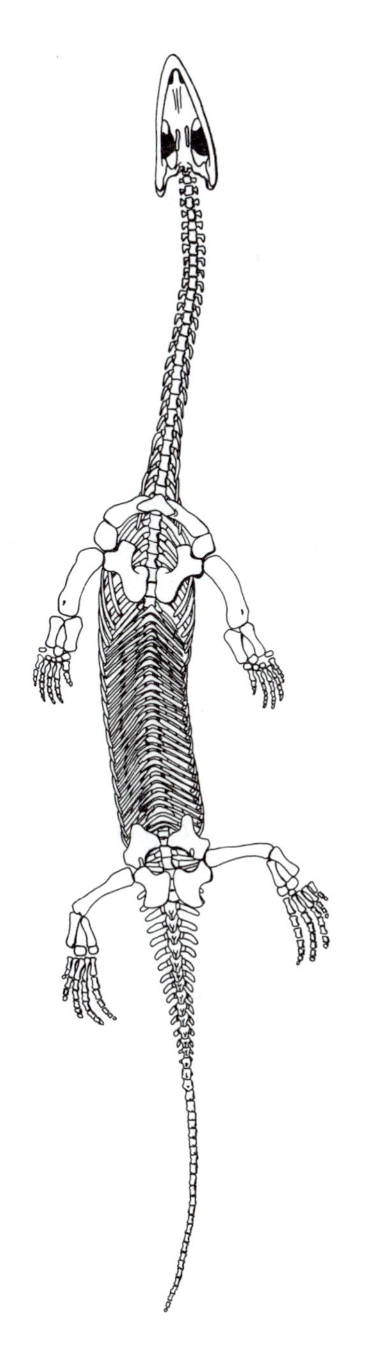

Figure 12-7. SKELETON OF THE NOTHOSAUR CERESIOSAURUS IN VENTRAL VIEW. This genus reached 4 meters in length. *From Kuhn-Schnyder, 1963. By permission of the Archivio Storico Ticinese.*

The pachypleurosaurids are primitive in the structure of the skull roof, with small upper temporal openings like those of early eosuchians. Most pachypleurosaurids are less than 1 meter long, and the low neural spines and broad flat neural arches indicate that lateral undulation of the trunk region may have remained the most important force in aquatic locomotion.

Other nothosaurs, including *Nothosaurus*, are much larger—up to 4 meters long—and the skull is much wider posteriorly, with the upper temporal openings larger than the orbits (Figure 12-8). *Nothosaurus* has much taller neural spines and narrower neural arches, which suggest a more rigid trunk. The forelimbs are much more robust and more strongly differentiated from the rear limbs than in the smaller pachypleurosaurids, which suggests that they were more important in propulsion.

The ancestry of plesiosaurs

Nothosaurs appear to be plausible ancestors of the plesiosaurs in most skeletal features. However, such a relationship seems to be contradicted by the structure of the palate and the shoulder girdle. The palate of plesiosaurs is *less* specialized than that of nothosaurs in the retention of interpterygoid vacuities and exposure of the base of the braincase between the pterygoids. This configuration suggests that plesiosaurs may have evolved from more primitive diapsids rather than from any of the well-known nothosaurs.

One genus from the Middle Triassic, *Pistosaurus*, which was originally described as a nothosaur, retains a more primitive pattern of the palate (Figure 12-9) (Sues, 1987). The skull is similar to that of nothosaurs in retaining nasal bones that are lost in typical plesiosaurs, and it has been associated with postcranial remains that are similar to those of nothosaurs (Figure 12-10). Unfortunately, the phylogenetic significance of this material is

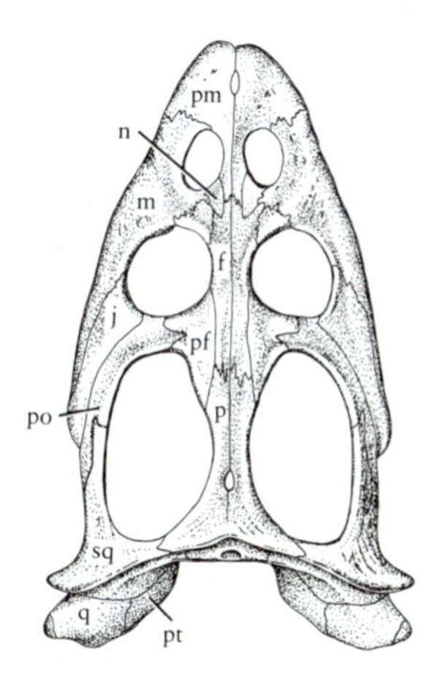

Figure 12-8. SKULL OF THE LARGE NOTHOSAUR *SIMOSAURUS*. The upper temporal openings are very large in contrast with those of *Pachypleurosaurus*. Length approximately 10 centimeters. Abbreviations as in Figure 8-3. *From von Meyer, 1847–1855.*

the elaboration of muscles that move the forelimb posteriorly. We can attribute the redevelopment of the anterior portion of the ventral surface of the pectoral girdle to a later stage in which more powerful protractors of the humerus evolved. The elaboration of the forelimbs in locomotion presumes a change in behavior and perhaps in the central nervous system control that emphasizes symmetrical, rather than asymmetrical, movement.

The pelvic girdle and rear limbs in nothosaurs are reduced and show little specialization indicative of an active role in propulsion.

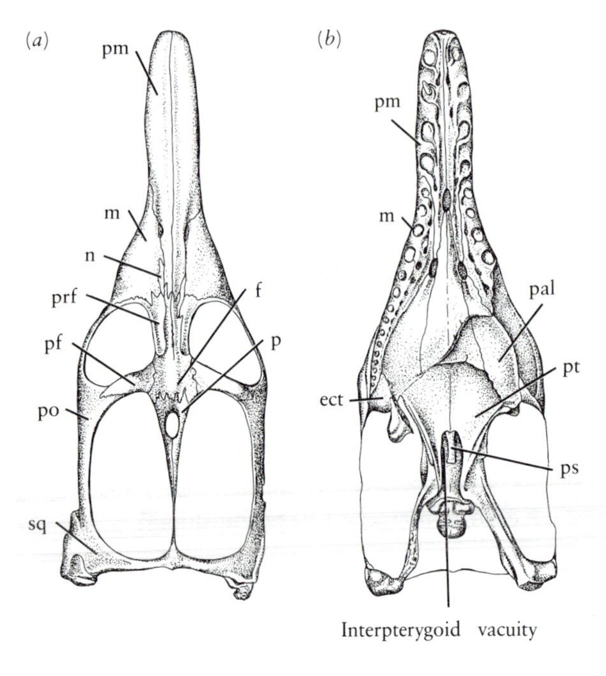

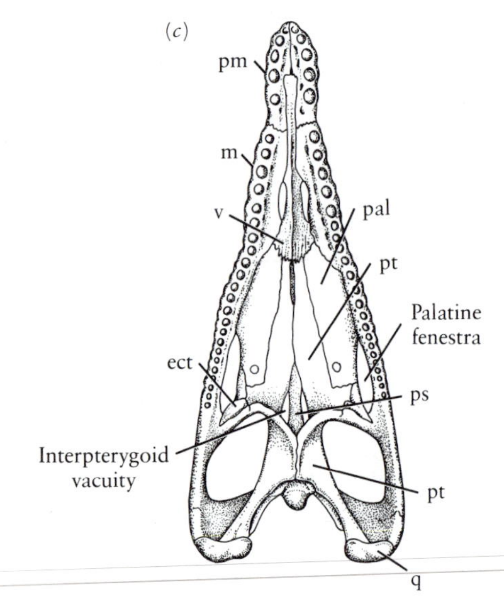

Figure 12-9. SKULL OF *PISTOSAURUS* FROM THE MIDDLE TRIASSIC. (*a*) Dorsal and (*b*) palatal views. The palate resembles that of plesiosaurs in retaining interpterygoid vacuities and having the braincase visible ventrally. The skull roof retains small nasal bones that are lost in Jurassic plesiosaurs. About 20 centimeters long. *From von Meyer, 1847–1855.* (*c*) Palate of the plesiosaur *Pliosaurus*. Length approximately 1 meter. *Modified from Saint-Seine, 1955.* Abbreviations as in Figure 8-3.

difficult to assess, since we are not certain that the skull and postcranial material belong to the same species. If they do, *Pistosaurus* provides a strong link between nothosaurs and plesiosaurs. The pattern of the palate of *Pistosaurus* may represent retention of a primitive condition and so indicate that the lineage leading to plesiosaurs had diverged earlier than the appearance of the oldest known typical nothosaurs. Alternately, it might reflect a later reversion to a primitive condition. Such a reversion could result from a change in developmental patterns, with the retention of an early onotogenetic stage in which the pterygoids had not yet extended to the midline beneath the braincase.

The other problem in accepting a nothosaurian origin for plesiosaurs is the great difference in the configuration of the shoulder girdle. Typical plesiosaurs have what appears to be a more primitive pattern, with both the scapula and coracoid greatly expanded ventrally, somewhat like those in primitive eosuchians. Andrews (1910) described growth stages in advanced plesiosaurs that show that the ventral portion of the girdle is initially open, somewhat as in nothosaurs, and the scapulae extend toward the midline only late in ontogeny. We can see a somewhat similar sequence phylogenetically from the nothosaurs through the most primitive Lower Jurassic plesiosaurs to the more advanced plesiosaurs (Figure 12-11). Unfortunately, the anatomy of the earliest plesiosaurs remains poorly known.

More detailed knowledge of the transition between nothosaurs and plesiosaurs is limited by the absence of articulated remains of either group in the late Triassic. Disarticulated vertebrae, girdles, and limb elements hint at a transition between the two groups, but so far there is no evidence of the manner of change between the markedly different limb proportions and pattern of the pelvic girdle that differentiate the two groups.

PLESIOSAURS

As a group, the plesiosaurs are distinguished from nothosaurs by the much greater size and increased similarity of the forelimbs and hind limbs. The pectoral and pelvic girdles are both greatly expanded ventrally, and there is only a short distance between them. The gastralia are limited in extent, with no more than nine rows, but they are extremely massive (Figure 12-12).

The means of locomotion among plesiosaurs has long been in dispute. Watson (1924) suggested that they rowed through the water, while Tarlo (1958) and Robinson (1975) proposed that plesiosaurs employed subaqueous flight in the manner of sea turtles and penguins.

Godfrey (1984) argued that the structure of the girdles differs from that of sea turtles and penguins in lacking any specialized accommodation for dorso-ventral movement of the limbs. The great anterior-posterior extent of both the pectoral and pelvic girdles in plesiosaurs would have permitted strong anterior and posterior strokes like those that are used for aquatic propulsion in the sea lion. The sea lion is propelled primarily by horizontal retraction of the forelimbs, but English (1976) demonstrated that the recovery stroke provides some anteriorly directed lift as well. Although the recovery stroke would not provide as much thrust as the dorso-ventral movements of the flippers in penguins and sea turtles, the great inertia of the large plesiosaurs would have allowed them to continue forward during the recovery stroke as is the case for large penguins between wing beats (Clark and Bemis, 1979).

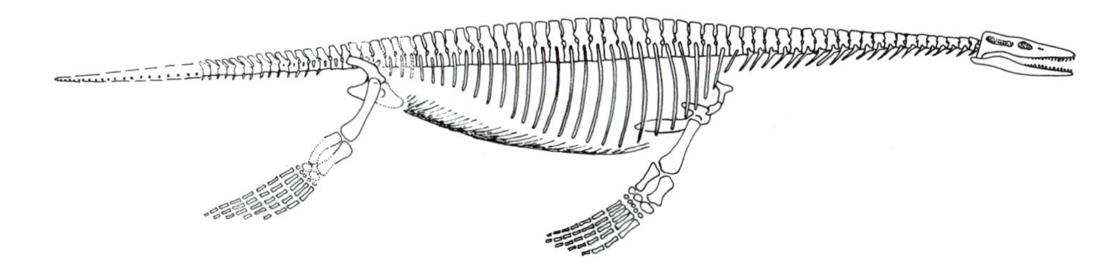

Figure 12-10. SKELETON OF *PISTOSAURUS*. The postcranial skeleton was found in the same locality as the skull but not in direct articulation. If the association is correct, this genus combines a postcranial skeleton like that of typical nothosaurs with a skull similar to that of plesiosaurs. Length approximately 3 meters. *From von Huene, 1948.*

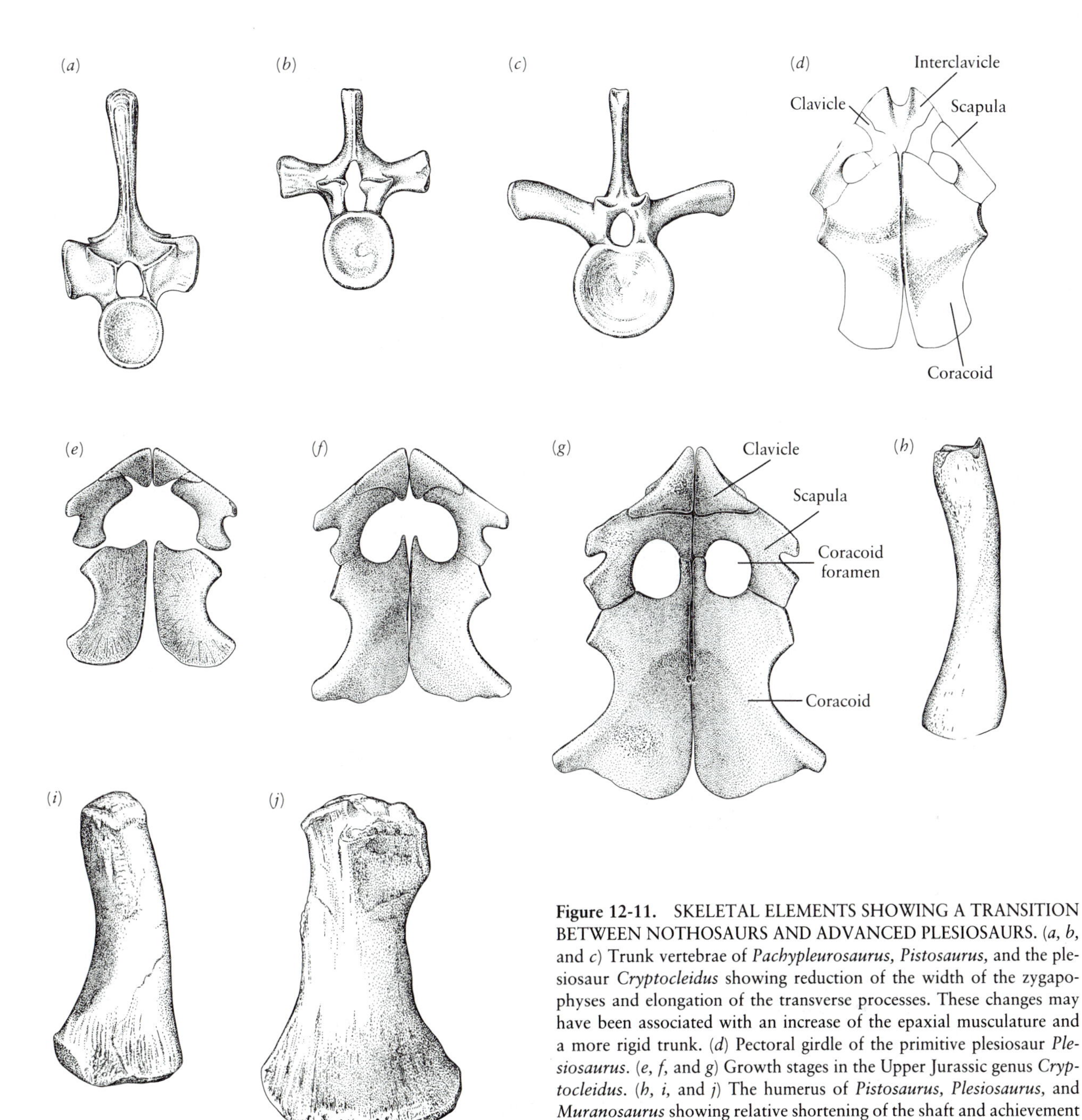

Figure 12-11. SKELETAL ELEMENTS SHOWING A TRANSITION BETWEEN NOTHOSAURS AND ADVANCED PLESIOSAURS. (*a, b,* and *c*) Trunk vertebrae of *Pachypleurosaurus, Pistosaurus,* and the plesiosaur *Cryptocleidus* showing reduction of the width of the zygapophyses and elongation of the transverse processes. These changes may have been associated with an increase of the epaxial musculature and a more rigid trunk. (*d*) Pectoral girdle of the primitive plesiosaur *Plesiosaurus.* (*e, f,* and *g*) Growth stages in the Upper Jurassic genus *Cryptocleidus.* (*h, i,* and *j*) The humerus of *Pistosaurus, Plesiosaurus,* and *Muranosaurus* showing relative shortening of the shaft and achievement of a more symmetrical appearance. *From Carroll and Gaskill, 1985.*

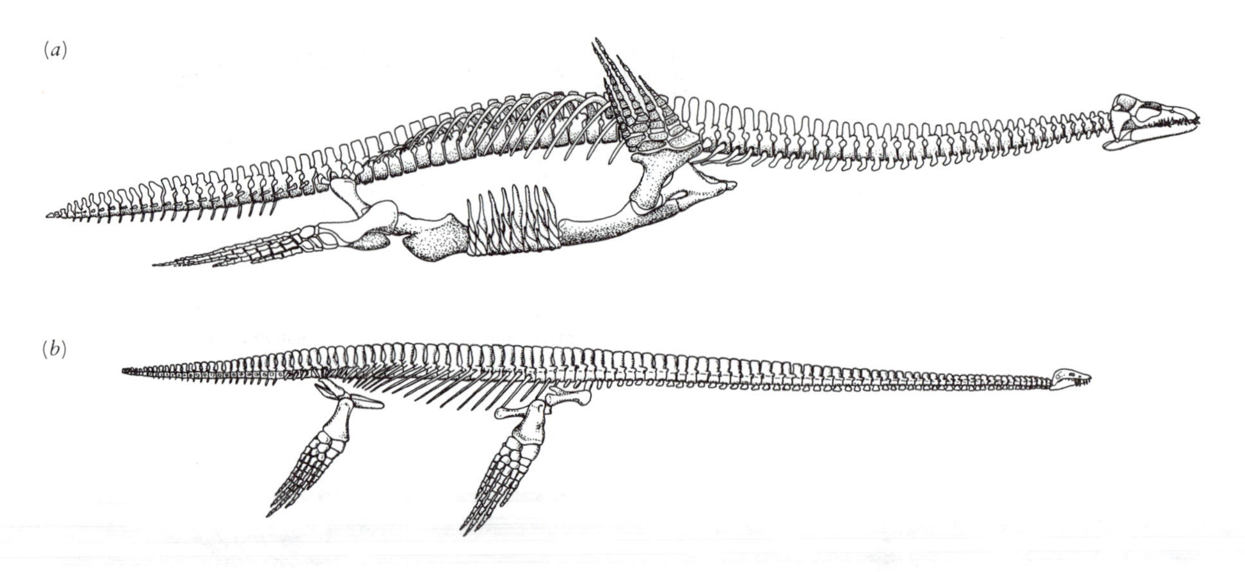

Figure 12-12. (*a*) Skeleton of the plesiosauroid *Cryptocleidus*, about 3 meters long. *From Brown, 1981.* (*b*) The elasmosaurid *Hydrothecrosaurus*, which reaches more than 12 meters in length. *From Saint-Seine, 1955.*

The lengthening of both girdles and the flattening of the puboischiadic plate into the horizontal plane, together with the increased bulk of the gastralia, would have contributed greatly to the rigidity of the trunk region in plesiosaurs. This rigidity was probably necessary to permit the effective use of both the forelimbs and hind limbs as paddles. The tail is short and probably served as a rudder rather than as an important element in propulsion. The stiffening of the trunk and the use of the rear limbs as paddles may be the most important changes between nothosaurs and plesiosaurs.

The proportions of the trunk and limbs remain relatively constant among plesiosaurs from the Lower Jurassic to the end of the Cretaceous. On the other hand, proportions of the head and neck vary extensively and progressive changes are evident in the details of the vertebrae, girdles, and limbs. In a recent review, Brown (1981) recognized approximately 40 plesiosaur genera. He divides them into four families. Like most other authors, including Welles (1962) and Persson (1963), Brown recognized two major groups of plesiosaurs, the Plesiosauroidea and the Pliosauroidea (Figure 12-13).

Among the Plesiosauroidea, the head is relatively small, as among the nothosaurs. The neck is relatively long, with up to 76 vertebrae. Pliosaurs have much larger skulls (in some genera they exceed 3 meters in length) with a much longer symphysis between the lower jaws, but the neck may have as few as 13 cervical vertebrae. All pliosaurs are included in a single family. The plesiosauroids are typically divided into two groups, the more primitive Plesiosauridae, which may be close to the origin of the entire assemblage and have as few as 28 cervical vertebrae, and the more specialized elasmosaurs, which have 32 to 76 cervical vertebrae.

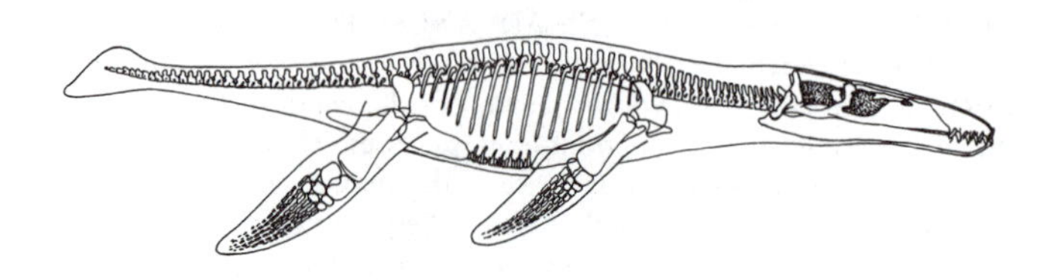

Figure 12-13. THE PLIOSAUR *LIOPLEURODON. From Newman and Tarlo, 1967.*

All advanced plesiosaurs, regardless of the families to which they belong, increase the number of phalanges in the paddles and develop single-headed, rather than double-headed, cervical ribs. Both elasmosaurs and pliosaurs persisted until the end of the Mesozoic.

PLACODONTS

The placodonts are a distinct aquatic group that was long thought to share a common ancestry with the nothosaurs. Placodonts have short, stout bodies, with the limbs only moderately specialized as paddles. The tail is long and laterally compressed but not in the form of a specialized propulsive organ. In most placodonts, the cheek and palatal teeth are large, flattened structures that would have been effective in crushing hard-shelled prey such as molluscs (Figure 12-14).

We know placodonts from the Middle and Upper Triassic of Europe, North Africa, and the Middle East. They were probably confined to shallow, coastal waters. Placodonts are quite diverse when they first appear in the lower beds of the Middle Triassic, and they are already so specialized that we have not been able to establish their ancestry (Sues, 1986).

Placodus, which Broili (1912) and Drevermann (1933) described, is characteristic of the group (Figure 12-15). The skull is massive and well consolidated for support of the large, flattened teeth on the palate, maxilla, and lower

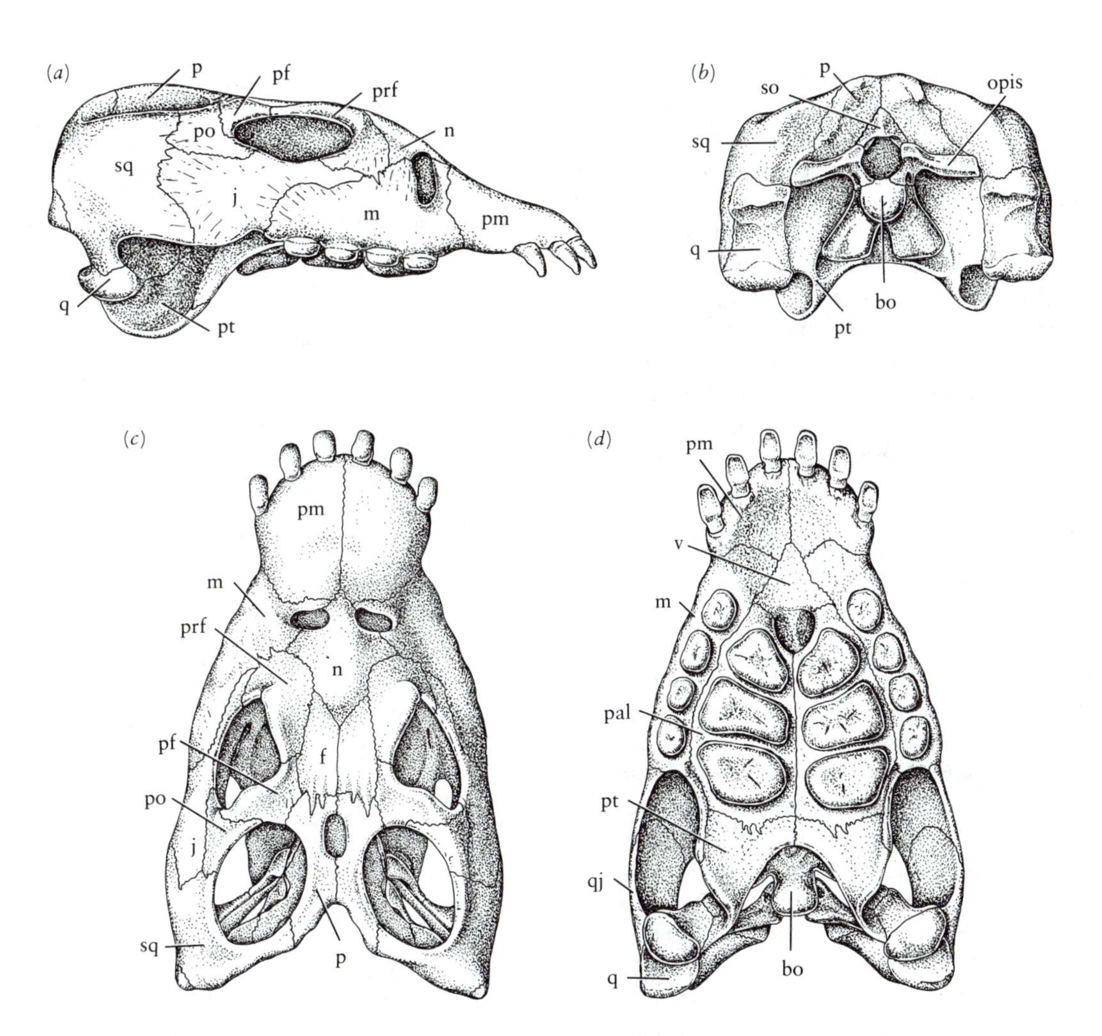

Figure 12-14. SKULL OF THE PLACODONT *PLACODUS*. (*a*) Lateral, (*b*) occipital, (*c*) dorsal, and (*d*) palatal views. Placodonts are characterized by enormous crushing teeth on the palate. The entire skull is strongly constructed. The palate is suturally attached to the braincase, and the occiput is a nearly solid plate of bone. The solid cheek may have evolved from the fenestrate condition that was evident in early diapsids, but there is no direct evidence for the specific origin of the placodonts. The lower jaw is characterized by the conspicuous coronoid process to which the adductor jaw muscles were attached. The skull is approximately 20 centimeters long. Abbreviations as in Figure 8-3. *From Peyer and Kuhn-Schnyder, 1955.*

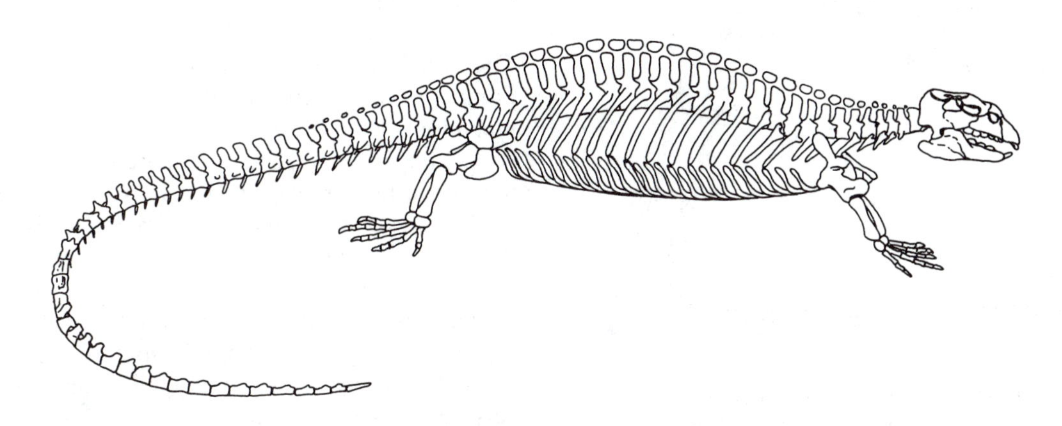

Figure 12-15. SKELETON OF THE PLACODONT *PLACODUS*, ABOUT 150 CENTIMETERS LONG. There is a single row of dermal ossicles above the neural spines. From *Peyer, 1950.*

jaw. The anterior teeth of the dentary and premaxilla are procumbent and spatulate at the tip. They may have served to dislodge attached prey. In contrast with most early reptiles, the dentary bone is elevated posteriorly as a large coronoid process for the insertion of jaw muscles. The palate is strengthened by the sutural attachment of the palatine and pterygoid at the midline, and there is no trace of interpterygoid vacuities. The suborbital fenestra is reduced to a narrow slit. The internal nares opens by a single passage behind the median vomer. The base of the braincase and the basipterygoid processes are still visible behind the palate, in contrast with the condition in nothosaurs. In *Placodus,* the braincase and dermal bones are integrated to form a nearly solid occipital surface, with greatly reduced posttemporal fenestrae. As in nothosaurs, the large upper temporal openings are surrounded by the parietal, squamosal, and postorbital. The cheek is solid, without a trace of a lower temporal opening, and shows little ventral emargination.

No features of the skull support close affinities with other aquatic reptiles or with any particular group of primitive terrestrial amniotes.

Placodus has approximately 28 presacral vertebrae, only slightly more than in most terrestrial reptiles. The neural spines are tall and the transverse processes extremely elongate. However, the centra are primitive in being deeply amphicoelous (Figure 12-16). This unusual combination of vertebral features is diagnostic of the placodonts. There are three sacral vertebrae and a long slender tail. The endoskeletal elements of the pectoral girdle are poorly ossified, like those in other members of the group; the coracoid is oval and the scapula is short and narrow. As in nothosaurs, the interclavicle underlies the blades of the clavicles (Figure 12-17).

The pelvic girdle is primitive in the platelike nature of the pubis and ischium, with no thyroid fenestration. The limbs, especially the carpals and tarsals, are poorly ossified and they have little specialization for aquatic propulsion.

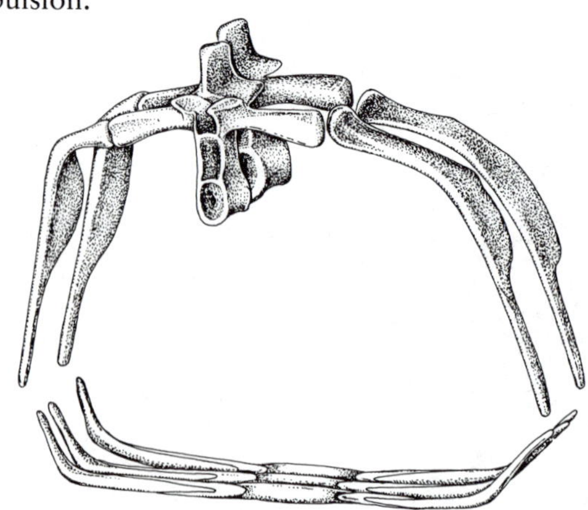

Figure 12-16. VERTEBRAE, RIBS, AND GASTRALIA OF THE PLACODONT *PARAPLACODUS.* This genus shows the long transverse processes and deeply amphicoelous centra that are characteristic of this group. *From Peyer, 1935.*

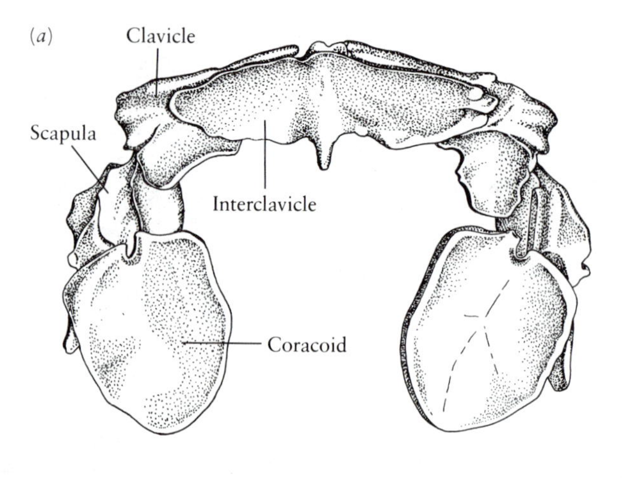

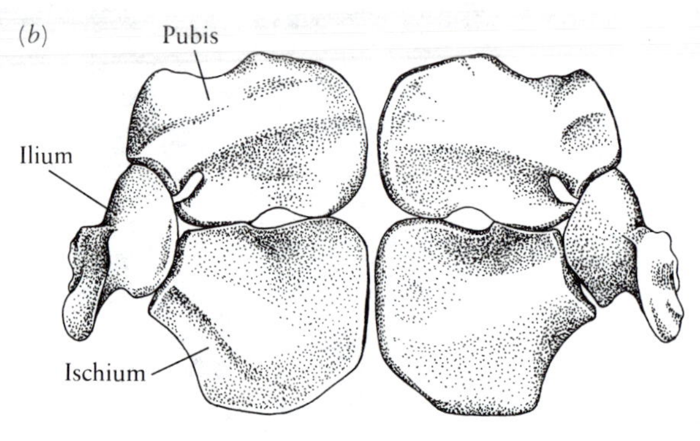

Figure 12-17. (*a*) Pectoral girdle of *Placodus* in ventral view. As in nothosaurs, the interclavicle broadly underlies the clavicle. (*b*) Pelvic girdle in dorsal view. It is primitive in lacking large thyroid fenestrae. *From Peyer and Kuhn-Schnyder, 1955.*

The gastralia are well developed in all placodonts. *Placodus* is characterized by the presence of a single row of dermal ossifications above the neural spines. However, the generally more primitive genus *Paraplacodus* lacks such ossification. In a closely related, but divergent, lineage (which is represented by the genera *Cyamodus, Placochelys,* and *Henodus*), the entire trunk region is covered by a dermal carapace that superficially resembles that of turtles (Figure 12-18). It is composed of a large number of polygonal ossicles. They are not at all comparable with the large plates of turtles, but in *Henodus* they were apparently covered by epidermal scutes (Westphal, 1976). The vertebrae are fused to the carapace and reduced in numbers. Most members of this assemblage of armored placodonts have cheek and palatal teeth like those of *Placodus,* but in some genera the anterior teeth are lost and may have been replaced by a horny beak. Most of the teeth are lost in *Henodus.* This genus is also peculiar in the closure of the upper temporal opening.

Helveticosaurus (Figure 12-19) apparently represents an early offshoot from the remainder of the placodonts. The vertebrae are comparable to those in *Placodus* in being deeply amphicoelous and in having very long transverse processes. The limbs and girdles show a similarly low degree of ossification. However, the trunk region is much longer, with approximately 43 presacral vertebrae.

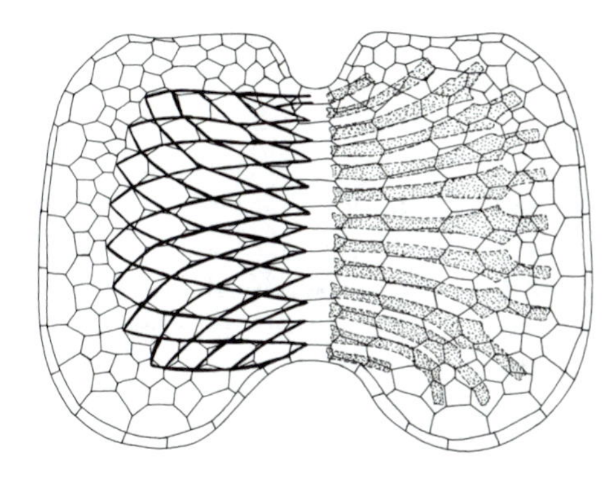

Figure 12-18. DORSAL ARMOR OF THE PLACODONT *HENODUS.* Irregular polygons show the pattern of the dermal armor. Heavier lines on the left show the pattern of overlying epidermal scutes. Stippled bars on the right represent the underlying ribs. The carapace reached approximately 1 meter in width. *Redrawn from Westphal, 1976.*

The cheek teeth are not flattened but resemble the more anterior teeth of other placodonts in being long and somewhat recurved. Palatal teeth are not evident. In the only two speciments we know, the skull is badly disarticulated, which precludes restoration of its original configuration. The interclavicle retains a long stem. The margins of the anterior portion of the interclavicle may be covered by the clavicular blades, in common with most primitive terrestrial reptiles. Despite some primitive features, *Helveticosaurus* does not help to establish the origin of placodonts or their relationships with other groups of aquatic reptiles.

One may argue that the absence of a lateral temporal opening is a specialized character of placodonts that evolved within the group to strengthen the skull in relationship to the crushing dentition. However, there is no direct evidence for this hypothesis. Other reptiles, including *Araeoscelis* (see Chapter 11) and *Trilophosaurus* (see Chapter 13), appear to have closed a primitively open cheek in relationship to a crushing dentition, but they show no other features in common with placodonts. The establishment of the affinities of this group must await the discovery of new fossils from the Lower Triassic or Upper Permian.

ICHTHYOSAURS

Throughout their known fossil record, ichthyosaurs were the most highly specialized of all marine reptiles. In advanced genera from the Jurassic and Cretaceous, the body is spindle shaped, the limbs are reduced to small steering fins, and the caudal fin is a large, lunate structure (Figure 12-20). The body form corresponds very closely with that of the modern teleost family, Carangidae, which includes the fastest swimming of all fish, the mackerals and tuna, which achieve speeds in excess of 40 kilometers an hour. The vertebrae are highly specialized, with the centra in the form of very short, deeply biconcave disks. The neural arches are separated from them by cartilage and do not bear transverse processes. The configuration of the skull is greatly modified, with a large orbit, a greatly reduced cheek, and a long snout. The teeth are of uniform shape and set in a long groove instead of distinct alveoli. The ichthyosaurs gave birth in the water to live young, thus avoiding the critical problem of going on land to lay eggs as must modern sea turtles, a practice that may have been necessary for mosasaurs and plesiosaurs as well.

Since specimens are present in many museums, it might be expected that the evolutionary history of ichthyosaurs would be well known. As McGowan (1983) emphasized, most fossils have come from a few extremely productive localities that represent only very limited periods of geological time; the remainder of ichthyosaur history is very poorly documented. Relatively few detailed anatomical studies have been undertaken during the past 50 years, and most of our knowledge still rests on preliminary work published in the nineteenth and early twentieth century.

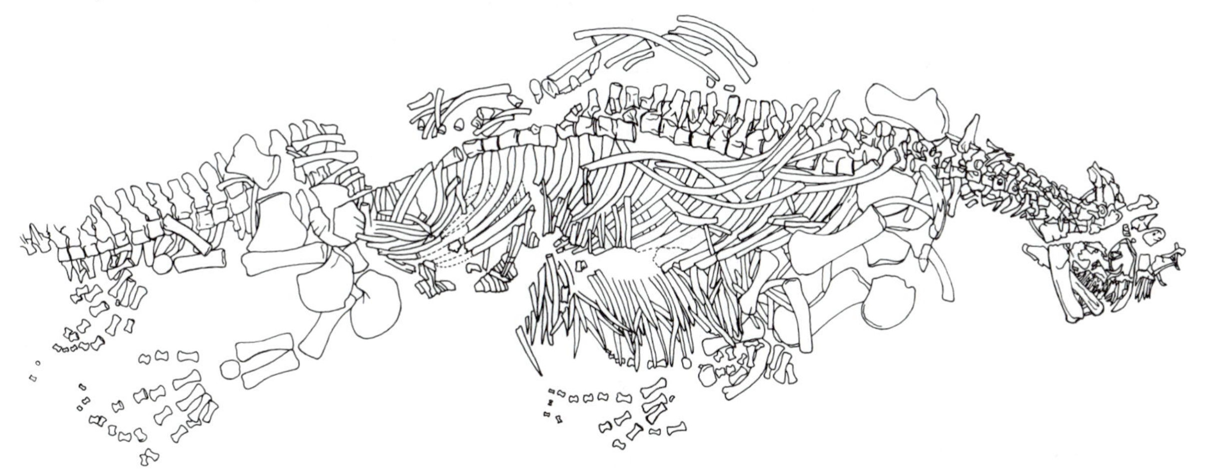

Figure 12-19. *HELVETICOSAURUS,* A PRIMITIVE PLACODONT IN WHICH THE MARGINAL TEETH ARE NOT CRUSHING PLATES. There is no trace of dermal armor. The specimen is approximately 2 meters long. *From Peyer, 1955.*

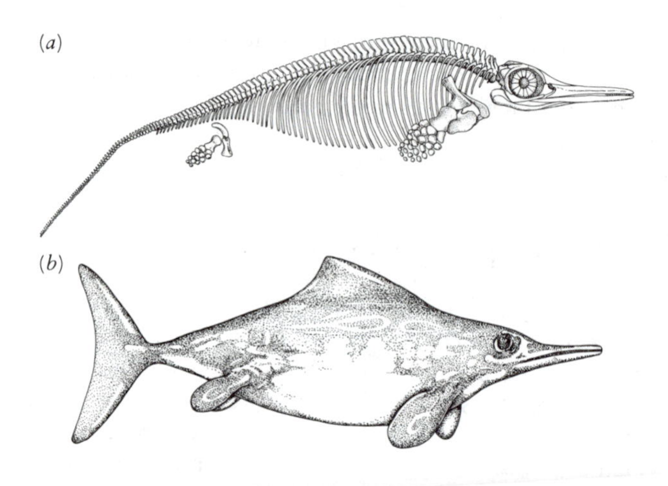

Figure 12-20. THE ADVANCED ICHTHYOSAUR *OPHTHALMOSAURUS* FROM THE UPPER JURASSIC. (*a*) Skeleton, approximately $3\frac{1}{2}$ meters long. *From Andrews, 1910.* (*b*) Restoration. *From McGowan, 1983.*

EARLY TRIASSIC ICHTHYOSAURS

The earliest ichthyosaurs are known from deposits at the top of the Lower Triassic in Spitsbergen and Japan and at the base of the Triassic sequence in China (Mazin, 1981, 1983; Shikama, Kamei, and Murata, 1978; Young and Dong, 1972). We know most of the skeleton from numerous specimens of the Japanese genus *Utatsusaurus* (Figure 12-21). The body form is broadly similar to Jurassic and Cretaceous ichthyosaurs, but most skeletal elements are notably more primitive. The entire skeleton is approximately $1\frac{1}{2}$ meters long. The skull is poorly known, but the lower jaws indicate that it was long and slender. The teeth are long and very slender, and the enamel is infolded at the tip. The neck is short, but the restored presacral column has approximately 40 presacral vertebrae. In contrast with later ichthyosaurs, the individual centra are approximately as long as they are high and not deeply amphicoelous. The neural arches appear to be only loosely attached to the centra. The tail is long and little specialized from the pattern of terrestrial reptiles, although the neural spines are somewhat modified and may have supported a low fin. The ribs in the trunk region are very long and slightly expanded distally. They are double headed, but both heads apparently attached to the centra—as in later ichthyosaurs—without the arch possessing a transverse process.

The scapula and coracoid are separately ossified, like those of the slightly modified aquatic diapsid *Askeptosaurus*. The interclavicle is a small triangular element, as in later Triassic ichthyosaurs. The clavicles are long narrow elements that resemble their counterparts in primitive terrestrial reptiles. The pubis and ischium are greatly reduced, as is the ilium, which was apparently not attached to the vertebral column. The forelimb and, even more so, the hind limb are reduced in size and degree of ossification. The humerus is short and broad. The ulna, radius, and metacarpals, in contrast, are little modified from the pattern in terrestrial reptiles. The carpus appears to have four elements in the proximal row and five in the distal. The manus has five distinct digits. There are four very short phalanges in each of the first three digits, with the hand as a whole giving the appearance of a fin.

The femur, tibia, and fibula are even more reduced than the forelimb, but the tarsus and pes are not preserved. A pattern of gastralia like that of primitive diapsids is retained.

Grippia, which is found in the uppermost beds of the Lower Triassic of Spitsbergen, provides the earliest evidence of the cranial structure in ichthyosaurs (Figure 12-22). The skull has a large upper temporal opening, but the cheek shows no evidence of a lateral fenestra, although it is slightly emarginated ventrally. The orbit is large, but the pre- and postfrontals meet above it. The snout, which is incompletely known, is somewhat elongate but probably not as greatly as in later ichthyosaurs. The pineal opening lies between the parietals in primitive fashion, in contrast with its position in later ichthyosaurs at the border of the frontals. The parietals, frontals, and nasals retain a relatively primitive configuration, whereas their pattern is greatly modified in later ichthyosaurs. The quadratojugal makes up a large portion of the cheek, as in later ichthyosaurs but in contrast with primitive amniotes. The squamosal forms much of the lateral border of the upper temporal opening. The external nares are high on the snout and relatively close to the orbit, like those in other aquatic forms.

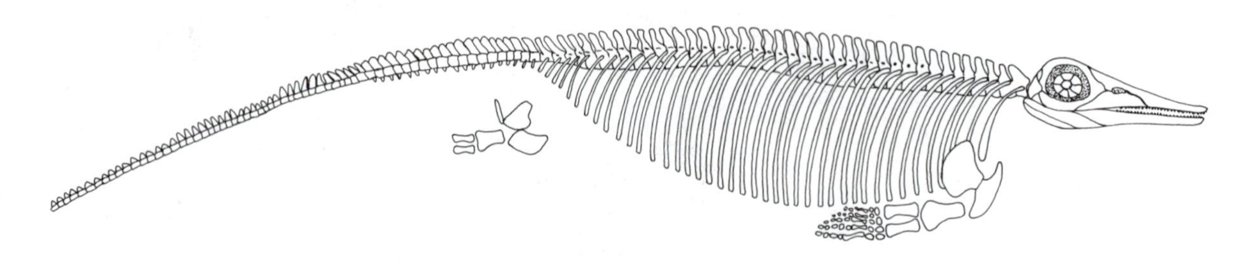

Figure 12-21. THE EARLIEST KNOWN ICHTHYOSAUR *UTATSUSAURUS* FROM THE LOWER TRIASSIC OF JAPAN. The skeleton is approximately $1\frac{1}{2}$ meters long. The form of the body resembles that of advanced ichthyosaurs, but the details of the vertebrae, limbs, and girdle are much more primitive. The skull is very poorly known and has been restored according to the pattern of later ichthyosaurs. *From Shikama, Kamei, and Murata, 1978.*

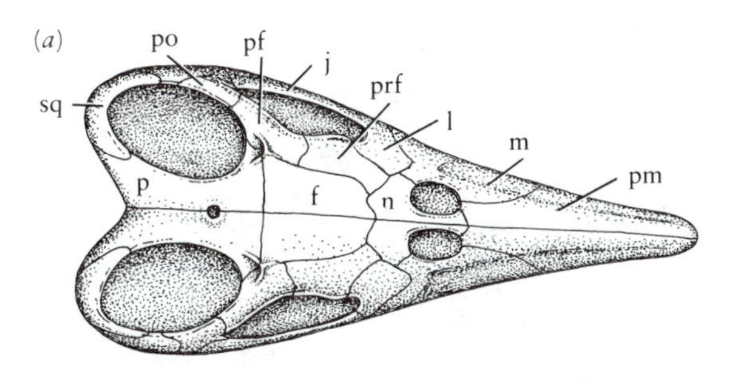

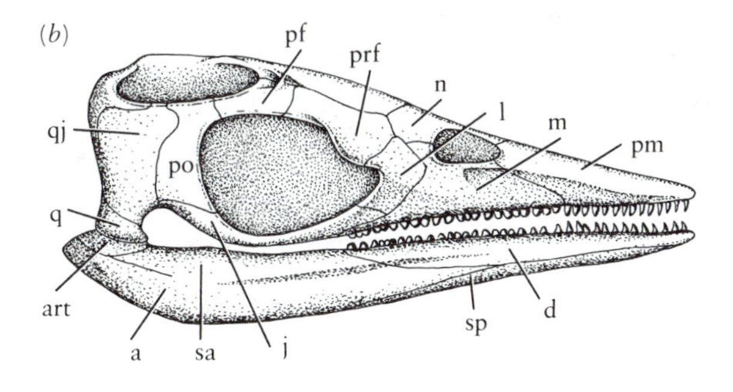

Figure 12-22. SKULL OF THE PRIMITIVE ICHTHYOSAUR *GRIPPIA* FROM THE LOWER TRIASSIC OF SPITSBERGEN. (*a*) Dorsal and (*b*) lateral views. Length about 15 centimeters. The tip of the snout has been restored. Abbreviations as in Figure 8-3. *From Mazin, 1981.*

Surprisingly, the teeth of the maxilla and back of the dentary are hemispherical, like those of the mosasaur *Globidens,* and set in two rows. The more anterior teeth are sharply pointed cones. All the teeth are set in distinct sockets. The Lower and Middle Triassic genus *Omphalosaurus,* which we find in California as well as Spitsbergen, has several rows of similar teeth in the cheek region but a much more advanced structure of the limbs than *Grippia.* Heterodonty and multiple rows of cheek teeth may be specializations of this group of early ichthyosaurs. We do not know the general body form of *Grippia,* but the individual vertebrae and elements of the girdles and limbs appear nearly as primitive as those of *Utatsusaurus.*

The earliest known ichthyosaur from China, *Chaohusaurus* (Figure 12-23), shares primitive features with *Utatsusaurus* and *Grippia.* The snout is somewhat elongated, the vertebrae are longer than high, and the forelimb is modified as a paddle, but the individual elements retain a configuration that is reminiscent of terrestrial reptiles, with a phalangeal count of 2, 3, 4, 4, 2. However, this limb is much larger than that of *Utatsusaurus.*

The problem of ichthyosaur origin

These early Triassic genera probably exemplify in a general way the pattern from which the later Mesozoic ichthyosaurs arose. In contrast with the sauropterygians, the early ichthyosaurs were already highly adapted to an aquatic way of life, although the anatomy of the skull, vertebrae, girdles, and limbs of the Lower Triassic genera is much more primitive than that of their Jurassic and Cretaceous counterparts. However, not even the most primitive features link them to any particular group of terrestrial or aquatic reptiles, although the presence of a dorsal temporal opening suggests affinities with early diapsids.

It was once thought that the presence of an upper temporal opening and a solid cheek constituted the basis for a large taxonomic group termed the Parapsida or Euryapsida that included (at various times in the history of the concept) ichthyosaurs, plesiosaurs, nothosaurs, placodonts, and such terrestrial forms as *Araeoscelis* (see Chapter 11), *Trilophosaurus, Protorosaurus, Prolacerta* (see Chapter 13), and even lizards (Carroll, 1981). It is now clear that lizards, *Protorosaurus,* and *Prolacerta* evolved from the primitive diapsid stock by loss of the lower temporal bar. *Araeoscelis* and *Trilophosaurus,* in which the cheek is solid, may have a similar origin. The postcranial skeleton of these strictly terrestrial forms shows no significant specialization in common with either ichthyosaurs or placodonts, and the common presence of a dorsal opening was either retained from an early diapsid condition or achieved convergently. In the later case, either or both of these groups might have arisen separately from primitive anapsid reptiles.

The anatomy of the skeleton of ichthyosaurs does point to a common origin within primitive amniotes, with marked similarities of the braincase to that of early captorhinomorphs. There is no support for von Huene's (1949) suggestion that ichthyosaurs arose directly from embolomerous labyrinthodonts.

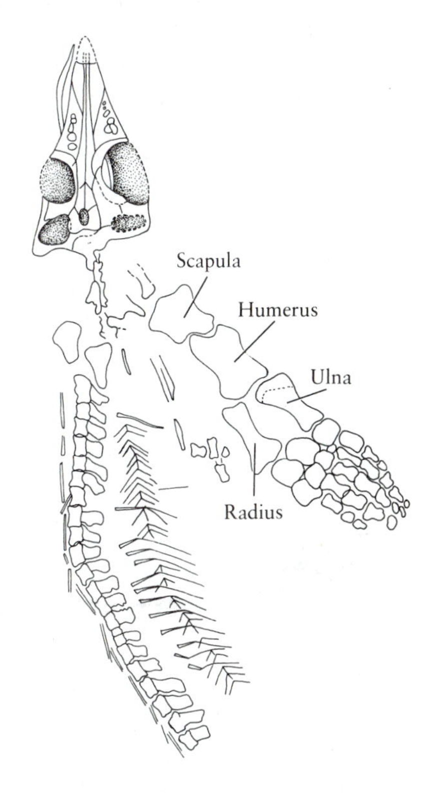

Figure 12-23. THE PRIMITIVE ICHTHYOSAUR *CHAOHUSAURUS* FROM THE BASE OF THE TRIASSIC SEQUENCE IN CHINA. Approximately 1 meter long. The pectoral limb is much larger than that of *Utatsusaurus. From Young and Dong, 1972.*

Figure 12-24. SKELETON OF THE MARINE REPTILE *NANCHANGOSAURUS [HUPEHSUCHUS]* FROM THE TRIASSIC OF CHINA. There are dermal ossifications along the vertebral column like those of *Placodus* and thecodont archosaurs. *From Young and Dong, 1972.*

Nanchangosaurus

Another early form that may have affinities with the primitive ichthyosaurs is *Nanchangosaurus [Hupehsuchus]* from China (Figures 12-24 and 12-25). The trunk is fusiform and the limbs are reduced but retain the characteristics of terrestrial forms. The skull has a long, toothless snout and an upper temporal opening. According to Young and Dong (1972), the skull also shows a lateral temporal opening and an antorbital fenestra. There is a row of dermal plates along the vertebral column. These features are characteristic of thecodont archosaurs (see Chapter 13). Young and Dong place this genus among the thecodonts, which suggests that the ichthyosaur habitus was achieved convergently in the two groups. Alternatively, one could interpret this form as a primitive ichthyosaur, showing evidence of the relationship of this group to the archosaurs. Much more remains to be learned of all the early Triassic forms before we can understand the origin of this group.

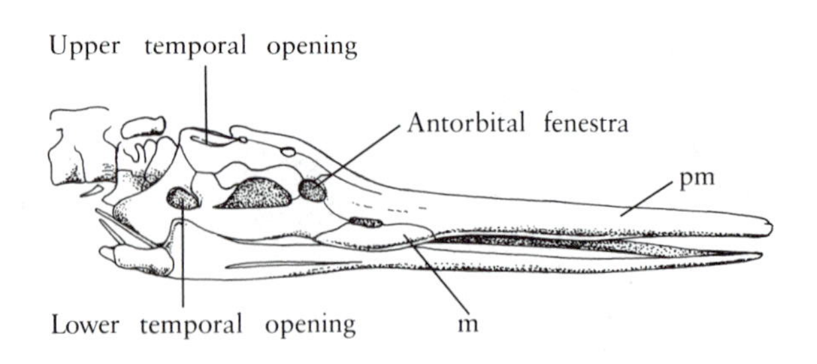

Figure 12-25. SKULL OF *NANCHANGOSAURUS [HUPEHSUCHUS]*. The presence of an antorbital fenestra is a feature it shares with archosaurs. In contrast with all orthodox ichthyosaurs, there is also a lateral temporal opening. The taxonomic position of this genus is uncertain. It may be an ichthyosaurlike archosaur or a link between primitive archosaurs and ichthyosaurs. Abbreviations as in Figure 8-3. *From Young and Dong, 1972.*

MIDDLE AND UPPER TRIASSIC ICHTHYOSAURS

Middle and Upper Triassic ichthyosaurs are even more widespread geographically and anatomically diverse than those of the Lower Triassic. Two very distinct forms are known from the Middle Triassic, *Cymbospondylus* and *Mixosaurus* (Figures 12-26 and 12-27). *Cymbospondylus* from Nevada was last described by Merriam in 1908. This genus was much larger than the Lower Triassic forms, reaching 10 meters in length. The trunk was very long, with approximately 60 presacral vertebrae. The tail was also very long and apparently straight, and the skull had a very long snout. The teeth are long, sharp, and set in deep sockets.

The proximal limb bones do not appear greatly modified from the terrestrial pattern. The carpals and tarsals are rounded elements; the distal bones are unknown. The pelvic girdle is strikingly primitive, with a platelike pubis and ischium; the ilium has a narrow blade (Figure 12-28). The bones of the pectoral girdle resemble those of *Utatasusaurus*.

Mixosaurus (Repossi, 1902; von Huene, 1949) is the most widely known of Middle Triassic ichthyosaurs, with fossils from Spitsbergen, the Alpine region, the western United States, China, and Indonesia. Superficially, *Mixosaurus* appears close to the pattern of later ichthyosaurs. The body is somewhat more than 1 meter long. The limbs are modified as paddles with five principle digits, which show considerable hyperphalangy. The pectoral fin is substantially larger than the pelvic. The tail is modified as a caudal fin, with the neural and haemal arches considerably elongated, but the end of the tail is not sharply downturned (see Figure 12-27).

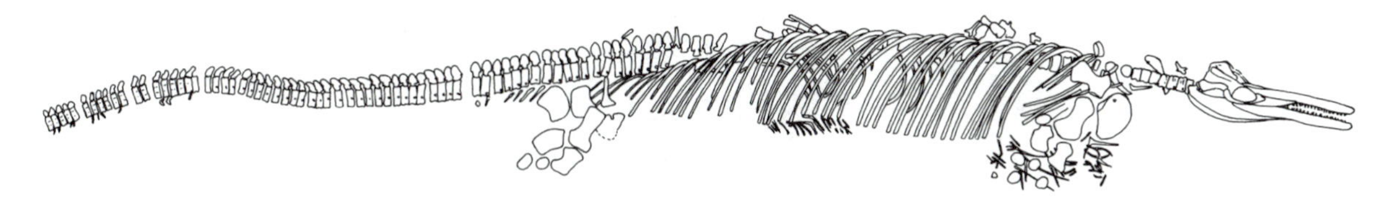

Figure 12-26. SKELETON OF THE ICHTHYOSAUR *CYMBOSPONDYLUS.* Skeleton is from the Middle Triassic of Nevada, approximately 10 meters long. *From Merriam, 1908.*

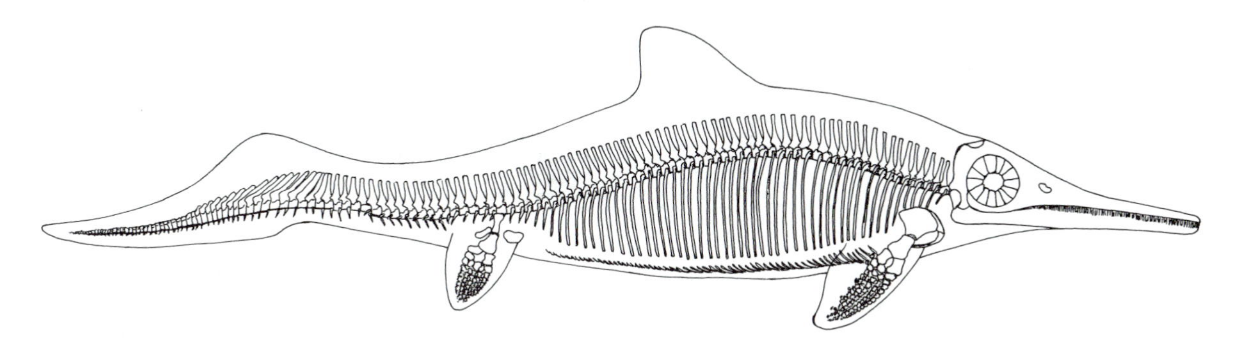

Figure 12-27. RESTORATION OF THE MIDDLE TRIASSIC ICHTHYOSAUR *MIXOSAURUS* FROM CENTRAL EUROPE. Length about 1 meter. *From Kuhn-Schnyder, 1963. By permission of the Archivio Storico Ticinese.*

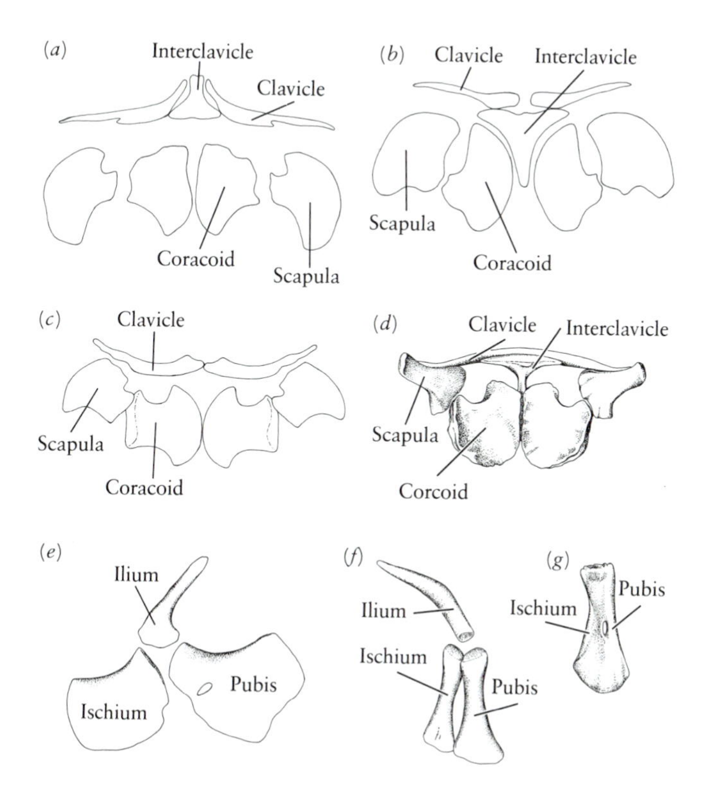

Figure 12-28. CHANGING PATTERN OF THE PECTORAL AND PELVIC GIRDLES AMONG ICHTHYOSAURS. (*a, b, c,* and *d*) Pectoral girdles of *Utatsusaurus* (Lower Triassic), *Mixosaurus* (Middle Triassic), *Shastasaurus* (Upper Triassic), and *Ophtalmosaurus* (Upper Jurassic). (*e, f,* and *g*) Pelvic girdles of *Cymbospondylus* (Middle Triassic), *Ichthyosaurus* (Lower Jurassic), and *Ophtalmosaurus* (Upper Jurassic). *(a) From Shikama, Kamei, and Murata, 1972, restored on the basis of the specimen drawing. (b, c, e, and f) From Merriam, 1908. (d and g) From Andrews, 1910.*

There are 45 to 55 presacral vertebrae. As in *Grippia,* the cheek teeth have wide, blunt tips, although the anterior teeth are sharply pointed. Unlike either earlier or later ichthyosaurs, the dorsal temporal opening is quite narrow.

The only Upper Triassic ichthyosaurs for which most of the skeleton is known, is *Shonisaurus,* which is represented by the remains of many specimens from Nevada that Camp (1980) recently described (Figure 12-29). *Shonisaurus* is the largest of all ichthyosaurs, reaching a length of 15 meters. In marked contrast with other genera, the manus and pes are both greatly elongated but there are only three rows of phalanges, which gives the appearance of paddles rather than fins. The ribs are expanded distally. The teeth are restricted to the front of the jaw, and the end of the tail is bent ventrally. In contrast with other ichthyosaurs, two vertebrae bear sacral ribs for attachment of the ilia. The other vertebrae approach the pattern of advanced ichthyosaurs. The centra are short, biconcave disks. They bear two rounded articulating surfaces for the attachment of rib heads.

Merriam (1902, 1904) named several genera of ichthyosaurs from the Upper Triassic of California, including *Shastasaurus, Delphinosaurus,* and *Toretocnemus,* none of which is represented by a complete skeleton. In an additional genus, *Merriamia,* the forelimb is much longer than the hind limb, as in *Mixosaurus* and Jurassic ichthyosaurs, but it has only three principle digits; the hind limb has four digits. The skull is advanced over that of most other Triassic forms in having the teeth set in a continuous groove rather than in separate sockets.

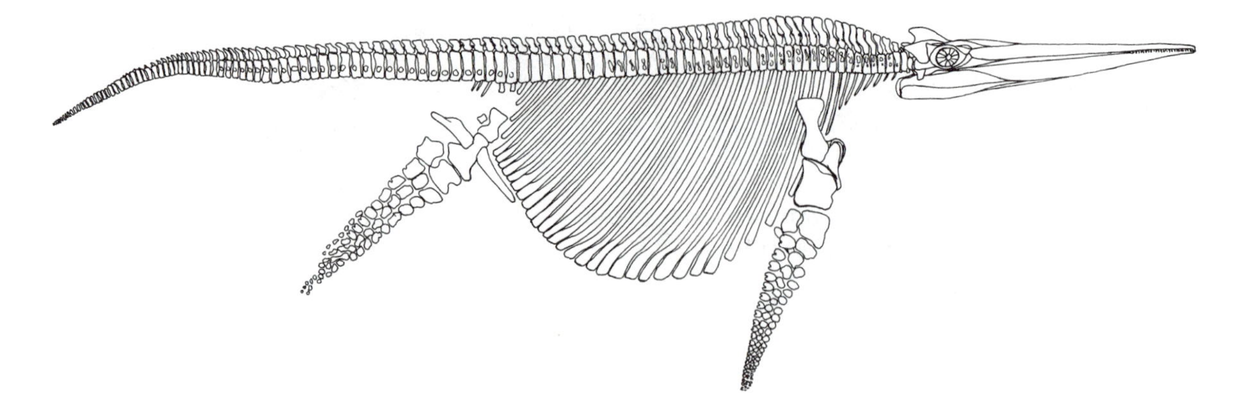

Figure 12-29. THE UPPER TRIASSIC ICHTHYOSAUR *SHONISAURUS* WHICH REACHED A LENGTH 15 METERS. The heads of the ribs are not drawn, in order to show the articulating surfaces of the vertebrae. *From Camp, 1980.*

JURASSIC AND CRETACEOUS ICHTHYOSAURS

Ichthyosaurs reached their greatest diversity in the early Jurassic, as documented by remains from England and Germany (McGowan, 1979). Post-Triassic ichthyosaurs appear to represent a new radiation from a single preexisting lineage, rather than being a continuation of the early Triassic radiation. Jurassic and Cretaceous ichthyosaurs appear more similar to one another than do the Triassic genera, and all show the same advanced characters.

The most important features relate to the mode of locomotion. Among the adequately known Triassic genera, the tail is straight or only gently curved. From the base of the Jurassic, all advanced ichthyosaurs show a sharp ventral bend in the tail that supports the lower portion of a high, lunate caudal fin. The caudal zygapophyses form a ball-and-socket joint to facilitate lateral flexion but limit mobility in a dorso-ventral direction. In Triassic genera, the zygopophyses are paired and relatively flat, but they become confluent and form a single, broadly curved surface by the early Jurassic.

A comparison with modern fish indicates that the change in body form between Triassic and Jurassic ichthyosaurs can be associated with a change in swimming mechanics from a moderately fast anguilliform mode to the much more effective advanced carangiform mode of modern tunas and the fastest-swimming sharks (Webb, 1982).

Other skeletal changes also distinguish the post-Triassic ichthyosaurs. In most Triassic genera, the ribs attach by a single head except in the cervical region. In Jurassic genera, the trunk ribs are double headed as well. The pectoral girdles of Jurassic and Cretaceous ichthyosaurs are characterized by a *T*-shaped rather than a triangular interclavicle. The pelvic girdle is greatly reduced and the ischium and pubis are fused to one another in the Upper Jurassic genus *Ophtalmosaurus* (see Figure 12-28).

The paired fins are both more highly specialized. The ulna, radius, tibia, and fibula are so modified that they form an unbroken series with the metapodials and phalanges (Figure 12-30). There is considerable variety in the configuration of the paddles in Jurassic and Cretaceous

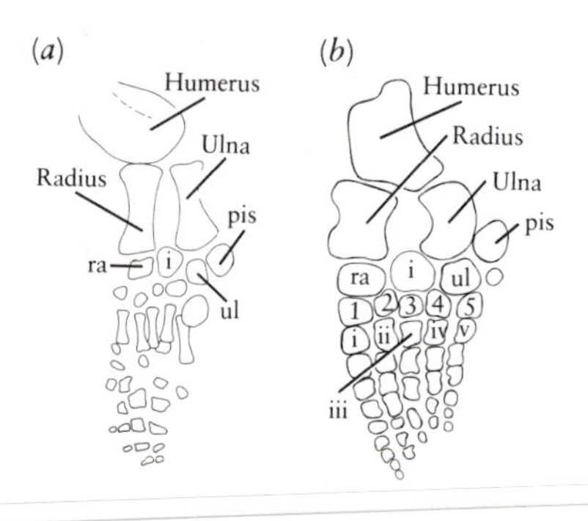

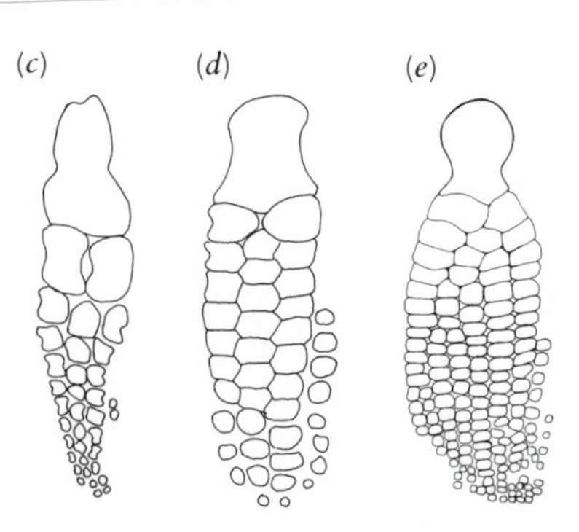

Figure 12-30. DIFFERENT PATTERNS OF THE FRONT LIMBS OF ICHTHYOSAURS. (*a*) *Utatsusaurus* (Lower Triassic). *From Shikama, Kamei, and Murata, 1978.* (*b*) *Mixosaurus* (Middle Triassic). *From McGowan, 1972.* (*c*) *Merriamia* (Upper Triassic). *From McGowan, 1972.* (*d*) *Proteosaurus,* showing the pattern that is characteristic of longipinnates (Lower Jurassic) *From McGowan, 1972.* (*e*) *Ichthyosaurus communis,* showing the pattern of latipinnates (Lower Jurassic). *From McGowan, 1972.* Abbreviations as follows: i, intermedium; pis, pisiform; ul, ulnare; 1–5, distal carpals; i–v, metacarpals. (*b–e*) *By permission of The Royal Ontario Museum.*

ichthyosaurs, and the number and arrangement of the carpals and phalanges has long been used to separate two taxonomic groups. In one group, termed the longipinnates, there are three distal carpals and three primary digits; in the other group, the latipinnates, there are four distal carpals and four or more primary digits. McGowan (1972) extended these groups back into the Triassic and recognized *Mixosaurus* as a latipinnate and *Cymbospondylus, Merriamia* and *Shonisaurus* as longipinnates. Appleby (1979) demonstrated that several early Jurassic species exhibited an intermediate pattern and suggested that the divergence of these two patterns began only after the end of the Triassic. He suggested that all the Jurassic ichthyosaurs arose from a single lineage of Triassic longipinnates rather than from the latipinnate *Mixosaurus,* which was thought to have such an ancestral position. The poorly known genus *Merriamia* might be the most closely related of the Upper Triassic genera.

Detailed descriptions of the skull of early Jurassic ichthyosaurs have only recently been published. They are based on specimens that have been acid etched from nodules in which the original three-dimensional character of the bone is retained (Romer, 1968; McGowan, 1973) (Figure 12-31). It was long thought that the supratemporal, a bone lost in many reptilian groups, formed the lateral margin of the temporal opening in Jurassic and Cretaceous genera and that the remainder of the cheek included both a squamosal and a large quadratojugal separated by a posterior extension of the postorbital bone. More complete preparation demonstrates that the areas identified as squamosal and quadratojugal are parts of a single bone that is simply overlapped by the postorbital. If the single bone is identified as the quadratojugal, the more dorsal element bordering the temporal opening can be recognized as the squamosal.

The occiput exhibits a common pattern in genera from the Upper Triassic through the Jurassic. The elements are not well integrated with one another but were presumably united by cartilage. The pattern broadly resembles that of primitive amniotes with a large supraoccipital, but the exoccipitals are small and do not contribute to the occipital condyle. The otic capsules are large and supported laterally against the squamosal like those in early archosaurs (see Figure 13-10). In the Jurassic genus *Ichthyosaurus,* the stapes is strikingly massive, unlike that of Mesozoic diapsids but resembling that of Carboniferous amniotes. It abuts ventrolaterally against the quadrate. It certainly would not have contributed to an impedance-matching system but may have been effective in transmitting vibrations from the water to the inner ear. The quadrate extends fairly far dorsally but was not embayed posteriorly for support of a tympanum.

The palate is specialized in the loss of the ectopterygoid and the transverse flange of the pterygoid. The basicranial articulation between the braincase and the palate is still retained in the early Jurassic genera, and the interpterygoid vacuities are open.

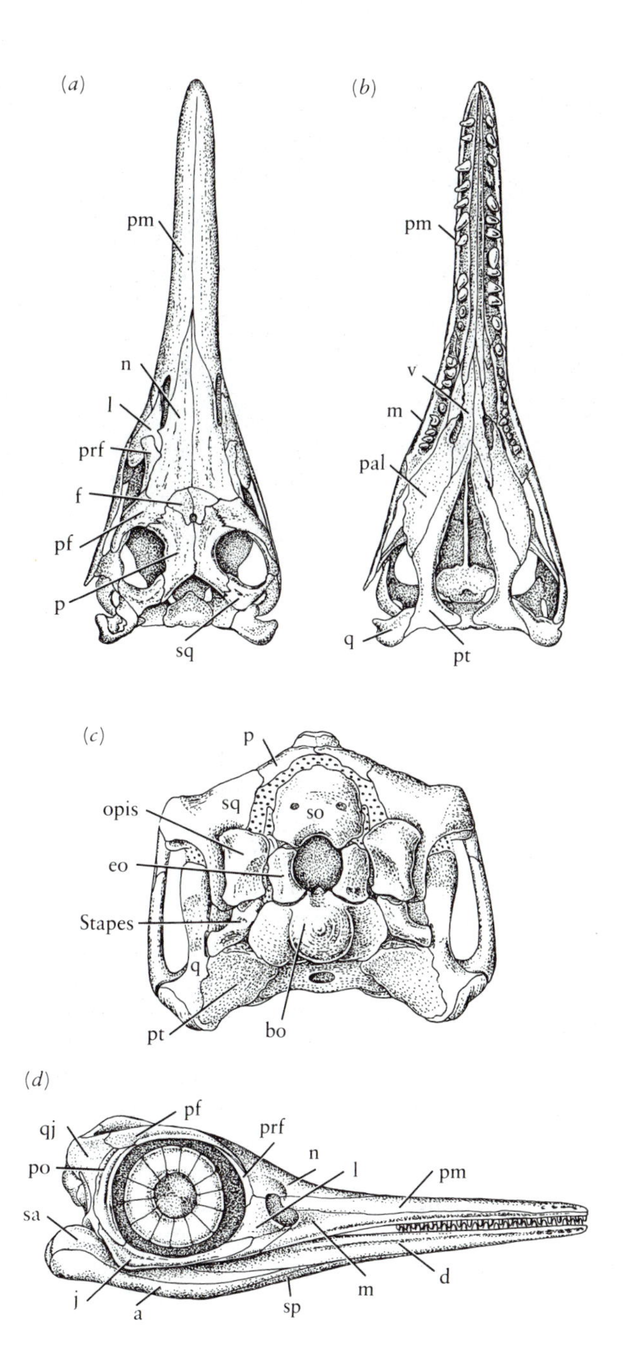

Figure 12-31. SKULL OF JURASSIC ICHTHYOSAURS. (*a, b,* and *c*) Dorsal, palatal, and occipital views of *Ichthyosaurus,* about 50 centimeters long. (*d*) Lateral view of *Ophtalmosaurus,* about 80 centimeters long. Abbreviations as in Figure 8-3. (*a and b*) *From Sollas, 1916.* (*c*) *From Romer, 1968.* (*d*) *From Andrews, 1910.*

The Upper Liassic (the lower division of the Jurassic) localities near Holzmaden in southern Germany provide the most striking record of ichthyosaur anatomy. In addition to several hundred completely articulated skeletons, many of the specimens show the outline of the body preserved as a carbonaceous film. These specimens demonstrate the configuration of the fleshy portion of the tail and a large, sharklike dorsal fin. Many specimens show the young in utero or in the process of emerging from the birth canal (Figure 12-32).

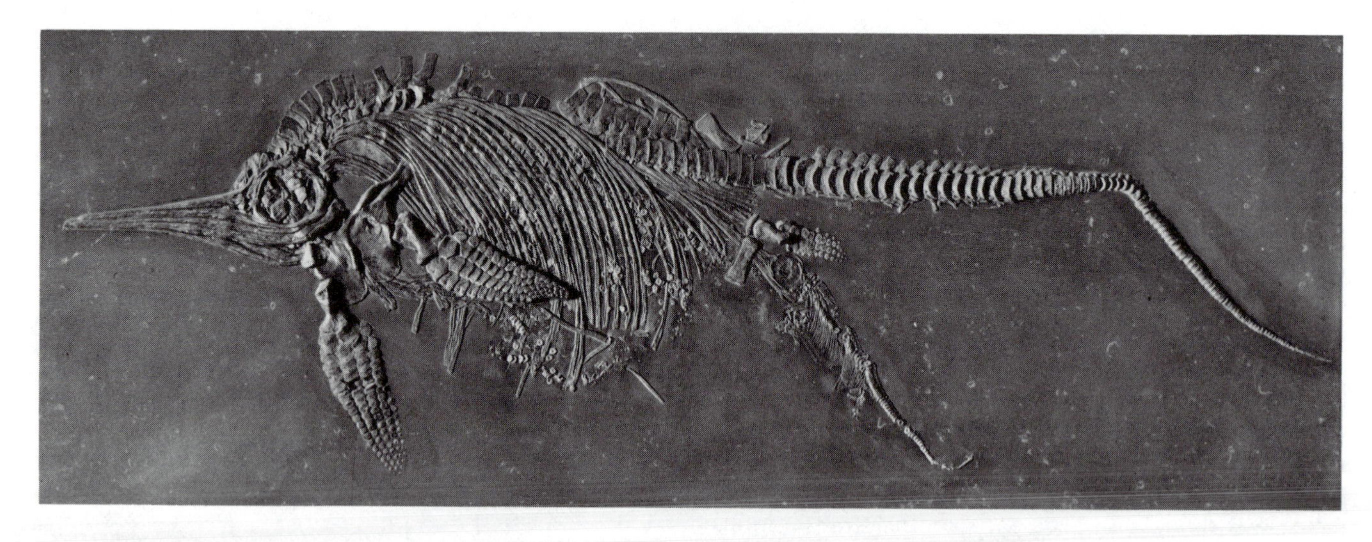

Figure 12-32. SKELETON OF A LIASSIC ICHTHYOSAUR IN THE PROCESS OF GIVING BIRTH. *Photograph courtesy of Dr. R. Wild, State Museum, Stuttgart.*

Despite the diversity of ichthyosaurs in the Lower Jurassic, the group diminishes rapidly in the later Mesozoic. We know little of ichthyosaurs in the middle Jurassic. In the late Jurassic and early Cretaceous, one of the most common and widespread genera is *Ophthalmosaurus* (see Figure 12-19), which is characterized by the enormous size of the orbits. The elements of the paddles are rounded rather than polygonal. The last ichthyosaur, *Playpterygius,* is known from scattered remains throughout the world that cover a span of approximately 45 million years and extend to the base of the Upper Cretaceous.

The last of the ichthyosaurs disappeared from the fossil record long before the late Mesozoic extinction of the mosasaurs and plesiosaurs. Despite their earlier appearance, more rapid diversification, and higher degree of aquatic specialization, the ichthyosaurs appear to have failed in competition with the other marine reptiles and the advanced sharks that were becoming dominant in the late Mesozoic.

SUMMARY

Aquatic adaptation has occurred among many amniotes lineages. Three major groups of aquatic reptiles are known in the Mesozoic, the sauropterygians, placodonts, and ichthyosaurs.

The sauropterygians, which are typified by the nothosaurs and plesiosaurs, probably originated among the primitive eosuchians. The Upper Permian genus *Claudiosaurus* is a primitive diapsid that combines features of terrestrial younginoids with those of primitive nothosaurs. It shows few aquatic adaptations, but the loss of the lower temporal bar and specializations of the palate are similar to those of later sauropterygians. Nothosaurs, which are known primarily from the Middle Triassic, are obligatorily aquatic but show relatively little skeletal specialization for aquatic locomotion. Primitive nothosaurs probably swam like modern aquatic lizards and crocodiles, relying primarily on lateral undulation of the trunk and tail for propulsion. The limbs would have been held close to the body to reduce drag.

Later nothosaurs evolved the ability to use the forelimbs in a symmetrical fashion and reelaborated the ventral portion of the shoulder girdle for more effective protraction and retraction. Larger nothosaurs have a more rigid trunk region and probably relied less on lateral undulation of the trunk and tail for propulsion.

Pistosaurus may represent a link between nothosaurs and plesiosaurs. The palate retains the primitive configuration of the pterygoid that is common to plesiosaurs, but the postcranial skeleton, if correctly associated, is typical of nothosaurs.

Plesiosaurs are advanced over nothosaurs in the greater relative size of the limbs, their greater specialization for aquatic propulsion, and the similarity of the pectoral and pelvic limbs. Like modern sea lions, plesiosaurs probably moved the pectoral fins primarily in the horizontal plane, with most of the force being delivered by their posterior movement. Nothosaurs are restricted to the Triassic, while both major groups of plesiosaurs, the Plesiosauroidea and Pliosauroidea, continued to the end of the Cretaceous.

Placodonts have long been allied with the nothosaurs and plesiosaurs, but there is no strong evidence that they are closely related. Even the earliest placodonts from the Middle Triassic are highly specialized in the possession of blunt crushing teeth, the high degree of consolidation of the skull, and the structure of the vertebrae, which have long transverse processes and deeply amphicoelous centra. Placodonts show little specialization for aquatic locomotion, but the limbs and girdles are poorly ossified. One lineage of placodonts evolved a carapace that resembles that of turtles in many features. Within this group,

Henodus further resembles turtles in the presence of a horny beak instead of teeth and the absence of a dorsal temporal opening.

Ichthyosaurs were the most highly specialized of all aquatic reptiles, with a body outline that resembles that of the fastest-swimming modern fish, the tunas. Several ichthyosaur genera have been described from the Lower Triassic, but they are already very highly specialized and provide little evidence of the origin of the group.

Several distinct types of ichthyosaurs have been described from the Middle and Upper Triassic, but we have not established their specific interrelationships. Jurassic and Cretaceous ichthyosaurs appear to represent a new radiation from a single lineage of earlier forms. They are characterized by a high lunate tail, double-headed ribs throughout the vertebral column, more highly specialized paddles, and teeth set in an open groove rather than in separate sockets.

The greatest diversity of the ichthyosaurs was in the Lower Jurassic. Specimens from Holzmaden in Southern Germany show the body outline, and many specimens have young in utero. The number and diversity of ichthyosaurs declines in the later Jurassic, and only a single genus, *Platypterygius* continues to the base of the Upper Cretaceous. Ichthyosaurs became extinct long before the end of the Cretaceous.

REFERENCES

Andrews, C. W. (1910). *Descriptive Catalogue of the Marine Reptiles of the Oxford Clay. Brit. Mus. (Nat. Hist.),* Part 1: 1–205.

Appleby, R. M. (1979). The affinities of Liassic and later ichthyosaurs. *Palaeontology,* **22**(4): 921–946.

Broili, F. (1912). Zur Osteologie des Schaedels von *Placodus. Palaeontographica,* **59**: 147–155.

Brown, D. S. (1981). The English Upper Jurassic Plesiosauroidea (Reptilia), and a review of the phylogeny and classification of the Plesiosauria. *Bull. Brit. Mus. (Nat. Hist.), Geol.,* **35**(4): 253–347.

Camp, C. L. (1980). Large ichthyosaurs from the Upper Triassic of Nevada. *Palaeontographica,* **170**: 139–200.

Carroll, R. L. (1981). Plesiosaur ancestors from the Upper Permian of Madagascar. *Phil. Trans. Roy. Soc., Lond. B,* **293**: 315–383.

Carroll, R. L., and Gaskill, P. (1985). The nothosaur *Pachypleurosaurus* and the origin of plesiosaurs. *Phil. Trans. Roy. Soc., B,* **309**: 343–393.

Clark, B. D., and Bemis, W. (1979). Kinematics of swimming penguins at the Detroit Zoo. *J. Zool., Lond.,* **188**: 411–428.

Dawson, W. R., Bartholomew, G. A., and Bennett, A. F. (1977). A reappraisal of the aquatic specializations of the Galapagos marine iguana (*Amblyrhynchus cristatus*). *Evolution,* **31**: 891–897.

Drevermann, F. (1933). Die Placodontier. 3. Das Skelett von *Placodus gigas* Agassiz im Senckenberg Museum. *Abh. Senckenberg Naturf. Ges.,* **38**: 319–364.

English, A. W. (1976). Limb movements and locomotor function in the California sea lion (*Zalophus californianus*). *J. Zool., Lond.,* **178**: 341–364.

Godfrey, S. (1984). Plesiosaur subaqueous locomotion: A reappraisal. *Neues Jahrb. Geol. Paläont. Mh.,* **11**: 661–672.

von Huene, F. (1948). *Pistosaurus,* a Middle Triassic plesiosaur. *Amer. J. Sci.,* **246**: 46–52.

von Huene, F. (1949). Ein Schaedel von *Mixosaurus* und die Verwandtschaft der Ichthyosaurier. *Neues Jahrb. Min. Geol. Pal. Mh., Abt. B*: 88–95.

Kuhn-Schnyder, E. (1963). I Sauri del Monte San Giorgio. *Comunicazioni dell 'Istitutio di Paleontologia dell 'Universita di Zurigo,* **20**: 811–854.

Mazin, J.- M. (1981). *Grippia longirostris* Wiman, 1929, un Ichthyopterygia primitif du Trias inferieur du Spitsberg. *Bull. Mus. natn. Hist. nat.,* **3**: 317–340.

Mazin, J.- M. (1983). *Omphalosaurus nisseri* (Wiman, 1910) un ichthyopterygien a denture broyeuse du Trias moyen du Spitsberg. *Bull. Mus. natn. Hist. nat.,* section C, no. 2 : 243–263.

McGowan, C. (1972). The distinction between latipinnate and longipinnate ichthyosaurs. *Roy. Ont. Mus., Life Sci. Occ. Papers,* **20**: 1–8.

McGowan, C. (1973). The cranial morphology of the Lower Liassic latipinnate ichthyosaurs of England. *Bull. Brit. Mus. (Nat. Hist.), Geol.,* **24**(1): 3–109.

McGowan, C. (1979). A revision of the Lower Jurassic ichthyosaurs of Germany with descriptions of two new species. *Palaeontographica,* **166**: 93–135.

McGowan, C. (1983). *The Successful Dragons. A Natural History of Extinct Reptiles.* Samuel Stevens, Toronto and Sarasota.

Merriam, J. C. (1902). Triassic Ichthyopterygia from California and Nevada. *Univ. Calif., Bull. Dept. Geol.,* **3**: 63–108.

Merriam, J. C. (1904). A new marine reptile from the Triassic of California. *Univ. Calif., Bull. Dept. Geol.,* **3**: 419–421.

Merriam, J. C. (1908). Triassic Ichthyosauria with special reference to the American forms. *Mem. Univ. Calif.,* **1**: 1–196.

von Meyer, H. (1847–1855). *Zur Fauna der Vorwelt. Die Saurier des Muschelkalkes, mit Ruecksicht auf die Saurier aus buntem Sandstein und Keuper.* Heinrich Keller, Frankfurt am Main.

Newman, B., and Tarlo, L. B. H. (1967). A giant marine reptile from Bedfordshire. *Animals, Lond.,* **10**(2): 61–63.

Persson, P. O. (1963). A revision of the classification of the Plesiosauria with a synopsis of the stratigraphical distribution of the group. *Lunds Univ. Arsskrift.,* **59**: 1–60, 9 figs.

Peyer, B. (1935). Die Triasfauna der Tessiner Kalkalpen. 8. Weitere Placodontierfunde. *Abh. Schweiz. Palaeont. Ges.,* **55**: 1–26, 5 plates.

Peyer, B. (1950). *Geschichte der Tierwelt.* Büchergilde Gutenberg, Zürich.

Peyer, B. (1955). Die Triasfauna der Tessiner Kalkalpen. 18. *Helveticosaurus zollingeri* n. g. n. sp. *Abh. Schweiz. Palaeont. Ges.,* **72**: 1–50.

Peyer, B., and Kuhn-Schnyder, E. (1955). Placodontia. In J. Piveteau (ed.), *Traite de Paleontologie,* **5**: 458–486. Masson S.A., Paris.

Repossi, E. (1902). Mixosauro degli strati triasici di Besano in Lombardia. *Atti della Soc. Ital. di Sci. Nat.*, **41**: 361–372.

Robinson, J. A. (1975). The locomotion of plesiosurs. *Neues Jahrbuch, Geol. Palaeont. Abh.*, **149**(3): 286–332.

Romer, A. S. (1968). An ichthyosaur skull from the Cretaceous of Wyoming. *Univ. Wyoming, Contrib. Geol.*, **7**: 27–41.

de Saint-Seine, P. (1955). Sauropterygia. In J. Piveteau (ed.), *Traite de Paleontologie, Volume* 5; pp. 420–428. Masson S.A., Paris.

Seymour, R. S. (1982). Physiological adaptations to aquatic life. In C. Gans and F. H. Pough (eds.), *Biology of the Reptilia, Vol. 13; pp. 1–51. Academic Press, New York.*

Shikama, T., Kamei, T., and Murata, M. (1978). Early Triassic Ichthyosaurus, *Utatsusaurus hataii* gen. et sp. nov., from the Kitakami Massif, Northeast Japan. *Tohoku Univ., Sci. Rept., 2nd ser., Geol.*, **48**: 77–97.

Sollas, W. J. (1916). The skull of *Ichthyosaurus* studied in serial sections. *Phil. Trans. Roy. Soc., Lond. (B)*, **208**: 63–126.

Sues, H.-D. (1986). On the skull of *Placodus gigas* and the relationships of the Placodontia. *Jour. Vert. Paleont.* (In press).

Sues, H.-D. (1987). The postcranial skeleton of *Pistosaurus* and the interrelationships of the Sauropterygia (Diapsida). *Zool. Jour. Linn. Soc.* (In press).

Tarlo, L. B. (1958). The scapula of *Pliosaurus macromerus* Phillips. *Palaeontology*, **1**: 193–199.

Watson, D. M. S. (1924). The elasmosaurid shoulder girdle and forelimb. *Proc. Zool. Soc. Lond.*: 885–917.

Webb, P. W. (1982). Locomotor patterns in the evolution of actinopterygian fishes. *Amer. Zool.*, **22**: 392–342.

Welles, S. P. (1962). A new species of elasmosaur from the Aptian of Columbia and a review of the Cretaceous plesiosaurs. *Univ. Calif. Publ. Geol. Sci.*, **44**: 1–89.

Westphal, F. (1976). The dermal armour of some Triassic placodont reptiles. In A. d'A. Bellairs and C. B. Cox (eds.), *Morphology and Biology of Reptiles. Linn. Soc. Symp. Ser.*, **3**: 31–41.

Young, C. C., and Dong, Z. M. (1972). On the aquatic reptiles of the Triassic in China. *Vert. Paleont. Mem.*, **9**: 1–34. In Chinese.

CHAPTER XIII

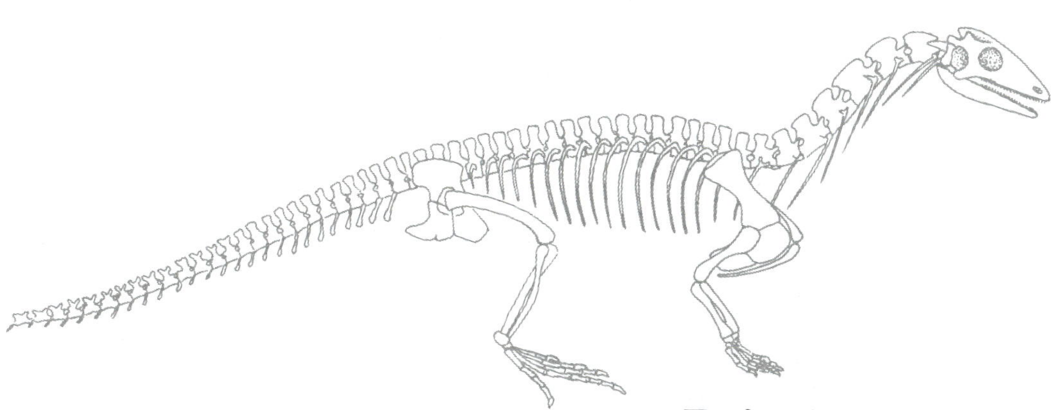

Primitive Archosauromorphs and Crocodiles

Archosaurs are the most spectacular reptiles. They include dinosaurs, which dominated the terrestrial environment throughout the Mesozoic; crocodiles, which are the largest of modern reptiles; and pterosaurs, which matched the birds in their degree of skeletal specialization for flight. Birds themselves arose from archosaurs and would be included in this group if a strictly phylogenetic classification were being used.

We include five orders in the Archosauria: the Crocodylia; two orders of dinosaurs, the Saurischia and Ornithischia; the Pterosauria; and an assemblage of primitive forms that are primarily restricted to the Triassic, the Thecodontia. Archosaurs almost certainly arose from a single ancestral stock that had diverged from other diapsid reptiles by the late Paleozoic (Figure 13-1).

As a group, the archosaurs are characterized by a host of skeletal specializations, many of which are associated with a more upright posture and more effective

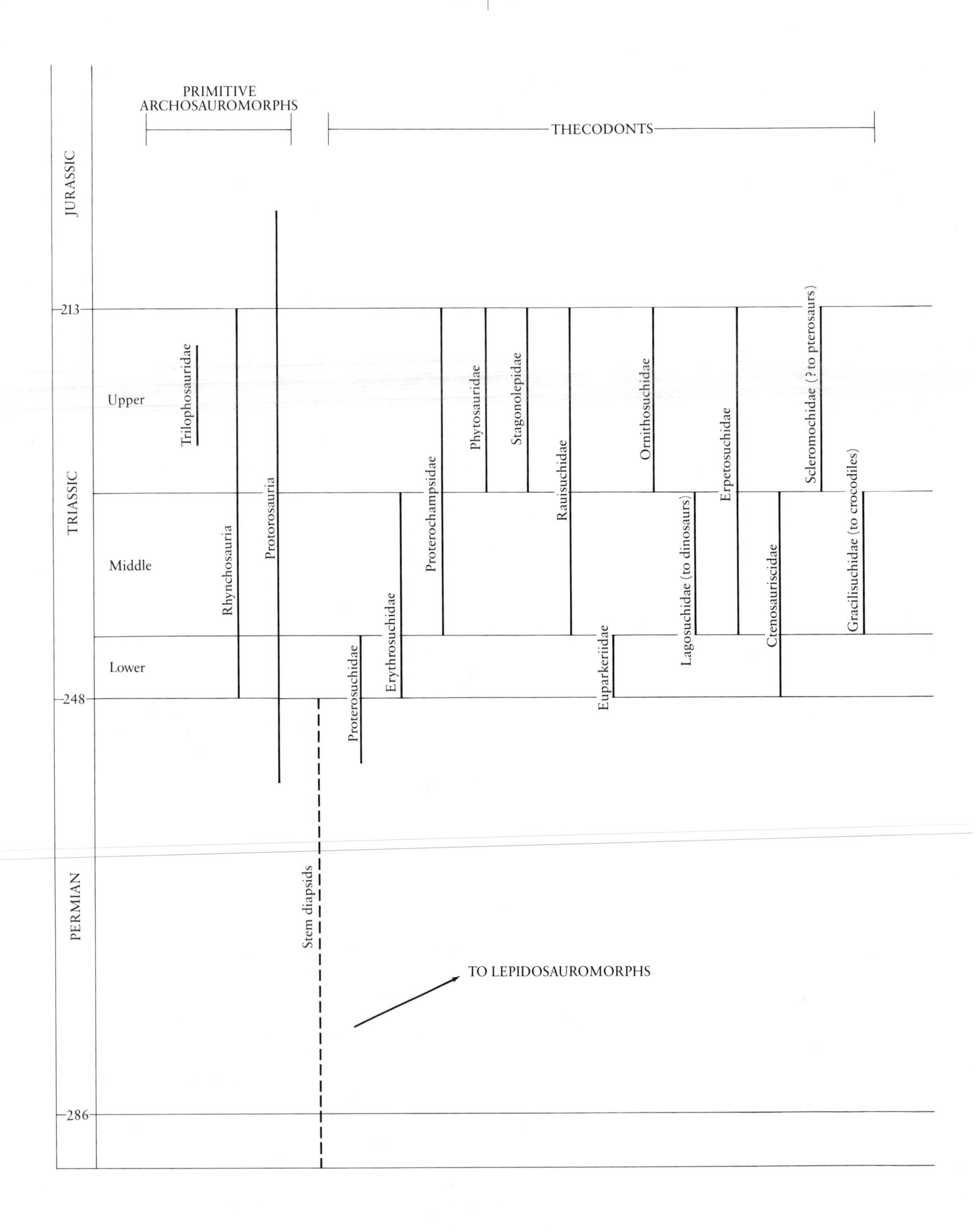

Figure 13-1. (*a* and *b*) Stratigraphic ranges of primitive archosauromorphs, thecodonts, and crocodiles. *Crocodile relationship based largely on Buffetaut, 1979.*

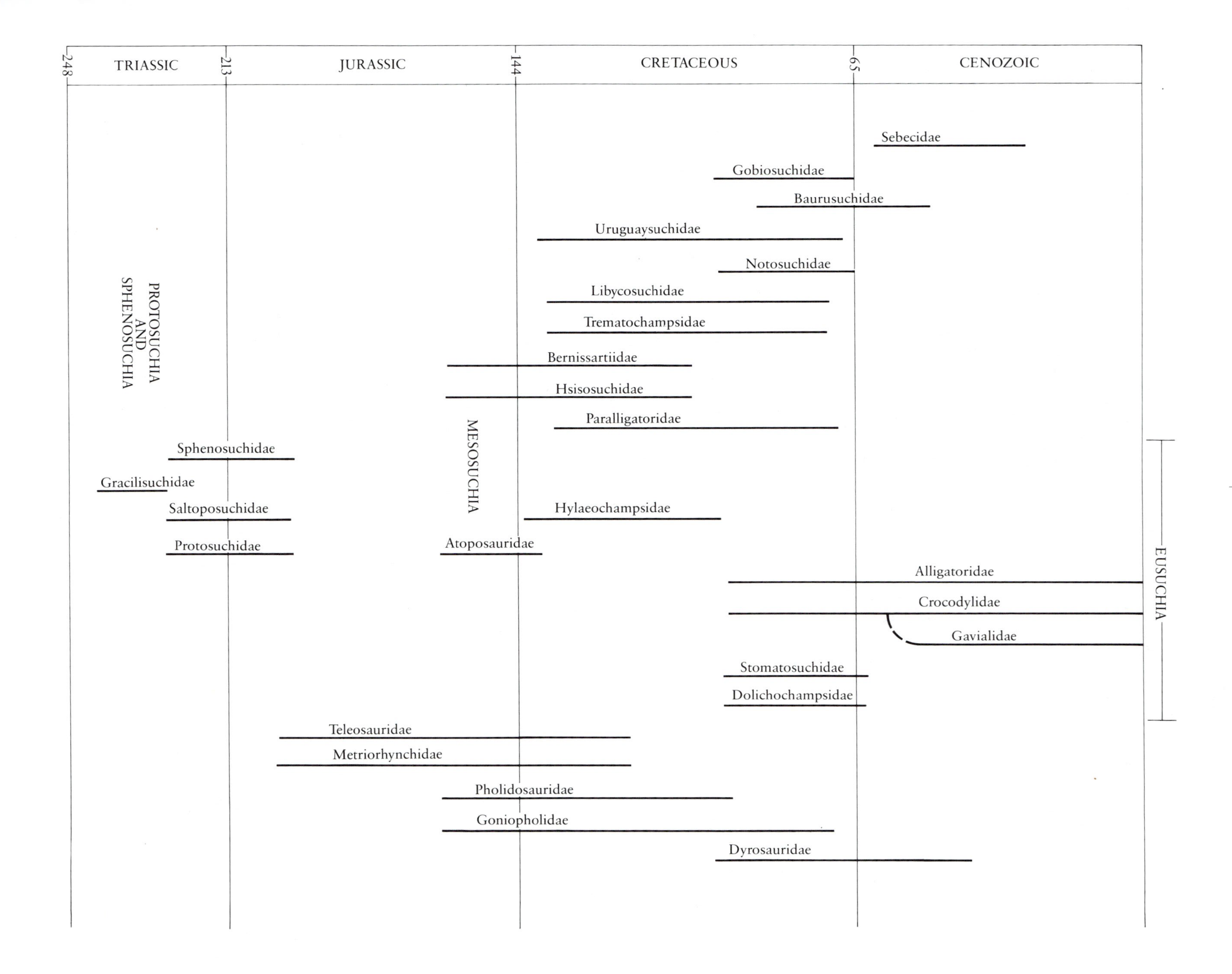
248
TRIASSIC
213
JURASSIC
144
CRETACEOUS
65
CENOZOIC
Sebecidae
Gobiosuchidae
Baurusuchidae
Uruguaysuchidae
Notosuchidae
Libycosuchidae
Trematochampsidae
Bernissartiidae
Hsisosuchidae
Paralligatoridae
PROTOSUCHIA
AND
SPHENOSUCHIA
MESOSUCHIA
Sphenosuchidae
Gracilisuchidae
Saltoposuchidae
Protosuchidae
Hylaeochampsidae
Atoposauridae
Alligatoridae
Crocodylidae
Gavialidae
EUSUCHIA
Stomatosuchidae
Dolichochampsidae
Teleosauridae
Metriorhynchidae
Pholidosauridae
Goniopholidae
Dyrosauridae

fore-and-aft movement of the limbs than is evident among the primitive diapsids and lepidosaurs. However, the most primitive archosaurs can be distinguished from other early diapsids only by a single, clearly definable skeletal character, the presence of a large opening anterior to the eye, the antorbital fenestra (see Figures 13-2*b* and 13-10).

Primitive Archosauromorphs

Within the last five years paleontologists have recognized that archosaurs probably shared a common ancestry with several other groups of early diapsids and may be included with them in a single larger assemblage, the Archosauromorpha (Benton, 1985). In addition to the archosaurs, we recognize three groups from the late Permian and Triassic as archosauromorphs: protorosaurs, rhynchosaurs, and trilophosaurids (Figure 13-2).

Many skeletal features distinguish the early members of this assemblage from lepidosauromorphs and primitive diapsids. The premaxilla extends dorsally behind the nares, separating the maxilla from the border of this opening. The teeth are set in sockets (the thecodont condition) rather than in a shallow groove. All may have had an impedance-matching middle ear. In the earliest adequately known members of each group, the quadrate is high and in contact with the paroccipital process. It is emarginated posteriorly and may have supported a tympanum. All early rhynchosaurs, archosaurs, and protorosaurs had a long, narrow stapes that was directed at right angles to the cheek.

The neck is elongate with seven or eight cervical vertebrae; throughout the column the centra lose their deeply amphicoelous nature. The humerus has lost the entepicondylar foramen. In the carpus, the pisiform is typically absent and the other bones are slow to ossify. The kind of sternum that was described in lepidosauromorphs does not occur in archosauromorphs.

The most important feature that unites the members of the archosauromorph assemblage is the structure of the tarsus and foot (Figure 13-3). In contrast with primitive diapsids, the astragalus and calcaneum articulate with one another via broad concave-convex articulating surfaces that are proximal and distal to the perforating foramen. The fifth distal tarsal is lost and the head of the

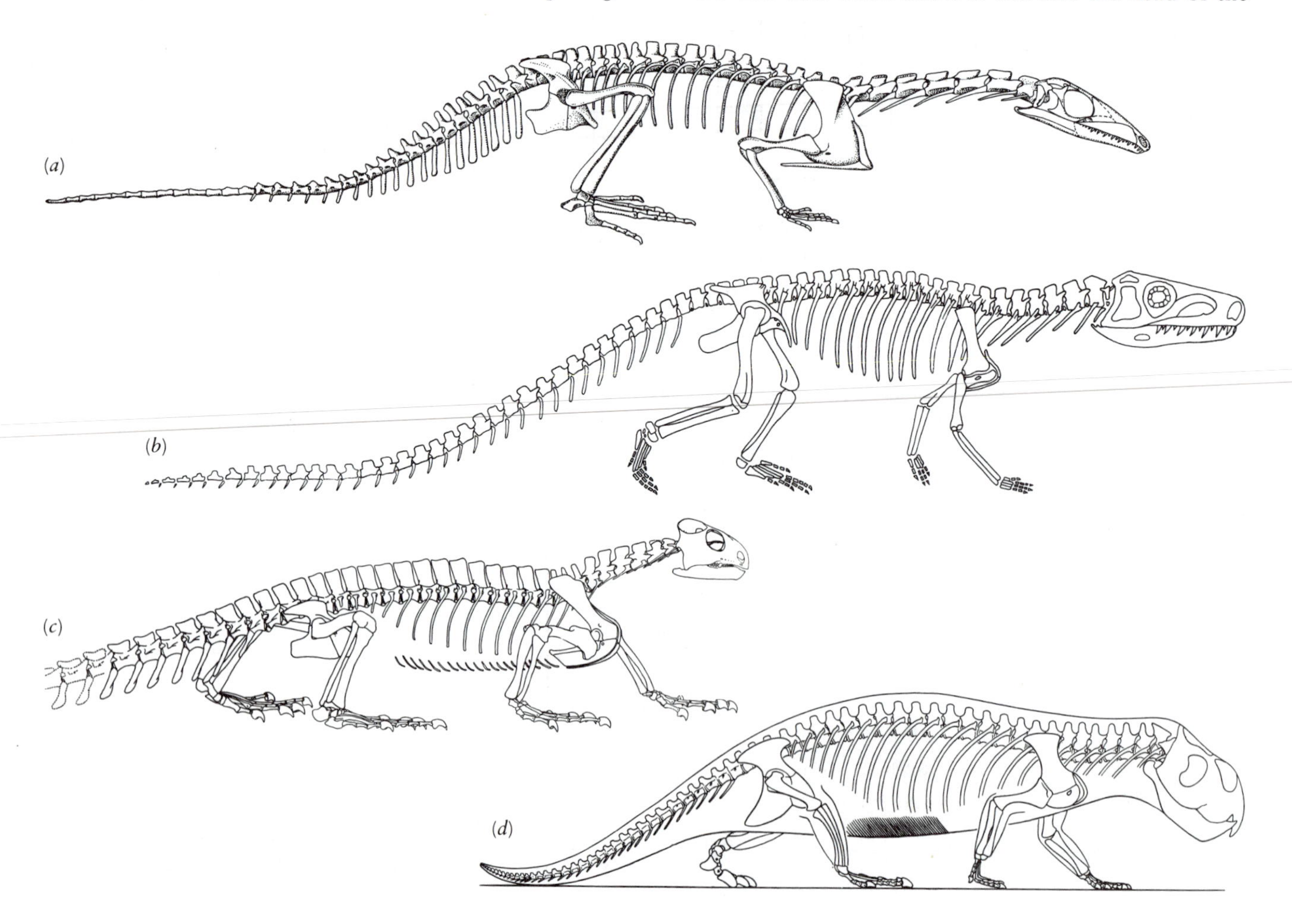

Figure 13-2. SKELETONS OF PRIMITIVE ARCHOSAUROMORPH REPTILES. (*a*) *Prolacerta*, about 1 meter long. *Modified from Gow, 1975.* (*b*) The primitive archosaur *Euparkeria*, about ½ meter long. *Modified from Ewer, 1965.* (*c*) *Trilophosaurus*, about 2 meters long. *From Gregory, 1945.* (*d*) The rhynchosaur *Paradapedon*, about 1½ meters long. *From Chatterjee, 1974.*

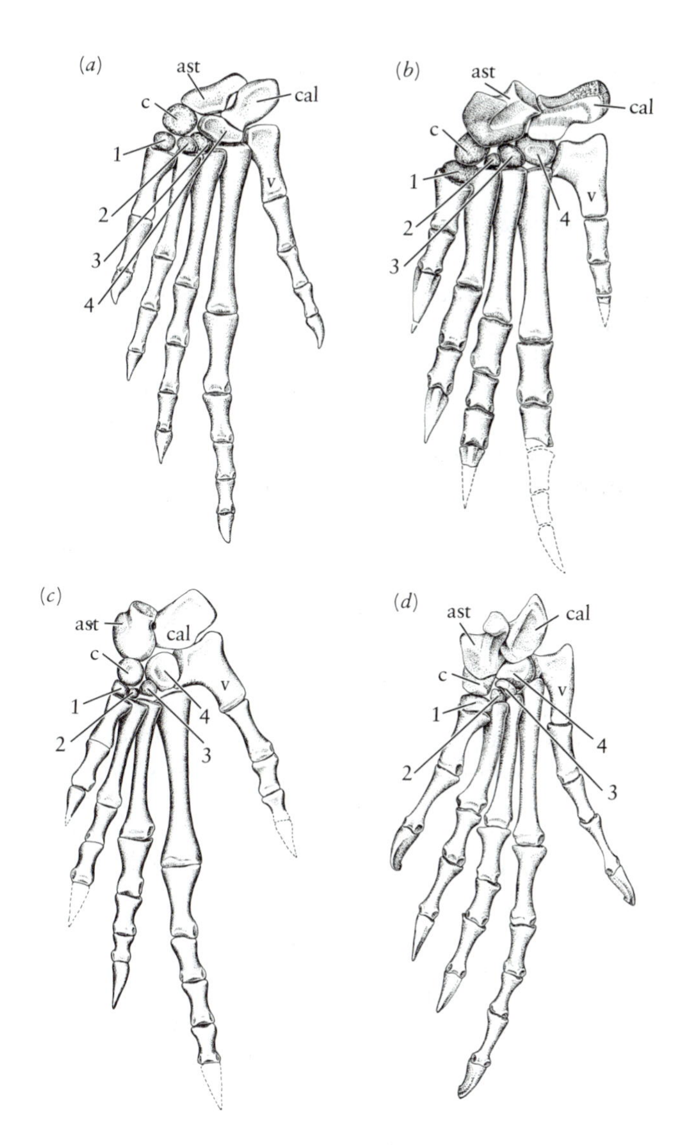

Figure 13-3. TARSUS AND FOOT OF PRIMITIVE ARCHOSAUROMORPHS. (*a*) *Protorosaurus. Restoration based on illustrations in von Meyer, 1856.* (*b*) The primitive archosaur *Chasmatosaurus. Modified from Cruickshank, 1972.* (*c*) The early rhynchosaur *Noteosuchus. From Carroll, 1976b.* (*d*) *Trilophosaurus. From Gregory, 1945.* Abbreviations as follows: ast, astragalus; c, centrale; cal, calcaneum; 1–4 distal tarsals; v, fifth metatarsal.

fifth metatarsal is bent medially, or "hooked," so that it articulates with the lateral surface of the fourth distal tarsal. The other metatarsals and phalanges follow the pattern of primitive diapsids and lepidosauromorphs. The phalangeal count is 2, 3, 4, 5, 4, and the fourth digit is the longest.

As in the lepidosauromorphs, the tarsus is consolidated and the hooked fifth metatarsal serves as a lever to ventroflex the foot (Brinkman, 1981), but these changes took place separately and at different rates in the two groups. A hooked fifth metatarsal does not appear among lepidosauromorphs until the Upper Triassic, but it appears in the archosauromorphs by the late Permian.

According to Brinkman (1979), the articulation between the astragalus and calcaneum that distinguishes early archosauromorphs is necessary to maintain contact between the calcaneum and the distal tarsals when the limb is brought into a more upright position.

Early archosauromorphs also differ from primitive diapsids in the lateral extension of the calcaneum beyond the articulating surfaces. This extension may have served as a pulley over which passed the tendon of the gastrocnemeus, which was attached to the plantar surface of the fifth metatarsal.

Among the earliest archosauromorphs, the structure of the remainder of the rear limb and pelvis does not differ significantly from that of primitive diapsids.

We have not found any fossils that link primitive diapsids with the earliest of archosauromorphs. Except for the elaboration of the sternum, the younginoids, including *Youngina* and *Thadeosaurus* (which were discussed in Chapter 11), may exemplify the skeletal pattern that was ancestral to the archosauromorphs. The long neck of *Petrolacosaurus* suggests an earlier initiation of the archosauromorph pattern, but this genus is otherwise too primitive for detailed comparison with the Upper Permian and early Triassic genera.

Aside from the shared specializations of the rear limb and middle ear, the four archosauromorph lineages are quite distinct from one another when they first appear in the fossil record, and none are close to the pattern that would be expected in the ancestors of the other three.

PROTOROSAURS (PROLACERTIFORMES)

The earliest member of the archosauromorph assemblage to appear in the fossil record is *Protorosaurus* from the early part of the Upper Permian of Europe (Figure 13-4). *Protorosaurus* is characterized by its very long neck. As in *Claudiosaurus* and *Petrolacosaurus,* this length is achieved by both a posterior displacement of the shoulder girdle and an elongation of the individual vertebrae. The skull appears superficially like that of early lizards in the reduction of the lower temporal bar, but the configuration of the quadrate does not resemble that of *Paliguana* (the earliest of lizards) and it is supported by a ventral process of the squamosal that precludes streptostylic movement.

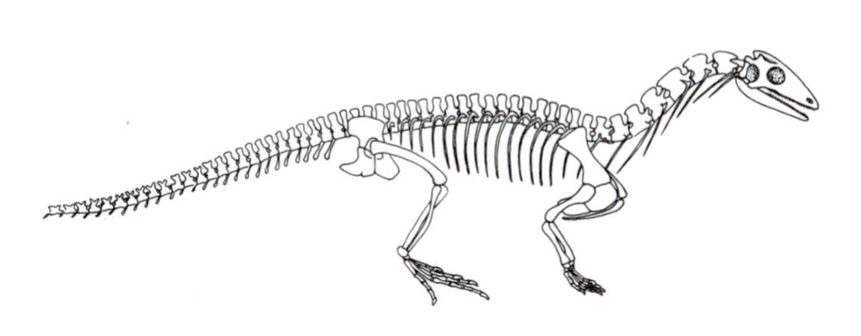

Figure 13-4. *PROTOROSAURUS,* THE EARLIEST-KNOWN ARCHOSAUROMORPH FROM THE UPPER PERMIAN OF EUROPE. 1 to 2 meters long. *Modified from Seeley, 1888.*

Nearly identical proportions of the head and postcranial skeleton occur in the Lower Triassic genus *Prolacerta* from southern Africa, which Gow (1975) described comprehensively. He was the first to recognize that the affinities of this group lie with the archosaurs rather than with lepidosaurs.

Protorosaurs are represented in the Middle Triassic of Central Europe by one of the most bizarre of the early reptiles, *Tanystropheus* (Figure 13-5). The neck, which is already distinctive in *Protorosaurus* and *Prolacerta,* is extended to more than twice the length of the trunk. The limbs are short and poorly ossified. Wild (1973) suggests that the adult life must have been spent in the water, for it is difficult to envisage how such a long neck could be supported on land. Interestingly, the teeth differ markedly between small and large individuals. In the young they are tricuspid, as in the most primitive mammals, whereas among the adults they are simple, sharp pegs. The last of the protorosaurs is *Tanytrachelus* from the late Triassic and early Jurassic of eastern North America (Olsen, 1979).

TRILOPHOSAURIDS

Although some material from the early Triassic has been assigned to the Trilophosauridae, we know this family primarily from a single genus in the Upper Triassic of western North America. Gregory (1945) thoroughly described its skeleton (Figure 13-6). The dentition is unique among archosauromorph reptiles. The cheek teeth are transversely expanded and form sharp, shearing surfaces. Teeth are absent on the premaxilla and front of the lower jaw and were probably replaced by a horny beak. Unlike other primitive archosauromorphs, the cheek is solidly constructed, without a trace of temporal fenestration or ventral emargination. We may attribute the absence of a lateral temporal opening to the necessity to strengthen the skull in relationship to the massive dentition, but no fossils are known from the early Triassic that demonstrate the origin of this condition. The dorsal temporal openings are separated by a high, narrow parietal crest that demonstrates the high degree of elaboration of the adductor jaw musculature. The postcranial skeleton of *Trilophosaurus* resembles that of primitive archosaurs, especially the shape of the ilium and the elongation of the transverse processes of the vertebrae (see Figure 13-2*c*).

RHYNCHOSAURS

Rhynchosaurs are the most common and widespread of the primitive archosauromorph groups. We find them throughout the Triassic, but they were most common in the middle and later parts of the period, with many fossils from Europe, South America, India, and East Africa. A few genera, which represent the oldest and youngest records of the group, occur in southern Africa and eastern North America.

The common middle and late Triassic rhynchosaurs are characterized by a highly specialized dentition, with many rows of teeth in the maxilla and dentary. The edentulous premaxillary bones form a long, overhanging "beak." The temporal arcade is greatly expanded to surround an extremely massive adductor musculature (Figure 13-7). Rhynchosaurs were almost certainly herbivorous, with heavy, short-limbed bodies that were roughly equivalent in size to a large pig. Fossils from the Lower, Middle, and Upper Triassic show a progressive achievement of cranial specialization, that Chatterjee (1974, 1980) and Benton (1983) recently described.

We can document their early history from the genera *Mesosuchus* and *Howesia* from the Lower Triassic of southern Africa (Figure 13-8). In these genera, the skull is broadly similar to that of *Youngina* but differs in the medial position of the external nares and the overhanging

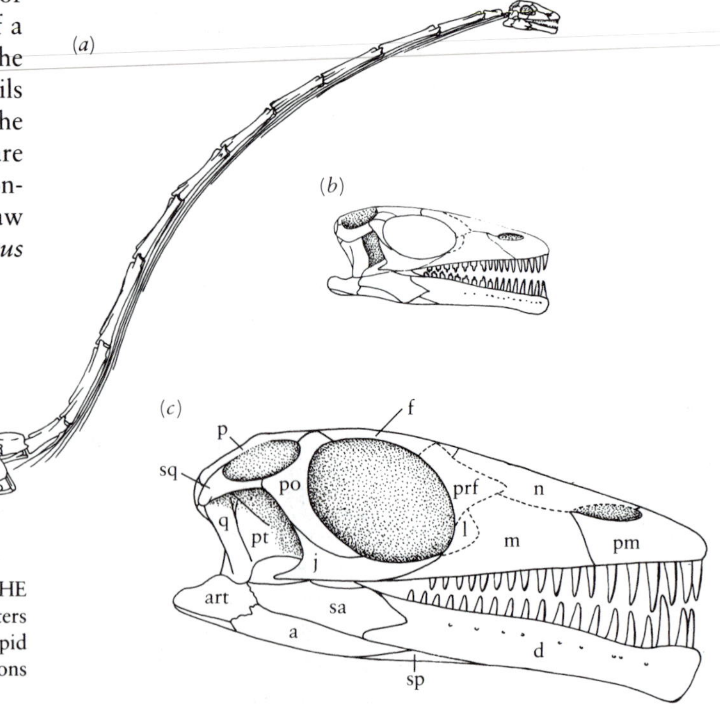

Figure 13-5. THE PROTOROSAUR *TANYSTROPHEUS* FROM THE MIDDLE TRIASSIC OF CENTRAL EUROPE. Approximately 3 meters long. (*a*) Skeleton. (*b*) Skull of immature specimen showing tricuspid cheek teeth. (*c*) Skull of an adult with simple cheek teeth. Abbreviations as in Figure 8-3. *From Wild, 1973.*

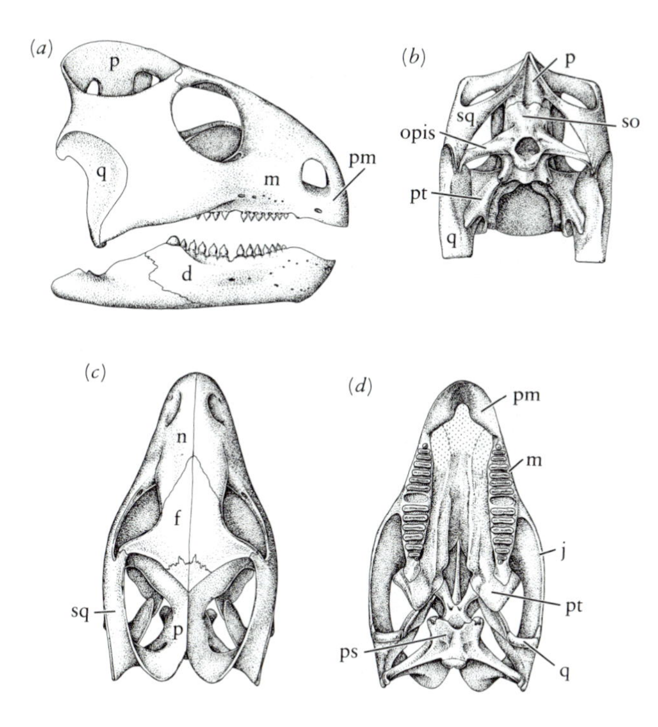

Figure 13-6. SKULL OF *TRILOPHOSAURUS*. (*a*) Lateral, (*b*) occipital, (*c*) dorsal, and (*d*) palatal views. The toothless premaxilla and front of the dentary may have been covered with a horny beak. Many of the sutures are closed in mature individuals. Approximately 10 centimeters long. Abbreviations as in Figure 8-3. *From Gregory, 1945.*

premaxilla, which in primitive species retains a few teeth. *Mesosuchus* is primitive in having a single row of marginal teeth. Unlike the eosuchians, these teeth are set in definite sockets. Teeth have been lost from the palatine bone and the transverse flange of the pterygoid. *Howesia* has developed additional rows of marginal teeth in the upper and lower jaw (Malan, 1963).

In later rhynchosaurs, the palatal surface of the maxilla is greatly widened and forms a broad tooth plate that is made up of many longitudinal rows of teeth and receives the toothed surface of the lower jaw in a longitudinal groove (Figure 13-9). These teeth are not regularly replaced as in most reptiles, and the worn teeth do not drop out, as in most genera with thecodont dentition, but are held in place by secondary bone of attachment. Chatterjee (1974) terms this type of tooth implantation **ankylothecodont.** Teeth are added posteriorly as the jaws grow. The upper and lower teeth wear against one another, and as they are worn flat, the bony surface itself occludes.

Benton (1983) states that the late Triassic rhynchosaur *Hyperodapedon* lacked a tympanum, but the early Triassic genus *Mesosuchus* has a long slender lizardlike stapes. The delicacy of this structure may explain its absence from the fossils of later genera.

Throughout the history of the rhynchosaurs the skull broadens, the body becomes larger, and the rear limbs

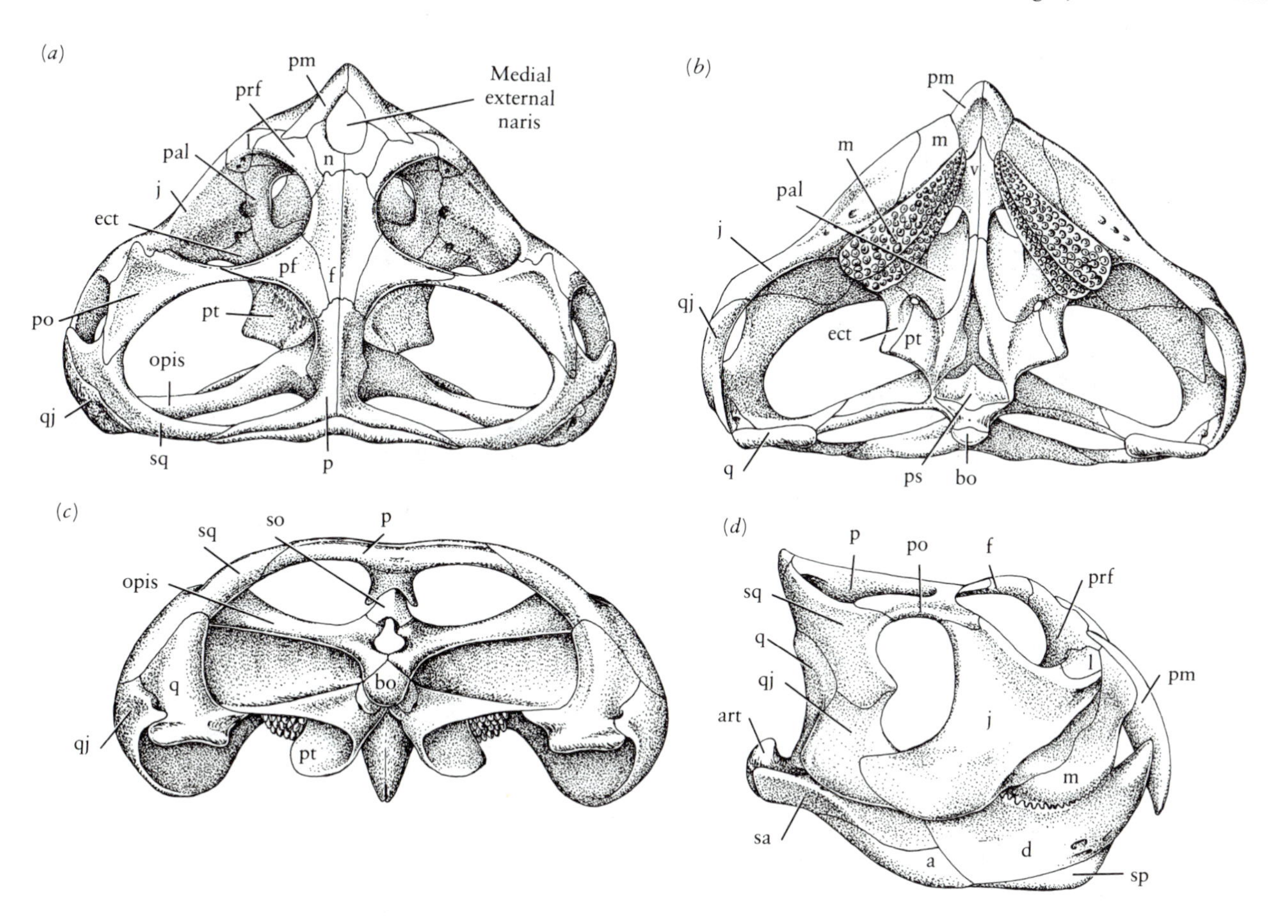

Figure 13-7. SKULL OF THE RHYNCHOSAUR *PARADAPEDON*. (*a*) Dorsal, (*b*) palatal, (*c*) occipital, and (*d*) lateral views. Approximately 20 centimeters wide. Abbreviations as in Figure 8-3. *From Chatterjee, 1974.*

are brought more directly under the body to facilitate a fore-and-aft gait. The forelimbs retain a more sprawling posture.

The tarsus of primitive rhynchosaurs is almost indistinguishable from those of protorosaurs and archosaurs, but the centrale takes on an increasingly important role and becomes attached to the astragalus and calcaneum in the proximal row. The astragalus and calcaneum lose their primitive articulating surfaces and become attached to one another. The tarsus is consolidated into a simple hinge joint as the limb becomes capable of fore-and-aft movement (Carroll, 1976b).

In the Middle Triassic, rhynchosaurs were a major component of the fauna in South America and East Africa. According to Benton (1983), all of the adequately known Upper Triassic rhynchosaurs may be included in two or three closely related genera. Together, they have a cosmopolitan distribution. *Hyperodapedon* occurs in Scotland and India, *Scaphonyx* in South America, and a form that is close to both *Scaphonyx* and *Hyperodapedon* comes from East Africa. Locally, there were very large populations of rhynchosaurs in the Norian (late Triassic), but no trace of the group survives into the Jurassic. Benton (1983) attributes the extinction of the group to the change of vegetation near the end of the Triassic. Chatterjee (1980) suggests that they succumbed to predation from the thecodont archosaurs and dinosaurs that became dominant at the end of the Triassic.

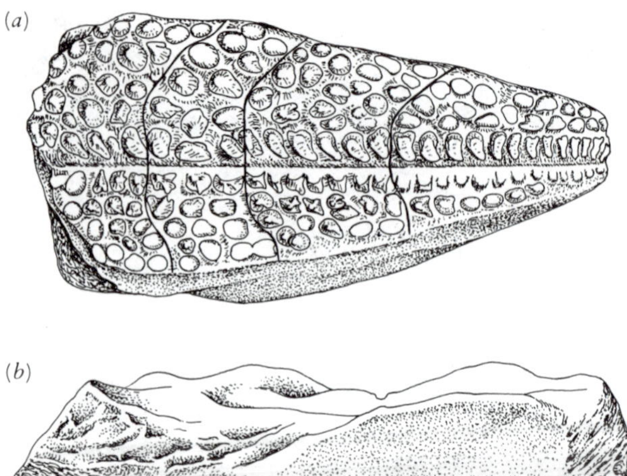

Figure 13-9. (*a*) Maxillary tooth plate of the rhynchosaur *Paradapedon* in ventral view showing groove for dentary and growth lines where teeth are added successively at the back of the jaw. (*b*) Maxillary tooth plate in dorsal view showing growth lines. *From Chatterjee, 1974.*

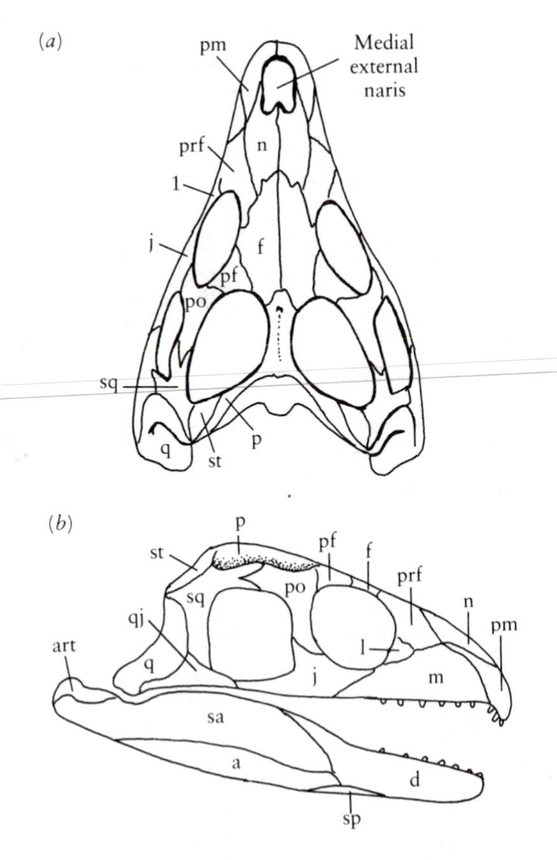

Figure 13-8. SKULL OF THE LOWER TRIASSIC RHYNCHOSAUR *MESOSUCHUS*, ABOUT 5 CENTIMETERS LONG. (*a*) Dorsal and (*b*) lateral views. The median external naris is a hallmark of rhynchosaurs. This primitive genus retains teeth in the premaxilla. Abbreviations as in Figure 8-3. *Drawings based on Broom, 1913, and photographs of specimen.*

It was long thought that rhynchosaurs were closely related to modern sphenodontids on the basis of general similarities of the skull and dentition. The common presence of primitive features such as the lower temporal bar only points to their common origin among early diapsids. Although the dentition appears to be vaguely similar, it is fundamentally different. Sphenodontids have only a single row of acrodont teeth in the maxilla, but rhynchosaurs have multiple rows of teeth set in sockets. Sphenodontids have a second row of teeth in the palatine, but this bone is edentulous in the rhynchosaurs. What appear to be long premaxillary teeth in rhynchosaurs are actually processes from the premaxillary bones. Sphenodontids have true premaxillary teeth.

THECODONTIA

The protorosaurs, trilophosaurids, and rhynchosaurs are limited primarily to the Triassic. The archosaurs show their first major radiation in that period but continue to expand throughout the Mesozoic. We include most Triassic archosaurs in a single order, the Thecodontia, a name that is based on the fact that the teeth are set in sockets. (It should be noted that this character is common to other archosauromorphs and has evolved convergently in mosasaurs among the lepidosaurs and among the mammal-like reptiles.) Thecodonts are unique among early archosauromorphs in having an antorbital fenestra, a character that unites them with later archosaurs. Additional characters that distinguish them from other archosauromorph groups include the retention of a complete lower temporal bar, a relatively short neck, and a single row of conical teeth in the premaxilla as well as in the maxilla and dentary.

Thecodonts were a dominant assemblage with a worldwide distribution throughout the Triassic. Their fossil record is extensive, and they have been the subject of intensive study by many authors. Nevertheless, the interrelationships within the assemblage are subject to continuing debate. They have customarily been grouped in four suborders: (1) the Proterosuchia, a primitive stock that may include the ancestors of all other thecodonts; (2) the Pseudosuchia, an assemblage of progressive forms including the probable ancestors of the dinosaurs; and (3 and 4) two specialized groups that were clearly distinct from the rest, the crocodilelike phytosaurs and the heavily armored Aetosauria, which were almost certainly herbivorous.

Within the last 15 years it has become apparent that the animals termed pseudosuchians are a heterogenous group, some of which may be related to the crocodiles and others to the dinosaurs. Various relationships between the proterosuchians and more advanced groups have been proposed (Romer, 1972b; Sill, 1974; Thulborn, 1982; Chatterjee, 1982; Bonaparte, 1984).

At least 10 families of thecodonts can be recognized. Although many well-preserved specimens are known, the fossil record is still very incomplete. A variety of poorly known forms may connect these families, but their interrelationships have not been clearly demonstrated. Much needs to be learned before we can establish a consistent phylogeny. For this reason the various families will be considered separately.

PROTEROSUCHIDAE

The proterosuchid thecodonts are unquestionably the most primitive archosaurs. The oldest known form is *Archosaurus,* from the Upper Permian of Russia, but the first adequately known genus is *Chasmatosaurus* [*Proterosuchus*] from the Lower Triassic of southern Africa and China, which Cruickshank (1972) recently reviewed. Proterosuchids also occur in India and possibly Australia. The skull is characteristic of archosaurs in the presence of a large antorbital opening that is surrounded by the maxilla, lacrimal, and jugal. Like the suborbital fenestra, this opening may have developed to lighten the skull and more effectively distribute mechanical forces, although it has also been suggested as accommodating a gland or permitting the expansion of the pterygoideus musculature (Walker, 1964) (Figure 13-10).

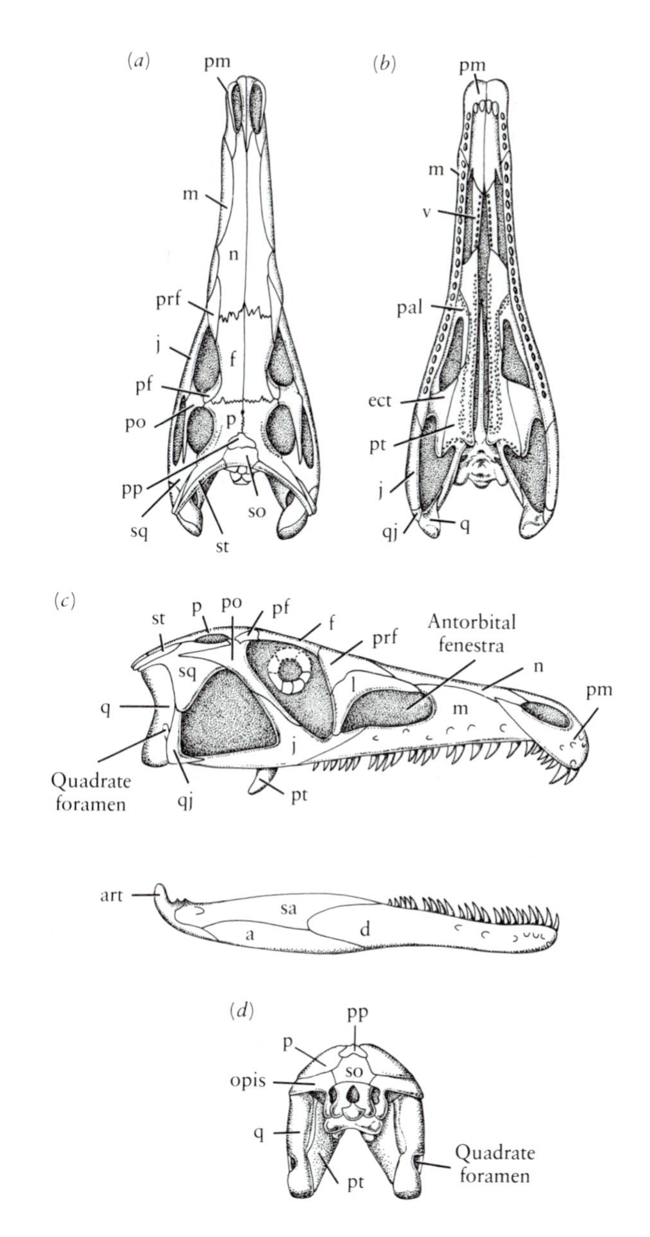

Figure 13-10. SKULL OF THE PRIMITIVE ARCHOSAUR *CHASMATOSAURUS* [*PROTEROSUCHUS*] FROM THE LOWER TRIASSIC OF SOUTH AFRICA. (*a*) Dorsal, (*b*) palatal, (*c*) lateral, including lower jaw, and (*d*) occipital views. Approximately 15 centimeters long. Abbreviations as in Figure 8-3. *From Cruickshank, 1972.*

The skull is characterized primarily by the possession of primitive features. The supratemporal and postparietal bones of early diapsids are retained, as is the pineal opening, and the pterygoid still bears teeth along a well-defined transverse flange. The marginal teeth are set in very shallow sockets.

As restored by Charig and Sues (1976), the body has a crocodiloid form. There are seven cervical vertebrae, but they are not greatly elongate. As in later archosaurs, the trunk vertebrae have long transverse processes. The shoulder girdle is clearly divided between the scapula and coracoid, in contrast with the unity in primitive diapsids and early lepidosauromorphs. These bones retain the primitive outline that we see in eosuchians. The humerus is primitive in having very large articulating surfaces at right angles with one another. The carpus is poorly ossified and appears to retain only two rows of elements, without a pisiform. Five digits are retained, but the phalangeal count is not known. The pelvis retains the primitive configuration of early diapsids without a trace of thyroid fenestration. The tarsus and pes (see Figure 13-3) are nearly identical with those of primitive protorosaurs, rhynchosaurs, and *Trilophosaurus*.

Among the currently known forms, we can clearly differentiate the family Proterosuchidae from all other thecodonts based on its retention of many primitive characters. These features do not preclude members of the family from being ancestral to any or all of the later

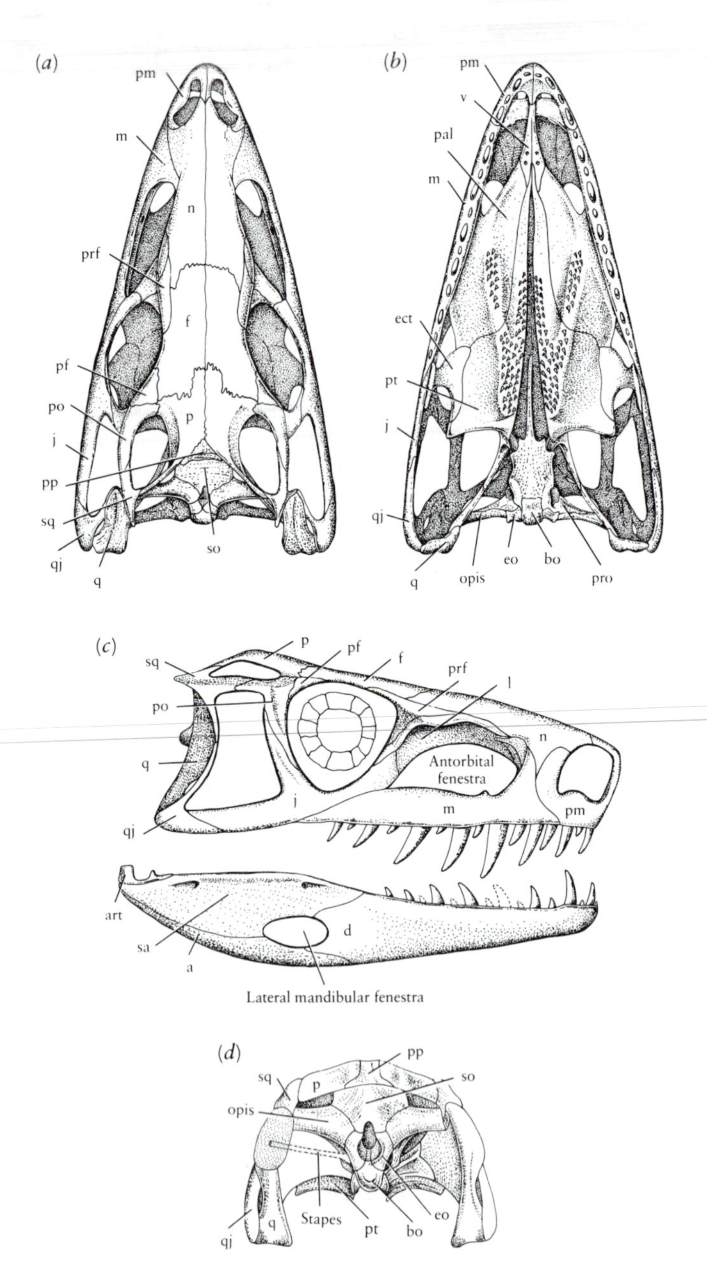

Figure 13-11. SKULL OF THE GRACILE THECODONT *EUPARKERIA*. (*a*) Dorsal, (*b*) palatal, (*c*) lateral, and (*d*) occipital views. About 10 centimeters long. Abbreviations as in Figure 8-3. *From Ewer, 1965.*

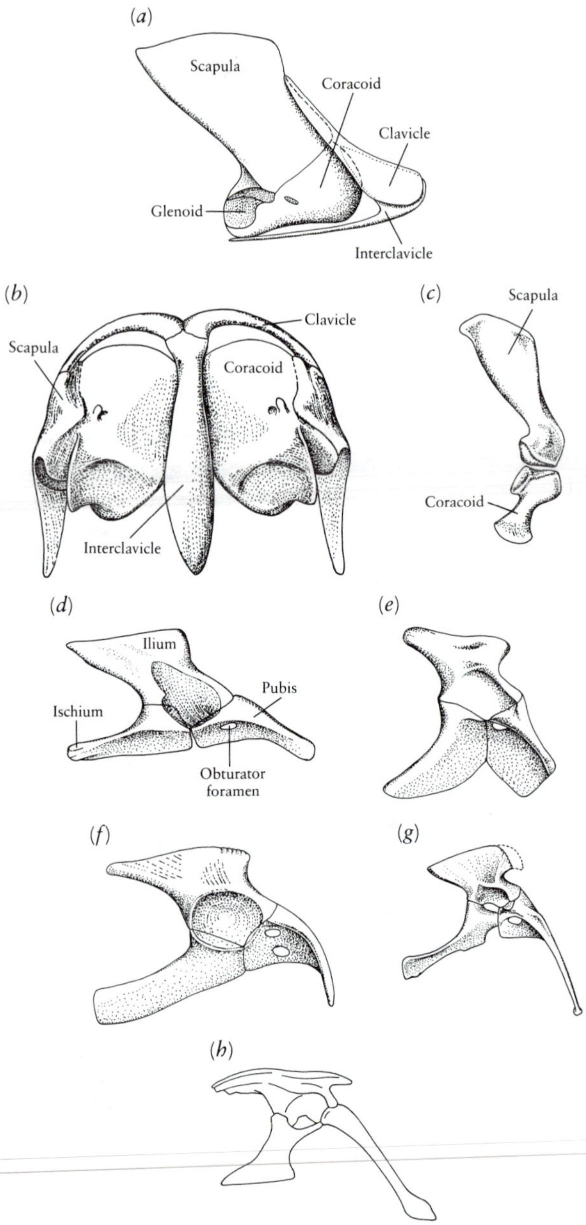

Figure 13-12. PECTORAL AND PELVIC GIRDLES OF PRIMITIVE ARCHOSAURS. (*a*) Pectoral girdle of *Chasmatosaurus* [*Proterosuchus*] in lateral view. In general, it resembles the shoulder girdle of eosuchians. The cleithrum is lost, but the clavicle and interclavicle remain large. The scapula and coracoid are not coossified in adults. *From Cruickshank, 1972.* (*b*) Ventral view of the shoulder girdle of the aetosaur *Stagonolepis*. *From Walker, 1961.* (*c*) Lateral view of the shoulder girdle of the protosuchian crocodile *Nothochampsa* [*Orthosuchus*] from the Lower Jurassic. The coracoid is clearly distinct and elongated, as in modern crocodiles. *From Nash, 1975.* (*d*) Pelvis of the proterosuchian *Chasmatosaurus*. This genus retains the flat puboischiadic plate common to primitive diapsids. *From Cruickshank, 1972.* (*e*) Pelvis of the phytosaur *Parasuchus*. A small thyroid fenestra has developed, but the pelvis remains primitive in the limited extension of the pubis and ischium. *From Chatterjee, 1978.* (*f*) Pelvis of *Euparkeria*. Both the pubis and ischium have become elongate. *From Ewer, 1965.* (*g*) Pelvis of the advanced "pseudosuchian" *Ornithosuchus*, in which the pubis and ischium approach the condition in dinosaurs. *From Walker, 1964.* (*h*) Pelvis of the Upper Triassic crocodile *Protosuchus*. As in modern crocodiles, the pubis is largely excluded from the acetabulum. *From Colbert and Mook, 1951.*

thecodonts, but they provide no basis for demonstrating specific relationships.

The adequately known proterosuchid genera are also differentiated by at least one specialized character, the marked down-turning of the premaxilla, which may preclude their occupying an ancestral position. Their very large size and robust limbs are further specializations that preclude close affinities with most later thecodonts.

All other thecodonts are advanced above the level of the proterosuchids in having a lateral mandibular fenestra in the lower jaw (Figure 13-11), in the loss of teeth from the transverse flange of the pterygoid, in the presence of dermal armor along the vertebral column, and in having at least some degree of specialization in the pelvis and rear limb (Figure 13-12).

The head of the femur in proterosuchids is terminal, and the bone was probably directed almost horizontally from the acetabulum, as in primitive tetrapods. This character is associated with the flattened, nearly horizontal puboischiadic plate, from which the major muscles that protracted and retracted the femur would have been oriented nearly horizontally. As the femur assumed a more vertical orientation in later thecodonts, the head became inclined medially toward the acetabulum and the pubis and ischium extend anteroventrally and posteroventrally to form a triradiate structure so that the associated muscle could exert a more effective force on the femur.

Charig (1972) has used the terms sprawling, semi-improved, and fully improved to describe the orientation of the rear limb in archosaurs. Only the proterosuchids retain the sprawling condition. Most other thecodonts and crocodiles illustrate the semi-improved posture, and the dinosaurs and their descendants, the birds, typify the fully improved stance.

As the limb is held more erect and comes to move in a more nearly parasagittal plane, the foot assumes a more symmetrical appearance with the third, rather than the fourth, digit becoming the longest.

The numerous derived thecodont families show different degrees of advancement and different structural details that are associated with attaining an upright posture. Two more advanced families accompany the Proterosuchidae in the Lower Triassic, the Erythrosuchidae and the Euparkeriidae.

ERYTHROSUCHIDAE

Erythrosuchids were first described from deposits in southern Africa and have since been recognized in China, Western Russia, Europe, and possibly North and South America. Erythrosuchids were the largest terrestrial vertebrates in the Lower Triassic, with a skull that was almost 1 meter long. They were heavy, quadrupedal animals that reached lengths of up to 5 meters. The glenoid appears to open more ventrally and the pelvis has initiated a triradiate structure that indicates that the limbs were held somewhat more vertically than in *Chasmatosaurus,* although the head of the femur does not appear to be inturned and the fourth digit of the pes is still the longest. The articulating surfaces of the limb bones and the carpals and tarsals are very poorly ossified, which has been attributed to semiaquatic adaptation.

EUPARKERIA AND THE ANKLE JOINT OF THECODONTS

Euparkeria, which occurs in a single locality in southern Africa, and three genera from China represent a further, very distinct lineage of early Triassic thecodonts. *Euparkeria* differs significantly from both proterosuchids and erythrosuchids in body form (see Figure 13-2). It is little more than one-half meter in length, including a long tail. The limbs are slender and it is usually illustrated in a bipedal pose (Ewer, 1965). *Euparkeria* is characterized by the presence of a dorsal row of dermal ossicles that run along the vertebral column of the trunk and tail, as is the case in many later thecodonts.

Although the body form is very different from that of *Chasmatosaurus,* the skull shows a fundamentally similar pattern (Figure 13-11). However, it is advanced in some features, including the loss of the pineal opening and the row of teeth along the transverse flange of the pterygoid. The teeth are sharp, compressed blades with serrated edges. The rear limbs of *Euparkeria* are one and one-half times as long as the forelimbs, a ratio that approaches that of some bipedal lizards, but the joint surfaces and the configuration of the femur and tarsus are not specialized for bipedal locomotion. Like modern lizards such as *Basiliscus* and *Crotaphytus, Euparkeria* may have been facultatively bipedal. The trunk is somewhat shortened with only 22 presacral vertebrae; the tail is long and would have served as an effective counterbalance if the animal were supported by its rear limbs.

Euparkeria probably did not evolve from any of the known proterosuchians. A single specimen, which Carroll (1976a) described from the Upper Permian of southern Africa, may indicate the earlier appearance of small-bodied thecodonts. *Heleosaurus,* like *Euparkeria,* has dermal armor along the vertebral column and a small number of serrated, bladelike teeth. Unlike contemporary eosuchians, it lacks a sternum and has a relatively long neck. The nature of the girdles suggests that *Heleosaurus* had a sprawling posture, but the limbs are poorly known and its affinities with the archosaurs cannot be firmly demonstrated. If *Heleosaurus* is a primitive thecodont, we must consider the absence of armor in proterosuchids to be a specialization.

Possibly because of the small size of the body, the tarsus of *Euparkeria* is not highly ossified but shows a pattern that may be interpreted as being intermediate between that of primitive archosaurs on one hand and both advanced thecodonts and dinosaurs on the other (Figure 13-13*e*).

In proterosuchids and erythrosuchids, the ankle joint passes between the proximal and distal tarsals to form a **mesotarsal** joint. Dinosaurs also have a mesotarsal joint, although the configuration of the individual elements is greatly altered and the proximal bones are integrated with the tibia and fibula.

Most thecodonts other than the proterosuchids and erythrosuchids evolve a new hinge joint between the astragalus and calcaneum; the astragalus becomes more closely associated with the tibia and fibula, and the calcaneum underlies the astragalus and is associated with the foot. This pattern is also shared by the modern crocodiles and is termed a **crocodiloid tarsus.** Because the ankle joint is made up of one bone of the crus (the fibula) as well as the tarsus it is also termed a **crurotarsal** joint (Figure 13-13*j*, *k*).

Among crocodiles as well as ornithosuchid and rauisuchid thecodonts, the tuber of the calcaneum extends directly posteriorly and serves as a lever to ventroflex the foot. The fifth metatarsal is reduced and loses its primitive role as the primary lever to flex the foot. In other thecodonts, including the generally more primitive aetosaurs and phytosaurs, the crurotarsal joint is developed but the fifth metatarsal remains large and may have shared the role of flexing the foot with the calcaneal tuber.

As Krebs (1963a) and Thulborn (1982) emphasized, a well-developed crurotarsal joint probably precludes the elaboration of a mesotarsal joint such as is characteristic of dinosaurs, which eliminates most thecodonts from the

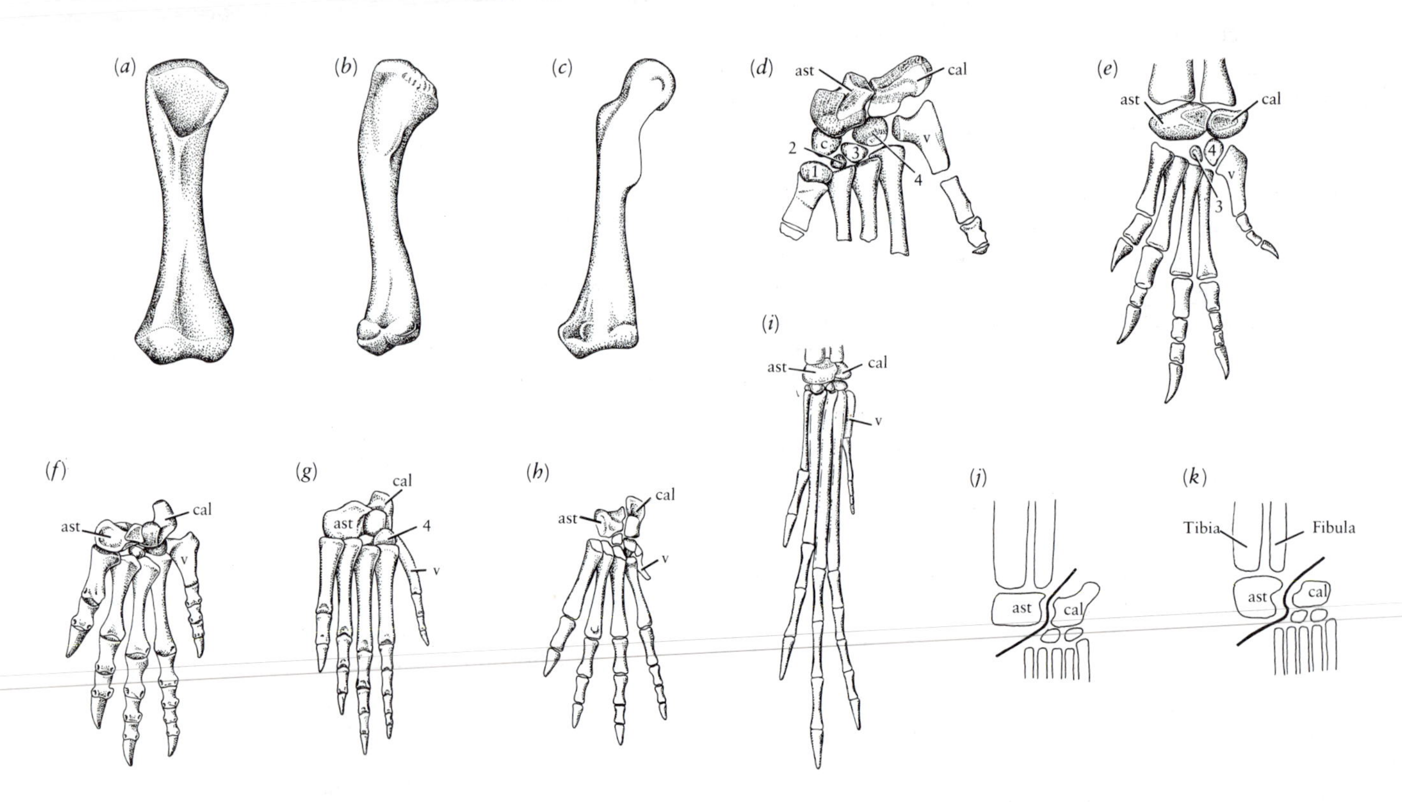

Figure 13-13. HIND LIMBS OF THECODONTS. (*a–c*) Femora, all in ventral view. (*a*) The proterosuchian *Chasmatosaurus* [*Proterosuchus*]. The shaft is straight, with the head terminal. These features, together with the retention of an intertrochanteric fossa and the ventral adductor crest, are characteristic of the primitive captorhinomorphs and early eosuchian reptiles. *From Cruickshank, 1972.* (*b*) The phytosaur *Parasuchus*. This genus illustrates a pattern that is common to more advanced thecodonts and crocodiles. The proximal head is offset from the shaft, the shaft is sigmoidal, and the intertrochanteric fossa and ventral ridge system are lost. *From Chatterjee, 1978.* (*c*) The advanced "pseudosuchian" *Riojasuchus*. The shaft is nearly straight, but the head is strongly angled medially, as is that of the human femur, to support the limb in a nearly vertical posture. *From Bonaparte, 1971.* (*d–k*) Tarsals and pes. (*d*) Tarsus of the proterosuchian *Chasmatosaurus* [*Proterosuchus*], which retains a primitive tarsal joint, with no clearly defined line of flexure. *From Carroll, 1976b.* (*e*) *Euparkeria*, in which the distal tarsals are reduced. Mobility is possible between the astragalus and calcaneum, but the main hinge was probably distal to these bones. *Original, based on specimens from the South African Museum and the Museum of Zoology, Cambridge University.* (*f*) The phytosaur *Parasuchus*, in which the heel of the calcaneum is directed obliquely posteriorly. The hooked fifth metatarsal serves as the primary level to ventroflex the foot. *From Chatterjee, 1978.* (*g*) The rauisuchid *Ticinosuchus*. The heel of the calcaneum is directed posteriorly, and the tarsal joint passes between the astragalus and the calcaneum as in crocodiles. *From Krebs, 1963b.* (*h*) The Lower Jurassic crocodile *Nothochampsa* [*Ornithosuchus*]. The astragalus functions as an extension of the crus, and the calcaneum is strongly integrated with the foot. *From Nash, 1975.* (*l*) *Lagosuchus*, a thecodont that may be closely related to the ancestry of dinosaurs. The calcaneum is reduced and lacks a heel. The major joint passes between the proximal and distal tarsals (a mesotarsal joint). The closest comparison is with *Euparkeria*. *From Bonaparte, 1978.* (*j*) Diagram of the crocodile-normal joint, in which a peg on the astragalus fits into a socket in the calcaneum. (*k*) The crocodile-reverse tarsus, in which a process on the calcaneum fits into a socket in the astragalus. Abbreviations as in Figure 13-3.

role of possible dinosaur ancestors. *Euparkeria* is exceptional in retaining a mesotarsal joint while developing a more symmetrical foot with a long third digit.

Most thecodonts are middle and late Triassic in age, during which time they made up a large portion of the reptilian fauna. They are especially well known in South America (Bonaparte, 1971, 1978; Romer, 1972b).

PROTEROCHAMPSIDAE

Many genera from the Middle and early Upper Triassic of South America appear to occupy a relatively primitive position among the thecodonts, although they are advanced above the level of *Chasmatosaurus* and *Erythrosuchus*. *Cerritosaurus, Chanaresuchus, Gualosuchus,* and *Proterochampsa* are grouped together in the family Proterochampsidae, which is characterized by having a relatively broad but low skull with small upper temporal openings that are visible primarily in dorsal view. Sill (1967) suggests that they are related to crocodiles, while Walker (1968) relates them to phytosaurs and others propose that they are part of the primitive proterosuchian radiation (Romer, 1972b; Bonaparte, 1982). None of these relationships are adequately substantiated. *Chanaresuchus* is specialized in the reduction of the fifth digit to a small remnant. The tarsus remains poorly known. The pelvis is very primitive in the platelike pubis and ischium.

PHYTOSAURS (PARASUCHIA)

Phytosaurs were abundant in the Upper Triassic of Europe, India, and western North America. They resemble modern crocodiles in their body form and probable way of life, but the external nares are situated far back on the long snout (Figure 13-14). The openings are typically elevated above the surrounding bones so that they could protrude from the water when the remainder of the head was submerged. The palate is arched behind the internal nares, and a soft secondary palate probably separated the air passage from the mouth. The long snout may be correlated with feeding in the water, as in modern crocodiles such as the gavial. Their carnivorous habits are clearly demonstrated in specimens that Chatterjee (1978) described, in which a variety of other reptiles are found among the stomach contents.

The lateral wall of the braincase is formed by a large laterosphenoid (or pleurosphenoid) anterior to the proötic (Figure 13-14). This bone is also present in aetosaurs and crocodiles and may be characteristic of all archosaurs, although it has not been described in ornithosuchids or rauisuchids. It is definitely not ossified in *Euparkeria,* but it is present in *Erythrosuchus* (Cruickshank, 1970).

The limb proportions resemble those of crocodiles, although their detailed anatomy is much more primitive. The pubis and ischium are still platelike. The astragalus and calcaneum have a crocodilianlike relationship, but the fifth metatarsal is hooked and the tuber of the calcaneum still faces laterally rather than posteriorly, which indicates the retention of the primitive leverage system. The trunk and tail were covered by an extensive dermal armor.

The six or seven recognized genera are very similar to one another in these and other skeletal features. The skull is clearly divergent from all other thecodont groups, but the postcranial skeleton suggests that phytosaurs evolved from a very primitive level that is comparable with the known Lower Triassic thecodonts. Surprisingly, there is no definite evidence of phytosaurs below the Upper Triassic. A single skull of a clearly phytosaurian pattern, *Mesorhinosuchus,* has been described as coming from the Lower Triassic, but there is continuing controversy over its actual age (Gregory, 1962; Westphal, 1976; Chatterjee, 1978). The specimen is now lost.

The phytosaurs were widely distributed in the northern continents and are represented by several specimens from Madagascar, but there is no record in Africa south of the Sahara or in South America or Australia.

AETOSAURS

Most thecodonts were clearly carnivorous with large, piercing teeth. The only exceptions are the aetosaurs, a distinct group that is known only from the Upper Triassic. They have small, leaf-shaped teeth, which suggests a herbivorous diet. Teeth were missing from the front of the snout and the lower jaw (Figure 13-15). Aetosaurs are limited to Europe (*Aetosaurus* and *Stagonolepis*), North America (*Typothorax, Stegomus,* and *Desmatosuchus*), and South America (*Aetosauroides, Argentinosuchus,* and *Neoaetosauroides*).

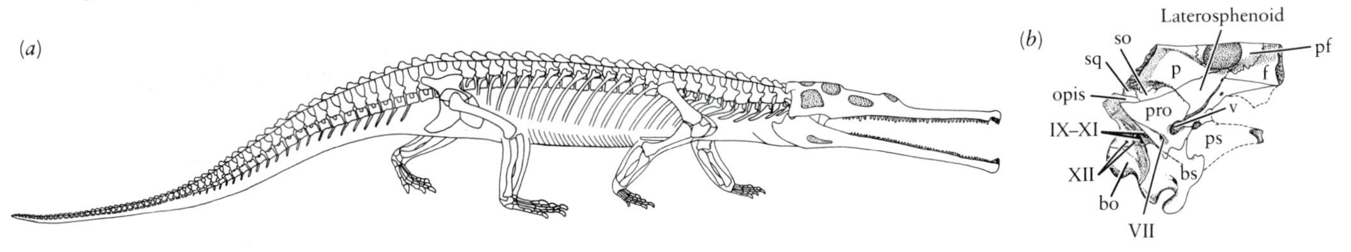

Figure 13-14. (*a*) Skeleton of the phytosaur *Parasuchus,* 3 meters long. (*b*) Braincase of *Parasuchus,* showing the ossified laterosphenoid. Abbreviations as in Figure 8-3. *From Chatterjee, 1978.*

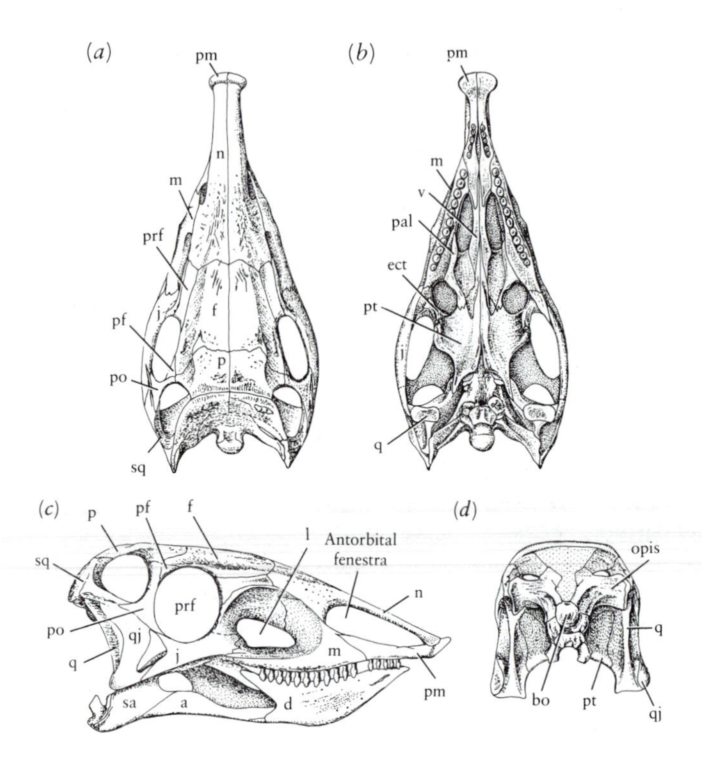

Figure 13-15. SKULL OF THE AETOSAUR *STAGONOLEPIS.* (*a*) Dorsal, (*b*) palatal, (*c*) lateral, and (*d*) occipital views. Approximately 25 centimeters long. Aetosaurs are the only group of herbivorous thecodonts. Abbreviations as in Figure 8-3. *From Walker, 1961.*

The head is small relative to the body, especially in the larger North American genera. The lower jaw has a shape that is strikingly similar to that of a lady's slipper and has a particularly large lateral fenestra.

The pelvis appears advanced in the great ventral projection of the pubis. The hands and feet resemble those of phytosaurs in the retention of primitive features. Aetosaurs have a crocodiloid tarsus with a particularly large calcaneal tuber (Walker, 1961). The body was extremely heavily armored, with large quadrangular plates along the back that extended down the sides, surrounded the tail, and covered the abdomen (Figure 13-16).

No fossils link aetosaurs with other thecodont groups.

RAUISUCHIDAE

The rauisuchids are the largest of the Middle and Upper Triassic thecodonts, reaching up to 6 meters long. Their footprints and skeletal evidence show that they were unquestionably quadrupedal. One of the most completely known forms, and also one of the oldest genera, is *Ticinosuchus* (Krebs, 1963b) from the early middle Triassic of Switzerland (Figure 13-17). The group is best represented in the Middle and Upper Triassic of South America but is also known from East Africa and North America.

The ankle and foot have advanced to the pattern that we find in modern crocodiles. The tuber of the calcaneum is oriented posteriorly to form a lever to which the tendon of the gastrocnemius was probably attached. The fifth metatarsal and, indeed, the entire fifth digit are reduced and probably have lost their role as a lever. The use of the calcaneal tuber as the primary lever may have transmitted the force to the foot more symmetrically as the limb was drawn into a more vertical orientation.

The pubis and ischium are extended obliquely anteriorly and posteriorly and a third pair of sacral ribs is incorporated in later members of the family.

Bonaparte (1984) discussed the function of the rear limbs in rauisuchids, which evidently attained a vertical posture independently of the lineage that led to dinosaurs. The femur retains a primitive configuration with an only slightly inturned head, but the acetabulum is reoriented, with the iliac portion extending laterally and nearly horizontally over the end of the femur (Figure 13-18). Thus, the weight of the body is supported directly by the vertical shaft of the limb.

The rauisuchids' armor consists of two rows of small plates that extend along the trunk with a single row above and below the tail. The individual pieces are linked by a narrow anterior process that fits into a groove beneath the preceding plate.

Both Sill (1974) and Bonaparte (1984) suggest that rauisuchids are closely related to the erythrosuchids, and Bonaparte places them in the same infraorder. Unfortu-

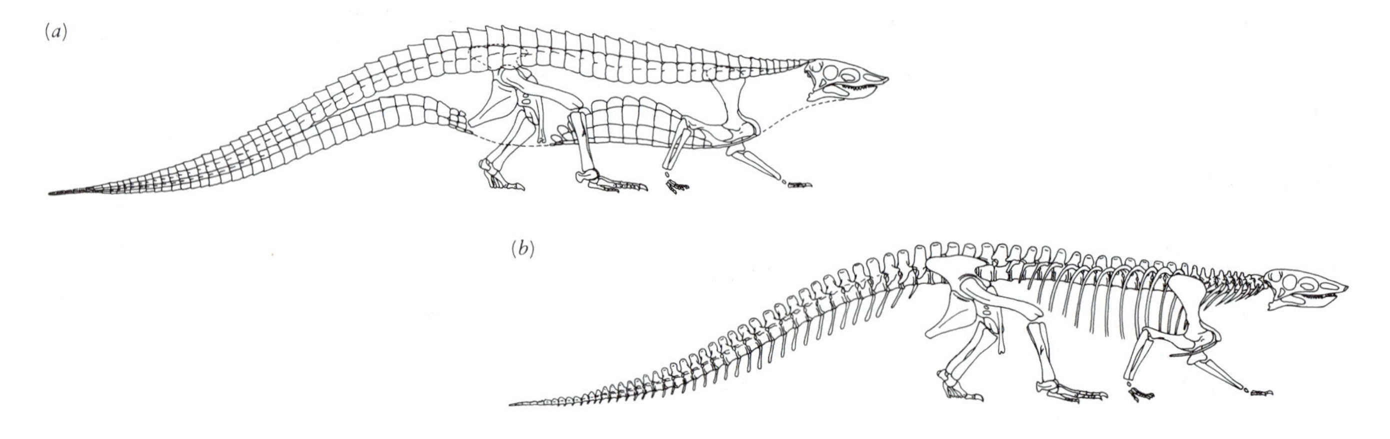

Figure 13-16. SKELETON OF THE AETOSAUR *STAGONOLEPIS.* (*a*) With and (*b*) without armor. Approximately 3 meters long. *From Walker, 1961.*

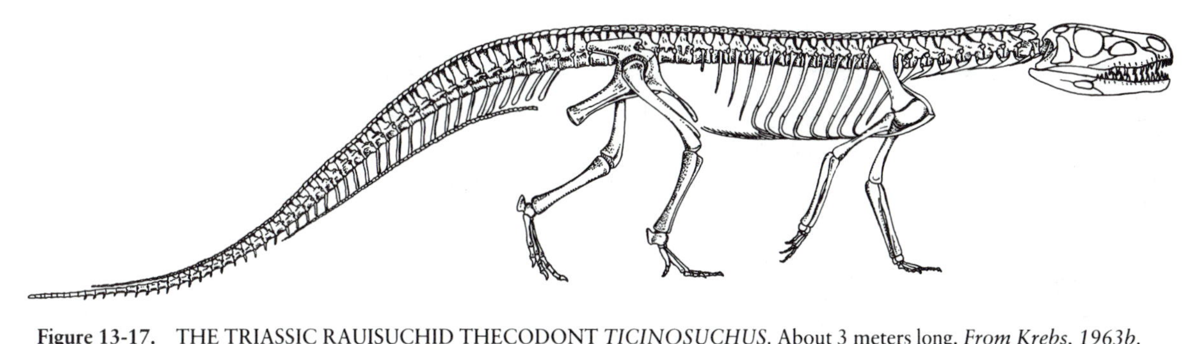

Figure 13-17. THE TRIASSIC RAUISUCHID THECODONT *TICINOSUCHUS*. About 3 meters long. *From Krebs, 1963b.*

nately, possible intermediate forms from the uppermost Lower Triassic and lowermost Middle Triassic are incompletely known, and the adequately known genera are quite distinct in their level of evolution. The only specific derived feature that has been cited as uniting these forms is the presence of an accessory antorbital fenestra or slitlike opening between the premaxilla and maxilla in some genera of both groups. Direct comparison between rauisuchids and erythrosuchids is complicated by the specializations of the well-known erythrosuchids that occurred in relationship to their great bulk and probably semiaquatic habits.

In a recent description of the late Triassic rauisuchoid *Postosuchus*, Chatterjee (1985) suggests that this genus might be close to the ancestry of the tyrannosaurid dinosaurs of the Cretaceous. The structure of the hind limb

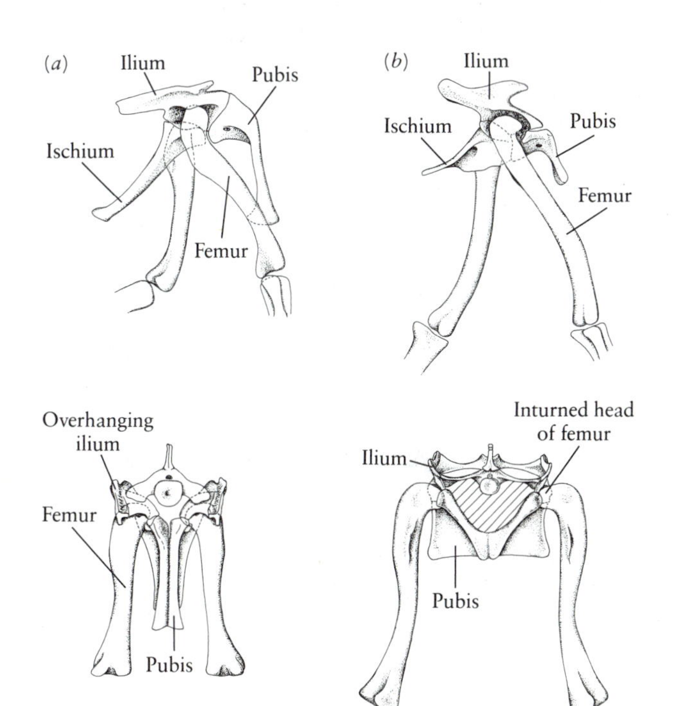

Figure 13-18. COMPARATIVE VIEWS OF THE PELVIS AND REAR LIMBS OF A RAUISUCHID AND A LAGOSUCHID. (*a*) The iliac surface of the acetabulum of the rauisuchid *Saurosuchus* is nearly horizontal, and the articulating surface of the femur is nearly terminal. (*b*) In *Lagerpeton*, the acetabulum is more nearly vertical and the head of the femur is inturned, as in primitive dinosaurs. *From Bonaparte, 1984.*

is much more primitive than that of any dinosaur and the hip and ankle joints both appear to be specialized in a different direction from the pattern of the dinosaur orders.

ORNITHOSUCHIDAE

The concept of the pseudosuchians as lightly built, bipedal ancestors of the dinosaurs is based primarily on members of the family Ornithosuchidae, of which *Ornithosuchus* from the Upper Triassic of Scotland is the best-known genus (Walker, 1964). *Riojasuchus* and *Venaticosuchus* are found in beds of roughly comparable age in South America. *Saltoposuchus*, which has been illustrated as a typical pseudosuchian, is now allied with the primitive crocodiles that are discussed in the following section.

Ornithosuchus (Figures 13-19 and 13-20) was a relatively large animal, with adult specimens reaching 4 meters in length. The forelimb is about two-thirds the length of the rear. Although *Ornithosuchus* is frequently illustrated in a bipedal pose, it may have walked in a quadrupedal manner. The skull is large relative to the trunk, and the teeth are large, laterally compressed blades.

Walker included *Ornithosuchus* among the carnosaurian dinosaurs because of the similarity of the skull to genera such as *Tyrannosaurus*, but the pelvic girdle and rear limb are far more primitive. The acetabulum is only slightly open, although the pubis extends far ventrally. The femoral head is not sharply inturned, and the fourth trochanter is poorly developed. The tarsal joint is crurotarsal rather than mesotarsal, and the calcaneum retains a prominent heel.

There are three pairs of sacral ribs. The dermal armor consists of a paired row of scutes along the trunk and a single row above the tail as in rauisuchids, but they lack a peg-and-socket articulation.

The closest affinities of the ornithosuchids are thought to lie among the euparkeriids.

LAGOSUCHIDAE

The lagosuchids from the Middle Triassic of South America provide the strongest evidence of the link between thecodonts and dinosaurs. We know only two genera,

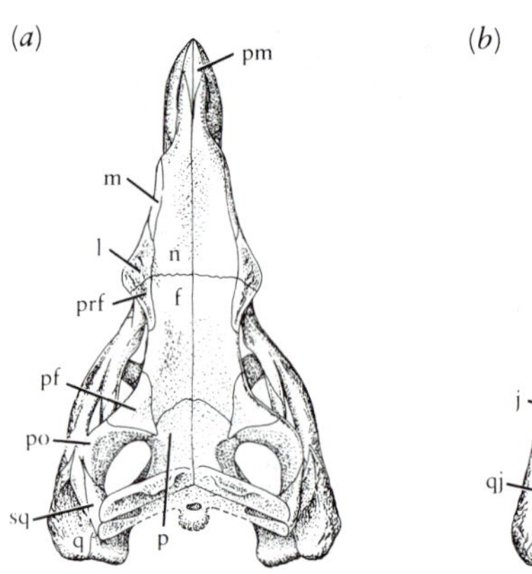

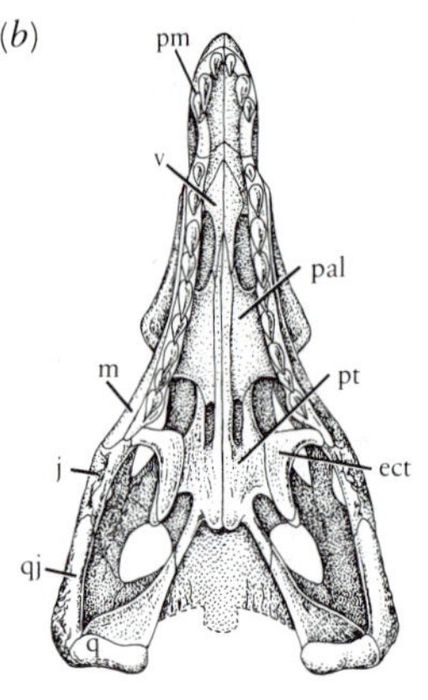

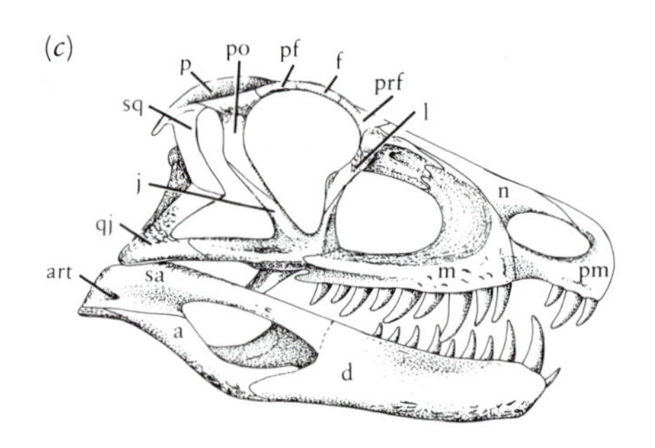

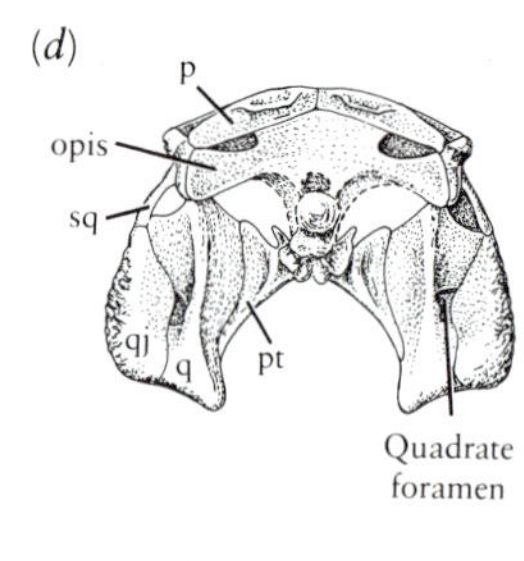

Figure 13-19. SKULL OF *ORNITHOSUCHUS*, AN ADVANCED CARNIVOROUS THECODONT. (*a*) Dorsal, (*b*) palatal, (*c*) lateral, and (*d*) occipital views. Approximately 25 centimeters long. Abbreviations as in Figure 8-3. *From Walker, 1964.*

Lagosuchus, for which most of the skeleton is described, and *Lagerpeton,* of which only the rear limb is known. *Lagosuchus* is about ⅓ meter in length with extremely long, slender limbs (Figure 13-21). The posterior limb is much longer than the anterior, and the tibia is longer than the femur. The relatively short length of the pubis and ischium in *Lagerpeton* is a primitive feature, but those of *Lagosuchus* are elongate. In both genera, the tarsus approaches that of dinosaurs in having a mesotarsal hinge between the astragalus and calcaneum proximally and the distal tarsals and metatarsals distally. The proximal tarsals are not yet closely integrated with the tibia and fibula. As in *Euparkeria,* the astragalus is considerably larger than the calcaneum and overlaps it slightly on the medial surface. The fifth metatarsal is not hooked.

The Lagosuchidae represent the most dinosaurlike of the known thecodonts. Their possible relationship with that group will be discussed in the next chapter, following consideration of a second major group of thecodont derivatives, the crocodiles.

CLASSIFICATION OF THECODONTS

Several authors have recently attempted to establish relationships among the thecodont families. Sill (1974), Brinkman (1981), and Chatterjee (1982) suggest that larger groups can be recognized on the basis of the configuration of the tarsus. Bonaparte (1971) recognized two different patterns among Middle and Upper Triassic thecodonts. In one, a process on the lateral surface of the astragalus fits into a recess on the medial surface of the calcaneum. This pattern is also recognized in the modern crocodile and is referred to as the **crocodile-normal** pattern. In the other pattern, the process is on the calcaneum and the recess is in the astragalus. This is referred to as the **crocodile-reverse** pattern. Crocodiles, phytosaurs, aetosaurs, and rauisuchids all have the crocodile-normal pattern and might share a common ancestry. Lagosuchids, ornithosuchids, and *Euparkeria* are all said to have the crocodile-reverse pattern. The tarsals in *Euparkeria* are not strongly ossified, and their configuration is not unequivically related to either pattern (see Figure 13-13*j*, *k*).

Although it may be true that neither pattern, when fully developed, is likely to have given rise to the other, both might have evolved separately in more than one lineage from a pattern like that exhibited by *Euparkeria.* The tarsals of erythrosuchids are too poorly ossified to support any particular affinities on this basis.

The characters by which each of the thecodont families are distinguished are far more obvious than those that support relationship between any of them.

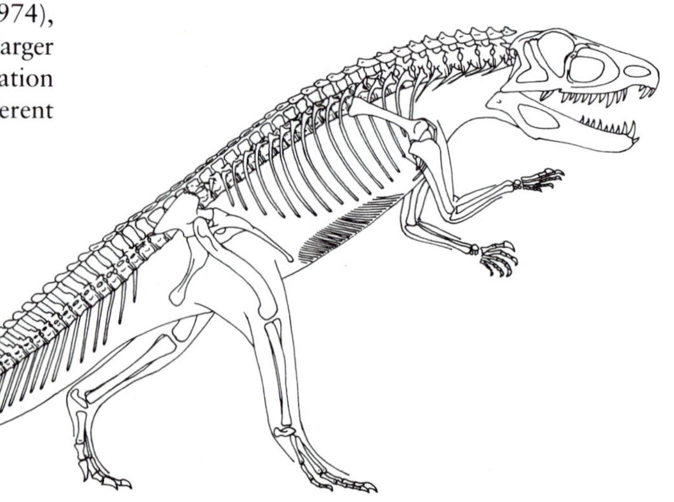

Figure 13-20. SKELETON OF *ORNITHOSUCHUS*. About 4 meters long. *From Walker, 1964.*

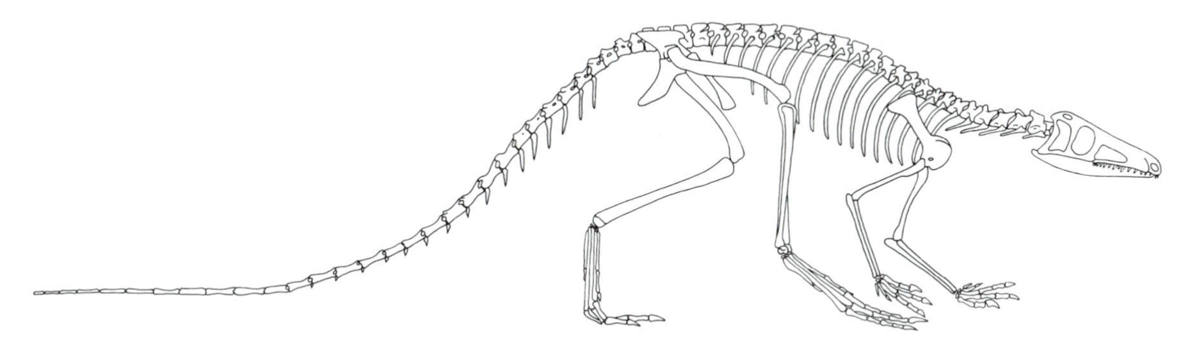

Figure 13-21. *LAGOSUCHUS*, A SMALL, LIGHTLY BUILT THECODONT FROM THE MIDDLE TRIASSIC OF SOUTH AMERICA. The structure of the rear limbs is similar to that of primitive dinosaurs. About 30 centimeters long. *From Bonaparte, 1978.*

CROCODYLIA

Crocodiles are the only surviving archosaurs. The history of this assemblage can be traced back into the Middle Triassic. Their fossil record is exceedingly rich and provides the potential for investigating many evolutionary problems. Crocodiles have undergone at least three major episodes of adaptive radiation that are the basis of a succession of suborders. We place the modern crocodiles, which have been known since the Upper Cretaceous, in the Eusuchia. We group an assemblage of mainly Jurassic and early Cretaceous genera as the Mesosuchia. The most primitive forms that are yet unquestionably crocodiles are placed in the suborder Protosuchia, which spans the late Triassic and earliest Jurassic.

It has long been thought that crocodiles arose from among the thecodonts, but their specific relationship with that group has not been established. It is now recognized that several genera that were previously classified as pseudosuchians share derived features with early crocodiles and should logically be included with that assemblage. These forms are placed in a distinct suborder, the Sphenosuchia.

SKELETAL ANATOMY OF CROCODILES

Since the late Triassic, crocodiles have been characterized by very consistent skeletal features. Iordansky (1973) and Langston (1973) provide useful reviews of the cranial anatomy. Although some structural details have changed over the last 180 to 200 million years, basic crocodilian features were established in the late Triassic and early Jurassic protosuchians.

The skull is massive and strongly buttressed throughout, in contrast with the fenestrate appearance of most thecodonts. The antorbital fenestrae are never large and are closed in all modern forms. The dorsal temporal openings are usually (but not always) small, and the configuration of the lateral openings is greatly altered by the anterior extension of the quadrate. The long snout is reinforced ventrally by a variably developed secondary palate that is more extensive in advanced crocodiles than in mammals. The presence of a secondary palate is customarily associated with separation of the nasal passages from the mouth, which may be especially important in aquatic forms. Langston argues that medial extensions of the premaxilla, maxilla, and palatine serve a primarily structural role to help support the elongate, flattened snout.

The occiput appears as a continuous plate of bone with no more than tiny posttemporal fossae. The prefrontals extend ventrally on either side of the midline to buttress the palate. Posteriorly, the palate and braincase are very strongly integrated so that there is not a trace of kinetism. The extremely well-reinforced skull appears to be specialized to resist forces generated by a very strong and rapid bite.

The function of the jaw muscles of modern species, as Schumacher (1973) described, may be applied to most fossil crocodiles that have a fundamentally similar skull shape. The largest of the muscles are the anterior and posterior pterygoideus. The fibers of these muscles are close to horizontal, and their force is greatest when the jaws are widely open. The anterior pterygoideus extends far anteriorly above the palate so that the long fibers can exert a strong constricting force over a wide range of jaw angles. The posterior pterygoideus wraps around the back of the jaw and attaches to the very long retroarticular process, which results in rapid jaw closure. The adductor fossa is angled posteriorly so that the adductor mandibulae posterior is also directed posteriorly rather than more nearly vertically as in thecodonts. Only the adductor mandibulae externus and pseudotemoporalis are vertically oriented. The resolved force of all the jaw muscles is obliquely anterodorsal. This force is resisted by the great hypertrophy of the quadrate in the same direction.

The quadrate has an extremely complex relationship with the other bones at the back of the skull. The primary head has contact with the proötic and laterosphenoid of the lateral braincase wall and the undersurface of the squamosal anterior to the opisthotic. The quadrate secondarily contacts the squamosal and opisthotic behind the middle ear cavity and the pterygoid and basisphenoid anteriorly and ventrally.

The proportions of the skull and the orientation of the quadrate in even the late Triassic ancestors of the modern crocodile genera suggest a feeding strategy that is distinct from other archosaurs and may explain their survival as contemporaries of the dinosaurs throughout the Mesozoic long after the extinction of the more dinosaurlike thecodonts. All the muscles insert near the back of the jaw to give a wide gape and rapid closure of the long jaws. Together with the massive jaw musculature, this arrangement implies that the late Triassic ancestors of the modern crocodile fed on large, powerful prey.

The complex pneumatization of nearly all of the skull bones is another highly distinctive character of crocodiles. There are two major systems of pneumatic ducts; one is elaborated from the eustachian tubes and the second is associated with the nasal passages (Iordansky, 1973).

The ducts associated with the eustachian tube develop ontogenetically by expansion of the diverticula of the middle ear cavity. Three passages open into the pharynx at the base of the basioccipital—the median and paired lateral eustachian ducts. All extend to the middle ear cavities and connect with sinuses in the supraoccipital, proötic, laterosphenoid, quadrate, and even into the articular bone of the lower jaw. Ducts that lead from the nasal passages ramify within the premaxilla, the maxilla, the palatine, and the pterygoid.

In his review of hearing in reptiles, Wever (1978) provides no explanation for the complexity of the pneumatic system in crocodiles. Crocodiles' ears do show a much greater sensitivity to airborne sounds than do those of other reptiles, and they compare well with those of birds and mammals.

Other characters that distinguish the skull of crocodiles from more primitive archosaurs include the low and flat skull table and the loss of the postfrontal, postparietal, and epipterygoid bones. The frontals and usually the parietals are fused at the midline.

Except for some marine genera, all crocodiles have 24 presacral vertebrae, 2 sacrals, and a tail with 30 to 40 segments. In modern crocodiles, the first 9 vertebrae are designated as cervicals since their ribs do not reach the sternum. The ribs associated with the atlas and axis are simple rods with slightly expanded heads. The next 6 have widely separated heads. The dorsal head is attached to the diapophysis at the end of the transverse process and the capitular head attaches to the parapophysis low on the centrum. Between them, they encircle the cervical artery. The bladelike shaft of the rib lies parallel to the vertebral column. The pointed anterior end lies beneath the posterior part of the preceding rib. This same pattern is evident in late Triassic and early Jurassic crocodiles. *Protosuchus,* from the Lower Jurassic, differs only in having a longer shaft on the eighth vertebra, which is intermediate in configuration between the other cervicals and the anterior trunk ribs.

In genera from the Triassic to the Recent, the diapophysis gradually rises starting with the eighth vertebra and reaches the level of the zygapophysis by about the eleventh or twelfth vertebra. The parapophysis rises as well and by the twelfth vertebra becomes associated with the base of the transverse process, where it occupies a position considerably medial to that of the parapophysis in the anterior vertebrae but becomes progressively more lateral in position posteriorly. The heads may become confluent at the end of the trunk.

The ribs of the most posterior trunk vertebrae become progressively shorter, and the more posterior are missing or fused to the transverse process. Only the last presacral is modified in *Protosuchus*. The last three trunk ribs are missing in gavials, and the last four or five are missing in crocodiles and alligators.

Among the thecodonts, there is little specialization of the pectoral girdle and forelimb. The dermal girdle remains primitive, with large clavicles and interclavicles. The scapula is high and relatively slim, and the coracoid is a small oval plate. The carpus is poorly known and apparently slow to ossify.

Among the early crocodiles, the clavicles are lost and the interclavicle is reduced to a longitudinal rod. There is no evidence of a sternum in thecodonts, but modern crocodiles have a complex cartilaginous structure that integrates the ends of the ribs, the base of the coracoids, and the interclavicle. The posteromedial portion of the structure is calcified in primitive crocodiles and resembles an extension of the interclavicle.

The coracoids extend posteriorly and ventrally. They tend to be elongate and somewhat resemble the coracoids of nothosaurs, but they are separated at the midline by the sternum. In modern crocodiles, the coracoids are approximately as long as the scapulae and are similarly shaped.

The changes in the shoulder girdle in crocodiles suggest that the forelimbs were more important in supporting the body than they were in the advanced thecodont groups, which may have been facultatively bipedal. The ventral and posterior extension of the coracoid would have provided a more effective angle for the muscles that retract the limb and lift the body. The loss of the clavicles in crocodiles, as in cursorial mammals, allows the glenoid to move anteriorly and posteriorly, which extends the length of the stride.

As in advanced thecodonts, the humerus has a distinct deltoid crest; the ulna and radius are long and slender. The carpus is significantly altered in the elongation of the ulnare and radiale and the retention or reelaboration of the pisiform, a bone that is only rarely preserved in other archosaurs. There are one or two distal carpals. The manus is not significantly different from that of advanced thecodonts.

The pelvic girdle and hind limbs of crocodiles are not substantially modified from the level of Middle Triassic thecodonts. The two sacral vertebrae are never fused to form a solid sacrum, and the blade of the ilium is not greatly expanded above the acetabulum. The acetabulum is perforate in some early forms but not in the modern genera. The most distinctive feature is the exclusion of the pubis from the acetabulum in all genera since the beginning of the Jurassic. The pubes remain large elements that extend forward and meet in a horizontally expanded plate. The ischium in modern crocodiles extends almost directly ventrally.

The femur has the gently sigmoidal shape of primitive thecodonts, with only a moderate inturning of the head and little development of trochanters. The tibia supports most of the weight of the body, and the fibula is slender. Both the tibia and fibula are supported primarily by the astragalus. The underlying calcaneum forms a stout lever to ventroflex the foot. In contrast with most thecodonts, the fifth digit is reduced to a splint of the metatarsal. This bone articulates with the fourth distal tarsal but lacks the distinct hooking of the head that is typical of the primitive thecodonts.

Long slender gastralia are retained behind the extensive sternal area. Most crocodiles have two rows of dermal plates that extend the length of the vertebral column dorsally, and some have dermal plates ventrally between the girdles as well.

With minor exceptions, this description holds not only for all genera of living crocodiles but for nearly all of the fossil groups back to the end of the Triassic. Like the lungfish, most of the skeletal specializations occurred in the earliest stages of crocodilian evolution, and the last 200 million years have seen only minor variations on a basic theme.

All modern crocodiles are amphibious, a pattern that has been characteristic of many members of the group since the early Jurassic. In contrast, the protosuchian crocodiles and the sphenosuchids had very long slim limbs and were almost certainly agile, terrestrial forms. A striking feature of crocodile evolution is that most of the specialized features that we associate with the modern aquatic genera were evolved in the late Triassic in a highly terrestrial assemblage.

SPHENOSUCHIA

The basic crocodilian pattern is well established in the Lower Jurassic protosuchians. The Upper Triassic genera that belong to the suborder Sphenosuchia are more primitive but still recognizably crocodilian. *Gracilisuchus* from the Middle Triassic of South America is the earliest genus that we may include within this group (Figure 13-22). *Gracilisuchus* was originally classified among the Ornithosuchidae (Romer, 1972a; Bonaparte, 1975). Brinkman (1981) pointed out that it differed from all members of that family in the configuration of the tarsus and the cheek region. He suggested that it might be related to *Sphenosuchus* and other members of the Sphenosuchia.

Gracilisuchus resembles the pattern of moderately advanced thecodonts in most skeletal features. The tarsus is crocodilian, with the lateral surface of the astragalus overriding the calcaneum. The fifth digit is reduced, and the proximal end of the fifth metatarsal does not retain a primitive "hooked" configuration. The pelvis has a long pubis that does not appear to contribute to the acetabulum. However, the coracoid is short. Unfortunately, the carpus and manus are not known. As in most thecodonts, the antorbital fenestra is large and surrounded by a broad depression.

The configuration of the cheek region strongly supports crocodilian affinities. The dorsal end of the quadrate is angled forward and is overhung by a broad rim of the squamosal. Unlike typical thecodonts, the squamosal does not extend ventrally along the quadrate. The parietals are fused posteriorly, and there is no pineal foramen. The pterygoids meet at the midline at the back of the palate. Romer assumed that they were fused to the braincase, but this fusion cannot be demonstrated in the available material. The front of the palate is not known, so there is no evidence of the formation of a secondary palate.

Dermal armor extends along the entire length of the vertebral column. There are roughly two units per segment. The elements of the two rows interdigitate at the midline. The cervical ribs show a striking resemblance to

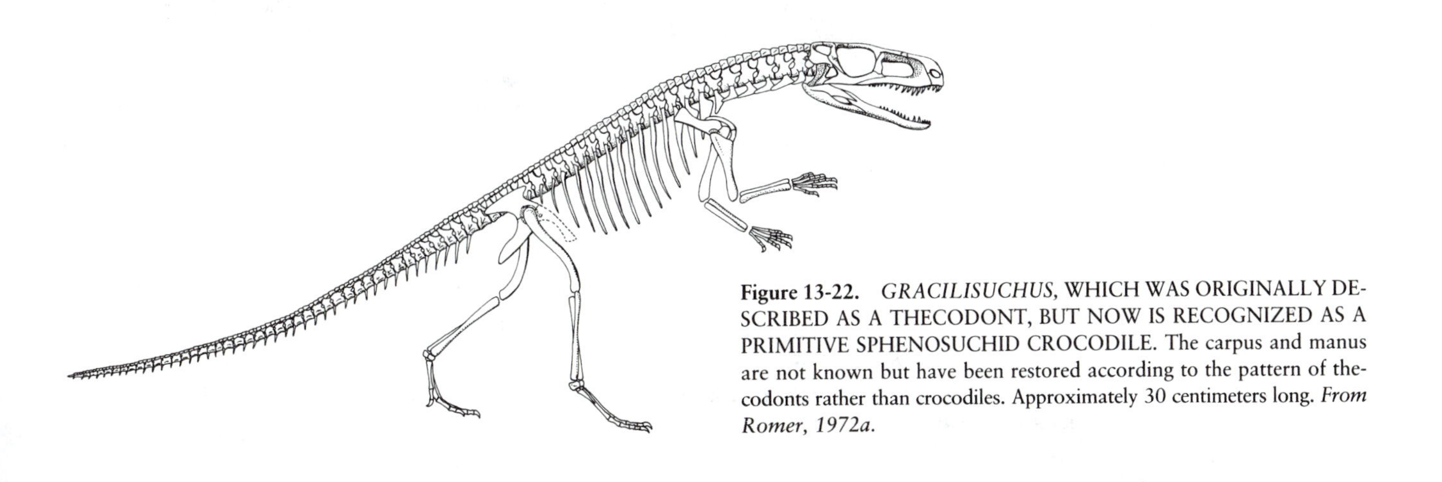

Figure 13-22. *GRACILISUCHUS*, WHICH WAS ORIGINALLY DESCRIBED AS A THECODONT, BUT NOW IS RECOGNIZED AS A PRIMITIVE SPHENOSUCHID CROCODILE. The carpus and manus are not known but have been restored according to the pattern of thecodonts rather than crocodiles. Approximately 30 centimeters long. *From Romer, 1972a.*

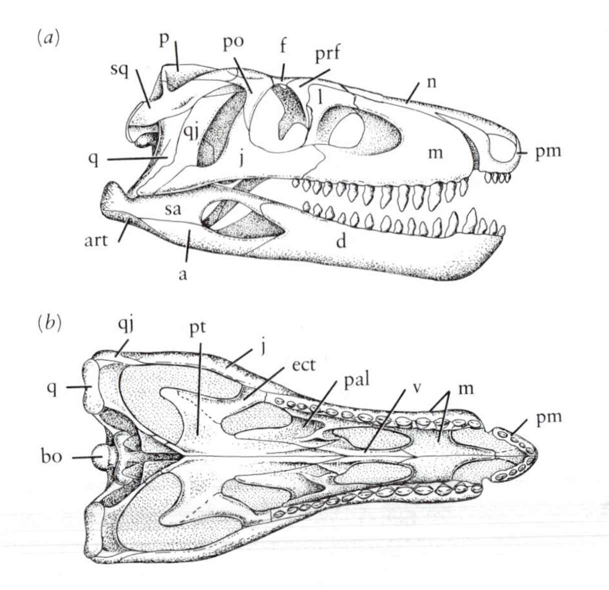

Figure 13-23. SKULL OF THE PRIMITIVE CROCODILE *SPHENOSUCHUS* FROM THE LOWER JURASSIC OF SOUTHERN AFRICA. (*a*) Lateral and (*b*) palatal views. Abbreviations as in Figure 8-3. *From Walker, 1972. Redrawn from* Nature. *Copyright © 1972; Macmillan Journals, Ltd.*

those of modern crocodiles, with two clearly separated heads and a short, posteriorly directed shaft that is overlapped anteriorly by the next preceding element. There are seven cervical vertebrae and a further one that bears ribs of intermediate character.

Despite some crocodilian features of the skull, most skeletal characteristics link *Gracilisuchus* with the thecodonts. However, it cannot be placed in any of the recognized thecodont families. The absence of a jugular process anterior to the orbit is a more primitive condition than that described in ornithosuchids, rauisuchids, or even the Lower Triassic genus *Euparkeria. Gracilisuchus* may represent a separate lineage that has evolved from near the base of the thecodont stock, above the level of proterosuchids and distinct from erythrosuchids, aetosaurs, or phytosaurs.

Most previously recognized members of the Sphenosuchia are from the Upper Triassic, including *Saltoposuchus* and *Terrestrisuchus* from Europe (Crush, 1984), *Pseudhesperosuchus* from South America (Bonaparte, 1971), and *Hesperosuchus* from North America (Colbert, 1952). *Sphenosuchus* is from the Lower Jurassic of southern Africa (Walker, 1968).

The skull of *Sphenosuchus* (Figure 13-23) shows a clearly crocodiloid appearance, with a low profile and elongate snout. The quadrate and quadratojugal slant forward, and the antorbital fenestra is reduced. The skull is akinetic, and the prefrontal extends ventrally to brace the palate. The parietals are united posteriorly, but the postorbital and postfrontal remain distinct. Crocodiloid pneumatization of the occipital area is initiated, and the cochlea of the inner ear is already long. As in later crocodiles, the laterosphenoid forms the lateral wall of the braincase anterior to the proötic.

The skull is more primitive than in later crocodiles in the absence of connections between the quadrate and the lateral wall of the braincase. The quadrate reaches the proötic anterior to the opisthotic. The upper temporal openings are long. The palate appears open as in thecodonts, although the internal nares are posterior, and the maxillae form a short secondary palate that is anterior to them.

The postcranial skeleton approaches the level of later crocodiles in the length of the coracoid, but the clavicle is retained. The long, anteriorly directed pubis contributes little to the acetabulum, which is perforate in the related genus *Terrestrisuchus*. As in modern genera, the carpus is elongate.

The sphenosuchids are clearly crocodiles but more primitive than the long-accepted members of this group. They appear divergent from the crocodilian habitus in their extremely light skeleton and presumably cursorial habits, especially in the genus *Terrestrisuchus* (Figure 13-24), but these features may be primitive ones that were inherited from the thecodonts.

PROTOSUCHIA

The Protosuchia retain long limbs and probably were basically terrestrial in habit, but they resemble modern croc-

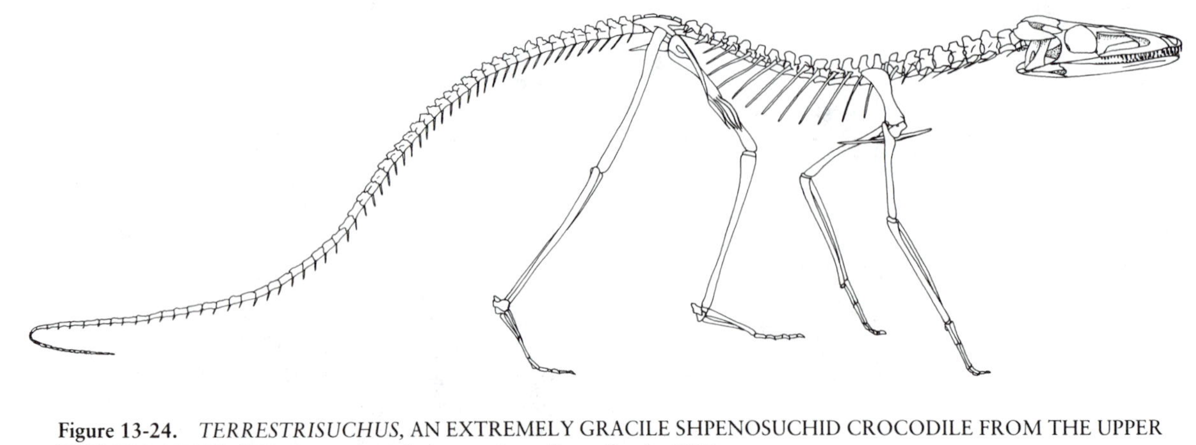

Figure 13-24. *TERRESTRISUCHUS*, AN EXTREMELY GRACILE SHPENOSUCHID CROCODILE FROM THE UPPER TRIASSIC OF EUROPE. Approximately ½ meter long. *From Crush, 1984.*

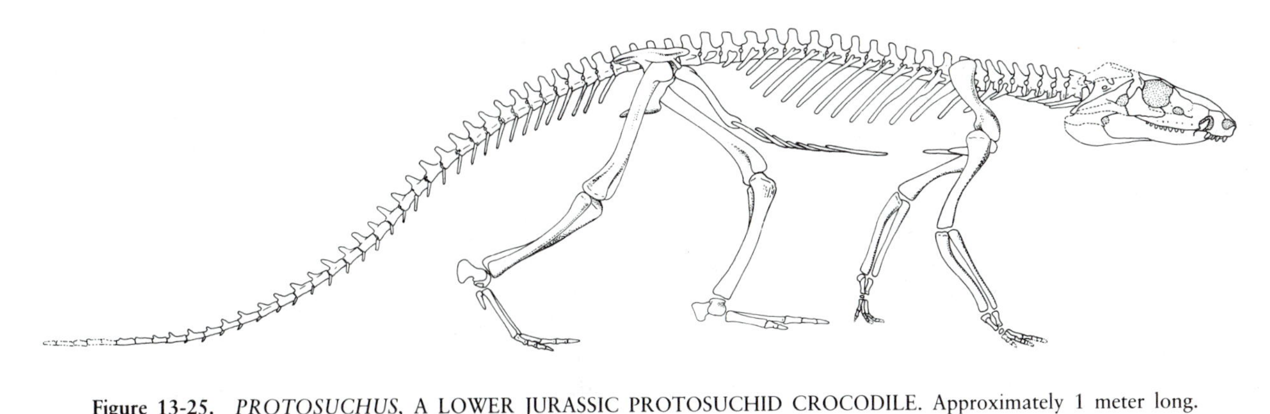

Figure 13-25. *PROTOSUCHUS*, A LOWER JURASSIC PROTOSUCHID CROCODILE. Approximately 1 meter long. *From Colbert and Mook, 1951. Skull after Crompton and Smith, 1980.*

odiles more closely than sphenosuchids in their general appearance and configuration of the skull (Figures 13-25 and 13-26). The upper temporal opening and antorbital fenestra are much reduced. The parietals are fused along the entire midline, and the postorbital and postfrontal form a single ossification. Like the sphenosuchids, a caninelike tooth in the lower jaw fits into a recess between the premaxilla and maxilla.

The palate is more platelike than that of sphenosuchians, and the palatine bone forms part of the secondary palate. In *Protosuchus* (Crompton and Smith, 1980), the pterygoids are flat and solidly fused along the midline, but in *Notochampsa* [*Orthosuchus*] (Nash, 1975), they are grooved, which suggests that the nasal passages extended between them to the level of the throat and were covered by a fleshy secondary palate. The quadrate is strongly integrated with the lateral wall of the braincase. In *Eopneumatosuchus* (Crompton and Smith, 1980), we see an essentially modern pattern of pneumatic ducts.

Postcranially, there is little that separates protosuchids from modern crocodiles except for the vertebrae, which are shallowly amphicoelous throughout the column. The fifth metatarsal is reduced to a splint, the coracoid is elongate although not quite to the extent of modern crocodiles, and the pubis is excluded from the acetabulum.

The protosuchids had a worldwide distribution with genera from South America (*Hemiprotosuchus*) (Bonaparte, 1971), southern Africa (*Notochampsa* [*Orthosuchus Erythrochampsa*] and *Pedeticosaurus*) (Nash, 1975), North America (*Stegomosuchus* and *Protosuchus*) (Colbert and Mook, 1951) and from China (*Dianosuchus*).

Based on the available fossil record, the protosuchians appear as a clearly distinct assemblage that was probably more terrestrial than modern crocodiles but less specialized for cursorial locomotion than the sphenosuchids.

It was long thought that the protosuchians were all Upper Triassic, which separated them clearly in time from the mesosuchians that appear in the upper portion of the Lower Jurassic. We now recognize that all the protosuchids except *Hemiprotosuchus* come from Lower Jurassic deposits (Olsen, McCune, and Thomson, 1982). Protosuchids are from older deposits than are the mesosuchians but, more importantly, these groups are distinguished by marked habitat differences.

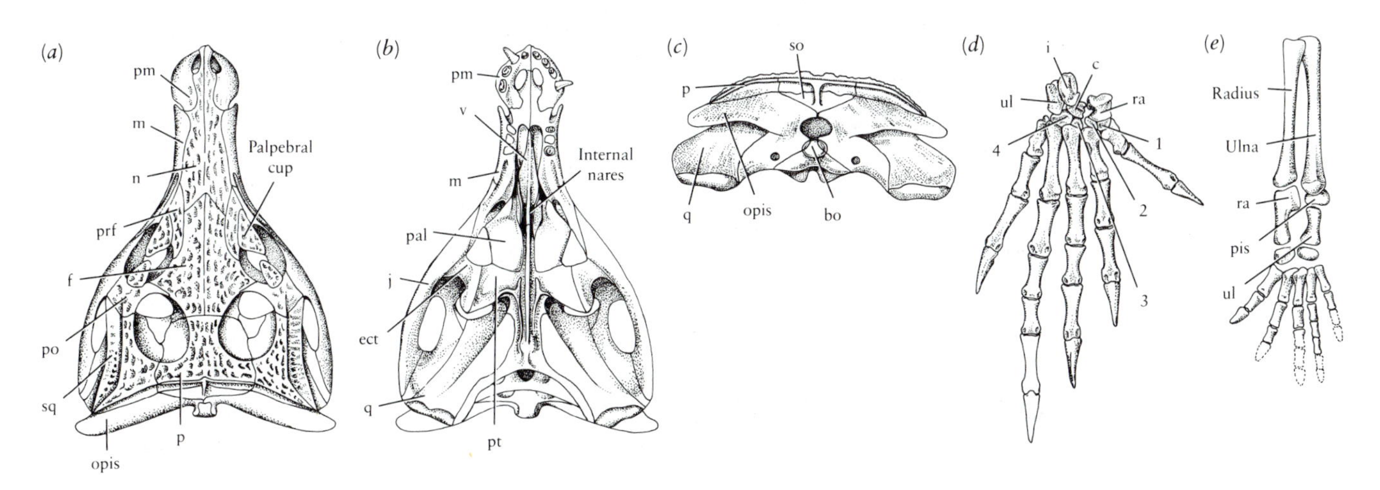

Figure 13-26. Skull of the Lower Jurassic Protosuchid crocodile *Nothochampsa* [*Orthosuchus*], 10 centimeters long. (*a*) Dorsal, (*b*) palatal, and (*c*) occipital views. *From Nash, 1975.* (*d*) Forelimb of the primitive archosauromorph *Trilophosaurus. From Gregory, 1945.* (*e*) Forelimb of *Nothochampsa. From Nash, 1975.* Abbreviations as in Figure 8-3, plus: c, centrale; i, intermedium; pis, pisiform; ra, radiale; ul, ulnare; 1–4, distal carpals.

All the protosuchians come from deposits that clearly indicate a terrestrial environment and are closely associated with the remains of dinosaurs. The earliest mesosuchians come from clearly marine deposits and are accompanied by ichthyosaurs and plesiosaurs. All are specialized for aquatic locomotion and feeding.

MESOSUCHIAN CROCODILES

Mesosuchians constitute the largest assemblage of crocodiles, including approximately 70 genera that occupy an intermediate position between protosuchians and eusuchians. They are advanced over the protosuchians in the posterior extension of the palatine bones that form the anterior margin of the internal nares at a point well behind the end of the tooth row. In contrast, the pterygoids form the anterior margin of the choana in eusuchians. The centra are still slightly amphicoelous in most mesosuchians, in contrast with the procoelous configuration of the eusuchians. Changing depositional environments throughout the Mesozoic result in an incomplete fossil record that makes it difficult to establish relationships among the approximately 16 recognized mesosuchian families (Buffetaut, 1982).

The transition between the primarily terrestrial protosuchians and the specialized marine mesosuchians may not have occurred before the late Liassic. Lower and upper Liassic deposits in Great Britain, Germany, and France record a similar fauna of ichthyosaurs and plesiosaurs. Crocodiles are almost unknown in the lower Liassic but are abundant and diverse in the upper Liassic.

AQUATIC MESOSUCHIANS

Four families of mesosuchians were predominantly or obligatorily aquatic. Two related families appear in the Upper Liassic, the Teleosauridae and the Metriorhynchidae. The teleosaurs are characterized by short forelimbs, which were only half the length of the hind limbs. This reduction corresponds with the limb reduction seen in various groups of aquatic reptiles that were discussed in Chapter 12. However, the elements of the limbs and girdles are only slightly altered from those of terrestrial crocodiles. The snout is greatly elongated, as in modern gavials, probably in relationship to feeding in the water. The teleosaurs retain a continuous covering of heavy armor.

The metriorhynchids, which continue into the Cretaceous, are the most specialized of all crocodiles and the only archosaurs that became highly adapted to life in the water. They are known primarily from Europe, with some representatives in North and South America. The forelimbs are transformed into paddles and, as in the ichthyosaurs, the vertebral column is bent ventrally to support a large caudal fin (Figure 13-27). The neck is one or two segments shorter than in modern crocodiles, and the total number of presacral vertebrae is increased to 26. Dermal armor is lost within the group. In teleosaurs and metriorhynchids, the upper temporal openings are much larger than in semiaquatic and terrestrial genera. Wide temporal openings may be associated with long jaw muscles and a very wide gape. The early metriorhynchid *Pelagosaurus*, which retains some dermal armor and has only a slightly downturned tail, provides a possible link between metriorhynchids and teleosaurs.

Other aquatic mesosuchians apparently had a separate origin. The pholidosaurids extend from the Upper Jurassic into the Upper Cretaceous. They have the long snout that is associated with aquatic feeding and the anterior end is bent sharply downward, but the upper temporal openings are somewhat smaller than in teleosaurs and metriorhynchids.

Dyrosaurids are known from the Upper Cretaceous into the Eocene. Elongation of the snout apparently developed separately within this group rather than as an inheritance from other aquatic forms. They are known primarily from north and west Africa, Saudi Arabia, and

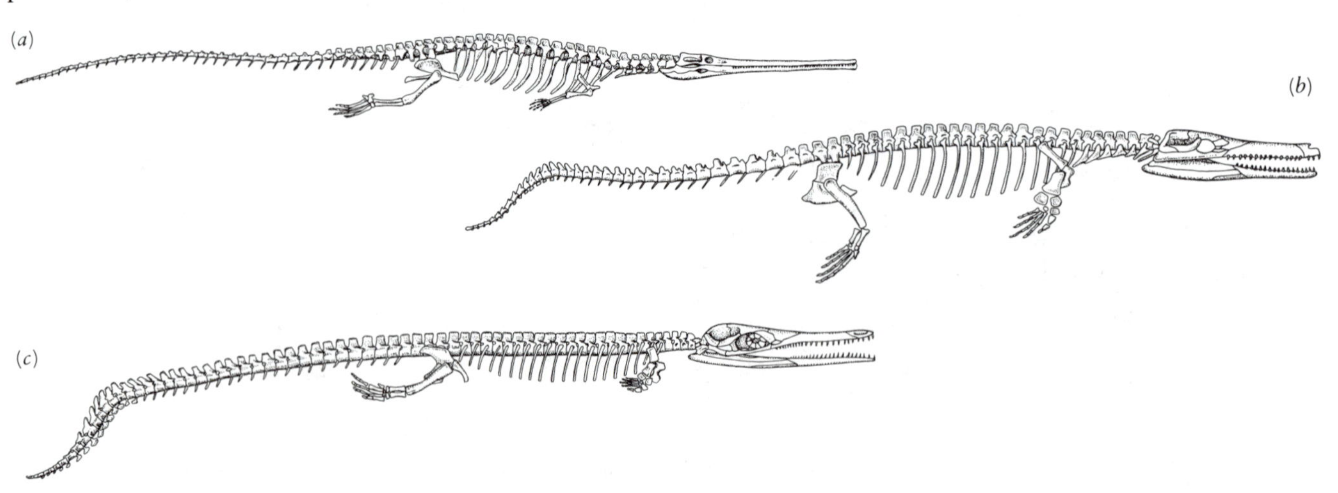

Figure 13-27. MARINE MESOSUCHIAN CROCODILES. (*a*) *Teleosaurus* from the Lower Jurassic. *Modified from Owen, 1849–1884.* (*b*) *Metriorhynchus* from the Upper Jurassic. *From Steel, 1973.* (*c*) *Geosaurus*, from the Lower Jurassic. *Modified from Fraas, 1902.*

eastern Asia, which corresponds with the margins of the Mesozoic Tethys Sea. The genus *Hyposaurus* also occurs in North and South America.

SEMIAQUATIC AND TERRESTRIAL MESOSUCHIANS

The Goniopholidae from the Upper Jurassic into the Upper Cretaceous and the Atoposauridae from the Upper Jurassic and Lower Cretaceous, have a worldwide distribution and closely approach the skeletal anatomy of modern crocodiles. The skull broadly resembles the pattern of conservative crocodiles living today, without the elongate snout that characterizes the specialized aquatic forms. Although they continue later in time, the goniopholids retained amphicoelous vertebrae and the palate is primitive. Among the atoposaurids, *Theriosuchus* has somewhat procoelous vertebrae and the pterygoids form the lateral, but not yet the anterior, margin of the choana. This genus may provide a link with the eusuchians.

Two other families occupy an isolated position among the mesosuchians, the Baurusuchidae (Upper Cretaceous to Eocene) and the Sebecidae (Paleocene to Miocene) (Figure 13-28). They occur primarily in South America, with possible European and North African relatives. Most known remains are of the skull, which has a high narrow snout, in contrast with most other crocodiles. The teeth are long, laterally compressed, serrate blades, a shape designated as **ziphodont.** Sebecid teeth so closely resemble those of carnivorous dinosaurs that they were once used as evidence that dinosaurs had lingered into the Tertiary in South America. Similarly shaped teeth are also present in *Pristichampsus,* an early Tertiary eusuchian crocodile.

Based on their dentition, paleontologists think that sebecids were active terrestrial predators, perhaps occupying the position in the early Tertiary of South America that was filled by large carnivorous mammals in other parts of the world (Langston, 1965). In the absence of adequate postcranial evidence, this hypothesis remains speculative.

Buffetaut (1982) discusses several other, less-well-known families of uncertain taxonomic position in a review of all the mesosuchian groups.

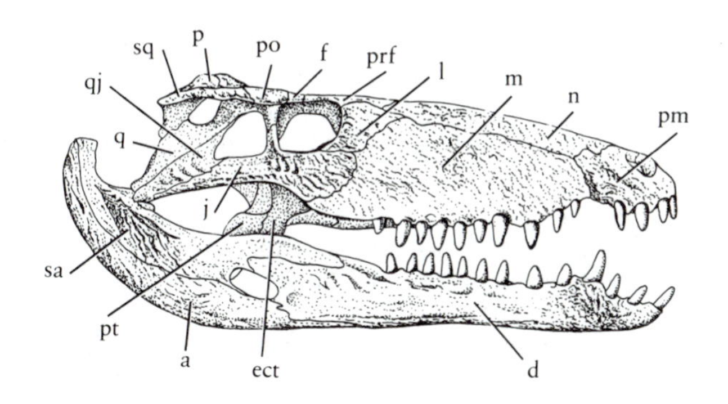

Figure 13-28. SKULL OF THE CROCODILE *SEBECUS.* From the Eocene of South America. Lateral view. *From Colbert, 1946.* Abbreviations as in Figure 8-3.

EUSUCHIANS

The modern crocodilians appear to represent the culmination of a lineage of semiaquatic forms that were common since at least the late Jurassic. We can distinguish the eusuchians from the mesosuchians by technical features of the palate and vertebrae. The internal nares are completely surrounded by the pterygoids, and the centra of all the vertebrae, except the atlas, the second sacral, and the first caudal, are procoelous, like those of advanced lizards and snakes.

Aside from some poorly known and rather specialized forms from the Lower Cretaceous (the Hylaeochampsidae), modern crocodiles first occur in the Upper Cretaceous, and we can place nearly all of the genera in the living families Alligatoridae, Crocodylidae, and Gavialidae.

We know gavials from the Eocene; the modern genus *Gavialis* is already present in the Miocene. Alligators appear in the Upper Cretaceous and the genera *Alligator* and *Caiman* both occur in the Oligocene. *Melanosuchus* and *Paleosuchus* appear in the Pliocene. The Crocodylidae also appear in the Upper Cretaceous; *Crocodylus* in the Paleocene, *Tomistoma* from the Eocene. *Osteolaemus,* the African dwarf crocodile, is the only living crocodilian genus without a fossil record.

Crocodiles that are essentially similar to modern genera were contemporaries of the dinosaurs in the late Mesozoic. They appear to show no effects of the general extinction that overtook so many other reptiles at the end of the Cretaceous. The eusuchians were much more numerous, diverse, and widespread in the early Tertiary than they are today. Their decline probably resulted from climatic deterioration since the early Cenozoic.

SUMMARY

Archosaurs include the living crocodiles, dinosaurs, pterosaurs, and a primitive assemblage from the Triassic, the thecodonts. Archosaurs may share a common ancestry in the late Paleozoic with rhynchosaurs, protorosaurs, and trilophosaurs. Together, these groups constitute the Archosauromorpha. They share a similar specialization of the ankle that facilitates a more upright posture.

Protorosaurs first occur in the Upper Permian and continue into the Lower Jurassic. They are characterized by the loss of the lower temporal bar and a long neck. In *Tanystropheus,* the neck is considerably longer than the trunk. The trilophosaurs are known primarily from a single late Triassic genus in which the cheek teeth are transversely expanded to form a wide, shearing surface. The cheek is not fenestrate.

Rhynchosaurs are the most common and widespread of primitive archosauromorphs. Middle and Upper Triassic genera are characterized by a broadly expanded skull, multiple tooth rows, and an overhanging toothless premaxilla. Rhynchosaurs were a common herbivorous group throughout the Triassic.

Primitive archosaurs, which are distinguished by the presence of an antorbital fenestra, are placed in the order Thecodontia, which includes four groups of Triassic reptiles, the Proterosuchia, Pseudosuchia, Phytosauria, and Aetosauria. Proterosuchians were large, very primitive carnivores. Pseudosuchians include more advanced forms, some of which were bipedal and may be close to the ancestry of dinosaurs. Phytosaurs were crocodilelike forms that were restricted to the Upper Triassic, and aetosaurs were heavily armored, quadrupedal herbivores. The interrelationships of these groups remain in dispute.

The proterosuchid *Chasmatosaurus,* from the Lower Triassic has very primitive girdles and limbs, which indicate a sprawling posture that was not very advanced above the primitive diapsids. Other thecodonts show changes in the pelvis, femur, and tarsus that enable the rear limbs to be held more erect and move in a fore-and-aft direction. The small, possibly facultatively bipedal genus *Euparkeria* represents the initiation of a more advanced tarsal structure within the Lower Triassic.

Most thecodonts have a crurotarsal or crocodiloid tarsus in which the main joint is between the astragalus and the calcaneum and the calcaneal tuber forms the main lever for ventroflexion of the foot. The pattern is initiated in proterochampsids, phytosaurs, and aetosaurs and is fully developed in the pseudosuchians. The lagosuchids, which have a mesotarsal joint, may be close to the ancestry of dinosaurs.

Crocodiles are conservative descendants of the early thecodont radiation. Most skeletal features that characterize modern genera were established by the late Triassic. The ancestry of crocodiles can be found among the sphenosuchids of the Middle and Upper Triassic, which have been classified among the ornithosuchid thecodonts. What is known of the postcranial skeleton is typical of advanced thecodonts, but the quadrate and cheek region have the characteristics of primitive crocodiles.

Sphenosuchids and Protosuchids from the Upper Triassic and Lower Jurassic were small, agile terrestrial animals. The mainly Jurassic mesosuchians and modern eusuchians include many semiaquatic and some fully marine derivatives. The modern crocodilian families are known from the Upper Cretaceous.

REFERENCES

Benton, M. J. (1983). The Triassic reptile *Hyperodapedon* from Elgin: Functional morphology and relationships. *Phil. Trans. Roy. Soc. Lond., B* **302**: 605–720.

Benton, M. J. (1985). Classification and phylogeny of the diapsid reptiles. *Zool. Journ. Linn. Soc.,* **84**: 97–164.

Bonaparte, J. F. (1971). Los tetrapodos del sector superior de la formacion Los Colorados, La Rioja, Argentina (Triasico Superior). I Parte. *Opera Lilloana,* **22**: 1–183.

Bonaparte, J. F. (1975). The Family Ornithosuchidae (Archosauria: Theocodontia). *Cent. Natl. Rech. Sci. Intl. Problemes Actuels de Paleontologie (Evolution de Vertebres),* **218**: 485–502.

Bonaparte, J. F. (1978). El Mesozoico de America del Sur y sus Tetrapodos. *Opera Lilloana,* **26**: 1–596.

Bonaparte, J. F. (1982). Faunal replacement in the Triassic of South America. *J. Vert. Paleont.,* **2**(3): 362–371.

Bonaparte, J. F. (1984). Locomotion in rauisuchid thecodonts. *J. Vert. Paleont.,* **3**(4): 210–218.

Brinkman, D. (1979). The structural and functional evolution of the diapsid tarsus. *Unpublished Ph. D. thesis.* McGill University, Montreal.

Brinkman, D. (1981). The origin of the crocodiloid tarsi and the interrelationships of thecodontian archosaurs. *Breviora,* **464**: 1–23.

Broom, R. (1913). Note on *Mesosuchus browni,* Watson, and on a new South African Triassic pseudosuchian (*Euparkeria capensis*). Records Albany Mus., **2**: 394–396.

Buffetaut, E. (1979). The evolution of crocodilians. *Scientific American,* **241**(4): 130–144.

Buffetaut, E. (1982). Radiation evolutive, Paleoecologie et Biogeographie des crocodiliens mesosuchiens. *Memoires de la Societe Geologique de France,* **142**: 1–88.

Carroll, R. L. (1976a). Eosuchians and the origin of archosaurs. In C. S. Churcher (ed.), *Athlone Essays in Palaeontology in Honour of Loris Shano Russell. Roy. Ont. Mus. Life Sci. Misc. Pub.*: 58–79.

Carroll, R. L. (1976b). *Noteosuchus*—The oldest known rhynchosaur. *Ann. S. Afr. Mus.,* **72**: 37–57.

Charig, A. J. (1972). The evolution of the archosaur pelvis and hindlimb: An explanation in functional terms. In K. A. Joysey and T. S. Kemp (eds.), *Studies in Vertebrate Evolution,* pp. 121–155. Oliver and Boyd, Edinburgh.

Charig, A. J., and Sues, H.-D. (1976). Suborder Proterosuchia Broom 1906b. In O. Kuhn (ed.), *Handbuch der Palaeoherpetologie. 13. Thecodontia,* pp. 11–39. Gustav Fischer Verlag, Stuttgart.

Chatterjee, S. (1974). A rhynchosaur from the Upper Triassic Maleri Formation of India. *Phil. Trans. Roy. Soc. Lond., B,* **267**: 209–261.

Chatterjee, S. (1978). A primitive parasuchid (phytosaur) reptile from the Upper Triassic Maleri Formation of India. *Palaeontology,* **21**: 83–127.

Chatterjee, S. (1980b). The evolution of rhynchosaurs. *Mem. Soc. geol. Fr., N.S.,* **139**: 57–65.

Chatterjee, S. (1982). Phylogeny and classification of thecodont reptiles. *Nature,* **295**: 317–320.

Chatterjee, S. (1985). *Postosuchus,* a new thecodontian reptile from the Triassic of Texas and the origin of tyrannosaurs. *Phil. Trans. Roy. Soc. Lond., B,* **309**: 395–460.

Colbert, E. H. (1946). *Sebecus,* representative of a peculiar suborder of fossil Crocodilia from Patagonia. *Bull. Am. Mus. Nat. Hist.,* **87**: 217–270.

Colbert, E. H. (1952). A pseudosuchian reptile from Arizona. *Bull. Am. Mus. Nat. Hist.,* **99**: 565–592.

Colbert, E. H., and Mook, C. C. (1951). The ancestral crocodilian *Protosuchus. Bull. Am. Mus. Nat. Hist.,* **97**: 147–182.

Crompton, A. W., and Smith, K. K. (1980). A new genus and species of crocodilian from the Kayenta Formation (Late Triassic?) of Northern Arizona. In L. L. Jacobs (ed.), *Aspects of Vertebrate History,* pp. 193–217. Museum of Northern Arizona Press, Flagstaff.

Cruickshank, A. R. I. (1970). Early thecodont braincases. *Internat. Union of Geol. Sci. Commission on Stratigraphy. Second Gondwana Symposium. Proceedings and Papers,* **1970**: 683–685.

Cruickshank, A. R. I. (1972). The proterosuchian thecodonts. In K. A. Joysey and T. S. Kemp (eds.), *Studies in Vertebrate Evolution,* pp. 89–119. Oliver and Boyd, Edinburgh.

Crush, P. J. (1984). A late Upper Triassic sphenosuchid crocodilian from Wales. *Palaeontology,* **27**: 131–157.

Ewer, R. F. (1965). The anatomy of the thecodont reptile *Euparkeria capensis* Broom. *Phil. Trans. Roy. Soc. Lond., B,* **248**: 379–435.

Fraas, E. (1902). Die Meer-Crocodilier (Thallattosuchia) des oberen Jura unter specieller Beruecksichtigung von *Dracosaurus* und *Geosaurus. Palaeontographica,* **49**: 1–72.

Gow, C. E. (1975). The morphology and relationships of *Youngina capensis* Broom and *Prolacerta broomi* Parrington. *Palaeont. afr.,* **18**: 89–131.

Gregory, J. T. (1945). Osteology and relationships of *Trilophosaurus. U. of Texas Publ.,* no. 4401: 273–359.

Gregory, J. T. (1962). The genera of phytosaurs. *Amer. J. Sci.,* **260**: 652–690.

Iordansky, N. N. (1973). The skull of the Crocodilia. In C. Gans and T. S. Parsons (eds.), *Biology of the Reptilia,* 4, pp. 201–262. Academic Press, New York.

Krebs, B. (1963). Bau und Funktion des Tarsus eines Pseudosuchiers aus der Trias des Monte San Giorgio (Kanton Tessin, Schweiz). *Palaeont. Zeitschrift,* **37**: 88–95.

Krebs, B. (1963b). *Ticinosuchus ferox nov. gen. nov. sp. Ein neuer Pseudosuchier aus der Trias des Monte San Giorgio. Schweiz. Paläont. Abhandl.,* **81**: 1–140.

Langston, W. (1965). Fossil crocodilians from Columbia and the Cenozoic history of the Crocodylia in South America. *Univ. Calif. Publ. Geol. Sci.,* **52**: 1–157.

Langston, W. (1973). The crocodilian skull in historical perspective. In C. Gans and T. S. Parson (eds.), *The Biology of the Reptilia. 4,* pp. 263–284. Academic Press, New York.

Malan, M. E. (1963). The dentitions of the South African Rhynchocephalia and their bearing on the origin of the rhynchosaurs. *South Afr. J. Sci.,* **59**(5): 214–220.

Meyer, H. von, (1856). *Zur Fauna der Vorwelt. Saurier aus dem Kupferschiefer der Zechstein-Formation.* Frankfurt-am-Main.

Nash, D. S. (1975). The morphology and relationships of a crocodilian, *Orthosuchus stormbergi,* from the Upper Triassic of Lesotho. *Ann. South Afr. Mus.,* **67**(7): 227–329.

Olsen, P. E. (1979). A new aquatic eosuchian from the Newark Supergroup (Late Triassic-Early Jurassic) of North Carolina and Virginia. *Postilla,* **176**: 1–14.

Olsen, P. E., McCune, A. R., and Thomson, K. S. (1982). Correlation of the early Mesozoic Newark Supergroup by vertebrates, principally fishes. *Am. J. Sci.,* **282**: 1–44.

Owen, R. (1849–1884). *History of British Fossil Reptiles.* Cassell, London.

Romer, A. S. (1972a). The Chanares (Argentina) Triassic reptile fauna. 13. An early ornithosuchid pseudosuchian, *Gracilisuchus stipanicicorum,* gen. et sp. nov. *Breviora,* **389**: 1–24.

Romer, A. S. (1972b). The Chanares (Argentina) Triassic reptile fauna. 16. Thecodont classification. *Breviora,* **395**: 1–24.

Schumacher, G. H. (1973). The head muscles and hyolaryngeal skeleton of turtles and crocodilians. In C. Gans and T. S. Parson (eds.), *Biology of the Reptilia,* 4, pp. 101–199. Academic Press, New York.

Seeley, K. (1888). Researches on the structure, organization and classification of the fossil Reptilia. 1. On *Protorosaurus speneri* (von Meyer). *Phil. Trans. Roy. Soc. Lond., B,* **178**: 187–213.

Sill, W. D. (1967). *Proterochampsa barrionuevoi* and the early evolution of the Crocodilia. *Bull. Mus. Comp. Zool.,* **135**(8): 415–446.

Sill, W. D. (1974). The anatomy of *Saurosuchus galilei* and the relationships of the rauisuchid thecodonts. *Bull. Mus. Comp. Zool.,* **146**(7): 317–362.

Steel, R. (1973). Crocodylia. In O. Kuhn (ed.), *Handbuch der Paläoherpetologie,* 16. Gustav Fischer Verlag, Stuttgart.

Thulborn, R. A. (1982). Significance of ankle structures in archosaur phylogency. *Nature,* **299**: 657.

Walker, A. D. (1961). Triassic reptiles from the Elgin area: *Stagonolepis, Dasygnathus* and their allies. *Phil. Trans. Roy. Soc. Lond., B,* **244**: 103–204.

Walker, A. D. (1964). Triassic reptiles from the Elgin area: *Ornithosuchus* and the origin of carnosaurs. *Phil. Trans. Roy. Soc. Lond., B,* **248**: 53–134.

Walker, A. D. (1968). *Protosuchus, Proterochampsa* and the origin of phytosaurs and crocodiles. *Geol. Mag.,* **105**: 1–14.

Walker, A. D. (1972). New light on the origin of birds and crocodiles. *Nature,* **137**: 257–263.

Westphal, F. (1976). Phytosauria. In O. Kuhn (ed.), *Handbuch der Paläoherpetologie. 13. Thecodontia,* pp. 99–120. Gustav Fischer Verlag, Stuttgart.

Wever, E. G. (1978). *The Reptile Ear.* Princeton University Press, Princeton, New Jersey.

Wild, R. (1973). Die Triasfauna der Tessiner Kalkalpen. 23. *Tanystropheus longobardicus* (Bassani) (Neue Ergebnisse). *Schweiz. Palaeontol. Abh.,* **95**: 1–162.

C H A P T E R **XIV**

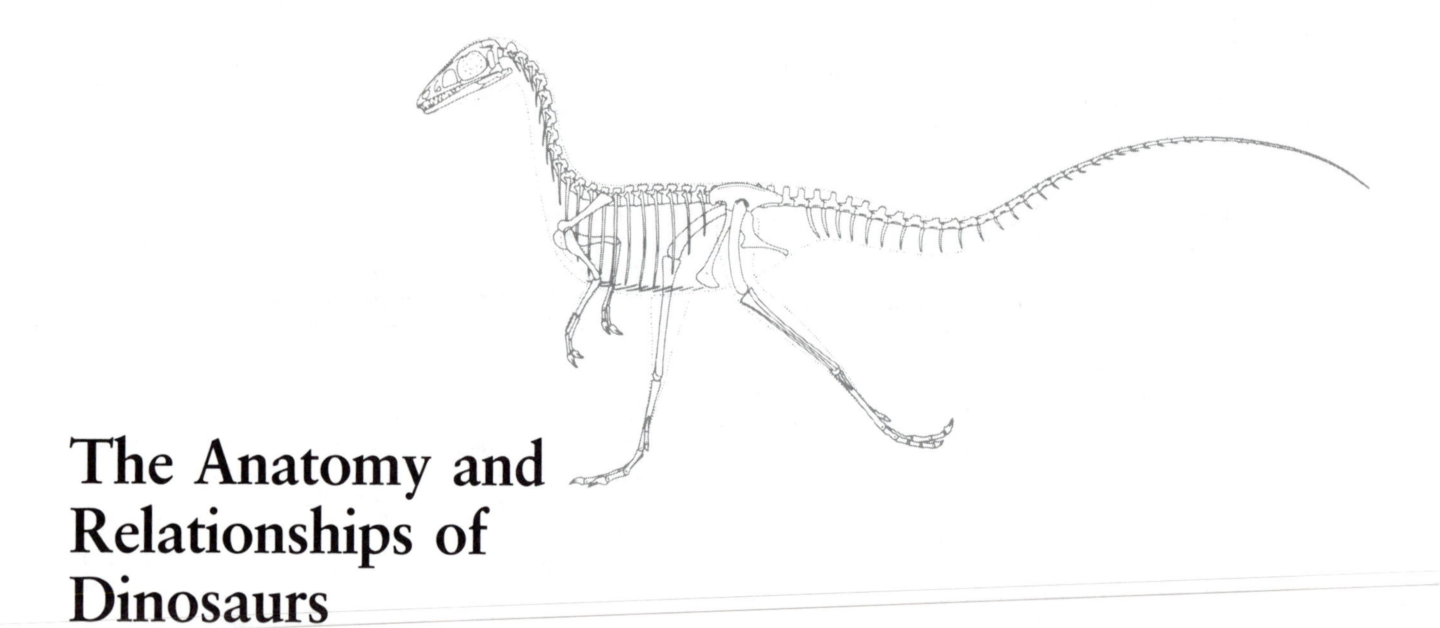

The Anatomy and Relationships of Dinosaurs

For 150 million years, from the Late Triassic to the end of the Cretaceous, dinosaurs dominated the terrestrial environment (Figure 14-1). They were unquestionably the most successful of all reptiles and rivaled the Cenozoic mammals in their structural diversity and wide distribution. Dinosaurs included the largest terrestrial animals that have ever lived, as well as species that may have equaled many mammals in their high metabolic rate and relative brain size.

Dinosaurs first appear in the fossil record at the end of the Middle Triassic in South America (Bonaparte, 1982). Several lineages are evident by the end of the Triassic, all of which share a number of advanced features that were not present in most thecodonts. The most important advances are associated with posture and locomotion.

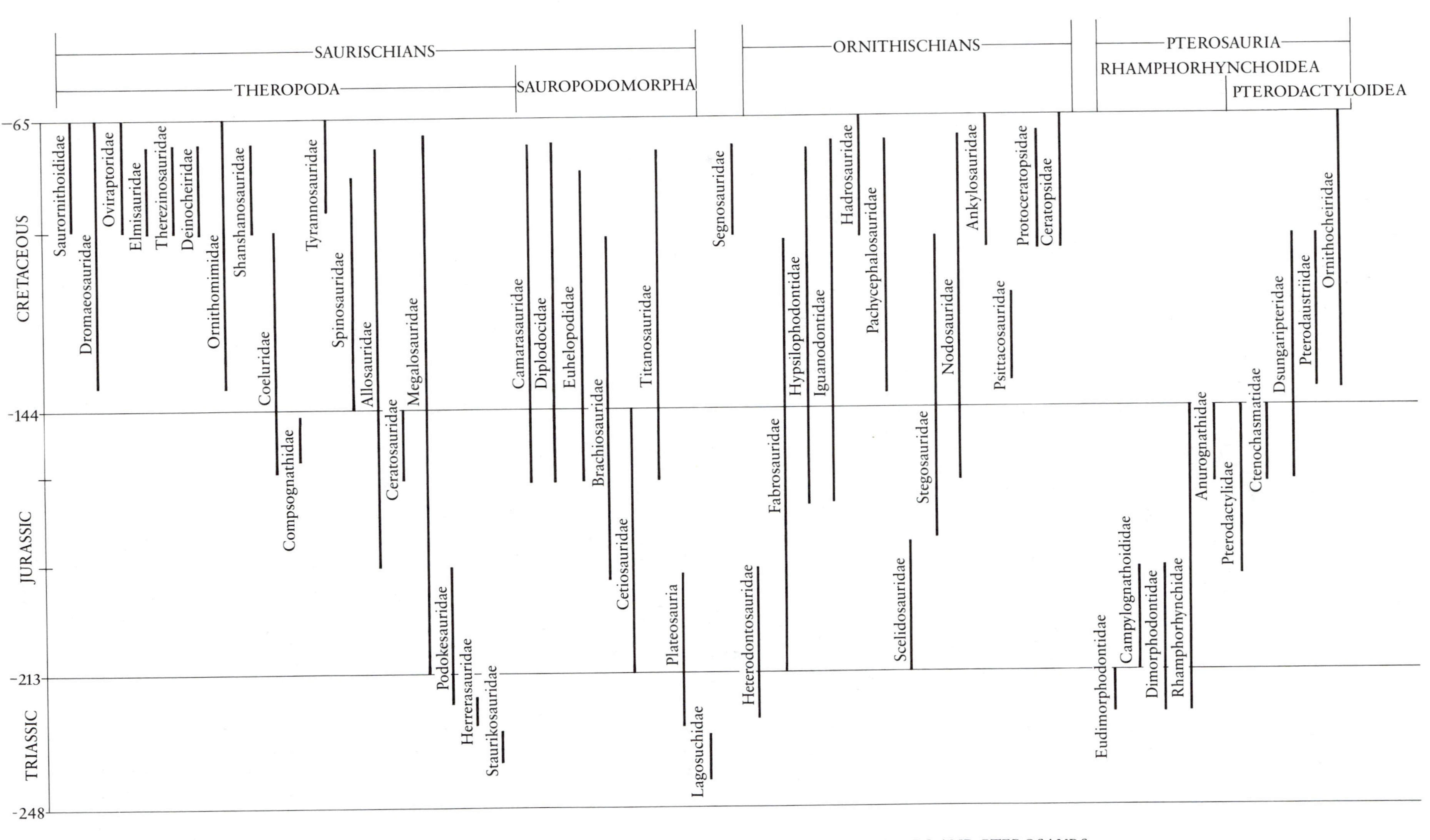

Figure 14-1. STRATIGRAPHIC RANGES OF THE FAMILIES OF DINOSAURS AND PTEROSAURS.

THE POSTURE AND ANCESTRY OF DINOSAURS

All thecodont groups above the level of proterosuchians tend to bring the limbs under the body and to move them in a more strictly fore-and-aft direction, without the large degree of lateral extension that is evident in primitive tetrapods. However, most retain vestiges of a sprawling posture in the structure of the hip and ankle joints. In the earliest members of all dinosaur groups, these joints show specialization that resulted in a nearly vertical posture.

The head of the femur is angled anteromedially so that it articulates with the dorsal margin of the acetabulum. This margin is strongly buttressed to accommodate the entire weight of the body. The central portion of the acetabulum is not ossified but appears fenestrated in the fossils. In contrast with thecodonts, it no longer needs to resist the medially directed force generated by the femur, which was angled toward the midline (Figure 14-2).

The shaft of the femur is straight or bowed slightly anteriorly, in contrast with the gently sigmoidal configuration of thecodonts and crocodiles. Just distal to the head are two tuberosities, the greater and lesser trochanters, that serve for insertion of the iliofemoralis internus and the iliofemoralis externus muscles. Further distally, on the posterior surface, is the fourth trochanter, which serves as the area of insertion for the major retractor of the femur, the caudifemoralis. The head of the tibia is twisted to produce a nearly vertical orientation of the shaft with respect to both its proximal and distal articulating surfaces.

A mesotarsal ankle joint is established, with the line of flexion passing transversely between the proximal and distal tarsals. The astragalus and calcaneum are reduced to relatively simple elements that are closely integrated with the ends of the tibia and fibula. The metatarsals are elongate and function as an additional unit of the lower limb. The digits form the main surface that contacts with the ground. The posture in dinosaurs may hence be considered to be digitigrade.

These changes result in a nearly vertical limb, as viewed anteriorly, with both the ankle and knee functioning primarily as hinge joints. Structurally and functionally, the closest modern homologue is provided by birds.

Among the thecodonts, the greatest similarity can be found among the Lagosuchidae, in which the rear limb achieved many features that are common to the earliest dinosaurs (see Figure 13-21). *Lagosuchus* is *more* specialized than early dinosaurs in the greater relative length of the metatarsals, but this is a character that might have been reversed in descendants with a larger body size. An animal that was very similar to *Lagosuchus* may have given rise to all the dinosaur groups. However, *Lagosuchus* is more primitive than any dinosaurs in the small degree of fenestration of the acetabulum. We do not know the skull well enough to make a detailed comparison with the major dinosaur groups. The exact phylogenetic position of this genus cannot yet be established.

The upright posture of the limbs in dinosaurs has long been confused with the question of their bipedality. A clear distinction between these features was made by Charig (1972). Among the thecodonts, the ornithosuchids and rauisuchids brought the rear limbs close to a vertical orientation. It was once thought that they were habitually bipedal, but information from footprints as well as skeletal anatomy indicates that they were primarily quadrupedal. *Lagosuchus* may also have been facultatively, rather than habitually, bipedal.

Many Triassic and early Jurassic dinosaurs show strong skeletal evidence for bipedality in the very great disparity in the length and structure of the forelimbs and hind limbs (see Figures 14-6 and 14-26). On the other hand, some early dinosaurs that approached the Jurassic sauropods in their great body size were almost certainly quadrupedal, although their rear limbs were also longer than the front limbs (see Figure 14-19). Romer (1956, 1966) argued that this limb disparity reflects a reversion to a quadrupedal stance from a primitive bipedal condition. If this were the

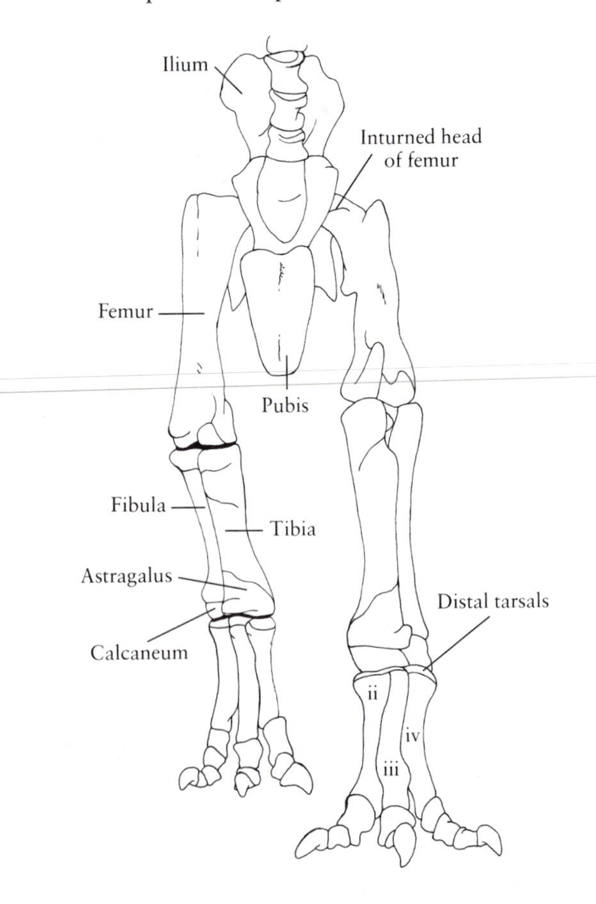

Figure 14-2. ANTERIOR VIEW OF THE REAR LIMBS OF THE DINOSAUR *TYRANNOSAURUS*. Note the vertical posture and the transverse knee and ankle joints that are characteristic of both saurischians and ornithischians. *Modified from Osborn, 1916.*

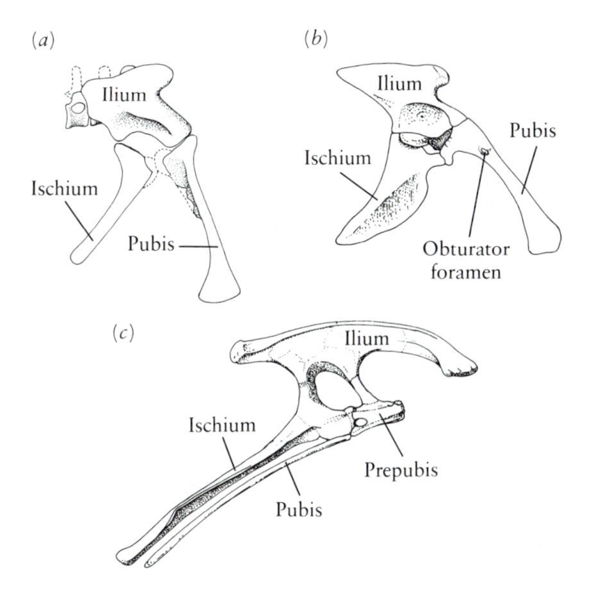

Figure 14-3. THE CONFIGURATION OF THE PELVIS, THE PRIMARY CRITERION FOR DISTINGUISHING THE ORDERS SAURISCHIA AND ORNITHISCHIA. The saurischian pelvis, exemplified by (*a*) *Staurikosaurus* is similar to that of thecodonts such as (*b*) *Lagosuchus* in which the elongate pubis and ischium together with the ilium form a triradiate structure. The pelvis of ornithischians, as exemplified by the early genus (*c*) *Heterodontosaurus* is advanced in having the pubis oriented posteriorly, alongside the ischium. In some advanced ornithischians (see Figure 14-28), a new structure, the prepubis, has developed anterior to the pubis. This results in a tetraradiate configuration. (*a*) *From Galton, 1977.* (*b*) *From Bonaparte, 1978.* (*c*) *From Santa Luca, 1980.*

case, it would provide further evidence that all dinosaurs shared a close common ancestry above the level of the thecodonts. Unfortunately, the fossil record does not demonstrate unequivocally that the early quadrupedal dinosaurs descended from fully bipedal ancestors. The early dinosaurs may have diverged from a primitive stock that was facultatively, but not habitually, bipedal. The clear distinction between these patterns of locomotion may have evolved after the occurence of the changes in the structure of the rear limb and girdle by which we recognize dinosaurs and after the initial divergence of the major dinosaur lineages.

Dinosaurs have long been divided into two major groups, the ornithischians and the saurischians, primarily on the basis of the structure of the pelvis (Figure 14-3). The configuration of the pelvis and most other features that distinguish early saurischians are primitive, as demonstrated by their similarity with those of thecodonts. In contrast, the ornithischian pattern of the pelvis and many features of the skull are clearly derived relative to those of the saurischians. Ornithischians may have evolved from primitive saurischians, but there is no fossil evidence to support the long-held assumption (which Charig reiterated in 1976) that saurischians and ornithischians had separate origins from distinct groups of thecodonts. No thecodonts other than the lagosuchids share significant derived features with either of the two major groups of early dinosaurs.

SAURISCHIANS

The ornithischians are a diverse assemblage and their interrelationships are not fully known, but they unquestionably have a common ancestry, established on the basis of a host of shared derived characters. We cannot state with equal assurance that the several groups of saurischians share a unique common ancestry since the features they share are primarily primitive. Two major groups have long been recognized, the carnivorous and obligatorily bipedal theropods and the typically quadrupedal and herbivorous sauropods. The theropods have been further separated into carnosaurs and coelurosaurs, primarily on the basis of size. Typical sauropods are only known from the Jurassic and Cretaceous. In the late Triassic and early Jurassic, a distinctive group of more primitive herbivorous saurischians are common. These are termed the plateosaurs or prosauropods; their relationships with other dinosaurs are subject to continuing speculation.

Unfortunately, the late Triassic record of dinosaurs remains very incomplete; the few fossils that we know share a number of primitive features, but the nature of their relationships with later forms is difficult to establish.

THEROPODS

PRIMITIVE TRIASSIC CARNIVORES

The only dinosaur that we recognize from the Middle Triassic is *Staurikosaurus* (Figure 14-4) from southern Brazil. It is missing the skull, the forelimbs, and the tarsus and foot, but the lower jaw, vertebrae, pelvis, and rear limb are comparable to those of later saurischian dinosaurs. The lower jaw is as long as the femur, with sharp piercing teeth. The trunk vertebrae are short and the rear limb is fully erect. Galton (1977) suggests that the posture may have been fully bipedal. The tibia is longer than the femur, as is common in cursorial animals, and the limb bones are hollow. Together with the size of the jaw, these characters suggest that *Staurikosaurus* was an active predator. As restored, the skeleton is slightly more than 2 meters long.

A possibly related genus from the Upper Triassic is *Herrerasaurus,* which was about 4 meters in length. The ankle and foot are typical of primitive dinosaurs in the integration of the astragalus and calcaneum with the tibia and fibula, the reduction of the distal tarsals, and the elongation of the metatarsals (Figure 14-5). The fifth digit is reduced and the weight is supported evenly on the other four digits.

The large size of the skull and the expansion of the distal end of the pubes into a vertical plate are features

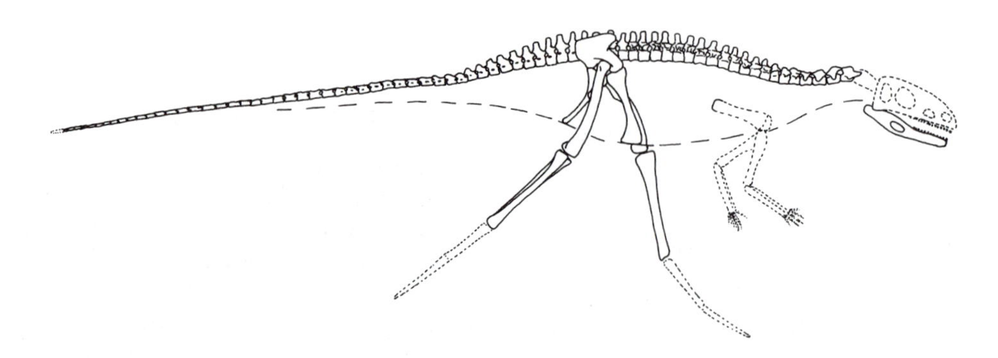

Figure 14-4. RESTORATION OF *STAURIKOSAURUS*. This is the earliest-known dinosaur from the late Middle Triassic of South America. Total length is 2.1 meters. *From Galton, 1977.*

that are shared with the large carnivorous dinosaurs of the Jurassic and Cretaceous. However, these features also evolved independently in large carnivorous thecodonts.

Because these Triassic genera are so incompletely known and primitive, we cannot ally them with certainty with later dinosaur groups. What we know of their anatomy and general habitus suggests that they may be related to the large carnivorous dinosaurs of the Jurassic and Cretaceous.

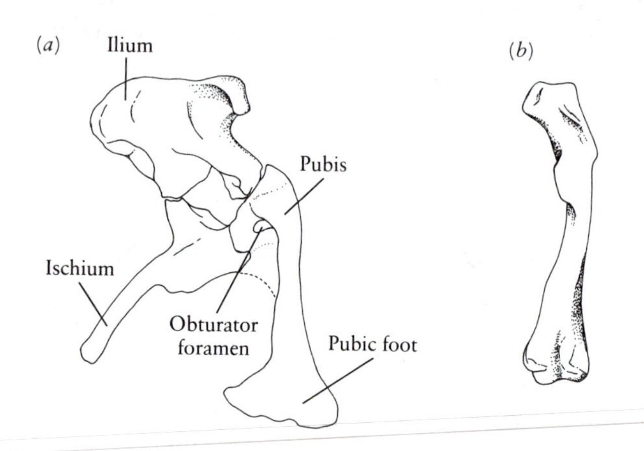

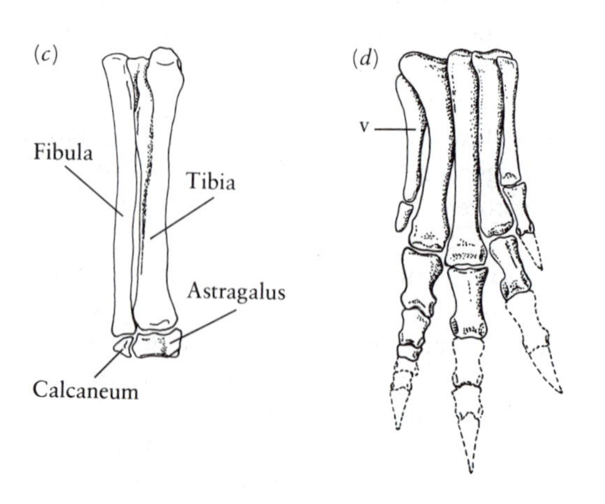

Figure 14-5. (*a*) Pelvis, (*b* and *c*) rear limb, and (*d*) foot of *Herrerasaurus*, a carnivorous dinosaur from the Upper Triassic of South America. *From Galton, 1977.*

"COELUROSAURS" AND "CARNOSAURS"

Late Triassic, Jurassic and Cretaceous carnivorous dinosaurs have customarily been divided into two infraorders, the Carnosauria and the Coelurosauria, that are based primarily on differences in size and proportions. The term carnosaur has been applied to the largest carnivorous dinosaurs, which are further characterized by the disproportionately large skull and relatively short forelimbs. Most genera that have been recognized as coelurosaurs are smaller, with a skull that is typically smaller relative to the length of the trunk and relatively larger forelimbs.

Genera that have been described within the past 15 years show a combination of features that were previously thought to distinguish these two groups. *Deinocheirus* and *Therizinosaurus* from the Cretaceous of Mongolia are still incompletely known but demonstrate the existence of very large theropods with extremely long forelimbs (Osmolska and Roniewicz, 1970; Barsbold, 1976). Nearly complete skeletons of *Dilophosaurus* (Welles, 1984) and *Deinonychus* (Ostrom, 1969) combine a nearly equal number of "carnosaur" and "coelurosaur" traits. Use of these two infraorders to classify the theropods has become difficult to justify. Unfortunately, we have not been able to establish an alternative classification that encompasses all the theropod families in a consistent phylogenetic scheme. The better-known theropod families are each readily distinguished by unique features, but shared, derived characters that unite them with one another are more difficult to recognize. The "carnosaur" families may each have evolved separately from different groups that have been classified as coelurosaurs. Or perhaps several separate lineages of both large and small theropods evolved independently from among primitive forms such as *Staurikosaurus* and *Herrerasaurus*.

Podokesaurids

A number of relatively completely known theropods have been described from the late Triassic and early Jurassic. *Coelophysis* (Colbert, 1972) and *Syntarsus* (Raath, 1969) are small, lightly built forms known from western North America and eastern Africa, respectively. Similar genera, which are placed in the family Podokesauridae, are present in Europe, South America, and Asia.

Coelophysis (Figure 14-6*a*) is a small form that has been considered typical of the coelurosaurs. The skeleton is approximately $2\frac{1}{2}$ meters long and extremely lightly built. The limbs are long and slender, and the bones are hollow. The skull is low and slender with small but sharp piercing teeth. The neck is longer than the trunk, and the sacrum is formed by five pairs of sacral ribs that are fused to the vertebrae and ilia in mature individuals. The ilium is very long and low, and the pubes extend far forward. They are narrow flattened rods that are fused at the midline, but, in contrast with larger theropods, they are not expanded into a vertical plate distally. The ischium, as in the other early dinosaurs, is a long, posteroventrally directed rod.

The tail is long, to counterbalance the weight of the presacral region. The scapula is long and narrow, and the coracoid is a small oval plate. The clavicle is present but reduced, which frees the shoulder girdle to allow greater maneuverability of the forelimb. The forelimbs are slim and show no evidence that they supported the body. The manus, which is reduced to three functional digits and a trace of a fourth, is well developed for grasping. The tibia and fibula are 20 percent longer than the femur, and the tarsus is greatly reduced. In the related genus *Syntarsus* from the Lower Jurassic of Africa, the astragalus and calcaneum are fused to one another and the distal tarsals are fused to the metatarsals, which strictly limits the joint to a fore-and-aft hinge. The pes of these podokesaurids is extremely birdlike, with three principal digits. The first digit is very small and does not reach the ground; it is not reversed in the manner of larger theropods.

Coelurids and *Compsognathus*

We have not found any fossils of small theropods in the Middle Jurassic. In the late Jurassic and early Cretaceous several distinct lineages may be recognized. *Compsognathus,* which is represented by two specimens from the Upper Jurassic of Europe, is the smallest known dinosaur (Figure 14-6*b*). Ostrom (1978) estimates that the body weight was 3 to $3\frac{1}{2}$ kilograms. In contrast with all other small theropods, the manus is reduced to two digits, with a phalangeal count of 2, 2, 0. The tibia is considerably longer than the femur, and the metatarsus is elongated as well, which suggests that *Compsognathus* was an effective cursorial predator. The type specimen has the bones of an agile lizard in its abdominal cavity. In contrast with other small theropods, interdental plates, which once were considered diagnostic of "carnosaurs," are present in both the upper and lower jaws (Figure 14-7).

The family Coeluridae represents a continuation of the pattern that was established by small podokesaurids. *Ornitholestes* (Figure 14-8) is the best-known member of the group. The restored skeleton is about 2 meters long. The skull is relatively high; the teeth are small and limited

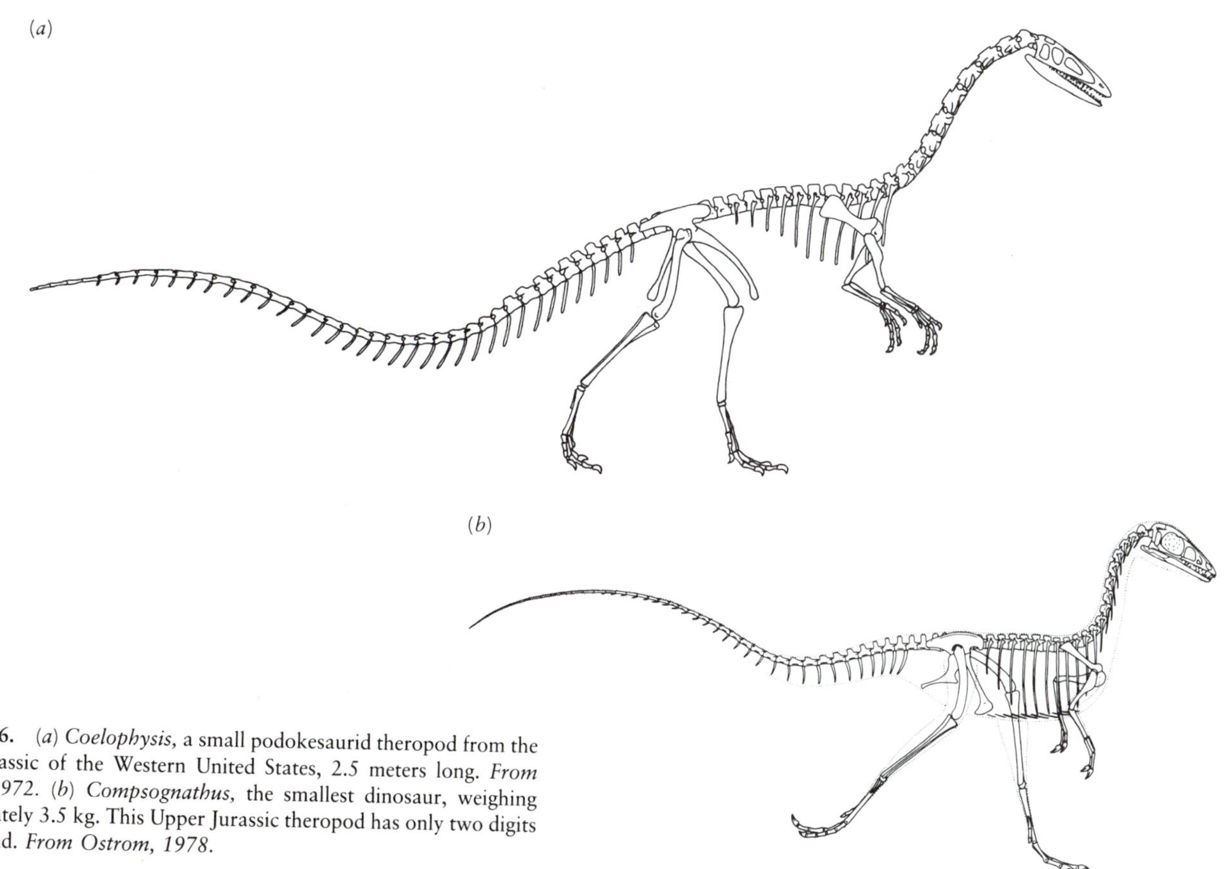

Figure 14-6. (*a*) *Coelophysis,* a small podokesaurid theropod from the Upper Triassic of the Western United States, 2.5 meters long. *From Colbert, 1972.* (*b*) *Compsognathus,* the smallest dinosaur, weighing approximately 3.5 kg. This Upper Jurassic theropod has only two digits on the hand. *From Ostrom, 1978.*

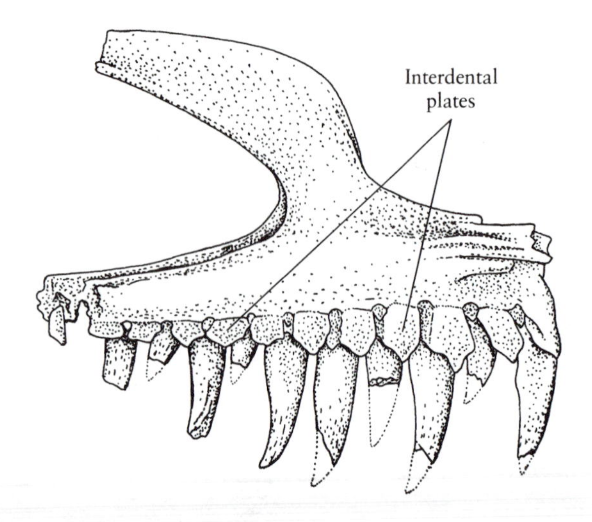

Figure 14-7. MEDIAL SURFACE OF THE UPPER JAW OF THE EARLY MEGALOSAUR *MEGALOSAURUS* FROM ENGLAND. Length of fragment 30 cm. This jaw clearly shows the interdental plates that were once thought to be characteristic of "carnosaurs." They are now known to have an irregular distribution among both "coelurosaurs" and "carnosaurs." *From Charig, 1979.*

to the front of the jaws. The cervical vertebrae are relatively short and are recessed posteriorly to form a more effective articulating surface, a pattern termed **opisthocoelous.** They are further characterized by the presence of large depressions on the lateral surface called **pleurocoels,** which would reduce the weight of the vertebrae without lessening their strength. There are four sacral vertebrae and a long tail, which is characterized by the great length of the prezygapophyses and anterior extension of the haemal arches.

The metacarpals are very long, as is the manus, which is strictly tridactyl, with the total loss of the fourth digit. The tibia is shorter than the femur, in contrast with other small theropods.

The family Coeluridae is not known with certainty above the Lower Cretaceous.

Ornithomimids

Several more specialized lineages have evolved from animals with the general anatomy of the coelurids. Among the most distinctive are the ornithomimids or ostrich dinosaurs, which are found primarily in the Cretaceous of North America and East Asia. Well-known genera such as *Ornithomimus* and *Struthiomimus* have the size and proportions of an ostrich (Figure 14-9). They are distinguished from all other dinosaurs by the total absence of teeth (Figure 14-10). The skull is smaller relative to the vertebrae and trunk than in other coelurosaurs. The bones are extremely thin, and the skull is thought to be kinetic in the manner of birds (Bock, 1964). The orbits and the braincase are both very large, with the estimated ratio of brain size to body weight approaching that of some birds (Russell, 1972). The cervical centra are elongated and fused to the cervical ribs. There are six coossified sacral vertebrae, which include elements that are part of the trunk and tail in more primitive dinosaurs. The forelimbs are very long, but the carpals are reduced and allow little mediolateral movement of the digits. In contrast with most other small theropods, there is a large pubic boot. The tibia is 20 percent longer than the femur and the metatarsals are very long. The first digit of the pes is lost, which results in a strictly tridactyl foot.

Specimens of the Upper Jurassic genus *Elaphrosaurus* from East Africa lack a skull, but the postcranial skeleton suggests a transitional position between early coelurids and ornithomimids (Galton, 1982a). Ornithomimids were poorly represented in the early Cretaceous but became widespread later in that period; only a few fragmentary specimens are found in the uppermost Mesozoic.

Dromaeosaurs and the Saurornithoididae

Small podokesaurids, coelurids, *Compsognathus*, and ornithomimids fit the old concept of coelurosaurs. In contrast, dromaeosaurs and saurornithoidids appear to be structurally intermediate between the primitive members

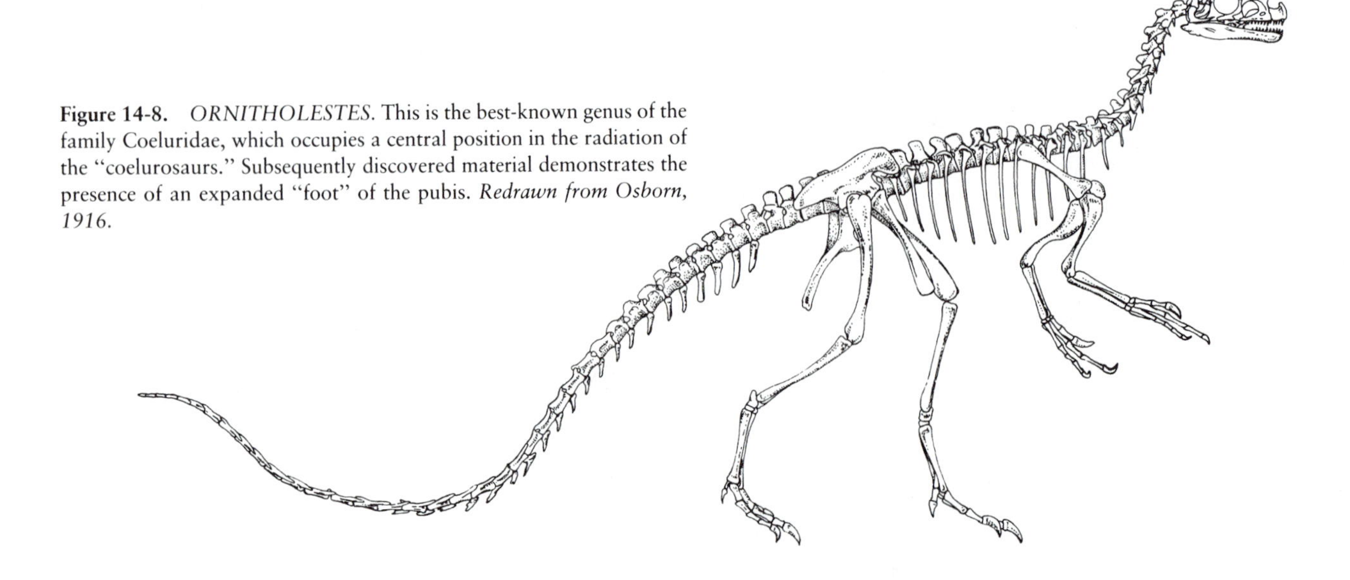

Figure 14-8. *ORNITHOLESTES.* This is the best-known genus of the family Coeluridae, which occupies a central position in the radiation of the "coelurosaurs." Subsequently discovered material demonstrates the presence of an expanded "foot" of the pubis. *Redrawn from Osborn, 1916.*

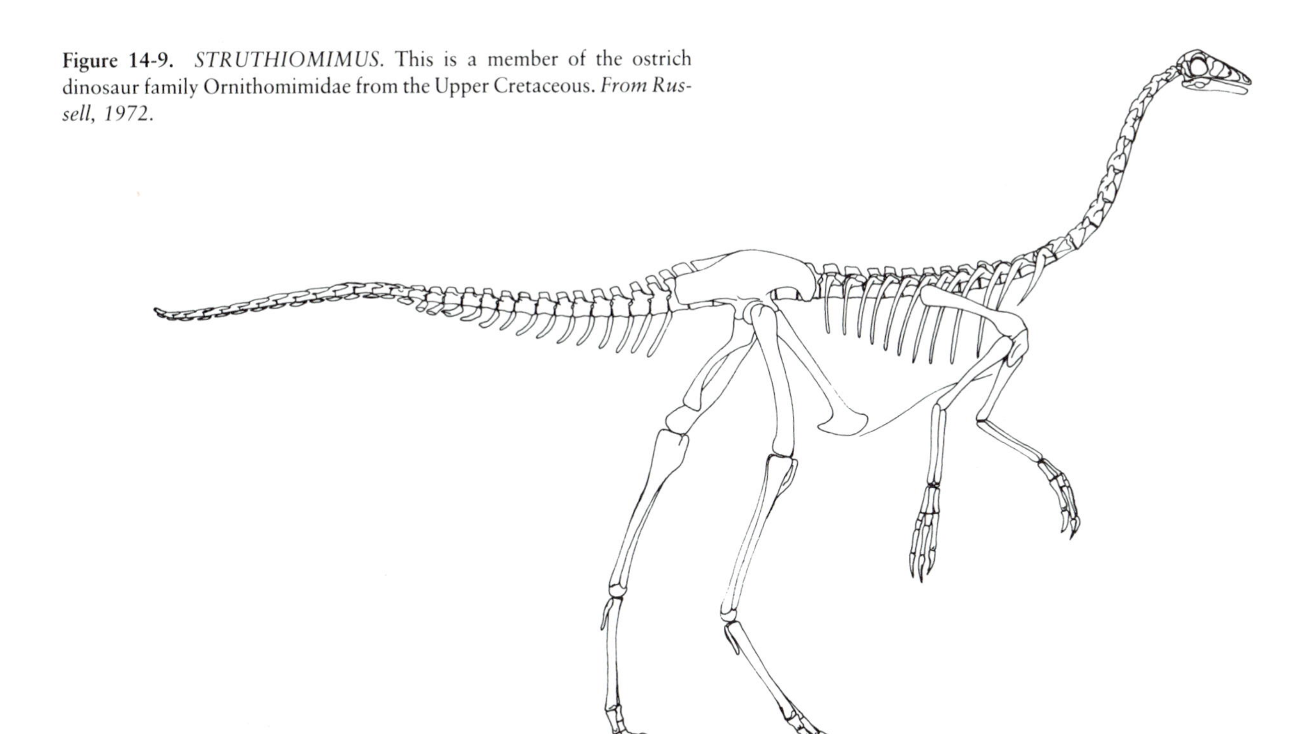

Figure 14-9. *STRUTHIOMIMUS.* This is a member of the ostrich dinosaur family Ornithomimidae from the Upper Cretaceous. *From Russell, 1972.*

of that assemblage and the large "carnosaurs." One of the most thoroughly described of all dinosaurs is *Deinonychus* from the Lower Cretaceous of Montana (Ostrom, 1969, 1976) (Figure, 14-11). It is a lightly built animal, about 3 meters long, with an estimated weight of 60 to 75 kilograms. It appears to combine almost equally characters of carnosaurs and coelurosaurs. The skull is relatively large, with a deep maxilla and lower jaw to accommodate large, bladelike teeth. However, the interdental plates that are characteristic of most large theropods are missing.

The centra of the cervical vertebrae are not opisthocoelous, as in almost all other advanced dinosaurs, but flattened on both ends and beveled to produce a fixed curvature that maintains the large head above the level of the trunk.

The trunk was held nearly horizontal and supported by five sacral vertebrae. The posterior caudal vertebrae are highly distinctive in having the prezygapophyses and anterior processes of the haemal arches elongated as narrow rods that run the length of 8 to 10 vertebrae and form supporting bundles around the centra for much of the length of the tail. They would have stiffened the tail so that it would move more as a unit than in other reptiles, but the persistence of the articulating surfaces of the zygapophyses indicates that some flexibility was retained. The tail was probably vital in maintaining balance. Coelurids and ornithomimids show a slight elongation of the caudal prezygapophyses as well.

In contrast with the well-known large theropods, the forelimbs are very long. There are three digits, each bearing long, recurved claws, that indicate rapacious habits.

Deinonychus differs from other carnivorous dinosaurs in the relatively great size of the coracoids, which would have served for the origin of large muscles that manipulated the forelimbs. The pubis extends directly ventrally and ends in a large "foot."

The rear limbs are very long, with a tibia that was 20 percent longer than the femur. These proportions are common to small theropods but not to the large genera. However, the metatarsals are not greatly elongated, and they and the foot are heavily built to support the body weight.

One of the most striking features of this genus is the enormous size of the claw on the second pedal digit, which must have served as a fearsome predatory weapon. It is

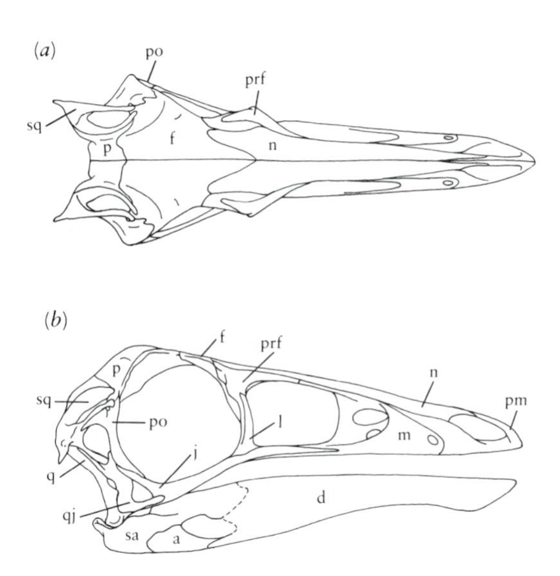

Figure 14-10. SKULL OF THE OSTRICH DINOSAUR *DROMICEIOMIMUS.* (*a*) Dorsal and (*b*) lateral view. Abbreviations as in Figure 8-3. *From Russell, 1972.*

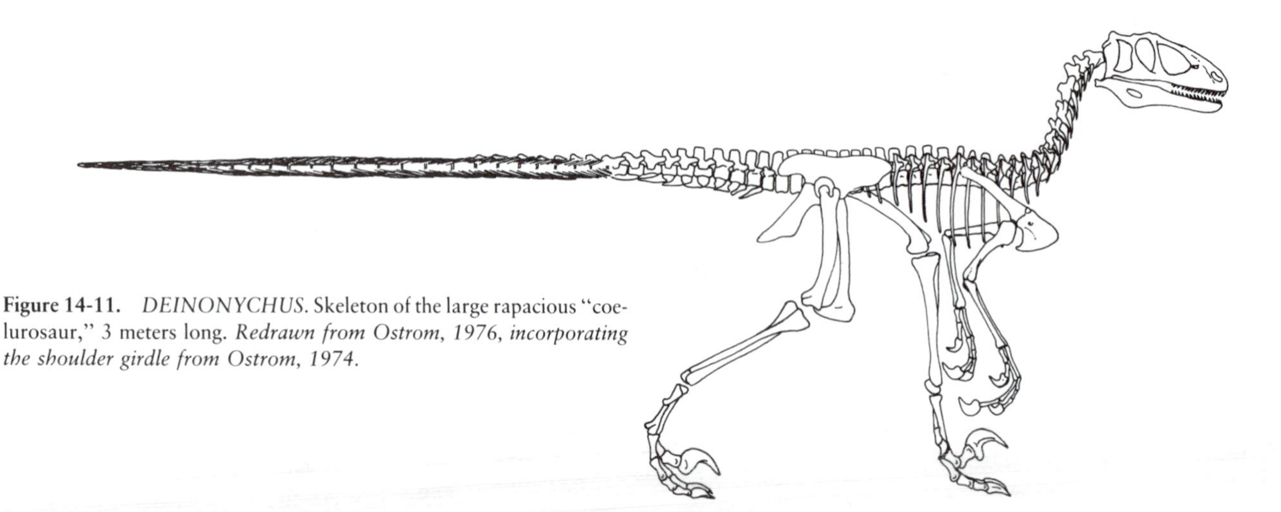

Figure 14-11. *DEINONYCHUS.* Skeleton of the large rapacious "coelurosaur," 3 meters long. *Redrawn from Ostrom, 1976, incorporating the shoulder girdle from Ostrom, 1974.*

so large that the digit was probably normally held above the remainder of the foot, which was functionally didactyl. The phalangeal count is 2, 3, 4, 5, 0.

The fourth trochanter, which is typical of most dinosaurs, is not evident on the femur, but a separate proximal process, the posterior trochanter, is conspicuous and may have served for the attachment of the ischiotrochantericus muscle. Ostrom (1969) suggests that this muscle may have served to retract the leg when the great claw dug into the prey. The use of this muscle, which originated on the ischium, would be less likely to upset the balance of the animal as it stood on one foot than would the use of the principal retractor, the caudifemoralis, which moved the tail laterally as it was contracted.

Ostrom (1969) tabulated an almost equal number of "carnosaur" and "coelurosaur" characters in *Deinonychus* and argued that these groups could not be clearly distinguished. On the other hand, he accepted that the ancestry of *Deinonychus* probably lay among the Coeluridae and hence that this genus was phylogenetically allied with the "coelurosaurs" despite the presence of many "carnosaur" features. The hands and feet in particular point to an ancestry that was shared with *Ornitholestes*. *Deinonychus* belongs to the family Dromaeosauridae, other members of which are known primarily from the Upper Cretaceous of North America and central Asia.

The Saurornithoididae constitute a related family of rapacious coelurosaurs from the Upper Cretaceous. They are somewhat smaller than dromaeosaurs, with lighter limbs but a similar enlargement of the ungual phalanx of the second pedal digit. They lack pleurocoels in the cervical vertebrae, which are present in dromaeosaurs and coelurids. The skull is smaller and relatively lower, but the braincase is relatively large and bulbous.

To judge by the structure of the skull, both dromaeosaurs and saurornithoidids had much larger brains than most reptiles, reaching about seven times the volume relative to body weight of crocodiles. This ratio is the same as that found in many birds and primitive living mammalian groups. The configuration of the orbits suggests that the eyes may have had a large area of overlap for binocular vision (Russell, 1969; Russell and Seguin, 1982). In their sensory acuity, degree of cerebral integration, and agility, the saurornithoidids represent a high point of dinosaur evolution.

The remaining theropods are all considerably larger and include the largest terrestrial predators that have ever lived. They culminate in the Upper Cretaceous in animals 15 meters long with a bulk of 7000 kilograms. The skull of *Tyrannosaurus* reached well over 1 meter in length, with individual teeth that were 15 centimeters long.

Many genera of large theropods are known, from as early as the late Triassic, but only a few have been adequately described and their relationships are not firmly established. As few as two or as many as five families may be recognized. The tyrannosaurids are a clearly defined group from the Upper Cretaceous. Most other large theropods may be placed in the Megalosauridae, or this name may be restricted to the most primitive large theropods, with separate families recognized for a series of more derived lineages: the Ceratosauridae, the Spinosauridae, and the Allosauridae.

The genus *Dilophosaurus* from the Lower Jurassic of North America (Welles, 1984) occupies a particularly isolated position among early theropods. With a length of over 6 meters, its bulk rivals animals that were previously classified as carnosaurs. The skull is large relative to the length of the cervical vertebrae but is lightly constructed. The forelimbs are relatively large (Figure 14-12).

The skull is distinctive in having a pair of narrow, parasagittal crests above the nasals and frontals, and the premaxilla is only weakly attached. The cervical vertebrae are short and planoconcave rather than amphicoelous, as they are in small podokesaurids. They are lightened by pleurocoels. There are four sacral vertebrae that remain unfused. The manus consists of three large and one reduced digit. In contrast with other large theropods, the pubis lacks a boot. The femur is slightly longer than the tibia, and the metatarsals are not greatly elongated. Three large toes face forward and a small one turns to the rear. *Dilophosaurus* shows no clear affinities with any later theropods.

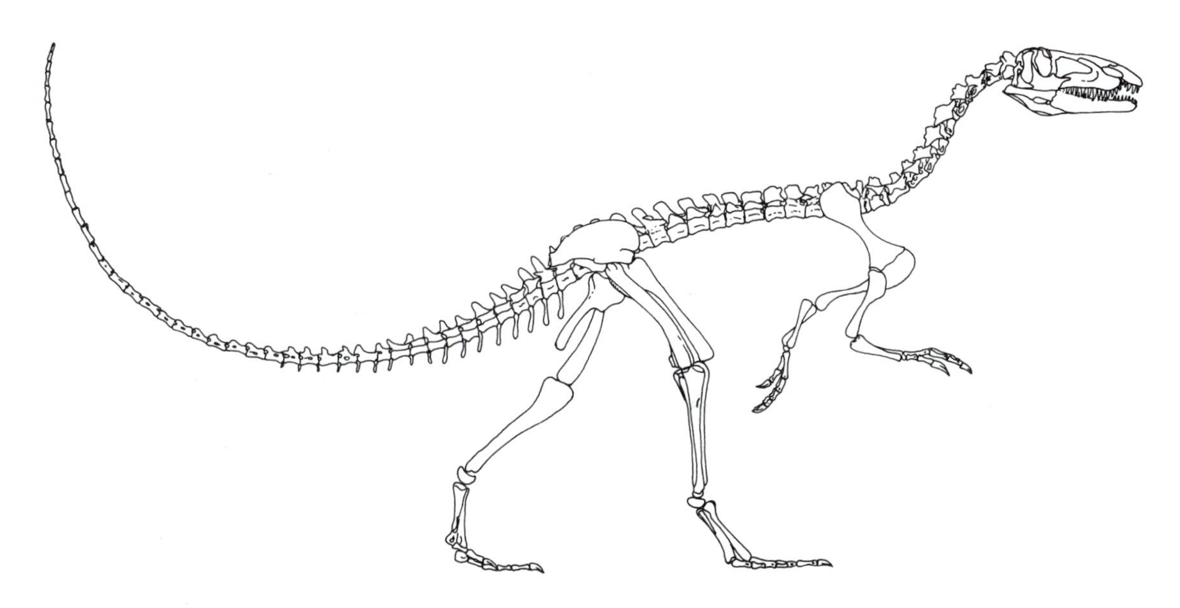

Figure 14-12. *DILOPHOSAURUS.* This is a large theropod from the Lower Jurassic, exhibiting a mosaic of "coelurosaur" and "carnosaur" traits. Approximately 6 meters long. *From Welles, 1984.*

In relationship to their large size, all tyrannosaurids and megalosaurs (if one uses that term in the more general sense) share a massively built skeleton, with a femur that is typically longer than the tibia and short, massive metatarsals. In addition, the forelimbs are always relatively short. There is as yet no strong evidence that these features were uniquely derived within the ancestry of these families. Tyrannosaurids and megalosaurs may be no more closely related to one another than either are to early members of the "coelurosaur" assemblage.

Megalosaurs

We find megalosaur remains from as early as the base of the Jurassic and possibly even in the late Triassic, but few animals earlier than the late Jurassic are represented by relatively complete remains and fewer yet have been adequately described. Jaw and limb elements that are placed in the genus *Megalosaurus* are known from the Lower and Middle Jurassic of England. The femur is almost 1 meter long, and the jaws and teeth are very large. Interdental plates are present at the base of the inside surface of the teeth of the upper and the lower jaw (see Figure 14-7).

Other material shows that early megalosaurs were fully bipedal, reached a length of approximately 6 meters, and that the distal end of the pubis, as in *Herrerasaurus,* was expanded into a large "foot" or "boot." As in other large theropods, the centra of the cervical vertebrae are opisthocoelous. Four digits are retained in both the front and hind limb.

The most completely known of the early megalosaurs is *Eustreptospondylus* from the Middle Jurassic of England, a specimen of which is on display at Oxford University. Von Huene (1926) described it under the name *Streptospondylus.* It consists of a nearly complete skeleton, although the skull is fragmentary. In general, it resembles the Upper Jurassic *Allosaurus* (Madsen, 1976), although it is more primitive in having only three sacral

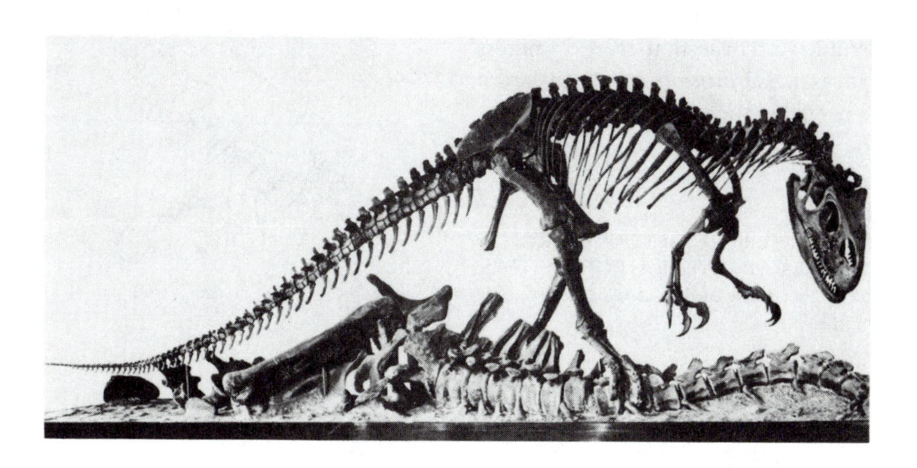

Figure 14-13. THE UPPER JURASSIC THEROPOD *ALLOSAURUS.* Adults of this genus reached lengths of 4 to 5 meters. *From Colbert, 1983.*

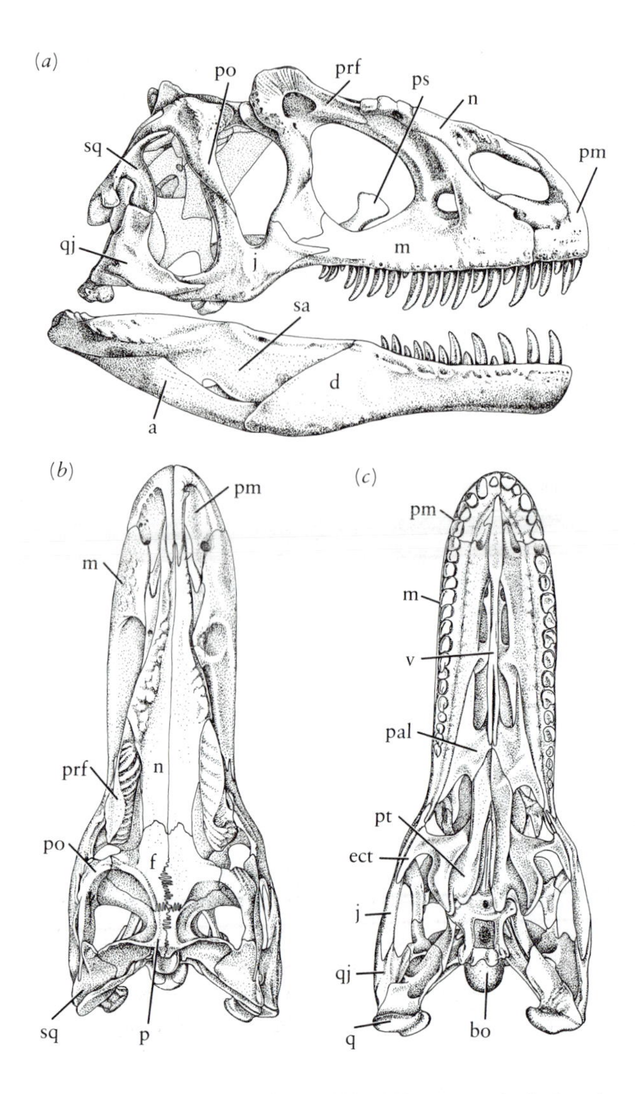

Figure 14-14. SKULL OF *ALLOSAURUS*. (*a*) Lateral, (*b*) dorsal, and (*c*) palatal views. Abbreviations as in Figure 8-3. *From Madsen, 1976.*

The upper Jurassic genus *Allosaurus* is the best-known representative of a conservative lineage that extends into the Upper Cretaceous. *Allosaurus* reached 12 meters in length and was among the most powerful carnivores of its time (Figure 14-13; see page 295). The body was strengthened to support the extra weight, and the sacrum had five fused vertebrae.

The skull is high and laterally compressed; the orbital opening is triangular and smaller than the principal antorbital fenestra (Figure 14-14). The teeth are long, laterally compressed, and recurved. In contrast with more primitive megalosaurs, the cervical vertebrae have well-ossified anterior condyles that contribute to well-defined ball-and-socket joints between the vertebrae. The more posterior trunk vertebrae retain a shallowly amphicoelous configuration. The tail is long, and the posterior prezygapophyses are considerably elongated.

The forelimb is considerably shorter than the rear limb and could not possibly have supported the body. Only three digits of the manus are retained. Each has a long recurved claw. Metatarsals II, III, and IV are closely integrated elements. The first digit, like that of birds, was oriented posteriorly.

Tyrannosaurids

The gigantic tyrannosaurids of the Upper Cretaceous were the largest of all theropods (Figure 14-15). Specializations of this family include the further reduction in the length of the forelimbs (which could not have reached the mouth) and the retention of only two functional digits, the first and second, which bear respectively two and three phalanges.

The skull is distinguished by the shape of the parietals, which form a sharp sagittal crest between the upper

vertebrae and a single antorbital opening. The femur is slightly longer than the tibia.

The early megalosaurs radiated during the Jurassic and led to specialized forms that included *Spinosaurus* from the Upper Cretaceous and its relatives, which had neural spines of the trunk vertebrae that were 2 meters long, and the Upper Jurassic *Ceratosaurus*, which had a short "horn" that was borne on the nasal bones (Stromer, 1915; Gilmore, 1920).

Figure 14-15. THE GIGANTIC UPPER CRETACEOUS THEROPOD *TYRANNOSAURUS*, WHICH REACHED A HEIGHT OF 6 METERS. The forelimbs are much shorter than those of *Allosaurus* and have only two digits. *From Osborn, 1916.*

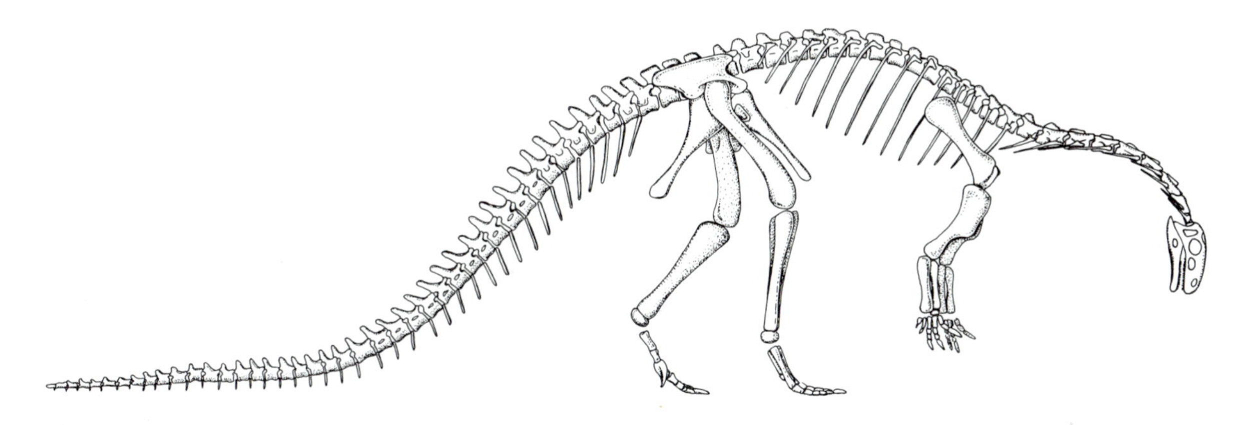

Figure 14-16. *PLATEOSAURUS.* Skeleton of the large plateosaur from the Upper Triassic of Germany. Length is about 7 meters. Plateosaurs are among the most primitive dinosaurs. Their dentition indicates that they were herbivores. They may have been facultatively, but not habitually, bipedal. *From Galton, 1971.*

temporal openings. In megalosaurs, the parietals are flat in this area. Little, if any, movement was possible between any of the elements of the skull in tyrannosaurids. In contrast, the lower jaw had some mobility between the dentary and more posterior elements, which allowed the gape to be increased so that the teeth could bite into extremely large prey.

Russell (1970a) recognizes three genera in North America: *Albertosaurus, Daspletosaurus,* and *Tyrannosaurus. Tyrannosaurus* also occurs in central Asia, where it was originally described under the name *Tarbosaurus,* together with *Alioramus* and *Alectrosaurus* (Kurzanov, 1976). The genus *Indosuchus* is recognized in India (Chatterjee, 1978).

We assume that tyrannosaurids evolved from among the megalosaurs, but no specific ancestor has been recognized. Chatterjee (1985) proposes that they may have evolved directly from Triassic thecodonts.

SAUROPODOMORPHS

PLATEOSAURS

The theropods were the first dinosaurs to be described because their anatomy could be readily derived from the pattern of the carnivorous thecodonts. These large predators probably fed on comparable-sized prey. The major food sources for the megalosaurs and tyrannosaurids probably included the giant sauropods of the Jurassic and Cretaceous. These sauropods were preceded in the late Triassic by an assemblage of more primitive herbivorous saurischians, the plateosaurs or prosauropods.

The best known plateosaur is *Plateosaurus* (Figures 14-16 and 14-17) from the Upper Triassic of Germany, but we find other genera in eastern and western North America, South America, southern Africa, and China. Plateosaurs were 1 to 7 meters long and of moderately heavy

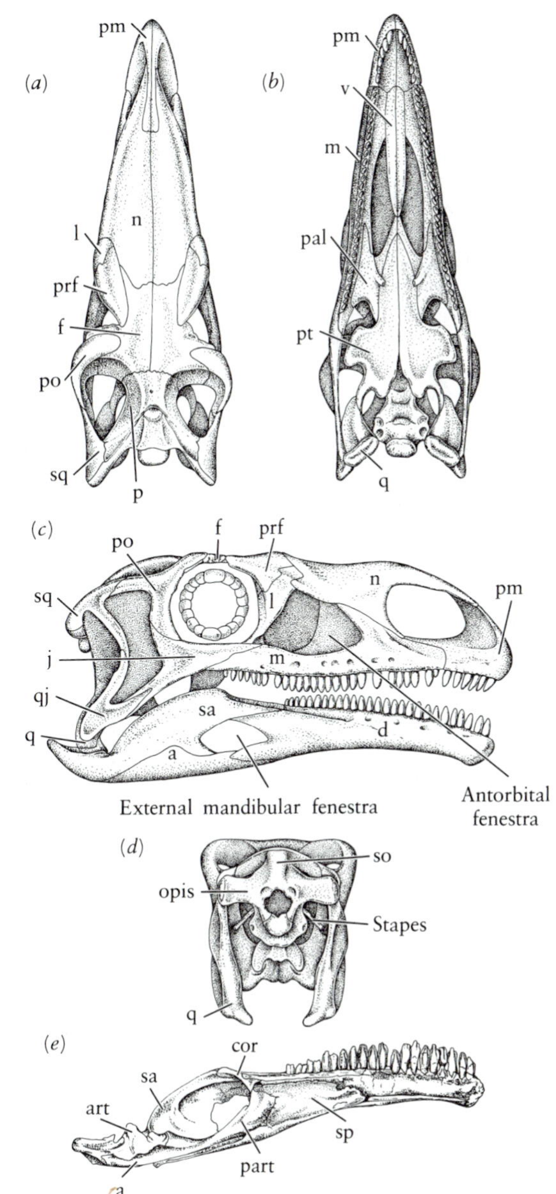

Figure 14-17. THE SKULL OF THE UPPER TRIASSIC HERBIVOROUS DINOSAUR *PLATEOSAURUS.* (*a*) Dorsal, (*b*) palatal, (*c*) lateral, and (*d*) occipital views, and (*e*) the medial surface of the lower jaw. *Courtesy of Peter Galton.*

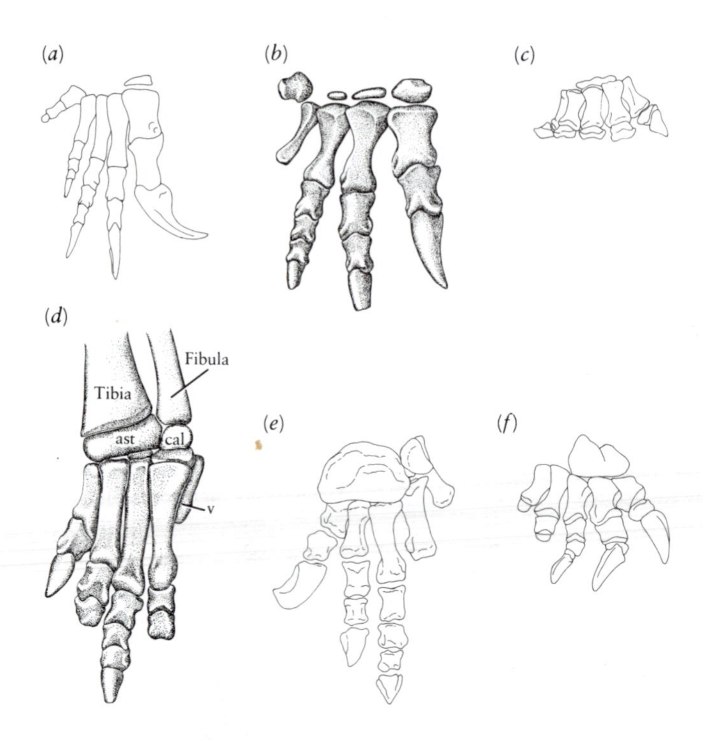

Figure 14-18. HANDS AND FEET OF SAUROPODOMORPHS. (*a*) Manus of the plateosaur *Anchisaurus*, Upper Triassic, ×¼. *From Galton, 1976.* (*b*) Manus of *Riojasaurus*, ×¼, a form that is structurally transitional between plateosaurs and sauropods, from the Upper Triassic. *From Bonaparte, 1971.* (*c*) The Upper Jurassic sauropod *Apatosaurus. From Coombs, 1975.* (*d*) The foot of *Riojasaurus*, ×$\frac{1}{12}$. *From Bonaparte, 1971.* (*e*) Foot of *Vulcanodon,* a form that is structurally intermediate between plateosaurs and sauropods, from the Lower Jurassic of East Africa. *From Raath, 1972.* (*f*) Foot of the Upper Jurassic sauropod *Apatosaurus. From Coombs, 1975.* Abbreviations as follows: ast, astragalus; cal, calcaneum; v, fifth metatarsal.

build. They were initially restored in a bipedal pose because of the greater length of their hind limbs, but footprints demonstrate that they were usually quadrupedal.

The skull is small relative to the trunk, and the neck is long. The teeth have laterally compressed crowns with crenulated edges that resemble those of herbivorous lizards. In advanced genera, the jaw articulation is well below the level of the tooth row, a feature that frequently developed in herbivorous groups to provide stronger leverage for the jaw muscles.

Plateosaurs were once thought to include carnivorous genera because of the discovery of large, bladelike teeth in association with their skeletons, but these teeth were probably left by other dinosaurs or large thecodonts that preyed upon them or fed on their carcasses.

Plateosaurs had three sacral vertebrae. Like *Staurikosaurus* and *Herrerasaurus,* the iliac blade is very short; the pubes are long narrow bones that fused at the midline but were not expanded vertically at the distal end. Proximally, each surrounds a large obturator foramen.

The forelimb is approximately two-thirds the length of the hind limb. The humerus, ulna, and radius are stout, and the manus is short and retains all five digits. It probably served for support, although the first digit is larger than the remaining ones and carries a huge claw (Figure 14-18).

The femur broadly resembles that of other early dinosaurs but is relatively thicker. The tarsus shows a well-developed mesotarsal hinge. The foot has four strongly developed, but short, digits. The fifth is reduced to a splint. According to van Heerden (1979), the posture was not fully improved, with the femur angled at about 25 degrees from the vertical.

The plateosaurs are among the earliest and most primitive of all dinosaurs. They probably shared a common ancestry with forms such as *Staurikosaurus* and *Herrerasaurus.* As in the case of the protosuchian crocodiles, we now recognize many of the plateosaurs that were once thought to be from the Triassic as being early Jurassic in age. They represent an early experiment in herbivory among the dinosaurs.

In the later Mesozoic, plateosaurs were replaced as effective herbivores by the sauropods and ornithopods. Already in the Upper Triassic, more advanced saurischian herbivores had evolved, as evidenced by *Riojasaurus* (Figure 14-19).

Galton (1977, 1985) and van Heerden (1979) recognized four families of prosauropods. The fully quadrupedal Melanorosauridae from the Late Triassic and early Jurassic appear close to the ancestry of the sauropods.

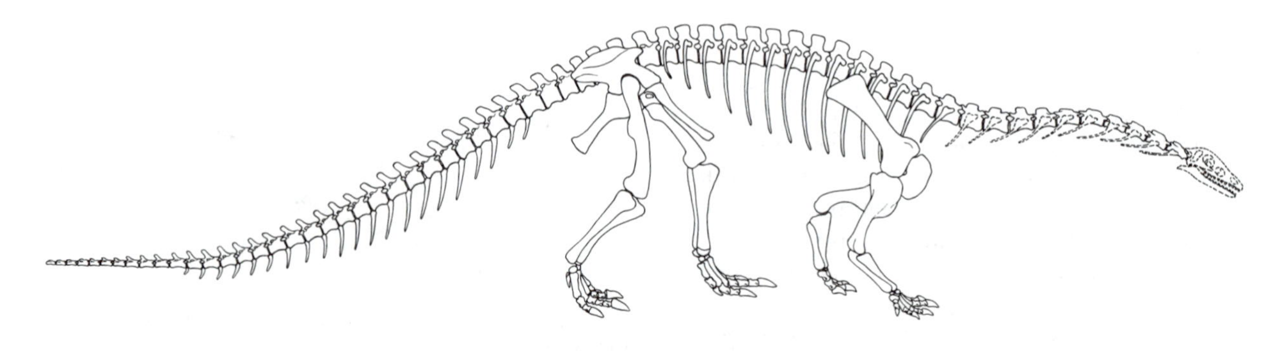

Figure 14-19. SKELETON OF *RIOJASAURUS.* This primitive sauropodomorph, from the Upper Triassic of South America, is 10 meters long. *From Bonaparte, 1978.*

SAUROPODS

Sauropods, which are exemplified by the Upper Jurassic genera *Diplodocus, Camarasaurus* and *Brachiosaurus,* are characterized by their enormous size, with weights as great as 80,000 kilograms (Colbert, 1962) and body lengths approaching 30 meters. They were obligatorily quadrupedal, and the limb bones are greatly thickened, solidly ossified, and held in a nearly vertical position. Like elephants, they did not bend much at the elbow and knee, and the stride would have been short. The feet were graviportal, with five short digits on both the manus and pes.

Surprisingly, the joint surfaces of the limbs are poorly defined, and there must have been a good deal of cartilage in the carpus and tarsus as well. The capacity for the cartilage to yield under pressure and conform to a shape that would most effectively distribute the force produced by the weight of the body was apparently more important than the greater per unit strength of bone.

The skull is strikingly small, and the teeth are simple, incisorlike structures that form a single functional row. The teeth show wear that could be associated with cropping vegetation, but they lack the grinding or shearing surfaces that are provided by the cheek teeth of ornithischians. The teeth are set at the margins of the jaw, which precludes the presence of a fleshy cheek that would help to retain the food within the mouth cavity. Processing of food in sauropods may have occurred primarily within the digestive tract, either with the help of gizzard stones, which are reported in a few sauropods, or through a symbiotic relationship with intestinal microorganisms (Coombs, 1975).

All adequately known sauropods have long necks, with 12 to 19 cervical vertebrae, each of which are elongated relative to those of the trunk. There are usually about 12 trunk vertebrae and from four to six sacrals.

We recognize nearly 50 genera of sauropods that are grouped in 6 subfamilies or families, but few are well known. Because of their great size, sauropod remains are easily found but difficult to collect and prepare. Nevertheless, we have an excellent record in the Upper Jurassic from geographically diverse localities: the Tendagaru in East Africa, the Szechwan basin in China, and the Morrison Formation in Utah, Colorado, and Wyoming. In contrast, their history in the early Jurassic and for most of the Cretaceous is poorly known.

Because of their gigantic size, the habitat of sauropods has long been cause for speculation. Some authors have suggested that their enormous bulk could only have been supported if they lived in the water. An aquatic way of life also seems to be suggested by the dorsal position of the nostrils in several genera. On the other hand, there is nothing in the skeleton that can be associated with aquatic locomotion, and the limbs and girdles have many attributes that are found in large terrestrial mammals such as elephants. Some strictly terrestrial mammals have dorsal narial openings, including genera with a trunk or complex nasal structures.

Coombs (1975) concludes that sauropods may have been quite varied in their habits, both as a group and seasonally within individual species. The presence of five or six well-defined genera within a single fossil locality suggests that they must have differed considerably in their specific ways of life to avoid competition for particular food sources. Most sauropods may have lived near streams, swamps, and lakes to judge by footprints and depositional evidence, but it is unlikely that they habitually lived in deep water, as has been suggested in some restorations (Dodson, Behrensmayer, Bakker, and McIntosh, 1980). No direct evidence is available regarding their diet. With such small heads, they must have spent most of their lives eating.

Sauropods have long been thought to have evolved from the plateosaurs. In addition to their common herbivorous habits, they are united by similar features of the postcranial skeleton. Some of these similarities may be attributed to common specializations for a massive body. The presence of a very large first digit of the manus with a strong claw may have been a unique specialization of the common ancestor of both groups. In most sauropods, especially the most primitive forms, the forelimbs are significantly shorter than the rear limbs, which suggests an ancestry among facultatively bipedal animals.

Most of the well-known plateosaurs are too late to be the actual ancestors of the sauropods, which appear to have diverged from the ancestral sauropodomorph lineage before the end of the Triassic.

Primitive sauropods

The oldest known genus that can be closely associated with the later Mesozoic sauropods is *Riojasaurus* (Bonaparte, 1978) from the Upper Triassic of South America (Figures 14-19 and 14-20). The skeleton is only about 6 meters long, but the girdles and limbs are much more massive than those of the plateosaurs, which indicate an

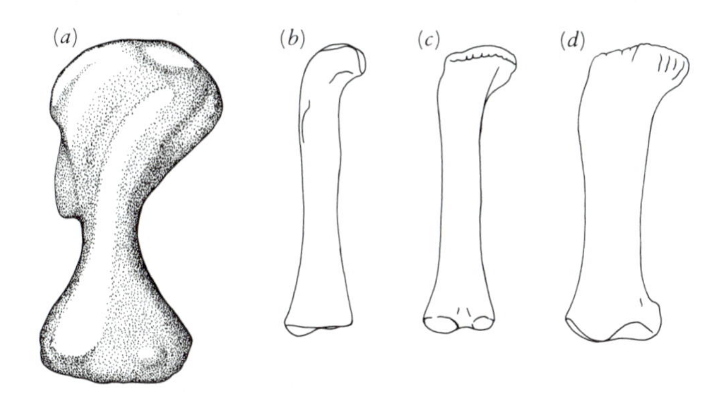

Figure 14-20. LIMB BONES OF SAUROPODOMORPHS. (*a*) Humerus of *Riojasaurus* from the Upper Triassic, $\times \frac{1}{10}$. *From Bonaparte, 1971.* (*b, c,* and *d*) Femora of *Plateosaurus* (Upper Triassic), *Barapasaurus* (Lower Jurassic) and *Apatosaurus* (Upper Jurassic). *From Jain, Kutty, Roychowdhury, and Chatterjee, 1977.*

obligatorily quadrupedal stance. The articulating surfaces of the humerus are enormously expanded. The ulna is massive, with little development of the olecranon. These features contrast with the quadrupedal ankylosaurs and stegosaurs that are discussed later in this chapter, both of which have a well-developed olecranon that may have been associated with a more sharply angled elbow joint. As in the plateosaurs, all five digits of the forelimb are retained. The first digit bears a very large claw, but the other phalanges of all the digits are relatively short.

The ilium is massive but short from back to front and is associated with only three sacral vertebrae. The pubis and ischium are short relative to those of theropods and generally resemble those of the plateosaurs (Figure 14-21). The puboischiac plate is extensive, as in primitive archosaurs. The femur is massive and considerably longer than the tibia. The tarsus and pes closely resemble those of the plateosaurs, with wide but flat astragalus and calcaneum and two small distal tarsals. The metatarsals of digits I through IV are long and fairly slender, and the fifth metatarsal is much shorter. The digits are all short but retain claws.

The skull is not known. There are estimated to be 10 or 11 cervical and 15 trunk vertebrae. In contrast with later sauropods, the vertebrae remain primitive in their general form but, like plateosaurs and advanced sauropods, they have extra articulating surfaces—the hyposphene and hypantrum, which are medial to the zygapophyses—that would contribute to the rigidity of the column.

Vulcanodon from the early Jurassic of Zimbabwe is even more similar to later sauropods (Cooper, 1984). Much of the girdles, limbs, and posterior axial skeleton is known but, unfortunately, not the head or neck. The trunk and tail would have been approximately 6 meters long. Most skeletal features are close to, but slightly more primitive than, typical sauropods and similar to, but advanced over the condition in robust prosauropods. The pelvis and rear limbs exhibit a clear mosaic of prosauropod and sauropod features. The sacrum is advanced over prosauropods in incorporating four fused vertebrae. However, the shape of the pubes, which form an anteriorly facing "apron," is typical of prosauropods. The femur is straight rather

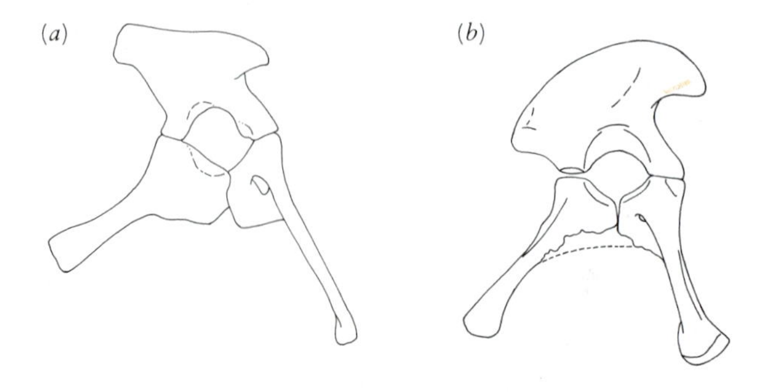

Figure 14-21. PELVES OF SAUROPODOMORPHS. (*a*) Pelvis of *Plateosaurus* from the Upper Triassic. *From Raath, 1972.* (*b*) Pelvis of *Barapasaurus* from the Lower Jurassic. *From Jain, Kutty, Roychowdhury, and Chatterjee, 1977.*

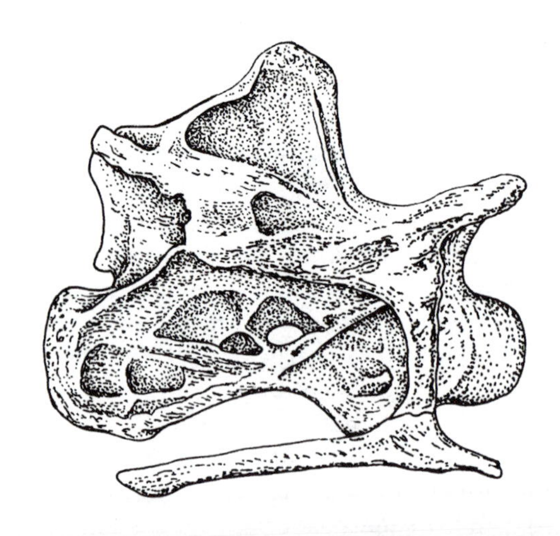

Figure 14-22. CERVICAL VERTEBRA OF THE UPPER JURASSIC SAUROPOD *DIPLODOCUS*, $\times \frac{1}{10}$. The very open structure conserves weight. *From Hatcher, 1901.*

than sigmoidal, and the distal tarsals are not ossified. In contrast, the large size of the ungual of the hallux resembles that of prosauropods.

The structure of both the forelimbs and hind limbs indicates that *Vulcanodon* was strictly quadrupedal and shifting from digitigrade to semiplantigrade posture. Cooper emphasized particularly close similarities in primitive features to *Plateosaurus* and the melanorosaurid *Euskelosaurus* combined with advanced characters that presage the pattern in diplodocid and camarasaurid sauropods.

Cetiosauridae

Barapasaurus from the Liassic of India is contemporary with *Vulcanodon* (Jain, Kutty, Roychowdhury, and Chatterjee, 1977). This genus is represented by a great deal of disarticulated material but no reconstruction has yet been attempted. It has already reached sauropod dimensions, approaching the size of *Diplodocus*. The individual bones are generally similar to those of the later sauropods but are primitive in some respects (Figures 14-20 and 14-21). Unlike those of *Riojasaurus* and *Vulcanodon*, the vertebrae resemble those of advanced sauropods in the development of large cavities in the centra and neural arches that serve to lighten these structures (Figure 14-22). The cervical vertebrae are opisthocoelous and elongate, but most of the trunk vertebrae remain shallowly amphicoelous. Four fused sacral vertebrae are attached to a long and high iliac blade of typical sauropod proportions.

Barapasaurus is placed in the family Cetiosauridae, which spans the length of the Jurassic. Another member of this family was briefly described on the basis of numerous skeletons from the Middle Jurassic of Szechwan Province, China (Dong and Tang, 1984). *Shunosaurus* is approximately 7 meters long, with 13 cervicals (which are one and one-half times as long as the dorsal), 12 dorsals, and 4 sacral vertebrae. The skull has spatulate teeth and large, paired narial openings that are located high on the snout—a configuration that is similar to that

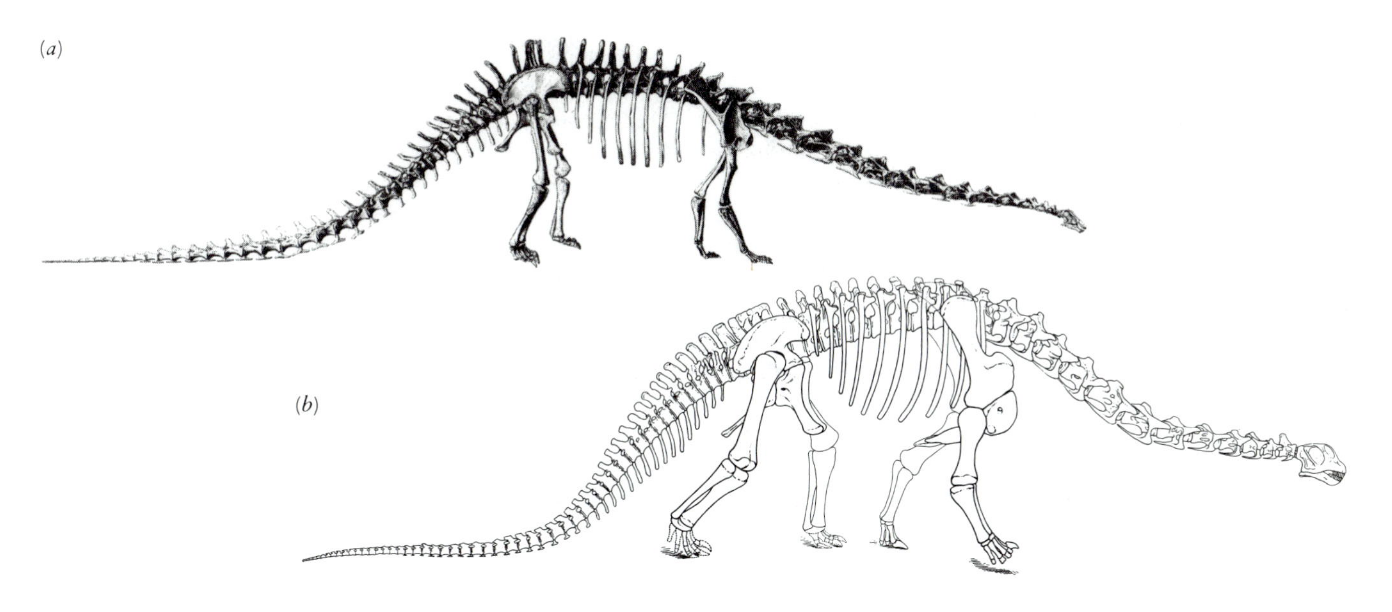

Figure 14-23. (*a*) Skeleton of the Upper Jurassic sauropod *Diplodocus*, 30 meters long. *From Hatcher, 1901.* (*b*) Skeleton of *Camarosaurus*, from the Upper Jurassic. *From Osborn and Mook, 1921.*

of the Upper Jurassic genus *Camarasaurus*. Within this family, the forelimbs range from two-thirds to four-fifth the length of the hind limbs.

The Cetiosauridae remains a poorly known and incompletely described assemblage that probably represents a basal stock that is ancestral to the more specialized forms of the late Jurassic and Cretaceous. The later forms are all advanced in the excavation of the vertebrae but vary in the number of sacral vertebrae; the proportions of the neck, tail, and limbs; and the nature of the skull and its dentition.

Diplodocidae

The Diplodocidae are found throughout the world in the Upper Jurassic, but this family is limited to eastern Asia in the Upper Cretaceous. Both *Diplodocus* and *Apatosaurus* [*Brontosaurus*] are known from complete, thoroughly described skeletons (Figure 14-23). Berman and McIntosh (1978) most recently reviewed this group. The body reaches approximately 30 meters in length, with the skull about 55 centimeters long at the end of a very long neck. Within the family, the number of cervical vertebrae varies from 12 to 19, with the individual vertebrae two and one-half times the length of the trunk vertebrae. There are five coossified sacrals, and the tail has a maximum of over 80 vertebrae; the last are long, narrow, and form a thin whiplash.

The skull (Figure 14-24) has large orbits with a median narial opening between them at the top of the skull. The teeth are long slender structures that are limited to the front of the mouth. They would be effective for cropping food and are rapidly replaced, if we judge by the number of replacement teeth that are present in the jaws.

The vertebrae, especially the cervicals, are huge. They are not formed of solid bone but by a complex lattice

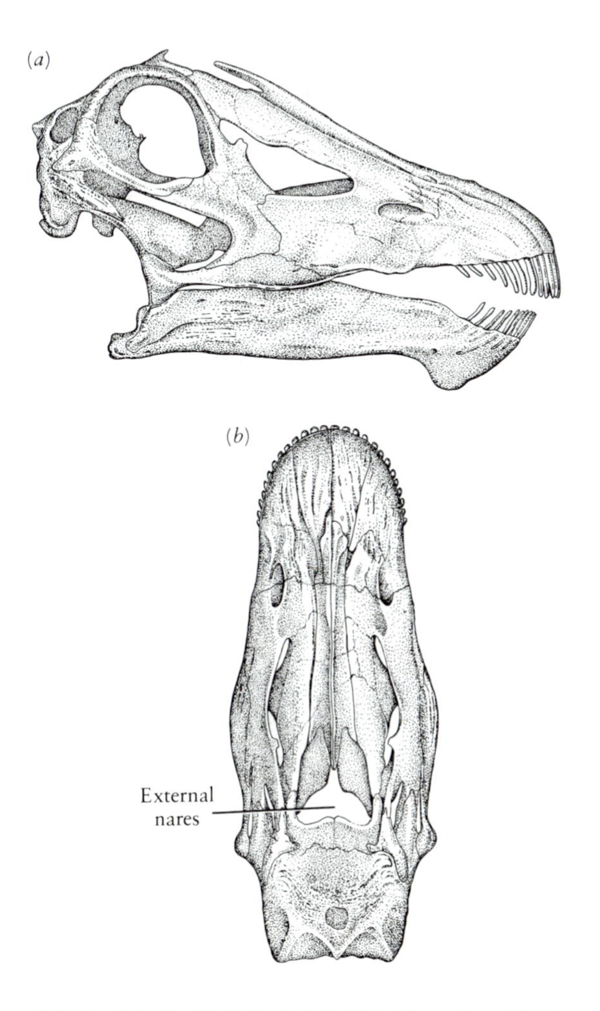

Figure 14-24. SKULL OF *DIPLODOCUS*. (*a*) Lateral and (*b*) dorsal views, 55 centimeters long. *From Ostrom and McIntosh, 1966.*

work that surrounds large openings. Some authors suggest that these spaces might have been occupied by air sacs that were connected with the lungs, as is the case among birds. The cervical and anterior trunk vertebrae are strongly opisthocoelous. There are only 10 trunk vertebrae that have tall neural spines.

The rib cage is high and narrow. Compared with other sauropods, the limbs of *Diplodocus* are relatively slender. The front limbs are two-thirds to three-fourths the length of the rear limbs. In *Diplodocus,* there are only two carpals, and the rest of the carpus is formed entirely of cartilage. In *Apatosaurus,* there is only a single carpal. In both genera, the only bone of the tarsus to ossify is the astragalus. The weight is borne by five metacarpals and five metatarsals of the hand and foot. The phalanges are greatly shortened. The first toe in the hand bears a large claw, like that of the plateosaurs. Two or three of the toes on the foot end with claws (see Figure 14-18).

Camarasauridae

The Camarasauridae is represented by *Camarasaurus* from the Upper Jurassic of North America and *Opisthocoelocaudia* from the Upper Cretaceous of Mongolia. *Camarasaurus* has a short neck with only about 12 cervical vertebrae that are only a little longer than the dorsals (which also number 12). The tail is short. The skull is high, and the teeth are spoon shaped and extend for much of the length of the jaw margin, in contrast with those of the Diplodocidae. The nasal openings are high but paired (Figure 14-25*a*). The front limbs are two-thirds to three-fourth the length of the rear. The caudal centra of *Camarasaurus* are procoelous, as in most sauropods, but they are opisthocoelous in the later, Asian genus.

In camarasaurids, as in all other Upper Jurassic sauropods, the sacrum is composed of five vertebrae: one dorsosacral, three true sacrals, and one caudosacral.

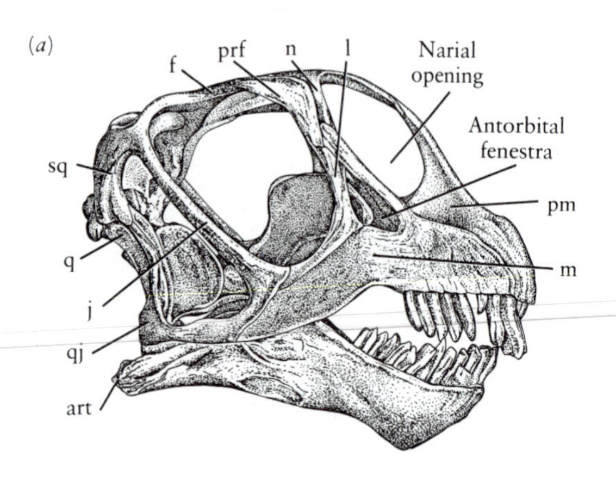

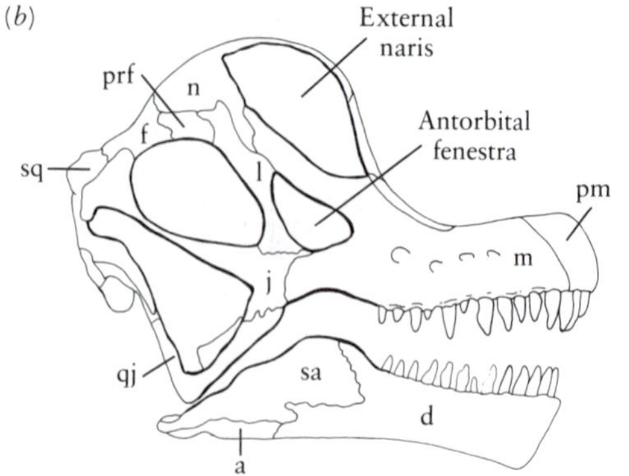

Figure 14-25. (*a*) Skull of *Camarasaurus,* $\times \frac{1}{10}$. *From Osborn and Mook, 1921.* (*b*) *Brachiosaurus. From Lapparant and Lavocat, 1955.* Abbreviations as in Figure 8-3.

Brachiosauridae

The Brachiosauridae are among the largest of all sauropods, with estimated weights of 80,000 kilograms. The skull is high, with the nasal openings bulging up in front of the orbits. The neck is extremely long, and the individual vertebra are three times as long as the dorsals. The front limbs are as long or longer than the rear limbs. We find brachiosaurs from the Middle Jurassic into the Lower Cretaceous in North America, Africa (including Madagascar), Europe, and perhaps in Australia but not in Asia.

Euhelopodidae

In contrast, the Euhelopodidae are known only in eastern Asia from the Upper Jurassic and Lower Cretaceous (Dong, Zhou, and Zhaug, 1983). These genera exhibit 17 to 19 cervicals that range in size from one and one-third to two and one-third times the length of the dorsals. There are 14 dorsals, which results in a very long presacral column. The teeth are spatulate and the skull of *Euhelopus* is similar to that of *Camarasaurus.*

Titanosauridae

The Titanosauridae are a primarily Cretaceous assemblage that had a possible forerunner in the Upper Jurassic of East Africa. Although occurring in North America and Europe, they are more common in the southern continents. The most important defining character is the strongly procoelous nature of the anterior caudal vertebrae, which allowed the tail to move more freely from side to side. There are six sacral vertebrae, and the neck is comparatively short. The skull of the South American genus *Antarctosaurus* is decidedly like that of *Diplodocus* in character (von Huene, 1929).

ORNITHISCHIANS

We have difficulty establishing the specific interrelationships of the major groups of saurischian dinosaurs because of the incomplete fossil record in the late Triassic. The early history of the ornithischians is also incompletely known, but this assemblage shares a number of derived features that clearly demonstrate a common origin.

The most significant feature is the presence of a unique bone at the symphysis of the lower jaws, the predentary.

This element does not bear teeth but may have had a horny covering like the beak of turtles. The predentary is not firmly attached to the more posterior bones of the jaws in primitive genera and may have allowed them to rotate slightly on their long axis to help manipulate food.

The teeth of all ornithischians have laterally compressed crowns that are crenulated along the edges. As with the plateosaurs, this pattern is associated with feeding on plant material. In contrast with plateosaurs and sauropods, the tooth rows of most ornithischians are medial to the margins of the dentary and maxilla, which indicates the presence of fleshy cheeks that would have assisted in retaining the food in the mouth as it was chewed. In most genera, a specific occlusal relationship is achieved between the upper and lower teeth. According to Galton (1973), these advances in the feeding apparatus were probably the main reason for the success of the ornithischians throughout the Jurassic and Cretaceous.

Crocodiles and some other tetrapods have bony plates in the eyelid or along the dorsal margin of the orbit, while in primitive ornithischians, a narrow bone, the supraorbital, attaches to the anterior margin of the orbit. A comparable bone is not present in saurischians. Two supraorbitals are present in several ornithischian groups, including the iguanodontids and pachycephalosaurs.

The character that is most commonly used to distinguish ornithischians from saurischians, and that Seeley (1888) used as the basis for recognizing the two dinosaur orders, is the configuration of the pelvis (see Figure 14-3). Saurischians retain the primitive, "lizardlike" pelvic pattern that is modified beyond that of early archosaurs by the extension of the pubis and ischium anteriorly and posteriorly to form, with the ilium, a triradiate structure. Ornithischians are specialized in having the pubis lie alongside the ischium in a posterior orientation. It is commonly said that the pubis has rotated posteriorly, but without any fossils that show an intermediate condition leading to that of the early ornithischians, we cannot say with assurance why the change occurred or how it was achieved.

Romer (1956), Galton (1969), Charig (1972), Walker (1977), and Santa Luca (1980) have all tried to explain why the pubis is in a posterior position. But their explanations are not fully satisfactory since it is very difficult to understand why the shift would have occurred in ornithischians but not among the obligatorily bipedal saurischians, such as *Coelophysis*, which resemble them in many other skeletal features. This change is particularly difficult to understand since many later ornithischians, including the small bipedal hypsilophodontids, redevelop an anterior pubic process that compares topographically with the primitive saurischian pubis.

Charig (1972) suggests that all dinosaurs had to modify the configuration of the pelvis from the pattern of advanced thecodonts because of the greater stride developed as a result of a fully upright rear limb. In primitive archosaurs, the swing of the femur was limited by the anterior extent of the pubis because the major muscles that protract the limb originated on that bone. If the femur moved anteriorly to the level of the pubis or even beyond, the efficiency of these muscles would be progressively reduced.

Several solutions to this problem are possible. If the body were to assume a habitually bipedal stance, the pubis would be tipped upward and out of the way of the femur. This solution was achieved in the theropods. The sauropods, which are graviportal animals, would not have moved the limbs in a wide arc, so that the configuration of the primitive pelvis would not have been a problem.

Ornithischians may have solved this problem by shifting the site of origin of the femoral protractors to other bones. The ilium of primitive ornithischians extends far forward of the acetabulum, which would provide space for the origin of the protractors of the femur, including the sartorius (iliotibialis) and the puboischiofemoralis. As in modern crocodiles, the latter muscle could also have originated from the posterior trunk vertebrae.

If the pubis were not functionally necessary for the origin of these muscles, it would not need to have remained in its primitive position. Selection may have occurred to reverse its position, possibly to allow the abdominal cavity to extend posteriorly and increase the volume of the gut in the herbivorous ornithischians (but not in the carnivorous theropods) or to shift the center of gravity posteriorly to facilitate a bipedal stance.

Two major groups of ornithischians have a well-developed pattern of dermal armor that might have been inherited from the thecodonts. In this one feature ornithischians may be more primitive than saurischians. Ornithischians, but not saurischians, have ossified tendons lateral to the vertebral column in the posterior trunk and the sacral and caudal regions.

We have not found any fossils that link early ornithischians with any of the saurischian groups. Bakker and Galton (1974) pointed out features of the dentition in which ornithischians resemble plateosaurs and aspects of the rear limb and feet that they share with early coelurosaurs, but none of the known saurischians combine the skull and postcranial features that we would expect to find in the immediate ancestors of ornithischians. Similar derived features of the rear limb and tarsus suggest that ornithischians and saurischians share a common ancestry among animals that are similar to *Lagosuchus*.

The earliest known ornithischian is *Pisanosaurus* from the Upper Triassic of South America (Bonaparte, 1976). It is very incompletely known but shows the specializations of the lower jaw and dentition that characterize the much-better-known early Jurassic genera.

FABROSAURIDS

We find several distinct ornithischian lineages in the lowest Jurassic from beds that were previously thought to be

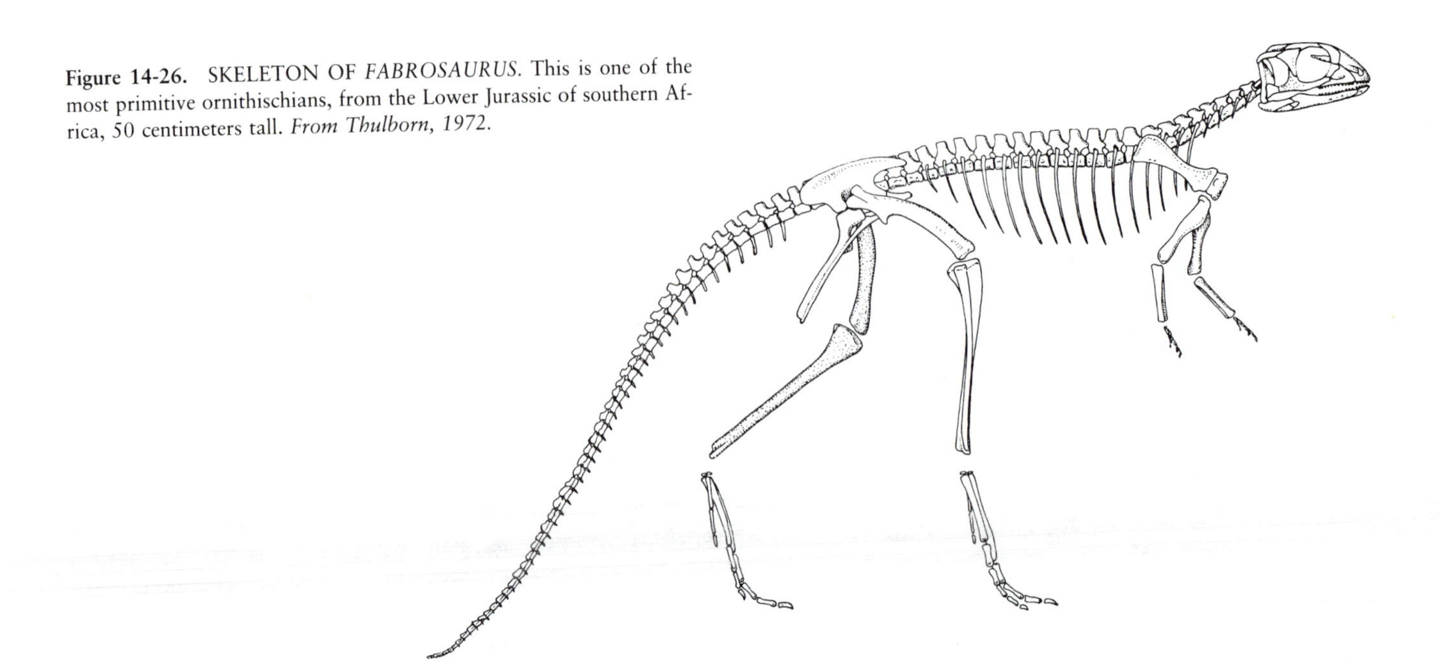

Figure 14-26. SKELETON OF *FABROSAURUS*. This is one of the most primitive ornithischians, from the Lower Jurassic of southern Africa, 50 centimeters tall. *From Thulborn, 1972.*

late Triassic in age. Although they are distinct from one another in important features of the dentition, their basic skeletal anatomy is very similar. The fabrosaurids show a pattern that could be close to that from which most other ornithischians evolved (Figure 14-26).

The best-known genus is *Fabrosaurus* [*Lesothosaurus*] from southern Africa (Thulborn, 1972; Galton, 1978; Gow, 1981). Like the podokesaurid *Coelophysis*, the skeleton is light and has exceedingly slender limbs. The forelimbs are even shorter, indicating obligatory bipedality. The total length is under 1 meter.

The skull is small and short with a large orbit that is crossed by a slim supraorbital bone. The antorbital opening, as in most ornithischians, is small and partially occluded by the maxilla (Figure 14-27). The premaxilla bears a continuous row of small incisiform teeth; the teeth in the maxilla and dentary extend along the margin of the bones. The tooth shape is common to that of all ornithischians. Like those of plateosaurs, the crowns are laterally compressed and leaf shaped. The upper and lower teeth occlude alternately between one another and have enamel on both the medial and lateral surfaces (see Figure 14-30*a*).

The cervical centra are short, which results in a shorter neck than is found in the early theropod dinosaurs. The trunk centra remain shallowly amphicoelous. There are approximately 22 presacral vertebrae and 5 sacrals. Alongside the posterior trunk and sacral vertebrae are numerous rod-shaped, ossified tendons that stiffened the column. They may have evolved in relationship to a habitually bipedal stance in primitive ornithischians but are retained in later quadrupedal forms as well.

There is a long, slender tail; the zygapophyses at its base are nearly vertical, which would have limited movement of the tail to the vertical plane. This limitation may have facilitated balance on the rear limbs. The high angle of the zygopophyses would also have reduced the tendency for the tail to be bent laterally during contraction of the caudifemoralis musculature, which served as a major retractor of the rear limbs.

The scapula is thin and slender as in small theropods. The coracoid is not known in fabrosaurids, but in other primitive ornithischians it is a small oval bone. Primitive ornithischians retain a slender clavicle like the early saurischians. The humerus, like that of other small, primitive dinosaurs, has a slender shaft but a large deltopectoral crest. In *Fabrosaurus*, the carpus is poorly ossified and the manus has four functional digits, with the fifth greatly reduced.

As in early "coelurosaurs," the ilium is long and accommodates five pairs of sacral ribs. The ischium, like that of small theropods, is slender and extends posteroventrally. Proximally it bears a ventral projection, the

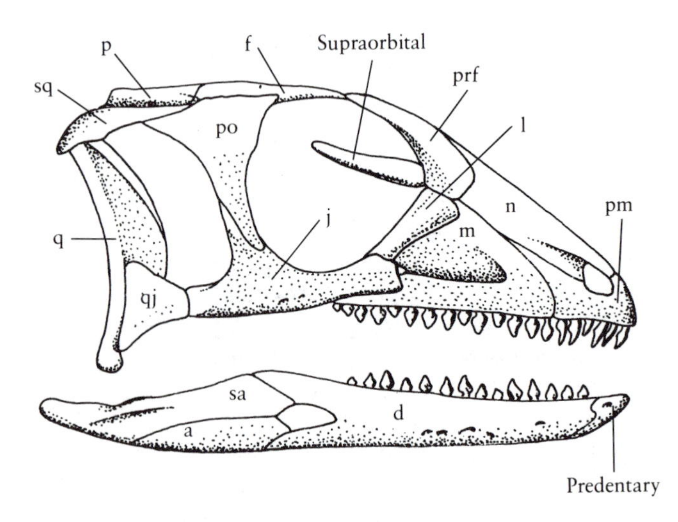

Figure 14-27. SKULL OF *FABROSAURUS*. Abbreviations as in Figure 8-3. *From Thulborn, 1970.*

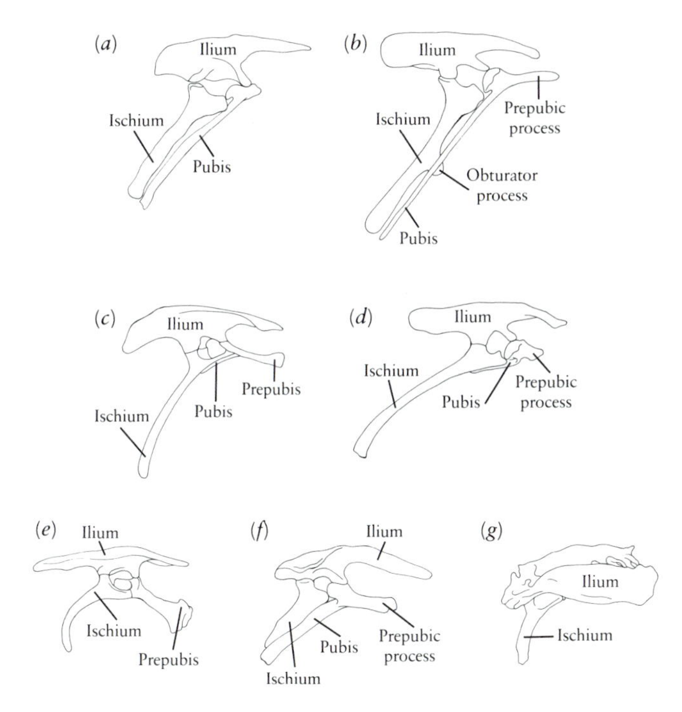

Figure 14-28. PELVES OF ORNITHISCHIANS. (*a*) *Scelidosaurus*, a primitive ornithischian without a prepubic process. (*b*) *Hypsilophodon*, a typical ornithopod with a prominent obturator process. (*c*) *Homoalcephale*, a pachycephalosaurid. (*d*) *Protoceratops*. (*e*) *Triceratops*. (*f*) *Stegosaurus*. (*g*) *Euoplocephalus*, an ankylosaur. *From Coombs, 1979.*

obturator process, that is also present in a major assemblage of later ornithischians, the ornithopods. The very slender pubis lies just below the ischium. In contrast with most later ornithischians, there is only a short anterior pubic process (Figure 14-28).

The head of the femur is *Fabrosaurus* is not strongly inturned. The shaft is slightly bowed anteriorly, and there is a long pendant fourth trochanter for insertion of the strong retractor muscles that originated at the base of the tail. The tibia is considerably longer than the femur and bears most of the weight of the lower limb. The fibula is reduced to a narrow splint. As in other ornithischians, the astragalus and calcaneum are so strongly integrated with the ends of the tibia and fibula that they are functionally an extension of the crus. The distal tarsals are separate but closely associated with the heads of the metatarsals. As in early "coelurosaurs," the pes is functionally tridactyl, with the fifth digit lost and the first reduced and turned slightly to the rear.

Fabrosaurus shows no trace of dermal armor. In contrast, *Scutellosaurus*, a possibly related form from North America, has an extensive covering along the back and onto the flanks (Colbert, 1981).

Well-known fabrosaurids are restricted to the Lower Jurassic, but Galton (1978) described jaws and teeth from the Jurassic-Cretaceous boundary that appear to represent a continuation of this group and similar teeth are known from both the Upper Triassic and Upper Cretaceous. The teeth in later genera are slightly inset, which demonstrates the initial development of cheeks within this group.

HETERODONTOSAURIDAE

Another group of primitive ornithischians, the heterodontosaurids, accompany the fabrosaurids in the early Jurassic. This family probably did not give rise to any later ornithischians, but they are important in establishing the primitive anatomy of this assemblage since we know their skeleton in greater detail than that of the fabrosaurids.

Heterodontosaurus is known from two complete skeletons from southern Africa (Charig and Crompton, 1974; Santa Luca, 1980). The body is slightly more than 1 meter long. The skull differs from that of *Fabrosaurus* in several significant features (Figure 14-29). There are conspicuous caniniform teeth in both the upper and lower jaw. The upper "canine" originates from the back of the premaxilla (and not the front of the maxilla, as do the canine teeth of primitive amniotes and mammals). The lower canine originates at the front of the dentary, just behind the predentary bone, and fits into a notch in the upper jaw that is part of a diastema between the premaxillary teeth and the cheek teeth. The other very important feature is the fact that the cheek teeth are not located at the margin of the skull and lower jaw but are inset, leaving a space lateral to the tooth row. The bone flares outward and is ridged above and below the tooth row in the maxilla and dentary, as if to support a lateral sheet of tissue that would have functioned like the mammalian cheek to retain food within the oral cavity.

Heterodontosaurus shows tooth wear that results from a specific occlusal pattern between the chisel-shaped upper and lower teeth. The lateral surface of the lower teeth wear against the medial surface of the maxillary teeth (Figure 14-30). Enamel is present only on the lateral surface of the upper teeth and the medial surface of the dentary teeth to provide a resistant cutting edge. The rate

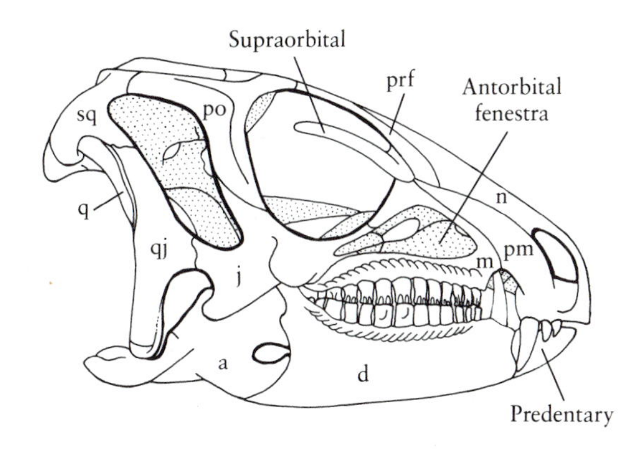

Figure 14-29. SKULL OF *HETERODONTOSAURUS*. From the Lower Jurassic of Southern Africa. Abbreviations as in Figure 8-3. *From Charig and Crompton, 1974.*

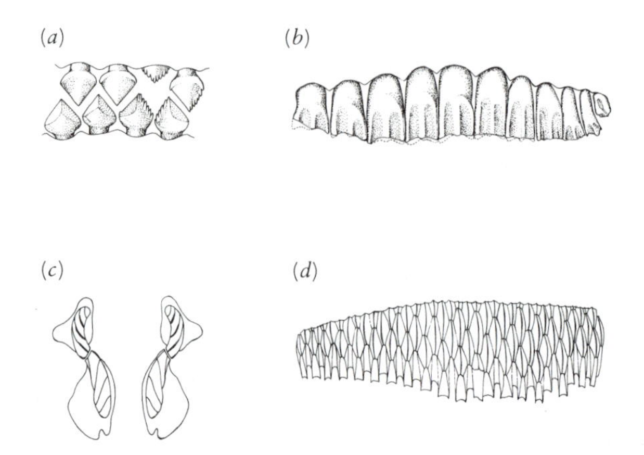

Figure 14-30. TOOTH PATTERN AND OCCLUSION IN ORNITHISCHIAN DINOSAURS. (*a*) *Fabrosaurus,* in which the upper and lower teeth alternate with one another. *From Thulborn, 1971. By permission of the Zoological Society of London.* (*b*) *Heterodontosaurus. From Charig and Crompton, 1974.* (*c*) Diagrammatic cross-section to show occlusal pattern in hadrosaurs. The same basic pattern also applies to most other ornithopods. *From Ostrom, 1961.* (*d*) Hadrosaur tooth battery in medial view showing many ranks of replacement teeth. *From Ostrom, 1961.*

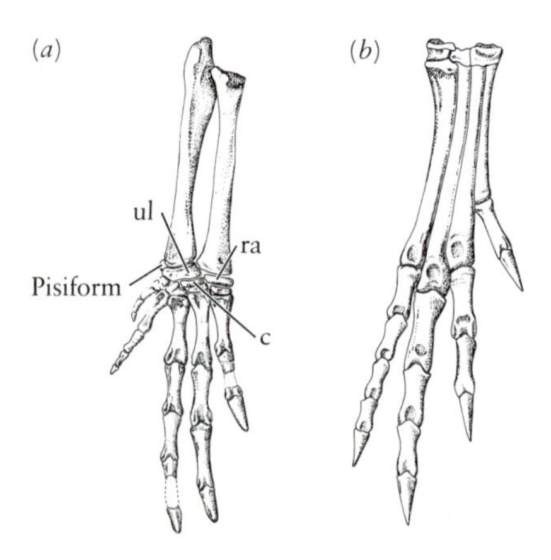

Figure 14-32. *HETERODONTOSAURUS.* (*a*) Forelimb and (*b*) hindlimb. Abbreviations as follows: c, centrale; ra, radiale; ul, ulnare. *From Santa Luca, 1980.*

and pattern of tooth replacement is modified relative to more primitive dinosaurs to maintain an even cutting edge. The late Triassic genus *Pisanosaurus* has a similar pattern of tooth wear and may belong to this family, although it does not have a lower caniniform tooth (the anterior part of the upper dentition is not known).

The forelimbs of *Heterodontosaurus* are considerably longer and more powerfully built than those of *Fabrosaurus* and may have been used for support in quadrupedal locomotion (Figure 14-31). The nature of the elbow joint, with a prominent olecranon, indicates that the manus was capable of powerful grasping. The carpus is more fully ossified than in other ornithischians and may provide a model for the primitive pattern in the group (Figure 14-32). Interestingly, there is a small pisiform, that is missing or unossified in most other archosauromorphs. Digits four and five are much smaller than the first three.

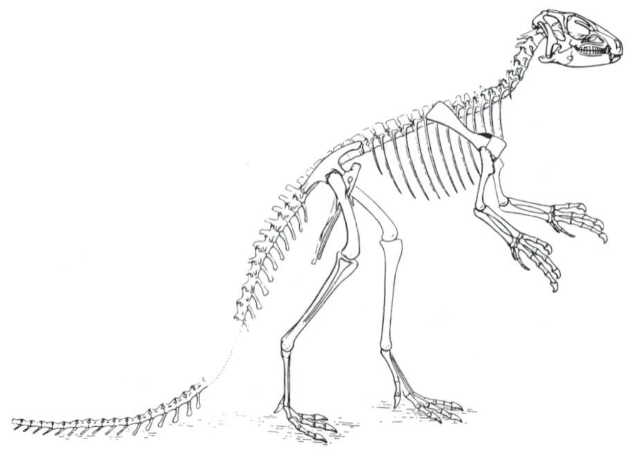

Figure 14-31. RESTORATION OF THE SKELETON OF *HETERODONTOSAURUS.* 50 centimeters tall. *From Santa Luca, 1980.*

The pelvis resembles that of fabrosaurids but lacks an obturator process. The joint surfaces between the femur and the fused tibiofibula indicate that the limb was not held absolutely vertically but, as in modern birds, the femur was slightly abducted. The knee is angled well forward. The astragalus and calcaneum are almost indistinguishably fused to the ends of the crus, and the distal tarsals are fused to the ends of the metatarsals that are, in turn, fused to one another proximally in a nearly avian pattern. The fifth digit is lost. The first is shorter than the next three, and the phalanges are somewhat divergent.

HYPSILOPHODONTIDS

The fabrosaurids may have given rise directly to the late Jurassic and Cretaceous hypsilophodontids, which in turn gave rise to the iguanodontids and hadrosaurs (Figure 14-33). The hypsilophodontids form a conservative group that is widespread but neither common nor diverse. They appear as a continuation of the fabrosaurid habitus as modest-sized, light-bodied, cursorial bipeds (Figures 14-34*a, b*).

All are advanced over the early fabrosaurids in the medial position and regular occlusion between the upper and lower teeth. In *Hypsilophodon,* the teeth were replaced in groups of three and maintained a functional shearing surface more effectively than did the alternate replacement of primitive reptiles, although the resulting wear pattern appears less regular than that of *Heterodontosaurus.*

The upper temporal openings of hypsilophodontids are larger than in fabrosaurids. The fused parietals form a medial crest. The lower jaw exhibits a high coronoid process. Teeth are retained in the premaxilla.

The best-known genus is *Hypsilophodon* from the Lower Cretaceous of England (Galton, 1974), which is less than 5 meters long. In contrast with the fabrosaurids, an anterior prepubic process extends upward toward the

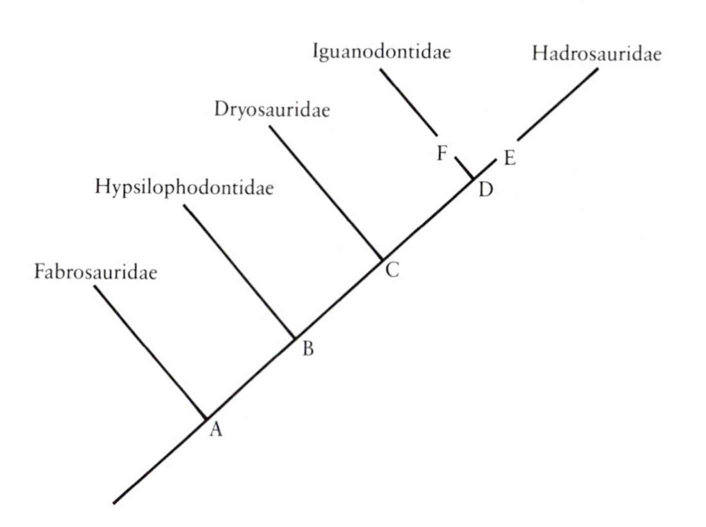

Figure 14-33. CLADOGRAM OF THE ORNITHOPODA.
Node A—The Ornithopoda: spinelike anterior pubic process; obturator process on ischium; bipedal; six teeth in premaxilla.
Node B—Hooked posterior margin to coracoid; thumbprint scar on femoral shaft; skull kinetic and pleurokinetic; adjacent maxillary and dentary teeth interlock; maxillae meet in midline in recess in premaxillae, coronoid with dorsal expansion; median ventral recess in premaxillae; median ridge on dentary teeth.
Node C—Premaxilla contacts prefrontal and lacrimal; dentary forms elevated coronoid process; reduced antorbital fenestra; median ridge on maxillary teeth; premaxillary teeth absent; posterior alveoli of dentary lie medial to coronoid process; quadrate is notched and forms part of the quadrate foramen; ischium curved and footed; bilobed ventral process on predentary.
Node D—Metatarsal V lost; fourth trochanter of femur low and crested, femur longer than tibia; anterior pubic process laterally flattened and deep; anterior intercondylar groove on femur; snout elongated (extended premaxillae, maxillae and nasals).
Node E—Dental magazines; palpebral lost/fused; maxillary and dentary teeth lozenge-shaped; quadrate foramen lost; manus digit I lost; metatarsals I and V lost; antitrochanter on ilium.
Node F—Metacarpal I in carpus; carpus massive; robust forelimb and shoulder girdle; shaft of femur curved; broad asymmetrical teeth; postacetabular blade deep.
Fabrosauridae = *Lesothosaurus* (= *Fabrosaurus*?), *Echinodon, Nanosaurus. Scutellosaurus, Alocodon* and *Trimucrodon* are considered provisionally *incertae sedis.*
Hypsilophodontidae = *Hypsilophodon, Zephyrosaurus, Thescelosaurus, Tenontosaurus.*
Dryosauridae = *Dryosaurus* (= *Dysalotosaurus*), *Valdosaurus* and ? *Parksosaurus* and *Mochlodon.*
Iguanodontidae = *Camptosaurus, Iguanodon, Ouranosaurus, Vectisaurus* and (?) *Probactrosaurus* + *Muttaburrasaurus* and *Craspedodon.*
Hadrosauridae = *Bactrosaurus* and all the other hadrosaurine and lambeosaurine hadrosaurs. *From Milner and Norman, 1984.*

tip of the ilium. The ischium has a strongly developed obturator process near the midpoint of its length. Ossified tendons sheath the end of the tail.

IGUANODONTIDS

Two other ornithopod groups that are characterized by larger body size and a graviportal posture evolved from animals that resembled the hypsilophodontids. The more primitive group, the iguanodontids, appear in the Middle Jurassic, reach their greatest diversity at the end of the Lower Cretaceous, and continue to the end of the Mesozoic.

Camptosaurus (Figure 14-35) is a well-known early iguanodontid. Its body length of approximately 6 meters exceeds that of most hypsilophodontids. The preorbital region of the skull is considerably elongated, the size of the antorbital fenestra is greatly reduced, and the premaxillary teeth are lost. The single row of marginal teeth resemble those of hypsilophodontids, except for the complete loss of the enamel on the wear surface.

The neck is long and the centra of the cervical vertebrae are opisthocoelous, as in advanced saurischians. There are 26 to 28 presacral vertebrae, an increase of 4 to 6 over primitive ornithischians.

The limbs are more massive, the tibia is shorter than the femur, and the metatarsals are short and stout. The manus ends in blunt claws that were probably capable of supporting the body in awkward quadrupedal locomo-

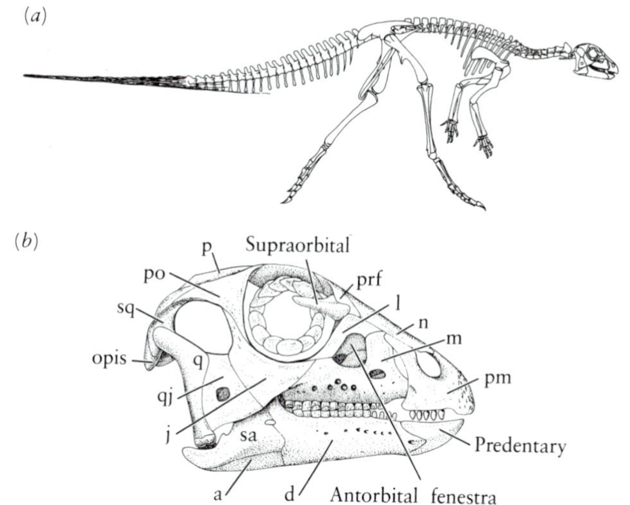

Figure 14-34. (*a*) Skeleton of *Hypsilophodon,* a small ornithischian that may represent the main lineage of ornithopod evolution, 3 meters long. (*b*) Skull of *Hypsilophodon.* Abbreviations as in Figure 8-3. *From Galton, 1974.*

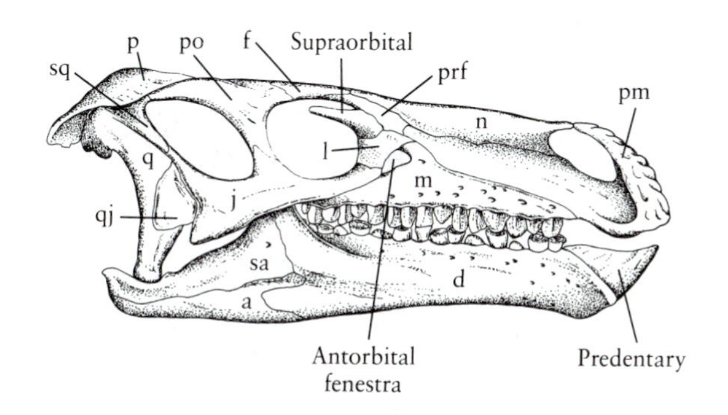

Figure 14-35. *CAMPTOSAURUS.* The skull of the primitive Upper Jurassic iguanodontid. Abbreviations as in Figure 8-3. *From Ostrom, 1961.*

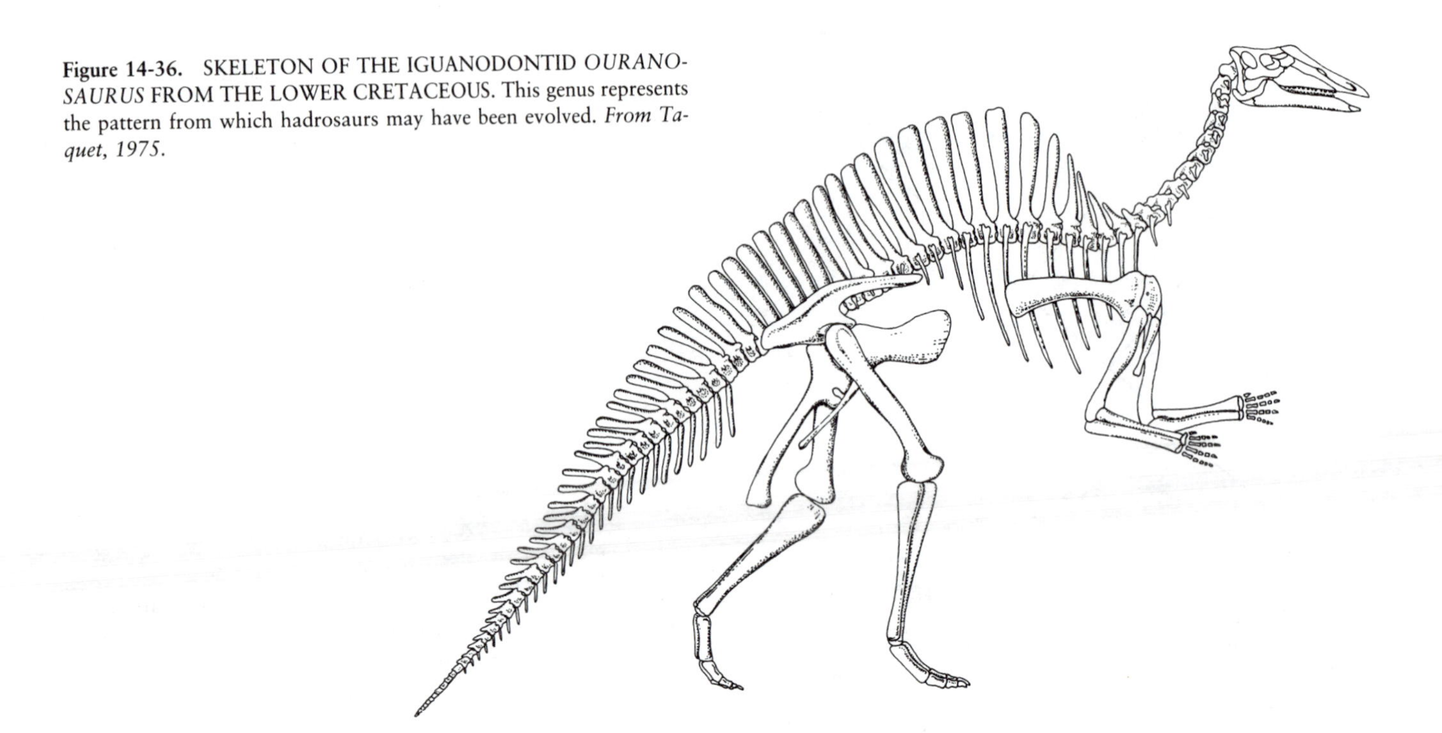

Figure 14-36. SKELETON OF THE IGUANODONTID *OURANOSAURUS* FROM THE LOWER CRETACEOUS. This genus represents the pattern from which hadrosaurs may have been evolved. *From Taquet, 1975.*

tion. In the Lower Cretaceous genus *Iguanodon,* which was one of the first dinosaurs to be described, the first digit of the manus ends in a huge spike set at right angles to the other digits (Norman, 1980).

The late Lower Cretaceous genera *Ouranosaurus* (Figure 14-36) and *Probactrosaurus* provide appropriate ancestors for the next advance in ornithopod evolution, the family Hadrosauridae.

HADROSAURS

The hadrosaurs or "duck-billed" dinosaurs were common and varied in the Upper Cretaceous of North America and Eurasia, and a single species has been reported from South America. The presacral vertebral column was elongated relative to more primitive ornithopods. Within hadrosaurs, the number of cervical vertebrae increased from 12 to 15. There were 15 to 19 trunk vertebrae, as well as 8 sacrals. The anterior trunk as well as the cervical centra are opisthocoelous. The tail is laterally compressed, and the neural and haemal spines of the tail are greatly elongated (Figure 14-37).

The appendicular skeleton resembled that of the iguanodontids but was somewhat heavier and had longer forelimbs that bore small hoofs on the digits rather than claws. Nevertheless, we think that they were habitually bipedal.

Much more important changes were evident in the skull. Instead of a single row of functional teeth in each jaw ramus, a series of replacement teeth were exposed simultaneously. There were 45 to 60 tooth positions in each jaw, each with several replacement teeth in sequence, giving a total of as many as 700 teeth that were exposed

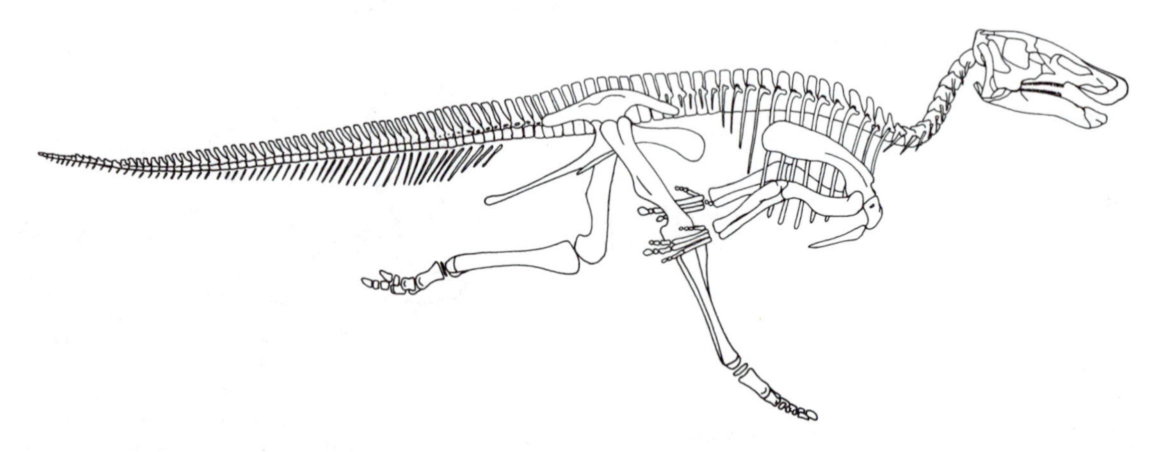

Figure 14-37. *ANATOSAURUS.* The primitive, noncrested hadrosaur is shown in running pose. *From Galton, 1970a.*

at once. Ostrom (1961) argued that the upper and lower tooth batteries sheared past one another, with the pterygoideus acting as a protractor and the posterior adductor acting as an antagonist to retract the jaws (Figure 14-38). In contrast, Weishampel (1983) provided evidence that the jaws were closed primarily vertically but that chewing also involved transverse movements of the dentition that were produced by rotating the maxillae.

In many, but not all, hadrosaur genera, the skull was greatly elaborated with crests of various sizes and shapes (Figure 14-39). Hopson (1975) and Dodson (1975) argued that the crests and other specializations of the nasal region were associated with species recognition. Weishampel (1981) provided evidence that the hollow crests of lambeosaurine hadrosaurs functioned as vocal resonators to produce species-specific call notes.

The habitat of hadrosaurs has long been in dispute. Skin impressions show that the hands were webbed as would be expected in aquatic forms. The lateral compression and vertical expansion of the tail would have provided an effective swimming organ. These factors have led to the conclusion that hadrosaurs spent much of their time in the water and fed on aquatic vegetation. On the other hand, Ostrom (1964a) argued that the highly specialized dentition would have been well suited to hard, fibrous terrestrial plants. The joint surfaces are highly ossified to support the body on land, and the ossified tendons would make the tail too rigid to serve as an effective paddle.

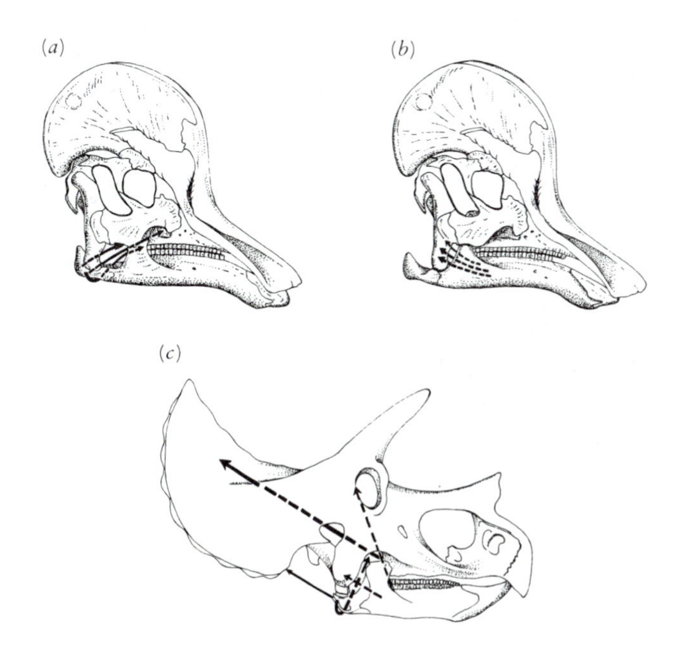

Figure 14-38. OCCLUSAL PATTERN OF THE HADROSAUR *CORYTHOSAURUS*. Shown is propalineal movement of the lower jaw (*a*) protracted and (*b*) retracted. (*a, b*) *From Ostrom, 1961*. Weishampel (1983) presents evidence that jaw closure was primarily vertical in this group. (*c*) Jaw mechanics in the ceratopsian *Triceratops*. Ceratopsians had a scissorslike jaw closer with no propalineal movement. Arrows show direction of force of major jaw muscles. (*c*) *From Ostrom, 1964b.*

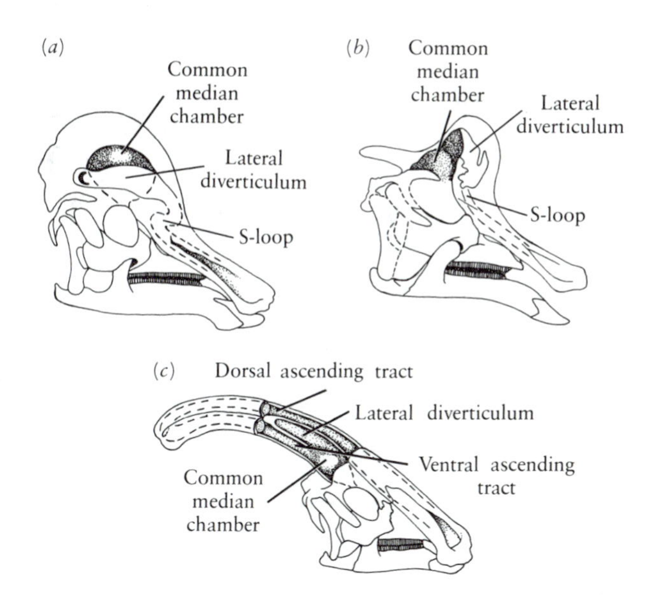

Figure 14-39. INTERNAL ANATOMY OF THE CREST OF LAMBEOSAURINE HADROSAURS. Weishampel described the acoustic properties which would provide for species recognition. (*a*) *Corythosaurus*. (*b*) *Lambeosaurus*. (*c*) *Parasaurolophus*. *From Weishampel, 1981.*

Most of the remains of hadrosaurs are from lowlying deposits of coastal planes and the margins of rivers. The absence of immature individuals led to the hypothesis that hadrosaurs migrated to higher land to reproduce. This assumption has been strikingly confirmed by the discovery of nesting sites in the foothills of the ancestral Rocky Mountains (Horner, 1984). The great success of hadrosaurs may have been the result of their capacity to inhabit a range of different environments. We may associate the change in dental patterns and the subsequent success of hadrosaurs with the proliferation of the angiosperms at the beginning of the Upper Cretaceous.

There were at least 26 genera of hadrosaurs in the Upper Cretaceous. Of two subfamilies, only the Hadrosaurinae, which lacked a crest, continued to the very end of the period.

We can include fabrosaurids, hypsilophodontids, iguanodontids, and hadrosaurs in the single suborder Ornithopoda. They are united by a unique derived character, the presence of an obturator process on the ischium (see Figure 14-28), and are also linked by a series of morphological intermediates. The remaining ornithischian groups lack an obturator process and may have evolved separately from the base of this assemblage (Sereno, 1986).

CERATOPSIANS

Our concept of the ceratopsians is based primarily on the well-known Upper Cretacous genera from North America, as epitomized by *Triceratops*. These animals are stocky, quadrupedal herbivores. The posterior portion of the skull is extended as a huge frill that spreads over the neck region. These forms are further distinguished by a variable

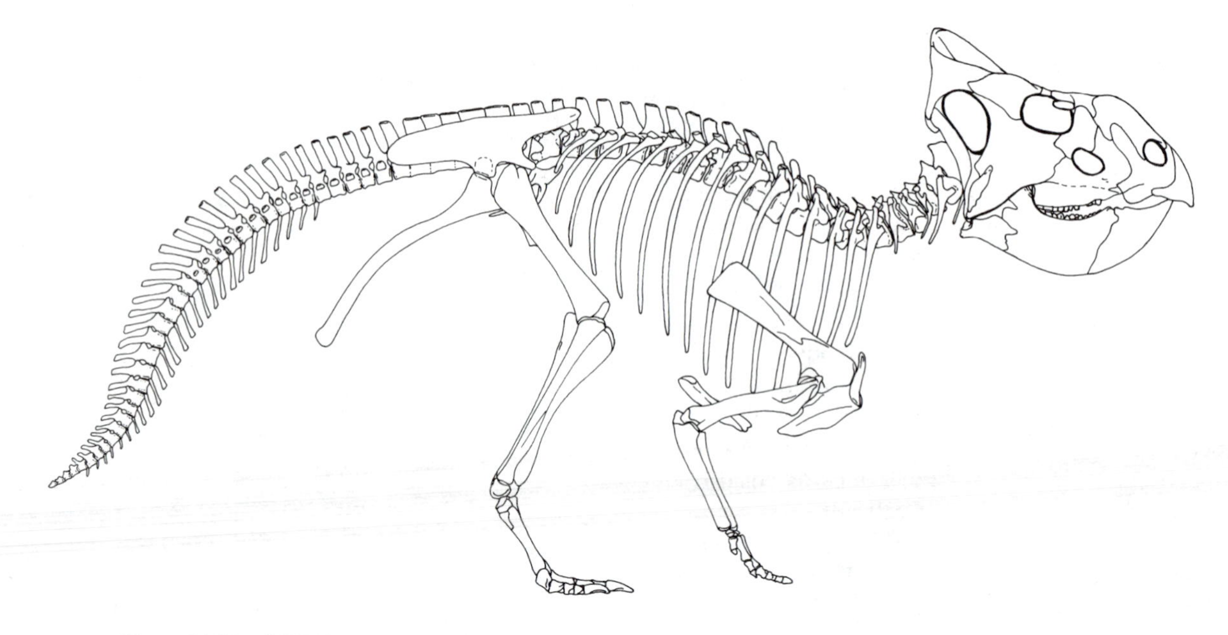

Figure 14-40. *LEPTOCERATOPS*, the North American Upper Cretaceous ceratopsian. *From Russell, 1970b.*

pattern of "horns" in the nasal region and above the eyes. The end of the snout is in the form of a laterally compressed "beak" that overhangs the lower jaw and terminates in a unique rostral bone that is comparable to the predentary in the lower jaw. We find the most advanced of the ceratopsians, the family Ceratopsidae, only in western North America.

Ceratopsians with somewhat less-specialized cranial anatomy, placed in the family Protoceratopsidae, are known in both North America and eastern Asia in the Upper Cretaceous (Figure 14-40). A further Asian genus, *Psittacosaurus* (Figure 14-41), from the Lower Cretaceous is thought to have been bipedal rather than quadrupedal and may provide a link with the base of the ornithischian stock. In both *Psittacosaurus* and the protoceratopsids, the tibia is much longer than the femur, which suggests cursorial locomotion. The skull of *Psittacosaurus* is comparable with that of later ceratopsians in having a beaklike nasal area, including a rostral bone, and one species has a nasal horn, but there is no evidence of the frill that characterizes the Upper Cretaceous families. This genus cannot be directly ancestral to the later families since it is more advanced in the closure of the antorbital fenestrae that are present in protoceratopsids and the reduction of

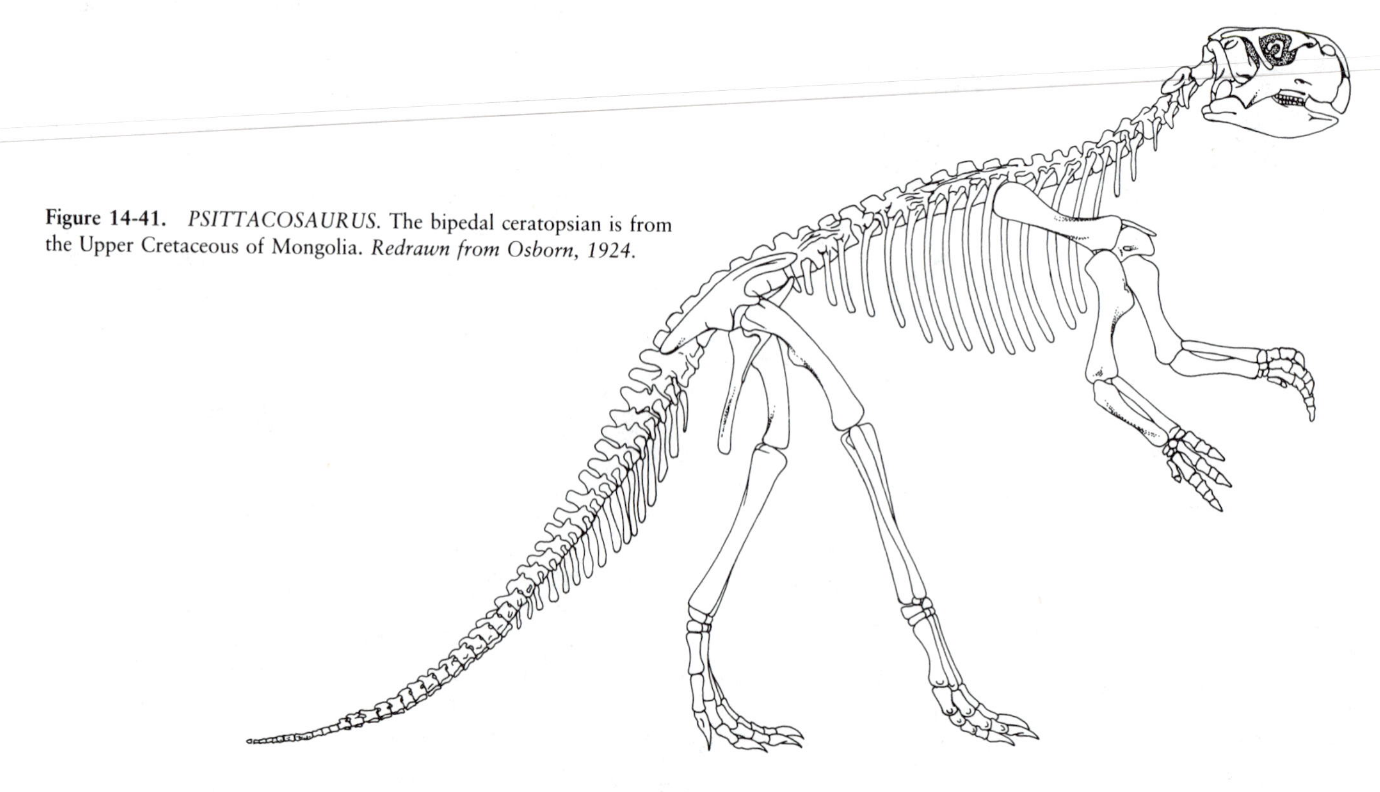

Figure 14-41. *PSITTACOSAURUS*. The bipedal ceratopsian is from the Upper Cretaceous of Mongolia. *Redrawn from Osborn, 1924.*

the fourth and fifth digits of the manus. These three groups appear to represent distinct radiations from an earlier ceratopsian stock.

The oldest putative ceratopsian is a single specimen of *Stenopelix* from the base of the Cretaceous in Germany (Sues and Galton, 1982). Its specific affinities with later ceratopsians are difficult to establish because the skull is missing, but the structure of the pelvis and rear limbs suggests that it is a member of this suborder. Despite its early appearance, it differs from psittacosaurids in having a femur that is longer than its tibia. The pelvis closely resembles that of the other members of the suborder in having a short prepubic process and a short posterior pubic process but no obturator process. As in other ceratopsians, the feet remain primitive. The metatarsals are slender but not greatly elongated, and the phalangeal formula is 2, 3, 4, 5, 0.

The pelvis of these ornithischians does not possess the one key feature that unites the ornithopods, an obturator process, but this absence might be expected in forms that have a short posterior portion of the pubis. The short prepubis together with the retention of premaxillary teeth in primitive ceratopsians suggest an early divergence from the ornithischian assemblage. We have not found any possible ceratopsian ancestors in the Jurassic; no fossils of this suborder are known in the southern hemisphere.

The success of ceratopsians in the Upper Cretaceous may be attributed to a highly specialized feeding mechanism (Ostrom, 1966). This apparatus involves the dentition, the configuration of the lower jaw, and the long frill formed by the squamosal and parietal. The teeth are arranged in a single functional row in each jaw ramus. The individual teeth resemble those of hadrosaurs in having leaf-shaped crowns in which the enamel is restricted to the medial surface of the lower teeth and the lateral surface of the upper teeth.

Although only a single row of teeth was functional at a given time, the teeth were apparently replaced very rapidly so that the ones at the jaw margin were always sharp. Tooth occlusion was produced by vertical shear that resulted from a scissorslike closure of the jaw. Leverage on the lower jaw was increased by the high coronoid process. Much of the force of jaw closure was provided by the adductor mandibulae posterior that extended posterodorsally behind the jaw, through the upper temporal fenestra, and out over the dorsal surface of the squamosal-parietal frill (see Figure 14-38*c*).

Although the primary function of the frill apparently was for muscle attachment, Rowe, Colbert, and Nations (1983) pointed out that the edge of the skull in some genera extended well beyond the area of muscle attachment. In genera such as *Triceratops,* in which the skull forms a nearly continuous bony surface, it would have served for protection from attack. In other genera such as *Pentaceratops,* there are large openings that would be vulnerable to injury by either predators or members of the same species. The large size of the frill would have made it important for species recognition and sexual selection, as is the case in many large mammalian herbivores (Farlow and Dodson, 1975).

PACHYCEPHALOSAURS

It was long thought that all advanced ornithischian groups might trace their origin to the primitive ornithopods. However, the remaining groups apparently evolved from an even more primitive stage at the very base of ornithischian evolution, because they share no derived features with even the most primitive ornithopods that are not present in all ornithischians.

All retain a simple dentition that is composed of a single row of teeth with leaf-shaped crowns that have enamel on both surfaces. The predentary is small, and the upper teeth continue onto the premaxilla in primitive members of each of the groups. They are all advanced over the early fabrosaurids in having cheeks, but all lack the obturator process that characterizes that group and other ornithopods.

The pachycephalosaurids are relatively small, bipedal forms that are known primarily from the Upper Cretaceous of Asia and North America, with one genus from Madagascar (Sues, 1980; Sues and Galton, 1987). The postcranial skeleton resembles that of primitive ornithopods except in having a very short posterior process on the pubis and a long prepubis. The acetabulum is formed entirely by the ilium and ischium.

The most striking feature of pachycephalosaurids is the domed appearance of the skull (Figure 14-42). The very high forehead is not a reflection of a large brain, however, but an enormously thickened frontoparietal. A knobby occipital frill is variably developed and may be ornamented with short spikes.

Galton (1970b) suggested that the skull was used as a battering ram in intraspecific combat, like the horns of sheep and goats. The pattern of the thickening would most effectively resist blows directed against the top of the skull, which suggests that they ran at each other with the trunk held horizontally. The internal structure of the skull is also specially reinforced with extra ossifications around the brain and by closing the interorbital spaces. In advanced genera, the dorsal temporal openings are also closed.

The Asian genera *Goyocephale* and *Homalecephale* have essentially the same pattern of skeletal organization as other pachycephalosaurs but lack a frontoparietal dome. The frontoparietal is thickened and ornamented with either pits or tubercles, which implies a mode of fighting that is comparable to that observed in the marine iguana *Amblyrhynchus*.

The oldest known pachycephalosaur is the genus *Yaverlandia* from the Lower Cretaceous of England, which is represented by a distinctively thickened skull cap. Maryanska and Osmolska (1974) described several genera

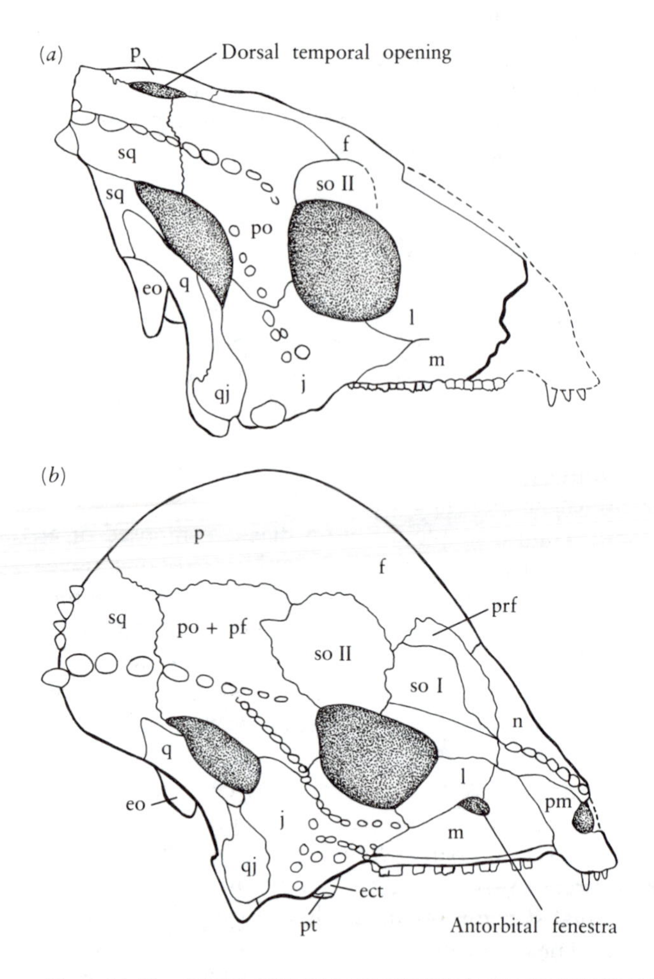

Figure 14-42. SKULL OF THE PACHYCEPHALOSAURIDS *HOMOALCEPHALE* AND *PRENOCEPHALE*. From the Upper Cretaceous of Mongolia representing (*a*) flat and (*b*) domeheaded genera. Abbreviations as in Figure 8-3 plus: so I and so II, supraorbitals. *From Maryanska and Osmolska, 1974.*

from the Upper Cretaceous of Mongolia. They argue that pachycephalosaurs should be recognized as a distinct suborder of ornithischians because of their distinctive skeletal anatomy and the absence of derived characters that link them with any of the other advanced ornithischian groups.

ARMORED ORNITHISCHIANS

All the ornithischian groups that have been discussed thus far have lost the armor that was present in their thecodont ancestry. Other ornithischians retained or elaborated dermal protection (Figure 14-43). Extensive armor is present in *Scutellosaurus* from the Lower Jurassic of North America, which otherwise resembles the early ornithopods, and in *Scelidosaurus* from Europe. Better-known armored groups, the stegosaurs and ankylosaurs, are found in the Middle Jurassic through the Cretaceous. Both are strictly quadrupedal, although the rear limbs remain much larger than the front. The toes bear hoofs, rather than claws, and the posture is graviportal.

Stegosaurs

Stegosaurs are known from the Middle Jurassic into the Upper Cretaceous. The best-known form is *Stegosaurus* from the Upper Jurassic of North America, which has a double row of dermal plates that are oriented vertically above the vertebral column (Figure 14-44). Farlow, Thompson, and Rosner (1976) argued that the orientation

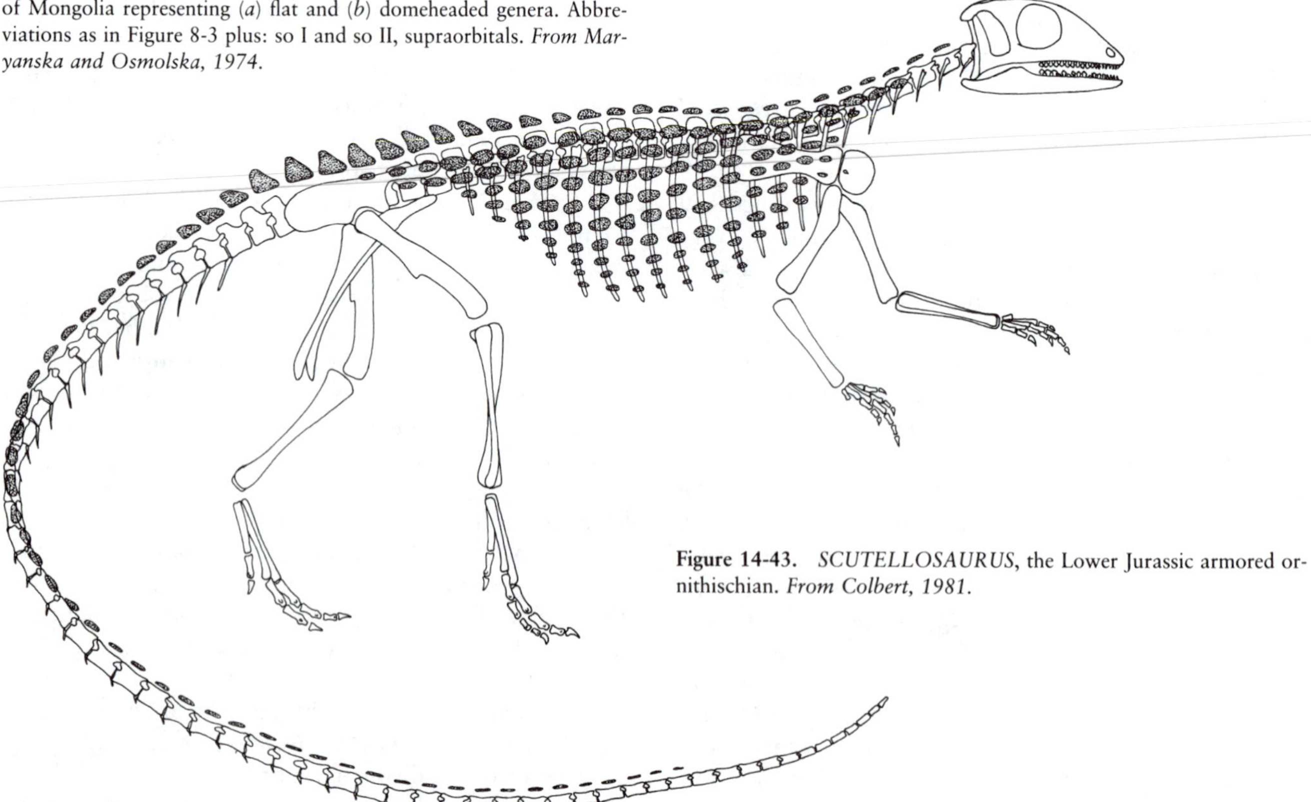

Figure 14-43. *SCUTELLOSAURUS*, the Lower Jurassic armored ornithischian. *From Colbert, 1981.*

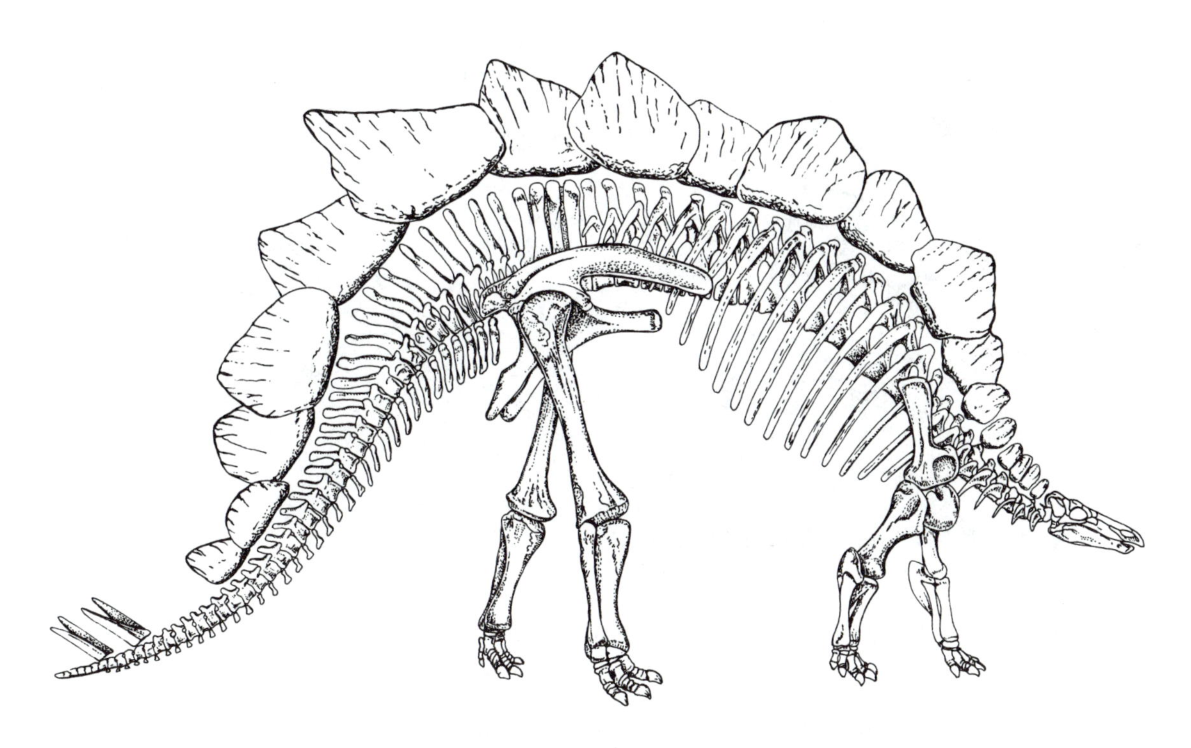

Figure 14-44. *STEGOSAURUS,* the Upper Jurassic armored dinosaur. *Redrawn from Gilmore, 1914.*

and alternating pattern of these plates would have made them admirably suited to radiate heat from their large body. *Stegosaurus* also has two pairs of bony spikes near the end of the tail. The skull is long and low. *Kentrosaurus* is a well-known form of similar morphology from the Upper Jurassic of East Africa (Galton, 1982b). *Huayangosaurus,* whose complete skeletons are known from the Middle Jurassic of China, may be close to the ancestry of stegosaurs. The group is also represented in the Middle Jurassic of Europe (Galton, Brun, and Rioult, 1980).

A few scattered remains of this group are known in the Lower Cretaceous with possibly attributable fragments from the Upper Cretaceous of India (Galton, 1981).

Ankylosaurs

Ankylosaurs were the most effectively armored of all dinosaurs (Figure 14-45). Their entire trunk was covered by a continuous mosaic of small, flat, interlocking, bony plates. Within this mosaic were larger, keeled plates. In the neck region, the armor was formed in half rings. The low, massive skull was covered by osteoderms that vaguely resemble those of scincomorph lizards (Figure 14-46). These plates were firmly attached to the underlying dermal bones of the skull roof and contributed to the closure of the dorsal temporal opening. The lateral opening was also covered in some genera, in which further plates extended from the cheek and skull table over the front of the neck.

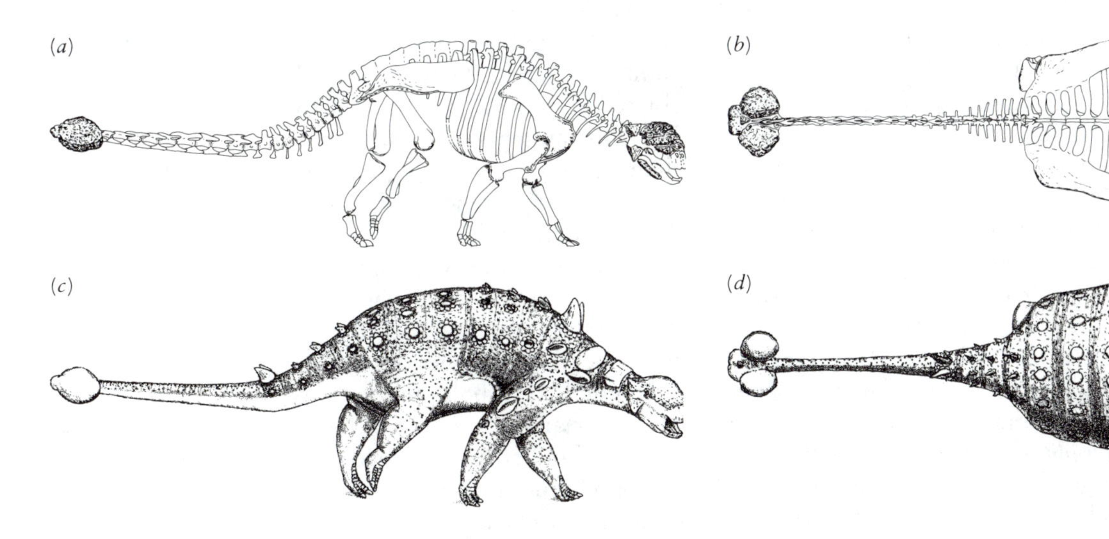

Figure 14-45. SKELETON AND ARMOR OF THE ANKYLOSAUR *EUOPLOCEPHALUS.* (*a*) Lateral and (*b*) dorsal views of skeleton. Restoration in (*c*) lateral and (*d*) dorsal views, showing pattern of armor. *From Carpenter, 1982.*

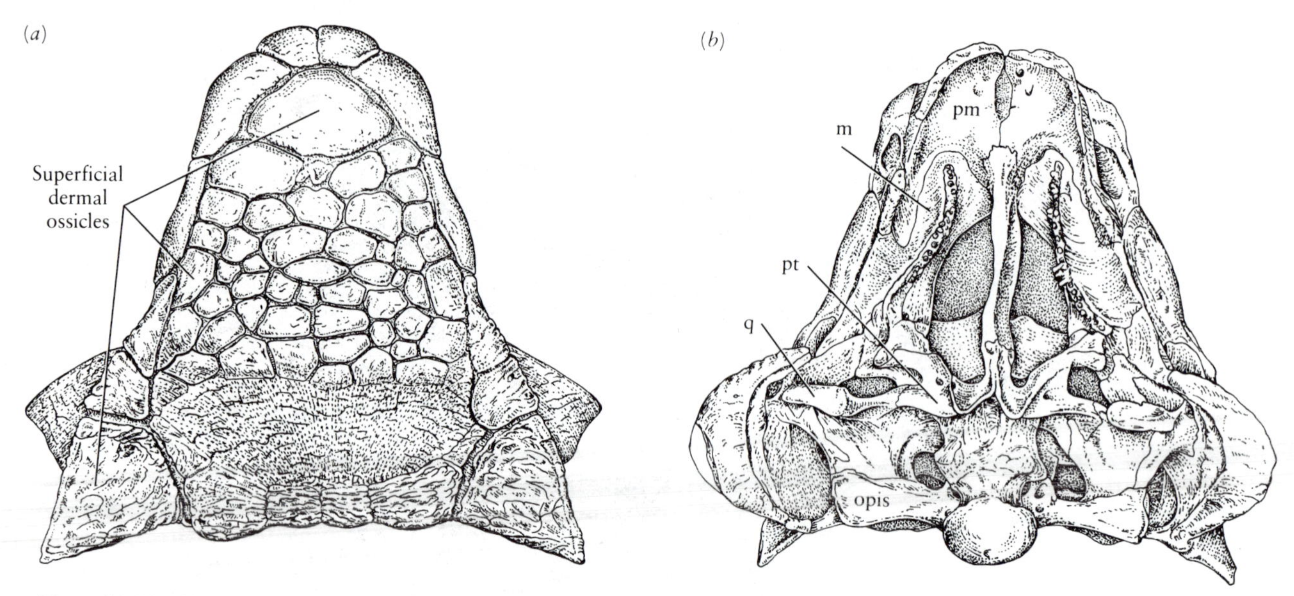

Figure 14-46. SKULL OF *ANKYLOSAURUS*. (*a*) Dorsal and (*b*) palatal views. Abbreviations as in Figure 8-3. *From Coombs, 1978.*

Coombs (1978) recognized two families in his recent review of the Ankylosauria. The Ankylosauridae are distinguished by the elaboration of the end of the tail into a massive bony club, which integrates additional dermal plates with the caudal vertebrae. Members of the Nodosauridae lack the caudal expansion but are distinguished by a solid fusion between the braincase and the palate. Some nodosaurs have long bony spines on the sides of the body that resemble the tail spines of the stegosaurs.

Both families have short heavy limbs and an obligatorily quadrupedal stance. The ilium is peculiar in that it broadly overhangs the femur. It is greatly expanded anteriorly to support the armor and surround the posterior ribs and greatly enlarged abdominal region. There is no prepubic process. The pubis is small and is virtually excluded from the acetabulum; the acetabulum is imperforate. At least eight vertebrae are incorporated in the sacrum. In advanced ankylosaurs, the head of the femur is terminal, rather than angled medially.

Adequately known ankylosaurs are limited to the Cretaceous of North America, Europe, Asia, and possibly Australia. Galton (1980a,b) recognized several fragmentary specimens from the Middle through Upper Jurassic of Europe as probable ancestors. *Sarcolestes* from the middle Jurassic of England is represented by a nearly complete lower jaw in which the diagnostic ankylosaur dermal plates are fused to the external surface. The dentition extends to the distal end of the bone, with space for only a very small predentary. *Cryptodraco* from the Upper Jurassic is based on a femur that is similar to the North American nodosaurid *Hoplitosaurus*. *Dracopelta* from the Upper Jurassic of Portugal is known from a trunk region that is covered with ankylosaurlike armor. We have not found any forms that might link ankylosaurs with other ornithischians.

Surprisingly, ankylosaurs are not known from the very rich Upper Jurassic deposits from North America, East Africa, and China. Galton suggests that they may have evolved in Europe and only reached the other continents in the Cretaceous. Ankylosaurs are widespread in North America and Asia in the Upper Cretaceous, but only a single genus appears in the latest Mesozoic beds.

SEGNOSAURIA

In 1980, Barsbold and Perle recognized a new infraorder of dinosaurs based on recently discovered material from the Upper Cretaceous of Mongolia. The Segnosauria is represented by two genera, *Segnosaurus* and *Erlikosaurus* (Figure 14-47), much of whose skeletons are known. They are distinguished from all other dinosaurs by a combination of specializations of the pelvis and skull. The pubis, like that of ornithischian dinosaurs, has rotated posteriorly to lie parallel with the ischium, which has an ornithopodlike obturator process. However, the ilium is very short and gives the dorsal portion of the pelvis the appearance of primitive theropod dinosaurs. The skull resembles that of ornithischians in having the tooth row of the maxilla and dentary inset from the skull margin. As in that order, the premaxilla is toothless. Segnosaurs lack the predentary bone that is diagnostic of ornithischians, but the anterior extremity of the dentary is toothless. Barsbold and Perle suggest that the end of the upper and lower jaws was covered by a horny beak.

Barsbold and Perle place the Segnosauria among the theropod dinosaurs, but the known genera show no derived characters that support that assignment. Paul (1984) argues that the segnosaurs belong to an otherwise unrepresented lineage that diverged from the base of dinosaur

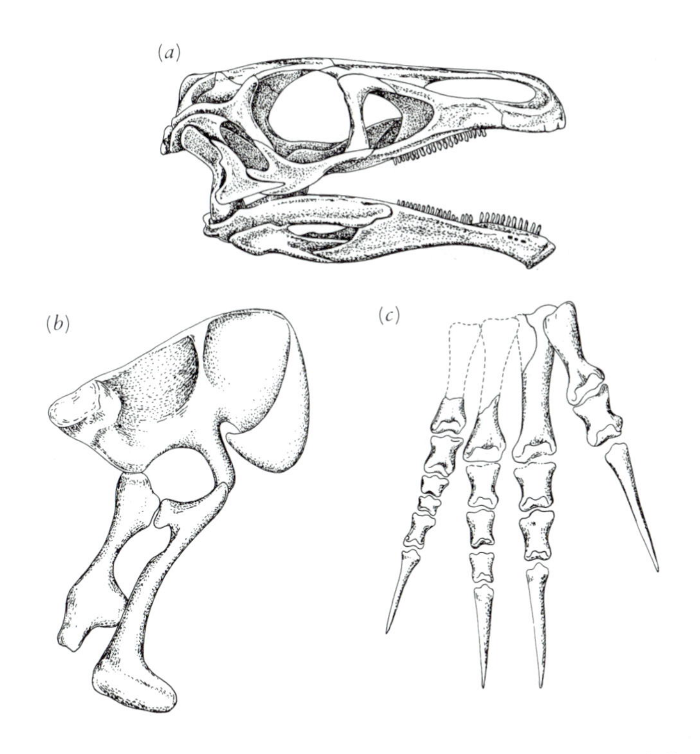

Figure 14-47. REPRESENTATIVES OF THE SEGNOSAURIA. (*a*) Skull of *Erlikosaurus*. *From Paul, 1984.* (*b*) Pelvic girdle of *Segnosaurus*. *From Barsbold, 1983.* (*c*) Dorsal view of right pes of *Erlikosaurus*. *From Barsbold and Perle, 1980.*

radiation close to the point of divergence of prosauropods and the ornithischians. Aside from the specializations of the dentition and pelvis, most of the characters of this group reflect a primitive level of evolution that more closely resembles that of Upper Triassic than other Upper Cretaceous dinosaurs. The foot, with four complete digits and an unconsolidated metatarsus, resembles that of plateosaurs and is far more primitive than that of even the earliest theropods. Paul points out that the combination of prosauropod and ornithischian features exhibited by these genera further supports the common ancestry of the two long-recognized dinosaur orders above the level of the thecodonts. The large number of derived characters that are shared by prosauropods, segnosaurs, and ornithischians suggests that all the herbivorous dinosaurs may belong to a single, monophyletic assemblage.

SUMMARY

Dinosaurs were the dominant terrestrial vertebrates from the Late Triassic until the end of the Mesozoic. They are characterized by the vertical posture of the rear limbs and their movement in a parasagittal plane. Early dinosaurs are differentiated from thecodonts by perforation of the acetabulum, inturning of the head of the femur, and the establishment of a mesotarsal joint between the proximal and distal tarsals. The calcaneum has lost the posterior tuber and the metatarsals are elongate. The structure and function of the rear limb are closely comparable with those of modern birds. Many early dinosaurs were bipedal, but we are not certain that this was a common heritage of all lineages.

Dinosaurs are divided into two large groups, the saurischians and the ornithischians, that probably shared a common ancestry at about the level of the lagosuchid thecodonts. The saurischians are further divided into the carnivorous theropods and the herbivorous sauropodomorphs.

The oldest-known dinosaur is *Staurikosaurus* from the latest middle Triassic of Brazil. It may be related to the large theropods of the Jurassic and Cretaceous.

Jurassic and Cretaceous theropods include the largest terrestrial carnivores that have ever lived. The customary subdivision into smaller coelurosaurs and larger carnosaurs can no longer be justified, but we have not yet established an alternative classification.

Podokesaurids include a variety of small- and medium-sized carnivores of the Upper Triassic and Lower Jurassic. The coelurids and compsognathids represent a central stock of small theropods that were common in the Upper Jurassic. Ornithomimids, or ostrich dinosaurs, were widespread in the Cretaceous. They had a large bulbous braincase like birds and lacked teeth. Dromaeosaurs include one of the most striking of carnivorous dinosaurs, *Deinonychus,* which bears an enormously enlarged claw on its rear foot. The dromaeosaurs and saurornithoidids are moderate-sized carnivores from the Cretaceous with the largest brain size to body weight ratio of any dinosaurs. Large theropods are represented throughout the Jurassic and the Cretaceous by the megalosaurs. The even larger tyrannosaurids were restricted to the Upper Cretaceous.

Prosauropods are among the most primitive of dinosaurs. These herbivores were common and widespread in the late Triassic and early Jurassic. Primitive members may have been ancestral to the sauropods, which were the largest of all dinosaurs and weighed up to 80,000 kilograms. Six families of sauropods are common in the Jurassic and Cretaceous.

Ornithischians form a clearly defined monophyletic group that is characterized by the common presence of a predentary bone and posteriorly directed pubis. All ornithischians are herbivores, and in all but the earliest genera the teeth are inset from the margins of the jaws, which indicates the presence of fleshy cheeks.

The earliest ornithischian is *Pisanosaurus* from the Upper Triassic of South America. Fabrosaurids and heterodontosaurids appear in the early Jurassic; both are small, obligatorily bipedal forms. Fabrosaurids, hypsilophodontids, iguanodontids, and hadrosaurs constitute a monophyletic assemblage, the suborder Ornithopoda, that is united by the presence of an obturator process on the ischium. In many hadrosaurs, the nasal passages were greatly elaborated with crests extending above the skull. They may have served for species recognition and sexual selection.

The Pachycephalosaurids, the ceratopsians, and the armored dinosaurs each evolved separately from the base of the ornithischian assemblage. The pachycephalosaurs are bipedal forms with a thickened frontoparietal region of the braincase that may have been used for intraspecific combat.

Ceratopsians are represented by three families, the bipedal psittacosaurids and the horned, quadrupedal protoceratopsids and ceratopsids. All are restricted to the Cretaceous. Stegosaurs and ankylosaurs are armored dinosaurs. Stegosaurs have a double row of alternating dermal plates that are oriented vertically along the back and spines on the tail. This group is known primarily from the Jurassic. Ankylosaurs, which we find primarily in the Cretaceous, have nearly continuous armor over the entire trunk region, and one family has an armored club at the end of the tail.

The newly discovered segnosaurs from the Upper Cretaceous of Asia represent a distinct lineage combining features of primitive Upper Triassic dinosaurs with dental and pelvic characteristics of ornithischians.

REFERENCES

Bakker, R. T., and Galton, P. M. (1974). Dinosaur monophyly and a new class of vertebrates. *Nature,* **248**: 168–172.

Barsbold, R. (1976). Novyye dannyye o terizinozavre (Therizinosauridae, Theropoda). (New data on *Therizinosaurus;* Therizinosauridae, Theropoda). In N. N. Kramarenko et al., *Sovmestnaya Sov.-Mongol. Paleontol. Eksped., Trudy,* **3**: 76–92.

Barsbold, R. (1983). Carnivorous dinosaurs from the Cretaceous of Mongolia. *Joint Soviet-Mongolian Expeditions, Transactions,* **19**: 1–116.

Barsbold, R., and Perle, A. (1980). Segnosauria, a new infraorder of carnivorous dinosaurs. *Acta Palaeontologica Polonica,* **25**(2): 187–195.

Berman, D. S., and McIntosh, J. S. (1978). Skull and relationships of the Upper Jurassic sauropod *Apatosaurus* (Reptilia, Saurischia). *Bull. Carnegie Mus. Nat. Hist.,* **8**: 1–35.

Bock, W. J. (1964). Kinetics of the avian skull. *J. Morphol.,* **114**: 1–42.

Bonaparte, J. F. (1971). Los tetrapodos del sector superior de la formacion Los Colorados, La Rioja, Argentina (Triasico Superior). *Opera Lilloana,* **22**: 1–83.

Bonaparte, J. F. (1976). *Pisanosaurus mertii* Casamiquela and the origin of the Ornithischia. *J. Paleont.,* **50**(5): 808–820.

Bonaparte, J. F. (1978). El Mesozoico de America del Sur y sus tetrapodos. *Opera Lilloana,* **26**: 1–596.

Bonaparte, J. F. (1982). Faunal replacement in the Triassic of South America. *J. Vert. Paleont.,* **2**(3): 362–371.

Carpenter, K. (1982). Skeletal and dermal armor reconstruction of *Euoplocephalus tutus* (Ornithischia: Ankylosauridae) from the Late Cretaceous Oldman Formation of Alberta. *Can. J. Earth Sci.,* **19**(4): 689–697.

Charig, A. J. (1972). The evolution of the archosaur pelvis and hindlimb: An explanation in functional terms. In K. A. Joysey and T. S. Kemp (eds.), *Studies in Vertebrate Evolution,* pp. 121–155. Oliver and Boyd, Edinburgh.

Charig, A. J. (1976). "Dinosaur monophyly and a new class of vertebrates": A critical review. In A. d' A. Belairs and C. B. Cox (eds.), *Morphology and Biology of Reptiles. Linn. Soc. Symp. Ser.,* **3**: 65–104.

Charig, A. J. 1979. *A New Look at the Dinosaurs.* Heinemann (in association with the British Museum of Natural History), London.

Charig, A. J., and Crompton, A. W. (1974). The alleged synonymy of *Lycorhinus* and *Heterodontosaurus. Ann. S. Afr. Mus.,* **64**: 167–189.

Chatterjee, S. (1978). *Indosuchus* and *Indosaurus,* Cretaceous carnosaurs from India. *J. Paleont.,* **52**(3): 570–580.

Chatterjee, S. (1985). *Postosuchus,* a new thecodontian reptile from the Triassic of Texas and the origin of tyrannosaurs. *Phil. Trans. Roy. Soc. Lond., B,* **309**: 395–460.

Colbert, E. H. (1962). The weights of dinosaurs. *Amer. Mus. Novitates,* **2076**: 1–16.

Colbert, E. H. (1972). Vertebrates from the Chinle Formation. In S. Carol and W. J. Breed (eds.), *Investigations in the Triassic Chinle Formation. Mus. Northern Arizona, Bull.,* **47**: 1–11.

Colbert, E. H. (1981). A primitive ornithischian dinosaur from the Kayenta Formation of Arizona. *Mus. Northern Arizona Press, Bull. Series,* **53**: 1–61.

Colbert, E. H. (1983). *Dinosaurs. An Illustrated History.* Hammond, New York.

Coombs, W. P. (1975). Sauropod habits and habitats. *Paleogeography, Paleoclimatology, Paleoecology,* **17**: 1–33.

Coombs, W. P. (1978). The families of the ornithischian dinosaur order Ankylosauria. *Palaeontology,* **21**: 143–170.

Coombs, W. P. (1979). Osteology and myology of the hindlimb in the Ankylosauria (Reptilia, Ornithischia). *J. Paleont.,* **53**(3): 666–684.

Cooper, M. R. (1984). A reassessment of *Vulcanodon karibaensis* Raath (Dinosauria: Saurischia) and the origin of the Sauropoda. *Palaeont. afr.,* **25**: 203–231.

Dodson, P. (1975). Taxonomic implications of relative growth in lambeosaurine hadrosaurs. *Syst. Zool.,* **24**: 37–54.

Dodson, P., Behrensmayer, A. K., Bakker, R. T., and McIntosh, J. S. (1980). Taphonomy and paleoecology of the dinosaur beds of the Jurassic Morrison Formation. *Paleobiology,* **6**(2): 208–232.

Dong, Z., Zhou, S., and Zhang, Y. (1983). The dinosaurian remains from Sichuan Basin, China. *Palaeontologia Sinica,* **162**, ser. C, (23): 1–145.

Dong, Z., and Tang, Z. (1984). Note on a Mid-Jurassic sauropod (*Datousaurus bashanensis* gen. et sp. nov.) from Sichuan Basin, China. *Vertebrata Palasiatica,* **22**: 69–75.

Farlow, J. O., and Dodson, P. (1975). The behavioral significance of frill and horn morphology in ceratopsian dinosaurs. *Evolution,* **29**(2): 353–361.

Farlow, J. O., Thompson, C. V., and Rosner, D. E. (1976). Plates of the dinosaur *Stegosaurus:* Forced convection heat loss fins? *Science,* **192**: 1123–1125.

Galton, P. M. (1969). The pelvic musculature of the dinosaur

Hypsilophodon (Reptilia: Ornithischia). *Postilla, Yale Peabody Mus.* **131**: 1–64.
Galton, P. M. (1970a). The posture of hadrosaurian dinosaurs. *J. Paleont.*, **44**: 464–473.
Galton, P. M. (1970b). Pachycephalosaurids—Dinosaurian battering rams. *Discovery*, **6**: 23–32.
Galton, P. M. (1971). The prosauropod *Ammosaurus*, the crocodile *Protosuchus*, and their bearing on the age of the Navajo sandstone of northeastern Arizona. *J. Paleont.*, **45**: 781–795.
Galton, P. M. (1973). The cheeks of ornithischian dinosaurs. *Lethaia*, **6**: 67–89.
Galton, P. M. (1974). The ornithischian dinosaur *Hypsilophodon* from the Wealden of the Isle of Wight. *Bull. Brit. Mus. (Nat. Hist.)* Geol., **25**: 1–152.
Galton, P. M. (1976). Prosauropod dinosaurs (Reptilia: Saurischia) of North America. *Postilla, Yale Peabody Mus.* **169**: 1–98.
Galton, P. M. (1977). On *Staurikosaurus pricei*, an early saurischian dinosaur from the Triassic of Brazil, with notes on the Herrerasauridae and Poposauridae. *Palaeont. Z.*, **51**: 234–245.
Galton, P. M. (1978). Fabrosauridae, the basal family of ornithischian dinosaurs (Reptilia: Ornithopoda). *Palaeont. Z.*, **52**: 138–159.
Galton, P. M. (1980a). Partial skeleton of *Dracopelta zbyszewskii* n. gen. and n. sp., an ankylosaurian dinosaur from the Upper Jurassic of Portugal. *Geobios*, **13**(3): 451–457.
Galton, P. M. (1980b). Armored dinosaurs (Ornithischia: Ankylosauria) from the Middle and Upper Jurassic of England. *Geobios*, **13**(6): 825–837.
Galton, P. M. (1981). *Craterosaurus pottonensis* Seeley, a stegosaurian dinosaur from the Lower Cretaceous of England, and a review of Cretaceous stegosaurs. *Neues Jahrb. Geol. Palaeont. Abh.*, **161**: 28–46.
Galton, P. M. (1982a). *Elaphrosaurus*, an ornithomimid dinosaur from the Upper Jurassic of North America and Africa. *Palaeont. Z.*, **56**: 265–275.
Galton, P. M. (1982b). The postcranial anatomy of the stegosaurian dinosaur *Kentrosaurus* from the Upper Jurassic of Tanzania, East Africa. *Geologica et Palaeontologica*, **15**: 139–160.
Galton, P. M. (1985). Notes on the Melanorosauridae, a family of large prosauropod dinosaurs (Saurischia: Sauropodomorpha). *Geobios*, **18**: 671–676.
Galton, P. M., Brun, R., and Rioult, M. (1980). Skeleton of the stegosaurian dinosaur *Lexovisaurus* from the lower part of Middle Callovian (Middle Jurassic) of Argences (Calvados), Normandy. *Bull. trim. Soc. geol. Normandie et Amis Museum du Havre*, **67**(4): 39–53.
Gilmore, C. W. (1914). Osteology of the armored Dinosauria in the United States National Museum, with special reference to the genus *Stegosaurus*. *Bull. U.S. Natl. Mus.*, **89**: 1–143.
Gilmore, C. W. (1920). Osteology of the carnivorous Dinosauria in the United States National Museum, with special reference to the genera *Antrodemus* [*Allosaurus*] and *Ceratosaurus*. *Bull. U.S. Natl. Mus.*, **110**: 1–154.
Gow, C. E. (1981). Taxonomy of the Fabrosauridae (Reptilia, Ornithischia) and the *Lesothosaurus* myth. *S. Afr. J. Sci.*, **77**: 1–43.
Hatcher, J. B. (1901). *Diplodocus* (Marsh): Its osteology, taxonomy, and probable habits, with a restoration of the skeleton. *Mem. Carnegie Mus.*, **1**(1): 1–64.
Heerden, J. van, 1979. The morphology and taxonomy of *Euskelosaurus* (Reptilia: Saurischia; Late Triassic) from South Africa. *Navorsinge van die Nazionale Museum*, **4**: 21–84.
Hopson, J. A. (1975). The evolution of cranial display structures in hadrosaurian dinosaurs. *Paleobiology*, **1**(1): 21–43.
Horner, J. R. (1984). The nesting behavior of dinosaurs. *Scientific American*, **250**(4): 130–137.
von Huene, F. (1926). The carnivorous Saurischia in the Jura and Cretaceous formations principally in Europe. *Rev. Mus. La Plata*, **29**: 35–167.
von Huene, F. (1929). Los Saurisquios y Ornitisquios del Cretaceo Argentino. *An. Mus. La Plata*, ser. 2, **3**: 1–196.
Jain, S. L., Kutty, T. S., Roychowdhury, T., and Chatterjee, S. (1977). Some characteristics of *Barapasaurus tagorei*, a sauropod dinosaur from the Lower Jurassic of Deccan, India. *Fourth International Gondwana Symposium*, pp. 204–216.
Kurzanov, S. M. (1976). A new Late Cretaceous carnosaur from Nogon-Tsava, Mongolia. In N. N. Kramarenko (ed.), *Paleontologiya i biostratigrafiya Mongolii. Sovmestnaya Sov.-Mong. Paleontol. Eksped.*, **3**: 93–104.
Lapparent, A.-F. de, and Lavocat, R. (1955). Dinosauriens. In J. Piveteau (ed.), *Traité de Paléontologie*, **5**: 785–962.
Madsen, J. H. (1976). *Allosaurus fragilis*: A revised osteology. *Utah Geol. and Mineral. Survey, Bull.*, **109**: 1–163.
Maryanska, T., and Osmolska, H. (1974). Pachycephalosauria, a new suborder of ornithischian dinosaurs. *Palaeontologia Polonica*, **30**: 45–102.
Milner, A. R., and Norman, D. B. (1984). The biogeography of advanced ornithopod dinosaurs (Archosauria: Ornithischia)—A cladistic-vicariance model. In W. E. Reif and F. Westphal (eds.), *Third Symposium on Mesozoic Terrestrial Ecosystems, Short Papers*, pp. 145–150. ATTEMPTO Verlag, Tübingen.
Norman, D. B. (1980). On the ornithischian dinosaur *Iguanodon bernissartensis* from the Lower Cretaceous of Bernissart (Belgium). *Inst. R. Sci. Natl. Belg., Mem.*, **178**: 1–103.
Osborn, H. F. (1916). Skeletal adaptations of *Ornitholestes, Struthiomimus, Tyrannosaurus*. *Bull. Am. Mus. Nat. Hist.*, **35**: 733–771.
Osborn, H. F. (1924). *Psittacosaurus* and *Protiguanodon*: Two Lower Cretaceous iguanodonts from Mongolia. *Amer. Mus. Novitates*, **127**: 1–16.
Osborn, H. F., and Mook, C. C. (1921). *Camarasaurus, Amphicoelias* and other sauropods of Cope. *Mem. Am. Mus. Nat. Hist., N.S.*, 3(3): 245–387.
Osmolska, H., and Roniewicz, E. (1970). Deinocheiridae, a new family of theropod dinosaurs. In Z. Kielan-Jaworowska (ed.), *Results of the Polish-Mongolian Palaeontological Expeditions. Part II. Palaeontologia Polonica*, **21**: 5–19.
Ostrom, J. H. (1961). Cranial morphology of the hadrosaurian dinosaurs of North America. *Bull. Am. Mus. Nat. Hist.*, **122**: 33–186.
Ostrom, J. H. (1964a). A reconsideration of the paleoecology of hadrosaurian dinosaurs. *Am. J. Sci.*, **262**: 975–997.
Ostrom, J. H. (1964b). A functional analysis of jaw mechanics in the dinosaur *Triceratops*. *Postilla, Yale Peabody Mus.* **88**: 1–35.
Ostrom, J. H. (1966). Functional morphology and evolution of

the ceratopsian dinosaurs. *Evolution*, **20**: 290–308.

Ostrom, J. H. (1969). Osteology of *Deinonychus antirrhopus*, an unusual theropod from the Lower Cretaceous of Montana. *Bull. Peabody Mus. Nat. Hist.* (*Yale Univ.*), **30**: 1–165.

Ostrom, J. H. (1974). The pectoral girdle and forelimb function of *Deinonychus* (Reptilia: Saurischia): A correction. *Postilla, Yale Peabody Mus.* **165**: 1–11.

Ostrom, J. H. (1976). On a new specimen of the Lower Cretaceous theropod dinosaur *Deinonychus antirrhopus. Breviora,* **439**: 1–21.

Ostrom, J. H. (1978). The osteology of *Compsognathus longipes* Wagner. *Zitteliana,* **4**: 73–118.

Ostrom, J. H., and McIntosh, J. S. (1966). *Marsh's Dinosaurs. The Collections from Como Bluff.* Yale University Press, New Haven and London.

Paul, G. S. (1984). The segnosaurian dinosaurs: Relics of the prosauropod-ornithischian transition? *J. Vert. Paleont.*, **4**(4): 507–515.

Raath, M. A. (1969). A new coelurosaurian dinosaur from the forest sandstone of Rhodesia. *Arnoldia,* **4**(28): 1–25.

Raath, M. A. (1972). Fossil vertebrate studies in Rhodesia: A new dinosaur (Reptilia: Saurischia) from near the Trias-Jurassic Boundary. *Arnoldia,* **5**(30): 1–37.

Romer, A. S. (1956). *Osteology of the Reptiles.* University of Chicago Press, Chicago.

Romer, A. S. (1966). *Vertebrate Paleontology* (3d ed.). University of Chicago Press, Chicago.

Rowe, T., Colbert, E. H., and Nations, J. D. (1983). The occurrence of *Pentaceratops* with a description of its frill. In S. G. Lucas, J. K. Rigby, Jr., and B. S. Kues (eds.), *Advances in San Juan Basin Paleontology.* University of New Mexico Press, Albuquerque.

Russell, D. A. (1969). A new specimen of *Stenonychosaurus* from the Oldman Formation (Cretaceous) of Alberta. *Can. J. Earth Sci.,* **6**: 595–612.

Russell, D. A. (1970a). *Tyrannosaurus* from the Late Cretaceous of Western Canada. *Natl. Mus. Nat. Sci., Publ. Palaeont.,* **1**: 1–30.

Russell, D. A. (1970b). A skeletal reconstruction of *Leptoceratops gracilis* from the Upper Edmonton Formation (Cretaceous) of Alberta. *Can. J. Earth. Sci.,* 7: 181–184.

Russell, D. A. (1972). Ostrich dinosaurs from the Late Cretaceous of Western Canada. *Can. J. Earth Sci.*, **9**(4): 375–402.

Russell, D. A., and Seguin, R. (1982). Reconstruction of the small Cretaceous theropod *Stenonychosaurus inequalis* and a hypothetical dinosauroid. *Syllogeus,* **37**: 1–43.

Santa Luca, A. P. (1980). The postcranial skeleton of *Heterodontosaurus tucki* (Reptilia, Ornithischia) from the Stormberg of South Africa. *Ann. S. Afr. Mus.,* **79**(7): 159–211.

Seeley, H. G. (1888). The classification of the Dinosauria. *Rep. Br. Ass. Advmt. Sci., 57th Meeting at Manchester,* **1887**: 698–699.

Sereno, P. C. (1986). Phylogeny of the bird-hipped dinosaurs (Order Ornithischia). *National Geographic Research* **2**: 234–256.

Stromer, E. (1915). Ergebnisse der Forschungen Prof. E. Stromersin den Wuesten Aegyptens. II. Wirbeltier-Reste der Baharije-Stufe (unterstes Cenoman). 3. Das Original des Theropoden *Spinosaurus aegyptiacus* nov. gen., nov. spec. *Abh. Bayer. Akad. Wiss.,* **28**(3): 1–32.

Sues, H.-D. (1980). A pachycephalosaurid dinosaur from the Upper Cretaceous of Madagascar and its paleobiogeographical implications. *J. Paleont.*, **54**(5): 954–962.

Sues, H.-D., and Galton, P. M. (1982). The systematic position of *Stenopelix valdensis* (Reptilia:Ornithischia) from the Wealden of North-Western Germany. *Palaeontographica, Abt. A,* **178**: 183–190.

Sues, H.-D., and Galton, P. M. (1987). Anatomy and classification of the North American Pachycephalosauria (Dinosauria: Ornithischia). *Palaentographica* (*A*).

Taquet, P. (1975). Remarques sur l'evolution des iguanodontides et l'origine des hadrosaurides. *Coll. Internatl. C.N.R.S., Probleèmes Actuels des Paléontologie—Evolution des Vertèbres,* **218**: 503–511.

Thulborn, R. A. (1970). The skull of *Fabrosaurus australis,* a Triassic ornithischian dinosaur. *Palaeontology,* **13**: 414–432.

Thulborn, R. A. (1971). Tooth wear and jaw action in the Triassic ornithischian dinosaur *Fabrosaurus. J. Zool., Lond.,* **164**: 165–179.

Thulborn, R. A. (1972). The postcranial skeleton of the Triassic ornithischian dinosaur *Fabrosaurus australis. Palaeontology,* **15**: 29–60.

Walker, A. D. (1977). Evolution of the pelvis in birds and dinosaurs. In S. M. Andrews, R. S. Miles, and A. D. Walker (eds.), *Problems in Vertebrate Evolution. Linn. Soc. Symp. Ser.,* **4**: 319–358.

Weishampel, D. B. (1981). Acoustic analysis of potential vocalization in lambeosaurine dinosaurs (Reptilia: Ornithischia). *Paleobiology,* 7(2): 252–261.

Weishampel, D. B. (1983). Hadrosaurid jaw mechanics. *Acta Palaeontologica Polonica,* **28**: 271–280.

Welles, S. P. (1984). *Dilophosaurus wetherilli* (Dinosauria, Theropoda), osteology and comparison. *Palaeontographica, Abt. A,* **185**: 85–180.

CHAPTER XV

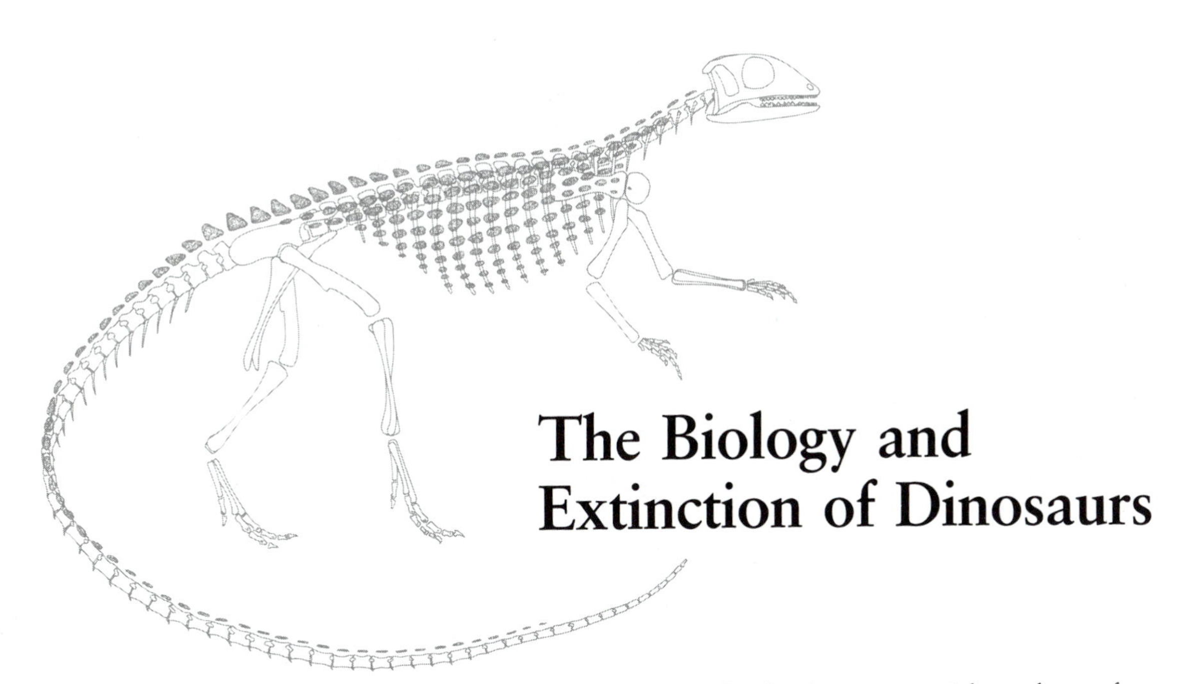

The Biology and Extinction of Dinosaurs

Dinosaurs were the dominant terrestrial vertebrates from the late Triassic to the end of the Cretaceous. During this time, they occupied a wide spectrum of adaptive zones that were broadly comparable to those of large herbivorous and carnivorous mammals in the Tertiary. Despite their long period of dominance, dinosaurs became extinct during a very short period at the end of the Mesozoic, when they were still near the peak of their diversity with a worldwide distribution and without evident competitors.

How did dinosaurs achieve such a position of dominance and why did they suddenly become extinct, only to be replaced by the descendants of the Mesozoic mammals with which they had lived since the close of the Triassic?

Were the dinosaurs wiped out by a catastrophic event at the end of the Cretaceous that mysteriously spared most other groups of terrestrial vertebrates, or did dinosaurs have a unique physiology that was particularly well suited to conditions during the Jurassic and Cretaceous but that was not sufficiently sophisticated to cope with climatic changes that led to the success of Tertiary mammals?

To answer these questions, it is necessary to consider first the evidence for the possible physiological specializations of the dinosaurs and then to evaluate the nature of the extinction event at the end of the Mesozoic.

BIOLOGY

Soon after the recognition of dinosaurs in the early nineteenth century, it was realized that their closest skeletal resemblance to living animals lay with the lizards and crocodiles. Since these modern reptiles have a low metabolic rate and limited capacity for sustained activity, it was long assumed that dinosaurs had a similarly sluggish way of life. During the past 15 years, much evidence has been presented to suggest that dinosaurs differed significantly from modern reptiles physiologically and may have more closely resembled modern mammals and birds. This subject is discussed most comprehensively in *A Cold Look at the Warm-Blooded Dinosaurs* (Thomas and Olson, 1980). Several lines of evidence suggest that dinosaurs were more active animals than any living reptiles.

POSTURE AND GAIT

All dinosaurs had a more upright posture than any living reptiles. This fact is demonstrated by the anatomy of the limbs and girdles and by evidence from footprints that show narrow trackways like those of the cursorial mammals and birds. The limbs of modern reptiles are normally sprawled to the side. They can raise themselves on the limbs when moving rapidly but spend most of the time resting with their bellies on the ground.

The absence of skeletal specializations to facilitate a fully erect posture in reptiles other than dinosaurs can be associated with their basic metabolic limitations. Modern reptiles rely primarily on fermentative metabolism for muscle contraction, which leads to a rapid build up of lactic acid; strenuous activity can only be sustained for a few minutes and must be followed by a long period of recovery. No living reptiles are capable of sustained running, in contrast with most mammals, which rely mainly on oxidative metabolism for continued muscle contraction (Bennett and Dawson, 1976; Bennett, 1982).

Selection probably would never lead to the evolution of limbs and girdles that were capable of habitually upright posture in animals with a typically reptilian physiology. Hence, the upright posture of all dinosaurs suggests that they had the capacity for more continuous activity than do any living reptiles. Coombs (1978) argued that the limb proportions of some dinosaurs are specifically comparable to those of cursorial mammals.

One can argue that an upright posture was necessary for most dinosaurs since their bones would not have had the strength to support their great weight unless they were oriented vertically, nor could the muscles lift or support the body if the limbs had the sprawling attitude of lizards and crocodiles. However, this argument does not apply to the many small dinosaurs, such as the early ornithischians and many coelurosaurs. In fact, these small dinosaurs have fully erect rear limbs and their obligatorily bipedal stance argues for sustained muscle activity to maintain balance and sustain locomotion. It seems particularly significant that the earliest known dinosaurs and their probable ancestors among the lagosuchid thecodonts were both comparatively small and had an erect posture. This fact suggests that the emergence of dinosaurs was associated with advances in their metabolic capacities that enabled them to have sustained periods of activity, in contrast with most primitive amniotes.

Sustained periods of muscle activity would have required a shift to mainly oxidative metabolism and a constant supply of both oxygen and nutrients to the body tissues. These modifications, in turn, would have required changes in the digestive, circulatory, and respiratory systems toward a pattern approaching that of mammals or birds. Unfortunately, little is preserved in the fossils that provides evidence of these changes.

Although the limbs and girdles of all dinosaurs are advanced over the pattern of other reptiles, they have not achieved a wholly mammalian character. Hotton (1980) pointed out that the large amounts of cartilage in the joints of dinosaurs would preclude the rapid acceleration common to cursorial mammals and that the shape of the joint surfaces suggests the short strides of a walking gait, rather than the fleetness of a gazelle or a jaguar. Nevertheless, the upright posture of all dinosaurs is closer to the pattern of large mammals than it is to more typical reptiles.

It is more difficult to establish whether dinosaurs were truly close to the mammals and birds in their metabolic physiology, or whether they were simply exceptional reptiles.

ECTOTHERMY AND ENDOTHERMY

The metabolism of living mammals and birds differs from that of most fish, amphibians, and reptiles in several respects: the possible duration of strenuous activity, the resting metabolic rate, the metabolic scope (the difference between resting metabolism and the rate when undergoing strenuous activity), the degree of temperature control, and the primary source of body heat.

Metabolic rate is measured in terms of the amount of oxygen used per gram of body weight per unit time (Figure 15-1). It is significantly influenced by the total weight of the animal. In mammals, which have been most

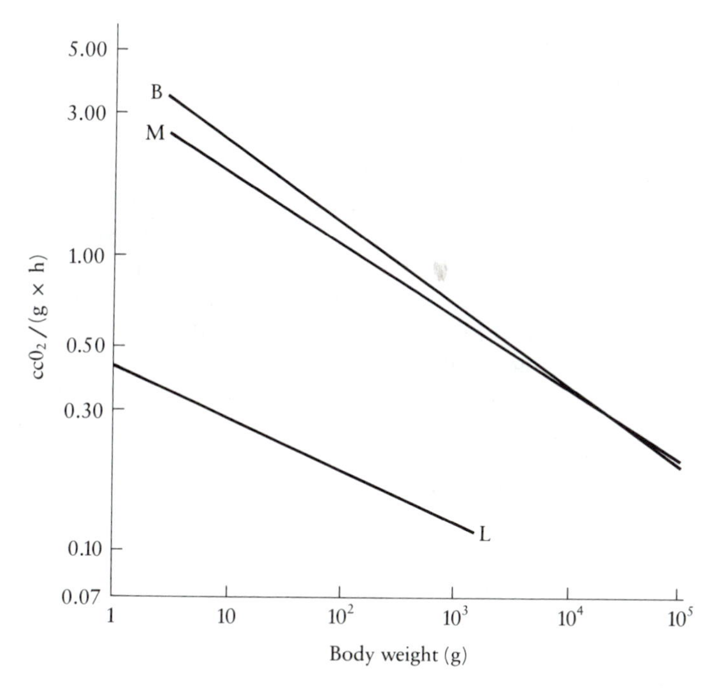

Figure 15-1. Comparison of the relationship of weight-specific metabolic rate to body height for lizards (L) resting at 37° C with the relationships of basal metabolic rate to body weight for birds (B) and mammals (M). *From Bennett and Dawson, 1976.*

extensively studied, the relationship is expressed by the formula $M_{std} = 17(W)^{-0.25}$, where M_{std} is the resting metabolic rate and W is the weight of the animal.

For a given body weight, the resting metabolic rate for mammals is approximately 10 times that of modern reptiles. During strenuous activity, this rate in reptiles may increase by six- to fifteenfold and in mammals by 20 to 30 times. The metabolic scope in mammals is approximately twice that of modern reptiles, but the difference is much less than that between the standard metabolic rates.

Except when hibernating or in a torpid state, all mammals and birds maintain a constant body temperature that is substantially higher than that of the environment. Because of the constancy of their body temperature, they are termed **homoiotherms.** Until about 40 years ago, it was thought that the body temperature of all modern reptiles fluctuated directly with the environmental temperature, which is why they were termed **poikilotherms.** It is now recognized that a great many reptiles control their body temperature by behavioral means and maintain a nearly constant body temperature during their time of maximum activity that is in the range of the birds and mammals.

A much more important factor in distinguishing birds and mammals from most reptiles is the source of their body heat. Reptiles that maintain a constant body temperature above that of the environment do so mainly through the absorption of radiant heat from the sun. Since this heat source is external to the body, they are termed **ectotherms.** In contrast, mammals and birds rely primarily on their higher metabolic rate to raise their body temperature. Since this is an internal heat source, they are designated **endotherms.**

Can the apparently greater degree of activity in dinosaurs be correlated with a significantly higher metabolic rate than that of modern reptiles? This conclusion may be supported by their bone histology.

BONE HISTOLOGY

In an extended review of the histology of vertebrate skeletal tissue, Enlow and Brown (1958) observed that the pattern seen in some dinosaurs closely resembles that of large mammals rather than that of other reptiles or amphibians. Ricqlès (1969, 1978, 1980) used their observations to argue that the metabolism of dinosaurs probably exceeded the typical reptilian rate.

Two major differences are evident between the bones of dinosaurs and larger mammals on one hand and other reptiles on the other. Mammals and dinosaurs rarely show distinct growth lines, which are common in typical reptiles. The absence of growth lines points to a more constant internal environment such as is maintained by warm-blooded animals. Dinosaurs and mammals also show a much higher proportion of remodeled secondary bone, which is associated with numerous Haversian canals that carry large blood vessels. This feature suggests accommodation to more rapid metabolic processes, including rapid sustained growth, elevated temperature, and rapid cycling of minerals between the bones and the body fluid.

These factors all point to metabolic activity that is higher than that of typical reptiles, but there is as yet no way of establishing how close it may have been to the mammalian level nor is it possible to distinguish whether the bone histology is related to an elevated or constant body temperature or to an elevated metabolic rate per se.

Lanyon (1981) argued that continued and extensive reorganization of bone tissue is particularly important in animals of great body weight, regardless of metabolic rate. The dinosaurs whose bone histology is most similar to the mammalian pattern are the gigantic sauropods, which on other grounds seem to show the least evidence for mammalian physiology. On the other hand, a relatively small specimen of *Allosaurus* shows a much more typically reptilian bone histology, although the skeletal anatomy and brain size among the theropods strongly suggest a mammalian level of physiology (Madsen, 1976). In contrast, tritylodonts, which were very advanced mammal-like reptiles that almost certainly approached the primitive mammals in their metabolic rate, retain a bone histology similar to that of typical reptiles (Sues, 1985).

The histological evidence for endothermy is further compromised by the fact that some crocodilian bone is extensively remodeled, although these animals are typically ectothermic.

COMMUNITY STRUCTURE

Bakker (1971, 1972, 1975, 1980) attempted to use ecological arguments to demonstrate that dinosaurs had an essentially mammalian metabolic level. If dinosaurs, like mammals, need at least 10 times the food resources required by ectotherms of comparable body size, it might be possible to detect this difference in the relative number of fossils of predators and prey species. In communities of ectothermic vertebrates, such as Paleozoic amphibians and reptiles, the biomass of predator and prey species is approximately equal. Among modern species of mammals, in contrast, the proportion of predators is about 3 percent that of prey. Bakker found essentially mammalian ratios in several dinosaur communities, which supports his contentions that the dinosaurs were endothermic at an essentially mammalian level. This argument assumes that predatory dinosaurs were food limited.

However, there are several problems with this conclusion. Clearly, it applies only to the predators and says nothing about the metabolism of the prey species. Extensive studies of the komodo dragon, a large varanid lizard of Indonesia, demonstrate that it is present in a ratio of 3 percent or less of its prey species, and yet it is unquestionably an ectothermic reptile (Hotton, 1980). Similar ratios are found in communities where the top predator is the garter snake.

Predator/prey ratios cannot of themselves demonstrate that dinosaurs were endotherms of a mammalian grade, although no dinosaur community has been described in which the predator/prey ratio is at the level that would be expected if the predator species were ectothermic.

BRAIN SIZE

Hopson (1977, 1980) argued that there might be a general correspondence between the brain size of dinosaurs and their metabolic rate. Animals with a metabolic rate approaching that of mammals would be expected to have a similarly effective sensory apparatus and to require a high level of integration between sensory input and motor activities.

As with bone histology, there is no simple and complete correlation between brain size and metabolic rate among living vertebrates, but there is a broad relationship. In general, mammals have approximately ten times the brain size of modern reptiles of similar body weight (Figure 15-2).

Dinosaurs have long been stigmatized by the fact that their brain-size-to-body-weight ratio was extremely small. As in the case of their metabolic requirements, this ratio is directly related to their great absolute size. Among mammals, for which the greatest amount of data is available, brain size is roughly proportional to the two-thirds power of body size. In a group such as modern carnivores or artiodactyls, brain size conforms to the expression $E = KP^{2/3}$, where E is brain size, P is body size, and K is a constant equal to the value of E when $P = 1$ (Jerison, 1973). For mammals as a group, $K = 0.12$. From this equation, we can see that all large vertebrates will have relatively small brains. Since all dinosaurs are fairly large, it is not surprising that their brain size appears very small. Bauchot (1978) showed that there is a further diminution in relative brain size in the largest members of groups with very large body size, such as the whales and ungulates. Applying this factor to the largest dinosaurs shows that their brain size would be expected to be approximately one-half the volume calculated from the expression $E = KP^{2/3}$.

Jerison (1973) elaborated a further procedure for comparing the relative size of the brain between species. He uses the term **encephalization quotient** (EQ) to indicate the ratio of measured brain size divided by the expected brain size within a particular group: $EQ = E_i/E_e$. The expected brain size (corrected for body size) is determined empirically as the average size for all known members of a particular group. The average mammal is established as having a EQ of 1. On this scale, higher apes have an EQ of 4 and that of humans is 7. Modern reptiles as a group, compared with mammals, have an average EQ of 0.12.

Hopson used Jerison's and Bauchot's work to reevaluate the relative brain size of representatives of all the major dinosaur groups. If modern crocodiles are assigned an EQ of 1, dinosaurs show a spread from 0.17 to 5.8, with an average not far from the only living archosaurs. The sauropods score the lowest, with ankylosaurs and stegosaurs in the range of 0.55 to 0.60 and ceratopsians 0.65 to 0.90. Only the ornithopods and theropods score substantially above the crocodiles. The highest EQ is possessed by the "coelurosaur" *Stenonychosaurus,* which alone overlaps with birds and mammals (Figure 15-3).

A similar range of EQ is exhibited by modern mammals, from 0.2 for some marsupials and insectivores to 7 for humans. (Note that these animals are compared with the mammalian standard, which is approximately 10 units above the crocodiles that were used as the standard for comparing dinosaurs.) *Within* mammals, there is not a good correlation between EQ and metabolic rate, but all mammals have both a high EQ and a high metabolic rate compared with all living amphibians and reptiles, which suggests that there is a generally strong relationship between these properties.

The fact that the average EQ of dinosaurs is similar to that of modern reptiles certainly does not support the contention that this group as a whole was endothermic. On the other hand, it might be argued that the very large-brained "coelurosaurs" did have a metabolic rate approaching that of mammals and birds. This contention is further supported by the close relationship that probably exists between "coelurosaurs" and birds, as discussed in Chapter 16. Hopson concludes that dinosaurs in general may have had a somewhat higher metabolic rate than that

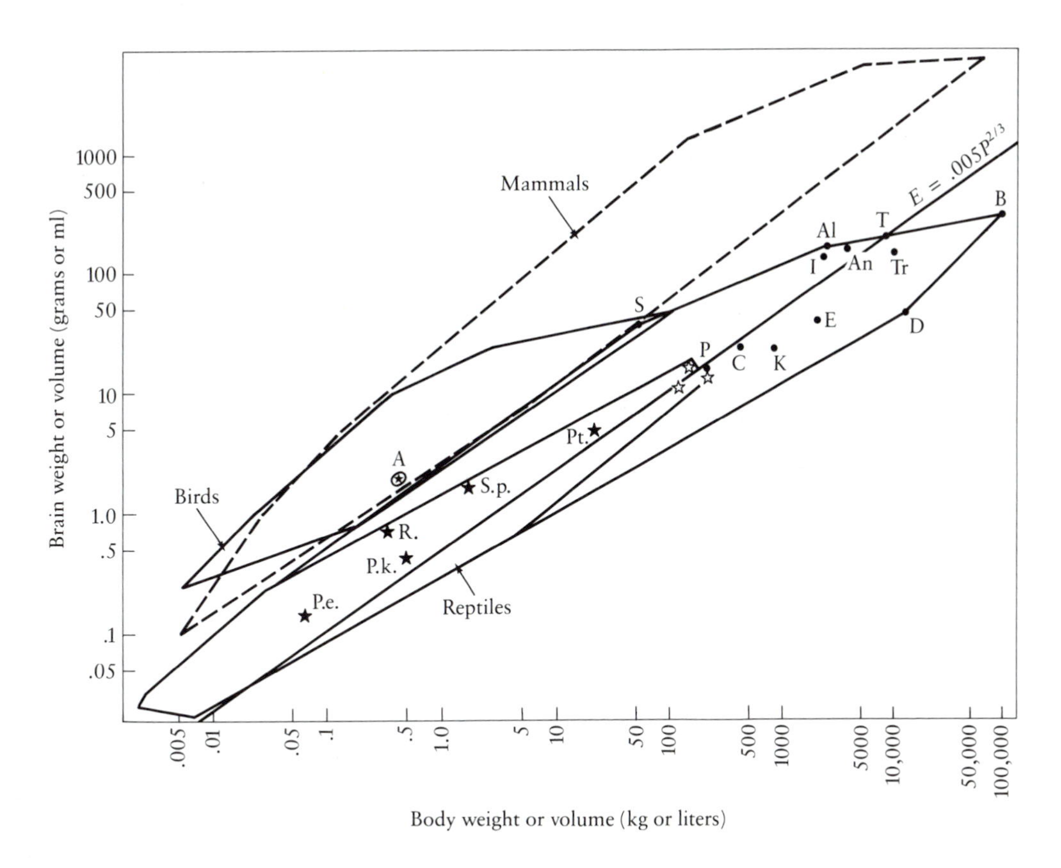

Figure 15-2. Brain-to-body-size relations in dinosaurs, pterosaurs, and the Jurassic bird *Archaeopteryx* superimposed on minimum convex polygons for living reptiles, birds, and mammals. Scale is log-log. The line with slope $\frac{2}{3}$ passes through the approximate center of points for living crocodilians (open stars). New data for the endocast of *Brachiosaurus* indicates that the point for this genus should lie the same distance below the line of slope $\frac{2}{3}$ as that for *Diplodocus*. Abbreviations: Pterosaurs (solid stars)—P.e., *Pterodactylus elegans;* P.k., *Pterodactylus kochi;* Pt., *Pteranodon* sp.; R., *Rhamphorhynchus;* S.p., *"Scaphognathus" purdoni.* Bird (starred circle)—A, *Archaeopteryx.* Dinosaurs—Al, *Allosaurus;* An, *Anatosaurus;* B, *Brachiosaurus;* C, *Camptosaurus;* D, *Diplodocus;* E, *Euoplocephalus;* I, *Iguanodon;* K, *Kentrosaurus;* P, *Protoceratops;* S, *Stenonychosaurus;* Tr, *Triceratops;* T, *Tyrannosaurus. From Hopson, 1980.*

of modern reptiles, but only a few genera may have approached true endothermy.

UNIQUE FACTORS OF DINOSAUR PHYSIOLOGY

Recent work suggests that dinosaurs may have had a metabolic regime that was not simply intermediate between that of typical ectotherms and endotherms but one that differed significantly from both (Regal and Gans, 1980; Spotila, 1980).

The most conspicuous feature of dinosaurs was their gigantic size. Only a few species weighed less than 5 kilograms, and many weighed a ton or more. In contrast, nearly all birds, mammals, and modern reptiles are much smaller. One of the most important effects of body size is its relationship to heat loss and gain. Because of the large surface-to-volume ratio and lack of insulation, most modern ectotherms, such as amphibians, lizards, and snakes, rapidly loose body heat. Rapid heat transfer renders them effective ectotherms but makes endothermy difficult to attain.

In contrast, large animals such as giant land tortoises, adult crocodiles, and the komodo dragon can gain body heat from external sources more rapidly than it is lost; even during the night, their temperature does not drop to that of the surroundings. Spotila and his coworkers (1980) have demonstrated that a reptile with a body diameter of 1 meter living in a subtropical climate would have a mean body temperature of approximately 34°C with a daily fluctuation of less than ± 1°. This figure is based entirely on the use of external heat sources, with no contribution from the animal's own metabolism. Animals of these dimensions that have a mammalian metabolic rate would need special structures or behavioral patterns to dissipate body heat, as is the case for large mammals such as the elephant.

Animals the size of most dinosaurs could achieve a mammalian level of temperature control with no more than a typically reptilian metabolic rate. This pattern is termed **inertial homoiothermy** (Hotton, 1980). Such a high and constant body temperature would provide many of the advantages of mammalian endothermy without the cost.

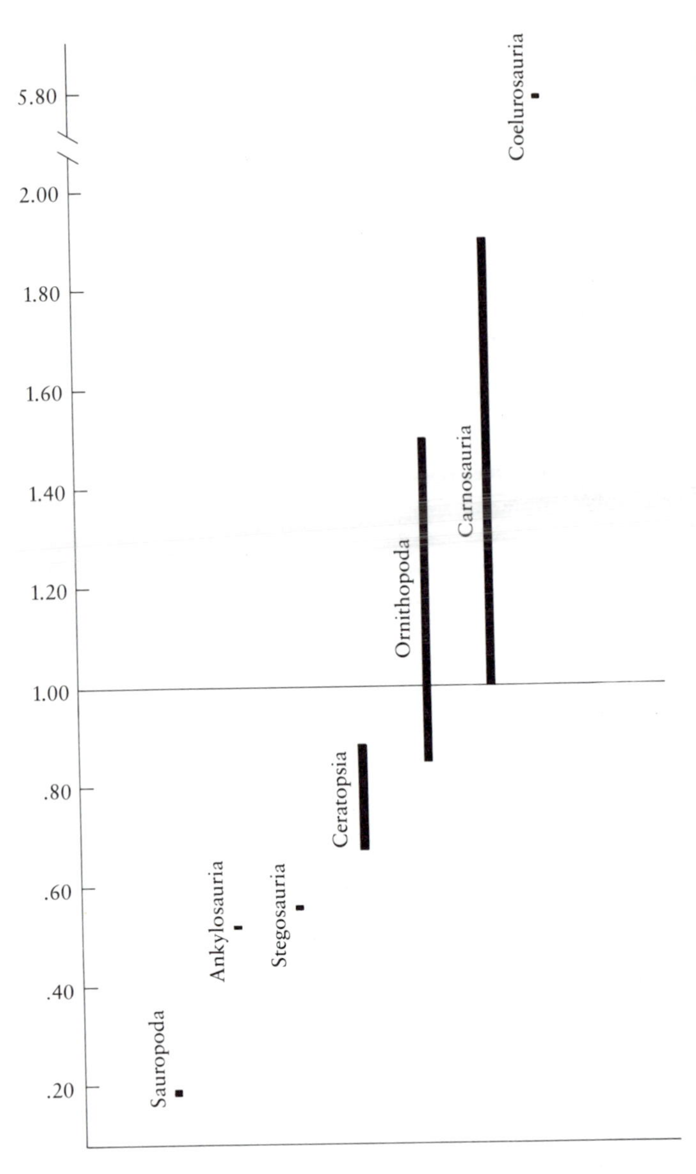

Figure 15-3. Encephalization quotients (EQ) for the suborders of dinosaurs (infraorders Carnosauria and Coelurosauria in the suborder Theropoda) calculated with reference to the line with the equation $E = .005\ P^{2/3}$ in Figure 15-2. Living crocodiles lie at 1.00. Note the break in the scale between 2.00 and 5.80. *From Hopson, 1980.*

Approximately 90 percent of the food consumed by small mammals is used simply for maintaining a constant high body temperature. Endothermy allows mammals to exploit a variety of environments that are not available to reptiles, including those that are continuously cold or dark, and enables them to achieve much smaller body size than did any dinosaurs. These advantages must be balanced by the fact that small endotherms must consume at least 10 times the amount of food required by an ectotherm of comparable body weight.

Dinosaurs almost certainly had a higher rate of oxygen consumption than do any modern reptiles if they relied primarily on oxidative metabolism to sustain long periods of muscle activity, but this would not necessitate the high rate of oxygen and food consumption required for a mammalian type of temperature control.

Large dinosaurs may have had a much lower metabolic rate than most mammals and still have been capable of sustained activity well above the level of typical reptiles as long as they lived in environments where the temperature did not have extremes of either hot or cold. Such a metabolic pattern accords with our interpretation of climatic conditions in the Mesozoic. Evidence from the ratio of oxygen isotopes in marine carbonates suggests a relatively high and constant temperature for much of the Mesozoic. This temperature pattern is associated with the low relief of most of the continents and large areas of continental seas, which resulted in maritime rather than continental climates over much of the earth. Under such conditions, dinosaurs would not have needed to cope with temperature extremes or rapid changes in temperature. Hotton (1980) suggests that dinosaurs may have migrated extensively to follow the most advantageous climatic conditions. Migration would explain the presence of their bones in the Arctic, above 70° latitude, where it was dark for much of the year even if temperatures were not below freezing.

Large size appears to be a simple way to maintain a high and constant body temperature at low metabolic cost, but it would have rendered dinosaurs extremely vulnerable to any climatic deterioration. Extremes of temperature, both high and low, would have been incapacitating or fatal if they were widespread or long lasting.

Small dinosaurs and immature individuals of all species could not have been inertial homoiotherms. They would have been subject to rapid heat loss and would not be able to maintain high body temperature except by the use of environmental sources unless they had a relatively high metabolic rate or insulation. A few dinosaurs are known from skin impressions and none show evidence of insulation, although this lack cannot be demonstrated for most species. Baby dinosaurs were probably ectotherms but rapidly reached a size where they could maintain a high body temperature as a result of their bulk.

SUMMARY OF METABOLIC PHYSIOLOGY

The problem in reaching a definitive conclusion regarding the physiology of dinosaurs is that there are no closely comparable animals living today. In the modern vertebrate fauna, there is a clear distinction between ectotherms and endotherms. There are no vertebrates other than mammals that reach a size comparable to that of the large dinosaurs. We have no way of directly studying any animals that might be inertial homoiotherms.

The predominance of forms with weights above 5 kilograms makes dinosaurs unique among major groups of vertebrates and clearly contrasts them with both birds and mammals. Their generally large size certainly had a great effect on their metabolic processes. Although some small dinosaurs may have had metabolic rates that approached those of mammals, there is no direct evidence that this attribute was common for the group. Evidence from brain size and the structure and degree of ossification of the girdles and limbs suggests that dinosaurs as a group were distinct from both typical ectotherms and endotherms but may have had an intermediate metabolic rate approaching, but not necessarily reaching, that of birds and mammals.

BIRTH AND GROWTH

Eggs that include embryos or are closely associated with recently hatched young are known from sauropods, ornithischians, and ceratopsians. Recent discoveries described by Horner (1984) show that particular nesting patterns are characteristic of different taxonomic groups and that some hadrosaurs apparently tended the young in the nest for some time after their hatching. Dinosaur eggs reached sizes of 5 to 7 kilograms, which is small relative to the size of a 20-ton sauropod and may reflect the mechanical limits of eggshell and the relatively high surface-to-volume ratio that was necessary for the transmission of respiratory gases.

Unlike birds and mammals, dinosaurs did not have determinant growth but continued to increase in size, at least slowly, as long as they lived. The limb bones show no evidence of separate epiphyseal ossifications, and the joint surfaces are more poorly defined than those of mammals and birds, which indicates the persistence of a thick layer of cartilage.

By analogy with modern crocodiles and carnivorous lizards, dinosaurs probably took advantage of a changing suite of prey items during growth. These factors may explain the relative rarity of small adult dinosaurs (Callison and Quimby, 1984). The adaptive zones that in mammals are occupied by small-sized adults may have been occupied, among dinosaurs, by juveniles of large species.

THE EXTINCTION AT THE END OF THE MESOZOIC

Dinosaurs appear in the middle Triassic and by the beginning of the Jurassic they had reached a position of dominance in the terrestrial biota. They continued to diversify throughout the late Jurassic and Cretaceous. The fossil record is too incomplete to judge diversity continuously throughout the Jurassic and Cretaceous, but in the late Cretaceous, approximately 70 million years ago, dinosaurs appear to have been near their peak in the number of reported genera, inhabited all continents except Antarctica, and lived as far north as the Arctic Circle. Then, at the end of the Mesozoic, approximately 65 million years ago, dinosaurs became extinct, and few, if any, fossils are found in beds of Cenozoic age (but see Sloan, Rigby, Van Valen, and Gabriel, 1986). The demise of the dinosaurs is but one element in a period of extinction that affected almost all groups of organisms.

Among the reptiles, pterosaurs, plesiosaurs, and mosasaurs also became extinct at the very end of the Mesozoic. The extinction of marine plants and invertebrates was even more impressive. The most dramatic extinction occurred among unicellular planktonic organisms. Apparently only a single species of calcareous planktonic foramenifera survived to give rise to the subsequent Tertiary radiation. Coccolithophores (unicellular algae with calcareous tests) were almost as severely affected. Diatoms, silicoflagellates, and radiolarians with silicious, rather than calcareous, skeletons were also significantly reduced. In some localities, Cretaceous species are common and diverse up to the last millimeter before the Mesozoic-Cenozoic boundary.

Multicellular invertebrates show an irregular pattern of extinction. Belemnites and ammonites become extinct at the boundary. Ammonites had been declining in numbers and diversity in the later Cretaceous, but nine genera persisted to the very top of the Mesozoic (Alvarez, Kauffman, Surlyk, Alvarez, Asaro, and Michel, 1984). Rudists and trigoniid bivalves, calcareous nannoplankton, and scleractinian hermatypic corals became extinct near the peak of their radiation (Archibald and Clemens, 1982). In general, planktonic organisms and sea-bottom filter feeders, including crinoids and bryozoans, were greatly reduced, but sea-bottom predators and detritus feeders were less affected (Van Valen, 1984).

The marked change in the Earth's biota at the end of the Mesozoic was recognized as early as the eighteenth century, but the probable cause of this phenomenon has been debated with increasing rigor during the past decade.

TIME

To determine the nature of the extinction event at the end of the Mesozoic, it is necessary to establish the length of its duration and the degree to which it was synchronous in different environments throughout the world. This period of extinction has formed part of the basis for recognizing the Cretaceous-Tertiary boundary (abbreviated as either C-T or K-T in many papers). It may also be referred to as the Maastrichtian-Danian boundary, to use the names of the latest stage within the Cretaceous and the earliest stage of the Tertiary.

This boundary is defined on a worldwide basis by the first appearance of species of plants and animals that are typical of the Tertiary. However, such biological criteria are not consistent in all localities. A good correlation can be established among widespread marine sequences, but it is difficult to correlate between marine, freshwater, and terrestrial biotas. Because of physical differences and geographical distances between distinct environments, biological criteria are unlikely to help us establish a consistent chronological marker. Only nonbiological factors can provide a basis for dating that can be used as a test for the synchrony of the extinctions at the end of the Mesozoic.

Radiometric dating is accurate only within about a million years for events at the time of the K-T boundary. Somewhat greater accuracy is possible using the sequence of reversals of the earth's magnetic field. Periodically, the polarity reverses so that the magnetic pole is south, rather than north. The direction of polarity is recorded in iron minerals of both sedimentary and igneous rocks. The specific cause of these reversals is unknown; they occur at extremely irregular intervals, with periods of normal or reversed polarity being of significantly different duration. Comparable time periods in different parts of the world can be recognized because of the similarity in relative duration of the intervals of normal and reversed polarity in a manner that is closely analogous with dating by tree rings. The time of greatest faunal change at the end of the Mesozoic occurs within a period of reversed polarity, interval 39. This entire interval is estimated as lasting for somewhat less than 500,000 years. (Hsü, 1982).

The most precise basis for dating the K-T boundary may be the presence of an abnormally high concentration of iridium, a metal of the platinum group. It was first discovered in a thin layer of clay that marks the faunal boundary between the Cretaceous and Tertiary near Gubbio, Italy (Alvarez, Alvarez, Asaro, and Michel, 1980). Iridium is extremely rare in most rocks of the earth's crust, making up only about 0.3 parts per billion. At Gubbio, the concentration is approximately 6.3 parts per billion—more than 20 times the background level. Similar clays in Denmark showed an even higher concentration—160 times higher than expected for crustal rocks. Since this initial work, an iridium "anomaly" or "spike" has been found in nearly 50 localities throughout the world (Alvarez, 1983). Most of these determinations have been made from drill cores taken from beneath the ocean where there is nearly continuous sedimentation across the K-T boundary. Samples have also been taken from several localities in western North America where nearly continuous deposition of terrestrial sediments occurred across the boundary.

In nearly every locality tested, an iridium spike has been detected. In all cases, this spike occurs at or extremely close to the horizon that had previously been accepted as indicating the K-T boundary on the basis of the fossil record. This iridium anomaly appears to be as close to synchronous as any known geological marker on a worldwide basis. The sediments containing the high concentration of iridium were deposited during a very short period of time to judge by their thickness. Exactly how short a period is difficult to establish by other means, but it was probably no more than a few thousand years.

THE ALVAREZ EXTINCTION HYPOTHESIS

The iridium anomaly not only serves as an extraordinarily precise means of establishing the time of the major extinction at the end of the Cretaceous, but it may also provide evidence of the cause of this extinction. Iridium is thought to be considerably more abundant in the solar system as a whole than it is in the crust of the earth. It is soluble in iron, and most of the iridium that was present when the earth coalesced was probably taken up by the iron in its core. The higher concentration of iridium in the solar system is based on proportions in meteorites, some types of which have as much as 0.5 parts per million. Since it is difficult to hypothesize any geological factors that would operate on a worldwide basis to concentrate the iridium at the Cretaceous-Tertiary boundary, Alvarez suggested that it resulted from the impact of a large meteorite or asteroid whose contents were then distributed over the surface of the earth. To account for the observed amount of iridium requires an extraterrestrial body approximately 10 kilometers in diameter that would have weighed 10^{16} grams. The impact of such a body would have created a crater approximately 100 kilometers in diameter.

Alvarez and his colleagues hypothesized that the impact of such a large object would have thrown debris up to 60 times its own mass into the atmosphere, where it would have circled the globe for several years before falling back to the earth. This amount of material, in the form of fine particles, would have formed a cloud so dense that very little sunlight would have reached the surface of the earth. The resulting darkness would have halted photosynthesis, killing first the plants and then the animals that feed on them. This disaster would certainly have been a sufficient cause for the extinction at the end of the Mesozoic.

Several additional effects of the impact have been noted in subsequent papers (Alvarez, Alvarez, Asaro, and Michel, 1982; Alvarez, 1983). Blocking of the sunlight would have led to a rapid chilling of the earth, since heat in the infrared end of the spectrum would have been able to escape the earth through the dust cloud. With photosynthesis much reduced, CO_2 would accumulate in both the atmosphere and in seawater. The buildup of carbon dioxide in the water would lead to higher acidity and the solution of the shells of carbonate-secreting organisms, which would account for the absence of calcareous fossils near the K-T boundary in some marine sections. Additional carbon dioxide in the atmosphere would produce a greenhouse effect once the dust cloud had gone. A general cooling across the K-T boundary has long been suspected, but the time interval for the cooling and subsequent heating predicted by the Alvarez model may have been too rapid to be detected by current methods of analysis of seawater temperatures that use the relative abundance of oxygen and carbon isotopes.

The Alvarez hypothesis has created an enormous interest in the problem of extinction at the end of the Mesozoic. Countless individual papers have been written and there have been several major symposia on the subject. The most comprehensive publication dealing with this question is *Geological Implications of Impacts of Large Asteroids and Comets on the Earth*, which was edited by Silver and Schultz (1982). This book and other important articles have been reviewed by Van Valen (1984) in a short summary of the problem. A series of papers on this subject also appeared in the April 1, 1984, issue of *Nature*.

Physical scientists tend to accept the major conclusions reached by Alvarez but contest some details. Scientists who have studied the fossil record have generally been highly critical. Officer and Drake (1983) present evidence that the period of greatest marine extinction was not instantaneous but extended over at least 10,000 and perhaps as much as 100,000 years. The sequence of extinction of different marine microorganisms differs from locality to locality, and pronounced extinction events may vary in their relationship to the time of the iridium anomaly by tens of thousands of years.

Studies of the behavior of particles in the high atmosphere indicate that the dust cloud would settle to the earth after a few months, rather than remaining for several years (Toon, Pollack, Ackerman, Turco, McCay, and Liv, 1982). The longer period of darkness first proposed by Alvarez would probably have killed all plant life and might have left the earth permanently covered with a thick layer of snow and ice.

Hörz (1982) raised another physical problem. He pointed out that most of the ejecta from impact craters on the earth and moon is confined to within a few crater diameters of the impact, with only small amounts distributed over greater distances. Several models are discussed in the Silver and Schultz volume to explain how a great mass of debris might have been forced into the upper atmosphere.

No large crater has yet been identified as coincident with the iridium anomaly, but it would be difficult to detect if it had been formed on the ocean floor. What is more difficult to explain is the lack of great extinctions at the other times when craters 100 kilometers or more in diameter are known to have been formed (Shoemaker, 1983).

Iridium anomalies have been discovered at other horizons that are not associated with mass extinctions and are missing from horizons, such as the Permo-Triassic boundary, when there were mass extinctions. Locally high concentrations of iridium are associated with reducing environments and the presence of iron pyrite. In these cases, the abundance of iridium probably resulted from a concentration from normal terrestrial sources.

On the other hand, it appears inescapable that the iridium anomaly at the end of the Mesozoic was both synchronous and worldwide. Since it characterizes both terrestrial and marine sediments, it is very difficult to hypothesize any single way in which it could have been concentrated from sediments of the earth's crust. McLean (1982) proposed a volcanic source for the iridium, but it is difficult to accept that a period of volcanic activity would have been of such great magnitude and yet last for such a short period of time.

Although an extraterrestrial source for the iridium marker may be difficult to prove unequivocally, it currently appears to be the most plausible hypothesis. It is also difficult to escape the conclusion that the synchrony between the iridium spike and the mass extinction of marine organisms indicates a causal relationship.

Surprisingly, the record of the terrestrial biota does not appear to support this hypothesis but suggests a different, apparently gradual pattern of extinction.

THE TERRESTRIAL BIOTA

Despite the emphasis that is typically placed on the demise of the dinosaurs at the end of the Mesozoic, this group actually provides little information regarding the nature of the extinction. One of the most serious problems is that there are very few places where terrestrial sedimentation is continuous across the K-T boundary. The only area in which the terrestrial fauna is well known at the very end of the Maastrichtian is in western North America along the foothills of the ancestral Rocky Mountains. Rich dinosaur faunas are known from the earlier Cretaceous in Central Asia, China, and South America, but none are known from the very end of the Mesozoic. Specific information regarding the last years of the dinosaurs is limited to North American localities.

Another problem is the emphasis that different authors place on different taxonomic categories. It is frequently stated that five orders of reptiles became extinct at the end of the Mesozoic: the Saurischia, Ornithischia, Pterosauria, Ichthyosauria, and the Plesiosauria. This appears to represent a very dramatic change in the fauna, but it actually involves a relatively small number of genera. In fact, ichthyosaurs are not known in the late Cretaceous. Their apparent Maastrichtian record was based on a misidentified fragment of a shoulder girdle. Twenty-two genera of plesiosaurs have been named from the Upper Cretaceous, but only four are known at the end of the Maastrichtian (Welles, 1952). We know only three pterosaur genera at the end of the Cretaceous, although *Pteranodon* is common and widely distributed. Only about a dozen dinosaur genera continue to the end of the Maastrichtian (Padian and Clemens, 1985). There is some indication that these genera were becoming progressively less common within the Maastrichtian and there may have been a progressive reduction from the north to the south, although this phenomenon has not been fully documented (Sloan, 1976; Van Valen and Sloan, 1977; Archibald and Clemens, 1982; Clemens, 1984; Padian and Clemens, 1985; Sloan, Rigby, Van Valen, and Gabriel, 1986).

In order to evaluate the significance of particular extinction events, we must compare them with the normal, or average, rates of extinction within the group throughout its history (Raup and Sepkoski, 1982). Among Mesozoic tetrapods, nearly all period and stage boundaries are marked by very large-scale extinctions, if we judge by the rarity of genera that continue from one stage or period to the next. Only a few dinosaur genera are known from more than a single formation or lithological unit. Throughout the Mesozoic, new genera continued to evolve at least as rapidly as others became extinct. What marks the K-T boundary as a unique event is not the number of extinctions, but the fact that no new genera evolved to replace them. Throughout the Mesozoic, the fossil record of dinosaurs is so incomplete that we rarely, if ever, have a record of the particular genera from one period that were ancestral to those in the next. Nor can we say with any degree of accuracy what percentage of the total number of genera living in one stage or period became extinct before the beginning of the next stage or period. What is known suggests that this percentage is always quite high (probably more than 50 percent and perhaps approaching 90 percent). The fact that it reached 100 percent at the end of the Cretaceous, with no surviving lineages, indicates a significant change in the world in which they lived. In contrast, an extinction rate of 50 to 90 percent within a particular stage or period might be normal and related only to progressive evolutionary changes, competition, and predation.

In many localities, the last dinosaurs occur about 3 meters below the iridium anomaly and not coincident with it. As observed in North America, one can attribute the final extinction of the dinosaurs to gradual changes in climate (Sloan, 1976). Unfortunately, for most of the world there is no fossil evidence that bears on this question.

Other reptiles living with the dinosaurs—turtles and lizards—show no significant extinction across the K-T boundary. Other important survivors were the crocodiles and the champsosaurs, which were as large as many dinosaurs and may have had a comparable physiology. They were distinguished ecologically as primarily aquatic forms.

Mammals also show a mixed pattern of extinction and survival across the K-T boundary. In North America, only one of thirteen described species of marsupials survived the end of the Cretaceous. Clemens (1984) points out that this extinction was not instantaneous but occurred gradually and progressively over approximately 200,000 years, coincident with the extinction of dinosaurs. Presumably, marsupials did not experience such a strong decline in the southern continents, if we judge by their subsequent radiation in both South America and Australia.

On the other hand, placental mammals appear to have crossed the boundary without significant reduction in numbers of species. The major radiation that culminated in their dominance in the early Tertiary had already begun in the late Cretaceous. Archibald (1982) described the gradual evolution of mammalian communities that were contemporaries of the dinosaurs in the late Maastrichtian and their successors in the early Paleocene.

Although we would expect that land plants would have undergone the same mass extinction as the phytoplankton if there had been a protracted period of darkness, this was not obviously the case. The great revolution in the history of vascular plants occurred early in the Cretaceous with the origin and rapid radiation of the angiosperms, which became the dominant land plants well before the end of the Cretaceous (Doyle, 1977). The greatest period of extinction among land plants near the K-T boundary is 3 to 6 meters above the level of the iridium anomaly or 50,000 to 90,000 years later (Hickey, 1981).

In different floral provinces, 50 to 75 percent of Cretaceous species became extinct by the early Tertiary. This figure would appear to be very high, but it is not greater than that observed in the Paleocene-Eocene boundary, which has never been thought of as a time of major floral extinction. Hickey points out that northern floras would be expected to be more resistant to catastrophic extinction than those in the tropics because they probably had already evolved the capacity to remain dormant for part of the year and may have had greater resistance to cold. The pattern of extinction observed is just the reverse, with greater extinction occurring in the north than the south. There are also marked regional differences in the extent of extinction that do not fit readily with the expectations of a worldwide catastrophic event.

One dramatic and short-term change suggests at least a local catastrophe at the K-T boundary. This change is the sudden dominance of fern spores rather than angiosperm pollen in the Raton Basin in New Mexico (Orth, Gilmore, Knight, Pillmore, Tschudy, and Fassett, 1982). Such flora changes occur as a result of forest fires, floods, and other local catastrophies. However, such a marked floral change has not been established in other parts of the world.

HYPOTHESIS OF GRADUAL CHANGE

Changes in the terrestrial community have long been attributed to a gradual climatic deterioration that culminated at the end of the Cretaceous and early in the Tertiary. It has been recognized that the end of the Cretaceous was marked by a major regression. The sea level dropped markedly throughout the world, resulting in the drying up of the many large epicontinental seas. The wide extent of seas within the continents during the Cretaceous would have resulted in maritime rather than continental climates over wide areas and the absence of sharp seasonality.

For much of the Mesozoic, the level of the land was low and without major mountain chains, which also contributed to an equable climate worldwide. Toward the end of the Mesozoic, the modern system of mountain ranges began to form as a result of movements of major oceanic and continental plates. This tectonic activity led to a breakup of the previous global patterns of circulation, bringing about regional differences in climate and greater seasonality. The tectonic activity coincided with and was probably causally related to the late Cretaceous regression (Hallam, 1984). Together they were responsible for a major deterioration of climate that continued into the Cenozoic. This climatic change is most directly reflected in changes in the flora, which indicate a reduction of the annual mean temperature in western North America of almost 10°C between the late Cretaceous and the early Paleocene (Hickey, 1981). This gradual lowering of the temperature and particularly the loss of equability of temperature worldwide would be sufficient to account for the observed floral and faunal changes in the North Temperate zone. According to the model of dinosaur physiology discussed earlier in this chapter, they would have been much more vulnerable to climatic deterioration than would the placental mammals. Van Valen and Sloan (1977) hypothesize that dinosaurs may have survived into the Tertiary in the tropics, but there is little fossil evidence yet available in those regions.

The processes of gradual geological change appear to provide an adequate explanation for the known evolutionary events among terrestrial vertebrates across the K-T boundary. A catastrophic event does not seem necessary or even likely on the basis of this evidence, but one must remember that our knowledge comes from only a very small fraction of the areas that may have been inhabited by dinosaurs and early mammals at the very end of the Mesozoic.

On the other hand, the sudden extinction of marine organisms, especially planktonic microorganisms, does appear most readily explicable by a catastrophic event. Although the marine extinctions were not all simultaneous and instantaneous, they could certainly be considered catastrophic as measured by the number of species and lineages involved and in their relatively rapid occurrence. The major extinctions all occurred within a fraction of the time represented by magnetic reversal 39, which itself lasted slightly less than 500,000 years.

The instantaneous appearance of the iridium anomaly throughout the world provides evidence of a simultaneous catastrophic event. It seems logical that these two events were causally related, although the exact relationship remains debatable. The Alvarez hypothesis of a long-lasting dust cloud that would have halted photosynthesis appears to be contradicted by the specific patterns of extinction of both marine and terrestrial plants. Hsü (1982) suggested both physical and chemical factors that might have had more immediate and wide-ranging effects in the sea than on land. His hypothesis might explain, at least partially, the apparent contradiction between the patterns of extinction on land and in the ocean.

W. Alvarez and his coauthors (1984) suggest that many of the apparent contradictions regarding the K-T extinction might be resolved by recognizing two distinct causes, one short term and catastrophic and the other long term. Many of the changes that happened in the terrestrial community and among marine vertebrates may have resulted from the gradual changes, which have been long recognized by geologists, in the pattern of the continents and seas toward the end of the Mesozoic. Superimposed on this pattern of gradual extinction were the sudden effects of a collision with an extraterrestrial body that caused the instantaneous termination of many groups but may have, for geographical and environmental reasons, produced less-profound extinction in others.

Much more evidence may eventually be available regarding the temporal distribution of terrestrial vertebrates across the K-T boundary in areas other than North America. Such findings may permit a final decision to be reached as to whether the major extinction at the end of the Cretaceous was the result of terrestrial or extraterrestrial causes.

REFERENCES

Alvarez, L. W. (1983). Experimental evidence that an asteroid impact led to the extinction of many species 65 million years ago. *Proc. Natl. Acad. Sci. U.S.A.*, **80**: 627–642.

Alvarez, L. W., Alvarez, W., Asaro, F., Michel, H. V. (1980). Extraterrestrial cause of the Cretaceous-Tertiary extinction. *Science*, **208**: 1095–1108.

Alvarez, W., Alvarez, L. W., Asaro, F., and Michel, H. V. (1982). Current status of the impact theory for the terminal Cretaceous extinction. In L. T. Silver and P. H. Schultz (eds.), *Geological Implications of Impacts of Large Asteroids and Comets on Earth. Geol. Soc. America, Special Paper*, **190**: 305–315.

Alvarez, W., Kauffman, E. G., Surlyk, F., Alvarez, L. W., Asaro, F., and Michel, H. V. (1984). Impact theory of mass extinctions and the invertebrate fossil record. *Science*, **223**: 1135–1141.

Archibald, J. D. (1982). A study of Mammalia and geology across the Cretaceous-Tertiary boundary in Garfield County, Montana. *Univ. Calif. Publ. Geol. Sci.*, **122**: 1–286.

Archibald, J. D., and Clemens, W. A. (1982). Late Cretaceous extinctions. *American Scientist*, **70**(4): 377–385.

Bakker, R. T. (1971). Dinosaur physiology and the origin of mammals. *Evolution*, **25**: 636–658.

Bakker, R. T. (1972). Anatomical and ecological evidence of endothermy in dinosaurs. *Nature*, **238**: 81–85.

Bakker, R. T. (1975). Dinosaur renaissance. *Scientific American*, **232**(4): 58–78.

Bakker, R. T. (1980). Dinosaur heresy—Dinosaur renaissance. Why we need endothermic archosaurs for a comprehensive theory of bioenergetic evolution. In R. D. K. Thomas and E. C. Olson (eds.), *A Cold Look at the Warm-Blooded Dinosaurs. AAAS Selected Symposium*, **28**: 351–462.

Bauchot, R. (1978). Encephalization in vertebrates: A new mode of calculation of allometry coefficients and isoponderal indices. *Brain Behav. Evol.*, **15**: 1–8.

Bennett, A. F. (1982). The energetics of reptilian activity. In C. Gans (ed.), *Biology of the Reptilia*. Vol. 13, pp. 155–199. Academic Press, London and New York.

Bennett, A. F., and Dawson, W. R. (1976). Metabolism. In C. Gans (ed.), *Biology of the Reptilia*. Vol. 5, pp. 127–223. Academic Press, London and New York.

Callison, G., and Quimby, H. M. (1984). Tiny dinosaurs: Are they fully grown? *J. Vert. Paleont.*, **3**: 200–209.

Clemens, W. A. (1984). Evolution of marsupials during the Cretaceous-Tertiary transition. In W. E. Reif and F. Westphal (eds.), *Third Symposium on Mesozoic Terrestrial Ecosystems*, pp. 47–53. ATTEMPTO Verlag, Tübingen.

Coombs, W. P. (1978). Theoretical aspects of cursorial adaptations in dinosaurs. *Quart. Rev. Biol.*, **53**: 393–418.

Doyle, J. A. (1977). Patterns of evolution in early angiosperms. In A. Hallam (ed.), *Patterns of Evolution, as Illustrated by the Fossil Record*, pp. 501–546. Elsevier, Amsterdam.

Enlow, D. H., and Brown, S. O. (1958). A comparative histological study of fossil and Recent bone tissues. Part III. *Texas J. Sci.*, **10**: 187–230.

Hallam, A. (1984). Pre-Quarternary sea-level changes. *Ann. Rev. Earth Planet. Sci.*, **12**: 205–243.

Hickey, L. J. (1981). Land plant evidence compatible with gradual, not catastrophic change at the end of the Cretaceous. *Nature*, **292**: 529–531.

Hopson, J. A. (1977). Relative brain size and behaviour in archosaurian reptiles. *Ann. Rev. Ecol. Syst.*, **8**: 429–448.

Hopson, J. A. (1980). Relative brain size in dinosaurs. Implications for dinosaurian endothermy. In R. D. K. Thomas and E. C. Olson (eds.), *A Cold Look at the Warm-Blooded Dinosaurs. AAAS Selected Symposium*, **28**: 287–310.

Horner, J. R. (1984). The nesting behaviour of dinosaurs. *Scientific American*, **250**: 130–137.

Hörz, F. (1982). Ejecta of the Ries Crater, Germany. In L. T. Silver and P. H. Schultz (eds.), *Geological Implications of Impacts of Large Asteroids and Comets on the Earth. Geol. Soc. America Special Paper*, **190**: 39–56.

Hotton, N., III. (1980). An alternative to dinosaur endothermy: The happy wanderers. In R. D. K. Thomas and E. C. Olson (eds.), *A Cold Look at the Warm-Blooded Dinosaurs. AAAS Selected Symposium*, **28**: 311–350.

Hsü, K. J. (1982). Evolutionary and environmental consequences of a terminal Cretaceous event. In D. A. Russell and G. Rice (eds.), *Extinctions and Possible Terrestrial and Extraterrestrial Causes. Syllogeus*, **39**: 140–142.

Jerison, H. J. (1973). *Evolution of the Brain and Intelligence*. Academic Press, New York.

Lanyon, L. E. 1981. Locomotor loading and functional adaptation in limb bones. In M. H. Day (ed.), *Vertebrate Locomotion. Symp. Zool. Soc. London*, **48**: 305–329.

Madsen, J. H., Jr. (1976). *Allosaurus fragiles:* A revised osteology. *Utah Geol. Min. Surv. Bull.*, **109**: 1–163.

McLean, D. M. (1982). Deccan volcanism and the Cretaceous-Tertiary Transition. In D. A. Russell and G. Rice (eds.), *Cretaceous-Tertiary Extinction and Possible Terrestrial Extraterrestrial Causes. Syllogeus*, **39**: 143–144.

Officer, C. B., and Drake, C. L. (1983). The Cretaceous-Tertiary Transition. *Science*, **219**: 1383–1390.

Orth, C. J., Gilmore, J. S., Knight, J. D., Pillmore, C. L., Tschudy, R. H., and Fassett, J. E. (1982). Iridium abundance measurements across the Cretaceous-Tertiary boundary in the San Juan and Raton Basins of northern New Mexico. In L. T. Silver and P. H. Schultz (eds.), *Geological Implications of Impacts of Large Asteroids and Comets on the Earth. Geol. Soc. America, Special Paper*, **190**: 423–433.

Padian, K., and Clemens, W. A. (1985). Terrestrial Vertebrate Diversity: Episodes and insights. In J. W. Valentine (ed.), *Phanerozoic Diversity Patterns*, pp. 41–96. Princeton University Press, Princeton.

Raup, D. M., and Sepkoski, J. J. (1982). Mass extinctions in the marine fossil record. *Science*, **215**: 1501–1503.

Regal, J., and Gans, C. (1980). The revolution in thermal physiology. Implications for dinosaurs. In R. D. K. Thomas and E. C. Olson (eds.), *A Cold Look at the Warm-Blooded Dinosaurs. AAAS Selected Symposium*, **28**: 167–188.

Ricqlès, A. J. (1969). L'histologie osseuse envisagée comme indicateur de la physiologie thermique chez les tétrapodes fossiles. *C.R. Hebd. Séanc. Acad. Sci. Paris, D.*, **268**: 782–785.

Ricqlès, A. J. (1978). Recherches paléohistologiques sur les os longs des tétrapodes. VII. Idem. Troisième partie: évolution (fin). *Ann. Paléont. (Vertébrés)*, **64**: 153–176.

Ricqlès, A. J. (1980). Tissue structures of dinosaur bone. Functional significance and possible relation to dinosaur physiology. In R. D. K. Thomas and E. C. Olson (eds.), *A Cold Look at the Warm-Blooded Dinosaurs. AAAS Selected Symposium*, **28**: 103–139.

Shoemaker, E. M. (1983). Asteroid and comet bombardment of the earth. *Ann. Rev. Earth Planet. Sci.*, **11**: 461–494.

Silver, L. T., and Schultz, P. H., eds. (1982). *Geological Implications of Large Asteroid and Comets on the Earth. Geological Society of America, Special Paper*, **190**.

Sloan, R. E. (1976). The ecology of dinosaur extinction. In C. S. Churcher (ed.), *Athlon-Essays on Paleontology in Honour of Loris Shano Russell. Roy. Ont. Mus. Life Sci. Misc. Publ.*, pp. 134–153.

Sloan, R. E., Rigby, J. R., Jr., Van Valen, L. M., Gabriel, D. (1986). Gradual dinosaur extinction and simultaneous ungulate radiation in the Hell Creek Formation. *Science*, **232**: 629–633.

Spotila, J. R. (1980). Constraints of body size and environment on the temperature regulation of dinosaurs. In R. D. K. Thomas and E. C. Olson (eds.), *A Cold Look at the Warm-Blooded Dinosaurs. AAAS Selected Symposium*, **28**: 233–252.

Sues, H. D. (1985). Personal communication.

Thomas, R. D. K., and Olson, E. C., eds. (1980). *A Cold Look at the Warm-Blooded Dinosaurs. AAAS Selected Symposium*, **28**. Westview Press, Colorado.

Toon, O. B., Pollack, J. B., Ackerman, T. P., Turco, R. P., McCay, C. P., and Liu, M. S. (1982). Evolution of an impact-generated dust cloud and its effects on the atmosphere. In L. T. Silver and P. H. Schultz (eds.), *Geological Implications of Impacts of Large Asteroids and Comets on the Earth. Geol. Soc. America, Special Paper*, **190**: 187–200.

Van Valen, L. M., and Sloan, R. E. (1977). Ecology and the extinction of the dinosaurs. *Evol. Theory*, **2**: 37–64.

Van Valen, L. M. (1984). Catastrophes, expectations, and the evidence. Review of L. T. Silver and P. H. Schultz (eds.), *Geological Implications of Impacts of Large Asteroids and Comets on the Earth. Paleobiology*, **10**: 121–137.

Welles, S. P. (1952). A review of the North American Cretaceous elasmosaurs. *Univ. Calif. Publ., Geol. Sci. Bull.*, **29**: 47–143.

C H A P T E R **XVI**

Flight

Active flight has evolved three times among vertebrates: in bats among the mammals and in the ancestors of birds and pterosaurs, both of which evolved from early archosaurs. Pterosaurs first appeared in the fossil record in the late Triassic; the oldest bird is known from the late Jurassic. These groups share many skeletal features, but flight was certainly achieved separately in each.

Pterosaurs

Pterosaurs are characterized by membranous wings that are supported by a single greatly enlarged fourth digit. They are divided into two suborders. The primitive rhamphorhynchoids appear in the late Triassic and are common throughout the Jurassic (Figure 16-1). They retain a long tail, short neck, and short face. Some were as

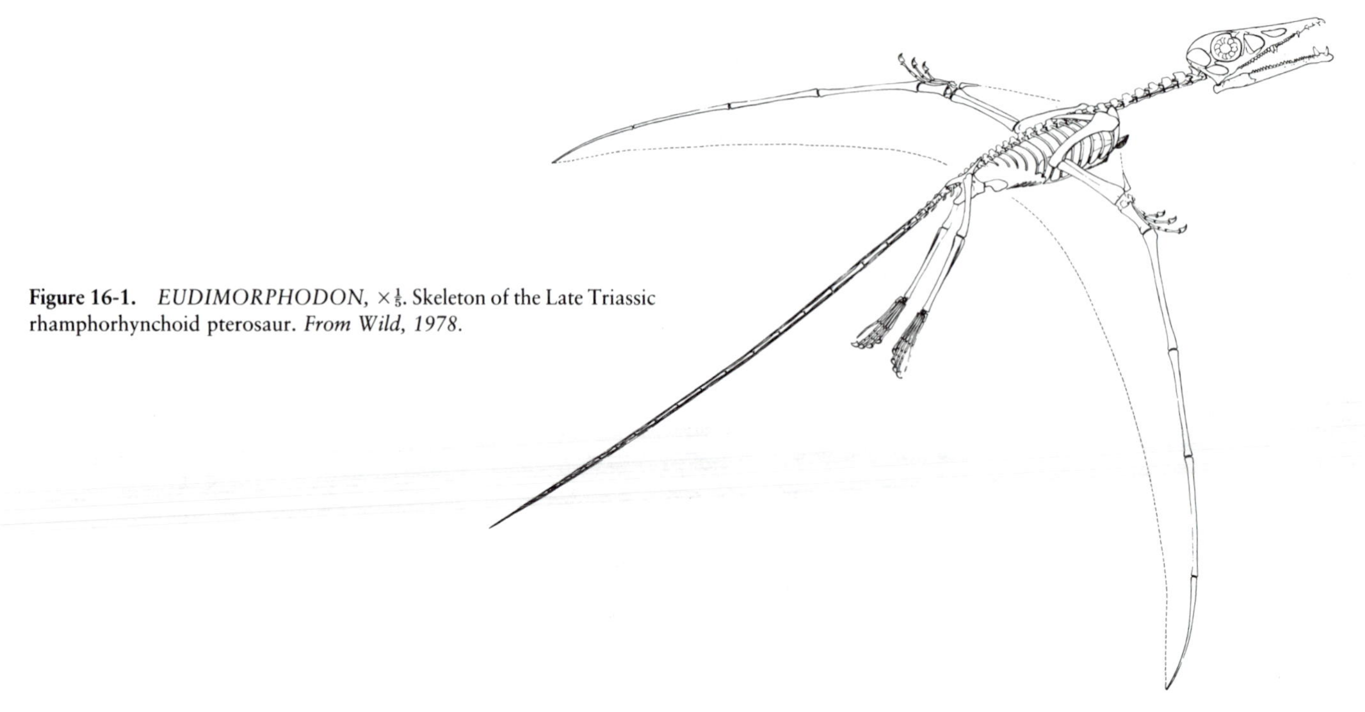

Figure 16-1. *EUDIMORPHODON*, $\times\frac{1}{5}$. Skeleton of the Late Triassic rhamphorhynchoid pterosaur. *From Wild, 1978.*

small as a sparrow. The pterodactyloids, which appeared in the late Jurassic, have shortened the tail while increasing the length of the metacarpals, neck, and skull. They attained the greatest size of any flying vertebrates, with *Quetzalcoatlus* from the late Cretaceous having an estimated wing span of 11 to 12 meters (Langston, 1981). Wellnhofer (1978) published the most recent review of the pterosaurs.

EARLY PTEROSAURS

The earliest pterosaurs occur in the Upper Triassic (Norian) of northern Italy (Wild, 1978). Nearly the entire skeleton of *Eudimorphodon* is known. It resembles later rhamphorhynchoids very closely and provides an excellent basis for characterizing the anatomy of all other pterosaurs. The contemporary genus *Peteinosaurus* is more primitive in having wings that are approximately two-thirds the length of those in other comparable-size pterosaurs. A third Norian genus, *Preondactylus* (Wild, 1984b), is known from only incomplete remains but is apparently the most primitive of all pterosaurs.

The skull is proportionately very large in all pterosaurs, with a latticework of narrow rods between the large orbital, antorbital, and temporal openings. The cranial bones tend to fuse, which makes it difficult to determine the position of sutures, but their general pattern resembles that of primitive archosaurs. The quadrate is streptostylic. The dentition of *Eudimorphodon* is conspicuously dimorphic, with long peglike teeth in the premaxilla and anterior extremity of the dentary, as well as two caninelike teeth in the maxilla. The other teeth are shorter and multicuspate, superficially resembling those of primitive Mesozoic mammals (see Chapter 18).

Eudimorphodon has 8 cervical, 12 thoracic, 2 lumbar, and 4 fused sacral vertebrae and nearly 35 caudals. *Peteinosaurus* has only 3 sacral vertebrae that remain unfused. The cervicals of *Eudimorphodon* are clearly procoelous, and as in other archosaurs, the trunk vertebrae bear long transverse processes. Like other rhamphorhynchoids, the caudal vertebrae are supported by bundles of ossified tendons that develop as greatly elongated processes from the pre- and postzygapophyses and haemapophyses, in comparable fashion to those in the theropod dinosaur *Deinonychus*.

The shoulder girdle is highly modified in a manner that is closely analogous with that of advanced birds to provide enlarged surfaces for the origin of flight muscles and to resist the forces generated by their contraction (Figure 16-2). The most conspicuous element is the broad sternum that bears a low keel. The sternum is notched laterally to receive the ends of the thoracic ribs. Lepidosaurs also possess a sternum, but it does not articulate with the coracoids in the manner of pterosaurs and almost certainly evolved independently.

No elements of the dermal shoulder girdle are known among pterosaurs. The absence of clavicles is a significant difference from birds. The scapula and coracoid are fused to form a simple L-shaped structure. The coracoid is elongate, with the extremity braced against the anterior margin of the sternum. The glenoid articulation is located far dorsally and the narrow scapular blade extends over the back.

The humerus is short relative to the length of the ulna and radius. The proximal head bears a very large deltopectoral crest for attachment of the muscles to depress the forelimb. The articulation with the glenoid is saddle shaped, so that the forelimb is rotated forward as it is moved ventrally. There is a hingelike articulation between

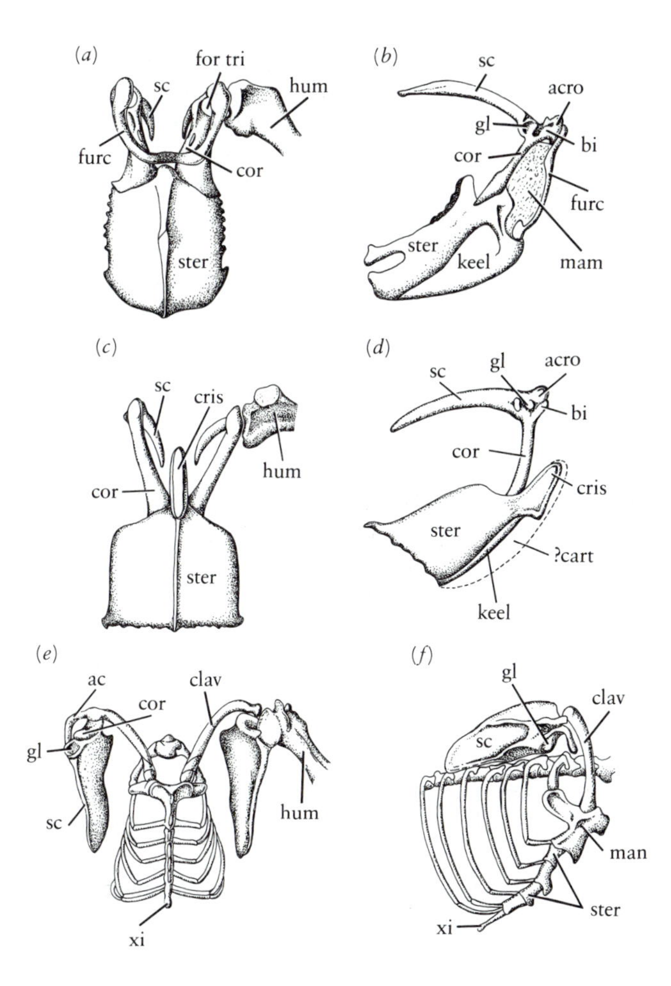

Figure 16-2. PECTORAL GIRDLES OF FLYING VERTEBRATES IN VENTRAL AND LATERAL VIEWS. (*a–b*) Bird, (*c–d*) pterosaur, and (*e–f*) bat. Abbreviations as follows: ac, acromion process; acro, acrocoracoid process; bi, biceps tubercle; ?cart, possible cartilagineous extension of sternal keel; clav, clavicle; cor, coracoid; cris, cristospine; for tri, foramen triosseum (junction of clavicle, coracoid, and scapula); furc, furcula; gl, glenoid fossa; hum, humerus; keel, sternal keel; man, manubrium; mem, membrane connecting sternum, coracoid, and clavicle; sc, scapula; ster, sternum; xi, xiphoid process of sternum. Not to scale. *From Padian, 1983b.*

the distal end of the humerus and the long narrow ulna and radius.

There are five or six carpals in primitive pterosaurs, but these are fused into one proximal and one distal element in later genera that form a simple hinge joint between the ulna and radius at one end and the remainder of the limb at the other. The first three metacarpals and digits are of comparable proportions to those in other early archosaurs and bear sharp, recurved claws. The fourth metacarpal is greatly enlarged to form the base of the finger to which is attached the flight membrane. Four thickened and elongate phalanges follow. There is no trace of the fifth finger. The pteroid, a bone that is unique to pterosaurs, extends medially from the anterior margin of the wrist. In later pterosaurs in which impressions of the soft anatomy are preserved this bone serves as the distal anchor for a narrow membrane that extends to the base of the neck.

The pelvic girdle of *Eudimorphodon* has a strange construction for an archosaur but is typical of later pterosaurs, with the acetabulum far dorsal in position above the broad puboischiadic plate (Figure 16-3). The acetabulum is imperforate. A large paired ossification, termed a prepubis but of a form that is unique to pterosaurs, articulates with the anterior margin of the short pubis.

It has long been thought that the rear limbs of pterosaurs had a sprawling posture, but Padian (1983a,b) argued that they were probably held in an upright position. He postulated that the pubis and ischium met at the midline, as in other diapsids, so that the acetabulum faced ventrolaterally rather than dorsolaterally as in previous reconstructions. In all pterosaurs, the head of the femur is bent medially, as in dinosaurs and birds, and the shaft was oriented in a parasagittal plane (Figures 16-4 and 16-5). The articulation with the tibia resembles that of birds, in which the femur projects well forward. The tibia is conspicuously longer than the femur, and the fibula is reduced to a splint of bone that is fused to the tibia proximally and is incomplete distally.

In *Peteinosaurus* and several later rhamphorhynchoids, there are four tarsal bones, the more proximal of which can be compared with the astragalus and calca-

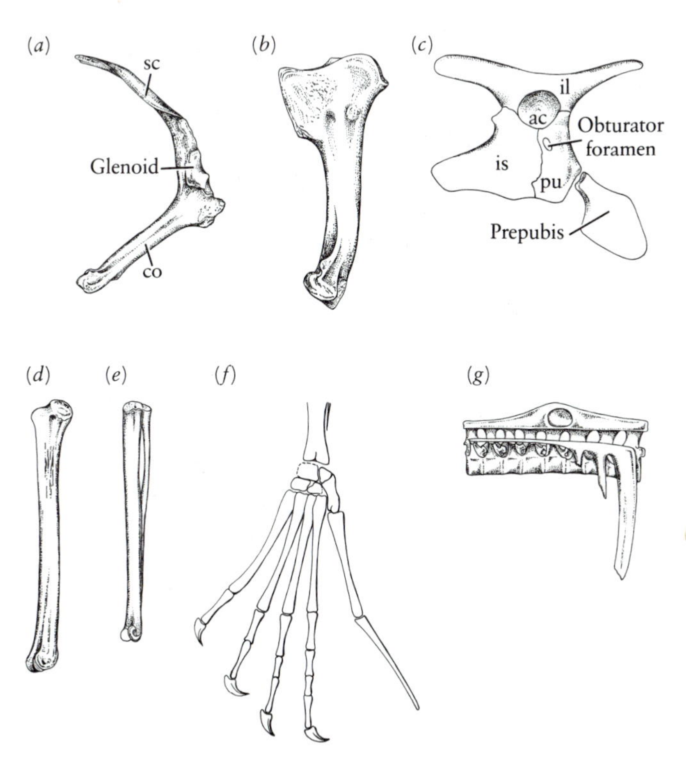

Figure 16-3. LIMB AND GIRDLE ELEMENTS OF PTEROSAURS. (*a*) The coossified scapula and coracoid of *Eudimorphodon*, $\times\frac{1}{2}$. (*b*) Humerus of *Eudimorphodon*, $\times\frac{1}{2}$. (*c*) Pelvis of *Eudimorphodon*, $\times 1$. (*d*) Femur of *Eudimorphodon*, $\times\frac{3}{4}$. (*e*) Tibia and fibula of *Peteinosaurus*, $\times\frac{1}{2}$. (*f*) Foot of *Peteinosaurus*, $\times\frac{3}{4}$. Note retention of four tarsals. (*g*) Notarium of *Pteranodon*, much reduced. Anterior is to the left. Abbreviations as in Figure 16-2, plus: ac, acetabulum; co, coracoid; il, ilium; is, ischium; pu, pubis. (*a–f*) *From Wild, 1978;* (*g*) *from Wellnhofer, 1978.*

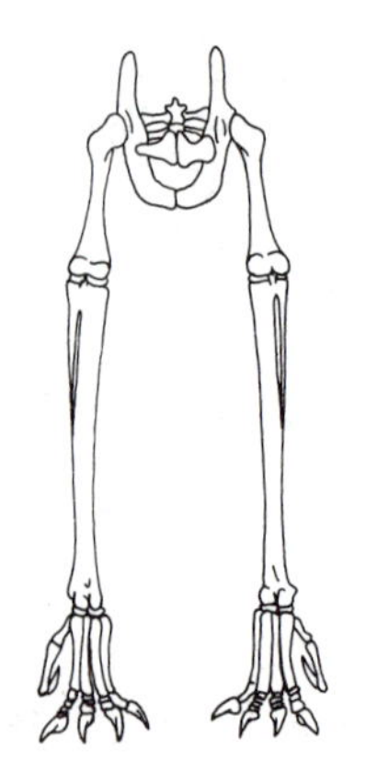

Figure 16-4. THE HIND LIMBS OF A RHAMPHORHYNCHOID PTEROSAUR. Anterior view showing their erect posture. *From Padian, 1983b.*

neum of other reptiles. In *Eudimorphodon* and most other pterosaurs, the tarsal joint is formed by clearly defined double condyles at the end of the tibia and two tarsal bones that are functionally incorporated with the foot. The astragalus and calcaneum are fused to the end of the crus, as in birds and many dinosaurs, so that a simple mesotarsal joint is formed between them and the distal tarsals (Figure 16-4).

In contrast with birds, the distal tarsals remain as independent centers of ossification in pterosaurs, rather than becoming fused with the metatarsals. In pterosaurs, all five metatarsals are retained, as in thecodonts. The first four are closely appressed, but do not fuse with one another as do metatarsals 2 to 4 in birds. Digits 1 to 4 retain the general pattern of primitive diapsids, with a typically reptilian phalangeal count and well-developed claws. The posture was digitigrade.

The fifth metatarsal is a short divergent element in all pterosaurs that superficially resembles the hooked fifth metatarsal of primitive archosaurs and and squamates (see Figure 16-3*f*). In rhamphorhynchoids, the fifth digit retains two long phalanges; these are lost in pterodactyloids. The function of this digit is subject to controversy. It superficially resembles the calcar of bats, which supports the flight membrane between the rear limbs, but is on the lateral rather than the medial surface of the foot. It had been thought to serve for attachment of the posterior portion of the main wing membrane, but Padian (1983a,b) showed that the membrane did not extend onto the surface of the rear limb. It might be a relict of an early stage in the development of flight in which a separate membrane was attached in this position.

FLIGHT APPARATUS

The first recognized pterosaurs were discovered in fine-grained lithographic limestone of late Jurassic age in the Solenhofen area of southern Germany. They show impressions of membranous wings that were supported by the greatly elongated fourth digit of the forelimb. They were initially thought to be related to bats, and Soemmering (1817) reconstructed the membrane as including the rear limbs and tail in the fashion of modern bats.

Based on comprehensive studies of *Rhamphorhynchus* and *Pterodactylus* from Solenhofen, Padian (1983b) demonstrated that the wing membrane attached to the trunk posteriorly and did not reach the rear limbs. Nor do these genera provide evidence for a membrane between the rear limbs. Padian reconstructs the wing as a narrow structure that resembles a gull's wing (Figure 16-6).

A separate membrane is present anterior to the proximal portion of the forelimb and runs from the base of the pteroid to the base of the neck. A tendon on its margin would have extended the limb when in flight. Several genera have membranes between the toes of the rear limb that may have been used in swimming. Species of *Rhamphorhynchus* had a short triangular or diamond-shaped vertical rudder at the end of the long tail.

The wing is formed by a continuous membrane. Microscopic examination shows a consistent pattern of narrow structural fibers radiating posteriorly to provide internal supports that are comparable in orientation to the

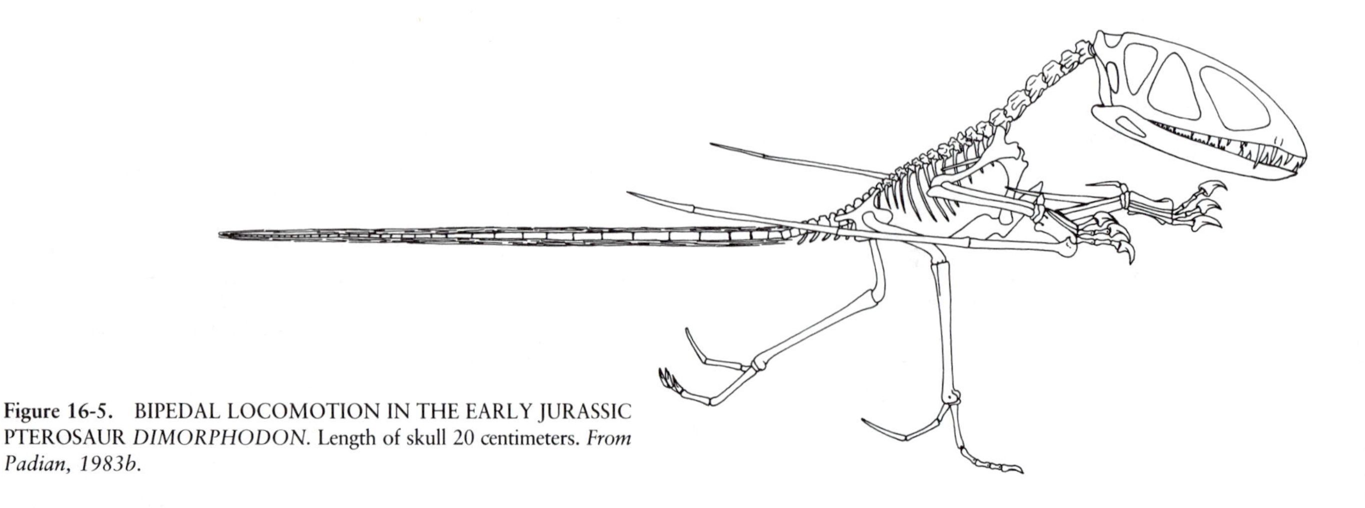

Figure 16-5. BIPEDAL LOCOMOTION IN THE EARLY JURASSIC PTEROSAUR *DIMORPHODON*. Length of skull 20 centimeters. *From Padian, 1983b.*

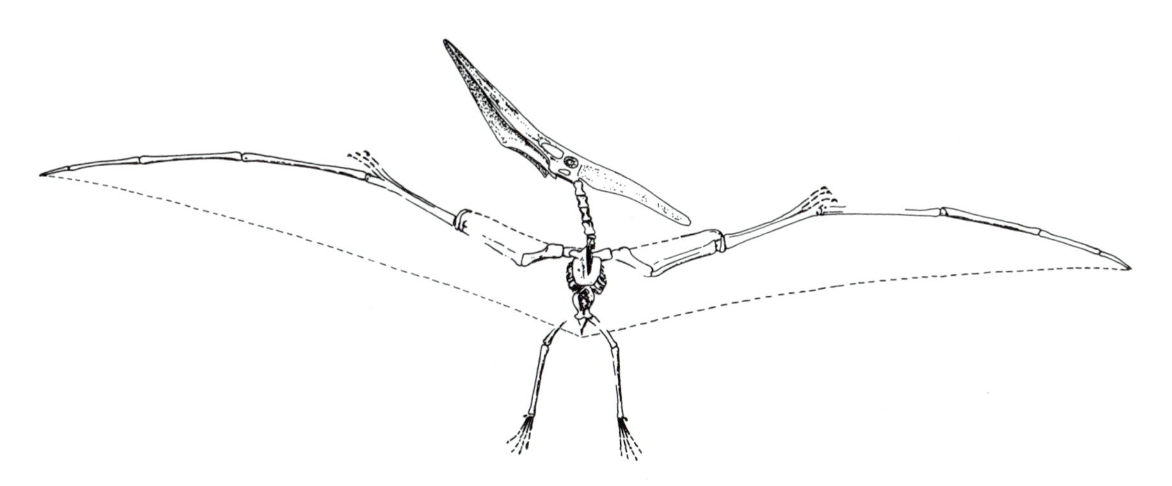

Figure 16-6. THE CRETACEOUS PTEROSAUR *PTERANODON*. This restoration shows the extent of the wing membrane. *From Padian, 1983b.*

additional digits that support the wings of bats and the feathers of birds. They are thought to be cartilaginous or collagenous, and they contribute to the camber of the airfoil.

The species *Sordes pilosus* (Sharov, 1971) has been described as having a "furry" covering over the body. Such a covering is not evident in the very well-preserved pterosaurs from Solenhofen, and Feduccia (1985) doubts that this material represents the natural body surface in *Sordes*. This pigeon-sized rhamphorhynchoid appears to have had a flight membrane between the rear limbs, in contrast with the species that Padian studied.

As a result of the recognition that pterosaurs were reptiles related to crocodiles and dinosaurs, it was long thought that they lacked the high metabolic rate of birds and bats and probably relied greatly on the wind to provide lift, rather than actively flapping the wings. They were typically portrayed as passive gliders rather than active flyers (see, for example, Romer, 1966, and Stahl, 1985).

Padian (1983b, 1985) showed that the skeleton of even the earliest pterosaurs was adapted toward active flight to a degree very similar to that of advanced birds and much greater than that of *Archaeopteryx*. The sternum and anteriorly directed cristospine would have provided a large area for the origin of the pectoralis muscle that in birds and bats is the largest muscle in the body and is primarily responsible for the power stroke in flight. The acromial process of the scapulocoracoid, like that of advanced birds, presumably served as a pulley over which the tendon of the supracoracoideus muscle passed so that it could elevate the wing. As in birds and bats, the deltopectoral crest of pterosaurs is greatly expanded to form a large area for the insertion of flight muscles. The sternum is strongly braced by the coracoid.

As in birds, the bones of pterosaurs are hollow and thin walled. They also have openings comparable to the pneumatic foramina that in birds lead to air sacs within the bones. These openings suggest that the respiratory system of pterosaurs may have been modified in a manner analogous with that of birds and in association with a very high metabolic rate.

Cranial endocasts of pterosaurs demonstrate that the brain was also advanced in a manner comparable with that of birds. Even in the early Jurassic genus *Parapsicephalus* (Figure 16-7), the relative size of the brain was greater than that of most comparable-sized reptiles and approached that of birds. The optic lobes are very large, but the olfactory lobes are greatly reduced, in common with birds in which the sense of sight is enhanced but that of smell is reduced. The cerebellum is greatly elaborated and the floccular lobes are even larger than in birds, which suggests that the pterosaurs had a high degree of aerial maneuverability.

It is improbable that all these features, which are common to advanced birds, would have evolved in pterosaurs if they had not been capable of active, flapping flight.

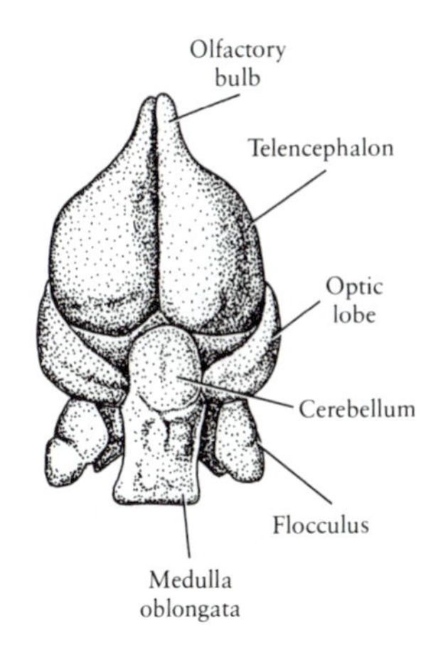

Figure 16-7. *PARAPSICEPHALUS*. Reconstruction of an endocast of the Lower Jurassic pterosaur. *From Wellnhofer, 1978.*

THE ORIGIN OF PTEROSAURS

The most recently described member of the Upper Triassic pterosaur fauna, *Preondactylus,* shows some of the most primitive features known in this group (Wild, 1984b). The humerus is shorter than the femur; the pteroid is short; the scapula and coracoid are not fused. The entire rear limb is longer than in any other pterosaur and the forelimb is shorter. The proportions of the wing finger indicate that elongation of the wing proceeded from distal to proximal elements.

Nevertheless, all the Triassic pterosaurs were highly specialized for flight and were very similar to later rhamphorhynchoids in most features. They provide little evidence of their specific ancestry and no evidence of earlier stages in the origin of flight. The presence of an antorbital opening has long been used to support affinities with archosaurs. The structure and probable posture of the rear limbs suggest close relationship with the lagosuchids and the immediate ancestors of the two dinosaur orders. In his description of the earliest known pterosaurs, Wild (1978, 1984a) argued for affinities with the eosuchians. The relatively few similarities between early pterosaurs and eosuchians are primarily primitive features that do not preclude affinities with archosaurs. His primary argument is that there was not enough time for the origin of flight between the appearance of relatively advanced thecodonts in the middle Triassic and the appearance of pterosaurs in the late Triassic. Certainly the early pterosaurs were much more advanced in many skeletal features than were late Triassic dinosaurs. On the other hand, 10 to 20 million years may have been available for evolution from lagosuchid-level thecodonts to the early pterosaurs, which is comparable to the time available for the origin of bats (see Chapter 20).

Padian (1984) pointed out numerous shared derived skeletal features, especially in the structure of the skull and rear limbs, that unite pterosaurs, lagosuchids, and dinosaurs but exclude eosuchians. He concluded that available evidence is not sufficient to resolve the specific interrelationships among these three groups. One feature, the retention of a hooked fifth metatarsal, appears more primitive than the condition in lagosuchids and dinosaurs. Unless this character has been reelaborated in relationship to the large fifth digit in early pterosaurs, its presence suggests that pterosaurs diverged from the thecodont stock at an earlier stage than did lagosuchids and dinosaurs. The advanced structure of the rear limbs nevertheless points to closer affinities with lagosuchids than with any other adequately known group of thecodonts.

Padian suggests that the closest affinities of pterosaurs may lie with *Scleromochlus,* a small form from the Upper Triassic of Scotland (Figure 16-8). Like early pterosaurs, the skull is very large with extensive fenestrae, and the posture of the rear limbs suggests that they moved in a parasagittal plane. The scapula is strap shaped and the coracoid is elongated. The fibula and tarsals are reduced and the fifth metatarsal is much shorter than the remainder. Unfortunately, this animal is very poorly preserved. The bones are represented only as natural casts in a coarse sandstone matrix and few morphological details are known with certainty. *Scleromochlus* was a contemporary of early pterosaurs but might be a descendant of the same ancestral stock, although the forelimbs are much shorter than the rear limbs and show no specializations for flight.

Padian (1985) suggests that pterosaurs evolved flight directly from terrestrial ancestors. He argues that they show no skeletal evidence for arboreality but were effective terrestrial bipeds. Padian emphasizes that all pterosaurs show skeletal specializations that are common to active flying birds and none that are similar to gliding vertebrates. Soaring was probably an important component of flight in the large Cretaceous species, as in the largest modern birds, but resulted from a reduction of the emphasis on flapping flight that was much more important in the smaller, earlier species. In contrast, Wild (1984b) argued for an origin among climbing ancestors.

PTEROSAUR DIVERSITY

Three families of rhamphorhynchoids are represented in the Upper Triassic. The best-known genus, *Eudimorphodon,* is placed in a monotypic family, but Wild suggests that it may be related to the Upper Triassic genus *Campylognathoides. Peteinosaurus* may be close to the ancestry of the well-known Lower Jurassic genus *Dimorphodon. Preondactylus* resembles Jurassic members of the Rhamphorhynchidae in the proportions of the phalanges of the flight finger. Despite being placed in different families, all three Triassic genera are basically similar to one another. Rhamphorhynchoids continued to radiate throughout the Jurassic and reached their greatest diversity toward the end of that period (Wellnhofer, 1975). *Rhamphorhynchus* is the best-known late Jurassic genus.

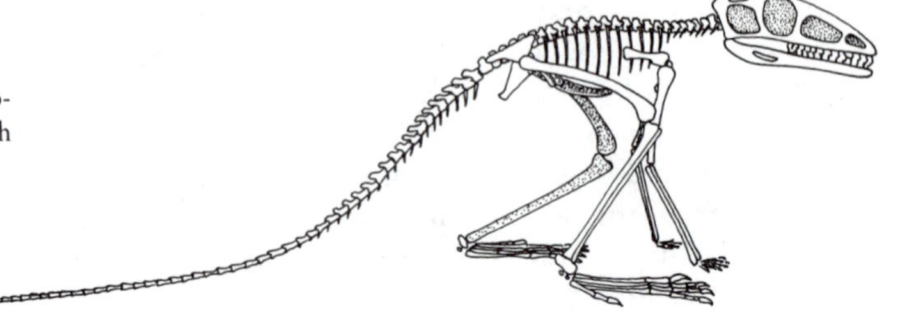

Figure 16-8. *SCLEROMOCHLUS.* This small archosaur from the Upper Triassic might be related to the ancestry of pterosaurs. Skull length approximately 30 millimeters. *From Padian, 1984.*

Anurognathus is a small, short-faced form whose peglike teeth suggest a diet of insects. The long face and piercing dentition of *Rhamphorhynchus* and the discovery of fish remains in the crop and gastric region of this and other genera suggest that many pterosaurs may have been piscivorous. There is no direct evidence of the reproductive habits of any pterosaurs. No rhamphorhynchoid is known to have survived into the Cretaceous.

Pterodactyloids were already numerous and diverse in the Upper Jurassic, as we can see by fossils from Solenhofen (Wellnhofer, 1970). These remains include some of the smallest pterosaurs and the ancestors of the largest flying vertebrates. No forms are known that are intermediate between rhamphorhynchoids and pterodactyloids, although all features of the skeleton confirm their close relationship.

The difference in proportions between these two suborders, especially the great reduction in the tail of pterodactyloids, probably changed their flight properties greatly (see Figure 16-6). The presence of a tail behind the center of lift in aircraft contributes to stability. If the airspeed is lowered, the nose automatically rises and increases the angle of attack, contributing to greater lift. In aircraft, stability results in automatic compensation for changes in direction and altitude, which is advantageous in providing a comfortable flight but reduces maneuverability. Stability is reduced as far as possible in combat aircraft so that changes in direction can occur rapidly. Selection for increased maneuverability may lay behind the changes in proportions among pterosaurs. Pterodactyloids more closely resemble modern birds in the shortness of the bony tail. However, the lift provided by the tail feathers of birds is missing. Pterodactyloids strengthened the area of the shoulder girdle by fusion of the anterior trunk vertebrae into a notarium with which the end of the scapula articulated (see Figure 16-3).

The dentition varied greatly among pterodactyloids. Genera such as *Pterodaustro* had hundreds of long narrow teeth in the lower jaw that formed a pattern like the baleen of whales and may have been used to strain invertebrates from the water (Figure 16-9). The common Cretaceous genus *Pteranodon* completely lost its teeth.

The late Cretaceous pterosaurs include the largest of all flying vertebrates. We think that *Pteranodon,* with a wing span of 7 meters, weighed approximately 17 kilograms. Scattered remains assigned to the genus *Quetzalcoatlus* suggest that it may have had a wing span of 11 to 12 meters and a weight of approximately 65 kilograms (Padian, 1984). The large pterosaurs, like large flying birds in the modern fauna, must have relied extensively on the wind, rather than continuous flapping flight, for remaining aloft (Brower, 1983). Young (1981) cited 12 to 16 kilograms as the limit for body weight in forms with sustained flapping flight. *Pteranodon* was approximately at the limit of this weight, and *Quetzalcoatlus* was far heavier. Unfortunately, impressions of the flight membrane have not been reported from the larger species.

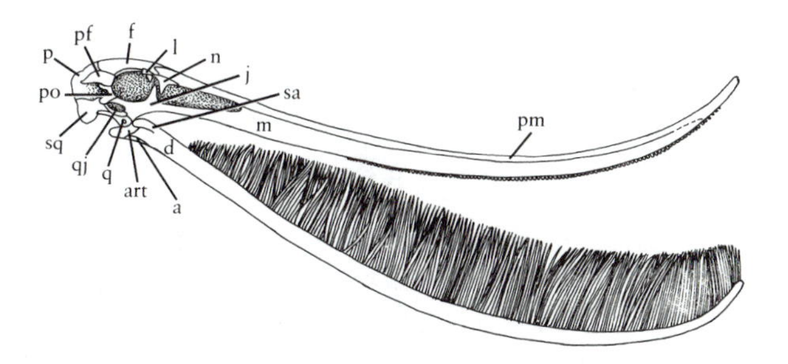

Figure 16-9. SKULL OF THE PTERODACTYLOID *PTERODAUSTRO* FROM THE UPPER CRETACEOUS OF ARGENTINA. The slender but extremely numerous teeth in the lower jaw probably functioned like the baleen of whales to strain small invertebrates from the water. Length 23 centimeters. Abbreviations as in Figure 8-3. *From Wellnhofer, 1978.*

The fossil record of pterosaurs extends for approximately 150 million years, and nearly 90 species have been recognized. They have been reported from every continent except Antarctica. Most are known from shallow marine deposits, but *Quetzalcoatlus* comes from 400 kilometers inland. Most genera were short-lived and restricted to small geographical areas.

It can be said that the pterosaurs became extinct at the end of the Cretacious, coincident with the extinction of the dinosaurs, since they are known in the Maastrichtian but are unknown from the Tertiary. However, this gives a misleading impression, for only *Pteranodon* and *Nyctosaurus* are common in the late Cretaceous, accompanied by a few remains of *Quetzalcoatlus.* The end of the Cretaceous did not record a catastrophic termination of a diverse assemblage but only the extinction of three members of an order that had been gradually diminishing in numbers throughout the Cretaceous.

The large size of *Pteranodon* and *Quetzalcoatlus* probably limited their distribution to areas that had regular, gentle winds throughout much of the year (Bramwell and Whitfield, 1974). The tectonic activity at the end of the Cretaceous would almost certainly have restricted their geographical range and may have directly resulted in their extinction.

A more dramatic decline had occurred at the end of the Jurassic with the termination of all the rhamphorhynchoid lineages. This event coincides with the emergence of birds, but the fossil record of that group is too poorly known to document competition between the two groups.

Much of the success of birds may be attributed to aspects of the physiology and soft anatomy that cannot be established in pterosaurs; therefore, it is difficult to determine why pterosaurs became extinct while birds continued to radiate throughout the late Mesozoic and Cenozoic. The flexibility of the bird wing and the ability to replace component feathers probably provided a more effective flight mechanism. Bats have a pterosaurlike membraneous wing but avoid competition with birds almost completely as a result of their nocturnal habits.

MESOZOIC BIRDS

In many respects, birds may be considered the most advanced vertebrates. Because of the energy requirements for sustained powered flight, their body temperature and metabolic rate are higher than those of even the most advanced mammals, and the structure and function of their respiratory system are unique among chordates. The number of modern species (nearly 9000) exceeds that of all other vertebrate groups except the higher bony fish. The physical and metabolic requirements of flight result in a basically uniform anatomy, even among secondarily flightless genera.

Feduccia (1980) recently reviewed their evolutionary history. Olson (1985) provided a more technical discussion of their fossil record. The radiation of modern birds apparently began in the late Mesozoic and most of the major modern lineages had become differentiated by the early Cenozoic. Their Mesozoic history remains very poorly known.

ARCHAEOPTERYX AND THE ORIGIN OF BIRDS

In contrast with pterosaurs, the fossil record of birds provides evidence of a crucial stage in their early evolution. *Archaeopteryx,* from the Upper Jurassic (Portlandian) of the Solenhofen region in southern Germany, remains the prime example of a genus that occupies the position of a "missing link" uniting two major vertebrate groups. Three nearly complete skeletons, two partial skeletons, and an isolated feather have been described. Every bone in the body is represented, and one specimen includes a natural endocast of the brain. The most important feature of these specimens is the preservation of flight feathers on the wing and along the tail in nearly their natural position. These are most spectacularly displayed in the Berlin specimen (Figure 16-10).

Were it not for these feathers, *Archaeopteryx* would not have been recognized as a bird, as is demonstrated by the fact that one nearly complete skeleton in which the feathers were not recognized was initially identified as a dinosaur. In fact, there are no features of the bony skeleton of *Archaeopteryx* that are uniquely avian. All

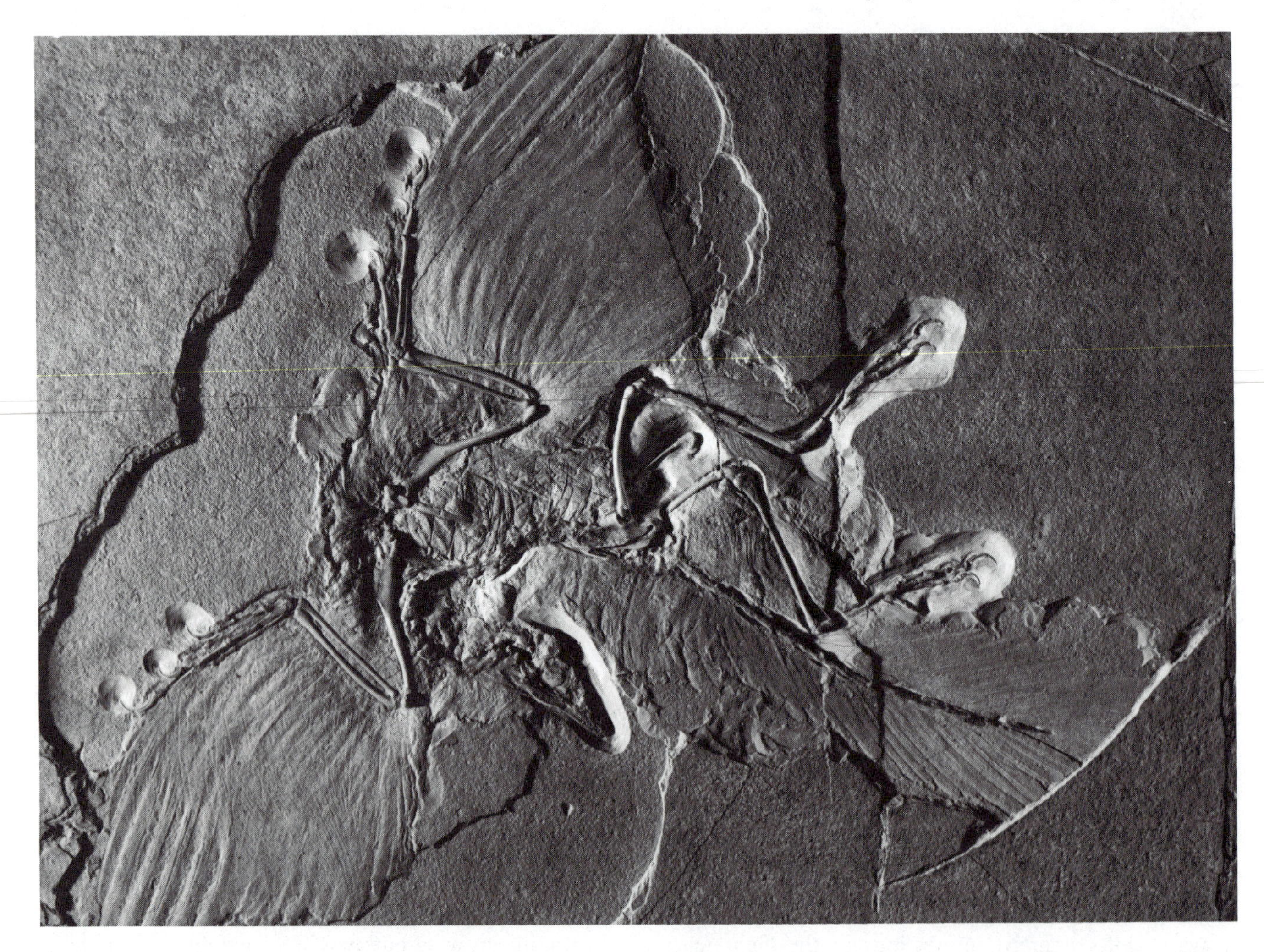

Figure 16-10. SKELETON OF *ARCHAEOPTERYX. From the Humboldt Museum, Berlin. Photograph courtesy of Dr. Hermann Jaeger.*

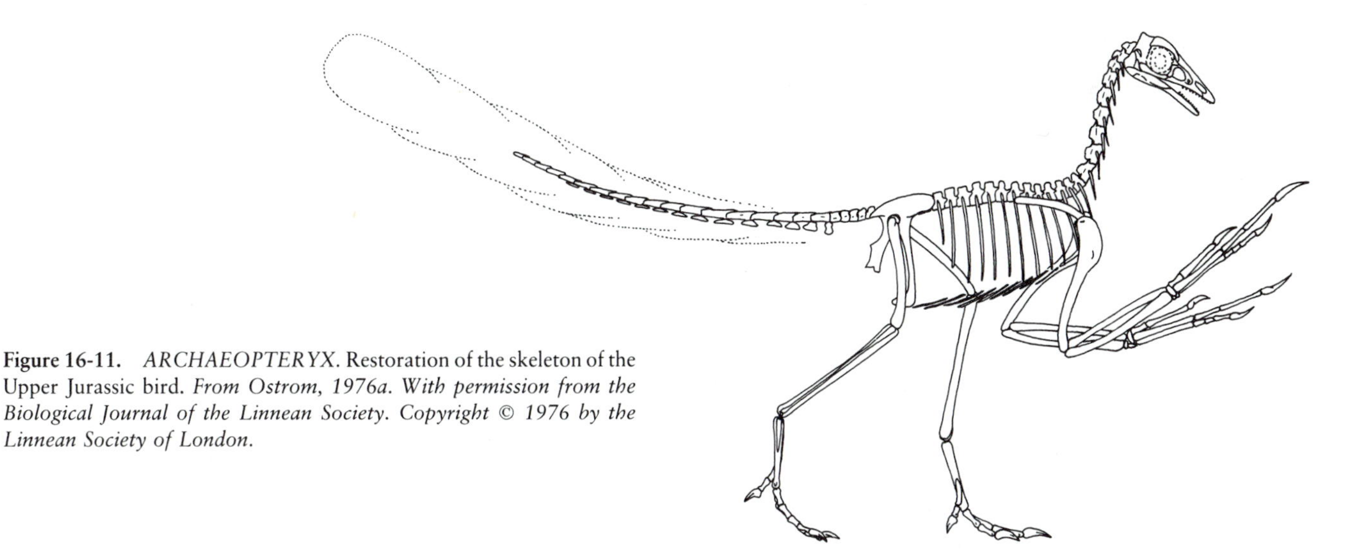

Figure 16-11. *ARCHAEOPTERYX.* Restoration of the skeleton of the Upper Jurassic bird. *From Ostrom, 1976a. With permission from the Biological Journal of the Linnean Society. Copyright © 1976 by the Linnean Society of London.*

have been described in genera that are classified among the dinosaurs.

If all elements of the skeleton were considered of equivalent value in classification, *Archaeopteryx* would certainly be considered a feathered dinosaur. Most paleontologists consider *Archaeopteryx* to be a bird because of the presence of a single, albeit complex, character—feathers—that are uniquely shared with modern birds. The structure and arrangement of the feathers are given a uniquely important position in classification because they provide almost unequivocal evidence of the capacity for powered flight.

Ostrom (1974, 1975a,b, 1976a,b) thoroughly discussed the anatomy and phylogenetic position of *Archaeopteryx.* He argues that this genus is both the most primitive known bird and that it provides the best available evidence of the origin of birds and the origin of avian flight. Certainly, there is no other adequately known genus that occupies such a central position in the origin of a major group. However, it seems probable that additional knowledge of small theropod dinosaurs and other late Jurassic and early Cretaceous birds may demonstrate a phylogenetically more complex transition between dinosaurs and birds (Thulborn, 1984).

Reconstruction of the skeleton of *Archaeopteryx* (Figures 16-11 and 16-12) shows a pattern that is very similar to that of small coelurosaurian dinosaurs such as *Compsognathus* (see Figure 14-6). *Archaeopteryx* was about the size of a pigeon but was markedly more primitive than modern birds in a number of important respects. None of the specimens shows a trace of the sternum, which provides the main area of origin for the major flight muscles in modern birds. The area that it would have occupied is largely covered by a chevron pattern of dermal scales, as in primitive reptiles, which indicates that the pectoralis muscles were not significantly more massive than in dinosaurs. The bones are thick walled, without the pneumatic ducts common to both modern birds and pterosaurs. Unlike all other birds, *Archaeopteryx* has a long bony tail, a feature that is also characteristic of primitive pterosaurs.

Despite the presence of feathers, the forelimb of *Archaeopteryx* shows no specifically avian features (Figure 16-13). It retains three fully developed digits, a flexible carpus, and unaltered ulna and radius, as in dinosaurs. The skull is primitive in the retention of teeth and the limited fusion of the bones.

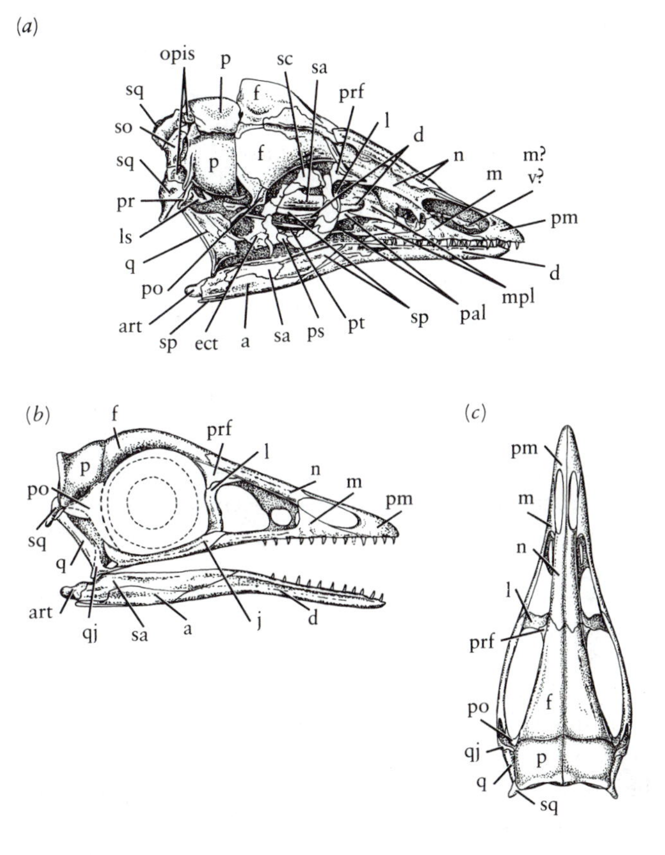

Figure 16-12. SKULL OF *ARCHAEOPTERYX.* (*a*) Drawing based on the Eichstätt specimen. (*b* and *c*) Reconstructions in lateral and dorsal view. Abbreviations as in Figure 8-3, plus: ls, laterosphenoid; mpl, maxillo-palatine; pr, prötic; sc, sclerotic ring. *From Wellnhofer, 1974.*

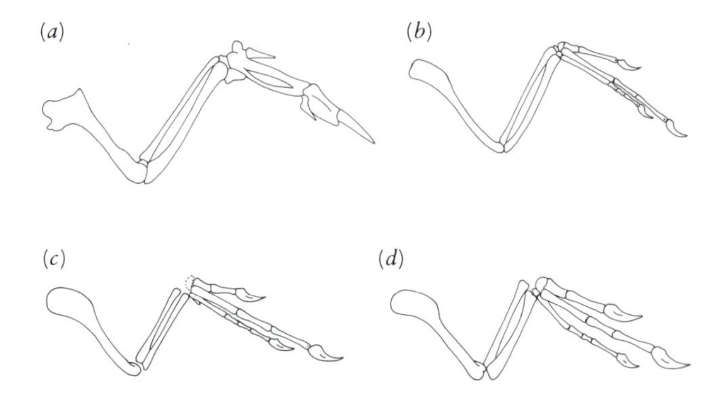

Figure 16-13. FORELIMBS OF BIRDS AND THEROPOD DINOSAURS. (*a*) Modern pigeon. (*b*) *Archaeopteryx*. (*c*) *Ornitholestes*. (*d*) *Deinonychus*. *From Ostrom, 1975a.*

It was once thought that the configuration of the pelvis and the presence of a furcula (wishbone) were uniquely avian features of the skeleton of *Archaeopteryx,* but both have been described in dinosaurs. Several theropods are known to have fused the clavicles medially to form a wishbonelike structure, and some Asian genera have rotated the pubis posteriorly in a birdlike pattern. The exact orientation of the pubis in *Archaeopteryx* is still debated.

The portion of the skeleton that is most like that of modern birds is the rear limb (Figure 16-14). The head of the femur is inturned; the knee and the ankle form simple hinge joints, which indicate that movement was in the parasagittal plane; and the fibula is reduced. The proximal tarsals are integrated with the end of the tibia and the distal tarsals are fused to the ends of the metatarsals. The metatarsals are elongate and partially fused with one another. Three digits face forward and the first is turned to the rear. These features, which are very close to the modern avian pattern, are also very similar to those of small theropod dinosaurs and so must be considered to form a common dinosaur-avian heritage.

Ostrom argues that to establish the origin of *Archaeopteryx* is to establish the origin of birds. *Archaeopteryx* has the most primitive structure that one could conceive for the immediate ancestors of birds, and only a few minor features of the skeleton have been demonstrated as being more specialized than those of later birds (Martin, 1984).

When *Archaeopteryx* was first described, it was thought to be closely related to dinosaurs. As that group became better known, it appeared that all dinosaurs were too specialized to be directly ancestral to birds. In his very influential book, Heilmann (1926) argued that *Archaeopteryx* must have evolved from more primitive archosaurs, the Triassic thecodonts. There are no features of primitive thecodonts that preclude them from being the ultimate ancestors of both dinosaurs and birds, but no thecodonts can be demonstrated as sharing a unique, common ancestry with birds. The features that they share are all primitive for archosaurs in general. Walker (1972), Martin, Stewart, and Whetstone (1980), and Martin (1983) showed that there are some characters of birds that are shared with crocodiles, but there is no evidence that they are uniquely shared with a common ancestor of these groups. The high degree of specialization of the entire skull of early crocodiles makes it difficult to accept that similarities of particular aspects of the quadrate and middle ear were the result of an immediate common ancestry.

Ostrom (1975b) demonstrated that many specialized aspects of the skeleton of *Archaeopteryx* are uniquely shared with small theropods and with no other groups of archosaurs. These characters include the structure of the forelimbs and hind limbs, the shoulder girdle, and skull. There is a great overall similarity to *Compsognathus,* but this genus is too late in time (as a direct contemporary of *Archaeopteryx*) and too specialized in the reduction of the manus to two digits. No other adequately known theropod appears to be an appropriate ancestor.

The divergence of the line leading to *Archaeopteryx* from that of primitive theropods may have occurred at any time during the late Triassic or early Jurassic. We have no more specific limits on the period of time during which the unique flight adaptations of this group evolved.

FEATHERS AND THE ORIGIN OF FLIGHT

Flight in *Archaeopteryx* is associated with a single character that is capable of fossilization, feathers. The preservation of feathers (or rather their impressions) is a rare phenomenon. They are known in *Archaeopteryx* only because the remains of this genus were preserved in fine-grained, lithographic limestone. No dinosaurs are known to have had feathers. Some, including hadrosaurs and ceratopsians, are known from mummified remains showing the surface of the body that demonstrate that these

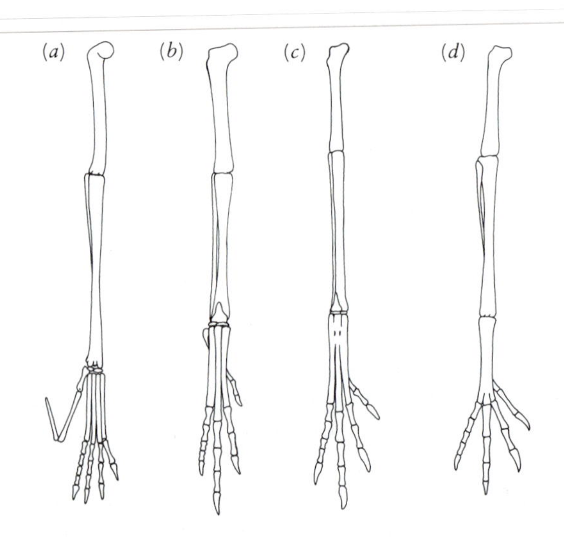

Figure 16-14. Rear limbs of (*a*) the pterosaur *Dimorphodon,* (*b*) the theropod dinosaur *Compsognathus,* (*c*) *Archaeopteryx,* and (*d*) a pigeon. Most of the birdlike features of the limb in *Archaeopteryx* were already achieved within dinosaurs. *From Padian, 1983b.*

genera lacked feathers, but feathers might have been present in other dinosaur groups without having been preserved. It is more significant that feathers are not associated with the one very well-preserved skeleton of *Compsognathus* from Solenhofen (Ostrom, 1978).

Developmentally, feathers and reptilian scales are homologous structures. In contrast with hair, which is strictly epidermal in origin, feathers and scales incorporate mesodermal tissue as well. Either feathers or scales can develop in a particular area of the skin but not both. This phenomenon is readily seen on the legs of birds such as domestic fowl, where the relative extent of the two tissues vary considerably.

Avian flight depends on large flight feathers, but feathers are also vital for insulation. It is difficult to imagine how selection could have acted directly to produce the large structures that were necessary for flight. On the other hand, relatively slight changes in the size and structure of reptilian scales would have been sufficient for them to function in insulating the body. Such insulation would have been important in any endotherm the size of *Archaeopteryx*. In contrast, Feduccia (1985) argues that feathers arose initially for flight rather than for insulation.

The primary and secondary wing feathers of *Archaeopteryx* are already arranged in the same manner as in modern flying birds (Figure 16-15). It is nearly impossible to attribute selection for this arrangement to factors other than flight. Moreover, each feather has the characteristic of an airfoil as a result of the asymmetrical position of the shaft (Feduccia and Tordoff, 1979). The absence of any uniquely avian characters of the forelimb and pectoral girdle suggests that *Archaeopteryx* must have been a weak flyer, but the nature of the feathers argue that it did fly. The proportions of the endocast are significantly advanced over the pattern of typical reptiles and approach the avian condition (Whetstone, 1983).

Figure 16-15. Arrangement of feathers in (*a*) *Archaeopteryx* and (*b*) a modern bird. (*c*) The wing of a nestling hoatzin showing the presence of claws on the first and second digits. The rudiment of the first digit in modern birds supports the alula or bastard wing. The number and arrangement of the primary flight feathers correspond almost exactly between *Archaeopteryx* and modern birds. The shaft of the feather is close to the anterior margin so that each feather acts as an airfoil. *From Heilmann, 1926.*

The specific evolutionary sequence leading to the origin of avian flight remains subject to controversy (Hecht, Ostrom, Viohl, and Wellnhofer, 1985). Two major hypotheses have been advanced. Heilmann (1926) and many other authors argued that the immediate ancestors of *Archaeopteryx* were arboreal. Feathers were elaborated to facilitate gliding from branch to branch and from trees to the ground. Active, flapping flight is thought to have evolved secondarily (Bock, 1985). Ostrom (1985 and references cited therein) suggests that flapping flight evolved in cursorial terrestrial ancestors. He specifically argues that flight evolved from a behavioral pattern involving anteroventral movement of the forelimbs associated with the capture of prey.

Ostrom, as well as Padian (1985) and Caple, Balda, and Willis (1983) point out that specializations for gliding and flapping flight are antithetical and that a gliding ancestry, as characterized by "flying" squirrels and "flying lemurs" (see Chapter 20), is very unlikely for birds or pterosaurs. In gliders, the wing membrane is attached to the body and extends between the front and rear limbs. The proximal, rather than the distal, elements of the limbs are elongated. The membrane of squirrels and flying lemurs have no internal support. In pterosaurs and birds, the distal elements, which would be most effective in flapping flight, are enlarged, but the proximal elements are not; the hind limbs are not involved in support of the flight membrane.

The structure of the rear limbs of *Archaeopteryx* and most later birds is clearly an adaptation to rapid terrestrial locomotion. There is no evidence for arboreal specialization, although the claw structure does not preclude climbing in trees. The great length of the distal elements of the forelimbs indicates that flapping was important for *Archaeopteryx*, whereas gliders would have had shorter limbs if the membrane continued all along the trunk.

Caple, Balda, and Willis (1983) argue persuasively that landing on small branches, as in truly arboreal birds, requires extremely highly developed abilities of flight and maneuverability that are achieved in only the most advanced groups of living birds. They contend that this factor indicates that the early birds were primarily terrestrial in habits. Caple, Balda, and Willis maintain that the immediate ancestors of birds were rapidly running terrestrial predators that caught insects in their mouths. These protobirds used the forelimbs to maintain stability while leaping after prey. Changing the distribution of weight at the ends of the limbs would be most effective in maintaining lateral stability. If the ends of the limbs provided even a small amount of lift, it would greatly increase stability. The movements of the limbs that would be used for maintaining stability in small leaping dinosaurs are comparable with those used by birds in flapping flight. Such reasoning suggests a continuous functional transition between the two groups, without an arboreal intermediate.

THE ORIGIN OF ADVANCED AVIAN FEATURES

Leaving aside the feathers, all uniquely avian skeletal features of modern birds were achieved subsequent to the level of *Archaeopteryx*. The most conspicuous osteological changes were the reduction of the tail to a short pygostyle and the evolution of a large keeled sternum. The coracoid is greatly elongated and forms an articulating brace with the sternum. Where the coracoid, clavicle, and scapula join one another is a conspicuous groove, the triosseal canal, which in living birds serves for passage of the tendon of the supracoracoideus muscle. This tendon extends dorsally from the muscle, passes through the canal (which acts as a pulley), and inserts on to the dorsal surface of the proximal end of the humerus (Figure 16-16). Although this muscle retains its primitive position ventral to the humerus, it lifts rather than depresses the forelimb. In modern birds, it is the largest muscle that raise the wing. As such, it is an important antagonist of the pectoralis and together they produce the flapping flight of modern birds.

The configuration of the coracoid in *Archaeopteryx* indicates that the supracoracoideus muscle retained the primitive function, common to other tetrapods, and would have served to pull the humerus anteriorly and ventrally. Ostrom (1976b) traced the change in orientation of the force of this muscle through a series of hypothetical in-

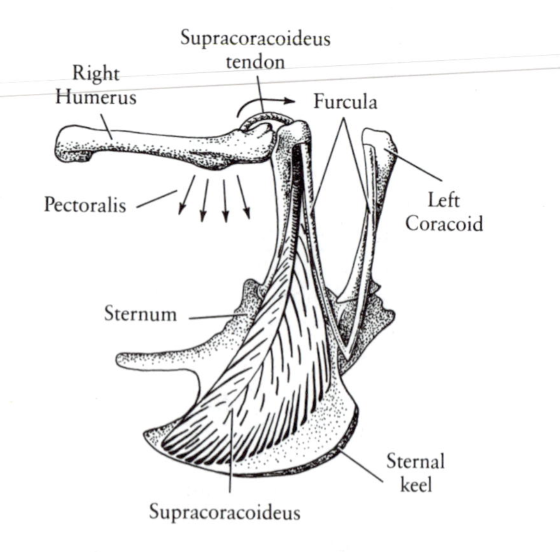

Figure 16-16. ANTEROLATERAL VIEW OF THE PECTORAL GIRDLE AND STERNUM OF A PIGEON. Drawn to show the function of the supracoracoideus muscle. The upper arrow indicates the course and action of the supracoracoideus tendon from the insertion toward the triosseal canal. The lower arrows indicate the location of the pectoralis, which is not illustrated. *From Ostrom, 1976b.*

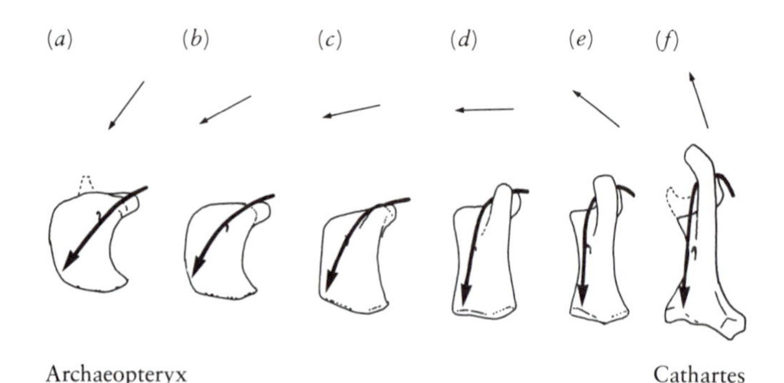

Figure 16-17. HYPOTHETICAL STAGES IN THE EVOLUTION OF THE AVIAN CORACOID. From the *Archaeopteryx* stage to that of a modern vulture, *Cathartes*. The arrows indicate the hypothesized course of the supracoracoideus fibers in each stage and their progressive deflection resulting from evolutionary elevation and expansion of the biceps tubercle (= acrocoracoid). Upper arrows indicate the line of action of the supracoracoideus at each stage. Dashed lines indicate the acromion and adjacent regions of the scapula. All stages are of a left coracoid viewed from the front. *From Ostrom, 1976b.*

termediate stages to the condition in modern birds (Figure 16-17). He argues that the primitive function of the supracoracoideus in *Archaeopteryx* together with the absence of a sternum indicate that this genus did not fly. Feduccia (1980) and others pointed out that bats retain a primitive supracoracoideus and, as in other tetrapods, the deltoid and coracobrachialis lift the forelimb. This pattern could have been operative in *Archaeopteryx*. The change in the function of the supracoracoideus was a refinement of flight but not a necessity.

The structure of the forelimb has also changed significantly between *Archaeopteryx* and modern birds, with fusion of the carpals to form a simple hinge joint, fusion of the digits and metacarpals, and development of joint surfaces to facilitate folding of the wing (Figure 16-18).

In contrast with the close skeletal similarities between *Archaeopteryx* and its theropod ancestors, a very large morphological gap separates this genus from all other known birds. For this reason, *Archaeopteryx* is placed in a subclass on its own, the Archaeornithes, with all other birds classified as Neornithes.

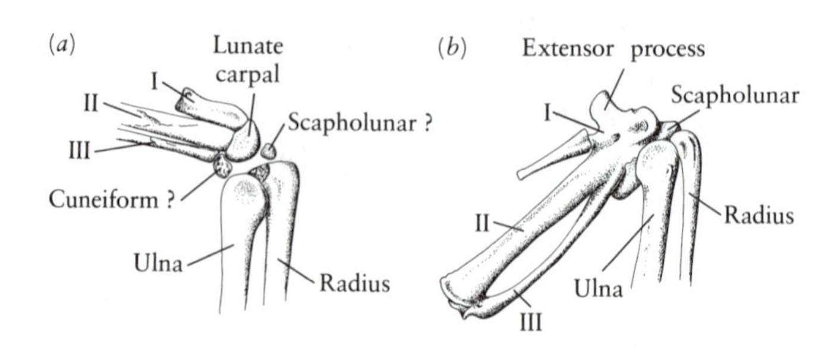

Figure 16-18. Comparison of the wrist area between (*a*) *Archaeopteryx* and (*b*) a modern bird, represented by *Cathartes*, showing suggested homology of elements. The pattern in *Archaeopteryx* retains all the features of coelurosaur dinosaurs. Abbreviations as follows: I, II, III; metacarpals. *From Ostrom, 1976b.*

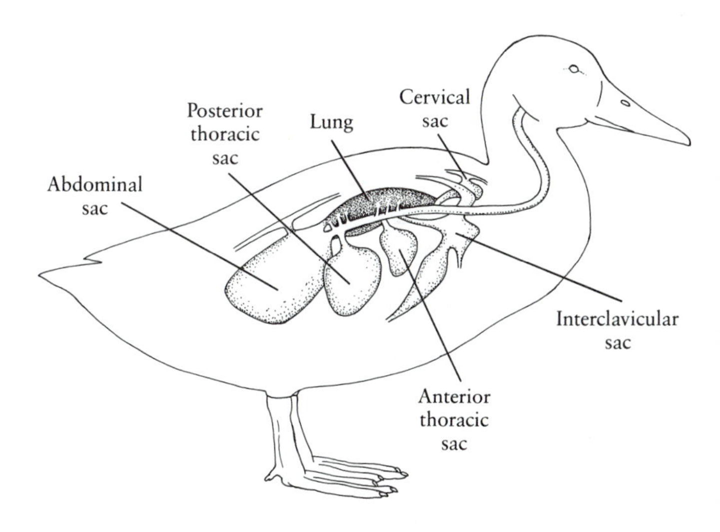

Figure 16-19. THE AVIAN RESPIRATORY SYSTEM, REPRESENTED BY A MALLARD DUCK. One of its most distinctive features is the presence of numerous air sacs that are connected to the bronchial passages and the lungs. Additional air sacs that enter into the bones are not shown. Most of the air inhaled goes directly into the posterior sacs. As the respiratory cycle continues, the air passes through the lungs and into the anterior sacs. This mechanism provides a continuous flow of air through the lung, in contrast with all other tetrapods, in which the air is pumped into and out of blind sacs. *From Schmidt-Nielsen, 1971.*

CRETACEOUS BIRDS

No avian remains other than those of *Archaeopteryx* have been definitely identified from the Jurassic, although Jensen (1981) argued that vertebral, pelvic, and limb elements from the Upper Jurassic of Utah belong to more advanced birds (see Ostrom, 1986).

All Cretaceous birds were much more advanced than *Archaeopteryx* and had already diverged along several distinctive adaptive pathways, which suggests that the appearance and initial diversification of the neornithes may have occurred within the Upper Jurassic among forms contemporary with *Archaeopteryx*. Considering the extremely incomplete record of birds in the later Mesozoic, it would not be surprising if several lineages were present in the late Jurassic without their remains having been found. An animal very like *Archaeopteryx* might still have been their ultimate ancestor.

We know two very different kinds of birds from the Lower Cretaceous. *Ambiortus* (Kurochkin, 1982), from the Neocomian of Central Asia, has a keeled sternum, elongate coracoids, and fused carpometacarpals resembling those of modern flying birds. It retains a third phalanx on the major digit of the wing, in contrast with all more advanced birds, and is placed in an order of its own, the Ambiortiformes. In contrast, *Enaliornis*, from the Lower Cretaceous of England, is an early member of a highly specialized group of flightless diving birds, the Hesperornithiformes. Further evidence of the distribution of Lower Cretaceous birds is provided by the discovery of feathers in Australia (Rich, 1976) and footprints in western Canada (Currie, 1981).

In the Upper Cretaceous, the Hesperornithiformes are among the most common and well-known birds not only throughout North America but in South America as well (Martin, 1984). Their remains are typically found in near-shore marine deposits. In all members of this group, the sternum lacked a keel and the forelimbs were greatly reduced. However, the feet were large as in loons and grebes, and they were probably foot-propelled divers. They are primitive in the retention of teeth and amphicoelous vertebrae. The nonpneumatic nature of the limb bones may have been a specialization for increasing their weight for more effective diving. The largest members of the genus *Hesperornis* were more than 1 meter high (Figures 16-20 and 16-21). *Baptornis* was smaller and the forelimbs were somewhat less reduced (Martin and Tate, 1976).

A second order of Cretaceous toothed birds, the Ichthyornithiformes, is also well represented in shallow marine deposits in North America, but the strongly keeled sternum indicates that they were proficient flyers. In proportions and possibly in habits *Ichthyornis* (Figure 16-22) resembled gulls, but it is not thought to be related to any of the modern orders. Both orders of toothed birds became extinct near the end of the Cretaceous.

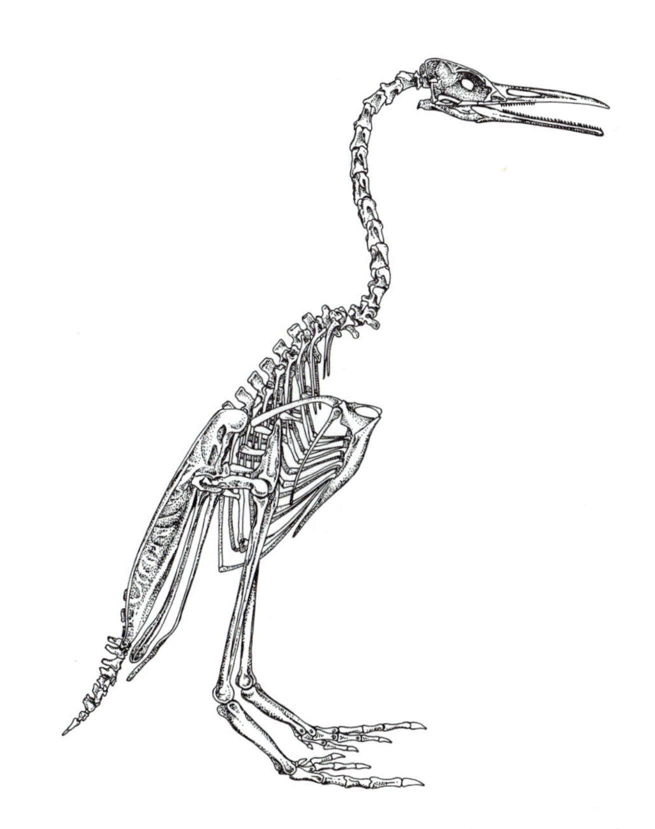

Figure 16-20. SKELETON OF THE UPPER CRETACEOUS DIVING BIRD *HESPERORNIS*, WHICH REACHED MORE THAN 1 METER TALL. The forelimb is reduced to a slender humerus and the sternum lacks a keel. *From Marsh, 1880.*

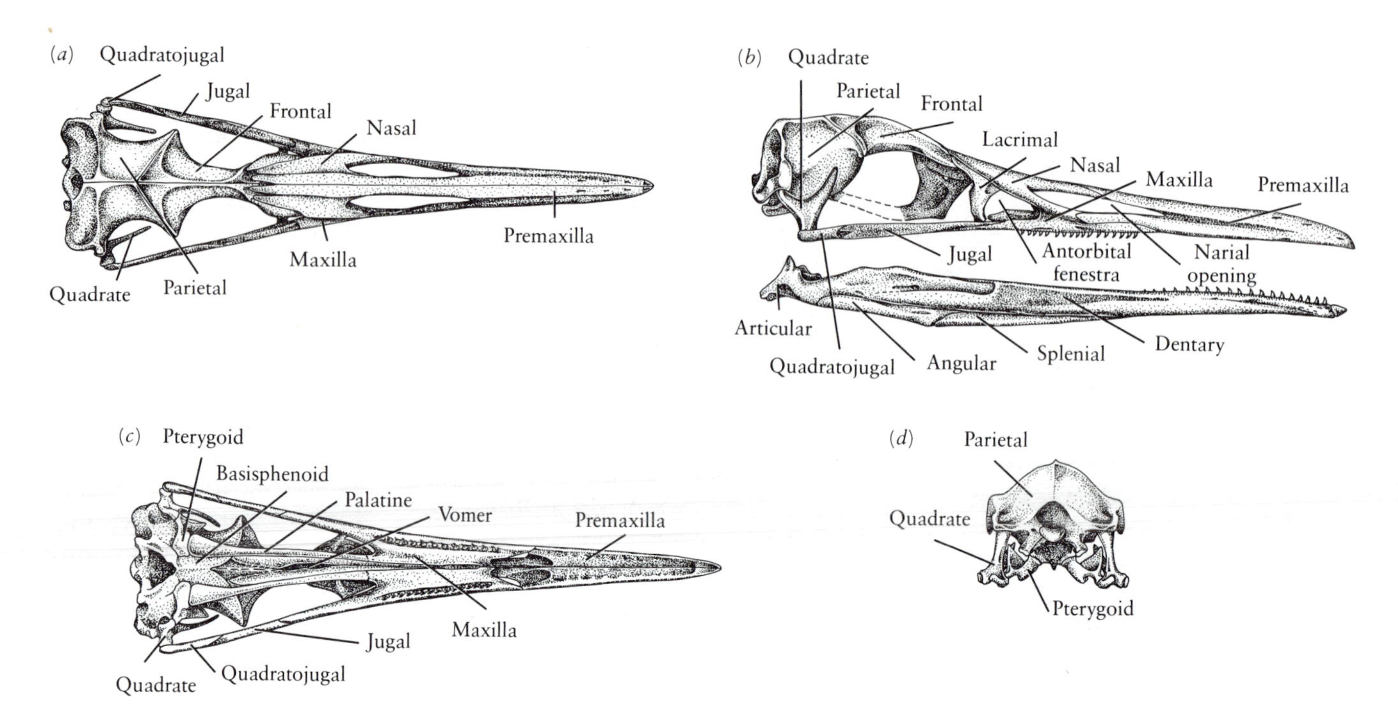

Figure 16-21. RESTORATION OF THE SKULL AND LOWER JAW OF *HESPERORNIS*. (*a*) Dorsal, (*b*) lateral, (*c*) palatal, and (*d*) occipital views. As in advanced birds, teeth have been lost from the premaxilla but they are retained on the maxilla and dentary. The palate retains primitive features common to coelurosaur dinosaurs, including a movable basicranial joint and a large vomer. *Courtesy of Larry Martin.*

Figure 16-22. SKELETON OF THE UPPER CRETACEOUS BIRD *ICHTHYORNIS*. A well-developed keel is on the sternum. There had been some question as to whether this genus had teeth, but this fact is now firmly established. *From Marsh, 1880.*

Other, less-well-known late Cretaceous birds include *Enantiornis* from Argentina, whose pectoral girdle and wing elements indicate well-developed flight capabilities, and *Gobipteryx* from Mongolia (Walker, 1981; Elzanowski 1981). Neither closely resembles other Cretaceous birds or any genera from the Cenozoic. *Alexornis,* which Brodkorb (1976) described from Baja California, was originally thought to be related to the Coraciiformes and Piciformes, but Olson (1985) now considers that it may be related to the Enantiornithiformes.

Only two modern orders are currently recognized from the Upper Cretaceous (Olsen, 1985). Remains from New Jersey and Wyoming appear to be members of the Procellariiformes (represented in the modern fauna by the albatrosses and petrels) and the Charadriiformes, whose closest living representatives are the shorebirds, gulls, and auks. These fossils demonstrate that the diversification of the modern orders began prior to the Cenozoic but provide little evidence of their interrelationships.

Cenozoic birds

All birds that we know from the Cenozoic appear to share a common ancestry above the level of the toothed Hesperornithiformes and Ichthyornithiformes. The fossil record demonstrated that at least some of the modern orders had already differentiated by the end of the Mesozoic, but their remains are so incomplete that it is not possible to establish the common skeletal pattern from which they evolved or to establish their interrelationships.

Within the modern fauna, birds are the most extensively studied class of vertebrates other than mammals. Taxonomy at the species level is very completely known; Welty (1982) estimates that no more than 2 percent of the living species remain to be described. Parkes (1975) listed approximately 8900 living species that are grouped in 166 families and approximately 27 orders. In contrast with our knowledge at the level of the species and family, little evidence is available to establish the interrelationships of the many orders.

The species, genera, and families of modern birds are for the most part readily recognized and distinguished from one another on the basis of the feathers, soft anatomy, and behavior. All of these features are associated with particular ways of life that differentiate the long-accepted orders. On the other hand, very few specialized (or derived) characters have been recognized that demonstrate specific sister-group relationships between any of these orders. Nearly all modern texts classify the living orders in a linear sequence based on ideas that were formulated in the last century regarding what were then thought to be primitive and advanced characters of living birds (Table 16-1). By modern criteria, some aspects of this sequence appear quite arbitrary, such as the placement of penguins in a position that implies that they are among the most primitive of Cenozoic birds, and the classification of diurnal predators, the Falconiformes, between the groups including ducks and swans (Anseriformes) and that represented by the chickens (Galliformes).

Nearly all of the living orders have some fossil record, but it has so far contributed little to an understanding of their interrelationships (Figure 16-23). There are several reasons for this problem. Only a few bones or parts of bones are both sufficiently robust to be preserved and sufficiently diagnostic for identification. In addition, powered flight requires that most elements of the skeleton be constrained to a particular pattern. The potential for change is very limited. Families that have adapted to particular ways of life may appear extremely similar to one another, even if they evolved from different ancestral groups. Diurnal predators, perching birds, foot-propelled divers, and large soaring birds may each have very similar skeletal specializations, but this does not necessarily indicate close common ancestry.

Among most orders, a few remains document the appearance of modern genera as early as the middle or late Miocene. We can identify many living families from the Late Eocene or Oligocene on the basis of isolated skeletal elements. Earlier remains typically differ so much from modern genera that it is difficult to establish to which family they belong, and many specimens have been placed in a succession of orders by different authors because the differences in the configuration of particular bones may be greater between genera within orders than between members of different orders.

Nevertheless, some fossils from the early Cenozoic do provide clear evidence to support relationships between modern orders, and current study of both the anatomy and biochemistry of modern birds suggests grouping of orders in a sequence that is far different from that provided in most texts. Feduccia (1980) and Olson (1985) used this evidence to propose a new arrangement of the Cenozoic orders. Many of their suggestions are followed in this chapter.

However, one very serious problem remains to be resolved, namely, the classification of a large assemblage of flightless birds termed the ratites.

TABLE 16-1. **Classification of Birds That Appears in Most Modern Texts**

Class Aves
- Subclass Archaeornithes
  - Archaeopterygiformes
- Subclass Neornithes
  - Superorder Odontognathae
    - Hesperornithiformes
    - Ichthyornithiformes
  - Superorder Neognathae
    - Struthioniformes
    - Rheiformes
    - Casuariiformes
    - Aepyornithiformes
    - Dinornithiformes
    - Apterygiformes
    - Tinamiformes
    - Sphenisciformes
    - Gaviiformes
    - Podicipediformes
    - Procellariiformes
    - Pelecaniformes
    - Ciconiiformes
    - Anseriformes
    - Falconiformes
    - Galliformes
    - Gruiformes
    - Charadriiformes
    - Columbiformes
    - Psittaciformes
    - Cuculiformes
    - Strigiformes
    - Caprimulgiformes
    - Apodiformes
    - Coliiformes
    - Trogoniformes
    - Coraciiformes
    - Piciformes
    - Passeriformes

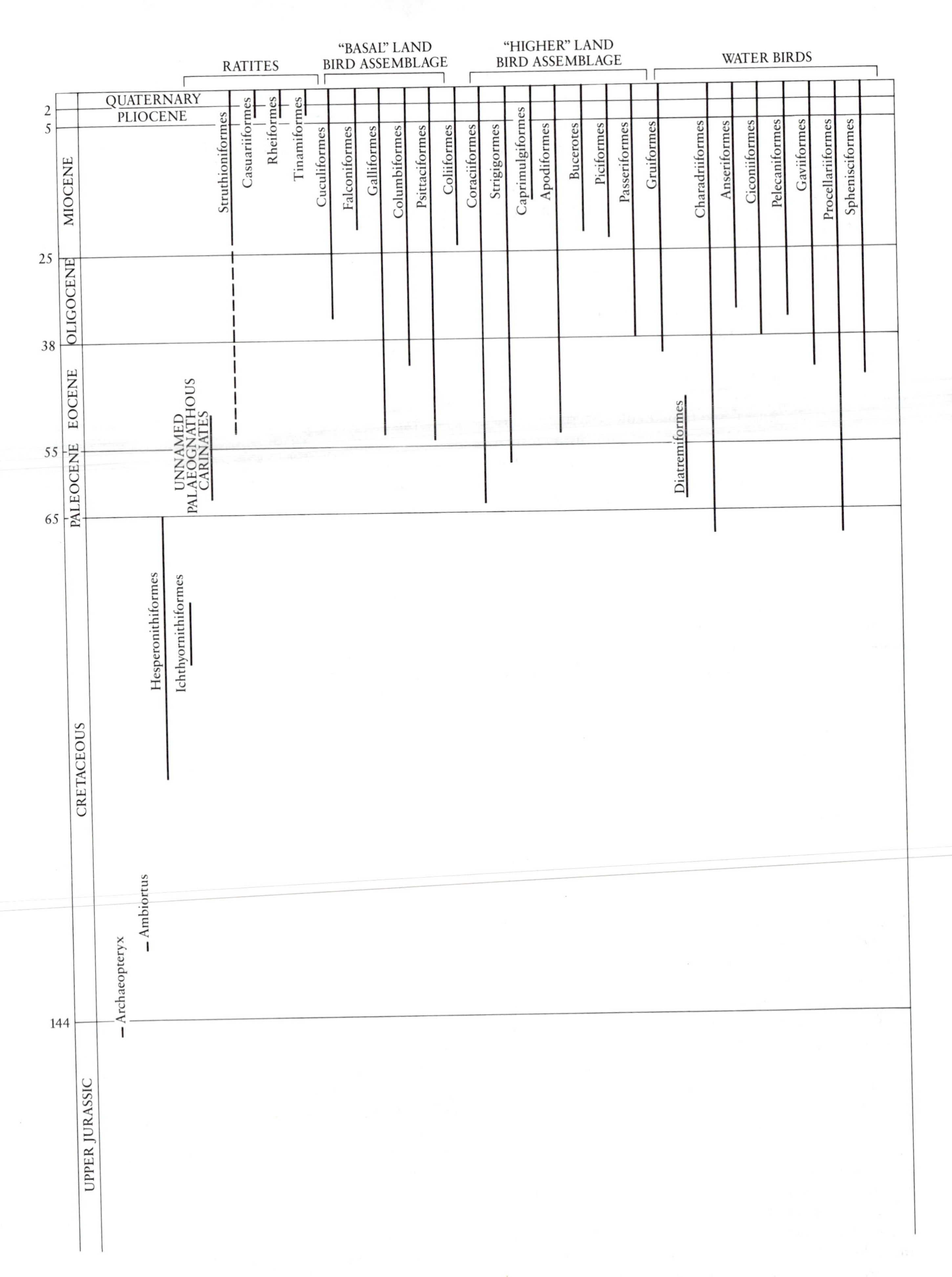

Figure 16-23. TEMPORAL DISTRIBUTION OF AVIAN ORDERS. Orders without a fossil record are not included.

PALAEOGNATHOUS BIRDS

Although the most important characteristic of birds is their ability to fly, flightlessness has evolved in many lineages of both fossil and modern bird groups. In general, it can be attributed to situations in which selection favored attributes such as large body size that were incompatible with flight. Flightlessness has also evolved in birds that were isolated on oceanic islands where the absence of terrestrial predators made it unnecessary for them to escape by flight.

The ancestry of many flightless birds can be traced to the well-known orders of flying birds. The recently extinct dodo from the island of Mauritius and the solitaire from Reunion Island are closely related to pigeons. Numerous other examples from the fossil record include a giant flightless goose on the Hawaiian islands (Olson and Wetmore, 1976), the giant auk of the North Atlantic, and giant analogues of the penguin that lived around the margin of the North Pacific Ocean.

Other flightless birds are so distinct from all orders of flying birds that their relationships have not yet been established. One of the most perplexing problems is the phylogenetic position of a number of flightless orders, including the kiwis, ostriches, rheas, cassowaries, and emus. They are grouped as ratites, in reference to the flat "raft-shaped" sternum that lacks a keel. The forelimbs and flight feathers are reduced to a variable extent.

Since flight has been lost in some members of other avian orders, the ratites may not have had a common ancestry but evolved their distinctive features convergently from a number of different ancestral groups as a result of similar selective pressures for a common way of life. On the other hand, the presence of similar characters suggests that they may share a common ancestry.

As early as 1867, Huxley recognized distinctive features in the palate of all ratites that he suggested were indicative of a common ancestry (Figure 16-24). These features include a large vomer firmly attached to the pterygoid, the absence of a joint between the pterygoid and palatine, and the presence of a movable joint between the base of the braincase and the pterygoid. Huxley recognized that these were primitive features and termed the ratite palate **palaeognathous.**

The palate in most flying birds (termed **neognathous**) has become much more mobile. The basipterygoid articulation with the braincase is lost, a joint is present between the pterygoid and palatine, and the vomers are reduced and do not reach the pterygoid or are lost entirely.

We now recognize that the presence of primitive features is not sufficient to establish that groups have a monophyletic origin (Hennig, 1966; Wiley, 1981). If the structure of the palate in ratites is truly primitive, it only indicates that they evolved from near the base of the radiation that gave rise to other Cenozoic birds, prior to the origin of the modern orders with neognathous palates. However,

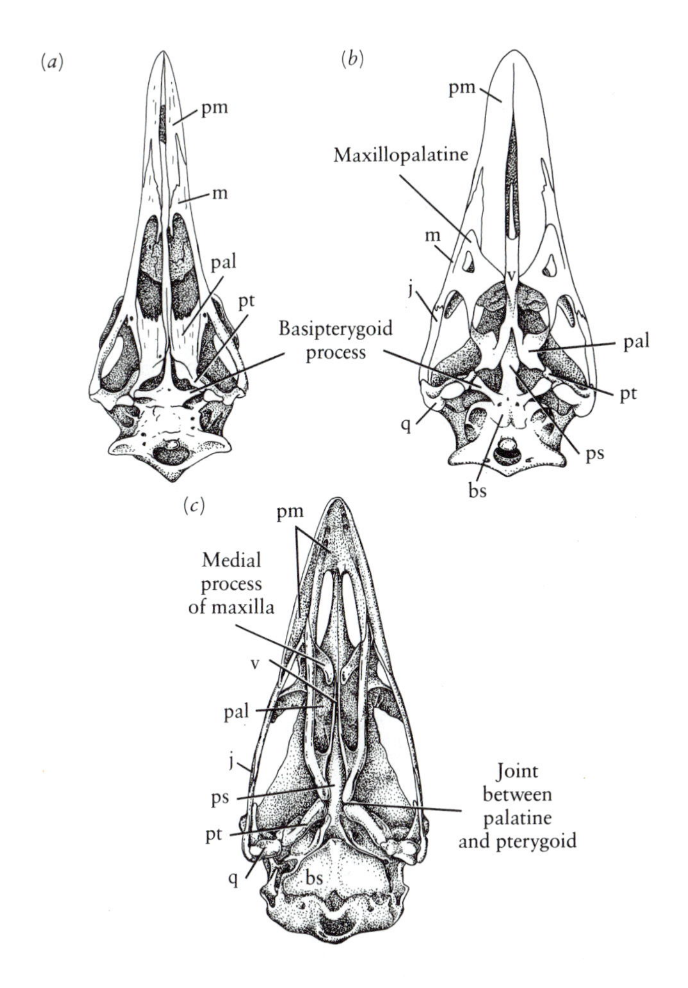

Figure 16-24. PALAEOGNATHOUS AND NEOGNATHOUS PALATES. (*a* and *b*) Palaeognathous palates of the cassowary and the rhea. *From Feduccia, 1980. Reprinted by permission of Harvard University Press.* (*c*) Neognathous palate of the Bronze Turkey. *From Lucas and Stettenheim, 1972.* Abbreviations as in Figure 8-3.

this evidence does not *preclude* ratites from having had a common ancestry.

One group of early Cenozoic birds, which Houde and Olson (1981) described but did not name, demonstrates the combination of a palaeognathous palate with the body of a typical flying bird that had a large keeled sternum (Figure 16-25). One or more of the ratite orders may have evolved from this group, although there are no specific features that demonstrate such affinities.

It is also possible that the palaeognathous features of the palate in ratites are not strictly primitive but have evolved as a result of neoteny by elaboration of primitive features that are evident in early developmental stages of neognathous birds (Feduccia, 1980; Olson, 1985). The loss of the sternal keel and the reduction of the wing in flightless birds can be attributed to a delay in developmental processes so that the adult retains an essentially embryonic condition. Changes in developmental patterns affecting the pectoral girdle and forelimbs may also influence the skull, although the changes in the palate would also have to be selectively advantageous. There is no obvious reason why a palaeognathous palate would be se-

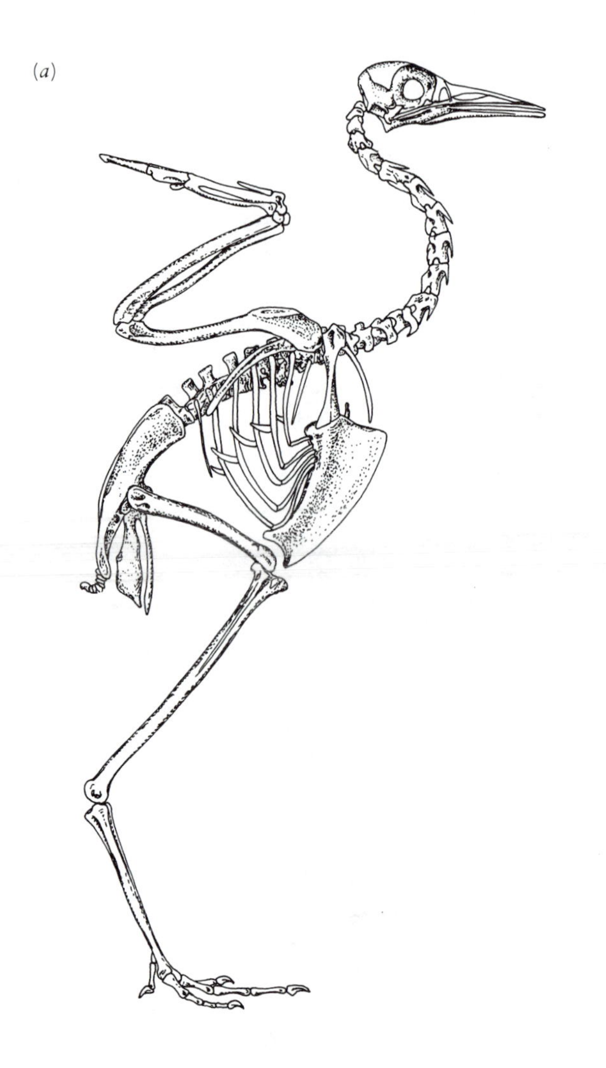

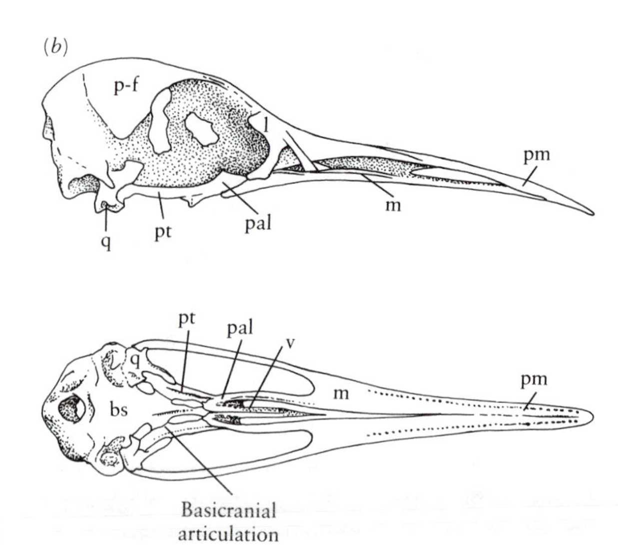

Figure 16-25. (*a*) Reconstruction of the skeleton of an unnamed genus from the Paleocene of Montana that has a keeled sternum but a palaeognathous palate. The tail, like that of a ratite, is very short. Approximately 40 centimeters tall. *Courtesy of Peter Houde.* Skull in lateral and palatal views. Abbreviations as in Figure 8-3. *From Houde and Olson, 1981. With permission from* Science. *Copyright 1981 by The American Association for the Advancement of Science.*

lected for in ratites and not in other modern flightless birds, nor is there any fossil evidence for such a reversal.

With the possible exception of ostriches, there is no fossil evidence linking any of the ratite orders with any of the other avian orders. Although there is no strong evidence that they share a unique common ancestry, they may still be more closely related to one another than they are to any of the modern orders of flying birds.

The palaeognathous palate is also present in the tinamous (Order Tinamiformes), which are common members of the modern fauna of South and Central America. They resemble the quails and partridges but are not thought to be closely related to them. All members of this group forage for food on the ground but are capable of strong flight for short distances and have a strongly keeled sternum. Their ground-dwelling habits are what might be expected in the ancestors of ratites, and they have often been linked with the origin of rheas and, possibly, other flightless birds (Cracraft, 1974). Tinamous are known no earlier than the late Pliocene, and their fossil record provides no evidence of their affinities.

Ratites

All other late Cenozoic birds with a palaeognathous palate are flightless. The reduction of the forelimb and loss of the keel are certainly specializations relative to the condition in primitive flying birds. They also resemble one another in the presence of a broad unossified region between the ilium and ischium, the ilioischiadic fenestra. This feature is certainly primitive since it is evident in Mesozoic birds as well.

The current ranges of kiwis, rheas, cassowaries, emus, and ostriches, as well as the now extinct moas and elephantbirds, are restricted to the Southern Hemisphere. Cracraft (1976) used this distribution to suggest that they must have shared a common ancestry among forms, such as the tinamous, that were restricted to the Gondwanaland continents. However, the fossil record of the ostriches indicates that they were common in Eurasia in the late Tertiary.

Elephantbirds and Moas

Two extinct groups that are customarily allied with the living ratites are the elephantbirds (Order Aepyornithiformes) from Madagascar and the moas (Order Dinornithiformes) from New Zealand. Neither has a fossil record earlier than the Pleistocene, and both became extinct following the appearance of man on these islands. We think that both groups were browsers, a way of life that was permitted by the absence of large herbivorous mammals. In contrast with ostriches, rheas, and cassowaries, which must be fleet to escape continental predators, the moas and elephantbirds had short, thick metatarsals and a graviportal stance. Cracraft (1976) recognized five genera and thirteen species of moas that ranged in size from that of a turkey to *Dinornis*, which reached 3 meters in height.

The elephantbirds were not as tall as the moas, but *Aepyornis maximus* (one of seven species) may have reached a weight close to 500 kilograms. Elephantbirds retained vestiges of wings, but these were lost entirely in the moas.

In the absence of fossils of elephantbirds or moas outside of Madagascar and New Zealand, it is probable that their ancestors flew to these islands and then became flightless. There is no strong evidence to link them to any specific group of flying birds.

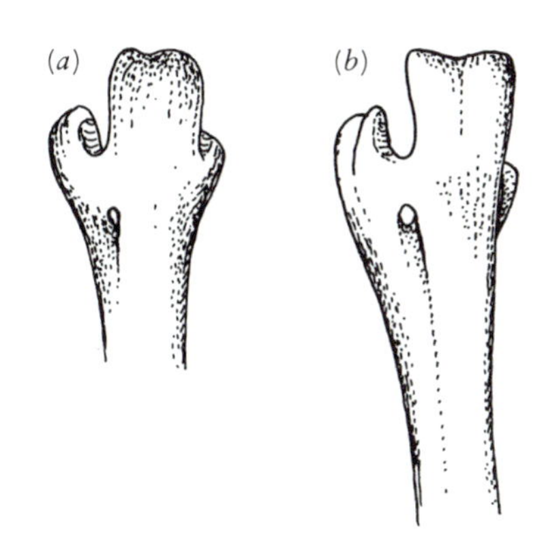

Figure 16-26. The distal ends of the left tarsometarsi of (*a*) *Proergilornis*, showing the beginning of reduction of the inner trochlea, and (*b*) *Struthio*, in which the inner trochlea is reduced to a stub and the inner toe is completely lost. *From Feduccia, 1980. Redrawn by permission of Harvard University Press.*

Rheas

Members of the Rheiformes have been restricted to South America throughout their known history. The only Tertiary genus now assigned to the Rheidae is *Heterorhea* from the late Pliocene of Argentina that is based on the distal portion of a tarsometatarus (Olson, 1985). Leg and wing bones from the late Paleocene of Brazil demonstrate the earlier appearance of other flightless birds in South America, but these remains cannot be assigned to any specific order.

Cassowaries and Emus

The Casuariiformes are confined to Australia, New Guinea, and adjacent islands. Their only Tertiary record is that of an emu from the Lower Pliocene of South Australia. Cassowaries are distinguished by the presence of a large bony casque that extends from the forehead like a helmet. Unlike the other large ratites, they inhabit forests rather than open country.

Kiwis

The chicken-sized kiwis are the smallest of the ratites. They are confined to thick, swampy forests in New Zealand. They use their long slender bill to probe for worms and are aided by a sense of smell that is uncommonly acute among birds. Kiwis nest in underground burrows and lay eggs that are more than 15 centimeters long. Their fossil record, which is limited to the Pleistocene, casts no light on their origin. Except for the structure of the palate and common attributes of flightlessness, they share few features with other ratites.

Ostriches

The Struthioniformes are the largest living birds. Like the other large ratites, they feed primarily on plant food but also eat insects and other small animals. The last ostriches in Arabia were killed during World War II. This group is now restricted primarily to central and southern Africa.

The ostriches are the only group of ratites in which the fossil record contributes significantly to establishing their prior history. Remains of the modern genus *Struthio* are known as early as the Upper Miocene in Turkey (Sauer, 1979). Kurochkin and Lungu (1970) assigned most Pliocene and Pleistocene specimens, ranging from Greece to China, to the species *S. asiaticus*. They suggest that it is a direct ancestor of the living species.

The remains of ostriches are readily recognized by the fact that they alone among modern birds have only two toes (Figure 16-26). On this basis, many other fossils ranging back to the Eocene have been included within the Struthioniformes (Olson, 1985). Several families have been named to include these forms, but all might be synonymized with the Struthionidae. The Geranoididae, which include several superficially cranelike birds from the early and middle Eocene of North America, are known primarily from fragmentary tibiotarsi and tarsometatarsi. The Eogruidae, which are based on similar material, occur in the late Eocene and early Oligocene of Central Asia. We know the Ergilornithidae from the early Oligocene of Asia and Europe (Kurochkin, 1981).

On the basis of these fossils, which demonstrate the gradual reduction of the lateral digit among large, cranelike birds, Olson suggests that the ostriches should be included within the Gruiformes, which are otherwise exemplified by cranes and rails. This affiliation assumes that the palaeognathous palate of ostriches was derived from the neognathous pattern of other gruiformes through neoteny, independently of other ratites.

THE BASAL LAND BIRD ASSEMBLAGE

The neognathous palate is certainly a specialized structure relative to that of Mesozoic birds and indicates that all orders in which it is present share a common ancestry.

In most modern classifications, the ratites are followed by a series of aquatic groups, then by the birds of prey, and finally a sequence of terrestrial birds, culminating with the passerines. Placement of the avian orders in such a linear sequence is partially an artifact of the Linnean system of classification and partially a reflection of our general ignorance of the actual relationship of most orders.

The fossil record shows that most modern families were present by the end of the Eocene, which indicates a rapid radiation within the late Mesozoic and earliest Cenozoic. Many of the orders may have arisen from a common ancestral stock with few clearly definable sister-group relationships.

Feduccia (1980) and Olson (1985) maintain that there was a major dichotomy between two large groups, the water bird assemblage on one hand and more terrestrial forms on the other. Olson suggests that they shared a close common ancestry and that each large assemblage can be divided into primitive and advanced representatives.

Olson (1985) groups the following orders among the primitive land bird assemblage:

Cuculiformes, including the hoatzin, cuckoos, and roadrunners
Falconiformes, diurnal predators, excluding the Vulturidae
Galliformes, including game birds and domestic fowl
Columbiformes, pigeons, and the dodo
Psittaciformes, parrots

Of all these groups, the Cuculiformes may be the most primitive. Olson places the hoatzin (Family Opisthocomidae) near the base of all land birds. The hatchlings of the genus *Opisthocomus* are characterized by the presence of clawed digits on the wing (see Figure 16-15). This feature resembles that of *Archaeopteryx* but is almost certainly not the result of direct inheritance, since the adult wing resembles that of other advanced birds. The Opisthocomidae shares no unique derived characters with any of the more advanced land bird orders and may represent an ancestral stock. However, the modern genus is specialized in being strictly herbivorous, with an enormous crop to assist in digesting arum leaves that are the principal component of its diet.

The Cuculiformes are widespread throughout the tropical and temporate regions of the world in a pattern that suggests an early Tertiary radiation. The feet are specialized in having the outer toe reversed to better grasp twigs. This pattern is termed **zygodactylous.**

The Galliformes, which include grouse, quail, pheasants, domestic chickens, guineafowls, turkeys, peafowl, and some less familiar families such as the Australian mound builders and the South American curassows, have a fossil record going back to the Lower Eocene. Most features of the skeleton are primitive. They are basically ground birds but can fly well for short distances. Some authorities place the hoatzin within the Galliformes because of its large crop. Olson suggests that this order may share a close common ancestry with the Cuculiformes.

The Columbiformes, which are represented in the modern fauna by the sand grouse, pigeons, and doves, are known from the fossil record as early as the late Eocene. The dodos and solitaires, which inhabited the islands of Mauritius, Rodriquez, and Reunion prior to their extinction in the seventeenth and eighteenth centuries, are usually placed in a separate family because of their loss of flight, but Olson includes them with the pigeons in the family Columbidae.

Parrots (order Psittaciformes) are known from the Lower Miocene. They are specialized in the evolution of a zygodactylous foot structure but are otherwise similar in some respects to pigeons and may share a common ancestry with them.

Vultures, eagles, hawks, and falcons have long been classified together because of their similar predatory adaptations. All have a sharply hooked beak with a soft mass called a cere across its top through which the nostrils pass and powerful feet with long claws and an opposable hind toe. Jollie (1976–1977) argued that this is probably not a natural, monophyletic grouping. The Falconidae has a fossil record going back to the early Miocene in South America. Olson (1985) suggests that the Falconidae may have evolved from the basal radiation of land birds within that continent.

There is currently no firm hypothesis as to the specific relationships of the hawks and eagles (family Accipitridae), despite their long fossil record. Brodkorb (1964) recognizes 62 extinct species, the earliest of which is from the late Eocene or early Oligocene of France. We find modern forms in the Middle Miocene. The skull resembles that of the secretary bird, family Sagittariidae, another large predator that has customarily been placed in the Falconiformes. This family has only a single living species confined to Africa, but extinct genera are recognized from the Middle and Upper Oligocene and Lower Miocene of France. Included within the Accipitridae are the vultures of the subfamily Aegypiinae, which are now confined to the Old World. This group was common in North America during the Tertiary, and only became extinct there during the Quarternary (Rich, 1980). Fossils of ospreys (Pandionidae) are known in the Old World as early as the Oligocene, but the nature of the relationships of this group with other members of the Falconiformes is uncertain.

THE HIGHER LAND BIRD ASSEMBLAGE

Olson (1985) concludes that the advanced land birds belong to a monophyletic assemblage, including the order Coraciiformes (within which the rollers and kingfishers are well-known forms), Strigiformes (owls), Caprimulgiformes (oil birds, frogmouths, and potoos), Apodiformes (swifts and hummingbirds), Piciformes (woodpeckers), and the Passeriformes (including some 5000 species of songbirds). It corresponds to the group Anomalogonatae as constituted by Beddard (1898). This assemblage is characterized by the consistent absence of the ambiens muscle, which is present in nearly all families of other avian groups.

Most of the higher land birds have specialized the articulation between the tarsometatarsus and the base of the digits to permit a better grip on small branches. However, the specific way in which the elements articulate differs from family to family, and similar patterns may have evolved separately in members of different orders.

Olson suggests that the Coliiformes may represent a link between the Anomalogonatae and the basal land bird assemblage. The colies, also termed mousebirds in reference to their rodentlike scurrying in the foliage, include only two modern genera that are restricted to Africa. This group is peculiar in having both the outer and hind toes reversible so that two, three, or four toes may face forward. Ballmann (1969) described members of this group from the Upper Eocene and Miocene of Europe, some of which may belong to the modern genera.

Coraciiformes

The most primitive members of the higher land bird assemblage are the Coraciiformes, within which the rollers (named for their aerial acrobatics) may be considered an ancestral assemblage. Olson (1985) identified a beautifully preserved specimen from the early Eocene of North America as very similar to the genus *Eurystomus,* which would provide the earliest known record of any extant avian genus. Rollers and kingfishers are both characterized by having two front toes (III and IV) joined part way in their length. We know fossils of modern families of Coraciiformes from the Eocene and Oligocene of Europe and North America.

Olson removes the hoopoes (family Upupidae) and the hornbills (Bucerotidae) from the Coraciiformes and places them closer to the stock that includes the woodpeckers and passerines. On the other hand, he uses the common presence of specialized features of the stapes, which Feduccia (1975) described, to include the Trogonidae within the Coraciiformes, close to the kingfishers. The trogons are a group of brightly colored birds common in the tropics of both the New and Old World, that are unique in having the inner, rather than the outer, toe reversible.

Olson (1982) also includes the suborder Galbulae in this order. This group, which includes jacamars and puffbirds, is usually considered close to the woodpeckers because of the zygodactylous nature of the feet, although the configuration in the two groups is quite different. Jacamars are iridescent birds that are currently centered in the New World tropics.

As the Coraciiformes are defined by Olson, this group was the dominant order of arboreal perching birds in North America and Europe during the early Tertiary. He suggests that the early members of this assemblage may have been part of a stem group that gave rise to several more derived orders. Both the Strigiformes and Caprimulgiformes may have evolved from early rollers, and the Caprimulgiformes in turn may have given rise to the Apodidae (swifts) and the hummingbirds.

Strigiformes and Caprimulgiformes

Rich and Bohaska (1981) recognized the oldest-known owl, *Ogygoptynx,* on the basis of a tarsometatarsus from the Paleocene of Colorado. It appears to belong to a group that could include the ancestry of both living owl families. Several other early Tertiary owls have been placed in a separate group, the Protostrigidae (Mourer-Chauvire, 1981). Complete skeletons of owls are known from the Oligocene. Modern genera appear in the Miocene. The largest of all owls are from the Pleistocene of Cuba (Kurochkin and Mayo, 1973). The limbs are relatively small and the sternum is reduced, but Olson doubts that they were flightless.

The Caprimulgiformes encompass a diverse group of families including the echo-locating oil bird, frogmouths, nightjars, and goat suckers, all of which are mainly nocturnal and have gaping mouths for catching insects. All have been reported from the late Eocene or the Oligocene of France (Mourer-Chauvire, 1982) and the owlet-frogmouths (Aegothelidae) are known from the middle Miocene of Australia (Rich and McEvey, 1977).

Apodiformes and Trogoniformes

We know primitive swiftlike birds from as early as the Lower Eocene in England and North America. *Primapus,* which is based on several humeri, is placed in an extinct family Aegialornithidae. Early swifts are very similar to members of the Caprimulgiformes, and Olson suggests that these groups might be combined into a single order. The earliest true swifts appear in the early Miocene of France. There is no Tertiary record of the hummingbirds (Trochilidae). They are usually placed with the swifts, but this association is not well documented. Cohn (1968) suggested that they may be closer to the Passeriformes.

Bucerotiformes

Olson suggests that the higher land birds may be divided into two lineages, one centered about the Coraciiformes and a second including the passerines and the woodpeckers. Three families, the Upupidae, the Phoeniculidae and the Bucerotidae, that are typically included in the Coraciiformes appear to be closer to the pico-passere lineage. The Bucerotidae (hornbills) are first represented in the middle Miocene of Morocco by a very distinctive tarsometarsus. Among the Upupidae and Phoeniculidae (hoopoes and wood hoopoes) only the latter have a Tertiary fossil record, which goes back to the early Miocene of Europe (Ballmann, 1969).

Piciformes

Woodpeckers, barbets, and toucans are characterized by their zygodactylous foot structure, with the outermost toe reversed to parallel the hind toe. This structural specialization is accompanied by characteristic changes in the muscles and tendons of the hind limb. Toucans, which are distinguished by their enormous bills, have no significant fossil record. We know barbets from the early Miocene. True woodpeckers, Picidae, which are specifically adapted for tree trunk foraging, appear in the middle Miocene.

Passeriformes

Passerine birds include over 5000 species and are dominant elements of the avian fauna on all continents except Antarctica. However, their fossil record is still very sketchy. Fifty to seventy families are recognized on the basis of species diversity, but they may differ from one another osteologically less than do genera of other avian orders.

Ornithologists recognize two large groups of passerines on the basis of the complexity of the syrinx and the structure of the stapes. Four-fifths of the modern species, including the songbirds of the temperate zone, are grouped among the oscines (suborder Passeres), which are characterized by a complex syrinx with more than three pairs of intrinsic muscles. Ames (1971) argued that the specializations of this structure demonstrate that all members of this huge assemblage share a close common ancestry.

The suboscines are now the dominant passerines of South America, where they total about 1000 species. They have a more primitive but variable pattern of the syrinx than do the oscines, but, according to Feduccia (1980), they can be accepted as a monophyletic group on the basis of the specialized structure of the stapes, which is characterized by a large, bulbous footplate.

Only a single fossil of a suboscine has been reported from the Tertiary, an unidentified species of the family Eurylaimidae from the early Miocene of Europe (Ballmann, 1969). This family, commonly referred to as broadbills, includes brightly colored birds of the Old World tropics.

The earliest-known passerines come from deposits in France that are latest Oligocene in age. No fossils of passerines are known in North America until the Miocene, despite a relatively good record of other bird groups. This absence supports the contention that they did not become important elements of the avian fauna of the Northern Hemisphere until late in the Tertiary, when they replaced the previously dominant Coraciiformes. However, they rapidly achieved dominance. In the early Miocene deposits at Wintershof West in Bavaria, they outnumber all other birds combined (Ballmann, 1969).

Oscines appear to have originated in the Old World, probably in the tropics, and spread into North America by the middle to late Miocene. They apparently became established in South America during the late Tertiary. The suboscines came north with the rise of the Panama land bridge in the late Pliocene and are now represented in North America by the flycatchers, Tyrannidae, many members of which migrate between North and South America.

Feduccia (1980) summarized the major groups of passerines. Sibley (1970) and Raikow (1978) have provided detailed classifications.

THE WATER BIRD ASSEMBLAGE

Gruiformes

According to Olson (1985), all remaining birds may be included in a single assemblage that arose late in the Mesozoic from the base of the land bird radiation. Most of this group is specialized for life in or over the surface of the water. The most primitive modern order within the assemblage, the Gruiformes, includes cranes, limpkins, and rails. Most inhabit an aquatic or swampy environment, but a few live on dry land and some even in deserts. Some genera are strong flyers, but others have a limited tendency to fly, and still others are flightless. Within the Gruiformes, many families have disjunct distributions and a higher percentage of recently extinct genera than any other major order of birds. Both features suggest great antiquity for the group.

Olson recognizes the family Cariamidae as the most primitive living gruiforms, and early Tertiary fossils link them to the base of the land bird assemblage. The cariamids are relatively large cursorial predators and scavengers that are now restricted to South America. They are capable of flight but are not strong flyers. Their greatest skeletal resemblance is to the hoatzin, which may be close to the base of the land bird assemblage.

The living members of the Cariamidae provide a model for the skeletal pattern of a group that could have given rise to the diverse forms making up the water bird assemblage, although their fossil record goes back no further than the late Pliocene. We know closely related genera, which are placed in the families Bathornithidae and Idiornithidae, from the early Tertiary of North America and Europe. The Cariamidae are classified within the suborder Cariamae, which also includes one or more groups of giant, early Tertiary predators.

During the early Cenozoic, there were relatively few large mammalian carnivores. At that time, South America was isolated from the other continents and no cursorial carnivores evolved among the native marsupials until late in the Cenozoic. Their role was taken by large predaceous birds of the family Phorusrhacidae (commonly referred to as Phororhachidae). We know the phorusrhacids from the early Oligocene to the late Pliocene in South America. All are flightless. Among the dozen or more species, some reached nearly 3 meters tall. Their huge skulls, up to 48 centimeters long, would have made them extremely formidable predators (Figure 16-27). This group was long thought to have been restricted to South America, but

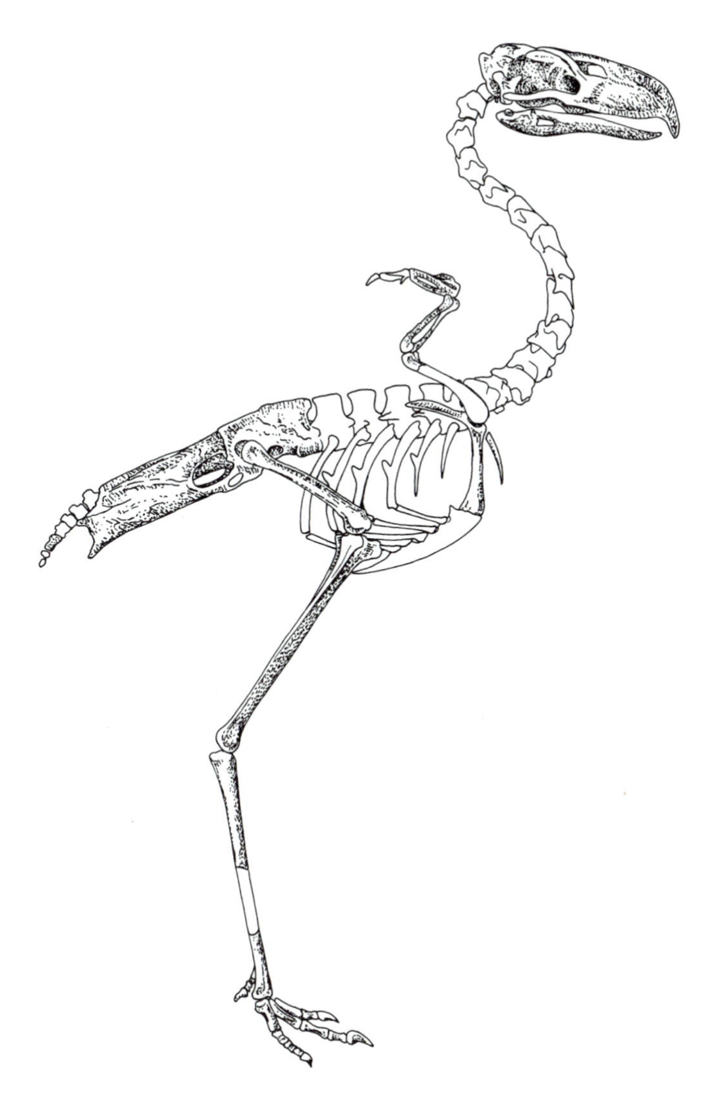

Figure 16-27. PHORUSRHACUS [*PHORORHACOS*]. A gigantic nonflying predaceous bird from the Tertiary of South America, approximately 1.5 meters tall. *From Andrews, 1901.*

Brodkorb (1963) described a distinct genus from the Pliocene of Florida that probably entered North America over the Panama land bridge in the late Pliocene. Even more striking was the discovery of a phorusrhacid from the Eo-Oligocene of France, which Mourer-Chauvire (1981) reported.

A second family of large, flightless, predatory birds is known from the northern continents in the early Tertiary. The Diatrymidae are frequently allied with the Gruiformes, although Olson (1985) considers that they are best placed in a separate order, the Diatrymiformes. They are restricted to the Paleocene and Eocene but have been described from China as well as North America and Europe. They reached more than 2 meters in height and, like the phorusrhacids, had huge skulls (Figure 16-28). The legs were shorter and more massive, with large claws.

Among the living gruiform families that have a fossil record, the Rallidae are known from the late Oligocene. Rails (including coots) are of interest because they include a great number of flightless species. Many are known on isolated oceanic islands. Some species include both flying and flightless races, which suggests that the loss of flight may occur very rapidly. Unfortunately, the fossil record tells us little about the history of the family (Olson, 1977a).

The cranes (Gruidae) are recognized from the late Eocene on the basis of a coracoid with a large pneumatic fossa on the sternal end of the dorsal surface (Cracraft, 1973). Nearly complete skeletons are known from the early Oligocene in North America.

Herons and bitterns are typically classified with the storks in the Ciconiiformes, but Olson argues that their relationships cannot be established on the basis of presently available evidence and he places them adjacent to the Gruiformes. Their fossil record extends back to the Eo-Oligocene.

Olson also allies the grebes (usually placed in their own order, Podicipediformes) close to the Gruiformes. They are strong-swimming, foot-propelled divers. Association with the Gruiformes is suggested by the neck musculature (Zusi and Storer, 1969) and by cranial similarities between grebes and rails. The fossil record, which goes back to the Lower Miocene, does not help to establish their affinities. Some grebes can fly well, while others are completely flightless.

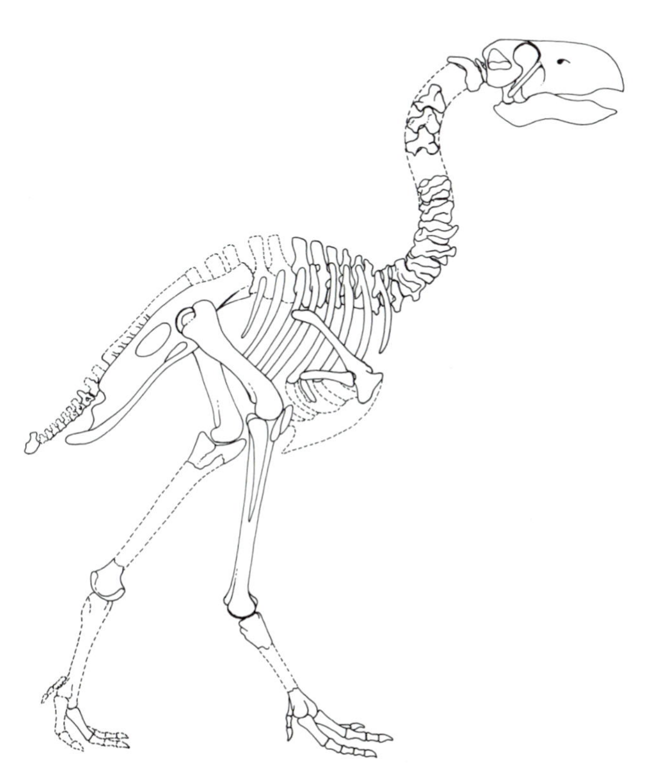

Figure 16-28. SKELETON OF THE GIANT PREDATORY BIRD *DIATRYMA*. This represents a family that was common in Asia, North America, and Europe during the Early Tertiary. This genus was more than 2 meters tall. *From Matthew and Granger, 1917.*

Charadriiformes

The Charadriiformes are a diverse group with approximately 20 modern families, including the plovers, sandpipers, avocets, phalaropes, snipes, gulls, terns, and auks. Olson provides evidence that the flamingos (Phoenicopteridae), which were formerly allied with storks, and the plains wanderer (Pedionomidae) and the bustards (Otididae), which are typically considered among the Gruiformes, should also be placed in this order.

The Charadriiformes are represented by a complex array of late Cretaceous fossils that are included in the family Graculavidae. These long-legged wading birds bridge the gap between primitive gruiforms and advanced charadriiforms. Within this primitive array, the early Tertiary shore bird *Presbyornis* appears close to the ancestry of ducks and swans (the Anseriformes). Like later Charadriiformes, members of the Graculavidae are distinguished by a four-notched sternum.

Olson (1979) suggested that the ibises (Plataleidae) evolved from the base of the charadriiform assemblage rather than being close to the storks. The earliest unquestioned members of this family are known from the middle Eocene (Peters, 1983). Flightless ibises have been described from the Quarternary of the Hawaiian islands and Jamaica.

All but six of the modern families of the Charadriiformes have a fossil record, which typically extends back into the middle Tertiary, but the fossils contribute little to establishing specific affinities. Flamingos are exceptional. Remains from the middle Eocene show a number of features in common with stilts that indicate their close relationship with the Recurvirostridae (Olson and Feduccia, 1980a).

The auks (Alcidae) are of particular interest because of their rich fossil record in the middle and late Tertiary (Olson, 1985). They are wing-propelled diving birds of the Northern Hemisphere that parallel the penguins in the south. They are known from the late Eocene of the eastern Pacific and later radiated around the North Atlantic. Simpson (1974) discussed the death of the last great auk in 1844.

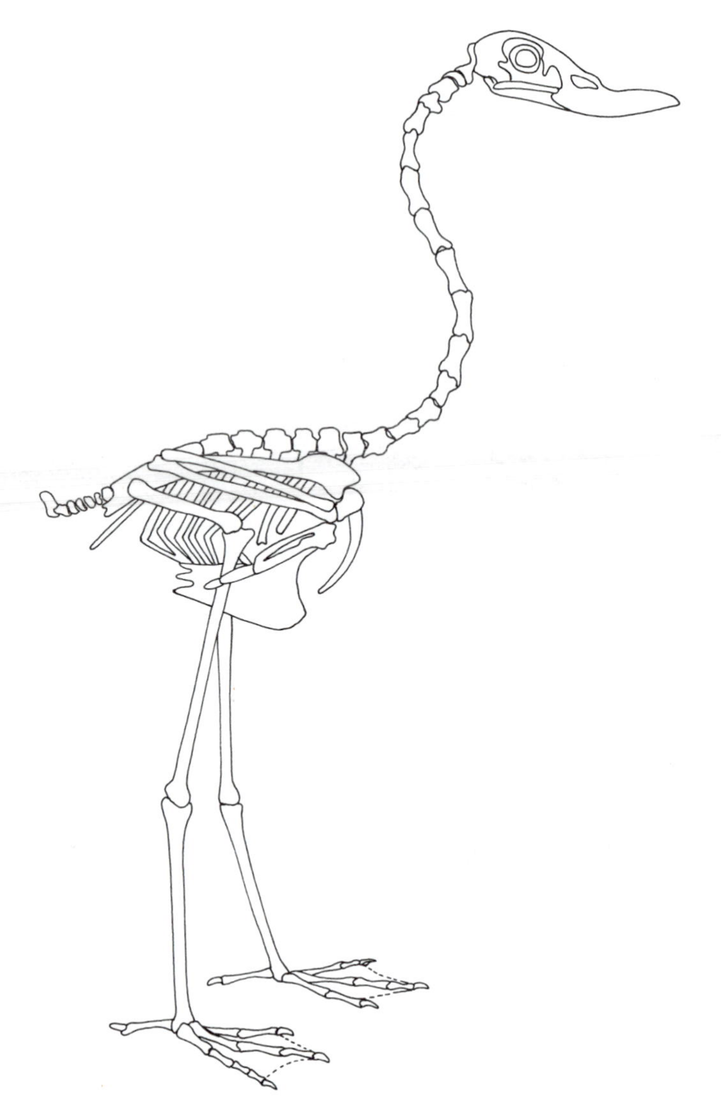

Figure 16-29. *PRESBYORNIS.* Provisional reconstruction of the skeleton of the Paleocene-Eocene shore bird. *From Olson and Feduccia, 1980b.*

Anseriformes

The origin of the ducks and swans is almost certainly to be found in the late Cretaceous and early Tertiary families Graculavidae and Presbyornithidae (Olson and Feduccia, 1980b). The genus *Presbyornis* shows a mosaic of postcranial characters that indicate a shore bird ancestry and derived cranial characters that are unique to ducks and swans. *Presbyornis* was a highly colonial bird whose bones are found in extremely dense concentrations, suggesting a behavior similar to that of modern flamingos. Almost the entire skeleton is known, although a complete description has not been published (Figures 16-29 and 16-30). Both the skull and the hyoid apparatus are unmistakably ducklike, showing the uniquely derived double-piston filter feeding apparatus that characterizes that group. The body, with its very long legs, reflects its shore bird ancestry.

Olson (1985) does not include the Presbyornithidae in the Anseriformes but considers that true ducks do not appear until the Oligocene. The early Oligocene genera *Romainvillia* and *Cygnopterus* were the size of geese, but their affinities with modern families are uncertain. By the middle Miocene, most of the modern genera had appeared, but they did not become dominant elements in the freshwater fauna until the Pliocene (Howard, 1973).

Ciconiiformes

The removal of flamingos, ibises, and herons from the Ciconiiformes leaves only the storks (Ciconiidae), the whalehead stork (Balaenicipitidae), and the hammerhead (Scopidae) of the groups typically included in this order. The fossil record of these forms goes back to the early Oli-

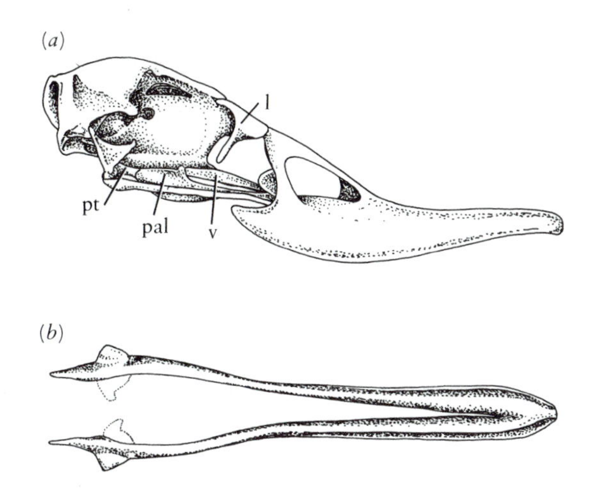

Figure 16-30. (*a* and *b*) Lateral view of the skull and ventral view of the mandible of *Presbyornis*. The structure is nearly identical with that of primitive living ducks. Abbreviations as in Figure 8-3. *From Olson and Feduccia, 1980b.*

gocene of Egypt. Olson (1985) also includes two other families, the Teratornithidae and the Vulturidae.

Members of the Vulturidae, which are now confined to the New World, appear superficially similar to the Old World vultures, which have close affinities with the hawks and eagles. It has long been recognized that the "New World" vultures are actually more closely related to storks. Their toes are not strongly hooked for grasping. The hind toe is somewhat elevated and the front toes have rudimentary webs. Members of the Vulturidae are termed the "New World" vultures, but their fossil record in North America goes back no earlier than the Pliocene. They were much more common in the Old World earlier in the Tertiary. Mourer-Chauvire (1982) recognized the genera *Plesiocathartes* and *Diatropornis* from the Eo-Oligocene. Members of the Vulturidae are the most accomplished land-soaring birds. The condor's wing span of 3 meters is the greatest of any modern land birds. However, this span is greatly exceeded by that of a separate but related family of extinct birds, the Teratornithidae.

The Teratornithidae include the largest flying birds ever reported. A gigantic vulturelike species, *Argentavis magnificens* from the late Miocene of Argentina, had a wing span of 7 to 7.6 meters and an estimated weight of 120 kilograms (Campbell and Tonni, 1983), which is far larger and heavier than it was thought possible for a flying bird. This group is known as recently as the late Pleistocene in North America. The unfused jaw symphysis, which would have allowed them to engulf very large prey, suggests that they were predators rather than scavengers.

Pelecaniformes

The remaining birds are progressively more highly adapted to an aquatic existence. The pelicaniformes are the most diverse, with six living families and three known only as fossils. All are relatively large birds that feed on fish. The group is united by the uniquely extensive webbing on the foot. The hind toe is turned partly forward and webbed to the innermost front toe. They are strong fliers but poor walkers. All have long beaks with throat pouches. Most have small nostrils or lack external openings entirely. Their salt glands are located within the orbit. The salt glands in other marine birds, including the Procellariiformes, are located in furrows on top of the skull.

Among the living families, the frigatebirds (Fregatidae) have the earliest fossil record, which goes back to the Lower Eocene (Figure 16-31). The boobies and gannets (Sulidae) and cormorants (Phalacrocoracidae) first appear in the early Oligocene. We find pelicans (Pelecanidae) and anhingas (Anhingidae) in the Lower Miocene. Although they are primitive in some respects, even the earliest members of these groups are readily recognized as characteristic of their respective families.

The Phaethontidae, tropic birds, are known from the middle Miocene and may be linked to an extinct family, the Prophaethontidae, which is known from the early Eocene. They are primitive in retaining long open nostrils, but the salt glands are within the orbit as in other members of the Pelecaniformes.

The most striking of the extinct Pelecaniformes were certainly the Pelagornithidae, which include gigantic soaring birds with a wing span of 5 to 6 meters. The bones were extremely thin and light, but much of the skeleton is known in some species. The long bills are distinguished by the presence of numerous, pointed, toothlike, bony projections. The humeral articulation was restricted so that the wing could not have been raised much above the horizontal, which made them almost incapable of sustained flapping flight. They probably relied primarily on the wind to provide lift.

Fossils of the Pelagornithidae are known from the early Eocene to at least the end of the Miocene, with a distribution that encompasses Europe, North and South America, Antarctica, New Zealand, and Japan (Olson, 1985).

Some characters of the pelagornithids are also present among the albatrosses (Procellariiformes), but the loca-

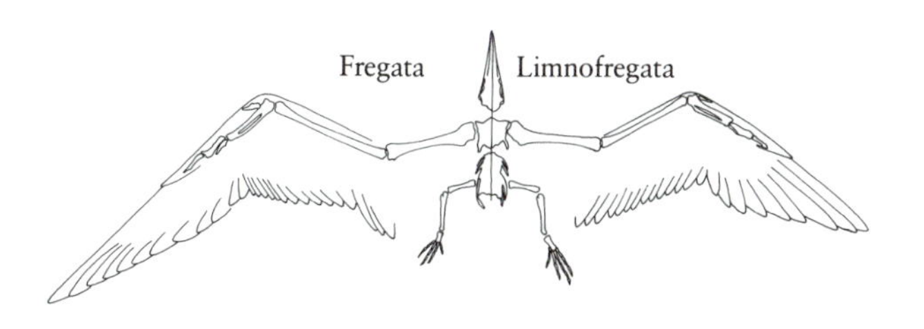

Figure 16-31. OUTLINE COMPARING THE OVERALL SKELETAL PROPORTIONS OF MODERN AND FOSSIL FRIGATEBIRDS. *Fregata* is depicted on the left half of this composite individual and *Limnofregata* from the Lower Eocene is pictured on the right. The body size of the two species is virtually the same, as demonstrated by the length of the sterna and pelves. *From Olson, 1977b. By permission of Smithsonian Institution Press, Smithsonian Institution, Washington, D.C., 1977.*

tion of the salt gland within the orbit is strong evidence of pelecaniform affinities. Further study of this group may provide evidence of the nature of the relationships between these two orders.

Closely related to the Sulidae and Anhingidae is a strikingly specialized family of large flightless diving birds, the Plotopteridae, which Olson and Hasegawa (1979) described. This group is known from both the North American and Japanese shores of the North Pacific in rocks ranging in age from late Oligocene to middle Miocene. Olson (1980) emphasized their similarities to both flightless auks and penguins in the modification of the forelimb to a paddlelike flipper (Figure 16-32). The largest of the five species far exceeds the size of the living penguins, with an estimated length of 2 meters. Olson speculated that their extinction may be correlated with the radiation of seals and porpoises. Despite some wing features that are similar to either penguins or auks, the sternum is typical of pelecaniforms and elements of the hind limb show affinities with the Anhingidae.

Procellariiformes

The procellariiforms include the albatrosses, shearwaters, storm petrels, and diving petrels (families Diomedeidae,

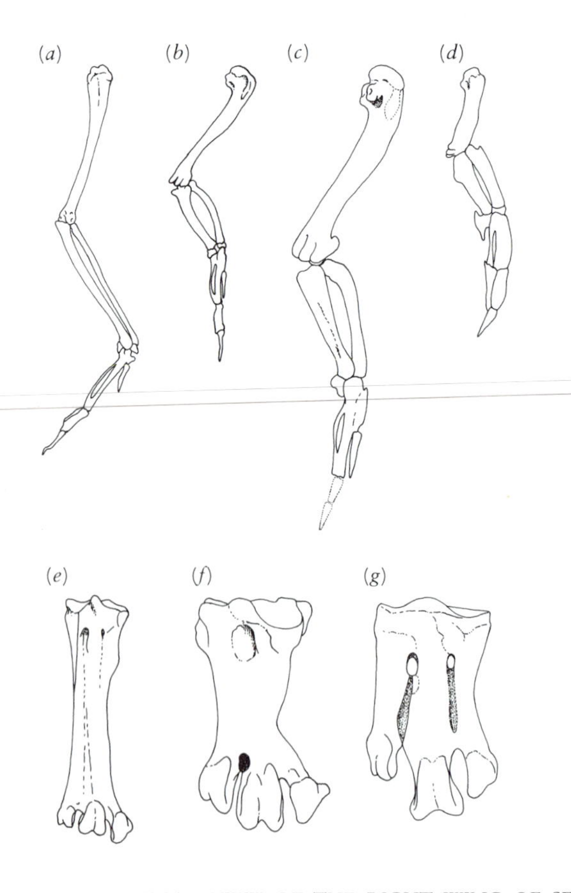

Figure 16-32. DORSAL VIEW OF THE RIGHT WING OF SEABIRDS. (*a*) *Anhinga* (Pelecaniformes). (*b*) Great auk (Charadriiformes). (*c*) Plotopterid (Pelecaniformes). (*d*) Penguin. Anterior view of right tarsometatarsus. (*e*) *Anhinga*. (*f*) Plotopterid. (*g*) Penguin. *From Olson and Hasegawa, 1979. With the permission of* Science. *Copyright 1979 by The American Association for the Advancement of Science.*

Procellariidae, Hydrobatidae or Oceanitidae, and Pelecanoididae). They are recognized by nostrils that extend onto the bill in short tubes. Members of this group spend most of their lives over the open ocean and only come to land to nest. The albatrosses are champion gliders but are almost helpless without fairly strong winds. The wing span of the wandering albatross, which is up to 3.5 meters, is the greatest of any living bird.

We know diving petrels from the early Pliocene. A modern genus of storm petrel is known from the late Miocene. Shearwaters are known from the early Oligocene. Living members of this family show very little change during the past 15 million years. The oldest albatrosses are known from the Upper Oligocene of South Carolina. They continue to be common in the North Atlantic into the late Cenozoic but disappear from this ocean in the Quarternary.

Gaviiformes

The loons (or divers) occupy a peculiarly isolated position phylogenetically. They resemble the grebes (Podicipedidae) in being foot-propelled divers. They are almost tailless, with the feet placed to the rear so that they are almost helpless on land. The muscles of the trunk encase the rear limb down to the ankle joint. Divers have fully webbed feet, in contrast to the paddlelike lobes on the toes that are a feature of grebes. Loons also superfically converge on the Hesperornithiformes, but unlike that group, they retain well-developed powers of flight. They may also use their wings for underwater locomotion.

Many workers support affinities between the Gaviiformes and the Procellariiformes and Sphenisciformes, although Storer (1956) suggested affinities with the Charadriiformes. Like penguins and the order including the albatross, loons are specialized in having two successive coats of down preceding the juvenal plumage. The earliest loon is *Colymboides anglicus* from the late Eocene of England. The single modern genus *Gavia* is known from the Lower Miocene.

Sphenisciformes

Penguins are typically placed at the beginning of the classification of carinate birds, implying a very primitive phylogenetic position. In fact, penguins are among the most highly specialized birds in their total loss of aerial flight and the high degree of specialization for subaqueous locomotion. Penguins differ from all ratites and other nonflying terrestrial birds in the retention of a large keel on the sternum and massive flight muscles. They have retained the basic structure and functions of flying birds, but their flight occurs underwater rather than in the air. Unlike other birds, upward movement of the wing requires as much force as downward movement because of the density of the medium and buoyancy of the body. The scapula provides a much larger surface area for the origin of muscles to lift the wing than that of aerial flyers. The bones have completely lost their pneumatization. The bones of the forelimb are flattened and do not bend at the elbow. The hind limbs are short and placed to the rear so that they have an upright terrestrial posture resembling that of advanced hominoids.

Ornithologists have not established the specific relationships of penguins, but most evidence points to affinities with the Gaviiformes and Procellariiformes. The pterygoid bone in early penguins is significantly more primitive than that of modern genera and closely resembles that of the Procellariiformes. Sibley and Ahlquist's (1972) electrophoretic analysis of proteins indicates that the closest affinities are with the petrels. The bill of one Eocene penguin, ?*Paleoeudyptes*, is long and sharply pointed, resembling that of loons, which may indicate a close relationship with the Gaviiformes as well (Olson, 1985).

Penguins probably evolved from flying ancestors like the auks and diving petrels, which used their wings for both aerial and subaqueous flight. Stonehouse (1975) argued that these functions are compatible as long as the body weight does not exceed 1 kilogram. Increase in size, comparable to that of the great auk and most penguins facilitates sustained diving and allows wider scope for underwater hunting but leads to excessive wing loading and loss of maneuverability in flight.

The fossil record of penguins was summarized by Simpson (1975, 1979), who also has written a very interesting popular account of both fossil and living genera (1974). Penguins have been restricted to the Southern Hemisphere throughout their history, although some modern genera range as far north as the equator. Even the earliest known fossils from the Upper Eocene of Seymour Island, Antarctica, are highly specialized for underwater flight and provide no evidence of their origin from primitive flying ancestors.

Although no complete or even nearly complete skeletons of fossil genera have been described, individual bones are massive and show distinguishing features that permit unquestioned identification as penguins and serve to differentiate individual genera. A total of 17 fossil genera are recognized, ranging from the Upper Eocene into the Pliocene. Only six genera are recognized today, all of which appear to represent a recent radiation with fossils that are known no earlier than the Pliocene. Zusi (1975) suggests that the flattened bill of modern genera may be more suitable for feeding on plankton as well as fish, in contrast with the narrow bill of at least some early penguins.

The fossil record demonstrates that penguins were more diverse in the Tertiary than in the modern fauna. The largest of all penguins, which was approximately 1½ meters high, is known from the Oligocene and early Miocene. The extinction of these large genera, like that of the large wing-propelled divers of the North Pacific, the Plotopteridae, may be attributed to the spread of pinnipeds and small cetaceans in the Miocene.

SUMMARY

Pterosaurs were the first vertebrates to achieve active, powered flight. They are already taxonomically diverse when they first appear in the fossil record in the late Triassic. The presence of an antorbital fenestra, the parasagittal orientation of the rear limbs, and the mesotarsal ankle joint indicate their close affinities with advanced thecodonts such as *Lagosuchus*. The retention of a hooked fifth metatarsal and a long fifth digit suggests their divergence from thecodonts prior to the emergence of the dinosaur orders.

The earliest known pterosaurs were already highly specialized for active flight, as demonstrated by the presence of a large keeled sternum and an elongate coracoid fused to the scapula. Functionally, the pectoral girdle and humerus are comparable to those of modern flying birds.

Pterosaurs differ markedly from birds in having a membraneous wing that superficially resembles that of bats and is supported by a greatly elongated fourth digit. In contrast with bats, the main membrane did not attach to the rear limb but was a relatively narrow structure. It was reinforced by a radiating pattern of collagenous or cartilaginous fibers.

Two suborders of pterosaurs are recognized, the rhamphorhynchoids, which appeared in the late Triassic and were dominant throughout the Jurassic, and the pterodactyloids, which originated in the late Jurassic and continued to the end of the Cretaceous. The pterodactyloids are distinguished by their very short tails and longer neck and metacarpals. The late Cretaceous pterodactyloid *Quetzalcoatlus* was the largest vertebrate ever to fly, with an estimated wing span of 11 to 12 meters.

The earliest known fossil bird is *Archaeopteryx* from the Upper Jurassic of Germany. Its flight feathers were extremely similar to those of modern birds in their structure and distribution, but the skeleton was almost identical with those of small theropod dinosaurs such as *Compsognathus*. Birds arose from small, bipedal, insectivorous dinosaurs. Feathers may have evolved initially for insulation. The feathered distal ends of the forelimbs may initially have served as airfoils to provide stability and lift in cursorial animals in association with leaping capture of insects. There is no evidence that the antecedents of birds were arboreal gliders.

No birds other than *Archaeopteryx* are known from the Jurassic. All later birds are much more advanced in the loss of the tail and evolution of a keeled sternum, a long coracoid, and a triosseal canal for passage of the tendon of the supracoracoideus muscle. Cretaceous birds include the toothed orders Hesperornithiformes and Ichthyornithiformes as well as incomplete remains of two modern orders.

The ratites, which include the living kiwis, ostriches, rheas, cassowaries, and emus, as well as the extinct moas and elephantbirds, may share a common ancestry among early carinate genera that had retained a palaeognathous palate. All other Cenozoic birds have a neognathous palate and make up a separate, monophyletic group.

Neognathous birds may be divided into two large assemblages, those that live primarily on and over the land and those that live habitually in or over the water. Olson includes the Cuculiformes, Falconiformes, Galliformes, Columbiformes, and Psittaciformes within the primitive land bird assemblage. The Coraciiformes, Strigiformes, Caprimulgiformes, Apodiformes, Piciformes, and Passeriformes constitute the higher land bird assemblage.

The water bird assemblage include the Gruiformes as a basal stock from which may have evolved the Charadriiformes, Anseriformes, and Ciconiiformes. The Pelecaniformes, Gaviiformes, Procellariiformes, and Sphenisciformes are successively more specialized aquatic birds.

Many of the modern bird orders are represented by fossils of modern genera as early as the Miocene, and modern families are known from the late Eocene or early Oligocene. We know fossils of passerines in the late Oligocene, but this order did not become dominant in Europe and North America until the Miocene.

REFERENCES

Ames, P. L. (1971). The morphology of the syrinx in passerine birds. *Bull. Peabody Mus. Nat. Hist., Yale Univ.*, **37**: 1–194.

Andrews, C. W. (1901). On the extinct birds of Patagonia. I. The skull and skeleton of *Phororhacos inflatus* Ameghino. *Trans. Zool. Soc. Lond.*, **15**: 55–86.

Ballmann, P. (1969). Die Vögel aus der altburdigalen Spaltenfüllung von Wintershof (West) bei Eichstätt in Bayern. *Zitteliana*, **1**: 5–60.

Beddard, F. E. (1898). *The Structure and Classification of Birds.* Longmans, London.

Bock, W. (1985). The arboreal theory of the origin of birds. In M. K. Hecht, J. H. Ostrom, G. Viohl, and P. Wellnhofer (eds.), *The Beginning of Birds*, pp. 199–207. Freunde des Jura-Museums Eichstätt, Willibaldsburg, Eichstätt.

Bramwell, C. D., and Whitfield, G. R. (1974). Biomechanics of *Pteranodon. Phil. Trans. Roy. Soc. London, B*, **267**: 503–581.

Brodkorb, P. (1963). A giant flightless bird from the Pleistocene of Florida. *Auk*, **80**: 111–115.

Brodkorb, P. (1964). Catalogue of fossil birds. Part 2 (Anseriformes through Galliformes). *Bull. Florida State Mus., Biol. Sci.*, **8**: 195–335.

Brodkorb, P. (1976). Discovery of a Cretaceous bird apparently ancestral to the Orders Coraciiformes and Piciformes. In S. L. Olson (ed.), *Collected Papers in Avian Paleontology Honoring the 90th Birthday of Alexander Wetmore. Smithsonian Contr. Paleobiol.*, **27**: 67–73.

Brower, J. C. (1983). The aerodynamics of *Pteranodon* and *Nyctosaurus*, two large pterosaurs from the Upper Cretaceous of Kansas. *J. Vert. Paleont.*, **3**(2): 84–124.

Campbell, K. E., and Tonni, E. P. (1983). Size and locomotion in teratorns (Aves: Teratornithidae). *Auk*, **100**: 390–403.

Caple, G. R., Balda, R. T., and Willis, W. R. (1983). The physics of leaping animals and the evolution of pre-flight. *Am. Nat.*, **121**: 455–467.

Cohn, J. M. W. (1968). *The Convergent Flight Mechanism of Swifts (Apodi) and Hummingbirds (Trochili) (Aves).* Ph. D. dissertation, University of Michigan.

Cracraft, J. (1973). Systematics and evolution of the Gruiformes (Class Aves). 3. Phylogeny of the Suborder Grues. *Bull. Am. Mus. Nat. Hist.*, **151**: 1–127.

Cracraft, J. (1974). Continental drift, paleoclimatology, and the evolution and biogeography of birds. *J. Zool.*, **169**: 455–545.

Cracraft, J. (1976). The species of moas (Aves: Dinornithidae). *Smithsonian Contr. Paleobiol.*, **27**: 189–205.

Currie, P. J. (1981). Birds footprints from the Gething Formation (Aptian, Lower Cretaceous) of northeastern British Columbia, Canada. *J. Vert. Paleont.*, **1**(3–4): 257–264.

Elzanowski, A. (1981). Results of the Polish-Mongolian palaeontological expedition. Part IX. Embryonic bird skeletons from the late Cretaceous of Mongolia. *Palaeontologia Polonica*, **42**: 147–179.

Feduccia, A. (1975). Morphology of the bony stapes (columella) in the Passeriformes and related groups: Evolutionary implications. *Univ. Kansas Mus. Nat. Hist., Misc. Publ.*, **63**: 1–34.

Feduccia, A. (1980). *The Age of Birds.* Harvard University Press, Cambridge.

Feduccia, A. (1985). On why the dinosaurs lacked feathers. In M. K. Hecht, J. H. Ostrom, G. Viohl, and P. Wellnhofer (eds.), *The Beginning of Birds*, pp. 75–79. Freunde des Jura-Museums Eichstätt, Willibaldsburg, Eichstätt.

Feduccia, A., and Tordoff, H. B. (1979). Feathers of *Archaeopteryx:* Asymmetric vanes indicate aerodynamic function. *Science*, **203**: 1021–1022.

Gingerich, P. D. (1971). Skull of *Hesperornis* and early evolution of birds. *Nature*, **243**: 70–73.

Hecht, M. K., Ostrom, J. H., Viohl, G., and Wellnhofer, P. (eds.) (1985). *The Beginning of Birds.* Freunde des Jura-Museums Eichstätt, Willibaldsburg, Eichstätt.

Heilmann, G. (1926). *The Origin of Birds.* Appleton, New York.

Hennig, W. (1966). Phylogenetic systematics. *Ann. Rev. Entomol.*, **10**: 97–116.

Houde, P., and Olson, S. L. (1981). Paleognathous carinate birds from the Early Tertiary of North America. *Science*, **214**: 1236–1237.

Howard, H. (1973). Fossil Anseriformes, corrections and additions. In J. Delacour, *Waterfowl of the World* (2d ed.), Vol. 4, pp. 371–378. Country Life, London.

Huxley, T. H. (1867). On the classification of birds and on the taxonomic value of the modifications of certain of the cranial bones observable in that class. *Proc. Zool. Soc. Lond.*, **1867**: 415–472.

Jensen, J. A. (1981). Another look at *Archaeopteryx* as the "oldest" bird. *Encyclia*, **58**: 109–128.

Jollie, M. (1976–1977). A contribution to the morphology and phylogeny of the Falconiformes. *Evol. Theory*, **1**: 285–298; **2**: 285–300; **3**: 1–141.

Kurochkin, E. N. (1981). Novyye predstaviteli i evolyutsiya dvukh semeystv arkhaichnykh zhuravleobraznykh v Yevrazii (New representatives and evolution of two archaic gruiform families in Eurasia). *Sovmestn. Sov.-Mong. Paleontol. Eksped., Trudy,* **15**: 59–85.

Kurochkin, E. N. (1982). Novyy otryad ptits iz nizhnego mela Mongolii (A new order of birds from the Lower Cretaceous of Mongolia). *Dokl. Akad. Nauk SSSR,* **262**: 452–455.

Kurochkin, E. N., and Lungu, A. N. (1970). A new ostrich from the Middle Sarmatian of Moldavia. *Paleont. Jour.,* **1970**: 103–111. (English translation of *Paleont. Zhurnal,* **1**: 118–126.)

Kurochkin, E. N., and Mayo, N. (1973). Las lechuzas gigantes Pleistoceno Superior de Cuba (The giant owls of the Upper Pleistocene of Cuba). *Acad. Cienc. Cuba, Inst. Geol., Actas,* **3**: 56–60.

Langston, W. (1981). Pterosaurs. *Sci. Amer.,* **244**(2): 122–136.

Lucas, A. M., and Stettenheim, P. R. (1972). Avian Anatomy. Integument. Part 1. *U.S. Dept. Agriculture, Handbook,* **362**: 1–340.

Marsh, O. C. (1880). *Odontornithes: A Monograph on the Extinct Toothed Birds of North America.* United States Geological Exploration of the Fortieth Parallel. Clarence King, geologist-in-charge. Government Printing Office, Washington.

Martin, L. D. (1983). The origin and early radiation of birds. In A. H. Brush and G. A. Clark, Jr. (eds.), *Perspectives in Ornithology,* pp. 291–338. Cambridge University Press, Cambridge.

Martin, L. D. (1984). A new hesperornithid and the relationships of the Mesozoic birds. *Trans. Kansas Acad. Sci.,* **87**: 141–150.

Martin, L. D., and Tate, J., Jr. (1976). The skeleton of *Baptornis advenus* (Aves: Hesperornithiformes). In S. L. Olson (ed.), *Collected Papers in Avian Paleontology Honoring the 90th Birthday of Alexander Wetmore. Smithsonian Contr. Paleobiol.,* **27**: 35–66.

Martin, L. D., Stewart, J. D., and Whetstone, K. N. (1980). The origin of birds: Structure of the tarsus and teeth. *Auk,* **97**: 86–93.

Matthew, W. D., and Granger, W. (1917). The skeleton of *Diatryma,* a gigantic bird from the Lower Eocene of Wyoming. *Bull. Am. Mus. Nat. Hist.,* **37**: 307–326.

Mourer-Chauvire, C. (1981). Première indication de la présence de phorusracidés, famille d'oiseaux géants d'Amerique du Sud, dans le Tertiaire Européen: *Ameghinornis* nov. gen. (Aves, Ralliformes) des Phosphorites du Quercy, France. *Géobios,* **14**: 637–647.

Mourer-Chauvire, C. (1982). Les oiseaux fossiles des Phosphorites du Quercy (Eocène Supérieur a Oligocène Supérieur): Implications paléobiogeographiques. *Géobios, Mem. Spec.,* **6**: 413–426.

Olson, S. L. (1977a). A synopsis of the fossil Rallidae. In S. D. Ripley, *Rails of the World: A Monograph of the Family Rallidae,* pp. 509–525. David R. Godine, Boston.

Olson, S. L. (1977b). A Lower Eocene frigatebird from the Green River Formation of Wyoming (Pelecaniformes: Fregatidae). *Smithsonian Contr. Paleobiol.,* **35**: 1–33.

Olson, S. L. (1979). Multiple origins of the Ciconiiformes. *Proc. Colonial Waterbird Group,* **1978**: 165–170.

Olson, S. L. (1980). A new genus of penguin-like pelecaniform bird from the Oligocene of Washington (Pelecaniformes: Plotopteridae). *Nat. Hist. Mus. Los Angeles County, Contr. Sci.,* **330**: 51–57.

Olson, S. L. (1982). A critique of Cracraft's classification of birds. *Auk,* **99**(4): 733–739.

Olson, S. L. (1985). The fossil record of birds. In D. Farner, J. King, and K. Parkes (eds.), *Avian Biology,* Vol. 8, pp. 79–238. Academic Press, Orlando. Reviewed by Hoffman, *Auk,* **104**(4), 1986.

Olson, S. L., and Feduccia, A. (1980a). Relationships and evolution of flamingos (Aves: Phoenicopteridae). *Smithsonian Contr. Zool.,* **316**: 1–73.

Olson, S. L., and Feduccia, A. (1980b). *Presbyornis* and the origin of the Anseriformes (Aves: Charadriomorphae). *Smithsonian Contr. Zool.,* **323**: 1–24.

Olson, S. L., and Hasegawa, Y. (1979). Fossil counterparts of giant penguins from the North Pacific. *Science,* **206**: 688–689.

Olson, S. L., and Wetmore, A. (1976). Preliminary diagnoses of two extraordinary new genera of birds from Pleistocene deposits in the Hawaiian Islands. *Proc. Biol. Soc. Wash.,* **89**: 247–258.

Ostrom, J. H. (1974). *Archaeopteryx* and the origin of flight. *Quart. Rev. Biol.,* **49**: 27–47.

Ostrom, J. H. (1975a). The origin of birds. *Ann. Rev. Earth Planet. Sci.,* **3**: 55–77.

Ostrom, J. H. (1975b). On the origin of *Archaeopteryx* and the ancestry of birds. *Coll. Int. C.N.R.S,* **218**: 519–532.

Ostrom, J. H. (1976a). *Archaeopteryx* and the origin of birds. *Biol. J. Linn. Soc.,* **8**: 91–182.

Ostrom, J. H. (1976b). Some hypothetical anatomical stages in the evolution of avian flight. In S. L. Olson (ed.), *Collected Papers in Avian Paleontology Honoring the 90th Birthday of Alexander Wetmore. Smithsonian Contr. Paleobiol.,* **27**: 1–21.

Ostrom, J. H. (1978). The osteology of *Compsognathus longipes* Wagner. *Zitteliana, Abh. Bayer. Staatss. Paläont. und hist. Geol.,* **4**: 73–118.

Ostrom, J. H. (1985). The meaning of *Archaeopteryx.* In M. K. Hecht, J. H. Ostrom, G. Viohl, and P. Wellnhofer (eds.), *The Beginning of Birds,* pp. 161–176. Freunde des Jura-Museums Eichstätt, Willibaldsburg, Eichstätt.

Ostrom, J. H. (1986). The Jurassic "bird" *Laopteryx priscus* re-examined. In K. M. Flanagan and J. A. Lillegraven (eds.), *Vertebrates, Phylogeny, and Philosophy. Cont. Geol. Univ. Wyo. Spec. Pap. 3,* Laramie.

Padian, K. (1983a). Description and reconstruction of new material of *Dimorphodon macronyx* (Buckland) (Pterosauria: Ramphorhynchoidea) in the Yale Peabody Museum. *Postilla, Yale Peabody Museum,* **189**: 1–44.

Padian, K. (1983b). A functional analysis of flying and walking in pterosaurs. *Paleobiology,* **9**(3): 218–239.

Padian, K. (1984). The origin of pterosaurs. In W.-E. Reif and F. Westphal (eds.), *Third Symposium on Terrestrial Mesozoic Ecosystems Proceedings,* pp. 163–168. Attempto Verlag, Tübingen.

Padian, K. (1985). The origins and aerodynamics of flight in extinct vertebrates. *Palaeontology,* **28**(3): 413–433.

Parkes, K. C. (1975). Special review. *Auk,* **92**: 818–830.

Peters, D. S. (1983). Die "Schnepfenralle" *Rynchaeites messelensis* Wittich 1898 ist ein Ibis. *J. Ornithol.,* **124**: 1–27.

Raikow, R. J. (1978). Appendicular myology and relationships of the New World nine-primaried oscines (Aves: Passeriformes). *Bull. Carnegie Mus. Nat. Hist.,* **7**: 1–52.

Rich, P. V. (1976). The history of birds on the island continent Australia. *Proc. 16th Intl. Ornithol. Congr.:* 53–65.

Rich, P. V. (1980). "New World vultures" with Old World

affinities? *Contr. Vert. Evol.*, **5**: 1–115.
Rich, P. V., and Bohaska, D. J. (1981). The Ogygoptyngidae, a new family of owls from the Paleocene of North America. *Alcheringa*, **5**: 95–102.
Rich, P. V., and McEvey, A. R. (1977). A new owlet-nightjar from the early to mid-Miocene of eastern New South Wales. *Mem. Natl. Mus. Victoria*, **38**: 247–253.
Romer, A. S. (1966). *Vertebrate Paleontology* (3d. ed.). University of Chicago Press, Chicago.
Sauer, E. G. F. (1979). A Miocene ostrich from Anatolia. *Ibis*, **121**: 494–501.
Schmidt-Nielsen, K. (1971). How birds breathe. *Sci. Am.*, **225**(6): 72–79.
Sharov, A. G. (1971). New flying reptiles from the Mesozoic of Kazakhstan and Kirgizia. *Akad. Nauk SSSR Trudy Paleont. Inst.*, **130**: 104–113. (In Russian.)
Sibley, C. G. (1970). A comparative study of the egg-white proteins of passerine birds. *Bull. Peabody Mus. Nat. Hist., Yale Univ.*, **32**: 1–131B.
Sibley, C. G., and Ahlquist, J. E. (1972). A comparative study of the egg white proteins of non-passerine birds. *Bull. Peabody Mus. Nat. Hist., Yale Univ.*, **39**: 1–276.
Simpson, G. G. (1974). *Penguins, Past and Present, Here and There.* Yale University Press, New Haven.
Simpson, G. G. (1975). Fossil penguins. In B. Stonehouse (ed.), *The Biology of Penguins*, pp. 19–41. Macmillan, London.
Simpson, G. G. (1979). Tertiary penguins from the Duinefontein site, Cape Province, South Africa. *Ann. S. Afr. Mus.*, **79**: 1–7.
Soemmering, S. T. von. (1817). Über einen *Ornithocephalus brevirostris* der Vorwelt. *Denkschrift Kgl. Bayer, Akad. Wiss., math.-phys. Cl.*, **6**: 89–104.
Stahl, B. J. (1985). *Vertebrate History. Problems in Evolution.* Dover, New York.
Stonehouse, B. (1975). Introduction: The Spheniscidae. In B. Stonehouse (ed.), *The Biology of Penguins*, pp. 1–15. Macmillan, London.
Storer, R. W. (1956). The fossil loon, *Columboides minutus. Condor*, **58**: 413–426.
Thulborn, R. A. 1984. The avian relationships of *Archaeopteryx*, and the origin of birds. *Zool. J. Linn. Soc.*, **82**: 119–158.
Walker, A. D. (1972). New light on the origin of birds and crocodiles. *Nature*, **237**: 257–263.
Walker, C. A. (1981). New subclass of birds from the Cretaceous of South America. *Nature* (London), **292**: 51–53.
Wellnhofer, P. (1970). Die Pterodactyloidea (Pterosauria) der Oberjura-Plattenkalke Süddeutschlands. *Abh. Bayer. Akad. Wiss., math.-naturwiss. Klasse, N. F.*, **141**: 1–133.
Wellnhofer, P. (1974). Das fünfte Skelettexemplar von *Archaeopteryx. Palaeontographica, Abt. A*, **147**: 169–216.
Wellnhofer, P. (1975). Die Rhamphorhynchoidea der Oberjura Plattenkalke Süddeutschlands. *Palaeontographica Abt. A*, **148**: 1–33, 132–186; **149**: 1–30.
Wellnhofer, P. (1978). Pterosauria. In P. Wellnhofer (ed.), *Handbuch der Paläoherpetologie.* Part 19; pp. 1–82. Gustav Fischer Verlag, Stuttgart.
Welty, J. C. (1982). *The Life of Birds.* Saunders, Philadelphia.
Whetstone, K. N. (1983). Braincase of Mesozoic birds. I. New preparation of the "London" *Archaeopteryx. J. Vert. Paleont.*, **2**: 439–452.
Wild, R. (1978). Die Flugsaurier (Reptilia, Pterosauria) aus der Oberen Trias von Cene bei Bergamo, Italien. *Boll. Soc. Pal. Ital.*, **17**: 176–256.
Wild, R. (1984a). Flugsaurier aus der Obertrias von Italien. *Naturwissenschaften*, **71**: 1–11.
Wild, R. (1984b). A new pterosaur (Reptilia, Pterosauria) from the Upper Triassic (Norian) of Friuli, Italy. *Gortania-Atti Museo Fruilano di Storia Naturale*, **5**: 45–62.
Wiley, E. O. (1981). *Phylogenetics.* Wiley, New York.
Young, J. Z. (1981). *The Life of Vertebrates* (3d ed.). Clarendon Press, Oxford.
Zusi, R. L. (1975). An interpretation of skull structure in penguins. In B. Stonehouse (ed.), *The Biology of Penguins*, pp. 59–84. Macmillan, London.
Zusi, R. L., and Storer, R. W. (1969). Osteology and myology of the head and neck of the Pied-billed grebes (*Podilymbus*). *Misc. Publ. Mus. Zool. Univ. Michigan*, **139**: 1–49.

CHAPTER XVII

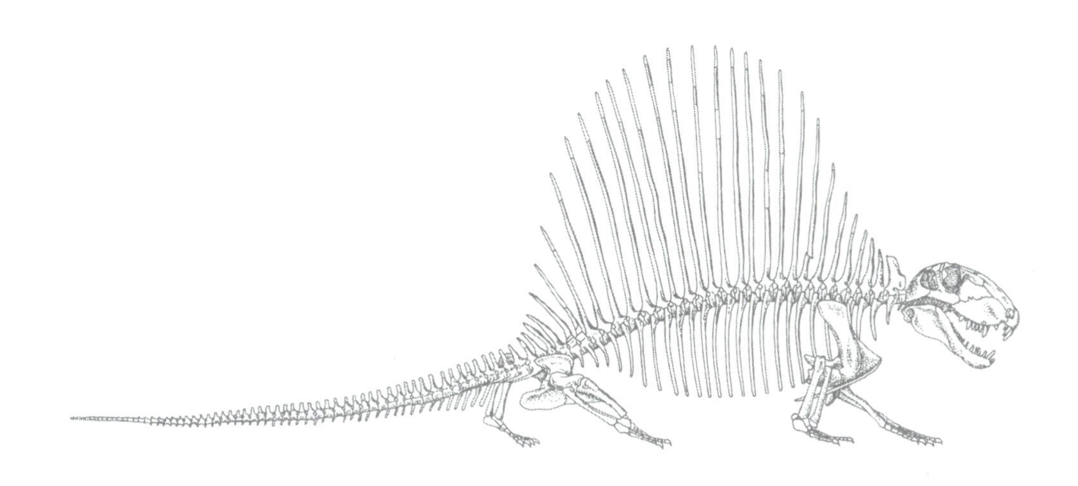

The Origin of Mammals

Mammals have been the dominant terrestrial vertebrates for the past 65 million years. Their history, however, extends over a much longer period. Primitive mammals are recognized as early as the late Triassic, more than 200 million years ago, and their ancestors are distinguishable from other amniotes as long ago as the Carboniferous.

The ancestors of mammals are identified from the same horizon and locality as the earliest conventional reptile, *Hylonomus,* in the early Pennsylvanian of Joggins, Nova Scotia (Carroll, 1964). The sequence from the early amniotes to the early mammals is the most fully documented of the major transitions in vertebrate evolution. The entire skeleton was modified, as was the soft anatomy, behavior, and physiology down to the level of cellular metabolism. Many of these changes are demonstrable, either directly or indirectly, through the fossil record.

The transition begins with animals that broadly resembled modern lizards in their skeletal anatomy and probably had a typically reptilian physiology. We can trace the transformation over a period of 150 million years from small, cold-blooded, scaly reptiles to tiny, warm-blooded, furry mammals.

The ancestors of mammals are usually classified in the subclass Synapsida within the Reptilia. However, these early genera can be recognized as being members of the same monophyletic assemblage as living mammals. One might formalize this relationship by including the synapsids within the Mammalia or by recognizing a single larger taxonomic grouping for both. This procedure is not followed here because it obscures the very important structural and physiological differences that distinguish mammals, as that group is usually recognized, from their ancestors that retain features generally regarded as being typical of reptiles.

The synapsids are included in two successive orders, the Pelycosauria, which are known from the base of the Pennsylvanian into the Upper Permian, and the Therapsida, which appears in the middle Permian and extends into the Mid-Jurassic. Pelycosaurs and therapsids represent sequential radiations, each of which originated from a single ancestral stock (Figure 17-1). Pelycosaurs retain many of the skeletal features of primitive reptiles and show little evidence of the attainment of mammalian attributes. Most skeletal features that are common to mammals evolved among the therapsids.

PELYCOSAURS

THE ANATOMY OF EARLY MAMMAL-LIKE REPTILES

Pelycosaurs are recognized by fragmentary remains as early as the Lower Pennsylvanian (Reisz, 1972). By the Upper Pennsylvanian, they constituted approximately 50 percent of the described reptilian genera. *Archaeothyris,* from the middle Pennsylvanian (Figure 17-2), provides the earliest evidence of the configuration of the pelycosaur skull and gives some information regarding body proportions and details of vertebral and limb structure.

In their general appearance, early pelycosaurs would have resembled large living lizards such as iguanids and

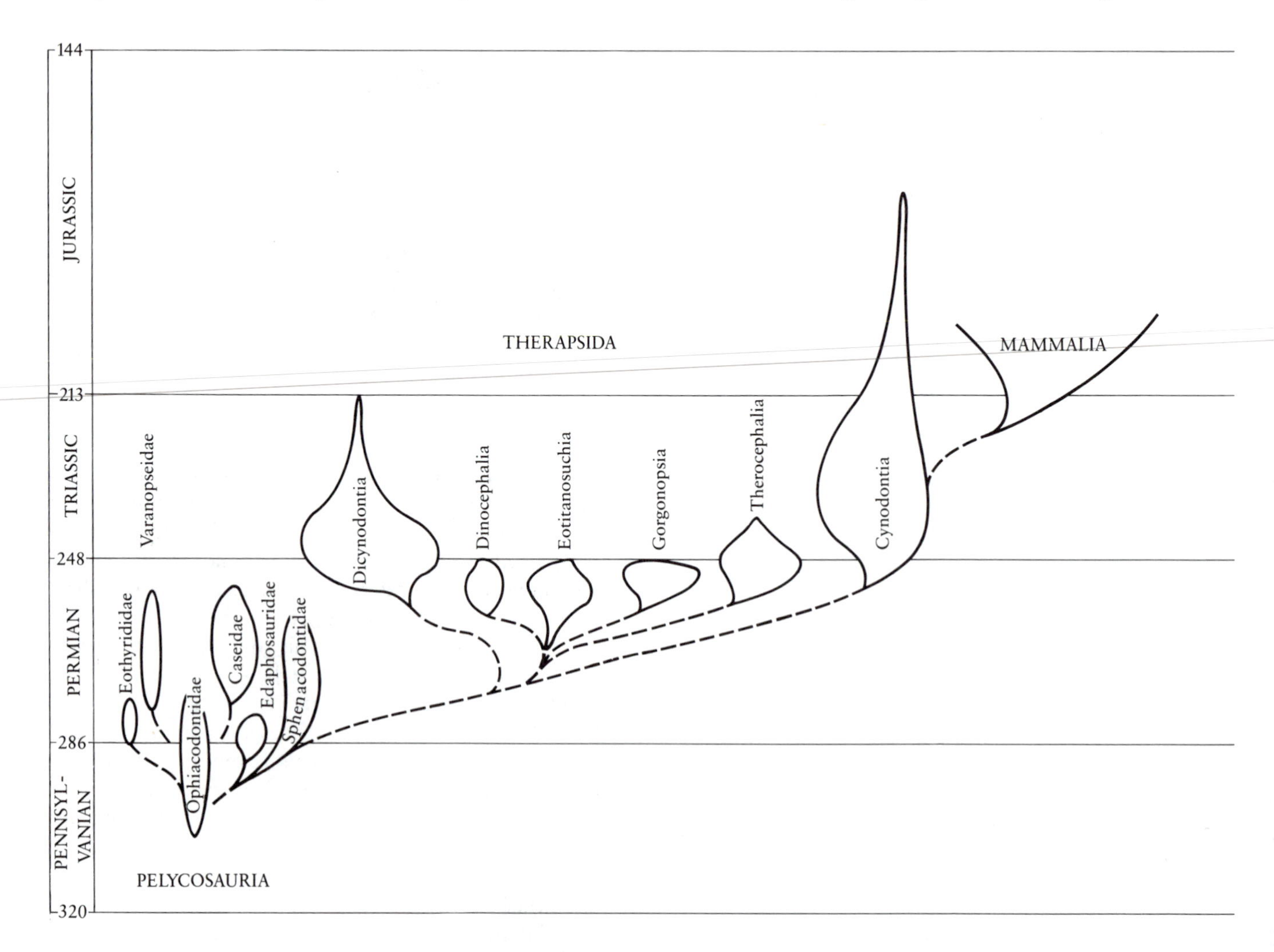

Figure 17-1. STRATIGRAPHIC RANGES OF THE MAJOR GROUPS OF MAMMAL-LIKE REPTILES.

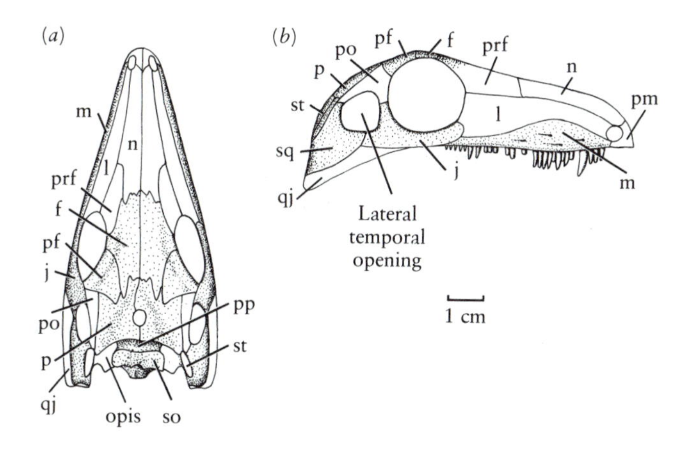

Figure 17-2. *ARCHAEOTHYRIS.* This is the oldest-known pelycosaur in which the skull can be restored, from the Middle Pennsylvanian. (*a*) Dorsal and (*b*) lateral views, ×½. The skull of this mammal-like reptile is distinguished from that of other early amniotes by the presence of a lateral temporal opening that is bordered ventrally by the squamosal and jugal. Unshaded areas are reconstructed. Abbreviations as in Figure 8-3. *From Reisz, 1972.*

varanids, although the pelycosaur's limbs were relatively shorter. Middle Pennsylvanian pelycosaurs were at least twice the length of contemporary protorothyrids, and they show progressive size increase into the early Permian.

The original divergence of the pelycosaurs from among the primitive amniotes may have resulted from an adaptation toward feeding on large prey. Not only is the entire body much larger, but the size of the skull increases disproportionately. The canine teeth, which are present in the most primitive amniotes, are accentuated, and the remaining teeth are also enlarged relative to those of the insectivorous protorothyrids. In relation to feeding on larger prey, the postorbital region of the skull is shorter but higher than in other primitive amniotes. There is a shorter area for the adductor muscles to attach to the lower jaw, but the individual muscle fibers were longer. Both features permit a wider gape and a rapid, but less powerful, jaw-closing action. Pelycosaurs were the first carnivorous (as opposed to insectivorous) amniotes.

Pelycosaurs also appear to be the first major group to diverge from the primitive amniote stock, as can be seen by the retention of several primitive features that are lost in all other groups of primitive reptiles. These features include the presence of two, rather than a single, coronoid bone in the lower jaw and the large size of the medial centrale of the tarsus.

On the other hand, even the earliest pelycosaurs are advanced in a number of cranial features that characterize later synapsids and provide the basis for the origin of mammalian anatomy. The most important of these features is a lateral temporal opening where the postorbital, squamosal, and jugal met in early anapsids. In contrast with early diapsids, the quadratojugal does not form the ventral margin of this opening in primitive pelycosaurs. In contrast with protorothyrids and early diapsids, the supratemporal and postorbital meet above the temporal opening and separate the squamosal and parietal superficially. The arrangement of bones is somewhat similar to that seen in anthracosaurian labyrinthodonts, which may be close to the ancestry of reptiles. The pattern in pelycosaurs might represent the primitive condition for amniotes, but the great posterior extension of the postorbital can also be interpreted as a specialized feature that was developed to strengthen the skull as the temporal opening developed. All the bones of the skull seen in other early amniotes are retained, but the postparietals fuse at the midline to form a single median element.

The skull of early pelycosaurs is also distinguished by the configuration of the occiput. It is essentially vertical in other groups of early amniotes. In pelycosaurs, it slopes posteriorly from the skull table to the area of the jaw articulation. As in protorothyrids, the braincase in primitive pelycosaurs is not solidly integrated with the dermal bones that surround the occiput. The middle Pennsylvanian genus *Archaeothyris* retains a discrete supraoccipital that was not suturally attached to the skull table or firmly connected to either the tabular or the otic capsule. The otic capsule is small and does not contact the cheek. This pattern is retained in the primitive Lower Permian genus *Ophiacodon* (see Figure 10-3). Other pelycosaurs integrate the bones of the occiput into a platelike structure that incorporates the supraoccipital, broad paroccipital processes, and the dermal bones of the margin of the skull (Figure 17-3). The stapes, which may have assisted in the support of the braincase in early pelycosaurs, remains a large element throughout the group, although it no longer served a structural role after the solidification of the occiput.

Postcranially, primitive pelycosaurs differ from the early protorothyrids primarily in features that may be associated with their greater body weight. The limb bones

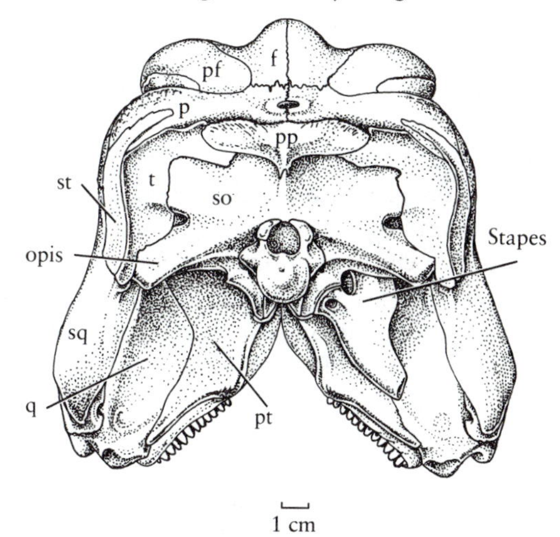

Figure 17-3. OCCIPUT OF THE LOWER PERMIAN PELYCOSAUR *DIMETRODON.* This genus shows the platelike nature of the supraoccipital, which is fused to the opisthotic and suturally attached to the tabular. The stapes is a massive element. Abbreviations as in Figure 8-3. *From Romer and Price, 1940.*

are more massive, with large and more complex areas for muscle attachment, and the metapodials and phalanges are relatively shorter (Carroll, 1986).

Like the protorothyrids, the pelycosaurs retain narrow neural arches, with the zygapophyses close to the midline. In primitive species, the neural spines are low and triangular, but they become greatly elongated in several families (Figure 17-4) and the transverse processes also become longer. These changes may be related to an early stage in reorganization of the trunk musculature. In mammals, the elaboration of axial musculature above and below the transverse processes helps limit lateral undulation of the trunk and support the vertebral column as an arch above the girdles. Lateral undulation of the column remained an important attribute of locomotion in most synapsids, but the elongation of the transverse processes may indicate more active support of the column on the girdles than was the case in other early amniotes. The number of trunk vertebrae ranges from 24 to 27.

In pelycosaurs, we can identify three centers of ossification in the scapulocoracoid: an anterior and a posterior coracoid as well as the scapula (Figure 17-4*d*). In adult protorothyrids, the scapulocoracoid is ossified without a trace of subdivision, but there are three distinct elements in young individuals of several Pennsylvanian genera. Coossification appears to be progressively delayed in pelycosaurs so that sutures are evident between the elements in more mature specimens.

Primitive pelycosaurs have two pairs of sacral ribs, and the blade of the ilium is narrow and directed posteriorly. Several advanced pelycosaur groups have one or two additional pairs of sacral ribs and a broad iliac blade to support their greater body size.

The configuration of the articulating surfaces of the girdles and limbs of all pelycosaurs indicate a sprawling posture like that of the most primitive tetrapods, which were discussed in Chapter 9.

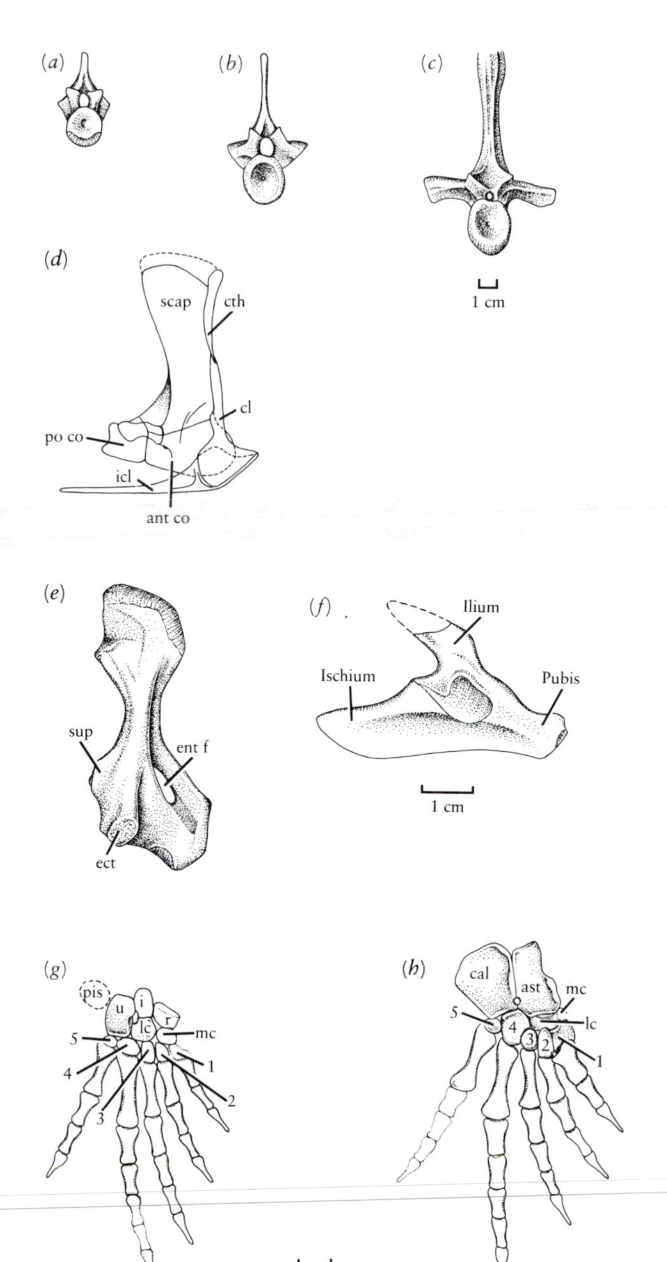

Figure 17-4. POSTCRANIAL ELEMENTS OF PELYCOSAURS. Anterior views of the trunk vertebrae of (*a*) *Protoclepsydrops*, (*b*) *Archaeothyris*, and (*c*) *Dimetrodon* that show the increasing length of the transverse processes from the early Pennsylvanian to the early Permian. (*a*) *From Reisz, 1972.* (*b*) *From Carroll, Belt, Dineley, Baird, and McGregor, 1972.* (*c*) *From Romer and Price, 1940.* (*d*) Shoulder girdle of the Lower Permian pelycosaur *Dimetrodon*, showing the three-fold division of the scapulocoracoid. *From Romer and Price, 1940.* (*e*) Humerus of *Archaeothyris*. The presence of a prominent supinator process and an entepicondylar ridge are typical of primitive pelycosaurs. As in all early reptiles, there is a large entepicondylar foramen. In some pelycosaurs, the supinator process extends distally to surround an entepicondylar foramen. (*f*) Pelvis of *Archaeothyris*. *From Reisz, 1972.* (*g*) Manus and (*h*) pes of the sphenacodont pelycosaur *Haptodus*. *From Currie, 1977.* Abbreviations as follows: ant co, anterior coracoid; ast, astragalus; cal, calcaneum; cl, clavicle; cth, cleithrum; ect, ectepicondylar ridge; ent f, entepicondylar foramen; i, intermedium; icl, interclavicle; lc, lateral centrale; mc, medial centrale; pis, pisiform; po co, posterior coracoid; r, radiale; scap, scapula; sup, supinator process; u, ulnare; 1–5, distal carpals and tarsals.

THE DIVERSITY OF PELYCOSAURS

By the early Permian, pelycosaurs made up 70 percent of the known amniote genera and had diversified into a number of distinct families. Unfortunately, remains from the Carboniferous, when these groups underwent initial differentiation, are still poorly known, and specific interrelationships of these families continue to be in dispute.

Romer and Price (1940) provided the most comprehensive anatomical and taxonomic review of the pelycosaurs. They recognized three suborders: the Ophiacodontia, which include a primitive ancestral assemblage and large piscivorous genera from the Lower Permian; the Sphenacodontia, which include large carnivorous genera; and the Edaphosauria, which consist of two groups of herbivores. In their recent taxonomic revisions, Reisz (1980, 1986) and Brinkman and Eberth (1983) questioned the validity of these suborders but recognize most of the family units that Romer and Price had described.

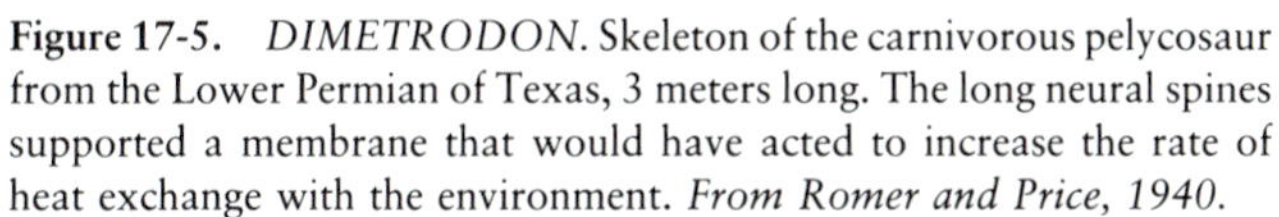

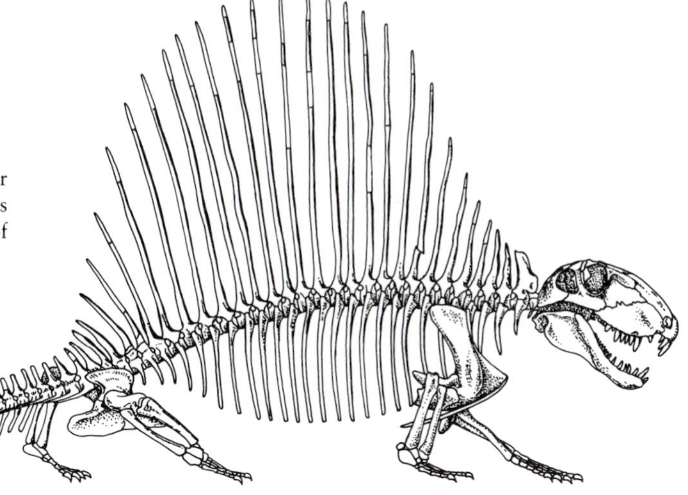

Figure 17-5. *DIMETRODON.* Skeleton of the carnivorous pelycosaur from the Lower Permian of Texas, 3 meters long. The long neural spines supported a membrane that would have acted to increase the rate of heat exchange with the environment. *From Romer and Price, 1940.*

Ophiacodontidae

The oldest-known and most primitive pelycosaurs are included in the family Ophiacodontidae. Aside from primitive features, this family is characterized by a long narrow snout and relatively low skull table. It is represented by several genera in the Pennsylvanian and remains common in the Lower Permian. The largest and best-known genus is *Ophiacodon,* which was more than 4 meters long. The carpals and tarsals were poorly ossified, which led Romer and Price to suggest that this group was semiaquatic and that the long narrow snout was an adaptation to feeding on fish.

Sphenacodontidae

Phylogenetically, the most important of the pelycosaur families are the sphenacodonts. They appear in the middle or late Pennsylvanian and are the dominant terrestrial carnivores in the Lower Permian. At some stage in their history they gave rise to the therapsids. The best-known sphenacodonts are the Lower Permian genera *Dimetrodon* (Figure 17-5) and *Sphenacodon,* which exceed 3 meters in length. The skull is typically much deeper than in the ophiacodonts and the canine teeth are greatly enlarged. To support the dentition, the maxilla extends far up on the snout and separates the lacrimal bone from the margin of the external nares.

The jaw articulation is well below the level of the tooth row in sphenacodonts, the articular is greatly enlarged, and the articulating surface extends far down on the medial face of the lower jaw. A feature that is unique to sphenacodonts is the presence of a reflected lamina of the angular (Figure 17-6), which can be associated with the development of the middle ear in later synapsids and mammals.

Because of their position close to the origin of the therapsids, the feeding apparatus of the sphenacodontids is of special interest. As in the case of the early diapsids, the presence of a temporal opening does not appear to have had much effect on the pattern of the jaw muscles in pelycosaurs, although it is highly important in therapsids. The opening in the cheek probably developed in relationship to the pattern of mechanical stresses on the back of the skull.

When compared to other early amniotes, the changes in the shape of the back of the skull and the configuration of the adductor chamber suggest that the jaw muscles

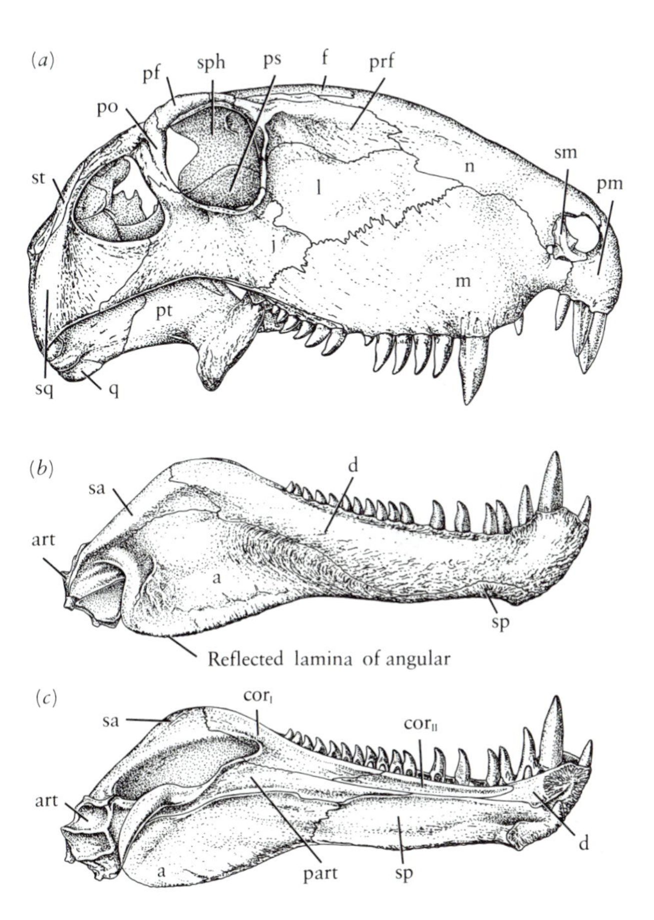

Figure 17-6. (*a*) Skull of *Dimetrodon,* in lateral view. (*b*) Lateral and (*c*) medial views of the lower jaw, $\times\frac{1}{4}$. Abbreviations as in Figure 8-3. *From Romer and Price, 1940.*

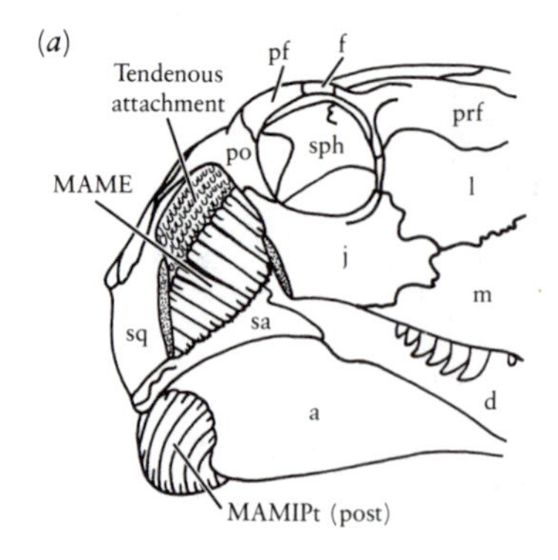

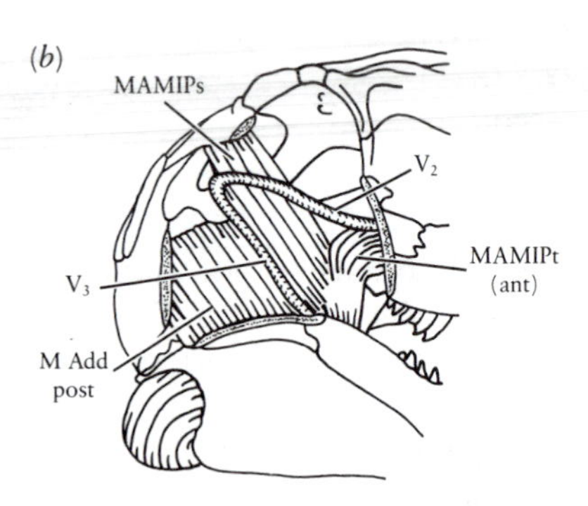

Figure 17-7. RECONSTRUCTION OF THE JAW MUSCULATURE OF THE CARNIVOROUS PELYCOSAUR *DIMETRODON.* (*a*) Superficial view and (*b*) deeper view. The anteroventral orientation of the major adductor muscles may have acted to resist anteriorly directed forces produced by struggling prey. Abbreviations as in Figure 8-3, plus: M Add post, adductor mandibulae posterior; MAME, adductor mandibulae externus; MAMIPs, adductor mandibulae internus, pseudotemporalis; MAMIPt (ant), adductor mandibulae internus, pterygoideus (anterior); MAMIPt (post), adductor mandibulae internus, pterygoideus (posterior); V_2 maxillary branch of Vth nerve; V_3 mandibular branch of Vth nerve. *From Barghusen, 1973.*

were differently oriented. Instead of extending vertically from the skull roof to the dorsal margin of the jaw, they were angled anteriorly (Figure 17-7). Barghusen (1973) argued that this orientation would serve to better resist struggling prey that might dislocate the lower jaw anteriorly. A similarly directed force would also result if the pelycosaur were attempting to tear chunks of meat from large prey.

The jaw articulation lies well below the level of the tooth row, so that the jaw muscles are attached to the sloping dorsal margin of the jaw, which rises to form a prominent coronoid process. The heightening of the coronoid process provides a longer lever arm for closing the jaw that is more nearly perpendicular to the fibers of the adductor muscles.

Among the larger pelycosaurs, sphenacodonts have the most gracile limbs. All sphenacodonts have tall neural spines, but they are spectacularly elongated in the genus *Dimetrodon.* Their function will be discussed with a consideration of the biology of the pelycosaurs.

The genus *Haptodus* (Figure 17-8) from the late Pennsylvanian and early Permian shows the initiation of structural features that are elaborated in the later sphenacodonts, while retaining many primitive features of the early ophiacodonts.

Edaphosauridae

Among the most striking of pelycosaurs is the genus *Edaphosaurus,* which is known from the late Pennsylvanian and early Permian (Figure 17-9). Like *Dimetrodon,* it has greatly elongated neural spines, but they are further characterized by the presence of short, transversely directed cross bars. The skull is markedly different from that of sphenacodonts in being very short and broad. The teeth

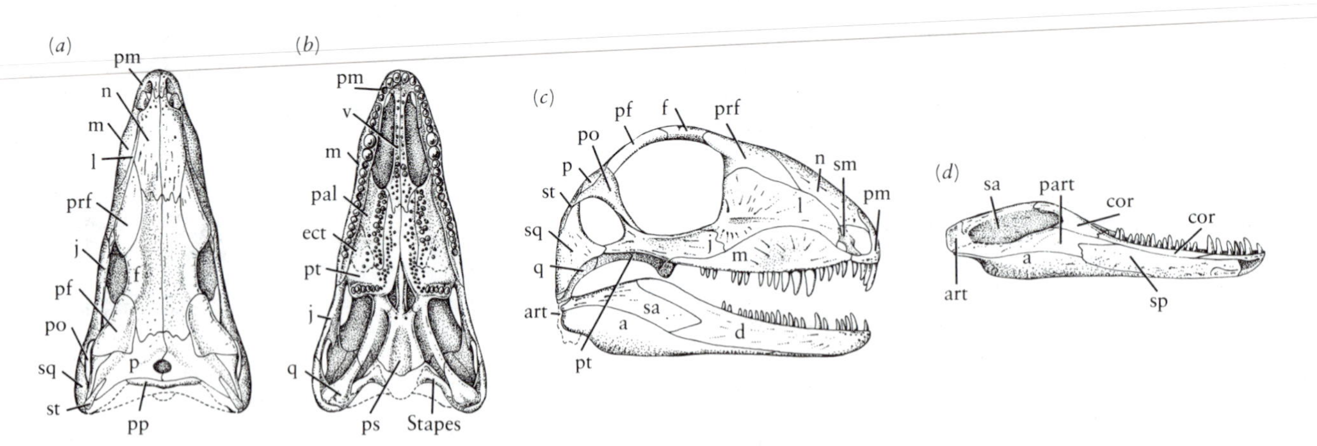

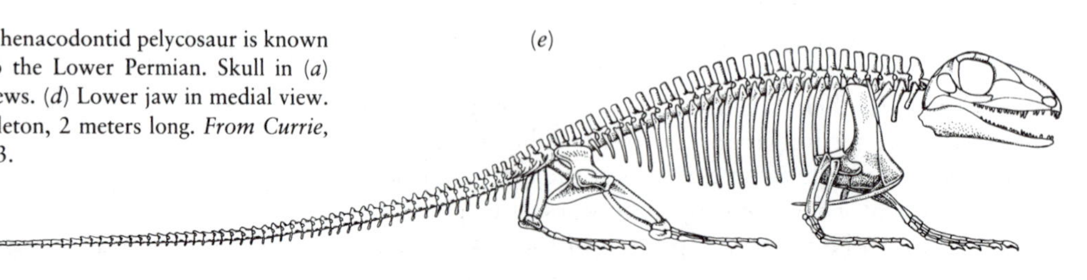

Figure 17-8. *HAPTODUS.* This sphenacodontid pelycosaur is known from the Upper Pennsylvanian into the Lower Permian. Skull in (*a*) dorsal, (*b*) palatal, and (*c*) lateral views. (*d*) Lower jaw in medial view. (*a* to *d*) *From Currie, 1979.* (*e*) Skeleton, 2 meters long. *From Currie, 1977.* Abbreviations as in Figure 8-3.

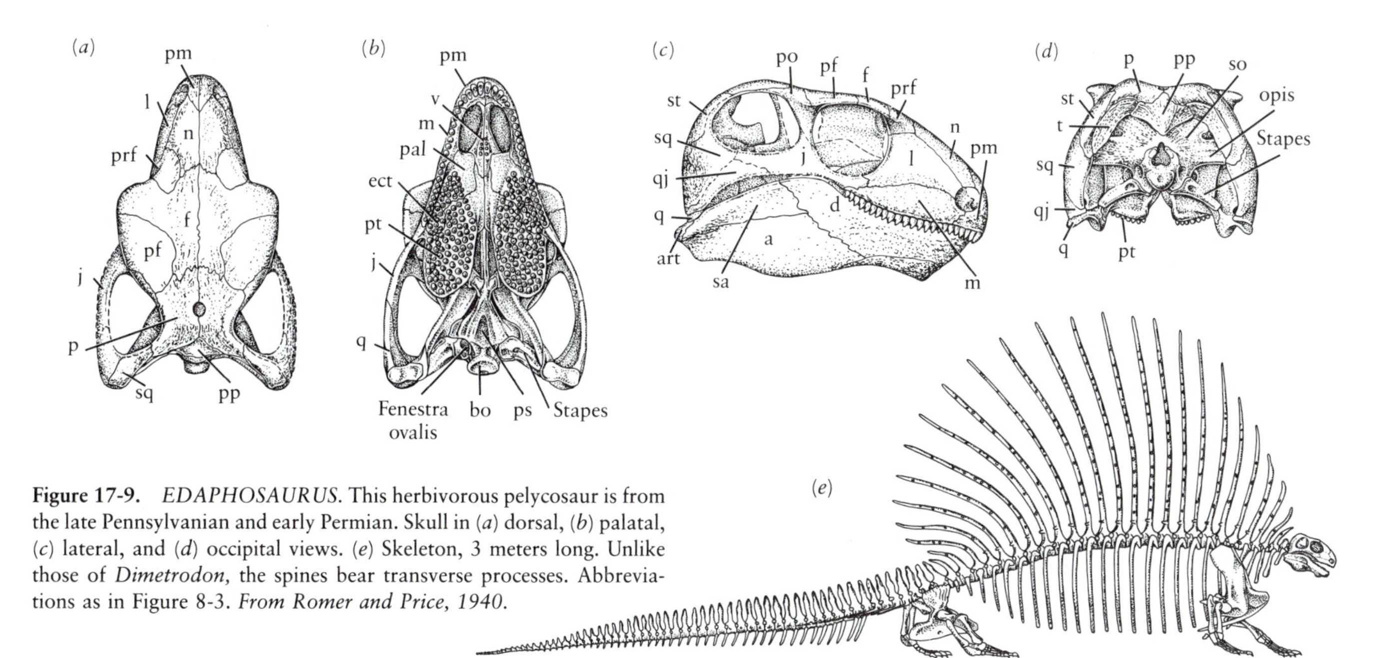

Figure 17-9. *EDAPHOSAURUS.* This herbivorous pelycosaur is from the late Pennsylvanian and early Permian. Skull in (*a*) dorsal, (*b*) palatal, (*c*) lateral, and (*d*) occipital views. (*e*) Skeleton, 3 meters long. Unlike those of *Dimetrodon,* the spines bear transverse processes. Abbreviations as in Figure 8-3. *From Romer and Price, 1940.*

are not long piercing and slashing structures but short pegs of nearly uniform length. The palate and inside surface of the lower jaw are covered with additional teeth of similar size that form a large biting surface. The individual teeth are not expanded or otherwise specialized for crushing hard food and Romer and Price suggest that they may have fed on relatively soft plant material. The limbs are shorter than those of sphenacodonts, and the rib cage is long and laterally expanded as one might expect in a herbivore.

A newly described edaphosaur from the Upper Pennsylvanian, *Ianthosaurus* (Figure 17-10), lacks tooth plates on the palate and lower jaw, and the marginal dentition resembles that of primitive sphenacodonts. Reisz and Berman (1986) argue that *Ianthosaurus* may demonstrate a close link between these families. The separation of these two groups must have occurred before the evolution of the distinctive sphenacodont reflected lamina of the angular that is not observed in the edaphosaurs.

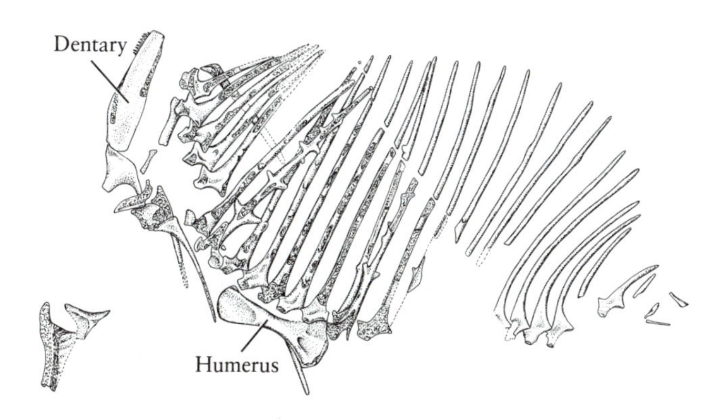

Figure 17-10. INCOMPLETE SKELETON OF *IANTHASAURUS,* $\times\frac{1}{3}$. This primitive edaphosaur is from the Upper Pennsylvanian. The dentition resembles that of early carnivorous pelycosaurs. *From Reisz and Berman, 1986.*

Eothyrids and Varanopseids

The phylogenetic position of two families of smaller pelycosaurs, the eothyrids and varanopseids, remains in doubt. Their skulls appear primitive relative to those of other pelycosaurs in having a short antorbital region and a lacrimal bone that reaches the narial opening (Figure 17-11). The tooth row is horizontal and at the same level as the jaw articulation. They seem advanced in the large size of the temporal opening, which is bordered ventrally by a narrow bar made up of a long quadratojugal that meets the maxilla and separates the jugal from the margin of the skull. The occiput is formed by a large platelike supraoccipital that is fused to the enlarged paroccipital processes. This pattern is quite distinct from that seen in early pelycosaurs and the genus *Ophiacodon.* We do not recognize any members of these families before the Lower Permian. Romer and Price considered that the eothyrids were among the most primitive pelycosaurs but allied the varanopseids with the sphenacodontids. No postcranial material is known of the eothyrids. The varanopseids (Langston and Reisz, 1981), which are typically about 1 meter long, are among the most agile pelycosaurs of the late Lower Permian and may have survived into the Upper Permian in southern Africa. Within this group, the jaw articulation extended posteriorly to lie well behind the occipital condyle.

Caseidae

Caseids were the most diverse and widespread group of herbivores among the pelycosaurs. They appear in the latter part of the Lower Permian, at the end of the period of abundance of *Edaphosaurus.* They lack long spines and have a very different type of dentition. The palatal denticles are much smaller than the marginal teeth, and there are no tooth plates on the lower jaws. The marginal teeth have laterally compressed, more-or-less spatulate

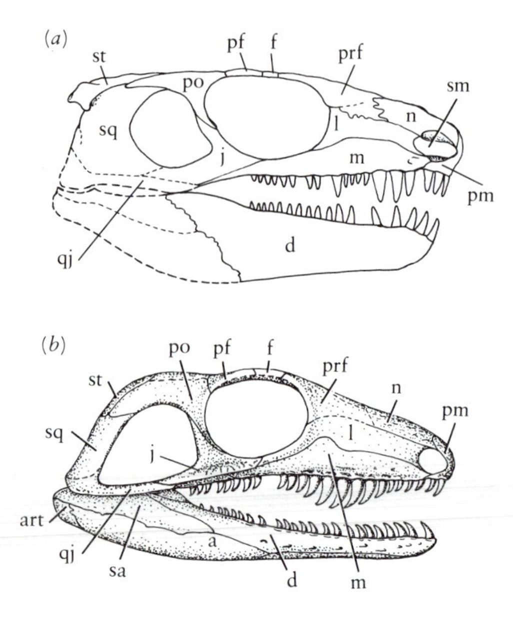

Figure 17-11. SKULLS OF PRIMITIVE CARNIVOROUS PELYCOSAURS. (*a*) The eothyrid *Oedaleops. From Langston, 1965.* (*b*) The varanopseid *Aerosaurus. From Langston and Reisz, 1981.* Both are 10 centimeters long. Abbreviation as in Figure 8-3.

crowns that are crenulated along the edge, which give them a general resemblance to those of modern herbivorous lizards, pareiasaurs, and plateosaurs.

The skull is distinctive in the large size of the temporal opening and the relatively enormous external narial opening (Figure 17-12). The surface of the skull is sculptured with a scattering of rounded pits. The oldest-known genus, *Casea*, is little more than 1 meter long, but *Cotylorhynchus* reached 3 meters and weighed at least 600 kilograms. In most genera, the skull is extremely small relative to the trunk length, and the rib cage is enormously expanded to surround a great mass of food undergoing digestion (Figure 17-12*e*). The number of sacral ribs increases from three to four pairs within the group. As in edaphosaurs, the supinator process has expanded distally to surround completely the ectepicondylar foramen. The hands and feet are short and wide (a feature elaborated progressively within the group). The phalangeal count is always less than in other pelycosaurs and may be as little as 2, 2, 3, 3, 2.

Romer and Price placed edaphosaurs and caseids in the same suborder. Both share a number of derived features, but most can be associated with their large size. The dental specialization toward herbivory is entirely different in the two groups. If they did share a common ancestry, it must have been at the level of primitive, carnivorous pelycosaurs. Reisz (1981) pointed out that eothyrids and caseids are unique among pelycosaurs in the possession of an anteroposteriorly elongated external narial opening, a pointed rostrum, and the entrance of the maxilla into the ventral margin of the orbit. Both also possess a short face and a low platelike occiput, although these features are not unique to these groups.

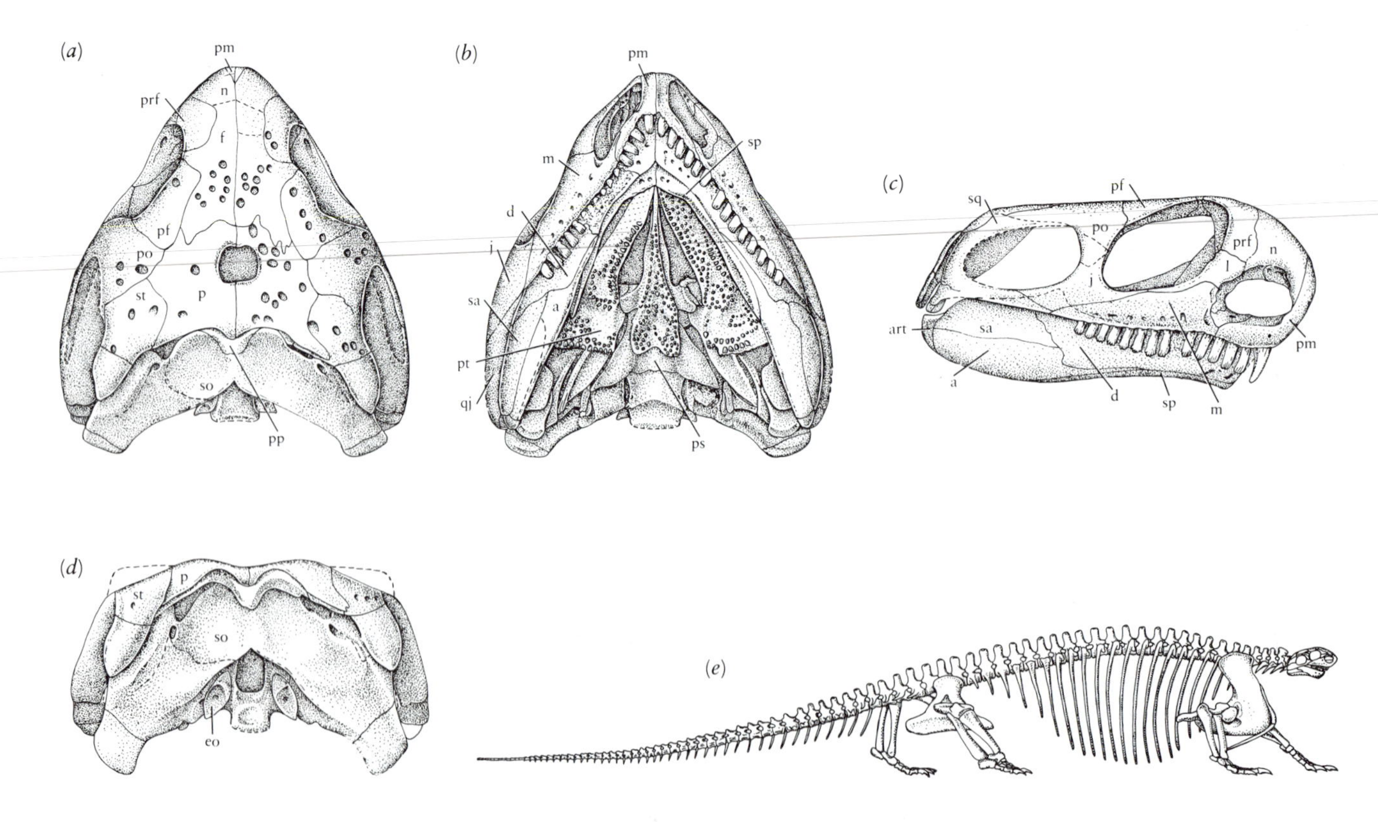

Figure 17-12. CASEID PELYCOSAURS. Skull of *Casea* from the Middle Permian of France in (*a*) dorsal, (*b*) palatial, (*c*) lateral, and (*d*) occipital views. Original is 11 centimeters long. Abbreviations as in Figure 8-3. *From Sigogneau-Russell and Russell, 1974.* (*e*) *Cotylorhynchus* skeleton from the Lower Permian of North America, 3 meters long. *From Stovall, Price, and Romer, 1966.*

As Olson (1968) pointed out, the eothyrids do not closely approach the caseids in the nature of their dentition. When they first appear in the fossil record, caseids are very distinct from all other groups of pelycosaurs. Fossils of caseids do not occur with the normal Permian fauna in Texas and Oklahoma and they may have lived in a distinct environment. Caseids are represented in the middle Permian of France and Russia but are otherwise limited to North America (Sigogneau-Russell and Russell, 1974).

THE BIOLOGY OF PELYCOSAURS

There is only one skeletal feature of pelycosaurs that shows unequivocally that they are related to the ancestry of mammals—the presence of a lateral temporal opening of a pattern known in no other early tetrapods. There is no feature that suggests any tendency in the direction of a high metabolic rate or other physiological characteristic that typifies mammals. Pelycosaurs probably had a physiology that was very similar to modern reptiles such as lizards, turtles, and crocodiles.

One feature that indicates temperature control is the tall neural spines of the sphenacodonts. Their high degree of articulation in the fossils, even when individual spines have been broken, shows that they were embedded in a single sheet of tissue or "sail" in the living animal. Romer (1948) calculated that the length of the spines in *Dimetrodon* increased disproportionately relative to other linear dimensions, so that the area of the sail varied in proportion to the *volume* of the body. Grooves at the base of the spines probably accommodated blood vessels. A vascular supply to the sail would permit it to function as a rapid means of absorbing or radiating heat from the body.

According to Bramwell and Fellgett (1973), a specimen of *Dimetrodon* weighing about 200 kilograms could warm up from 26°C to 32°C in 205 minutes without the sail or 80 minutes with the sail. The sail would have enabled them to be active much earlier in the day than were other predators or prey of comparable size. While the presence of a sail argues strongly for a selection regime favoring temperature control, it also demonstrates that these animals functioned primarily as ectotherms rather than endotherms.

THERAPSIDS

THE ORIGIN OF THERAPSIDS

Pelycosaurs are known primarily from North America and Europe, with a few late-surviving genera in southern Africa and Russia. The most complete fossil record is from the Lower Permian Redbeds sequence of the southwestern United States. These deposits continue into the Upper Permian, but the fauna becomes progressively depauperate as a result of increasing aridity. Olson (1962, 1974) described fossils from the early Upper Permian that have some features in common with therapsids, but their remains are too fragmentary to provide information regarding the transition between the two groups. The upper beds in the southwestern United States are sometimes referred to informally as middle Permian in age, as are the Russian beds in which we find the earliest therapsids. However, only Lower and Upper Permian are recognized as formal geological units.

The oldest unquestioned therapsids come from deposits in European Russia at the base of the Upper Permian. Several distinct groups can be recognized, which suggests a significant period of prior evolution. The early therapsids overlap in time with the youngest caseid pelycosaur from Russia, *Ennatosaurus,* but their remains come from different localities, which leads us to believe that they lived in different environments. On the basis of anatomical differences and possible environmental separation, the early therapsids probably had adapted to a way of life that was distinct from that of the pelycosaurs, perhaps because of differences in their physiology. These differences may account for the rapid radiation of therapsids that quickly reached a level of diversity far greater than that of their ancestors.

The closest affinities of the therapsids lie among the sphenacodont pelycosaurs, based specifically on the common presence of a reflected lamina of the angular. *Dimetrodon* and other members of the subfamily Sphenacodontinae appear too specialized in the great length of the neural spines and the reduced number of premaxillary teeth to be the direct ancestors of therapsids. However, the more primitive genus *Haptodus* could have filled this role. The lineage leading to therapsids may have diverged from animals that were similar to *Haptodus* at any time between the late Pennsylvanian and the middle Permian, a period of at least 25 million years.

The therapsids are clearly advanced over the pelycosaurs when they appear in the Upper Permian, particularly in the specializations of the postcranial skeleton.

EARLY THERAPSIDS

Eotitanosuchia (Phthinosuchia)

Therapsids are typically grouped in two major assemblages, the carnivorous theriodonts and the herbivorous anomodonts. They share a common ancestry among primitive therapsids, which also include both carnivorous and herbivorous genera.

Among the earliest and most primitive therapsids are the biarmosuchids from the Ocher (Ezhovo) locality at the base of the Russian sequence (Sigogneau and Chudinov, 1972). The skull of *Biarmosuchus* (Figure 17-13) resembles that of the sphenacodonts in most features. The temporal musculature is largely confined to the inner surface of the skull. The occiput is inclined slightly anteroventrally, the reverse of the orientation seen in pelycosaurs, but has a similar platelike configuration. The supratemporal bone is lost. The canine tooth is especially prominent, but there is not an anterior "step" in the maxilla, as in *Dimetrodon*. Like the advanced sphenacodontids, the maxilla extends dorsally to the nasal and separates the lacrimal from the narial opening. Unlike any pelycosaur, the septomaxilla has a long exposure on the surface of the skull.

The palate retains most of the primitive features of sphenacodontids, including the presence of teeth on the transverse flange of the pterygoid, but the vomers are recessed above the level of the internal nares and are partially fused. The palatines and anterior portion of the pterygoids are arched above the midline, which suggests the presence of a narrow air passage above the remainder of the denticulate palate. The stapes is a large bone that is oriented transversely and has a very large stapedial foramen. The reflected lamina of the angular is clearly separated from the remainder of the bone and is characterized by radiating ridges.

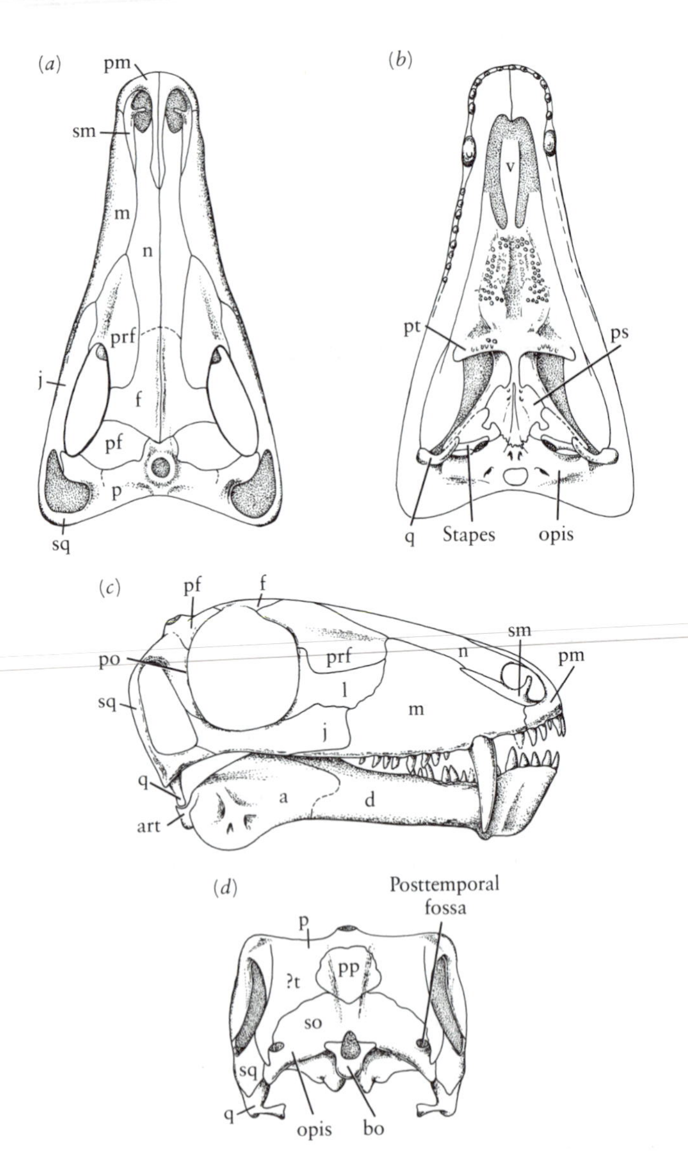

Figure 17-13. SKULL OF THE PRIMITIVE THERAPSID *BIARMOSUCHUS*. (*a*) Dorsal, (*b*) palatal, (*c*) lateral, and (*d*) occipital views, $\times \frac{1}{3}$. Abbreviations as in Figure 8-3. *From Sigogneau and Chudinov, 1972.*

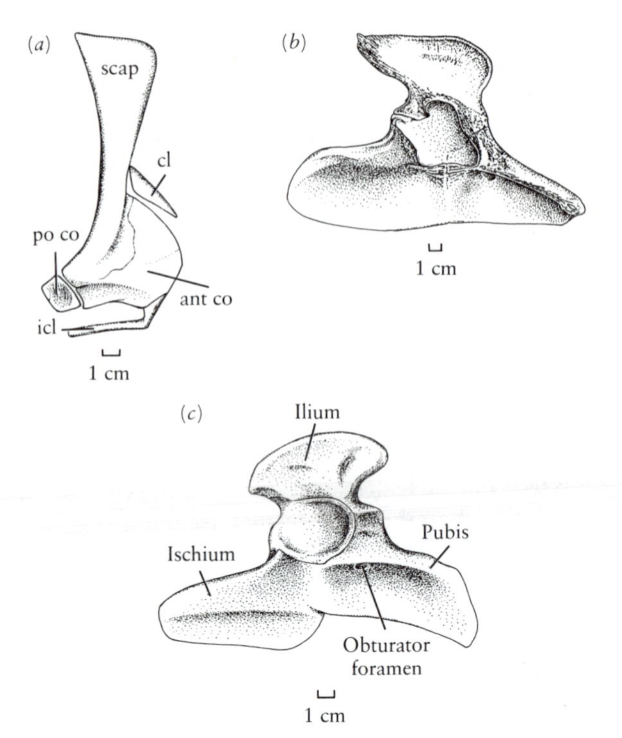

Figure 17-14. GIRDLES OF PELYCOSAURS AND THERAPSIDS. (*a*) Shoulder girdle of the therapsid *Biarmosuchus*. Abbreviations as in Figure 17-4. *From Sigogneau and Chudinov, 1972.* (*b*) Pelvis of the pelycosaur *Dimetrodon*. *From Romer and Price, 1940.* (*c*) Pelvis of the therapsid *Biarmosuchus*. *From Sigogneau and Chudinov, 1972.*

Much of the postcranial skeleton is known, although no restoration has been published. The vertebrae resemble those of primitive sphenacodontids, without the exaggerated neural spines of the Sphenacodontinae. The structure of the girdles and limbs indicates a posture much advanced above the level of the pelycosaurs. The glenoid and acetabulum both open more ventrally, and the femur may have assumed a semierect posture that was comparable to moderately advanced thecodonts (Figure 17-14). The blade of the scapula is much narrower than that of pelycosaurs, but the clavicle and interclavicle remain large elements. The outline of the pelvis remains primitive in the presence of platelike pubis and ischium.

The humerus retains a primitive configuration with both ends widely expanded, but the femur has a sigmoid curvature and inturned head that is comparable to that of a crocodile (Kemp, 1982). The hands and feet are more symmetrical than those of pelycosaurs, which indicates that they faced more directly forward throughout the stride, and the length of some phalanges is greatly reduced, which presages the reduction in phalangeal number that is typical of later therapsids and mammals (Figure 17-15).

Dinocephalians

The dinocephalians were contemporary with the primitive carnivorous biarmosuchids and include some of the earliest herbivorous therapsids. They are confined to the Upper Permian but are known in southern Africa as well as Russia. They left no descendants.

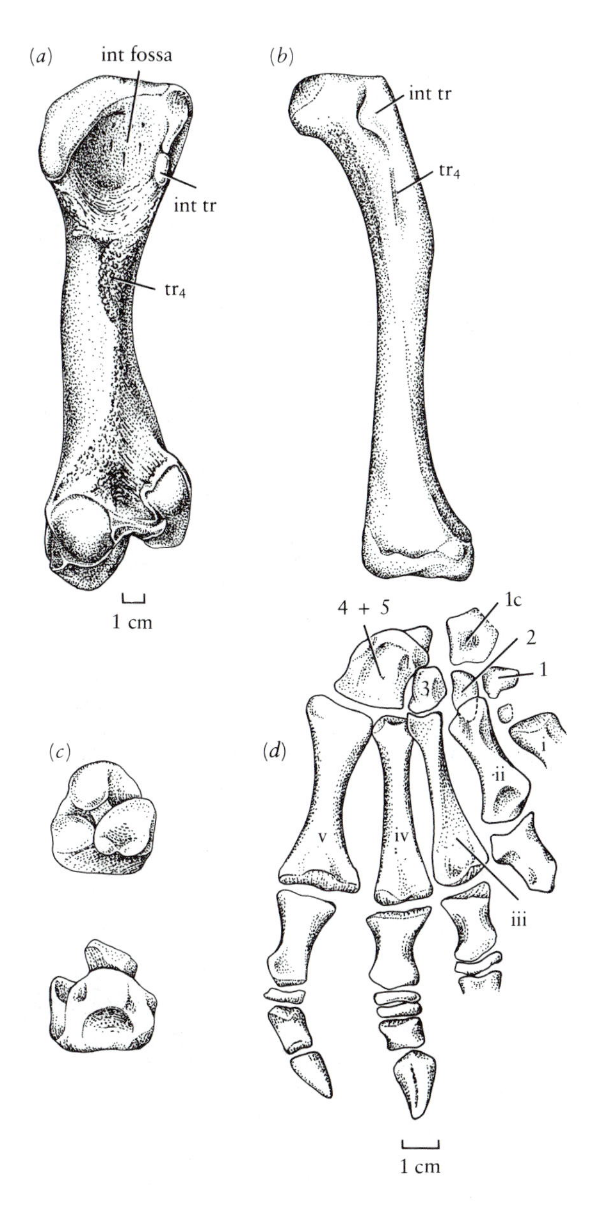

Figure 17-15. LIMB ELEMENTS OF PELYCOSAURS AND THERAPSIDS. Ventral views of the femora of (*a*) *Dimetrodon* and (*b*) a biarmosuchid. Note inturning of head in the therapsid. (*c*) Medial and lateral views of the complex astragalus of a biarmosuchid. (*d*) Foot of a biarmosuchid. Note reduction in the length of the phalanges relative to primitive pelycosaurs. Abbreviations: int fossa, intertrochanteric fossa; int tr, internal trochanter; lc, lateral centrale; tr_4, fourth trochanter; 1 to 5 distal tarsals; i to v, metatarsals. (*a*) *From Romer and Price, 1940.* (*b, c,* and *d*) *From Sigogneau and Chudinov, 1972.*

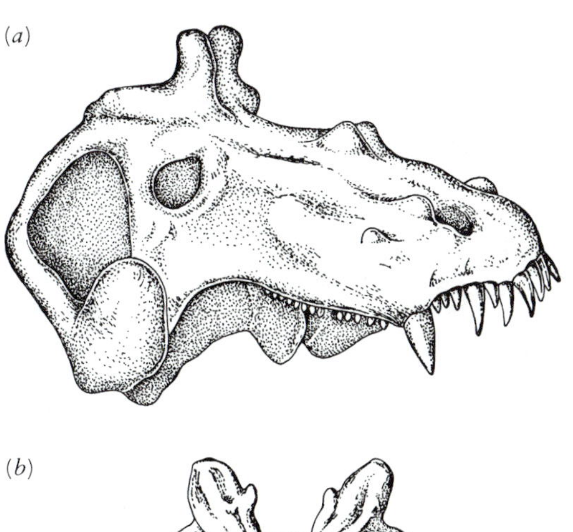

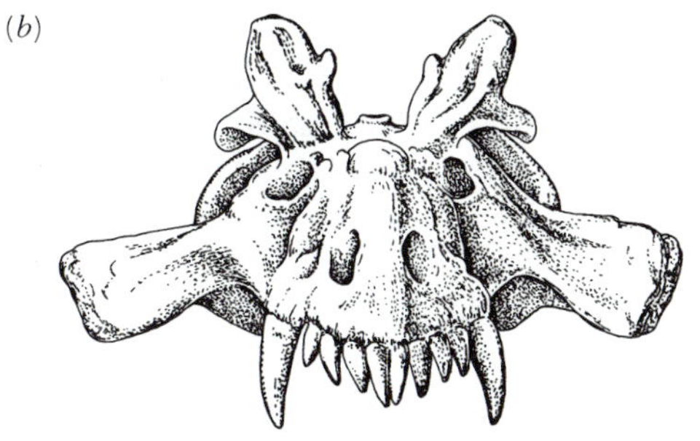

Figure 17-16. THE SKULL OF *ESTEMMENOSUCHUS.* (*a*) Lateral and (*b*) anterior views. It is one of the oldest-known therapsids, coming from the lowermost Upper Permian of Russia. The small size of the cheek teeth suggests that it was a herbivore. Its specific affinities are unknown. *From Chudinov, 1965. By permission of the University of Chicago Press.* Skull 80 cm long.

Among the most specialized of the dinocephalians is the genus *Estemmenosuchus* from the lowest Upper Permian zone in Russia (Figure 17-16). The skull is large and massive, with hornlike protuberances on the maxillae, frontals, parietals, and jugals. The estemmenosuchids had large canine teeth but tiny cheek teeth. Like other dinocephalians, the incisors are large and have a specific interdigitating arrangement.

Better known are the brithopodids from the Russian Zone II, including *Titanophoneus* (Figure 17-17). The limbs show a more upright posture, but the postcranial skeleton retains the general pattern of sphenacodont pelycosaurs with a very long tail. The digital formula is reduced to 2, 3, 3, 4, 5 in the manus and 2, 3, 3, 3, 3 in the pes. The temporal opening is greatly enlarged dorsoventrally compared with the pattern of biarmosuchids. Dorsally, there are sharp crests near the midline that indicate the great dorsal extent of the jaw musculature. The canines are long, and the piercing cheek teeth suggest that the brithopodids had a primarily carnivorous diet. The incisors are large and interdigitating; they bear a narrow shelf at their base, which ensures a close occlusion that would have enabled them to cut pieces of flesh neatly from their prey.

We find more advanced dinocephalians in southern Africa. They include the titanosuchids and tapinocephal-

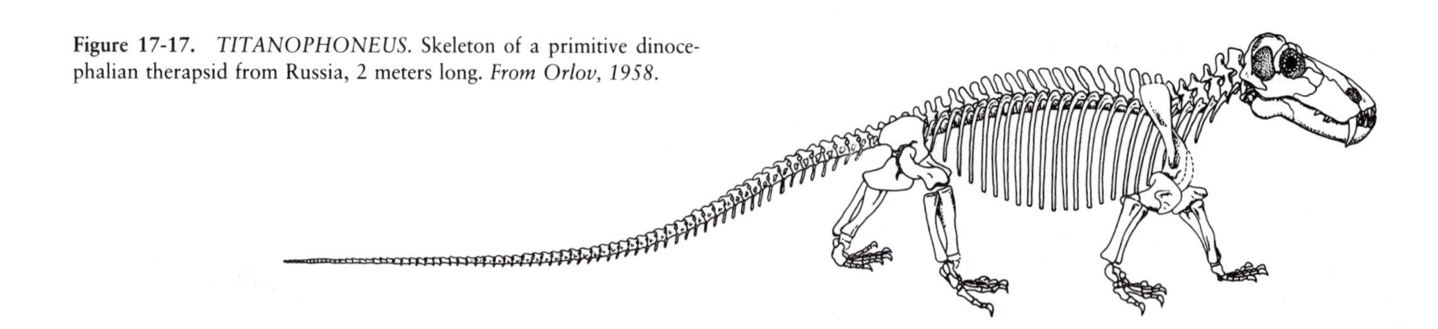

Figure 17-17. *TITANOPHONEUS.* Skeleton of a primitive dinocephalian therapsid from Russia, 2 meters long. *From Orlov, 1958.*

ids, in which the cheek teeth are reduced in size but increased in number and evolved a chisel shape like that of the incisors. The canine teeth are indistinguishable from the remainder of the dentition. These animals were surely herbivores. In advanced genera, the snout region is broad and low.

Among the titanosuchids and especially the tapinocephalids, the bones at the back of the skull are extremely thick (up to 10 centimeters) and pachyostotic. Barghusen (1975) showed that the structure of the skull and its articulation with the vertebral column are ideally suited for head-butting behavior such as was suggested for pachycephalosaurids and is observed today among sheep and goats (Figure 17-18). The upper temporal openings are greatly reduced as a result of the extensive development of surrounding bones. The occiput is broad and slopes far forward, with the jaw articulation displaced well anteriorly. The postcranial skeleton of the tapinocephalid *Moschops*, which reached a length approaching 3 meters, resembles that of an ox (Figure 17-19). The forelimbs retain a sprawling posture, but the hind limbs would have been held nearly erect. The phalangeal count is reduced to a mammalian 2, 3, 3, 3, 3.

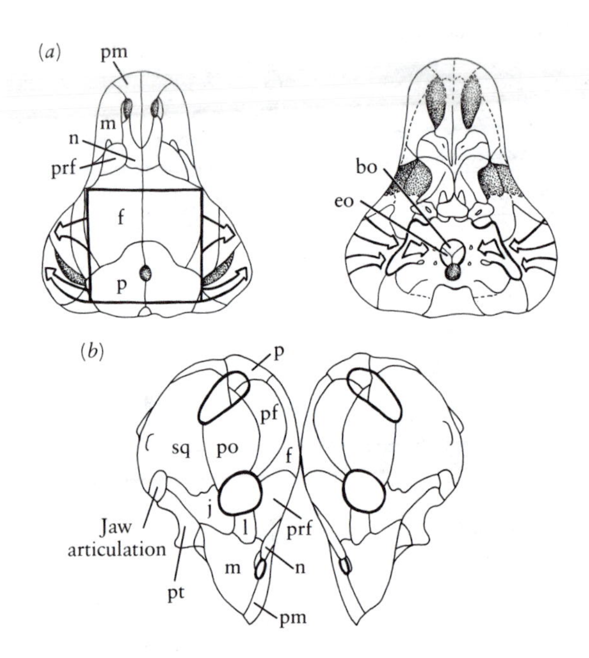

Figure 17-18. (*a*) Skull of the dinocephalian *Moschops* showing areas of dorsal thickening and distribution of forces produced when head butting. (*b*) Orientation of skulls at time of contact. Abbreviations as in Figure 8-3. *From Barghusen, 1975.*

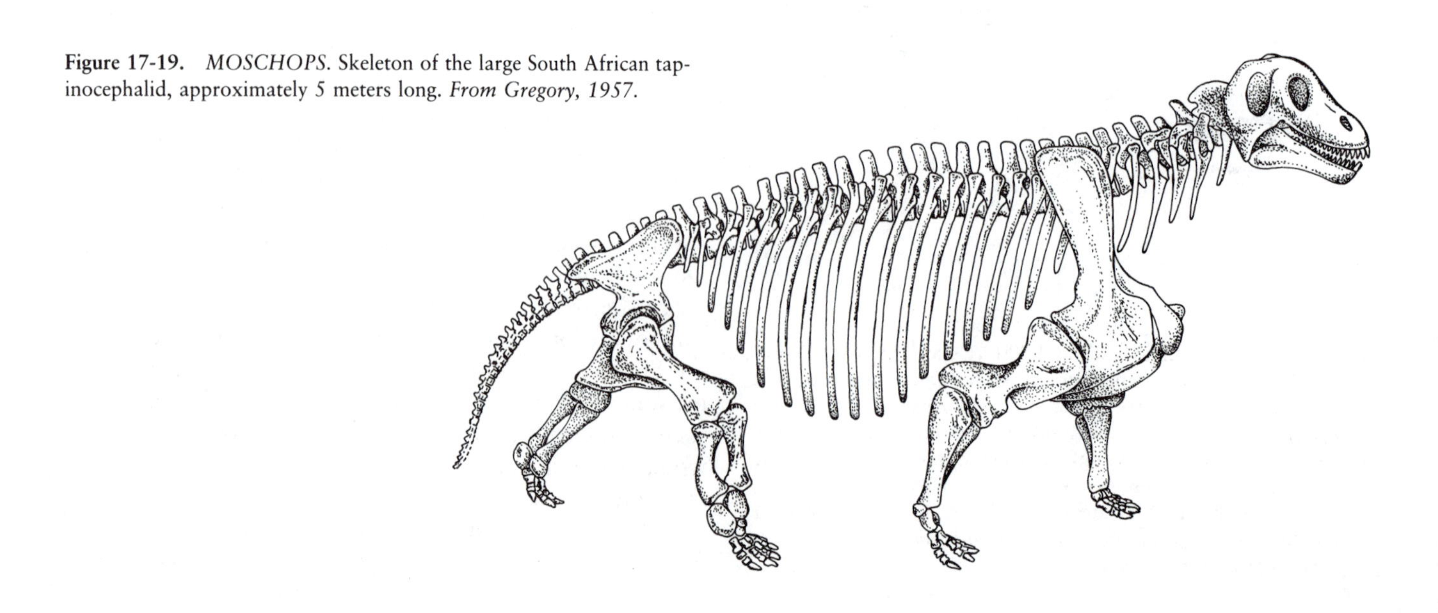

Figure 17-19. *MOSCHOPS.* Skeleton of the large South African tapinocephalid, approximately 5 meters long. *From Gregory, 1957.*

ADVANCED HERBIVOROUS THERAPSIDS: THE ANOMODONTS

Venjukoviamorphs

The dinocephalians did not survive the end of the Permian. Most herbivorous therapsids belong to a distinct assemblage, the Anomodontia. The earliest members of this group are known in the early Russian beds and are represented by the genus *Otsheria* in Zone I and *Venjukovia* in Zone II (Figure 17-20). The skulls of these genera are small and short, superficially resembling the herbivorous pelycosaurs. The zygomatic arch is located high on the cheek and exposes the adductor musculature widely between the cheek and the lower jaw. The temporal openings have expanded posteriorly beyond the level of the occipital condyle, but they do not extend as far dorsally as those of early dinocephalians. The canine teeth are reduced and the cheek teeth are short and blunt.

The premaxillae form a broad shelf of bone anterior to the internal nares. The palate is primitive in retaining a movable basicranial articulation and a distinct transverse flange of the pterygoid, but it has lost the denticles that occupied the margin of the flange in the more primitive therapsids.

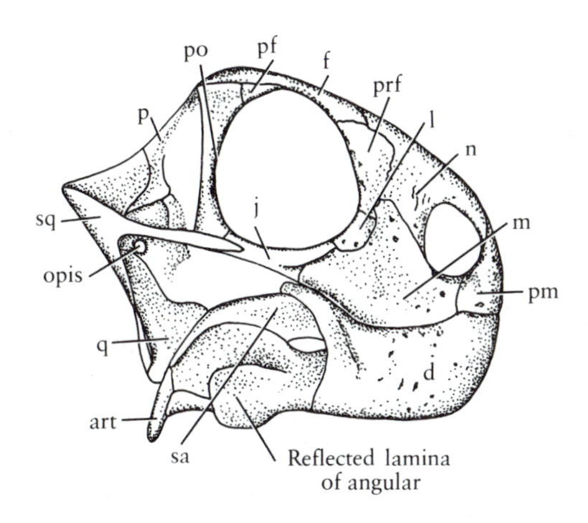

Figure 17-21. SKULL OF THE DROMASAUR *GALEOPS*. This small, agile herbivore is related to the dicynodonts. Natural size. Abbreviations as in Figure 8-3. *From Brinkman, 1981.*

Like the latter anomodonts, the coronoid bone is lost from the lower jaw and a large mandibular foramen is evident. The dentary bones in *Venjukovia* are fused at the symphysis.

As herbivorous forms, it was long thought that the venjukoviamorphs might form a link between the primitive dinocephalians and later anomodonts. Barghusen (1976) pointed out that *Venjukovia* is more primitive than the earliest known dinocephalians in not having the adductor musculature expanded over the dorsal skull table behind the orbits. In contrast, *Venjukovia* initiated the spread of a lateral sheet of the adductor mandibulae externus, which originates on the lateral surface of the squamosal and inserts on the lateral surface of the lower jaw. Many of the specializations for a herbivorous diet are clearly different in these two groups and evolved separately. There is no evidence to ally the anomodonts with any other particular group of primitive therapsids.

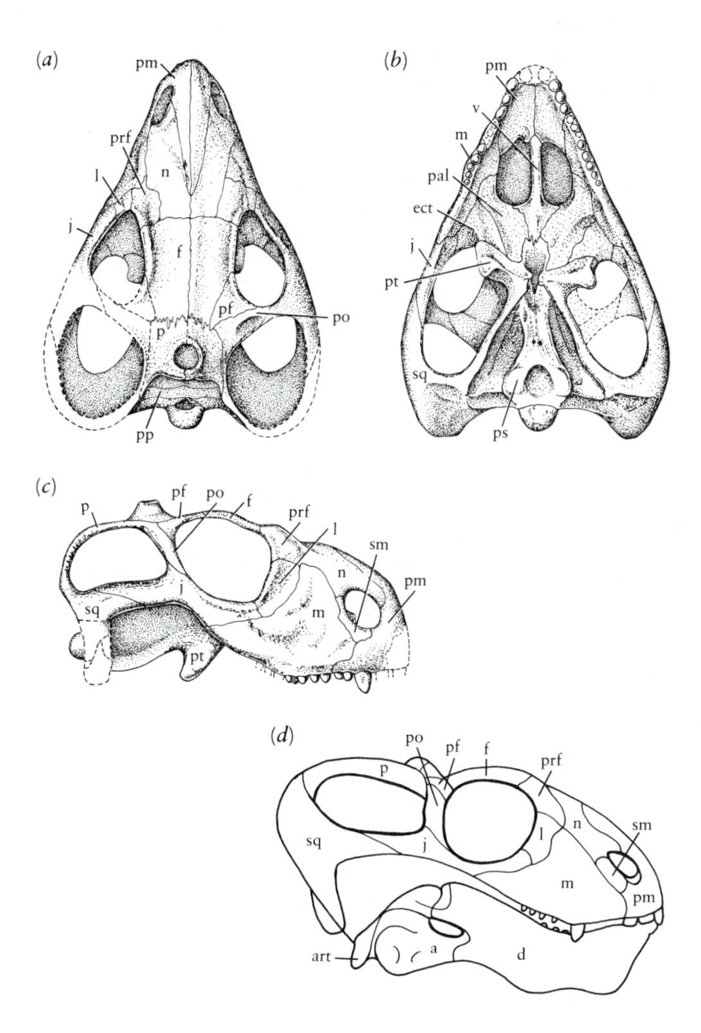

Figure 17-20. SKULLS OF PRIMITIVE UPPER PERMIAN ANOMODONTS. *Otsheria* in (*a*) dorsal, (*b*) palatal, and (*c*) lateral views, $\times \frac{1}{2}$. *From Chudinov, 1960.* (*d*) *Venjukovia* in lateral view, $\times \frac{1}{2}$. *From Barghusen, 1976.* Abbreviations as in Figure 8-3.

Dromasaurs

Three genera that are represented by only four specimens are the sole members of a second group of anomodonts, the dromasaurs, which are known only from the Upper Permian of southern Africa. The skull of dromasaurs (Figure 17-21) appears superficially like that of the venjukoviamorphs, with a high and narrow zygomatic arch, narrow postorbital bar, and narrow squamosal. It is relatively short and there is a partial secondary palate that is formed by the premaxillae. Canine teeth are not prominent and premaxillary teeth are progressively lost within the group. In contrast with venjukoviamorphs and dicynodonts, the postorbital does not extend back to the squamosal. The adductor musculature expands out of the temporal opening dorsally, and the lateral surface of the lower jaw is grooved to suggest an area for insertion of a superficial layer of the external adductor.

The limbs are long and slender, in contrast with the dicynodonts, and dromasaurs lack the key changes in the jaw apparatus that characterize that group.

Dicynodonts

The most numerous, diverse, and long-lived anomodonts were the dicynodonts, which appeared at the very base of the Upper Permian sequence in southern Africa, became worldwide in distribution in the Lower Triassic, and continued to the end of the Triassic. They were by far the most abundant of terrestrial vertebrates in the late Permian and early Triassic but lost ground by the mid-Triassic, perhaps in competition with rhynchosaurs and herbivorous cynodonts.

The early dicynodonts were small in comparison with the contemporary dinocephalians and pareisaurs of the Upper Permian, but later genera ranged up to the size of an ox. The body of dicynodonts is short and stout. In large genera, the rear limbs were held erect, but the fore limbs were sharply bent at the elbow. In all members, the phalangeal count was reduced to 2, 3, 3, 3, 3. The postcranial skeleton varied in relationship to body weight and to different ways of life that ranged from semiaquatic to subterranean, but the skull shows a remarkable consistency in its proportions that is related to a unique specialization of the feeding apparatus.

The genus *Dicynodon* (Figure 17-22) shows the cranial pattern that is typical of the group. Dicynodonts differ from venjukoviamorphs in lengthening the temporal region and elaboration of the squamosal, which results in a broad plate of bone on the lateral surface. The teeth are greatly reduced, typically leaving only a pair of canines, each fitting into a massive canine boss at the front of the maxilla. Even these teeth are lost in some genera. Their presence or absence in some species may reflect sexual dimorphism.

The premaxillae as well as the dentaries are fused at the midline in all but the most primitive genus, *Eodicynodon*. The bone at the front of the upper and lower jaws is marked by tiny nutritive foramina that resemble the surfaces that underlie the horny beak of birds and turtles. It is logical to assume that dicynodonts had a similar covering that formed a beaklike structure.

The palate is advanced in having the braincase sutured to the pterygoids, rather than retaining a movable basicranial joint. The premaxillae, maxillae, and palatine form a long secondary palate, with the internal nares opening behind the midpoint of the skull. The temporal fossae are enormous, which indicates the great mass of the adductor musculature. The area of the pterygoid that served for origin of the pterygoideus musculature extends anteriorly rather than laterally, so that the conspicuous transverse flange of the pterygoid, which is present in *Otsheria*, is no longer evident.

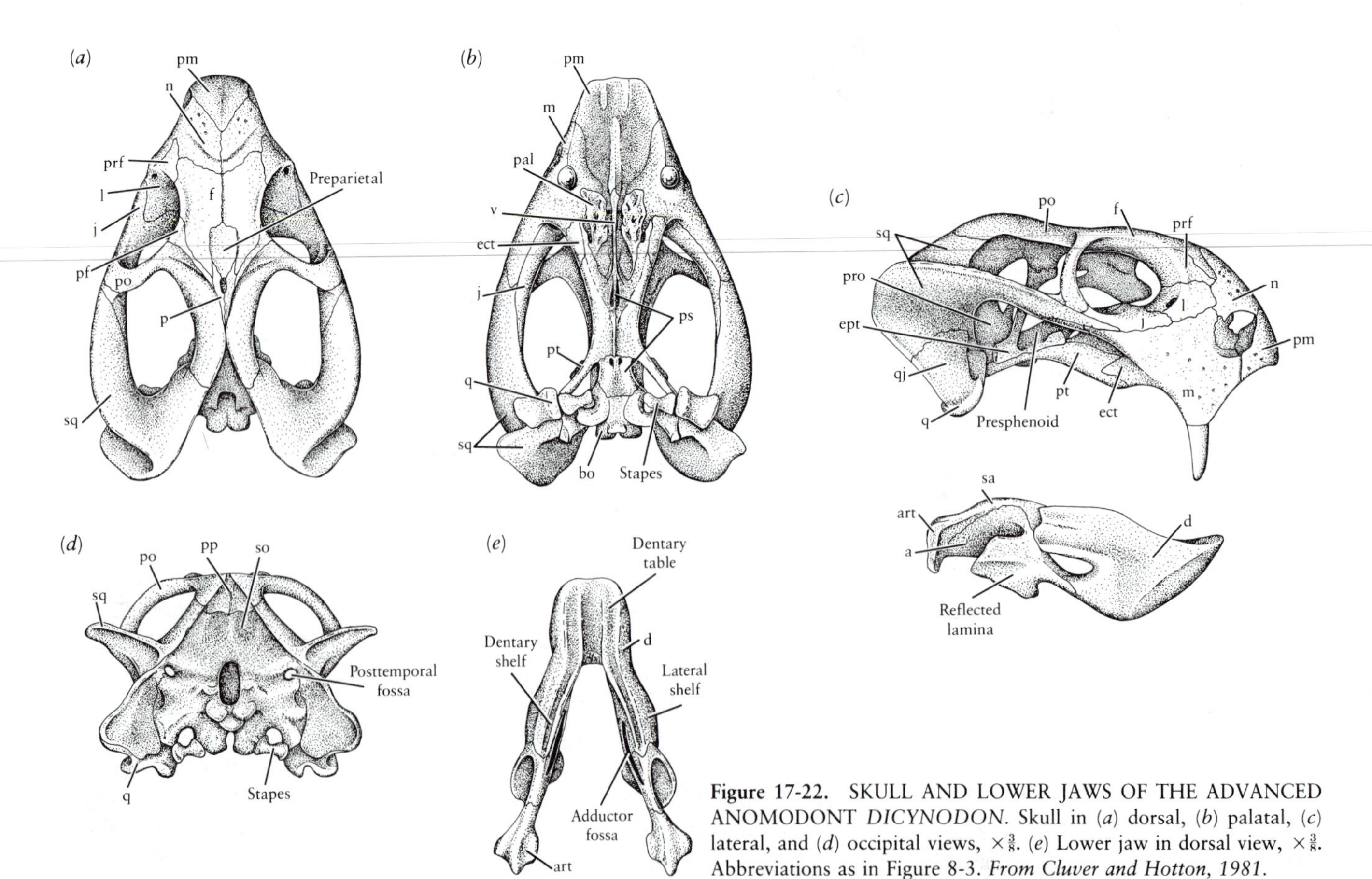

Figure 17-22. SKULL AND LOWER JAWS OF THE ADVANCED ANOMODONT *DICYNODON*. Skull in (*a*) dorsal, (*b*) palatal, (*c*) lateral, and (*d*) occipital views, ×⅜. (*e*) Lower jaw in dorsal view, ×⅜. Abbreviations as in Figure 8-3. *From Cluver and Hotton, 1981.*

The configuration of the jaw articulation differs significantly from that of venjukoviamorphs. The articulating surface of the articular is approximately twice as long as that of the quadrate. The surface of both bones is convex in lateral view but grooved longitudinally, which indicates that the articulating surface of the lower jaw was translated across that of the quadrate, in contrast with the simple hinge joint of most tetrapods.

The shape of the jaws and temporal region shows that the major muscles that opened and closed the mouth were oriented nearly horizontally, which suggests that anterior and posterior movements of the jaw were of great importance (Figure 17-23).

Within the anomodonts, a major new muscle mass has evolved. Already in the venjukoviamorphs, the adductor mandibulae externus was widely exposed laterally. Barghusen (1976) identified a depression on the lateral surface of the lower jaw into which inserted a superficial portion of this muscle. The origin may have been in the process of shifting from the medial surface of the back of

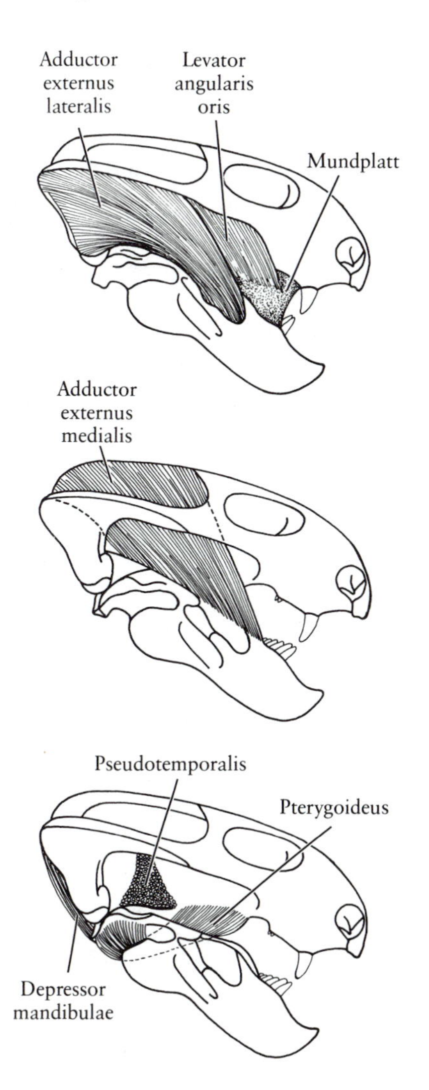

Figure 17-23. *EMYDOPS.* Reconstruction of the jaw-closing muscles in the primitive dicynodont at progressively deeper levels. *From Crompton and Hotton, 1967.*

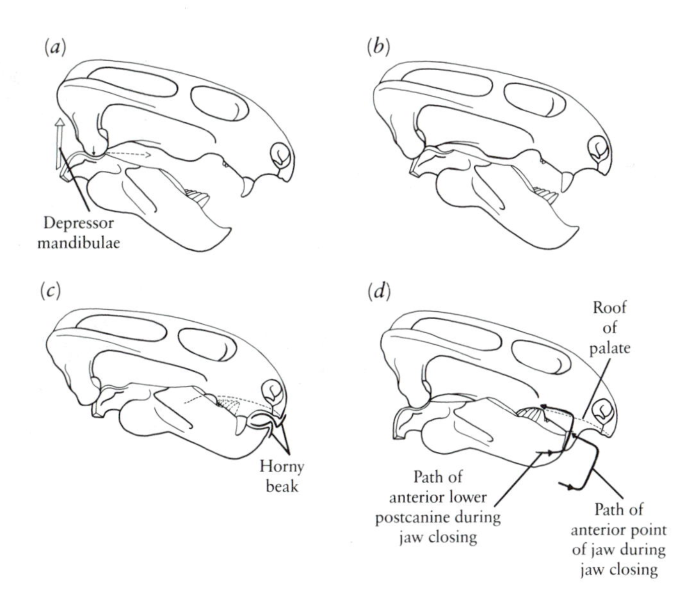

Figure 17-24. LATERAL VIEW OF THE SKULL IN THE PRIMITIVE DICYNODONT *EMYDOPS.* Stages in the masticatory cycle are illustrated. (*a*) Depression and beginning of protraction. Dashed arrow indicates direction of jaw movement. (*b*) Full protraction and beginning of elevation. (*c*) Beak bite and beginning of retraction. (*d*) Complete retraction. *From Crompton and Hotton, 1967.*

the zygomatic arch onto the lateral surface of the squamosal. In dicynodonts, the squamosal has greatly expanded laterally to provide a very large surface for the origin of the newly expanded muscle. The lateral surface of the lower jaw has evolved a specialized area for its insertion, the lateral shelf.

The presence of this lateral muscle would have balanced the more medially directed force of the remaining elements of the adductor mass. This balancing force may have made it possible for the reduction of the transverse flange of the pterygoid, which in more primitive synapsids prevented the lower jaw from moving medially. The change in orientation of the transverse flange of the pterygoid increased the area of the subtemporal fenestra and permitted the anterior expansion of the adductor muscles.

Crompton and Hotton (1967) demonstrated that the major force for breaking up food resulted from retraction of the lower jaw when it was nearly closed (Figure 17-24). The jaw opened from its maximally retracted position. As the jaw began to close, it was drawn forward by the pterygoideus muscle. It then closed and retracted simultaneously, with the force of the adductors originating from the back of the temporal region.

In primitive dicynodonts, this jaw action produced shearing at two points, posteriorly between the remaining cheek teeth of the lower jaw, maxilla, and palate and anteriorly between the lateral edge of the lower jaw and the canine boss. These surfaces were covered with horny material that is assumed to have formed sharp cutting edges. Advanced dicynodonts loose the cheek teeth entirely and shearing is limited to the more anterior region, which is increased in surface area. Neither grinding nor

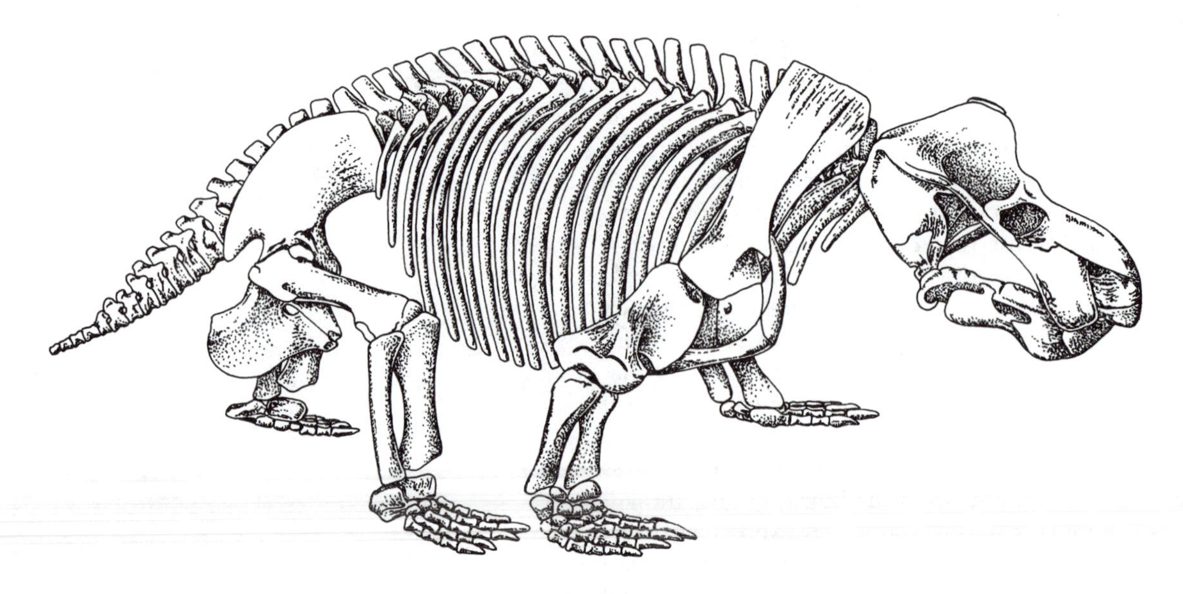

Figure 17-25. SKELETON OF THE LARGE DICYNODONT *KANNEMAYERIA*. From the Lower Triassic of southern Africa, approximately 3 meters long. *From Pearson, 1924. By permission of the Zoological Society of London.*

chewing occurred in this group; the food was comminuted entirely by shearing.

It is unusual to find a group as diverse in postcranial features and as long lived as the dicynodonts that has maintained such a stereotyped feeding pattern. We assume that all dicynodonts depended primarily on plant material, but the specific diet must have been different in genera such as *Lystrosaurus*, which is thought to have been semi-aquatic, huge terrestrial animals such as *Kannemeyeria* from the Lower Triassic (Figure 17-25) and *Placerias* from the Upper Triassic, which have the body proportions of large grazing or browsing mammals, and small forms with wedged-shaped skulls such as *Cistecephalus* and *Kawingasaurus*, in which the limbs and girdles suggest a burrowing habitus (Cox, 1972, Cluver, 1978).

Relationships among the dicynodont lineages are reviewed by Kemp (1982) and Cluver and King (1983).

ADVANCED CARNIVOROUS THERAPSIDS: THE THERIODONTS

Gorgonopsians

Although the origin of anomodonts is lost in the obscurity of ancestral therapsids, advanced carnivorous therapsids may all trace their origin to forms like the biarmosuchids, which lead with little question to the gorgonopsians, first known from the Upper Permian of southern Africa. The ictidorhinids—known primarily from cranial remains—provide a close link between the two groups. The skull closely resembles that of primitive carnivorous therapsids except for the prominence of lower canine teeth, reduction in the number of cheek teeth, and the presence of a separate median bone in the skull table, the preparietal, between the parietal and frontal (Figure 17-26). A bone is present in a similar position in dicynodonts but probably evolved separately in the two groups.

Gorgonopsids are advanced over the ictidorhinids in the expansion of the adductor chamber and reduction in the relative size of the orbit. The canine teeth are further emphasized and the cheek dentition reduced. The skull is massively constructed and reaches a length of 45 centimeters in one genus. We find gorgonopsids primarily in southern Africa, but they also appear in Russia toward the end of the Permian. Twenty-two genera have been described; they were the dominant carnivores in the later Permian and probably preyed on the large pareiasaurs and dinocephalians.

The postcranial skeleton, which is well known in the genus *Lycaenops* (Figure 17-27), superficially resembles that of cursorial mammals. The posture of the forelimb remains primitive, with the humerus held essentially horizontal. According to Kemp (1982), the structure of the femur is comparable with that of crocodiles and indicates a similar capacity for two styles of locomotion—one in which the limb retains a sprawling posture like that of primitive amniotes and the other in which the femur was held at about a 45-degree angle, which enabled the lower limb to move in a parasagittal plane. Changes in the structure of the ankle and foot that distinguish advanced theriodonts are little evident in gorgonopsids, which retain the primitive phalangeal count of 2, 3, 4, 5, 3, with only a slight diminution in the lengths of some phalanges.

Gorgonopsians did not survive past the end of the Permian and are not closely related to the origin of either of the more advanced groups of carnivorous therapsids. Sigogneau (1970) published the most recent review of this group.

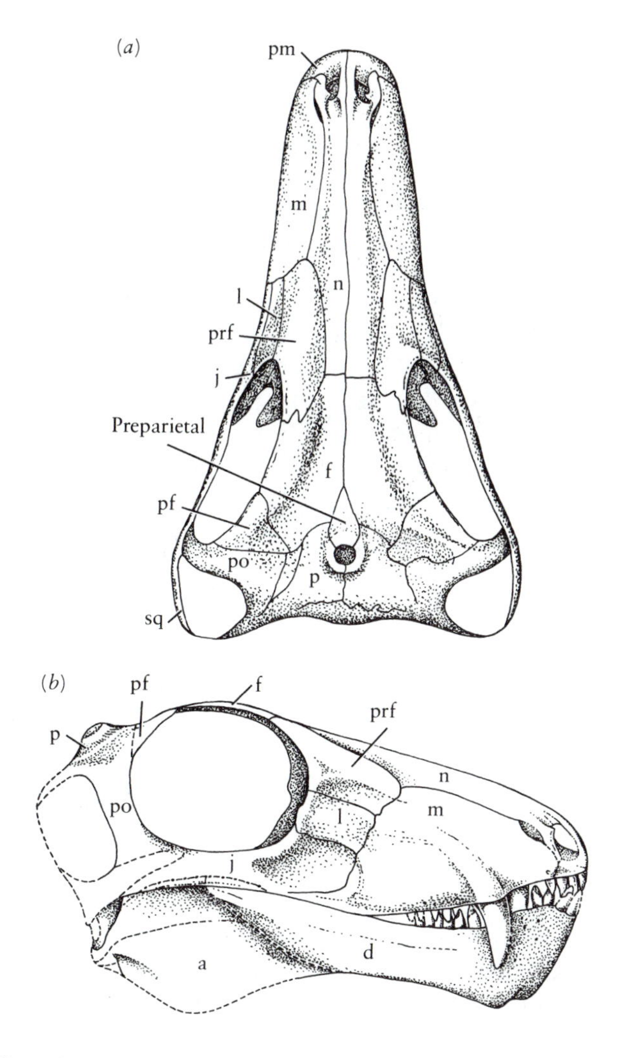

Figure 17-26. SKULL OF THE PRIMITIVE ICTIDORHINID GORGONOPSIAN *RUBIDGINA*. From the Upper Permian of Southern Africa. (*a*) Dorsal and (*b*) lateral views. As in *Biarmosuchus*, the orbit is much larger than the temporal opening, in contrast with more advanced gorgonopsians such as *Lycaenops* (see Figure 17-27). This skull is 10 centimeters long. Abbreviations as in Figure 8-3. *From Sigogneau, 1970.*

Therocephalians

Two much more advanced groups of carnivorous therapsids, the therocephalians and cynodonts, appear in the Upper Permian of Russia and southern Africa. We have not established the specific origin and interrelationships of these groups. They may have evolved separately from primitive carnivorous therapsids.

The therocephalians were much more diverse than the gorgonopsians, including small, possibly insectivorous forms (the scaloposaurids) (Figure 17-28), large carnivores (the pristerognathids), and the herbivorous bauriids. They range from the base of the Upper Permian sequence to the end of the Lower Triassic and are known in China, Antarctica, Russia, and southern and eastern Africa. Mendrez (1972, 1974, 1975, 1979) published detailed descriptions of several genera.

In contrast with the gorgonopsians, the jaw musculature in early therocephalians expanded dorsally over the braincase, leaving only a narrow sagittal crest between the adductor chambers. Both therocephalians and cynodonts developed a secondary palate, but this structure evolved in different ways in the two groups. In primitive therocephalians (Figure 17-28*b*), the vomer participates along with the premaxillae and maxillae, but the palatine remains dorsal in position. In primitive cynodonts (see Figure 17-30*b*), the vomer remains dorsal to the secondary palate, which incorporates the palatine in its posterior border. In both groups, the epipterygoid is expanded as a plate of bone lateral to the braincase.

Kemp's (1986) description of a nearly complete skeleton of a small therocephalian shows that some members of this group had very mammalian proportions, with relatively long limbs, attenuated lumbar ribs, and a highly reduced tail.

Some of the therocephalians competed with the gorgonopsians as the large carnivores of the late Permian. Like that group, these early therocephalians had large canines and tended to lose the cheek teeth.

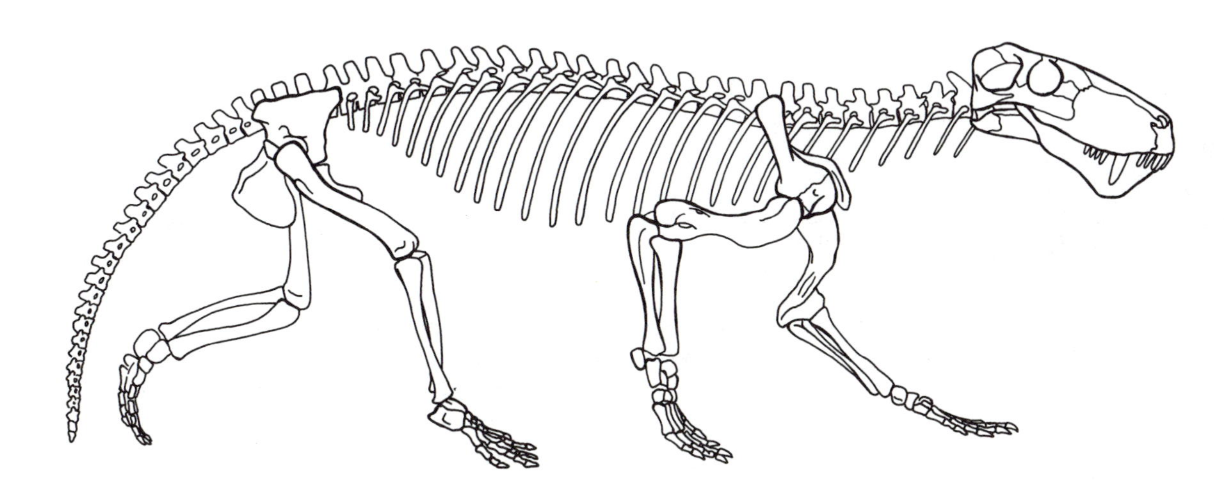

Figure 17-27. SKELETON OF THE GORGONOPSID *LYCAENOPS*, ORIGINAL 1 METER LONG. *From Colbert, 1948.*

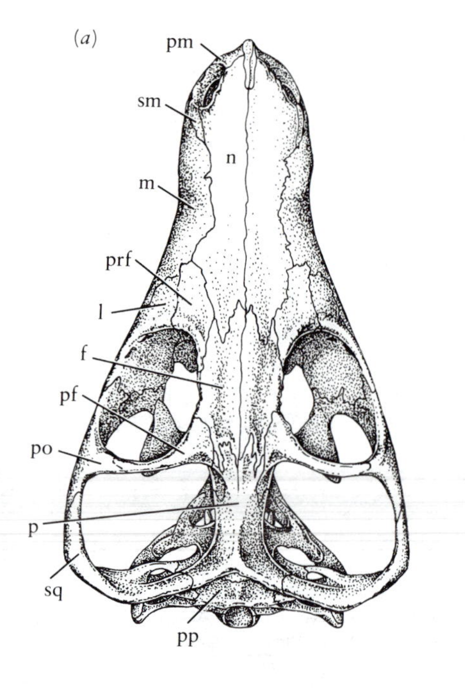

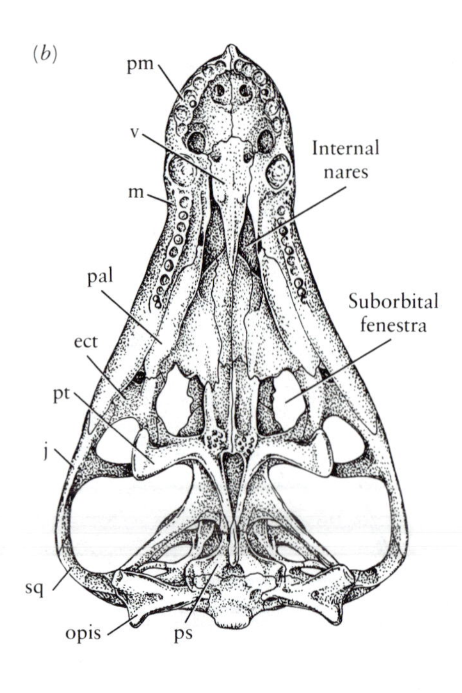

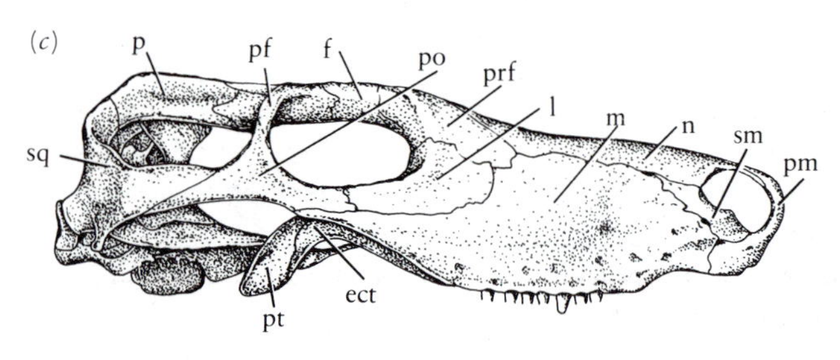

Figure 17-28. SKULL OF THE LOWER TRIASSIC THEROCEPHALIAN *REGISAURUS*. (*a*) Dorsal, (*b*) palatal, and (*c*) lateral views, $\times \frac{2}{3}$. Abbreviations as in Figure 8-3. *From Mendrez, 1972.*

We find a variety of small and medium-sized therocephalians in the Lower Triassic. Among them are the bauriids, which parallel the more mammal-like cynodonts in several features (Figure 17-29). A nearly complete secondary palate is developed and the cheek teeth have expanded crowns to crush and grind food. The postorbital bar is no longer complete. The dentary bone is enlarged, but the other bones of the lower jaw are not significantly reduced. Paleontologists once thought that therocephalians related to *Bauria* might be close to the ancestry of at least some mammals, but increased knowledge of both therocephalians and cynodonts indicates that only the latter groups shows the specialized features of the dentition, braincase, and lower jaw that are expected in mammalian ancestors.

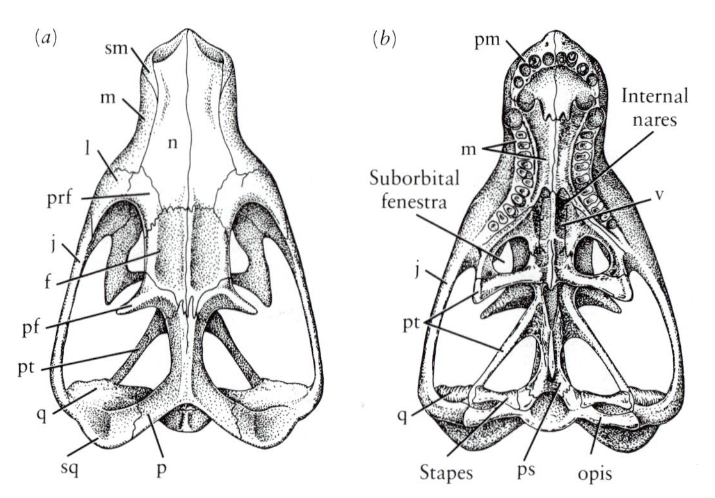

Figure 17-29. SKULL OF THE HERBIVOROUS THEROCEPHALIAN *BAURIA*. (*a*) Dorsal and (*b*) palatal views, $\times \frac{1}{2}$. Abbreviations as in Figure 8-3. *Drawn on the basis of sketches of Mendrez-Carroll.*

CYNODONTS AND THE ORIGIN OF MAMMALS

Therocephalians and gorgonopsians were the dominant reptilian carnivores in the late Permian. Within the early Triassic they were supplanted by the most advanced theriodonts, the cynodonts. Although some features that we associate with mammals were achieved in other therapsid groups, only the cynodonts show a significant approach to the mammalian condition in their general morphology. Cynodonts are represented in the latest Permian of both Russia and southern Africa by the family Procynosuchidae. The skull of *Procynosuchus* (Figure 17-30) is ad-

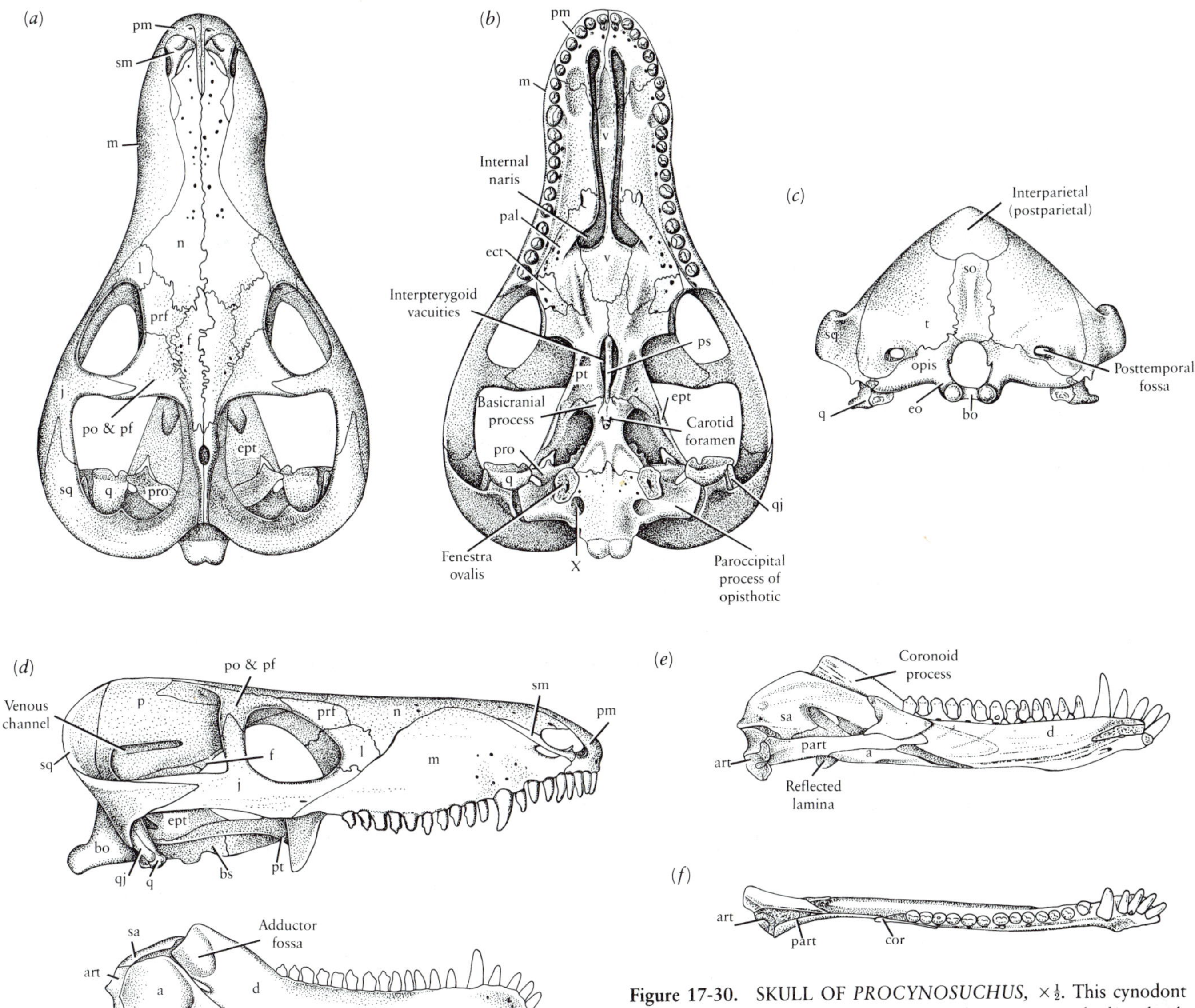

Figure 17-30. SKULL OF *PROCYNOSUCHUS*, $\times \frac{1}{2}$. This cynodont is from the Upper Permian of southern Africa. (*a*) Dorsal, (*b*) palatal, (*c*) occipital, and (*d*) lateral views. (*e* and *f*) Lower jaw in medial and occlusal view. Abbreviations as in Figure 8-3. *From Kemp, 1979.*

vanced over other early therapsids in a complex of features that presages the mammalian condition. To appreciate the importance of these features it is necessary to consider some of the basic differences between primitive reptiles and mammals.

We tend to think of large brain size and live birth as being particularly important attributes of modern mammals, but both were achieved relatively late in mammalian evolution. The placental pattern of live birth was certainly not achieved until late in the Cretaceous and large brain size evolved only within the Tertiary.

A much more fundamental mammalian feature is their high metabolic rate. Modern mammals require approximately ten times more food and oxygen than do reptiles of comparable size. The higher metabolic rate allows mammals to be active more continuously and to maintain a high, constant body temperature that is independent of the environment. Such a radical difference in metabolic rate affects nearly all the systems of the body and is responsible, directly or indirectly, for nearly all the differences we observe between reptiles (be they modern or Paleozoic) and mammals.

The initial change may have been a shift in the quantitative importance of fermentative and oxidative metabolism of the voluntary muscles. Reptiles rely primarily on fermentative, anaerobic metabolism of glucose to power muscle contractions. Glucose is present within the muscles and the response can be immediate and is little affected by temperature. By relying on anaerobic metabolism, reptiles can run short distances as rapidly as mammals. A critical drawback of anaerobic metabolism is the accumulation of lactic acid, which results in muscle fatigue after 1 or 2 minutes of vigorous activity. After a few short dashes, a lizard will be almost incapable of further exertion. The breakdown of the lactic acid requires several hours before the muscles are again capable of active contraction (Bennett and Dawson, 1976).

Mammals also make use of anaerobic metabolism for short bursts of speed, but for sustained locomotion they rely on oxidative metabolism. This reaction yields substantially larger amounts of energy and produces only water and carbon dioxide as waste products. Muscles can continue to contract vigorously for hours without significant fatigue.

The more upright posture and structural adaptations for more effective fore-and-aft movement of the limbs in therapsids, as in dinosaurs, may be related to an enhanced capacity to maintain muscle activity for long periods of time without fatigue. Judging by the structure of the limbs and girdles, we can see that even the eotitanosuchids may have had a significantly higher metabolic rate than the pelycosaurs and other primitive amniotes. It is difficult to explain the upright posture of the gorgonopsians and later theriodonts if they were not capable of sustained locomotor activity.

When viewed from our position as warm-blooded mammals, we can easily appreciate the value of a high metabolic rate. However, its benefits are costly. Modern reptiles require much less food and can survive long periods without eating. They much more readily adapt to conditions of marked seasonal and diurnal temperature change since their body temperature can fluctuate widely without ill effects (Pough, 1980).

A high metabolic rate requires a large and dependable food supply. All early amphibians and reptiles were carnivores. The latest pelycosaur fauna, dominated by caseids, included for the first time a majority of herbivores. The herbivorous dicynodonts far outnumbered the carnivores in the latest Permian and early Triassic. The presence of a large number of primary consumers may have provided a sufficiently stable food source that it became practical for carnivorous therapsids to attain a higher metabolic rate.

A high metabolic rate also requires a constant supply of oxygen and rapid removal of CO_2, which, in turn, requires more effective circulation and respiration than occur in modern reptiles. One may assume that the capacity for gas exchange within the lungs was increased in the ancestor of mammals and that a complete separation of the ventricles was achieved so that mixing of oxygenated and deoxygenated blood did not occur. Neither of these changes leaves any direct evidence in the skeleton and we can only speculate on the degree of their attainment in the therapsids.

Other aspects of a higher metabolic rate are more directly apparent in the skeletal anatomy. A constant supply of oxygen to the lungs in mammals is ensured by the separation of the air passage from the mouth by the evolution of an effective secondary palate. The requirement for additional food is related to more effective feeding strategies and more rapid digestion. Modern reptiles swallow their food whole or in large pieces and digestion is slow. Birds use the gizzard to break up their food into small particles so that it can be more rapidly digested. In the advanced therapsids and mammals, food is broken up in the mouth by the use of a specialized dentition and more prolonged chewing. Complex teeth, specialized jaw mechanics, and modification of the jaw musculature are characteristic of the line leading to mammals.

None of these changes are evident in the gorgonopsians. Some therocephalians have an extensive secondary palate and complex cheek teeth, but none show evidence of significant changes in the distribution of the jaw musculature. Such changes are evident in even the earliest known cynodonts, and additional modifications occur progressively in this group throughout the Triassic.

PROCYNOSUCHIDS AND THE ORIGIN OF THE MAMMALIAN JAW MUSCULATURE

The Upper Permian procynosuchid cynodonts show marked advances in the development of a secondary palate, complex cheek teeth, and changes in the jaw and temporal region that indicate the initial steps in reorganizing the jaw musculature toward a mammalian pattern. The secondary palate in *Procynosuchus* (Figure 17-30*b*) is formed by ventral and medial extensions of the premaxilla, maxilla, and palatine. They do not meet at the midline and do not continue to the back of the tooth row. However, a fleshy secondary palate probably completed the separation of the air passage from the mouth.

The cheek teeth are multicuspate and superficially resemble the molars of late Triassic mammals. The crowns of the teeth are elongate, with a series of linearly arranged cusps. The upper and lower teeth do not occlude with one another and are continuously replaced, as in most reptiles.

The temporal opening is enormous, which indicates the presence of very large muscles to close the jaws. In contrast with other groups of therapsids, the lower temporal bar, or zygomatic arch, extends laterally beyond the level of the dentary. This extension allows the insertion of jaw muscles on the lateral surface of the dentary. This also occurred in dicynodonts, but as a result of raising the zygomatic arch above the level of the tooth row rather than by its lateral extension. The dentary of cynodonts has a high coronoid process that is recessed laterally for muscle insertion. In mammals, the muscle that occupies this position is termed the deep masseter and the recess for its insertion is termed the masseteric fossa.

In Chapter 9, we noted that primitive reptiles and amphibians have three major jaw adductors, the adductor mandibulae posterior, the adductor mandibulae externus, and the adductor mandibulae internus (including the pseudotemporalis and the pterygoideus). Judging by the

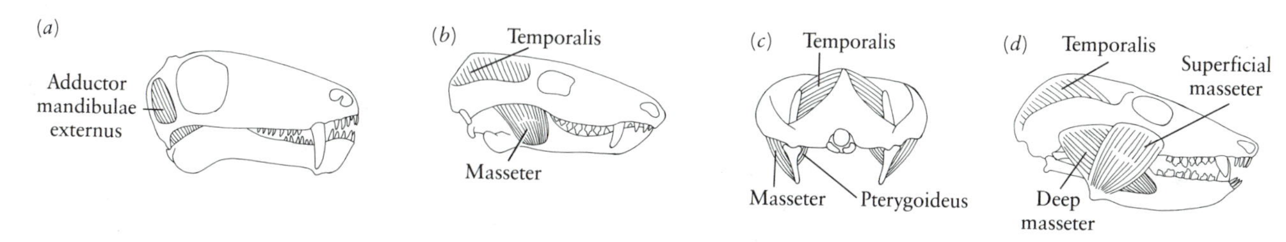

Figure 17-31. EVOLUTION OF THE MAMMALIAN JAW-CLOSING MUSCULATURE. Of the three major elements of the adductor musculature in primitive tetrapods described in Chapter 9, the adductor mandibulae internus and the adductor mandibulae posterior are both very much reduced in the ancestry of mammals. The major muscles that originate from the adductor chamber in mammals are all derived from the adductor mandibulae externus. In primitive therapsids, as illustrated by *Biarmosuchus*, (*a*), the externus, as in primitive amniotes, consists of a single functional unit that is entirely contained within the adductor chamber and inserts on the dorsal and medial surface of the back of the lower jaw. In primitive cynodonts such as *Thrinaxodon*, (*b*), the zygomatic arches are bowed laterally and the superficial portion of the externus inserts on the lateral surface of the lower jaw. Its origin has shifted to the inside surface of the zygomatic arch. With the separation of the adductor mandibulae into two functional units, the medial portion, which originates on the skull roof and the lateral surface of the braincase, is now termed the temporalis. The lateral portion is designated the masseter. In posterior view, (*c*), we see that the force of the masseter acts to pull the lower jaw laterally as well as dorsally, thus balancing the medially directed force produced by the temporalis. (*d*) In more advanced cynodonts such as *Probainognathus*, the masseter has split into two units. The fibers of the deep masseter, like the single element in *Thrinaxodon*, are oriented nearly parallel with those of the temporalis. The fibers of the newly elaborated superficial masseter are oriented obliquely anteriorly.

configuration of the skull and lower jaw, we can see that the primitive pattern is retained in pelycosaurs and early therapsids. In later synapsids, the adductor mandibulae posterior and the pseudotemporal portion of the adductor mandibulae internus are greatly reduced. Most of the mammalian jaw musculature and all of that which occupies the temporal region is derived from the reptilian adductor mandibulae externus. In cynodonts, the outer layer of this muscle extends its insertion onto the lateral surface of the dentary and thus becomes recognizable as the masseter. Its origin switches to the inner surface of the zygomatic arch. The muscle that retains its insertion on the medial and dorsal surface of the coronoid process and that originates on the skull roof and lateral wall of the braincase is distinguished as the temporalis (Figure 17-31).

The increase in size of the temporal opening in early cynodonts provides more space for jaw musculature, thus producing a stronger bite. The development of a masseter also redistributes the forces that are applied to the jaw and allows mediolateral control of jaw movement that was not possible in more primitive reptilian groups. This control becomes very important in establishing the precise occlusion of the molar teeth that occurs in mammals.

THRINAXODON AND THE EARLY STAGES IN THE DEVELOPMENT OF THE MAMMALIAN SKELETON

Skull

The procynosuchids of the late Permian were succeeded in the *Lystrosaurus* zone at the base of the Triassic by the galesaurids, of which *Thrinaxodon* is the best-known example. Both the skull and the postcranial skeleton have been the subject of detailed study and provide a thorough understanding of a representative therapsid that is close to the ancestry of mammals (Jenkins, 1971; Fourie, 1974; Crompton and Jenkins, 1979).

Thrinaxodon was a lightly built, active carnivore, approximately 50 centimeters in length. The trunk is long, with 27 presacral vertebrae, and the limbs are relatively short. The shortness of the limbs is accentuated by the plantigrade posture of the feet. The general appearance brings to mind the proportions of a mustelid carnivore.

The skull is about 10 centimeters long, and the adductor chamber occupies nearly half of its length (Figure 17-32). The two sides are separated by a very narrow braincase that forms a sharp sagittal crest that still retains a pineal opening. The squamosal expands broadly around the back of the adductor chamber and forms the lateral surface of the occiput. As in *Procynosuchus*, a single area of ossification occupies the position of the postorbital and postfrontal of more primitive therapsids. This bone meets the prefrontal above the orbit.

Thrinaxodon is advanced over the procynosuchids in the formation of a solid secondary palate, with sutural attachment of the maxillae and palatines at the midline beneath the nasal passage. The pterygoids meet at the midline and close the interpterygoid vacuity in adults. The dentition has a nearly mammalian appearance. There are four upper and three lower incisors, a number that is retained in most more advanced cynodonts. Canine teeth are present in both upper and lower jaws, although they are not as large as those of more primitive theriodonts. There are seven to nine cheek teeth. The crowns are laterally compressed and marked by a series of linearly arranged cusps. However, in contrast with mammals, the teeth are still replaced regularly and do not show a specific pattern of occlusion.

The lower jaw is dominated by the dentary, with a coronoid process that extends dorsally above the zygo-

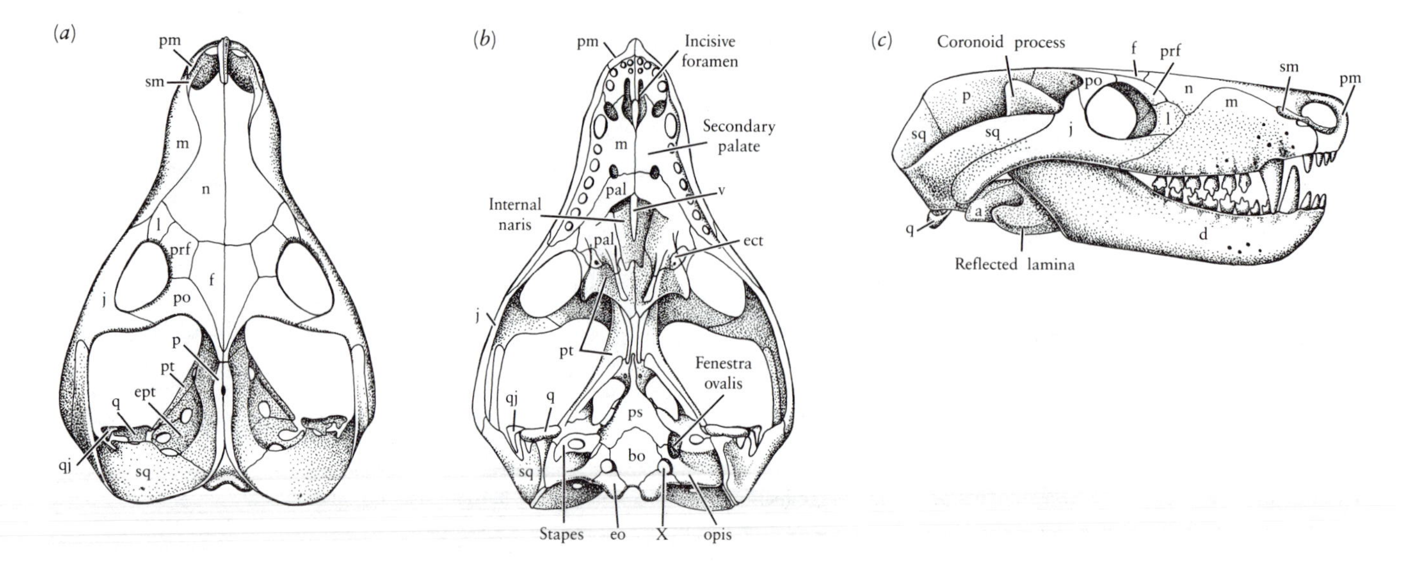

Figure 17-32. SKULL OF THE LOWER TRIASSIC CYNODONT *THRINAXODON*. (*a*) Dorsal, (*b*) palatal, and (*c*) lateral views, ×¾. Abbreviations as in Figure 8-3. *By permission of the Zoological Society of London. From Parrington, 1946.*

matic arch. The masseteric fossa reaches to its base. The postdentary bones are not greatly reduced but, like those of *Procynosuchus,* lose sutural contact with the dentary and can move separately. The reflected lamina of the angular is still a large structure that extends laterally from the remaining bone surface. As in primitive reptiles, the articular forms the entire surface for articulation with the skull. The quadrate and closely linked quadratojugal are greatly reduced and fit loosely into adjacent sockets at the base of the squamosal.

An early commitment to a higher metabolic rate among therapsids led to selection for a more sophisticated postcranial skeleton to facilitate rapid and agile locomotion. Throughout the evolution of cynodonts and Mesozoic mammals, there is progressive change in the morphology of the vertebrae, girdles, and limbs. The initial stages in these changes are clearly evident in *Thrinaxodon.*

The Origin of the Mammalian Atlas-Axis Complex

Among therapsids, there is a marked change in the nature of the mobility of the head on the trunk. The pattern in primitive amniotes (see Chapter 10) is retained with little modification throughout the pelycosaurs and into the primitive therapsids (Figure 17-33 and 17-34). The occipital condyle is a roughly hemispherical structure that is situated beneath the foramen magnum. The atlas arches and intercentrum form an incomplete ring that fits about it like the socket of a ball-and-socket joint. The mobility of this joint was limited by the proatlas, whose paired elements formed a link between articulating surfaces on the exoccipitals and the paired atlas arches. This linkage restricted both rotation and dorsoventral flexion between the skull and the trunk.

Among primitive therapsids, mobility is achieved by a limited amount of rotation and flexion at each point of articulation between the skull and the anterior elements of the cervical series. In mammals, the two movements are concentrated between different elements of the atlas-axis complex. Dorsoventral flexion in an arc as great as 90 degrees occurs primarily between the occipital condyle and the atlas. This mobility is achieved among the cynodonts and early mammals by division of the originally single occipital condyle into two articulating surfaces that shift progressively from a position beneath the foramen

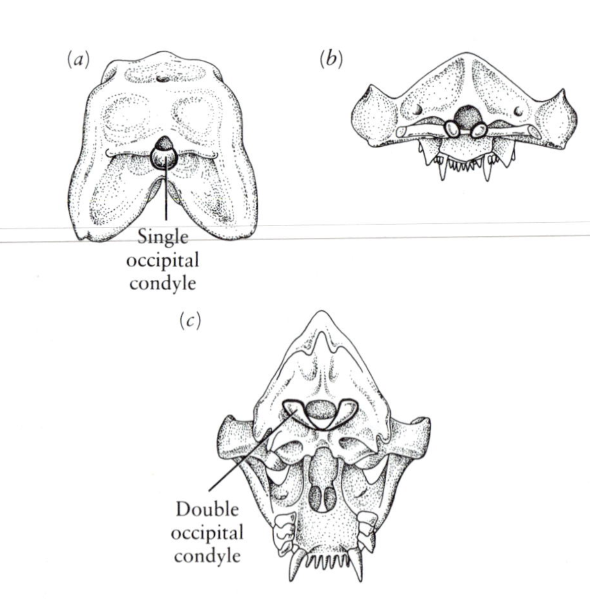

Figure 17-33. The articulation between the skull and the cervical vertebrae changes significantly between pelycosaurs and mammals. (*a*) In primitive pelycosaurs, the occipital condyle is a single hemispherical structure located directly below the foramen magnum. (*b*) Within the therapsids, the condyle splits to form a paired structure. (*c*) The area of articulation moves dorsally in mammals, so that the hinge line passes midway in the height of the foramen magnum to minimize tension on the nerve cord when the skull is flexed dorsally or ventrally. Pairing of the occipital condyle precludes lateral flexion and rotation at this point. *Redrawn from Jenkins, 1971.*

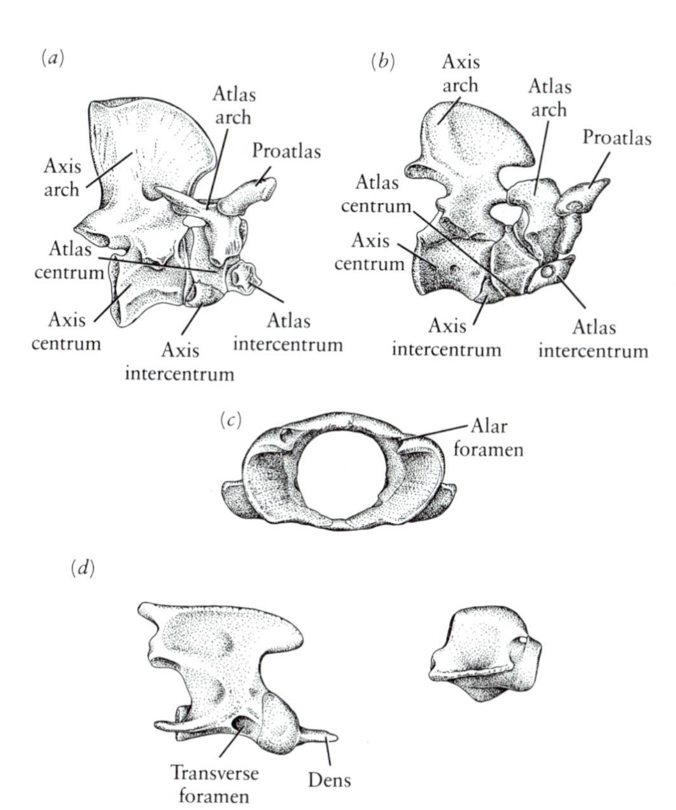

Figure 17-34. CHANGES IN THE ATLAS-AXIS COMPLEX BETWEEN PELYCOSAURS AND MAMMALS. (*a*) The pelycosaur *Ophiacodon*. Limited flexion and rotation are possible at a series of joints between the occipital condyle and the cervical vertebrae. The link between the proatlas and the occiput limits dorsoventral flexion and rotation between the skull and the atlas. The atlas arch restricts rotation between the atlas and the axis. *From Romer and Price, 1940.* (*b*) The cynodont *Thrinaxodon*. The zygopophyseal articulation between the atlas and axis is simplified to allow rotation between these elements. The pleurocentra of the atlas and axis are closely integrated to form an axis around which the atlas arch and intercentrum can rotate. *Modified from Jenkins, 1970.* (*c* and *d*) Modern mammals. (*c*) Anterior view of atlas of a cat. (*d*) Lateral view of atlas-axis complex separated to show articulating surfaces. The atlas arch and intercentrum have fused to form a ring-shaped structure. The anterior surface restricted movement by the skull to flexure in the vertical plane. Rotation and lateral flexure occur between the atlas and the axis.

magnum to positions on either side of this opening. In primitive amniotes, flexion at this joint would cause a considerable stretching at the top of the spinal cord and compression at its base. As the condyles moved dorsally to reach the midpoint of the height of the foramen magnum, these forces on the spinal cord were minimized. The paired nature of the condylar surface effectively precludes rotation and lateral bending at this joint. In *Thrinaxodon,* the condyle is clearly paired, but the proatlas is retained. The articulating surfaces are still in a relatively ventral position throughout the cynodonts.

The restriction of rotation between the head and the atlas is compensated for by specialization of the atlas-axis articulation. In primitive therapsids, rotation between these elements is limited by the well-developed zygapophyseal joints between the atlas and axis arches. The zygapophyses are reduced in early cynodonts. In addition, the centrum of the atlas becomes suturally attached and later fused to the centrum of the second vertebra to form an axislike structure around which the arches of the atlas rotate. The atlas intercentrum retains its connection with the atlas arches. They are not coossified in either advanced cynodonts or in the earliest mammals, but they must have been closely united by cartilage and ligaments. They form a functional ring in living mammals.

In primitive synapsids and other early amniotes, the cervical vertebrae extend horizontally toward the skull. In cynodonts, there is a progressive trend for them to angle dorsally toward the foramen magnum, as in mammals. As in most modern mammals, *Thrinaxodon* already has seven cervical vertebrae, which are distinguished by the retention of intercentra that are lost in the trunk region, the relatively low angle of the zygapophyses, and the simplicity of the ribs.

Trunk and Tail

The configuration of the trunk vertebrae and ribs in therapsids and early mammals shows progressive changes to restrict lateral flexion and enable the trunk to be flexed in the sagittal plane. Throughout the trunk in *Thrinaxodon,* the zygapophyses are very steeply angled to reduced lateral flexion. For the first time there is a clear distinction between the thoracic and lumbar region of the trunk as shown by the configuration of the ribs. There are 13 thoracic, 7 lumbar, and 5 sacral vertebrae. The most striking feature of the trunk region is the expansion of the proximal portion of the ribs into a broad costal plate (Figure 17-35). The ribs in the lumbar region consist of only the costal plate, without any shaft. The costal plates are logically associated with establishing greater rigidity of the column so that the trunk region could be held persistently off the ground. Their presence also suggests a reduction in the degree of lateral undulation of the column (Kemp, 1982). The costal plates are progressively reduced in several lineages of more advanced cynodonts, which may be attributed to a progressive elaboration of the epaxial musculature to support the trunk, without the need to rely on the expanded ribs. It may be recalled that the earliest terrestrial vertebrates, the ichthyostegids, had extremely broad ribs to support the trunk, but they were reduced in subsequent amphibian lineages.

In contrast with more primitive therapsids, the tail is quite short in *Thrinaxodon,* with only 10 to 15 segments. This feature seems trivial, but in fact the shortness of the tail is related to a fundamental change in the control of posture and locomotion. Both therapsids and thecodonts modified their stance from the sprawling posture of primitive amniotes, but they did so in different ways. Archosaurs retain the long tail that is characteristic of primitive amniotes. As the limbs achieved a more habitually upright posture among advanced thecodonts, the maintenance of an arched trunk was achieved by the counterbalancing of the long heavy tail. The weight of the tail also enabled some thecodonts and dinosaurs to achieve

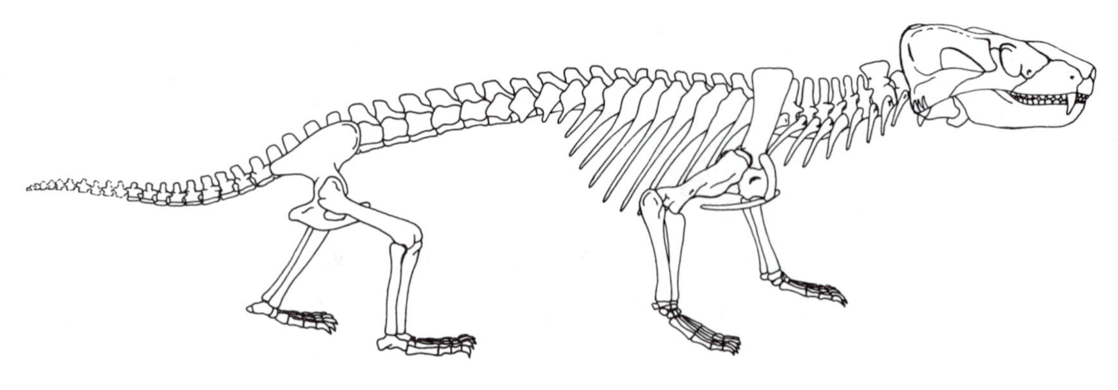

Figure 17-35. SKELETON OF THE LOWER TRIASSIC CYNODONT *THRINAXODON*, ½ METER LONG. Greatly widened ribs probably served to support the trunk and to limit lateral undulation. *From Jenkins, 1984.*

facultative and even habitual bipedality. In contrast, the tail was progressively shortened in cynodonts. The elaboration of costal plates and more effective trunk musculature may have enabled the cynodonts to shorten the tail early in their evolution, but this shortening precluded the development of bipedality at the therapsid level. Some late therapsids did redevelop a long tail, but it was a slender appendage like that of modern rodents, without the heavy musculature of their reptilian ancestors.

Changes in the pelvic girdle related to the evolution of more fore-and-aft movement of the limbs also were markedly different in the two groups. Thecodonts and primitive dinosaurs extend the pubis and ischium to increase the distance between the protractors and retractors of the femur. In cynodonts, the pubis and ischium were reduced rather than elaborated, but the iliac blade was expanded, especially in an anterior direction, to increase the area of origin for both protractors and retractors that have a much different mechanical arrangement than those of dinosaurs.

The changes in cynodonts were already well underway at the base of the Triassic, at which time the early archosaur *Chasmatosaurus* retained a pelvis that was similar to that of the most primitive diapsids.

Appendicular Skeleton

Throughout the cynodonts, changes in the structure of the girdles and limbs are associated with a more erect posture. As the limbs approached a vertical orientation, they could support the trunk without the necessity for the extensive musculature originating from the ventral portion of the girdles that was common to more primitive amniotes. There is a gradual reduction in the coracoid and puboischiadic plates and changes in the glenoid and acetabulum to resist more vertically directed forces from the limbs. The pelvic girdle achieves a structure within the cynodonts similar to that of primitive living marsupials and placentals. The pectoral girdle is slower to evolve and the forelimbs retain a sprawling posture throughout the Triassic.

In primitive amniotes, the broad ventral blade of the clavicle and the robust T-shaped interclavicle served to resist the strong medially directed force that resulted from the laterally oriented limbs. These dermal elements remain large throughout the cynodonts and into primitive mammals, which indicates that the forelimbs continued to exert a substantial medial force as a result of their sprawling posture. In the living monotremes, these bones have retained an essentially cynodont pattern.

Throughout the cynodonts, there is a gradual reduction in the extent of the coracoid and procoracoid. In *Thrinaxodon,* the coracoid and scapula share about equally in the formation of the glenoid. The small procoracoid forms a narrow portion of the anterior margin. The scapula is tall and narrow; it is concave medially to conform with the configuration of the trunk. The anterior margin is reduced relative to that of pelycosaurs, but a short process, the acromion, extends anteriorly to articulate with the stem of the clavicle. The scapular portion of the glenoid faces ventrally as well as laterally and would have received most of the weight of the body on the forelimb.

The humerus remains a heavy, complex bone, like that of pelycosaurs. The head is deflected dorsally to translate more effectively the force of the limbs to the vertical scapula. Movement of the humerus would have still been primarily in the horizontal plane and the function of the lower limb would have been similar to that in pelycosaurs. The opening of the glenoid and the development of a more distinct head would have allowed the humerus to be moved closer to the body when it was retracted, a necessary initial step in the development of the mammalian fore-and-aft swing of the limb.

Thrinaxodon had a phalangeal count of 2, 3, 4, 4, 3, with one disk-shaped phalanx in digits 3 and 4. Later cynodonts reduced the number to 2, 3, 3, 3, 3.

The pelvic girdle and rear limb of early cynodonts show a series of important changes from the configuration in pelycosaurs that presage the mammalian pattern. Knowledge of modern reptiles and mammals shows that there were also major changes in the functional arrangement of the muscles during this transition (Figure 17-36). In pelycosaurs, the acetabulum faces directly laterally. The shaft of the femur is straight and the head terminal, which indicates that it extended nearly horizontally and at right angles from the trunk. The surfaces for articulation with the tibia and fibula demonstrate that the distal end was actually elevated slightly above the acetabulum. The tibia and fibula were not held vertically but angled laterally, as is shown by their articulation with the astragalus and calcaneum.

The puboischiadic plate was extensive and expanded symmetrically anterior and posterior to the acetabulum. From its surface originated two major muscles to depress the femur or to lift the body on the limb. The puboischiofemoralis externus inserted primarily on the internal

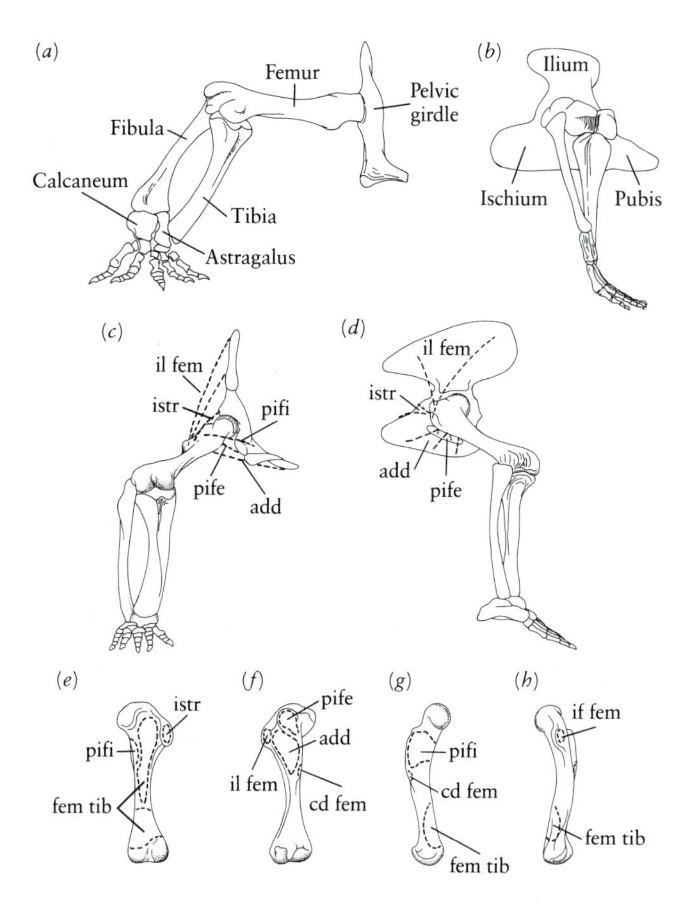

Figure 17-36. THE PELVIC GIRDLE AND REAR LIMB. Changes in their structure and function between pelycosaurs and cynodonts. (*a*) Anterior and (*b*) lateral views of the pelvis and rear limb of the pelycosaur *Dimetrodon*. (*c*) Anterior and (*d*) lateral views of the pelvis and rear limb of the cynodont *Thrinaxodon*. (*e*) Dorsal, (*g*) ventral, (*g*) anterior, and (*h*) posterior views of the femur of *Thrinaxodon* showing the position of muscle attachments. The orientation of the major muscles changes to facilitate more effective fore-and-aft movement of the limb, which is drawn closer to the body. The posture of the foot also changes in this transition from digitigrade to plantigrade. In *Thrinaxodon*, the calcaneum serves as a lever to ventroflex the foot. Abbreviations as follows: add, adductor musculature; cd fem, caudofemoralis; fem tib, femorotibialis; il fem, iliofemoralis; istr, ischiotrocantericus; pife, puboischiofemoralis externus; pifi, puboischiofemoralis internus. *From Jenkins, 1971.*

trochanter, a conspicuous ridge on the anteroventral margin of the femur just distal to the head. The adductors inserted more distally along the adductor crest.

The main retractor of the femur, which served to move the body forward on the rear limb, was the caudifemoralis. This muscle originated at the base of the tail and extended horizontally forward to insert on the prominent fourth trochanter, about one-third of the way down the femur on its ventral surface. A second retractor was the ischiotrochantericus, which originated along the dorsal surface of the ischium and inserted on the posterodorsal surface of the femur just distal to the head.

The puboischiofemoralis internus, which originated on the dorsal (or internal) surface of the puboischiadic plate anterior to the ilium and inserted on the dorsal surface of the femur, served as a major protractor, returned the femur to its anterior position following the power stroke. The femur could be lifted by the iliofemoralis, which originated on the dorsal surface of the femur just distal to its head. The orientation of the major protractors and retractors indicate that the femur moved primarily in a horizontal plane.

In Lower and Middle Triassic cynodonts, the acetabulum was more open anteriorly and ventrally. The head of the femur angles dorsally and anteriorly from the shaft, so that the bone could extend anteriorly at about a 55-degree angle to the sagittal plane and move more effectively in a dorsoventral arc. The tibia and fibula were oriented vertically.

The pelvic girdle was modified to accommodate a shift in the orientation of the major muscles, which also modified their areas of insertion on the femur. The puboischiadic plate is reduced and rotated somewhat posteriorly, which reflects a reduction in the size of the puboischiofemoralis externus and the adductor muscles and their importance in lifting the body on the rear limb. As the pubis and ischium were rotated posteriorly, the adductor and puboischiofemoralis externus could serve as retractors of the limb. The area of insertion of both muscles shifted proximally. The point of origin of the caudifemoralis was smaller, and the importance of this muscle as a retractor was certainly diminished.

The orientation and function of the ischiotrochantericus in cynodonts (termed the obturator internus in mammals) remains the same as in pelycosaurs, but its area of insertion is elaborated as a conspicuous new structure, the greater trochanter, that was just distal to the head of the femur on the posterior margin.

The origin of the iliofemoralis shifts anteriorly in cynodonts with the anterior extension of the iliac blade. Although its origin is partially anterior to the acetabulum, it acts as a retractor because the area of insertion is posterior and ventral to the inturned head of the femur. This muscle, which forms part of the glutei in mammals, serves as one of the major retractors of the femur.

The puboischiofemoralis internus retains its role as a protractor of the femur. Its insertion moves from the dorsal to the anterior surface of the femur onto the area that is known as the lesser trochanter in mammals. This tuberosity is in the same position as the internal trochanter in reptiles, and they may be considered structurally homologous although they serve for the insertion of different muscles. In early cynodonts, the puboischiofemoralis externus moved posteriorly and proximally to occupy the intertrochanteric fossa, while the puboischiofemoralis internus came to occupy the lesser trochanter. The shift in insertion of the internus to a more ventral and anterior position was necessary because of the change in the orientation of the femoral head. If the primitive dorsal insertion were retained in cynodonts and mammals, the muscle would have rotated the femur and swung the limb laterally, rather than moving it forward.

In primitive cynodonts, the origin of the puboischiofemoralis internus remained as in pelycosaurs, along the anterior surface of the puboischiadic plate. In advanced cynodonts and mammals, it migrated anteriorly and dorsally to occupy the anterior margin of the iliac blade and lumbar vertebrae and adjacent area of the body wall. In this position, the muscle is termed the iliopsoas.

A nearly mammalian configuration of the pelvic girdle and rear limb is achieved in the advanced cynodonts through the continuation of changes that were initiated at the beginning of the Triassic.

The articulating surfaces of the tibia and fibula in *Thrinaxodon* are modified to accommodate their more vertical orientation. A major functional change occurred in the ankle joint. As in the squamates and primitive archosaurs, a simple joint evolved between the lower limb and the foot that was activated through a lever system. The pattern in mammals, which is already evident in the early cynodonts, is analogous with that of crocodiles. The astragalus is closely integrated with the tibia and the main joint occurs between it and the calcaneum, which is closely integrated with the foot and develops a posterior heel. As in crocodiles, the gastocnemius muscles inserts on the calcaneal tuber. The posture of the foot is plantigrade and the phalangeal formula is reduced to 2, 3, 3, 3, 3 in middle Triassic cynodonts, but additional phalanges may have been present in *Thrinaxodon*.

ADVANCED CYNODONTS

Cynognathidae

From an ancestry among the galesaurids, the later cynodonts underwent an adaptive radiation that led to a diverse assemblage of herbivores and large and moderate-sized carnivores that were common in the middle and late Triassic and extended into the Jurassic.

Cynognathus was a large, heavily built carnivore that is diagnostic of the *Cynognathus* zone in the upper part of the Lower Triassic of southern Africa and is known from South America as well. With a skull that was as long as 40 centimeters, *Cynognathus* must have been one of the most formidable carnivores in the early Triassic (Figure 17-37). The snout is long and constricted behind the large canines. The adductor chamber is expanded laterally and posteriorly; the surrounding bones are wider and thicker than in *Thrinaxodon*, which suggests a very strong bite. The chamber does not extend as far anteriorly relative to the total length of the skull so that the gape would be wider to accommodate very large prey. The cheek teeth are laterally compressed and coarsely serrated. The occiput was widely expanded, probably for insertion of very massive trunk musculature, and confluent laterally with the ends of the very high zygomatic arches.

The dentary bone makes up a substantially larger contribution to the lower jaw than does that of the galesaurids. Superficially, it resembles the dentary of large

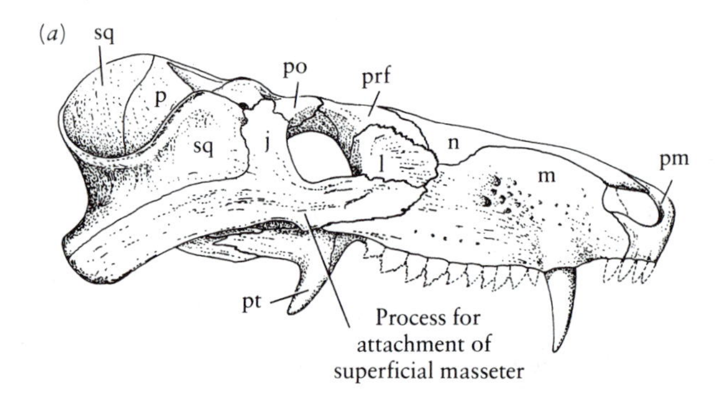

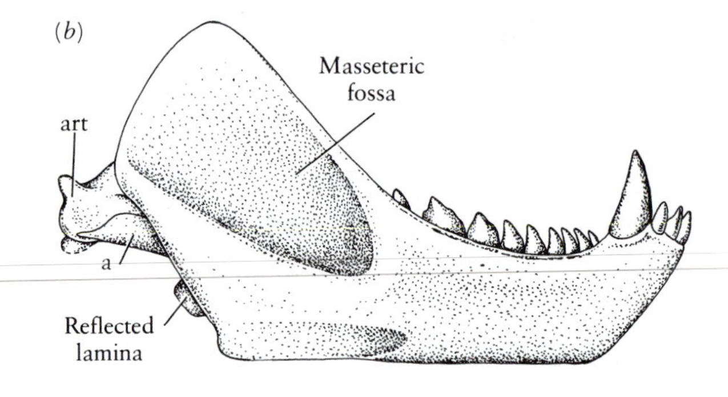

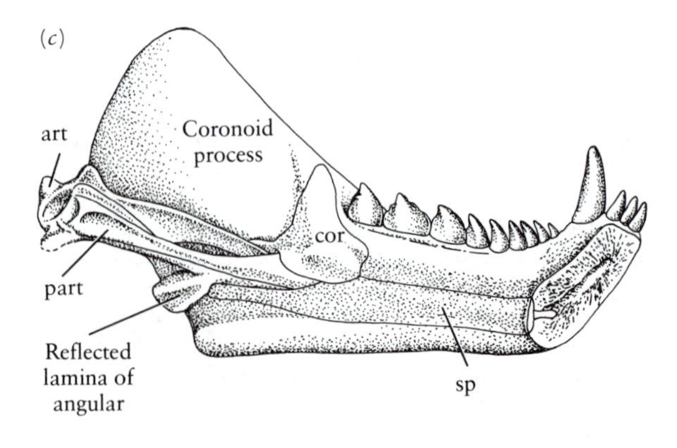

Figure 17-37. (*a*) Skull and (*b* and *c*) lower jaw of the large cynodont *Cynognathus*, maximum length 40 centimeters. Abbreviations as in Figure 8-3. (*a*) *From Broili and Schröder, 1934.* (*b* and *c*) *From Kermack, Mussett, and Rigney, 1973.*

mammalian carnivores, with a high coronoid process. The masseteric fossa is deep and extends far forward. The ventral margin of the dentary extends posteriorly as the angular process. In modern mammals, this process serves as the area of insertion for the superficial masseter, a muscle that originates on the lateral surface of the anterior portion of the zygomatic arch. In *Cynognathus*, a masseteric process associated with the origin of this muscle extends ventrally from the arch. The superficial masseter split off from the anterior margin of the earlier developed deep masseter, which is present in primitive cynodonts. The fibers of the superficial masseter are oriented posteroventrally, at nearly right angles to those of the deep masseter. Elaboration of the superficial masseter completes the major changes in the elaboration of the mammalian jaw musculature.

Although the lower jaw has a superficially mammalian appearance, closer examination shows that the dentary is accompanied by a number of smaller bones that are united in a narrow bar, which fit into a groove on its medial surface. This bar is made up of the articular, prearticular, angular, and surangular. The coronoid remains a flat plate of bone that overlaps the anterior end of the rod. The jaw articulation is formed in primitive reptilian fashion by the articular and quadrate.

The postcranial skeleton was very similar to that of *Thrinaxodon*, except for proportional differences that were associated with the greater weight of the body.

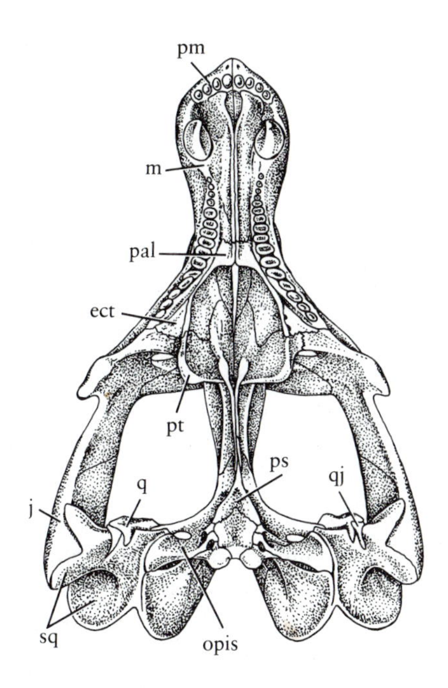

Figure 17-38. PALATE OF THE LOWER TRIASSIC GOMPHODONT *DIADEMODON*, ×¼. Abbreviations as in Figure 8-3. *From Brink, 1963.*

Gomphodonts

Accompanying *Cynognathus* in the early Triassic were primitive members of two families of herbivorous cynodonts, the Diademodontidae and the Traversodontidae, which are together referred to as gomphodonts. The primitive genus *Diademodon* (Figure 17-38) closely resembled *Cynognathus* in most cranial and postcranial features, but the dentition was markedly different in the presence of transversely expanded cheek teeth.

Like *Cynognathus*, but in contrast with galesaurids, the posterior end of the squamosal is flared laterally to form a groove that extends down toward the area of the jaw articulation. This structure remains prominent throughout the gomphodonts but is less conspicuously developed in other cynodonts. The possible function for this structure will be discussed in relationship to the jaw apparatus.

The adductor chamber of the gomphodonts was considerably larger than in *Cynognathus*, with a relatively narrow postorbital bar.

In diademodontids, the snout is narrow behind the large canines, as in *Cynognathus*. The most anterior of the cheek teeth are conical and the most posterior are laterally compressed sectorial teeth, as in primitive cynodonts. The remaining teeth are transversely expanded and relatively flat at the tip. The upper teeth are much wider than the lower.

When the teeth first erupt, the enamel is elaborated in a series of small cusps around the periphery, with a large cusp near the middle of the external margin that connects with a weak transverse crest. The enamel is lost with wear, leaving a cylinder of dentine rimmed with enamel. Galesaurids and cynognathids show some tooth wear, but there is no regular pattern of occlusion between the upper and lower teeth. In gomphodonts, the middle of the crown of the lower teeth fits between the adjacent crowns of the upper teeth.

Diademodontids are known from China and east Africa as well as southern Africa, and they extend into the Middle Triassic.

The traversodontids are a more specialized family of herbivores that shares a close common ancestry with the diademodontids. Genera from the Lower Triassic of South America show an intermediate structure between the two families. Primitively, the posterior teeth are still sectorial, but in later genera all the postcanines are transversely widened with flat occlusal surfaces. The middle Triassic genus *Massetognathus* (Figures 17-39 and 17-40) is typical of the group. The canines are greatly reduced and there is a short diastema between them and the cheek teeth. The snout is not constricted as in diademodontids but is expanded beyond the tooth row, which may indicate the presence of fleshy cheeks to retain the food, as in ornithischian danosaurs. The upper teeth are even more expanded than in diademodontids, with a strong transverse ridge on the posterior edge. This ridge would have sheared against an anterior ridge on the lower teeth. Both upper and lower teeth have an external ridge. Between the ridges is a basin where the food was crushed.

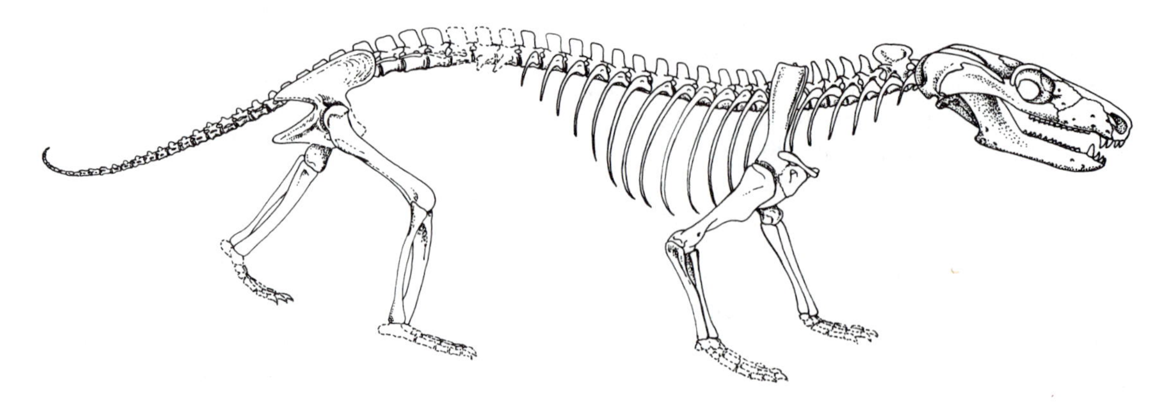

Figure 17-39. SKELETON OF THE SOUTH AMERICAN GOMPHODONT CYNODONT *MASSETOGNATHUS*, ×⅕. *From Jenkins, 1970.*

Primitive diademodontids retain broad costal plates in the lumbar region like those of the galesaurids, but the plates are absent from the cervical and anterior thoracic ribs. In contrast, advanced traversodontids have reevolved the slim ribs of typical reptiles. The vertebral column is specialized in the presence of accessory articulating surfaces that may have helped to support the trunk. The epaxial musculature, situated between the neural spines and the transverse processes, has probably become further elaborated. The ilium is elongated anteriorly and supported by up to four sacral ribs.

Traversodontids survive nearly to the end of the Triassic (Hopson, 1984). We know them from the Lower to the Upper Triassic in South America, the Middle Triassic of East Africa, and the Upper Triassic of North America, India (Chatterjee, 1983), and southern Africa.

Tritylodonts

The tritylodonts are certainly the most specialized of the herbivorous cynodonts. They do not appear until the Upper Triassic but persist into the late Middle Jurassic as the last surviving therapsids. The skull is very mammalian in general appearances (Figure 17-41). The temporal opening is huge and confluent with the orbit. The postorbital and prefrontal bones are both lost. The dentition gives them a rodentlike appearance. One pair of incisors is greatly enlarged, but the canine teeth are not developed, which leaves a long diastema between the incisors (one to three pairs) and the molariform cheek teeth. In contrast with all other therapsids, but like the mammals, the cheek teeth have multiple roots. However, the dentary retains a strictly reptilian pattern, in which the condylar process does not reach the squamosal.

The upper teeth have a squarish appearance in occlusal view, with three rows of cusps arranged longitudinally; the lower teeth have two rows of cusps. The individual cusps are crescentic; those of the upper teeth are concave anteriorly and those of the lower teeth are concave posteriorly. The lower teeth are drawn posteriorly along the grooves between the upper teeth as the jaw is closed. The pattern of tooth wear indicates that similar jaw movements occurs in traversodontids. *Massetognathus* has a pattern of cusps that could have given rise to that of the tritylodonts, but the morphological gap between the two groups is significant. Only fragmentary remains of tritylodonts have been described from the latest Triassic in South America, but they are common in the early Jurassic of China, western Europe, southern Africa, and western North America (Roth and Hopson, 1986; Sues, 1983, 1986). *Stereognathus* is known from the Middle Jurassic of England. The anterior portion of the postcranial skeleton is best known in *Kayentherium* (Figure 17-42). The limbs are short and there are no costal plates.

The generally mammalian appearance of the skull and the similarity of the dentition to that of gnawing mammals led in the nineteenth century to the identifica-

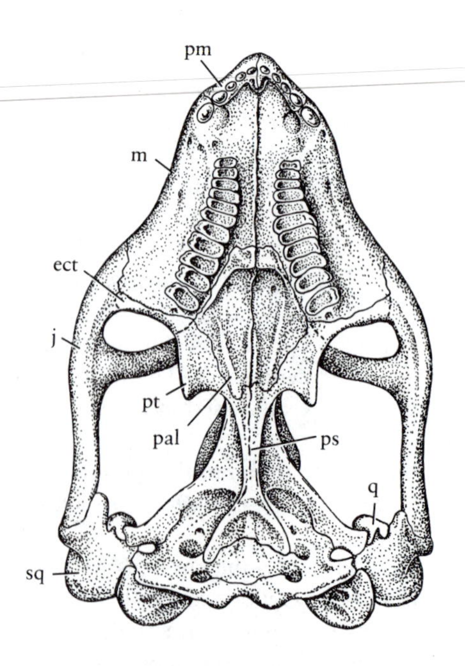

Figure 17-40. PALATE OF THE HERBIVOROUS CYNODONT *MASSETOGNATHUS*. Abbreviations as in Figure 8-3. *From Romer, 1967.*

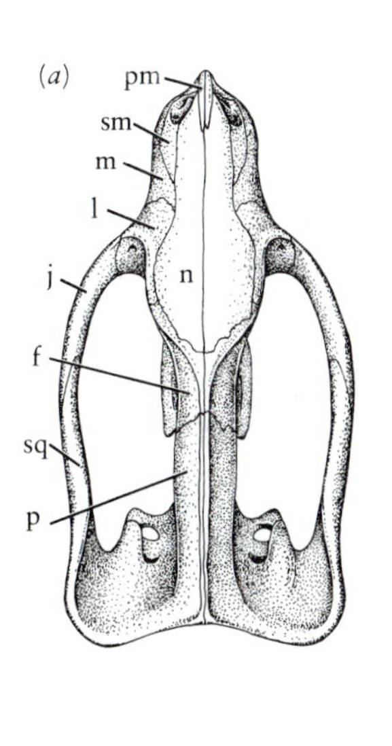

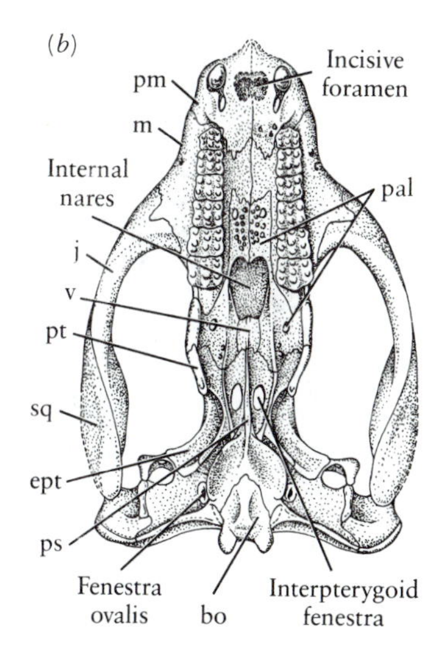

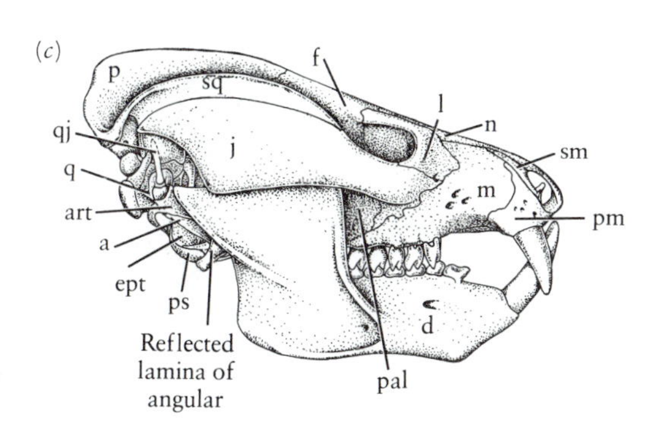

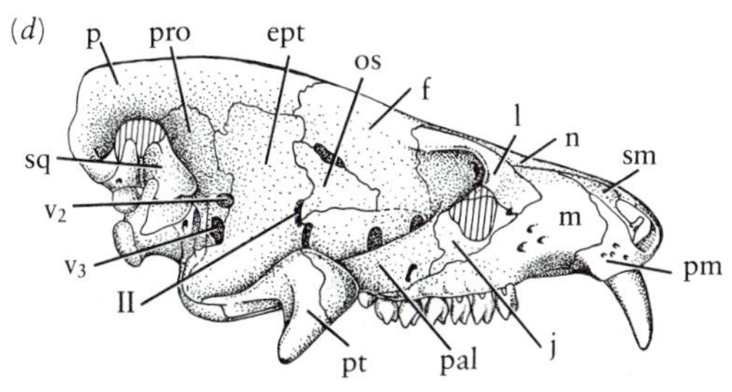

Figure 17-41. THE LOWER JURASSIC TRITYLODONT *KAYENTATHERIUM*. (*a*) Dorsal, (*b*) palatal, and (*c*) lateral views, $\times\frac{1}{4}$. (*d*) Lateral view of skull with zygomatic arch removed to show braincase, $\times\frac{1}{4}$. Abbreviations as in Figure 8-3, plus: os, orbitosphenoid; v_2, v_3, openings for branches of the Vth nerve. *From Sues, 1983.*

tion of the tritylodonts as mammals. We now know that the postcranial skeleton is extremely mammalian in many features as well. Kemp (1983) suggested that they may be the most closely related of any of the therapsids to the ancestry of mammals.

The specialization of the dentition, with the loss of canines and the complex molarization of all cheek teeth, certainly precludes all known tritylodonts from the ancestry of mammals. Kemp proposed that the ancestors of mammals would lack all the dental specializations of the tritylodonts but would have the derived postcranial features of this group. As yet, no such therapsids have been described. Alternatively, one must accept that a great deal of convergence in the structure of the girdles and limbs occurred between the ancestors of mammals and the ancestors of tritylodonts (Sues, 1985).

Chiniquodontidae

In addition to the large cynognathids and the herbivorous gomphodonts and tritylodonts, a more conservative line of small to medium-sized cynodonts is known from the Middle and Upper Triassic, the Chiniquodontidae. We recognize five genera—*Aleodon* from the Middle Triassic of East Africa and the remainder from the Middle and Upper Triassic of South America.

In all members of this group, the temporal opening is very large, extending nearly half the length of the skull (Figure 17-43). Nevertheless, it is still separated from the orbit by a narrow postorbital bar and, in contrast with tritylodonts, the postorbital and prefrontal bones are retained. However, the pineal opening is lost. The secondary palate extends back to the end of the tooth row with a major contribution from the palatine bones. The dentition

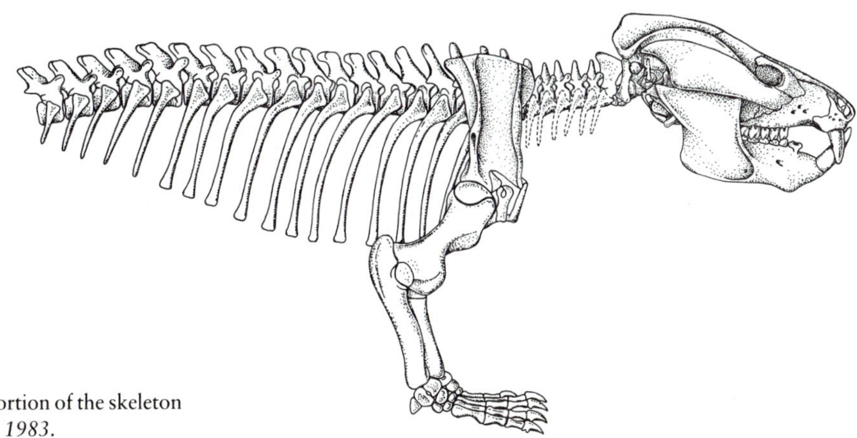

Figure 17-42. *KAYENTATHERIUM*. Anterior portion of the skeleton of the Lower Jurassic tritylodont, $\times\frac{1}{7}$. *From Sues, 1983.*

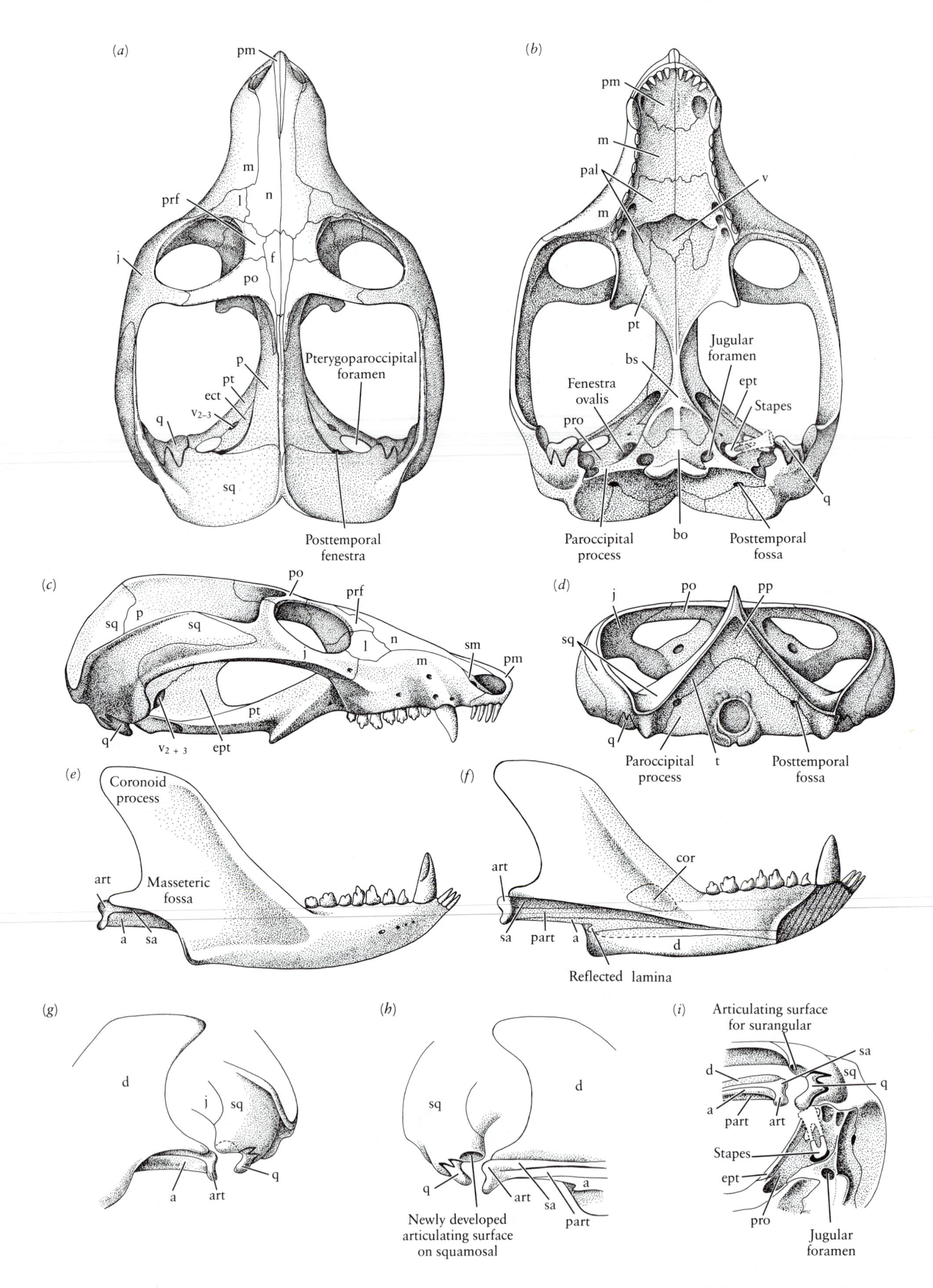

Figure 17-43. SKULL OF THE ADVANCED CARNIVOROUS CYNODONT *PROBAINOGNATHUS*. From the Middle Triassic of South America, 9½ centimeters long. (*a*) Dorsal, (*b*) ventral, (*c*) lateral, and (*d*) occipital views. (*e*) Lateral and (*f*) medial view of lower jaw. (*g*, *h*, and *i*) Details of jaw articulation. Abbreviations as in Figure 8-3, plus: V_{2+3}, openings for maxillary and mandibular branches of Vth nerve. *From Romer, 1970.*

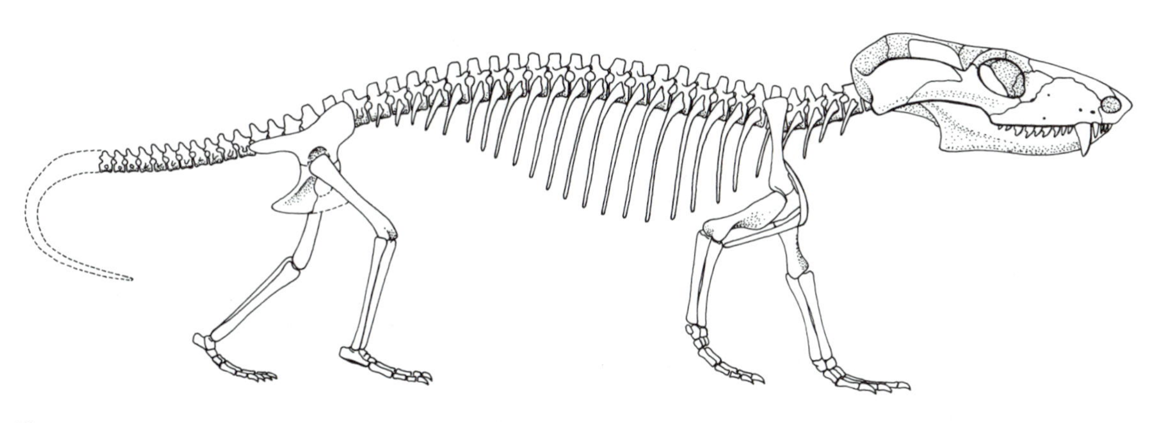

Figure 17-44. *PROBELESODON.* Skeleton of the advanced carnivorous cynodant from the Middle Triassic of South America, 65 centimeters long. *From Romer and Lewis, 1973.*

closely resembles that of the galesaurids, with four upper and three lower incisors, moderately long canine teeth, and seven laterally compressed cheek teeth with a longitudinal arrangement of cusps. In most chiniquodonts, there is only a single row of cusps, but in *Probainognathus,* a cingulum bearing cusps is present on the inner side of the teeth. The teeth show wear but they did not have a regular pattern of occlusion. One of the most notable features of *Probainognathus* from the middle Triassic of Argentina is the posterior extension of the dentary. The adjacent surangular articulates with the squamosal, forming a new jaw joint lateral to the persistent reptilian joint between the articular and quadrate.

We know most of the postcranial skeleton of the middle Triassic genus *Probelesodon* (Romer and Lewis, 1973) (Figure 17-44). It is clearly advanced over the pattern of the galesaurids in the loss of costal plates on all the ribs. The scapula is narrow and the coracoid reduced. The anterior process of the ilium is long, but the posterior portion remains long as well.

Probelesodon is less mammalian in appearance than the tritylodonts *Oligokyphus* and *Kayentherium,* but we should note that *Probelesodon* is Middle Triassic, while *Oligokyphus* is Lower Jurassic and is significantly *younger* than the earliest mammals.

Tritheledonts (Ictidosaurs)

The assemblage of genera that is grouped as the tritheledonts or ictidosaurs are among the most tantalizing, or frustrating, cynodonts. Like the tritylodonts, they are known only from the very end of the Triassic and beginning of the Jurassic. Five genera have been assigned to the group, but all are known from fragmentary remains and few have been adequately described. The skull ranges from 3 to 6 centimeters long, the temporal opening (as in tritylodonts) is confluent with the orbit, and the postorbital and prefrontal bones are lost. In *Diarthrognathus,* the frontal extends ventrally anterior to the orbit to reach a dorsal process of the palatine that extends up from the palatal surface. The dentary appears to have made contact with the squamosal. As Gow (1980) described, the dentition is unlike that of chiniquidontids and mammals in the relatively great width of the cheek teeth (Figure 17-45). In *Pachygenelus* there are only two pairs of upper and lower incisors, although *Chaliminia* has three uppers, behind which is a diastema. *Theioherpeton* from South

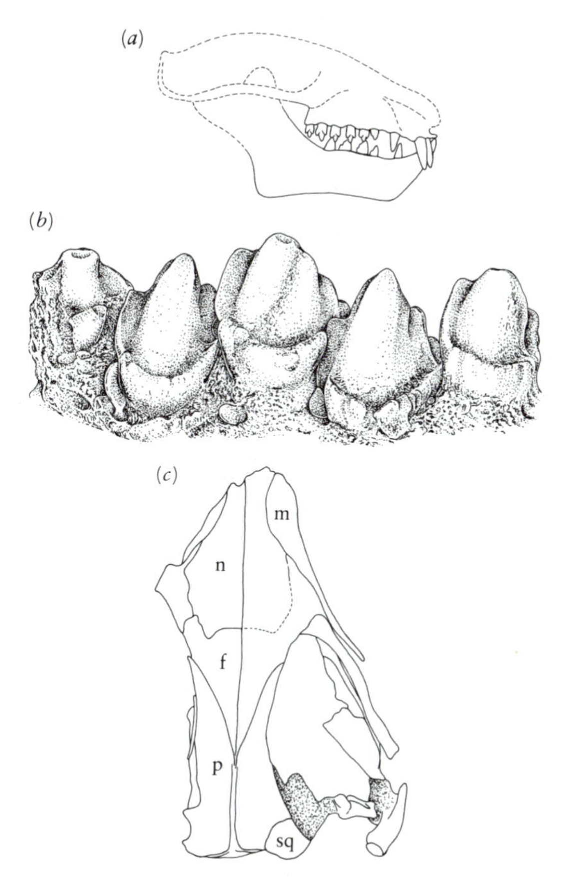

Figure 17-45. TRITHELODONTS. These advanced carnivorous cynodonts are from the Upper Triassic and Lower Jurassic. (*a*) Skull of *Pachygenelus* from the Upper Triassic of North America. *From Chatterjee, 1983. With permission from* Science. *Copyright 1983 by The American Association for the Advancement of Science.* (*b*) Teeth of a specimen of *Pachygenelus* from southern Africa. *From Gow, 1980.* (*c*) Dorsal view of the skull of *Theioherpeton* from the Upper Triassic of South America. Abbreviations as in Figure 8-3. *From Bonaparte and Barberena, 1975.*

America appears to be the most mammalian of them all with a small, low skull that has very narrow zygomatic arches and narrow, crowned cheek teeth with partially divided roots. *Diarthrognathus* and *Chaliminia* appear to have retained an interpterygoid vacuity, a feature that was present only in very primitive cynodonts, but its expression in the trithelodonts may be a result of their very small size, which may favor the retention of juvenile characters.

Because of the incomplete nature of most genera in this group, we are not certain that they are closely related. Nevertheless, they appear to show the closest approach to a mammalian morphology among small carnivorous cynodonts, which are otherwise the most appropriate ancestors of mammals.

The Origin of the Mammalian Skeleton

We see the initial stages in the origin of most of the features that characterize the mammalian skeleton in the galesaurids and chiniquidontids of the Lower to Middle Triassic. Unfortunately, the record of the immediate ancestors of mammals becomes less complete in the Upper Triassic. The trithelodonts provide only a tantalizing glimpse of small forms that may be derived from the chiniquodontids. The tritylodonts provide good evidence of the final stages in the evolution of the therapsids, but their highly specialized dentition indicates that they are not close to the ancestry of mammals.

SKULL

Most of the major features of the mammalian skull evolved among the carnivorous cynodonts. The chiniquodontids still retain the postorbital and prefrontal bones, but they are lost in both trithelodonts and tritylodonts. With the loss of the postorbital bar, the temporal opening becomes confluent with the orbit in these groups, as in the early mammals.

Among the chiniquodontids, the secondary palate extended posteriorly to essentially the same extent as in primitive mammals. Important changes in the dentition between cynodonts and mammals will be discussed in the following chapter.

The major features of the mammalian braincase were already established in the carnivorous cynodonts (Figure 17-46). In modern mammals, the skull is dominated by a large, bulbous braincase that is formed largely by dermal bones, notably the frontal, parietal, and squamosal. In contrast, the braincase of primitive amniotes is formed primarily of cartilage and cartilage replacement bones. The changes that led to the mammalian type of braincase began among the early therapsids not as a means of housing a larger brain but to provide more extensive areas of origin for the jaw muscles and to strengthen the skull to resist the forces of mastication. Among the therocephalians, there is already a tendency for the dermal bones of the skull table to extend ventrally and anteriorly from the dorsal and posterior margins of the adductor chamber, but this process is much more fully elaborated among the cynodonts. In *Thrinaxodon,* the squamosal forms the posterior border of the adductor chamber and extends ventrally to the level of the otic capsule. The parietal extends laterally and ventrally over the original, cartilaginous lateral wall of the braincase. Posteriorly, the endoskeletal bones of the otic-occipital portion of the primitive amniote braincase are retained. Throughout the cynodonts, the opisthotic is exposed at the base of the occipital surface as a rodlike paroccipital process, and the postparietal, tabular, and supraoccipital are separated by persistent sutures, as are the exoccipitals and basioccipital. These bones become fused into a unitary occipital plate in advanced mammals.

In pelycosaurs, as in primitive amniotes, the proötic is a simple bone that forms the anterior portion of the otic capsule and surrounds the anterior margin of the fenestra ovalis. In cynodonts, the proötic, which is now termed the periotic, becomes elaborated anteriorly. Dorsally, it forms sutural contacts with the ventral lamellae of the parietal and squamosal.

In early synapsids, the epipterygoid was a narrow rod of bone that extended dorsally from the pterygoid at the level of the basicranial articulation. In cynodonts, it expands into a broad plate that is suturally connected to the parietal dorsally and the periotic posterodorsally to form a bony wall that is lateral to the primitive braincase and anterior to the otic capsule. With its incorporation into the braincase, the epipterygoid may now be designated by the name used in mammals, the alisphenoid. The alisphenoid differs from all other elements of the mammalian braincase in tracing its origin to a visceral arch component.

The posteroventral portion of the epipterygoid, or alisphenoid, in cynodonts lies lateral to the periotic. Between these bones is a space termed the cavum epiptericum that contains the semilunar and geniculate ganglia of the Vth nerve in modern reptiles. This area is open ventrally in cynodonts and is not yet really within the braincase.

In primitive cynodonts, the foramen for the mandibular branch of the trigeminal nerve opens between the periotoic and the alisphenoid. The ophthalmic and probably the maxillary branches ran forward medial to the alisphenoid and emerged along its anterior margin. The VIIth (facial nerve) passed out through the periotic behind the foramen for the mandibular branch of the Vth.

Anteriorly, the periotic in *Thrinaxodon* is divided into a superficial sheet of tissue that extends lateral to the alisphenoid and a more medial structure, the anterior lamina (Figure 17-46*d*). The relative extent of these two parts of the periotic are important in distinguishing early members of the major mammalian lineages in the later Me-

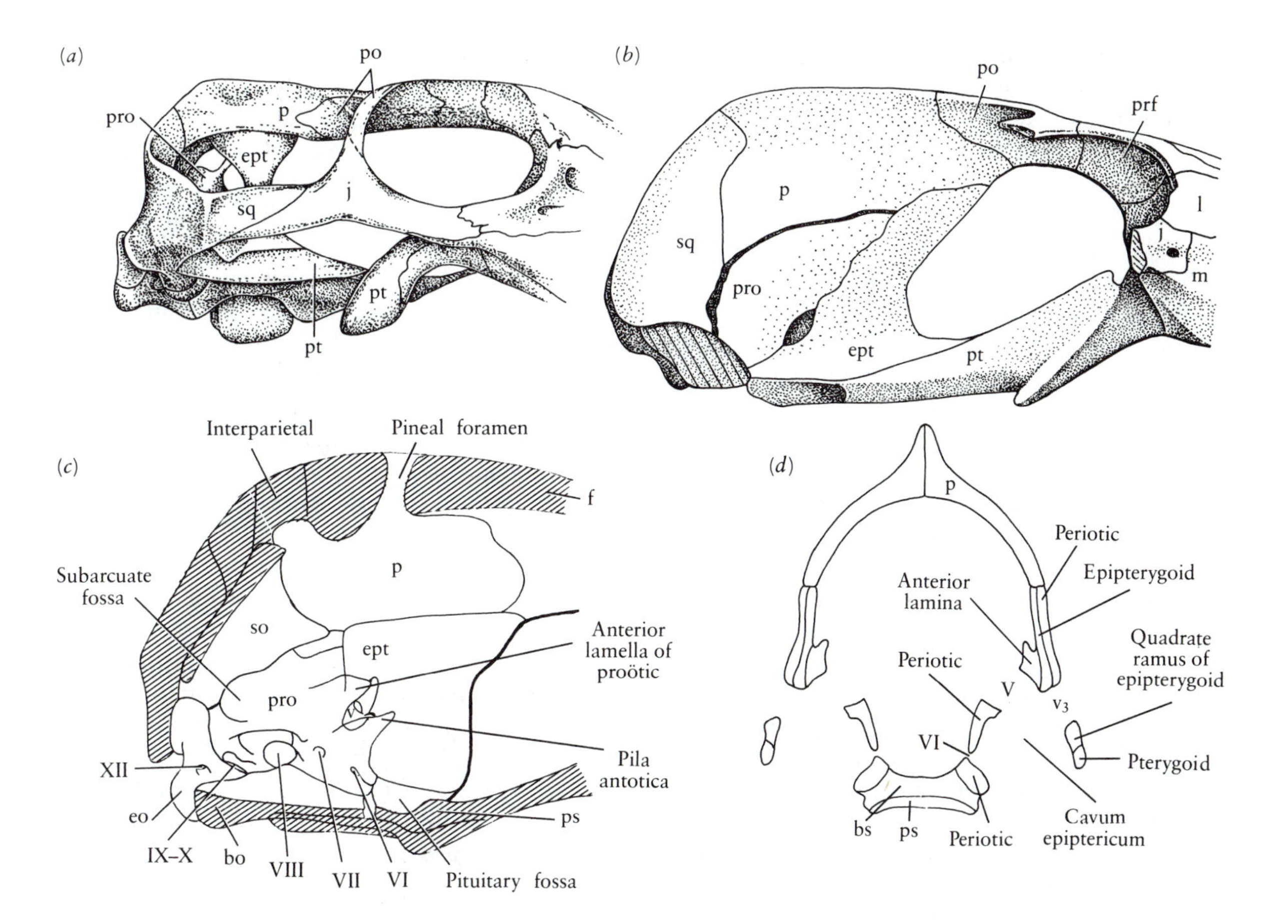

Figure 17-46. EVOLUTION OF THE MAMMALIAN BRAINCASE. (*a*) Posterior portion of the skull in the primitive therapsid *Regisaurus*. As in primitive amniotes, the braincase is still formed primarily by endochondral bones and cartilage. The parietal and squamosal are primarily superficial elements. The epipterygoid forms a broad vertical pillar that is lateral to the brain. *From Mendrez, 1972.* (*b*) Lateral view of the braincase in *Probaignathus*, a Middle Triassic cynodont. The parietal and squamosal have spread ventrally to provide additional surface for the origin of the jaw musculature. They reach the otic capsule to form a continuous wall of the braincase posteriorly. The epipterygoid is suturally attached to both the otic capsule and the parietal to form a partial wall, lateral to the original cartilagenous braincase. The area of the braincase medial to the orbit remains unossified. *From Romer, 1970.* (*c*) Medial view of the lateral wall of the braincase in the Lower Triassic cynodont *Thrinaxodon*. *From Crompton and Jenkins, 1979.* (*d*) Transverse section through the back of the right side of the braincase in *Thrinaxodon* to show the position of the cavum epiptericum between the periotic and the epipterygoid. *From Crompton and Jenkins, 1979.* (*c and d*) *By permission of the University of California Press.* Abbreviations as in Figure 8-3, plus: V_3 opening for mandibular ramus of Vth nerve.

sozoic. Anterior to the alisphenoid and medial to the orbit, the lateral wall of the braincase remains cartilaginous.

Endocasts of the braincase of cynodonts indicate that the brain size is not significantly larger than that of reptiles of comparable body weight. The brain of cynodonts is distinguished by the presence of a floccular area adjacent to the inner ear that is concerned with balance in modern mammals and may be related to the more continuously upright posture of cynodonts (Ulinski, 1986).

THE ORIGIN OF THE MAMMALIAN MIDDLE EAR

In chiniquodontids and tritylodonts, the lower jaw approaches the mammalian pattern in the large size of the dentary. Throughout the cynodonts, the coronoid process expands dorsally to provide additional area for attachment of the adductor muscles and to provide better leverage to close the jaw. At the level of the cynognathids, the angular process expanded to accommodate the origin of the superficial masseter. Chiniquidontids also elaborate the condylar process, which in mammals forms the jaw articulation. Bramble (1978) pointed out that selection to increase the length of the condylar process must have acted initially to support the extradentary bones, since there would have been no selective advantage for developing the dentary-squamosal joint until contact was actually made between these bones.

Factors that are associated with mastication do not serve to explain the concomitant changes in the extradentary bones. If strengthening of the lower jaw for more effective resistance to the forces of mastication were the primary selective factor, one would expect that the extradentary bones, including the articular, would remain solidly attached to the dentary in all therapsids. In contrast, the earliest cynodonts (and also the therocephalians) have lost the sutural connection between the dentary and the other jaw bones, which appear to be able to move independently to at least a limited degree. The progressive reduction in size and loose attachment of the postdentary bones has long been an enigmatic aspect of the evolution of mammals. The most plausible explanation is that elaborated by Allin (1975) in relationship to the origin of the mammalian middle ear.

All modern mammals differ from other amniotes in the presence of three ear ossicles, the malleus and incus in addition to the stapes which is homologous with the single ossicle of reptiles and birds. As Manley (1972) demonstrated, the acuity of hearing in birds with a single ossicle is as great as that in mammals, so that incorporation of the additional bones has no obvious selective advantage but bespeaks a very different evolutionary history. The configuration of the back of the skull and the orientation and large size of the stapes in early pelycosaurs (see Figure 17-3) indicate that it could not have participated in an impedance-matching system like that of lizards, crocodiles, or turtles. Pelycosaurs may have heard loud, low-frequency sounds through the general surface of the cheek and lower jaw, as do modern earless lizards and snakes that have a maximum sensitivity between 200 and 500 hertz. Like the ancestors of other amniote groups that integrated the stapes into the middle ear, pelycosaurs solidified the attachment of the braincase to the skull roof, which freed the stapes from the supporting role it had in early amniotes. For some reason, the stapes remained massive in synapsids and there is no evidence for the presence of a reptilian type of tympanum attached to the back of the cheek. The failure to develop a tympanum in this position in the ancestors of mammals may be related to the configuration of the back of the skull and lower jaws in pelycosaurs, specifically sphenacodonts, which differs from that of the ancestors of lizards, turtles, and crocodiles. In sphenacodonts and early therapsids, the jaw articulation and the distal end of the stapes are below the level of the tooth row, behind the back of the lower jaw. If sound were to be transmitted along the length of the stapes to the inner ear, the most appropriate area for its reception would be the surface of the lower jaw rather than the back of the cheek as in conventional reptiles.

The first evidence of a specialized area for the reception of airborne vibrations in pelycosaurs is the reflected lamina of the angular in early sphenacodont pelycosaurs. It is difficult to accept that this structure was effective in sound transmission at this stage of synapsid evolution, but the subsequent role of the angular in supporting the tympanum is nearly indisputable if one examines its structure in later cynodonts and the earliest mammals (Figure 17-47). Allin (1975) suggested that the reflected lamina in pelycosaurs may have acted somewhat as a tympanum. To be effective, the underlying space must have been air filled, which could have occurred if a diverticulum from the eustachian tube were to extend into this area along the surface of the pterygoideus muscle.

The significance of the rear of the jaw to hearing is much easier to accept in early cynodonts, in which there is unquestionable mobility of the postdentary bones. In *Procynosuchus* and *Thrinaxodon,* the reflected lamina of the angular is extensive and well separated from the body of the bone. Allin suggested that all the postdentary elements vibrated as a unit. The movement of the articular would be translated to the quadrate, which was only loosely attached to the squamosal. The loose attachment of the quadrate is only explicable in relationship to its mobility in consort with the bones in the lower jaw. In turn, the quadrate would have activated the stapes, which is still a large element but is lightened by the presence of a very large opening for the stapedial artery.

In later cynodonts—both the herbivorous gomphodonts and tritylodonts and the carnivorous chiniquodontids—the postdentary bones are further reduced. The angular is reduced to a narrow rod that lies parallel with the prearticular and surangular. The reflected lamina consists of a narrow process that extends posteroventrally toward the retroarticular process of the articular. Together, these processes form an incomplete ring. According to Allin, the large bony surface of the reflected lamina in early cynodonts was replaced by a membrane (the tympanum) that stretched to the articular. These elements form a structure that is comparable, except for its larger size, with the tympanic ring and malleus in modern mar-

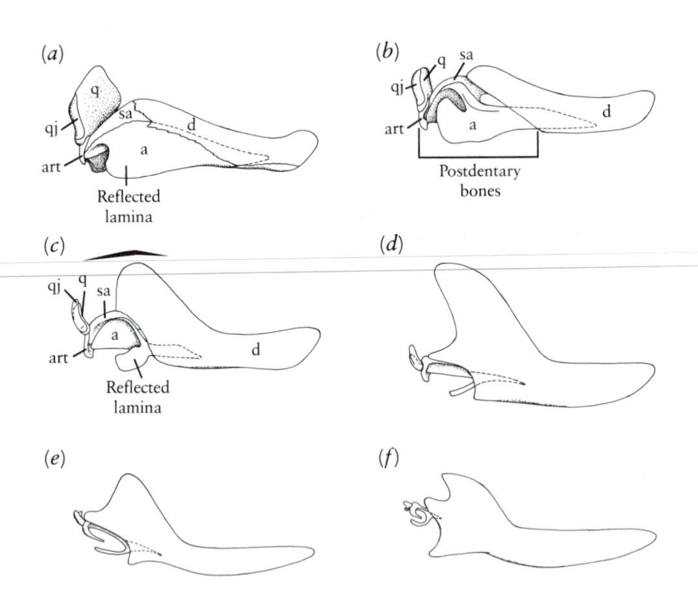

Figure 17-47. PROGRESSIVE CHANGES IN THE STRUCTURE OF THE JAW AND ELEMENTS OF THE MIDDLE EAR FROM PELYCOSAURS TO MAMMALS. All in lateral view. (*a*) The sphenacodont pelycosaur *Dimetrodon*. All the elements of the lower jaw are suturally attached. The angular bears a reflected lamina. (*b*) Condition of an advanced therocephalian. Postdentary bones are no longer suturally attached to the dentary, but they remain very large. (*c*) *Thrinaxodon*, a primitive cynodont. (*d*) The advanced cynodont *Probainognathus*. (*e*) An early Jurassic mammal, *Morganucodon*. (*f*) Hypothetical reconstruction of a Jurassic panthothere. The ear ossicles are not known in the immediate ancestors of placentals and marsupials, but they would have been about this size. Abbreviations as in Figure 8-3. *From Allin, 1975.*

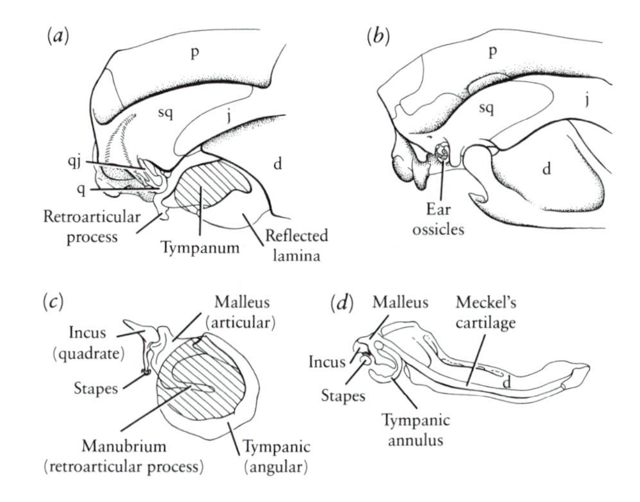

Figure 17-48. (*a*) Jaw and ear region in a mammal-like reptile, *Thrinaxodon*, and (*b*) a living opossum. (*c*) Enlarged view of ear ossicles in the opossum. (*d*) Medial view of the jaw in a fetal mammal to show the parallel between the ontogenetic development of the middle ear and jaw joint complex with the phylogenetic origin of this structure. Abbreviations as in Figure 8-3. *From Crompton and Jenkins, 1979. By permission of the University of California Press.*

supials (Figure 17-48). The malleus articulates with the incus in exactly the same way as the articular articulates with the quadrate in advanced therapsids and the quadrate (incus) articulates with the stapes.

The ear ossicles in adult mammals are functionally very distinct, since they do not serve a role as elements in the jaw articulation, as was the case in all therapsids. Indicative of this evolutionary transformation, the function of these bones changes in every generation of marsupials. During their development in the pouch, the malleus and incus retain the reptilian role of the articular and quadrate. Only when the young leave the pouch do these bones separate from the lower jaw and enter the middle ear.

In *Probainognathus*, the surangular and dentary extend back to the squamosal to form a second jaw articulation. As the dentary is further elaborated, we can recognize two functional jaw joints, a medial reptilian joint, which consists of the articular and quadrate, and a lateral mammalian joint, which is formed by the dentary and squamosal. Both are retained in early mammals (see Chapter 18). It is not yet certain when the malleus and incus became incorporated into the middle ear, but the grooves on the medial surface of the dentary that indicate their position of attachment in early Jurassic mammals are missing in Upper Jurassic genera.

JAW MECHANICS

Selection for improved reception of airborne vibrations must have remained strong throughout the evolution of the cynodonts, even against what must have been strong pressure for greater solidification of the elements of the lower jaw.

It is difficult to understand how the jaws of advanced cynodonts resisted the forces placed on them by the increasing volume of jaw musculature. Bramble (1978) elaborated an elegant model to explain how the forces of mastication were distributed.

We have long recognized that even in modern mammals the area of jaw articulation is small and probably cannot resist the full force of the jaw muscles. This problem would have been much greater in the advanced cynodonts. Although the forces of the various components of the jaw musculature are in different directions, all would appear to place pressure on the jaw articulation if it is viewed as the only fulcrum about which the jaw rotates.

The most important feature of Bramble's model is the recognition that the position at which the food is held in the mouth will act as a second fulcrum. This point is most clearly evident if we consider a moderately advanced cynodont such as *Trirachodon* and concentrate on a single muscle, the temporalis (Figure 17-49). The force of the muscle acts posterodorsally on the coronoid process. One may continue this line anteroventrally to intersect the tooth row. This line is termed the projected line of action. If the food being bitten is anterior to this intersection, the force of the muscle lifts the back of the jaw and exerts a

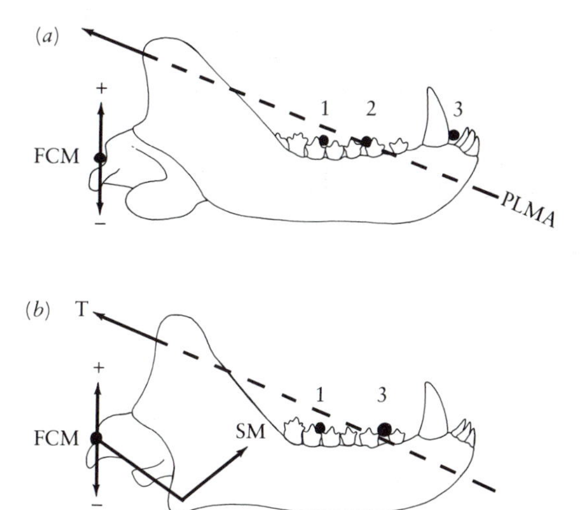

Figure 17-49. RESOLUTION OF FORCES ON THE JAW JOINT OF A MAMMAL-LIKE REPTILE. *Based on the bi-fulcral model of Bramble, 1978.* Abbreviations as follows: PLMA, projected line of muscle action, extended anteriorly from the temporalis; FCM, fulcrum of craniomandibular joint; T, temporalis; SM, superficial masseter; 1, 2, 3, occlusal fulcra. (*a*) When the occlusal fulcrum is at 3, near the front of the mouth, contraction of the temporalis places a positive force on the jaw articulation. When the occlusal fulcrum is at 1, near the back of the jaw, the force on the jaw articulation is negative. When the occlusal fulcrum is directly in line with the projected line of muscle action, the force on the jaw articulation is neutral. (*b*) Representation of the counterbalancing of forces produced by the temporalis and the superficial masseter. The superficial masseter, whose force is represented by the arrow SM, acts at nearly right angles to the temporalis, T. When the occlusal fulcrum is near the front of the jaw, 3, the force on the jaw joint is negative, and when it is at 1, the force is positive. There will be a position near the middle of the jaw where the forces of the temporalis and superficial masseter neutralize one another.

positive pressure on the jaw articulation. If the food, and so the fulcrum, is posterior to the intersection, the back of the jaw is lowered and produces a negative force on the jaw joint. The further posterior the point of bite, the greater the negative force. The deep masseter acts in the same direction as does the temporalis.

In contrast, the superficial masseter is oriented so as to give a balancing effect. If that muscle is considered alone, the greatest positive force is applied when the food is at the back of the jaw, and the greatest negative force is when the food is in an anterior position.

When food is held at the back of the jaw, the strong negative force produced by the temporalis and deep masseter are balanced by the positive force of the superficial masseter. It would be possible for the forces to be balanced all along the jaw by reducing the force of one or the other muscle masses. In practice, the negative force of the temporalis dominates in the posterior portion of the jaw and the force is positive for food held anteriorly. In no area is the force on the jaw articulation as great as it would be if only the temporalis or the superficial masseter were active.

The balance between the forces differs in relationship to the height of the coronoid process and the orientation of the temporalis. During the course of cynodont evolution, the projected line of action intersects the jaw at a progressively more anterior position (Figure 17-50). In *Probainognathus,* it passes over the end of the jaw so that the force on the jaw articulation is negative no matter where along the jaw the bite may be.

Allin suggested that this negative force had to be compensated for by the depressor mandibulae, which originated on the back of the skull and inserted on the articular to prevent the dislocation of the lower jaw. He placed the origin of the muscle within the trough that was formed by the lateral flaring of the squamosal. Other workers, including Sues (1983), reconstructed an external auditory meatus within the groove and argued that the retroarticular process of the articular was too fragile to serve as the origin for a strong depressor.

POSTCRANIAL SKELETON

The latest of the chiniquodontids in which the postcranial skeleton is well known is *Probelesodon* (see Figure 17-44). The limbs are considerably more gracile than in primitive cynodonts, but the structure of the girdles and limbs have not advanced far from the pattern of the galesaurids. As Kemp (1982, 1983) emphasized, the postcranial skeleton of the late Triassic and early Jurassic tritylodonts is far closer to the mammalian pattern than is that of the Middle Triassic chiniquodontids.

In both groups, the costal plates, which are so conspicuous in primitive cynodonts, are lost and the ribs have an appearance that is common to most reptiles and early mammals. The axial musculature probably assumed a much more important role in supporting the trunk. Even the extra zygapophyseal articulations, which are common in more primitive forms, are lost. Surprisingly, the distinction between the thoracic and lumbar ribs that is evident in *Thrinaxodon* is lost in the most advanced cynodonts, although tritylodonts show a definite distinction between thoracic and lumbar vertebrae.

The axis centrum of tritylodonts assumes a mammalian configuration with the elaboration of a large anterior dens, which is only an incipient feature in early cynodonts. However, the atlas arch and intercentrum are not yet fused to one another. The ends of the centra are flat or platycoelous, as in mammals, with the loss of the notochordal remnants between the centra that were reflected in the amphicoelus configuration in more primitive synapsids. In tritylodonts, there are just two sacral vertebrae, as in the early mammals, and the more posterior caudal vertebrae are elongated to form a long, but slender, tail.

Although it is still far more primitive than that of early placental and marsupial mammals, the shoulder girdle closely approaches the pattern of the primitive mammals of the early Jurassic. The coracoid elements are further reduced than in earlier cynodonts, and the glenoid is very open laterally (Figure 17-51). The acromion and the

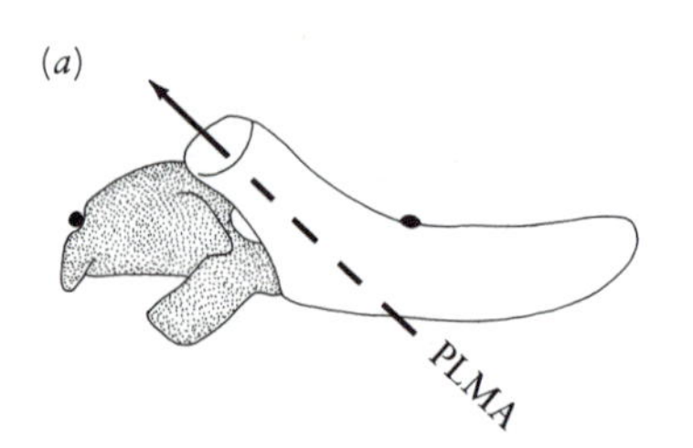

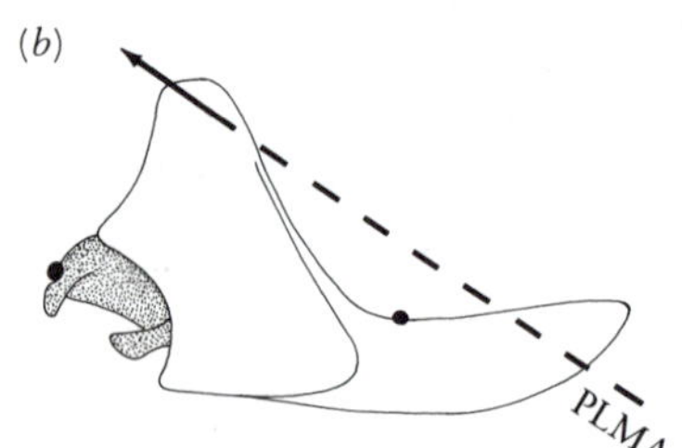

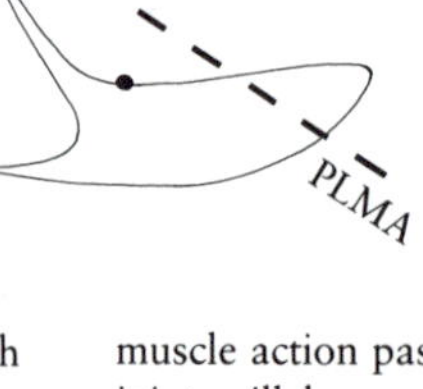
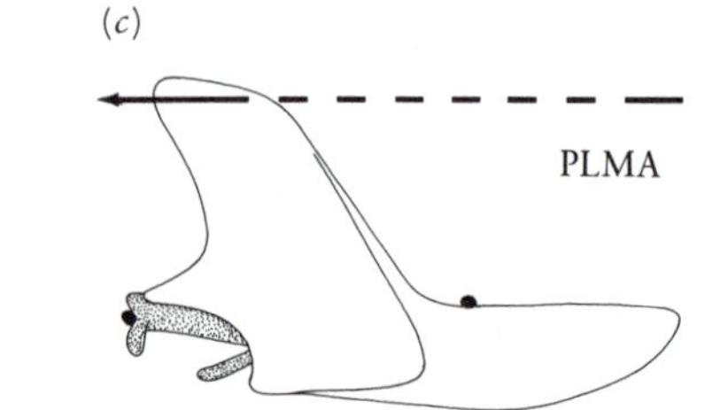

Figure 17-50. JAWS OF THREE CYNODONTS. The way in which the progressive heightening of the coronoid process changes the forces acting on the jaw joint is shown. Only the temporalis is considered. (*a*) In *Procynosuchus,* the projected line of muscle action crosses the lower jaw behind the teeth row, and the force on the jaw joint is positive no matter where the food is bitten. (*b*) In *Trirachodon,* the projected line of muscle action intersects the middle of the tooth row, so that a more anterior bite will produce a positive force but a more posterior bite will produce a negative force. (*c*) In *Probainognathus,* the projected line of muscle action passes over the end of the jaw, and the force on the jaw joint will be negative no matter where the food is bitten. With the evolution of the superficial masseter, cynodonts were able to exert a progressively stronger bite on the food while the jaw articulation was stressed to a smaller extent. This process explains how the extradentary elements could have been reduced in late therapsids in relationship to the evolution of the mammalian middle ear. Abbreviation as in Figure 17-49. *From Bramble, 1978.*

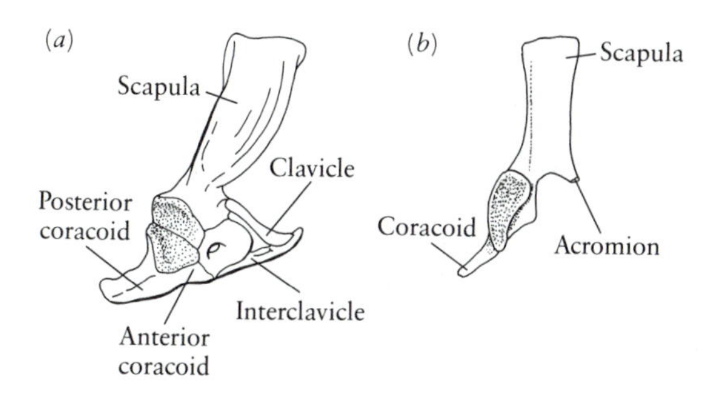

Figure 17-51. CHANGES IN THE SHOULDER GIRDLE IN ADVANCED THERAPSIDS. (*a*) A cynodont. *From Jenkins and Parrington, 1976.* (*b*) The tritylodont *Oligokyphus. From Kemp, 1982.*

base of the anterior margin of the blade of the scapula are angled laterally away from the anterior margin of the coracoid region and apparently provided passage for the extension of a portion of the supracoracoideus muscle, which is thought to have originated on the medial surface of the scapular blade. This muscle is homologous with the supraspinatus in advanced mammals and helps to stabilize the shoulder joint when the humerus is held in a more upright position.

The humerus has a long shaft and a large hemispherical head. According to Kemp (1982), the forelimb had achieved a primitive mammalian posture within the tritylodonts.

The pelvic girdle and rear limb also achieved an essentially mammalian appearance in the Lower Jurassic tritylodonts (Figure 17-52). The ilium is long and directly anterodorsally, with the loss of the posterior expansion that was common to early synapsids. A ridge running the length of the lateral surface of the ilium separates the area of origin of the dorsal gluteus (iliofemoralis) from the ventral iliacus and psoas (puboischiofemoralis internus). The pubis and ischium have rotated posteriorly so that they are largely posterior to the acetabulum. The three bones remain as separate areas of ossification in all cynodonts but become indistinguishably fused in mammals. A large obturator foramen, whose initial development occurred in early cynodonts, is evident.

The femur of tritylodonts is essentially straight, with a large, medially inflected, hemispherical head. The posture and gait of the rear limb may have been comparable to those of primitive modern mammals.

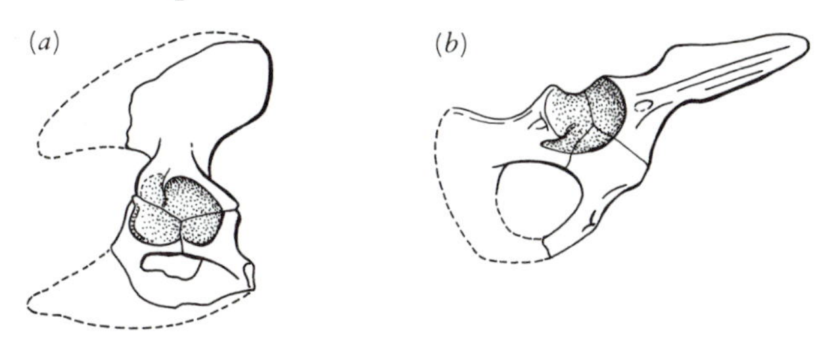

Figure 17-52. CHANGES IN THE PELVIC GIRDLE BETWEEN PRIMITIVE AND ADVANCED THERAPSIDS. (*a*) The cynodont *Thrinaxodon.* (*b*) The tritylodont *Oligokyphus. From Jenkins and Parrington, 1976.*

SUMMARY

The ancestry of mammals can be traced via the mammal-like reptiles to the base of the amniote radiation in the early Pennsylvanian. The mammal-like reptiles are grouped in two orders, the Pelycosauria and the Therapsida within the subclass Synapsida. Early pelycosaurs are differentiated from other primitive amniotes by their larger size, dentition and skull proportions (which indicate a carnivorous diet), and the presence of a lateral temporal opening. Pelycosaurs were the most common amniotes in the late Pennsylvanian and early Permian. We recognize several families, including the primitive ophiacodonts, the herbivorous edaphosaurs and caseids, and the large, carnivorous sphenacodontids. The large "sail" of the sphenacodontid *Dimetrodon,* which was supported by very elongate neural spines, probably functioned to absorb and radiate heat, indicating that the most advanced pelycosaurs were ectothermic, as are most reptiles, and had a low metabolic rate.

The transition between pelycosaurs and therapsids has not been documented. It may have involved an environmental shift, as well as changes in morphology and physiology. The therapsids are already quite diverse when they first appear in the Upper Permian of Russia. The skull of the primitive eotitanosuchids resembles that of sphenacodont pelycosaurs, especially in the presence of a reflected lamina of the angular, but the limbs could be held in a more upright position and moved in a more nearly fore-and-aft direction. Primitive therapsids include both carnivores and large herbivorous genera that were common in both Russia and southern Africa in the late Permian. The most common therapsids were the herbivorous dicynodonts, in which the dentition was functionally replaced with a horny beak. They sheared their food with a posteriorly directed, nearly horizontal stroke. Dicynodonts were extremely varied in habitat and body size, but all had a very similar jaw structure.

We recognize three major groups of carnivorous therapsids, the primitive gorgonopsians, which were restricted to the Upper Permian; the more diverse therocephalians; and the cynodonts, which led to the mammals.

The most significant factor that differentiated advanced therapsids from primitive amniotes was an increased metabolic rate, which is indicated in primitive cynodonts by expansion of the adductor chamber, more complex cheek teeth, and formation of a secondary palate, which can be associated with prolonged mastication. Among cynodonts, the structure and function of the atlas-axis complex approaches the mammalian condition, as does the pattern of the girdles and limbs.

Advanced cynodonts include both carnivorous and herbivorous lineages. The herbivorous tritylodonts approached the mammalian pattern of the postcranial skeleton very closely but were strongly divergent in the loss of the canines and anterior cheek teeth and the elaboration of crushing and grinding posterior teeth.

The carnivorous chiniquodontids are the most mammal-like of adequately known cynodonts. The secondary palate is highly developed and the cheek teeth are laterally compressed with a longitudinal series of cusps, as in early mammals. In contrast with the condition in tritylodonts, the dentary approaches the squamosal as an initial stage in the formation of the mammalian jaw joint. The braincase in cynodonts assumed a mammalian configuration by the ventral elaboration of the parietal and frontal and integration of the epipterygoid (or alisphenoid) into the skull wall anterior to the otic capsule. These changes can be attributed to extension and strengthening of the areas of origin for the jaw musculature. The brain itself remained relatively small.

Mammal-like reptiles evolved an impedance-matching middle ear in an entirely different way than did the ancestors of modern reptiles and birds. Among pelycosaurs, the reflected lamina of the angular in sphenacodonts probably functioned as a tympanum. Its vibrations were transmitted to the inner ear via the articular, quadrate, and stapes. Among therapsids, the bones at the back of the jaw are not suturally attached to the dentary but were able to move separately, which facilitated transmission of vibrations. The extradentary bones were gradually reduced but remained part of the lower jaw even among early mammals. The configuration and relative position of these bones indicate without any doubt that the quadrate, articular, and angular of reptiles are homologous with the incus, malleus, and tympanic of modern mammals.

Carnivorous cynodonts became progressively rare toward the end of the Triassic. We cannot yet recognize the specific lineage that led to mammals.

REFERENCES

Allin, E. F. (1975). Evolution of the mammalian middle ear. *J. Morph.*, **147**: 403–438.

Allin, E. F. (1986). The auditory apparatus of advanced mammal-like reptiles and early mammals. In N. Hotton III, P. D. MacLean, J. J. Roth, and E. C. Roth (eds.), *The Evolution and Ecology of Mammal-Like Reptiles*, pp. 283–294. Smithsonian Institution Press, Washington, D.C.

Barghusen, H. R. (1973). The adductor jaw musculature of *Dimetrodon* (Reptilia, Pelycosauria). *J. Paleont.*, **47**: 823–834.

Barghusen, H. R. (1975). A review of fighting adaptations in dinocephalians (Reptilia, Therapsida). *Paleobiology*, **1**: 295–311.

Barghusen, H. R. (1976). Notes on the adductor jaw musculature of *Venjukovia*, a primitive anomodont therapsid from the Permian of the U.S.S.R. *Ann. S. Afr. Mus.*, **69**: 249–260.

Bennett, A. F., and Dawson, W. R. (1976). Metabolism. In C. Gans and W. R. Dawson (eds.), Vol. 5, *Biology of the Reptilia*, pp. 127–211. Academic Press, London.

Bonaparte, J. F., and Barberèna, M. C. (1975). A possible mammalian ancestor from the Middle Triassic of Brazil (Therapsida-Cynodontia). *J. Paleont.*, **49**(5): 931–936.

Bramble, D. M. (1978). Origin of the mammalian feeding complex; models and mechanisms. *Paleobiology*, **4**: 271–301.

Bramwell, C. D., and Fellgett, P. B. (1973). Thermal regulation in sail lizards. *Nature*, **242**: 203–205.

Brink, A. S. (1963). Notes on some new *Diademodon* specimens in the collection of the Bernard Price Institute. *Palaeontologia afr.*, **8**: 97–111.

Brinkman, D. (1981). The structure and relationships of the dromasaurs (Reptilia: Therapsida). *Breviora*, **465**: 1–34.

Brinkman, D. and Eberth, D. A. (1983). The interrelationships of pelycosaurs. *Breviora*, **473**: 1–35.

Broili, F., and Schröder, J. (1934). Zur Osteologie des Kopfes von *Cynognathus*. *Bayer. Akad. Wissenschaft München, Sitzungsberichte, Math.-Naturw. Abt.*, **1934**: 95–128.

Carroll, R. L. (1964). The earliest reptiles. *Zool. Jour. Linn. Soc.*, **45**: 61–83.

Carroll, R. L. (1986). The skeletal anatomy and some aspects of the physiology of primitive reptiles. In N. Hotton III, P. D. MacLean, J. J. Roth, and E. C. Roth (eds.), *The Evolution and Ecology of Mammal-Like Reptiles*, pp. 25–45. Smithsonian Institution Press, Washington, D.C.

Carroll, R. L., Belt, E. S., Dineley, D. L., Baird, D., and McGregor, D. C. (1972). Vertebrate paleontology of eastern Canada. *24th Intl. Geol. Congress, Canada 1972, Field Excursion A59, Guidebook:* 1–113.

Chatterjee, S. (1983). An ictidosaur fossil from North America. *Science*, **220**: 1151–1153.

Chudinov, P. K. (1960). Upper Permian therapsids from the Ezhovo locality. *Pal. Zhurnal*, **4**: 81–94. (In Russian.)

Chudinov, P. K. (1965). New facts about the fauna of the Upper Permian of the U.S.S.R. *J. Geol.*, **73**: 117–130.

Cluver, M. A. (1978). The skeleton of the mammal-like reptile *Cistecephalus* with evidence for a fossorial mode of life. *Ann. S. Afr. Mus.*, **76**: 213–246.

Cluver, M. A., and Hotton, N., III. (1981). The genera *Dicynodon* and *Diictodon* and their bearing on the classification of the Dicynodontia (Reptilia, Therapsida). *Ann. S. Afr. Mus.*, **83**: 99–146.

Cluver, M. A., and King, G. M. (1983). A reassessment of the relationships of Permian Dicynodontia (Reptilia, Therapsida) and a new classification of dicynodonts. *Ann. S. Afr. Mus.*, **91**(3): 195–273.

Colbert, E. H. (1948). The mammal-like reptile *Lycaenops*. *Bull. Am. Mus. Nat. Hist.*, **89**: 353–404.

Cox, C. B. (1972). A new digging dicynodont from the Upper Permian of Tanzania. In K. A. Joysey and T. S. Kemp (eds.), *Studies in Vertebrate Evolution*, pp. 173–189. Oliver and Boyd, Edinburgh.

Crompton, A. W., and Hotton, N., III. (1967). Functional morphology of the masticatory apparatus of two dicynodonts (Reptilia, Therapsida). *Postilla, Yale Peabody Museum*, **109**: 1–51.

Crompton, A. W., and Jenkins, F. A. (1979). Origin of mammals. In J. A. Lillegraven, Z. Kielan-Jaworowska, and

W. A. Clemens (eds.), *Mesozoic Mammals: The First Two-Thirds of Mammalian History,* pp. 59–73. University of California Press, Berkeley.

Currie, P. J. (1977). A new haptodontine sphenacodont (Reptilia: Pelycosauria) from the Upper Pennsylvanian of North America. *J. Paleont.,* **51**(5): 927–942.

Currie, P. J. (1979). The osteology of haptodontine sphenacodonts (Reptilia: Pelycosauria). *Palaeontographica, Abt. A,* **163**: 130–168.

Fourie, S. (1974). The cranial morphology of *Thrinaxodon liorhinus* Seeley. *Ann. S. Afr. Mus.,* **65**(10): 337–400.

Gow, C. E. (1980). The dentitions of the Tritheledontidae (Therapsida: Cynodontia). *Proc. Roy. Soc. Lond., B,* **208**: 461–481.

Gregory, W. K. (1957). *Evolution Emerging,* Vol. 2. Macmillan, New York.

Hopson, J. A. (1984). Late Triassic traversodont cynodonts from Nova Scotia and Southern Africa. *Paleont. Afr.,* **25**: 181–201.

Jenkins, F. A. (1970). The Chanares (Argentina) Triassic reptile fauna VII. The postcranial skeleton of the traversosontid *Massetognathus pascuali* (Therapsida, Cynodontia). *Breviora,* **352**: 1–28.

Jenkins, F. A. (1971). The postcranial skeleton of African cynodonts. *Bull. Peabody Mus. Nat. Hist.,* **36**: 1–216.

Jenkins, F. A. (1984). A survey of mammalian origins. In P. D. Gingerich and C. E. Badgley (eds.), *Mammals. Notes for a Short Course. Univ. Tenn., Studies Geol.* **8**: 32–47.

Jenkins, F. A., and Parrington, F. R. (1976). The postcranial skeletons of the Triassic mammals *Eozostrodon, Megazostrodon* and *Erythrotherium. Phil. Trans. Roy. Soc., Lond., B,* **173**: 387–431.

Kemp, T. S. (1979). The primitive cynodont *Procynosuchus:* Functional anatomy of the skull and relationships. *Phil. Trans. Roy. Soc., Lond., B,* **285**: 73–122.

Kemp, T. S. (1982). *Mammal-like Reptiles and the Origin of Mammals.* Academic Press, London.

Kemp, T. S. (1983). The relationships of mammals. *Zool. J. Linn. Soc. Lond.,* **77**: 353–384.

Kemp, T. S. (1986). The skeleton of a baurioid therocephalian therapsid from the Lower Triassic (*Lystrosaurus* zone) of South Africa. *Jour. Vert. Paleon.* **6**: 215–232.

Kermack, K. A., Mussett, F., and Rigney, H. W. (1973). The lower jaw of *Morganucodon. Zool. J. Linn. Soc.,* **53**: 87–175.

Langston, W., Jr. (1965). *Oedalepos campi* (Reptilia: Pelycosauria). A new genus and species from the Lower Permian of New Mexico, and the family Eothyrididae. *Bull. Texas Mem. Mus.,* **9**: 1–47.

Langston, W., Jr., and Reisz, R. R. (1981). *Aerosaurus wellesi,* new species, a varanopseid mammal-like reptile (Synapsida: Pelycosauria) from the Lower Permian of New Mexico. *J. Vert. Pal.,* **1**(1): 73–96.

Manley, G. A. (1972). A review of some current concepts of the functional evolution of the ear in terrestrial vertebrates. *Evolution,* **26**: 608–621.

Mendrez, C. H. (1972). On the skull of *Regisaurus jacobi,* a new genus and species of Bauriamorpha Watson and Romer 1956 (= Scaloposauria Boonstra 1953), from the *Lystrosaurus*-zone of South Africa. In K. A. Joysey and T. S. Kemp (eds.), *Studies in Vertebrate Evolution,* pp. 191–219. Oliver and Boyd, Edinburgh.

Mendrez, C. H. (1974). Etude du crâne d'un jeune specimen de *Moschorhinus kitchingi* Broom 1920 (? *Tigrisuchus simus* Owen, 1876), Therocephalia Pristerosauria Moschorhinidae d'Afrique australe. *Ann. S. Afr. Mus.,* **64**: 71–115.

Mendrez, C. H. (1975). Principales variations du palais chez les thérocéphales Sud-Africains (Pristerosauria et Scaloposauria) au cours du Permien Supérieur et du Trias Inférieur. *Coll. Intl. C.N.R.S. 218. Problèmes actuels de paléontologie-évolution des vertébrés*: 379–408.

Mendrez-Carroll, C. H. (1979). Nouvelle étude du crâne du type de *Scalaposaurus constrictus* Owen, 1876, de la zone à *Cistecephalus* (Permien Supérieur) d'Afrique australe. *Bull. Mus. Natn. Hist. nat. 4e ser.,* **1** (Section C, No. 3): 155–201.

Olson, E. C. (1962). Late Permian terrestrial vertebrates, U.S.A. and U.S.S.R. *Trans. Am. Phil. Soc.,* **52**: 3–224.

Olson, E. C. (1968). The family Caseidae. *Fieldiana Geol.,* **17**: 225–349.

Olson, E. C. (1974). On the source of the therapsids. *Ann. S. Afr. Mus.,* **64**: 27–46.

Orlov, J. A. (1958). Carnivorous dinocephalians from the fauna of Isheev (Titanosuchia). *Trudy Pal. Inst. Akad. Nauk Sci. U.S.S.R.,* **72**: 1–114. (In Russian.)

Parrington, F. R. (1946). On the cranial anatomy of cynodonts. *Proc. Zool. Soc. Lond.,* **116**: 181–197.

Pearson, H. S. (1924). A dicynodont reptile reconstructed. *Proc. Zool. Soc. Lond.,* **1924**: 827–855.

Pough, F. H. (1980). The advantages of ectothermy for tetrapods. *Am. Naturalist,* **115**: 92–112.

Reisz, R. (1972). Pelycosaurian reptiles from the Middle Pennsylvanian of North America. *Bull. Mus. Comp. Zool. Harv.,* **144**: 27–62.

Reisz, R. (1980). The Pelycosauria: A review of phylogenetic relationships. In A. L. Panchen (ed.), *The Terrestrial Environment and the Origin of Land Vertebrates,* pp. 553–591. Systematics Association Special Volume No. 15.

Reisz, R. R. (1986). Pelycosauria. In P. Wellnhofer (ed.), *Handbuch der Paläoherpetologie,* Part 17A, pp. 1–102. Gustaf Fisher Verlag, Stuttgart.

Reisz, R. R., and Berman, D. S. (1986). *Ianthosaurus hardestii* n. sp., a primitive ephadosaur (Reptilia, Pelycosauria) from the Upper Pennsylvanian Rock Lake Shale near Garnett, Kansas. *Can. Journ. Earth Sci.,* **23**: 77–91.

Romer, A. S. (1948). Relative growth in pelycosaurian reptiles. *Roy. Soc. S. Afr., Spec. Publ., Robert Broom Comm. Vol.*: 45–55.

Romer, A. S. (1967). The Chanares (Argentina) Triassic reptile fauna III. Two new gomphodonts, *Massetognathus pascuali* and *M. teruggii. Breviora,* **264**: 1–25.

Romer, A. S. (1970). The Chanares (Argentina) Triassic reptile fauna. VI. A chiniquodontid cynodont with an incipient squamosal-dentary jaw articulation. *Breviora,* **344**: 1–18.

Romer, A. S., and Lewis, A. D. (1973). The Chanares (Argentina) Triassic reptile fauna 19. Postcranial material of the cynodonts *Probelesodon* and *Probainognathus. Breviora,* **407**: 1–26.

Romer, A. S., and Price, L. I. (1940). Review of the Pelycosauria. *Geol. Soc. Am., Spec. Paper,* **28**: 1–538.

Sigogneau, D. (1970). *Révision systématique des gorgonopsiens Sud-Africains. Cahiers de Paléontologie.* Editions C.N.R.S., Paris.

Sigogneau, D., and Chudinov, P. K. (1972). Reflections on some Russian eotheriodonts (Reptilia, Synapsida, Therapsida). *Palaeovertebrata,* **5**: 79–109.

Sigogneau-Russell, D., and Russell, D. E. (1974). Etude du premier Caséidé (Reptilia, Pelycosauria) d'Europe occidentale. *Bull. Mus. Natl. Hist. Nat.*, Ser. 3, No. 230, Sci terre, **38**: 145–215.

Stovall, J. W., Price, L. I., and Romer, A. S. (1966). The postcranial skeleton of the giant Permian pelycosaur *Cotylorhynchus romeri. Bull. Mus. Comp. Zool.*, **135**: 1–30.

Sues, H.-D. (1983). Advanced mammal-like reptiles from the Early Jurassic of Arizona. *Doctoral thesis.* Harvard University, Cambridge.

Sues, H.-D. (1985). The relationships of the Tritylodontidae (Synapsida). *Zool. J. Linn. Soc.*, **85**: 205–217.

Sues, H.-D. (1986). The skull and dentition of two tritylodontid synapsids from the Lower Jurassic of western North America. *Bull. Mus. Comp. Zool.*, **151**: 215–266.

Ulinski, P. S. (1986). Neurobiology of the therapsid-mammal transition. In N. Hotton III, P. D. MacLean, J. J. Roth, and E. C. Roth (eds.), *The Evolution and Ecology of Mammal-Like Reptiles*, pp. 149–171. Smithsonian Institution Press, Washington, D.C.

C H A P T E R XVIII

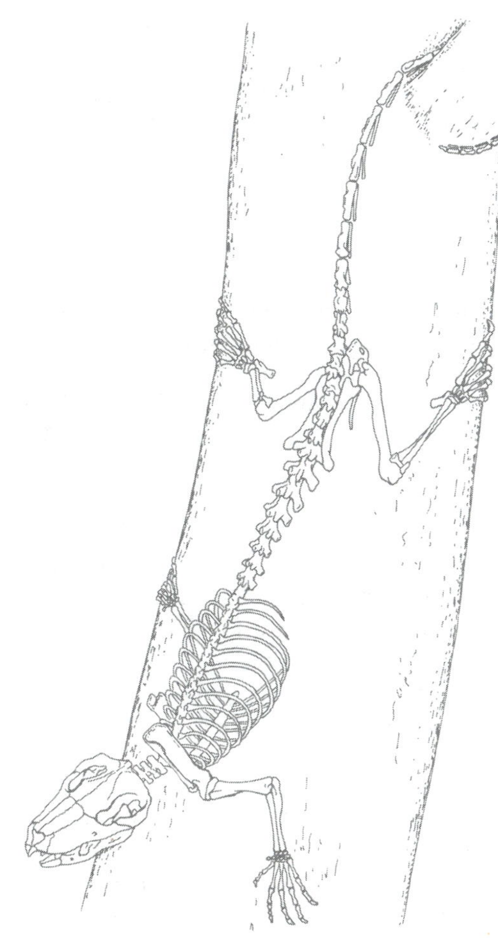

Primitive Mesozoic Mammals and Monotremes

Many mammalian features were achieved within the therapsids, but the two groups can still be clearly distinguished. We recognize the oldest fossil mammals from the late Triassic and early Jurassic. They were small forms that probably resembled living shrews in their general appearance and way of life (Figure 18-1). Their remains are identified as mammalian primarily on the basis of characters of the jaws and teeth. In contrast with therapsids, the dentary and squamosal form the principal jaw joint. The incisors, canines, and premolars are replaced only once, and the molar teeth are not replaced at all. The molars have two roots and show a specific pattern of occlusion.

Early mammals represent a new radiation that was clearly separate from that of the therapsids and stemmed from a distinct lineage of small carnivorous or insectivorous cynodonts that had achieved a relatively larger brain size and, probably, a higher metabolic rate than

(a)

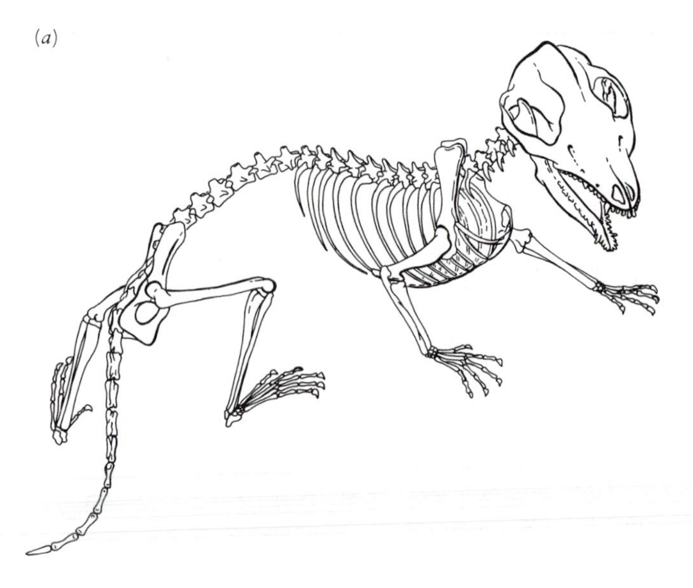

(b)

Figure 18-1. (*a*) Skeleton of a primitive mammal based on disarticulated elements of *Morganucodon* and a partial skeleton of *Megazostrodon*. The skeleton was 10 centimeters long to the base of the tail. *From Jenkins and Parrington, 1976.* (*b*) Restoration of a primitive mammal based on *Morganucodon* and *Megazostrodon*. *From Crompton, 1968.*

other mammal-like reptiles. Several distinct groups of mammals are known from the Upper Triassic and Jurassic, but the morganucodontids are by far the best known. In most features of their skeletal anatomy, they are almost ideal ancestors for later mammals.

THE ANATOMY OF *MORGANUCODON*

SKULL

Morganucodon is known from many specimens that provide detailed knowledge of most of the skeleton. Kermack, Mussett, and Rigney (1981) described the skull in detail on the basis of articulated material from China and thousands of disarticulated elements from fissure fillings in Great Britain (Figure 18-2). All the material appears to be early Jurassic in age (Sinemurian).

The skull is approximately 3 centimeters long and appears large relative to the body. It is long and slender compared with those of *Thrinaxodon* and *Probainognathus*. The zygomatic arch is slender and arched dorsally but less expanded laterally. The temporal opening and the orbit are confluent. The braincase appears much larger than in any well-known cynodonts, which can in part be attributed to the smaller size of the skull (less than half the length of *Probainognathus*).

The parietals are fused, and their line of fusion forms a narrow sagittal crest. With the loss of the postorbital and prefrontal, the frontal appears as a much larger element than in cynodonts and extends ventrolaterally over the dorsal portion of the braincase, as does the parietal. Anterior to the orbit, the frontal meets a dorsal extension of the palatine. In contrast with therian mammals (marsupials and placentals), the squamosal forms no part of the braincase.

The external nares are apparently confluent, although the tip of the snout is not well known. Ventrally, the premaxillae and maxillae form the border of the incisive foramen, just lateral to the midline. In living mammals, this opening is associated with the Jacobson's organ, an accessory nasal pouch for smelling the contents of the mouth.

A large nasal cavity occupies the snout above the secondary palate. It was probably lined by extensive olfactory epithelium. In modern mammals, this tissue is supported by cartilaginous turbinal bones that have a scroll-like configuration. Longitudinal ridges on the internal surfaces of the maxillary and nasal bones of both advanced cynodonts and early mammals indicate where the turbinals were attached.

The tooth row extends more than half the length of the skull, as does the secondary palate. The area between the back of the skull and the palate is telescoped relative to cynodonts and is further shortened in later mammals. The pterygoid is reduced in extent but still retains a lateral flange that limited medial movements of the lower jaw. The ectopterygoid, which is lost in most later mammals, may have been retained in *Morganucodon*. The quadrate ramus of the pterygoid extends posteriorly as a narrow process lateral to the cavum epiptericum, which remains open ventrally, lateral to the basisphenoid.

The braincase can be seen in lateral view in the intact skull, and we can study the internal surface and details of the individual bones from those collected from the fissure fillings. The external surface of the back of the braincase broadly resembles that of carnivorous cynodonts, but the petrosal extends relatively further forward than in *Probainognathus* and surrounds the opening for the maxillary and mandibular branches of the trigeminal nerve (Figure 18-2). These nerves passed between the petrosal and the alisphenoid in cynodonts and penetrate the alisphenoid in modern therian mammals (Figure 18-3). In marsupials and placentals, these opening in the alisphenoid are termed the foramen rotundum (for the maxillary branch) and the foramen ovale (for the mandibular branch). Since these openings pass through the petrosal rather than the alisphenoid in *Morganucodon,* they are designated the pseudorotundum and pseudoovale. Kermack and his colleagues (1981, 1984) attributed great taxonomic significance to the different relative extent of the alisphenoid and petrosal in *Morganucodon* compared with both cynodonts and therian mammals. Developmental studies of modern mammals by Griffiths (1978) and Presley (1981) show that ossification of the wall of the braincase is quite opportunistic, especially in monotremes. The relative extent of ossification of the petrosal and alisphenoid might have been quite variable in evolution.

Posteriorly, the alisphenoid extends as a narrow quadrate ramus beneath the petrosal nearly to the posterior end of the anterior lamella. The large opening in the wall of the skull medial to the orbits, which was retained in carnivorous cynodonts, is partially closed by the formation of a new area of ossification, the orbitosphenoid, through which emerged the optic nerve (II). Some tritylodonts are more advanced than primitive mammals in ossifying this area completely.

The base of the braincase forms a broad, triangular area. The basisphenoid is solidly sutured to the pterygoid, but the shape of the once movable basipterygoid process is evident in dorsal view (Figure 18-4*b*) and where the bone is isolated from the rest of the skull. The basisphenoid is fused anteriorly to the narrow process of the parasphenoid. In contrast, the parasphenoid appears as a separate ossification in most mammals. The slender, median vomer continues anteriorly from the parasphenoid.

The basisphenoid is pierced ventrally by foramina for the carotid arteries. The course of these arteries is quite variable in mammals and the position of foramina for their passage is an important basis for classification. In *Cynognathus,* gomphodonts, and tritylodontids, the carotids enter the sella turcica from the side.

The otic capsule in *Morganucodon,* as in all mammals, is ossified as a single unit, the petrosal, which represents the fusion of the opisthotic and proötic of more primitive tetrapods. Ventrally, the petrosal forms a broad floor of the braincase that is in close association with the basisphenoid and basioccipital.

An important distinction from cynodonts is the anterior extension of the base of the petrosal to form a partial floor of the cavum epiptericum, medial to the vertically oriented anterior lamina of the petrosal. A large recess is evident dorsally (Figure 18-4*c*) for the semilunar ganglion of the Vth nerve. Posteriorly and dorsally we can see a second conspicuous depression in the petrosal, the subarculate fossa, which houses the flocculus, a portion of the brain that is important in muscular coordination.

The cochlea of the inner ear is much larger than in cynodonts, and its position is evident externally by a conspicuous promontory. As in reptiles and monotremes, the cochlea is not coiled, as it is in marsupials and placentals. Ventrally, the otic capsule is penetrated by the cochlear foramen and the vestibular foramen (or fenestra ovalis), which receives the footplate of the stapes.

This area of the braincase differs markedly from that of advanced placental mammals, since there is no auditory bulla, which could form only after the articular and quadrate left the jaw joint. Primitive marsupials also lack a bulla and so retain the primitive condition (Figure 18-3).

The dorsal margin of the parietal and squamosal form a transverse lambdoidal crest that separates the occipital surface from the temporal opening. In contrast with advanced mammals, sutures separating the bones of the occiput are still clearly evident. The occipital condyles are formed entirely by the exoccipitals, with the basioccipital being only narrowly exposed posteriorly. The condyles remain relatively low in comparison with modern therian mammals. Just anterior to the condyle is a tiny anterior condylar foramen for the XIIth cranial nerve. Between the exoccipital and the petrosal is a foramen for the internal jugular vein, glossopharyngeal nerve (IX), vagus nerve (X), and spinal accessory nerve (XI). This opening is termed the jugular foramen in reptiles but the posterior lacerate foramen in mammals.

The posttemporal fossa, which is lost in later mammals, opens between the squamosal and petrosal. The tabulars and supraoccipital are not present in any of the available specimens, but we think that they remained as distinct centers of ossification. The postparietal also appears as a separate center of ossification during early development in living mammals.

The jaw articulation is just anterior to the level of the occipital condyles in primitive mammals but assumes a more anterior position in advanced forms. The primitive, posterior position of the jaw articulation may be associated with the close relationship between the extradentary jaw bones and the otic capsule, which is inherited from cynodonts. The squamosal forms a distinct but shallow glenoid for articulation with the condyle of the dentary in *Morganucodon* (Figure 18-4*d*). Immediately medial to this joint is a recess for the quadratojugal and quadrate, which articulate with the articular. The quadrate is supported medially by the paroccipital process of the otic capsule. The jaw articulation is slightly above the level of the tooth row rather than more ventral, as was the case in carnivorous cynodonts.

The stapes is distinguished by a large stapedial foramen and articulates with the medial surface of the quadrate. These elements probably already functioned as ear ossicles, but they are only slightly reduced in size relative to their proportions in cynodonts. Judging by the size of the angular and the retroarticular process of the articular, the tympanum must have remained extremely large in comparison with modern mammals. Kermack and his colleagues showed that the size relationship of the tympanum and the footplate of the stapes would have enabled them

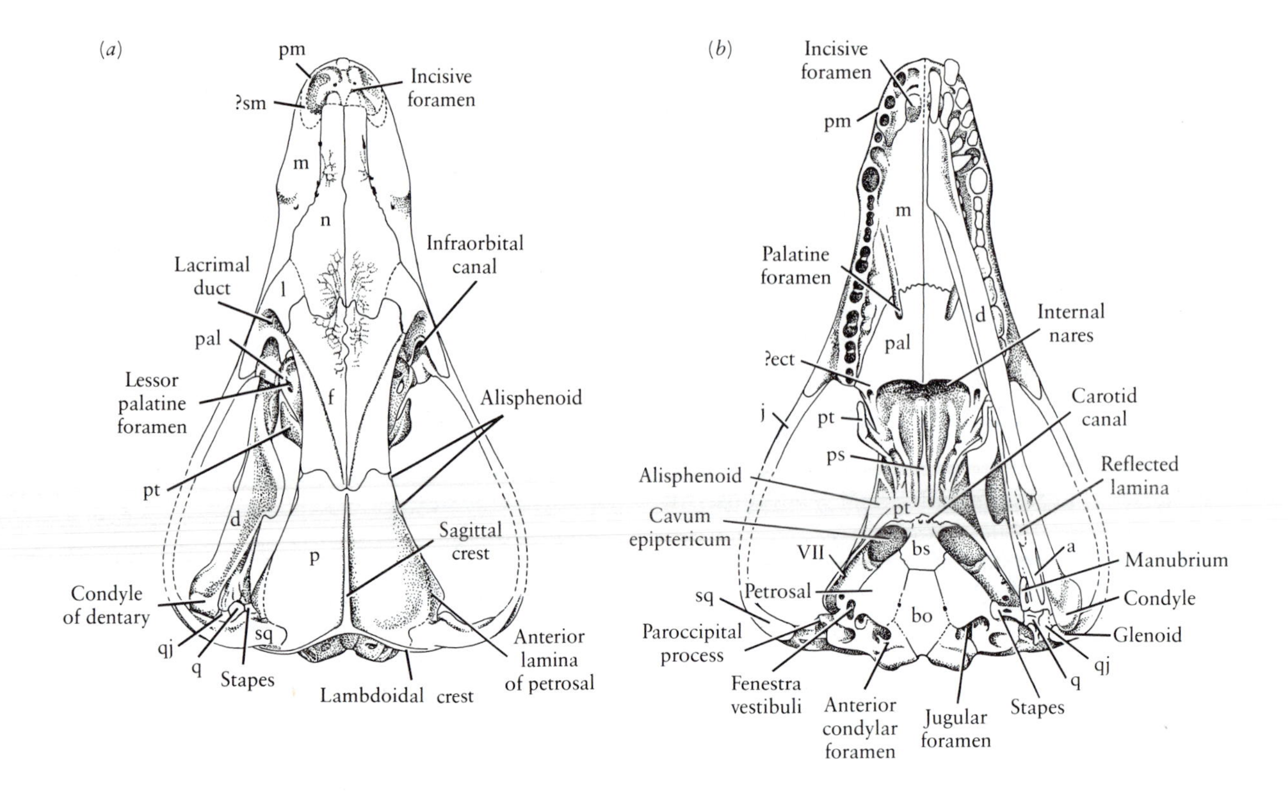

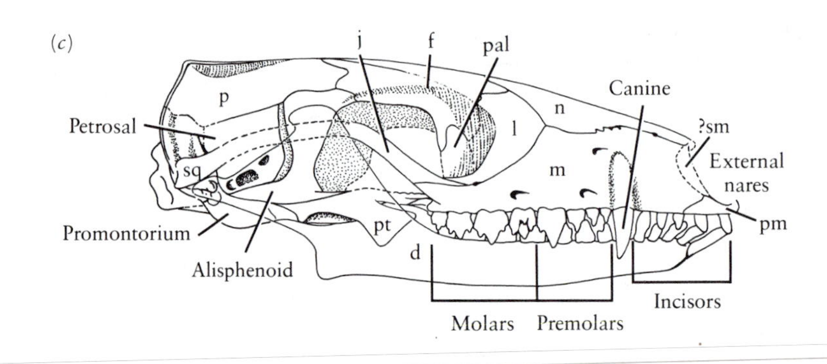

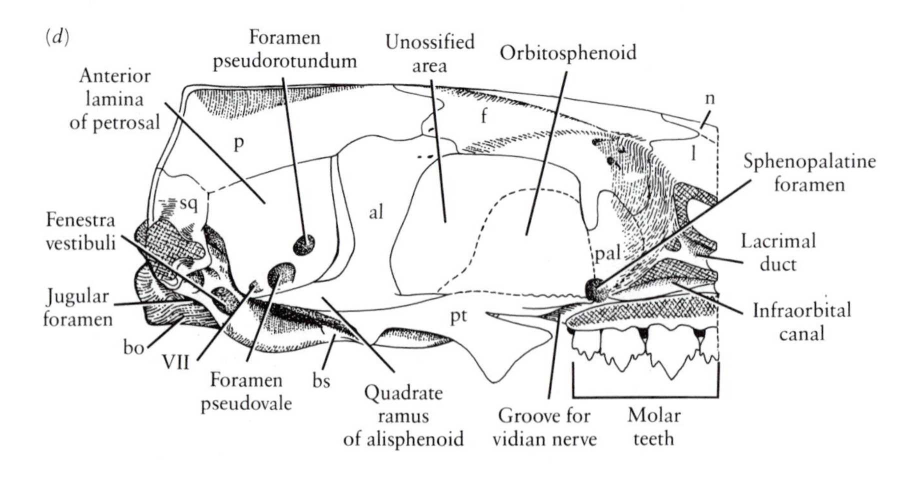

Figure 18-2. SKULL OF THE EARLIEST WELL KNOWN MAMMAL *MORGANUCODON*. (*a*) Dorsal, (*b*) palatal, and (*c*) lateral views. (*d*) Lateral view of the braincase with zygomatic arch removed. Original 2 centimeters long. Alisphenoid in specimen is broken, so that it appears unnaturally narrow. *From Kermack, Mussett, and Rigney, 1981.* Abbreviations as in Figure 8-3.

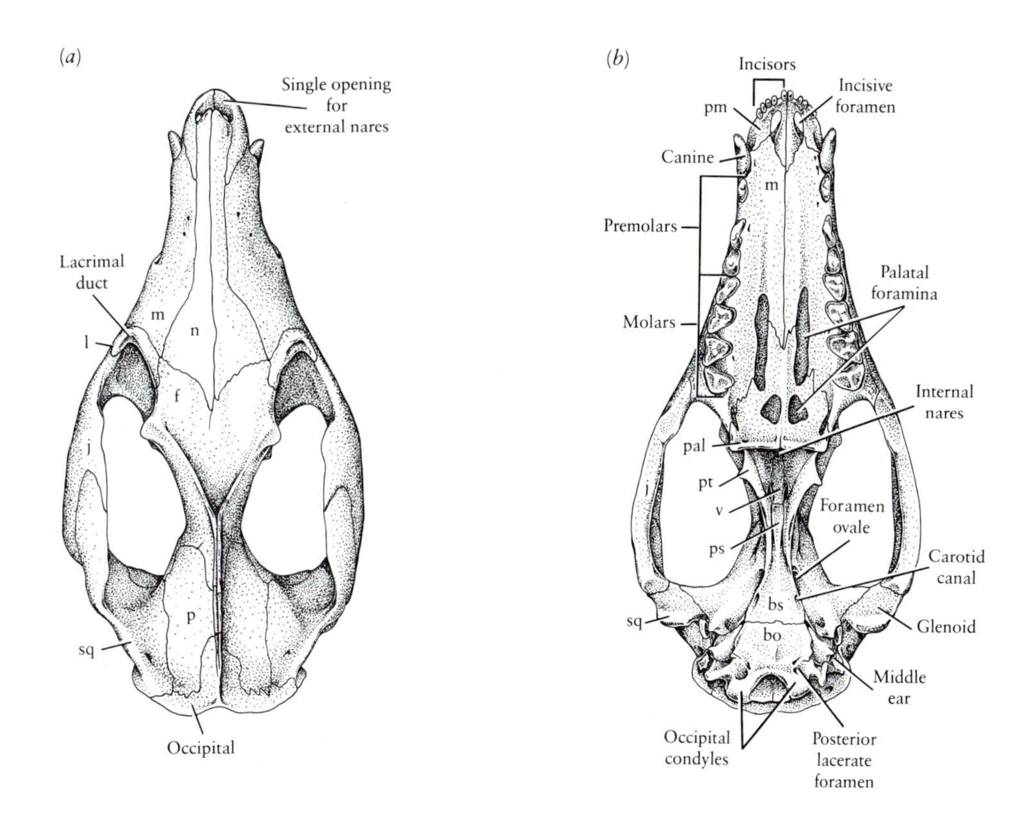

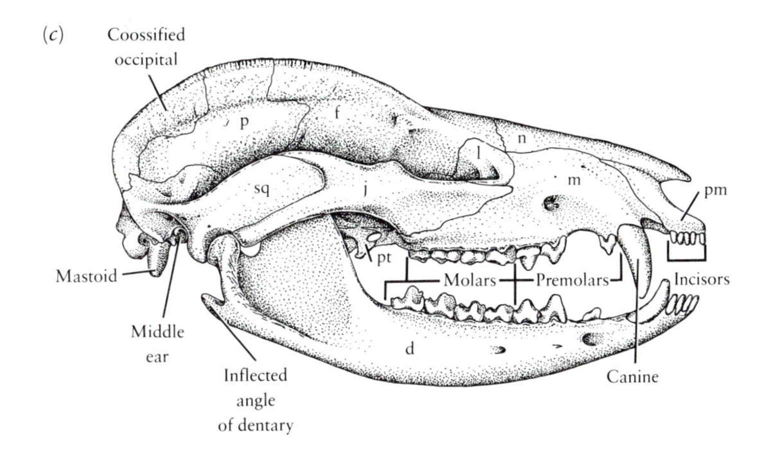

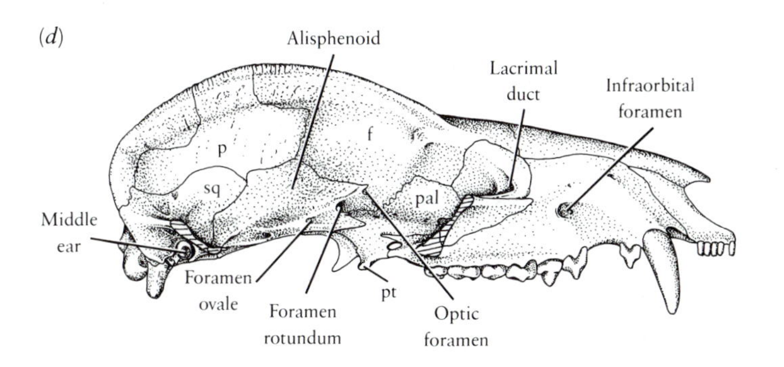

Figure 18-3. SKULL OF A PRIMITIVE LIVING MAMMAL, THE MARSUPIAL *DIDELPHIS.* (*a*) Dorsal, (*b*) palatal, and (*c*) lateral views. (*d*) Lateral view of the skull with the zygomatic arch removed. Abbreviations as in Figure 8-3.

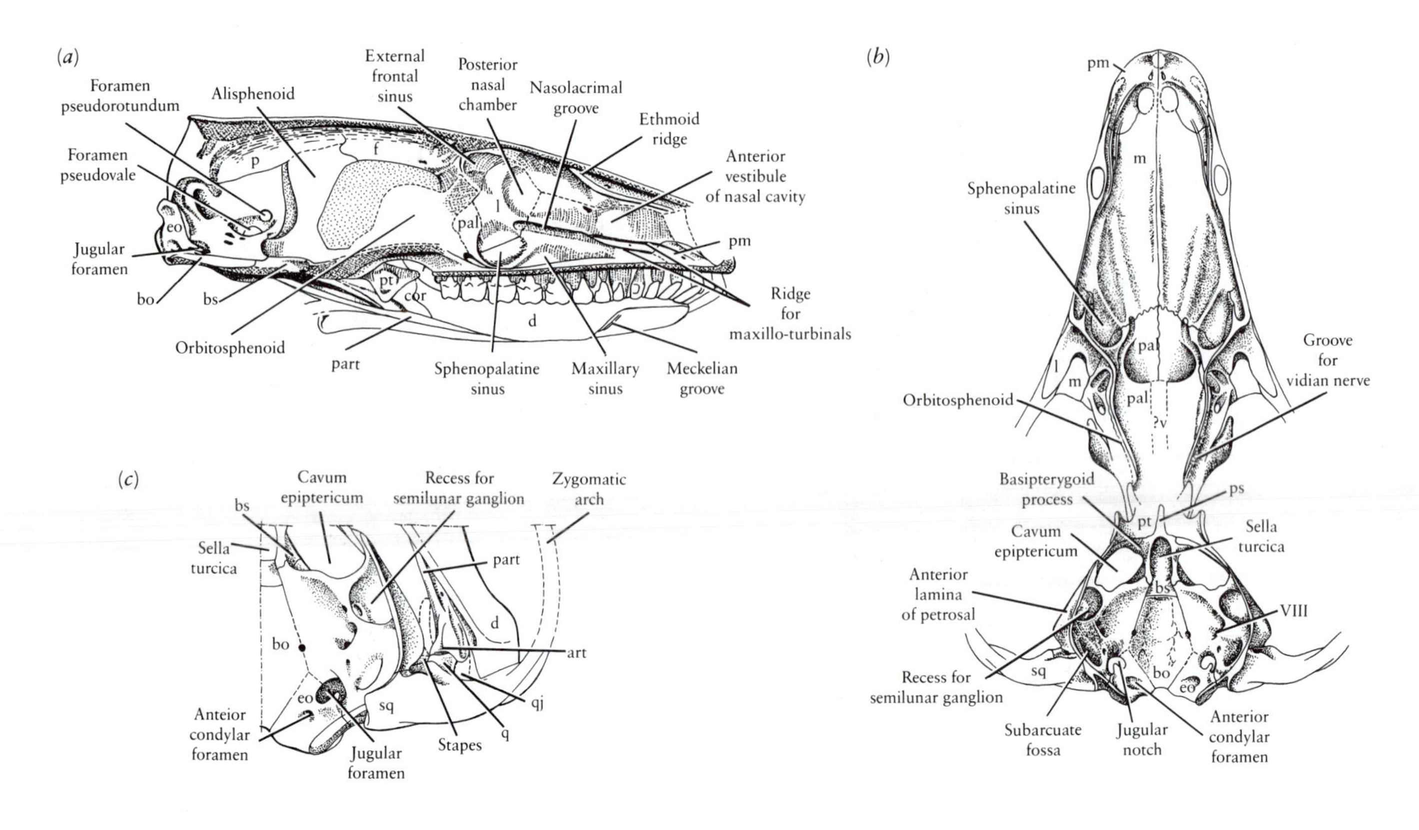

Figure 18-4. *MORGANUCODON.* (*a*) Medial view of the skull as seen from a sagittal section. (*b*) Floor of the skull in dorsal view. (*c*) Dorsal view of the back of the braincase and area of jaw articulation. (*d*) Ventral view of the area of the middle ear and jaw articulation. Abbreviations as in Figure 8-3. *From Kermack, Mussett, and Rigney, 1981.*

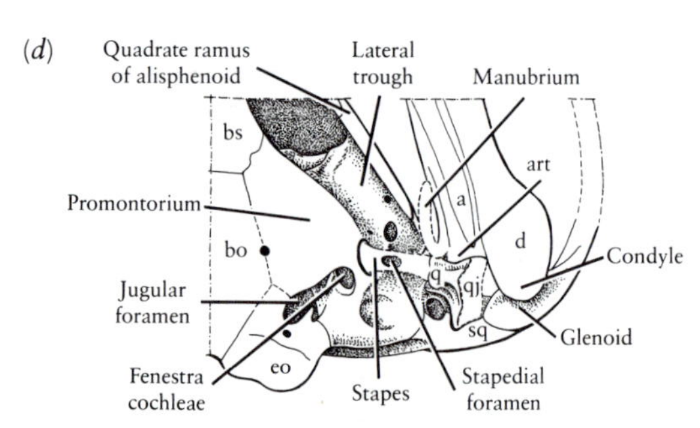

to function as an effective impedence-matching structure, but their great size and persistent association with the jaw apparatus would have made them much less responsive to sound than are the elements in modern mammals.

LOWER JAW

The lower jaw of morganucodontids appears typically mammalian in lateral view, with the large dentary forming a high coronoid process and a distinct condylar process for articulation with the skull (Figure 18-5). As in advanced cynodonts, the lower margin of the dentary ends posteriorly in a rounded process, which has been compared with the angular process in therian mammals, despite its more anterior position relative to the condyle. The pattern in other early mammals, especially *Dinnetherium* (see Figure 18-12), suggests that this structure is more primitive and may better be termed a pseudangular process. Between the pseudangular process and the condyle we can see the narrow reflected lamina of the angular

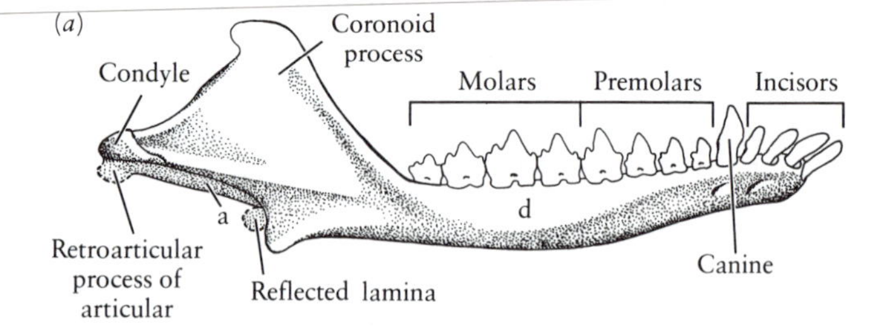

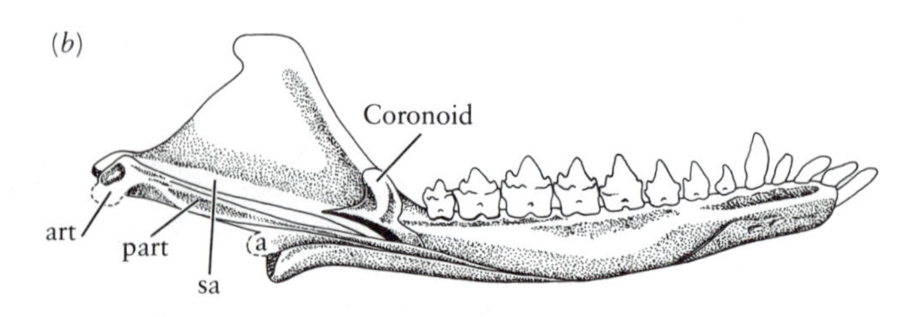

Figure 18-5. LOWER JAW OF *MORGANUCODON.* (*a*) Medial and (*b*) lateral views. Abbreviations as in Figure 8-3. *From Kermack, Mussett, and Rigney, 1973.*

and the manubrium of the articular, which would have supported the very large tympanum.

The medial surface shows that all the extradentary bones of the cynodonts are still retained, although their size is greatly reduced. The rodlike prearticular, angular, and surangular fit into a groove on the inside surface of the dentary. The coronoid is a flat, irregular plate of bone at the base of the coronoid process. The articulating surface of the articular is concave and faces almost directly posterior to surround the anteriorly facing base of the quadrate.

The symphysis between the jaws is not strongly developed and the two sides would have been able to move somewhat independently of one another.

DENTITION

The dentition is one of the most important features for understanding the evolution and classification of mammals. The shape of the teeth and their manner of occlusion is much more highly correlated with particular diets and ways of life among mammals than is the case in other vertebrates. Dental enamel is the hardest part of the skeleton and so the most likely to be fossilized and recovered. The potential for evolution of different patterns of the teeth is so great that most mammalian species can be distinguished by the nature of the cusps of a single molar tooth.

Functional and developmental constraints do limit the potential for different patterns, so that each major taxonomic group can be characterized by a limited range of tooth morphology that correlates extremely well with classification based on other characters.

The establishment of a complex suite of dental characters is coincident with what is considered the origin of mammals. Many other features of the skeleton show more-or-less continuous change through the advanced mammal-like reptiles and early mammals. The establishment of a particular tooth morphology and pattern of occlusion correlate well with the time of the initial radiation that led to the divergence of the major groups of Mesozoic mammals. The changes in the dentition may also be correlated with significant size reduction from the cynodont pattern, which suggests important differences in metabolism and reproduction as well (Hopson, 1973).

The mammalian dentition as exemplified by *Morganucodon* show differentiation into incisors, canines, premolars, and molars. The specific number of teeth in each category is important in relationship to feeding habits and classification of mammals and is indicated by the **tooth formula,** which shows the number in both the upper and lower jaws. The tooth formula of *Morganucodon* is I 5/4 C 1/1 P 4/4 M 4/4 or

$$\frac{5\quad 1\quad 4\quad 4}{4\quad 1\quad 4\quad 4}$$

The incisors, canines and premolars of primitive mammals are replaced only once, and the molars are not replaced at all. The cheek teeth have two roots and show a specific pattern of occlusion that is associated with a particular arrangement of the cusps. Some of these characters are encountered among both herbivorous and carnivorous cynodonts, but no known cynodont exhibits all of them. Both tritylodonts and carnivorous cynodonts have reduced the amount of tooth replacement, but the molar teeth are replaced at least once. Tritylodonts have multiple-rooted cheek teeth but have lost the canines and anterior cheek teeth. Occlusion is achieved in gomphodonts and tritylodonts, but it results exclusively from movements of the lower jaw in the vertical plane, whereas occlusion in most early mammals includes a transverse component as well.

The teeth of galesaurids and chiniquodontids have fewer mammalian features than those of tritylodonts, but they have no specialized features that are lacking in the early mammals. Although all cheek teeth are replaced more than once, teeth that are structurally similar to the premolars and molars are evident in both *Thrinaxodon* and *Probainognathus.* These teeth show wear, but the teeth of the lower jaw are completely medial to those in the upper jaw, and neither the shape nor the position of the upper and lower teeth are specifically related to one another.

We can associate the absence of specific tooth occlusion in carnivorous cynodonts with the retention of the primitive pattern of continuous tooth replacement. As long as the teeth were repeatedly replaced and many were not fully functional at a particular time, there could not be strong selection for a precise pattern of occlusion. With only minor exceptions, mammals have eliminated replacement of the molar teeth no matter how long their life span may be, and the more anterior teeth are replaced only once.

Crompton (1971, 1972, 1974) and his colleagues (Crompton and Jenkins, 1968; Crompton and Kielan-Jaworowska, 1978; Crompton and Hiiemäe, 1970) have admirably traced the evolution of molar occlusion from advanced therapsids into therian mammals. In chiniquodontids, some individuals show wear between the lateral surface of the lower and the medial surface of the upper molars, but this wear was apparently not an important component of mastication. In contrast, *Morganucodon* shows a very specific pattern of wear that is highly correlated with the structure of the teeth (Figure 18-6). In common with the galesaurids and chiniquodontids, the crowns are laterally compressed, with a series of apical cusps arranged anteroposteriorly. The principal cusps are consistent in their position and function. The cusps on the upper teeth are identified by capital letters and those of the lower teeth by lower case letters. The highest cusp, near the middle of the tooth's length, is termed **A** (upper) and **a** (lower). A more anterior cusp is designed **B** and **b.** Cusps **C, c** and **D, d** are posterior to **A** and **a.** Galesaurids and some chiniquodontids have a low ridge, or cingulum, along the internal or lingual (toward the tongue) margin of the molars. *Morganucodon* has added a lateral or buccal (toward the cheek) cingulum on the upper teeth. Small cusps are present on the cingula.

The cusps are sharp and could be used for piercing prey as soon as the teeth erupt, but newly erupted teeth have only limited areas that can be used for shearing. More effective shearing surfaces develop during the life of the animal as a result of the specific wear pattern between the upper and lower teeth. The medial surface of **A** (the principal cusp of the upper molar) wears against the lateral surface of the lower tooth between **a** and **c**. The lateral surface of **a** wears against the medial surface of the upper tooth between **A** and **B**. As wear proceeds, a zig-zag pattern of wear facets is produced, which maximizes the shearing function of the tooth. This pattern can be most clearly seen in occlusal view (Figure 18-6*c*).

The geometry of the wear facets indicates that the teeth in the lower jaw must have moved medially as well as dorsally during occlusion. The pattern of jaw movements in morganucodontids was probably very similar to that of some primitive modern mammals, specifically the opposum. Crompton and Hiiemäe (1970) studied mastication in this animal as a guide to interpreting the pattern

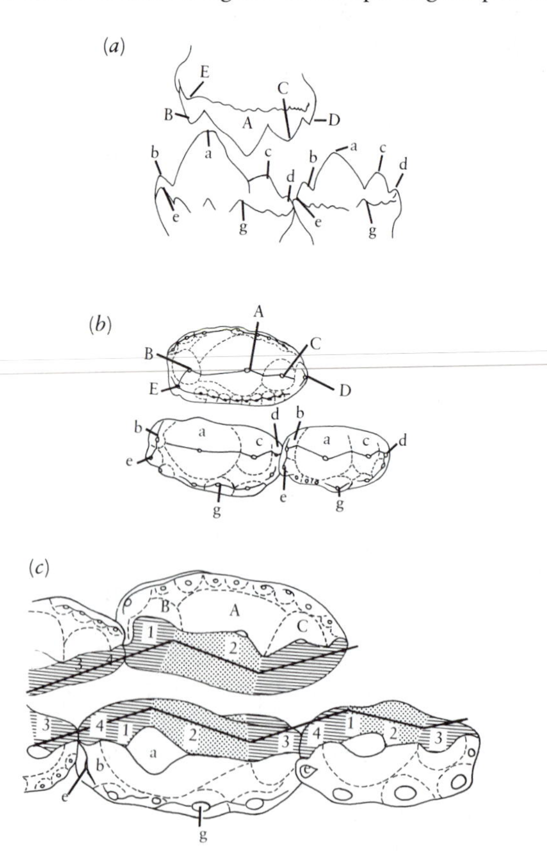

Figure 18-6. (*a* and *b*) Upper and lower molar teeth of *Morganucodon* in internal and occlusal views. (*c*) Occlusal views of two upper and three lower molars. The wear facets form two sides of "wide-angled" reversed triangles. The orientation of these wear facets is illustrated by the thick black lines. *From Crompton and Jenkins, 1968.* Anterior to left.

in early mammals. The jaws are only loosely attached at the symphysis, which allows the two sides to move somewhat independently. One of the most significant features that they discovered is that *occlusion occurs on only one side of the jaw at a time.* As seen in anterior view, the occluding jaw moves first laterally and then medially and slightly anteriorly as the teeth come into contact. The teeth move directly ventrally as the jaw opens. Hence, the occluding teeth trace a roughly triangular path as the jaw is opened and closed. The transverse movement of the teeth during occlusion and the independence of the two sides of the jaw are among the most important features that characterize early mammals and distinguish them from all cynodonts.

POSTCRANIAL SKELETON

Jenkins and Parrington (1976) described the postcranial skeleton of morganucodonts on the basis of a nearly complete skeleton of *Megazostrodon* from the Upper Triassic or Lower Jurassic of southern Africa and well-preserved but disarticulated elements of *Morganucodon* [*Eozostrodon*] (Figures 18-1 and 18-7). These early mammals were approximately 10 centimeters long to the base of the tail and may have weighed between 20 and 30 grams.

The head appears large relative to the trunk, and the limbs are slender. Perhaps the most significant difference from all therapsids is in the structure of the vertebral column. There is a clear distinction between the thoracic and lumbar vertebrae. The latter lack ribs, as do those of modern mammals, and the zygapophyses are tilted at a 35- to 45-degree angle to resist torsion, in contrast with the nearly flat zygapophyses in the thoracic region. The neural spines are low, but they show the initiation of the mammalian pattern in which those of the anterior vertebrae angle posteriorly, while the more posterior spines angle anteriorly. Three transitional, or anticlinal, vertebrae are evident in the posterior thoracic region. This pattern of the trunk vertebrae in mammals is associated with the elaboration of flexion-extension in the sagittal plane, in contrast with the lateral flexion that is possible in more primitive amniotes. The exact number of presacral vertebrae cannot be established, but there were probably about 27. There are 2 or possibly 3 sacrals and approximately 12 caudals, which formed a long, but slender, tail. As in later mammals, the cervical vertebrae are characterized by the presence of a very large neural canal, which may be associated with the elaboration of a larger brachial nerve plexus than in reptiles. In contrast with more advanced mammals, the cervical ribs of *Morganucodon* are not fused to the vertebrae. In later mammals, the fused bases of the cervical ribs surround the transverse foramen for the cervical artery.

The atlas arches remain paired and have not fused with the intercentrum, but the proatlas and the zygapophyseal articulations with the axis have been lost, thus

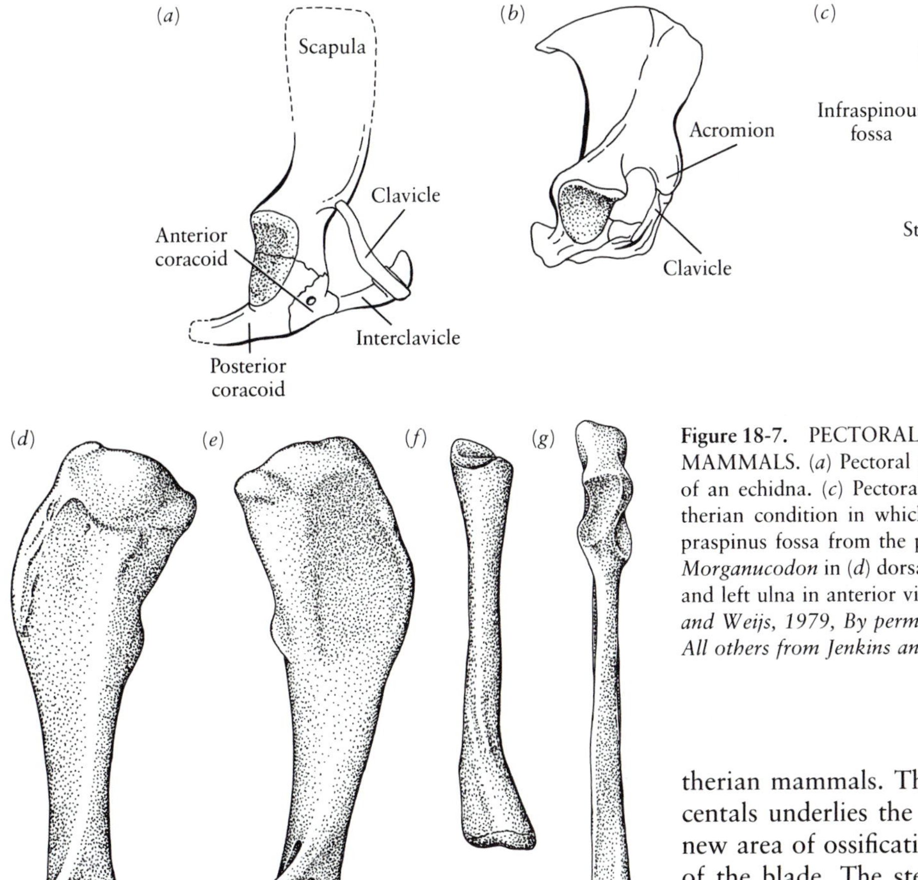

Figure 18-7. PECTORAL GIRDLE AND FORELIMB OF PRIMITIVE MAMMALS. (*a*) Pectoral girdle of *Morganucodon*. (*b*) Pectoral girdle of an echidna. (*c*) Pectoral girdle of an opossum showing the typical therian condition in which a strong spine separates the anterior supraspinus fossa from the posterior infraspinus fossa. Left humerus of *Morganucodon* in (*d*) dorsal and (*e*) ventral views. (*f* and *g*) right radius and left ulna in anterior view. All limb elements ×6. (*c*) *From Jenkins and Weijs, 1979, By permission of the Zoological Society of London. All others from Jenkins and Parrington, 1976.*

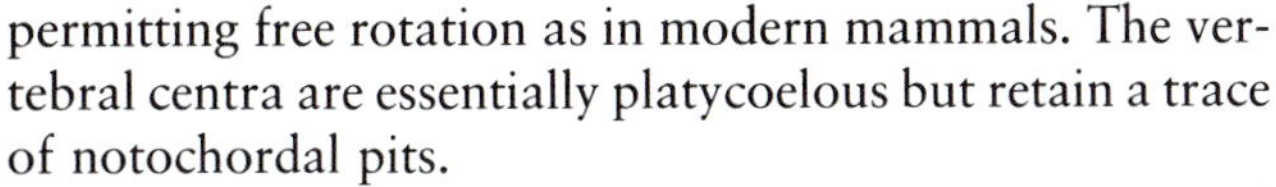

permitting free rotation as in modern mammals. The vertebral centra are essentially platycoelous but retain a trace of notochordal pits.

The limbs and girdles resemble those of advanced cynodonts, especially *Oligokyphus,* and show few new, uniquely mammalian, features. However, there are some differences in detail. In the shoulder girdle, the procoracoid is still retained, although it is completely excluded from the glenoid (Figure 18-7*a*). As in cynodonts and monotremes, the posterior coracoid is characterized by a strong, posteriorly directed process. As in advanced cynodonts, the scapula is still clearly more primitive than that of modern marsupials and placentals. In therian mammals, it is in the form of a broad plate of bone that is divided by the vertical scapular spine into the anterior supraspinus fossa and the posterior infraspinus fossa. These surfaces are occupied by the infraspinatus and supraspinatus muscles, which stabilize the shoulder joint in modern mammals in which the glenoid is very open (Jenkins and Weijs, 1979).

In advanced cynodonts, Triassic mammals, and monotremes, the supraspinatus muscle originates on the anteromedial surface of the blade. The anterior edge of the scapular blade is homologous with the scapula spine in therian mammals. The area that in marsupials and placentals underlies the supraspinatus muscle evolved as a new area of ossification anterior to the primitive margin of the blade. The stem of the clavicle retains a similar articulation with the acromion throughout this transition.

Jenkins and Parrington do not refer to a sternum, although its presence is implied in their restoration. Sternal elements have been described in tritylodonts and were presumably present in all early mammals. The clavicle and interclavicle are still large in early mammals, and the interclavicle is retained in living monotremes, although it is lost in therian mammals.

Surprisingly, there is no evidence of separate epiphyseal ossifications in *Morganucodon*. These are characteristic of all modern mammals and would be expected in the early forms, which already have well-defined joint surfaces and, by their small size, demonstrate closely regulated growth.

The humerus retains a generally primitive appearance, with widely expanded ends that are oriented at an angle of approximately 50 degrees to one another. The ectepicondylar foramen of cynodonts is lost, but a large entepicondylar foramen is retained. The hemispherical head is tilted strongly dorsally from the end of the shaft.

In the pelvis, the obturator foramen is further enlarged and the pubis further reduced from the pattern in cynodonts (Figure 18-8). The acetabulum retains vestiges of the separate articulating surfaces formed by the pubis, ilium, and ischium, which become confluent in later mammals. The lower margin of the acetabulum is incomplete.

Marsupial bones are not recognized, but since they occur in both tritylodonts and several groups of primitive mammals, they were probably present in *Morganucodon*.

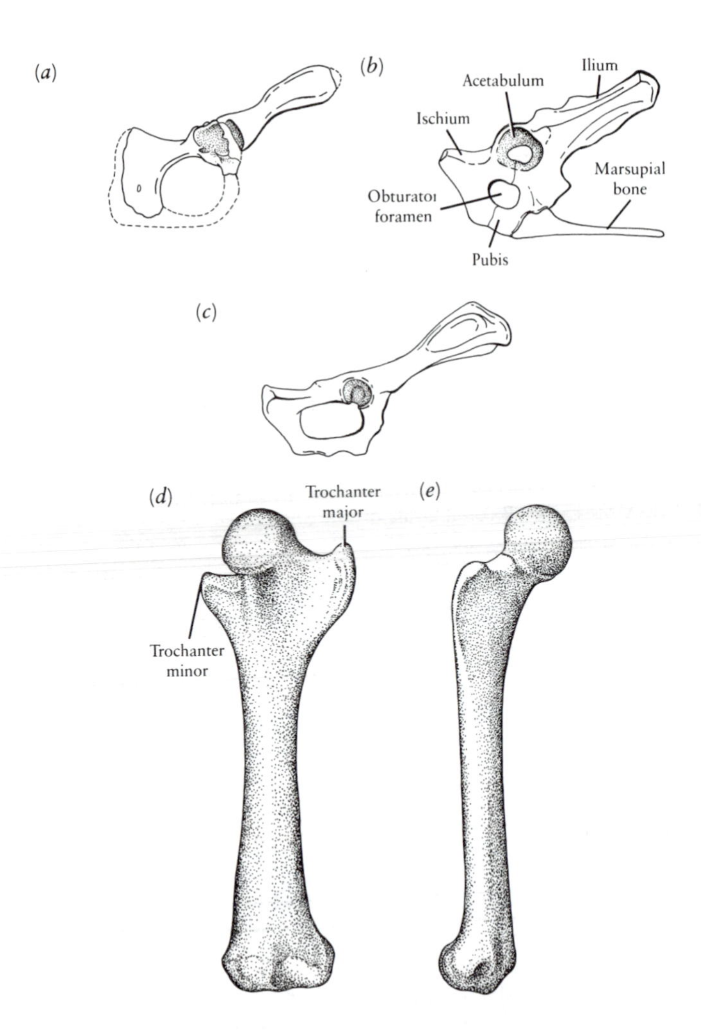

Figure 18-8. PELVIC GIRDLE AND FEMUR OF PRIMITIVE MAMMALS. (*a*) Pelvis of *Morganucodon*. (*b*) Pelvis of echidna. (*c*) Pelvis of the primitive placental *Tupaia*, a tree shrew. Left femur of *Morganucodon* in (*d*) dorsal and (*e*) medial views, ×4. *From Jenkins and Parrington, 1976.*

The femur appears much more typically mammalian than does the humerus, with the nearly spherical head well distinguished from the long narrow shaft. A depression near the center of the head indicates the presence of the ligamentum capitis femoris, which in modern mammals attaches the bone to the acetabulum. The greater and lesser trochanters extend medially and laterally from the proximal end of the shaft in the same plane as the distal articulating surface.

The configuration of the joint surfaces of the lower limbs indicates a posture and mode of limb movement that were similar to that of primitive, living, noncursorial mammals, such as the marsupial *Didelphis* and the primitive placental *Tupaia*. The ankle joint retains a pattern that is little advanced from that of cynodonts, with the astragalus not yet fully dorsal to the calcaneum, as it is in modern mammals.

Aside from the astragalus and calcaneum, the specific pattern of the carpals and tarsals can be traced only incompletely between cynodonts and mammals because of the small size of the elements and their tendency to become disarticulated. The number of centralia of the carpus is reduced from two to one. The fifth distal carpal has fused with the fourth within the therapsids, and a comparable change has occurred in the tarsus.

Different names are applied to many of the elements in mammals. The radiale, intermedium, and ulnare of the reptilian wrist are termed the scaphoid, lunar, and cuneiform in mammals. However, the centrale and pisiform retain their reptilian names. Distal carpals 1 to 4 are called the trapezium, trapezoid, magnum, and unciform. In the foot, the centrale is now termed the navicular, and the first three distal tarsals are named internal, middle, and external cuneiform (or entocuneiform, mesocuneiform, and ectocuneiform). The fourth distal tarsal is called the cuboid.

The hallux (big toe) of morganucodontids may have been somewhat divergent. This feature and the presence of well-developed claws suggest an arboreal habit. As Jenkins and Parrington emphasized, the small size of early mammals would result in any environment appearing extremely irregular and requiring much grasping, scrambling, and climbing, so that the distinction between terrestrial and arboreal habits would have little functional significance.

THE BIOLOGY OF PRIMITIVE MAMMALS

Remains of *Morganucodon* and closely related genera have been described from Europe, China, Africa, and North America in beds of late Triassic and early Jurassic age. There are remains of thousands of individuals in the fissure fillings of Wales.

Except for some specializations of the dentition, *Morganucodon* appears to represent an almost ideal pattern for the origin of the skeletal features of all later mammals. Skeletal evidence may also be used to reconstruct the soft anatomy and physiology of primitive mammals.

One of the most significant features of early mammals was their extremely small body size, which is equivalent to that of the smallest living mammals. It is reasonable to believe that the ancestors of mammals can be found among cynodonts such as the chiniquodontids or galesaurids that reduced their body size, probably in relationship to an insectivorous diet. Small size would have facilitated pursuit of tiny prey and probably involved life close to the ground and within the vegetation, where they would be relatively free from predation by the dinosaurs that were gaining dominance at that time.

It is generally accepted that early mammals were probably nocturnal, which may have been but an extension of life among rocks and in thick vegetation. This idea is based on the predominance of rods rather than cones in the eyes of primitive living mammals (Walls, 1942).

METABOLIC RATE

The body proportions, size, and dentition of early mammals resemble those of modern insectivores, which suggests a broadly similar way of life and physiology. Their general similarity with primitive modern mammals suggests a comparable metabolic rate and body temperature. Crompton, Taylor, and Jagger (1978) argued that the body temperature of early mammals may have been lower than even the most primitive of living mammals, the monotremes, which have a temperature of only 30 to 32°C. Nevertheless, if they were endotherms and maintained a relatively constant body temperature, they would have had a fairly high metabolic rate simply because of their small size.

HAIR

There is an inverse relationship between body size and metabolic rate (as measured per gram of body weight) in all endotherms (Figure 18-9). The metabolic rate increases very rapidly in animals that weigh less than 100 grams largely because of rapid heat loss due to the increased surface-to-volume ratio. Small mammals are able to maintain their body heat only because of effective insulation. If the morganucodontids were endotherms, as is implied by the specialization of the dentition and presence of a secondary palate, it is impossible to escape the conclusion that they must have evolved hair at an earlier stage in evolution when they were somewhat larger. Unfortunately, hair does not fossilize, and no one has recognized any skeletal features that are specifically and directly associated with the presence of hair. One might expect hair to be visible in footprints, but no such evidence has been described from the Mesozoic.

Hair is a uniquely mammalian tissue that is not directly homologous with any derivative of the skin present in other amniotes. Although we think of hair primarily in terms of its importance in insulation, it also has a sensory function, as is most clearly shown by the vibrissae of carnivores and rodents. Maderson (1972) suggested that hair originated as short keratinous projections extending from between the scales that were derived from structures resembling stretch receptors in modern lizards. Such extended sensory structures could warn the animal of sharp objects and tight places without risking damage to the surface of the body. These hairlike processes might logically develop all over the body, particularly in small animals living close to the ground. If these processes were to extend far beyond the scales, they would break up the flow of air over the surface and thus serve as insulation.

The presence of hair implies the existence of sebaceous and sweat glands to lubricate the hair and provide controlled heat loss. The embryological similarity of mammary glands and sweat glands suggests that the development of mammary glands may have followed closely

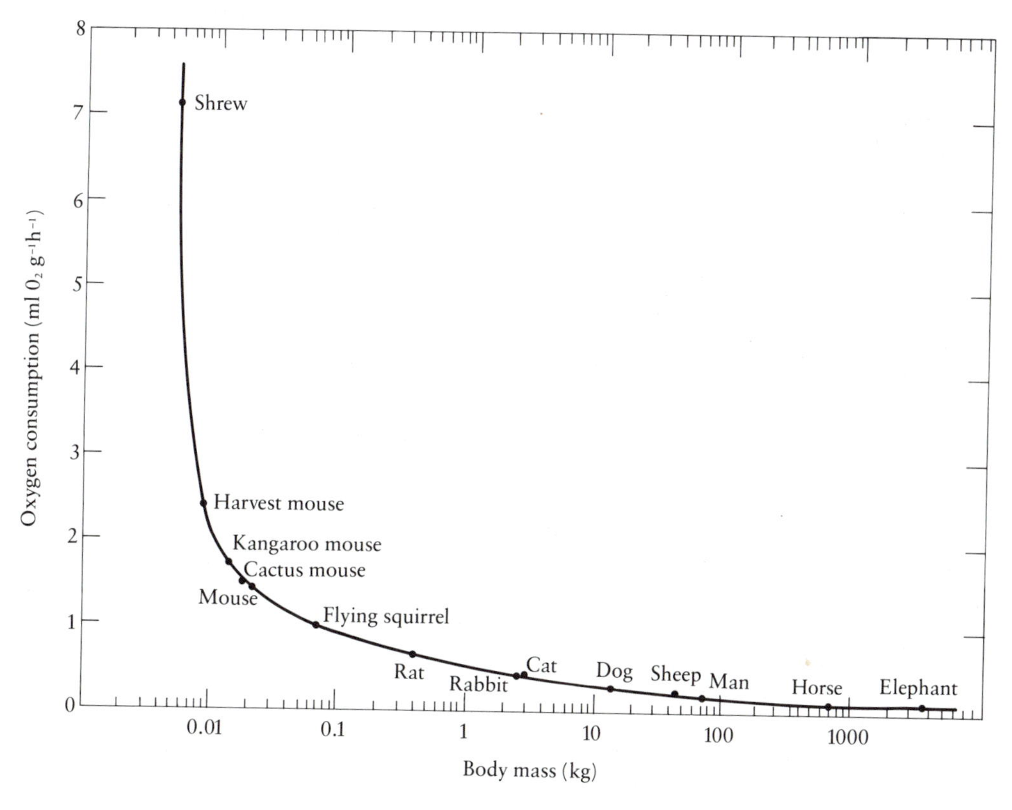

Figure 18-9. OBSERVED RATES OF OXYGEN CONSUMPTION OF VARIOUS MAMMALS. The oxygen consumption per unit body mass increases rapidly with decreasing body size. Note that the abscissa is logarithmic while the ordinate has an arithmetic scale. *From Schmidt-Nielsen, 1975.*

the evolution of hair. The presence of mammary glands in monotremes, marsupials, and placentals suggests that they must have evolved in the common ancestors of these groups, possibly before the end of the Triassic.

Hopson (1973) suggested that the evolution of mammary glands, in turn, may have been a requisite for the development of the precise pattern of tooth occlusion that is evident in the late Triassic mammals. In conventional reptiles, the young must be able to feed as soon as they are hatched and so must have a functional dentition at that time. Because of their small size at hatching, the teeth must be lost and replaced several times before adult size is reached. Since the teeth are repeatedly replaced and their relative position changes as the jaw increases in size, specific tooth occlusion cannot evolve.

Among mammalian ancestors, feeding of the young on maternal milk would have allowed tooth eruption to be delayed until considerable growth had occurred. Growth would be far advanced by the time the single generation of molar teeth had erupted behind the adult premolars. Specific tooth occlusion would also have been facilitated by the development of the definitive mammalian jaw joint between the squamosal and dentary and the more solid attachment of the molar teeth afforded by multiple roots. These factors may have been particularly important in permitting controlled transverse movement of the jaws during occlusion.

REPRODUCTION

A high, internally maintained body temperature and the evolution of mammary glands in the late therapsids set the stage for the development of the primitive mammalian reproductive pattern. The early Mesozoic mammals probably reproduced like living monotremes—hatching from eggs at a very immature, essentially embryonic level of development (Griffiths, 1978). It is somewhat surprising that primitive mammals should be born or hatched in a totally helpless condition. A major feature of reptiles, as opposed to amphibians, is their active, nearly adult condition at birth. Yet their successors, if we can judge on the basis of modern marsupials and monotremes, weighed less than a gram at birth and were blind and totally dependent on the adult for protection, warmth, and food.

As Hopson (1973) pointed out, the immaturity of ancestral mammals at birth may be explained by the small size of the adults and the physiological constraints that were imposed by a high metabolic rate. A mouse or a shrew requires approximately 100 times as much food per unit of body weight as does an elephant. The food and oxygen requirements of a warm-blooded mammal weighing less than 2 or 3 grams would theoretically approach infinity. This physiological limitation precludes the existence of mammals of smaller size, but some shrews closely approach this limit and we estimate that the late Mesozoic mammals were no larger than 20 to 30 grams. As adults, such small mammals must feed nearly constantly and can starve to death within only a few hours. It would be physiologically impossible for the young of such small mammals to maintain a high body temperature by their own metabolic activities. In fact, they do not. The young of small shrews and rodents are not endothermic but ectothermic. They are completely dependent on their parents to maintain a high body temperature. Their nakedness allows them to absorb heat more effectively. The nest, as well as the parents, provide insulation. Approximately 90 percent of the food consumed by small mammals is normally used to maintain a high body temperature. By relieving the young of the need to maintain their own body temperature, nearly all of what they are fed can go to growth and development.

BRAIN SIZE

The braincase of the Lower Jurassic mammal *Morganucodon* appears to be three to four times as large as that expected for therapsids of equivalent size (Crompton and Jenkins, 1979). This is a very substantial change and approaches the brain size of primitive Cenozoic mammals.

The increase in brain size between the advanced therapsids and the earliest mammals may be attributed to several factors. In the night or in dark tunnels and passageways, the early mammals would have needed extremely acute vision and would have placed more reliance on other sensory systems. The senses of smell, hearing, and touch were all augmented in late cynodonts. The distinctly larger size of the cochlea in early mammals suggests that they probably had greater acuity of hearing and a larger frequency response than the late cynodonts. The sense of smell, which is very important to most primitive mammals, was probably already well developed in the early cynodonts, if we are to judge by their large and complex nasal cavities. The presence of hair, which almost certainly evolved prior to the appearance of true mammals in the late Triassic, would have provided a wealth of tactile stimuli from the entire body surface.

The greater range of movements at the joint surfaces of the limbs and girdles in early mammals, in contrast with primitive reptiles, must have developed together with a more complex feedback system from the sensory receptors in the muscles, which led to the elaboration of integrative centers in the cerebellum.

The very small size of the Triassic mammals required a very high metabolic rate to compensate for rapid heat loss. In turn, the high metabolic rate necessitated very efficient feeding behavior, based on better integration of sensory input and motor function than was common in the mammal's reptilian predecessors. This improved integration led to further elaboration of the higher brain centers.

THE DIVERSITY OF PRIMITIVE MAMMALS

OTHER RHAETO-LIASSIC MAMMALS

In addition to *Morganucodon,* a number of less well-known mammals have been described from the late Triassic and early Jurassic of southern Africa, Europe, China, western North America, and India. The specific age and precise correlation of these deposits are difficult to establish and they are often referred to as Rhaeto-Liassic, a term that combines the names of the latest geological stage of the Triassic with the earliest stage in the Jurassic.

Sinoconodon from Yunnan, China, appears to be a relict of an earlier stage in the evolution of mammals. It is larger than other early genera, with a skull more than 5 centimeters long (Figure 18-10). *Sinoconodon* appears to be more derived than *Morganucodon* in the size and depth of the glenoid fossa and the degree of reduction of the postdentary bones. The jaw articulation resembles that of carnivorous cynodonts in being well below the level of the tooth row, in contrast with the more dorsal position in *Morganucodon.* The cheek teeth resemble those of *Morganucodon* in being laterally compressed and bearing a linear series of cusps, but they show *no* evidence of precise occlusion. This condition is clearly primitive and can be associated with the solid attachment of the jaws at the symphysis, which would not permit the two sides to move independently or to undergo rotation along the long axis. The pattern of tooth replacement is also primitive. The anterior cheek teeth are lost without being replaced. This phenomenon sometimes occurs in *Morganucodon,* but no skull of *Sinoconodon* retains anterior cheek teeth. A comparable situation occurs in advanced herbivorous cynodonts.

On the basis of these features, it is difficult to include *Sinoconodon* among the mammals. However, the structure of the petrosal resembles that of *Morganucodon* in its extension beneath the area of the semilunar ganglion, and the anterior lamina surrounds the mandibular and maxillary branches of the trigeminal nerve. Crompton and Sun (1985) classify *Sinoconodon* among the mammals but suggest that it diverged at the very base of the assemblage, at a more primitive level than the morganucodontids.

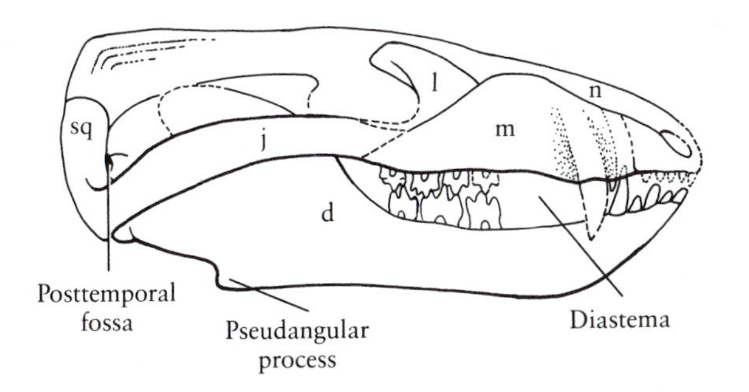

Figure 18-10. SKULL OF *SINOCODON* FROM THE LIASSIC OF CHINA. Original 5 centimeters long. The gap between the canines and the cheek teeth results from the nonreplacement of the anterior teeth. This genus is primitive in lacking precise tooth occlusion. Abbreviations as in Figure 8-3. *From Crompton and Sun, 1985. With permission from the Zoological Journal of the Linnean Society. Copyright 1985 by the Linnean Society of London.*

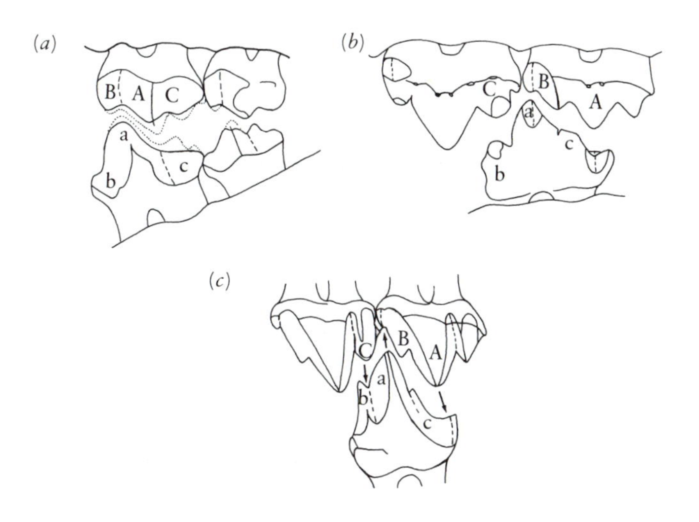

Figure 18-11. TOOTH OCCLUSION IN MESOZOIC MAMMALS. *(a) Morganucodon. (b) Megazostrodon. (c) Kuehneotherium. From Crompton, 1974. Redrawn with permission by the Trustees of the British Museum (Natural History).* Anterior is to the left.

The genus *Megazostrodon* from southern Africa is very similar to *Morganucodon* in the structure of the postcranial skeleton, but the teeth occlude in a slightly different manner (Figure 18-11). The upper and lower teeth alternate with one another, with cusp **A** occluding behind **c**, while **a** occludes in front of **B**. The difference from the pattern in *Morganucodon* is not great, but it is comparable with the initial changes that separate major groups of later Mesozoic mammals (Crompton, 1974; Gow, 1986).

Dinnetherium, from the western United States, also differs in its occlusal pattern, with **a** wearing between **B** and **A**, and **A** wearing against the external face of **c.** A more important distinction is the extensive transverse component in the movement of the teeth, which is apparently produced by a significant degree of jaw rotation along its long axis during occlusion. The degree of rotation may be related to the elaboration of a true angular process that is lateral and ventral to the extradentary bones, in addition to a more anterior pseudangular process (Figure 18-12).

Morganucodon, Sinoconodon, Megazostrodon, and *Dinnetherium* retain the basic tooth structure of their cynodont ancestors, in which the cusps are arranged in a strictly linear fashion. A significantly different pattern occurs in *Kuehneotherium,* in which the main cusps are arranged in the pattern of an obtuse-angle triangle. The central cusp of the upper tooth is situated lingually (toward the tongue) and that of the lower tooth faces buccally or labially (toward the cheek). In contrast with *Mor-*

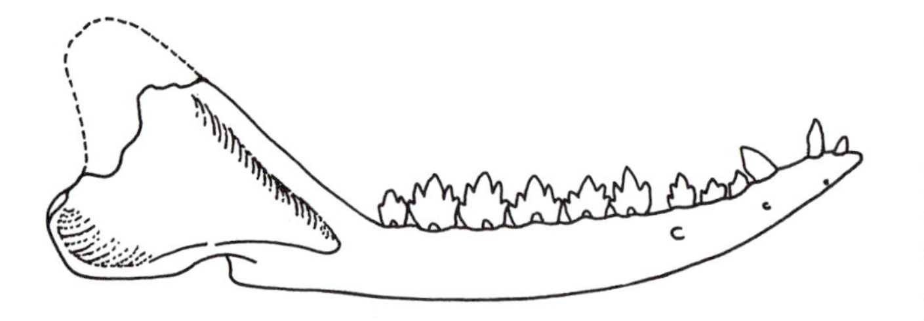

Figure 18-12. LOWER JAW OF *DINNETHERIUM.* This primitive mammal from the Lower Jurassic of Western North America shows the beginning of a true angular process. *From Jenkins, 1984.*

ganucodon, the principal cusp of the lower tooth fits between the teeth in the upper jaw (Figure 18-11*c*). In occlusal view, the upper and lower teeth form a series of reversed triangles, which is the basis for the pattern of occlusion in both marsupials and placentals. The initiation of this pattern would require but a slight shift in the relative position of the upper and lower teeth from a pattern like that of *Megazostrodon.* In *Megazostrodon* and *Morganucodon,* a triangular pattern of occlusal surfaces is produced by wear of the teeth after they erupt. In *Kuehneotherium,* the triangular pattern of occlusion is established genetically by the triangular arrangement of the cusps and is further accentuated by wear.

The lower jaw of *Kuehneotherium* retains a groove for the extradentary bones like that of *Morganucodon,* which indicates that it had a similar double-jaw articulation. We know *Kuehneotherium* only from isolated teeth and jaw fragments from the same fissure fillings in Wales from which *Morganucodon* came. The apparent absence of other remains of *Kuehneotherium* suggests that the remainder of the skeleton may have been so similar to that of the much more common genus that they cannot be differentiated.

The relationships of *Kuehneotherium* and other therian mammals will be discussed in the next chapter.

We see an even more divergent cusp pattern in the haramiyids (Figure 18-13). This group is only known by isolated teeth from a variety of deposits in Europe. They have two or more roots, like those of both tritylodonts and early mammals. The crowns are broad, with multiple cusps arranged around the margin. In some teeth, the cusps are arranged in two anteroposteriorly oriented rows, which suggests a similar function to that of the tritylodonts. Wear would have been produced primarily by movements of the jaw in the vertical plane.

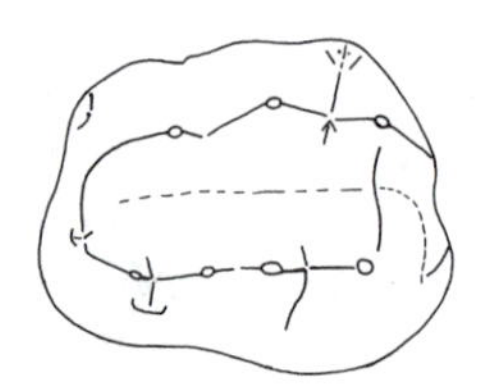

Figure 18-13. CHEEK TOOTH OF *HARAMIYA.* Lateral and occlusal views, ×10. *From Clemens, 1980.*

Hahn (1973) and Clemens (1980) discussed the anatomy and taxonomic position of these teeth. With no knowledge of the remainder of the skeleton, there is no solid evidence that haramiyids are mammals and not separate cynodont derivates. Their possible affinities with later Mesozoic multituberculates has led to their inclusion within the Mammalia.

Teeth of haramiyids, morganucodontids, and relatives of *Kuehneotherium* are all known as early as the base of the Rhaetic, approximately 225 million years ago, from the locality of Saint-Nicolas-de-Port near Nancy, France (Sigogneau-Russell, 1983). A single tooth of *Kuehneotherium* has been reported from the Norian of Great Britain (Fraser, Walkden, and Stewart, 1985).

The Upper Triassic and Lower Jurassic provide an exceptionally rich record of early mammals. For the remainder of the Mesozoic, a period of more than 100 million years, most remains are limited to little more than jaws and teeth. During this period the skeletal and physiological characters that distinguish marsupial and placental mammals evolved. In addition to the ancestors of the modern therian mammals, we know a variety of nontherian mammals from the Jurassic and Cretaceous. Most were small, insectivorous forms, but they also include the herbivorous multituberculates and carnivorous amphilestids that reached the size of the modern opossum. The fossil record of Mesozoic mammals is reviewed in detail in the book *Mesozoic Mammals. The First Two-Thirds of Mammalian History,* which was edited by Lillegraven, Kielan-Jaworowska, and Clemens (1979).

The record of fossils from the Rhaeto-Liassic covers approximately 15 million years. In contrast, there is no record of fossil mammals for the following 20 to 30 million years, until the Middle Jurassic (Bathonian) (Figure 18-14).

We know Middle Jurassic mammals from several localities in Great Britain and one in China. These remains include isolated teeth of morganucodontids and possible haramiyids. In addition, nearly complete jaws are known that represent the first appearance of at least four other lineages that diversified during the later Mesozoic (Freeman, 1979).

DOCODONTS

The docodonts are the shortest-lived and least diverse of all groups of Mesozoic mammals. They are known only from the Middle and Upper Jurassic and include only four genera from North America and Europe. Their specific ancestry has not been recognized and they apparently had no descendants (Kron, 1979; Krusat, 1980).

The pattern of the molar teeth clearly distinguishes docodonts from all other Mesozoic mammals (Figure 18-15). The lower molars are rectangular, with a row of high

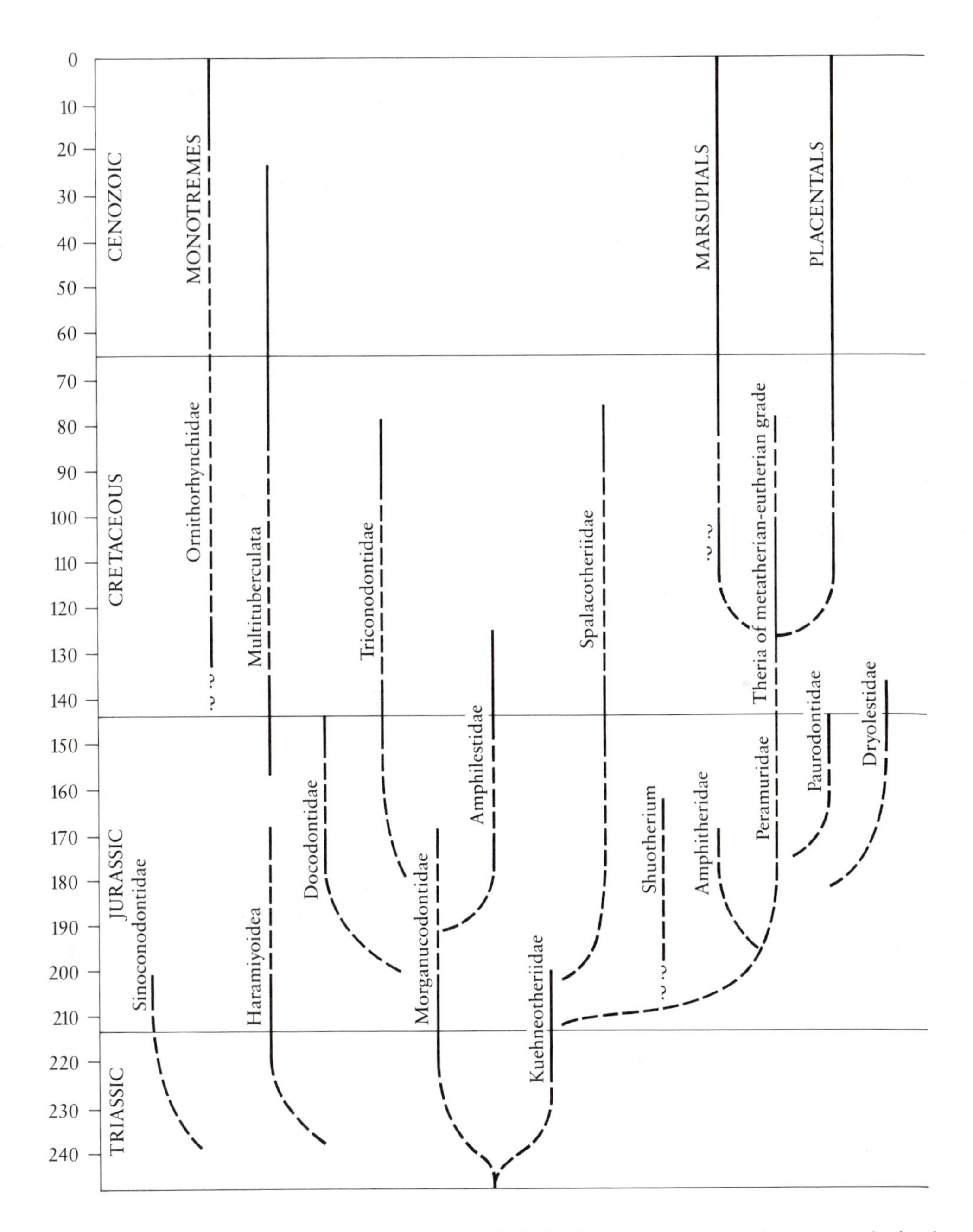

Figure 18-14. PHYLOGENY OF MESOZOIC MAMMALS. Note that there are very long gaps in the fossil record. *Data from Crompton and Jenkins, 1979, modified to include subsequent discoveries.*

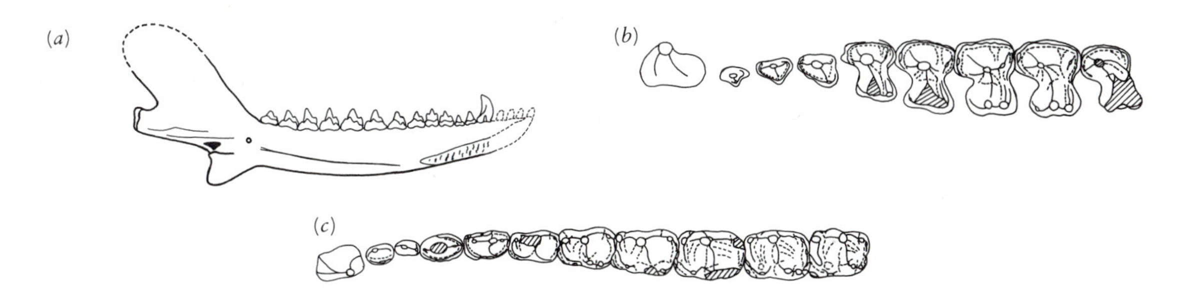

Figure 18-15. (*a*) Medial view of the lower jaw of *Docodon*, ×2.5. *From Kron, 1979. By permission of the University of California Press.* (*b*) Upper and (*c*) lower teeth of *Docodon*. *From Jenkins, 1969.*

buccal cusps and somewhat lower lingual cusps that probably evolved by elaboration of the lingual cingulum of more primitive mammals. The principal lateral cusp is joined to the lingual row by a transverse ridge. The upper molars are much more expanded transversely, with a widely extended medial portion that is distinguished from the more lateral surface by a waisted midsection. Wear produced a complex pattern of shearing surfaces. In older individuals, the cusps became blunt and they would have served to crush and grind the food.

If we judge from their jaws, docodonts were the size of small mice and had elongated snouts. The number of cheek teeth appears to increase over time, with Upper Jurassic species having as many as four premolars and eight molars. The jaws are grooved medially for the attachment of extradentary bones, which indicates that they still retained a reptilian jaw joint. An extension from the lower margin of the dentary is in the position of the pseudangular process of *Dinnetherium*. The middle Jurassic docodont *Borealestes* and the late Jurassic *Haldanodon* are known from considerable postcranial material that has not yet been described. (Kermack et al., 1987).

AMPHILESTIDS

Amphilestids retained the cusp pattern of the morganucodontids, with a single row of linearly arranged cusps. The occlusal pattern more specifically resembles that of *Megazostrodon* and *Dinnetherium,* in which the upper and lower teeth more nearly alternate in position. Amphilestids are represented in the Middle Jurassic by the genera *Amphilestes* and *Phascolotherium. Amphilestes* has a tooth count of

$$\frac{? \quad ? \quad ? \quad ?}{3\text{–}4 \quad 1 \quad 4 \quad 5}$$

They have neither an angular nor a pseudangular process (Figure 18-16).

Amphilestids also occur in the late Jurassic and early Cretaceous. Much of the skeleton is known of *Gobiconodon* from the early Cretaceous of the Mongolian People's Republic and the western United States (Jenkins, 1984).

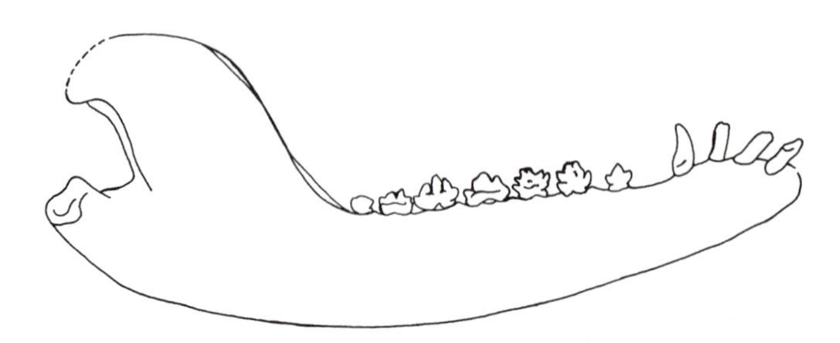

Figure 18-16. LOWER JAW OF THE AMPHILESTID *PHASCOLOTHERIUM. From Jenkins and Crompton, 1979. By permission of the University of California Press.*

The body was 35 centimeters long (excluding the tail) and more heavily built than the living opossum. An important difference from other primitive mammals is the development of a typically therian supraspinous fossa of the scapula (Jenkins and Crompton, 1979). The alisphenoid is relatively larger than in *Morganucodon* and approaches the therian condition. Freeman (1979) suggested that the amphilestids be placed within the therian order Symmetrodonta.

Surprisingly, *Gobiconodon* shows replacement of the molar teeth (Jenkins, 1984), which may be related to the large size and presumably longer life span of this genus. One of the reasons that tooth replacement was limited in the earliest mammals may have been because they were so small and, short-lived that there was not time for several generations of teeth to erupt.

Other groups of mammals that appear first in the Middle Jurassic are clearly related to the ancestry of marsupials and placentals. They will be discussed in the next chapter.

TRICONODONTIDS

Another long hiatus in the fossil record, which lasted approximately 25 million years, separates the Middle and Upper Jurassic faunas. Among the animals first known in the Upper Jurassic are the triconodontids, which extend into the Upper Cretaceous. Their closest affinities appear to lie with *Morganucodon.* This relationship is based primarily on dental similarities, since little of the remainder of the skeleton of triconodontids has been described.

The fact that the principal cusps of triconodontids are arranged in a linear pattern is certainly primitive. The particular pattern of occlusion shared by triconodontids and morganucodontids, in which the upper and lower teeth are aligned more or less one to one, may be primitive or derived relative to that exemplified by *Megazostrodon* in which the upper and lower molars are opposed more or less two to one. The teeth of triconodontids are definitely specialized in having the major cusps of nearly uniform height, which gives the teeth the appearance of a saw blade (Figure 18-17).

A partial braincase described by Kermack (1963) demonstrates the presence of an uncoiled cochlea, as in early Jurassic mammals, and the cavum epiptericum is still open anteriorly. Newly discovered skeletal remains from the Lower Cretaceous being studied by Jenkins (1984) show the possible association of a scapula with a supraspinous fossa, as in the amphilestids and therian mammals.

The only triconodontid in which the tooth count is known has a dental formula of

$$\frac{2 \quad 1 \quad 4 \quad 5}{1 \quad 1 \quad 4 \quad 5}$$

No clearly defined angular process is present.

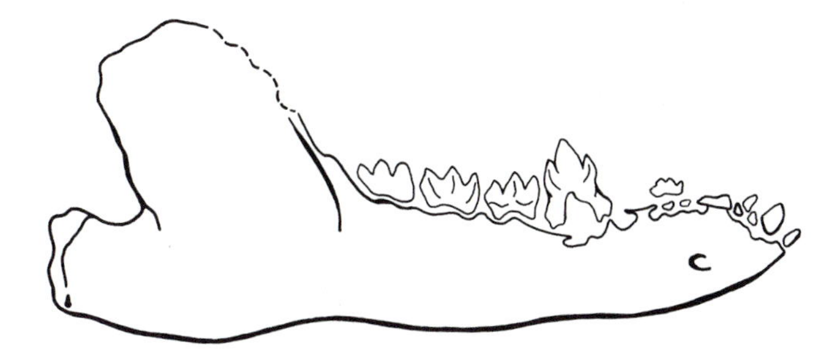

Figure 18-17. LOWER JAW OF THE TRICONODONT *TRICONODON*. *From Jenkins and Crompton, 1979. By permission of the University of California Press.*

MULTITUBERCULATES

The most diverse and numerous of Mesozoic mammals, and the only group that were omnivorous rather than insectivorous or carnivorous, were the multituberculates (see Figure 18-21). They are first known with certainty in the Upper Jurassic but have long been thought to show affinities with the haramiyids, which are recognized as early as the late Triassic.

Multituberculates range in size from the dimensions of a small mouse to those of a woodchuck. The skull has a superficially rodentlike appearance, which results from the presence of a single pair of long, procumbent, lower incisors followed by a diastema due to the loss of the canines. There may be one to three pairs of upper incisors, and the upper canine may be retained in primitive genera. In most multituberculates, the anterior lower premolars are specialized as laterally compressed shearing blades that bear a series of vertical ridges. The molar teeth have two or three rows of linearly arranged cusps that may have been used to crush or grind food (Figure 18-18).

Tooth wear indicates that jaw motion was entirely in a sagittal plane, with no trace of the transverse component that was common to the most primitive mammals. In most genera, the glenoid was open anteriorly, which permitted propalinal movement of the lower jaw. Krause (1982) demonstrated that the power stroke involved retraction of the mandible during closure (a broad analogy may be drawn with mastication in dicynodonts, which were discussed in the previous chapter). The symphysis is not solidly fused, which indicates the possibility of unilateral movement of the jaw rami. There is no angular process.

Like modern rodents, multituberculates may have been largely herbivorous, but the shearing lower premolars show a pattern that is particularly close to that of modern omnivorous marsupials, as Clemens and Kielan-Jaworowska (1979) pointed out.

The skull of multituberculates is low and broad. The frontal overhangs the orbit. The eyes faced primarily laterally, in contrast with their more anterior orientation in early therian mammals. The jugal is missing, and the zyg-

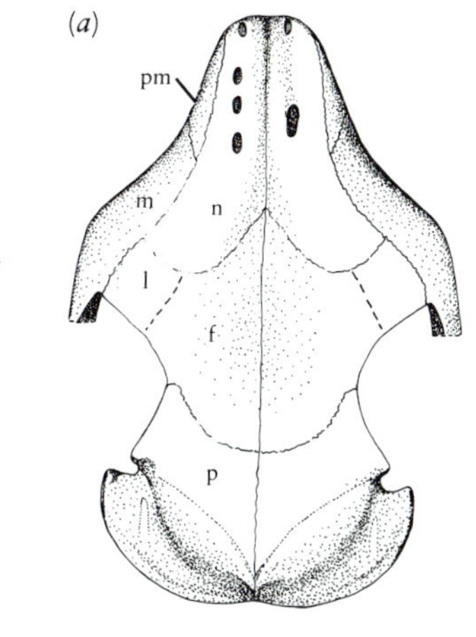

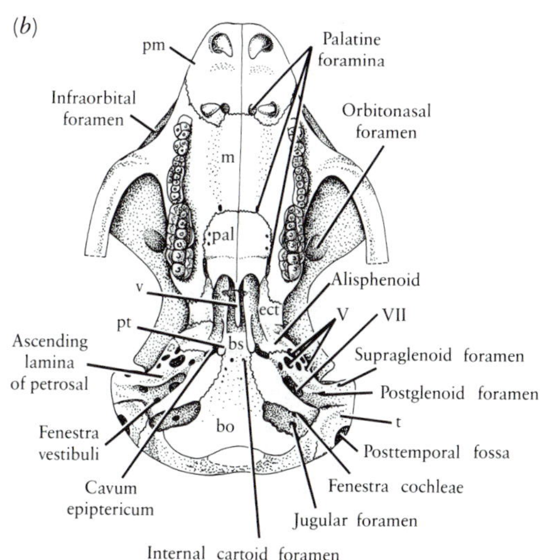

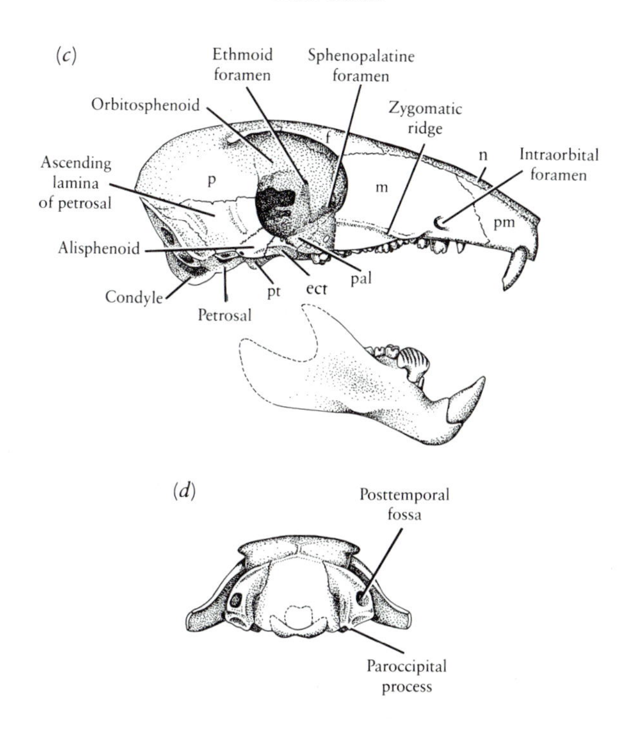

Figure 18-18. (*a*) Dorsal, (*b*) palatal, (*c*) lateral, and (*d*) occipital views of the skull of the late Cretaceous multituberculate *Kamptobaatar* from Mongolia, ×3. (Occiput is drawn slightly smaller). Abbreviations as in Figure 8-3. *From Kielan-Jaworowska, 1971.*

omatic arch is formed entirely by the squamosal and a long posterior process from the maxilla.

The braincase of Upper Cretaceous multituberculates described by Kielan-Jaworowska (1971) resembles those of *Morganucodon* and triconodontids in having a relatively small alisphenoid but a large ascending lamina of the petrosal (Figure 18-19). The squamosal does not contribute substantially to the wall of the braincase. The occiput is primitive in the retention of a posttemporal fossa. As in primitive mammals, the cochlea is uncoiled. Desui and Lillegraven (1986) have described the incorporation of the malleus and incus into the middle ear of a Paleocene multituberculate. The cranial vascular system is described by Kielan-Jaworowska, Presley, and Poplin (1986).

The most primitive of the multituberculates are included in the family Paulchoffatiidae (suborder Plagiaulacoidea) from the Upper Jurassic (Hahn 1969, 1977) (Figure 18-20). They retain three upper incisors, a possible upper canine, five upper premolars, and two upper molars. Three lower premolars and three lower molars are recognized, although the specific homologies of the molar and premolar teeth of multituberculates have not been satisfactorily established. The lower premolars have not yet developed the shearing function that was common to later members of the group, and like the haramiyids, at least one lower molar has a central basin rather than a linear pattern of cusps.

In the Cretaceous, the plagiaulacoids are succeeded by the Ptilodontoidea (a primarily North American group) and the predominantly Asian Taeniolabidoidea. Both groups continued to radiate into the early Tertiary. The taeniolabidoids are the most rodentlike in the development of continuously growing incisors in which the enamel is limited to the anterior margin. As the teeth wear, the enamel forms a self-sharpening edge. In some genera, the bladelike lower premolars are lost. The tooth formula may be reduced to as little as

$$\frac{2\ 0\ 1\ 2}{1\ 0\ 0\ 3}$$

Although the teeth of multituberculates are common and varied in size and configuration, skeletal remains are rare. Several articulated specimens have been reported from Asia but have not yet been fully described. Krause and Jenkins (1983) reviewed the postcranial anatomy of all the North American multituberculates. They found that adequately known elements of the skeleton are very similar throughout the suborders Ptilodontoidea and Taeniolabidoides from the late Cretaceous into the early Tertiary. A specimen of *Ptilodus* from the late Paleocene of Saskatchewan provides evidence of much of the skeleton. This genus shows strong evidence of arboreal adaptation. Comparison with living mammals that are known to be specifically adapted to life in trees demonstrates that the tail of *Ptilodus* is specialized as a prehensile organ. The joints of the tarsus and pes are modified to permit the

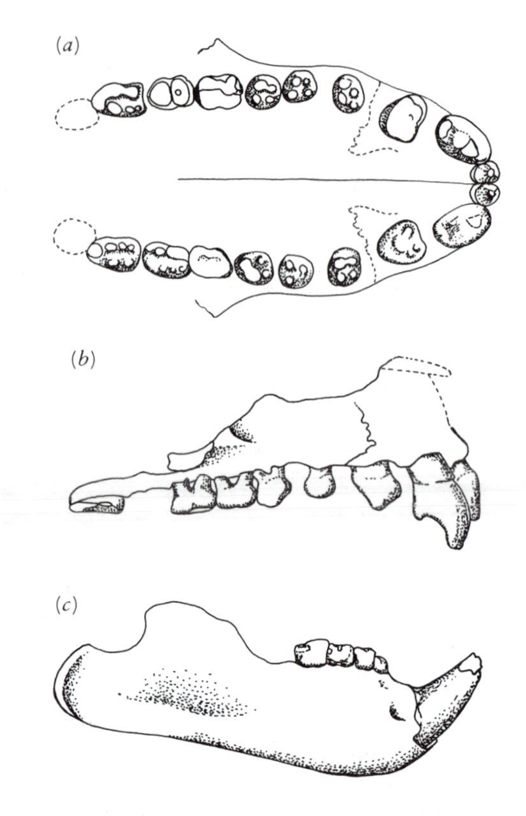

Figure 18-20. PRIMITIVE PLAGIAULACOID MULTITUBERCULATES. (*a* and *b*) *Kuehneodon* palate and lateral view of cheek teeth, ×4. (*c*) Lower jaw of *Paulchoffatia*, ×2.5. *From Clemens and Kielan-Jaworowska, 1979. By permission of the University of California Press.*

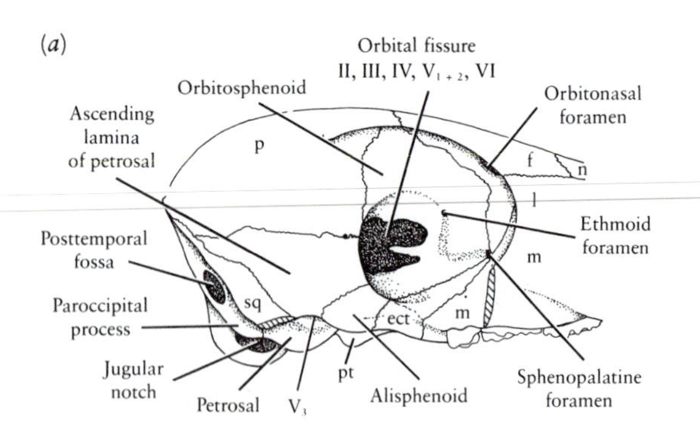

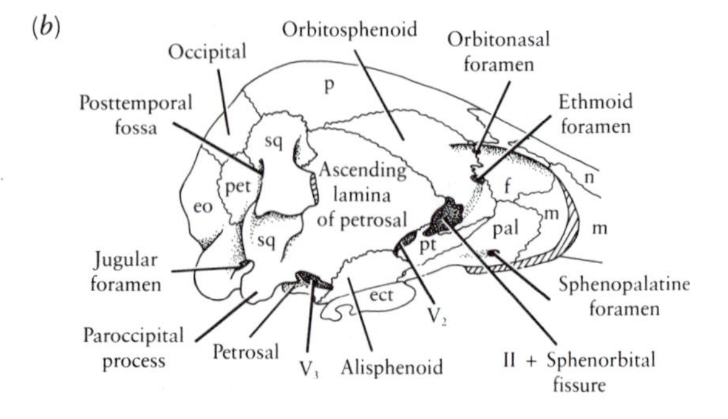

Figure 18-19. (*a*) Lateral view of the braincase of *Kamptobaatar*. (*b*) Lateral view of the braincase of *Ornithorhynchus*. Abbreviations as in Figure 8-3. *From Kielan-Jaworowska, 1971.*

foot to be reversed in the manner of squirrels, which allows them to descend the trunk of a tree headfirst (Figure 18-21).

The scapula is primitive in lacking the supraspinous fossa that characterizes therian mammals, but the glenoid is advanced in being widely open ventrally. The acetabulum is shallow and open dorsally, which would have contributed to the extensive range of limb movements that are necessary in an arboreal mammal. The area for articulation of the tibia extends primarily along the posterior aspect of the femoral condyles, which indicates that the rear limb was held in a crouched position, as in other primitive mammals.

An extra bone, the parafibula, is associated with the knee joint. This element is also present in monotremes, where it is fused to the fibula, and in a few therian mammals. Its function has not been fully investigated in living genera.

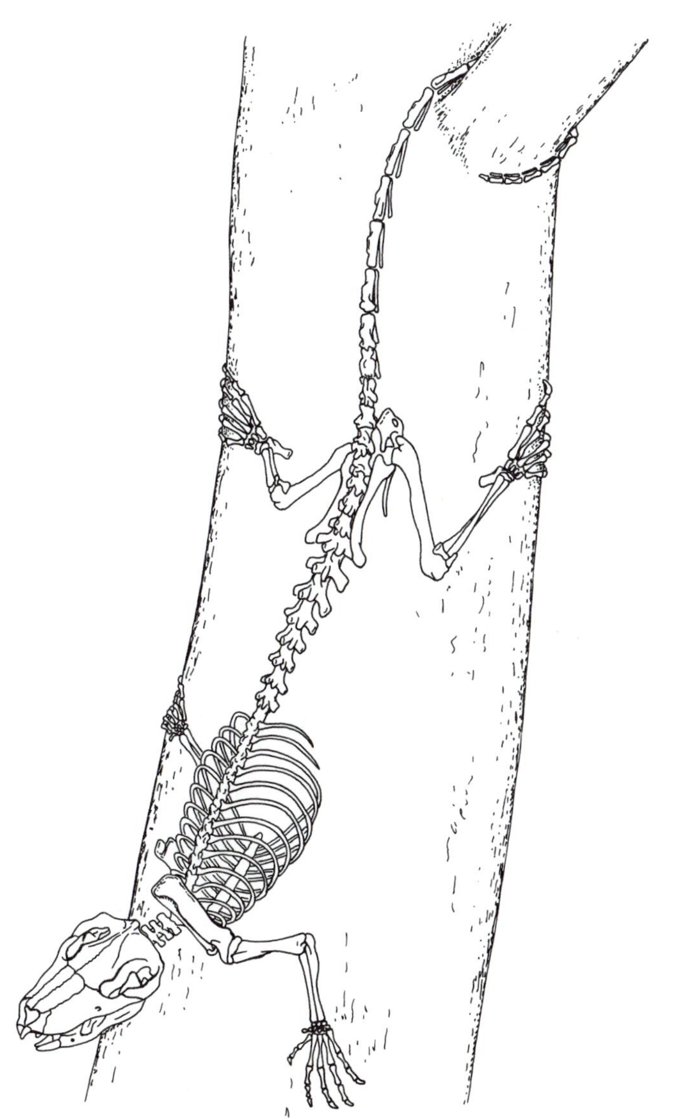

Figure 18-21. A MULTITUBERCULATE SKELETON. Reconstruction is based principally on a specimen of *Ptilodus*. *From Krause and Jenkins, 1983.*

Kielan-Jaworowska (1969) described a pelvis with epipubic (marsupial) bones in a specimen from the Upper Cretaceous of Mongolia. She suggested that the very narrow space between the halves of the pelvis might indicate that they gave birth to very small, but live, young, as in the case of marsupials.

The evolutionary pattern of the multituberculates is unique among mammals in several respects. They are the longest living order. Even if we exclude the haramiyids, they extend from the Upper Jurassic into the Oligocene, a period of over 100 million years (Krishtalka, Emry, Storer, and Sutton, 1982). They are the most diverse of the primarily Mesozoic mammalian orders, with 11 families and nearly 50 genera, and the only one to survive into the Tertiary. They were also the first group of mammalian herbivores to gain dominance. Clemens and Kielan-Jaworowska (1979) attribute the great success of the multituberculates in the late Mesozoic and early Cenozoic to the radiation of angiosperms, which may have provided a new food source that they were uniquely adapted to utilize.

Several multituberculate genera became extinct at the end of the Cretaceous, but the remaining lineages continued to differentiate and the order apparently reached its greatest diversity in the Paleocene. Their subsequent decline and extinction has been attributed to competitive inferiority relative to a series of herbivorous placentals, first the condylarths, later the primates, and finally the rodents (Hopson, 1967; Van Valen and Sloan, 1966; Krause, 1986). Krause and Jenkins argued that they had no skeletal features that were obviously inferior to the pattern in contemporary placental mammals. They might have been inferior in one or more aspects of their physiology or behavior, but these characters cannot be judged directly from the fossil record.

Because of their considerable diversity, distinct anatomy, and long period of independent evolution, the multituberculates have been placed in a separate subclass, the Allotheria.

INTERRELATIONSHIPS OF MESOZOIC MAMMALS

Kuehneotherium and other early therian mammals (which will be discussed in the next chapter) can be recognized as a monophyletic assemblage on the basis of shared derived features of the dentition. It is much more difficult to establish interrelationships among the remaining, nontherian mammals of the Mesozoic. It has been thought that they represent a distinct group that was characterized by a more primitive level of evolution. The entire assemblage was placed within the subclass Prototheria, which also includes the monotremes. However, it is presently impossible to establish that this is a natural group. Most of the features that the included families share may be primitive for all mammals, rather than being derived relative to those of the ancestors of the therians.

Morganucodontids, amphilestids, and triconodontids have been united in the order Triconodonta on the basis of having the major cusps of the molar teeth arranged in a linear sequence (Jenkins and Crompton, 1979). This pattern is surely primitive (plesiomorphic) for all mammalian lineages, since it is basically similar to the pattern of carnivorous cynodonts and is also shared by *Dinnetherium* and probably by the immediate ancestors of the therian mammals. Presley (1981) and Kemp (1983) argued that the pattern of the braincase that is common to morganucodontids, multituberculates, and triconodontids is probably also primitive for all mammals.

We can define the families Triconodontidae, Amphilestidae, and Dinnetheriidae on the basis of apomorphies, but we may not be able to do so for the Morganucodontidae. The Morganucodontidae may occupy a position among early mammals that is comparable to that of the Protorothyridae among early amniotes. Animals with the basic morphology of *Morganucodon* and *Megazostrodon* may have given rise to all mammals, with the probable exception of *Sinocondon* and the haramiyids.

The early appearance and very distinct morphology of the haramiyid teeth suggest that they are phylogenetically distinct from all other mammals. Clemens and Kielan-Jaworowska (1979) pointed out that these teeth are very much smaller than those of the contemporary tritylodonts, which suggests that they may have been more similar to the other tiny mammals in their physiology. If haramiyids are related to the multituberculates, the many derived features of the skeleton that the latter shares with morganucodonts would support a common origin at the mammalian level. If multituberculates are not closely related to the haramiyids, they may have evolved from among the late Triassic or early Jurassic morganucodontids, but there is no evidence of specific affinities. Docodonts may have evolved independently from a similar level.

MONOTREMES

Among living mammals, the monotreme genera *Ornithorhynchus*, *Tachyglossus*, and *Zaglossus* are unique in the primitive features of their reproductive system. They alone lay eggs and have an essentially reptilian configuration of the reproductive tracts. They retain a cloaca, rather than separating the aperture of the digestive system and the urogenital opening, and the male lacks a scrotum.

In contrast, most other features of their soft anatomy and physiology support the inclusion of the platypus and echidnas within the Mammalia. They have hair, suckle their young, and maintain a relatively constant body temperature by metabolic means that is well above that of the environment. The basically mammalian pattern of a wide range of physiological processes that Griffiths (1968, 1978) discussed supports the assumption that the monotremes share a much closer common ancestry with the marsupials and placentals than they do with any reptilian group.

However, the phylogenetic position of monotremes remains subject to debate. The skull of the platypus and echidnas are highly specialized in a manner divergent from those of all other groups of mammals, fossil or living. The snout is drawn out as a long tubular structure in the echidnas and appears as a long flattened beak in the platypus. The echidnas have no trace of teeth; the platypus develops two pairs of upper teeth and three pairs of lower molariform teeth, but these are shed before maturity and are functionally replaced by a leathery beak (Figure 18-22).

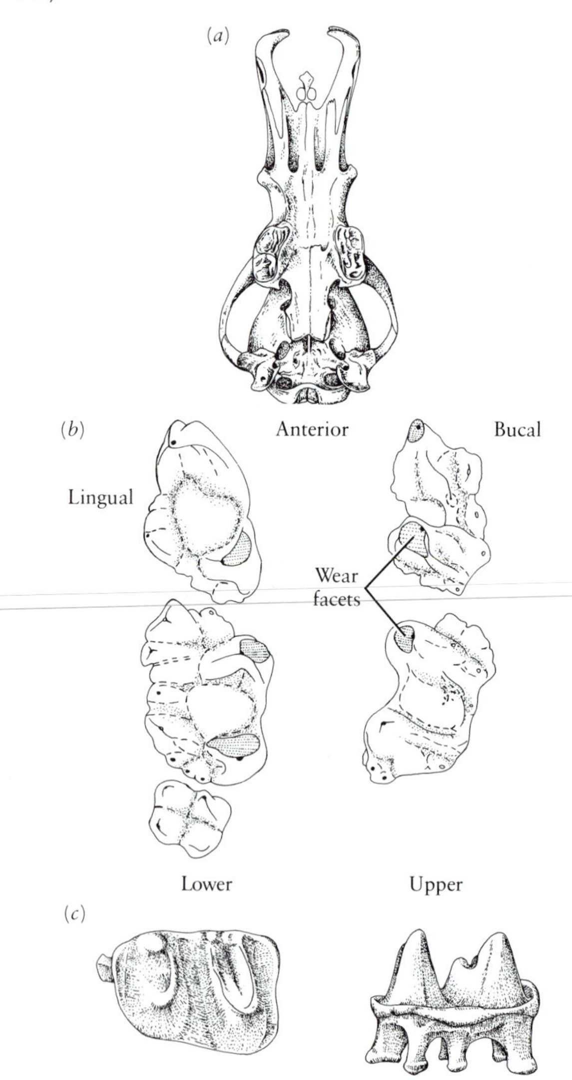

Figure 18-22. MONOTREMES. (*a*) Skull of the duck-billed platypus. *From Gregory, 1951.* (*b*) Molar teeth of the platypus. These are not functional but are replaced by a leathery beak in the adult. *From Hopson and Crompton, 1969.* (*c*) Teeth of the Miocene monotreme *Obdurodon*. *From Clemens, 1979. By permission of the University of California Press.*

The postcranial skeleton, specifically the shoulder girdle, retains the pattern of advanced cynodonts and the most primitive Mesozoic mammals. Both coracoids are retained, and the interclavicle, which is lost in therian mammals, is still a large element. The anterior margin of the scapula is formed by the spine, with no elaboration of a supraspinous fossa. As in cynodonts, the forelimbs have a primitive structure and sprawling posture, but this may be exaggerated in relationship to their specializations for burrowing. The cervical ribs have not fused to the vertebrae. These primitive features of the skeleton have been used to suggest that monotremes diverged from the remainder of mammals before the end of the Triassic.

The fossil record of monotremes provides little help in establishing their specific affinities. Remains from the Miocene include molar teeth, a partial lower jaw, and an ilium (Archer, Plane, and Pledge, 1978). The teeth differ from those of the modern platypus in having two strong transverse ridges but are not readily compared with those of any other mammals (Figure 18-23). Equally distinctive teeth have evolved separately among many groups of Tertiary mammals.

Archer, Flannery, Ritchie, and Molnar (1985) have described a lower jaw from the Lower Cretaceous of Australia as a possible monotreme. There is a strongly transverse component to the arrangement of the cusps, but the pattern could be derived from that of primitive tribosphenic molars, suggesting that monotremes may be much more closely related to therian mammals than had been previously thought. The phylogenetic significance of this specimen is further discussed by Kielan-Jaworowska, Crompton, and Jenkins (1987).

Kermack and Kielan-Jaworowska (1971) and Kielan-Jaworowska (1971) argued that the structure of the braincase of monotremes is fundamentally similar to that of multituberculates and other nontherian Mesozoic mammals in the great extent of the anterior lamella of the petrosal and the relatively limited extent of the alisphenoid when compared with therian mammals.

There are two problems with this line of reasoning. Since *Morganucodon* has a large petrosal and relatively small alisphenoid, one can argue that this condition is primitive for mammals and thus of limited significance in establishing relationships between any particular groups of mammals. Presley (1981) provided evidence from developmental studies that the composition of the wall of the braincase is not fundamentally different in therian and nontherian mammals and that the condition seen in monotremes may have evolved from the same pattern as that of the ancestors of marsupials and placentals.

Kemp (1982, 1983) suggested that monotremes shared a common ancestry with the therian mammals as late as the Upper Jurassic. He argues that the incorporation of the quadrate and articular into the middle ear as the incus and malleus and the final loss of the reptilian jaw articulation occurred only once in the common ancestor of marsupials, placentals, and monotremes. Support of this

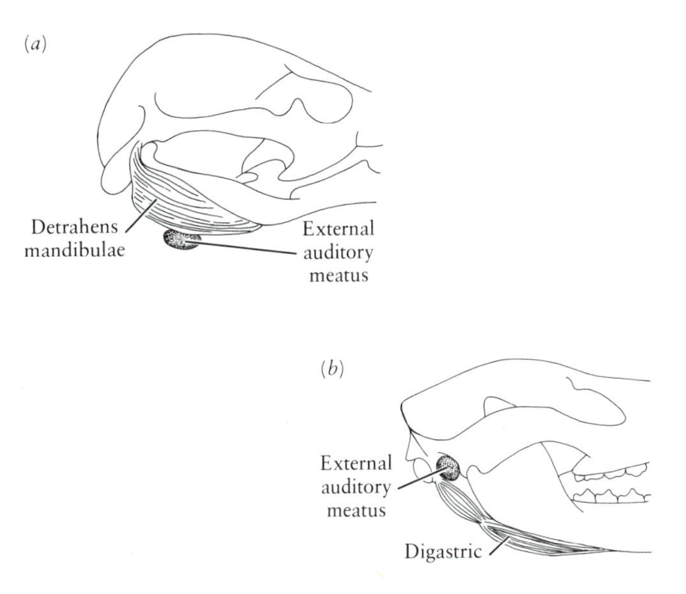

Figure 18-23. EXTERNAL AUDITORY MEATUS AND JAW DEPRESSOR MUSCLES. Relationship between them in monotreme and therian mammals is shown. (*a*) *Ornithorhynchus*. The external auditor meatus lies below the detrahens mandibulae. (*b*) *Didelphis*, the auditory meatus lies above the digastic. *From Hopson, 1966.*

hypothesis requires the demonstration that all the primitive features of the skeleton of monotremes were still retained in the ancestors of placentals and marsupials in the late Jurassic. Such evidence seems unlikely on the basis of Henkel and Krebs' (1977) preliminary description of the complete skeleton of an Upper Jurassic therian mammal.

Multituberculates almost certainly lost the reptilian jaw articulation independently, as shown by the incorporation of the malleus and incus into the middle ear in a Paleocene species. Unfortunately, there is no evidence before the end of the Cretaceous to establish when this occurred in marsupials or placentals. There is indirect evidence from the nonhomologous nature of the jaw depressors in monotremes and therian mammals that structures closely associated with the middle ear evolved separately in these groups (Hopson, 1966) (Figure 18-23).

The anatomy of the postcranial skeleton of living monotremes still suggests that their ancestors diverged from the remainder of the Mammalia at a comparatively early stage, probably at a level little advanced from the Rhaeto-Liassic morganucodontids. There is no strong evidence that they share a close common ancestry with any of the later groups of therian or nontherian mammals. Their ancestors may have become isolated in the Australian region by the early Jurassic. There is no record of any other fossil mammals there until the late Oligocene, but there is also no evidence in any other part of the world of fossil monotremes or their plausible ancestors since the early Jurassic.

Summary

The earliest mammals are recognized in the Upper Triassic and Lower Jurassic. They are distinguished from advanced therapsids by their smaller body size (as little as 20 grams), the relatively larger size of the cranial cavity, and the nature of the dentition. The incisors, canines, and premolars were replaced only once, and there is only a single generation of molar teeth. In most groups, there is a specific pattern of tooth occlusion in which the lower jaw moves medially as well as dorsally. Occlusion occurs on only one side of the skull at a time. The dentary and squamosal form the primary jaw joint, but the articular and quadrate remain part of the jaw apparatus.

The vertebrae are differentiated from those of therapsids and resemble those of later mammals in the structure of the atlas-axis complex, the expansion of the neural canal of the cervical vertebrae, and in the distinction between the thoracic and lumbar region. The configuration of the neural spines indicates that vertebral flexure was primarily in the vertical plane.

The skeleton of the best-known early Mesozoic mammals, the morganucodontids, is sufficiently similar to those of the most primitive living marsupials and placentals to suggest that they were physiologically similar as well. They were almost certainly endothermic, which would have necessitated the presence of hair and implies the development of sebaceous, sweat, and mammary glands. Their very small body size would have forced them to give birth to young that were at a very immature stage of development and totally dependent on maternal milk, warmth, and protection.

The relatively larger brain size of early mammals, compared with similar-sized reptiles, may be attributed to the integration of enhanced auditory, tactile, and olfactory input and the need to coordinate sophisticated locomotor and feeding activities associated with their high metabolic requirements.

Several other lineages of primitive mammals accompany the morganucodontids in the late Triassic and early Jurassic. *Sinoconodon* may be a relict of a more primitive stage of evolution if we judge by the lack of tooth occlusion and the primitive pattern of tooth replacement. The teeth of *Megazostrodon* and *Dinnetherium* resemble those of *Morganucodon,* but the upper and lower teeth more nearly alternate with one another at occlusion. In occlusal view, the molar cusps of the upper and lower teeth of *Kuehneotherium* form a pattern of reversed triangles, which provides a very effective shearing structure as soon as they erupt. This pattern is elaborated in the later therian mammals, the marsupials and placentals.

We only know the haramiyids from their tiny teeth. The pattern of their cusps indicates that they occluded by propalinal movements of the lower jaw without the transverse movements that were common to other early Mesozoic mammals.

After the Lower Jurassic, the fossil record of mammals is very incomplete until the very end of the Cretaceous. Among the later Mesozoic mammals are the docodonts and triconodontids, which we know primarily from teeth and isolated jaws. Amphilestids were among the largest of Mesozoic mammals, reaching the size of an opossum. *Gobiconodon* has a scapula like that of therian mammals with a supraspinus fossa but replaces its molar teeth.

The most common and diverse of Mesozoic mammals were the omnivorous multituberculates, which ranged into the Oligocene. They appear first in the Upper Jurassic but may have evolved from the Rhaeto-Liassic haramiyids. The dentition is superficially rodentlike, with large incisors and grinding cheek teeth. One of the lower premolars is typically highly developed as a shearing blade. As in the haramiyids, jaw movements were propalinal. As in most primitive mammals, the cochlea is uncoiled. The postcranial skeleton of multituberculates appears to have been quite uniform. *Ptilodus* shows arboreal specializations. An Upper Cretaceous genus had marsupial bones.

Morganucodonts, triconodonts, amphilestids, docodonts, and multituberculates have been grouped in the subclass Prototheria primarily because of the possession of primitive characters relative to therian mammals; they have not been shown to form a natural, monophyletic assemblage.

The monotremes are the most primitive of living mammals if we judge by their low metabolic rate, archaic features of the skeleton, and egg-laying habits. The high degree of specialization of the skull and the loss or great reduction of the dentition makes it difficult to compare them with any other mammals. A recently discovered fossil from the Lower Cretaceous suggests affinities with primitive therian mammals, in contrast with the long-held opinion that monotremes had diverged from the ancestors of marsupials and placentals by the early Jurassic.

References

Archer, M., Flannery, T. F., Ritchie, A., and Molnar, R. E. (1985). First Mesozoic mammal from Australia—An early Cretaceous monotreme. *Nature* **318**: 363–366.

Archer, M., Plane, M. D., and Pledge, N. S. (1978). Additional evidence for interpreting the Miocene *Obdurodon insignis* Woodburne and Tedford, 1975, to be a fossil platypus (Ornithorhynchidae: Monotremata) and a reconsideration of the status of *Ornithorhynchus agilis* De Vis, 1885. *The Australian Zoologist*, **20**: 9–27.

Clemens, W. A. (1979). Notes on the Monotremata. In J. A. Lillegraven, Z. Kielan-Jaworowska, and W. A. Clemens (eds.), *Mesozoic Mammals. The First Two-Thirds of Mammalian History*, pp. 309–311. University of California Press, Berkeley.

Clemens, W. A. (1980). Rhaeto-Liassic mammals from Switzerland and West Germany. *Zitteliana*, **5**: 51–92.

Clemens, W. A., and Kielan-Jaworowska, Z. (1979). Multituberculata. In J. A. Lillegraven, Z. Kielan-Jaworowska, and W. A. Clemens (eds.), *Mesozoic Mammals. The First Two-Thirds of Mammalian History*, pp. 99–149. University of California Press, Berkeley.

Crompton, A. W. (1968). The enigma of the evolution of mammals. *Optima*, **18**(3): 137–151.

Crompton, A. W. (1971). The origin of the tribosphenic molar. In D. M. Kermack and K. A. Kermack (eds.), *Early Mammals. Zool. J. Linn. Soc.*, **50** (suppl. 1): 65–87.

Crompton, A. W. (1972). Postcanine occlusion in cynodonts and tritylodontids. *Bull. Brit. Mus. (Nat. Hist.) Geol.*, **21**: 27–71.

Crompton, A. W. (1974). The dentitions and relationships of the southern African Triassic mammals, *Erythrotherium parringtoni* and *Megazostrodon rudnerae*. *Bull. Brit. Mus. (Nat. Hist.), Geol.*, **24**: 397–437.

Crompton, A. W., and Hiiemäe, K. M. (1970). Molar occlusion and the mandibular movements during occlusion in the American opossum, *Didelphis marsupialis*. *Zool. J. Linn. Soc.*, **49**: 21–47.

Crompton, A. W., and Jenkins, F. A., Jr. (1968). Molar occlusion in Late Triassic mammals. *Biol. Rev.*, **43**: 427–458.

Crompton, A. W., and Jenkins, F. A., Jr. (1979). Origin of mammals. In J. A. Lillegraven, Z. Kielan-Jaworowska, and W. A. Clemens (eds.), *Mesozoic Mammals. The First Two-Thirds of Mammalian History*, pp. 59–73. University of California Press, Berkeley.

Crompton, A. W., and Kielan-Jaworowska, Z. (1978). Molar structure and occlusion in Cretaceous therian mammals. In P. M. Butler and K. A. Joysey (eds.), *Studies in the Development, Function and Evolution of Teeth*, pp. 249–287. Academic Press, London.

Crompton, A. W., and Sun, A.-L. (1985). Cranial structure and relationships of the Liassic mammal *Sinoconodon*. *Zool. J. Linn. Soc.*, **85**: 99–119.

Crompton, A. W., Taylor, C. R., and Jagger, J. A. (1978). Evolution of homeothermy in mammals. *Nature*, **272**: 333–336.

Desui, M. and Lillegraven, J. A. (1986). Discovery of three ear ossicles in a multituberculate mammal. *National Geographic Research*, **2**: 500–507.

Fraser, N. C., Walkden, G. M., and Stewart, V. (1985). The first pre-Rhaetic therian mammal. *Nature*, **314**: 161–163.

Freeman, E. F. (1979). A Middle Jurassic mammal bed from Oxfordshire. *Palaeontology*, **22**: 135–166.

Gow, C. E. (1986). A new skull of *Megazostrodon* (Mammalia, Triconodonta) from the Elliot Formation (Lower Jurassic) of southern Africa. *Palaeont. afr.*, **26**: 13–23.

Gregory, W. K. (1951). *Evolution Emerging*. Vol. 2. Macmillan, New York.

Griffiths, M. (1968). *Echidnas*. Pergamon Press, Oxford and New York.

Griffiths, M. (1978). *The Biology of the Monotremes*. Academic Press, London.

Hahn, G. (1969). Beiträge zur Fauna der Grube Guimarota Nr. 3, die Multituberculata. *Palaeontographica, Abt. A*, **133**: 1–100.

Hahn, G. (1973). Neue Zäne von Haramiyiden aus der deutschen Ober-Trias und ihre Beziehungen zu den Multituberculaten. *Palaeontographica, Abt. A*, **142**: 1–15.

Hahn, G. (1977). Das Coronoid der Paulchoffatiidae (Multituberculata; Ober Jura). *Paläont. Z.*, **51**: 234–245.

Henkel, S., and Krebs, B. (1977). Der erste Fund eines Säugetier-Skelettes aus der Jura Zeit. *Umschau der Wissenschaft und Technik*, **77**: 217–218.

Hopson, J. A. (1966). The origin of the mammalian middle ear. *Am. Zool.*, **6**: 437–450.

Hopson, J. A. (1967). Comments on the competitive inferiority of the multituberculates. *Syst. Zool.*, **16**: 352–355.

Hopson, J. A. (1973). Endothermy, small size and the origin of mammalian reproduction. *Amer. Naturalist*, **107**: 446–452.

Hopson, J. A., and Crompton, A. W. (1969). Origin of mammals. In T. Dobzhansky, M. K. Hecht, and W. C. Steere (eds.), *Evolutionary Biology*, Vol. III., pp. 15–72. Appleton, New York.

Jenkins, F. A., Jr. (1969). Occlusion in *Docodon* (Mammalia, Docodonta). *Postilla, Yale Peabody Museum*. **139**: 1–24.

Jenkins, F. A., Jr. (1984). A survey of mammalian origins. In P. D. Gingerich and C. E. Badgley (eds.), *Mammals. Notes for a Short Course. Univ. Tennessee, Studies in Geol.*, **8**: 32–47.

Jenkins, F. A., Jr., and Crompton, A. W. (1979). Triconodonta. In J. A. Lillegraven, Z. Kielan-Jaworowska, and W. A. Clemens (eds.), *Mesozoic Mammals. The First Two-Thirds of Mammalian History*, pp. 74–90. University of California Press, Berkeley.

Jenkins, F. A., Jr., and Parrington, F. R. (1976). The postcranial skeletons of the Triassic mammals *Eozostrodon, Megazostrodon* and *Erythrotherium*. *Phil. Trans. Roy. Soc., Lond., B*, **273**: 387–431.

Jenkins, F. A., Jr., and Weijs, W. A. (1979). The functional anatomy of the shoulder in the Virginia opossum (*Didelphis virginiana*). *J. Zool.*, **188**: 379–410.

Kemp, T. S. (1982). *Mammal-like Reptiles and the Origin of Mammals*. Academic Press, London.

Kemp, T. S. (1983). The relationships of mammals. *Zool. J. Linn. Soc.*, **77**: 353–384.

Kermack, K. A. (1963). The cranial structure of the triconodonts. *Phil. Trans. Roy. Soc. Lond., B*, **246**: 83–103.

Kermack, K. A., and Kielan-Jaworoska, Z. (1971). Therian and nontherian mammals. In D. M. Kermack and K. A. Kermack (eds.), *Early Mammals. Zool. J. Linn. Soc.*, **50** (suppl. 1): 103–115.

Kermack, K. A., Lee, A. J., Lees, P. M., and Mussett, F. (1987). A new docodont from the Forest Marble. *Zool. J. Linn. Soc.*, **89**: 1–39.

Kermack, K. A., Mussett, F., and Rigney, H. W. (1973). The lower jaw of *Morganucodon*. *Zool. J. Linn. Soc.*, **53**: 87–175.

Kermack, K. A., Mussett, F., and Rigney, H. W. (1981). The skull of *Morganucodon*. *Zool. J. Linn. Soc.*, **71**: 1–158.

Kielan-Jaworowska, Z. (1969). Discovery of a multituberculate marsupial bone. *Nature*, **222**: 1091–1092.

Kielan-Jaworowska, Z. (1971). Skull structure and affinities of the Multituberculata. *Palaeontologia Polonica*, **25**: 5–41.

Kielan-Jaworowska, Z., Crompton, A. W., and Jenkins, F. A., Jr. (1987). Monotreme relatives. *Nature* (in press).

Kielan-Jaworowska, Z., Presley, R., and Poplin, C. (1986). The cranial vascular system in taeniolabidoid multituberculate mammals. *Phil. Trans. Roy. Soc.* B., **313**: 525–602.

Krause, D. W. (1982). Jaw movement, dental function, and diet in the Paleocene multituberculate *Ptilodus*. *Paleobiology*, 8: 265–281.

Krause, D. W. (1986). Competitive exclusion and taxonomic displacement in the fossil record: The case of rodents and multituberculates in North America. In R. M. Flanagan and J. A. Lillegraven (eds.), *Vertebrates, Phylogeny, and Philosophy. Cont. Geol. Univ. Wyo. Spec. Pap. 3.*

Krause, D. W., and Jenkins, F. A., Jr. (1983). The postcranial skeleton of North American multituberculates. *Bull. Mus. Comp. Zool.*, **150**: 199–246.

Krishtalka, L., Emry, R. J., Storer, J. E., and Sutton, J. F. (1982). Oligocene multituberculates (Mammalia: Allotheria): Youngest known record. *J. Paleont.*, **56**(3): 791–794.

Kron, D. G. (1979). Docodonta. In J. A. Lillegraven, Z. Kielan-Jaworowska, and W. A. Clemens (eds.), *Mesozoic Mammals. The First Two-Thirds of Mammalian History*, pp. 91–98. University of California Press, Berkeley.

Krusat, G. (1980). Contribuicao para o conhecimento da fauna do Kimeridgiano da mina de lignito Guimarota (Leiria, Portugal). IV parte. *Haldanodon expectatus* Kuehne and Krusat 1972 (Mammalia, Docodonta) [The Kimmeridgian fauna of the Guimarota lignite mine, Leiria, Portugal; Part 4, *Haldanodon expectatus;* Mammalia, Docodonta]. *Memorias dos Servicos Geologicos de Portugal*, **27**: 1–79.

Lillegraven, J. A., Kielan-Jaworowska, Z., and Clemens, W. A. (eds.). (1979). *Mesozoic Mammals. The First Two-Thirds of Mammalian History.* University of California Press, Berkeley.

Maderson, P. F. A. (1972). When? Why? and How?: Some speculations on the evolution of the vertebrate integument. *Am. Zoologist*, **12**: 159–171.

Presley, R. (1981). Alisphenoid equivalents in placentals, marsupials, monotremes and fossils. *Nature*, **294**: 668–670.

Schmidt-Nielsen, K. (1975). *Animal Physiology. Adaptation and Environment.* Cambridge University Press, Cambridge.

Sigogneau-Russell, D. (1983). A new therian mammal from the Rhaetic locality of Saint-Nicolas-de-Port (France). *Zool. J. Linn. Soc.*, **78**: 175–186.

Van Valen, L., and Sloan, R. E. (1966). The extinction of the multituberculates. *Syst. Zool.*, **15**: 261–278.

Walls, G. L. (1942). The vertebrate eye and its adaptive radiation. *Bull. Cranbrook Inst. Sci.*, (Bloomfield Hills), **19**: 1–785.

CHAPTER XIX

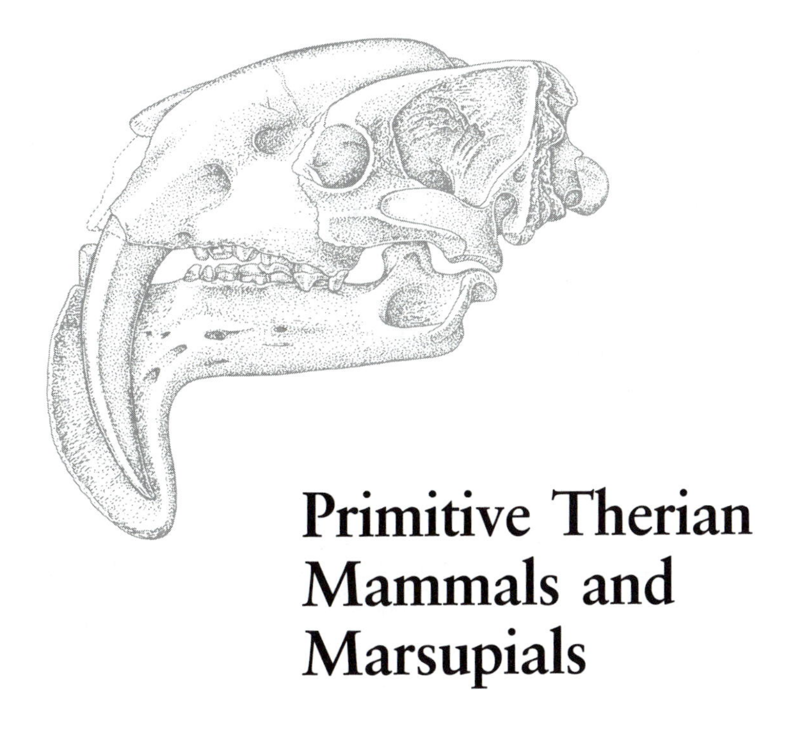

Primitive Therian Mammals and Marsupials

Aside from the echidnas and platypus, all living mammals can be included in a single monophyletic assemblage, the Theria. The modern therian groups—marsupials and placentals—diverged from a common ancestor early in the Cretaceous. More primitive therians are known from the late Triassic into the Cretaceous.

THE EVOLUTION OF THERIAN MOLAR TEETH

The fossil record of early therian mammals consists primarily of teeth and jaws; hence, the structure and function of these teeth is of paramount importance in establishing relationships. The molar teeth of primitive therian mammals are characterized by a triangular arrangement of the principal cusps. The inception of this pattern is recognized in the Rhaeto-Liassic genera *Woutersia* and *Kuehneotherium*. *Woutersia* is known only from isolated teeth from the base of the Rhaetic (Sigogneau-Russell, 1983). *Kuehneotherium* is represented by many teeth and jaw fragments from fissure fillings of Lower Jurassic (Sinemurian) age (Kermack, Mussett, and Rigney, 1973). The number of incisors is not known. There is one canine, from five to six premolars and three to six molars, with a maximum postcanine count of ten to eleven (Figure 19-1). In contrast with *Morganucodon,* the teeth are not strongly attached to the jaws. Based on differences in the shape of isolated teeth, Mills (1984) suggests that there were five constant molars and a sixth that was inconstant in size and not always present.

As outlined in the previous chapter, the molar teeth of *Kuehneotherium* occlude with one another like those of later therian mammals. The edges joining the principal cusps shear past one another like the sides of reversed triangles. *Kuehneotherium* also approaches the therian configuration in the presence of an incipient talonid (or heel) at the rear of the lower molars that is formed by an expansion of the cingulum and bears a single cusp. This area of the cingulum overlaps that of the anterior edge of the succeeding tooth so that they are locked into place, which ensures greater precision of occlusion. This overlap also prevents food from being wedged between the teeth during mastication.

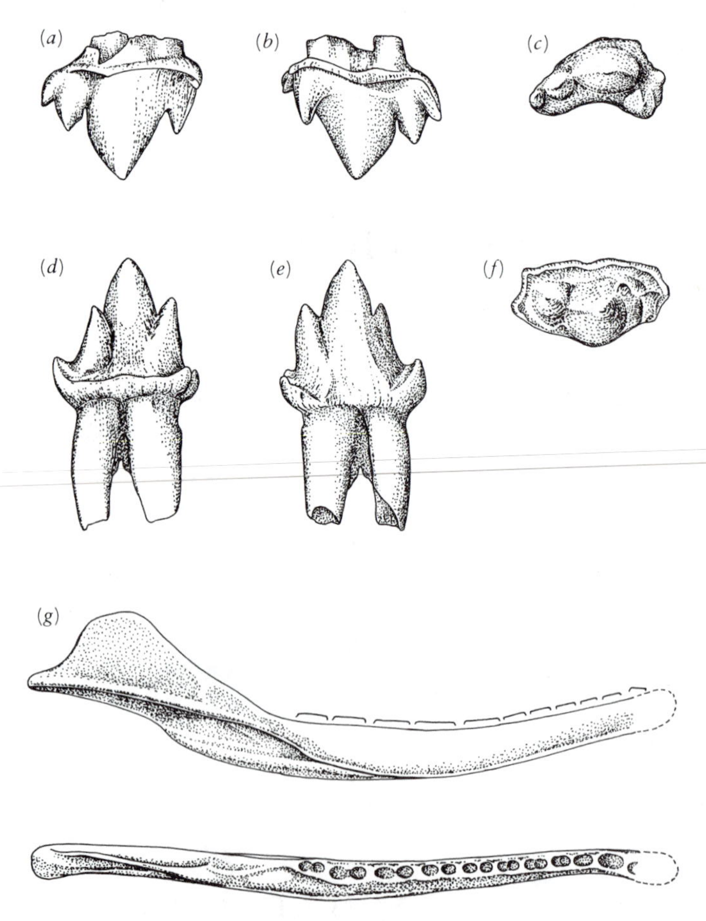

Figure 19-1. JAWS AND TEETH OF *KUEHNEOTHERIUM*, AN EARLY THERIAN MAMMAL FROM THE LOWER JURASSIC. Left upper molar in (*a*) lingual, (*b*) buccal, and (*c*) occlusal views. Left lower molar in (*d*) lingual, (*e*) buccal, and (*f*) occlusal views. Nature of occlusion is illustrated in Figure 18-11. (*g*) Lower jaw of *Kuehneotherium* in medial and dorsal views. *From Kermack, Kermack, and Mussett, 1968.*

In the previous chapter, the principal cusps were identified by the same alphabetical scheme as those of nontherian mammals. The relationship of *Kuehneotherium* to later therian mammals makes it possible to use the system of cusp designation that is applied to marsupials and placentals (Figure 19-2).

Cope and Osborn (Osborn, 1888) established that the molar teeth of all marsupials and placental mammals evolved from a common ancestral pattern. Osborn referred to this pattern as tritubercular, but this term was later changed to **tribosphenic.** In the characteristic tribosphenic tooth, the occlusal surface of the upper molar is in the shape of a triangle, or trigon, with the apex toward the tongue. The lower teeth have a similar (but reversed) triangle (the trigonid) anteriorly and, in addition, a posterior heel or talonid. Early students of fossil mammals thought that the apical cusps of the upper and lower triangles were comparable to the single cusp of primitive reptiles. For this reason, this cusp was designated the **protocone** in the upper tooth and the **protoconid** in the lower tooth. The names **paracone** (anteriorly) and **metacone** (posteriorly) are used for the other major cusps of the upper tooth, and **paraconid** and **metaconid** for the anterior and posterior cusps of the lower trigonid. The talonid is outlined by the **entoconid** (buccally), **hypoconid** (lingually), and **hypoconulid** (posteriorly), which surround a depression or basin. Additional features are shown in Figure 19-2.

In primitive therians such as *Kuehneotherium,* the cusps of both the upper and lower teeth are arranged in the pattern of an obtuse triangle that may be conceived as having evolved by the slight displacement of the linearly arranged cusps in *Morganucodon* and other more primitive mammals. When Mesozoic mammals with this tooth pattern were first discovered, it seemed natural to apply the same terminology to the major cusps as had been established for marsupials and placentals. As more emphasis was placed on the significance of wear patterns, it was recognized that the protocone was specifically associated with the presence of a fully developed talonid with a deep basin. Primitive therians that had only an incipient talonid would have had no place for the protocone to occlude, and so it was suggested that the apical cusp in genera such as *Kuehneotherium* was not a protocone. Further work has established its homology with the paracone of marsupials and placentals. The posterior cusp can be accepted as homologous with the metacone, but the anterior cusp (B) appears comparable to the stylocone of later therians (Cassiliano and Clemens, 1979). The cusps of the lower trigonid of *Kuehneotherium* are directly

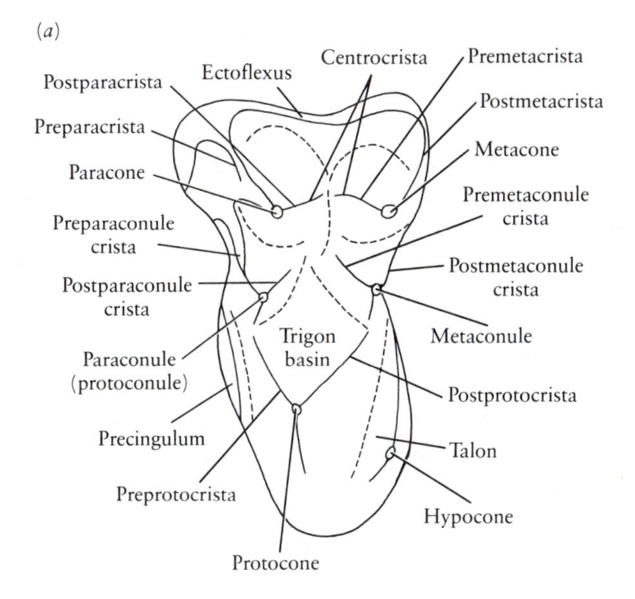

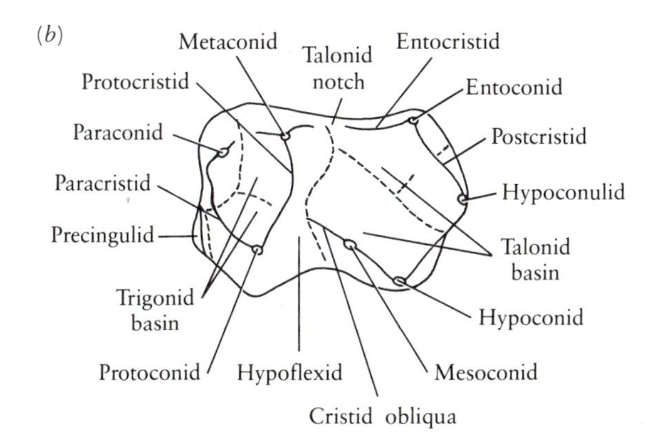

Figure 19-2. Pattern of the cusps and other important features of tribosphenic molars that are common to primitive therian mammals. (*a*) Upper molar and (*b*) lower molar, both in occlusal view. *From Bown and Kraus, 1979. By permission of the University of California Press.*

comparable with those of modern therians. The talonid cusp is identified as the hypoconulid.

SYMMETRODONTS

Among other Mesozoic mammals, the molar teeth of *Kuehneotherium* are most similar to those of the Middle and Upper Jurassic genera *Cyrtlatherium* and *Tinodon* [*Eurylambda*]. They are grouped within a larger assemblage, the Symmetrodonta, which is characterized by the broadly symmetrical appearance of upper and lower molar teeth, both of which are dominated by a trigonid, with only a small talonid on the lower teeth (Figure 19-3).

Kuehneotherium and *Tinodon* are designated obtuse angle symmetrodonts since the angle between the lines joining the major cusps at the apex is greater than 90 degrees. In the Spalacotheriidae, which is known from the Middle Jurassic to the Upper Cretaceous, the angle is less than 90 degrees. Members of this family are termed the acute angle symmetrodonts. The Spalacotheriidae appear to be advanced in the direction of marsupials and placentals in the arrangement of the cusps, but the talonid is reduced or lost.

Despite the persistence of symmetrodonts from the late Triassic to almost the end of the Cretaceous, their remains are exceedingly rare. The best-known genus is *Spalacotherium* from the Middle Jurassic, which is represented by associated jaws and teeth. It has a large canine, three premolars, and seven molar teeth. The jaw lacks a distinct angle.

We assume that all the early therian mammals broadly resembled the morganucodontids in their skeletal anatomy and would have appeared superficially similar to the smallest and most primitive living placentals and marsupials.

Shuotherium from the Middle or Upper Jurassic of China may be derived from the base of the symmetrodont radiation. It is represented by a single lower jaw that shows most of the premolar and molar dentition (Figure 19-4). The principal cusps are arranged in the pattern of an acute triangle, as in spalacotheriids, but there is a small talonid on the *anterior* rather than the posterior end of the molars. This is opposite to the pattern that is elaborated in marsupials and placentals and must have evolved independently. The jaw has a groove for the postdentary bones and probably retains a reptilian jaw articulation. There is no evidence of an angular process (Chow and Rich, 1982). Kermack, Lee, Lees, and Mussett (1987) suggest that *Shuotherium* may be a docodont.

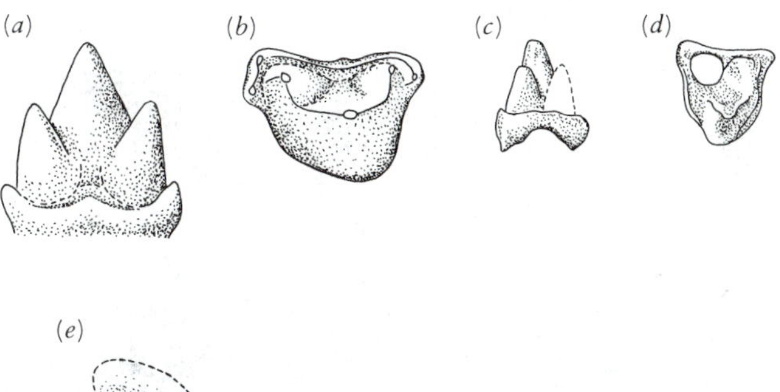

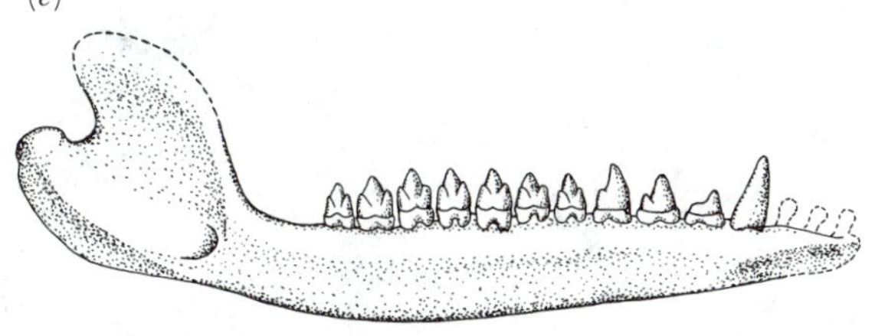

Figure 19-3. TEETH AND JAW OF SYMMETRODONTS. (*a*) Lingual and (*b*) occlusal views of the lower molar of the obtuse angle symmetrodont *Tinodon*. Note similarity with *Kuehneotherium* (Figure 19-1). (*c*) Lingual and (*d*) occlusal views of the lower molar of the acute angle symmetrodont of *Spalacotheroides*. (*e*) Lingual view of the lower jaw of *Spalacotherium*, ×2.5. *From Cassiliano and Clemens, 1979. By permission of the University of California Press.*

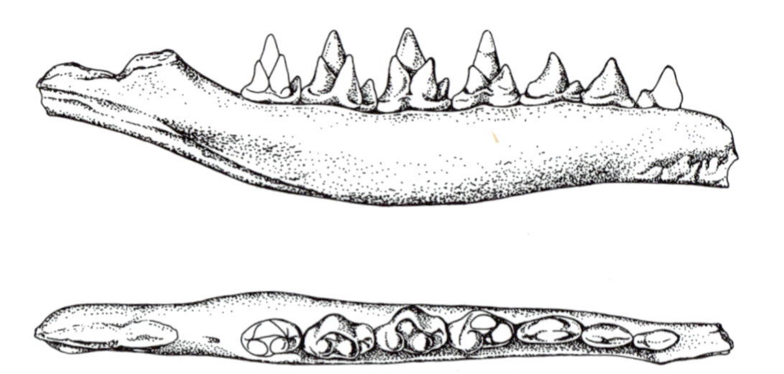

Figure 19-4. MEDIAL AND OCCLUSAL VIEWS OF THE LOWER JAW OF *SHUOTHERIUM*. From the Middle Jurassic of China, ×7. It has a structure similar to the talonid of pantotheres on the anterior end of the molar teeth. *From Chow and Rich, 1982.*

EUPANTOTHERIA

The main line of evolution toward marsupials and placentals is represented by the Eupantotheria. Except for one undescribed skeleton (Figure 19-5), this group is known primarily from jaws and teeth. The talonid is somewhat better developed than in *Kuehneotherium*, but it still has only a single cusp. The upper and lower teeth are both in the shape of acute triangles, but the upper is significantly wider than the lower. A protocone has not yet evolved. The lower jaw has developed a strong angular process, in contrast with the symmetrodonts.

Figure 19-5. SKELETON OF AN UNDESCRIBED PANTOTHERE. From the Upper Jurassic of Portugal, ×2. *Photograph courtesy Dr. Krebs.*

Kraus (1979) recognized four families in a recent review. At least three may have been present in the Middle Jurassic. Prothero (1981) discussed possible interrelationships within the group.

We know the Amphitheriidae from only a single Middle Jurassic genus. *Amphitherium* has an elongated talonid, but it is not basined. There are at least four incisors, a canine, and at least eleven postcanine teeth, of which five are considered to be premolars (Figure 19-6).

The Paurodontidae are known only from the Upper Jurassic; they typically have no more than eight postcanine teeth, but some genera have at least ten.

The Dryolestidae may be represented by isolated teeth as early as the Middle Jurassic; numerous genera are known in the Upper Jurassic, and they are the only family of eupantotheres to continue into the early Cretaceous (Krebs, 1971). In contrast with the line leading to marsupials and placentals, the talonid is reduced and the molars are anteroposteriorly compressed (Figure 19-6). Some Upper Jurassic genera retain the coronoid, splenial, and a per-

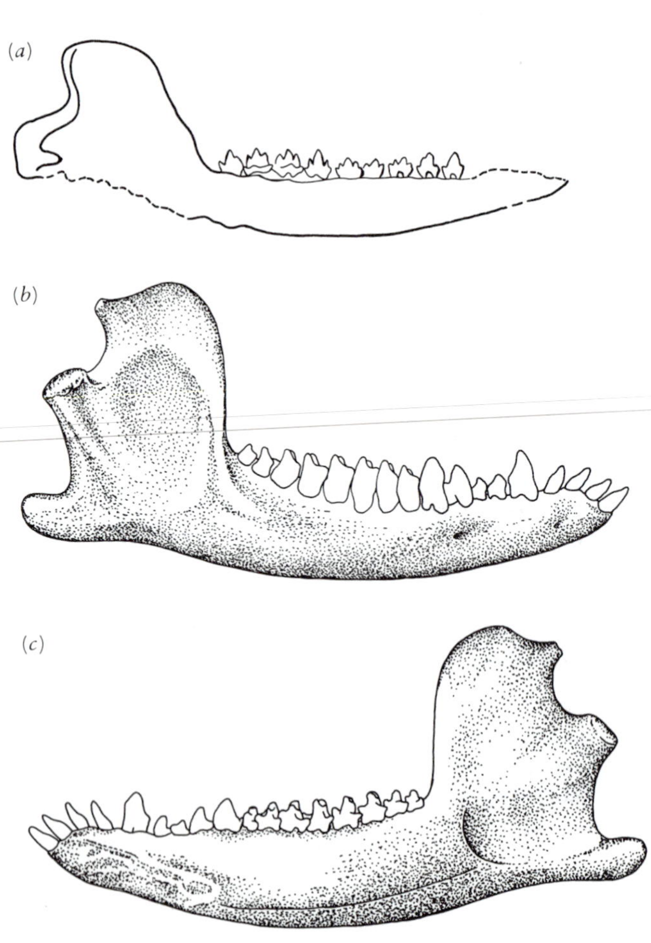

Figure 19-6. JAWS OF PANTOTHERES. (*a*) *Amphitherium* from the middle Jurassic of England, ×2. *From Jenkins and Crompton, 1979. By permission of the University of California Press.* (*b*) Lateral and (*c*) medial views of the lower jaw of a dryolestid pantothere from the Upper Jurassic of Portugal. *From Krebs, 1971.*

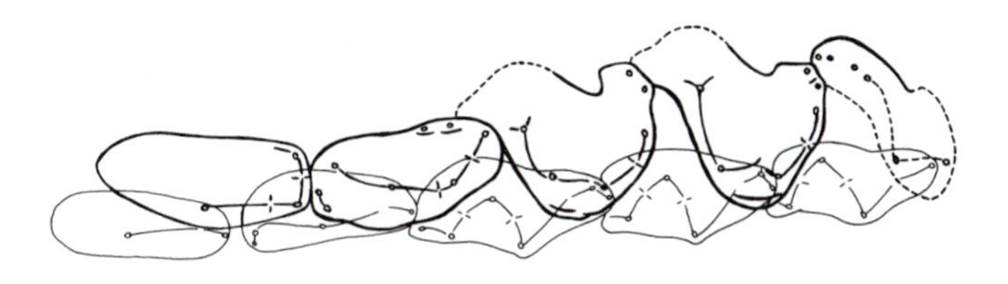

Figure 19-7. DISTAL PREMOLARS AND MOLARS OF *PERAMUS*. Occlusal relationship is shown. Upper teeth are drawn with heavy lines. Front is to the left, lingual is to the bottom. *From Clemens and Mills, 1971. Redrawn with permission by the Trustees of the British Museum (Natural History).*

sistent Meckel's cartilage, but these features are lost by the Lower Cretaceous. None of the genera show any trace of the reptilian jaw joint. There are typically twelve upper and lower postcanine teeth, but some genera have only eight.

The family Peramuridae, which is known with certainty only from a single Upper Jurassic genus *Peramus*, appears most closely related to the ancestry of the later therian mammals (Figure 19-7). The talonid has a hypoconid and an entoconid in addition to the hypoconulid of *Amphitherium* and it is incipiently basined. As Clemens and Mills (1971) showed, the upper molars have not yet developed a protocone. There are eight postcanine teeth in both the upper and lower dentition. This is less than the number in later therians, but other features of *Peramus* seem close to the pattern that is expected in the ancestors of marsupials and placentals (McKenna, 1975).

Crompton (1971) admirably outlined the changes leading to the establishment of the tribosphenic molar. These changes involve progressive addition of shearing surfaces to exploit the transverse component of jaw motion (Figure 19-8). In *Kuehneotherium,* there are three major shearing surfaces; in the fully developed tribosphenic molar there are six. Shearing surfaces are added primarily by medial extension of the upper molars, which become approximately twice the width of the lowers. Because of the medial movement of the lower jaw during mastication, the shearing surfaces of the lower teeth are moved across those of the upper. The protocone develops as a new structure near the apex of the medial extension to form a crushing and grinding surface against the basined talonid.

Fully developed tribosphenic molars, which are recognized by the establishment of a definite talonid basin and the elaboration of a protocone, are first known from the early Cretaceous (Neocomian through Albian) of Europe, North America, and Asia (Dashzeveg and Kielan-Jaworowska, 1984). The oldest is *Aegialodon* (Figure 19-8*d*), which we know from a single lower molar with a clearly basined, three-cusped talonid. *Pappotherium* from the Albian of Texas is among the earliest forms in which the upper molars are known to have a fully developed protocone (Crompton and Kielan-Jaworowska, 1978).

THERIANS OF METATHERIAN-EUTHERIAN GRADE

We can classify all Cenozoic mammals with tribosphenic molars as either marsupials or placentals on the basis of different patterns of tooth replacement, the number of molars and premolars, and the detailed morphology of the individual molars. The earliest known tribosphenic teeth, and some from the later Cretaceous, cannot be definitely assigned to either of these groups. The earliest may belong to an ancestral stock that existed before the divergence of the modern infraorders. Others represent distinct therian lineages that coexisted with the early marsupials and placentals but cannot be classified with either group (Clemens and Lillegraven, 1986). These primitive therians have been lumped together in an informal assemblage "Theria of metatherian-eutherian grade."

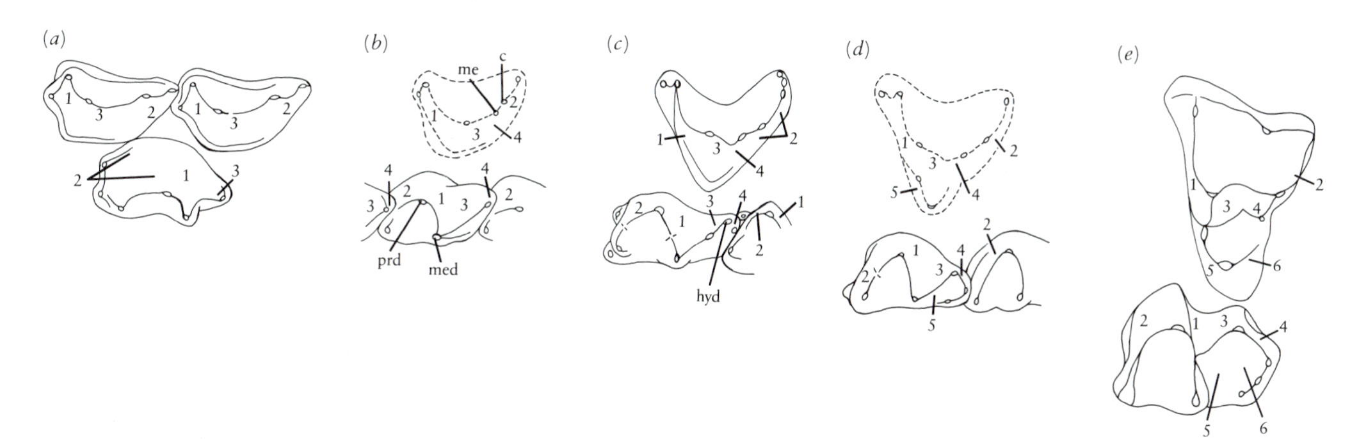

Figure 19-8. OCCLUSAL VIEW OF THERIAN MOLAR TEETH. These genera show important stages in the evolution of tribosphenic molars. For each drawing, upper teeth are above and lower are below. Anterior is to the left, buccal is to the top. Numbers identify shearing surfaces, which increase from 3 to 6. (*a*) *Kuehneotherium.* (*b*) *Amphitherium* (upper molar hypothetical). (*c*) *Peramus.* (*d*) *Aigialodon* (upper molar hypothetical). (*e*) *Didelphodus,* an early Cenozoic eutherian. Abbreviations as follows: hyd, hypoconulid; me, metacone; med, metaconid; prd, protoconid. *From Bown and Kraus, 1979. By permission of the University of California Press.*

Deltatherium is one of the most completely known members of this assemblage. It is represented by nearly complete skulls from the Campanian of Mongolia (Figure 19-9). It was originally thought to belong to the basal stock of placental mammals, the Proteutheria. Kielan-Jaworowska (1975) recognized that the tooth formula

$$\frac{?4 \quad 1 \quad 3 \quad 3\text{–}4}{1\text{–}2 \quad 1 \quad 3 \quad 4\text{–}3}$$

was in fact closer to the pattern of metatherian mammals, except for the variable loss of the upper fourth molar (M^4). Compared with other early therians, the skull is relatively large (4 centimeters long), but the snout is short (compare with Figure 20-3). Palatine foramina (which are common to marsupials) are lacking.

Kielantherium from the Aptian or Albian of Mongolia is closer in time to the marsupial-placental dichotomy and may be closer to the morphological pattern of the common ancestors of these groups (Figure 19-9*b*). Dashzeveg and Kielan-Jaworowska (1984) describe a nearly complete jaw with five premolars and four molars. The presence of five premolars corresponds with the highest number in primitive placentals. Early marsupials have only three.

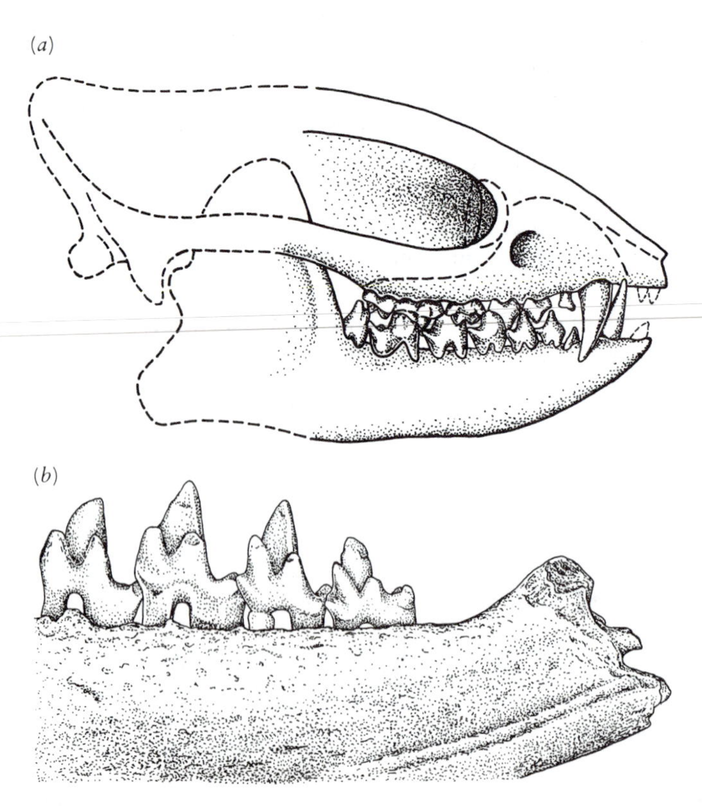

Figure 19-9. THERIANS OF THE METATHERIAN-EUTHERIAN GRADE. (*a*) Skull of *Deltatheridium* from the Upper Cretaceous of Mongolia, ×2. *From Kielan-Jaworowska, Eaton, and Bown, 1979. By permission of the University of California Press.* (*b*) Lingual view of lower jaw of *Kielantherium* from the Lower Cretaceous of Mongolia, ×12. *From Dashzeveg and Kielan-Jaworowska, 1984.*

Different authors have classified isolated teeth from the Albian of Texas as either therians of metatherian-eutherian grade (Patterson, 1956; Butler, 1978; Kielan-Jaworowska, Eaton, and Bown, 1979) or as belonging specifically to one or another of the advanced infraclasses. Slaughter (1981) and Fox (1980) assigned *Holoclemensia* to the Metatheria and *Pappotherium* to the Eutheria. These groups probably had differentiated by this time, but it remains difficult to assign isolated teeth to any particular group of subsequent therians.

A gap of approximately 20 million years separates the rare, early therians of metatherian-eutherian grade from the comparatively rich fossil record of the Upper Cretaceous. By that time (Santonian and Campanian), unquestioned marsupials and placentals can be recognized, and they occur in considerable diversity.

MARSUPIALS

We assume that marsupials and placentals diverged essentially simultaneously from a common ancestry that is represented by the early therians of metatherian-eutherian grade. In the late Cretaceous, the geographical distribution of the two groups shows significantly different patterns. Marsupials and placentals are of nearly equal abundance in North America and are known in lesser numbers in South America. Placentals are relatively common in the Upper Cretaceous of central Asia, but no Cretaceous marsupials have been described from Asia. Neither group is known from the Cretaceous of Africa or Australia.

Marsupials can be distinguished from placentals by unique derived characteristics of the dentition. Tooth replacement is limited to the third premolar in the upper and lower jaw (P^3_3). Most marsupials have only three premolars and four molars, while placentals have four premolars (five in primitive species) and three molars. In marsupials with an ossified tympanic bulla, the alisphenoid is typically an important component. The jugal extends posteriorly to the glenoid fossa. The presence of a reflected angular process of the jaw distinguishes marsupials from all living placentals, but this structure is present in some early eutherian mammals.

The very immature state of the newborn in marsupials is almost certainly a primitive feature for therians, as is the presence of marsupial bones.

We can distinguish the individual teeth of primitive marsupials from those of placentals by the following features (Figure 19-10). The upper molar have a broad buccal cingulum with several large stylar cusps. The crowns of the upper molars are not as wide, relative to their length. The hypoconulid and entoconid are typically close together (twinned) and well separated from the hypoconid.

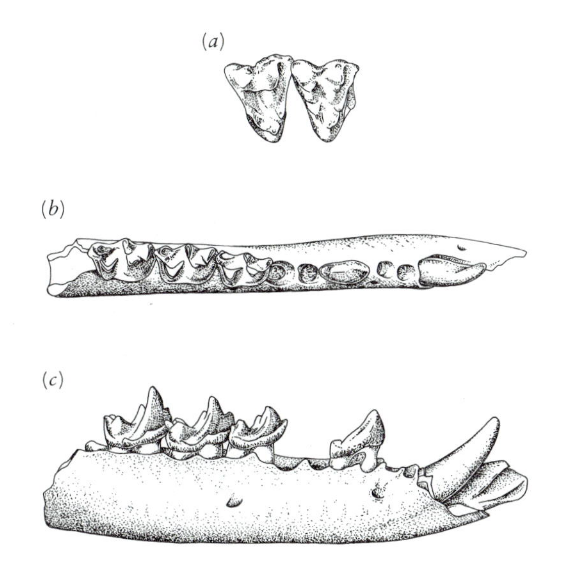

Figure 19-10. THE PRIMITIVE MARSUPIAL *ALPHADON* FROM THE UPPER CRETACEOUS OF NORTH AMERICA. (*a*) Upper molars in occlusal view and lower jaw in (*b*) occlusal and (*c*) lateral views, ×6. *From Clemens, 1979. By permission of the University of California Press.*

GEOGRAPHICAL AND TEMPORAL DISTRIBUTION

The earliest record of marsupials is in North America. The marsupial affinities of the isolated teeth from the Lower Cretaceous remain in question, but there is diverse material from the Upper Cretaceous (Campanian and Maastrichtian) of southern Canada (Fox, 1979) and the western United States (Clemens, 1966). The earliest comes from the early Campanian, nearly 20 million years before the end of the Mesozoic (Fox, 1986). Clemens recognized three families of Upper Cretaceous marsupials, all within the superfamily Didelphoidea, which includes the living opossum. Remains from the Upper Cretaceous are restricted to teeth, partial jaws, and cranial fragments. Members of the Didelphidae are known in North America as late as the Miocene, approximately 25 million years ago, after which they became extinct. The modern genus *Didelphis* reentered North America from South America approximately 3 million years ago, after the emergence of the isthmus of Panama.

Didelphids migrated from North America into Europe in the early Eocene. The genus *Peratherium* is known there into the Miocene. Closely related forms are known from the Eocene and Oligocene of North Africa (Crochet, 1984; Bown and Simons, 1984), but they apparently became extinct there without becoming more widely distributed.

The only marsupial remain so far reported from Asia is a single molar tooth from the Lower Oligocene, similar to that of European and North American didelphids (Benton, 1985).

Marsupials very similar to North American genera are reported from Peru and Bolivia in the late Cretaceous, together with primitive placentals (Marshall, de Muizon, and Sigé, 1983). Marsupials underwent a major radiation in the late Cretaceous or early Cenozoic in South America, which culminated in the differentiation of 10 families.

One member of a typical South American family, the Polydolopidae, is known from the late Eocene of Seymour Island on the Antarctic Peninsula. This discovery provides the first solid evidence for a possible route of dispersal between South America and Australia, across Antarctica (Woodburne and Zinsmeister, 1984).

The fossil record of marsupials in the Australian region begins in the late Oligocene (Tedford, Banks, Kemp, McDougal, and Sutherland, 1975). By this time, most of the major groups had already differentiated. There is no direct evidence to document when marsupials first entered Australia.

The place of origin and direction of dispersal of marsupials in the southern continents is subject to continuing debate (Figure 19-11). Keast (1977), Marshall (1980b, 1982a), Szalay (1982), Woodburne (1984), and Woodburne and Zinsmeister (1984), have discussed this problem. Marshall emphasizes that there is still no definite evidence. Woodburne argues that the great diversity of marsupials in South America in the early Paleocene suggests that their radiation may have been well underway in the late Cretaceous. Cretaceous fossils are as yet very

Figure 19-11. MODEL OF MARSUPIAL RADIATION. The group first appeared in the fossil record in the New World, and may have originated in either North or South America as shown on this map of the distribution of continents in the late Cretaceous. They spread into Europe and North Africa in the Eocene. They are not known in Australia until the end of the Oligocene. *From Marshall, 1980b.*

poorly known, but it is possible that the main radiation of early marsupials may have occurred here rather than in North America. The radiation of marsupials may have preceded that of the placentals in the New World, although placentals are reported in Asia as early as the end of the Lower Cretaceous. Marsupials may have been the only mammals in South America at the time that migration to Australia via Antarctica was possible in the late Cretaceous. This would explain why placentals did not contribute to the early Tertiary fauna of Australia. Alternately, the diverse omnivorous, opportunistically feeding marsupials may have been at an advantage in making the journey to Australia, in comparison with the relatively few, primarily herbivorous placentals that were in South America at the end of the Cretaceous and the beginning of the Cenozoic.

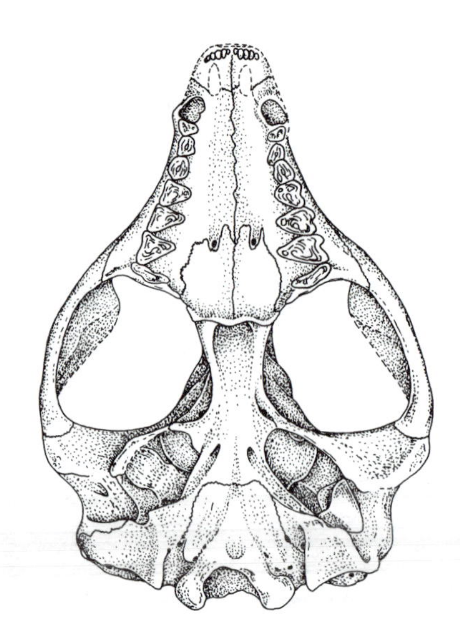

Figure 19-12. *SPARASSOCYNUS.* Palatal view of the didelphid from the Pliocene of Argentina, ×1. *From Reig and Simpson, 1972. By permission of the Zoological Society of London.*

SOUTH AMERICAN MARSUPIALS

The earliest and most primitive South American marsupials are closely related to those that are known from the Upper Cretaceous of North America. Clemens (1979) classified the genera that occur on both continents within the Didelphidae, but Szalay (1982) suggested that they belong to the closely related family, Pediomyidae.

North and South America became separated in the late Cretaceous, and faunal interchange between these continents was greatly restricted until the very end of the Cenozoic. Marsupial evolution in South America proceeded for most of the Cenozoic without influence from the rest of the world. This geographical separation, as well as anatomical evidence, clearly demonstrate that South American marsupials constitute a monophyletic assemblage.

Most of the major groups of South American marsupials were recently restudied by Marshall (1977, 1978, 1979, 1980a, b, 1981, 1982b, c), who has also published a summary review (1982a) with extensive citation of previous literature. Marsupials form a relatively minor element in the modern fauna, with only 17 genera, but 76 additional genera have been described from the Tertiary.

Didelphidae

The didelphids remain important elements in the South American fauna throughout the Cenozoic. Thirteen genera are present in the living fauna and approximately 25 genera have been described from fossils. Didelphids are recognized primarily by primitive features. The dental formula is

$$\frac{5\ 1\ 3\ 4}{4\ 1\ 3\ 4}$$

The configuration of the teeth typically resembles that of modern genera that are opportunistic feeders on invertebrates, small vertebrates, and plant material (Figure 19-12). Most are at least facultatively arboreal, with a prehensile tail and opposable big toe. Members of the family range from the size of a mouse to that of a large domestic cat. Some had the proportions of a weasel and the living genus *Chironectes* is specialized for aquatic locomotion, with webbed feet and a watertight marsupium.

At least 12 genera are recognized from the Paleocene, which suggests that marsupials had an even larger late Cretaceous radiation in South America than that known in North America. There is considerable diversity among the early didelphids, but the various phylogenetic lineages within this family have not yet been worked out.

Beginning in the late Cretaceous and continuing in the early Cenozoic, a broad spectrum of more specialized marsupials evolved from the base of the didelphid assemblage. This group includes a number of lineages that resemble insectivores and small rodents and others that paralleled the placental carnivores of other continents.

Microbiotheriidae

The Microbiotheriidae is known from only two genera, *Dromiciops,* a small, mouselike form in the living fauna, and *Microbiotherium* from the Upper Oligocene and Lower Miocene. The dental formula is primitive

$$\frac{5\ 1\ 3\ 4}{4\ 1\ 3\ 4}$$

but the stylar shelves on the upper molars are small and the posterior molars are reduced. Microbiotheres are distinguished from all other marsupials by the enormous inflation of the auditory bulla. The posterior two-thirds is formed by a new ossification, the entotympanic, and the anterior one-third by the alisphenoid (Figure 19-13).

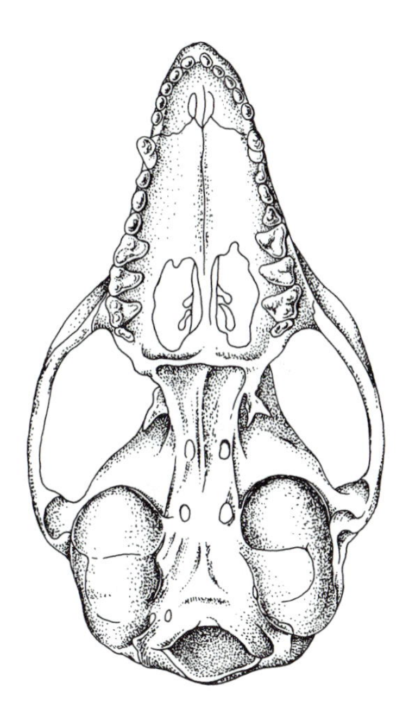

Figure 19-13. PALATAL VIEW OF *DROMICIOPS*. The microbiothere is from the recent fauna of South America. The auditory bulla is greatly expanded. The foot structure resembles that of Australian marsupials. *From Marshall, 1982b.*

These genera are customarily grouped among the Didelphoidea. Szalay (1982) recognized specializations of the tarsus in the modern genus that are very similar to those of the Australian marsupials and suggests that microbiotheres should be separated taxonomically from all other South American groups as the order Dromiciopsia.

During the Lower Eocene, two other didelphoid families are recognized on the basis of single genera, Bonapartheriidae and Prepidolopidae. Other groups are sufficiently distinct that they are placed in separate superfamilies.

Borhyaenoidea

The Borhyaenoidea broadly resemble the medium-to-large-sized placental carnivores of the northern continents. They encompass approximately 35 genera and range in age from the early Paleocene to the end of the Pliocene. They almost certainly evolved from early didelphids, but a specific ancestral stock has not been recognized. The Paleocene genus *Patene* is an almost ideal intermediate between the two groups. The dentition is somewhat specialized in the reduction of the incisors to give a dental formula of

$$\frac{4\ \ 1\ \ 3\ \ 4}{3\ \ 1\ \ 3\ \ 4}$$

The molar teeth increase in size posteriorly. The stylar cusps are reduced, which gives the upper molars a more bladelike appearance. In genera in which the postcranial skeleton is adequately known, the limbs are short and show no cursorial specializations (Figure 19-14). From the Oligocene into the Pliocene, the role of cursorial carnivores in South America seems to have been taken by the phoruschacoids, a group of large terrestrial birds (see Chapter 16).

Among the Oligocene borhyaenids, *Proborhyaena* was the size of a large bear, with molar teeth that were specialized in a manner analogous with the carnassial teeth of placental carnivores. Borhyaenids are not known later than the Lower Pliocene.

Special affinities have been repeatedly postulated between the doglike carnivores of South America and Australia, but recent work on both the osteology and the biochemistry refutes this notion (Kirsch, 1984) (Figure 19-15).

From among the borhyaenids evolved the thylacosmilids, which appear in the Upper Miocene and continue to late Pliocene. They have a highly distinctive dentition with no incisors but enormous, ever-growing upper canines that are comparable to those of the placental "saber-toothed tiger." There is a distinct symphyseal flange of the lower jaw underlying these teeth (Figure 19-16). Thylacosmilids do not survive into the Pleistocene, by which time North American placental carnivores, including the true saber tooths, had entered South America.

Caenolestoids

The caenolestoids are an assemblage of small insectivorous and omnivorous forms. They were most diverse in the Lower Miocene, when they were the most abundant of all small marsupials. A total of thirteen genera are recognized, of which three are present in the modern fauna.

Figure 19-14. *PROTHYLACYNUS*. Skeleton of the South American borhyaenoid $\times \frac{1}{8}$. *From Sinclair, 1906.*

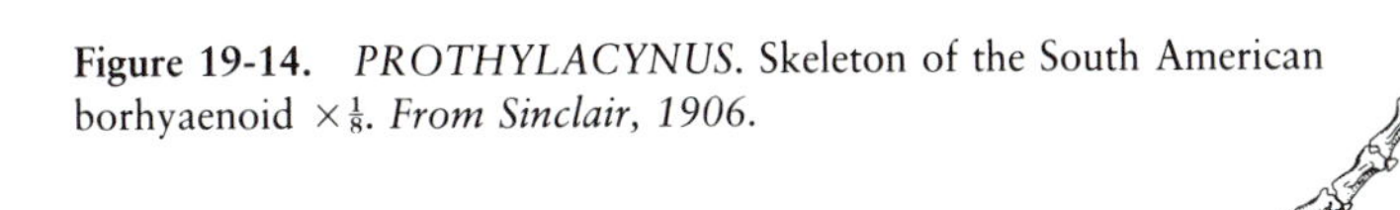

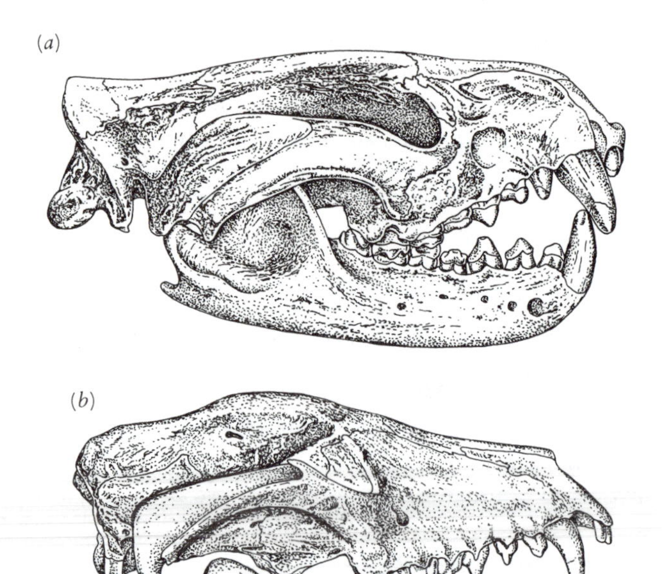

Figure 19-15. CARNIVOROUS MARSUPIALS. (*a*) The South American genus *Borhyaena* from the Miocene, $\times \frac{1}{4}$. (*b*) The recently extinct Australian genus *Thylacynus*, $\times \frac{1}{4}$. *From Sinclair, 1906.*

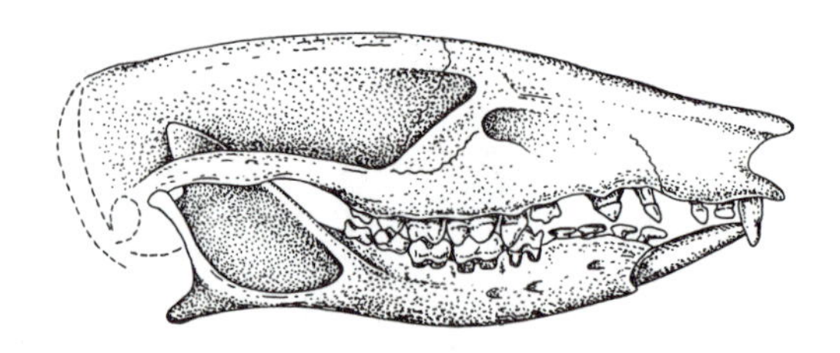

Figure 19-17. *PALAEOTHENTES*, a shrewlike caenolestoid marsupial from the Oligocene of South America. *From Marshall, 1980a. By permission of the Field Museum of Natural History, Chicago.*

The earliest reported fossil is Lower Eocene. The dental formula is

$$\frac{3\text{–}4 \quad 1 \quad 3 \quad 4}{2\text{–}4 \quad 1 \quad 2\text{–}3 \quad 4}$$

The most distinctive feature is the structure of the first molar, which is customarily enlarged and bladelike, closely resembling the bladelike premolar of the multituberculates (Figure 19-17). The first and/or second lower incisors are long and procumbent. A similar pattern is achieved in some Australian marsupials, but in those forms the third lower incisor is specialized. Caenolestoids have long been recognized as being very distinct from other South American marsupials, but they share with them a highly distinctive pattern of the spermatozoa, which become paired within the epididymis. Paired sperm are not known in any placental groups or among the Australian marsupials.

Polydolopoidea

A totally extinct group, the polydolopoids, which are known from the Paleocene into the Lower Oligocene, somewhat resemble the caenolestoids but achieved similar dental specialization earlier and in a different way. They too have a multituberculatelike enlarged cheek tooth, but it is the last premolar, not the first molar. The anterior teeth in the lower jaw resemble the procumbent incisors of caenolestoids but include the canines, as well as the small incisors (Figure 19-18). A single polydolophid genus, *Antarctodolops*, is the only marsupial known to have reached Antarctica, although the portion of the Antarctic peninsula where it was found was probably a part of South America in the Eocene (Woodburne and Zinsmeister, 1984).

Groeberioidea and Argyrolagoidea

Two other groups that are known only from fossils evolved a superficially rodentlike dentition. *Groeberia* is known from two incomplete specimens from the Upper Eocene. The dental formula is reduced to

$$\frac{2 \quad 0 \quad 0 \quad 4}{1 \quad 0 \quad 0 \quad 4}$$

The enamel on the ever-growing incisors is restricted to the labial side, as in modern rodents, to give a self-sharpening surface. The facial region is very short but deep. Placental rodents entered South America in the Oligocene and may have restricted the opportunities for their marsupial analogues.

A later group of highly specialized rodent analogues, the Argyrolagoidea (Figure 19-19), are reported from the Oligocene into the Pliocene. They show a remarkable con-

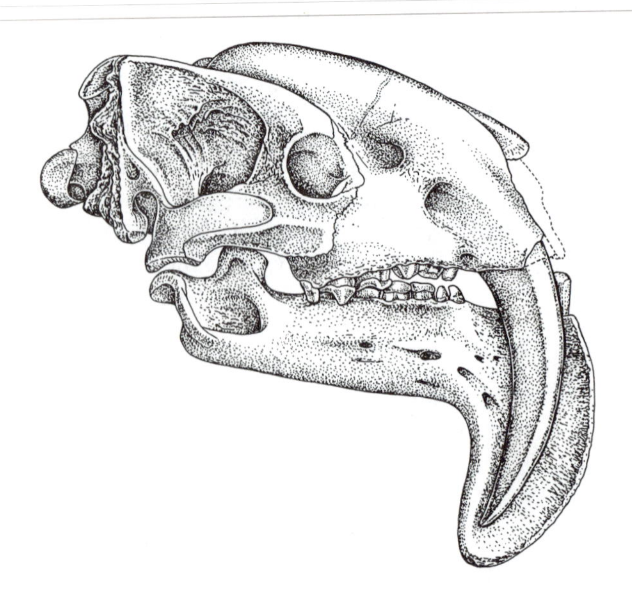

Figure 19-16. *THYLACOSMILUS*, a marsupial "saber tooth tiger" from South America. *From Riggs, 1934.*

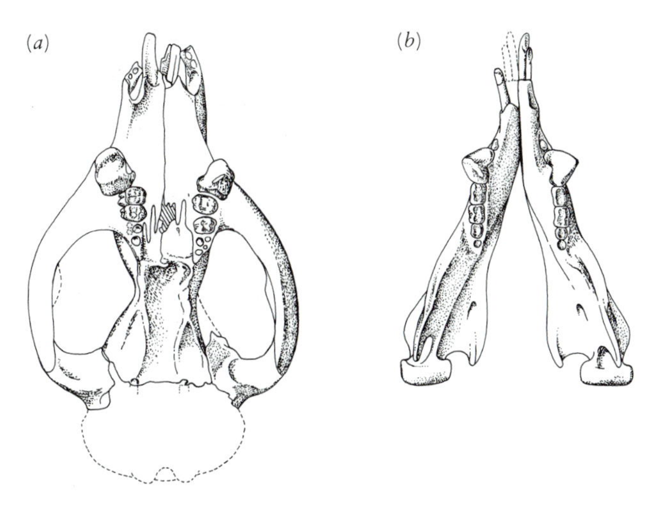

Figure 19-18. (*a*) Palate and (*b*) lower jaws of the polydolopid marsupial *Epidolops* from the Upper Paleocene of Brazil, ×1. *From Paula Couto, 1952.*

vergence with kangaroo rats and jerboas. The dental formula is reduced to

$$\frac{2\ 0\ 1\ 4}{2\ 0\ 1\ 4}$$

The remaining premolars and molars are ever growing. As in modern jumping mice, the auditory bulla is greatly inflated (although composed, in typical marsupial fashion, of the alisphenoid), the front legs are short, the hind legs are elongate, and the tibia and fibula are fused distally. The pes has only two digits. The tail is heavy to counterbalance the weight of the front of the body. As in other marsupials, there are large palatal vacuities and a reflected angular process.

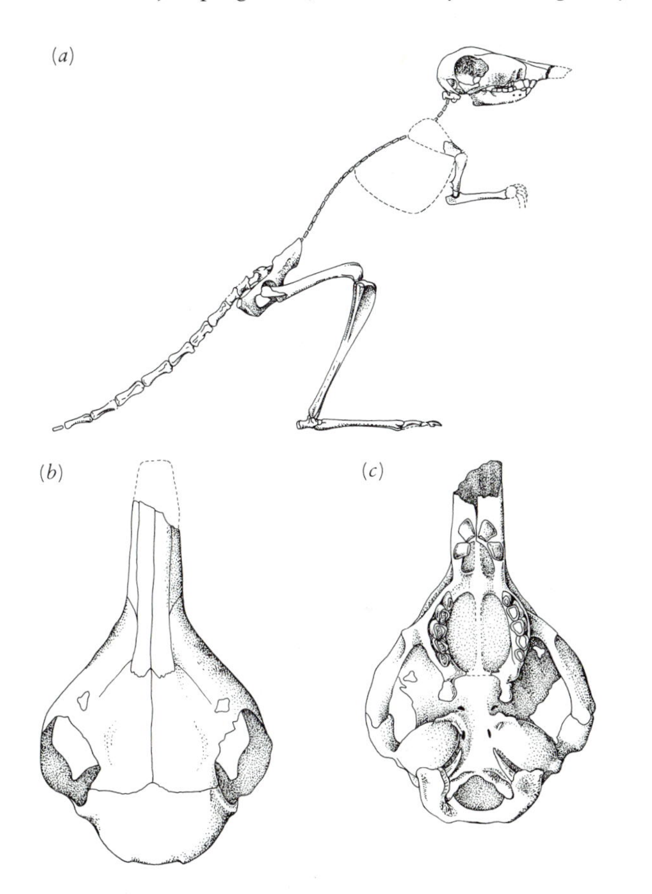

Figure 19-19. (*a*) Skeleton of *Argyrolagus*, a riochetal marsupial from the Pliocene of Argentina, ×⅓. Skull of *Argyrolagus* in (*b*) dorsal and (*c*) palatal views. *From Simpson, 1970.*

AUSTRALIA

The earliest-known marsupials from the Australian region come from the late Oligocene of Tasmania (Tedford, Banks, Kemp, McDougal, and Sutherland, 1975). This may be at least 40 million years after marsupials first entered Australia. A diverse fauna is known from the Miocene, by which time most of the modern families had become differentiated. Woodburne (1984) and Rich (1982) recently reviewed the diversity of Australian marsupials. The carnivorous groups are considered in detail in the book *Carnivorous Marsupials*, which Archer (1982a) edited.

Dasyuroidea

The relationships and classification of the Australian marsupials continue to be hotly debated. At least two major groups are recognized that may have evolved separately from an essentially didelphoid level. The Dasyuroidea, as small-to-medium-sized insectivores, carnivores, and omnivores, occupy a position in Australia that is comparable to the Didelphoidea in South America. Archer (1976) emphasized that the early Miocene dasyurid *Akotarinja* would certainly have been identified as a didelphid had it been found in South America rather than Australia. All dasyuroids can be distinguished from primitive didelphoids by the reduction in the number of incisors, to give a dental formula of

$$\frac{4\ 1\ 2\text{–}3\ 4}{3\ 1\ 2\text{–}3\ 4}$$

but the configuration of the individual teeth remains basically similar (Figure 19-20).

The Dasyuridae includes 14 living genera of "mouse"- and "cat"-like forms that range in size from 5 grams to 10 kilograms. Another six genera are known only as fossils. Archer (1982b) discussed relationships within the group. The smaller forms have the feeding habits of the placental insectivores. *Sarcophilus*, the Tasmanian devil, has large, blunt molars that are used for crushing bones, which make it an effective scavenger on large animals.

Thylacinus, the Tasmanian "wolf," is placed in a separate family. The general body form as well as details of the dentition provide a strikingly close parallel with the placental canids (see Figure 19-15*b*). This genus is known from the Miocene and became extinct in 1934. It has repeatedly been suggested that the Australian thyla-

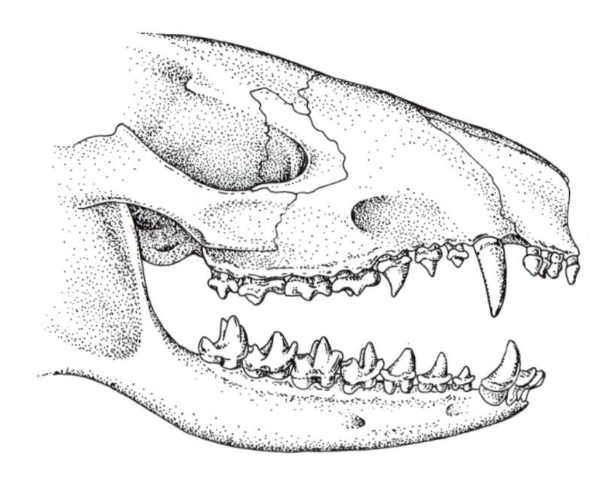

Figure 19-20. *SMINTHOPSIS*. Dentition of the modern dasyurid from Australia. *From Archer, 1976.*

cinids had a close relationship with the South American borhyaenids (for example, see Kirsch, 1977). Archer (1982c) and Sarich and Cronin (1980) discussed both the anatomical and seriological evidence and concluded that *Thylacinus* is more closely related to the dasyurids than to any other group of marsupials.

A more distant relative of the dasyurids is *Myrmecobius* (family Myrmecobiidae), the numbat or Australian anteater. The snout is elongate in this genus, as in placental anteaters, and the tongue is very long and extensible. The size of the teeth is greatly reduced, but the number of molars is increased to five or six. Despite the absence of a fossil record prior to the Pleistocene, there is no question of the affinity of the one modern species with the dasyuroids.

Perameloidea

Aside from the dasyuroids, all other adequately known Australian marsupials share a specialization of the rear foot in which digits 2 and 3 are greatly reduced and incorporated in a single sheath of tissue, a configuration termed **syndactyly** (Figure 19-21). These specialized digits are used in grooming. Genera with syndactyly include two very distinct groups, the Perameloidea, or bandicoots, and the Diprotodontoidea, a much more diverse assemblage.

The Perameloidea includes only eight recent genera, which are placed in two closely related families, the Peramelidae and the Thylacomyidae (the latter including only the rabbit-eared bandicoot and the Pliocene fossil *Ischnodon*). Bandicoots retain an even more primitive dental formula than the dasuyrids:

$$\frac{4\text{–}5 \quad 1 \quad 3 \quad 4}{3 \quad 1 \quad 3 \quad 4}$$

The molars differ in the rectangular outline of the crowns, which results from an addition of cusps to the primitively triangular trigonid. This pattern is achieved separately in the two families. In peramelids, it results primarily by the addition of hypocone, but in the thylacomyids, the metacone shifts lingually and the stylar cusps are elaborated. The enlarged crown is associated with a more omnivorous diet than the dasyurids. Perameloids vary in size from that of a mouse up to nearly 5 kilograms. Many are burrowers.

Perameloids are exceptional among metatherians in having a chorioallantoic placenta, as do eutherian mammals, but their young are born at the same immature stage as those of other marsupials. The fossil record of the Peramelidae is scanty but goes back to the Miocene.

Opinion differs as to whether the perameloids are close to the ancestry of the diprotodonts or have a separate origin at the level of dasyurids or didelphids.

Phalangeroids

The diprotodonts include three superfamilies, the Vombatoidea (including the wombats of the modern fauna and a variety of extinct groups), the Phascolarctoidea (koalas), and the Phalangeroidea, which encompasses the modern phalangerids, kangaroos, and gliding and pygmy possums. The diprotodonts are united by the derived condition of the lower incisors, which are reduced to two procumbent teeth. The group consists of primarily herbivorous forms.

Among the diprotodonts, the phalangeroids are relatively primitive in their dentition and the arboreal way of life of the phalangerids may be close to the primitive pattern for the entire group. The three living genera are the size of squirrels and retain five toes on both hands and feet, with the first toe opposable. The dental formula is

$$\frac{3 \quad 1 \quad 1\text{–}3 \quad 4}{1\text{–}3 \quad 0\text{–}1 \quad 1\text{–}2 \quad 4}$$

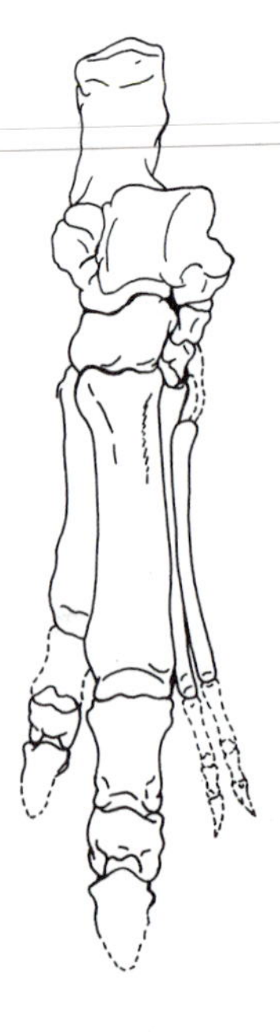

Figure 19-21. FOOT OF THE PLIO-PLEISTOCENE KANGAROO *PROTEMNODON*. This genus shows the syndactyl condition, in which digits 2 and 3 are greatly reduced and inclosed in a single sheath of tissue. *From Flannery, 1982.*

The molars have square crowns with semilophodont cusps. They are basically herbivorous but are known to eat insects and young birds. The fossil record of phalangerids goes back to the Upper Oligocene; the Middle Miocene genera are already modern in appearance.

The family Ektopodontidae is an extinct group limited to the Miocene and Pliocene that, although yet poorly known, probably arose from primitive phalangers (Woodburne and Clemens, 1984). The teeth are bilophodont, with the occlusal surface subdivided by longitudinal or radiating structures (Figure 19-22). The teeth of *Ektopodon* are so unusual that it was originally thought to be a monotreme (Stirton, Tedford, and Woodburne, 1967).

Other phalangeroid families, the Petauridae, including five genera of gliding possums, and the pygmy possums, Burramyidae, have a fossil record going back to the Miocene.

One of the most striking of extinct marsupials groups, the Thylacoleonidae, may also be related to the phalangerids. *Wakaleo* from the Miocene and *Thylacoleo* of the Pliocene and Pleistocene are large forms with a highly specialized dentition (Figure 19-23). The first of the upper three incisors and the lower incisor are greatly enlarged and have the appearance of canine teeth. The upper canine is retained, but it is very small, as are upper premolars 1 and 2. The third premolars of the upper and lower jaws are very greatly expanded in length to form gigantic shearing blades that appear to have functioned like the carnassial teeth of placental carnivores. In the lower jaw, the premolar is continuous with a smaller, bladelike first molar. The second lower molar and the first upper molar are much smaller. The absence of crushing or grinding molars and the caniniform configuration of the incisors give these genera the appearance of highly specialized carnivores. The probable diet of the thylacoleonids has long been debated. It is unusual for highly specialized carnivores to evolve among basically herbivorous groups, but groups such as the rodents and phalangers, in which the dentition appears specialized for feeding on plant material, include members that make extensive use of animal prey. Wells, Horton, and Rogers (1982) presented considerable evidence that the teeth of *Thylacoleo* were used in the manner of modern placental carnivores.

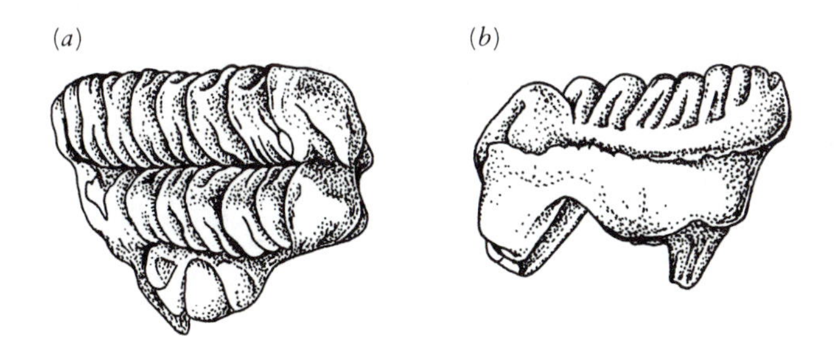

Figure 19-22. (*a*) Occlusal and (*b*) side views of molar tooth of *Ektopodon,* a peculiar phalangeroid from the Miocene and Pliocene of Australia, ×4. *From Stirton, Tedford and Woodburne, 1967.*

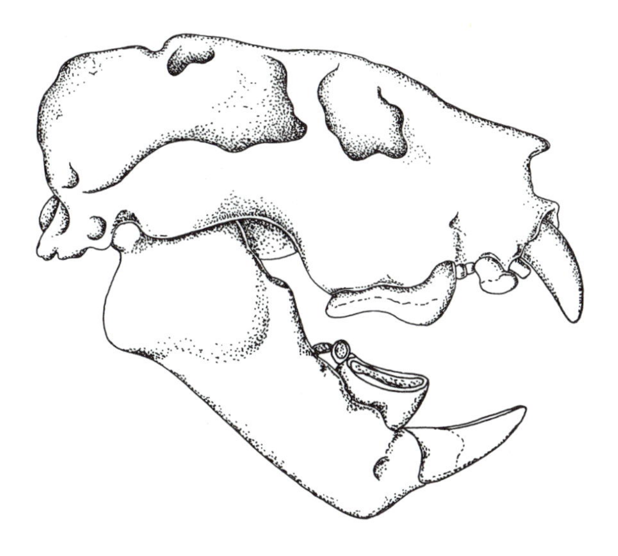

Figure 19-23. SKULL OF THE PLEISTOCENE PHALANGEROID *THYLACOLEO*, ×⅓. There is continuing dispute as to whether the greatly elongated cheek tooth is indicative of a carnivorous diet. *From Finch and Freedman, 1982.*

The postcranial skeleton follows the phalangerid pattern and is thought to be well adapted for climbing. Wells and his colleagues suggested a leopardlike habitus.

The most dramatic of the living Australian marsupials are certainly the kangaroos, with 19 living genera and approximately 60 species. All may be placed in a single family, the Macropodidae, or the smaller, more primitive rat-kangaroos may be placed in their own family, the Potoroinidae.

Except for two genera of rat-kangaroos, all genera have a rear limb that is significantly longer than the forelimb and hop bipedally when moving rapidly. The tooth formula is

$$\frac{3 \quad 0\text{–}1 \quad 2 \quad 4}{3 \quad 0 \quad 2 \quad 4}$$

The dentition of the rat-kangaroos resembles that of the phalangerids, with relatively low-crowned bunodont or sublophodont molar teeth, although most differ in having a very long, shearing, lower third premolar. The enamel is restricted to the lateral surface of the lower incisors. In true kangaroos, the molars are higher crowned and bilophodont. The third lower premolar is less specialized. Another feature that distinguishes the two groups is the presence of an opening in the lower jaw in rat-kangaroo through which passes a medial slip of the masseter that penetrates into the body of the horizonal ramus of the lower jaw (Figure 19-24).

We recognize both families (or subfamilies) in the middle Miocene. Rat-kangaroos have not changed significantly since the Miocene and most modern genera extend back into the late Cenozoic.

In contrast, the true kangaroos and wallabies show a major radiation in the Pliocene, possibly associated with

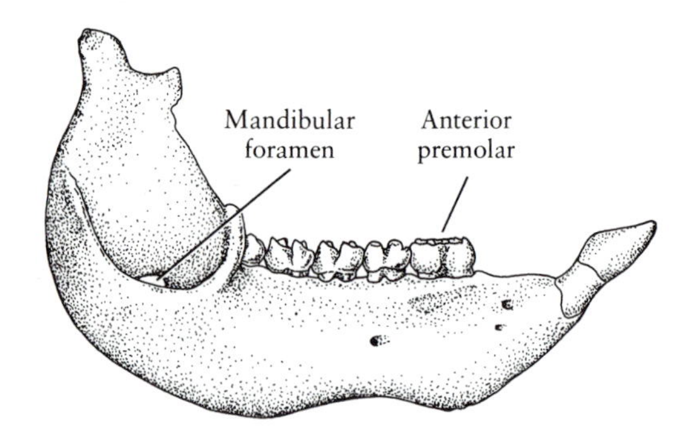

Figure 19-24. MANDIBLE OF THE RAT-KANGAROO *STHENURUS*. This genus shows the long anterior premolar and the mandibular foramen which distinguishes this group from the "true" kangaroos. *From Rich, 1982.*

climatic changes that led to the spread of grassland. The 28 fossil and living species of *Macropus* are thought to have differentiated within the last 5 million years (Rich, 1982, p. 444). Like the placental ruminates, the kangaroos have a symbiotic relationship with bacteria that assists in the digestion of cellulose.

The taxonomy of early macropodids is discussed in three papers in the *Journal of Paleontology* for July 1984. Flannery (1982) outlined the currently limited knowledge of the evolution and mechanics of kangaroo locomotion. The most primitive genera, like phalangers, move on the ground by a quadrupedal bound. This pattern may be the result of secondarily terrestrial locomotion that evolved from the primarily arboreal mode of their phalangeroid ancestors. The presence of syndactyly may have constrained the possible pattern of evolution of the foot. Dawson (1977) showed that bidepal hopping in kangaroos is energetically more economical than walking when they exceed a speed of 15 kilometers per hour.

Phascolarctoidea and Vombatoidea

The Phascolarctidae (koalas) were included among the phalangeroids but are now thought to share a closer relationship with the wombats. This group has but a single living genus, but its fossil record goes back to the Miocene. The dentition is not as extremely reduced as that of wombats but retains three upper incisors and a canine:

$$\frac{3 \quad 1 \quad 1 \quad 4}{1 \quad 0 \quad 1 \quad 4}$$

The teeth are low crowned, with a dual V-shaped array of crests. They feed almost exclusively on eucalyptus leaves and almost never leave the trees.

Living vombatoids include only two genera, but the group has a substantial fossil record that goes back to the Miocene. Wombats are burrowers and exceed 1 meter in length. They have the appearance of small bears and feed on grass and roots. From the Pliocene on, the group is characterized by the presence of rootless, ever-growing, prismatic, bilobed cheek teeth and only a single pair of upper and lower incisors. In the modern genera, the tooth formula is reduced to

$$\frac{1 \quad 0 \quad 1 \quad 4}{1 \quad 0 \quad 1 \quad 4}$$

Such a rodentlike specialization is unique to wombats among living marsupials but is closely paralleled by the South American Argyrolagidae.

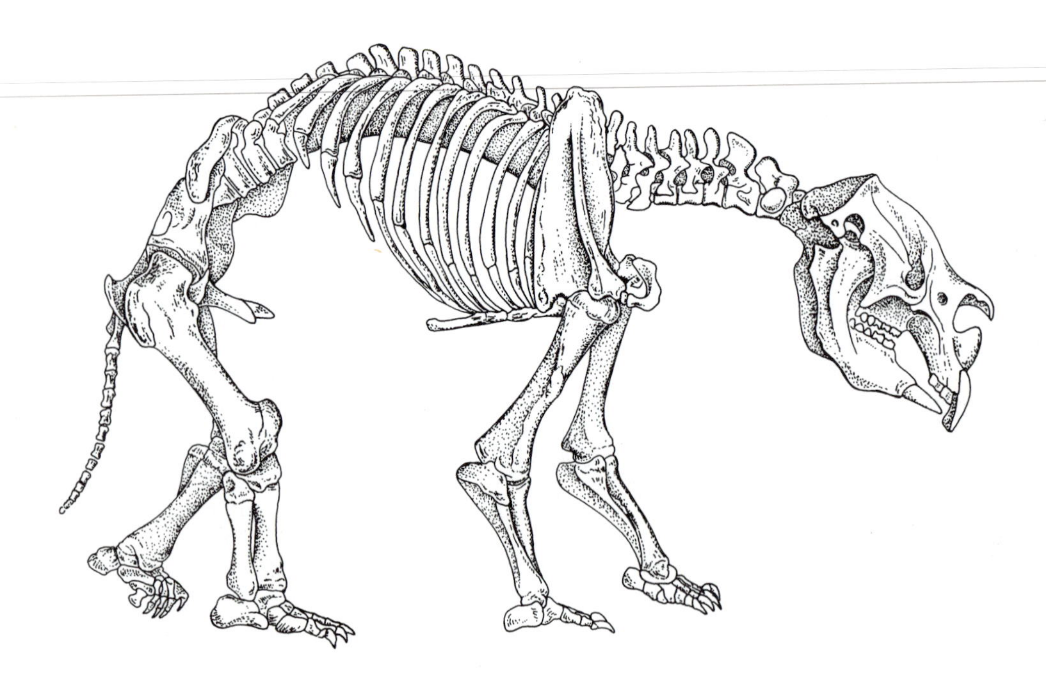

Figure 19-25. *DIPROTODON*, a giant relative of the wombat from the Pleistocene of Australia. *From Gregory, 1951.*

Wynyardia from the early Miocene is a poorly known form placed in a family of its own, which may have affinities with the wombats. It illustrates the early development of the diprotodont condition and shows some characteristics of phalangers and kangaroos.

The Diprotodontidae and Palorchestidae are related groups that are known from the Upper Oligocene to the Pleistocene. They are the largest of all known marsupials. Their sizes range from that of a sheep to a hippopotamus (Figure 19-25). They are more primitive than wombats in retaining three upper incisors. The cheek teeth bear two transverse crests. The nasal bones are short, which may indicate the presence of a proboscis (Figure 19-26). Twelve genera have been described. Their extinction may have been related to climatic changes at the end of the Pleistocene or the appearance of man in Australia.

Two final groups whose taxonomic position awaits clarification must be mentioned. *Notorcytes,* the marsupial "mole," is highly specialized in the reduction of the eyes, the loss of the pinnna of the ear, the configuration of the skull, and the fusion of the cervical vertebrae. It has no fossil record, and its specific relationship with other Australian marsupials has not been established. Dental evidence suggests alliance with dasyurids or peramelids, but the foot structure resembles that of diprotodonts.

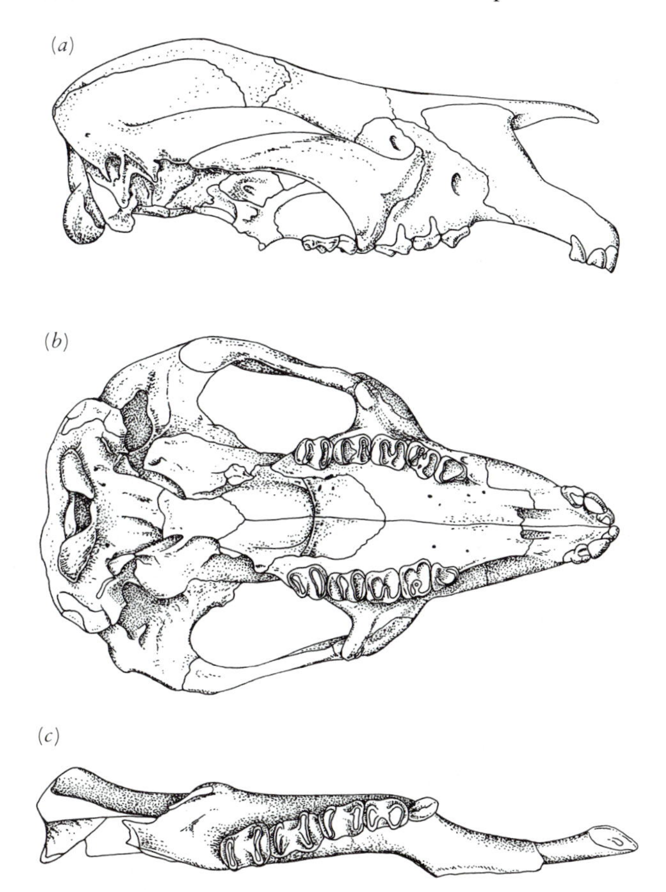

Figure 19-26. THE PALORCHESTID DIPRODODONT *NGAPAKALDIA* FROM THE LATE CENOZOIC OF AUSTRALIA. Skull in (*a*) lateral and (*b*) palatal views. (*c*) Occlusal view of lower jaw, × ⅓. *After Stirton, 1967.*

The Tarsipedidae, or honey possum, is known from a single species with no fossil record. It was once included among the phalangeroids but is now recognized as being very distinct anatomically and probably had a long, separate evolutionary history. The only well-developed teeth are the upper canines and lower incisors. The cheek teeth are reduced to pegs. It feeds on the honey and pollen in flowers and weighs less than 20 grams.

MARSUPIAL CLASSIFICATION

Marsupials have long been placed in a single order, Marsupialia, as opposed to the twenty-odd orders of living placentals. With approximately 200 genera, one could argue that this assemblage is much more diverse than many of the smaller placental orders. Recently, a spate of papers have been published supporting a subdivision into four or more orders (Kirsch, 1977; Szalay, 1982; Archer, 1976). However, Marshall (1981) retained a single order, although he implied that there was a major dichotomy between New World and Australian groups. These various schemes differ significantly in the way in which the Australian groups are subdivided. A continuing problem is the absence of any fossil evidence in Australia before the end of the Oligocene, by which time all the major groups had become well differentiated from one another.

Where the fossil evidence is lacking, we may look more closely at serological comparison and the soft anatomy of living forms (Kirsch, 1977, 1984). Whatever the evidence, the strongest support is for a division between the New World and Australian forms.

SUMMARY

Marsupials and placentals constitute a monophyletic assemblage whose common ancestors are recognized among the Lower Cretaceous aegialodontids. More primitive therians appear as early as the late Triassic. All Mesozoic therians were small and may have resembled modern shrews in their general structure and way of life. Early therian mammals are known primarily from isolated teeth and jaws. The cusps of the molar teeth of *Kuehneotherium* are arranged in the pattern of an obtuse triangle. The apex of the upper molar faces medially and that of the lower faces laterally. The edges of the teeth shear transversely past one another as the jaws are closed. The lower molar has a small posterior heel or talonid.

Early marsupials and placentals are characterized by tribosphenic molars in which the upper molar has three principal cusps—the apical protocone, an anterior paracone, and posterior metacone—arranged as an acute angle triangle or trigon. The lower molar has a similarly triangular pattern of cusps anteriorly that is dominated by the protoconid, paraconid, and metaconid. The posterior talonid is outlined by the entoconid, hypoconid, and metaconid. In *Kuehneotherium*, the upper molar lacks a protocone and the trigon is outlined by the paracone (in the apical position), the metacone, and an anterior stylocone. The cusps of the lower trigonid are the same as in later therians, but there is only a single talonid cusp, the hypoconulid.

Kuehneotherium belongs to a group known as symmetrodonts that extended, with little diversity, into the Upper Cretaceous. The Jurassic pantotheres have sharply triangular trigonids and a large talonid that approaches the condition in marsupials and placentals. The genus *Peramus* from the Upper Jurassic appears close to the ancestry of the later therians.

In the ancestors of marsupials and placentals, the upper molar teeth became enlarged transversely and progressively increase the number of shearing surfaces. The protocone developed on a newly formed apical surface by the Lower Cretaceous. The most primitive mammals that have a large-basined talonid for occlusion with the protocone are placed in a group termed therians of metatherian-eutherian grade, since they cannot be classified with either of the modern groups.

Marsupials and placentals probably diverged from a common ancestor in the early Cretaceous. Marsupials are first known in the Upper Cretaceous of North America but may have originated in Central or South America and then migrated to Australia. Primitive didelphid marsupials reached Europe from North America in the early Tertiary, and a few spread into Northern Africa and as far east as central Asia. These lineages apparently became extinct without having radiated significantly.

Marsupials radiated extensively in both South America and Australia during the Cenozoic. Didelphoids represent a central stock in South America from which numerous other families evolved, including the microbiotherids, which are characterized by greatly enlarged auditory bullae; the carnivorous borhyaenoids; and the insectivorous and rodentlike caenolestoids, polydolopoids, and argyrolagoids.

When marsupials first appeared in the fossil record of Australia in the late Oligocene, they were already very diverse and the interrelationships of the various lineages have not been satisfactorily established. The dasyuroids appear to form an ancestral stock that is comparable to the didelphoids in the New World. The carnivorous Tasmanian wolf and Tasmanian devil converge closely on the pattern of the South American borhyaenoids. All other Australian marsupials are specialized in having digits 2 and 3 greatly reduced and incorporated in a single sheath of tissue. This condition is termed syndactyly. The perameloids or bandicoots retain three to five incisors and are unique among marsupials in having a chorioallantoic placenta. The diprotodonts include three groups that are represented in the modern fauna by the wombats, koalas, phalangeroids, and kangaroos. The small, arboreal phalangeroids may most closely resemble the ancestral pattern of this assemblage. We know most of the modern families from the early Miocene. Several important groups were common in the middle and late Tertiary but are now extinct. Among the phalangeroids, the thylacoleonids were the size and proportions of a leopard, with premolar teeth in the form of gigantic shearing blades. Other diprotodonts are primarily herbivorous, but *Thylacoleo* appears to have been a carnivore. The diprotodontids and palorchestids were sheep- to hippopotamus-sized relatives of the wombats and were common from the Upper Oligocene to the late Pleistocene. Kangaroos are known from the middle Miocene, but the radiation of the large modern species appears to have occurred within the past 5 million years.

REFERENCES

Archer, M. (1976). The dasyurid dentition and its relationships to that of didelphids, thylacinids, borhyaenids (Marsupicarnivora) and peramelids (Peramelina: Marsupialia). *Austral. J. Zool. Suppl. Ser.*, **39**: 1–34.

Archer, M. (ed.). (1982a). *Carnivorous Marsupials.* Royal Zoological Society, New South Wales.

Archer, M. (1982b). Review of the dasyurid (Marsupialia) fossil record, integration of data bearing on phylogenetic interpretation and suprageneric classification. In M. Archer (ed.), *Carnivorous Marsupials,* pp. 397–443. Royal Zoological Society, New South Wales.

Archer, M. (1982c). A review of Miocene thylacinids (Thylacinidae, Marsupialia), the phylogenetic position of the Thylacinidae and the problem of apriorisms in character analysis. In M. Archer (ed.), *Carnivorous Marsupials,* pp. 445–476. Royal Zoological Society, New South Wales.

Benton, M. J. (1985). The first marsupial fossil from Asia. *Nature,* **318**: 313.

Bown, T. M., and Kraus, M. J. (1979). Origin of the tribosphenic molar and metatherian and eutherian dental formulae. In J. A. Lillegraven, Z. Kielan-Jaworowska, and W. A. Clemens (eds.), *Mesozoic Mammals. The First Two-Thirds of Mammalian History,* pp. 172–181. University of California Press, Berkeley.

Bown, T. M., and Simons, E. L. (1984). First record of marsupials (Metatheria: Polyprotodonta) from the Oligocene in Africa. *Nature,* **308**: 447–449.

Butler, P. M. (1978). Insectivores and Chiroptera. In V. J. Maglio and H. B. S. Cooke (eds.), *Evolution in African Mammals,* pp. 56–68. Harvard University Press, Cambridge.

Cassiliano, M. L., and Clemens, W. A. (1979). Symmetrodonta. In J. A. Lillegraven, Z. Kielan-Jaworowska, and W. A. Clemens (eds.), *Mesozoic Mammals. The First Two-Thirds of Mammalian History,* pp. 150–161. University of California Press, Berkeley.

Chow, M., and Rich, T. H. V. (1982). *Shuotherium dongi,* n. gen. and sp., a therian with pseudo-tribosphenic molars from the Jurassic of Sichuan, China. *Aust. Mammal.*, **5**: 127–142.

Clemens, W. A. (1966). Fossil mammals of the type Lance Formation, Wyoming. Part II. Marsupialia. *Univ. Calif. Publs. Geol. Sci.*, **62**: 1–122.

Clemens, W. A. (1979). Marsupialia. In J. A. Lillegraven, Z. Kielan-Jaworowska, and W. A. Clemens (eds.), *Mesozoic Mammals. The First Two-Thirds of Mammalian History,* pp. 192–220. University of California Press, Berkeley.

Clemens, W. A. and Lillegraven, J. A. (1986). New Late Cretaceous, North American advanced therian mammals that fit neither the marsupial nor eutherian molds. In R. M. Flanagan and J. A. Lillegraven (eds.), *Vertebrates, Phylogeny, and Philosophy. Cont. Geol. Univ. Wyo. Spec. Pap. 3.*

Clemens, W. A., and Mills, J. R. E. (1971). Review of *Peramus tenuirostris* Owen (Eupantotheria, Mammalia). *Bull. Brit. Mus. Nat. Hist. (Geol.)*, **20**: 87–113.

Crochet, J.-Y. (1984). *Garatherium mahboubii* nov. gen., nov. sp., marsupial de l'Eocène inférieur d'el Kohol (Sud-Oranais, Algérie). *Ann. Paléont. (Vert.-Invert.)*, **70**: 275–294.

Crompton, A. W. (1971). The origin of the tribosphenic molar. In D. M. Kermack and K. A. Kermack (eds.), *Early Mammals J. Linn. Soc.*, **50** (suppl. 1): 65–87.

Crompton, A. W., and Kielan-Jaworowska, Z. (1978). Molar structure and occlusion in Cretaceous therian mammals. In P. M. Butler and K. A. Joysey (eds.), *Studies in the Development, Function and Evolution of Teeth*, pp. 249–287. Academic Press, London.

Dashzeveg, D., and Kielan-Jaworowska, Z. (1984). The lower jaw of an aegialodontid mammal from the Early Cretaceous of Mongolia. *Zool. J. Linn. Soc.*, **82**: 217–227.

Dawson, T. J. (1977). Kangaroos. *Scientific American*, **237**(2): 78–89.

Finch, M. E., and Freedman, L. (1982). An odontometric study of the species of *Thylacoleo* (Thylacoleonidae, Marsupialia). In M. Archer (ed.), *Carnivorous Marsupials,* pp. 553–561. Royal Zoological Society, New South Wales.

Flannery, T. F. (1982). Hindlimb structure and evolution in the Kangaroos (Marsupialia: Macropodoidea). In P. V. Rich and E. M. Thompson (eds.), *The Fossil Vertebrate Record of Australasia,* pp. 507–524. Monash University Press, Clayton.

Fox, R. C. (1979). Mammals from the Upper Cretaceous Oldman Formation, Alberta. II. *Pediomys* Marsh (Marsupialia). *Can. J. Earth Sci.*, **16**: 103–113.

Fox, R. C. (1980). *Picopsis pattersoni,* n. gen. and sp., an unusual therian from the Upper Cretaceous of Alberta, and the classification of primitive tribosphenic mammals. *Can. J. Earth Sci.*, **17**: 1489–1498.

Fox, R. C. (1985). Upper molar structure in the Late Cretaceous symmetrodont *Symmetrodontoides* Fox, and a classification of the Symmetrodonta (Mammalia). *J. Paleont.*, **53**(1): 21–26.

Fox, R. C. (1986). Paleontology and the early evolution of marsupials. In M. Archer (ed.), *The Evolution of Possums and Opossums.* Royal Zoological Society, New South Wales.

Gregory, W. K. (1951). *Evolution Emerging.* Macmillan, New York.

Jenkins, F. A., and Crompton, A. W. (1979). Triconodonta. In J. A. Lillegraven, Z. Kielan-Jaworowska, and W. A. Clemens (eds.), *Mesozoic Mammals. The First Two-Thirds of Mammalian History,* pp. 74–90. University of California Press, Berkeley.

Keast, A. (1977). Historical biogeography of the marsupials. In B. Stonehouse and D. Gilmore (eds.), *The Biology of Marsupials,* pp. 69–95. Macmillan, London.

Kermack, D. M., Kermack, K. A., and Mussett, F. (1968). The Welsh pantothere *Kuehneotherium praecursoris. Zool. J. Linn. Soc.*, **47**: 407–423.

Kermack, K. A., Mussett, F., and Rigney, H. W. (1973). The lower jaw of *Morganucodon. Zool. J. Linn. Soc.*, **71**: 1–158.

Kielan-Jaworowska, Z. (1975). Evolution of the therian mammals in the Late Cretaceous of Asia. Part I. Deltatherididae. In Z. Kielan-Jaworowska (ed.), Results Polish-Mongol. Paleont. Exped. Part VI. *Palaeontologia Polonica*, **33**: 103–132.

Kielan-Jaworowska, Z., Eaton, J. G., and Bown, T. M. (1979). Theria of metatherian-eutherian grade. In J. A. Lillegraven, Z. Kielan-Jaworowska, and W. A. Clemens (eds.), *Mesozoic Mammals. The First Two-Thirds of Mammalian History,* pp. 182–191. University of California Press, Berkeley.

Kirsch, J. A. W. (1977). The comparative serology of the Marsupialia, and a classification of marsupials. *Austral. J. Zool., Suppl.*, **52**: 1–152.

Kirsch, J. A. W. (1984). Living mammals and the fossil record. In P. D. Gingerich and C. E. Badgley (eds.), *Mammals. Notes for a Short Course. Univ. Tennessee, Studies Geol.*, **8**: 17–31.

Kraus, M. J. (1979). Eupanthotheria. In J. A. Lillegraven, Z. Kielan-Jaworowska, and W. A. Clemens (eds.), *Mesozoic Mammals. The First Two-Thirds of Mammalian History,* pp. 162–171. University of California Press, Berkeley.

Krebs, B. (1971). Evolution of the mandible and lower dentition in dryolestids (Pantotheria, Mammalia). In D. M. Kermack and K. A. Kermack (eds.), *Early Mammals. Zool. J. Linn. Soc.*, **50** (Suppl. 1): 89–102.

Marshall, L. G. (1977). Cladistic analysis of borhyaenoid, dasyuroid, and thylacinid (Marsupialia: Mammalia) affinity. *Syst. Zool.*, **26**: 410–425.

Marshall, L. G. (1978). Evolution of the Borhyaenidae, extinct South American predaceous marsupials. *Univ. Calif. Publ. Geol. Sci.*, **117**: 1–89.

Marshall, L. G. (1979). Evolution of metatherian and eutherian (mammalian) characters: A review based on cladistic methodology. *Zool. J. Linn. Soc.*, **66**: 369–410.

Marshall, L. G. (1980a). Systematics of the South American marsupial family Caenolestidae. *Fieldiana (Geol.), New Ser.*, **5**: 1–145.

Marshall, L. G. (1980b). Marsupial paleobiogeography. In L. L. Jacobs (ed.), *Aspects of Vertebrate History,* pp. 345–386. Museum of Northern Arizona Press, Flagstaff.

Marshall, L. G. (1981). The families and genera of Marsupialia. *Fieldianan (Geol.), New Ser.*, **8**: 1–65.

Marshall, L. G. (1982a). Evolution of South American Marsupialia. In M. A. Mares and H. H. Genoways (eds.), *Mammalian biology in South America. Pymatuning Laboratory of Ecology, Univ. Pittsburgh, Spec. Publ. Ser.*, **6**: 251–272.

Marshall, L. G. (1982b). Systematics of the South American marsupial family Microbiotheriidae. *Fieldiana (Geol.), New Ser.*, **10**: 1–75.

Marshall, L. G. (1982c). Systematics of the extinct South American marsupial family Polydolopidae. *Fieldiana Geol., New Ser.*, **12**: 1–109.

Marshall, L. G., de Muizon, C., and Sigé, B. (1983). Late Cretaceous mammals (Marsupialia) from Bolivia. *Géobios*, **16**: 739–745.

McKenna, M. C. (1975). Toward a phylogenetic classification of the Mammalia. In W. P. Luckett and F. S. Szalay (eds.), *Phylogeny of the Primates*, pp. 21–46. Plenum, New York.

Mills, J. R. E. (1984). The molar dentition of a Welsh pantothere. *Zool. J. Linn. Soc.*, **82**: 189–205.

Osborn, H. F. (1888). The evolution of mammalian molars to and from the tritubercular type. *Amer. Nat.*, **22**: 1067–1079.

Patterson, B. (1956). Early Cretaceous mammals and the evolution of mammalian molar teeth. *Fieldiana (Geol.)*, **13**: 1–105.

Paula Couto, C. de (1952). Fossil mammals from the beginning of the Cenozoic in Brazil. Marsupialia: Polydolopidae and Borhyaenidae. *Amer. Mus. Novit.*, **1559**: 1–27.

Prothero, D. R. (1981). New Jurassic mammals from Como Bluff, Wyoming, and the interrelationships of non-tribosphenic Theria. *Bull. Am. Mus. Nat. Hist.*, **167**: 281–325.

Reig, O. A., and Simpson, G. G. (1972). *Sparassocynus* (Marsupialia, Didelphidae), a peculiar mammal from the late Cenozoic of Argentina. *J. Zool., Lond.*, **167**: 511–539.

Rich, T. H. (1982). Monotremes, placentals, and marsupials: Their record in Australia and its biases. In P. V. Rich and E. M. Thompson (eds.), *The Fossil Vertebrate Record of Australasia*, pp. 385–488. Monash University Press, Clayton.

Riggs, E. S. (1934). A new marsupial saber-tooth from the Pliocene of Argentina and its relationships to other South American predacious marsupials. *Trans. Am. Phil. Soc., New Ser.*, **24**: 1–32.

Sarich, V. M., and Cronin, J. E. (1980). South American mammal systematics, evolutionary clocks, and continental drift. In R. L. Ciochon and A. B. Chiarelli (eds.), *Evolutionary Biology of the New World Monkeys and Continental Drift*, pp. 399–421. Plenum, New York.

Sigogneau-Russell, D. (1983). A new therian mammal from the Rhaetic locality of Saint-Nicolas-du-Port (France). *Zool. J. Linn. Soc.*, **78**: 175–186.

Simpson, G. G. (1970). The Argyrolagidae, extinct South American marsupials. *Bull. Mus. Comp. Zool.*, **139**(1): 1–86.

Sinclair, W. J. (1906). Mammalia of the Santa Cruz beds. Marsupialia. *Rep. Princeton Univ. Expeditions Patagonia, 1896–1899*, **4**(3): 1–110.

Slaughter, B. H. (1981). The Trinity therians (Albian, Mid-Cretaceous) as marsupials and placentals. *J. Paleont.*, **55**(3): 682–683.

Stirton, R. A. (1967). The Diprotodontidae from the Ngapakaldi Fauna, South Australia. In R. A. Stirton and M. O. Woodburne (eds.), *Tertiary Diprotodondidae from Australia and New Guinea. Bull. Austral. Bureau Min. Res.*, **85**: 1–44.

Stirton, R. A., Tedford, R. H. and Woodburne, M. O. (1967). A new Tertiary formation and fauna from the Tirari Desert, South Australia. *Rec. S. Austral. Mus.*, **15**: 427–462.

Szalay, F. S. (1982). A new appraisal of marsupial phylogeny and classification. In M. Archer (ed.), *Carnivorous Marsupials*, pp. 621–640. Royal Zoological Society, New South Wales.

Tedford, R. H., Banks, M. R., Kemp, N., McDougal, I., and Sutherland, F. L. (1975). Recognition of the oldest known fossil marsupials from Australia. *Nature*, **255**: 141–142.

Wells, R. T., Horton, D. R., and Rogers, P. (1982). *Thylacoleo carnifex* Owen (Thylacoleonidae, Marsupialia): Marsupial carnivore? In M. Archer (ed.), *Carnivorous Marsupials*, pp. 573–576. Royal Zoological Society, New South Wales.

Woodburne, M. O. (1984). Families of marsupials: Relationships, evolution and biogeography. In P. D. Gingerich and C. E. Badgley (eds.), *Mammals. Notes for a Short Course, Univ. Tennessee, Studies in Geol.*, **8**: 48–71.

Woodburne, M. O., and Clemens, W. A. (eds.), (1984). Revision of the Ektopodonditae (Mammalia, Marsupialia, Phalangeroidea) of the Australian Neogene. *Univ. Calif. Publ. Geol. Sci.*, **131**: 1–114.

Woodburne, M. O., and Zinsmeister, W. J. (1984). The first land mammal from Antarctica and its biogeographic implications. *J. Paleont.*, **58**: 913–948.

CHAPTER XX

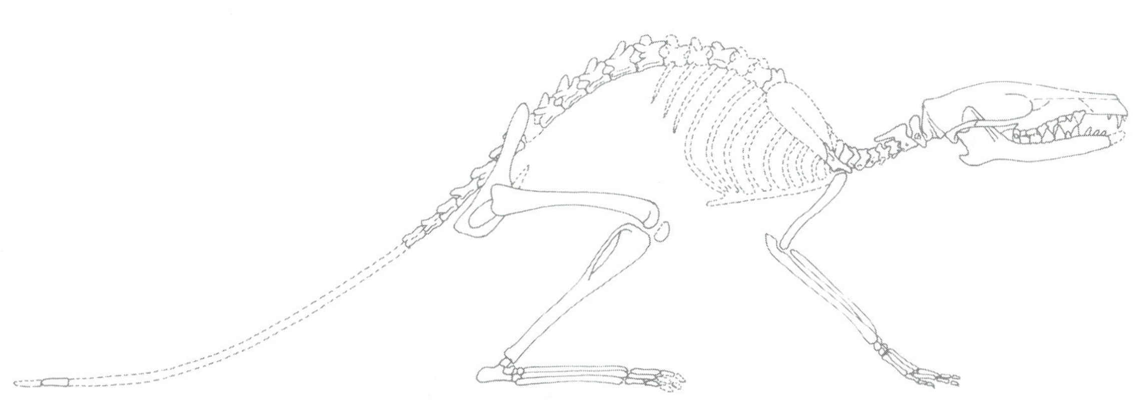

The Radiation of Placental Mammals

PLACENTAL REPRODUCTION

Since the end of the Mesozoic, placental mammals have been the dominant terrestrial vertebrates on all continents except Australia and Antarctica. In most features of the skeleton, soft anatomy, and physiology, marsupial and placentals appear to be very similar, and many species have adapted to nearly identical ways of life (Hunsaker, 1977; Stonehouse and Gilmore, 1977; Tyndale-Biscoe, 1973). Nevertheless, the pattern of reproduction is fundamentally different in the two groups. Despite the loss of the shell, all marsupial young are born at a very immature stage of development that is comparable to that of monotremes (Figure 20-1). The young of the giant kangaroo weigh less than 1 gram and are no more developed than the embryo of a comparably sized placental 12 days after fertilization. Extremely small placentals such as shrews and many rodents also give birth to hairless, blind, initially ectothermic young because of the impossibility of maintaining a high body temperature by very

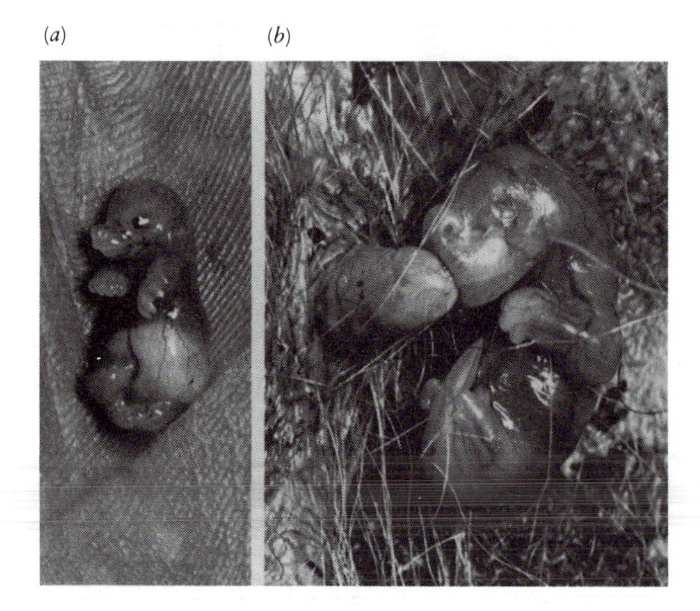

Figure 20-1. NEONATES OF MONOTREMES AND MARSUPIALS 4 HOURS AFTER BIRTH. (*a*) The echidna *Tachyglossus*. (*b*) The kangaroo *Macropus*. *From Kirsch, 1977. By permission of the Zoological Society of London.*

small animals, but the young of most placentals are much more highly developed and many are able to move about and fend for themselves soon after birth, although they may depend on their mothers' milk for a significant period of time.

Subjectively, it seems obvious that the pattern of reproduction of placental mammals is more advanced than that of marsupials and may be related to the dominance of placentals in most parts of the world for the last 65 million years. It is more difficult to understand what specific selective advantages there may have been for the initial stages in the origin of the placental pattern of reproduction and how a long gestation period became possible (Rodger and Drake, 1987).

Lillegraven (1969, 1975, 1979, 1984) has treated this subject in detail. It is hazardous to extrapolate aspects of reproductive and metabolic physiology from modern mammals back to their early Cretaceous antecedents, but several significant advances that are evident in all placentals must have been initiated before the differentiation of the modern orders in the Upper Cretaceous.

Lillegraven points out two important features that provide a selective advantage for nearly all placentals over their marsupial counterparts. Although the energetic cost of gestation is much lower in marsupials because of the small size of the neonate, the *total* cost of raising the young to the point of independence is higher because the metabolic cost of the long period of lactation is higher in marsupials. In addition, the time required for maturation is longer in most marsupials, so that their capacity for reproductive increase is lower than that of placentals. Among animals of similar body size and living in the same environment, placentals are more efficient in making use of resources and have the potential to reproduce more rapidly. Lillegraven (1984) and Parker (1977) suggest that a long period of intrauterine development may also be necessary for the elaboration of the higher brain centers that are a conspicuous feature of placentals but are much less evident in marsupials.

One of the reasons for more rapid maturation among placentals is that development occurs more quickly in the uterus than it does in the pouch of marsupials. In addition, there are significant differences in intrauterine metabolism. In both groups, the first ten days to two weeks of embryonic development are relatively slow because of the reliance on nonoxidative metabolism. Birth follows soon after this period in marsupials. Among placentals, the remainder of development is more rapid as a result of a shift to oxidative metabolism. This shift is made possible by the more intimate association with the maternal circulation that is provided by the chorioallantoic placenta. Most marsupials have a choriovitelline placenta that is much less effective in the transfer of oxygen from the mother to the embryo. Perameloids are exceptional in having an allantoic placenta, but it lacks the villae that characterize that of placentals. Marsupials probably could not maintain such intimate contact between the mother and the embryo even if they had more effective placentation because of the problem of tissue incompatability.

As placentals, we take for granted our long period of intrauterine development, which, in fact, depends on the evolution of entirely new developmental processes. Since all mammalian embryos combine the genetic material of both parents, their proteins will be recognized as foreign tissue and would be expected to be rejected by the mother. This problem does not occur among marsupials since their period of active development is so short that there is not time for the rejection process to occur.

How do placentals solve this problem?

Early developmental stages in placental mammals differ from those of all other vertebrates. The egg is very small and yolk poor in both marsupials and placentals (Figure 20-2). In placentals, the fertilized egg divides repeatedly to form a mass of cells. It then separates into an outer layer of cells and an inner cell mass. The inner cell mass forms all the tissue of the developing embryo. The outer layer, termed the **trophoblast**, is unique to placentals. It has many functions. It digests its way into the uterine mucosa; it secretes the hormone chorionic gonadotrophin, which signals the corpus luteum and pituitary that implantation has occurred; and, most important, its cells form an active barrier between the maternal and embryonic tissue to prevent rejection (Kaufman, 1983). Lillegraven (1975, p. 720) states "The 'invention' of the trophoblast tissue by primaeval eutherians was probably the single most important evolutionary event in the history of the infraclass." The evolution of the trophoblast would allow longer and more intimate placentation, even if the structure of the placenta were of a marsupial type. Once rejection was not a problem, selection could act to im-

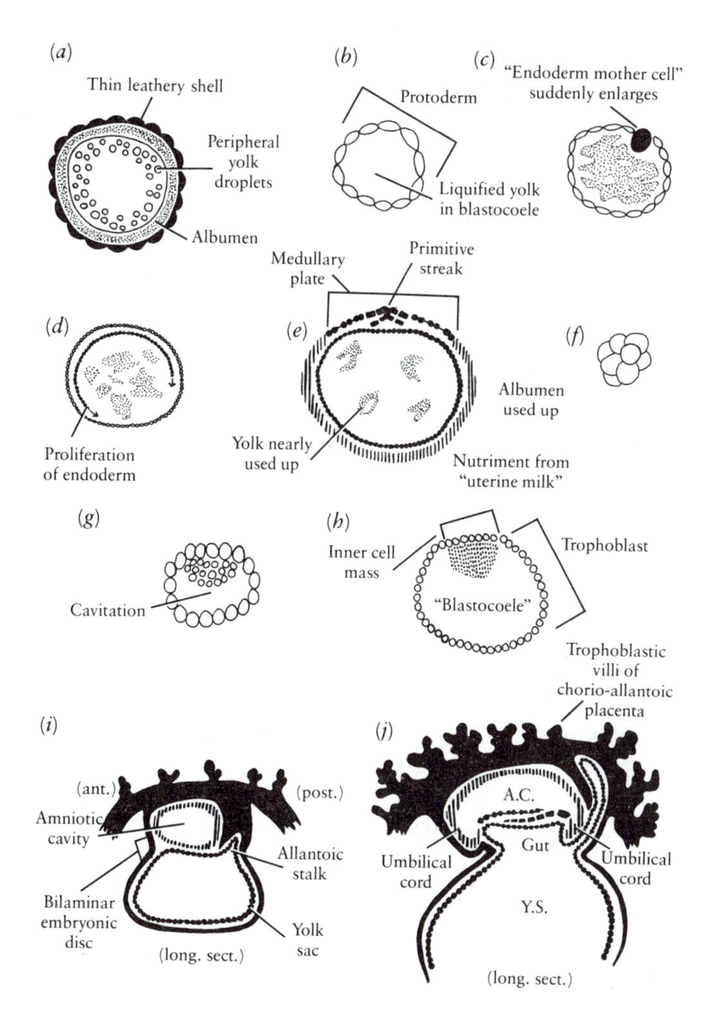

Figure 20-2. EARLY DEVELOPMENTAL STAGES OF MARSUPIALS AND PLACENTALS. (*a* to *e*) Marsupial development from fertilized egg, proliferation of endoderm, and formation of blastocyst. (*f* to *j*) Placental development including elaboration of trophoblast, which acts to prevent rejection. *From Lillegraven, 1969.*

prove the efficiency of the placentation and allow more rapid and complete development of the embryo.

There is no evidence of how the trophoblast may have evolved, nor is there any way in which its presence might be reflected in the skeletal remains of Mesozoic mammals. We find fossils in the Cenozoic that show embryos within the mother's body, but this is long after the original diversification of placental mammals.

LOWER CRETACEOUS EUTHERIANS

Some authors recognize isolated teeth from the Lower Cretaceous of Texas (placed in the genus *Pappotherium*) as belonging to primitive placentals (Fox, 1975). Eutherians that are informally designated as "*Prokennalestes*" and "*Prozalambdalestes*," have been listed as part of a fauna of early Cretaceous mammals from Mongolia, but they have not yet been figured or described (Beliajeva, Trovimov, and Reshetov, 1974).

UPPER CRETACEOUS EUTHERIANS

SANTONIAN AND CAMPANIAN MAMMALS FROM MONGOLIA

Placentals probably became phylogenetically distinct from ancestral marsupials within the Lower Cretaceous, but the earliest well-dated and positively identified eutherians come from the Upper Cretaceous (Santonian and Campanian) of Mongolia. They have been described in a series of papers by Kielan-Jaworowska (1984c and references therein). The earlier discoveries were reviewed by Kielan-Jaworowska, Bown, and Lillegraven (1979).

Four genera are known from nearly complete skulls. *Kennalestes* and *Asioryctes* (Figure 20-3) appear to be almost ideal structural ancestors for later eutherian mammals. Among living genera, the closest overall resemblance lies with the tree shrew *Tupaia*. The genera *Barunlestes* and *Zalambdalestes* represent a more specialized, divergent lineage. The skulls of *Kennalestes* and *Asioryctes* are slender and about 3 centimeters long; the braincase is narrow and there is no separation between the orbit and the temporal region. In contrast with *Morganucodon,* the squamosal contributes significantly to the wall of the braincase. The zygomatic arch is long and slender. As in more primitive mammals, the auditory bulla is not ossified. The ectotympanic (reptilian angular) forms a simple ring to support the ear drum that is oriented at

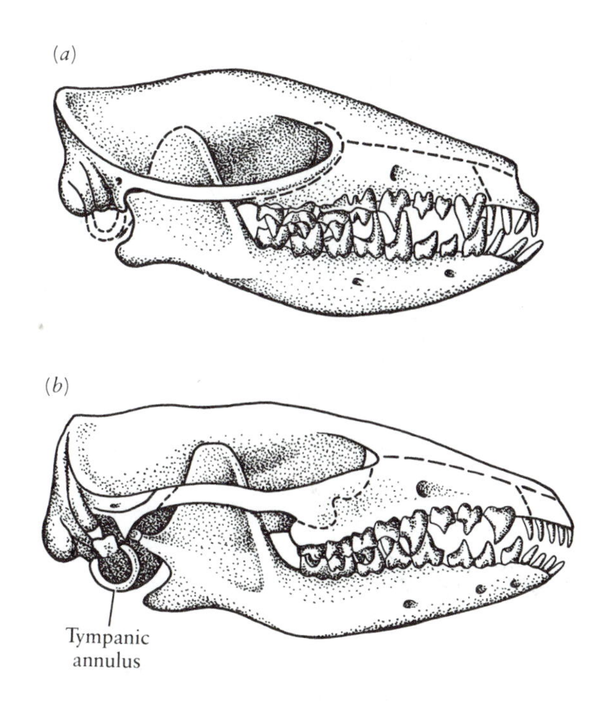

Figure 20-3. UPPER CRETACEOUS PLACENTALS FROM MONGOLIA. (*a*) *Kennalestes.* (*b*) *Asioryctes. From Kielan-Jaworowska, 1980.*

only a slight angle from the horizontal. Ear ossicles are not preserved; they were probably much smaller than in Rhaeto-Liassic mammals. The vestibular foramen opens at the base of the external auditory meatus, a groove formed in the squamosal just behind the glenoid. The posterior wall of the glenoid forms a ridged postglenoid process, behind which opens the postglenoid foramen, through which drained the external jugular vein. The posterior portion of the otic capsule is visible ventrally and posteriorly as the mastoid process.

The base of the braincase is well preserved in *Barunlestes* and *Zalambdalestes* (Figure 20-4) and is readily

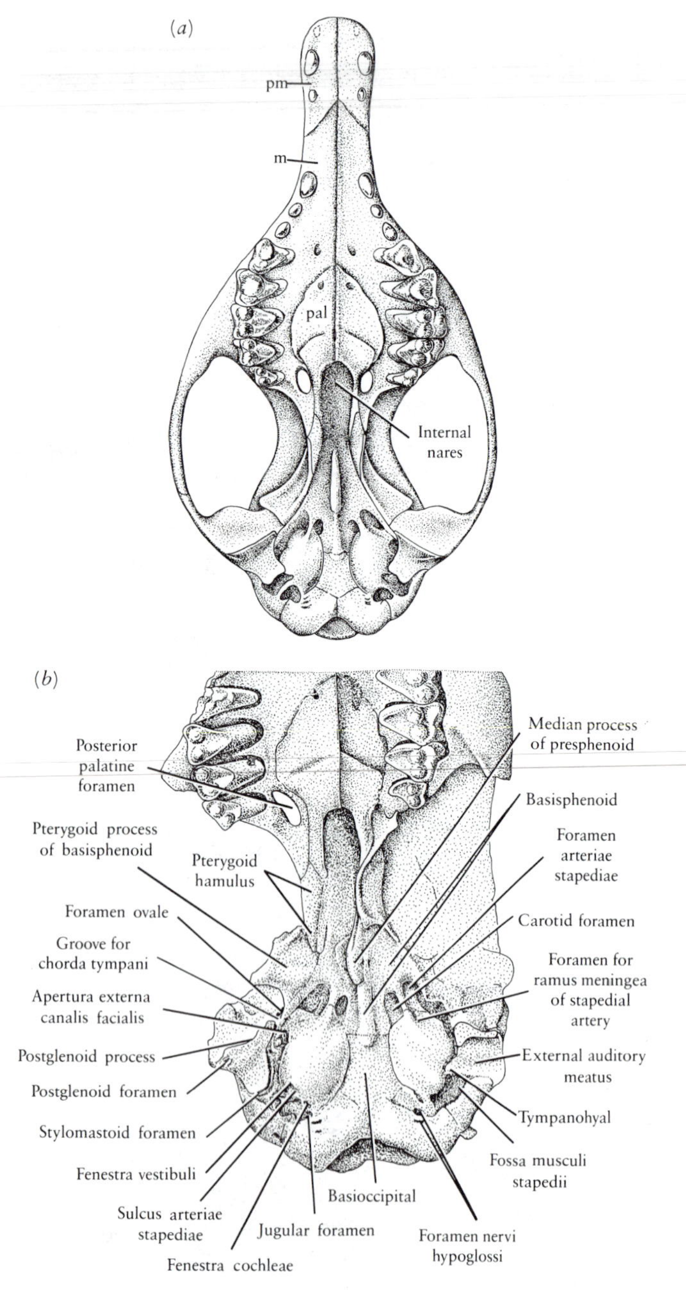

Figure 20-4. (*a*) Ventral view of the skull of *Zalambdalestes* and (*b*) the base of the braincase of *Barunlestes*, both from the Upper Cretaceous of Mongolia ×2. Abbreviations as in Figure 8-3. (*a*) *From Kielan-Jaworowska, 1984a;* (*b*) *from Kielan-Jaworowska and Trofimov, 1980.*

compared with that of *Morganucodon*. The skull proportions and dentition of *Barunlestes* have diverged somewhat from the pattern of the most primitive eutherians, but the braincase appears to have retained a primitive configuration. A progressive shortening of the skull between the otic capsule and the back of the palate has been evident from advanced therapsids through the early Jurassic mammals. This area is further shortened in the early eutherians, but it is still significantly longer than in modern placentals.

The basisphenoid extends laterally to form the floor of the anterior portion of the cavum epiptericum anterior to the otic capsule. It is pierced posteriorly by foramina for the carotid and stapedial arteries. Just posterior to these openings, the floor of the otic capsule forms a conspicuous promontory below the cochlea. In many eutherians, this structure is grooved for the passage of the promontory artery, a branch of the internal carotid. There is no groove in this position in the earliest eutherians and the artery probably passed medial to the otic capsule. The presence of a promontory artery in later placentals is thought to result from a lateral shift of the medial branch of the internal carotid (Presley, 1979).

Kielan-Jaworowska (1984b) described endocasts of *Kennalestes* and *Asioryctes* that show that the brain was primitive for therian mammals, with very large olfactory bulbs, cerebral hemispheres that were widely separated posteriorly, large midbrain exposure, and a comparatively short and wide cerebellum (Figure 20-5). If the rhinal fissure is correctly identified in *Asioryctes*, the neocortex is very small. The encephalization quotient is 0.36 for *Kennalestes* and 0.56 for *Asioryctes*. The cochlea has the shape of a crescent consisting of only one whorl.

The dentition of *Kennalestes* and *Asioryctes* appears to illustrate the most primitive condition for placentals. *Asioryctes* retains five upper and four lower incisors, as in early marsupials, but the number is reduced to four and three in *Kennalestes*. The canine is double rooted, a specialization relative to more primitive therians but retained in some later eutherians. The adults of both genera have four premolars and three molars, but the deciduous dentition of *Kennalestes* retains five premolars. McKenna (1975) argued that this is probably the primitive condition for placentals, although reduction to four or fewer premolars occurs in almost all later members of this group.

In contrast with marsupials, the last premolar in early placentals may primitively be semimolariform. In *Kennalestes*, the general outline of this tooth somewhat resembles the first molar, but there is no metacone on P^4 or metaconid on P_4. The last lower premolar has an abbreviated talonid without a basin and only a single trigonid cusp. The upper molars are wider than those of early marsupials but tend to have narrower stylar shelves and less conspicuous stylar cusps. A lingual cingulum is present, but it is not prominent. In the lower molars, the paraconid is markedly smaller than the protoconid and metaconid, in contrast with their more equal proportions

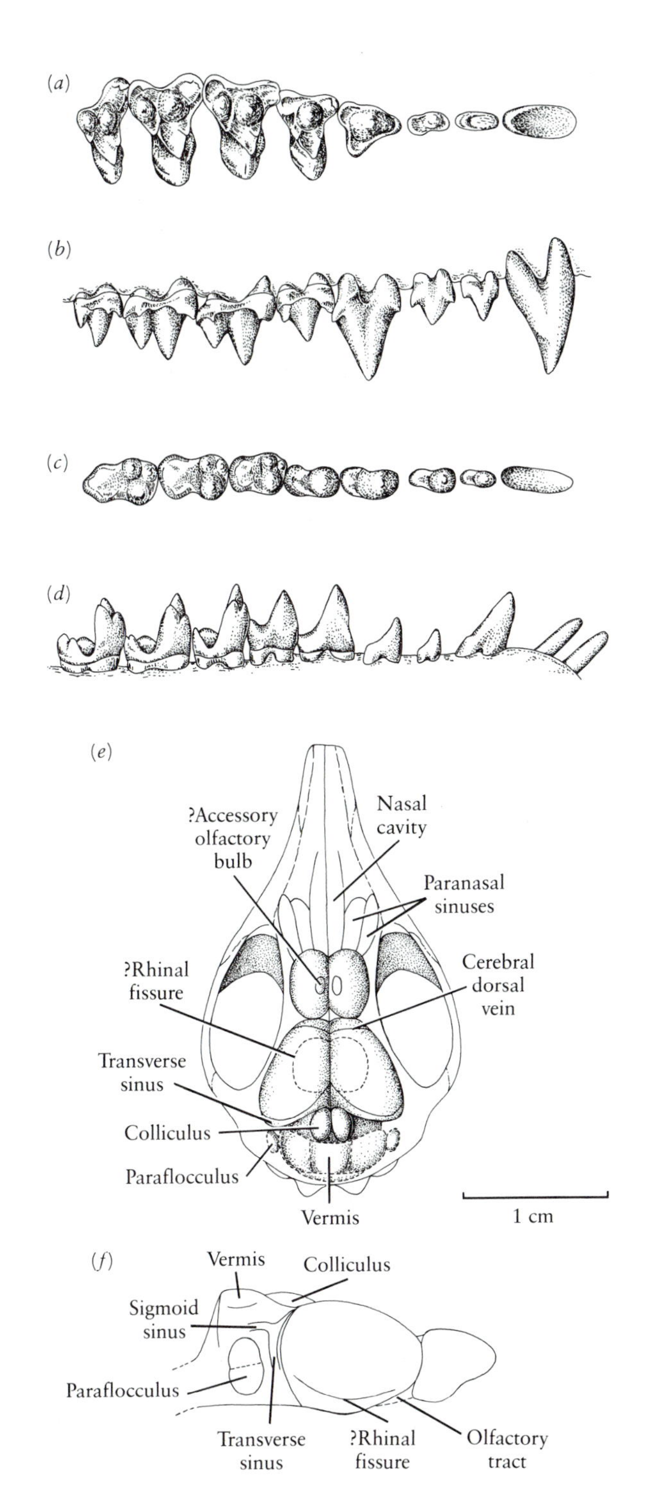

Figure 20-5. TEETH OF *KENNALESTES.* (*a*) Upper teeth in occlusal view. (*b*) Upper teeth in buccal view. (*c*) Lower teeth in occlusal view. (*d*) Lower teeth in lingual view. (*e*) Dorsal view of the endocast of *Asioryctes.* (*f*) Lateral view of the endocast of *Zalambdalestes.* (*a* to *d*) *From Kielan-Jaworowska, 1968;* (*e* and *f*) *from Kielan-Jaworowska, 1984.*

in marsupials. The trigonid appears anteroposteriorly compressed relative to the long talonid.

The angular process of the dentary is inflected as in marsupials, and there is still a trace of the coronoid bone at the base of the high coronoid process.

The entire skeleton can be reconstructed in *Zalambdalestes* (Figure 20-6), but the fusion of the limb bones and elongation of the metapodials are specializations that are not encountered in other primitive placentals. Less is known of the postcranial material of *Asioryctes,* but it appears to represent more closely the primitive condition for placentals. In *Asioryctes* (Figure 20-7), the atlas arches are fused dorsally, but the intercentrum is no more than suturally attached and may remain a distinct ossification. The carpus is compaiable to the most primitive living eutherians and lacks the grasping specializations seen in living didelphids. The scaphoid, lunar, and centrale, which fuse in some later placentals, remain separate in *Asioryctes.* The astragalus is primitively lateral rather than directly dorsal to the calcaneum, and the hallux is not opposable. The shoulder girdle in *Zalambdalestes* is typical of therians in the presence of a scapular spine and the loss of distinct coracoid elements and the interclavicle.

Animals with an anatomy like *Kennalestes* and *Asioryctes* could have given rise to nearly all subsequent placentals. They have been placed in an ill-defined assemblage, the "Proteutheria," which is considered a stem group that includes the ancestors of most, if not all of the later placentals.

In *Zalambdalestes* and *Barunlestes* (Figure 20-8), the snout is much more elongate than that of *Kennalestes* and *Asioryctes,* the upper incisors are reduced to three, and the first lower incisor is procumbent and greatly enlarged, grossly resembling the pattern of South American caenolestoids and polydolopids. The postcranial skeleton shows a number of specializations in common with living ricochetal rodents. The mobility of the cervical vertebrae is limited by the great posterior extension of the neural spine of the axis. The hind limb is much longer than the front and the fibula is extensively fused to the tibia. Other skeletal features are advanced over the condition in *Asioryctes* but are closer to the pattern in later eutherians. The intercentrum of the atlas is suturally attached to the neural arch. The astragalus is fully dorsal to the calcaneum and has a well-developed tibial trochlea. Kielan-Jaworowska (1975) suggested that *Zalambdalestes* may have possessed marsupial bones.

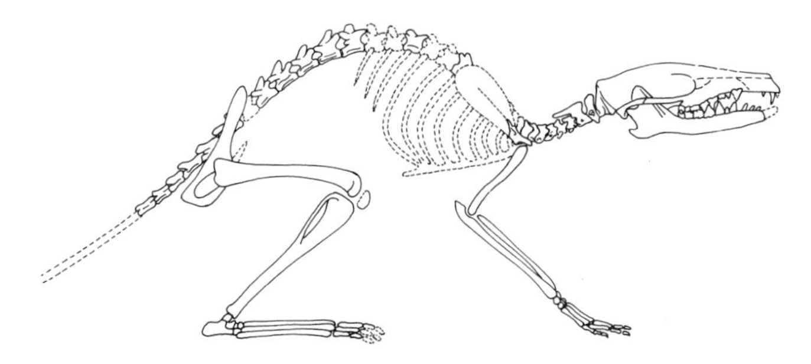

Figure 20-6. POSTCRANIAL SKELETON OF THE UPPER CRETACEOUS PLACENTAL *ZALAMBDALESTES.* Restoration is partly based on skeletal elements of *Barunlestes,* $\times\frac{1}{2}$. *From Kielan-Jaworowska, 1978.*

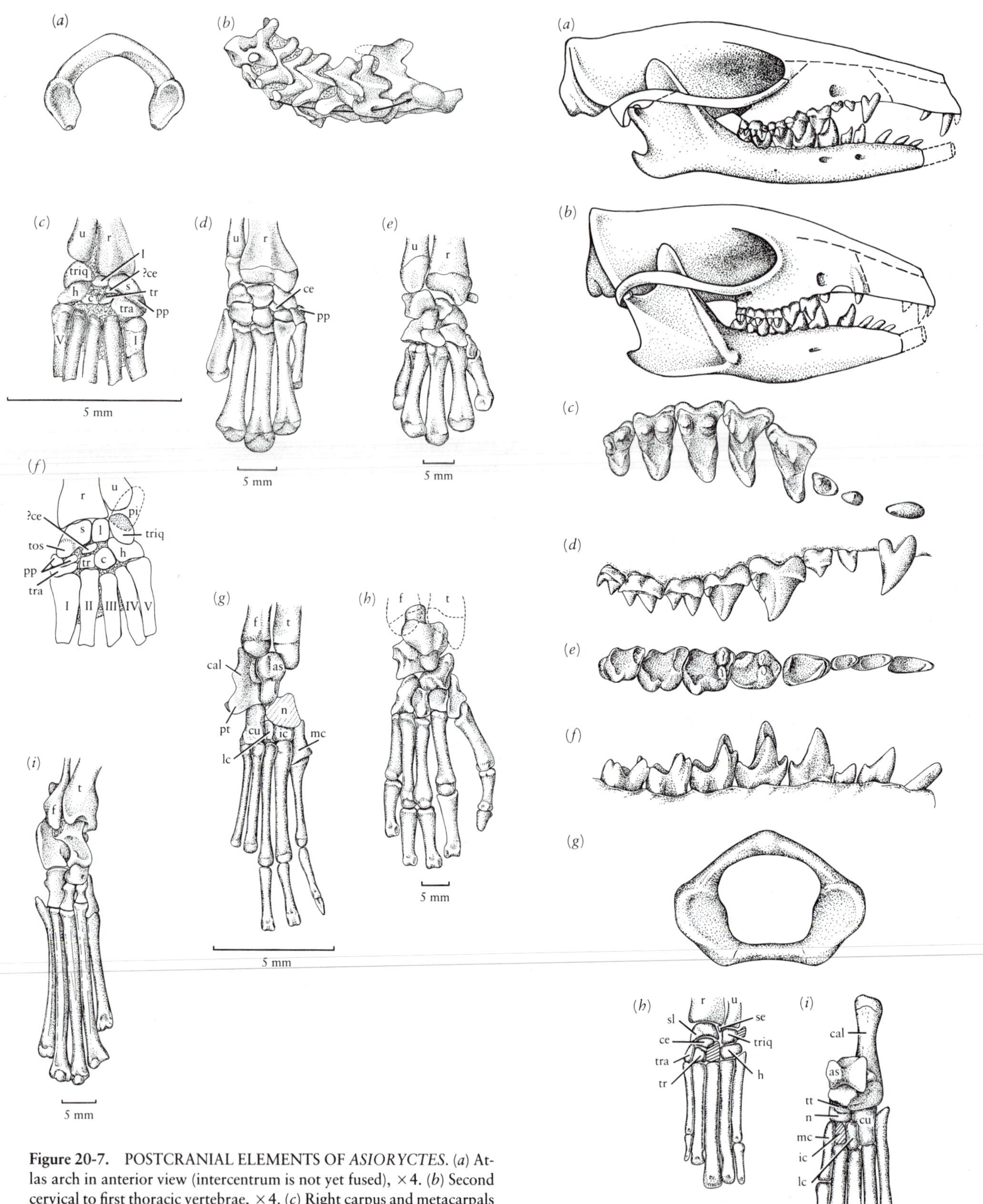

Figure 20-7. POSTCRANIAL ELEMENTS OF *ASIORYCTES*. (*a*) Atlas arch in anterior view (intercentrum is not yet fused), ×4. (*b*) Second cervical to first thoracic vertebrae, ×4. (*c*) Right carpus and metacarpals in dorsal view. (*d* and *e*) Comparable view of the carpus of *Tenrec* and *Didelphis*, ×1.5. (*f*) Ventral view of *Asioryctes*, ×5. Pisiform is reconstructed. (*g*) Right tarsus and metatarsals of *Asioryctes* in dorsal view, ×4. (*h* and *i*) Comparable views of *Didelphis* (×.5) and *Tupaia* (×.6). Abbreviations as follows: as, astragalus; c, capitatum; cal, calcaneum; ce, centrale; ct, calcaneal tuberosity; cu, cuboideum; cuf, cuboid facet; f, fibula; h, hamatum; ic, intermedial cuneiform; l, lunatum; lc, lateral cuneiform; mc, medial cuneiform; n, naviculare; paf, plantar astragalar foramen; pi, pisiform; pp, praepollex; pt, peroneal tubercle; r, radius; s, scaphoideum; st, sustentacular facet; triq, triquetrum; t, tibia; tos, tuber ossi scaphoidei; tr, trapezoideum; tra, trapezium; u, ulna; I, II, III, IV, V, metacarpals and metatarsals. *From Kielan-Jaworowska, 1977.*

Figure 20-8. Lateral view of the skulls of (*a*) *Zalambdalestes* and (*b*) *Barunlestes* from the Upper Cretaceous of Mongolia, ×2. *From Kielan-Jaworowska, 1980.* Upper dentition of *Zalambdalestes* in (*c*) occlusal and (*d*) labial views. *From Kielan-Jaworowska, 1968.* (*e* and *f*) Lower dentition of *Zalambdalestes* in occlusal and lingual views. *From Kielan-Jaworowska, 1968.* (*g*) Atlas vertebra of *Barunlestes* in anterior view. (*h*) Hand of *Barunlestes* in dorsal view. (*i*) Reconstruction of tarsus of *Zalambdalestes.* (*g* to *i*) *From Kielan-Jaworowska, 1978.* Abbreviations as in Figure 20-7, plus: se, sesamoid bone; sl, scapholunatum; tt, tuber tibialis.

MAASTRICHTIAN MAMMALS

We find a relatively diverse placental fauna in the western parts of Canada and the United States at the end of the Maastrichtian that continues without a break into the early Cenozoic (Archibald, 1982; Clemens, 1973; Johnston and Fox, 1984; Sloan and Van Valen, 1965; Van Valen 1978). The remains are almost entirely limited to jaws and teeth, but they demonstrate the first stages in the diversification of Cenozoic mammals.

The most conservative genera are *Cimolestes* (Figure 20-9), *Procerberus,* and *Gypsonictops,* whose teeth retain the general pattern seen in *Kennalestes* and *Asioryctes. Paranyctoides* (Fox, 1984) shows the early appearance of true insectivores. The greatest variety occurs among the condylarths, an archaic group of herbivores that is akin to the hoofed mammals of the Tertiary. The most completely known genus is *Protungulatum,* whose remains include a complete lower jaw with most of the teeth in place (see Chapter 21). The teeth are clearly distinct from those of the "proteutherians" in the blunt nature of the cusps, the more rectangular appearance of the crown, and the more equal height of the trigonid and talonid of the lower molars. All of these features are associated in living mammals with feeding on plant material. Perhaps the most surprising discovery within the Upper Cretaceous is the presence of a single tooth of a very primitive primate *Purgatorius.*

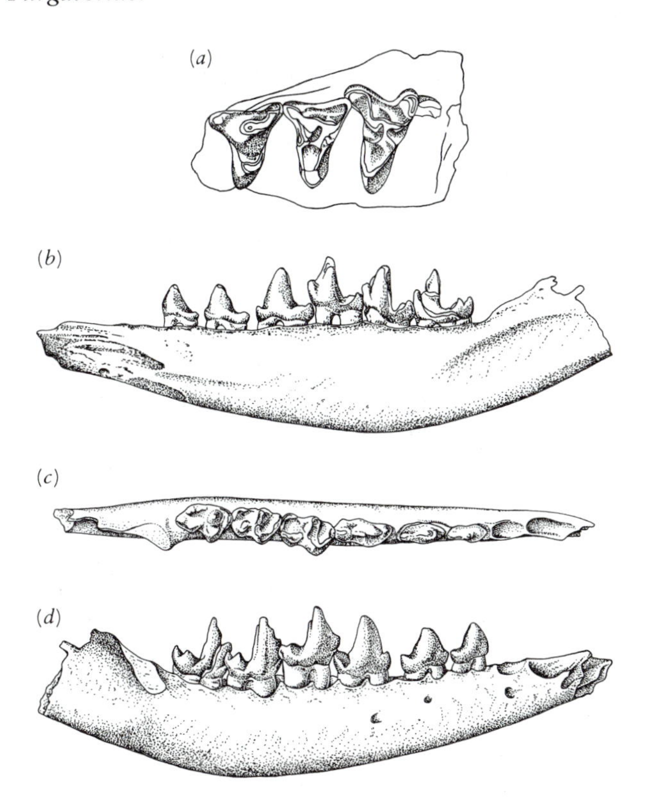

Figure 20-9. JAWS AND DENTITION OF *CIMOLESTES,* a primitive placental from the Upper Cretaceous of North America. (*a*) Occlusal view of maxilla with P^4 and M^{1-2}, ×3. (*b*) Medial, (*c*) occlusal, and (*d*) lateral views of lower jaw, ×2½. *From Clemens, 1973.*

A single tooth of questionable affinities provides the only evidence of Cretaceous eutherians in Europe. In South America, they are represented by *Perutherium,* which is known by a jaw with two teeth that may belong to the ungulate order Notoungulata (Marshall, de Muizon, and Sigé, 1983).

THE BEGINNING OF THE CENOZOIC MAMMALIAN RADIATION

The extinction of dinosaurs left vacant a broad range of adaptive zones that were subsequently occupied by therian mammals. The early record is best documented in North America, with considerable information available from Europe and much less from the rest of the world (Savage and Russell, 1983). The earliest Tertiary eutherians were small forms that resembled the many early insectivorous and omnivorous marsupial groups that are known in South America.

The fossil record at the very base of the Cenozoic (Lower Paleocene) remains relatively incomplete, and as in the Maastrichtian, most species are known only from jaws and teeth. The record rapidly improves in the Middle and Upper Paleocene, and by the beginning of the Eocene most of the living orders, as well as many extinct groups, are fairly well represented (Gingerich and Badgley, 1984).

The incomplete fossil record in the latest Cretaceous and early Cenozoic makes it very difficult to establish the nature of the interrelationships among the many groups of eutherians found in the later Tertiary. This problem is complicated by the fact that the early members of many groups differed only slightly in their dentition. If the known fossil record of the Upper Cretaceous accurately reflects the diversity of placentals at that time, the initial differentiation of most of the placental orders must have occurred within approximately 10 million years.

At least 30 distinct families are recognized by the Middle Paleocene. Very few specific interrelationships can be documented at present. The limited evidence provided by late Cretaceous mammals suggests that there was a series of successive radiations in the late Cretaceous and early Tertiary (McKenna, 1975; Novacek, 1982, 1986). The first radiation occurred prior to the appearance of the earliest adequately documented eutherian fossils in the Upper Santonian or Lower Campanian of Mongolia and resulted in two major lineages, represented by *Kennelestes* and *Asioryctes* on the one hand and *Zalambdalestes* and *Barunlestes* on the other. By the end of the Cretaceous, we recognize three additional groups that are represented by an ancestral primate, an insectivore, and a variety of condylarths.

During the latest Cretaceous and early Tertiary, each of these lineages underwent further diversification. The descendants of the condylarths include the modern orders Artiodactyla, Perissodactyla, Tubulidentata (the aardvark), Proboscidia, Sirenia, Hyracoidea, and the Cetacea, as well as a host of extinct groups. The Carnivora, Creodonta (an extinct group of carnivores), and an assemblage including primates, tree shrews, insectivores, bats, and the flying lemur may have diverged from an earlier radiation whose ancestors closely resemble *Asioryctes* and *Kennelestes*.

McKenna (1982) postulated that the lagomorphs, rodents and elephant shrews might share a common ancestry with the Upper Cretaceous genera *Zalambdalestes* and *Barunlestes*, but this connection has not been fully documented. Of even greater uncertainty is the ancestry of the Xenarthra and Pholidota (pangolins or scaly anteaters), which may have been the first of all eutherian groups to differentiate. The characters of the early members of these groups are so specialized that their affinities remain subject to controversy.

On the basis of a large number of cranial features of Cenozoic mammals, Novacek (1986) hypothesized a hierarchical arrangement involving all major placental groups (Figure 20-10). This work provides an extremely valuable basis for establishing supraordinal relationships. Unfortunately, this publication was received too late for effective integration in this text.

PROBLEMS OF CLASSIFICATION

A better understanding of the radiation of early placentals remains a major challenge to paleontologists. Many of the conspicuous characteristics of early eutherians are either primitive in nature (and so of little use in establishing specific interrelationships) or are unique to each group.

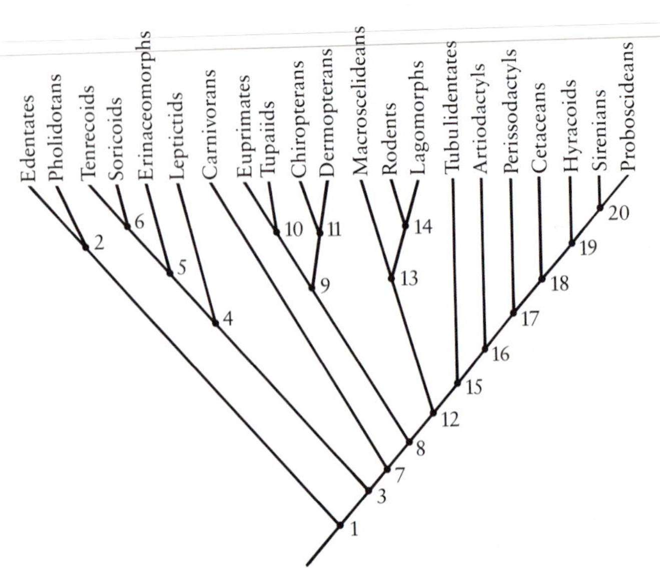

Figure 20-10. COMPUTER-GENERATED CLADOGRAM OF LIVING EUTHERIAN GROUPS PLUS LEPTICTIDS. Cladogram is based on 104 cranial characters described in detail by Novacek, 1986.

To establish relationships among these groups, it is necessary to recognize derived features that are common to two or more distinct lineages.

The Upper Cretaceous placentals, especially *Asioryctes* and *Kennalestes*, provide a strong basis for establishing the polarity of character transformation. Nearly all of the skeletal characters of these genera are primitive for eutherians, if we judge by out-group comparison with early therians and the character distribution seen in later placentals.

Novacek (1980) tabulated a number of character states that are recognized as primitive for eutherians and show a more derived condition in various early Tertiary groups (Table 20-1).

Some structural complexes seem especially important in evaluating possible relationships among early eutherians. Among these are features of the auditory bulla and carotid circulation. In most mammals, the main artery that brings blood to the brain is the internal carotid. Matthew (1909) hypothesized that in primitive mammals this vessel diverged into two branches, one that passed medial to the otic capsule and a more lateral branch that passed through a groove near the crest of the promontory beneath the cochlea. These are termed the medial and promontory branches of the internal carotid. More derived mammals were thought to have lost one or the other of these vessels. More recent evidence from both the fossil record of early eutherians (Kielan-Jaworowska, 1981) and developmental studies of modern mammals (Presley, 1979) indicates that the presence of the carotid in a medial position was the primitive condition and that the promontory position resulted from a lateral shift during development. There is no evidence that any mammal had both branches. Another major vessel in the area of the auditory capsule is the stapedial artery, which diverges from the internal carotid. The relative size of these two branches is important in the classification of primates (Szalay and Delson, 1979).

Even more variable, and so potentially more useful in classification, is the way in which the auditory bulla develops. In primitive therians, the bulla is not ossified. The middle ear chamber is covered by a thin membrane of connective tissue that is attached to the medial surface of the dermal tympanic ring. The primitive condition is retained in some marsupials and primitive placentals. Most eutherians develop an ossified structure that may be formed by outgrowths from any of a number of surrounding bones or by completely separate areas of ossification. Novacek (1977b) outlined these patterns. MacPhee (1981) discussed in detail the pattern in primates and possibly related groups.

The bulla may form from any of the following bones that are adjacent to the otic capsule: the alisphenoid (a common contributor in marsupials), the tympanic ring (a dermal ossification frequently termed the ectotympanic), the petrosal, or the basisphenoid. Alternatively, the bulla may form from one or more separate centers of ossifi-

TABLE 20-1 **Primitive Traits of Early Placentals**

Cranial
- Carotid artery in medial position
- Occiput not posteriorly expanded
- Auditory bulla not ossified
- Ectotympanic annular and inclined at an acute angle to the horizontal plane of skull
- Jaw condyle distinctly lower than coronoid process
- Postglenoid foramen retained
- Origin of temporalis muscle extending anteriorly over frontal
- Palatal fenestrae absent
- Nasals broad posteriorly
- Infraorbital canal long
- Zygomatic arch complete, with large jugal
- Postorbital process of frontal absent
- Anterior opening of alisphenoid canal confluent with foramen rotundum
- Supraorbital foramen absent
- Foramen rotundum confluent with sphenorbital foramen
- Optic foramen small
- Suboptic foramen present
- Maxilla orbital wing small or absent
- Fronto-maxillary contact absent
- Palatine orbital wing extensive
- Lacrimal-palatine contact extensive
- Lacrimal facial wing large
- Single lacrimal foramen exits par facialis
- Orbitosphenoid large
- Alisphenoid orbital wing small

Postcranial
- Scapula fossae narrow, shallow, subequal in area
- Manubrium of sternum not enlarged
- Deltopectoral crest on humerus strong
- Entepicondylar foramen present
- Ulna robust, with sigmoid lateral curvature
- Pelvic-sacral fusion limited
- Greater trochanter prominent, exceeding height of femoral head
- Lesser trochanter large, lamelliform
- Tibia-fibula broadly separate
- Distal elements of forelimbs and/or hind limbs not markedly elongate
- Scaphoid and lunate unfused
- Os central present
- Metatarsals not greatly elongated
- No metatarsals significantly reduced
- Calcaneal fibular facet pronounced
- Superior astragalar foramen present

Data primarily from Novacek, 1980.

cation termed entotympanics, which are frequently differentiated as rostral or caudal, depending on whether they are anterior or posterior in position.

The bulla is cartilaginous in a few genera. This pattern might appear to be primitive but its presence in only a few genera that belong to distantly related groups indicates that this condition is derived rather than primitive. The distribution of different patterns of ossification of the auditory bulla is shown in Table 20-2. A particular pattern, once evolved, appears to be retained with little change and so may be important in confirming relationships within each major group. However, particular orders may show more than one pattern. There is a broad correlation with the relationships of the orders based on their dentition, but the absence of a bulla in primitive members of many lineages indicates that the ossification of this structure has occurred separately many times. Similar elements may have been incorporated independently in many separate groups.

The detailed structure of the cheek teeth remains the primary basis for establishing relationships among the early placentals. The structure and function of the teeth in *Kennalestes, Asioryctes,* and *Cimolestes* appear to represent a generally primitive pattern for eutherians. A few Cenozoic genera retain the fifth premolar, but in most orders, the primitive dental formula is

$$\frac{3 \quad 1 \quad 4 \quad 3}{3 \quad 1 \quad 4 \quad 3}$$

McKenna (1975) argued that the loss of the fifth premolar may have occurred in different ways in different groups of placentals, but this pattern is not clearly established. In contrast with the Upper Cretaceous genera, the canine is almost always single rooted. In primitive genera, the last premolar is clearly distinguished from the first molar but may become increasingly molariform in derived groups. The molar cusps are initially tall and sharp. The upper molars are triangular in outline with little development of cingula, and the talonid of the lower molar is much lower than the trigonid cusps.

In derived groups, this pattern is modified in various ways. Animals that specialize in feeding on larger prey typically develop longer shearing surfaces on the molars and/or premolars. The posterior molars may be reduced or lost but, where present, retain sharp cusps. The incisors are retained and the canines may be much enlarged.

The molar teeth of omnivores and herbivores become bulbous and more nearly square in occlusal view and frequently develop a new cusp from the cingulum, the hypocone, posterior to the protocone. The talonid and trigonid of the lower molars become more nearly equal in height.

The teeth of herbivores may become further specialized by the development of lophs and crests to break up the food and increase their height to compensate for wear (**hypsodonty**). The most specialized condition is the capacity for continuous growth of the teeth, which is achieved by rodents and rarely by other groups. This condition is judged in fossils by the failure of the roots to close.

TABLE 20-2 **Major Components of the Auditory Bulla in the Mammalian Orders**

	No Bulla	Ectotympanic	Entotympanic	Petrosal	Alisphenoid	Basisphenoid
Marsupialia	X		Rare		X	
Primitive placentals: *Kennalestes, Asioryctes*	X					
Leptictids			X			
Insectivora	Soricids, some talpids	X	X	Tenrecs, erinaceids	X	Tenrecs, erinaceids
Dermoptera		X	X			
Macroscelidea			X	X	X	X
Scandentia (tupaiids)			X			
Primates				X		
Chiroptera			X			
Creodonta	X		Rare			
Carnivora (canids, phocids, procyonids, felids, viveriids, hyaenids, most mustelids)			X			
Carnivora (ursids, odobenids, *Ailurus*, lutrine, and mephitine mustilids)		X				
Anagalida			X			
Rodents		X				
Lagomorphs		X				
Artiodactyla		X				
Perissodactyla		X				
Mesonychia			X			
Cetacea		X				
Proboscidea		X				
Hyracoidea			X			
Toxodonta		X				
Notoungulata		X				
Pholidota			X			
Palaeanodontidae			X			
Edentata			X			

The following genera have cartilaginous bulla: the megachiropteran bats *Pteropus, Acerodon,* and *Boneia;* the edentate *Dasypus;* and the viveriid carnivore *Nandinia.*

Data from Novacek, 1977.

Herbivores tend to reduce or lose the canine and frequently the lateral incisors and anterior premolars, leaving a long gap, or diastema, between the anterior incisors and the remaining cheek teeth.

These changes can occur in a variety of ways and in different sequences, which allows us to recognize many different lineages of insectivores, carnivores, and herbivores. Characteristics of the dentition that form the basis for recognition of different genera and families can be readily documented, but it remains much more difficult to determine what basic similarities can be used to demonstrate the relationships between groups.

Teeth once lost are rarely regained, but loss of particular teeth (especially the posterior molars) has occurred many times in unrelated groups. Cusps once reduced or blunted are not likely to reappear or become sharp and conspicuous, but new cusps and extensions of cingula may evolve separately in different lineages and yet appear very similar in position and function. A pattern that superficially resembles that of the simple triangular upper molar of the most primitive eutherians apparently evolved separately in several different lineages of insectivorous mammals, and many groups of early eutherians evolved very large procumbent incisors in the lower jaw. Establishing specific homologies of dental patterns may require a more-or-less continuous sequence of fossils that represent a particular lineage.

Although the postcranial skeleton is stressed in characterizing each of the major groups of Cenozoic mammals, it has not been used in a systematic way to establish their interrelationships. Szalay and Decker (1974) pointed out that many paleontologists feel that the postcranial skeleton is inherently less useful in establishing relationships because of its greater plasticity relative to the dentition and the greater tendency to exhibit convergence. There is no consistent evidence to support this assumption, and the taxonomic patterns elucidated from the two elements of the anatomy appear generally congruent.

Given its greater complexity, the entire skeleton exhibits more features that could be used in classification than have been recognized in the dentition. The use of the postcranial skeleton depends on associated remains that are known for only a relatively small number of species. Articulated skeletons are especially rare in the early Cenozoic, when the major radiation of placentals occurred.

Nevertheless, elements of the postcranial skeleton are certainly useful in distinguishing the major groups throughout most of the Cenozoic. The carpals and tarsals have proven especially valuable because of the complex interrelationship of articulating surfaces that are associated with particular modes of locomotion.

Szalay (1977) used the structure of the astragalus and calcaneum to hypothesize supraordinal relationship among the eutherian orders and for determining their primitive locomotor patterns. Carpals and tarsals may fuse with one another or become lost. Heavy graviportal mammals, such as the elephant, retain all the toes, but smaller, more agile forms tend to loose the lateral digits. Either the middle toe or digits 3 and 4 may be emphasized, which leads to a **mesaxonic** or **paraxonic** condition.

As our knowledge of the fossil record improves and as more detailed anatomical studies are made, it should be possible to clarify the interrelationships of the placental orders. On the other hand, recent work on the relationships of tree shrews (Luckett, 1980a) and primates (MacPhee, Cartmill, and Gingerich, 1983; Novacek, McKenna, Neff, and Cifelli, 1983) suggests that none of the criteria now recognized are adequate to establish specific relationships between these and other placental orders, which may reflect a more general problem.

We know that the initial divergence of most eutherian groups occurred within a time span of 10 to 15 million years. During this time, at least 30 lineages, which we recognize as being distinct at the family level, became differentiated. Within this assemblage, common ancestry of sister groups could not have lasted, on the average, for more than 2 or 3 million years.

If the successive divergence of these groups occurred so rapidly, there may have been little time for the evolution of significant changes in the skeleton to have occurred *between* each point of divergence. It may never be possible to establish specific sister-group relationships among some groups on the basis of skeletal characteristics.

Much attention has recently been focused on the use of immunological reactions and direct study of the proteins and nucleic acids for establishing interrelationships among the placental groups (e.g., Ciochon and Corruccini, 1983; Luckett, 1980a; Miyamoto and Goodman, 1986). There is a general correspondence between the phylogenetic patterns established from the fossil record and those derived from molecular evidence. Groups with well-established affinities are easily recognized by both methods, and groups whose specific affinities have long been contested on the basis of fossil and anatomical evidence show a similarly conflicting pattern on the basis of molecular data. Both methods reflect the same difficulties. Closely related groups that are undergoing rapid, successive dichotomous branching have little time to evolve either structural or chemical changes that can be detected in their descendants.

GEOGRAPHY

Another factor that can be used in judging possible relationships among Cenozoic mammals is the geographical position of the major land masses and seaways that may have facilitated or prevented the migration of land mammals. In the early Mesozoic, there was a nearly continuous world continent, which helps to explain the wide distribution of synapsids and the earliest mammals.

By the late Jurassic, there is evidence of separation of a large northern continent, Laurasia (composed of North America, Greenland, and Eurasia) from a southern land mass, Gondwanaland (made up of South America, Africa, India, Antarctica, and Australia). They were separated by the Tethys seaway, which extended from the area of the present Caribbean through the Mediterranean region and across the Middle East to southern Asia.

During the later Jurassic and early Cretaceous, the southern continents began to separate from one another. India and Madagascar moved away from Africa approximately 160 million years ago, followed by the separation of India from Antarctica about 150 million years ago, and the initial separation of southern Africa and South America 130 million years ago (Tarling, 1980). During the middle Cretaceous, contact was still maintained between northeastern North America and northern Europe and between Brazilian South America and the bight of Africa (see Figure 19-11).

By the late Cretaceous, the Atlantic Ocean was continuous from the Arctic to Antarctica, with at least a 500-kilometer gap between South America and Africa. Asia and North America established contact across the area of the Bering Strait. A chain of islands may have provided tenuous connections between North and South America. To the south, South America, Antarctica, and Australia remained in contact. India was a large island that was moving slowly northward toward Asia.

During the Cenozoic, North America and Europe were intermittently in contact (especially to the north) (McKenna, 1983a, b). A degree of contact was also maintained between Alaska and Siberia. Although some groups were endemic, Asia, Europe, and North America had a broadly similar mammalian fauna throughout the Cenozoic (Simpson, 1965).

In contrast, South America was effectively isolated from North America and Africa throughout the Cenozoic, although North American and/or African immigrants entered South America during the late Eocene and the Miocene. A solid contact with North America was achieved in the late Pliocene. South America was separated from Antarctica by the formation of the Drake Passage approximately 30 million years ago (Woodburne and Zinsmeister, 1984).

Australia became separated from Antarctica during the Eocene and was isolated from the rest of the world until it approached Asia at the end of the Cenozoic. Rodents and bats then entered Australia, but the influence of the Australian fauna never spread further than Indonesia.

Africa had tenuous connections with Europe during the early Cenozoic, but several groups evolved there in apparent isolation until the end of the early Miocene, when contact with Asia was established via the Arabian peninsula.

EUTHERIAN MAMMALS *incertae sedis*

We can recognize nearly all the modern eutherian orders by the early Eocene. Most of the early Cenozoic mammals can be related to these groups or to the major orders of extinct mammals, but a few families remain more difficult to classify. These families were once allied with the modern members of the order Insectivora. Butler (1972) argued that the families that include the shrews, moles, hedgehogs, etc., can be readily defined as a monophyletic assemblage and distinguished from these more primitive placentals. Following Romer (1966) and Butler (1972), other authorities have used a separate suborder or order Proteutheria to accommodate both the most primitive eutherians and a number of early Cenozoic families that cannot be allied with the major orders. The Proteutheria is clearly an unnatural assemblage. It seems more judicious to admit our ignorance of the specific phylogenetic position of these families and refer to them as Eutheria, *incertae sedis*. Within this assemblage, the apatemyids, pantolestids, and leptictids appear as distinct as early members of the different eutherian orders. Only their lack of diversity and longevity precludes their recognition as taxa of equal rank.

The taxonomic position of the Leptictidae and Palaeoryctidae has been particularly difficult to assess. They were originally based on early Cenozoic genera, but the concept of these families has been extended to embrace most or all of the Upper Cretaceous placentals (Kielan-Jaworowska, Bown, and Lillegraven, 1979). The content and taxonomic position of these families is under review by Novacek and McKenna. McKenna and his colleagues (1984) classified the palaeoryctids within the order Insectivora, as defined by Butler (1972), based on evidence of the basicranium. There is no strong evidence that either of these families has a uniquely close relationship with *Kennalestes* and *Asioryctes*. These Upper Cretaceous genera, together with *Cimolestes, Procerberus,* and *Gypsonictops,* appear to occupy a position that is close to the ancestry of all later eutherians. As such, they are especially difficult to classify in an orthodox manner. They too may be listed as "Eutheria, *incertae sedis,*" but a quite different phylogenetic position is implied than for the apatemyids, pantolestids, and Cenozoic leptictids.

LEPTICTIDAE

Work by Novacek (1977a) indicates that leptictids were a short-lived group of relatively conservative early Tertiary forms that left no descendants. They were relatively diverse throughout the Paleocene and Eocene and extended into the middle Oligocene in North America. Several genera have been described from Europe as well, and others may occur in Asia.

Complete skeletons of several genera have been mentioned in the literature, but none are fully described (Guth, 1962). The skull is primitive in having an elongate snout and retaining the jugal (unlike many true insectivores), and the carotid is medial in position (Figure 20-11). They are advanced in the incorporation of an entotympanic bone in the auditory bulla, but this bone does not form a complete covering. Leptictids are unique among placentals in having a conspicuous triangular exposure of the parietal on the occipital surface.

Among Tertiary genera the primitive tooth count is

$$\frac{2\ 1\ 4\ 3}{3\ 1\ 4\ 3}$$

Novacek accepts McKenna's suggestion that the third premolar has been lost from a primitive count of 5. The canine and first premolar are single rooted and the last premolar is molariform. The trigonid is anteroposteriorly compressed and the talonid has a large shallow basin. The teeth retain the primitively sectorial pattern with sharp cusps on transversely widened upper molars.

Where known, the hind limb, especially the tibia, and the foot are elongate. In the Oligocene *Leptictis,* the tibia and fibula are fused, which suggests convergence with the pattern of *Zalambdalestes.*

Gypsonictops from the Upper Cretaceous is the only pre-Cenozoic genus to be assigned to the Leptictidae. It is primitive in the retention of five lower premolars and an inflected angle. It is not thought to be directly related to the ancestry of the later genera but links them, in a general way, with *Kennalestes,* which has been included in the superfamily Leptictoidea.

In the most recent appraisal of leptictids, Novacek (1986) recognized them as members of a distinct order, Leptictida, but placed them in the superorder Insectivora,

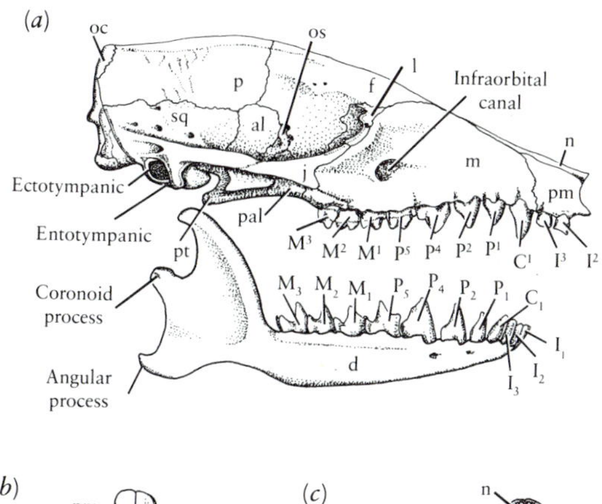

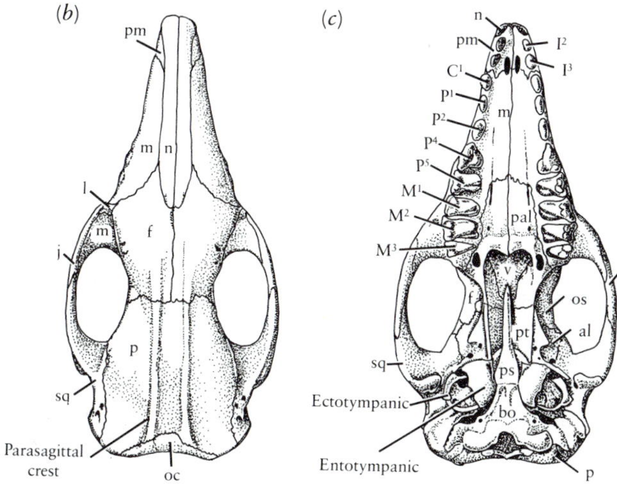

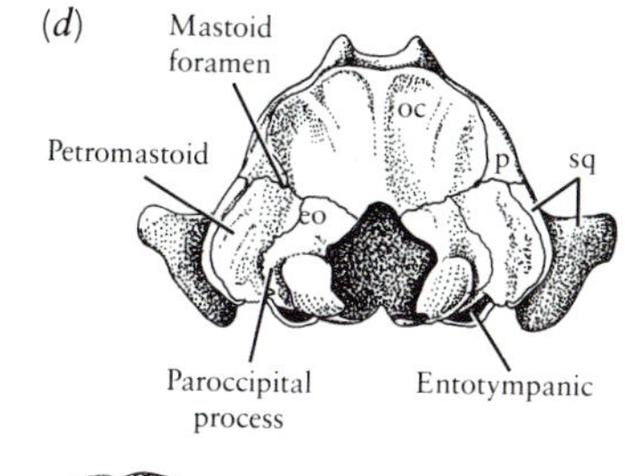

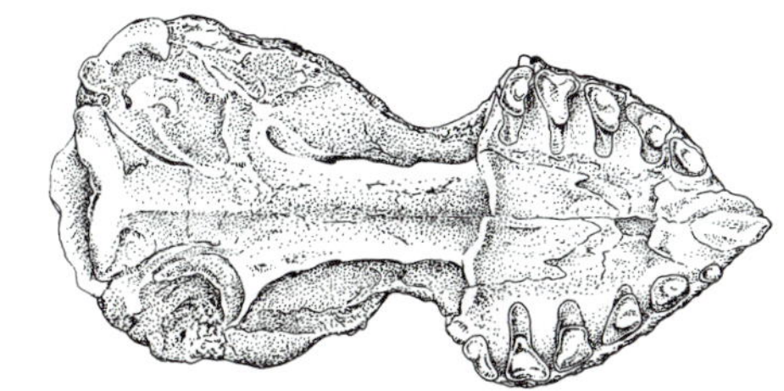

Figure 20-11. (*a*) Lateral, (*b*) dorsal, (*c*) palatal, and (*d*) occipital views of the Oligocene leptictid *Leptictis dakotensis*, $\times\frac{3}{4}$. This species retains many primitive placental features. *From Novacek, 1986.* (*e*). Palate of *Palaeoryctes*, ×1. *From McDowell, 1958.*

which also includes the families that are grouped here within the order Insectivora.

APATEMYIDAE

The aptemyids are a rare, but clearly distinct, group that is known from the Lower Paleocene into the Oligocene of North America and Europe. Their remains consist primarily of jaws and teeth. The only adequately known cranial remains are those of *Sinclairella* from the Oligocene (Figure 20-12). No postcranial elements have been attributed to the family.

Apatemyids are characterized primarily by the highly specialized dentition of the lower jaw, which was already established by the early Paleocene. The most distinctive feature is the extremely large, procumbent anterior incisor, which has a curved, spoon-shaped crown and a long root. Lateral incisors and the lower canine are lost. The most anterior premolar is a large, laterally compressed and procumbent tooth that superficially resembles the bladelike lower premolar of multituberculates and caenolestoids. The more anterior of the two pairs of upper incisors is also greatly enlarged, which gives a generally rodentlike appearance to the skull. The upper canine is lost, resulting in a short diastema. West (1973) suggested that apatemyids may have used their enlarged lower teeth like those of the phalangerid *Dactylopsila* to pierce and slice insect cuticle.

During the evolution of this group, the lower jaw became progressively shorter and deeper. Two of the original four premolars were lost, and the molars reduced in size to accommodate the posterior extension of the incisor root. The trigonid became lower relative to the talonid, and all of the molar cusps were rounded. Apatemyids have been allied with ungulates, rodents, primates, carnivores, taeniodonts, tillodonts, condylarths, a variety of "insectivores," and "proteutherians." McKenna (1975) associated them with the Carnivora in the "Grandorder" Ferae, based on similarities of the molar dentition to *Cimolestes*, as both Szalay (1968) and Lillegraven (1969) previously suggested.

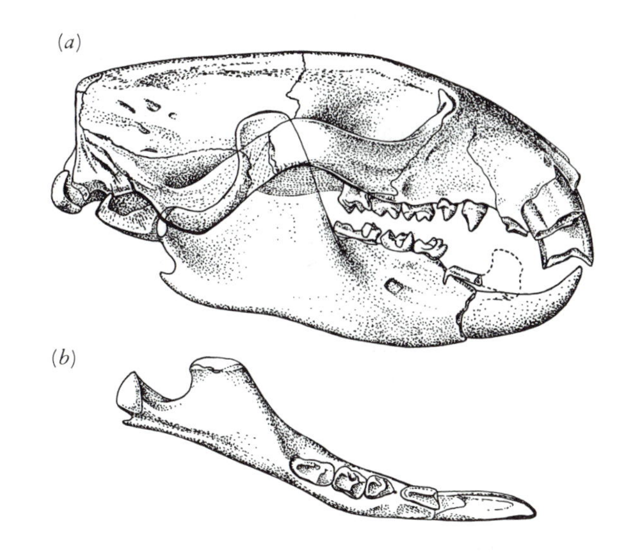

Figure 20-12. SKULL OF *SINCLAIRELLA*, AN APATEMYID FROM THE OLIGOCENE. (*a*) Skull in lateral view and (*b*) lower jaw in occlusal view, ×1. *From Scott and Jensen, 1936.*

PANTOLESTIDAE

Pantolestids, which Koenigswald (1980) recently discussed, are among the largest of the "proteutherians," with skulls approaching 15 centimeters long. We know this group from the Middle Paleocene to the Lower Oligocene in North America and from the late Paleocene to middle Eocene in Europe. The poorly known Oligocene family Ptolemaiidae (Butler, 1978) may represent an extension of the group into northern Africa. *Pantolestes* from the Middle Eocene of North America shows skeletal specializations comparable with those of the sea otter toward an aquatic or semiaquatic way of life. Fish remains are found within the stomach of the European genus *Buxolestes.* The broad, low skull also resembles that of the seals and sea otters. The dentition of the earliest pantolestids can be derived from that of the Upper Cretaceous and Lower Paleocene "proteutherians," such as *Procerberus.* The canines are large, and the third upper molar is expanded transversely with a hooklike parastylar extension. Within the pantolestids, the molars become specialized for crushing with lower cusps and a quadrangular occlusal surface. The temporal region of the skull becomes elongate, probably to accommodate massive jaw musculature. These features are rapidly accentuated between the late Paleocene and middle Eocene. These changes are generally associated with feeding on molluscs.

In addition to these taxonomically isolated families, there are several extinct orders, which are limited to the lower Cenozoic, that appear to have diverged directly from the ancestral eutherians. Their specific origin and interrelationships are subject to continuing dispute, but they show no derived features in common with any of the surviving placental groups.

TAENIODONTA

The taeniodonts were among the most highly specialized terrestrial placentals of the late Paleocene and early Eocene, but the early Paleocene representatives have a postcranial skeleton that retains most of the features of primitive eutherians (Figure 20-13).

Onychodectes, which Schoch (1982) recently described, broadly resembles the living opossum *Didelphis,* except for its slightly greater size and more robust skeleton. The limbs are moderately long but without evidence of cursorial adaptations. They may have been adept at climbing and had some capacity for digging. The ulna and radius are unreduced and unfused. The carpals are unreduced, unfused, and alternating. One specialization is the elaboration of the astragalocalcaneal complex for extreme plantar flexion, a character shared with the lepticids, *Cimolestes* and *Procerberus.* The feet are plantigrade, with five digits each bearing unfissured claws. The tail is long and heavy.

The skull is primitive in the absence of a postorbital bar. The dentition is complete, without trace of a diastema, and the canine is only moderately enlarged. The cheek teeth are primitive in their basically tritubercular configuration. The premolars are sectorial, but the molar cusps are reduced, as would be appropriate in an animal that grinds its food. More importantly, the cheek teeth are higher crowned than in primitive eutherians, with the enamel extending lingually on the upper molars and labially on the lowers to provide the greater wearing surfaces necessary for an omnivorous to herbivorous diet. The upper molars are transversely narrow, with the protocones, protoconules, and metaconules small and lingually placed; the stylar shelves are reduced. In contrast with most other herbivorous groups, neither the upper nor the lower teeth have cingula primitively, and the hypocone is absent.

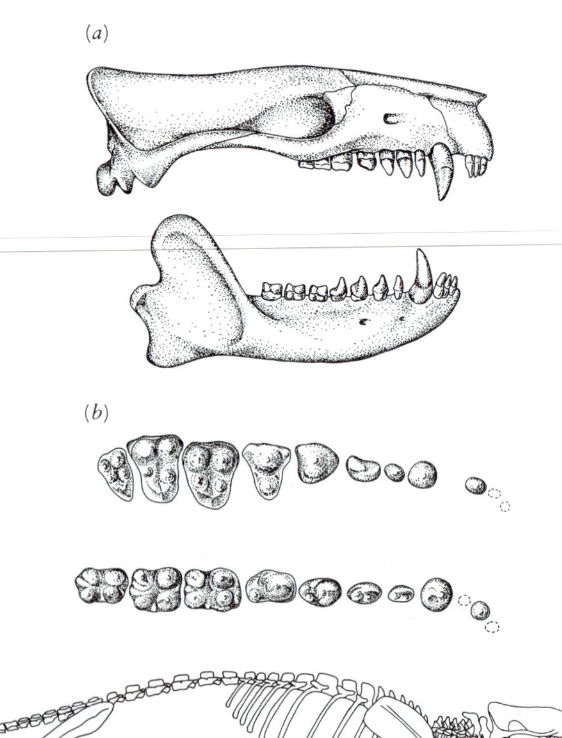

Figure 20-13. THE PRIMITIVE TAENIODONT *ONYCHODECTES.* (*a*) Skull and lower jaw in lateral view. (*b*) Upper and lower dentition in occlusal view. (*c*) Skeleton in lateral view. Approximately the size of an opossum. *From Schoch, 1982.*

By the early Eocene, genera such as *Stylinodon* (Figure 20-14) had become highly specialized in their dentition and postcranial skeleton and reached the size of a bear. Patterson (1949) suggested that the rate of evolution of the dentition, illustrated through a series of progressive stages, may have been among the most rapid in any mammalian group. In *Stylinodon,* all the teeth are rootless and ever growing. One upper and one lower incisor have been lost. The canine is by far the largest tooth. Like the incisors, it bears enamel only on the anterior surface, which forms a sharp, chisel-like blade. The canine is greatly expanded posteriorly in a manner that is unique among mammals to provide a large crushing surface. The premolars are molariform and all the cheek teeth rapidly lose their crowns through wear, so they appear as cylinders of dentine with only a thin surrounding edge of enamel.

The diet of taeniodonts was tough and highly abrasive. The well-developed claws, especially on the forelimbs, and the elaboration of large areas for muscle attachment suggest that they were used for digging food from the ground. Schoch suggested that the advanced taeniodonts may have resembled the aardvark in digging abilities.

The fossil record of taeniodonts is entirely limited to western North America. Previous reports from other areas are based on misidentified specimens. They are not known after the middle Eocene. Schoch and Lucas (1981) suggest that their rarity in most localities may be due to their living in a poorly sampled environment, such as the uplands, away from the typical areas of deposition.

Lillegraven (1969) and McKenna (1973) suggest that taeniodonts may be the sister group of *Cimolestes*.

TILLODONTS AND PANTODONTS

The tillodonts constitute a second group of rare herbivores that were once thought to be closely related to the taeniodonts. They too have chisel-shaped anterior teeth, but the most highly specialized teeth are the second incisors, not the canines. It is clear that herbivorous adaptations occurred separately in these two groups and not from a similarly specialized common ancestor.

No tillodonts are completely enough known for a skeletal restoration, but enough has been collected to indicate that they were large, massive animals with clawed, five-toed, plantigrade feet. The skull of the Eocene genus *Trogosus* (Figure 20-15) was approximately 30 centimeters long. The second incisors are greatly enlarged, rootless, and ever growing. Like the incisors of taeniodonts, the enamel is limited to the anterior surface. The

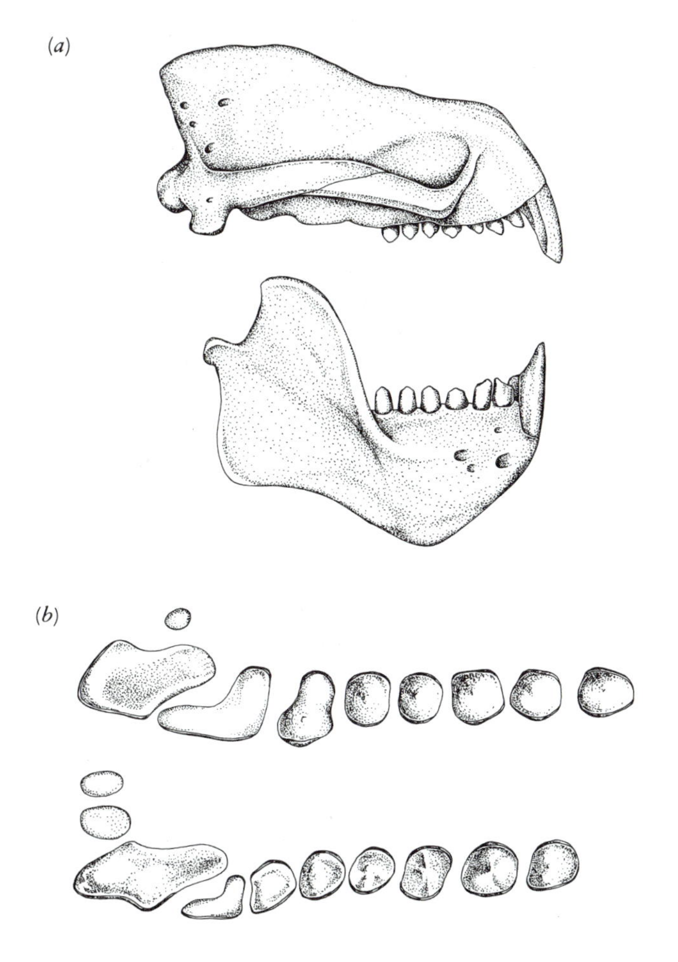

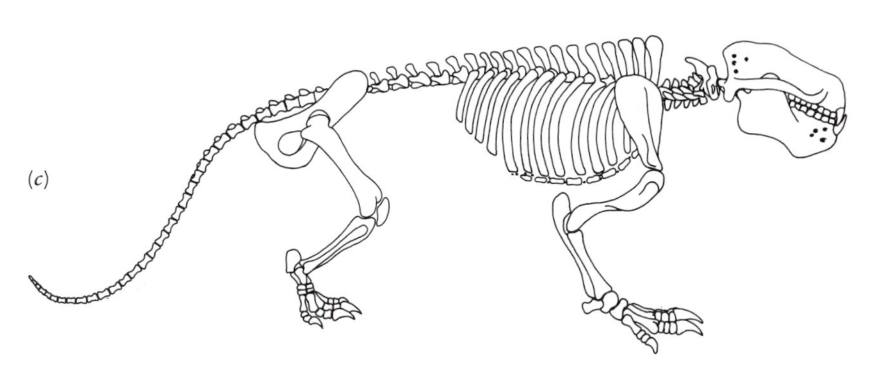

Figure 20-14. THE ADVANCED TAENIODONT *STYLINODON.* (*a*) Skull in lateral view. (*b*) Upper and lower dentition in occlusal view. (*c*) Skeleton the size of a bear. *From Schoch, 1982.*

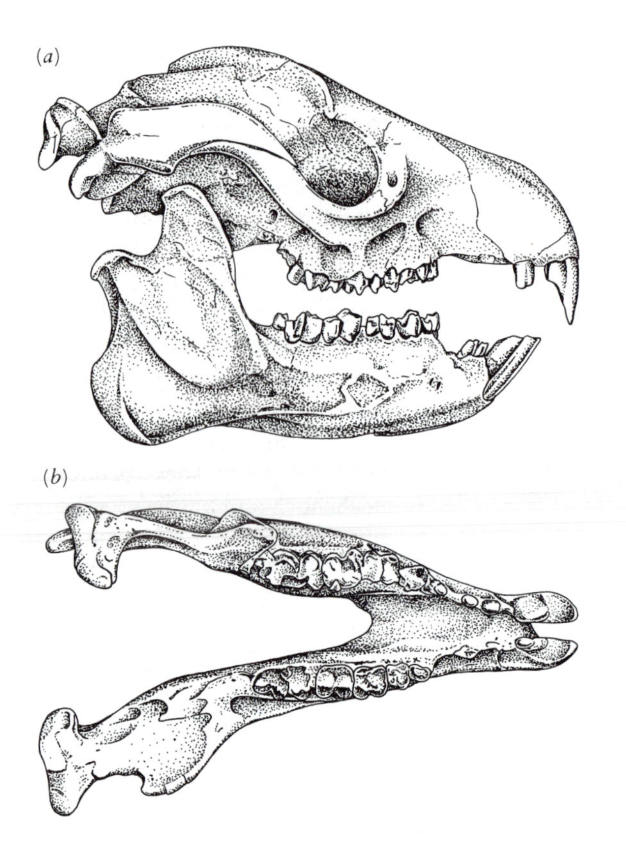

Figure 20-15. (*a*) Skull of the Eocene tillodont *Trogosus,* 30 centimeters long. (*b*) Lower jaw of *Tillodon* in occlusal view. *From Gazin, 1953. By permission of Smithsonian Institution Press, Smithsonian Institution, Washington, D.C.*

other anterior teeth are reduced or lost, leaving a wide diastema anterior to the cheek teeth. The buccal side of the lower cheek teeth and the lingual side of the uppers are arcuately columnar, while the opposite side bears low cusps. They are rapidly worn away, which suggests a highly abrasive diet. Gazin (1953) hypothesized a rodentlike anterior-posterior excursion of the mandible.

The skull is primitive in the small size of the braincase and absence of a postorbital bar, but the basicranium is shorter then in primitive eutherians.

The primitive Upper Paleocene and Lower to Middle Eocene genus *Esthonyx* (Figure 20-16) (Gingerich and Gunnell, 1979; Stucky and Krishtalka, 1983) shows the initial elaboration of the second incisors, but a diastema has not yet developed. When unworn, the cheek teeth show a primitive pattern that is not far removed from Upper Cretaceous eutherians, except for greatly expanded hypoconal and periconal shelves.

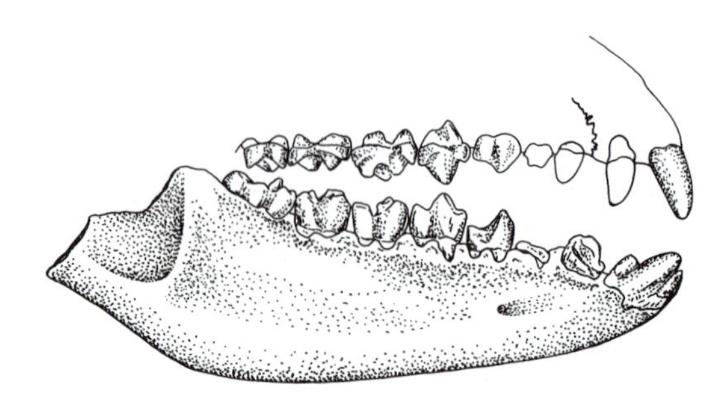

Figure 20-16. *ESTHONYX.* The dentition of the Lower Eocene tillodont, $\times\frac{2}{3}$. *From Gazin, 1953. By permission of Smithsonian Institution Press, Smithsonian Institution, Washington, D.C.*

Tillodonts first appear in the Lower or Middle Paleocene of Asia, where they may have originated (Chow, Chang, Wang, and Ting, 1973; Zhou and Wang, 1979). They linger there into the Upper Eocene. They are known in North America from the Upper Paleocene into the Middle Eocene, and in Europe, only in the Lower Eocene.

Their origin has been sought among condylarths, anagalids, and with the pantodonts (Gingerich and Gunnell, 1979). Cifelli (1983), following Gazin (1953), suggested that they may have originated from a genus such as *Deltatherium* (Figure 20-17). This genus is typically placed within the condylarths, but Cifelli pointed out that it is more primitive in many features of its dentition and corresponds more closely with ancestral eutherians. The upper molar stylar shelf is still wide, with large parastylar and metastylar lobes. The protocone is anterior in position, and the trigonid is still substantially higher than the talonid. Derived features shared by tillodonts and *Deltatherium* include reduction of the entoconid and upper

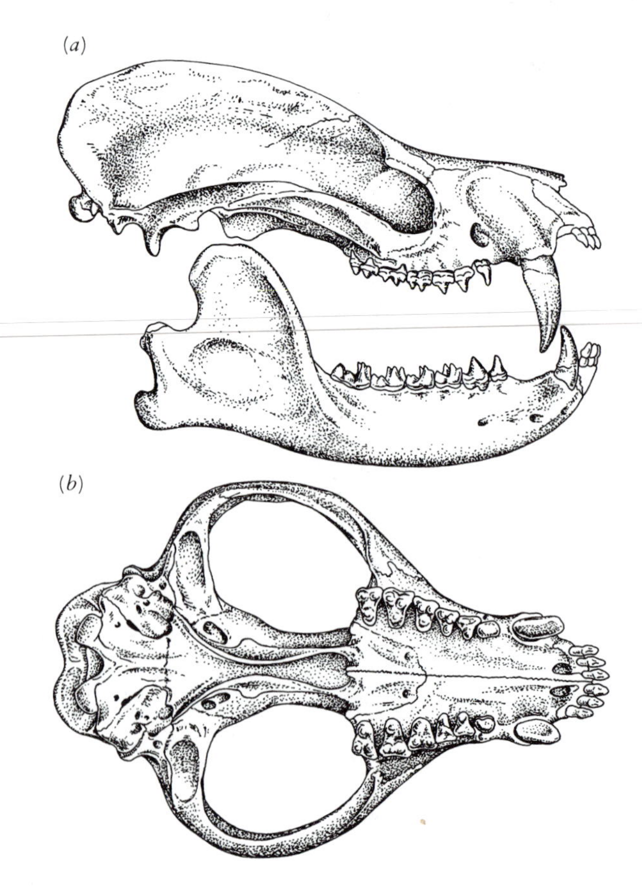

Figure 20-17. SKULL OF THE PRIMITIVE EUTHERIAN *DELTATHERIUM.* (*a*) Lateral and (*b*) palatal views. This genus has long been classified among the ungulates, but the dentition retains the features of "proteutherians." *From Matthew, 1937.*

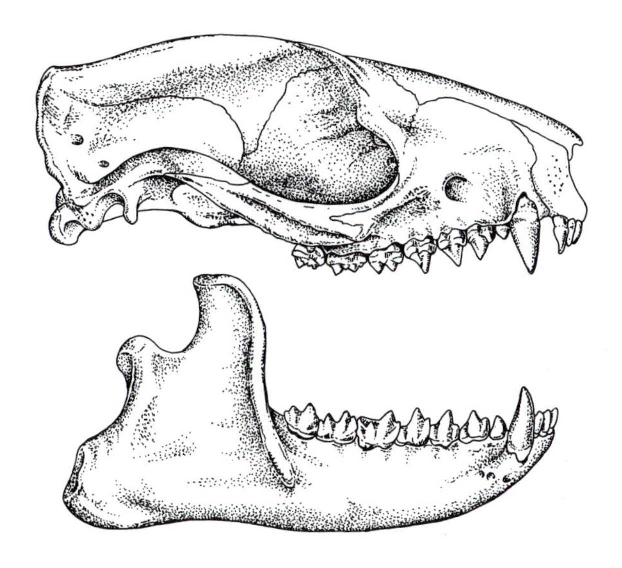

Figure 20-18. *PANTOLAMBDA.* Skull of the primitive pantodont is 15 centimeters long. *From Matthew, 1937.*

molar conules and hypertrophy of the upper molar parastylar and metastylar shear surfaces.

Zhou and Wang (1979) also contend that tillodonts had an ancestry that was distinct from condylarths and other ungulates. They argued strongly for close affinities with a second group of archaic herbivores, the pantodonts. Cifelli (1983) also suggested pantodont affinities.

The pantodonts were a much more diverse and well-known assemblage. Eleven families are recognized from the Paleocene and they extend into the Lower Oligocene in Asia. Early knowledge was based primarily on large forms from North America, including the sheep-sized *Pantolambda* (Figure 20-18) and the rhinoceros-sized *Titanoides* (Figure 20-19) and *Coryphodon.* The limbs are short and stout. The feet are generally primitive, with five spreading digits. *Titanoides* bears claws, but other genera have small hoofs.

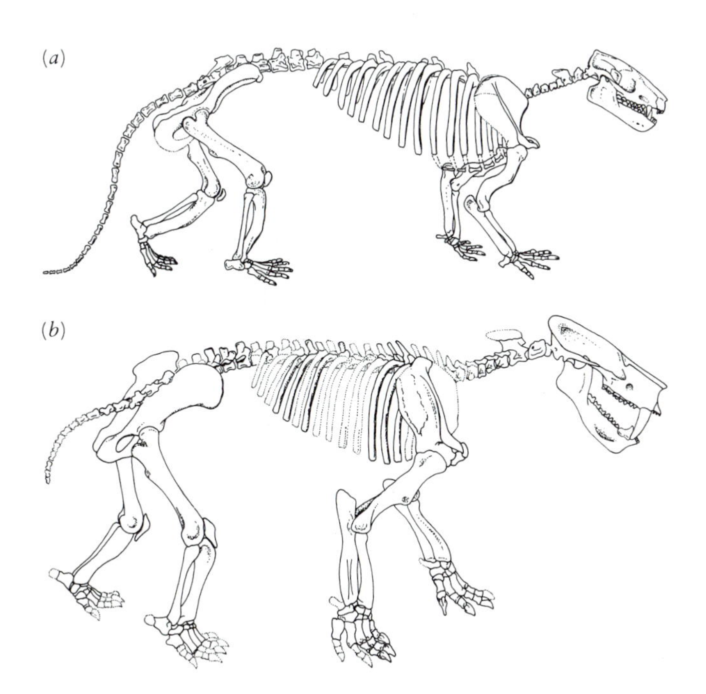

Figure 20-19. PANTODONTS. (*a*) Skeleton of *Pantolambda,* $\times\frac{1}{12}$. (*b*) Skeleton of the clawed pantodont *Titanoides,* $\times\frac{1}{16}$. *From Simons, 1960.*

Our knowledge of pantodonts has recently been greatly expanded by the discovery in Asia of a number of smaller forms, some as little as a rat (Chow, Chang, Wang, and Ting, 1977). *Bemalambda* is one of the best-known members of the order. Another genus may have had a tapirlike proboscis (Ting, Schiebout, and Chow, 1982).

The dentition is primitively complete without a diastema. All genera can be recognized by the V-shaped lophs on upper premolars 3 and 4, and the W-shaped pattern of the molar lophs. The teeth show relatively little wear, which suggests a nonabrasive diet (Coombs, 1983).

Cifelli suggested that the pantodonts and tillodonts evolved separately from forms like *Deltatherium.* Zhou and Wang (1979) suggested a common ancestry in the Upper Cretaceous. McKenna (1975) traced pantodonts directly to *Cimolestes,* in common with taeniodonts, but associates tillodonts with primitive ungulates.

DINOCERATA (UINTATHERES)

The uintatheres include approximately 15 genera of large, graviportal herbivores that have also been allied with the pantodonts. They were known initially from North America (Wheeler, 1961) but much material has recently been recognized from Asia (Tong and Lucas, 1982). They did not occur in Europe.

Uintatheres first appeared near the end of the Paleocene and were common in the Eocene but did not survive the end of that epoch. The size and general proportions of the postcranial skeleton resemble those of the pantodonts; the carpals and tarsals are alternating and the fibula does not articulate with the calcaneum (Figure 20-20).

The skull is very distinctive. In most genera, it bears a number of bony protuberances and has greatly elongated canine teeth. The upper incisors are typically missing. The cheek teeth are distinguished by the presence of a V-shaped crest on P^3-M^3, with the protocone at the apex (Figure 20-21). The lower molars have high trigonids with a prominent metalophid. There is no hypolophid, but this crest is primitively present in pantodonts.

Van Valen (1978) proposed that the Dinocerata might be derived from loxolophine arctocyonids (which are close to the ancestry of perissodactyls). In contrast, Tong and Lucas suggested affinities with the anagalids, which are otherwise thought to be related to the ancestry of rodents, lagomorphs, and elephant shrews. McKenna (1980) pointed out resemblances to the Paleocene xenungulates among the South American ungulates. Xenungulates (which are discussed in Chapter 21, see Figure 21-82) have bilophodont molars but retain nonmolariform premolars.

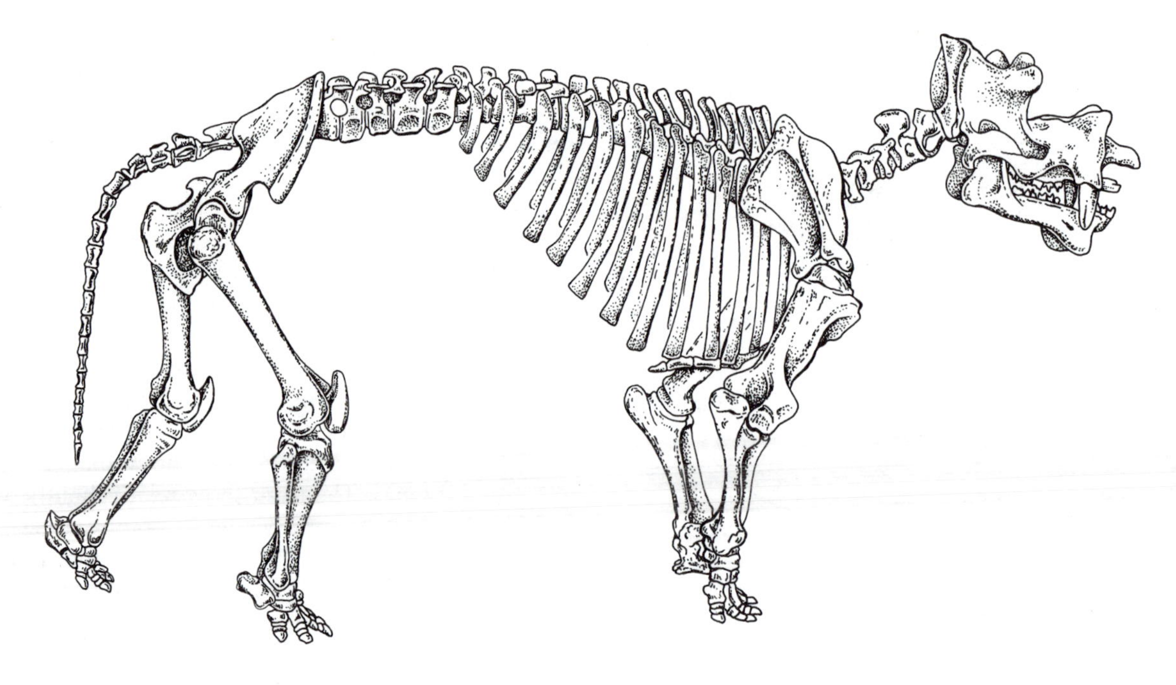

Figure 20-20. SKELETON OF THE UINTATHERE *DINOCERAS*. *From Gregory, 1957.*

Wheeler accepted that no known uintathere could be derived from known pantodonts but argued that the similarity of the feet, especially the astragalus, and the dentition could be attributed to common ancestry at a more primitive level. He also noted specific similarities to the Xenungulata.

We hope that fossils from the early and middle Paleocene will eventually provide substantial evidence of a link between uintatheres and some other group of early placentals.

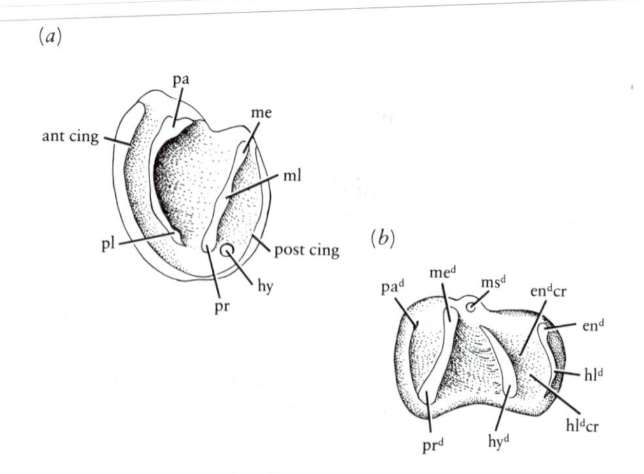

Figure 20-21. CUSP PATTERN OF THE MOLAR TEETH OF A UINTATHERE. (*a*) Left upper molar. (*b*) Left lower molar. Abbreviations: ant cing, anterior cingulum; en^d, entoconid; en^d cr, entoconid crest; hl^d, hypoconulid; hl^d cr, hypoconulid crest; hy, hypocone; hy^d, hypoconid; me, metacone; me^d, metaconid; ml, metaloph; ms^d, metastylid; pa, paracone; pa^d, paraconid; post cing, posterior cingulum; pr, protocone; pl, protoloph; pr^d, protoconid. *From Wheeler, 1961.*

The remaining orders that are considered in this chapter, which contribute to the modern fauna, also appear to have their origin near the base of the initial placental radiation.

INSECTIVORA

Biologists have long recognized that shrews, moles, hedgehogs, and their less-well-known relatives—the tenrecs and African golden mole—are among the most primitive living placentals. The brain is small, and the cerebral hemispheres are smooth and do not expand over the cerebellum. The testes are abdominal, inguinal, or in a sac in front of the penis. Some genera retain a cloaca. The skull is primitive in the absence of a postorbital bar and the auditory bulla is rarely ossified. The number and configuration of the teeth in primitive genera resemble the pattern in the early "Proteutheria." The feet are usually plantigrade and pentadactyl, with the pollex and hallux not opposable.

Butler (1972) and Novacek (1980) demonstrated that these modern groups belong to a monophyletic order that is clearly distinct from the primitive "Proteutheria" and from tree shrews and elephant shrews, which were once included within the Insectivora as the Menotyphla.

The following are shared derived characters of the Insectivora as so restricted. The pattern of the skull wall medial to the orbit is modified in association with the small size of the eye and the relatively large nasal capsule. The maxilla is widely expanded, but the palatine is much reduced. The lacrimal has no facial wing. The orbitosphenoid is anterior to the braincase. The jugal is absent or markedly reduced. In several groups, the zygomatic arch is lost or incomplete. A postglenoid foramen is usually present for passage of the external jugular vein. There is a large pyriform fenestra between the otic capsule and the basisphenoid and the bony dorsum sellae is absent. The pubic symphysis is reduced.

Butler (1972) recognized four suborders of insectivores: the Erinaceomorpha (hedgehogs); the Soricomorpha, including *Solenodon* (a modern genus that is known only in the West Indies), moles (Talpidae), and shrews (Soricidae); the Tenrecomorpha (the tenrecs); and the Chrysochlorida, (the golden mole). These groups are also known as the Lipotyphla, in reference to the absence of an intestinal caecum. McKenna (1975) and Dawson and Krishtalka (1984) united all families other than the hedgehog in a single suborder, Soricomorpha.

These authors recognize *Batodon* (Figure 20-22) from the Upper Cretaceous of North America as the earliest-known soricomorph on the basis of its dentition. Three or four families from the early Cenozoic are placed in this suborder. The Lower to Mid-Paleocene genus *Prosarcodon* (family Micropternodontidae) provides the oldest known cranial remains of this group (McKenna, Xue, and Zhou, 1984). It shows affinities with later lipotyphlans in having a large pyriform fenestra separating the basisphenoid from the petrosal, and the glenoid fossa of the squamosal faces forward as in soricoids. The Micropternodontidae do not show affinities with later groups of soricomorphs, but McKenna and associates (1984) suggested that they are closely related to the palaeoryctids.

The closest affinities of *Batodon* among the early Cenozoic soricomorphs apparently lie with the Geolabididae, a family that extends into the Miocene. According to Lillegraven, McKenna, and Krishtalka (1981), the geolabidids are close to the soricoid (other than the talpids). This relationship is supported by the transverse expansion of the condyle of the dentary, which is not seen in either

(a)

(b)

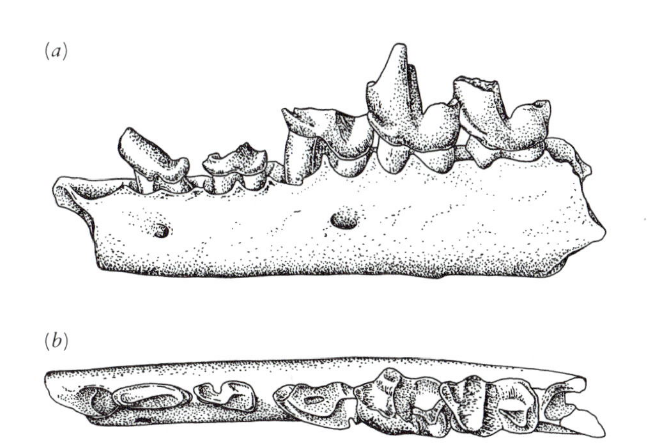

Figure 20-22. *BATODON.* Lower jaw of the possible soricomorph insectivore from the Upper Cretaceous. (*a*) Lateral and (*b*) occlusal views, ×3. *From Clemens, 1973.*

(a)

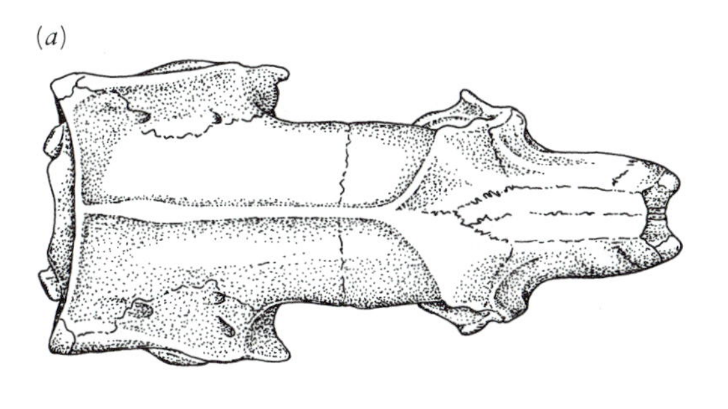

(b)

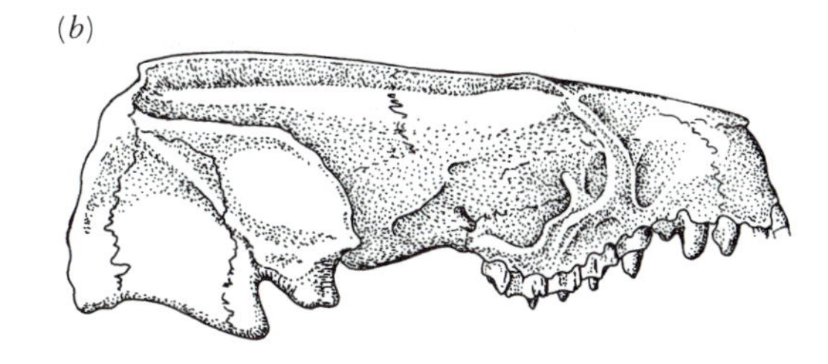

(c)

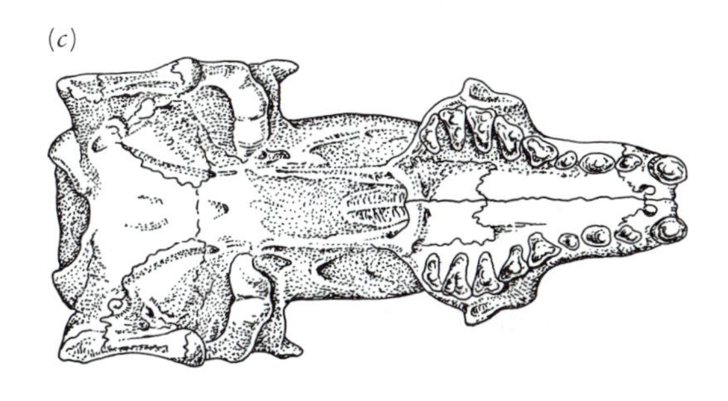

Figure 20-23. SKULL OF *APTERNODUS,* A PRIMITIVE INSECTIVORE. (*a*) Dorsal, (*b*) lateral, and (*c*) palatal views. *From McDowell, 1958.*

moles or erinaceids. As in soricids and *Solenodon,* the deciduous dentition appears to be very rapidly replaced.

The Eocene to Oligocene family Apternodontidae (Figure 20-23) may have risen from, or in common with, the Geolabididae. According to Dawson and Krishtalka (1984), apternodontids include the ancestors of the tenrecs and the golden mole, which are known as early as the Lower Miocene in East Africa (Butler, 1978). In contrast with Lillegraven and his coauthors, Dawson and Krishtalka suggest that *Solenodon* (which lacks a fossil record) may have arisen from the Apternodontidae rather than from the Geolabididae.

Tenrecids, chrysochlorids, and *Solenodon* are all characterized by a distinctive pattern of the upper molars, termed zalambdodonty, in which the crowns are in the shape of a narrow V and the metacone is lost (Figure 20-24a). A vaguely similar tooth shape is also present in *Zalambdalestes, Palaeoryctes,* and in the marsupials *Necrolestes* and *Notoryctes.* McDowell (1958) provided strong evidence that this pattern evolved independently several times among the Insectivora.

We find the oldest members of the modern families Soricidae and Talpidae in the late Eocene (Krishtalka and Setoguchi, 1977; Sigé, Crochet, and Iusole, 1977). The

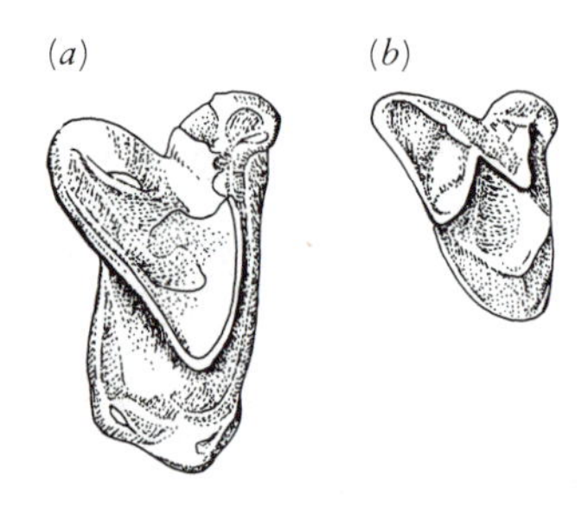

Figure 20-24. ZALAMBDONTY AND DILAMBDONTY AS ILLUSTRATED BY THE UPPER MOLARS OF INSECTIVORES. (*a*) The zalambdont molar of *Solenodon* and (*b*) the dilambdont molar of *Nesophontes*. *From McDowell, 1958.*

shrews are first represented by *Domnina* and the moles by *Eotalpa* (Figure 20-25). In contrast with the views of Lillegraven and his colleagues, Dawson and Krishtalka (1984) suggest that both shrews and moles can trace their origin to the family Nyctitheriidae, which is known only from dental remains of mid-Paleocene to early Oligocene age. The molars of shrews, moles and nyctitheriids have a pattern termed dilambdodont, based on the W-shaped configuration of crests formed from the paracone and metacone on the lateral edge of the tooth crown (Figure 20-24*b*). Repenning (1967) and more recently George (1986) reviewed later Tertiary shrews and their relationships with living species.

The earliest known genus that may be included among the erinaceomorphs is *Paranyctoides* from the mid-Upper Cretaceous (Fox, 1984). *Litolestes* from the late Paleocene illustrates features that characterize the modern family Erinaceidae (Novacek, Bown, and Schankler, 1985). The molars are rectangular in occlusal outline and decrease in size posteriorly. In contrast with soricomorph insectivores, the cusps are low, which indicates their use in crushing and grinding rather than piercing. Von Koenigswald and Storch (1983) described complete and beautifully preserved skeletons of the erinaceomorph *Pholidocercus* from the Middle Eocene of Germany. The well-known Miocene genus *Brachyerix* (Figure 20-26) shows the basic cranial structure of this family.

Yates (1984) discussed the distribution of modern insectivore families.

Early erinaceomorphs and soricomorphs have a primitive pattern of the ear region, without ossification of the bulla, and a median position of the internal carotids as in the most primitive eutherians (Novacek, McKenna, Neff, and Cifelli, 1983; McKenna, Xue, and Zhou, 1984). The dentition of the oldest-known insectivore, *Paranyctoides,* is similar to that of *Cimolestes incisus,* which McKenna (1975) conjectured was close to the ancestry of carnivores, primates, and ancestral ungulates, although he was not able to establish more specific interrelationships among these groups. No more recently discovered material has clarified this problem.

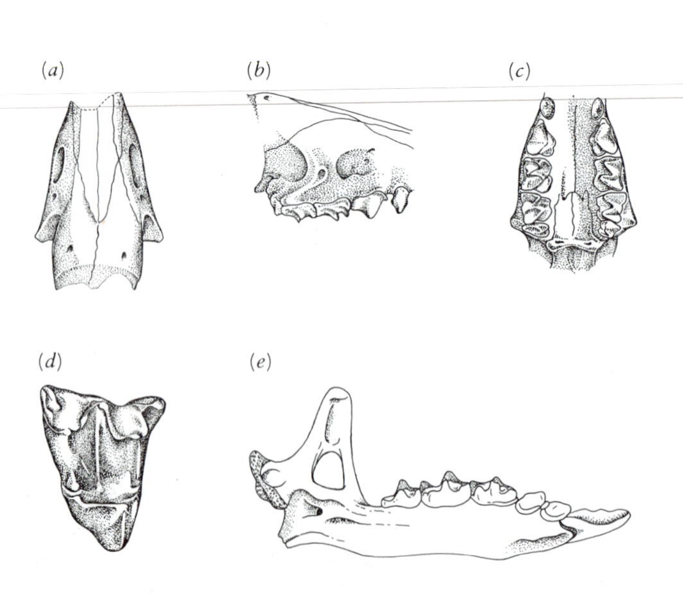

Figure 20-25. SKULL OF *DOMNINA*, A LOWER OLIGOCENE SHREW. (*a*) Dorsal, (*b*) lateral, and (*c*) palatal views. *From McDowell, 1958.* (*d*) Molar tooth of the oldest recognized mole *Eotalpa* from the late Eocene of Europe. *From Sigé, Crochet, and Insole, 1977.* (*e*) Jaw of *Limnoecus*, a Miocene shrew. Note long procumbent incisor. *From Repenning, 1967.*

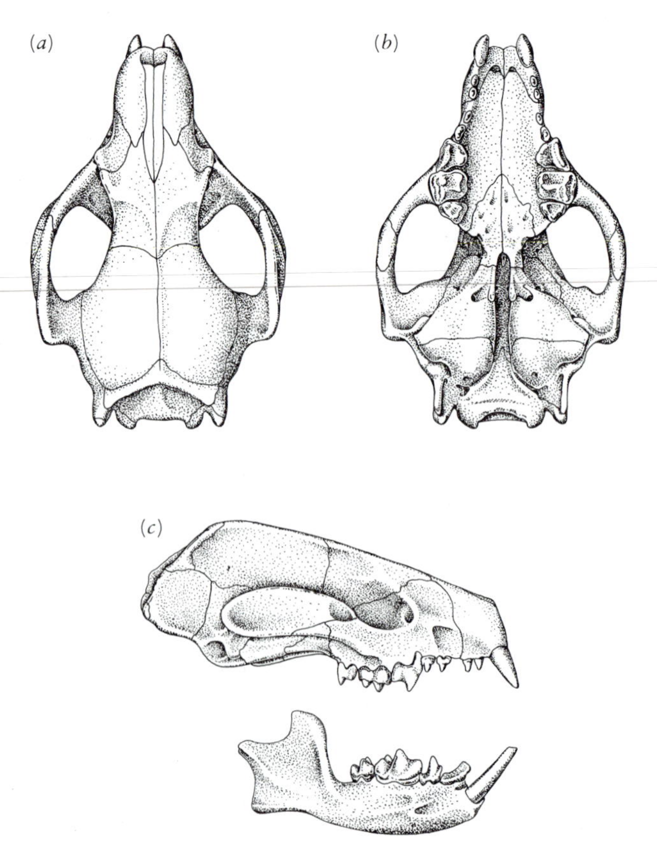

Figure 20-26. *BRACHYERIX,* A MIOCENE HEDGEHOG FROM NORTH AMERICA. Skull in (*a*) dorsal, (*b*) palatal, and (*c*) lateral views, 3 centimeters long. *From Rich, 1981.*

BATS

Bats are among the most specialized of modern mammals. All are accomplished flyers, and the insectivorous microchiropterans have a highly developed sonar that enables them to hunt insects in the dark. Like the pterosaurs, the flight structure of bats was already highly evolved when they first appeared in the fossil record. The oldest skeleton of a bat, *Icaronycteris* from the early Eocene, appears almost indistinguishable from living bats (Figure 20-27).

The teeth of earlier bats, from the Paleocene of France, which Russell, Louis, and Savage (1973) described, show a pattern that could be readily derived from that of Upper Cretaceous eutherians such as *Cimolestes,* with transversely expanded upper molars and the talonid well lower than the trigonid. The presence of a W-shaped ectoloph corresponds closely to the pattern of mid-Paleocene to late Eocene nyctitheriid soricomorphs (Figure 20-28). This finding provides the strongest evidence for specific insectivore-bat affinities and is not contradicted by other skeletal features.

In the absence of adequate descriptions of the skeletons of primitive insectivores or early Paleocene "proteutherians," we cannot establish the specific skeletal pattern from which bats arose. Jepsen (1970) argued that it is impossible to specify when the lineage leading to bats diverged from primitive eutherians, but a maximum age of 75 million years might be hypothesized on the basis of the level of dental specialization known in the Asian and North American Upper Cretaceous eutherians. On the other hand, the divergence might have been as late as the Middle or Upper Paleocene, if the dental similaries of early bats and primitive soricomoph insectivores reflect a close common ancestry. The evolution of the distinctive chiropteran skeleton may have occurred over a period as long as 25 million years or as short as 8 million years.

Whenever the divergence of bats began, they reached the most highly specialized and most modern grade of evolution of any Eocene eutherians. They had nearly completed their skeletal evolution when primates and horses had only just begun.

Although it is not possible to state how rapid the rate of change was during the early stages of bat evolution, it is certain that changes since the early Eocene have been extremely slow. *Icaronycteris* had already evolved an essentially modern forelimb, with great elongation of the humerus, partial fusion of the ulna with the radius, and great elongation of digits 2 through 5. The scapula has assumed a dorsal position. As in modern bats, most of

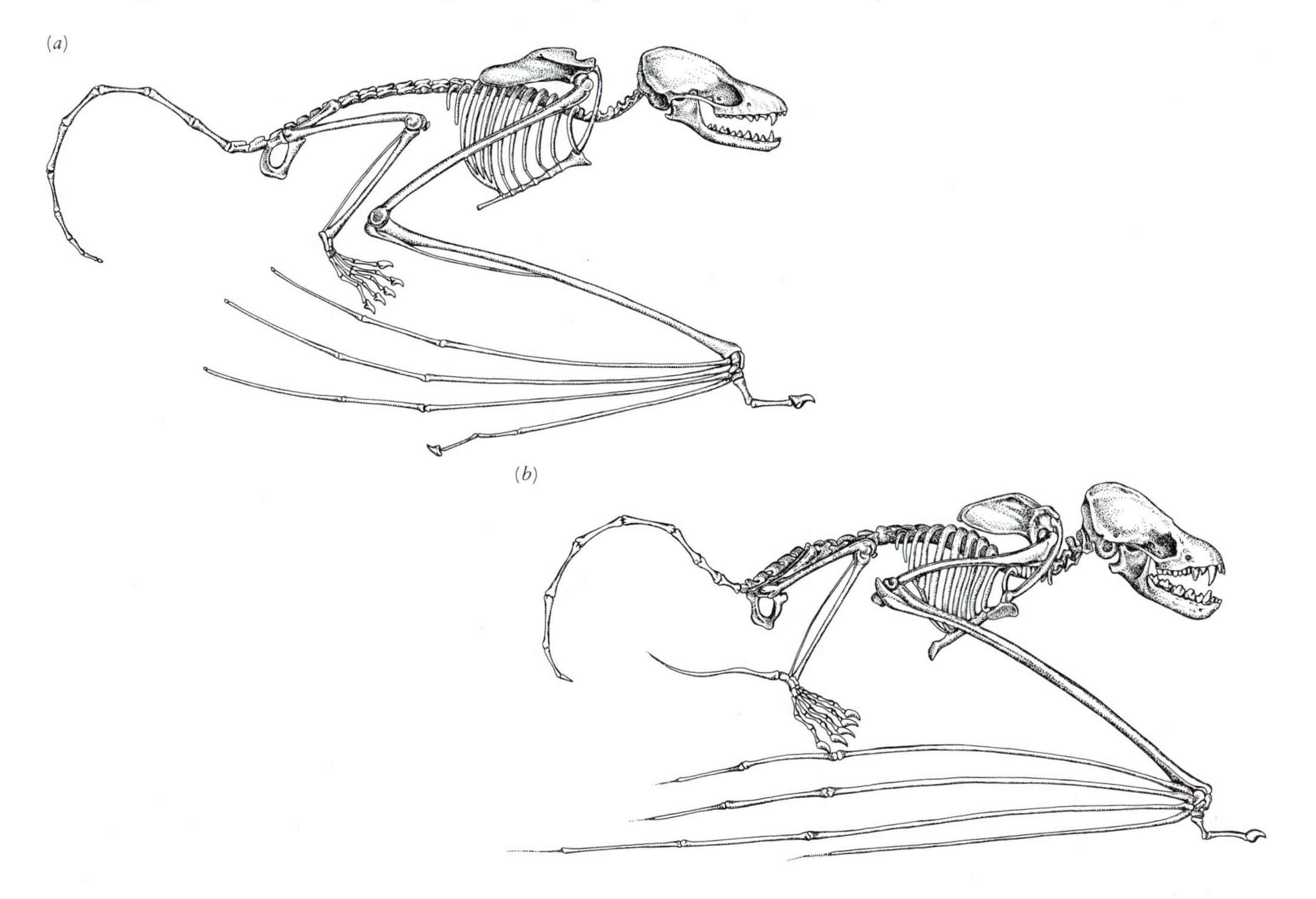

Figure 20-27. BATS. (*a*) The early Eocene bat *Icaronycteris.* (*b*) The modern bat *Myotis.* Both ×1. *From Jepsen, 1970.*

(a)

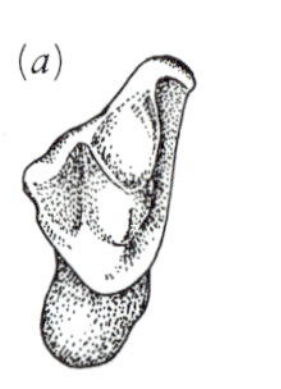

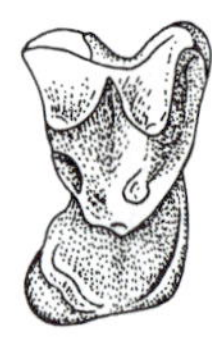

(b)

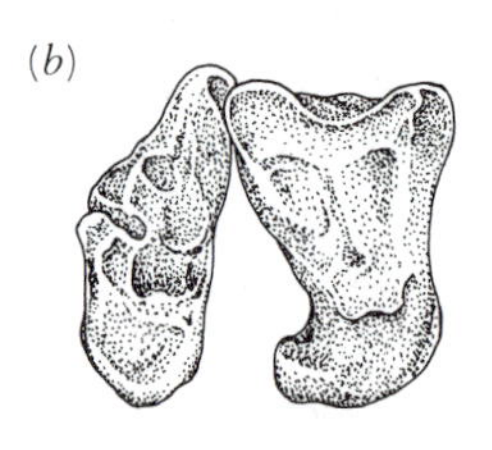

Figure 20-28. (*a*) Second and third upper right molars of the Eocene bat *Icaronycteris.* (*b*) Second and third upper right molars of the Eocene insectivore *Nyctitherium.* (*a*) *From Russell, Louis, and Savage, 1973.* (*b*) *From Krishtalka and Setoguchi, 1977.*

the cranial sutures are obliterated. A relatively small number of features are less well developed than in modern bats. The dental formula of

$$\frac{2\ \ 1\ \ 3\ \ 3}{3\ \ 1\ \ 3\ \ 3}$$

includes one or more teeth that are lost in all modern bats. The ribs have not coalesced and the elements of the sternum are neither fused nor keeled. A primitive phalangeal count of 2, 3, 3, 3, 3 is retained in both the forelimbs and rear limbs. The terminal phalanges of digits 2 through 5 of the manus are very short and blunt, whereas they are more slender in modern bats.

In most features of its anatomy, *Icaronycteris* resembles the modern microchiropteran bats. Work by Novacek (1985) indicates that the structure of the auditory region already shows specializations associated with echo locating.

Jepson originally classified *Icaronycteris* within the Microchiroptera. Some features of this genus, including the presence of a suture separating the premaxillae and a claw on the index finger, are shared with megachiropterans, but these features are clearly primitive and would be expected in the ancestors of all bats. In contrast, the structure of the shoulder girdle seems advanced in the direction of the microchiropterans.

A host of derived features of both the skeleton and soft anatomy establishes that the two major groups of modern bats share a close common ancestry, but they may have diverged prior to the appearance of the earliest adequately known bats in the early Eocene.

The fossil record of the modern bat families remains very incomplete prior to the Pleistocene. Only two fossil genera are recognized as belonging to the Megachiroptera: *Archaeopteropus* from the early Oligocene of Europe and *Propotto* from the early Miocene of Africa (Butler, 1978).

The known ranges of the modern families are shown in Figure 20-29. The living genera *Myotis, Rhinolophus, Hipposideros, Tadarida,* and *Taphozous* are known from the Oligocene, and *Vespertilio, Minioptera* and *Eptesicus* appear in the Miocene (Dawson and Krishtalka, 1984).

Primates

The primates, including the lemurs, tarsioids, monkeys, apes, and man, have a rich and intensively studied fossil record going back to the early Cenozoic. The modern assemblage can be traced with little question to the base of the Eocene, at which time the members can be readily distinguished from other early eutherians by the presence of a postorbital bar, the relatively large size of the braincase, formation of an auditory bulla by elaboration of the petrosal, and details of dental anatomy.

More primitive genera, the plesiadapiforms, were common in the Paleocene. Their molar teeth closely resemble those of later primates, but the remainder of the anatomy is quite distinct. There is continued debate as to whether they should be included within the same order as the modern primates (e.g., MacPhee, Cartmill, and Gingerich, 1983), but there is no definite evidence that precludes early plesiadapiforms from being ancestral to later primates and no other group shares such close dental similarities.

Szalay and Nelson (1979) provided the most comprehensive and detailed review of the skeletal anatomy of both Recent and fossil primates. Gingerich (1984) and Badgley (1984) briefly covered more recent literature. Many recent articles on the anatomy of fossil primates have appeared in *The American Journal of Physical Anthropology.* See also Wood, Martin, and Andrews (1986).

Several different schemes for taxonomic subdivision of the primates are used by current authors. Many authors recognize a major subdivision into two units, the Prosimii, including the tarsioids, lemuroids, and their extinct relatives, and the Anthropoidea—monkeys, apes, and man. The Plesiadapiformes may be included within the Prosimii or treated as a third group of equivalent rank. Other authors, notably Szalay and Delson (1979), group the lemuroids and their extinct relatives in the suborder Strepsirhini and include the tarsioids and anthropoids in the suborder Haplorhini. There is general agreement that the South American monkeys should be placed in an infraorder Platyrrhini of equal taxonomic rank with the Old World monkeys, apes, and man—the Catarrhini (Figure 20-30).

The specific origin of primates among more primitive eutherians has not been established, although they could have evolved from animals with a molar pattern like that of *Cimolestes* (see Figure 20-9) (McKenna, 1975; Kielan-Jaworowska, Bown, and Lillegraven, 1979). No specific derived characters have been demonstrated as being

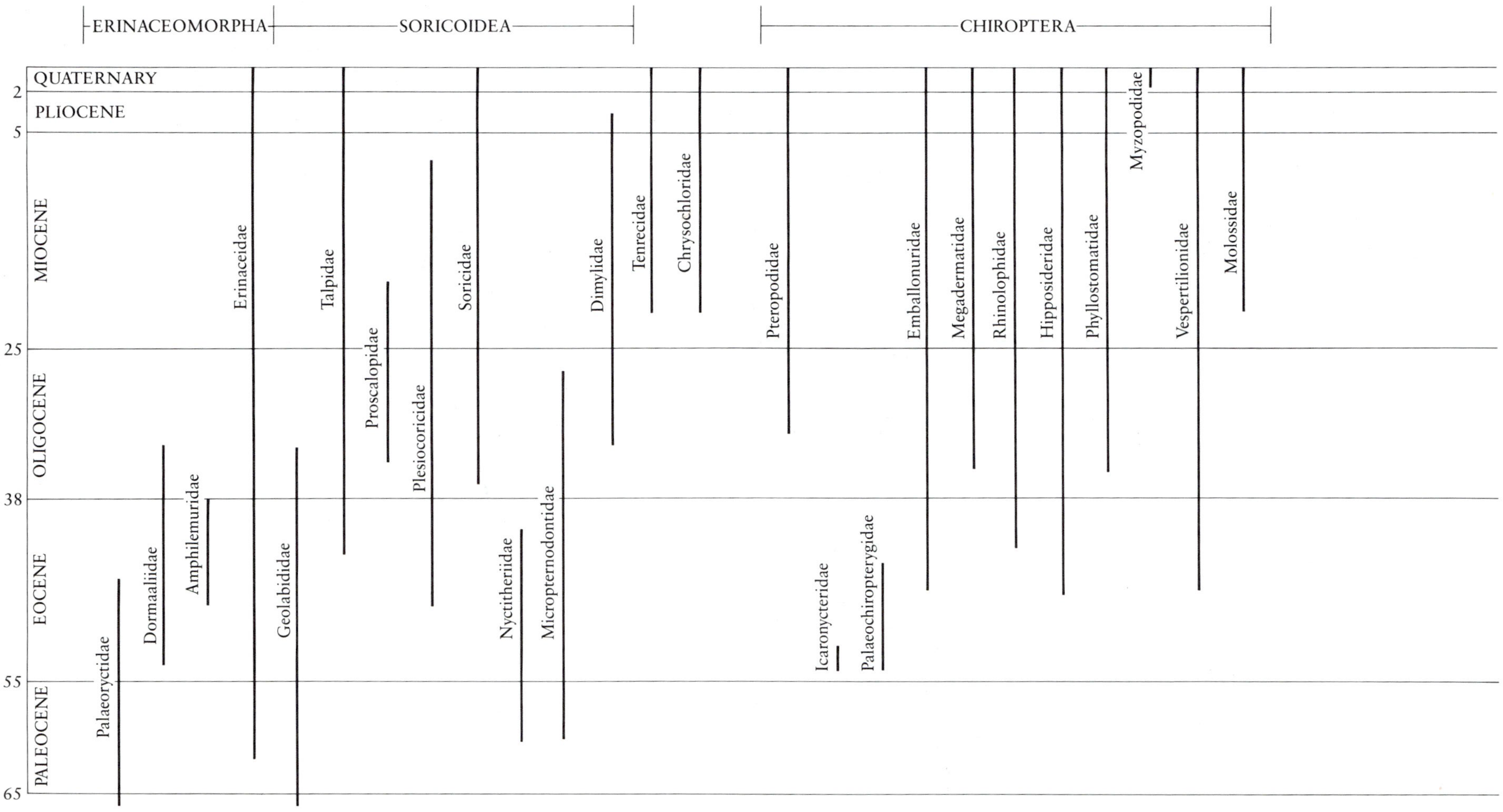

Figure 20-29. FAMILY RANGES OF INSECTIVORES AND BATS.

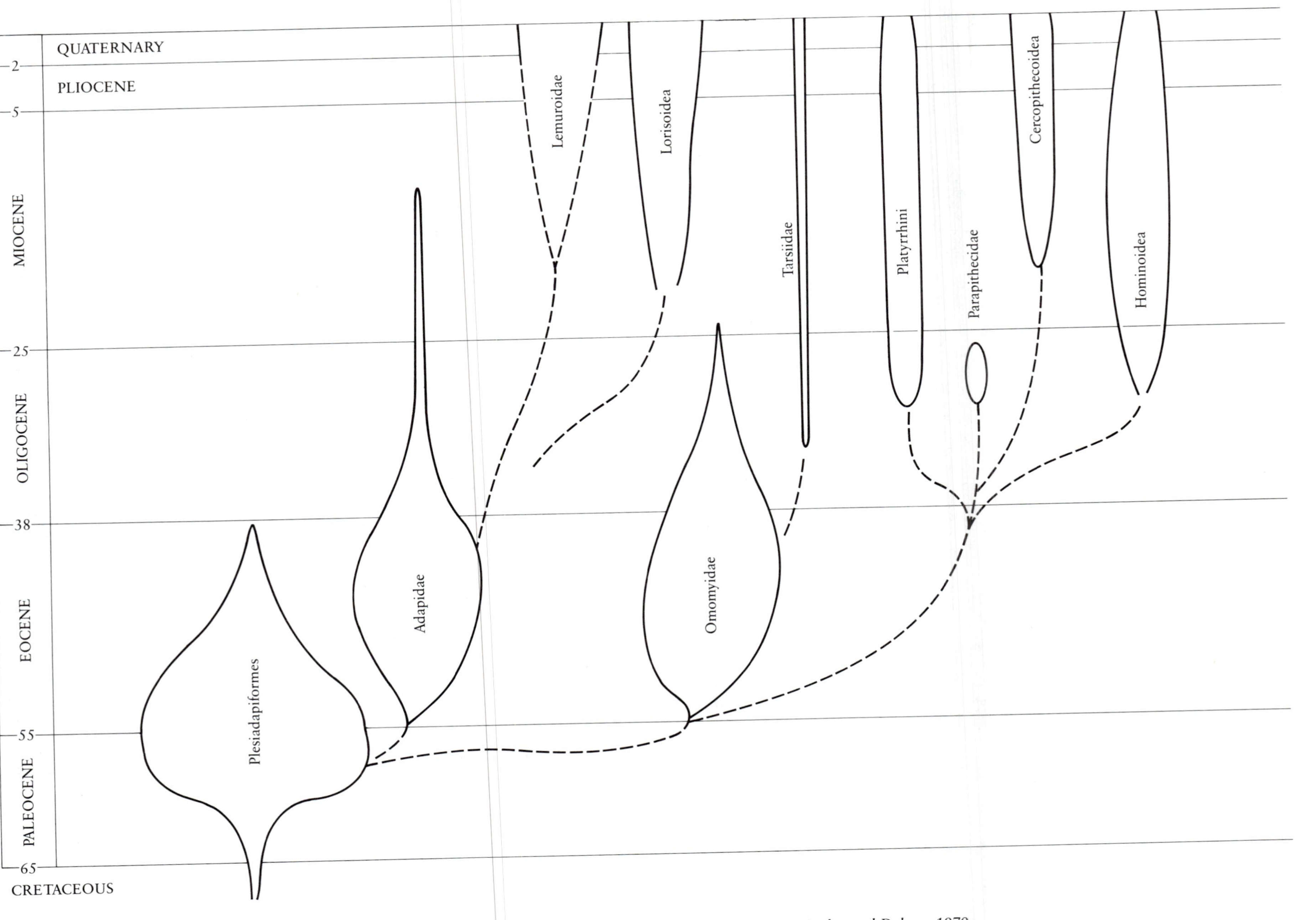

Figure 20-30. LONGEVITY OF PRIMATE GROUPS. *Data taken from Szalay and Delson, 1979.*

uniquely shared between early primates and the early members of any other order (Novacek, McKenna, Neff, and Cifelli, 1983).

PLESIADAPIFORMES

A single lower molar from the late Cretaceous of Montana provides the oldest fossil evidence of the primates. This tooth is distinguished from those of other primitive eutherians by the relatively blunt cusps and its squarish outline (Figure 20-31). It is almost identical with the molars of the genus *Purgatorius* from the Lower Paleocene, which is known from a nearly complete dentition with a formula of

$$\frac{3\ 1\ 4\ 3}{3\ 1\ 4\ 3}$$

Aside from the very small size (estimated body weight of 20 grams), we have little evidence of the overall form of these early primates.

Only incomplete remains are known from the early Paleocene, but genera with a similar pattern of the molar teeth become very important elements of the fauna in North America and Europe from the Middle Paleocene into the Middle Eocene. We recognize four or five quite distinct families, all of which are grouped within the suborder Plesiadapiformes. This assemblage may include the ancestors of all higher primates. The most primitive genus in which the skull can be reconstructed is *Palaechthon,* a member of the superfamily Paromomyoidea from the middle Paleocene (Figure 20-32). It is approximately 4 centimeters long. The posterior margin of the orbit is constricted, but there is not a complete postorbital bar. The tooth row is continuous, without a diastema, but the third incisor and the first premolar are lost, giving a dental formula of

$$\frac{2\ 1\ 3\ 3}{2\ 1\ 3\ 3}$$

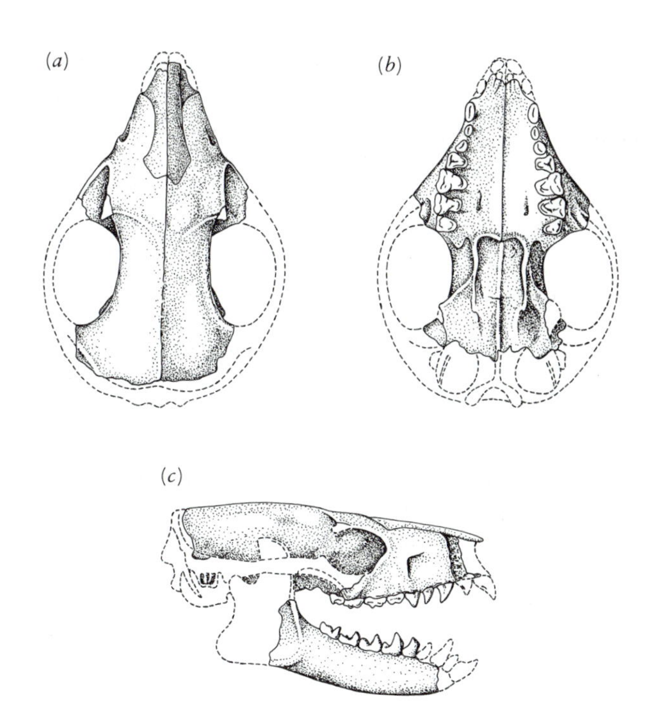

Figure 20-32. *PALAECHTHON.* Reconstruction of the skull of the paromomyoid primate from the Middle Paleocene. (*a*) Dorsal, (*b*) palatal, and (*c*) lateral views, 4 centimeters long. *From Kay and Cartmill, 1977.*

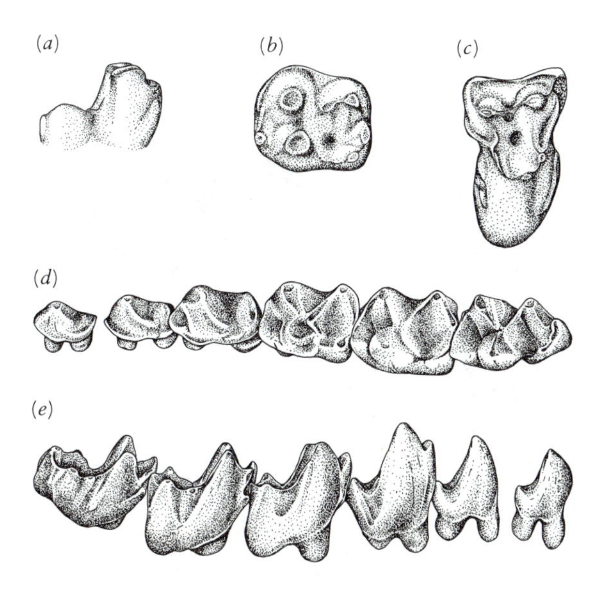

Figure 20-31. TEETH OF *PURGATORIUS.* (*a*) Buccal and (*b*) occlusal view of lower molar from the Upper Cretaceous. (*c*) Occlusal view of upper molar from the Lower Paleocene. (*d*) Occlusal and (*e*) buccal views of lower dentition from the Lower Paleocene. *Redrawn from Szalay and Delson, 1979.*

The canine is relatively short. The wider molars with blunt cusps imply an omnivorous diet, although the small body size suggests that they were probably insectivorous by analogy with small living species. In the late Paleocene-early Eocene genus *Ignacius,* the basicranial region is preserved, which provides a basis for comparison with the pattern in early members of other eutherian orders. The bulla is ossified but it is not possible to determine whether this ossification is developed from the petrosal, as in advanced primates, or from some other center of ossification (MacPhee, Cartmill, and Gingerich, 1983). The petrosal is also an important element in the bulla of most erinaceomorph insectivores but is not a primitive character of that group (Novacek, McKenna, Neff, and Cifelli, 1983). More certain is the absence of a groove for the promontory artery, which clearly distinguishes *Ignacius* and, by inference, other primitive primates from more advanced members of this assemblage.

Ignacius may be primitive in the configuration of the basicranial region, but the anterior dentition is highly specialized, with greatly enlarged and procumbent lower incisors and a long diastema between the upper canine and the cheek teeth. In these features, *Ignacius* resembles the second group of primates that was common in the Paleocene and early Eocene, the Plesiadapoidea. They are characterized by a superficially rodentlike dentition, with a long diastema between the procumbent incisors and the grinding molar teeth that suggests a herbivorous diet.

Nearly the entire skeleton is known in the advanced genus *Plesiadapis* (Figure 20-33). It would have weighed approximately 4 to 5 kilograms. The skeleton is primitive, showing neither loss nor fusion of elements. This feature is characteristic of primates, as opposed to advanced members of other orders. Gingerich (1976, 1984) argued that *Plesiadapis* was predominantly terrestrial, but Szalay and Delson (1979) considered that the skeleton was adapted to arboreal locomotion. Some derived features of the post-cranial skeleton are shared with more advanced primates, but they are probably the result of convergent specialization for arboreality.

STREPSIRHINI

Of the known Paleocene genera, only *Purgatorius* has a sufficiently primitive dentition to be considered as a possible ancestor of primates other than the Plesiadapiformes. Two major groups of more advanced primates appear at the base of the Eocene in Europe and North America, the Adapidae and the Omomyidae. Omomyids are also known at this time in Asia. Both groups exhibit advanced characteristics, including the presence of a postorbital bar and a higher degree of arboreal adaptation, which is indicated by the grasping hallux and pollex. The auditory capsule is marked by a grove for the promontory artery, which indicates that the internal carotid has shifted from the medial position common to primitive therians. In both groups, the stapedial artery is enlarged. In the adapids *Adapis*, *Notharctus*, and *Smilodectes*, the foramen for the stapedial artery is larger than that for the promontory artery, while in the omomyids *Necrolemur* and *Tetonius*, the reverse is true. Szaley and Delson (1979) suggest that the adapid condition is the more primitive of the two.

The adapids resemble the living Malagasy lemurs in size and skeletal morphology. The most primitive genus, *Cantius*, is represented by very common fossils of teeth and jaws in the lower Eocene of North America and is also known in Europe. The dental formula of

$$\frac{2\ \ 1\ \ 4\ \ 3}{2\ \ 1\ \ 4\ \ 3}$$

is more primitive in the retention of all the premolars than is that of the earliest paromomyoid or plesiadapoid known from cranial remains. *Cantius* shows the beginning of development of the hypocone to square the upper molars. Elements of the postcranial skeleton show specific similarities with modern lemurs in their specialization for arboreal leaping and grasping (Rose and Walker, 1984).

The skeleton of the middle Eocene adapid *Smilodectes* gives a good idea of this early stage in the evolution of lemuriform primates (Figure 20-34 and 20-35). The

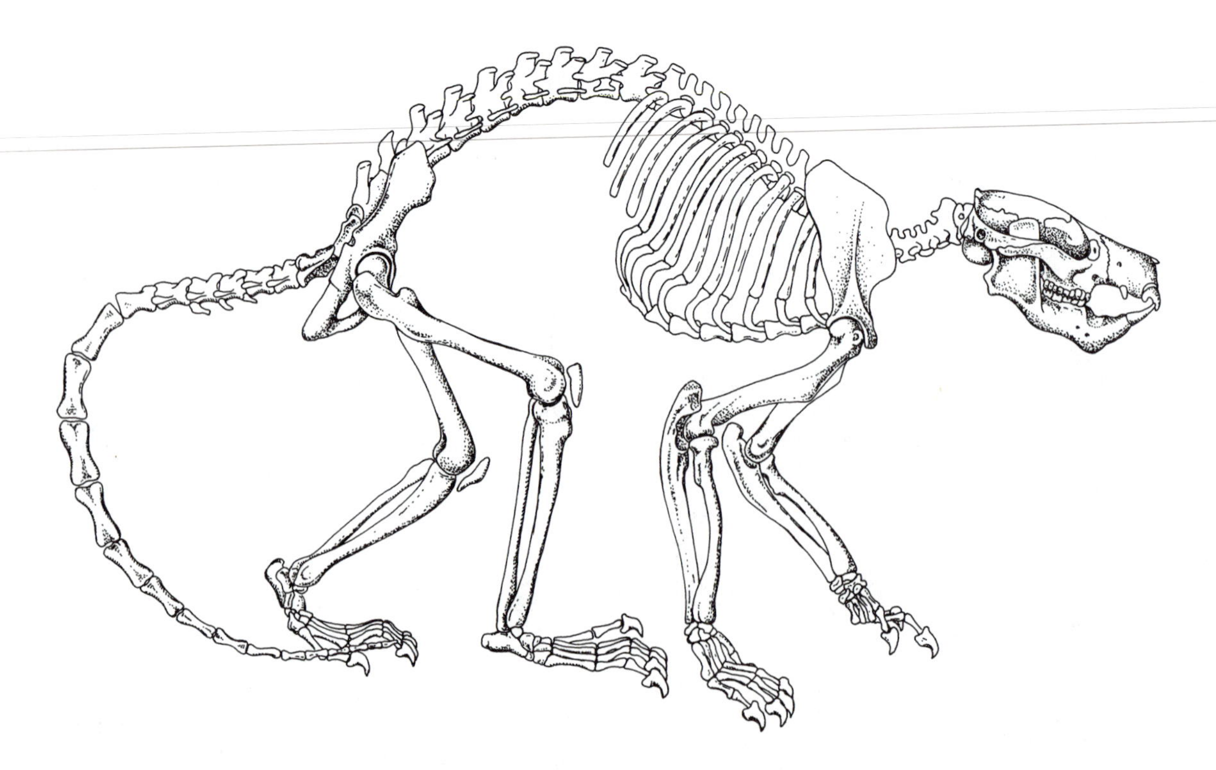

Figure 20-33. SKELETON OF THE COMMON LATE PALEOCENE PRIMATE *PLESIADAPIS*. *From Simons, 1964.*

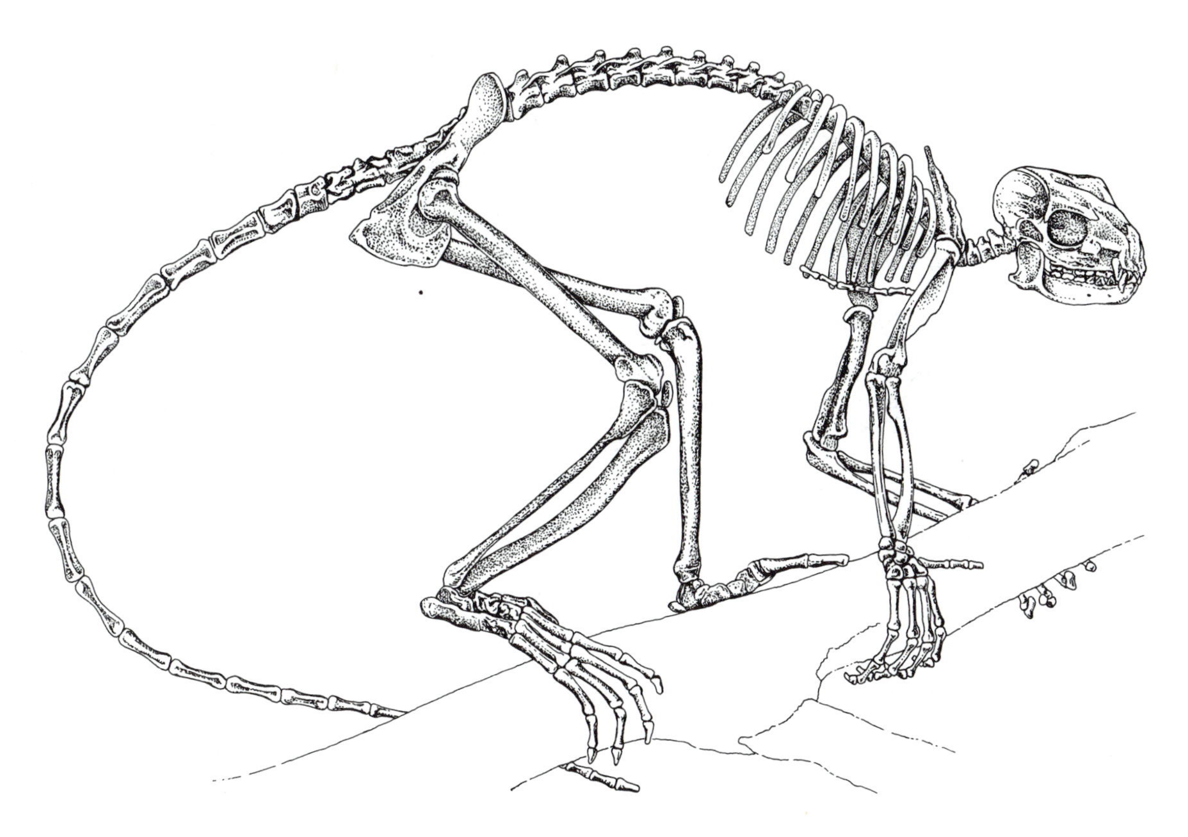

Figure 20-34. SKELETON OF THE MIDDLE EOCENE ADAPID *SMILODECTES.* This genus is close to the ancestry of the modern lemuriform primates. *From Simons, 1964.*

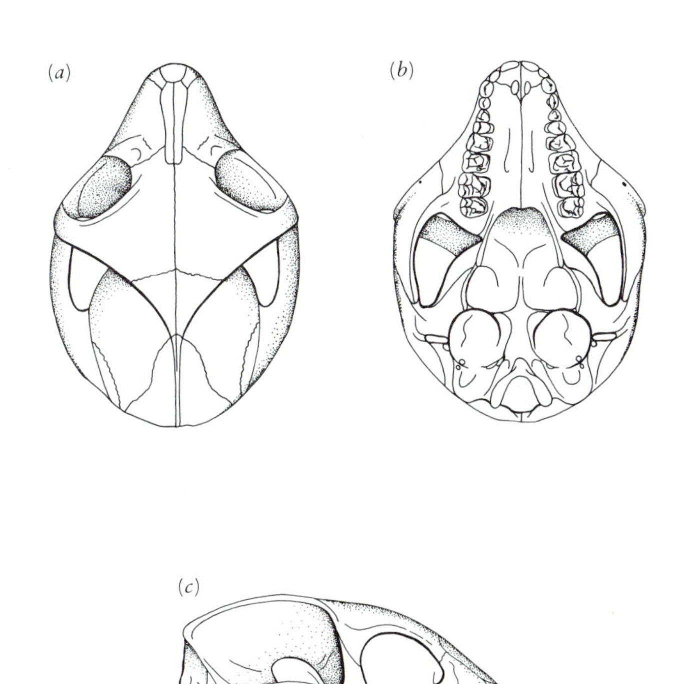

Figure 20-35. SKULL OF *SMILODECTES.* (*a*) Dorsal, (*b*) palatal, and (*c*) lateral views. *From Szalay and Delson, 1979.*

face is moderately long, and the temporal openings are divided by a sagittal crest over the small braincase. Adapids were most common in the Eocene, and approximately 20 genera have been described from North America. The last member of this family was *Sivaladapis,* which was known as recently as 7 to 8 million years ago in southern Asia (Gingerich and Sahni, 1984; Gingerich, 1986).

Modern lemuroids are clearly distinguished from the adapids by specializations of the procumbent lower incisors and incisiform canines to form a tooth comb (Figure 20-36). The only group of modern prosimians with a significant fossil record are the Lorisidae, which are known from the Miocene of India (MacPhee and Jacobs, 1986) and East Africa (Walker, 1978). *Progalago* and *Komba* resemble the living bush babies, while *Mioeuoticus* may be included among the pottos.

The Lemuroidea and Indrioidea are today confined to Madagascar and the Comora Islands. The postcranial skeleton shows little change from the adapid pattern, but the relative size of the braincase has increased. All are herbivorous, some are terrestrial, but most are arboreal. Many recently extinct forms are known from subfossils. All are larger than their living counterparts and include species that were convergent on cercopithecoid monkeys and baboons; one had very long forelimbs and hooklike hands and feet and may have hung upside down like a tree sloth.

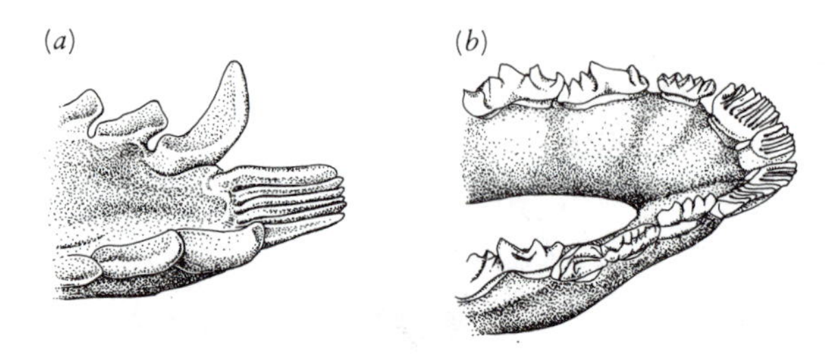

Figure 20-36. TOOTH COMBS OF LEMURS. (*a*) Tooth comb of the modern lemur *Euoticus.* (*b*) Tooth comb of the "flying lemur" *Cynocephalus. From Rose, Walker, and Jacobs, 1981. Reprinted by permission from* Nature. *Copyright © 1981, Macmillan Journals Limited.*

TARSIIFORMES

Omomyids were common in the Eocene, with twenty genera in North America, four in Europe, and one in Asia, and extend into the late Oligocene or earliest Miocene. According to Szalay and Delson, the dental pattern of the primitive adapid *Pelycodus* could have given rise to that of the omomyids. Most members of this group are smaller than the adapids and weigh less than 500 grams. They have pointed and enlarged central incisors and a tubular external auditory meatus.

Teilhardina from the early Eocene of the Paris Basin still retains a dental formula of

$$\frac{2\ \ 1\ \ 4\ \ 3}{2\ \ 1\ \ 4\ \ 3}$$

The much-better-known North American contemporary, *Tetonius* (Figure 20-37), has lost the first premolar and greatly reduced the second. Its large brain occupies much of the small, globular skull. The eyes are large and close together, with the facial region much reduced. The auditory bulla is much enlarged.

It has long been thought that the omomyids included the ancestors of the modern genus *Tarsius,* although no fossil tarsiids had been described. A recently discovered jaw from the Lower Oligocene of North Africa demonstrates a close link between these groups. Simons and Bown (1985) classify *Afrotarsius* within the Tarsiidae with some hesitation, while emphasizing similarities with the omomyid *Pseudoloris* from the Upper Eocene of Europe.

Tarsius, which is known today from the Philippines and Indonesia, is characterized by a reduced rostrum and enormous eyes, with a partial closing of the orbit posteriorly. The foramen magnum is oriented ventrally in association with vertical leaping and clinging. The calcaneum and navicular are greatly elongated, the tibia and fibula are partially fused, and the big toe is enlarged for grasping. The postcranial skeleton is not well known among the omomyids, but there is some hint of tarsioid specialization in the Eocene, including elongation of the tarsals, and fusion of the tibia and fibula.

ANTHROPOIDEA

During the Paleocene and Eocene, when primitive primates were common in Europe and North America, these areas had a moist, tropical climate. The late Eocene and early Oligocene marks the beginning of a sharp increase in seasonality and reduced temperatures in the North Temperate region. The number of primate fossils drops markedly, with only two genera reported in the Oligocene and Lower Miocene of North America. In the later Cenozoic, the fossil record of primates is almost totally limited to southern Asia, Africa, and South America and consists almost entirely of members of the Anthropoidea.

The earliest remains that have been assigned to the Anthropoidea are teeth and jaw fragments from the late Eocene of Burma, which are assigned to the genera *Amphipithecus* and *Pondaungia* (Ba Maw, Ciochon, and Savage, 1979) (Figure 20-38). Unfortunately, they are too incomplete to establish their specific affinities or to characterize the primitive nature of this group. Unquestioned members of the catarrhine radiation are common in North Africa at the beginning of the Oligocene, and a single genus marks the first appearance of the Platyrrhini in South America in the middle to late Oligocene. The early

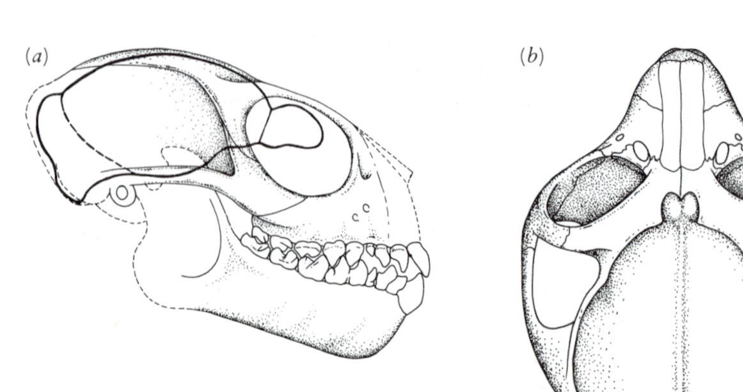

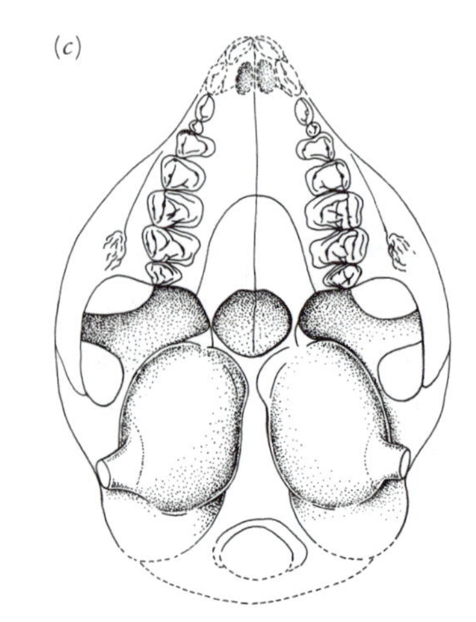

Figure 20-37. SKULL OF THE EOCENE OMOMYID *TETONIUS.* (*a*) Lateral view showing outline of endocast. (*b*) Dorsal view. (*c*) Palatal view. *From Szalay, 1976.*

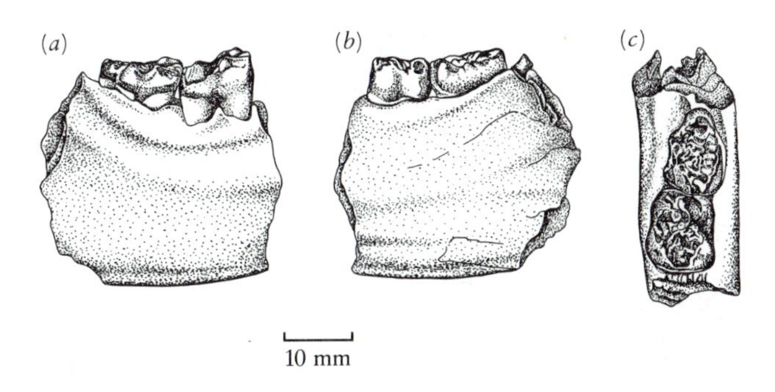

Figure 20-38. RIGHT DENTARY OF *PONDAUNGIA* FROM THE UPPER EOCENE OF BURMA. M_{2-3} in (*a*) labial, (*b*) lingual, and (*c*) occlusal views. This genus may belong to the Anthropoidea. *From Ba Maw, Ciochon, and Savage, 1979. Reprinted by permission from* Nature. *Copyright © 1979, Macmillan Journals Limited.*

anthropoids are clearly distinguished from all prosimians by a relatively larger and more convoluted brain with reduced olfactory lobes. The rostrum is shortened and the amount of olfactory epithelium is reduced. The orbits are more forwardly directed and are separated from the temporal fossa by a postorbital septum. The mandibular rami and frontal bones are typically fused at the midline.

The early anthropoids are already too advanced to make close comparison with any of the early Cenozoic prosimian groups. Many authors favor affinities with the Omomyidae (e.g., Luckett and Szalay, 1975; Ciochon and Chiarelli, 1980a; Szalay and Delson, 1979), but Gingerich (1980a, 1981) argues for closer affinities with the Adapidae. Rosenberger and Szalay (1980) admit that all adequately known omomyids possess some specialized anatomical features that preclude their being ancestral to the anthropoids. Anthropoid affinities are based primarily on the small body size and the correspondingly large apparent brain size and related modifications of the facial region and dentition of the omomyids. Relatively large brain size may also be reflected in the relative size of the branches of the carotid artery that supply the brain. Packer and Sarmiento (1984) argue that many of the similarities between the living genus *Tarsius* and anthropoids may be the result of convergent increase in relative brain size and acuity of hearing. Nevertheless, the general similarities of anthropoids and Eocene omomyids seem to suggest closer affinities than with any of the known adapids.

PLATYRRHINI

The Platyrrhini are restricted to South and Central America throughout their history. The earliest known fossil is *Branisella,* a member of the modern family Cebidae from the middle to late Oligocene of Bolivia. This genus and all South American monkeys retain a dental formula of

$$\frac{2\ \ 1\ \ 3\ \ 3}{2\ \ 1\ \ 3\ \ 3}$$

whereas most adequately known Catarrhini have lost one of the premolars. The Ceboidea can also be distinguished by the derived character of a greatly inflated auditory bulla and the flattened face and widespread nostrils that give the group its name. Sixteen living genera and eight that are known only as fossil represent this group during the Cenozoic (Rose and Fleagle, 1981).

Hoffstetter (1980) suggested that the Platyrrhini may be most closely related to the family Parapithecidae, among the Old World Anthropoidea. This family is represented by the genera *Parapithecus* and *Apidium* from the Oligocene of Egypt, which are primitive in retaining three premolars. They may represent remnants of the stem group of all anthropoids but are generally considered to be more closely related to the later Catarrhini.

The geographical origin of the South American monkeys is also subject to continuing debate, which is considered in detail in *Evolutionary Biology of the New World Monkeys and Continental Drift* (Ciochon and Chiarelli, 1980a). Several authors in this volume argued that Old World monkeys originated from precatarrhine anthropoids that were established in Africa by the late Eocene. Hoffstetter suggested that they reached South America by chance rafting on floating vegetation at a time when Africa and South America were significantly closer than at present. There is no evidence that there was more than a single migrational event, since all South American primates appear to have evolved from a single stock.

Gingerich (1980a) subscribes to the view that was earlier expressed by Simpson (1945) that platyrrhines dispersed from North America. This hypothesis would also require crossing a large water body. More significantly, this hypothesis requires a long migration across the northern continents to account for their fairly well-documented common ancestry with the Old World anthropoids. There are no fossils of anthropoids from the Cenozoic of North America, during which time the fossil record of other groups is exceedingly rich. Nor is there any record of prosimian primates in South America at any time during the Cenozoic, although there is a moderately good record of other mammals. These facts argue strongly against the origin of New World monkeys within South America.

CATARRHINI

With the possible exception of the late Eocene fossils from Burma and the Parapithecidae, all the Old World anthropoids certainly have a unique common ancestry. This is most simply demonstrated by the consistent dental formula of

$$\frac{2\ \ 1\ \ 2\ \ 3}{2\ \ 1\ \ 2\ \ 3}$$

which represents a loss of one premolar relative to that of the New World monkeys. They also lack the inflation of the bulla that characterizes the Platyrrhini.

A number of early catarrhine genera are known from the Oligocene of the Fayum Basin, west of Cairo. These sediments were deposited between 28 and 34 million years ago in a swampy, well-forested area. Early work by Simons (1964) suggested that these genera might represent a number of lineages, including the specific ancestors of Old World monkeys, the gibbon, and the great apes and humans. Paleontologists now agree that the Old World monkeys, the Cercopithecoidea, did not differentiate until later in the Cenozoic. Fleagle and Kay (1983) suggest that the Oligocene genera *Propliopithecus* and *Aegyptopithecus* might belong to an assemblage that existed prior to the major diversification of the catarrhines.

Simons (1967) recognized that the early catarrhines resemble the modern apes move than they do the Old World monkeys. Especially in their dentition, the Cercopithecoidea are more specialized than these early apelike forms, with high-crowned, bilophodont cheek teeth and a protruding muzzle (Figure 20-39). The earliest-known cercopithecoids are the early Miocene *Prohylobates* from North Africa and *Victoriapithecus* from East Africa. Szalay and Delson (1979) suggest that the nearest ancestry of these forms may lie with *Propliopithecus*.

The major radiation of the Old World monkeys did not occur until 7 to 8 million years ago. This assemblage may also includes *Oreopithecus,* a superfically apelike genus from the late Miocene or early Pliocene of Italy that was once thought to have close affinities with man.

The Oligocene genus *Aegyptopithecus* may belong to the stem group of all later catarrhines, but its anatomy more closely resembles that of later hominoids than it does the Cercopithecoidea. The skull has a globular braincase and a relatively short face (Figure 20-40). The configuration of the teeth suggests that it was a fruit eater. The dentition is strongly dimorphic, with large canine teeth in the males. The postcranial skeleton is badly crushed and has not been reconstructed, but what is known suggests quadrupedal arboreal locomotion—climbing and running through the branches of the forest canopy (Simons, 1984).

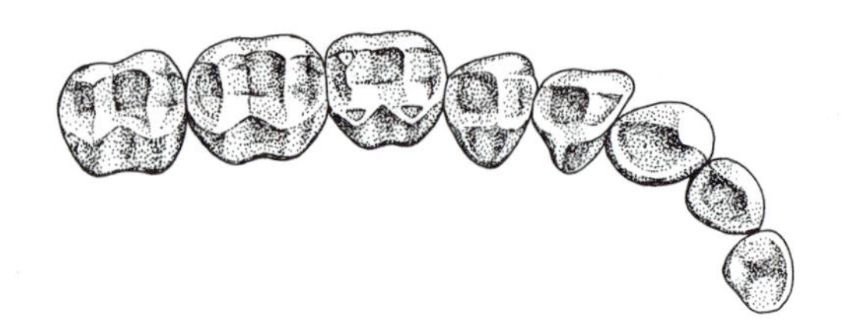

Figure 20-39. *DOLICHOPITHECUS.* Lower dentition of the cercopithecoid from the Pliocene of Europe. The squared molars are dominated by four cusps. *From Szalay and Delson, 1979.*

HOMINOIDEA

Members of the superfamily Hominoidea are known from the early Miocene of East Africa, some 22 million years

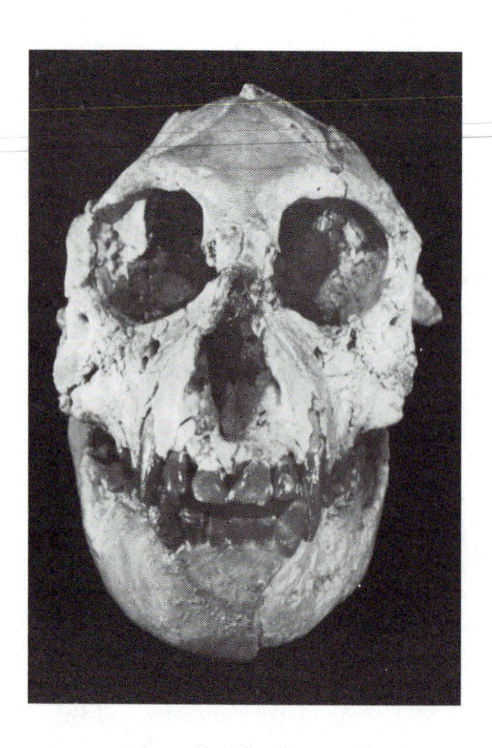

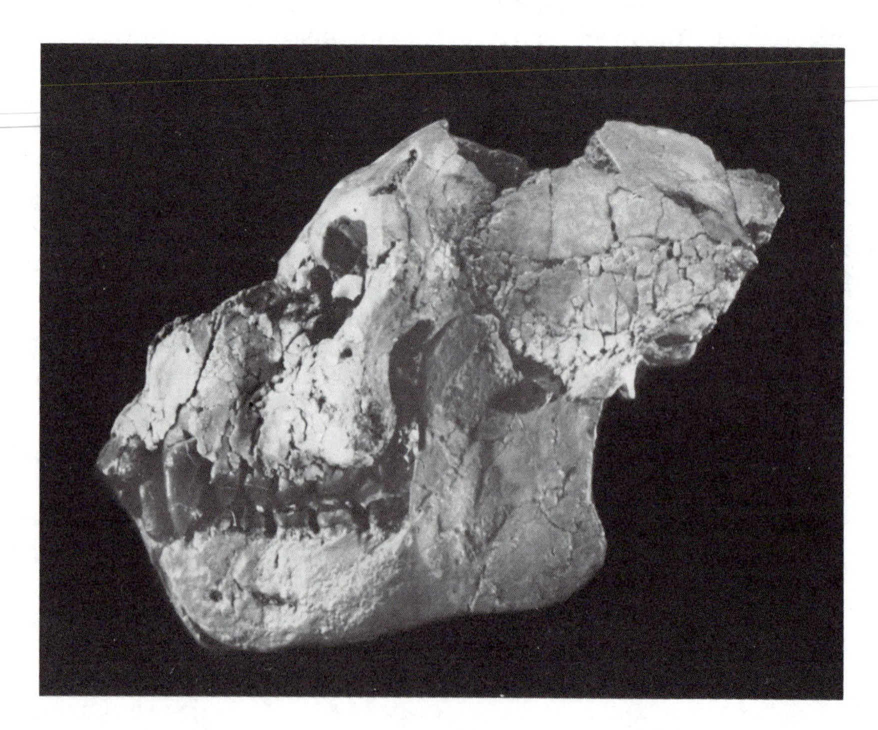

Figure 20-40. SKULL OF THE EARLY APE *AEGYPTOPITHECUS* FROM THE LOWER OLIGOCENE OF EGYPT. Anterior and lateral views. *From Simons, 1967.*

ago. This superfamily is the most extensively studied of all mammalian groups, yet the nature and interrelationships of the fossil species remain subject to continuing debate. Many of these problems are discussed in the recent books *New Interpretations of Ape and Human Ancestry* (Ciochon and Corruccini, 1983) and *Ancestors: The Hard Evidence* (Delson, 1985). Pilbeam (1984) recently provided a brief but authoritative summary.

The earliest adequately known hominoid is *Proconsul africanus*, a baboon-sized species from Kenya (Figures 20-41 and 20-42). Like *Aegyptopithecus*, it was a sexually dimorphic, fruit-eating, arboreal quadruped. The general anatomy is sufficiently primitive to be ancestral to all the later apes and humans. Details show a mosaic of features that illustrate more specific similarities with later forms. The elbow, shoulder joint, and feet resemble those of a chimpanzee, the wrist resembles that of monkeys, and the lumbar vertebrae are like those of a gibbon. Other feature are not closely similar to those of any other primates.

During the Miocene, there was an extensive African radiation of genera that were similar to *Proconsul*. At the end of the early Miocene, approximately 17 million years ago, connection was achieved between Africa and Asia via the Arabian Penninsula, and primitive hominoids migrated into Europe and Asia. A degree of specialization was achieved in both these areas, which resulted in the recognition of two informal groups of early hominoids: the dryomorphs (based on the genus *Dryopithecus*), which include most of the European and African species, and the ramamorphs (based on the genus *Ramapithecus*), which were primarily Asian, where they are known from 12 to 7 million years ago. During the 1960s and 1970s, it was widely accepted that *Ramapithecus* was closely related to the ancestry of hominids and that this group had separated from the great apes (Pongidae) approximately 17 million years ago. At that time, the ramamorphs were known almost exclusively from teeth and jaws. The teeth were recognized as having very thick enamel, like those of primitive hominids, and they were arranged in a parabolic arc around the palate. In contrast, the teeth of the great apes are typically arranged in a U shape and the enamel is thinner. With Ward and Pilbeam's (1983) description of cranial material (Figure 20-43), we now recognize that *Ramapithecus* is very closely related to the extant orangutan *Pongo*. This conclusion is based on shared derived features of the facial region and some characters of the limbs. Ramamorphs are clearly not closely related

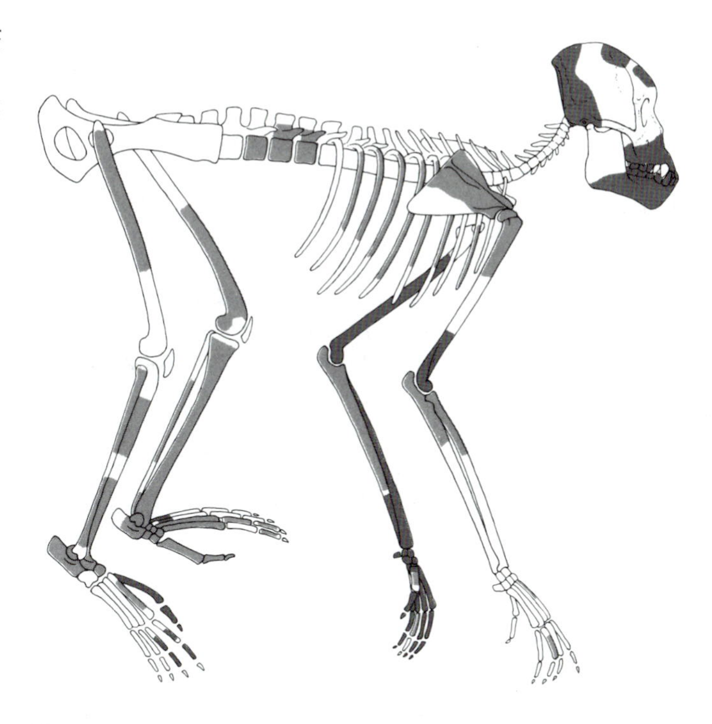

Figure 20-42. SKELETON OF *PROCONSUL*. *From Pilbeam, 1984.*

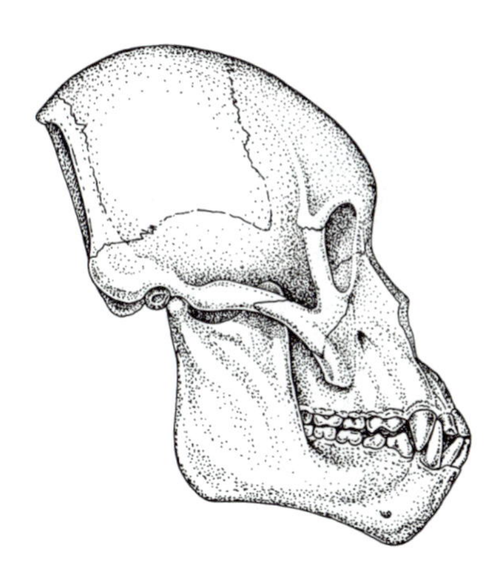

Figure 20-41. *PROCONSUL*. Skull of the early homonoid from the Miocene of Africa. *From Pilbeam, 1984.*

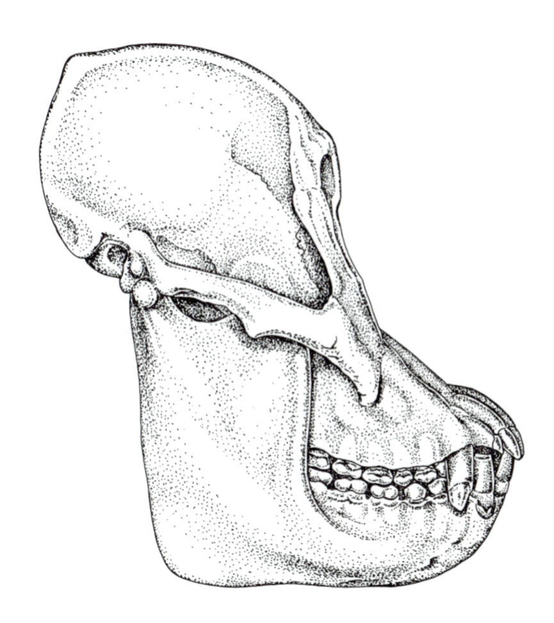

Figure 20-43. SKULL OF *SIVAPITHECUS* [*RAMAPITHECUS*]. Orangutan is from the Late Miocene of India. *From Pilbeam, 1984.*

to hominids, and the dental similarities must be attributed to convergence (Ciochon and Corruccini, 1983). *Ramapithecus* is no longer thought to be a separate genus but is probably referable to *Sivapithecus*, which exhibits marked sexual dimorphism. *Gigantopithecus* from the Pliocene of India and the Pleistocene of China, which was once thought to be close to the hominids, is now also allied with *Sivapithecus*.

These new findings eliminate the Asian ramamorph complex from close affinities with the hominids, whose early history is now thought to have been confined to Africa. *Kenyapithecus*, which is known from strata approximately 16 million years old in East Africa, may be a member of a stem group that was ancestral to both African apes and humans (Wolpoff, 1983).

Unfortunately, there are no fossil hominoids known in Africa between 4 and 14 million years ago. The African apes *Gorilla* and *Pan* have no fossil record. Based on anatomical evidence, they might have diverged from the ancestors of humans any time from about 5 to 14 million years ago.

Molecular evidence, which is now based on a host of different techniques, places the dichotomy between man and the African great apes as having occurred between 6 and 10 million years ago (Ciochon and Coruccini, 1983; Sibley and Ahlquist, 1984) (Figure 20-44).

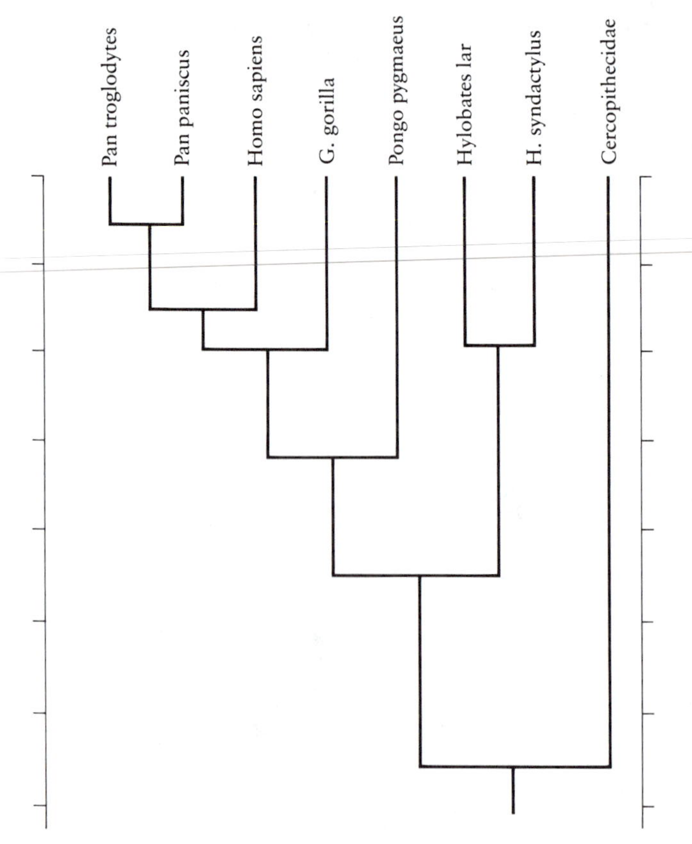

Figure 20-44. HOMINOID PHYLOGENY, BASED ON MOLECULAR DATA. *From Sibley and Ahlquist, 1984.*

HOMINIDAE

The hominoid fossil record in Africa begins again approximately 4 million years ago, with primitive hominids of the genus *Australopithecus* in Ethiopia and Tanzania (*American Journal of Physical Anthropology*, 1982, Vol. 57(4)). The earliest-known species is *Australopithecus afarensis*, which is represented by several specimens, including much of the skeleton of an individual named "Lucy" (Johanson and Edey, 1981; Kimbel, White, and Johanson, 1984). *Australopithecus afarensis* was fully bipedal if we judge by the configuration of the joints of the hip, knee, and ankle (Stern and Susman, 1983). Numerous foot prints of a hominid of this type are known from strata 3.75 million years old. *Australopithecus afarensis* ranged from 25 to 50 kilograms, with the males 50 to 100 percent larger than the females. This strong sexual dimorphism suggests that, like many modern primates, they traveled in troops rather than forming permanent family groups. The brain ranged from approximately 450 to 550 cubic centimeters, which is 20 to 30 percent larger than that of a chimpanzee with a similar-sized body.

The general configuration of the face and palate resembles that of a chimpanzee, but the canines are reduced and do not extend much above the height of the surrounding teeth, which suggests that they were not used for defense (Figure 20-45). It has been argued that this indicates the initial use of simple tools for defense and obtaining food. The cheek teeth are large and have thick enamel, which enables them to feed on hard plant food. Compared with humans, the arms are relatively longer and the legs relatively shorter. The hand may have been capable of more precise manipulation than is possible for the chimpanzee.

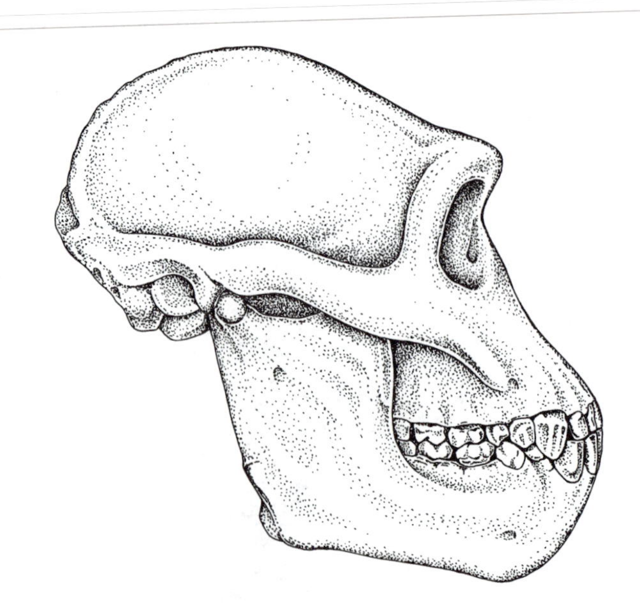

Figure 20-45. SKULL OF THE MOST PRIMITIVE HOMINID *AUSTRALOPITHECUS AFARENSIS.* This species is known as early as 4 million years ago in East Africa. *From Pilbeam, 1984.*

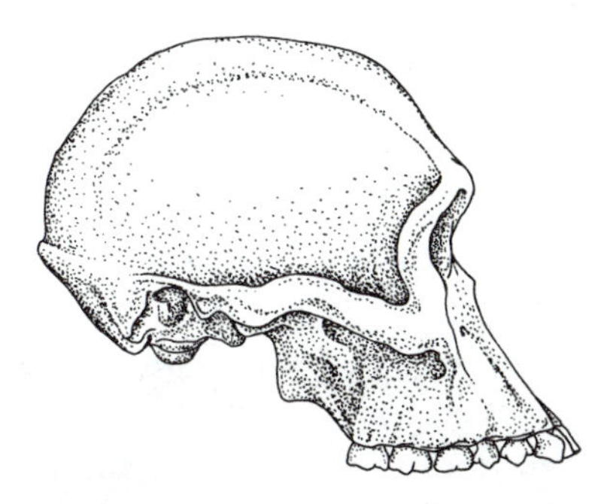

Figure 20-46. *AUSTRALOPITHECUS AFRICANUS.* Skull in lateral view. *From Howell, 1978. Reprinted by permission of Harvard University Press.*

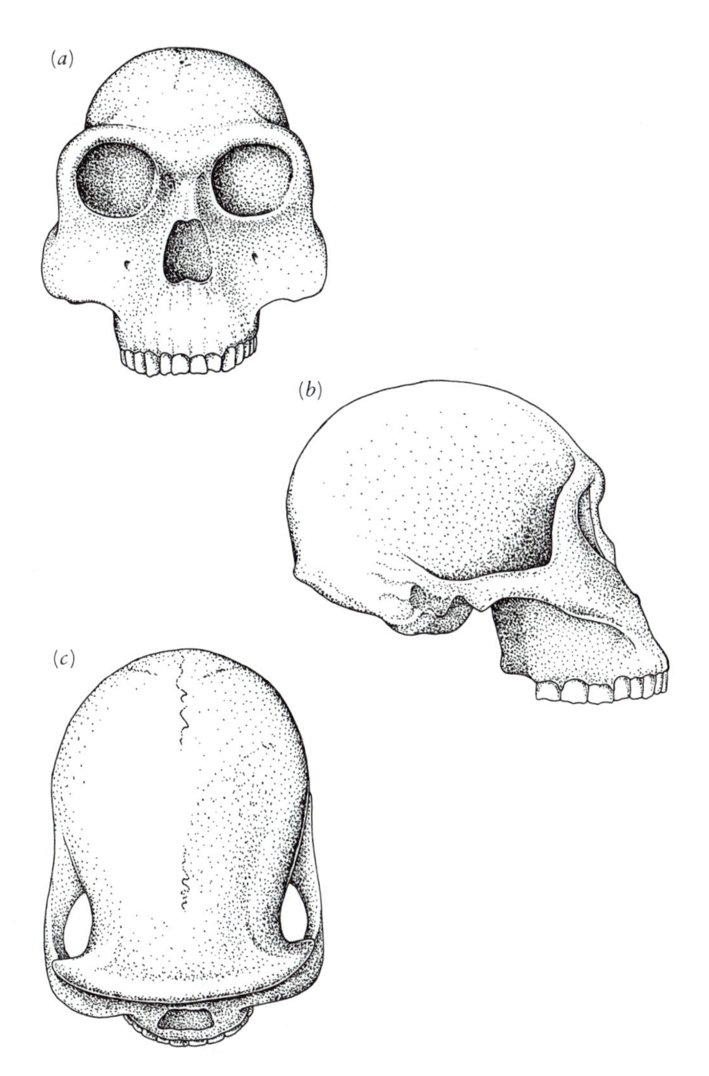

Figure 20-48. *HOMO HABILIS.* Skull in (*a*) anterior, (*b*) lateral, and (*c*) dorsal views. *From Howell, 1978. Reprinted by permission of Harvard University Press.*

The emergence of hominids from earlier apes may have been associated with a shift in environment from life in the wet forests to dryer grassland and savanna. The necessity for rapid movement in the open may have provided the selective advantage for the development of bipedality. The fossil record clearly demonstrates that bipedality was achieved among hominids with a brain that was little advanced above that of the great apes.

One of the most surprising aspects of the early evolution of hominids is the fact that there was not just one, but a number of species living at the same time in southern and eastern Africa (Walker, Leakey, Harris, and Brown, 1986). There is still considerable controversy over the exact identity of many remains as well as their ages and specific relationships, but there appear to have been at least three coexisting lineages between 2 and 3 million years ago. In addition to *Australopithecus afarensis,* these lineages include the larger, but lightly built, *Australopithecus africanus* (Figure 20-46), *Australopithecus boisei* (Figure 20-47), which had extremely massive teeth, and, by about 2 million years ago, the first member of the modern genus, *Homo habilis.*

Primitive members of our own genus *Homo* are differentiated from *Australopithecus* by further increase in cranial capacity and modifications of the pelvis for improved bipedal locomotion and to accomodate the large size of the head of the neonate. The brain of *Homo habilis* was approximately 700 cubic centimeters in volume, the face was shortened, and the teeth reduced in size (Figure 20-48).

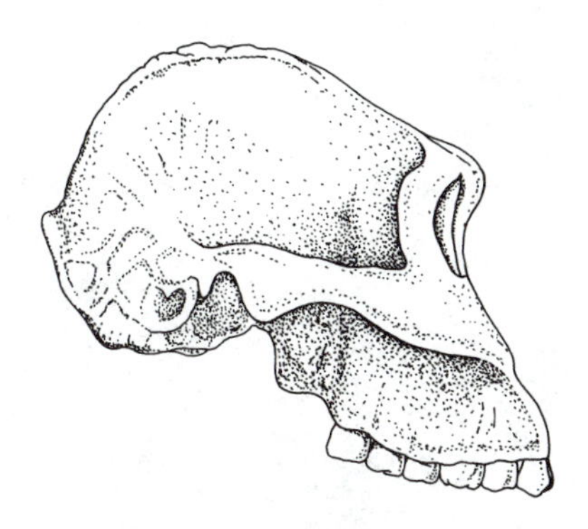

Figure 20-47. *AUSTRALOPITHECUS BOISEI,* A ROBUST AUSTRALOPITHICINE, IN LATERAL VIEW. *From Howell, 1978. Reprinted by permission of Harvard University Press.*

Homo habilis is not recognized after 1.75 million years ago but appears to have been replaced by *Homo erectus,* which first appears at that time. This species has a more robust skeleton and a cranial capacity of 850 to 1000 cubic centimeters. The face and teeth are further reduced (Figure 20-49*a*). A nearly complete skeleton of *Homo erectus* has recently been discovered from beds 1.6 million years old in east Africa (Figure 20-50).

Approximately 1 million years ago, *Homo erectus* spread out of Africa into eastern and southern Asia. There is considerable debate regarding the pattern and rate of evolutionary change between *Homo erectus* and *Homo sapiens* (see Chapter 22), but by approximately 300,000 years ago, populations with an anatomy that is considered

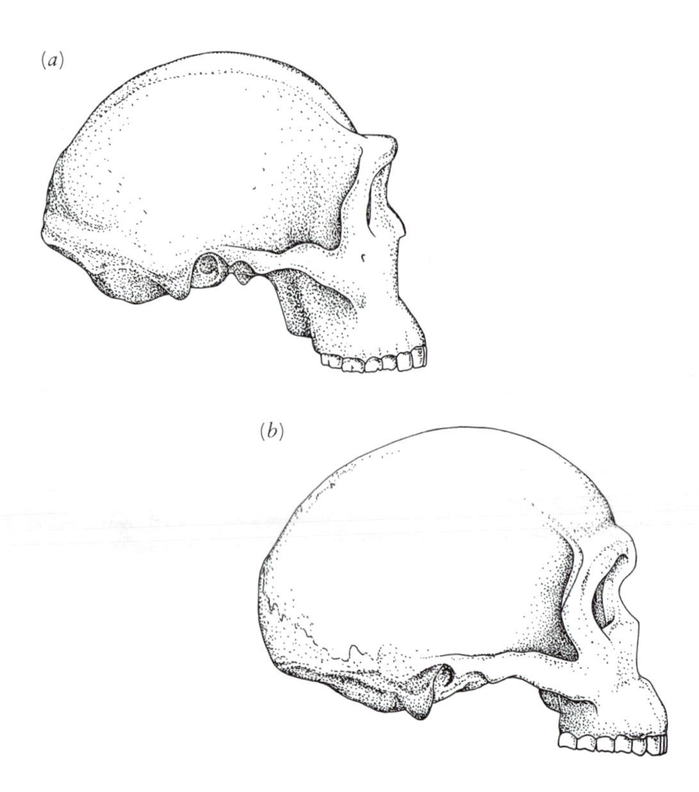

Figure 20-49. (*a*) *Homo erectus.* (*b*) *Homo sapiens neanderthalaenis. From Howell, 1978. Reprinted by permission of Harvard University Press.*

typical of *Homo erectus* were replaced by primitive members of our own species. *Homo sapiens* evolved a nearly modern appearance, with the brain size approaching 1400 cubic centimeters, over the next 100,000 to 200,000 years. The term "archaic" *Homo sapiens* is used from members of our species that lived from about 300,000 years ago to approximately 40,000 years ago. Among the archaic *Homo sapiens* were the Neanderthals, who were common in Europe and southwest Asia between 70,000 and 30,000 years ago. The Neanderthals were shorter and more heavily built than modern humans but had a comparable cranial capacity. The skull was characterized by the presence of prominent brow ridges and strong anterior teeth (Figure 20-49*b*). Many features of the skeleton show analogies with modern hominids living in arctic regions, which may be correlated with the glacial conditions that existed in Europe during the late Pleistocene.

The degree to which the Neanderthals were related to modern humans in Europe is still a matter of contention (Smith and Spencer, 1984). *Homo sapiens* of more modern aspect are known in Africa as long as 100,000 years ago. They apparently invaded Europe between 30,000 and 40,000 years ago and may have replaced the Neanderthals with a minimum of interbreeding.

Within the genus *Homo*, we cross the line from paleontology to archaeology. Badgley (1984) cited the following dates for the achievement of important cultural advances: manufacture of stone tools, 2 million years ago; control of fire, 500,000 years ago; the origin of agriculture, 10,000 years ago. The Neanderthals buried their dead in a ritualistic manner, and the earliest art is known from about 35,000 years ago.

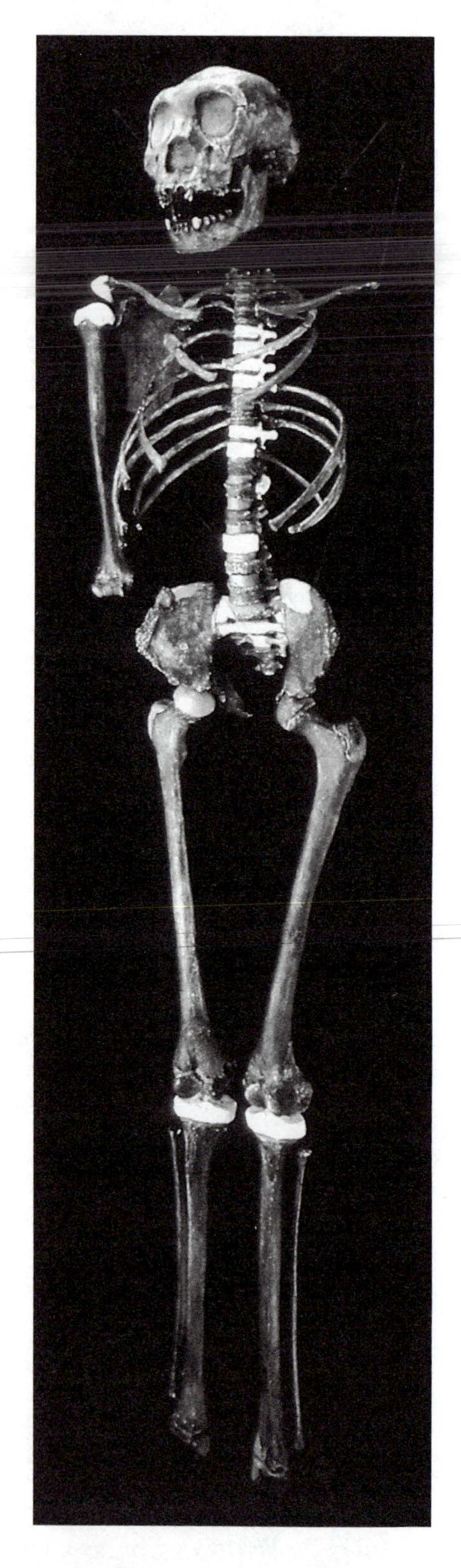

Figure 20-50. SKELETON OF *HOMO ERECTUS*, 1.6 MILLION YEARS OLD. *Courtesy of Richard Leakey.*

Scandentia (Tree Shrews)

It has been suggested that several other eutherian groups might have close affinities with primates. These groups include the tree shrews, elephant shrews, bats, and Dermoptera or "flying lemurs." Tree shrews are superficially squirrel-like animals that are common in southeast Asia. Their phylogenetic position has long been debated. They were once considered members of the Insectivora but have repeatedly been suggested as having close affinities with the primates.

The skull proportions and dentition are little different from the Upper Cretaceous eutherians. The bulla is formed by the entotympanic, and the internal carotid is in the promontory position (Figure 20-51). The fossil record consists of fairly complete skulls from the Miocene that are essentially modern (Jacobs, 1980).

Their taxonomic position has been considered from many angles in the recent book, *Comparative Biology and Evolutionary Relationships of Tree Shrews* (Luckett, 1980a). The consensus is that tree shrews cannot be associated closely with any other groups but are better placed in a separate order, the Scandentia (which was earlier proposed by Butler, 1972).

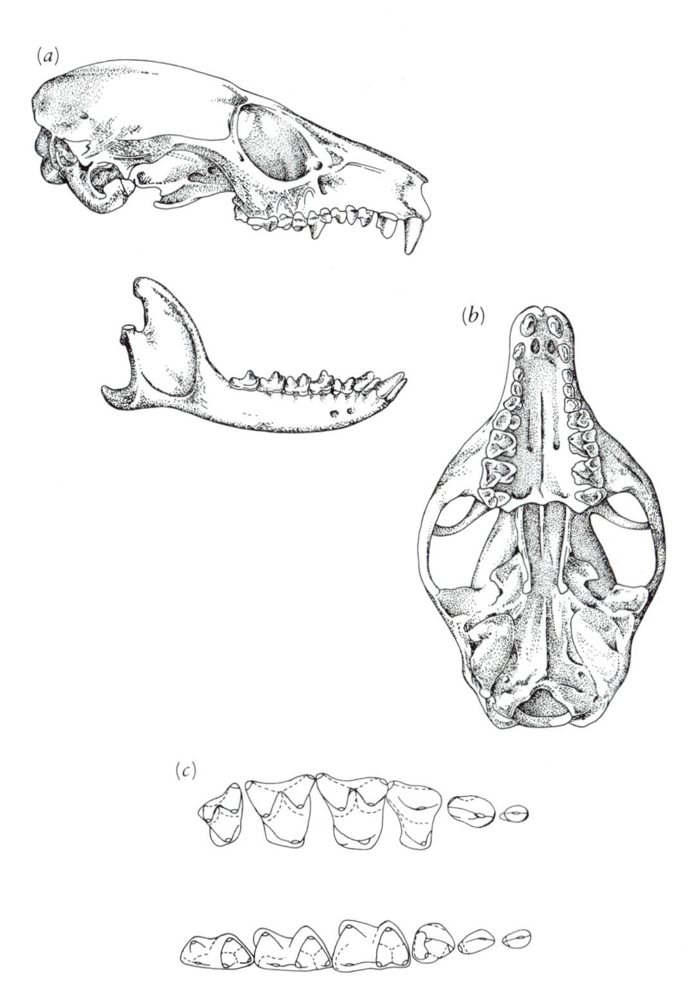

Figure 20-51. SKULL OF THE TREE SHREW *PTILOCERCUS*. (*a*) Lateral and (*b*) palatal view. *From Gregory, 1957.* (*c*) Dentition of *Tupaia*. *From Evans, 1942.*

Nevertheless, there is some evidence, especially from the molecular side, that tree shrews are more closely related to primates and insectivores than they are to other orders. This evidence is not contradicted by the dental anatomy (Butler, 1980) or the configuration of the tarsals (Szalay and Drawhorn, 1980).

Modern tupaiids appear to represent the retention of a very primitive stage in the evolution of eutherians, and their general appearance may differ little from that of their Upper Cretaceous ancestors.

Dermoptera

A single modern genus, *Cynocephalus* from tropical Asia, Indonesia, and the Philippines is the only living representative of the order Dermoptera. The most striking feature of this animal is the presence of a gliding membrane that extends between the limbs and onto the tail. *Cynocephalus* spends most its time in trees and is nearly helpless on the ground. The skull is vaguely lemurlike, but the postcranial skeleton is highly distinctive in the great length of the slender limbs. The ulna is reduced and fused to the radius. The fibula does not articulate with the calcaneum.

The cheek teeth are broad and the enamel is wrinkled so that wear produces many shearing surfaces to deal with a strictly vegetarian diet. Dermopterans have a large intestinal caecum filled with bacteria that are capable of breaking down cellulose.

The lower incisors are spatulate, with the margins extended like the teeth of a comb. We find similarly specialized incisors in members of the family Plagiomenidae from the mid-Paleocene to late Eocene of North America (Figure 20-52). This group is known primarily from jaws and teeth. The lower dentition has the primitive formula

$$\frac{\quad}{3\ \ 1\ \ 4\ \ 3}$$

The cheek teeth also show the structure noted in *Cynocephalus* (Rose and Simons, 1977; Rose, 1982). Plagiomenids are especially common in the early Eocene of Ellesmere Island (West and Dawson, 1978).

In turn, the plagiomenids may be related to the mixodectids. This group, which Szalay (1969) last reviewed, includes five genera restricted to the Paleocene of North America. Their remains consist primarily of jaws and teeth. The upper and lower incisors, like those of apatemyids, are greatly enlarged and procumbent, but the lower anterior premolar is not bladelike and the lower incisor does not extend so far back in the mandible as in that group. Rose and Simons (1977) noted similarities between mix-

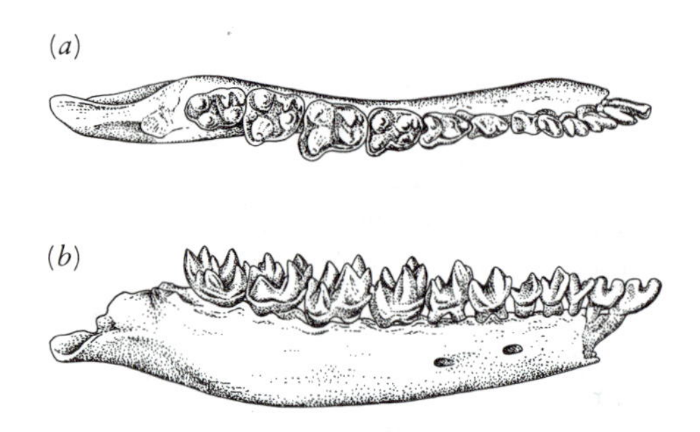

Figure 20-52. LOWER JAW OF *PLAGIOMENE*. From the early Eocene, thought to be related to the modern dermopteran genus *Cynocephalus*. (*a*) Occlusal and (*b*) lateral views, ×2. *From Rose and Simons, 1977.*

odectids and plesiadapiform primates but suggest that they may result from convergence.

Bats, primates, tree shrews, and dermopterans have long been thought to be closely related to one another and to have special affinities with the insectivores. McKenna (1975) classified them together in the superorder Archonta. Luckett (1980b) considered the relationships of these groups in detail. Few derived features of the skeleton can be shown to unite these orders, but macromolecular studies suggest that they are more closely related to one another than any are to other eutherians (Cronin and Sarich, 1980).

Szalay and Drawhorn (1980) provided evidence that the tarsal structure of tree shrews, dermopterans, and archaic primates show a number of derived features in common that may be attributed to a common factor of arboreality. However, arboreality might have been achieved separately in the ancestors of each group. Their specific interrelationships remain unresolved.

CARNIVORES

TERRESTRIAL CARNIVORES

The dentition of early eutherians appears suitable for piercing and shearing small prey, and it is generally accepted that they fed primarily on insects, as do the most primitive living mammals. It would seem a short step from such small insectivores to larger predaceous carnivores. Surprisingly, this change was relatively slow; it is not until near the end of the Paleocene that any moderate-sized mammals that are clearly recognizable as carnivores appeared in the fossil record.

In contrast with the plethora of herbivorous orders, there are only two major groups of terrestrial carnivores in the early Cenozoic, the ancestors of the modern order Carnivora and an archaic group, the Creodonta. In both groups, the posterior cheek teeth are modified to form specialized shearing surfaces. Such teeth are termed **carnassials.** Different teeth are modified as carnassials in the two groups, which indicates that a carnivorous habit was elaborated separately. In creodonts, this modification involved upper molars 1 and/or 2 and lower molars 2 or 3. In the early members of the Carnivora, the last upper premolar and the first lower molar are modified (Figure 20-53).

MacIntyre (1966) pointed out that specialization of the molars as carnassials in the creodonts led to the reduction of crushing and grinding surfaces that were maintained in true carnivores. The dentition in early members of both groups remained basically very primitive, however, and both were originally included among the insectivores as ancestral eutherians (Cope, 1875).

MacIntyre showed that relatively minor changes would have been necessary to modify the molars and posterior premolars of the late Cretaceous and early Paleocene genus *Cimolestes* to the pattern seen in early creodonts and Carnivora (Figure 20–54). Since *Cimolestes* also appears to be close to the ancestry of other groups of Cenozoic mammals, we have been unable to demonstrate that creodonts and the Carnivora had a unique common ancestry, which would be necessary to justify their inclusion in a single supraordinal taxon. Despite their limited diversity and longevity, it seems appropriate to place the creodonts in a separate order.

The two families of creodonts, the Oxyaenidae and Hyaenodontidae, appear in the late Paleocene and are common throughout the Eocene. Oxyaenids, which may have originated in North America, were generally small, with long bodies and short limbs. In contrast with the hyaenodontids, they have lost the third molar. The posture of the feet was plantigrade with spreading toes. Hyaenodontids, which were more common in the Old World, include genera in which the body proportions were comparable with those of felids, canids, and hyaenids (Figure 20-55). Some reached the size of large bears. Like other primitive eutherians, the creodonts lacked an auditory bulla and five digits were retained on both the forelimbs and hind limbs (Figure 20-56).

Romer (1966) and Jerison (1973) stigmatized the creodonts as archaic and small brained, but Radinsky (1977) demonstrated that relative brain size increased as rapidly among the creodonts as it did in the early members

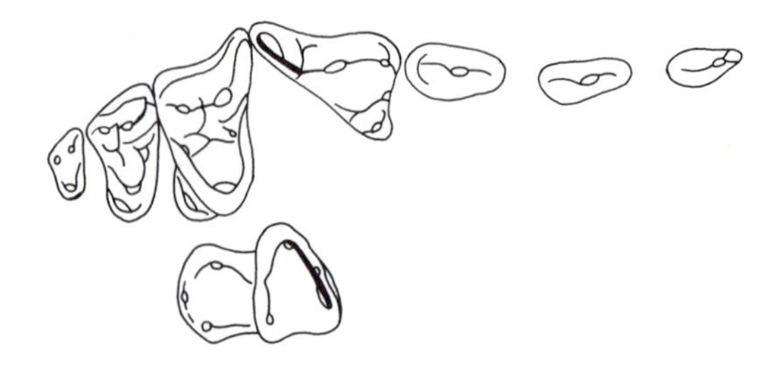

Figure 20-53. CARNASSIAL TEETH IN THE ORDER CARNIVORA. Heavy line indicates position of carnassial blades on last upper premolar and first lower molar of the primitive carnivore *Miacis*. *From Savage, 1977.*

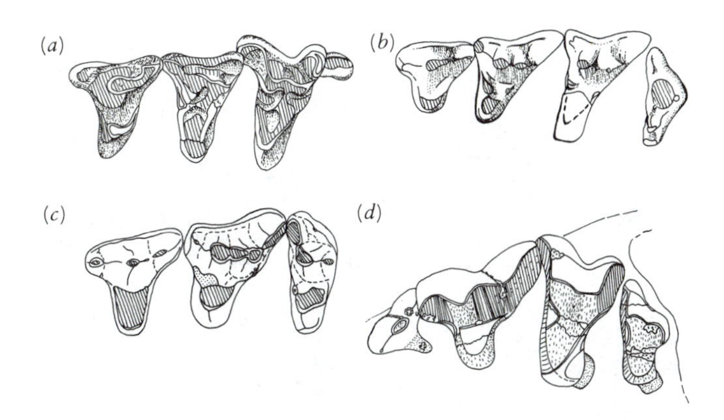

Figure 20-54. UPPER CHEEK TEETH OF PRIMITIVE CARNIVOROUS PLACENTALS. Development of carnassial blades is shown. (*a*) Last upper premolar and first two molars of the Upper Cretaceous "proteutherian" *Cimolestes*. *From Clemens, 1973*. (*b*) Last premolar and all three molars of the Lower Eocene hyaenodont creodont *Prototomus*. Shearing blades are developed on both the first and second molars. *From Matthew and Granger, 1915*. (*c*) Last premolar and first two molars of the oxyaenid creodont *Oxyaena* from the Upper Paleocene. The postmetacrista of the first molar is elaborated as a carnassial blade. *From Gingerich, 1980b*. (*d*) Fourth premolar and first two molars of the viverravid carnivore *Ictidopappus* from the Middle Paleocene. The major shearing surface has developed on the last premolar. *From MacIntyre, 1966*. Anterior is to the left.

of the Carnivora, together with an increase in the extent of the neocortex. In North America and Europe, the creodonts became greatly reduced after the end of the Eocene, but hyaenodontids remained the dominant carnivores in Africa through the Oligocene, and one genus survived in Asia into the Pliocene.

We first recognize members of the order Carnivora in the early and middle Paleocene. They are characterized by the specialization of the last upper premolar and first lower molar for shearing. This specialization is first shown by the presence of a well-developed metastyle blade at the posterior angle of the premolar, which sheared against the anterior surface of the trigonid of the first lower molar. The protocone of P^4 is located far forward of the paracone. The posterior molars are reduced. All of the early carnivores were small—comparable in size to a weasel or small cat. They are primitive in lacking ossification of the auditory bulla; the scaphoid, lunar, and centrale of the carpus remain separate, in contrast with their fusion in more advanced carnivores.

Among the primitive Carnivora, two families are recognized, the Viverravidae from the Lower Paleocene into the Oligocene and the Miacidae, which appears first in the earliest Eocene. Both are Holarctic in distribution. Although appearing earlier in time, the Viverravidae are more specialized in the loss of the third molar in both upper and lower jaws. The viverravid *Protictis* (Figure 20-57) has a small skull and extremely high, sharp cusps, which suggest an insectivorous diet.

Miacids are clearly distinct when they first appear in the Lower Eocene of Europe and North America. Gingerich (1980b) suggests that they may have immigrated from Asia where they are also known in the early Eocene. The miacids are generally more similar to the derived carnivore families and assumed dominance over the Viverravidae by the end of the Eocene. Most are poorly known, but a complete articulated skeleton was described from

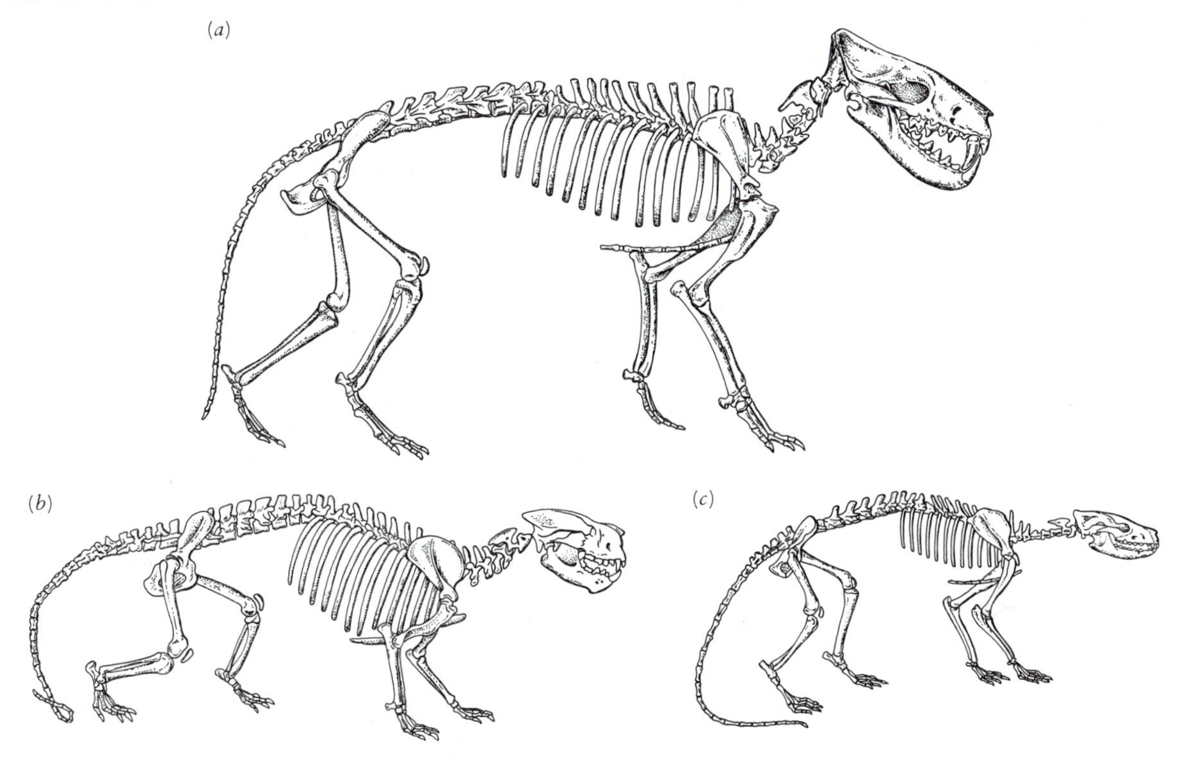

Figure 20-55. SKELETAL RESTORATION OF OXYAENIDS AND HYAENODONTS. (*a*) *Hyaenodon*. (*b*) *Patriofelis*. (*c*) *Sinopa*. *From Gregory, 1957*.

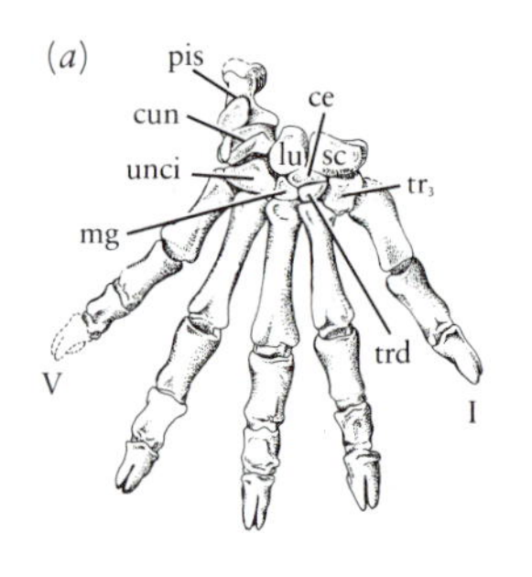

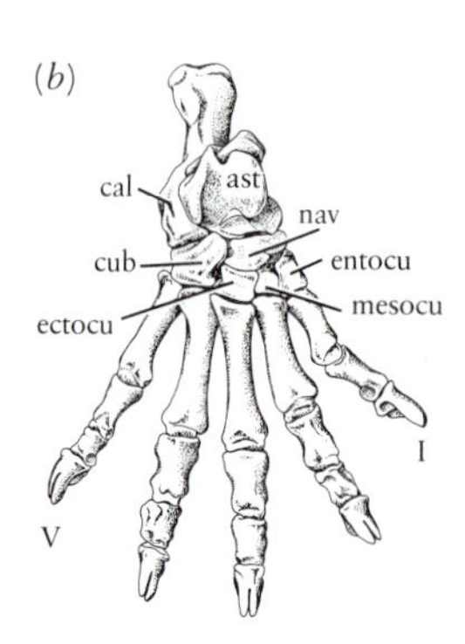

Figure 20-56. FEET OF THE CREODONT *PATRIOFELIS*. (*a*) Front. (*b*) Rear. Abbreviations as follows: ast, astragalus; cal, calcaneum; ce, centrale; cub, cuboid; cun, cuneiform; ectocu, ectocuneiform; entocu, entocuneiform; lu, lunar; mesocu, mesocuneiform; mg, magnum; nav, navicular; pis, pisiform; sc, scaphoid; trd, trapezoid; trz, trapezium; unci, unciform; I, V, metacarpals and metatarsals. *From Gregory, 1957.*

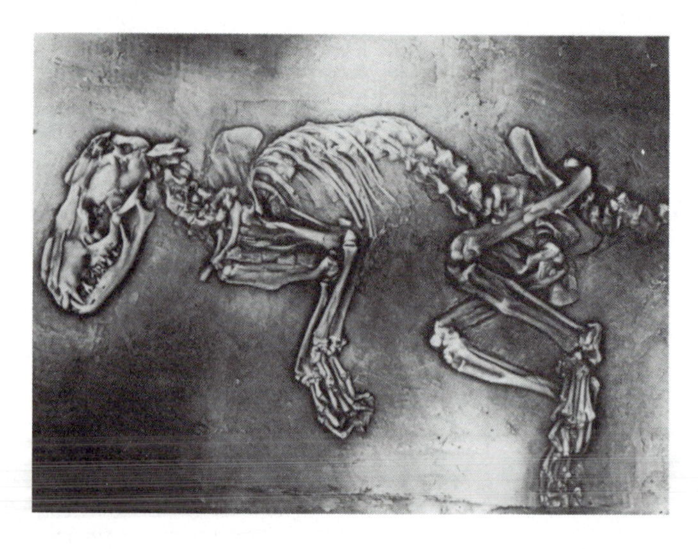

Figure 20-58. *PAROODECTES*. Skeleton of the miacid carnivore from the Middle Eocene of Germany, $\times \frac{3}{8}$. *From Springhorn, 1980.*

the middle Eocene of Germany (Springhorn, 1980) (Figure 20-58). The limbs are relatively short, the trunk is elongate, and the feet are plantigrade.

The modern carnivore families may have begun to diverge from among these primitive genera by the end of the Eocene, but few remains are known until the early Oligocene, by which time many of the modern groups were clearly established.

All the advanced families are distinguished from the miacids and viverravids by ossification of the auditory bulla, but this structure shows different patterns of evolution in each family. According to Hunt (1974), all living carnivores have at least three elements of the bulla: the dermal ectotympanic and both a rostral and caudal entotympanic. The ossification of the bulla was probably associated with an increase in hearing ability. The greater acuity of other mammals with large bullae has been established physiologically (Lay, 1972).

Two major groups of advanced carnivores are recognized on the basis of differences in the nature of the bulla together with changes in the carotid circulation that can be observed in the bones of the basicranial region. In the Aeluroidea (or Feloidea), including viverrids, felids, and hyaenids, the main branch of the internal carotid is reduced or lost, with the arterial circulation to the brain coming primarily from the external carotid in conjunction with the evolution of a countercurrent exchanger in the vicinity of the orbit, which apparently cools the blood entering the brain.

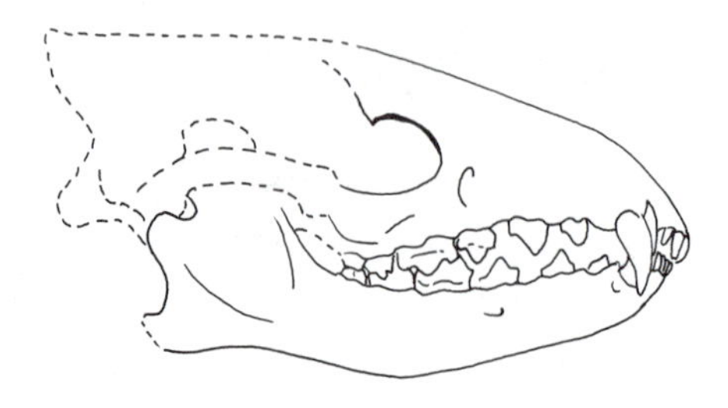

Figure 20-57. *PROTICTIS*. Skull of the Paleocene carnivore, $\times \frac{2}{3}$. *From MacIntyre, 1966.*

In felids and viverrids, both the ectotympanic and the caudal entotympanic contribute to the formation of a septum that divides the bulla into anterior and posterior chambers. In hyaenids, the septum is formed primarily by the ectotympanic, and the chamber of the middle ear is increased by expansion into the mastoid.

In the arctoids (also termed canoids), which include the Canidae, Procyonidae, Ursidae, and Mustelidae, as well as the marine carnivores, the internal carotid remains an important artery and the auditory bulla is not clearly divided into two chambers. Canids have an incomplete division of the bulla, and there is no septum in procyonids and mustelids.

Flynn and Galiano (1982) suggest that the origin of the arctoids and aeluroids can be traced to separate groups among the earlier carnivores. They argued that the canoids can be traced to the miacids and the aeluroids to genera that were previously included among the viverravids. They divide all carnivores into the suborders Caniformia and Feliformia and distinguish early members of these groups by small differences in the degree of development of cusps on the last upper premolar and the absence of the third molar in all Feliformia.

In contrast, Gingerich (1983) argues that there is no evidence of evolutionary continuity between the Viverravidae, which radiated in the Paleocene and early Eocene, and the aeluroid families, which did not begin to differentiate until the late Eocene. He contends that the loss of M^3_3 occurred independently in these two groups, as it did in the oxyaenid creodonts.

Most post-Eocene carnivores can be assigned to living families. The major exception are the amphicyonids.

The amphicyonids are an assemblage of medium-to-very-large-sized genera that are known from the early Oligocene to the early Pliocene. They are common in North America and also known in Africa and Eurasia. The bulla shows a primitive arctoid pattern and they were previously placed in the Canidae.

Hunt (1977) and Ginsburg (1977) suggested that amphicyonids are closely related to the ursids, because both of these groups are specialized in having an enlarged inferior petrosal sinus with an excavation in the basioccipital. In bears, this opening houses a loop of the internal carotid that forms a countercurrent exchanger with an adjacent vein to cool the blood going into the brain. The structure of the opening in amphicyonids suggests a comparable function.

Radinsky (1980) described the endocasts of 10 amphicyonid genera that showed a progressive expansion and increased infolding of the neocortex and an increase in relative braincase from an EQ of 0.6 to 0.7 in the Oligocene to 1.19 to 1.4 in the late Miocene, as well as other features that show strong parallels with changes observed in the canids.

We find fossils of clearly distinguishable canids, viverrids, mustelids, and felids in the early Oligocene. Radinsky (1982) showed that the basic cranial proportions of these families were already established when they first appear in the fossil record, although their degree of distinction is clearly lower than that between the modern members of these families (Figure 20-59). He saw no obvious "key" characters that have enabled these groups to expand into the distinctive adaptive zones of their living counterparts, which are based on the method of prey capture and jaw mechanics. He suggested that they may initially have partitioned the carnivore adaptive zones primarily on the basis of size. He attributed the initial success of the modern carnivore families not to any general or particular structural advances but rather to the extinction of the archaic carnivores toward the end of the Eocene. An extensive faunal turnover at this time may have resulted from climatic deterioration. On the other hand, all the modern carnivore families are advanced in the development of an ossified auditory bulla and at least some increase in relative brainsize (Radinsky, 1973, 1975, 1977). All the modern families have fused the scaphoid and lunar

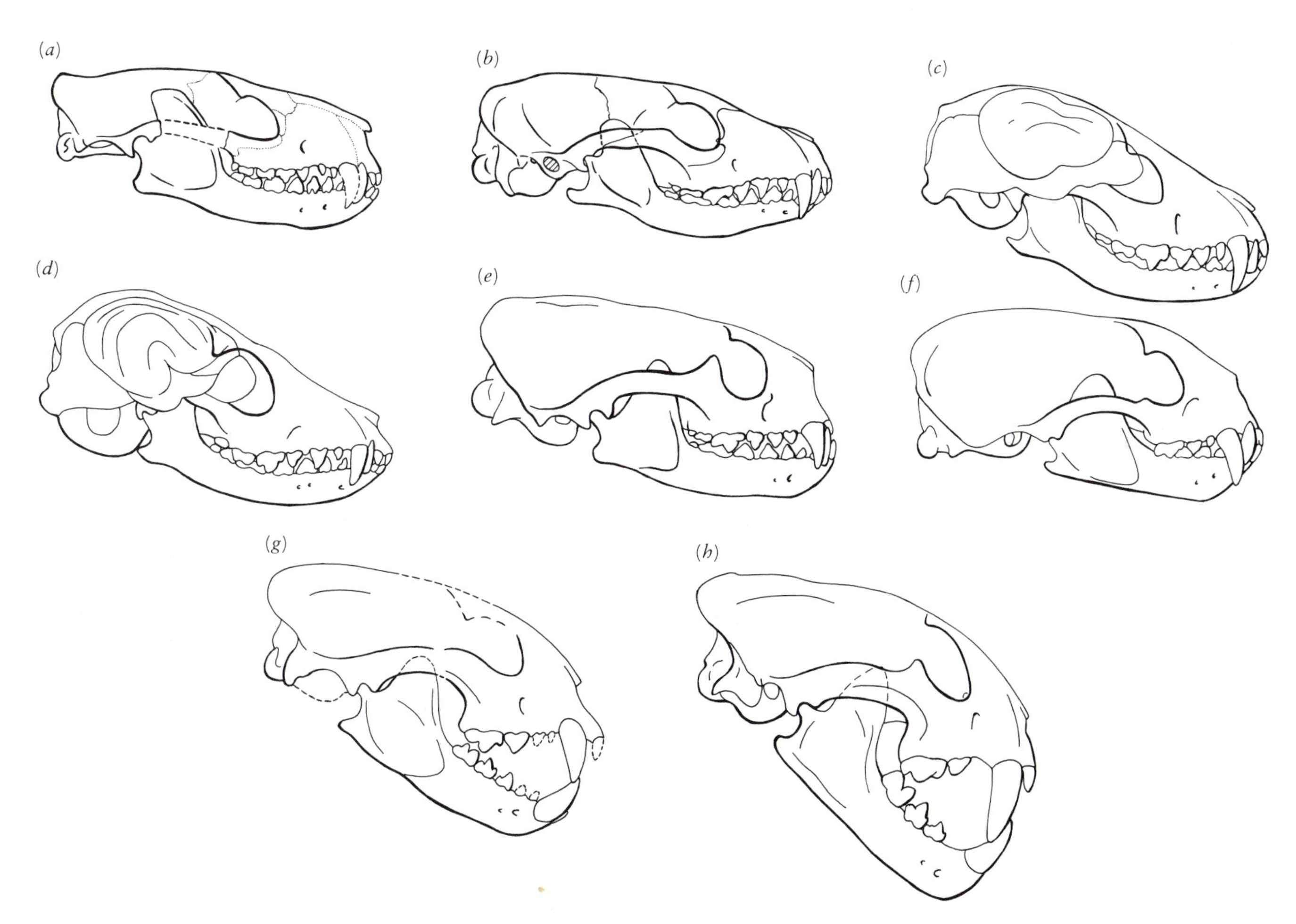

Figure 20-59. COMPARATIVE ILLUSTRATIONS OF THE SKULLS OF MODERN AND EARLY CENOZOIC CARNIVORA. (*a*) *Vulpavus*, a middle Eocene miacid. (*b*) A modern viverrid, *Viverra*. (*c*) The early canid *Hesperocyon*. (*d*) The modern fennec. (*e*) The Miocene mustelid *Potamotherium*. (*f*) The modern South American mustelid *Tayra*. (*g*) The early felid *Proailurus* from the early Miocene. (*h*) The modern clouded leopard. *From Radinsky, 1982.*

bones in the carpus, which suggests more effective running and support of their generally larger body size, although this fusion did occur in at least one miacid.

The Canidae is represented in the early Oligocene by *Hesperocyon.* The main line of canid evolution can be traced in North America. This family appears in Africa and South America only in the Pleistocene (Berta, 1987).

Procyonids (raccoons and their allies) are questionably identified from the early Oligocene, although their remains are easily confused with those of mustelids. The skull proportions of this group vary greatly, but they can be recognized by the blunting of the cusps and the squaring of the molars, which are associated with an omnivorous diet. The lesser panda *Ailurus* (which is sometimes placed in a separate family) apparently arose from the early procyonids and appears in the middle and late Miocene of Asia and Europe.

Bears appear in the middle Oligocene in Europe. *Ursavus* from the middle Miocene appears close to the ancestry of modern bears. The giant Panda *Ailuropoda* is probably an early offshoot of the Ursidae and is first known in the late Miocene of Europe (Thenius, 1979).

Hyaenids appear in the Miocene of Africa. This group apparently evolved from the early viverrids (Savage, 1978).

Cats are recognized in the early Oligocene by the shortness of the skull, the large size of the sectorial carnassials, reduction of other cheek teeth, and the presence of retractile claws. The early cats are further distinguished by the great length of the upper canines, which led to their description as "saber-toothed" cats. These cats are also distinguished by the pattern of the auditory bulla, which may not be completely ossified (Martin, 1980). The division of the bulla that characterizes modern cats is not developed. Because of these primitive features, the early cats have been referred to in an informal way as paleofelids, and some authors have placed them in a distinct family, the Nimravidae (Baskin, 1981; Martin, 1980). Such primitive cats are known until about 7 million years ago. Following this scheme, the term Felidae should be restricted to cats with a more modern type auditory bulla. Such a modern cat, *Proailurus,* is known from the base of the Miocene. The specific relationships of these groups remain uncertain. The monophyly of the cats as a whole seems more firmly established than is the exact nature of the relationships between the two subgroups. Hence, it seems judicious to retain the family name Felidae for all cats while recognizing a distinction between Nimravinae and Felinae at the subfamily level.

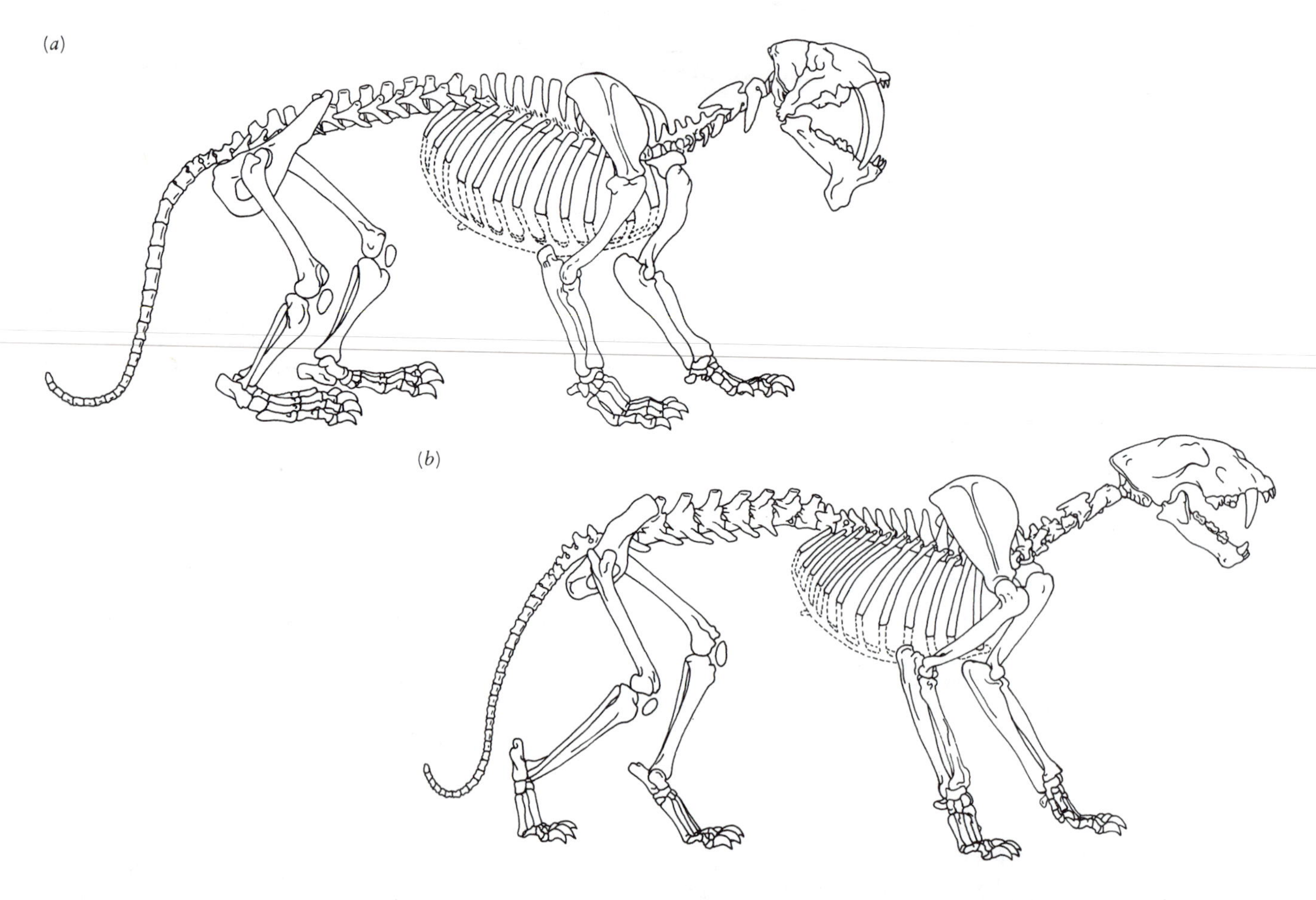

Figure 20-60. SABER-TOOTHED CATS. (*a*) The dirk-toothed cat *Barbourofelis.* (*b*) The scimitar-toothed cat *Machairodus.* *From Martin, 1980.*

The "saber-tooth" specialization is not restricted to the Nimravinae but has evolved independently in several different lineages among both subfamilies. For example, the well-known Pleistocene genus *Smilodon* is a neofelid. Martin (1980) distinguished between two types of saber tooths: the scimitar-toothed cats have short, broad canines, usually with very coarse serrations, and the dirk-toothed cats have long, broad canines, usually with fine serrations. The dirk-toothed cats are short limbed and probably slow running. The scimitar-toothed cats are long limbed and may have been better adapted for pursuit of prey. Both of these patterns developed in both subfamilies (Figure 20-60). The saber-toothed adaptation appears to have been a specialization for feeding on especially large prey. Many of the largest Cenozoic herbivores became extinct at the end of the Pleistocene and with them vanished the last of the saber tooths.

Flynn and Galiano (1982) argued that nimravids are not closely related to felids but have closer affinities with the canids. They base this conclusion on differences in the structure of P^4 and the absence of a septate bulla in the nimravids. The nimravids share with the Caniformes a specialization of the tarsus in which the fibula does not articulate with the calcaenum. Unfortunately, they did not make detailed comparisons with non-nimravid felids, which would be necessary to demonstrate that these two groups did not share a close common ancestor.

Neff (1983) argues on the basis of the basicranial anatomy that nimravids are distinct from both felids and canids and are instead a separate lineage that evolved from primitive miacids independent of both aeluroids and arctoids.

AQUATIC CARNIVORES

In addition to their position as the dominant terrestrial predators, the Carnivora also successfully invaded marine habitats in the middle Tertiary. The seals, sea lions, and walruses have typically been classified together as pinnipeds. However, there is considerable evidence that seals (Phocoidea) evolved from a different group of terrestrial carnivores than that which gave rise to the sea lions and walruses, the Otarioidea (Tedford, 1976).

The Otarioidea is the better-known group, with a fossil record that goes back to the latest Oligocene (Mitchell and Tedford, 1973, Barnes, 1979). The oldest-known genus, *Enaliarctos* (Figure 20-61) shows affinities with the Oligocene ursids of the subfamily Amphicyonodontinae (not to be confused with the family Amphicyonidae). They still retain features of terrestrial carnivores, such as clearly defined carnassial teeth with an enlarged, anteriorly placed protocone. The tympanic bulla shows primitive features that are typical of ursids, but the ectotympanic has become the major element. The mastoid and paroccipital processes are conjoined. The foramina for venous drainage of the skull are enlarged as in modern

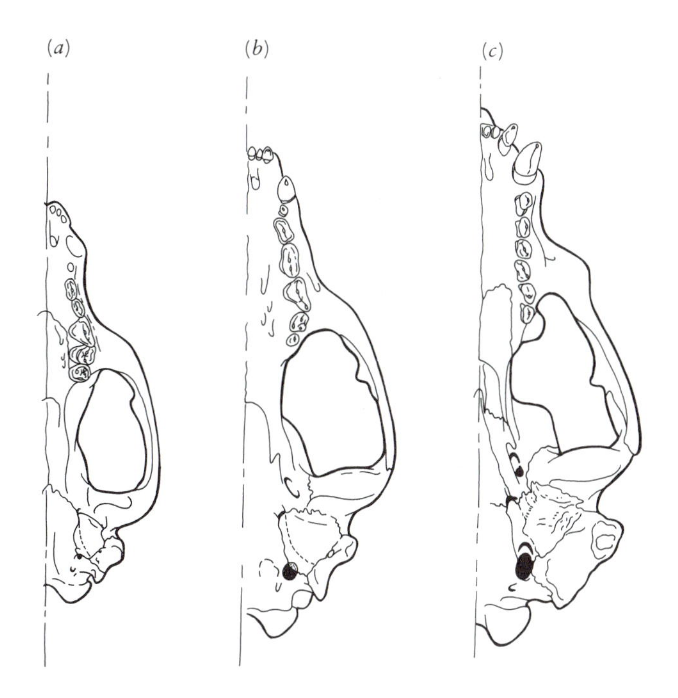

Figure 20-61. COMPARATIVE VENTRAL VIEWS OF THREE ARCTOID CARNIVORES. These skulls show the morphological steps in the evolution of the Otarioidea. (*a*) *Pachycynodon*, Oligocene, France. (*b*) *Enaliarctos*, early Miocene, California. (*c*) *Arctochephalus*, Recent, South Atlantic. *From Tedford, 1976.*

marine mammals, and the dentition is somewhat simplified. The limbs are already modified as flippers.

Three more advanced families have evolved in succession from the primitive enaliarctid stock. All have developed a homodont dentition, probably an adaptation to feeding on small prey that was swallowed whole. The desmatophocids are a second, totally extinct group that was widespread between 10 and 17 million years ago. *Allodesmus* is a well-known member (Figure 20-62). As in *Enaliarctos*, the eyes are large and the otic region shows some specialization for determination of the directionality of sound under water.

The walruses (Odobenidae) succeeded the desmatophocids. They are first known approximately 14 million years ago, represented by the genus *Neotherium*. *Imagotaria*, which lived adjacent to the west coast of North America from 9 to 12 million years ago, probably includes the ancestors of the living species (Figure 20-63). In these

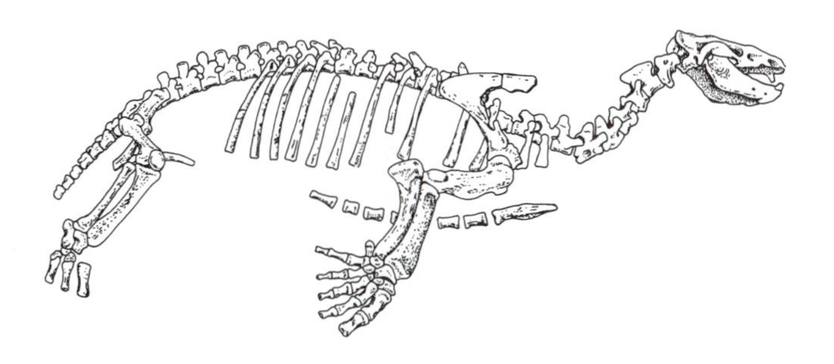

Figure 20-62. SKELETON OF *ALLODESMUS*. *From Mitchell, 1975.*

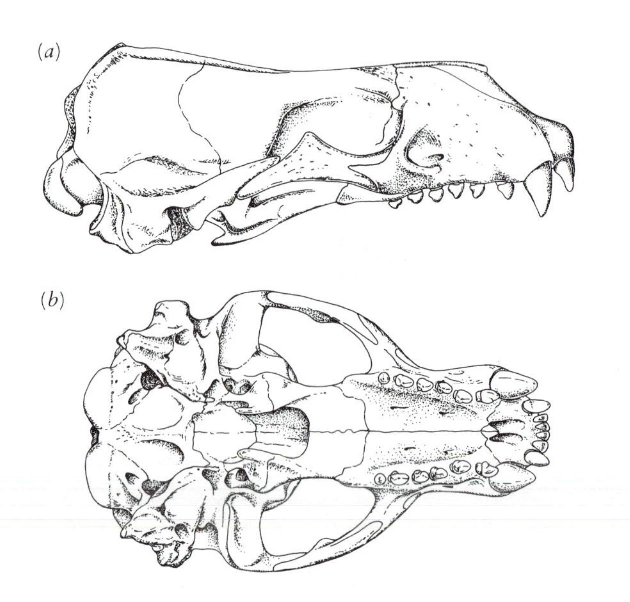

Figure 20-63. SKULL OF *IMAGOTARIA*. This genus is an ancestor of the modern walrus from the Miocene of western North America. (*a*) Lateral and (*b*) palatal views. Original 30 centimeters long. *From Repenning and Tedford, 1977.*

forms, the simplified cheek teeth have only a single root. Most of the history of the Odobenidae is recorded around the Pacific Basin. The walruses established themselves in the North Atlantic between 8 and 5 million years ago and the reentered the North Pacific about 1 million years ago.

The Otariidae (fur seals and sea lions) are first represented by *Pithanotaria* from beds that are 11 million years old. Members of this genus are small and primitive in some respects, but they are already unmistakable otariids in the form of the postcranial skeleton. *Thalassoleon* is a well-known genus from the late Miocene (Figure 20-64). The modern genus *Eumetopias* is known from Japan at least 2 million years ago.

The oldest phocids are *Leptophoca* and *Montherium*, which are based on very incomplete remains from the middle Miocene, (Ray, 1976). In contrast with the Pacific origin of the otarioids, we know the phocoids primarily from the Atlantic Basin and the areas adjacent to the Mediterranean Sea. The origin of the phocoids may be traced to the mustelids. The early Miocene genus *Potamotherium* (Figure 20-65*b*) represents the pattern from which they may have evolved (Savage, 1957). This genus retains primitive features that are characteristic of the mustelids together with evidence of aquatic adaptation parallel to that of the marine otter. In contrast with the early otarioids, the mastoid and paroccipital processes of *Potamotherium* are widely separated. Both phocoid and otarioids show modification of the ear region for determining the direction of sound, but this facility developed separately in the two groups (Repenning, 1972). In contrast with the previous discussion, work being carried out by Wyss (1987) and Berta suggests that phocoids and odobenids share a close common ancestry among the desmatophocids.

Although too late to be actually ancestral, *Semantor* from the Upper Miocene of the Caspian Basin shows anatomical features intermediate between *Potamotherium* and true phocids.

The modern phocid fauna became established approximately 4 million years ago with the beginning of the

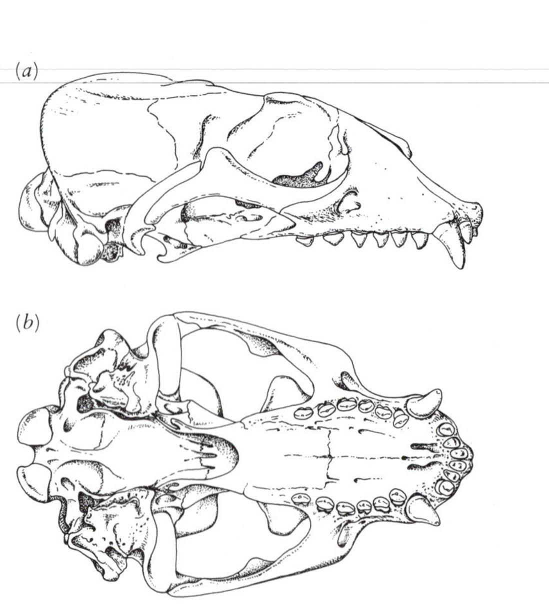

Figure 20-64. SKULL OF *THALASSOLEON*. An ancestor of the fur seals and sea lions from the late Miocene of western North America. (*a*) Lateral and (*b*) palatal views. 25 centimeters long. *From Repenning and Tedford, 1977.*

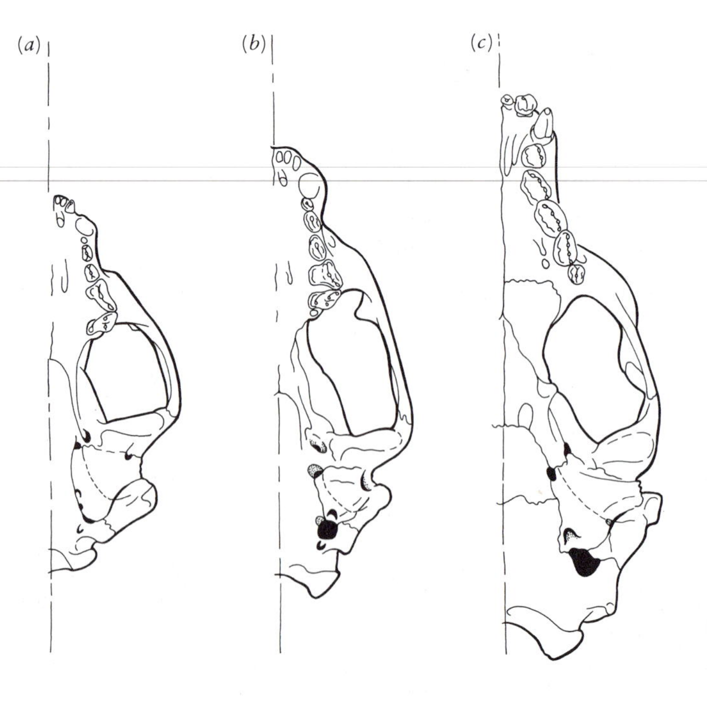

Figure 20-65. COMPARATIVE VENTRAL VIEWS OF THREE ARCTOID CARNIVORES. These genera show the possible morphological steps in the evolution of the Phocoidea. (*a*) *Paragale*, early Miocene, France. (*b*) *Potamotherium*, early Miocene, France. (*c*) *Monachus*, Recent, Caribbean. *From Tedford, 1976.*

climatic deterioration that heralded the Pleistocene. In the late Cenozoic, they extended into the North Pacific and one group spread into the southern hemisphere.

RODENTS, LAGOMORPHS, AND ELEPHANT SHREWS

Most orders of herbivorous placentals consist of medium-to-large-sized animals that probably had a common ancestry among the condylarths in the earliest Paleocene. The small herbivore adaptive niches are filled primarily by members of the single order Rodentia, with the remainder from the order Lagomorpha, the rabbits. Opinion has varied through the years as to whether the general similarities of rabbits and rodents are due to close relationship or convergence. Recently discovered fossils indicate that they probably shared a common ancestry among a group of early Cenozoic families from Asia that are united in the order Anagalida (McKenna, 1975, 1982; Dawson, Li, and Qi, 1984; Hartenberger, 1980). Although they are primarily insectivorous in habit, the elephant shrews may be closely related. The question of the origin of lagomorphs and rodents and many problems of classification within the rodents are discussed in a recent symposium volume edited by Luckett and Hartenberger (1985).

Although there was considerable migration of mammals between Eurasia and North America during some intervals of the Cenozoic, eastern Asia was effectively isolated from western Asia and Europe during the early Tertiary by the Obik Sea and the Turgai Straits, which crossed the west Siberian lowlands and connected the Tethy Sea to the south with the Arctic Ocean (McKenna, 1983b). During the Paleocene, the Bering Land Bridge was at such a high latitude that eastern Asia was also effectively separated from North America. Within this continental area evolved a rich and varied fauna that is little represented in the rest of the world and lacks many common elements of the North American-west European fauna (Li and Ting, 1983).

ANAGALIDA

Members of the order Anagalida are among the most important elements of the early Cenozoic east Asian fauna. Although more than a score of genera have been included in this group (Li and Ting, 1983), only a few have been adequately described and illustrated and their specific interrelationships have not been established.

The Oligocene genus *Anagale* remains one of the best-known forms and, despite its age, appears to represent a generally primitive pattern (Figure 20-66). The skull is small and the temporal opening is confluent with the orbit, although there is a strong postorbital process.

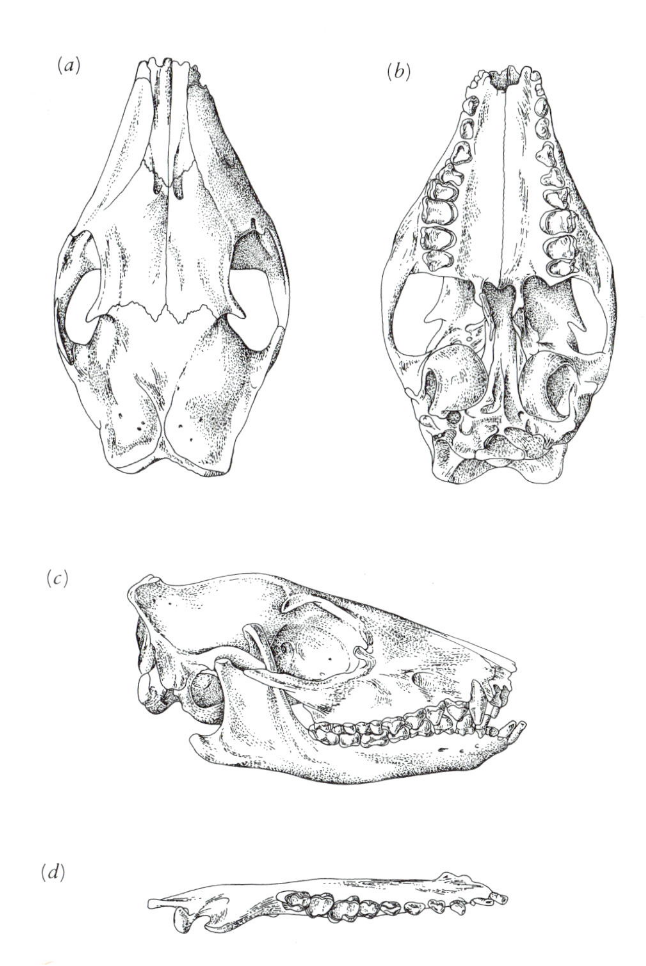

Figure 20-66. *ANAGALE* FROM THE OLIGOCENE OF MONGOLIA. Skull in (*a*) dorsal, (*b*) palatal, and (*c*) lateral views. (*d*) Lower jaw in occlusal view. *From Simpson, 1931.*

The dentition is complete. Primitive features led Simpson (1931) to classify *Anagale* with the tree shrews, but McKenna (1963) showed that it possessed no unique derived features in common with the Tupaiidae.

The upper molars are distinctive in the position of the paracone and metacone near the buccal margin. The posterior premolars are molariform and the cheek teeth are lingually high crowned. The prismatic lower cheek teeth resemble those of lagomorphs. The dentition is clearly distinct from the *Cimolestes* morphotype, which may be ancestral to all other groups discussed in this chapter, and shows some resemblance to that of *Zalambdalestes* (see Figure 20-8).

The jaw is unusual in the great height of the condyle above the tooth row and the short recurved coronoid process. Another advanced feature is the complete ossification of the auditory bulla, which is a compound structure formed from the ectotympanic and entotympanic.

The postcranial skeleton is generally primitive. In contrast with *Zalambdalestes*, the limb proportions are not specialized for saltatory locomotion, and the tibia and fibula are not fused. The astragalus and calcaneum resemble those of lagomorphs. The feet are specialized in having large fissured claws on the manus and distally

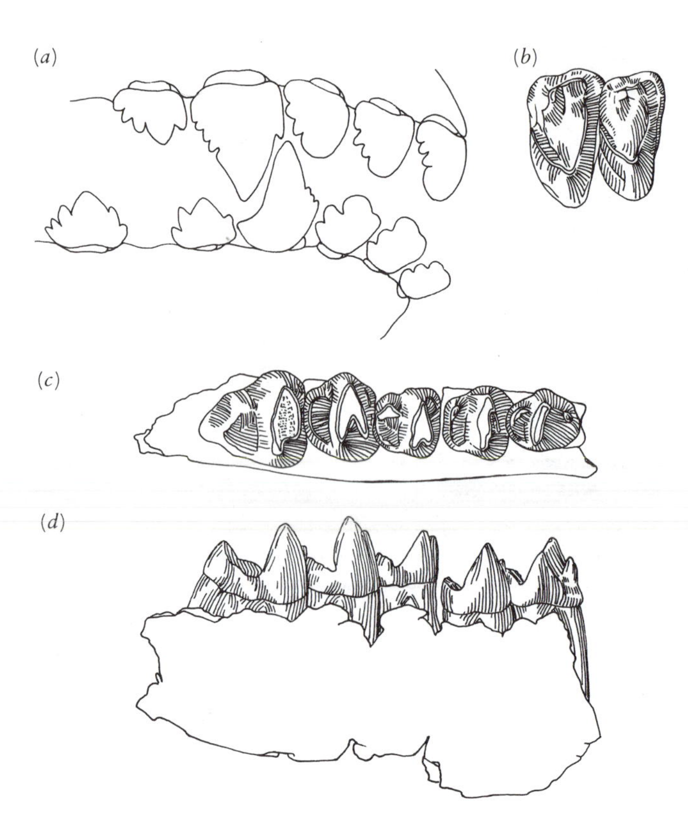

Figure 20-67. DENTITION OF THE ANAGALID *PSEUDICTOPS*. (*a*) Schematic drawing of the anterior dentition. *From Sulimski, 1968.* (*b*) Upper cheek teeth in occlusal view. *From Matthew, Granger, and Simpson, 1929.* (*c* and *d*) Lower cheek teeth in occlusal and lateral views. *From Matthew, Granger, and Simpson, 1929.*

spatulate unguals of the pes. McKenna suggested that they may have dug for food. Ingestion of dirt with the food would account for the high degree of tooth wear.

Pseudictops from the Upper Paleocene of Mongolia represents another family of anagalids in which the cheek teeth have a superficially rabbitlike appearance, with the trigon greatly extended lingually. The anterior dentition, which Sulimski (1968) described, is complete and without a diastema, and the teeth have laterally compressed lobate crowns that show no resemblance to those of either lagomorphs or rodents (Figure 20-67).

MACROSCELIDEA—ELEPHANT SHREWS

Elephant shrews are an assemblage of mouse-to rabbit-sized insectivorous species that are confined today and throughout their known history to Africa. Superficially, they are distinguished by a long flexible snout. Skeletally, they are characterized by the great length of their rear limbs and feet, which enable them to hop rapidly. The ulna and radius are closely appressed and the tibia and fibula are fused. The pollex and hallux are reduced or absent. The auditory bulla is a uniquely compound structure with variable contributions from the ectotympanic, rostral and caudal entotympanics, squamosal, petrosal, alisphenoid, basisphenoid, and pterygoid.

The skull is otherwise primitive in the absence of a postorbital bar and the retention of nearly all of the primitive eutherian complement of teeth

$$\frac{1\text{–}3 \quad 1 \quad 4 \quad 2}{3 \quad 1 \quad 4 \quad 2\text{–}3}$$

The last molar is small or vestigial and the upper fourth premolar is large and molariform. The cheek teeth are specialized for grinding in a similar manner to those of lagomorphs. They are initially low crowned, but the Miocene genus *Myohyrax* (which was originally described as a hyracoid) has high-crowned prismatic molars like those of herbivorous rodents.

We recognize four living genera and five are known only as fossils. Novacek (1984) suggested that the earliest-known genus, *Metoldobotes* (Figure 20-68) from the mid-

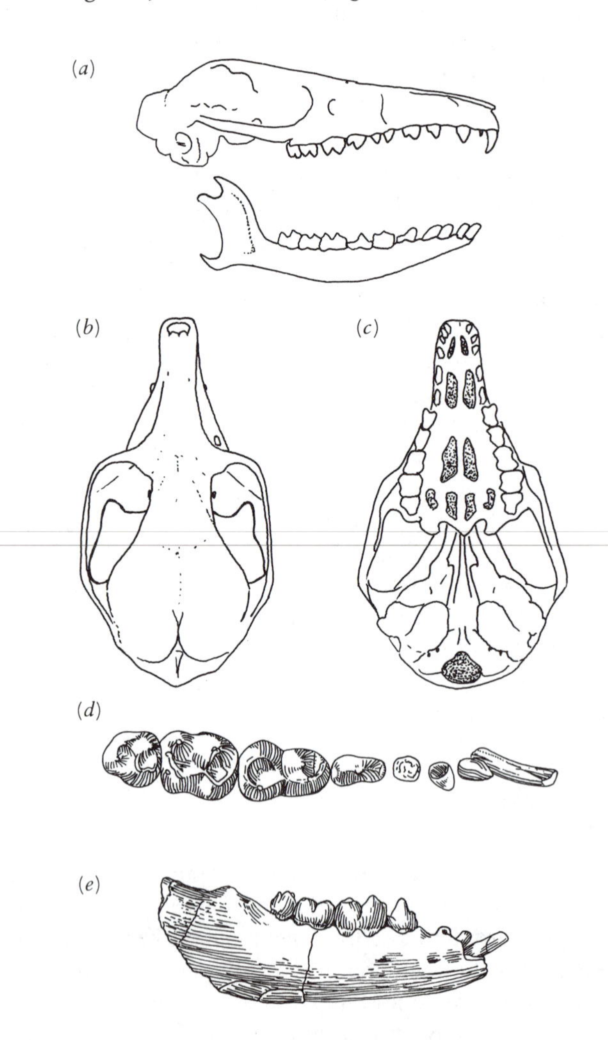

Figure 20-68. SKULL OF THE MODERN ELEPHANT SHREW *ELEPHANTULUS*. (*a*) Lateral, (*b*) dorsal, and (*c*) palatal views. (*d*) Occlusal view of dentition and (*e*) lateral view of the lower jaw of the mid-Oligocene elephant shrew *Metoldobotes*, jaw 2½ centimeters long. (*a*) *From Lawlor, 1979;* (*b* and *c*) *from Novacek, 1984;* (*d* and *e*) *from Patterson, 1965.*

Oligocene, may represent an early radiation of the family rather than having close affinities with any modern genera. The modern genus *Rhynchocyon* is known as early as the Miocene.

Elephant shrews were long allied with the Insectivora because of their generally primitive anatomy and were grouped with the tree shrews as menotyphlans because of the presence of an intestinal caecum. Recent authors do not recognize any shared derived features that unite them with either the Tupaiidae or any insectivore families.

McKenna (1975), followed by Szalay (1977), Hartenberger (1980), and Novacek (1982, 1984) suggested close affinities with anagalids, rodents, lagomorphs, and zalambdalestids. McKenna united these groups on the basis of a common loss of the original P_3^3 from among the five premolars (which are thought to be primitive for eutherians), the height of the jaw condyle above the occlusal surface, and the hook-shaped coronoid process. All show a tendency toward unilaterally hypsodont or prismatic teeth with reduced stylar shelves and a lagomorph-like chewing action, in which the wear surfaces of the lower molar trigonids continue posterodorsally the wear surfaces of the talonids of the teeth in front of them. Stressing tarsal similarities, Szalay placed elephant shrews with the lagomorphs. Evans (1942) made the strongest arguments for a specific relationship between anagalids and elephant shrews. The auditory bulla of elephant shrews is quite unlike that of *Anagale,* but both may have developed an ossified bulla subsequent to their divergence, even if they had an immediate common ancestry in the early Cenozoic. Novacek (1984) drew attention to the similarities of elephant shrews to members of the order Rhinogradentia (Stümpke, 1967).

Despite the structural similarities observed between elephant shrews and the Asian anagalids and their relatives, the rodents and lagomorphs, it should be emphasized that there is no evidence of modern or fossil macroscelids outside of Africa. Since they are not known prior to the Oligocene, elephant shrews may have evolved within Africa from a primitive eutherian stock that was quite distinct from the Asian groups.

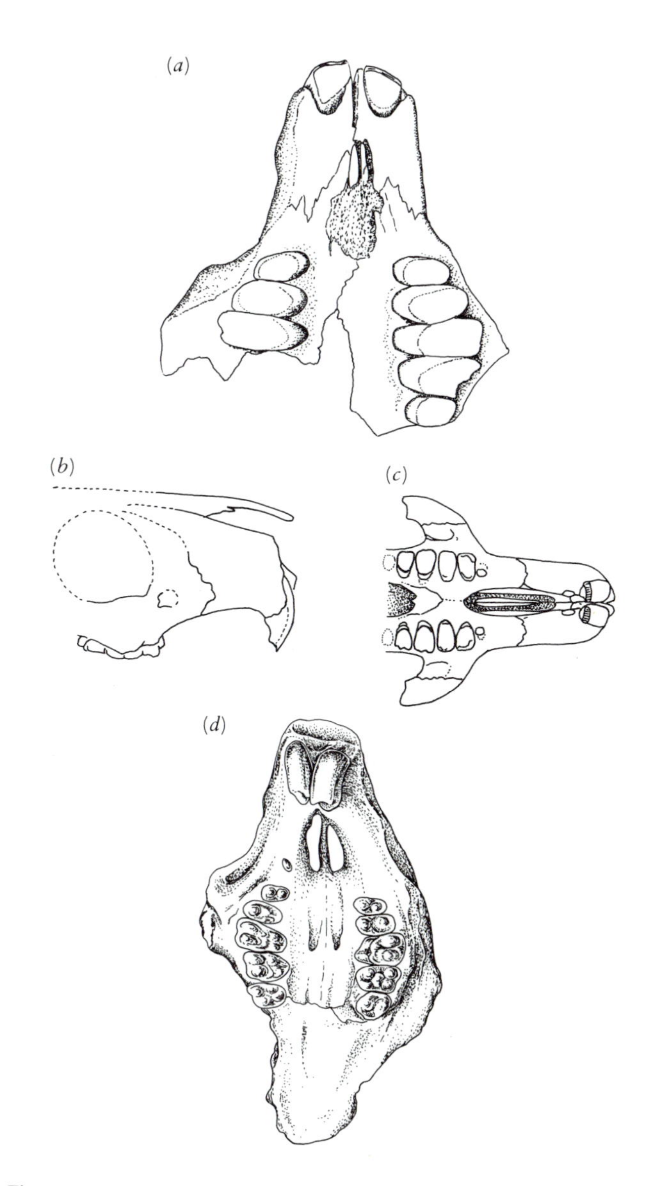

Figure 20-69. (*a*) View of the anterior end of the palate of the anagalid *Eurymylus,* ×3. *From Sych, 1971.* Snout of *Mimolagus* from the Oligocene of Mongolia—which is thought to be one of the most primitive lagomorphs—in (*b*) lateral and (*c*) palatal views, ×2. *From Bohlin, 1951.* (*d*) Palate of *Heomys,* known from the Middle and Late Paleocene of China, ×3. This genus may be closer to the ancestry of rodents. *From Li and Ting, 1985.*

EURYMYLOIDEA

The groups that have just been discussed are characterized by a primitive tooth pattern that lacks a diastema. Among the families grouped within the Anagalida, the eurymyloids more closely resemble rodents and rabbits in the presence of a long diastema between the anterior incisors and the cheek teeth. *Eurymylus* from the Upper Paleocene of Mongolia combines characteristics of both lagomorphs and rodents but is clearly distinct from both (Figure 20-69*a*). The upper and lower incisors are ever growing and have a double layer of enamel, as in rodents. The lower incisors extend posteriorly to the base of M_3. As in rodents, there is a long palatal bridge anterior to the internal nares. In contrast, the cheek teeth resemble the simple triangular pattern of lagomorphs and show no resemblance to the complex multicuspate teeth of early rodents. The talonid of one tooth and the trigonid of the next form a common wear surface, and the wear pattern indicates that horizontal transverse movements of the mandible were significant. On the other hand, *Eurymylus* is clearly distinct from lagomorphs in the absence of the second pair of upper incisors and the loss of P^2. The angular process is reflected, as in rodents and some very primitive placentals and marsupials, but in contrast with rabbits.

According to Li and Ting (1985), the eurymylids may be especially close to the ancestry of rodents. In this family, the functional separation between gnawing and chewing, which is characteristic of rodents, has already been established to judge from the elongation of the glenoid and the difference in length between the diastema of the upper and lower jaws. The dental formula of

$$\frac{1\ 0\ 2\ 3}{1\ 0\ 2\ 3}$$

is comparable to that of rodents, except for the retention of one extra lower premolar. The molar teeth of *Heomys* from the middle Paleocene (Figure 20-69*d*) closely resemble those of the earliest ctenodactyloid rodents such as *Cocomys*.

Li and Ting point out dental similarities between eurymylids and a lower jaw from the later Cretaceous that was described as *Barunlestes* (Kielan-Jaworowska and Trofimov, 1980) in the great elongation of the second lower incisor and the anteroposterior compression of the trigonids. They argue that this specimen is not a true zalambdalestid but is closer to other Asian "proteutherians" than to leptictids. These arguments imply that eurymylids and other anagalids may have diverged from the basal placental stock earlier than other groups discussed in this chapter.

Li and Ting suggested that the lagomorphs may be closely related to a further eurymyloid family, the Mimotonidae. *Mimotoma* from the late Paleocene and *Mimolagus* from the Oligocene have a dental formula of

$$\frac{2(?)\ 0\ 3\ 3}{2\text{–}1\ 0\ 3\ 3}$$

and have only one layer of enamel on the incisors.

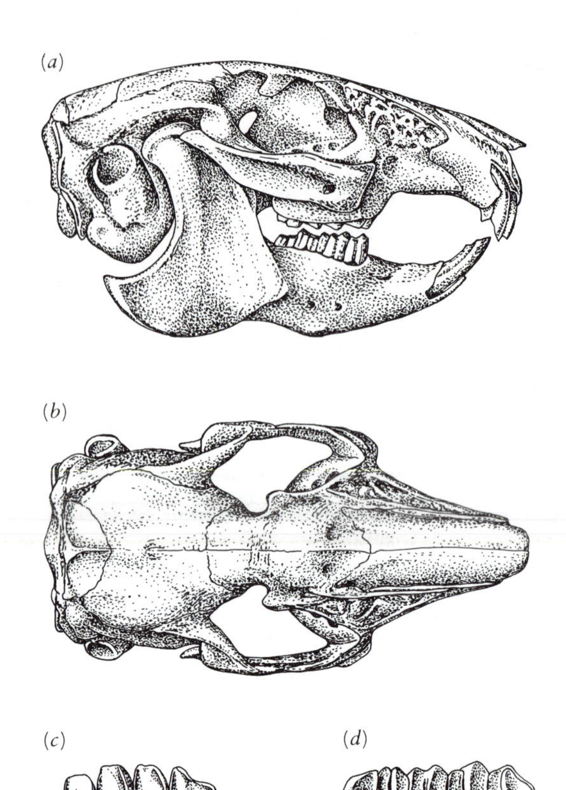

Figure 20-70. *PALAEOLAGUS.* Skull and dentition of the Oligocene lagomorph. *From Wood, 1940.*

LAGOMORPHA

Modern lagomorphs are a clearly defined group consisting of rabbits, hares, and pikas. Rabbits and hares are placed in the family Leporidae, which includes 11 living and 21 fossil genera. *Ochotona* is the only surviving member of the Ochotonidae, but this family was considerably more diverse in the Tertiary, with 23 fossil genera.

The general pattern of the skull and dentition are very similar throughout this order. All are clearly distinguished from rodents by the presence of a second pair of incisors immediately behind the first (Figure 20-70). There is only a single layer of enamel on their anterior surface, in contrast with two layers in rodents. The pattern of mastication in the two orders is fundamentally different. Lagomorphs are characterized by a great deal of transverse grinding, while propalinal movement is extremely important among rodents since their first appearance in the fossil record. In most lagomorphs, the maxilla is conspicuously fenestrated. There is only a very short secondary palate behind the long incisive foramen. In contrast with *Anagale*, the auditory bulla is made up entirely of the ectotympanic without an entotympanic.

The postcranial skeleton is also very stereotyped in lagomorphs, with the tail short or rudimentary. The tibia and fibula are fused distally, and there are five digits on the fore limb and four on the rear.

McKenna (1982) recognized *Hsiuannania* and *Mimotona* from the Paleocene of China as the earliest lagomorphs. As in later members of this order, the molar teeth exhibit lingual extension of the anterior and posterior cingula to form lingually coalescing columns, the pericone and hypocone (Figure 20-71). A comparison with their anagalid ancestors demonstrates that the central cusp, whose homology has long been questioned, is the protocone. The paracone and metacone are far lateral in position. The talonid basin also shifts buccally to become a mere hypoflexid. In advanced lagomorphs, the crown of the teeth is crossed by a thin band of enamel, the hypostria, each end of which is marked by a deep reentrant, which gives the teeth the shape of a much depressed figure 8 in occlusal view.

Among early lagomorphs, the upper teeth develop lingual hypsodonty as the enamel border crosses the cheek teeth at an angle. In advanced genera, the teeth become rootless.

Connections between the Paleocene mimotonids and later lagomorphs are not well known. Lagomorphs from

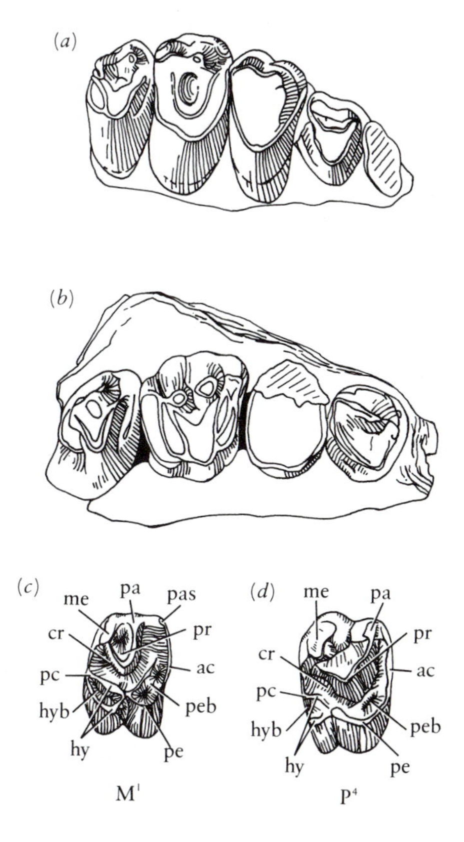

Figure 20-71. HOMOLOGIES OF THE ANAGALID AND LAGOMORPH UPPER-CHEEK-TOOTH PATTERN. The lagomorph P^4 and upper molars have a central protocone. In *Hsiuannania, Huaiyangale, Linnania,* eurymylids, mimotonids, and pseudictopids, the protocone is still lingual but a hypoconal shelf and a pericone are developing. The basic lagomorph upper-cheek-tooth pattern involves a progressive buccal shift of the primordeal tooth-crown cusps. The talonid basin of the lower molars also shifts buccally to become a mere hypoflexid. (*a*) *Hyaiyangale chianshanensis,* right P^4–M^3, Paleocene, Anhui. (*b*) *Hsiuannania tabiensis,* right P^4–M^3, Paleocene, Anhui. (*c*) *Palaeolagus tennodon,* right M^1, early Oligocene. (*d*) *Palaeolagus temnodon,* right P^4, early Oligocene. Abbreviations as follows: ac, anterior cingulum; cr, crescent (posterior part indicated); hy, hypocone (subdivided by secondary basin); myb, secondary hypocone basin; pc, posterior cingulum; mc, metacone; pa, paracone; pas, parastyle; pe, pericone; peb, pericone basin; pr, protocone. Drawings are not to scale. *From McKenna, 1982.*

the middle to late Eocene of Asia and the late Eocene of North America have a dental structure from which both the Leporidae and the Ochotonidae could have evolved. Leporids are known from the late Eocene in North America; ochotonids appear in the Oligocene. McKenna (1982) listed many features that distinguish these groups and differentiate them from more primitive lagomorphs.

Lagomorphs spread into Europe by the beginning of the Oligocene. Ochotonids reached Africa in the early Miocene, but the leporids did not arrive until the Pliocene. Rabbits reached South America only in the Pleistocene.

Lagomorphs and rodents apparently diverged from a common ancestor in the early Cenozoic. Both groups are clearly defined and relatively stereotyped in their anatomy, but they show very different patterns of radiation (Simpson, 1959). We recognize only two or three families of lagomorphs during the entire Cenozoic, and only 12 genera with 46 species are living today. Rodents radiated into approximately 50 families, with over 400 genera recognized in the modern fauna. In a detailed analysis of mastication, Weijs and Dantuma (1981) suggested that the particular type of tooth occlusion that is common to all lagomorphs enabled them to utilize a wide variety of vegetable matter so that each species could occupy a very broad adaptive zone. The geographical ranges of many lagomorph species and genera are extremely wide and their numbers are prodigious. White and Keller (1984) argued that the pattern of tooth occlusion is related to a particular configuration of the skull and jaw musculature that was not readily amenable to change. In contrast, the dental apparatus and pattern of mastication in rodents has certainly been much more variable, and dietary specializations allowed them to become partitioned into a host of separate niches.

RODENTS

Rodents may be considered the most successful of all mammals in terms of their worldwide distribution, taxonomic diversity, and number of individuals. We recognize more than 1700 species in the recent fauna. About 50 families evolved during the Cenozoic, of which approximately one-fourth are now extinct (Carleton, 1984). Rodents vary considerably in size, with *Eumegamys* from the Pliocene of South America having a skull nearly 60 centimeters long (Dawson and Krishtalka, 1984), but most are small and their basic anatomy remains quite stereotyped.

Rodents are all characterized by a particular pattern of the dentition and jaw mechanics that is probably the basis for their enormously successful radiation into a wide range of habitats. From their first appearance in the fossil record in the early Tertiary, they had reduced their dental formula to

$$\frac{1\quad 0\quad 2\quad 3}{1\quad 0\quad 1\quad 3}$$

As in lagomorphs, the upper and lower incisors are ever growing. The enamel is limited to the labial surface, and the lower incisors wear across the uppers so that both retain a chisel-like surface. All rodents have two layers of enamel, in contrast with one in lagomorphs. Gnawing with the incisors occurs independently of chewing with the cheek teeth. The masseter and pterygoideus muscles move the jaw forward to gnaw and backward to chew. This propalinal movement of the lower jaw is utilized for the chewing stroke in some advanced families.

Differences in the jaw musculature and the histology of the enamel on the incisors form the basis for rodent classification. The most fundamental subdivision is based on the configuration of the lower jaw that reflects the length of the pterygoideus muscle (Figure 20-72). In the primitive condition, termed **sciurognathous** (and typified by squirrels, mice, and rats), the internal pterygoideus is relatively short and the angle of the jaw originates below the plane of the lower incisors. In the more derived **hystricognathous** condition (which is typified by porcupines, a few African forms, and all native South American rodents), the internal pterygoid is longer and the angle originates lateral to the plane of the incisors. The hystricognathous condition probably evolved only once.

All members of the Hystricognathi are also characterized by a wavy pattern of the hydroxyapatite crystals making up the inner layer of enamel. This enamel pattern is termed **multiserial.** Primitive rodents, including all Eocene species, have poorly organized bands of inner enamel, called the **pauciserial** condition. A second derived pattern, termed **uniserial,** with alternating bands of similarly oriented crystals, occurs in advanced rodents that retain a sciurognathous jaw pattern (Wahlert, 1968).

The importance of propalinal movement of the lower jaw in rodents has resulted in modifications of the origin of the masseter above and anterior to the tooth row. The superficial masseter maintains the position that is common to other mammals. The deep masseter becomes divided into two layers, a middle layer beneath the superficial masseter and a deep layer. Among primitive rodents, the origin of the deep masseter remains associated with the zygomatic arch, as in primitive eutherians. This pattern is termed **protrogomorphous** (Figure 20-73). In the **sciuromorphous** pattern, the middle layer of the masseter originates anterior to the zygomatic arch, lateral and dorsal to the infraorbital foramen. In other groups, the infraorbital foramen is enlarged and some portion of the masseter passes through it. In the **hystricomorphous** condition, which is common to both Old and New World porcupines, the infraorbital foramen is very large and the deep masseter passes through it and originates on the snout. In the **myomorphous** condition, which is exemplified by rats and mice, the infraorbital foramen is keyhole-shaped and a portion of the deeper masseter passes through the enlarged dorsal area. The middle layer of the masseter (masseter lateralis) extends forward onto a platelike surface, anterior to the zygomatic arch.

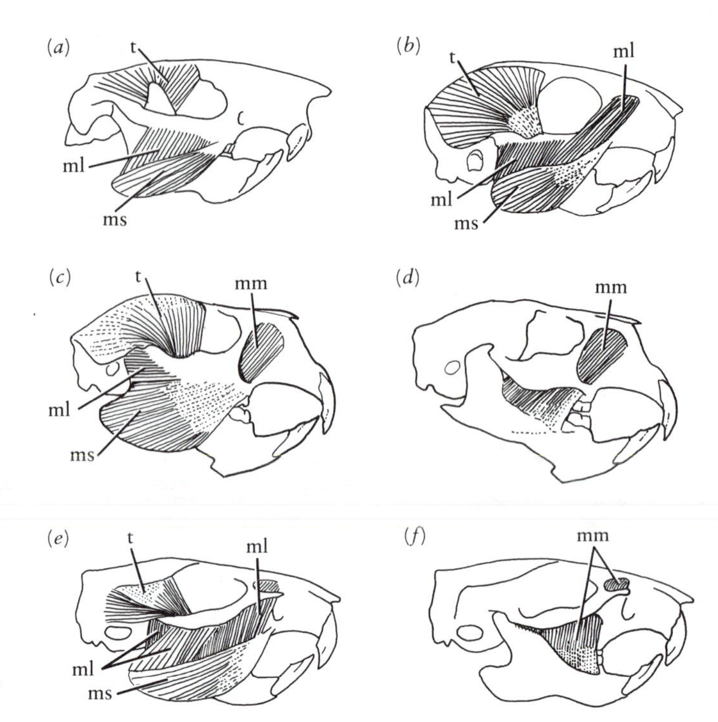

Figure 20-73. ZYGOMASSETERIC PATTERNS IN RODENTS. (*a*) *Ischryotomus* (Paramyidae), primitive Eocene rodent. The masseter muscles originate entirely on the zygomatic arch. (*b*) Abert's squirrel (*Sciurus aberti*, Sciuridae); the anterior part of the masseter lateralis originates on the rostrum and the zygomatic plate. (*c*) Porcupine (*Erethizon dorsatum*, Erethizontidae); the anterior part of the masseter medialis originates largely on the rostrum and passes through the enlarged infraorbital foramen. (*d*) Porcupine, showing the attachments of the masseter medialis. (*e*) A cotton rat (*Sigmodon hispidus*, Cricetidae); the masseter superficialis originates on the rostrum, and the anterior part of the masseter lateralis originates on the anterior extension of the zygomatic arch. (*f*) The superficial muscles shown in (*e*) have been removed; the masseter medialis originates partly on the rostrum and passes through the narrow infraorbital foramen. Abbreviations as follows: ml, masseter lateralis; mm, masseter medialis; ms, masseter superficialis; t, temporalis. *From Vaughan, 1972.*

The sciuromorph, hystricomorph, and myomorph conditions were once used as the basis for a threefold subordinal division of all advanced rodents. We now recognize that all three patterns have been achieved more than once and that transitions from one condition to another are possible as well. They remain important descriptive terms, but they have lost their original taxonomic significance.

Like the lagomorphs, rodents appear to have evolved from eurymyloid anagalids within the Paleocene of Asia (Dawson, Li, and Qi, 1984). The genus *Heomys* resembles rodents in many features, but the teeth are already more high crowned than are those of the earliest-known unquestioned rodents.

The earliest-known rodent is *Acritoparamys* [*Paramys*] *atavus,* which is represented by isolated teeth from

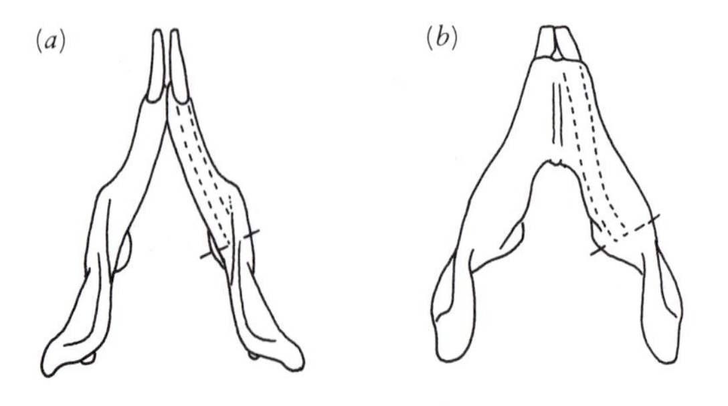

Figure 20-72. (*a*) Sciurognathy. Demonstrated by a marmot in which the angle of the jaw arises below the incisor and (*b*) hystricognathy, as demonstrated by a New World porcupine in which the angle originates lateral to the plane of the incisor. Dashed lines represent the incisor in the socket. *From Jacobs, 1984.*

the Upper Paleocene of North America. Dawson and her colleagues argue that the Lower Eocene *Cocomys* from China is more primitive in the nonmolariform pattern of P^4_4 (Figure 20-74). Wood (1962) had earlier argued that this condition would be expected in the immediate ancestors of known rodents. The upper and lower fourth premolars also remain nonmolariform in the eurymyloid *Heomys*.

Cocomys and *Acritoparamys* apparently represent two diverging lines of early rodents. *Cocomys* is a member of a relatively restricted assemblage, the Ctenodactyloidea, that was common in Asia in the Eocene and is represented today in Africa by four genera that superficially resemble the pikas. The structure of the lower jaw and dentition is primitive, but the masseter pattern is hystricomorphous. The family Ischyromyidae, including *Acritoparamys*, may be close to the ancestry of all other rodents.

The ischyromyids (including the paramyids of Wood, 1962, 1977) appear first in North America but may have evolved in Asia. They provide good examples of the primitive rodent morphotype (Figure 20-75). *Paramys* resembled a squirrel in its general appearance, although the limbs were relatively shorter and no specifically arboreal adaptations are evident. The jaw was sciurognathous, and the arrangement of the masseter protrogomorphous. The dental formula is primitive for the order and the cheek teeth are rooted and retain vestiges of their bunodont heritage (Figure 20-76).

The ischyromyids may include the ancestry of a number of conservative modern lineages, but the specific origin of several more advanced groups remains controversial. In the nature of the jaw, distribution of the masseter muscle, and the tooth formula, the most conservative living descendants of the primitive rodent assemblage are the Aplodontidae, which is represented by the sewellel or mountain beaver that is restricted to the west coast of North America (Rensberger, 1983). It is tentatively linked to the ischyromyids via the late Eocene genus *Eohaplomys*. The middle and late Tertiary Mylagaulidae is a related family that includes genera that developed "horn"-like extensions from the nasal bones.

The modern family Sciuridae shows the best-documented transition from the ischyromyids. The early Oligocene squirrel *Protosciurus* retains an essentially protrogomorphous pattern of the masseter and a very primitive dentition, but the otic region has already achieved the unique derived pattern of modern genera, in which the enlarged septate tympanic bulla is tightly fused to the periotic. The postcranial skeleton is virtually identical to that of the living fox squirrel, *Sciurus niger*, which indicates that it was fully arboreal. Miocene members of the genus *Sciurus* are so similar to living squirrels that Emry

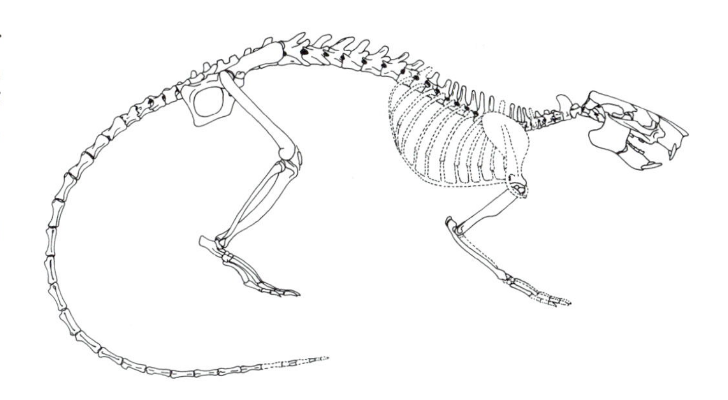

Figure 20-75. SKELETON OF THE LOWER EOCENE RODENT *PARAMYS*. *From Wood, 1962.*

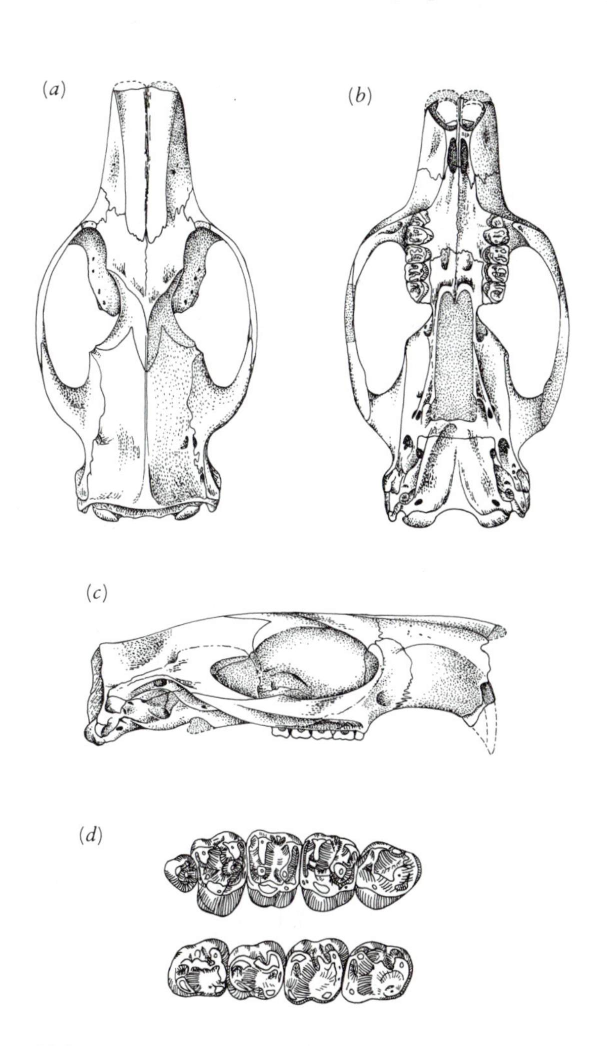

Figure 20-76. SKULL OF *PARAMYS*. (*a*) Dorsal, (*b*) palatal, and (*c*) lateral views. (*d*) Occlusal views of upper and lower cheek teeth. *From Wood, 1962.*

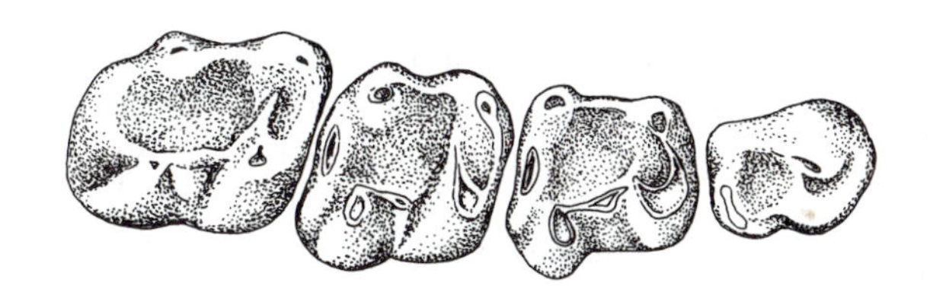

Figure 20-74. TEETH OF THE PRIMITIVE ASIAN RODENT *COCOMYS*. Occlusal view of P_4–M_3. *From Dawson, Li, and Qi, 1984.*

and Thorington (1984) considered this genus to be a "living fossil," despite its worldwide distribution and great species diversity in the modern fauna.

The beavers and their relatives, the Castoroidea, have been classified with squirrels as members of the conservative suborder Sciuromorpha, but the sciuromorphous pattern of the masseter may have been achieved separately in the two groups. The Castoridae has been restricted to the Holarctic region since its appearance in the Oligocene. The Pleistocene genus *Castoroides* reached the size of a bear.

Most modern dormice, placed in the Gliroidea, have achieved a myomorphous pattern of the masseter independent of the genera classified as the Myomorpha. However, one modern African genus *Graphiurus*, is virtually protrogomorphous. The oldest-known glirid, *Eogliravus* from the middle Eocene of Europe, has an enlarged infraorbital foramen, as in the hystricomorphs, but in more advanced genera the lateral portion of the masseter has extended anteriorly to produce the myomorphous pattern. The gliroids can be traced to the *Microparamys* lineage within the Ischyromyidae.

While hystricomorphy and myomorphy were achieved among the gliroids, which are classified within the Sciuromorpha, the earliest members of the Myomorpha, the Geomyoidea, have an essentially sciuromorphous pattern. This superfamily is classified with the muroids and dipodoids on the basis of dental similarities and knowledge of intermediate forms from the fossil record.

According to Korth (1984), the enormous myomorph assemblage may have evolved from the family Sciuravidae, which diverged from the base of the ischyromyid radiation in the early Eocene. Eomyids from the later Eocene can be united with the geomyoids based on the common presence of a long infraorbital canal that is depressed into the rostrum and a tooth pattern with transverse lophs (Wahlert, 1985). Geomyoids are represented in the modern fauna by the heteromyids (kangaroo rats and pocket mice), which appear in the early Oligocene of North America, and the Geomyidae (pocket gophers).

Muroids and dipodoids may share a common ancestry in the late Eocene and early Oligocene. Myomorphy in muroids, as in the gliroids, appears to have been derived via an intermediate hystricomorphous stage. The Dipodoidea, which include jumping mice (Zapodidae) and jerboas (Dipodidae), appear in the late Oligocene of Eurasia.

The muroids are the most diverse assemblage of rodents, with approximately 1135 living species. Because of their basically similar anatomy, all have been classified in a single family. On the other hand, Chaline and his coauthors (1977) recognized eight families. Carleton and Musser (1984) divided the assemblage into 15 living and 11 extinct subfamilies. Dawson and Krishtalka (1984) recognized four families, but nearly all species can be placed either in the Muridae, which includes 460 species of Old World rats and mice, or the Cricetidae, which includes a host of fossil forms as well as the modern hamsters and voles. The cricetids are known from the late Eocene of China and the early Oligocene of North America and Europe. The murids first appear in the Middle Miocene of southern Asia. They are restricted to the Old World and are the only group of rodents to reach Australia.

The complexity of muroid radiation is daunting (Figure 20-77) but the potential for evolutionary studies is enormous. They have a rich fossil record that is easily recovered through bulk washing and screening. The teeth show a complex morphology with a great variety of patterns. The group shows a succession of radiations throughout the late Cenozoic that has led to a great many new species within the past 3 million years. According to Chaline (1977), *Microtus* gave rise to 217 species and subspecies in North America in fewer than 1.5 million years. Jacobs (1984) cited the appearance of 30 genera and 180 species of cricetids in South America within the past 3.5 million years and over 100 species of murids in Africa during the past 10.5 million years.

Two families from Africa occupy very isolated positions among the rodents. Both the Anomaluridae (scaly tailed flying squirrels, with three living genera) and the Pedetidae (a single species, the spring hare) are known from the Miocene, but the early fossils resemble the living species too closely to help in establishing their ancestry. Both are hystricomorphous but sciurognathous.

All remaining rodents have been included in the Hystricognathi, or Hystricomorpha, used as a suborder. The structure of the lower jaw and the multiserial pattern of the inner layer of enamel are generally accepted as demonstrating that all had a common ancestry. Most, but not all, are hystricomorphous. However, common ancestry has proven difficult to document from the fossil record.

This assemblage is clearly divided into two groups on the basis of their distribution—the primarily South American caviomorphs and several groups that are predominantly African. No member of this assemblage is definitely known before the middle to late Oligocene, at which time they first appear in both Africa and South America. This pattern is exactly the same as that of the anthropoid primates and exactly the same phylogenetic and zoogeographical problems are being argued (Ciochon and Chiarelli, 1980b).

It has long been assumed that all members of the Hystricognathi shared a common origin among northern hemisphere rodents, but no particular genus or family has been convincingly demonstrated as occupying an ancestral position. Lavocat (1978, 1980) contended that African hystricognathous genera arose from the Eurasian thryonomyoids, which are characterized by the presence of five transverse crests on the cheek teeth. He suggested that the South American caviomorphs arrived from Africa by rafting across the then much narrower South Atlantic during the Eocene. On the other hand, Wood (1980, 1983) and Patterson and Wood (1982) argued that the South American caviomorphs are more primitive when they first appear in the fossil record with, among other features,

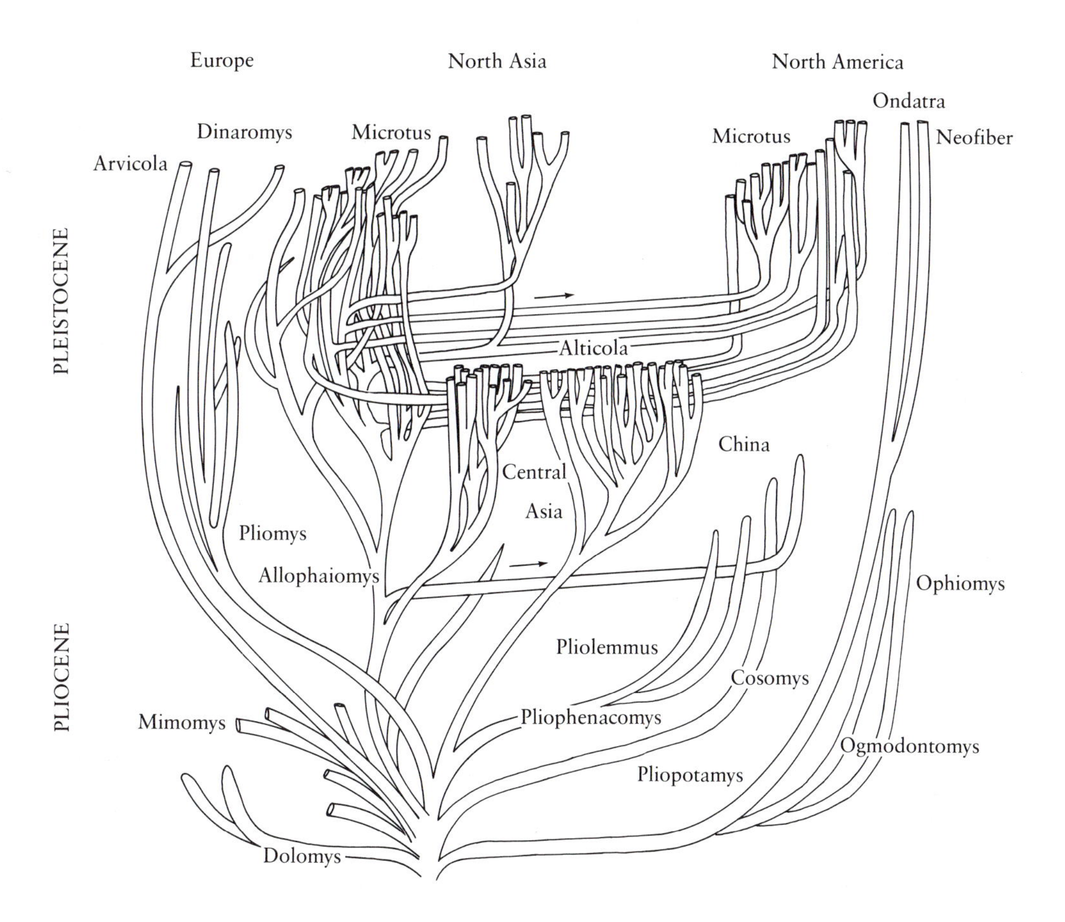

Figure 20-77. RADIATION OF VOLES. *From Chaline, 1977. By permission.*

only four crests on the cheek teeth and so could not have evolved from the African genera (Figure 20-78). Rather, they suggested, the Caviomorpha evolved from a Middle American assemblage termed the Franimorpha, which they considered to have had an incipiently hystricognathous lower jaw. The ancestral caviomorphs would have had to cross the water gap between North and South America by waif dispersal or rafting, but the distance was shorter than that between Africa and South America. Korth (1984) argued that all the genera that Wood included in the Franimorpha should be classified with other families of primitive rodents and that none show any trace of hystricognathous jaw structure. This problem remains unresolved.

Lavocat (1973) united the Thyronomyoidea, Bathyergidae, Hystricidae, the Oligocene family Phiomyidae, and several other extinct groups in the infraorder Phiomorpha to include all the Old World hystricognaths. Among the modern families the Bathyergidae, or mole rats, have only five living genera. They are exceptional in not having an hystricomorphous arrangement of the masseter. However, Miocene fossils indicate that this condition was present in early members of the family. They can be traced back to the late Oligocene of Asia. The Hystricidae, or Old World porcupines, include three living genera; their fossil record goes back to the late Miocene in India, Africa, and Europe. The Pertromyidae (dassie rats) and Thryonomyidae (cane rats), both of which include only a single living genus, are placed in the superfamily Thyronomyoidea with a record from the Oligocene.

The South American native rodents are a much more closely knit group and are all included in the infraorder Caviomorpha. The early evolution of this assemblage was most recently discussed by Patterson and Wood (1982).

Seven families of caviomorph rodents are recognized in the middle Oligicene of Bolivia (Figure 20-79). In all, the dentition is reduced to

$$\frac{1\ \ 0\ \ 1\ \ 3}{1\ \ 0\ \ 1\ \ 3}$$

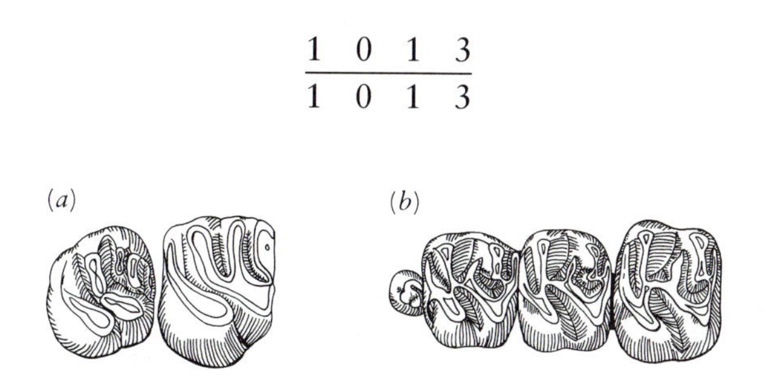

Figure 20-78. DENTITION OF OLD WORLD AND SOUTH AMERICAN MEMBERS OF THE HYSTRICOGNATHI. (*a*) The caviomorph *Branisamys* from the Middle Oligocene, LP^4–M^1. (*b*) The Eurasian thryonomyoid *Metaphiomys*, Ldm^{3-4}, M^{1-2}. *From Patterson and Wood, 1982.*

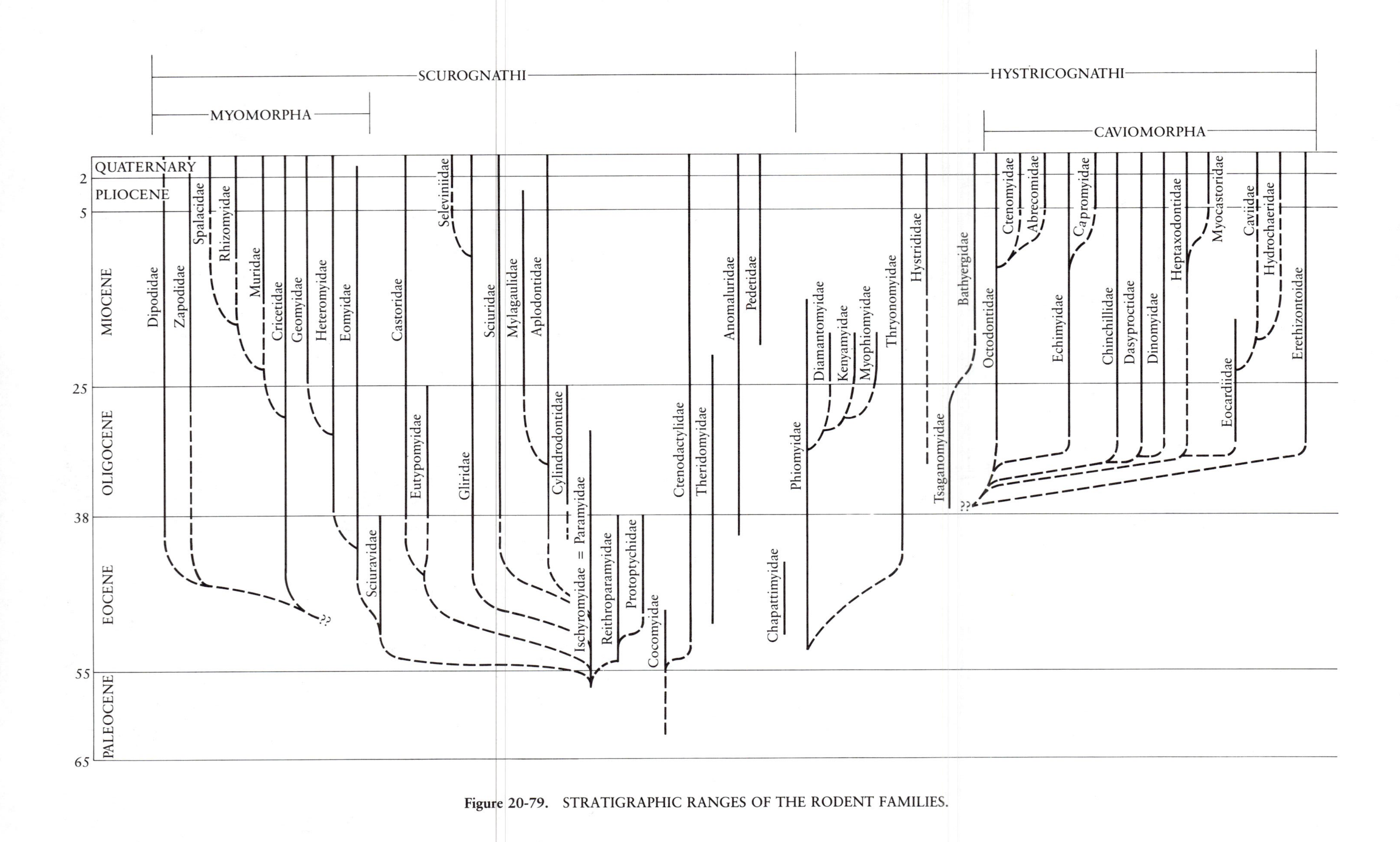

Figure 20-79. STRATIGRAPHIC RANGES OF THE RODENT FAMILIES.

They are actually quite similar to one another in dental morphology but are classified according to the divergence of their descendants. Early members of the Erethizontoidea, the ancestors of the New World porcupine, are the most divergent, but the remainder may have shared a recent common ancestry. Members of the Chinchilloidea, Cavioidea, and Octodontoidea are all present in this early fauna. The Octodontidae may be closest to the ancestry of the remaining groups. Only two of the sixteen families of caviomorphs recorded in the Tertiary are extinct. The capybaras extended into southern North America during the Pleistocene and the porcupine has spread throughout the forested portion of this continent.

Cricetid rodents from North America expanded greatly in South America at the end of the Cenozoic, into niches that were apparently not occupied by any of the native rodents.

SUMMARY

The long gestation period of placental mammals is possible due to the evolution of a new embryonic tissue, the trophoblast, which is responsible for preventing rejection of the embryo by the maternal immune system.

Fossils of eutherian mammals are known as early as the Lower Cretaceous. Many specimens have been described from the Upper Cretaceous of Mongolia. *Kennalestes* and *Asioryctes* may be close to the ancestry of most later placental mammals. *Zalambdalestes* and *Barunlestes* represent a divergent lineage that shares some similarities with rodents and lagomorphs.

A diverse fauna of placentals is known from the latest Cretaceous in North America, but most species are known from little more than jaws and teeth. A continuous depositional sequence preserves fossils across the Maastrichtian-Paleocene boundary and records the early stages in the radiation of Cenozoic mammals. Most of the major orders of mammals are known by the Eocene, but their specific interrelationships remain difficult to establish. Their classification is based primarily on the pattern of molar cusps. Each major group can be distinguished by specialized characters, but it is more difficult to discover shared derived features that demonstrate specific interrelationships between the major orders.

Eurasia and North America were intermittently in contact throughout the Cenozoic. Africa was separated from Eurasia for much of the early Tertiary and South America was separated from North America except at the very end of the Cenozoic.

Apatemyids, pantolestids, and leptictids are early Tertiary mammal groups that show only a limited radiation and whose specific affinities with the major placental orders have not been established.

Taeniodonts, tillodonts, pantodonts, and uintatheres were medium-to-large-sized herbivores that were common in the early Tertiary but left no descendants. Their origins and interrelationships have not been established. Advanced members are clearly distinguished by divergent specializations of their dentition. Taeniodonts are known only in western North America, but the pantodonts and uintatheres (Dinocerata) are known in both North America and Asia. Only the tillodonts are known in Europe as well.

Living members of the order Insectivora retain features of the skull, dentition, and soft anatomy that may be little altered from the pattern of late Mesozoic eutherians. Shrews, moles, hedgehogs, and tenrecs share a common ancestry that is distinct from that of other placental orders as indicated by the pattern of the bones around the orbit, the absence of a bony dorsum sellae, and the reduced pubic symphysis.

Fossil bats, which are very similar to living species, are known from the early Eocene. The structure of the ear region indicates that they already had the capacity for echolocation. Their dentition resembles that of Paleocene insectivores related to shrews.

Primates appear at the very end of the Cretaceous. Most Paleocene genera belong to the Plesiadapiformes, which are characterized by a dentition resembling that of rodents. More typical primates, the adapids and omomyids, are common in the Eocene. They include the ancestors of the modern prosimians. Members of the Anthropoidea appear in the Lower Oligocene of Africa and the middle to late Oligocene of South America. Their specific origin has not been established. Hominids diverged from African apes between 6 and 10 million years ago. The ancestry of modern humans may be traced through the species *Australopithecus afarensis, Homo habilis,* and *Homo erectus* to *Homo sapiens.*

Bats, primates, tree shrews, and the Dermoptera (flying lemurs) have been grouped as the Archonta, which suggests that they had a close common ancestry. This assumption is supported by molecular evidence but has not been confirmed by unique skeletal characters that are common to the primitive members of these groups.

The creodonts were an archaic order of carnivores that were common in the late Paleocene and Eocene. The order Carnivora is represented in the Paleocene and early Eocene by the primitive miacids and viverravids. The modern families emerged in the late Eocene and early Oligocene. The fossil record of marine carnivores demonstrates that the sea lions and walruses evolved from relatives of primitive bears. Seals may share a common ancestry with walruses.

Rodents and lagomorphs both evolved from early Cenozoic anagalids, which were common in eastern Asia. The elephant shrews, although known only in Africa, may have emerged from a similar ancestral stock.

REFERENCES

American Journal of Physical Anthropology. (1982). Pliocene Hominid Fossils from Hadar, Ethiopia. Vol. 57(4).

Archibald, J. D. (1982). A study of Mammalia and geology across the Cretaceous-Tertiary boundary in Garfield County, Montana. *Univ. Calif., Publ. Geol. Sci.,* **122**: 1–286.

Ba Maw, Ciochon, R. L., and Savage, D. E. (1979). Late Eocene of Burma yields earliest anthropoid primate, *Pondaungia cotteri. Nature,* **282**: 65–67.

Badgley, C. E. (1984). Human Evolution. In P. D. Gingerich and C. E. Badgley (eds.), *Mammals. Notes for a Short Course. Univ. Tennessee, Studies Geol.,* **8**: 182–198.

Barnes, L. G. (1979). Fossil enaliarctine pinnipeds (Mammalia: Otariidae) from Pyramid Hill, Kern County, California. *Contr. Sci., Nat. Hist. Mus. Los Angeles County,* **318**: 1–41.

Baskin, J. A. (1981). *Barbourofelis* (Nimravidae) and *Nimravides* (Felidae), with a description of two new species from the late Miocene of Florida. *J. Mammal.,* **62**: 122–139.

Beliajeva, E. E., Trovimov, B. A., and Reshetov, V. J. (1974). General stages in evolution of late Mesozoic and early Tertiary mammalian faunas in Central Asia. In H. H. Kpamapehko (ed.), *Mesozoic and Cenozoic Faunas and Biostratigraphy of Mongolia,* pp. 19–45. [In Russian.]

Berta, A. (1987). Origin, diversity, and zoogeography of South American Canidae. In B. D. Patterson and R. M. Timm (eds.), *Studies in Neotropical Mammalogy: Essays in Honor of Philip Hershkovitz. Fieldiana Zoology,* New Series. (in press).

Bohlin, B. (1951). Some mammalian remains from Shi-ehr-ma-ch'eng, Hui-hui-p'u area, Western Kansu. VI. Vertebrate Palaeontology 5. Reports from the Scientific Expedition to the North-Western Provinces of China under Leadership of Dr. Sven Hedin. *The Sino-Swedish Expedition Publ.,* **35**: 1–47.

Butler, P. M. (1972). The problem of insectivore classification. In K. A. Joysey and T. S. Kemp (eds.), *Studies in Vertebrate Evolution,* pp. 253–265. Oliver and Boyd, Edinburgh.

Butler, P. M. (1978). Insectivores and Chiroptera. In V. J. Maglio and H. B. S. Cooke (eds.), *Evolution of African Mammals,* pp. 56–68. Harvard University Press, Cambridge.

Butler, P. M. (1980). The tupaiid dentition. In W. P. Luckett (ed.), *Comparative Biology and Evolutionary Relationships of Tree Shrews,* pp. 171–204. Plenum Press, New York.

Carleton, M. D. (1984). Introduction to rodents. In S. Anderson and J. K. Jones, Jr., (eds.), *Orders and Families of Recent Mammals of the World,* pp. 255–265. Wiley, New York.

Carleton, M. D., and Musser, G. G. (1984). Muroid rodents. In S. Anderson and J. K. Jones, Jr., (eds.), *Orders and Families of Recent Mammals of the World,* pp. 289–379. Wiley, New York.

Chaline, J. (1977). Rodents, evolution and prehistory. *Endeavour, New Ser.,* **1**: 44–51.

Chaline, J., Mein, P., and Petter, F. (1977). Les grandes lignes d'une classification évolutive des Muroidea. *Mammalia,* **41**: 245–252.

Chow Min-Chen, Chang Yu-Ping, Wang Ban-Yue, and Ting Su-Yin. (1973). New mammalian genera and species from the Paleocene of Nanhsiung, N. Kwantung. *Vert. Pal. Asiat.,* **11**(1): 31–35 [In Chinese.]

Chow Min-Chen, Chang Yu-Ping, Wang Ban-Yue, and Ting Su-Yin. (1977). Mammalian fauna from the Paleocene of Nanxiong Basin, Guangdong. *Lab. Vert. Pal. Peking, Mem.,* **20**: 1–100.

Cifelli, R. L. (1983). The origin and affinities of the South American Condylarthra and early Tertiary Litopterna (Mammalia). *Am. Mus. Novitates,* **2772**: 1–49.

Ciochon, R. L., and Chiarelli, A. B. (eds.). (1980a). *Evolutionary Biology of the New World Monkeys and Continental Drift.* Plenum Press, New York.

Ciochon, R. L., and Chiarelli, A. B. (1980b). Concluding remarks. In R. L. Ciochon and A. B. Chiarelli (eds.), *Evolutionary Biology of the New World Monkeys and Continental Drift,* pp. 495–501. Plenum Press, New York.

Ciochon, R., and Corruccini, R. S. (eds.). (1983). *New Interpretations of Ape and Human Ancestry.* Plenum Press, New York.

Clemens, W. A. (1973). Fossil mammals of the type Lance Formation, Wyoming. Part III. Eutheria and Summary. *Univ. Calif. Publ. Geol. Sci.,* **94**: 1–102.

Coombs, M. C. (1983). Large mammalian clawed herbivores: A comparative study. *Trans. Am. Phil. Soc.,* **73**(7): 1–96.

Cope, E. D. (1875). Systematic catalogue of Vertebrata of the Eocene of New Mexico collected in 1874. *Report to Engineers Dept., U.S. Army. Geographical Explorations and Surveys West of the 100th Meridian. Wheeler Wash., Govt. Printing Office,* pp. 1–37.

Cronin, J. E., and Sarich, V. M. (1980). Tupaiid and Archonta phylogeny: The macromolecular evidence. In W. P. Luckett (ed.), *Comparative and Evolutionary Relationships of Tree Shrews,* pp. 293–312. Plenum Press, New York.

Dawson, M. R., and Krishtalka, L. (1984). Fossil history of the families of recent mammals. In S. Anderson and J. Knox Jones (eds.), *Orders and Families of Recent Mammals of the World,* pp. 11–57. Wiley, New York.

Dawson, M. R., Chuan-Kuei Li, and Tao Qi. (1984). Eocene ctenodactyloid rodents (Mammalia) of Eastern and Central Asia. *Carnegie Mus. Nat. Hist., Spec. Pub.,* **9**: 138–150.

Delson, E. (ed.). (1985). *Ancestors: The Hard Evidence.* Liss, New York.

Emry, R. J., and Thorington, R. W., Jr. (1984). The tree squirrel *Sciurus* (Sciuridae, Rodentia) as a living fossil. In N. Eldredge and S. M. Stanley (eds.), *Living Fossils,* pp. 23–31. Springer-Verlag, New York, Berlin, Heidelberg, Tokyo.

Evans, F. G. (1942). The osteology and relationships of the elephant shrews (Macroscelididae). *Bull. Am. Mus. Nat. Hist.,* **80**: 85–125.

Fleagle, J. G., and Kay, R. F. (1983). New interpretations of the phyletic position of Oligocene hominoids. In R. Ciochon and R. S. Corruccini (eds.), *New Interpretations of Ape and Human Ancestry,* pp. 181–210. Plenum Press, New York.

Flynn, J., and Galiano, H. (1982). Phylogeny of the Early Tertiary Carnivora, with a description of a new subgenus of *Protictis* from the Middle Eocene of northwestern Wyoming. *Am. Mus. Novitates,* **2725**: 1–64.

Fox, R. C. (1975). Molar structure and function in the early Cretaceous mammal *Pappotherium:* Evolutionary implications for Mesozoic Theria. *Can. J. Earth Sci.,* **12**: 412–442.

Fox, R. C. (1984). *Paranyctoides maleficus* (New Species), an early eutherian mammal from the Cretaceous of Alberta. In R. M. Mengel (ed.), *Papers in Vertebrate Paleontology Honoring Robert Warren Wilson. Carnegie Mus. Nat. Hist., Spec. Papers,* **9**: 9–20.

Gazin, C. L. (1953). The Tillodontia: An early Tertiary order of mammals. *Smithsonian Misc. Coll.,* **121**(10): 1–110.

George, S. B. (1986). Evolution and biogeography of soricine shrews. *Syst. Zool.,* **35**: 153–162.

Gingerich, P. D. (1976). Cranial anatomy and evolution of early Tertiary Plesiadapidae (Mammalia, Primates). *Univ. Michigan, Papers on Paleont.,* **15**: 1–140.

Gingerich, P. D. (1980a). Eocene Adapidae, paleobiogeography, and the origin of South American Platyrrhini. In R. L. Ciochon and A. B. Chiarelli (eds.), *Evolutionary Biology of the New World Monkeys and Continental Drift,* pp. 123–138. Plenum Press, New York.

Gingerich, P. D. (1980b). *Tytthaena parrisi,* oldest known oxyaenid (Mammalia, Creodonta) from the Late Paleocene of Western North America. *J. Paleont.,* **54**(3): 570–576.

Gingerich, P. D. (1981). Early Cenozoic Omomyidae and the evolutionary history of tarsiiform primates. *J. Human Evol.,* **10**: 345–374.

Gingerich, P. D. (1983). Systematics of early Eocene Miacidae (Mammalia, Carnivora) in the Clark's Fork Basin, Wyoming. *Univ. Mich., Contr. Mus. Paleont.,* **26**: 197–225.

Gingerich, P. D. (1984). Primate evolution. In P. D. Gingerich and C. E. Badgley (eds.), *Mammals. Notes for a Short Course. Univ. Tennessee, Studies Geol.,* **8**: 167–181.

Gingerich, P. D. (1986). Early Eocene *Cantius torresi*—oldest primate of modern aspect from North America. *Nature,* **319**: 319–321.

Gingerich, P. D., and Badgley, C. E. (eds.). (1984). *Mammals. Notes for a Short Course. Univ. Tennessee, Studies Geol.,* **8**: 1–234.

Gingerich, P. D., and Gunnell, G. F. (1979). Systematics and evolution of the genus *Esthonyx* (Mammalia, Tillodontia) in the early Eocene of North America. *Univ. Mich., Mus. Paleont. Contr.,* **25**: 125–153.

Gingerich, P. D., and Sahni, A. (1984). Dentition of *Sivaladapis nagrii* (Adapidae) from the late Miocene of India. *Intl. J. Primatology,* **5**: 63–79.

Ginsburg, L. (1977). *Cynelos lemanensis* (Pomel), carnivore ursidé de l'Aquitanien d'Europe. *Ann. Paléont. (vert.),* **63**: 57–104.

Gregory, W. K. (1957). *Evolution Emerging.* Vol. 2. Macmillan, New York.

Guth, C. (1962). Un insectivore de Menat. *Ann. Paléont.,* **48**: 1–10.

Hartenberger, J.-L. (1980). Données et hypothéses sur la radiation initiale des rongeurs. *Palaeovertebrata, Mem. Jubilée R. Lavocat,* 285–301.

Hoffstetter, R. (1980). Origin and deployment of New World monkeys emphasizing the southern continents route. In R. L. Ciochon and A. B. Chiarelli (eds.), *Evolutionary Biology of the New World Monkeys and Continental Drift,* pp. 103–122. Plenum Press, New York.

Howell, F. C. (1978). Hominidae. In V. J. Maglio and H. B. S. Cooke (eds.), *Evolution of African Mammals,* pp. 154–248. Harvard University Press, Cambridge.

Hunsaker, D. (ed.). (1977). *The Biology of Marsupials.* Academic Press, New York.

Hunt, R. M. (1974). The auditory bulla in carnivora: An anatomical basis for reappraisal of carnivore evolution. *Jour. Morph.,* **143**: 21–76.

Hunt, R. M. (1977). Basicranial anatomy of *Cynelos* Jourdan (Mammalia: Carnivora) an Aquitanian amphicyonid from the Allier Basin, France. *J. Paleont.,* **51**: 826–843.

Jacobs, L. L. (1980). Siwalik fossil tree shrews. In W. P. Luckett (ed.), *Comparative Biology and Evolutionary Relationships of Tree Shrews,* pp. 205–216. Plenum Press, New York.

Jacobs, L. L. (1984). Rodentia. In P. D. Gingerich and C. E. Badgley (eds.), *Mammals. Notes for a Short Course. Univ. Tennessee, Stud. Geol.,* **8**: 155–166.

Jepsen, G. L. (1970). *Biology of Bats. Vol. 1. Bat Origins and Evolution.* Academic Press, New York and London.

Jerison, H. (1973). *Evolution of the Brain and Intelligence.* Academic Press, New York.

Johanson, D., and Edey, M. (1981). *Lucy. The Beginnings of Humankind.* Simon and Schuster, New York.

Johnston, P. A., and Fox, R. C. (1984). Paleocene and Late Cretaceous mammals from Saskatchewan, Canada. *Palaeontographica, Abt. A.,* **186**: 163–222.

Kaufman, M. H. (1983). The origin, properties and fate of trophoblast in the mouse. In Y. W. Loke and A. Whyte (eds.), *Biology of Trophoblast,* pp. 23–68. Elsevier, Amsterdam.

Kay, R. F., and Cartmill, M. (1977). Cranial morphology and adaptations of *Palaechthon nacimienti* and other Paromomyidae (Plesiodapoidea, ? Primates), with a description of a new genus and species. *J. Human Evol.,* **6**: 19–53.

Kielan-Jaworowska, Z. (1968). Preliminary data on the Upper Cretaceous eutherian mammals from Bayn Dzak, Gobi Desert. In Z. Kielan-Jaworowska (ed.), Results of the Polish-Mongolian Expeditions. Part I. *Palaeontologia Polonica,* **19**: 171–191.

Kielan-Jaworowska, Z. (1975). Possible occurrence of marsupial bones in Cretaceous eutherian mammals. *Nature,* **255**: 698–699.

Kielan-Jaworowska, Z. (1977). Evolution of the therian mammals in the Late Cretaceous of Asia. Part II. Postcranial skeleton in *Kennalestes* and *Asioryctes.* In Z. Kielan-Jaworowska (ed.), Results of the Polish-Mongolian Palaeontological Expeditions. Part VII. *Palaeontologia Polonica,* **37**: 65–83.

Kielan-Jaworowska, Z. (1978). Evolution of the therian mammals in the Late Cretaceous of Asia. Part III. Postcranial skeleton in Zalambdalestidae. In: Z. Kielan-Jaworowska (ed.), Results of the Polish-Mongolian Palaeontological Expeditions. Part VIII. *Palaeontologia Polonica,* **38**: 3–41.

Kielan-Jaworowska, Z. (1980). Les premieres mammiferes. *Recherche,* **11**(108): 146–155.

Kielan-Jaworowska, Z. (1981). Evolution of the therian mammals in the Late Cretaceous of Asia. Part IV. Skull structure in *Kennalestes* and *Asioryctes.* In Z. Kielan-Jaworowska (ed.), Results of the Polish-Mongolian Palaeontological Expeditions. Part IX. *Palaeontologia Polonica,* **42**: 25–78.

Kielan-Jaworowska, Z. (1984a). Evolution of the therian mammals in the Late Cretaceous of Asia. Part V. Skull structure in Zalambdalestidae. In Z. Kielan-Jaworowska (ed.), Results of the Polish-Mongolian Palaeontological Expeditions. Part X. *Palaeontologia Polonica,* **46**: 107–117.

Kielan-Jaworowska, Z. (1984b). Evolution of the therian mammals in the Late Cretaceous of Asia. Part VI. Endocranial casts of eutherian mammals. In Z. Kielan-Jaworowska (ed.), Results of the Polish-Mongolian Palaeontological Expeditions. Part X. *Palaeontologia Polonica,* **46**: 157–171.

Kielan-Jaworowska, Z. (1984c). Evolution of the therian mammals in the Late Cretaceous of Asia. Part VII. Synopsis. In Z. Kielan-Jaworowska (ed.), Results of the Polish-Mongolian Palaeontological Expeditions. Part X. *Palaeontologia Polonica,* **46**: 173–183.

Kielan-Jaworowska, Z., Bown, T. M., and Lillegraven, J. A. (1979). Eutheria. In J. A. Lillegraven, Z. Kielan-Jaworowska, and W. A. Clemens (eds.), *Mesozoic Mammals. The First Two-Thirds of Mammalian History,* pp. 221–258. University of California Press, Berkeley.

Kielan-Jaworowska, Z., and Trofimov, B. A. (1980). Cranial morphology of the Cretaceous eutherian mammal *Barunlestes. Acta Palaeontol. Polonica,* **25**(2): 167–186.

Kimbel, W. H., White, T. D., and Johanson, D. C. (1984). Cranial morphology of *Australopithecus afarensis:* A comparative study based on a composite reconstruction of the adult skull. *Am. J. Phys. Anthrop.,* **64**: 336–388.

Kirsch, J. A. (1977). The six-percent solution: Second thoughts on the adaptedness of the Marsupialia. *Am. Scientist,* **65**(3): 276–288.

Koenigswald, W. von (1980). Das Skelett eines Pantolestiden (Proteutheria, Mammalia) aus dem mittleren Eozän von Messel bei Darmstadt. *Paläontol. Z.,* **54**(3-4): 267–287.

Koenigswald, W. von, and Storch, G. (1983). *Pholidocercus hassiacus,* ein Amphilemuride aus dem Eozän der "Grube Messel" bei Darmstadt (Mammalia, Lipotyphyla). *Senckenbergiana lethaea,* **64**: 447–495.

Korth, W. W. (1984). Earliest Tertiary evolution and radiation of rodents in North America. *Bull. Carnegie Mus. Nat. Hist.,* **24**: 1–71.

Krishtalka, L., and Setoguchi, T. (1977). Paleontology and geology of the Badwater Creek area, central Wyoming. Part 13. The late Eocene Insectivora and Dermoptera. *Annals Carnegie Mus. Nat. Hist.,* **46**: 71–99.

Lavocat, R. (1973). Les rongeurs du Miocene d'Afrique orientale—I. Miocene inferieur. Trav. Mem. Inst. Ecole Pratique des Hautes Etudes. *Inst. de Montpellier Mem.,* **1**: 1–284.

Lavocat, R. (1978). Rodentia and Lagomorpha. In V. J. Maglio and H. B. S. Cooke (eds.), *Evolution of African Mammals,* pp. 69–89. Harvard University Press, Cambridge.

Lavocat, R. (1980). The implications of rodent paleontology and biogeography to the geographical sources and origin of platyrrhine primates. In R. L. Ciochon and A. B. Chiarelli (eds.), *Evolutionary Biology of New World Monkeys and Continental Drift,* pp. 93–102. Plenum Press, New York.

Lawlor, T. E. (1979). *Handbook to the Orders and Families of Living Mammals.* Mad River Press, Eureka, California.

Lay, D. M. (1972). The anatomy, physiology, functional significance and evolution of specialized hearing organs of gerbilline rodents. *J. Morph.,* **138**: 41–120.

Li Chuan-Kuei, and Ting Su-Yin. (1983). The Paleogene mammals of China. *Bull. Carnegie Mus. Nat. Hist.,* **21**: 1–98.

Li Chuan-Kuei, and Ting Su-Yin. (1985). Possible phylogenetic relationship of Asiatic eurymylids and rodents, with comments on mimotonids. In W. P. Luckett and J.-L. Hartenberger (eds.), *Evolutionary Relationships among Rodents,* pp. 35–58. Plenum Press, New York.

Lillegraven, J. A. (1969). Latest Cretaceous mammals of upper part of Edmonton Formation of Alberta, Canada, and review of marsupial-placental dichotomy in mammalian evolution. *Paleont. Contr. Univ. Kansas,* **50**: 1–122.

Lillegraven, J. A. (1975). Biological considerations of the marsupial-placental dichotomy. *Evolution,* **29**: 707–722.

Lillegraven, J. A. (1979). Reproduction in Mesozoic mammals. In J. A. Lillegraven, Z. Kielan-Jaworowska, and W. A. Clemens (eds.), *Mesozoic Mammals. The First Two-Thirds of Mammalian History* pp. 259–276. University of California Press, Berkeley.

Lillegraven, J. A. (1984). Why *was* there a "Marsupial-Placental Dichotomy"? In P. D. Gingerich and C. E. Badgley (eds.), *Mammals. Notes for a Short Course. Univ. Tennessee, Stud. Geol.,* **8**: 72–86.

Lillegraven, J. A., McKenna, M. C., and Krishtalka, L. (1981). Evolutionary relationships of middle Eocene and younger species of *Centetodon* (Mammalia, Insectivora, Geolabididae) with a description of the dentition of *Ankylodon* (Adapisoricidae). *Univ. Wyoming Publ.,* **45**: 1–115.

Luckett, W. P. (ed.). (1980a). *Comparative Biology and Evolutionary Relationships of Tree Shrews.* Plenum Press, New York.

Luckett, W. P. (1980b). The suggested evolutionary relationships and classification of tree shrews. In W. P. Luckett (ed.), *Comparative and Evolutionary Relationships of Tree Shrews,* pp. 3–31. Plenum Press, New York.

Luckett, W. P., and Hartenberger, J.-L. (1985). *Evolutionary Relationships among Rodents.* Plenum Press, New York.

Luckett, W. P., and Szalay, F. S. (eds.). (1975). *Phylogeny of the Primates, a Multidisciplinary Approach: July 6–14, 1974, Burg Wartenstein, Austria.* Plenum Press, New York.

MacIntyre, G. T. (1966). The Miacidae (Mammalia, Carnivora). Part 1. The systematics of *Ictidopappus* and *Protictis. Bull. Am. Mus. Nat. Hist.,* **131**: 115–210.

MacPhee, R. D. E. (1981). Auditory regions of primates and eutherian insectivores. Morphology, ontogeny, and character analysis. *Contr. Primatol.,* **18**: 1–282.

MacPhee, R. D. E., Cartmill, M., and Gingerich, P. D. (1983). New Palaeogene primate basicrania and the definition of the order primates. *Nature,* **301**: 509–511.

MacPhee, R. D. E., and Jacobs, L. J. (1986). *Nycticeboides simpsoni* and the morphology, adaptations and relationships of Miocene Siwalik Lorisidae. In R. M. Flanagan and J. A. Lillegraven (eds.), *Vertebrates, Phylogeny, and Philosophy. Cont. Geol. Univ. Wyo. Spec. Pap. 3.*

Marshall, L. G., de Muizon, C., and Sigé, B. (1983). *Perutherium altiplanense,* un notongulé du Crétacé Supérieur du Pérou. *Palaeovertebrata,* **13**(4): 145–155.

Martin, L. D. (1980). Functional morphology and the evolution of cats. *Trans. Nebraska Acad. Sci.,* **8**: 141–154.

Matthew, W. D. (1909). The Carnivora and Insectivora of the Bridger Basin, Middle Eocene. *Am. Mus. Nat. Hist. Mem.,* **9**: 291–567.

Matthew, W. D. (1937). Paleocene faunas of the San Juan Basin, New Mexico. *Trans. Am. Phil. Soc., New Ser.,* **30**: 1–510.

Matthew, W. D., and Granger, W. (1915). A revision of the Lower Eocene Wasatch and Wind River Faunas. *Bull. Am. Mus. Nat. Hist.,* **34**: 1–103.

Matthew, W. D., Granger, W., and Simpson, G. G. (1929). Additions to the fauna of the Gashato Formation of Mongolia. *Amer. Mus. Novitates,* **376**: 1–12.

McDowell, S. B., Jr. (1958). The greater Antillean insectivores. *Bull. Am. Mus. Nat. Hist.,* **115**: 113–214.

McKenna, M. C. (1963). New evidence against tupaioid affinities of the mammalian family Anagalidae. *Amer. Mus. Novitates,* **2158**: 1–16.

McKenna, M. C. (1973). Sweepstakes, filters, corridors, Noah's arks, and beached Viking funeral ships in palaeogeography. In D. H. Tarling and S. K. Runcorn (eds.), *Implications of Continental Drift to the Earth Sciences,* Vol. 1, pp. 295–308. Academic Press, London.

McKenna, M. C. (1975). Toward a phylogenetic classification of the Mammalia. In W. P. Luckett and F. S. Szalay (eds.), *Phylogeny of the Primates,* pp. 21–46. Plenum Press, New York.

McKenna, M. C. (1980). Early history and biogeography of South America's extinct land mammals. In R. L. Ciochon and A. B. Chiarelli (eds.), *Evolutionary Biology of the New World Monkeys and Continental Drift,* pp. 43–77. Plenum Press, New York.

McKenna, M. C. (1982). Lagomorph interrelationships. *Geobios, mem. spec.,* **6**: 213–223.

McKenna, M. C. (1983a). Cenozoic paleogeography of North Atlantic land bridges. In M. H. P. Bott, S. Saxon, M. Talwani, and J. Thiede (eds.), *Structure and Development of the Greenland-Scotland Ridge,* pp. 351–399. Plenum Press, New York.

McKenna, M. C. (1983b). Holarctic landmass rearrangement, cosmic events, and Cenozoic terrestrial organisms. *Ann. Missouri Bot. Gard.,* **70**: 459–489.

McKenna, M. C., Xue Xiangxu, and Zhou Mingzhen. (1984). *Prosarcodon lonanensis,* a new Paleocene micropternodontid palaeoryctoid insectivore from Asia. *Am. Mus. Novitates,* **2780**: 1–17.

Mitchell, E. D. (1975). Parallelism and convergence in the evolution of Otariidae and Phocidae. Conseil international de l'Exploration de la rev. Rapports et Procés-Verbaux de Réunions, **169**: 12–26.

Mitchell, E., and Tedford, R. H. (1973). The Enaliarctinae, a new group of extinct aquatic carnivora and a consideration of the origin of the Otariidae. *Bull. Am. Mus. Nat. Hist.,* **151**: 201–248.

Miyamoto, M. M., and Goodman, M. (1986). Biomolecular systematics of eutherian mammals: Phylogenetic patterns and classification. *Sys. Zool.* **35**: 230–240.

Neff, N. A. (1983). The basicranial anatomy of the Nimvavidae (Mammalia: Carnivora): Character analyses and phylogenetic inferences. Ph. D. thesis, The City University of New York.

Novacek, M. J. (1977a). A review of Paleocene and Eocene Leptictidae (Eutheria; Mammalia) from North America. *Paleo Bios,* **24**: 1–42.

Novacek, M. J. (1977b). Aspects of the problem of variation, origin and evolution of the eutherian auditory bulla. *Mammal Rev.,* **7**: 131–149.

Novacek, M. J. (1980). Cranioskeletal features in tupaiids and selected Eutheria as phylogenetic evidence. In W. P. Luckett (ed.), *Comparative Biology and Evolutionary Relationships of Tree Shrews,* pp. 35–93. Plenum Press, New York.

Novacek, M. J. (1982). Information for molecular studies from anatomical and fossil evidence on higher eutherian phylogeny. In M. Goodman (ed.), *Macromolecular Sequences in Systematic and Evolutionary Biology,* pp. 3–41. Plenum Press, New York.

Novacek, M. J. (1984). Evolutionary stasis in the elephant shrew, *Rhynchocyon.* In N. Eldredge and S. M. Stanley (eds.), *Living Fossils,* pp. 4–22. Springer-Verlag, New York, Berlin, Heidelberg, Tokyo.

Novacek, M. J. (1985). Evidence for echolocation in the oldest known bats. *Nature,* **315**: 140–141.

Novacek, M. J. (1986). The skull of leptictid insectivorans and the higher-level classification of eutherian mammals. *Bull. Am. Mus. Nat. Hist.,* **183**: 1–112.

Novacek, M. J., McKenna, M. C., Neff, N. A., and Cifelli, R. L. (1983). Evidence from earliest known crinaceomorph basicranum that insectivorans and primates are not closely related. *Nature,* **306**: 683–684.

Novacek, M. J., Bown, T. M., and Schankler, D. (1985). On the classification of the early Tertiary Erinaceomorpha (Insectivora, Mammalia). *Amer. Mus. Novit.,* **2813**: 1–22.

Packer, D. J., and Sarmiento, E. E. (1984). External and middle ear characteristics of primates, with reference to tarsier-anthropoid affinities. *Am. Mus. Novitates,* **2787**: 1–23.

Parker, P. (1977). An evolutionary comparison of placental and marsupial patterns of reproduction. In B. Stonehouse and D. Gilmore (eds.), *The Biology of Marsupials,* pp. 273–286. Macmillan, London.

Patterson, B. (1949). Rates of evolution in taeniodonts. In G. L. Jepsen, G. G. Simpson, and E. Mayr (eds.), *Genetics, Paleontology and Evolution,* pp. 243–278. Princeton University Press, Princeton.

Patterson, B. (1965). The fossil elephant shrews (Family Macroscelidae). *Bull. Mus. Comp. Zool.,* **133**: 297–335.

Patterson, B., and Wood, A. E. (1982). Rodents from the Deseadan Oligocene of Bolivia and the relationships of the Caviomorpha. *Bull. Mus. Comp. Zool.,* **149**: 371–543.

Pilbeam, D. (1984). The descent of hominoids and hominids. *Scientific American,* **250**: 84–96.

Presley, R. (1979). The primitive course of the internal carotid artery in mammals. *Acta Anat.,* **103**: 238–244.

Radinsky, L. B. (1973). Evolution of the canid brain. *Brain, Behav., Evol.,* 7: 169–202.

Radinsky, L. B. (1975). Evolution of the felid brain. *Brain, Behav., Evol.,* **11**: 214–254.

Radinsky, L. (1977). Brains of early carnivores. *Paleobiology,* **3**: 33–349.

Radinsky, L. B. (1980). Endocasts of amphicyonid carnivores. *Am. Mus. Novitates,* **2694**: 1–11.

Radinsky, L. B. (1982). Evolution of skull shape in carnivores. 3. The origin and early radiation of the modern carnivore families. *Paleobiology,* **8**(3): 177–195.

Ray, C. E. (1976). Geography of phocid evolution. *Syst. Zool.,* **25**: 391–406.

Rensberger, J. M. (1983). Successions of meniscomyine and allomyine rodents (Aplodontidae) in the Oligo-Miocene John Day Formation, Oregon. *Univ. Cal. Pub. Geol. Sci.,* **124**: 1–157.

Repenning, C. A. (1967). Subfamilies and genera of the Soricidae. *U.S. Dept. Interior, Geol Survey Prof. Paper,* **565**: 1–74.

Repenning, C. A. (1972). Underwater hearing in seals: Functional morphology. In R. J. Harrison (ed.), *Functional Anatomy of Marine Mammals,* pp. 307–331. Academic Press, London.

Repenning, C. A., and Tedford, R. H. (1977). Ontarioid seals of the Neogene. *U.S. Dept. Interior, Geol. Survey Prof. Paper,* **992**: 1–93.

Rich, T. H. V. (1981). Origin and history of the Erinaceinae and Brachyericinae (Mammalia, Insectivora) in North America. *Bull. Am. Mus. Nat. Hist.,* **171**: 1–116.

Romer, A. S. (1966). *Vertebrate Paleontology* (3d ed.). University of Chicago Press, Chicago.

Rose, K. D. (1982). Anterior dentition of the early Eocene plagiomenid dermopteran *Worlandia. J. Mamm.,* **63**(1): 179–183.

Rose, K. D., and Fleagle, J. G. (1981). The fossil history of nonhuman primates in the Americas. In A. F. Coimbra-Filho and R. A. Mittermeier (eds.), *Ecology and Behavior of Neotropical Primates,* pp. 111–167. Academia Brasileira de Ciécias, Rio de Janeiro.

Rose, K. D., and Simons, E. L. (1977). Dental function in the Plagiomenidae: Origin and relationships of the mammalian order Dermoptera. *Univ. Mich., Contr. Mus. Paleont.,* **24**: 221–236.

Rose, K. D., and Walker, A. (1984). The skeleton of Early Eocene *Cantius,* oldest lemuriform primate. *Am. J. Phys. Anthrop.,* **66**: 73–89.

Rose, K. D., Walker, A., and Jacobs, L. L. (1981). Function of the mandibular tooth comb in living and extinct mammals. *Nature,* **289**: 583–585.

Rosenberger, A. L., and Szalay, F. S. (1980). On the tarsiiform origins of Anthropoidea. In R. L. Ciochon and A. B. Chiarelli (eds.), *Evolutionary Biology of the New World Monkeys and Continental Drift,* pp. 139–157. Plenum Press, New York.

Russell, D. E., Louis, P., and Savage, D. E. (1973). Chiroptera and Dermoptera of the French Early Eocene. *Univ. Calif. Publ. Geol. Sci.,* **95**: 1–57.

Savage, D. E., and Russell, D. E. (1983). *Mammalian Paleofaunas of the World.* Addison-Wesley, London.

Savage, R. J. G. (1957). The anatomy of *Potamotherium* an Oligocene lutrine. *Proc. Zool. Soc. Lond.,* **129**: 151–244.

Savage, R. J. G. (1977). Evolution of carnivorous mammals. *Palaeontology,* **20**: 237–271.

Savage, R. J. G. (1978). Carnivora. In V. J. Maglio and H. B. S. Cooke (eds.), *Evolution of African Mammals,* pp. 249–267. Harvard University Press, Cambridge.

Schoch, R. M. (1982). Phylogeny, classification and palaeobiology of the Taeniodonta (Mammalia: Eutheria). *Third North Am. Paleont. Conv. Proc., Montreal,* **2**: 465–470.

Scott, W. B., and Jepsen, G. L. (1936). Insectivora and Carnivora. In W. B. Scott and G. L. Jepsen (eds.), The Mammalian Fauna of the White River Oligocene. Part I. *Trans. Am. Phil. Soc., New Ser.,* **28**: 1–153.

Sibley, C. G., and Ahlquist, J. E. (1984). The phylogeny of the hominoid primates as indicated by DNA-DNA hybridization. *J. Molecular Evol.,* **20**: 2–15.

Sigé, B., Crochet, J.-Y., and Insole, A. (1977). Les plus vieilles taupes. *Géobios, Mém. spéc.,* **1**: 141–157.

Simons, E. L. (1960). The Paleocene Pantodonta. *Trans. Am. Phil. Soc.,* **50**(6): 1–80.

Simons, E. L. (1964). The early relatives of man. In *Human Ancestors,* pp. 22–42. W. H. Freeman and Company, New York.

Simons, E. L. (1967). The earliest apes. *Scientific American,* **217**: 28–35.

Simons, E. L. (1984). Dawn ape of the Fayum. *Nat. Hist.,* **93**: 18–20.

Simons, E. L., and Bown, T. M. (1985). *Afrotarsius chatrathi,* first tarsiiform primate (? Tarsiidae) from Africa. *Nature,* **313**: 475–477.

Simpson, G. G. (1931). A new insectivore from the Oligocene, Ulan Gochu Horizon, of Mongolia. *Amer. Mus. Novitates,* **505**: 1–22.

Simpson, G. G. (1945). The principles of classification and classification of mammals. *Bull. Am. Mus. Nat. Hist.,* **85**: 1–350.

Simpson, G. G. (1959). The nature and origin of supraspecific taxa. *Cold Spring Harbor Symp. Quant. Biol.,* **24**: 255–271.

Simpson, G. G. (1965). *The Geography of Evolution.* Chilton, Philadelphia and New York.

Sloan, R. E., and Van Valen, L. M. (1965). Cretaceous mammals from Montana. *Science,* **148**: 220–227.

Smith, F. H., and Spencer, F. (eds.). (1984). *The Origin of Modern Humans. A World Survey of the Fossil Evidence.* Liss, New York.

Springhorn, R. (1980). *Paroodectes feisti,* der erste Miacide (Carnivora, Mammalia) aus dem Mittel-Eozän von Messel. *Paläont. Z.,* **54**: 171–198.

Stern, J. T., and Susman, R. L. (1983). The locomotor anatomy of *Australopithecus afarensis. Am. J. Phys. Anthrop.,* **60**: 279–317.

Stonehouse, B., and Gilmore, D. (eds.). (1977). *The Biology of Marsupials.* Macmillan, London.

Stucky, R., and Krishtalka, L. K. (1983). Paleocene and Eocene marsupials of North America. *Ann. Carnegie Mus.,* **52**: 229–263.

Stümpke, H. (1967). *Bau und Leben der Rhinogradentia.* Gustav Fischer Verlag, Stuttgart.

Sulimski, A. (1968). Paleocene genus *Pseudictops* Matthew, Granger and Simpson, 1929 (Mammalia) and its revision. Results Polish-Mongol. Paleont. Exped., I. *Palaeont. Polonica,* **19**: 101–129.

Sych, L. (1971). Mixodontia, a new order of mammals from the Paleocene of Mongolia. *Palaeontologia Polonica,* **25**: 147–158.

Szalay, F. S. (1968). The beginnings of primates. *Evolution,* **22**: 19–36.

Szalay, F. S. (1969). Mixodectidae, Microsyopidae and the insectivore-primate transition. *Bull. Am. Mus. Nat. Hist.,* **140**: 193–330.

Szalay, F. S. (1976). Systematics of the Omomyidae (Tarsiiformes Primates), taxonomy, phylogeny and adaptations. *Bull. Am. Mus. Nat. Hist.,* **156**: 157–450.

Szalay, F. S. (1977). Phylogenetic relationships and a classification of the eutherian Mammalia. In M. K. Hecht, P. C. Goody, and M. B. Hecht (eds.), *Major Patterns in Vertebrate Evolution,* pp. 315–174. Plenum Press, New York.

Szalay, F. S., and Decker, R. L. (1974). Origins, evolution and function of the tarsus in late Cretaceous eutherians and Paleocene primates. In F. A. Jenkins, Jr. (ed.), *Primate Locomotion,* pp. 223–259. Academic Press, New York.

Szalay, F. S., and Delson, E. (1979). *Evolutionary History of the Primates.* Academic Press, New York.

Szalay, F. S., and Drawhorn, G. (1980). Evolution and diversification of the Archonta in an arboreal milieu. In W. P. Luckett (ed.), *Comparative Biology and Evolutionary Relationships of Tree Shrews,* pp. 133–169. Plenum Press, New York.

Tarling, D. H. (1980). The geologic evolution of South America with special reference to the last 200 million years. In R. L. Ciochon and A. B. Chiarelli (eds.), *Evolutionary Biology of the New World Monkeys and Continental Drift,* pp. 1–41. Plenum Press, New York.

Tedford, R. H. (1976). Relationship of pinnipeds to other carnivores (Mammalia). *Syst. Zool.,* **25**(4): 363–374.

Thenius, E. (1979). Die Ahnen des grossen Panda. *Kosmos,* **75**(7): 478–480.

Ting Su-Yin, Schiebout, J. A., and Chow Min-Chen. (1982). Morphological diversity of early Tertiary pantodonts: A new tapir-like pantodont from China. *Third North. Am. Paleont. Conv., Proc.,* **2**: 547–550.

Tong Yongsheng, and Lucas, S. G. (1982). A review of Chinese uintatheres and the origin of the Dinocerata (Mammalia, Eutheria). *Third North Am. Paleont. Conv. Proc.,* **2**: 551–556.

Tyndale-Biscoe, H. (1973). *Life of Marsupials.* Edward Arnold, London.

Van Valen, L. M. (1978). The beginning of the age of mammals. *Evol. Theory,* **4**: 45–80.

Vaughan, T. A. (1972). *Mammalogy.* Saunders, Philadelphia, London, Toronto.

Wahlert, J. H. (1968). Variability of rodent incisor enamel as viewed in thin section, and the microstructure of the enamel in fossil and Recent rodent groups. *Breviora,* **309**: 1–18.

Wahlert, J. H. (1985). Skull morphology and relationships of geomyoid rodents. *Amer. Mus. Novitates,* **2812**: 1–20.

Walker, A. (1978). Prosimian primates. In V. J. Maglio and H. B. S. Cooke (eds.), *Evolution of African Mammals,* pp. 90–99. Harvard University Press, Cambridge.

Walker, A., Leakey, R. E., Harris, J. M., and Brown, F. H. (1986). 2.5 Myr *Australopithecus boisei* from west of Lake Turkana, Kenya. *Nature,* **322**: 517–522.

Ward, S. C., and Pilbeam, D. R. (1983). Maxillofacial morphology of Miocene hominoids from Africa and Indo-Pakistan. In R. Ciochon and R. S. Corruccini (eds.), *New Interpretations of Ape and Human Ancestry,* pp. 211–238. Plenum Press, New York.

Weijs, W. A., and Dantuma, R. (1981). Functional anatomy of the masticatory apparatus in the rabbit (*Oryctolagus cuniculus* L.). *Netherland J. Zool.,* **31**(1): 99–147.

West, R. M. (1973). Review of the North American Eocene and Oligocene Apatemyidae (Mammalia, Insectivora). *Spec. Publ. Texas Tech Univ. Mus.,* **3**: 1–42.

West, R. M., and Dawson, M. R. (1978). Vertebrate paleontology and the Cenozoic history of the North Atlantic region. *Polarforschung,* **48**: 103–119.

Wheeler, W. H. (1961). Revision of the uintatheres. *Peabody Mus. Nat. Hist. Bull.* (*Yale*), **14**: 1–126.

White, J. A., and Keller, B. L. (1984). Evolutionary stability and ecological relationships of morphology in North American Lagomorpha. *Carnegie Mus. Nat. Hist., Spec. Pub.,* **9**: 58–66.

Wolpoff, M. H. (1983). *Ramapithecus* and human origins. An anthropologist's perspective of changing interpretations. In R. L. Ciochon and R. S. Corruccini (eds.), *New Interpretations of Ape and Human Ancestry,* pp. 651–676. Plenum Press, New York.

Wood, A. E. (1940). Lagomorpha. In W. B. Scott and G. L. Jepsen (eds.), The Mammalian Fauna of the White River Oligocene. Part III. *Trans. Am. Phil. Soc., New Ser.,* **28**: 271–362.

Wood, A. E. (1962). The early Tertiary rodents of the family Paramyidae. *Trans. Am. Phil. Soc., New Ser.,* **52**: 1–261.

Wood, A. E. (1977). The evolution of the rodent family Ctenodactylidae. *J. Palaeont. Soc. India,* **20**: 120–137.

Wood, A. E. (1980). The origin of the caviomorph rodents from a source in Middle America: A clue to the area of origin of the platyrrhine primates. In R. L. Ciochon and A. B. Chiarelli (eds.), *Evolutionary Biology of the New World Monkeys and Continental Drift,* pp. 79–91. Plenum Press, New York.

Wood, A. E. (1983). The radiation of the order Rodentia in the southern continents: The dates, numbers and sources of the invasions. *Schriftenreihe für geologische Wissenschaften, Berlin,* **19/20**: 381–394.

Wood, B., Martin, L., and Andrews, P. (eds.) (1986). *Major Topics in Primate and Human Evolution.* Cambridge University Press, Cambridge.

Woodburne, M. O., and Zinsmeister, W. J. (1984). The first land mammal from Antarctica and its biogeographic implications. *J. Paleont.,* **58**: 913–948.

Wyss, A. B. (1987). Comments on the walrus auditory region and the monophyly of pinnipeds. *Am. Mus. Novitates.* (in press).

Yates, T. L. (1984). Insectivores, elephant shrews, tree shrews, and dermopterans. In S. Anderson and J. Knox Jones, Jr. (eds.), *Orders and Families of Recent Mammals of the World,* pp. 117–144. Wiley, New York.

Zhou, M.-Z., and Wang, B.-Y. (1979). Relationships between the pantodonts and tillodonts and classification of the Order Pantodonta. *Vert. PalAsiatica,* **17**: 37–48. [In Chinese.]

CHAPTER XXI

Ungulates, Edentates, and Whales

Condylarths

Although the earliest placentals had a dentition that was suitable for feeding on insects, the early Cenozoic fauna was dominated by small omnivores and herbivores. Nearly 70 percent of the Puercan fauna (at the base of the Paleocene) can be included in the archaic order Condylarthra (Van Valen, 1978; Rose, 1981). The Condylarthra, as that term is generally used, consists of a vast and diverse assemblage of primarily herbivorous placentals.

The earliest adequately known genus is *Protungulatum* from the latest Cretaceous (Figure 21-1). The lower jaw and dentition are well known and provide evidence of a significant dietary shift from the pattern of other Upper Cretaceous placentals. The cusps are relatively blunt and bulbous, which improves their capacity for crushing and grinding, and the significance of transverse shear is reduced; the trigonid and talonid of the lower molars are nearly equal in height. As in most early Tertiary euthe-

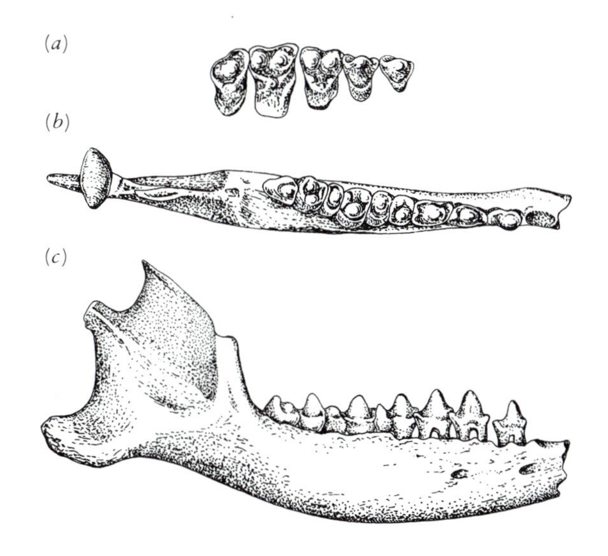

Figure 21-1. *PROTUNGULATUM,* AN UPPER CRETACEOUS CONDYLARTH. (*a*) Posterior teeth from the upper jaw. (*b* and *c*) Lower jaw in occlusal and lateral view. *From Sloan and Van Valen, 1965. With permission from* Science. *Copyright 1965 by the American Association for the Advancement of Science.*

rians, the teeth are advanced over those of the Cretaceous "proteutherians" in the reduction of the stylar shelf. Those of *Protungulatum* are broadened by the elaboration of strong pre- and postcingula, and the pattern of cusps is squared up through the elaboration of a hypocone from the postcingulum. The protocone is shifted posteriorly and the paraconule and metaconule are enlarged.

The pattern of the dentition and structure of the lower jaw are what would be expected in the common ancestor of many medium- and large-sized, herbivorous placentals of the Cenozoic, including the dominant large herbivores of the Northern Hemisphere, the artiodactyls and perissodactyls; a host of extinct groups common in the Tertiary of South America; a radiation of African orders including the elephants; and the terrestrial ancestors of the whales and sirenians.

Condylarths are considered to represent an evolutionary grade. This assemblage is differentiated from more derived ungulates primarily on the basis of the retention of primitive characters. Condylarths include many recognizable lineages, several of which may be specifically related to one or more descendant orders.

It is clear that the condylarths are due for a major taxonomic revision that would recognize the affinities of their constituent genera with the major orders of the later Cenozoic. However, such a revision will not be attempted here. Since there is still little evidence regarding specific interrelationships among the condylarths, the taxonomic usage in this text will follow the pattern of most recent papers. It also remains convenient to conceive of the condylarths as a single taxonomic unit, since all the members are clearly related to one another by immediate common descent, even if many are also related to the ancestry of later groups. Although it may be possible to demonstrate relationships between particular condylarth genera and primitive members of orders that have differentiated in the later Cenozoic, most condylarth genera are clearly recognized on the basis of the retention of a host of primitive characters.

Protungulatum is placed in the family Arctocyonidae, which is accepted as including the most primitive condylarths and the ultimate ancestors of all later ungulates. In contrast with the specialization of the molar teeth for a diet that required crushing and grinding, the dentition is primitive in retaining a tooth count of

$$\frac{3\ 1\ 4\ 3}{3\ 1\ 4\ 3}$$

without a diastema, and the premolars remain clearly distinct from the molars. The skull, which is known in early Tertiary genera, retains a narrow braincase and lacks a postorbital bar and the auditory bulla. A postglenoid foramen is retained.

Following the classification of Van Valen (1978) and Cifelli (1983b), one can divide the Arctocyonidae into four subfamilies. The Oxyclaeninae is the most primitive of all and includes the ancestry of the other three arctocyonid subfamilies and may be directly ancestral to one or two other condylarth families as well as the artiodactyls.

The Arctocyoninae is considered typical of primitive condylarths but is not thought to be directly ancestral to any later groups. The skull of *Arctocyon* is well known and illustrates the general pattern from which the later ungulates have arisen (Figure 21-2). Compared with later therians, we can see that it combines features that might be considered characteristic of both carnivores and herbivores, which reflects the fact that later condylarths have specialized in both directions. The cheek teeth seem to be specialized for crushing and grinding and the incisors for cropping, as in herbivores, but the canines are enlarged and pointed. The great size of the adductor chamber and the relatively low position of the dentary condyle are similar to those of advanced carnivores, not herbivores, but these features are probably primitive and were retained from earlier, insectivorous placentals. The Triisodontinae

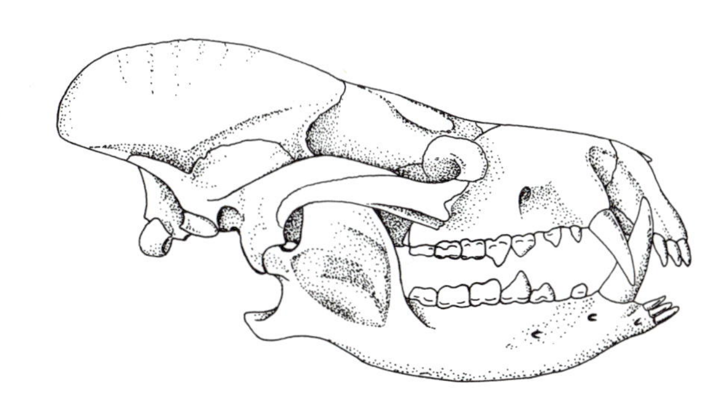

Figure 21-2. SKULL OF *ARCTOCYON,* A PRIMITIVE CONDYLARTH, ×⅓. *From Russell, 1964.*

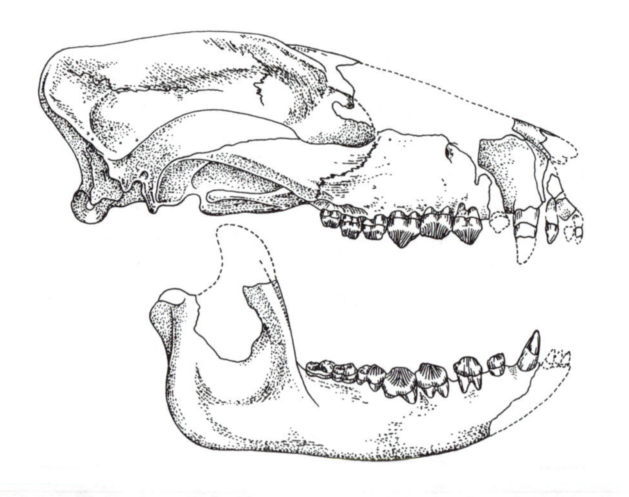

Figure 21-3. SKULL OF THE PRIMITIVE CONDYLARTH *PERIPTYCHUS*, ×⅓. The Periptychidae is characterized by the wrinkling of the molar enamel. *From Matthew, 1937.*

are thought to have given rise to the mesonychid condylarths, which in turn include the ancestors of cetaceans. The Loxolophinae may include the ancestry of most other ungulate groups.

These subfamilies occur in the early Paleocene, within 1 or 2 million years of the appearance of *Proungulatum*. The distinguishing characters of the individual arctocyonid subfamilies will be discussed with the derived groups with which they show close affinities.

A second family of primitive condylarths, the Periptychidae, also appears in the late Cretaceous and is common in the Paleocene. The molars are typically distinguished by a wrinkling of the enamel (Figure 21-3). Van Valen (1978) places *Perutherium*, the only Cretaceous eutherian from South America, in the Periptychidae, but Cifelli (1983b) doubts that this genus is even a eutherian.

The periptychid *Ectoconus* is one of the very few early Paleocene mammals that is known from a nearly complete skeleton (Matthew, 1937; Simpson, 1941). Although the dental anatomy may be somewhat divergent, the remainder of the skeleton is very primitive and may represent the pattern from which all later ungulates evolved (Figure 21-4). The skull resembles that of *Arctocyon* in its primitive features. The vertebral count is 7 cervicals, 14 dorsals, 6 lumbars, 4 sacrals, and 24 caudals. The ulna and fibula, which are reduced and variably fused in later ungulates, are stout and separate. The manus and pes are fully pendactyl and retain all the carpals and tarsals common to primitive placentals. The pattern of the carpals and tarsals would be considered alternating (Figure 21-5). One feature of *Ectoconus* that is clearly not primitive is its great size, with a length of nearly 2 meters including the tail.

The Meniscotheriidae include some of the most specialized condylarths with molarized premolars and lophodont cheek teeth (Gazin, 1955). They occur in the late Paleocene and early Eocene of North America and the late Paleocene of Europe. Despite their precocious achievement of an advanced dentition, they are not implicated in the ancestry of any of the modern ungulate groups (Figure 21-6).

The hyopsodontids are common condylarths in the early Eocene and can be traced to the base of the Paleocene. *Hyopsodus* is a small form, with short, unspecialized limbs (Figure 21-7). The crescentic molar cusps of hyopsodontids resemble those of primitive artiodactyls and they have frequently been suggested as being closely related. However, no more specific features have been identified that might ally the groups. Cifelli (1983b) considered that hyopsodontids include the ancestry of some, if not all, of the South American ungulates.

Phenacodonts are among the best-known condylarths. Their anatomy will be discussed with that of their probable descendants, the perissodactyls. The phenacolophids are an, as yet, poorly known Asian family of con-

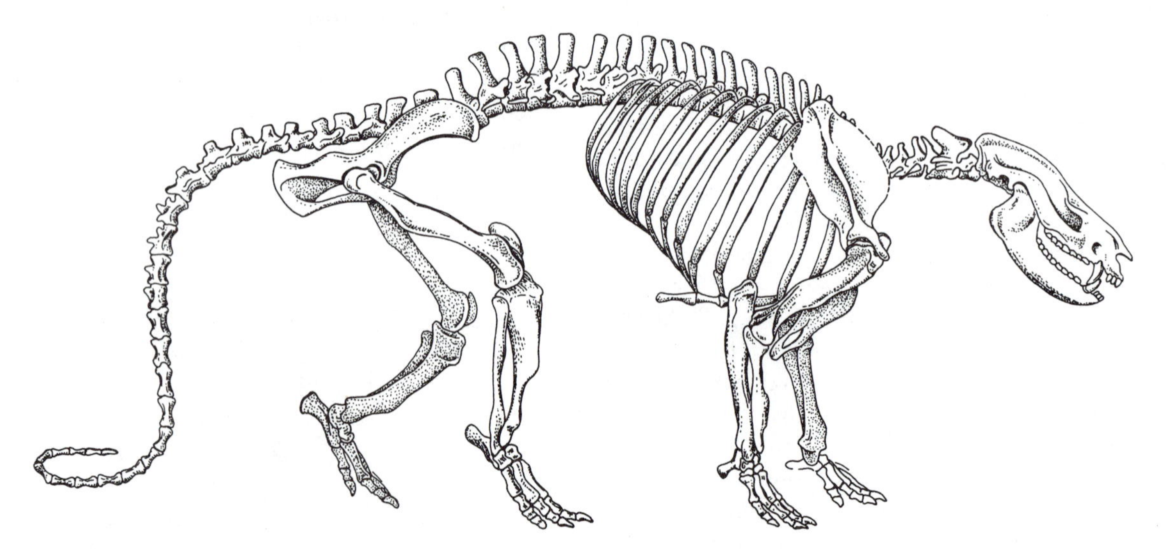

Figure 21-4. SKELETON OF THE PALEOCENE CONDYLARTH *ECTOCONUS*. This genus exemplifies the primitive pattern for ungulates. Length about 2 meters. *From Gregory, 1951 and 1957.*

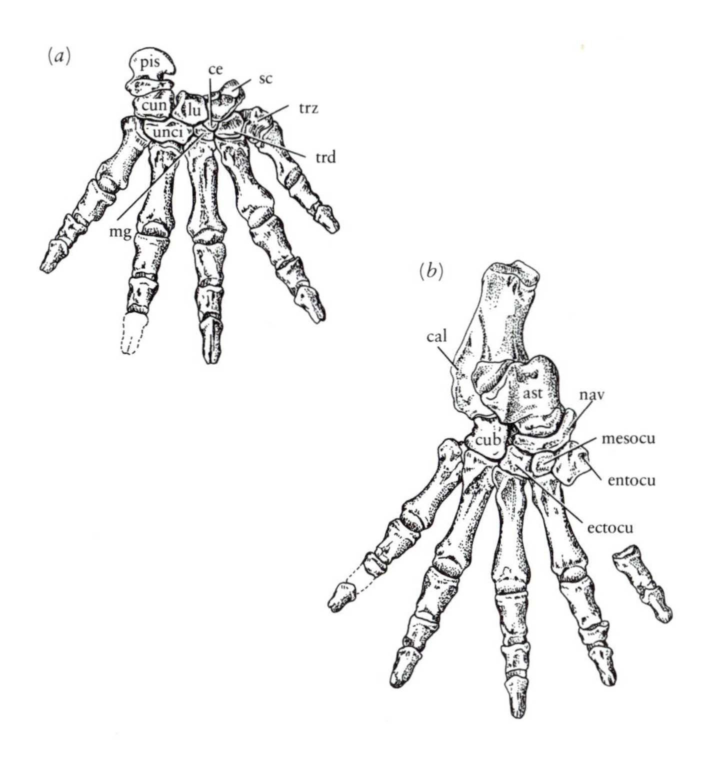

Figure 21-5. MANUS AND PES OF THE PRIMITIVE CONDYLARTH *ECTOCONUS*, ×⅓. The carpals and tarsals are alternating or overlapping, which is a primitive condition for ungulates. Abbreviations as in Figure 20-56. *From Matthew, 1937.*

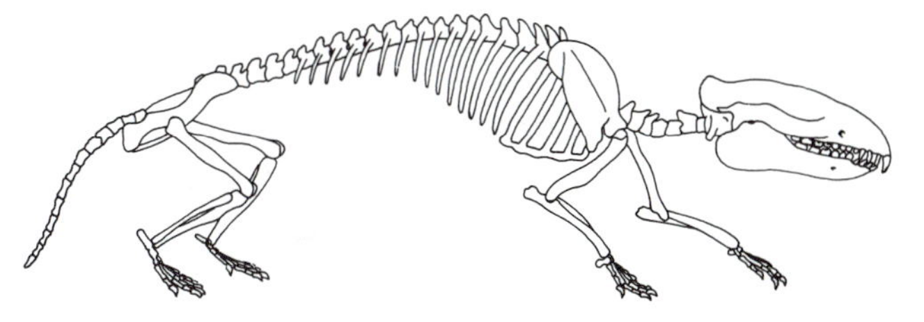

Figure 21-7. THE SMALL, SHORT-LIMBED EOCENE CONDYLARTH *HYOPSODUS*, ×⅓. *From Gazin, 1968. By permission of Smithsonian Institution Press, Smithsonian Institution, Washington, D.C.*

dylarths that may be close to the ancestry of the African ungulate groups.

The radiation of ungulates provides an important example of the different viewpoints that may be gained by different approaches to classification. The orders of Cenozoic mammals have been recognized primarily on the basis of distinct anatomical and adaptational patterns that are exhibited by typical members of the major groups (see Van Valen, 1971b). For example, cetaceans are recognized on the basis of aquatic adaptation, and the artiodactyls and perissodactyls are defined on the basis of marked advances in the structure of the tarsus. As such, these groups are clearly recognized, and we may speak of the major evolutionary changes that are evident in their origins. The fossil record of both artiodactyls and perissodactyls shows a significant radiation following the achievement of these anatomical changes.

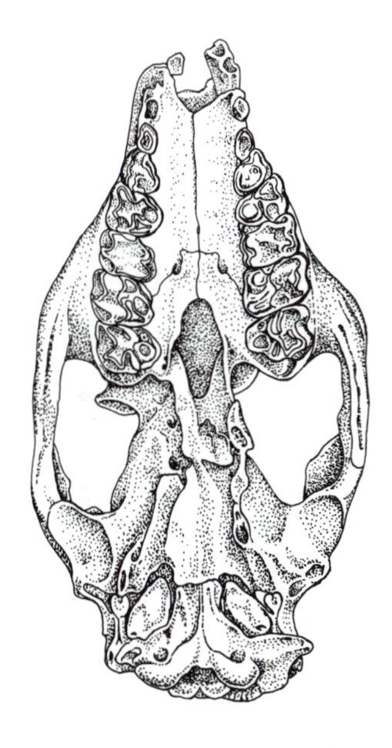

Figure 21-6. PALATE OF THE CONDYLARTH *MENISCOTHERIUM*. This genus shows a complex pattern of the cusps, which is advanced over that of all other condylarths. The solenodont condition resembles that of artiodactyls but was separately evolved. The posterior premolars are molarized to a higher degree than in any other early Tertiary ungulates. *From Gazin, 1965. By permission of Smithsonian Institution Press, Smithsonian Institution, Washington, D.C.*

As our knowledge of condylarths increases, our way of viewing these evolutionary advances may change significantly (Figure 21-8). In the case of the cetaceans and perissodactyls, their origin among the condylarths has been clearly documented. With this increase in our knowledge of their phylogenetic relationships, we can modify the classification of early ungulates to better reflect these affinities. The phenacodontids can be included among the perissodactyls and the mesonychids among the whales.

Such a system of classification gives a much different picture of the origin of these mammalian orders (Van Valen, 1971b). If, as seems likely, it may eventually be possible to trace the ancestry of most of the placental orders back to the early Paleocene, or even the latest Cretaceous, the differences between the earliest ancestral forms will be very small—potentially no more than those that distinguish species or even populations within species. The origin of orders will become synonymous with the origin of species or geographical subspecies. In fact, this pattern is what one would expect from our understanding of evolution going back to Darwin. The selective forces related to the origin of major groups would be seen as no different than those leading to adaptation to very slightly differing environments and ways of life.

On the basis of a better understanding of the anatomy and relationships of the earliest ungulates, we can see that the origin of the Cetacea and the perissodactyls resulted not from major differences in their anatomy and ways of life but from slight differences in their diet and mode of locomotion, as reflected in the pattern of the tooth cusps and details of the bones of the carpus and tarsus.

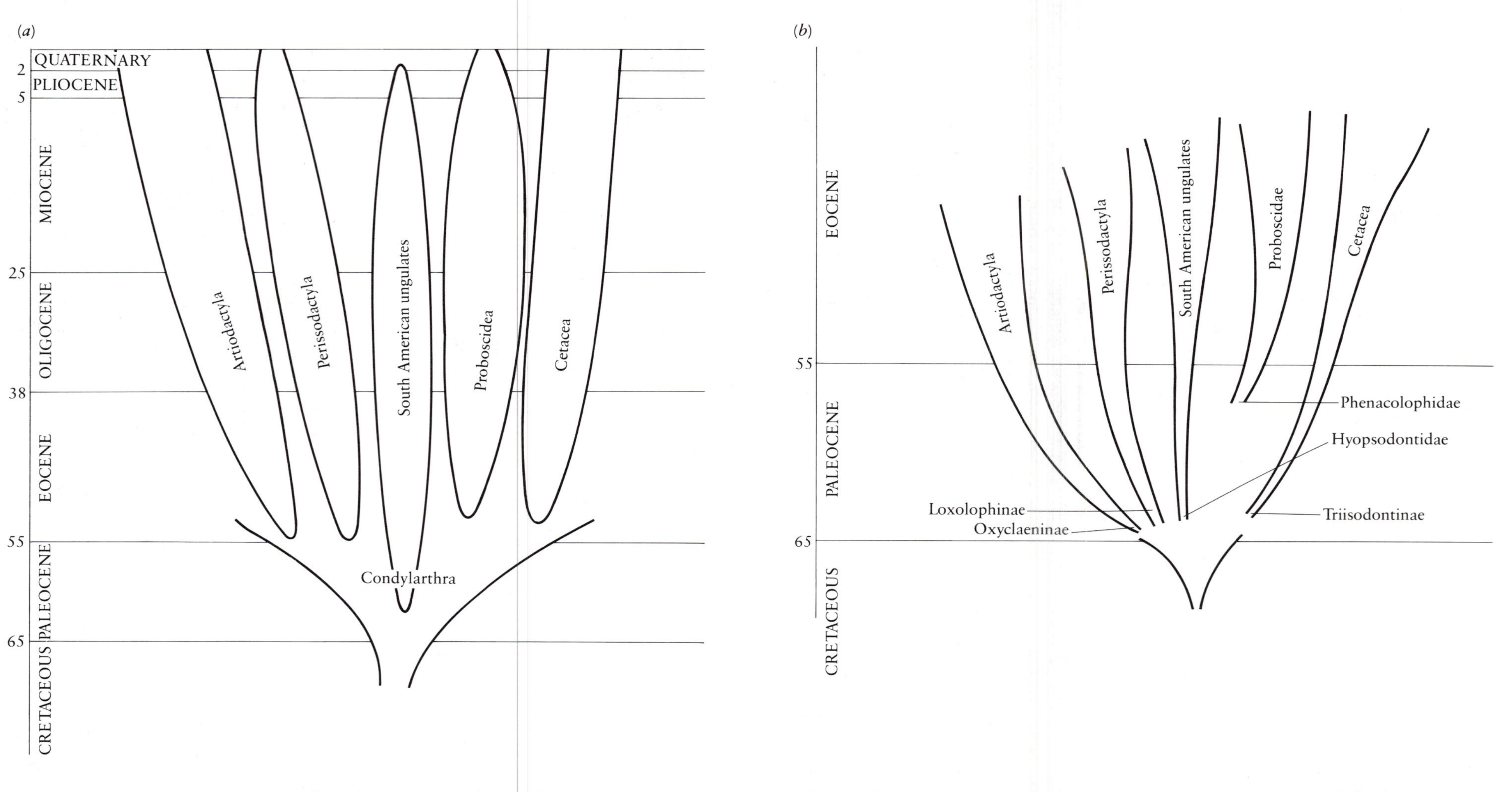

Figure 21-8. ALTERNATE PHYLOGENIES OF UNGULATE GROUPS. (*a*) Phylogeny showing the customary range of the major ungulate groups based on the achievement of new structural or adaptational features. (*b*) Phylogeny based on recognized affinities with particular families or subfamilies that are typically classified as condylarths. The Oxyclaeninae, Loxolophinae, and Triisodontinae are all classified within the condylarth family Archocyonidae.

By the use of a strictly phylogenetic method of classification, we see that most of the major anatomical and adaptational changes occur *within* rather than *between* major taxonomic groups (in this case orders).

Radinsky (1982) commented that the pattern of evolution seen in carnivores between the miacids and the modern families is qualitatively different from that seen in the origin of the artiodactyls and perissodactyls in that the latter is based on marked morphological changes and the former is not. To some degree, this may be a justifiable observation (considering the limitations of the fossil record), but the validity of this comparison may be questioned if we consider that the origin of the perissodactyls may be recognized as having occurred not at the emergence of *Hyracotherium* with its specialized calcaneal-navicular joint but rather at the time that the ancestral phenacodontids diverged from other early condylarths, which involved (as far as we know) only subtle differences in the dentition.

ARTIODACTYLA

The artiodactyls are the most diverse ungulate group, with 79 living genera and a rich fossil record that goes back to the earliest Eocene. Twenty-seven families are recognized during the Cenozoic, of which ten survive in the modern fauna.

The pattern of radiation resembles that of the ungulates as a whole. From a clearly differentiated ancestral stock, they rapidly radiated into many distinct lineages, most of which had a relatively long subsequent history (Figure 21-9). The early evolution of the artiodactyls is fairly well documented by both the dentition and considerable skeletal material and provides the basis for fairly detailed analysis of evolutionary patterns. Numerous papers describing Eocene artiodactyls have appeared recently (Rose, 1982, 1985; Golz, 1976; Sudre, 1977, 1978; Wilson, 1971, 1974; Black, 1978; Webb and Taylor, 1980; Krishtalka and Stucky, 1985). The most important conclusion that can be drawn from this work is that the origin of nearly all the recognized families can be traced to the late Middle Eocene or the Upper Eocene and that specific interrelationships between these families are very difficult to document.

PRIMITIVE ARTIODACTYLS

The oldest genus that is recognized as an artiodactyl is *Diacodexis*, a rabbit-sized animal known from the Lower Eocene of North America, Europe and Asia (Rose, 1982, 1985; Thewissek, Russell, Gingerich, and Hussain, 1983) (Figure 21-10). The entire skeleton presages the pattern of later artiodactyls, but it is allied with this group specifically on the basis of the configuration of the astragalus, in which, uniquely among mammals, there are pulley-shaped articulating surfaces both proximally with the tibia and distally with the navicular and cuboid (Figure 21-11). This pattern limits movement of the foot to the vertical plane, enabling much more effective translation of force across the joints.

Comparison of limb proportions of *Diacodexis* with living ungulates indicates that it was the most highly cursorial of Lower Eocene mammals. Like other small cursorial mammals, it was probably capable of leaping and bounding as well. The hind limb is substantially longer than the front, and the tibia is markedly longer than the femur. The fibula is reduced to a splint and in some specimens is fused to the tibia distally. The elbow as well as the ankle joint are specialized so as to restrict movement to the vertical plane. The metacarpals and metatarsals are very elongate. They are not fused but are closely integrated with one another to form a single functional unit. As is characteristic of other artiodactyls, the third and fourth metapodials are the largest, and the third and fourth digits bear most of the weight of the body. The toes probably bore small hoofs.

In contrast with advanced artiodactyls, the clavicle is still retained, the ulna is complete and unfused, and the cuboid and navicular remain separate. The manus, which is more completely known in other early artiodactyls, retains five digits. As in condylarths and perissodactyls, the femur retains the third trochanter, which is lost in later artiodactyls. The first metatarsal is greatly reduced but the pes, in contrast with all other artiodactyls, retains all five digits.

In marked contrast with the limb proportions and structure of the ankle, the skull and teeth of *Diacodexis* remain very primitive. The basicranium retains an essentially condylarth pattern (Coombs and Coombs, 1982). All the teeth of primitive ungulates are retained without gaps. The cusps are blunt and the upper molars retain a primitively triangular configuration (Figure 21-12). The extremely rapid radiation of artiodactyls within the Eocene must be attributed primarily to advances in their locomotor apparatus, not to changes in their dentition.

Van Valen (1971a) recognized five genera of early Eocene artiodactyls: *Diacodexis, Wasatchia, Bunophorus, Hexacodus,* and *Protodichobune*, all of which are placed within the Dichobunidae. The postcranial anatomy of the dichobunids appears significantly advanced relative to that of all condylarths and no intermediates are known. Schaeffer (1947) discussed the important changes in the tarsus that distinguish artiodactyls. The pattern of the astragalus can be derived from that of primitive condylarths, but they are distinguished by a significant morphological and functional gap.

The dentition retains a pattern that suggests derivation from among the most primitive arctocyonids. Van Valen (1971a, 1978) indicated that their closest affinities may lie with *Chriacus* [*Tricentes*] in the subfamily Oxyclaeninae from the Lower Paleocene. Among the derived features shared by *Chriacus* and *Diacodexis* are the mod-

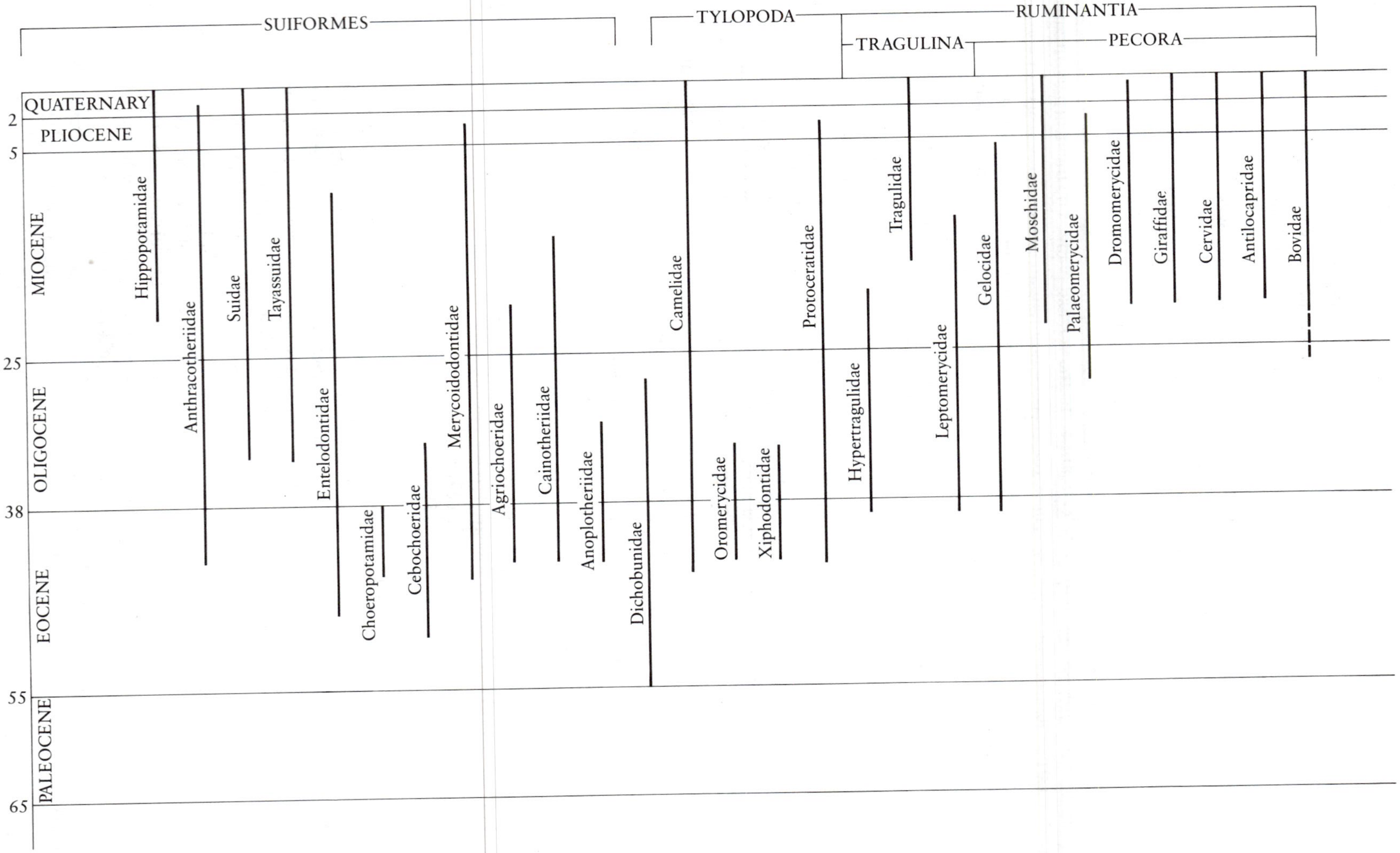

Figure 21-9. STRATIGRAPHIC RANGES OF ARTIODACTYL FAMILIES.

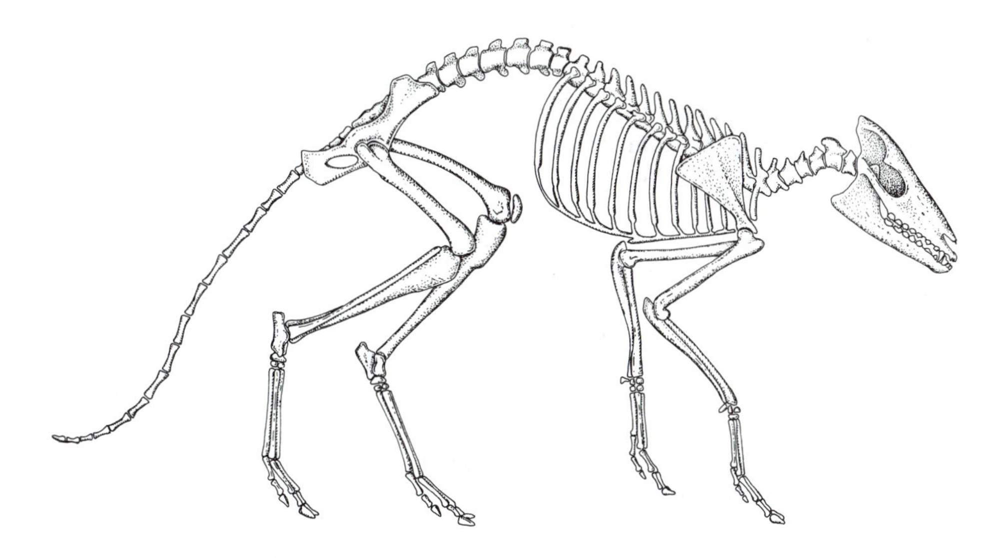

Figure 21-10. SKELETON OF THE LOWER EOCENE ARTIODACTYL *DIACODEXIS,* ABOUT THE SIZE OF A RABBIT. The limb structure is very specialized for cursorial locomotion. *From Rose, 1982. With permission from* Science. *Copyright 1982 by the American Association for the Advancement of Science.*

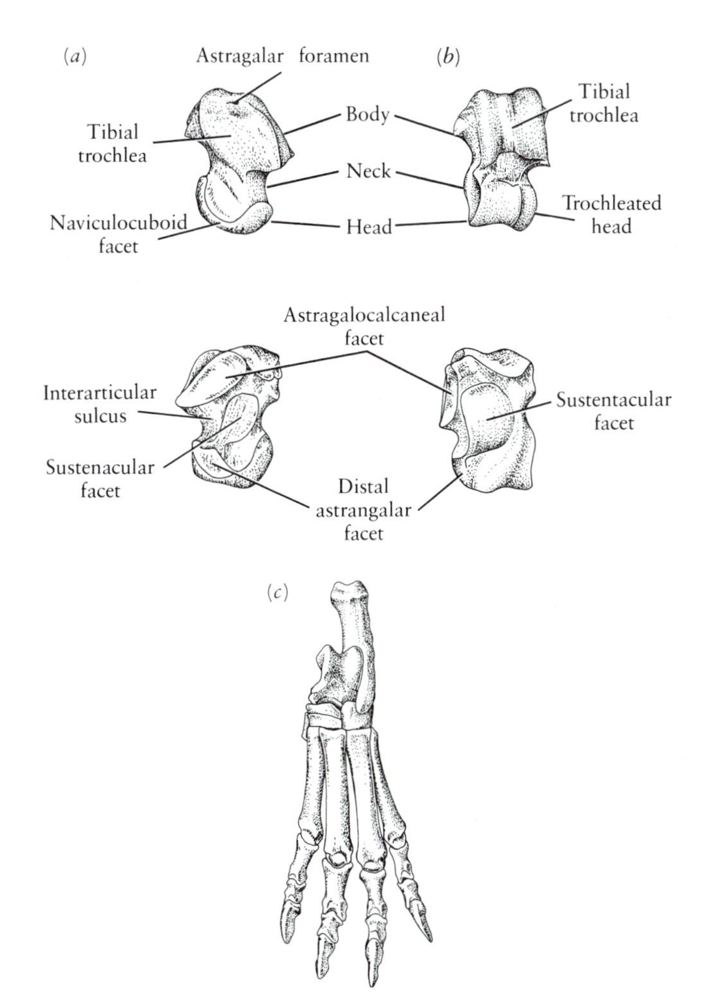

Figure 21-11. Comparison of the astragalus of (*a*) the hyopsodont *Choeroclaenus* and (*b*) the primitive artiodactyl *Diacodexis*. The proximal and distal articulating surfaces of *Diacodexis* are in the shape of a pulley, which allows controlled flexion in the vertical plane between the lower leg and the foot. *From Schaeffer, 1947.* (*c*) Foot of a primitive artiodactyl, the oreodon *Agriochoerus. From Scott, 1940.*

erately strong paracristid on P_4 and the fact that the talonid of M_1 and M_2 is wider than the trigonid; both have relatively large lingual cingula. As importantly, these genera share primitive features that are lost in other groups of early ungulates but are retained by other early artiodactyls. Of particular importance is the fact that the upper molars of the earliest artiodactyls still lack a hypocone. The third lower premolar is not reduced and the canine is still moderately large. The condyle of the lower jaw is not greatly elevated above the tooth row. In contrast with the skull, the postcranial skeleton of *Chriacus* remains primitive, with no evidence for the inception of a paraxonic manus or pes.

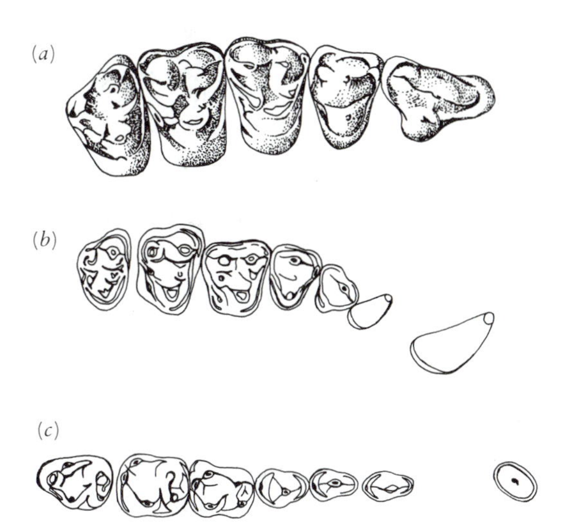

Figure 21-12. (*a*) Upper molars and posterior premolars of the earliest artiodactyl *Diacodexis,* ×3. *From Sinclair, 1914.* (*b*) Upper dentition of the oxyclaenine arctocyonid *Chriacus* [*Tricentes*], which is close to the pattern of primitive artiodactyls, ×1. *From Matthew, 1937.* (*c*) Lower dentition of *Chriacus. From Matthew, 1937.*

It is customary to use the achievement of the specialized astragalus as a basis for defining the origin of artiodactyls. Alternatively, one might define them on the basis of their initial phylogenetic divergence from the arctocyonid stock, at the point when they branched from a sister group that includes the ancestors of one of the other ungulate orders.

By the middle Eocene, we can recognize several artiodactyls lineages. Three or four subfamilies can be included within the Dichobunidae, which are united by the retention of primitive characters. Members of the Diacodexinae, which are known from the Lower Eocene, are considered the most primitive of all artiodactyls. They are thought to include the ancestry of the Middle and Upper Eocene Dichobuninae and Homacodontinae (Figure 21-13). The latter subfamilies are distinguished by the presence of a consistently well-developed hypocone, while this cusp was variably present in the Diacodexinae. The hypoconulid of M_1 and M_2 develops from a cingular position in a saddle between the hypoconid and the entoconid.

The Helohyinae is included within the Dichobunidae by most authors, but Coombs and Coombs (1977) consider that it represents a distinct family. The molars are inflated and bulbous but the cusp pattern appears primitive; the premolars are simple and trenchant and the hypocone is absent (West, 1984). The Diacodexinae and Homacodontinae are common in North America but the Helohyinae (idae) is more common in Asia, where it may have originated.

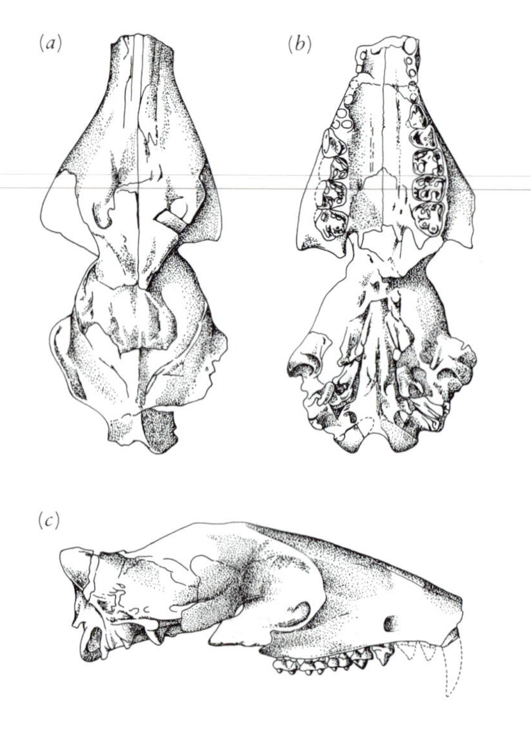

Figure 21-13. SKULL OF THE PRIMITIVE ARTIODACTYL *HOMACODON*. (*a*) Dorsal, (*b*) palatal, and (*c*) lateral views, $\times \frac{3}{4}$. This genus may be close to the ancestry of several advanced artiodactyl families. *From Sinclair, 1914.*

Artiodactyls radiated extensively in all the northern continents in the early and middle Eocene; approximately 20 families appeared by the late Eocene. By the end of the Eocene, all of the major groups of artiodactyls had emerged. Modern members of the order can be conveniently classified in three major groups: the suborder Suiformes, including the modern pigs, peccaries, and the hippopotamus; the Tylopoda (camels and llamas); and the Ruminantia, including tragulids, giraffes, deer, and the diverse assemblage of antelopes and cattle.

An informal twofold division may be recognized on the basis of the dentition between the primitive artiodactyls and the suiforms on one hand, which generally retain a **bunodont** dentition with gently rounded cusps, and the remaining genera, in which the cusps assume the shape of a crescent or half moon, the **selenodont** condition. Limb structure and proportions remain primitive in the suiforms but are more advanced in the other two groups. In their dentition and limb structures, the suiforms are characterized by primitive features, but they also share one readily recognized cranial apomorphy. Like other eutherians, the posterior surface of the otic capsule, the mastoid, is exposed posteriorly in camels and early ruminants. In contrast, the squamosal extends over the mastoid to reach the exoccipital in the modern suiform groups.

A variety of primitive artiodactyls have been placed in a separate, clearly paraphyletic suborder, the Palaeodonta (Romer, 1966). Several of these groups may now be included within more derived taxa, but the Dichobunidae retains the position of a stem group that may include the ancestors of all other artiodactyls.

SUIFORMS

Diacodexis is the earliest and generally most primitive artiodactyl, but the cursorial specializations indicated by the great length of the distal limb elements are more advanced than the condition seen in early members of the suiforms. According to Rose (1982, 1985), the short limbs of the suiforms may be interpreted as indicating either that they had secondarily reduced the degree of cursorial specialization or that this group had diverged from the basal stock of artiodactyls at a more primitive stage than that represented by *Diacodexis*. Both Guthrie (1968) and Rose (1985) believe that the great elongation of the distal limb elements and the reduction of the lateral toes in *Diacodexis* are specializations that are reduced in later dichobunids and perhaps in the immediate ancestors of suiforms. The partial fusion of the ecto- and mesocuneiform is a derived feature that may specifically exclude this genus from the ancestry of groups other than the ruminants and camels.

Anthracotheriids and Hippopotami

Among the Suiformes, the earliest group that became abundant was the Anthracotheriidae. Coombs and Coombs

(1977) trace this family back to the middle Eocene helohyids. Primitive members of these two families are distinguished from the diacodexids by the more bulbous cusps of the molar teeth, the presence of a strong continuous cingulum, and larger body size. Anthracotheriids are a primarily Old World group; they may have originated in Asia. They were diverse in the late Eocene of Burma and later became widespread in Europe. They were common from the early Oligocene into the early Miocene in North America and in Africa from the early Oligocene into the Pliocene (Black, 1978). They lingered into the Pleistocene in Asia.

The early anthracotheriids were terrier sized, but later genera were as large as a hippopotamus (Figure 21-14). In some advanced genera, the molars were squared and the cusps selenodont. The limbs were short and stout, without cursorial adaptations. The metapodials were short and unfused; five digits were retained in the front foot and four in the rear. Advanced members of the group are found in deposits that suggest that they were amphibious in habit. *Merycopotamus* from the late Miocene to Pliocene of Africa and southern Asia links this family with the Hippopotamidae. More than other anthracotheriid, it shares with them a deep flange at the angle of the jaw. Both the upper and lower canines and the lateral lower incisors are considerably enlarged, especially among the males.

Only two genera are recognized among fossil and recent members of the Hippopotamidae—the amphibious form *Hippopotamus*, which is common today in wet areas throughout Africa south of the Sahara, and the slightly more terrestrial pygmy hippo *Hexaprotodon* [*Choeropsis*[, which is restricted to forest and coastal plains from Guiana to the Ivory Coast (Figure 21-15).

Fossils are known throughout Eurasia in the Pliocene and Pleistocene. Fragmentary remains of early members of the Hippopotamidae are known as early as the early Miocene, some 18 million years ago, in Africa. They retain the primitive ungulate tooth formula of

$$\frac{3 \quad 1 \quad 4 \quad 3}{3 \quad 1 \quad 4 \quad 3}$$

but the upper canines are diagnostic in the presence of a deep posterior groove, and the molars wear so as to show a triangular trefoil pattern that clearly distinguishes them from the anthracotheriids.

The modern species have reduced the formula to

$$\frac{2 \quad 1 \quad 3\text{–}4 \quad 3}{2 \quad 1 \quad 3\text{–}4 \quad 3}\ (\textit{Hexaprotodon})$$

and

$$\frac{2 \quad 1 \quad 3 \quad 3}{1 \quad 1 \quad 3 \quad 3}\ (\textit{Hippopotamus})$$

These genera are also distinguished by skull proportions, with the orbit more posterior and dorsal in position in *Hippopotamus* and the snout broader and longer relative to the extent of the cheek teeth.

The Hippopotamidae became common between 4 and 6 million years ago. The first complete skull is that of *Hexaprotodon*. The postcranial skeleton shows slender limbs that suggest an animal more agile and less amphibious than *Hippopotamus*. The rise of the hippos, which were first recorded in East Africa, seems to be correlated with a local decline in the anthracotheriids.

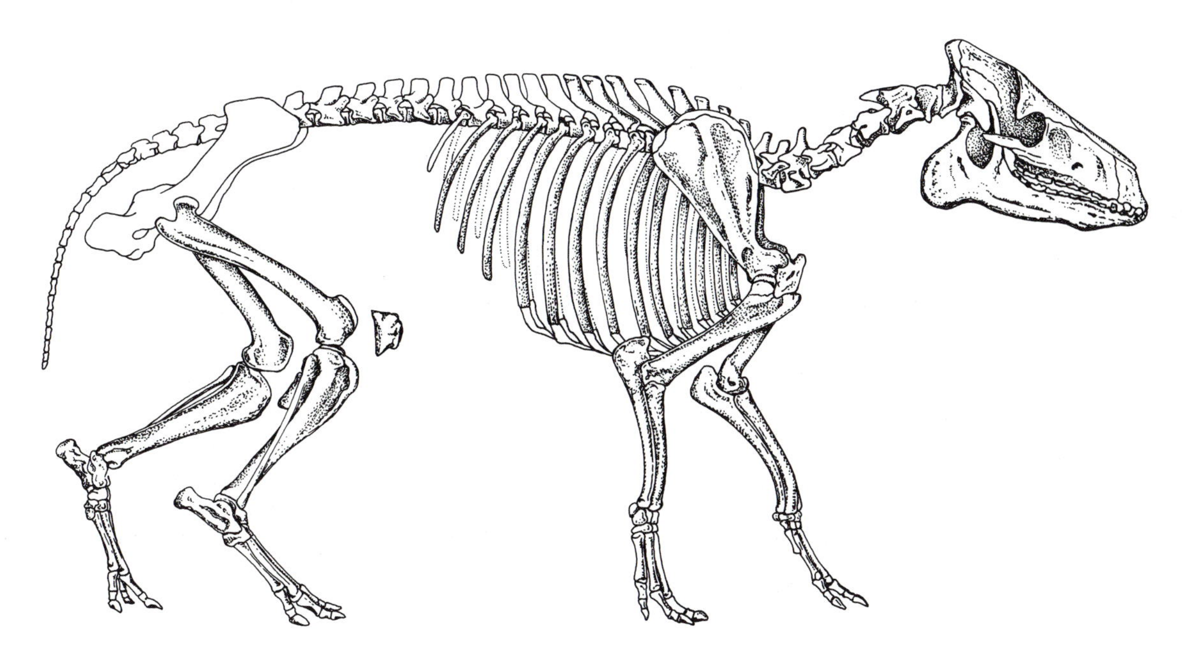

Figure 21-14. THE ANTHRACOTHERID *ANCODUS*. From the Oligocene of western North America, $\times\frac{1}{10}$. *From Scott, 1894.*

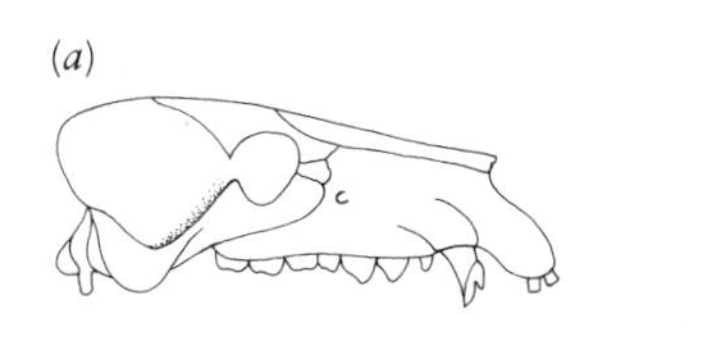

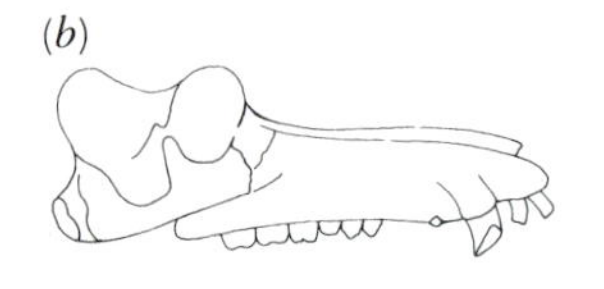

Figure 21-15. DIAGRAMMATIC COMPARISON OF THE CRANIUM OF THE LIVING HIPPOPOTAMUS GENERA. (*a*) *Hexaprotodon* and (*b*) *Hippotamus. From Coryndon, 1978. Reprinted by permission of Harvard University Press.*

The genus *Hippopotamus*, with only two pairs of lower incisors and a larger, heavier body, appears between 3 and 4 million years ago. Coryndon (1978), who most recently reviewed the evolution of this family, attributes the restriction of the terrestrial forms and the aquatic specialization of the genus *Hippopotamus* to competition with the terrestrial Bovidae that became increasingly abundant in the late Cenozoic. The Anthracotheriidae and Hippopotamidae are combined in the superfamily Hippopotamoidea.

Primitive piglike forms

Several families of primitive piglike forms are known in the Eocene and Oligocene, including the Cebochoeridae and Choeropotamidae, which Coombs and Coombs (1977) suggest may share an ancestry among the Helohyidae. The entelodonts, which extended into the early Miocene, were common in North America and Europe. They were large animals, with the skull reaching nearly 1 meter long. They are characterized by the presence of bony processes from the lower jaw and zygomatic arch (Figure 21-16). Gazin (1955) suggested an origin directly from the helohyines of the Dichobunidae, although the likelihood of this relationship is questioned by Coombs and Coombs.

Suidae and Tayassuidae

Members of the Suidae are known as early as the Lower Oligocene in Europe; they have been restricted to the Old World throughout their history. Within the group, the first digit of the manus is lost but there are few other changes in the postcranial skeleton. Although the cheek teeth remain bunodont, the cusp pattern may become very complicated (Figure 21-17). The upper canines curve outward and upward.

The Tayassuidae may have had a common ancestry with the suids, but peccaries are first known in the early Oligocene of North America. They appear in Europe from the Middle Oligocene to the Miocene, extend into Asia and Africa in the Pliocene, and enter South America in the Pleistocene. The molar teeth are less complex than those of the Suidae and the upper canines are straight. All three incisors are retained in both upper and lower jaws. Some genera within the group developed a more advanced postcranial skeleton. In *Platygonus*, the metatarsals and metacarpals are fused and there are only two toes on the fore feet and hind feet. The living genus *Tayassu* is more primitive in retaining four toes on the forefoot and three on the hind.

PRIMITIVE SELENODONT ARTIODACTYLS

While the suiforms retained a relatively conservative dentition and locomotor apparatus throughout their history, many other artiodactyl lineages achieved a more advanced dentition within the late Eocene, with squared molars and crescentic cusps (Figure 21-19). The genera included within the Dichobunidae may have given rise to all the advanced families, but it is difficult to establish specific interrelationships. Wilson (1974) demonstrated five different ways

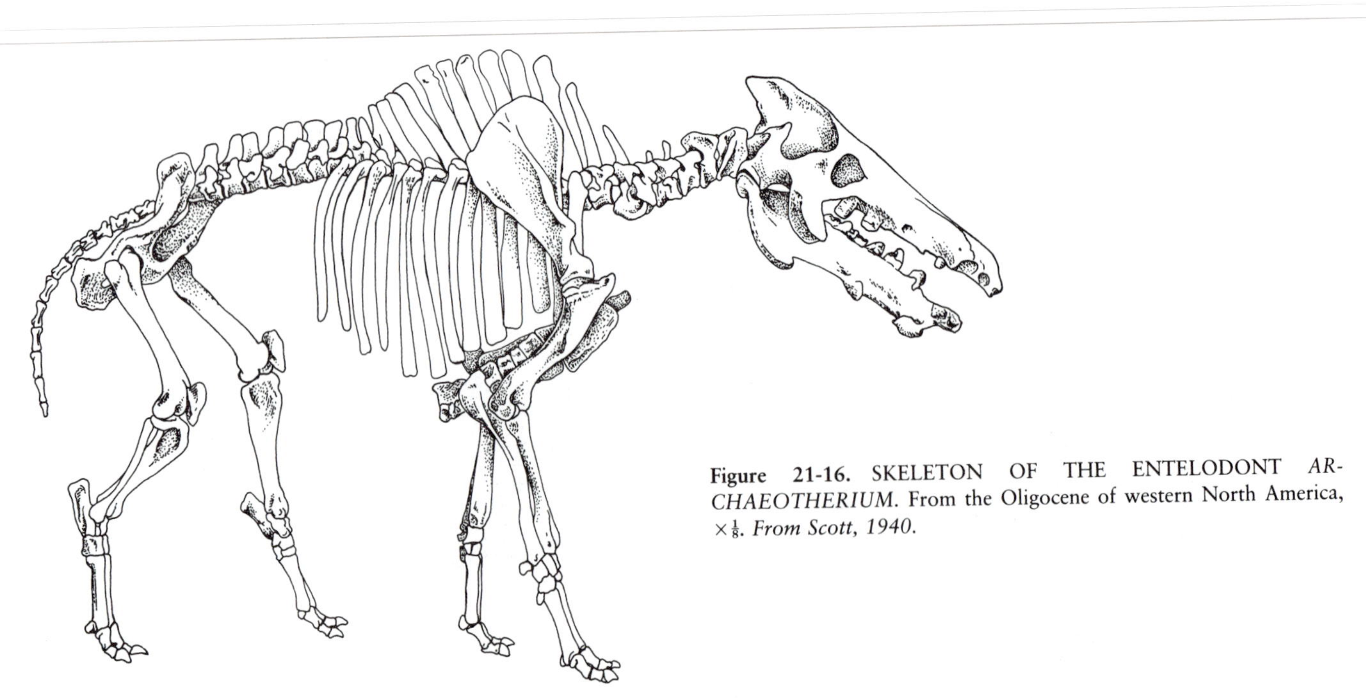

Figure 21-16. SKELETON OF THE ENTELODONT *ARCHAEOTHERIUM*. From the Oligocene of western North America, $\times \frac{1}{8}$. *From Scott, 1940.*

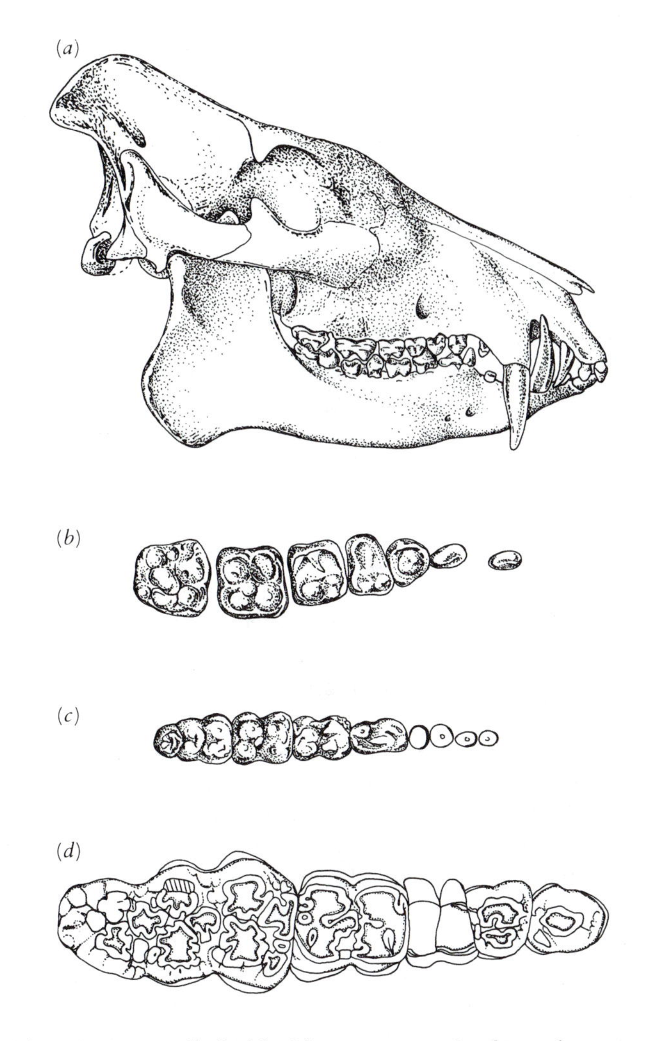

Figure 21-17. (*a*) Skull of the Oligocene peccary *Perchoerus* from western North America. *From Scott, 1940.* (*b* and *c*) Upper and lower dentition of *Perchoerus. From Scott, 1940.* (*d*) Complex cheek teeth of the suid *Nyanzachoerus. From Harris and White, 1979.*

in which a selenodont pattern evolved in different groups of Upper Eocene artiodactyls, all of which may have been derived from a pattern like that of the homacodontine dichobunids (see Figure 21-13). In association with changes in the molar teeth, the anterior end of the skull is elongated in many groups, which results in gaps behind the incisors and between the canines and anterior premolars. The lower incisors become procumbent, probably for more effective cropping of vegetation. In several groups, the lower canine teeth come to resemble the incisors and the first lower premolar becomes large and caniniform. This specialization appears to have occurred separately in several groups, if we judge by relationships based on the molar teeth and the postcranial skeleton.

Among primitive members of the derived groups of selenodont artiodactyls, the radius and ulna are not initially fused, the metapodials remain separate, and five toes are retained in the manus and four in the pes. The orbit is only partially closed posteriorly, and the mastoid is exposed laterally.

Merycoidodontoidea

The Merycoidodontoidea were a common and diverse assemblage that we know from the late Middle Eocene into the Pliocene; they were entirely restricted to North America. Most were the size and proportions of pigs and sheep, with relatively short limbs and primitive feet (Figure 21-18).

Throughout the group, the face remained short and gaps did not develop between the anterior teeth (Figure 21-19). The labial crests of the upper molars were the first to become highly crescentic, but in most members of the group both the upper and lower molars bear four large crests. The upper incisors are reduced, the lower

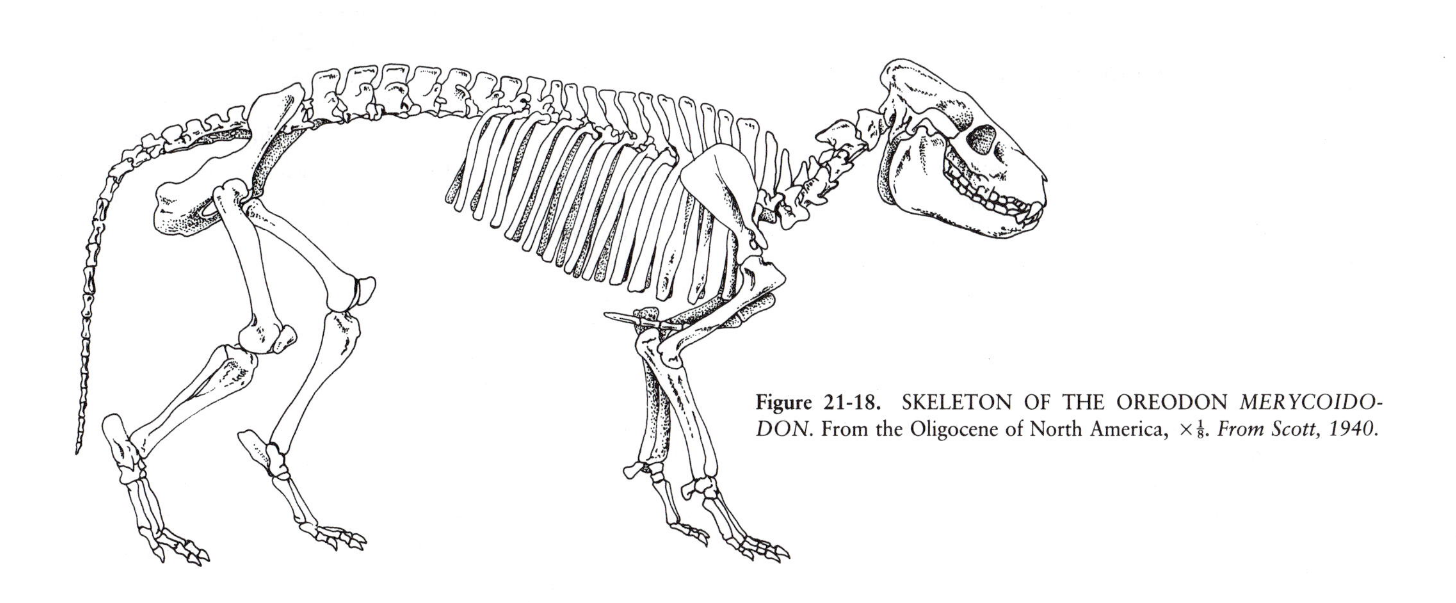

Figure 21-18. SKELETON OF THE OREODON *MERYCOIDODON.* From the Oligocene of North America, $\times \frac{1}{8}$. *From Scott, 1940.*

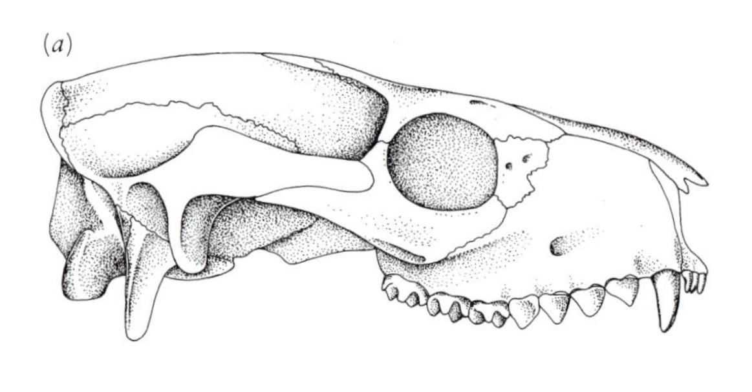

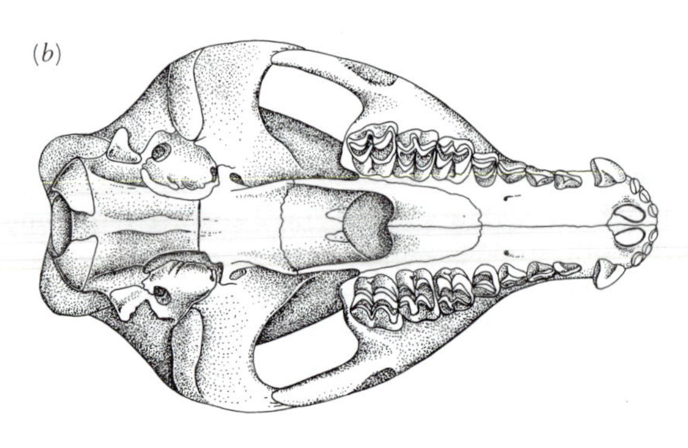

Figure 21-19. SKULL OF THE OREODON *BATHYGENYS*. (*a*) Lateral and (*b*) palatal view. *From Wilson, 1971.*

incisors and canines are procumbent, and the first lower premolar assumes the function of the canine.

The early genera *Protoreodon, Diplobunops,* and *Agriochoerus* are placed in the family Agriochoeridae, which Golz (1976) suggests probably evolved from a homacodontine dichobunid near *Bunomeryx*. Early agriochoerids are differentiated by a progressive development of selenodonty, molarization of the premolars, and reduction of the hypocone. The ungual phalanges of *Diplobunops* and *Agriochoerus* are long and laterally compressed—more like claws than the hoof-bearing feet of other artiodactyls (see Figure 21-11). It has been suggested that they climbed trees, although the hands and feet show little adaptation for grasping (Coombs, 1983). Elongation of the unguals may have been less extreme in the early agriochoerids, which are known primarily from skulls. Wilson (1971) described the gradual evolution of the late middle Eocene genus *Protoreodon* into the diverging lineages of *Agriochoerus* and the oreodon *Merycoidodon*. The agriochoerids continue in diminishing numbers into the Lower Miocene. In contrast, the Merycoidodontidae were extremely common in the Oligocene and Miocene and might have formed large herds. Schultz and Falkenbach (1968) recognized 11 subfamilies.

Anoplotheroidea

A second but less diverse assemblage of primitive selenodont artiodactyls are known in Europe as contemporaries of the merycododontoids. Sudre (1977) combined the cainotheriids and anoplotherids in the superfamily Anoplotheroidea. Their specific affinities are uncertain. Romer (1966) placed them among the Tylopoda, but no specific affinities with the Camelidae have been established. Rose (1985) pointed out that the postcranial morphology and limb proportions of the well-known *Cainotherium* show detailed resemblances to *Diacodexis*, which suggest origin from the basal artiodactyl assemblage.

TYLOPODA

Camelidae

The first modern family of selenodont artiodactyls to appear in the fossil record is the Camelidae, which is known in the Upper Eocene. The oldest recognized genus is *Poebrodon* (Gazin, 1955). Like most early members of this family, it is from North America. Golz (1976) was unable to identify a specific ancestry for camels but suggested that they may have arisen from the homacodontine dichobunids. The lower molars of primitive camels are distinguished from those of other early selenodont artiodactyls by the deflection of the anterior crest of the hypoconid inward and away from the posterior crest of the protoconid and its joining instead with the entoconid (Gazin, 1955). In contrast with the Merycoidodontoidea, the lower first premolar and canine are of equal size.

We have complete skeletons of the Lower Oligocene genus *Poebrotherium* (Figures 21-20 and 21-21). Early camels are notable for the great elongation of the neck and limbs and the consolidation of the metapodials in advance of other early artiodactyls. The ulna is coossified with the radius. The proximal portion of the fibula is fused to the tibia. The distal portion, termed the malleolar, serves the function of a tarsal element and moves with the calcaneum and the astragalus. By the Oligocene, metapodials II and V were reduced to tiny splints and lacked phalanges. All the weight was supported by digits III and IV (Figure 21-22*a*).

Webb (1972) outlined the evolution of the locomotor apparatus in later camels. Primitive genera bore small hooves, but they are reduced in the Miocene and functionally replaced by pads as in the modern camel. This is confirmed by footprints of a modern pattern from the Upper Miocene. These footprints also demonstrate that early genera had already achieved the specialized pattern of limb movements that characterize the living camels. In contrast with the pattern in horses and most running mammals, the front and hind limbs on one side of the body are moved in unison; as both limbs on the left side are moved forward, those on the right remain in place. This pattern is termed **pacing** and enables the limbs to have a very long stride without interfering with one another. This gait is most effective in open terrain where there is little need for maneuverability, but even there, lateral stability is drastically reduced. To compensate for this instability, the limbs are brought close to the midline so there is as little lateral movement as possible. Camels have reversed the trend toward more extreme unguligrady that is seen in other ungulates and returned to a digitigrade

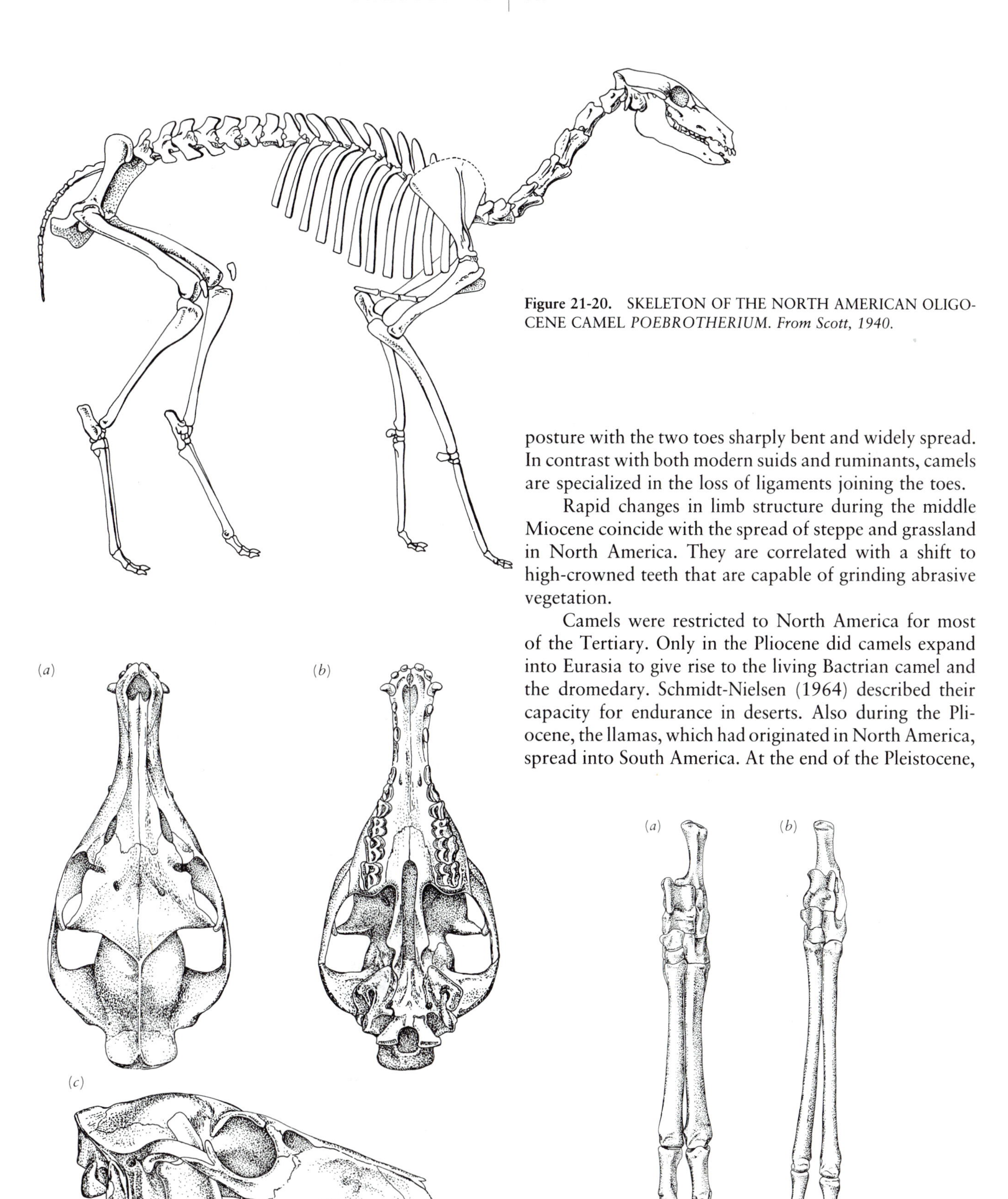

Figure 21-20. SKELETON OF THE NORTH AMERICAN OLIGOCENE CAMEL *POEBROTHERIUM. From Scott, 1940.*

posture with the two toes sharply bent and widely spread. In contrast with both modern suids and ruminants, camels are specialized in the loss of ligaments joining the toes.

Rapid changes in limb structure during the middle Miocene coincide with the spread of steppe and grassland in North America. They are correlated with a shift to high-crowned teeth that are capable of grinding abrasive vegetation.

Camels were restricted to North America for most of the Tertiary. Only in the Pliocene did camels expand into Eurasia to give rise to the living Bactrian camel and the dromedary. Schmidt-Nielsen (1964) described their capacity for endurance in deserts. Also during the Pliocene, the llamas, which had originated in North America, spread into South America. At the end of the Pleistocene,

Figure 21-21. (*a* to *c*) Skull of the camel *Poebrotherium*. In (*a*) dorsal, (*b*) palatal and (*c*) lateral views. *From Scott, 1940.*

Figure 21-22. REAR FEET OF SELENODONT ARTIODACTYLS. (*a*) The Oligocene camelid *Poebrotherium*. (*b*) The protocerid *Protoceras*. *From Scott, 1940.*

camels became extinct in North America, which coincided with the extinction of many other large mammals (Martin and Klein, 1984).

In the past, several other families have been grouped with the camels, but their phylogenetic position remains uncertain.

Oromerycidae

The Oromerycidae (which were most recently discussed by Prothero, 1986) include six genera limited to the late Eocene and early Oligocene of North America. The metapodials are not fused and the manus is still four toed, but the lateral toes of the pes are greatly reduced. Within the group the limbs become elongate, the ulna and radius are fused, and the fibula is reduced.

The oromerycids are distinguished from all other early North American selenodontids in having a bifurcate protocone; the entoconid is joined by a crest to the metaconid. Selenodonty is further elaborated within this group.

Many features of the limbs and feet advanced toward the pattern of camels but they were certainly achieved separately within each group. Golz (1976) suggests that they could have shared a common ancestry only at the level of the dichobunids. Their molars share general similarities with those of the agiochoerids, but the P_1 is not caniniform.

Xiphodontidae

Xiphodonts are a further camel-like group common in Europe in the late Eocene and early Oligocene. They early achieved a didactyl limb with only splints of side metapodials. The dentition appears camel-like, but primarily in primitive features. Since the early evolution of camelids is entirely limited to North America, and they appear to have diverged directly from primitive dichobunids, it seems unlikely that they share a unique common ancestry with the xiphodontids, although they have long been considered to be closely related. Viret (1961) suggested that the Amphimerycidae may be allied with the Xiphodontidae rather than with the Pecora, as has previously been suggested.

Protoceratidae

The protoceratids are yet another group of long-limbed, selenodont artiodactyls that is known from the late Eocene into the Pliocene in North America. They have been classified with camelids, hypertragulids, and leptomerycids on the basis of similarity of limb structure (Figure 21-22*b*). Late Eocene fossils do not demonstrate specific sister-group relationships of any of these families with the early Protoceratidae. Golz (1976) proposed that they, like the camelids, hypertragulids, and agriochoerids, may have evolved from among the homacodontine dichobunids. Webb and Taylor (1980) cite several shared derived features, most importantly the structure of the vertebral arterial canal that unites the camelids and protoceratids.

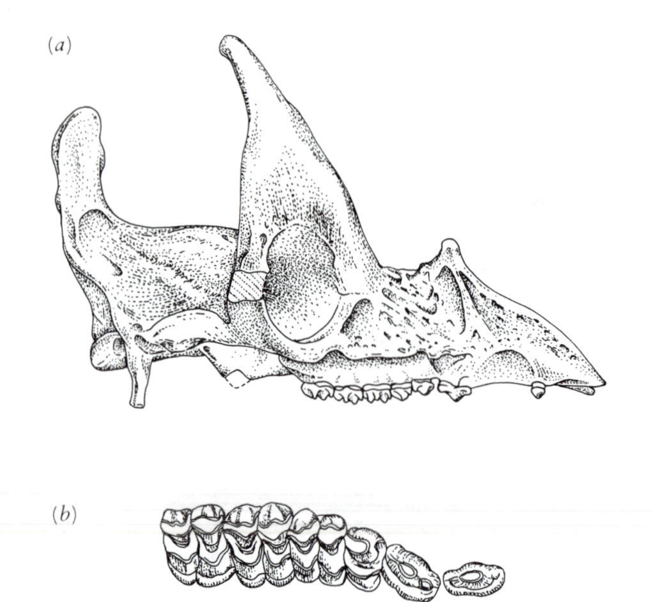

Figure 21-23. (*a*) Skull of the protocerid *Paratoceras* from the Miocene of North America. (*b*) Teeth of the same genus. *From Patton and Taylor, 1973.*

Early protoceratids are primitive in retaining four toes in the front and hind limbs, but the side toes of the pes are reduced. In contrast with early camels, they lack coossification of the ulna and radius and the fibula and tibia and the metapodials are not fused. The forelimbs are significantly shorter than the hind limbs. The dentition is advanced relative to that of contemporary hypertragulids, agriochoerids, and oromerycids in developing a tetra-selenodont pattern of the upper molars. In contrast with early camels, the upper molars are widened in occlusal view, rather than narrowed. In advanced genera, as in camelids and ruminants, the upper incisors are lost. Like the merycoidodontoids, the first lower premolar is caniniform and the lower canine is incisiform (Patton and Taylor, 1973).

Advanced protoceratids, such as *Paratoceras* (Figure 21-23) are characterized by the development of bony protuberances in a variety of bizarre patterns. They may extend from the temporal crests of the parietal, the supraorbital border of the frontal, and the dorsal border of the maxilla (Webb, 1981).

RUMINANTS

Although the differentiation of early artiodactyls from primitive ungulates and their large-scale radiation in the Middle and Upper Eocene may be attributed primarily to more sophisticated locomotor apparatus, the large-scale radiation in the Upper Eocene is accompanied by the evolution of a selenodont dentition that presumably enabled them to process coarse plant food much more effectively than the primitive bunodont genera. The continued success of the hippos, camelids, and ruminants may be attributed to advances in the structure and function of the digestive tract. These changes have left no direct fossil record, but the distribution of features among modern genera makes it possible to estimate the time of their origin in relationship to the pattern of phylogenetic divergence.

Among modern artiodactyl families, only the suids retain a primitive stomach that shows little difference from that of nonungulate mammals. The tayassuids and hippopotamids both possess expansions from the fore part of the stomach that store food undergoing digestion, but differences in structural details suggest that they evolved separately. The camels and the Ruminantia have more complex stomachs, which include large chambers for the breakdown of cellulose by bacteria and protozoa. Langer (1974) suggested that the structure of the stomachs in these groups is homologous in a general way but that much specialization must have occurred separately since their divergence in the Eocene. The pecorans have the most complex stomach of all, including the omasum, which is missing in tragulids.

Vaughan (1978) emphasized the importance of the slow but thorough extraction of nutrients from plant food that characterize the ruminants and place them at a considerable advantage over other ungulates in times of food shortage.

Primitive ruminants

Most modern artiodactyls, a total of approximately 65 genera, belong to a single group, the Pecora, including deer, giraffes, cattle, sheep, goats, and antelopes. Their ancestry can be traced back to the early artiodactyls, but their major radiation apparently occurred somewhat later than that which gave rise to pigs, hippos, and camels.

Rose (1985) emphasized that the limb proportions of *Diacodexis* from the early Eocene already approached the pattern of modern ruminants such as the tragulids (small deerlike genera living today in Africa and southern Asia), although details of limb structure are much more primitive.

The ancestors of the pecorans can be recognized in the late Eocene by the achievement of specializations of the foot structure that unite all members of this assemblage as a monophyletic group. The first uniquely pecoran character that can be recognized is the fusion of the cuboid and navicular of the tarsus (Figure 21-24). This character is evident in three families that appear first in the Upper Eocene: the Hypertragulidae, which are restricted to North America; the Leptomerycidae, which are known first in central Asia; and the Eurasian family Gelocidae. The early hypertragulids are the most primitive in other features of the skeleton, which suggests that this family is close to the origin of the entire assemblage. The earliest recognized member of the Hypertragulidae is *Simimeryx*, which we know primarily from the dentition that indicates a close relationship to the homocodont dichobunid *Mesomeryx*

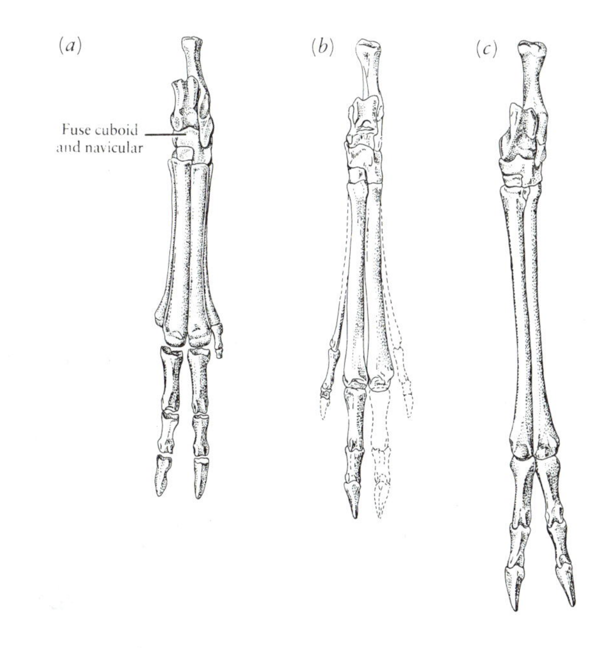

Figure 21-24. (*a*) Rear foot of the primitive ruminant *Hypertragulus* from the Oligocene of North America showing fusion of cuboid and navicular. *From Scott, 1940.* (*b*) Foot of *Archaeomeryx*. *From Colbert, 1941b.* (*c*) Foot of *Leptomeryx*. *From Scott, 1940.*

(Golz, 1976). To the extent that these fossils are known, the original divergence of the pecoran stock from the primitive artiodactyls involved reduction of the cingula, increased selenodonty and hypsodonty, and loss of the paraconule of the upper molars.

Hypertragulus from the early Oligocene is a much-better-known member of this family (Figure 21-25). It was a small animal with long limbs that was capable of rapid running and bounding. Details of limb structure indicate the retention of many primitive features. There are still five toes on the manus and four on the pes, although the lateral toes were somewhat reduced. The ulna and radius are coossified and the fibula is complete but much reduced and fused at both ends with the tibia. Metapodals II and IV remain separate.

The postorbital bar remains incomplete and the mastoid bone is extensively exposed laterally. It is progressively covered by the squamosal in later ruminants. The upper incisors are reduced, which foreshadows their loss in later pecorans. The upper canine retains its primitive configuration, but the lower canine is incisiform and its functional role is taken by the caniniform first premolar, as in merycoidodontoids and protoceratids.

The two other families that appear in the Eocene, the Leptomerycidae (Figure 21-26) and the Gelocidae, are more advanced than the hypertragulids in the fusion of the magnum and trapezoid and the loss of the trapezium and metacarpal I (Figure 21-27). The shaft of the fibula is no longer fully ossified; the distal end (termed the malleolar) functions like a tarsal element. A similar change occurred among the camelids.

These advances are also evident in the fossil and living tragulids. The fossil record of the Tragulidae goes back

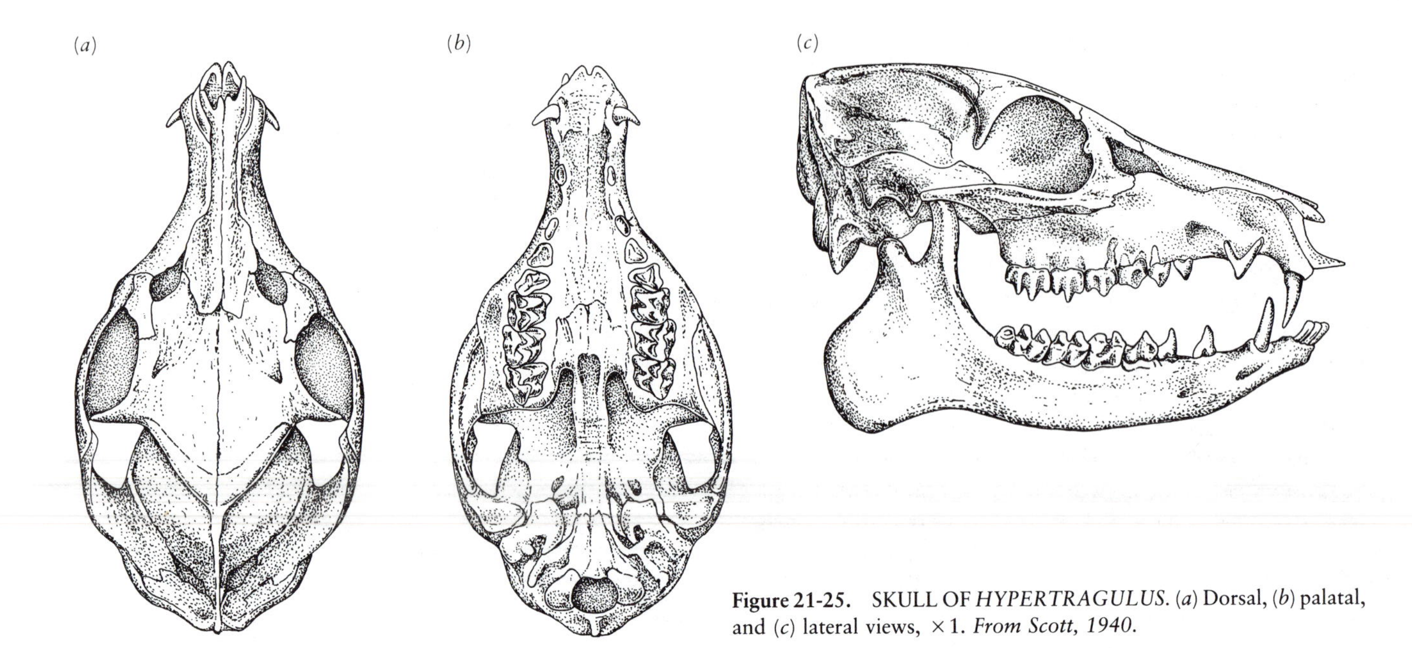

Figure 21-25. SKULL OF *HYPERTRAGULUS*. (*a*) Dorsal, (*b*) palatal, and (*c*) lateral views, $\times 1$. *From Scott, 1940.*

no further than the early Miocene when they appear in Europe, Africa, and Asia, but the living genera have long been considered the most primitive ruminants. Webb and Taylor (1980) suggest that they represent a level of evolution just above that of the hypertragulids, since they are more primitive than the leptomerycids and other ruminants in the nature of the articulation of the malleolar with the calcaneum. The postorbital bar in tragulids is complete but it is formed primarily by the jugal rather than by the frontal as in other advanced ruminants, suggesting that this feature was achieved separately in tragulids. The side metapodials are retained, although their size is reduced. Gentry (1978) contends that the living and Miocene tragulids might be classified within a single genus.

The gelocids are the most advanced of the Upper Eocene ruminants in the elongation of the principal metatarsals. The astragalus is compact and parallel sided, as in the Pecora, with the proximal and distal articulation surfaces in the same plane. The lateral metapodials are further reduced. The premolar teeth are more complex, with four lingual crests, and the mastoid has only a narrow posterior exposure. All these features suggest that they are close to the ancestry of the living pecorans.

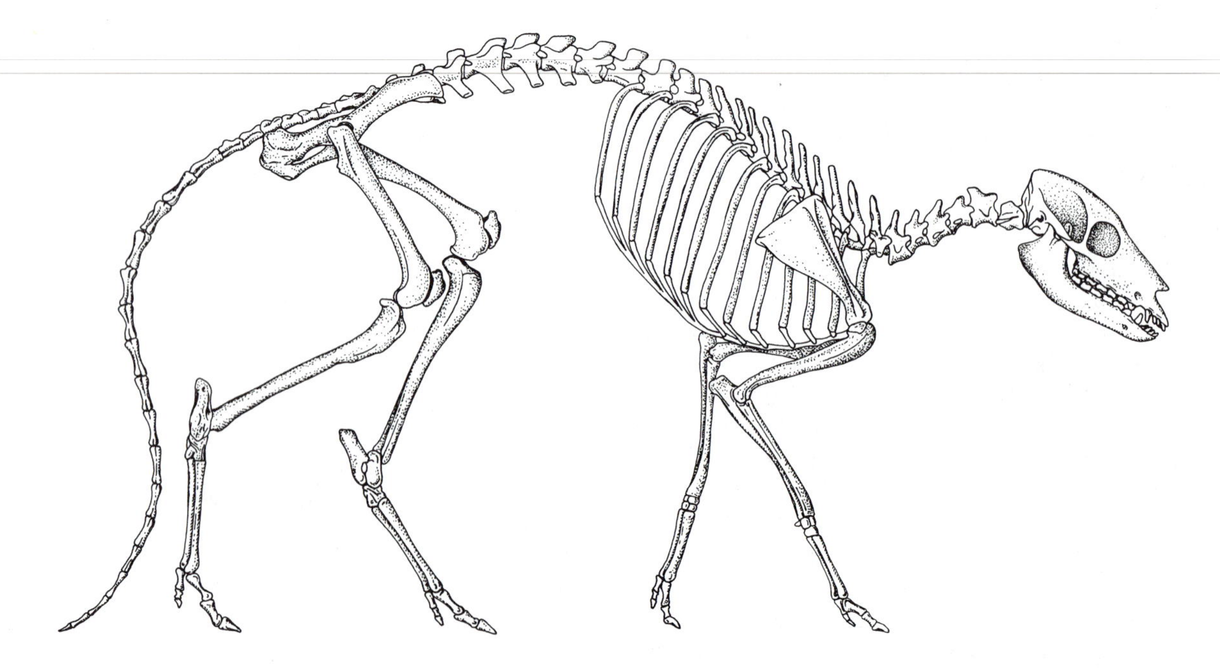

Figure 21-26. *ARCHAEOMERYX*. Skeleton of this member of the primitive ruminant family Leptomerycidae, $\times \frac{1}{3}$. *Colbert, 1941b.*

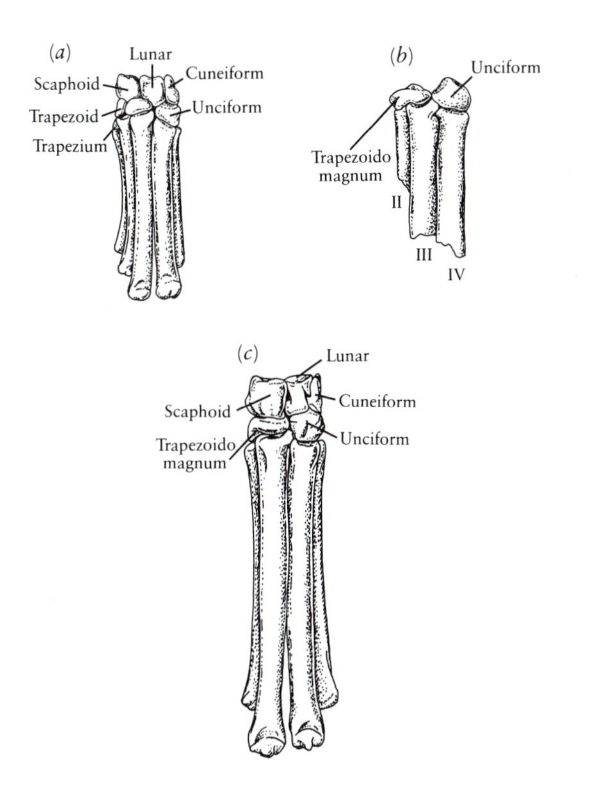

Figure 21-27. MORPHOLOGICAL CHANGES IN THE FRONT FEET OF HORNLESS RUMINANTS. (*a*) *Hypertragalus*. (*b*) *Archaeomeryx*. (*c*) *Leptomeryx*. *From Webb and Taylor, 1980.* I to V, metacarpals.

Webb and Taylor (1980) suggest that the hypertragulids, tragulids, leptomerycids, and gelocids represent a sequence of families that progressively approached the ancestry of modern pecorans. Study of the gelocids by Viret (1961) and Hamilton (1973) suggests that the Pecora probably diverged from this family in the Upper Eocene or Lower Oligocene.

Webb and Taylor consider the extant family Moschidae, the musk deer, to be a sister group of other living pecorans. Members of this family are notable for their enlarged canines and absence of antlers. The exclusion of the Moschidae from the remaining pecorans allows that group to be defined on the basis of the presence of frontal appendages.

Pecora

Four families of pecorans are recognized in the modern fauna—the Cervidae, Giraffidae, Antilocapridae, and Bovidae—all of which can be distinguished by the nature of their frontal appendages. The Bovidae alone are said to have true **horns** consisting of a conical bony horn core surrounded by a keratin sheath; neither are shed. The Antilocapridae, the pronghorned "antelope" of North America, has the same elements but the sheath is shed annually and both core and sheath may be forked. Giraffes have a bony horn core but it is covered with skin rather than a keratin sheath. The term **ossione** is applied to this structure. Ossicones may be in the shape of a simple spike, as in the living giraffe and okapi, or they may be forked or palmate as in diverse late Cenozoic genera of the Old World. Cervids characteristically have forked bony structures, termed **antlers** that are shed annually. They are attached to a pedicel extending from the skull that is comparable to the ossicone of giraffes. The horn cores of bovids and giraffids develop separately and later fuse to the skull. The antlers of cervids develop as outgrowths of the frontals (Bubenik, 1966).

The Antilocapridae and Bovidae are customarily united in the superfamily Bovoidea, and the Cervidae and Giraffidae are joined in the Cervoidea. However, the interrelationships of these families are not firmly established (Janis and Scott, 1987). Hamilton (1978a,b) believes that frontal appendages have evolved separately within the Cervidae and suggests that this family should be considered as the sister group of giraffids, antilocaprids, and bovids. Janis (1982) argues that frontal appendages have evolved several times within the Pecora. These differences result from our incomplete knowledge of the immediate ancestors of the modern pecoran families that appear in the early to middle Miocene.

There is a fairly rich record of primitive pecorans from the early Miocene into the Pleistocene. The North American family Dromomerycidae and the Old World Palaeomerycidae include animals with frontal appendages that are similar to those of living giraffes (Frick, 1937). The Giraffidae is an Old World group that is typically thought of as an extension of the palaeomerycids on the basis of the similarity of the ossicones. Hamilton (1978b) recognized three main types, the short-limbed, heavy bodied Sivatheriinae, which retain palmate horns similar to those of the palaeomerycids; the long-necked Giraffinae; and an assemblage that includes *Okapia,* which he considered the most primitive of these groups. He recognized two extinct families from the Miocene as more primitive members of the superfamily Giraffoidea, which he defined on the basis of the presence of a lobate lower canine tooth.

Clearly identifiable deer appear in the Lower Miocene of Europe, Asia, and Africa. In contrast with giraffids and bovids, the cervids are limited primarily to the North Temperate region. They spread into North America at the close of the Cenozoic and extended into South America after the emergence of the Panama Land Bridge.

The North American antilocaprids also emerged in the early Miocene and were moderately diverse until the late Cenozoic when, perhaps as a result of competition from the immigrant bison, they were reduced to a single genus, *Antilocapra.* Leinders and Heintz (1980) cite shared derived characters of the lacrimal duct that suggest a close relationship to the cervids.

Isolated teeth of possible bovids are reported from the late Oligocene of central Asia (Trofimov, 1958), but the earliest well-documented occurrence of the family is *Eotragus* from the late early Miocene of Europe (Ginsburg and Heintz, 1968). The major radiation of bovids occurred in Africa, where their fossil record goes back to the late early Miocene. Gentry (1978) and Solounias (1982) discuss the radiation of the subfamilies and tribes that make up this group. Their radiation may be associated with the spread of grassland and savanna. Their remains are notably missing in the Oligocene deposits from North Africa, which record a damp, forested environment.

In contrast with their dominance in the Old World, only a few bovids have entered North America. They include the bison, mountain sheep, mountain goat, and musk oxen, all of which are exceptional among bovids in their tolerance of cold conditions. The bison reached Central America but did not enter South America.

MESONYCHIA

There were no large mammalian carnivores in the early Paleocene. Oxyaenids and hyaenodontids appeared only in the late Paleocene, and members of the Carnivora were still relatively small until the Oligocene. In the absence of other large carnivores in the early Cenozoic, this role was taken by a family within the ungulate assemblage, the Mesonychidae.

Most early ungulates and their descendants were herbivores, but the skull proportions and dentition of the earliest condylarths show only limited specialization for feeding on plant material and may have retained the primitive capacity for dealing with animal food.

The large size of the temporal fossae and the strong canine teeth give *Arctocyon* (Figure 21-2) the appearance of a carnivore despite its blunt cheek teeth. Szalay (1969b) suggested that such genera, like members of the Ursidae, might have been omnivorous or even carnivorous rather than strictly herbivorous. A carnivorous role seems even more likely for the mesonychids. Specializations of the molar dentition indicate that they had reversed the evolutionary trend of the early ungulates and become primarily carnivorous through increasing vertical shear, rather than crushing and grinding.

Mesonychids appear in the middle Paleocene and extend into the early Oligocene, a span of approximately 20 million years. They are typified by the Eocene genera *Mesonyx* and *Harpagolestes*. *Mesonyx* was the size and proportions of a wolf and, perhaps, had a similar way of life. The limbs were specialized for cursorial locomotion (Figure 21-28). The metapodials were closely integrated and the posture of the foot was digitigrade. However, all mesonychids are thought to have retained hoofs rather than claws, and the terminal digits, like those of condylarths, are deeply fissured.

The middle Paleocene genus *Dissacus*, which is known from Europe, Asia, and North America, was more primitive, retaining a plantigrade posture and five digits in the manus. Even this genus was fairly large, with the skull nearly 20 centimeters long. The large temporal fossa and high sagittal crest in *Harpagolestes* and *Mesonyx* and the low position of the mandibular condyle and its hingelike action are typical of carnivores and suggest a very strong bite (Figure 21-29). The dentition has some specialized features in common with members of the Carnivora as well. The lower molars are laterally compressed blades and both the upper and lower teeth have carnassial notches that in modern carnivores serve to hold the flesh as it is torn from the bone. However, in occlusal view, one can see that the upper molars retain a basically triangular pattern, with no development of carnassials.

In squaring the upper molars to increase the surface area to crush and grind plant material, the ancestral condylarths significantly reduced the postvallum-prevallid shear between the back of the last upper premolar and the an-

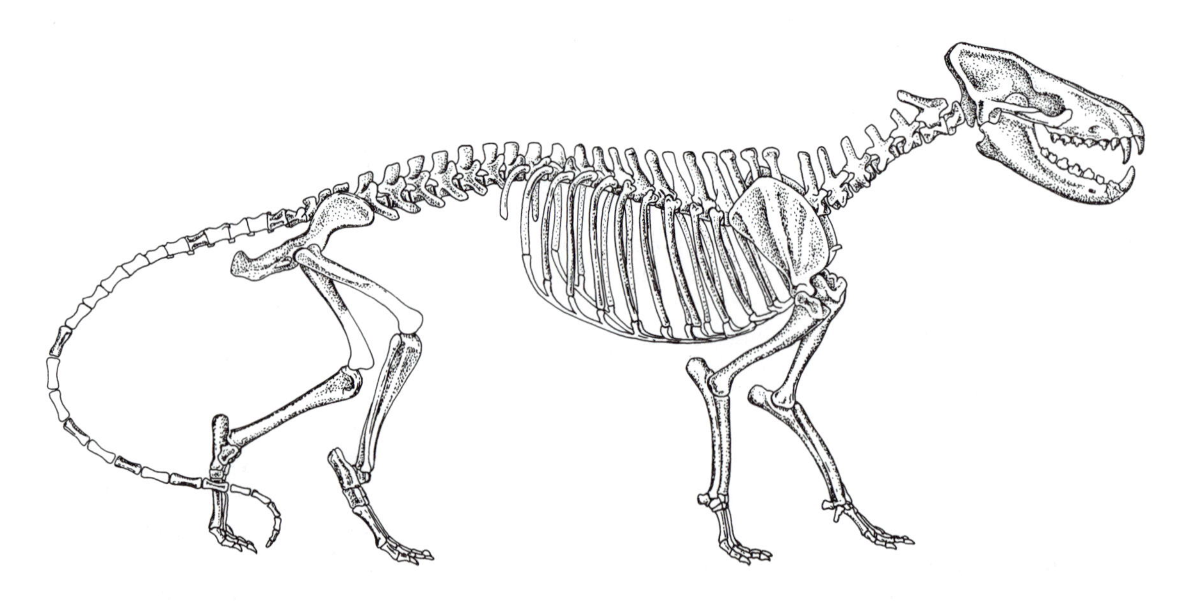

Figure 21-28. *MESONYX*. Skeleton of the carnivorous mesonychid from the Eocene of North America. *From Scott, 1888.*

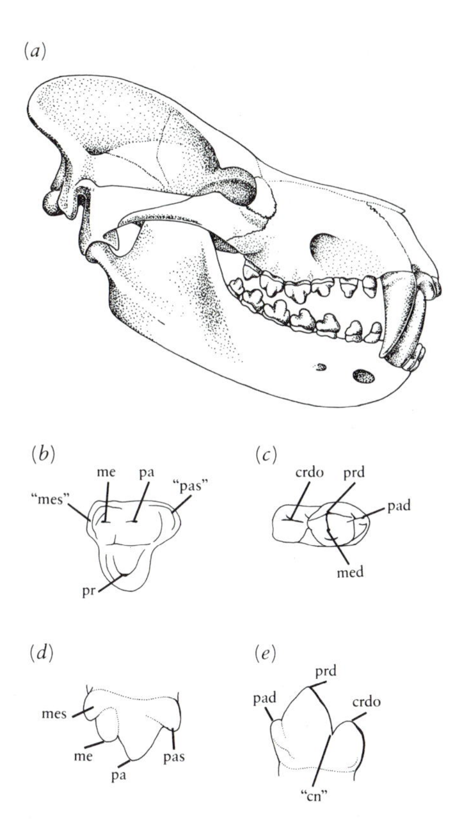

Figure 21-29. (*a*) Skull of the mesonychid *Harpagolestes.* (*b* to *e*) Molar teeth of the middle Paleocene mesonychid *Dissacus.* (*b*) and (*d*) are occlusal and buccal views of right upper molar. (*c*) and (*e*) are occlusal and buccal views of left lower molar. Abbreviations as follows: "cn," analogous feature to the carnassial notches of the carnivoran and hyaenodontan trigonids; crdo, cristid obliqua, the crest formed by the buccal wall of the talonid; me, metacone; med, metaconid; "mes," mestastyle; pa, paracone; pad, paraconid; "pas," parastyle; pr, protocone; prd, protoconid. *From Szalay, 1969b.*

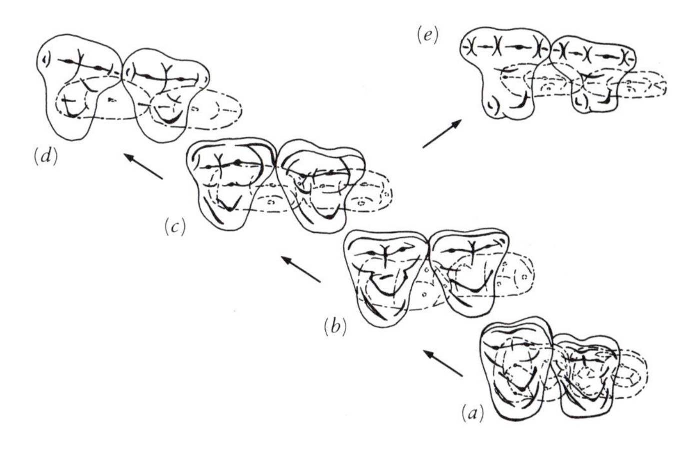

Figure 21-30. THE CONFIGURATION OF MOLAR TEETH. Progressive changes from triisodont arctocyonids to mesonychids. (*a*) The early Eocene triisodont *Eoconodon.* (*b*) *Microclaenodon,* an intermediate form. (*c*) The primitive mesonychid *Dissacus.* (*d*) *Mesonyx* of the middle Eocene. (*e*) *Hapalodectes* from the early Eocene. Darkly outlined upper molars are drawn in occlusion with dashed lower molars. *From Szalay, 1969b.*

terior molars and the front of the lower trigonids that in creodonts and the Carnivora became elaborated to form the carnassial blades. Instead, the ancestors of the mesonychids developed different surfaces of the teeth for shearing that were never as efficient.

Matthew (1937), Szalay (1969b), and Van Valen (1978) identified the ancestors of the mesonychids among the triisodontine arctocyonids. Szalay traced the evolution of the molars between these groups (Figure 21-30). In the sequence from early Paleocene triisodontines, the lower molars became progressively laterally compressed. The protoconid became the largest cusp, the paraconid shifted laterally to lie directly in front, and the metaconid was reduced. The talonid remained low, lost the hypoconulid and entocondulid, and eliminated the basin. The hypoconid and cristid obliqua remained as the primary elements. As a result of these changes, the lower molars developed a close resemblance to the premolars. In contrast with the common trend in other ungulates to molarize the premolars, the triisodontines and mesonychids premolarized the molars.

The upper molars also became more linear in function, with anterior and posterior extension of the parastyle and metastyle. The paracone and metacone remained conspicuous. The paraconule and metaconule were lost, leaving a valley between the buccal cusps and the protocone. The lateral surface of the lower molars sheared against the medial surface of the buccal cusps. The apex of the lower molars crushed against the surface lateral to the protocone.

Only a relatively few, small mesonychids had sharp cusps. In most genera, the teeth are massive and the cusps blunt, which gives the general appearance of the teeth in hyaenids and suggests that they were bone-crushing scavengers. Others may have been bearlike omnivores. *Andrewsarchus,* the largest genus among the mesonychids, is also the largest-known terrestrial mammalian carnivore. The skull from the Upper Eocene of Mongolia is 83 centimeters long and 56 centimeters wide across the zygomatic arches (Figure 21-31).

Mesonychids were common and diverse throughout the Eocene and continued into the early Oligocene in Asia (Szalay and Gould, 1966; Szalay, 1969a; Li and Ting, 1983). Their extinction may be associated with the rise of large creodonts and modern families of Carnivora in the late Eocene and early Oligocene.

The appropriate taxonomic rank of the mesonychids is difficult to judge. They were less diverse and long lived than most of the placental orders, but they are anatomically very distinct from the remainder of the early ungulates. They are often classified as a separate suborder of the Condylarthra, but sometimes as a distinct order. The problem is complicated by the fact that early mesonychids were almost certainly close to the ancestry of whales. Despite the extreme difference in habitus, it is logical from the standpoint of phylogenetic classification to include the mesonychids among the Cetacea.

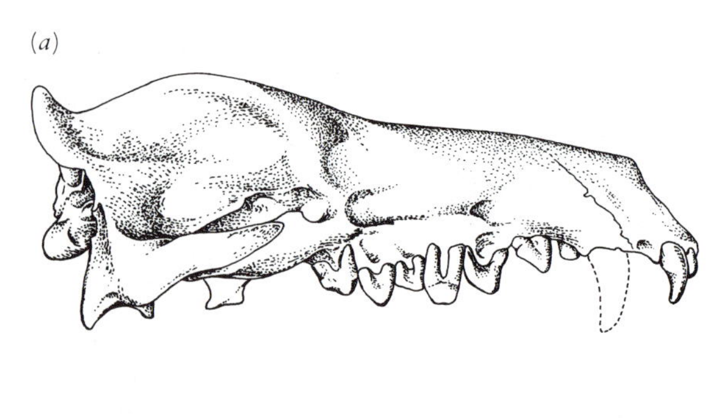

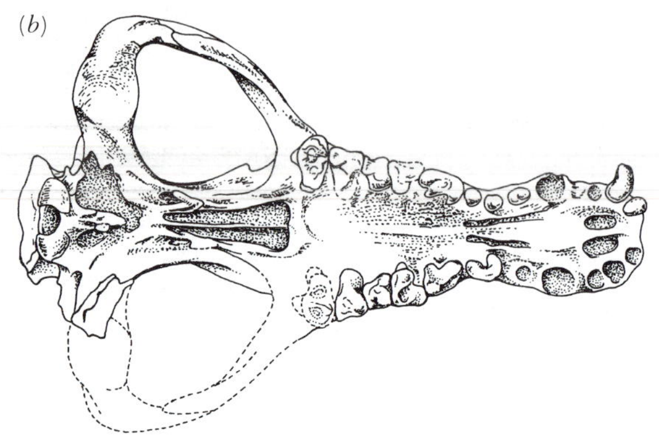

Figure 21-31. (*a*) Lateral and (*b*) palatal views of the skull of *Andrewsarchus,* a giant omnivorous mesonychid from the Upper Eocene of Mongolia, 83 centimeters long. *From Osborn, 1924.*

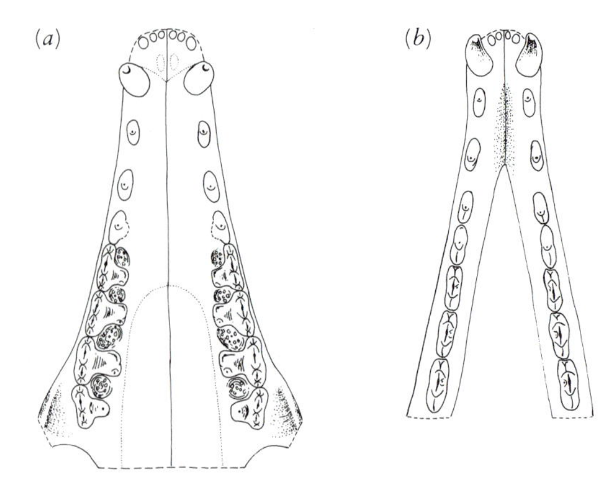

Figure 21-32. (*a* and *b*) Upper and lower dentition of the early Eocene mesonychid *Hapalodectes,* which approaches the pattern of early whales. *From Szalay, 1969b.*

CETACEANS

Whales are among the most spectacular of all vertebrates. They include the largest animals that have ever lived, which far exceed the bulk of the largest dinosaurs. The brains of some modern species exceed those of man in absolute size, and they have an exceedingly elaborate, but as yet poorly understood, sound communication system. Toothed whales have evolved a system of sonar that enables them to hunt prey in the depth of the ocean.

As descendants of primitive terrestrial mammals, whales have become secondarily adapted to a completely aquatic way of life to a degree rivaled by no other vertebrate groups. Their entire skeleton, physiology, and behavior are modified for feeding, communication, locomotion, and reproduction within the water.

Whereas the locomotor and feeding apparatus of most advanced groups of terrestrial ungulates evolved progressively throughout the Cenozoic, whales had become highly adapted to life in the water by the Middle Eocene. The brains of whales had evolved the size and degree of surface convolutions comparable to those of advanced hominids 30 million years ago (Lilly, 1977).

The skulls of Eocene whales bear unmistakable resemblances to those of primitive terrestrial mammals of the early Cenozoic. Early genera retain a primitive tooth count with distinct incisors, canines, premolars, and multirooted molar teeth. Although the snout is elongate, the skull shape resembles that of the mesonychids, especially *Hapalodectes,* a small Eocene genus with particularly narrow shearing lower molars (Figure 21-32). Although the lateral portion of the upper molars forms a similarly narrow shearing blade, the medial portion is divergent in the presence of a large hypocone. As in whales, the zygomatic arches turn ventrally at their point of origin on the maxilla. Also like the early whales, *Hapalodectes* has vascularized areas between the medial portions of the upper molars. *Hapalodectes* is probably too late to be an actual ancestor of whales, which are known from the end of the lower Eocene, but according to Szalay (1969a) they may share a close common ancestry with this genus.

EARLY PROTOCETID WHALES

The oldest whales have been described from latest early Eocene deposits in Pakistan (Gingerich and Russell, 1981; Gingerich, Wells, Russell, and Shah, 1983). The cranial remains are intermediate between those of well-known late Eocene whales and mesonychids. Some incomplete material was initially assigned to the Mesonychidae, but more recently discovered remains show that the basicranium is unquestionably that of a cetacean. The most complete of these remains belong to the genus *Pakicetus* (Figure 21-33). The skull is incomplete but may have been 30 to 35 centimeters long. The braincase is very narrow, with a high sagittal crest and a lambdoidal crest much as in late Eocene whales.

The dentition and elongate jaws are very similar to those of well-known Upper Eocene whales. Like them, the auditory bulla is a massive structure formed from the ectotympanic. In more advanced whales, the ear region is significantly altered in relationship to directional hearing under water and for protection when diving. These

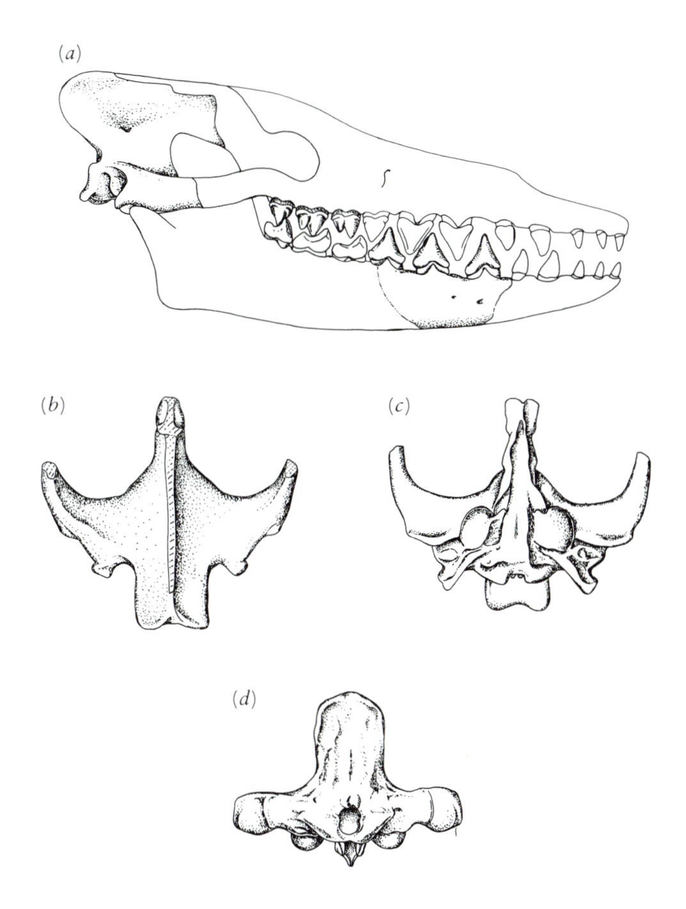

Figure 21-33. SKULL OF *PAKICETUS*, A CETACEAN FROM THE LATEST EARLY EOCENE OF PAKISTAN. (*a*) Restoration of the skull in lateral view. (*b*) Dorsal, (*c*) ventral, and (*d*) occipital views of back of braincase, $\times \frac{3}{4}$. *From Gingerich, Wells, Russell, and Shah, 1983. With permission from* Science. *Copyright 1983 by the American Association for the Advancement of Science.*

changes are not evident in *Pakicetus*. The auditory bullae in modern whales are isolated by sinuses so that sound can be detected independently on the two sides of the skull. In *Pakicetus,* the bulla is still attached to the squamosal, basioccipital, and paroccipital, whereas in modern whales it is attached only to the periotic. The presence of a well-developed fossa for the tensor tympani indicates that *Pakicetus* retained a functional tympanic membrane, which is lost in advanced whales. There is no evidence of vascularization of the middle ear to maintain pressure while diving.

The remains of *Pakicetus* are found together with terrestrial mammals, which indicates that it spent at least some of its life on land. Unfortunately, no postcranial remains have been associated with this genus. Gingerich and his colleagues hypothesize that this genus may have been amphibious, spending much of its time in the water feeding on fish, but the rest of the time on land.

The premolar teeth of *Pakicetus* are serrated triangular blades that superficially resemble the teeth of large sharks. The molars retain the cusp pattern of mesonychids. The upper molars are triple rooted, with a distinct protocone, paracone, and metacone. The lowers retain a distinct protoconid and hypoconid, exactly as in the mesochynids.

Whales are separated into three suborders, the Archaeoceti, which are limited to the Eocene, and the Odontoceti and Mysticeti, the toothed and baleen whales that both appear in the early Oligocene (Fordyce, 1980).

The earliest whales in the archaeocete family Protocetidae come from the margins of the Tethys Sea in the Indian subcontinent, North Africa, Nigeria, and possibly from the southeastern United States. They are all short bodied—less than 3 meters in length. They retain a primitive dentition, with the first two upper premolars retaining two roots and the remaining cheek teeth keeping all three. The nostrils are at the anterior end of the snout. *Protocetes* has facets on the sacral ribs for a large pelvis.

UPPER EOCENE WHALES

More advanced Upper Eocene archeocetes are included in the families Basilosauridae and Dorudontidae, which Barnes and Mitchell (1978) considered subfamilies of the Basilosauridae. Members of the Basilosaurinae (Figure 21-34) approached 25 meters in length by the end of the Eocene and had a worldwide distribution. The upper third molar was lost and both molars and premolars had lost the third root.

The dorudontinae were smaller and retained all the molar teeth (Figure 21-35). Both this group and the Protocetidae have been suggested as ancestors of advanced whales.

The Upper Eocene archaeocetes look superficially modern with an elongate, streamlined body, paddle-shaped forelimbs, and (to judge by similarities of the caudal vertebrae) a horizontal tail fluke supported by fibro-cartilage. However, fairly large elements of the rear limbs were retained in some species. A fleshy dorsal fin may have been present, since it appears to be a primitive feature of the major groups of living whales. The ear structure was coupled acoustically with seawater, but there is no evidence for the development of sonar. Fordyce (1980) suggests that the archaeocetes might still have been capable of coming out on land to breed, as do modern seals. This stage of aquatic adaptation may have required about 15 million years to achieve from fully terrestrial Middle Paleocene mesonychids.

MODERN WHALES

The odontocetes and mysticetes of the modern fauna differ in so many features that some authors suggest that they evolved separately from distinct terrestrial ancestors. Barnes (1984) points out a number of derived characters shared by both groups that demonstrate that they had a unique common ancestry above the level of the known archaeocetes.

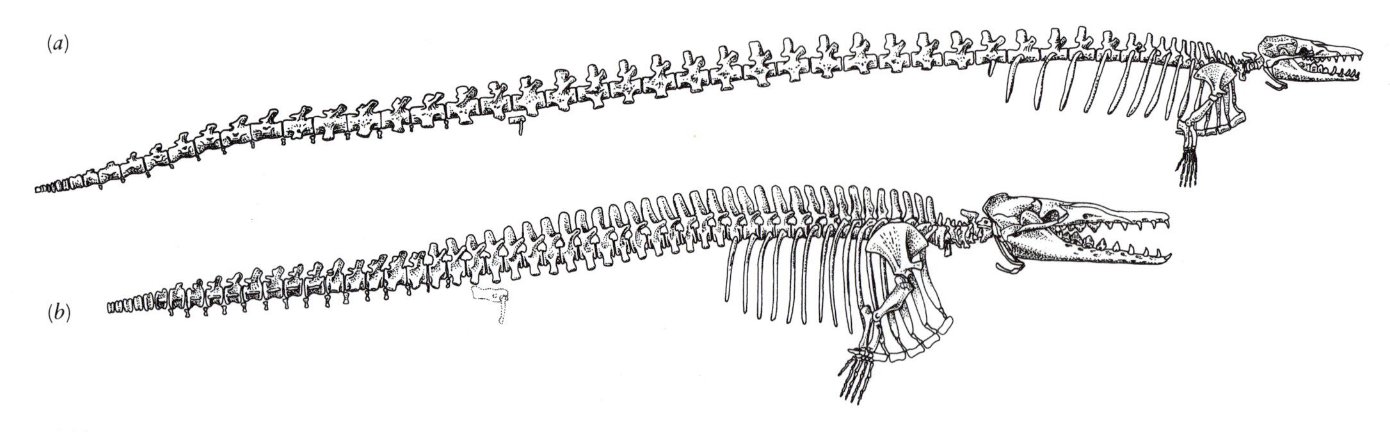

Figure 21-34. UPPER EOCENE ARCHEOCETE WHALES. (*a*) *Basilosaurus*, which reached 25 meters in length. (*b*) *Zygorhiza* (Dorudontinae), which reached 5 meters in length. *From Kellogg, 1936. With permission of Carnegie Institution of Washington.*

The skull is telescoped in association with the posterior movement of the external nares. The rostral and occipital elements have both extended over what was primitively the top of the skull. On the palatal surface, the vomer is exposed on the basicranium and covers the basioccipital-basisphenoid suture. The orbit remains confluent with the temporal opening, but the zygomatic process of the squamosal contacts the postorbital process of the frontal or is connected to it by a ligament.

The elbow joint is nonrotational and the olecranon fossa of the humerus is lost. The number of phalanges in the forelimb is increased, but the number of digits never increases.

All living whales are covered with a thick layer of blubber to maintain their body heat, but this probably evolved within the most primitive whales.

Odontocetes and mysticetes are clearly distinct when they first appear in the Southern Hemisphere in the early Oligocene, but the most primitive families show a transition from the earlier archaeocetes.

The fossil record of cetaceans is less complete in the Oligocene than in the Eocene or Miocene. This dearth has been considered a result of ecological factors that limited their original numbers (Lipps and Mitchell, 1976) or geological factors that biased their preservation (Dawson and Krishtalka, 1984). Fordyce (1980) outlined changes

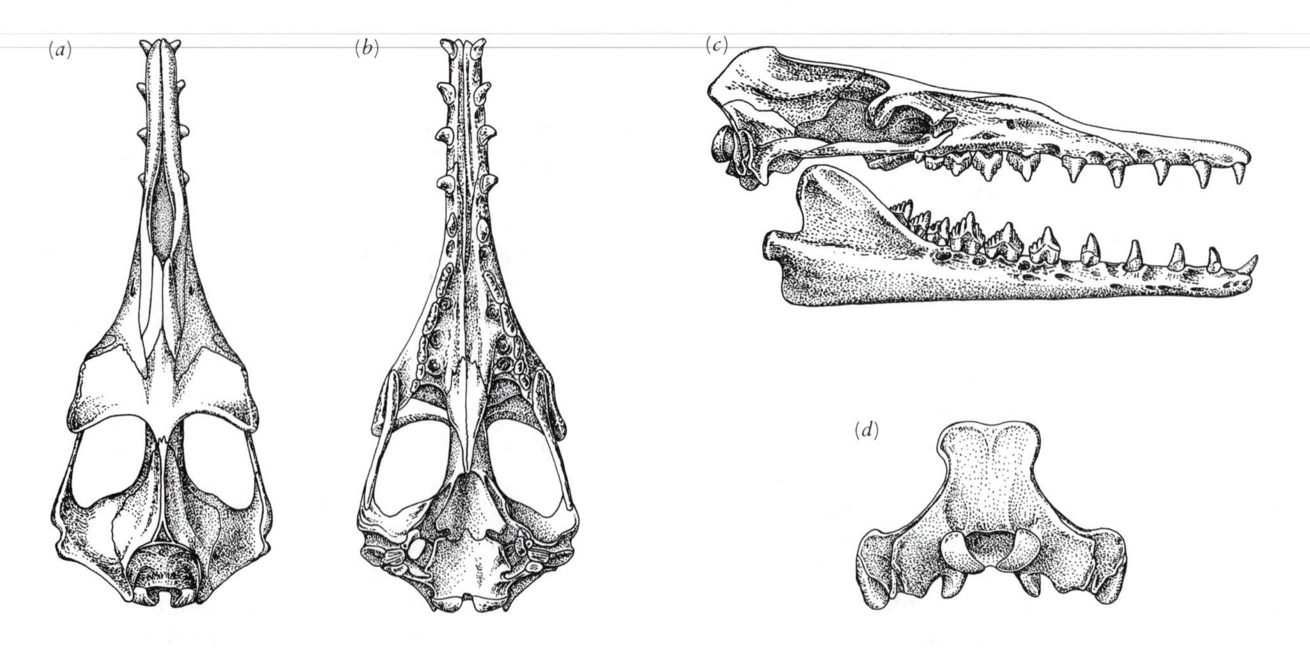

Figure 21-35. SKULL OF THE UPPER EOCENE WHALE *ZYGORHIZA*. (*a*) Dorsal, (*b*) palatal, (*c*) lateral, and (*d*) occipital views. The original was 84 centimeters long. *From Kellog, 1936. With permission of Carnegie Institution of Washington.*

in circulation in the southern oceans in the early Oligocene and subsequent establishment of the circum-antarctic current that probably triggered odontocete and mysticete evolution.

Odontocete radiation

Odontocetes and mysticetes have diverged considerably since the Oligocene. Although both suborders underwent telescoping of the skull, it was accomplished differently in the two groups. In odontocetes, the most conspicuous change is the backward extension of the rostral bones toward the occiput. In mysticetes, by contrast, telescoping is achieved primarily by the forward extension of the occipital bones onto the skull table.

Odontocetes are primitive in the retention of teeth but they are modified progressively within the group. The teeth loose their multiple roots and become reduced to simple pointed pegs. In some toothed whales, the number of teeth is greatly increased, while other families have greatly reduced dental counts.

One of the most significant features of the odontocetes, and one not developed in mysticetes, is the capacity to echolocate. As Barnes (1984, p. 140) described:

> High frequency sound, in the form of clicks, is produced by movement of recycled air within complex diverticula, sacs and valves of the nasal passages, is focused through a fatty "melon" on the face that acts as an acoustic lens, then projected into the environment. A flat surface on the posterior part of each premaxilla in most fossil and extant odontocetes indicates the site of a premaxillary sac, and presumably even the earliest odontocetes of the Oligocene actively echolocated or were at least preadapted to do this. The external auditory meatus is closed. Sound waves are reflected off objects in the water, and are transmitted by the lower jaw via a thin bony area in its posterior part to the ear region. The ear bones are isolated by fat bodies and air sacs allowing directional hearing.

The directionality of the sound is controlled by the structure of the head in the vicinity of the narial opening. In modern odontocetes, this opening is single rather than double as in other whales, and the modification of the opening results in a marked asymmetry of the surrounding bones. Asymmetry in this region evolved at least six and possibly as many as ten times among the different odontocete families.

We can group the odontocetes into five superfamilies, each representing a major radiation (Figure 21-36). Each can be characterized by the development of a different pattern of asymmetry. The Squalodontoidea includes four families from the Oligocene and early Miocene that may encompass the ancestry of all modern groups of toothed whales. The most primitive family is the Agorophiidae, which is both temporally and morphologically intermediate between archaeocetes and more advanced odontocetes. The skull shows only a limited degree of telescoping, and the cheek teeth are still multirooted and have accessory denticles. Within the Squalodontidae, the skull is fully telescoped with the nostrils between the orbits. The parietals are eliminated from the top of the skull and the maxillae make contact with the supraoccipital. The cheek teeth, although increased in number, retain their primitive triangular configuration. Squalodontids may be ancestral to most later toothed whales. The extinct Rhabdosteidae (Eurhinodelphidae) were long-snouted whales that had reduced all their teeth to simple pegs. The Squalodelphidae had already achieved an asymmetrical skull by the early Miocene.

The Delphinoidea include most of the living species. The Kentriodontidae, which are known from the Lower Miocene into the early Pliocene, link them to the first radiation of toothed whales. They were less than 2 meters long and retained a primitively symmetrical skull. Their descendants, the living families Delphinidae (dolphins and killer whales), Phocaenidae (Porpoises), and Monodontidae (belugas and narwhales), all appear in the late Miocene, some 10 to 11 million years ago. All may have separately evolved cranial asymmetry.

The Physeteroidea, which include sperm whales and pygmy sperm whales, are first known from the early Miocene, as are the Platanistoidea, the modern Asiatic freshwater dolphin, and the Amazon dolphin. The beaked whales, Ziphioidea, appear in the Middle Miocene.

Mysticeti

Mysticetes are characterized by the possession of baleen, which enables them to feed on the enormous quantities of zooplankton that are available in the world's oceans. Baleen develops from specialized epithelial tissue in the roof of the mouth that becomes keratinized. Pivorunas (1979) outlined a variety of feeding techniques practiced by different species. As much as 70 tons of water can be taken into a huge pouch beneath the mouth, throat, and chest by the giant blue whale.

Remains of baleen whales are known as early as the Lower Oligocene of New Zealand (Fordyce, 1977). Although baleen does not fossilize, the skulls show a pattern of vascularization similar to that which supplies the baleen in living whales. *Aetiocetus* from the Upper Oligocene of Oregon is the most primitive known mysticete (Figure 21-37). It already shows the pterygoid air sinus and the loose articulation of the lower jaw that are characteristic of this group, although it retains a full complement of marginal teeth (Emlong, 1966). The Aetiocetidae, which are limited to the Oligocene, appear to be a stem group that includes the ancestry of all other mysticete families. The Cetotheriidae from the late Oligocene to late Pliocene are an extremely diverse assemblage with 60 recognized species, some as small as 3 meters long. The Balaenidae, the right whales and bowheads, are known from the early Miocene. The pygmy right whale is placed in a separate family, the Neobalaenidae. The Balaenopteridae, called rorquals, are

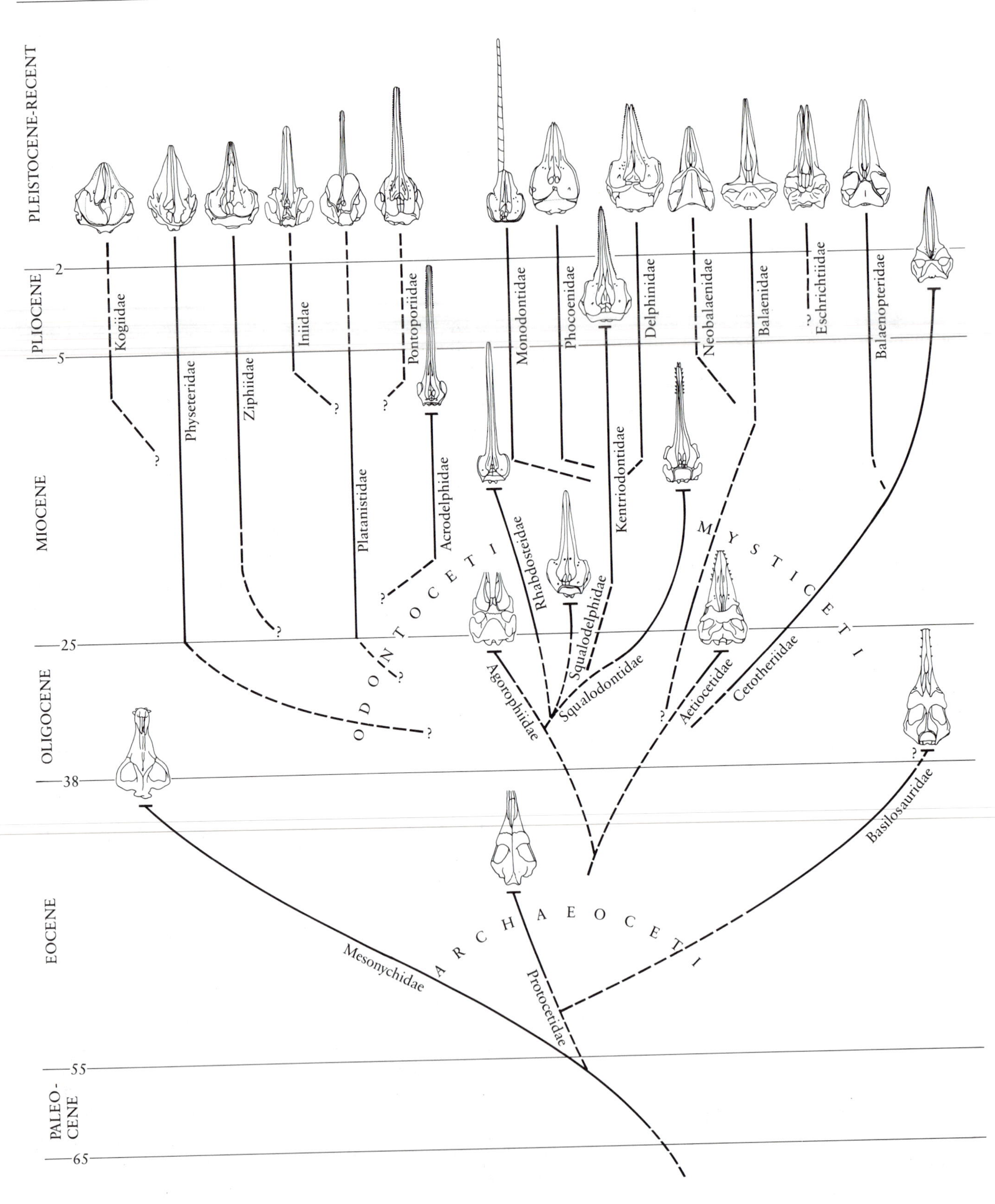

Figure 21-36. PHYLOGENY OF THE CETACEA. Included are all currently recognized fossil and living families. *From Barnes, Domning, and Ray, 1985.*

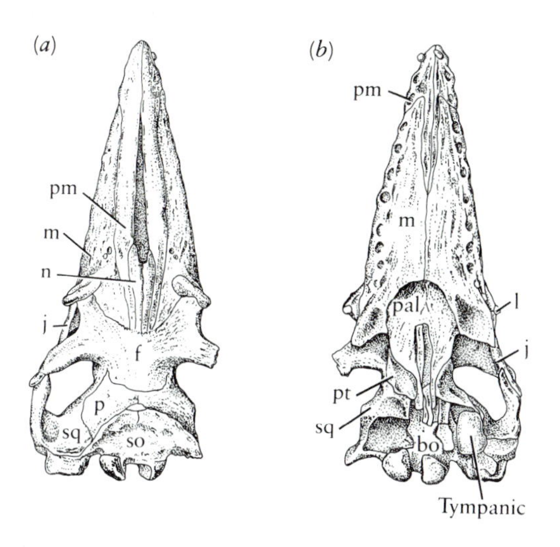

Figure 21-37. SKULL IN DORSAL AND PALATAL VIEWS OF THE PRIMITIVE OLIGOCENE WHALE *ACTIOCETUS.* Although teeth are retained, the pattern of the skull is otherwise typical of primitive mysticete whales. Abbreviations as in Figure 8-3. *From Emlong, 1966.*

the most abundant living family and include the minke and giant blue. This group is known from the late Miocene. The Eschrichtiidae, which include the gray whale, has a fossil record going back to the late Miocene, but the small size of the skull is a primitive feature that may suggest an even earlier origin. The number of families gives an inflated impression of the diversity of modern mysticetes, for there are only 8 living genera with fewer than 15 species.

PERISSODACTYLA

THE ORIGIN OF PERISSODACTYLS

Only six genera of perissodactyls remain in the modern fauna. The living horses, tapirs, and rhinos represent but a small fraction of the numerous lineages that were common in the early Cenozoic (Figure 21-38).

Perissodactyls are known from the earliest Eocene and their ancestry can be traced with little question to the condylarth family Phenacodontidae. Van Valen (1978) suggested that phenacodontids may in turn be derived from the loxolophine arctocyonids at the base of the Paleocene. He distinguished early members of the Loxolophinae from other arctocyonid subfamilies by the possession of relatively low-crowned and transverse lower molars; the trigonid basins lack a central crest, and the M_1 paraconids do not project forward. *Loxolophus* (Figures 21-39 and 21-40) has a rather doglike skull, with large canines and a short diastema behind the first premolar.

Phenacodus from the late Paleocene and early Eocene has been considered a typical condylarth (Figures 21-41). The limbs are relatively long but unspecialized. Both manus and pes are pendactyl but digits I and V are reduced. The middle Paleocene genus *Tetraclaenodon* is less completely known but occupies a position closer to the ancestry of the perissodactyls. Although it is the most advanced phenacodont that is not too specialized to have given rise to perissodactyls, it is still significantly more primitive than the earliest members of that group.

Hyracotherium, which was common in the Lower Eocene, is the best-known primitive perissodactyl. Radinsky (1966) described a number of changes between *Tetraclaenodon* and *Hyracotherium.* The most important involve the dental and locomotor apparatus.

The molars of *Hyracotherium* have relatively higher and more acute cusps and ridges and the crests connecting the cusps are better developed (Figure 21-42). The protocone-metaconule connection has been lost, but a crest has developed that joins the hypocone and metaconule. These changes result in the elaboration of two oblique transverse crests, an anterior protoloph and posterior metaloph. In the lower teeth, the hypoconulid has been displaced posteriorly, leaving the posterior side of the hypoconulid and entoconid clear for shear against the anterior side of the metaloph.

The third upper molar is enlarged to the size of the second and has added a hypocone. The lower third molar becomes larger than the second. These changes would have resulted in an increase in the amount of shear and a decrease in crushing.

Elaboration of the angle of the jaw suggests increased importance of the masseter-pterygoid complex relative to the temporalis, as in most mammalian herbivores. This complex contributes to transverse movements of the jaws, in contrast to the vertical movement produced by the temporalis, which is emphasized in carnivores.

In *Tetraclaenodon* (Figure 21-43), the mobility of the forelimb is already restricted relative to that of more primitive mammals. The elements of the carpus alternate with one another, which adds strength to the carpus but reduces its mobility. The scaphoid rests partly on the magnum and the lunar partly on the unciform. In *Hyracotherium,* the length of the forearm relative to the humerus is increased and the carpals have more extensive areas of articulation. There is increased overlap between the cuneiform and the scaphoid and between the unciform and magnum. The first digit is lost and the fifth is reduced; the third digit has assumed the major role in support of the foot. The metacarpals are much longer than those of *Phenacodus.*

Similar changes in proportions can be seen in the rear limb. The tarsus is narrow and more compact. The astragalus, calcaneum, and navicular are radically modified to eliminate the possibility of lateral movement of the foot. The astragalo-navicular articulation is saddle shaped, which is a unique and diagnostic feature of the perissodactyls. The first and fifth toes are lost. In these features, early perissodactyls are advanced over the contemporary

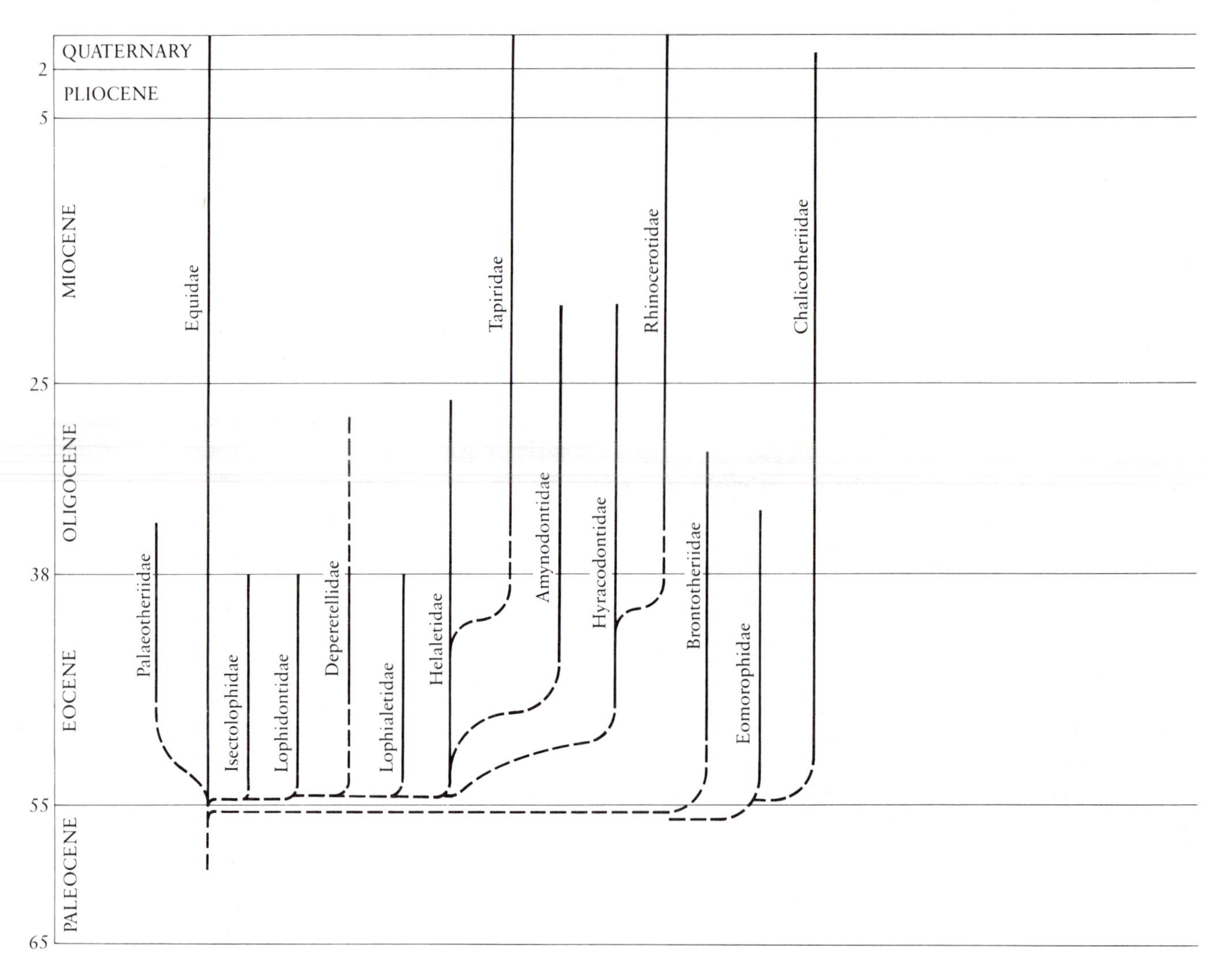

Figure 21-38. STRATIGRAPHIC RANGES OF PERISSODACTYL FAMILIES.

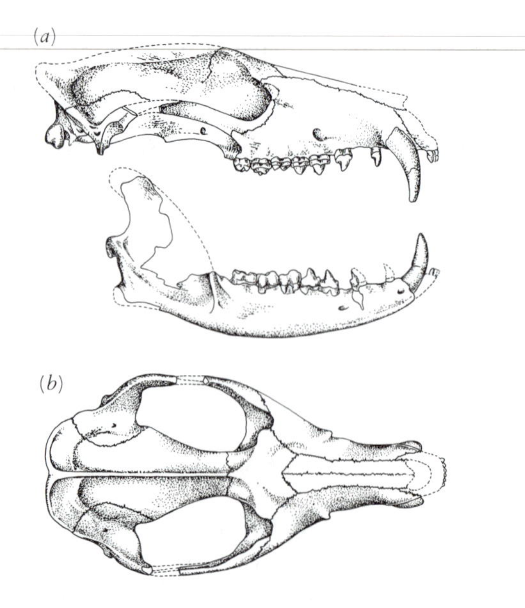

Figure 21-39. (*a*) Lateral and (*b*) dorsal views of the skull of the arctocyonid *Loxolophus,* which may be close to the ancestors of the phenacodonts and so to the perissodactyls. *From Matthew, 1937.*

artiodactyls and distinguished from them in the emphasis that is placed on the third digit, a pattern termed **mesaxonic.**

Radinsky estimated that the changes from *Tetraclaenodon* to *Hyracotherium* took place in less than 5 million years—considerably more rapidly than other changes within the order in the subsequent 55 million years. As in the case of the artiodactyls, Radinsky considers that the change in foot structure and other modifications for cursorial locomotion were more significant

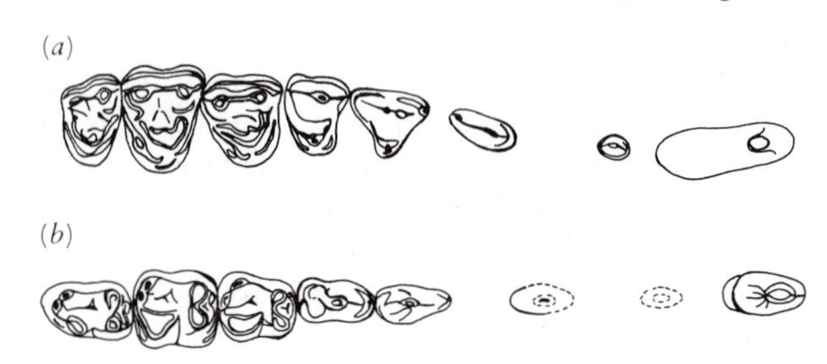

Figure 21-40. (*a* and *b*) Upper and lower dentition of *Loxolophus. From Matthew, 1937.*

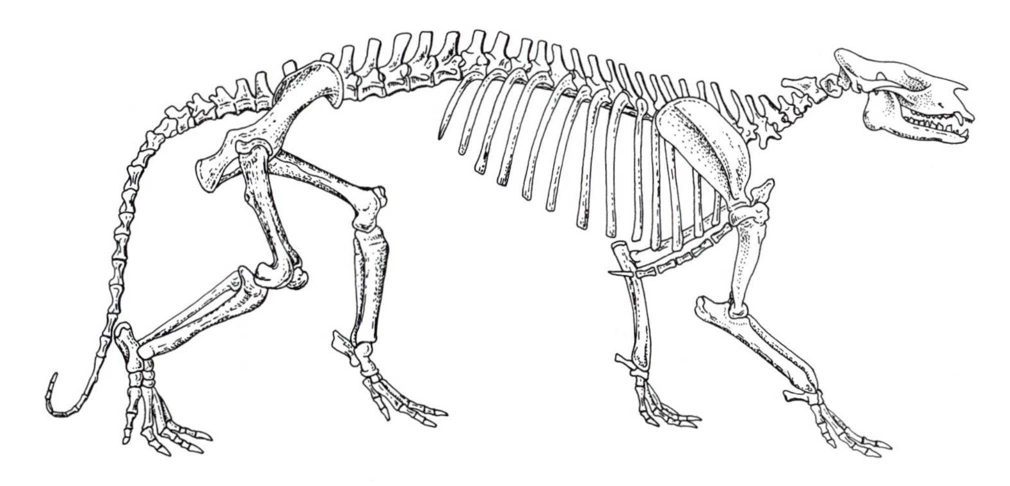

Figure 21-41. SKELETON OF THE CONDYLARTH *PHENACODUS*. *From Gregory, 1951 and 1957.*

than changes in the dentition in accounting for the great success of perissodactyls in the early Eocene.

The absence of intermediates between *Tetraclaenodon* and *Hyracotherium* in North America suggests that the actual transition may have taken place in some other part of the world. As in the case of the artiodactyls, one might debate whether it is more useful to consider that the origin of perissodactyls occurred when the definitive ankle morphology evolved or at the earlier time when the ancestral phenacondont lineage first diverged from the ancestors of other ungulate groups.

Hyracotherium varied from 25 to 50 centimeters in height at the shoulder. The orbit is midway in the length of the skull and lacks a postorbital bar. The teeth are low crowned (brachydont) and basically bunodont, despite the initial development of cross lophs. The premolars are not molariform. A short diastema is present but all the incisors are retained, as is the case in modern equids. Endocasts demonstrate that the brain is significantly advanced over that of condylarths in its relative size and expanded neocortex (Radinsky, 1976). Most skeletal features of *Hyracotherium* may be close to the ancestral pattern for all perissodactyls. MacFadden (1976) showed that the confluence of the foramen ovale and the middle lacerate foramen and the migration of the optic foramen close to, or confluent with, a group of posteroventral foramina are derived features that demonstrate that *Hy-*

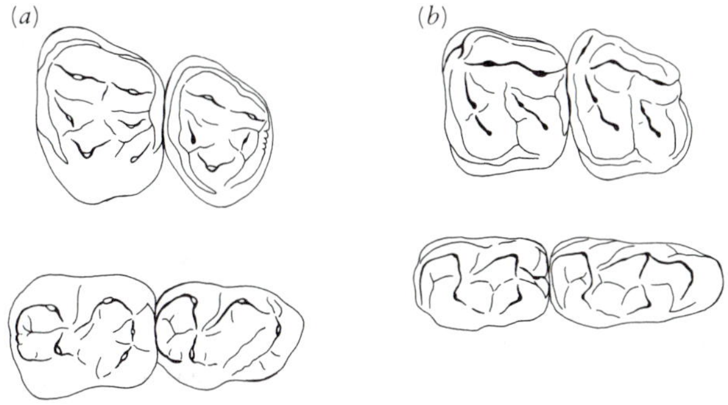

(*a*) (*b*)

Figure 21-42. Upper and lower second and third molars of (*a*) *Tetraclaenodon* and (*b*) *Hyracotherium*. These teeth show the transition from the condylarth to the primitive perissodactyl pattern. *From Radinsky, 1966.*

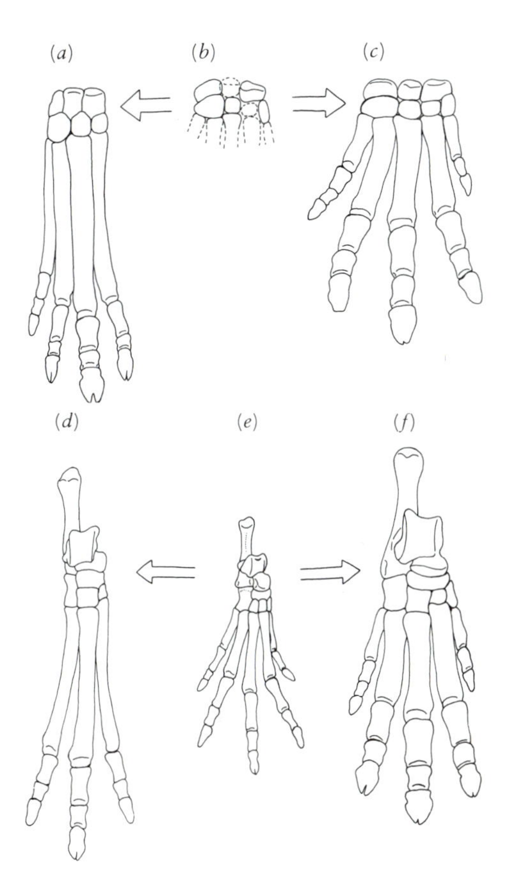

Figure 21-43. CHANGES IN THE LIMB STRUCTURE BETWEEN PHENACODONTS AND PRIMITIVE PERISSODACTYLS. (*a*) Front feet of *Hyracotherium*, (*b*) *Tetraclaenodon*, and (*c*) *Phenacodus*. (*d*) Hind feet of *Hyracotherium*, (*e*) *Tetraclaenodon*, and (*f*) *Phenacodus*. *From Radinsky, 1966.*

racotherium belongs to the same monophyletic group as other members of the family Equidae.

Representatives of three superfamilies of perissodactyls are known in the Lower Eocene. In addition to the equid *Hyracotherium,* we find the tapiroid *Homagalax* and the chalicotheroid *Paleomoropus* (a group whose later members are characterized by the elaboration of long claws) (Radinsky, 1969). The early brontotheroids (rhinolike forms with large bony protuberances on the skull) appeared in the later part of the early Eocene, and the Rhinocerotoidea is known from the beginning of the late Eocene. Fourteen perissodactyl families had differentiated by the end of the Eocene. In strong contrast with the Artiodactyla, the subsequent history of this group shows a progressive reduction in diversity.

Although later members of the perissodactyl superfamilies are very distinct from one another, the early Eocene forms are differentiated primarily by small differences in the pattern of molar cusps (Figure 21-44). Equids retain a distinct protoconule and metaconule. In the early tapiroids, these cusps join the protoloph and metaloph and the protolophid and hypolophid are elaborated in the lower molars. The chalicotheroids retain the protoconule and protocone as distinct cusps, but the metaconule merges with the hypocone to form a high, unbroken metaloph. Brontotheroids have a strong mesostyle and a W-shaped ectoloph; the metaconule is absent and the protoconule is reduced. The functional significance of these changes in cusp pattern is discussed by Butler (1952) in relationship to different patterns of mastication and particular types of food. The early perissodactyl genera were also differentiated to some degree by differences in size.

The three initial groups underwent subsequent radiation within the Eocene.

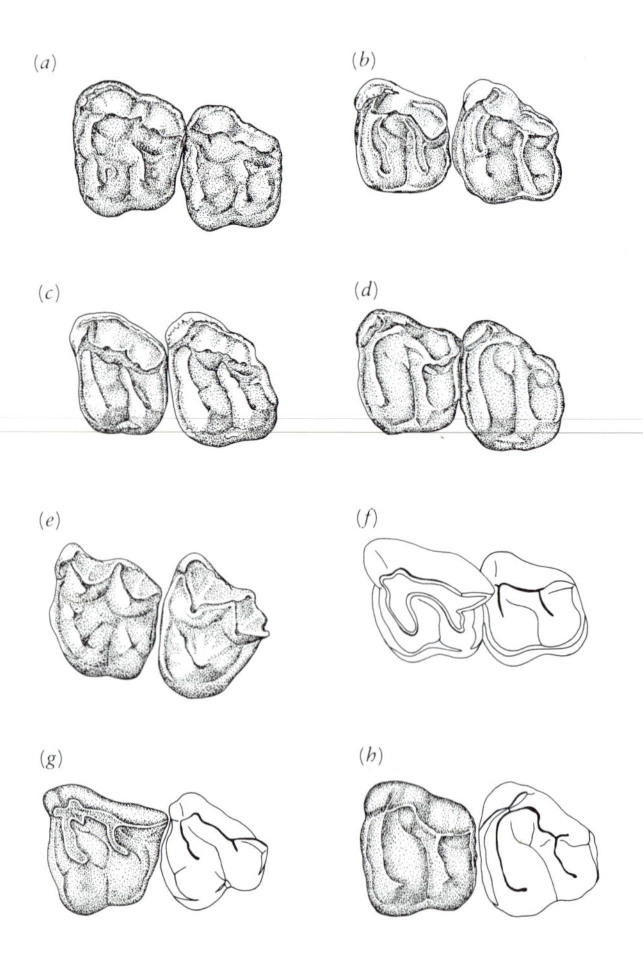

Figure 21-44. SECOND AND THIRD UPPER MOLARS OF EARLY PERISSODACTYLS. (*a*) *Hyracotherium.* (*b*) *Paleomoropus,* a chalicotherioid. (*c* and *d*) *Homogalax* and *Heptodon,* both tapiroids. (*e*) *Eotitanops,* a brototherioid. (*f, g,* and *h*) The rhinocerotoids *Amynodon, Hyracodon,* and *Hyrachyus. From Radinsky, 1969.*

TAPIROIDS

Although tapirs are the most conservative of living perissodactyls, they are but one surviving lineage of a diverse assemblage in the Eocene. The early tapiroids are distinguished from the equid *Hyracotherium* by elaboration of the cross lophs and some molarization of the premolars. They were common in all the northern continents in the early Eocene. Subsequently, separate families differentiated in North America, Europe, and Asia. Short-lived, little-differentiated groups include the Isectolophidae in North America, the Lophidontidae in Europe, and the Deperetellidae and Lophialetidae in Asia. In the Asian families, the manus became tridactyl and the metapodials were relatively long and slender.

Both isectolophids and a second family common in North America in the middle to late early Eocene, the Helaletidae, may be traced to *Homagalax.* The early helaletid genus *Heptodon* (Figure 21-44*d*) has a sharp-crested transverse protoloph and metaloph that meet the ectoloph, a pattern that is common to all later tapiroids. It is also advanced in the presence of a postcanine diastema. *Heptodon* appears to be at the base of a dichotomy leading in one direction to the Rhinocerotoidea and in another to the modern family Tapiridae. *Helaletes,* which is known from the middle Eocene of North America and the early late Eocene of Asia, is close to the origin of modern tapirs. As in *Tapirus,* this genus shows a deep nasal incision, which indicates development of a short proboscis. Other helaletids survived until near the end of the Oligocene. All other early tapiroid families became extinct at the end of the Eocene.

Radinsky (1965) detailed the changes in the skeleton between the Eocene tapiroid *Heptodon* and the Recent genus *Tapirus.* The main advances in the skull are associated with the elaboration of the proboscis and were essentially completed by the Oligocene (Figure 21-45). Postcranially, most changes can be attributed to the larger size of the modern genus, but we can see some cursorial adaptations, including the fusion of the radius and ulna and the loss of the acromium, which in noncursorial mammals supports the clavicle. As in all other tapiroids, the modern tapirs retain brachydont cheek teeth and a rela-

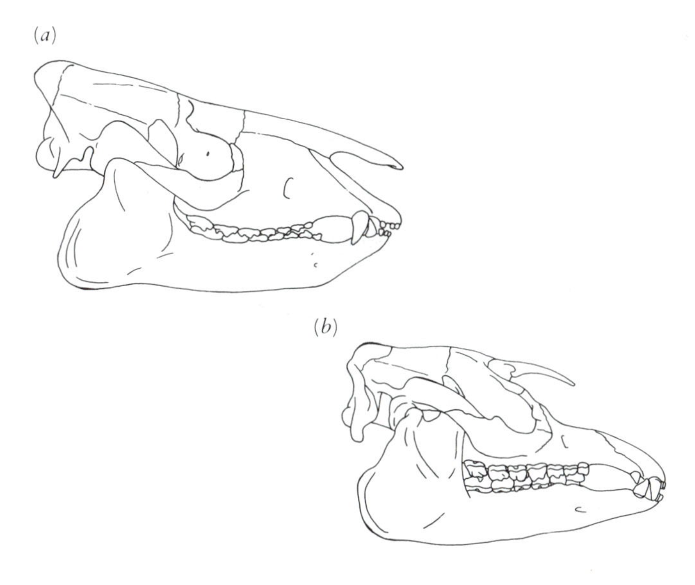

Figure 21-45. SKULLS OF EOCENE TAPIROIDS. (*a*) *Heptodon* and (*b*) the modern genus *Tapirus*. *From Radinsky, 1965.*

tively primitive pattern of the feet. The earliest tapirid, *Prototapir*, appears in the early Oligocene in Europe and the Middle Oligocene in North America. The modern genus *Tapirus* is known first in the late Miocene of China and is common in North and South America during the Pleistocene (Dawson and Krishtalka, 1984).

RHINOCEROTOIDEA

Ancestral rhinoceroses can be recognized from the beginning of the late Eocene in North America and Asia. Primitive genera were tapirlike in size and body form, but the teeth show an increase in shearing function that, as in the case of early selenodont artiodactyls, is associated in some groups with subsequent increase in size and greater cursorial adaptation. The teeth of the early rhinos are differentiated from those of the tapiroids in being higher crowned; the ectolophs of the first and second molars are long and flat and the paralophids and metalophids are high (Figure 21-44*f*, *g*, and *h*). Shearing along the ectoloph is enhanced and horizontal shearing becomes important. The morphology of the molar teeth among the Rhinocerotoidea is fairly conservative, but there is considerable variation of the pattern of the incisors and canines, which Radinsky (1969) used to characterize each of the families (Figure 21-46).

Advanced characters of the molars are initiated in some variants of a late middle Eocene species of the helaletid tapiroid *Hyrachyus*. They are elaborated in two late Eocene families, the Hyracodontidae and the Amynodontidae. Radinsky (1969) suggests that this pattern may have evolved separately from different lineages, which would make the Rhinocerotoidea, as it is generally understood, polyphyletic. One might solve this problem by referring the Amynodontidae to the Tapiroidea and restricting the Rhinocerotoidea to the Hyracodontidae and their descendant, the modern family Rhinocerotidae.

The structure and general habitus of the amynodontids are like those of other rhinoceroses and so will be considered in this section. This family is distinguished by increasing hypsodonty and further elaboration of ectoloph shear; the canines are usually large and erect and the incisors are small and pointed. The molars are transversely compressed. The skull is massive and the antorbital portion shortened. As in other rhinos, there is progressive size increase in most lineages, although some genera remain small. The limbs remain conservative, with the manus remaining tetradactyl; no cursorial adaptations are evident beyond those already present in tapirs. Amynodontids were common and diverse in the early Oligocene of North America and especially in Asia, but they did not survive to the end of that epoch.

The hyracodontids, which Radinsky (1967) reviewed, were small- to medium-sized forms that show some cursorial specializations. The pes is long and slender and the manus tridactyl. They are distinguished from the amynodonts dentally by the fact that the third upper molar is triangular in outline rather than squared, and the ectoloph is confluent with the metaloph. Within the family, most lineages show an increase in hypsodonty and the premolars become more molariform. The canines are smaller and the incisors are spatulate or pointed. A variety of lineages, some of which were highly cursorial, evolved during the late Eocene and early Oligocene. Two genera

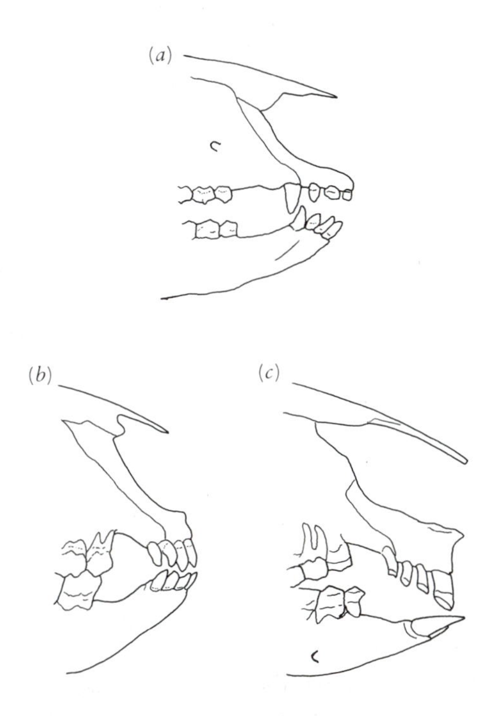

Figure 21-46. SPECIALIZATION OF THE ANTERIOR TEETH IN RHINOCEROTOIDS. (*a*) *Hyrachyus*, a tapiroid, showing a primitive perissodactyl condition. (*b*) *Hyracodon*, a hyracodontid rhinocerotoid. (*c*) *Trigonias*, a primitive rhinocerotid. *From Radinsky, 1969.*

were common in both North America and Asia in the late Eocene, but subsequent evolution proceeded separately in both areas. In the Oligocene, the Asian genus *Indricotherium* was gigantic. This family did not survive the Oligocene.

The ancestry of the modern family Rhinocerotidae is apparently to be found among primitive hyracodontids in which the manus retained four digits. Early members are recognized by the chisel-shaped upper incisor and the enlarged, procumbent second lower incisor. The other incisors and canines are reduced or lost. The Rhinocerotidae was common and diverse throughout the northern hemisphere in the Oligocene and Miocene. The central Asian genus *Baluchitherium* from the late Oligocene and early Miocene was the largest known land mammal, standing 5 meters tall at the shoulder, with a skull that was 1.2 meters long. Like other primitive rhinos this genus lacked horns. *Teleoceras* was a common late Miocene and Pliocene form from North America that had the proportions of a hippopotamus. Rhinos became extinct in North America after the Pliocene but remained widespread and diverse in the Old World into the late Pleistocene.

The fossil record of rhinos since the Miocene is well documented in Africa (Hooijer, 1978). The black and white rhinos, *Diceros bicornis* and *Ceratotherium simum*, are among the most long-lived modern ungulate species. Their transition from more primitive African species occurred approximately 4 million years ago. The divergence between these rhinoceros genera occurred at least 12 million years ago.

Dicerorhinus, which is now restricted to southeast Asia and Indonesia, was present in Europe in the early Miocene and appeared in Africa more than 20 million years ago. The genus *Rhinoceros* is known from the late Pliocene. The now extinct woolly rhinoceros *Coelodonta*, which is illustrated in neolithic cave paintings in Europe, was widespread in Eurasia during the Pleistocene.

Throughout their history, rhinos have included both browsing and grazing forms that were capable of exploiting a fairly wide range of diets. The actions of humans, both in direct predation and by widespread destruction of habitats, have placed all five of the surviving rhino species in danger of extinction. A revised phylogeny of the Rhinocerotoidea is proposed by Prothero, Manning and Hanson (1986).

BRONTOTHERIOIDEA

During the early Cenozoic, the most spectacular perissodactyls in North America and Asia were the brontotheres or titanotheres (Figure 21-47). From dog-sized forms in the late early Eocene, they reached the size of a rhinoceros by the time of their extinction in the middle Oligocene. Accompanying the great increase in overall size was an inordinate increase in the size of bony projections from the front of the skull. Stanley (1974) described the

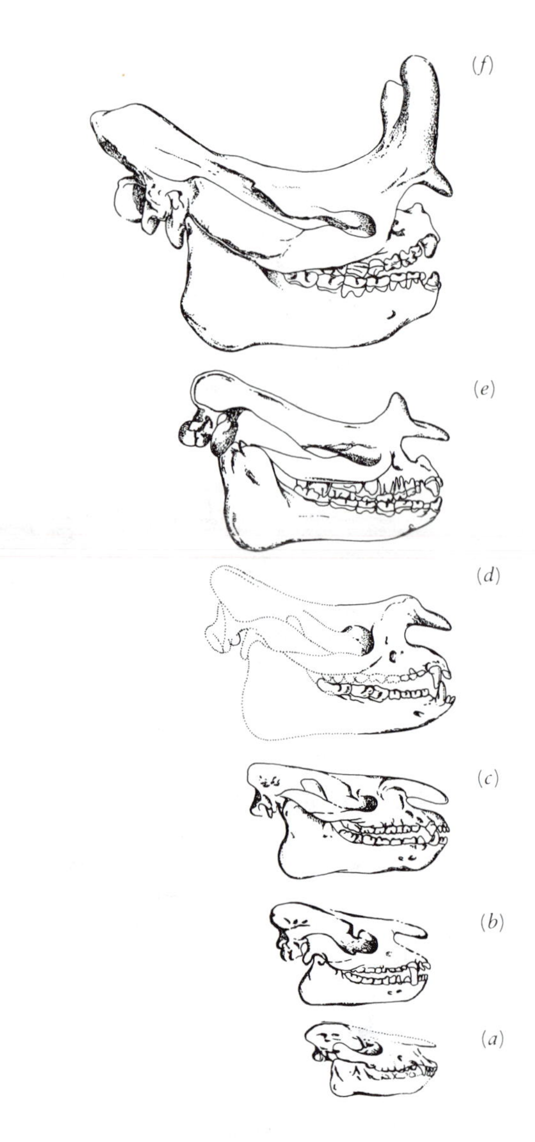

Figure 21-47. ELABORATION OF THE "HORNS" OF TITANOTHERES. (*a*) *Eotitanops*, Lower Eocene. (*b*) *Limnohyops*, Middle Eocene. (*c*) *Manteoceras*, Middle Eocene. (*d*) *Protitanotherium*, Upper Eocene. (*e*) *Brontotherium leidyi*, Lower Oligocene. (*f*) *Brontotherium gigas*, Lower Oligocene. *From Stanley, 1974.*

evolution of these structures for intraspecific combat in relationship to progressive adjustment to the increase in total body size.

Eotitanops from the latter part of the early Eocene had already evolved the W-shaped ectoloph that was characteristic of this group throughout its history. The body size increased by 50 percent by the middle Eocene. As in the Rhinocerotoidea, brontotheres show a progressive shortening of the anterior portion of the skull. By the late middle Eocene, the bones were elaborated as hornlike structures. The phylogeny of titanotheres established by Osborn (1929) indicates that large size and elaboration of the "horns" occurred independently in three related subfamilies between the middle Eocene and the early Oligocene. In relationship to their large body size, the limbs remained little modified and four digits were retained in the manus.

CHALICOTHEROIDEA

Advanced chalicotheroids include some of the most bizarre ungulates. *Chalicotherium* from the Miocene of Europe had limb proportions and specializations of the pelvic girdle that would have enabled it to assume a semibipedal stance comparable to that of a gorilla (Figure 21-48). Zapfe (1979) and Coombs (1983) suggested that it may have been adapted to browsing on tall trees.

Moropus, in which the body proportions were more horselike, is characterized by the presence of large claws on both the forelimbs and hind limbs that were retractable like those of a cat (Figure 21-49). *Chalicotherium* also had claws on the rear limbs. It has been suggested that their claws were used to dig up tubers or roots, but the teeth do not show extensive wear as would be expected if they ate food from the ground. No other elements of the skeleton are specialized like those of known digging forms. The fusion of the ulna and radius precludes twisting of the lower arm. The forelimbs may have been used to bring vegetation to the mouth.

Tylocephalonyx from the Miocene of North America resembled *Moropus* in general proportions, but the skull was greatly thickened posteriorly like the dome of *Pachycephalosaurus* and may have been used in intraspecific head butting (Coombs, 1979).

Chalicotheres are represented in the early Eocene by members of the primitive family Eomoropidae, which are known from both North America and Asia (Radinsky, 1964). Their dentition resembles that of later chalicotheres in possessing a W-shaped ectoloph that enhances vertical shear. The lower molars have a double-V pattern

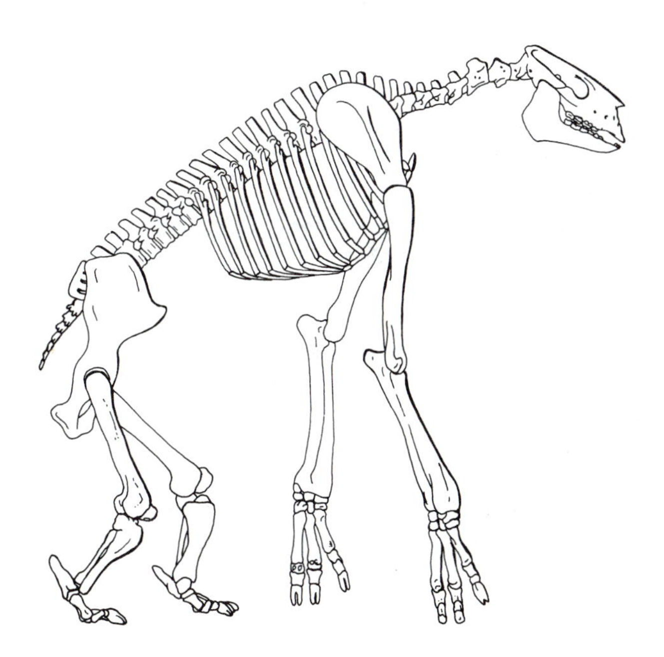

Figure 21-48. *CHALICOTHERIUM* FROM THE MIOCENE OF EUROPE. This genus is thought to have been capable of a partially bipedal posture and browsed from trees. *From Coombs, 1983.*

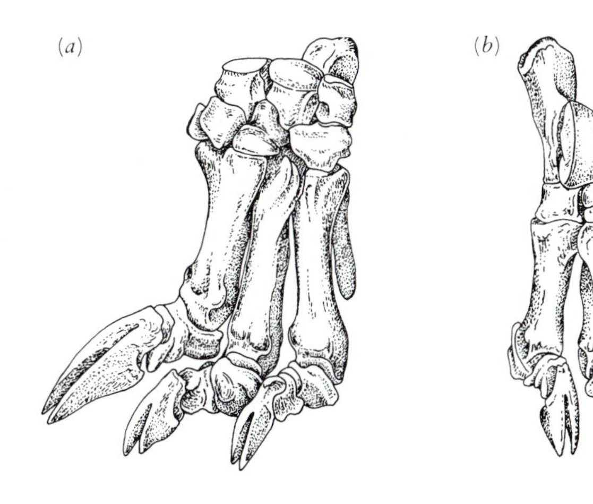

Figure 21-49. (*a*) Forefoot and (*b*) hind foot of the chalicothere *Moropus. From Gregory, 1951 and 1957.*

of crests and surfaces for crushing against the protocone. Chalicotheroids remain conservative in not evolving molariform premolars. It is not known whether or not the eomoropids had yet developed claws.

The family Chalicotheriidae is recognized from the latest Eocene and continued into the Pleistocene in Asia and Africa (Butler, 1978). After the Eocene, most of the history of this group was restricted to the Old World. Only a few genera close to *Moropus* were represented in North America during the Miocene (Coombs, 1978).

EQUOIDEA

The extensive fossil record of the family Equidae provides an excellent example of long-term, large-scale evolutionary change. Changes in body size, skull proportions, dentition, limb structure, and relative brain size have all been thoroughly documented (Simpson, 1961; Edinger, 1948; Radinsky, 1976, 1984).

Early work suggested that horses constituted a single assemblage that progressed relatively steadily from the small-sized *Hyracotherium* [*Eohippus*], with low-crowned teeth and four toes on the front feet and three on the rear, to the modern genus *Equus*, which has high-crowned teeth and whose manus and pes are reduced to a single toe. Subsequent research has demonstrated a much more complex radiation, with many divergent lineages of browsers and grazers overlapping one another in time (Figure 21-50).

Hyracotherium is known from the base of the Eocene in both North America and Europe. The equids were among the most abundant mammals during the early Eocene in North America, where the evolution of the horse is best documented. Several early genera were also common in Europe where a divergent family (or subfamily), the Palaeotheriidae, is recognized. *Palaeotherium* is distinguished by a W-shaped ectoloph and oblique, sharp-crested cross lophs. With wear, the molars functioned like

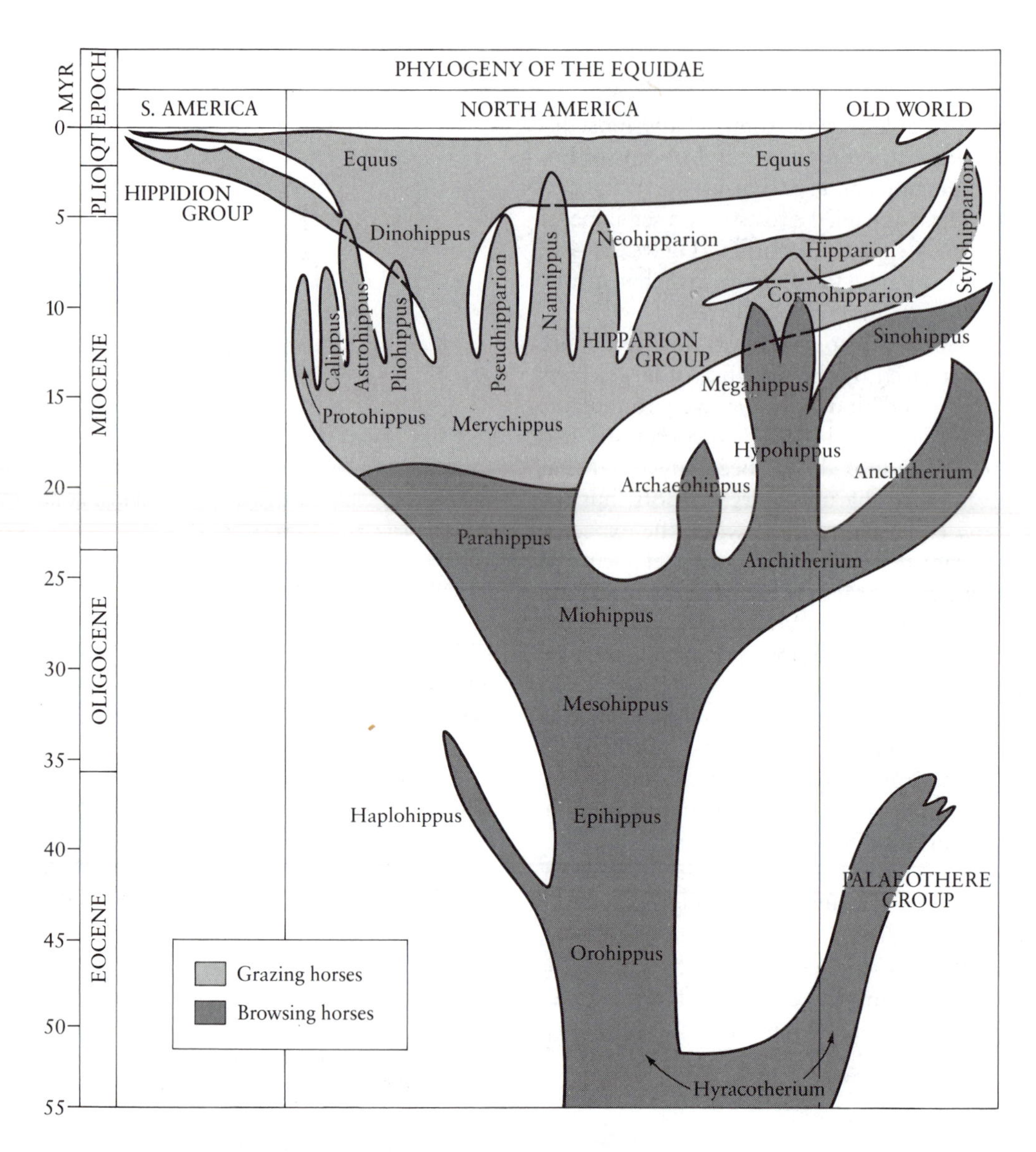

Figure 21-50. PHYLOGENY OF THE HORSE. *From MacFadden, 1985.*

those of solenodont artiodactyls. *Palaeotherium,* which was the largest equoid of the Eocene, also molarized its premolars. European equoids became extinct early in the Oligocene, leaving only the family Equidae to represent this group for the remainder of the Cenozoic.

Only a single lineage of equids is known in the Eocene of North America, represented by *Orohippus* and *Epihippus,* which are successors of *Hyracotherium* in the Middle and Upper Eocene. They show moderate size increase and progressive molarization of the premolars.

Mesohippus was the characteristic equid of the North American Oligocene. It was the size of a sheep and had lost the fourth toe of the manus. The teeth were now clearly lophodont. The ectoloph is prominent but does not have the W-shape of chalicotheres and brontotheres. All but the small first premolar had become molariform (Figure 21-51). However, the cheek teeth remained low crowned and so were better suited for browsing than grazing. From *Mesohippus,* one line, including the genus *Anchitherium,* migrated to Eurasia in the early Miocene and gave rise to species that are known in China as late as the Pliocene. All members of this lineage retained conservative features of the dentition and limbs.

A more progressive line led via *Parahippus* to the characteristic Miocene genus *Merychippus.* Within the 5 to 10 million years of the range of *Parahippus,* most of the major changes in cranial proportions leading to the modern horse were accomplished (Radinsky, 1984) (Figure 21-52). The teeth became high crowned to resist the greater wear of a diet of hard grasses. To accommodate the long roots of the cheek teeth (about 60 percent their length in modern horses), the jaws and face became deepened and the tooth row was displaced anteriorly relative to the orbit and the jaw articulation. The zygomatic arch was strengthened by the completion of the postorbital bar. The high-crowned teeth were supported by a newly elaborated tissue, **cement,** which is soft but tough and serves to support the hard but brittle columns of enamel. The pattern of the molar and premolar lophs approached those of the modern horse. The muzzle was elongated to

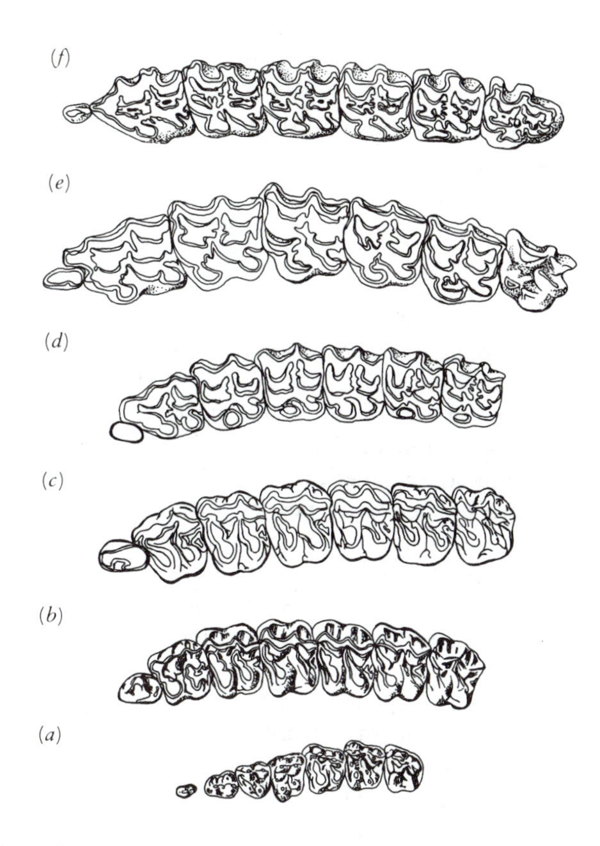

Figure 21-51. CHANGES IN THE DENTITION WITHIN THE EQUIDAE. (*a*) *Orohippus*, middle Eocene. (*b*) *Mesohippus*, Oligocene. (*c*) *Miohippus*, Miocene. (*d*) *Merychippus*, Miocene. (*e*) *Pliohippus*, Pliocene. (*f*) *Equus*, Pleistocene. *From Gregory, 1951 and 1957.*

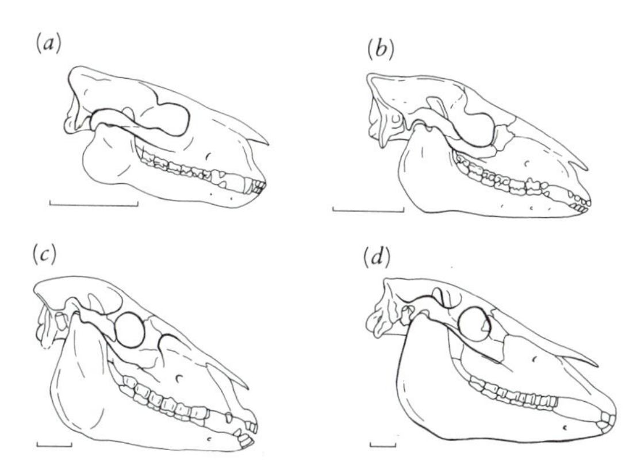

Figure 21-52. SKULLS OF REPRESENTATIVE EQUIDS THAT SPAN THE EVOLUTIONARY HISTORY OF THE FAMILY. (*a*) *Hyracotherium*. (*b*) *Mesohippus*. (*c*) *Merychippus*. (*d*) *Equus*. Note progressive changes in proportions. The greatest change occurs between *Mesohippus* and *Merychippus*. Scale bars equal 5 centimeters. *From Radinsky, 1984.*

extend the reach of the enlarged incisors. In contrast with the pecorans, the upper as well as lower incisors are retained in the most progressive perissodactyls.

Three toes are retained in *Merychippus*, but the larger central one evidently bore most of the weight (Thomason, 1986). The configuration of the manus and pes indicates that a strong elastic ligament, like that of modern horses, passed behind the central digit. The ligament stretched as the foot struck the ground, and its elastic energy assisted in ventroflexion of the foot to increase the force of the next stride (Camp and Smith, 1942). These specializations of the dentition and limbs point to *Merychippus* as a rapidly running, grazing animal of the newly expanded North American prairies. *Merychippus* was overshadowed in the late Miocene by a diversity of grazing genera that are divisible into the three-toed hipparions and the monodactyl equines. *Hipparion* is distinguished by the isolated protocone in its upper cheek teeth and by a deep depression or fossa in front of the orbit. *Hipparion* spread from North America into Eurasia and Africa. The appearance of *Hipparion* has been used as an indicator of the Miocene-Pliocene boundary in Europe and much of Asia, but this history is complicated by the probability that more than one lineage migrated to the Old World (MacFadden and Skinner, 1981; MacFadden, 1984a). Hipparions persisted in Africa into the Pleistocene.

A separate branch from *Merychippus* led to the late Miocene and Pliocene genus *Pliohippus*, in which the lateral toes became vestigial (Figure 21-53) and hypsodonty was further increased. One descendant lineage of *Pliohippus* with relatively short limbs, exemplified by *Hippidion*, diversified in South America in the late Pliocene. Another line gave rise to the dominant Pleistocene genus *Equus*, which appeared approximately 3.5 million years ago.

Equus quickly spread to Europe, Asia, Africa, and South America (Eisenmann, 1980). Surprisingly, this genus became extinct in the New World at the end of the Pleistocene, whereas nine species still live in the Old World including the wild Asian horse, four species of asses, and four zebras.

Among the perissodactyls, only the horses have been common and widespread in the late Cenozoic. Janis (1976) discussed the dietary limitations that might have been imposed on perissodactyls by the character of their digestive system. In contrast with most living artiodactyls, they do not ruminate and their efficiency in extracting nutrients from plant material is significantly lower. Perissodactyls all possess a fermentation chamber, the caecum, but it is posterior rather than anterior to the stomach, opening between the small intestine and the colon. In contrast with

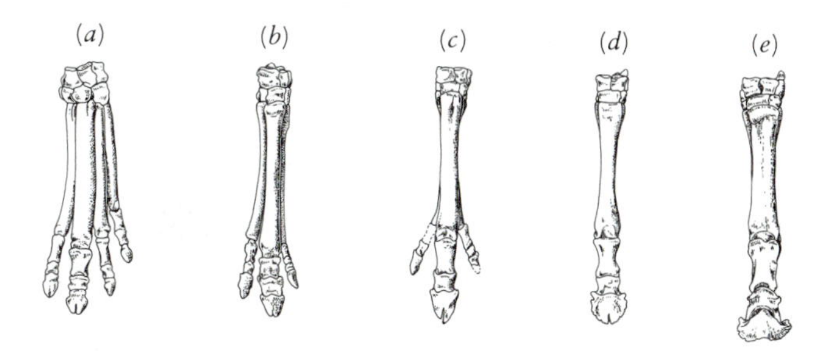

Figure 21-53. EVOLUTION OF THE FRONT FOOT IN THE EQUIDAE. (*a*) *Hyracotherium*. (*b*) *Miohippus*. (*c*) *Parahippus*. (*d*) *Pliohippus*. (*e*) *Equus*. *From Gregory, 1951 and 1957.*

artiodactyls, food passes relatively rapidly through the digestive tract. This enables them to eat very large amount of fibrous foods that are low in nutritive value, but they are relatively less efficient in handling foods of lower fiber content than are the artiodactyls. Bell (1969) described the ecological relationship in which horses feed on highly fibrous grasses and thereby make other food more accessible to artiodactyls. Janis argued that the greater efficiency of feeding on fibrous material by modern horses restricted the number of species that might have evolved to make use of this type of food. In contrast, artiodactyls are much more selective feeders, and so many species have been able to evolve and coexist in the same general area.

AFRICAN UNGULATES

Africa was never as completely isolated as Australia or South America, but in the early Tertiary there was certainly much more continuous movement of mammals between North America and Eurasia than between either of these continents and Africa (Maglio, 1978).

As yet, little is known of Paleocene mammals in Africa (Cappetta, Jaeger, Sabatier, Sigé, Sudre, and Vianey-Liaud, 1978). The late Eocene and Oligocene fauna shows a considerable diversity of forms, including some highly specialized groups that indicate a significant period of prior evolution.

A limited number of lineages appear to have entered Africa in the early Tertiary; these included prosimian primates, creodont carnivores, and possibly several lineages of early condylarths. Orders typical of the northern continents such as artiodactyls, perissodactyls, insectivores, and the Carnivora were almost certainly later immigrants that were highly differentiated before they entered Africa.

Other groups appear to have differentiated in Africa. The elephants and their close relatives within the order Proboscidea are the best known and are most specifically associated with Africa. The hyraxes or conies differentiated there, as did the short-lived, rhinoceroslike embrithopods. Although early fossils are not known, the aardvarks have been confined to Africa for most of their history. Fossils of early sirenians are common in northern Africa, which suggests that their origin may be associated with the margins of that continent.

Because of this geographical association, it has long been thought that many of these African groups shared a close ancestry. Simpson (1945) used the term Subungulate to unite proboscideans, hyraxes, embrithopods, and sirenians. When the desmostylians, a marine group from the Pacific Basin, were recognized as a distinct order, they too were included in this assemblage. McKenna (1975) grouped the Proboscidea, Sirenia, and Desmostylia in the Mirorder Tethytheria. This classification has been further substantiated by Novacek (1986) and Domning, Ray, and McKenna (1986). The history of mammals in Africa has been reviewed comprehensively by Maglio and Cooke (1978). Recent discoveries have extended the fossil record of some of these groups and demonstrated probable affinities with earlier Asian forms.

PROBOSCIDEA

The Proboscidea includes only two surviving species, the African elephant, *Loxodonta africana,* and the Asian elephant *Elephas maximus.* Proboscideans were an important and very widespread group for most of the Cenozoic, and until the end of the Pleistocene they were common in North and South America and throughout Eurasia, as well as Africa. Their early history—until the end of the early Miocene—was confined to Africa.

The oldest currently known proboscideans are from the early Eocene of southern Algeria (Mahboubi, Ameur, Crochet, and Jaeger, 1984). The skull of this (as yet unnamed) genus shows the high profile common to elephants. The nasal opening is posterior in position, indicating the initiation of a trunk, and the second upper incisors are enlarged (although not yet of elephantine proportions, (see Figure 21-54*a*)). The upper canines are re-

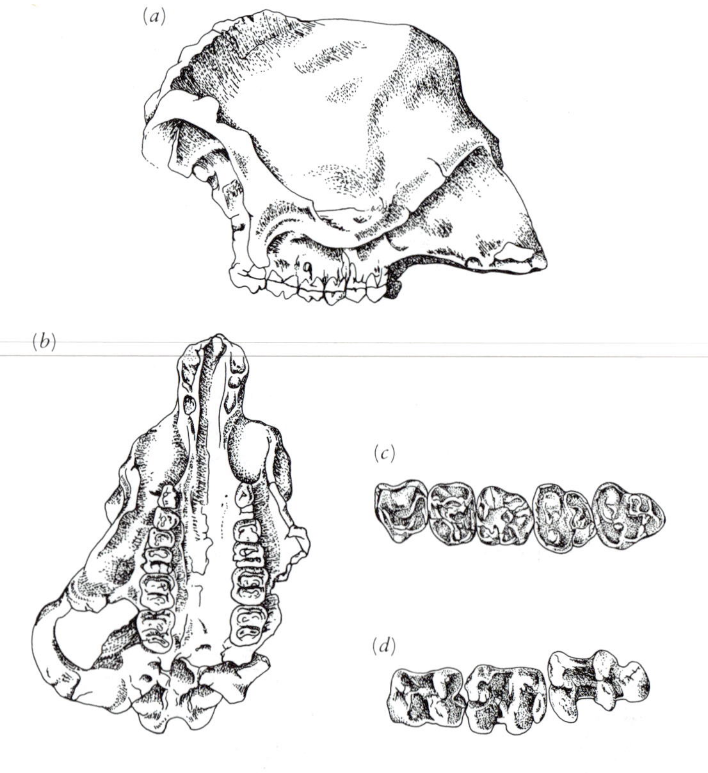

Figure 21-54. SKULL OF A NEWLY DISCOVERED PROBOSCIDEAN FROM THE EARLY EOCENE OF ALGERIA. (*a*) Lateral and (*b*) palatal views. The entire skeleton was less than 1 meter tall. *From Mahboubi, Ameur, Crochet, and Jaeger, 1984. Reprinted by permission from* Nature. *Copyright © 1984. Macmillan Journals Ltd.* (*c* and *d*) Teeth of members of the family Anthracobunidae, which may be related to the ancestry of elephants. (*c*) Posterior upper teeth of *Anthracobune. From West, 1983.* (*d*) Lower molars of *Jozaria,* $\times \frac{1}{2}$. *From Wells and Gingerich, 1983.*

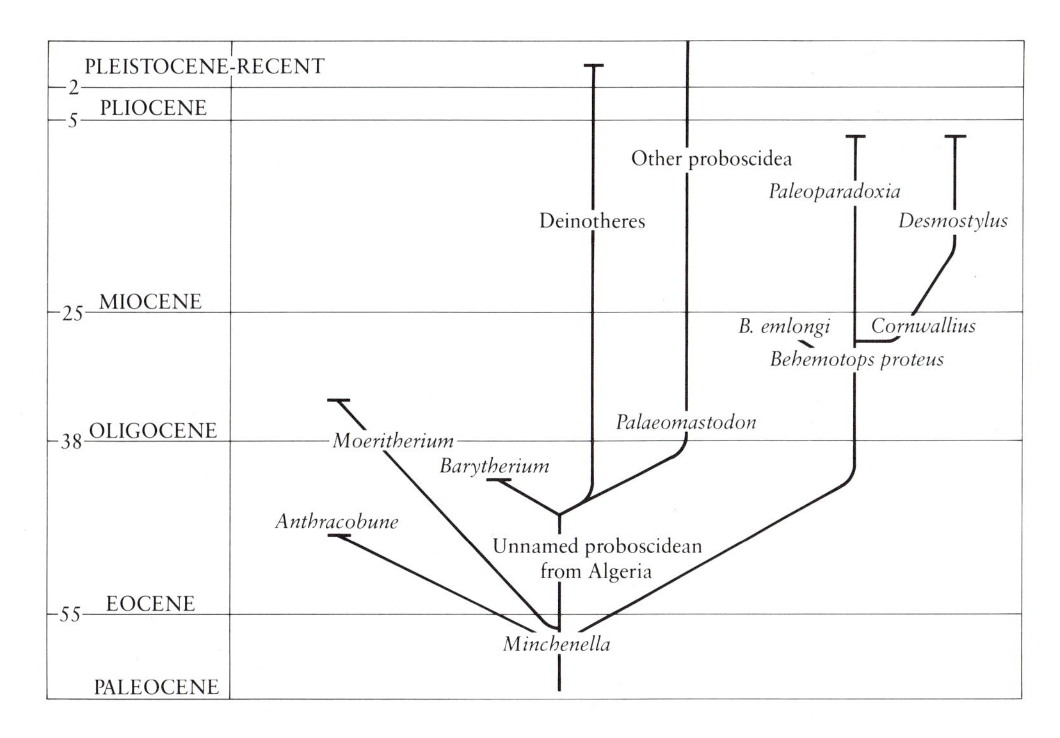

Figure 21-55. SIMPLIFIED PHYLOGRAM OF PROBOSCIDEA AND DESMOSTYLIA. *From Domning Ray and McKenna 1986. By permission of Smithsonian Institution Press, Smithsonian Institution, Washington, D.C., 1986.*

duced and separated from the premolars by a long diastema; the first premolar is lost. The molars are distinguished by two well-developed transverse lophs. As in other primitive proboscideans, the second lower incisors are also significantly enlarged, spatulate, and procumbent. The third lower incisor is reduced and the canine is lost.

The occiput is high, with a broad nuchal crest for muscles to support a heavy skull. The bones of the head are already pneumatized and as in later proboscideans, the auditory meatus is high.

Although it is only about 1 meter tall, the postcranial skeleton shows the graviportal adaptations common to later proboscideans. The forelimb is extremely robust and the radius is fixed in a pronating position. The femur is longer than the humerus and much longer than the tibia. The astragalus and calcaneum are closely similar to those of later Eocene and early Oligocene proboscideans.

The only feature that seems to remove this genus from direct ancestry to later proboscideans is the presence of a deep submaxillary fossa, which is not reported in other early members of the order.

These Lower Eocene fossils are definitely close to the ancestry of later proboscideans. Much less-well-known genera from the early to middle Eocene of the Indian subcontinents have molar teeth of a similar pattern, which West (1983) and Wells and Gingerich (1983) suggest may link them to more primitive ungulates from the Paleocene. Members of the family Anthracobunidae (Figure 21-54*c*) have molars that are dominated by four massive, blunt cusps arranged so as to form two transverse ridges. The addition of a small entoconid II presages the proboscidean condition. However, the anterior teeth give no clue of proboscidean affinities. The canines were large and the incisors unspecialized. The dental count is typical of primitive eutherians. Some genera may have been as large as a pig or tapir. The anthracobunids might be relicts of the ancestral stock that gave rise to the proboscideans, but they are clearly too late to be directly ancestral and do not show the initiation of any typically proboscidean characters. The ancestry of the anthracobunids may lie with genera such as *Minchenella* from the late Paleocene of China, which Zhang (1978, 1980) placed in the condylarth family Phenacolophidae. Although *Minchenella* is known only from lower jaws, Domning, Ray, and McKenna (1986) argue that the transversly broadened hypoconulid shelf of the lower third molar with a small entoconid II is a sufficiently significant derived feature to establish its affinities with both the Proboscidea and a second order of aquatic mammals, the Desmostylia (Figure 21-55).

Order Proboscidea
- Suborder Gomphotherioidea
  - Family Gomphotheriidae: late Eocene to middle Pleistocene
  - Family Elephantidae: late Miocene to Recent
- Suborder Mammutoidea
  - Family Mammutidae: early Miocene to sub-Recent
  - Family Stegodontidae: middle Miocene to late Pleistocene

There is a rich record of proboscideans from the Upper Eocene and Lower Oligocene of northern Africa. These are already large animals and have most of the postcranial specializations that are common to the modern species. The limbs are columnar, with the distal elements short. The bones lack medullary cavities, and the manus and pes are short and pentadactyl.

These early proboscideans have been classified in a single genus, *Palaeomastodon* within the Gomphotheriidae, although they show extensive variability and may include the ancestors of several subsequent lineages.

The Gomphotheriidae is considered a long-living ancestral stock that gave rise to a succession of other groups. Primitively, they have long spatulate incisors in the elongate lower jaws, as well as tusks in the upper jaws (Figure 21-56). The teeth are bunodont, with up to seven pairs of rounded cusps. In contrast with their early Eocene ancestors, they have lost all trace of canine teeth. The fossil record of proboscideans is poor for most of the Oligocene but rich in the Miocene, by which time all the major lineages had emerged. At the end of the early Miocene, Africa joined Asia and the proboscideans began their migrations to the ends of the earth (Figure 21-57). Descendants of early gomphotheriids reached all continents

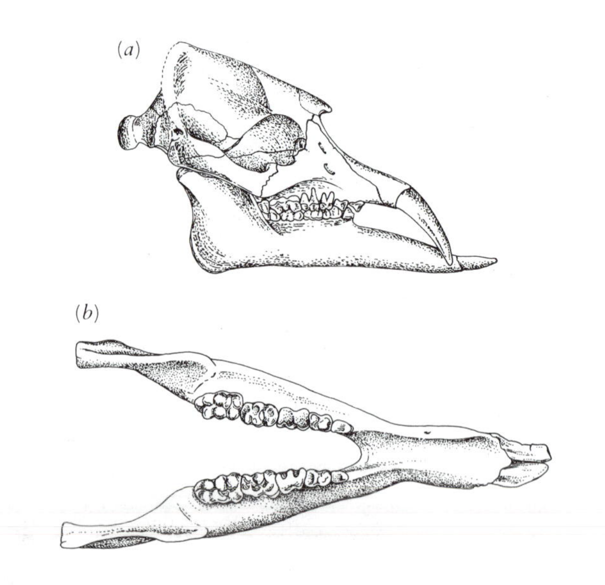

Figure 21-56. (*a*) Skull and (*b*) lower jaws of the gomphotheriid proboscidean *Palaeomastodon* from the Upper Eocene of North Africa. *From Andrews, 1906.*

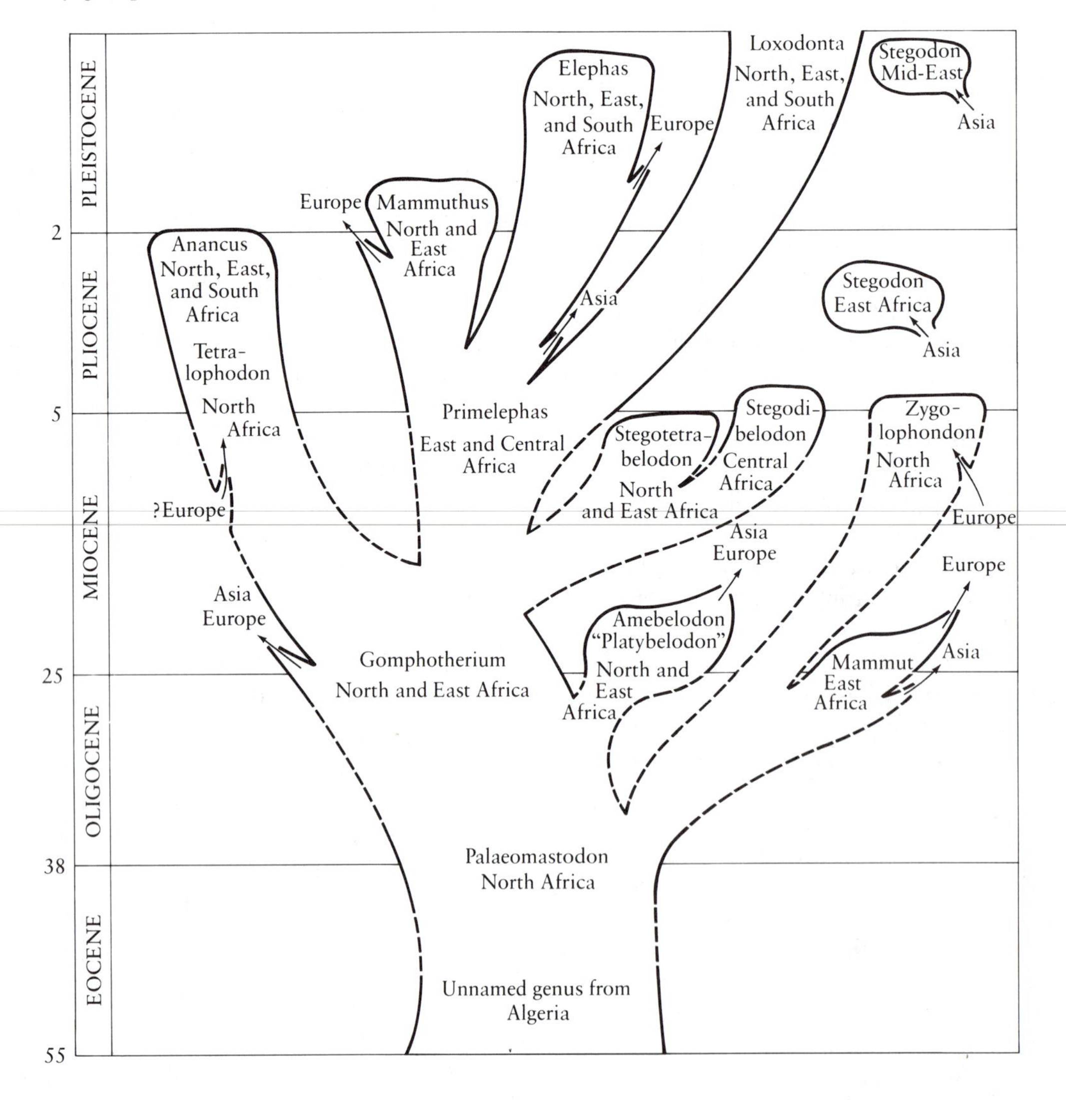

Figure 21-57. PHYLOGENY OF THE PROBOSCIDEA SHOWING THEIR GEOGRAPHICAL DISTRIBUTION. *From Coppens, Maglio, Madden, and Beden, 1978. Reprinted by permission of Harvard University Press.*

except Australia and Antarctica. The Gomphotheriinae extended into Asia, Europe, and North America; the Cuvierienoninae originated in North America and spread into South America; and the Anancinae rose in Africa and spread into Europe and parts of Asia. The latter two subfamilies approached the pattern of modern elephants in the loss of their lower tusks, elongation of their molars, and deepening of their skulls.

The surviving family Elephantidae is recognizable as early as the late Miocene. It is characterized by a unique pattern of tooth structure and succession that is not evident in other proboscideans. The molar teeth are so long that the anterior portion erupts and becomes functional while the posterior portion is still undergoing development. As the teeth form at the back of the jaw, they move forward and replace the earlier formed teeth that are lost at the front. In early genera, which are placed in the subfamily Stegotetrabelodontinae, the premolars are functional, as in other proboscideans, but in advanced elephants they are suppressed. Instead, the deciduous cheek teeth tend to become molariform and function as the molars develop. Only one and one-half pairs of teeth are functional at a time in their relatively short jaws.

The teeth of the primitive genus *Stegotetrabelodon* retain evidence of two two longitudinal rows of cusps that were common to primitive proboscideans (Figure 21-58). The cusps are anteroposteriorly compressed. With wear, the two halves become confluent, as in modern elephants. In the most derived elephant species, the wooly mammoth, there may be as many as 30 plates in the last molar. Each plate is composed of a compressed loop of enamel surrounding the dentine and separated by cementum. The hard enamel forms a series of shearing blades that, with wear, stand above the softer dentine and cementum.

In early genera, each molariform tooth bears only as many plates as there were rows of cusps in advanced gomphotheres (five to seven in the last molar). They were also primitive in retaining very long tusks in the lower jaws. The subfamily Elephantinae appeared in the latest Miocene, represented by the genus *Primelephas* from east and central Africa, which is distinguished by the short symphysis of the lower jaw and reduction or loss of the premolars. This primitive lineage gave rise to three derived genera by the middle Pliocene, *Elephas, Loxodonta,* and *Mammuthus. Loxodonta* was highly advanced when it first appeared, but unlike other late Tertiary proboscideans, it showed little change within the genus and never left Africa. *Loxodonta africana* is now confined to the area south of the Sahara but lived in Egypt during predynastic times. *Elephas* spread out of Africa in the mid-Pliocene and is now restricted to southern Asia. Its African record is very complete during the Pliocene and Pleistocene and shows a sequence of rapidly evolving forms within a single lineage. *Mammuthus,* which is distinguished by spirally twisted tusks and a wrinkled or wavy pattern of the enamel, was common in North America and Eurasia, where it persisted into the late Pleistocene.

The suborder Mammutoidea paralleled the Elephantidae in many features of the dentition but did not evolve their specialized pattern of serial tooth replacement. Like the Elephantidae, they lost the lower tusks and shortened the jaw symphysis while elaborating molars that showed a platelike structure. However, the entire tooth is formed prior to eruption. They appear in Africa in the early Oligocene. The Mammutidae were common in North America and Eurasia as well as Africa and persisted into the sub-Recent. The Stegodontidae were a primarily Asian group that reentered Africa several times during their history. Their teeth remained low crowned, with thick enamel.

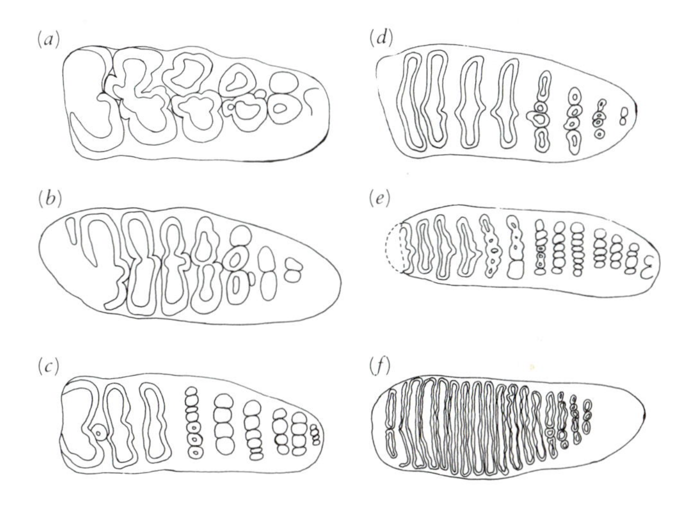

Figure 21-58. CHANGES IN THE MOLAR CUSPS OF PROBOSCIDEANS. These species show progressive consolidation of the gomphothere cone-pairs, loss of the median cleft, fusion of accessory columns, increase in plate number, and thinning of enamel. (*a*) *Gomphotherium.* (*b*) *Stegotetrabelodon.* (*c*) *Primelephas.* (*d*) *Mammuthus subplanifrons.* (*e*) *Mammuthus africanavus.* (*f*) *Mammuthus primigenius.* Similar changes occurred in the genus *Elephas. From Maglio, 1973.*

DEINOTHERIOIDEA AND MOERITHERIOIDEA

Two distinct groups, the Deinotherioidea and the Moeritherioidea, may have diverged before the Gomphotherioidea and Mammutoidea but may nevertheless be included within the Proboscidea (Figure 21-55).

The Deinotheriodea, which were present in the Old World from the early Miocene to the end of the Pleistocene, have long been considered a divergent suborder within the Proboscidea. They were specifically excluded by Maglio (1973) and by Coppens, Maglio, Madden, and Bedey, (1978), but Mahboubi, Ameur, Crochet, and Jaeger (1984) and Domning, Ray, and McKenna (1986) argued that they shared a close common ancestry.

The deinotherioids were elephantine in size and limb structure. They are clearly distinguished from other proboscideans by the absence of upper tusks and the presence of large, recurved tusks in the lower jaws (Figure 21-59).

The posterior position of the narial opening indicates the presence of a proboscis, but the shape of the skull is very distinct from that of typical elephants. The occiput is extended far ventrally to permit attachment of large muscles to ventroflex and rotate the skull. As in the gomphotheriids, the molar teeth are bilophodont. They remain low crowned throughout the history of the group. The teeth show considerable variability in both of the recognized genera but no progressive evolutionary change except for general increase in size. Harris (1976) showed that the anterior teeth were used for crushing and the posterior for shearing.

Typical members of this group are not known in Africa or elsewhere until the early Miocene and no possible ancestors are recognized in the Oligocene, although typical proboscideans are fairly well known during this time. Harris (1978) suggested that the Upper Eocene genus *Barytherium,* which is known from North Africa, is a plausible relative of this group on the basis of the similarity of the cheek teeth. This genus was assigned a more distant relationship by Mahboubi and his colleagues on the basis of the discovery of Lower Eocene proboscideans.

Deinotherioids extended into Europe and Asia in the late Miocene and Pliocene but were never as common or widespread as the elephants.

Moeritherium, which we know from the Upper Eocene and Oligocene of northern Africa (from Senegal to Egypt), has long been associated with the proboscideans, but it also shares some features with a later group of marine mammals, the Desmostylia.

Moeritherium was less than 1 meter tall with fairly lightly built limbs. Their structure and the discovery of many specimens in marine deltaic deposits suggests that *Moeritherium* was amphibious in habits. The dentition is similar to that of the earliest proboscideans in having both the upper and lower second incisors greatly enlarged and, like them, there are upper but no lower canines (Figure 21-60). The six pairs of cheek teeth are bilophodont.

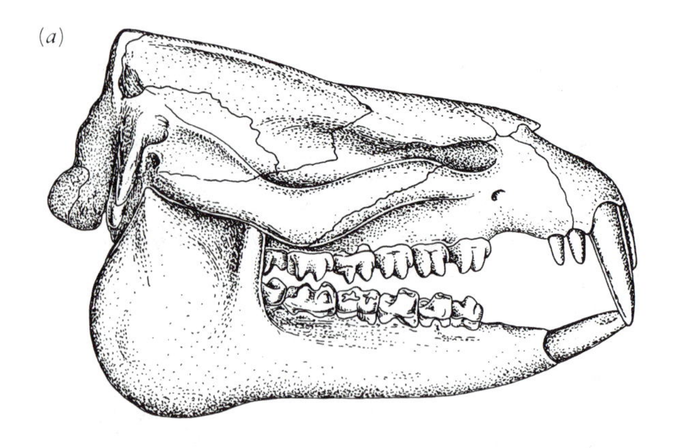

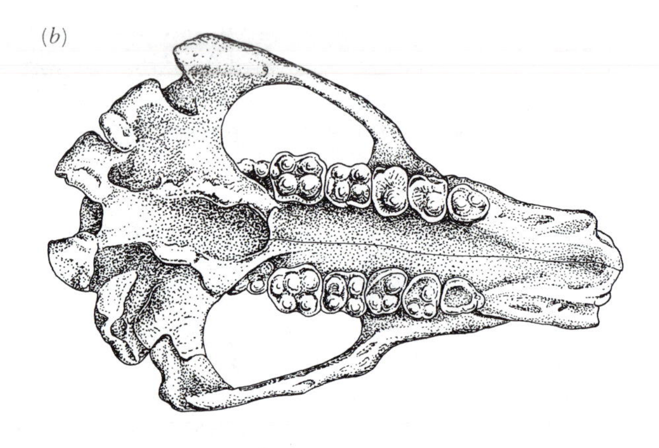

Figure 21-60. THE EARLY PROBOSCIDEAN *MOERITHERIUM.* (*a*) Restoration of skull in lateral view. (*b*) Palate. *From Andrews, 1906.*

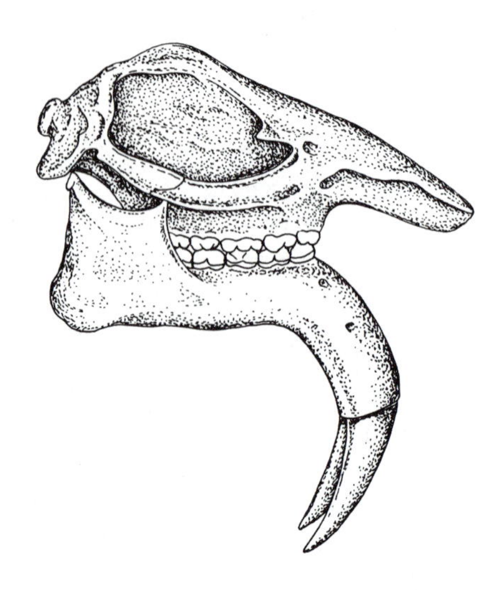

Figure 21-59. SKULL OF *DEINOTHERIUM. From Harris, 1978. Reprinted by permission of Harvard University Press.*

In contrast with other early proboscideans, the skull is low and only weakly pneumatized but the external auditory meatus is high. There is no evidence of a trunk.

Moeritheres may be relicts of an early stage in the evolution of the proboscideans. Domning, Ray, and McKenna (1986) point out features in which they are divergent from later proboscideans, including loss of the third lower incisor, loss of the lacrimal bone, and the shortening of the anterior portion of the skull. They retain primitive characters that indicate that they diverged from early tethytheres close to the point of origin of the desmostylians.

DESMOSTYLIA

Desmostylians were long known only from isolated teeth consisting of a number of cusps that developed as closely placed enamel cylinders (Figure 21-61*c*). They were found in marine deposits and, as they somewhat resemble the teeth of some sirenians, it was assumed that they were closely related. The discovery of associated postcranial

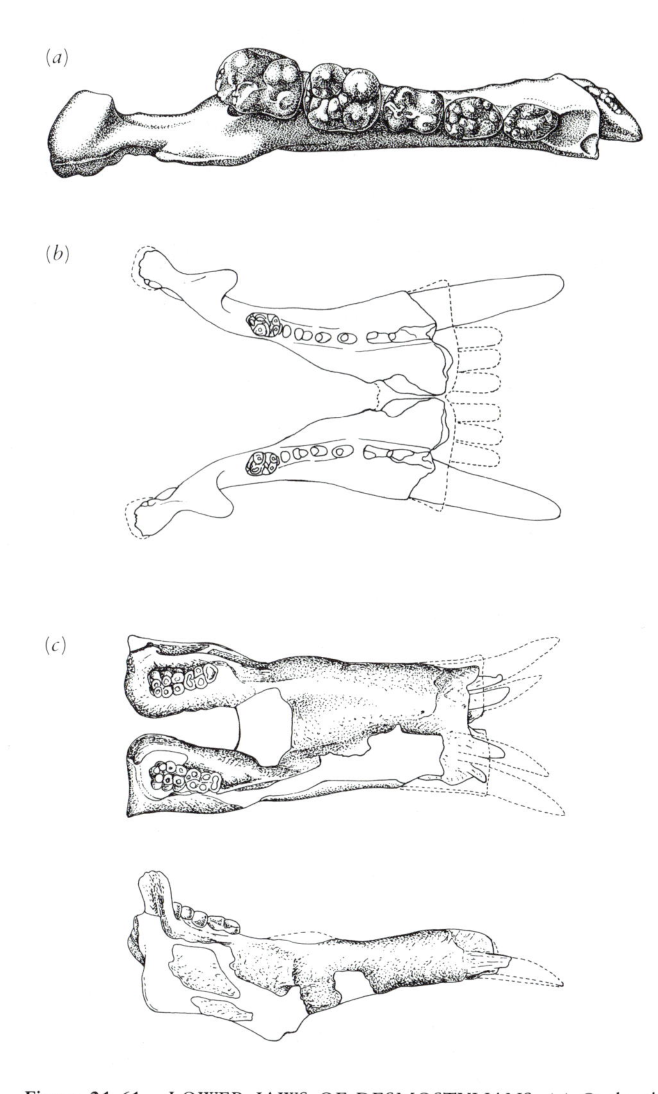

Figure 21-61. LOWER JAWS OF DESMOSTYLIANS. (*a*) Occlusal view of the lower jaw of the primitive Upper Oligocene desmostylian *Behemotops proteus.* Note similarity with the teeth of the primitive proboscidean *Moeritherium,* 25 centimeters long. *From Domning, Ray and McKenna, 1986.* (*b*) Occlusal view of the lower jaw of *Behemotops emlongi.* Anterior cheek teeth are represented by empty alveoli. The anterior dentition resembles that of *Desmostylus;* the relative length of the jaw is much shorter, but the symphyseal region broader. These proportions resemble the modern hippopotamus, 50 centimeters long. *From Domning, Ray and McKenna, (1986.)* (*a* and *b*) *By permission of Smithsonian Institution Press. Smithsonian Institution, Washington, D.C., 1986.* (*c*) Dorsal and lateral views of the lower jaws of *Desmostylus,* from the Middle Miocene. The cheek teeth are restricted to the back of the jaw, leaving a long diastema behind the tusklike canines. *From Reinhart, 1959.*

material demonstrated that they had a totally different body form, with both front and hind limbs that were well developed but with hands and feet that were somewhat specialized as paddles (Figure 21-62). These animals may have been amphibious or perhaps somewhat seal-like in habits. Elaboration of the incisors, elongation of both the upper and lower jaws, and the long diastema anterior to the cheek teeth give them the look of primitive elephants. However, the fact that the canine teeth are also elaborated as tusks rules out close relationship with typical proboscideans. Desmostylians are known only in the Upper Oligocene and Miocene and are restricted to the margins of the North Pacific ocean. *Palaeoparadoxia* and *Desmostylus* were common in the Miocene. The Upper Oligocene genus *Behemotops,* which was recently described by Domning, Ray, and McKenna (1986), appears close to the ancestry of later desmostylians and demonstrate close affinities with the base of the proboscidean assemblage. Domning, Ray, and McKenna suggest that desmostylians were amphibious herbivores that fed on marine algae and that the earlier genera depended to a large extent on plants exposed in the intertidal zone.

It is fairly easy to conceive of a common, Paleocene ancestor for typical proboscideans, moeritherioids, and desmostylians that was already distinct in the elaboration of the anterior dentition and the bilophodont arrangement of the molar cusps. It is more difficult to envisage that common ancestor also showing significant derived features in common with other African groups such as embrithopods, sirenians, and hyraxes.

EMBRITHOPODA

Embrithopods are known primarily from a single locality in the Oligocene of Egypt. *Arsinoitherium* is known from the entire skeleton (Figure 21-63). Although it is elephantine in its general form, the skull is entirely different. It is dominated by two gigantic bony processes that arise from the nasals and a much smaller pair that is medial to the orbits. The teeth form a nearly uniform series without tusks, conspicuous canines, or a significant diastema. They retain a full primitive dental formula of

$$\frac{3\quad 1\quad 4\quad 3}{3\quad 1\quad 4\quad 3}$$

The molar teeth are conspicuously bilophodont and high crowned.

Until recently, no remains of embrithopods had been described outside Africa. In 1977, McKenna and Manning suggested that the Upper Paleocene or Lower Eocene Mongolian genus *Phenacolophus* might be related to the ancestry of *Arsiniotherium. Phenacolophus* is represented by remains of a skull that is approximately 15 centimeters long and some postcranial elements. Both the upper and lower teeth are bilophodont (Figure 21-64*a*). Unlike *Arsiniotherium,* the canines are enlarged. Limb elements are short, stocky, and primitive.

Sen and Heintz (1979) described material from several localities in north central Turkey and Romania that can be assigned to the Arsiniotheriidae with much greater assurance (Figure 21-64*d*). The specimens are smaller than the Egyptian genus and somewhat more primitive in details of the dental anatomy. The most significant difference is the absence of horns.

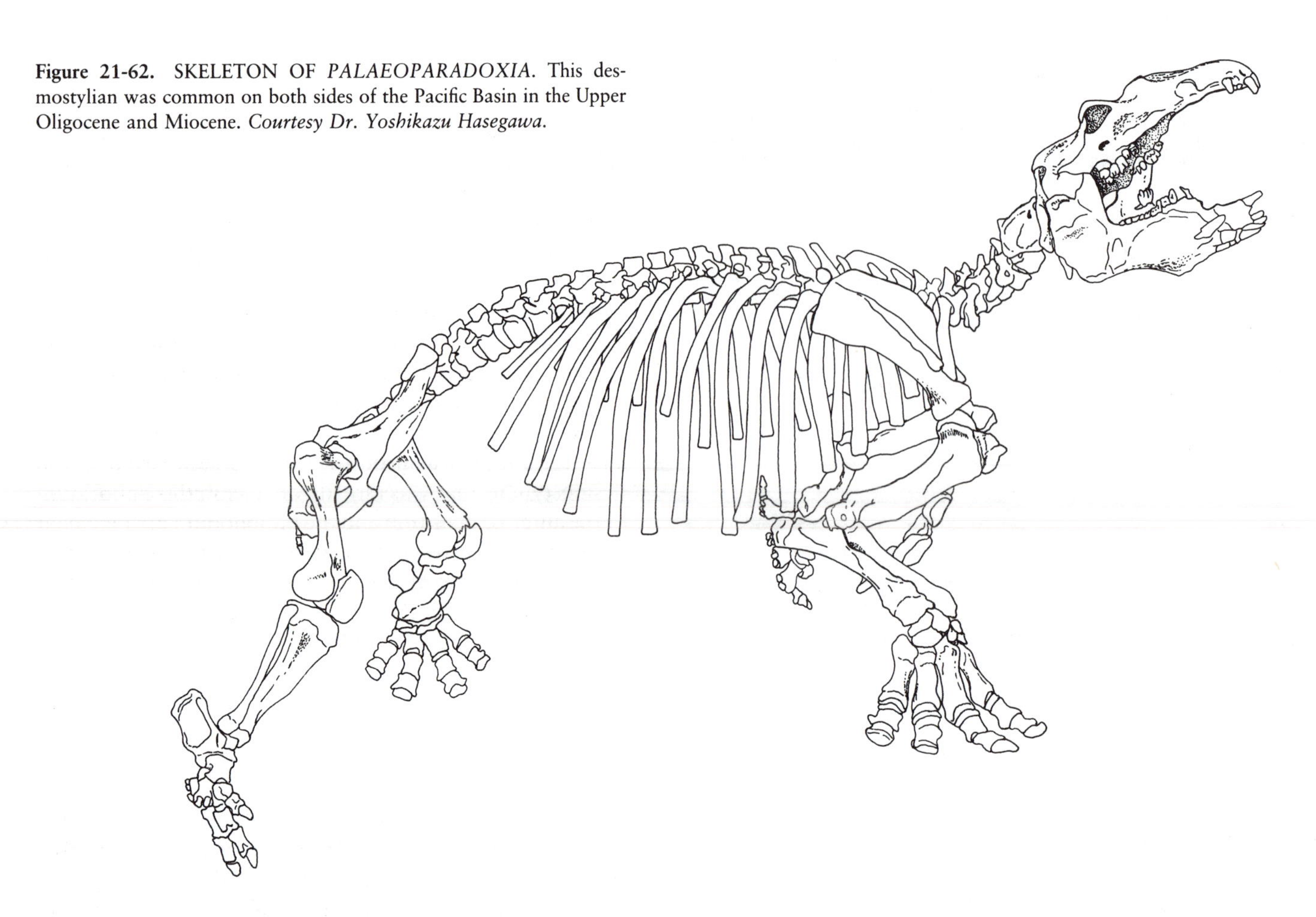

Figure 21-62. SKELETON OF *PALAEOPARADOXIA*. This desmostylian was common on both sides of the Pacific Basin in the Upper Oligocene and Miocene. *Courtesy Dr. Yoshikazu Hasegawa.*

On the basis of available evidence, one could argue that the Embrithopoda originated in Asia and migrated to Africa, or that the group originated in Africa and spread into Asia at an early stage in evolution, before the emergence of horned forms. Bearing in mind the rarity of fossils from the Paleocene in Africa, an Asian origin seems a simpler explanation. Both the Turkish and African forms come from deposits that suggest amphibious habits.

Both embrithopods and proboscideans are suggested as being derived from members of the condylarth family Phenacolophidae, which have large blunt molar cusps arranged so as to form transverse lophs.

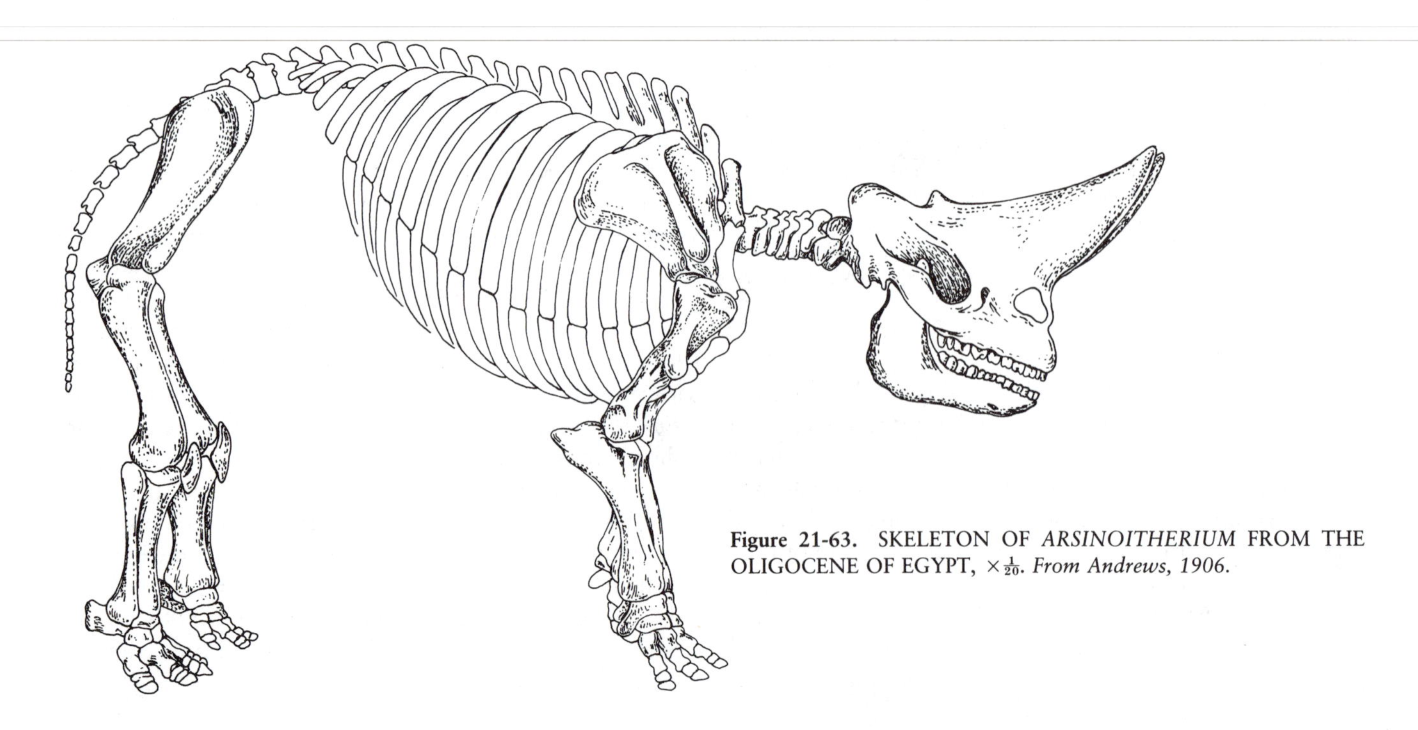

Figure 21-63. SKELETON OF *ARSINOITHERIUM* FROM THE OLIGOCENE OF EGYPT, $\times\frac{1}{20}$. *From Andrews, 1906.*

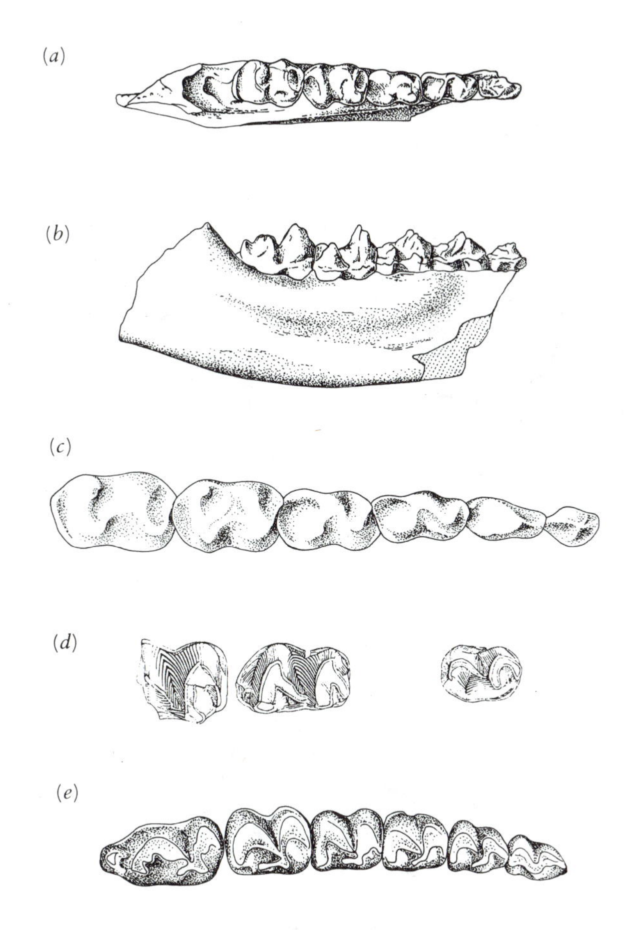

Figure 21-64. POSSIBLE RELATIVES OF *ARSINOITHERIUM.* (*a*) Occlusal and (*b*) lateral views of the lower jaw of *Phenacolophus*, a condylarth from the Upper Paleocene or Lower Eocene of Mongolia. *From Matthew and Granger, 1925.* (*c*) Composite reconstruction of lower cheek teeth of *Phenacolophus. From McKenna and Manning, 1977.* (*d*) Lower cheek teeth of the late Eocene of Oligocene Romanian arsinoithere *Crivadiatherium. From McKenna and Manning, 1977.* (*e*) Lower cheek teeth of *Arsinotherium. From Andrews, 1906.*

SIRENIANS

Sirenians have long been thought to have some affinities with proboscideans and embrithopods, but the current evidence does not seem very convincing.

Sirenians superficially resemble whales in having a fusiform body, forelimbs that are specialized as paddles, and a horizontal tail fluke. They differ from all other specialized marine mammals in adhering to a strictly herbivorous diet. They are primarily coastal in distribution. Only four species survive today, *Dugong dugon* of the Indopacific basin, and three species of manatee: *Trichechus inunguis* in the Amazon basin, *T. manatus* in the Caribbean and along the northern coast of South America, and *T. senegaliensis* on the west coast of Africa. Stellar's sea cow, *Hydrodamalis,* was a dugong that lived along the northern margin of the Pacific ocean until its extinction in the eighteenth century.

Sirenians lack sonar, although the periotic, like that of whales, is not solidly fused to the skull and is dense and expanded. All sirenians going back to the Lower Eocene are characterized by massive, pachyostotic ribs that significantly increase the body weight. Horizontal stability is increased by elongation of the lungs and a horizontal diaphragm. Compression of the thoracic cavity permits sirenians to sink with a minimum of effort. Unlike whales, they are not active divers.

In all modern sirenians, the rostrum and the end of the lower jaw are deflected ventrally. This feature is most conspicuous in the dugong, which feeds on bottom-hugging sea grasses. Living dugongs have two or three pairs of cheek teeth that are open rooted but lack enamel. The manatee has up to eight teeth that are functional at a time in each jaw ramus. These teeth are steadily replaced by new teeth that erupt at the back of the jaw; as many as 20 cheek teeth may erupt in each jaw. In some genera, there are tusklike upper teeth that appear to be comparable to the first incisors in contrast to the tusks in elephants, which form from the second incisors.

Sirenians first appear in the fossil record in the Lower Eocene of Hungary (Kretzoi, 1953). Their remains are fragmentary but include the pachyostotic ribs that are a hallmark of the group.

The Middle Eocene *Prorastomus* from Jamaica, the only genus in the family Prorastomidae, shows the most primitive cranial pattern yet known. The lower jaw is straight, and there is only a slight deflection of the rostrum. The skull is pachyostotic. The dental count of

$$\frac{3\ \ 1\ \ 5\ \ 3}{3\ \ 1\ \ 5\ \ 3}$$

is striking in the retention of a fifth premolar, which is lost in all other post-Cretaceous groups of placentals. Aside from pachyostotic ribs and vertebrae, no definitely attributable postcranial remains are known to indicate the degree of aquatic adaptation attained at this stage of sirenian evolution.

Apart from the primitive nature of the skull, with its general similarities to condylarths, these early sirenids do not show any features that demonstrate specific relationships with other placental orders. The cheek teeth are bilophodont like those of early proboscideans and members of the Anthracobunidae, but this feature evolved separately in several other orders. The retention of five premolars in all Eocene species raises the possibility of derivation from a very primitive stock of late Cretaceous or early Tertiary eutherians, without close affinities with any other advanced mammalian orders. The early distribution of sirenians suggests an origin along the shores of the ancestral Tethys Sea rather than specifically from African progenitors.

Most Middle and Upper Eocene sirenians are placed in the genus *Protosiren,* which is reported from Java, India, Europe, and southeastern North America (Domning, Morgan, and Ray, 1982). They argue that the family Protosirenidae is ancestral to both living families. The postcranial skeleton is already essentially modern.

Dugongids are represented in the middle Eocene by *Eotheroides* from North Africa (Figure 21-65); like other Eocene genera, it still retains five premolars. It is placed in the subfamily Halitheriinae, which has a scattered record throughout the early Tertiary in Africa, Europe, and both the Atlantic and Pacific shores of the New World (Domning, 1978). The Dugonginae, which is represented today by a single species of the genus *Dugong* in the Indo-Pacific basin, has no fossil record.

Although most sirenians were tropical in distribution throughout their history, one lineage, the Hydrodamalinae, became adapted to life around the shores of the North Pacific in the late Cenozoic. The origin of this group lies with halitherine dugongids, which are known along the western shores of North America since the early Miocene (Domning, 1978). The earliest hydrodamalinae, *Dusisiren* (Figures 21-66 and 21-67) appeared about 19 million years ago. A series of species provide a nearly continuous morphological sequence leading to the modern genus *Hydrodamalis,* which appeared 7 million years ago. In this sequence, the size increases from 2 to 3 meters to over 9 meters (probably an accommodation to living in cold waters) and the cheek teeth are completely lost as are the phalanges of the front limb. Domning attributes the change in dentition to feeding on softer plants, including kelp and other brown-red algae, as was observed by early explorers in the Arctic. Heavy insulation with blubber made them so bouyant that they apparently did not dive at all.

In the late Pliocene, *Hydrodamalis* was known from the Japanese archipelago around the North Pacific to northern Mexico. The range of the group has apparently been progressively reduced over the last 20,000 years. The last animals were killed about 1768.

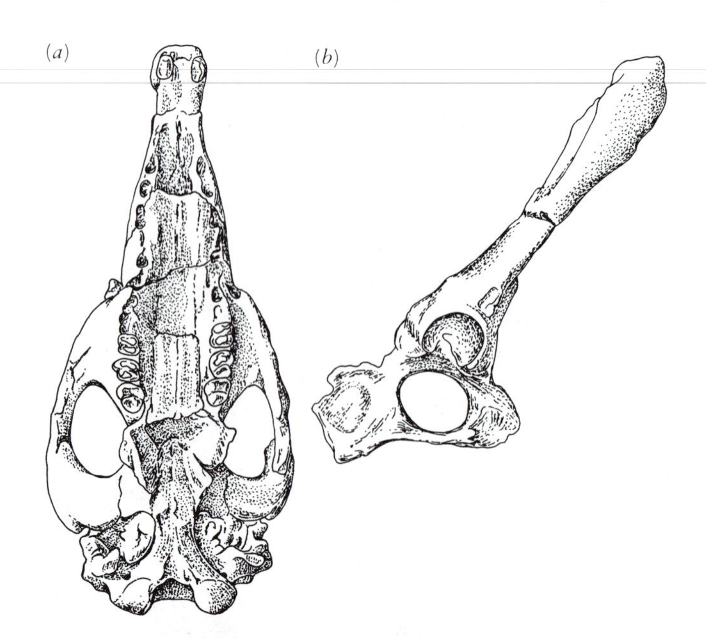

Figure 21-65. (*a*) Palate of the sirenid *Eotheroides* from the middle Eocene of North Africa. (*b*) Pelvic girdle of *Eotheroides. From Andrews, 1906.*

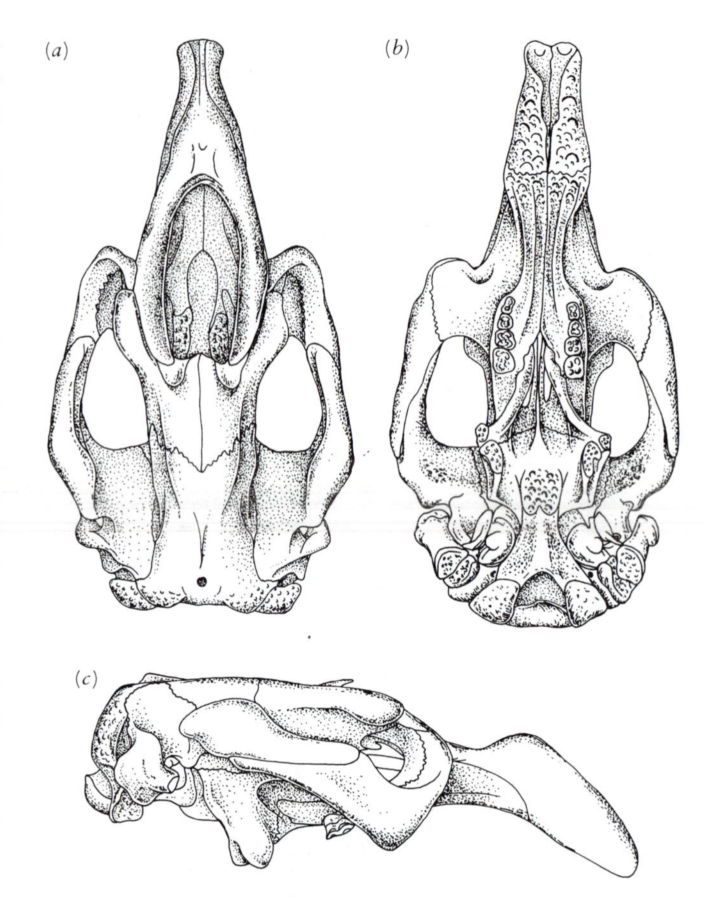

Figure 21-66. (*a*) Dorsal, (*b*) palatal, and (*c*) lateral views of the skull of *Dusisiren,* an early Miocene ancestor of Stellar's sea cow. *From Domning, 1978.*

Domning (1982) recently reviewed the evolution of the manatees (Trichechidae). Their ancestry probably lies in primitive dugongids or possibly protosirenids that were isolated in the South American area in the early Tertiary, at which time the Amazon Basin was open to both the Atlantic and Pacific Oceans. The first possible member of the modern family is *Potamosiren* from the Middle Miocene. This genus has reduced the cheek teeth to molars 1 to 3, as in most dugongids. The feature of continuous cheek tooth replacement, which is fed by indefinite tooth-germ division, is established in *Ribodon* by the late Miocene. Domning attributed selection for the changing dental pattern to feeding on siliceous aquatic grasses in the Amazon Basin, in contrast with the less-abrasive sea grasses eaten by the dugongs.

Movement of the teeth from the back to the front of the jaw and the loss of permanent premolars are features in common with members of the Elephantidae, but they clearly evolved independently in the two groups, and the sirenians are distinct in the continuous replacement of the molars.

Dugongids preceded the manatees in the Caribbean, where they were present for most of the Tertiary. They became extinct in the Atlantic and Mediterranean at the end of the Cenozoic, possibly as a result of climatic de-

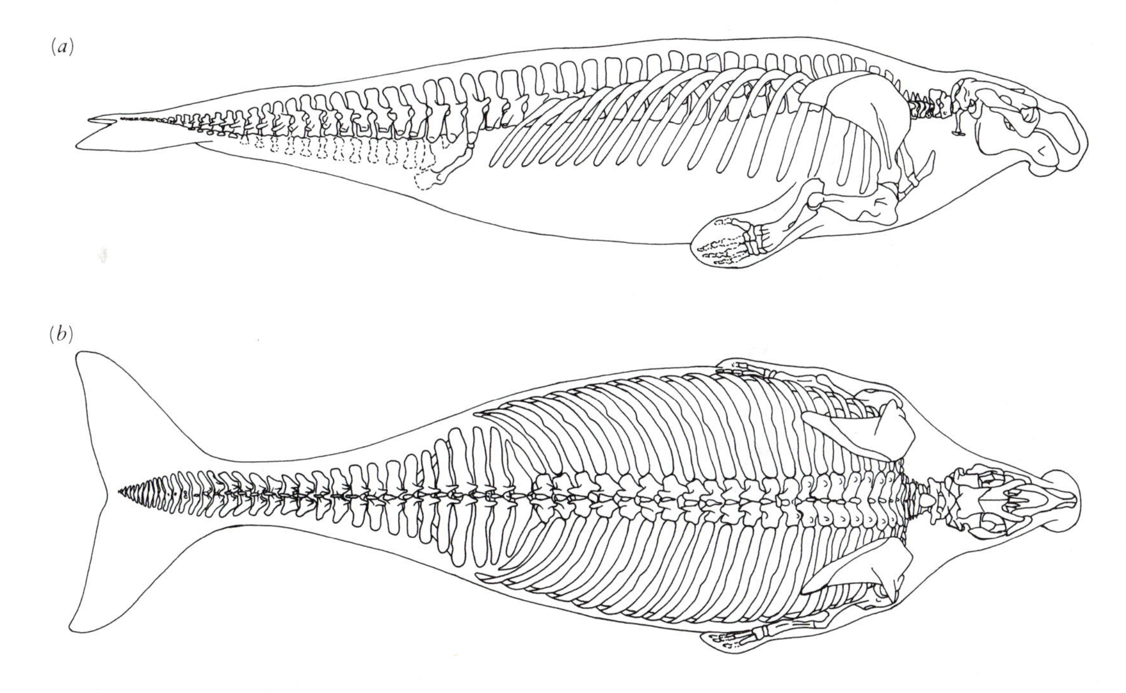

Figure 21-67. SKELETON OF *DUSISIREN*. (*a*) Lateral and (*b*) dorsal views. Original was 4 meters long. *From Domning, 1978.*

terioration leading to the Pleistocene glaciation. The last record of dugongids in the Atlantic Basin is in the late Miocene, whereas manatees evidently entered that ocean from South America in the Pliocene.

HYRACOIDEA

The hyracoids or conys are superficially rabbitlike forms whose history has been confined largely to Africa, although they are also known in limited areas of Europe and Asia.

They appear first in the Upper Eocene of North Africa (Sudre, 1979). In the early Oligocene, Lower Fossil Wood Zone of the Fayum, Egypt, they make up approximately half the fauna, but by the Upper Fossil Wood Zone, they account for only 16 percent of the specimens. Their importance continues to dwindle through the Cenozoic (Meyer, 1978). Today, there are three genera that are restricted to Africa and the near East.

The current fossil record provides little evidence of the specific origin of the group and little to confirm affinities with the other "African" orders.

The dentition of hyraxes for most of their history maintained nearly the primitive eutherian formula, with loss of only the lateral lower incisors. The most important specialization is that the central upper incisor is a triangular, recurved tusk that grows from a persistant pulp cavity. The enamel on the lingual side is present when the tooth first erupts but is rapidly worn down. As in rodents, this tusklike medial incisor is sharpened by contact with the lower incisor. The upper incisor is always larger in the male than in the female. There may be a short diastema behind the incisors but the rest of the dentititon is of uniform height. The cusps are bunodont to lophodont, with a distinctly perissodactyl appearance in some genera (Figure 21-68).

The modern genera are omnivorous. Some are arboreal and others live among the rocks. All climb well with the help of moist foot pads.

The postcranial skeleton is distinctive in the great length of the vertebral column, with 20 to 23 thoracic and 4 to 9 lumbar vertebrae. In contrast, the tail is exceedingly short. There are five toes on the manus and three on the pes. All bear flat nails except the inside digit of the pes, which has a claw. The posture is plantigrade, with the rear limbs kept in a crouch.

The abundant early Tertiary fossils are included in the extinct family Pliohyracidae. We first find the modern Procaviidae in the Miocene. The lateral incisors and canines are lost in the Pleistocene and Recent species. Sudre agrees with the frequently suggested idea that hyraxes originated from the Condylarthra but does not support affiliation with proboscideans and sirenians. Current work by Fischer (Prothero, Manning, and Fischer, 1986) supports close affinities between hyraxes and perissodactyls.

TUBULIDENTATA—THE AARDVARK

The aardvark, or earth pig, is known from a single modern species, *Orycteropus afer*, which is widespread in Africa south of the Sahara. It has a heavily built, generally archaic skeleton, with limbs that are specialized for digging. There are four toes on the manus, five on the pes, and

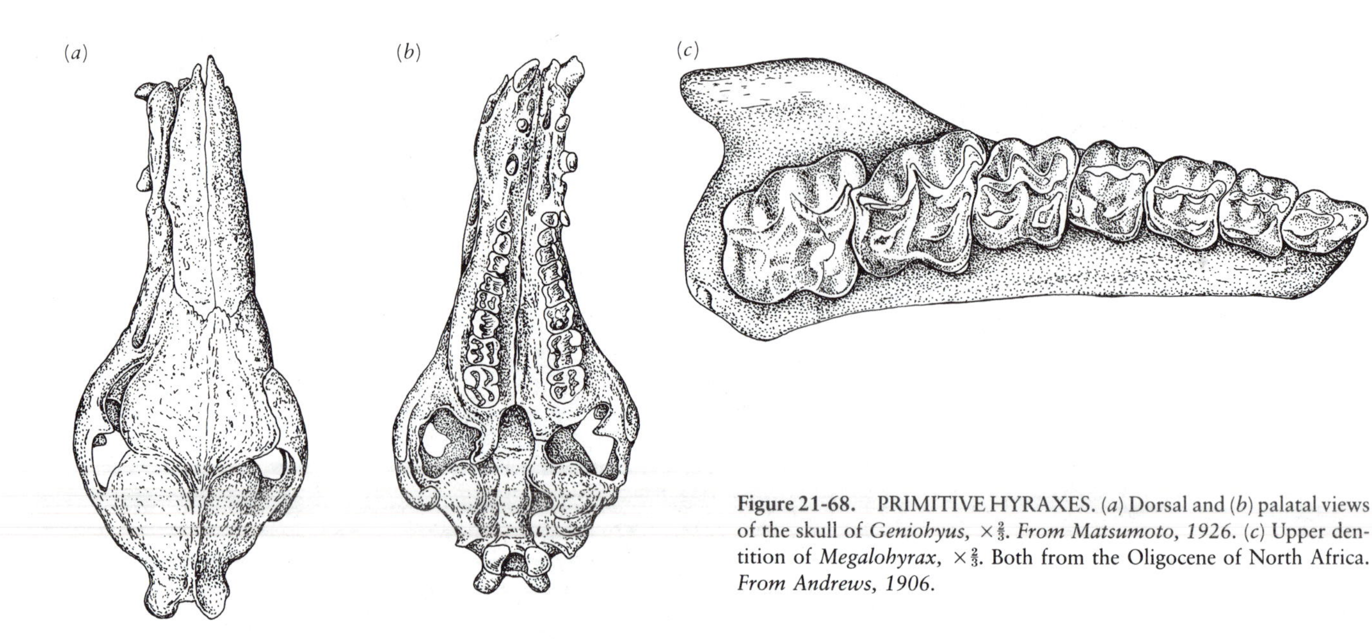

Figure 21-68. PRIMITIVE HYRAXES. (*a*) Dorsal and (*b*) palatal views of the skull of *Geniohyus*, × ⅔. *From Matsumoto, 1926.* (*c*) Upper dentition of *Megalohyrax*, × ⅔. Both from the Oligocene of North Africa. *From Andrews, 1906.*

the clavicle is retained. The limb bones show conspicuous tuberosities for muscle attachment.

The aardvark lives in burrows and feeds primarily on termites that it digs from the ground. Its sense of smell is extremely acute. The incisors and canines are lost, as is the enamel from all the teeth. The cheek teeth are large and unique among mammals in being composed of numerous hexagonal prisms of dentine that form around tubular pulp cavities. The absence of enamel is compensated for by continuous growth of the cheek teeth. The teeth and lower jaws are much better developed than in ant-eating specialists in other orders.

Aardvarks are of special phylogenetic interest in being almost certainly African in origin, but without special skeletal resemblance to the other groups whose origin or early evolution has been associated with Africa: sirenians, elephants, and hyraxes. They are also clearly distinct phylogenetically from the other groups that are highly committed to feeding on ants—the South American Myrmecophagidae and the pangolins.

Aardvarks nevertheless show a similar pattern of enamel loss to that of South American xenarthrans and the North American palaeanodonts (which will be discussed in a later section). Patterson (1975, 1978) discussed

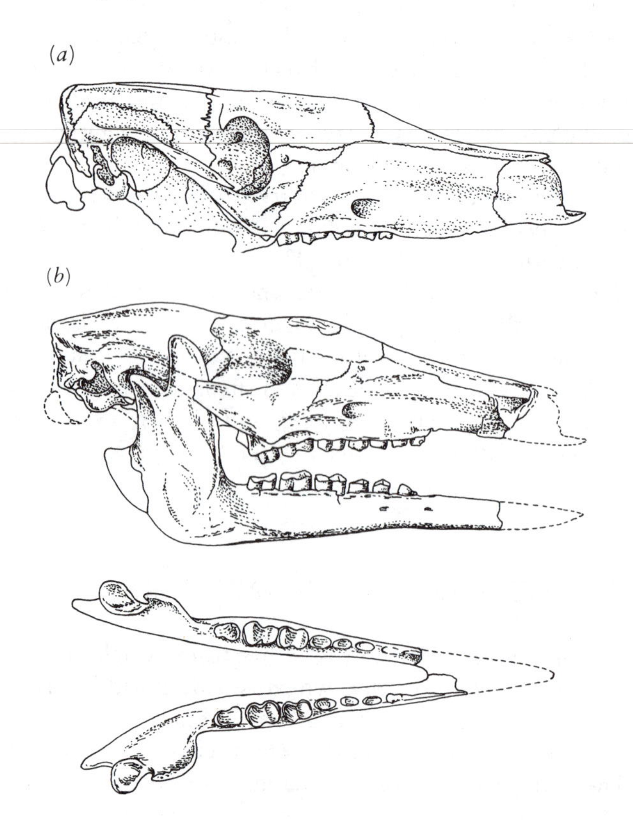

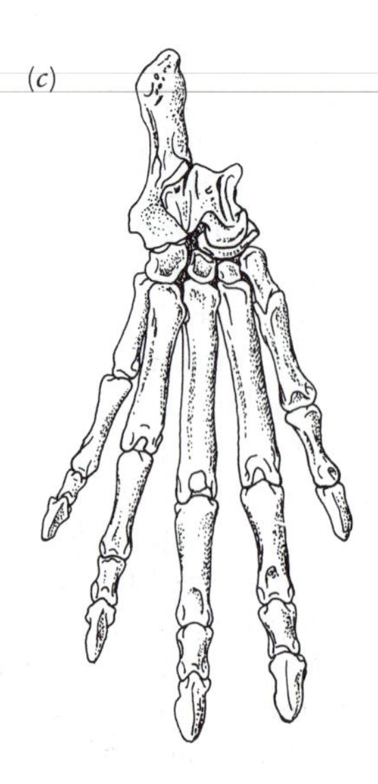

Figure 21-69. AARDVARKS, ORDER TUBULIDENTATA. (*a*) Skull of the modern species *Orycteropus afer*. (*b*) The Miocene species *Orycteropus gaudryi*, skull in lateral view and lower jaw in occlusal view. (*c*) Foot of *Orycteropus gaudryi*. Note similarity to the condylarth *Ectoconus* (see Figure 21-5). *From Colbert, 1941a.*

the fossil record in two recent papers. He recognized four distinct genera, all from the late Cenozoic. The modern genus *Orycteropus* can be traced back to the late Miocene. Colbert (1941a) demonstrated that the modern species differs from the extremely well-known late Miocene form *O. gaudryi* by little more that its larger size and more robust skeleton (Figure 21-69). The late Miocene species retains all seven cheek teeth, which are reduced to five or six in the modern species. The specimens described by Colbert were found on the island of Samos, adjacent to the coast of Turkey; other fossils referable to the modern genus are known from France and western Russia. The earlier record is confined to Africa. The oldest-recognized aardvard, *Myorycteropus* from the Lower Miocene has a similar dentition to the living genus, but the limbs show somewhat different adaptations to digging. *Leptorycteropus* from the middle Pliocene, despite its late appearance, is the least-specialized member of the group. The snout is not elongate, but the jaw symphysis is long and large canine teeth are retained. The skeleton is much more lightly built and the zygapophyses are less extensively interlocking than in other aardvards, which suggests less commitment to digging. The skull of *Plesiorycteropus* from the sub-Recent of Madagascar is very similar to that of the pangolin, suggesting a much higher degree of specialization for ant eating than we see in other aardvarks.

Unlike the other anteaters, there is little controversy regarding the ultimate ancestry of the aardvarks. Most recent authors have allied them closely with the condylarths. Patterson (1975) compared various skeletal elements with individual condylarth genera but did not specify a particular point of origin within that order. The absence of normal cusps in all the known genera precludes specific comparison of the teeth, which form the basis for the recognition of the condylarth families. Thewissen (1985) calls attention to similarities with members of the Insectivora.

SOUTH AMERICAN UNGULATES

South America was much more effectively isolated during the Tertiary than was Africa. From the late Cretaceous until the end of the Pliocene, there was no direct land connection between South America and any other continent. There was a variably emergent chain of islands that may have allowed rare waif dispersal across the Caribbean from North America. The alternative possibility of access from Africa via ephemeral mid-Atlantic islands and rafting remains controversial (Ciochon and Chiarelli, 1980; Simpson, 1978).

In the late Cretaceous or early Tertiary, at least two major lineages entered South America, one giving rise to the South American edentates and a second to a series of ungulate orders.

The fossil history in South America is not as continuously recorded as that of North America, but most major time intervals are represented. The most significant hiatus is the absence of any remains from the early Paleocene. As in Africa, by the time groups appeared in the later Paleocene and Eocene, they were already so specialized that their specific interrelationships are very difficult to establish. We may be missing 15 to 20 million years of their early evolution.

Because of its isolation, correlation between the fossil-bearing beds in South America and those of the rest of the world was subject to doubt until absolute dating came into use. As in Europe and North America, provincial age terms were long used. The modern correlation, which is based on radiometric dating and magnetic stratigraphy, is shown in Figure 21-70, but provincial terms are also cited in the text to facilitate use of the earlier literature.

The only Cretaceous mammal from South America that may be assigned to the ungulates, *Perutherium*, is known from a single jaw with fragments of two teeth. Van Valen (1978) places it in the family Periptychidae, but Cifelli (1983b) questions whether it is even a eutherian.

Recent authors recognize six uniquely South American ungulate orders: Notoungulata, Astrapotheria, Trigonostylopoidea, Xenungulata, Pyrotheria, and Litopterna. Many families show specializations in common with well-known orders from the northern continents and Africa. Ameghino (1906), who was the first to describe many of the South American groups, thought that these similarities indicated close relationships, but all are now attributed to convergence (Simpson, 1967, 1980).

The litopterns include both horselike and camel-like forms. The pyrotheres have features common to elephants and the astrapotheres somewhat resemble rhinos. The notoungulates included the rodentlike typotheres and hegetotheres and the sheeplike to rhinolike toxodonts. The trigonostylopoids are presently included within the Astrapotheria, and the Xenungulata are allied with the pyrotheres.

The origin and affinities of the South American ungulates were most recently reviewed by Cifelli (1983b). Their ultimate ancestry has long been assumed to lie with the condylarths of the northern continents. One family of conservative condylarths, the Didolodontidae, is known in South America from the late Paleocene (Riochican) into the middle Miocene (Friasian) (Figure 21-71).

The ancestry of the didolodonts can be traced to the hyopsodont subfamily Mioclaeninae, particularly genera such as *Litaletes* from the Middle Paleocene. Cifelli characterized the Miocleaninae by the possession of inflated premolars, with the loss of accessory cusps and cuspules and by reduction of M^3_3. The mioclaenines and the didolodonts share a similar pattern of molarization of the upper premolars; in contrast with phenacodonts, which they otherwise closely resemble, the protocone of P^4 is

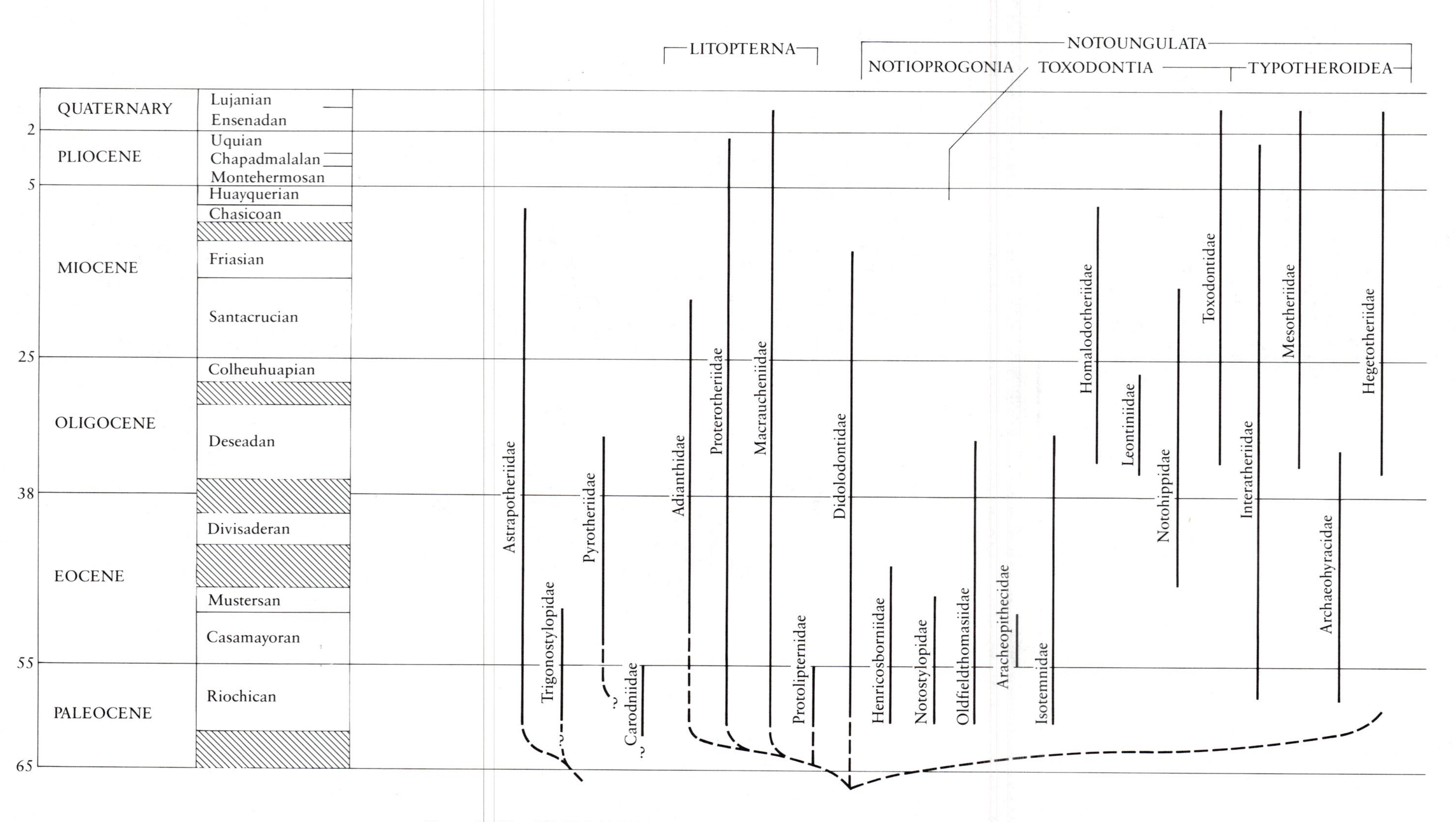

Figure 21-70. TIME RANGES OF SOUTH AMERICAN UNGULATE FAMILIES.

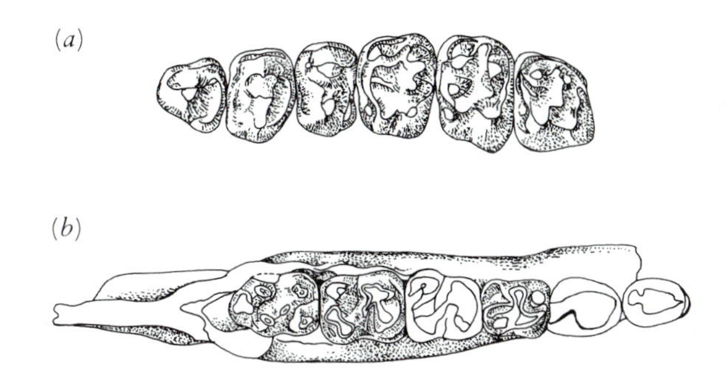

Figure 21-71. (*a* and *b*) Upper and lower teeth of the South American condylarth *Didolodus*, ×1. *From Simpson, 1948.*

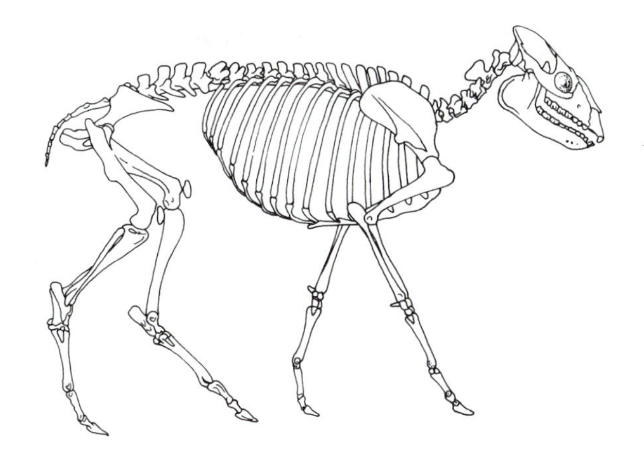

Figure 21-72. THE HORSELIKE LITOPTERN *DIADIAPHORUS. From Gregory, 1951 and 1957.*

large and transversely aligned with the paracone. The early South American condylarths are distinguished in having slightly more bunodont cusps, with the talonid cusps somewhat more distinct.

The didolodonts, which are known primarily from jaw fragments and isolated teeth, may be ancestral to all other South American ungulates, but they show particularly close affinities with the Litopterna. Based on the anatomy of primitive didolodonts and early litopterns such as *Asmithwoodwardia*, the immediate common ancestors of these groups would be expected to have been small, comparable with the better-known hyopsodontids, with a complete dental series. The anterior teeth were unspecialized and a diastema had not developed. The P^3 has a small metacone close to the paracone and a low protocone. The P^4 had both paracone and metacone, but the styles were reduced. In general, the upper molars had low cusps. A hypocone had developed on M^{1-2}, but not on M^3. In the lower jaw, the P_3 was simple and uninflated. P_4 had a metaconid and paraconid as well as a protoconid and a unicuspid talonid. The trigonid and talonid on the lower molars were of nearly the same height. This morphotype provides a basis for establishing the degree of relationship of all the other South American ungulate groups to the didolodontids and litopterns.

Figure 21-73. THE CAMEL-LIKE LITOPTERN *THEOSODON. From Gregory, 1951 and 1957.*

LITOPTERNA

The litopterns are best known by the proterotheres, which were superficially horselike (Figure 21-72). *Thoatherium* from the Miocene achieved a reduction of the lateral digits that was even greater than that of the modern horse at the end of the Tertiary. The limbs were short and the dentition was much less specialized. A diastema was little developed, and the teeth remained low crowned but lophodont. They appear to have been suitable for browsing but not grazing. The macraucheniids had the general build of camels, but the retracted nasal bones suggest the presence of a proboscis (Figure 21-73). The Adianthidae were small, delicately built animals that are known only in the middle Tertiary (Cifelli and Soria, 1983).

NOTOUNGULATA

The largest assemblage of South American ungulates belongs to the order Notoungulata, which includes 13 families that exhibit a wide range of body forms.

Eight families had already appeared by the Riochican and Casamayoran (Middle Paleocene to Lower Eocene). We can recognize their close relationship by a particular pattern of the molar cusps. In early members of the order, the teeth are low crowned and a trace of the primitive pattern of distinct cusps is still evident, but the cusps have begun to join to form a pattern of lophs with a straight ectoloph on the outer surface, a long protoloph, and a metaloph, which may have a separate forward-directed process termed a crochet. There may be accessory cusps in the central valley of the upper molars (Figure 21-74). In the lower teeth, the entoconid is isolated in a crest of the posterior lophid. An anterior lophid is formed from the protoconid and metaconid.

The complete eutherian tooth complement is present in early genera, and only a few forms reduce the tooth count significantly or develop a long diastema.

Patterson (1934) and Simpson (1948) described a distinctive pattern of the middle ear that characterizes notoungulates. In addition to the normal auditory bulla formed from an expanded ectotympanic, there are additional chambers both above and below the normal middle ear cavity that may have enhanced the acuity of hearing. The external auditory meatus is an ossified tube (Figure 21-75).

Postcranially, early notoungulates, as characterized by *Thomashuxleya* (Figure 21-76), retain primitive features little advanced above the level of the North American amblypods and dinocerata. There are five toes on both the front feet and hind feet, and the posture is plantigrade.

When the notoungulates are first recognized in the late Paleocene and early Eocene, they were in the initial stages of an explosive radiation, which is reflected in an astounding degree of individual variability. The species *Henricosbornia lophodonta* is represented by the dentitions of hundreds of specimens. Simpson's (1948) thorough statistical study of both qualitative and quantitative traits indicates that all belong to a single biological unit that shows continuous variability. Previous descriptions by Ameghino (1906), which were based on a typological species concept, led to the recognition of sixteen species that were placed in eight genera, four families, and three orders.

Four suborders are recognized within the notoungulates, all of which had diverged by the end of the Paleocene. The most primitive forms are included in the Notioprogonia, which are restricted to the Paleocene and Eocene. The Toxodontia is the most diverse group and includes the largest genera, which reached the size and proportions of hippos and rhinos. The typothers and the

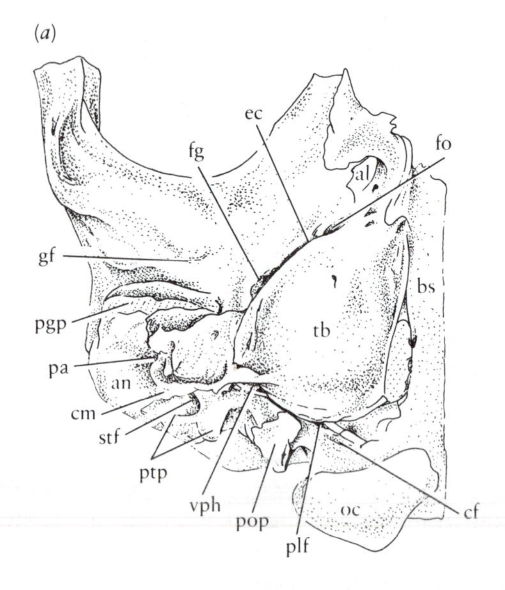

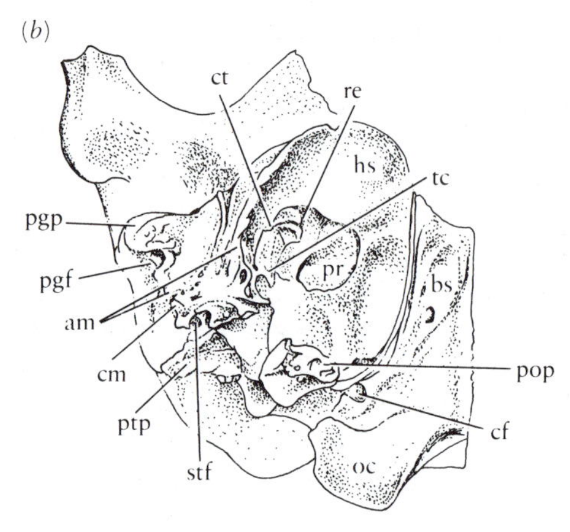

Figure 21-75. (*a* and *b*) Detail of ear region of the notoungulate *Notostylops.* Abbreviations as follows: al, alisphenoid; am, auditory meatus; an, acoustic notch; bs, basisphenoid; cf, condylar foramen; cm, crista meati; ct, crista tympanica; ec, eustachian canal; fg, fissura glaseri; fo, foramen ovale; gf, glenoid fossa; hs, hypotympanic sinus; oc, occipital condyle; pa, porus acusticus; pgf, postglenoid foramen; pgp, postglenoid process; plf, posterior lacerate foramen; pop, paroccipital process; pr, promontorium of the petrosal; ptp, posttympanic process; re, recessus epitympanicus; stf, stylomastoid foramen; tb, tympanic bulla; tc, tympanic cavity; vph, vagina processus hyoidei. *From Simpson, 1948.*

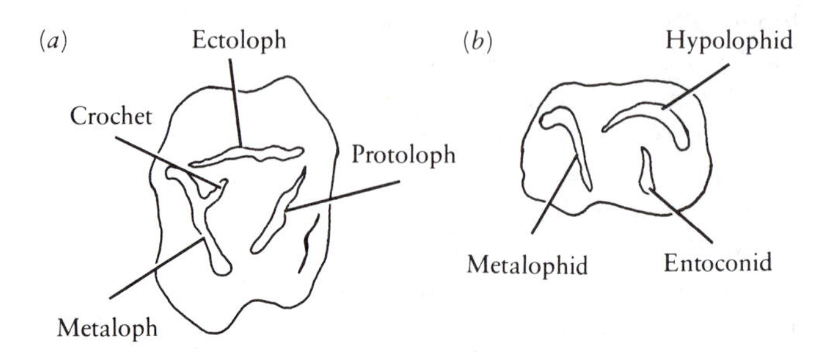

Figure 21-74. (*a*) Upper and (*b*) lower molars of notoungulate showing pattern of the lophs. *From Simpson, 1980.*

hegetotheres included rabbitlike forms and large rodentlike genera.

Early members of all the suborders are characterized by primarily primitive features when they first appear in the fossil record, including a complete dentition with low-crowned teeth and without a diastema. In each group, advanced genera achieved high-crowned and ever-growing teeth.

Within the Notioprogonia, the family Notostylopidae shows early dental specialization in the elaboration of large nipping incisors that are separated from the cheek teeth by a long diastema (Figure 21-77). As in the case of *Henricosbornia,* the genus *Notioprongia* shows ex-

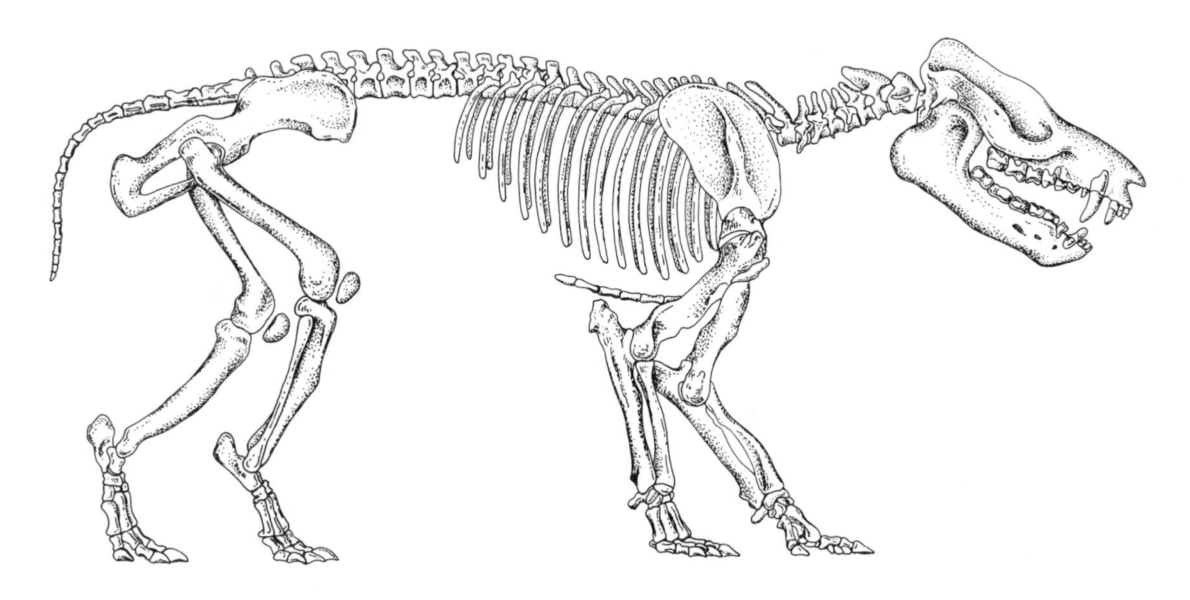

Figure 21-76. SKELETON OF THE PRIMITIVE NOTOUNGULATE *THOMASHUXLEYA*, $\times \frac{1}{10}$. *From Simpson, 1967.*

treme variability in tooth count (Simpson, 1980), with a single species ranging from

$$\frac{3\ 1\ 4\ 3}{3\ 1\ 4\ 3}$$

to

$$\frac{2\ 0\ 3\ 3}{2\ 0\ 3\ 3}$$

Both typotheres and hegetheres include advanced members in which the incisors were ever-growing structures like the gnawing incisors of rodents (Figure 21-78). The Pliocene and Pleistocene typothere *Mesotherium* reached the size of a black bear. Within the typotheres, the number of digits is reduced to a greater degree than in other notoungulates. Among Northern Hemisphere ungulates, there is a very clear distinction between the mesaxonoic pattern of the perissodactyls, in which the main axis of the foot runs through the central digit and the number of digits is reduced to three or one, as opposed to the paraxonic pattern of the artiodactyls, in which the main axis runs between digits III and IV and the number of digits is reduced to four or two. In the typothere *Miocochilius,* the forefoot is three toed, with the largest being the second and third. The hind foot has only two toes, the third and fourth. All other notoungulates place most of their weight on the middle toes but retain digits II and IV as well. The degree of developmental constraint governing digital evolution would appear to be different among various ungulate orders.

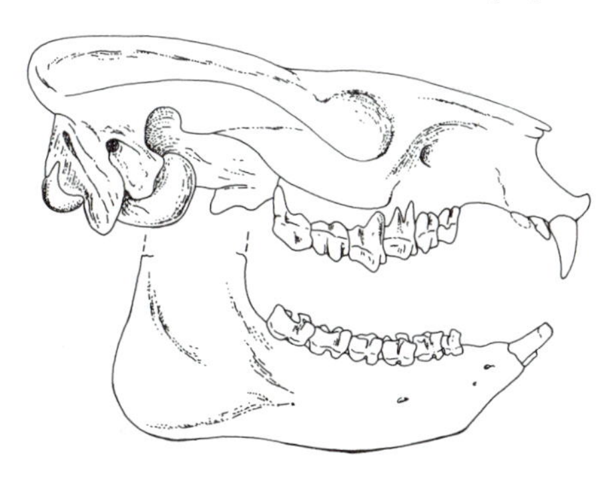

Figure 21-77. SKULL OF *NOTOSTYLOPS,* A PRIMITIVE NOTOUNGULATE. Note elaboration of incisors and retention of canines and anterior premolars, $\times \frac{3}{8}$. *From Simpson, 1948.*

Among the hegetotheres, *Pachyrukhos* achieved a very rabbitlike anatomy, with long rear limbs and feet that rested flat on the ground that could have served to spring the animal up in a rabbitlike leaping action. Both the cheek teeth and the incisors were evergrowing, and the dental formula was reduced to

$$\frac{1\ 0\ 3\ 3}{2\ 0\ 3\ 3}$$

The Toxodonta include five families. *Thomashuxleya,* which was described as a representative early notoungulate, is a primitive toxodont. The Santacrucian (early Miocene) genus *Homalodotherium* (Figure 21-79) has converged on the pattern of the perissodactyl superfamily Chalicotheroidea in the elaboration of claws, in contrast with all other notoungulates. *Toxodon* is a late-surviving member of the group that had the build and size of a rhinoceros and even bore a stubby horn on its nasals. The dentition shows a short diastema, with the loss of the last upper incisor and canine and the first lower premolar. The other teeth were all high crowned and ever growing. This genus was common to the end of the Pleistocene. Typotheres and hegetotheres also survived in reduced numbers into the Pleistocene before becoming extinct.

The notoungulates are the only South American ungulate order that may have a fossil record in other con-

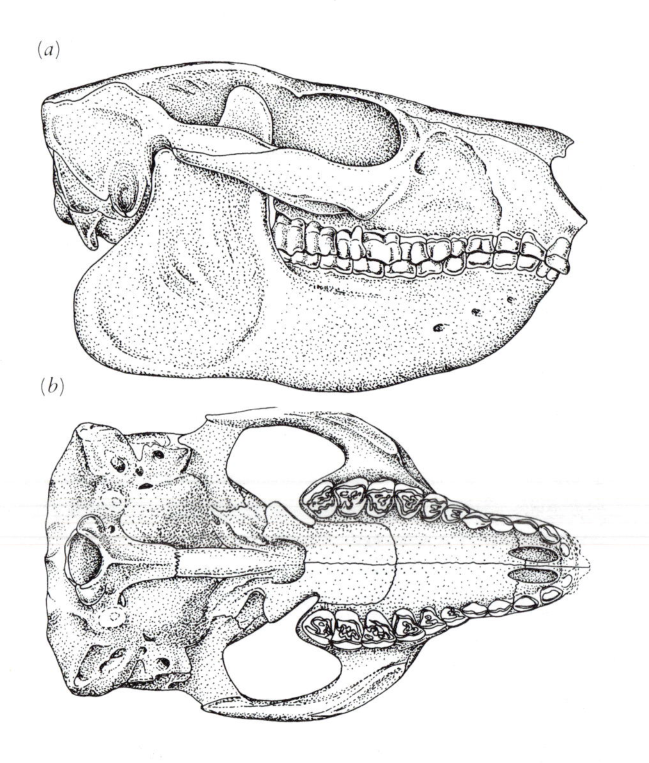

Figure 21-78. THE TYPOTHERE *NOTOPITHECUS*. (*a*) Lateral and (*b*) palatal views, ×1. *From Simpson, 1967.*

tinents. *Arctostylops* from North America has been assigned to the Notioprogonia, as have numerous Asian genera, including *Palaeostylops* (Figure 21-80). These remains have been interpreted as indicating either that notoungulates originated in the northern continents and later entered South America or that some notoungulates emigrated from South America early in their history. Advanced arctostylopids have cheek teeth that closely resemble those of South American notoungulates, but Cifelli (1983b) argues that these similarities are the result of convergence and that the most primitive Asian species lack important characteristics of notoungulates and have specializations that are not encountered in the South

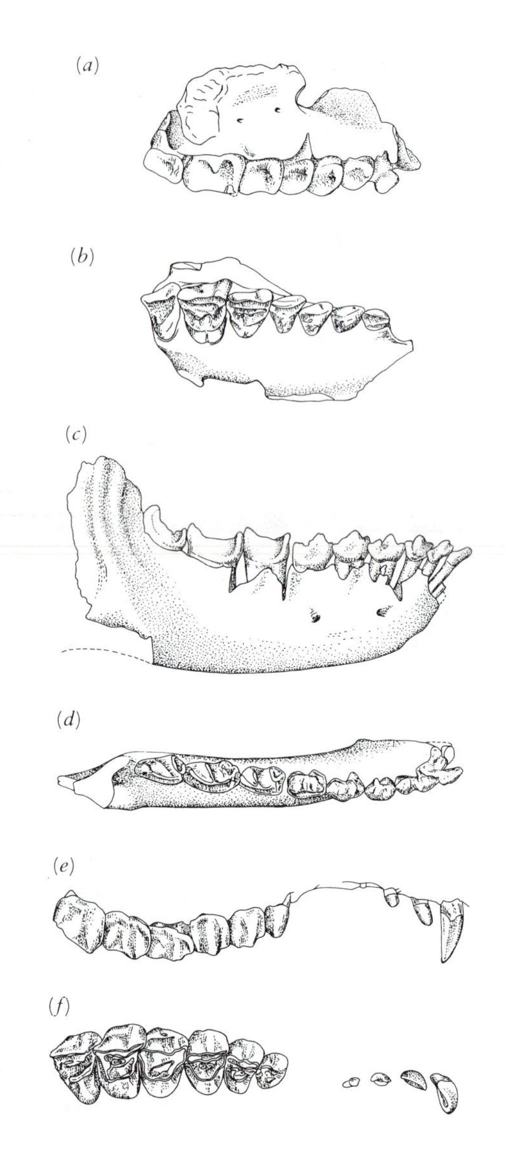

Figure 21-80. (*a* to *d*) Upper and lower dentition of *Palaeostylops*, a possible notoungulate from the Upper Paleocene or lower Eocene of Mongolia. *From Matthew and Granger, 1925.* (*e* and *f*) Upper dentition of *Notostylops*, a primitive South American notoungulate, in lateral and occlusal view. *From Simpson, 1948.*

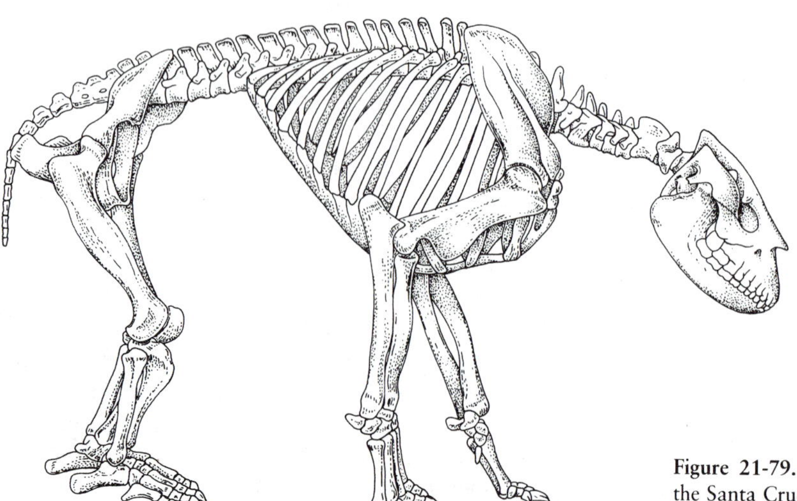

Figure 21-79. *HOMALODOTHERIUM*. A clawed toxodont from the Santa Crucian, $\times\frac{1}{15}$. *From Riggs, 1937. By permission of the Field Museum of Natural History, Chicago.*

American genera. The premolars are exceedingly simple and the upper molars primitively lack a hypocone. What later develops in the position of the hypocone is actually a displaced metaconule. The arctostylopids lack the transverse entolophid on the lower molar and have a simpler trigonid in the upper molar. They are divergent in the great development of anteroposterior vertical shearing surfaces on the upper and lower molars. Unfortunately, the ear structure of the Asian forms has not been investigated to see if it resembles the highly distinctive pattern of the South American genera. Cifelli suggests that the arctostylopids may have evolved directly from the atypical Chinese arctocyonid *Lantianius*.

The teeth of early notoungulates are already clearly characteristic of that order but could have been derived from the pattern of the primitive didolodont-litoptern stock.

PYROTHERES

The pyrotheres were large animals with long bodies and short columnar limbs that are known from the late Paleocene into the Deseadan (Lower and Middle Oligocene) (MacFadden and Frailey, 1984). *Pyrotherium* from the Deseadan is the only genus in which the skull is known. Two pairs of upper and one pair of lower incisors were specialized as short tusks; the cheek teeth are bilophodont. The nasal region suggests a proboscis (Figure 21-81). Like proboscideans and embrithopods, the pyrotheres had a serial tarsus with modified calcaneofibular contact.

Patterson (1977) reported that the ear region of *Pyrotherium* resembles that of notoungulates, but the dental specialization is so different that an immediate common ancestry seems very unlikely. Cifelli (1983b) suggested that the teeth could be derived from those of the common didolodont-litoptern ancestors, although he stressed the problem of accounting for such a drastic change in the dentition prior to the Upper Eocene.

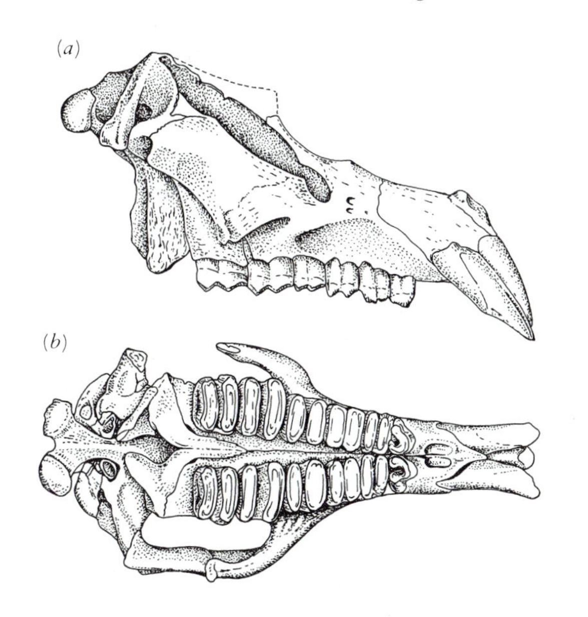

Figure 21-81. SKULL OF THE ELEPHANTLIKE GENUS *PYROTHERIUM*. (*a*) Lateral and (*b*) palatal views. *From Lavocat, 1958.*

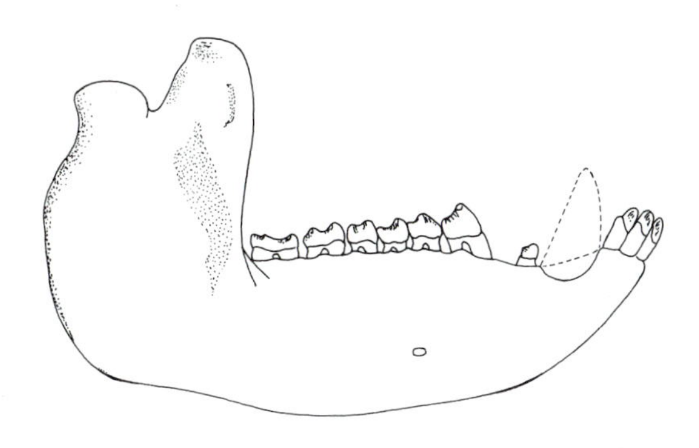

Figure 21-82. LOWER JAW OF *CARODNIA*, THE ONLY GENUS INCLUDED IN THE ORDER XENUNGULATA. The molar teeth are strongly bilophodont. *From Paula Couto, 1952.*

XENUNGULATA

The order Xenungulata was named by Paula Couto (1952) for the reception of a single genus, *Carodnia*, from the Upper Paleocene (Riochican). It was a large form that broadly resembles the pantodonts, unitatheres, and embrithopods of the other continents. The limbs were slender, but the feet retained all five digits. The incisors were chisel-like, the canines were large and sharp, and the cheek teeth were partially bilophodont (Figure 21-82). This combination of dental characters distinguishes this genus from members of all other South American orders. Wheeler (1961) and McKenna (1980) suggested affinities with the Dinocerata based on dental resemblances. This hypothesis is elaborated by Schoch and Lucas (1985). Cifelli (1983a) suggested that *Carodnia* may share a common ancestry with the pyrotheres, based on tarsal similarity.

ASTRAPOTHERIA

We know astrapotheres from the Upper Paleocene to the end of the Miocene. Advanced genera, such as *Astrapotherium* from the Oligocene, were 3 meters or more in length (Figure 21-83). The forelimbs were stout, but the rear ones were more slender. Like that of the Dinocerata, the tarsus is strongly alternating, with great development of the medial malleolus of the tibia as a weight-supporting area. The skull was highly specialized, with very short, toothless premaxillae; the frontal sinuses give the forehead a domed appearance. The upper canines were large and ever growing. The lower canines were also elongate, but the anterior premolars were vestigial; the last two molars were enormous. The narial opening was posterior in position, which suggests the presence of a proboscis. The lower jaw was apparently longer than the upper.

The Trigonostyloidea, which are known primarily from the Lower Eocene, have long been allied with the astrapotheres, but Simpson (1967) placed them in a dis-

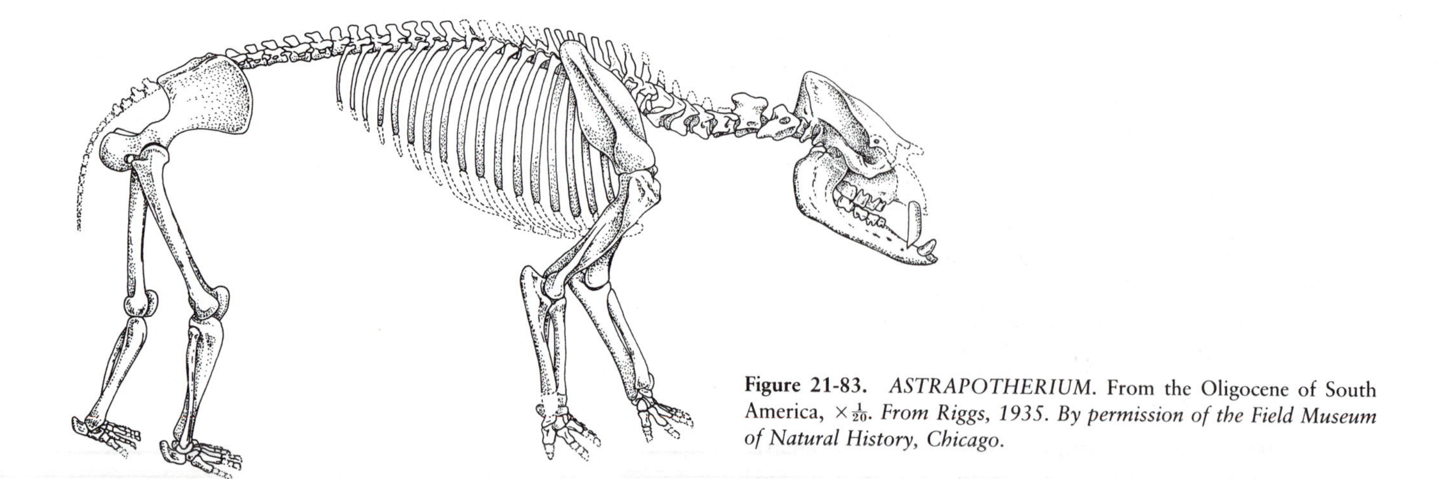

Figure 21-83. *ASTRAPOTHERIUM.* From the Oligocene of South America, ×$\frac{1}{20}$. *From Riggs, 1935. By permission of the Field Museum of Natural History, Chicago.*

tinct order. Cifelli (1983a,b) returned them to the Astrapotheria, arguing that they were distinguished primarily by primitive features. The postcranial skeleton is not known. The skull (Figure 21-84) resembles that of *Astrapotherium* in the great size of the lower canines and the reduction of the anterior premolars. The third and fourth upper premolars are molarized. The lower molars lack a paraconid, and the labial talonid cusps are joined by a crest that is continuous with the cristid obliqua.

A striking feature of the trigonostylopids is the fact that the hypocone is almost or completely absent in the primitively triangular upper molars. Since this cusp develops in later astrapotheres, its early absence cannot be considered a simplification but is almost certainly a primitive feature of the group. In this feature, the trigonostylopids are more primitive than the early litopterns and didolodonts.

This evidence suggests that astrapotheres had evolved prior to the emergence of the immediate common ancestor of the didolodonts and litopterns. Cifelli (1983b) uses this evidence to argue that the South American ungulates may not have had a unique common ancestry. Such a common ancestry could be imagined in forms only slightly more primitive than the early didolodonts, but they would have very few dental characters that would differentiate them from the most primitive condylarths.

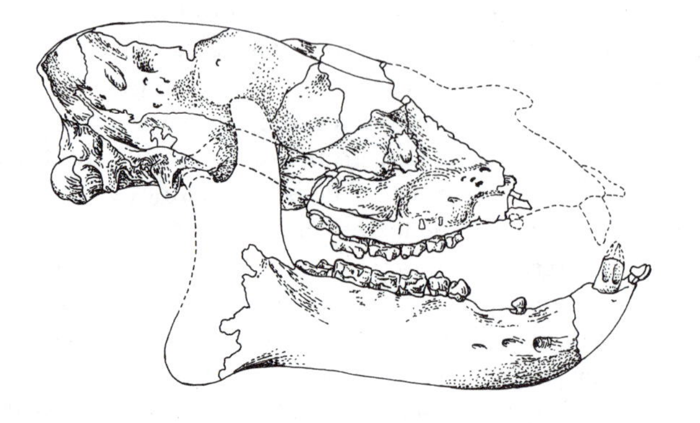

Figure 21-84. *TRIGONOSTYLOPS.* Skull in lateral view, ×$\frac{1}{4}$. *From Simpson, 1967.*

A late Cretaceous or early Paleocene migration of primitive ungulates (still lacking a hypocone) into South America would give time for the divergence and specialization of all the known orders. In contrast, Cifelli (1983b) argues that there may have been two ancestral lineages, one that gave rise to the astrapotheres and pyrotheres and the other that led to the litopterns and notoungulates. Only the later groups are securely tied to the northern condylarths, but both must trace their ultimate origin to that group. Schoch and Lucas (1985) argue that xenungulates and pyrotheres are not condylarth derivatives but arose from Asian anagalids.

The ungulates of South America appear as a copy in miniature of the diversity seen in the rest of the world, with their own models of hippos, rhinos, horses, camels, elephants, and conies. They diversified and flourished to the end of the Tertiary. By the end of the Pleistocene, they had all become extinct. Their final demise may be attributed to competition and predation by the placentals from the rest of the world that entered South America 3 million years ago, when a land connection was established with North America.

Edentates

Several very different sorts of mammals have been grouped together in a single assemblage, the Edentata. They include the Asian and African pangolins or scaly anteaters (Order Pholidota); a diverse but phylogenetically coherent group of South American forms, the Order Xenartha, which includes the armadillos (suborder Cingulata), anteaters (Vermilingua), and sloths (Pilosa); and the palaeanodonts from the early Tertiary of North America. All are characterized by the reduction or complete loss of the dentition. If teeth are retained, they lack enamel. Most edentates have highly developed claws, especially on the forelimbs, which in many genera are used for digging. In most modern edentates, this combination of features is associated with a diet that includes ants, termites, and other small insects that are dug from the ground or from termite nests. The living tree sloths and a number of extinct Xenarthrans are notable exceptions in being strictly herbivorous.

In contrast with primitive mammals and modern shrews, insectivorous edentates do not use their teeth to pierce and shear their prey. Instead, they are crushed or, more commonly, swallowed whole to be broken up by a muscular, gizzardlike portion of the stomach. Their highly developed sense of smell enables them to detect prey buried deep in the ground.

The great reduction of the dental enamel and elaboration of the skeleton for digging are features of the earliest known fossil edentates and suggest that feeding on subterranean colonies of social insects led to the initial divergence of this assemblage.

The evolution of each of the major edentate groups has been confined largely, if not entirely, to a single continental area. It is not yet possible to establish if they originated from a single ancestral stock whose descendants became isolated in each of these areas or whether these groups arose from two or three separate ancestral lineages that were not closely related to one another.

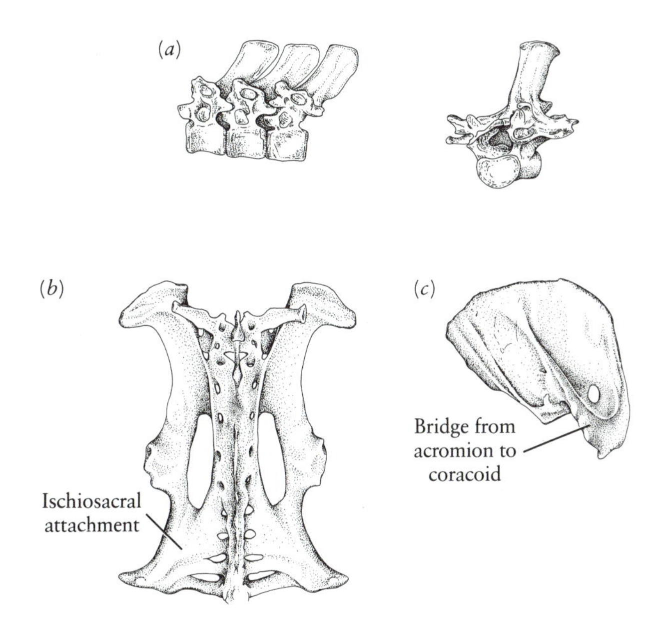

Figure 21-85. (*a*) Vertebrae of the South American anteater *Myrmecophaga* showing extra articulating surfaces. *From Gregory, 1951 and 1957.* (*b*) Pelvic girdle of an armadillo showing accessory articulation between the ischia and the caudal vertebrae. (*c*) Scapula of the ground sloth *Nothrotherium. From Stock, 1925. With permission of Carnegie Institution of Washington.*

XENARTHRA

The South American order Xenarthra is the most diverse of the edentate groups, with a rich fossil record throughout the Cenozoic (Simpson, 1948, 1980; Patterson and Pascual, 1968). The term xenarthra refers to the "strange" or accessory articulations between the vertebrae (Figure 21-85*a*). Their elaboration may be associated with the initial specialization of the postcranial skeleton for digging and may have enabled divergent groups such as the armadillos and glyptodonts to support a heavy carapace and the ground sloths to support their massive body in a near vertical posture. All members have an accessory area of attachment for the pelvic girdle between the ischium and the transverse processes of the caudal vertebrae (Figure 21-85*b*). In most genera, the scapula bears a second spine that is parallel but posterior to the first. In the sloths, the acromion extends anteriorly to fuse with the coracoid area.

Cingulata

Despite the obvious specialization of armadillos in possessing an external bony carapace, the remainder of the skeleton of early genera is primitive relative to other xenarthrans, and they appear close to the ancestry of the entire order. Some sloths retain isolated elements of dermal armor within their skin. Armor is completely absent in South American anteaters. It may have been lost in an early stage of their evolution, or they may have diverged from the common xenarthran stock before the elaboration of armor.

We find armadillo scutes in the lowest horizon that bears Cenozoic mammals in South America, the Riochican (Middle to Upper Paleocene). Much of the skeleton of the primitive armadillo *Utaetus* is known from the Casamayoran (Lower Eocene). It expresses all the important characteristics of xenarthrans but in a primitive form. Only the back of the skull is preserved, and it shows that the glenoid is fairly far forward, low, and nearly flat. The teeth are simple pegs, without roots, and were probably ever growing. Unlike later xenarthrans, the teeth still have a thin covering of enamel on the internal and external surfaces, but not elsewhere. There are ten teeth in each jaw ramus that are plausibly identified as two incisors, a slightly larger canine, and seven cheek teeth, which are circular to eliptical in outline (Figure 21-86). The absence of enamel results in a relatively soft crown, without cusps or a specific occlusal pattern. Rapid wear of the teeth allows them to conform to a pattern that is appropriate for the available food. This wear is compensated for by continuous growth of the teeth. The possibility for the teeth to adjust to different foods may be a major factor that is responsible for the wide range of diets of modern armadillos.

The lower jaw in *Utaetus* has a very short process anterior to the dentition, in contrast with its much greater elaboration in later armadillos. The cervical vertebrae are all separate, whereas the axis and the succeeding cervicles fuse in lower Miocene and later genera. The trunk vertebrae show distinct xenarthrous articulations, which are characteristic of the order. The scapula has a strong ac-

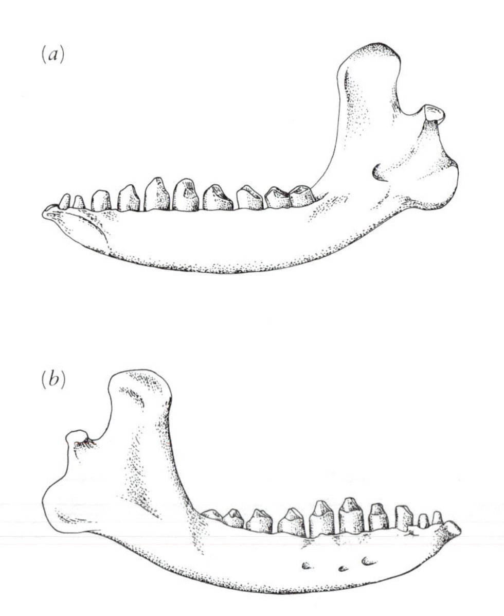

Figure 21-86. (*a*) Medial and (*b*) lateral views of the lower jaw of the Eocene armadillo *Utaetus*, $\times\frac{3}{4}$. *From Simpson, 1948.*

romion and a thickening of the posterior margin of the blade in the position where the second scapular spine develops in later genera. The ischium has already elaborated the specialized area of attachment to the anterior caudal vertebrae. The ulna has a long olecranon, which suggests digging capabilities.

The carapace has already evolved most of the features of modern genera, although the individual scutes were not as tightly integrated. There was a separate head shield and at least 12 flexible bands over the shoulder area. The more anterior of these became consolidated in post-Eocene genera to form a solid unit. Scutes are tentatively associated with the tail, but they may not have formed well-developed rings. As in modern genera, the scutes were covered with epidermal scales; they do not fossilize, but their presence is indicated by impressions in the underlying dermal elements.

Many lineages of armadillos are known throughout the Cenozoic in South America. Despite their readily preserved and easily identified scutes, their fossil record remains spotty. Several of the modern genera have a fossil record going back to the Pliocene or Pleistocene. At the end of the Pliocene, several genera entered North America. The only species now remaining on that continent is *Dasypus novemcinctus*, with a wide distribution in the southern states.

Simpson (1948) pointed out that the evolutionary rate of armadillos since the early Eocene appears slow if we compare *Utaetus* with primitive modern forms. On the other hand, much more significant changes are evident in the lineages that lead to more specialized living genera such as *Priodontes* and *Chlamyphorus*. Even more dramatic modifications in the skeleton are seen in a separate cingulate lineage, the family Glyptodontidae.

Glyptodonts are distinguished from other armored edentates by the integration of the carapace into a single, inflexible unit. Glyptodonts are also distinguished by the pattern of the cheek teeth, which are bilobate or trilobate in occlusal view (Figure 21-87). We find teeth with this pattern as early as the Mustersan (middle Eocene). The pampatheriine armadillos may provide a link between the two families. Four or five subfamilies of glyptodonts are known in the Cenozoic of South America. One genus, *Glyptotherium*, entered North America in the early Pleistocene, and several species are recognized in the Gulf states. None survived to the end of the Pleistocene. Gillette and Ray (1981) thoroughly described this genus, which provides a good example of the terminal members of this family (Figures 21-87 and 21-88).

The skeleton is approximately 2.5 meters long, and almost all of it is covered by dermal armor. The skull has a small cephalic shield, as in armadillos, and the tail is

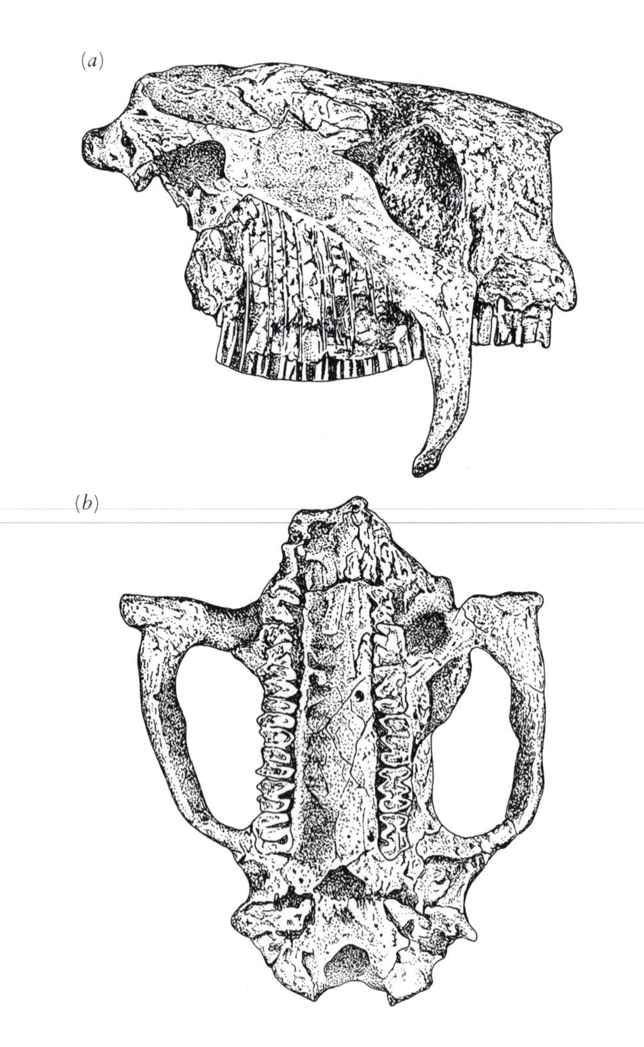

Figure 21-87. SKULL OF THE GLYPTODONT *GLYPTOTHERIUM*. (*a*) Lateral and (*b*) palatal views. Note extremely high-crowned molar teeth revealed by damage of the cheek. *From Gillette and Ray, 1981. By permission of Smithsonian Institution Press, Smithsonian Institution, Washington, D.C.*

Figure 21-88. (*a*) Skeleton of *Glyptotherium*. Armor is shown in dotted lines. (*b*) Reconstruction of *Glyptotherium*. Original was 2 meters long. *From Gillette and Ray, 1981. By permission of Smithsonian Institution Press, Smithsonian Institution, Washington D.C.*

surrounded by overlapping cylinders. The tail is tapered in the North American genus but ends in an expanded mace in the South American subfamily Doedicurinae. The skeleton is highly modified to support the carapace. Most of the trunk vertebrae and the enormous pelvis are fused into a single, immobile unit. The limbs are massive to accommodate the extra weight of the carapace, with the tibia and fibula no more than 50 percent the length of the femur. The feet are stout and broad. Presumably only very slow locomotion was possible.

The skull is very high, narrow, and short and is dominated by enormously high-crowned teeth. As in other xenarthrans, they lack enamel but accommodate for its absence by the potential for continuous growth. There are eight nearly identical molariform teeth in each jaw quadrant. It is not known whether the most anterior are supernumerary cheek teeth or modified canines. As in armadillos, no incisiform teeth are retained. The distribution of the jaw muscles indicates that jaw movement was primarily in the horizontal plane. Gillette and Ray suggest that *Glyptotherium* probably fed on relatively soft vegetation along the banks of water courses during the late Pleistocene. The final extinction of this genus was probably the result of the restriction during the Pleistocene of the warm, damp environment that it required.

Pilosa

The strictly arboreal modern tree sloths, which lack dermal armor, appear very distinct from armadillos. The now extinct ground sloths bridge this gap to some degree. Three or four major lineages of sloths are recognized. The mylodontoids appear in the late Eocene. In addition to showing the characteristic edentate features of the girdles and vertebrae, they retain a mosaic of ossicles within their skin. Like other sloths, they are distinguished from the armadillos by the reduction in the number of cheek teeth to four or five, and the reduction of the zygomatic arch. The acromion extends forward to reach the coracoid region. Like the armadillos, the mylodontoids were strictly terrestrial and most were fairly large, at least the size of a black bear. Their forelimbs were shorter than the rear limbs, and their claws were subcircular in cross section and nearly straight. The cheek teeth were enlarged.

The megalonychoids and megatherioids were common and diverse in the early Miocene. The early genera were smaller than the mylodontoids, lacked armor, and had claws that were laterally compressed and curved. The forelimbs were nearly as long as the rear limbs. The teeth retained the primitive peg shape. In their small size, limb proportions, and claw configuration, these early genera resembled the modern tree sloths and they are thought to have been arboreal. The best-known descendants of these groups are the large ground sloths, including *Megatherium*, *Nothrotherium*, and *Megalonyx*. The ancestors of *Megalonyx* extended into the Carribean and North America in the late Miocene and early Pliocene (Figure 21-89). The genus *Megalonyx* evolved in North America and by the late Pleistocene had reached as far north as Alaska. The late Pleistocene genera were almost certainly terrestrial, but this adaptation appears to have been secondary. They accommodated to a renewed life on the ground in a manner vaguely like that of the gorilla. Their hands and feet were turned inward so that they would have walked on their outer digits. The configuration of the pelvis and rear limb suggests that they might have been capable of a semibipedal stance, which would have enabled them to feed from high trees like the chalicotheres (Coombs, 1983).

The living tree sloths are very distinct from the ground sloths in general structure and habitus. In the modern genera *Bradypus* and *Choleopus*, the hands and feet appear like hooks with which the animals can support themselves upside down in trees. About 10 percent of their time is spent in this position. Neither genus is able to walk on the ground but must drag itself along during the rare times when it leaves the trees. Both are strictly vegetarians and have a complex stomach containing microorganisms that can break down cellulose. The body temperature in tree sloths is both low and irregular, ranging from 18 to 35°C.

The two genera are differentiated by the number of toes in the forelimb: *Bradypus* has three and *Choleopus*

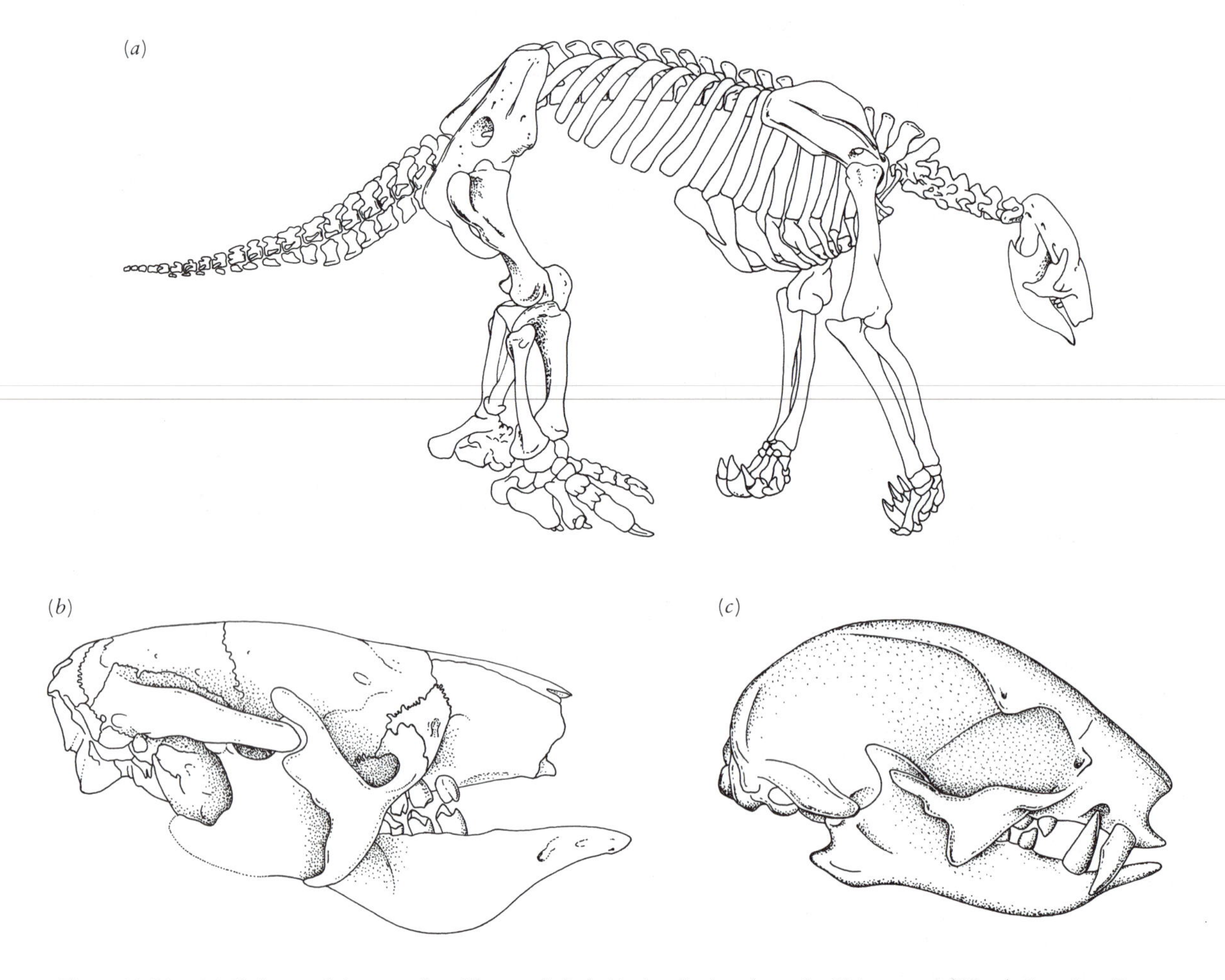

Figure 21-89. (*a*) Skeleton of the megatheroid ground sloth *Nothrotherium* from the Pleistocene of North America. *From Stock, 1925.* (*b*) Skull of *Nothrotherium. From Stock, 1925. With permission of Carnegie Institution of Washington.* (*c*) Skull of the modern tree sloth *Choloepus. From Radinsky and Ting, 1984.*

has only two. *Bradypus* has eight or nine cervical vertebrae, while *Choleopus* has five to seven and rarely eight.

The two modern genera have long been united in a single family Bradypodidae. In contrast, Patterson and Pascual (1968) suggested that each genus might be more closely related to families that were common in the early Cenozoic than they are to each other. This suggestion was further elaborated by Webb (1985). He points out a number of features in which *Bradypus* resembles the megatherioids and *Choloepus* resembles the megalonychoids. *Choloepus* and megalonychoids have anteriorly displaced caniniform teeth that are not distinguishable in *Bradypus* and megatheroids. Both groups have reelaborated the zygomatic arch, but in different ways. *Bradypus* and megatherioids have a hemispherical auditory bulla, while *Choloepus* and other megalonychoids, like the primitive mylodontoids, lack a bulla.

We do not think that the very similar arboreal adaptations of the two tree sloth genera evolved by convergence, but rather they reflect a common origin in primitive arboreal sloths of the late Oligocene or early Miocene, which lived prior to the divergence of the large terrestrial megatherioids and megalonychoids.

The late Cenozoic antellian genera *Synocnus* and *Acratocnus* may be closely related to *Choloepus*. No close relative of *Bradypus* has been recognized.

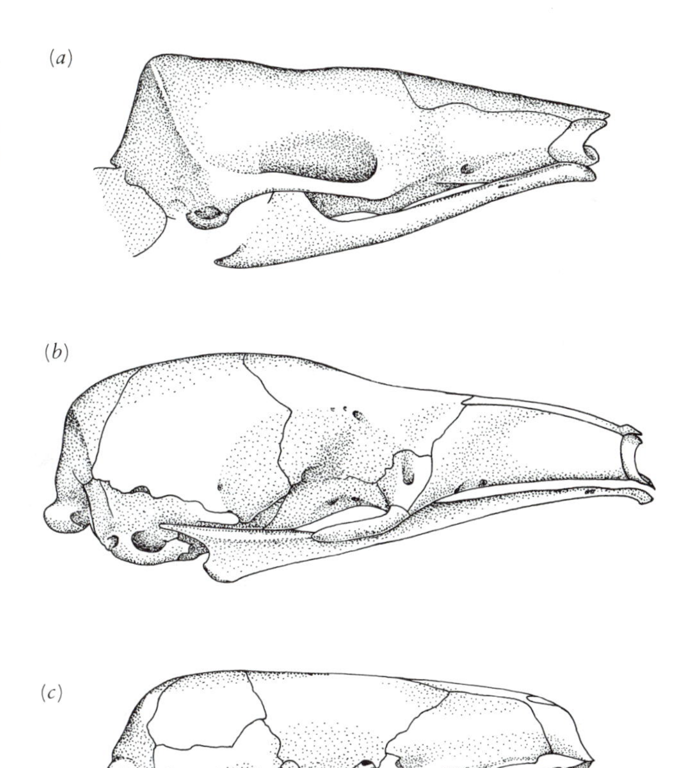

Figure 21-90. SKULLS OF ANTEATERS. (*a*) *Eurotamandua* from the middle Eocene of Europe. (*b*) *Tamandua*, a modern member of the Myrnecophagidae from South America. (*c*) *Manis*, an Asian pangolin. *From Storch, 1981.*

Vermilingua

South American anteaters (Myrmecophagidae) have the most highly specialized skull of all xenarthrans, with a long snout and an exceedingly slender lower jaw with no trace of teeth. The zygomatic arch is lost in modern genera (Figure 21-90). The tongue can be extended far beyond the mouth and is covered by a sticky secretion to which the prey becomes attached. Of the three living genera, the arboreal *Cyclopes* and *Tamandua* have prehensile tails, a feature that is absent in the larger terrestrial genus *Myrmecophaga.*

Myrmecophagids, which appear in the fossil record in South America in the Santacrucian (Middle Miocene), show an anatomy similar to that of modern genera. Surprisingly, Storch (1981) described an even older and yet essentially modern appearing species from the Middle Eocene of Europe. The skull of *Eurotamandua* (Figure 21-90*a*) is very similar to those of the living genera except for the retention of a slender zygomatic arch. It also resembles that of African and Asian pangolins. On the basis of the geographical distribution of the modern genera, one might consider *Eurotamandua* as a possible pangolin, but the postcranial skeleton shows definite xenarthran features, including the second scapular spine, sutural attachment between the ilium and the caudal vertebrae, and extra articulating surfaces of the vertebrae, none of which are seen in the pangolins. The Vermilingua probably shared a common ancestry with sloths and armadillos, but no intermediates are known.

PHOLIDOTA

The pangolins or scaly anteaters are placed in a separate order, Pholidota. Pangolins are today limited to seven species, four in Africa south of the Sahara and three in southeast Asia and adjacent islands. All may be included in a single genus *Manis*, or the African species may be distinguished as members of the genus *Phataginus* (Patterson, 1978).

Some pangolins are strictly arboreal, but most are at least partially subterranean. They are fully committed to feeding on subterranean social insects, including ants and termites, having completely lost the teeth and greatly reduced the lower jaw. The forelimbs and hind limbs are stout for digging. An essentially modern-looking pangolin *Eomanis*, complete with epidermal scales, is known from the Middle Eocene of Europe (Storch, 1978) (Figure 21-91). The striking modernity of this early genus makes it difficult to establish its ancestry.

In describing the only North American pangolin, the Lower Oligocene genus *Patriomanis*, Emry (1970) argued that they were not closely related to the xenarthrans but may have originated from the palaeanodonts (which are discussed in the next section). Patterson (1978) accepted

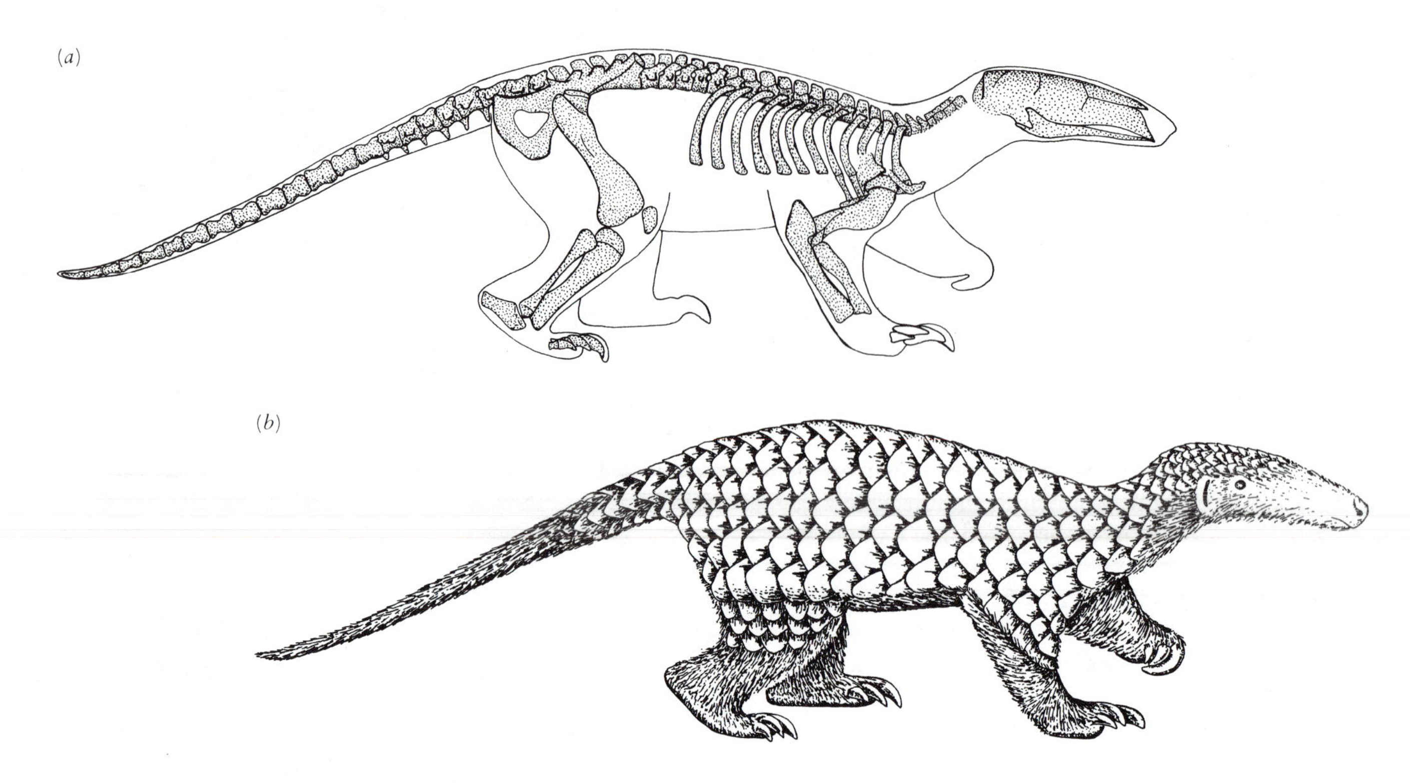

Figure 21-91. THE PANGOLIN *EOMANIS* FROM THE MIDDLE EOCENE OF EUROPE. (*a*) Skeleton. (*b*) Restoration (scales are preserved with the skeleton). *From Storch, 1978.*

Emry's identification of *Patriomanis* as a pholidotan but questioned its close affinities with the palaeanodonts. What is known of the skull is certainly much closer to that of *Manis* than to that of the palaeanodonts. Currently, there is no convincing evidence for the specific relationships of the Pholidota. The skull is strikingly convergent with that of the South American Myrmecophagidae, but they lack all the definitive xenarthran postcranial specializations. On the basis of biomolecular studies, Miyamoto and Goodman (1986) place the Pholidota close to the Carnivora and Insectivora.

PALAEANODONTS

With the exception of the myrmecophagid *Eurotamandua* in the Eocene of Europe and the late Cenozoic migration of glyptodonts, ground sloths, and armadillos into North America, the xenarthrans were restricted to South America for most of their history. The establishment of most of the definitive features of the order and the origin of all the major groups certainly occurred there, in isolation from the rest of the world.

Their ultimate origin and relationship with other eutherians remain unresolved. Matthew (1918) and Simpson (1931, 1948) emphasized many resemblances to the early Tertiary North American Palaeanodonta. The best-known member of this group is the Eocene genus *Metacheiromys* (Figure 21-92). It resembles the xenarthrans in the loss of enamel and the reduction of the number of cheek teeth. The limbs are primitive but characterized by tuberosities which suggests that they were used in digging. Simpson pointed out a number of features that he believed showed incipient development toward the xenarthrous condition, including extra areas of vertebral articulation, a thickening of the posterior margin of the scapular spine, and the structure of the pelvis. None of these reach the stage of development of the earliest-known South American xenarthran, and Emry (1970) contended that they do not indicate close relationships.

In any case, *Metacheiromys* is too late and too specialized in the loss of all but two of the cheek teeth to be considered directly ancestral. The most primitive member of the Metacheiromyidae is *Propalaeanodon* from the Upper Paleocene (Rose, 1979) (Figure 21-93*a*). It consists of a lower jaw with seven teeth and a possibly associated humerus that suggests a digging habit. The enamel is already all but lost and the crowns of the teeth have no cusps, in common with the early South American armadillos. The ramus of the jaw is also relatively slender.

The Epoicotheriidae is a second family of palaeanodonts that also appears in the Upper Paleocene. The Eocene and Oligocene members are highly specialized as subterranean burrowers, showing a degree of specialization of the skull, vertebrae, and limbs that is comparable to that of the African golden mole and some burrowing rodents (Figure 21-94) (Rose and Emry, 1983). The dentition is less reduced in this group than in the Metach-

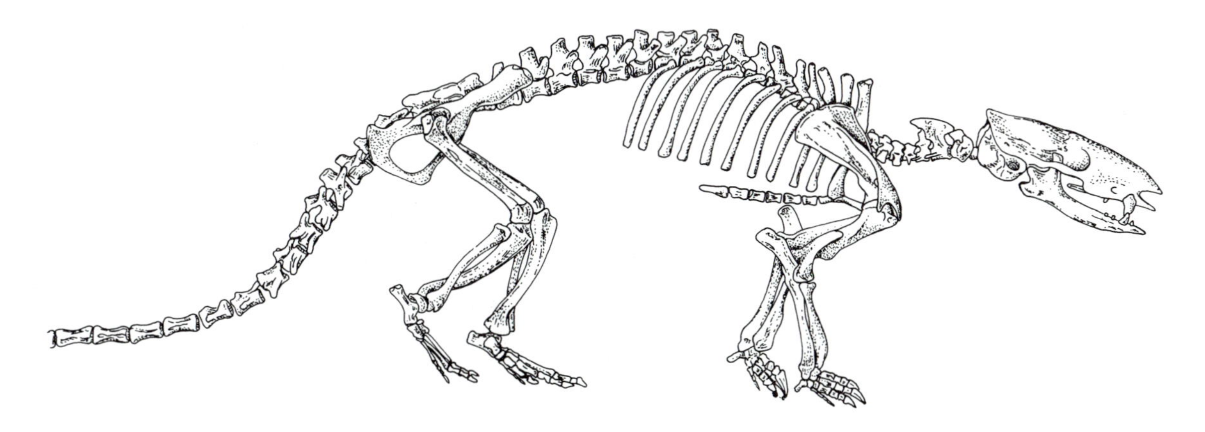

Figure 21-92. SKELETON OF THE EOCENE PALAEANODONT *METACHEIROMYS*. The original was 2 meters long. *From Simpson, 1931.*

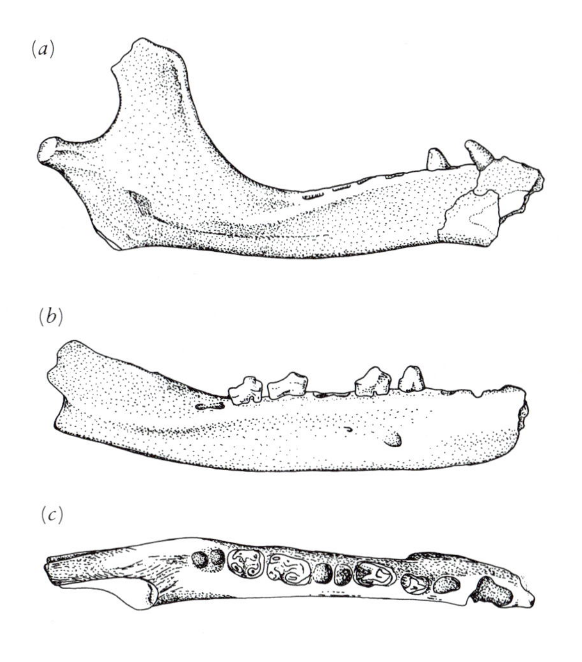

eiromyidae, retaining four to five cheek teeth and a fairly large canine. It seems possible that the otherwise isolated Upper Paleocene genus from China *Ernanodon* (Radinsky and Ting, 1984) may be related to the epoicotheres (Figure 21-95). The skull superficially resembles that of some sloths in the presence of both upper and lower caniniform teeth, but the lower tooth bites in front of the canine in both *Ernanodon* and the epoicothere, as in most mammals, whereas it bites behind the upper tooth in the sloth.

The early epoicotheriid *Amelotabes* from the Upper Paleocene is known from a jaw with seven cheek teeth, a large canine, and at least one incisor. Like that of *Propalaeanodon* and the early armadillos, the jaw ramus is slender and the coronoid process somewhat reduced. The cheek teeth have very thin enamel, but when the teeth first erupt they retain a cusp pattern that is common to primitive placentals. Rose (1978) pointed out similarities

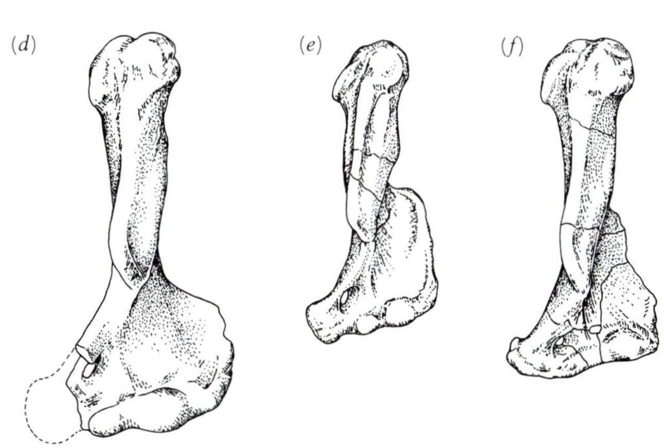

Figure 21-93. PALAEANODONTS. (*a*) Lower jaw in medial view of *Propalaeanodon*. (*b*) Medial and (*c*) occlusal view of *Amelotabes*, ×2. (*d*, *e*, and *f*) Humeri of the metacheiromyids *Palaeanodon* and *Propalaeanodon*, and the epoicotheriid *Pentapassalus*, ×1. (*a*, *b*, *d*, *e*, and *f*) *From Rose, 1979.* (*c*) *From Rose, 1978.*

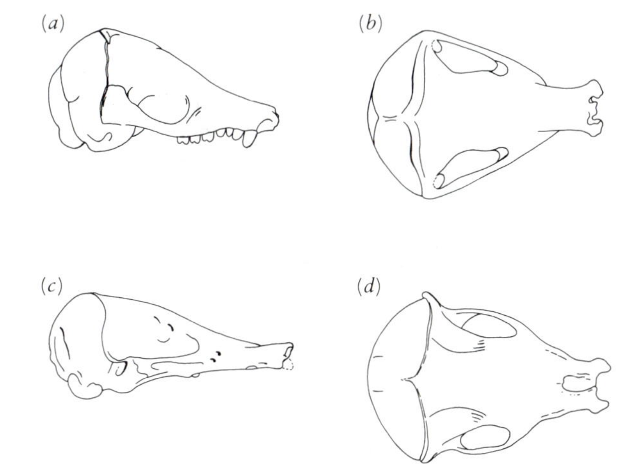

Figure 21-94. SUBTERRANEAN BURROWERS. (*a* and *b*) Skull of the modern African mole, ×1½. Skull of *Epoicotherium* in (*c*) dorsal and (*d*) lateral views. *From Rose and Emry, 1983.*

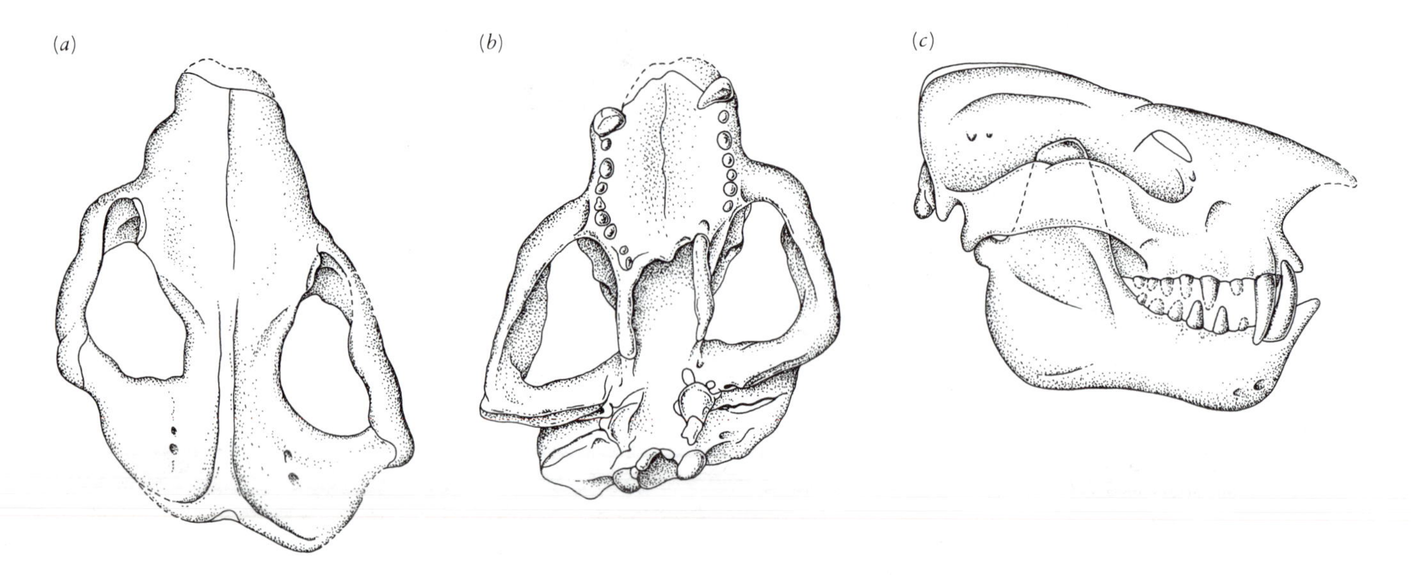

Figure 21-95. SKULL OF *ERNANODON* FROM THE LATE PALEOCENE OF CHINA. (*a*) Dorsal, (*b*) palatal, and (*c*) lateral views, $\times\frac{2}{3}$. *From Radinsky and Ting, 1984.*

with pantolestids, one of the small, taxonomically isolated groups of early placentals that were once included in the Proteutheria.

One can make a fairly strong case for the origin of the palaeanodonts from primitive eutherians and the subsequent divergence of the Epoicotheriidae (including *Ernanodon*) and the Metacheiromyidae. Emry (1970) argued strongly that palaeanodonts were closely related to the pangolins, but a very wide morphological gap separates these forms, which is further emphasized by the presence of an essentially modern pangolin in the middle Eocene.

Close relationship of palaeanodonts with xenarthrans was denied by Emry but tentatively supported by Rose (1979) on the basis of earlier and more primitive members of the Metacheiromyidae.

Among the very limited remains of the earliest metacheiromyids, there are no features that are specialized so as to preclude common ancestry with the xenarthrans. They do show two derived features in common—great reduction of the enamel and specialization of the forelimb for digging. The high degree of specialization of the postcranial skeleton in Lower Eocene armadillos and the presence of their armor in the Upper Paleocene suggest a significant period of prior evolution during the early history of South American mammals.

Whether the common ancestors of palaeanodonts and xenarthrans would be recognizable as such prior to their entry into South America can only be revealed by the discovery of fossils from the earliest Cenozoic.

McKenna (1975) suggested that xenarthrans have no close affinities with any of the placentals that emerged in the latest Cretaceous and earliest Cenozoic. He argued that xenarthrans occupy a unique position as the primitive sister group of all other eutherians. He based this conclusion on the presence of primitive characters that were lost in all other placentals. These characters include the retention of septomaxillary bones, ossified ribs that reach the sternum, low and variable body temperature, and primitive features of the reproductive system. As Novacek (1982) pointed out, the ossified ventral portion of the ribs is not a primitive feature among early placentals but is almost certainly a specialization of xenarthrans.

Among living edentates, only the tree sloths have a low and variable body temperature, which may be attributed to their restriction to the tropics where they can depend on a high, nearly constant, external heat source. The occurrence of the sloth *Megalonyx* in Alaska during the Pleistocene brings into question whether a low, poorly controlled body temperature was a general feature of xenarthrans. We find equally primitive features of the reproductive system in some living insectivores.

The only character cited by McKenna that supports an early divergence of the xenarthrans is the presence of a septomaxillary bone in the armadillos. This bone is not known in the early members of the other placental orders, although our knowledge of the skulls of late Cretaceous and early Paleocene placentals is not sufficiently complete to demonstrate its absence with great assurance. The fact that armadillos, like the anteaters, have greatly elaborated turbinals in relationship to their highly elaborated sense of smell might have resulted in the retention or reexpression of this bone adjacent to the nasal chamber.

The question of edentate affinities was most recently discussed by Novacek (1986), who cites the columnar shape of the stapes as a particularly important primitive feature of both xenarthrans and pangolins that suggests that these groups diverged prior to all other placental orders. Discovery of fossils from the late Cretaceous and early Tertiary that document the initial stages in the evolution of each of these groups is necessary before we can firmly establish the specific phylogenetic position of the edentates.

THE GREAT INTERCHANGE

South America was isolated from the northern continents from some time in the late Cretaceous to the late Pliocene, approximately 3 million years ago. Some immigration occurred during this period either from North America or conceivably from Africa. Gingerich (1985) argues that there may have been a brief period in the Upper Paleocene (late Tiffanian) when typically South American groups migrated to North America. The Arctostylopidae, Uintatheriidae and palaeanodonts, all of which *may* have affinities with South American orders, first appear in North America at this time. However, none of these groups have been firmly allied with South American genera, and all might have entered North America from Asia. Primates and rodents entered South America in the middle Oligocene. The procyonid *Cyonasua* reached there during the late Miocene, and two groups of ground sloths moved north during the same time interval.

At the end of the Pliocene, a solid land route was established across the isthmus of Panama, and there was a great interchange of mammals between North and South America. This interchange has been chronicled in a series of papers by Webb (1976, 1977, 1978), Marshall, Webb, Sepkoski, and Raup (1982), and Webb and Marshall (1982) and most comprehensively by Stehli and Webb (1985). The fauna of the intervening area of Central America was discussed in a symposium published in the *Journal of Vertebrate Paleontology* (MacFadden, 1984).

Following the interchange, the number of genera in both continents increased and then returned to nearly their former value. At the generic level, the number of forms that moved north and south was comparable with 16 southern genera going north and 23 northern genera moving south. However, the long-term effect on the fauna of the two continents was very different. The overall effect on North America was of little significance. The modern fauna includes only five southern genera, an opossum, a porcupine, an armadillo, and two sigmadontine rodents. On the other hand, more than 50 percent of the genera in the modern fauna of South America are descendants of the North American invaders. This number is strongly affected by the presence of some 40 genera of cricetid rodents that radiated in South America from northern immigrants but is also influenced by diverse canids, felids, and cervids. The most conspicuous change was the gradual decline of South American ungulates during the interchange and, finally, the extinction of the last three genera in the late Pleistocene.

SUMMARY

Nearly all medium- to large-sized placental herbivores belong to a single assemblage, the ungulates, which share a common ancestry among late Cretaceous and early Tertiary condylarths. We have not yet established the specific interrelationships among the many condylarth lineages.

Artiodactyls are the most diverse ungulate order. The most primitive known genus, *Diacodexis* from the earliest Eocene, was the size of a rabbit, with very long slender limbs. Specialization for cursorial locomotion appears to be the prime factor in the emergence of artiodactyls. They radiated extremely rapidly in the Eocene, with most major groups appearing by the end of that epoch. Artiodactyls are divided into three major groups: Suiformes, including pigs and the hippopotamus; the Tylopoda, including camels and llamas; and the Ruminantia, including tragulids, giraffes, deer, antelopes, and cattle. The position of a number of extinct groups remains contentious. The modern families of pigs and peccaries appeared in the early Oligocene, and the hippopotamus appears in the Miocene. Camels are known from the Upper Eocene. Ancestors of the ruminants are recognized in the late Eocene. The modern families differentiated within the Oligocene, with deer, giraffes, and bovids appearing in the Miocene.

Mesonychids evolved from the base of the ungulate assemblage but specialized their dentition to fill the role of large carnivores and scavangers in the Paleocene and Eocene. Mesonychids almost certainly share a close common ancestry with the cetaceans. Primitive whales appear at the end of the early Eocene in Pakistan. Eocene archaeocetes gave rise to the odontocetes and mysticetes by the early Oligocene, when both groups appear in the Southern Hemisphere.

We can trace the ancestry of perissodactyls to loxolophine arctocyonids at the base of the Paleocene. *Hyracotherium*, one of the earliest-known perissodactyls, is distinguished from phenacodontid condylarths by the unique saddle-shaped astragalo-navicular articulation. The ancestors of horses, tapirs, chalicotheres, brontotheres, and rhinoceroses all diverged within the Eocene. In contrast with artiodactyls, the diversity of perissodactyls decreased in the later Cenozoic.

Africa was partially isolated from the northern continents in the early Cenozoic when a number of unique groups evolved, including elephants, hyraxes, and embrithopods, which may be traced to Asian condylarths of the families Anthracobunidae and Phenacolophidae. The extinct aquatic order Desmostylia may have evolved from the elephantlike Moeritherioidea. Eocene sirenians are known from the margins of the Tethys Sea in northern Africa, Hungary, and Jamaica. The retention of five premolars suggests a very early divergence from primitive eutherians rather than close affinities with African ungulates. Aardvarks are known in Africa since the Lower Miocene, but they are thought to have evolved from the early condylarths without strong affinities with any of the other orders.

South America was effectively isolated from other continents throughout the Tertiary. Early condylarths gave rise to a large radiation of ungulate orders that are almost, if not entirely, restricted to that continent; these include the Notoungulata, Astrapotheria, Pyrotheria, and Litopterna. All may have risen from the didolodonts, which are closely related to the mioclaenine condylarths. The arctostylopids of Asia and North America may be related to the notoungulates.

At the end of the Cenozoic, the South American ungulate fauna became extinct, probably as a result of competition and predation from North American herbivores and carnivores that invaded across the Panama Land Bridge. A few South American rodent and edentate genera became established in North America, but many more North American forms are now present in the South.

The xenarthrans are a second group of primarily South American placentals. Their affinities with other orders remain unresolved. We know fairly advanced armadillos from the late Paleocene. Both armadillos and the closely related glyptodonts spread into North America with the emergence of the Panama Land Bridge in the late Pliocene.

Sloths appear in the late Eocene. The early genera were arboreal, but some retained a mosaic of ossicles within their skin. The modern tree sloths appear to have evolved separately from two groups of ground sloths: *Bradypus* from the megatherioids and *Choloepus* from the megalonychoids.

A member of the South American family of anteaters, the Myrmecophagidae, is known in Europe in the Middle Eocene. The ancestry of the Old World anteaters, the pangolins, is not known. The North American Paleocene and Eocene palaeanodonts may have evolved from the same ancestral groups as the xenarthrans, but this connection has not been firmly established.

REFERENCES

Ameghino, F. (1906). Les formations sedimentaires du Crétacé Supérieur et du Tertiare de Patagonie. *An. Mus. Nac. Buenos Aires*, **15**: 1–568.

Andrews, C. W. (1906). *A Descriptive Catalogue of the Tertiary Vertebrata of the Fayum, Egypt. British Museum (Natural History)*, London.

Barnes, L. G. (1984). Whales, dolphins and porpoises: Origin and evolution of the Cetacea. In P. D. Gingerich and C. E. Badgley (eds.), *Mammals. Notes for a Short Course. Univ. Tennessee Dept. Geol. Sci. Stud. Geol.*, **8**: 139–154.

Barnes, L. G., Domning, D. P., and Ray, C. E. (1985). Status of studies in fossil marine mammals. *Marine Mammal Science*, **1**: 15–53.

Barnes, L. G., and Mitchell, E. D. (1978). Cetacea. In V. J. Maglio and H. B. S. Cooke (eds.), *Evolution of African Mammals*, pp. 582–602. Harvard University Press, Cambridge.

Bell, R. H. V. (1969). The use of the herb layer by grazing ungulates in the Serengeti. In A. Watson (ed.), *Animal Populations in Relation to their Food Resources*, pp. 111–128. Symp. Brit. Ecol. Soc. Blackwell, Oxford and Edinburgh.

Black, C. C. (1978). Paleontology and geology of the Badwater Creek area, central Wyoming. Part 14. The artiodactyls. *Ann. Carnegie Mus.*, **47**(10): 223–259.

Bubenik, A. B. (1966). *Das Geweih*. P. Parey Verlag, Hamburg and Berlin.

Butler, P. N. (1952). Molarisation of the premolars in the Perissodactyla. *Proc. Zool. Soc. Lond.*, **121**: 819–843.

Butler, P. M. (1978). Chalicotheriidae. In V. J. Maglio and H. B. S. Cooke (eds.), *Evolution of African Mammals*, pp. 368–370. Harvard University Press, Cambridge.

Camp, C. L., and Smith, N. (1942). Phylogeny and functions of the digital ligaments of the horse. *Mem. Univ. Calif.*, **13**: 69–124.

Cappetta, H., Jaeger, J.-J., Sabatier, M., Sigé, B., Sudre, J., and Vianey-Liaud, M. (1978). Découverte dans le Paleocene du Maroc des plus anciens mammifères autheriens d'Afrique. *Géobios*, **11**: 257–263.

Cifelli, R. L. (1983a). Eutherian tarsals from the Late Paleocene of Brazil. *Amer. Mus. Novit.*, **2761**: 1–31.

Cifelli, R. L. (1983b). The origin and affinities of the South American Condylarthra and early Tertiary Litopterna (Mammalia). *Amer. Mus. Novit.*, **2772**: 1–49.

Cifelli, R. L., and Soria, M. F. (1983). Systematics of the Adianthidae (Litopterna, Mammalia). *Amer. Mus. Novit.*, **2771**: 1–25.

Ciochon, R. L., and Chiarelli, A. B. (eds.). (1980). *Evolutionary Biology of the New World Monkeys and Continental Drift*. Plenum, New York.

Colbert, E. H. (1941a). A study of *Orcyteropus gaudryi* from the Island of Samos. *Bull. Am. Mus. Nat. Hist.*, **78**: 305–351.

Colbert, E. H. (1941b). The osteology and relationships of *Archaeomeryx*, an ancestral ruminant. *Amer. Mus. Novit.*, **1135**: 1–24.

Coombs, M. C. (1978). Reevaluation of early Miocene North American *Moropus* (Perissodactyla, Chalicotheriidae, Schizotheriinae). *Bull. Carnegie Mus. Nat. Hist.*, **4**: 1–62.

Coombs, M. C. (1979). *Tylocephalonyx*, a new genus of North American dome-skulled chalicotheres (Mammalia, Perissodactyla). *Bull. Am. Mus. Nat. Hist.*, **164**: 1–64.

Coombs, M. C. (1983). Large mammalian clawed herbivores: A comparative study. *Trans. Am. Phil. Soc.*, **73**(7): 1–96.

Coombs, M. C., and Coombs, W. P. (1977). Dentition of *Gobiohyus* and a reevaluation of the Helohyidae (Artiodactyla). *J. Mamm.*, **58**: 291–308.

Coombs, M. C., and Coombs, W. P. (1982). Anatomy of the ear region of four Eocene artiodactyls: *Gobiohyus*, ? *Helohyus*, *Diacodexis* and *Homacodon*. *J. Vert. Paleont.*, **2**(2): 219–236.

Coppens, Y., Maglio, V. J., Madden, C. T., and Beden, M.

(1978). Proboscidea. In V. J. Maglio and H. B. S. Cooke (eds.), *Evolution of African Mammals*, pp. 336–367. Harvard University Press, Cambridge.

Coryndon, S. C. (1978). Hippopotamidae. In V. J. Maglio and H. B. S. Cooke (eds.), *Evolution of African Mammals*, pp. 483–495. Harvard University Press, Cambridge.

Dawson, M. R., and Krishtalka, L. (1984). Fossil history of the families of Recent mammals. In S. Anderson and J. Knox Jones, Jr. (eds.), *Orders and Families of Recent Mammals of the World*, pp. 11–57. Wiley, New York.

Domning, D. P. (1978). Sirenian evolution in the North Pacific Ocean. *Univ. Calif. Publ. Geol. Sci.*, **118**: 1–178.

Domning, D. P. (1982). Evolution of manatees: A speculative history. *J. Paleont.*, **56**(3): 599–619.

Domning, D. P., Morgan, G. S., and Ray, C. E. (1982). North American Eocene sea cows. *Smithsonian Contr. Paleo.*, **52**: 1–69.

Domning, D. P., Ray, C. E., and McKenna, M. C. (1986). Two new Oligocene Desmostylians and a discussion of tethytherian systematics. *Smithsonian Contr. Paleo.*, **59**: 1–56.

Edinger, T. (1948). Evolution of the horse brain. *Mem. Geol. Soc. Am.*, **25**: 1–177.

Eisenmann, V. (1980). Les chevaux (*Equus sensu lato*) fossiles et actuels: cranes et dents jugales supérieurs. 180 pp. *Cah. Paléont.* CNRS, Paris.

Emlong, D. (1966). A new archaic cetacean from the Oligocene of northwest Oregon. *Univ. Oregon, Bull. Mus. Nat. Hist.*, **3**: 1–51.

Emry, R. J. (1970). A North American Oligocene pangolin and other additions to the Pholidota. *Bull. Am. Mus. Nat. Hist.*, **142**: 455–510.

Fordyce, R. E. (1977). The development of the Circum-Antarctic Current and the evolution of the Mysticeti (Mammalia: Cetacea). *Paleogeography, Paleoclimatology, Paleoecology*, **21**: 265–271.

Fordyce, R. E. (1980). Whale evolution and Oligocene southern ocean environments. *Paleogeography, Paleoclimatology, Paleoecology*, **31**: 319–336.

Frick, C. (1937). Horned ruminants of North America. *Bull. Am. Mus. Nat. Hist.*, **69**: 1–669.

Gazin, C. L. (1955). A review of the upper Eocene Artiodactyla of North America. *Smithsonian Misc. Coll.*, **128**(8): 1–96.

Gazin, C. L. (1965). A study of Early Tertiary condylarthran mammal *Meniscotherium*. *Smithsonian Misc. Coll.*, **149**(2): 1–98.

Gazin, C. L. (1968). A study of the Eocene condylarthran mammal *Hyopsodus*. *Smithsonian Misc. Coll.*, **153**(4): 1–90.

Gentry, A. W. (1978). Bovidae. In V. J. Maglio and H. B. S. Cooke (eds.), *Evolution of African Mammals*, pp. 540–572. Harvard University Press, Cambridge.

Gillette, D. D., and Ray, C. E. (1981). Glyptodonts of North America. *Smithsonian Contr. Paleobiol.*, **40**: 1–255.

Gingerich, P. D. (1985). South American mammals in the Paleocene of North America. In F. G. Stehli and S. D. Webb (eds.), *The Great American Biotic Interchange*, pp. 123–137. Plenum, New York.

Gingerich, P. D., and Russell, D. E. (1981). *Pakicetus inachus*, a new archaeocete (Mammalia, Cetacea) from the early-middle Eocene Kuldana Formation of Kohat (Pakistan). *Univ. Mich. Contr. Mus. Paleont.*, **25**: 235–246.

Gingerich, P. D., Wells, N. A., Russell, D. E., and Shah, S. M. I. (1983). Origin of whales in epicontinental remnant seas: New evidence from the early Eocene of Pakistan. *Science*, **220**: 403–406.

Ginsburg, L., and Heintz, E. (1968). La plus ancienne antilope, *Eotragus artenensis* du Burdigalien d'Artenay. *Bull. Mus. Natn. Hist. Nat. Paris*, **40**: 837–842.

Golz, D. J. (1976). Eocene Artidactyla of southern California. *Nat. Hist. Mus. Los Angeles County, Sci. Bull.*, **26**: 1–85.

Gregory, W. K. (1951 and 1957). *Evolution Emerging*. Macmillan, New York.

Guthrie, D. A. (1968). The tarsus of early Eocene artiodactyls. *J. Mamm.*, **49**(2): 297–302.

Hamilton, W. R. (1973). The lower Miocene ruminants of Gebel Zelten, Libya. *Bull. Brit. Mus. (Nat. Hist.) Geol.*, **21**: 73–150.

Hamilton, W. R. (1978a). Cervidae and Palaeomerycidae. In V. J. Maglio and H. B. S. Cooke (eds.), *Evolution of African Mammals*, pp. 496–508. Harvard University Press, Cambridge.

Hamilton, W. R. (1978b). Fossil giraffes from the Miocene of Africa and a revision of the phylogeny of the Giraffoidea. *Phil. Trans. Roy. Soc., B*, **283**: 165–229.

Harris, J. M. (1976). Evolution of feeding mechanisms in the family Deinotheriidae (Mammalia: Proboscidea). *Zool. J. Linn. Soc.*, **56**: 331–362.

Harris, J. M. (1978). Deinotherioidea and Barytherioidea. In: V. J. Maglio and H. B. S. Cooke (eds.), *Evolution of African Mammals*, pp. 315–332. Harvard University Press, Cambridge.

Harris, J. M., and White, T. D. (1979). Evolution of the Plio-Pleistocene African Suidae. *Trans. Am. Phil. Soc.*, **69**(part 2): 1–128.

Hooijer, D. A. (1978). Rhinocerotidae. In V. J. Maglio and H. B. S. Cooke (eds.), *Evolution of African Mammals*, pp. 371–378. Harvard University Press, Cambridge.

Janis, C. (1976). The evolutionary strategy of the Equidae and the origins of rumen and cecae digestion. *Evolution*, **30**: 757–774.

Janis, C. (1982). Evolution of horns in ungulates: Ecology and paleoecology. *Biol. Rev.*, **57**: 261–318.

Janis, C. and Scott, K. (1987). The interrelationships of higher ruminant families with special emphasis on the members of the Cervoidea. *Bull. Amer. Mus. Nat. Hist.* (in press).

Kellogg, R. (1936). A review of the Archaeoceti. *Carnegie Inst. Wash. Publ.*, **482**: 1–366.

Kretzoi, M. (1953). A legidösebb magyar ösemlös-lelet. *Földtani Közlöny*, **83**(7–9): 273–277. (In Hungarian; Russian and French abstracts.)

Krishtalka, L., and Stucky, R. K. (1985). Revision of the Wind River fauna, early Eocene of Central Wyoming. Part 7. Revision of *Diacodexis* (Mammalia, Artiodactyla). *Ann. Carnegie Mus. Art.*, **14**: 413–486.

Langer, P. (1974). Stomach evolution in the Artiodactyla. *Mammalia*, **38**: 295–314.

Lavocat, R. (1958). Pyrotheria. In J. Piveteau (ed.), *Traité de Paléontologie*, 6, pp. 181–186. Masson, S.A., Paris.

Leinders, J. J. M., and Heintz, E. (1980). The configuration of the lacrimal orifices in pecorans and tragulids (Artiodactyla, Mammalia) and its significance for the distinction between Bovidae and Cervidae. *Beaufortia*, **30**(7): 155–160.

Li, C.-K., and Ting, S.-Y. (1983). The Paleogene mammals of China. *Bull. Carnegie Mus. Nat. Hist.*, **21**: 1–93.

Lilly, J. C. (1977). The cetacean brain. *Oceans*, **10**: 4–7.

Lipps, J. H., and Mitchell, E. D. (1976). Trophic model for the adaptive radiations and extinctions of pelagic marine mammals. *Paleobiology,* 2: 147–155.

MacFadden, B. J. (1976). Cladistic analysis of primitive equids, with notes on other perissodactyls. *Syst. Zool.,* 25: 1–14.

MacFadden, B. J. (1984a). Systematics and phylogeny of *Hipparion,* and *Cormohipparion* (Mammalia, Equidae) from the Miocene and Pliocene of the New World. *Bull. Am. Mus. Nat. Hist.,* 179(1): 1–196.

MacFadden, B. J. (ed.). (1984b). Origin and evolution of the Cenozoic vertebrate fauna of Middle America. *J. Vert. Paleont.,* 4(2): 169–283.

MacFadden, B. J. (1985). Patterns of phylogeny and rates of evolution in fossil horses: Hipparions from the Miocene and Pliocene of North America. *Paleobiology,* 11: 245–257.

MacFadden, B. J., and Frailey, C. D. (1984). *Pyrotherium,* a large enigmatic ungulate (Mammalia, *incertae sedis*) from the Deseadan (Oligocene) of Salla, Bolivia. *Palaeontology,* 27: 867–874.

MacFadden, B. J., and Skinner, M. F. (1981). Earliest holarctic hipparion, *Cormohipparion goorisi* n. sp. (Mammalia, Equidae), from the Barstovian (Medial Miocene) Texas Gulf coastal plain. *J. Paleont.,* 55(3): 619–627.

Maglio, V. J. (1973). Origin and evolution of the Elephantidae. *Trans. Am. Phil. Soc., New Ser.,* 63: 1–149.

Maglio, V. J. (1978). Patterns of faunal evolution. In V. J. Maglio and H. B. S. Cooke (eds.), *Evolution of African Mammals,* pp. 603–619. Harvard University Press, Cambridge.

Maglio, V. J., and Cooke, H. B. S. (eds.). (1978). *Evolution of African Mammals.* Harvard University Press, Cambridge.

Mahboubi, M., Ameur, R., Crochet, J.-Y., and Jaeger, J.-J. (1984). Earliest known proboscidean from early Eocene of northwest Africa. *Nature,* 308: 543–544.

Marshall, L. G., Webb, S. D., Sepkoski, J. J., Jr., and Raup, D. M. (1982). Mammalian evolution and the great American interchange. *Science,* 215: 1351–1357.

Martin, P. S., and Klein, R. G. (1984). *Quaternary Extinctions, a Prehistoric Evolution.* University of Arizona Press, Tucson.

Matsumoto, H. (1926). Contribution to the knowledge of the fossil Hyracoidea of the Fayum, Egypt, with description of several new species. *Bull. Am. Mus. Nat. Hist.,* 56(4): 253–351.

Matthew, W. D. (1918). Edentata. In W. D. Matthew and W. Granger (eds.), A Revision of the Lower Eocene Wasatch and Wind River Faunas. Part V. Insectivora (continued), Glires, Edentata. *Bull. Am. Mus. Nat. Hist.,* 38: 565–657.

Matthew, W. D. (1937). Paleocene faunas of the San Juan Basin, New Mexico. *Trans. Am. Phil. Soc.,* 30: 1–510.

Matthew, W. D., and Granger, W. (1925). Fauna and correlation of the Gashato Formation of Mongolia. *Amer. Mus. Novitates,* 186: 1–12.

McKenna, M. C. (1975). Toward a phylogenetic classification of the Mammalia. In W. P. Luckett and F. S. Szalay (eds.), *Phylogeny of the Primates: A Multidisciplinary Approach,* pp. 21–46. Plenum, New York.

McKenna, M. C. (1980). Early history and biogeography of South America's extinct mammals. In R. L. Ciochon and A. E. Chiarelli (eds.), *Evolutionary Biology of the New World Monkeys and Continental Drift,* pp. 43–77. Plenum, New York.

McKenna, M. C., and Manning, E. (1977). Affinities and palaeobiogeographic significance of the Mongolian Paleogene genus *Phenacolophus. Gébios, Mem. spec.,* 1: 61–85.

Meyer, G. E. (1978). Hyracoidea. In V. J. Maglio and H. B. S. Cooke (eds.), *Evolution of African Mammals,* pp. 284–314. Harvard University Press, Cambridge.

Miyamoto, M. M., and Goodman, M. (1986). Biomolecular systematics of eutherian mammals: Phylogenetic patterns and classification. *Sys. Zool.,* 35: 230–240.

Novacek, M. J. (1982). Information for molecular studies from anatomical and fossil evidence on higher eutherian phylogeny. In M. Goodman (ed.), *Macromolecular Sequences in Systematic and Evolutionary Biology,* pp. 3–41. Plenum, New York.

Novacek, M. J. (1986). The skull of leptictid insectivorans and the higher-level classification of eutherian mammals. *Bull. Amer. Mus. Nat. Hist.,* 183: 1–111.

Osborn, H. F. (1924). *Andrewsarchus,* giant mesonychid of Mongolia. *Amer. Mus. Novitates,* 146: 1–5.

Osborn, H. F. (1929). The titanotheres of ancient Wyoming, Dakota and Nebraska. *U.S. Geol. Surv., Mon.,* 55: 1–953.

Patterson, B. (1934). The auditory region of an upper Pliocene typotherid. *Field Mus. Nat. Hist., Geol. Ser.,* 6(5): 83–89.

Patterson, B. (1975). The fossil aardvarks (Mammalia: Tubulidentata). *Bull. Mus. Comp. Zool.,* 147(5): 185–237.

Patterson, B. (1977). A primitive pyrothere (Mammalia, Notoungulata) from the early Tertiary of northwestern Venezuela. *Fieldiana Geol.,* 33: 397–422.

Patterson, B. (1978). Pholidota and Tubulidentata. In V. J. Maglio and H. B. S. Cooke (eds.), *Evolution of African Mammals,* pp. 268–277. Harvard University Press, Cambridge.

Patterson, B., and Pascual, R. (1968). The fossil mammal fauna of South America. *Quart. Rev. Biol.,* 43: 409–451.

Patton, T. H., and Taylor, B. E. (1973). The Protoceratinae (Mammalia, Tylopoda, Protoceratidae) and the systematics of the Protoceratidae. *Bull. Am. Mus. Nat. Hist.,* 150: 347–414.

Paula Couto, C. de., (1952). Fossil mammals from the beginning of the Cenozoic in Brazil. Condylarthra, Litopterna, Xenungulata, and Astrapotheria. *Bull. Am. Mus. Nat. Hist.,* 99: 355–394.

Pivorunas, A. (1979). The feeding mechanisms of baleen whales. *Am. Scientist,* 67: 432–440.

Prothero, D. R. (1986). A new oromerycid (Mammalia, Artiodactyla) from the early Oligocene of Montana. *Jour. Paleon.,* 60: 458–465.

Prothero, D. R., Manning, E., and Fischer, M. (1986). The phylogeny of the ungulate mammals: Evidence from Paleontology and molecular biology. (Manuscript in preparation).

Prothero, D. R., Manning, E., and Hanson, C. B. (1986). The phylogeny of the Rhinocerotoidea. *Zool. Jour. Linn. Soc.,* 87: 341–366.

Radinsky, L. B. (1964). *Paleomoropus,* a new early Eocene chalicothere (Mammalia, Perissodactyla), and a revision of Eocene chalicotheres. *Amer. Mus. Novitates,* 2179: 1–28.

Radinsky, L. B. (1965). Evolution of the tapiroid skeleton from *Heptodon* to *Tapirus. Bull. Mus. Comp. Zool.,* 134: 69–106.

Radinsky, L. B. (1966). The adaptive radiation of the phenacodontid condylarths and the origin of the Perissodactyla. *Evolution,* 20: 408–417.

Radinsky, L. B. (1967). *Hyrachyus, Chasmotherium,* and the early evolution of helaletid tapiroids. *Amer. Mus. Novitates,* 2313: 1–13.

Radinsky, L. B. (1969). The early evolution of the Perissodactyla. *Evolution,* **23**: 308–328.

Radinsky, L. B. (1976). Oldest horse brains: More advanced than previously realized. *Science,* **194**: 626–627.

Radinsky, L. B. (1982). Evolution of skull shape in carnivores. 3. The origin and early radiation of the modern carnivore families. *Paleobiology,* **8**: 177–195.

Radinsky, L. B. (1984). Ontogeny and phylogeny in horse skull evolution. *Evolution,* **38**: 1–15.

Radinsky, L. B., and Ting, S. (1984). The skull of *Ernanodon,* an unusual fossil mammal. *J. Mamm.,* **65**: 155–158.

Reinhart, R. H. (1959). A review of the Sirenia and Demostylia. *Univ. Calif. Publ. Geol. Sci.,* **36**: 1–145.

Riggs, E. S. (1935). A skeleton of *Astrapotherium. Geol. Ser. Field. Mus. Nat. Hist.,* **6**: 167–177.

Riggs, E. S. (1937). Mounted skeleton of *Homalodotherium. Geol. Ser. Field Mus. Nat. Hist.,* **6**: 233–243.

Romer, A. S. (1966). *Vertebrate Paleontology* (3d ed). University of Chicago Press, Chicago.

Rose, K. D. (1978). A new Paleocene epoicotheriid (Mammalia), with comments on the palaeanodonts. *J. Paleont.,* **52**: 658–674.

Rose, K. D. (1979). A new Paleocene palaeanodont and the origin of the Metacheiromyidae (Mammalia). *Breviora,* **455**: 1–14.

Rose, K. D. (1981). Composition and species diversity in Paleocene and Eocene mammal assemblages: An empirical study. *J. Vert. Paleont.,* **1**(3–4): 367–388.

Rose, K. D. (1982). Skeleton of *Diacodexis,* oldest known artiodactyl. *Science,* **216: 621–623.**

Rose, K. D. (1985). Comparative osteology of North American dichobunid artiodactyls. *J. Paleont.,* **59**: 1203–1226.

Rose, K. D., and Emry, R. J. (1983). Extraordinary fossorial adaptations in the Oligocene palaeanodonts *Epoicotherium* and *Xenocranium* (Mammalia). *J. Morph.,* **175**: 33–56.

Russell, D. E. (1964). Les mammifères Paléocènes d'Europe. *Mem. Mus. Natl. Hist. Nat. Paris, Ser. C,* **13**: 1–324.

Schaeffer, B. (1947). Notes on the origin and function of the artiodactyl tarsus. *Amer. Mus. Novitates,* **1356**: 1–24.

Schmidt-Nielsen, K. (1964). *Desert Animals. Physiological Problems of Heat and Water.* Oxford University Press, New York.

Schoch, R. M., and Lucas, S. G. (1985). The phylogeny and classification of the Dinocerata (Mammalia, Eutheria). Bull. Geol. Inst. Univ. Uppsala, **11**: 31–58.

Schultz, C. B., and Falkenbach, C. H. (1968). The phylogeny of the oreodonts. Parts 1 and 2. *Bull. Am. Mus. Nat. Hist.,* **139**: 1–498.

Scott, W. B. (1888). On some new and little known creodonts. *J. Acad. Nat. Sci. Philadelphia,* **9**: 155–185.

Scott, W. B. (1894). The structure and relationships of *Ancodus. J. Acad. Nat. Sci. Philadelphia,* **9**: 461–497.

Scott, W. B. (1940). Artiodactyla. In W. B. Scott and G. L. Jepsen (eds.) *The Mammalian Fauna of the White River Oligocene. Part IV. Trans. Am. Phil. Soc., New Ser.,* **28**: 363–746.

Sen, S., and Heintz, E. (1979). *Palaeoamasia kansui* Ozansoy 1966, embrithopode (Mammalia) de l'Eocene d'Anatolie. *Ann. Paléont. (Vert.),* **65**: 73–91.

Simpson, G. G. (1931). *Metacheiromys* and the relationships of the Edentata. *Bull. Am. Mus. Nat. Hist.,* **59**: 295–381.

Simpson, G. G. (1941). Mounted skeleton and restoration of an early Paleocene mammal. *Amer. Mus. Novitates,* **1155**: 1–5.

Simpson, G. G. (1945). Principles of classification and a classification of mammals. *Bull. Am. Mus. Nat. Hist.,* **85**: 1–360.

Simpson, G. G. (1948). The beginning of the age of mammals in South America. Part 1. *Bull. Am. Mus. Nat. Hist.,* **91**: 1–232.

Simpson, G. G. (1961). *Horses.* Anchor Books, Garden City.

Simpson, G. G. (1967). The beginning of the age of mammals in South America. Part 2. *Bull. Am. Mus. Nat. Hist.,* **137**: 1–259.

Simpson, G. G. (1978). Early mammals in South America: Fact, controversy, and mystery. *Proc. Am. Phil. Soc.,* **122**: 318–328.

Simpson, G. G. (1980). *Splendid Isolation, the Curious History of South American Mammals.* Yale University Press, New Haven.

Sinclair, W. J. (1914). A revision of the bunodont Artiodactyla of the Middle and Lower Eocene of North America. *Bull. Am. Mus. Nat. Hist.,* **33**: 267–295.

Sloan, R. E., and Van Valen, L. (1965). Cretaceous mammals from Montana. *Science,* **148**: 220–227.

Solounias, N. (1982). Evolutionary patterns of the Bovidae (Mammalia). *Third North American Paleont. Conv., Proc.,* **2**: 495–499.

Stanley, S. M. (1974). Relative growth of the titanothere horn: A new approach to an old problem. *Evolution,* **28**: 447–457.

Stehli, F. G., and Webb, S. D. (eds.). (1985). *The Great American Biotic Interchange.* Plenum, New York.

Stock, C. (1925). Cenozoic gravigrade edentates of western North America with special reference to the Pleistocene Megalonychinae and Mylodontidae of Rancho la Brea. *Carnegie Inst. Wash., Publ.,* **331**: 1–206.

Storch, G. (1978). *Eomanis waldi,* ein Schuppentier aus dem Mittel-Eozän der "Grube Messel" bei Darmstadt (Mammalia: Pholidota). *Senckenbergiana lethaea,* **59**: 503–529.

Storch, G. (1981). *Eurotamandua jorensi,* ein Myrmecophagide aus dem Eozän der "Grube Messel" bei Darmstadt (Mammalia, Xenarthra). *Senckenbergiana lethaea,* **61**: 247–289.

Sudre, J. (1977). L'evolution du genre *Robiacina* Sudre 1969, et l'origine des Cainotheriidae; implications systématiques. *Géobios, Mém. spéc.,* **1**: 213–231.

Sudre, J. (1978). Les artiodactyles de l'Eocene moyen et Supérieur d'Europe occidentale (systematique et évolution). *Mem. et Travaux, Ecole Prat. Hautes Etudes, Inst. Montpellier,* **7**: 1–229.

Sudre, J. (1979). Nouveaux mammiferes Eocene du Sahara occidental. *Palaeovertebrata,* **9**: 83–115.

Szalay, F. S. (1969a). The Hapalodectinae and a phylogeny of the Mesonychidae (Mammalia, Condylarthra). *Amer. Mus. Novitates,* **2361**: 1–26.

Szalay, F. S. (1969b). Origin and evolution of function of the mesonychid condylarth feeding mechanism. *Evolution,* **23**: 703–720.

Szalay, F. S., and Gould, S. J. (1966). Asiatic Mesonychidae (Mammalia, Condylarthra). *Bull. Am. Mus. Nat. Hist.,* **132**: 127–174.

Thewissen, J. G. M. (1985). Cephalic evidence for the affinities of Tubulidentata. *Mammalia,* **49**: 257–284.

Thewissen J. G. M., Russell, D. E., Gingerich, P. D., and Hussain, S. T. (1983). A new dichobunid artiodactyl (Mammalia) from the Eocene of north-west Pakistan: Dentition and classification. *Proc. Koninklijke Nederlandse Akad. van Wetenschappen, ser. B,* **86**: 153–180.

Thomason, J. J. (1986). The functional morphology of the manus in the tridactyl equids *Merychippus* and *Mesohippus:* Paleontological inferences from neontological models. *Jour. Vert. Pal.,* **6**: 143–161.

Trofimov, B. A. (1958). New Bovidae from the Oligocene of Central Asia. *Vert. Palasiatica,* **2**: 243–247.

Van Valen, L. (1971a). Toward the origin of artiodactyls. *Evolution,* **25**: 523–529.

Van Valen, L. (1971b). Adaptive zones and the orders of mammals. *Evolution,* **25**: 420–428.

Van Valen, L. (1978). The beginning of the age of mammals. *Evol. Theory,* **4**: 45–80.

Vaughan, T. A. (1978). *Mammology.* Saunders, Philadelphia.

Viret, J. (1961). Artiodactyla. In J. Piveteau (ed.) *Traité de Paléont.,* 6, pp. 887–1021. Masson, S. A., Paris.

Webb, S. D. (1972). Locomotor evolution in camels. *forma et functio,* **5**: 99–112.

Webb, S. D. (1976). Mammalian faunal dynamics of the great American interchange. *Paleobiology,* **2**: 220–234.

Webb, S. D. (1977). A history of Savanna vertebrates in the New World. Part I. North America. *Ann. Rev. Ecol. Syst.,* **8**: 355–380.

Webb, S. D. (1978). A history of Savanna vertebrates in the New World. Part II. South America and the Great Interchange. *Ann. Rev. Ecol. Syst.,* **9**: 393–426.

Webb, S. D. (1981). *Kyptoceras amatorum,* new genus and species from the Pliocene of Florida, the last protoceratid artiodactyl. *J. Vert. Paleont.,* **1**(3–4): 357–365.

Webb, S. D. (1986). On the interrelationships of tree sloths and ground sloths. In G. G. Montgomery (ed.), *The Evolution and Ecology of Armadillos, Sloths, and Vermilinguas,* pp. 105–112. Smithsonian Institute Special Publications, Washington, D.C.

Webb, S. D., and Marshall, L. G. (1982). Historical biogeography of Recent South American land mammals. In M. A. Mares and H. H. Genoways (eds.), *Mammalian Biology in South America. Pymatuning Symp. Ecol., Spec. Publ. Ser.,* **6**: 39–52.

Webb, S. D., and Taylor, B. E. (1980). The phylogeny of hornless ruminants and a description of the cranium of *Archaeomeryx. Bull. Am. Mus. Nat. Hist.,* **167**: 121–157.

Wells, N. A., and Gingrich, P. D. (1983). Review of Eocene Anthracobunidae (Mammalia, Proboscidea) with a new genus and species, *Jozaria palustris,* from the Kuldana Formation of Kohat (Pakistan). *Univ. Michigan, Contr. Mus. Paleont.,* **26**: 117–139.

West, R. M. (1983). South Asian middle Eocene moeritheres (Mammalia: Tethytheria). *Ann. Carnegie Mus. Nat. Hist.,* **52**: 359–373.

West, R. M. (1984). Paleontology and geology of the Bridger Formation, Southern Green River Basin, southwestern Wyoming. Part 7. Survey of Bridgerian Artiodactyla, including description of a skull and partial skeleton of *Antiacodon pygmaeus. Milwaukee Public Museum, Contr. Biol. Geol.,* **56**: 1–47.

Wheeler, W. H. (1961). Revision of the uintatheres. *Bull. Peabody Mus. Nat. Hist.,* **14**: 1–93.

Wilson, J. A. (1971). Early Tertiary vertebrate faunas, Vieja Group, Trans Pecos-Texas: Agriochoeridae and Merycoidodontidae. *Texas Mem. Mus. Bull.,* **18**: 1–83.

Wilson, J. A. (1974). Early Tertiary vertebrate faunas, Vieja Group and Buck Hill Group, Trans-Pecos Texas: Protoceratidae, Camelidae, Hypertragulidae. *Texas Mem. Mus. Bull.,* **23**: 1–34.

Zapfe, H. (1979). *Chalicotherium grande* (Blainv.) aus der miozänen Spaltenfüllung von Neudorf an der March (Devinska Nova Ves) Tschechoslowakei. *Neue Denkschriften Naturhist. Mus. Wien,* **2**: 1–282.

Zhang, Y. (1978). Two new genera of condylarthran phenacolophids from the Paleocene of Nanxiong Basin, Guangdong. *Vert. Palasiatica,* **16**: 267–274.

Zhang, Y. (1980). *Minchenella,* a new name for *Conolophus Zhang. Vert. Palasiatica,* **18**(3): 257.

CHAPTER XXII

Evolution

This book has been devoted primarily to an historical account of the pattern of vertebrate evolution. The fossil record also provides the basis for establishing the nature of long-term processes of evolution. Knowledge of fossil vertebrates makes it possible to determine the rates of evolutionary change and evaluate what factors are involved in major adaptational changes, the evolution of new structures, and the origin of major taxonomic groups.

EVOLUTION AT THE SPECIES LEVEL

The nature of evolutionary patterns and processes is currently being debated more vigorously than at any other time since the publication of Darwin's *The Origin of Species*. These arguments involve both the rate of evolution and the significance of natural selection as a determining factor.

Darwin (1859) argued that the course of evolution resulted primarily from natural selection acting on variations within populations. He believed that this process led to gradual and progressive changes that could account for all organic diversity.

Mendelian and population genetics have since provided knowledge of the way in which variability is inherited and a means for establishing the relative significance of mutation rates, selection coefficients, and population size in determining the direction and rate of evolution. Fisher (1930), Wright (1931), Dobzhansky (1937), Huxley (1942), Mayr (1942), and Simpson (1944, 1953) integrated this information with knowledge of the fossil record to establish a comprehensive or synthetic theory that has dominated evolutionary thought since the 1940s. The modern evolutionary synthesis has since incorporated the recent advances in molecular biology but still emphasizes the role of selection acting at the population level initially elaborated by Darwin (Dobzhansky, Ayala, Stebbins, and Valentine, 1977; Stebbins and Ayala, 1981; Mayr, 1982).

PHYLETIC EVOLUTION AND PUNCTUATED EQUILIBRIUM

Most detailed studies of evolutionary processes have been carried out in the laboratory and through field studies of modern populations. In rapidly reproducing species, laboratory studies may involve hundreds of generations and field studies may cover several decades of observations. At these time scales, biologists can show that selection is significant in producing progressive change in allele frequencies of characters that are important for the survival of populations under differing environmental conditions (Endler, 1986). Well-established examples include the spread of alleles for dark pigmentation among moths in industrial areas of Europe during the past 150 years and differing frequencies of alleles for color patterns in the snail genus *Cepaea,* which lives in a variety of environments that are subject to seasonal change (*Biological Journal of the Linnean Society,* nos. 3 and 4, 1980).

Observations from contemporary populations have been extrapolated to longer time scales, and most textbooks argue that evolution over millions and tens of millions of years probably followed a pattern of gradual, progressive change. Evidence from the fossil record, especially that of marine invertebrates, appears to show a much different pattern. Eldredge and Gould (1972) pointed out that there are very few adequately documented cases among fossil invertebrates in which species undergo significant change through time. Most common species appear to retain a particular morphological pattern for millions of years without significant change. When change does occur, it appears to require only a relatively short period of time compared with the total longevity of the species.

Eldredge and Gould coined the term **punctuated equilibrium** to describe a pattern of evolution in which most change occurs extremely rapidly at the time of the initiation of new species, while most of their history passes with very little change (Figure 22-1*a*). This pattern has been further discussed in a series of subsequent papers (Gould and Eldredge, 1977; Eldredge, 1984, 1985; Stanley, 1975, 1981, 1982; Eldredge and Stanley, 1984b; Gould, 1980, 1982a,b).

Most biologists consider that evolutionary change has two components—**phyletic evolution** and **speciation.** Phyletic evolution (also termed **anagenesis**) is the process envisaged by Darwin in which a single lineage changes over time, gradually becoming so distinct that ancestral and descendant portions of the lineage are recognized as different species (Figure 22-1*b*). Darwin believed that this process could also account for the origin of higher categories—families, orders, and classes. Although new species may be produced through phyletic evolution, the term speciation is applied to a distinct process in which a single species divides into two lineages that become reproductively isolated from one another. Phyletic evolution accounts for change, but the process of speciation (or **cladogenesis**) is necessary to explain the diversity of organisms.

Neither Darwin nor most authors of the modern synthesis emphasize speciation as a particularly important factor in accounting for progressive morphological change. In contrast, Eldredge, Gould, and Stanley, who find little evidence for significant modification within species, argue that speciation plays a dominant role in evolutionary change.

If one could demonstrate that most species change very little for most of their history but retain an essentially static morphology, it would imply a very different pattern of long-term evolution than that envisaged by Darwin and accepted by most modern biologists.

The work of Eldredge, Gould, and Stanley has provided a great impetus for reevaluation of evolutionary patterns in many groups of organisms (Bergren and Casey, 1983; Cope and Skelton, 1985). The fossil record of vertebrates has the potential to test this hypothesis, while providing additional insights into other evolutionary processes.

THE NATURE OF THE FOSSIL RECORD

It would seem a simple matter to establish whether species change progressively or remain stable for long periods of time. In fact, only rarely are geological processes sufficiently regular and continuous to lead to the preservation of numerous fossils that are evenly distributed throughout

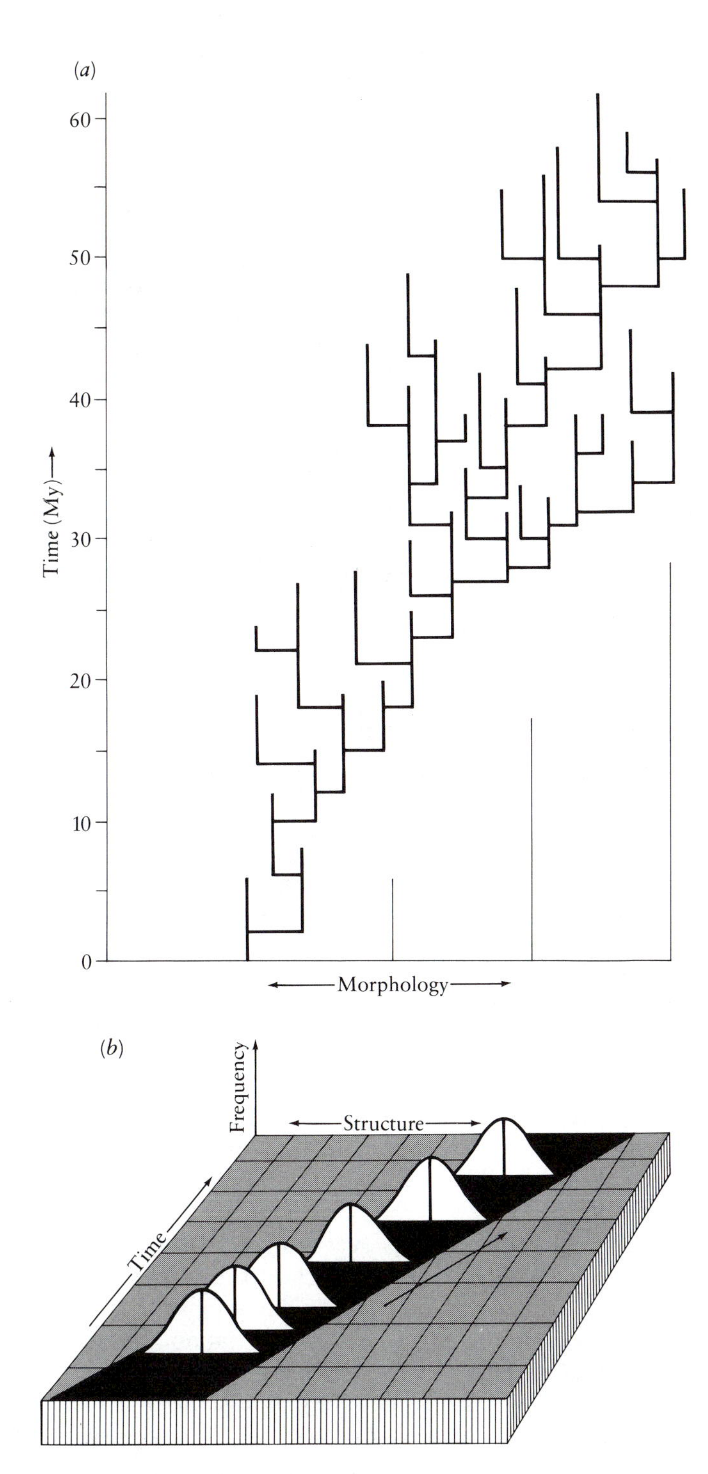

Figure 22-1. (*a*) The pattern of evolution termed "punctuated equilibrium" by Eldredge and Gould, in which most morphological change occurs during the process of speciation and most of the duration of a species is characterized by stasis. *From Stanley, 1979.* (*b*) The pattern of morphological change termed "phyletic evolution," as envisaged by Darwin, in which gradual change occurs throughout the duration of a species. *From Moore, Lalicker, and Fischer, 1952. With permission of McGraw-Hill Book Company.*

the time range of species. Most fossiliferous horizons record only an extremely short period of time and are separated by long gaps from other productive zones. Because of this problem, few paleontologists have been concerned with change within species and most simply assumed that it followed the pattern hypothesized by Darwin.

One of the major reasons for the continuing debate regarding the rates and patterns of evolution is certainly the incompleteness of the fossil record. In Darwin's time, no evolving sequences had been discovered. At the present time, the fossil record provides a good framework of evolutionary patterns, but significant gaps remain between many of the major groups and between most well-known species and genera. Where dating is possible, most depositional sequences show significant gaps (Dingus and Sadler, 1982; Schindel, 1982; Dingus, 1984), which may give the impression of sudden appearance of taxa but provides no evidence of evolutionary rates or processes. This problem is clearly evident even in the best-known and most thoroughly studied sequences of Cenozoic mammals (Gingerich, 1977, 1982). The problem of gaps is particularly evident in the fossil record of Mesozoic mammals (see Figure 18-14), but we should recognize that a similar, if less extreme, condition holds for all Mesozoic tetrapods. The problems are even greater in the Paleozoic, where most species and genera are reported from only a single locality (Carroll, 1977, 1984). Such a record does not demonstrate anything about evolutionary processes at the level of the genus and species.

There are exceptions. In some lake deposits, sedimentation has been extremely regular, recording what appear to be yearly or seasonal sequences. A notable example is provided by the lakes of the Newark supergroup in the late Triassic and early Jurassic of eastern North America (McCune, Thomson, and Olsen, 1984). In individual basins of deposition, each year of sedimentation may be recorded, but the total time interval for each lake is only about 20,000 years. There is little evidence of gradual, progressive evolution in these lakes, but this may be attributed to the relatively short length of the total cycle or the lack of sediments from the earliest horizons, when the original colonization and adaptation would have occurred. Bell, Baumgarten, and Olson (1985) described gradualistic patterns of evolution in six characters of the three-spined stickleback, *Gasterosteus doryssus,* over a period of approximately 110,000 years in a middle-to-late Miocene saline lake in Nevada.

The fossil record is incomplete in space as well as time. In modern species, clinal variation is well documented in many wide-ranging species that occur in areas with progressively changing environmental conditions (Endler, 1977). There are few individual fossiliferous areas that are so extensive as to demonstrate geographical variation. The fossil record may include deposits of similar ages in different continents, but rarely are there enough intervening deposits known to document clinal variation.

Even in well-documented sequences, species and genera commonly appear suddenly in the fossil record. This pattern may be attributed to sudden evolution within the area being sampled, but it can almost always be accounted for by migration from some other part of the world.

The rarity of relatively complete stratigraphic sequences makes it difficult to document the nature of evolutionary change within species. On the other hand, it should be possible to demonstrate long-term stasis even if the stratigraphic record is relatively incomplete. However, even this requires relatively complete fossil remains.

In addition to the total absence of evidence from many horizons and geographical areas, the fossils that are known are often very incomplete. Among mammals, a single tooth may be sufficient to document the occurrence of a species, which explains the apparently excellent record of late Cenozoic mammals (Kurtén, 1968; Kurtén and Anderson, 1980). However, such scanty evidence is not sufficient to demonstrate evolutionary rates. The remainder of the skeleton from different horizons or localities may differ dramatically or not at all.

A particularly important bias of the fossil record is expressed against organisms that are rare and/or geographically restricted. There is obviously a direct correlation between the total numbers, longevity, and geographical distribution of species and the likelihood of their being represented in the fossil record. Most fossils that are known were probably members of species that were common, widespread, and long-lived. Eldredge, Gould, and Stanley, as well as Mayr (1982), have argued that evolutionary change is most likely to occur through rapid change within small, geographically isolated populations. If this thesis is correct, the most important portion of the biota for establishing the patterns and processes of evolution is the least likely to be represented in the fossil record.

In order to establish evolutionary patterns and rates at the species level, some requirements must be met:

1. The sedimentary record must be relatively continuous, with gaps of no more than 10,000 to 20,000 years.
2. The total sequence must be fairly long, probably in excess of 100,000 years.
3. The sequence must be adequately dated, ideally by radiometric methods.
4. If one is attempting to establish change within a species, the entire species range should be represented. This can be established only if both ancestral and descendant species are known in the same geographical area.
5. The geographical range of the species should also be known with some assurance.
6. Ideally, a significant proportion of the skeletal anatomy should be known.

Not surprisingly, these conditions are almost never completely met. Only under exceptional circumstances is it possible to measure evolutionary rates at the species level, so as to test empirically the hypothesis elaborated by Eldredge, Gould, and Stanley.

EVIDENCE FROM THE FOSSIL RECORD

Because of the importance to evolutionary theory of establishing whether or not significant change does occur within species, numerous studies have been carried out in recent years. The difficulty of analyzing data from the fossil record is illustrated by a series of studies of fossil hominids from the Pliocene and Pleistocene. Using essentially the same information, Rightmire (1981) concluded that there was almost no change in the size of the braincase within the species *Homo erectus,* while Cronin and his colleagues (1981) concluded that there was progressive change from *Australopithecus* to *Homo sapiens.* Wolpoff (1984) undertook a further study of *Homo erectus.* From analysis of 13 characters of the skull and lower jaw, he concluded that they showed a variety of different rates, but most exhibited gradual and progressive change linking *Homo habilis* and *Homo sapiens.*

The most detailed and extensive study, which was designed specifically to test the model of punctuated equilibrium among fossil vertebrates, is that of Gingerich (1976) in the early Cenozoic of the Big Horn Basin in the Western United States. The genera *Hyopsodus, Haplomylus,* and *Pelycodus* have a rich fossil record in the Lower Eocene Willwood Formation, where approximately 3.5 million years of deposition are represented by 1500 meters of fluvial sediments. Morphometric analysis of molar teeth demonstrates both phyletic evolution and speciation. All three genera are known from sequential series of species showing gradual increase in size over time. Most of the species had previously been recognized on the basis of a few, clearly distinct specimens from widely separated horizons. Study of hundreds of specimens from intervening levels demonstrated several continuous morphological series that can be divided into separate species only arbitrarily (Figure 22-2).

Divergence of lineages over time is observed in both *Pelycodus* and *Hyopsodus.* Evidence from a single geographical area makes this appear like sympatric speciation. Mayr (1963) emphasized that speciation almost always requires geographical isolation of populations over a significant period of time. We may assume that these fossils record a later period of evolution, subsequent to allopatric speciation, when the species had again come into contact. The morphological divergence can be attributed to selection pressure for changes that enabled them to coexist without having to compete for the same resources.

The geographical ranges of the fossil species studied by Gingerich were almost certainly much larger than the localities in which their remains were found. Populations of a species may have been isolated in other areas for thousands or tens of thousands of years and then reappeared in the same area as the parent species. The time required for the development of mechanisms to promote reproductive isolation may have been shorter than the

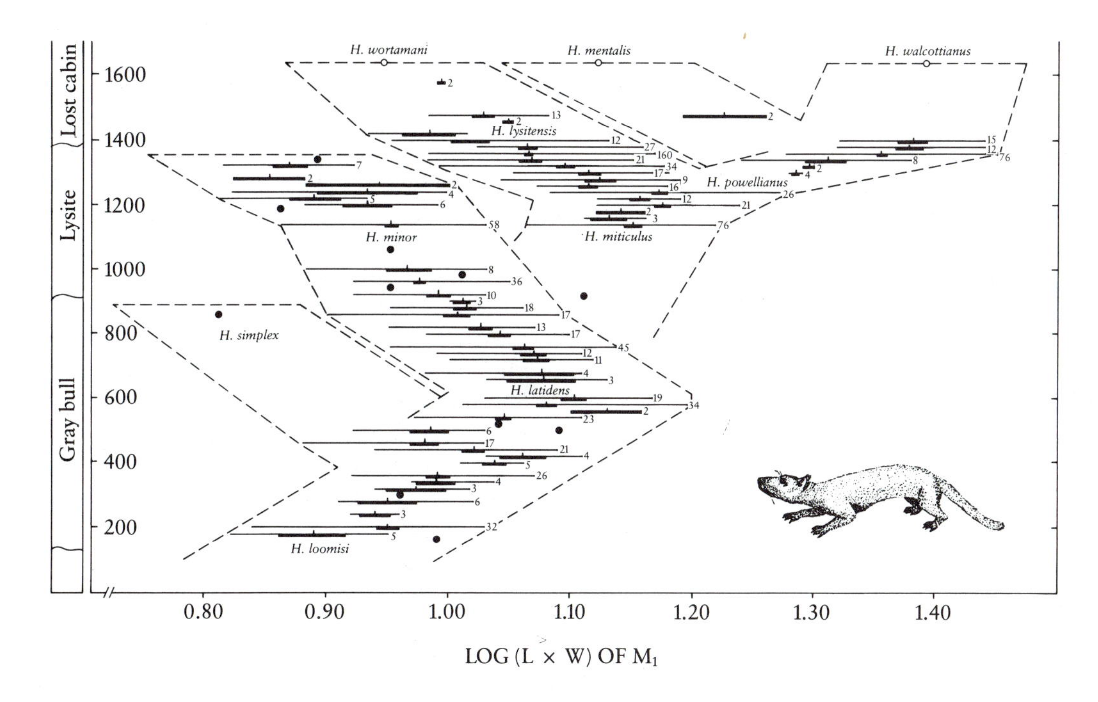

Figure 22-2. STRATIGRAPHIC RECORD OF EARLY EOCENE *HYOPSODUS* IN NORTHWESTERN WYOMING. Figure shows variation and distribution of tooth size in samples from many levels spanning most of the early Eocene. Specimens come from localities at approximately 6-meter intervals in or near a measured stratigraphic section totaling about 500 meters in thickness. The small species *Hyopsodus loomisi* became larger gradually through time, until it differed sufficiently to be recognized as a different species *H. latidens*. *H. "simplex"* was another early species derived from *H. loomisi*. *H. latidens* apparently gave rise to both *H. minor* and *H. miticulus,* which in turn gave rise to *H. lysitensis* and *H. powllianus*. Note the regular pattern of divergence in tooth size (and by inference body size) in pairs of sympatric sister lineages. *From Gingerich, 1976.*

intervals recorded in the sedimentary record or led to little morphological change.

The rate of evolutionary change for specific characters in diverging species does not appear higher than that for characters changing within a single lineage. In these examples, speciation does not appear to introduce factors of morphological change that are different from those that occur through phyletic evolution (Gingerich, 1982).

Additional work by Gingerich (1980) in a wider area of western North America that combines both late Paleocene and early Eocene sediments deposited over approximately 12.5 million years documents 24 species that have arisen by phyletic evolution and 14 cases in which species appear suddenly in the fossil record; the latter events are attributed to immigration. We should note that not all species showed progressive change; some did exhibit stasis for variable periods of time (see also West, 1979).

Unfortunately, there are inherent limits to the degree of resolution in fluvial sediments from which come most fossils of Cenozoic mammals. Gingerich (1982) observed that sedimentation does not provide a consistent record over intervals of less than 5000 to 6000 years duration. Lake deposits have the potential for analysis of changes occuring from year to year, although their total duration is usually much shorter than the average longevity of individual species (McCune, Thomson, and Olsen, 1984; Bell, Baumgarten, and Olson, 1985).

At a higher taxonomic level, Rose and Brown (1984) documented evidence of progressive change between genera on the basis of over 600 specimens of primates collected from a 700-meter-thick sequence representing approximately 4 million years of the Eocene. One example is provided by marked changes in the dentition between *Tetonius homunculus* and *Pseudotetonius ambiguus* (Figure 22-3). There is no evidence of a change in evolutionary rate across the transition; it is smoothly gradational. Numerous character changes were noted, not all of which were synchronous.

Other well-documented examples of progressive morphological change within species and genera are provided by Hürzeler (1962, mid-Cenozoic rabbits), Chaline and Laurin (1986, Plio-Pleistocene rodents), Fahlbusch (1983, Miocene rodents), Harris and White (1979, African suids from the Plio-Pleistocene), MacFadden (1985, hipparion horses), and Krishtalka and Stucky (1985, early Eocene artiodactyls). Harris and White (p. 89) observed: "Many changes in size and morphological attributes in the various suid taxa appear to conform to simple Darwinian directional selection processes resulting in gradualistic change in population means and ranges through time."

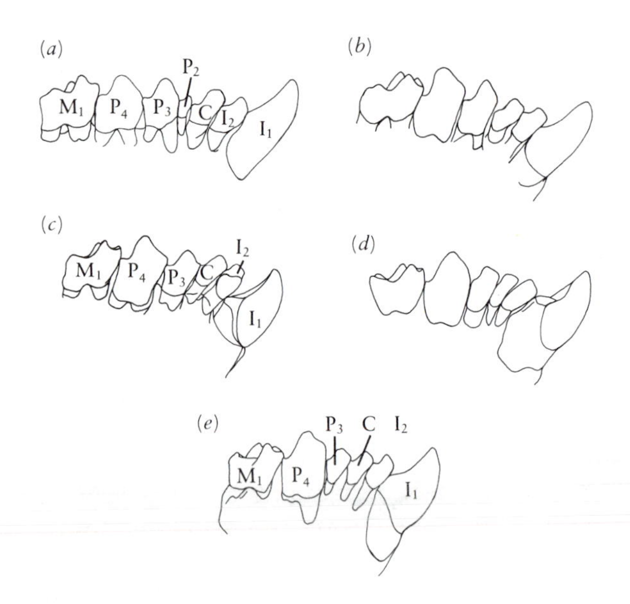

Figure 22-3. GRADUAL EVOLUTION OF ANTERIOR LOWER DENTITION. From the *Tetonius homunculus–P. ambiguus* lineage from the Central and Southern Bighorn Basin. Stages in the transition are arbitrarily delimited by slight morphological differences and/or intervals with little or no data, and each is a composite dentition (M_{2-3} omitted). Based on all known material from the indicated interval. *From Rose and Bown, 1984. Reprinted by permission from* Nature. *Copyright 1984, Macmillan Journals Ltd.*

LONGEVITY OF SPECIES AND GENERA AS A MEASURE OF EVOLUTIONARY RATES

For most species, accurate establishment of rates and patterns of morphological change is not possible on the basis of the current knowledge of the fossil record. For this reason, Gould, Eldredge, and Stanley attempted to determine evolutionary rates on the basis of published data on the longevity of genera and species. Stanley (1982) accepts an average species longevity of Cenozoic mammals of 2 to 3 million years and an average longevity of 8 million years for genera. These data are based primarily on Romer (1966), Van Valen, (1973), Kurtén (1968), and Kurtén and Anderson (1980). Stanley assumed that the recognition of species and genera depends on their morphological stability through time. If genera and species do not undergo change during their recognized longevity, evolution can only have occurred during the short, unrecorded intervals between species and genera.

This interpretation of these data appears to demonstrate punctuated equilibrium at the level of the genus and species for the Class Mammalia throughout the Cenozoic. In fact, it simply reflects the conventions of classification.

All identifiable fossils must be placed in one species or another. There is no formal way to indicate the position of specimens that are intermediate between the typical members of two related species. For the purpose of classification, all biological information is digitized. This procedure has led to a generally useful system of classification, but it cannot reflect gradual evolutionary change. It is inevitable that Stanley would find a punctuational pattern on the basis of the Linnean classification of Cenozoic mammals.

The problems of using taxonomic units to demonstrate a pattern of punctuated equilibrium is graphically illustrated by Stanley's comments on evolution within the family Elephantidae, based on the work of Maglio (1973). Stanley states (1981, p. 100): "What stands out in the phylogeny of the elephants is a pattern strongly indicative of punctuated evolution. All three advanced genera descended from the ancestral genus *Primelephas,* and all three appear abruptly and almost simultaneously in the fossil record." This observation appears to have been based on Maglio's phylogeny, which is reproduced here in Figure 22-4.

It may be noted that the "sudden" appearance of the genera *Loxodonta, Elephas,* and *Mammuthus* occurs after a 500,000-year gap in the fossil record. Maglio (1973, p. 64) clearly viewed the differences between the early members of these lineages as being at the level of the species: "*Primelephas* was succeeded by three species, each at the base of a distinct phyletic complex. For this reason they have been placed in three genera, *Loxodonta, Elephas* and *Mammuthus.*" The generic distinction is made for the sake of taxonomic convenience and does not demonstrate major morphological differences. It is clear from the text and several illustrations showing changes in dental features that the three generic lineages diverged from

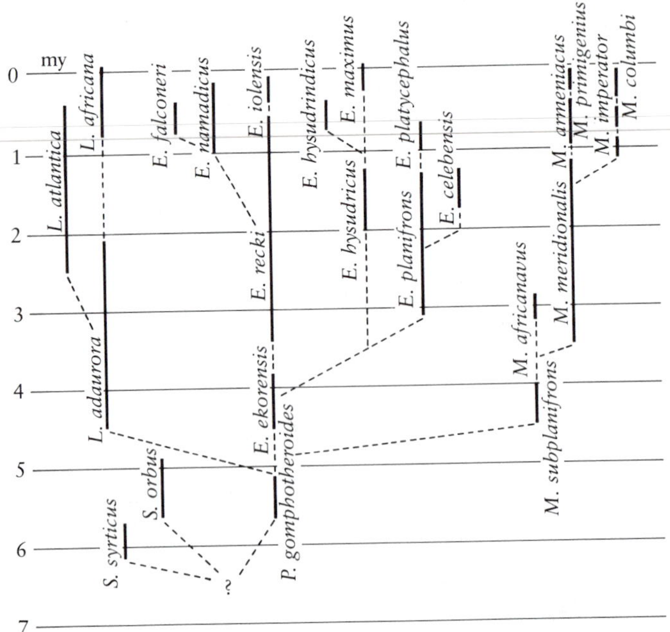

Figure 22-4. PHYLOGENY OF THE ELEPHANTINAE. Generic abbreviations: *E—Elephas; L—Loxodonta; M—Mammuthus; P—Primelephas; S—Stegotetrabelodon.* Vertical alignment of chronospecies depicts apparent phyletic transition. *From Maglio, 1973.*

a common morphological pattern represented by the putative ancestor *Primelephas*.

The subsequent evolution of the three genera illustrates different patterns. The early species of *Loxodonta*, *L. adaurora*, shows little change over a period of 2 million years. In general, the dentition of *Loxodonta* is very conservative, but the length of the lower jaw is much reduced within the span of the genus. Within the genus *Elephas*, species demonstrate continuous change over a period of 4.5 million years. Maglio (1973, p. 81) stated:

> *Elephas recki* is the best known species of *Elephas* from Africa; it can be traced through a progressive series of stages from an early form in the middle Pliocene of Kikagati to a rather progressive form in the middle Pleistocene of Bed IV at Olduvai Gorge. These stages pass almost imperceptibly into each other, as can be seen in the successive populations known from Kaiso, Omo, Laetolil, Koobi Fora and Illeret, Ouadi Derdemi, Olduvai, Kanjera and Ologesailie, among others. The molar teeth show progressive increase in the number of plates, their relative height and spacing, reduction in enamel thickness, increased folding of enamel and loss of the median loop on the wear figures. The skull shows all of the modifications already begun in *E. ekorensis* but here they are developed even further. On stratigraphic and faunal evidence the earliest stage of *E. recki* (Kikagati) would have been later than the latest record of *E. ekorensis*. At present, a direct phyletic relationship between *E. ekorensis* and *E. recki* seems certain.
>
> In the late Pleistocene of Africa a more progressive elephant appears which I retain as a distinct species, *Elephas iolensis*, only as a matter of convenience. Although as a group, material referred to *E. iolensis* is distinct from that of *E. recki*, some intermediate specimens are known and *E. iolensis* seems to represent a very progressive, terminal stage in the *E. recki* specific lineage.

Attention to anatomical detail, rather than to taxonomic designation, shows that the elephants provide excellent evidence of significant morphological change within species, through species within genera, and through genera within a family.

Clearly, it is not possible to estimate rates of morphological change on the basis of the longevity of genera and species as those data are usually presented. The rate of morphological change can only be established through detailed study of unusually complete fossil sequences. There is considerable evidence from Tertiary mammals that significant change does occur during the duration of species, as they are typically recognized, and this change can account for the emergence of new species and genera.

There are no cases among vertebrates that clearly document extremely rapid evolution from one species to another, although most transitions are without an adequate fossil record.

The most misleading aspect of this debate is the assertion that evolutionary rates have been predominantly either gradual *or* punctuated. As these terms are used by Eldredge, Gould, and Stanley, one is presented with a choice between two extremes—either evolutionary change is slow and continuous (a gradual process at a constant rate) or evolutionary rates alternate between very short episodes of rapid change and long periods of stasis. There is no evidence from the study of modern populations or the fossil record of vertebrates that evolutionary rates conform to any such stereotyped patterns.

One of the most obvious features demonstrated by the fossil record is that the rate of evolutionary change has been extremely irregular at all taxonomic levels. Simpson (1944) coined the terms **bradytely, horotely,** and **tachytely** to refer to slow, medium, and rapid rates of evolution and elaborated the concept of quantum evolution for extremely rapid shifts between one adaptive zone and another.

As study of the fossil record is concentrated on smaller and smaller time units, it is possible to measure rates of change both within and between species that reveal a similarly broad spectrum of rates at the species level (Gingerich, 1982; Maglio, 1973; Wolpoff, 1984).

The writings of Gould, Eldredge, and Stanley emphasize the consistency of evolutionary patterns at the level of the species as if most of the characters of the species were evolving at the same rate. As Wolpoff (1984) emphasized in the case of the genus *Homo*, various characters are evolving at significantly different rates. This is a common feature throughout the history of vertebrates. It is clearly evident among the common ungulates of the northern continents. As was outlined in Chapter 21, rates of evolution are different for cranial features, the dentition, and locomotor apparatus within single lineages, between different lineages, and at different times within a single lineage.

If a single word is to be used to describe evolutionary rates, it is certainly neither *gradual* nor *punctuational*, but *irregular* or *opportunistic*.

QUANTITATIVE MEASURES OF EVOLUTIONARY RATES

It may eventually be possible to clarify this problem by more extensive quantitative comparison of rates. Haldane (1949) established the **darwin** as a unit for measuring evolutionary change. A darwin (*d*) is defined as a change by a factor of *e* per million years, where *e* is the base of natural logarithms. The logarithmic scale is appropriate since proportional rather than absolute change is of interest.

Gingerich (1983) emphasized that the apparent rate of evolution differs significantly over different time ranges. A rate of approximately 0.1 darwin is common for fossil vertebrates measured over a period of approximately 1 million years (for example, the early Tertiary mammals studied by Gingerich, 1982, and fossil hominids studied by Wolpoff, 1984). For change over a 10-million-year period, Gingerich calculates that the average rate is close to 0.02 *d*. Measures of change during the Pleistocene glaciation average close to 4 *d*, and for historical colonization, approximately 400 *d*. At this rate, a mouse could give rise to an elephant in 10,000 years. He cites rates up to 60,000 darwins for changes resulting from laboratory selection experiments. These situations are unnatural from the standpoint of selection, but they demonstrate the genetic potential for extremely rapid change.

These drastically different rates result from averaging factors over long periods of evolution. In living species, the frequency of alternative polymorphic traits may change dramatically during a single season as a result of environmental changes (many examples are discussed in the *Biological Journal of the Linnean Society*, Vol. 14 (3 and 4), 1980). Averaged over several years, the frequencies remain essentially constant. Size change is common in many groups and may be rapid for short periods of time, but it will appear much slower if periodic increases and decreases are averaged over longer periods. A particular structural feature, like the astragalus of perissodactyls, may change significantly over a period of less than 5 million years and then remain essentially unaltered for 55 million years. The rate of change will appear much slower if it is measured over the whole time span of the order.

Evolutionary rates that have been measured for the Pleistocene (Kurtén, 1959) are far faster than most of the changes that are seen in the earlier fossil record and would account for the most rapid morphological transformations that have occurred in the emergence of new groups.

Objective measures of change in complex shapes have been described by Benson and his coauthors (1982) and Bookstein and his coworkers (1985) but have not yet been widely applied to the fossil record.

One of the most pervasive problems in appreciating the rates of evolution is the difference of scaling. As Maynard-Smith (1981) pointed out, what would appear as a gradual change to a population biologist working with the modern fauna would appear to a paleontologist as an instantaneous event occurring over such a short interval of time that it would not be represented in the geological record. All the changes that occurred as a result of the domestication of plants and animals during human history occurred in a time interval too short to be recorded in most sedimentary sequences.

It should also be emphasized that nearly all published measures of the rates of evolution refer to a small number of characters and may not be representative of the entire organism. Molecular evidence suggests that only a very small portion of the genome is concerned with anatomical change. This is demonstrated by the findings of King and Wilson (1975) that the amino acid sequence of 12 varied proteins differ by only 1 percent between humans and chimpanzees.

MACROEVOLUTION

It is fairly simple to envisage directional selection resulting in changes at the species level and speciation leading to the proliferation of distinct species within genera—for example, the emergence of the modern felid and canid species from ancestral members of these families. Even the evolution of hominids from earlier, apelike forms seems to be a fairly simple process conceptually.

On the other hand, it is much more difficult to imagine evolution between groups with significantly different ways of life. The differentiation of groups as distinct as the carnivores, primates, cetaceans, and bats from a common ancestor seems to require quite different mechanisms.

A distinction has long been recognized between these processes, which is emphasized by the terms **microevolution** and **macroevolution.**

It has been repeatedly suggested that special factors that are not evident at the species level are responsible for macroevolutionary events. The most recent mechanism to be emphasized is that of **species selection,** which Stanley (1975, 1979, 1981) proposed and Gould and Eldredge have supported. The claim by these authors that evolutionary processes differ between species and higher taxonomic levels is emphasized by their belief that little change occurs within species.

SPECIES SELECTION

The term species selection may be applied to any process that produces differential survival among a number of species that descended from a common ancestor. It may result from the greater longevity of individual species or from different rates of speciation. According to Eldredge, Gould, and Stanley, groups that are capable of rapid speciation will have a greater capacity for morphological change and an enhanced chance for long-term survival.

They contend that selection operates among species much as other biologists have argued for the importance of natural selection at the level of individuals *within* a species. It is logical to think that selection acts on a series of species and leads to the extinction of some and the survival of others. The characters of the successful species are perpetuated, just as are those of the successful individuals under Darwinian selection.

It is much more difficult to accept the further assumption that rapid, progressive morphological change can be attributed to selection among species. We will first examine this process where only a single character is involved. Several related species may have different numbers of vertebrae and selection may favor the species with the most. Subsequent speciation will occur in the surviving lineage with the most vertebrae. Further speciation and subsequent selection will increase both the number of vertebrae in each species and the number of species with higher vertebral number.

On the other hand, similar results can be achieved by individual selection within species. If, as is certainly the case, vertebral number is variable and inherited and a higher vertebral number is advantageous in certain circumstances, the number of individuals with higher vertebral numbers will increase. Following speciation, there will be a larger number of species with higher vertebral numbers. Variability, selection, and speciation occur in both models but in different sequences. In the Darwinian model, variation occurs within populations and species. Since there are thousands or tens of thousands of individuals within species, the chance for advantageous variants to occur is quite high. If variation is only considered to be of importance *between* species, the changes of the appearance of new advantageous variants must be governed by the number of new species. The relative amount of variability that will be available according to the two models is roughly proportional to the difference in numbers between individuals within species and the number of species. As an added quantitative factor, the time required for the origin of new species is four or five orders of magnitude greater than the generation time of individuals within species.

This problem becomes much more significant if we consider all the anatomical characters that may be undergoing evolutionary change within a lineage. For example, the evolution of advanced hominids involves features of the dentition, size and configuration of the cranial vault, the facial area, vertebral column, hands, pelvis, and rear limbs. Change in any one of these characters might be attributed to species selection, but to account for all of them evolving independently but at the same time would require an extremely large number of coexisting species.

The fossil record in the late Cenozoic is certainly sufficiently complete to demonstrate that most mammalian groups have never included the vast number of contemporary species that would be necessary for the process of species selection to lead to continued progressive morphological change. Other arguments regarding species selection are elaborated by Charlesworth and his colleagues (1982), Maynard-Smith (1983), and Gilinsky (1986).

Species selection may be important in a few particular circumstances (as outlined by Stanley in 1982), but it cannot be considered a probable cause of progressive change in most vertebrate groups.

Eldredge, Gould, and Stanley also suggest that there are aspects of the speciation process itself that facilitate rapid change. This argument had previously been made by Mayr (1954, 1963, 1970, 1982). He proposed that rapid change is much more likely to occur in small populations restricted to limited geographical areas than in large, widespread species. Environmental differences within small areas are likely to be limited, so that it is not necessary for the population to retain a wide range of alternative alleles to cope with geographically variable conditions. In small, isolated populations, inbreeding will lead to homozygosity, so that recessive as well as dominant traits will commonly be exposed to selection. We can expect directional selection to operate much more rapidly under these conditions than it does in widespread species with large population sizes. Population isolation can be considered a prelude to speciation.

It is generally accepted that speciation among vertebrates requires a significant period of geographical separation between populations during which time they develop physical and behavioral characters that preclude interbreeding if they subsequently inhabit the same geographical area. If newly formed species become secondarily sympatric, they may compete strongly with one another for the same resources. If they are to survive in the same geographical area, they must diverge behaviorally and/or physically so that they can take advantage of different resources.

Natural selection can thus be expected to act especially strongly during speciation, first when the incipient species are represented by small, isolated populations and later, if they subsequently become sympatric.

These processes may occur over only a very short period of time relative to the total longevity of a species and would thus produce a punctuated pattern in the fossil record. These processes explain the clear distinction we see between contemporary species, but it is difficult to understand how they can account for the major structural changes that typify vertebrate evolution.

Mayr (1982) and Gould (1982b) both argued that speciation in some way releases inherited genetic and developmental constraints so that, for a short time subsequent to speciation, species are much more susceptible to significant change through natural selection. No specific mechanism has been proposed to account for this process, and the fossil record of vertebrates does not demonstrate its occurrence. Many of Mayr's arguments are challenged by Barton and Charlesworth (1984).

Species selection and other theories of macroevolution have been proposed because of the perceived differences between changes at the species level and larger-scale evolutionary events.

We can recognize at least three different aspects of macroevolution:

1. The origin and radiation of major taxonomic groups.
2. The evolution of distinct new structures or physiological processes.
3. Major adaptive shifts.

These features may or may not be correlated in time. Macroevolutionary events may appear to be relatively rapid, although they are frequently associated with a significant gap in the fossil record.

We may look at each of these aspects separately and consider whether they can be understood on the basis of what we know of evolution at the species level.

THE ORIGIN OF MAJOR TAXONOMIC GROUPS

Most species may be grouped into a relatively small number of higher categories. Among placental mammals there are some 30 orders, most of which include several families. If one accepts that they are natural (monophyletic) groups, does this mean that special evolutionary processes are necessary to account for their origin?

Taxonomic procedures play a very important role in how we look at evolution. Linnean systematics is basically a hierarchical system. Species are grouped in genera, genera into families, families into orders and so on. This system leads to the impression that species in different categories differ from one another in proportion to differences in taxonomic rank. Species in different families are presumed to be more different than species of different genera, species of different orders are more different than species of different families, and so on. This impression may be correct if we consider only members of the modern fauna, but it is clearly not so at the beginning of a major radiation. As more and more emphasis is placed on vertical or cladistic classification and as the fossil record improves, it will potentially be possible to allocate populations within a single species to different genera and even families or orders of descendant forms. There are already some instances among Cenozoic mammals where such subdivisions of ancestral species may be possible. Wilson (1971) described progressive evolution among primitive selenodont artiodactyls in which populations of early species give rise successively to members of different genera and different families.

The largest-scale radiation that is relatively well documented is that of the latest Cretaceous and early Paleocene ungulates, in which as many as 16 orders differentiated within 2 to 3 million years from a single family of early condylarths (Van Valen, 1978). As yet, we know little of these animals except their dentition, but the pattern of radiation is not refuted on the basis of much more complete skeletal remains from later in the Cenozoic.

What we see in the fossil record of these ancestral groups is not obviously different from the pattern of radiation within many other families during the Cenozoic in terms of the pattern of variation and speciation. The most significant factor is that many of the species and genera went on to subsequent periods of radiation that are recognized as distinct families and orders. At the level of individuals and species, the evolution of the early condylarths does not seem significantly different from typical microevolutionary events. The differences are not in the species themselves, but rather in what opportunities there were for their descendants.

Van Valen did note that the rate of generation of new taxa was considerably faster than the average for the Cenozoic, but this difference can be associated with the very low number of initiating lineages and the almost total absence of competition.

The subsequent radiation of successful ungulate orders such as the artiodactyls and perissodactyls in North America and the notoungulates in South America follow comparable patterns, as described by Krishtalka and Stucky (1986), Radinsky (1966), and Simpson (1948, 1967). Features that are initially variable within species become stabilized and amplified and come to be recognized as characters of families and orders.

If we examine evolutionary events at the level of populations and species, there is no phenomenon that can be associated with the origin of genera or other higher categories. The processes leading to speciation are no different whether the daughter species become extinct or give rise to a new order. Higher taxonomic categories can only be recognized subsequently, after a host of later speciation events have given rise to many descendant taxa.

The search for special evolutionary factors must be centered on the conditions that facilitate the subsequent successful radiation of a group, not on the initial appearance of a new species.

THE EMERGENCE OF NEW STRUCTURES AND WAYS OF LIFE

The most spectacular evolutionary changes involve the emergence of new structures and ways of life, for example, the appearance of the vertebrate body plan and the origin of amphibians, birds, and bats. Can these events be explained by the same processes that account for minor changes observed at the level of the genus and species? Can the foot of a tetrapod or the wing of a bird evolve slowly and by small increments from the fin of a fish or the forelimb of a dinosaur, or must we look for significantly different mechanisms?

Table 22-1 lists many large-scale structural and physiological changes that have occurred in vertebrate history, with an emphasis on those that are related to the emergence of new groups or ways of life.

None of the changes in soft anatomy and physiology can be studied directly from the fossil record. We know that endothermy evolved among the ancestors of both the mammals and the birds, but we cannot establish exactly when the change was initiated or the degree to which the modern condition was approached during the Mesozoic.

The time of phylogenetic divergence can be used to judge the minimum time of establishment of features such as the vertebrate sense organs, swimbladder, amniote egg, avian respiratory system, mammary glands, hair, and the

TABLE 22-1 **Major Structural and Physiological Changes in the History of Vertebrates**

I. Features associated with the origin of vertebrates

- The emergence of the chordate body plan
- Development of the neural crest and its derivatives
- Origin of brain and paired sense organs
- Origin of bone (which occurred separately in several lineages)

II. Features associated with the origin of gnathostomes

- Origin of jaws
- Origin of teeth with regular replacement pattern
- Origin of paired fins
- Origin of swimbladder in osteichthyes

III. Features associated with the origin of tetrapod groups

- Origin of limbs in amphibians
- Origin of the amniote egg
- Origin of an impedance-matching ear in squamates, archosaurs, turtles, and mammals
- Loss of limbs in aïstopods, caecilians, and snakes
- Secondary aquatic adaptation of mesosaurs, ichthyosaurs, nothosaurs, placodonts, mosasaurs, whales, and sirenians

IV. Features associated with the origin of flight

- Origin of wings in pterosaurs, birds, and bats
- Elaboration of the avian respiratory system
- Echolocation in bats

V. Features associated with the origin of mammals

- Endothermy
- Hair
- Mammary glands
- Precise tooth occlusion and the tribosphenic molar
- Trophoblast of placentals

trophoblast, but it is unlikely that the fossil record will ever provide much information regarding the nature of their origin. Study of embryological development has the potential for establishing how these structures may have evolved but cannot specify the time frame.

Currently, we know neither the time scale nor the specific ancestral condition for the emergence of the chordate body plan, the origin of bone, jaws, and teeth, or the paired limbs of gnathostomes, but the fossil record has the potential to yield this information.

In order to have an objective basis for evaluating the factors involved in the emergence of new structures, we must have unequivocal evidence of the ancestral condition and at least a minimum limit on the time during which the transition occurred. We are left with a shorter list of major structural changes for which we now have some evidence from the fossil record.

STRUCTURAL CHANGES ASSOCIATED WITH ENVIRONMENTAL SHIFTS

The most dramatic evolutionary events are those that combine a major change in habitus with the appearance of a new structural-functional complex. Examples include the origin of amphibians, birds, pterosaurs, and ichthyosaurs. The appearance of these groups is made all the more striking by the rarity of intermediate forms in the fossil record.

The origin of amphibians, which is discussed in Chapter 9, is certainly one of the most significant episodes in vertebrate history. The group from which they arose, the rhipidistian sarcopterygians, were common throughout the Devonian and lingered into the Lower Permian. Their structure is known in considerable detail (Andrews and Westoll, 1970). Genera in which the skeleton is appropriate for ancestors of amphibians are known from the middle Devonian, approximately 15 million years prior to the appearance of amphibians at the end of the Devonian (Jarvik, 1980). These early amphibians are also known in considerable detail. Their limbs have already achieved a typically tetrapod pattern.

The remains of rhipidistian fish are common in Middle and Upper Devonian deposits that might be expected to yield the remains of ancestral amphibians, but no fossil is known that could be considered intermediate between these two groups.

Although the general body form of rhipidistians and amphibians appears very distinct, most of the individual skeletal units are very similar and can be readily homologized. Except for proportions, the skulls are basically similar, and we can directly compare nearly all the bones present in early amphibians with those in the fish. In both groups, the vertebrae consist of a number of distinct central and arch elements; the only difference is the elaboration of zygapophyses in the amphibian. The most striking change in the axial skeleton is the great elaboration of the ribs, but they were already present in rhipidistians.

The most significant change is the modification of fish fins into tetrapod limbs. Even here, strong skeletal similarities link the two groups. All the proximal limb elements were already present in the fish, including the joint structures of the elbow and knee. The only portion of the skeleton to undergo fundamental change was the distal end of the fin. The distal carpals and tarsals, the metapodials, and the phalanges have no homologues in rhipidistians.

If we consider the skeleton as a whole, we see that the only changes to occur were those directly associated with the requirements of terrestrial support and locomotion. The presence of large axial elements of the paired fins in rhipidistians indicates that they could already have been used to support and propel the body against the substrate when it was only partially supported by the buoyancy of the water. The presence of internal nares suggests that rhipidistians already made use of atmospheric oxygen, despite the retention of gill supports that are typical of fish. Most of the features that distinguish later Paleozoic tetrapods from rhipidistians evolved after the shift from an aquatic to a terrestrial habitat.

At least some rhipidistians may have been adapted to life in very shallow water with occassional limited excursions onto the land. The early ichthyostegid amphibians, on the other hand, retain a fishlike tail and almost certainly reproduced in the water, as do many modern amphibians. Nevertheless, the general body form shows an essential difference in habitus. Rhipidistians were basically and obligatorily aquatic animals that were incapable of effective terrestrial locomotion or supporting their body out of the water. In contrast, ichthyostegids were unquestionably capable of terrestrial locomotion.

If intermediates between rhipidistians and amphibians had been common or widespread, their remains would almost certainly have been discovered in the well-known deposits of middle and late Devonian age. We may attribute their absence to the limited adaptive opportunities available for animals in intermediate environments. A partially evolved tetrapod limb would be effective only in forms that lived in very shallow water, where selection was most stringent for moving the body when the trunk and tail could no longer provide effective thrust against an aquatic medium. Stranded fish such as actinopterygians, which did not have large muscular fins, may have provided a periodically rich food source for rhipidistians living in this narrow adaptive zone. This stage in the transition between fish and amphibians may have occurred quite rapidly in small, scattered populations that were also strongly susceptible to extinction. On the other hand, evolution may have been slowed by counter selection, which continued to act to retain a fin structure that was effective in the water until ancestral amphibians were behaviorly committed and structurally capable of spending a significant amount of time on land.

The origin of flight in birds provides a striking example of a major change in the way of life that was accompanied by very limited modification of skeletal features. As a result of exceptional preservation of several skeletons of the late Jurassic genus *Archaeopteryx*, we have evidence of an extremely early stage in the evolution of avian flight. Skeletally, *Archaeopteryx* is as primitive as a bird could possibly be, and yet the structure of the individual feathers and their arrangement on the wing are strictly avian. The only advance in the skeleton over the condition in small theropod dinosaurs is a proportional increase in the size of the forelimb. Most changes in the skeleton that characterize all modern birds occurred *after* the change in habitus.

The presence of feathers in association with the skeleton of most specimens of *Archaeopteryx* is a fortuitous result of their preservation in extremely fine-grained lithographic limestone. Feathers might have been present in some theropod dinosaurs without being preserved. Feathers are unquestionably homologous with reptilian scales and probably evolved for insulation long before their use as elements of a wing. The key to the origin of birds as flying animals was probably a primarily behavioral change in the use of the forelimbs. The anatomical pattern common to *Archaeopteryx* had already evolved among theropod dinosaurs. The only structural change that must have accompanied the behavioral shift from a small, terrestrial, or arborial dinosaur to a flying bird was an increase in the size of the feathers associated with the incipient flight surfaces.

Other transitions, like that between advanced thecodonts and pterosaurs and those leading to the origin of ichthyosaurs and bats, seem much more dramatic because we lack knowledge of immediately ancestral forms. Like the origin of tetrapods, these transitions may have occurred over relatively long periods of time (in the order of 10 million years).

In all these examples, evolution involved primarily those elements of the skeleton that were specifically associated with the change in habitat. Homologues of structures that characterize the derived groups are recognizable among their ancestors, but the functions of the elements may be fundamentally different.

In the case of amphibians, birds, and ichthyosaurs, most of the anatomical features that are characteristic of advanced members of the groups evolved after the actual habitat change. On the other hand, early Eocene bats were nearly as advanced as many living species.

These examples were chosen to illustrate some of the largest-scale skeletal changes involving habitat shifts for which there is some fossil evidence. Although one might hypothesize some special mechanism or process to account for the morphological or temporal gaps that do exist, the currently available evidence does not require any special factors. These changes did not necessarily occur any faster than those that did under other conditions. Most morphological changes involve structures that could function either in an intermediate state or in different ways in both habitats. The absence of numerous intermediate forms can be attributed to the limited possibilities for adaptation to intermediate habitats.

Other examples of habitat shifts are seen in the fossil record in which at least some intermediate stages are known. Many different lineages of aquatic crocodiles evolved from terrestrial or semiaquatic ancestors, and numerous intermediate forms are known. The transitions between terrestrial sphenodontids and aquatic pleurosaurs and between varanoid lizards and mosasaurs involved little more than progressive changes in proportions (see Chapter 11). Aigialosaurs provide a good link to the more specialized marine mosasaurs. Although all mosasaurs are more advanced in having paddle-shaped fins, greater changes occur in the number of phalanges and the number of trunk and caudal vertebrae within the mosasaurs than in the origin of this group from aigialosaurs.

In fact, adaptation to an aquatic way of life can occur without any perceptible skeletal changes. This phenomenon would be difficult to demonstrate from the fossil record but is evidenced by the modern marine iguanids, which depend on underwater algae for their diet and spend much of their time in the water. Bartholomew and his colleagues (1976) showed that they do not differ significantly from terrestrial iguanids in either their anatomy or physiology.

This example suggests that major shifts in habitat can be initiated primarily by modification of behavior without any changes in anatomy or physiology. Animals such as crocodiles and iguanids that are capable of swimming and feeding in the water may do so to take advantage of more readily available food sources or to avoid terrestrial predators or competitors. Once feeding and locomotion in the water have become important factors in their survival, selection will act to perpetuate structural and behavioral features that are more advantageous in that habitat. Such changes can occur without speciation and do not require genetic revolutions.

CHANGES WITHIN ADAPTIVE ZONES

Further support for an essentially Darwinian explanation of large-scale structural change can be gained from skeletal modifications that are not associated with major adaptive shifts. The changes that occur in the structure of the skull, including the braincase, lower jaw, and especially the middle ear, in therapsids and early mammals are as great as the skeletal changes associated with the origin of amphibians, birds, and aquatic reptiles. The changes in the ancestry of mammals occurred over a period of approximately 100 million years. This transition is not without temporal gaps, but the structural changes are essentially continuous. The fossil record is not sufficiently complete for a thorough quantitative analysis, but there are no clearly defined episodes of especially rapid evolution. We can explain the entire sequence by persistent directional selection associated with perfection of hearing and mastication. A similar progressive sequence can be noted in the evolution of feeding and locomotion among neopterygian fish in the Mesozoic, which preceded the great radiation of higher teleosts in the late Cretaceous. Changes in the locomotor apparatus among Cenozoic ungulates were somewhat episodic but apparently represent a continuous process in many lineages that requires only a moderately strong selection pressure for its continuation.

The clearly documented cases of large-scale progressive morphological change in groups within a particular adaptive zone and for many features in groups that have entered new adaptive zones make it very difficult to accept any hypotheses that require special processes to account for the remaining gaps in the fossil record.

CONSTRAINTS AND RADIATIONS

Although detailed knowledge of the fossil record is not available for all groups, well-documented examples of evolution at the level of the species and genus as well as the emergence of new structures, ways of life, and major taxonomic groups demonstrate that natural selection acting on variations within populations is sufficient to explain most evolutionary changes.

The amount of variability that we observe in fossil and modern populations, together with moderately strong directional selection, is capable of causing morphological changes over short periods of time that are far more rapid than the long-term changes responsible for the macroevolutionary advances observed in the fossil record.

Establishing a mechanism to explain evolutionary change does not in itself fully explain the observed *pattern* of vertebrate evolution. On the basis of an understanding of variation and selection, one might expect that the history of evolution would exhibit an essentially uniform pattern of continuous radiation and adaptation (Figure 22-5). In fact, the phylogenies of all major vertebrate groups show an irregular, episodic history of occasional large-scale radiations followed by the long-term survival of a relatively small number of basically distinct structural and/or adaptive types. The potential universe of organisms is occupied in a clearly heterogeneous fashion.

We should now look for ways to explain how the factors of variability and/or selection may be directed or constrained to produce this large-scale pattern.

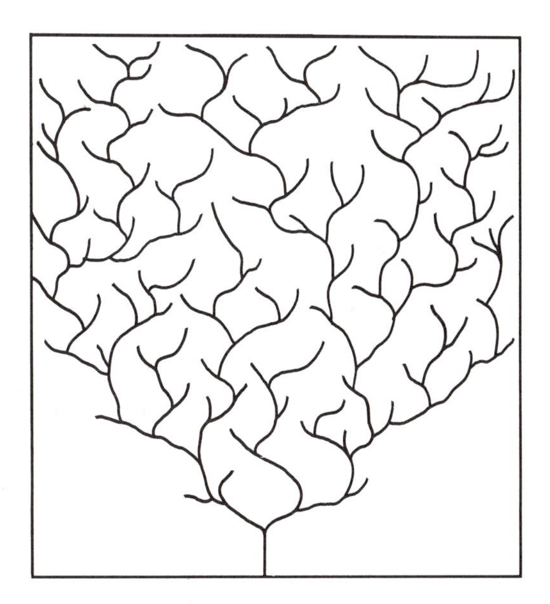

Figure 22-5. HYPOTHETICAL RADIATION OF A GROUP IN WHICH THERE ARE NO CONSTRAINTS. All potential morphotypes are produced by natural selection acting on random variation. Morphological change proceeds at a uniform rate.

ENVIRONMENTAL CONSTRAINTS

The most obvious explanation for the discontinuous distribution of organisms is found in the heterogeneity of the environment. Throughout vertebrate history, most organisms have adapted to particular environments or ways of life. Almost all can be described as being primarily aquatic, terrestrial, or aerial. Within these broad environments, further habitat subdivisions, such as pelagic, benthonic, nectonic, burrowing, arboreal, and so on, can be recognized.

The physical and biological requirements of any particular habitat result in selection for features that are nearly always disadvantageous in other environments or ways of life, which leads to the perpetuation and intensification of particular adaptive types. For example, most mammalian orders have retained a basically similar morphology and way of life for 50 to 60 million years.

Within particular adaptive zones, competition with other species restricts the number and nature of species and frequently leads to progressively higher degrees of specialization in diet, locomotor patterns, and other aspects of structure and behavior. Since the biological environment, including both other species and other members of the same species, is continually changing, evolution in most groups results in progressive modification within broad adaptive zones. There may be an optimal solution for a particular aspect of a species at a particular time, but it may never be realized before it is altered by other, conflicting selection pressures.

In contrast, physical aspects of the environment may set more nearly absolute limits on adaptive change. Optimal designs among biological systems may be recognized by the convergent attainment of closely similar structures in groups with distinct ancestry. Perhaps the best example is supplied by the body form of rapidly swimming aquatic vertebrates. Among modern fish, the most rapidly swimming species are members of the teleost family Carangidae, including the tuna. They are characterized by a laterally compressed, spindle-shaped trunk and a high, lunate tail. This configuration provides the greatest thrust and the least amount of drag (Lighthill, 1970; Webb, 1982). An essentially similar body form has been achieved among sharks, including some genera as early as the late Devonian, by ichthyosaurs among Mesozoic reptiles, and by Cenozoic whales (although in the whales, the caudal fin

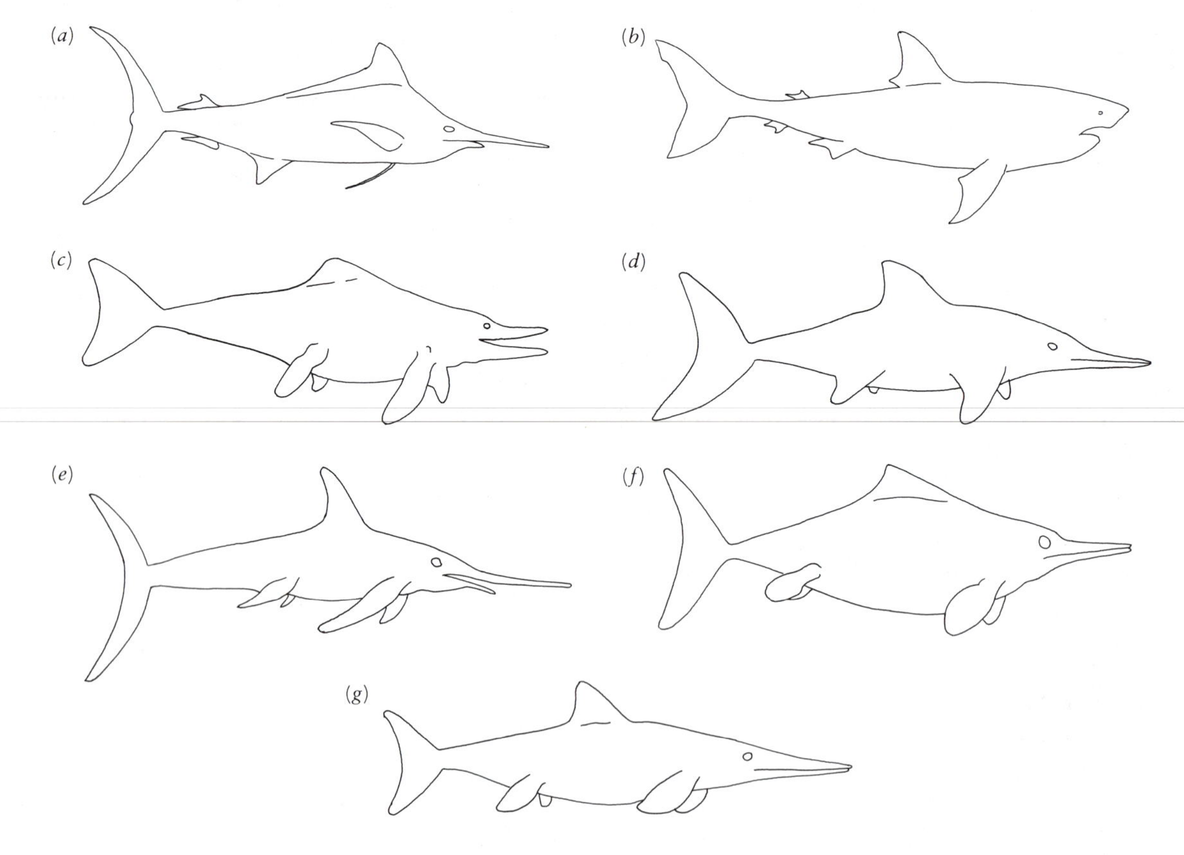

Figure 22-6. BODY OUTLINE OF RAPIDLY SWIMMING AQUATIC VERTEBRATES. (*a*) The marlin, a living teleost bony fish. (*b*) A predaceous shark. (*c* to *g*) A variety of ichthyosaurs: (*c*) *Ichthyosaurus*, Lower Liassic, Lower Jurassic. (*d*) *Stenopterygius*, Upper Liassic, Lower Jurassic. (*e*) *Eurhinosaurus*, Upper Liassic. (*f*) *Ophthalmosaurus*, Upper Jurassic and Lower Cretaceous. (*g*) *Platypterigius*, Upper Cretaceous. These ichthyosaurs differ somewhat in fin proportions and the configuration of the skull and jaws, which reflect different degrees of maneuverability and feeding habits, but the general body form has remained nearly constant for 120 million years in response to the constraints of rapid swimming. *Outlines of ichthyosaurs adapted from McGowan, 1983.*

is oriented horizontally rather than vertically, since lateral flexure of the vertebral column is much restricted in the mammals.)

This body form is clearly optimal for rapid aquatic locomotion. Other factors also affect the shape of individual species, such as the nature of their prey, requirements for maneuverability as well as speed, and structural differences of their ancestors. Between groups, specialized aquatic locomotion provides an excellent example of convergence. Within groups, it demonstrates changing evolutionary rates and stasis. From the early Jurassic into the late Cretaceous (approximately 120 million years) ichthyosaurs exhibit a nearly constant body form (Figure 22-6). Their fossil record in the Triassic is incomplete, but it appears to have required about 40 million years to achieve the pattern of advanced ichthyosaurs.

Active flight similarly constrains the body form to a particular pattern, and nearly identical structures were achieved separately by flying reptiles and birds (see Figure 16-2). Rose and Emry (1983) describe striking similarities of the skull in a variety of burrowing genera that evolved from different groups of reptiles and mammals.

Physical constraints of the environment are clearly sufficient to explain why particular structural patterns may be achieved fairly quickly and persist for long periods of time.

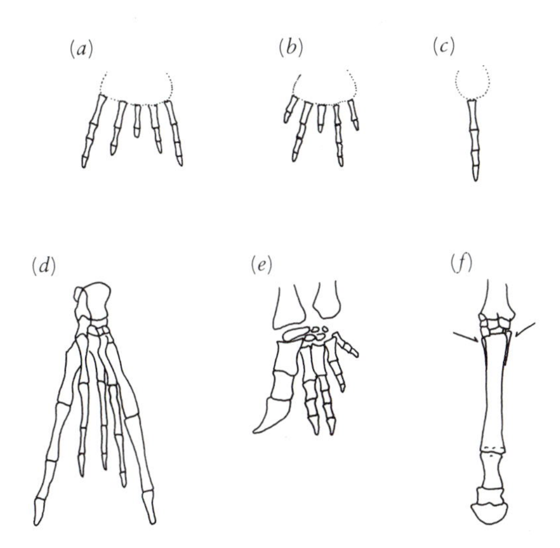

Figure 22-7. (*a* to *c*) Examples of digit patterns that are forbidden in terms of the Stock and Bryant version of the polar coordinate model for digit formation. Digit complexity is shown as phalangeal number for convenience. (*d* to *f*) Identified forbidden morphologies. (*d*) The fin of the elephant seal (*Macrorhinus leoninus*). Although the phalangeal formula is acceptable, the lateral and medial digits are clearly more complex than the three central digits. (*e*) The forelimb pattern of the reptile *Massospondylus*. Although the phalangeal formula is acceptable, the most complex digit in terms of phalangeal number (the central digit) is separated by one digit from the lateral digit, which is adjudged complex on structural grounds. (*f*) The single digit of the foot of the horse (*Equus caballus*). This clearly symmetrical digit is flanked proximally by two much reduced metatarsals (arrows). *From Holder, 1983.*

DEVELOPMENTAL CONSTRAINTS

It has been suggested that developmental constraints are very important factors in limiting the potential for evolutionary change. Alberch (1980), Gould (1980), and Williamson (1981) argue that inherent limits on developmental processes canalize the direction that change may take. Alberch and Gale (1985) provide as an example different patterns of digit reduction that characterize frogs and salamanders over more than 200 million years of evolution. Stock and Bryant (1981) argue that there are certain patterns of the digits that can never be expected to evolve due to the basic processes of limb formation. Holder (1983) shows that most of these supposedly forbidden designs do occur, although they are relatively uncommon (Figure 22-7).

The pattern of ossification of the carpals and tarsals in early diapsid reptiles provides a useful example of the limits of developmental constraints. Among primitive diapsids, several genera show a consistent pattern in the sequence of ossification of the bones of the wrist and ankle. In the carpus, the first bone to appear is the ulnare, followed by the fourth distal carpal and the intermedium, then the lateral centrale, distal carpals 1 and 3, followed by the radiale, medial centrale and distal carpals 2 and 5, and, last of all, the pisiform. In the tarsus, the astragalus and calcaneum appear first, followed by the fourth distal tarsal, the centrale, distal tarsals 3, 1, and 2, and finally the fifth distal tarsal, which may then fuse with the fourth.

These early genera were near the base of a large radiation that lead to all modern reptiles (other than turtles) and a number of extinct aquatic groups. Early members of the aquatic lineages show a reduction in the number of carpals and tarsals to provide greater flexibility of the wrist and ankle and to reduce weight. The first elements to be lost were the last elements to be ossified in the primitive group. It is a general rule of development that the last elements to appear are the most subject to change and loss. This rule may be considered a developmental constraint that predicts, in a general way, the pattern of the carpus and tarsus in many groups of secondarily aquatic reptiles, as is clearly shown by the nothosaurs (Figure 22-8). It is also evident in primitive members of more advanced groups, including plesiosaurs and ichthyosaurs. However, it is not an absolute constraint. Advanced plesiosaurs and ichthyosaurs have clearly different patterns of the limbs that overcome this constraint with the elaboration of new elements and significant changes in proportions and configuration. The primitive distinction between elements of the lower limb and the hand and foot are lost, and some ichthyosaurs loose the serial pattern of the digits (see Figure 12-30) (Carroll, 1985).

Another developmental "rule" that is broken among secondarily aquatic reptiles is the one that states that limb reduction begins with the most distal elements and pro-

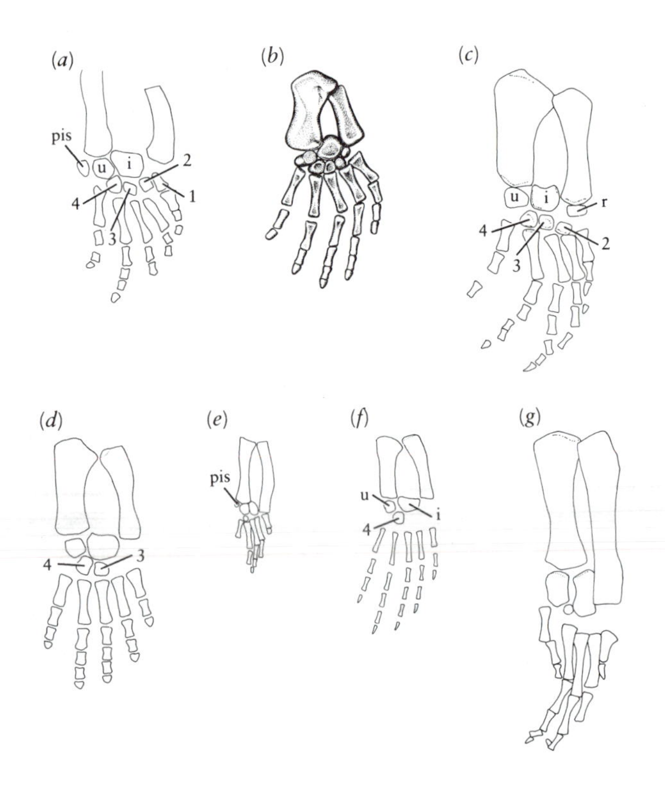

Figure 22-8. FORELIMB OF NOTHOSAURS SHOWING DIFFERENT PATTERNS OF THE CARPUS. They correspond in general to the sequence of ossification of the carpals in primitive eosuchians. (*a*) *Proneusticosaurus*. (*b*) *Lariosaurus*. (*c*)*Ceresiosaurus*. (*d*)*Paranothosaurus*. (*e*)*Dactylosaurus*. (*f*)*Nothosaurus*. (*g*)*Pachypleurosaurus;* the number of phalanges is reduced from the primitive reptilian formula of 2, 3, 4, 5, 3 to 1, 2, 3, 4, 3. Abbreviations as follows: i, intermedium; pis, pisiform; r, radiale; u, ulnare; 1 to 4, distal carpals. The medial and lateral centrale, which are present in primitive reptiles, are missing in all these genera.

ceeds proximally (Wolpert, 1983). This pattern is followed in nothosaurs by reduction of the digits and then the distal limb elements, while the humerus and femur remain large. In plesiosaurs and ichthyosaurs, the number of phalanges and the total length of the hand *increase*, while the distal limb elements are reduced. The humerus and femur remain large in plesiosaurs but are much reduced in ichthyosaurs.

Hinchliffe and Griffiths (1983) used examples from amphibians, birds, and mammals to show that both general patterns and details of carpal formation can be influenced by specific mutations and that there are no overriding developmental rules that limit evolutionary potentials.

On the basis of currently available evidence, it is difficult, if not impossible, to determine whether conservative developmental patterns are a constraint on evolutionary change or, conversely, reflect long-term stabilizing selection.

Unfortunately, little is yet known of the general way in which developmental processes are controlled among vertebrates (Bonner, 1982; Goodwin, Holder, and Wylie, 1983; Raff and Kaufman, 1983; Maynard-Smith et al., 1985). One may attribute the overall consistency of the vertebrate body plan to some inherent aspects of development. Such pervasive features of vertebrate anatomy might also be termed historical constraints. Among tetrapods, the paired limbs always show the same general pattern (Stubin and Alberch, 1986), but this is not a basic vertebrate constraint since cartilaginous and bony fish show many different arrangements of the fin skeleton. Long-term selection can probably eliminate some developmental pathways and thus forever eliminate some originally possible directions of adaptation. Limbs and other major features that are lost are probably never regained.

STRUCTURAL AND PHYSIOLOGICAL CONSTRAINTS

There are other aspects of organisms that show little if any significant variability and are hence not amenable to change in response to selection. They may be considered the most clear-cut examples of evolutionary constraints.

Most vertebrate activities are associated directly with the skeletal and muscular systems whose fundamental units show extremely limited variability. Although the configuration and distribution of bone and cartilage differ greatly among the vertebrate groups, their strength per unit area is essentially constant. Selection may act to make the most effective use of their physical properties, but their strength under tension, compression, or shear cannot be significantly altered (Hildebrand, 1982).

The distribution of skeletal tissue can be adjusted to take advantage of the relative strength of bone and the relative lightness and compressibility of cartilage. The elaboration of cartilage as the only skeletal material in sharks and their allies probably precludes the evolution within that group of the complex feeding structures that are common to bony fish, which rely on the greater strength of bone.

The strength of muscle contraction is also restricted within narrow limits. The arrangement of the fibers and some aspects of their physiology are variable, but all exert a force of approximately 2 to 3 kilograms per square centimeter of cross section, measured at right angles to the orientation of the fibers. The degree by which a muscle contracts and the amount that it can be stretched without damage are also closely limited. Muscles can contract only about 30 percent of their resting length and are capable of stretching by approximately 50 percent of their resting length. Changes in the size and proportions of the skull and appendicular skeleton are closely constrained by these properties of the muscles.

Perhaps the most general structural constraints are those associated with the relationships between linear dimensions, surface area, and volume. If one considers the vertebrate body as a roughly cubical structure, doubling of linear dimensions results in squaring the surface area, while the volume increases as the cube of linear dimensions. Since the weight-bearing capacity of bone and the force that is generated by muscles are both proportional to cross-sectional area, the muscles and bones of the limbs must increase faster than linear dimensions to support and move a heavier body.

This factor also affects processes such as heat gain and loss and exchange of respiratory gases and nutrient molecules across membranes, all of which occur at an essentially constant rate that is proportional to surface area. Increased body size requires a disproportional elaboration of the area of the gills or lungs for gas exchange and the area of the intestinal surface for absorption of food. The rate of heat loss and gain and water loss are correspondingly reduced in animals with larger bodies.

Changes in proportions related to changes in absolute size are termed **allometric** and are very significant in constraining the general body form of vertebrates (Gould, 1966, 1974, 1975b).

The relative size of the sense organs of the head provides a striking example of structural constraints. Both the eye and the otic capsule are significantly larger, relative to other features of the head, in small vertebrates. This phenomenon is related to physical factors that restrict their practical size within fixed limits. The flow of fluids is restricted in small tubes, and hence the diameter of the semicircular canal varies only slightly over a great range of body weights (Jones and Spells, 1963). The dimensions of the rods and cones of the eye are limited by the interference pattern of the waves of light, which sets bounds

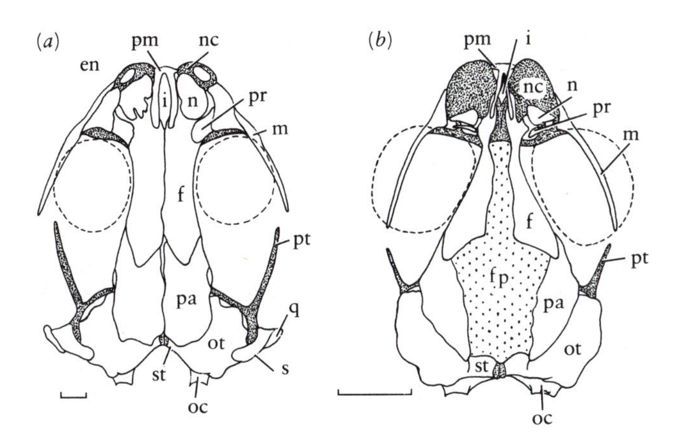

Figure 22-9. SKULLS OF TWO CLOSELY RELATED SALAMANDERS. These species show striking differences in the pattern of the dermal bones to accommodate the great relative increase in size of the sense organs as a result of the smaller absolute size of the skull. (*a*) *Pseudoeurycea goebeli.* (*b*) The much smaller species, *Thorius narisovalis.* Scale bars = 1 millimeter. Abbreviations as follows: en, external naris; f, frontal; fp, frontoparietal fontanelle; i, internasal fontanelle; m, maxilla; n, nasal; nc, nasal capsule; oc, occipital condyle; ot, otic capsule; pa, parietal; pm, premaxilla; pr, prefrontal; pt, pterygoid process; q, quadrate; s, squamosal; st, synotic tectum. *From Hanken, 1984. With permission from the Biological Journal of the Linnean Society, copyright 1984.*

to the production of a clear retinal image (D'Arcy Thompson, 1966). In tiny vertebrates, the eye and otic capsule dominate the structure of the skull (Figure 22-9), which greatly influences the configuration of the skeletal elements. The development of the bones conforms to the earlier established pattern of the sense organs (Hanken, 1984). The bone can thus change very significantly in shape without any corresponding genetic change in the mechanism that controls bone formation. We can assume that the manifest changes in the details of skull structure are governed by one factor, selection for small body size.

Reorganization of the skull in relationship to reduced body size has been an important factor in the early evolutionary stages of several groups: lepospondyls and modern amphibians (Carroll and Holmes, 1980), primitive reptiles (Carroll, 1970), modern lizards, and possibly ancestral snakes (Rieppel, 1984).

Physiological factors may also be considered among the evolutionary constraints. Low metabolic rates strongly constrain the adaptive potential of fish, amphibians, and early amniotes. None of these groups could attain active flight or adaptation to cold terrestrial environments with little sunlight. None are as active in the dark as are many mammals. On the other hand, a high metabolic rate makes it difficult for birds and mammals to adapt to environments with limited or irregular food supplies, unless they are capable of intermittant periods of dormancy.

Metabolic rates within each group of vertebrates increase in proportion to approximately the 0.75 power of the body weight (Schmidt-Nielsen, 1975). Requirements for food, oxygen, and elimination of metabolic wastes increase faster than linear dimensions, although not at the rate of volume. This increase requires compensating changes in various anatomical elements. For example, Gould (1975a) demonstrated that the occlusal area of the cheek teeth in many mammals increased significantly faster than linear measures of the skull.

"LIVING FOSSILS"

Constraints resulting from physical factors of the environment, structural limitations of vertebrate tissue, and developmental processes would be expected to act in a similar manner in all lineages within taxonomic groups that have relatively similar structural and behavioral characteristics. It is hence difficult to explain why a few members of most major groups have evolved much more slowly and show significantly less diversity than other related forms.

Darwin coined the term "living fossil" to apply to species that appeared to retain very primitive morphological patterns and were assumed to have had very slow evolutionary rates throughout their history.

Schopf (1984) points out that several different categories of "living fossils" are recognized by some recent authors.

1. Species that are the only living remnants of groups that have otherwise become extinct. The coelacanth *Latimeria* is a good example of this category. Although recognizably distinct, it retains most of the skeletal characters of its Cretaceous ancestors.
2. Persistently primitive members of modern groups that have otherwise changed dramatically. The tree shrews and the opossum *Didelphis* retain a skeletal anatomy similar to that of Upper Cretaceous therians, while the remaining placental and marsupial lineages underwent dramatic adaptive radiations. The tapir among the perissodactyls and the tragulids among the ungulates are less dramatic examples in this category that have been discussed in a recent review by Eldredge and Stanley (1984a).
3. Species or genera that have persisted over a long interval of geological time. The squirrel genus *Sciurus* and the elephant shew *Rhynchocyon* are both known from the Miocene and have shown little morphological change during the past 15 to 20 million years (Emry and Thorington, 1984; Novacek, 1984). In addition, elephant shrews retain many features of even more primitive placental mammals and so they might be included in category 2 as well. On the other hand, tree squirrels are common and widespread today and do not have a conspicuously archaic skeletal anatomy. They hence do not fit other criteria of "living fossils," despite their relative longevity.

We assume that "living fossils" evolved very slowly but relatively few examples have an adequate fossil record, so that it is difficult to establish actual rates of evolution. For example, *Sphenodon* and *Latimeria* have no fossil record and *Didelphis* is not known from fossils earlier than the Pleistocene.

Schopf notes that "living fossils" are generally recognized on the basis of a limited number of features of the skeletal anatomy. When their soft anatomy and physiology have been studied, these species are frequently found to be as advanced as other more uniformly progressive genera. The duck-billed platypus and echidna illustrate the reverse condition. They are considered the most primitive living mammals on the basis of their reproductive pattern, and yet the skulls are extremely highly specialized.

The various examples of "living fossils" that have been recognized show that this is not a well defined concept. Schopf (1984) provides a definition that may help to clarify this problem: "A relatively little morphologically modified representative of a relatively archaic lineage with little modern representation."

Simpson (1944) coined the term **bradytely** to apply to very slow morphological change and considered that this represented a distinct category of evolutionary rates. However, current authors argue that slow rates of evolution are not statistically separable from the broad spectrum of moderate and rapid rates. Nevertheless, there may be some features in common that do differentiate more slowly evolving lineages.

Eldredge (1984) notes that it is not infrequent for the most primitive living members of a group to represent a little-diversified lineage from the base of the radiation. We might expect that the lineage that first invades a new adaptive zone would remain preeminent within that zone for a long period of time during which descendant species would be forced to specialize to utilize smaller subsections of the environment. *Didelphis*, as an apparent remnant of the first radiation of the marsupials, may be an example of this phenomenon, as might the tree shrews and perhaps some insectivores among the placentals.

Wake and his coauthors (1983) explain the long-term stasis of the salamander *Plethodon* by its capacity to accommodate to environmental perturbation without responding by structural change. They attribute this capacity to behavioral, physiological, and developmental plasticity. The ability of *Plethodon* to adapt behaviorally and physiologically to a wide range of diets and specific habitats without the necessity to change morphologically has permitted this genus to persist for more than 60 million years. They point out that limited morphological change in *Plethodon* is not correlated with a low rate of speciation, as Stanley (1984) argued to explain other examples of stasis. In fact, 26 species of *Plethodon* are recognized in the modern fauna, in contrast with 5 species of the morphologically diverse derivative *Aneides*.

Further study of long-lived, conservative lineages is desirable to establish just how limited their change has been and to determine whether or not they share significant features in common.

ADAPTIVE RADIATION

Environmental, structural, and physiological factors all tend to constrain the potential for evolutionary change. If any of these constraints are removed, the way may be open for large-scale adaptive radiation. Organisms may have the opportunity to invade a new adaptive zone simply because it is empty without any significant changes in their adaptive ability. Alternatively, changes in structure, physiology, or behavior may give organisms the facility to evolve within a previously unavailable adaptive zone.

The extinction of dinosaurs provided the opportunity for an extremely wide-scale radiation of mammals throughout the world in the early Cenozoic. There is no evidence for any significant structural or physiological change in either placentals or marsupials that coincided with the end of the Mesozoic. The placentals evolved explosively to fill a wide range of adaptive zones within a period of 1 to 5 million years (Van Valen, 1978). Most of the terrestrial adaptive zones were occupied by the end of the Paleocene, after which radiation was largely restricted within each zone. The rapid rate of adaptive and taxonomic evolution at the beginning of the Cenozoic can be explained by the lack of either competitors or predators. There is no need to assume that mutation rates were greater than normal or that speciation was more common or rapid than in the late Mesozoic. For a period of at least 1 million years, most lineages had a much higher potential for survival and gave rise to groups that were to persist for tens of millions of years.

The fossil record is not sufficiently complete in the latest Mesozoic and earliest Cenozoic to establish the rate of anatomical change, but it would not have had to be any higher than in the later Cenozoic to account for the amount of morphological diversity. Van Valen suggests that all the early Cenozoic groups could be included in a single family on anatomical grounds, although they are classified in several orders on the basis of the subsequent evolutionary history of their descendants.

The radiation of marsupials in South America probably followed a similar pattern, although the fossil record is not sufficiently complete to document its detailed characteristics. Even less is known of the early radiation of marsupials in Australia.

Some of the most dramatic examples of short-term radiation in newly available habitats are seen in the evolution of fish species flocks in both ancient and modern lakes (Echell and Kornfield, 1984). Although we lack information on critical periods of their evolution, evidence from the geological record demonstrates that hundreds of species may become differentiated within fewer than 10,000 years. Speciation and structural changes have occurred more rapidly than in any other well-documented sequences. Unfortunately, there is no evidence that this phenomenon has ever led to large-scale evolutionary change, since nearly all of these lakes have had a very short history, with most drying up after no more than 25,000 years. Lake Baikal is an obvious exception, but it is so isolated that its fauna is unlikely to contribute to the rest of the world.

Other dramatic radiations are associated with new structural, physiological, or behavioral patterns. The Paleozoic amphibians were able to invade the land only after the achievement of skeletal changes that permitted locomotion and support outside an aquatic environment. The amphibians radiated extensively during the Carboniferous, although the fossil record is too incompletely known to establish how rapid this process was.

The achievement of active flight in pterosaurs and birds similarly opened up an entirely new way of life and led to large-scale radiations. On a smaller scale, the achievement of more effective tarsal structure appears to have triggered the radiation of artiodactyls and perissodactyls in the early Cenozoic, as did the evolution of ever-growing incisors among the ancestral rodents.

However, there is not always a close correlation between significant structural and physiological changes and major radiations. Several other groups of Cenozoic mammals developed ever-growing incisors but without the success of rodents. Many of the physiological features of modern therian mammals had almost certainly evolved before the divergence of marsupials and placentals in the early Cretaceous, but the major radiation of these groups only occurred in the early Cenozoic. The definitive carnivore carnassial pattern evolved in the early Paleocene, but the major radiation leading to modern families did not occur until the late Eocene. In these cases, competition, predation, or other environmental factors may have been more important in influencing evolution than the attainment of new structural or physiological features.

On the other hand, the radiation of amniotes in the early Pennsylvanian and their dominance by the early Permian was achieved in the face of an earlier established amphibian fauna. However, the rate of early amniote radiation was not nearly as rapid as that of early Cenozoic placentals. Similarly, archosaurs appear to have achieved dominance in the middle and late Triassic at the expense of the therapsids, which had been much more widespread and diverse in the earlier Triassic. Successive radiations of both cartilaginous and bony fish appear to have resulted from the evolution of more effective patterns of feeding and locomotion, which led to the demise of more archaic groups.

The capacity of a group to undergo a major radiation requires both the facility to invade a new adaptive zone as a result of inherent structural, physiological, and behavioral factors and the opportunity to do so in relationship to available food supplies and a relatively low level of competition and predation.

EXTINCTION

It seems appropriate to end a book on vertebrate history with a discussion of extinction. Among mammals, for which we have the most complete record, the average species longevity is 2 to 3 million years. The average for other groups appears to be higher, but it may be biased by the better fossil record of longer-lived species. Schopf (1984) suggests that if we could apply the same criteria to fossil groups that are used for modern species, average species longevity might be as short as 200,000 years. No vertebrate species has escaped extinction for more than a few million years.

Two distinct events are seen as extinction: the termination of a lineage without descendants and its transition into a recognizably different species by phyletic evolution. The latter process may be termed **pseudoextinction.** Taxa may become extinct either locally (as in the case of the genus *Equus* in the late Pleistocene in North America) or throughout their range.

Extinction at the species level is clearly as natural and inevitable a result of evolution as is change. The nature of extinction has been studied extensively among modern species, which provide a model for local, short-term events. In a review of Pleistocene and recent extinctions of reptiles, birds, and mammals, Diamond (1984) shows that the probability of extinction is related most directly to population size and the area of its distribution. Extinction is most likely in species with small population size that occupy limited geographical areas. This phenomenon is most clearly demonstrated among animals occupying groups of islands of varying size but has been measured in other physically and biologically restricted areas. Population size is associated with trophic level and body size, so that large carnivores are the most liable to extinction and small herbivores the least. These observations suggest that over short time spans in limited geographical areas, stochastic factors of population dynamics are more important in governing the probability of extinction than are predation, competition, or physical changes in the environment.

Stochastic factors must have played an important role in producing a relatively constant level of extinction throughout vertebrate history. We may be able to establish the relative importance of stochastic and deterministic factors through study of the relative longevity of herbivorous and carnivorous species and species with large and small body size.

Prior to the Cenozoic, it is very difficult to establish species ranges and population sizes, and emphasis has been placed on other possible causes of extinction. The extinction of a number of genera with a similar morphology suggests some common environmental cause. Two categories are typically considered—interaction with other species, especially competition, and changes in the physical environment.

The successive replacement of groups inhabiting the same geographical area and making use of the same resources is logically explained by competition. The difficulty of demonstrating competition, even among living groups, is discussed in a symposium published by the *American Naturalist* (November 1983). It is even more difficult to demonstrate competition in a convincing manner from the fossil record. In the Paleozoic and Mesozoic, we rarely have sufficiently detailed evidence of time ranges to show that two species or larger taxonomic groups actually occupied the same habitat and succeeded one another immediately in time. Even in the case of the Plio-Pleistocene species *Australopithecus robustus* and *Homo habilis,* such evidence is very difficult to establish (Walker, 1984).

However, in some cases, the inference of extinction through competition and predation is fairly strong. Perhaps the most convincing example is provided by the extinction of all South American ungulates following the invasion of a host of competitors and predators at the close of the Tertiary (Figure 22-10) (Webb, 1976).

Large-scale extinction involving many elements of a fauna may be the result of modifications of the physical environment. The extinction of the rich fauna of primates in the North Temperate Zone following the Eocene has been attributed to gradual cooling, which led to the loss of extensive tropical forests. Changes in the terrestrial fauna in North America during the later Tertiary are associated with further replacement of the forests by savanna (Webb, 1984), which provided a new environment for grazing mammals.

The fossil record in the Cenozoic is sufficiently complete, at least in Europe and North America, to show the true spatial and temporal distribution of most genera. From this evidence, we can attribute most extinctions to causes similar to those that are evident in the modern fauna, although changes in climate and sea level were much greater over longer periods of time. The most dramatic recent change in the environment—caused by the repeated advances of continental ice sheets during the Pleistocene—modified the ranges of many species. Many others became extinct at the close of the Pleistocene because of the geographical reduction of arctic conditions (Martin and Klein, 1984).

Systematic consideration of extinctions during the Paleozoic and Mesozoic is much more difficult. The incompleteness of the stratigraphic record gives the appearance of a succession of large-scale extinctions. During the Mesozoic, for example, no dinosaur species are known to extend beyond a single stage boundary. Since most organisms are commonly, if not entirely, restricted to particular environments, their remains are typically found in similar sediments. When geological conditions change, groups frequently vanish from the fossil record. Their disappearance may reflect local extinction but does not necessarily indicate total extinction, since individual species or entire faunas may be able to migrate to more favorable areas that may not be represented in the fossil record.

The longevity and the turnover rates of species and genera measured from our current knowledge of the fossil record of the Paleozoic and Mesozoic may be largely a reflection of geological factors and may have only limited biological significance. Because of the large number of species in families, the presence or absence of families should be much less influenced by chance and so should give a more accurate, if less detailed, view of extinction events.

In addition to the more-or-less constant rate of extinction that results from stochastic processes and normal environmental change (which may be referred to as background extinction), the fossil record of marine invertebrates shows several much larger-scale extinction events. Raup and Sepkoski (1984) argue that these major extinction events are recurrent, with a periodicity of approximately 26 million years, suggesting a single, probably extraterrestrial, cause. However, their paper has been seriously criticized by Hallam (1984) and Hoffman (1985). Major extinctions in the marine environment are said to have peaked at the following times:

	Peak observed, million years before present
Tertiary	
Middle Miocene	11.3
Late Eocene	38
Cretaceous	
Maastrichtian	65
Cenomanian	91
Hauterivian	125
Jurassic	
Tithonian	144
Callovian	163
Bajocian	175
Pliensbachian	194
Triassic	
Norian	219
Olenekian	243
Permian	
Dzhulfian	248

Except for the extinction at the end of the Cretaceous, which is discussed in Chapter 15, there is not much confirmation for this pattern from the vertebrate fossil record. Thomson's (1977) study of rates of extinction of fish showed no significant coincidence with the Raup-Sepkoski time table. The most dramatic extinction in the marine environment occurred at the end of the Permian, wiping out 95 percent of the nonvertebrate species and more than half the families. Surprisingly, there was not a correspondingly large extinction of either terrestrial or aquatic vertebrates (Schaeffer, 1973). The fossil record of terrestrial vertebrates is very poor in the Lower Triassic, but the late Permian and early Triassic appear to be the time of origination of a host of lineages that dominated the Mesozoic.

The problems of evaluating the vertebrate fossil record across the Cretaceous-Tertiary boundary were discussed in Chapter 15. The present evidence is not sufficient to support a sudden, catastrophic extinction at that time. McKenna (1983) and Prothero (1985) also cast doubt on the catastrophic nature of extinctions near the Eocene-Oligocene boundary.

Although some periodically recurring factor may have led to large-scale extinction events among marine invertebrates, there is little in the fossil record of either vertebrates or vascular plants (Knoll, 1984) to support this hypothesis. With the possible exception of the extinction at the end of the Cretaceous, there is no strong evidence for a significant effect of catastrophic extraterrestrial forces on the history of vertebrates (Stanley, 1987).

A new agent of large-scale extinction was added in the Pleistocene with the emergence and wide distribution of modern man. There is continued dispute as to how significant a role humans had in the last dramatic extinction at the end of the Pleistocene. This extinction involved primarily large mammals weighing more than 44 kilograms (Martin and Wright, 1967; Martin and Klein, 1984). Seventy-eight species, 37 genera, and 7 families became extinct in North America. Martin (1984) argues that these extinctions occurred approximately 11 thousand years ago and can be attributed directly to the activities of human hunters. In contrast, Webb (1984) argues that these extinctions can be accounted for primarily by climatic factors.

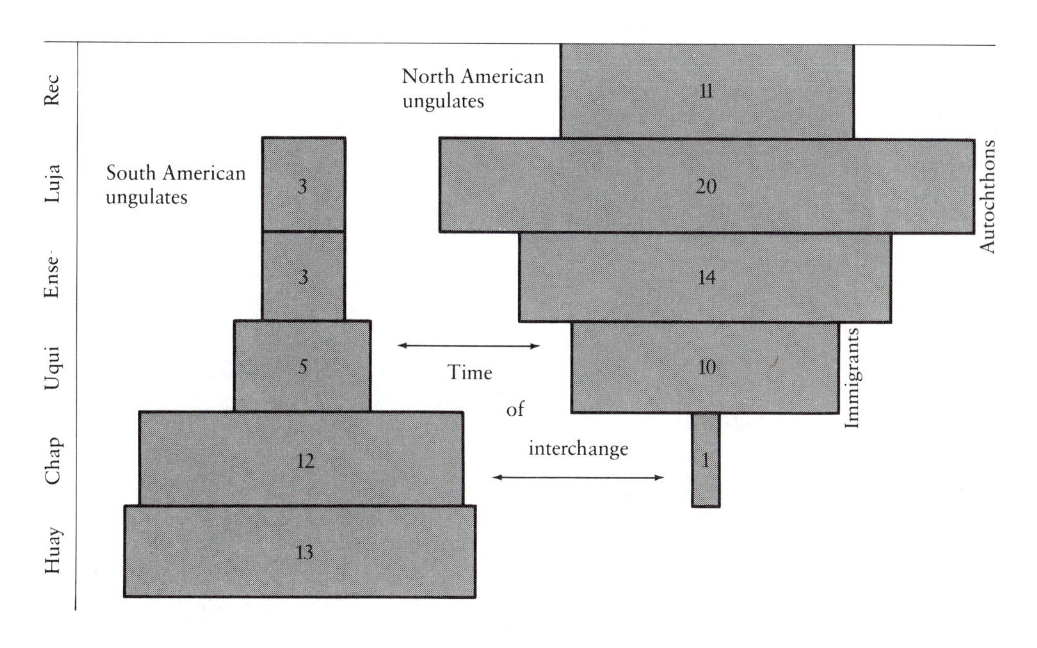

Figure 22-10. TURNOVER OF UNGULATE GENERA IN SOUTH AMERICA. Numbers in boxes represent number of genera. Scale at left shows late Tertiary and Quaternary stages recognized in South America. *From Webb, 1976.*

There is no debate concerning the human responsibility for a host of more recent extinctions. Day (1981) stated that more than 150 vertebrate species have become extinct during the past 300 years. There is no reason to think that mankind is exempt from the same processes that have led to the extinction of all other species that have inhabited this planet.

SUMMARY

The fossil record of vertebrates provides an important basis for determining large-scale evolutionary processes and rates. In contrast with the ideas of Darwin and the modern evolutionary synthesis, Eldredge, Gould, and Stanley argue that morphological change occurs primarily at the time of speciation and that little significant change occurs during their subsequent duration. Because of the incomplete nature of the fossil record, it is difficult to test this hypothesis, but many examples of progressive change within species and between species and genera have been described. The fossil record shows that rates of evolution are extremely variable, rather than consistently following either a gradualistic pattern or a punctuated pattern.

Evolutionary rates can be measured in darwins, a change by a factor of *e* per million years. A rate of approximately 0.1 darwin is common for vertebrates measured over a period of 1 million years. Lower rates are recorded over longer periods of time because of the averaging of variable factors.

Macroevolution involves the origin and radiation of major groups, the evolution of distinct new structures and physiological processes, and major adaptive shifts. Many theories have been proposed to explain these processes, but all can be attributed to variation and selection acting at the level of populations and species. The following features are characteristic of transitions between major adaptive zones:

1. A gap of 10 to 20 million years frequently separates ancestral and descendant forms.
2. Intermediates are rare.
3. Only those parts of the skeleton that are specifically associated with the habitat shift are modified.
4. The transition may involve functional changes in elements that remain structurally similar.
5. Many features that are characteristic of typical members of the derived group evolve subsequent to the adaptive shift.

The absence of intermediate forms may be explained by the difficulty of adapting to intermediate habitats. The entry into a new adaptive zone may be initiated by behavioral changes, with modification of structures and physiology occurring subsequently.

Physical and biological aspects of the environment, as well as inherent limitations of structural elements and physiology, constrain vertebrate evolution to a relatively few major adaptive zones and explain why vertebrate history is characterized by occasional large-scale radiations followed by the long-term survival of a relatively small number of adaptive types. It is not clear whether developmental patterns are important in constraining the course of evolution or are themselves controlled by stabilizing selection.

Major radiations require both the structural, physiological, and behavioral facility to invade a new adaptive zone and the opportunity to do so with respect to available resources and a relatively low level of competition and predation.

Species have an estimated longevity ranging from 200,000 to several million years. In Pleistocene and Recent populations, the probability for extinction is directly proportional to body size and trophic level and inversely proportional to population size and the extent of its range. Competition and changes in the physical environment are probably important factors that lead to extinction over the longer scale of the fossil record, although they are difficult to demonstrate. The depositional record in the Paleozoic and Mesozoic is too incomplete to provide accurate measures of the duration of species and genera. The fossil record of vertebrates does not show convincing evidence of recurrent catastrophic extinction as a result of extraterrestrial factors.

REFERENCES

Alberch, P. (1980). Ontogenesis and morphological diversification. *Amer. Zool.*, **20**: 653–667.

Alberch, P., and Gale, E. A. (1985). A developmental analysis of an evolutionary trend: Digital reduction in amphibians. *Evolution*, **39**: 8–23.

American Naturalist. (1983). **122**(5): 601–738.

Andrews, S. M., and Westoll, T. S. (1970). The postcranial skeleton of rhipidistian fishes excluding *Eusthenopteron*. *Trans. Roy. Soc. Edinb.*, **68**: 391–489.

Bartholomew, G. A., Bennet, A. F., and Dawson, W. R. (1976). Swimming, diving and lactate production of the marine iguana *Amblyrhynchus cristatus*. *Copeia*; 709–720.

Barton, N. H., and Charlesworth, B. (1984). Genetic revolutions, founder effects, and speciation. *Ann. Rev. Ecol. Syst.*, **15**: 133–164.

Bell, M. A., Baumgarten, J. V., and Olson, E. C. (1985). Patterns of temporal change in single morphological characters of a Miocene stickleback fish. *Paleobiology*, **11**: 258–271.

Benson, R. H., Chapman, R. E., and Siegel, A. F. (1982). On the measurement of morphology and its change. *Paleobiology*, **8**: 328–339.

Bergren, W. A., and Casey, R. E. (1983). Introduction to the symposium on the tempo and mode of evolution from micropaleontological data. *Paleobiology,* **9**: 326.

Biological Journal of the Linnean Society. (1980). **14**(3–4): 259–434.

Bonner, J. T. (1982). *Evolution and Development.* Springer-Verlag, Berlin, Heidelberg, New York.

Bookstein, F., Chernoff, B., Elder, R., Humphries, J., Smith, G., and Strauss, R. (1985). Morphometrics in Evolutionary Biology. *Special Publication* **15**. The Academy of Natural Sciences of Philadelphia. 277 pages.

Carroll, R. L. (1970). Quantitative aspects of the amphibian-reptilian transition. *Forma et functio,* **3**: 165–178.

Carroll, R. L. (1977). Patterns of amphibian evolution: An extended example of the incompleteness of the fossil record. In A. Hallam (ed.), *Patterns of Evolution,* pp. 405–437. Elsevier, Amsterdam, Oxford, New York.

Carroll, R. L. (1984). Problems in the use of terrestrial vertebrates for zoning of the Carboniferous. *Neuvième Congrès Internationale de Stratigraphie et de du Géologie du Carbonifère. Compte rendu,* **2**: 135–147.

Carroll, R. L. (1985). Evolutionary constraints in aquatic diapsid reptiles. In J. C. W. Cope and P. W. Skelton (eds.), Evolutionary case histories from the fossil record. *Special Papers in Palaeontology,* **33**: 1–203. The Palaeontological Association, London.

Carroll, R. L., and Holmes, R. (1980). The skull and jaw musculature as guides to the ancestry of salamanders. *Zool. J. Linn. Soc.,* **68**(1): 1–40.

Chaline, J., and Laurin, B. (1986). Phyletic gradualism in a European Plio-Pleistocene *Mimomys* lineage (Arvicolidae, Rodentia). *Paleobiology,* **12**: 203–216.

Charlesworth, B., Lande, R., and Slatkin, M. (1982). A neo-Darwinian commentary on macroevolution. *Evolution,* **36**(3): 474–498.

Cope, J. C. W., and Skelton, P. W. (eds.). (1985). Evolutionary case histories from the fossil record. *Special Papers in Palaeontology,* **33**: 1–203. The Palaeontological Association, London.

Cronin, J. E., Boaz, N. T., Stringer, C. B., and Rak, Y. (1981). Tempo and mode in hominid evolution. *Nature,* **292**: 113–122.

Darwin, C. (1859). *The Origin of Species.* John Murray, London.

Day, D. (1981). *The Doomsday Book of Animals. A Natural History of Vanished Species.* Viking Press, New York.

Diamond, J. M. (1984). "Normal" extinctions of isolated populations. In M. H. Nitecki (ed.), *Extinctions,* pp. 191–246. University of Chicago Press, Chicago.

Dingus, L. W. (1984). Effects of stratigraphic completeness on interpretations of extinction rates across the Cretaceous-Tertiary boundary. *Paleobiology,* **10**: 420–438.

Dingus, L. W., and Sadler, P. M. (1982). Expected completeness of sedimentary sections: Estimating a time-scale dependent, limiting factor in the resolution of the fossil record. *Third North Am. Paleont. Conv. Proc.,* **2**: 461–464.

Dobzhansky, T. (1937). *Genetics and the Origin of Species.* Columbia University Press, New York.

Dobzhansky, T., Ayala, F. J., Stebbins, G. L., and Valentine, J. W. (1977). *Evolution.* W. H. Freeman and Company, New York.

Echell, A. A., and Kornfield, I. (eds.). (1984). *Evolution of Fish Species Flocks.* University of Maine at Orono Press, Orono, Maine.

Eldredge, N. (1984). Simpson's Inverse: Bradytely and the phenomenon of living fossils. In N. Eldredge and S. M. Stanley (eds.), *Living Fossils,* pp. 272–277. Springer-Verlag, Berlin, Heidelberg, New York.

Eldredge, N. (1985). *Time Frames.* Simon and Schuster, New York.

Eldredge, N., and Gould, S. J. (1972). Punctuated equilibria: An alternative to phyletic gradualism. In T. J. M. Schopf (ed.), *Models in Paleobiology,* pp. 82–115. Freeman, Cooper, San Francisco.

Eldredge, N., and Stanley, S. M. (eds.). (1984a). *Living Fossils.* Springer-Verlag, Berlin, Heidelberg, New York, Tokyo.

Eldredge, N., and Stanley, S. M. (1984b). Living fossils: Introduction to the casebook. In N. Eldredge and S. M. Stanley (eds.), *Living Fossils,* pp. 1–3. Springer-Verlag, Berlin, Heidelberg, New York, Tokyo.

Emry, R. J., and Thorington, R. W., Jr. (1984). The tree squirrel *Sciurus* (Sciuridae, Rodentia) as a living fossil. In N. Eldredge and S. M. Stanley (eds.), *Living Fossils,* pp. 23–31. Springer-Verlag, Berlin, Heidelberg, New York, Tokyo.

Endler, J. A. (1977). Geographic variation, speciation, and clines. *Monographs in Population Biology,* no. 10. Princeton University Press, Princeton.

Endler, J. A. (1986). *Natural Selection in the Wild.* Princeton University Press, Princeton.

Extinctions in Vertebrate History. (1987). *Memoires de la Société géologique de France, N. S.,* no. 150.

Fahlbusch, V. (1983). Makroevolution. Punktualismus. Ein Diskussionsbeitrag am Beispiel miozäner Eomyiden (Mammalia, Rodentia). *Paläont. Z.,* **57**: 213–230.

Fisher, R. A. (1930). *The Genetical Theory of Natural Selection.* Clarendon Press, Oxford.

Gilinsky, N. L. (1986). Species selection as a causal process. *Evol. Biol.* **20**: 249–273.

Gingerich, P. D. (1976). Paleontology and phylogeny: Patterns of evolution at the species level in early Tertiary mammals. *Am. J. Sci.,* **276**: 1–28.

Gingerich, P. D. (1977). Patterns of evolution in the mammalian fossil record. In A. Hallam (ed.), *Patterns of Evolution, as Illustrated by the Fossil Record,* pp. 469–500. Elsevier, Amsterdam.

Gingerich, P. D. (1980). Evolutionary patterns in early Cenozoic mammals. *Ann. Rev. Earth Planet. Sci.,* **8**: 407–424.

Gingerich, P. D. (1982). Time resolution in mammalian evolution: Sampling, lineages, and faunal turnover. *Third North Am. Paleont. Conv., Proc.,* **1**: 205–210.

Gingerich, P. D. (1983). Rates of evolution: Effects of time and temporal scaling. *Science,* **222**: 159–161.

Goodwin, B. C., Holder, N., and Wylie, C. C. (eds.). (1983). *Development and Evolution. Sixth Symp. Brit. Soc. Dev. Biol.* Cambridge University Press, Cambridge.

Gould, S. J. (1966). Allometry and size in ontogeny and phylogeny. *Biol. Rev.,* **41**: 587–640.

Gould, S. J. (1974). The origin and function of "bizarre" structures: Antler size and skull size in the "Irish Elk," *Megaloceros gigantus. Evolution,* **28**: 191–220.

Gould, S. J. (1975a). On the scaling of tooth size in mammals. *Am. Zool.,* **15**: 351–362.

Gould, S. J. (1975b). Allometry in primates, with emphasis on scaling and the evolution of the brain. In H. Kuhn, W. P.

Luckett, C. R. Nobach, A. H. Schultz, D. Stark, and F. S. Szalay (eds.), *Contributions to Primatology,* Vol. 5, pp. 244–292. S. Karker, Basel.

Gould, S. J. (1980). The evolutionary biology of constraint. *Daedalus,* **109**: 39–52.

Gould, S. J. (1982a). Darwinism and the expansion of evolutionary theory. *Science,* **216**: 380–387.

Gould, S. J. (1982b). The meaning of punctuated equilibrium and its role in validating a hierarchical approach to macroevolution. In R. Milkman (ed.), *Perspectives on Evolution,* pp. 83–104. Sinauer Associates Inc., Sunderland, Mass.

Gould, S. J., and Eldredge, N. (1977). Punctuated equilibria: The tempo and mode of evolution reconsidered. *Paleobiology,* **3**: 115–151.

Haldane, J. B. S. (1949). Suggestions as to quantitative measurement of rates of evolution. *Evolution,* **3**: 51–56.

Hallam, T. (1984). Asteroids and extinction—No cause for concern. *New Scientist,* **8**: 30–33.

Hanken, J. (1984). Miniaturization and its effect on cranial morphology in plethodontid salamanders, genus *Thorius* (Amphibia: Plethodontidae). I. Osteological variation. *Biol. J. Linn. Soc.,* **23**: 55–75.

Harris, J., and White, T. D. (1979). Evolution of Plio-Pleistocene African Suidae. *Trans. Am. Phil. Soc.,* **69**: 1–128.

Hildebrand, M. (1982). *Analysis of Vertebrate Structure* (2d. ed.). Wiley, New York.

Hinchliffe, J. R., and Griffiths, P. J. (1983). The prechondrogenic patterns in tetrapod limb development and their phylogenetic significance. In B. C. Goodwin, N. Holder, and C. C. Wylie (eds.), *Development and Evolution. Sixth Symp. Brit. Soc. Dev. Biol.,* pp. 99–121. Cambridge University Press, Cambridge.

Hoffman, A. (1985). Patterns of family extinction depend on definition and geological time scale. *Nature,* **315**: 659–662.

Holder, N. (1983). Developmental constraints and the evolution of vertebrate digit patterns. *J. Theoret. Biol.,* **104**: 451–471.

Hürzeler, J. (1962). Kann die biologische Evolution, wie sie sich in der Vergangenheit abgespielt hat, exakt erfasst werden? *Stud. Kath. Akad. Bayern,* **16**: 15–36.

Huxley, J. S. (1942). *Evolution and the Modern Synthesis.* Allen and Unwin, London.

Jarvik, E. (1980). *Basic Structure and Evolution of Vertebrates.* Vols. 1 and 2. Academic Press, London.

Jones, G. M., and Spells, K. E. (1963). A theoretical and comparative study of the functional dependence of the semicircular canal upon its physical dimensions. *Proc. Roy. Soc. Lond., B,* **157**: 403–419.

King, M. C., and Wilson, A. C. (1975). Evolution at two levels in humans and chimpanzees. *Science,* **188**: 107–116.

Knoll, A. H. (1984). Patterns of extinction in the fossil record of vascular plants. In M. H. Nitecki (ed.), *Extinction,* pp. 21–68. University of Chicago Press, Chicago.

Krishtalka, L., and Stucky, R. K. (1985). Revision of the Wind River Faunas. Early Eocene of Central Wyoming. Part 7. Revision of *Diacodexis* (Mammalia, Artiodactyla). *Ann. Carnegie Mus.,* **54**: 413–486.

Kurtén, B. (1959). Rates of evolution in fossil mammals. In *Genetics and Twentieth Century Darwinism,* pp. 205–215. *Cold Spring Harbor Symposia on Quantitative Biology,* vol. 24.

Kurtén, B. (1968). *Pleistocene Mammals of Europe.* Weidenfels and Nicolson, London.

Kurtén, B., and Anderson, E. (1980). *Pleistocene Mammals of North America.* Columbia University Press, New York.

Lighthill, M. J. (1970). Aquatic animal propulsion of high hydromechanical efficiency. *J. Fluid Mech.,* **44**: 265–301.

MacFadden, B. J. (1985). Patterns of phylogeny and rates of evolution in fossil horses: Hipparions from the Miocene and Pliocene of North America. *Paleobiology,* **11**: 245–257.

Maglio, V. J. (1973). Origin and evolution of the Elephantidae. *Trans. Am. Phil. Soc., New Ser.,* **63**: 1–149.

Martin, P. S. (1984). Catastrophic extinctions and Late Pleistocene Blitzkrieg: Two radiocarbon tests. In M. H. Nitecki (ed.), *Extinctions,* pp. 153–189. University of Chicago Press, Chicago.

Martin, P. S., and Klein, R. G. (1984). *Quaternary Extinctions, a Prehistoric Evolution.* University of Arizona Press, Tucson.

Martin, P. S., and Wright, H. E. (eds.). (1967). *Pleistocene Extinctions: A Search for a Cause.* Yale University Press, New Haven.

Maynard-Smith, J. (1981). Macroevolution. *Nature,* **289**: 13–14.

Maynard-Smith, J. (1983). Current controversies in evolutionary biology. In M. Greene (ed.), *Dimensions of Darwinism,* pp. 273–286. Cambridge University Press, Cambridge.

Maynard-Smith, J., Burian, R., Kauffman, S., Alberch, P., Campbell, J., Goodwin, B., Lande, R., Raup, D., and Wolpert, L. (1985). Developmental constraints and evolution. *Quart. Rev. Biol.,* **60**: 265–287.

Mayr, E. (1942). *Systematics and the Origin of Species.* Columbia University Press, New York.

Mayr, E. (1954). Change of genetic environment and evolution. In J. Huxley, A. C. Hardy, and E. B. Ford (eds.), *Evolution as a Process,* pp. 157–180. Allen and Unwin, London.

Mayr, E. (1963). *Animal Species and Evolution.* Harvard University Press, Cambridge, Mass.

Mayr, E. (1970). *Populations, Species, and Evolution.* Harvard University Press, Cambridge.

Mayr, E. (1982). Speciation and macroevolution. *Evolution,* **36**: 1119–1132.

McCune, A. R., Thomson, K. S., and Olsen, P. E. (1984). Semionotid fishes from the Mesozoic Great Lakes of North America. In A. A. Echelle and I. Kornfield (eds.), *Evolution of Fish Species Flocks,* pp. 27–44. University of Maine at Orono Press, Orono, Maine.

McGowan, C. (1983). The Successful Dragons. *A Natural History of Extinct Reptiles.* Samuel Stevens, Toronto.

McKenna, M. C. (1983). Holarctic landmass rearrangement, cosmic events, and Cenozoic terrestrial organisms. *Ann. Missouri Bot. Gard.,* **70**: 459–489.

Moore, R. C., Lalicker, C. G., and Fischer, A. G. (1952). *Invertebrate Fossils.* McGraw-Hill, New York.

Novacek, M. (1984). Evolutionary stasis in the elephant shrew, *Rhynchocyon.* In N. Eldredge and S. M. Stanley (eds.), *Living Fossils,* pp. 4–22. Springer-Verlag, Berlin, Heidelberg, New York, Tokyo.

Prothero, D. R. (1985). North American mammalian diversity and Eocene-Oligocene extinctions. *Paleobiology,* **11**: 389–405.

Radinsky, L. B. (1966). The adaptive radiation of the phenacodontid condylarths and the origin of the Perissodactyla. *Evolution,* **20**: 408–417.

Raff, R. A., and Kaufman, T. C. (1983). *Embryos, Genes and Evolution. The Developmental-Genetic Basis of Evolution-*

ary Change. Macmillan, New York.
Raup, D. M., and Sepkoski, J. J. (1984). Periodicity of extinctions in the geologic past. *Proc. Natl. Acad. Sci. USA,* **81**: 801–805.
Rieppel, O. (1984). Miniaturization of the lizard skull: Its functional and evolutionary implications. In M. W. J. Ferguson (ed.), The Structure, Development and Evolution of Reptiles. *Symposia Zool. Soc. Lond.,* **52**: 503–520.
Rightmire, G. P. (1981). Patterns in the evolution of *Homo erectus. Paleobiology,* **7**: 241–246.
Romer, A. S. (1966). *Vertebrate Paleontology* (3d ed.). University of Chicago Press, Chicago.
Rose, K. D., and Bown, T. M. (1984). Gradual phyletic evolution at the generic level in early Eocene omomyid primates. *Nature,* **309**: 250–252.
Rose, K. D., and Emry, R. J. (1983). Extraordinary fossorial adaptations in the Oligocene palaeanodonts *Epoicotherium* and *Xenocranium* (Mammalia). *J. Morph.,* **175**: 33–56.
Schaeffer, B. (1973). Fishes and the Permian-Triassic boundary. *Pub. Can. Soc. Pet. Geol. Calgary,* 1973: 493–497.
Schindel, D. E. (1982). Resolution analysis: A new approach to the gaps in the fossil record. *Paleobiology,* **8**: 340–353.
Schmidt-Nielsen, K. (1975). *Animal Physiology. Adaptation and Environment.* Cambridge University Press, Cambridge.
Schopf, T. J. M. (1984). Rates of evolution and the notion of "Living Fossils". *Ann. Rev. Earth Planet. Sci.,* **12**: 245–292.
Shubin, N. H., and Alberch, P. (1986). A morphogenetic approach to the origin and basic organization of the tetrapod limb. *Evol. Biol.,* **20**: 319–387.
Simpson, G. G. (1944). *Tempo and Mode in Evolution.* Columbia University Press, New York.
Simpson, G. G. (1948). The beginning of the age of mammals in South America. Part 1. *Bull. Am. Mus. Nat. Hist.,* **91**: 1–232.
Simpson, G. G. (1953). *The Major Features of Evolution.* Columbia University Press, New York.
Simpson, G. G. (1967). The beginning of the age of mammals in South America. Part 2. *Bull. Am. Mus. Nat. Hist.,* **137**: 1–259.
Stanley, S. M. (1975). A theory of evolution above the species level. *Proc. Natl. Acad. Sci. USA,* **72**: 646–650.
Stanley, S. M. (1979). *Macroevolution: Pattern and Process.* W. H. Freeman and Company, New York.
Stanley, S. M. (1981). *The New Evolutionary Timetable. Fossils, Genes, and the Origin of Species.* Basic Books, New York.
Stanley, S. M. (1982). Macroevolution and the fossil record. *Evolution,* **36**: 460–473.
Stanley, S. M. (1984). Does bradytely exist? In N. Eldredge and S. M. Stanley (eds.), *Living Fossils,* pp. 278–280. Springer-Verlag, Berlin, Heidelberg, New York, Tokyo.
Stanley, S. M. (1987). *Extinction.* Scientific American Library, New York.
Stebbins, G. L., and Ayala, F. J. (1981). Is a new evolutionary synthesis necessary? *Science,* **213**: 967–971.
Stock, G. B., and Bryant, S. V. (1981). Studies of digit regeneration and their implications for theories of development and evolution of vertebrate limbs. *J. Exp. Zool.,* **216**: 423–433.
Thompson, D'Arcy, W. (1966). *On Growth and Form* (Abridged ed.), J. T. Bonner (ed.). Cambridge University Press, Cambridge.
Thomson, K. S. (1977). The pattern of diversification among fishes. In A. Hallam (ed.), *Patterns of Evolution, as Illustrated by the Fossil Record,* pp. 377–404. Elsevier, Amsterdam.
Van Valen, L. (1973). A new evolutionary law. *Evol. Theory,* **1**: 1–30.
Van Valen, L. (1978). The beginning of the age of mammals. *Evol. Theory,* **4**: 45–80.
Wake, D. B., Roth, G., and Wake, M. H. (1983). On the problem of stasis in organismal evolution. *J. Theoret. Biol.* **101**: 211–224.
Walker, A. (1984). Extinction in hominid evolution. In M. H. Nitecki (ed.), *Extinctions,* pp. 119–152. University of Chicago Press, Chicago.
Webb, P. W. (1982). Locomotor patterns in the evolution of actinopterygian fishes. *Amer. Zool.,* **22**: 329–342.
Webb, S. D. (1976). Mammalian fauna dynamics of the great American interchange. *Paleobiology,* **2**: 220–234.
Webb, S. D. (1984). Neogene faunas of North America. In P. D. Gingerich and C. E. Badgley (eds.), *Mammals. Notes for a Short Course. Univ. Tennessee Stud. Geol.,* **8**: 128–138.
West, R. M. (1979). Apparent prolonged evolutionary stasis in the middle Eocene hoofed mammal *Hyopsodus. Paleobiology,* **5**: 252–260.
Williamson, P. G. (1981). Palaeontological documentation of speciation in Cenozoic molluscs from Turkana Basin. *Nature,* **293**: 437–443.
Wilson, J. A. (1971). Early Tertiary vertebrate faunas, Vieja Group. Trans-Pecos Texas: Agrichoeridae and Merycoidodontidae. *Texas Mem. Mus. Bull.,* **18**: 1–83.
Wolpert, L. (1983). Constancy and change in the development and evolution of pattern. In B. C. Goodwin, N. Holder, and C. C. Wylie (eds.), *Development and Evolution. Sixth Symp. Brit. Soc. Dev. Biol.,* pp. 47–57. Cambridge University Press, Cambridge.
Wolpoff, M. (1984). Evolution in *Homo erectus:* The question of stasis. *Paleobiology,* **10**: 389–406.
Wright, S. (1931). Evolution in Mendelian populations. *Genetics,* **16**: 97–159.

A P P E N D I X

VERTEBRATE CLASSIFICATION

The following tabulation is intended to include most vertebrate genera that are known from fossil remains. They are arranged in an essentially Linnean system of classification that groups together closely related taxa in a hierarchical order. This classification is based on the most recent general reviews, plus the incorporation of subsequently described genera.

Agnatha	Reviewed by Dr. David Elliott
Placodermi	(Denison, 1978)
Chondrichthyes	Paleozoic genera (Zangerl, 1981)
	Modern elasmobranchs (Compagno, 1977)
Acanthodii	(Denison, 1979)
Osteichthyes	General classification from Nelson (1984), based on the cladistic analysis of Lauder and Liem (1983)
	Sarcopterygians (Thomson, 1969)
Amphibians	(Carroll and Winer, 1977; Estes, 1981)
Reptiles	Anapsida (Carroll, 1982, and references cited)
	Chelonia (Młynarski, 1976)
	Lepidosaurs (Estes, 1983; Rage, 1984)
	Archosauromorpha and Synapsida (tabulation by Dr. Hans-Dieter Sues)
Birds	(Olson, 1985, and references cited; Gruson, 1976)
Mammals	Nontherian mammals (Lillegraven, Kielan-Jaworowska, and Clemens, 1979)
	Marsupials (Marshall, 1981)
	Placental mammals (Savage and Russell, 1983)
	Classification of modern families (Anderson and Jones, 1984)

For Mesozoic and Cenozoic elasmobranchs and actinopterygian groups that have not been systematically reviewed during the past 20 years, listing at the generic level is emended from Romer (1966). (References for these citations appear at the end of the appendix.)

It should be recognized that such a compendium is subject to many sources of error. A particularly serious problem is the omission of species-level diversity. For this information, the reader must turn to the primary literature. For most groups, there is still significant disagreement as to specific phylogenetic relationships, the validity of particular taxa, and the name and rank of higher taxonomic units. What appears here is a compromise among several, sometimes quite divergent, approaches.

It should also be recognized that such a listing is not an accurate reflection of the longevity of taxa. They are cited as occurring within a particular geological stage but may be known from only a single horizon or may have lived throughout the stage. Uncritical analysis of data of this nature may give a very misleading impression of evolutionary processes.

The methodology of classification is currently in a state of instability greater than at any period since the time of Linnaeus. Together with the great increase in knowledge of many aspects of vertebrate phylogeny, a major modification of the taxonomy of all vertebrates may be possible before the end of the century. However, it is certainly not practical at the present time.

As discussed in Chapter 1, systems of classification can be used to show specific phylogenetic relationships, but this approach requires special notations and such a great number of taxonomic ranks that it is not practical when applied to a group as diverse as the vertebrates.

This classification has the more limited goal of indicating what genera and families are grouped in higher taxonomic categories. It is not meant to show the specific nature of the interrelationships of either the genera or the larger groups. In fact, these relationships are only rarely known with sufficient assurance to justify their inclusion in a system of formal classification.

ABBREVIATIONS

Geological

Camb., Cambrian; Carb., Carboniferous; Cret., Cretaceous; Dev., Devonian; Eoc., Eocene; Jur., Jurassic; L., Lower; M., Middle; Mioc., Miocene; Miss., Mississippian; Olig., Oligocene; Ord., Ordovician; Paleoc., Paleocene; Penn., Pennsylvanian; Perm., Permian; Pleist., Pleistocene; Plioc., Pliocene; R., Recent; Sil., Silurian; Quat., Quaternary; Tert., Tertiary; Trias., Triassic; U., Upper.

Geographical

Af., Africa; Ant., Antarctica; Arc., Arctic; As., Asia; Atl., Atlantic; Aus., Australia; CA., Central America; CAs., Central Asia; Cos., Cosmopolitan; EAf., East Africa; EAs., Eastern Asia; EInd., East Indies; EEu., Eastern Europe; Eu., Europe; Gr., Greenland; Ind., India; Indo-Pac. Oc., Indo-Pacific Ocean; Mad., Madagascar; Maurit., Mauritius; Med., Mediterranean; NA., North America; NAf., North Africa; NAs., Northern Asia; NAtl., North Atlantic; NNA., Northern North America; NOc., Northern Oceans; NPac., North Pacific; NZ., New Zealand; Oc., Ocean(s); Pac., Pacific; SA., South America; SAf., South Africa; SAs., Southern Asia; SAtl., South Atlantic; SOc., Southern Oceans; SPac., South Pacific; Spits., Spitsbergen; SWAs., Southwest Asia; Tas., Tasmania; Trop., Tropics; WAf., West Africa; WAs., Western Asia; WInd., West Indies.

CLASS AGNATHA

SUBCLASS PTERASPIDOMORPHI (DIPLORHINA)

ORDER HETEROSTRACI (PTERASPIDIFORMES) **?Arandaspididae** *Anatolepis* U. Camb. NA. L. Ord. Spits. Gr. NA. *Arandaspis* M. Ord. Aus. *Porophoraspis* M. Ord. Aus. **Astraspidae** *Astraspis* [*Pychaspis*] M. Ord. NA. **Eriptychiidae** *Eriptychius* M. Ord. NA. **Cyathaspididae** *Alainaspis* U. Sil. NNA. *Allocryptaspis* [*Cryptaspis*] L. Dev. NA. *Americaspis* [*Palaeaspis*] U. Sil. NA. *Anglaspis* [*Frankelaspis*] L. Dev. Eu. Spits. *Archegonaspis* [*Eoarchegonaspis Lauaspis*] U. Sil. ? L. Dev. Eu. U. Sil. NA. *Ariaspis* U. Sil. NA. *Asketaspis* L. Dev. NNA. *Boothiaspis* U. Sil. NNA. *Ctenaspis* [*Bothriaspis*] L. Dev. Eu. Spits. *Cyathaspis* [*Diplaspis*] U. Sil. NA. U. Sil. L. Dev. Eu. L. Dev. NAs. *Davelaspis* L. Dev. Spits. *Dikenaspis* L. Dev. NA. *Dinaspidella* [*Dinaspis*] L. Dev. NA. Spits. *Homalaspidella* [*Homalaspis Hemaspis*] L. Dev. NA. Spits. *Irregulareaspis* [*Dictyaspidella Dictaspis*] L. Dev. Eu. Spits. *Liliaspis* U. Sil. NAtl. *Listraspis* L. Dev. NA. *Nahanniaspis* U. Sil. L. Dev. NNA. *Pionaspis* L. Dev. NA. *Poraspis* [*Holaspis*] L. Dev. Eu. Spits. NNA. *Ptomaspis* M.-U. Sil. NNA. *Seretaspis* L. Dev. Eu. *Steinaspis* L. Dev. Eu. *Tolypelepis* [*Tolypaspis*] U. Sil. L. Dev. NNA. Eu. *Torpedaspis* U. Sil. L. Dev. NNA. *Vernonaspis* [*?Anatiftopsis Eoarchegonaspis*] U. Sil. L. Dev. NNA. **Amphiaspidae** *Amphiaspis Angaraspis Aphataspis Argyriaspis Edaphaspis Eglonaspis Empedaspis Gabreyaspis Gerronaspis* L. Dev. NAs. *Gunaspis* L. Dev. Eu. *Hibernaspis Kureykaspis Lecaniaspis Litotaspis Pelurgaspis Olbiaspis Pelaspis Prosarctaspis Putoranaspis Sanidaspis Siberiaspis Tareyaspis Tuxeraspis* L. Dev. NAs. **Corvaspidae** *Corvaspis* U. Sil. L. Dev. EEu. NNA. Spits. **Traquairaspididae** *Traquairaspis* [*Lophopiscis Lophaspis Orthaspis Phialaspis Yukonaspis*] U. Sil. Eu. U. Sil. L. Dev. NA. L. Dev. Spits. *Weigeltaspis* L. Dev. NNA. Eu. **Pteraspidae** *Althaspis* [*Podolaspis*] L. Dev. Eu., EEu. *Anchipteraspis* U. Sil. NNA. *Brachipteraspis* L. Dev. Eu. *Canadapteraspis Cosmaspis* L. Dev. NA. *Doryaspis* [*Lyktaspis Scaphaspis*] L. Dev. Spits. *Errivaspis* L. Dev. Eu. *Escharaspis* L. Dev. NNA. *Eucyclaspis* L. Dev. NA. *Europrotaspis* [*Glossoidaspis*] L. Dev. Eu. *Grumantaspis* L. Dev. Spits. *Lampraspis* L. Dev. NA. *Larnovaspis* L. Dev. Eu. *Loricopteraspis* [*Brotzenaspis*] L. Dev. EEu. *Miltaspis* L. Dev. Spits. *Mylopteraspidella Mylopteraspis* L. Dev. Eu. *Oreaspis* L. Dev. NA. *Protaspis* [*Cyrtaspidichthys Cyrtaspis Eucyrtaspis Europrotaspis Gigantaspis*] L. Dev. NA. Eu. Spits. *Protopteraspis* [*Simopterspis*] U. Sil. L. Dev. NNA., L. Dev. Eu. Spits. *Psephaspis* L. Dev. NA. *Pteraspis* [*Archaeoteuthis Brachipteraspis Cymripteraspis Lerichaspis Paleoteuthis Parapteraspis Penygaspis Plesiopteraspis Podolaspis Pseudopteraspis Simopteraspis Steganodictyum*] L. Dev. Eu., EEu. Spits. NAs. NA. *Rhachiaspis* U. Sil. NNA. *Rhinopteraspis* [*Belgicaspis*] L. Dev. Eu. EEu. *Stegobranchiaspis* L. Dev. NNA. *Ulutitaspis* U. Sil. NNA. *Unarkaspis* L. Dev. NNA. *Zascinaspis* L. Dev. EEu. NA. **Cardipeltidae** *Cardipeltis* L. Dev. Spits. NA. **Drepanaspidae (Psammosteidae)** *Aspidosteus* [*Aspidophorus Obruchevia*] U. Dev. EEu. *Crenosteus* U. Dev. Eu. *Drepanaspis* L. Dev. Eu. *Ganosteus* M. Dev. Eu. *Guerichosteus* L. Dev. Eu. *Hariosteus* L. Dev. Eu. *Karelosteus* U. Dev. Eu. *Obruchevia* U. Dev. Eu. *Psammolepis* M. Dev. Eu. M. Dev. Gr. U. Dev. NNA. CAs. *Psammosteus* [*Dyptychosteus Megalopteryx Placosteus*] U. Dev. EEu. NA. CAs. *Psephaspis* L. Dev. NA. *Pyc-*

nolepis M. Dev. EEu. *Pycnosteus Schizosteus* [*Cheirolepis Microlepis*] M. Dev. Eu. Spits. *Rohonosteus* U. Dev. NNA. *Stroshiperus* U. Sil. Eu. *Tartuosteus* M. Dev. Eu. *Traquairosteus* U. Dev. Eu. *Yoglinia* M. Dev. Eu.

ORDER HETEROSTRACI FAMILY INCERTAE SEDIS *Kallostrakon* L. Dev. Eu. *Natlaspis* U. Sil. L. Dev. NNA. *Oniscolepis* U. Sil. Eu. *Tesseraspis* L. Dev. Eu.

ORDER THELODONTIDA **Katoporidae** *Goniporus* U. Sil. Dev. Eu. EEu. *Helenolepis* U. Sil. CAs. *Katoporus* U. Sil. L. Dev. Eu. EEu. *Lanarkia* U. Sil. Eu. *Phlebolepis* U. Sil. NNA. Eu. EEu. **Loganiidae** *Coelolepis* U. Sil. L. Dev. Eu. *Logania* U. Sil. NA. Eu. EEu. CAs. NA. L. Dev. EEu. *Sigurdia* L. Dev. NNA. Spits. *Thelodus* [*Thelolepis Thelolepoides Thelyodus*] L. Sil. NNA. U. Sil. L. Dev. EEu. Eu. NA. Spits. CAs. **Turiniidae** *Turinia* L. Dev. Eu. EEu. Spits. Aus. CAs. NAs. **Apalolepididae** *Apalolepis* L. Dev. EEu. *Skamolepis* L. Dev. EEu. **Nikoliviidae** *Amaltheolepis* L.-M. Dev. Spits. L. Dev. NAs. EEu. *Gompsolepis* L. Dev. EEu. *Nikolivia* L. Dev. EEu. Eu.

SUBCLASS CEPHALASPIDOMORPHA

ORDER OSTEOSTRACI **Tremataspidae** *Timanaspis* U. Sil. EEu. L. Dev. Eu. *Tremataspis* [*Odontododus Stigmolepis*] U. Sil. Eu. *Witaaspis* U. Sil. EEu. **Dartmuthiidae** *Dartmuthia* [*Lophosteus*] U. Sil. Eu. *Oeselaspis* [*?Trachylepis*] U. Sil. Eu. *Saaremaaspis* [*Dasylepis ?Dictyolepis Rotsikuellaspis*] U. Sil. L. Dev. Eu. *Tyriaspis* U. Sil. Eu. **Ateleaspidae** *Aceraspis* [*Hemiteleaspis*] *Hirella* [*Micraspis*] L. Dev. Eu. *Hemicyclaspis* L. Dev. Eu. NNA *Tuvaspis* L. Dev. NAs. **Sclerodontidae** *Sclerodus* [*Eukeraspis*] L. Dev. Eu. **Cephalaspidae** *Benneviaspis* L. Dev. Eu. Spits. *Boreaspis* L. Dev. Spits. *Cephalaspis* [*Alaspis Camptaspis Escuminaspis Eucephalaspis Minestaspis Pattenaspis Scolenaspis Zenaspis*] U. Sil. L. Dev. Eu. L.-M. Dev. Spits. L. Dev. EAs. L.-U. Dev. NA. *Ectinaspis* [*Hoelaspis*] L. Dev. Spits. *Procephalaspis* U. Sil. Eu. *Securiaspis* L. Dev. Eu. Spits. *Stensiopelta* L. Dev. Eu. *Tannuaspis* L. Dev. NAs. *Tegaspis* L. Dev. Spits. *Thyestes* [*Auchenaspis*] U. Sil. L. Dev. Eu. **Kiaeraspididae** *Acrotomaspis Axinaspis* L. Dev. Spits. *Ilemoraspis* L. Dev. EEu. *Kieraspis* L. Dev. Spits. *Nectaspis* L.-M. Dev. Spits. *Turinea* [*Cephalopterus*] L. Dev. Eu.

ORDER GALEASPIDA **Hanyangaspidae** *Hanyangaspis Latirostraspis* M. Sil. EAs. **Galeaspidae** *Galeaspis* L. Dev. EAs. **Eugaleaspidae** *Eugaleaspis Sinogaleaspis Yunnanogaleaspis* L. Dev. EAs. **Nanpanaspidae** *Nanpanaspis* L. Dev. EAs. **Polybranchiaspidae** *Cyclodiscaspis Damaspis Diandongaspis Dongfangaspis Kwangnanaspis Laxaspis Polybranchiaspis Siyingia* L. Dev. EAs. **Hunanaspidae** *Antquisagittaspis Asiaspis Huananaspis Sanchaspis Sangiaspis Szechuanaspis* L. Dev. EAs. **Duyunolepidae** *Duyunolepis Neoduyunaspis Paraduyunaspis* L. Dev. EAs. **Lungmenshanaspidae** Lungmenshanaspis Sinoszechuanaspis Qingmenaspis L. Dev. EAs. **Tridensaspidae** *Tridensaspis* L. Dev. EAs. **Dayongaspidae** *Dayongaspis* L. Sil EAs.

ORDER ANASPIDA **?Jaymoytiidae** *Jaymoytius* M. Sil. Eu. **Birkeniidae** *Birkenia* M.-U. Sil. L. Dev. Eu. *Ctenopleuron* U. Sil. NA. *Pharynogolepis Pterygolepis* [*Pterolepidops Pterolepis*] *Saarolepis* U. Sil. Eu. *Rhyncholepis* L. Dev. Eu. **Euphaneropsidae** *Euphanerops* U. Dev. NA. **Endeiolepidae** *Endeiolepis* U. Dev. NA. **Lasaniidae** *Lasanius* U. Sil. L. Dev. E.U.

ORDER PETROMYZONTIFORMES **Petromyzontidae** R. Oc. Eu. Aus. NZ. NA. SA. **Mayomyzontidae** *Hardistiella* U. Miss. NA. *Mayomyzon* M. Penn. NA.

AGNATHA INCERTAE SEDIS

Cyclostomis Gilpichthys Pipiscius Penn. NA.

ORDER MYXINIFORMES **Myxinidae** (**Bdellostomatidae**) R. Oc. Unnamed fossil, Penn. NA.

CLASS PLACODERMI

ORDER STENSIOELLIDA **Stensioellidae** *Stensioella* L. Dev. Eu.

ORDER PSEUDOPETALICHTHYDA **Paraplesiobatidae** *Paraplesiobatis* L. Dev. Eu. *Pseudopetalichthys* [*Parapetalichthys*] L. Dev. Eu.

ORDER RHENANIDA **Asterosteidae** *Asterosteus* M. Dev. NA. SWAs. *Bolivosteus* M. Dev. SA. *?Brindabellaspis* L. Dev. Aus. *Gemuendina* [*Broiliina*] L. Dev. Eu. *Jagorina* U. Dev. Eu. *?Ohioaspis* M. Dev. NA. Aus.

ORDER PTYCTODONTIDA **Ptyctodontidae** *Campbellodus* U. Dev. Aus. *Chelyophorus* [*Cheliophorus*] U. Dev. EEu. *Ctenurella* U. Dev. Eu. EEu. Aus. *Desmoporella* M. Dev. Eu. *Eczematolepis* [*Acantholepis Oracanthus Phlyctaenacanthus*] M.-U. Dev. NA. *Goniosteus* M. Dev. Eu. *Palaeomylus* [*Rhynchodus*] M.-U. Dev. NA. Eu. *Ptyctodopis* M. Dev. NA. *Ptyctodus* [*Aulacosteus Paraptyctodus Rinodus*] M.-U. Dev. ?L. Miss. NA. M.-U. Dev. Eu. EEu. U. Dev. NAs. NAf. Aus. *Rhamphodopsis* M. Dev. Eu. *Rhynchodus* [*Ramphodus Rhampodontus Rhampodus Rhynchodontus Rhynchognathis Rhynchosteus Ringinia*] M.-U. Dev. Eu. NA.

ORDER ACANTHOTHORACI **Palaeacanthaspidae** *Dobrowlania* L. Dev. EEu. *Kimaspis* Dev. CAs. *Kolymaspis* L. Dev. NAs. *Kosoraspis* L. Dev. EEu. *Palaeacanthaspis* L. Dev. EEu. *Radotina* [*Holopetalichthys*] L. Dev. Eu. *Romundina* L. Dev. NNA. **Weejasperaspididae** *Weejasperaspis* L. Dev. Aus.

ORDER PETALICHTHYIDA **Macropetalichthyidae** *Ellopetalichthys* M. Dev. NNA. *Epipetalichthys* U. Dev. Eu. *Lunaspis* L. Dev. Eu. *Macropetalichthys* [*Acanthaspis Agassichthys Heintzaspis Ohiodorulites Physichthys*] M. Dev. NA. Eu. *Notopetalichthys* L. Dev. Aus. *Quasipetalichthys* M. Dev. EAs. *Shearsbyaspis* L. Dev. Aus. *Wijeaspis* L. Dev. SWAs. M. Dev. Spits. NAs.

ORDER PHYLLOLEPIDA **Antarctaspidae** *Antarctaspis* M. or U. Dev. Ant. **Phyllolepididae** *Austrophyllolepis* U. Dev. Aus. *Phyllolepis* [*Pentagonolepis*] U. Dev. Eu. EEu. Gr. NA. Aus. *Placeolepis* U. Dev. Aus.

ORDER ARTHRODIRA

SUBORDER ACTINOLEPINA **Actinolepidae** *Actinolepis* M. Dev. EEu. L.-M. Dev. Spits. *Aethaspis* L. Dev. NA. *Ailuracantha* L. Dev. Eu. *Anarthraspis* [*Svalbardaspis*] L. Dev. NA. *Baringaspis* L. Dev. NNA. *Bryantolepis* [*Bryantaspis Euraspis Euryaspidichthys*] L. Dev. NA. *Heightingtonaspis* L.

Dev. Eu. NA. *Kujdanowiaspis* L. Dev. EEu. *Lataspis* [*Lehmanosteus Plataspis*] L. Dev. Spits. *Mediaspis* M. Dev. Spits. *Proaethaspis* L. Dev. NA. *Sigaspis* L. Dev. Spits. *Simblaspis* L. Dev. NA. *Steurtzaspis* L. Dev. Eu. *Szeaspis* L. Dev. EAs. **Goodradigbeeonidae** *Goodradigbeeon* L. Dev. Aus.

SUBORDER WUTTAGOONASPINA **Wuttagoonaspidae** *Wuttagoonaspis* M. Dev. Aus.

SUBORDER PHLYCTAENIINA **Phlyctaeniidae** *Aggeraspis* L. Dev. Eu. *Arctaspis* L. Dev. Spits. *Arctolepis* [*Acanthaspis Jaekelaspis*] L. ? M. Dev. Spits. *Cartieraspis* M. Dev. NA. *Diadsomaspis* L. Dev. Eu. *Dicksonosteus* L. Dev. Spits. *Elegantaspis* L. Dev. Spits. *Gaspeaspis* M. Dev. NA. *Heintzosteus* L. Dev. Spits. *Heterogaspis* [*Monaspis*] L. Dev. Spits. *Huginaspis* M. Dev. Spits. *Kolpaspis* M. Dev. NA. *Kueichowlepis* L. Dev. EAs. *Neophlyctaenius* U. Dev. NA. *Pageauaspis* [*Quebecaspis*] M. Dev. NA. *Phlyctaenius* [*Phlyctaenaspis Batteraspis*] L. Dev. NA. *Prosphymaspis* L. Dev. Eu. **Holonematidae** *Belemnacanthus* M. Dev. Eu. *Deirosteus* M. Dev. EEu. U. Dev. NA. SWAs. *Deveonema* U. Dev. EEu. *Groenlandaspis* U. Dev. Eu. Gr. SWAs. Aus. Ant. *Gyroplacosteus* [*Operchallosteus*] U. Dev. Eu. EEu. *Holonema* M. Dev. NAs. Ind. M.-U. Dev. Eu. EEu. NA. U. Dev. EEu. SWAs. Spits. NAf. NNA. Aus. *Megaloplax* U. Dev. EEu. *Rhenonema* M. Dev. Eu. *Tiaraspis* L. Dev. Eu. Spits. *Tropidosteus* M. Dev. Eu.

SUBORDER BRACHYTHORACI

SUPERFAMILY BUCHANOSTEOIDEA **Buchanosteidae** *Arenipiscis Buchanosteus Burrinjucosteus Errolosteus Parabuchanosteus Taemasosteus Toombsosteus* L. Dev. Aus. **?Williamsaspidae** *Williamsaspis* L. Dev. Aus.

SUPERFAMILY HETEROSTEOIDEA **Heterosteidae** *Herasmius* M. Dev. Spits. *Heterosteus* [*Chelonichthys Heterostius Ichthyosauroides*] M. Dev. Eu. EEu. Gr. Spits.

SUPERFAMILY HOMOSTEOIDEA **Homosteidae** *Angarichthys* M. Dev. NAs. *Euleptaspis* [*Leptaspis*] L. Dev. Eu. Spits. *Homosteus* [*Homostius*] *Trionyx* L.-M. Dev. Spits. M. Dev. Eu. EEu. Gr. NA. *?Lophostracon* L. Dev. Spits. *Luetkeichthys* M. Dev. NAs. *Tityosteus* L. Dev. Eu.

SUPERFAMILY BRACHYDEIROIDEA **Brachydeiridae** *Brachydeirus* [*Brachydirus Auchenosteus*] U. Dev. Eu. *Oxyosteus* [*Platyosteus*] U. Dev. Eu. EEu. *Synauchenia* [*Synosteus*] U. Dev. Eu. **Leptosteidae** *Leptosteus* U. Dev. Eu. NA.

SUPERFAMILY COCCOSTEOIDEA **Gemuendenaspidae** *Gemuendenaspis* L. Dev. Eu. **Coccosteidae** *Belgiosteus* M. Dev. Eu. *Clarkeosteus* M. Dev. NA. ?U. Dev. Gr. *Coccosteus* M. ?U. Dev. Eu. EEu. ?NA. *Dickosteus* M. Dev. Eu. *Eldenosteus* U. Dev. NA. *Harrytoombsia* U. Dev. Aus. *Livosteus* M.-U. Dev. EEu. *Millerosteus* M. Dev. Eu. EEu. *Plourdosteus* [*Pelycophorus Tomaiosteus*] U. Dev. Eu. NA. NAs. *Protitanichthys* M. Dev. NA. *Watsonosteus* M. Dev. Eu. *Woodwardosteus* [*Liognathus Lispognathus*] M. Dev. NA. **Camuropiscidae** *Camuropiscis* U. Dev. Aus. *?Rolfosteus Simosteus Tubonasus* U. Dev. Aus. **Pholidosteidae** *Malerosteus* U. Dev. EEu. *Pholidosteus* U. Dev. Eu. *Tapinosteus* U. Dev. Eu. **Incisoscutidae** *Incisoscutum* U. Dev. Aus.

SUPERFAMILY DINICHTHYLOIDEA **Dinichthyidae** *Bruntonichthys* U. Dev. Aus. *Bullerichthys* U. Dev. Aus. *Dinichthys* [*Ponerichthys*] U. Dev. NA. *Dunkleosteus* U. Dev. Eu. EEu. NAf. NA. *Eastmanosteus* M. Dev. NAf. M.-U. Dev. NA. U. Dev. Eu. EEu. SWAs. Aus. *Gorgonichthys* U. Dev. NA. *Hadrosteus* U. Dev. Eu. *Heintzichthys* [*Stenognathus*] U. Dev. NA.? EEu. *Holdenius* U. Dev. NA. *Hussakofia* [*Brachygnathus*] U. Dev. NA. *Kianyousteus* M. Dev. EAs. **Leiosteidae** *Erromenosteus* [*Leiosteus*] U. Dev. Eu. **Trematosteidae** *Belosteus* U. Dev. Eu. *Brachyosteus* U. Dev. Eu. *Cyrtosteus* U. Dev. Eu. *Parabelosteus* U. Dev. Eu. *Trematosteus* U. Dev. Eu.

SUBORDER OR SUPERFAMILY INCERTAE SEDIS **Rachiosteidae** *Rachiosteus* U. Dev. Eu. **Selenosteidae** *Braunosteus* U. Dev. Eu. *Enseosteus* [*Ottonosteus Walterosteus*] U. Dev. Eu. *Gymnotrachelus* U. Dev. Eu. *Microsteus* [*Parawalterosteus*] U. Dev. Eu. *Paramylostoma* U. Dev. NA. *Rhinosteus* U. Dev. Eu. *Selenosteus* U. Dev. NA. *Stenosteus* U. Dev. NA. EEu. **Pachyosteidae** *Pachyosteus* U. Dev. Eu. EEu. **Mylostomatidae** *Dinomylostoma* U. Dev. NA. Aus. *Kendrickichthys* U. Dev. Aus. *Mylostoma* U. Dev. NA. *?Tafilalichthys* U. Dev. NAf. **Titanichthyidae** *Titanichthys* [*Brontichthys*] U. Dev. EEu. NAf. NA. **Bungartiidae** *Bungartius* U. Dev. NA.

ARTHRODIRA INCERTAE SEDIS *Antarctolepis* M. Dev. Ant. *Aspidichthys* [*Anomalichthys Aspidophorus*] U. Dev. Eu. NA. NAf. SWAs. *Atlantidosteus* L. Dev. NAf. *Callognathus* U. Dev. NA. *Copanognathus* U. Dev. NA. *Cosmacanthus* U. Dev. Eu. *Diplognathus* U. Dev. NA. *Glyptaspis* U. Dev. NA. *Grazosteus* ?M. Dev. Eu. *Hollardosteus* M. Dev. NAf. *Laurentaspis* M. Dev. NA. *Machaerognathus* U. Dev. NA. *Murmur* [*Euptychaspis Ptychaspis*] *Overtonaspis* L. Dev. Eu. *Prescottaspis* L. Dev. Eu. *Qataraspis* L. Dev. SWAs. *Taemasosteus* L. Dev. Aus. *Taunaspis* L. Dev. Eu. *Timanosteus* U. Dev. CAs. *Trachosteus* U. Dev. NA. *Wheathillaspis* L. Dev. Eu.

ORDER ANTIARCHI **Bothriolepidae** *Bothriolepis* [*Bothryolepis Glyptosteus Homothorax Macrobrachius Parmphractus Phoebammon Placothorax Shurcabroma Stanacanthus*] U. Dev. EEu. Eu. Ant. NA. Gr. CAs. M. Dev. NAs. Aus. SWAs. *Dianolepis* M. Dev. EAs. *Grossilepis* U. Dev. EEu. Eu. Aus. *Hillsaspis* U. Dev. Aus. *Wudinolepis* M. Dev. EAs. *Yunnanolepis* L. Dev. EAs. **Asterolepidae** *Asterolepis* M.-U. Dev. EEu. NA. Eu. Gr. Spits. EAs. M. Dev. Aus. *Byssacanthus* M.-U. Dev. EEu. Eu. *Gerdalepis* M. Dev. Eu. *Microbachius* M. Dev. Eu. *Pambulaspis* U. Dev. Aus. *Pterichthyodes* M. Dev. Eu. L. Dev. Aus. *Remigolepis* U. Dev. Aus. Eu. *Stegolepis* U. Dev. CAs. **Sinolepidae** *Sinolepis* U. Dev. EAs.

ANTIARCHI INCERTAE SEDIS *Grossaspis* M. Dev. EEu. *Ledapolepis* U. Dev. Eu. *Taeniolepis* U. Dev. EEu.

PLACODERMI INCERTAE SEDIS

Asiacanthus L. Dev. EAs. *Changyonophyton* U. Dev. EAs. *Deinodus* M. Dev. NA. *Hybosteus* [*Coelosteichthys Grossosteus*] M. Dev. EEu. *Neopetalichthys* L. Dev. EAs. *Nessariostoma* L. Dev. Eu. *Oestophorus* M. Dev. NA. *Sedowichthys* M. Dev. NAs. *Tollichthys* M. Dev. NAs. *Yunnanacanthus* Dev. CAs.

CLASS CHONDRICHTHYES

SUBCLASS ELASMOBRANCHII

SUPERORDER UNDESIGNATED

ORDER CLADOSELACHIDA **Cladoselachidae** *Cladoselache* [*Cladodus*] U. Dev. NA. *Monocladodus* U. Dev. NA.

ORDER CORONODONTIA **Family not designated** *Coronodus* U. Dev. NA. *Diademodus* [*Tiarodontus*] U. Dev. NA.

ORDER SYMMORIIDA **Symmoriidae** *Cobelodus* [*Styptobasis*] *Denaea* U. Carb. NA. *Symmorium* [*Cladodus*] Miss. Eu. Penn. NA. **Stethacanthidae** *Orestiacanthus* Penn. NA. *Stethacanthus* [*Cladodus Lambdodus Physonemus*] U. Dev. Penn. NA. Miss. Eu.

ORDER EUGENEODONTIDA (EDESTIDA, HELICOPRIONIDA)

SUPERFAMILY CASEODONTOIDEA **Caseodontidae** *Caseodus* [*Orodus*] Penn. NA. *Erikodus* [*Agassizodus Copodus*] U. Perm. Gr. *Fadenia* Penn. NA. U. Perm. Gr. *Ornithoprion* Penn. NA. *Romerodus* Penn. NA. **Eugeneodontidae** *Bobbodus* Penn. NA. *Eugeneodus* Penn. NA. *Gilliodus* Penn. NA. **Caseodontoidea incertae sedis** *Campodus* Penn. Eu. *Chiastodus* Penn. EEu.

SUPERFAMILY EDESTOIDEA (EDESTIDA) **Agassizodontidae (Helicoprionidae)** *Agassizodus* [*Lophodus*] Miss. NA. *Arpagodus* Penn. As. *Campyloprion* ?Penn. EEu. *Helicoprion* [*Lissoprion*] L. Perm. Eu. ?WAs. EAs. Aus. NA. *Sarcoprion* Perm. Gr. *Toxoprion* ?Penn. NA. **Edestidae** *Edestus* [*Edestes Edestodus Protospirata*] Penn. NA. Eu. EEu. *Helicampodus* U. Perm. Ind. L. Trias. WAs. *Lestrodus* [*Edestus*] Penn. Eu. *Parahelicampodus* L. Trias. Gr. *Syntomodus* U. Perm. EAs.

ORDER ORODONTIDA **Orodontidae** *?Hercynolepis* L. Dev. Eu. *?Leiodus* Miss. NA. *?Mesodmodus* Miss. NA. *Orodus* [*Hybodopsis*] Miss. Eu. Penn. NA.

ORDER SQUATINACTIDA **Family not designated** *Squatinactics* Miss. NA.

SUPERORDER EUSELACHII

ORDER CTENACANTHIFORMES

SUPERFAMILY CTENACANTHOIDEA **Ctenacanthidae** *Amelacanthus Ctenacanthus* [*Eunemacanthus Sphenacanthus*] U. Dev. NA. Miss. Eu. *Cratoselache* Miss. Eu. *Goodrichthys* [*Moythomasina*] Miss. Eu. *Wodnika* [*Radamas*] U. Perm. Eu. NA. **Bandringidae** *Bandringa* Penn. NA. **Phoebodontidae** *Phoebodus* M. Dev. Penn. NA. U. Dev. or L. Carb. EAs. **Incertae sedis** *Carinacanthus* Trias. NA. *Acandylacanthus* Miss. NA. *Anaclitacanthus* Miss. NA. *Asteroptychius* Miss. Eu.

SUPERFAMILY HYBODONTOIDEA **Family not designated** *Arctacanthus* [*Ancistriodus Dolophonodus Hamatus Homacanthus*] M. Perm. NA. U. Perm. Gr. *Dabasacanthus* Penn. NA. *Moyacanthus* Miss. Eu. *Onychoselache* Miss. Eu. *Tristychius* Miss. Eu. **Hybodontidae** *Acrodonchus* M.-U. Trias. Eu. *Acrodus* [*Adiapneustes ?Psilacanthus Sphenonchus Thectodus*] L. Trias.-U. Cret. Eu. Trias. SAs. M. Trias.-U. Cret. NA. U. Trias. Spits. U. Cret. SA. *Arctacanthus* [*Dolophonodus Hamatus*] M. Perm. NA. Gr. *Asteracanthus* [*Curtodus Strophodus*] U. Trias.-L. Cret. Eu. U. Trias. NA. Jur. EAs. Mad. Jur.-Paleoc. NAf. *Bdellodus* L. Jur. Eu. *Carinacanthus* U. Trias. NA. *Coelosteus* Miss. NA. *Dicrenodus* [*Carchariopsis Pristicladodus*] Miss.-Penn. Eu. Miss. NA. *Doratodus* M.-U. Trias. Eu. *Echinodus* Penn. Eu. *Eoörodus* U. Dev. NA. *Hybocladodus* Miss. NA. *Hybodonchus* M.-U. Trias. Eu. *Hybodus* [*Leiacanthus Meristodon Orthybodus Parhybodus ?Selachidea*] ?U. Perm. M. Trias.-U. Cret. Eu. L. Trias. Spits. L.-U. Trias. Gr. L. Trias.-U. Cret. NA. U. Trias. EAs. U. Trias.-Paleoc. Af. Jur.-Cret. Aus. *?Lambdodus* Miss. Eu. NA. *Lissodus* L. Trias. SAf. *Lonchidion* U. Cret. NA. *Mesodmodus* Miss. NA. *?Monocladodus* Penn. NA. *Nemacanthus* [*Desmacanthus Nematacanthus*] ?L. Trias. Gr. M. Trias.-U. Jur. Eu. U. Trias. Spits. NA. *Orthacodus* U. Jur.-L. Cret. Eu. *Palaeobates* M.-U. Trias. Eu. *?Petrodus* [*Octinaspis Ostinaspis*] Miss.-Penn. Eu. Penn. NA. *Polyacrodus* L. Trias. Gr. M.-U. Trias. Eu. *Priorybodus* U. Jur.-L. Cret. Af. *Pristacanthus* U. Jur. Eu. *Prohybodus* L. Cret. NAf. *Protacrodus* U. Dev. Eu. *Scoliorhiza* U. Trias. NA. *Sphenacanthus* Miss.-Perm. Eu. *Styracodus* [*Centrodus*] Penn. Eu. *Symmorium* Penn. NA. *Xystrodus* U. Perm. Eu.

SUPERFAMILY PROTACRODONTOIDEA **Tamiobatidae** *Holmesella* Penn. NA. *Protacrodus* M.-U. Dev. Eu. NA. U. Dev. or L. Carb. EAs. *Tamiobates* Miss. NA.

ORDER XENACANTHIDA (PLEUROCANTHODII) **Diplodoselachidae** *Diplodoselache* Miss. Eu. **Xenacanthidae (Xenacanthi Xenacanthini)** *Orthacanthus* [*Aganodus Compsacanthus Diplodus Diploctus Dissodus Dittodus Eucompsacanthus Ochlodus Pternodus*] Carb.-Perm. Eu. Perm. NA. U. Dev or L. Carb. EAs. *Pleuracanthus* Penn. Eu. *Xenacanthus* [*Hypospondylus Orthacanthus*] L. Perm. Eu. NA. **Incertae sedis** *Anodontacanthus* U. Carb. Eu. *Bransonella* Penn. NA. *Diacranodus* L. Perm. NA. *Iriodus* [*Expleuracanthus*] L. Perm. Eu. *Phricacanthus* Penn. Eu. *Platyacanthus* L. Perm. Eu.

ORDER GALEOMORPHA **Palaeospinacidae** *Palaeospinax* U. Trias. L. Jur. Eu. **Family undesignated** *Synechodus* U. Jur. U. Cret. Paleoc. Eu.

SUBORDER HETERODONTOIDEA **Heterodontidae** R. Indo-Pac. Oc. *Heterodontus* [*Cestracion Drepanephorus Gyropleurodus Platyacrodus Pseudacrodus Tropidotus*] U. Jur. Mioc. Eu. U. Cret. Mioc. SA. U. Cret. Eoc. Af. Mioc. Aus. NZ. R. Indo-Pac. Oc. *Paracestracion* U. Jur. Eu. *Strongyliscus* Mioc. WNA.

SUBORDER ORECTOLOBOIDEA **Orectolobidae** R. Oc. *Brachaelurus* U. Cret. NA. Eu. *Cantioscyllium* U. Cret. Eu. *Corysodon Crossorhinops* U. Jur. Eu. *Ginglymostoma* [*Plicodus*] U. Cret. Mioc. Eu. Af. U. Cret. WInd. U. Cret. Mioc. NA. Eoc. As. R. Oc. *Mesiteia* U. Cret. SWAs. M. Eoc. Eu. *Crossorhinus* [*Palaeocrossorhinus*] U. Jur. Eu. *Orectoloboides* L. Cret. Eu. *Palaeocarcharias* U. Jur. Eu. *Phorcynus* U. Jur. Eu. *Squatirhina* U. Cret. NA. U. Cret. Eoc. Eu. Paleoc. Eoc. Af. *Squatirhynchus* U. Cret. SWAs. **Rhinocodontidae** R. Oc. **Hemiscyllidae** *Acanthoscyllium Almascyllium* U. Cret. SWAs. *Chiloscyllium* U. Cret. NAf. WInd. Mioc. Eu. R. Oc. **Parascyllidae** *Pararhincodon* U. Cret. Eu. SWAs.

SUBORDER LAMNOIDEA **Carchariidae** (**Odontaspidae**) R. Oc. *?Anotodus* Olig. Plioc. Eu. Af. NA. *Carcharias* [*Hypotodus ?Iekelotodus Odontaspis ?Palaeohypotodus Parodontaspis Priodontaspis Striatolamia Synodontaspis Triglochis*] L. Cret. Plioc. Eu. U. Cret. Mioc. As. U. Cret. Plioc. SA. NZ. Af. U. Cret. Pleist. NA. Pleist. EInd. R. Oc. **Cretoxyrhinidae** *Cretolamna* U. Cret. Eu. EAs. NAf. WAf. Mad. *Cretoxyrhina* U. Cret. Eu. NA. *Paraisurus* U. Cret. Eu. NA. Af. *Plicatolamna* U. Cret. SWAs. NA. Eu. **Orthacodontidae** *Orthacodus* [*Parorthacodus Sphenodus*] L. Jur. Eoc. Eu. U. Jur. Af. U. Cret. NA. **Lamnidae** (**Carcharodontidae Isuridae**) R. Oc. *Carcharoides* Mioc. Aus. SA. *Carcharodon* [*Agassizodon Carchariolamna Eocarcharodon Macrohizodus Megaselachus Palaeocarcharodon*] ?L. Cret. Paleoc. Pleist. Eu. Paleoc. Plioc. Af. Eoc. Pleist. NA. Olig. Pleist. Aus. Mioc. SAs. SA. EInd. Mioc. Plioc. NZ. WInd. Mioc. Pleist. Aus. R. Oc. *Isurolamna* L. Eoc. Eu. *Isurus* [*Carcharocles Cosmopolitodus Isuropsis Oxyrhina*] L. Cret. Pleist. Eu. U. Cret. Mioc. WInd. As. Aus. SA. U. Cret. Pleist. NA. Mioc. Mad. Mioc. Plioc. NZ. Pleist. EInd. R. Oc. *Lamiostoma* Mioc. Eu. R. Oc. *Lamna* [*?Euchlaodus Jekelotodus Leptostyrax Otodus*] L. Cret. Plioc. Eu. U. Cret. Aus. As. U. Cret. Eoc. Af. U. Cret. Pleist. NA. Tert. NZ. *Palaeocorax* U. Cret. Eu. WAs. *Parisurus Procarcharodon* Paleoc. Eoc. Af. *Pseudocorax* U. Cret. Eu. NA. U. Cret. Paleoc. Af. *Pseudoisurus* U. Cret. Eu. NAs. *Squalicorax* [*Anacorax Corax ?Xenolamia*] U. Cret. Eu. U. Cret. Paleoc. NA. **Mitsukurinidae** R. Oc. *Anomotodon* U. Cret. NAf. SWAs. U. Cret. Eoc. Eu. *Scapanorhynchus* L. Cret. Paleoc. Eu. U. Cret. SWAs. Aus. NZ. NA. SA. U. Cret. Paleoc. Af. R. NAtl. NPac. **Cetorhinidae** R. Oc. *Cetorhinus* [*Hannoveria*] Olig. Plioc. Eu. Mioc. Plioc. NA. R. Oc. **Alopiidae** R. Oc. *Alopias* [*Alopecias Vulpecula*] Eoc. NA. Af. Olig. Mioc. Eu. Mioc. WInd. Plioc. SA. R. Oc.

SUBORDER CARCHARHINOIDEA **Scyliorhinidae** (**Scylliidae**) R. Oc. *Galeus* U. Jur. Plioc. Eu. Eoc. NA. Af. R. Oc. *Palaeoscyllium* U. Jur. U. Cret. Eu. *Pararhincodon* U. Cret. Eoc. Eu. U. Cret. NA. *Pristiurus* U. Jur. Eu. R. Oc. *Protogaleus* Paleoc. Eoc. NAf. *Pteroscyllium* U. Cret. SWAs. *Scyliorhinus* [*Scyllium Thyellina*] U. Cret. SWAs. U. Cret. Mioc. NA. U. Cret. Plioc. Eu. Paleoc. Eoc. Af. R. Oc. *?Scylliodus* U. Cret. Eu. *?Trigonodus* Eoc. Eu. **Pseudotriakidae** R. Oc. *Archaeotriakis* U. Cret. NA. **Triakidae** (**Mustelidae**) R. Oc. *Mustelus* [*Galeus*] Olig. Plioc. Eu. R. Oc. *Paratriakis* U. Cret. Eu. SWAs. *Triakis* ?U. Cret. Eu. Paleoc. WInd. **Carcharhinidae** R. Oc. *Alopiopsis* [*Pseudogaleus*] Eoc. NAf. Eoc. Olig. Eu. R. Oc. *Aprionodon* [*Aprion*] Eoc. Af. Eoc. Plioc. Eu. Mioc. As. R. Oc. *Carcharhinus* Mioc. Plioc. Eu. Af. Mioc. Pleist. NA. ?Olig. Mioc. SA. Mioc. Aus. Plioc. EInd. R. Oc. *Eogaleus* M. Eoc. Eu. *Galeocerdo* Eoc. Mioc. NAf. Eoc. Plioc. NA. Eu. Mioc. WInd. SA. Mioc. Plioc. As. Aus. Pleist. EInd. R. Oc. *Galeorhinus* U. Cret. Eoc. NAf. Eoc. Pleist. Eu. NA. R. Oc. *Hemipristis* Eoc. Mioc. NA. Af. Eoc. Plioc. Eu. Mioc. As. Aus. SA. Mioc. Pleist. EInd. R. Red Sea Indian Oc. *Hypoprion* Eoc. NA. Mioc. Eu. Af. R. Oc. *Negaprion* Eoc. Mioc. NA. R. Oc. *Physodon* Eoc. Af. Eoc. Mioc. Eu. R. Oc. *Prionace* Pleist. NA. R. Oc. *Prionodon* [*Glyphis*] Eoc. Af. Eoc. Plioc. Eu. NA. Olig. Mioc. SA. Mioc. As. WInd. R. Oc. *Scoliodon* [*Loxodon Rhizoprinodon*] Eoc. Af. Eoc. Mioc. Eu. Eoc. Pleist. NA. R. Oc. **Sphyrnidae** *Sphyrna* [*Zygaena*] Eoc. Plioc. Eu. Eoc. Pleist. NA. Mioc. Af. Aus. Plioc. EInd. SA. R. Oc.

SUBORDER HEXANCHOIDEA **Hexanchidae** (**Notidanidae**) R. Oc. *Hexanchus* R. Oc. U. Cret. SWAs. U. Eoc. Aus. U. Cret. Tert. Eu. NAf. *Heptranchias* [*Notidanodon*] R. Oc. Cret. Mioc. NA. Olig. EAs. U. Eoc. Aus. L. Cret. Eu. *Notorynchus* R. Oc. Mioc. Aus. Cret. Mioc. NA. "*Notidanus*" L.-U. Jur. Eu. **Chlamydoselachidae** R. NAtl. NPac. *Chlamydoselache* Mioc. WInd. Plioc. Eu. R. NAtl. NPac.

ORDER SQUALOMORPHA

SUBORDER SQUALOIDEA **Squalidae** R. Oc. *Centrophoroides* U. Cret. SWAs. *Centrophorus* ?U. Cret. Mioc. Eu. R. Oc. *Centropterus* U. Cret. Eu. *Centroscymnus* Mioc. WInd. R. Oc. *Centrosqualus* U. Cret. SWAs. *Cheirostephanus* Mioc. WInd. *Cretascymnus* U. Cret. SWAs. *Etmopterus* [*Spinax*] ?U. Cret. Mioc. Eu. R. Oc. *Oxynotus* [*Centrina*] Mioc. Plioc. Eu. R. Oc. *Protospinax* U. Jur. Eu. *Protosqualus* L. Cret. Eu. *Squalus* [*Acanthias*] U. Cret. Plioc. Eu. Paleoc. Eoc. NAf. Olig. SA. Mioc. Aus. Mioc. Pleist. NA. R. Oc. **Dalatiidae** (**Scymnorhinidae**) R. Oc. *Dalatias* [*Scymnus Scymnorhinus*] U. Cret. SWAs. NA. Eoc. Plioc. Eu. R. Oc. *Isitius* U. Cret. Eoc. Af. Eoc. Mioc. Eu. R. Oc. *Somniosus* [*Laemargus*] Eoc. NAf. R. Oc. **Echinorhinidae** R. Oc. *Echinorhinus* [*?Goniodus*] Eoc. Mioc. NA. Plioc. Eu. R. Oc.

SUBORDER PRISTIOPHOROIDEA **Pristiophoridae** R. Oc. *Pliotrema* Tert. NZ. R. Oc. *Pristiophorus* U. Cret. SWAs. Mioc. Eu. NA. Mioc. Plioc. Aus. R. Oc. *Propristiophorus* U. Cret. SWAs.

SUBORDER SQUATINOIDEA **Squatinidae** (**Rhinidae**) R. Oc. *Squatina* [*Rhina Thaumas Trigenodus*] U. Jur. Plioc. Eu. U. Cret. SWAs. U. Cret. Pleist. NA. Eoc. Af. Mioc. Aus. R. Oc.

ORDER BATOIDEA

SUBORDER TORPEDINOIDEA **Torpedinidae** R. Oc. *Eotorpedo* Paleoc. Eoc. Af. M. Eoc. Eu. *Narcine* Eoc. Eu. R. Oc. *Narcopterus* Eoc. Eu. *Torpedo* [*Narcobatus*] ?Eoc. Eu. R. Oc. **Narkidae Temeridae** R. Oc.

SUBORDER PRISTOIDEA **Pristidae** R. Oc. *Anoxypristis* [*Oxypristis*] Eoc. Eu. *Ctenopristis* U. Cret. SWAs. Paleoc. Af. *Marckgrafia* L. Cret. Paleoc. NAf. *Onchopristis* L.-U. Cret. NAf. U. Cret. NA. *Onchosaurus* [*Dalpiazia Gigantichthys*] U. Cret. Eu. NA. SA. U. Cret. Paleoc. NAf. *Peyeria* U. Cret. NAf. *Pristis* [*Myripristis Pristibatus*] ?U. Cret. Eoc. NA. Eoc. Mioc. Eu. As. Eoc. Af. Mioc. Plioc. Aus. *Propristis* [*Amblypristis Eopristis*] Eoc. NAf. NA. *Schizorhiza* U. Cret. SWAs. NA. SA. U. Cret. Paleoc. Af.

SUBORDER RHINOBATOIDEA **Rhynchobatidae** R. Oc. *Rhynchobatus* Cret. Af. Eoc. Mioc. Eu. R. Oc. **Rhinobatidae** R. Oc. *Aellopos* [*Euryarthra Spathobatis*] *Asterodermus Belemnobatis* U. Jur. Eu. *Cyclarthrus* L. Jur. Eu. *Platyrhina* Eoc. Eu. R. NPac. *Protoplatyrhina* U. Cret. NA. *Rhinobatos* [*Rhinobatus*] L. Cret. Mioc. Eu. U. Cret. SWAs. U. Cret. Eoc. NAf. R. Oc. *Rhombopterygia* U. Cret. SWAs. *Trygonorrhina* Eoc. Mioc. Eu. R. Oc.

SUBORDER RAJOIDEA **Rajidae** R. Oc. *Acanthobatis* Mioc. Eu. *Cyclobatis* U. Cret. SWAs. *?Dynatobatis* ?Plioc. SA. *On-*

cobatis Plioc. NA. *Pararaja* U. Cret. SWAs. *Platyspondylus* L.-U. Cret. NAf. *Raja* [*Actinobatis Raia*] ?U. Cret. SWAs. WInd. U. Cret. Plioc. Eu. U. Cret. Eoc. NAf. Eoc. Mioc. NA. R. Oc. **Sclerorhynchidae** *Ankystrorhynchus* U. Cret. Eu. NA. *Ischyrhiza* U. Cret. NA. *Libanopristis Micropristis* U. Cret. SWAs. *Pucapristis* U. Cret. SA. *Sclerorhynchus* [*Ganopristis*] U. Cret. Paleoc. Eu. Af. U. Cret. Eoc. SWAs. **Pseudorajidae Anacanthobatidae** R. Oc.

SUBORDER MYLIOBATOIDEA **Urolophidae** R. Oc. *Urolophus* [*Leiobatis*] M. Eoc. Eu. Pleist. NA. R. Oc. **Dasyatidae** (**Trygonidae**) R. Oc. *Dasyatis* [*Dasibatus Dasybatus Heliobatis Palaeodasybatis Pastinachus Trygon Xiphotrygus*] ?L. Cret. M. Eoc. Eu. U. Cret. NAf. M. Eoc. Pleist. NA. U. Cret. SA. Mioc. As. Aus. Pleist. EInd. R. Oc. *Gryphodobatis* Plioc. NA. *Hypolophites* U. Cret. Eoc. Af. *Parapalaeobates* U. Cret. SWAs. NAf. *Rhombodus* U. Cret. Eu. SWAs. NAf. SA. *Taeniura* M. Eoc. Plioc. Eu. R. Oc. **Potamotrygonidae** R. Af. SA. *Potamotrygon* Pleist. R. Af. R. SA. **Gymnuridae** R. Oc. *Gymnura* [*Pteroplatea*] Mioc. NA. R. Oc. **Mobulidae** R. Oc. *Burmhamia* L. Eoc. Eu. *Eomanta* L. Olig. Eu. *Manta* Mioc. Plioc. NA. Mioc. Eu. R. Oc. *Paramobula* L. Olig. Eu. **Rhinopteridae** R. Oc. *Rhinoptera* [*Zygobatis*] U. Cret. Plioc. Eu. U. Cret. Mioc. Af. U. Cret. Mioc. Plioc. SA. Eoc. Mioc. NA. Mioc. As. **Myliobatidae** R. Oc. *Aetobatus* [*Plinthicus*] Paleoc. Plioc. Eu. Eoc. Mioc. Af. Eoc. Pleist. NA. Mioc. SAs. SA. R. Oc. *Apocopodon* U. Cret. NA. SA. *Brachyrhizodus* U. Cret. NA. *Hypolophites* Paleoc. WAf. *Igdabatis* U. Cret. WAf. *Mesibatis* Plioc. NA. *Myliobatis* [*Ichthyaetus Ptychopleurus*] U. Cret. Plioc. Eu. NA. Paleoc. Eoc. Af. Eoc. Mioc. As. Mioc. Pleist. Aus. Mioc. Plioc. SA. Tert. NZ. R. Oc. *Promyliobatis* M. Eoc. Eu. *Pucabatis* U. Cret. SA. **Incertae sedis** *Ptychotrigon* M. Eoc. Eu. U. Cret. NA.

PRESUMED ELASMOBRANCHII REMAINS NOT IDENTIFIABLE TO ORDER

ISOLATED NEUROCRANIA
"*Cladodus*" *hassiacus* U. Dev. Eu. "*Cladodus*" *wildungensis* U. Dev. Eu. "*Tamiobatis*" ?L. Carb. NA.

ISOLATED TEETH OR PARTS OF DENTITIONS
Ageleodus [*Callopristodus Ctenoptychius*] Miss. Eu. *Carcharopis* [*Dicrenodus*] Miss. Eu. *Cladodus* ubiquitous in Paleozoic *Crassidonta* L. Perm. NA. *Cynopodius* Miss. Eu. *Eoorodus* U. Dev. NA. *Euglossodus* Miss. Eu. *Hybocladodus* Miss. NA. *Lambdodus* [*Dicentrodus*] Penn. NA. "*Lophodus*" *Mesodmodus* Miss. NA. *Pleurodus* L. Carb. Eu. *Pristicladodus Rhamphodus* Miss. Eu. *Centrodus* [*Styracodus*] Penn. Eu. *Venustodus* Miss. NA.

MUCOUS MEMBRANE DENTICLES
Echinodus Miss. Eu. *Multidentodus* L. Carb. NA. "*Scolopodus*" *Thrinacodus* Miss. NA.

DERMAL DENTICLES
Cladolepis M. Dev. NA. *Deirolepis* [*Deviolepis*] M. Dev. NA. *Elegestolepis* U. Sil. EEu. *Ellesmereia* L. Dev. NNA. *Listracanthus* Penn. NA. *Maplemillia* U. Dev. NA. *Ohiolepis* M. Dev. NA. *Petrodus Ostinaspis* Miss. EEu.

ICHTHYODORULITES (SPINES)
Aganacanthus Miss. Eu. *Amacanthus* [*Homacanthus*] Miss. NA. *Antacanthus* Miss. Eu. *Batacanthus* Miss. NA. *Bulbocanthus* L. Dev. NA *Bythiacanthus* Miss. NA. *Chalazacanthus* Miss. Eu. *Cosmacanthus* [*Leptacanthus*] Dev. Eu. *Drepanacanthus* [*Drepanocanthus*] Penn. NA. *Euctenius* Miss. Eu. *Euctenodopis* Miss. Eu. *Euphyacanthus* Miss. Eu. *Glymmatacanthus* Miss. NA. *Gnathacanthus* Miss. Eu. *Homacanthus* Dev. EEu. *Lispacanthus* Miss. Eu. *Lophacanthus* Penn. Eu. *Margaritacanthus* [*Euacanthus*] Miss. EEu. *Metaxyacanthus* [*Dactylodus*] M. Carb. EEu. *Ostracanthus* U. Carb. Eu. *Physonemus* Miss. Eu. *Ptychacanthus* Miss. Eu. *Stichacanthus* Miss. Eu. *Thaumatacanthus* U. Perm. Ind. *Tubulacanthus* L. Perm. EEu. *Xystracanthus* Penn. NA.

SUBCLASS HOLOCEPHALI

ORDER CHONDRENCHELYIFORMES **Chondrenchelyidae** *Chondrenchelys* Miss. Eu. *Harpagofututor* U. Miss. NA. ?*Platyxystrodus* Carb. NA. ?*Solenodus* Penn. Eu.

ORDER COPODONTIFORMES **Copodontidae** *Acmoniodus* U. Dev. NA. *Copodus* [*Characodus* ?*Dimyleus Labodus Mesogomphus Mylacodus Mylax Pinacodus Pleurogomphus Rhymodus*] Miss. Eu. NA.

ORDER PSAMMODONTIFORMES **Psammodontidae** *Lagarodus* Miss. Penn. Eu. *Mazodus* Miss. NA. *Psammodus* [*Arachaeobatis Astrobodus Homalodus*] ?U. Dev. Miss. NA. Miss. Eu.

ORDER INCERTAE SEDIS

SUBORDER COCHLIODONTOIDEI **Cochliodontidae** *Cochliodus* [*Chitinodus Cyrtonodus*] Miss. Penn. Eu. Miss. NA. ?*Cranodus Crassidonta* L. Perm. Aus. ?*Cymatodus* Penn. NA. *Deltodus* [*Deltodopsis Stenopterodus Taeniodus*] Miss. Eu. Miss. Penn. NA. Perm. Aus. *Dichelodus Diplacodus* Miss. Eu. ?*Erismacanthus* [*Cladacanthus Dipriacanthus Gampsacanthus Lecracanthus*] Miss. Eu. NA. ?Aus. *Helodopsis* Perm. SAs. *Icanodus* [*Enniskillen Eutomodus*] Miss. Eu. NA. Perm. Aus. ?*Macrodontacanthus* L. Perm. NA. ?*Menaspacanthus* Miss. Eu. *Platyodus* Miss. NA. *Poecilodus* Miss. Eu. Miss. Penn. NA. ?Perm. Aus. *Psephodus* [*Aspidodus*] Miss. Penn. Eu. Miss. L. Perm. NA. Penn. SAs. *Sandalodus* [?*Orthopleurodus Trigonodus Vaticinodus*] U. Dev. Penn. NA. Miss. Eu. *Synthetodus* U. Dev. NA. *Thoralodus* U. Dev. Eu. *Xenodus* [*Goniodus*] U. Dev. NA.

SUBORDER HELODONTOIDEI **Helodontidae** *Helodus* [*Diclitodus Pleurodus Pleuroplax*] U. Dev. L. Perm. NA. Miss. Penn. Eu. Perm. Aus. *Venustodus* [*Lophodus Oxytomodus* ?*Rhampodus Tomodus*] Miss. Eu. NA.

SUBORDER MENASPOIDEI **Menaspidae** *Deltoptychius* [?*Antacanthus Listracanthus* ?*Lophocanthus Phigeacanthus* ?*Phricacanthus Platacanthus Platycanthus Streblodus*] Miss. Penn. Eu. NA. *Menaspis* [?*Asima* ?*Radamus*] U. Perm. Eu.

SUBORDER SQUALORAJOIDEI **Squalorajidae** *Squaloraia* [*Spinocorhinus*] L. Jur. Eu.

SUBORDER MYRIACANTHOIDEI **Acanthorhinidae** *Acanthorhina* L. Jur. Eu. **Chimaeropsidae** *Chimaeropsis* L. U. Jur. Eu. **Myriacanthidae** *Myriacanthus* [*Metopacanthus Prognathodus*] L.-M. Jur. Eu.

ORDER CHIMAERIFORMES

SUBORDER AND FAMILY UNNAMED *Delphyodontos* U. Miss. NA.

SUBORDER ECHINOCHIMAEROIDEI **Echinochimaeridae** *Echinochimaera* U. Miss. NA. *Marracanthus* Miss. NA.

SUBORDER CHIMAEROIDEI **Chimaeridae** R. Oc. *Brachymylus* [*Aletodus*] L.-U. Jur. Eu. *Chimaera* [*Plethodus*] U. Cret. Aus. Eoc. Plioc. Eu. Tert. EInd. NZ. R. Oc. *Edaphodon* [*Bryactinus Dipristis Driphrissa Eumylodus ?Isotaenia ?Leptomylus Loxomylus Mylognathus Passalodon Psittacodon ?Sphagepoea*] L. Cret. Plioc. Eu. U. Cret. Eoc. NA. Mioc. Plioc. Aus. *Ganodus* [*Leptacanthus*] U. Jur. Eu. *?Ichthypriapus* U. Cret. NA. *Ischyodon* ?Mioc. Aus. *Ischyodus* [*Auluxacanthus Chimaeracanthus*] M. Jur. Paleoc. Eu. Cret. NZ. *Myledaphus* U. Cret. NA. *Pachymylus* M. Jur. Eu. *Psaliodus* Eoc. Eu. *Similihariotta* U. Carb. NA. **Rhinochimaeridae** R. Oc. *Amylodon* Olig. Eu. *Elasmodectes* [*Elasmognathus*] U. Jur. U. Cret. Eu. *Elasmodus* U. Cret. Eoc. Eu. **Callorhinchidae** R. Oc. *Callorhinchus* [*Callorhynchus*] U. Cret. NZ. Mioc. SA. R. Oc.

CHONDRICHTHYES SUBCLASS INCERTAE SEDIS

ORDER INIOPTERYGIFORMES **Iniopterygidae** *Iniopteryx* Penn. NA. **Sibyrhynchidae** *Sibyrhynchus* Penn. NA. *Iniopera* Penn. NA. *Inioxyele* Penn. NA.

ORDER PETALODONTIDA **Family not designated** *Antiodus* [*Chomatodus*] Miss. NA. *Chomatodus* [*Chromatodus Palaeobatis Petalodus Psammodus*] Miss. Eu. Carb. NA. *Ctenoptychius* [*Petalodus*] Penn. Eu. *Fissodus* [*Cholodus*] Miss. Eu. NA. Carb. NA. *Glyphanodus* [*Glyphanodon*] Miss. Eu. *Harpacodus* [*Ctenopetalus Ctenoptychius*] Miss. Eu. *Janassa* [*Acrodus Byzenos Climaxodus Cymatodus Dictea Ianassa ?Peltodus Strigilina Thoracodus Trilobites*] Miss. L. Perm. NA. Miss. U. Perm. Eu. L. Perm. Gr. Miss. Perm. EEu. *Lisogodus* Miss. NA. *Paracymatodus* [*Cymatodus Cimatodus*] Miss. EEu. *Petalodus* [*Ctenopetalus Getalodus Sicarius*] Miss. Eu. U. Dev. or L. Carb. EAs. *Petalorhynchus* [*Petalodus*] Miss. NA. Eu. EEu. U. Perm. Ind. *Polyrhizodus* [*Ctenoptychius Dacatylodus Petalodus Rhomboderma*] Miss. Eu. EEu. Carb. NA. *Serratodus* Miss. Eu. *Tanaodus* [*Chomatodus Tonaodus*] Miss. NA. Eu. **Pristodontidae** *Pristodus* [*Diodontopsodus Hoplodus Petalorhynchus Pristicladodus*] Miss. Eu. *Megactenopetalus* [*Megactenopectalodus*] L.-M. Perm. NA. M.-U. Perm. EAs. U. Perm. WAs. *Peripristis* [*Ctenoptychius*] Miss. Perm. NA.

CLASS INCERTAE SEDIS ACANTHODII

ORDER CLIMATIIFORMES **Climatiidae** *Brachyacanthus* L. Dev. Eu. *?Brochoadamones* L. Dev. NNA. *Cheiracanthoides* [*Helolepis*] M. Dev. NA. *Climatius* U. Sil. L. Dev. Eu. NA. EEu. *Eiffellepis* M. Dev. Eu. *Erriwacanthus* U. Sil. Eu. L. Dev. EEu. *Euthacanthus* L. Dev. Eu. *Latviacanthus* L. Dev. EEu. *Lupopsyrus* L. Dev. NNA. *Nostolepis* [*Dendracanthus Diplacanthoides Dontacanthus Rhabdiodus*] U. Sil. L. Dev. Eu. EEu. Gr. Spits. *Parexus* L. Dev. Eu. *Pruemolepis* M. Dev. Eu. *Ptomacanthus* L. Dev. Eu. EEu. *Sabrinacanthus* L. Dev. Eu. *Vernicomacanthus* L. Dev. Eu. *Welteldorifa* M. Dev. Eu. **Diplacanthidae** *Diplacanthus* [*Rhadinacanthus*] M. Dev. Eu. EEu. U. Dev. NA. *Gladiobranchus* L. Dev. NNA. **Gyracanthidae** *Agnacanthus* Miss. Eu. *Antacanthus* Miss. Eu. *Gyracanthus* L. Dev. Penn. NA. ?L. Dev. Penn. Eu. *Oracanthus* Miss. NA. Eu.

ORDER ISCHNACANTHIFORMES **Ischnacanthidae** *Acanthodopsis* Penn. Eu. *Apateacanthus* U. Dev. NA. *Atopacanthus* U. Dev. Eu. EEu. *Doliodus* ?M. Dev. NA. *Gomphonchus* [*Gomphodus Poracanthodes*] U. Sil. Eu. EEu. L. Dev. NA. Spits. Gr. ?SAs. *Helenacanthus* L. Dev. NA. *Ischnacanthus* L. Dev. Eu. NA. *Marsdenius* Miss. Eu. *Persacanthus* U. Dev. SWAs. *?Plectrodus* U. Sil. Eu. *Rockycampacanthus* L. Dev. Aus. *Taemasacanthus* L. Dev. Aus. *Uraniacanthus* L. Dev. Eu. *Xylacanthus* L. Dev. Spits.

ORDER ACANTHODIFORMES **Acanthodidae** *Acanthodes* Miss. L. Perm. Eu. Miss. Aus. SAf. Penn. L. Perm. Eu. U. Dev. or L. Carb EAs. *Carycinacanthus* Miss. EEu. *Cheiracanthus* M. Dev. Eu. EEu. ?Ant. *Homalacanthus* U. Dev. NA. Miss. EEu. *Mesacanthus* L.-M. Dev. Eu. *Protogonacanthus* U. Dev. Eu. *Pseudacanthodes* [*Protacanthodes*] Penn. Eu. *Traquairichthys* [*Traquairia*] Penn. Eu. ?L. Perm. ?NA. *Triazeugacanthus* U. Dev. NA.

ACANTHODIANS INCERTAE SEDIS

Antarchtonchus M. or U. Dev. Ant. *Archaeacanthus* M. Dev. EEu. *Byssacanthoides* M. or U. Dev. Ant. *Campylodus* U. Sil. EEu. L.-M. Dev. Eu. *Devononchus* U. Dev. EEu. *Eupleurogmus* Miss. Aus. *Gemuendolepis* L. Dev. Eu. *Haplacanthus* M.-U. Dev. EEu. ?M. Dev. ?Gr. *Holmesella* Penn. NA. *Homacanthus* M.-U. Dev. EEu. U. Dev. Penn. Eu. U. Dev. NA. *Machaeracanthus* [*Dinacanthodes*] L.-M. Dev. Eu. L. Dev. NAf. L.-M. Dev. NA. *Monopleurodus* U. Sil. EEu. *Nodacosta* L.-M. Dev. EEu. *Nodonchus* L. Dev. Eu. *Onchus* [*Leptocheles*] L. Sil. L. Dev. NA. U. Sil. L. Dev. Eu. L.-U. Dev. EEu. *Pinnacanthus* L. Dev. NA. *Protodus* L.-M. Dev. NA. *Ptychodictyon* M. Dev. EEu. *Sinacanthus* M. Dev. EAs. *Striacanthus* M.-U. Dev. Aus.

CLASS OSTEICHTHYES

SUBCLASS ACTINOPTERYGII

INFRACLASS CHONDROSTEI

ORDER PALAEONISCIFORMES

SUBORDER PALAEONISCOIDEA **Cheirolepidae** *Cheirolepis* M. Dev. Eu. U. Dev. NA. **Stegotrachelidae** *?Borichthys* Miss. Eu. *Kentuckia* Miss. NA. *Mimia* U. Dev. Aus. *Moythomasia* M.-U. Dev. Eu. U. Dev. Aus. *Orvikuina* M. Dev. Eu. *Stegotrachelus* Dev. Eu. ?U. Dev. NA. Ant. **Tegeolepidae** *Tegeolepis* [*Actinophorus*] U. Dev. NA. **Rhabdolepidae** *Osorioichthys* [*Stereolepidella Stereolepis*] U. Dev. Eu. *Rhabdolepis* L. Perm. Eu. **Rhadinichthyidae** *Aetheretomon* Miss. Eu. *Cycloptychius* Miss. NAs. Miss. Penn. Eu. *?Eurylepidoides* L. Perm. NA. *Rhadinichthys* Miss. Af. Miss. Penn. Eu. Penn. ?M. Trias. SA. *Rhadinoniscus Strepheoschema* Miss. Eu. **Carbovelidae** *Carboveles* Miss. Eu. *Phanerosteon* [*Gymnoniscus Sceletophorus*] Miss. L. Perm. Eu. *?Sphaerolepis* [*Trissolepis*] L. Perm. Eu. **Canobiidae** *Canobius Mesopoma* Miss. Eu. *Whiteichthys* Penn. Gr. **Cornuboniscidae** *Cornuboniscus* Miss. Eu. **Styracopteridae** *Benedenius* [*Benedenichthys*] *Styracopterus* [*?Fouldenia*] Miss. Eu. **Cryphiolepidae** *Cryphiolepis* Miss. Eu. **Holuriidae** *Holuropsis* U. Perm. Eu. *Holurus* Miss. Eu. **Cosmoptychiidae** *Cosmoptychius* Miss. Eu. *Watsonichthys* Miss. Eu. L.

Perm. SAf. **Pygopteridae** *Nematoptychius* Miss. Eu. *Pygopterus* M. Perm. Gr. U. Perm. L. Trias. Eu. L. Trias. SAf. Spits. **Elonichthyidae** *Drydenius* Miss. Perm. Eu. *Elonichthys* [*Ganacrodus ?Pariostegus ?Propalaeoniscus*] Miss. U. Perm. Eu. Miss. Penn. ?U. Trias. NA. Miss. U. Perm. ?U. Trias. Aus. L. Perm. SAf. M. Perm. Gr. U. Perm. SA. ?Trias. WAf. *Ganolepis* Perm. NAs. *Gonatodus* Miss. NA. Miss. Penn. EU. **Acrolepidae** *Acrolepis* Miss. L. Perm. NA. Miss. U. Perm. Eu. Penn. SA. Penn. L. Perm. NAs. ?L.-U. Perm. SAf. ?Trias. Aus. *Acrorhabdus* L. Trias. Spits. Gr. *Boreosomus* [*Diaphorognathus*] L. Trias. Gr. Spits. Mad. ?M. Trias. NA. *Hyllingea* U. Trias. Eu. *Leptogenichthys* M. Trias. Aus. *Mesonichthys* Miss. Eu. *Namaichthys* L. Perm. SAf. *Plegmolepis* M. Perm. Gr. *Reticulolepis* ?Miss. U. Perm. Eu. *Tholonotus* L. Perm. SA. **Coccocephalichthyidae** *Coccolcephalichthys* [*Coccocephalus Cocconiscus*] Penn. Eu. **Amblypteridae** *Amblypterina* U. Perm. Eu. *Amblypterus* L.-U. Perm. Eu. ?L. Perm. NA. As. ?M. Trias. SA. *?Lawnia* L. Perm. NA. **Aeduellidae** *Aeduella* L. Perm. Eu. **Commentryidae** *Commentrya* [*Elaveria*] *Paramblypterus* [*Amblypterops Cosmopoma Dipteroma*] L. Perm. Eu. **Palaeoniscidae** *?Aegicephalichthys* M. Trias. Aus. *Cosmolepis* [*Oxygnathus Thrissonotus*] L. Jur. Eu. *?Gyrolepidoides* M. Trias. SA. *Gyrolepis* L. Trias. EAs. M.-U. Trias. Eu. U. Trias. NA. *Palaeoniscum* [*Eupalaeoniscus Geomichthys Palaeoniscus Palaeothrissum*] ?L. Perm. NA. ?L. Perm. U. Perm. ?L. Trias. Eu. M. Perm. Gr. L. Trias. Spits. ?EAs. ?U. Trias. Aus. *Progyrolepis* L. Perm. Eu. NA. *Pteronisculus* [*Glaucolepis*] L. Trias. Spits. Gr. Mad. Aus. *?Trachelacanthus* L. Perm. Eu. *Turseodus* [*Eurecana Gwyneddichthys*] U. Trias. NA. *Westollia* [*Lepidopterus*] L. Perm. Eu. **Dicellopygidae** *?Aneurolepis* [*Urolepis*] M.-U. Trias. Eu. *?Brachydegma* L. Perm. NA. *Dicellopyge* L. Trias. SAf. **Boreolepidae** *Boreolepis* U. Perm. Gr. **Birgeriidae** *Birgeria* [*Xenestes*] L. Trias. Spits. Gr. NA. Mad. L.-U. Trias. Eu. *Ohmdenia* L. Jur. Eu. *?Psilichthys* Trias. or Jur. Aus. **Scanilepidae** *Scanilepis* L. Trias. Spits. U. Trias. Eu. **Centrolepidae** *Centrolepis* L. Jur. Eu. **Coccolepidae** *Browneichthys* L. Jur. Eu. *Coccolepis* [*?Palaeoniscionotus*] L. Jur. L. Cret. Eu. M. Jur. As. Jur. Aus.

SUBORDER PLATYSOMOIDEI **Platysomidae** *Mesolepis* [*Pododus*] Miss. Penn. Eu. *Paramesolepis* Miss. Eu. *Platysomus* [*?Tonipoichthys Uropteryx*] Miss. U. Perm. Eu. Miss. L. Perm. L. Trias. NA. M. Perm. M. Trias. Gr. ?L. Trias Spits. U. Trias. Aus. **Chirodontidae** (**Amphicentridae**) *Cheirodopsis* Miss. Eu. *Chirodus* [*Amphicentrum Cheirodus Hemicladodus*] Miss. Penn. Eu. NA. *Eurynothus* [*Eurynotus Notacmon Plectrolepis*] Miss. Penn. Eu. NA. Penn. NAs. ?M. Trias. SA. *Globulodus* [*Eurysomus*] *Lekanichthys* U. Perm. Eu. *Paraeurynotus* L. Perm. Eu. *Proteurynotus* Miss. Eu. **Bobastraniidae** *Bobastrania* L. Trias. Spits. NA. Gr. Mad. *?Caruichthys* L. Trias. SAf. *Ebenaqua* U. Perm. Aus. *Ecrinesomus* L. Trias. Mad. *Lambeichthys* L. Trias. NA.

ORDER HAPLOLEPIFORMES **Haplolepidae** *Haplolepis* [*Eurylepis Mecolepis Mekolepis Parahaplolepis*] *Pyritocephalus* [*Teleopterina*] Penn. Eu. NA.

ORDER DORYPTERIFORMES **Dorypteridae** *Dorypterus* L. Perm. EAs. U. Perm. Eu.

ORDER TARRASIIFORMES **Tarrasiidae** *Tarrasius* Miss. Eu. *Paratarrasius* Miss. NA.

ORDER PTYCHOLEPIFORMES **Ptycholepidae** *Ptycholepis* M. Trias. L. Jur. Eu. U. Trias. NA.

ORDER PHOLIDOPLEURIFORMES **Pholidopleuridae** *Arctosomus* [*Neavichthys*] L. Trias. NAs. *Australosomus* L. Trias. Af. Mad. Gr. M. Trias. Spits. *Macroaethes* M. Trias. Aus. *Pholidopleurus* M.-U. Trias. Eu.

ORDER LUGANOIIFORMES **Luganoiidae** *Besania Luganoia* M.-U. Trias. Eu. **Habroichtyidae** *Habroichthys Nannolepis* U. Trias. Eu. **Thoracopteridae** *Gigantopterus* U. Trias. Eu. *Thoracopterus* [*Pterygopterus*] ?L. Trias. NA. M. Trias. Aus. U. Trias. Eu.

ORDER REDFIELDIIFORMES **Redfieldiidae** (**Catopteridae Dictyopygidae**) *Atopocephala* L. Trias. SAf. *?Beaconia Brookvalia* M. Trias. Aus. *Cionichthys* U. Trias. NA. *Daedalichthys* L. Trias. SAf. *?Dictyopleurichthys* M. Trias. Aus. *Dictyopyge* U. Trias. NA. *Geitonichthys* M. Trias. Aus. *Helichthys* L. Trias. SAf. *Ischnolepis* L. Trias. SAf. *Lasalichthys* U. Trias. NA. *Mauritanichthys* U. Trias. NA. *Molybdichthys Phlyctaenichthys* M. Trias. Aus. *Pseudobeaconia* M. Trias. SA. *Redfieldius* L. Jur. NA. *Schizurichthys* M. Trias. Aus. *Synorichthys* U. Trias. NA.

ORDER PERLEIDIFORMES **Perleididae** (**Colobodontidae**) *Chrotichthys* L. Trias. Aus. *Colodus* [*Asterodon ?Cenchrodus ?Charitodon ?Charitosaurus Eupleurodus ?Hemilopus ?Nephrotus ?Omphalodus*] L. Trias. SAf. NAs. EAf. L.-U. Trias. Aus. M.-U. Trias. Eu. *Crenolepis* [*Crenilepis Cernilepoides*] M. Trias. Eu. *Dimorpholepis* Trias. WAf. *Dollopterus* L. Trias. NA. M. Trias. Eu. *Engycolobodus* M. Trias. Eu. *?Helmolepis* L. Trias. Gr. *Manlietta* M. Trias. Aus. *Meidiichthys* L. Trias. SAf. *Mendocinichthys* [*Mendocinia*] M. Trias. SA. *Meridensia* M.-U. Trias. Eu. *Perleidus* L. Trias. EAs. Mad. Gr. Spits. M.-U. Trias. Eu. Trias. WAf. *Pristisomus* L. Trias. Aus. Mad. *Procheirichthys* M. Trias. Aus. *Tripelta* L. Trias. Aus. *Zeuchthiscus* L. Trias. Aus. **Cleithrolepidae** *Cleithrolepis* L. Trias. SAf. L.-U. Trias. Aus. M. Trias. SA. U. Trias. Eu. *Dipteronotus* U. Trias. Eu. *Hydropessum* L. Trias. SAf. **Platysiagidae** *Platysiagum* M. Trias. L. Jur. Eu. **Cephaloxenidae** *Cephaloxenus* M.-U. Trias. Eu. **Aethodontidae** *Aethodontus* M.-U. Trias. Eu.

ORDER PELTOPLEURIFORMES **Peltopleuridae** *Peltopleurus* M. Trias. EAs. M.-U. Trias. Eu. *Placopleurus* M.-U. Trias. Eu. **Polzbergiidae** *Polzbergia* U. Trias. Eu.

ORDER PHANERORHYNCHIFORMES **Phanerorhynchidae** *Phanerorhynchus* Penn. Eu.

ORDER SAURICHTHYIFORMES **Saurichthyidae** *Brevisaurichthys* M. Trias. Eu. *Saurichthys* [*Acidorhynchus Belonorhynchus Giffonus Gymnosaurichthys Ichthyorhynchus*] L. Trias. EEu. NA. NNA. Spits. Mad. Gr. SAf. SAs. EAs. L.-M. Trias. Aus. L.-U. Trias. Eu. U. Trias. CAs. *Saurorhynchus* L. Jur. Eu. *Systolichthys* M. Trias. Eu.

CHONDROSTEANS NOT ASSIGNED TO FAMILIES *Aldingeria* Penn. Gr. *?Anaglyphys* L. Perm. Eu. *Anatoia* L. Trias. SA. *Apateolepis* M. Trias. Aus. *?Atherstonia* [*Hypterus*] U. Perm. Eu. L. Trias. SAf. Mad. U. Trias. NA. *Belichthys* M. Trias. Aus. *Broometta* L. Trias. SAf. *Canin-*

chaia Cenechoia L. Trias. SA. *Challaia* L. Trias. SA. *Diphyodus* Miss. NA. *Disichthys* U. Perm. SAf. *Echentaia* L. Trias. SA. *Elpisopholis* U. Trias. Aus. *Eurynotoides* U. Perm. Eu. *Evenkia* L. Trias. NAs. *Guaymayenia* L. Trias. SA. *Isodus* Miss. Eu. *Leighiscus Megapteriscus Mesembroniscus* M. Trias. Aus. *Luederia* L. Perm. NA. *Myriolepis* L. U. Trias. Aus. ?L. Trias. SA. *Oxypteriscus* Miss. NAs. *Palaeobergeria* Miss. NAs. *Pasamhaya* L. Trias. SA. *Peleichthys* U. Perm. SAf. *Pteroniscus* Jur. WAs. *?Schizospondylus* U. Cret. Eu. *Tanaocrossus* U. Jur. WAf.

ORDER POLYPTERIFORMES (CLADISTIA) **Polypteridae** R. Af. *Polypterus* Eoc. R. Af.

ORDER ACIPENSERIFORMES

SUBORDER CHONDROSTEOIDEI **?Errolichthyidae** *Errolichthys* L. Trias. Mad. **Chondrosteidae** *Chondrosteus Gyrosteus* [*?Strongylosteus*] L. Jur. Eu. *?Stichopterus* L. Cret. As.

SUBORDER ACIPENSEROIDEI **Acipenseridae** *Acipenser* U. Cret. R. NA. Eu. R. As. *Huso* Plioc. R. Eu. R. NAs. *Protoscaphirhynchus* U. Cret. NA.

SUBORDER POLYODONTOIDEI **Polyodontidae** *Crossopholis* Eoc. NA. *Paleopsephurus* U. Cret. NA.

INFRACLASS NEOPTERYGII

ORDER LEPISOSTEIFORMES (GINGLYMODI) **Lepisosteidae** *Atractosteus* U. Cret. R. NA. Eoc. Eu. ?U. Cret. WAf. *Lepisosteus* U. Cret. SAs. U. Cret. R. NA. *?Paralepidosteus* L. Cret. WAf.

ORDER SEMIONOTIFORMES **Semionotidae** (**Lepidotidae Dapediidae**) *Acentrophorus* U. Perm. Eu. U. Trias. NA. *Aetheolepis* Jur. Aus. *Alleiolepis* [*Leiolepis*] M. Trias. Eu. *Angolaichthys* Trias. WAf. *Aphelolepis* Trias. Eu. *Asialepidotus* M. Trias. EAs. *Corunegenys* U. Trias. Aus. *Dapedium* [*Aechmodus Amblyurus Dapedius Omalopleurus Pholidotus*] U. Trias. L. Jur. Eu. SAs. *Enigmatichthys* M. Trias. Aus. *Eosemionotus* L.-M. Trias. Eu. *Hemicalypterus* U. Trias. NA. *Heterostrophus* [*Heterostichus*] U. Jur. Eu. *Lepidotes* [*Lepidosaurus Lepidotus Plesiodus Prolepidotus Scrobodus Sphaerodus*] U. Trias. L. Cret. ?U. Cret. Eu. U. Trias. L. Cret. Af. NA. U. Trias. U. Cret. As. ?Trias. L. Jur. U. Cret. SA. L. Jur. Mad. *?Orthurus* U. Trias. Eu. *Paracentrophorus* L. Trias. Mad. *Paralepidotus* U. Perm. U. Trias. Eu. *Pericentrophus* L. Trias. Eu. *Prionopleurus* [*Pantelion*] Jur. Eu. *Pristiosomus* U. Trias. Aus. *Sargodon* U. Trias. Eu. *Semionotus* [*?Archaeosemionotus Ischypetrus*] L.-U. Trias. Eu. ?L. Trias. SA. U. Trias. Aus. SAf. NA. *Serrolepis* M.-U. Trias. Eu. *Sinosemionotus* M. Trias. EAs. *Tetragonolepis* [*Homoeolepis Pleurolepis*] L.-U. Jur. Eu. L. Jur. As.

ORDER PYCNODONTIFORMES **Pycnodontidae** *Acrotemus* U. Cret. Eu. WAf. *Anomaeodus* L.-U. Cret. Eu. NA. U. Cret. Af. *Athrodon* U. Jur. U. Cret. Eu. *Coccodus* U. Cret. SWAs. *Coelodus* [*Anomiophthalmus Cosmodus Glossodus*] U. Jur. U. Cret. Eu. L.-U. Cret. NA. Af. U. Cret. Mad. U. Cret. Eoc. As. *Ellipsodus* L. Cret. Eu. *Eomesodon* U. Trias. U. Jur. Eu. *Grypodon* [*Ancistrodon Ankistrodus*] U. Cret. Eu. NAf. NA. *Gyrodus* [*Stromateus*] M. Jur. U. Cret. Eu. U. Jur. WInd. U. Jur. ?U. Cret. NA. U. Cret. Af. *Gyronchus* [*Gynonchus ?Gyroconchus Macromesodon Mesodon Scaphodus Typodus*] M. Jur. L. Cret. Eu. *Ichthyoceros* U. Cret. SWAs. *Mesturus* U. Jur. Eu. *Micropycnodon* [*Pycnomicrodon*] U. Cret. NA. *Palaebalistum* [*Palaeobalistes*] U. Cret. As. SA. U. Cret. Eoc. Eu. Af. *Polygyrodus* U. Cret. Eu. *Proscinetes* [*Microdon Polysephis*] M. Jur. L. Cret. Eu. L. Cret. NA. *Pycnodus* [*Periodus Pychnodus*] ?M. Jur. L. Cret. Eoc. Eu. U. Jur. WInd. L. Cret. Aus. L.-U. Cret. NA. Eoc. Af. As. *Stemmatodus* ?U. Jur. L. Cret. Eu. *Tibetodus* U. Jur. CAs. *Trewavasia* [*Xenopholis*] U. Cret. SWAs. *Uranoplosus* L. Cret. Eu. NA. *Woodthropea* U. Trias. Eu.

ORDER MACROSEMIIFORMES **Macrosemiidae** *?Enchelyolepis* U. Jur. Eu. *Histionotus* U. Jur. L. Cret. Eu. *Legnonotus* U. Trias. Eu. *Macrosemius* [*Disticholepis*] U. Jur. Eu. WAf. *Notagogus* [*Blenniomogeus Callignathus*] U. Jur. L. Cret. Eu. *?Orthurus* U. Trias. Eu. *Petalopteryx* [*Aphanepygus*] L.-U. Cret. Eu. U. Cret. SWAs. *Propterus* U. Jur. L. Cret. Eu. **Uarbyichthyidae** *Uarbyichthys* Jur. Aus. **Incertae sedis** *Ophiosis* M. Trias. L. Cret. Eu. U. Jur. WAf. *Songanella* U. Jur. WAf.

ORDER AMIIFORMES **Parasemionotidae** *Archaeolepidotus* L. Trias. Eu. *Broughia ?Helmolepis* L. Trias. Gr. *Jacobulus* L. Trias. Mad. *Ospia* L. Trias. Gr. *Paracentrophus* L. Trias. Mad. *Parasemionotus* L. Trias. Mad. Gr. *Phaidrosoma* U. Trias. Eu. *Praesemionotus* L. Trias. Eu. *Promecosomina* M.-U. Trias. Aus. *Stensionotus Thomasinotus* L. Trias. Mad. *Tungusichthys* L. Trias. NAs. *Watsonulus* [*Watsonia*] L. Trias. Mad. Gr. **Caturidae** (**Eugnathidae, Furidae**) *Allolepidotus* [*Plesiolepidotus*] U. Trias. Eu. *Caturus* [*?Amblysemius Conodus Ditaxiodus Endactis Strobilodus Thlattodus Uraeus*] M. Jur. L. Cret. Eu. U. Jur. WAf. *Eoeugnathus* M.-U. Trias. Eu. *Furo* [*Eugnathus Isopholis Lissolepis*] ?U. Trias. U. Jur. Eu. Jur. As. *Heterolepidotus* [*Brachyichthyes Eulepidotus*] L.-U. Jur. Eu. Jur. CAs. *Macrepistus* L. Cret. NA. *Neorhombolepis* L.-U. Cret. Eu. L. Cret. SA. *Osteorachis* [*Harpactes Harpactira Isocolum*] L.-U. Jur. Eu. *Otomitla* L. Cret. NA. *Sinoeugnathus* M. Trias. EAs. **Amiidae** *Amia* [*Amiatus Cyclurusypamia ?Kindleia Notaeus Pappichthys Paramiatus Protamia Stylomuleodon*] U. Cret. R. NA. Paleoc. Mioc. Eu. Eoc. As. Spits. *Amiopsis* U. Jur. L. Cret. Eu. *Enneles* L. Cret. SA. *Ikechaoamia* ?L. Cret. EAs. *Liodesmus* [*Lophiurus*] U. Jur. Eu. *Platacodon* U. Cret. NA. *Pseudamiatus* [*Pseudamia*] Eoc. Spits. *Sinamia* ?U. Jur. EAs. *Urocles* [*Megalurus Synergus*] U. Jur. Eu. ?L. Cret. ?SA. *Vidalamia* U. Jur. Eu.

ORDER PACHYCORMIFORMES **Pachycormidae** *Asthenocormus* [*Agassizia*] U. Jur. Eu. *?Eugnathides* U. Jur. NA. *Euthynotoides Euthynotus* [*Cyclospondylus Heterothrissops Parathrissops Pseudothrissops*] L. Jur. Eu. *Hypsocormus* U. Trias. U. Jur. Eu. *?Leedsichthys* [*Leedsia*] *Orthocormus* U. Jur. Eu. *Pachycormus* [*Cephenoplosus ?Pachylepis ?Lycodus*] *Prosauropsis* [*Protosauropsis Saurostomus*] L. Jur. Eu. *?Protosphyraena* [*Erisichthe Pelecopterus*] U. Cret. Eu. As. NA. SA.

ORDER ASPIDORHYNCHIFORMES **Aspidorhynchidae** *Aspidorhynchus* M.-U. Jur. Eu. Cret. SA. *Belonostomus* [*Belonostmus Dichelospondylus Hemirhynchus Ophirachis Vinctifer*] U. Jur. U. Cret. Eu. L.-U. Cret. NA. As. SA. Aus. U. Cret. NAf.

DIVISION TELEOSTEI

ORDER PHOLIDOPHORIFORMES **Pholidophoridae** *Eurycormus* U. Jur. Eu. *?Flugopterus* [*Megalopterus*] U. Trias. Eu. *Hungki-*

ichthys U. Jur. EAs. *Pholidophoretes* U. Trias. Eu. *Pholidophorides* L. Jur. Eu. *Pholidophoristion* U. Jur. Eu. *Pholidophorus* [*Baleiichthys ?Microps ?Nothosomus Phelidophorus Poreirgus*] ?L. Trias. U. Jur. SA. M. Trias. U. Jur. Eu. U. Trias. U. Jur. WAf. L. Jur. NA. U. Jur. As. *?Prohalecites* M.-U. Trias. Eu. **Ichthyokentemidae** *Elpistoichthys* U. Trias. Eu. *Ichthyokentema* U. Jur. Eu. *?Catervanolus* U. Jur. WAf. **Majokidae** *Majokia* U. Jur. WAf. **?Lingulellidae** *Lingulella* U. Jur. **Pleuropholidae** *Austropleuropholis Parapleuropholis* U. Jur. WAf. *Pleuropholis* U. Jur. L. Cret. Eu. U. Jur. WAf. **Archaeomenidae** *Archaeomaene Aetheolepis Aphnelepis Madariscus* Jur. Aus. *Wadeichthys* L. Cret. Aus. **Oligopleuridae** *?Calamopleurus* U. Jur. L. Cret. Eu. L. Cret. SA. *?Callopterus* U. Jur. Eu. *Ionoscopus* [*Opsigonus*] U. Jur. L. Cret. Eu. *Oligopleurus* U. Jur. L. Cret. Eu. U. Cret. SA. *Spathiurus* [*Amphilaphurus*] U. Cret. SWAs. **Incertae sedis** *Ceramurus* U. Jur. Eu. *Galkinia* U. Jur. CAs. *Hulettia* M. Jur. NA. *Lombardina Signeuxella* U. Jur. WAf. *Lophiostomus* U. Cret. Eu.

ORDER LEPTOLEPIFORMES **Leptolepidae** *Ascalabos* U. Jur. Eu. *Carsothrissops* U. Cret. Eu. *?Clupavus* U. Jur. L. Cret. Eu. Cret. Af. NA. *Ctenielepis* U. Jur. Eu. *Eurystichthys* [*Eurystethus*] U. Jur. Eu. *Leptolepides* U. Jur. Eu. *Leptolepis* [*Liassolepis Megastoma Sarginites*] U. Jur. U. Cret. Eu. NA. Jur. Spits. Aus. Jur. Cret. As. ?Jur. L.-U. Cret. SA. L.-U. Cret. Af. *Luisichthys* U. Jur. WInd. *Proleptolepis* L. Jur. Eu. *Tharrias* U. Cret. SA. *Tharsis* U. Jur. Eu. *Todiltia* M. Jur. NA. *Varasichthys* U. Jur. SA.

ORDER ICHTHYODECTIFORMES **Allothrissopidae** *Allothrissops* U. Jur. L. Cret. Eu. **Ichthyodectidae** *Chirocentrites* L. Cret. Eu. *Cladocyclus* L. Cret. SA. *Eubiodectes* U. Cret. WAs. *Gillicus* L.-U. Cret. Eu. NA. *Ichthyodectes* L.-U. Cret. Eu. NA. *Proportheus* L. Cret. WAf. *Spathodactylus* L. Cret. Eu. *Thrissops* U. Jur. U. Cret. Eu. NAf. *Xiphactinus* [*Portheus*] L.-U. Cret. NA. Aus. Eu. **Saurodontidae** *Saurocephalus* L.-U. Cret. Eu. U. Cret. NA. ?NAf. *Saurodon* [*Daptinus*] U. Cret. Eu. ?NAf. **?Thryptodontidae (Plethodonidae)** *Bananogmius* [*Ananogmius Anogmius*] U. Cret. Eu. NA. *Martinichthys Niobrara* U. Cret. NA. *Paranogmius* U. Cret. NAf. *Plethodus* U. Cret. Eu. NAf. *Syntengmodus Thryptodus* [*?Pseudothryptodus*] *Zanclites* U. Cret. NA. **Incertae sedis Tselfatiidae** *Occithrissops* M. Jur. NA. *Protobrama Tselfatia* U. Cret. NAf.

SUBDIVISION OSTEOGLOSSOMORPHA

ORDER OSTEOGLOSSIFORMES

SUBORDER OSTEOGLOSSOIDEI **Osteoglossidae** R. Af. SAs. EInd. Aus. SA. *?Brychaetus* [*Platops Pomphractus*] Paleoc. NAf. Eoc. Eu. *?Eurychir* U. Cret. NA. Paleoc. Eoc. NAf. *?Genartina* Eoc. NA. *Musperia* L. Tert. EInd. *Phareodus* [*Dapedoglossus*] M. Eoc. NA. Olig. Aus. *Scleropages* Olig. R. EInd. Mioc. R. Aus. *Sinoglossus* Eoc. EAs. **Singididae** *Singida* ?Paleoc. EAf.

SUBORDER NOTOPTEROIDEI **Lycopteridae** *Lycoptera* [*Asiatolepis*] *Paralycoptera* U. Jur. EAs. *Manchurichthys* L. Cret. EAs. **Hiodontidae** R. NA. *Eohiodon* Paleoc. Eoc. NA. **Notopteridae** *Nopterus* L. Olig.-R. EInd. R. SAs.

SUBORDER MORMYROIDEI **Mormyridae** R. Af. **Gymnarchidae** R. Af.

OSTEOGLOSSOMORPH INCERTAE SEDIS *Chandlerichthys* M. Cret. NNA.

SUBDIVISION ELOPOMORPHA

ORDER ELOPIFORMES

SUBORDER UNDESIGNATED **Anaethaliontidae** *Anaethalion* [*Aethalion*] U. Jur. L. Cret. Eu. **Crossognathidae** *Apsopelix* [*Helmintholepis Leptichthys Palaeoclupea Pelecorapis Syllaemus*] U. Cret. Eu. NA. *Crossognathus* L. Cret. Eu.

SUBORDER ELOPOIDEI **Elopidae** *Davichthys* U. Cret. SWAs. *Elops* Eoc. Eu. R. Oc. **Megalopidae** *Elopoides* L. Cret. Eu. *?Pachythrissops* U. Jur. L. Cret. Eu. *Promegalops* L. Eoc. Eu. *Protarpon* L. Eoc. Eu. *Sedenhorstia* U. Cret. Eu.

SUBORDER ALBULOIDEI **Osmeroididae** *Dinelops* U. Cret. Eu. *Osmeroides* L.-U. Cret. Eu. ?U. Cret. ?EAs. **Pterothrissidae** *Hajulia* U. Cret. SWAs. *Istieus* U. Cret. Eu. SWAs. *Pterothrissus* ?M. Jur. U. Jur. U. Cret. Eoc. Eu. Olig. Aus. R. Atl. Pac. **Albulidae** *Albula* ?U. Cret. ?NA. L.-M. Eoc. Eu. R. Oc. *Coriops Kleinpellia* U. Cret. NA. *Lebonichthys* U. Cret. SWAs. **?Phyllodontidae** *?Casierius* L. Cret. Eu. NA. *Egertonia* U. Paleoc. L. Eoc. Eu. *?Eodiaphyodus* U. Cret. WAf. *Paralbula* U. Cret. Mioc. NA. L. Eoc. Eu. U. Eoc. NAf. *Phyllodus* L. Paleoc. U. Eoc. Eu. L. Paleoc. Mioc. NA *Pseudoegertonia* L. Paleoc. WAf.

SUBORDER PACHYRHIZODONTOIDEI **Notelopidae** *Notelops* L. Cret. SA. **Pachyrhizodontidae** *Elopopsis* U. Cret. Eu. NAf. *Pachyrhizodus* [*Raphiosaurus Thrissopater*] U. Cret. Eu. NA. Aus. *Rhacolepis* L. Cret. SA.

ORDER ANGUILLIFORMES (APODES) **Anguillavidae** R. Cos. *Anguillavus* U. Cret. SWAs *Enchelurus* U. Cret. Eu. *Eoanguilla* M. Eoc. Eu. *Mastygocerus* Mioc. R. EInd. **Paranguillidae** *Dalpiazella* M. Eoc. Eu. *Paranguilla* M. Eoc. Eu. **Moringuidae** R. Oc. **Myrocongridae** R. Atl. Oc. **Xenocongidae** R. Oc. *Echelus Eomyrophis* [*Eomyrus*] M. Eoc. Eu. R. Oc. *Mylomyrus* Eoc. NAf. *Rhynchorinus* Eoc. Eu. *?Whitapodus* M. Eoc. Eu. **Muraenidae** R. Oc. *?Deprandus* Mioc. NA. *Muraena* Plioc. NAf. R. Oc. **Heterenchelyidae** R. Oc. *Heterenchelys* Mioc. Aus. R. Oc. **Dyssomminidae** R. Oc. **Proteomyridae** *Proteomyrus* M. Eoc. Eu. **Anguilloididae** *Anguilloides* M. Eoc. Mioc. Eu. *Veronanguilla* M. Eoc. Eu. **Milananguillidae** *Milananguilla* M. Eoc. Eu. **Muraenesocidae** R. Oc. *Muraenesox* Eoc. Eu. Mioc. Aus. R. Oc. **Neenchelyidae** R. Ind. Oc. **Nettastomatidae** R. Oc. *Nettastoma* Eoc. Eu. R. Oc. **Nessorhamphidae** R. Oc. **Congridae** R. Oc. *Ariosoma* [*Congermuraena*] Olig. Plioc. Eu. R. Oc. *Astroconger* Olig. Aus. *Bolcyrus* M. Eoc. Eu. *Conger* Eoc. Olig. NA. Mioc. NZ. R. Oc. *Enchelion* U. Cret. SWAs. *Paracongroides* M. Eoc. Eu. *Parbatmya* Eoc. NA. *Uroconger* Mioc. Plioc. Aus. R. Oc. *Voltaconger* M. Eoc. Eu. **Ophichthidae** R. Oc. *Caecula* Mioc. Eu. R. Oc. *Goslinophis* M. Eoc. Eu. *Mystriophis* ?Mioc. NZ. R. Oc. *Ophichthus* Eoc. Eu. R. Oc. **Todaridae Synaphobranchidae** R. Oc. **Simenchelyidae** R. NAtl. NPac. Oc. **Dysommidae Derichthyidae Macrocephenchelyidae Serrivomeridae** R. Oc. **Patavichtidae** *Patavichthys* M. Eoc. Eu. **Nemichthyidae** R. Oc. *Nemichthys* Eoc. Olig. WAs. R. Oc. **Cyemidae Aoteidae** R. Oc. **Urenchelyidae** *Urenchelys* U. Cret. Eu. NA. SWAs. **Incertae sedis** *Bolcanguilla* M. Eoc. Eu. *Gazolapodus* M. Eoc. Eu.

SUBORDER SACCOPHARYNGIDAE (LYOMERI) **Saccopharyngidae Eurypharyngidae Monognathidae** R. Oc.

ORDER NOTACANTHIFORMES (LYOPOMI AND HETEROMI) **Halosauridae** R. Oc. *Echidnocephalus* U. Cret. Eu. ?NA. *Enchelurus* U. Cret. Eu. SWAs. *Halosaurus* Mioc. NA. R. Oc. *Laytonia* ?Cret. Olig. Plioc. NA. **Lipogenyidae** R. Oc. **Notacanthidae** R. Oc. *Pronotacanthus* U. Cret. SWAs. Olig. Eu.

SUBDIVISION CLUPEOMORPHA

ORDER ELLIMMICHTHYIFORMES **Ellimmichthyidae** *Diplomystus* [*Copeichthys Histurius Hyperlophus ?Oncochetos*] U. Cret. SWAs. U. Cret. Eoc. Af. NA. SA. U. Cret. Mioc. Eu. *Ellimmichthys* L. Cret. SA.

ORDER CLUPEIFORMES

SUBORDER DENTICIPITOIDEI **Denticipitidae** R. Af. *Paleodenticeps* Olig. or Mioc. EAf.

SUBORDER CLUPEOIDEI **Clupeidae + Engraulidae** R. Cos. *Alisea* Mioc. NA. *Alosa* [*Alausa Caspialos Clupeonella*] ?Eoc. NA. Olig. Plioc. Eu. Plioc. NAf. Plioc. R. WAs. R. Oc. *Austroclupea* Plioc. SA. *Bramlettia* U. Cret. NA. *Brevoortia* Plioc. NAf. R. Oc. *?Chanopsis* L. Cret. WAf. *Clupea* [*Alonsina Clupeops Sahelinia*] Eoc. Pleist. Eu. Olig. SA. Mioc. WAs. EInd. Plioc. NAf. R. Oc. *Clupeopsis* Eoc. Eu. *Crossognathus* L. Cret. Eu. *Domeykos* U. Jur. SA. *Driverius* U. Cret. NA. *Engraulis* Mioc. Eu. R. Oc. *Engraulites Epelichthys Etringus* Mioc. NA. *Etrumeus* [*Halecula Parahalecula*] Olig. WAs. Mioc. Plioc. Eu. Plioc. NAf. R. Oc. *Ganoessus Ganolytes* [*Diradias*] Mioc. NA. *Gasteroclupea* U. Cret. SA. *Hacquetia Halecopsis* Eoc. Eu. *Hayina* [*Jobertina Smithites*] Mioc. NA. *Histiothrissa* U. Cret. Eu. SWAs. Af. *Iquius* U. Tert. EAs. *Knightia* [*Ellimma Ellipos*] Eoc. SA. *Lembacus* Mioc. NA. *Lygisoma* Mioc. NA. *Melettina Neohalecopsis* Olig. Eu. *Opisthonema* Mioc. NA. R. Oc. *?Ostariostoma* U. Cret. NA. *Paraclupavus* U. Jur. Cret. Af. *Paraclupea* L. Cret. EAs. *Pateroperca* U. Cret. SWAs. *Pomolobus* Mioc. Eu. R. Oc. *Protoclupea* U. Jur. SA. *Pseudoberyx* U. Cret. SWAs. *Pseudochilsa* Mioc. Eu. R. Oc. *Pseudoetringus* Eoc. NA. *Quisque* Mioc. NA. *Rhomarus* Mioc. NA. *Sardinella* Mioc. Eu. R. Oc. *Sarmatella* Mioc. R. Eu. *Spratus* [*Meletta*] Olig. Eu. R. Oc. *Steinbergia* Mioc. NA. *Stolephorus* [*Spratteloides*] Mioc. NA. Plioc. NAf. R. Oc. *Syllaemus* U. Cret. Eu. NA. *Wisslerius* U. Cret. NA. *Xenothrissa Xyne Xyrinus* Mioc. NA.

SUBDIVISION EUTELEOSTEI

ORDER INCERTAE SEDIS *Pharmacichthys* U. Cret. SWAs.

ORDER SALMONIFORMES

SUBORDER ESOCOIDEI **Esodidae** *Esox* [*Trematina*] Olig. R. Eu. R. As. Paleoc. R. NA. **Umbridae** *Novumbra* Olig. R. NA. *Palaeosox* M. Eoc. Eu. *Proumbra* Paleoc. NEAs. *Umbra* Olig. R. Eu.

SUBORDER ARGENTINOIDEI **Argentinidae + Bathylagidae** R. NOc. *Argentina* L. Olig. EEu. R. NOc. *Azalois* Mioc. NA. *Bathylagus* [*Auesita*] Mioc. NA. R. NOc. *Lygisma* Mioc. NA. *Proargentina* Mioc. Eu. *Sternbergia* Mioc. NA. **Opisthoproctidae** R. NOc. **Alepocephalidae** R. Oc. *Palaeotroctes* L. Olig. EEu. *Xenodernichthys* Mioc. Eu. R. Oc. **Family undesignated** *Gaudryella Humbertia* U. Cret. WAs.

SUBORDER OSMEROIDEI **Osmeridae** R. NOc. *Mallotus* Pleist, Eu. NA. Gr. R. NAtl. *Osmerus* Mioc. Eu. R. Oc.

SUBORDER GALAXIODEI **Salangidae** R. EAs. **Retropinnidae** R. Aus. NZ. **Galaxiidae** R. SAf. Aus. NZ. SA. *Galaxias* Plioc. R. NZ.

SUBORDER SALMONOIDEI **Salmonidae** R. Oc. Eu. As. NA. *Beckius* Eoc. NA. *Coregonus* Pleist. Eu. R. As. NA. *Cyclolepis* U. Cret. NA. *Cyclolepoides* Eoc. NA. *?Goudkoffia ?Leucichthyops ?Natlandia* U. Cret. NA. *Oncorhynchus* Plioc. R. NA. *Paleolox* Mioc.-Plioc. NA. *Parastenodus* Eoc. NA. *Procharacinus Prohydrocyon* Eoc. Eu. *Prosopium* Plioc. Pleist. R. NA. *Protohymallus* [*Prothymallus*] Mioc. Eu. *Rhabdofario Salmo* Mioc. R. Eu. NA. R. Oc. *?Salmodium* Eoc. Eu. *Salvelinus* Pleist. R. Eu. ?Pleist. NA. *Thaumaturus* M. Eoc. L. Mioc. Eu. *Thymallus* Eoc. Mioc. Eu. R. NOc.

SUPERORDER UNDESIGNATED

Pyrenichthys U. Cret. Eu.

SUPERORDER OSTARIOPHYSI

ORDER GONORHYNCHIFORMES

SUBORDER GONORHYNCHOIDEI **Gonorhynchidae** R. Indo-Pac. Oc. *Charistosomus* [*Solenognathus*] U. Cret. Eu. SWAs. NA. *Gonorhynchops* U. Cret. Eu. *Notogoneus* [*Colpopholis Phalacropholis Protocatostomus Sphenolepis*] Eoc. Olig. Eu. ?Paleoc. NA. Olig. Aus. **Judeichthyidae** *Judeichthys* U. Cret. SWAs.

SUBORDER CHANOIDEI **Chanidae** R. Indo-Pac. Oc. *?Chanopsis* L. Cret. WAf. *Chanos* Eoc. Mioc. Eu. R. Indo-Pac. Oc. *Dartbile* U. Cret. SA. *Parachanos* L. Cret. Af. SA. U. Cret. Eu. *Prochanos* U. Cret. Eu. *Rubiesichthys* L. Cret. Eu. **Kneriidae Phractolaemidae** R. Af.

ORDER INCERTAE SEDIS **Unnamed family** *Chanoides* M. Eoc. Eu.

ORDER CHARACIFORMES **Characidae** R. Af. SA. CA. *Alestes* Tert. R. Af. *Astyanax Brycon* Pleist. R. SA. *Characilepis* Mioc. SA. *Eobrycon* L. Tert. Plioc. SA *?Eurocharax* L. Eoc. Eu. *Pareobasis* Pleist. R. SA. *Procharax* Plioc. SA. *Triportheus* Pleist. R. SA. **Erythrinidae Xiphostomatidae** R. SA. **Hepsetidae** R. Af. **Cynodontidae** R. SA. **Lebiasinidae** R. SA. CA. **Parodontidae Gasteropelecidae Prochilodontidae** R. SA. **Curimatidae** R. SA. *Curimatus* Pleist. R. SA. **Anostomidae Hemidontidae Chilodontidae** R. SA. **Distichodontidae Citharinidae Ichthyoboridae** R. Af.

ORDER CYPRINIFORMES **Cyprinidae** R. Eu. As. Af. NA. *Abramis* [*Ballerus*] Mioc. NAs. Mioc. R. Eu. *Acrocheilus* Mioc.-Plioc. R. NA. *Alburnoides* Mioc. R. As. R. Eu. *Alburnus* Mioc. R. Eu. Plioc. R. As. *Alisodon* Pleist. NA. *Amblypharyngodon* U. Tert. NAs. *Aspius* Mioc. R. Eu. *Barbus* Olig. R. Eu. Mioc. R. Af. As. U. Tert. R. EInd. *Blicca* L. Eoc. R. Eu. Mioc. R. NAs. *Campostoma* Pleist. R. NA. *Capitodus* Mioc. Eu. *Carassius* Plioc. R. As. R. Eu. *Chela* Eoc. Eu. R. As. EInd. *Chondrostoma* Olig. R. Eu.

?Mioc. As. *Chrosomus* Pleist. R. NA. *Ctenopharyngodon* Plioc. R. As. *Cyprinus* Mioc. R. Eu. Plioc. R. As. *Daunichthys* Tert. R. EInd. *Diastichus* Pleist. NA. *Dionda* Pleist. R. NA. *Enoplophthalmus* Mioc. Eu. *Eocyprinus* L. Tert. EInd. *Evomus* Plioc. NA. *Gila* [*?Anchypopsis Siphateles*] Mioc. R. NA. *Gobio* Mioc. R. Eu. *Hemiculturella* Plioc. R. As. *Hemitrichas* Olig. Eu. *Hexapsephus* Tert. R. EInd. *Hybognathus* Pleist. R. NA. *Hybophthalmichthys* Plioc. R. EAs. *Hybopis Ictiobus* Pleist. R. NA. *Idadon* Mio.-Plioc. NA. *Leuciscus* Olig. R. Eu. Mioc. R. As. R. Af. NA. *Leucus* Pleist NA. *Mylocheilus* Mioc.-Plioc. R. NA. *Mylocyprinus* Plioc. Pleist. NA. *Mylopharodon* Plioc. R. NA. *Mylopharyngodon* Plioc. NA. Plioc. R. EAs. *Nocomis* Plioc. R. NA. *Notemigonus* Pleist. R. NA. *Notropis* Mioc. Pleist. R. NA. *Oligobelus* Tert. R. EInd. *Orthodon* Mioc.-Plioc. R. NA. *Osteochilus* Tert. R. EInd. *Paraleuciscus* Mioc. Eu. *Pelecus* [*Culter*] Plioc. R. *Pimephales ?Proballostomus* Pleist. R. NA. *Pseudorasbora* Mioc. R. As. *Puntius Rasbora* Tert. R. EInd. *Ptychocheilus* [*Squalius*] Mioc.-Plioc. R. NA. *Rhinichthys* L. Pleist. NA. *Richardsonius* Plioc. R. NA. *Rodeus* Mioc. R. Eu. *Rutilus* Eoc. R. Eu. *Scardinius* Olig. R. Eu. Mioc. NAs. *Semotilus* L. Pliest. R. NA. *Sigmopharyngodon* Pleist. NA. *Soricidens* Mioc. Eu. *Thynnichthys* Tert. R. EInd. *Tinca* [*Tarsichthys*] Olig. R. Eu. *Varicorhinus* Plioc. R. Eu. As. *Xenocypris* Plioc. R. As. **Gyrinocheillidae** R. EAs. **Psilorhynchidae** R. SAs. **Catostomidae** R. EAs. NA. *Amyzon* Eoc. Olig. NA. *Carpoides* Pleist. R. NA. *Catostomus* Eoc. R. As. Mioc.-Plioc. Pleist. R. NA. *Chasmites* Mioc.-Plioc. Pleist. R. NA. *Deltistes* Pleist. R. NA. *Minytrema* L. Pleist. NA. *Moxostoma* Pleist. R. NA. *Pantosteus* Plioc. R. NA. **Homalopteridae** R. As. **Cobitidae** R. Eu. As. EInd. NAf. NA. *Cobitis* [*Acanthopsis*] *Nemacheilus* Olig. R. Eu. R. As. NAf.

ORDER SILURIFORMES (NEMATOGNATHI) **Diplomystidae** R. SA. **Ictaluridae** (**Ameiuridae**) R. As. NA. CA. *Ictalurus* [*Ameiurus*] Olig. As. Mioc. R. NA. *Plyodictis* Mioc. Plioc. Pleist. R. NA. **Bagridae** (**Porcidae Mystidae**) *Bagre* [*Bagrus Felichthys Porcus*] Pleist. R. Af. *Bucklandium* [*Glyptocephalus*] Eoc. Eu. *Chrysichthys* Pleist. R. Af. *?Clairbornichthys* Eoc. NA. *Eaglesoma* Eoc. Af. *Eomacronas* [*Macronoides*] Eoc. WAf. *Fajumia* Eoc. NAf. *Heterobagrus* Plioc. R. As. *Mystus* [*Macrones Macronichthys*] Pleist. R. EInd. *Nigerium* Eoc. WAf. *Rita* Plioc. R. As. *Socnopaea* Eoc. NAf. **Craniglanidae** R. Eas. **Siluridae** R. Eu. As. Af. *?Bachmannia* Tert. SA. *Parasilurus* Plioc. R. EAs. *Pliosilurus* Plioc. R. As. *Silurus* Eoc. R. Eu. Plioc. R. As. **Schilbidae + Pangasiidae** R. SAs. EInd. Af. *Pangasius* L. Tert. R. As. *Pseudeutropius* [*Brachyspondylus*] Tert. R. EInd. **Amblycipitidae Amphiliidae** R. Af. **Akysidae** R. SAs. **Sisoridae** (**Bagariidae**) R. SAs. EInd. *Bagarius* Plioc. R. SAs. Tert. R. EInd. **Clariidae** R. SAs. EInd. Af. *Clarias* Plioc. R. SAs. Af. Pleist. R. EInd. *Clarotes* Pleist. R. Af. *Heterobranchus* Pleist. Eu. Pleist. R. Af. **Heteropneustidae** R. SAs. **Chacidae** R. SAs. EInd. **Olyridae** R. SAs. **Malapteruridae** R. SA. **Mochokidae** (**Synodontidae**) R. Af. *Synodontis* Mioc. Eu. Plioc. R. Af. **Arriidae** (**Tachysuridae**) R. tropical coasts *Arius* [*Tachysurus*] Eoc. R. Af. SAs. EInd. WInd. SA. *Auchenoglanis* Pleist. R. Af. *Eopeyeria* [*Ariopsis Peyeria*] Eoc. NA. *Osteogeneiosus* Eoc. Af. R. EInd. *Rhineastes* [*Astephas*] Eoc. Olig. NA. R. As. **Doradidae Auchenipteridae Aspredinidae** (**Bunocephalidae**) R. SA. **Plotosidae** R. coasts As. Aus. EAf. *Tandanus* Olig. R. Aus. **Pimelodontidae** R. CA. SA. *Pimelodus* Tert. R. SA. ?Eoc. Af. **Ageneiosidae Hypopthalmidae Helogeneidae Cetopsidae** R. SA. **Stegophilidae** (**Pygidiidae Eretmophilidae**) R. SA. **Trichomycteridae** R. SA. *Propygidium* Eoc. SA. **Callichthyidae** R. SA. *Corydoras* Mioc. R. SA. **Loricariidae** R. SA.

SUPERORDER STENOPTERYGII

ORDER STOMIIFORMES (STOMIATIFORMES) **Gonostomidae** R. Oc. *Cyclothone* [*Regenius*] Mioc. NA. R. Oc. *?Gonorhynchops* U. Cret. Eu. NA. *Gonostoma* Mioc. Eu. Plioc. NAf. *Indrissa* U. Cret. NAf. *Paravinciguerria* U. Cret. NAf. *Photichthys* Mioc. Eu. Plioc. NAf. R. Oc. *?Protostomias* U. Cret. NAf. *Scopeloides* [*Mrazecia*] *Vinciguerria* [*Zalarges*] Olig. Mioc. Eu. **Sternoptychidae** R. Oc. *Argyropelecus* Olig. Mioc. Eu. Mioc. NAf. NA. R. Oc. *Polyipnoides* Eoc. Eu. *Polyipnus Sternoptyx* Olig. Eu. R. Oc. **Astronesthidae Melanostomiidae Malacosteidae** R. Oc. **Chauliondontidae** *Chauliodus* Mioc. NA. R. Oc. **Protostomiatidae** *Pronotacanthus* U. Cret. SWAs. *Protostomia* U. Cret. NAf. **Stomiatidae** R. Oc. *Eostomias* Mioc. NA. **Idacanthidae** R. Oc.

SUPERORDER SCOPELOMORPHA

ORDER AULOPIFORMES

SUBORDER ENCHODONTOIDEI **Enchodontidae** *Diplolepis* U. Cret. Eu. *Enchodus* [*Enurygnathus Holcodon Ischyrocephalus Isodon Phasganodus Solenodon Tetheodus*] U. Cret. SWAs. NA. U. Cret. Paleoc. Eu. Af. U. Cret. Eoc. SA. *?Halecodon ?Leptecodon ?Luxilitos* U. Cret. NA. *Palaeolycus* U. Cret. Eu. *Panthophilus* U. Cret. SWAs. *Rharbichthys* U. Cret. NAf. *Volcichthys* U. Cret. Eu. **?Tomognathidae** *Tomognathus* U. Cret. Eu. **Eurypholidae** *Eurypholis* U. Cret. Eu. NAf. SWAs. *Saurorhamphus* U. Cret. SWAs.

SUBORDER CIMOLICHTHYOIDEI **Cimolichthyidae** *Cimolichthys* [*Empo*] U. Cret. NA. Eu. SA. U. Cret. Paleoc. NAf. *Prionolepis* U. Cret. Eu.

SUBORDER HALECOIDEI **Halecidae** *Halec* [*Archaeogadus Pomognathus*] U. Cret. Eu. SWAs. *Hemisaurida* U. Cret. Eu. *Phylactocephalus* U. Cret. SWAs.

SUBORDER ALEPISAUROIDEI **Paralepidae** (**Sudidae**) *Holosteus* [*Pavlovichthys*] Eoc. Olig. Eu. *Iniomus* Eoc. Olig. NA. *Lestichthys* Mioc. NA. *Paralepis* [*Anapterus Tydeus*] Mioc. Eu. NAf. R. Oc. *Parascopelus* Mioc. Eu. *Sudis* Mioc. Plioc. Eu. NAf. R. Oc. *Trossulus* Mioc. NA. **Synodontidae** R. Atl. Ind. Pac. Oc. *Sardinius* U. Cret. Eu. **Giganturidae Paralepididae Anotopteridae Evermannellidae Omosudidae Alepidauridae Pseudotrichonotidae** R. Oc. **Incertae sedis**-*Apateodus* U. Cret. Eu. U. Cret. Paleoc. NAf.

SUBORDER AULOPOIDEI **Aulopodidae Chlorophthalmidae Scopelarchidae Notosudidae** (**Scopelosauridae**) R. Oc.

SUBORDER ICHTHYOTRINGOIDEI **Ichthyotringidae** *Ichthyotringa* [*Rhinellus*] U. Cret. Eu. SWAs. NA. NAf. **Cheirothricidae** *Cheirothrix* [*Megapus Megistopuspholis*] U. Cret. Eu. SWAs. *Exocoetoides* U. Cret. SWAs. **Dercetidae** *Dercetis* U. Cret. Eu. As. *Pelargorhyncus* U. Cret. Eu. *Rhynchodercetis* U. Cret. Eu. SWAs. *Stratodus* U. Cret. SWAs. NA. U. Cret. Paleoc. NAf. *?Trianaspis* U. Cret. NA.

ORDER MYCTOPHIFORMES **Myctophidae** (**Scopelidae**) R. Oc. *Acrognathus* U. Cret. Eu. SWAs. *Cassandra* [*Leptosomus*] U. Cret. Eu. As. *Ceratoscopus* ?Mioc. NA. *Diaphus* Olig. Mioc. Eu. R. Oc. *Eomyctophum* Olig. Mioc. Eu. Olig. WAs.

Hakelia U. Cret. SWAs. *Lampanyctus* Mioc. NZ. NA. Mioc. Plioc. Eu. *Myctophum* [*Hygophum Scopellus*] Eoc. Plioc. Eu. Mioc. NZ. NA. Plioc. NAf. R. Oc. *?Nematonotus* U. Cret. SWAs. *Nyctophus* Mioc. Plioc. Eu. NA. *Omniodon* Eoc. Eu. *Opistopteryx* U. Cret. SWAs. *?Palimphemis* Mioc. Eu. *?Rhamphornimia* U. Cret. SWAs. *Sardinius* U. Cret. Eu. **Sardinoididae** *Sardinoides* U. Cret. Eu. SWAs. *Nematonotus* U. Cret. Eu. **Neoscopelidae** R. Oc.

SUPERORDER UNDESIGNATED

ORDER PATTERSONICHTHYIFORMES **Pattersonichthyidae** *Humilichthys Pattersonichthys Phoenicolepis* U. Cret. SWAs.

ORDER CTENOTHRISSIFORMES **Ctenothrissidae** *Ctenothrissa* [*Aeothrissa*] U. Cret. Eu. SWAs. *Heterothrissa* U. Cret. SWAs. **Aulolepidae** *Aulolepis* U. Cret. Eu.

SUPERORDER PARACANTHOPTERYGII

ORDER PERCOPSIFORMES

SUBORDER SPHENOCEPHALOIDEI **Sphenocephalidae** *Sphenocephalus* U. Cret. Eu.

SUBORDER APHREDODEROIDEI **Aphredoderidae** R. NA. *Amphiplaga Asineops Erismatopterus* M. Eoc. NA. *Trichophanes* U. Mioc. NA. **Amblyopsidae** R. NA.

SUBORDER PERCOPSOIDEI **Percopsidae** R. NA.

ORDER BATRACHOIDIFORMES (HAPLODOCI) **Batrachoididae** R. Oc. CA. SA. *Batrachoides* U. Mioc. NAf. R. Oc.

ORDER GOBIESOCIFORMES (XENOPTERI) **Gobiesocidae** R. Oc. *?Bulbiceps* Mioc. NA.

ORDER LOPHIIFORMES (PEDICULATI)

SUBORDER LOPHIOIDEI **Lophiidae** R. Oc. *Lophius* Eoc. Eu. Plioc. NAf. R. Oc.

SUBORDER ANTENNARIOIDEI **Brachionichthyidae** R. Oc. **Antennariidae** R. Oc. *Histionotophorus* [*Histiocephalus*] M. Eoc. Eu. **Chaunacidae Ogcocephalidae** R. Oc.

SUBORDER CERATIOIDEI **Melanocetidae Diceratiidae Himantolophidae Oneirodidae Gigantacinidae Neoceratidae Centrophrynidae Ceratiidae Caulophrynidae Linophrynidae** R. Oc.

ORDER GADIFORMES

SUBORDER MURAENOLEPIDOIDEI **Muraenolepididae** R. SOc.

SUBORDER GADOIDEI **Moridae + Bregmacerotidae + Gadidae + Merulucciidae** R. Oc. Eu. As. NA. *Arnoldites* [*Arnoldina*] Plioc. NA. *Bregmacernia* Mioc. Eu. *Bregmaceros* [*Podopteryx*] ?Eoc. NZ. Eoc. Mioc. NA. Eoc. Plioc. Eu. Olig. WAs. Mioc. Aus. Plioc. NAf. R. Oc. *Brosmius* [*Brosme*] Olig. Plioc. Eu. Plioc. NAf. R. Oc. *Eclipes* [*Merriamina*] Mioc. NA. *Gadus* [*Morhus*] Paleoc. Pleist. Eu. Olig. Aus. Pleist. Gr. R. Oc. *Lepidion* Mioc. EAs. R. Oc. *Lota* Plioc. R. Eu. R. As. NA. *Lotella* Mioc. R. Eu. *Melanogrammus Melanotus* Olig. Eu. R. Oc. *Merlangus* Mioc. Eu. R. Oc. *Merluccius* [*Spinogadus*] Eoc. Mioc. Eu. Mioc. Aus. NZ. R. Oc. *Molva* Plioc. Eu. R. Oc. *Neopythites* Eoc. Eu. R. Oc. *Odontogadus* Mioc. Eu. *Onobrosmius* Mioc. WAs. Mioc. Plioc. Eu. *Palaeogadus* [*Lotimorpha Megalolepis Nemopteryx Palaeobrosmius Pseucolota Ruppelianus*] Eoc. WAs. Olig. Mioc. Eu. *Palaeomolva* Mioc. Eu. *?Paractichthys* U. Cret. NA. *Petalolepis* U. Cret. Eu. *Progadius* Plioc. NA. *Physiculus* Mioc. Eu. NZ. R. Oc. *Promerluccius* Olig. WAs. *Raniceps Strinsia* Mioc. Eu. R. Oc. *Rhinocephalus* L. Eoc. Eu. *Urophycis* [*Phycis*] Eoc. Mioc. Eu. R. Oc.

SUBORDER MACROUROIDEI **Macrouroidei** R. Oc. *Amblygoniolepidus* Olig. NA. *Bolbocara* Mioc. NA. *Calilepidus* Olig. Mioc. NA. *Coelorincus* Mioc. Aus. R. Oc. *Homeocoryphaenoides* Mioc. NA. *Homeomacrurus* Plioc. NA. *Homeonezumia* Olig. NA. *Hymenocephalus* Olig. Mioc. Eu. R. Oc. *Leptacantholepidus* Mioc. NA. *Macrourus* [*Coryphaenoides*] Eoc. Mioc. Eu. Mioc. NA. NZ. R. Oc. *Oxygoniolepidus* Olig. NA. *Palaeobathygadus* Olig. Mioc. NA. *Promacrurus* Olig. Mioc. NA. *Pyknolepidus* Olig. NA. *Rankinian* U. Cret. NA. *Trachyrincus* Mioc. NZ. R. Oc. *Trichiurichthys* U. Mioc. Eu.

ORDER OPHIDIIFORMES **Ophidiidae (Brotulidae)** R. Oc. WInd. *?Bauzaia* Eoc. NA. *Brotula* Mioc. Plioc. Eu. R. Oc. *Glyptophidium* Mioc. EAs. *Neobythites* Eoc. Eu. R. Oc. *Ophidion* [*Ophidium*] ?Paleoc. M. Eoc. Eu. Eoc. WInd. Mioc. Aus. R. Oc. *Preophidion* Eoc. NA. *Protobrotula* L. Olig. Eu. **Carapidae (Fierasferidae)** R. Oc. *Carapus* [*Fierasfer Jordaniscus*] Olig. Mioc. NA. Mioc. Aus. R. Pac. Oc. **Pyramodontidae** R. Oc.

SUPERORDER ACANTHOPTERYGII

SERIES ATHERINOMORPHA

ORDER ATHERINIFORMES

SUBORDER EXOCOETOIDEI (SYNNENTOGNATHI) **Exocoetidae (Hemiramphidae)** *Chirodus* Mioc. Eu. *Cobitopsis* U. Cret. SWAs. Olig. Eu. *Derrhias* Mioc. NA. R. Oc. *Exoxoetus* [*Hemiexocoetus*] Eoc. NA. *Euleptorhamphus* [*Beltion*] Mioc. NA. R. Oc. *?Hemilampronites* U. Cret. NA. *Hemiramphus* M. Eoc. Eu. R. Oc. *Rogenites* [*Rogenia*] *Zelosis* Mioc. NA. **Belonidae** R. Oc. *Belone* Plioc. Eu. Mioc. NAf. R. Oc. **Scomberesocidae** R. Oc. *Praescomberesox* Eoc. Olig. NA. *Scombersus* Mioc. NA. *Scomberesox* Mioc. Eu. NA. Plioc. NAf. R. Oc. **Forficidae** *Forfex* Mioc. NA. *Zelotichthys* [*Selota Zelotes*] Mioc. NA.

ORDER CYPRINODONTIFORMES

SUBORDER CYPRINODONTOIDEA **Oryziatidae Adrianichthyidae** R. EInd. **Horaichthyidae** R. As. **Cyprinodontidae** R. Eu. Af. NA. SA. *Aphanius* Mioc. R. As. *Brachylebias* Mioc. WAs. *Carrionellus* L. Tert. SA. *Cyprinodon* Olig. R. Eu. L. Tert. R. NA. *Empetrichthys* Plioc. R. NA. *Fundulus* [*Gephyrura Parafundulus*] ?Mioc. Plioc. R. NA. Mioc. R. Eu. *Haplochilus* Olig. R. Eu. *Lithofundulus* Tert. SAs. *Pachylebias* [*Anelia ?Physocephalus*] Olig. Mioc. Eu. *Prolebias* [*Ismene Pachystetus*] L. Olig. Mioc. Eu. **Goodeidae** R. NA. **Anablepidae Nynsiidae** R. CA. SA. **Poeciliidae** R. NA. SA.

SUBORDER ATHERINOIDEI **Melanotaeniidae** R. Oc. **Atherinidae** R. As. Aus. NA. Oc. *Atherina* M. Eoc. R. Eu. Mioc. R. As. R. NA. *Menidia* Plioc. R. NA. *Prosphyraena Rhamphognathus* [*Mesogaster*] M. Eoc. Eu. *Zanteclites* Mioc. NA. **Isonidae** R. Oc. **Neostethidae Phallostethidae** R. NPac. EAs.

SERIES PERCOMORPHA

ORDER BERYCIFORMES

SUBORDER STEPHANOBERYCOIDEI **Stephanoberycidae** R. Oc. **Melamphaeidae** R. Oc. *Scopelogadus* U. Mioc. NA. **Gibberichthyidae** R. Oc.

SUBORDER POLYMIXIODEI **Polymixiidae** R. Oc. *Berycopsis* [*Platycormus*] *Homonotichthys* [*Homonotus Stenostoma*] U. Cret. Eu. *Omosoma* U. Cret. Eu. SWAs. NAf. *Parapolymyxia* Eoc. Olig. NA. *Polymixia* Eoc. Eu. R. Oc. *Pycnosterinx* [*Imogaster*] U. Cret. SWAs.

SUBORDER DINOPTERYGOIDEI **Dinopterygidae** *Dinopteryx* U. Cret. SWAs. **Aipichthyidae** *Aipichthys* U. Cret. SWAs. **Pycnosteroididae** *Pycnosteroides* U. Cret. SWAs. **Stichocentridae** *Stichocentrus* U. Cret. SWAs.

SUBORDER BERYCOIDEI **Diertmidae** R. Oc. *?Absalomichthys* [*Abantis*] Mioc. NA. **Trachichthyidae** *Acrogaster* [*Acanthophoria*] U. Cret. Eu. SWAs. NAf. *Gephyroberyx* Olig. EEu. R. Oc. *Gnathoberyx* U. Cret. SWAs. *Hoplopteryx* [*Goniolepis Hemicyclolepis Hemigonolepis ?Priconolepis*] U. Cret. Eu. SWAs. NAf. NA. *Hoplostethus* Plioc. Eu. R. Oc. *Libanoberyx Lissoberyx* U. Cret. SWAs. *Trachichthodes* Tert. Aus. *Tubantia* U. Cret. Eu. **Kosorogasteridae Anoplogasteridae** R. Oc. **Berycidae** R. Oc. *Beryx* ?Cret. Eoc. Eu. Eoc. Olig. WAs. R. Oc. *?Costaichthys* [*Heterolepis*] U. Cret. Eu. *Echinocephalus* Eoc. Eu. *?Electrolepis* U. Cret. Eu. *?Kemptichthys* Tert. NAf. *Lobopterus* [*Dictynopterus*] U. Cret. Eu. *Platylepis* Eoc. Eu. *?Spinacites* U. Cret. Eu. **Monocentridae** R. Oc. *Brazosiella* Eoc. NA. *Cleidopus* Olig. R. Aus. *Monocentris* [*Lepisacanthus*] Paleoc. Eoc. Eu. Eoc. WInd. Eoc. Plioc. Aus. R. Oc. **Anomalopidae** R. Oc. **Holocentridae** R. Oc. *Adriocentrus* U. Cret. Eu. *Africentrum* [*Microcentrum*] Mioc. NAf. *Alloberyx* U. Cret. SWAs. *Berybolcensis* M. Eoc. Eu. *Caproberyx* U. Cret. Eu. NAf. *Ctenocephalichthys* U. Cret. SWAs. *Eohollocentrum* M. Eoc. Eu. *Eugocentrus* U. Cret. Eu. *Holocentrites* Eoc. Olig. NA. *Holocentroides* Olig. Eu. *Holocentrus* Eoc. Mioc. Eu. R. Oc. *Kansius* U. Cret. NA. *Myripritis* Eoc. Mioc. Eu. R. Oc. *Paraberyx* U. Cret. Olig. NA. *Parospinus* U. Cret. SWAs. *Stichoberyx* U. Cret. NAf. *?Stintonia* Eoc. Olig. NA. Eoc. Mioc. Eu. *Tenuicentrum* M. Eoc. Eu. *Trachichtyoides* U. Cret. Eu. ?Olig. Aus. *Weileria* Eoc. Olig. NA. Eoc. Mioc. Eu. **Family Incertae Sedis** *Cryptoberyx Pattersonoberyx Plesioberyx* U. Cret. SWAs.

ORDER ZEIFORMES **Parazenidae Macrurocyttidae** R. Oc. **Zeidae** R. Oc. *Cylloides* Olig. Eu. *Zenopsis* Olig. Mioc. Eu. *Zenus* Olig. Plioc. Eu. ?Plioc. NAf. R. Oc. **Grammicolepidae** R. Oc. WInd. SAf. EInd. **Oreosomidae** R. Oc. **Caproidae** R. Oc. *Capros* [*Glyphisoma Metapomichthys Proantigonia*] Olig. Mioc. Eu. Plioc. NAf. R. Oc. *Caprovesposus* Olig. Mioc. Eu. *Microcapros* U. Cret. SWAs.

ORDER LAMPRIFORMES

SUBORDER LAMPRIDOIDEI **Lampridae** R. Oc. *Lampris* [*Diatomoeca*] U. Mioc. NA. R. Oc.

SUBORDER VELIFEROIDEI **Veliferidae** R. Oc. *Palaeocentrotus* L. Eoc. Eu.

SUBORDER TRACHIPTEROIDEI **Lophotidae** R. Oc. *Protolophotes* Olig. SWAs. R. Oc. **Trachipteridae Regalecidae** R. Oc.

SUBORDER STYLEOPHOROIDEI **Styleophoridae** R. Oc.

ORDER GASTEROSTEIFORMES **Gasterosteidae** R. Eu. As. NAf. NA. *Gasterosteus* [*Meriamella*] U. Mioc. R. NA. R. Eu. U. Mioc. R. As. NAf. *Pungitius* [*Gastrosteops*] Plioc. R. NA. R. As. NAf. **Aulorhynchidae** R. Oc. *Aulorhynchus* [*Protosyngnathus*] Tert. SAs. EInd. R. Oc. *Protaulopsis* M. Eoc. Eu.

ORDER SYNGNATHIFORMES

SUBORDER AULOSTOMOIDEI **Aulostomidae** R. Oc. *Eoaulostomus Jungersenichthys Macraulostomus Urosphen* Eoc. Olig. Eu. *Synhypuralis* M. Eoc. Eu. **Fistulariidae** R. Oc. *Fistularia* Olig. Plioc. Eu. R. Oc. **Centriscidae** R. Oc. *Aeoliscus* Olig. Mioc. Eu. *Centriscus* [*Amphisyle*] Olig. WAs. Olig. ?Plioc. Eu. R. Oc. *Paramphisile* M. Eoc. Eu. **Parasynarcualidae** *Parasynarcualis* M. Eoc. Eu. **Macrorhamphosidae** R. Oc. *?Aulorhamphus* M. Eoc. Eu. *Gasterorhamphosus* U. Cret. Eu. *Protorhamphosus* U. Paleoc. CAs. **Paraeoliscidae** *Aeoliscoides Paraeoliscus* M. Eoc. Eu.

SUBORDER SYNGNATHOIDEI **Solenostomidae** R. Oc. *Solenorhynchus* M. Eoc. Eu. *Prosolenostomus* M. Eoc. Eu. **Syngnathidae** R. Oc. *Calamostoma* M. Eoc. Eu. *Dunckerocampus* [*Acanthognathus*] Olig. Mioc. Eu. R. Oc. *Hipposyngnathus* Mioc. Eu. *Pseudosyngnathus* M. Eoc. Eu. *Syngnathus* [*Siphonostoma*] M. Eoc. Plioc. Eu. Mioc. NA. Plioc. NAf. R. Oc.

ORDER SYNBRANCHIFORMES **Synbranchidae** R. SAs. Af. Aus. NA. SA. Oc.

ORDER INDOSTOMIFORMES **Indostomidae** R. SAs.

ORDER PEGASIFORMES **Pegasidae** R. Ind. Pac. Oc.

ORDER DACTYLOPTERIDAE **Dactylopteriformes** R. Ind. Pac. Atl. Oc.

ORDER SCORPAENIFORMES

SUBORDER SCORPAENOIDEI **Scorpaenidae** R. Oc. *Ampheristius* [*Goniognathus*] L. Eoc. Eu. *Ctenopomichthys* [*Ctenopoma Jemelka*] *Eosynanceja* L. Eoc. Eu. Mioc. Eu. *Rhomarchus Rixator* Mioc. NA. *Scorpaenodes* [*Sebastodes*] Mioc. Plioc. NA. Plioc. EAs. R. Oc. *Scorpaenoides* Eoc. Olig. Eu. *Scorpaenopterus* Mioc. Eu. *Sebastavus Sebastinus Sebastoessus* Mioc. NA. **Triglidae** R. Oc. *Peristedion* Plioc. Eu. R. Oc. *Trigla* [*?Trigloides*] Eoc. Mioc. Eu. Mioc. WAs. Plioc. NAf. Pleist. NA. R. Oc. **Pterygocephalidae** *Pterygocephalus* M. Eoc. Eu. **Caracanthidae Aploactinidae** R. Oc. *Eocynanceja* Eoc. Eu. **Pataecidae** R. Aus. **Synanceiidae** R. Ind. Pac. Oc. **Congiopodidae** R. Oc.

SUBORDER HEXAGRAMMOIDEI **Hexagrammidae** R. Oc. *Achrestogrammus* ?U. Mioc. NA. *Zemigrammatus* Mioc. NA. **Zaniolepidae** R. NPac. Oc.

SUBORDER PLATYCEPHALOIDEI **Platycephalidae** R. Oc. **Hoplichthyidae** R. Ind. Pac. Oc. **?Anoplopomatidae** R. Oc. *?Aenoscorpius* [*Eoscorpius*] Mioc. NA.

SUBORDER COTTOIDEI **Cottidae** (**Icelidae**) R. Eu. As. NA. Oc. *Cottopsis* Olig. SWAs. *Cottus* R. As. Mioc. NZ. L. Olig. R. Eu. Mioc. Plioc. R. NA. *Kerocottus* Mioc.-Plioc. NA. *Lirosceles* Mioc. NA. *Myoxocephalus* Mioc.-Plioc. R. NA. *?Paraperca* Olig. Eu. **Cottocomephoridae** R. EAs. **Ereuniidae** R. Pac. Oc. **Normanichteryidae** R. SPac. **Psychrolutidae** R. Oc. **Cyclopteridae** R. Oc. *Cyclopterus* Pleist. NA. R. Oc. **Agonidae** R. Oc. **Comephoridae** R. EAs.

ORDER PERCIFORMES

SUBORDER PERCOIDEI **Centropomidae** R. Eu. As. NAf. Aus. NA. *Centropomus* M. Eoc. R. Eu. R. NA. *Cyclopoma* M. Eoc. Eu. *Eolates* M. Eoc. Eu. Eoc. R. Af. R. As. Aus. *Paralates Platylates* Olig. Eu. *Psammoperca* Eoc. Eu. R. Oc. **Percichthyidae** *Morone* L. Pleist. R. NA. *Percichthys* U. Paleoc. U. Eoc. R. SA. *Santosius* U. Tert. SA. **Serranidae** R. Oc. *Acanthroperca* Eoc. Eu. *Acanus* Olig. Eu. *Allomorone* Eoc. NA. *Amphiperca* Eoc. Mioc. Eu. *Anthias* Mioc. Eu. R. Oc. *Arambourgia* [*Apogonoides*] Plioc. NAf. *Aritolabrax* Mioc. EAs. *Blabe* Eoc. NAf. *Centropristis* Olig. Mioc. Eu. R. Oc. *Dapalis* [*Smerdis*] ?Eoc. NAf. Paleoc. Mioc. Eu. *Dicentrarchus* [*Labrax*] *Eoserranus* Tert. SAs. *Epinephelus* [*Emmachaere*] Mioc. NA. Plioc. NAf. R. Oc. *Maccullochella* Tert. R. Aus. *Morone* Eoc. Mioc. Eu. R. Oc. *Niphon* Olig. Eu. R. Oc. *Paracentropristis* Tert. Eu. *Paramorone* L. Tert. NA. *Percalates* Olig. R. Aus. *Percilia* Eoc. Eu. R. SA. *Phosphichthys* Eoc. NAf. NA. *Prolates* [*Pseudolates*] Paleoc. Eu. *Properca* Eoc. Mioc. Eu. *?Proserranus* Paleoc. Eu. *Protanthias* Mioc. NA. *Serranus* M. Eoc. Mioc. Eu. Plioc. NAf. R. Oc. **Plesiopidae** Indo-Pac. Oc. **Pseudoplesiopidae** R. Indo-Pac. Oc. **Anisochromidae** R. WInd. Oc. **Acanthoclinidae** R. Indo-Pac. Oc. **Glaucosomidae** R. Pac. Oc. **Therapomidae** R. Indo. Pac. Oc. **Banjosidae** R. NPac. Oc. **Kuhliidae** R. Oc. As. EInd. Polynesia **Gregoryinidae** R. Oc. **Centrarchidae** R. NA. *Ambloplites* Plioc. R. NA. *Archoplites* Mioc.-Plioc. R. NA. *Borescentranchus* Mioc. NA. *Centrarchites* Eoc. NA. *Chaenobryttus* ?Mioc. Plioc. R. NA. *Lepomis* Mioc. R. NA. *Micropterus Mioplarchus* Mioc. NA. *Oligoplarchus* Olig. NA. *Pomoxis* Mioc. Plioc. R. NA. **Priacanthidae** R. Oc. *Priacanthus* [*Apostasis*] Olig. Mioc. Eu. R. Oc. *Pseudopriacanthus* Eoc. Mioc. Eu. **Apogonidae** (**Cheilodipteridae**) R. Oc. *Apogon* [*?Eretima*] M. Eoc. Mioc. Eu. Mioc. NA. R. Oc. *Apogonoides* Plioc. NAf. *Praegalegra* Olig. NA. **Percidae** R. Eu. As. NA. *Acerina* U. Tert. R. As. R. Eu. *Anthracoperca Carangopis* M. Eoc. Eu. *Cristigerina* Eoc. Eu. *?Dasceles ?Erisceles* Mioc. NA. *?Guoyquichthys* Tert. SA. *Leobergia* Plio. NAs. *Lucioperca* Pleist. R. Eu. *Mioplosus* Eoc. NA. *?Pachygaster ?Paralates* Olig. Eu. *Perca* [*?Coeloperca Eoperca ?Percostoma ?Plioplarchus ?Sandroserrus*] Eoc. R. Eu. Olig. Pleist. R. NA. Mioc. WAs. *Percaletes* Mioc. NZ. *Percarina* Olig. R. Eu. *Podocys Propercarina* Olig. Eu. *Sandar* Pleist. R. Eu. *Stizostedion* Plioc. R. NA. **Sillaginidae** R. Indo-Pac. Oc. *Sillago* Olig. Plioc. Aus. R. Oc. **Branchiostegidae** (**Latilidae**) R. Oc. *Branchiostegus* [*Latilus*] Plioc. NAf. R. Oc. **Labracoglossidae** R. Pac. Oc. **Lactariidae** R. Indo-Pac. Oc. SAs. EInd. *Lactarius* Mioc. Aus. R. SAs. EInd. **Pomatomidae** (**Scombropsidae**) R. Oc. *Lophar* Mioc. NA. *Scombrops* Plioc. EAs. R. Oc. **Rachycentridae** R. Indo-Pac. Oc. **Echeneidae** R. Oc. *Echeneis* Olig. Mioc. Eu. R. Oc. *Opisthomyzon* Olig. Eu. **Carangidae** (**Seriolidae**) R. Oc. *Acanthonemopsis* Mioc. Eu. *Aliciola* Mioc. NA. *Archaeus* [*Archaeoides*] Olig. Mioc. Eu. *Carangopsis* M. Eoc. Eu. *Caranx* [*Citula Parequula*] M. Eoc. Plioc. Eu. Eoc. Pleist. NA. Plioc. NAf. R. Oc. *Ceratoichthys* M. Eoc. Eu. *Decapterus* [*Lompochites*] Mioc. NA. Plioc. NAf. R. Oc. *?Desmichthys* Mioc. Eu. *Ductor* M. Eoc. Eu. *Hypacanthus* [*Lichia*] Olig. Pleist. Eu. R. Oc. *Irifera* Mioc. NA. *Oligoplites* [*Palaeoscomber*] Mioc. Eu. R. Oc. *Paratrachinotus* M. Eoc. Eu. *Pseudoseriola* Mioc. NA. *?Pseudovomer* Mioc. Eu. *Seriola* [*Micropteryx*] M. Eoc. Eu. Mioc. NA. Plioc. NAf. R. Oc. *Trachurus* M. Eoc. Eu. Mioc. Plioc. NAf. R. Oc. *Vomer* Tert. Eu. R. Oc. *Vomeropsis* M. Eoc. Eu. **Coryphaenidae** R. Oc. **Formionidae** (**Apolectidae**) R. Indo-Pac. Oc. **Menidae** R. Indo-Pac. Oc. *Mene* [*?Gasteracanthus*] Paleoc. Plioc. NAf. M. Eoc. Eu. R. Oc. *Bathysoma* Paleoc. Eu. **Leiognathidae** (**Equulidae**) R. Oc. *Leiognathus* [*Equula*] Olig. Mioc. Eu. R. Oc. **Bramidae Caristiidae** R. Oc. **Arripididae** R. SPac. Oc. **Emmelichthyidae** R. Indo-Pac. Oc. **Lutjanidae** R. Oc. *Caesio* Eoc. Eu. *Lednevia* Mioc. Eu. *Lutjanus* [*Lutianus*] Mioc. Eu. Olig. NA. Mioc. Aus. R. Oc. **Lobotidae** R. Oc. *Protolobotus* Olig. SWAs. **Gerridae** R. Oc. **Pomadasyidae** (**Pristipomidae**) R. Oc. *Orthopristis* Plioc. NAf. R. Oc. *Parapristopoma* Plioc. NAf. *Pomadasys* [*Pristipoma*] Eoc. Mioc. Eu. NAf. R. Oc. **Sparidae** (**Nemipteridae Leithrinidae**) R. Oc. *Atkinsonella* Mioc. NA. *Boops* [*Box*] Mioc. Eu. Plioc. NAf. R. Oc. *Crednidentex* Plioc. NAf. R. Oc. *Crommyodus* [*Pliacodus*] Mioc. NA. *Ctenodentex* Eoc. Eu. NAf. *Dentex* Eoc. Plioc. Eu. Mioc. NZ. Plioc. NAf. R. Oc. *Kreyenhagenius* Eoc. NA. *Pagellus* Paleoc.-Mioc. Eu. Olig.-Mioc. NZ. R. Oc. *Pagrosomus* Plioc. Aus. R. Oc. *Pagrus* Paleoc. Mioc. Eu. Plioc. NAf. R. Oc. *Paracalanus* Plioc. WAf. *Plectrites Rhytmias* Mioc. NA. *Sargus* [*Diplodus*] Eoc. Plioc. Eu. ?Eoc. Mioc.-Plioc. NAf. Mioc. Aus. Plioc. NA. R. Oc. *Sparnodus* Eoc. Mioc. Eu. *Sparosoma* [*Rhamnubia*] Olig. Eu. *Sparus* [*Aurata Chrysophrys*] Eoc. WInd. Olig. Plioc. Eu. Plioc. NAf. R. Oc. *Spondylisoma* [*Cantharus*] Eoc. Eu. R. Oc. **Sciaenidae** R. Oc. Eu. NA. SA. *Aplodinotus* Plioc. R. NA. *Cynoscion* [*Aristocion*] Mioc. NA. R. Oc. *Eocilophyodus* U. Cret. Eoc. WAf. *Eokokemia* Eoc. NA. *Ioscion* Mioc. NA. *Jefitchia* Eoc. NA. *Larium* Mioc. Eu. R. Oc. *Lompoquia* Mioc. NA. *Otolithes* Eoc. Eu. R. Oc. *Pogonias* Mioc. Pleist. NA. R. Oc. *Sciaena* [*Sciaenops*] Olig. Mioc. Eu. NA. R. Oc. *Pseudoumbrina* Plioc. Eu. *Umbrina* Plioc. Eu. R. Oc. **Mullidae** R. Oc. *Mullus* Mioc. Eu. R. Oc. **Monodactylidae** (**Psettidae**) R. Indo-Pac. Oc. coasts *Pasaichthys Psettopsis* M. Eoc. Eu. **Pempheridae** R. Oc. **Bathyclupeidae** R. Indo-Pac. Oc. Carribbean **Toxotidae** *Toxotes* L. Tert. R. EInd. **Coracinidae** (**Dichistiidae**) R. SOc. **Kyphosidae** (**Scorpididae Givellidae**) R. Oc. **Amphistiidae** *Amphistium* [*?Macrostoma ?Woodwardichthys*] M. Eoc. Eu. **Ephippidae** (**Chaetodipteridae Platacidae**) R. Oc. *Archaephippus* M. Eoc. Eu. *Ephippites* Olig. Eu. *Exellia* [*Semiophorus*] M. Eoc. Eu. **Platacidae** R. Oc. *Eoplatax* M. Eoc. Eu. *?Paraplatax* Olig. EEu. **Scatophagidae** R. Indo-Pac. Oc. coasts *Scatophagus* M. Eoc. Mioc. Eu. R. Oc. **Chaetodontidae** (**Pomacanthidae**) R. Oc. *Chaetodon* Olig. Mioc. Eu. Plioc. NAf. R. Oc. *Chelmo* Mioc. EInd. R. Oc. *Holacanthus* Eoc.

Mioc. Eu. R. Oc. *Pomacanthus* Eoc. Eu. R. Oc. **Enoplosidae** R. SPac. *Enoplosus* Eoc. Eu. R. Oc. **Histiopteridae** R. Indo-Pac. Oc. **Nandidae (Polycentridae Pristolepidae)** R. SAs. Af. EInd. SA. **Oplegnathidae** R. EAs. Aus. SAf. SA. *Oplegnathus* Plioc. R. Aus. **Embiotocidae (Ditremidae)** R. NPac. *Ditrema* Eoc. Eu. R. NPac. *Eriquius* Mioc. NA. **Cichlidae** R. Af. SAs. CA. SA. WInd. *Acara* Tert. R. SA. *Acaronia* Pleist. R. SA. *Aequideus* Plioc. R. SA. *Cichlaurus* [*Chilasoma*] Mioc. WInd. R. SA. *Macracara* Eoc. SA. *?Palaeochromis* Tert. NAf. *Tilapia* Tert. R. Af. R. WAs. **Pomacentridae (Abudefdufidae)** R. Oc. Af. *Chromis* Mioc. R. Af. *Cockerellites Izuus* Mioc. EAs. *Odonteus* Eoc. Eu. *?Palaeochromis* Tert. NAf. *Priscacara* Eoc. NA. **Gadopridae** R. Aus. **Cirrhitidae** R. Indo-Pac. Oc. **Chironemidae** R. SOc. **Aplodactylidae** R. Oc. **Cheilodactylidae Latridae** R. Indo-Pac. Oc. **Owstoniidae** R. Oc. *Owstonia* Plioc. Eu. EAs. R. Oc. **Cepolidae** R. Oc. *Cepola* Eoc. Plioc. Eu. Plioc. NAf. R. Oc. **Arambourgellidae** *Arambourgella* M. Eoc. Eu. **Canctidae** R. Oc. *Eozanclus* M. Eoc. Eu.

SUBORDER MUGILOIDEI **Mugilidae** R. Oc. *Mugil* Eoc. Plioc. Eu. Plioc. NAf. R. Oc.

SUBORDER SPHYRAENIDEI **Sphyraenidae** R. Oc. *Sphyraena* Eoc. Af. M. Eoc. Eu. Mioc. EInd. Pleist. As. NA. R. Oc.

SUBORDER POLYNEMOIDEI **Polynemidae** R. Oc. *Polydactylus* Tert. EEu.

SUBORDER LABROIDEI **Labridae** R. Oc. *Coris* Mioc. Eu. R. Oc. *Eolabroides* M. Eoc. Eu. *Julis* Mioc. Eu. R. Oc. *Labrodon* [*Diaphyodus Nummopalatus Pharyngophilus*] L. Eoc. Plioc. Eu. Eoc. NAf. NZ. Mioc. NA. Mioc. Pleist. As. R. Oc. *Labrus* ?M. Eoc. Plioc. Eu. R. Oc. *Platylaemus* Eoc. Eu. NAf. *Pseudosphaerodon* L. Eoc. Eu. *Pseudostylodon* Eoc. Eu. *Pseudovomer Stylodus* Mioc. Eu. *Symphodus* [*Bodianus Crenilabrus*] Mioc. Eu. Plioc. NAf. *Taurinichthys* Mioc. Plioc. Eu. **Odacidae** R. Oc. Aus. NZ. **Scaridae** R. Oc. As. NA. SA. *Pseudoscaris* Eoc. Eu. R. As. NA. SA. *Scaroides* Tert. CA. R. Oc. *Scarus* [*Callyodon*] Eoc. Eu. Mioc. SAs. R. Oc.

SUBORDER ZOARCOIDEI **Zoarcidae Bathymasteridae Cryptacanthodidae Pholididae Ptilichthyidae Zaproridae Scytalinidae** R. Oc. **Anarhichadidae** R. NAtl. NPac. *Anarhichus* Plioc. Eu. R. Oc. **Xenocephalidae** R. EInd. **Stichaeidae** R. NOc. *Stichaeus* Tert. EAs. R. Oc.

SUBORDER TRACHINOIDEI **Trichonotidae** R. Indo-Pac. Oc. **Opisthognathidae** R. Oc. **Congrogadidae** R. Pac. Ind. Oc. **Bathymasteridae** R. NPac. **Notograptidae** R. WPac. Oc. **Mugiloididae** R. Indo-Pac. Oc. **Pholidichthyidae** R. WPac. Oc. **Cheimarrhychthyidae** R. NZ. **Trichonotidae** R. WPac. Oc. **Trachinidae** *Callipterys* M. Eoc. Eu. *Trachinopsis* Plioc. Eu. **Percophididae** R. SAtl. **Trichodontidae** R. NPac. **Creediidae** R. Aus. Coasts. **Uranoscopidae** R. Oc. **Leptoscopidae** R. SPac. Aus. NZ. **Champsodontidae** R. Indo-Pac. Oc. *Myersichthys* Olig. Eu. *Pseudoscopelus* Mioc. Eu. R. Oc.

SUBORDER NOTOTHENOIDEI **Bovichthyidae** R. SOc. **Nototheniidae** R. SOc. *Notothenia* Mioc. NZ. R. Oc. **Bathydreconidae Channichthyidae Harpagiferidae** R. SOc.

SUBORDER BLENNIOIDEI **Blenniidae** *Blennius* Mioc. Eu. R. Oc. *Oncolepis* M. Eoc. Eu. *Problennius* Eoc. WAs. **Labrisomidae** R. Oc. **Tripterygiidae** R. Oc. *Tripterygion* Plioc. NAf. R. Oc. **Clinidae** *Clinus* Mioc. Eu. Plioc. NAf. R. Oc. *Pterygocephalus* Eoc. Eu. R. Oc. **Chaenopsidae** R. Oc. **Dactyloscopidae** R. SAtl. SPac. Oc.

SUBORDER ICOSTEOIDEI **Icosteidae** R. Oc.

SUBORDER SCHINDLEROIDEI **Schindleriidae** R. Oc.

SUBORDER AMMODYTOIDEI **Ammodytidae** *Ammodytes* Olig. Mioc. Eu. R. Oc. *?Rhamphosus* M.-U. Eoc. Eu.

SUBORDER CALLIONYMOIDEI **Callionymidae** *Callionymus* M. Eoc. Plioc. Eu. R. Oc. **Draconettidae** R. Oc.

SUBORDER GOBIOIDEI **Gobiidae** *Eogottus* M. Eoc. Eu. *Gobiopsis* Eoc. WAs. R. Oc. *Gobius* ?M. Eoc. Plioc. Eu. Mioc. NA. WAs. Plioc. NAf. R. Oc. *Lepidogobius* Mioc. R. NA. *?Pirskenius* U. Olig. Eu. **Rhyacichthyidae Gobioididae Microdesmidae Eleotridae Kraemeriidae** R. Oc.

SUBORDER KUROIDEI **Kurtidae** R. EInd.

SUBORDER ACANTHUROIDEI **Acanthuridae** *Acanthonemus* M. Eoc. Eu. *Acanthurus* M. Eoc. Mioc. Eu. R. Oc. *Apostasella* [*Apostasis*] Olig. Mioc. Eu. *Eolabroides* M. Eoc. Eu. *Eozanclus* M. Eoc. Eu. *Naso* [*Naseus*] M. Eoc. Eu. R. Oc. *Parapygaeus* U. Eoc. Eu. *Protautoga* Mioc. NA. U. Tert. SA. *Pseudosphaerodon Pygaeus* Eoc. Eu. *Tylerichthys* M. Eoc. Eu. *Zauchus* M. Eoc. Eu. R. Oc. *Zebrastoma* Tert. WInd. R. Oc. **Siganidae** R. Oc. *Archaeteuthis* [*Protosigana*] Olig. Eu.

SUBORDER SCOMBROIDEI **Gempylidae** *Acanthonotos* [*Hemithyrsites*] Mioc. Plioc. Eu. Plioc. NAf. R. Oc. *?Bathysoma* U. Cret. Eu. *Eothyrsites* Olig. NZ. *Eutrichiurides* L.-M. Eoc. Eu. *Euzaphleges* [*Zaphleges*] Mioc. NA. *Gempylus* Olig. Eu. SWAs. R. Oc. *Progempylus* L. Eoc. Eu. *Thyrsites* Mioc. NA. R. Oc. *Thyrsitocephalus* Olig. Eu. *Thyrsocles Zaphlegus* Mioc. NA. **Trichiuridae** *Eutrichiurides* Paleoc. Eu. Paleoc. Eoc. Af. *Lepidopus* [*Acanthonotus Anenchelum Lepidopides*] Eoc. WAf. WInd. Olig. SA. Olig. Plioc. Eu. Plioc. NAf. *Trichiurides* Paleoc. Mioc. Eu. *Trichiurus* Eoc. Af. Eoc. Plioc. Eu. Pleist. NA. R. Oc. **Scombridae** *Amphodon* [*Scombramphodon*] Eoc. Olig. Eu. *Aramichthys* Eoc. WAs. *Ardiodus* L. Eoc. Eu. *Auxides* Mioc. NA. *Auxis* Eoc. Mioc. Eu. R. Oc. *Eocoelopoma* L. Eoc. Eu. *Eoscombrus* Eoc. Olig. NA. *Eothynnus* [*?Cariniceps ?Coelocephalus Phalacrus ?Rhonchus*] L. Eoc. Eu. Plioc. NAf. *Euthynnus* [*Katsuwonus*] Pleist. WInd. R. Oc. *Gymnosarda* [*Cybiosarda*] Eoc. Olig. Eu. R. Oc. *Isurichthys* Olig. Eu. Oc. *Landanichthys* L. Paleoc. WAf. *Matarchia* Mioc. Eu. R. Oc. *Miothunnus* Mioc. Eu. *Megalolepis* Olig. Eu. *Neocybium* Olig. Mioc. Eu. *Ocystias Ozymandias* Mioc. NA. *Palaeothunnus* U. Paleoc. SWAs. *Palimphyes* [*Krambergeria*] Olig. Eu. *Pelacybium Sarda* [*Pelamys*] Paleoc. Olig. Eu. Mioc. NA. Plioc. NAf. R. Oc. *Sarmata* Mioc. Eu. *Scomber* [*Pneumatophorus*] Eoc. Plioc. Eu. Af. Mioc. NA. Tert. EAs. R. Oc. *Scomberomurus* [*Cybium*] Paleoc. Af. Eoc. Mioc. Eu. Olig. SWAs. R. Oc. *Scombramphodon* L. Eoc. Olig. Eu. *Scombrinus* L. Eoc. Eu. *Scombrosarda* Eoc. Eu. EEu. *Sphyraenodus* [*Dictyodus*] Eoc. NAf. L. Eoc. Mioc. Eu. *Starrias* Mioc. NA. *Stereodus* Mioc. Eu. *Tamesichthys* L. Eoc. Eu. *Thunnus* [*Orcynus Thynnus*] Eoc. Pleist. Eu. Mioc. NA. Af. R. Oc. *Tunita* Mioc. NA. *Turio* Mioc. NA. *Wetherellus Woodwardiella*

L. Eoc. Eu. *Xestias* Mioc. NA. *Xiphopterus* Eoc. Eu. **Xiphiidae** *Acestrus Aglyptorhynchus* L. Eoc. Eu. *Blochius* U. Eoc. Eu. *Brachyrhynchus* Mioc. Eu. *Coelorhynchus* [*Cylindracanthus Glyptorhynchus*] U. Cret. Eoc. As. NAf. U. Cret. Olig. Eu. Eoc. NA. *Congorhynchus* U. Cret. Eoc. WAf. *Hemirhabdorhynchus* Eoc. Eu. WAf. *Xiphias* Eoc. WAf. Eoc. Olig. Eu. Pleist. NA. R. Oc. *Xiphiorhynchus* [*?Ommatolampes*] Paleoc. Plioc. NAf. Eoc. WAf. L. Eoc. Mioc. Eu. **Luvariidae** R. Oc. **Istiophoridae** *Istiophorus* [*Histiophorus*] U. Cret. Mioc. NA. Eoc. Plioc. Eu. R. Oc. *Tetrapterus* Eoc. Eu. R. Oc. **Palaeorhynchidae** *Enniskillenius* L. Eoc. Eu. *Homorhynchus* [*Hemirhynchus*] Eoc. Mioc. Eu. *Palaeorhynchus* Eoc. Mioc. Eu. Olig. WAs. *Pseudotetrapterus* Olig. Mioc. Eu.

SUBORDER STROMATEOIDEI **Amarsipidae Centrolophidae Ariommatidae** R. Oc. **Nomeidae** R. Oc. *Carangodes* Eoc. Eu. **Stromateidae** R. Oc. *Seserimus* [*Aspidolepis*] ?U. Cret. Eu. R. Oc. **Tetragonuridae** R. Oc.

SUBORDER ANABANTOIDEI **Anabantidae** *Anabas* Pleist. R. EInd. **Belontiidae Helostomidae** R. EInd. **Osphronemidae** *Osphronemus* U. Eoc. or L. Olig. R. EInd. R. As.

SUBORDER LUCIOCEPHALOIDEI **Luciocephalidae** R. EInd.

SUBORDER CHANOIDEI **Chanidae** R. Indo-Pac. Oc. *?Ancylostylus* U. Cret. Eu. *Chanos* Eoc. Mioc. Eu. R. Indo-Pac. Oc. *Dartbile* U. Cret. ?Eoc. SA. *Parachanos* L. Cret. Af. SA.

SUBORDER MASTACEMBELOIDEI **Mastacembelidae** R. SAs. EAf. **Chaudhuriidae** R. SEAs.

ORDER PLEURONECTIFORMES (HETEROSOMATA)

SUBORDER PSETTODOIDEI **Psettodidae** R. Oc. *Joleaudichthys* M. Eoc. NAf.

SUBORDER PLEURONECTOIDEI **Citharidae** *Citharus* [*Eucitharus*] Mioc. Plioc. Eu. R. Cos. *Citharichthys* Olig. Eu. Plioc. NAf. R. Oc. **Bothidae** *Arnoglossus* Mioc. Eu. R. Oc. *Bothus* Eoc. Olig. Eu. R. Oc. *Eobothus* M. Eoc. As. Eu. *Evesthes* Mioc. NA. *Imhoffius* M. Eoc. Eu. *Paralichthys* [*Vorator*] Mioc. NA. R. Oc. *Scophthalmus* [*Rhombus*] Eoc. Plioc. Eu. Mioc. WAs. R. Oc. **Pleuronectidae** *Cleisthenes* [*Protopsetta*] Tert. EAs. R. Oc. *Hippoglossoides* Mioc. NA. R. Oc. *Limanda* Plioc. Eu. R. Oc. *Pleuronectes* [*Platessa*] Mioc. Aus. R. Oc. *Pleuronichthys* [*Zoropsetta Zororhombus*] Mioc. NA. Plioc. Eu. R. Oc. *?Propsetta* Mioc. Eu.

SUBORDER SOLEIOIDEI **Soleidae** *Achirus* Plioc. NAf. R. Oc. *Anoterisma* Mioc. NAf. *Arambourgichthys* Mioc. NAf. *Eosolea* Eoc. NA. *Eobuglossus* M. Eoc. NAf. *Microchirus* [*Monochir*] Mioc. Eu. Plioc. NAf. R. Oc. *Solea* Paleoc. Mioc. Eu. Eoc. Plioc. NAf. Mioc. WAs. R. Oc. *Turahbuglossus* M. Eoc. NA. **Cynoglossidae** R. Oc. *Cynoglossus* Mioc. Eu. R. Oc.

ORDER TETRAODONTIFORMES

SUBORDER BALISTOIDEI **Aracanidae** *Proaracana* M. Eoc. Eu. **Ostraciodontidae** *Eolactonia* M. Eoc. Eu. **Triacanthidae** *Cryptobalistes Protacanthodes* Eoc. Eu. **Triacanthodidae** *Eoplectus Zignoichthys* M. Eoc. Eu. **Balistidae** R. Oc. **Ostraciidae** R. Oc.

SUBORDER TETRAODONTOIDEI **Tetraodontidae** *Tetraodon* [*Ovoides*] Eoc. Plioc. Eu. R. Oc. **Triodontidae** *Triodon* ?M. Eoc. Eu. Af. R. Oc. **Diodontidae** *Chilomycterus* Mioc. Eu. R. Oc. *Diodon* [*Enneodon Gymnodus Heptadiodon Megalurites Progymodon*] Eoc. Plioc. Eu. As. Af. Eoc. Pleist. NA. Mioc. EInd. WInd. Mioc. Plioc. Aus. R. Oc. *Kyrtodymnodon* Plioc. Eu. *Oligodiodon* Mioc. Eu. **Molidae** *Mola* [*Orthagoriscus*] Mioc. Plioc. Eu. Tert. SA. R. Oc. **Incertae sedis** *Eotrigodon* L.-M. Eoc. Eu. M. Eoc. WAf. NAf. *Plectocretacicus* U. Cret. SWAs.

SUBCLASS SARCOPTERYGII

ORDER CROSSOPTERYGII

SUBORDER RHIPIDISTIA

SUPERFAMILY OSTEOLEPIDOIDEA (OSTEOLEPIFORMES) **Osteolepidae** *Bogdanovia* U. Dev. Eu. *Canningius* M. Dev. Gr. *Chrysolepis* U. Dev. EEu. *Ectosteorhachis* L. Perm. NA. *Geiserolepis* Miss. EEu. *Glyptopomus* [*Glyptognathus Glyptolaemus Pennagnathus Platygnathus*] M.-U. Dev. Eu. NA. *Gogonasus* U. Dev. Aus. *Gyroptychius* [*Diplopterax Diplopterus Diptopterus*] M.-?U. Dev. Eu. M. Dev. Gr. *Latvius* U. Dev. Eu. NA. *Lohsania* L. Perm. NA. *Megalichthys* [*Carlukeus Centrodus Parabatrachus ?Plintholepis Rhomboptychius ?Sporelepis*] ?Miss, Penn. Eu. Penn. NA. *Megapomus* U. Dev. EEu. *Megistolepis* U. Dev. *Metaxygnathus* U. Dev. Aus. *Osteolepis* [*Pleiopterus Pliopterus Triplopterus Tripterus*] M. Dev. NAs. Ant. M.-U. Dev. Eu. *Shirolepis* M. Dev. EEu. *Sterropterygion* U. Dev. NA. *Thaumatolepis* U. Dev. Eu. NAs. *Thursius* M. Dev. Eu. ?U. Dev. NA. *Thysanolepis* ?Miss. EEu. **Eusthenopteridae** *Eusthenodon* U. Dev. NA. Gr. *Eusthenopteron* U. Dev. NA. Eu. *Hyenia* U. Dev. NA. *Jarvikina* U. Dev. EEu. *Marsdenichthys* U. Dev. Aus. *Platycephalichthys* U. Dev. EEu. *Tristichopterus* M. Dev. Eu. **?Eusthenopteridae** *Devonosteus* M.-U. Dev. Eu. *Litoptychius* U. Dev. NA. **Panderichthyidae** *Elpistostege* U. Dev. NA. *Panderichtys* [*?Cricodus ?Polyplocodus*] M.-U. Dev. EEu. *Obruchevichthys* U. Dev. EEu. **Rhizodontidae** *?Canowindra* U. Dev. Aus. *Rhizodus* [*Archichthys ?Coelosteus ?Colonodus Dendroptychius Labyrinthodontosaurus Polyporites Sigmodus*] *Strepsodus* Miss. Aus. Miss. Penn. Eu. NA. *Sauripteris* [*Sauripterus*] U. Dev. Eu. NA. *Spodichthys* U. Dev. Gr. **Incertae sedis** *Canowindra* U. Dev. Aus.

SUPERFAMILY HOLOPTYCHOIDEA (POROLEPIFORMES) **Powichthyidae** *Powichthys* L. Dev. NA. *Youngolepis* L. Dev. EAs. **Holoptychidae** *Glyptolepis* [*?Hamodus Plyphlepis ?Sclerolepis*] M. Dev. Spits. Gr. M.-U. Dev. Eu. EEu. *Holoptychus* [*?Apedodus ?Apendulus ?Dendrodus Holoptychius Lamnodus*] M. Dev. Miss. Eu. U. Dev. NA. Gr. Aus. Ant. NAs. *Laccognathus* U. Dev. EEu. *Pseudosauripterus* U. Dev. Eu. *?Ventalepis* U. Dev. EEu. **Porolepidae** *Heimenia* M. Dev. Spits. EAs. *Porolepis* [*Gyrolepis*] L.-M. Dev. Eu. Spits. NAs.

SUBORDER ONYCHODONTIFORMES **Onychodontidae** *Grossius* M. Dev. Eu. *Onychodus* [*Protodus*] L. Dev. NAs. L.-M. Dev. Spits. M.-U. Dev. Eu. NA. U. Dev. Gr. *Quebecius* U. Dev. NA. *Strunius* U. Dev. Eu.

SUBORDER COELACANTHIFORMES (ACTINISTIA) **Euporosteidae** *Euporosteus* U. Dev. Eu. **Diplocercidae** *Chagrinia* U. Dev. NA. *Dictyonosteus* U. Dev. Spits. *Diplocercides Nesides* U. Dev. Eu. **Rhabdodermatidae** *Cardiosuctor* Miss. NA. *?Dumfregia* Miss. Eu. *Rhabdoderma* [*Conchiopsis Holopygus*] Miss. NAf. Miss. Penn. Eu. EEu. NA. *Synaptolus* Penn. NA. **Coelacanthidae** *Axelia* L. Trias. Spits. *Bunoderma* Jur. SA. *Chinlea* U. Trias. NA. *Coccoderma* [*Dokkoderma*] U. Jur. Eu. *?Coelocanthropsis* Miss. Eu. EEu. *Coelacanthus* [*Hoplopygus*] Penn. U. Perm. Eu. L. Trias. Mad. *Cualabaea* U. Jur. WAf. *Diplurus* [*?Holophagoides Osteopleurus Pariestegus Rhabdiolepis*] U. Trias. NA. *Graphiurichthys* [*Graphiurus*] U. Trias. Eu. *Heptanema* M.-U. Trias. Eu. *Libys* U. Jur. Eu. *Macropoma* [*Eurypoma Lophoprionolepis*] L.-U. Cret. Eu. *Macropomoides* U. Cret. SWAs. *Mawsonia* L. Cret. SA. WAs. L.-U. Cret. Af. *Miguashaia* U. Dev. NA. *Moenkopia* L. Trias. NA. *Mylacanthus* L. Trias. Spits. *Piveteauvia* L. Trias. Mad. *Rhipis* Jur. Af. *Sassenia* L. Trias. Spits. Gr. *Scleracanthus* L. Trias. Spits. *Sinocoelacanthus* L. Trias. EAs. *Spermatodus* L. Perm. NA. *Whiteia* L. Trias. Mad. Gr. ?U. Trias. NA. *Wimania* [*Leioderma*] L. Trias. Spits. ?Gr. **Laugiidae** *?Holophagus* [*Trachymetopon Undina*] U. Trias. U. Jur. Eu. ?Jur. Aus. *Laugia* L. Trias. Gr. **Hadronectoridae** *Allenypterus Hadronector Polyosteorhynchus* Miss. NA. **Latimeriidae** R. Ind. Oc. **Insertae sedis** *Lochmocercus* Miss. NA.

ORDER DIPNOI *Diabolichthyes* L. Dev. EAs. **Uranolophidae** *Uranolophus* L. Dev. NA. **Dipnorhynchidae** *Dipnorhynchus* L. Dev. Eu. M. Dev. Aus. *Ganorhynchus* M.-U. Dev. Eu. Na. *Griphognathus Holodipterus* [*Archaeotylus Holodus*] U. Dev. Eu. Aus. **Dipteridae** *Chirodipterus* U. Dev. Eu. Aus. *Conchodus* [*Cheirodus*] M.-U. Dev. Eu. U. Dev. NA. *Dipteroides* U. Dev. Eu. *Dipterus* [*Catopterus Eoctenodus Paradipterus Polyphractus*] L.-U. Dev. Eu. M.-U. Dev. NA. U. Dev. NAs. Aus. *Grossipterus* U. Dev. Eu. *?Palaedaphus* [*?Archaeonectes ?Heliodus ?Paleodahus*] U. Dev. Eu. NA. *Pentlandia* M. Dev. Eu. *Rhinodipterus* M.-U. Dev. Eu. *?Stomiakykus* M. Dev. NA. **Rhynchodipteridae** *Rhynchodipterus* U. Dev. Eu. Gr. **Phaneropleuridae** *Fleurantia* U. Dev. NA. *Jarvikia Nielsenia Oervigia* U. Dev. Gr. *Phaneropleuron* U. Dev. Eu. ?Gr. *Scaumenacia* [*Canadiptarus Canadipterus*] U. Dev. NA. ?Eu. *Soederberghia* U. Dev. Gr. Aus. NA. **Ctenodontidae** *Ctenodus* [*Campylopleuron Proctenodus Rhadamista*] Miss. Penn. Eu. NA. Miss. Aus. *Tranodis* Miss. NA. **Sagenodontidae** *Megapleuron* Penn. NA. *?Proceratodus* Penn. L. Perm. NA. *Sagenodus* [*Petalodopsis Yonodus*] Miss. Penn. Eu. Miss. L. Perm. NA. *Straitonia* Miss. Eu. **Uronemidae** *Uronemus* [*Ganopristodus*] Miss. Eu. ?NA. **Conchopomidae** *Conchopoma* [*Conchiopsis Peplorhina*] Penn. L. Perm. NA. L. Perm. Eu. **Ceratodontidae** *Arganodus* U. Trias. NAf. *Ceratodus* [*Hemictenodus Metaceratodus Scropha*] L. Trias. U. Jur. Eu. L. Trias. L. Cret. As. L. Trias. U. Cret. Mad. NA. Aus. L. Trias. Paleoc. Af. L. Trias. Spits. *Gosfordia* L. Trias. Aus. *Microceratodus* Trias. NAf. L. Trias. Mad. *Paraceratodus* L. Trias. Mad. *Ptychoceratodus* U. Trias. Eu. *Neoceratodus* [*Epiceratodus ?Ompax*] U. Cret. R. Aus. **Lepidosirenidae** *Gnathorhiza* Penn. L. Perm. NA. U. Perm. Eu. *Lepidosiren* [*Amphibichthys*] Mioc. R. SA. *Protopterus* [*Protomalus Rhinocryptis*] Eoc. R. Af.

DIPNOI INCERTAE SEDIS *Osteoplax* Miss. Eu. *?Palaeophichthys* Penn. NA.

CLASS AMPHIBIA

SUBCLASS LABYRINTHODONTIA

ORDER ICHTHYOSTEGALIA **Acanthostegidae** *Acanthostega* U. Dev. Gr. **Ichthyostegidae** *Ichthyostega* U. Dev. Gr. *?Ichthyostegopsis* U. Dev. Gr.

ORDER INCERTAE SEDIS *Crassigyrinus* U. Miss. Eu.

SUPERFAMILY LOXOMMATOIDEA **Loxommatidae** *Baphetes* L.-M. Penn. Eu. M. Penn. NA. *Loxomma* U. Miss. M. Penn. Eu. *Megalocephalus* L.-M. Penn. Eu. M. Penn. NA. **Family to be named** *Spathicephalus* U. Miss. Eu. NA.

ORDER TEMNOSPONDYLI

SUPERFAMILY COLOSTEOIDEA **Colosteidae** *Colosteus* M. Penn. NA. *Greererpeton* U. Miss. NA. *Pholidogaster* U. Miss. Eu.

SUPERFAMILY TRIMERORHACHOIDEA **Saurerpetontidae** *Acroplous* L. Perm. NA. *Erpetosaurus* M. Penn. NA. *Isodectes* L. Perm. NA. *Saurerpeton* M. Penn. NA. **Trimerorhachidae** *?Doragnathus* U. Miss. Eu. *Lafonius* U. Penn. NA. *Nannospondylus* L. Perm. NA. *Neldasaurus* L. Perm. NA. *Slaughenhopia* U. Perm. NA. *Trimerorhachis* L. Perm. NA.

SUPERFAMILY INCERTAE SEDIS **Caerorhachidae** *Caerorhachis* U. Miss. Eu. **Dendrerpetontidae** *Dendrerpeton* L. Penn. NA. *?Erpetocephalus* L. Penn. Eu. *?Eugyrinus* L. Penn. Eu.

SUPERFAMILY EDOPOIDEA **Cochleosauridae** *Chenoprosopus* L. Perm. NA. *Cochleosaurus* M. Penn. Eu. NA. *Gaudrya* M. Penn. Eu. *Macrerpeton* M. Penn. NA. **Edopidae** *Edops* L. Perm. NA.

SUPERFAMILY ERYOPOIDEA **Eryopidae** *Actinodon* L. Perm. Eu. Ind. *Chelyderpeton* L. Perm. Eu. *Clamorosaurus* L. Perm. EEu. *Eryops* U. Penn. L. Perm. NA. *Intasuchus* U. Perm. EEu. *Onchiodon Osteophorus Sclerocephalus* L. Perm. Eu. *Syndyodosuchus* U. Perm. Eu. **Dissorophidae** *Alegeinosaurus* L. Perm. NA. *Amphibamus* M. Penn. NA. Eu. *Arkanserpeton* ?L. Penn. or M. Penn. NA. *Aspidosaurus* L. Perm. NA. *Astreptorhachis* U. Penn. NA. *Brevidorsum Broiliellus Cacops Conjunctio Dissorophus Ecolsonia Fayella* L. Perm. NA. *Iratusaurus Kamacops* U. Perm. EEu. *Longiscitula* L. Perm. NA. *Micropholis* L. Trias. SAf. *Platyhystrix* L. Perm. NA. *Tersomius* L. Perm. NA. *Zygosaurus* U. Perm. EEu. **Branchiosauridae** *Apateon* L. Perm. Eu. *Branchiosaurus* M. Penn. L. Perm. Eu. U. Penn. NA. *Schoenfelderpeton* L. Perm Eu. **Micromelerpetontidae** *Branchierpeton* L. Perm. Eu. *Limnerpeton* M. Penn. Eu. *Micromelerpeton* L. Perm. Eu. **Doleserpetontidae** *Doleserpeton* L. Perm. NA. **Trematopsidae** *Acheloma* L. Perm. NA. *?Actiobates* U. Penn. NA. *Trematops* L. Perm. NA. **Parioxyidae** *Parioxys* L. Perm. NA. **Zatracheidae** *Dasyceps* L. Perm. NA. *Stegops* M. Penn. NA. *Zatrachys* [*Acanthostoma ?Acanthostomatops*] L. Perm. Eu. NA. **Archegosauridae** *Archegosaurus* L. Perm. Eu. *Bashkirosaurus Platyoposaurus* U. Perm. EEu. *Prionosuchus* ?L. Perm. SA. **Melosauridae** *Melosaurus* U. Perm. EEu.

SUPERFAMILY RHINESUCHOIDEA **Rhinesuchidae** *Laccocephalus* L. Trias. SAf. *Muchocephalus* U. Perm. SAf. *Rhineceps* U. Perm. SAf. *Rhinesuchoides* U. Perm. SAf. *Rhinesuchus* U. Perm. SAf. *Uranocentrodon* L. Trias. SAf. **Lydekkerinidae** *?Broomulus* L. Trias. SAf. *Chomatobratrachus* L. Trias. Aus. *Cryobatrachus* L. Trias. Ant. *?Deltacephalus* L. Trias. Mad. *Limnoiketes* L. Trias. SAf. *Lydekkerina* L. Trias. SAf. *Luzocephalus* L. Trias. EEu. Gr. *?Putterillia* L. Trias. SAf. **Sclerothoracidae** *Sclerothorax* L. Trias. Eu. **Peltobatrachidae** *Peltobatrachus* U. Perm. EAf.

SUPERFAMILY CAPITOSAUROIDEA **Benthosuchidae** *Benthosuchus* L. Trias. EEu. ?Mad. *?Gondwanosaurus* L. Trias. Ind. *Odenwaldia* L. Trias. Eu. *?Pachygonia* L. Trias. Ind. *Parabenthosuchus* L. Trias. Eu. *Thoosuchus* L. Trias. EEu. **Capitosauridae** *Bukobaja* U. Trias. EEu. *Cyclotosaurus* M.-U. Trias. Eu. *Eocyclotosaurus* L. Trias. Eu. *Eryosuchus* L.-M. Trias. EEu. *Kestrosaurus* L. Trias. SAf. *Paracyclotosaurus* U. Trias. Aus. *Parotosuchus* [*Parotosaurus Archotosaurus*] L. Trias. SAf. EAf. Ind. M. Trias. NA. L.-M. Trias. Aus. Trias. Eu. EEu. NAf. *Promastodonsaurus* M. Trias. SA. *Sassenisaurus* L. Trias. Spits. *Volgasaurus* L. Trias. EEu. *Volgasuchus* L. Trias. EEu. *Wellesaurus* Trias. NAf. *Wetlugasaurus* L. Trias. EEu. **Mastodonsauridae** *Mastodonsaurus* U. Trias. Eu. EEu.

SUPERFAMILY RHYTIDOSTEOIDEA **Rhytidosteidae** *Arcadia* L. Trias. Aus. *Boreopelta* L. Trias. NAs. *Deltasaurus* L. Trias. Aus. SAf. *Derwentia* L. Trias. Aus. *Laidleria* L. Trias. SAf. *Peltostega* L. Trias. Eu. Spits. *Pneumatostega* L. Trias. SAf. *Rhytidosteus* L. Trias. SAf. **Indobrachyopidae** *Indobrachyops* L. Trias. SAs. *Mahavisaurus* L. Trias. Mad. *Rewana* L. Trias. Aus.

SUPERFAMILY TREMATOSAUROIDEA **Trematosauridae** *Aphaneramma* L. Trias. Spits. NA. Aus. *Erythrobatrachus* L. Trias. Aus. *Gonioglyptus* L. Trias. Ind. *Ifasaurus* L. Trias. Mad. *Inflectosaurus* L. Trias. EEu. *?Latiscopus* U. Trias. NA. *Lyrocephaliscus* [*Lyrocephalus*] L. Trias. Spits. *Lyrosaurus* L. Trias. Mad. *Mahavisaurus* L. Trias. Mad. *Microposaurus* L. Trias. SAf. *Platystega* L. Trias. Spits. *Stoschiosaurus* L. Trias. Gr. *Tertrema* L. Trias. Spits. *Tertremoides* L. Trias. Mad. *Trematosaurus* L. Trias. Eu. SAf. *Trematosuchus* L. Trias. SAf. *Wantzosaurus* L. Trias. Mad.

SUPERFAMILY BRACHYOPOIDEA **Kourerpetontidae** *Kourerpeton* ?Perm. NA. **Dvinosauridae** *Dvinosaurus* U. Perm. EEu. **Brachyopidae** *Austrobrachyops* L. Trias. Ant. *Batrachosaurus* L. Trias. SAf. EEu. *Blinasaurus* L. Trias. Aus. *Boreosaurus* L. Trias. Spits. *Bothriceps* U. Perm. Aus. *Brachyops* L. Trias. Aus. Ind. *Hadrokkosaurus* M. Trias. NA. *Notobrachyops* U. Trias. Aus. *Siderops* L. Jur. Aus. *Sinobrachyops* M. Jur. EAs. *Trucheosaurus* [*Bothriceps*] U. Perm. Aus. *Tupilakosaurus* L. Trias. Gr. Eu. **Chigutisauridae** *Keratobrachyops* L. Trias. Aus. *Pelorocephalus* [*Chigutisaurus Icarosaurus*] L. Trias. SA.

SUPERFAMILY METOPOSAUROIDEA **Metoposauridae** *Anaschisma* U. Trias. NA. *?Dictyocephalus* U. Trias. NA. *Metoposaurus* U. Trias. NA. Eu. Ind. NAf.

SUPERFAMILY ALMASAUROIDEA **Almasauridae** *Almasaurus* U. Trias. NAf.

SUPERFAMILY PLAGIOSAUROIDEA **Plagiosauridae** *Gerrothorax* U. Trias. Eu. *?Melanopelta* L. Trias. EEu. *Plagiobatrachus* L. Trias. Aus. *Plagiorophus* M. Trias. Eu. *Plagiosaurus* U. Trias. Eu. *Plagiosternum* M. Trias. Eu. *Plagiosuchus* M.-U. Trias. Eu. Spits.

ORDER ANTHRACOSAURIA

SUBORDER EMBOLOMERI **Eoherpetontidae** *Eoherpeton* U. Miss. Eu. **Proterogyrinidae** *?Papposaurus* U. Miss. Eu. *Proterogyrinus* [*Mauchchunkia*] U. Miss. NA. Eu. *Tulerpeton* U. Dev. EEu. **Eogyrinidae** *Calligenethlon* L. Penn. NA. *Carbonerpeton* M. Penn. NA. *Diplovertebron* M. Penn. Eu. *Eogyrinus* L. Penn. Eu. *Leptophractus* M. Penn. NA. *Neopteroplax* U. Penn. L. Perm. NA. *Palaeoherpeton* L. Penn. Eu. *Pholiderpeton* ?U. Miss. NA. L. Penn. Eu. *Pteroplax* L. Penn. Eu. **Archeriidae** *Archeria* L. Perm. NA. *Cricotus* U. Penn. NA. *Spondylerpeton* M. Penn. NA. **Anthracosauridae** *Anthracosaurus* L. Penn. Eu.

SUBORDER GEPHYROSTEGIDA **Gephyrostegidae** *Bruktererpeton* U. Miss. Eu. *Eusauropleura* M. Penn. NA. *Gephyrostegus* M. Penn. Eu.

SUBORDER SEYMOURIAMORPHA **Discosauriscidae** *Ariekanerpeton* L. Perm. CAs. *Discosauriscus* L. Perm. Eu. *Letoverpeton* L. Perm. EEu. *Urumgia* L. Perm. EAs. *Utegenia* L. Perm. CAs. **Kotlassiidae** *Buzulukia Bystrowiana ?Enosuchus Karpinskiosaurus Kotlassia* U. Perm. EEu. **Seymouriidae** *Gnorhinosuchus* L. Perm. EEu. *Nyctiboetus Rhinosauriscus* U. Perm. EEu. *Seymouria* L. Perm. NA.

SUBORDER INCERTAE SEDIS **Lanthanosuchidae** *Lanthaniscus Lanthanosuchus Chalcosaurus* U. Perm. EEu.

Chroniosuchidae *Chroniosaurus Chroniosuchus* U. Perm. EEu.

Limnoscelidae *Limnoscelis* U. Penn. L. Perm. NA. *Limnosceloides Limnoscelops* L. Perm. NA. *Limnostygis* M. Penn. NA. *Romeriscus* L. Penn. NA.

Solenodonsauridae *Solenodonsaurus* M. Penn. Eu.

Tseajaiidae *Tseajaia* L. Perm. NA.

Tokosauridae *Tokosaurus* U. Perm. EEu.

Nycteroleteridae *Nycteroleter Macroleter* U. Perm. EEu.

CLASS INCERTAE SEDIS

Diadectidae *Diadectes* U. Penn. L. Perm. NA. *Desmatodon* U. Penn. NA. *Stephanospondylus* L. Perm. Eu.

SUBCLASS LEPOSPONDYLI

ORDER AÏSTOPODA **Ophiderpetontidae** *Coloraderpeton* U. Penn. NA. *Ophiderpeton* L.-U. Penn. Eu. L.-M. Penn. NA. **Lethiscidae** *Lethiscus* L. Miss. Eu. **Phlegethontiidae** *Dolichosoma* M. Penn. Eu. *Phlegethontia* L.-M. Penn. NA. Eu. L. Perm. NA.

ORDER NECTRIDEA **Keraterpetontidae** *Batrachiderpeton* L. Penn. Eu. *Diceratosaurus* M. Penn. NA. *Diplocaulus* L.-U. Perm. NA. *Diploceraspis* U. Penn. NA. *Keraterpeton* L.-M. Penn. NA. Eu. **Scincosauridae** *Sauravus* U. Penn. L. Perm. Eu. *Scincosaurus* M. Penn. Eu. **Urocordylidae** *Crossotelos* L. Perm. NA. *Ctenerpeton* M. Penn. NA. *Lepterpeton* L. Penn. Eu. *Ptyonius* ?L.-M. Penn. NA. *Sauropleura* M. Penn. Eu. M. Penn. L. Perm. NA. *Urocordylus* L. Penn. Eu. **Family incertae sedis** *Arizonerpeton* L. Penn. NA.

ORDER MICROSAURIA

SUBORDER TUDITANOMORPHA **Tuditanidae** *Asaphestera* L. Penn. NA. *Boii* U. Penn. Eu. *Crinodon* U. Penn. Eu. *Tuditanus* M. Penn. NA. **Hapsidopareiontidae** *Hapsidopareion* L. Perm. NA. *Llistrofus* L. Perm. NA. *Ricnodon* L. Penn. NA. M. Penn. Eu. *Saxonerpeton* L. Perm. Eu. **Pantylidae** *Pantylus* L. Perm. NA. *Trachystegos* L. Penn. NA. **Gymnarthridae** *Cardiocephalus* L. Perm. NA. *Elfridia* L. Penn. NA. *Euryodus* L. Perm. NA. *Hylerpeton* L. Penn. NA. *Leiocephalikon* L. Penn. NA. *Pariotichus* L. Perm. NA. *Sparodus* M. Penn. Eu. **Ostodolepididae** *Micraroter* L. Perm. NA. *Ostodolepis* L. Perm. NA. *Pelodosotis* L. Perm. NA. **Trihecatontidae** *Trihecaton* U. Penn. NA. **Goniorhynchidae** *Rhynchonkos* [*Goniorhynchus*] L. Perm. NA.

SUBORDER MICROBRACHOMORPHA **Microbrachidae** *Microbrachis* M. Penn. Eu. **Hyloplesiontidae** *Hyloplesion* M. Penn. Eu. **Brachystelechidae** *Brachystelechus* L. Perm. Eu. *Carrolla* L. Perm. NA. **Odonterpetontidae** *Odonterpeton* M. Penn. NA.

ORDER LYSOROPHIA **Lysorophidae** *Cocytinus* M. Penn. NA. *Lysorophus* L. Perm. NA. *?Megamolgophis* L. Perm. NA. *Molgophis* M. Penn. NA.

SUBCLASS LEPOSPONDYLI ORDER INCERTAE SEDIS

Adelogyrinidae *Adelogyrinus* U. Miss. Eu. *Adelospondylus* U. Miss. Eu. *Dolichopareias* L. Miss. Eu. *Palaeomolgophis* L. Miss. Eu.
Acherontiscidae *Acherontiscus* U. Miss. Eu.

MODERN AMPHIBIAN ORDERS

ORDER GYMNOPHIONA **Caeciliidae** R. CA. Ind. NAf. SA. *Apodops* U. Paleoc. SA. **Ichthyophiidae** R. WInd. Ind. SAs. SPac. **Scolecomorphidae** R. NAf. **Typhlonectidae** R. SA. **Rhinatrematidae** R. SA.

ORDER URODELA

SUBORDER KARAUROIDEA **Karauridae** *Karaurus* U. Jur. WAs.

SUBORDER PROSIRENOIDEA **Prosirenoidea** *Albanerpeton* [*Heteroclitotriton*] U. Cret. NA. M. Jur. M. Mioc. Eu. *Prosiren* L. Cret. NA. *?Ramonellus* L. Cret. SWAs.

SUBORDER CRYPTOBRANCHOIDEA **Hynobiidae** R. EAs. **Cryptobranchidae** *Andrias* [*Hydrosalamandra Plicagnathus Proteocordylus Sieboldia Tritogenius Tritomegas Zaissanurus*] U. Olig. U. Plioc. Eu. M. Mioc. U. Mioc. NA. Pleist. R. EAs. *Cryptobranchus* U. Paleoc. M. Pleist. R. NA.

SUBORDER PROTEOIDEA **Proteidae** *Mioproteus* M. Mioc. EEu. *Necturus* U. Paleoc. R. NA. *Orthophyia* U. Mioc. Eu. *Proteus* Pleist. Eu. R. **Batrachosauroididae** *Batrachosauroides* L. Eoc. M. Mioc. NA. *Opisthotriton* U. Cret. U. Paleoc. NA. *Palaeoproteus* U. Paleoc. M. Eoc. Eu. *Peratosauroides* L. Plioc. NA. *Prodesmodon* [*Cuttysarkus*] U. Cret. L. Paleoc. NA.

SUBORDER AMPHIUMOIDEA **Amphiumidae** *Amphiuma* U. Paleoc. M. Mioc. R. NA. *Proamphiuma* U. Cret. NA.

SUBORDER AMBYSTOMATOIDEA **Dicamptodontidae** *Ambystomichnus* [*Ammobatrachus*] Paleoc. NA. *Bargmannia* U. Mioc. EEu. *Chrysotriton* L. Eoc. NA. *Dicamptodon* L. Plioc. R. NA. *Geyeriella* U. Paleoc. Eu. *Wolterstorffiella* U. Paleoc. Eu. **Ambystomatidae** *Ambystoma* [*Ogallalabatrachus Plioambystoma*] L. Olig. R. NA. *Amphitriton* U. Plioc. NA. **Scapherpetontidae** *Lisserpeton* U. Cret. U. Paleoc. NA. *Piceoerpeton* U. Paleoc. L. Eoc. NA. *Scapherpeton* [*Hedronchus Hemitrypus*] U. Cret. U. Paleoc. NA.

SUBORDER PLETHODONTOIDEA **Plethodontidae** *Aneides* L. Mioc. R. NA. *Batrachoseps* L. Plioc. R. NA. *Desmognathus* Pleist. R. NA. *Gyrinophilus* L. Pleist. R. NA. *Plethodon* L. Mioc. R. NA. *Pseudotriton* L. Pleist. R. NA.

SUBORDER SALAMANDROIDEA **Salamandridae** *Archaeotriton* U. Olig. L. Mioc. EEu. *Brachycormus* [*Molge Oligosema Triton Triturus Tylotriton*] L. Mioc. Eu. *Chelotriton* [*Epipolysemial Grippiella Heliarchon Palaeosalamandrina Polysemia Salamandra Tischleviella Tylototriton*] M. Eoc.-U. Mioc. Eu. *Chioglossa* U. Olig.-R. Eu. *Euproctus* U. Pleist.-R. Eu. *Koalliella* U. Paleoc. Eu. *Megalotriton* U. Eoc. or L. Olig. Eu. *Mertensiella* L. Mioc. EEu. R. WAs. *Notophthalmus* L. Mioc.-R. NA. *Oligosemia* U. Mioc. Eu. *Palaeopleurodeles* U. Olig. Eu. *Pleurodeles* U. Mioc. or L. Plioc.-R. NAf. *Procynops* U. Mioc. EAs. *Salamandra* [*Dehmiella Heteroclitotriton Palaeosalamandra Voigtiella*] U. Eoc. or L. Olig.-R. EEu. Eu. *Salamandrina* L. Mioc.-R. Eu. *Taricha* [*Palaeotaricha*] U. Olig.-R. NA. *Triturus* U. Olig.-R. Eu. U. Mioc.–R. WAs. *Tylototriton* M. Eoc. Eu. R. EAs. **Sirenidae** *Siren* M. Eoc.-R. NA. *Habrosaurus* [*Adelphesiren*] U. Cret. Paleoc. NA. *Pseudobranchus* Plioc.-R. NA.

URODELA INCERTAE SEDIS *Comonecturoides* U. Jur. NA. *Hylaeobatrachus* L. Cret. Eu. *?Triassurus* U. Trias. WAs.

?ORDER PROANURA **Protobatrachidae** *Triadobatrachus* [*Protobatrachus*] L. Trias. Mad.

ORDER ANURA **Ascaphidae** *Notobatrachus* U. Jur. SA. *Vieraella* L. Jur. SA. **Discoglossidae** *Discoglossus* M. Eoc. L. Mioc. Eu. *Eodiscoglossus* U. Jur. Eu. *Gobiates*, U. Cret. EAs. *Latonia* Mioc. ?Plioc. Eu. *Pelophilus* M. Mioc. Eu. *Prodiscoglossus* U. Olig. Eu. *Scotiphryne* U. Cret. M. Paleoc. NA. *Zaphrissa* M.-U. Olig. Eu. **Pipidae** *Cordicephalus* L. Cret. SWAs. *Eoxenopoides* L. Olig. SAf. *Saltenia* U. Cret. SA. *Shelania* L. Paleoc. SA. *Thoraciliacus* L. Cret. SWAs. *Xenopus* Paleoc. SA. Mioc. R. Af. NAf. **Rhinophrynidae** *Eorhinophrynus* U. Paleoc. M. Eoc. NA. *Rhinophrynus* L. Olig. Pleist. R. NA. **Palaeobatrachidae** *Neusibatrachus* U. Jur. Mioc. Eu. *Palaeobatrachus* Eoc. Mioc. Eu. *Pliobatrachus* Plioc. Pleist. Eu. **Pelobatidae** *Eopelobates* U. Cret. Olig. NA. M. Eoc. L. Mioc. Eu. *Macropelobates* L. or M. Olig. As. M. Mioc. EAs. *Miopelobates* Mioc. L. Plioc. Eu. *Pelobates* Plioc. R. Eu. *Scaphiopus* L. Olig. R. NA. **Pelodytidae** *Miopelod-*

ytes M. Mioc. NA. **Leptodactylidae** *Caudiverbera* L. Eoc. Mioc. SA. *Eleutherodactylus* Mioc. SA. Pleist. NA. *Eusophus* L. Olig. SA. *Indobatrachus* Eoc. As. *Leptodactylus* L. Mioc. R. NA. Pleist. R. SA. *Syrrhopus* Pleist. NA. **Bufonidae** *Bufo* Paleoc. Pleist. SA. Mioc. Pleist. NA. Plioc. As. Eu. **Ceratophrynidae** *Ceratophrys* Plioc. Pleist. SA. *Wawelia* Mioc. SA. **Hylidae** *Acris* L. Mioc. Plioc. R. NA. *Hyla* L. Olig. Pleist. R. NA. L. Mioc. Eu. *Proacris* L. Mioc. NA. *Pseudacris* U. Mioc. R. NA. **Microhylidae** *Gastrophryne* L. Mioc. R. NA. **Rhacophoridae** *Rhacophorus* Pleist. R. As. **Ranidae** *Asphaerion* L. Mioc. Eu. *Ptychadena* Mioc. *Rana* M. Eoc. R. Eu. Mioc. R. NA. Plioc. R. As. *?Theatonius* U. Cret. NA.

CLASS REPTILIA

SUBCLASS ANAPSIDA

ORDER CAPTORHINIDA

SUBORDER CAPTORHINOMORPHA **Protorothyrididae** *Anthracodromeus* M. Penn. NA. *Archerpeton* L. Penn. NA. *Brouffia* M. Penn. Eu. *Cephalerpeton* M. Penn. NA. *Coelostegus* M. Penn. Eu. *Hylonomus* [*Fritschia*] L. Penn. NA. *Paleothyris* M. Penn. NA. Eu. *Protorothyris* [*Melanothyris*] L. Perm. NA. **Captorhinidae** *Captorhinikos Captorhinus* [*Ectocynodon Hypopnous*] *Captorhinoides Eocaptorhinus Labidosaurikos Labidosaurus ?Peurcosaurus Rhiodenticulatus Romeria* L. Perm. NA. *Protocaptorhinus* L. Perm. NA. U. Perm. EAf. *Kahneria Rothaniscus* [*Rothia*] M. Perm. NA. *Hecatogomphius* M. Perm. EEu. *Moradisaurus* U. Perm. NAf. **Bolosauridae** *Bolosaurus* L. Perm. NA. EEu. **?Batropetidae** *Batropetes* L. Perm. Eu. **Acleistorhinidae** *Acleistorhinus* L. Perm. NA.

SUBORDER PROCOLOPHONIA

SUPERFAMILY PROCOLOPHONOIDEA **Nyctiphruretidae** *Barasaurus* U. Perm. Mad. *Nyctiphruretus* U. Perm. EEu. *Owenetta* U. Perm. SAf. **Procolophonidae** *Anomoiodon* L. Trias. Eu. *Burtensia* L. Trias. EEu. *Candelaria* M. Trias. SA. *Contritosaurus* L. Trias. EEu. *?Estheriophagus* U. Perm. NAs. *Eumetabolodon* L. Trias. EAs. *Hypsognathus* U. Trias. NA. *Kapes* L. Trias. EEu. *Koiloskiosaurus* L. Trias. Eu. *Leptopleuron* [*Telerpeton*] M. Trias. Eu. *Macrophon Microphon* L. Trias. EEu. *?Microcnemus* L. Trias. EEu. *Myocephalus Myognathus* L. Trias. SAf. *Neoprocolophon* L. Trias. EAs. *Orenburgia* L.-M. Trias. EEu. *Paoteodon* U. Trias. EAs. *Procolophon Microthelodon Thelegnathus* L. Trias. SAf. *Procolophonoides* L. Trias. SAf. *?Santaisaurus* L. Trias. EAs. *Sphodrosaurus* U. Trias. NA. *Spondylolestes* L. Trias. SAf. *Tichvinskia Vitalia* L. Trias. EEu. **Sclerosauridae** *Sclerosaurus* [*Aristodesmus*] L. Trias. Eu. *Basileosaurus* L. Trias. Eu.

SUBORDER PAREIASAUROIDEA **Rhipaeosauridae** *Leptoropha Rhipaeosaurus* M. Perm. EEu. **Pareiasauridae** *Anthodon* U. Perm. SAf. EAf. EEu. *Bradysaurus* [*Brachypareia Bradysuchus Koalemasaurus Platyoropha*] M. Perm. SAf. *Elginia* U. Perm. Eu. *Embrithosaurus* [*Dolichopareia Nochelesaurus*] M. Perm. SAf. *Nanoparia* U. Perm. SAf. *Parasaurus* U. Perm. Eu. *Pareiasaurus* [*Pareiasuchus Propappus*] U. Perm. SAf. EAf. EEu. *Scutosaurus* [*Amalitzkia Proelginia*] U. Perm. EEu. *Shihtienfenia* U. Perm. EAs.

SUBORDER MILLEROSAUROIDEA **Millerettidae** *Broomia ?Heleophilus Milleretta* [*Millerina*] *Milleretoides Millerettops Milleropsis Millerosaurus Nanomilleretta* U. Perm. SAf.

ORDER MESOSAURIA **Mesosauridae** *Brasileosaurus Mesosaurus* [*Ditrichosaurus Noteosaurus Notosaurus*] *Stereosternum* L. Perm. SAf. SA.

ORDER INCERTAE SEDIS *Eunotosaurus* M. Perm. SAf.

SUBCLASS TESTUDINATA

ORDER CHELONIA

SUBORDER PROGANOCHELYDIA **Proganochelyidae** *Proganochelys* [*Chelytherium Psammochelys Stegochelys Triassochelys*] U. Trias. Eu. SAs. **Proterochersidae** *Proterochersis ?Saurischiocomes* U. Trias. Eu.

SUBORDER PLEURODIRA **Pelomedusidae** *Apoidochelys* U. Cret. SA. *Bothremys* U. Cret. Mioc. NA. *Carteremys* Eoc. Ind. *Dacquemys* L. Olig. NAf. *Neochelys Palaeaspis* [*Palaeochelys Palemys*] Eoc. Eu. *Paralichelys* Olig. Plioc. Eu. *Pelomedusa* ?Olig. R. Af. Mad. *Podocnemis* [*Erymnochelys*] Cret. R. SA. Af. Eu. As. *Polysternon* Eoc. Olig. Eu. *Potamochelys* L. Cret. WAf. *Rosasia* U. Cret. Eu. *Roxochelys* U. Cret. WAf. SA. *Shweboemys* Eoc. NAf. Mioc. Plioc. SAs. *Sokotochelys* U. Cret. WAf. *Stereogenys* Eoc. Olig. Af. Eoc. Eu. *Stupendemys* L. Plioc. SA. *Taphrosphys* [*Prochonias*] U. Cret. Mioc. NA. SA. Eoc. Mioc. Af. **Chelidae** *Chelodina* Plioc. R. Aus. *Chelus* [*Chelydra Chelys*] U. Mioc. R. SA. *Emydura* Pleist. R. Aus. *Hydromedusa* Eoc. R. SA. *Parahydraspis* Plioc. SA. *Pelocomastes* Pleist. Aus. *Phrynops* [*Acrohydraspis Rhinemys*] Plioc. R. SA. *Platemys* R. SA. **Platychelyidae** *Platychelys* U. Jur. Eu. **Eusarkiidae** *Eusarkia* Eoc. NAf.

SUBORDER CRYPTODIRA

SUPERFAMILY BAENOIDEA **Glyptopsidae** *Glyptops* U. Jur. NA. ?Eu. *Mesochelys* L. Cret. Eu. **Baenidae** *Baena* U. Cret. NA. *Chisternon* Eoc. NA. *Dorsetochelys* L. Cret. Eu. *Eubaena* U. Cret. NA. *Hayemys* L. Cret. NA. *Palatobaena* U. Cret. NA. *Plesiobaena* Jur. Paleoc. NA. *Stygichelys* U. Cret. NA. *Trinitichelys* U. Jur. NA. **Neurankylidae** *Boremys Neurankylus Thescelus* U. Cret. NA. *Compsemys* U. Cret. Paleoc. NA. **Meiolaniidae** *Meiolania* [*Ceratochelys Miolania*] Pleist. Aus. *Crossochelys* Eoc. SA. *Niolamia* U. Cret. SA.

SUPERFAMILY TRIONYCHOIDEA **Kinosternidae** *Kinosternon* [*Cinosternum Sternotherus*] ?Plioc R. NA. SA. *Staurotypus* R. NA. *Xenochelys* Olig. NA. **Dermatemydidae** R. CA. *Adocus* U. Cret. NA. Eoc. EAs. M. Eoc. WAs. *Agomphus* [*Amphiemys*] U. Cret. Olig. NA. *Baptemys* Eoc. NA. *Basilemys* U. Cret. NA. *Dermatemys* R. NA. *?Heishanemys* ?U. Cret. EAs. *Hoplochelys* Paleoc. Olig. NA. *Lindholmemys* U. Cret. NAs. *Mongolemys* U. Cret. Paleoc. EAs. *Peshanemys* L. Cret. SAs. *Sinochelys* ?Cret. EAs. *Trachyaspis* Eoc. Mioc. Eu. NAf. *Tretosternon* [*Heolochelydra ?Peltochelys*] ?U. Jur. Cret. Eu. *Tsaotanemys* U. Cret. EAs. *Zangerlia* U. Cret. **Carettochelyidae** R. New Guinea *Akrochelys Allaeochelys* Eoc. Eu. *Anosteira* [*?Apholidemys Castresia Pseudotrionyx*] Eoc. Mioc. NA. EAs. Eu. *Carettochelys* Mioc. R. New Guinea

Hemichelys ?Eoc. Ind. *Pseudoanosteira* Eoc. NA. **Trionychidae** R. As. EInd. Af. NA. *Aspideretes* U. Cret. EAs. *Chitra* Pleist. EInd. Pleist. R. SAs. *Cyclanorbis* ?Plioc. R. Af. *Cycloderma* ?Mioc. R. Af. *Eurycephalochelys* L. Eoc. Eu. *Lissemys* ?Plioc. R. SAs. Ind. *Plastomenus* U. Cret. Eoc. NA. M. Eoc. WAs. *Trionyx* [*Amyda Asperidites Axestemys Conchochelys Paleotrionyx Plastomenus Platypeltis Temnotrionyx*] Cret. R. As. Af. NA. Eu. *Palaeotrionyx* Paleoc. NA. *Sinaspideretes* U. Jur. EAs.

SUPERFAMILY CHELONIOIDEA **Plesiochelyidae** *Craspedochelys* U. Jur. Eu. *Plesiochelys* [*Brodiechelys Hylaeochelys Parachelys Tholemys Wincania*] U. Jur. EAs. U. Jur. L. Cret. Eu. *Portlandemys* U. Jur. Eu. *Tienfuchelys* U. Jur. EAs. **Protostegidae** *Archelon Calcarichelys Chelosphargis Protostega* U. Cret. NA. *Pneumatoarthrus* U. Cret. NA. *Rhinochelys* Cret. Eu. **Toxochelyidae** *Ctenochelys Lophochelys* U. Cret. NA. *Dollochelys* U. Cret. NA. L. Eoc. Eu. *Erquelinnesia* Eoc. Olig. Eu. *Osteopygis* [*Euclastes ?Lytoloma ?Propleura Rhetechelys*] L. Cret. NA. *Glossochelys* Paleoc. Olig. Eu. *Peritresius Prionochelys* U. Cret. NA. *Protochelys* U. Cret. NA. *Thinochelys* U. Cret. NA. *Toxochelys* [*Phyllemys*] U. Cret. NA. **Dermochelyidae** R. Oc. *Cosmochelys* Eoc. WAf. *Dermochelys* [*Sphargis*] ?Mioc. R. Oc. *Eosphargis* Eoc. Eu. *Psephophorus* [*Macrochelys*] Eoc. NAf. Eoc. Plioc. Eu. *Protosphargis* U. Cret. Eu. Med. **Cheloniidae** R. Oc. *Argillochelys* Eoc. Eu. *Caretta* [*?Pliochelys ?Proganosaurus Thalassochelys*] ?U. Cret. Eoc. R. Oc. *Chelonia* [*Chelone*] Paleoc. R. Oc. *Eochelone* M. Olig. Eu. *Glarichelys* L. Olig. Eu. *Procolpochelys* Mioc. NA. *Puppigerus* [*Erquelinnesia ?Glossochelys ?Pachyrhynchus*] Eoc. Eu. *Rhinochelys* L.-U. Cret. Eu. *Allopleuron* U. Cret. Eu. *Corsochelys* U. Cret. NA. *Desmatochelys* U. Cret. NA. *Syllomus* Mioc. NA. EAs. NAf. **Thalassemyidae** *Desmemys* L. Cret. Eu. *Eurysternum* [*Achelonia Acichelys Aplax Euryaspis Hydropelta Palaeomedusa*] *Padiochelys* [*?Chelonemys*] U. Jur. Eu. *Thalassemys* U. Jur. Eu. *Tropidemys* U. Jur. Eu. *Yaxartermys* U. Jur. WAs.

SUPERFAMILY TESTUDINOIDEA **Chelydridae** R. NA. SA. EInd. *Acherontemys* Mioc. NA. *Chelydra* Olig. R. NA. CA. Olig. Plioc. ?Eu. *Chelydrops* U. Mioc. NA. *Chelydropsis* Olig. Mioc. Eu. ?As. *Emarginachelys* U. Cret. NA. *Macrocephalochelys* Plioc. EEu. *Macroclemys* [*Macrochelys*] Mioc. R. NA. Mioc. ?Eu. *Planiplastron* M. Olig. WAs. *Protochelydra* L. Paleoc. NA. **Emydidae** R. Eu. As. Af. NA. SA. *Batagur* R. SAs. *Chinemys* Olig. EEu. Mioc. EAs. Eu. Pleist. R. EAs. *Chrysemys* Eoc. Mioc. ?As. Eoc. R. NA. CA. SA. Eu. *Clemmydopsis* Plioc. Mioc. *Clemmys* [*Geoliemys ?Paralichelys*] R. NA. *Cuora* (*Cyclemys*) Plioc. R. EAs. *Echmatemys* Eoc. NA. WAs. *Emydoidea* Plioc. R. NA. *Emys* [*Platemys*] ?Plioc. R. NAf. WAs. Eu. *Epiemys* Plioc. EAs. *Geoclemys* [*Polyechmatemys*] R. Ind. *Geoemyda* [*Nicoria*] Eoc. R. EAs. CA. SA. Eu. *Grayemys* Eoc. WAs. *Hardella* Plioc. R. Ind. *Hokouchelys* Eoc. EAs. *Kachuga* Plioc. R. Ind. *Mauremys* Olig. R. As. Eu. NAf. *Ocadia* Eoc. Mioc. ?Eu. R. EAs. ?Af. *Ptychogaster* Eoc. Mioc. Eu. *Sakya* Plioc. Pleist. EEu. WAs. *Sharemys* Olig. EAs. *Temnoclemmys* Mioc. Eu. *Terrapene* [*Cistudo*] Plioc. R. NA. *Broilia* Olig. Eu. Mioc. **Testudinidae** *Geochelone* Eoc. R. NA. SA. As. Eu. Af. R. Ind. *Gopherus* Olig. R. NA. *Kinixys* [*Cinixys Cinothorax*] ?Olig. R. Af. ?Eu. *Stylemys* Olig. Mioc. NA. As. ?Eu. *Testudo* [*Caudochelys Colossochelys Ergilemys Eupachemys Geochelone Hadrianus Hesperotestudo Megalochelys*] Eoc. R. Eu. As. NAf. *Floridemys* [*Bystra*] Mioc. NA. *Impregnochelys* L. Mioc. EAf. *Kansuchelys* U. Mioc. or ?Olig. EAs. *?Sinohadrianus* Eoc. EAs.

CHELONIA INCERTAE SEDIS **Sinemydidae** *Manchurochelys* ?U. Jur. EAs. *Sinemys* U. Jur. L. Cret. EAs. **Kallokibotiidae** *Kallokibotium* U. Cret. Eu. **Pleurosternidae** *Chengyuchelys* U. Jur. L. Cret. SAs. *Helochelys* U. Cret. Eu. *Pleurosternon* [*Digerrhum Megasernon Megasternum*] U. Jur. L. Cret. Eu. As. **Chelycarapookidae** *Chelycarapookus* L. Cret. Aus. **Family Undesignated** *Apertotemporalis* U. Cret. NAf. *Chitracephalus* L. Cret. Eu. *Dinochelys* U. Jur. NA. *Hangaiemys* L. Cret. EAs. *Macrobaena* Eoc. EAs. *Nanhsiungchelys* U. Cret. EAs. *Scutemys* ?U. Cret. EAs. *Solnhofia* U. Jur. Eu.

SUBCLASS DIAPSIDA

ORDER ARAEOSCELIDA **Petrolacosauridae** *Petrolacosaurus* U. Penn. NA. **Araeoscelididae** *Araeoscelis* [*Ophiodeirus*] *Zarcasaurus* L. Perm. NA. *?Kadaliosaurus* L. Perm. Eu.

ORDER INCERTAE SEDIS
Mesenosauridae *Mesenosaurus* ?M. Perm. EEu.

Coelurosauravidae *Coelurosauravus* [*Daedalosaurus*] U. Perm. Mad. *Weigeltisaurus* [*Palaeochamaeleo*] U. Perm. Eu.

Drepanosauridae *Drepanosaurus* U. Tris. Eu.

Endennasauridae *Endennasaurus* U. Trias. Eu.

ORDER CHORISTODERA **Champsosauridae** *Champsosaurus Simoedosaurus* U. Cret. Eoc. Eu. NA. *Eotomistoma* U. Cret. EAs. *Tchoiria Khurendukhosaurus* L. Cret. EAs.

ORDER THALATTOSAURIA **Thalattosauridae** *Thalattosaurus* [*Scenodon*] M. Trias. NA. **Askeptosauridae** *Askeptosaurus* M. Trias. Eu. **Claraziidae** *Clarazia* [*Heschelería*] M. Trias. Eu.

INFRACLASS LEPIDOSAUROMORPHA

ORDER EOSUCHIA **Acerosodontosauridae** *Acerosodonosaurus* U. Perm. Mad. **Younginidae** *Heleosuchus Youngina* [*Youngoides Youngopsis*] U. Perm. SAf. **Tangasauridae** *Hovasaurus Thadeosaurus* U. Perm. Mad. *Tangasaurus* U. Perm. EAf. *Kenyasaurus* L. Trias. EAf. **Galesphyridae** *Galesphyrus* U. Perm. SAf.

SUPERORDER LEPIDOSAURIA

ORDER SPHENODONTA **?Gephyrosauridae** *Gephyrosaurus* L. Jur. Eu. **Sphenodontidae** R. NZ. *Brachyrhinodon* U. Trias. Eu. *?Chometokadmon* U. Cret. Eu. *Eilenosaurus* U. Jur. NA. *Glevosaurus* U. Trias. Eu. *Homeosaurus* [*Leptosaurus Stelliosaurus*] *Meyasaurus* U. Jur. Eu. *Monjurosuchus* U. Jur. EAs. *Opisthias* [*Theretairus*] U. Jur. NA. *?Palacrodon* L. Trias. SAf. *?Pachystropheus Pelecymala Polysphenodon Planocephalosaurus Simila* U. Trias. Eu. *Sapheosaurus* [*Piocormus Sauranodon*] U. Jur. Eu. *?Scharschengia* L. Trias. EEu. *Sigmala* U. Trias. Eu. *Toxolophosaurus* L. Cret. NA. **Pleurosauridae** *Palaeopleurosaurus* L. Jur. Eu. *Pleurosaurus* U. Jur. L. Cret. Eu.

ORDER SQUAMATA

SUBORDER LACERTILIA

INFRAORDER EOLACERTILIA **Paliguanidae** *Palaeagama Paliguana Saurosternon* U. Perm. or L. Trias. SAf. **Kuehneosauridae** *Cteniogenys* U. Jur. Eu. NA. *Icarosaurus* U. Trias. NA. *Kuehneosaurus Kuehneosuchus Perparvus* U. Trias. Eu. *?Rhabdopelix* U. Trias. NA. **Fulengidae** *Fulengia* L. Jur. SAs. **Eolacertilia incertae sedis** *Colubrifer* L. Trias. SAf. *Lacertulus* U. Perm. or L. Trias. SAf. *Litakis* U. Cret. NA.

INFRAORDER IGUANIA **Euposauridae** *Euposaurus* U. Jur. Eu. **Arretosauridae** *Arretosaurus* U. Eoc. EAs. **Iguanidae** *Aciprion* M. Olig. NA. *Anolis* U. Olig. Pleist. R. NA. R. WInd. *Corytophanes* Pleist. NA. R. CA. SA. *Crotaphytus* [*Gambelia*] ?L. Olig. Pleist. R. NA. *Ctenosaura* R. NA. CA. *Cyclura* U. Pleist. R. WInd. *Cypressaurus* L. Olig. NA. *Diposaurus* L. Mioc. R. NA. *Geiseltaliellus* M. Eoc. Eu. *Harrisonsaurus* L. Mioc. NA. *Holbrookia* Pleist. R. NA. ?M. Mioc. NA. *Iguana* U. Pleist. R. WInd. R. CA. SA. U. Pleist. R. SA. *Laemanctus* R. CA. *Leiocephalus* Mioc. NA. Pleist. R. WInd. *Leiosaurus* U. Pleist. R. SA. *Oplurus* R. Mad. *Paradipsosaurus* Eoc. or Olig. NA. *Parasauromalus* L.-M. Eoc. NA. *Phrynosoma* [*Eumecoides*] M. Mioc.-R. NA. *Pristiguana* U. Cret. SA. *Sauromalus* Pleist. R. NA. *Sceloporus* Mioc. R. NA. *Swainiguanoides* M. Paleoc. NA. *Urosaurus* R. NA. *Uta* Pleist. R. NA. **Agamidae** R. EEu. As. Af. EInd. Aus. *Agama* ?U. Eoc. or L. Olig. U. Mioc. Eu. Af. Paleoc. SAs. *Clamydosaurus* Pleist. R. Aus. *Mimeosaurus* U. Cret. EAs. *Stellio* Pleist. SWAs. R. Eu. WAs. NAf. *Tinosaurus* Eoc. EAs. NA. Eu. SAs. Paleoc. SAs. *Uromastyx* U. Eoc. or L. Olig. U. Pleist. Eu. **Chameleontidae** R. Eu. Af. Mad. *Anquingosaurus* ?M. Paleoc. SAs. *Chamaeleo* Mioc. EEu. Eu. Mioc. R. EAf.

INFRAORDER NYCTISAURIA (GEKKOTA) **Ardeosauridae** *Ardeosaurus* [*Eichstattosaurus*] U. Jur. Eu. *Eichstaettisaurus* [*Broilisaurus*] U. Jur. Eu. *Yabeinosaurus* U. Jur. EAs. ?U. Jur. SAs. **Bavarisauridae** *Bavarisaurus* U. Jur. Eu. *Palaeolacerta* U. Jur. Eu. **Gekkonidae** R. As. Af. EInd. Aus. CA. WInd. SA. Oceania. *Aristelliger* Pleist. R. WInd. R. CA. *Cadurcogekko* U. Eoc. or L. Olig. Eu. *Coleonyx* U. Pleist. R. NA. *Cyrtodactylus* R. SAs. EInd. SPac. Oc. Indo-Pac. Oc. *Geckolepis* U. Pleist. R. Mad. *Gekko* R. ?EAf. *Gerandogekko* Mioc. Eu. *Hemidactylus* R. Af. SAs. Pac. Oc. Maurit. *Lygodactylus* R. WAf. *Pareodura* U. Pleist. R. Mad. *Phelsuma* U. Pleist. Mad. R. Maurit. *Phyllodactylus* M. Mioc. EEu. R. Med. NA. SA. Af. Mad. Aus. *Rhodanogekko* U. Eoc. Eu. *Sphaerodactylus* U. Pleist. R. WInd. R. CA. *Tarentola* U. Pleist. R. WInd. R. Med. NAf. SWAs. *Thecadactylus* U. Pleist. R. WInd. R. CA. SA. **Pygopodidae** R. Aus.

INFRAORDER LEPTOGLOSSA (SCINCOMORPHA) **Paramacellodidae** *Becklesius* [*Becklesisaurus*] U. Jur. Eu. *Paramacellodus* U. Jur. Eu. NA. *Pseudosaurillus* U. Jur. Eu. *Saurillodon* U. Jur. Eu. *Saurillus* U. Jur. Eu. **Xantusiidae** R. NA. CA. WInd. *Lepidophyma* [*Impensodens*] Pleist. NA. R. CA. *Palaeoxantusia* M. Paleoc. L. Olig. NA. *Xantusia* Pleist. R. NA. **Teiidae** R. NA. CA. WInd. SA. *Adamisaurus* U. Cret. EAs. *Ameiva* Pleist. SA. U. Pleist. R. WInd. R. CA. SA. *Callopistes* U. Plio. SA. *Chamops* [*Alethesaurus Lanceosaurus*] U. Cret. NA. *Cherminsaurus* U. Cret. EAs. *Cnemidophorus* Plioc. R. NA. R. CA. SA. WInd. *Darchansaurus* U. Cret. EAs. *Dicrodon* U. Pleist. R. SA. *Dracaena* Mioc. R. SA. *Erdenetesaurus* U. Cret. EAs. *Haptosphenus Leptochamops Meniscognathus* U. Cret. NA. *Macrocephalosaurus* U. Cret. EAs. *Paraglyphanodon Polyglyphanodon Peneteius* U. Cret. NA. *Tupinambis* Olig. Mioc. R. SA. R. WInd. **Scincidae** R. SEu. As. Af. EInd. Aus. Oceania NA. CA. WInd. SA. *Ablepharus* L. Pleist. CAs. *Chalcides* Pleist. Med. *Contogenys* U. Cret. M. Paleoc. NA. *Egernia* U. Mioc. R. Aus. *Eumeces* Olig. R. NA. M. Mioc. NAf. Pleist. NAtl. *Gongylomorphus* R. Maurit. *Mabuya* U. Pleist. WInd. Mad. R. WInd. SA. Af. Mad. SAs. EInd. Aus. CA. *Mimobecklesisaurus* U. Jur. EAs. *Paracontogenys* U. Eoc. NA. *Sauriscus* U. Cret. NA. *Trachydosaurus* ?Cret. R. Aus. **Lacertidae** R. Eu. As. Af. *Dracaenosaurus* Olig. (?U. Eoc.) *Eolacerta* M. Eoc. Eu. *Eremias* M. Mioc. NAf. R. Af. As. *Lacerta* Mioc. R. Eu. Plioc. R. Eu. Med. Plioc. EInd. R. NAf. CAs. *Plesiolacerta* ?Paleoc. Olig. Eu. *Pseudeumeces* U. Eoc. L. Olig. Eu. **Cordylidae** (**Gerrhosauridae Zonuridae**) R. Af. *Gerrhosaurus* Mioc. R. Af. *Pseudolacerta* L. Eoc. U. Eoc. or L. Olig. Eu. **Dibamidae** R. EAs. EInd. NA.

INFRAORDER ANNULATA (AMPHISBAENIA) **Oligodontosauridae** *Oligodontosaurus* U. Paleoc. NA. **Amphisbaenidae** R. Af. NA. CA. SA. WInd. SWAs. *Changlosaurus* Olig. EAs. *Omoiotyphlops* U. Eoc. or L. Olig. (?Mioc.) Eu. **Rhineuridae** *Dyticonastis* U. Olig. L. Mioc. NA. *Jepsibaena* Eoc. NA. *Lestophis* Eoc. ?L. Olig. NA. *Macrorhineura* L. Mioc. NA. *Ototriton* L. Eoc. NA. *Plesiorhineura* M. Paleoc. NA. *Pseudorhineura* M. Olig. NA. *Rhineura* [*Platyrhachis*] Olig. R. NA. *Spathorhynchus* M. Eoc. L. Olig. NA. **Hyporhinidae** *Hyporhina* Olig. NA. **Bipedidae** R. CA. **Trogonophidae** R. NAf. SWAs.

INFRAORDER DIPLOGLOSSA (ANGUIMORPHA)

SUPERFAMILY UNCERTAIN **Paravaranidae** *Paravanus* U. Cret. EAs. **Bainguidae** *Bainguis* U. Cret. EAs.

SUPERFAMILY ANGUOIDEA **Anguidae** R. Eu. As. NAf. NA. SA. WInd. *Anguis* Olig. R. Eu. R. WAs. SWAs. Mioc. R. EEu. Pleist. R. SAs. *Anniella* U. Mioc. NA. *Apodosauriscus* L. Eoc. NA. *Arpadosaurus* L. Eoc. NA. *Celestus* U. Pleist. R. WInd. R. CA. *Diploglossus* U. Pleist. R. WInd. R. SA. CA. *Eodiploglossus* L. Eoc. NA. *Eoglyptosaurus* L. Eoc. NA. *Gerrhonotus* U. Cret. R. NA. R. CA. *Glyptosaurus* Eoc. NA. *Helodermoides* ?Eoc. L.-M. Olig. NA. *Machaerosaurus* Paleoc. Olig. NA. *Melanosaurus* Eoc. NA. ?L. Olig. ?Eu. *Odaxosaurus* U. Cret. ?Paleoc. NA. *Ophipseudopus* M. Eoc. Eu. *Ophisauriscus* M. Eoc. Eu. *Ophisaurus* [*Propseudopus Pseudopus Sauromorus*] Eoc. R. Eu. Mioc. NA. EEu. U. Plioc. SWAs. R. SAs. NAf. EInd. *Paragerrhonotus* L. Plioc. NA. *Paraglyptosaurus* L.-M. Eoc. NA. *Parapseudopus Ophipseudopus* Eoc. Eu. *Parodaxosaurus* M. Paleoc. NA. *Peltosaurus* Olig. NA. *Placosaurus* [*Helodermoides Loricotherium Necrodasypus Placotherium Proiguana Protrachysaurus*] EAs. SAs. Eoc. L. Olig. Eu. *Proxestops* Paleoc. NA. *Xestops* [*Oreosaurus*] Eoc. NA. Eu. **Anniellidae** R. NA. **Xenosauridae** R. EAs. CA. *Exostinus*

[*Harpagosaurus Prionosaurus*] U. Cret. Olig. NA. *Restes* U. Paleoc. L. Eoc. NA. **Dorsetisauridae** *Dorsetisaurus* U. Jur. Eu. NA.

SUPERFAMILY VARANOIDEA (PLATYNOTA) **Necrosauridae** *Colpodontosaurus* U. Cret. NA. *Eosaniwa* M. Eoc. Eu. *Necrosaurus* [*Melanosauroides Odontomophis Palaeosaurus Palaeovaranus*] Paleoc. Olig. Eu. *Parasaniwa* U. Cret. ?M. Eoc. NA. *Parviderma* U. Cret. EAs. *?Provaranosaurus* Paleoc. NA. **Helodermatidae** R. NA. *Eurheloderma* U. Eoc. or L. Olig. Eu. *Heloderma* Olig. Mioc. R. NA. *?Paraderms* U. Cret. NA. **Varanidae** R. SAs. Af. EInd. Aus. *Chilingosaurus* U. Cret. EAs. *Iberovaranus* M. Mioc. Eu. *Megalania* [*Notiosaurus*] Pleist. Aus. *?Pachyvanarus* U. Cret. NAf. *Palaeosaniwa* [*Megasaurus*] U. Cret. NA. *Saniwa* [*Thinosaurus*] Eoc. Olig. NA. Eoc. Eu. *Saniwides* U. Cret. EAs. *Telmasaurus* U. Cret. EAs. *Varanus* Mioc. ?Pleist. Eu. Mioc. EAf. CAs. Plioc. EEu. SWAs. Med. Plioc. R. Ind. Pleist. Aus. Pleist. R. SAs. **Lanthanotidae** R. EInd. *Cherminotus* U. Cret. EAs. **Aigialosauridae** *Aigialosaurus Carsosaurus* [*Opetiosaurus Mesoleptos*] *?Coniasaurus* M. Cret. Eu. *Proaigialosaurus* U. Jur. Eu. **Dolichosauridae** *Acteosaurus* [*Adriosaurus*] M. Cret. Eu. *Dolichosaurus* M. Cret. Eu. *Eidolosaurus Pontosaurus* [*Hydrosaurus*] L. Cret. Eu. **Mosasauridae** *Amphekepubis* U. Cret. NA. *Angolosaurus* U. Cret. WAf. *Clidastes* U. Cret. NA. *Compressidens Dollosaurus* U. Cret. Eu. *Ectenosaurus* U. Cret. NA. *Globidens* U. Cret. Eu. NA. SWAs. NAf. ?EInd. *Hainosaurus* U. Cret. Eu. *Halisaurus* U. Cret. NA. *Liodon* U. Cret. Eu. NA. *Mosasaurus* [*Baseodon Batrachiosaurus Batrachotherium Drepanodon Lesticodus Macrosaurus Nectoportheus Pterycollosaurus*] U. Cret. Eu. NA. *Platecarpus* [*Holosaurus Lestosaurus Sironectes*] U. Cret. Eu. NA. *Plesiotylosaurus* U. Cret. NA. *Plioplatecarpus* U. Cret. Eu. NA. *Plotosaurus* [*Kolposaurus*] *Prognathodon* [*Ancylocentrum Brachysaurus Prognathosaurus*] U. Cret. Eu. EEu. NA. *Taniwhasaurus* U. Cret. NZ. *Tylosaurus* [*Rhamposaurus Rhinosaurus*] U. Cret. NA. **Anguimorpha incertae sedis** *Gobiderma Proplatynotia* U. Cret. EAs.

LACERTILIA INCERTAE SEDIS *Anhuisaurus* Paleoc. SAs. *Araeosaurus* U. Cret. Eu. *Changjiangosaurus* Paleoc. SAs. *Conicodontosaurus* U. Jur. or ?L. Cret. SAs. U. Cret. EAs. *Costasaurus* U. Cret. Eu. *Cteniogenys* U. Jur. NA. Eu. *Dibolosodon* L. Mioc. SA. *Durotrigia* U. Jur. Eu. *Haplodontosaurus* Paleoc. NA. *Iguanavus* M. Eoc. NA. *Ilerdaesaurus* U. Jur. or L. Cret. Eu. *Isodontosaurus* U. Cret. EAs. *Lisboasaurus* U. Jur. Eu. *Litakis* U. Cret. NA. *Meyasaurus* U. Jur. or L. Cret. Eu. *Paraprionosaurus* ?Paleoc. M. Eoc. NA. *Qianshanosaurus* Paleoc. SAs. *Teilhardosaurus* U. Jur. EAs.

SUBORDER SERPENTES

INFRAORDER SCOLECOPHIDIA **Typhlopidae** *Typhlops* M. Mio. Eu. R. Eu. SAs. Af. SA. EInd. WInd. Aus. **Leptotyphlopidae** R. SA. WInd. Af. WAs.

INFRAORDER HENOPHIDIA

SUPERFAMILY SIMOLIOPHEOIDEA **Lapparentopheidae** *Lapparentophis* L. Cret. NAf. **Simoliopheidae** *Simoliophis* U. Cret. Eu.

SUPERFAMILY ANILIOIDEA **Aniliidae** *Colombophis* M. Mioc. SA. *Coniophis* U. Cret. M. Eoc. NA. U. Eoc. Eu. *Eoanilius* U. Eoc. Eu. **Uropeltidae** R. Ind. Ceylon

SUPERFAMILY BOOIDEA **Dinilysiidae** *Dinilysia* U. Cret. SA. **Xenopeltidae** R. CA. SEAs. **Boidae** R. Af. Mad. Aus. SAs. EEu. NA. CA. SA. *Albaneryx* M. Mioc. Eu. *Anilioides* L. Mioc. NA. *Boavus* R. Af. Mad. Aus. SAs. EEu. NA. CA. SA. *Bransateryx* U. Olig. Eu. *Cadurceryx* U. Eoc. Eu. *Cadurcoboa* U. Eoc. Eu. *Calamagras* M. Eoc. ?M.-U. Olig. L. Mioc. NA. L. Eoc. Eu. *Charina* M. Mioc. R. NA. *Cheilophis* M. Eoc. NA. *Daunoplis* Plioc. SAs. *Dawsonophis* U. Eoc. NA. *Dunnophis* L.-U. Eoc. Eu. M. Eoc. NA. *Eunectes* M. Mioc. R. SA. *?Geringophis* M. Olig. L. Mioc. NA. *Gigantophis* U. Eoc. NAf. *Helagras* Paleoc. NA. *Huberophis* U. Eoc. NA. *?Lithophis* M. Eoc. NA. *Madtsoia* L. Eoc. SA. *Ogmophis* U. Eoc. L.-M. Olig. L.-M. Plioc. NA. *Paleopython* ?M.-U. Eoc. ?Olig. Eu. *Paleryx* U. Eoc. Eu. *Paraepicrates* M. Eoc. NA. *Platyspondylia* U. Olig. Eu. *Plesiotortrix* U. Eoc. Olig. Eu. *Pseudopicrates* L. Mioc. NA. *Pterygoboa* L.-M. Mioc. NA. *Python* L. Mioc. EAf. *Tregophis* U. Mioc. NA. *Wonambi* Pleist. SAf. **?Palaeophidae** *Archaeophis* L. Eoc. Eu. WAs. *Palaeophis* [*Dinophis Titanophis*] M. Eoc. WAf. U. Cret. L. Eoc. NAf. *Pterophenus* M.-U. Eoc. NA. M. Eoc. NAf. U. Eoc. SA.

SUPERFAMILY ACROCHORDOIDEA **Acrochordidae** *Acrochordus* M. Mioc. Ind. R. SAs. Aus. **Nigeropheidae** L. Paleoc. WAf. *Woutersophis* M. Eoc. Eu.

INFRAORDER CAENOPHIDIA

SUPERFAMILY COLUBROIDEA **Anomalopheidae** *Anomalophis* L. Eoc. Eu. **Russellopheidae** *Russellophis* L. Eoc. NA. **Colubridae** *Ameiseophis* L. Mioc. NA. *Coluber* M.-U. Olig. M.-U. Mioc. L.-M. Pleist. Eu. U. Mioc. U. Plioc. R. EEu. Plioc. R. NA. R. As. NAf. EInd. *Dakotaophis* L. Mioc. NA. *Diadophis* M. Mioc. R. NA. *Dolniceophis* L. Mioc. Eu. *Dryinoides* M. Mioc. NA. *Elaphe* M.-U. Mioc. Plioc. R. NA. U. Mioc. EEu. R. Eu. EInd. As. *Heterodon* U. Mioc. Plioc. R. NA. *Lampropeltis* M. Mioc. Plioc. L. Pleist. R. NA. *Malpolon* L. Plioc. R. Eu. R. SWAs. NAf. *Mionatrix* M. Mioc. EAs. *Natrix* M. Mioc. R. Eu. R. NA. As. Af. EInd. *Nebraskophis* M. Mioc. NA. *Neonatrix* M. Mioc. NA. *Nerodia* U. Mioc. Plioc. NA. *Palaeomalpolon* M. Pleist. EEu. *Palaeonatrix* M. Mioc. EEu. *Paleofarancia* U. Mioc. NA. *Paleoheterodon* M. Mioc. NA. *Paracoluber* M. Mioc. NA. *Paraoxybelis* L. Mioc. NA. *Protropidonotus* M. Mioc. Eu. *Pseudocemophora* L. Mioc. NA. *Regina* L. Pleist. NA. *Salvadora* M. Mioc. NA. *Stilosoma* U. Mioc. R. NA. *Texasophis* M. Olig. M.-U. Mioc. NA. Mioc. Eu. **Elapidae** [including **Hydropheidae**] *Naja* M. Mioc. NAf. R. SAs. EInd. *Palaeonaja* M. Mioc. L. Plioc. Eu. **Viperidae** [including **Crotalidae**] *Bitis* L. Pleist. EAf. R. Eu. *Vipera* U. Mioc. EEu. R. Eu. As. EInd. *Crotalus* U. Pleist. R. NA.

SERPENTES INCERTAE SEDIS

Goinophis L. Mioc. NA. *Ophidium* U. Mioc. Plioc. SA. *Mesophis Pachyophis* U. Cret. Eu.

SUPERORDER SAUROPTERYGIA

ORDER INCERTAE SEDIS **Claudiosauridae** *Claudiosaurus* U. Perm. Mad.

ORDER NOTHOSAURIA **Pachypleurosauridae** *Anarosaurus Dactylosaurus* M. Trias. Eu. *Keichousaurus* M. Trias. EAs. *Neusticosaurus* M.-U. Trias. Eu. *Pachypleurosaurus* [*Pachypleura*] *?Psilotrachelosaurus Phygosaurus* M. Trias. Eu. **Simosauridae** *Simosaurus* [*?Opeosaurus*] M. Trias. Eu. **Nothosauridae** *Ceresiosaurus Lariosaurus* [*Macromerosaurus*] M. Trias. Eu. *Nothosaurus* [*Conchiosaurus Dracosaurus Oligolycus*] M.-U. Trias. Eu. M. Trias. SWAs. *Paranothosaurus Proneusticosaurus ?Rhaeticonia* M. Trias. Eu. **Cymatosauridae** *Cymatosaurus* [*Germanosaurus Micronothosaurus*] M. Trias. Eu. SWAs. **Pistosauridae** *Pistosaurus* M. Trias. Eu.

NOTHOSAURIA INCERTAE SEDIS *Corosaurus* U. Trias. NA. *Elmosaurus* M. Trias. Eu. *Kwangsisaurus Metanothosaurus* L. Trias. EAs. *Parthanosaurus* M. Trias. Eu.

ORDER PLESIOSAURIA

SUPERFAMILY PLESIOSAUROIDEA **Plesiosauridae** *Plesiosaurus* L. Jur. Eu. **Cryptoclididae** *Cryptoclidus* [*Cryptocleidus*] *Kimmerosaurus* U. Jur. Eu. *?Aristonectes* U. Cret. SA. **Elasmosauridae** *Alzadasaurus* L. Cret. NA. SA. *Aphrosaurus* U. Cret. NA. *Brancasaurus* L. Cret. Eu. *Elasmosaurus Fresnosaurus Hydralmosaurus Hydrotherosaurus Leurospondylus* U. Cret. NA. *Mauisaurus* U. Cret. NZ. *Microcleidus* L. Jur. Eu. *Muraenosaurus* U. Jur. Eu. WInd. *Morenosaurus* U. Cret. NA. *Styxosaurus Thalassomedon* U. Cret. NA. *Tricleidus* U. Jur. Eu. *Woolungasaurus* U. Cret. Aus.

SUPERFAMILY PLIOSAUROIDEA **Pliosauridae** *Archaeonectrus Eretmosaurus Eurycleidus ?Eurysaurus Macroplata Rhomaleosaurus* L. Jur. Eu. *Bishanopliosaurus* L. Jur. EAs. *Brachauchenius Dolichorhynchops* [*Trinacromerum*] U. Cret. NA. *Kronosaurus* L. Cret. Aus. *Leptocleidus* L. Cret. Eu. *Liopleurodon* [*Ischyrodon*] U. Jur. Eu. *?Megalneusaurus* U. Jur. NA. *Peloneustes* U. Jur. Eu. *Peyerus* L. Cret. SAf. *Pliosaurus* M.-U. Jur. Eu. *Polyptychodon* U. Cret. Eu. NA. *Simolestes Stretosaurus* U. Jur. Eu. *Sinopliosaurus* U. Jur. EAs. *Strongylokrotaphus* U. Jur. EEu. *Yuzhoupliosaurus* M. Jur. EAs.

PLESIOSAURIA INCERTAE SEDIS *Aptychodon* U. Cret. Eu. *Brimosaurus Discosaurus Piptomerus Piratosaurus* U. Cret. NA. *Sthenarosaurus* L. Jur. Eu.

INFRACLASS ARCHOSAUROMORPHA

ORDER PROTOROSAURIA **Protorosauridae** *Protorosaurus* U. Perm. Eu. **Prolacertidae** *Boreopricea Cosesaurus Macrocnemus Megacnemus* M. Trias. Eu. *?Kadimakara* L. Trias. Aus. *?Malerisaurus* U. Trias. SAs. *Prolacerta* [*Pricea*] L. Trias. SAf. *Prolacertoides* L. Trias. EAs. **Tanystropheidae** *Tanystropheus* L.-M. Trias. Eu. *Tanytrachelos* U. Trias. NA. **Incertae sedis** *?Trachelosaurus* L. Trias. Eu.

ORDER TRILOPHOSAURIA **Trilophosauridae** *?Anisodontosaurus* L. Trias. NA. *Tricuspisaurus Variodens* U. Trias. Eu. *Trilophosaurus* U. Trias. NA.

ORDER RHYNCHOSAURIA **Rhynchosauridae** *?Eifelosaurus* L. Trias. Eu. *Howesia Mesosuchus* L. Trias. SAf. *Hyperodapedon* [*Stenometopon Paradapedon*] U. Trias. Eu. SAs. *Mesodapedon* U. Trias. SAs. *Rhynchosaurus* M.-U. Trias. Eu. *Scaphonyx* [*Cephalonia Cephalastron Cephalastronius Scaphonychimus*] M. Trias. SA. *Stenaulorhynchus* M. Trias. EAf.

SUPERORDER ARCHOSAURIA

ORDER THECODONTIA

SUBORDER PROTEROSUCHIA **Proterosuchidae** *Archosaurus* U. Perm. EEu. *Chasmatosaurus* [*?Ankistrodon ?Elaphrosuchus ?Proterosuchus*] L. Trias. SAf. EAs. SAs. *Chasmatosuchus* L. Trias. EEu. *Kalisuchus Tasmaniosaurus* L. Trias. Aus. *Xilousuchus* ?M. Trias. EAs. **Erythrosuchidae** *?Arizonasaurus* M. Trias. NA. *Erythrosuchus* L. Trias. SAf. *?Cuyosuchus* L. Trias. SA. *Fugusuchus* L. Trias. EAs. *Garjainia* L. Trias. EEu. *?Kalisuchus* L. Trias. Aus. *Shansisuchus* L. Trias. EAs. *Vjushkovia* L. Trias. EEu. EAs. **?Proterochampsidae** *Cerritosaurus* [*?Rhadinosuchus*] *Chanaresuchus Gualosuchus* M. Trias. SA. *Proterochampsa* U. Trias. SA.

SUBORDER ORNITHOSUCHIA **Euparkeriidae** *Euparkeria* [*Browniella*] L. Trias. SAf. *?Halazaisuchus Turfanosuchus ?Wangisuchus* L. Trias. EAs. **Ornithosuchidae** *Ornithosuchus* [*Dasygnathus*] U. Trias. Eu. *Riojasuchus* U. Trias. SA. *Venaticosuchus* U. Trias. SA. **Lagosuchidae** *Lagosuchus Lagerpeton* M. Trias. SA.

SUBORDER RAUISUCHIA **Rauisuchidae** *Fasolasuchus* U. Trias. SA. *Heptasuchus* U. Trias. NA. *?Hoplitosaurus ?Procerosuchus Luperosuchus* M. Trias. SA. "*Mandasuchus*" *Stagonosuchus* M. Trias. EAf. *Prestosuchus* M. Trias. SA. *Rauisuchus* M. Trias. SA. *Saurosuchus* U. Trias. SA. *Ticinosuchus* M. Trias. Eu. *?Vjushkovisaurus* M. Trias. NA. **Poposauridae** *Poposaurus Postosuchus* U. Trias. NA. *?Teratosaurus* U. Trias. Eu. *?Sinosaurus* L. Jur. EAs.

SUBORDER AETOSAURIA **Stagonolepididae** *Aetosaurus ?Ebrachosaurus Stagonolepis* U. Trias. Eu. *Aetosauroides Argentinosuchus Neoaetosauroides* U. Trias. SA. *Desmatosuchus Stegomus Typothorax* U. Trias. NA.

SUBORDER INCERTAE SEDIS **Erpetosuchidae** *Erpetosuchus ?Dyoplax* U. Trias. Eu. *?Parringtonia* M. Trias. EAf. **Ctenosauriscidae** *Ctenosauriscus* L. Trias. Eu. *Hypselorhachis* M. Trias. SA. *Lotosaurus* M. Trias. EAs. **Gracilisuchidae** *Gracilisuchus ?Lewisuchus* M. Trias. SA. **Scleromochlidae** *Scleromochlus* U. Trias. Eu. Unnamed family *Megalancosaurus* U. Trias. Eu.

SUBORDER PHYTOSAURIA **Phytosauridae** *Angistorhinus* [*Brachysuchus*] U. Trias. NA. NAf. *Belodon Mystriosuchus* U. Trias. Eu. *Nicrosaurus* [*Heterodontosuchus Lophoprosopus Lophorhinus*] U. Trias. Eu. NA. SAs. *Parasuchus* [*Ebrachosuchus Francosuchus ?Mesorhinosuchus Paleorhinus ?Promystriosuchus*] U. Trias. Eu. SA. NA. Mad. *Rutiodon* [*Angistorhinopsis ?Clepsysaurus Leptosuchus Machaeroprosopus Metarhinus Pseudopalatus Rhytidodon*] U. Trias. Eu. NA. ?SAs.

THECODONTIA INCERTAE SEDIS *Clarencea* U. Trias. SAf. *Dongusia* M. Trias. EEu. *?Doswellia* U. Trias. NA. *Fenhosu-*

chus L. Trias. EAs. *Heleosaurus* U. Perm. SAf. *?Megalancosaurus* U. Trias. Eu. *?Microchampsa* L. Jur. EAs. "*Pallisteria*" "*Teleocrater*" M. Trias. EAf. *Strigosuchus* L. Jur. EAs.

ORDER CROCODYLIA

?SUBORDER TRIALESTIA **Trialestidae** *Trialestes* [*Triassolestes*] U. Trias. Eu.

SUBORDER SPHENOSUCHIA **Saltoposuchidae** *Saltoposuchus Terrestrisuchus* U. Trias. Eu. **Sphenosuchidae** *?Dibothrosuchus* L. Jur. EAs. *Hesperosuchus* U. Trias. NA. *Pseudhesperosuchus* U. Trias. SA. *Sphenosuchus* U. Trias. SAf.

SUBORDER PROTOSUCHIA **Platyognathidae** *Platyognathus* L. Jur. EAs. **Protosuchidae** *Baroqueosuchus* L. Jur. SAf. *Dianosuchus* L. Jur. EAs. *Eopneumatosuchus* L. Jur. NA. *Hemiprotosuchus* U. Trias. SA. *Lesothosuchus Notochampsa* [*Orthosuchus*] *Erythrochampsa Pedeticosaurus* L. Jur. SAf. *Protosuchus Stegomosuchus* L. Jur. NA.

SUBORDER HALLOPODA **Hallopidae** *Hallopus* U. Jur. NA.

SUBORDER MESOSUCHIA **Teleosauridae** *Aeolodon* [*Engyonimasaurus Glaphyrorhynchus*] U. Jur. Eu. *Gavialinum* M. Jur. Eu. *?Haematosaurus* U. Jur. Eu. *Machimosaurus* [*Madrimosaurus*] U. Jur. L. Cret. Eu. *Mycterosuchus* U. Jur. Eu. *Mystriosaurus Platysuchus* L. Jur. Eu. *Steneosaurus* L. Jur. SA. L.-U. Jur. Eu. L.-M. Jur. Mad. *Teleosaurus* M. Jur. Eu. **Metriorhynchidae** *Capellineosuchus Enaliosuchus* L. Cret. Eu. *Dakosaurus* U. Jur. L. Cret. Eu. *Geosaurus* [*Brachytaenius Cricosaurus Itlilimnosaurus Neustosaurus*] U. Jur. L. Cret. Eu. U. Jur. SA. *Metriorhynchus* [*Rhachaeosaurus Purranisaurus*] M.-U. Jur. Eu. M.-U. Jur. SA. *Pelagosaurus* L. Jur. Eu. *Teleidosaurus* M. Jur. Eu. **Pholidosauridae** *Anglosuchus Crocodilaemus* U. Jur. Eu. *?Meridiosaurus* L. Cret. SA. *Peipehsuchus* U. Jur. EAs. *?Petrosuchus Pholidosaurus* U. Jur. L. Cret. Eu. *Sarcosuchus* L. Cret. SA. NAf. *Suchosaurus* L. Cret. Eu. *Teleorhinus* [*Terminonaris*] L.-U. Cret. NA. ?Eu. **Atoposauridae** *Alligatorellus Alligatorium Atoposaurus* U. Jur. Eu. *Hoplosuchus* U. Jur. NA. *Karatausuchus* U. Jur. CAs. *Shantungosuchus* U. Jur. EAs. *Theriosuchus* U. Jur. Eu. **Goniopholididae** *?Coelosuchus ?Dakotasuchus* U. Cret. NA. *Goniopholis* [*Nannosuchus ?Amphicotylus ?Eutetrauranosuchus ?Diplosaurus*] U. Jur. L. Cret. Eu. U. Jur. EAs. ?U. Jur. U. Cret. ?NA. *?Itasuchus ?Microsuchus* U. Cret. SA. *Kansajsuchus* U. Jur. CAs. *Oweniasuchus* [*Brachydectes*] U. Jur. Eu. *?Pinacosuchus ?Pliogonodon ?Polydectes* U. Cret. NA. *?Sunosuchus* U. Jur. EAs. SEAs. *?Symptosuchus* U. Cret. SA. *Vectisuchus* L. Cret. Eu. **Dyrosauridae** *Atlantosuchus* Paleoc. NAf. *Dyrosaurus* Paleoc. NAf. *Hyposaurus* [*Congosaurus Sokotosaurus Wurnosaurus*] U. Cret. NA. SA. Paleoc. Eoc. NAf. WAf. *Phosphatosaurus* Eoc. NAf. *Rhabdognathus* [*Rhabdosaurus*] Eoc. NAf. WAf. *Sokotosuchus* U. Cret. WAf. *Tilemsisuchus* Eoc. NAf. **Paralligatoridae** *Paralligator* U. Cret. EAs. *Shamosuchus* L. Cret. EAs. **Hsisosuchidae** *?Doratodon* U. Cret. Eu. *Hsisosuchus* U. Cret. Eu. **Bernissartiidae** *Bernissartia* U. Jur. L. Cret. Eu. **Trematochampsidae** *?Baharijodon* U. Cret. NAf. *Trematochampsa* L. Cret. NAf. U. Cret. Mad. **Libycosuchidae** *Libycosuchus* L.-U. Cret. NAf. **Notosuchidae** *Notosuchus* U. Cret. SA. **Uruguaysuchidae** *Araripesuchus* L. Cret. SA. NAf. *?Peirosaurus* U. Cret. SA. *Uruguaysuchus* U. Cret. SA. **Baurusuchidae** *Baurusuchus Cynodontosuchus* U. Cret. SA. *?Bergisuchus Iberosuchus* Eoc. Eu. **Sebecidae** *Ilchunaia* Eoc. SA. *Sebecus* ?Paleoc. Eoc. Mioc. SA. **?Gobiosuchidae** *Gobiosuchus* U. Cret. EAs. **?Edentosuchidae** *Edentosuchus* L. Cret. EAs.

SUBORDER EOSUCHIA **?Hylaeochampsidae** *Hylaeochampsa* [*Heterosuchus*] L. Cret. Eu. **Stomatosuchidae** *?Aegyptosuchus Stomatosuchus* [*Stromerosuchus*] U. Cret. NAf. *?Chiayusuchus* U. Cret. EAs. **Dolichochampsidae** *Dolichochampsa* U. Cret. SA. **Gavialidae** *Eogavialis* Eoc. Olig. NAf. *Gavialis* Mioc. R. SAs. Pleist. EAs. *Gavialosuchus* Mioc. Eu. *Gryposuchus* [*Rhamphostopsis Rhamphostoma*] Olig. Plioc. SA. *Ikanogavialis Hesperogavialis* Mioc. SA. *Rhamphosuchus* Plioc. SA. **Alligatoridae** *Akanthosuchus* Paleoc. NA. *Alligator* [*Caimanoidea*] Olig. R. NA. R. EAs. *Allognathosuchus* [*Arambourgia Hassiacosuchus*] Paleoc. Eoc. NA. Eu. *Balanerodus* Olig. SA. *Caiman* [*Brachygnathosuchus Dinosuchus Proalligator Purrusaurus Xenosuchus*] Olig. R. SA. R. CA. *Ceratosuchus* Paleoc. NA. *Diplocynodon* [*Orthosaurus Boverisuchus Caimanosuchus Eocenosuchus*] Eoc. NA. Eoc. Plioc. Eu. *Eoalligator* ?Paleoc. EAs. *Hispanochampsa* Olig. Eu. *Melanosuchus* Plioc. R. SA. *Paleosuchus* Plioc. R. SA. R. CA. *Procaimanoidea* Eoc. NA. *?Prodiplocynodon* U. Cret. NA. *Wannangosuchus* Paleoc. NA. **Crocodylidae** *Aigialosuchus* U. Cret. Eu. *Allodaposuchus* U. Cret. Eu. *Asiatosuchus* Eoc. EAs. Eu. ?NA. *Brachyuranochampsa* Eoc. NA. *Charactosuchus* Mioc. SA. Eoc. EInd. *Crocodylus* [*Crocodilus Champsa Thecachampsa*] ?Eoc. Plioc. Eu. Paleoc. R. Af. NA. As. Pleist. R. Aus. EInd. CA. WInd. Mad. *Deinosuchus* [*Phobosuchus*] U. Cret. NA. *Dollosuchus* Eoc. Eu. *Eosuchus* Eoc. Eu. *Euthecodon* Mioc. Pleist. NAf. *Holopsisuchus* U. Cret. NA. *Kentisuchus* Eoc. Eu. *Leidyosuchus* U. Cret. Eoc. NA. *Lianghusuchus* Eoc. EAs. *Megadontosuchus* Eoc. Eu. *Mourasuchus* [*Nettosuchus*] Mioc. Plioc. SA. *?Navajosuchus* Paleoc. NA. *Necrosuchus* Paleoc. SA. *Orthogenysuchus* Eoc. NA. *Osteolaemus* R. Af. *Pallimnarchus* Pleist. Aus. *Planocrania* Paleoc. EAs. *Pristichampus* [*Limnosaurus Boverisuchus Weigeitisuchus*] Eoc. Eu. NA. SAs. *Quinkana* Pleist. Aus. *Thoracosaurus* [*Sphenosaurus*] U. Cret. NA. Eu. *?Tienosuchus* Eoc. EAs. *Tomistoma* Mioc. Eu. NA. Eoc. R. As.

ORDER PTEROSAURIA

SUBORDER RHAMPHORHYNCHOIDEA **Dimorphodontidae** *Dimorphodon* L. Jur. Eu. *Peteinosaurus* U. Trias. Eu. **Eudimorphodontidae** *Eudimorphodon* U. Trias. Eu. **Campylognathoididae** *Campylognathoides* L. Jur. Eu. **Rhamphorhynchidae** *Preondactylus* U. Trias. Eu. *Dorygnathus ?Comodactylus Nesodactylus* U. Jur. NA. *Parapsicephalus* U. Jur. Eu. *Rhamphorhynchus* [*?Odontorhynchus*] U. Jur. Eu. Af. *Scaphognathus* U. Jur. Eu. *Sordes* U. Jur. CAs.

SUBORDER PTERODACTYLOIDEA **Dsungaripteridae** *Dsungaripterus* ?U. Jur. EAf. L. Cret. EAs. *Putanipterus* L. Cret. SA. *Noripterus* L. Cret. EAs. **Ctenochasmatidae** *Ctenochasma* U. Jur. Eu. *Gnathosaurus* U. Jur. Eu. *Huanhepterus* U. Jur. EAs. **Pterodaustriidae** *Pterodaustro* ?L. Cret. SA. **Pterodactylidae** *?Dermodactylus* U. Jur. NA. *Germanodactylus Gallodactylus* U. Jur. Eu. *?Herbstosaurus* M. Jur. SA. *Pterodactylus* [*Cycnorhamphus Diopecephalus Ornithocephalus Ptenodracon*] U. Jur. Eu. ?EAs. **Ornithocheiridae**

Araripesaurus Araripedactylus Santanadactylus L. Cret. SA. *Criorhynchus* [*Coloborhynchus*] U. Cret. Eu. *Ornithocheirus* [*Cimoliornis Lonchodectes*] U. Cret. Eu. *Ornithodesmus* U. Cret. Eu. *Pteranodon* [*Nyctodactylus Nyctosaurus Occidentalia Geosternbergia*] U. Cret. NA. *Titanopteryx* U. Cret. SWAs. "*Quetzalcoatlus*" U. Cret. NA.

ORDER SAURISCHIA

SUBORDER STAURIKOSAURIA **Staurikosauridae** *Staurikosaurus* M.-U. Trias. SA. *?Spondylosoma* M. Trias. SA. **Herrerasauridae** *Herrerasaurus ?Ischisaurus* U. Trias. SA.

SUBORDER THEROPODA **Podokesauridae** *Avipes Halticosaurus* [*Dolichosuchus*] *Procompsognathus* [*?Pterospondylus*] *Saltopus Velocipes* U. Trias. Eu. *Coelophysis Podokesaurus* U. Trias. NA. *Dilophosaurus* L. Jur. NA. *?Lukousaurus* L. Jur. EAs. *Segisaurus* L. Jur. NA. *Syntarsus* L. Jur. SAf. **Coeluridae** *Aristosuchus* [*Calamospondylus Thecocoelurus Thecospondylus*] L. Cret. Eu. *Coelurus Ornitholestes* U. Jur. NA. *Inosaurus* L. Cret. NAf. *Kakuru* L. Cret. Aus. *Microvenator* L. Cret. NA. *?Sinocoelurus* U. Jur. EAs. *?Teinurosaurus* [*Caudocoelus*] U. Jur. Eu. **Shanshanosauridae** *Shanshanosaurus* U. Cret. EAs. **Compsognathidae** *Compsognathus* U. Jur. Eu. **Ornithomimidae** *Archaeornithomimus* L. Cret. NA. U. Cret. EAs. *?Betasuchus* U. Cret. Eu. *Dromiceiomimus Coelosaurus Ornithomimus Struthiomimus* U. Cret. NA. *Elaphrosaurus* U. Jur. EAf. ?NA. *Gallimimius* U. Cret. EAs. *Garudimimus* U. Cret. EAs. *Tugulusaurus* L. Cret. EAs. **Deinocheiridae** *Deinocheirus* U. Cret. EAs. **Therezinosauridae** *Therezinosaurus* U. Cret. EAs. **Elmisauridae** *Chirostenotes* [*Macrophalangia*] U. Cret. NA. *Elmisaurus* U. Cret. EAs. **Oviraptoridae** *Caenagnathus* U. Cret. NA. *Oviraptor Ingenia* U. Cret. EAs. **Dromaeosauridae** *Deinonychus* L. Cret. NA. *?Hulsanpes Velociraptor* U. Cret. EAs. *?Paronychodon Dromaeosaurus Saurornitholestes* U. Cret. NA. *Phaedrolosaurus* L. Cret. EAs. **Saurornithoididae** *Pectinodon Stenonychosaurus ?Troodon* U. Cret. NA. *Saurornithoides* U. Cret. EAs. **Megalosauridae** *Bahariasaurus Carcharodontosaurus* U. Cret. NAf. *?Chingkankousaurus* U. Cret. EAs. *?Embasasaurus* L. Cret. CAs. *Erectopus* L. Cret. Eu. ?U. Cret. NAf. *Eustreptospondylus* N. Jur. Eu. *Kelmayisaurus* L. Cret. EAs. *Majungasaurus* U. Cret. Mad. *Megalosaurus* [*Magnosaurus Sarcosaurus*] L.-U. Jur. Eu. NAf. ?EAf. *Metriacanthosaurus* U. Jur. Eu. *Poekilopleuron* M. Jur. Eu. *Szechuanosaurus* U. Jur. EAs. *Torvosaurus* U. Jur. NA. *Xuanhanosaurus* M. Jur. EAs. **Allosauridae** *Allosaurus* [*?Antrodemus*] U. Jur. NA. ?EAf. L. Cret. Aus. *Indosaurus* [*?Orthogoniosaurus*] U. Cret. SAs. *Piatnitzkysaurus* M. Jur. SA. *Piveteausaurus* U. Jur. Eu. *Yangchuanosaurus* U. Jur. EAs. **Spinosauridae** *Altispinax* L. Cret. Eu. *?Acrocanthosaurus* L. Cret. NA. *Spinosaurus* U. Cret. NAf. **Ceratosauridae** *Ceratosaurus* U. Jur. NA. ?EAf. *?Chienkosaurus* U. Jur. EAs. *?Proceratosaurus* M. Jur. Eu. **Dryptosauridae** *?Dryptosauroides* U. Cret. SAs. *Dryptosaurus* U. Cret. NA. **Tyrannosauridae** *Albertosaurus* [*Gorgosaurus*] *Daspletosaurus* U. Cret. NA. *Alioramus Alectrosaurus* U. Cret. EAs. *Genyodectes* U. Cret. SA. *Indosuchus* U. Cret. SAs. *Tyrannosaurus* [*Tarbosaurus*] U. Cret. NA. EAs.

THEROPODA INCERTAE SEDIS *Aublysodon* U. Cret. NA. *Avimimus* U. Cret. EAs. *Avisaurus* U. Cret. NA. ?SA. *Bradycneme Heptasteornis* U. Cret. Eu. *Chilantaisaurus* U. Cret. EAs. *Coeluroides Compsosuchus Jubbulpuria Laevisuchus Ornithomimoides* U. Cret. SAs. *Iliosuchus* M. Jur. Eu. *Itemirus* U. Cret. CAs. *Labocania* U. Cret. CA. *Marshosaurus Stokesosaurus* U. Jur. NA. *Noasaurus* U. Cret. SA. *Rapator Walgettosuchus* U. Cret. Aus.

SUBORDER SAUROPODOMORPHA

INFRAORDER PLATEOSAURIA **Anchisauridae** *Ammosaurus* L. Jur. NA. *Anchisaurus* [*Amphisaurus Gyposaurus Megadactylus Yaleosaurus*] L. Jur. NA. SAf. *Azendohsaurus* U. Trias. NAf. *Coloradia* U. Trias. SAf. *Euskelosaurus* [*Euskelesaurus Gigantoscelis Plateosauravus*] U. Trias. SAf. *Lugengosaurus* [*?Yunnanosaurus*] L. Jur. EAs. *Massospondylus* [*Aetonyx Aristosaurus Dromicosaurus Gryponyx Leptospondylus*] L. Jur. SAf. NA. *?Mussaurus* U. Trias. SA. *Plateosaurus* [*Dimodosaurus Gresslyosaurus Pachysauriscus Pachysaurus*] U. Trias. Eu. ?SA. *Sellosaurus* [*Efraasia*] U. Trias. Eu. **Melanorosauridae** *Camelotia* U. Trias. Eu. *Melanorosaurus* U. Trias. SAf. *Riojasaurus* [*Strenusaurus*] U. Trias. SA. *Vulcanodon* L. Jur. SAf. **Blikanasauridae** *Blikanasaurus* U. Trias. SAf.

INFRAORDER SAUROPODA **Cetiosauridae** *Amygdalodon* M. Jur. SA. *Barapasaurus* L. Jur. SAs. *Cetiosaurus* [*Cardiodon*] M.-U. Jur. Eu. *Haplocanthosaurus* U. Jur. NA. *Rhoetosaurus* L. Jur. Aus. *Shunosaurus* M. Jur. EAs. *Volkheimeria Patagosaurus* M. Jur. SA. *?Zizhongosaurus* ?L. Jur. EAs. **Diplodocidae** *Apatosaurus* [*Atlantosaurus Brontosaurus Elosaurus*] *Barosaurus* U. Jur. NA. EAf. *Cetiosauriscus* U. Jur. Eu. *Dicraeosaurus* U. Jur. EAf. *Diplodocus* U. Jur. NA. *Mamenchisaurus* U. Jur. EAs. *Nemegtosaurus Questosaurus* U. Cret. EAs. **Brachiosauridae** *?Astrodon* [*?Pleurocoelus*] L. Cret. NA. ?Eu. *?Austrosaurus* L. Cret. Aus. *Bothriospondylus* M.-U. Jur. Eu. Mad. *Brachiosaurus* U. Jur. NA. EAf. *Pelorosaurus* [*Dinodocus Gigantosaurus Oplosaurus Ornithopsis*] L. Cret. Eu. *Rebbachisaurus* L. Cret. NAf. **Titanosauridae** *Aegyptosaurus* U. Cret. NAf. *?Aepisaurus* L. Cret. Eu. *Alamosaurus* U. Cret. NA. *?Algoasaurus* L. Cret. SAf. *Antarctosaurus Argyrosaurus ?Campylodoniscus* U. Cret. SA. *?Chubutisaurus* L. Cret. SA. *Hypselosaurus* U. Cret. Eu. *Laplatasaurus* U. Cret. SA. Mad. *Loricosaurus* U. Cret. SA. *?Macrurosaurus* U. Cret. Eu. *Saltasaurus* U. Cret. SA. *Titanosaurus* [*Magyarosaurus*] ?L.-U. Cret. Eu. U. Cret. SAs. SA. *Tornieria* [*Gigantosaurus*] U. Jur. EAf. **Camarasauridae** *Camarasaurus* [*Uintasaurus Morosaurus*] U. Jur. NA. *Opisthocoelocaudia* U. Cret. EAs. **Euhelopodidae** *?Chiayusaurus* U. Cret. EAs. *Euhelopus* [*Helopus*] U. Jur. or L. Cret. EAs. *Omeisaurus* U. Jur. EAs. *Tienshanosaurus* ?U. Jur. EAs.

SAUROPODA INCERTAE SEDIS *Asiatosaurus* L. Cret. EAs. *Epanterias Dystrophaeus* U. Jur. NA. *Microcoelus* U. Cret. SA. *Mongolosaurus* L. Cret. EAs.

DINOSAURIA INCERTAE SEDIS **Segnosauridae** *Erlikosaurus Segnosaurus* U. Cret. EAs.

ORDER ORNITHISCHIA

SUBORDER ORNITHOPODA **Fabrosauridae** *Alocodon Trimucrodon* U. Jur. Eu. *Echinodon* L. Cret. Eu. *Fabrosaurus* [*Lesothosaurus*] L. Jur. SAf. *Gongbusaurus* U. Jur. EAs. *Tawasaurus* L. Jur. EAs. *Technosaurus* U. Trias. NA.

Heterodontosauridae *Abrictosaurus Geranosaurus Heterodontosaurus Lanasaurus Lycorhinus* L. Jur. SAf. *Dianchungosaurus* L. Jur. EAs. *?Pisanosaurus* U. Trias. SA. **Dryosauridae** *Dryosaurus* [*Dysalatosaurus*] U. Jur. NA. EAf. *?Kangnasaurus* ?Cret. SAf. *Valdosaurus* L. Cret. Eu. NAf. **Hypsilophodontidae** *?Fulgurotherium* L. Cret. Aus. *Hypsilophodon* ?U. Jur. L. Cret. Eu. ?NA. *Othnielia ?Nanosaurus* U. Jur. NA. *Parksosaurus* U. Cret. NA. *Phyllodon* U. Jur. Eu. *Tenontosaurus* L. Cret. NA. *Thescelosaurus* U. Cret. NA. *Zephyrosaurus* L. Cret. NA. **Iguanodontidae** *?Anoplosaurus Craspedodon* U. Cret. Eu. *Callovosaurus* U. Jur. Eu. *Camptosaurus* [*Camptonotus Cumnoria*] U. Jur. NA. Eu. L. Cret. NA. *Iguanodon* L. Cret. Eu. ?EAs. NA. NAf. *Muttaburrasaurus* L. Cret. Aus. *Ouranosaurus* L. Cret. NAf. *Probactrosaurus* L. Cret. EAs. *Rhabdodon* [*Mochlodon Oligosaurus*] U. Cret. Eu. *Vectisaurus* L. Cret. Eu. **Hadrosauridae** *Anatosaurus Brachylophosaurus Cheneosaurus Claosaurus Corythosaurus Edmontosaurus Hadrosaurus* [*Kritosaurus*] *Lambeosaurus Lophorhothon Maiasaura Parasaurolophus Prosaurolophus Saurolophus* U. Cret. NA. *Aralosaurus* U. Cret. SAs. *Bactrosaurus Barsboldia Gilmoreosaurus Jaxartosaurus Mandschurosaurus Nipponosaurus Shantungosaurus Tanius Tsintaosaurus* U. Cret. EAs. *?Notoceratops* U. Cret. SA. *Orthomerus* [*Telmatosaurus Limnosaurus*] U. Cret. Eu. *Secernosaurus* U. Cret. SA.

SUBORDER PACHYCEPHALOSAURIA **Pachycephalosauridae** *Gravitholus Ornatotholus Pachycephalosaurus Stegoceras Stygimoloch* U. Cret. NA. *Majungatholus* U. Cret. Mad. *Prenocephale Tylocephale* U. Cret. EAs. *Yaverlandia* L. Cret. Eu. **Homalocephalidae** *Goyocephale Homalocephale ?Micropachycephalosaurus Wannanosaurus* U. Cret. EAs.

SUBORDER STEGOSAURIA **?Scelidosauridae** *?Lusitanosaurus Scelidosaurus* L. Jur. Eu. *?Scutellosaurus* L. Jur. NA. **Stegosauridae** *Chialingosaurus Chungkingosaurus Tuojiangosaurus* U. Jur. EAs. *Craterosaurus* L. Cret. Eu. *Dacentrurus* [*Omosaurus*] M.-U. Jur. Eu. *Dravidosaurus* U. Cret. EAs. *Huayangosaurus* M. Jur. EAs. *Kentrosaurus* [*Kentrurosaurus Doryphorosaurus*] U. Jur. EAf. *Lexovisaurus* U. Jur. Eu. *Paranthodon* L. Cret. SAf. *Stegosaurus* [*Diracodon Hypsirophus*] U. Jur. NA. *Wuerhosaurus* L. Cret. EAs.

SUBORDER ANKYLOSAURIA **Nodosauridae** *Acanthopholis* U. Cret. Eu. *Brachypodosaurus* U. Cret. SAs. *?Cryptodraco* U. Jur. Eu. *?Dracopelta* U. Jur. Eu. *Hylaeosaurus* [*Polacanthoides Polacanthus*] L. Cret. Eu. *Nodosaurus* [*Hierosaurus Stegopelta*] *Palaeoscinus Panoplosaurus* [*Edmontonia*] U. Cret. NA. *?Priodontognathus* ?U. Jur. Eu. *Hoplitosaurus Sauropelta Silvisaurus* L. Cret. NA. *Sarcolestes* U. Jur. Eu. *Struthiosaurus* [*Crataeomus Danubiosaurus Leipsanosaurus Pleuropeltus Onychosaurus*] U. Cret. Eu. **Ankylosauridae** *Amtosaurus Pinacosaurus* [*Syrmosaurus*] *Saichania Talarurus Tarchia* U. Cret. EAs. *Ankylosaurus Euoplocephalus* [*Anodontosaurus Dyoplosaurus Scolosaurus*] U. Cret. NA. *?Heishansaurus ?Sauroplites ?Stegosaurides* U. Cret. EAs.

SUBORDER CERATOPSIA **Psittacosauridae** *Psittacosaurus* [*Protiguanodon*] L. Cret. EAs. **Protoceratopsidae** *Bagaceratops Microceratops Protoceratops* U. Cret. EAs. *Leptoceratops Montanoceratops* U. Cret. NA. **Ceratopsidae** *Anchiceratops Arrhinoceratops Centrosaurus Chasmosaurus Eoceratops Monoclonius* [*Brachyceratops*] *Pachyrhinosaurus Pentaceratops Styracosaurus Torosaurus Triceratops* U. Cret. NA.

CERATOPSIA INCERTAE SEDIS *Stenopelix* L. Cret. Eu.

DIAPSIDA INCERTAE SEDIS

ORDER PLACODONTIA **?Helveticosauridae** *Helveticosaurus* M. Trias. Eu. **Placodontidae** *Paraplacodus* M. Trias. Eu. *Placodus* [*Anomosaurus Crurosaurus*] L.-M. Trias. Eu. M. Trias. SWAs. **Cyamodontidae** *Cyamodus* M. Trias. Eu. *Placochelys* [*Placochelyanus*] U. Trias. Eu. *Psephoderma* U. Trias. Eu. *Psephosaurus* ?M. Trias. ?SWAs. U. Trias. Eu. *Saurosphargis* M. Trias. Eu. **Henodontidae** *Henodus* U. Trias. Eu. **Incertae sedis** *Chelyoposuchus* U. Trias. Eu.

ORDER OR SUBCLASS ICHTHYOPTERYGIA (ICHTHYOSAURIA)

?Hupehsuchidae *Nanchangasaurus* [*Hupehsuchus*] M. Trias. EAs. **Utatsusauridae** *Utatsusaurus* L. Trias. EAs. **Omphalosauridae** *Chaosaurus* L. Trias. EAs. *Omphalosaurus* [*Pessopteryx*] M. Trias. Spits. *Grippia* L. Trias. Spits. *Phalarodon* M. Trias. NA. **Mixosauridae** *Mixosaurus* M. Trias. Eu. ?EInd. Spits. SWAs. **Shastasauridae** *Cymbospondylus* M. Trias. NA. *Delphinosaurus* U. Trias. NA. *Himalayosaurus* U. Trias. EAs. *Merriamia* U. Trias. NA. *?Pessosaurus* M. Trias. Spits. *Shastasaurus* U. Trias. NA. *Shonisaurus* U. Trias. NA. *Toretocnemus* [*Californosaurus*] U. Trias. NA. **Ichthyosauridae** *Baptanodon* U. Jur. NA. *Ichthyosaurus* [*Proteosaurus Eurypterygius*] L. Jur. Eu. Gr. *Ophthalmosaurus* U. Jur. Eu. SA. **Stenopterygiidae** *Stenopterygius* L. Jur. Eu. **Protoichthyosauridae** *Protoichthyosaurus* L. Jur. Eu. **Leptopterygiidae** *Eurhinosaurus* L. Jur. Eu. *Grendelius* U. Jur. Eu. *Leptopterygius* L. Jur. Eu. *Platypterygius* [*Myopterygius Myobradypterygius*] U. Cret. NA. U. Cret. EEu. L.-U. Cret. Aus. L. Cret. Eu. *?Nannopterygius* U. Jur. Eu. *Temnodontosaurus* L. Jur. Eu.

SUBCLASS SYNAPSIDA

ORDER PELYCOSAURIA **Ophiacodontidae** *Archaeothyris* M. Penn. NA. *Baldwinonus* L. Perm. NA. *Clepsydrops* [*Archaeobelus*] U. Penn. NA. *Ophiacodon* [*Arribasaurus Theropleura Poliosaurus Winfieldia Therosaurus Diopeus*] U. Penn. L. Perm. NA. *?Protoclepsydrops* L. Penn. NA. *Stereophallodon* L. Perm. NA. *Stereorhachis* U. Penn. Eu. **Varanopseidae** *Aerosaurus ?Basicranodon Mycterosaurus* [*Eumatthevia*] *Varanodon Varanops* L. Perm. NA. **Eothyrididae** *Eothyris Oedaleops* L. Perm. NA. **Sphenacodontidae** *?Bathygnathus Ctenospondylus Dimetrodon* [*Bathyglyptus Embolophorus Theropleura*] L. Perm. NA. *Haptodus* [*Callibrachion Cutleria Datheosaurus Palaeohatteria Palaeosphenodon Pantelosaurus*] U. Penn. L. Perm. NA. L. Perm. Eu. *Macromerion* U. Penn. Eu. *Neosaurus* L. Perm. Eu. *Secodontosaurus Sphenacodon* [*Oxyodon*] L. Perm. NA. **Edaphosauridae** *Edaphosaurus* [*Brachycnemius Naosaurus*] U. Penn. L. Perm. NA. Eu. *Ianthosaurus* U. Penn. NA. **Caseidae** *Angelosaurus* U. Perm. NA. *Casea* L. Perm. NA. Eu. *Caseopsis Cotylorhynchus* U. Perm. NA. *Ennatosaurus* U. Perm. EEu. *?Caseoides* U. Perm. NA. *?Phreatophasma* U. Perm. Eu.

PELYCOSAURIA INCERTAE SEDIS *Colobomycter Delorhynchus* L. Perm. NA. *Echinerpeton* M. Penn. NA. *Elliotsmithia* U. Perm. SAf. *Glaucosaurus Lupeosaurus* L. Perm. NA. *Milosaurus* U. Penn. NA. *Nitosaurus Scoliomus Tetraceratops Thrausmosaurus Trichasaurus Varanosaurus* L. Perm. NA. *Xyrospondylus* U. Penn. NA.

ORDER THERAPSIDA

SUBORDER EOTITANOSUCHIA **Biarmosuchidae** *Biarmosuchus* [*Biarmosaurus*] U. Perm. EEu. **Eotitanosuchidae** *Eotitanosuchus Ivantosaurus* U. Perm. EEu. **Phthinosuchidae** *Phthinosuchus Phthinosaurus* U. Perm. EEu. **Incertae sedis** *Gorgodon Knoxosaurus Steppesaurus Watongia* U. Perm. NA.

SUBORDER DINOCEPHALIA

INFRAORDER TITANOSUCHIA **Brithopodidae** *Admetophoneus Archaeosyodon Brithopus Chthomaloporus Doliosauriscus* [*Doliosaurus*] *Mnemeiosaurus Notosyodon Syodon Titanophoneus* U. Perm. EEu. *Eosyodon* U. Perm. NA. **Deuterosauridae** *Deuterosaurus* U. Perm. EEU. **Estemmenosuchidae** *Anoplosuchus Estemmenosuchus Molybdopygus Parabradysaurus Rhopalodon Zopherosuchus* U. Perm. EEu. **Anterosauridae** *Anteosaurus* [*Broomosuchus Dinosuchus Titanognathus Pseudanteosaurus*] *Micranteosaurus Paranteosaurus* U. Perm. SAf. **Titanosuchidae** *Jonkeria* [*Dinophoneus*] *Titanosuchus* [*Scapanodon Parascapanodon*] U. Perm. SAf.

INFRAORDER TAPINOCEPHALIA **Tapinocephalidae** *Moschops* [*Agnosaurus Moschognathus Moschoides Pnigalion Ulemosaurus*] U. Perm. SAf. EEu. *Avenantia Delphinognathus Keratocephalus Mormosaurus Moschosaurus Phocosaurus Riebeeckosaurus Struthiocephaloides Struthiocephalus Struthionops Styracocephalus Tapinocephalus Taurocephalus* U. Perm. SAf. **?Incertae sedis** *Dimacrodon Driveria Mastersonia* U. Perm. NA.

SUBORDER DICYNODONTIA

INFRAORDER VENJUKOVIAMORPHA **Venjukoviidae** *Otsheria Venjukovia* [*Myctosuchus*] U. Perm. EEu.

INFRAORDER DROMASAURIA **Galeopsidae** *Galechirus Galeops Galepus* U. Perm. SAf.

INFRAORDER EODICYNODONTIA **Eodicynodontidae** *Eodicynodon* U. Perm. SAf.

INFRAORDER ENDOTHIODONTIA **Endothiodontidae** *Chelyodontops Endothiodon* [*Esoterodon Emydochampsa Endogomphodon*] *Pachytegos* U. Perm. SAf.

INFRAORDER PRISTERODONTIA **Aulacocephalodontidae** *Aulacephalodon* [*Aulacocephalodon*] *?Digalodon Pelanomodon* U. Perm. SAf. **Dicynodontidae** *Dicynodon* [*Dinanomodon Daptocephalus*] U. Perm. SAf. EAf. EEu. ?SEAs. EAs. *Geikia Gordonia* U. Perm. Eu. **Kannemeyeriidae** *Angonisaurus* M. Trias. EAf. *Barysoma Chanaria Dinodontosaurus Dolichuranus* [*Rhopalorhinus*] M. Trias. SWAf. *?Elephantosaurus* M. Trias. EEu. *Ischigualastia* U. Trias. SA. *Kannemeyeria Rechnisaurus* [*Proplacerias*] L. Trias. SAf. M. Trias. EAf. SAs. SA. *Parakannemeyeria Sinokannemeyeria* L. Trias. EAs. *Placerias* U. Trias. NA. *Rhinocerocephalus ?Rhadiodromus Rhinodicynodon* M. Trias. EEu. *Sangusaurus Zambiasaurus* M. Trias. EAf. *Shansiodon* L. Trias. EAs. *Stahleckeria* M.-U. Trias. SA. *Tetragonias* M. Trias. EAf. *Uralokannemeyeria Rabidosaurus* M. Trias. EEu. *Vinceria Jacheleria* L. Trias. SA. *Wadiasaurus* M. Trias. SAs. **Lystrosauridae** *Lystrosaurus* [*Prolystrosaurus*] L. Trias. SAf. SAs. EEu. EAs. ?SEAs. Ant. **Oudenodontidae** *Cteniosaurus Oudenodon Rhachiocephalus* [*Eocyclops Kitchingia Megacyclops Neomegacyclops Odontocyclops Pelorocyclops Platycyclops*] *Tropidostoma* U. Perm. SAf. **Pristerodontidae** *Emyduranus Eurychororhinus Pristerodon Storthyggnathus Synostocephalus* U. Perm. SAf.

INFRAORDER DIICTODONTIA **Emydopidae** *Emydops* [*Emydopsoides*] *Myosauroides Palemydops* U. Perm. SAf. *Myosaurus* L. Trias. SAf. Ant. **Cistecephalidae** *Cistecephaloides Cistecephalus* [*Kistecephalus*] U. Perm. SAf. *Kawingasaurus* U. Perm. EAf. **Robertiidae** *Robertia* U. Perm. SAf. **Diictodontidae** *?Anomodon* U. Perm. SAf. *Diictodon* U. Perm. SAf. EAs.

INFRAORDER KINGORIAMORPHA **Kingoriidae** *Kingoria* [*?Dicynodontoides*] U. Perm. EAf. SAf. *Kombusia* L. Trias. SAf.

DICYNODONTIA INCERTAE SEDIS *Brachyprosopus Brachyuraniscus Broilius Cerataelurus Compsodon Cryptocynodon Emydorhinus Eosimops Eumantellia Haughtoniana Heuneus Koupia Newtonella Parringtoniella Prodicynodon Taognathus* U. Perm. SAf.

SUBORDER GORGONOPSIA **?Ictidorhinidae** *Ictidorhinus Lycaenodon Lemurosaurus Rubidgina* U. Perm. SAf. **?Hipposauridae** *Hipposaurus ?Pseudhipposaurus* U. Perm. SAf. **?Burnetiidae** *Burnetia* U. Perm. SAf. *Proburnetia* U. Perm. EEu. **Gorgonopsidae** *Aelurognathus* [*Dixeya*] *Aelurosaurus* [*Aelurosauroides Aelurosauropsis*] *Aloposaurus* [*Aloposauroides*] *Arctognathus* [*Lycosaurus Lycaenodontoides Arctognathoides*] *Arctops* [*?Pardocephalus Smilesaurus*] *Broomicephalus Broomisaurus Cephalicustroidus Cerdorhinus* [*?Galerhynchus*] *Clelandina* [*Dracocephalus*] *Cyanosaurus* [*Cyniscopoides*] *Dinogorgon Eoarctops Galesuchus Gorgonops* [*Pachyrhinos Gorgonognathus Leptotrachelus Chiwetasaurus*] *Leontocephalus Lycaenops* [*Tigricephalus ?Tangagorgon*] *Paragalerhinus Prorubidgea Rubidgea Scylacognathus Scylacops Sycosaurus* U. Perm. SAf. *Inostrancevia Pravoslavleria Sauroctonus* U. Perm. EEu.

SUBORDER THEROCEPHALIA **Crapartinellidae** *Crapartinella* U. Perm. SAf. **Pristerognathidae** *Pristerognathus* U. Perm. SAf. **Hofmeyriidae** *Hofmeyria Ictidostoma* [*Ictidognathus*] U. Perm. SAf. **Lycideopsidae** *Lycideops* U. Perm. SAf. **Ictidosuchidae** *Ictidosuchus Ictidosuchoides Ictidosuchops* U. Perm. SAf. *Silphoictidoides* U. Perm. EAf. *Regisaurus* L. Trias. SAf. *?Nanictops* U. Perm. SAf. **Whaitsiidae** *Moschowhaitsia* U. Perm. EEu. *Theriognathus* [*Alopecopsis Aneugomphius Hyenosaurus Notaelurops Notosollasia Whaitsia*] U. Perm. SAf. EAf. **Moschorhinidae** *Annatherapsidus* [*Anna*] *Chthonosaurus* U. Perm. EEu. *Akidnognathus Euchambersia* U. Perm. SAf. *Moschorhinus* [*?Tigisuchus Cerdops Hewittia*] U. Perm. L. Trias. SAf. *Promoschorhynchus* U. Perm. SAf. **Ericiolacertidae** *Ericio-*

lacerta L. Trias. SAf. **Scaloposauridae** *Nanicticephalus Scaloposaurus* U. Perm. SAf. *Tetracynodon* [*?Homodontosaurus*] U. Perm. L. Trias. SAf. *Scalopolacerta ?Zorillodontops* L. Trias. SAf. **Simorhinellidae** *Simorhinella* U. Perm. SAf. **Bauridae** *Bauria* [*Baurioides Melinodon Microgomphodon Sesamodon Watsoniella*] L. Trias. SAf.

SUBORDER CYNODONTIA

INFRAORDER PROCYONSUCHIA **Procynosuchidae** *Parathrinaxodon* U. Perm. EAf. *Procynosuchus* [*Galecranium Galeophrys Leavachia Mygalesaurus Nanictosuchus Protocynodon*] U. Perm. SAf. **Dviniidae** *Dvinia* [*Permocynodon*] U. Perm. EEu. **Galesauridae** *Cromptodon* L. Trias. SA. *Galesaurus Cyanosaurus* [*Cynosuchus*] U. Perm. SAf. *Thrinaxodon* [*?Nythosaurus Notictosaurus*] L. Trias. SAf. Ant. *?Nanocynodon* U. Perm. EEu. *Platycraniellus* [*Platycranion*] *Tribolodon* L. Trias. SAf.

INFRAORDER EUCYNODONTIA

SUPERFAMILY CYNOGNATHOIDEA **Cynognathidae** *Cynognathus* [*Cynidiognathus Lycaenognathus Lycochampsa ?Karoomys*] L. Trias. SAf. SA.

SUPERFAMILY TRITYLODONTOIDEA **Diademodontidae** *Diademodon* [*Cragievanus ?Cynochampsa Gomphognathus Protacmon Sysphinctostoma*] L. Trias. SAf. *?Ordosiodon* L. Trias. EAs. *Titanogomphodon* M. Trias. SWAf. **Trirachodontidae** *Cricodon* M. Trias. EAf. *?Sinognathus* L. Trias. EAs. *Trirachodon* [*Triachodontoides*] L. Trias. SAf. **Traversodontidae** *Andescynodon Pascualgnathus Rusconiodon* L. Trias. SA. *Colbertosaurus* M. Trias. SA. *Exaeretodon* [*Proexaeretodon Theropsis*] M.-U. Trias. SA. U. Triassic. SAs. *Ischignathus* U. Trias. SA. *Luangwa* M. Trias. SAf. *Massetognathus* [*Megagomphodon*] M. Trias. SA. *Scalenodon* M. Trias. EAf. *Scalenodontoides* U. Trias. SAf. *Traversodon Gomphodontosuchus* M.-U. Trias. SA. **Tritylodontidae** *Bienotherium* L. Jur. EAs. *Bienotheroides* M. Jur. EAs. *Bocatherium* ?M. Jur. CA. *Dinnebitodon* L. Jur. NA. *Kayentatherium* [*Nearctylodon*] L. Jur. NA. *Lufengia* L. Jur. EAs. *Oligokyphus* [*?Chalepotherium Mucrotherium Uniserium*] L. Jur. Eu. NA. *Stereognathus* M. Jur. Eu. *Tritylodon* [*?Triglyphus Likhoelia Tritylodontoideus*] L. Jur. SAf. ?Eu. *Yunnanodon* L. Jur. EAs.

SUPERFAMILY CHINIQUODONTOIDEA **Chiniquodontidae** *?Aleodon* M. Trias. EAf. *Belesodon Chiniquodon* M.-U. Trias. SA. *?Dromatherium ?Microconodon* U. Trias. NA. *Probainognathus Probelesodon* M. Trias. SA. *?Pseudotriconodon* U. Trias. Eu. *?Therioherpeton* M. Trias. SA. **Tritheledontidae** *Chalimia* U. Trias. SA. *Diarthrognathus Tritheledon* L. Jur. SAf. *Pachygenelus* L. Jur. SAf. NA. **Incertae sedis** *Tricuspes* L. Jur. Eu. *Eoraetia* U. Trias. Eu. *Kunminia* L. Jur. EAs.

AVES

This list of the genera of fossil birds is taken primarily from S. L. Olson (1985) and emphasises first occurrences. This listing is by no means complete. Many more genera are cited by Brodkorb (1978, and references cited therein), but the taxonomic position of many of these genera is now subject to question.

CLASS AVES

SUBCLASS ARCHAEORNITHES

ORDER ARCHAEOPTERYGIFORMES **Archaeopterygidae** *Archaeopteryx* [*Archaeornis Griphosaurus Gryphornis Juravis*] U. Jur. Eu.

SUBCLASS NEORNITHES

Ambiortidae *Ambiortus* L. Cret. EAs.

SUPERORDER ODONTOGNATHAE

ORDER HESPERORNITHIFORMES **Enaliornithidae** *Enaliornis* L. Cret. Eu. **Baptornithidae** *Baptornis* U. Cret. NA. *Neogaeornis* U. Cret. SA. **Hesperornithidae** *Hesperornis Parahesperornis ?Coniornis* U. Cret. NA. ?SA.

ORDER ICHTHYORNITHIFORMES **Ichthyornithidae** *Ichthyornis* [*Plegadornis*] U. Cret. NA.

SUPERORDER INCERTAE SEDIS

ORDER GOBIPTERYGIFORMES **Gobipterygidae** *Gobipteryx* U. Cret. EAs.

ORDER ENANTIORNITHIFORMES **Enantiornithidae** *?Alexornis* U. Cret. NA. *Enantiornis* U. Cret. SA. **?Zhyraornithidae** *Zhyraornis* U. Cret. As.

SUPERORDER PALAEOGNATHAE

ORDER UNNAMED **Lithornidae** *Lithornis Paracathartes* Paleoc. Eoc. NA. Eu. *Promusophaga* L. Eoc. Eu.

ORDER TINAMIFORMES **Tinamidae** *Eudromia* [*Roveretornis Tinamisornis*] *Nothoprocta* [*Cayetornis*] *Nothura Querandiornis* U. Plioc. R. SA. *Crypturellus Nothoprocta Tinamus* Pleist. R. SA.

ORDER STRUTHIONIFORMES **Struthionidae** *Palaeotis* M. Eoc. Eu. *Struthio* [*Megaloscelornis Pachystruthio Palaeostruthio Struthiolithus*] U. Mioc. Plioc. Pleist. EEu. As. NAf. R. WAs. NAf. Plioc. Pleist. R. SAf.

ORDER RHEIFORMES **Opisthodactylidae** *Diogenornis* U. Paleoc. SA. *Opisthodactylus* L. Mioc. SA. **Rheidae** *Heterorhea* U. Plioc. SA.

ORDER CASUARIIFORMES **Casuariidae** *Casuarius* Plioc. New Guinea R. Aus. **Dromaiidae** *Dromaius* [*Dromiceius*] Mioc. Plioc. R. Aus. **Dromornithidae** R. Aus.

ORDER AEPYORNITHIFORMES **Aepyornithidae** *Aepyornis* Pleist. R. Mad.

ORDER DINORNITHIFORMES **Dinornithidae** *Anomalpterys Dinornis Emeus Euryapteryx Megalapteryx Pachyornis* Pleist. R. NZ.

ORDER APTERYGIFORMES **Apterygidae** *Apteryx* Pleist. R. NZ.

SUPERORDER NEOGNATHAE

ORDER CUCULIFORMES **Opisthocomidae** *Hoazinoides* U. Mioc. SA. **Musophagidae** *Crinifer* Olig. NAf. R. Af. *Musophaga* [*Apopempsis*] U. Mioc. Eu. L. Mioc. EAf. **Cuculidae** *Cur-*

soricoccyx L. Mioc. NA. *Dynamopteryx* Eoc. Olig. Eu. *Neococcyx* L. Olig. NA.

ORDER FALCONIFORMES **Falconidae** *Badrostes* L. Mioc. SA. *Falco* L. Mioc. NA. U. Pleist. Aus. R. Cos. *Polyborus* Pleist. R. SA. R. NA. **Sagittariidae** *Pelargopappus* M.-U. Olig. L. Mioc. Eu. **Accipitridae** *Aquilavus* U. Eoc. L. Olig. L. Mioc. Eu. *Buteo* M.-U. Olig. R. Cos. *Palaeohierax* L. Mioc. Eu. *Neophrontops* M. Mioc. NA. **Pandionidae** *Pandion* M.-U. Mioc. Plioc. NA. R. Cos.

ORDER GALLIFORMES **Cracidae** *Boreortalis* L. Mioc. NA. *Gallinuloides* L. Eoc. NA. *Procrax* L. Olig. NA. **Megapodiidae** *Progura* [*Palaeopelargus*] U. Eoc. Eu. Quat. Aus. *?Sylviornis* Quat. EInd. **Numididae** *Telecrex* U. Eoc. EAs. **Phasianidae** *Meleagris* [*Agriocharis Parapavo*] U. Mioc. U. Plioc. Pleist. R. NA. *Miophasianus* M. Mioc. Eu. *Miortyx* M. Olig. L. Mioc. NA. *Palaeocryptonyx* L.-U. Mioc. Eu. *Palaeortyx* U. Olig. L.-U. Mioc. Eu. *Rhegminornis* L. Mioc. NA. **?Turnicidae** *Turnix* L. Plioc. R. SAf.

ORDER COLUMBIFORMES **Pteroclidae** R. Eu. As. Af. *Pterocles* U. Eoc. L. Mioc. R. Eu. R. As. *Syrrhaptes* M. Plioc. EAs. Pleist. R. As. **Columbidae** *Columba* U. Mioc. R. Eu. R. Cos. *Pezophaps* Pleist. Rodriguez *Raphus* [*Didus*] R. Maurit.

ORDER PSITTACIFORMES **Psittacidae** *Conuropsis* L. Mioc. R. NA. *Palaeopsittacus* L. Eoc. Eu. *Psittacus* [*Archaeopsittacus*] L.-M. Mioc. Eu. R. Af.

ORDER INCERTAE SEDIS **Zygodactylidae** *Zygodactylus* L. Mioc. Eu.

ORDER COLIIFORMES **Coliidae** U. Eoc. Eu. R. Af. *Colius* [*Necrornis*] L.-U. Mioc. Eu. R. Af. *Limnatornis* L. Mioc. Eu.

ORDER CORACIIFORMES (INCLUDING TROGONIFORMES AND GALBULAE)

SUBORDER INCERTAE SEDIS **?Halcyornithidae** *Halcyornis* L. Eoc. Eu.

SUBORDER CORACII **Atelornithidae Leptosomidae** R. Mad. **Galbulidae Bucconidae** R. SA. **Coraciidae** *Geranopterus* U. Eoc. Olig. Eu. *Eurystomus* ?L. Eoc. R. NA. **Primobucconidae** *Primobucco Neanis* L. Eoc. NA. *Uintornis Botauroides Eobucco* M. Eoc. NA.

SUBORDER HALCYONES (ALCEDINI) **Alcedinidae** Eoc. Olig. R. Eu. R. Cos. **Meropidae** Eoc. NA. Eoc.-R. Eu. **Todidae** R. WInd. *Palaeotodus* M. Olig. NA. **Momotidae** R. WInd. *Protornis* L. Olig. Eu. **Trogonidae** L. Olig. R. As. Af. NA. WInd. SA. *Paratrogon* L. Mioc. Eu. *Trogon* Pleist. R. SA. R. NA. **Archaeotrogonidae** *Archaeotrogon* U. Eoc. U. Olig. Eu.

ORDER STRIGIFORMES **Ogygoptyngidae** *Ogygoptynx* U. Paleoc. NA. **Protostrigidae** *Minerva* [*Aquila Protostrix*] M.-U. Eoc. NA. *Eostrix* L.-M. Eoc. NA. ?Eu. **Strigidae** R. Cos. *Strix* L. Mioc. Eu. ?NA. *Ornimegalonyx* Pleist. Cuba *Otus* L. Mioc. Eu. **Tytonidae** R. Cos. *Prosybris* L. Mioc. Eu. *Tyto* M.-L. Mioc. Plioc. L. Pleist. Eu. Quat. WInd.

ORDER CAPRIMULGIFORMES **Aegothelidae** Eoc. Olig. Eu. R. Aus. SPac. *Quipollornis* L.-M. Mioc. Aus. *Megaegotheles* Pleist. Sub-Recent NZ. **Podargidae** U. Eoc. Eu. R. SAs. Aus. **Steatornithidae** L. Eoc. NA. ?U. Olig. Eu. R. SA. **Caprimulgidae** Eoc. Olig. Eu. R. Cos.

ORDER APODIFORMES

SUBORDER APODI **Aegialornithidae** *Aegialornis* U. Eoc. Eu. *Cypselavus* U. Eoc. Olig. Eu. *Primapus* [*Procuculus*] L. Eoc. Eu. ?L. Eoc. NA. **Apodidae** *Apus* M.-U. Mioc. Eu. R. Cos. *Cypseloides* L. Mioc. Eu. R. SA.

SUBORDER TROCHILI **Trochilidae** R. NA. WInd. SA.

ORDER BUCEROTIFORMES **Bucerotidae** R. As. Af. SPac. *Bucorvus* M. Mioc. NA. R. Af. **Upupidae** Eoc. Olig. Eu. R. Eu. As. Af. Mad. *Upupa* Quat. SAtl. **Phoeniculidae** L.-U. Mioc. Eu. R. Af.

ORDER PICIFORMES **Indicatoridae** L. Plioc. SA. R. As. Af. **Capitonidae** L. Mioc. NA. Eu. R. As. Af. CA. SA. *Capitonides* L.-M. Mioc. Eu. **Picidae** M. Mioc. NA. R. Eu. As. Af. NA. SA. *Campephilus* U. Plioc. R. NA. R. SA. CA. *Pliopicus* L. Plioc. NA.

ORDER PASSERIFORMES **?Palaeoscinidae** *Palaeoscinis* M.-U. Mioc. NA. **Alaudidae** L. Mioc. Eu. R. Cos. **Corvidae** *Corvus* M. Mioc. Eu. R. Cos. *Miocitta* U. Mioc. NA. **Sittidae** *Sitta* M.-U. Mioc. Eu. R. Cos. (except SA.) **Fringillidae** *Ammodramus* U. Mioc. R. NA. R. Eu. As. Af. WInd. SA. *Passerina* L. Plioc. R. NA. R. CA. SA. **Eurylaimidae** L. Mioc. R. Af. As. A more complete record of late Cenozoic passerines can be found in Brodkorb (1978).

ORDER GRUIFORMES

SUBORDER CARIAMAE **Cariamidae** L. Olig. R. SA. *Chunga* U. Plioc. SA. *?Riacama* L. Olig. SA. **?Cunampaiidae** *Cunampaia* U. Eoc. SA. **Phorusrhacidae** *Ameginornis* Eoc.-Olig. Eu. *Andalgalornis* L.-M. Plioc. SA. *Andrewsornis* L. Olig. SA. *Devincenzia* ?U. Mioc. SA. *Onactornis* L.-?M. Plioc. SA. *Phororhacos* L.-M. Mioc. SA. *Physornis* L. Olig. SA. *Psilopterus* L. Plioc. SA. *Titanis* U. Plioc. NA. *Titanornis* Pleist. NA. *Tomodus* [*?Palaeocicomia*] M. Mioc. SA. **Bathornithidae** *Bathornis* [*Neocathartes*] U. Eoc. L.-U. Olig. L. Mioc. NA. *?Eutreptornis* U. Eoc. NA. *Paracra* Olig. NA. **Idiornithidae** *Elaphrocnemus* [*Filholornis*] U. Eoc. L. Olig. Eu. *?Gypsornis* U. Eoc. Eu. *Idiornis* U. Eoc. L. Olig. Eu. *Oblitavis* U. Eoc. L. Olig. Eu. *Occitaniavis* U. Eoc. L. Olig. Eu. *Propelargus* L. Olig. Eu.

SUBORDER GRUES **Geranoididae** *Eogeranoides ?Geranodornis Geranoides Palaeophasianus Paragrus* M. Eoc. NA. **Eogruidae** *Eogrus* U. Eoc. ?U. Mioc. CAs. *Sonogrus* L. Olig. CAs. **Ergilornithidae** *Amphipelargus* U. Mioc. M. Plioc. Eu. As. *Ergilornis* [*Proergilornis*] L. Olig. EAs. **Eleutherornithidae** *Eleutherornis* M. Eoc. Eu. **Gruidae** NA. CA. Ind. As. Aus. Af. *Aramornis* [*Probalearica*] L. Mioc. NA. ?U. Mioc. EEu. *Geranopsis* U. Eoc. Eu. *Grus* Mioc. L. Plioc. Eu. U. Mioc. SAs. L. Plioc. NA. R. Cos. (except SA.) *Palaeogrus* L. Mioc. Eu. *Pliogrus* L. Plioc. Eu. EEu. **Aramidae** R. NA. SA. CA. *Badistornis* M. Olig. NA. **Psophiidae** R. SA. **Heliornithidae** R. CA. SA. **Rhynochetidae** R. SPac. **Eurypygidae** R. SA. CA. **Mesitornithidae** R. Mad.

SUBORDER RALLI

Rallidae *Aphanapteryx* Quat. Ind. Oc. *Capelirallus* Quat. NZ. *Coturnicops* U. Plioc. NA. *?Creccoides* L. Pleist. NA. *Diaphorapteryx* Quat. NZ. *Eocrex* L. Eoc. NA. *Euryonotus* U. Pleist. SA. *Fulica* L. Plioc. M.-U. Pleist. R. NA. R.

As. Eu. SA. CA. M. Pleist. WAs. *Fulicaletornis* M. Eoc. NA. *Gallinula* U. Plioc. U. Pleist. R. NA. M. Pleist. WAs. R. Eu. As. Af. NA. SA. *Ibidopsis* U. Eoc. Eu. *Laterallus* U. Plioc. NA. *?Ludiortyx* U. Eoc. Eu. *Miofulica* M. Mioc. Eu. *Miorallus* U. Mioc. Eu. *Nesotrochis* Quat. WInd. *Palaeogramides* U. Olig. or L. Mioc. U. Mioc. Eu. U. Mioc. EEu. *Paleorallus* L. Eoc. NA. *Pararallus* U. Mioc. Eu. *Paraortygometra* U. Olig. or L. Mioc. ?U. Mioc. Eu. *Pardirallus* U. Plioc. R. NA. *Quereyrallus* U. Eoc. or L. Olig. Eu. *Rallicrex* U. Olig. or L. Mioc. EEu. *Rallus* L. Plioc. Eu. M.-U. Plioc. NA. R. Cos. *Youngornis* M. Mioc. EAs. **Apterornithidae** *Apterornis* [*Aptornis*] Quat. NZ.

SUBORDER INCERTAE SEDIS **Ardeidae** *Ardea* M. Mioc.-Plioc. NA. U. Plioc. NAf. R. Cos. *Ardeagrandis* U. Mioc. EEu. *Botaurus* U. Plioc. R. NA. R. Cos. *Butorides* Pleist. R. NA. *Gnotornis* U. Olig. NA. *?Nyctanassa* M. Plioc. CAs. *Nycticorax* Plioc. R. NA. R. Cos. *Proardea* [*Proardeola*] Eoc. Olig. Mioc. Eu. *Zeltornis* L. Mioc. NAf.

ORDER PODICIPEDIFORMES **Podicipedidae** *Podiceps* L. Mioc. L. Plioc. NA. R. Cos.

ORDER DIATRYMIFORMES **Diatrymatidae (Gastornithidae)** *Diatryma* U. Paleoc. M. Eoc. NA. Eu. *Gastornis Remiornis* U. Paleoc. Eu. *Omorhamphus* U. Paleoc. Eu. *Zhonguanus* L. Eoc. EAs.

ORDER CHARADRIIFORMES **Burhinidae** *Burhinus* L. Mioc. Pleist. NA. Pleist R. CA. R. Eu. As. Af. Aus. SA. WInd. **Plataleidae** *Apteribis* Quat. SPac. *Eudocimus* Plioc. NA. *Milnea* Mioc. Eu. *Plegadis* L. Mioc. Eu. U. Plioc. R. NA. R. As. Af. SA. CA. As. Aus. *Rhynchaeites* M. Eoc. Eu. *Xencibis* Quat. WInd. **Chionididae** R. Ant. **Graculavidae** *Graculavus Telmatornis Laornis* U. Cret. NA. **Cimolopterygidae** *Palintropus* U. Cret. NA. **Dakotornithidae** *Dakotornis* Paleoc. NA. **Rostratulidae** R. SA. Af. SAs. **Dromadidae** R. Ind. Oc. Af. **Thinocoridae** R. SA. **Pedionomidae** R. SA. **Jacanidae** *Jacana* Olig. NAf. R. Af. Aus. Ind. As. NA. SA. CA. **Scolopacidae** R. Cos. *Paractitis* L. Olig. NA. *Totanus* U. Eoc. L. Olig. Eu. **Charadriidae** R. Cos. **Haematopodidae** *Haematopus* [*Palostralegus*] L.-M. Plioc. R. NA. **Recurvirostridae** *Recurvirostra* U. Eoc. L. Olig. Eu. M. Mioc. NA. R. Cos. **Phoenicopteridae** *Juncitarsus* M. Eoc. NA. *Megapaloelodus* L. Mioc. Eu. L.-M. Mioc. L. Plioc. NA. *Palaelodus* U. Olig. L. Mioc. Eu. *Phoeniconotius* Mioc. Aus. *Phoenicopterus* [*Leakeyornis Phoeniconaias Phoenicoparrus*] U. Olig. L. Mioc. R. Eu. L. Mioc. R. EAf. R. NA. SA. As. **Glareolidae** R. As. Af. Ind. SEAs. Aus. Mad. *Glareola* M. Mioc. R. Eu. *Mioglareola* M. Mioc. Eu. *Paractiornis* L. Mioc. NA. **Otididae** R. Af. SEAs. Ind. *Otis* M. or U. Mioc. R. Eu. *Gryzaja* L. Plioc. EEu. **Stercorariidae** R. Circumpolar *Stercorarius* M. Mioc. Pleist. NA. ?L. Mioc. Eu. **Laridae** R. Cos. *Larus* L.-U. Mioc. Eu. Plioc. NA. U. Mioc. EEu. R. Cos. *Gaviota* U. Mioc. NA. **Alcidae** R. NA. Eu. SEAs. CA. *Aethia* U. Mioc. U. Plioc. R. NA. R. Eu. *Alca* L. Plioc. R. NA. R. Eu. *Alcodes* U. Mioc. NA. *Alle* L. Plioc. R. NA. R. Eu. *Australca* U. Mioc. L. Plioc. NA. *Brachyramphus* U. Mioc. U. Plioc. R. NA. R. Eu. *Cepphus* U. Mioc. Pleist. R. NA. R. Eu. *Cerorhinca* U. Mioc. U. Plioc. R. NA. R. CA. SEAs. *?Endomychura* U. Mioc. U. Plioc. NA. *Fratercula* L. Plioc. R. NA. R. Eu. *Mancalla* Plioc. NA. *Miocepphus* M. Mioc. NA. *Pinguinus* L. Plioc. R. NA. R. Eu. *Praemancalla* U. Mioc. NA. *Ptychoramphus* U. Mioc. U. Plioc. R. NA.

ORDER ANSERIFORMES **Presbyornithidae** *Presbyornis* [*Nautilornis Telmabates*] Paleoc. NA. EAs. L. Eoc. NA. SA. U. Cret. NA. *?Telmatornis* U. Cret. NA. **Anatidae** R. Cos. *Cygnopterus* L. Olig. Eu. *Dendrochen* L. Mioc. Eu. *Mergus* M. Mioc. R. NA. R. As. Eu. CA. Ind. SEAs. Aus. *Paranyroca* L. Mioc. NA. *Romainvillia* L. Olig. Eu. *Sinanas* M. Mioc. EAs.* *Tadorna* M. Mioc. R. Eu. R. NA. As. Af. Ind. Aus. **Anhimidae** R. SA.

ORDER CICONIIFORMES **Ciconiidae** *?Ciconiopsis* L. Olig. SA. *Grallavis* L. Mioc. Eu. *Palaeoephippiorhynchus* L. Olig. NAf. **Scopidae** *Scopus* L. Plioc. SAf. R. Af. **Balaenicipitidae** R. Af. *Goliathia* U. Eoc. L. Olig. NAf. *Paludavis* U. Mioc. SAs. NAf. **Teratornithidae** *Argentavis* U. Mioc. SA. *Cathartornis* U. Pleist. NA. *Teratornis* L.-U. Pleist. NA. **Vulturidae** R. NA. SA. *Breagyps* Pleist. NA. *Diatopornis* U. Eoc. L. Olig. Eu. *Plesiocathartes* U. Eoc. L. Olig. ?L. Mioc. Eu. *Pliogyps* Plioc. NA. *Sarcoramphus* Plioc. NA. *Vultur* Pleist. SA. R. SA. NA.

ORDER PELECANIFORMES

SUBORDER PHAETHONTES **Prophaethontidae** *Prophaethon* L. Eoc. Eu. **Phaethontidae** R. Tropics Atl. Pac. Ind. Oc.

SUBORDER ODONTOPTERYGIA **Pelagornithidae** *Caspiodontornis* M. Olig. SWAs. *Cyphornis* [*Osteodontornis*] L. Mioc. NA. *?Dasornis* L. Eoc. Eu. *Gigantornis* M. Eoc. WAf. *Odontopteryx* L. Eoc. Eu. *Osteodontornis* U. Mioc. NA. *Palaeochenoides* U. Olig. NA. *Pelagornis* M. Mioc. Eu. ?NA. *Pseudodontornis* U. Olig. NA. ?L. Mioc. ?U. Plioc. NZ. *?Tympanoneisiotes* U. Olig. NA.

SUBORDER FREGATAE **Fregatidae** R. Oc. *Limnofregata* L. Eoc. NA.

SUBORDER PELECANI **Pelecanidae** R. Cos. (except SA.) *Pelecanus* Mioc. R. Aus. L. Mioc. L. Plioc. R. Eu. L. Plioc. SAs. WAs. L.-U. Plioc. R. NA. R. SA. WInd. *?Protopelicanus* U. Eoc. Eu.

SUBORDER SULAE **Sulidae** *Microsula* U. Olig. M. Mioc. NA. M. Mioc. Eu. *Miosula* Mioc. Plioc. NA. *Morus* Mioc. U. Pleist. R. NA. *Palaeosula* Mioc. Plioc. NA. *Sarmatosula* U. Mioc. EEu. *Sula* L.-U. Olig. Eu. Mioc. Plioc. NA. R. Cos. **Plotopteridae** *Plotopterum* L. Mioc. NA. U. Olig. L. Mioc. EAs. *Tonsala* U. Olig. NA. **Anhingidae** *Anhinga* U. Mioc. EEu. NAf. SAs. NA. U. Plioc. L. Pleist. EAf. ?U. Pleist. Aus. R. NA. Af. SA. Mad. EAs. Aus. *Protoplotus* U. Eoc. SAs. **Phalacrocoracidae** R. Cos. *Phalacrocorax* L. Mioc. Plioc. Eu. WAs. NA. Pleist. NA. U. Mioc. EEu. M. Plioc. CAs. L. Plioc. SAf. R. Cos. *Pliocarbo* L. Plioc. WAs. *?Valenticarbo* U. Plioc. L. Pleist. SAs.

ORDER PROCELLARIIFORMES **Diomedeidae** *Diomedea* M. Mioc. Plioc. NA. U. Mioc. Aus. U. Mioc. SA. U. Plioc. L. Pleist. Eu. R. Trop. Oc. *Plotornis* M. Mioc. Eu. U. Olig. M. Mioc. NA. **Procellariidae** *?Argyrodyptes* L. Mioc. SA. *Calonectris* L. Plioc. SAf. *Fulmarus* M.-U. Mioc. NA. R. Oc. *Pachyptila* L. Plioc. SAf. *Procellaria* L. Plioc. SAf. R. Southern Oc. *Pterodroma* L. Mioc. Eu. R. Oc. *Puffinus* ?L. Olig. Eu. L. Mioc. Plioc. Eu. NA. M. Mioc. L. Plioc. SAf. R. Oc. **Pelecanoididae** *Pelecanoides* L. Plioc. SAf. R. Southern Oc. **Oceanitidae (Hydrobatidae)** *Oceanodroma* U. Mioc. NA. R. Oc. *Oceanites* L. Plioc. SAf. R. Oc.

ORDER GAVIIFORMES **Gaviidae** *Colymboides* U. Eoc. L. Mioc. Eu. *Gavia* [*Colymbus*] L. Mioc. EEu. M.-U. Mioc. Plioc. Pleist. R. NA. R. Eu. CA. Ind. SEAs.

ORDER SPHENISCIFORMES **Spheniscidae** *Anthropodyptes* U. Eoc. L.-U. Olig. M. Mioc. Aus. *Anthropornis* U. Eoc. L. Olig. Ant. *Aptenodytes* U. Plioc. NZ. R. Ant. *Archaeospheniscus* U. Eoc. L. Olig. NZ. Ant. *Arthrodytes* U. Olig. L. Mioc. SA. *Chubutodyptes* U. Olig. L. Mioc. SA. *Dege* U. Mioc. or L. Plioc. SAf. *Delphinornis* U. Eoc. L. Olig. Ant. *Duntroonornis* U. Eoc. L. Olig. NZ. *Eudyptes* U. Plioc. R. NZ. R. Aus. *Eudyptula* U. Plioc. R. NZ. R. Aus. *Insuza* U. Mioc. or L. Plioc. SAf. *Korora* U. Olig. NZ. *Marplesornis* U. Plioc. NZ. *Megadyptes* Pleis. R. NZ. R. Ant. *Nucleornis* ?Mioc. SAf. *Pachydyptes* U. Eoc. L. Olig. NZ. *Palaeondyptes* U. Eoc. L.-U. Olig. Aus. NZ. Ant. *Palaeospheniscus* U. Olig. L. Mioc. SA. *Paraptenodytes* U. Olig. L. Mioc. SA. *Platydyptes* U. Eoc. L.-U. Olig. L. Mioc. NZ. *Pseudaptenodytes* U. Mioc. Aus. *Pygoscelis* U. Plioc. NZ. R. Ant. *Spheniscus* U. Plioc. R. Af. R. SA. *Wimanornis* U. Eoc. L. Olig. Ant.

CLASS MAMMALIA

SUBCLASS PROTOTHERIA

ORDER MONOTREMATA **Ornithorhynchidae** R. Aus. *Obdurodon* M. Mioc. Aus. *Ornithorhynchus* [*Platypus*] Plioc. R. Aus. *Steropodon* L. Cret. Aus. **Tachyglossidae** R. Aus. New Guinea *Tachyglossus* [*Echidna*] *Zaglossus* Pleist. R. Aus.

ORDER TRICONODONTA **Sinoconodontidae** *Sinoconodon* L. Jur. EAs. **Morganucodontidae** *Brachyzostrodon* U. Trias. Eu. *Erythrotherium* L. Jur. SAf. *Klamelia* M. or U. Jur. EAs. *Megazostrodon* L. Jur. SAf. *Morganucodon* ?U. Trias. L. Jur. Eu. L. Jur. EAs. **Amphilestidae** *Amphilestes* M. Jur. Eu. *Aploconodon* U. Jur. NA. *?Gobiconodon* L. Cret. EAs. NA. *Phascolodon* U. Jur. NA. *Phascolotherium* M. Jur. Eu. **Triconodontidae** *Alticonodon* U. Cret. NA. *Astroconodon* L. Cret. NA. *Priacodon* U. Jur. NA. *Triconodon* U. Jur. Eu. *Trioracodon* U. Jur. Eu. NA. **Incertae sedis** *Dinnetherium* L. Jur. NA. *Hallautherium Helvetiodon* ?U. Trias. Eu.

ORDER DOCODONTA **Docodontidae** *Borealestes* M. Jur. Eu. *Docodon* [*Dicrocynodon Diplocynodon Ennacodon*] U. Jur. NA. *Haldanodon Peraiocynodon* U. Jur. Eu.

SUBCLASS ALLOTHERIA

ORDER MULTITUBERCULATA

SUBORDER PLAGIAULACOIDEA **Arginbaataridae** *Arginbaatar* L. Cret. EAs. **Paulchoffatiidae** *Bolodon* U. Jur. L. Cret. Eu. *Guimarotodon* U. Jur. Eu. *Henkelodon* U. Jur. Eu. *Kuehneodon* U. Jur. L. Cret. Eu. *?Parendotherium* L. Cret. Eu. *Paulchoffatia* U. Jur. L. Cret. Eu. *Plioprion* U. Jur. Eu. *Pseudobolodon* U. Jur. Eu. **Plagiaulacidae** *Ctenacodon* [*Allodon*] U. Jur. Eu. NA. *Loxaulax* L. Cret. Eu. *Plagiaulax* [*?Bolodon*] U. Jur. L. Cret. Eu. *Psalodon* U. Jur. NA.

SUBORDER PTILODONTOIDEA **Boffiidae** *Boffia* M. Paleoc. Eu. **Neoplagiaulacidae** *Cimexomys* U. Cret. L. Paleoc. NA. *Ectypodus* L. Paleoc. U. Eoc. NA. U. Paleoc. Eu. *Gobiaatar* U. Cret. CAs. *Mesodma* [*Parectypodus*] U. Cret. U. Paleoc. NA. *Mimetodon* M.-U. Paleoc. NA. *Neoplagiaulax* L.-U. Paleoc. NA. U. Paleoc. Eu. *Parectypodus* L. Paleoc. U. Eoc. NA. L. Eoc. Eu. *Xanclomys* M. Paleoc. NA. **Cimolodontidae** *Anconodon* M.-U. Paleoc. NA. *Cimolodon* U. Cret. NA. *Liotomus* [*Neoctenacodon*] U. Paleoc. Eu. **Ptilodontidae** *Kimbetohia* U. Cret. L. Paleoc. NA. *Prochetodon* U. Paleoc. L. Eoc. NA. *Ptilodus* M.-U. Paleoc. NA.

SUBORDER TAENIOLABIDOIDEA **Taeniolabididae** *Catopsalis* U. Cret. NA. CAs. L.-U. Paleoc. NA. *Kamptobaatar* U. Cret. CAs. *Lambdopsalis* U. Paleoc. EAs. *Prionessus* U. Paleoc. L. Eoc. As. *Sphenopsalis* U. Paleoc. EAs. *Taeniolabis* [*Polymastodon*] L. Paleoc. NA. **Eucosmodontidae** *Acheronodon* U. Cret. NA. *Buginbaatar* U. Cret. or ?Paleoc. CAs. *Eucosmodon* L. Paleoc. NA. *Bulganbaatar Kryptobaatar* [*Gobibaatar*] U. Cret. CAs. *Microcosmodon* L.-U. Paleoc. NA. *Nemegtbaatar* U. Cret. CAs. *Neoliotomus* M.-U. Paleoc. L. Eoc. NA. *Pentacosmodon* U. Paleoc. NA. *Stygimys* U. Cret. M. Paleoc. NA. *Tugrigbaatar* U. Cret. CAs. *Xironomys* M. Paleoc. NA. **Chulsanbaataridae** *Chulsanbaatar* U. Cret. CAs. **Sloanbaataridae** *Sloanbaatar* U. Cret. CAs.

SUBORDER INCERTAE SEDIS **Cimolomyidae** *Cimolomys Meniscoessus* U. Cret. NA. *Essonodon* U. Cret. NA.
Incertae sedis *Allacodon Viridomys* U. Cret. NA. *Cimexomys* U. Cret. L. Paleoc. NA.

SUBORDER HARAMIYOIDEA **Haramiyidae** *Haramiya* U. Trias. L. Jur. Eu. *Thomasia* U. Trias. L. Jur. Eu.

SUBCLASS THERIA

INFRACLASS TRITUBERCULATA

ORDER SYMMETRODONTA **Kuehneotheriidae** *Kuhneon* U. Trias. L. Jur. Eu. *Kuehneotherium* U. Trias. L. Jur. Eu. *Woutersia* U. Trias. Eu. **Spalacotheriidae** *Eurylambda* U. Jur. NA. *Peralestes* U. Jur. Eu. *Spalacotherium* U. Jur. L. Cret. Eu. *Spalacotheroides* L. Cret. NA. *Symmetrodontoides* U. Cret. NA. *Tinodon* U. Jur. NA. **Amphidontidae** *Amphidon* U. Jur. NA. *Manchurodon* L. Cret. EAs. *Nakunodon* L. Jur. SAs.

ORDER INCERTAE SEDIS **Family unnamed** *Shuotherium* M. Jur. EAs.

ORDER EUPANTOTHERIA **Amphitheriidae** *Amphitherium* M. Jur. Eu. **Peramuridae** *Peramus* U. Jur. Eu. *Brancatherulum* U. Jur. EAf. *Palaeoxonodon* M. Jur. Eu. **Paurodontidae** *Araeodon Archaeotrigon* U. Jur. NA. *Paurodon Tathiodon* U. Jur. NA. **Dryolestidae** *Amblotherium* [*Stylacodon Stylodon*] U. Jur. Eu. NA. *Crusafontia* L. Cret. Eu. *Dryolestes Herpetairus Kepolestes* U. Jur. NA. *Euthlastus* U. Jur. NA. *Kurtodon* U. Jur. Eu. *Laolestes Malthacolestes* U. Jur. NA. *Melanodon* U. Jur. NA. L. Cret. Eu. *Miccylotyrans Pelicopsis* U. Jur. NA. *Peraspalax Phascolestes* U. Jur. Eu. **Incertae sedis** *Butlerigale Guimarota Simpsonodon* U. Jur. Eu.

THERIA OF METATHERIAN-EUTHERIAN GRADE

Aegialodontidae *Aegialodon* L. Cret. Eu. *Kermackia* L. Cret. NA. *Kielantherium* L. Cret. CAs. **Deltatherididae** *Deltatheridium Deltatheroides Hyotheridium* U. Cret. CAs. **Incertae sedis** *Beleutinus* U. Cret. SWAs. *Holoclemensia Pappotherium*

Slaughteria Trinititherium L. Cret. NA. *Potamotelses Falepetrus Bistius* U. Cret. NA.

INFRACLASS METATHERIA

ORDER MARSUPIALIA

(NEW WORLD AND EUROPEAN MARSUPIALS)

SUBORDER DIDELPHOIDEA **Didelphidae** *Albertatherium* U. Cret. NA. *Alphadon* U. Cret. NA. ?SA. *Amphiperatherium* [*Oxygomphius*] L. Eoc. U. Mioc. Eu. *Bobbschaefferia* [*Schaefferia*] U. Paleoc. SA. *Caluromys* [*Mallodelphys Philander*] L. Pleist. R. SA. R. CA. *Caroloameghina* L. Eoc. SA. *Chironectes* [*Cheironectes Gamba Memina*] L. Plioc. R. SA. R. CA. *Coona* L. Eoc. SA. *Derorhynchus* U. Paleoc. SA. *Didelphis* [*Dasyurotherium Didelphys Gambatherium Leucodelphis Lucodidelphys Opossum Sarigua Thylacotherium*] Pleist. R. NA. L. Pleist. R. SA. R. CA. *Didelphopsis* U. Paleoc. SA. *Entomacodon* [*Centracodon*] M. Eoc. NA. *Gaylordia* [*Xenodelphis*] *Guggenheimia* U. Paleoc. SA. *Glasbius* U. Cret. NA. *Herpetotherium* L. Eoc. L. Mioc. NA. *Hondadelphys* M. Mioc. SA. *Hyperdidelphys* [*Cladodidelphys Paradidelphys*] U. Mioc. U. Plioc. SA. *Lestodelphys* [*Notodelphys*] L. Pleist. R. SA. *Lutreolina* [*Peramys*] U. Mioc. R. SA. *Marmosa* [*Asagis Grymaeomys Notogogus Quica*] M. Mioc. R. SA. R. CA. *Marmosopsis* U. Paleoc. SA. *Micoureus* U. Pleist. R. SA. *Mimoperadectes* L. Eoc. NA. *Minusculodelphis* U. Paleoc. SA. *Mirandatherium* [*Mirandaia*] Mioc. R. SA. *Monodelphis* [*Hemiurus Microdelphys Minuania Monodelphiops Peramys*] U. Mioc. R. SA. *Monodelphopsis* U. Paleoc. SA. *Nanodelphys* [*Didelphidectes*] M. Eoc. M. Olig. NA. *Pachybiotherium* U. Olig. SA. *Paradidelphys* U. Mioc. U. Plioc. SA. *Peradectes* [*Thylacodon*] U. Cret. L. Eoc. NA. L. Eoc. Eu. ?U. Cret. SA. *Peratotherium* [*Alacodon*] L. Eoc. U. Olig. Eu. *Philander* [*Holothylax Metacherius Metachirops*] L. Plioc. R. SA. R. CA. *Protodidelphis* U. Paleoc. SA. *Sparassocynus* [*Gerazoyphus Perazoyphium*] U. Mioc. L. Plioc. SA. *Sternbergia* U. Paleoc. SA. *Thylamys* L. Plioc. R. SA. *Thylatheridium* U. Mioc. U. Plioc. SA. *Thylophorops* U. Plioc. L. Pleist. SA. *Zygolestes* L. Plioc. SA. **Pediomyidae** *Aquiladelphis* U. Cret. NA. *Pediomys* [*Synconodon Protolambda*] U. Cret. NA. ?U. Cret. SA. **Microbiotheriidae** *Microbiotherium* [*Clenia Clenialites Eodidelphys Hadrorhynchus Microbiotheridion Oligobiotherium Phonocdromus Prodidelphys Proteodidelphys Stylognathus*] U. Olig. L. Mioc. SA. **Stagodontidae** *Boreodon Delphodon* U. Cret. NA. *Didelphodon* [*Diaphorodon Didelphops Ectoconodon Stagodon Thlaeodon*] *Eodelphis* U. Cret. NA. **Borhyaenidae** *Acrocyon* U. Olig. L. Mioc. SA. *Anatherium* [*Acyon*] U. Olig. L. Mioc. SA. *Angelocabrerus* L. Eoc. SA. *Arctodictis* U. Olig. L. Mioc. SA. *Arminiheringia* [*Dilestes*] L. Eoc. SA. *Argyrolestes* L. Eoc. SA. *Borhyaena* [*Conodonictis Dynamictis Pseudoborhyaena*] U. Olig. L. Mioc. SA. *Borhyaenidium* U. Mioc. L. Plioc. SA. *Chasicostylus* U. Mioc. SA. *Cladosictis* [*Agustylus Cladictis Hathliacymus Ictioborus*] U. Olig. L. Mioc. SA. *Cladosictis Lycopsis Prothylacynus* [*Napodonictis Prothylacocyon*] L.-M. Mioc. SA. *Eobrasilia* U. Paleoc. SA. *Eutemodus* [*Apera*] U. Mioc. L. Plioc. SA. *Nemolestes* ?U. Paleoc. ?L. Eoc. SA. *Notictis* U. Mioc. SA. *Notocynus* U. Plioc. SA. *Notogale* U. Olig. SA. *Parahyaenodon* M. Plioc. SA. *Patene* [*Ischyrodidelphis*] U. Paleoc. L. Eoc. SA. *Peratherеutes* L. Mioc. SA. *Pharsophorus* U. Olig. SA. *Plesiofelis* M. Eoc. SA. *Proborhyaena* U. Olig. SA. *Procladosictis* M. Eoc. SA. *Prothylacynus* L. Mioc. SA. *Pseudolycopsis* U. Mioc. SA. *Pseudonotictis* L. Mioc. SA. *Pseudothylacynus* U. Olig. SA. *Sipalocyon* [*Amphiproviverra Amphithereutes Protoproviverra Thylacodictis*] U. Olig. L. Mioc. SA. *Stylocinus* U. Mioc. SA. **Thylacosmilidae** *Achlysictis* [*Acrohyaenodon*] Plioc. SA. *Hyaenodontops* U. Plioc. SA. *Notosmilus* U. Plioc. SA. *Thylacosmilus* U. Mioc. L. Plioc. SA. **Argyrolagidae** *Argyrolagus* Plioc. SA. *Microtragulus* U. Mioc. L. Pleist. SA. *Proargyrolagus* L. Olig. SA.

SUBORDER CAENOLESTOIDEA **Caenolestidae** *Abderites* [*Homunculites*] *Acdestis* [*Callomenus Dipilus*] U. Olig. L. Mioc. SA. *Eomanodon Halmarhippus Micrabderites* U. Olig. SA. *Palaeothentes* [*Cladoclinus Essoprion Halmadromus Halmaselus Metaepanorthus Metriodromus Palaepanorthus Paraepanorthus Pilchenia Prepanorthus*] *Parabderites* [*Mannodon Tidaeus Tideus*] U. Olig. L. Mioc. SA. *Phonocdromus* L. Mioc. SA. *Pichipilus* U. Olig. L. Mioc. SA. *Pilchenia Pitheculites* [*Eomannodon Micrabderites*] U. Olig. SA. *Pliolestes* U. Mioc. L. Plioc. SA. *Pseudhalmarhiphus* U. Olig. SA. *Stilotherium* [*Garzonia Halmarhiphus Parhalmarhipus*] Mioc. SA. **Polydolopidae** *Amphidolops* [*Anadolops*] L. Eoc. SA. *Antarctodolops* U. Eoc. Ant. *Epidolops* U. Paleoc. SA. *Eudolops* [*Promysops Propolymastodon*] L. Eoc. SA. *Polydolops* [*Anissodolops Archaeodolops Orthodolops Pliodolops Pseudolops*] U. Paleoc. M. Eoc. SA. *Prepidolops* L. Eoc. SA. *Seumadia* U. Paleoc. SA.

SUBORDER INCERTAE SEDIS **Groeberiidae** *Groeberia* U. Eoc. SA.

INCERTAE SEDIS *Eobrasilia Gashternia Ischyrodidelphis Xenodelphis* U. Paleoc. SA. *Ideodelphys Progarzonia* L. Eoc. SA. **Bonapartheriidae** *Bonapartherium* L. Eoc. SA. **Necrolestidae** *Necrolestes* L. Mioc. SA.

AUSTRALASIAN MARSUPIALIA

SUBORDER DASYUROIDEA **Dasyuridae** *Ankotarinja* M. Mioc. Aus. *Antechinomys* Pleist. R. Aus. *Antechinus* [*Parantechinus Pseudantechinus*] ?U. Mioc. Pleist. R. Aus. *Dasycercus* Pleist. R. Aus. *Dasylurinja* M. Mioc. Aus. *Dasyuroides* Pleist. R. Aus. *Dasyurus* [*Dasyurinus Dasyurops Nasira Notoctonus Satanellus Stictophonuss*] Plioc. R. Aus. R. EInd. *Glaucodon* L. Plioc. Aus. *Keeuna* M. Mioc. Aus. *Phascogale* [*Ascogale Phascolictis Phascologale Tapoa*] Pleist. R. Aus. *Phascolosorex* M. Plioc. R. EInd. *Planigale* Plioc. R. Aus. EInd. *Sarcophilus* [*Diabolus Ursinus*] Pleist. R. Aus. *Sminthopsis* [*Podabrus*] Pleist. R. Aus. R. EInd. *Wakamatha* ?M. Mioc. Aus. **Thylacinidae** *Thylacinus* [*Paracyon Peralopex*] U. Mioc. R. Aus. Plioc. Pleist. EInd. **Myrmecobiidae** *Myrmecobius* U. Pleist. R. Aus. **Notoryctidae** R. Aus.

SUBORDER PERAMELOIDEA **Peramelidae** *Chaeropus* [*Choeropus*] L. Pleist. R. Aus. *Echymiper* [*Anuromeles Brachymelis Suillomeles*] M.-U. Mioc. R. Aus. *Isoodon* [*Thylacis*] ?M.-U. Mioc. Plioc. R. Aus. R. EInd. *Perameles* [*Thylacis Thylax*] ?M.-U. Mioc. Plioc. R. Aus. *Peroryctes* [*Ornoryctes*] ?M.-U. Mioc. Aus. *Thylacis* Pleist. R. Aus. **Thylacomyidae** *Ischnodon* Plioc. Aus. *Macrotis* [*Peragale* [*Thalacomys Thylacomys*] Pleist. R. Aus.

SUBORDER DIPROTODONTA

SUPERFAMILY PHALANGEROIDEA **Phalangeridae** *Cercaertus* Pleist. Aus. *Palaeoptaurus* Pleist. Aus. *Petaurus* Plioc. Pleist. R. Aus. R. EInd. *Phalanger* [*Ailurops Balantia Ceonyx Coescoes Cuscus Encuscus Sipalus Phalangista Spilocuscus Strigocuscus*] ?M. Mioc. L. Plioc. R. Aus. R. EInd. *Trichosurus* [*Ceraertus Psilogrammurus Trichurus*] Plioc. R. Aus. **Ektopodontidae** *Ektopodon* M. Mioc. Aus. **Petauridae** *Gymnobelideus* [*Palaeopetaurus*] Pleist. R. Aus. *Petaurus* [*Belideus Petaurella Ptilotus Xenochirus*] Pleist. R. Aus. R. EInd. *Pseudocheirops* U. Mioc. R. Aus. *Pseudocheirus* [*Hemibelideus Hepoona Petropseudes Pseudocheirops*] Plioc. R. Aus. R. Tas. EInd. *Pseudokoala* Plioc. Aus. *Schinobates* Pleist. R. Aus. **Burramyidae** *Acrobates* Pleist. R. Aus. *Burramys* Plioc. R. Aus. *Cercartetus* [*Dromicia Dromiciella Dromiciola Endromicia*] M. Mioc. R. Aus. EInd. Tas. **Thylacoleonidae** *Thylacoleo* Plioc. Pleist. Aus. *Wakaleo* M. Mioc. Aus. **Macropodidae** *Aepyprymnus* Pleist. R. Aus. *Bettongia* [*Bettongiops*] M. Mioc. R. Aus. *Brachalletes* Pleist. Aus. *Caloprymnus* Pleist. R. Aus. *Dendrolagus* Plioc. Aus. *Dorcopsis* Plioc. Aus. R. EInd. *Dorcopsoides* M.-U. Mioc. Aus. *Fissuridon* Pleist. Aus. *Gumardee* M. Mioc. Aus. *Hadronomas* U. Mioc. Aus. *Hypsiprymnodon* [*Pleopus*] Plioc. R. Aus. *Lagorchestes* Pleist. R. Aus. *Macropus* [*Boriogale Dendrodorcopsis Gerboides Gigantomys Halmaturus Kangurus Leptosiagon Megaloia Osphranter Phascolagus*] Plioc. R. Aus. R. EInd. *Nambaroo* M. Mioc. Aus. *Onychogale* Pleist. R. Aus. *Osphranter* Plioc. Aus. *Palaeopotorous* M. Mioc. Aus. *Petrogale* Plioc. Pleist. R. Aus. *Potorous* [*Hypsiprymnus Potoroops*] Pleist. R. Aus. *Prionotemnus* Plioc. Pleist. Aus. *Procoptodon* [*Pachysiagon*] Pleist. Aus. *Propleopus* [*Triclis*] Plioc. Pleist. Aus. *Protemnodon* Plioc. Aus. EInd. *Setonix* ?M. Mioc. Pleist. R. Aus. *Sthenurus* [*Simosthenurus*] Plioc. Pleist. Aus. *Synaptodon* Pleist. Aus. *Thylogale* Plioc. R. Aus. R. EInd. *Tropsodon* Plioc. Pleist. Aus. *Wabularoo* U. Mioc. Aus. *Wakiewakie* M. Mioc. Aus. *Wallabia* Pleist. R. Aus.

SUPERFAMILY PHASCOLARCTOIDEA **Phascolarctidae** *Koobor* Plioc. Aus. *Litokoala* M. Mioc. Aus. *Periokoala* M. Mioc. Aus. *Phascolarctos* Pleist. R. Aus.

SUPERFAMILY VOMBATOIDEA **Vombatidae** *Lasiorhinus* [*Wombatula*] Pleist. R. Aus. *Phascolonus* [*Sceparnodon*] Plioc. Pleist. Aus. *Ramsayia* Pleist. Aus. *Rhizophascolonus* M. Mioc. Aus. *Vombatus* [*Phascolomis*] Plioc. Pleist. R. Aus. **Diprotodontidae** *Bematherium* M. Mioc. Aus. *Diprotodon* [*Diarcodon*] Pleist. Aus. *Euryzygoma* Plioc. Aus. *Euowenia* [*Owenia*] Plioc. Pleist. Aus. *Kolopsis* U. Mioc. Aus. Plioc. EInd. *Kolopsoides* Plioc. EInd. *Meniscolophus* Plioc. Aus. *Neohelos* M. Mioc. Aus. *Nototherium* Plioc. Pleist. Aus. Plioc. EInd. *Plaisiodon* U. Mioc. Aus. *Pyramios* U. Mioc. Aus. *Raemeotherium* M. Mioc. Aus. *Zygomaturus* U. Mioc. Pleist. Aus. *?Brachalletes* Plioc. Aus. *?Koalemus* Plioc. Aus. *?Sthenmerus* Pleist. Aus. **Palorchestidae** *Ngapakaldia* M. Mioc. Aus. *Palorchestes* U. Mioc. Pleist. Aus. Pleist. Tas. *Pitikantia* M. Mioc. Aus. **Wynyardiidae** *Namilamadeta* M. Mioc. Aus. *Wynyardia* L. Mioc. Aus.

SUBORDER INCERTAE SEDIS **Tarsipedidae** R. Aus.

INFRACLASS EUTHERIA

(A possible supraordinal classification of Eutherian mammals is seen in Table A-1.)

ORDER INCERTAE SEDIS **Kennalestidae** *Asioryctes Kennalestes* U. Cret. CAs. **Zalambdalestidae** *Barunlestes Zalambdalestes*

TABLE A-1. **Supraordinal Classification of Eutheria, Proposed by Novacek (1986)**

Class Mammalia
- Subclass Theria
  - Infraclass Eutheria
    - Cohort Edentata
      - Order Xenarthra
      - Order Pholidota
    - Cohort Epitheria
      - Superorder Insectivora
        - Order Leptictida
        - Order Lipotyphla
          - Suborder Erinaceomorpha
          - Suborder Soricomorpha
            - Superfamily Tenrecoidea
            - Superfamily Soricoidea
      - Superorder Volitantia
        - Order Dermoptera
        - Order Chiroptera
      - Superorder Anagalida
        - Order Macroscelidae (may include Anagalidae)
        - Grandorder Glires
          - Order Rodentia
          - Order Lagomorpha
      - Superorder Ungulata
        - Order Arctocyonia
        - Order Dinocerata
        - Order Embrithopoda
        - Order Artiodactyla
        - Order Cetacea
        - Order Perissodactyla
        - Grandorder Meridiungulata (most South American ungulates)
        - Grandorder Paenungulata
          - Order Hyracoidea
        - Mirorder Tethytheria
          - Order Sirenia
          - Order Proboscidea
          - Order Desmostylia
    - Cohort Epitheria incertae sedis
      - Order Tubulidentata
      - Order Carnivora
      - Order Primates
      - Order Scandentia
      - Order Tillodontia
      - Order Taeniodonta

U. Cret. EAs. **Family unnamed** *Batodon Gallolestes* U. Cret. NA. *Cimolestes Procerberus* U. Cret. L. Paleoc. NA.

ORDER APATOTHERIA **Apatemyidae** *Apatemys [Teilhardella]* U. Paleoc. U. Eoc. NA. L. Eoc. Eu. *Eochiromys* L. Eoc. EU. ?NA. *Heterohyus [Amphichiromys Heterochiromys ?Necrosorex]* L.-U. Eoc. Eu. *Jepsenella* M. Paleoc. NA. ?U. Paleoc. Eu. *Labidolemur* U. Paleoc. NA. *Sinclairella* L.-M. Olig. NA. *Stehlinella [Stehlinius]* U. Eoc. NA. *Unuchinia [Apator]* U. Paleoc. NA.

ORDER LEPTICTIDA **Gypsonictopidae** *Gypsonictops [Euangelistes]* U. Cret. NA. **Leptictidae** *Diacodon* M. Paleoc. NA. *Diaphyodectes* U. Paleoc. NA. *Hypictops* M. Eoc. NA. *Ictopidium* U. Eoc. EAs. *Leptictis* U. Eoc. Olig. NA. *Myrmecoboides* M.-U. Paleoc. NA. *Palaeictops [Parictops]* ?U. Paleoc. L.-M. Paleoc. NA. *Prodiacodon* L. Paleoc. L. Olig. NA. *Protictops* U. Eoc. NA. **Pseudorhyncocyonidae** *Leptictidium* M. Eoc. Eu. *Pseudorhyncocyon* U. Eoc. Eu.

ORDER PANTOLESTA **Pantolestidae** *Amaramnis* L. Eoc. NA. *Bogdia* M. Eoc. EAs. *Buxolestes* M. Eoc. Eu. *Chadronia* L. Olig. NA. *Cryptopithecus* U. Eoc. L. Olig. Eu. *Dyspterna* U. Eoc. L. Olig. Eu. *?Kelba* L. Mioc. EAf. *Niphredil* U. Paleoc. NA. *Pagonomus* U. Paleoc. Eu. *Palaeosinopa* M. Paleoc. L. Eoc. NA. ?L. Eoc. EAs. *Paleotomus* L.-U. Paleoc. NA. *Pantolestes* M. Eoc. NA. *Pantomimus* L. Paleoc. NA. *Propalaeosinopa* M.-U. Paleoc. NA. *Simidectes* U. Eoc. NA. **Pentacodontidae** *Aphronorus* M. Paleoc. NA. Eu. *Bisonalveus* U. Paleoc. NA. *Coriphagus [Mixoclaenus] Pentacodon* M. Paleoc. NA. *Protentomodon* U. Paleoc. NA. **?Ptolemiidae** *Ptolemaia* L. Olig. NAf. *Qarunavus* Olig. NAf.

ORDER SCANDENTIA **Tupaiidae** *Tupaia* Plioc. SAs. R. SEAs.

ORDER MACROSCELIDEA **Macroscelididae** *Elephantulus [Elephantomys]* U. Plioc. R. Af. *Macroscelides* U. Plioc. R. Af. *Metoldobotes* L. Olig. NAf. *Mylomygale* U. Plioc. L. Pleist. SAf. *Myohyrax* L. Mioc. SAf. EAf. *Miorhynchocyon* L. Mioc. EAf. *Palaeothentoides* L. Plioc. SAf. *Protypotheroides* L. Mioc. SAf. *Rhynchocyon* L. Mioc. R. Af. *Pronasilio Hiwegicyon* M. Mioc. EAf.

ORDER DERMOPTERA

SUPERFAMILY PLAGIOMENOIDEA **Plagiomenidae** *Elpidophorus* L.-U. Paleoc. NA. *Plagiomene* U. Paleoc. L. Eoc. NA. *Planetetherium* U. Paleoc. NA. *?Thylacaelurus* U. Eoc. NA. *Worlandia* U. Paleoc. L. Eoc. NA. **Galeopithecidae (Cynocephalidae)** R. SEAs. **?Mixodectidae** *Eudaemonema* M. Paleoc. NA. *Mixodectes [Indrodon Oldobotes]* M.-U. Paleoc. NA. *?Remiculus* U. Paleoc. Eu. **Placentidentidae** *Placentidens* L. Eoc. Eu.

ORDER INSECTIVORA **Family unnamed** *Paranyctoides* U. Cret. NA.

SUBORDER ERINACEOMORPHA (LIPOTYPHLA)

SUPERFAMILY ERINACEOIDEA **Dormaaliidae** *Ankylodon* M. Eoc. L. Olig. NA. *Crypholestes* M. Eoc. NA. *Dormaalius* L. Eoc. Eu. *Macrocranion [Messelina]* L.-M. Eoc. Eu. NA. *Proterixoides* M. Eoc. NA. *Scenopagus* L.-M. Eoc. NA. *Sespedectes* M. Eoc. NA. **Amphilemuridae** *Alsaticopithecus Amphilemur* M. Eoc. Eu. *Gesneropithex* U. Eoc. Eu. *Pholidocercus* M. Eoc. Eu. **Erinaceidae** *Adapisorex* M.-U. Paleoc. Eu. *Amphiechinus Brachyerix* L.-M. Mioc. NA. *Cedrochoerus* U. Paleoc. NA. *Dartonius* L. Eoc. NA. *Dimylechinus* L. Mioc. Eu. *Entomolestes* M. Eoc. NA. *Eolestes* L. Eoc. NA. *Erinaceus* U. Mioc. R. Eu. R. As. Af. *Galerix [Parasorex Pseudogalerix]* U. Mioc. Eu. EAf. *Gymnurechinus* M. Mioc. EAf. *Lanthanotherium* M.-U. Mioc. Eu. U. Mioc. L. Plioc. NA. *Leipsanolestes* U. Paleoc. L. Eoc. NA. *Litolestes* U. Paleoc. NA. *Metechinus* M.-U. Mioc. NA. *Miochinus* L.-U. Mioc. Eu. *Neomatronella* L. Eoc. Eu. *Neurogymnurus* U. Eoc. L. Olig. Eu. *Ocajila* L. Mioc. NA. *Parvechinus* U. Olig. As. L.-M. Mioc. NA. *Postpalerinaceus* L. Plioc. Eu. *Protechinus* U. Mioc. EAf. *Proterix* M. Olig. NA. *Stenoechinus* L. Mioc. NA. *Untermannerix* M.-U. Mioc. NA. **Incertae sedis** *Adunator* M. Paleoc. L. Eoc. Eu. *Diacochoerus* U. Paleoc. NA. *Exallerix* M. Olig. EAs. *Mckennatherium* M. Paleoc. NA. *Talpavus* L.-U. Eoc. NA. *Talpavoides* L. Eoc. NA. *Tupaiodon* U. Eoc. As. *Xenacodon* U. Paleoc. NA.

SUBORDER SORICOMORPHA **Palaeoryctidae** *Aaptoryctes* U. Paleoc. NA. *?Aboletylestes* U. Paleoc. Eu. *Acmeodon Avunculus* M. Paleoc. NA. *Didelphodus [Didelphyodus Phenacops]* L.-M. Eoc. NA. *Gelastops* M. Paleoc. NA. *Leptonysson* M. Paleoc. NA. *Naranius* L. Eoc. EAs. *Palaeoryctes* L. Paleoc. L. Eoc. NA. *Pararyctes* U. Paleoc. NA. *Stilpnodon* M. Paleoc. NA. *?Thelysia* L. Eoc. NA. *Tsaganius* L. Eoc. EAs.

SUPERFAMILY SORICOIDEA **Geolabididae** *Batodontoides* M. Eoc. NA. *Centetodon [Embassis Geolabis Hypacodon Metacodon]* L. Eoc. L. Mioc. NA. **Talpidae** R. Eu. As. NA. *Ankyloscapter* L.-M. Mioc. NA. *Asthenoscapter* L.-M. Mioc. Eu. *Condylura* Pleist. R. NA. *Cryporyctes* L. Olig. NA. *Desmana [Desmagale Mygale Myogale]* L. Plioc. R. Eu. R. As. *Desmanella* M.-U. Mioc. Eu. M. Mioc. L. Plioc. As. *Domninoides* Mioc. NA. *Eotalpa* U. Eoc. Eu. *Gaillardia* U. Mioc. NA. *Galemys* L.-U. Plioc. R. Eu. *Geotrypus* U. Olig. L. Mioc. Eu. *Hesperoscalops* L. Plioc. NA. *Mygalea* U. Mioc. Eu. *Mygalinia* L. Plioc. Eu. *Mygatalpa* M.-U. Olig. Eu. *Mystipterus* L.-U. Mioc. NA. *Myxomygale* L.-M. Olig. Eu. *Neurotrichus* U. Mioc. L. Plioc. NA. *Parascalops* Pleist. R. NA. *Paratalpa* M. Olig. L. Mioc. Eu. *Proscapanus* L.-U. Mioc. Eu. *Quadrodens* U. Olig. NA. *Scalopoides* L.-U. Mioc. NA. ?M. Mioc. Eu. *Scalopus [Scalops]* U. Mioc. R. NA. *Scapanus [Xeroscapheus]* L. Plioc. R. NA. *Scaptochirus* Pleist. Eu. Pleist. R. As. *Scaptogale [Echinogale]* L. Mioc. Eu. *?Scaptonyx* M.-U. Mioc. Eu. R. EAs. *Talpa [Mogera]* L. Mioc. R. Eu. Pleist. R. As. *Teutonotalpa* L. Mioc. Eu. *Urotrichus* ?M. Mioc. Eu. Pleist. R. EAs. **Proscalopidae** *Cryptoryctes* L. Olig. NA. *Mesoscalops* L. Mioc. NA. *Oligoscalops* L.-M. Olig. NA. *Proscalops* U. Olig. L. Mioc. NA. **Plesiosoricidae** *?Butselia* L. Olig. Eu. *Plesiosorex* M. Olig. U. Mioc. NA. *?Saturninia* M. Eoc. L. Olig. Eu. **Soricidae** *Adeloblarina* M. Mioc. NA. *Allosorex* U. Mioc. U. Plioc. Eu. *Alluvisorex* M.-U. Mioc. NA. *Amblycoptus* L. Plioc. Eu. *Amphisorex* L. Olig. Eu. *Anchiblarinella* U. Mioc. NA. *Anourosorex [Shikamainosorex]* M. Plioc. R. As. U. Mioc. ?NA. Eu. *Antesorex Augustidens* L. Mioc. NA. *Beckiasorex* L.-M. Mioc. NA. *Beremendia* L. Plioc. L. Pleist. Eu. ?EAs. *Blarina* U. Plioc. R. NA. *Blarinella* U. Plioc. R. EAs. *Blarinoides* U. Plioc. Eu. *Carposorex* L. Mioc. Eu. *Clapasorex* L. Mioc.

Eu. *Crocidosorex* U. Olig. L. Mioc. Eu. *Crocidura* L. Plioc. R. Eu. Pleist. R. As. L. Mioc. L. Pleist. R. Af. L. Pleist. R. EInd. *Cryptotis* U. Mioc. R. NA. Pleist. R. SA. *Dinosorex* U. Olig. Mioc. Eu. *Diplomesodon* U. Plioc. Af. R. EEu. *Domina* [*Miothen Protosorex*] L. Olig. Eu. L. Mioc. NA. *Episoriculus* L. Plioc. U. Plioc. Eu. *Hesperosorex* ?U. Mioc. L. Plioc. NA. *Ingentisorex* U. Mioc. NA. *Limnoecus* L.-U. Mioc. Eu. L. Mioc. L. Plioc. NA. *Microsorex* Pleist. R. NA. *Myosorex* U. Plioc. R. SAf. *Neomys* [*Crossopus*] U. Pleist. R. Eu. Pleist. R. As. *Nesiotites* U. Pleist. Eu. *Notiosorex* U. Mioc. R. NA. *Oligosorex* L.-M. Mioc. Eu. *Paenelimnoecus* M. Mioc. Eu. *Paranourosorex* U. Mioc. Eu. *Siwalikosorex* U. Mioc. SAs. *Srinitium* M. Olig. Eu. *Sorex* [*Drepanosorex*] L. Mioc. R. NA. U. Mioc. R. Eu. *Soricella* L. Mioc. Eu. *Soriculus* L. Plioc. R. As. ?Plioc. R. NAf. *Suncus* [*Pachura*] L. Plioc. R. Eu. Pleist. R. Af. R. As. *Sylvisorex* U. Plioc. R. Af. *Tregosorex* U. Mioc. NA. *Trimylus* [*Heterosorex*] M. Olig. M.-U. Mioc. NA. U. Olig. L. Plioc. Eu. *Wilsonosorex* L. Mioc. NA. *Zelceina* L. Plioc. L. Pleist. Eu. **Nyctitheriidae** *Bumbanius* L. Eoc. EAs. *Leptacodon* M. Paleoc. L. Eoc. NA. *Nyctitherium* L.-U. Eoc. NA. *Oedollus* L. Eoc. EAs. *Plagioctenodon* U. Paleoc. L. Eoc. NA. *Plagioctenoides* L. Eoc. NA. *Pontifactor* U. Paleoc. M. Eoc. NA. **Micropternodontidae** *Micropternodus* [*Kentrogomphius*] M. Eoc. U. Olig. NA. *Prosarcodon* M. Paleoc. EAs. *Sarcodon* [*Opisthopsalis*] U. Paleoc. EAs. *Sinosinopa* M. Eoc. EAs. **Dimylidae** *Cordylodon* U. Olig. L. Mioc. Eu. *Dimyloides* U. Olig. Eu. *Dimylus* L. Mioc. Eu. *Exodaenodus* M. Olig. Eu. *Metacordylodon* U. Mioc. Eu. *Plesiodimylus* M. Mioc. L. Plioc. Eu. *Pseudocordylodon* L. Mioc. Eu. **Incertae sedis** *Clinopternodus* L. Olig. NA. *Nycticonodon* L. Eoc. Eu. *Parapternodus* L. Eoc. NA. *Scraeva* M.-U. Eoc. Eu.

SUBORDER ZALAMBDODONTA

SUPERFAMILY TENRECOIDEA **Tenrecidae** (**Centetidae**) *Erythrozootes* L. Mioc. EAf. *Parageogale* M. Mioc. EAf. *Protenrec* L. Mioc. EAf.

SUPERFAMILY CHRYSOCHLOROIDEA **Chrysochloridae** *Amblysomus* U. Plioc. Af. *Chlorotalpa* [*?Chrystotricha*] Pleist. R. SAf. *Chrysochloris* L. Plioc. R. Af. *Proamblysomus* U. Plioc. Pleist. SAf. *Prochrysochloris* L. Mioc. EAf.

ORDER INSECTIVORA INCERTAE SEDIS *Aethomylus* M. Eoc. NA. *Arctoryctes* L. Mioc. NA. *Creotarsus* L. Eoc. NA. *Adapisoriculus* U. Paleoc. Eu. *Hyracolestes* U. Paleoc. EAs.

ORDER TILLODONTIA **Esthonychidae** *Adapidium* U. Eoc. As. *Esthonyx* [*Plesesthonyx*] U. Paleoc. L. Eoc. NA. L. Eoc. Eu. *Lofochaius* L. Paleoc. EAs. *Kuanchuanius* M. Eoc. EAs. *Megalesthonyx* L. Eoc. NA. *Meiostylodon* L. Paleoc. EAs. *Tillodon* M. Eoc. NA. *Trogosus* [*Tillotherium*] L.-M. Eoc. NA. **Incertae sedis** *Basalina* M. Eoc. SAs. *Dysnoetodon* Paleoc. EAs.

ORDER PANTODONTA **Archaeolambdidae** *Archaeolambda* U. Paleoc. L. Eoc. EAs. *Nanlingilambda* U. Paleoc. EAs. **Bemalambdidae** *Bemalambda* M.-U. Paleoc. EAs. *Hypsilolambda* M. Paleoc. EAs. **Pantolambdidae** *Caenolambda* M.-U. Paleoc. NA. *Pantolambda* M. Paleoc. NA. **Barylambdidae** *Barylambda* U. Paleoc. NA. *Haplolambda* [*Archaeolambda*] U. Paleoc. NA. EAs. *Ignatiolambda Leptolambda* U. Paleoc. NA. **Titanoideidae** *Titanoides* [*Sparactolambda*] M.-U. Paleoc. NA. **Coryphodontidae** *Asiocoryphodon* L. Eoc. EAs. *Coryphodon* [*Letalophodon Loxolophodon*] U. Paleoc. L. Eoc. NA. L. Eoc. Eu. L.-M. Eoc. EAs. *Eudinoceras* M.-U. Eoc. EAs. *Hypercoryphodon* L. Olig. EAs. *Manteodon* L. Eoc. EAs. *Metacoryphodon* M. Eoc. EAs. **Harpyodidae** *Harpyodus* U. Paleoc. EAs. **Pantolambdodontidae** *Dilamba* U. Eoc. EAs. *Oroklambda* M. Eoc. EAs. *Pantolambdodon* M.-U. Eoc. EAs. **Pastoralodontidae** *Altilambda* U. Paleoc. EAs. *Convallisodon* U. Paleoc. EAs. *Pastoralodon* U. Paleoc. L. Eoc. EAs. **Cyriacotheriidae** *Cyriacotherium* U. Paleoc. NA.

ORDER DINOCERATA **Uintatheriidae** *Bathyopsis* L.-M. Eoc. NA. *Bathyopsoides* U. Paleoc. NA. *Eobasileus* [*Uintacolotherium*] U. Eoc. NA. *Ganatherium* L. Eoc. As. *Gobiatherium* ?L. Eoc. M.-U. Eoc. EAs. *Houyanotherium* U. Paleoc. EAs. *Jiaoluotherium* U. Paleoc. EAs. *Mongolotherium* U. Paleoc. EAs. *Phenaceras* L. Eoc. As. *Probathyopsis* [*Prouintatherium*] U. Paleoc. L. Eoc. NA. ?L. Eoc. EAs. *Prodinoceras* U. Paleoc. EAs. *Pyrodon* L. Eoc. EAs. *Tetheopis* M.-U. Eoc. NA. *Uintatherium* [*Dinoceras Ditetrodon Elachoceras Laoceras Octotomus Paroceras Platoceras Tinoceras Uintamastix*] ?M. Eoc. As. U.-M. Eoc. NA. **Gobiatheriidae** *Gobiatherium* M. Eoc. EAs.

ORDER TAENIODONTIA **Stylinodontidae** *Chungchienia* U. Eoc. EAs. *Conoryctella Conoryctes* M. Paleoc. NA. *Ectoganus* [*Camalodon*] L. Eoc. NA. *Huerfanodon* L. Paleoc. NA. *Lampadophorus* U. Paleoc. NA. *Onychodectes* L. Paleoc. NA. *Psittacotherium* M. Paleoc. NA. *Stylinodon* L.-U. Eoc. NA. ?U. Eoc. As. *Wortmania* L. Paleoc. NA.

ORDER CHIROPTERA

SUBORDER MEGACHIROPTERA **Pteropodidae** *Propotto* L. Mioc. Af. *Archaeopteropus* M. Olig. Eu.

SUBORDER MICROCHIROPTERA

SUPERFAMILY ICARONYCTEROIDEA **Icaronycteridae** *Icaronycteris* L. Eoc. NA. ?Eu. **Palaeochiropterygidae** *Archaeonycteris* L.-M. Eoc. Eu. *Cecilionycteris Matthesia Palaeochiropteryx* M. Eoc. Eu.

SUPERFAMILY EMBALLONUROIDEA **Emballonuridae** *Taphozous* ?Eoc. Olig. Eu. L. Mioc. R. Af. R. SAs. Aus. *Vespertiliavus* M. Eoc. M. Olig. Plioc. Eu.

SUPERFAMILY RHINOLOPHOIDEA **Megadermatidae** *Cardioderma* U. Plioc. L. Pleist. R. Af. *Megaderma* Mioc. Eu. Pleist. R. As. R. Aus. *Necromantis* L. Olig. Eu. **Rhinolophidae** *Palaeonycteris* U. Olig. Eu. *Rhinolophus* U. Eoc. R. Eu. Pleist. R. As. Aus. M. Mioc. U. Plioc. R. Af. **Hipposideridae** *Asellia* M. Mioc. Eu. Af. R. NAf. SAs. *Hipposideros* M. Eoc. M. Mioc. Eu. L. Mioc. NAf. Pleist. R. EInd. R. As. Af. Aus. *Palaeophyllopora* U. Eoc. M. Olig. Eu. *Paraphyllophora* U. Eoc. L. Olig. ?Mioc. Eu. *Rhinonycteris* M. Mioc. Aus.

SUPERFAMILY PHYLLOSTOMATOIDEA **Phyllostomatidae** *Desmodus* Pleist. NA. R. SA. *Leptonycteris* U. Pleist. R. CA. NA. *Mormoops* Pleist. R. WInd. Pleist. R. NA. R. SA. *Notonycteris* U. Mioc. SA. *?Vampyravus* L. Olig. NAf.

SUPERFAMILY VESPERTILIONOIDEA **Myzopodidae** *Myzopoda* L. Pleist. Af. R. Mad. **Vespertilionidae** *Antrozous*

Plioc. R. NA. *Anzanycteris* U. Plioc. NA. *Barbastella* Pleist. R. Eu. *Chamtwaria* M. Mioc. EAf. *Eptesicus* [*Hesperoptenus Histiotus*] M. Mioc. R. Eu. ?L. Plioc. As. Pleist. R. NA. R. SA. WInd. As. Af. Aus. *Histiotus* L. Pleist. NA. *Lasionycterus* M. Plioc. R. NA. *Lasiurus* [*Atalapha Dasypterus*] U. Plioc. R. NA. Pleist. R. SA. R. WInd. Hawaii *Miniopterus* Mioc. R. Eu. Pleist. R. As. U. Plioc. U. Pleist. Af. *Myotis* M. Olig. R. Eu. Pleist. R. As. Af. R. SA. Aus. U. Mioc. R. NA. *Nyctalus* U. Pleist. R. Eu. *Nycticeius* U. Pleist. R. NA. *Paleptesicus* M. Mioc. Eu. *Pipistrellus* [*Nyctalus Vesperugo*] Pleist. R. Eu. R. As. NA. Af. EInd. *Plecotus* [*Corynorhinus*] L. Plioc. R. Eu. As. R. NAf. Pleist. NA. *Scotophilus* ?M. Mioc. Eu. *Stehlinia* [*Nycterobius Revilliodia*] M. Eoc. U. Olig. Eu. *Tadarida* [*Chaerophon Nyctinomus*] U. Olig. SA. ?U. Olig. R. Eu. Pleist. R. NA. WInd. R. SAs. Af. Aus. SA. *Vespertilio* U. Plioc. R. Eu. ?Pleist. NA. Pleist. R. Af. R. As. **Molossidae** *Eumops* U. Pleist. NA. R. SA. CA. *Nyctinomus* L. Mioc. Eu.

SUPERFAMILY INCERTAE SEDIS *Afropterus* M. Mioc. Af.

ORDER PRIMATES

SUBORDER PLESIADAPIFORMES

SUPERFAMILY PARAMOMYOIDEA **Paromomyidae** *Berruvius* U. Paleoc. L. Eoc. NA. *Ignacius* U. Paleoc. U. Eoc. NA. *Micromomys* U. Paleoc. NA. *Navajovius* U. Paleoc. L. Eoc. NA. *Palaechthon Palenochtha* M. Paleoc. NA. *Paromomys* M. Paleoc. NA. *Phenacolemur* U. Paleoc. M. Eoc. NA. Eu. *Plesiolestes* M.-U. Paleoc. NA. *Purgatorius* U. Cret. L. Paleoc. NA. *Tinimomys* U. Paleoc. L. Eoc. NA. **Picrodontidae** *Picrodus* [*Megapterna*] M.-U. Paleoc. NA. *Zanycteris* U. Paleoc. NA. **?Microsyopidae** *Alveojunctus* M. Eoc. NA. *Microsyops* [*Cynodontomys*] U. Paleoc. M. Eoc. NA. *Niptonomys* U. Paleoc. L. Eoc. NA. *Uintasorex* M. Eoc. NA.

SUPERFAMILY PLESIADAPOIDEA **Plesiadapidae** *Chiromyoides* U. Paleoc. NA. Eu. L. Eoc. NA. *Platychoerops* L. Eoc. Eu. *Pronothodectes* M. Paleoc. NA. *Plesiadapis* [*Ancepsoides Nothodectes*] U. Paleoc. L. Eoc. Eu. NA. **Saxonellidae** *Saxonella* U. Paleoc. Eu. NA. **Carpolestidae** *Carpodaptes* M. Paleoc. L. Eoc. NA. *Carpolestes* U. Paleoc. NA. *Elphidotarsius* M. Paleoc. NA.

SUBORDER PROSIMII

INFRAORDER ADAPIFORMES **Adapidae** *Adapis* [*Aphelotherium Palaeolemur*] *Agerinia* L. Eoc. Eu. *Anchomomys* M. Eoc. L. Olig. Eu. *Caenopithecus* M. Eoc. Eu. *Copelemur* L. Eoc. NA. *Europolemur* M. Eoc. Eu. *Huerzeleris* M.-U. Eoc. Eu. *Indraloris* U. Mioc. Ind. *Leptadapis* M.-U. Eoc. Eu. *Lushius* U. Eoc. EAs. *Mahgarita* U. Eoc. NA. *Microadapis* M. Eoc. Eu. *Notharctus* [*Hipposyus Limnotherium Prosinopa Telmalestes Telmatolestes Thinolestes Tomitherium*] L.-M. Eoc. NA. *Pelycodus* L. Eoc. NA. Eu. *Periconodon* M. Eoc. Eu. *?Petrolemur* U. Paleoc. EAs. *Pronycticebus* U. Eoc. Eu. *Protoadapis* [*Europolemur Megatarsius*] L.-U. Eoc. Eu. *Smilodectes* [*Aphanolemur*] M. Eoc. NA.

INFRAORDER LEMURIFORMES

SUPERFAMILY LEMUROIDEA **Lemuridae** R. Mad. Comora Islands. **Megalapidae** *Megalapis* R. Mad.

SUPERFAMILY LORISOIDEA **Lorisidae** *Komba* L. Mioc. EAf. *Mioeuoticus* L. Mioc. EAf. *Progalago* L. Mioc. EAf. **Cheirogaleidae** R. Mad.

SUPERFAMILY INDRIOIDEA **Indriidae** *Mesopropithecus* [*Neopropithecus*] subfossil Mad. **Daubentoniidae** R. Mad. **Archaeolemuridae** *Archaeolemur Hadropithecus* subfossil Mad. **Palaeopropithecidae** *Palaeopropithecus Archaeoindris* subfossil Mad.

INFRAORDER TARSIIFORMES **Omomyidae** *Absarokius* L.-M. Eoc. NA. *Altanius* L. Eoc. CAs. *Anaptomorphus* M. Eoc. NA. *Anemorhysis* L. Eoc. NA. *Chlororhysis* L. Eoc. NA. *Chumashius* U. Eoc. NA. *Donrussellia* L. Eoc. Eu. *Dyseolemur* M. Eoc. NA. *Ekgmowechashala* L. Mioc. NA. *Hemiacodon* M. Eoc. NA. *Hoanghonius* M. or U. Eoc. EAs. *Kohatius* L. Eoc. SAs. *Loveina* L. Eoc. NA. *Macrotarsius* M. Eoc. L. Olig. NA. *Mckennamorphus* L. Eoc. NA. *Microchoerus* U. Eoc. L. Olig. Eu. *Nannopithex* M. Eoc. Eu. *Necrolemur* M.-U. Eoc. Eu. *Omomys* [*Euryacodon Palaeacodon*] L.-M. Eoc. NA. *Ourayia* M. Eoc. NA. *Pseudoloris* U. Eoc. Eu. *Rooneyia* L. Olig. NA. *Shoshonius* L. Eoc. NA. *Stockia* M. Eoc. NA. *Teilhardina* L. Eoc. Eu. *Tetonius* [*Paratetonius*] L. Eoc. NA. *Trogolemus* M.-U. Eoc. NA. *Uintanius* L.-M. Eoc. NA. *Utahia* L.-M. Eoc. NA. *Washakus* [*Yamanius*] M. Eoc. NA. **Tarsiidae** *?Afrotarsius* Olig. NAf.

SUBORDER ANTHROPOIDEA

INFRAORDER INCERTAE SEDIS *Amphipithecus Pondaungia* U. Eoc. SAs.

INFRAORDER PLATYRRHINI **Cebidae** *Branisella* U. Olig. SA. *Dolichocebus* U. Olig. SA. *Neosaimiri* M. Mioc. SA. **Atelidae** *Cebupithecia* M. Mioc. SA. *Homunculus* L. Mioc. SA. *Stirtonia* M. Mioc. SA. *Tremacebus* U. Olig. SA. *Xenothrix* Pleist. WInd.

INFRAORDER CATARRHINI

SUPERFAMILY PARAPITHECOIDEA **Parapithecidae** *Apidium Parapithecus* M. Olig. NAf.

SUPERFAMILY CERCOPITHECOIDEA **Cercopithecidae** *Cercocebus* U. Plioc. L. Pleist. R. Af. *Cercopithecus* U. Plioc. R. Af. *Cercopithecoides* U. Plioc. M. Pleist. Af. *Colobus* U. Mioc. R. Af. *Dinopithecus* U. Plioc. Af. *Dolichopithecus* ?U. Mioc. Plioc. Eu. *Gorgopithecus* U. Plioc. SAf. *Libypithecus* U. Mioc. NAf. *Macaca* [*Aulacinus Macacus Rhesus*] U. Mioc. R. NAf. L. Plioc. L. Pleist. Eu. U. Plioc. R. As. ?M. Pleist. R. EInd. *Mesopithecus* U. Mioc. U. Plioc. Eu. WAs. *Papio* [*Brachygnathopithecus Choeropithecus Cynocephalus Dinopithecus Gorgopithecus Parapapio Simopithecus*] Plioc. R. Af. R. SAs. *Paracolobus* Plioc. EAf. *Paradolichopithecus* Plioc. Eu. EEu. CAs. *Parapapio* ?U. Mioc. L. Pleist. Af. *Presbytis* [*Paradolichopithecus Semnopithecus*] Pleist. R. SAs. EInd. Ind. ?U. Mioc. SAs. *Procynocephalus* U. Plioc. EAs. EInd. *Prohylobates* [*Victoriapithecus*] L. Mioc. NAf. *Pygathrix* L. Pleist. R. EAs. R. SAs. *Theropithecus* M. Plioc. R. Af. *Victoriapithecus* L.-M. Mioc. EAf. **Oreopithecidae** *Oreopithecus* U. Mioc. Eu.

SUPERFAMILY HOMINOIDEA **Pliopithecidae** *Dendropithecus* L.-M. Mioc. EAf. *Limnopithecus* M. Mioc. EAf. *Pliopithecus* [*Epipliopithecus Plesiopliopithecus*] M.-U. Mioc.

Eu. *Propliopithecus* M. Olig. NAf. **Hylobatidae** *?Aeolopithecus* Olig. NAf. *Hylobates* Pleist. R. EInd. R. SEAs. **Pongidae** *Dryopithecus* [*Adaetontherium Ankarapithecus Anthropodus Griphopithecus ?Hylopithecus Indopithecus Neopithecus Paidopithex Palaeopithicus Paleosimia Rhenopithicus Sugrivapithecus Udabnopithecus Xeonpithecus*] L.-M. Mioc. EAs. Mioc. EAf. M.-U. Mioc. Eu. *Gigantopithecus* [*Gigantanthropus*] U. Mioc. M. Pleist. EAs. SAs. *Proconsul* Mioc. EAf. *Oligopithecus* L. Olig. NAf. *Pongo* [*Simia*] L. Pleist. SAs. Pleist. R. EInd. *Sivapithecus* [*Bramapithecus ?Kenyapithecus Ramapithecus*] M.-U. Mioc. Eu. SAs. Ind. EAs. EAf. **Hominidae** *Australopithecus* [*Hemanthropus ?Meganthropus Paranthropus Plesianthropus Zinjanthropus*] ?M. Plioc. L. Pleist. SAf. EAf. *Homo* [*Atlanthropus Cyphanthropus Nipponanthropus Palaeanthropus Pithecanthropus Protanthropus Sinanthropus Telanthropus*] Pleist. R. Eu. Af. As. Aus. NA.

ORDER CREODONTA

SUBORDER HYAENODONTIA **Hyaenodontidae** *Alienetherium* M. Eoc. Eu. *Allopterodon* M. Eoc. Eu. *Anasinopa* L. Mioc. Af. *Apataelurus* U. Eoc. NA. *Apterodon* L. Olig. L. Mioc. Af. L. Olig. Eu. ?As. *Arfia* L. Eoc. NA. *Consobrinus* L. Olig. Eu. *Cynohyaenodon* [*?Pseudosinopa*] M.-U. Eoc. Eu. *Dissopsalis* L. Mioc. Af. M.-U. Mioc. SAs. *Francotherium* L. Eoc. Eu. *Hemipsalodon* L. Olig. NA. *Hessolestes* U. Eoc. NA. *Hyaenodon* [*Neohyaenodon Pseudopterodon Taxotherium*] U. Eoc. M. Olig. Eu. NA. Olig. L. Mioc. NAf. U. Eoc. U. Olig. As. *Hyaenodontipus* U. Eoc. Eu. *Hyainailourus* [*Hyaenaelurus*] L. Mioc. Eu. NAf. L.-M. Mioc. As. *Isohyaenodon* L. Mioc. Af. *Leakitherium* L. Mioc. EAf. *Limnocyon* [*Telmatocyon*] M.-U. Eoc. NA. *Machaeroides* M. Eoc. NA. *Masrasector* Olig. NAf. *Megistotherium* L. Mioc. NAf. *Metapterodon* L. Mioc. Af. *Metasinopa* U. Eoc. Eu. L. Olig. NAf. *Oxyaenodon* U. Eoc. NA. *Oxyaenoides* M. Eoc. Eu. *Paenoxyaenoides* L. Olig. Eu. *Paracynohyaenodon* U. Eoc. Eu. EAs. *Parapterodon* U. Eoc. Eu. *Paratritemnodon* M. Eoc. SAs. *Paroxyaena* U. Eoc. Eu. *Praecodens* M. Eoc. Eu. *Prodissopsalis* M. Eoc. Eu. *Prolaena* U. Eoc. EAs. *Prolimnocyon* M. Paleoc. L. Eoc. NA. *Propterodon* M. Eoc. Eu. U. Eoc. As. *Prototomus* [*Prolimnocyon*] L. Eoc. NA. L.-U. Eoc. Eu. *Proviverra* [*Geiselotherium Leonhardtina Prorhyzaena*] L.-M. Eoc. Eu. M. Eoc. NA. *Proviverroides* M. Eoc. NA. *Pterodon* U. Eoc. As. NA. U. Eoc. L. Olig. L. Mioc. NAf. *Quercytherium* U. Eoc. Eu. *Schizophagus* U. Eoc. Eu. *Sinopa* [*Stypolophus ?Triacodon*] ?U. Eoc. As. *Sivapterodon* M. Mioc. As. *Teratodon* L. Mioc. Af. *Thereutherium* M. Olig. Eu. *Thinocyon* [*Entomodon*] M. Eoc. NA. ?EAs. *Tritemnodon* ?L. Eoc. Eu. L.-M. Eoc. NA. ?U. Eoc. EAs. **Oxyaenidae** *Ambloctonus* [*Amblyctonus*] L. Eoc. NA. *Dipsalidictides* L. Eoc. NA. *Dipsalodon* U. Paleoc. NA. *Galethylax* U. Eoc. Eu. *Oxyaena* [*Dipsalidictis Argillotherium*] U. Paleoc. L. Eoc. NA. L. Eoc. EAs. *Palaeonictis* U. Paleoc. L. Eoc. NA. L. Eoc. Eu. *Patriofelis* [*Aelurotherium Limnofelis Oreocyon*] L.-M. Eoc. NA. *Sarkastodon* U. Eoc. EAs. *Tytthaena* U. Paleoc. NA.

ORDER CARNIVORA

SUPERFAMILY MIACOIDEA **Miacidae** *Eostictis* U. Eoc. NA. *Miacis* [*Mimocyon Xinyuictis*] L.-U. Eoc. NA. Eu. U. Eoc. As. *Oodectes* L.-M. Eoc. NA. *Palaearctonyx* M. Eoc. NA. *Pappictidops* M. Paleoc. EAs. *Paroodectes* M. Eoc. Eu. *Plesiomiacis* [*Pleurocyon*] U. Eoc. NA. *Procynodictis* U. Eoc. NA. *Prodaphaenus* U. Eoc. NA. *Simamphicyon* U. Eoc. Eu. *Tapocyon* U. Eoc. NA. *Uintacyon* L.-U. Eoc. NA. ?L. Eoc. ?Eu. *Vassacyon* L. Eoc. NA. *Vulpavus* L.-M. Eoc. NA. **Viverravidae** *Didymictis* M. Paleoc. L. Eoc. NA. L. Eoc. Eu. *Ictidopappus* M. Paleoc. NA. *Protictis* M.-U. Paleoc. NA. *Quercygale* M.-U. Eoc. Eu. *Viverravus* U. Paleoc. U. Eoc. NA. ?U. Eoc. Eu. ?L. Olig. As.

SUPERFAMILY AELUROIDEA (FELOIDEA) **Viverridae** R. Eu. As. Af. *Anictis* L. Olig. Eu. *Atilax Crossarchus* Pleist. R. Af. *Civettictis* ?U. Mioc. U. Plioc. L. Pleist. R. Af. *Cynictis* Pleist. R. Af. *Genetta* M. Mioc. R. Af. R. WAs. Eu. *Helogale* U. Plioc. U. Pleist. R. Af. *Herpestes* [*Calogale*] L. Mioc. R. Eu. L. Plioc. R. As. Af. *Herpestides* L.-M. Mioc. Eu. *Kichechia* L. Mioc. Af. *Macrogalidia* Pleist. R. EInd. *Mungos* Pleist. R. Af. *Paleoprionodon* L.-M. Olig. Eu. M. Olig. As. *Progenetta* [*Miohyaena*] L.-U. Mioc. Eu. U. Mioc. As. *Pseudocivetta* U. Plioc. L. Pleist. Af. *Semigenetta* L.-U. Mioc. Eu. M. Mioc. EAs. *Stenoplesictis* L.-U. Olig. Eu. *Suricata* Pleist. R. Af. *Tunguristis* U. Mioc. EAs. *Vishnuictis* U. Mioc. M. Plioc. L. Pleist. As. *Viverra* [*Anictis*] M. Mioc. R. Eu. L. Plioc. R. As. Pleist. R. EInd. Af. *?Viverricula* U. Pleist. R. EAs. **Hyaenidae** *Adcrocuta* U. Mioc. Eu. As. *Allohyaena* [*Dinocrocuta Xenohyaena*] U. Mioc. Eu. *Chasmaporthetes* [*Ailurena*] U. Plioc. L. Pleist. NA. Af. Plioc. Eu. U. Plioc. SWAs. *Hyaena* [*Plesiocrocuta*] L. Plioc. Pleist. Eu. L. Plioc. R. As. L. Plioc. R. Af. *Hyaenictis* M.–?U. Mioc. L. Plioc. Eu. L. Pleist. As. L. Plioc. Af. *Ictitherium* [*Galeotherium Hyaenalopex Palhyaena Thalassictis*] M. Mioc. L. Plioc. Eu. As. M. Mioc. NAf. *Leecyaena* L. Pleist. EAs. Pleist. SAf. *Lycyaena* U. Mioc. L. Plioc. Eu. M.-U. Mioc. As. *Miohyaena* M. Mioc. As. *Pachycrocuta* U. Mioc. L. Pleist. As. U. Mioc. U. Plioc. Eu. U. Plioc. Af. *Palinhyaena* U. Mioc. As. *Percrocuta* M.-U. Mioc. As. U. Mioc. Eu. Af. *Plioviverrops* L.-U. Mioc. Eu. *Protictitherium* M.-U. Mioc. SWAs. *Thalassictis* U. Mioc. Eu. **Felidae** *Acinonyx* [*Cynaelurus Schaubia*] U. Plioc. U. Pleist. Eu. As. U. Plioc. R. Af. Pleist. NA. *Adelphailurus* U. Mioc. or L. Plioc. NA. *Aelurogale* [*Ailurictis Ictidailurus Nimraviscus*] L. Olig. EAs. U. Eoc. M. Olig. Eu. *Aeluropsis* U. Mioc. As. *Afrosmilus* L. Mioc. Af. *Barbourofelis* U. Mioc. NA. *Dinaelurus* L. Mioc. NA. *Dinictis* L. Olig. L. Mioc. NA. *Dinobastis* Pleist. Eu. As. NA. *Dinofelis* U. Mioc. U. Plioc. As. L. Plioc. Eu. L. Plioc. L. Pleist. Af. U. Plioc. NA. *Eofelis* L.-M. Olig. Eu. *Epimachairodus* U. Mioc. EAs. ?U. Mioc. U. Plioc. Eu. *Eusmilus* [*Paraeusmilus*] U. Eoc. EAs. U. Eoc. L. Olig. Eu. U. Eoc. U. Olig. NA. *Felis* [*Leptailurus Lynx Neofelis Panthera Printinofelis*] U. Mioc. R. Eu. As. NA. M. Mioc. R. Af. Pleist. R. EInd. SA. *Homotherium* [*Epimachairodus*] U. Plioc. As. EInd. Eu. Af. NA. *Hoplophoneus* L. Olig. As. L.-U. Olig. NA. *Lynx* L. Plioc. U. Pleist. R. Eu. Af. U. Plioc. U. Pleist. R. NA. As. *Machairodus* [*Heterofelis Therailurus*] U. Mioc. L. Plioc. Eu. L. Plioc. NA. As. L. Plioc. U. Pleist. Af. *Megantereon* [*Toscanius*] U. Mioc. U. Plioc. As. L. Pleist. Eu. U. Pleist. Af. EInd. *Mellivorodon* U. Mioc. As. *Metailurus* M.-U. Mioc. As. U. Mioc. Eu. *Miomachairodus* U. Mioc.

SWAs. *Nimravides* U. Mioc. L. Plioc. NA. *Nimravus* [*Nimravinus*] U. Olig. L. Mioc. NA. L.-M. Olig. Eu. M. Olig. L. Mioc. As. *Paramachairodus* [*Propontosmilus Sivasmilus*] L. Plioc. Eu. L.-U. Plioc. As. *Pratifelis* U. Mioc. NA. *Proailurus* [*Brachictis Stenogale*] L. Olig. L. Mioc. Eu. M. Olig. As. *Propontosmilus* U. Mioc. L. Plioc. As. *Prosansanosmilus* L. Mioc. Eu. *Pseudaelurus* [*Ailuromachairodus Metailurus Sansanailurus Schizailurus*] L. Mioc. Af. L. Mioc. L. Plioc. Eu. NA. L. Mioc. U. Mioc. As. M. Mioc. EAs. *Quercyailurus* L. Olig. Eu. *Sansanosmilus* [*Albanosmilus Grivasmilus*] M. Mioc. Eu. M. Mioc. As. ?U. Mioc. NA. *Sivaelurus* M. Mioc. As. *Sivafelis* U. Plioc. As. *Sivasmilus* M. Mioc. As. *Smilodon* [*Smilodontopsis Trucifelis*] Pleist. SA. NA. *Smilodontidion* L. Pleist. SA. *Syrtosmilus* L. Mioc. NAf. *Vinayakia* M.-U. Mioc. As. *Viretailurus* U. Plioc. Eu. *Vishnufelis* M. Mioc. As.

SUPERFAMILY ARCTOIDEA (CANOIDEA) **Mustelidae** *Aelurocyon* L. Mioc. NA. *Ailurictis* M. Mioc. Eu. *Amphictis* U. Olig. L. Mioc. NA. *Anatolictis* M. Mioc. Eu. *Aonyx* ?U. Mioc. As. Pleist. R. Af. Eu. *Arctomeles* U. Plioc. Eu. *Arctonyx* ?Pleist. R. As. *Baranogale* U. Mioc. U. Plioc. Eu. *Beckia* U. Mioc. NA. *Brachyprotoma* L.-U. Pleist. NA. *Brachypsalis* L.-U. Mioc. NA. *Brachypsaloides* U. Mioc. NA. *Brachypsigale* L.-M. Plioc. NA. *Buisnictis* L.-U. Plioc. NA. *Canimartes* L. Pleist. NA. *Cernictis* U. Mioc. NA. *Circamustela* U. Mioc. Eu. *Conepatus* U. Plioc. R. SA. Pleist. R. NA. *Craterogale* U. Mioc. NA. *Dinogale* L. Mioc. NA. *Enhydra* L. Pleist. Eu. Pleist. NA. R. NPac. *Enhydrictis* [*Pannonictis*] L. Plioc. EAs. U. Plioc. Pleist. Eu. *Enhydriodon* M. Mioc. ?U. Plioc. Eu. M. Plioc. Af. U. Mioc. L. Pleist. As. *Enhydritherium* U. Mioc. L. Plioc. NA. *Eomellivora* [*Sivamellivora*] U. Mioc. Eu. As. NA. *Ferinestrix* M.-U. Plioc. NA. *Galictis* Pleist. R. SA. *Gulo* Pleist. R. Eu. NA. As. *Hadrictis Hydrictis* U. Mioc. Eu. *Ictonyx* U. Pleist. R. Af. *Ischyrictis* L.-U. Mioc. Eu. M. Mioc. SWAs. *Laphyctis* [*Ischyrictis*] ?L. Mioc. Eu. *Leptarctus* L. Mioc. L. Plioc. NA. U. Mioc. EAs. *Limnonyx* M. Mioc. Eu. *Lutra* [*Basarabictis Plesiolatax*] U. Mioc. R. Eu. Pleist. R. As. SA. NAf. NA. *Lutravus* U. Mioc. L. Plioc. NA. *Lyncodon* Pleist. R. SA. *Marcetia* U. Mioc. Eu. *Martes* ["*Hydrocyon*" ?*Paramartes Sansanictis*] M. Mioc. R. Eu. As. NA. *Martinogale* U. Mioc. L. Plioc. NA. *Megalictis* L. Mioc. NA. *Meles* [*Heterictis Iranictis*] L. Plioc. R. As. L. Pleist. R. Eu. *Mellalictis* M. Mioc. Af. *Mellidelavus* M. Mioc. Af. *Mellivora* [*Ursitaxus*] U. Plioc. As. L. Plioc. R. Af. *Melodon* ?U. Mioc. L. Plioc. As. *Mephitis* M. Plioc. R. NA. *Mephititaxus* L. Mioc. NA. *Mesomephitis* U. Mioc. Eu. *Miomephitis* L. Mioc. Eu. *Miomustela* L.-U. Mioc. NA. *Mionictis* M. Mioc. Eu. SWAs. M.-U. Mioc. NA. *Mustela* [*Paratanuki Putorius*] U. Mioc. R. Eu. NA. L. Plioc. R. As. Pleist. R. SA. Af. *Mustelictis* L. Olig. Eu. *Oligobunis* ?L. Mioc. NA. *Osmotherium* [*Pelycictis*] Pleist. NA. *Palaeogale* [*Bunaelurus*] L. Olig. L. Mioc. Eu. NA. M. Olig. As. *Palaemeles Paralutra* M. Mioc. Eu. *Pannonictis* L.-U. Plioc. Eu. U. Plioc. SWAs. *Paragale* L. Mioc. Eu. *Parataxidea* L. Plioc. Eu. As. *Paroligobunis* L. Mioc. NA. *Perunium* U. Mioc. Eu. *Plesiogale* L. Mioc. Eu. *Plesiogulo* U. Mioc. SWAs. Eu. L. Plioc. As. NA. *Plesiomeles* M. Mioc. Eu. *Pliogale* L. Plioc. NA. *Plionictis* M.-U. Mioc. NA. *Pliotaxidea* U. Mioc. L. Plioc. NA. *Potamotherium* U. Olig. L. Mioc. Eu. L.-U. Mioc. NA. *Presictis* M. Mioc. Eu. *Promartes* L. Mioc. NA. *Promeles* [?*Polgardia*] L. Plioc. Eu. As. *Promellivora* U. Mioc. As. *Promephitis* [?*Nannomephitis*] U. Mioc. L. Plioc. Eu. M.-U. Mioc. SWAs. *Proputorius* L. Mioc. Eu. M. Mioc. SWAs. L. Plioc. As. *Protarctos?* L. Plioc. Eu. *Sabadellictis* L. Plioc. Eu. *Satherium* M.-U. Plioc. NA. *Sinictis* L. Plioc. Pleist. As. L. Plioc. Eu. *Sivalictis* M. Mioc. Eu. As. *Sivaonyx* U. Mioc. Eu. U. Mioc. L. Plioc. As. L. Plioc. NAf. *Sminthosinus* M.-U. Plioc. NA. *Spilogale* U. Plioc. R. NA. *Stenogale* Olig. Eu. *Sthenictis* M. Mioc. L. Plioc. NA. *Stipanicicia* L. Pleist. SA. *Taxidea* U. Mioc. R. NA. *Taxodon* M. Mioc. Eu. *Tisisthenes* L. Pleist. NA. *Trigonictis* Plioc. L. Pleist. NA. *Trocharion* M. Mioc. Eu. *Trochictis* M. Mioc. SWAs. M. Mioc. L. Plioc. Eu. U. Mioc. As. *Trochotherium* M. Mioc. Eu. *Vishnuonyx* M. Mioc. As. *Vormela* U. Plioc. R. As. Plioc. R. Eu. *Xenictis* L.-U. Plioc. Eu. **Phocidae** *Acrophoca Pisciphoca* L. Plioc. SA. *Callophoca* U. Mioc. Eu. *Gryphoca* U. Mioc. Eu. *Leptophoca* M. Mioc. NA. *Monotherium* U. Mioc. Eu. *Palaeophoca* U. Mioc. Eu. *Phoca* [*Monachopsis ?Pontophoca ?Praepusa*] M. Mioc. R. Eu. NA. Plioc. R. NAs. NAtl. NPac. *Phocanella Platyphoca* U. Mioc. Eu. *Prophoca* U. Mioc. Eu. **Canidae** *Aelurodon* M. Mioc. U. Mioc. NA. *Aletocyon* L. Mioc. NA. *Alopex* [?*Xenalopex*] Pleist. R. Eu. NA. *Bassariscops* L.-M. Mioc. NA. *Borocyon* L. Mioc. NA. *Borophagus* [*Hyaenognathus*] U. Mioc. Pleist. NA. *Canis* [*Aenocyon Dinocynops Theriodictis Thos*] L. Plioc. R. Eu. As. NA. Aus. U. Plioc. R. Af. Pleist. SA. *Carpocyon* M.-U. Mioc. NA. *Cerdocyon* Pleist. SA. *Chailicyon* U. Eoc. As. *Cuon* [*Crassicuon Cyon Semicuon Sinocuon Xenocyon*] L. Pleist. Eu. U. Pleist. NA. Pleist. R. As. R. EInd. *Cynarctoides* U. Olig. L. Mioc. NA. *Cynarctus* M.-U. Mioc. NA. *Cynodesmus* U. Olig. L. Mioc. NA. *Dusicyon* Pleist. R. SA. *Epicyon* M.-U. Mioc. NA. *Fennecus* U. Pleist. R. Af. *Gobicyon* U. Mioc. As. Eu. *Hesperocyon* [*Pseudocynodictis*] L. Olig. L. Mioc. NA. *Leptocyon* M.-U. Mioc. NA. *Lycaon* Pleist. R. Af. *Mesocyon* ?L.-U. Olig. L. Mioc. NA. *Neocynodesmus* L. Mioc. NA. *Nothocyon* U. Olig. L. Mioc. NA. *Nyctereutes* U. Plioc. R. As. L. Plioc. Eu. *Osteoborus* U. Mioc. L. Plioc. NA. *Otocyon* [*Prototocyon*] Pleist. R. Af. *Pachycynodon* M. Olig. Eu. ?L. Eoc. L. Olig. EAs. *Philotrox* L. Mioc. NA. *Phlaocyon* L.-M. Mioc. NA. *Protocyon* [*Palaeocyon Palaeospeothos*] Pleist. SA. NA. *Proturocyon* U. Mioc. NA. *Strobodon* U. Mioc. NA. *Sunkahetanka* U. Olig. L. Mioc. NA. *Theriodictis* M. Pleist. SA. *Tomarctus* L. Mioc. L. Plioc. NA. *Urocyon* U. Plioc. R. NA. R. SA. *Vulpes* ?U. Mioc. L. Plioc. R. NA. U. Plioc. R. Eu. Af. Plioc. R. As. **Procyonidae** *Bassariscus* U. Mioc. R. NA. *Brachnasua* Pleist. SA. *Chapalmalania* U. Plioc. SA. *Cyonasua* [*Amphinasua ?Chapalmalania Pacynasua*] U. Mioc. U. Plioc. SA. *Edaphocyon* M. Mioc. NA. *Nasua* Pleist. R. SA. R. CA. M.-U. Plioc. NA. *Parailurus* L.-U. Plioc. Eu. L. Plioc. NA. *Plesictis* [*Mustelavus*] L. Olig. L. Mioc. Eu. NA. U. Olig. As. *Procyon* U. Mioc. R. NA. Pleist. R. SA. *Sivanasua* [*Ailuravus Schlossericyon*] L.-U. Mioc. Eu. M.-U. Mioc. As. *Zodiolestes* L. Mioc. NA. **Amphicyonidae** *Afrocyon* L. Mioc. NAf. As. *Agnotherium* [*Tomocyon*]

L. Mioc. L. Plioc. Eu. *Amphicyanis* L. Olig. Eu. *Amphicyon* [*Amphicyonops Ictiocyon*] M. Olig. L. Plioc. Eu. L.-U. Mioc. NA. ?M. Olig. M. Mioc. As. *Arctamphicyon* U. Mioc. As. *Brachycyon* L. Olig. Eu. *Cynelos* L. Olig. L. Mioc. Eu. L. Mioc. NA. *Cynodictis* U. Eoc. ?M. Olig. EAs. U. Eoc. L. Olig. Eu. *Daphoenictis* L. Olig. NA. *Daphoenocyon* L. Olig. NA. *Daphoenodon* L. Mioc. NA. *Daphoenus* L. Olig. L. Mioc. NA. *Euoplocyon* M. Mioc. NA. *Goupilictis* U. Olig. Eu. *Hadrocyon* U. Mioc. NA. *Haplocyon* L. Olig. L. Mioc. Eu. *Haplocyonoides* L. Mioc. Eu. *Haplocyonopsis* U. Olig. L. Eoc. Eu. *Harpagophagus* L.-M. Olig. Eu. *Hebucides* L. Mioc. Af. *Ischyrocyon* M.-U. Mioc. NA. *Mammacyon* L. Mioc. NA. *Paradaphoenus* L. Mioc. NA. *Pericyon* U. Olig. L. Mioc. NA. *Pliocyon* U. Mioc. NA. *Proamphicyon* L.-M. Olig. NA. *Protemnocyon* M. Olig. NA. *Pseudamphicyon* L.-M. Olig. Eu. *Pseudarctos* M.-U. Mioc. Eu. *Pseudocyon* U. Olig. M. Mioc. Eu. M.-U. Mioc. NA. *Pseudocyonopsis* U. Eoc. L. Mioc. Eu. *Sarcocyon* L.-M. Olig. Eu. *Symplectocyon* L. Olig. Eu. *Temnocyon* U. Olig. L. Mioc. NA. *Vishnucyon* M. Mioc. As. *Ysengrinia* U. Olig. L. Mioc. Eu. L. Mioc. ?NA. **Ursidae** *Absonodaphoenus* L. Mioc. NA. *Adelpharctos* M. Olig. Eu. *Agriotherium* [*Agriarctos Hyaenarctos Lydekkerion*] ?U. Mioc. L. Plioc. Pleist. Eu. U. Mioc. L. Plioc. NA. U. Plioc. Pleist. As. U. Mioc. L. Plioc. Af. *Ailuropoda* [*Aelureidopus Ailuropus*] U. Pleist. R. As. *Allocyon* U. Olig. L. Mioc. NA. *Alopecocyon* [*Alopecodon ?Galecynus Viretius*] M. Mioc. Eu. *Amphicticeps* L.-U. Olig. EAs. *Amphicynodon* [*Cynodon Paracynodon Plesiocyon*] L.-M. Olig. Eu. As. L. Mioc. NA. *Arctodus* [*Arctoidotherium Arctotherium Dinarctotherium Pararctotherium Proarctotherium Pseudarctotherium Tremarctotherium*] ?U. Plioc. Pleist. SA. Pleist. NA. *Broiliana* L. Mioc. Eu. *Campylocynodon* L. Olig. NA. *Cephalogale* M. Olig. L. Mioc. Eu. ?U. Eoc. L. Mioc. As. L. Mioc. NA. *Dinocyon* M.-U. Mioc. Eu. U. Mioc. As. *Enhydrocyon* U. Olig. L. Mioc. NA. *Harpaleocyon* M. Mioc. Eu. *Helarctos* U. Mioc. Eu. Pleist. R. As. *Hemicyon* [*Harpaleocyon Phoberocyon Plithocyon*] L.-U. Mioc. Eu. U. Mioc. As. L.-M. Mioc. NA. *Indarctos* U. Mioc. Eu. As. NA. *Melursus* Pleist. R. SA. ?As. *Pararctotherium* U. Pleist. SA. *Parictis* L. Olig. L. Mioc. NA. *Phoberocyon* L. Mioc. Eu. *Plionarctos* L. Plioc. NA. *Plithocyon* L.-M. Mioc. Eu. M. Mioc. SWAs. U. Mioc. NA. *?Selenarctos* U. Pleist. EAs. *Simocyon* [*Araeocyon Metarctos*] U. Mioc. L. Plioc. Eu. L. Plioc. NA. *Stromeriella* L. Mioc. Eu. *Thaumastocyon* M. Mioc. Eu. *Tremarctos* Pleist. NA. CA. R. SA. *Ursavus* M. Mioc. SWAs. EAs. NA. *Ursulus* U. Plioc. Eu. *Ursus* [*Euarctos Selenarctos Thalarctos*] L. Plioc. R. Eu. NA. Pleist. NAf. Pleist. R. As. *?Metarctos* L. Mioc. As.

SUPERFAMILY OTARIOIDEA **Enaliarctidae** *Enaliarctos Pinnarctidion* L. Mioc. NA. *Kamtschatarctos* M. Mioc. NAs. **Desmatophocidae** *Allodesmus* L. Mioc. NA. M. Mioc. EAs. *Desmatophoca* L. Mioc. NA. **Otariidae** *Arctocephalus* Pleist. SAf. R. SOc. *Callorhinus* Plioc. R. Pac. *Eumetopias* Plioc. R. NA. SA. *Pithanotaria* M.-U. Mioc. NA. *Thalassoleon* U. Mioc. Plioc. NA. **Odobenidae** *Alachtherium* Plioc. Eu. *Aivukus* U. Mioc. Eu. NA. *Dusignathus* U. Mioc. NA. *Imagotaria* M.-U. Mioc. NA. *Neotherium* M. Mioc. NA. *Odobenus* Plioc. Pleist. Eu. Pleist. R. NAtl. NPac. *Pliopedia* U. Mioc. NA. *Pontolis* U. Mioc. NA. *Prorosmarus* U. Mioc. NA. *Valenictis* L. Plioc. NA.

CARNIVORA INCERTAE SEDIS *Adracon* L. Olig. Eu. *Kolponomus* L. Mioc. NA.

ORDER ANAGALIDA **Anagalidae** *Anagale* L. Olig. EAs. *Anagalopsis* ?L. Olig. EAs. *Anaptogale* M. Paleoc. EAs. *Chianshania* M. Paleoc. As. *Diacronus* M. Paleoc. EAs. *Hsiuannania* U. Paleoc. EAs. *Huaiyangale* M.-U. Paleoc. EAs. *Kashanagale* U. Paleoc. As. *Linnania Stenanagale Wangogale* M. Paleoc. EAs. **Pseudictopidae** *Allictops* U. Paleoc. EAs. *Anictops* M. Paleoc. EAs. *Cartictops* M. Paleoc. EAs. *Halictops* U. Paleoc. As. *Paranictops* U. Paleoc. L. Eoc. EAs. **Eurymylidae** *Eurymylus* U. Paleoc. EAs. *Heomys* M.-U. Paleoc. EAs. *Matutinia Rhombomylus* L. Eoc. EAs. **Mimotonidae** *Gomphos* U. Paleoc. EAs. *Hypsimylus* U. Eoc. EAs. *Mimolagus* ?L. Olig. EAs. *Mimotona* M.-U. Paleoc. EAs. **Family incertae sedis** *Anchilestes* M. Paleoc. EAs. *Ardynictis* L. Olig. EAs. *Praolestes* U. Paleoc. As. L. Eoc. NA.

ORDER RODENTIA

SUBORDER SCIUROGNATHI

INFRAORDER PROTROGOMORPHA

SUPERFAMILY ISCHYROMYOIDEA **Paramyidae** *Ailuravus* [*Megachiromyoides*] L.-U. Eoc. Eu. *Decticadapis* L. Eoc. Eu. *Franimys* U. Paleoc. L. Eoc. NA. *Ischyrotomus* M. Eoc. L. Olig. NA. *Leptotomus* L. Eoc. L. Olig. NA. *Lophiparamys* L.-U. Eoc. NA. *Manitsha* U. Eoc. L. Olig. NA. *?Masillamys* M.-U. Eoc. Eu. *Microparamys* L. Eoc. Eu. U. Paleoc. L. Olig. NA. *Myronomys* U. Eoc. L. Olig. NA. *Paramys* U. Paleoc. M. Eoc. NA. L.-M. Eoc. Eu. *Pseudotomus* L.-M. Eoc. NA. *Rapamys* U. Eoc. NA. *Reithroparomys Thisbemys* L.-U. Eoc. NA. **Sciuravidae** *Knightomys* L. Eoc. NA. *Pauromys* M. Eoc. NA. *Sciuravus* L.-U. Eoc. NA. U. Eoc. EAs. *Taxymys* M.-U. Eoc. NA. *Tillomys* M. Eoc. NA. **Cylindrodontidae** *Ardynomys* L. Olig. EAs. NA. M. Olig. CAs. *Cyclomylus* M.-U. Olig. As. *Cylindrodon* L. Olig. NA. *Dawsonemys* L. Eoc. NA. *Jaywilsonomys* L. Olig. NA. *Mysops* L.-U. Eoc. NA. *Pareumys Presbymys* U. Eoc. NA. *Pseudocylindrodon* U. Eoc. L. Olig. NA. M.-U. Olig. EAs. *Sespemys* U. Olig. NA. **Protoptychidae** *Protoptychus* M.-U. Eoc. NA. **Ischyromyidae** *Hulgana* L. Olig. EAs. *Ischyromys* ?U. Eoc. Olig. NA. *Plesiarctomys* M.-U. Eoc. Eu. *Titanotheriomys* L. Olig. NA.

ISCHYROMYOIDEA INCERTAE SEDIS *Floresomys* L. Olig. NA. *Meldimys* L. Eoc. Eu. *Metkamys* M. Eoc. SAs. *Paracitellus* U. Olig. L. Mioc. Eu. *Pseudoparamys* L. Eoc. Eu.

SUPERFAMILY APLODONTOIDEA **Aplodontidae** *Allomys* U. Olig. L. Mioc. NA. *Ameniscomys* M. Mioc. Eu. *Aplodontia* Pleist. R. NA. *Cedromus* L.-M. Olig. NA. *Crucimys* L. Mioc. NA. *Downsimus* U. Olig. NA. *Eohaplomys* U. Eoc. NA. *Haplomys* U. Olig. L. Mioc. NA. *Liodontia* M.-U. Mioc. NA. *Meniscomys* L. Mioc. NA. *Niglarodon* U. Olig. L. Mioc. NA. *Oligopetes* L. Olig. Eu. *Pelycomys* L.-M. Olig. NA. *Pipestoneomys* L. Olig. NA. *Pleisispermophilus* L.-U. Olig. Eu. U. Olig. CAs. ?L. Olig. NA. *Prosciurus* L. Olig. L. Mioc. NA. *Pseudaplo-*

don L. Plioc. As. NA. *Sciurodon* L. Olig. M. Mioc. Eu. *Selenomys* U. Olig. EAs. *Spurimus* U. Eoc. L. Olig. NA. *Tardontia* M. Mioc. L. Plioc. NA. *Trigonomys* L. Olig. Eu. *Wellelodon* L. Mioc. NA. **Mylagaulidae** *Epigaulus* U. Mioc. NA. *Mesogaulus* L.-U. Mioc. NA. *Mylagaulodon* L. Mioc. NA. *Mylagaulus* L. Mioc. L. Plioc. NA. *Promylagaulus* L. Mioc. NA.

INFRAORDER SCIUROMORPHA

SUPERFAMILY SCIUROIDEA **Sciuridae** *Albanensia* Mioc. Eu. *Aliveria* L. Mioc. Eu. *Ammospermophilus* U. Mioc. R. NA. *Atlantoxerus* Mioc. L. Plioc. Eu. M. Mioc. SWAs. R. NAf. *Blackia* L. Mioc. L. Plioc. Eu. L. Mioc. NA. *Cryptopterus* L. Plioc. Eu. M.-U. Plioc. NA. *Cynomys* L. Plioc. R. NA. R. As. *Eutamias* U. Mioc. R. As. NA. L. Plioc. R. Eu. *Forsythia* M. Mioc. Eu. *Getuloxerus* U. Mioc. NAf. *Glaucomys* Pleist. R. NA. *Heteroxerus* L. Olig. U. Mioc. Eu. *Miopetaurista* L.-U. Mioc. Eu. *Miosciurus* L. Mioc. NA. *Miospermophilus* L.-M. Mioc. NA. *Paenemarmota* U. Mioc. L. Pleist. NA. *Palaearctomys* L.-U. Mioc. NA. *Palaeosciurus* L. Olig. L. Mioc. Eu. *Petaurista* U. Pleist. R. As. EInd. *Petauristodon* M.-U. Mioc. NA. *Petinomys?* L. Plioc. Eu. *Pliopetaurista* U. Mioc. L. Plioc. Eu. *Pliopetes* U. Mioc. U. Plioc. Eu. *Pliosciuropterus* L.-U. Plioc. Eu. *Protosciurus* U. Olig. L. Mioc. NA. *Protospermophilus* L.-U. Mioc. NA. *Ratufa* L. Mioc. Eu. ?L. Mioc. Pleist. R. EInd. R. SAs. *?Sciuropterus* U. Plioc. Eu. *Sciurotamias* Pleist. R. As. *Sciurus* L. Olig. R. Eu. L. Pleist. R. NA. Pleist. R. As. SA. R. NAf. *Similisciurus* L. Mioc. NA. *Spermophilinus* L. Mioc. L. Plioc. Eu. *Spermophilus* M. Mioc. R. NA. U. Plioc. R. Eu. Pleist. R. As. *Tamias* [*Neotamias*] L. Mioc. Eu. L. Mioc. R. NA. Pleist. EAs. *Tamiasciurus* Pleist. R. NA. *Vulcanisciurus* L. Mioc. Af. *Xerus* ?Mioc. Eu. Pleist. R. Af.

INFRAORDER CASTORIMORPHA **Castoridae** *Agnotocastor* U. Eoc. L. Olig. NA. M. Olig. As. *Anchitheriomys* [*Amblycastos*] L.-U. Mioc. NA. M. Mioc. Eu. As. *Boreofiber* L. Plioc. Eu. *Capacikala Capatanka* L. Mioc. NA. *Castor* L. Plioc. R. NA. Eu. As. *Castoroides* U. Plioc. Pleist. NA. *Chalicomys* U. Mioc. As. *Dipoides* U. Mioc. L. Plioc. Eu. L. Plioc. L. Pleist. NA. U. Mioc. As. *Eucastor* [*Sipmogomphius*] U. Mioc. L. Plioc. NA. Plioc. As. *Euhapsis* L. Mioc. NA. *Hystricops* U. Mioc. NA. *Monosaulax* M.-U. Mioc. NA. Mioc. EAs. U. Mioc. Eu. *Palaeocastor* U. Olig. WAs. Mioc. NA. *Palaeomys* [*Chalicomys*] U. Mioc. L. Plioc. Eu. *Paradipoides* Pleist. NA. *Procastoroides* M. Plioc. L. Pleist. NA. *Propalaeocastor* L. Olig. L. Mioc. As. *Romanocastor* U. Plioc. Eu. *Sinocastor* Plioc. L. Pleist. EAs. *Steneofiber* [*Steneotherium*] L. Olig. L. Plioc. Eu. U. Olig. As. *Trogontherium* [*Conodontes*] M. Mioc. U. Pleist. Eu. Pleist. As. *Youngifiber* M. Mioc. As. *Zamolxifiber* U. Plioc. Eu. **Eutypomyidae** *Eutypomys* L. Olig. L. Mioc. NA. *Janimus* U. Eoc. NA.

INFRAORDER UNNAMED

SUPERFAMILY GLIROIDEA **Gliridae** (**Myoxidae**) *Armantomys* L.-M. Mioc. Eu. *Branssatoglis* U. Olig. M. Mioc. Eu. *Dryomimus* L. Plioc. Eu. *Dryomys* [*Dyromys*] U. Pleist. R. Eu. M. Mioc. Af. R. As. *Eliomys* Mioc. R. Eu. *Eogliravus* L.-M. Eoc. Eu. *Eomuscardinus* M. Mioc. Eu. *Gliravus* U. Eoc. U. Olig. Eu. *Glirudinus* M. Olig. M. Mioc. Eu. *Glirulus* U. Mioc. U. Plioc. Eu. R. As. *Glis* [*Myoxus*] L. Mioc. R. Eu. R. WAs. *Graphiurops* U. Mioc. Eu. *Heteromyoxus* L.-M. Mioc. Eu. *Hyponomys* L. Plioc. U. Pleist. Eu. *Leithia* Pleist. Eu. Malta *Microdyromys* L.-U. Mioc. Eu. M. Mioc. SWAs. M. Mioc. EAs. *Mioglis* L. Mioc. Eu. *Muscardinus* M. Mioc. R. Eu. R. WAs. *Myoglis* M.-U. Mioc. Eu. *Myomimus* [*Dryomimus Philistomys*] L. Mioc. L. Plioc. Eu. U. Mioc. R. As. *Nievella* L. Mioc. Eu. *Oligodromys* M. Olig. Eu. *Paraglirulus Pentaglis* M. Mioc. Eu. *Peridyromys* L. Olig. L. Mioc. Eu. *Praearmantomys* L. Mioc. Eu. *Pseudodyromys* L.-M. Mioc. Eu. M. Mioc. SWAs. *Tempestia* M.-U. Mioc. Eu. *Vasseuromys* L.-U. Mioc. Eu. **Seleviniidae** R. As. *Plioselevinia* L. Plioc. Eu.

INFRAORDER MYOMORPHA

SUPERFAMILY GEOMYOIDEA **Eomyidae** *Adjidaumo* [*Gymnoptychus*] L. Olig. L. or M. Mioc. NA. *Apeomys* L. Mioc. Eu. *Aulolithomys Centimanomys* L. Olig. NA. *Cupressimus* L. Olig. Eu. *Eomyops* M. Mioc. Eu. *Eomys* L.-U. Olig. Eu. L.-M. Olig. NA. *Kansasimys* U. Mioc. NA. *Keramidomys* L. Mioc. L. Plioc. Eu. *Leptodontomys* M.-U. Mioc. NA. Af. *Ligerimys* L. Mioc. Eu. *Meliakrouniomys* L. Olig. NA. *Metadjidaumo* L. Olig. NA. *Namatomys* L. Olig. NA. *Paradjidaumo* Olig. NA. *Protoadjidaumo* U. Eoc. NA. *Pseudadjidaumo* M. Mioc. NA. *Pseudotheridomys* M. Olig. L. Mioc. Eu. L.-M. Mioc. NA. *Rhodanomys* M. Olig. L. Mioc. Eu. *Ritteneria* L. Mioc. Eu. *Ronquillomys* U. Mioc. NA. *Viejadjidaumo* L. Olig. NA. *Yoderimys* L. Olig. NA. **Florentiamyidae** *Florentiamys* U. Olig. L. Mioc. NA. *Kirkomys* U. Olig. NA. *Sanctimus* U. Olig. L. Mioc. NA. **Geomyidae** *Cratogeomys* L. Plioc. R. NA. *Dikkomys* L. Mioc. NA. *Diplolophus* [*Gidleumys*] M. Olig. NA. *Entoptychus* U. Olig. L. Mioc. NA. *Geomys* [*Nerterogeomys Parageomys*] U. Mioc. R. NA. *Grangerimus* U. Olig. L. Mioc. NA. *Gregorymys* L. Mioc. NA. *Griphomys* U. Eoc. L. Olig. NA. *Heterogeomys* U. Pleist. R. NA. *Jimomys* L.-M. Mioc. NA. *Lignimus* M.-U. Mioc. NA. *Mojavemys* M. Mioc. NA. *Nerterogeomys* M. Plioc. L. Pleist. NA. *Nonomys* [*Subsumus*] L. Olig. NA. *Parapliosaccomys* M.-U. Mioc. NA. *Pleurolicus* L. Mioc. NA. *Pliogeomys* U. Mioc. U. Plioc. NA. *Pliosaccomys* U. Mioc. NA. *Prodipodomys* M. Mioc. L. Pleist. NA. *Prothomomys* U. Mioc. NA. *Schizodontomys* L. Mioc. NA. *Tenudomys* ?U. Olig. L. Mioc. NA. *Thomomys* L. Plioc. R. NA. *Ziamys* L. Mioc. NA. **Heteromyidae** *Akmaiomys Apletotomeus* M. Olig. NA. *Cupidinimus* M.-U. Mioc. NA. M. Mioc. Eu. *Dipodomys* U. Mioc. L. Pleist. R. NA. *Diprionomys* M.-U. Mioc. NA. *Etadonomys* M. Plioc. L. Pleist. NA. *Heliscomys* L. Olig. L. Mioc. NA. *Hitonkala* L. Mioc. NA. *Liomys* L. Pleist. R. NA. *Mookomys* L.-M. Mioc. NA. *Peridiomys* M.-U. Mioc. NA. *Perognathoides* U. Mioc. L. Plioc. NA. *Perognathus* U. Mioc. R. NA. *Proheteromys* L.-U. Mioc. NA.

SUPERFAMILY DIPODOIDEA **Dipodidae** *Alactaga* U. Plioc. R. As. Pleist. R. Eu. R. NAf. *Dipus* Pleist. R. As. EEu. *Paralactaga* U. Mioc. L. Plioc. EAs. EEu. *Sminthoides* L. Pleist. As. **Zapodidae** *Eozapus* U. Mioc. Eu. U. Mioc. R. As. *Heterosminthus* M. Mioc. As. *Macrognathomys* M.

Mioc. L. Plioc. NA. *Megasminthus* M. Mioc. NA. *Napaeozapus* Pleist. R. NA. *Plesiosminthus* [*Parasminthus Schaubenemys*] U. Eoc. L. Mioc. As. M. Olig. L. Mioc. Eu. U. Olig. U. Mioc. NA. *Pliozapus* U. Mioc. L. Plioc. NA. *Proalactaga* M. Mioc. As. *Protozapus* U. Mioc. Eu. *Sicista* U. Plioc. R. Eu. As. *Sminthozapus* U. Mioc. L. Plioc. Eu. *Zapus* Plioc. R. NA. R. As. **Simimyidae** *Simimys* U. Eoc. L. Olig. NA.

SUPERFAMILY MUROIDEA **Cricetidae** *Adelomyarion* U. Olig. Eu. *Afrocricetodon* Mioc. Af. *Akodon* U. Plioc. R. SA. *Allocricetus* U. Plioc. Eu. WAs. *Allophaiomys* U. Plioc. L. Pleist. Eu. *Allospalax* U. Mioc. Eu. *Alticola* Pleist. R. As. *Anomalomys* Mioc. Eu. *Aralomys Argyromys* Olig. WAs. *Aratomys* L. Plioc. As. *Arvicola* Pleist. R. As. L. Plioc. Pleist. R. Eu. *Atopomys* L.-M. Pleist. NA. *Auliscomys* M. Plioc. SA. *Baiomys* L. Plioc. R. NA. *Baranomys* Plioc. Eu. *Bensonomys* U. Mioc. Plioc. NA. *Blancomys* L. Plioc. Eu. *Bolomys* U. Plioc. R. SA. *Byzantinia* U. Mioc. Eu. *Calomys* U. Mioc. U. Plioc. NA. R. SA. *Calomyscus* L. Plioc. Eu. R. As. *Cimarronomys* L. Plioc. NA. *Clethrionomys* [*Evotomys*] Pleist. R. As. NA. ?L.-U. Plioc. R. Eu. *Coloradoeumys* M. Olig. NA. *Copemys* L. Mioc. L. Plioc. NA. M.-U. Mioc. EAs. *Cotimus* U. Mioc. Eu. *Cricetinus* Pleist. EAs. *Cricetodon* [*Pseudoruscionomys*] U. Eoc. L. Olig. EAs. L.-U. Mioc. Eu. M. Olig. NA. U. Mioc. NAf. *Cricetops* U. Olig. As. *Cricetulodon* U. Mioc. Eu. *Cricetulus* L. Plioc. R. As. Pleist. R. Eu. *Cricetus* L. Plioc. R. Eu. ?Plioc. L. Pleist. R. As. *Dakkamys* M. Mioc. SWAs. *Dankomys* U. Plioc. SA. *Democricetodon* L.-U. Mioc. Eu. M. Mioc. SWAs. ?*Desmodillus* L. Pleist. Af. *Discrostonyx* Pleist. R. Eu. As. R. NA. *Dolomys* [*Pliomys Apistomys*] L.-U. Plioc. R. Eu. *Ellobius* Pleist. R. Eu. As. NAf. *Eoemys* L.-M. Olig. NA. *Eolagurus* L. Pleist. NAs. *Eothenomys* Pleist. R. As. *Epimeriones* L. Plioc. Eu. *Eucricetodon* U. Eoc. L. Mioc. Eu. As. *Eumyarion* L.-U. Mioc. Eu. *Eumys* L. Olig. L. Mioc. NA. ?U. Olig. As. *Eumysodon* U. Olig. L. Mioc. WAs. *Fahlbuschia* [*Pararuschinomys*] L.-U. Mioc. Eu. *Galushamys* U. Mioc. NA. *Gerbillus* Pleist. R. As. U. Plioc. R. Af. ?U. Mioc. Eu. *Geringia* U. Olig. NA. *Germanomys* L. Plioc. Eu. *Graomys* U. Plioc. R. SA. *Heterocricetodon* M.-U. Olig. Eu. *Holochilius* U. Pleist. R. SA. *Ischymomys* U. Mioc. Eu. *Ishimomys* U. Mioc. As. *Kanisamys* M.-U. Mioc. EAs. *Kislangia Lagurodon* U. Plioc. L. Pleist. Eu. *Kowalskia* U. Mioc. L. Plioc. Eu. *Lagurus* L. Plioc. Pleist. Eu. Pleist. R. As. U. Pleist. R. NA. *Lartertomys* L. Mioc. Eu. *Laugaritiomys* L. Pleist. Eu. *Leakeymys* M. Mioc. Af. *Leidymys* M. Olig. L. Mioc. NA. *Lemmus* [*Myodes*] Pleist. R. Eu. As. NA. *Leukaristomys* L. Plioc. Eu. *Megacricetodon* L.-U. Mioc. Eu. M. Mioc. As. *Melissiodon* L. Olig. M. Mioc. Eu. *Mesocricetus* Pleist. R. As. L. Plioc. R. Eu. *Microcricetus* U. Mioc. Eu. *Micromys* L. Plioc. R. As. Eu. *Microtodon* L. Plioc. As. U. Plioc. NA. L. Plioc. Eu. *Microtoscoptes* [*Goniodontomys*] L. Plioc. As. U. Mioc. NA. *Microtus* [*Chionomys Pedomys Phaiomys*] U. Plioc. R. Eu. L. Pleist. R. NA. Pleist. R. As. Pleist. R. NAf. *Mimomys Cosomys Cseria* L. Plioc. Pleist. Eu. NA. As. *Miochomys* U. Mioc. NA. *Myocricetodon* U. Mioc. NAf. As. *Myospalax* [*Miotalpa Siphneus*] U. Plioc. R. As. *Mystromys* Pleist. R. Af. *Nebraskomys* Plioc. NA. *Necromys* Pleist. SA. *Neocometes* L.-M. Mioc. Eu. *Neofiber* L. Pleist. R. NA. *Neondatra* M.-U. Plioc. NA. *Neotoma* L. Plioc. R. NA. *Notocricetodon* L. Mioc. Af. *Ochrotomys* U. Pleist. R. NA. *Ogmodontomys* U. Plioc. L. Pleist. NA. *Ondatra* [*Anaptogonia Sycium*] M. Plioc. R. NA. *Onychomys* L. Plioc. R. NA. *Oryzomys* U. Mioc. R. NA. Pleist. R. SA. *Paciculus* L.-M. Mioc. NA. *Palaeocricetus* M. Mioc. Eu. *Paracricetodon* Olig. Eu. ?L.-U. Olig. NA. *Paracricetulus* M. Mioc. EAs. *Paramicrotoscoptes* U. Mioc. NA. *Paratarsomys* L. Mioc. Af. *Paronychomys* U. Mioc. NA. *Peromyscus* [*Haplomylomys*] U. Mioc. R. NA. *Phenacomys* L. Pleist. R. NA. *Pitymys* Pleist. R. Eu. As. *Plesiocricetodon* M. Mioc. As. *Pliolemmus* L. Plioc. L. Pleist. NA. *Pliomys* L. Plioc. U. Pleist. Eu. U. Plioc. SWAs. L. Pleist. NA. *Pliophenacomys* Plioc. Pleist. NA. *Pliopotamys* M.-U. Plioc. NA. *Pliospalax* U. Mioc. L. Plioc. Eu. *Pliotomodon* U. Mioc. L. Plioc. NA. *Poamys* U. Mioc. NA. *Praesynaptomys* U. Plioc. Eu. *Prediscrostonyx* L.-M. Pleist. NA. *Promimomys* U. Mioc. NA. L. Plioc. As. U. Plioc. Eu. *Proneofiber* L. Pleist. NA. *Propliomys* L. Plioc. Eu. *Propliophenacomys* U. Mioc. NA. *Prosiphneus* U. Mioc. L. Plioc. EAs. *Prosomys* L. Plioc. Eu. *Prospalax* U. Mioc. L. Pleist. Eu. *Protalactaga* U. Olig. or L. Mioc. As. *Protatera* L. Plioc. As. *Prototomys* Pleist. Af. *Pseudocricetodon* L. Olig. L. Mioc. Eu. *Pseudomeriones* U. Mioc. L. Plioc. EAs. Eu. U. Mioc. SWAs. *Pterospalax* U. Mioc. or L. Plioc. Eu. *Ptyssophorus* Pleist. SA. *Reithrodon* Pleist. R. SA. *Reithrodontomys* M. Plioc. R. NA. R. SA. *Repomys* U. Mioc. NA. *Rhinocricetus* L. Plioc. L. Pleist. Eu. *Rhizospalax* M.-U. Olig. Eu. *Rotundomys* U. Mioc. Eu. *Ruscinomys* U. Mioc. L. Plioc. Eu. U. Plioc. Af. *Scottimus* M. Olig. L. Mioc. NA. *Sigmodon* U. Plioc. R. SA. U. Plioc. NA. *Spalax* L. Plioc. Pleist. R. Eu. Pleist. R. SWAs. NAf. *Spanocricetodon* M. Mioc. As. *Stachomys* L. Plioc. Eu. *Symmetrodontomys* M.-U. Plioc. NA. *Synaptomys* [*Microtomys Mictomys*] U. Plioc. R. NA. *Tatera* Plioc. Af. Pleist. R. Af. As. *Tregomys* U. Mioc. NA. *Trilophomys* L.-U. Plioc. Eu. *Turkomys* M. Mioc. Eu. *Tyrrhenicola* Pleist. Eu. *Ungaromys* L. Plioc. U. Plioc. Pleist. Eu. *Wilsoneumys* M.-U. Olig. NA. *Yatkolamys* L. Mioc. NA. *Zetamys* U. Olig. NA. *Zramys* U. Mioc. Af. **Muridae** *Acomys* U. Plioc. L. Pleist. R. Af. *Aethomys* U. Plioc. R. Af. *Antemus* M.-U. Mioc. As. *Anthracomys* L. Plioc. Eu. *Apodemus* U. Mioc. R. Eu. As. *Arvicanthus* Pleist. R. Af. L. Plioc. As. *Castillomys* U. Mioc. L. Plioc. Eu. *Dasymys Dendromus Grammomys* Pleist. R. Af. *Gerboa* U. Mioc. Eu. *Hydromys* Pleist. Aus. *Karnimata* U. Mioc. L. Plioc. As. *Leggada* ?L. Pleist. Af. *Limnioscomys* U. Plioc. Af. *Malacothrix* Pleist. R. Af. *Mastacomys* Pleist. R. Aus. *Mastomys* U. Plioc. Af. *Mus* L. Plioc. U. Pleist. As. Pleist. R. Eu. Af. R. Cos. *Myotomys* U. Pleist. Af. *Nesokia* Pleist. R. SAs. R. NAf. *Occitanomys* U. Mioc. L. Plioc. Eu. *Orientalomys* L. Plioc. Eu. U. Plioc. As. *Otomys* Pleist. R. Af. *Paraethomys* U. Mioc. L. Plioc. Eu. U. Mioc. SWAs. U. Plioc. Af. *Parapelomys* U. Mioc. L. Plioc. As. *Parapodemus* U. Mioc. Pleist. Eu. U. Mioc. L. Plioc. As. *Pelomys* L. Plioc. Eu. As. Pleist. R. Af. *Proceromys* L. Plioc. As. *Progonomys* U. Mioc. L. Plioc. Eu. As. *Rattus* [*Epimys*] Pleist. R. As. Pleist. R. Eu. Aus. Af. EInd. R. Cos. *Rhabdomys* Pleist. R. Af. *Rhagamys* Pleist. Eu. *Rhagapodemus* U. Mioc. U.

Plioc. Eu. *Saccostomus* Pleist. R. Af. *Saidomys* L. Plioc. Af. *Steatomys* U. Plioc. U. Pleist. R. Af. *Stephanomys* U. Mioc. U. Plioc. Eu. ?Plioc. Pleist. As. *Thallomys* ?L.-U. Pleist. R. Af. *Valermys* U. Mioc. Eu. *Zelotomys* Pleist. R. Af.

SUPERFAMILY SPALACOIDEA **Rhizomyidae** *Brachyrhizomys* U. Mioc. EAs. *Parrarhizomys* L. Plioc. EAs. *Rhizomyoides* M. Mioc. L. Plioc. As. *Rhizomys* U. Mioc. R. As. *Tachyoryctoides* M. Olig. L. Mioc. EAs.

INFRAORDER INDETERMINATE

SUPERFAMILY CTENODACTYLOIDEA **Ctenodactylidae** *Advenimus* L.-U. Eoc. As. *Africanomys Dubiomys* Mioc. NAf. *Irhoudia* U. Mioc. U. Plioc. Af. *Karakormys* M.-?U. Olig. As. *Leptotataromys* U. Olig. EAs. *Metasayimys* M. Mioc. As. Af. *Petrokozlovia* L.-U. Eoc. As. *Saykanomys* L.-U. Eoc. As. *Tamquammys* L.-M. Eoc. EAs. *Tataromys* M. Olig. L. Mioc. EAs. *Terrarboreus* M. Olig. CAs. *Testouromys* M.-U. Mioc. NAf. *Tsinlingomys* U. Eoc. EAs. *Woodomys* M. Olig. As. *Yindirtemys* U. Olig. EAs. *Yuomys* U. Eoc. EAs. **Chapattimyidae** *Chapattimys* M. Eoc. Ind. **Cocomyidae** *Cocomys* L. Eoc. EAs.

SUPERFAMILY PEDETOIDEA **Pedetidae** *Megapedetes* L. Mioc. EAf. Mioc. Eu. *Parapedetes* L. Mioc. SWAf. *Pedetes* Pleist. R. Af.

SUPERFAMILY ANOMALUROIDEA **Anomaluridae** R. Af. *?Nementchamys* U. Eoc. NAf. *Paranomalurus* L. Mioc. Af. *Zenderella* L. Mioc. Af.

SUPERFAMILY THERIDOMYOIDEA **Theridomyidae (Peudosciuridae)** *Columbomys* U. Olig. Eu. *?Diatomys* M. Mioc. As. *Elfomys* M. Eoc. M. Olig. Eu. *Estellomys* U. Eoc. Eu. *Issiodoromys* M. Olig. L. Mioc. Eu. *Microsuevosciurus* U. Eoc. L. Olig. Eu. *Neosciuromys* L. Mioc. Af. *Oltinomys* U. Eoc. Eu. *Paradelomys* M.-U. Eoc. Eu. *Patriotheridomys* U. Eoc. Eu. *Protadelomys* L.-M. Eoc. Eu. *Protophiomys* U. Eoc. NAf. *Pseudosciurus* U. Eoc. L. Olig. Eu. *Remys* M.-U. Eoc. Eu. *Sciuroides* M.-U. Eoc. Eu. *Sciuromys* L.-M. Olig. Eu. *Suevosciurus* U. Eoc. M. Olig. Eu. *Tarnomys* L. Olig. Eu. *Thalerimys* U. Eoc. Eu. *Theridomys* [*Trechomys*] U. Eoc. M. Olig. Eu. *Treposciurus* M.-U. Eoc. Eu.

SUBORDER HYSTRICOGNATHI

INFRAORDER BATHYGEROMORPHA **Bathygeridae** *Bathygeroides* L. Mioc. SWAf. *Cryptomys* Pleist. R. Af. *Gypsorhychus* U. Plioc. L. Pleist. Af. *Heterocephalus* Pleist. R. Af. *Paracryptomys* L. Mioc. Af. *Proheliophobius* L. Mioc. Af. **Tsaganomyidae** *Tsaganomys* M.-U. Olig. As.

INFRAORDER HYSTRICOMORPHA **Hystricidae** *Hystrix* [*Xenohystrix*] M. Mioc. R. Eu. L. Plioc. R. Af. U. Mioc. R. As. Pleist. EInd. *Sivacanthion* M. Mioc. As. *Xenohystrix* U. Plioc. L. Pleist. Af.

INFRAORDER PHIOMORPHA

SUPERFAMILY THRYONOMYOIDEA **Phiomyidae** *Andrewsimys* L. Mioc. Af. *Gaudeamus Metaphiomys* Olig. Af. *Paraphiomys* Olig. U. Mioc. Af. *Phiocricetomys* Olig. Af. *Phiomys* L. Olig. L. Mioc. NAf. **Thryonomyidae** *Epiphiomys* L. Mioc. Af. *Paraulacodus* M. Mioc. As. U. Mioc. Af. *Petromus* [*Petromys*] Pleist. R. Af. *Sayimys* M. Olig. U. Mioc. As. M. Mioc. Af. *Thryonomys* [*Aulacodus*] Pleist. R. Af. **Diamantomyidae** *Diamantomys* L. Mioc. Af. *Promonomys* L. Mioc. SAf. **Kenyamidae** *Kenyamys* L. Mioc. Af. *Simonimys* L. Mioc. Af. **Myophiomyidae** *Elmerimys* L. Mioc. Af. *Myophiomys* L. Mioc. Af. *Phiomyoides* L. Mioc. EAf.

INFRAORDER CAVIOMORPHA

SUPERFAMILY OCTODONTOIDEA **Octodontidae** *Acaremys* U. Olig. L. Mioc. SA. *Chasicomys* U. Mioc. SA. *Eoctodon* U. Olig. SA. *Eucelophorus* U. Plioc. SA. *Eucoleophorus* U. Plioc. SA. *Phthoramys* L. Plioc. SA. *Pithanotomys* U. Plioc. ?Pleist. SA. *Platypittamys* U. Olig. SA. *Pseudoplataeomys* Plioc. SA. *Sciamys* L. Mioc. SA. *Xenodontomys* M. Plioc. Pleist. SA. **Echimyidae** *?Acarechimys* L. Mioc. SA. *Adelphomys* L. Mioc. SA. *Cercomys* U. Mioc. L. Pleist. R. SA. *Chasichimys* U. Mioc. SA. *Deseadomys* U. Olig. SA. *Eumysops* [*Proaguti Proatherura*] U. Mioc. L. Pleist. SA. *Paradelphomys* U. Olig. SA. *Pattersomys* U. Mioc. SA. *Prospaniomys* U. Olig. SA. *Protadelphomys* U. Olig. SA. *Spaniomys* [*Graphimys Gyrignophus*] L. Mioc. SA. *Stichomys* L.-?M. Mioc. SA. **Ctenomyidae** *Actenomys* [*Dicoelophorus*] L. Plioc. L. Pleist. SA. *Ctemys* [*Paractenomys*] U. Plioc. R. SA. *Megactenomys* L. Pleist. SA. **Abrocomidae** *Protabrocoma* L.-U. Plioc. SA. **Capromyidae** *Isomyopotamus* U. Plioc. SA. *Myocastor* [*Myopotamus*] U. Plioc. R. SA. *Tramycastor* L. Pleist. SA.

SUPERFAMILY CHINCHILLOIDEA **Chinchillidae** *Lagostomopsis* U. Mioc. U. Plioc. SA. *Lagostomus* U. Plioc. R. SA. *Perimys* U. Olig. L. Mioc. SA. *Prolagostomus* L.-?M. Mioc. SA. *Scotaeumys* U. Olig. L. Mioc. SA. *Scotamys* U. Olig. SA. **Dasyproctidae** *Cephalomys* [*Orchiomys*] U. Olig. SA. *Litodontomys* U. Olig. SA. *Lomomys* ?L. Mioc. SA. *Neoreomys* L.-U. Mioc. SA. *Olenopsis* L.-U. Mioc. SA. *Scleromys* [*Lomomys*] L.-U. Mioc. SA. **Dinomyidae** *?Colpostemma Dabbenea Diaphoromys* U. Mioc. SA. *Eusigmomys* M. Mioc. SA. *Gyriabrus* U. Mioc. SA. *Potamarchus* [*Discolomys*] U. Mioc. M. Plioc. SA. *Simplimus* Mioc. SA. *Telicomys* U. Mioc. U. Plioc. SA. *Terastylus* U. Mioc. L. Plioc. SA.

SUPERFAMILY CAVIOIDEA **Eocardiidae** *Asteromys* U. Olig. SA. *Chubutomys* U. Olig. SA. *Eocardia* [*Dicardia ?Hedymys Tricardia*] L. Mioc. SA. *Luanthus Phanomys* U. Olig. L. Mioc. SA. *Megastus* M. Mioc. SA. *Schistomys* [*Procardia*] L. Mioc. SA. **Caviidae** *Allocavia* U. Mioc. SA. *Cardiomys* [*Caviodon Diocartherium Lelongia Neoprocavia Pseudocerdiomys*] M. Mioc. U. Plioc. SA. *Cavia* Pleist. R. SA. *Caviodon* ?U. Mioc. U. Plioc. SA *Caviops Dolicavia* U. Plioc. SA. *Dolichotis* [*Paradolichotis*] L. Pleist. R. SA. *Microcavia* [*Nannocavia*] Pleist. R. SA. *Neocavia* M. Plioc. SA. *Orthomyctera* U. Mioc. M. Pleist. SA. *Palaeocavia* U. Mioc. L. Pleist. SA. *Pascualia* ?U. Plioc. SA. *Procardiomys* L. Plioc. SA. *Prodolichotis* U. Mioc. SA.*?Propediolagus* U. Plioc. SA. **Hydrochoeridae** *Anchimysops* U. Plioc. SA. *Cardiatherium* [*Cardiotherium*] U. Mioc. ?L. Plioc. SA. *Chapalmatherium* U. Plioc. SA. *Hydrochoerus* Pleist. NA. WInd. Pleist. R. SA. *Kiyutherium* U. Mioc. SA. *Neoanchimys* U. Plioc. SA. *Neochoerus* [*Palaeohydrochoerus Pliohydrochoerus*

Prohydrochoerus Protohydrochoerus] Pleist. SA. NA. CA. *Procardiatherium* U. Mioc. ?M. Plioc. SA. *Protohydrochoerus* M.-U. Plioc. SA.

SUPERFAMILY ERETHIZONTOIDEA **Erethizontidae** *Coendu* M. Plioc. M. Pleist. NA. R. SA. *Disteiromys* M. Mioc. SA. *Eosteriomys* U. Olig. SA. *Erethizon* L. Pleist. R. NA. *Hypsosteiromys* U. Olig. SA. *Neosteiromys* M. Plioc. SA. *Parasteiromys* U. Olig. SA. *Protacaremys* [*Archaeocardia Eoctodon "Palaeocardia"*] U. Olig. SA. *Protosteiromys* U. Olig. SA. *Steiromys* [*Parasteiromys*] U. Olig. L. Mioc. SA.

ORDER RODENTIA INCERTAE SEDIS *Branisamys* U. Olig. SA. *Ectropomys* U. Eoc. Eu. *Guanajuatomys* L. Olig. NA. *Luribayomys* U. Olig. SA. *Nonomys* L. Olig. NA. *Pairomys* U. Eoc. Eu. *Prolapsus* M.-U. Eoc. NA. *Quercymys* L. Olig. Eu.

ORDER LAGOMORPHA **Stem lagomorphs—no family designated** *Desmatolagus* U. Olig. EAs. U. Eoc. U. Olig. NA. *Gobiolagus* U. Eoc. L. Olig. EAs. *Hsiuannania* U. Paleoc. EAs. *Lushilagus* U. Eoc. EAs. *Megalagus* L. Olig. L. Mioc. NA. *Mimolagus* ?Olig. EAs. *Mimotona* L.-U. Paleoc. EAs. *Mytonolagus* U. Eoc. NA. *Shamolagus* U. Eoc. EAs. ?L. Olig. Eu. **Ochotonidae** R. As. NA. *Alloptox* [*Metochotona*] M.-U. Mioc. As. *Amphilagus* U. Olig. U. Mioc. Eu. *Austrolagomys* L. Mioc. Af. *Bellatona* U. Mioc. EAs. *Bohlinotona* M. Olig. EAs. *Cuyamalagus* L. Mioc. NA. *Eurolagus* M.-U. Mioc. Eu. *Gripholagomys* L. Mioc. NA. *Hesperolagomys* M.-U. Mioc. NA. *Kenyalagomys* L. Mioc. EAf. *Lagopsis* L.-M. Mioc. Eu. *Marcuinomys* U. Olig. Eu. *Ochotona* [*Lagomys Lagotona Pliochotona Prochotona*] U. Mioc. R. Eu. NA. As. *Ochotonoides* L. Plioc. Eu. Plioc. Pleist. As. *Ochotonolagus* U. Olig. EAs. *Oreolagus* L.-M. Mioc. NA. *Paludotona* L. Plioc. Eu. *Piezodus* U. Olig. M. Mioc. Eu. *Procaprolagus* L.-M. Olig. EAs. *Prolagus* M. Mioc. Pleist. Eu. M.-U. Mioc. SWAs. *Ptychoprolagus* M. Mioc. Eu. *Russellagus* M. Mioc. NA. *Sinolagomys* L. Olig. L. Mioc. EAs. *Titanomys* U. Olig. M. Mioc. Eu. **Leporidae** *Agispelagus* U. Olig. EAs. *Alilepus* U. Mioc. L. Plioc. Eu. ?NA. ?Pleist. NA. *Aluralagus* L. Plioc. NA. *Archaeolagus* L. Mioc. NA. *Brachylagus* U. Pleist. NA. *Caprolagus* Plioc. R. As. *Chadrolagus* L. Olig. NA. *Hypolagus* L. Mioc. U. Plioc. NA. L. Plioc. L. Pleist. Eu. L. Pleist. As. *Lagotherium* U. Plioc. Eu. *Lepus* L. Plioc. R. Eu. Af. Pleist. R. As. NA. *Litogalus* M. Olig. NA. *Nekrolagus* L.-U. Plioc. NA. *Notolagus* [*Dicea*] Plioc. Pleist. NA. *Ordolagus* M. Olig. EAs. *Oryctolagus* L. Pleist. R. Eu. NAf. *Palaeolagus* L. Olig. L. Mioc. NA. *Panolax* M. Mioc. NA. *Pliolagus* L. Plioc. L. Pleist. Eu. *Pliopentalagus* L. Plioc. Eu. As. *Pratilepus* M.-U. Plioc. NA. *Procaprolagus* L.-U. Olig. EAs. *Pronolagus* Pleist. R. Af. *Sylvilagus* [*Palaeotapeti*] L. Pleist. R. NA. Pleist. R. SA. *Trischizolagus* L.-U. Plioc. Eu. *Veterilepus* U. Mioc. Eu.

ORDER CONDYLARTHRA **Arctocyonidae** (**Oxyclaenidae**) *Anacodon* ?U. Paleoc. L. Eoc. NA. *Arctocyon* [*Arctotherium Heterobrous Hyodectes*] U. Paleoc. Eu. Paleoc. NA. *Arctocyonides* [*Creodapis Procyonictis*] U. Paleoc. Eu. *Baioconodon* L. Paleoc. NA. *Chriacus* [*Lipodectes*] L. Paleoc. L. Eoc. NA. *Colpoclaenus* M.-U. Paleoc. NA. *?Deltatherium Deuterogonodon* M. Paleoc. NA. *Desmatoclaenus* L.-M. Paleoc. NA. *Eoconodon* L. Paleoc. NA. *Goniacodon* L.-M. Paleoc. NA. *Lambertocyon* M.-U. Paleoc. NA. *Landenodon* U. Paleoc. L. Eoc. Eu. *Loxolophus* L. Paleoc. NA. *Mentoclaenodon* U. Paleoc. NA. Eu. *Mimotricentes* L.-M. Paleoc. NA. *Neoclaenodon* M. Paleoc. NA. *Oxyclaenus* L.-M. Paleoc. NA. *Oxyprimus* L. Paleoc. NA. *Paratriisodon* U. Eoc. EAs. *Prothryptacodon* M. Paleoc. NA. *Protungulatum* U. Cret. L. Paleoc. NA. *Ragnarok* L. Paleoc. NA. *Spanoxyodon Stelocyon* M. Paleoc. NA. *Thangorodrim* L. Paleoc. NA. *Thryptacodon* U. Paleoc. L. Eoc. NA. *Tricentes* M.-U. Paleoc. NA. *Triisodon* M. Paleoc. NA. **Paroxyclaenidae** *?Dulcidon* M. Eoc. NAs. *Kiinkerishella* U. Eoc. NAs. *Kochictis* M. Olig. Eu. *Kopidodon* M. Eoc. Eu. *Paroxyclaenus* U. Eoc. Eu. *Pugiodens* [*Vulpavoides*] *Russellites* M. Eoc. Eu. *Spaniella* L. Eoc. Eu. **Tricuspiodontidae** *Paratricuspiodon* U. Paleoc. Eu. *Tricuspiodon* [*Conaspidotherium Plesiphenacodus*] U. Paleoc. Eu. **Mioclaenidae** *Bomburia Choeroclaenus* L. Paleoc. NA. *Eliopsodon* Paleoc. NA. *Litaletes* M. Paleoc. NA. *Mioclaenus* M. Paleoc. NA. *Promioclaenus* L.-M. Paleoc. NA. *Protoselene* Paleoc. NA. *Phenacodaptes* M.-U. Paleoc. NA. **Hyopsodontidae** *Aletodon* [*Platymastus*] U. Paleoc. L. Eoc. NA. *Decoredon* L. Paleoc. EAs. *Dipavali* U. Paleoc. Eu. *Dorralestes* U. Paleoc. NA. *Haplaletes* L.-U. Paleoc. NA. *Haplomylus* U. Paleoc. L. Eoc. NA. ?M. Eoc. EAs. *Hyposodus* U. Paleoc. U. Eoc. NA. L. Eoc. EAs. *Lessnessina* L. Eoc. Eu. *Litomylus* M.-U. Paleoc. NA. *Louisina* U. Paleoc. Eu. *Microhyus* U. Paleoc. L. Eoc. Eu. *Palasiodon* M. Paleoc. As. *Paschatherium* U. Paleoc. L. Eoc. Eu. *Utemylus* U. Paleoc. NA. *Yuodon* M. Paleoc. EAs. **Meniscotheriidae** *Ectocion* [*Gidleyina*] M. Paleoc. L. Eoc. NA. *Meniscotherium* [*Hyracops*] U. Paleoc. L. Eoc. NA. *Orthaspidotherium Pleuraspidotherium* U. Paleoc. Eu. *Prosthecion* U. Paleoc. NA. **Periptychidae** *Anisonchus* L.-U. Paleoc. NA. *Carsioptychus* [*Plagioptychus*] L.-M. Paleoc. NA. *Conacodon Earendil* L. Paleoc. NA. *Ectoconus* L. Paleoc. NA. ?EAs. *Fimbrethil Gillisonchus Haploconus Hemithlaeus Maiorana Mimatuta Oxyacodon Periptychus* L. Paleoc. NA. *Pseudanisonchus* U. Paleoc. EAs. *Tinuviel* L. Paleoc. NA. **Phenacodontidae** *Almogaver* M. Eoc. Eu. *Eodesmatodon* U. Eoc. EAs. *Phenacodus* U. Paleoc. L. Eoc. NA. L.-M. Eoc. Eu. *Tetraclaenodon* [*Euprotogonia*] M.-U. Paleoc. NA. **Didolodontidae** *Argyrolambda* L. Eoc. SA. *Asmithwoodwardia* U. Paleoc. L. Eoc. SA. *Didolodus* [*Cephanodus Lonchoconus Nephacodus*] L. Eoc. SA. *Enneoconus* L. Eoc. SA. *Ernestokokenia* [*Notoprogonia Progonia*] U. Paleoc. L. Eoc. SA. *Lamegoia* U. Paleoc. SA. *Oxybunotherium* L. Eoc. SA. *Paulogervaisia Proectocion* L. Eoc. SA. *Protheosodon* L. Olig. SA. **Phenacolophidae** *Ganolophus Minchenella Phenacolophus Tienshanilophus Yuelophus* U. Paleoc. EAs.

ORDER ARTIODACTYLA

SUBORDER PALAEODONTA **Dichobunidae** *Achaenodon* U. Eoc. Af. NA. *Antiacodon* [*Sarcolemur*] M. Eoc. NA. *Apriculus* [*Protelotherium*] U. Eoc. NA. *Aumelasia* M. Eoc. Eu. *Auxontodon* U. Eoc. NA. *Bunomeryx* U. Eoc. NA. *Bunophorus* L. Eoc. NA. *Buxobune* M. Eoc. Eu. *Chorlakkia* M. Eoc. SAs. *Diacodexis* L. Eoc. NA. Eu. *Dichobune* [*?Thylacomorphus*] M. Eoc. L. Olig. Eu. ?U. Eoc. NAs. M. Eoc. SAs. *Lophiohys* M. Eoc. NA. *Hexacodus* L. Eoc. NA. *Homacodon* [*Nanomeryx*] M. Eoc. NA. *Hylomeryx* [*?Sphenomeryx*] *Mesomeryx* U. Eoc. NA. *Hyperdichobune* U. Eoc. L. Olig. Eu. *Lantianius* L. Olig. EAs. *Lophiobunodon Meniscodon Messelobunodon* M. Eoc. Eu. *Metriotherium* L.-U. Olig. Eu. *Microsus* M. Eoc. NA. *Mioachoerus* U. Eoc. Eu. *Mouillacitherium* M. Eoc. L.

Olig. Eu. *Neodiacodexis* M. Eoc. NA. *Mytonomeryx Pentacemylus Taphochoerus Parahyus* U. Eoc. NA. *Protodichobune* L. Eoc. Eu. U. Eoc. EAs. **Helohyidae** *Bunodentus* M. Eoc. SAs. *Gobiohyus* M.-U. Eoc. EAs. *Haqueina* M. Eoc. SAs. *Helohyus* M. Eoc. NA. *Indohyus* M.-?U. Eoc. SAs. *Khirtharia Kunmunella Raoella* M. Eoc. SAs.

SUBORDER SUINA

SUPERFAMILY ENTELODONTOIDEA **Choeropotamidae** *Choeropotamus* U. Eoc. Eu. **Cebochoeridae** *Acotherulum* U. Eoc. Eu. *Cebochoerus* [*Acotherulum Leptacoltherulum*] M. Eoc. L. Olig. Eu. *Leptotheridium* M.-U. Eoc. Eu. *Moiachoerus* U. Eoc. Eu. *Mixtotherium* M.-U. Eoc. Eu. L. Olig. NAf. **Entelodontidae** (**Elotheridae**) *Archaeotherium* [*Choerodon Megachoerus Pelonax Scaptohyus*] L. Mioc. L.-U. Olig. EAs. *Brachyhyops* U. Eoc. L. Olig. NA. *Dinohyus* [*?Ammodon ?Boochoerus ?Daeodon*] L.-U. Mioc. NA. *?Dyscritochoerus* U. Eoc. NA. *Entelodon* [*Elodon Elotherium*] L.-M. Olig. Eu. As. ?NA. *Eoentelodon* U. Eoc. L. Olig. EAs. *?Megachoerus* M. Olig. NA. *Paraentelodon* U. Olig. L. Mioc. EAs. **Leptochoeridae** *Nanochoerus* Olig. NA. *Stibarus* L.-U. Olig. NA.

SUPERFAMILY SUOIDEA **Suidae** *Auerliachoerus* L. Mioc. Eu. *Bugtitherium* L. Mioc. As. *Bunolistriodon* L. Mioc. As. L.-M. Mioc. Eu. *Chleuastochoerus* L. Plioc. As. *Dicoryphochoerus* [*Hyosus*] M. Mioc. U. Plioc. As. *Emichoerus* M. Olig. Eu. *Hippohyus* U. Mioc. U. Plioc. As. *Hylochoerus* U. Pleist. R. Af. *Hyosus* U. Mioc. As. *Hyotherium* [*Amphichoerus Choeromorus Choerotherium Palaeochoerus*] L.-M. Mioc. Eu. L. Mioc. Af. M. Mioc. As. *Kolpochoerus* U. Plioc. Pleist. Af. *Korynochoerus* M.-U. Mioc. Eu. U. Mioc. SWAs. *Kubanochoerus* L. Mioc. NAf. M. Mioc. WAs. M. Mioc. Eu. *Listriodon* L.-M. Mioc. Af. L. Mioc. L. Plioc. Eu. L.-U. Mioc. As. *Lophochoerus* L. Plioc. As. *Lopholistriodon* M. Mioc. Af. *Mabokopithecus* L. Mioc. Af. *Mesochoerus* L. Pleist. Af. *Metridiochoerus* U. Plioc. U. Pleist. Af. *Microstonyx* M.-U. Mioc. Eu. U. Mioc. As. *Notochoerus* U. Plioc. U. Pleist. Af. *Nyanzachoerus* U. Mioc. Eu. Mioc. L. Plioc. Af. *Phacochoerus* [*Afrochoerus Gerontochoerus Kolpochoerus Metridiochoerus Notochoerus Potamochoeroides Potamochoerops Prontochoerus Synaptochoerus Tapinochoerus*] L. Pleist. R. Af. Pleist. WAs. *Potamochoeroides* U. Plioc. L. Pleist. Af. *Potamochoerus* [*Choiropotamus Eostopotamochoerus Koiropotamus Postpoamochoerus Propotamochoerus*] L. Pleist. R. Af. L. Plioc. Eu. M. Mioc. Pleist. As. *Promesochoerus* U. Plioc. Af. *Propalaeochoerus* U. Olig. Eu. *Propotamochoerus* M. Mioc. L. Plioc. As. L. Plioc. Eu. *Sanitherium* L.-M. Mioc. Af. M. Mioc. L. Plioc. Eu. As. *Schizochoerus* M. Mioc. SWAs. U. Mioc. Eu. As. *Sivachoerus* U. Mioc. SWAs. U. Mioc. U. Plioc. SAs. L. Pleist. NAf. *Sivahyus* U. Mioc. As. *Sonohyus* L.-M. Mioc. As. M.-U. Mioc. Eu. *Stylochoerus* U. Pleist. Af. *Sus* [*Microstonyx*] U. Mioc. R. Eu. As. L. Pleist. R. Af. Pleist. R. EInd. *Tetraconodon* L.-U. Plioc. As. *Xenochoerus* [*Diamantohyus*] U. Mioc. Eu. *Xenophyus* L. Mioc. Eu. **Tayassuidae** (**Dicotylidae**) *Albanohyus* Mioc. Eu. *Argyrohyus* M.-U. Plioc. SA. *Catagonus* Pleist. SA. *Chaenohyus* U. Olig. NA. *Cynorca* L. Mioc. NA. *Desmahyus* Mioc. NA. *Doliochoerus* M.-U. Olig. Eu. *Dyseohyus* L.-U. Mioc. NA. *Hesperhyus* [*Desmathyus Pediohyus*] L.-U. Mioc. NA. *Mylohyus* M. Plioc. Pleist. NA. *Palaeochoerus* L. Olig. M. Mioc. Af. L.-U. Mioc. As. *Pecarichoerus* U. Mioc. SAs. L. Plioc. SAf. *Perchoerus* [*Bothrolabis Thinohyus*] L. Olig. L. Mioc. NA. *Platygonus* [*Parachoerus*] Plioc. U. Pleist. NA. U. Plioc. Pleist. SA. *Prosthennops* M.-U. Mioc. NA. *Taucanamo* L.-M. Mioc. Eu. M. Mioc. SWAs. *Tayassu* [*Dicotyles Pecari*] Pleist. R. SA. Pleist. R. NA. *Thinohyus* L.-M. Olig. ?L. Mioc. NA.

SUPERFAMILY HIPPOPOTAMOIDEA **Anthracotheriidae** *Aepinacodon* L.-M. Olig. NA. *Anthracochoerus* L. Olig. Eu. *Anthracokeryx* U. Eoc. L. Olig. As. *Anthracosenex* U. Eoc. As. *Anthracothema* U. Eoc. As. *Anthracotherium* U. Eoc. L. Mioc. Eu. U. Eoc. M. Mioc. As. *Arretotherium* U. Olig. L. Mioc. NA. *Bothriodon* [*Aepinacodon Ancodon Ancodus Hypotamus*] U. Eoc. U. Olig. As. L. Olig. Eu. *Bothriogenys* L. Olig. L. Mioc. Af. *Brachyodus* L. Olig. L. Mioc. Af. L. Olig. M. Mioc. Eu. As. *Choeromeryx* U. Mioc. As. *Diplopus* U. Eoc. Eu. *Elomeryx* U. Eoc. U. Olig. Eu. U. Olig. NA. *Gelasmodon* L. Mioc. As. ?NAf. *Gonotelma* L. Mioc. As. *Hemimeryx* M. Olig. M. Mioc. As. *Heothema* U. Eoc. L. Olig. As. *Heptacodon* L.-U. Olig. NA. *Huananothema* U. Eoc. L. Olig. EAs. *Hyoboops* [*Merycops*] M. Olig. M. Mioc. As. L. Mioc. Af. *Masritherium* L. Mioc. Af. *Merycopotamus* U. Mioc. L. Plioc. NAf. M. Mioc. U. Plioc. SAs. *Microbunodon* [*Microselenodon*] M. Olig. Eu. *Octacodon* U. Olig. NA. *Parabrachyodus* L. Mioc. SAs. *Probrachyodos* U. Eoc. L. Olig. EAs. *Prominatherium* U. Eoc. Eu. *Telmatodon* [*Gonotelma*] L.-M. Mioc. As. *Thaumastognathus* U. Eoc. Eu. *Ulausuodon* U. Eoc. As. **Haplobunodontidae** *Anthracobunodon* M. Eoc. Eu. *Haplobunodon* M.-U. Eoc. Eu. *Massilabune* M. Eoc. Eu. *Rhagatherium* [*Amphirhagatherium*] M. Eoc. L. Olig. Eu. L. Olig. NAf. ?M. Mioc. As. **Hippopotamidae** *Hexaprotodon* [*Choeropsis*] U. Mioc. ?Eu. As. Af. L. Plioc. R. Af. *Hippopotamus* [*Hexaprotodon Hippoleakius Prochoeropsis Tetraprotodon*] U. Plioc. R. As. L. Plioc. R. Af. Pleist. Eu. EInd.

SUBORDER TYLOPODA

SUPERFAMILY MERYCOIDODONTOIDEA **Agriochoeridae** *Agriochoerus* L. Olig. L. Mioc. NA. *Diplobunops* [*Agriotherium Chorotherium Eomeryx Mesagriochoerus Protagriochoerus*] U. Eoc. NA. *Protoreodon* U. Eoc. L. Olig. NA. **Merycoidodontidae** (**Oreodontidae**) *Aclistomycter* U. Eoc. L. Olig. NA. *Bathygenys* L. Olig. NA. *Brachycrus* [*Pronomotherium*] M. Mioc. NA. *Cyclopidius* L.-U. Mioc. NA. *Dayohyus* L. Mioc. NA. *Desmatochoerus* [*Hypselochoerus Paradesmatochoerus*] L.-M. Olig. NA. *Epigenetochoerus* L. Mioc. NA. *Eporeodon* [*?Eucrotaphus*] M. Olig. L. Mioc. NA. *Generochoerus* M.-U. Olig. NA. *Hadroleptauchenia* M.-U. Olig. NA. *Hypsiops* U. Olig. L. Mioc. NA. *Leptauchenia* U. Olig. L. Mioc. NA. *Limnenetes* L. Olig. NA. *Mediochoerus* M. Mioc. NA. *Megabathygenys* L. Olig. NA. *Megasespia* U. Olig. NA. *Megoreodon* U. Olig. L. Mioc. NA. *Merychyus* [*Metoreodon*] L. Mioc. NA. *Merycochoerus* L.-M. Mioc. NA. *Mercoides* L. Mioc. NA. *Merycoidodon* [*Oreodon*] L.-U. Olig. NA. *Mesoreodon* L. Mioc. NA.

Metoreodon M. Mioc. NA. *Miniochoerus* [*Paraminiochoerus*] M.-U. Olig. NA. *Oreodontoides* [*Paroreodon*] L. Mioc. NA. *Oreonetes* L. Olig. NA. *Otionohyus* L.-U. Olig. NA. *Parabathygenys* L. Olig. NA. *Paramerychyus* L. Mioc. NA. *Paramerycoidodon* M.-U. Olig. NA. *Parastenopsochoerus* M. Olig. NA. *Phenacocoelus* L.-M. Mioc. NA. *Pithecistes* U. Olig. NA. *Platyochoerus* M.-U. Olig. NA. *Prodesmatochoerus* L.-M. Olig. NA. *Promerycochoerus* [*Paracotylops Paramerycochoerus Parapromerycochoerus Pseudopromerycochoerus*] U. Olig. L. Mioc. NA. *Promesodreodon Pseudocyclopidius* U. Olig. NA. *Pseudodesmatochoerus Pseudomesoreodon Pseudogenetochoerus* L. Mioc. NA. *Pseudoleptauchenia* M. Olig. NA. *Sespia* U. Olig. NA. *Stenopsochoerus* [*Pseudostenopsochoerus*] *Subdesmatochoerus* M.-U. Olig. NA. *Submerycochoerus Superdesmatochoerus* L. Mioc. NA. *Ticholeptus* [*Poatrephes*] M. Mioc. NA. *Ustatochoerus* U. Mioc. L. Plioc. NA.

SUPERFAMILY ANOPLOTHEROIDEA **Cainotheriidae** (**Caenotheriidae**) *Caenomeryx* L.-M. Olig. Eu. *Cainotherium* [*Caenotherium Procaenotherium*] M. Olig. M. Mioc. Eu. *Oxacron* U. Eoc. L. Olig. Eu. *Paroxacron* U. Eoc. L. Olig. Eu. *Plesiomeryx* M.-U. Olig. Eu. *Procaenotherium* M. Olig. Eu. **Anoplotheriidae** *Anoplotherium* U. Eoc. L. Olig. Eu. *Diplartiopus* U. Eoc. Eu. *Diplobune* U. Eoc. M. Olig. Eu. *Ephelcomenus* [*Hyracodontherium*] M. Olig. Eu. *Hyracodontherium* M. Olig. Eu.

SUPERFAMILY CAMELOIDEA **Camelidae** *Aepycamelus* Mioc. NA. *Aguascalientia* L. Mioc. NA. *Alforjas* U. Mioc. NA. *Blanococamelus* M. Plioc. NA. *Blickomylus* L.-M. Mioc. NA. *Camelodon* U. Eoc. NA. *Camelops* L. Plioc. Pleist. NA. *Camelus* [*Cameliscus*] U. Pleist. R. As. Pleist. R. NAf. *Delahomeryx* L. Mioc. NA. *Dyseotylopus* U. Olig. NA. *Eulameops* U. Pleist. SA. *Floridatragulus* L.-M. Mioc. NA. *Gentilicamelus* L. Mioc. NA. *Hemiauchenia* U. Mioc. L. Pleist. NA. *Hesperocamelus* M. Mioc. NA. *Hidrosotherium* L. Olig. NA. *Homocamelus* M.-U. Mioc. NA. *Lama* [*Auchenia Hemiauchenia Palaeolama Vicugna*] U. Plioc. R. SA. *Oromeryx Protylopus* U. Eoc. NA. *Megatylops* U. Mioc. U. Plioc. NA. *Michenia* L.-U. Mioc. NA. *Miolabis* M. Mioc. NA. *Miotylopus Nothokemas Oxydactylus* L. Mioc. NA. *Nothotylopus* U. Mioc. NA. *Palaeolama* U. Mioc. U. Plioc. NA. Pleist. SA. *Paracamelus* [*Megatylopus Neoparacamelus*] L.-U. Plioc. NA. Eu. U. Plioc. Pleist. As. *Paratylopus* M. Olig. NA. *Pliauchenia* M.-U. Mioc. NA. *Peobrodon* U. Eoc. NA. *Poebrotherium* M. Olig. NA. *Priscocamelus* L. Mioc. NA. *Procamelus* U. Mioc. L. Plioc. NA. *Protolabis* M. Mioc. L. Plioc. NA. *Pseudolabis* U. Olig. L. Mioc. NA. *Rakomylus* M. Mioc. NA. *Stenomylus Tanymykter* L. Mioc. NA. *Titanotylopus* Plioc. Pleist. NA. *Vicugna* M. Pleist. R. SA. **Oromerycidae** *Eotylopus* U. Eoc. M. Olig. NA. *Malaquiferus Merycobunodon* U. Eoc. NA. *Montanatylopus* L. Olig. NA. *Oromeryx Protylopus* M.-U. Eoc. NA.

SUPERFAMILY INCERTAE SEDIS **Xiphodontidae** *Dichodon* [*Tetraselenodon*] M. Eoc. L. Olig. Eu. *Haplomeryx Paraxiphodon* M.-U. Eoc. Eu. *Xiphodon* U. Eoc. L. Olig. Eu. **Amphimerycidae** *Amphimeryx* U. Eoc. L. Olig. Eu. *Pseudamphimeryx* L. Eoc. L. Olig. Eu. **Protoceratidae** *Heteromeryx* L. Olig. NA. *Leptoreodon* [*Camelomeryx Hesperomeryx Merycodesmus*] *Leptotragulus* [*Parameryx*] U. Eoc. NA. *Paratoceras* L.-U. Mioc. NA. *Poabromylus* U. Eoc. NA. *Prosynthetoceras* L.-M. Mioc. NA. *Protoceras* [*Calops Pseudoproceras*] U. Olig. L. Mioc. NA. *Pseudoprotoceras* L. Olig. NA. *Syndoceras* L. Mioc. NA. *Synthetoceras* [*Prosynthetoceras*] L. Mioc. L. Plioc. NA. *Toromeryx* U. Eoc. NA.

SUBORDER RUMINANTIA

INFRAORDER TRAGULOIDEA **Hypertragulidae** *Andegameryx* L.-M. Mioc. Eu. *Hypertragulus* [*Allomeryx*] L. Olig. L. Mioc. NA. *Hypisodus* Olig. NA. *Indomeryx* U. Eoc. SAs. *Nanotragulus* U. Olig. L. Mioc. NA. *Notomeryx* U. Eoc. EAs. *Parvitragulus* L. Olig. NA. *Simimeryx* U. Eoc. NA. **Tragulidae** R. As. Af. *Dorcabune* M. Mioc. L. Plioc. As. *Dorcatherium* L.-M. Mioc. Af. M. Mioc. L. Plioc. Eu. U. Mioc. L. Pleist. As. *Gobiomeryx* L.-M. Olig. EAs. *Tragulus* L. Mioc. R. As. Pleist. EInd. **Leptomerycidae** *Archaeomeryx* U. Eoc. L. Olig. As. *Bachitherium* L.-M. Olig. Eu. *Gobiomeryx* L. Olig. EAs. *Hendryomeryx* U. Eoc. NA. *Leptomeryx* L. Olig. L. Mioc. NA. *Lophiomeryx* L.-U. Olig. As. Eu. *Miomeryx* L. Olig. EAs. *Prodremotherium* L.-M. Olig. Eu. M.-U. Olig. WAs. L. Mioc. Af. *Pronodens* L. Mioc. NA. *Pseudomeryx* M. Olig. As. *Pseudoparablastomeryx* L.-U. Mioc. NA. *Rutitherium* U. Olig. Eu. *Xinjiangmeryx* U. Eoc. EAs. **Gelocidae** *Cryptomeryx* L. Olig. Eu. *Gelocus* [*Paragelocus Pseudogelocus*] U. Eoc. M. Olig. Eu. L. Mioc. As. Af. *Paragelocus* U. Eoc. Eu. *Pseudoceras* U. Mioc. L. Plioc. NA. *Walangania* L. Mioc. EAf.

INFRAORDER PECORA

SUPERFAMILY CERVOIDEA **Palaeomerycidae** *Aletomeryx* [*Dyseomeryx Sinclairomeryx*] L.-M. Mioc. NA. *Barbouromeryx* L.-M. Mioc. NA. *Climacoceras* Mioc. EAf. *Cranioceras Drepanomeryx* M. Mioc. NA. *Dromomeryx* M.-U. Mioc. NA. *Lagomeryx* [*Heterocemas*] L.-U. Mioc. Eu. Mioc. L. Plioc. EAs. *Palaeomeryx* L. Mioc. Af. As. M. Mioc. L. Plioc. Eu. *Pediomeryx* [*Procoileus*] U. Mioc. L. Plioc. NA. *Procervulus* L.-U. Mioc. Eu. *Rakomeryx* M.-U. Mioc. NA. *Yumaceras* ?U. Mioc. U. Plioc. L. Pleist. NA. **Moschidae** *Amphitragulus* U. Olig. L. Mioc. Eu. L. Mioc. As. Af. *Blastomeryx* [*Parablastomeryx Problastomeryx Propalaeoryx Pseudoblastomeryx*] L. Mioc. U. Plioc. NA. *Dremotherium* U. Olig. L. Mioc. Eu. *Iberomeryx* U. Olig. L. Mioc. As. *Longirostromeryx* M.-U. Mioc. NA. *Machaeomeryx* L. Mioc. NA. *Micromeryx* [*Orygotherium*] M. Mioc. L. Plioc. Eu. *Moschus* U. Mioc. R. As. **Cervidae** *Alces* [*Libralces*] Pleist. R. Eu. As. NA. *Anoglochis* Plioc. Eu. *Antifer* [*Paraceros*] U. Plioc. Pleist. SA. *Arvernoceros* U. Plioc. Eu. *Axis* U. Plioc. R. As. Pleist. Eu. EInd. *Blastocerus* Pleist. R. SA. NA. *Bretzia* L. Plioc. NA. *Capreolus* ?U. Mioc. R. Eu. U. Plioc. R. As. *Cervalces* Pleist. NA. *Cervavitulus* U. Mioc. Eu. *Cervavitus* U. Mioc. As. *Cervocerus* [*Cervavitus Damacerus Procervus*] L. Plioc. Eu. As. *Cervulus* U. Plioc. As. *Cervus* [*Cervodama Deperetia Elaphus Epirusa Euctenoceros Nipponicervus ?Procoileus Pseudaxis Rucervus Rusa Sika*] U. Plioc. R. Eu. U. Mioc. Pleist. NAf. M. Plioc. R. NA. U. Mioc. Pleist. As. *Dama* U. Plioc. R. As. Pleist. NAf. Pleist. R.

Eu. *Dicrocerus* [*Euprox Heteroprox*]. L. Mioc. Eu. As. ?U. Mioc. As. *Elaphurus* L. Pleist. R. EAs. *Eostylocerus* U. Mioc. EAs. *Eucladocerus* [*Polycladus*] U. Mioc. U. Plioc. Eu. *Euctenoceros* U. Plioc. Eu. *Eumeryx* L.-U. Olig. EAs. *Habromeryx* L.-U. Pleist. SA. *Hippocamelus* Pleist. R. SA. *Hydropotes* Pleist. R. EAs. *Libralces* U. Plioc. Eu. *Megaloceros* [*Dolichodoryceros Megaceros Megaceroides Orthogonoceros Sinomegaceroides Sinomegaceras*] L.-U. Pleist. Eu. Pleist. NAf. As. *Metacervulus* [*Paracervulus*] U. Mioc. EAs. L. Pleist. Eu. *Muntiacus* [*Cervulus*] L. Pleist. R. As. U. Mioc. L. Plioc. EEu. Pleist. EInd. *Navahoceros* U. Pleist. NA. *Odocoileus* [*Palaeoödocoileus Protomazama*] M. Plioc. R. NA. *Ozotoceras* Pleist. R. SA. *Paracervulus* U. Mioc. L. Plioc. Eu. *Praemegaceros* U. Plioc. L. Pleist. Eu. *Procapreolus* L. Plioc. Eu. As. *Pseudaxis* ?U. Mioc. L. Pleist. As. *Rangifer* Pleist. R. Eu. As. NA. *Sangamona* U. Pleist. NA. *Stephanocemas* M.-U. Mioc. Eu. As. **Giraffidae** *Birgerbohlinia* U. Mioc. Eu. *Bohlinia* [*Orasius*] U. Mioc. L. Plioc. Eu. *Bramatherium* U. Mioc. SAs. *Decennatherium* U. Mioc. L. Plioc. Eu. SWAs. *Giraffa* [*Camelopardalis*] L. Plioc. Eu. M. Mioc. U. Plioc. As. U. Mioc. R. Af. *Giraffokeryx* U. Mioc. L. Plioc. SAs. L. Mioc. Eu. *Helladotherium* [*Panotherium*] L. Plioc. EEu. SWAs. *Honanotherium* U. Mioc. EAs. Eu. *Hydaspitherium* U. Mioc. SAs. *Indratherium* U. Plioc. As. *Injanatherium* U. Mioc. EAs. *Libytherium* U. Mioc. Af. *Macedonitherium* U. Plioc. SWAs. *Palaeotragus* [*Achtiaria*] L. Mioc. U. Plioc. Af. U. Mioc. L. Plioc. EEu. As. *Prolibytherium* L. Mioc. NAf. *Propalaemeryx* U. Mioc. SAs. *Samotherium* [*Akicephalus Cheronotherium Shansitherium*] U. Mioc. L. Plioc. Eu. As. M. Mioc. Af. *Sivatherium* [*Griquatherium Indratherium ?Orangiatherium*] U. Plioc. SAs. L. Plioc. U. Pleist. Af. *Triceromeryx* [*Hispanocervus*] L. Mioc. Eu. As. *Vishnutherium* L. Plioc. SAs. *Zarafa* L. Mioc. NAf. *Progiraffa* L. Mioc. As.

SUPERFAMILY CERVOIDEA INCERTAE SEDIS *Amphiprox* U. Mioc. Eu. *Canthumeryx* L. Mioc. NAf. *Euprox* L.-U. Mioc. Eu. *Heterocemas* U. Mioc. EAs. *Heteroprox* M. Mioc. Eu. *Pliocervus* U. Mioc. Eu. *Pseudoblastomeryx* L. Mioc. NA.

SUPERFAMILY BOVOIDEA **Antilocapridae** *Antilocapra* [*Neomeryx*] U. Mioc. R. NA. *Capromeryx* [*Breameryx Dorcameryx*] U. Plioc. Pleist. NA. *Ceratomeryx* L.-M. Plioc. NA. *Hayoceros* L. Pleist. NA. *Hexameryx* U. Mioc. NA. *Hexobelomeryx* [*Hexameryx*] U. Mioc. NA. *Meryceros* [*Submeryceros*] M.-U. Mioc. NA. *Merycodus* [*Cosoryx Paracosoryx Subcosoryx Subparacosoryx*] L.-M. Mioc. NA. *Osbornoceros Ottoceros Plioceros Proantilocapra* U. Mioc. NA. *Ramoceros* [*Merriamoceros Paramoceros*] L.-U. Mioc. NA. *Sphenophalos* [*Plioceros*] U. Mioc. or L. Plioc. NA. *Stockoceros* U. Pleist. NA. *Tetrameryx* [*Hayoceros Stockoceros*] U. Plioc. U. Pleist. NA. *Texoceros* U. Mioc. L. Plioc. NA. **Bovidae** *Aepyceros* U. Mioc. R. Af. *Alcelaphus* [*Bubalis Pelorocerus*] L. Pleist. R. Af. *Antidorcas* U. Mioc. Eu. As. U. Plioc. Pleist. Af. *Antilope* L. Pleist. R. As. U. Mioc. L. Pleist. Af. *Antilospira* L.-U. Plioc. ?L. Pleist. EAs. *Beatragus* L. Pleist. R. Af. *Benicerus* M. Mioc. Af. *Bibos* Pleist. R. As. Pleist. Eu. Pleist. R. EInd. *Bison* [*Gigantobison Parabison Platycerobison Simobison Superbison Stelabison*] U. Plioc. U. Pleist. As. Pleist. R. Eu. NA. *Bootherium* Pleist. NA. *Bos* [*Poephagus*] Pleist. R. Eu. As. Af. NA. *Boselaphus* U. Mioc. Pleist. R. As. *Bubalus* L. Pleist. Eu. L. Pleist. R. As. *Bucapra* L. Pleist. SAs. *Budorcas* U. Pleist. R. As. *Cambayella* U. Mioc. SAs. *Capra* [*Aegoceros Ibex*] U. Pleist. R. Eu. Pleist. R. NAf. *Cephalophus* U. Mioc. R. Af. *Connochaetes* Pleist. R. Af. *Criotherium* L. Plioc. EEu. *Damalavus* L. Plioc. NAf. Eu. *Damaliscus* U. Mioc. R. Af. *Damalops* U. Mioc. L. Pleist. EAs. SAs. *Deperetia* U. Plioc. Eu. *Dorcadoryx* U. Plioc. As. *Dorcadoxa* U. Mioc. As. *Eotragus* [*Eocerus Murphelaphus*] Mioc. Af. As. L.-U. Mioc. Eu. *Euceratherium* [*Aftonius Preptoceras*] Pleist. NA. *Eusyncerus* L. Plioc. Eu. *Gallogoral* U. Plioc. Eu. *Gangicobus* L. Pleist. SAs. *Gazella* [*Antidorcas Gazelloportax Procapra*] U. Mioc. L. Pleist. Eu. M. Mioc. R. As. Af. *Gobicerus* M. Mioc. EAs. *Helicoportax* M.-U. Mioc. SAs. *Helicotragus* [*Helioceras Helicophora*] U. Mioc. EEu. SWAs. *Hemibos* [*Amphibos Peribos ?Probunalis*] U. Plioc. As. *Hemistrepsiceros* L. Plioc. EEu. *Hemitragus* ?Plioc. Pleist. As. L. Plioc. Pleist. Eu. R. As. *Hesperidoceras* U. Plioc. Eu. *Hippotragus* U. Mioc. Eu. ?U. Plioc. As. L. Pleist. R. Af. *Hydaspicobus* U. Plioc. L. Pleist. SAs. *Hypsodontus* M. Mioc. EEu. *Indoredunca* U. Plioc. L. Pleist. SAs. *Kabulicornis* L.-U. Plioc. As. *Kobikeryx* U. Mioc. L. Plioc. SAs. *Kobus* [*Cobus*] U. Mioc. As. Plioc. R. Af. *Kubanotragus* M. Mioc. As. *Leptobos* [*Epileptobos*] U. Mioc. U. Plioc. Af. L. Pleist. Eu. As. U. Pleist. EInd. *Leptotragus* L. Plioc. EEu. SWAs. *Lyrocerus* U. Mioc. L. Plioc. EAs. *Madoqua* U. Mioc. U. Plioc. Af. *Makapania* U. Plioc. L. Pleist. Af. *Megalotragus* Pleist. Af. *Megalovis* U. Plioc. L. Pleist. EAs. U. Plioc. Eu. *Menelikia* L. Pleist. Af. *Mesembriacus* U. Mioc. SWAs. *Mesembriportax* L. Plioc. Af. *Microtragus* L. Plioc. EEu. SWAs. *Mitragocerus* [*Dystychoceras*] U. Mioc. L. Plioc. Eu. As. *Myotragus* Pleist. Med. *Naemorhaedus* ?U. Pleist. Eu. Pleist. R. As. *Neotragocerus* U. Mioc. L. Plioc. NA. *Nisidorcas* U. Mioc. Eu. *Oioceros* L.-U. Mioc. As. M. Mioc. Af. U. Mioc. Eu. *Olonbulukia* L. Plioc. EAs. *?Orasius* U. Mioc. SWAs. *Orchonoceros* L. Plioc. As. *Oreamnos* U. Pleist. NA. *Oreonager* U. Plioc. NAf. *Oreotragus* U. Plioc. R. Af. *Orygotherium* M. Mioc. Eu. *Oryx* U. Plioc. U. Pleist. Af. SWAs. *Ovibos* [*Parovibos Praeovibos*] Pleist. Eu. As. Pleist. R. NA. *Ovis* [*Ammotragus Caprovis Pachyceros*] U. Mioc. R. As. Eu. L. Pleist. R. NA. NAf. *Pachygazella* L. Plioc. EAs. *Pachyportax* L. Plioc. L. Pleist. SAs. *Pachytragus* M. Mioc. L. Plioc. EEu. SWAs. Plioc. NAf. *Palaeohyopsodontus* M. Olig. EAs. *Palaeoreas* U. Mioc. L. Plioc. Eu. L. Plioc. As. NAf. *Palaeoryx* L. Plioc. As. Eu. Af. *Pantholops* Pleist. R. As. *Parabos* L.-U. Plioc. Eu. Af. *Paraprotoryx* U. Mioc. EAs. *Parmularius* U. Plioc. L. Pleist. Af. *Parurmiatherium* L. Plioc. As. EEu. *Pelea* Pleist. R. Af. *Pelorovis* L.-U. Pleist. Af. *Perimia* L. Plioc. SAs. *Platybos* L. Pleist. SAs. *Platycerabos* [*Parabos*] Pleist. R. NA. *Plesiaddax* L. Plioc. EAs. Eu. *Pliotragus* [*Deperetia*] L. Pleist. Eu. *Praeovibos* L. Pleist. As. U. Pleist. NA. *Proamphibos* L. Plioc. L. Pleist. SAs. *Procamptoceras* L. Plioc. L. Pleist. Eu. *Procapra* U. Pleist. R. As. *Procobus* L. Plioc. EEu. *Prodamaliscus* L. Plioc. EEu. SWAs. *Proleptobos* L. Plioc. SAs. *Prosinotragus* L. Plioc. EAs. Eu. *Prostrepsiceros* U. Mioc. L. Plioc. EEu. SWAs. U. Mioc. EAs. Af. *Protoryx* ?M. Mioc. SWAs. L. Plioc.

EEu. As. *Protragelaphus* L. Plioc. EEu. SWAs. *Protragocerus* [*Paratragoceros*] L.-U. Mioc. Af. M. Mioc. As. U. Mioc. Eu. *Pseudobos* U. Mioc. As. *Pseudotragus* M. Mioc. EAs. ?Af. L. Plioc. EEu. SWAs. *Pultiphagnoides* L. Plioc. As. L. Pleist. Af. *Qurlignoria* L. Plioc. EAs. *Rabaticeras* Pleist. NAf. *Raphicerus Redunca* [*Cervicapra*] L. Plioc. Pleist. R. Af. *Rhyncotragus* ?U. Mioc. Af. *Rhynotragus* Pleist. Af. *Rupicapra* L. Pleist. R. Eu. SWAs. *Ruticeros* U. Mioc. As. Pleist. Af. *Saiga* Pleist. R. As. EEu. NA. *Samokeras* U. Mioc. Eu. *Selenoportax* L. Plioc. SAs. Eu. *Shensispira* U. Mioc. EAs. *Simatherium* ?L. Plioc. L. Pleist. Af. *Sinoreas* L. Plioc. EAs. *Sinoryx* L. Plioc. EAs. *Sinotragus* L. Plioc. EAs. Eu. *Sivacapra* L. Pleist. SAs. *Sivaceros* L. Plioc. SAs. *Sivacobus Sivadenota* L. Pleist. SAs. *Sivaportax* L. Plioc. SAs. *Sivatragus* U. Plioc. L. Pleist. As. *Sivoreas* M. Mioc. L. Plioc. SAs. *Sivoryx* L. Pleist. SAs. *Soergelia* Pleist. Eu. ?As. NA. *Spirocerus* U. Plioc. U. Pleist. As. *Sporadotragus* U. Mioc. Eu. *Strepsiportax* U. Mioc. As. *Strogulognathus* U. Mioc. Eu. *Symbos* [*Scaphoceros*] Pleist. NA. *Syncerus* ?U. Plioc. Eu. Pleist. R. Af. *Tetracerus* Pleist. R. SAs. *Thaleroceros* L. Pleist. EAf. *Toribos* L. Plioc. Eu. *Torticornis* U. Mioc. EAs. *Tossunnoria* M. Mioc. SWAs. L. Plioc. EAs. *Tragelaphus* U. Mioc. U. Pleist. Af. *Tragocerus* [*Austroportax Dystichoceras Graecoryx Indotragus Pikermicerus Pontoportax Tragoportax*] U. Mioc. L. Plioc. Eu. L. Plioc. NAf. As. *Tragoportax* U. Mioc. Eu. As. *Tragoreas* U. Mioc. L. Plioc. EEu. As. *Tragospira* U. Plioc. L. Pleist. Eu. *Tsaidamotherium* L. Plioc. EAs. *Ugandax* U. Mioc. L. Pleist. Af. *Urmiatherium* U. Mioc. L. Plioc. As. Eu. *Vishnucobus* L. Pleist. SAs.

SUPERFAMILY BOVOIDEA INCERTAE SEDIS *Amphimoschus* L. Mioc. Eu.

ORDER ARTIODACTYLA INCERTAE SEDIS *Aksyria* L. Eoc. As. *Dulcidon* M. Eoc. SAs. *Raphenacodus* M. Eoc. EAs. *Tragulohyus* L. Olig. Eu.

ORDER MESONYCHIA (ACREODI) **Mesonychidae** *Andrewsarchus* U. Eoc. EAs. *Ankalagon* L. Paleoc. NA. *Dissacus* [*Hyaenodictis*] M. Paleoc. L. Eoc. NA. U. Paleoc. M. Eoc. Eu. ?M.-U. Paleoc. EAs. *Dissacusium* L. Paleoc. As. *Guilestes* U. Eoc. EAs. *Hapalodectes* L. Eoc. NA. ?U. Paleoc. Eoc. As. *Harpagolestes* M.-U. Eoc. NA. U. Eoc. L. Olig. As. *Honanodon* M.-U. Eoc. As. *Hukoutherium* L.-M. Paleoc. EAs. *Ichthyolestes* M. Eoc. SAs. *Jiangxia* U. Paleoc. EAs. *Lihoodon* U. Eoc. EAs. *Lophoodon* U. Eoc. As. *Mesonyx* M.-U. Eoc. NA. U. Eoc. EAs. *Metahapalodectes* U. Eoc. As. *Microclaenodon* M. Paleoc. NA. *Mongolestes* L. Olig. As. *Mongolonyx* M.-U. Eoc. EAs. *Pachyaena* ?U. Paleoc. L.-U. Eoc. EAs. L. Eoc. Eu. NA. *Plagiocristodon* U. Paleoc. L. Eoc. EAs. *Synoplotherium* [*Dromocyon*] M. Eoc. NA. *Wyolestes* L. Eoc. NA. *Yantanglestes* M.-U. Paleoc. EAs.

ORDER CETACEA

SUBORDER ARCHAEOCETI **Protocetidae** *Eocetus* U. Eoc. NAf. *Gandakasia Ichthyolestes Pakicetus* M. Eoc. SAs. *Pappocetus* M. Eoc. WAf. *Protocetus* M. Eoc. NAf. NA. SAs. **Basilosauridae** (**Zeuglodontidae**) *Basilosaurus* [*Hydrargos Zeuglodon*] U. Eoc. NA. NAf. *Dorudon* U. Eoc. NAf. NA. *?Platyosphys* L. Olig. Eu. *?Pontogeneus* U. Eoc. NA. *Prozeuglodon* M.-U. Eoc. NAf. *Zygorhiza* U. Eoc. NA. Eu.

SUBORDER ARCHAEOCETI INCERTAE SEDIS *Microzeuglodon* U. Olig. Eu. *Kekenodon* L. Olig. or L. Mioc. NZ. *Pachycetus* M. Eoc. Eu.

SUBORDER ODONTOCETI **Kentriodontidae** *Delphinodon Kampholophus Kentriodon* M. Mioc. NA. *Leptodelphis* U. Mioc. EEu. *Liolithax Lophocetus* M. Mioc. NA. *Microphocaena* U. Mioc. EEu. *Pithanodelphis* U. Mioc. Eu. *Sarmatodelphis* U. Mioc. EEu. **Squalodontidae** *Australosqualodon* Olig. Aus. *Eosqualodon* U. Olig. Eu. *Metasqualodon* L. Mioc. Aus. *Microcetus* U. Olig. Eu. *Neosqualodon* L. Mioc. Eu. *Oligodelphis* M. Olig. SWAs. *Parasqualodon* U. Olig. Aus. *Patriocetus* U. Olig. Eoc. Eu. *Phoberodon* L. Mioc. SA. *Prosqualodon* Olig. Aus. L. Mioc. SA. NZ. *Sachalinocetus* L. Mioc. EAs. *Saurocetus* U. Mioc. NA. *Squalodon* [*Phocodon Rhytisodon*] L.-U. Mioc. Eu. M.-U. Mioc. NA. Mioc. NZ. *Sulakocetus* U. Olig. SWAs. *Tangaroasaurus* L. Mioc. NZ. **Platanistidae** *Allodelphis* Mioc. NA. *Anisodelphis* U. Mioc. SA. *Hesperocetus* U. Mioc. NA. *Ischyrorhynchus* U. Mioc. SA. *Pachyacanthus* U. Mioc. Eu. *Pontistes Pontivaga* L. Plioc. SA. *Proinia* L. Mioc. SA. *Rhabdosteus* M. Mioc. NA. *Saurodelphis* [*Pontoplanodes Saurocetes*] M.-U. Mioc. SA. *Zarhachis* M. Mioc. NA. **Ziphiidae** *Anoplonassa* M.-U. Mioc. NA. *Belemnoziphius* U. Mioc. Eu. NA. *Cetorhynchus* M.-U. Mioc. Eu. *Choneziphius* U. Mioc. U. Plioc. Eu. U. Mioc. NA. *Eboroziphius* U. Mioc. NA. *Incacetus* M. Mioc. SA. *Mesoplodon* U. Mioc. U. Plioc. Eu. U. Mioc. NA. Plioc. Aus. R. Oc. *Palaeoziphius* U. Mioc. Eu. *Proroziphius* U. Mioc. NA. *Squalodelphis* L. Mioc. Eu. *Ziphioides* M. Mioc. Eu. *Ziphirostrum* [*Mioziphius*] U. Mioc. Eu. **Delphinidae** *Agabelus* M. Mioc. NA. *Anacharsis* U. Mioc. SWAs. *Araeodelphis* M. Mioc. NA. *Belosphys Cetorhinops* M. Mioc. NA. *Delphinavus* L.-U. Mioc. NA. *Delphinopsis* U. Mioc. Eu. *Delphinus* L. Plioc. Pleist. Eu. R. Oc. *Globicephala* Pleist. NA. R. Oc. *Goniodelphis* M. Mioc. NA. *Hadrodelphis* M. Mioc. *Imerodelphis* U. Mioc. SWAs. *Iniopsis* L.-U. Mioc. Eu. *Ixacanthis* M. Mioc. NA. *Machrochirifer* U. Mioc. Eu. *Megalodelphis* M. Mioc. NA. *Orcinus* U. Mioc. R. Oc. *Pelodelphis* M. Mioc. NA. *Phocageneus* M. Mioc. *Protodelphinus* L. Mioc. Eu. *Stenella* ?M. Mioc. Plioc. *Steno* L. Plioc. Eu. R. Oc. *Stereodelphis* M. Mioc. Eu. *Tretosphys* M. Mioc. NA. *Tursiops* [*Tursio*] L. Pleist. Eu. NA. R. Oc. **Rhabdosteidae** (**Eurhinodelphidae**) *Argyrocetus* L. Mioc. SA. *Eurhinodelphis* M. Mioc. NA. U. Mioc. Eu. EAs. *Hemisynthachelus* U. Mioc. Eu. *Macrodelphinus* L. Mioc. NA. *Ziphiodelphis* L. Mioc. Eu. **Albireonidae** *Albireo* U. Mioc. NA. **Acrodelphidae** *Acrodelphis* L.-U. Mioc. Eu. M. Mioc. NA. *Champsodelphis* M. Mioc. Eu. *Cyrtodelphis* L.-U. Mioc. Eu. *Heterodelphis* M.-U. Mioc. Eu. *Potamodelphis* U. Mioc. NA. *?Schizodelphis* [*Cyrtodelphis*] L. Mioc. Af. L. Mioc. L. Plioc. Eu. M. Mioc. L. Plioc. NA. **Monodontidae** (**Delphinapteridae**) *Delphinapterus* U. Mioc. Eu. NA. R. NOc. *Denebola* U. Mioc. NA. *Monodon* Pleist. Eu. NA. **Phocaenidae** *Loxolithax* M. Mioc. NA. *Palaeophocaena* U. Mioc. Eu. *Phocaena* [*Phocaenoides*] U. Mioc. EAs. R. Oc. *Phocaenopsis* L. Mioc. NZ. *Piscolithax* L. Plioc. SA. U. Mioc. NA. *Protophocaena* U. Mioc. Eu. *Salumniphocaena* U. Mioc. NA. **Pontoporiidae** *Parapontoparia* U. Mioc. U. Plioc. NA. *Pliopontos* L. Plioc. SA. *Pontistes* Plioc. SA. *Prolipotes* ?Mioc. EAs. **Physeteridae** *Apenophyseter* L.

Mioc. SA. *Auiophyseter* M. Mioc. NA. *Balaenodon* U. Mioc. Eu. *Diaphorocetus* L. Mioc. SA. *Dinoziphius* M.-U. Mioc. NA. *Haplocetus* U. Mioc. NA. *Hoplocetus* M. Mioc. Eu. ?L. Plioc. NA. *Idiophyseter* M. Mioc. NA. *Idiorophus* L. Mioc. SA. *Kogiopsis* L. Plioc. NA. *Ontocetus* M. Mioc. EAs. M.-U. Mioc. NA. *Orycterocetus* M. Mioc. NA. *Physeter* L. Mioc. Eu. U. Mioc. NA. R. Oc. *Physeterula* L.-U. Mioc. Eu. *Physetodon* L. Plioc. Aus. *Plesiocetopsis* [*Plesiocetus*] Mioc. SA. U. Mioc. NA. L. Plioc. Eu. *Praekogia* U. Mioc. NA. *Priscophyseter* U. Mioc. Eu. *Prophyseter* U. Mioc. Eu. *Scaldicetus* ?L. Mioc. U. Mioc. Eu. ?L. Plioc. Aus. *Thalassocetus* U. Mioc. Eu. **Agorophiidae** *Agorophius* U. Olig. NA.

SUBORDER ODONTOCETI INCERTAE SEDIS *Agriocetus* U. Olig. Eu. *Eoplatanista* L. Mioc. Eu. *Lamprolithax* M. Mioc. NA. *Lonchodelphis* U. Mioc. NA. *Miodelphis* L. Mioc. NA. *Nannolithax Oedolithax Platylithax* M. Mioc. NA. *Xenorophus* U. Olig. NA.

SUBORDER MYSTICETI **Aetiocetidae** *Aetiocetus* U. Olig. NA. **Cetotheriidae** *Amphicetus* L. Plioc. Eu. *Cephalotropis* U. Mioc. NA. *Cetotheriomorphis* M.-U. Mioc. Eu. *Cetotheriopsis* U. Olig. Eu. L. Mioc. SA. *Cetotherium* U. Mioc. L. Plioc. Eu. *Cophocetus* M. Mioc. NA. *Diorocetus* M. Mioc. NA. *Eucetotherium* M.-U. Mioc. Eu. *Herpetocetus* U. Mioc. Eu. *Heterocetus* M.-U. Mioc. Eu. *Imerocetus* U. Mioc. SWAs. *Isocetus* ?M. Mioc. NA. U. Mioc. Eu. *Mauicetus* M. Olig. L. Mioc. NZ. *Mesocetus* ?M. Mioc. NA. M.-U. Mioc. Eu. *Metopocetus* U. Mioc. NA. Eu. *Mixocetus Nanisocetus Parietobalaena* U. Mioc. NA. *Pelocetus Peripolocetus* M. Mioc. NA. *Rhegnopsis Siphonocetus* U. Mioc. NA. *Tiphyocetus* M. Mioc. NA. **Eschrichtiidae** (**Rhachianectidae**) *Eschrichtius* U. Pleist. NA. R. NPac. **Balaenopteridae** R. Oc. *Balaenoptera* [*Cetotheriophanes*] U. Mioc. Pleist. Eu. L. Plioc. Pleist. NA. Pleist. As. R. Oc. *Burtinopsis* U. Mioc. Eu. *Idiocetus* Mioc. EAs. U. Plioc. Eu. *Megaptera* L. Plioc. Pleist. NA. U. Plioc. Eu. R. Oc. *Megapteropsis* L. Plioc. Eu. *Mesoteras* U. Mioc. NA. *Notiocetus* L. Plioc. SA. *Palaeocetus* U. Mioc. Eu. *Plesiocetus* Mioc. SA. U. Mioc. NA. L. Plioc. Eu. **Balaenidae** R. Oc. *Aglaocetus* L. Mioc. SA. *Balaena* Plioc. Aus. L. Plioc. Pleist. Eu. R. Arc. Oc. *Balaenotus Balaenula* U. Mioc. Plioc. Eu. *Eschrictius* [*Rhachianectes*] M. Mioc. Eu. R. NPac. *Morenocetus* L. Mioc. SA. *Protobalaena* L. Plioc. Eu.

ORDER CETACEA INCERTAE SEDIS *Archaeodelphis* U. Olig. NA. *Chonecetus* Olig. NA. *Ferecetotherium* Olig. WAs. *Mirocetus* Olig. SWAs. *Nannocetus* U. Mioc. NA.

ORDER PERISSODACTYLA

SUBORDER HIPPOMORPHA

SUPERFAMILY EQUOIDEA **Equidae** *Anchitherium* [*Kalobatippus Paranchitherium*] L.-M. Mioc. NA. M. Mioc. Eu. M. Mioc. L. Plioc. As *Archaeohippus* L.-M. Mioc. NA. *Astrohippus* U. Mioc. NA. *Calippus* M. Mioc. NA. *Cormohipparion* M.-U. Mioc. As. Eu. NA. *Dinohippus* U. Mioc. NA. *Epihippus* [*Duchesnehippus*] U. Eoc. NA. *Equus* [*Allohippus Allozebra Amerhippus Asinus Dolichohippus Hemionus Hesperohippus Hippotigris Kolpohippus Kreterofhippus Neohippus Onager Plesippus etc.*] M. Plioc. U. Pleist. NA. Plioc. R. As. Af. Eu. Pleist. SA. *Gobihippus* U. Eoc. NAs. *Haplohippus* L. Olig. NA. *Hipparion* [*Hemihipparion Hippotherium Notohipparion Proboscidohipparion*] M.-U. Mioc. NA. U. Mioc. L. Pleist. Eu. Af. As. M. Pleist. Af. *Hippidion* [*Hippidium Plagiohippus Stereohippus*] Pleist. SA. *Hippodon* L.-M. Mioc. NA. *Hypohippus* M. Mioc. NA. *Hyracotherium* [*Eohippus Protorohippus Xenicohippus*] L. Eoc. NA. Eu. As. *Lophiopus* U. Eoc. Eu. *Lophiotherium* M.-U. Eoc. Eu. *Megahippus* M. Mioc. NA. *Merychippus* [*Eoequus*] M. Mioc. Plioc. NA. *Mesohippus* [*Pediohippus*] L.-M. Olig. NA. *Miohippus* M. Olig. L. Mioc. NA. *Nannippus* M. Mioc. L. Pleist. NA. *Neohipparion* M.-U. Mioc. NA. *Onohippidium* [*Hyperhippus*] U. Mioc. NA. Pleist. SA. *Orohippus* M. Eoc. NA. *Parahippus* L. Mioc. NA. *Pliohippus* M. Mioc. NA. *Protohippus* L.-U. Mioc. NA. *Pseudhipparion* M. Mioc. U. Mioc. NA. *Sinohippus* M. Mioc. As. *Stylohipparion* Plioc. Pleist. Af. **Palaeotheriidae** *Anchilophus* M.-U. Eoc. Eu. *Leptolophus* M. Eoc. Eu. *Pachynolophus* M.-U. Eoc. Eu. *Palaeotherium* U. Eoc. L. Olig. Eu. *Paraplagiolophus* M. Eoc. Eu. *Plagiolophus* [*Paloplotherium*] M. Eoc. L. Olig. Eu. *Propachynolophus* L.-M. Eoc. Eu. *Propalaeotherium* M. Eoc. L. Olig. Eu. ?L. Eoc. M. Eoc. As. *Pseudopalaeotherium* L. Olig. Eu.

SUPERFAMILY BRONTOTHERIOIDEA **Brontotheriidae** (**Titanotheriidae**) *Acrotitan Arctotitan* U. Eoc. EAs. *Brachydiastematherium* U. Eoc. EEu. *Brontops* [*Diploclonus Megacerops*] *Brontotherium* L. Olig. NA. *Desmatotitan* M.-U. Eoc. EAs. *Dianotitan* U. Eoc. EAs. *Diplacodon Dolichorhinus* U. Eoc. NA. *Dolichothinoides* U. Eoc. EAs. *Duchesneodus* U. Eoc. NA. *Embolotherium* L.-M. Olig. As. *Eotitanops* L. Eoc. NA. M. Eoc. As. *Eotitanotherium* U. Eoc. NA. *Epimanteoceras* U. Eoc. L. Olig. EAs. *Gnathotitan* U. Eoc. As. *Heterotitanops* U. Eoc. NA. *Hyotitan* M. Olig. EAs. *Lambdotherium* L. Eoc. NA. ?EAs. *Limnohyops* M. Eoc. NA. *Manteoceras* M.-U. Eoc. NA. U. Eoc. ?NAs. *Megacerops* [*Symborodon*] L. Olig. NA. *Menodus* [*Allops Titanotherium*] L. Olig. NA. Eu. EAs. *Mesatirhinus Metarhinus* M.-U. Eoc. NA. *Metatelmatherium* U. Eoc. EAs. NA. *Metatitan* L.-M. Olig. As. *Microtitan* M.-?U. Eoc. EAs. *Pachytitan* U. Eoc. EAs. *Pakotitanops* M. Eoc. SAs. *Palaeosyops* L. Eoc. As. L.-M. Eoc. NA. *Parabrontops* U. Eoc. L. Olig. EAs. *Protembolotherium* U. Eoc. L. Olig. EAs. *Protitan* U. Eoc. ?L. Olig. As. *Protitanops* L. Olig. NA. *Protitanotherium* U. Eoc. NA. EAs. *Rhadinorhinus* U. Eoc. NA. *Rhinotitan* U. Eoc. L. Olig. EAs. *Sivatitanops* U. Eoc. SAs. *Sphenocoelus Sthenodectes* U. Eoc. NA. *Teleodus* U. Eoc. L. Olig. NA. *Telmatherium* M. Eoc. NA. ?EAs. *Titanodectes* U. Eoc. L. Olig. EAs.

SUBORDER ANCYLOPODA **Eomoropidae** *Eomoropus* U. Eoc. NA. EAs. L. Olig. As. *Grangeria* U. Eoc. EAs. ?NA. *Lunania* U. Eoc. EAs. *Paleomoropus* L. Eoc. NA. **Chalicotheriidae** *Ancylotherium* [*Circotherium Nestoritherium*] M. Mioc. L. Plioc. Eu. U. Mioc. EAs. L. Pleist. Af. *Borissiakia* U. Olig. WAs. *Chalicotherium* [*Macrotherium*] L. Mioc. L. Plioc. As. Eu. L. Mioc. Af. *Chemositia* U. Mioc. Af. *Limognitherium* L. Olig. Eu. *Litolophus* U. Eoc. EAs. *Lophiaspis* L.-M. Eoc. Eu. *Macrotherium* M.-U. Mioc. EAs. *Moropus* L. Mioc. As. L.-M. Mioc. NA. *Nestoritherium* M. Mioc. L. Pleist. As. *Olsenia* U. Eoc. EAs. *Phyllotillon*

[*Metaschizotherium*] U. Olig. L. Mioc. As. L. Mioc. Eu. ?L. Mioc. L. Pleist. Af. *Schizotherium* Olig. Eu. U. Eoc. L. Mioc. As. *Tylocephalonyx* L.-M. Mioc. NA.

SUBORDER CERATOMORPHA

SUPERFAMILY TAPIROIDEA **Isectolophidae** *Homogalax* [*Systemodon*] L. Eoc. NA. EAs. *Isectolophus* [*Parisectolophus Schizolophodon*] M.-U. Eoc. NA. M. Eoc. EAs. *Paralophidon* M. Eoc. Eu. *Sastrilophus* M. Eoc. SAs. **Helaletidae** (**Hyrachyiidae**) *Atalonodon* M. Eoc. Eu. *Chasmotherium* L.-U. Eoc. Eu. M. Eoc. Ind. *Colodon* [*Paracolodon*] L.-U. Olig. NA. M. Eoc. U. Olig. EAs. *?Cymbalophus* U. Paleoc. Eu. *Dilophodon* [*Heteraletes*] M.-U. Eoc. NA. *Helaletes* [*Chasmotheroides Desmatotherium Veragromovia*] M.-U. Eoc. NA. L.-U. Eoc. EAs. *Heptodon* L. Eoc. EAs. NA. *Hyrachyus* U. Eoc. Eu. NA. M.-U. Eoc. As. *Paracolodon* L. Olig. As. *Selenaletes* L. Eoc. NA. *Veragromovia* M. Eoc. As. **Lophialetidae** *Breviodon Lophialetes* U. Eoc. EAs. *Eoletes* M. Eoc. As. *Parabreviodon* U. Eoc. EAs. *Pataecops* M. Eoc. EAs. *Schlosseria* M.-U. Eoc. As. *Simplaletes* U. Eoc. NAs. *Zhongjianoletes* U. Eoc. EAs. **Deperetellidae** *Deperetella* [*Cristidentinus Diplophodon*] M.-U. Eoc. ?M. Olig. EAs. L. Pleist. Eu. *Diplolophodon* U. Eoc. As. *Teleolophus* M. Eoc. M. Olig. As. **Lophiodontidae** *Lophiodon* L.-U.Eoc. Eu. ?U. Eoc. EAs. *Rhinocerolophiodon* M. Eoc. Eu. **Tapiridae** *Miotapirus* L. Mioc. NA. *Palaeotapirus* [*Paratapirus*] L. Mioc. Eu. L.-M. Mioc. As. *Protapirus* U. Olig. Eu. M. Olig. L. Mioc. NA. *Selenolophodon* M. Mioc. As. *Tapiravus* L.-U. Mioc. NA. *Tapirus* [*Megatapirus Tapiriscus*] L. Olig. Pleist. Eu. L. Olig. R. NA. Mioc. R. SAs. Pleist. R. SA.

TAPIROIDEA INCERTAE SEDIS *Euryletes* M. Eoc. EAs. *Indolphus* U. Eoc. As.

SUPERFAMILY RHINOCEROTOIDEA **Hyracodontidae** *Allacerops* L. Olig. As. *Aprotodon* U. Olig. L. Mioc. As. *Ardynia* [*Ergilia*] Olig. EAs. *Benaritherium* U. Olig. As. *Caenolophus* M. Eoc. L. Olig. As. *Dzungariotherium* M.-U. Olig. As. *Eggysodon* M.-U. Olig. Eu. *Epitriplopus* U. Eoc. NA. *Forstercooperia* [*Cooperia*] M. Eoc. L. Olig. As. U. Eoc. NA. *Guixia* U. Eoc. L. Olig. As. *Hyracodon* L. Olig. L. Mioc. NA. *Ilianodon* U. Eoc. L. Olig. EAs. *Imeguincisoria* U. Eoc. As. *Indricotherium* U. Eoc. L. Mioc. As. L. Mioc. Eu. *Juxia* U. Eoc. As. *Pappaceras* U. Eoc. EAs. *Paraceratherium* [*Aralotherium*] M. Olig. L. Mioc. As. *Prohyracodon* M. Eoc. Eu. L. Eoc. L. Olig. EAs. *Prothyracodon* U. Eoc. NA. ?As. *Rhodopagus* L.-U. Eoc. EAs. *Symphyssorrhachis* L. Olig. EAs. *Triplopus* L.-M. Eoc. ?M. Olig. As. M.-U. Eoc. NA. *Urtinotherium* M. Eoc. L. Olig. EAs. **Amynodontidae** *Amynodon* [*Sharamynodon*] U. Eoc. L. Olig. NA. EAs. *Amynodontopsis* U. Eoc. NA. *Cadurcodon* U. Eoc. U. Olig. As. M. Olig. Eu. *Cadurcotherium* Olig. Eu. L. Mioc. As. *Euryodon* M. Eoc. As. *Gigantamynodon* U. Eoc. L. Olig. EAs. *Hypsamynodon* L. Olig. EAs. *Huananodon* U. Eoc. L. Olig. EAs. *Lushiamynodon* ?M.-U. Eoc. EAs. *Metamynodon* [*Cadurcopsis*] U. Eoc. U. Olig. NA. ?U. Eoc. L. Mioc. EAs. *Paracadurcodon* ?U. Eoc. L. Olig. EAs. *Paramynodon Procadurcodon* U. Eoc. As. *Sharamynodon* U. Eoc. L. Olig. As. *Sianodon* U. Eoc. L. Olig. EAs. *Teilhardia* U. Eoc. EAs. *Toxotherium* L. Olig. NA. *Zaisanamynodon* U. Eoc. As. **Rhinocerotidae** *Aceratherium* [*Acerorhinus Turkanatherium*] M. Olig. L. Plioc. Eu. U. Olig. L. Plioc. EInd. ?M. Olig. Mioc. Af. U. Plioc. EInd. *Amphicaenopus* L.-U. Olig. NA. *Aphelops* Mioc. NA. *Baluchitherium* U. Olig. L. Mioc. As. *Beliajevina* M. Mioc. SWAs. *Brachydiceratherium* L. Mioc. Eu. *Brachypotherium* [*Indotherium Thaumastotherium*] U. Olig. L. Plioc. As. Mioc. Af. L. Mioc. M. Mioc. NA. *Caenopus* L.-U. Olig. NA. M. Olig. EAs. *Ceratotherium* U. Mioc. Pleist. Af. *Chilotherium* L. Mioc. U. Mioc. As. Af. Eu. *Coelodonta* [*Tichorhinus*] Pleist. Eu. L. Mioc. U. Pleist. As. *Diceratherium* [*Metacaenopus Menoceras*] U. Olig. Eu. U. Olig. M. Mioc. NA. M. Mioc. CA. L.-U. Mioc. As. *Dicerorhinus* [*Ceratorhinus*] U. Olig. R. As. L. Mioc. Pleist. Eu. L. Mioc. U. Plioc. NAf. *Diceros* [*Atelodus Ceratotherium Opsiceros Pliodiceros ?Serenageticeros*] M.-U. Mioc. Eu. U. Plioc. R. Af. U. Mioc. As. *Didermocerus* M. Mioc. Eu. *Dromaceratherium* L. Mioc. Eu. *Elasmotherium* Pleist. Eu. L.-U. Pleist. NAs. *Floridaceras* L. Mioc. NA. *Gaindatherium* M.-U. Mioc. As. L. Mioc. Eu. *Hispanotherium* M. Mioc. Eu. As. *Huangotherium* U. Mioc. As. *Iranotherium* L. Plioc. Eu. *Lartertotherium* L.-M. Mioc. Eu. *Menoceras* L.-M. Mioc. NA. *Mesaceratherium* U. Olig. M. Mioc. Eu. M. Mioc. SWAs. *Meschotherium* U. Olig. L. Mioc. As. *Moschoedestes* L. Mioc. NA. *Paradiceros* M. Mioc. Af. *Penetrigonias* L. Olig. NA. *Peraceras* U. Mioc. L. Plioc. NA. *Plesiaceratherium* M.-U. Mioc. EAs. U. Mioc. Eu. *Pleuroceros* U. Olig. ?L. Plioc. Eu. L. Mioc. ?L. Plioc. As. *Preaceratherium* M. Olig. Eu. *Proaceratherium* L. Mioc. Eu. *Prosantorhinus* L. Mioc. Eu. As. *Rhinoceros* [*Procerorhinus*] M. Mioc. Eu. U. Mioc. U. Plioc. L. Pleist. R. As. EInd. *Ronzotherium* [*Paracaenopus*] L.-U. Olig. Eu. *Stephanorhinus* L. Plioc. ?As. L. Plioc. U. Pleist. Eu. U. Plioc. SWAs. *Subhyracodon* L.-M. Olig. NA. *Teleoceras* [*Aprotodon*] M. Mioc. L. Plioc. NA. Plioc. EAs. *Tesselodon* M. Mioc. As. *Trigonias* L. Olig. NA. Eu. **Ceratomorpha incertae sedis** *Schizotheroides* U. Eoc. NA.

ORDER PROBOSCIDEA

?SUBORDER MOERITHERIOIDEA **Anthracobunidae** *Anthracobune* [*?Ishatherium*] M. Eoc. SAs. *Ishatherium* L. Eoc. SAs. *Jozaria* M. Eoc. SAs. *Lammidhania* M. Eoc. Ind. *Pilgrimella* M. Eoc. SAs. **Moeritheriidae** *Moeritherium* U. Eoc. L. Olig. NAf. ?SAs.

SUBORDER EUELEPHANTOIDEA **Gomphotheriidae** (**Trilophodontidae**) *Amebelodon* U. Mioc. NA. *Anacus* [*Dibunodon Pentalophodon*] U. Mioc. U. Plioc. Eu. Af. As. L. Pleist. Eu. Af. *Choerolophodon* M.-U. Mioc. Eu. As. Af. *Cuvieronius* [*Cordillerion Teleobunomastodon*] Mioc. U. Pleist. NA. *Eubelodon* U. Mioc. NA. *Gomphotherium* [*Bunolophodon Cheorolophodon ?Geisotodon Genomastodon ?Hemilophodon Megabelodon ?Protanancus Tatabelodon Tetrabelodon Trilophodon*] L. Mioc. L. Plioc. Eu. L.-U. Mioc. As. L.-U. Mioc. Af. U. Mioc. L. Plioc. NA. *Haplomastodon* ?M. Plioc. NA. *Megabelodon* U. Mioc. NA. *Notiomastodon* U. Pleist. SA. *Palaeomastodon Phiomia* L. Olig. ?U. Olig. NAf. *Pentalophodon* U. Plioc. As. *Platybelodon* [*Torynobelodon*] L. Mioc. Af. U. Mioc. As. NA. *Protanancus* L.-M. Mioc. Af. *Rhynchotherium* [*Aybelodon*

Blickotherium Dibelodon] L. Mioc. Af. M. Mioc. Eu. U. Mioc. L. Pleist. NA. U. Mioc. L. Plioc. As. *Stegomastodon* [*Aleamastodon Haplomastodon Rhabdobunus*] U. Plioc. Pleist. NA. Pleist. SA. *Stegotetrabelodon* U. Mioc. Eu. U. Mioc. M. Plioc. Af. *Synconolophus* U. Mioc. U. Plioc. As. Eu. *Tetralophodon* [*Lydekkeria Morrillia*] M.-U. Mioc. Eu. U. Mioc. L. Plioc. As. NA. **Elephantidae** *Elephas* [*Hypselephas Platelephas*] Pleist. R. As. L. Plioc. U. Pleist. Af. Pleist. Eu. *Loxodonta* [*Hesperoloxodon Omoloxodon Palaeoloxodon Phanagaroloxodon Pilgrimia Sivalikia*] L. Pleist. R. Af. *Mammuthus* [*Archidiskodon Dicyclotherium Metarchidoskodon Parelephas Stegoloxodon*] L.-U. Plioc. Af. L.-U. Pleist. Eu. As. EInd. Pleist. NA. *Primelephas* U. Mioc. L. Plioc. Af. *Stegodibelodon* L.-M. Plioc. NAf.

SUBORDER MAMMUTOIDEA **Stegodontidae** *Stegodon* [*Parastegodon Platystegodon*] L. Plioc. Pleist. Af. U. Plioc. Pleist. As. Pleist. EInd. *Stegolophodon* [*?Eostegodon*] L. Mioc. Pleist. Af. U. Mioc. U. Plioc. As. U. Plioc. Eu. L.-U. Pleist. EInd. **Mammutidae** *Mammut* U. Mioc. L. Plioc. U. L. Plioc. U. Pleist. NA. *Miomastodon* M.-U. Mioc. NA. *Palaeomastodon* L. Olig. NAf. *Pliomastodon* U. Mioc. NA. *Zygolophodon* L.-U. Mioc. Eu. Af. As.

SUBORDER DEINOTHERIOIDEA **Deinotheriidae** *Deinotherium* L. Mioc. U. Mioc. Eu. As. L. Mioc. M. Pleist. Af. *Prodeinotherium* L. Mioc. Af. Eu. SAs.

SUBORDER BARYTHERIOIDEA **Barytheriidae** *Barytherium* U. Eoc. NAf.

ORDER SIRENIA **Prorastomidae** *Prorastomus* M. Eoc. W.Ind. **Dugongidae** (**Halicoridae**) R. WPac. Ind. Oc. Red Sea *Anisosiren* M. Eoc. *Anomotherium* M.-U. Olig. EAs. *Caribosiren* M. Olig. WInd. *Dioplotherium* L.-M. Mioc. Eu. *Dusisiren* M.-U. Mioc. L. Plioc. NA. *Eotheroides* [*Archaeosiren Eosiren Eotherium*] M.-U. Eoc. NAf. *Felsinotherium* [*Cheirotherium Halysiren*] L. Plioc. NAf. NA. L.-U. Plioc. Eu. *Halitherium* [*Manatherium*] L. Olig. L. Mioc. Eu. Olig. Mad. ?Mioc. NA. *Hesperosiren* M. Mioc. NA. *Hydromalis* U. Mioc. R. NPac. Arc. *Indosiren* U. Mioc. EInd. *Manatus* [*Trichechus*] Pleist. R. NA. WInd. R. WAf. *Metaxytherium* L.-M. Mioc. Eu. *Prototherium* [*Mesosiren Paraliosiren*] U. Eoc. Eu. *Rytiodus* [*Thytiodus*] U. Olig. Eu. *Sirenavus* M. Eoc. Eu. *Thalattosiren* L. Mioc. Eu. **Manatidae** (**Trichechidae**) *Potamosiren* L. Mioc. SA. *Ribodon* U. Mioc. SA. **Protosirenidae** *Protosiren* M. Eoc. Eu. NAf.

ORDER DESMOSTYLIA **Desmostylidae** *Behemotops* U. Olig. NA. *Cornwallius* Olig. L. Mioc. WNA. EAs. *Desmostylus* [*Desmostylella Kronokotherium*] L. Mioc. L. Plioc. WNA. EAs. *Paleoparadoxia* L. Mioc. L. Plioc. EAs. M. Mioc. L. Plioc. WNA.

ORDER HYRACOIDEA **Procaviidae** *Gigantohyrax* U. Plioc. Af. *Procavia* U. Plioc. R. Af. SWAs. *Prohyrax* L. Mioc. Af. *Saghatherium* L. Olig. NAf. **Pliohyracidae** *Bunohyrax* [*Mixohyrax*] L. Olig. L. Mioc. NAf. *Geniohyus* L. Olig. Af. *Kvabebihyrax* U. Plioc. EEu. *Megalohyrax* L. Olig. ?L. Mioc. Af. *Meroehyrax* Mioc. EAf. *Pachyhyrax* L. Olig. NAf. L. Mioc. EAf. *Pliohyrax* U. Mioc. Eu. As. *Postschizotherium* U. Plioc. Eu. As. *Thyrohyrax* Olig. Af. *Titanohyrax* L. Olig. Af.

ORDER EMBRITHOPODA **Arsinoitheriidae** *Arsinotherium* L. Olig. NAf. *Crivadiatherium* U. Eoc. L. Olig. EEu. *Palaeoamasia* L. Eoc. SWAs.

ORDER TUBULIDENTATA **Orycteropodidae** *Leptorycteropus* U. Plioc. Af. *Myorycteropus* L. Mioc. EAf. *Orycteropus* U. Mioc. R. Af. M. Mioc. L. Plioc. EEu. SWAs. *Plesiorycteropus* subfossil Mad.

ORDER NOTOUNGULATA

SUBORDER NOTOPRONGONIA **?Arctostylopidae** *Allostylops* U. Paleoc. EAs. *Anastylops* L. Eoc. As. *Arctostylops* U. Paleoc. NA. *Asiostylops* U. Paleoc. As. *Palaeostylops* U. Paleoc. L. Eoc. EAs. *Sinostylops* U. Paleoc. As. **Henricosborniidae** *Henricosbornia* [*Hemistylops Microstylops Monolophodon Pantostylops Polystylops Prohyracotherium Selenoconus*] U. Paleoc. M. Eoc. SA. *Othneilmarshia* [*Postpithecus*] *Peripantostylops* U. Paleoc. L. Eoc. SA. *Simpsonotus* U. Paleoc. SA. **Notostylopidae** *Edvardotrouessartia* L. Eoc. SA. *Homalostylops* [*Acrostylops*] U. Paleoc. L. Eoc. SA. *Otronia* M. Eoc. SA. *Seudenius* U. Paleoc. SA. *Notostylops* [*Anastylops Catastylops Entelostylops Eostylops Isostylops Pliostylops*] L. Eoc. SA.

SUBORDER TOXODONTIA **Oldfieldthomasiidae** (**Acoelodidae**) *Acoelodus* L. Eoc. SA. *Allalmeia* U. Olig. SA. *Brachystephanus* U. Eoc. SA. *Carmargomendesis Colbertia Kibenikhoria Itaboraitherium* U. Paleoc. SA. *Maxschlosseria* [*Paracoelodus Oldfieldthomasia*] L. Eoc. SA. *Paginula* L. Eoc. SA. *Tsmanichoria* M. Eoc. SA. *Ultrapithercus* L. Eoc. SA. *Xenostephanus* U. Olig. SA. **Archaeopithecidae** *Acropithecus Archaeopithecus* L. Eoc. SA. **Isotemnidae** *?Acoelohyrax* L. Eoc. SA. *Anisotemnus* L. Eoc. SA. *?Calodontotherium Distylophorus* M. Eoc. SA. *Coelostylodon* L. Eoc. SA. *Isotemnus* [*Prostylops*] U. Paleoc. L. Eoc. SA. *Periphragnis* [*Proasmodeus*] M. Eoc. SA. *Pleurocoelodon* U. Olig. SA. *Pleurostylodon* [*Pleurotemnus*] L. Eoc. SA. *Rhyphodon* M. Eoc. SA. *Thomashuxleyia* L. Eoc. SA. *Trimerostephanos* M. Eoc. U. Olig. SA. **Homalodotheriidae** *Asmodus* U. Olig. SA. *Chasicotherium* [*Puntanotherium*] U. Mioc. SA. *Homalodotherium* ?U. Olig. L.-U. Mioc. SA. **Leontiniidae** *Ancylocoelus* U. Olig. SA. *Colpodon* U. Olig. SA. *Henricofilholia Leontinia Scarrittia* U. Olig. SA. **Notohippidae** *Argyrohippus* U. Olig. SA. *Eomorphippus* M. Eoc. SA. *Eurygenium* U. Olig. SA. *Interhippus* M. Eoc. U. Olig. SA. *Morphippus Nesohippus* U. Olig. SA. *Notohippus* L. Mioc. SA. *Perhippidium* U. Olig. SA. *Rhynchippus* U. Olig. SA. *Stilhippus* U. Olig. SA. **Toxodontidae** *Adinotherium* L.-M. Mioc. SA. *Alitoxodon* U. Plioc. SA. *Hemioxotodon* U. Mioc. SA. *Hyperoxotodon* U. Mioc. SA. *Nesodon* L.-U. Mioc. SA. *Nesodonopsis* M. Mioc. SA. *Ocnerotherium* L. Plioc. SA. *Palaeotoxodon* ?U. Plioc. SA. *Palyeidodon* M. Mioc. SA. *Paratrigodon* U. Mioc. SA. *Pisanodon* U. Mioc. L. Plioc. SA. *Oriadubitherium* U. Olig. SA. *Stereotoxodon* U. Mioc. SA. *Toxodon* [*?Posnanskytherium*] U. Plioc. Pleist. SA. *Trigodon* [*Eutrigodon*] ?M.-U. Plioc. SA. *Xotodon* L.-U. Plioc. SA.

SUBORDER TYPOTHEROIDEA

SUPERFAMILY TYPOTHEROIDEA **Interatheriidae** *Antepithecus* L. Eoc. SA. *Archaeophylus* [*Progaleopithecus*] U. Olig. SA. *Caenophilus* U. Mioc. SA. *Cochilus* U. Olig. SA. *Epipatriarchus* L.-U. Mioc. SA. *Guilielmoscottia* M. Eoc. SA. *Interatherium* [*Icochilus*] L.-M. Mioc. SA. *Medistylus* L.-M. Olig. SA. *Miocochilius* U. Mioc. SA. *Notopithecus* [*?Pseudadiantus*] ?U. Paleoc. L. Eoc. ?L. Plioc. SA. *Paracochilius* U. Olig. SA. *Plagiarthus* [*Argy-*

rohyrax] U. Olig. SA. *Protypotherium* [*Patriarchus*] U. Olig. L. Plioc. SA. *Transpithecus* ?U. Paleoc. L. Eoc. SA. **Mesotheriidae** *Eutypotherium* [*Tachtypotherium Typothericulus*] U. Mioc. SA. *Mesotherium* L.-M. Pleist. SA. *Proedium* [*Isoproedrium Proedrium*] U. Olig. SA. *Pseudotypotherium* L.-U. Plioc. SA. *Trachytherus* [*Eutrachytherus*] ?U. Eoc. U. Olig. SA. *Typothericulus* M. Mioc. SA. *Typotheriopsis* [?*Acrotypotherium*] U. Mioc. ?L. Plioc. SA.

SUPERFAMILY HEGETOTHEROIDEA **Archaeohyracidae** *Archaeohyrax* ?U. Eoc. U. Olig. SA. *Bryanpattersonia* L. Eoc. SA. *Eohegotherium* M. Eoc. SA. *Eohyrax* ?U. Paleoc. L. Eoc. SA. *Pseudhyrax* M. Eoc. SA. **Hegetotheriidae** *Ethegotherium* U. Eoc. SA. *Hegetotherium* [*Selatherium*] U. Olig. M. Mioc. SA. *Hemihegetotherium* U. Mioc. SA. *Pachyrukhos* [*Pachyrucos*] U. Olig. U. Mioc. SA. *Paedsotherium* U. Mioc. L. Pleist. SA. *Prohegetotherium Propachyrucos Prosotherium* [*Medistylus Phanophilus*] U. Olig. SA. *Pseudohegetotherium* L. Plioc. SA. *Raulringueletia* U. Mioc. or L. Plioc. SA. *Tremacyllus* L. Plioc. Pleist. SA.

NOTOUNGULATA INCERTAE SEDIS *Acama* U. Eoc. SA. *Brandmayria* U. Paleoc. SA.

ORDER ASTRAPOTHERIA **Trigonostylopidae** *Albertogaudryia* [*Scabellia*] L. Eoc. SA. *Shecenia* U. Paleoc. SA. *Teragonostylops* L. Eoc. SA. *Trigonostylops* [*Chiodon Staurodon*] U. Paleoc. L. Eoc. SA. **Astrapotheriidae** *Astraponotus* [*Notamynus*] U. Eoc. SA. *Astrapothericulus* L. Mioc. SA. *Astrapotherium* U. Olig. U. Mioc. SA. *Parastrapotherium* U. Olig. SA. *Scaglia* L. Eoc. SA. *Synastrapotherium* U. Olig. SA. *Xenastrapotherium* U. Mioc. SA.

ORDER LITOPTERNA **Proterotheriidae** *Anisolambda* [*Eulambda Josepholeidya Ricardolydekkeria*] U. Paleoc. SA. *Brachytherium* M.-U. Plioc. SA. *Deuterotherium* U. Olig. SA. *Diadiaphorus* L. Mioc. L. Plioc. SA. *Diplasiotherium* U. Plioc. SA. *Eoauchenia* U. Mioc. U. Plioc. SA. *Eoproterotherium* U. Olig. SA. *Epecuenia* U. Mioc. SA. *Guilielmofloweria* L. Eoc. SA. *Heteroglyphis* M. Eoc. SA. *Licaphrium* L.-?M. Mioc. SA. *Licaphrops* U. Olig. L. Mioc. SA. *Phoradiadus* U. Eoc. SA. *Polyacrodon* [*Decaconus Oroacrodon Periacrodon*] U. Eoc. SA. *Polymorphis* [*Megacrodon*] U. Eoc. SA. *Prolicaphrium* U. Olig. SA. *Proterotherium* L. Mioc. SA. *Prothoatherium* U. Olig. SA. *Thoatherium* L. Mioc. SA. *Wainka* U. Paleoc. SA. *Xesmodon* [*Glyphodon*] U. Eoc. SA. **Protolipternidae** *Asmithwoodwardia* U. Paleoc. SA. *Miguelsoria Protolipterna* U. Paleoc. SA. **Macraucheniidae** *Cramauchenia* U. Olig. SA. *Cullinia* L. Plioc. SA. *Ernestohaeckelia* L. Eoc. SA. *Macrauchenia Macraucheniopsis* M.-U. Pleist. SA. *Notodiaphorus* M.-U. Olig. SA. *Paramacrauchenia* U. Olig. SA. *Phoenixauchenia* M. Mioc. SA. *Promacrauchenia* M.-U. Plioc. SA. *Theosodon* U. Olig. L. Plioc. SA. *Victorlemoinea* U. Paleoc. L. Eoc. SA. **Adianthidae** *Adianthus* [*Adiantus*] U. Olig. L. Mioc. SA. *Adiantoides* U. Eoc. SA. *Indalecia* L. Eoc. SA. *Proheptoconus* U. Olig. SA. *Thadanius* U. Olig. SA. *Tricoelodus* [*Proadiantus*] U. Olig. SA.

ORDER XENUNGULATA **Carodniidae** *Carodnia* [*Ctalecarodnia*] U. Paleoc. SA.

ORDER PYROTHERIA **Pyrotheriidae** *Carolozittelia* L. Eoc. SA. *Griphodon* M. Eoc. L. Olig. SA. *Propyrotherium* U. Eoc. SA. *Pyrotherium* U. Olig. SA. **Colombitheriidae** *Colombitherium* ?U. Eoc. SA. *Proticia* L. Eoc. SA.

ORDER XENARTHRA

INFRAORDER LORICATA (CINGULATA)

SUPERFAMILY DASYPODOIDEA **Dasypodidae** *Astegotherium* L. Eoc. SA. *Cabassous* [*Xenurus*] Pleist. R. SA. *Chaetophractus* U. Plioc. R. SA. *Chasicotatus* U. Mioc. SA. *Chlamyphorus* [*Chlamydophorus*] U. Pleist. R. SA. *Chorobates* U. Plioc. SA. *Coelutateus* L. Eoc. SA. *Dasypus* [*Paropus Tatu Tatusia*] Pleist. R. SA. M. Plioc. R. NA. *Doellotatus* [*Eutatopsis*] M.-U. Pleist. R. SA. M. Plioc. SA. *Epipeltephilus* U. Mioc. ?L. Plioc. SA. *Euphractus* [*Scleropleura*] Pleist. R. SA. *Eutatus* Pleist. SA. *Holmesina* Pleist. NA. *Kraglievichia* U. Mioc. U. Plioc. SA. M. Plioc. L. Pleist. NA. *Machlydotherium* L.-U. Eoc. SA. *Macroeuphractus* U. Mioc. U. Plioc. SA. *Meteutatus* [*Sadypus*] L. Eoc. U. Olig. SA. *Paleuphractus* L. Plioc. SA. *Pampatherium* [*Chalamytherium Chlamydotherium Holmesina*] Pleist. SA. NA. *Paraeuphractus* ?U. Plioc. SA. *Parapeltocoelus* U. Olig. SA. *Peltcoelus* U. Olig. SA. *Peltephilus* U. Olig. L. Mioc. SA. *Proeuphractus* U. Olig. SA. *Proetus* U. Olig. U. Mioc. SA. *Propraopus* ?U. Plioc. Pleist. SA. Pleist. NA. *Prostegotherium* L. Eoc. SA. *Prozaedius* U. Olig. M. Mioc. SA. *Pseudeutatus* [*Anutaetus Isutaetus Pachyzedys*] M. Eoc. SA. *Pseudostegotherium* ?L. Eoc. U. Olig. SA. *Ringueletia* U. Plioc. SA. *Stegotheriopsis* U. Olig. SA. *Stegotherium* L. Mioc. SA. *Stenotatus* [*Prodasypus*] U. Olig. M. Mioc. SA. *Tolypeutes* ?U. Plioc. Pleist. R. SA. *Utaetus* [*Anteutatus Orthutaetus Parataetus Posteutatus ?Coelutaetus*] ?U. Paleoc. L.-M. Eoc. SA. *Vassallia* M. Mioc. SA. *Zaedyus* U. Plioc. R. SA. **Palaeopeltidae** *Palaeopeltis* U. Eoc. M. Olig. SA.

SUPERFAMILY GLYPTODONTOIDEA **Glyptodontidae** (**Hoplophoridae**) *Asterostemma* L.-U. Mioc. SA. *Cochlops* L. Mioc. SA. *Doedicuroides* L. Pleist. SA. *Doedicurus* Pleist. SA. *Ecinepeltus* L. Mioc. SA. *Eleutherocercus* M.-U. Plioc. SA. *Glyptatelus* ?M. Eoc. U. Olig. SA. *Glyptodon* [*Glyptocoileus Glyptopedius Glyptostracon Stromatherium Xenoglyptodon*] Pleist. SA. NA. *Glyptotherium* U. Plioc. L. Pleist. NA. *Hoplophorus* [*Sclerocalyptus*] Pleist. SA. *Lomaphorus* Pleist. SA. *Metopotoxus* L. Mioc. SA. *Neothoracophorus Neuryurus* Pleist. SA. *Nopachtus* U. Plioc. SA. *Palaeodoedicurus* U. Plioc. SA. *Palaeohoplophorus* M. Mioc. L. Plioc. SA. *Panochthus* Pleist. SA. *Paraglyptodon* U. Plioc. ?L. Pleist. SA. *Plaxhaplous* Pleist. SA. *Propalaeohoplophorus* U. Olig. U. Mioc. SA. *Sclerocalyptus* Pleist. SA. *Trachycalyptus Urotherium* ?M. Plioc. U. Plioc. SA.

INFRAORDER PILOSA

SUPERFAMILY MEGALONYCHOIDEA **Megalonychidae** *Analcimorphus* L. Mioc. SA. *Diheterocnus* [*Heterocnus*] U. Plioc. ?L. Pleist. SA. *Eucholoeops* L.-M. Mioc. SA. *Hyperleptus* L. Mioc. SA. *Megalonychops* ?Pleist. SA. *Megalonychotherium* L. Mioc. SA. *Megalonyx* [*Onychotherium*] L. Plioc. L.-U. Pleist. NA. *Pelecyodon* L. Mioc. SA. *Pliometanastes* U. Mioc. NA. *Proplatyarthrus* M. Eoc. SA. **Megatheriidae** *Diellipsodon* M. Mioc. SA. *Eomegatherium* M. Mioc. SA. *Eremotherium* [*Pseudere-*

motherium Schaubia Schaubitherium] Pleist. NA. SA. *Essonodontherium* U. Pleist. SA. *Hapaloides* U. Olig. SA. *Hapalops* [*Parhapalops Pseudhapalops Xyophorus*] L.-M. Mioc. SA. *Megathericulus* U. Mioc. SA. *Megatherium* [*Paramegatherium*] Pleist. SA. U. Plioc. Pleist. NA. *Nothrotheriops* Pleist. NA. *Nothrotherium* [*Coelodon Nothrotheriops*] U. Pleist. SA. *Planops* [*Prepotherium*] L.-U. Mioc. SA. *Plesiomegatherium* U. Mioc. L. Plioc. ?M. Plioc. SA. *Prepotherium* L.-M. Mioc. SA. *Promegatherium* [*Eomegatherium*] M.-U. Mioc. SA. *Pronothrotherium* [*Senetia*] U. Mioc. U. Plioc. SA. *Proschizmotherium* U. Olig. SA. *Pseudhapalops* M. Mioc. SA. *Pyramiodontherium* [*Megatheriops*] L. Plioc. SA. *Schismotherium* L. Mioc. SA.

SUPERFAMILY MYLODONTOIDEA **Mylodontidae** *Analcitherium* L. Mioc. SA. *Elassotherium* U. Mioc. SA. *Glossotheridium* U. Plioc. SA. *Glossotheriopsis* M. Mioc. SA. *Glossotherium* [*Eumylodon Mylodon Oreomylodon Pseudolestodon*] M. Plioc. Pleist. NA. Pleist. SA. *Lestodon* [*Prolestodon*] *Mylodon* [*Glossotherium Grypotherium Neomylodon*] M.-U. Pleist. SA. *Nematherium* L. Mioc. ?M. Mioc. SA. *Neonematherium* U. Mioc. SA. *Octodontotherium* U. Olig. SA. *Octomylodon* U. Mioc. SA. *Orophodon* M.-U. Olig. SA. *Proscelidodon* U. Plioc. SA. *Pseudoprepotherium* M. Mioc. SA. *Scelidodon* Pleist. SA. *Scelidotherium* [*Catonyx*] U. Plioc. Pleist. SA. *Sphenotherus* M. Plioc. SA. *Thinobadistes* U. Mioc. NA. **Entelopidae** *Entelops* L. Mioc. SA.

INFRAORDER VERMILINGUA **Myrmecophagidae** *Eurotamandua* M. Eoc. Eu. *Myrmecophaga* [*Neotamandua*] M. Plioc. R. SA. R. CA. *Neotamandua* M.-U. Mioc. SA. *Palaeomyrmedon* M. Plioc. SA. *Protamandua* L. Mioc. SA.

ORDER INCERTAE SEDIS

SUBORDER PALAEANODONTA **Metacheiromyidae** *Metacheiromys* M. Eoc. NA. *Palaeanodon* U. Paleoc. L. Eoc. NA. *Propalaeanodon* U. Paleoc. NA. **Epoicotheriidae** *Alocodontulum* [*Alocodon*] L. Eoc. NA. *Amelotabes* U. Paleoc. NA. *Epoicotherium* [*Xenotherium*] L. Olig. NA. *Pentapassalus* L. Eoc. NA. *Tetrapassalus* M. Eoc. NA. *Tubulodon* L. Eoc. NA. *Xenocranium* L. Olig. NA. **?Ernanodontidae** *Ernanodon* U. Paleoc. As.

ORDER PHOLIDOTA **Manidae** *Eomanis* M. Eoc. Eu. *Manis* [*Phataginus*] L.-U. Plioc. ?Af. EInd. Pleist. R. As. R. Af. *Necromanis* L. Olig. L. Mioc. Eu. *Patriomanis* L. Olig. NA. ?*Teutmanis* Olig. Mioc. Eu.

MAMMALIA INCERTAE SEDIS

Acamana U. Eoc. SA. *Florentinoameghinia* L. Eoc. SA. *Ithygrammodon* M. Eoc. NA. *Obtusudon* Paleoc. As. *Pakilestes* M. Eoc. SAs. *Wannotherium* U. Paleoc. As. **Didymoconidae (Tshelkariidae)** *Archaeoryctes* U. Paleoc. As. *Didymoconus* [*Tshelkaria*] M. Olig. EAs. *Hunanictis* L. Eoc. As. *Kennatherium* U. Eoc. As. *Mongoloryctes* U. Eoc. EAs. *Zeuctherium* M. Paleoc. As.

REFERENCES

Anderson, S., and Jones, J. K. (1984). *Orders and Families of Recent Mammals of the World.* Wiley, New York.

Brodkorb, P. (1978). Catalogue of fossil birds. Part 5. Passeriformes. *Bull. Fla. State Mus., Biol. Sci.*, **23**: 139–228.

Carroll, R. L. (1982). The early evolution of reptiles. *Ann. Rev. Ecol. Syst.*, **13**: 87–109.

Carroll, R. L., and Winer, L. (1977). Appendix to Chapter 13. Patterns of amphibian evolution: An extended example of the incompleteness of the fossil record. In A. Hallam (ed.), *Patterns of Evolution.* Elsevier, Amsterdam.

Compagno, L. J. V. (1977). Phyletic relationships of living sharks and rays. *Amer. Zool.*, **17**(2): 303–322.

Denison, R. H. (1978). Placoderms. In H.-P. Schultze (ed.), *Handbook of Paleoichthylogy. Vol. 2.* Gustav Fischer Verlag, Stuttgart.

Denison, R. H. (1979). Acanthodii. In H.-P. Schultze (ed.), *Handbook of Paleoichthyology.* Vol. 5. Gustav Fischer Verlag, Stuttgart.

Estes, R. (1981). Gymnophiona, Caudata. In P. Wellnhofer (ed.), *Handbuch der Paläoherpetologie.* Part 2. Gustav Fischer Verlag, Stuttgart.

Estes, R. (1983). Sauria terrestria, Amphisbaenia. In P. Wellnhofer (ed.), *Handbuch der Paläoherpetologie.* Part 10. Gustav Fischer Verlag, Stuttgart.

Gruson, E. S. (1976). *Checklist of the World's Birds.* Quadrangle, New York.

Lauder, G. V., and Liem, K. F. (1983). The evolution and interrelationships of the actinopterygian fishes. *Bull. Mus. Comp. Zool.*, **150**(3): 95–197.

Lillegraven, J. A., Kielan-Jaworowska, Z., and Clemens, W. A. (eds.), (1979). *Mesozoic Mammals. The First Two-Thirds of Mammalian History.* University of California Press, Berkeley.

Marshall, L. G. (1981). The families and genera of Marsupialia. *Fieldiana (Geol.), New Ser.*, **8**: 1–65.

Młynarski, M. (1976). Testudines. In P. Wellnhofer (ed.), *Handbuch der Paläoherpetologie.* Part 7. Gustav Fischer Verlag, Stuttgart.

Nelson, J. S. (1984). *Fishes of the World.* Wiley, New York.

Olson, S. L. (1985). The fossil record of birds. In D. Farner, J. King, and K. Parkes (eds.), *Avian Biology*, Vol. 8, pp. 79–238. Academic Press, New York.

Rage, J.-C. (1984). Serpentes. In P. Wellnhofer (ed.), *Handbuch der Paläoherpetologie.* Part 11. Gustav Fischer Verlag, Stuttgart.

Romer, A. S. (1966). *Vertebrate Paleontology* (3d ed.). University of Chicago Press, Chicago.

Savage, D. E., and Russell, D. E. (1983). *Mammalian Paleofaunas of the World.* Addison-Wesley, London.

Thomson, K. S. (1969). The biology of the lobe-finned fishes. *Biol. Rev.*, **44**: 91–154.

Zangerl, R. (1981). Chondrichthyes. I. Paleozoic Elasmobranchii. In H.-P. Schultze (ed.), *Handbook of Paleoichthyology.* Vol. 3A. Gustav Fischer Verlag, Stuttgart.

Index*

*Letters L and R following page numbers refer to left and right columns in the classification that appears in the Appendix.